HANDBOOK OF MODERN HOSPITAL SAFETY

Second Edition

HANDBOOK OF

MODERN HOSPITAL SAFETY

Second Edition

Edited by

WILLIAM CHARNEY

CRC Press
Taylor & Francis Group
Boca Raton London New York

CRC Press is an imprint of the
Taylor & Francis Group, an **informa** business

CRC Press
Taylor & Francis Group
6000 Broken Sound Parkway NW, Suite 300
Boca Raton, FL 33487-2742

Printed in the United States of America on acid-free paper
10 9 8 7 6 5 4 3 2 1

International Standard Book Number-13: 978-1-4200-4785-1 (Hardcover)

Library of Congress Cataloging-in-Publication Data

Handbook of modern hospital safety / edited by William Charney. -- 2nd ed.
p. ; cm.
Includes bibliographical references and index.
ISBN-13: 978-1-4200-4785-1 (hardcover : alk. paper)
ISBN-10: 1-4200-4785-X (hardcover : alk. paper)
1. Hospitals--Security measures--Handbooks, manuals, etc. 2. Hospitals--Safety measures--Handbooks, manuals, etc. 3. Hospitals--Employees--Health and hygiene--Handbooks, manuals, etc. I. Charney, William, 1947- II. Title.
[DNLM: 1. Hospitals. 2. Safety Management--methods. 3. Health Personnel. 4. Occupational Health. WX 185 H2365 2010]

RA969.95.H37 2010
362.11068'4--dc22 2008051637

Visit the Taylor & Francis Web site at
http://www.taylorandfrancis.com

and the CRC Press Web site at
http://www.crcpress.com

Contents

Preface ix

Introduction xiii

Editor xvii

Contributors xix

1 Overview of Health and Safety in the Health Care Environment **1**-1
David A. Sterling

2 Epidemiology **2**-1
James M. Boiano, John P. E. Sestito, Sara E. Luckhaupt, Cynthia F. Robinson, and James T. Walker

3 Back Injury Prevention in Health Care **3**-1
William Charney, Anne Hudson, Susan Gallagher, John D. Lloyd, Andrea Baptiste, Audrey Nelson, Mary Matz, Bernice D. Owen, Arun Garg, Jocelyn Villeneuve, Justin Gilmore, and Willis Martin

4 Emerging Infectious Diseases **4**-1
William Charney, J. H. Lange, G. Mastrangelo, Jacques Lavoie, Yves Cloutier, Jaime Lara, Geneviève Marchand, Jeanne Anderson, Andrew Geeslin, and Andrew Streifel

5 Disinfection and Infection Control **5**-1
Erica J. Stewart, Russell N. Olmsted, Pier-George Zanoni, Laurence Lee, Jules Millogo, and Kathryn M. Duesman

6 Progress in Preventing Sharps Injuries in the United States **6**-1
Jane Perry, Ahmed E. Gomaa, Janine Jagger, and Emergency Care Research Institute

7 Violence in the Health Care Industry **7**-1
Jane Lipscomb, Jonathan Rosen, Matthew London, Kathleen McPhaul, Joyce A. Simonowitz, and Service Employees International Union

8 Prevention of Slip, Trip, and Fall Hazards for Workers in Hospital Settings **8**-1
James W. Collins and Jennifer L. Bell

9 Ergonomics 9-1
Margaret Wan, Jocelyn Villeneuve, and Guy Fragala

10 Chemodrugs 10-1
Thomas P. Fuller

11 Tuberculosis Engineering Controls 11-1
Lorraine M. Conroy, John E. Franke, Byron S. Tepper, John Mehring, Derrick Hodge, Daniel Kass, William Charney, June Fisher, Patrice M. Sutton, Mark Nicas, and Robert J. Harrison

12 Electrocautery Smoke: Reasons for Scavenging 12-1
Stackhouse, Incorporated; Jacob D. Paz; Charles J. Bryant; Richard Gorman; John Stewart; Wen-Zong Whong; Lindsey V. Kayman; Association of Operating Room Nurses; William Charney; Scott Miller; Shawn Campbell; and Bob Reynolds

13 Safe Use of Ethylene Oxide in the Hospital Environment 13-1
Arthur R. Reich

14 Monitoring Aldehydes 14-1
Edward W. Finucane and William Charney

15 Pentamidine 15-1
William Charney, Lawrence Rose, and Patricia Quinlan

16 Ribavirin 16-1
Jim Bellows, Robert J. Harrison, William Charney, Kevin J. Corkery, and Lee Wugofski

17 Trace Anesthetic Gas 17-1
Edward W. Finucane, Anthony Schapera, William Charney, Patricia A. Heinsohn, and D. L. Jewett

18 Cost–Benefit 18-1
William Charney and Donald Maynes

19 Clinical Approach to Glove Dermatitis and Latex Allergy 19-1
Dennis Shusterman

20 Latex Glove Use: Essentials in Modern Hospital Safety 20-1
Wava M. Truscott

21 Health and Safety Hazards of Shiftwork: Implications for Health Care Workers and Strategies for Prevention 21-1
Joseph Schirmer

22 Radiation Protection in Hospitals 22-1
Jean-Pierre Gauvin

23 Reproductive Hazards in Hospitals 23-1
Lindsey V. Kayman

24 Medical Waste 24-1
Michael L. Garvin, Herb B. Kuhn, and Elaine Peters

25 The Occupational Hazards of Home Health Care 25-1
Elaine Askari and Barbara DeBaun

26 Caring Until It Hurts: How Nursing Work Is Becoming the Most Dangerous Job in America 26-1
Service Employees International Union

27 Laboratory Safety 27-1
Wayne Wood

28 Biological Exposure Index Testing: Two Case Studies 28-1
William Charney, Martin Lipnowski, and Lee Wugofski

29 Functions and Staffing of a Hospital Safety Office 29-1
Lindsey V. Kayman and John F. Clemons

30 Education for Action: An Innovative Approach to Training Hospital Employees 30-1
Merri Weinger and Nina Wallerstein

31 Stress Factors 31-1
Laura A. Job, Wendy E. Shearn, Kathleen Kahler, Jolie Pearl, and Marian McDonald

32 Case Studies of Health Care Workers in the Compensation System 32-1
Lenora S. Colbert

Index **I**-1

Preface

We intend this book to address the major occupational health issues in hospitals. In this book, we have assembled the best available current information about well-known hazards such as back injuries, antineoplastic agents, anesthetic gases, and ethylene oxide as well as laboratory safety and some newer concerns. The increasing number of patients with human immunodeficiency virus has heightened public awareness of infectious waste hazards and needlestick injuries. Health care personnel need to be aware of the risks associated with new treatments such as pentamidine, ribavirin, and lasers. The broad scope of chapters on staffing needs, safety committees, and employee training provides a context for the chapters on the more specific hazards.

We hope that the various authors' extensive experiences in the health care industry will be used to prevent hazardous exposures, reducing the incidences of occupational injuries, disabilities, and diseases associated with employment in health care. We thank Jon Lewis, Brian Lewis, and Janet Tarolli, RN, of Lewis Publishers for their support and encouragement in making this book possible.

The health care industry has multiple potential injury and disease hazards.[1,2] Hospital workers' injuries tend to be both severe and costly. For example, back injuries are one of the most expensive hazards in all U.S. industry.[3] The health care industry has multiple problems; occupational health issues are but one symptom of these growing crises. A recently released study by the Joint Commission on Accreditation of Healthcare Organizations (JCAHO) highlights the convergence of occupational health and patient care issues. More than 40% of 5208 hospitals surveyed from 1986 to 1988 by the JCAHO had less than significant compliance with published safety standards such as electrical wiring and hazardous waste.[4] The same survey found serious problems with quality of patient care issues. For example, 51% did not have adequate procedures for reviewing surgery to determine whether the surgery should be done or if it was done correctly, and 35% did not always review blood transfusion requests and procedures to evaluate the objective need or whether whole blood or blood components were needed. This last problem is important because blood components carry less risk of infection from viruses. Why are all these critical problems surfacing in the health care industry? In explaining these findings, the president of the Joint Commission, Dennis O'Leary, cited lack of resources and financial pressures.[5]

Many of the chapters discuss U.S. rules and regulations governing occupational safety and health. It is not comprehensive in this regard. The reader with responsibilities for hospital environmental and occupational health should make an independent effort to determine the latest applicable federal and state regulations. Only a continuous search will offer the reader any security in the rapidly changing legal context for this work. New and rapidly changing federal and state regulations governing medical and infectious wastes are an example.

There are solutions to all the problems described in this book; however, some may be more easily arrived at than others. It may be easier to persuade administrators to adopt solutions requiring capital expenditures than to adopt solutions requiring increased labor costs, such as more staff or more education and training programs. Most hazards described can be solved with engineering controls. The technology to create improvements in ventilation and to engineer out technological hazards is fairly well developed. These often can be considered as one-time capital investments. Other hazards that include back injuries, shift work, and understaffing are not so easily solved with technical fixes and capital improvements. These problems will require additional staff costs, either directly through new employees or indirectly through increased training time.

Many hospital administrators can be expected to resist increased labor expenditures as a means to resolve the hazards of chronic understaffing, back injuries, and rotational shift work. To appreciate the reasons for this apparent stubbornness, we need to develop an understanding of the economic context in which these health care industry managers operate. Himmelstein and Woolhandler have written a quantitative, historical introduction to the economics of the U.S. health care industry, highlighting the tension between the economic forces shaping the organization of health care and a rational health care delivery system.[6] They report that the production of goods for the health care industry is more profitable than providing health care services. In 1983, the profits of proprietary hospitals ($1.2 billion) were less than those of the drug industry ($5.6 billion) and the suppliers of medical equipment ($2.8 billion). This may be not only because the growth of proprietary hospitals is a recent phenomenon, but also because there may be some intrinsic economic advantages to creating value-added, marketable products for the ill when compared with providing care for the ill, however that care may be organized.

Through participation in collective bargaining for health care workers in Wisconsin, I recognize employers' financial incentive to resist increased labor costs. From the labor perspective, it is often difficult to convince employers that hiring and retaining qualified workers will save money and human resources in the long run. These labor cost issues will remain difficult to resolve until the true impacts of economic decisions on employee health and patient care are more widely appreciated. Recently, for example, an analysis of 3100 U.S. hospitals found that private not-for-profit hospitals were associated with lower patient mortality rates than for-profit hospitals. (Other institutional factors associated with lower mortality rates include higher percentages of board-certified physicians and registered nurses, higher occupancy rates, technological sophistication, and greater size.) Lower mortality rates were also associated with higher payroll expenses per hospital bed.[7] Quality care costs money. If resources are available to hire and retain qualified staff (resources that, in a for-profit setting, might be required for investors' dividends), then patient care seems to benefit.

Bailar and Smith have evaluated statistical data on cancer in the United States from 1950 to 1982 from an epidemiologic perspective and found that the "war on cancer" has largely failed to reduce cancer mortality.[8] They interpret these data to mean that more emphasis needs to be placed on prevention and less on treatment. Since the associations between exposures and health outcomes are clearer for occupational injuries and illnesses than for most cancers, there is an even greater need for prevention to dominate treatment. Unhappily, under the current system of health care delivery in the United States, most routine payment mechanisms are established for treatment, not for prevention. It is our hope that any light shed by these chapters on hazard identification and control will also help to illuminate the path toward prevention of injury and disease.

References

1. Patterson WB, Craven DE, Schwartz DA, Nardell EA, Kasmer J, and Noble J. Occupational hazards to hospital personnel. *Ann Int Med* 102:658–680, 1985.
2. Gun RT, Mullan RJ, Brown D, Clapp D, Dubrow R, Gordon J, Liscomb J, Pedersen DH, Salg J, and Schnor TT. *Report of the Division of Surveillance, Hazard Evaluations and Field Studies Task Force on Hospital Worker Health.* U.S. Department of Health and Human Services, National Institute of Occupational Safety and Health, 1985.
3. Office of Technology Assessment. *Preventing Illness and Injury in the Workplace*, OTA-H-256. US Congress, OTA, Washington DC, 1985.

4. Joint Commission on Accreditation of Healthcare Organizations. *Background on the Accreditation Process and Data for Surveys Conducted January 1, 1986–December 31, 1988*. Cited by Michael Millenson in the *Chicago Tribune*, December 3, 1989, p. 3.
5. Hospitals fall short on quality. *Wis State J*, December 4, 1989.
6. Himmelstein DU and Woolhandler S. The corporate compromise: A Marxist view of health maintenance organizations and prospective payment. *Ann Int Med* 109:494–501, 1988.
7. Hartz AJ, Krakauer H, Kuhn EM, Young M, Jacobsen SJ, Gay G, Muenz L, Katzoff M, Bailey RC, and Rimm AA. Hospital characteristics and mortality rates. *N Eng J Med* 321:1720–1725, 1989.
8. Bailar JC and Smith EM. Progress against cancer? *N Eng J Med* 314:1226–1232, 1986.

Joseph Schirmer

Introduction

Working in health care is certainly dangerous to your health. Every day in the United States, 9000 health care workers sustain disabling injuries on the job.[1] Every 30 seconds a health care worker is stuck by a needle, and one out of every 10 health care workers has developed an allergy to latex. This is just the beginning of the roll call of exposures. In many states, health care workers incur more lost time due to a violent act committed against them than law enforcement personnel. According to another book just published, *The Epidemic of Health Care Work Injury: An Epidemiology*, working in health care is more dangerous than working in a coal mine.[2] And as a study conducted by the Minnesota Nurses Association has recently proved,[3] managed care with its associated downsizing and de-skilling is turning an already bad/epidemic situation worse.

The *Handbook of Modern Hospital Safety* has been compiled to not only outline the problems that health care workers face, but also to show that there are solutions to the problems. This book covers the major exposures health care workers face; however, since the medical technology is always evolving, new problems have surfaced that did not make it into this book. For example, sonographers (ultrasound technicians), according to anecdotal information, have developed high rates of ergonomic injuries. In fact, the majority of the medical and clinical methods employed today were developed without any thought or *a priori* testing for occupational effects. And now, as there seems to be ever-increasing peer review evidence defining the occupational threats to health care workers, there are ongoing obstacles to the implementation of the solutions that are both economic and political.

Economics of Health Care Safety

Over the past century, medical care has evolved from a small cottage industry and almost charitable public service to an enormously profitable and increasingly private business.[4] Medicine has become one of the largest industries in the United States, and economics now readily competes with science and humanitarian concerns in shaping medical care.[4] This paradigm is transferable to the problem of funding for hospital health and safety programs. Too often safety is made to compete for budgets with other capital funding needs. For example, a lead abatement program will not be funded because the system needs another ambulance. It is one or the other.

One of the major economic blockades is the workers' compensation system and its downstream effects on the implementation process. Health care institutions are either self-insured or part of an overall external compensation system. Therefore, the real cost of the injuries suffered by health care workers is not seen by the CEOs. CEOs are so insulated from the cost of the injuries and the medical costs incurred that to persuade them to allocate budgets for prevention when they are not paying the costs of the injuries is extremely difficult and sometimes impossible. Concurrently, since the health care provider system is an integral part of the "free market" system, "profitability" and "cost–benefit" are two important considerations that have to be met by the safety design criteria. This creates a dynamic dilemma for the safety officers of any health care system: to not only design a safety system but also design one that either pays for itself or is profitable in order to meet the criteria of the "master system."

Hospitals and their systems do not perform a professional cost–benefit analysis when a question of safety is involved. The purchasing systems compare only the cost of one item against the cost of the safety item. Often the safety item is more expensive. But this is a superficial analysis because without an evaluation of the injury cost data, the downstream savings cannot be calculated. A good example of this is replacing latex gloves with nonlatex gloves. There is an initial changeover expense, but the cost savings of preventing health care worker allergies and lost time is not calculated into the initial assessment by most institutions. The same is true with the needle safety device. A safety device is at this stage more expensive than a traditional nonsafety device. However, the real cost of a needlestick is not integrated into the cost–benefit analysis. Often, hospital systems do not even have the capability to retrieve or assess their injury costs.

Another type of example of this conflict is the back injury and ergonomic problems faced by health care workers and their institutions. On the national average, back injuries account for approximately 33%–50% of workers' compensation costs in each facility. Depending on the size of the facility, when hidden costs and replacement costs are included, this can amount to an expenditure of half a million dollars to millions of dollars for each hospital system. Prevention of these injuries includes either funding a "lift team," which traditionally costs only 10%–20% of the injury cost, or providing the proper amount of mechanical equipment so all the total body transfers can be done mechanically. However, getting the funding for either of these proven programs is acutely difficult as the institutions are insulated against the injury cost dollars.

Political Obstacles

Political problems for the implementation of safety programs in hospitals also abound. The average "shelf life" of a hospital CEO is 5 years. The priorities of the CEO, in order to maintain career advantage, are often short term and most often do not include safety concerns. The average CEO is bureaucratically protected from the safety probes by layers of managers and departments and rarely even knows that safety exists unless a safety problem becomes a disaster.

Politically, health care worker injury is not on the agenda. Less than 2% of all inspections done by state or federal OSHAs are done in the hospital sectors. There are few, if any, lobbyists for health care workers on a national level in Washington, and it was not until June 1998, when the governor of California signed into law a bill mandating the use of needle safety devices, that a health care safety problem reached such a high political plateau on the state level. On a national level, only 10% of health care workers are organized in unions, and those unions that have been successful in organizing the health care sector have a difficult time pushing a health and safety agenda in an atmosphere of downsizing and corporate control.

As the atmosphere in hospitals becomes more corporate due to mergers and "big" being seen as "better" and more difficult, so too have the occupational health and safety professionals in hospitals become more corporate in their thinking and approaches. Hospital safety professionals function within specified roles largely created and shaped to meet the needs of the corporate class.[5] I have often had to argue more strenuously with occupational health and safety professionals setting themselves up as "gatekeepers" than I have had to argue with management to get a certain program implemented. This class relationship between occupational health and safety professionals in the hospital sector and management makes it more difficult to put forward another form of political thinking within the structure of health and safety.

The *Handbook of Modern Hospital Safety* does not offer a panacea for the problems stated above, but a reference point. Its 1000-plus pages are all points of demarcation. The information contained in these chapters, each written by knowledgeable people in the field, is ammunition that can help to begin to correct hazardous conditions that have been taken for granted for too long. The handbook is a reference book that has compiled the first three volumes of *Essentials of Modern Hospital Safety* and then added a fourth volume, in one single volume. It contains resources on the major areas of health care delivery, acute care, long-term nursing home care, and the growing field of home health care.

Health care worker safety is in a crisis, parallel to the crisis of the health care delivery system in the United States. Saving money seems to be the "bottom line." Janitors are now being asked in some facilities to do some patient bedside care. De-skilling of the workforce is a myopic formula that puts the health care patient as well as the health care worker at risk, which in the long run will drive up costs and not save money. In the last 10 years, health care worker injuries have doubled in the nursing home industry and have risen by 40% in the acute care setting. Not many people are paying attention to this. The handbook is designed to assist in the overall analysis, not create the actual analysis. It contains enough information to allow the reader to shape a system of intervention and to provide enough data to persuade hospital systems to implement safety programs.

References

1. Slattery M. The epidemic hazards of nursing. *Am J Nursing* 98(11), 1998.
2. Charney W and Fragala G (eds.). *The Epidemic of Health Care Worker Injury: An Epidemiology*. CRC Press, Boca Raton, FL, 1998, Chapter 1.
3. Shogren B. Findings of Minnesota Nurses Association study between 1990 and 1994. In: W Charney and G Fragala (eds.), *Epidemic of Health Care Worker Injury: An Epidemiology*, CRC Press, Boca Raton, FL, 1998, Appendix 1, pp. 215–225.
4. Himmelstein D. The corporate compromise. *Ann Intern Med* 109:494–501, 1988.
5. Lax M. Workers and occupational safety and health professionals. *New Solutions* 8(1):99–116, 1998.

William Charney
DOH

Editor

William Charney, DOH, is the president of Healthcare Safety Consulting based in Brattleboro, Vermont. He was formerly the safety coordinator for Washington Hospitals Workers' Compensation Program in Seattle, a position he held for 5 years. Prior to this, he was the director of environmental health at San Francisco General Hospital, Department of Public Health, for 10 years. Charney also served as director of safety at the Jewish General Hospital in Montreal, Quebec, before this.

Charney is the author/editor of nine books with topics ranging from occupational health to health care workers including three volumes of *Essentials of Modern Hospital Safety*, published by Lewis Publishers, a division of Taylor & Francis; *The Epidemic of Health Care Worker Injury: An Epidemiology*, published by CRC Press, also a division of Taylor & Francis; and *The Handbook of Modern Hospital Safety Emerging Infections Diseases and the Threat to Occupational Health and Back Injury among Healthcare Workers*. Charney is also the author of many peer reviewed articles on hospital safety systems, including his research on lifting teams, published in the *American Journal of Occupational Health Nursing*; research on needle safety devices, published in the *Journal of Healthcare Safety, Compliance and Infection Control*; research on ribavirin, published in the *Journal of Respiratory Care*; and many others. Charney has spoken on health care safety issues at major conferences and received the Environmental Health Award for 1998 by the State Department of Health of California and the Advocacy Award in 2007 from the Safe Patient Handling Conference.

Contributors

Jeanne Anderson
Office of Emergency Preparedness
Minnesota Department of Health
St. Paul, Minnesota

Elaine Askari
School of Public Health
University of California, Berkeley
Berkeley, California

Association of Operating Room Nurses
Denver, Colorado

Andrea Baptiste
Safe Patient Handling Center
James A. Haley Veterans' Hospital
Tampa, Florida

Jennifer L. Bell
National Institute for Occupational Safety and Health
Centers for Disease Control and Prevention
Cincinnati, Ohio

Jim Bellows
California Department of Health Services
Berkeley, California

James M. Boiano
Division of Surveillance Hazard Evaluation and Health Studies
National Institute for Occupational Safety and Health
Centers for Disease Control and Prevention
Cincinnati, Ohio

Charles J. Bryant
National Institute for Occupational Safety and Health
Centers for Disease Control and Prevention
Cincinnati, Ohio

Shawn Campbell
Biomedical Engineering
San Francisco General Hospital
San Francisco, California

William Charney
Healthcare Safety Consulting
Brattleboro, Vermount

John F. Clemons
Boston University Medical Center
Boston, Massachusetts

Yves Cloutier
Institute of Research for Health and Safety
Montreal, Quebec, Canada

Lenora S. Colbert
National Health & Human Service Employees Union
New York, New York

James W. Collins
Division of Safety Research
National Institute for Occupational Safety and Health
Centers for Disease Control and Prevention
Cincinnati, Ohio

Lorraine M. Conroy
School of Public Health Environmental and Occupational Health Sciences
University of Illinois at Chicago
Chicago, Illinois

Kevin J. Corkery
San Francisco General Hospital
San Francisco, California

Barbara DeBaun
California Pacific Medical Center
San Francisco, California

Kathryn M. Duesman
Retractable Technologies
Little Elm, Texas

Emergency Care Research Institute
Plymouth Meeting, Pennsylvania

Edward W. Finucane
High Technology Enterprises
Stockton, California

June Fisher
University of California, San Francisco
San Francisco, California

and

Training for Development of Innovative Control Technology
San Francisco, California

Guy Fragala
Patient Safety Center of Inquiry
James A. Haley Veterans' Hospital
Tampa, Florida

John E. Franke
School of Public Health, Environmental and Occupational Health Sciences
University of Illinois at Chicago
Chicago, Illinois

Thomas P. Fuller
Illinois State University
Normal, Illinois

Susan Gallagher
Celebration Institute
Houston, Texas

Arun Garg
Industrial and Systems Engineering
University of Wisconsin–Milwaukee
Milwaukee, Wisconsin

Michael L. Garvin
University of Iowa Hospitals and Clinics
Iowa City, Iowa
and
Garvin Consulting Services
Iowa City, Iowa

Jean-Pierre Gauvin
Radiation Protection/Occupational Hygiene Service
Montreal Joint Hospital Institute
Montreal, Quebec, Canada

Andrew Geeslin
Office of Emergency Preparedness
Minnesota Department of Health
St. Paul, Minnesota

Justin Gilmore
LIKO Incorporated
Franklin, Massachusetts

Ahmed E. Gomaa
Division of Surveillance Hazard Evaluation and Health Studies
National Institute for Occupational Safety and Health
Centers for Disease Control and Prevention
Cincinnati, Ohio

Richard Gorman
National Institute for Occupational Safety and Health
Centers for Disease Control and Prevention
Cincinnati, Ohio

Robert J. Harrison
California Department of Health Services
Berkeley, California

Patricia A. Heinsohn
San Francisco Department of Public Health
San Francisco, California

Derrick Hodge
FOJP Service Corporation
New York, New York

Anne Hudson
Coos County Public Health Department
North Bend, Oregon

Janine Jagger
International Healthcare Worker Safety Center
University of Virginia Health System
Charlottesville, Virginia

D. L. Jewett
Department of Orthopaedic Surgery
School of Medicine
University of California, San Francisco
San Francisco, California

Laura A. Job
New York State Department of Labor
New York, New York

Kathleen Kahler
Kaiser Foundation Hospital
San Francisco, California

Daniel Kass
The Center for Occupational and Environmental Health
Hunter College
New York, New York

Lindsey V. Kayman
Department of Environmental and Occupational Health Support Services
University of Medicine and Dentistry of New Jersey
Piscataway, New Jersey

Herb B. Kuhn
American Hospital Association
Washington, District of Columbia

J. H. Lange
Enviro-Safe Training and Consulting
Pittsburgh, Pennsylvania

Jaime Lara
Institute of Research for Health and Safety
Montreal, Quebec, Canada

Jacques Lavoie
Institute of Research for Health and Safety
Montreal, Quebec, Canada

Laurence Lee
Pacific Industrial Hygiene
Oakland, California

Martin Lipnowski
Jewish General Hospital
Montreal, Quebec, Canada

Jane Lipscomb
School of Nursing
University of California
San Francisco, California

John D. Lloyd
Patient Safety Center of Inquiry
James A. Haley Veterans' Hospital
Tampa, Florida

Matthew London
Professional Employees Federation
Albany, New York

Sara E. Luckhaupt
Division of Surveillance, Hazard Evaluation, and Field Studies
National Institute for Occupational Safety and Health
Centers for Disease Control and Prevention
Cincinnati, Ohio

Geneviève Marchand
Institute of Research for Health and Safety
Montreal, Quebec, Canada

Willis Martin
Martin Technologies
Rocky Mount, North Carolina

G. Mastrangelo
Department of Environmental Medicine and Public Health
University of Padua
Padua, Italy

Mary Matz
Patient Safety Center of Inquiry
James A. Haley Veterans' Hospital
Tampa, Florida

Donald Maynes
Equitable Safety Solutions
Des Moines, Iowa

Marian McDonald
School of Public Health and Tropical Medicine
Tulane University
New Orleans, Louisiana

Kathleen McPhaul
Work and Health Research Center
University of Maryland School of Nursing
Baltimore, Maryland

John Mehring
Western Region Health and Safety Department
Service Employees International Union
Seattle, Washington

Scott Miller
Biomedical Engineering
San Francisco General Hospital
San Francisco, California

Jules Millogo
Retractable Technologies
Little Elm, Texas

Audrey Nelson
Patient Safety Center of Inquiry
James A. Haley Veterans' Hospital
Tampa, Florida

Mark Nicas
School of Public Health
University of California, Berkeley
Berkeley, California

Olmstead Environmental Services
Garrison, New York

Russell N. Olmsted
Infection Control Service
Saint Joseph Mercy Health System
Ann Arbor, Michigan

Bernice D. Owen
School of Nursing
University of Wisconsin–Madison
Madison, Wisconsin

Jacob D. Paz
J&L Environmental Services, Incorporated
Las Vegas, Nevada

Jolie Pearl
Epidemiology and Prevention Interventions Center
San Francisco General Hospital
San Francisco, California

Jane Perry
International Health Care Worker Safety Center
University of Virginia Health System
Charlottesville, Virginia

Elaine Peters
KPMG Peat Marwick
Chicago, Illinois

Patricia Quinlan
School of Public Health
University of California, Berkeley
Berkeley, California

Arthur R. Reich
Office of Environmental Health and Safety
University of California, San Francisco
San Francisco, California

Bob Reynolds
Biomedical Engineering
San Francisco General Hospital
San Francisco, California

Cynthia F. Robinson
Division of Surveillance, Hazard Evaluation, and Field Studies
National Institute for Occupational Safety and Health
Centers for Disease Control and Prevention
Cincinnati, Ohio

Lawrence Rose
Division of Occupational Safety and Health
California Department of Industrial Relations
Oakland, California

Jonathan Rosen
Occupational Safety & Health Department
New York State Public Employees Federation
Albany, New York

Anthony Schapera
Department of Anesthesiology
San Francisco General Hospital
San Francisco, California

Joseph Schirmer
Service Employees International Union
Washington, District of Columbia

and

American Federation of Labor and Congress of Industrial Organizations
Washington, District of Columbia

Service Employees International Union
Washington, District of Columbia

John P. E. Sestito
Division of Surveillance, Hazard Evaluation, and Field Studies
National Institute for Occupational Safety and Health
Centers for Disease Control and Prevention
Cincinnati, Ohio

Wendy E. Shearn
Kaiser Permanente Medical Center
San Francisco, California

Dennis Shusterman
Division of Occupational and Environmental Medicine
University of California, San Francisco
San Francisco, California

Joyce A. Simonowitz
Division of Occupational Safety and Health
California Department of Industrial Relations
Los Angeles, California

Stackhouse, Incorporated
Riverside, California

David A. Sterling
School of Public Health
Saint Louis University Health Sciences Center
Saint Louis University
Saint Louis, Missouri

Erica J. Stewart
National Environmental Health and Safety
Kaiser Permanente
Oakland, California

John Stewart
National Institute
for Occupational Safety
and Health
Morgantown, West Virginia

Andrew Streifel
Office of Emergency
Preparedness
Minnesota Department of
Health
St. Paul, Minnesota

Patrice M. Sutton
Public Health Institute
Berkeley, California

Byron S. Tepper
Office of Safety and
Environmental Health
Johns Hopkins University
and Johns Hopkins Hospital
Baltimore, Maryland

Wava M. Truscott
Safeskin Corporation
San Diego, California

Jocelyn Villeneuve
Association for Health
and Safety Sector
Social Affairs
Montreal, Quebec, Canada

James T. Walker
Division of Surveillance,
Hazard Evaluation, and
Field Studies
National Institute
for Occupational Safety
and Health
Cincinnati, Ohio

Nina Wallerstein
Department of Family,
Community, and Emergency
Medicine
University of New Mexico
Albuquerque, New Mexico

Margaret Wan
National Environmental Health
and Safety
Kaiser Permanente
Oakland, California

Merri Weinger
Environmental Epidemiology
and Toxicology Section
California Department
of Health Services
Emeryville, California

Wen-Zong Whong
National Institute
for Occupational Safety
and Health
Morgantown, West Virginia

Wayne Wood
Safety Office
McGill University
Montreal, Quebec, Canada

Lee Wugofski
Center for Municipal
Occupational Safety
and Health
San Francisco General Hospital
San Francisco, California

Pier-George Zanoni
Michigan Department
of Community Health
Lansing, Michigan

1

Overview of Health and Safety in the Health Care Environment

David A. Sterling
Saint Louis University

Introduction 1-1
Selected Hazards in the Health Care Profession 1-3
Government and Professional Association Impact on the Health Care Profession 1-4
Governmental Agencies • Hospital Accreditation and the JCAHO • Professional Associations • Unions
Growth of the Health Care Profession 1-6
Conclusions 1-7
References 1-8

Introduction

Health care workers, by the very nature of their work, have both unique and common health hazards in the workplace environment. Many of these hazards are obvious while others are difficult to recognize and include physical risk factors and agents such as communicable diseases, exposure to chemical and biological toxins, carcinogens, ionizing and nonionizing radiation, and ergonomic/human factor hazards; as well as psychological risks, stress-induced disorders, chemical dependency, marital dysfunction, and suicide. Paradoxically, these same workers are responsible for the health care of others. Many of these health problems may also be brought home and be transmitted to their families. Meanwhile, the cost of health insurance for health care practitioners and their employers in the United States is increasing.

Health care facilities themselves represent a situation, where the facility and operational needs to satisfy health care requirements serve as the source of occupational hazards. The health care environment is diverse and includes hospitals of all types, outpatient clinics and other provider services, emergency medical treatment centers (stationary and mobile), dental clinics, pharmacies, testing and research laboratories, patient rehabilitation (physical therapy), as well as veterinary facilities.

Selected examples of health care facility operations that may have a health impact on personnel include operating rooms, patient and treatment areas, sterilization areas, pharmacies, and support laboratory facilities.[37] These areas may have sources of contaminants such as toxic gases and vapors (anesthetic gases—nitrous oxide, ethane, halothane; sterilant gases—ethylene oxide; and antineoplastics); physical hazards (radiation, noise, ergonomic); and infectious microorganisms.[25] Investigations of operating room, sterilization, and other specific exposed personnel have shown symptoms ranging from acute effects of fatigue, headache, and skin irritation, to adverse reproductive and cancer outcomes.[1–13,29,30,35,77]

Maintenance and deterioration of the facility and equipment can expose non-health care staff to toxic[69] and infectious agents.[14] Disposal of hazardous chemicals, radioactive, and infectious waste is a proven potential health hazard to health care personnel, non-health care staff, the community, and the environment.[15,19]

The stress of providing health care also has psychological impacts, which may be associated with increased chemical dependency, lifestyle, and family difficulties.[16] High stress situations may also lead to increased accident rates as well as physiological upsets, which may reduce tolerance to other toxic exposures.[16,31,36] Violence toward health care workers is well documented in psychiatric wards, but is becoming more prevalent under other routine health care settings such as the emergency room, pediatric clinics, medical surgery units, and long-term care facilities.[78]

An aspect, which in the past has not typically been a concern, although it is now often a major consideration, is the building design decisions in new buildings and renovations. Proper venting of work areas such as laboratories,[28,65] operating rooms, and pharmacies, as well as specific equipment, is crucial to the health of the general patient population and the employees. Improper design, installation, operation, or maintenance may be responsible for insufficient removal or spread of toxic and infectious agents. With ongoing energy conservation and cost-cutting measures, building ventilation and maintenance are often the first ones to suffer, but may have one of the greatest consequences.[26,27,65]

Energy conservation in buildings, hospitals included, has concentrated on reducing ventilation. Hospitals, because of special potentially hazardous conditions, are even more prone to problems than other buildings.[25] Any energy conservation strategy must seek to guarantee reasonable air quality in hospitals. Other design aspects of growing concern are properly designed, or ergonomically acceptable, work stations, and appropriate lighting.[22,23]

The issue of health and safety for health care providers has been growing in not only professional concerns, as shown by the abundance of articles in the professional and scientific journals[2–11,14,15,25–27,29–33,35–37] and books,[16–20,28,34] but also by the increasing regulatory requirements.[21]

Yet, despite the ubiquitousness documentation concerning hazards of work in hospitals, hospital workers, especially those considered well educated, have been identified as one of the workforces ranked with the greatest difference between proportion of workers exposed to workplace hazards and the number perceiving risk to themselves and coworkers.[44] Hospital workers, who have been shown to be highly exposed to various hazards, such as radiologic technicians, have low awareness of their own exposure.[45] The diversity and complexity of hospital exposures may be one reason why awareness of hazard is low. Hospital workers may be well informed concerning their specific responsibilities, but may be unaware of other exposures that may be directly related to their work.

Although continuing education and training awareness are requirements for most workers either under the Occupational Safety and Health Administration (OSHA) Hazard Communication Standard (HCS) or other specific standards that various health care facilities must comply with, the level and quality is variable and, unfortunately, is typically ineffective. Programs are most often brief and infrequent and tend to be routine and without hands-on components. Individuals in technically oriented skills are often, by virtue of their training, considered already educated in the recognition of such hazards, and instruction is passive. In certain hospital services, participants may have poor education, be minimally literate, and/or speak English as a second language. The utilization of education and training programs not designed or does not address appropriately the problems to specific worker populations is of questionable value.[46]

The early years of hospital administration in the 1900s centered on ways to attract middle class (paying) patients, keeping the census high and the costs low. To this end, visual appearances were important and led to health and safety improvements primarily in infectious control by improved housekeeping, use of flooring materials in wards and operating rooms, which were easy to clean and, by chance alone, were aseptic. On the other hand, cost-cutting techniques contrary to health and safety were also promoted, such as cleaning and reuse of soiled gauze and reduction in surgical instrument use to reduce cleaning and sterilization costs.[47,61] The cost–benefit approach to hospital management has not changed, particularly in these times of rapidly rising health care costs. Continuing education and training of employees in a manner that would ensure appropriate awareness of hazards and engineering methods of control are examples of numerous areas that have suffered.

Selected Hazards in the Health Care Profession

Health care workers have traditionally been viewed as accepting certain risks as part of patient care. The worker compensation acts, which all states have, were to eliminate the concept of acceptance of risk as part of employment, and because the health care services are thought of as part of the service industry, health care is considered less hazardous than the manufacturing industry. In contrast, the incidence of lost workdays per 100 full-time workers per year for health service workers in 1987 was 66.7, compared with 45.8 for all service workers combined.[24] In addition, the Centers for Disease Control and Prevention (CDC) report that among the 10 highest statistically significant proportionate mortality ratios (PMRs) by occupation, industry and cause of death are malignant lymphomas (PMR of 216 with $p < 0.01$) for males over 20 years of age employed in health diagnosis and treatment occupations; and malignant melanoma of the skin (PMR of 188 with $p < 0.05$) for females over 20 years of age employed in health service occupations.[40]

The highest incidents of injuries are reported in custodial/housekeeping personnel,[67] followed by food services/nutrition employees, nurses, and laboratory technicians, as compared to all job classifications. Needlestick injuries, typically by used needles, were the most prevalent type of injury followed by strains and sprains, over half of which are related to the back from lifting and twisting; lacerations; and contusions. Chemical and biologic exposures are also a large component of reported injuries.[38–49]

Needlestick injuries expose personnel to bloodborne pathogens and other body fluids with the increased potential for such serious infections as human immunodeficiency virus (HIV), hepatitis C virus (HCV), hepatitis B virus (HBV), and others. It is also estimated that there may be underreporting of injuries ranging from 40% to 60%.[39,49,62,66,73]

Housekeeping and custodial personnel are routinely exposed to cleaning solvents and disinfectant agents of industrial strength, as well as refuse that may contain biologically contaminated items such as needles, gauze, and broken glass improperly disposed. Because of the work environment and varied tasks, typical industrial engineering controls are rarely feasible, resulting in regular exposure to dermal irritants as well as inhalation of suspect and known systemic toxins, placing this group at high risk to chemically induced skin diseases and chronic illnesses.

Maintenance workers are also found to have similar exposures as housekeeping and custodial personnel and are at increased risk.[38,39] For example, the simple repair of broken blood pressure machines, which is seldom performed, has been shown to lead to elevated exposure to mercury of maintenance workers and those in the vicinity.

Musculoskeletal injuries, the sprains, and strains rank second among all work-related injuries, with the greatest prevalence, according to the National Institute for Occupational and Safety and Health (NIOSH), among those employed in the health care industry.[63] This is no surprise given the requirements of lifting, pulling, sliding and turning of patients, transfer of patients, moving of equipment, and standing for long hours.

Potential exposures and adverse health outcomes of personnel in the operating rooms have been an area of concern and controversy. Although exposure to anesthetics has been recognized for years, the methods of control can be costly, and exposure to anesthetics cannot be regarded as a problem solved.[34] Percutaneous injuries during surgical procedures occur regularly and increase the risk of infection to surgical personnel.[73] Electrocautery surgery produces aerosols and smoke, which may be mutagenic,[74] as well as blood-containing aerosols,[75] and pose a potential respiratory infection hazard to operating personnel. The electrosurgical devices, as well as other operating room equipment, produce microwave and other nonionizing radiation, which may be of concern, particularly for ocular exposure to surgeons.[70]

Antineoplastics, although of therapeutic value to patients, have mutagenic, carcinogenic, and teratogenic potential. Health care personnel such as pharmacists, physicians, nurses, and others handling these drugs have been shown to be potentially at risk.[71,72,81] Often used laminar flow hoods are more useful in protecting the integrity of the drug than personnel. Antineoplastics have been measured within rooms outside of the hoods where they are handled,[71,80] posing a risk to the users and other occupants in the room.

Elevated levels of noise capable of producing noise-induced hearing loss are common in large hospital facilities. Noise levels over 80 dBA and up to 110 dBA have been measured in food preparation areas, medical laboratories, mechanical and power plant rooms, medical records offices, floor nursing units, print shops, and maintenance areas.[76]

Use of high-technology medical devices for diagnosis and treatment is common. These devices, depending on the conditions of use, have potential health risks associated. For example, the use of magnetic resonance imaging (MRI) as a diagnostic procedure is increasing in use. MRI devices produce magnetic fields, radio frequency radiation, noise, and vibration. Although at this time there is no evidence on adverse health effects of technicians associated with the proper use of MRI devices, it has been suggested that surveillance continue with the continued increased use.[68]

Government and Professional Association Impact on the Health Care Profession

Government regulatory agencies are in a large part responsible for the manner in which health care is managed and priorities are set.[47] Acceptance of federal funds allowed for acquisition and use of state-of-the-art diagnostic equipment, financing for uninsured patient care, and other services. However, also acquired were regulatory standards with which to comply, ranging from burdensome documentation and reporting paperwork to requirements associated with building construction and equipment specifications for hospitals and medical facilities.[48] Selected governmental agencies and standards are discussed below.

Governmental Agencies

Diversity of the health care environment brings it under the compliance requirements of many governmental regulatory agencies. The primary agency responsible for employee health and safety is the OSHA, established in 1970 in the U.S. Department of Labor. OSHA is responsible for the promulgation and enforcement of workplace safety and health standards. Almost all of the OSHA standards are not workplace specific, but apply to all workplaces, except where exempt, to protect the majority of healthy workers. Examples of relevant standards associated with health care include the list of permissible exposure levels (PELs) to gases, vapors, and particles, as well as specific requirements for ethylene oxide, mercury, formaldehyde, noise, bloodborne pathogens, hazard communication, and laboratory standard.

Probably the OSHA standard that has had the greatest impact in all work places has been the OSHA HCS, which came into full effect as of March 1989. Among the requirements, employers are responsible for performing inventory and labeling all hazardous chemicals, maintaining, reviewing, and updating Material Safety Data Sheets of all hazardous chemicals, and training/educating all employees as to the hazards, health effects, and protective measures for all hazardous chemicals they may contact. The requirement alone of educating all employees, when performed appropriately, has brought forth both an increase in awareness that many workers did not have of the hazards associated with their work, as well as a decrease in anxiety over work conditions wrongly perceived as hazardous due to lack of information.

A related standard that also applies to most hospital facilities is the Laboratory Standard, which was specifically designed to deal with the use of small quantities of multiple chemicals and procedures that are not part of a production process. In the hospital environment, this would apply to the medical laboratory diagnostic support facilities, pharmacy, and all chemical research activities.

The Bloodborne Pathogen Standard is the first standard OSHA has promulgated specifically for the protection of health care workers. Compliance with this standard became effective as of March 1992. OSHAs Bloodborne Pathogen requirements are based on the guidelines from the CDC, which were first available in 1982 and consolidated with recommended practices for universal precautions in 1987.[49,50]

Discharges to the environment such as through incineration, wastewater discharge, hazardous and biological waste disposal, and fugitive (uncontrolled) emissions are regulated under standards by the U.S. Environmental Protection Agency (EPA). The Resource, Conservation, and Recovery Act (RCRA) is administered by the EPA and is basically a permit system requiring inventorying and tracking of all hazardous chemicals from the time they are considered waste products through their storage, treatment, and disposal. The OSHA HCS and Laboratory Standard requiring hazardous chemical documentation and worker education, along with the relevant EPA standards such as RCRA, have been instrumental in the recognition of potentially hazardous substances and situations. Regulation is the first step in hazard control.

The CDC is charged with the surveillance and investigation of infectious disease outbreaks in hospitals, and as such, the regulations are mostly in the form of reporting requirements. The CDC is also responsible for preparing recommendations and guidelines for the control of infectious disease.[51] NIOSH is a similar agency within the U.S. Department of Health and Human Services (USDHHS). This agency conducts research and investigates workplace hazards toward the preparation of recommendations and guidelines for their evaluation and control.[52]

The Nuclear Regulatory Commission (NRC) regulates sources and isotopes that produce ionizing radiation. These impact on areas such as diagnostic x-rays, nuclear medicine, and research-utilizing radioactive sources. The Federal Drug Agency (FDA) impacts in the hospital environment through regulation of drug products, medical devices, on food services handling and certain food products related to infectious disease.

State and local governments are typically involved with the enforcement of federal regulations through equivalent state programs, such as a state OSHA, state and local health departments, and other state and local agencies with varying titles in areas such as radiation, infectious disease control, biological and hazardous waste, and food handling. State and local governments may adopt other agency and association guidelines and standards and/or generate their own in regard to regulatory standards and licensing requirements.[52,60]

Hospital Accreditation and the JCAHO

Accreditation is a voluntary procedure and, on its own, it is an overhead few administrators would put forth the time and expense to comply with. With the passage of the Medicare Act in 1965, accreditation by the Joint Commission on Accreditation of Hospitals (JCAH) (known as the Joint Commission on Accreditation of Healthcare Organizations, JCAHO, as of 1987), among others, was promoted as a means of compliance with the Medicare and Medicaid program reimbursement requirements.[53–57] In addition, many state licensing programs accept accreditation by the JCAHO for hospital certification. Insurance companies have also added incentive toward accreditation. Certain insurers require accreditation for reimbursement certification requirements and offer discounts for physician, and other staff, for malpractice, liability, and other hazard insurance.[55]

Requirements for accreditation have been, and still are, however, primarily in support of improving and maintaining the quality of health services that hospitals and other health care organizations provide. Health and safety concerns of the patient are at the forefront, and those of the employees (the health care workers and supportive services personnel) are a secondary concern.

Professional Associations

Professional associations play a large role in the development of recommendations and guidelines, which often are adopted as regulatory requirements and/or accreditation and licensing standards by

TABLE 1.1 Selected Professional Associations with Health Care Worker Health and Safety Interest

American Association of Occupational Health Nurses
American College of Occupational and Environmental Medicine
American College of Physicians
American Conference of Governmental Industrial Hygienists
American Hospital Association
American Industrial Hygiene Association
American Medical Association
American Nursing Association
American Society for Heating, Refrigeration, and Air Conditioning Engineers
American Society for Safety Engineers
American Standards for Testing and Materials
NFPA
NSC

federal, state, and local agencies. For example, sections of the National Fire Protection Association (NFPA) code for safety to life from fire in buildings and structures, and other NFPA codes, have been adopted by OSHA, JCAHO, and HRSA for hospitals. The first OSHA PELs were mostly adopted from the ACGIH 1968 TLVS, and much of the OSHA safety standards sections were adopted from National Safety Council (NSC) recommendations. Table 1.1 contains a short list of selected associations, among many more, which have a relevant impact on hospital and health care worker health and safety.

Unions

Employee unions are not typically given credit for health and safety reform. However, in many cases, they have been a driving force through membership on health and safety committees, through grievances and grievance committees and contract negotiations. The Bloodborne Pathogen Standard was initiated by petitioning of OSHA to issue an emergency temporary standard to protect workers from bloodborne pathogens by several hospital employee unions.[50] Although an emergency standard was never issued, this led ultimately to the revision of the CDC guidelines of 1988 and development of the present OSHA Bloodborne Pathogen Standard. Probably the greatest union activities with the most success have come from working women, who make up approximately 46% of the U.S. work force and 80% of all hospital employees, represented by such unions as the Service Employees International Union and the Coalition of Labor Union Women, which advocates on behalf of women's issues and rights.[44,58]

Growth of the Health Care Profession

The demographics of health care workers are unclear. A 1988 report by the Bureau of Labor Statistics (BLS) found that there were approximately 4.5 million hospital employees out of 8 million health care workers in the United States, about 4% of the total U.S. workforce.[41] Another report from NIOSH only gave the number of health care workers in the United States to be 6.5 million in 1983.[64] More recent data from the BLS using only Standard Industrial Code (SIC) 806 for hospitals (Table 1.2) indicate approximately 3.8 million employees in hospitals as of 1993, where 80% of these individuals are women and 92% of all employees serving in nonsupervisory positions.[42] This may leave out up to 1 million support employees who fall under different SIC numbers. Also, according to these statistics, in 1981, there were only 2.9 million individuals employed in hospitals. This is an increase of almost 1 million people over the past 10 years employed in hospitals alone.

TABLE 1.2 Number of Employees in U.S. Hospitals for Selected Years[a]

Year	Number of Employees (×10³)	Number of Females Employed (×10³)	Nonsupervisory Positions (×10³)	Number in General Med. and Surg. Hospital (×10³)
1991	2904.2	2348.7	2661.9	2856.4[b]
1991	3655.1	2956.1	3352.6	3359.2
1993 (Feb.)	3806.7	3056.2	3492.8	3500.2

Source: BLS. Employment Hours, and Earnings, United States, 1981–93: March 1992 Benchmark Revisions and Historical Corrections. US Department of Labor, Bureau of Labor Statistics, Washington, DC, 1993.
[a] Derived from BLS 1993 Employment, Hours, and Earnings for SIC number 806.[42]
[b] For year 1982.

TABLE 1.3 Number of Hospitals, Employees, and Total Expenses of Selected Years for Member Hospitals of the American Hospital Association[a]

Year	Number of Hospitals	Full-Time Equivalent Employees (×10³)	Expenses (×10⁶)	Number of RNs	Number of LPNs
1950	6,788	1,058	3,651	—	—
1960	6,876	1,598	8,421	—	—
1970	7,123	2,537	25,556	—	—
1980	6,965	3,492	91,886	—	—
1991	6,634	4,165	258,508	925,947	193,507

Source: AHA. American Hospital Association hospital health statistics. *AHA 1991 Annual Survey of Hospitals*, Chicago, IL, 1993.
[a] Derived from AHA 1991 Annual Survey of Member Hospitals.[43]
—, indicates information is not available.

The American Hospital Association (AHA), using statistics only from member hospitals (Table 1.3), estimated approximately 4.2 million hospital employees as of 1991, with 22% as registered nurses and 4.6% as licensed practical nurses. A total expenditure for hospital operations of over 250 billion dollars per year for 1991 was estimated.[43] The data from the AHA over the previous four decades show an increase of almost 1 million additional hospital employees every 10 years since 1950. This corresponds with the statistics from the BLS. The hospital expenditures indicate an increase of 30%–40% each decade. It is expected that the health care workforce in the United States may exceed 10 million by the year 2000.[42]

Conclusions

Health care work and the facilities in which the work is performed pose an adverse health risk to the health care workers and support personnel. These individuals deserve, and should demand, healthy and safe conditions in which to work. Necessarily, because of the type of work performed, flexible controls need to be instituted to achieve and maintain a healthy and safe work environment.

The majority of health and safety controls and management procedures observed in the health care profession and facilities are for the protection of the patient, not the worker. In many instances, the health and safety practices serve to protect both the patient and worker. However, this is not always the case. For example, surgical masks, intended to protect the patient, have been used by the health care workers to protect against airborne aerosols, such as droplet nuclei tuberculosis, for which it is not effective.[84] The CDC and NIOSH have proposed guidelines for more effective respirator use. However, the respirators recommended are not designed for health care worker–patient interaction in terms of size, convenience and looks, and have met resistance.[83] This is one example of the numerous issues, only some of which have been included in this chapter, which must be addressed properly to protect workers in the health care environment.

Other issues of concern that need to be addressed as well include the lack of health and safety education and training within the certificate and degree-granting programs for doctors, nurses, medical laboratory technicians, and others in the health care environment who will be exposed to potential hazards on a daily basis; the impact of the Americans for Disability Act; and health care reform. These issues need to be addressed on a proactive, not reactive, basis.

References

1. NIOSH. *Criteria for a Recommended Standard: Occupational Exposure to Waste Anesthetic Gases and Vapors*. DBHS(NIOSH), Publ. No. 77-140, Cincinnati, OH, 1977.
2. Knill-Jones R, Newman B, and Spence A. Anesthetic practice and pregnancy: Controlled survey of male anesthetists in the United Kingdom. *Lancet* 25:1326–1328, 1975.
3. Knill-Jones R, Rodrogues L, Moir D, and Spence A. Anesthetic practice and pregnancy: Controlled survey of women anesthetists in the United Kingdom. *Lancet* 17:1326–1328, 1972.
4. Layzer R. Myeloneuropathy after prolonged exposure to nitrous oxide. *Lancet* 9:1227–1230, 1978.
5. Haas J and Schottenfeld D. Risk of the offspring from parental occupational exposure. *J Occup Med* 21(9):21–31, 1979.
6. Cohen E, et al. Occupational disease in dentistry and chronic exposure to trace anesthetic gases. *J Am Dent Assoc* 101:21–31, 1980.
7. Tarmenbaum T and Goldbery R. Exposure to anesthetic gases and reproductive outcomes. *J Occup Med* 27(9):659–668, 1985.
8. Choi-Lao A. Trace anesthetic vapors in hospital operating room environments. *J Nurs Res* 30(3): 156–161, 1981.
9. Vainio H. Inhalation anesthetics, anticancer drugs and sterilants as chemical hazards in hospitals. *Scand J Work Environ Health* 8:94–107, 1982.
10. Nguyen T, Theiss J, and Matney T. Exposure of pharmacy personnel to mutagenic antineoplastic drugs. *Cancer Res* 42:4792–4796, 1982.
11. Anderson R, et al. Risk of handling injectable antineoplastic agents. *Am J Hosp Pharm* 39:1881–1887, 1982.
12. USDOL(OSHA). *Guidelines for Cytotoxic (Antineoplastic) Drugs*. OSHA Instruction Publishing, No. 8-1.1, Washington, DC, 1986.
13. *Proceedings of the 26th Interscience Conference on Antimicrobial Agents and Chemotherapy*, Washington, DC, American Society Microbiology, 1986.
14. Goldberg M, et al. Mercury exposure from the repair of blood pressure machines in medical facilities. *Appl Occup Environ Hyg* 5(9):604, 1990.
15. Block S and Netherton J. Infectious hospital wastes: Their treatment and sanitary disposal. In: S. Block (ed.), *Disinfection, Sterilization, and Preservation,* 2nd edn. Lea & Febiger, Philadelphia, PA, 1977.
16. Pelleier K. *Healthy People in Unhealthy Places: Stress and Fitness at Work*. Dell Publishing Co., Inc., New York, 1985.
17. Brune D and Edling C (eds.). *Occupational Hazards in the Health Profession*. CRC Press LLC., Boca Raton, FL, 1989.
18. Charney W and Shirmer J (eds.). *Essentials of Modern Hospital Safety,* Vol. 1. Lewis Publishers, Boca Raton, FL, 1990.

19. Reinhardt P and Gordan J. *Infectious and Medical Waste Management*. Lewis Publishers, Boca Raton, FL, 1990.
20. Lewy R. *Employees at Risk: Protection and Health of the Health Care Worker*. Van Nostrand Reinhold, Florence, KY, 1990.
21. 29CFR 1910. 1450. Occupational exposure to hazardous chemicals in laboratories. DOL(OSHA). *Fed Regis* 55(21):3327–3335, 1990.
22. *2nd International Conference on AIDS*, Paris, France, June 23–25, 1986.
23. *3rd International Conference on AIDS*, Paris, France, June 1–5, 1987.
24. BLS. Occupational injuries and illnesses in US by industry, 1987. USDOL, Bureau of Labor Statistics, Bulletin No. 2328, Washington, DC, 1989.
25. Sterling E and Sterling D. Air quality in hospitals and health care facilities. *Proceedings of the 3rd International Conference on Indoor Air Quality*, Stockholm, Sweden, August 20–24, Vol. 5, pp. 209–213, 1984.
26. Bleckman J, Albrecht R, and Bertz E. Hospital air quality. *Proceedings of the 3rd International Conference on Indoor Air Quality*, Stockholm, Sweden, August 20–24, Vol. 5, pp. 215–220, 1984.
27. Sterling D, Clark C, and Bjornson S. The effects of air control systems on the indoor distribution of viable particles. *Environ Int* 8:559–571, 1982.
28. USDHHS. *Biosafety in Microbiological and Biomedical Laboratories*, 2nd edn. U.S. Department of Health and Human Services (CDC/NIH), Washington, DC, 1988.
29. Rubin R, et al. Neurobehavioural effects of the on-call experience in housestaff physicians. *J Occup Med* 33(1):13–18, 1991.
30. Sarri C, Eng E, and Rungan C. Injuries among medical laboratory housekeeping staff: Incidence and workers perception. *J Occup Med* 33(1):52–56, 1991.
31. Gauch R, Feeney K, and Brown J. Attitudes and behaviors of medical technologists as a result of AIDS. *J Occup Med* 33(10):74–79, 1991.
32. Putz-Anderson V (eds.). *Cumulative Trauma Disorders: A Manual for Musculoskeletal Disease of the Upper Limbs*. Taylor & Francis, Inc., Philadelphia, PA, 1988.
33. Rodgers S (ed.). *Ergonomic Design for People at Work*, Vols. 1 and 2. Van Nostrand Reinhold, New York, 1986.
34. Halsey M. Occupational health and pollution from anesthetics. *Anesthesia* 46:486–488, 1991.
35. Keleher K. Occupational health: How work environment can affect reproductive capacity and outcome. *Nurse Pract J* 16(l):23–34, 1991.
36. Sechrist S and Frazer G. Identification and ranking of stressors in nuclear medicine technology. *J Nucl Med* 18(l):44–48, 1990.
37. Schwartz J, et al. The risk of radiation exposure to laboratory personnel. *J Lab Med* 22(2):114–119, 1991.
38. Weaver V, McDiarmid M, Guidera J, Humphrey F, and Schaefer J. Occupational chemical exposures in an academic medical center. *J Occup Med* 35(7):701–705, 1993.
39. Wilkinson W, Salazar M, Uhl J, Koepsell T, Dekoos R, and Long R. Occupational injuries: A study of health care workers at a Northwestern health science center and teaching hospital. *Am Assoc Occup Health Nurse J* 40(6):287–293, 1992.
40. NCHS. Monthly vital statistics report, 42(4). Final draft from the U.S. Department of Health and Human Services, Centers for Disease Control and Prevention, and National Center for Health Statistics. 1993.
41. BLS. Employment and earnings, 35(3). U.S. Department of Labor, Bureau of Labor Statistics, Office of Employment and Unemployment Statistics, Washington, DC, 1988.
42. BLS. Employment hours, and earnings, United States, 1981–93: March 1992 benchmark revisions and historical corrections. U.S. Department of Labor, Bureau of Labor Statistics, Washington, DC, 1993.

43. AHA. American Hospital Association hospital health statistics. *AHA 1991 Annual Survey of Hospitals*, Chicago, IL, 1993.
44. Sexton P. *The New Nightingales: Hospital Workers, Unions, New Women Issues*. Enquiry Press, New York, 1982.
45. Behrens V and Brackbill R. Worker awareness of exposure: Industry and occupations with low awareness. *Am J Ind Med* 23:695–701, 1993.
46. LaMontagne A, Kelsey K, Ryan C, and Christiani D. A participatory workplace health and safety training program for ethylene oxide. *Am J Ind Med* 22:651–664, 1992.
47. Granshaw L and Porter R (eds.). Managing medicine: Creating a profession of hospital administration in the U.S., 1895–1915. In: M. Vogel (ed.), *The Hospital in History*. Routledge, New York, 1989.
48. HRA. Minimum requirements of construction and equipment for hospitals and medical facilities. U.S. Department of Health and Human Services, Publication No: 79-14500, Hyattsville, MD, 1979.
49. Behling D and Gay J. Industry profile: Hazards of the healthcare profession. *Occup Health Saf* 62(2):54–57, 1993.
50. Goldstein L and Johnson S. OSHA Bloodbome pathogen standard: Implications for the occupational health nurse. *Am Assoc Occup Health Nurse* 39(4):182–188, 1991.
51. CDC. Universal precautions for prevention of transmission of human immunodeficiency virus, hepatitis B virus, and other bloodborne pathogens in health care settings. *Morb Mortal Wkly Rev* 37:377–382, 387–388, 1988.
52. NIOSH. *Guidelines for Protecting the Safety and Health of Health Care Workers*. U.S. Department of Health and Human Services (NIOSH), Publication No. 88–119, 1988.
53. Martin F. *Fifty Years of Medicine and Surgery: An Autobiographical Sketch*. Lakeside Press, Chicago, IL, 338 p, 1934.
54. Roberts J, et al. A history of the Joint Commission on Accreditation of Hospitals. *J Am Med Assoc* 258:936–940, 1987.
55. Yodaiken R and Zeitz P. Accreditation policies in occupational health care. *J Occup Med* 35(6): 562–567, 1993.
56. Lynch JR and Pendergrass JA. Occupational health and safety program accreditation commission. *Am Ind Hyg Assoc J* 7:387–390, 1976.
57. JCAHO. Accreditation manual for hospitals, Vol. 1, Standards. Joint Commission on Accreditation of Healthcare Organizations, 645 N. Michigan Ave, Chicago, IL, 1992.
58. Stellman J, Stellman S, et al. The role of the union health and safety committee in evaluating the health hazards of hospital workers: A case study. *Prev Med* 7(3):332–337, 1978.
59. Hansen KS. Occupational dermatosis in hospital cleaning women. *Contact Dermatitis* 9:343–351, 1983.
60. Monagle J. *Risk Management: A Guide for Healthcare Professionals*. Aspen Systems Corporation, Rockville, MA, 1985.
61. Vogel M. *The Invention of the Modern Hospital: Boston 1870–1930*. The University of Chicago Press, Chicago, IL, 1980.
62. Sellick J, Hazany P, and Mylotte J. Influence of an educational program and mechanical opening needle disposal boxes on occupational needlestick injuries. *Infect Control Epidemiol* 12(12):725–731, 1991.
63. Allen A. On the job injury: A costly problem. *J Post Anesth Nurs* 5(5):367–368, 1993.
64. Martin L. Prevention of occupational transmission of bloodborne pathogens. *Abstract from the American Occupational Health Conference*, April 28–30, Atlanta, GA, p. 434, 1993.
65. Burton J. Choice, location of lab fume hood win effect performance, protection. *Occup Health Saf* 61(9):50, 1992.
66. Hoffman K, Weber D, and Rutala W. Infection control strategies relevant to employee health. *Am Assoc Occup Health Nurse* 39(4):167–181, 1991.

67. Toivanen H, Hehn P, and Hanninien O. Impact of regular relaxation training and psychosocial working factors on neck-shoulder tensions and absenteeism in hospital cleaners. *J Occup Med* 35(11):1123–1130, 1993.
68. Evans J, et al. Infertility and pregnancy outcomes among magnetic resonance imaging workers. *J Occup Med* 35(12):1191–1195, 1993.
69. Goldberg M, et al. Mercury exposure from the repair of blood pressure machines in medical facilities. *Appl Occup Environ Hyg J* 5(9):604, 1990.
70. Paz J, et al. Potential ocular damage from microwave exposure during electrosurgery: Dosimetric survey. *J Occup Med* 29(7):580–583, 1987.
71. McDevitt J, Lees P, and McDiarmid M. Exposure of hospital pharmacists and nurses to antineoplastic agents. *J Occup Med* 35(1):57–60, 1993.
72. McDiarmid M, Garley H, and Arrington D. Pharmaceuticals as hospital hazards: Managing the risk. *J Occup Med* 33(2):155–158, 1991.
73. Tokars J, et al. Percutaneous injuries during surgical procedures. *J Am Med Assoc* 267(21):2899–2904, 1992.
74. Gatti J, et al. The mutagenicity of electrocautery smoke. *Plast Reconstr Surg* 89(5):781–786, 1992.
75. Heinsohn P and Jewett D. Exposure to blood-containing aerosols in the operating room: A preliminary study. *Am Ind Hyg Assoc J* 54(8):446–453, 1993.
76. Yassi A, Gaborieau D, Gillespie I, and Elias J. The noise hazard in a large health care facility. *J Occup Med* 33(10):1067–1070, 1991.
77. Schulte P, et al. Biologic markers in hospital workers exposed to low levels of ethylene oxide. *Mutat Res* 278:237–251, 1992.
78. Lipscomb J and Love C. Violence toward health care workers. *Am Assoc Occup Health Nurs J* 40(5):219–227, 1992.
79. Saurel-Cubizolles M, et al. Neuropsychological symptoms and occupational exposure to anesthetics. *Brit J Ind Med* 49:276–281, 1992.
80. Harlow V. Occupational exposure to antineoplastics. Thesis for Master of Science in Community Health, Emphasis in Industrial Hygiene, Old Dominion University, Norfolk, VA, May, 1990.
81. Balanis B, et al. Antineoplastic drug handling protection after OSHA guidelines: Comparison by profession, handling activity, and work site. *J Occup Med* 34(2):149–155, 1992.
82. Harber P, Hsu P, and Fedoruk M. Personal risk assessment under the Americans with disability act. *J Occup Med* 35(10):1000–1010, 1993.
83. Nelson H. Objections to protective respirators. *Lancet* 340(8827):1088, 1992.
84. Charney W. The inefficiency of surgical masks for protection against droplet nuclei tuberculosis. *J Occup Med* 33(9):943–944, 1991.

2

Epidemiology

James M. Boiano
National Institute for Occupational Safety and Health

John P.E.Sestito
National Institute for Occupational Safety and Health

Sara E. Luckhaupt
National Institute for Occupational Safety and Health

Cynthia F. Robinson
National Institute for Occupational Safety and Health

James T. Walker
National Institute for Occupational Safety and Health

Overview of Workplaces and Occupations in Health Care 2-1
Employment and Forecast • Occupations and Forecast
References 2-9
Demographic Characteristics of Workers in Health Care 2-10
By Workplace • By Occupation
Reference 2-14
Burden of Injury and Illness in Health Care as Documented by Surveillance Systems 2-14
Bureau of Labor Statistics National Surveillance Systems Data • Total Nonfatal Occupational Injuries and Illnesses • Total Nonfatal Occupational Injuries • Nonfatal Occupational Illnesses • Nonfatal Occupational Injuries and Illnesses Involving Days away from Work • Major Injuries/Illnesses, Exposures, and Sources • Special Populations • Fatal Occupational Injuries • Other Key Facts • Reported BLS Cases Underestimate Magnitude of Occupational Injuries and Illnesses • Infectious Disease Data • Disease and Cause-Specific Mortality Data • Data Source Limitations • Summary
References 2-45

Overview of Workplaces and Occupations in Health Care

James M. Boiano and John P.E. Sestito

Health care is part of Health Care and Social Assistance (HCSA) sector, defined by North American Industry Classification System (NAICS) Sector 62.[1] Industries in this sector are arranged on a continuum starting with those establishments providing medical care exclusively, continuing with those providing health care and social assistance, and finally those providing only social assistance. Many of the industries in the sector are defined based on the educational degree held by the practitioners included in the industry. This sector is comprised of four subsectors: ambulatory health care services (621), hospitals (622), nursing and residential care facilities (623), and social assistance (624) and their respective constituent four digit industries; the first three subsectors collectively represent Health Care and are the focus of this section.

Health Care includes both employer (i.e., with paid employees) and nonemployer (i.e., without paid employees or self-employed) establishments, with the exception of hospitals that does not include the latter. Based on the most recent Economic Census figures (2002), there are over 564,000 employer establishments and over 739,000 self-employed establishments in Health Care, representing 8.2% and 4.2%

TABLE 2.1 Number of Employer and Self-Employed Establishments by Health Care Subsector and Industry, 2002

		Number of Establishments	
2002 NAICS	Industry	Employer[a]	Self-Employed[b] (Nonemployer)
621	Ambulatory health care services	489,021	697,239
6211	Physician offices	203,118	171,497
6212	Dental offices	118,305	33,234
6213	Offices of other health practitioners	104,222	284,314
6214	Outpatient care centers	25,750	7,717
6215	Medical and diagnostic laboratories	11,079	16,461
6216	Home health care services	17,666	132,685
6219	Other ambulatory health care services	8,881	51,331
622	Hospitals	6,411	NA[c]
6221	General medical and surgical	5,193	NA
6222	Psychiatric and substance abuse	603	NA
6223	Specialty (other than 6,222)	615	NA
623	Nursing and residential care facilities	69,342	42,571
6231	Nursing care facilities	16,568	—
6232	Residential mental retardation, mental health, and substance abuse facilities	28,508	—
6233	Community care facilities for the elderly	17,988	—
6239	Other residential care facilities	6,278	—
62, except 624	Health care	564,774	739,810
	All industry sectors[d]	6,891,382	17,646,062

Note: Dash (—) indicates that estimates are unavailable.

[a]*Source:* From Economic Census 2002, www.census.gov/econ/census02/data/us/US000_62.HTM. Includes only establishments of firms with paid employees and subject to payroll tax.

[b]*Source:* From Economic Census Nonemployer Statistics 2002, www.census.gov/epcd/nonemployer/2002/us/US000_62. HTM. Includes establishments of firms with no paid employees and not subject to payroll tax (typically self-employed individuals). Each distinct business income tax return filed by a nonemployer business is counted as an establishment. Nonemployer businesses may operate from a home address or a separate physical location.

[c] Not applicable (NA), there are no self-employed establishments in this subsector.

[d] Excludes public administration, i.e., federal, state, and local government agencies.

of all establishments, respectively (Table 2.1). Ambulatory health care services represents over 91% of all establishments in Health Care. Establishments of the self-employed far outnumber employed establishments in home health care services, offices of other health care practitioners, and other ambulatory health services. Offices of physicians, dentists, or other health practitioners represent 75% of employer and 66% of self-employed establishments. Hospitals represent less than 2% of all establishments in Health Care; over 80% of hospitals provide general medical and surgical services.

Employment and Forecast

There are over 14 million private, government, and self-employed workers in Health Care, representing over 10% of all employment (Table 2.2). Ambulatory health care services and hospitals account for over 82% of employed persons. Hospitals account for nearly 40% of employed persons and the largest number of government workers (0.74 million). Over half of the employed persons in ambulatory health care services represent offices of health practitioners (physicians, dentists, etc.).

TABLE 2.2 Number of Private and Government Wage and Salary Workers and Self-Employed Workers by Health Care Subsector and Industry, 2006 (Numbers in Thousands)

			Wage and Salary Workers						
					Government				Self-Employed
2002 NAICS	Industry	Total[a]	Total[a]	Private	Total	Federal	State	Local	Workers
621	Ambulatory health care services[b]	6,133	5,733	5,457	275	22	104	148	393
6211	Physician offices	1,785	1,685	1,661	23	7	11	5	99
6212	Dental offices	852	784	777	7	1	4	2	63
6213	Offices of other health practitioners	553	427	423	4	0	1	3	125
6214	Outpatient care centers	919	894	788	106	7	34	65	25
6216	Home health care services	928	866	801	65	1	34	29	62
6219	Other ambulatory health care services	1,096	1,077	1,007	70	6	20	44	19
622	Hospitals	5,712	5,703	4,963	740	206	330	204	9
623	Nursing and residential care facilities	2,507	2,478	2,304	173	11	85	78	28
6231	Nursing care facilities	1,807	1,798	1,696	101	6	40	55	9
6232, 6233, and 6239	Residential care facilities (without nursing)	700	680	608	72	5	45	23	19
62, except 624	Health care	14,352	13,813	12,724	1,190	240	519	429	429
	All industry sectors (16 years and over)	144,427	133,736	113,347	20,389	3,362	6,099	10,927	10,586

Source: From Current Population Survey, Table 16, Employed and experienced unemployed persons by detailed industry and class of worker, Annual Average 2006, http://www.bls.gov/cps/cpsaat16.pdf

[a] Totals may not add up due to exclusion of unpaid family workers.

[b] Excludes medical and diagnostic laboratories (NAICS 6215), where data is unavailable.

TABLE 2.3 Employment and Forecast by Health Care Subsector and Industry, 2004–2014

		Employment (in Thousands)		
2002 NAICS	Industry	2004	2014	Percent Growth
621	Ambulatory health care services	4,946	7,031	42
6211, 6212, and 6213	Offices of health practitioners	3,337	4,561	37
6216	Home health care services	773	1,310	69
6214, 6215, and 6219	Outpatient, laboratory, and other ambulatory health care services	836	1,160	38
622	Hospitals (private)	4,294	4,982	16
623	Nursing and residential care facilities	2,815	3,597	28
6231	Nursing care facilities	1,575	1,757	11
6232, 6233, and 6239	Residential care facilities	1,240	1,840	48
62, except 624	Health care	12,055	15,610	30
	All industry sectors	145,612	164,540	13

Source: From Berman, J.M., *Mon. Labor Rev.*, 128(11), 45, 2005, http://www.bls.gov/opub/mlr/2005/11/art4full.pdf
Note: Employment data for wage and salary workers are from the BLS establishment-based Current Employment Statistics Survey.

By 2014, Health Care is expected to grow by over 3.5 million jobs (30%), or about one out of every five new jobs (Table 2.3). Ambulatory health care services leads all Health Care subsectors with a projected 42% growth, adding nearly 2.1 million new jobs. Over 1.2 million (59%) of these new jobs will be in offices of health practitioners. Home health care services is leading the growth in this subsector (nearly 70%), and has the distinction of being the nation's fastest growing employer by 2014. Residential care facilities represent the second largest projected growth (48%), adding 0.60 million new jobs by 2014. Although projected growth in hospitals is a modest 16% by 2014, over 0.68 million new jobs will be added, ranking second only to offices of health practitioners. Health Care accounts for 4 of the 20 fastest growing industries (Table 2.4) and 5 of the 20 largest growing industries (Table 2.5).

Occupations and Forecast

Workers in Health Care represent a diverse group of professional, technical, and service occupations. Table 2.6 lists the 10 largest health care occupations by occupational title, as defined by the standard occupational code (SOC), for each of the three subsectors. Registered nurses constitute the largest occupation with nearly 2 million, of which over 70% are employed in hospitals. Other predominant occupations include nursing aides, orderlies, and attendants (1.2 million); licensed practical nurses and licensed vocational nurses (0.6 million); home health aides (0.5 million); medical assistants (0.3 million); and personal and home care aides (0.3 million). Collectively, these five occupational groups represent over 20% of all Health Care workers. Within hospitals, the three largest occupations are related to nursing care (registered nurses; nursing aides, orderlies, and attendants; and licensed practical nurses and licensed vocational nurses), and account for over 2 million jobs in 2005.

Rapid growth is projected for many occupations in Health Care from 2004 to 2014. Thirteen of the 30 fastest growing occupations are in Health Care, 6 of which are in the top 10 (Table 2.7). Home health aides, with an expected growth of 56%, are the nation's fastest growing occupation. Health Care accounts for 5 of the 30 largest growing occupations by 2014 (Table 2.8). The number of registered nurses

TABLE 2.4 Industries with Fastest Growing Employment, 2004–2014

2002 NAICS	Industry	Employment (in Thousands) 2004	Employment (in Thousands) 2014	Percent Growth 2004–2014
6216	**Home health care services**	**773**	**1,310**	**69**
5112	Software publishers	239	400	67
5416	Management, scientific, and technical consulting services	779	1,250	60
6232, 6233, and 6239	**Residential care facilities**	**1,240**	**1,840**	**48**
5612	Facilities support services	116	170	46
5613	Employment services	3,470	5,050	45
7115	Independent artists, writers, and performers	42	61	45
5611	Office administrative services	319	450	41
5415	Computer systems design and related services	1,147	1,600	39
6214, 6215, and 6219	**Outpatient, laboratory, and other ambulatory health care services**	**836**	**1,160**	**38**
6244	Child day care services	767	1,062	38
6114–17	Other educational services	475	650	37
6211, 6212, and 6213	Offices of health practitioners	3,337	4,561	37
5412	Accounting, tax preparation, bookkeeping, and payroll services	816	1,100	35
6112 and 6113	**Junior colleges, colleges, universities, and professional schools**	**1,462**	**1,965**	**34**
6241, 6242, and 6243	Individual, family, community, and vocational rehabilitation services	1,365	1,810	33
487	Scenic and sightseeing transportation	27	35	30
5622 and 5629	Waste treatment and disposal and waste management services	206	268	30
5419	Other professional, scientific, and technical services	503	646	28
5414	Specialized design services	121	155	28

Source: From Berman, J.M., *Mon. Labor Rev.*, 128(11), 45, 2005, http://www.bls.gov/opub/mlr/2005/11/art4full.pdf
Note: Bold indicates industries in Health Care.

is expected to grow by 0.7 million, the second largest increase across all industry sectors and ranking first among Health Care occupations. The number of home health aides is expected to grow by 0.35 million, the ninth largest increase across all industries and representing the second largest growing occupation in Health Care.

Combined establishment and employment figures yield the following salient facts[2] about Health Care:

- Hospitals account for less than 2% of the health care establishments, but employ 40% of all health care workers.
- More than 70% of hospital employees are in establishments with 1000 or more workers.
- Over 85% of nonhospital health care establishments employ fewer than 20 workers, and about 50% employ 5 or fewer workers.
- Nearly 70% of nonhospital employees are employed in establishments with 20 or more workers.

TABLE 2.5 Industries with Largest Employment Growth, 2004–2014

2002 NAICS	Industry	Employment (in Thousands) 2004	Employment (in Thousands) 2014	Change 2004–2014
44 and 45	Retail trade	15,034	16,683	1,649
5613	Employment services	3,470	5,050	1,580
722	Food services and drinking places	8,850	10,301	1,451
6211, 6212, and 6213	**Offices of health practitioners**	**3,337**	**4,561**	**1,224**
23	Construction	6,965	7,757	792
NA	Local government educational services	7,762	8,545	783
622	**Hospitals (private)**	**4,294**	**4,982**	**688**
6232, 6233, and 6239	**Residential care facilities**	**1,240**	**1,840**	**600**
6216	**Home health care services**	**773**	**1,310**	**537**
6112 and 6113	Junior colleges, colleges, universities, and professional schools	1,462	1,965	503
NA	Local government enterprises except passenger transit	4,216	4,699	483
42	Wholesale trade	5655	6131	476
5416	Management, scientific, and technical consulting services	779	1,250	471
5415	Computer systems design and related services	1,147	1,600	453
6241, 6242, and 6243	Individual, family, community, and vocational rehabilitation services	1,365	1,810	445
NA	State government educational services	2,249	2,691	442
713	Amusement, gambling, and recreation industries	1,351	1,710	359
5617	Services to buildings and dwellings	1,694	2,050	356
6214, 6215, and 6219	**Outpatient, laboratory, and other ambulatory health care services**	**836**	**1,160**	**324**
721	Accommodation	1,796	2,100	304

Source: From Berman, J.M., *Mon. Labor Rev.*, 128(11), 45, 2005, http://www.bls.gov/opub/mlr/2005/11/art4full.pdf
Note: Bold indicates industries in Health Care.

TABLE 2.6 Ten Largest Occupations by Health Care Subsector, 2005

SOC[a]	Occupation	Total Employed[b]
Ambulatory Health Care Services (NAICS 621)		
29-1111	Registered nurses	408,180
31-9092	Medical assistants	306,010
43-4171	Receptionists and information clerks	282,980
43-6013	Medical secretaries	263,710
31-9091	Dental assistants	259,810
31-1011	Home health aides	232,500
39-9021	Personal and home care aides	203,150
43-9061	Office clerks (general)	184,520
29-2061	Licensed practical nurses and licensed vocational nurses	164,420
29-2021	Dental hygienists	157,150
Hospitals (NAICS 622)		
29-1111	Registered nurses	1,424,860
31-1012	Nursing aides, orderlies, and attendants	403,500
29-2061	Licensed practical nurses and licensed vocational nurses	187,420
43-9061	Office clerks (general)	123,420
37-2012	Maids and housekeeping cleaners	121,850
29-2034	Radiological technologists and technicians	110,710
29-2070	Medical records and health information technicians	97,270
43-6013	Medical secretaries	93,450
11-9111	Medical and health services managers	92,650
43-4111	Interviewers (except eligibility and loan)	81,820
Nursing and Residential Care Facilities (NAICS 623)		
31-1012	Nursing aides, orderlies, and attendants	759,650
31-1011	Home health aides	254,340
29-2061	Licensed practical nurses and licensed vocational nurses	234,090
29-1111	Registered nurses	157,870
37-2012	Maids and housekeeping cleaners	116,590
39-9021	Personal and home care aides	102,180
35-2012	Cooks (institutional and cafeteria)	78,130
35-2021	Food preparation workers	68,570
39-9011	Child care workers	50,500
39-9032	Recreation workers	47,970

Note: Estimates do not include self-employed workers.

[a] *Source:* From 2000 SOC, BLS.

[b] *Source:* From National industry-specific occupational employment and wage estimates, May 2005: http://www.bls.gov/oes/current/naics3_621000.htm, http://www.bls.gov/oes/current/naics3_622000.htm, and http://www.bls.gov/oes/current/naics3_623000.htm

TABLE 2.7 Thirty Fastest-Growing Occupations, All Industry Sectors, 2004–2014

SOC	Occupation	Employment (in Thousands) 2004	Employment (in Thousands) 2014	Percent Growth
31-1011	**Home health aides**	624	974	**56.0**
15-1081	Network systems and data communications analysts	231	357	54.6
31-9092	**Medical assistants**	387	589	**52.1**
29-1071	**Physician assistants**	62	93	**49.6**
15-1031	Computer software engineers (applications)	460	682	48.4
31-2021	**Physical therapist assistants**	59	85	**44.2**
29-2021	**Dental hygienists**	158	226	**43.3**
15-1032	Computer software engineers (systems software)	340	486	43.0
31-9091	**Dental assistants**	267	382	**42.7**
39-9021	Personal and home care aides	701	988	41.0
15-1071	Network and computer systems administrators	278	385	38.4
15-1061	Database administrators	104	144	38.2
29-1123	**Physical therapists**	155	211	**36.7**
19-4092	Forensic science technicians[a]	10	13	36.4
29-2056	**Veterinary technologists and technicians**	60	81	**35.3**
29-2032	**Diagnostic medical sonographers**	42	57	**34.8**
31-2022	**Physical therapist aides**	43	57	**34.4**
31-2011	**Occupational therapist assistants**	21	29	**34.1**
19-1042	Medical scientists (except epidemiologists)[a]	72	97	34.1
29-1122	**Occupational therapists**	92	123	**33.6**
25-2011	Preschool teachers (except special education)	431	573	33.1
29-2031	**Cardiovascular technologists and technicians**	45	60	**32.6**
25-1000	Postsecondary teachers	1628	2153	32.2
19-2043	Hydrologists	8	11	31.6
15-1051	Computer systems analysts	487	640	31.4
47-4041	Hazardous materials removal workers	38	50	31.2
17-2031	Biomedical engineers[a]	10	13	30.7
13-1071	Employment, recruitment, and placement specialists	182	237	30.5
7-2081	Environmental engineers	49	64	30.0
23-2011	Paralegals and legal assistants	224	291	29.7

Source: From Hecker, D.E., *Mon. Labor Rev.*, 128(11), 70, 2005, http://www.bls.gov/opub/mlr/2005/11/art5full.pdf

Notes: Employment data for wage and salary workers are from the BLS establishment-based Current Employment Statistics Survey. Bold indicates health care occupations as defined by two SOC major groups: Health Care Practitioner and Technical Occupations (29-0000) and Health Care Support Occupations (31-0000).

[a] Health-care-related occupations.

TABLE 2.8 Thirty Largest-Growing Occupations, All Industry Sectors, 2004–2014

SOC	Occupation	Employment (in Thousands) 2004	Employment (in Thousands) 2014	Change 2004–2014
41-2031	Retail salespersons	4256	4992	736
29-1111	**Registered nurses**	**2394**	**3096**	**703**
25-1000	Postsecondary teachers	1628	2153	524
43-4051	Customer service representatives	2063	2534	471
37-2011	Janitors and cleaners (except maids and housekeepers)	2374	2813	440
35-3031	Waiters and waitresses	2252	2627	376
35-3021	Combined food preparation and serving workers	2150	2516	367
31-1011	**Home health aides**	**624**	**974**	**350**
31-1012	**Nursing aides, orderlies, and attendants**	**1455**	**1781**	**325**
11-1021	General and operations managers	1807	2115	308
39-9021	Personal and home care aides	701	988	287
25-2021	Elementary school teachers (except special education)	1457	1722	265
13-2011	Accountants and auditors	1176	1440	264
43-9061	Office clerks (general)	3138	3401	263
53-7062	Laborers and freight, stock, and material movers (hand)	2430	2678	248
43-4171	Receptionists and information clerks	1133	1379	246
37-3011	Landscaping and groundskeeping workers	1177	1407	230
53-3032	Truck drivers (heavy and tractor-trailer)	1738	1962	223
15-1031	Computer software engineers (applications)	460	682	222
49-9042	Maintenance and repair workers (general)	1332	1533	202
31-9092	**Medical Assistants**	**387**	**589**	**202**
43-6011	Executive secretaries and administrative assistants	1547	1739	192
41-4012	Sales representatives	1454	1641	187
47-2031	Carpenters	1349	1535	186
25-9041	Teachers' assistants	1296	1478	183
39-9011	Child care workers	1280	1456	176
35-2021	Food preparation workers	889	1064	175
37-2012	Maids and housekeeping cleaners	1422	1587	165
53-3033	Truck drivers (light or delivery services)	1042	1206	164
15-1051	Computer systems analysts	487	640	153

Source: From Hecker, D.E., *Mon. Labor Rev.*, 128(11), 70, 2005, http://www.bls.gov/opub/mlr/2005/11/art5full.pdf

Notes: Employment data for wage and salary workers are from the BLS establishment-based Current Employment Statistics Survey. Bold indicates health care occupations as defined by two SOC major groups: Health Care Practitioner and Technical Occupations (29-0000) and Health Care Support Occupations (31-0000).

References

1. North American Industry Classification System. Executive Office of the President, Office of Management and Budget, United States, 2002.
2. Career Guide to Industries. Health care. www.bls.gov/oco/cg/cgs035.htm
3. Berman JM. Industry output and employment projections to 2014. *Mon Labor Rev* 128(11):45–69, November 2005. http://www.bls.gov/opub/mlr/2005/11/art4full.pdf
4. Hecker DE. Occupational employment projections to 2014. *Mon Labor Rev* 128(11):70–101, 2005. http://www.bls.gov/opub/mlr/2005/11/art5full.pdf

Demographic Characteristics of Workers in Health Care

James M. Boiano

By Workplace

Selected demographic characteristics of employed persons in Health Care are provided by subsector, constituent industries (where available) and for all industries, for comparison purposes (Tables 2.9 and 2.10). These characteristics include gender, race and ethnicity, and age.

About 78% (11.2 million) of the workers in Health Care are women, more than any other industry sector and about 70% higher than the average for all industries (Table 2.9). Among industries (four digit codes), home health care employs the greatest percentage of women (90%), followed by nursing care facilities (85%) and dental offices (80%). The lowest percentage of women are employed by other ambulatory health care services and offices of other health practitioners (68.6% and 69.3%, respectively).

Available data on sex, race, and ethnicity show that there is a greater percentage of Blacks and Asians in Health Care when compared to all industries, while the percentage of Hispanics is somewhat less (Table 2.9). Blacks represent about 16% (2.3 million) of the workers in Health Care, leading all industry sectors, and 1.5 times the industry average of nearly 11%. Hospitals employ the largest number of Blacks in Health Care (0.94 million), followed by ambulatory health care services (0.70 million). Home health

TABLE 2.9 Percent Distribution of Employment within Health Care by Sex, Race, and Hispanic or Latino Ethnicity, 2006

			Percent of Total			
2002 NAICS Code[a]	Industry	Total Employed[1] (in Thousands)	Women	Black or African American	Hispanic or Latino	Asian
621	Ambulatory health care services[b]	6,133	77.2	11.5	9.9	6.6
6211	Physician offices	1,785	76.4	6.9	8.7	6.5
6212	Dental offices	852	79.7	3.6	9.6	6.3
6213	Offices of other health practitioners	553	69.3	3.2	5.4	4.6
6214	Outpatient care centers	919	78.4	12.2	11.0	5.4
6216	Home health care services	928	90.2	27.0	16.7	4.1
6219	Other ambulatory health care services	1,096	68.6	15.8	8.0	6.9
622	Hospitals	5,712	76.6	16.4	7.6	7.0
623	Nursing and residential care facilities	2,507	82.0	24.6	8.6	4.1
6231	Nursing care facilities	1,807	85.5	26.7	8.3	4.5
6232, 6233, and 6239	Residential care facilities (without nursing)	700	73.0	19.1	9.3	3.0
62, except 624	Health care	14,352	77.8	15.7	8.7	6.3
	All industry sectors (16 years and over)	144,427	46.3	10.9	13.6	4.5

Source: From Current Population Survey, Household Data Annual Averages, Table 18 Employed persons by detailed industry, sex, race, and Hispanic or Latino ethnicity, 2006, ftp://ftp.bls.gov/pub/special.requests/lf/aat18.txt

[a] NAICS, U.S. Census Bureau, www.census.gov/epcd/www/naics.html

[b] Excludes medical and diagnostic laboratories (NAICS 6215); data are unavailable.

TABLE 2.10 Percent Distribution of Employment in Health Care by Age, 2006

			Percent Distribution of Employment by Age Group						
2002 NAICS Code[a]	Industry	Total Employed[b] (in Thousands)	16–19	20–24	25–34	35–44	45–54	55–64	65 and Over
621	Ambulatory health care services[a]	6,133	1.5	6.9	21.9	25.1	26.1	14.6	3.9
6211	Physician offices	1,785	1.2	6.0	20.9	25.2	27.8	14.3	4.5
6212	Dental offices	852	2.8	8.8	21.4	24.5	24.5	13.8	4.0
6213	Offices of other health practitioners	553	1.9	7.3	21.1	22.4	25.0	16.9	5.4
6214	Outpatient care centers	919	1.3	7.4	23.4	22.1	27.5	15.5	2.7
6216	Home health care services	928	1.2	5.6	18.9	28.5	25.3	15.9	4.5
6219	Other ambulatory health care services	1,096	1.2	7.3	25.4	26.1	24.9	12.5	2.3
622	Hospitals	5,712	0.8	7.0	21.4	25.6	27.7	14.9	2.5
623	Nursing and residential care facilities	2,507	4.5	10.0	20.0	21.9	24.5	15.0	4.0
6231	Nursing care facilities	1,807	4.7	9.5	20.1	22.3	24.9	14.6	3.8
6232, 6233, and 6239	Residential care facilities (without nursing)	700	4.1	11.2	20.0	20.6	23.6	15.7	4.8
62, except 624	Health care	14,352	1.7	7.5	21.4	24.7	26.4	14.8	3.4
	All industry sectors (16 years and over)	144,427	4.2	9.6	21.5	23.9	23.6	13.4	3.7

Source: From Current Population Survey, NIOSH IST Data Mart Query, Query 1: column variable, age; row variable, NORA sector (primary job) and Query 2: column variable, age; row variable, industry code (primary job), http://isx-morg1/dsr/IST/ISTDM2/cps/cpsestimatesresults2.aspx

Note: Values may not sum to the totals due to rounding.

[a] NAICS, U.S. Census Bureau, www.census.gov/epcd/www/naics.html

[b] Excludes medical and diagnostic laboratories (NAICS 6215); data are unavailable.

care services (27%) employ the largest percentages of any industry group within Health Care, while dental offices (3.6%) and offices of other health care practitioners (3.2%) employ the smallest percentages.

Hispanics represent about 8.7% (1.2 million) of the workforce in Health Care (Table 2.9). Ambulatory health care services employ the largest number of Hispanics of any Health Care subsector (0.61 million), followed by hospitals (0.43 million). Home health care services (16.7%) and outpatient care centers (11.0%) employ the greatest percentages of any Health Care industry group.

Asians represent over 6.3% of the workforce in Health Care, with an employment of 0.90 million (Table 2.9). Hospitals employ the greatest percentage of Asians of any Health Care subsector (7.0%); ambulatory health care services employ the largest number (0.4 million) slightly more than in hospitals.

Percent distribution of employed persons in Health Care by age is provided for each of the three Health Care subsectors and four digit constituent industries, where available (Table 2.10). When compared to all industries, percents within each of the 7 age groups were very similar, with exception of the 16–19 age group where the industry average was about 2.5 times that in Health Care. Among the three Health Care subsectors, the highest percent employed within the 16–19, 20–24, 55–64, and 65 and over age groups were for nursing and residential care facilities; in the 35–44 and 45–54 age groups, hospitals employed the largest percent; and in the 25–34 age group, ambulatory health care services employed the largest percent. When focusing on industries (four digit codes), nursing care facilities and residential care facilities without nursing had the highest percent distributions for young workers (16–19 and 20–24 age groups), while offices of other health care practitioners had the highest percent distributions for older workers (55–64 and 65 and over).

By Occupation

Table 2.11 presents sex, race, and ethnicity demographics of the top 20 Health Care occupations with the largest number of workers. For 16 of these occupations, the percents of women are higher than the industry average. Dental hygienists represent the highest concentration of women (98.6%), whereas dentists represent the lowest (22.6%). Registered nurses account for the largest number of women of any health care occupation (2.5 million).

TABLE 2.11 Percent Distribution of Women, Blacks, Hispanics, and Asians in 20 Largest Occupations in Health Care, 2006

		Percent of Total			
Occupation	Total Employed (in Thousands)	Women	Black or African American	Hispanic or Latino	Asian
Registered nurses	2,529	91.3	10.9	4.2	7.5
Nursing, psychiatric, and home health aides	1,906	88.9	34.8	13.1	4.0
Physicians and surgeons	863	32.2	5.2	5.7	17.0
Personal and home care aides	703	87.3	22.4	14.9	5.8
Licensed practical nurses and licensed vocational nurses	556	94.2	23.2	7.0	3.1
Health diagnosing and treating practitioner support technicians	425	80.1	11.8	8.2	5.6
Clinical laboratory technologists and technicians	321	78.1	14.2	7.8	9.6
Diagnostic related technologists and technicians	281	72.9	7.5	6.3	2.9
Dental assistants	274	95.4	5.4	14.9	4.2
Pharmacists	245	48.9	6.0	5.6	19.5
Physical therapists	198	62.7	5.8	5.0	13.7
Dentists	196	22.6	3.1	4.3	11.4
Emergency medical technicians and paramedics	156	31.9	11.9	7.4	2.2
Dental hygienists	144	98.6	1.4	4.6	4.2
Speech language pathologists	114	95.3	8.1	3.6	1.4
Medical records and health information technicians	98	92.0	20.5	15.1	1.4
Dieticians and nutritionists	96	91.0	21.2	4.6	7.6
Respiratory therapists	85	66.0	15.3	6.2	4.6
Occupational therapists	78	90.3	3.1	2.0	4.7
Chiropractors	69	23.1	3.3	1.9	1.8
Total (16 years and older)	144,427	46.3	10.9	13.6	4.5

Source: From Current Population Survey (2006), Household Data Annual Averages, Table 11, Employed persons by detailed occupation, sex, race, and Hispanic or Latino ethnicity, http://www.bls.gov/cps/cpsaat11.pdf

Note: Data for occupations with fewer than 50,000 employed are not published.

For half of the listed occupations, the percents of Blacks meet or exceed the industry average of 10.9% (Table 2.11). Nursing, psychiatric, and home health care aides represent the occupational group with the highest percentage (34.8%); dental hygienists comprise the lowest percentage (1.4%). Nursing, psychiatric, and home health aides account for the largest number of Blacks of any health care occupation (0.66 million).

The percents of Hispanics are higher than the industry average (13.6%) for only 3 of the 20 occupations, with medical records and health information technicians representing the highest concentration (15.1%). By comparison, chiropractors represent the lowest percentage (1.9%). Nursing, psychiatric, and home health aides account for the largest number of Hispanics of any health care occupation (0.25 million).

The percents of Asians are higher than and exceed the industry average of 4.5% for 11 of the top 20 occupations. Pharmacists represent the highest concentration of individuals of Asian decent (19.5%); medical records and health information technicians and speech language pathologists represent the lowest (1.4%). Registered nurses account for the largest number of Asians of any health care occupation (0.19 million).

Very little data are available on age distribution for most of the health care occupations listed in Table 2.11, with the exception of registered nurses. Findings from a 2004 national sample survey[1] show continual movement to more registered nurses in older age groups and a general decline in the numbers of registered nurses in younger age groups (Figure 2.1). Based on data from seven quadrennial surveys conducted from 1980 to 2004, the average age of the registered nurse population continued to climb, increasing to 46.8 years of age in 2004 compared to 44.3 years in 1996. The largest age group of registered nurses in 1980 was 25–29 years of age, 35–39 years in 1992, 40–44 years in 2000, and 45–49 years in 2004. By contrast, the numbers of registered nurses in the two youngest age groups (less than 25 and 25–29 years of age) continued to decline over this 24 year period.

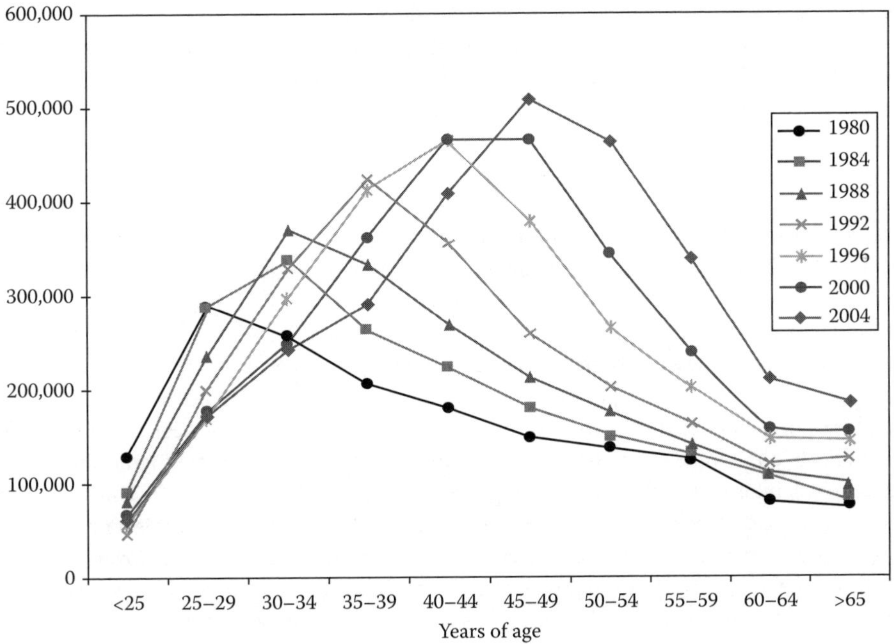

Note: The total numbers in each survey, across age ages, may not equal the estimated total of all registered nurses due to incomplete information provided by respondents. Only those who provided age information are included in the calculations used for this chart.

FIGURE 2.1 Age distribution of registered nurse population, 1980–2004.

Reference

1. The Registered Nurse Population: Findings from the 2004 National Sample of Registered Nurses. Health Resources and Services Administration, U.S. Department of Health and Human Services. http://bhpr.hrsa.gov/healthworkforce/rnsurvey04/2.htm

Burden of Injury and Illness in Health Care as Documented by Surveillance Systems

James M. Boiano, John P.E. Sestito, Sara E. Luckhaupt, Cynthia F. Robinson, and James T. Walker

This section presents an overview illustrated by charts and tables that describe the national magnitude and trends of occupational injuries, illnesses, and fatalities in the HCSA sector,* as documented by various federal and state-based surveillance systems. The data focus on the three Health Care subsectors (i.e., ambulatory health care services, hospitals, and nursing and residential care facilities), although data for social assistance are also provided, where available. Data are presented for predominant health and safety hazards and issues including sprains and strains, overexertion/repetitive trauma, falls on same level, assaults and violent acts, sharps injuries (SIs), and infectious disease. Significantly elevated causes of death for each of the three Health Care subsectors and for the largest health care occupations are also provided.

Bureau of Labor Statistics National Surveillance Systems Data

National statistics on occupational injuries, illnesses, and fatalities are compiled by the Bureau of Labor Statistics (BLS) in conjunction with participating state agencies. National estimates of the numbers and rates of illnesses and injuries are compiled from the annual Survey of Occupational Injuries and Illnesses (SOII), which is based solely on private industry employer's OSHA logs.[1] The SOII exclude self-employed persons, public sector workers, and workers employed on small farms, representing 22% of the U.S. workforce, and thus may underestimate the true prevalence of injuries, illnesses, and fatalities. Numbers of cases and incidence rates are reported by year with 2005 being the most recent year data are presented.

For each recordable case, employer's are required to complete OSHA Form 300 (Log of Work-Related Injuries and Illnesses), OSHA Form 301 (Injuries and Illnesses Incident Report), and OSHA Form 300A (Summary of Work-Related Injuries and Illnesses).[2] Collectively, these forms are used to develop a picture of the extent and severity of work-related incidents. In Form 300, employers must record information about every work-related death and about every significant work-related injury or illness that involves loss of consciousness, restricted work activity or job transfer, days away from work, or medical treatment beyond first aid. Employers must also classify the case as either an injury or one of the following types of illness: skin disorder, respiratory condition, poisoning, hearing loss, or all other illnesses. In Form 301, employers must record information about the employee (name, address, date of birth, gender, etc.), information about the treating health care professional, and information about the case (date and time of incident, job activity at time of incident, part of body affected, object or substance that harmed the employee, and date of death if employee died). In Form 301A, employers must provide yearly totals for the following: deaths, total cases with days away from work, total cases with job transfer or restriction, and total other recordable cases—total days away from work, total days of job transfer or restriction, injuries, or one of the five aforementioned illness categories. Establishment information including name, industry code, annual average number of employees, and total hours worked by all employees in the last year is also required.

* 2002 NAICS code 62 (see Chapter 1 for more information on NAICS codes for HCSA sector).

The circumstances of each case are classified based on the BLS Occupational Injury and Illness Classification Manual.[3] The survey uses four case characteristics to describe each incident that led to an injury or illness that led to one or more days away from work. These characteristics include the following:

- *Nature*: The physical characteristics of the disabling injury or illness, such as lacerations, fractures, or sprains/strains
- *Part of body affected*: Part of body directly linked to the nature of the reported injury or illness, such as back, finger, or eye
- *Event or exposure*: The manner in which the injury or illness was produced or inflicted by the source, such as falls, overexertion, or repetitive motion
- *Source*: The object, substance, exposure, or bodily motion that directly produced or inflicted the disabling condition, such as chemicals, vehicles, or machinery

In this section, numbers of cases and incidence rates are reported by year, with 2005 being the most recent year where data are presented. These estimates are provided for the HCSA sector as a whole, by three digit subsector and four digit industry (where available), with comparisons to all private industries and service-providing industries.

National statistics on fatal occupational injuries are from another BLS surveillance system called the Census of Fatal Occupational Injuries (CFOI). Unlike the SOII, the CFOI is considered a complete census that uses multiple data sources for tracking traumatic workplace fatalities resulting from intentional and unintentional injuries.[4]

Total Nonfatal Occupational Injuries and Illnesses

Of the 4.2 million nonfatal occupational injuries and illnesses reported by private industry employers in 2005, the HCSA sector represents the second largest share of injuries and illnesses (668,000 or 15.9% of total recordable cases)[3] (Figure 2.2). In fact, three of the four HCSA subsectors—hospitals, nursing and residential care facilities, and ambulatory health care services—are ranked 1st, 2nd, and 12th, respectively, and are among the 14 industries with 100,000 or more injuries and illnesses in 2005 (Table 2.12). Hospitals have led this group for the past 3 years, ever since NAICS-based tabulations began in 2003.[5]

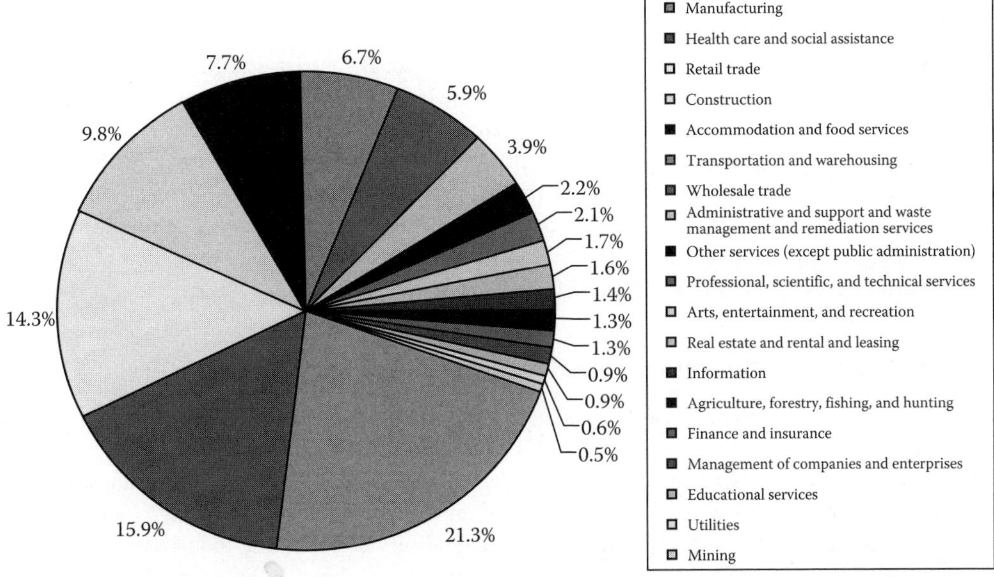

FIGURE 2.2 Percent distribution of nonfatal occupational injuries and illnesses by industry sector, 2005.

TABLE 2.12 Number of Cases and Incidence Rate of Nonfatal Occupational Injuries and Illnesses for Industries with 100,000 or More Cases, 2005

2002 NAICS Code[a]	Industry[b]	Total Cases (in Thousands)	Incidence Rate[c]
622	**Hospitals**	**281.51**	**8.1**
623	**Nursing and residential care facilities**	**209.1**	**9.1**
452	General merchandise stores	147.2	6.7
336	Transportation equipment manufacturing	146.8	8.3
561	Administrative and support services	141.1	3.4
332	Fabricated metal product manufacturing	121.8	8.0
423	Merchant wholesalers (durable goods)	119.5	4.1
2382	Building equipment contractors	117.8	6.7
311	Food manufacturing	114.2	7.7
7221	Full-service restaurants	111.7	3.9
44511	Supermarkets and other grocery (except convenience) stores	110.7	6.5
621	**Ambulatory health care services**	**110.6**	**2.8**
424	Merchant wholesalers (durable goods)	110.0	5.7
7222	Limited-service eating places	103.3	4.1
	Total (private industry[d])	4,214.2	4.6

Source: From Bureau of Labor Statistics, http://www.bls.gov/news.release/osh.t04.htm
Note: Bold indicates health care industry.
[a] NAICS, U.S. Census Bureau, www.census.gov/epcd/www/naics.html
[b] Totals include data for industries not shown separately.
[c] The incidence rates represent the number of injuries and illnesses per 100 full-time workers.
[d] Excludes farms with fewer than 11 employees.

Occupational illnesses account for only 7% of all total reportable injury and illness cases in the HCSA sector and was not different from private industry as a whole. Compared to injuries, illnesses are often difficult to relate to the workplace and more likely to be underreported due to the fact that many work-related diseases are associated with long latency periods (e.g., cancers, chronic respiratory ailments, etc.). The issue of underreporting is discussed later in section.

Figure 2.3 compares incidence rates of nonfatal occupational injuries and illnesses for the HCSA sector and by subsector for 2003–2005 to those for private industry and to service-providing industries. In 2005, the incidence rate of injuries and illnesses in the HCSA sector was 5.9 cases per 100 full-time workers, nearly 1.3 and 1.4 times higher than in private and service-providing industries, respectively. Injury and illness incidence rates in the sector were driven by nursing and residential facilities and hospitals. These rates declined for all HCSA subsectors, as well as in private and service-providing industries, for each year since 2003 (when NAICS-based tabulations began), with the exception of the rate in the social assistance subsector that increased from 2004 to 2005.

Total Nonfatal Occupational Injuries

In 2005, the overall incidence rate of nonfatal occupational injuries for the HCSA sector was 5.5 cases per 100 full-time workers, compared to 4.4 and 3.9 cases per 100 full-time workers in the private and service-providing industries, respectively (Figure 2.4). The number of nonfatal injuries for this sector (624,000) accounted for 15.7% of the total number of injury cases in private industry.[6] Incidence rates declined for all HCSA subsectors, as well as in private and service-providing industries, for each year since 2003, with the exception of the rate in the social assistance subsector that increased from 2004 to

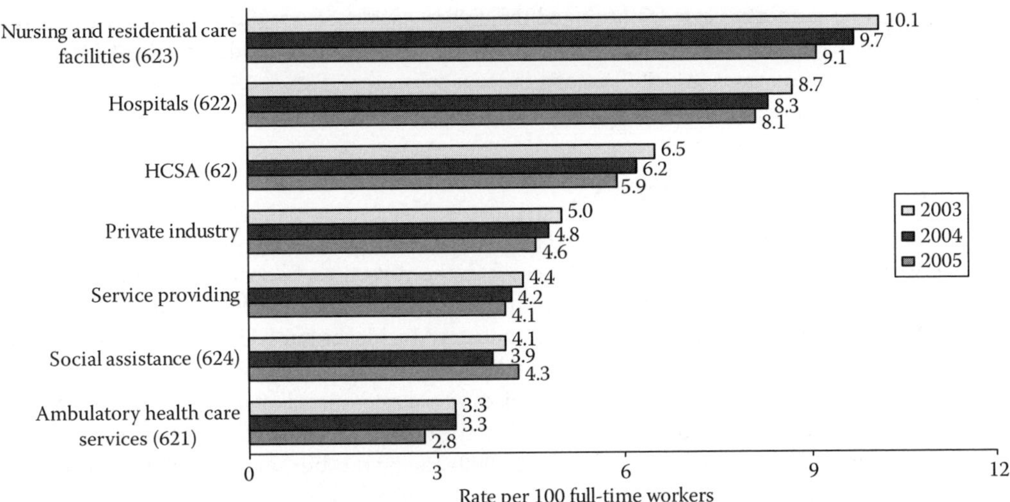

Note: NAICS codes in parentheses follow industry sector name.

FIGURE 2.3 Incidence rates of nonfatal occupational injuries and illnesses, HCSA sector and private industry, 2003–2005. (From Bureau of Labor Statistics, Table 1, Incidence rates for nonfatal occupational injuries and illnesses by industry and case type: 2005 data = http://www.bls.gov/iif/oshwc/osh/case/ostb1619.pdf, 2004 data = http://www.bls.gov/iif/oshwc/osh/case/ostb1487.pdf, and 2003 data = http://www.bls.gov/iif/oshwc/osh/case/ostb1355.pdf)

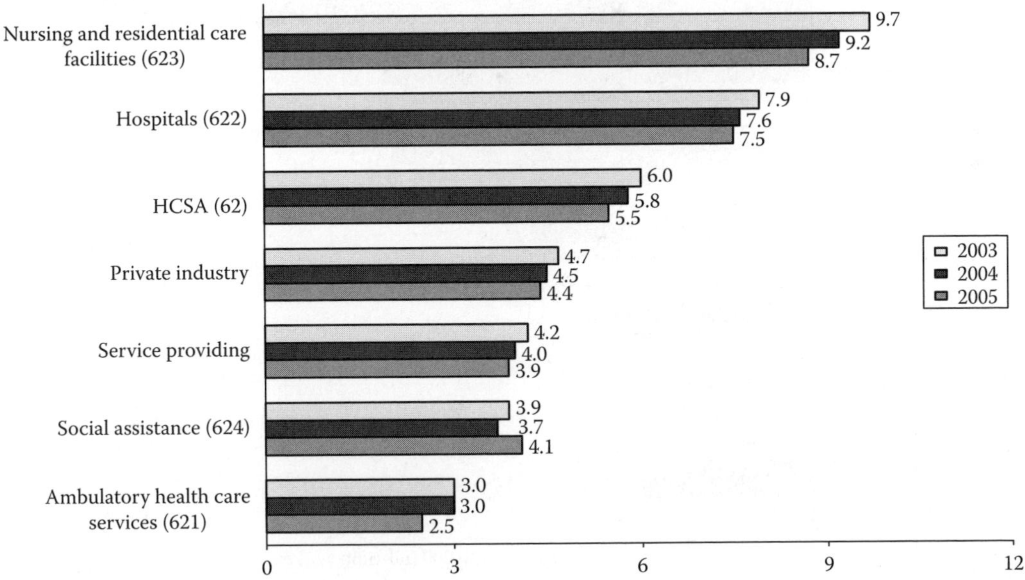

Note: NAICS codes in parentheses follow industry sector name.

FIGURE 2.4 Incidence rates of nonfatal occupational injuries, HCSA sector and private industry, 2003–2005. (From Bureau of Labor Statistics, Table SNR05, Incidence rate and number of nonfatal occupational injuries by industry, private industry, 2003–2005: 2005 data = http://www.bls.gov/iif/oshwc/osh/case/ostb1611.pdf, 2004 data = http://www.bls.gov/iif/oshwc/osh/case/ostb1479.pdf, and 2003 data = http://www.bls.gov/iif/oshwc/osh/case/ostb1347.pdf)

2005. Nursing and residential care facilities had the highest incidence rate (8.7 cases per 100 full-time workers) with nearly 200,000 injury cases, followed by hospitals with an incidence rate of 7.5 and the highest number of injury cases (259,000) among the four subsectors (Figure 2.4).[6] These two subsectors accounted for nearly three-quarters of the total injury cases for the sector.

Nonfatal Occupational Illnesses

In 2005, the incidence of nonfatal occupational illnesses for the HCSA sector was 39.9 cases per 10,000 full-time workers, compared to 26.7 and 19.6 cases in the private and service-providing industries, respectively (Figure 2.5).[7] The number of nonfatal illnesses for this sector (45,000) accounted for almost 20% of the total number of illness cases in private industry.[8] Incidence rates declined for all HCSA subsectors, as well as in private and service-providing industries for each year since 2003, with the exception of social assistance and ambulatory health care services whose rates increased and remained unchanged from 2004 to 2005, respectively. In 2005, hospitals had the highest incidence rate (66.2 cases per 10,000 full-time workers) and number of reported cases (22,900) among the four subsectors. Nursing and residential care facilities accounted for the second highest incidence rate (40 cases per 10,000 full-time workers) and the third highest number of reported cases (9,200) behind ambulatory health care services. Hospitals accounted for over half of the 45,000 total illness cases for the sector.[8]

In 2005, nonfatal occupational skin diseases and disorders and respiratory conditions represented the most frequently reported illness categories in HCSA, with overall incidence rates of 7.0 and 5.2 cases, respectively, per 10,000 full-time workers (Figures 2.6 and 2.7). By comparison, incidence rates

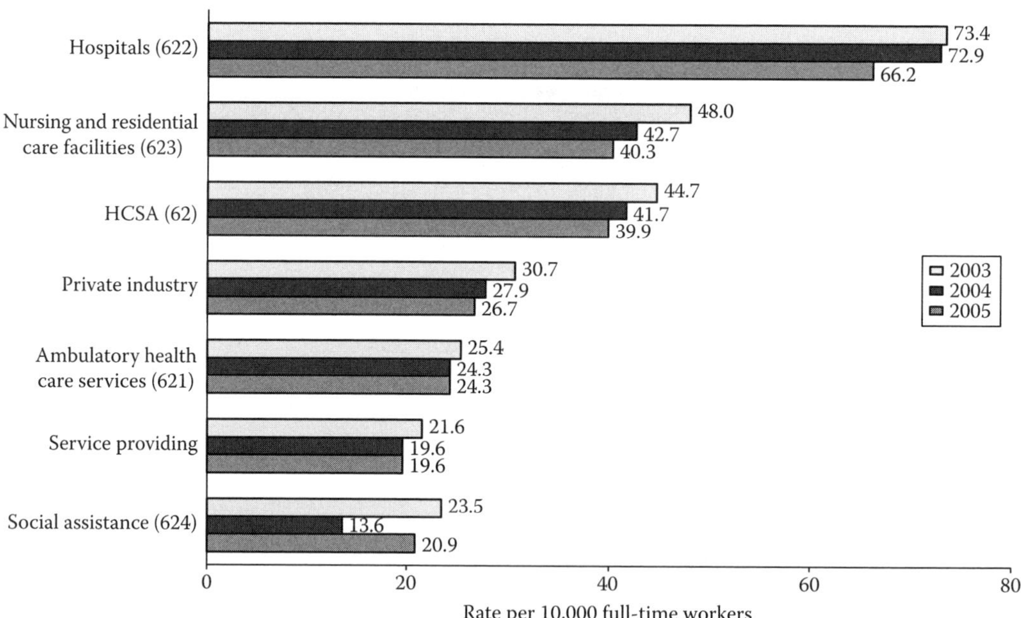

Note: NAICS codes in parentheses follow industry sector name.

FIGURE 2.5 Incidence rates of nonfatal occupational illnesses, HCSA sector and private industry, 2003–2005. (From Bureau of Labor Statistics, Table SNR08, Incidence rates of nonfatal occupational illness by industry and category of illness, private industry, 2003–2005: 2005 data = http://www.bls.gov/iif/oshwc/osh/case/ostb1614.pdf, 2004 data = http://www.bls.gov/iif/oshwc/osh/case/ostb1482.pdf, and 2003 data = http://www.bls.gov/iif/oshwc/osh/case/ostb1350.pdf)

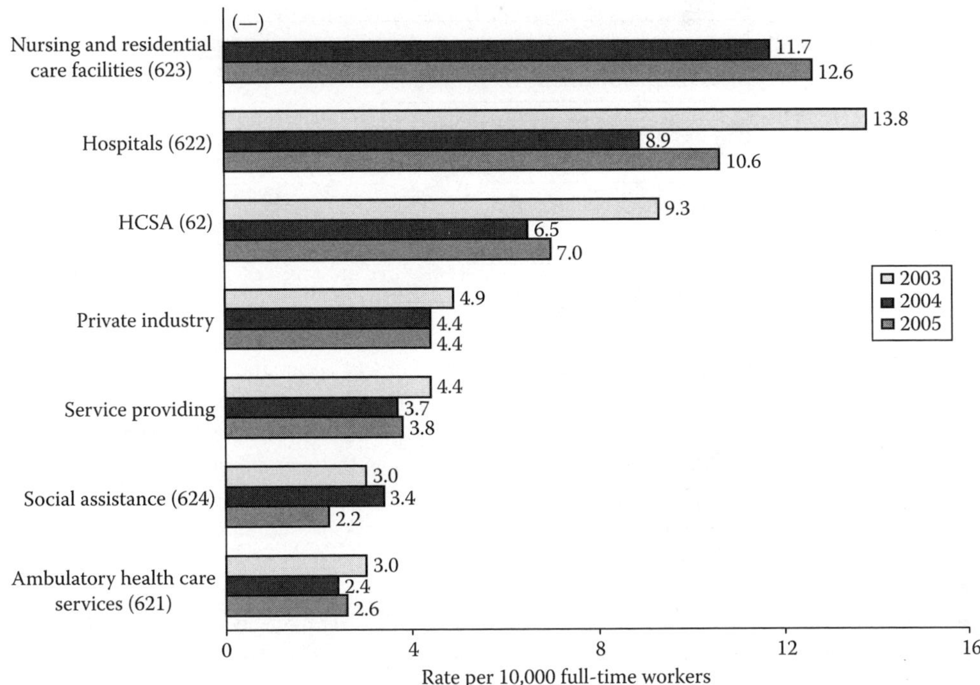

Note: NAICS codes in parentheses follow industry sector name.
(—) No publishable BLS data.

FIGURE 2.6 Incidence rates of nonfatal occupational skin diseases or disorders, HCSA sector and private industry, 2003–2005. (From Bureau of Labor Statistics, Table SNR08, Incidence rates of nonfatal occupational illness by industry and category of illness, private industry, 2003–2005: 2005 data = http://www.bls.gov/iif/oshwc/osh/case/ostb1614.pdf, 2004 data = http://www.bls.gov/iif/oshwc/osh/case/ostb1482.pdf, and 2003 data = http://www.bls.gov/iif/oshwc/osh/case/ostb1350.pdf)

in private industry were 4.4 and 2.2, respectively.[7] Eighty percent of the illness cases involving skin disorders and respiratory conditions were reported in nursing and residential care facilities and hospitals. Nursing and residential care facilities had the highest incidence rate (12.6) and second highest number of skin disease cases (2900) among the subsectors. Hospitals accounted for the second highest incidence rate (10.6) and the highest number of skin disease cases (3700). Hospitals had the highest incidence rate (8.0) and number of cases (2800) of nonfatal respiratory conditions.

The incidence rate of nonfatal occupational poisonings in the HCSA sector was 0.2 cases per 10,000 workers in 2005. The social assistance subsector had the highest incidence rate, four times higher than the HCSA sector average and the rate in 2004. The incidence rate for all other illnesses (primarily repetitive trauma cases) was 27.4 cases per 10,000 full-time workers, accounting for nearly 70% of the total illness cases in the sector (Figure 2.8).[8] Hospitals had an incidence rate of 47.3, nearly three times higher than in private industry, with over half of the "all other "illness cases for the sector.[8]

Nonfatal Occupational Injuries and Illnesses Involving Days away from Work

Of the 4.2 million nonfatal occupational injuries and illnesses reported by private industry employers in 2005, 1.2 million (28%) involved one or more days away from work.[9] The HCSA sector accounted for 175,900 (14.2%) of these 1.2 million cases. Within HCSA, health care accounted for 154,940 (88%) of the

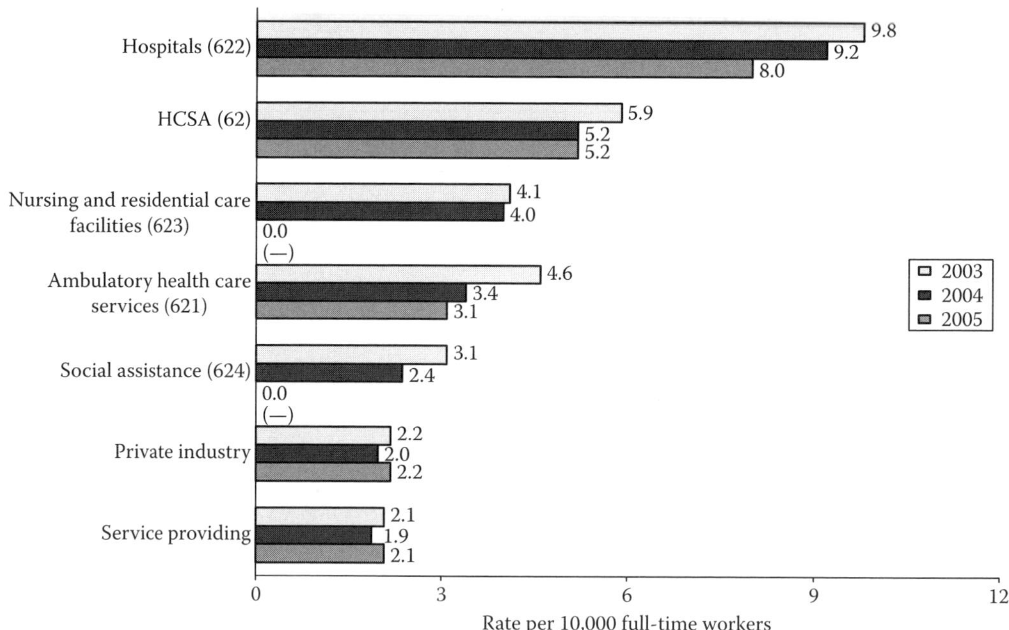

Note: NAICS codes in parentheses follow industry sector name.
(—) No publishable BLS data.

FIGURE 2.7 Incidence rates of nonfatal occupational respiratory conditions, HCSA sector and private industry, 2003–2005. (From Bureau of Labor Statistics, Table SNR08, Incidence rates of nonfatal occupational illness by industry and category of illness, private industry, 2003–2005: 2005 data = http://www.bls.gov/iif/oshwc/osh/case/ostb1614.pdf, 2004 data = http://www.bls.gov/iif/oshwc/osh/case/ostb1482.pdf, and 2003 data = http://www.bls.gov/iif/oshwc/osh/case/ostb1350.pdf)

175,900 cases, and social assistance the remainder (20,960 or 12%).[10] Nursing and residential care facilities (66,620 cases) and hospitals (62,930 cases) accounted for over 73% of the total number of injury and illness cases involving days away from work.

Figure 2.9 compares incidence rates of selected nonfatal occupational injuries and illnesses involving days away from work for the HCSA sector as a whole and by subsector for 2003–2005 to those for all private industry and to service-providing industries. In 2005, the incidence rate for the HCSA sector was 1.6 cases per 10,000 full-time workers, slightly higher than in private and service-providing industries. These data show that about one in four of the injury and illness cases involves days away from work. Nursing and residential care facilities and, to a lesser extent, hospitals were primary drivers for the increased incidence rates for the sector.

Cases involving days away from work are typically characterized by the nature of the injury or illness, the part of body affected, the source that caused the injury or illness, or the event that leads to exposure resulting in illness or injury. In 2005, sprains and strains (82.3 cases per 10,000 workers) were the most likely type of injury or illness in HCSA, nearly 1.5 times more likely to occur among workers in HCSA than in all private industry and about 5 times more likely than the next highest category of soreness and pain (Figure 2.10). The part of the body most affected was the trunk (66.8 cases per 10,000 workers), with an incidence rate nearly 1.5 times higher than in private industry, followed by lower extremities, upper extremities, and then multiple body parts (Figure 2.11). The health care patient (47.5 cases per 10,000 workers) was the most likely source of injury or illness for the health care worker, followed by floor/walkways/ground surfaces, and worker motion/position (Figure 2.12). Overexertion, falls on the same level, contact with object/equipment, and assaults/violent acts represent the top events or exposures leading to injury or illness in the sector (Figure 2.13).

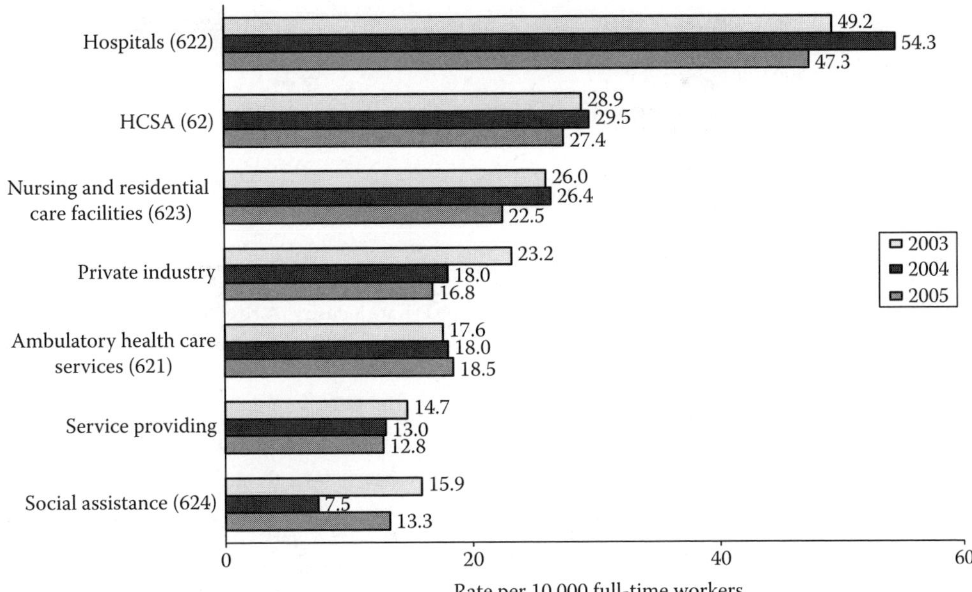

Note: NAICS codes in parentheses follow industry sector name.

FIGURE 2.8 Incidence rates of other nonfatal occupational illnesses, HCSA sector and private industry, 2003–2005. (From Bureau of Labor Statistics, Table SNR08, Incidence rates of nonfatal occupational illness by industry and category of illness, private industry, 2003–2005: 2005 data = http://www.bls.gov/iif/oshwc/osh/case/ostb1614.pdf, 2004 data = http://www.bls.gov/iif/oshwc/osh/case/ostb1482.pdf, and 2003 data = http://www.bls.gov/iif/oshwc/osh/case/ostb1350.pdf)

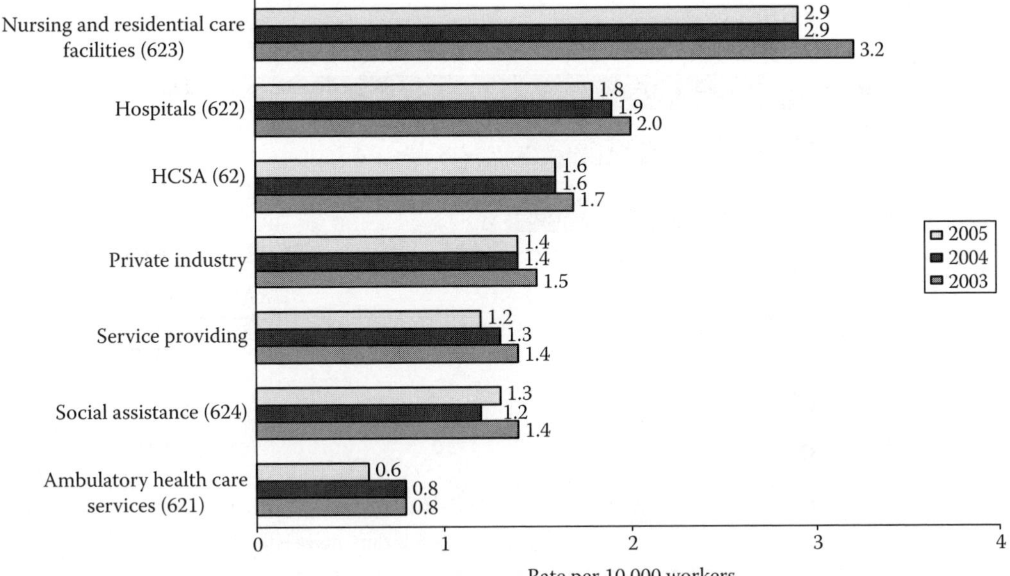

Note: NAICS codes in parentheses follow industry sector name.

FIGURE 2.9 Incidence rates of nonfatal occupational injury and illness cases involving days away from work, HCSA sector and private industry, 2003–2005. (From Bureau of Labor Statistics, Table R8, Incidence rates for nonfatal occupational injuries and illnesses involving days away from work per 10,000 full-time workers by industry and selected events or exposure leading to injury and illness, 2003–2005: 2005 data = http://www.bls.gov/iif/oshwc/osh/case/ostb1664.pdf, 2004 data = http://www.bls.gov/iif/oshwc/osh/case/ostb1518.pdf, and 2003 data = http://www.bls.gov/iif/oshwc/osh/case/ostb1386.pdf)

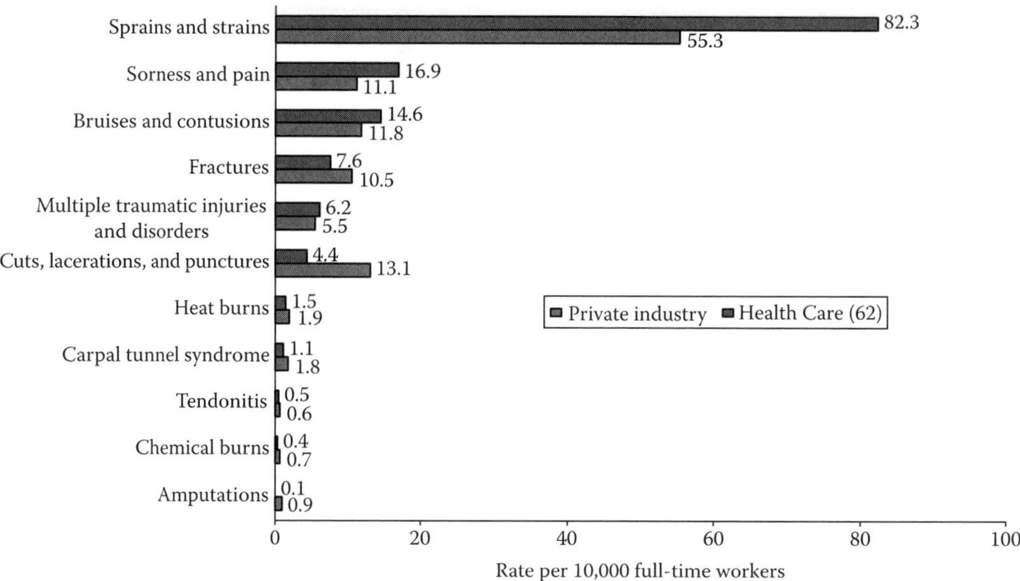

FIGURE 2.10 Incidence rates of nonfatal occupational injuries and illnesses involving days away from work by selected nature of injury or illness, HCSA sector (62)* and private industry, 2005. (* denotes NAICS codes in parentheses follow industry sector name.) (From Bureau of Labor Statistics, Table R5, Incidence rates for nonfatal occupational injuries and illnesses involving days away from work per 10,000 full-time workers by selected natures of injury or illness, 2005: 2005 data = http://www.bls.gov/iif/oshwc/osh/case/ostb1661.pdf)

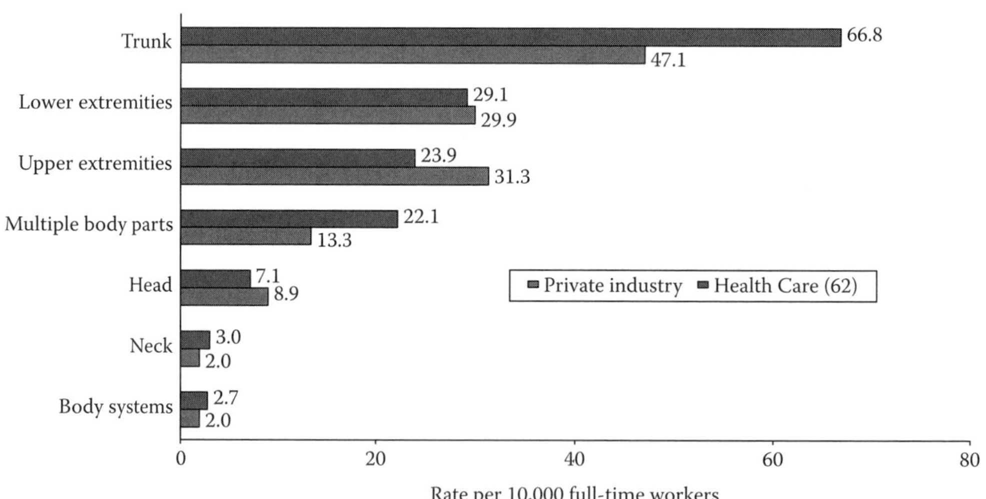

Note: NAICS code in parentheses follow industry sector name.

FIGURE 2.11 Incidence rates of nonfatal occupational injuries and illnesses involving days away from work by selected parts of body affected, HCSA sector and private industry, 2005. (From Bureau of Labor Statistics, Table R6, Incidence rates for nonfatal occupational injuries and illnesses involving days away from work per 10,000 full-time workers by selected parts of body affected by injury or illness, 2005: 2005 data = http://www.bls.gov/iif/oshwc/osh/case/ostb1662.pdf)

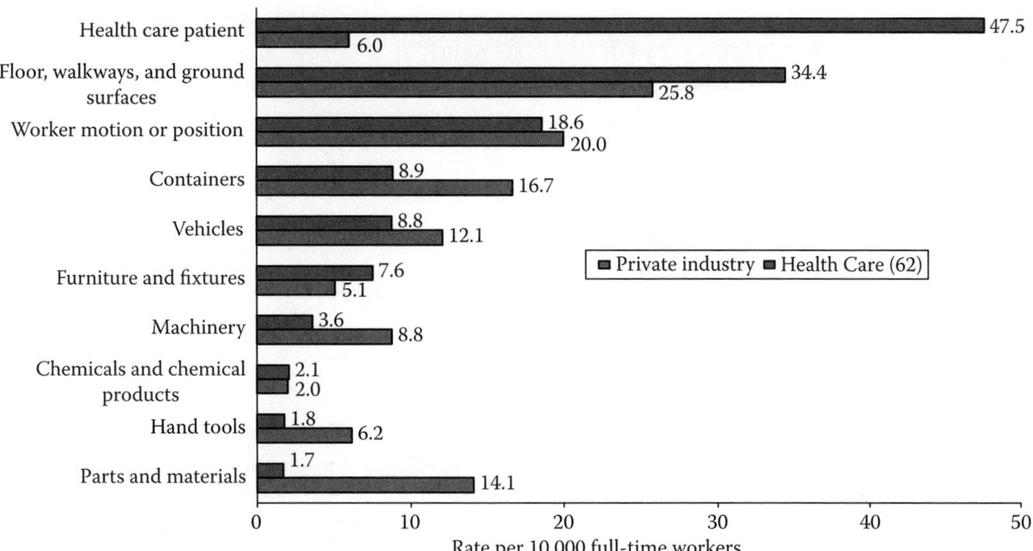

Note: NAICS code in parentheses follow industry sector name.

FIGURE 2.12 Incidence rates of nonfatal occupational injuries and illnesses involving days away from work by selected sources of injury or illness, HCSA sector and private industry, 2005. (From Bureau of Labor Statistics, Table R7, Incidence rates for nonfatal occupational injuries and illnesses involving days away from work per 10,000 full-time workers by industry and selected sources of injury or illness, 2005: 2005 data = http://www.bls.gov/iif/oshwc/osh/case/ostb1663.pdf)

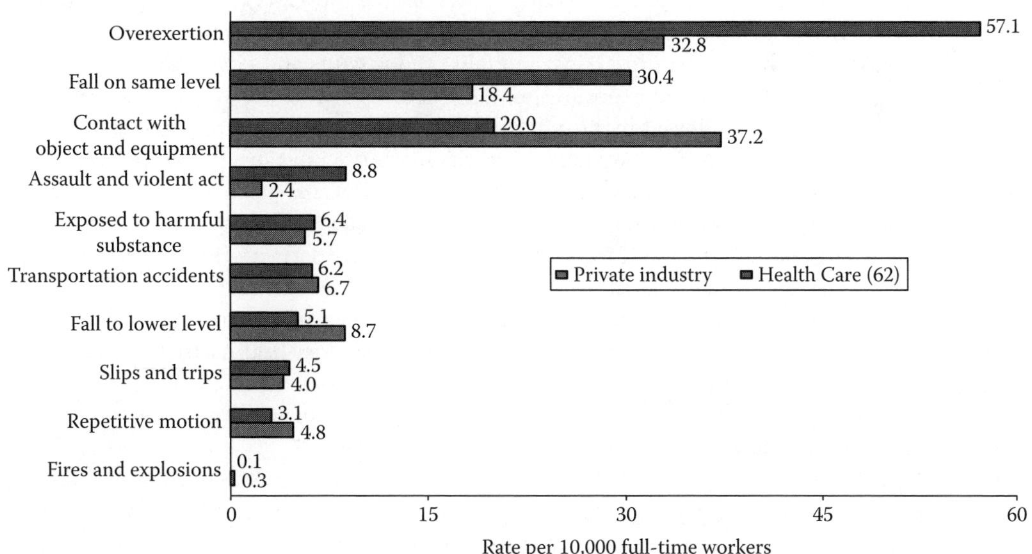

Note: NAICS code in parentheses follow industry sector name.

FIGURE 2.13 Incidence rates of nonfatal occupational injuries and illnesses involving days away from work by selected events of exposures leading to injury or illness, HCSA sector and private industry, 2005. (From Bureau of Labor Statistics, Table R7, Incidence rates for nonfatal occupational injuries and illnesses involving days away from work per 10,000 full-time workers by industry and selected events of exposures leading to injury or illness, 2005: 2005 data = http://www.bls.gov/iif/oshwc/osh/case/ostb1664.pdf)

The average incidence rate for assaults/violent acts in the HCSA sector (8.8 cases per 10,000 workers) was nearly four times higher than in all private industry.

In 2005, nursing and residential care facilities experienced the highest incidence rates (per 10,000 workers) for illnesses and injuries involving days away from work in the sector for

- Musculoskeletal disorders (MSDs) with an incidence rate of 131.4 cases, followed by other ambulatory health care services (89.0) and hospitals (82.7)[11]
- Overexertion including lifting with an incidence rate of 122.8, followed by other ambulatory health care services (90.2) and hospitals (71.9)[12]
- Falls on the same level with an incidence rate of 56.5, followed by social assistance (34.3) and hospitals (30.1)[12]
- Personal assaults and violent acts with an incidence rate of 20.1, followed by social assistance (9.7) and outpatient care centers (9.5)[12]

Incidence rates for MSDs, overexertion, falls on the same level, and personal assaults and violent acts in nursing and residential care facilities were 3.2, 3.7, 3.1, and 12.5 times higher in HCSA, respectively, than in private industry and 1.8–2.4 times higher than the sector average.[11,12]

In 2005, home health care services experienced the highest incidence rates (per 10,000 workers) of any one industry in the sector for

- Falls to lower level with an incidence rate of 10.9 cases, followed by nursing and residential care facilities (7.7) and social assistance (7.3)[12]
- Transportation (highway) accidents with an incidence rate of 22.4, followed by other ambulatory health care services (20.9) and social assistance (11.9)[12]

Major Injuries/Illnesses, Exposures, and Sources

Figures 2.14 through 2.18 present incidence rates of predominant lost workday occupational injuries and illnesses in Health Care for 2003–2005, including sprains and strains; overexertion/repetitive motion injuries; back injuries; slips, trips, and falls; and assaults and violent acts. Overall, for each of these injuries, illnesses, and exposures, nursing and residential care facilities experienced the highest incidence rates, followed by hospitals and ambulatory health care services. With the exception of injuries due to assaults and violent acts, the incidence rates for injuries, illnesses, and exposures showed similar trends in magnitude across the Health Care subsectors when compared to private industry: about 2–4 times higher for nursing and residential care facilities, 1.2–2 times higher for hospitals, and less than half for ambulatory health care services (Figures 2.14 through 2.17). Incidence rates for injuries and illnesses involving assaults/violent acts were about 10 and 4 times higher in nursing and residential care facilities and hospitals than in private industry, respectively, and were no different than private industry for ambulatory health care services (Figure 2.18). Incidence rates for the 3 year period were somewhat lower in 2005 when compared to 2003 for these outcomes/exposures, with the exception of slips, trips, and falls in nursing and residential care facilities that increased about 7%.

Incidence rates characterized by selected sources of injury and illness are presented by Health Care subsector for 2003–2005, with comparison to private industry, in Figures 2.19 and 2.20. For each source (i.e., health care patient, worker motion or position, and floor and ground surfaces), nursing and residential care facilities experienced the highest incidence rates, followed by hospitals and ambulatory health care services. The incidence rates for injuries associated with worker motion or position in ambulatory health care services showed marked declines from 2003 to 2005 (Figure 2.21). The rates in hospitals, by comparison, were similar and in 2005 slightly exceeded that in nursing and residential care facilities. Incidence rates for two of the three source categories (worker motion or position and floor and ground surfaces) showed similar trends in magnitude across Health Care subsectors when compared to

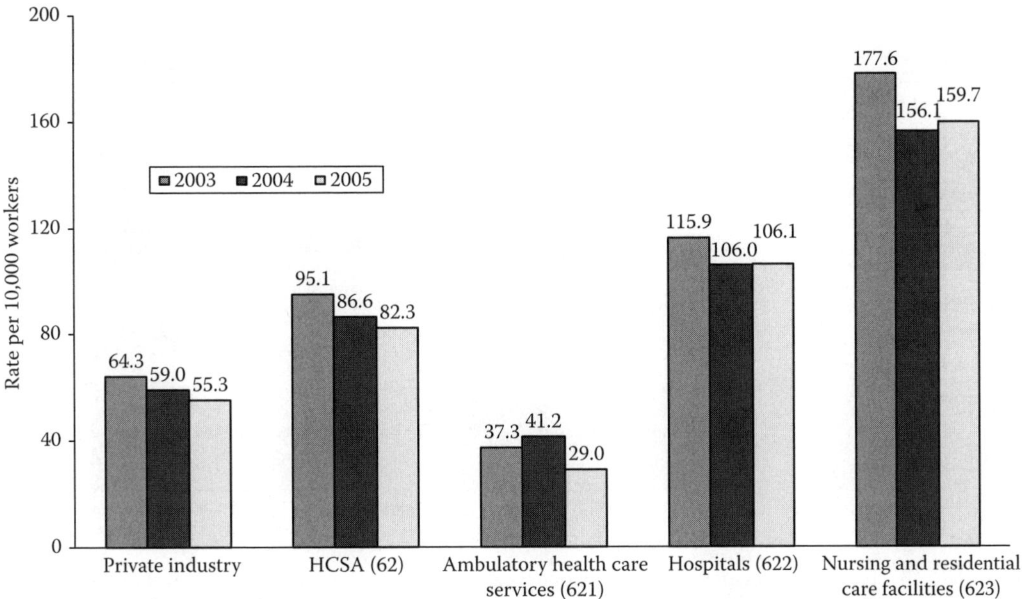

Note: NAICS codes in parentheses.

FIGURE 2.14 Incidence rates of nonfatal occupational sprains and strains involving days away from work, Health Care and private industry, 2003–2005.

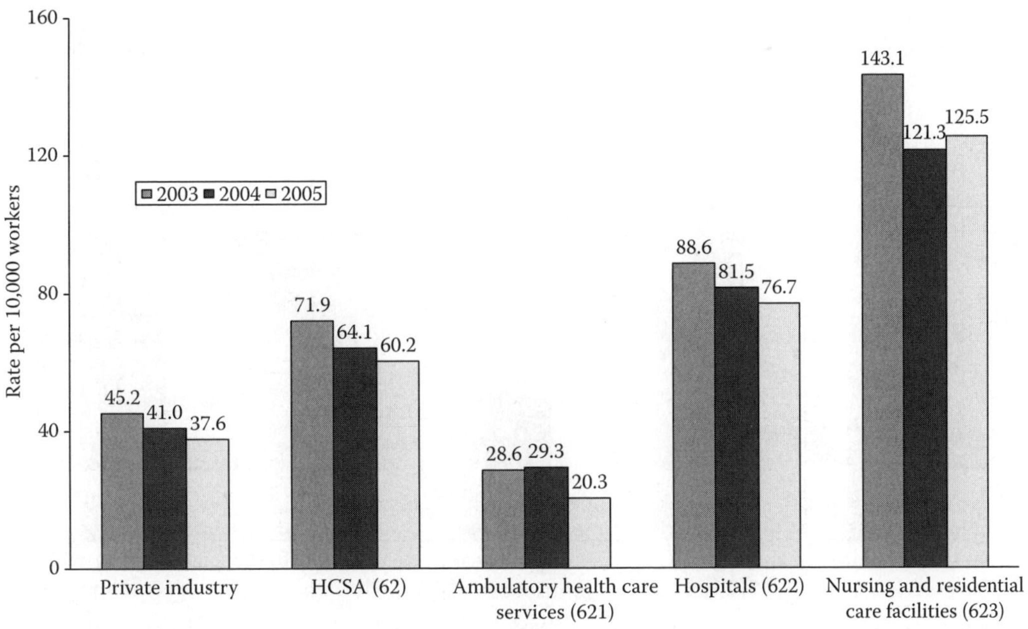

Note: NAICS codes in parentheses follow industry sector name.

FIGURE 2.15 Incidence rates of nonfatal occupational overexertion and repetitive motion injuries and illnesses involving days away from work, Health Care and private industry, 2003–2005.

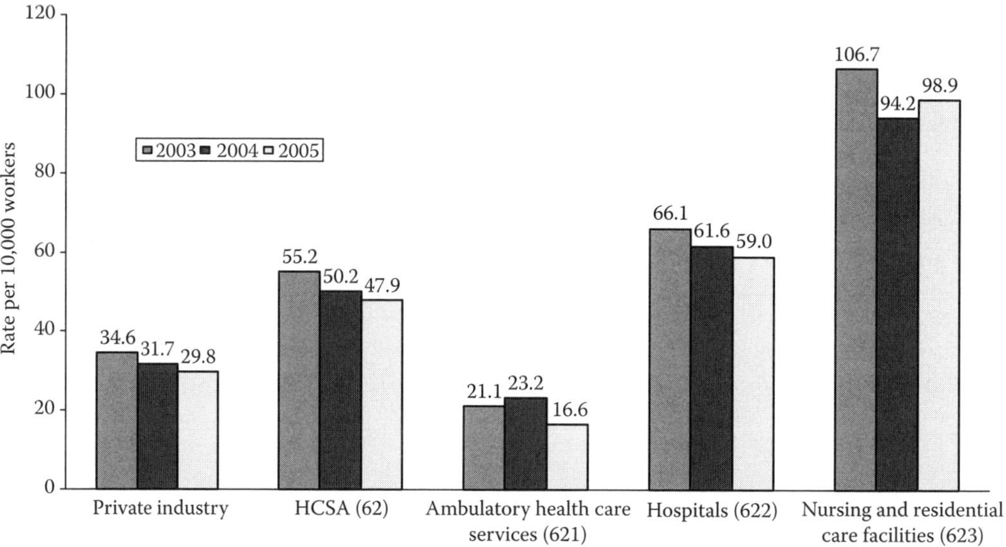

FIGURE 2.16 Incidence rates of nonfatal occupational back injuries involving days away from work, Health Care and private industry, 2003–2005. (NAICS codes in parentheses follow industry sector name.)

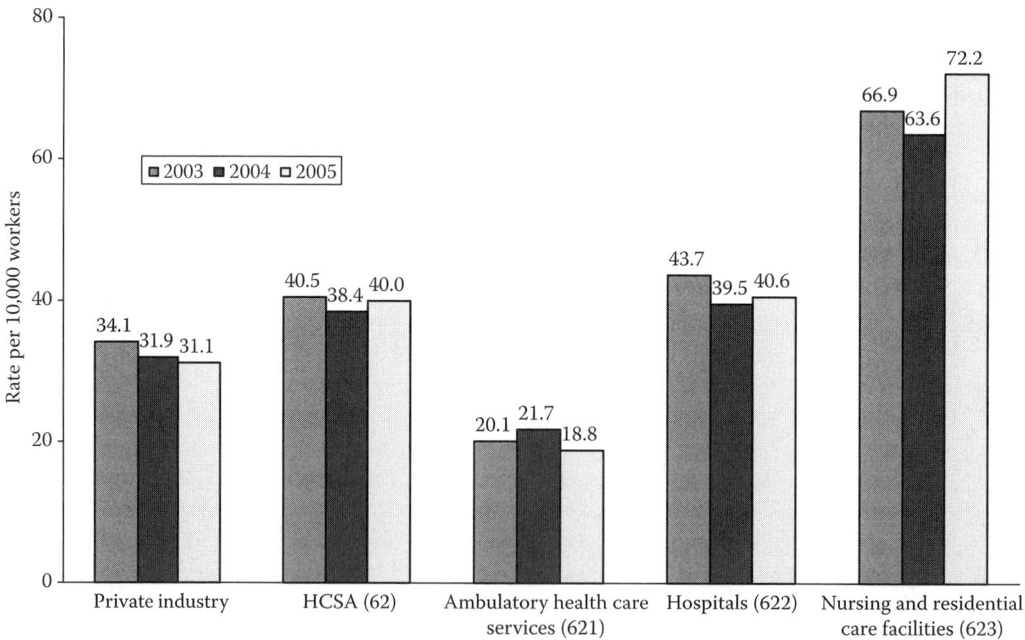

FIGURE 2.17 Incidence rates of nonfatal occupational injuries and illnesses involving days away from work due to slips, trips, and falls, Health Care and private industry, 2003–2005. (NAICS codes in parentheses follow industry sector name.)

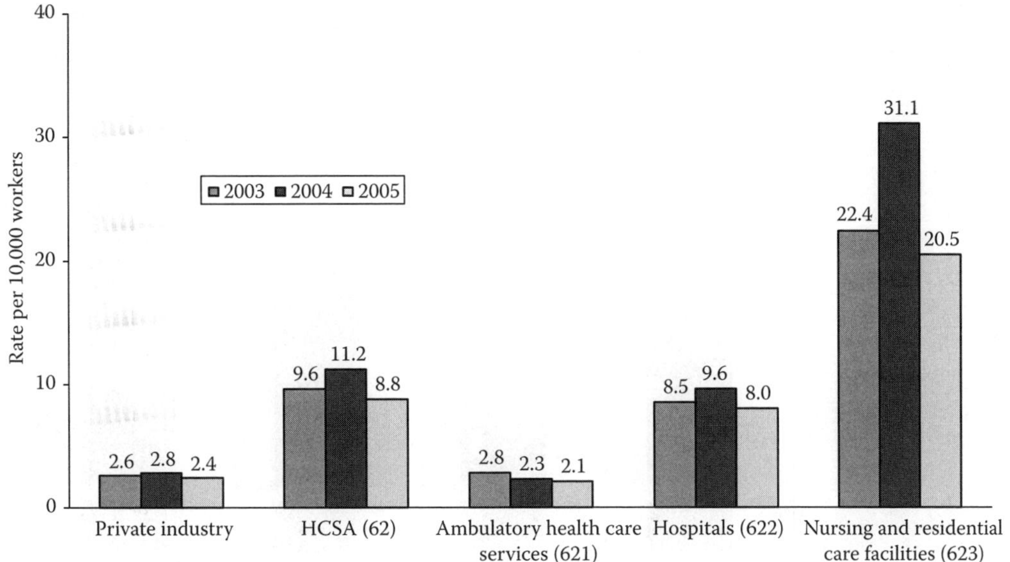

FIGURE 2.18 Incidence rates of nonfatal occupational injuries and illnesses involving days away from work due to assaults and violent acts, Health Care and private industry, 2003–2005. (NAICS codes in parentheses follow industry sector name.)

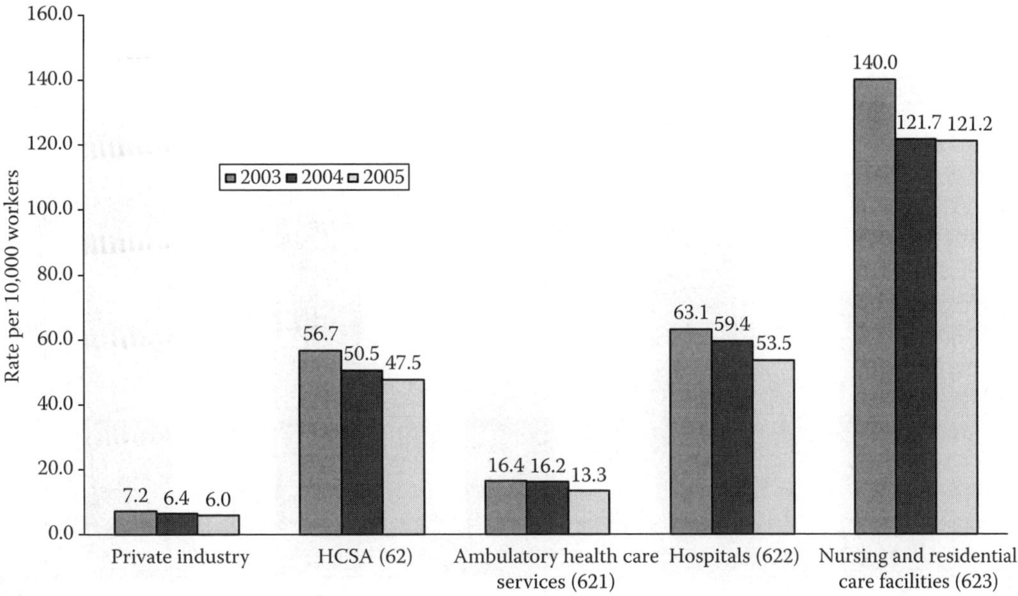

FIGURE 2.19 Incidence rates of nonfatal occupational injuries and illnesses involving days away from work, health care patient as source, Health Care and private industry, 2003–2005. (NAICS codes in parentheses follow industry sector name.)

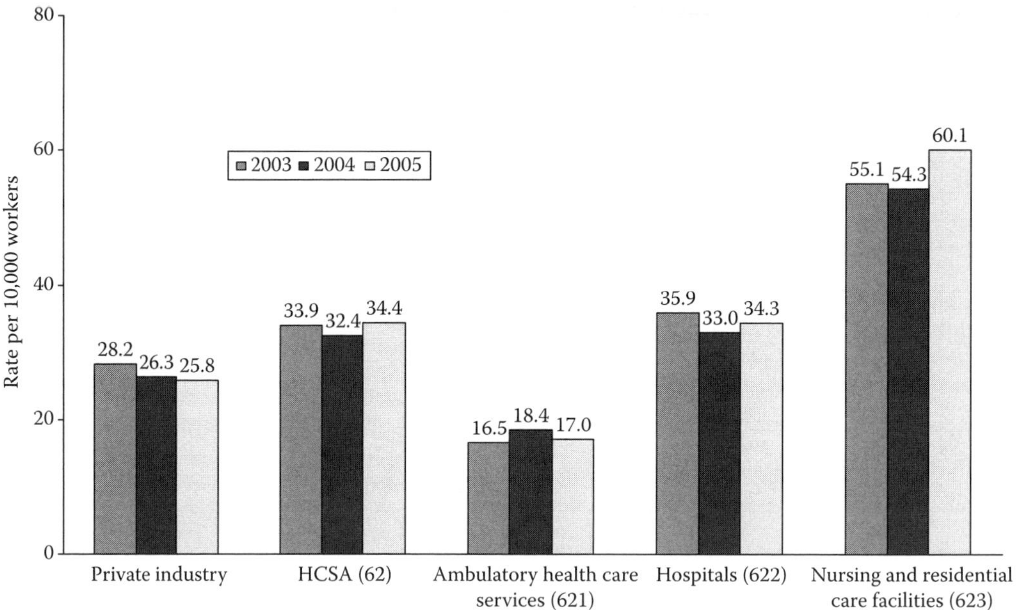

FIGURE 2.20 Incidence rates of nonfatal occupational injuries involving days away from work, floor and ground surfaces as source, Health Care and private industry, 2003–2005. (NAICS codes in parentheses follow industry sector name.)

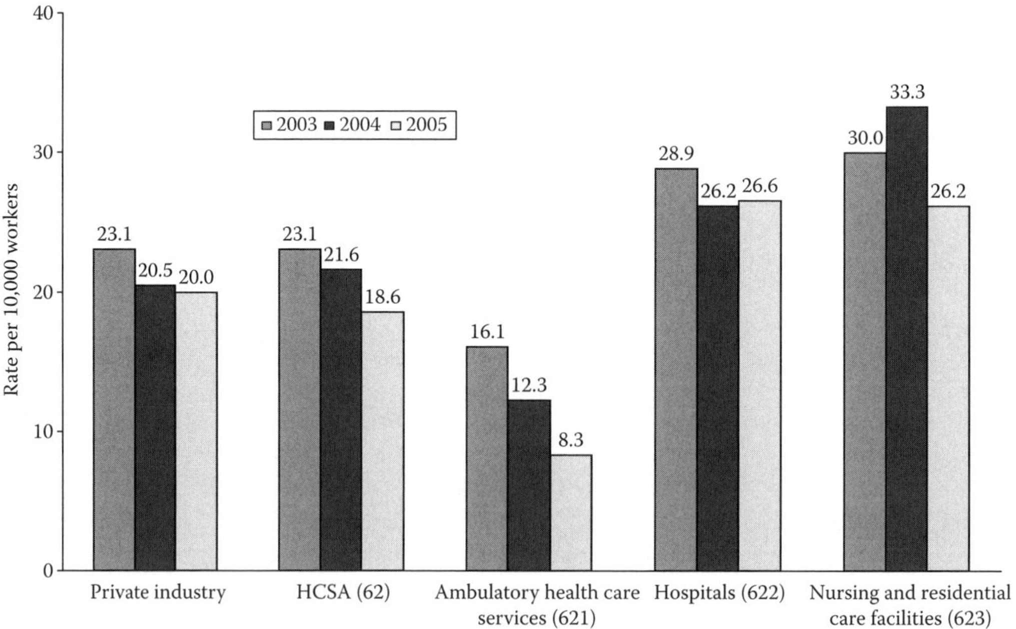

FIGURE 2.21 Incidence rates of nonfatal occupational injuries and illnesses involving days away from work, worker motion or position as source, Health Care and private industry, 2003–2005. (NAICS codes in parentheses follow industry sector name.)

private industry: about 2–3 times higher for nursing and residential care facilities, 1.2–1.3 times higher for hospitals, and 40%–60% less for ambulatory health care services (Figures 2.20 and 2.21). Incidence rates involving injuries where the health care patient was the source were about 20, 9, and 2 times higher in nursing and residential care facilities, hospitals, and ambulatory health care services, respectively, than in private industry, a finding directly attributable to the fact that patients are unique to Health Care (Figure 2.19).

Special Populations

This section provides data describing the distribution of nonfatal injuries and illnesses for selected special populations at risk within the HCSA sector (i.e., those workers who experience a disproportionate share of injury and disease due to sex, age, race, ethnicity, etc.).

The section focuses on women, young workers (16–19 year olds), minorities, and older workers (45 years and over). Reported figures are based on 2005 employment and injury and illness data.

Industry Level Data

Table 2.13 displays the 2005 percent distribution of nonfatal injury and illness cases involving days away from work by the sex, race, and ethnic origin of worker in the HCSA sector and private industry. Women experienced 80.7% of the lost workday injury and illness cases in this sector compared to 31%

TABLE 2.13 Percent Distribution of Nonfatal Injuries and Illnesses Involving Days away from Work by Industry, Sex, Race, and Ethnic Origin of Worker, 2005

			Percent of Nonfatal Injury and Illness Cases Involving Days away from Work			
2002 NAICS Code[a]	Industry	Total Cases	Women	Black or African American[b]	Hispanic or Latino[b]	Asian[b]
621	Ambulatory health care services[c]	25,390	84.2	17.6	7.7	1.1
6211	Physician offices	5,420	91.0	10.5	14.5	0.5
6212	Dental offices	1,010	100	—	—	—
6213	Offices of other health practitioners	900	96.7	30.9	7.1	—
6214	Outpatient care centers	4,380	75.8	29.9	4.2	0.2
6216	Home health care services	9,660	95.3	20.1	5.7	1.0
6219	Other ambulatory health care services	3,180	45.6	7.1	8.4	0.8
622	Hospitals	62,930	77.1	17.1	10.3	4.2
623	Nursing and residential care facilities	66,620	84.5	29.3	9.2	1.9
624	Social assistance	20,960	75.6	16.5	9.0	0.8
62	Health care and social assistance	175,900	80.7	23.0	9.6	2.4
	Total (private industry, 16 years and over)	1,234,680	33.7	11.8	19.0	1.5

Source: From Bureau of Labor Statistics, Table R39, http://www.bls.gov/iif/oshwc/osh/case/ostb1695.txt and Table R38, http://www.bls.gov/iif/oshwc/osh/case/ostb1694.txt

Notes: Dash (—) indicates data are unavailable. Because of rounding and data exclusion of nonclassifiable responses, data may not sum to the totals.

[a] NAICS, U.S. Census Bureau, www.census.gov/epcd/www/naics.html

[b] Includes women and men.

[c] Excludes medical and diagnostic laboratories (NAICS 6215) where data is unavailable.

private industry. Available data by four-digit industry group reveal that women represented the minority of cases in only one industry, other ambulatory health care services. Overall, Blacks (both women and men) experienced 23% of the lost workday injury and illness cases, nearly twice that of their counterparts in private industry. Blacks accounted for over 29% of the cases in nursing and residential care facilities, and about 17% of the cases for each of the remaining three subsectors. Hispanics of both sexes experienced 9.6% of the lost workday injury and illness cases in the sector, about half of that of their counterparts in private industry. The distribution of cases among Hispanics across the four subsectors ranged from 7.7% to 10.3%, with the highest in hospitals. Asians of both sexes experienced 2.4% of the lost workday injury and illness cases in this sector, about 1.6 times that of their counterparts in private industry. The percent distribution of cases among Asians across the four subsectors ranged from 0.8% to 4.2%, with the highest percent of cases experienced in hospitals.

Table 2.14 displays the 2005 percent distribution of nonfatal injuries and illnesses involving days away from work, by age category of workers in the HCSA sector and private industry. In general, HCSA workers in the 16–19, 20–24, 25–34, and 35–44 age groups experienced lower percent of lost workday injury and illness cases than their counterparts in private industry. However, the reverse was true for workers aged 45 and over. When comparing the proportions of cases within each subsector to one another, workers in nursing and residential care facilities experienced the highest proportion of cases in the age groups of 16–19, 20–24, and 25–34. Workers in hospitals experienced the highest proportions in the age groups of 35–44, 45–54, and 55–54 and workers in social assistance experienced the highest proportions in the 65 and over age group.

TABLE 2.14 Percent Distribution of Nonfatal Injuries and Illnesses Involving Days away from Work by Industry and Age of Worker, 2005

2002 NAICS Code[a]	Industry	Total Cases	Percent of Nonfatal Injury and Illness Cases Involving Days away from Work by Age Group						
			16–19	20–24	25–34	35–44	45–54	55–64	65 and Over
621	Ambulatory health care services[b]	25,390	0.6	7.4	21.4	25.4	26.7	15.8	2.7
6211	Physician offices	5,420	1.1	3.2	15.6	30.0	36.5	11.6	1.9
6212	Dental offices	1,010	—	18.8	13.9	45.5	16.8	—	—
6213	Offices of other health practitioners	900	—	—	18.9	44.4	26.7	7.8	—
6214	Outpatient care centers	4,380	1.1	10.3	21.9	19.6	24.6	16.7	5.5
6216	Home health care services	9,660	0.3	5.8	17.2	23.1	29.0	21.7	2.9
6219	Other ambulatory health care services	3,180	—	12.9	47.8	18.9	10.7	8.5	1.2
622	Hospitals	62,930	0.8	6.5	18.8	26.7	29.0	16.4	1.7
623	Nursing and residential care facilities	66,620	3.4	12.1	25.1	24.8	22.5	10.1	2.0
624	Social assistance	20,960	1.7	9.8	20.7	21.5	27.0	14.8	4.4
62	Health care and social assistance	175,900	1.8	9.2	21.8	25.2	25.9	13.7	2.3
	Total (private industry, 16 years and over)	1,234,680	3.4	10.9	23.8	25.5	23.1	11.1	2.2

Source: From Bureau of Labor Statistics, Table R37, http://www.bls.gov/iif/oshwc/osh/case/ostb1693.txt

Notes: Dash (—) indicates data are unavailable. Because of rounding and data exclusion of nonclassifiable responses, data may not sum to the totals.

[a] NAICS, U.S. Census Bureau, www.census.gov/epcd/www/naics.html

[b] Excludes medical and diagnostic laboratories (NAICS 6215) where data is unavailable.

Data by Selected Occupations

This section focuses on selected health care occupational groups, those which experienced the highest number of nonfatal injuries and illnesses involving days away from work in 2005. These include nursing aides, orderlies and attendants, registered nurses, licensed practical and vocational nurses, home health aides, personal and home care aides, and child care workers.

Tables 2.15 and 2.16 display the 2005 percent distribution of nonfatal injuries and illnesses involving days away from work by the sex, race, ethnic origin, and the age group of workers in these six occupations and private industry. Women in these occupations experienced most of the injury and illness burden, representing 84%–98% of the reported cases. With exception of registered nurses, Blacks in these occupations also experienced a disproportionately higher number of cases as compared to their counterparts in private industry. Registered nurses represented the only occupational group that experienced a disproportionately higher number of cases for Asian workers, as compared to private industry. The percents for Hispanics in each of the six occupations were less than their counterparts in private industry. When compared to private industry, there was a greater proportion of injury and illness cases in the older age groups for many of these occupations, with the exception of nursing aides, orderlies and attendants, and child care workers.

Fatal Occupational Injuries

In 2005, HCSA accounted for 104 work-related fatalities.[13] Fifty-six (56%) of these fatalities involved transportation accidents (mostly highway accidents). Assaults and violent acts accounted for 21% of the fatal occupational injuries within the HCSA sector, with about an equal number of homicides and suicides.

TABLE 2.15 Percent Distribution of Nonfatal Injuries and Illnesses Involving Days away from Work by Sex, Race, or Ethnic Origin of Worker for Occupations in HCSA Sector with Highest Number of Cases in 2005

		Percent Distribution of Nonfatal Injuries and Illnesses Involving Days away from Work			
Occupation	Lost Day Injury and Illness Cases	Women	Black or African American[a]	Hispanic or Latino[a]	Asian[a]
Nursing aides, orderlies, and attendants	52,150	89	31.6	8.6	2.1
Registered nurses	20,100	92	7.8	3.5	6.6
Licensed practical nurses and licensed vocational nurses	7,190	93	17.1	3.7	1.4
Home health aides	7,110	98	23.4	11.4	1.0
Personal and home care aides	4,420	84	33.2	10.2	—
Child care workers	2,560	86	33.9	10.2	—
Total (private industry, 16 years and older)		33.7	11.8	19.0	1.5

Source: From Bureau of Labor Statistics, Table R42, http://www.bls.gov/iif/oshwc/osh/case/ostb1698.txt and Table R43, http://www.bls.gov/iif/oshwc/osh/case/ostb1699.txt

Notes: Dash (—) indicates data are unavailable. Because of rounding and data exclusion of nonclassifiable responses, data may not sum to the totals.

[a] Includes women and men.

TABLE 2.16 Percent Distribution of Nonfatal Injuries and Illnesses Involving Days away from Work by Age for Occupations in HCSA Sector with Highest Number of Cases in 2005

Occupation	Lost Day Nonfatal Injury and Illness Cases	Percent of Nonfatal Injury and Illness Cases Involving Days away from Work by Age Group						
		16–19	20–24	25–34	35–44	45–54	55–64	65 and Over
Nursing aides, orderlies, and attendants	52,150	3.1	15.0	26.7	26.3	19.6	8.2	1.0
Registered nurses	20,100	0.6	2.2	0.7	26.6	34.8	17.2	2.6
Licensed practical nurses and licensed vocational nurses	7,190	0.3	2.8	21.3	27.9	30.5	15.8	1.1
Home health aides	7,110	0.4	8.0	18.5	20.3	30.8	19.2	2.7
Personal and home care aides	4,420	1.4	12.9	15.2	23.9	25.9	16.4	4.3
Child care workers	2,560	2.3	27.7	27.3	16.8	20.3	4.7	1.2
Total (private industry, 16 years and older)		3.4	10.9	23.8	25.5	23.1	11.1	2.2

Source: From Bureau of Labor Statistics, Table R41, http://www.bls.gov/iif/oshwc/osh/case/ostb1697.txt

In 2005, the incidence rate of fatal work-related injuries in the HCSA sector was 0.7 per 100,000 workers in 2005, compared to an incidence rate of 4.3 in private industry (Figure 2.22). Among the four subsectors, hospitals accounted for the lowest incidence rate (0.4) and ambulatory care services and social assistance represented the highest rate (0.9).

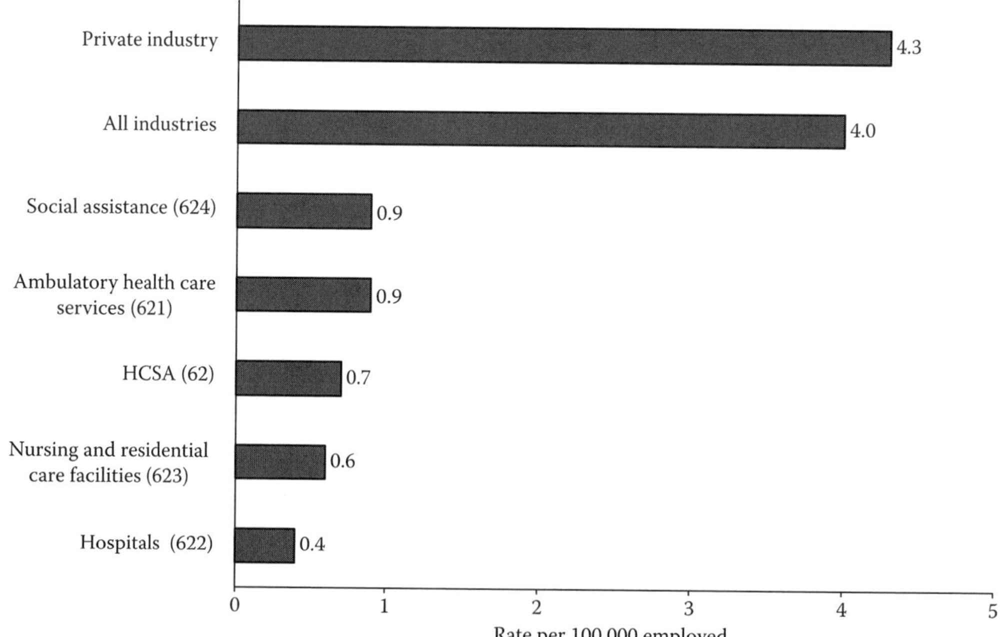

Note: NAICS codes in parentheses follow industry sector name.

FIGURE 2.22 Rate of fatal occupational injuries, HCSA sector and private industry, 2005. (From Bureau of Labor Statistics, Census of Fatal Occupational Injuries, Special tabulation—Number and rate of fatal occupational injuries by selected worker characteristics, 2003–2005, http://www.bls.gov/iif/)

Other Key Facts[9,14]

- In 2005, strains and sprains were the leading nature of injury in every major industry sector. HCSA accounted for nearly one in five cases of all sprains and strains.
- In 2005, HCSA accounted for one in five cases of all falls on the same level. Two-thirds of these cases were reported by nursing and residential care facilities and hospitals.
- Nursing aides, orderlies, and attendants experienced the third highest number of days away from work in 2005 (52,150 cases) among all occupations and the highest among health care occupations, with the majority (89%) of the cases involving women. Injuries to these workers were attributable to health care patients 58% of the time and were due to overexertion for 54% of the cases. The median number of days away from work for this occupation was 5.
- Registered nurses accounted for the 11th highest number of injuries and illnesses involving days away from work in 2005 (20,100) among all occupations and the 2nd highest among health care occupations. The median number of lost workdays for this occupation was 6.
- In 2005, the combined number of injury and illness cases involving days away from work for nursing aides, orderlies, and attendants; and registered nurses (72,250 cases) accounted for over 40% of all injuries and illnesses involving days away from work in the HCSA sector.
- The HCSA sector accounted for nearly 20% (n = 72,780) of all work-related MSDs involving days away from work in 2005, exceeding all industry sectors. Nursing aides, orderlies, and attendants had the highest number of MSD cases among health care occupations (28,920) and the second highest among all occupations. Registered nurses had the second highest number of cases among health care occupations (9,060) and the eighth highest number among all occupations. Home health aides and licensed practical and vocational nurses had the next highest number of MSD cases among health care occupations, ranking in the top 25 of all occupations.
- In 2005, two-thirds of personal assaults and violent acts occurred in the HCSA sector.

Reported BLS Cases Underestimate Magnitude of Occupational Injuries and Illnesses

Studies have shown that the BLS SOII fails to capture a large proportion of job-related injuries and illnesses of private sector employers.[15–17] A recent study of injury and illness reporting in Michigan found that the SOII missed more than two-thirds of job-related injuries and illnesses,[15] while another study estimated that the SOII missed between 33% and 69% of all injuries and illnesses.[16] Additionally, major changes in OSHA recordkeeping rules in 1995 and 2002 have been shown to correspond directly to substantial declines in the number of SOII recordable injuries and illnesses.[18] For example, starting in 2002, MSDs were recorded in the "all other illnesses" illness category on OSHA Form 300 that, in effect, lumped MSDs in with all reported illnesses not categorized as skin disorders, respiratory ailments, poisonings, or hearing loss. The change has been perceived by many to obscure the magnitude of MSD cases in Health Care and other industries where MSDs represent a major problem.

Apart from regulatory changes, causes of underreporting of nonfatal injuries and illnesses are many and diverse.[17] Causes for underreporting by employers include the following: neglect for or lack of knowledge of recordkeeping requirements, negative impact of injury records on management bonuses, control increase of workers' compensation/insurance rates, avoid targeted OSHA inspections, or maintain eligibility for contracts requiring a good safety record. Likewise, workers may not report safety or health problems to their employers for many reasons, such as fear of disciplinary action, not wanting supervisor to think worker was careless, injury too minor to report, unable to afford lost work time, lack of awareness that problem is work related (particularly true for diseases with long latency periods), injury is considered part of the job (particularly true of health care workers), frustration with workers' compensation procedures, or negative impact on company goal of a perfect safety record (especially when reinforced by incentive programs that inadvertently result in peer pressure and are perceived to offer large rewards for hiding injuries). Despite these limitations, the data from the SOII represent the only source of national data on the numbers and rates of occupational injuries and illnesses.

Infectious Disease Data

Sharps Injuries

Two surveillance systems have been developed to measure SIs among health care workers: the Exposure Prevention Information Network (EPINet) at the University of Virginia (http://www.healthsystem.virginia.edu/internet/epinet/about_epinet.cfm) and the CDC's National Surveillance System for Hospital Health Care Workers (NaSH) (http://www.cdc.gov/ncidod/dhqp/nash.html). Characteristics of these two systems and data derived from them are shown in Table 2.17. Data from EPINet and NASH, adjusted for underreporting, have been used to estimate that 384,325 percutaneous injuries are sustained annually by hospital-based health care personnel.[19] Since almost half of U.S. health care workers work outside of hospitals, as many as 600,000–800,000 SIs (Table 2.18) may occur annually among all health

TABLE 2.17 SI Surveillance Data from Four Sources

Characteristic	EPINet	NaSH	Massachusetts	California
Number of sites	48 hospitals	26 hospitals	99 hospitals	316 (of >3000 licensed acute care hospitals, home health care agencies, and skilled nursing facilities)
Most recent published data	2003	Summary report June 1995 to July 1999	2004	1998–1999
Number of SIs per year reported	1728	1380 (average)	3279	976 (average)
Rate of SIs in most recent year available	23.87 SIs per 100 occupied beds		18.3 per 100 licensed hospital beds	
Occupations associated with injuries	37.9% nurses 22.1% physicians 9.0% surgery attendants 5.4% phlebotomist/ venipuncture/ IV team 2.4% clinical laboratory workers	44% nurses 30% physicians 13% technicians	39% nurses 33% physicians 20% technicians (includes surgical, phlebotomists, and clinical laboratories)	49% nurses 9% physicians 10% technologists 9% aides, orderlies, and nursing assistants 8% phlebotomists needles per syringe 8% suture needles 7% butterfly needles
Devices associated with injuries	32% disposable syringes 21% suture needles	34% syringes 16% suture needles 13% butterfly needles	31% hypodermic needles 22% suture needles (56% hollow bore needle: hypodermic, butterfly, vacuum tube, and others)	32% disposable
Injuries with safety devices	32%	4.3 (195/4569)	33%	

Sources: From Perry, J., Parker, G., and Jagger, J., *Adv. Expo. Prev.*, 7, 42, 2005; National Surveillance System for Hospital Health Care Workers Summary Report for data collected from June 1995 through July 1999; Sharps Injuries among Hospital Workers in Massachusetts, 2004: Findings from the Massachusetts Sharps Injury Surveillance System, published April 2007; Cone, J., *Calif. Morbidity*, September 2000.

TABLE 2.18 Frequency Estimates for the United States, 1998

Health Outcome	Estimated Number
SIs	600,000–800,000
Occupationally acquired hepatitis B infection	461
Occupationally acquired acute hepatitis B	132
Occupationally acquired acute hepatitis C	70
Occupationally acquired HIV	1

Sources: From CDC, *NIOSH Alert: Preventing Needlestick Injuries in Health Care Settings*, Department of Health and Human Services, CDC, Cincinnati, OH, DHHS (NIOSH) Publication No. 2000-108, 1999; Personal communication with Ian Williams, Division of Viral Hepatitis (DVH), CDC, February 1, 2002 and Annemarie Wasley, DVH, CDC, April 20, 2007; Do, A.N., Ciesielski, C.A., Metler, R.P., Hammett, T.A., Li, J., and Fleming, P.L., *Infect. Control Hosp. Epidemiol.*, 24, 86, 2003.

care workers, but little data are available about the occurrence of SIs in outpatient settings.[20] A few states, including Massachusetts and California, have also developed their own SI surveillance systems. Based on data from these surveillance systems, it is known that most reported SIs are associated with hypodermic syringes or other hollow-bore needles and most reported SIs occur in nursing, medical, or laboratory staff, but housekeepers and other health care workers are also at risk (Table 2.17).[21–24]

Although reported SIs among health care workers are a common occurrence, fortunately, they rarely lead to infection with bloodborne pathogens (Table 2.18).[20,25,26] Using mathematical modeling, the World Health Organization estimated the incidence of infections attributable to percutaneous injuries and concluded that 39% of HCV, 37% of HBV, and 4.4% of HIV infections acquired among health care workers worldwide in 2000 were attributable to occupational exposure via SIs. The occupational attributable fractions for the United States were estimated to be substantially lower: 8%, 1%, and 0.5% for HCV, HBV, and HIV, respectively. The probability of acquiring an infection depends on the prevalence of infection among the patient population, the probability of health care worker exposure, the probability of infection occurring after exposure, and the proportion of health care workers that are susceptible to infection.[27] Sepkowitz and Eisenberg estimated annual death rates for U.S. health care workers from occupational events to be 17–57 per 1 million workers. They attributed more than half of these deaths (between 80 and 260 total deaths in 2002) to infection; 75–250 deaths from HBV, and 5–10 deaths from HIV, HCV, and tuberculosis (TB) combined. Their estimates were based on reported rates of needlestick injuries, infection prevalence among patients, reported infections among health care workers, and the risk of dying from infections once acquired.[28]

HIV

The average risk of HIV transmission after a percutaneous exposure from a known positive source is estimated to be 0.3%. Risk factors for transmission include exposure to a large quantity of blood from the source person (e.g., device visibly contaminated with patient's blood, procedure involving a needle being placed directly into a vein or artery, or deep injury), exposure to blood from a source person with terminal illness, hollow-bore needles, and, possibly, immunologic factors in the exposed worker.[29]

Data on HIV infection and AIDS among health care workers have been collected by the CDC through the HIV/AIDS Reporting System and the National Surveillance for Occupationally Acquired HIV Infection System.[30] Health care personnel with HIV/AIDS who are reported without any known risk for HIV infection are investigated by state and local health departments using the following case definitions. Documented cases of occupationally acquired HIV/AIDS are those in which HIV seroconversion is temporally related to an exposure to an HIV-positive source and in which the exposed worker has no nonoccupational risk factors for acquisition of HIV (e.g., male homosexual–bisexual contact or IV drug use). Possible cases of occupationally acquired HIV/AIDS are those in which a

TABLE 2.19 Occupations of Health Care Workers with Documented and Possible Occupationally Acquired HIV Infection, 1981–2006

Occupation	Documented	Possible
Nurse	24	35
Laboratory technician (clinical)	16	17
Physician (nonsurgical)	6	12
Laboratory technician (nonclinical)	3	—
Housekeeper/maintenance workers	2	13
Technician (surgical)	2	2
Embalmer/morgue technician	1	2
Health aide/attendant	1	15
Respiratory therapist	1	2
Technician (dialysis)	1	3
Dental worker (including dentist)	—	6
Emergency medical technician (paramedic)	—	12
Physician (surgical)	—	6
Other technician/therapist	—	9
Other health care occupation	—	6
Total	57	140

Source: From Do, A.N., Ciesielski, C.A., Metler, R.P., Hammett, T.A., Li, J., and Fleming, P.L., *Infect. Control Hosp. Epidemiol.*, 24, 86, 2003.

Note: Dash (—) indicates no reported cases.

worker is found to be HIV positive, has no nonoccupational risk factors for HIV/AIDS, and has opportunities for occupational exposure to blood, body fluids, or HIV-positive laboratory material. More than 90% of health care personnel infected with HIV have reported nonoccupational risk factors for acquiring their infection.

Between 1981 and 2006, the CDC received reports of 57 documented cases and 140 possible cases of occupationally acquired HIV in U.S. health care workers (Table 2.19). Thirty-one (54%) of the implicated exposures occurred prior to 1991. Eight of the documented HIV cases occurred despite antiviral postexposure prophylaxis. No documented occupationally acquired cases of HIV infection have been reported since 1999, and the most recent possible case of occupationally acquired HIV was reported in 2000.

HBV

CDC estimated an incidence of 17,000 HBV infections per year among health care workers in 1983, which declined to approximately 400 in 1995, after widespread immunization of health care workers, implementation of universal precautions, and adoption of the OSHA bloodborne pathogens standard. In 1983, the estimated incidence of HBV infections among health care workers was threefold higher than the incidence in the general U.S. population (386 per 100,000 vs. 122 per 100,000). By 1995, however, the estimated incidence of HBV infections among health care workers was more than fivefold lower than the incidence in the general U.S. population (9.1 per 100,000 vs. 50 per 100,000).[31] The CDC's Division of Viral Hepatitis estimates that 139 cases of acute HBV were occupationally acquired in 1995 (3.2 per 100,000 health care workers in patient care occupations), which declined to 87 in 2004 (1.6 per 100,000 workers) (Figure 2.23).[25,32]

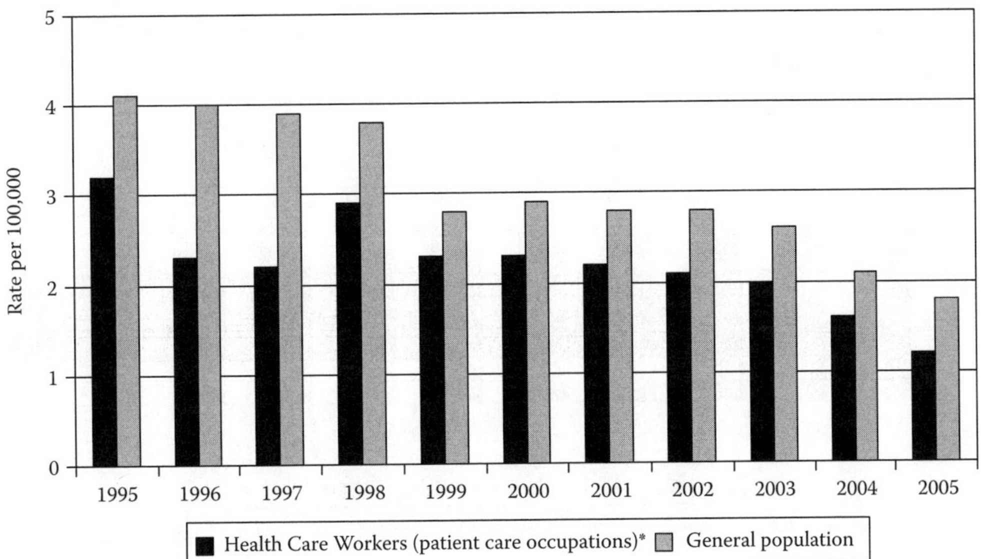

* Patient care occupations include physicians, dentists, nurses, physicians' assistants, and health technologists and technicians.

FIGURE 2.23 Incidence of acute hepatitis B among health care workers. (From personal communication with Ian Williams, Division of Viral Hepatitis (DVH), CDC, February 1, 2002 and Annemarie Wasley, DVH, CDC, April 20, 2007; CDC, *Morbidity Mortality Wkly Rep*, 56, 2007; Household Data Annual Averages, Current Population Survey, Bureau of Labor Statistics, U.S. Department of Labor, 1995–2005.)

HCV

From 1999 through 2004, the percentage of patients with acute hepatitis C case who reported being health care workers averaged about 2% (range: 1%–4%). This increased to 7.2% in 2005. It is unknown what proportion of these cases was occupationally acquired.[33] It has been estimated that percutaneous exposure leads to 50–150 transmissions of HCV among health care workers annually, assuming that hospitalized patients have the same HCV seroprevalence as the rest of the U.S. population.[28] Seroprevalence studies of HCV in health care workers suggest minimally increased risk compared with the general population.[34]

Tuberculosis

The TB incidence in health care workers declined from 4.9 per 100,000 workers in the health care industry in 1994 to 3.0 in 2005 (Figure 2.24).[32,35,36] In 2005, 3.1% (420) of nationally reported TB cases for whom occupational information was available occurred among health care workers.[36]

The prevalence of TB among health care workers in 2006 (3.2 per 100,000 population) was higher than the prevalence of TB among health care workers in 2005 (3.1 per 100,000 population), and the prevalence has been slightly increasing since 2001. The risk of occupational acquisition of TB among health care personnel has increased due to the emergence of multidrug-resistant and extensively drug-resistant TB and the need to hospitalize patients not responding to traditional outpatient antibiotic regimens. Even though the incidence of TB is decreasing in the U.S. population, health care personnel remain at risk without careful adherence to engineering and administrative controls.[37,38]

Other

A recent review found published case reports of occupationally acquired bloodborne infections for a total of 60 pathogens or species: 26 viruses, 18 bacteria/rickettsia, 13 parasites, and 3 yeasts.[39]

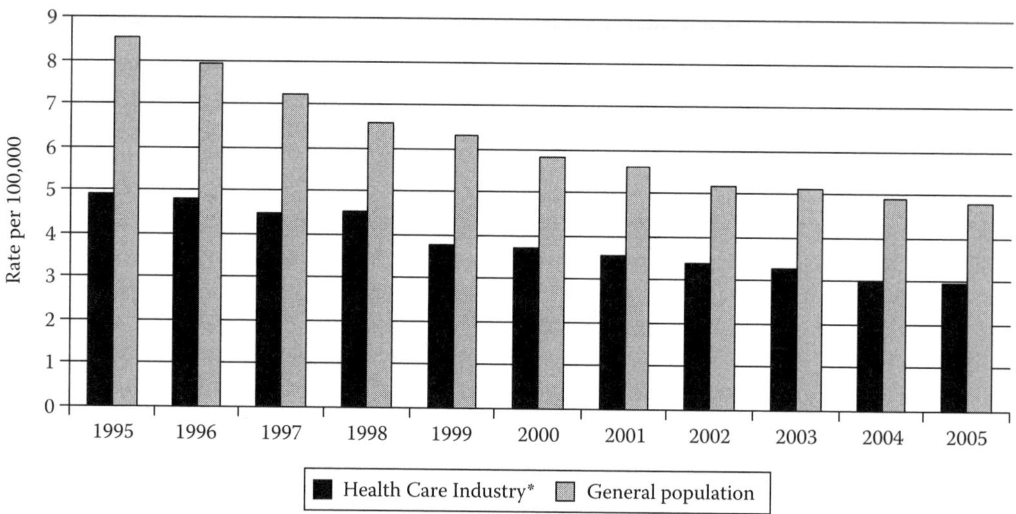

*Health care industry includes hospitals and nonhospital health services.

FIGURE 2.24 Incidence of TB among health care workers. (From Online Tuberculosis Information System (OTIS), National Tuberculosis Surveillance System, United States, 1993–2004, U.S. Department of Health and Human Services (US DHHS), Centers for Disease Control and Prevention (CDC), Division of TB Elimination, CDC WONDER Online Database, March 2006; CDC, Reported Tuberculosis in the United States, 2005, U.S. Department of Health and Human Services, CDC, Atlanta, GA, September 2006; Household Data Annual Averages, Current Population Survey, Bureau of Labor Statistics, U.S. Department of Labor, 1995–2005.)

Disease and Cause-Specific Mortality Data

Death certificate data information from the NOMS (National Occupational Mortality Surveillance) system, with multiple cause of death, coded usual or lifetime occupation, and industry information, were used to assess whether any associations exist between cause-specific mortality and occupation and industry. For this surveillance system, 28 states have provided coded data for this surveillance system for the years 1984–1998 and multiple cause analysis was conducted. The measure of association used most often was the proportionate mortality ratio (PMR), defined as the ratio of the proportion of deaths due to a specific cause for a specified occupation or industry during a specified time period divided by the proportion of deaths due to that cause for all occupations or industries during the same period, multiplied by 100. A PMR is considered to be significantly elevated when its value is greater than 100 and the lower 95% confidence interval (CI) exceeds 100. A significantly elevated PMR suggests that more deaths than expected are associated with a given cause of death in a specified occupation or industry. PMRs should be interpreted as flags or indicators that describe gaps, trends, and elevated risks for serious, acute and chronic disease, and fatal injuries. The NOMS system is available on the Web as an interactive query system for access to precalculated PMRs by occupation or industry.[40]

NOMS data were analyzed to produce mortality estimates for each of the four HCSA industry subsectors and for 18 major health care occupations (Tables 2.20 and 2.21, respectively). PMRs for the top 10 causes of death (i.e., most highly and significantly elevated causes of death), excluding those associated with small numbers of deaths, are reported in Table 2.20. Three causes of death (AIDS, non-A and non-B viral hepatitis, and various cancers) were observed in all four subsectors. Drug-related deaths were observed in all three health care subsectors. The following causes of death were observed in two of the subsectors: viral hepatitis B (ambulatory health care services and hospitals), sarcoidosis (hospitals and social assistance), and malignant melanoma of the skin and polyarteritis nodosa and allied conditions (ambulatory health care services and social assistance). Of the top 10 causes of death in

TABLE 2.20 Top 10 Significantly Elevated PMRs by HCSA Industry Subsector, 1984–1998

2002 NAICS Code[a]	CIC[b]	Industry	Cause of Death (ICD-9 Codes)[c]	PMR[d]	Number of Deaths	95% CI LCL, UCL
621	812, 820–822, 830, and 840	Ambulatory health care services	Air and space transport accidents (840–845)	380**	97	308, 464
			Other lung diseases due to external agents, excluding inhalation (506 and 5071–508)	219**	37	154, 302
			Viral hepatitis B (0701 and 0703)	186**	69	144, 235
			(AIDS) HIV infection (042–044)	176**	940	165, 187
			Non-A and non-B viral hepatitis (0704–0709)	159**	103	130, 193
			Myoneural disorders (358)	155**	53	116, 203
			Polyarteritis nodosa and allied conditions (446)	147**	97	120, 180
			Hodgkin's disease (201)	145**	127	121, 172
			Drug-related deaths (292, 304, 3052–3059, 850–858, 9500–9505, 9620, and 9800–9805)	143**	760	134, 154
			Malignant melanoma of skin (172)	142**	418	129, 156
622	831	Hospitals	(AIDS) HIV infection (042–044)	163**	2875	158, 167
			Sarcoidosis (135)	149**	269	132, 169
			Viral hepatitis B (0701, 0703)	145**	155	123, 169
			Non-A and non-B viral hepatitis (0704–0709)	121**	231	106, 137
			Drug-related deaths (292, 304, 3052–3059, 850–858, 9500–9505, 9620, and 9800–9805)	119**	1760	114, 124
			Cancer of small intestine, including duodenum (152)	119**	147	101, 140
			Asthma (493)	118**	1575	113, 123
			Disorders of the peripheral nervous system (350–357)	118*	249	104, 134
			Diffuse diseases of connective tissue (710)	115*	927	108, 123
			Acute myeloid leukemia	114**	681	105, 123
623	832 and 870	Nursing and residential care facilities	Accidents caused by storms, floods, and earth eruption (908–909)	209*	10	100, 384
			(AIDS) HIV infection (042–044)	148**	945	139, 158
			Cancer of unspecified female genital organs (184)	141**	83	112, 175
			Non-A and non-B viral hepatitis (0704–0709)	135*	74	106, 170

(continued)

TABLE 2.20 (continued) Top 10 Significantly Elevated PMRs by HCSA Industry Subsector, 1984–1998

2002 NAICS Code[a]	CIC[b]	Industry	Cause of Death (ICD-9 Codes)[c]	PMR[d]	Number of Deaths	95% CI LCL, UCL
			Cancer of cervix uteri (180)	123**	440	111, 135
			Diabetes mellitis (250)	118**	5700	115, 120
			Drug-related deaths (292, 304, 3052–3059, 850–858, 9500–9505, 9620, and 9800–9805)	113**	550	103, 122
			Endocrine, nutritional, metabolic, and immunity disorders (240–279)	111**	8080	109, 114
			Motor vehicle traffic accidents (810–819)	110**	1352	105, 115
			Cancer of pancreas (157)	109*	790	102, 117
624	871	Social assistance	Non-A and non-B viral hepatitis (0704–0709)	190**	42	137, 257
			(AIDS) HIV infection (042–044)	186**	514	170, 202
			Neurotic and personality disorders (300–301)	171*	27	113, 249
			Sarcoidosis (135)	163**	35	114, 227
			Polyarteritis nodosa and allied conditions (446)	153*	30	103, 219
			Malignant melanoma of skin (172)	143**	108	117, 172
			Mental disorders associated with solvent exposures (296–297, 300–301, and 3483)	136**	85	109, 169
			Lymphoid leukemia (204)	127*	78	100, 159
			Cancer of bone, connective tissue, skin, and breast (170–175)	122**	1263	117, 172
			Infectious and parasitic diseases (001–139)	120**	1716	115, 125

* = $p < 0.05$; ** = $p < 0.01$.

[a] NAICS, U.S. Census Bureau, www.census.gov/epcd/www/naics.html

[b] Census Industry Code, U.S. Census Bureau, http://www.census.gov/apsd/techdoc/cps/mar96/append_a.html

[c] Multiple cause PMR analysis was conducted using the ninth revision of the International Classification of Diseases (ICD-9) to code cause of death.

[d] PMR is defined as the ratio of the age-adjusted proportion of deaths from a specific cause of death for a particular occupation or industry during a specified time period compared to the proportion of that cause among all industries or occupations during the same period, multiplied by 100. To test for statistical significance, two-sided 95% CIs are calculated, based on the Poisson distribution for observed deaths, and using the normal approximation to the Poisson for large numbers. A statistically significantly elevated PMR must be interpreted as a flag that suggests elevated risk that should be further evaluated for confounding factors.

TABLE 2.21 Top Three Significantly Elevated PMRs for Largest Occupations in HCSA Sector, 1984–1998

Occupation	COC[a]	Total Employed (in Thousands)	Cause of Death (ICD-9 Codes)[b]	PMR[c]	Number of Deaths	95% CI LCL, UCL
Registered nurses	095	2,529	Air and space transport accidents (840–845)	211**	28	140, 305
			Viral hepatitis B (0701 and 0703)	192**	58	146, 248
			(AIDS) HIV infection (042–044)	180**	587	166, 195
Health aides (except nursing)	447	1,906[d]	Myoneural disorders (358)	245**	8	106, 483
			Hodgkin's disease (201)	216**	21	134, 330
			(AIDS) HIV infection (042–044)	154**	139	130, 182
Nursing aides, orderlies, and attendants	446		(AIDS) HIV infection (042–044)	172**	1416	166, 178
			Non-A and non-B viral hepatitis (0704–0709)	171**	120	142, 205
			Viral hepatitis B (0701 and 0703)	152**	60	116, 196
Child care workers	406 and 468	1,401	Systemic sclerosis (7101)	154*	27	102, 224
			Lymphatic cancer and multiple myeloma (202–203)	123**	223	108, 140
			Cancer of ovary and other uterine adnexa (183)	120*	207	104, 138
Physicians	084	863	Air and space transport accidents (840–845)	942**	65	727, 1201
			Viral hepatitis B (0701 and 0703)	304**	28	202, 439
			Drug-related deaths (292, 304, 3052–3059, 850–858, 9500–9505, 9620, and 9800–9805)	300**	241	263, 340
Personal service occupations, nec (includes personal and home care aides)	469	703	(AIDS) HIV infection (042–044)	199**	129	166, 236
			Chronic myeloid leukemia (2051)	171*	20	105, 264
			Cancer of urinary organs (188 and 1893–1899)	144**	73	113, 181
Licensed practical nurses	207	556 (includes licensed vocational nurses)	Sarcoidosis (135)	161*	34	112, 226
			Drug-related deaths (292, 304, 3052–3059, 850–858, 9500–9505, 9620, and 9800–9805)	146**	226	127, 166
			(AIDS) HIV infection (042–044)	146**	174	126, 170

(continued)

TABLE 2.21 (continued) Top Three Significantly Elevated PMRs for Largest Occupations in HCSA Sector, 1984–1998

Occupation	COC[a]	Total Employed (in Thousands)	Cause of Death (ICD-9 Codes)[b]	PMR[c]	Number of Deaths	95% CI LCL, UCL
Health diagnosing practitioners	089	425	Anterior horn cell disease including motor neurone disease (335)	333**	13	177, 569
			Asthma (493)	218*	13	116, 373
			Cancer of brain and nervous system (191–192)	201**	22	126, 305
Clinical laboratory technologists and technicians	203	321	Viral hepatitis B (0701 and 0703)	348**	18	206, 550
			Toxic encephalopathy (3483)	192**	24	123, 285
			Sarcoidosis (135)	191*	15	107, 315
Dental assistants	445	274	Hodgkin's disease (201)	264**	11	132, 472
			(AIDS) HIV infection (042–044)	163*	28	109, 236
			Alzheimer's disease (290, 3310, and 3311)	155**	125	129, 184
Pharmacists	096	245	Drug-related deaths (292, 304, 3052–3059, 850–858, 9500–9505, 9620, and 9800–9805)	242**	90	194, 297
			Cancer of ovary and other uterine adnexa (183)	179**	30	121, 255
			Cancer of brain and nervous system (191–192)	165**	79	130, 205
Physical therapists	103	198	(AIDS) HIV infection (042–044)	200**	35	140, 279
			Cancer of bone, connective tissue, skin, and breast (170–175)	179**	26	117, 262
			Non-Hodgkin's lymphomas (200, 2020–2022, 2028, and 2029)	151**	100	123, 184
Dentists	085	196	Air and space transport accidents (840–845)	643**	15	360, 1060
			Disorders of the peripheral nervous system (350–357)	218**	15	122, 359
			Chronic myeloid leukemia (2051)	213**	17	124, 342
Radiologic technicians	206	182	TB (010–018 and 137)	307*	12	159, 537
			(AIDS) HIV infection (042–044)	216**	73	170, 272
			Accidental poisonings (850–869 and 9292)	168*	25	109, 248

Emergency medical technicians (EMTs), paramedics, and other technologists	208	156 (EMTs alone)	(AIDS) HIV infection (042–044)	168**	199	146, 194
			Infectious and parasitic diseases (001–139)	124**	514	113, 136
			Alzheimer's disease (290, 3310, and 3311)	124*	99	101, 151
Dental hygienists	204	144	Accidental falls (880–888 and 9293)	174*	19	105, 272
			Cancer of ovary and other uterine adnexa (183)	167*	26	109, 244
			Cancer neoplasm of breast (174 and 175)	156**	92	126, 191
Speech therapists (speech language pathologists)	104	114	Mental disorders associated with solvent exposures (296–297, 300–301, and 3483)	360*	6	132, 783
			Multiple sclerosis and other demyelinating diseases of central nervous system (340–341)	279*	6	102, 607
			Cancer of brain and nervous system (191–192)	209*	11	104, 373
Health record technicians	205	98	(AIDS) HIV infection (042–044)	213**	27	140, 310
			Lymphatic cancer and multiple myeloma (202–203)	164*	31	112, 233
			Mental disorders related to substance abuse (291–292 and 303–305)	148*	33	102, 208

[a] Census Occupation Code (COC), U.S. Census Bureau, http://www.census.gov/hhes/www/ioindex/view.html

[b] Multiple cause PMR analysis was conducted using the ninth revision of the International Classification of Diseases (ICD-9) to code cause of death

[c] PMR is defined as the ratio of the age-adjusted proportion of deaths from a specific cause of death for a particular occupation or industry during a specified time period compared to the proportion of that cause among all industries or occupations during the same period, multiplied by 100. To test for statistical significance, two-sided 95% CIs are calculated, based on the Poisson distribution for observed deaths, and using the normal approximation to the Poisson for large numbers. A statistically significantly elevated PMR must be interpreted as a flag that suggests elevated risk that should be further evaluated for confounding factors.

[d] Separate employment figures for health aides (except nursing) and nursing aides, orderlies, and attendants are unavailable.

$*p < 0.05$; $**p < 0.01$.

ambulatory health care services, those unique to this subsector included air and space transport accidents, other lung diseases due to external agents, and myoneural disorders. Of the top 10 causes of death in hospitals, those unique to this subsector included asthma, disorders of the peripheral nervous system, and diffuse diseases of connective tissue and acute myeloid leukemia. Of the top 10 causes of death in nursing and residential care facilities, those unique to this subsector included accidents caused by nature; diabetes mellitus; endocrine, nutritional, metabolic, and immunity disorders; and motor vehicle traffic accidents. Of the top 10 causes of death in social assistance, those unique to this subsector included neurotic and personality disorders, mental disorders associated with solvent exposure, and infectious and parasitic diseases.

Table 2.21 presents the top 3 significantly elevated PMRs for each of the 18 largest HCSA occupations. Infectious diseases (AIDS, hepatitis, etc.) were among the top 3 significantly elevated PMRs for 11 of the 18 occupations, with AIDS being the most prevalent, accounting for thousands of deaths. Several cancers (small intestine, female genital organs, pancreas, bone, and Hodgkin's disease) were also among the top 3 significantly elevated PMRs for 11 of the 18 occupations. Other causes that were observed in more than one occupation included lymphatic cancer and multiple myeloma, drug-related deaths, air and space transport accidents, sarcoidosis, and mental disorders.

Data Source Limitations

The ability to describe the distribution and determinants of occupational injury, disease, and mortality has improved since the passage of the Occupational Safety and Health Act. However, occupational safety and health surveillance data remain fragmented and have substantial gaps. This section is illustrative of the fragmented nature of occupational health statistics, though providing the reader with information representative of the current statistical and surveillance data sources. As noted recently by Rosenman and colleagues, no comprehensive occupational injury and illness (acute and chronic) surveillance system exists in the United States, either at the national or state levels.[15]

Each of the data sources used herein has limitations, particularly those that attempt to quantify reports of occupational injury and illness (i.e., disease caused by exposures at work). For example, the design of SOII, conducted annually by the U.S. BLS, excludes ~22% of the labor force (self-employed workers, public sector workers, and individuals employed on farms employing 10 or fewer workers). In 1987, the National Academy of Sciences reported on deficiencies in the BLS's occupational health and safety statistical programs.[41] Despite improvements to the SOII, recent studies have reported that the SOII substantially undercounts acute nonfatal injuries, and is generally believed to undercount both acute and chronic illnesses from chronic chemical and other exposures. Recordkeeping and regulatory changes have been suggested as explaining recent declines in the magnitude of occupational injuries and illnesses.[18]

Selected data on occupational mortality are presented in this section. The strengths and limitations of death certificate data have been discussed at length.[42,43] The fact that many individuals do not die as a direct result of their work-related disease prompts NIOSH to maintain surveillance for all causes of death, underlying and contributing. Certifying physicians typically do not list all of a decedent's diseases on the death certificate. As with any analysis based on death certificate data, there is undoubtedly some misclassification of cause of death, and death certificates rely on usual industry and occupation, which may not reflect lifetime exposure histories.

Data that depend, either directly or indirectly, on physician reporting or recording of occupational or work-relatedness can be influenced significantly by the physician's ability or willingness to suspect and evaluate a relationship between work and health. These, in turn, are influenced by evolving medical/scientific information, and by the legal, political, and social environment. Some factors may lead to increased diagnosis and recording/reporting (e.g., the Occupational Safety and Health Act of 1970), while other factors may reduce occupational disease recognition or reporting by physicians (e.g., long latency between a work exposure and disease development, or concern about involvement in litigation).

Summary

The surveillance data described in this section provide information on the health status of workers in the HCSA sector. National surveillance systems show that this sector is particularly hazardous to workers in terms of nonfatal injuries and illnesses; the HCSA sector has the second largest share of all nonfatal injury and illness cases, as well as cases involving days away from work. Incidence rates for nonfatal injuries and illnesses in this sector are driven by nursing and residential care facilities and hospitals, with rates nearly double of those in all private industry. Predominant injuries and illnesses among health care workers include sprains and strains; injuries associated with slips, trips, and falls; overexertion/repetitive trauma injuries associated with patient lifting and assaults from patients; and SIs. Disease and cause-specific mortality data show significantly elevated causes of death in the sector for infectious disease, various cancers, and drugs. Limitations associated with the various reported data are also provided and underscore the importance of improved surveillance of acute and chronic illnesses and acute nonfatal injuries in this sector.

References

1. Bureau of Labor Statistics. Survey of Occupational Injuries and Illnesses. www.bls.gov/respondents/iif/home.htm
2. OSHA Recordkeeping Forms. www.osha.gov/recordkeeping/new-osha300form1-1-04.pdf
3. Bureau of Labor Statistics. Occupational Injuries and Illnesses Classification System Manual. www.bls.gov/iif/oshoiics.htm
4. Bureau of Labor Statistics. Census of Fatal Occupational Injuries. www.bls.gov/iif/oshcfoi1.htm
5. Bureau of Labor Statistics. Workplace Injuries and Illnesses in 2005, *USDOL News*, October 19, 2006. http://www.bls.gov/iif/oshwc/osh/os/osnr0025.txt
6. Bureau of Labor Statistics. Table SNR05, Incidence rate and number of nonfatal occupational injuries by industry, 2005. http://www.bls.gov/iif/oshwc/osh/os/ostb1611.txt
7. Bureau of Labor Statistics. Table SNR08, Incidence rates for nonfatal occupational illness by industry and category of illness, private industry, 2005. http://www.bls.gov/iif/oshwc/osh/os/ostb1614.txt
8. Bureau of Labor Statistics. Table SNR10, Numbers of nonfatal occupational illnesses by industry and category of illness, private industry, 2005. http://www.bls.gov/iif/oshwc/osh/os/ostb1616.txt
9. Bureau of Labor Statistics. Nonfatal occupational injuries and illnesses requiring days away from work, 2005. *USDOL News*, November 17, 2006. http://www.bls.gov/iif/oshcdnew.htm
10. Bureau of Labor Statistics. Table R-1, Number of nonfatal occupational injuries and illnesses involving days away from work by industry and selected natures of injury and illness, 2005. http://www.bls.gov/iif/oshwc/osh/case/ostb1657.txt
11. Bureau of Labor Statistics. Incidence rate of occupational injuries and illnesses involving days away from work with musculoskeletal disorders, private industry and health care and social assistance sector, United States, 2005. Special Request (unpublished) Analysis. USDOL, BLS, November 2006.
12. Bureau of Labor Statistics. Table R-8, Incidence rates for nonfatal occupational injuries and illnesses involving days away from work per 10,000 full-time workers by industry and selected events or exposures leading to injury or illness, 2005. http://www.bls.gov/iif/oshwc/osh/case/ostb1664.txt
13. Bureau of Labor Statistics. Fatal occupational injuries by selected worker characteristics and selected industry, all United States, private industry, 2005. http://data.bls.gov/GQT/servlet/RequestData
14. Bureau of Labor Statistics. Table 12, Number and median days away from work of occupational injuries and illnesses with days away from work involving musculoskeletal disorders by selected occupations, all United States, private industry, 2005. http://www.bls.gov/iif/oshcdnew.htm
15. Rosenman KD, Kalush A, Reilly MJ, Gardiner JC, Reeves M, and Luo Z. How much work-related injury and illness is missed by the current national surveillance system? *J Occup Environ Med* 48:357–365, 2006.

16. Leigh JP, Marcin JP, and Miller TR. An estimate of the US government's undercount of nonfatal occupational injuries. *J Occup Environ Med* 46:10–18, 2004.
17. Azaroff LS, Levenstein C, and Wegman DH. Occupational injury and illness surveillance: Conceptual filters explain underreporting. *Am J Public Health* 92:1421–1429, 2002.
18. Friedman LS and Forst LS. The impact of OSHA recordkeeping regulation changes on occupational injury and illness trends in the U.S.: A time-series analysis. *Occup Environ Med* 64:454–460, 2007.
19. Panlilio AL, Orelien JG, Srivastava PU, Jagger J, Cohn RD, Cardo DM, et al. Estimate of the annual number of percutaneous injuries among hospital-based healthcare workers in the United States, 1997–1998. *Infect Control Hosp Epidemiol* 25:556–562, 2004.
20. CDC. *NIOSH Alert: Preventing Needlestick Injuries in Health Care Settings.* Department of Health and Human Services, CDC, Cincinnati, OH, DHHS (NIOSH) Publication No. 2000-108, 1999.
21. Perry J, Parker G, and Jagger J. EPINet report: 2003 percutaneous injury rates. *Adv Expo Prev* 7:42–45, 2005.
22. The National Surveillance System for Hospital Health Care Workers Summary Report for Data Collected from June 1995 through July 1999. Available at http://www.cdc.gov/ncidod/hip/NASH/report99.PDF
23. Sharps Injuries among Hospital Workers in Massachusetts, 2004: Findings from the Massachusetts Sharps Injury Surveillance System. Published April 2007. Available at http://www.mass.gov/Eeohhs2/docs/dph/occupational_health/injuries_hospital_2004.pdf
24. Cone J. Needlestick injury surveillance in California, 1998–1999. *Calif Morb* September 2000.
25. Personal communication with Ian Williams, Division of Viral Hepatitis (DVH), CDC, February 1, 2002 and Annemarie Wasley, DVH, CDC, April 20, 2007.
26. Do AN, Ciesielski CA, Metler RP, Hammett TA, Li J, and Fleming PL. Occupationally acquired human immunodeficiency virus (HIV) infection: National case surveillance data during 20 years of the HIV epidemic in the United States. *Infect Control Hosp Epidemiol* 24:86–96, 2003.
27. Pruss-Usten A, Rapiti E, and Hutin Y. Estimation of the global burden of disease attributable to contaminated sharps injuries among health-care workers. *Am J Ind Med* 48:482–490, 2005.
28. Sepkowitz KA and Eisenberg L. Occupational deaths among healthcare workers. *Emerging Infect Dis* 11:1003–1008, 2005.
29. CDC. Updated U.S. public health service guidelines for the management of occupational exposures to HBV, HCV, and HIV and recommendations for postexposure prophylaxis. *Morb Mortal Wkly* Rep 50:RR-11, 2001.
30. Surveillance of Occupationally Acquired HIV/AIDS in Healthcare Personnel, as of December 2006. Available at http://www.cdc.gov/ncidod//dhqp/bp_hcp_w_hiv.html, accessed on January 22, 2006.
31. Mahoney FJ, Stewart K, Hu H, Coleman P, and Alter M. Progress toward the elimination of hepatitis B virus transmission among health care workers in the United States. *Arch Intern Med* 157:2601–2605, 1997.
32. Household Data Annual Averages, Current Population Survey, Bureau of Labor Statistics, U.S. Department of Labor, 1995–2005.
33. CDC. Surveillance for Acute Viral Hepatitis—United States, 2005. Surveillance Summaries, March 16. *Morb Mortal Wkly Rep* 56(SS-3), 2007.
34. Henderson DK. Managing occupational risk for hepatitis C transmission in the health care setting. *Clin Microbiol Rev* 16:546–568, 2003.
35. Online Tuberculosis Information System (OTIS), National Tuberculosis Surveillance System, United States, 1993–2004. U.S. Department of Health and Human Services (US DHHS), Centers for Disease Control and Prevention (CDC), Division of TB Elimination, CDC WONDER Online Database, March 2006.
36. CDC. Reported Tuberculosis in the United States, 2005. U.S. Department of Health and Human Services, CDC, Atlanta, GA, September 2006.

37. Rosenstock L. et al. *Clinical Occupational and Environmental Medicine*, 2nd edn. Elsevier Saunders, Philadelphia, PA, 2005, pp 145–262.
38. Maloney SA et al. Efficacy of control measures in preventing nosocomial transmission of multidrug-resistant tuberculosis to patients and healthcare workers. *Ann Intern Med* 122(2):90–95, 1995.
39. Tarantola A, Abiteboul D, and Rachline A. Infection risks following accidental exposure to blood and body fluids in health care workers: A review of pathogens transmitted in published cases. *Am J Infect Control* 34:367–375, 2006.
40. CDC/NIOSH. http://www.cdc.gov/niosh/topics/surveillance/NOMS/default.html
41. National Research Council. *Counting Injuries and Illness in the Workplace: Proposals for a Better System*, ES Pollack and DG Keimig (Eds.). National Academy Press, Washington DC, 1987.
42. Dubrow R, Sestito JP, Lalich NR, Burnett CA, and Salg JA. Death certificate-based occupational mortality surveillance in the United States. *Am J Ind Med* 11:329–342, 1987.
43. Burnett C, Maurer J, Rosenberg, HM, and Dosemeci M. *Mortality by Occupation, Industry and Cause of Death, 24 Reporting States (1884–1988)*. NIOSH, U.S. Department of Health and Human Services, Cincinnati, OH, DHHS (NIOSH) Publication No. 97–114, 1997.

3

Back Injury Prevention in Health Care

William Charney
Healthcare Safety Consulting

Anne Hudson
Coos County Public Health Department

Susan Gallagher
Celebration Institute

John D. Lloyd
James A. Haley Veterans' Hospital

Andrea Baptiste
James A. Haley Veterans' Hospital

Audrey Nelson
James A. Haley Veterans' Hospital

Mary Matz
James A. Haley Veterans' Hospital

Bernice D. Owen
University of Wisconsin–Madison

Arun Garg
University of Wisconsin–Milwaukee

Jocelyn Villeneuve
Association for Health and Safety Sector Social Affairs

Justin Gilmore
LIKO Incorporated

Willis Martin
Martin Technologies

Lift Teams: A Proven Method to Reduce Back Injury in Health Care Workers 3-2
Historical Background • Causes of Back Injury in Health Care Workers' Compensation • Goals: Mechanization • What Does Not Work? • Lifting Team Strategy • Lift Team Successes • Review of Some Data Published in Peer Review • Lift Team Advantages • Transfer Team Implementation Steps • Conclusions
Acknowledgment 3-12
References 3-12
Back Injury to Health Care Workers from Patient Lifting: The Legislative Solution 3-13
Overview of Back Injury from Manual Patient Lifting • No Manual Lift—Safer for Patients • Patient-Handling Policy in the United Kingdom • "No Lifting" in Victoria, Australia • U.S. Model for Back Injury Prevention in Nursing Homes • American Nurses Association (ANA) and Safe Patient Handling • Patient Handling in America: A Call for "Safe Patient Handling—No Manual Lift" Legislation • "Nurse and Patient Safety and Protection Act of 2007" Introduced
References 3-31
Disposable 3-34
The Meaning of Safety in Caring for the Larger, Heavier Patient 3-43
What Is Obesity? • The Relationship between Willingness to Accommodate and Sensitivity • Safe Practices in WLS • The Challenges of Unplanned Access • Creating a Safe Work Environment • Conclusion
References 3-61
Patient-Handling Technologies 3-64
Introduction • Technologies to Assist with Vertical Transfer of Patients • Technologies to Assist with Lateral Transfer of Patients • Bariatric Specialty Equipment Needs • Turning and Repositioning • Equipment and Use
References 3-75
Safe Patient Handling: Evidence-Based Solutions 3-76
Introduction • High-Risk Patient-Handling Tasks • Evidence-Based Solutions for High-Risk Patient-Handling Tasks • Conclusions • Recommendations
References 3-87
Supplemental Readings 3-91

An Ergonomic Approach to Reducing Back Stress in Nursing Personnel 3-92
Causes of Back Pain • Approaches to Prevention • Goal 1: Determining Stressful Tasks • Goal 2: Ergonomic Evaluation of Stressful Tasks • Goal 3: Testing Approaches to Decrease Stressfulness (Locating Assistive Devices) • Goal 4: Laboratory Study • Goal 5: Application of Findings to Clinical Areas • Ergonomic Approach within a Hospital Setting • Summary
References 3-108
The Ceiling Lift: An Efficient Way of Preventing Injuries among Nursing Staff 3-109
Introduction • Methodology • Results • Conclusion
Bibliography 3-114
Slings for Amputees 3-115
General Sling Usage • Removing Bariatric Patients from a Vehicle • Safe Lifts Outside of an Emergency Department • Maneuvering Floor Lifts on Carpeted Surfaces • Use of Ceiling Lifts with Hip Surgery Patients • Sling Washing Instructions • Mixing and Matching Equipment • Operating Room Transfers • Patient Assessment Tool • Transport vs. Transfer
Establishing the Standard of Care™ with the Martin Chair-A-Table® 3-122
Benefits to Institutions 3-124
Benefits to Risk Managers and Insurers 3-124
Benefits to Caregivers 3-124

Lift Teams: A Proven Method to Reduce Back Injury in Health Care Workers

William Charney

Historical Background

Back injuries remain one of the most significant injuries in the United States and according to the Bureau of Labor Statistics (BLS) data "health care work" is the leading industry for musculoskeletal disorders (MSDs) if all categories of health care worker exposure to patient lifting are added together. Transportation (truck drivers) is officially rated number 1 for MSDs at 45,327. However, CNAs, RNs, and LPNs are 44,660, 12,074, and 5,598 respectively for a total of 62,332 or 138% above truck drivers.[1] Nurses' annual prevalence rate for back injury is 40%–50%.[2]

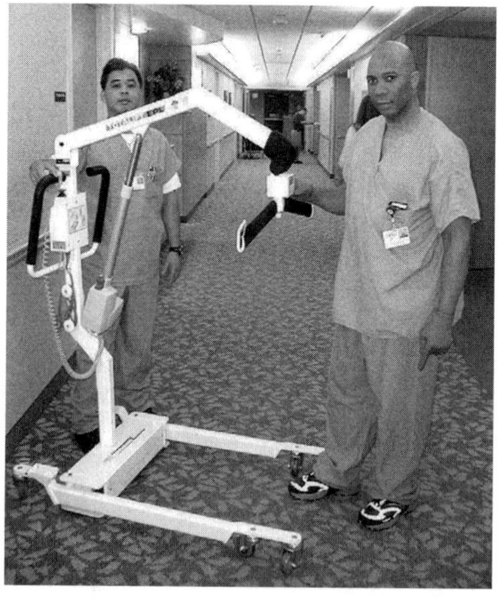

Lifting team

Manual lifting, transferring, and repositioning of patients are the activities and job tasks associated with MSDs in health care workers.[3–6] Marras[7] showed in biomechanical modeling that manual lifting and repositioning can put a health care worker's risk for injury at 75% for each task. Most of the lifting of patients, according to the Marras study, is in a compressive force range of between 4000 and 12,000 N, well above the recommended National Institute for Occupational Safety and Health (NIOSH) Guidelines of 3400 N and some are well above the upper limit of the NIOSH Guidelines of 6400 N.

Causes of Back Injury in Health Care Workers' Compensation

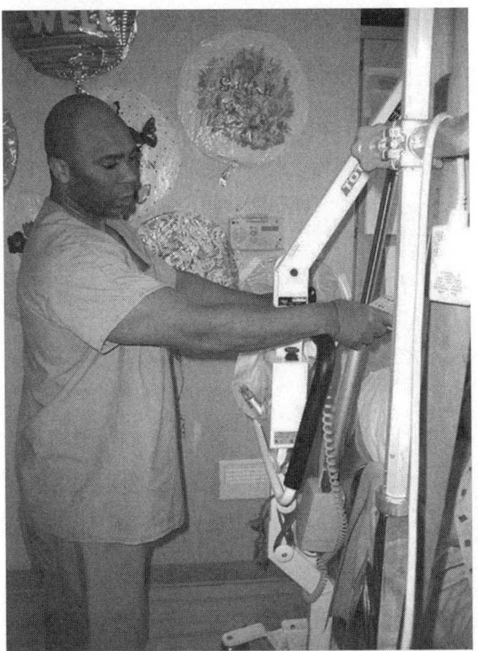

Lift team member adjusting lift

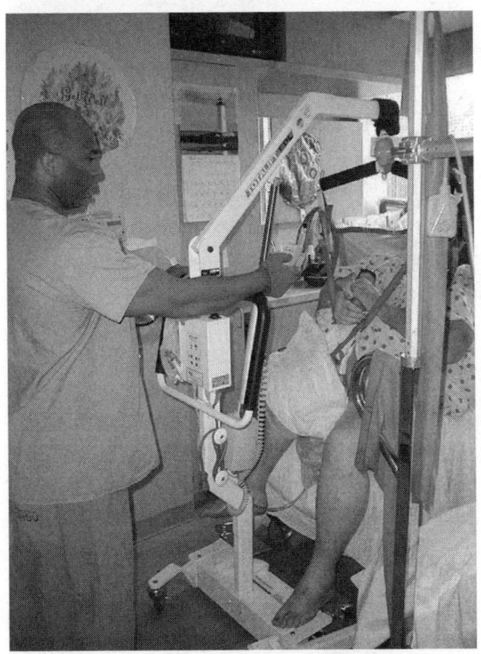

Mobilizing bariatric patient

According to Haiduven,[8] there are organizational factors and environmental factors that influence the incidence of back injury to health care workers. Organizational factors include time pressure to perform the task; lack of available lifting aids; and lack of personnel to assist with the lift task. Environmental factors include space restrictions; health care facilities not ergonomically designed so mechanical lifts, if available, do not fit into rooms, bathrooms, or bathing rooms.

Another factor that has led to these high rates of injury is lack of ergonomic legislation that would mandate use of mechanical equipment to reduce the compressive loads on the health care workers. Voluntary prevention programs are not keeping pace with the rates of injury. In nursing homes in particular, the rates of back injury have doubled in the last 10 years and in acute care the rates have increased by 40%.

A continuing problem is the lack of understanding of the cost of these injuries to the health care system.

Few CEOs understand the science of cost–benefit analysis for injury. In most health care facilities, back injury accounts for between 35% and 60% of workers' compensation costs. And that is computed in direct cost dollars, compensation premium costs that include workers' compensation costs, and medical costs. It is widely accepted now that indirect costs are four times the direct costs. These include replacement costs, recruitment costs, manager time costs, training and orientation costs, etc.

One can consider the back injury rates to which health care workers have been exposed to over the decades as part of the general malaise and problematic in occupational health generally. The United States is 27th per-capita on occupational health spending. A disservice has been perpetuated on the occupational health community in the United States in the form of a myth that occupational health prevention is a cost rather than part of the profit of the health care system.

All the savings generated as part of sound scientific prevention programs can be reinvested into the system to make it more efficient and profitable at the bedside.

Health care workers are continuously exposed to MSDs. One study[9] showed that the average ward nurse lifts 1.8 tons per shift. Another study[10] showed that 83% of nursing personnel work despite back pain and 60% fear a disabling back injury. An American Nurses Association Health and Safety Survey of 2001 which surveyed a cohort of 4826 nurses reported that 60% had a fear of developing a severe back injury.[11]

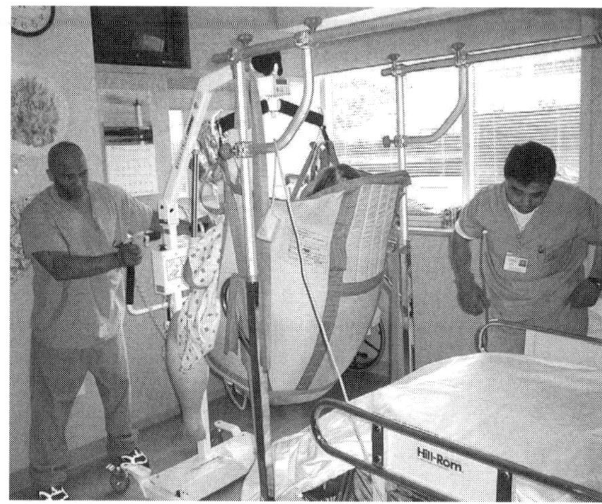

Transferring patient using vertical lift

Goals: Mechanization

The goal in the health care delivery system is to mechanize patient transfers and repositionings in order to reduce the compressive forces on the lumbar spine, and not turn the caregivers into patients. Other industries have mechanized to great advantage in reducing their injury claim rates. One element in the mechanization process is the Lift Team, as they are part of the overall model that mandates the use of mechanical equipment during patient transfers and removes nursing exposure to the manual loads. Lift Teams put "risk where it can be controlled" in a small group of teams rather than in large nursing departments.

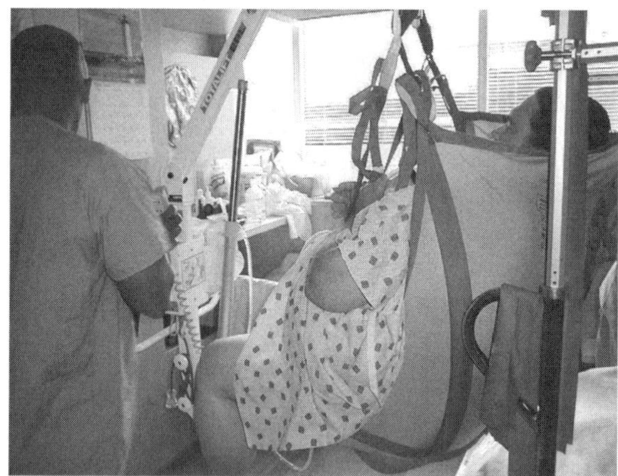
Transferring bariatric patient

The health care industry mechanizes for patient diagnostics and treatment; however, the second part of the equation of mechanization for health care worker safety is lagging far behind. The manufacturers of lifting equipment and patient movement technology have done a credible job in providing the industry with technical options to protect health care workers. The research and development phase of patient movement technology has produced advanced floor lift designs; ceiling mounted transfer designs; sit to stand models; lifts that remove patients from cars, baths, and showers mechanically; lateral transfer stretchers; bariatric lifts; beds that can reposition patients; and repositioning devices from slip sheets to handles on beds that can position patients (see chapter); and, despite peer review, science showing an extremely positive cost–benefit and payback within 12–15 months for equipment and other short payback periods for program implementation.[10–13] However, the industry is slow to implement.

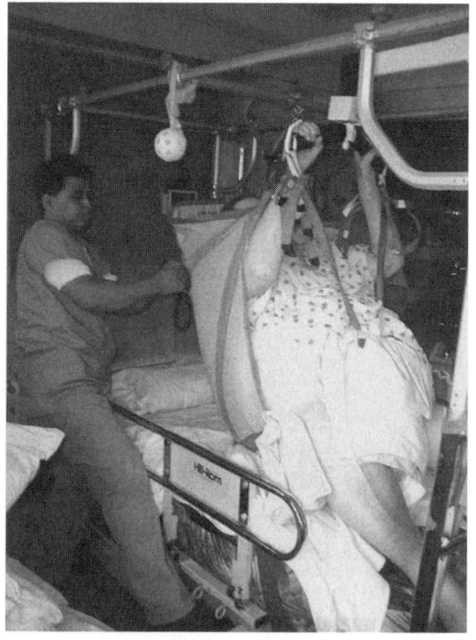
Lift team with patient

What Does Not Work?

1. **Manual lifting:** Manual lifting is the cause of low-back disorder risk during patient handling. According to different researchers simple two-dimensional models have indicated large loads during the various patient transfers (Garg and Owen, 1994). According to research, most, if not all manual lifts, exceed the recommended limits of 3400 N of force considered the safety limit by NIOSH. At 3400 N, we first begin to see vertebral and cartilage endplate microfractures. At between 3400 and 6400 N (upper NIOSH limit), up to 50% of population would be expected to develop endplate microfractures. And above 6400 N of force, more than 50% of people would be expected to develop endplate microfractures. A one-person reposition using the "hook" method approaches 9000 N of force. And according to Marras lower-back disorder probabilities ranged from 77% to 91% with compressive force ranges of between 3400 and 6400 N of force for manual transfer lifts.

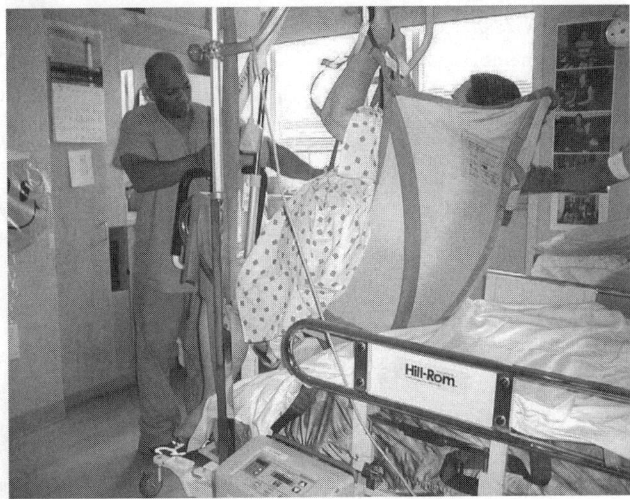

Lift team member directing patient

2. **Training in biomechanics**: Historically, it was widely accepted that classes in body mechanics and training in lifting techniques could prevent these job-related injuries. It was usually the first and last method attempted at controlling back injury in health care. However, 35 years of research disputes this belief.[14] These efforts have continuously failed at reducing rates of back injury in patient care settings. Modeling for biomechanics was based on research conducted with predominately male subjects who lifted boxes vertically from the floor. Unlike lifting a box with handles, a patient lift is much more problematic. A patient's weight is not evenly distributed and the mass is asymmetric, bulky, and cannot be held close to the body. Additionally, patient movement is unpredictable with combative patients, patients that lose balance, differentials in pain thresholds, etc. This vertical model does not account for all the lifting, turning, and repositioning of patients on the lateral plane which uses weaker muscles of the arms and shoulders instead of legs. Twisting, torquing, awkward positions during patient handling are 20%–30% of working time with patients, producing stressful compressive forces on the disks and muscles in the spine. Biomechanics training has never controlled for all these variables and, therefore, is not effective in reducing injuries.
3. Nursing schools are still teaching manual lifting as part of the curriculum. Owen pointed out that in her studies that over 95%[15] of nursing schools are still teaching this method. This creates a discordant paradigm with the new science conflicting with the current curriculum.

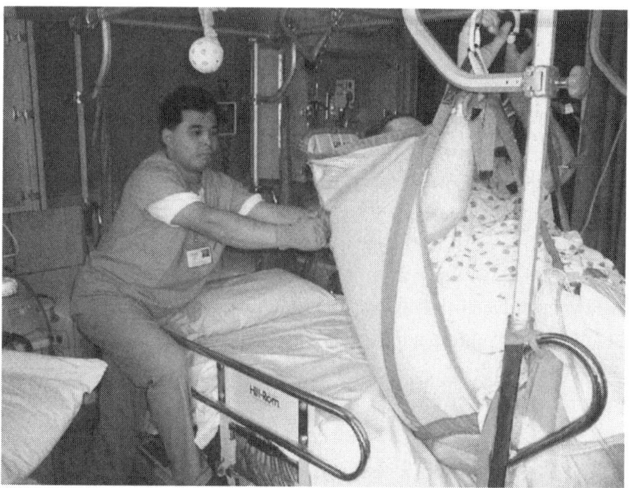

Lift team member assisting lift

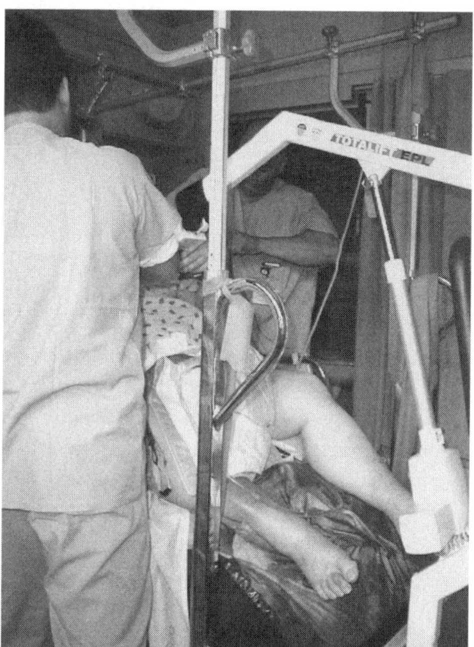

Lift team members assisting lift

Lifting Team Strategy

1. **Lifting is a skill rather than a random task.** Patient transfers and repositions are complex movements due to patient individuality, pain thresholds, room spacing, patient weight, etc. There are over 20 different types of patient transfers and repositionings that must be learned, from vertical transfers and lateral transfers; sit-to-stand paradigms; walking of patients; dressing patients, car transfers, toileting patients, bathing patients, turning patients, rolling patients; and ambulance transfers. All of which have been shown to create excessive compressive forces on the spine. Additionally, there are over 20 different types of mechanical equipment available that one must learn to manipulate; including, but not limited to vertical lifts, lateral lifts, sit-to-stand lifts, bath lifts, shower devices, ceiling lifts, toilet lifts, and car lifts. All with three or four different type of slings are to be used for different patients or type of lifts. All these maneuvers and equipment

require a specialized skill set to be performed in a safe manner for both patient and health care worker. Then there are even more subsets. Bariatric patients pose another risk-management challenge and they need to be handled safely and with dignity of care.

Patients have no handles and a 10 lb load lifted away from the body exerts a force of 100 lb on the lumbar spine.[16] The average age of U.S. nurses is 45 and many thousands have had a prior back episode, which increases their potential for an acute injury. Obesity in patients has increased the risk to caregivers who lift manually and the need for a higher level of training to reach the level of risk of exposure has never been mandated by the health care establishments. "Professionalizing" the task reduces all the variables due to the expertise of the professional, and this in turn contributes to the safe movement of patients. Concentrating a skill level in professional Lift Teams creates a safety factor for the staff previously exposed to random lifting. It guarantees that the lift will be done according to all the safest techniques, standards, and the use of equipment, applications of training and awareness of lift principles thereby reducing the variables that create injury potential.

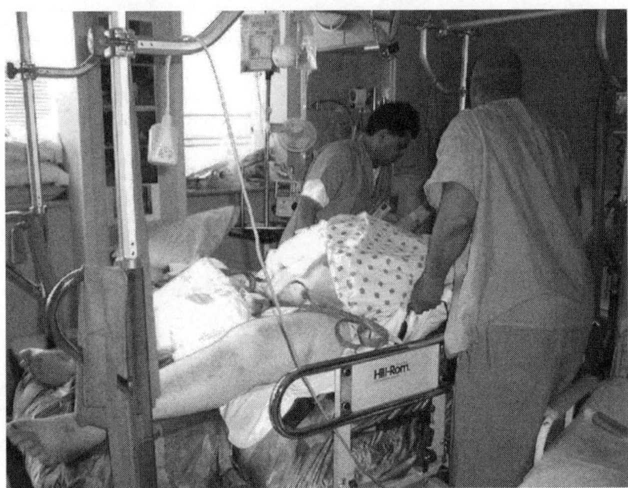

Saving nurses 1.5 h per shift

2. **Putting risk where it can be controlled**. This is a risk-management strategy that considers the mathematics of numbers of people exposed and the number of variables that contribute to incidence. In a large nursing department that includes RNs, LPNs, and others exposed to patient transfers, the challenges are formidable that the safety parameters will be followed in ways that reduce biomechanical forces. Training of 1000 employees exposed to patient transfers in an acute care setting, including per diems and others, on all the parameters of a safe patient handling program is a risk-management nightmare given turnover rates of 68% (American Hospital Association [AHA]), vacancy rates of 8.76% (AHA), per diems, release time for training, etc. The variables are more controllable in a set of teams that the criteria such as mandated use of mechanical equipment, previsualization of lift, and coordination of lift will be followed.

Lift Team Successes

Lifting teams have been studied for approximately 13 years. The first published study was by Charney.[17] This study was done at San Francisco General Hospital, a 250 bed acute care facility, and, after 1 year, the rate of injury on the Lift Team shift was reduced to zero. Since 1991, at least nine other studies by different authors have been published in peer review.[8,18–24] These studies have included approximately 30 different acute care facilities ranging in size from 200 to 1500 beds. There are also additional unpublished data from hospitals showing excellent results. For example, Tampa General Hospital (TGH), a 600-bed facility, has been running Lift Teams for 2 years and is showing pre-Lift Team Patient Handling injuries in 1999 of 105 injuries to 38 injuries 2 years post-Lift Team. In fact, all the published studies have all shown remarkable reductions in lost time injury, lost days, restricted days, compensation and medical dollars, and even one study (Hefti et al.[19]) shows reductions in turnover rates and vacancy rates attributed to the Lift Teams. There is now ample evidence proving the efficiency of Lift Teams in reducing injury rates and statistically significant data in peer review proving the effectiveness of Lift Teams in most if not all injury categories.

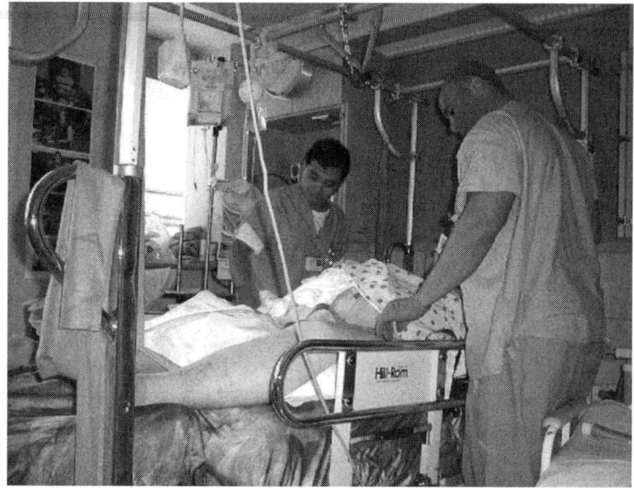

Lift team member preparing patient sling

Review of Some Data Published in Peer Review*

1. Charney, "Ten Hospital Study" 23.5 combined years of Lift Team Experience: 69% reduction of injury rates, 62% reduction of incident rates, 90% reduction in lost days, and 72% reduction in workers' compensation costs.
2. Charney, "18 Hospital Study" showed statistically significant reductions in all categories of risk in all hospitals studied, injury rates, compensation dollars, etc.
3. Kaiser Northern California Data: All hospitals' aggregate data showed a 7% overall reduction in workers' compensation costs in the facilities using Lift Teams and a 17% increase in workers' compensation costs in facilities not implementing Lift Teams. All Kaiser facilities in California are now required to have Lift Teams.
4. New data are now showing a 39% decrease in workers' compensation back claims in nursing department at Kaiser Lift Team Hospitals.
5. Harbor View Medical Center* 400 beds; Seattle Washington Anecdotal Data: Lift Teams reduced lost workdays from 1474 to 470 in the first year of implementation.
6. Tri-City Health Systems, San Diego, CA: 3 year annual average of workers' compensation data: $242,000 pre-Lift Team costs were reduced to $14,000 after 1 year of implementation. Lost days, prior to Lift Team, were reduced from 788 days to 0.
7. Sutter Health Systems in California: 6 year review showed a frequency reduction from 12.9

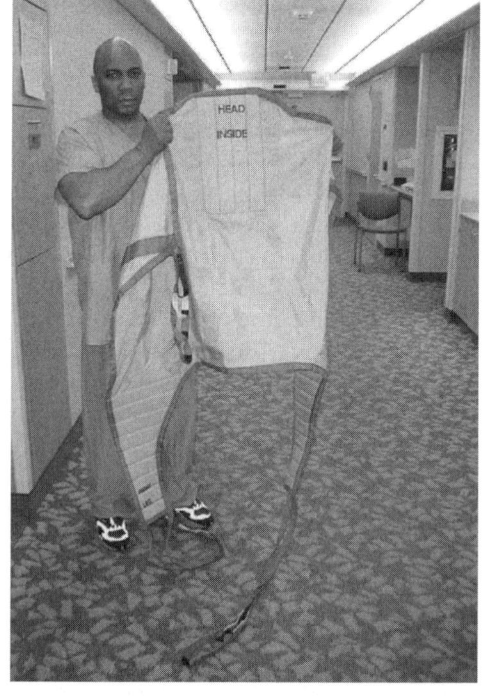

Lift team member showing sling

* Unless otherwise noted by "*" as anecdotal.

to 1.3 and an indemnity reduction from 9.7 to 0.8. Sutter also showed a reduction of workers' compensation claim dollars from a pre-Lift Team for patient transfer claims of $2,248,817 to post-Lift Team of $1,551,875, reduction of $696,942 over 2 years.

8. Woodland Healthcare, a member hospital of Catholic Healthcare West, crossed trained a Transport Team to a Lift Team. Restricted days were reduced from an average of 160 days pre-Lift Team to 2 days post.
9. The Mayo Clinic: A 1500-bed facility reduced restricted days by 361%; a $310,000 reduction in cost of back injury over 4 years and the Teams performed over 60,000 lifts without an injury to staff or Team members until the Teams were subject to budget cuts.
10. Alaska Native Medical Center: 5 year comparable data showed Lift Teams reduced lost days from 61/100 FTEs to 20/100 FTEs. The Teams perform 10,000 patient contacts per year.
11. Sioux Valley Medical Center: A 500-bed hospital with 2 year data reduced injury claims by 69%, reduced lost days by 95%, reduced restricted days by 88%, reduced average cost per injury due to patient handling from $9894 to $1099. Medical cost savings direct and indirect costs $690,504.
12. University of Chicago Hospital*: A 576-bed hospital performs 30–60 lifts per day: running two teams 7 days per week covering 8 a.m. to 9 p.m. Reduced injuries 40% for first year.
13. Tampa General Hospital: A 600-bed facility implemented Lift Teams on day and evening shifts 7 days per week and installed ceiling lifts as their technology. They showed a 62% reduction in workers' compensation claims in the first year and a reduction in the patient handling injury rate from pre-Lift Team of 2, to post-Lift Team in 2003 to 0.5. RN injuries went from a high pre-Lift Team of 47 in 1 year to 21 post in 2003.
14. Caska reported on Lift Teams in three nursing home units in one medical center. Zero injuries on units and shifts that had access to Lift Teams.
15. Oregon Health & Science University Medical Center*: 1700 lifts are performed each month, covering all shifts on all days. The team uses floor lifts for total body transfers. 75% of calls are for

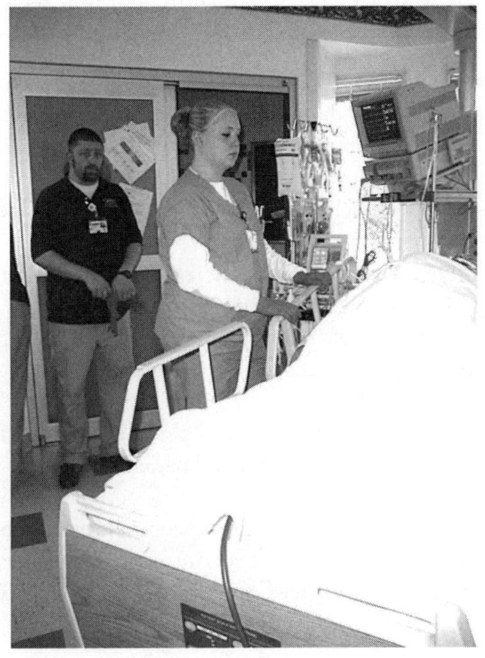

repositionings. Average response time for Team is 3 min with an average time spent performing the lift of 9.1 min. 62% of all services were in the intensive care units with medical/surgical accounting for the balance of usage.

16. Donaldson Study in a 296-bed facility showed a 60% reduction in claim frequency in 1 year and a 90% reduction in claim frequency over 6 years. A 98.6% decrease in incurred loses for the 6 year period.
17. Davis Study, University of Washington Medical Center: 350-bed showed a 62.7% reduction in sick calls due to back injury on Lift Team shift and an 85.5% reduction in lost workdays.

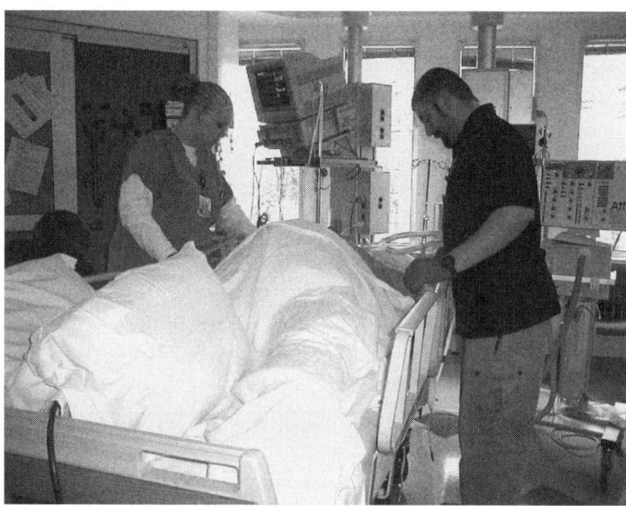

Lift Team Advantages

Lift Teams have several advantages that have been seen in both peer review science and in anecdotal information.

1. The use of Lift Teams allows the facilities to return injured nurses back to patient bed-side care without fear of being exposed to dangerous compressive loads and without fear of reinjury. In the United States, there are thousands of injured nurses who are not working due to severe or chronic back pain. This group would be able to return to their careers. Second, Lift Teams allow previously injured nurses who are still working but, who are at greater risk for a more acute episode, to work without fear of reinjury.
2. Lift Teams are inherently safer for patients. Lift Teams are professional patient transfer experts. In subjective findings (Oregon Health Science University Medical Center), it was found that patients have more confidence in Lift Teams to move them safely and with less trauma than ward nurses. Second, there are less patient falls during transfers with Lift Teams, which reduce hospital liability.
3. Lift Teams are saving nurses an average of 1.5 h per shift, which can be then used for patient bed-side care. This time saving is cost–benefit and patient benefit.
4. Lift Teams, due to their training, are able to ambulate complex or "dead weight" patients more often, which can amount to shorter patient stay outcomes for this cohort of patients.
5. Lift Teams, as a program design, assist in the recruitment of new nurses.
   a. Lift Teams have now shown in peer review science an ability to reduce turnover rates (Hefti et al.[19]).

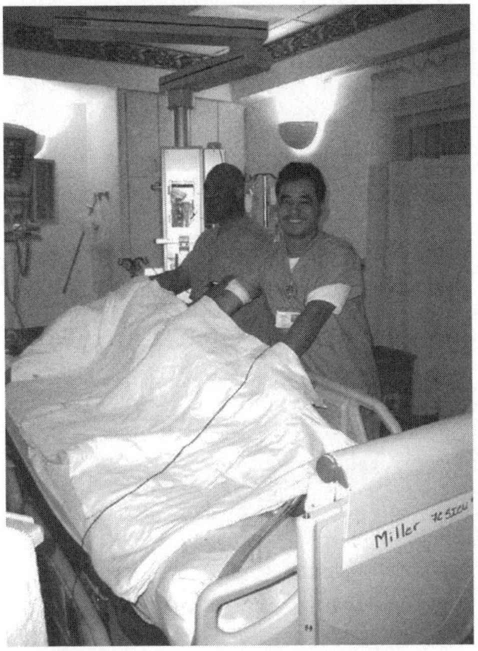

Relieving nurses of lifting

Some anecdotal data from a nurse survey taken at TGH in 2004 are as follows:

- For someone with a previous back injury, the Lift Team is a life saver.
- The Lift Team is extremely helpful in turning patients for skin checks.
- Nurses are less tired at the end of their shifts.
- We do not have to pull other nurses from patient care during lifts and repositions.
- My back no longer hurts at the end of the day.
- The Lift Team is the most valuable resource for the ICU nurses. Patients would not be turned as often and staff would experience more strain and fatigue.
- I love the Lift Team. It was one good asset when considering employment at TGH.
- Thanks for saving our backs.
- Patients get out of bed more.
- Increased patient satisfaction with transfers.
- More time for other nursing duties.

Transfer Team Implementation Steps

1. Task force committee: Oversees all aspects of implementation
   a. Consists of nursing, transportation, administration, safety, physical therapy, human resources, etc.
   b. Types of lifts to be performed
   c. Units to be covered
   d. Equipment purchases
   e. Policy and procedures
   f. Budgeting and business plan
   g. Outreach education for facility
2. Policy and procedures: Written P&P: Creating a commitment that is facility wide with disciplinary procedures for accountability
   a. Nursing must call the Team for all total body transfers
   b. The patient handling specialist (PHS) will mediate any problems that arise
   c. Creation of a bariatric safety system
3. Equipment inventory and process
   a. Selection of all patient handling equipment
   b. Selection of all slings
   c. Piloting of all new equipment
4. Appointment of a PHS
   a. Daily oversight and problem solving
   b. Coordinates patient screening and assessment
   c. Reports to Task Force Committee
5. Screening patients for lift handling status: Resident assessments
   a. All new admissions
   b. Updating of patient status for lift

6. Team administration
   a. Team management
   b. Team data log collection
   c. Distribution of Teams
7. Team communications systems
   a. On-call systems
   b. Hospital by rounds
   c. Prioritization of calls
8. Team hiring and selection process
   a. Hiring criteria
   b. Selection process
   c. Shifts
9. Team training modules
10. Hospital data analysis: Tracking data systems
11. Questionnaires for FAC performance evaluations: System evaluations of process
12. Roles and responsibilities of staff
13. Transfer team tracking documentation
14. Quality assurance

Conclusions

The Lift Team model has been working very successfully for the last decade in acute care hospitals. There is a wealth of data showing reductions in all categories of injury, as well as data showing the positive cost–benefit of having this model implemented. The risk reduction model of putting "risk where it can be controlled" and that "lifting and transferring patients is a skill rather than a random task" has a shown validity in the practice of busy acute care hospitals. Given all the variables within the paradigm of patient movement, the transferring of patients needs to be considered a profession with a specialized skill set that cannot be applied successfully by busy nursing personnel. Second, the physical design and layout of acute care facilities create hazardous conditions within, which to attempt safe lifting and transferring; rooms too small, bathrooms not large enough to accommodate mechanical equipment, doors too small for bariatric patients, sharp angles, inclines, rugs, etc.; all of which suggests a team approach to minimize the danger of exposing this population of nursing caregivers. Since it has been found in the American Nurses Association survey of 2001 that 88% of nurses think about leaving the profession due to the "physical demands" of the work, it is the responsibility of the health systems that employ them to reduce these physical work stresses. "One unique and proven method is the Lift Team."

Acknowledgment

Lift team pictures are courtesy of Oregon Health Science University Medical Center.

References

1. U.S. Department of Labor, Bureau of Labor Statistics, Non-fatal occupational injuries. Table 12. U.S. Department of Labor, Bureau of Labor Statistics, Washington, DC, 2002.
2. Edlich RF, Woodard CR, and Haines MJ. Disabling back injuries in nursing personnel. *J Emerg Nurs* 27(2):150–155, April 2001.
3. Caska BA, Patnode RE, and Clickner D. Feasibility of nurse staffed lift system. *J AAOHN* 46(6):283–288, 1998.
4. Leighton DJ and Reilly T. Epidemiological aspects of back pain. *J Occup Med* 45:263–267, 1995.

5. Owen BD and Garg A. Four methods for identification of most back stressing tasks. *Int J Econ* 9:213–220, 1992.
6. Stobbe TJ, Plummer RW, Jensen RC and Attfield MD. Incidence of low back pain. *J Safety Res* 19:21–28, 1988.
7. Marras W. A comprehensive analysis of low back disorder risk and spinal loading during the transferring and repositioning of patients. *Ergonomics* 42(7):904–926, 1999.
8. Haidavan D. Lift teams in healthcare facilities: A literature review. *AAOHN J* 51(5):210–218, 2003.
9. Tuohy-Main K. Why manual handling should be eliminated. *Geriaction* 15:10–14, 1997.
10. Worthington K. Stress and overwork top nurses concerns. *Am J Nurs* 101(12):96, 2001.
11. ANA Nursing World, Health and Safety Survey, September 2001.
12. Charney W. How to accomplish a responsible cost–benefit back injury analysis in the health care industry. In: W Charney and A Hudson (eds.), *Back Injury among Healthcare Workers: Causes, Solutions, and Impacts*, Chapter 5. CRC Press, Boca Raton, FL, 2003, pp. 41–47.
13. Nelson A. Equipment for safe patient handling and movement. In: W Charney and A Hudson (eds.), *Back Injury among Healthcare Workers: Causes, Solutions, and Impacts*, Chapter 9. CRC Press, Boca Raton, FL, 2003, pp. 121–135.
14. Ergonomics Technical Advisory Group. *Patient Care Ergonomics Resource Guide: Safe Patient Handling and Movement*, Patient Safety Center (Tampa, FL), Veterans Health Administration, October 2001, p. 6.
15. Owen BD. Magnitude of the problem. In: W Charney and A Hudson (eds.), *Back Injury among Healthcare Workers: Causes, Solutions, and Impacts*, Chapter 2. CRC Press, Boca Raton, FL, 2003, pp. 5–12.
16. Charney W, Zimmerman K, and Walara E. The lifting team. A design method to reduce lost time back injury in nursing. *AAOHN J* 39(5):231–234, 1991.
17. Charney W. Preventing back injuries to healthcare workers using lift teams: Data for 18 hospitals. *J Healthcare Safety* 1(2):21–29, Spring 2003.
18. Charney W. Lift team method for reducing back injury: A 10 Hospital Study. *AAOHN J* 45(6): 300–304, 1997.
19. Hefti K et al. Back injury prevention: A lift team success story. *AAOHN J* 51(6):246–251, 2003.
20. Charney W. Reducing back injury in nursing: A case study using mechanical equipment and a hospital transport team as a lift team. *J Healthcare Safety* 4(3):117–120, 2000.
21. Meiteneun E et al. Effects of focusing ergonomic risk factors on patient lifting teams. *J Healthcare Safety* 3(5), 1999.
22. Donaldson A. Lift team intervention: A six year picture. *J Healthcare Safety* 4(2):65–68, 2000.
23. Caska B. Implementing and using a nurse staff lift team. *AOHP*, Spring 2000.
24. Davis A. Birth of a lift team. *J Healthcare Safety* 1(2), 1999.

Back Injury to Health Care Workers from Patient Lifting: The Legislative Solution

Anne Hudson

Overview of Back Injury from Manual Patient Lifting

Throughout the country nurses continue suffering debilitating, and often career-ending and life-altering, back injuries from lifting and moving dependent patients.[1] Though modern patient-lift equipment is available, much of the health care industry continues to rely upon people to do the work of machines.[2] Although the Occupational Safety and Health Act of 1970 pledges "To assure so far as possible every working man and woman in the Nation safe and healthful working conditions and to preserve our human resources," there seem to be no enforced safeguards against hazardous lifting by health care workers. Nursing work can be compared to the hardest labor with nurses lifting weight estimated at 1.8 ton, or 3600 lb, per shift.[3] In 1992, the Occupational Safety and Health Administration (OSHA) investigated a chain of nursing homes in Pennsylvania finding nurse aides lifting over 5 ton, or 10,000 lb, per shift.[4] According to data provided by the U.S. Bureau of Labor Statistics (BLS), three to four times as many work-related musculoskeletal disorders (MSDs) occur among nursing aides than among registered nurses.

In 2006, the BLS reported 357,160 MSDs in a 12-page list of over 650 sources. "Health care patient or resident of health care facility" was listed as the source of 40,840 MSDs. Incredibly, patients and residents are the source of 11.4% of all MSDs occurring among U.S. workers![5] In 2000, "health care patient" was given as the cause of missed work time for nearly 11,000 registered nurses and nearly 45,000 nursing aides, orderlies, and attendants with "overexertion" and "overexertion in lifting" the primary events.[6] In 2002, health care workers, over 90% women, had twice the rate of injuries from overexertion than construction laborers.[7]

In 2006, BLS data showed that health care workers combined suffered 48,880 work-related MSDs requiring days away from work. Laborers were listed first with 28,860 MSDs. Nurse aides, orderlies, and attendants were listed second with 27,590 MSDs, and registered nurses were listed fifth with 9,200 MSDs. However, combining the number of MSDS to the nine categories of health care workers listed (nurse aides, orderlies, attendants—27,590; registered nurses—9,200; EMTs and paramedics—3,040; LPNs and LVNs—2,830; home health aides—2,660; personal and home care aides—1,770; radiologic technicians—670; psychiatric aides—580; and medical and clinical laboratory techs—540) totals 48,880 or 169% of "first place," and these are only the reported and accepted injuries![8]

It is important to note that only reported and accepted claims are reflected and that herniated spinal disks are not included (www.bls.gov). Not only is health care worker injury underreported at over 50%,[9] but "data on reported injuries...obscures the incidence and severity of injuries to U.S. healthcare workers."[10] In data on work-related MSDs, separating categories of health care workers, who all suffer similar injuries from lifting and moving patients and residents, is one way that the magnitude of injury related to manual patient handling is minimized.

A survey of nurses revealed that 83% work in spite of back pain; 60% fear suffering a disabling back injury; 88% consider health and safety concerns in choosing the type of nursing work and whether to remain in nursing.[11] An estimated 38% of nurses will require time off due to a back injury during their career.[12] An English survey found that 44% of injured nurses may never return, being either unemployed or medically retired.[13] Twelve percent of nurses leave nursing permanently due to back injury.[14]

"In the face of a nursing shortage that is fast reaching crisis proportions, injuries are a major contributing factor to nurses leaving the profession."[15] An important point is that many injured nurses have no choice about "leaving," as many employers terminate back-disabled nurses rather than restructure work assignments which do not require heavy lifting. Thus, 12% of nurses may be forced out after they have suffered largely preventable injuries.

With a diminishing supply of nurses and misguided efforts at cost-cutting, some employers are replacing registered nurses with unlicensed assistive personnel, which increases the number of patients per nurse. Research has shown that with each patient over four added to an RN's care assignment, risk of patient death within 30 days of admission increases by 7%.[16] Thus, serious risk to patients may increase as some employers force out nurses who have been needlessly disabled by manual patient lifting.

Back injury to health care workers from lifting patients has been studied for decades. Lifting the weight of adult patients greatly exceeds established tolerance limits of compressive force to spinal structures.[17] Body mechanics training has proven ineffective in preventing injury with lifting hazardous amounts of weight far above safe lifting limits set by the National Institute for Occupational Safety and Health (NIOSH) at 51 lb for men,[18] and by the "Snook Tables" at 46 lb for women.[19]

More recently, the Revised NIOSH Lifting Equation has yielded 35 lb as the maximum recommended weight limit for use with most patient-handling tasks. The weight limit for patient handling would be even less than 35 lb "when the task is performed under less than ideal circumstances, such as lifting with extended arms, lifting when near the floor, lifting when sitting or kneeling, lifting with the trunk twisted or the load off to the side of the body, lifting with one hand or in a restricted space, or lifting during a shift lasting longer than 8 h."[20]

With health care workers remaining at the top for disabling back injuries, the ready solution has been repeatedly demonstrated. From 1991, William Charney's pioneering work with "zero lift" policy calling for use of mechanical lift equipment either by nursing staff or by specially trained "lift teams," numerous studies have proven that safe patient handling is possible.[21] One study reported more than 60,000 patient transfers without injury to the transfer team using lift equipment or to nursing staff working with the team.[22] In a study of 18 hospitals, none of the members of lift teams were injured from the reporting facilities. In one facility averaging some 5500 lifts each year, the same lift team members had worked for 7 years without incident.[23] Overhead ceiling lifts, which are reportedly required in new hospital construction in Denmark and other European countries, have proven to be very successful with injury reduction.[24]

In spite of abundant evidence over many years that use of mechanical patient-lift equipment prevents injuries to nursing staff, many U.S. health care facilities continue requiring hazardous manual lifting. When nurses become back-disabled from lifting patients, many employers add insult to injury by terminating nurses when they can no longer lift people. The basis for such exploitation is often the mistaken belief that patient-lift equipment "costs too much" to provide for nursing staff. Direct costs, for medical and compensation alone, for one injured nurse requiring spinal surgery, or multiple surgical procedures, easily run into hundreds of thousands of dollars. Replacement costs for a medical/surgical nurse are estimated at $46,000 and for a critical care nurse at $64,000.[25] Outfitting one hospital room with a ceiling lift reportedly costs about $5,000; a transfer stretcher/chair costs about $11,000. Safety is a proven cost-saving measure, rather than an expense, for preserving valuable health care workers and the financial resources of employers and workers' compensation insurance carriers.

No Manual Lift—Safer for Patients

Manual turning to reposition has been reported by critically ill patients 18 years and older to be the most painful and distressing procedure, more painful than tracheal suctioning, tube advancement, and wound dressing changes![26] Thus, it seems that nursing staff themselves may unintentionally cause the greatest, most unnecessary pain to patients with manual turning and repositioning. Raising public awareness of available safe patient-handling equipment would allow patients and families to demand modern friction-reducing devices and repositioning aids such as sliding sheets, in addition to lift and transfer equipment.

Patients have reported feeling more secure and comfortable with mechanical lift equipment. Evidence is mounting for reduced pain and injury to patients with "no manual lift" policies. One study showed reduction of patient skin tears treated with dressing or medication from 16 preintervention to 5 postintervention.[27] Says Senior Industrial Officer Rob Bonner, in reporting on the Australian Nursing Federation (ANF) South Australian branch "No Lift, No Injury" program: "As well as the reduction in worker injuries, some of the important outcomes have been the improvement in resident care... There was a significant level of reporting of decreased skin tears and bruising among residents and significant reporting of residents being encouraged to take a greater role in their own movement. Some of these outcomes are as important as the implications for nurses and care workers."[28]

Measurable evidence is mounting for reduced skin tears, pain, and bruising to patients with "no manual lift" policies. Health care facility records are not readily available which document additional adverse patient events which sometimes occur with manual handling, though nursing staff may provide anecdotal accounts of such events including dislodgement of invasive tubes and lines, nerve damage to the brachial plexus, dislocation of shoulders, fractures of fragile bones, and patients being dropped. With availability of gentle repositioning, lift, and transfer equipment, which has proven to decrease risk of pain and injury to patients, manual handling of patients and residents without assistive equipment may come to be considered malpractice.

Patient-Handling Policy in the United Kingdom

Nurses in the United Kingdom can be disciplined by their employer if they lift patients manually, while nurses in many areas of the United States can be disciplined if they refuse to lift patients. Information follows on U.K. patient-handling practice highlighted from *The Guide to the Handling of Patients: Introducing a Safer Handling Policy.*[29]

Rather than a specific U.K. governmental "No Manual Lifting" document, codes of practice and guidelines have been set forth to comply with the Manual Handling Operations Regulations (MHOR) 1992,[30] derived from the European Council Directive (90/269/EEC) on health and safety requirements for the manual handling of loads where there is risk of back injury. Foundational for the Directive is the Health and Safety at Work Act (HSWA) of 1974.

Lifting and handling patients comes under the general banner of lifting and handling loads governed by the HSWA 1974, the MHOR 1992, and other European Community directives. These directives require employers to perform assessments on risks to the health and safety of employees. Where risks are identified relating to the manual handling of loads, employers are required to avoid hazardous manual handling as far as reasonably practicable. Where hazardous manual handling is unavoidable, the risk must be assessed according to a number of factors and must be reduced to the lowest reasonable practicable level.

Prosecution of employers by the Health and Safety Commission (HSC) may result when provisions in the regulations are breached. Employees are also responsible to follow the regulations in performance of their duties.

The *Health and Safety Executive (HSE) Guidance* accompanying MHOR 1992 lacks the force of law but is taken into account by HSE inspectors in determining the compliance with the regulations. The *HSE Guidance* contains guidelines on limits of weight for manual handling which preclude lifting adult patients, with lifting patients over 33.3 kg (73 lb) by two men, and over 22.2 kg (49 lb) by two women, considered unsafe. A significant advance of the MHOR 1992 is requiring assessment before lifting a load which carries a risk of injury.

Though "No Manual Patient Lifting" is not addressed specifically by distinct legislation, manually lifting adult patients is considered against U.K. law in compliance with the MHOR 1992, due to the weight of adults, the established risk of injury with lifting patients, and the requirement for employers to assess the risk of injury and to reduce that risk of injury to workers to the lowest practicable level.

The Royal College of Nursing (RCN), the leading professional nursing organization in the United Kingdom, is looked to for expert opinion on best practices for patient handling. The RCN gives their revised *Code of Practice for Patient Handling* as the framework for implementation of the MHOR 1992, stating: "The aim is to eliminate hazardous manual handling in all but exceptional or life-threatening situations." "There is no threshold below which handling may be regarded as safe."[31]

RCN's publication *Introducing a Safer Patient Handling Policy* asks, "Are you still lifting? No one working in a hospital, nursing home or community setting should need to lift patients manually any more. Hoists, sliding aids and other specialized equipment mean staff should no longer have to risk injury while doing their job." RCN's *Safer Patient Handling Policy* underscores that there is no acceptable weight limit for lifting adults; nurses should not lift any patient manually, whatever their weight, the only exception being babies and very young children.[32]

The RCN *Safer Patient Handling Policy* has become the gold standard for U.K. employers working to prevent back injuries among nursing staff. Health care facilities not working under such a policy are not in compliance with the MHOR 1992, which applies to all hospitals, nursing homes, residential homes, and the domestic environment, wherever dependent patients and residents require assistance with lifting and movement needs. Dozens of improvement notices have been issued and one prosecution with a £12,000 fine. Health care facilities in the United Kingdom are also vulnerable to large civil claims for negligence, ranging up to £800,000 to an ICU nurse who suffered a low-spine injury from lifting a patient when no mechanical hoist was available.[33–35] RCN attorneys represent nurse members with patient-handling injuries where failure in the duty of care by the employer can be proven, such as lack of appropriate lift equipment. It is reported that such cases are not difficult to win as the principles are so well established.

"Unsafe Lifting Practices," given in Chapter 22 of *The Guide to the Handling of Patients: Introducing a Safer Handling Policy* (the "Green Book"), details the specific danger to nurses and to patients for each of a number of manual lifting techniques. Reasons why adult patients must not be lifted are summarized: People weigh too much to safely lift and they can act unpredictably. It is difficult or impossible for caregivers to achieve a safe position to lift. All manual lifting techniques impose a risk of injury to nursing staff. Most manual lifts carry a risk of injury to the patient. Manual lifting is not therapeutic; it does not enhance patient mobility.

Suggested weight lifting limits are provided in the *HSE Guidance* to the MHOR 1992 which aid in identifying manual handling with risk of injury, as on page 96 of the "Green Book." The *HSE Guidance* itself is available online by subscription at http://baldwin.butterworths.co.uk/search/pages/firstime_index.htm. If the weight to be lifted presents risk of injury, which is given as 25 kg (55 lb) capacity for one man or 16.6 kg (36 lb) capacity for one woman, additional people may be required to safely lift the load. It is difficult for a load to be shared evenly and safely, especially with patient handling, when the patient's weight often falls onto one of the lifters, creating a high risk of injury. The *HSE Guidance* to the MHOR 1992 provides suggested capacities requiring more than one lifter, indicating that two people should not lift more than two-thirds of their combined lifting capacity and that three people should lift only one-half of their combined lifting capacity. With the weight lifting capacity of 25 kg (55 lb) for one man, the combined weight capacity is given for two men lifting together, in the ideal position, close to the lower body, as 33.3 kg (73 lb); for three men lifting together the weight limit is 37.5 kg (82 lb). With the weight lifting capacity of 16.6 kg (36 lb) for one woman, for two women lifting together, the combined weight capacity is given as 22.2 kg (49 lb); for three women lifting together, the weight limit is 25 kg (55 lb). A notation states that for safety three women lifting together close to the floor, and at a distance in front of their feet, should lift only 5 kg (11 lb).

"No Lifting" in Victoria, Australia

Australia's experience with legislation teaches that a general ergonomics standard, without specifically addressing the lifting of patients and residents, will not necessarily reduce nurse injury. Even with National Manual Handling Regulations in place for 10 years, Australian nurses still suffered the highest injury rate in the female workforce. Most injuries to nurses were to the back from lifting and moving patients and residents, accounting for more than 50% of WorkCover claims in the health industry, costing $26 million per year.[36]

Elizabeth Langford, Coordinator of the Injured Nurses' Support Group, authored the ground-breaking report, *Buried But Not Dead: A Survey of Occupational Illness and Injury Incurred by Nurses in the Victorian Health Service Industry*, detailing the extent and severity of nurse injury in the Australian state of Victoria.[37] Spurred by national media attention, with prime-time television, newspaper, and radio coverage of nurse injury exposed in *Buried but Not Dead*, and with efforts by ANF Victorian Branch, "No Lifting" for health care was ignited in Victoria, Australia.

Elizabeth Langford observed that her own injury and injuries to other nurses were serious life-altering spinal injuries, often ending the nurse's career, but that available data on nurse injury emphasized sprains and strains which may be expected to heal in a matter of days or weeks. Research given in *Buried but Not Dead* revealed that 73% of nurse back injury is to the lumbar spine, with the majority of 57% of the injuries to spinal disks. It appears that reports filed on nurse injury often list only the initial diagnosis of back, muscle, or lumbar strain, while diagnostic tests, which may be deferred during an initial period of weeks or months of conservative treatment, eventually verify that the majority of nurse back injury is actually to spinal disks. Thus, work injury data based on initial injury reports generally do not reflect the extremely serious back injuries of nurses, greatly minimizing the severity of injuries caused to nurses by manual patient lifting.

Only with a collaborative effort was nurse back injury reduced in the State of Victoria, when nursing, health care, and government worked together on the "Victorian Nurses' Back Injury Prevention Project." The $7.7 million costs for the "No Lift Program" were recovered within 1 year, with lost work days due to nurse back injury reduced 74%, WorkCover claims down 48%, and the cost of claims down 54%, saving $13 million per year by preventing back injuries to nurses.[38]

"Elizabeth Langford, who has worked tirelessly for over a decade to prevent what happened to her happening to other nurses, was rewarded with an Order of Australia on June 9, 2003. While Elizabeth is extremely pleased that her work in the area of injury prevention for nurses has been recognized, she is adamant the Award belongs to the whole team, from Gwyneth Evans who set up the Injured Nurses Support Group in 1988, to Jeanette Sdrinis, the current Occupational Health and Safety Officer at the Australian Nursing Federation Victorian Branch."[39] Jeanette Sdrinis said: "Prior to the ANF campaign for all Victorian health facilities to adopt a No Lifting Policy, nurses were expected to lift patients using outdated manual lifting techniques. Nurses were the hidden victims of preventable workplace injury; debilitating back pain, career ending injuries and daily chronic fatigue were considered just part of the job. **We've turned around a culture that's been there for more than 100 years**. Employers are now accepting their responsibilities and nurses now say it is unacceptable to be injured at work and know back pain can be prevented."[40]

U.S. Model for Back Injury Prevention in Nursing Homes

The U.S. Centers for Disease Control and Prevention (CDC) NIOSH and partners recognized the "Employee Back Injuries Prevention in Nursing Homes Project" with the National Occupational Research Agenda (NORA) Partnering Award for Worker Health and Safety 2003. From 10 nominated projects, the Partnering Award honors teamwork, innovative thinking, and strong science in advances for worker health and safety, which are established hallmarks of NORA. The biennial award goes to organizations collaborating in research partnerships for development of new practices, products, procedures, equipment, or policies for the protection of workers from job-related injury, illness, or death.

In 2003, the NORA Partnering Award went to a research project which focused on reduction of work-related back injuries to employees in a group of nursing homes operated by BJC Health Care. Comparison of injury rates was made for the 3 years before and after the intervention. Frequency of back injuries were successfully reduced by 57%, injury rates were lowered by 58%, and workers' compensation costs were reduced by 71%.[41]

The Employee Back Injuries Prevention in Nursing Homes project combined measures on reduction of causes of injury and evaluation of effectiveness of the measures. Employees helped to identify movements and postures which placed nursing assistants at risk of back strain, stress, and injury with the lifting and movement needs of residents. Evaluation was performed on the reduction of stress and strain with lifting by use of mechanical lift equipment. The project results, with employee input, led to implementation of a "best practices" program.

Project partners receiving the NORA Partnering Award at the NORA Symposium 2003 held June 23–24, 2003, in Arlington, VA, included BJC Health Care, BJC Occupational Health Nurse Council, Washington University, West Virginia University, Arjo Inc. (www.argo.com), EZ Way Inc. (www.ezlifts.com), and NIOSH.

Dr. John Howard, Director of NIOSH, stated: "Everyone benefits when partners work together to tackle demanding challenges in occupational safety and health." This model shows promise as the nursing home industry examines new ways to reduce the risk of painful, costly, work-related back injuries among employees.

"This project is a stellar example of NORA's value in stimulating new research partnerships to address serious work-related injury and illness concerns. Seven years in the national agenda, these results provide valuable data for planning our future course."

Following the successful project, NIOSH published a user's guide for broad distribution within U.S. health care to facilitate establishment of programs for reduction of back injury with patient and resident handling.[42] Development on the federal level of this proven model for reducing back injuries in health care is an important step toward a national policy banning the manual lifting of patients.

American Nurses Association (ANA) and Safe Patient Handling

Endorsement by nursing organizations is vital for the success of a national safe patient-handling policy. On June 21, 2003, ANA issued their landmark position statement calling for "Elimination of Manual Patient Handling to Prevent Work-Related Musculoskeletal Disorders."[43] This document underscores ANA's commitment to improve the health and safety of nurses, increase patient safety, and reduce health care costs.

ANA emphasizes that nurses suffer a disproportionate amount of MSDs from the cumulative effect of repeatedly lifting patients. Training in "proper" body mechanics does not apply to handling the human form and is not useful in reducing injuries caused by lifting and transferring patients. In addition to lifting a hazardous amount of weight, other factors that increase the risk of injury are outlined, such as awkward postures and positioning; the unpredictability of patients who might be agitated, combative, unresponsive, or with limited ability to assist; and limits within the patient care environment created by furniture, walls, treatment equipment, etc. An essential note is that the hazard of injury with patient handling persists even with the help of additional staff.

There are many types of specialized equipment for the lifting and transferring of patients. These include a wide variety of full-body rolling floor lifts, overhead sling lifts, sit/stand assist lifts, lateral transfer stretcher/chairs, repositioning aids, friction-reducing devices, etc. The health care industry must incorporate modern technology to reduce injuries caused by patient handling, just as technology has been applied to reduce exposure of health care workers to needlestick injury and communicable airborne disease.

Policies must be adopted by employers committing to implementation of the safest patient handling which prioritizes use of assistive equipment. Manual total body lifts would be eliminated apart from exceptional or life-threatening circumstances and in caring for infants and small children. Following involvement of staff with selection, adequate amounts of well-maintained readily available equipment is essential, as well as training on its use, and a nonpunitive approach so that staff are not fearful of reporting work-related injury.

ANA recognizes that using lift equipment prevents untoward events to patients that sometimes occur with manual handling such as pain with the underarm lift and injury such as falls, contusions, and skin tears. ANA reflects commitment to improving patient safety, comfort, and dignity by its endorsement of assistive patient-handling equipment. Employing use of modern technology for patient lifting reduces risk of injury to nursing personnel while improving the quality of care for patients.

Soon after release of ANA's position statement calling for elimination of manual patient handling was ANA's announcement on September 17, 2003, of their "Handle with Care" initiative designed to prevent potentially career-ending back, neck, and musculoskeletal injuries to nurses (www.ana.org/handlewithcare/).

ANA's "Handle with Care" brochure, which was sent to every hospital in the United States, includes the experiences of this writer and of Maggie Flanagan, RN, with work-related musculoskeletal injury. "The campaign is designed to support the individual nurse in the workplace with a profession-wide effort for greater education and training and for increased use of assistive equipment and patient-handling devices. The campaign also seeks to reshape nursing education and federal and state ergonomics policy by highlighting the ways technology-oriented safe-patient handling benefits patients and the nursing workforce."

ANA believes their "Handle with Care" campaign can dramatically improve the health and safety of American nurses, increase the safety of patient care, and reduce health care costs substantially.

ANA envisions their "Handle with Care" campaign to be similar to their "Safe Needles Save Lives" campaign from the late 1990s. "Safe Needles Saves Lives" was highly successful for passage of state and federal legislation and with institution of cultural changes to prevent needlestick injuries in the health care industry.[44]

Patient Handling in America: A Call for "Safe Patient Handling—No Manual Lift" Legislation

Nursing practice today is expected to be based on the best available scientific evidence. Pat Quigley, President of Florida Nurses Association, declared that the book *Back Injury among Healthcare Workers: Causes, Solutions, and Impacts*[45] "...should be in the library for all nursing faculty as you are teaching [manual patient lifting] skills that are not evidence-based."[46] With many nursing schools still teaching manual patient lifting, and many employers still requiring nurses to lift manually, an enormous gap exists between the "unsafe nursing practice" of manual lifting and an overabundance of irrefutable evidence on prevention of injuries to nurses and patients with mechanical patient-lift equipment.

Back Injury among Healthcare Workers: Causes, Solutions, and Impacts, coedited by William Charney and this writer, was released in July 2003, to expose the unchecked epidemic of back injuries from manual patient lifting and to promote legislation to assure the safe handling of patients. The book's dedication states, "This book is dedicated to the thousands of back-injured healthcare workers who have sacrificed their well-being, and often their careers, to painful injuries from manually lifting patients. It is the authors' hope that this book will lead to implementation of No Manual Lifting of patients in hospitals, nursing homes, and home health through the use of technology by nursing staff or specially-trained Lift Teams. We also hope that states will correlate the national nursing shortage with nursing injury and will pass Zero Lift for Healthcare legislation to halt the unnecessary loss of healthcare workers to preventable disabling injuries. Finally, we look forward to the day when nursing organizations will negotiate for retention of back-injured nurses, including, when necessary, provision by employers of Permanent Light Duty nursing work."[45]

Three decades of body mechanics training; education on "proper" lifting techniques; and time for volunteerism of the nearly 6,000 hospitals and 18,000 nursing homes in the country to implement policy requiring use of safe lift equipment, have proven ineffective in reducing injuries caused by lifting and moving dependent patients and residents. Workers' compensation insurance carriers are assisting some health care facilities reduce patient-handling injuries with financial incentives and with assistance establishing programs for safe patient handling. Still, these efforts have not substantially reduced injury caused by the manual lifting of patients and residents. Categories of health care workers combined still suffer more work-related MSDs than any other occupation in the nation. The hazard with manually lifting outrageous loads in awkward positions cannot be educated or trained away. People simply weigh too much for other people to be bending forward across hospital beds, down to the commode, etc., and lifting them. The burden of such dangerous lifting must be removed from the backs of health care workers and transferred to modern mechanical equipment designed to safely lift and transfer people. Enactment of industry-specific, safe patient-handling legislation is essential to halt the epidemic of devastating injuries caused by the manual lifting of patients and residents.

Our country's needlestick safety legislation may be used as a model for rapid enactment of legislation for health care worker safety. It took only 2 years from passage of California AB 1208 in September 1998 revising California's bloodborne pathogen standard, until the Needlestick Safety and Prevention Act of 2000 was signed into law by President William Clinton on November 6, 2000, directing OSHA to revise the national Bloodborne Pathogens Standard. OSHA published the revised standard in the *Federal Register* on January 18, 2001, and the Needlestick Safety and Prevention Act of 2000 went into effect on April 18, 2001 (http://www.oshaslc.gov/needlesticks/needlefaq.html).[47,48]

Needleless IV systems and safety needle devices had been available, and were widely advertised in nursing journals, leading nurses to wonder why their hospital did not have them. Meanwhile, though needleless systems were available, many nurses continued suffering preventable needlesticks, some being fatally infected with blood-borne pathogens. Legislation was required to force most employers to provide readily available safety equipment.

Likewise, a wide variety of modern, safe, gentle, patient-lift equipment and friction-reducing devices have been available for many years, with many peer-reviewed studies demonstrating nurse injury prevention by use of the equipment. A national mandate, which may be called "Safe Patient Handling—No Manual Lift," is necessary to force all health care facilities, with nursing schools following suit, to implement safe patient-handling programs with "no manual lift" policies requiring adequate amounts of readily available, well-maintained equipment, appropriate to the patient population, which may be used by either nursing staff or specially trained "lift teams."

Legislative efforts by states to eliminate hazardous manual patient lifting began with introduction of legislation for safe patient handling first by California in February 2004, followed by Massachusetts in December 2004.

"Massachusetts" continues pursuing legislation for safe patient handling with companion bills reintroduced in January 2007. Senator Richard T. Moore (D) introduced Senate Number 1294, "An act to require the use of evidence-based practices for safe patient handling and movement" on January 10, 2007. Representative Jennifer M. Callahan (D) introduced House Number 2052, "An act relative to safe patient handling in certain health facilities," on January 11, 2007, with the goal of ensuring that quality nursing staff are not lost to preventable injuries. SB 1294 and HB 2052 were referred to the Joint Committee on Public Health. A Public Hearing was held October 24, 2007, with the bills for the safe handling of patients in Massachusetts presently stalled in committee.

If passed, SN 1294 would require every licensed health care facility to implement an evidence-based policy for safe handling and movement of patients; and to provide training on use of patient-handling equipment and devices, patient care ergonomic assessment protocols, no lift policies, and patient-lift teams. The intent of the "No Lift Policy" is the elimination of manual handling in virtually every patient care situation, apart from all but exceptional or life threatening situations. Constituting a pledge from administrators that proper equipment, adequately maintained and in sufficient numbers, will be available to care providers, the "No Lift Policy" is an integral part of a comprehensive safe patient handling and movement program in acute care hospitals and long-term care facilities.

If passed, HB 2052 would require all health care facilities, including but not limited to acute care hospitals, psychiatric hospitals and nursing homes, to develop and implement safe patient-handling policies; and to identify, assess, and develop strategies to control risk of injury to patients and health care workers with lifting, transferring, repositioning, or movement of a patient or equipment. HB 2052 would regulate implementation of policies and would establish credits for costs with implementing a safe handling program. (History and text of MA SB 1294 and HB 2052: www.mass.gov/legis)

In "California," legislation for safe patient handling has passed the Senate and the Assembly every year since 2004 for 4 years running. But the legislative effort to protect health care workers and patients from needless pain and injuries caused by manual lifting has been vetoed every time by California Governor Arnold Schwarzenegger (R). Since his first veto message of Assembly Bill 2532, on September 22, 2004, Governor Schwarzenegger persists in maintaining that adequate safeguards are already in place and that the financial burden to hospitals of "zero lift" would be too great.

Most recently, Senate Bill 171 "Hospital Patient and Health Care Worker Injury Protection Act," introduced on February 5, 2007, by Senator Don Perata (D), passed the California Senate on June 4, 2007, and the Assembly on September 6, 2007, but was vetoed by Governor Schwarzenegger on October 13, 2007. The bill would have become effective on July 1, 2008, providing legal protection against preventable injuries from manual patient handling.

If Governor Schwarzenegger had not vetoed SB 171, all general acute care hospitals in California would have been required to establish a patient protection and health care worker back injury prevention plan; to conduct needs assessments to identify patients needing lift teams, and lift, repositioning, or transfer devices; to use lift teams, and lift, repositioning, and transfer devices; and to train health care workers on the appropriate use of lift, repositioning, and transfer devices. (For bill text and veto messages of AB 2532 introduced by Assembly Member Loni Hancock (D) in 2004, of SB 363 introduced by Senator Don Perata (D) in 2005, of SB 1204 introduced by Senator Don Perata (D) in 2006, and of SB 171 introduced by Senator Don Perata (D) in 2007, see www.leginfo.ca.gov)

Interestingly, companion bill AB 371 "An act…relating to health facility financing…" was introduced on February 14, 2007, by California Assemblyman Jared Huffman (D) and coauthor Assemblywoman Sally J. Lieber (D). AB 371 was amended on April 23, 2007, and was referred to the Assembly Committee on Appropriations. Per http://www.leginfo.ca.gov, the bill passed out of committee and on January 7, 2008, went to the Assembly with concurrence in Senate amendments pending. If AB 371 passes, hospitals applying for financing from issuance of tax-exempt bonds will be required to provide a copy of the hospital's injury and illness prevention program (IIPP) specifying how the hospital has implemented, or plans to implement, a hospital patient and health care worker injury prevention program, including a "zero lift policy" for use of powered patient-transfer devices, patient-lifting devices, or lift teams instead of manual lifting and transferring of patients.

Meanwhile, the belief stated by Governor Schwarzenegger and others, that purchasing safety equipment for employees "costs too much," keeps nursing personnel at the top for disabling back injuries and ignores greater costs resulting from the inevitable injuries. The ethics of such excuses, which essentially promote disabling injuries to health care workers, by permitting lack of safety equipment for lifting dangerous loads, certainly need to be examined.

Used as disposable human lift equipment, 90% of nursing staff injured lifting patients are women.[22] Is provision of forklifts, hoists, and cranes for use by predominantly male warehouse and construction workers considered too much of a financial burden? Nursing staff are often required to lift equivalent amounts of weight without such equipment and will likely be terminated by many employers when disabled by the lifting. The majority of back injuries suffered from lifting patients cannot rightfully be called "accidents." When nurses are required to repeatedly lift hazardous amounts of weight, often bending forward with the back in its most vulnerable position, spinal injury is the expected result rather than an "accident."

Since initial attempts by California and Massachusetts for industry-specific legislation for the safe handling of health care patients in the United States, at least 18 states, 36%, over one-third of states in the nation, have passed or have introduced legislation for or related to safe patient handling. Other states could be active on safe patient handling of which this writer is unaware. Nine states, nearly one-fifth of states in the nation, known at this writing to have passed legislation for or related to safe patient handling are Ohio, New York, Texas, Washington, Hawaii, Rhode Island, Maryland, Minnesota, and New Jersey. Nine other states known to have introduced or reintroduced legislation related to safe patient handling which has not yet passed are Massachusetts, California, Iowa, Nevada, Michigan, Florida, Vermont, Missouri, and Illinois.

States Which Have Introduced Legislation Related to Safe Patient Handling

Massachusetts reintroduced companion bills Senate Number 1294, "An Act to Require the Use of Evidence-Based Practices for Safe Patient Handling and Movement," on January 10, 2007, and House Number 2052, "An Act Relating to Safe Patient Handling in Certain Health Care Facilities," on February 19, 2007.

California reintroduced Senate Bill 171, "Hospital Patient and Health Care Worker Injury Protection Act," on February 5, 2007, which was vetoed for the fourth time on October 13, 2007. It is unknown if California has reintroduced again.

Iowa introduced House File 635, "An Act Relating to Manual Patient Handling by Nurses in a Hospital Setting," on March 8, 2005. It is unknown if Iowa has reintroduced.

Nevada introduced Assembly Bill 577, "Requires Certain Medical Facilities to Establish a Program for Safe Handling of Patients," on March 26, 2007.

Michigan introduced Senate Bill 377, to amend "Public Health Code" regarding each hospital's establishment of a safe patient-handling committee and implementation of a safe patient-handling program, on March 27, 2007.

Florida reintroduced companion bills SB 508, "Relating to Hospitalized Patients/Safe Lifting Policies," on October 15, 2007, and HB 471, "Relating to Hospitalized Patients/Safe Lifting Policies," on January 7, 2008.

Vermont introduced companion bills Senate Bill 141, "An Act Relating to Safe Patient Handling," on February 27, 2007, and House Bill 421, "An Act Relating to Safe Patient Handling," on February 28, 2007.

Missouri introduced House Bill 1940, "Hospital Patient Safety," on January 31, 2008.

Illinois reintroduced House Bill 5274, "Safe Patient Handling—Tech," "Creates the Safe Patient Handling Act," on February 14, 2008.

States Which Have Passed Legislation Related to Safe Patient Handling

Ohio House Bill 67 was signed into law on March 21, 2005, by Governor Bob Taft (R), with Section 4121.48 creating a program for interest-free loans to nursing homes for implementation of a no-manual-lift program. The law stipulates creation in the state treasury of the long-term care loan fund to be operated by the bureau of workers' compensation "to make loans without interest to employers that are nursing homes for the purpose of allowing those employers to purchase, improve, install, or erect sit-to-stand floor lifts, ceiling lifts, other lifts, and fast electric beds, and to pay for the education and training of personnel, in order to implement a facility policy of no manual lifting of residents by employees." (For text of HB 67 Section 4121.48 as enrolled, see http://www.legislature.state.oh.us/bills.cfm?ID = 126_HB_67_EN)

New York companion bills, Assembly Bill 7641 and Senate Bill 4929 were introduced in April 2005, and were signed into law on October 18, 2005, by Governor George Pataki (R). The law created a 2-year "Safe Patient Handling Demonstration Program" to establish safe patient-handling programs and collect data on the incidence of nursing staff and patient injury with patient-handling, manual versus lift equipment. Results will be used to describe best practices for improving health and safety of health care workers and patients during patient handling. (See http://assembly.state.ny.us/ and http://www.senate.state.ny.us)

In April 2007, New York introduced identical bills, A 7836 by Assembly Member Richard N. Gottfried (D) and S 5116 by Senator Kemp Hannon (R), "An act to amend chapter 738 of the laws of 2005, relating to establishing a safe patient handling demonstration program, in relation to the effectiveness thereof." In June 2007, S 5116 was substituted by A 7836 which was signed into law on July 3, 2007, by Governor Eliot Spitzer (D), extending the safe patient-handling demonstration program for 2 years to research the effect of safe patient-handling programs in health care facilities across New York State. The study is to build upon existing evidence-based data, with the ultimate goal of designing "best practices" for safe patient handling in New York State health care facilities. The bill also establishes specifications for safe patient-handling programs. (Summary text A 7836: http://assembly.state.ny.us/leg/?bn = A07836)

Texas was the first state in the nation to mandate implementation of policy for safe patient-handling and movement programs by hospitals and nursing homes. Senate Bill 1525, "An Act relating to safe patient handling and movement practices of nurses in hospitals and nursing homes," was introduced on March 10, 2005, by author Senator Judith Zaffirini (D) and Sponsor Representative Dianne Delisi (R). SB 1525 was signed into law by Governor Rick Perry (R) on June 17, 2005, and became effective January 1, 2006.

The Texas safe patient-handling law includes requiring hospitals and nursing homes to establish a policy to identify, assess, and develop methods of controlling the risk of injury to patients and nurses associated with lifting, transferring, repositioning, and movement of patients; evaluation of alternative methods from manual lifting to reduce the risk of injury from patient lifting, including equipment and patient care environment; restricting, to the extent feasible with existing equipment, manual handling of all or most of a patient's weight to emergency, life-threatening, or exceptional circumstances; and provision for refusal to perform patient-handling tasks believed in good faith to involve unacceptable risks of injury to a patient or nurse. (Enrolled text SB 1525: http://www.capitol.state.tx.us/tlodocs/79R/billtext/html/SB01525F.htm)[49]

Washington was the first state to pass legislation requiring hospitals to provide mechanical patient-lift equipment as part of their policy for safe patient handling, and was first to offer financial assistance with implementation of mechanical equipment to reduce injuries related to lifting and moving patients.

House Bill 1672 "An Act relating to reducing injuries among patients and health care workers" was originally introduced February 1, 2005, sponsored by Representatives Steve Conway (D), Zack Hudgins (D), Tami Green (D), Eileen Cody (D), Sherry Appleton (D), Dawn Morrell (D), Alex Wood (D), John McCoy (D), Phyllis Kenney (D), Jim Moeller (D), and Maralyn Chase (D). After being stalled in the House Committee on Commerce and Labor in 2005, HB 1672 was reintroduced on January 8, 2006. HB 1672 passed the House 85 to 13 on March 7, 2006, and passed the Senate 48 to zero on March 8, 2006. Washington State's law for the safe handling of patients was signed into law by Governor Christine Gregoire (D) on March 22, 2006, and went into effect June 7, 2006.

On a timeline between February 1, 2007, and January 30, 2010, Washington hospitals must establish a safe patient-handling committee, of which at least half of the members are to be frontline nonmanagerial employees who provide direct patient care, and must establish a safe patient-handling program, and implement a safe patient-handling policy for all shifts and units. Hospitals may choose among three options for implementation of lift equipment, either one readily available lift per acute care unit on the same floor, one lift for every 10 acute care inpatient beds, or lift equipment for use by specially trained lift teams. Financial assistance will be provided by reduced workers' compensation premiums for hospitals implementing a safe patient-handling program and by tax credits covering the cost of purchasing mechanical lifting or other patient-handling devices.

Washington hospitals are to develop procedures for employees to refuse to perform, without fear of reprisal, patient-handling activities which the employee believes in good faith would involve an unacceptable risk of injury to an employee or patient. An employee who in good faith follows the procedure shall not be the subject of disciplinary action for refusing to perform or be involved in the patient handling or movement in question.

Patient-handling hazard assessments are to be conducted and a process developed to identify the appropriate use of the safe patient-handling policy based on the patient's condition and the availability of lifting equipment or lift teams. An annual performance evaluation of the safe patient-handling program is to be conducted. Staffs are to be trained at least annually on safe patient-handling policies, equipment, and devices. With hospital construction or remodeling, feasibility is to be considered of incorporating patient-handling equipment, or of designing to incorporate at a later date.

Significantly, Washington State's law for the safe handling of patients was passed during National Patient Safety Awareness Week, which was March 5–11, 2006. HB 1672 states, "Patients are not at optimum levels of safety while being lifted, transferred, or repositioned manually. Mechanical lift programs can reduce skin tears suffered by patients by threefold. Nurses, thirty-eight percent of whom have previous back injuries, can drop patients if their pain thresholds are triggered." The BLS reports that the injury rate of hospital employees in Washington State exceeds that of construction, agriculture, manufacturing, and transportation. With passage of HB 1672, Washington hospital patients and health care workers will be protected from unintentional pain and injuries which sometimes occur with manual patient lifting and moving.

Among chief champions for safe patient handling in Washington State is William Charney whose article, "The need to legislate the health care industry in the state of Washington to protect health care workers from back injury," thoroughly documents the epidemic of back injury to health care workers in the state of Washington, which holds true throughout every state in the nation. Additionally outlined is the legislative solution calling for implementation of back injury prevention programs by either the "Zero Lift" model, with nursing staff use of lift equipment, or the "Lift Team" model, with use of lift equipment by a specially trained team.[50] (Enrolled text HB 1672: http://www.leg.wa.gov/pub/billinfo/2005-06/Pdf/Bills/House%20Passed%20Legislature/1672-S.PL.pdf)

Hawaii adopted a resolution on April 24, 2006, "Requesting appropriate safeguards be instituted in health care facilities to minimize the occurrence of musculoskeletal injuries suffered by nurses." House Concurrent Resolution No. 16 was introduced on February 3, 2006, by Representatives Marilyn B. Lee (D), Rida Cabanilla (D), Cindy Evans (D), Maile S. L. Shimabukuro (D), Roy M. Takumi (D), Clift Tsuji (D), Kirk Caldwell (D), Josh Green, MD (D), Robert N. Herkes (D), Ezra Kanoho (D), Bertha C. Kawakami (D), Bob Nakasone (D), Brian Schatz (D), Joseph M. Souki (D), Dwight Y. Takamine (D), and Kameo Tanaka (D). The report associated with HCR 16 is titled "American Nurses Association's 'Handle with Care' Campaign Support."

The resolution recognizes that work-related MSDs are the leading occupational health problem plaguing the nursing workforce; that of primary concern are back injuries, which can be severely debilitating for nurses, though musculoskeletal injuries can affect other parts of the body such as the neck, shoulders, wrists, and knees; that compared to other occupations, nursing personnel are among the highest at risk for MSDs; that the incidence rate for work days lost due to back injuries is twice as great for nursing home workers as it is for truck drivers; that nursing home workers and hospital workers incur more lost work days due to back injuries than construction workers, miners, and agriculture workers.

HCR 16 recognizes that the risk of nurse injury from manual patient handling crosses all specialty areas; that musculoskeletal injuries to nurses is particularly distressing in context of the current nurse shortage; and that injuries to nurses from patient-handling compound factors such as the aging nurse workforce, declining retention and recruitment rates, and lowering the social value of nursing to worsen the shortage.

HCR states that in response to the significant number and severity of work-related back injuries and other musculoskeletal injuries among nurses, ANA launched the "Handle with Care" campaign to build a better health care industry-wide effort for safe patient handling, "to prevent back and other musculoskeletal injuries." "This is being done through developing partnerships and coalitions, education and training, increasing use of assistive equipment and patient-handling devices, reshaping nursing education to incorporate safe patient handling, and pursuing federal and state ergonomics policy by highlighting technology-oriented and safe-patient handling benefits for patients and nurses."

Hawaii's resolution states that in 2005 the Council of State Governments' Health Capacity Task Force adopted and supported the policies contained in ANA's "Handle with Care" campaign, and asked member states to also support the campaign. With adoption of HCR 16, Hawaii says, "Be it resolved…that the Legislature of the State of Hawaii supports the policies contained in the American Nurses Association's 'Handle With Care' campaign; and… that certified copies of this Concurrent Resolution be transmitted to the Council of State Governments' Health Capacity Task Force and the American Nurses Association." (Text HCR 16: http://www.capitol.hawaii.gov/session2006/Bills/HCR16_.pdf)

Rhode Island addressed musculoskeletal injuries to health care workers with the "Safe Patient Handling Act of 2006" which became law on July 7, 2006, "to promote the safe handling of patients in health care facilities." Companion bills, Senate Bill 2760 and House Bill 7386, both entitled "An Act Relating to Health and Safety—Safe Patient Handling Legislation," were introduced in February 2006. SB 2760 was introduced on February 14, 2006, by Senators V. Susan Sosnowski (D), Beatrice A. Lanzi (D), Rhoda E. Perry (D), M. Teresa Paiva-Weed (D), and Juan M. Pichardo (D). HB 7386 was introduced on February 16, 2006, by Representatives Grace Diaz (D), Paul E. Moura (D), Amy G. Rice (D), Edith H. Ajello (D), and Raymond J. Sullivan, Jr. (D). The bills were transmitted on June 29, 2006, to Governor Donald L. Carcieri (R), and became law on July 7, 2006, without Governor Carcieri's signature. Rhode Island's Safe Patient Handling Act took effect on January 1, 2007.

Legislative findings listed in the law include greater risk of patient injury with manual lifting and moving, and that safe patient handling can reduce skin tears suffered by patients by threefold and can significantly reduce other injuries to patients as well. With nurses lifting an estimated 1.8 tons per shift, health care workers lead the nation in work-related MSDs. From 38% to 50% of nurses and other health care workers will suffer a work-related back injury; 44% will be unable to return to their preinjury position. Eighty-three percent of nurses work in spite of back pain, and 60% fear a disabling back injury. Twelve percent to 39% of nurses not yet disabled are considering leaving nursing due to back pain and injuries.

Case studies are cited showing reduction of injuries and costs with use of lifting equipment. In nine case studies, use of lift equipment resulted in injuries decreasing 60% to 95%, workers' compensation costs dropping 95%, and absenteeism related to lifting and handling injuries being reduced by 98%.

"Safe patient handling" is defined as "the use of engineering controls, transfer aids, or assistive devices whenever feasible and appropriate instead of manual lifting to perform the acts of lifting, transferring, and/or repositioning health care patients and residents." As a condition of licensure, health care facilities, defined as a hospital or a nursing facility, shall establish a safe patient-handling committee, chaired by a professional nurse or other appropriate licensed health care professional, and with at least half of the members hourly, nonmanagerial employees who provide direct patient care.

By July 1, 2007, each licensed health care facility shall develop a written safe patient-handling program, to prevent MSDs among health care workers and injuries to patients. The safe patient-handling program is to include implementation, by July 1, 2008, of a safe patient-handling policy for all shifts and units to achieve the maximum reasonable reduction of manual lifting, transferring, and repositioning of all or most of a patient's weight, except in emergency, life-threatening, or otherwise exceptional circumstances. Patient-handling hazard assessments are to be conducted; and a process developed to identify the appropriate use of the safe patient-handling policy based on the patient's physical and mental condition, the patient's choice, and the availability of lifting equipment or lift teams; and to address circumstances when it would be medically contraindicated to use lifting or transfer aids or assistive devices for particular patients.

A registered nurse or other appropriate licensed health care professional is to be trained to serve as an expert resource for training clinical staff on safe patient-handling policies, equipment, and devices before implementation, at least annually, and as needed. An annual performance evaluation of the safe patient-handling program is to be reported to the safe patient-handling committee, determining the extent of reduction in MSD claims and days of lost work caused by patient handling, with recommendations to increase the program's effectiveness. An annual report to the safe patient-handling committee of the facility shall be made available to the public upon request, on activities related to the identification, assessment, development, and evaluation of strategies to control risk of injury to patients, nurses, and other health care workers associated with the lifting, transferring, repositioning, or movement of a patient. Nothing in the law precludes lift team members from performing other assigned duties.

Protocols shall be established for an employee to report, without fear of discipline or adverse consequences, to the committee being required to perform patient handling that is believed in good faith to expose the patient and/or employee to an unacceptable risk of injury. These reportable incidents shall be included in the facility's annual performance evaluation.

The health services council shall consider, among other things, the proposed availability and use of safe patient-handling equipment in new or renovated space to be constructed, and input from the community to be served by the proposed equipment and services. Rhode Island's Safe Patient Handling Act of 2006 will increase health care safety by mandating use of modern technology to decrease injuries traditionally suffered by nursing staff, patients, and residents as the result of unsafe manual lifting and movement. (Wording: http://www.rilin.state.ri.us/BillText06/SenateText06/S2760A.pdf)

Maryland passed identical companion bills for safe patient handling on April 10, 2007. House Bill 1137 "Health Care Facilities and Regulation," sponsored by Delegates Joseline Pena-Melnyk (D), Aisha N.

Braveboy (D), Melony G. Griffith (D), Jolene Ivey (D), and Kris Valderrama (D), was introduced on February 19, 2007. After passing the House on March 21, 2007, and passing the Senate on April 1, 2007, HB 1137 was approved by Governor Marin O'Malley (D) on April 10, 2007. Senate Bill 879 "Hospitals—Safe Patient Lifting," sponsored by Senator Paul G. Pinsky (D), was introduced on February 21, 2007. SB 879 passed the Senate on March 24, 2007, passed the House on March 29, 2007, and was approved by Governor Martin O'Malley (D) on April 10, 2007. Maryland's "Act concerning Hospitals—Safe Patient Lifting" took effect on October 1, 2007.

HB 1137 and SB 879 define "safe patient lifting" as "use of mechanical lifting devices by hospital employees, instead of manual lifting, to lift, transfer, and reposition patients." The laws require Maryland hospitals to develop a safe patient lifting committee, with equal numbers of managers and employees by December 1, 2007, and for the committee to develop a safe patient lifting policy by July 1, 2008, with the goal of reducing employee injuries associated with patient lifting. Consideration is to be given to patient-handling hazard assessment processes; enhanced use of mechanical lifting devices; development of specialized lift teams; training programs for safe patient lifting required for all patient care personnel; incorporating space and construction design for mechanical lifting devices in architectural plans for hospital construction or renovation; and developing a process for evaluating effectiveness of the safe lifting policy. (Text of HB 1137: http://mlis.state.md.us/2007RS/chapters_noln/Ch_57_hb1137T.pdf. Text of SB 879: http://mlis.state.md.us/2007RS/chapters_noln/Ch_56_sb0879T.pdf)

Minnesota passed legislation for the safe handling of dependent health care patients and residents in May 2007. Companion bills were House File 712 introduced on February 8, 2007, by Representatives Patti Fritz (D), Erin Murphy (D), Maria Ruud (D), Karen Clark (D), Jim Abeler (R), and David Bly (D); and Senate File 828 introduced on February 15, 2007, by Senators Linda Higgins (D), Sharon L. Erickson Ropes (D), Kathy Sheran (D), Paul E. Koering (R), and John Marty (D). Minnesota's "Safe Patient Handling Act" was passed on May 20, 2007, within the second engrossment of a large omnibus bill, House File No. 122. HF 122 was approved by Governor Tim Pawlenty (R) on May 25, 2007.

Minnesota's Safe Patient Handling Act requires implementation of a safe patient-handling program by every licensed health care facility in the state, including hospitals, outpatient surgical centers, and nursing homes. By July 1, 2008, a written safe patient-handling policy will be adopted with the facility's plan to achieve by January 1, 2011, the goal of minimizing manual lifting of patients by nurses and other direct patient care workers by utilizing safe patient-handling equipment, rather than people, to transfer, move, and reposition patients and residents in all health care facilities. The program will address the assessment of hazards with patient handling, acquisition of an adequate supply of appropriate safe patient-handling equipment, initial and ongoing training of nurses and direct patient care workers on use of the equipment, procedures related to ensuring that remodeling and construction are consistent with program goals, and periodic evaluations of the safe patient-handling program.

By July 1, 2008, every licensed health care facility in Minnesota shall establish a safe patient-handling committee, with at least half the members nonmanagerial nurses and other direct patient care workers. In a health care facility where nurses and other direct patient care workers are covered by a collective bargaining agreement, the union shall select the committee members proportionate to its representation of nonmanagerial workers, nurses, and other direct patient care workers. Employees serving on the safe patient-handling committee must be compensated by their employer for all hours spent on committee business.

Duties of a safe patient-handling committee shall include completion of a patient-handling hazard assessment which includes identification of problems and solutions, highest risk areas for lifting injuries, and recommendations for a mechanism to report, track, and analyze injury trends. The committee is to make recommendations on the purchase, use, and maintenance of an adequate supply of appropriate safe patient-handling equipment, and on training of nurses and other direct patient care workers on use of safe patient-handling equipment, initially and periodically afterward. The committee will conduct annual evaluations of the safe patient-handling plan and progress toward goals in the safe patient-handling policy, and will recommend procedures to ensure that remodeling and construction of patient care areas incorporate safe patient-handling equipment or the design needed to accommodate such equipment at a later date.

The commissioner shall make training materials on implementation of the safe patient-handling program available to all health care facilities at no cost as part of the training and education duties of the commissioner. The safe patient-handling program shall be enforced by the commissioner with violations subject to penalties provided.

The commissioner may make dollar-for-dollar matching grants, not to exceed $40,000, to health care facilities to acquire and train on safe patient-handling equipment. Priority for grants may be given to facilities that demonstrate financial hardship; the 50% match requirement may be waived; and such facilities may be granted more than $40,000. Health care facilities experiencing hardship shall not be required to meet the safe patient-handling requirements until July 1, 2012.

Provisions are included for studying ways to require workers' compensation insurers to recognize compliance in premiums of health care and long-term care facilities, and for development of on-going funding sources, including, but not limited to, low-interest loans, interest-free loans, and federal, state, or county grants. The commissioner must make recommendations to the legislature regarding funding sources by January 15, 2008. A special workers' compensation fund was to provide $500,000 the first year and $500,000 the second year for safe patient-handling grants, but a line item veto reduced funding to $500,000 one time.

The Minnesota State Council on Disability shall convene a work group representing clinics, disability advocates, and direct care workers, to assess options for use of safe patient-handling equipment in unlicensed outpatient clinics, physician offices, and dental settings; identify barriers to use of safe patient-handling equipment in these settings; and define clinical settings that move patients to determine applicability of the Safe Patient Handling Act. The work group is to report findings to the legislature by January 15, 2008.

Omnibus bill HF 122 is very large, over 130 pages, with language pertinent to safe patient handling in three separate areas of the bill regarding 1. Initial grant funding in Article 1, Section 6, Subdivision 3, on pages 25–26; 2. The main body of wording for safe patient handling in Article 2, Section 23. 182.6551 to Section 25. 182.6553, on pages 48–51; and 3. Study of ways to require workers' compensation insurers to recognize compliance in premiums and for on-going funding in Article 2, Section 36, and work groups on safe patient handling and equipment in Section 37 on pages 58–59. (Wording of HF 122: http://www.leg.state.mn.us/leg/legis.asp)

On passage of Minnesota's Safe Patient Handling Statute, Elizabeth (Bettye) Shogren, RN, Health and Safety Specialist, Minnesota Nurses Association, stated, "We had many supporters in the community and a comprehensive strategic plan to advance the bill. Over-preparation and realistic expectations made this a success."

On January 17, 2008, the Minnesota Department of Labor and Industry announced awarding 67 grants totaling $500,000 to health care facilities to help purchase patient-lifting equipment in complying with the new state patient-handling regulations. For a list of the facilities in Minnesota receiving patient-handling grants see http://www.doli.state.mn.us/pdf/sph_approv_grants.pdf.

In this writing, *New Jersey* is the most recent state to pass safe patient-handling legislation. The new "Safe Patient Handling Act" was signed into law by New Jersey Governor Jon Corzine (D) on January 3, 2008. Identical bills, titled "Safe Patient Handling Act," were introduced as Senate Bill 1758 on March 21, 2006, and Assembly Bill 3028 on May 15, 2006. On December 13, 2007, A 3028 was substituted by S 1758. The Safe Patient Handling Act passed the New Jersey Senate 37 to 0 and the Assembly 77 to 3. Primary Sponsors of New Jersey's Safe Patient Handling Act were Senators Joseph F. Vitale (D) and Loretta Weinberg (D), Assemblymen Herb Conaway, MD (D), Vincent Prieto (D), and Gary S. Schaer (D), and Assemblywoman Joan M. Voss (D), plus 20 Co-Sponsors.

New Jersey's new law for reducing the risk of injury to patients and health care workers with patient-handling needs went into effect immediately upon signing, covering general and special hospitals, nursing homes, state developmental centers, and state and county psychiatric hospitals. Covered facilities are to establish within 12 months a safe patient-handling committee, comprised of at least 50% health care workers, and others trained in safe patient-handling procedures, for development, implementation, evaluation, and possible revision of the safe patient-handling program.

The safe patient-handling program is to include procedures for assessing and updating patient-handling requirements of each patient, and a plan for achieving prompt access to, and availability of, patient-handling equipment and handling aids.

Facilities shall establish a safe patient-handling policy on all units and for all shifts, posted in a location easily visible to staff, patients, and visitors, which minimizes unassisted patient handling, and includes a statement concerning the right of a patient to refuse the use of assisted patient handling.

"Assisted patient handling" means use of mechanical patient-handling equipment, including, but not limited to, electric beds, portable base and ceiling track-mounted full body sling lifts, stand assist lifts, and mechanized lateral transfer aids; and patient-handling aids, including, but not limited to, gait belts with handles, sliding boards, and surface friction-reducing devices.

Retaliatory action shall not be taken against any health care worker for refusing to perform a patient-handling task due to a reasonable concern about worker or patient safety, or the lack of appropriate and available patient-handling equipment. If a health care worker refuses to perform a patient-handling task, the worker shall promptly report to the supervisor the refusal and the reason for the refusal.

Covered facilities will have 36 months to establish a safe patient-handling program and begin training employees in safe patient-handling guidelines. Recommendations shall be included for a 3-year capital plan to purchase safe patient-handling equipment and patient-handling aids necessary to carry out the safe patient-handling policy, which is to take into account the financial constraints of the facility. (Text: http://www.njleg.state.nj.us/2006/Bills/PL07/225_.PDF)

"Nurse and Patient Safety and Protection Act of 2007" Introduced

With momentum building to legislate the safe handling of patients throughout the country, a national bill, House Resolution 6182, was first introduced into the U.S. House of Representatives on September 26, 2006, by U.S. Representative John Conyers (D-MI), "To amend the Occupational Safety and Health Act of 1970 to reduce injuries to patients, direct-care registered nurses, and other health care providers by establishing a safe patient handling standard" (HR 6182 history and text: http://thomas.loc.gov).

National legislation for safe patient handling was reintroduced by Representative John Conyers (D-MI) on January 10, 2007, with HR 378, the "Nurse and Patient Safety and Protection Act of 2007." HR 378 was referred to the Committee on Education and Labor and, also, to the Committee on Energy and Commerce where it was subsequently referred on February 2, 2007, to the Subcommittee on Health. The last action was May 9, 2007, when House Education and Labor referred to the Subcommittee on Workforce Protections.

Section 2 of HR 378 states that the "Nurse and Patient Safety and Protection Act of 2007—Requires the Secretary of Labor, acting through the Director of Occupational Safety and Health Administration, to establish a Federal Safe Patient Handling Standard to prevent musculoskeletal disorders for direct-care registered nurses and other health care providers working in health care facilities by requiring the elimination of manual lifting of patients through the use of mechanical devices, except during a declared state of emergency."

Section 5 of HR 378 defines "declared state of emergency" as "an officially designated state of emergency that has been declared by the Federal Government or the head of the appropriate State or local governmental agency having authority to declare that the State, county, municipality, or locality is in a state of emergency, but does not include a state of emergency that results from a labor dispute in the health care industry or consistent under staffing."

Findings of Congress in Section 1 of HR 378

1. Direct-care registered nurses rank 10th among all occupations for MSDs, sustaining injuries at a higher rate than laborers, movers, and truck drivers. In 2004, nurses sustained 8800 MSDs, most of which (over 7000) were back injuries. The leading cause of these injuries in health care are the result of patient-lifting, patient-transferring, and patient-repositioning injuries.
2. The physical demands of the nursing profession lead many nurses to leave the profession. Fifty-two percent of nurses complain of chronic back pain and 38% suffer from pain severe enough to require leave from work. Many nurses and other health care providers suffering back injury do not return to work.
3. Patients are not at optimum levels of safety while being lifted, transferred, or repositioned manually. Mechanical lift programs can substantially reduce skin tears suffered by patients, allowing patients a safer means to progress through their care.
4. The development of assistive patient-handling equipment and devices has essentially rendered the act of strict manual patient-handling unnecessary as a function of nursing care.
5. Application of assistive patient-handling technology fulfills an ergonomic approach within the nursing practice by designing and fitting the job or workplace to match the capabilities and limitations of the human body.
6. A growing number of health care facilities have incorporated patient-handling technology and have reported positive results. Injuries among nursing staff have dramatically declined since implementing patient-handling equipment and devices. As a result, the number of lost work days due to injury and staff turnover has declined. Cost-benefit analyses have also shown that assistive patient-handling technology successfully reduces workers' compensation costs for MSDs.
7. Establishing a safe patient-handling standard for direct-care registered nurses and other health care providers is a critical component in increasing patient safety, protecting nurses, and addressing the nursing shortage.

If HR 378 Nurse and Patient Safety and Protection Act of 2007 passes, a Federal Safe Patient Handling Standard will be established with which all health care facilities in the nation will be required to comply. "Health care facility" is defined in Section 5 as "an outpatient health care facility, hospital, nursing home, home health care agency, hospice, federally qualified health center, nurse managed health center, rural health clinic, or any similar health care facility that employs direct-care registered nurses."

With enactment of HR 378, requirements on all health care facilities will be to purchase, use, and maintain safe lift mechanical devices; input from direct-care registered nurses and organizations representing direct-care registered nurses in implementing the standard; a program to identify problems and solutions regarding safe patient handling; a system to report, track, and analyze trends in injuries, and to make injury data available to the public; training for staff on safe patient-handling policies, equipment, and devices at least annually, with training to also include hazard identification, assessment, and control of musculoskeletal hazards in patient care areas, including interactive classroom-based and hands-on training by a knowledgeable person or staff; and annual evaluations of safe patient-handling efforts, as well as new technology, handling procedures, and engineering controls with documentation of this process to include equipment selection and evaluation.

Nurse and Patient Safety and Protection Act of 2007 includes strong protection for a direct-care registered nurse or other health care provider who may refuse a work assignment if the assignment would violate the Federal Safe Patient Handling Standard established under the Act, or who in good faith may report a violation or a suspected violation of this Act or of the Standard established under this Act.

HR 378 also authorizes $50 million to be appropriated for a grant program for health care facilities that can prove a financial need to cover some or all of the costs of purchasing safe patient-handling equipment. The grant funds are to remain available until expended.

Following is the summary provided at http://thomas.loc.gov of requirements for the safe handling of patients which will become effective throughout the country if HR 378 is passed into law:

> Nurse and Patient Safety and Protection Act of 2007—Requires the Secretary of Labor, acting through the Director of Occupational Safety and Health Administration, to establish a Federal Safe Patient Handling Standard to prevent musculoskeletal disorders for direct-care registered nurses and other health care providers working in health care facilities by requiring the elimination of manual lifting of patients through the use of mechanical devices, except during a declared state of emergency.
>
> Requires health care facilities to (1) develop and implement a safe patient handling plan consistent with such standard; and (2) post a uniform notice that explains the standard and the procedures to report patient handling-related injuries. Requires the Secretary to direct the Occupational Safety and Health Administration to conduct audits of plan implementation and compliance.
>
> Authorizes health care providers to (1) refuse to accept an assignment in a health care facility if the assignment would violate the standard or if such provider is not prepared to fulfill the assignment without compromising the patient safety or jeopardizing the provider's license; and (2) file complaints against facilities that violate this Act. Requires the Secretary to investigate complaints and to prohibit retaliation if violations occur. Prohibits health care facilities from retaliating with respect to employment against providers for such refusal or against any individual who in good faith reports a violation, participates in an investigation or proceeding, or discusses violations.
>
> Authorizes health care providers who have been retaliated against in violation of this Act to bring a cause of action in a U.S. district court. Entitles providers that prevail to reinstatement, reimbursement of lost compensation, attorneys' fees, court costs, and/or other damages.
>
> Requires the Secretary of Health and Human Services to establish a grant program for purchasing safe patient handling equipment for health care facilities. (HR 378 status, text, summary: http://thomas.loc.gov)

More than availability of products, education, and training were required to move the massive health care industry to provide safety devices for prevention of needlesticks; it took legislation. The same holds true for prevention of serious musculoskeletal injuries caused by unsafe manual patient lifting. Modern equipment, methods, and abundant supportive research for injury prevention with patient handling are readily available. Yet, the on-going epidemic, of back, neck, shoulder, and other musculoskeletal injuries to health care workers, shows that much of the health care industry is unable to self-regulate for protection of employees against preventable injuries.

Industry-specific legislation will be required to force many hospitals and nursing homes to protect health care workers from needless physical, financial, and career ruin, and to protect patients and residents from avoidable pain and injury directly caused by manual patient handling. With the rising obesity of patients, with our most critical nurse shortage ever, and with an aging nurse population,[51] action must be taken to halt devastating injuries caused by the manual lifting and movement of patients. With increasing awareness of the extent and consequences of losing nurses to preventable injuries, the elimination of hazardous manual patient handling can be accomplished through legislation in our nation.

References

1. Foley ME. Letter to the editor. *US News and World Report*. Press releases. American Nurses Association. March 23, 2000. http://nursingworld.org/pressrel/2000/ltr0426.htm
2. Shogren E. The relationship between the nursing shortage and nursing injury. In: W Charney and A Hudson (eds.), *Back Injury Among Healthcare Workers: Causes, Solutions, and Impacts*. CRC Press/Lewis Publishers, Boca Raton, FL, 2004, p. 248.

3. Tuohy-Main K. Why manual handling should be eliminated for resident and carer safety. *Geriaction* 15:10–14, 1997.
4. U.S. Department of Labor Brief. OSH Review Commission. Secretary of Labor v. Beverly Enterprises, Inc. OSHRC Docket 91-3344, 92-0238, 0819, 1257, 93-0724. Filed May 1, 1996.
5. Bureau of Labor Statistics. U.S. Department of Labor. Number, incidence rate, median days away from work, and relative standard errors of occupational injuries and illnesses involving days away from work by selected sources with MSDs in private industry for All US, 2006, Table 5. Bureau of Labor Statistics. U.S. Department of Labor, November 2007. http://www.bls.gov
6. Bureau of Labor Statistics. U.S. Department of Labor. Number and percent of nonfatal occupational injuries and illnesses involving days away from work for the ten occupations with the largest number of cases by case and worker characteristics, 2000, Table 12. Bureau of Labor Statistics. U.S. Department of Labor, 2002, p. 12, 13. http://www.bls.gov
7. Bureau of Labor Statistics. U.S. Department of Labor. Number and percent of nonfatal occupational injuries and illnesses involving days away from work for the ten occupations with the largest number of cases by case and worker characteristics, 2002, Table 2. Bureau of Labor Statistics. U.S. Department of Labor, 2004, p. 3, 6. http://www.bls.gov
8. Bureau of Labor Statistics. U.S. Department of Labor. Number, incidence rate and median days away from work of occupational injuries and illnesses with days away from work involving musculoskeletal disorders by selected occupations, All US, private industry, 2006, Table 11. Bureau of Labor Statistics. U.S. Department of Labor, November 2007. www.bls.gov/iif/oshwc/osh/case/ostb1792.pdf
9. Owen BD. The magnitude of low-back problem in nursing. *West J Nurs Res* 11(2):234–242, 1989.
10. Hudson A. Government statistics particularly skewed against health care. Readers respond. *Ergonomics Today*, March 17, 2004. www.ergoweb.com/news/detail.cfm?id = 897
11. 2001 American Nurses Association Health and Safety Survey. Online survey conducted July 11 to August 15, 2001. http://www.NursingWorld.org
12. Owen BD. Preventing injuries using an ergonomic approach. *AORN J* 72(6):1031–1036, 2000.
13. Mitchelmore M. The psychosocial implications of back injury at work. *Nurs Stand* 10(38):33–38, 1996. http://www.nursing-standard.co.uk/archives/wk38/research.htm
14. Stubbs DA, Buckle PW, Hudson MP, Rivers PM, and Baty D. Backing out: Nurse wastage associated with back pain. *Intl J Nurs Stud* 23(4):325–336, 1986.
15. Foley ME. Hearing on ergonomics safety in the workplace, held July 20, 2001. U.S. Department of Labor. American Nurses Association report. Workplace Issues: Occupational Safety and Health. http://nursingworld.org/dlwa/osh/hearing.htm
16. Aiken LH, Clarke SP, Sloane DM, Sochalski J, and Silber JH. Hospital nurse staffing and patient mortality, nurse burnout, and job dissatisfaction. *J Am Med Assoc* 288(16):1987–1993, 2002.
17. Marras WS, Davis KG, Kirking BC, and Bertsche PK. A comprehensive analysis of low-back disorder risk and spinal loading during the transferring and repositioning of patients using different techniques. *Ergonomics* 42(7):904–926, 1999.
18. Waters TR, Putz-Anderson V, Garg A, and Fine LJ. Revised NIOSH equation for the design and evaluation of manual lifting tasks. *Ergonomics* 36(7):749–776, 1993.
19. Snook SH and Ciriello VM. The design of manual handling tasks: Revised tables of maximum acceptable weights and forces. *Ergonomics* 34:1197–1213, 1991.
20. Waters TR. When is it safe to manually lift a patient? *Am J Nurs* 107(8):53–58, 2007.
21. Charney W, Zimmerman K, and Walara E. The lifting team: A design method to reduce lost time back injury in nursing. *AAOHN J* 39(5):231–234, 1991.
22. Meittunen EJ, Matzke K, McCormack H, and Sobczak SC. The effect of focusing ergonomic risk factors on a patient transfer team to reduce incidents among nurses associated with patient care. *J Healthc Safety Compliance Infect Control* 3(7):305–312, 1999.

23. Charney W. The lift team method for reducing back injuries: A 10 hospital study. *AAOHN J* 45(6):300–304, 1997.
24. Villeneuve J. The ceiling lift: An efficient way to prevent injuries to nursing staff. *J Healthc Safety Compliance Infect Control* 2(1):19–23, 1998.
25. Kosel K and Olivo T. The business case for work force stability. Voluntary hospitals of America Center for Research and Innovation. *VHA Res Ser* 7:7, 2002. http://www.vha.com/research/public/stability.pdf
26. Pasero C and McCaffery M. Pain in the critically ill. *Am J Nurs* 102(1):59–60, 2002.
27. Garg A. *Long-Term Effectiveness of 'Zero-Lift Program' in Seven Nursing Homes and One Hospital.* U.S. Dept HHS, CDC, NIOSH, Cincinnati, OH, 1999, p. 64.
28. Australian Nursing Federation. *No Lift, No Injury Goes from Strength to Strength.* April 2002. http://www.anf.org.au/news_professional/news_professional_0204.html
29. The National Back Pain Association in collaboration with the Royal College of Nursing. *The Guide to the Handling of Patients: Introducing a Safer Handling Policy*, revised 4th edn. The National Back Pain Association, Teddington, Middlesex, 1997.
30. Manual Handling Operations Regulations 1992. At http://www.hmso.gov.uk/acts.htm, click "Search Engine," enter "manual handling regulations."
31. Royal College of Nursing. *RCN Code of Practice for Patient Handling.* Royal College of Nursing, London, April 1996, revised September 2000. http://www.rcn.org.uk/publications/pdf/code-practice-patient-handling.pdf
32. Royal College of Nursing. *RCN Introducing a Safer Patient Handling Policy.* Royal College of Nursing, London, April 1996, revised November 2000. http://www.rcn.org.uk/direct
33. Nurse Wins £800,000 for Back Injury. *BBC News.* February 15, 2000. http://news.bbc.co.uk/1/hi/health/642381.stm
34. Meikle J. Nurse Wins £345,000 for Back Pain. *The Guardian.* SocietyGuardian. Guardian Newspapers Limited. June 6, 2001. http://society.guardian.co.uk/print/0,3858,4199020-106678,00.html
35. Back Pain Nurse Awarded £420,000. *BBC News.* October 16, 2002. http://news.bbc.co.uk/1/hi/england/2333005.stm
36. Australian Nursing Federation Victorian Branch. No Lifting Policy Statement. Adopted March 1998, revised April 2000. http://www.anfvic.asn.au/services_pubs.htm
37. Langford E. *Buried But Not Dead: A Survey of Occupational Illness and Injury Incurred by Nurses in the Victorian Health Service Industry.* Australian Nursing Federation Victorian Branch Injured Nurses' Support Group, Melbourne, 1997, p. 10. http://www.anfvic.asn.au/services_pubs.htm
38. Policy and Strategic Projects Division, Victorian Government Dept of Human Services. Victorian nurses back injury prevention project evaluation report 2002. Policy and Strategic Projects Division, Victorian Government Dept of Human Services, Melbourne, Victoria, 2002. http://www.nursing.vic.gov.au
39. Nurse Awarded Order of Australia. Occupational Health and Safety Representatives Networking for Safe Workplaces. June 27, 2003. http://www.ohsrep.org.au/news/1056674467_31526.html
40. Australian Nursing Federation Victorian Branch. Australian Nursing Federation's No Lifting Policy Wins Outstanding Occupational Health and Safety Leadership Award. Australian Nursing Federation Victorian Branch media release. June 22, 2001. http://www.anfvic.asn.au/
41. Centers for Disease Control Recognizes Teamwork that Reduced Employee Back Injuries in Nursing Homes. CDC press release. June 23, 2003. http://www.cdc.gov/niosh/niosh-update-06-23-2003.html
42. Collins J, Nelson A, and Sublet V. *Safe Lifting and Movement of Nursing Home Residents.* U.S. Department HHS, CDC, NIOSH, Cincinnati, OH, 2006.
43. American Nurses Association Position Statement on Elimination of Manual Patient Handling to Prevent Work-Related Musculoskeletal Disorders. Effective June 21, 2003. Originated by Nurse Advocacy Programs Center for Occupational Health and Safety. Adopted by ANA Board of Directors. http://www.nursingworld.org

44. Blakeney B and Stierle L. *Welcome to 'Handle with Care.'* American Nurses Association, Silver Spring, MD. September 17, 2003. http://www.nursingworld.org/handlewithcare/bbltr.htm
45. Charney W and Hudson A (eds.), *Back Injury among Healthcare Workers: Causes, Solutions, and Impacts.* CRC Press/Lewis Publishers, Boca Raton, FL, 2004.
46. Quigley P. 2003 Health Care Legislation: Opportunities for unity. *Florida Nurse* 51(1):1–3, 2003.
47. US Dept of Labor. Occupational Safety and Health Admin. Occupational Exposure to Bloodborne Pathogens; Needlestick and Other Sharps Injuries; Final Rule. Rules and Regulations. US Dept of Labor. Occupational Safety and Health Admin. 29 CFR. Part 1910. Docket No. H370A, January 18, 2001. *Federal Register* 66(12):5318, 2001.
48. Needlestick Safety and Prevention Act (Enrolled Bill). Signed into law November 6, 2000. http://thomas.loc.gov. Under "Legislation," select "Bill Text 101st to 108th. Select "106." Under "Bill Number," enter HR 5178." Click "search." At #4, select HR 5178 ENR.
49. Hudson A. Texas passes first Law for safe patient handling in America: Landmark legislation protects healthcare workers and patients from injury related to manual patient lifting. *J Long Term Eff Med Implants* 15(5):559–566, 2005.
50. Charney W. The need to legislate the healthcare industry in the state of Washington to protect healthcare workers from back injury. *J Long Term Eff Med Implants* 15(5):567–572, 2005.
51. Buerhaus PI, Staiger DO, and Auerbach DI. Implications of an aging registered nurse workforce. *J Am Med Assoc* 283(22):2948–2954, 2000.

Disposable

Anne Hudson

This is my story, of how I became back-injured and learned that nurses are often disposed of when they can no longer be used to lift people. My aim is to help prevent the same from happening to others.

When the nursing instructor told our class, "Your job depends on your back," I, quite frankly, did not believe her. The startling announcement was made during a pediatrics nursing class, way off subject. Our instructors said nurses are especially prone to cumulative trauma injury to the back from lifting patients and to "take care of your back."

The student across from me appeared as alarmed as I was. Comparing notes later, we found that both of us saw our transcript pass before our eyes. We thought why are we sitting here taking all of these classes, if our job will depend on our back?! A college degree made no sense if being a nurse depended on your back. The instructor gave no explanation of "cumulative trauma injury," so, I naively chose to believe she meant muscle strain which should quickly heal. I thought surely hospitals would not require a nursing degree for work that could permanently damage nurses, and then toss them out. That is how little I knew about floor nursing before becoming a nurse myself.

We learned in nursing skills class how to lift and move people alone and with a partner. At 5 ft 3 in., and 115 lb, I learned that with rocking for momentum, and placing my arms beneath the person just so, I could move a supine 160 lb person up the bed by myself. And that was just for starters. We learned every manner of manual lifting, turning, and repositioning which was put to immediate use in clinicals at the nursing home.

"Elizabeth" was my first patient at the nursing home. I was to practice my new skills with giving her a bed bath. I arrived in my starched white dress with the blue and white "Student Nurse" patch hand-sewn onto the sleeve. My hair was put up just so, in what I thought was a cute little "do." I had not been prepared to find Elizabeth, emaciated and nonverbal, with severely contracted legs. She could use her arms though and, when I leaned over to introduce myself, she grabbed my hair, entwining a fistful in her bony fingers. There went my do and, with efforts to reassure poor Elizabeth, I struggled to bathe and turn a stiff scrawny body that could not unfold.

Somehow, I thought nursing would become more predictable from classroom to practice, but, actually, few things have gone as expected. One of the reasons I wanted to become a nurse was to demystify the whole hospital/medical arena, to learn more so I could help others and my own family. Missions accomplished. I learned plenty. And I know I helped a great many people, too.

Though it has been a while since my back injury in 2000 and leaving the bedside, I still see people in town who recognize me and say, "You were my nurse. You were a good nurse." That is a very nice compliment but seems all I painfully hear is "you were." "Leaving" hospital nursing was certainly not my choice. I had intended to work 25 or more years at the hospital building a nice retirement, but lasted only 10 years before herniated lumbar disks from years of lifting people forced me out. Or, rather, the hospital, which had no provision to retain injured nurses with lifting restrictions, forced me out.

I worked with acutely ill patients on medical/surgical, telemetry, and intermediate care units. The patients were generally very weak and many were also very heavy, requiring much strenuous lifting, repositioning, turning, and even lifting and carrying upon occasion. Once, a man had a stroke while seated on the toilet in his bathroom. Another nurse and I carried him back to his bed. The same nurse and I once carried a woman who became unconscious, from her chair across the room, back to her bed. Though I never saw a patient actually get dropped, I was involved in a number of "near misses" when nursing staff almost did not make it when lifting and carrying someone from the bedside chair or commode back to their bed.

On the units where I worked, the expectation was for each nurse to do as much lifting and repositioning of patients as possible on their own, before seeking help from others. Being able to do your own work, without asking for help, was perceived as being capable and efficient. Such independence was valued more highly than safety, with few nurses appearing to think much about risk to themselves with lifting patients. We had been taught how to perform the various lifts, were required to lift, and were given annual body mechanics training as "protection" from injury, if we would only lift "correctly." How could the vague warning from nursing school to "take care of your back" be applied, when nurses were required to lift dangerous amounts of weight with all sizes of patients throughout every shift?

There was one rolling floor lift in the hospital, with the slings kept in Central Supply on a lower level apart from the nursing unit floors. It was such a production to walk the halls and floors looking for the lift, or to call the other units to locate and go get the lift, and then to call the nursing supervisor to request the appropriate sling, and to wait for arrival of the sling from Supply, that the lift was used only when it was absolutely impossible for a team of nurses to do a manual lift. I saw the lift used about once every 6 or 8 months. I was involved in a number of patient lifts from the floor, including times when the nursing supervisor herself came to help manually lift, with no one suggesting to go get the lift. So, the lift was not used even when it would seem necessary.

All of the lifting and moving of patients did not trouble me for the most part. I felt strong and healthy and was eager to keep my patients well-positioned and mobilized. Rather then allowing frail elderly patients to experience incontinence, and then cleaning them up, I tried as much as possible to get dependent patients up to the bedside commode often, to help them maintain continence, skin integrity, and mobility along with their dignity. I was pleased to document measurable output on the "Intake and Output" record, hopefully indicating to the next shift that the patient could maintain continence if they were assisted up to the commode often enough. I was equally glad to help other nurses with repositioning and with lifting and transferring their patients up to the commode or to their chair for meals, too.

There were times, however, when I thought I might need to look for a position on a unit without so much lifting. Once, using the under-axilla lift to raise a short plump woman in room 2B from the commode to her feet, I felt a pop and fleeting sharp pain in my low spine. Another time, using the same under-axilla lift to raise a tall thin man in room 6A to his feet from the bedside, I felt another spinal pop with brief sharp pain. I was trained and required to use the under-axilla "drag lift" to lift patients to their feet or to pull them up in bed, the "cradle lift" with arms under the patient's shoulders and thighs to pull up in bed, the "hug lift" to raise the patient to their feet from a seated position with their arms around my neck, and so forth, not knowing that such manual lifts have been condemned in other countries for many years due to the proven danger to nurses and patients alike.

There was continual patient lifting and moving with many outrageous events. One of the last, before going out with my back injury, was with saving a patient from a fall. I rushed into the room when I saw from the hallway that he was about to fall headlong while rising from his bedside chair, pushing the wheeled over-bed table farther and farther away as he rose. The man grabbed my right shoulder and pushed me down hard into an "L" shape, now bearing his weight on my back. There was that familiar pop in my low spine. But this time, the sudden intense pain lasted and lasted and lasted until my screaming for help brought a nursing assistant and a nurse who had "floated" from another unit to my aid. They got the man off my back and assisted him to bed. When I stood upright, the pain wonderfully disappeared. I felt relieved that I had apparently once more escaped without injury, so I thought there was no need to complete an "Incident Report" and went about my work to complete my shift. I was busy and thought there was no time for extra paperwork.

Soon after, the day came, after thousands of patient lifts, with most of the patients heavier than I, that such outrageous lifting caught up with me. On a day off work, while walking through my kitchen, I had sudden, horrible, unrelenting pain across my low back and down my leg. The pain was incapacitating. I was unable to sit down. I could do nothing except barely shuffle around. The bed became my enemy. I could not get down onto the bed without great pain, could not turn over or sleep once down, and got "stuck" from pain part-way up trying to get out of bed. I kept thinking it would pass and that I would be able to work on my next scheduled day. I was in denial that I could be severely injured.

Well, the severe pain continued and I was not able to report for work. I reluctantly called in with back pain. I did not yet know that I had a cumulative trauma spinal injury, with two herniated lumbar disks from years of lifting people, and that my hospital career was over. I refused to think along those lines. When I had considered a possible back injury over the years, I had indulged in magical thinking, that, if injured, the hospital would help keep me employed. I believed I was a well-respected team member and a valued employee and, after all, it was "health care."

I went to the hospital to fill out the workers' compensation form. It asked, "What did you do?" "What will you do next time to keep this from happening again?" From nursing school on, I had occasionally heard that nurses were at risk of cumulative trauma back injury from lifting patients. Now I turn up with a back injury and it began to feel like it was my fault, that the hospital had no responsibility for a nurse being back-injured by thousands of patient lifts, that the nurse had done something else to her back apart from lifting patients.

My primary care doctor diagnosed lumbar strain and lumbar radiculopathy, gave me a prescription for pain medication, and sent me to "Back School" where I learned about exercises, hydration, etc. I was referred to a neurologist who ordered the first MRI, prescribed other medications, and suggested a chiropractor and an attorney for workers' compensation. I am grateful that I contacted the particular chiropractor and attorney suggested by the neurologist.

The chiropractor did manipulations and ultrasound, suggested a muscle relaxant, and said lumbar disks were the problem, not just lumbar strain, and that I needed a discogram. My primary care doctor referred me to an orthopedist who ordered physical therapy and, over time, offered fusion for bulging/herniation of L4/L5. Physical therapy provided heat, massage, ultrasound, and stretching and strengthening exercises. Two discograms confirmed internal disk disruption of L4/L5 and L5/S1 and identified both levels as pain generators.

Ultimately, after 2 years of conservative measures provided no relief, a neurosurgeon performed an anterior–posterior two-level lumbar inter-body fusion, replacing my damaged disks with cadaver bone grafts through my abdomen, and stabilizing my spine with titanium screws through my back.

During the 2 years leading to surgery, I was permitted only 6 months of light duty, including a total of 90 days per workers' compensation regulations, and a second 90 days obtained through negotiations with the hospital, with assistance from the nurses' bargaining unit.

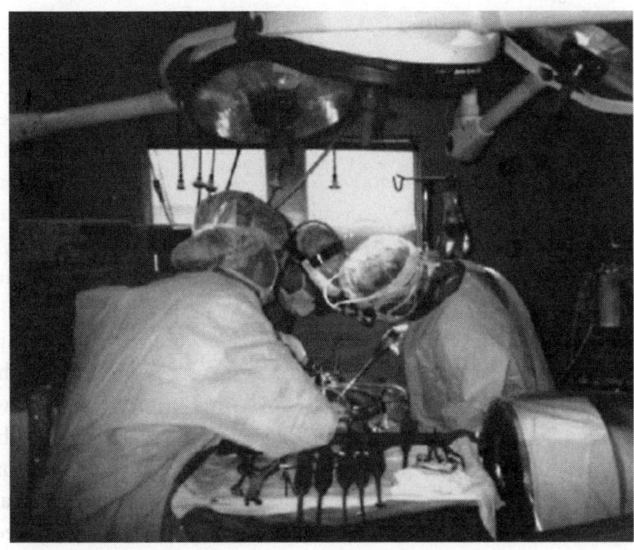

When I was initially injured, light duty was quickly arranged by the hospital's employee health nurse, and I was back working within 2 weeks. I had seen other injured nurses working light duty and thought it was wonderful that the hospital was helping to keep them working. Now they were doing the same for me, helping to keep me working. When other injured nurses disappeared, I thought they had chosen to go work someplace else. I knew nothing of the workers' compensation system, that workers' compensation, not the hospital, was paying my light duty wages, and that the clock was ticking off the number of light duty days I would be allowed.

Two months into my injury, about the time I first saw the neurologist and had my first MRI, a certified registered letter requiring my signature arrived stating that workers' compensation denied my claim of work-related back injury. I had worked 60 days of light duty in Nursing Education, training nurses on new medical products, outfitting the practice code cart, gathering and documenting data, etc. I had done well in light duty and appreciated and enjoyed working. I managed my back pain during work with non-narcotic analgesics, position changes with alternating sitting and standing, a rolled towel off the linen cart as a lumbar cushion in my chair, analgesic rubs, and an assortment of lumbar supports, including one with magnets and one which held a small ice pack. But when I reported for work the morning after my workers' compensation claim was denied, I was dismissed. The hospital sent me home to use accumulated sick leave and vacation leave until I could return to my position, or until my leave time and Family Medical Leave time ran out, and I would be terminated.

Several weeks later I had a consultation with a neurosurgeon and requested a prescription for a back brace and a release to return to work. I was afraid for my job and thought if I just tried harder I could do the lifting in spite of the pain. I had read conflicting reports on the pros and cons of back supports and braces, but I needed to find out if I could return to work with my back in its current shape.

Wearing a heavy brace with metal stays, I returned to my regular position in the Intermediate Care Unit, but lasted only 3 weeks. I found that the brace felt good to wear, supportive, and secure, preventing me from bending at the waist or twisting. But I also found, after only 3 weeks of wearing the brace, that I was weaker because I was not supporting my torso; the brace was doing it for me. I found, too, that as supportive and comforting as it was, the brace did nothing to diminish compressive force to spinal disks with lifting. My damaged low spine hurt badly with lifting, brace or no brace.

My last patient was a 425 lb woman about my same height. She did well after her gastric bypass surgery but still needed a lot of help lifting and shifting legs, raising and washing beneath her massive

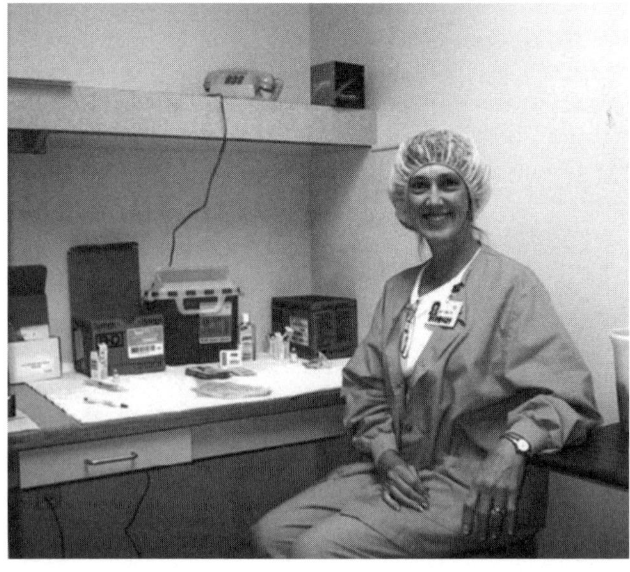

abdominal panus, repositioning and pulling up in bed, etc. Once she got upright onto her feet she could shuffle-walk into the bathroom, but after toileting her arms could not reach around her body to cleanse herself. Her torso was so large that it came to within one foot of the floor when she was standing. This made it necessary for me to go onto three points, low to the ground, and shove a cloth with much effort deep between ponderous flesh to reach the parts needing to be cleansed. She was working hard to get better and I do hope she is doing well today. No disrespect is intended in describing some of the strenuous care unique to the bariatric population.

The woman with the gastric bypass was my last patient in my position as a hospital floor nurse. Job or no job, the increased pain was too great to go on. I went off again with back pain and filed a second workers' compensation claim. Again I was placed into light duty right away, this time in the Business Office and Medical Records. This time it took only 30 days, after a total of 90 days light duty, to receive the workers' compensation denial, again per certified registered letter requiring my signature. Again, I was dismissed and sent home to use up my accumulated leave.

I was continuing with conservative treatment for my back and, at this point, through negotiations with the hospital, obtained the additional 90 days of light duty mentioned earlier. This period of light duty was in the pre-op area where it was wonderful to be back working with patients. I did well in pre-op and enjoyed helping prepare patients for their surgery. I also enjoyed working with the staff and learning about a different area of the hospital. When the 90 days were up, again I was dismissed to exhaust remaining leave time.

By this time I had researched back injuries to nurses from patient lifting and had learned a great deal. I wanted to help the hospital learn how they could prevent back injuries to other nurses. I learned that the hospital had a Back Injury Prevention Task Force and requested to speak at their meeting. I told the small group about research showing safety limits for lifting, research on spinal injury with patient lifting and preventing injuries with patient lift equipment, and evidence of tremendous cost savings with injury prevention.

They appeared bored with what I had to say and their reply stunned me. They said, "We know all that. We've been going for three years." Because the hospital had not implemented a "no lifting" policy, and had not provided the patient lifting equipment I had learned about, I believed they must be unaware. How could a hospital know that they could prevent staff injuries and not take the steps needed to do so? If they knew that preventing injuries could save money, why were they not aggressive in injury prevention? My injury, that of a single nurse, ultimately cost about $200,000 in direct costs for medical and compensation alone. Some nurses have far more serious injuries incurring far greater costs.

It felt awful to learn that the hospital had known for years how to prevent patient lifting injuries but had allowed me and others to be seriously injured and disabled from our jobs anyway. No remorse was expressed, no "we're sorry," never even a phone call to ask how I was doing. And it was becoming apparent that I, like others before me, would be terminated if unable to return to lifting.

By now I had an attorney to help me appeal the hospital's denials of my two workers' compensation claims. The hospital accepted as work-related the lesser "lumbar strain" injury, which should quickly heal, but denied the more serious disk herniations shown by MRI. My appeals of their denials first went to the workers' compensation board. When the hospital appealed the board's decisions in my favor, I requested hearings and appeared in workers' compensation court twice before an administrative law judge. Every workers' compensation board and court decision came back in my favor, confirming that my spinal disk injuries were from lifting hospital patients.

Winning my case was notable because workers' compensation typically accepts only single-point-in-time injuries and my case was based on cumulative trauma spinal injury from years of lifting patients. In court, my attorney used research showing the evidence between patient lifting and spinal injury to nurses, especially the review of research on the low back in the Federal Register on the "Ergonomics Standard Final Rule," November 11, 2000. Though the Ergonomics Rule itself was rescinded by President George Bush, the research stands and can be quite helpful in proving the relationship between patient lifting and cumulative trauma spinal injuries, or "degenerative disk disease," to nurses and other health care workers. Particularly helpful is the explanation of how microfractures to the disk nucleus and to vertebral endplates occur with the repetitive lifting of heavy loads without pain, due to the absence of pain receptors in those structures. Insidious extensive damage may have already resulted in degenerative disk disease before severe pain announces extension of damage from the center of the disk to nerves in the outer ring of the disk.

One of my physicians, who had treated back-injured nurses for 25 years, was uncertain that lifting patients by itself was the cause of spinal injury. But, after being convinced by my attorney to read the research, he found the evidence for spinal injury from lifting patients so compelling that he contacted the researchers to learn more. He then wrote a strong letter to the workers' compensation court on my behalf stating that there was no question that my spinal injury was caused by lifting patients as a nurse.

In court, two nurse coworkers testified about the type and amount of patient lifting we nurses routinely performed. I submitted line drawings of manual patient lifting techniques, along with a detailed description of the types of manual lifts, and amounts of weight I generally lifted, on a typical shift. This was estimated at 3,000 lb which I thought sounded excessive until I located an article estimating nurse lifting at 3,600 lb per shift. My estimate was low!

The same evening of my first court appearance, of fighting against the hospital to prove that my spinal injury was from lifting patients, was the annual hospital Employee Awards Banquet. I received my 10 year pin that night, along with some nice recognition in photo scenes displayed on a big screen. It was a lovely dinner affair with moving speeches by administration highlighting that "employees are our most valuable asset."

A special award that night was the hospital's first-ever "Team Spirit Award," which two other nurses and I accepted on behalf of the nurses' bargaining unit, for our fund-raising efforts to help with the critical health needs of a fellow nurse's child. There I was, on stage, accepting with others the hospital's "Team Spirit Award," while I was off work, without pay, and apparently barred from returning unless I could go back to the same heavy lifting. I was not feeling like a "valuable asset." In fact, I felt like taking the microphone and asking, "Can anyone use a good nurse?"

There had been a time when I felt I was a valuable team member, receiving Employee of the Month, and so forth. Now, I just felt rejected because I was "damaged goods"—damaged by lifting people and rejected by those who required me to do so. It felt like being used and discarded. Because I could not return to lifting, I was being processed out the door. The very thing I thought was too unethical to believe, that hospitals would allow, essentially cause, nurses to be disabled by lifting and then discard them, was now happening to me.

Before my injury, I tried to "lighten the load" by requesting female urinals, to reduce turning heavy women on/off the bedpan. Male patients were provided urinals. Could we provide female patients with a urinal, too? Objections to my request were that another unit had trialed female urinals several years ago and they did not like them and that turning for the bedpan was "good activity."

I also requested that our unit purchase "monkey poles" for patients to pull themselves up in bed. When available, nurses on our unit borrowed "bed frames" from the orthopedic unit for our patients who were alert and strong enough to pull themselves up. Could we provide more patients on our unit a device for pulling themselves up, to reduce strain on nurses? If I had known about friction-reducing "sliding sheets" at that time, I would have suggested them, too, but such devices were unknown to me and apparently to many other nurses as well.

A bed frame is a large, heavy, multiple-piece, metal unit, with horizontal pieces bolted by nurses onto the headboard and footboard of the bed to support a vertical bar extending the length of the bed overhead, from which dangles a metal triangle "trapeze" for patients to pull themselves up

ANNE HUDSON

Employee of the Month for February

Anne Hudson, R.N.

Anne Hudson, R N., an eight-year employee, was selected for her exceptional nursing skills and compassion for patients making her an outstanding role model.

in bed. Because a bed frame rises vertically head and foot, and extends horizontally the length of the bed, the bed becomes "enclosed." This makes it impossible to roll a patient's IV pole from one side of the bed to the other without removing the IV bag from the pump and passing it across the bed to the other side, then moving the pole and pump around the bed to be reunited with the bag, or, I have seen the pole, pump with bag and all, lifted and passed horizontally, through the bed frame, across the bed, and patient, to the other side.

A monkey pole is one bent bar attached vertically to the head of the bed extending over the upper portion of the bed with a metal triangle dangling on a chain. Monkey poles are far smaller and lighter than bed frames, requiring far less storage space, and do not "enclose" the bed, so there is no impediment to moving IV poles from side to side. Plus, our unit only had access to bed frames when the ortho unit had extras not in use on their unit. Much lifting and pulling of patients up in bed could be avoided if patients had equipment to move themselves.

The two objections to my suggestion for monkey poles were lack of storage space (How about on the bed?) and that some post-op patients should not pull themselves up (How about removing the triangle bar for such patients?). It was disappointing that my efforts to reduce the physical stress from lifting people were rejected. These were the only suggestions I knew to make to reduce lifting at the time. I had no idea of the tremendous variety of lift equipment and friction-reducing devices which were available, which likely could have prevented my injury, and the injuries of others, but which were not provided by the hospital.

At one point, workers' compensation evaluated me for eligibility for vocational rehabilitation. Voc rehab may be offered to those with accepted work-related injuries who are unable to perform the "essential functions" of their position, and those with no work available in their area providing at least 80% of their wage at injury. In my case, with permanent lifting restrictions, I could not return to my position as a floor nurse and there are no other jobs in my small town for which I am qualified which would provide 80% of my nurse wages.

Voc rehab determined that I was eligible for re-training though I discovered that nurses were typically retrained away from nursing into lower-paying non-nursing work. Voc rehab would not help me obtain an advanced nursing degree, but would likely approve only a certificate program for perhaps medical transcription, or other clerical work, which may provide half of my nurse wages.

The voc rehab evaluation determined my "transferable skills." That is, they determined that I maintained all of the following hospital-based nursing skills, apparently showing that with all of my nursing skills intact I could work as a nurse—someplace else.

"Transferable Skills"

"Ability to apply nursing and health care techniques which include administering injections, medications, and treatments.

Analyzing medical data, patient activity.

Applying clinical problem solving techniques.

Applying human anatomy, physiology, and biology knowledge.

Ability to apply infectious materials procedures, institutional care procedures, interpersonal communication techniques, life support procedures, medical lab techniques, nursing practices and procedures, patient observation and care procedures, personal care procedures.

Ability to apply sanitation practices to health care.

Apply uniform tests or procedures.

Assist in examining and treating patients.

Ability to chart medical data, collect blood samples, collect clinical data.

Ability to comprehend body response variations, comprehend composition of drugs.

Ability to conduct patient assessments, draw blood, maintain medical records, perform routine medical tests, routine medical treatments.

Ability to plan and organize work, prepare patient reports, prepare patients for exam and treatments, prepare patients for tests.

Ability to read and understand operating and technical manuals.
Ability to set up patient care equipment.
Ability to take vital signs.
Ability to comprehend and use medical terminology.
Ability to understand drug products, properties and composition of drugs.
Ability to utilize pharmacological terminology.
Ability to apply principles of gerontology.
Ability to make decisions based on response to care and treatment."

In my state of Oregon, workers' compensation operates to help employers return injured workers to their pre-injury position. If unable to return to the same duties due to their work injury, workers' compensation regulations do not provide for assisting injured persons to remain with their employer in another position. And the employer has no obligation to retain a worker who is unable to perform the "essential functions" of their job, in my case, lifting.

The employer is not prohibited from retaining injured workers, but it was not the general practice of my hospital to retain nurses with lifting restrictions in other non-lifting nursing work. Of course, they said I could apply for any position within my lifting restrictions. And, I did apply for several positions, but was never selected. "We wanted someone with recent experience." "We hired someone already in the department." "Some patients might need help transferring," etc.

They never focused on how they could use me, only on why they could not. I have since found the same has happened to many other back-injured nurses across the country—damaged goods don't sell. When I finally saw that they were not going to allow me to take any other position, I withdrew my last application. I did not want to feel another rejection. At one point I wondered, "Am I still a nurse?" and understood how feelings of self-worth can be tied to one's work.

Attempts at employing the Americans with Disabilities Act were unsuccessful. Apparently, employers are only required to allow persons with disabilities to discuss potential job modifications and accommodations, but can decline any such suggestions if the employer believes implementation would create an "undue hardship."

The lumbar fusion surgery relieved a great deal of my pain and I am glad I had it done. My legs no longer give out on me and most of the shooting pains are gone. I still have pain with sitting, with lifting any amount of weight, with bending forward, and with many other activities. A coccyx cut-out seat cushion helps a great deal with sitting and driving by eliminating direct pressure on the base of the spine. Some other helps are analgesics, heat, ice, exercises, massage, positioning with pillows for sleep, etc.

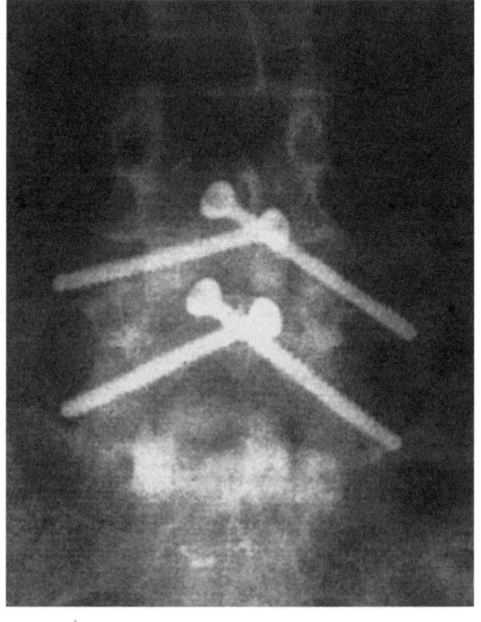

Before my injury manifested itself, I unknowingly practiced "harm without hurt" with thousands of patient lifts causing microfractures I did not feel. Now I must learn to live with "hurt without harm," with daily activities causing pain but which may not be causing further damage. The last x-ray, when I was declared "medically stationery" at closure of my workers' compensation claim, showed one graft well-fused and the other patchy looking. The future of my spine is uncertain, with potential collapse of the cadaver bone grafts, and high risk of "adjacent segment syndrome" with deterioration of the segments above and below the fused segments from assuming the extra stress. At times, increased back pain tells me

I should be seen for it again, but I dread the prospect of reopening my workers' compensation claim with all the related unpleasantries.

I am very fortunate that my husband has a solid career and that he and my family have been my mainstay in every way through all of this. Many nurses are their sole support and, when injured beyond returning to their position, some must sell, and some lose, possessions, vehicles, and even their homes, and some their marriages due to the multiple stressors with a disabling spinal injury.

After healing from spine surgery, a friend recruited me to work for the county public health department. Working as a public health nurse has expanded my horizons and allows me to help people and the community in new ways. I am grateful to be working as a nurse today, as many back-injured nurses are not able to find another employer who will accept a nurse who cannot lift. Still, tremendous losses will never be regained from losing my hospital career, with seniority and benefits of longevity, and financial planning and retirement. Just when I should be at the height of my earning power, I am starting over at the bottom, currently making almost $15 less per hour, now with a disabling spinal injury which likely could have been prevented by safe lift equipment. And I am one of the "lucky" ones.

I hope telling my story helps to expose and stop the exploitation of nurses by using them until "broken" and treating them as disposable. You may email me at anne@wingusa.org.

Anne Hudson, RN, BSN
12-1-07

The Meaning of Safety in Caring for the Larger, Heavier Patient

Susan Gallagher

Lifting, turning, and repositioning patients—an important part of health care in any clinical environment is a high-risk activity that most health care workers provide without consideration of the long-term impact.[1] A recent study indicated that during an 8 h shift a nurse may lift a total of 1.8 ton.[2] This may be one of the reasons why health care is considered one of the most dangerous jobs in the United States. Caregivers are not the only group at risk for injury in health care settings. Obese patients in the clinical environment are reportedly at a higher risk for certain common and predictable complications simply because of their body weight and size.[3] For instance, patients 45.4 kg (100 lb) or more above ideal body weight have exponential increases in mortality and serious morbidities as compared with their nonobese counterparts.[4] Because of both emotional and physical reasons, some obese people resist pursuing health care and frequently defer hospitalization until the last possible moment.[5] Therefore, in many cases, safe care for larger, heavier patients can be more complicated, and may be more difficult for health care clinicians.

Safe patient care among obese individuals encompasses all areas of patient safety. For example, as important as safe patient-handling and minimal-lift programs have become, when addressing the larger heavier patient this philosophy only addresses part of the safety equation. Airway management, pain control, intravenous (IV) access, urinary catheter placement, basic wound care, and more can pose safety risks not only to caregivers but patients as well. Additionally, patients having weight loss surgery (WLS) hold unique safety concerns throughout their lives such as the threat of vitamin and mineral depletion, emotional adaptation, and other long-term needs. The medical patient admitted unexpectedly through the emergency department poses a nearly unmanageable task for the unprepared facility. A holistic approach to safe management of the obese patient should include preventing caregiver injury and promoting patient safety through utilization of tools and resources including preplanning for expected and unexpected events.

What Is Obesity?

From an historical perspective, the word obesity originates from the Latin language and refers to the state of becoming fattened by eating.[6] Bariatrics is a term derived from the Greek expression "baros" and refers

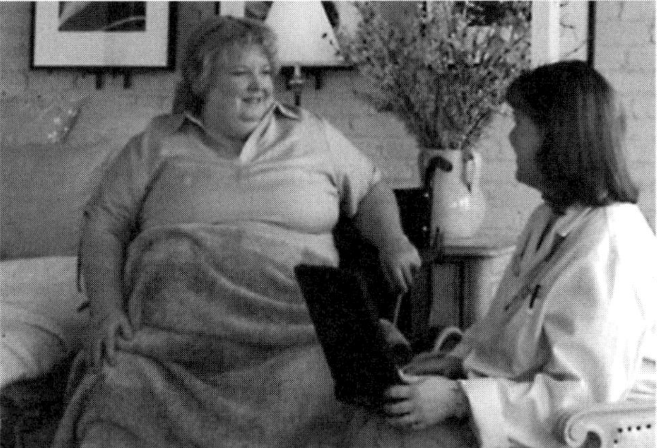

to issues pertaining to weight.[7] In some circles, the term "bariatrics" is thought to address WLS. However, the word comprises much broader meaning; currently, it refers to the practice of health care that relates to the treatment of weight and weight-related conditions. This includes weight-loss surgery as well as reconstruction after massive weight loss, medical weight management, safe patient-handling programs, and more.

Issues related to weight are of interest to the public for several reasons. Health and health-related concerns are at the forefront of the debate, along with the need for additional and specialized health services and overall access to health care.[8] Bariatrics as a specialty is becoming increasingly important in pace with the growing number of obese and overweight Americans. From a safety perspective, as the weights of patients and clinicians increase in line with the trend for the overall population, consider the effect on the incidence of occupational injuries.[9]

Obesity comes with a substantial economic burden: estimates of the total economic costs associated with the disease account for 5.5%–7.8% of all U.S. health care expenditures.[10,11] Obesity has been found to be associated with a 36% increase in inpatient and outpatient spending and a 28% increase in medications for obese smokers.[12] Obesity-related issues cost Americans nearly $150 billion annually—$117 billion is spent on health and health-related issues and $33 billion is spent on the largely unsuccessful weight-loss industry.

Recent estimates suggest that more than 67% of adults in the United States are overweight. Of all Americans between the ages of 26 and 75, 10%–25% are obese, 130 million are overweight, and more than 9 million are morbidly obese. From 1976 to 2000, the incidence of obesity increased from 14.4% to 30.9%. This increase has occurred regardless of age, gender, ethnicity, socioeconomic status, or race.[13] Overweight and obesity are not limited to the United States. Worldwide, nearly 2 billion individuals are overweight, equaling the number of individuals suffering from starvation.[14] To fully understand the meaning of these statistics it is important to know how overweight and obesity are defined and measured. Further, standardizing measurements and definitions ensures all stakeholders are speaking the same language. Safety guidelines and protocols may be defined by these standardized assessment tools.

Overweight simply refers to an excess of body weight compared to set standards. The excess weight may come from muscle, bone, fat, and/or water. Obesity refers specifically to the abnormal proportion of body fat. Many people who are overweight are also obese. Both obesity and overweight can be quantitatively defined using body mass index (BMI),[15] the most common and widely acceptable method of assessing overweight and obesity. BMI is a mathematical formula that describes relative height and weight; it is

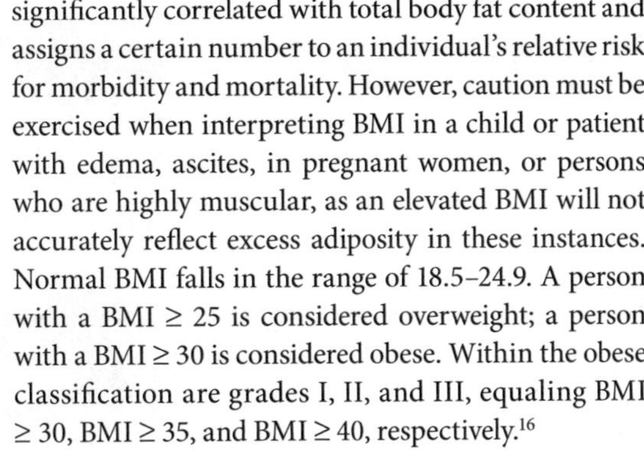

significantly correlated with total body fat content and assigns a certain number to an individual's relative risk for morbidity and mortality. However, caution must be exercised when interpreting BMI in a child or patient with edema, ascites, in pregnant women, or persons who are highly muscular, as an elevated BMI will not accurately reflect excess adiposity in these instances. Normal BMI falls in the range of 18.5–24.9. A person with a BMI $\geq$ 25 is considered overweight; a person with a BMI $\geq$ 30 is considered obese. Within the obese classification are grades I, II, and III, equaling BMI $\geq$ 30, BMI $\geq$ 35, and BMI $\geq$ 40, respectively.[16]

Other assessment tools establish certain risk factors for comorbidities, such as the waist-to-hip ratio.[17] A correlation study by Rexrode and others suggest that individuals with a high waist-to-hip ratio are at risk for certain cardiac and metabolic conditions; the presence of central obesity exacts greater tolls on this segment of the population.[18,19] The underlying issue with this is that patients with numerous comorbid conditions are likely to develop deconditioning even during short encounters; deconditioned patients pose threats to both patient and caregiver safety.

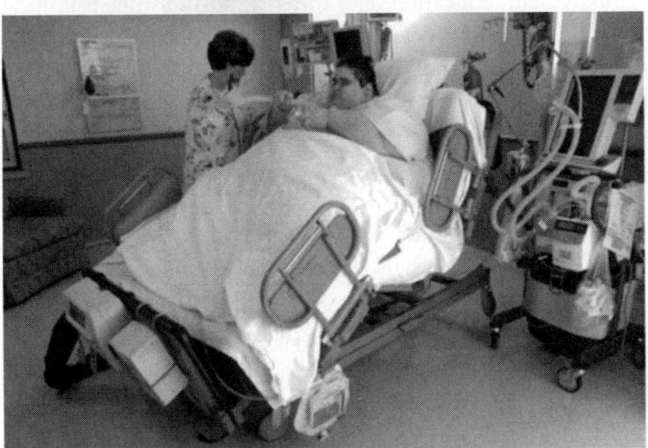

Understanding vocabulary and standardizing measurements and definitions help clinicians determine which patients may develop the common, predictable, and preventable complications related to weight issues and further anticipate what tools (equipment) and resources (clinical experts) can best prevent or manage these threats to safe patient care. A functional understanding of bariatrics serves clinicians and patients in this goal.

The Relationship between Willingness to Accommodate and Sensitivity

Modern culture idealizes thin and disparages obesity.[20] It is possible that in health care settings this could be more intense simply because of the challenges in lifting, repositioning, turning, or in the general care of a larger patient. Although, two-thirds of Americas are categorically considered overweight, most Americans, including health care workers hold bias toward those who are overweight.[21] This bias interferes with the willingness to accommodate the health care needs of this special patient population.

The American Society for Metabolic and Bariatric Surgery (ASMBS) defines the term obesity as a lifelong, progressive, life-threatening, genetically related mulitfactoral disease of excess fat storage with multiple comorbidities.[22] In contrast, the National Association for the Advancement for Fat Acceptance (NAAFA) does not consider a larger person obese—contending that this disapproving term medicalizes a very natural state. NAAFA reminds us that some people are born tall, some short, some skinny, and ultimately some fat, "We are tired of being labeled in a negative light simply for our God-given habitus."[23]

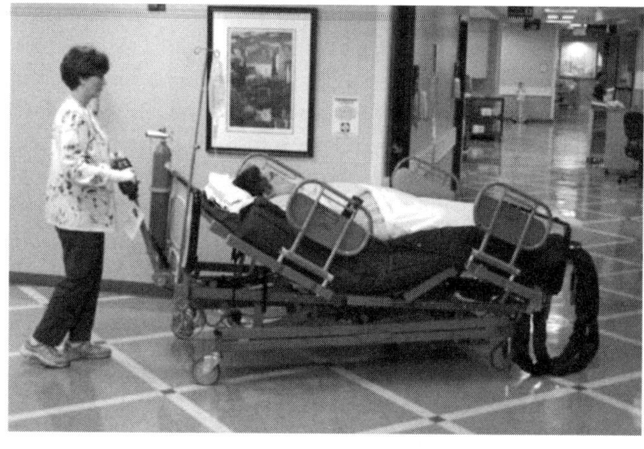

In one sense the cause of obesity is straightforward—the state of expending less energy than the amount consumed. But in another sense, obesity is intangible, involving the complex individual regulation of body weight, specifically body fat.[24] This individual regulation is the unknown factor in the weight management enigma.[25] The obese person, their friends and family members are likely to feel an economic, physical, and emotional effect in all areas of their lives. Despite efforts at weight loss, Americans continue to gain weight with obesity reaching pandemic proportions. An interviewee in the Public Broadcasting Service (PBS) *Frontline Fat* explains, "…I've tried Weight Watchers, all kinds of magazine diets, phen-fen…I have even had my jaw wired." Each weight loss experience ended with a regain of the lost weight plus more, creating bounding feelings of helplessness and hopelessness.[26] Over 175,000 Americans plan to have bariatric weight loss surgery (BWLS) this year. The preoccupation with weight control and the human body is pervasive, and some argue, for good reason.

Extensive investigation into biases toward obesity in general and the obese person specifically, suggests this prejudice and discrimination cross employment, education, and other important life-sustaining activities.[27] Research suggests prejudice toward obese people develops at a very young age. For example, children as young as 6 years old describe silhouettes of obese children as lazy, stupid, and ugly. According to this study, prejudice toward the obese child is observed regardless of race or socioeconomic status.[28] Children are not the only ones who hold a prejudice directed toward the overweight person, health care clinicians are also often biased against the larger patient.[29] In using the Bray Attitudes Toward Obesity Survey (BATOS) to measure attitudes of dieticians toward the obese patient, bias was even observed among obese persons themselves.[30] Health professionals are far from immune to these culturally driven views. In one study wherein mental health professionals were asked to evaluate identical case histories with corresponding photographs of either normal weight or obese women, the obese women were rated significantly higher on agitation, impaired judgment, inadequate hygiene, inappropriate behavior, intolerance of change, stereotyped behavior, suspiciousness, and total psychological dysfunction.[31] Another study incorporated use of the "FAT" suit, which was used to change the appearance of patients. The same patient was used for videotaped session, however, the person's physical appearance changed. Medical students reported that the patient, who was masquerading as the obese person, was more defensive, nervous, insincere, seductive, depressed, emotional, cold, and unlikable than the nonobese.[32] Despite the growing body of medical research that suggests otherwise, many health professionals continue to believe that obesity is a self-inflicted condition that results from a lack of will power.[33]

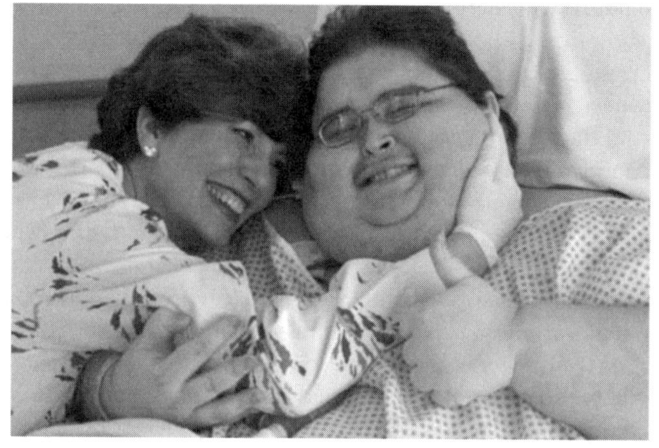

Historically, obesity has been perceived as a problem of self-discipline.[34] However, recent discoveries suggest that this may, in fact, be far from the truth. For example, consider the Pima Indian groups in the southwestern United States. They are thought to possess what James Neel refers to as the thrifty gene. Over centuries of evolution, this group survived in part by very efficiently metabolizing their calories, providing for long periods of famine. Today, the genetically homogenous group living south of the United States/Mexican border maintains an average of 22 h of intense activity each week. Most members of this community eat the indigenous diet, and have no problem with obesity. However those living north of the border, in the United States, have adopted the western diet—which is high fat, high calories, and high carbohydrate. To compound the problem, the group, in general, has adopted the sedentary lifestyle of many Americans. The Pima Indian people are experiencing profound levels of obesity, diabetes, hypertension, and all the complications inherent in morbid obesity.[35] There is no debate that weight gain occurs when intake, meaning food intake, exceeds output, meaning activity—the real mystery behind balancing body weight depends on a number of other factors. Genetics, gender, physiology, biochemistry, neuroscience, as well as cultural, environmental, and psychosocial factors influence weight and its regulation.[36] Health care clinicians best serve their patients when they recognize obesity as a multifactoral chronic condition.

Obese Americans neither chose to be overweight nor chose to experience widespread prejudice and discrimination.[37] Failure to provide safe, appropriate health care is often based on the premise that inadequate policies and procedures are justified by blaming the patient for his/her condition. A primary role of the health care clinicians in caring for this vulnerable population is to guarantee a safe haven from the obesity-related prejudice and discrimination, which often stems from misunderstandings. However, one study that examined caregiver attitudes toward the BWLS patient successful linked inadequate tools for providing care with negative attitudes toward the patient. Rather than simply exploring verbal prejudice, the focus of this study examined failure to provide reasonable accommodation in the form of medical equipment, comfortable surroundings, properly fitting attire, etc. This particular study suggested that bias exists toward the morbidly obese patient despite the degree of overweight. Although many issues were examined, the study also found that most hospital departments treated patients well, except for those with equipment that was insufficiently sized to meet the patient's needs safely.[38]

The concern of prejudice and discrimination is that these emotions pose barriers to health care regardless of practice setting or professional discipline. The overwhelming misunderstanding of obesity is likely to interfere with preplanning efforts, access to services, and resource allocation. Although this misunderstanding is not universal, it is pervasive enough to pose obstacles, and clinicians interested in making changes will need to acknowledge this barrier. However, quite interesting, is the fact that the department most ill-prepared to handle patients was thought to hold the greatest bias. Perhaps introduction of appropriate preplanning efforts could play a role in improving the level of compassion, sensitivity, and understanding in caring for the larger, heavier, and more complex patient. Clinicians across the country and across practice settings recognize that threats to compassion, sensitivity, and understanding will continue to pose barriers until caregiver-safe strategies are available. Even the most compassionate caregiver may be reluctant to provide adequate care because of the threat of caregiver injury. This realistic fear of injury, along with the failure to provide satisfactory care to a complex patient, further perpetuates discrimination toward the obese patient.

The value of recognizing these barriers is that before we can even begin the process to preplanning care for the heavier, more complex patient it is important to discuss the meaning of obesity in America. Health care clinicians, as members of society at large, are likely to carry the same bias toward obesity as the general population. Instead of investigating ways to provide the most comprehensive care possible, clinicians are faced with colleagues asking why patients allow themselves to become so heavy or shouldn't weight loss be the primary goal in care. For a well-meaning clinician to suggest weight loss to a person without an in-depth knowledge of the many tools available is like asking a carpenter to repair a cabinet with a single hammer. The etiologies of obesity are complex and multifactoral, and management options should reflect this. Misunderstandings such as this, suggest that as a culture we need to be more aware of the many dimensions of weight issues and recognize how this ultimately influences safe patient care.

Safe Practices in WLS

A number of intestinal surgeries in decades past held the secondary outcome of massive weight loss. This sparked interest in a surgical approach to weight management.[8] Today, two main categories of surgery are used to treat morbid obesity. These are gastric restrictive, and combined gastric restrictive and malabsorptive. The Roux-en-Y Gastric Bypass (RYGB) combines gastric restriction and malabsorption and is the most common procedure done in the United States today. A stomach pouch is reduced to about 15–30 mL to provide for the restriction, and a portion of the intestine is bypassed providing the malabsorptive component. Food the patient eats bypasses 90% of the stomach and duodenum, and a limb of jejunum in varying lengths, thus fewer calories are absorbed. When high-caloric foods are dumped into the limb of the small intestine, a feeling of satiety or even discomfort may result, helping to curb the appetite.[9] Reduced consumption and/or absorption of the food leads to weight loss. This procedure can be safely done laparoscopically in certain patients.

Circumgastric banding, a restrictive procedure, limits stomach size by placing an inflatable band around the fundus of the stomach. The band is connected to a subcutaneous port and can be inflated or deflated in the provider's office to meet the patient's need as he or she loses weight. Banding is performed laparoscopically.

The goal of WLS is to improve the patients' quality of life and health both short- and long-term.[39] Seventy-five percent of patients are expected to lose 75%–80% of their excess body weight. Well over 70% of patients with hypertension will no longer require medications, and more than 90% of patients with type 2 diabetes are expected to be free of their medications. Drastically improved if not completely resolved are asthma, reflux, fatigue, shortness of breath. Bi-level positive airway pressure (BiPAP) for sleep apnea and other supportive devices may no longer be required within just months of surgery. Patients report an overall sense of well-being. There is an expected reduction in the risks of heart disease, pulmonary disorders, and cancer. Patients can expect a significant improvement in quality of life and self-esteem.[40]

The numbers of obese people are increasing, and concurrently surgical options for weight loss are improving. Realistically, while some patients will do poorly after WLS, many are experiencing a dramatic improvement in quality of life. The key to safe WLS is an interdisciplinary team effort which is comprised of specially trained clinical experts, specially designed equipment, adequate resources, policies, procedures, and appropriate patient selection, education, and follow-up. The ASMBS, National Institutes of Health (NIH), and others have set forth guidelines and recommendations to improve safe WLS outcomes.

Candidates for surgery must meet certain criteria. For example, the patient must have a BMI of greater than 40, or greater than 35 with significant comorbidities such as sleep apnea, type 2 diabetes mellitus, hyperlipedemia, or hypertension.[41] Before WLS is considered for reimbursement, most insurance carriers require documentation of three unsuccessful attempts on medically supervised weight loss programs. The patient needs a thorough physical exam, including health history and weight history. Screening must determine absence of physical or emotional disorders that might be at the heart of weight gain. The screening process also includes multiple evaluations to assess whether a patient is able to comply with the many lifelong changes that are required postoperatively. Sometimes family members are involved with this evaluation as families can be a very important part of postoperative adaptation to change.

Patient teaching is a collaborative effort between the patient, family, and the health care team, and should begin on first contact. Patients must understand that the surgery itself does not guarantee long-term weight management. Commitment to lifelong behavioral changes is essential to the success of this strategy. Patients will need instruction preoperatively and reinforced postoperatively. Patients will need to understand the importance of early ambulation to decrease the likelihood of developing deep venous thrombosis (DVT), pneumonia, ileus, or other immobility-related complication. Pain management and medication administration should be discussed. Various medications, dosages, and routes can be used and patient instructions should be tailored to fit their situation.[42]

Dietary teaching will vary according to facility and surgeon preference. A dietician should be involved in the multidisciplinary approach to patient teaching. The patient must understand the need to consume at least three meals daily, chewing food completely before swallowing. Liquids cannot be taken at meal times, and at the first indication of fullness the patient needs to stop eating. High-calorie, sugar- and fat-dense foods and beverages need to be avoided. Each patient will need his or her own individualized dietary plan, with special attention to vitamin, mineral, and protein supplements for a lifetime.

Wound care will vary based on type of surgery performed. A small adhesive covering may be all that is necessary for a laparoscopic procedure. The left trocanteric site is at risk for infection because it is most likely to have contamination from the internal organs to the skin surface. The patient should be reminded to report any unusual tenderness at the site. A dry dressing over the surgical site and drain sites will be used after an open procedure and will need to be changed daily and as needed. Home health assistance may be necessary for patients if there is complicated wound care required. The patient needs to understand that discharge instructions will be provided and should include information about follow-up appointments, wound/dressing care, drain tube care, activity, diet, medications, and support groups. It is imperative that patients follow-up with their primary care provider soon after discharge to ensure adjustments to medications and procedures due to control of comorbidities.

Some patients will still not be candidates until underlying psychological issues are resolved. Preoperative psychological evaluations are required by many insurance carriers, and are thought to serve as a predictor of the candidates' postoperative success in the face of long-term lifelong changes. Some WLS programs and surgeons require a preoperative psychological evaluation of all surgical candidates in order to screen for psychopathology and prepare candidates for the many lifestyle changes expected after surgery. Bariatric surgery together with medical and behavioral follow-up is the most effective clinical approach to managing morbid obesity.[43] A psychological evaluation is comprised of two main parts, the psychological testing and a clinical interview. With regard to psychological testing, the Minnesota Multiphasic Personality Inventory-2 (MMMP-2) is a frequently used instrument.[44] Although initially designed to be administered to psychiatric patients, it is used extensively with other categories of patients. The clinical interview is comprised of a comprehensive assessment of the candidate's medical, surgical, and psychiatric history; drug or food allergies; history of eating disorders; alcohol or tobacco use; both prescription and over-the-counter medications; and more. The psychosocial history should also assess the candidate's family and social situation. Most obese patients are not the only obese person in the family or social support system and this may be important for the patient's success in the long term. The patient's knowledge of the surgery sought should also be assessed. It is important to make certain that the patient has the evaluation far enough in advance of the surgery such that any preoperative intervention can be well underway and not delay progress. For example, at the survey and the clinical interview, the results of the psychological tests are reviewed with the candidate. This is the time that additional questions and concerns are addressed. Any preoperative recommendations are made. Some programs require a written contract with the candidate and perhaps his/her support person. The written contract outlines any agreement, commitment, requirements, or specific responsibilities necessary before surgery is scheduled.[45]

Like all patients, immediate postoperative assessment and documentation of vital signs are imperative, especially if a change in the clinical condition occurs. Pain management is a priority. This may be accomplished by giving pain medications by various routes. Nonnarcotic implantable pain pumps can be used to decrease the pain at the incision site without the threat of an adverse respiratory consequence. Respiratory consequences such as postoperative atelectasis and pneumonia are common complications that can be minimized by early ambulation, repositioning, use of incentive spirometry, and coughing and deep breathing.

The patient may have a gastric tube; this should be monitored based strictly on the surgeon's expressed recommendations. All other tubes and catheter must be monitored for patency and fluid balance. DVT prevention is especially important among obese patients. Sequential compression devices and compression stockings must be size-appropriate.[46] For many post-op bariatric surgery patients, food had become a coping mechanism and part of their support system. Further, personal relationships can change. The evolution of one's lifestyle is dramatic and support groups are necessary to celebrate, and cope with the loss. Long-term follow-up to discuss the physical, social, and psychosocial implications of the new lifestyle are the basis of success.

The Challenges of Unplanned Access

Although nearly 3% of WLS patients experience a serious outcome postoperatively, in general this patient population has been well screened and therefore experience few adverse sequelae. Unlike the patient having WLS, the larger heavier individual who enters the health care facility for an unplanned event can be more complex and time consuming to care for, staffing seldom accommodates this difference, and reimbursement simply does not. Clinicians in home, hospital, out-patient, long-term care, and acute rehabilitation settings have concerns about safe care for the very large patient. Regardless of the practice setting, preplanning becomes an essential component of safe patient care.[47] Therefore, hospitals across the country are initiating bariatric task forces in hopes of designing processes to control or prevent some of the untoward complications associated with caring for the obese patient.[48] The challenge to these groups has been the limited availability of resources in which to build on. Adequate nutritional support, IV access, skin and wound considerations, appropriately sized equipment, airway and ventilatory management, resuscitation and diagnostic testing, pain control, social and emotional concerns, and the prevention of complications all present special and unique difficulties. Practical resources such as longer gloves, wider commodes, specialized tracheostomy tubes, heavy duty furniture, and many others are important to consider. Real-life challenges such as physical transfers, intubation, surgery, skilled nursing placement, home care present threats to safety.

Hospitalized patients complain of lack of privacy and loss of control. Those who are independent at home often become dependent because of unfamiliar surroundings. Many obese patients are embarrassed because of their size and personal appearance.[49] Because of problems related to immobility, dependence, and embarrassment these needs may be more intense once the patient is hospitalized.[50] Like all patients, the obese patient brings fears, expectations, and emotional needs to the health care experience.[51]

One paramedic confided that the real issue is the amount of time needed to transport a very large, critically ill patient. EMS services do not want to expend this degree of resources with one patient, and without appropriate equipment and preplanning these services can become very costly to the paramedic company. In privately run services, dispatchers learn the names and addresses of the very heavy, difficult patients and might make every effort to avoid providing service. This is the unspoken but very real concern, care may be delayed simply because EMS has become overwhelmed with the unreimbursed costs in managing a complex patient. It becomes more difficult to transport patients who cannot cooperate because of pain. Paramedics equipped with standard-sized equipment are not able to obtain accurate blood pressure readings, and starting an IV is difficult; however, the most significant challenge is lifting the patient into the transport van. In the past, patients have reportedly been placed directly onto the ambulance floor, but this causes skin trauma. When placed directly on the floor the patient is not secured, and in the unlikely event of a traffic collision, the patient and attendant are at risk for injury. Even if the patient was secured to a standard-sized gurney, problems could occur. Often the upper body weight can be so great that the head of the gurney cannot be elevated. In the presence of respiratory distress, failure to elevate the head of the gurney could result in respiratory arrest—compounding an already complex situation. A larger ambulance may be necessary to safely transport larger patients while maintaining the dignity of both patient and paramedic. Bariatric ambulances usually include both customized stretchers, modified suspension/loading systems, and a ramp-and-winch system. The unfortunate reality is that additional personnel, specialized equipment, and vehicles are costly. A properly outfitted bariatric ambulance could cost as much as $250,000. Retrofitting an existing ambulance with a ramp and winch could cost roughly $5000. For smaller EMS services, absorbing the costs to provide this type of service can be economically unrealistic. However, some EMS services recognize that providing safe patient-handling training and outfitting their ambulances with equipment needed to handle bariatric patients was an investment in safety for both their employees and patients. One innovative approach, at Southwest Ambulance, uses a ramp and winch system to pull the gurney into the ambulance. Stretchers that can hold patients up to 1000 lb are used for the safety and comfort of the patient. Design features include specialized air shock/lifts and heavy-duty suspension to lower the entire ambulance for a safer loading and unloading angle. A new loading ramp which extends for safer loading and unloading of patients with a motorized pulley system that attaches to the gurney, and a strengthened, lowered, and widened patient area.[52] Some communities are restricted by cost. Experts suggest that at a minimum, every EMS service should address employee concerns on the handling of patients at various weight limits. Policies should be developed to identify patient-movement strategies and set limits on the minimum number of people required to move a patient over a specified weight; and require staff to request lift assistance when confronted with a patient who exceeds the lifting limits of the crew on scene. EMS systems should consider creating a special response unit that could be shared as a regional resource. Instead of purchasing an ambulance already equipped for bariatric patients, a more cost-effective solution might be to equip a trailer with such supplemental items as a heavy-rated basket or scoop stretcher lined with layers of blankets to be used as cushions and supports to ensure a semifowlers position. Ramps, lifts or air displacement-type lateral transfer device for sliding a heavy patient off of a couch or bed or out of a building and/or into and out of the ambulance.[53]

Emergency department nurses report difficulty in assessment, diagnostic testing, and clinical intervention. Some hospital staff members report that the obese patient is placed directly onto a mattress on the floor—this is inappropriate for the bariatric patient who may need to be placed in a semifowlers' position to breathe. Once on the floor transferring to a gurney is impossible, not to mention the ergonomic challenges. A specially outfitted room with a lateral transfer device, gowns, blood pressure cuffs, exam table or bed frame, lift and transfer system, ceiling lift, and more should be considered based on the expressed needs of the clinicians.

Pain relief is often a consideration early in the emergency department experience. Excess body fat can alter drug absorption, depending on the medication. For example, drugs such as diazepam and carbamazepine are highly soluble in fat and are therefore absorbed mostly in adipose tissue. Dosage of these drugs must be calculated using the patient's actual body weight. Drugs that are absorbed mainly into lean tissue, such as acetaminophen, should be calculated using the patient's ideal body weight—what the patient should weigh.[54] Trying to remember which drugs fall into which category is almost impossible. A clinical pharmacist can be an important resource to ensure that the drug dose is accurate. Standard 1–1.5 in. needles may not be able to penetrate past the adipose tissue in a patient with especially thick hips. In this case, use a longer needle needs to deliver an intramuscular dosage of the intended medication or consider a drug that uses another route.

A thick layer of adipose tissue interferes with visualization and location of veins, which makes it technically more difficult to identify common landmarks; therefore, inserting an IV catheter can be difficult. Some nurses report difficulty gauging depth. If it requires more than two attempts, consider using a peripherally inserted central catheter (PICC) or a midline catheter instead of a standard peripheral catheter. Placement of lines, in a nonemergency situation, can be best achieved using portable bedside ultrasound guidance.[55] Both types of catheters can stay safely in place for weeks or months at a time, thereby eliminating the need to repeatedly stick the patient.[56]

Blood pressure cuffs must be the proper size, fitting the patients correctly. Cuffs that are too small and taped to keep them on the arm will display higher readings.[57] It is not only important to have the oversized blood pressure cuff, but staff needs to know where to locate it. Imagine the patient's embarrassment when hearing, "Where is that really big blood pressure cuff…you remember we used it on the huge blonde lady last week…." Equipment must be readily and discreetly accessible.

Selecting the proper advanced diagnostic tool will depend not only on the clinical situation but also the patient's weight and body circumference. Although a brain MRI may be the most appropriate technique for evaluating an acute stroke, if the patient cannot fit on the MRI scanner, it is of no practical use. Therefore, before scheduling an obese patient for a diagnostic imaging procedure, it is important to know the patient's weight and circumference, at the largest point, to assess whether the patient will be able to fit properly. Table weight limits and aperture diameters for fluoroscopy differ from those for CT and for MRI. Currently, industry standards exist for table weight limits and aperture diameters for each of the imaging techniques. In increasing order of cross-sectional diameter, according to current industry standards, the imaging techniques are fluoroscopy, CT, cylindric bore MRI, and vertical bore open MRI. Patients who exceed the weight limit of the table as defined by the manufacturer can potentially damage the table or its motor mechanics and injure themselves. The table and table motor are insured by manufacturers up to a certain weight. The cost to repair damage caused by a heavier patient is not likely covered under the manufacturer's warranty. In some cases, patients may meet the weight limit of a table but may exceed the gantry or bore diameter because of their girth. Typically, the industry-standard aperture in fluoroscopy is18 in.; the gantry diameter in CT, including MDCT, is nearly 28 in.; and the bore aperture in MRI is 24 in. Although the aperture diameters are accurate in the horizontal plane on CT and MRI, they do not account for the table thickness entering the gantry or bore and therefore overestimate the vertical distance. Typically, in the vertical plane, 6–7 in. must be subtracted from the gantry or bore diameter to account for the table thickness. Patients and caregivers should be spared the embarrassment of unsuccessful attempts to perform diagnostic examinations with inappropriately sized equipment.

Radiologists and technologists should be aware of weight and aperture limits of all imaging equipment at their diagnostic imaging facility. Weight and aperture limit data should be posted and be made easily available within the department and on nursing units. Despite these industry standard limits, some imaging vendors are now recognizing the issue of obesity and have increased the table weights and aperture dimensions of their newest imaging equipment.

Physical limitations on equipment are not the only safety challenges. The exams themselves may differ in larger, heavier patients with a high degree of adiposity. For example, radiographs are limited by x-ray beam attenuation that results in lower image contrast. Also, the increased body thickness through which the x-ray beam must travel results in increased exposure time and introduces motion artifact. The typical setting to obtain a chest radiograph is a kVp of 90–95 and mAs of 2–2.5. However, in obese patients these settings can result in inadequate penetration of the x-ray beams through the patient's body, along with more background scatter.[58] Care needs to be taken when performing and relying on diagnostic exams performed on the obese patient.

Depending on the patient's weight and distribution, a standard operating room (OR) table may not accommodate the patient properly. Hospitals and outpatient surgical centers need to factor in the accommodation of severely obese patients as part of their surgical table purchase decisions. Suppliers and manufacturers have responded to the need for tables that hold higher capacity weight, the OR tables will now hold 1200 lb. Manufacturers are also responding to the need for OR tables to be able to be lowered very low. When performing surgery on a morbidly obese patient, the incision site is often very high because the abdomen is so large. Surgeons need the OR table to be as low as possible so that staff members are able to ergonomically reach the patient. One such table can support, raise, and lower patients as heavy as 1100 lb. Patients as heavy as 600 lb can be fully positioned into various surgical postures, and can also be moved along the longitudinal slide of the table top. Another general surgical table can accommodate patients as heavy as 1000 lb in a normal orientation and 500 lb in reverse orientation. Complementing the tables are a complete line of table accessories including table-width extensions that expand from 20 to 28 in., and split-leg sections. Modular designs and an abundance of accessories means existing tables can be easily upgraded with bariatric capabilities. Bariatric restraints available with the some tables help assure secure positioning of the heaviest patients, while bariatric power-lift stirrups utilize gas-spring assistive technology to help users easily lift the legs of patients weighing up to 800 lb into lithotomy postures. Consider a pressure redistribution surface on the OR table to counter the stress of pressure on the soft tissue. Talk with vendors and seek resources as these products continue to evolve based on consumer need.[59]

Some hospitals reportedly have created an extension that is placed on the table; however, this could interfere with the complex functions of the table. Others report that an oversized bed is used. Although this is likely to accommodate size and weight, it too precludes use of the complex functions of a specially designed OR table. Care needs to be taken to prevent pressure related injury from straps, supports or simply the surgeon's body resting against any tissue that extends from the table. Prevention or support table pads must be considered. A lateral transfer device may help prevent caregiver injury and reduce shearing injury that develops from moving a heavy patient from one surface to another. In fact, the risk for caregiver injury is so great with lateral transfers that some hospitals have made the decision to move the patient from the OR table directly to frame/bed that will be used for the rest of the hospitalization. If the procedure has lasted over 2 h or if any discoloration of intact skin exists over the buttocks area, at the very least, a prevention (pressure redistribution) surface should be considered.

In the critical care settings, initial treatment is often aimed at managing the most life-threatening conditions. On-going treatment can extend over a prolonged period and prove difficult for the patient to tolerate. A comprehensive plan of care designed to address the specific needs of the larger patient in the critical care setting, sets the tone for the remainder of the hospitalization. As in all settings, caring for the patient must target his physical, emotional, and social needs.[60] A strong collaborative effort early on, that includes the patient and family members as team members, can provide the most successful outcome.

Many large people are healthy; however, obese patients who are hospitalized for prolonged periods of time have probably been affected by a cascade of events that occur from a number of comorbidities including cardiac or pulmonary complications.[61,62] Illness, even relatively minor, in the morbidly obese can result in major catastrophes and life-threatening situations requiring urgent medical intervention and physiological support.[63–65] For example, sleep apnea is found in normal weight individuals, but is more prevalent in the obese because of excess upper body weight.[66] A diagnosis of sleep apnea is made when the patient has five or more obstructive apneic episodes per hour of sleep.[67] Sleep apnea may be obstructive, central or mixed. Coexisting symptoms include restless sleep, snoring, and daytime somnolence.[68] Hypersomnolence is characteristic and is often associated with apneic pauses during sleep, a condition that increases with progressive weight gain.[69]

An overweight patient is not necessarily adequately nourished in that he or she may consume calorie dense foods that are nutritionally inadequate. Therefore, it is essential that nutritional assessment and appropriate interventions be accomplished in the critical care area and not erroneously given low priority. Because of the patient's body mass, nutritional needs are not easy to assess. Metabolic studies are very helpful in determining more specifically what level of nutritional support is most appropriate for the patient.[70] Choosing the right type of feeding is important. The goal for the patient in the critical care is to provide adequate nutritional needs taking care to keep the pCO_2 within normal limits.[71] A dietitian is instrumental in providing assessment and a nutritional plan of care tailored to meet the needs of the obese patient. Nutritional assessment includes diet history, BMI, physical exam, medical and social history, and lab values such as serum albumin, pre-albumin, serum transferrin, and lymphocyte count.[72] In the presence of obesity, the patient's large, heavy abdominal wall causes substantially greater pressures in the abdomen. In order to safely deliver enteric nutrition, care must be taken to ensure that the tube feeding is passing from the stomach and into the small intestines. The patient may be at significant risk for aspiration, especially in the presence of elevated intra-abdominal pressure and high gastric residuals. Feeding tubes passed through the pylorus can help to reduce this risk but can be very difficult to place in the obese patient.[73] Although use of the enteric route is preferred, sometimes this is not possible, and nutrition must be delivered parenterally.

Obese patients often present with atypical pressure ulcers. Pressure within skin folds can be sufficient to cause skin breakdown. Tubes and catheters burrow into skin folds, which can further erode the skin surface. The patient needs to be repositioned at least every 2 h, as do tubes and catheters. Tubes should be placed so that the patient does not rest on them. Tube/catheter holders may be helpful in this step. Commercially available securing devices that can be opened and closed several times, and remain in place, will reduce the likelihood of skin necrosis.

Wound healing can be problematic in some obese patients. Blood supply to fatty tissues may be insufficient to provide an adequate amount of oxygen and nutrients, which can interfere with wound healing. A delay in wound healing can occur if the patient has a diet that lacks protein or essential vitamins and nutrients. Wound healing can also be delayed if the wound is within a skin fold, where excess moisture and bacteria can accumulate. Furthermore, excess body fat increases tension at wound edges.[74] To reduce the occurrence of abdominal wound separation, some clinicians use a surgical binder to support the area. The binder will need to be large enough to comfortably fit the patient.

Maintaining a stable airway is a safety challenge. Providing adequate gas exchange and then weaning from mechanical ventilation is fraught with difficulties. A thick layer of fatty tissue can make breath sounds very hard to hear. Chest x-rays can be poorly penetrated, thus assessment of pulmonary function sometimes will rely more on blood gas analysis, ventilatory pressures, amount, appearance and culture and sensitivity results of sputum, and tracheal suctioning. Prior to extubation or full weaning from mechanical ventilation, the patient's ability to adequately take deep breaths should be assessed. Knowing whether the patient has a history of sleep apnea will be helpful in guiding this phase of pulmonary management. Optimizing pulmonary function during mechanical ventilation and weaning requires careful management. As weaning progresses, ventilatory support is reduced to allow greater patient participation in breathing. The patient's spontaneous breaths must be sufficient to provide a good tidal volume thus preventing or reducing atelectasis. Position changes while in bed are vital to help mobilize secretions and to improve gas exchange and reduce intrapulmonary shunting. Weaning may be expedited by placing the patient in a semifowler's position.[75,76] Many larger patients carry excessive weight in their abdominal region. When placed in a high-fowler's position the fatty tissue compresses against the diaphragm, therefore compromising respiratory function. Once hemodynamic stability is achieved, use of a chair specially designed for the obese patient is desirable. An oversized reclining wheelchair has been helpful in some critical care. Following extubation, increasing activity, and encouraging the patient to cough and take deep breaths to clear secretions is critical. These activities serve to maintain adequate gas exchange and to reduce atelectasis. He should avoid both the supine and high-fowlers' positions as both place undue pressure on the diaphragm. Nasal continuous positive airway pressure (CPAP) can be helpful as an adjunct in the long-term management of OSA; however, postextubation, many patients find this difficult to tolerate. BiPAP as a bridge after extubation and at night is especially helpful to prevent sleep apnea and hypoventilation.

During emergency intubation, it may be difficult to visualize anatomic landmarks, such as vocal cords, in the morbidly obese patient. An esophageal tracheal double lumen airway is recognized by the American Heart Association and the American Society of Anesthesiologists as an alternative to an endotracheal tube when obesity-related technical difficulties arise.[77] Even in a nonemergency situation, safe airway management can be challenging. Standard tracheostomy tubes can be too short for use in the presence of a very thick neck. Some clinicians use an endotracheal tube, others use a specially measured and ordered tracheostomy tube. Use of appropriate equipment can complement care, leading to safer and more effective care. Oversized walkers, wheelchairs, commodes, bed frames, specialty mattresses such as pressure relief or pressure reduction support surfaces, lateral rotation therapy, and lifts can reduce or eliminate complications related to prolonged immobility.[78] Equipment designed to assist in moving, transferring, or lifting obese patients creates a setting more sensitive to the safety of the patient and the nurse.

Following a long critical care experience, patients are often physically and emotionally exhausted. Preplanning is essential when transferring from the critical care area. Lack of appropriately sized equipment creates a multitude of care issues.[79] Numerous obesity-related complications are due to the inability to move deconditioned patients adequately or because patients are emotionally discouraged, afraid of falling or simply weak.[80,81] Challenges on the medical-surgical areas stem from this deconditioning. Generally, patients having WLS are placed on a specially designed bariatric surgery unit, but what about the non-WLS patient? More and more interest has emerged concerning centralizing care of the obese patient. Although reasons for admission vary, the question remains: Are there enough universal clinical needs to warrant aggregating patients? How about safety...patient and caregiver safety? The those who support the concept of centralized care argue that caregivers are more comfortable with not only clinical problems but the safe patient-handling concerns of larger patients. For example, staff members become comfortable with equipment selection and its use. There is speculation that a higher degree of sensitivity exists because staff members are more familiar with the complexities of care. On the other hand, is cohorting patients an unnecessary patient distinction based on weight—an already profoundly sensitive issue. The heart of this debate centers on the facility's ability to reconcile this concern.

Rooms designed to manage the needs of larger patients must accommodate for more personnel and larger equipment. There needs to be sufficient clearance at the bedside and in patient seating zones. The American Institute of Architects (AIA) *Guidelines for Design and Construction of Hospital and Health Care Facilities* calls for clearances around the patient bed in a single room to be a minimum of 3 ft. However, in a more recent publication addressing design of bariatric rooms, the organization indicates a need for at least 5 ft of clearance. This clearance is necessary to accommodate lifts and other equipment, but also patient safety in the event of a fall. Adequate spacing of furniture and equipment can mitigate the circumstance of the patient striking objects during a fall, and ensure adequate space for a lift to assist the patient. The door to the patient's room needs to be wide enough to allow larger equipment to pass easily. A width of 60 in. is considered to be sufficient to accomplish this goal. Options for a 60 in. opening include (1) a sliding door or (2) a pair of unequal-leaf swinging doors—one door 42 in. wide, the other 18 in. Along with a properly functioning climate control system, consider a ceiling mounted fan to help with cooling the room. Wider visitor chairs need to be available. A bariatric bed, wheelchair, and lift system need to be available at a minimum. The choice of a ceiling lift, floor lift, or combination system is entirely facility-specific. Regardless, it best serves planners to recognize the actual needs of their patient population and let that factor drive these choices. Additional bariatric equipment can be introduced as the patient's condition warrants. A walker, commode, air-displacement lateral transfer product, slide product, gown, overhead trapeze, power-driven product, and more can facilitate patient care in the patient's room. The bathroom should support a patient weighing 1000 lb. Consider a room of at least 45 ft^2, a 60 in. door, and waterproof walls and floor—the entire room becomes the shower. The floor-mounted toilet should be mounted with a minimum distance to the nearest wall of 21 in. Consider a bar or similar device accessible in front of the toilet, so the patient can reach forward and use both hands to pull himself up, and consider mounting the toilet tissue dispenser in front of the toilet to allow the patient ease of access. The sink must be structurally sufficient to resist pulling away from the wall if grabbed by a falling patient.[82]

The goals of discharge planning are to focus on ways to maximize the patient's physical and emotional transition toward independence in the home setting. In developing a plan of discharge care, the patient, family, and members of the interdisciplinary team should be involved. The discharge planning needs of the obese patient are not unlike needs of the nonobese patient, however, a prolonged hospitalization or numerous comorbidities can intensify complexities. Not all obese patients will require special accommodation at home; however, patients who have limited mobility are likely to have special needs and therefore require special accommodation. In a recent study, nurses reported five specific challenges in the home-care setting: equipment, reimbursement, access to resources, client motivation, and family/significant other support—the challenges cited most often pertained to issues of equipment.[83] Although there are currently no statistics to measure injuries at home, family members and caregivers may be at risk for injury when caring for the patient in the home, as fewer personnel are available to help.[84] Oversized wheelchairs and walkers with greater weight limitations than standard equipment are readily available for purchase or rent in major medical supply centers. Both items promote independence and dignity. Equipment that nurses find most helpful in the home are the wheelchair, walker, commode, electronically controlled bed frame, support surface, and lift. When planning for oversized equipment in the home, consider weight limits, width, and electrical needs. In other words, does the patient have a sliding glass door or extra wide doorway through which equipment can be delivered? Or, can the equipment be dissembled, so that it can be delivered through a standard-sized doorway?

Community resources could include physical therapists, weight control counselors, and others. Written instructions for treatments, medication schedules, and follow up appointments will be helpful in clarifying dates, times, and events for the patient.[85]

Creating a Safe Work Environment

The human element is at the heart of health care. Sufficient numbers of staff members, the health and safety of staff members and specialized training are the qualities that move the work place from good to great. In the face of caring for obese patients these qualities are essential. The questions become "Who will move patients? How will the patients be moved? and What equipment and training is needed to move patients?"[86] Baptiste suggests that at the organizational process level there is even more opportunity. Policies and procedures need to be in place from the time of admission through the care plan to include the patient's room—regardless of whether the admission is scheduled or unplanned. For example, there needs to be a plan in place for how the patient will be transported from the emergency department to their room, making sure everything is set up to accommodate the patient.[87] Drake agrees explaining that the first step to mobilizing bariatric patients involves analysis and preplanning of the event along with the understanding of goals. Taking the time to plan in advance will improve safety and success.[88] Preplanning is a common theme among experts. Despite support for preplanning efforts, these activities are generally not forthcoming. One recent study suggests that barriers to preplanning for obese patients may be related to the complex nature of the patient population. It has become difficult to determine how to begin to develop a criteria-based protocol.[89] Consider the departments that ought to be involved—it is likely that every department in the facility has contact with members of this patient group. For example, whether the patient is admitted through the emergency department or through the admitting office, special accommodation is necessary. Failure to preplan can hinder patient movement from one department to another therefore leading to further delays in necessary diagnostic and therapeutic intervention.

Although appropriate equipment is essential, other resources are equally valuable. Health care clinicians and hospital support personnel need to be involved in the preplanning process. Education is critical to planning individualized care that complements the criteria-based protocol. Competency tools can be a resource to set and maintain standards of care. In managing the complex needs of the bariatric patient, preplanning for equipment has been thought to be the first step for intervention; however, it simply is not enough. Rather, a comprehensive process for preplanning is necessary, and should include (1) a bariatric task force, (2) a criteria-based protocol, which includes preplanning for size appropriate equipment and experts, (3) a competencies/skill set, and (4) a surveillance.[90,91]

Getting started can be the most challenging aspect of developing a bariatric protocol. Little to no research exists to describe best-practice strategies for care. Clinical experts, teamwork, and group participation are critical to embracing a project of this complexity. Additionally, in striving for success, frontline employees must be part of the process. Staff members closest to the patient are in the best position to share actual safety concerns. For example, staff members in the emergency department might explain that they need a wider, smooth-surfaced ramp or walkway to aid in transporting the patient from their personal car to the department. Additionally, they likely could benefit from a bariatric chair (sometimes referred to as a cardiac chair, reclining chair, etc) that transforms a chair on wheels to an exam table, gurney, or recliner. This chair could also aid staff members in transporting the patient from their personal car to the department; however, it is also useful to safely move the patient throughout the hospital for tests or exams. While in the emergency department, the chair can be placed in the examination position. The goal of this type of product is to reduce unnecessary lateral transfers or lifts. For more independent patients, an oversized walker can be used to aid in walking to the bathroom or other partial weight bearing action.

Change requires strong leadership and administrative support.[92,93] In developing preplanning for care, the value of an interdisciplinary team cannot be overlooked. A team approach has become the standard of care for many organizations. The team can be helpful in two respects. One is to serve as the team charged with developing a Comprehensive Bariatric Plan of Care—which includes the preplanning tool, training, surveillance processes, etc. The second responsibility is to provide care conferences on individual patients. For example, some organizations require interdisciplinary patient care conferences within 8 h of admission if the patient meets certain criteria, such as a BMI greater than 50. Consider including the patient and/or his/her significant other, as this offers insight into the patient's special needs. Documentation of meetings, individual patient care goals, and corresponding intervention improves consistency and accountability. This level of accountability more fully defines each clinician's responsibilities.[94] (See What is an interdisciplinary team?)

However, this is important to an individual that the presence of a team does not necessarily ensure timely access to assessment and intervention.[95] A preplanning tool becomes the critical factor in caring for more complex patients.

Early in the process, the interdisciplinary team must identify equipment throughout the organization such as MRI scanners, OR, or x-ray tables, which typically have weight limitations. Identifying the weight limits of diagnostic and treatment equipment house-wide prevents misunderstandings as to what resources are actually available to clinicians. Further, talk with staff members in off-unit areas to determine safety concerns. Some facilities have introduced ceiling lifts or floor lifts to transfer patients, some have a dedicated air displacement/lateral transfer device on site. Either or all of these products serve as tools for clinicians.

Preplanning with manufacturers and vendors to provide equipment for the morbidly obese patient is essential. Institutional policies and procedures for obtaining oversized transportation and transfer devices, bed frames and support surfaces, wheelchairs, walkers and commodes, or furniture need to be instituted.[96] When selecting oversized equipment, it is essential to consider both the weight limits and the width of the equipment. For example, patients not exceeding the weight limit for a standard bedside commode may still be unable to use a standard device due to the size of their hips. Most medical equipment suppliers rent or sell extra wide wheelchairs, walkers, and commodes that accommodate patients weighing up to 1000 lb. Some rental companies provide a number of oversized bariatric items as a bundle, providing a price incentive.

Education provided to ensure basic skills or competencies is imperative, and has become a critical part of any care plan. When addressing patient safety, consider conducting a survey to determine the actual learning needs of clinicians. The value of a diverse, interdisciplinary bariatric task force is that it serves to provide a pool of experts to develop lesson plans/education addressing clinical needs. For example, assuming clinicians are seeking information pertaining to sensitivity—a social worker, chaplain, nurse expert, and patient member of the task force could develop a on-hour module to teach these skills. Training should include strategies for not only for patient safety but caregiver safety as well. Many authors talk about the need for a paradigm shift from requiring caregivers to learn body mechanic techniques to requiring the organization to provide a safe environment through ergonomic research, no lift policies, and unit-based education and mentoring.[97] Experts suggest than an annual safety program is simply not enough. For example, the National Institute for Occupational Health and Safety (NIOSH) has published lifting guidelines. According to these guidelines, an average woman should be able to safely lift about 46 lb. Therefore, the average individual is at risk when lifting only 50 static lb.[98] From a practical perspective, in working with the obese patient, is it feasible to think that 11 caregivers can truly lift a 550 lb woman? The goal simply is not achievable without proper training, equipment, and assessment. Part of appropriate training is to learn to utilize resources and tools. Consider an ergonomic assessment for obese patients who have challenges to mobility.

An ergonomic assessment is helpful when conducted on-unit to address the safety needs of bariatric patients and their caregivers. One issue is that the environment where caregivers work is often cramped, requiring the caregiver to reach across furniture to get to the patient. The patient may be uncooperative, sedated, or in pain adding to the physical forces against the caregiver.[99] Caregivers need to be aware of the risks and also specific ideas to mitigate the dangers. For example, consider a retractable bed frame—one which will allow expansion to enable the patient to turn and reposition himself yet can narrow to accommodate the ergonomic needs of caregivers who must reach the patient to perform tasks such as bathing, wound care or even resuscitation efforts. An ergonomist can share specific handling ideas. For example, a patient with a large panniculus may have difficulty moving from a supine to fowler's position. The ergonomist may suggest use of a 6-point lift to position the patient to the side of the bed and then use a walker to support the patient in ambulation. Experts can match available tools with the specific task to be accomplished. This sort of training is best done at the bedside on patient-to-patient basis.

In order to ensure long-term success of a comprehensive bariatric safety program, it is essential to understand and participate in surveillance and outcome studies. Cost, clinical, and satisfaction research can be conducted to measure the value of an organizational improvement effort. Studies examining time from admission to equipment availability, expert consultant, incidence of skin injury, and others document, from a quality perspective, the value of a comprehensive bariatric care plan.[100] Consider patient satisfaction as an opportunity, for instance, patients are concerned about the capabilities and confidence of caregivers. Edwards explains that patients may be thinking, "Are these two petite nurses able to help me move"? This causes stress for the patient. When a specially trained caregiver or a professional team comes in to help the patient move, it gives the patient a sense of security and thus improved satisfaction. On the other hand, when a facility sends the message that it is concerned about staff members as much as the patients, it dramatically improves the culture. Given the tools to provide excellent care to the patient, caregivers feel better and their job satisfaction rises. Savvy administrators know that the happier the caregiver is, the happier the patient is and this is reflected in satisfaction scores.[101]

Conclusion

With obesity on the rise, clinicians are increasingly responsible for managing the needs of this complex patient population. Although preplanning for equipment is a helpful adjunct to care, it is never a substitute for care. Numerous resources are available to clinicians across practice settings, and use of resources in a timely and appropriate manner are thought to improve measurable therapeutic, satisfaction, and cost outcomes, coordinating these resources in the form of a comprehensive bariatric care plan may ensure the most favorable outcome. The obese patient holds numerous care challenges, and it is in the interest of health care organizations to meet these care challenges in a clinically, ethically, and legally sound manner.[102]

WHAT IS THE INTERDISCIPLINARY TEAM?

Regardless of the reason for admission or practice setting injuries can occur. However, it is important to recognize that appropriately sized equipment is just one aspect of comprehensive outcome-driven care. In planning care for obese patients, it may best serve the institution, caregivers, and patients to consider the value of a timely interdisciplinary team conference, which identifies appropriate equipment and adequate resources. Many of the more common, predictable, and costly complications can be prevented by an interdisciplinary team conference designed to identify early in the admission what the patient's unique needs are. It is critical to patient and caregiver safety to have appropriate-sized equipment, for reasons described earlier. Resources within and outside the organization can be identified. This should also include staffing resources to help with the physical care of the patient. The purpose of the interdisciplinary team is to provide each discipline an opportunity to identify their specific goal for the patient, along with equipment and resources to meet these goals.

Typically, members of the interdisciplinary team would include the physician, physical therapist, nurse, occupational therapist, dietician, respiratory therapist, pharmacist, or counselor, among others. Some rehabilitation settings include the patient and family members as part of the team because this tends to formalize the commitment between the patient, caregivers, and the rehabilitation plan. Physical therapists help the patient to transfer, ambulate, and otherwise increase activity and subsequently endurance. Goals set by occupational therapists often entail greater independence in activities of daily living. Bathing can be a difficult function for obese patients. Skin folds can cause special hygiene problems that the occupational therapist can creatively address with the patient. Clinicians tend to collaborate with other disciplines to balance the patient's functional, emotional, physical, medical, and spiritual well-being.[103]

Many overweight patients are actually undernourished, especially if nutrition was neglected during the acute episode.[104] The goal of the dietician is likely to emphasize food choices within the context of health and well-being, taking into account any coexisting diagnoses, such as diabetes. One of the goals of the respiratory therapist is to help the patient achieve an adequate level of rest. This is thought to be difficult for larger patients because of sleep apnea, or OHS described earlier.

Some medications are absorbed primarily in adipose tissue, while others in the muscle. The pharmacist can be instrumental in tailoring a medication plan unique to a patient who may have a higher percentage of body fat than most.[105] Psychologists or other counseling specialists can guide the patient in identifying behaviors that may interfere with a healthy lifestyle. Many larger patients can be depressed, withdrawn, and express feelings of helplessness and hopelessness, especially after a prolonged hospitalization. This discipline is skilled at encouraging patients to be more accepting of themselves, and more active in meeting their emotional and physical needs.[103] Because of the range of skin injuries that can occur, the WOCN/ET can prove to be a valuable asset. Other disciplines may be added to the interdisciplinary team as indicated; however, the value of an interdisciplinary team is in its timing—the earlier the better as this provides a way for caregivers to prevent some of those common complications associated with caring for larger patients.

THE VOICE OF THE PATIENT REPRESENTATIVE ON THE BARIATRIC TASK FORCE

As various disciplines are beginning to acknowledge the rather special needs of an extremely obese population seeking adequate health care services, many are finding resources scant. Bariatric care, having much in common with America's waistline, is one of the fastest growing areas of specialty practice.

Though research is underway, many providers and institutions find a serious lack of information pertaining to everyday practicalities or even unique problems this population struggles with on a daily basis, some of them as mundane as to escape notice of those who have had little or no experience with massive body habitus. I think it is critical that a patient representative be present as a member of an interdisciplinary Bariatric Task Force, in that as a primary care provider in long-term care settings for over 20 years I have myself encountered numerous situations where staff had no idea of how to best approach and plan patient care. The patient representative brings this unique experience to the team, keeping ideas, and strategies appropriate to the patients' size and emotional well-being.

IDEAS FOR BARIATRIC EQUIPMENT

Ambulatory/mobility aids—Transfer bench, trapeze, hover-type lateral transfer, slide sheet, lift and transfer product, ceiling lift, walker, wheelchair, stretcher, reclining wheelchair

Bathing/bathroom—Commode, bedpan, shower chair, handheld shower

Beds—Expandable frame, support surface

Others—Blood pressure cuff, diapers, drape/gowns, diagnostic equipment, scales, tables, binders, surgical instruments, tracheostomy tubes, tube holders

TEN TIPS FOR SAFE PATIENT HANDLING

Recognize the danger

Recognize health care as a dangerous situation where caregivers forego their own safety for the safety of others

Ensure proper training

Annual safety education which includes training in body mechanics alone is not enough

Utilize tools and resources

Introduce tools (equipment) and resources (experts) in a timely, appropriate manner

Preplan for care

Coordinate a bariatric task force, preplanning document, training and outcomes tracking system to meet the actual needs of patients and their caregivers

Understand the environment

Consider an ergonomic assessment to reconcile risks with suitable solutions

Understand the patient

Assess the patient for pain, depression, fear, or sedation as these pose the threat of immobility

Frontline input

Involve individuals most closely involved with patients. Consider a patient representative who can explain best the lived experience of being an obese patient

Cohort patients

Group like patients when possible, but recognize the entire facility must be prepared to manage larger, heavier patients

Track injuries

Design and implement a unified surveillance system to measure cost, satisfaction, and therapeutic outcomes, including caregiver injuries

Recognize ongoing change

As products, equipment, and training evolve so does the nature of safe bariatric services

References

1. Humphreys SL. Obesity in patients and nurses increases the nurse's risk of injury lifting patients. *Bariatr Nurs Surg Patient Care* 2(1):3–6, 2007.
2. Siddharathan K, Nelson A, Tiesman H, and Chen F. Cost-effectiveness of a multifacted program for safe patient handling. *Adv Patient Safety* (3):347–358, 2005.
3. National Task Force on the Prevention and Treatment of Obesity. Medical care for obese patients: Advise for health care professional. *Am Fam Physician* 65:81–88, 2002.
4. Kral J. Morbid obesity and related health risks. *Ann Int Med* 103 (6 part 2):1043–1047, 1985.
5. Gallagher SM. Meeting the needs of the obese patient. *AJN* 96(8)supp:1s–12s, 1996.
6. Aronson SM. A physician's lexicon: The verbiage of obesity. *Med Health Rhode Island* 65(5): 154, 2003.
7. Deitel M and Melissas J. The origin of the word Bari. *Obes Surg* 15(7):1005–1008, 2005.
8. Camden SG. Nursing care of the bariatric patient. *Bariatr Nurs Surg Patient Care* 1(1):21–30, 2006.
9. Nelson A, Matz M, Chen F, Siddharathan K, Lloyd J, and Fragala F. Development and evaluation of a multifacted ergonomics program to prevent injuries associated with safe patient handling tasks. *Int J Nurs Stud* 43:717–733, 2006.
10. Kort MA, Langley PC, and Cox ER. A review of cost-of-illness studies on obesity. *Clin Ther* 20: 772–779, 1998.
11. Thompson D and Wolf AM. The medical cost burden of obesity. *Obes Rev* 2:189–197, 2001.
12. Strum R. The effects of obesity, smoking, and drinking on medical problems and costs. *Health Aff* 21:245–253, 2002.
13. Ogden CL, Carroll MD, Curtin LR et al. Prevalence of overweight and obesity in the United States 1999–2004. *JAMA* 295:1549–1555, 2006.
14. Buchwald H. Is morbid obesity a surgical disease? *General Surgery News* 9–15, June 2007.
15. NIDDK. Statistics related to overweight and obesity. Accessed June 1, 2007 at http://www.niddk.nih.gov
16. National Heart, Lung, and Blood Institute. Classification of overweight and obesity by BMI, weight circumference and associated disease risks. Accessed December 16, 2007 at http://www.nhlbi.nih.gov/health/public/heart/obesity/lose_wt/bmi_dis.htm
17. Gallagher S, Langlois C, Spacht D, Blackett A, and Henn T. Preplanning with protocols for skin and wound care in obese patients. *Adv Skin Wound Care* 17(8):436–441, 2004.
18. Rexrode KM et al. Abdominal adiposity and coronary heart disease in women. *JAMA* 280(21): 1843–1848, 1998.
19. Define Obesity and Overweight. Accessed June 21, 2007 at http://www.dshs.state.tx.us/phn/define.shtm
20. Flegal KM, Carroll MD, Ogden CL, and Johnson CL. Prevalence and trends in obesity among US adults, 1990–2000. *JAMA* 288:1723–1727, 2002.
21. Schwartz MB, Chambliss HO, Brownell KD, Blair SN, and Billington C. Weight bias among health professionals specializing in obesity. *Obes Res* 11(9):1033–1177, 2003.
22. American Society for Bariatric Surgery. Accessed June 20, 2004 at www.asbs.org
23. National Association for the Advancement of Fat Acceptance. Accessed August 4, 2007 at www.naafa.org
24. Gallagher S, Arzouman J, Lacovara J, Blackett A, McDonald P, Traver G, and Bartholomeaux F. Criteria-based protocols and the obese patient: Planning care for a high-risk population. *O/WM* 50(5):32–42, 2004.
25. Gallagher S. Taking the weight off with bariatric surgery. *Nursing* 34(4):58–64, 2004.
26. *Frontline Fat PBS Home Video*. Public Broadcasting Service, Seattle, Washington, 1998.

27. Faulcbaum L and Choban P. Surgical implications of obesity. *Ann Rev Med* 49:215–234, 1998.
28. Staffieri JR. A study of social stereotype of body image in children. *J Pers Soc Psychol* 7:101–104, 1967.
29. Thone RR. *Fat: A Fate Worse Than Death*. Harrington Park Press, New York, 1997.
30. Maiman LA, Wang VL, Becker MH, Finlay J, and Simonson M. Attitudes toward obesity and the obese among professionals. *J Am Diet Assoc* 74:331–336, 1979.
31. Young LM and Powell B. The effects of obesity on clinical judgments of mental health professionals. *J Health Soc Behav* 26:233–246, 1985.
32. Breytspraak LM, McGee J, Conger JC, Whatly JL, and Moore JT. Sensitizing medical students to impression formation processes in the patient interview. *J Med Ed* 52:47–54, 1977.
33. Falkner NH, French SA, Jeffrey RW, Newmark-Sztainer D, Sherwood NE, and Morton N. Mistreatment due to weight: prevalence and sources of perceived mistreatment in women and men. *Obes Res* 7:572–576, 1999.
34. Gallagher S. *A Tragic Case of Childhood Obesity*. UMC Press, MI, 2000.
35. *The Pima Indians: Pathfinders for Health*. Accessed August 14, 2007 at www.niddk.nih.gov/health/diabetes/pima/pathfind/pathfind.htm
36. Gustafson NJ. *Managing Obesity and Eating Disorders*. Western Schools Press, South Easton, MA, 1997, pp. 13–19.
37. Gustafson NJ. *Managing Obesity and Eating Disorders*. Western Schools Press, South Easton, MA, 1997, p. 2.
38. Kaminsky J and Gadaleta D. A study of discrimination within the medical community as viewed by obese patients. *Obes Surg* 12(1):14–18, 2002.
39. Buchwald H, Avidor Y, Braunwald E, Jensen MD, Pories W, Fahrbach K et al. Bariatric Surgery: A systematic review and meta-analysis. *JAMA* 292:1724–1737, 2004.
40. http://www.obesity-surgery-center.com/benefits_gastric_bypass_surgery.htm
41. Gastointestinal Surgery for Severe Obesity. *NIH Consensus Statement Online 1991 Mar 25–27*, 9(1):1–20.
42. Garza S. Bariatric weight loss surgery: Patient education, preparation, and follow-up. *Crit Care Nurs Quart* 26(2):101–104, 2003.
43. Buchwald H. Mainstreaming bariatric surgery. *Obes Surg* 9:462–470, 1991.
44. MMPI-2 The Minnesota Report. Regents of the University of Minnesota, 2001.
45. Woodward B. Bariatric surgery options. *Crit Care Nurs Quart* 26(2):89–100, 2003.
46. Deitel M and Shikora SA. The development of surgical treatment of morbid obesity. *J Am Coll Nutr* 21(5):365–371, 2002.
47. Gallagher S, Langlois C, Spacht D, Blackett A, and Henns T. Preplanning protocols for skin and wound care in obese patients. *Adv Skin Wound Care: J Prevent Healing* 17(8):436–443, 2004.
48. Gallagher S. Shedding weight with bariatric surgery. *Nursing* 34(3):58–64, 2004.
49. Gallagher SM. Meeting the needs of the obese patient. *AJN*, 96(8)supp:1s–12s, 1996.
50. Kramer K and Gallagher S. WOC nurses as advocates for patients who are morbidly obese: A case study promoting use of bariatric beds. *J Wound Ostomy Continence Nurs* 31(1):276–281, 2004.
51. Gallagher S. Understanding compassion, sensitivity, and the obese patient. *Bariatr Times* 1(1):1, 4–8, 2004.
52. http://www.swambulance.com/bariatric.html
53. Barishansky RM and O'Connor KE. Bariatric patients pose weighty challenges. *EMS Insider* 34(8):12, 2007.
54. Goodell TT. The obese trauma patient: Treatment strategies. *J Trauma Nurs* 3(2):36, 1996.
55. Gallagher SM. Caring for obese patients. *Nursing 98*, 28(3):32hn1–32hn3, 1998.
56. Gallagher SM. Caring for the obese patient. *Nursing 98*, 28(3):31hn–33hn, 1998.
57. Maxwell MH. Errors in blood-pressure measurement due to incorrect cuff size in obese patients. *Lancet* 2:33–36, 1982.
58. Uppot RU, Sahani DV, Hahn PF, Gervais D, and Mueller PR. Impact of obesity on medical imaging and image-guided intervention. *Am J Roentgenol* 188(2):433–440, 2007.

59. Akridge J. Maximum capacity: Facilities, manufacturers respond to soaring obesity rates with bariatric products and equipment for the continuum of care. *Healthcare Purchasing News* January 2007.
60. Lachet MF, Owen JW, and Ebel MD. Caring for the morbidly obese pregnant woman. *MCN* 20: 101–106, 1995.
61. Warner WA and Garrett LP. The obese patient and anesthesia. *JAMA* 205:92–93, 1968.
62. Catenacci AJ, Anderson JD, and Boersma D. Anesthetic hazards of obesity. *JAMA* 175:657–665, 1961.
63. Gould AB Jr. Effects of obesity on respiratory complications following general anesthesia. *Anesth Analg* 41:448–452, 1962.
64. Vaughn RN. Anesthesia and the Obese Patient. In: BR Brown Jr. (ed.), *Contemporary Anesthesia Practice*. F.A. Davis, Philadelphia, PA, 1982.
65. Postlewait RW and Johnson WD. Complications following surgery for duodenal ulcers in obese patients. *Arch Surg* 105:438–440, 1972.
66. Shinohara E, Kihara S, Yamashita S et al. Visceral fat accumulation as an important risk factor for obesity sleep apnea syndrome in obese subjects. *J Int Med* 24(1):11–18, 1997.
67. American Sleep Dissorders Association. *International Classification of Sleep Dissorders: Diagnostic and Coding Manual, Revised (ICSD-R)*. Westchester, IL: Amercian Academy of Sleep Medicine, 1997.
68. Gleason JM. Obese hypoventilation syndrome. *Crit Care Nurse* 7(6):74–78, 1987.
69. Shinohara E, Kihara S, Yamashita S et al. Visceral fat accumulation as an important risk factor for obesity sleep apnea syndrome in obese subjects. *J Int Med* 24(1):11–18, 1997.
70. AARC Clinical Practice Guidelines. Metabolic measurement using indirect calorimetry during mechanical ventilation. *Resp Care*, 39(12):1170–1175, 1994.
71. Salvi RJ. Metabolic monitoring. *Adv Manag Resp Care* 3:40–42, 1997.
72. Gallagher SM. Meeting the needs of the obese patient. *AJN* 96(8)supp:1s–12s, 1996.
73. Shikora SA. Enteral feeding tube placement in obese patients: Consideration for nutritional support. *Nutr Clin Pract* 12:9s–13s, 1997.
74. Gallagher SM. Morbid obesity: A chronic disease with an impact on wounds and related problems. *O/WM* 43(5):18–27, 1997.
75. Carpenter R, Burns SM, Egloff MB et al. Effects of body position on spontaneous respiratory rate and tidal volume in patients with obesity, abdominal distention, and ascites. *J Crit Care* 3(2):102–106, 1994.
76. Lasater-Erhad M. The effect of body position on arterial oxygen saturation. *Crit Care Nurs* 15(10): 31–36, 1995.
77. Banyai M, Falger S, Roggla M, et al. Emergency intubation with the Combitube in a grossly obese patient with bull neck. *Resusc* 26(3):271–276, 1993.
78. Gallagher SM. Tailoring care for obese patients. *RN* 62(5):43–48, 1999.
79. Gallagher S, Arzouman J, Lacovara J, Blackett A, McDonald P, Traver G, and Bartholomeaux F. Criteria-based protocols and the obese patient: Planning care for a high-risk population. *O/WM* 50(5):32–44, 2004.
80. Gallagher SM. Morbid obesity: A chronic condition with an impact on wounds and wound problems. *O/WM* 43(5):18–27, 1997.
81. Gallagher SM. Caring for the overweight patient in acute care setting. *Journal of Healthcare Safety Compliance Infect Control* 4(8):379–382, 2000.
82. Planning and Design Guidelines for Bariatric Healthcare Facilities. Accessed January 1, 2008 at http://www.aia.org/nwsltr_print.cfm?pagename=aah_jrnl_20061018_award_winner
83. Gallagher SM. Needs of the homebound morbidly obese patient: A descriptive survey of home health nurses. *O/WM* 44(4):23–29, 1998.
84. Humphreys SL. Obesity in patients and nurses increases the nurse's risk of injury lifting patients. *Bariatr Nurs Surg Patient Care* 2(1):3–6, 2007.

85. Gallagher SM. Needs of the homebound morbidly obese patient: A descriptive survey of home health nurses. *O/WM* 44(4):23–29, 1998.
86. Rowen L. Safety first. *Bariatr Nurs Surg Patient Care*, 2(1):1, 2007.
87. Baptiste A, Leffard B, Vieira ER, Rowen L, and Tyler RD. Caregiver injury and safe patient handling. *Bariatr Nurs Surg Patient Care* 2(1):7–16, 2007.
88. Drake D. Tips and tools for safe patient handling. *Bariatr Nurs Surg Patient Care* 2(1):83–34, 2007.
89. Gallagher S. Understanding barriers to protocol development. National WOCN Conference, Las Vegas, NV, 2002.
90. Gallagher S. Bariatrics: Considering mobility, patient safety, and caregiver injury. In: W Charney and A Hudson (eds.), *Back Injury among Healthcare Workers*. Lewis Publishers, Baton Rouge, LA, 2004.
91. Gallagher S. Restructuring the therapeutic environment to promote care and safety for the obese patient. *JWOCN* 26:292–297, 1999.
92. Gallagher SM. Outcomes in clinical practice: Pressure ulcers prevalence and incidence studies. *O/WM* 43(1):28–40, 1997.
93. Camden SG, Shaver J, and Cole K. Promoting dignity and preventing caregiver injury while caring for a morbidly obese woman with skin tears and a pressure ulcer. *Bariatric Nurs Surg Patient Care* 2(1):77–82, 2007.
94. DeRuiter H-P, Meitteunen E, and Sauder K. Improving safety for caregivers through collaborative practice. *J Healthc Safety Compliance Infect Control* 5(2):61–64, 2001.
95. Gallagher Morbid obesity: A chronic disease with an impact on wounds and related problems. *OWM* 45(5):18–27, 1997.
96. Gallagher SM. Restructuring the therapeutic environment to promote care and safety for the obese patient. *JWOCN* 26:292–297, 1999.
97. Humphries SL. Obesity in patients and nurses increases the nurse's risk of injury lifting patients. *Bariatr Nurs Surg Patient Care* 2(1):3–6, 2007.
98. Edlich R, Woodward C, and Haines M. Disabling back injuries in nursing personnel. *J Emerg Nurs* 27:150–1155, 2001.
99. Kneafsey R. The effect of occupational socialization on nurses' patient handling practices. *J Clin Nurs* 9:585–593, 2000.
100. Charney W. An epidemic of health care worker injury. In: W Charney and G Fragala (eds.), *An Epidemic of Health Care Worker Injury: An Epidemiology*. CRC Press, New York, 1998.
101. Edwards G. The experts speak. *Bariatr Nurs Surg Patient Care* 2(1):3–6, 2007.
102. Gallagher S. Morbid obesity: A chronic disease with an impact on wounds and related problems. *O/WM* 45(5):18–27, 1997.
103. Balters S. Reclaiming lives. Team Rehab, November 1998.
104. Gallagher SM. Morbid obesity: A chronic condition with an impact on wounds and wound problems. *O/WM* 43(5):18–27, 1997.
105. Gallagher SM. Tailoring care for obese patients. *RN* 62(5):43–50, 1999.

Patient-Handling Technologies

John D. Lloyd and Andrea Baptiste

Introduction

Patient handling is a common task in health care. With the steady rise in work-related injury costs in the nursing occupation, safe patient handling has become a major topic of discussion. The health care industry is gradually accepting that manually lifting and transferring physically dependent patients is a high-risk activity, for both the caregiver and the patient. The highest-risk element of most patient-handling tasks typically involves either a vertical or horizontal transfer. Vertical patient-handling activities include bed to chair, bed to commode, and chair to commode transfers. Horizontal or lateral patient-handling activities include bed to stretcher transfers and repositioning tasks.

Technological solutions are needed to address the rising prevalence and incidence rates of musculoskeletal injuries in nurses related to patient-handling and movement tasks. Direct and indirect costs associated with back injuries are estimated to be between $24 billion annually[5,13] and $64 billion with $20 billion of that attributed to the health care industry.[2] Preventative solutions are needed to control the hazards and financial burdens associated with patient-handling tasks. Research supports the use of assistive devices for nursing staff to care for patients.[1,6,7,9,10,14,15]

The purpose of this chapter is to identify, describe, and discuss technologies that make the workplace safer for nurses and patients. Key technologies are grouped according to principal function. Technologies to assist with vertical transfer of patients include powered full-body sling lifts, ceiling-mounted patient lifts, floor-based lifts, powered standing lifts, nonpowered standing aids, and gait belts. Technologies to assist with lateral transfer of patients include air-assisted systems, friction-reducing devices, mechanical lateral transfer aids, sliding boards, and transfer chairs. Other new and emerging technologies which stand to positively impact the nursing profession, such as powered transport devices, are also presented.

Technologies to Assist with Vertical Transfer of Patients

Powered Full-Body Sling Lifts

Perhaps the most commonly used patient-lifting technology is the powered full-body sling lift. A vast number of models and configurations are available and are typically used with patients who have physical and/or cognitive impairments. These lifting devices can be used for almost any type of lift transfer. Powered lifting devices offer many benefits over mechanical or manual alternatives, since caregivers do not have to physically lift or reposition patients. The patient transfer is accomplished with the powered advantage of the patient lift, so there is less risk of injury to the caregiver.

There is a wide variation in the types of slings available for full-body lifts. Newer sling designs are much easier to install beneath patients or residents. When using full-body sling lifts, patients need to be fitted with slings of the right size to ensure no skin shearing or pressure points exist during the transfer. Emerging patient-lift systems are exploring opportunities to integrate the sling into hospital bedding or the patient's clothing. Alternative new technologies for vertical transfer of patients propose sling-less patient handling, which poses enormous time and energy saving potential while affording appropriate considerations to patient safety.

The majority of powered full-body sling lifts are mounted on a portable base, however, use of ceiling-mounted patient lifts is growing dramatically. The portable base and the ceiling-mounted devices have differing advantages. Powered full-body sling lifts can be used to move patients out of beds, into and out of chairs, for toileting tasks, bathing tasks, and for any type of lift transfer. These lifts are available with many features and there is a wide variation in the types of slings available. The newer sling designs are much easier to install beneath the patient or resident.

While mechanical lifting devices of any type have been shown to be far safer for both nurses and patients, several limitations interfere with their use in practice, including difficulty using in confined spaces, the extra time required, lack of accessibility or availability, difficulty using and storing, and poor maintenance.[3] Many of these barriers can be eliminated through the use of ceiling-mounted lifts over each patient bed. Using a ceiling-mounted lift decreased internal forces significantly when compared to operation of a mechanical floor lift.[8]

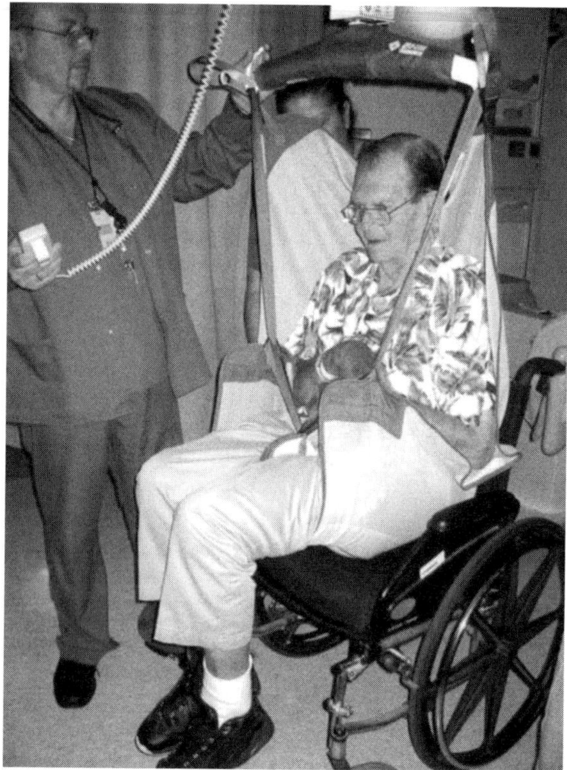

Ceiling-mounted patient lifts

Ceiling-mounted lifting systems provide solutions for reducing work-related nursing injuries and enhancing comfort and dignity for persons with spinal cord injury/disorders (SCI/D). Although they are not new in Europe and Canada, studies on the capability of these devices to reduce injury have been limited. Villenueve et al.[12] reported a Quebec study that showed a significant reduction of injuries attributed to the use of ceiling track lifting devices as opposed to floor-based devices. In the United States, the pioneering work of the Veterans Administration (VA) nurses in Florida on the use of the ceiling-mounted patient-lift system was first reported by Tiesman, Nelson, Charney, and Siddharthan at the VA Health Services Research Annual Meeting in 2002.[11] The report highlighted the impact of this new lifting and transfer technology in the reduction of musculoskeletal injuries among nurses as well as the ease and safety of its use in a 60-bed long-term care unit. It was also noted in this report that studies performed elsewhere have observed a reduction in occupational accidents by 75%–80% and an increase in personnel satisfaction.

A ceiling-mounted lift is installed with tracks that are secured into the beams to support the weight. It has a motor unit powered by battery that is attached to the track that raises and lowers the patient, who can then be moved along the track. There are two available ceiling lift configurations: single track and transverse track. A single-track system follows a dedicated path; therefore patient care activities involving vertical transfers are limited to this specific path. A transverse coverage system provides broader coverage within the room. The type of ceiling track installed depends on the need, frequency, patient population, and cost involved. A variety of sling designs and materials are available. According to Fragala,[4] "this lifting and transfer technology addresses two problems: the need to find a lift when and where you need one, and the need to push floor-based lifts from one place to another, sometimes over and around dangerous obstacles. Strategically positioned, the ceiling-mounted lift is designed to be available when and where you need it." Despite the myriad of attributes of this technology, the majority of health care facilities are not equipped with ceiling lifts.

Advantages of the ceiling-mounted patient lift include (1) ease in training nurses about use, (2) ease in maneuvering over floors and around furniture, (3) no need for storage, (4) conveniently located at bedside, (5) increased patient security and comfort, and (6) less strenuous on nurses than traditional floor-based lifting devices. These new devices have a few drawbacks, including (a) cost and (b) geographical restrictions of the lift based on where tracks are installed.

Tiesman et al. found that implementation of a ceiling-mounted patient-lift system decreased both the frequency and severity of injuries on a long-term care unit (in preparation). The number of musculoskeletal injuries on the unit decreased from 18 to 12, the number of restricted work days decreased from 16 to 6, and the number of lost work days decreased from 39 to 0 in a 12-month period. In addition, the nursing staff was extremely satisfied with the lifts measured via surveys and focus groups.

Portable bases can be used to suspend full-body sling lifts where overhead tracks are not available or practical. However, it is important to know the weight limit on these portable bases as they may not be able to handle the weight of a bariatric patient.

The benefit of using a ceiling-mounted lift should be weighed against the cost of alternative options, such as powered floor lifts.

Powered lifts offer many benefits when caring for patients. One key advantage is that caregivers do not have to manually move or reposition patients due to the mechanical nature of such lifts. The transfer becomes powered versus manual so there is less risk of injury to the caregiver. However, placement of a sling underneath the patient still proves to be a challenging task for caregivers. Many functional features of powered lifts include transferring patients from bed to chair or repositioning them by simply pressing a button. It is imperative that the motor of the powered lift is compatible with patients' weight, therefore lift capacities have increased considerably over the past decade. The present standard of 270 kg

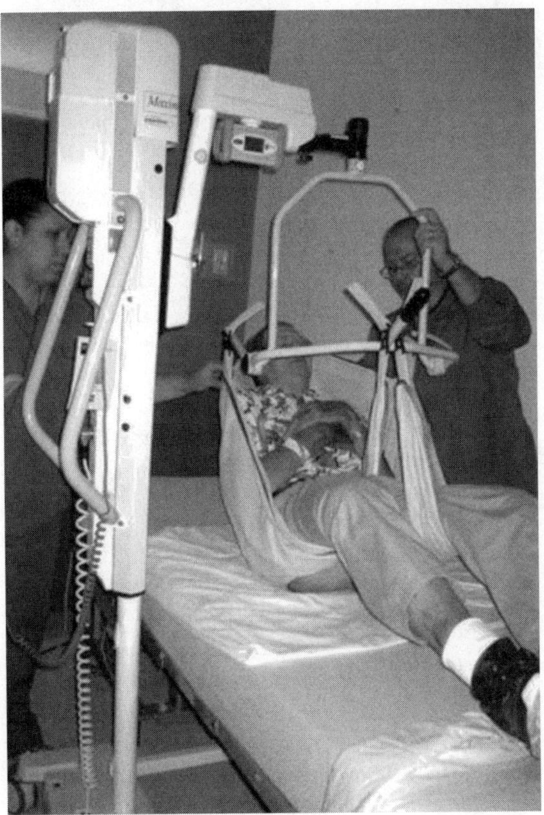

Floor-based lifts

(600 lb) for most floor-based and ceiling lift systems affords facility-wide coverage that should adequately address the needs of all by the most overweight patients. Lifts are now available with expanded capacity for morbidly obese patients up to 1000 lb. When using powered lifts, patients need to first be fitted with slings of the right size to ensure no skin shearing or pressure points exist during the transfer.

A new innovation in the market of powered full-body sling lifts is "powered positioning." Typically, while transferring a patient between a supine and seated posture, the caregiver manually directs the patient's position. This can be achieved by exerting a physical force against straps located on the sling, or using a positioning handle designed into the hanger bar. These forces can impose a biomechanical stress on the wrist, elbow, and shoulder joints of the caregiver in direct proportion to the weight of the patient. Powered positioning affords the nurse the facility to change a patient's posture using the powered advantage of the patient-lift system. This new technology is available for both floor-based and ceiling-mounted patient lifts.

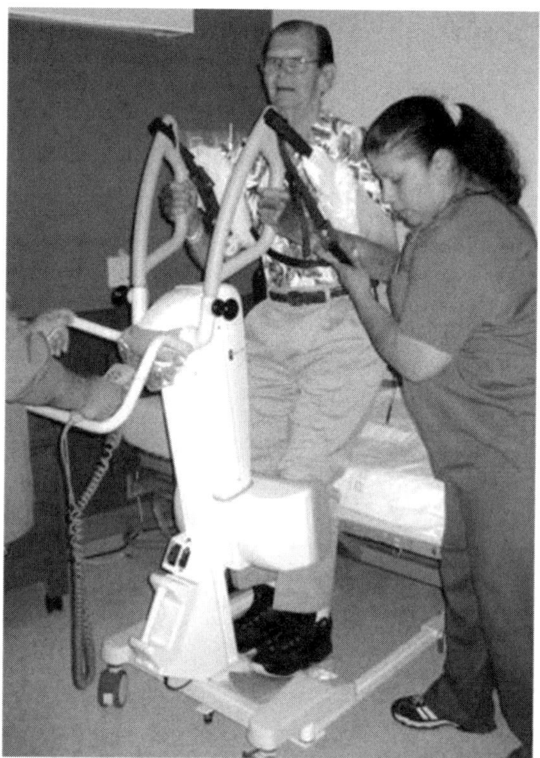

Powered standing lifts

Powered standing lifts provide an alternative to full-body sling lifts and are particularly useful for patients who are cognitively coherent, partially dependent and have some weight bearing capabilities. These lifts are excellent for moving patients in and out of chairs and for toileting tasks. Powered standing assist lifts have a relatively small base and are therefore easily maneuvered in restricted areas, such as small bathrooms. There are some variations in the sling design, but the basic concept is of simple design is very easy to place around the patient's torso. This type of sling allows unrestricted access to the patient's lower body for the purpose of toileting. Powered standing lifts should not be used with patients who are at times combative, unpredictable, or have cognitive deficits.

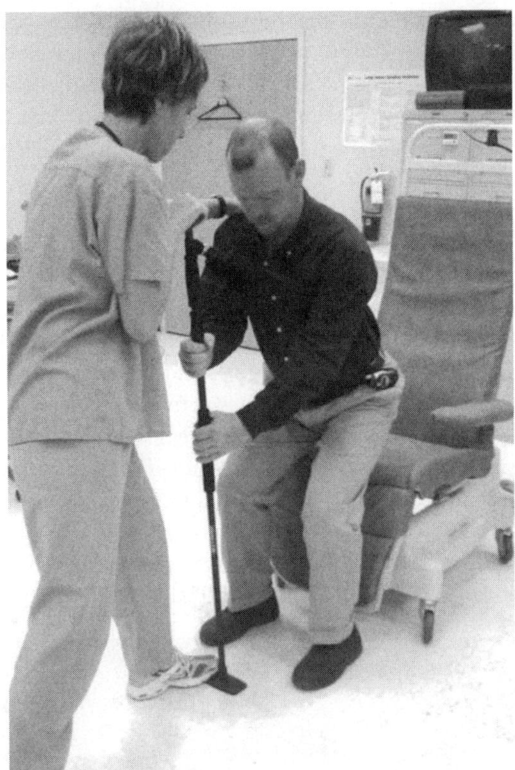

Standing assist and repositioning aids

Some patients or residents may only need a little support to stand. In this case, they can help themselves if they have a support to grasp. Various types of devices can be provided to assist a patient from a seated to standing position by allowing them to hold on to a secure device and pull themselves up, such as demonstrated in the figure. These devices may be freestanding or attached to beds.

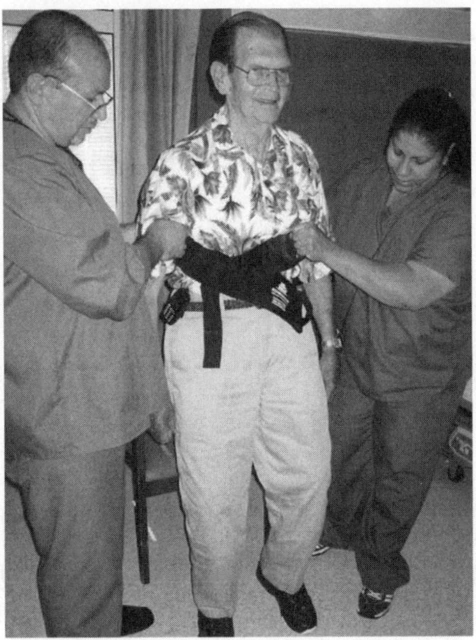

Gait/transfer belt with handles

An object with handles improves the grasp opportunity for the worker and thereby reduces the risk. Gait/transfer belts are installed on patients or residents, usually around the area of the waist providing handles for a worker to grasp when assisting or transferring a partially dependent patient or resident, as shown. Small handheld slings that go around the patient can also facilitate a transfer by providing handles. These options are available for patients with weight bearing capabilities that need only minimal assistance.

Technologies to Assist with Lateral Transfer of Patients

One of the highest risk tasks in nursing is to transfer a patient from a bed to a stretcher. This type of transfer involves the caregiver to reach over the stretcher to the bed where the patient is lying, to then pull the patient over to the stretcher. This task forces the caregiver into a poor, awkward posture. Lateral transfers are typically performed using a draw sheet, which has been proven to cause high shear forces on the nurse's spine. Furthermore, few nurses have adequate shoulder strength to perform this activity safely.[6] Use of lateral transfer aids have reduced the physical demands required in the horizontal transfer of patients.[1,6]

The use of lateral transfer aids can eliminate the need for poor postures, thus reducing the risk of injury to the caregivers. Lateral transfer aids are devices used for the lateral transfer of patients. Devices can be grouped into three categories: (1) friction-reducing lateral sliding aids, (2) air-assisted lateral sliding aids, and (3) mechanical lateral sliding aids. Each will be briefly described.

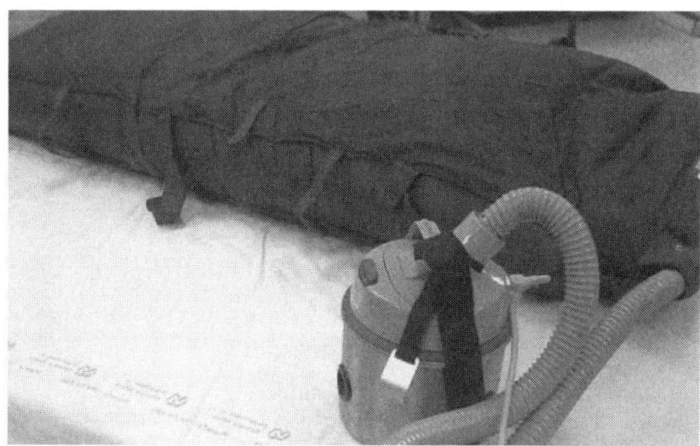

Air-assisted lateral transfer aids

These are devices where a flexible mattress is placed under a patient in the same manner as a transfer board. There is a portable air supply attached to the mattress that inflates the mattress. Air flows through perforations in the mattress and the patient is moved on a cushioned film of air allowing staff members to perform the task with much less effort. These technologies are particular suitable when performing lateral transfers involving patients with special medical conditions, such as pressure sores.

Friction-Reducing Lateral Sliding Aids

Friction-reducing lateral sliding aids can assist with bed to stretcher type transfers. These devices can be positioned beneath the patient or resident similar to a transfer board and provide a surface for the patient to be slid over more easily due to the friction-reducing properties of the device. These are simple low-cost devices, usually made of a smooth fabric that is foldable and very easy to store. Properly designed handles can reduce horizontal reach, as shown in the example.

Use of a friction-reducing device for the lateral transfer of patients significantly reduced muscle activity for the spine and the shoulders. While reduced muscle activity can reduce muscle fatigue, observation of intervention subjects revealed the friction-reducing device was not intuitive in its use, and despite training, subjects did not realize the true capability of this ergonomic intervention. Training and competency programs to assure appropriate use are needed to fully benefit from patient care equipment.[8]

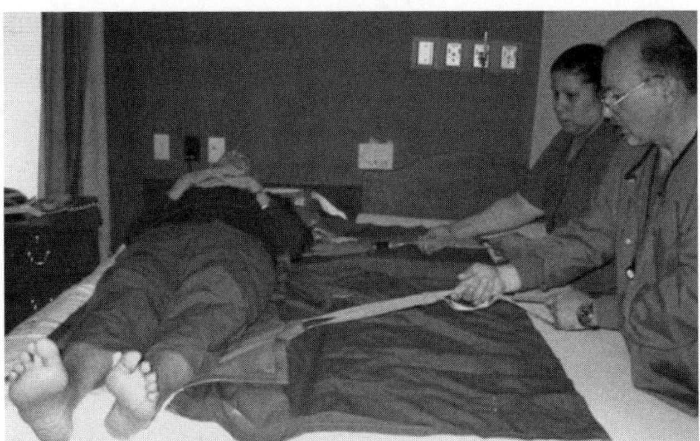

Mechanical lateral transfer aids

Friction-reducing devices should be used whenever there a dependent patient is transferred laterally. Such devices should be provided in sufficient number and in convenient locations to promote staff usage. Training in the correct use of any new device is imperative to successful implementation.

One of the highest risk tasks in nursing is to transfer a patient from a bed to a stretcher. This type of transfer involves the caregiver to reach over the stretcher to the bed where the patient is lying, to then pull the patient over to the stretcher. This task forces the caregiver into a poor, awkward posture. The use of lateral transfer aids can eliminate the need for poor postures, thus reducing the risk of injury to the caregivers. Lateral transfer aids are devices used for the lateral transfer of patients. These aids vary in types: (1) air-assisted aids, (2) mechanical lateral aids, and (c) manual lateral sliding aids.

An air-assisted device is a flexible mattress, which is placed under a patient and inflated by a portable air supply. The bottom side of the mattress has thousands of holes allowing the air to get through, while the patient is moved on a cushioned film of air, across to the other surface. This reduction in friction between the bed and bottom of mattress makes the transfer easier for the caregiver. Key strengths of air-assisted sliding aids are that they provide a good surface for patients with compromised skin integrity (pressure sores or burns) and that there is no weight limit. This type of device is especially useful for bariatric patients.

Stretchers are available that are height adjustable and have a mechanical means of transferring a patient on and off the stretcher. Some are motorized and some use a hand crank mechanical device. Mechanical means of mechanizing the lateral transfer are also available as independent options able to be used with most beds and stretchers, as shown. These devices eliminate the need to manually slide the patient, minimizing risk to the caregiver.

Mechanical lateral aids are those devices, which eliminate the need to manually slide patients, thus substantially reducing the risk of injury to caregivers. A mechanical lateral transfer aid is one, which may be mechanically assisted or motorized. The key benefit of these products is the reduction of risk of injury to the caregiver. Mechanical lateral aids are highly recommended when treating bariatric population, given the higher range of weight capacities. The mechanical features of these products cause them to be more costly than air-assisted and manual lateral sliding aids. However, the benefits of purchasing these devices far outweigh the cost.

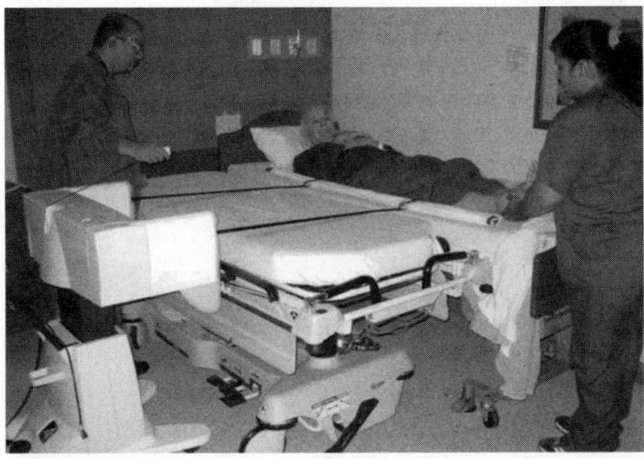

Sliding boards

For seated bed to chair or chair to toilet type transfers, low-cost sliding boards are available. Sliding boards are usually made of a smooth rigid material with a low coefficient to friction. The lower coefficient of friction allows for an easier sliding process. These boards act as a supporting bridge when seated slide transfers are performed. Some, but substantially reduced, manual lifting is still required to move the patient; however, sliding boards do offer considerable improvement at a minimal cost. The illustrated example is suitable for independent or assisted transfers from wheelchair to bed.

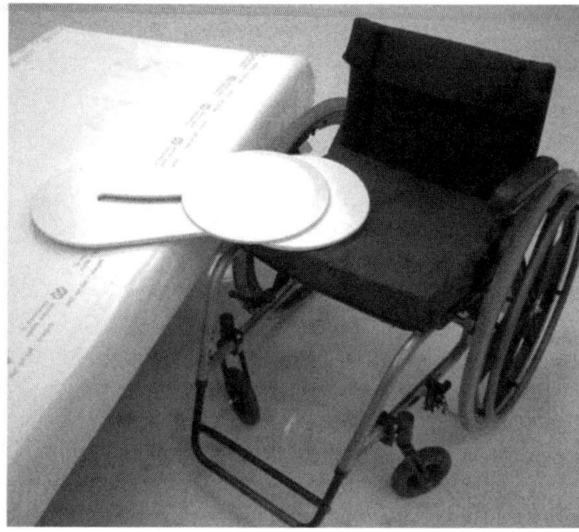

Repositioning devices

The traditional method of repositioning a patient in bed has been to hook under the patients' arms and pull them toward the head of the bed. This posture adopted by the caregiver places stress on backs and shoulders as they are required to reach excessively in a forward bent position, subjecting them to a high risk of back and shoulder injury.

Repositioning aids are available for patient positioning up in bed or can be used to turn a patient to their side. Repositioning aids are typically made with low-friction fabric either tubular in shape or consist of two pieces of material that slide one over the other. They are designed to reduce friction, thus making it easier to slide the patient around in bed or in a chair. There are devices that are used for seated repositioning and those used for supine. A benefit of some devices is the unidirectional feature, as when engaged by zipping a section, it prevents the patient from sliding down as friction is reintroduced.

Transfer Chairs

Some new wheelchairs and dependency chairs can convert into stretchers where the back of the chair pulls down and the leg supports come up to form a flat stretcher. These devices facilitate lateral transfer of the patient or resident and eliminate the need to perform lift transfer in and out of wheelchairs. There are wheelchair devices that convert to stretchers which also have a mechanical transfer aid built in for a bed to stretcher or stretcher to bed type transfer.

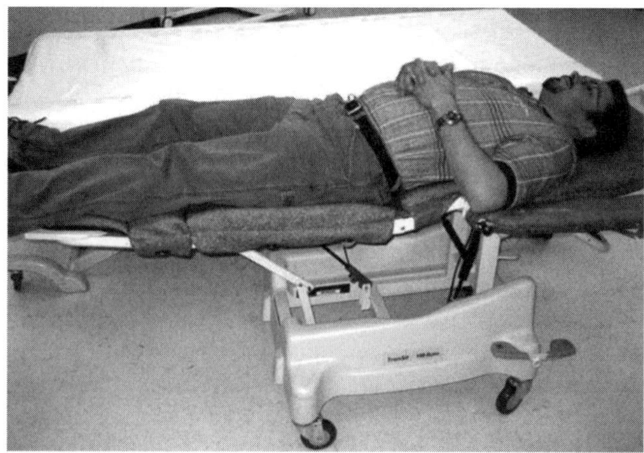

Patient transport devices

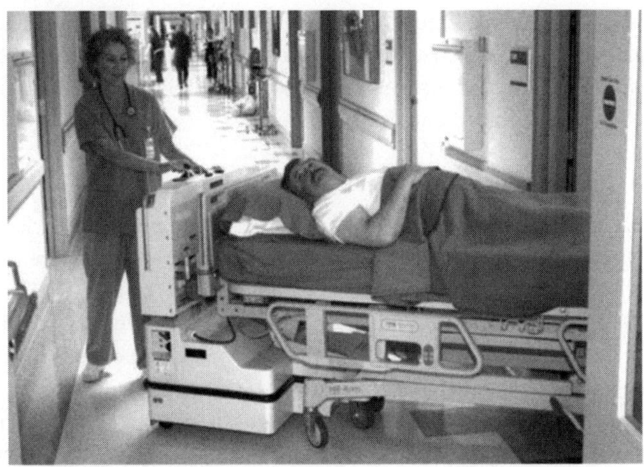

Patient transport devices

Transporting a patient in a bed, stretcher, or wheelchair requires significant effort, particularly over uneven terrain, carpeting, or long distances. This task typically requires two or more staff to perform safely.

Powered transport devices have recently been introduced to the market of patient-handling technologies to address this problem. These devices can be attached to the head of bed or stretcher and are motorized, thereby assisting in variable-speed propulsion of the patient. This low-cost device can be used for patient transport throughout a hospital or nursing home, requiring only one caregiver to perform the task. Some newer higher-end stretchers and hospital beds have integrated motorized capability. While more convenient and eliminating storage issues, which is otherwise an issue for the independent transport devices, the integrated systems can be considerably more expensive and serve only one patient at a time.

Use of powered transport devices such as bed and wheelchair movers and powered stretchers is becoming a popular choice as it reduces the risk of caregiver injury by reducing the push/pull forces involved, making patient transport a safer task for caregivers. This is especially important in bariatric patient care where the mass of the patient, in addition to the weight of the bed is excessive, thus demanding higher push or pull forces during patient transport.

Bariatric Specialty Equipment Needs

Managing obese patients provide special challenges to nursing care staff. Some of these difficulties including the inability to turn, transfer, or ambulate patients can contribute to pressure ulcers. Respiratory insufficiency can develop due to overweight hypoventilation syndrome. Colostomy care becomes complicated due to a large abdominal apron of fat. From the caregivers' perspective, treatment of bariatric patients subjects caregivers to an increased risk of injury. To minimize this risk, the appropriate use of technology is one solution in managing this special population.

Ambulatory/Mobility Aids

Ambulatory and mobility aids consist of assistive devices for patients who are unsteady in standing or walking. In this category, equipment includes but is not limited to walkers, canes, gait training devices, and crutches. A gait trainer is a frame on wheels used to provide support to a patient when walking or ambulating. The patient is placed in a specialized harness, which is affixed to the gait trainer. This enables the patients' weight to be fully supported if needed. Gait trainers are typically utilized in rehabilitation until the patient has developed enough strength and can walk independently.

When choosing a mobility aid, it is important to ensure that the device is adjustable in width or wide enough to accommodate the user. The structural integrity and weight limitation of each assistive device also needs to be taken into account to ensure safe use of the mobility aid. If storage is a concern, there are bariatric walkers that can be folded and easily carried.

Bathing Equipment

Many slips and falls occur in the shower or bathroom due to poor flooring and lack of proper equipment. By implementing the proper use of appropriate equipment in the bathroom, the probability of a fall is reduced. The bariatric patient is especially susceptible to falls given the lack of balance and weakness in the knees and ankles. Options to improve safety in the bathroom include use of grab bars, elevated toilet seats, tub transfer benches, bath benches, and shower trolleys. It is important that grab bars are strong enough to support the weight of the user and that they are strategically placed.

Tub transfer benches are devices that assist in the transfer of unsteady patients into the tub. They are seated surfaces that extend over tubs, and are designed to fit standard bath tubs. Weight capacity ranges from 650 to 850 lb. Another option for those residents steadier on the feet is security rails, which are attached to the rim of the tub.

Weight capacity of toilets should be considered as regular toilet capacities may have different weight accommodations due to the porcelain used. It is necessary to check with the manufacturer for the maximum weight. If the toilet is floor mounted or wall hung, that design difference would also have a significant impact on the weight capacity. Some Kohler models will handle weight capacities to 1000 lb if the commode is floor mounted and 500 lb for a wall hung model. Even at these capacities, extreme care should be taken not to underestimate the support points needed to support this weight.

Beds and Mattresses

New beds are designed to multitask and accommodate larger weight capacities up to 1000 lbs. Special features offered are turn assist up to 20° to aid in patient positioning, percussion therapy, pulsating air suspension therapy, pressure relief therapy, and cardiac chair positioning. Due to excessive skin folds, additional body weight, lack of mobility, difficulty to clean and treat certain areas, some bariatric patients are susceptible to pressure ulcers and thus require specialized surfaces. Low air loss mattresses greatly assist in the prevention of pressure sores and improve patient outcomes, comfort, and safety. Some treatment systems and beds are also specifically made for easier transport of patients.

Turning and Repositioning

Two tasks which prove very challenging and place caregivers at a high risk for injury include insertion of patient care slings under bariatric patients and repositioning a patient up in bed or turning to the side. Currently, manufacturers are working on designing slings that can be left under patients who have to be moved frequently. The advantages to this would be less strain on the caregivers to turn the patient to the side or log roll them to insert the sling, less time taken to perform the task of repositioning because the sling would already be there and most important, less risk and exposure to injury. Although there is no evidence or literature on leaving slings under patients and patient outcomes, this decision should be considered carefully.

Transportation

Transport of bariatric patients can be made easier by use of a powered system. Such powered systems may be built into the bed or via a detachable battery operated device.

The Intellidrive system is an example of a built-in powered device, which operates by unplugging the bed and releasing the brakes. There are two handles at the bed head, which are used to steer and the bed is moved by depressing the buttons and applying minimal force to initiate movement.

An alternative device used for patient is a detachable, battery-operated technology, capable of moving a patient in bed. It is typically attached to the bed head and powered by batteries. It has an adjustable steering angle and docking height and is operated by a toggle switch, similar to that of a scooter.

Inherent in the process of care and management of patients is transfers from bed to wheelchair or bed to toilet. There are several challenges involved in transferring bariatric patients, such as inadequate size of doorways, inappropriate-sized beds or commodes, and unavailable floor-based lifts with ability to accommodate weight upper limits. One way of reducing the number of transfers is the use of ceiling-mounted lifts.

Equipment and Use

Significant progress has been made in the area of patient-transfer equipment design in the past 15 years. Devices that were nonexistent are now prevalently available to meet the needs of nearly all-patient populations. Noted improvements include incremental increases in the comfort, lift capabilities, and style of patient transfer devices.

The collaborative effort among health care organizations, staff, and manufacturers has produced improvements to these devices resulting in a favorable impact on the quality of patient care. However continuous improvement is needed in areas of standardization of increased capacities in order to meet the needs of our changing patient demographics.

References

1. Bohannon R. Horizontal transfers between adjacent surfaces: Forces required using different methods. *Arch Phys Med Rehabil* 80:851–853, 1999.
2. Fragala G. Implementing ergonomic approach promotes workplace safety. *Provider* 18(12):31–34, 1992.
3. Fragala G. Injuries cut with lift use in ergonomics demonstration project. *Provider* 19(10): 39–40, 1993.
4. Fragala C. Mobility and transfer technology: Advances at three levels. *Nursing Homes Magazine* December 2001.
5. Garrett B, Singiser D, and Banks S. Back injuries among nursing personnel: The relationship of personal characteristics, risk factors and nursing practices. *AAOHN J* 40(11):510–516, 1992.
6. Lloyd JD and Baptiste A. Biomechanical evaluation of friction reducing devices for lateral patient transfers. *AAOHN J* 54(3):113–119, 2006.
7. Lynch RM and Freund A. Short-term efficacy of back injury intervention project for patient care providers at one hospital. *Am Ind Hyg Assoc J* 61(2):290–294, 2000.
8. Nelson A, Lloyd J, Gross C, and Menzel N. Preventing nursing back injuries. *AAOHN J* 51(3): 126–134, 2003.
9. Owen BD and Fragala G. Reducing perceived physical stress while transferring residents: An ergonomic approach. *AAOHN J* 47(7):316–323, 1999.
10. Owen BD. Preventing injuries using an ergonomic approach. *Assoc Oper Room Nurs J* 72(6): 1031–1036, 2000.
11. Tiesman HM, Nelson K, Charney B, and Siddharthan K. The impact of new lifting technology in a long term care unit. Paper presented at the VA Health Services Research 2002 Annual Meeting, Washington, DC, 2002.
12. Villenueve J, Goumain P, and Elabidi D. A comparative study of two types of patient-lifting devices for moving patients in long term care. Paper presented at the International Ergonomics Association Congress, Toronto, Ontario, Canada, 1994.
13. Williamson K, Turner J, Brown K, Newman K, Sirles A, and Selleck C. Occupational health hazards for nurses part. *J Nurs Schol* 2. 20(3):162–168, 1988.

14. Zhuang Z, Stobbe TJ, Hsiao H, Collins JW, and Hobbs GR. Biomechanical evaluation of assistive devices for transferring residents. *Appl Ergon* 30(4):285–294, 1999.
15. Zhuang Z, Stobbe TJ, Collins JW, Hsiao H, and Hobbs GR. Psychological assessment of assistive devices for transferring patients/residents. *Appl Ergon* 31(1):35–44, 2000.

Safe Patient Handling: Evidence-Based Solutions

Audrey Nelson, Andrea Baptiste, and Mary Matz

Disclaimer: The findings and conclusions in this report are those of the authors and do not necessarily represent the views of the Department of Veterans Affairs.

Introduction

Despite the high risk associated with many patient-handling tasks, patient care providers often rely on tradition, rather than evidence when deciding how best to perform these tasks. In the past decade, patient-handling technologies have progressed to a new level of safety and ease in use. Further, a significant body of science has emerged to guide clinicians toward safer patient handling. The purpose of this chapter is to summarize this new evidence toward supporting or refuting the safety of the most common approaches to patient handling.

High-Risk Patient-Handling Tasks

Most care providers identify patient lifting as a high-risk task, while minimizing the risk associated with other patient-handling tasks, such as bathing, feeding, or dressing. A high-risk patient-handling task is defined as any patient assignment that pushes the limits of human capabilities, including those that require lifting a heavy load, sustained awkward position, bending/twisting when performing the task, excessive reaching to get the task done, tasks of long duration that contribute to fatigue, tasks that require excessive force on one or more joints or body parts, and tasks that require standing for long periods of time.

The unique patient characteristics and physical environment of each patient care setting contributes to the frequency, type, and severity of the risk associated with patient-handling tasks. Table 3.1 outlines common high-risk tasks associated with various clinical settings.

Evidence-Based Solutions for High-Risk Patient-Handling Tasks

In the past decade, technology, techniques, processes, and tools have advanced to enhance the safety of many of the patient-handling tasks previously identified in Table 3.1. Evidence related to the most common approaches to patient handling will be described, including back belts, equipment/devices, patient transport aids, administrative controls (e.g., scheduling, assignments), minimal-lift policies, clinical decision-making tools (e.g., algorithms/patient care assessment protocols), staff training, manual patient lifting, unit-based peer leaders, facility-based champions, and lift teams.

Back Belts

In an attempt to protect workers from experiencing low-back discomfort or pain, some safety professionals have subscribed and recommended the use of back belts. However, the use of back belts has been controversial as to whether or not there is any real benefit of injury prevention. This debate has been one for many years across the manufacturing and health care industry. A recent article by Roelofs et al.[1] studied whether lumbar supports may prevent recurrent back pain in 360 health care workers in the Netherlands. This randomized controlled trial included an intervention where workers completed a

TABLE 3.1

Clinical Setting	Common High-Risk Tasks
Medical/surgical	• Transfer from bed to chair • Transfer from bed to stretcher • Moving occupied bed or stretcher • Making occupied bed • Bathing a confused or totally dependent patient • Moving a patient up from the floor • Weighing a patient • Applying antiembolism stockings • Repositioning in bed • Making occupied bed • Extensive dressing changes
Psychiatry	• Restraining a patient • Escorting a confused or combative patient • Toileting a confused or combative patient • Dressing a confused or combative patient • Picking a patient up from floor • Bathing/showering confused or combative patient • Bed-related care in beds that are not height adjustable
Critical care units	• Transporting patients (road trips) • Lateral transfers (bed to stretcher) • Lifting patient to the head of the bed • Repositioning patient in bed from side to side • Making occupied bed • Applying antiembolism stockings
Perioperative settings	• Standing long periods of time • Lifting and holding patient's extremities • Holding retractors for long periods of time • Transferring patients on and off operating room tables/beds • Reaching, lifting, and moving equipment • Repositioning patients on operating room beds
Long-term care/nursing homes	• Repositioning in bed • Making occupied bed • Transferring patient from bathtub to chair • Transferring patient from wheelchair to bed • Transferring patient from wheelchair to toilet • Lifting a patient up from the floor • Weighing a patient • Bathing a patient in bed • Bathing a patient in a shower chair /trolley • Undressing/dressing a patient • Repositioning patient in dependency chair • Making an occupied bed • Feeding bed-ridden patient • Changing absorbent pad
Home care	• Providing patient care in a bed that is not height adjustable • Providing care in crowded area, forcing awkward positions • Toileting and transfer tasks without proper lifting aids • No assistance for tasks

short course on healthy work methods including the use of four types of lumbar supports. Results indicated that there were 52.7 fewer days with workers wearing back belts than those who received the short course. There was no change in the number of sick days between the groups. There was a small significant difference in pain intensity and function of workers. It should be noted that this study was limited as participants were not blinded and there was missing data which required imputation. Researchers concluded that including patient directed use of lumbar supports in a short course on healthy work methods may decrease the number of days when low-back pain occurs, but not absenteeism among workers with prior low-back pain. In opposition, many studies present evidence which suggests back belts are not helpful.[2,3]

A back belt is a lightweight band, breathable in nature, with double-sided pulls that facilitate varying levels of tautness and pressure. Proponents of back belts claim the following: (1) reduction of internal spinal forces during forceful exertions of the back, (2) an increase in intra-abdominal pressure, (3) reinforcing the spine, (4) limiting bending, and (5) reminder to lift properly.[4] In opposition, some studies have shown that there is no relationship between intra-abdominal pressure and abdominal musculature.[5,6] Many studies have supported that intra-abdominal pressure does not play an important part in relieving intradiskal pressure or tension in the back extensors.[7–12] In addition to this evidence, The Centers for Disease Control and Prevention's (CDC's) National Institute for Occupational Safety and Health (NIOSH) found no evidence that back belts reduce back injury or back pain for retail workers who lift or move merchandise.[3]

Patient-Handling Equipment/Devices

Use of patient-handling technology has proven to be quite an effective solution in dealing with high-risk tasks in nursing. Evidence is mixed for two reasons. First, because technology is typically incorporated with educational programs, which make it difficult to attribute the success to one variable. Second, studies rely on retrospective data. More recently, there have been more clinical trials to try to parse out technology as an effective strategy without confounding variables such as education or training. Some examples of effective strategies within the patient-handling field are described in the next few paragraphs.

Research has demonstrated that using portable mechanical lifts have a positive effect on work-related injuries and health care workers.[13–16] Manufacturing companies are constantly developing and expanding their product line in an attempt to accommodate the growing population. There are more obese patients today than there ever was 20 years ago and with the increase in patient weight and more debilitating comorbidities, there is an increased injury risk to caregivers. Portable floor-based lifts can now accommodate higher weight capacities up to 1000 lb. However while increasing the weight capacity of the lift, the caregiver now has to maneuver a heavier lift, which requires more push force thus increasing risk (see Figure 3.1).

To avoid this push force use of ceiling lifts are encouraged. An additional benefit is that they are conveniently stored on an overhead track and take up minimal space in the room. Newer ceiling lifts have tracks and motors that can accommodate up to 1000 lb patients. Research has supported this technology because of its ability to facilitate many tasks by utilizing a variety of sling types, making it easier and safer for the caregiver and safer for the patient. Ceiling lifts can assist in transferring a patient from the chair/wheelchair to the bed, lift a patient up in bed so the nurse can change the linen, move a patient up in bed, turn a patient in bed or even perform lateral transfers (see Figure 3.2).

Some of these patient-handling tasks may seem to be or can be performed faster than if they were done manually but the benefits of these multifunctional devices far outweigh any perceived or actual additional time spent performing the activity. As well, the long-term savings in workers compensation injury costs are significant enough to justify purchasing such units. Many studies have evaluated ceiling lifts and have found that they are a good investment.[17–21]

In addition to floor-based, portable and ceiling lifts, beds now offer many features that assist caregivers in high-risk tasks. Manufacturing companies have improved surfaces and frames significantly to reduce risks to both the patient and the staff. Using height-adjustable beds and fast electric beds have reduced the strain of caregivers during tasks such as bathing.[22–24] New bed technology now provides more features such as bed egress, shearless pivot, lateral rotation, timed percussion, and vibration.

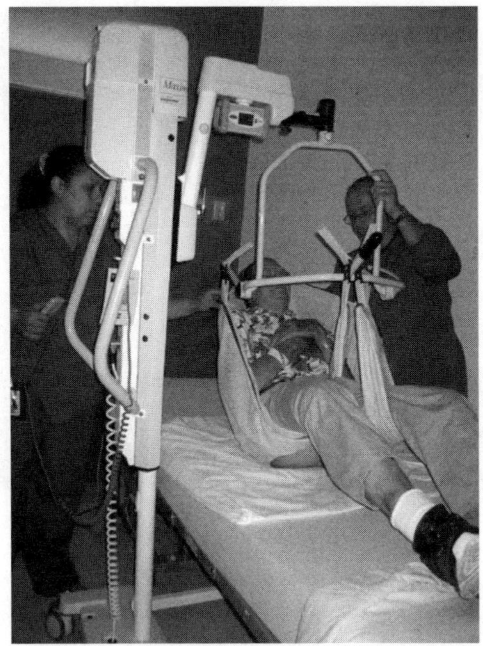

FIGURE 3.1 Floor-based lift.

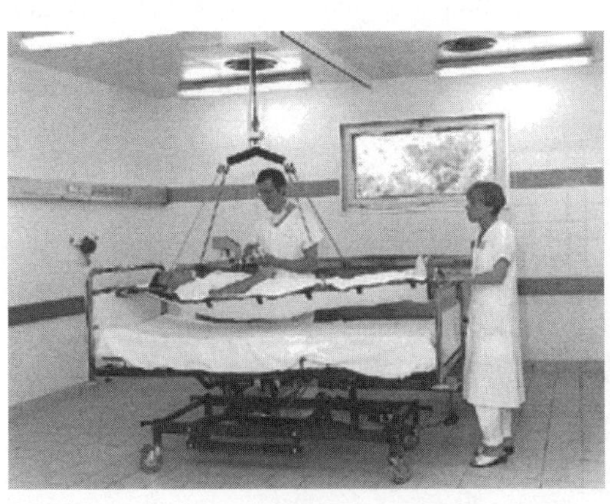

(a)

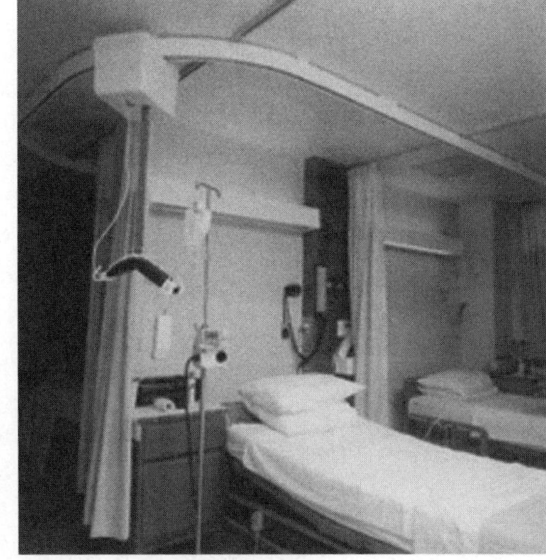

(b)

FIGURE 3.2 Ceiling lift.

One high-risk task that is currently being addressed through technology is repositioning a patient to the head of the bed and turning a patient on their side. Evidence-based studies concur that repositioning with a draw sheet (the traditional method) places caregivers at an increased risk of injuries due to high spinal loading.[25] There are two types of repositioning aids, seated and supine, that are commercially available but greatly underused (see Veterans Administration (VA), *Patient Safety Center Technology Resource Guide* for list of devices http://www.visn8.med.va.gov/patientsafetycenter/safePtHandling/).

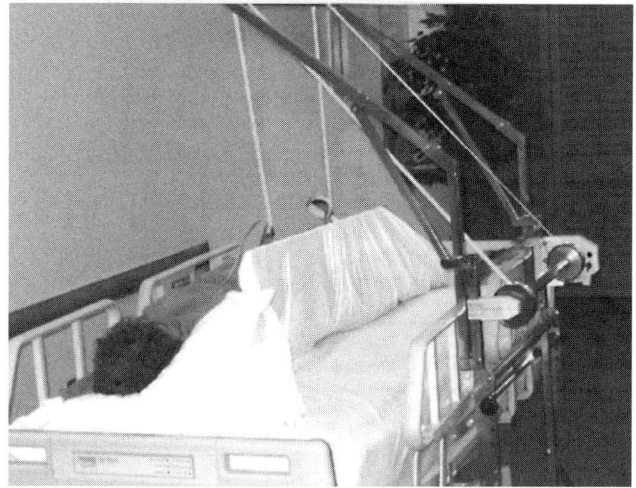

FIGURE 3.3 PRS system.

A clinical trial is in progress to evaluate a new multifunctional device capable of lateral transfers, repositioning to the head of the bed, and turning a patient on their side. Figure 3.3 illustrates this new technology turning a patient on their side. While the evidence to support this emerging technology is evolving, the device shows promise.

More clinical trials are needed to evaluate the usability, feasibility, and clinical issues associated with new technology. Turning a patient has been identified as a high-risk task as it is done quite frequently and places caregivers at a high injury risk.[26–28] A new solution to turning a patient on the side is the use of clips, which connect to the existing bed sheet by clamps. These clips are intended to assist caregivers to turn patients in bed in order to relieve pressure points and to allow access for linen or dressing changes. They are then attached to a spreader bar which is connected to either a floor-based lift or ceiling lift to facilitate turning a patient on the side as shown in Figure 3.4. A laboratory evaluation was performed and results and recommendations were provided to the manufacturing company for improvements to

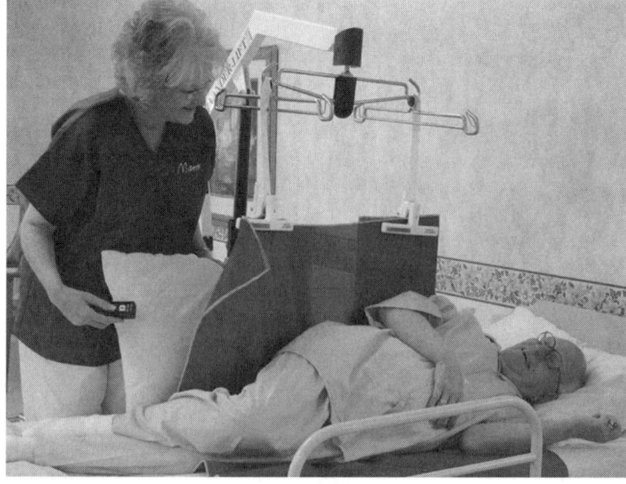

FIGURE 3.4 Clips for turning a patient in bed.

the design of the product. There is a need for more laboratory and clinical evaluations to add to the evidence that technology can make a positive difference in caregiver burden and reduce risk of injury associated with patient-handling tasks.

There has been more research done on the use of lateral transfer aids as a solution to reduce risk of injury (Zelenka et al., 1996; Baptiste, 2007).[29–32] Laterally moving a patient from a bed to a stretcher is a high-risk task due to the following key points: (a) awkward posture the caregiver has to adopt, (b) excessive horizontal reach to grab sheet before the transfer, (c) the weight of the patient, (d) the forces needed to exert sufficient pull force to move the patient, and (e) poor coupling/no handles.

A laboratory evaluation demonstrated that biomechanically, the best lateral transfer device is one made of a parachute-like material with extended pull straps and handles that provides reduction in friction thereby decreasing the forces on the spine. Close behind in rank order were the air-assisted lateral transfer devices (Lloyd and Baptiste, 2004).

There is a variety of lateral transfer solutions available: air-assisted devices, mechanical aids, slide sheets or friction-reducing devices, and slide boards. Powered mechanical devices like mechanical lateral transfer aids and ceiling lifts should be utilized as preferable first solutions.

Patient Transport

Patient transport is also a task that is done frequently by nurses and the physical demands are usually underestimated by most people. Pushing or maneuvering a bed or stretcher is difficult given the combined weight of the bed and patient, narrow hallways, flooring, and small doorways or elevators. There are several technological options to assist in this task.

Figure 3.5 shows an example of a solution which can significantly reduce the push forces involved in moving beds. This detachable device clamps onto the bed or cart and is battery operated. By simply pressing a button on the lever, the device now powers the bed, and the operator is only required to steer, not push the occupied bed. Bed movers are of great benefit as they reduce the need to push therefore reduce risk of strains by caregivers, however the downside to such devices is that they may not fit into the elevator.

Another solution to patient transport is integrated powered systems. This is where the power is built into the bed/stretcher making the device more expensive but less cumbersome as there is nothing to attach. Figure 3.6 illustrates an integrated system.

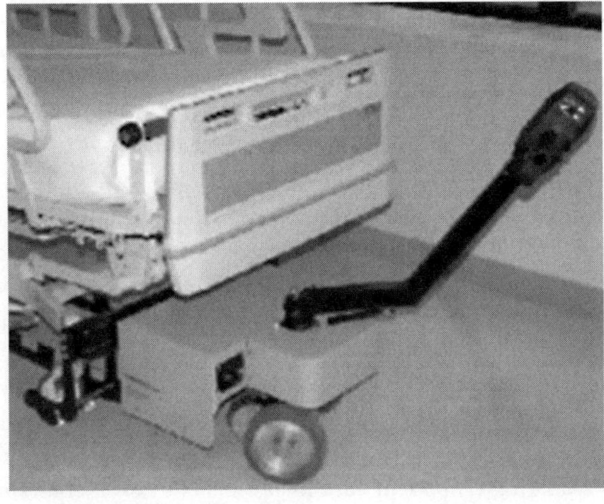

FIGURE 3.5 Bed mover.

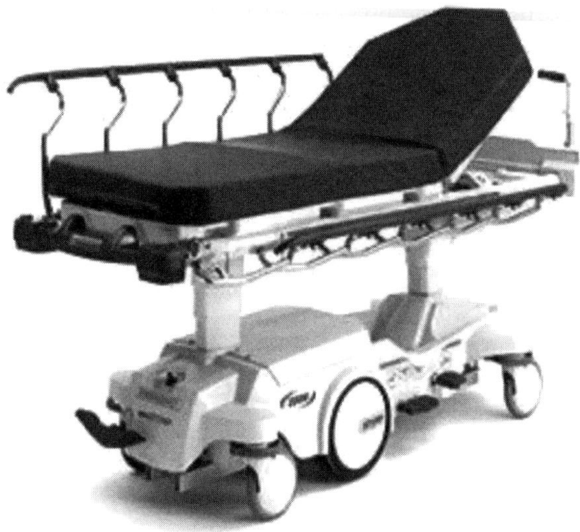

FIGURE 3.6 Powered stretcher.

Evidence has shown that patient-handling technology, such as portable lifts, ceiling lifts, lateral transfer devices, height adjustable and fast electric beds, repositioning devices, and patient transport aids are all effective solutions for minimizing caregiver risk.

Manual Patient Lifting

Lifting a patient manually has been taught in nursing schools for years and is still practiced in some institutions today. These techniques have been banned in Europe[33,34] and are unsafe for nurses and patients according to a 1993 study by Corlett and others. In the United Kingdom, manual handling techniques are outlawed by legislature nationally, but some nurses still handle patients manually and do not use lift equipment.[33,34] In a study by Owen et al., in 1995, 98% of nurses use the "hook and toss" method known also as the drag lift. This method involves reaching under a patients' armpits and carrying or transferring them this way. This lift is also used in the United States as it is taught by 83% of nurse educators in schools of nursing.[35] Although the drag lift is unsafe for the patient and for the caregiver, it is still being used today and new techniques and safer ways to transfer patients needs to included as part of the nursing curriculum in schools. The use of patient-handling equipment needs to be taught and supported by legislature nationally to protect caregivers and patients.

Administrative Controls

An administrative control is an intervention designed to decrease risk of injury by minimizing exposure to a specific hazard.[36] Examples of administrative controls designed to address awkward postures and loads involved in patient-handling tasks include use of policies, procedures, training, supervision, and communication,[37] as well as by scheduling and job rotation.[38] Despite the use of administrative controls, the hazard is still present and capable of inflicting injury,[37] and these interventions are rarely effective when used alone.

A common administrative control for safe patient handling would be distribution of workload away from traditional models that focus exclusively on patient acuity, toward models that also take into account patient dependency levels and/or exposure to high-risk patients. This can be accomplished in several ways, including changing (1) patient care assignments to even ergonomic exposure equitably among staff, (2) patient care delivery, by minimizing the number of times a very dependent patient needs to be transferred by scheduling multiple procedures during the same time period, or (3) reassessing whether to eliminate some traditional tasks, that are deemed high risk and not critical to quality of care (e.g., need for daily weights on all patients).

Another administrative control might include safe patient-handling policies, as well as policies and procedures for routine maintenance of patient-handling equipment and allocation of adequate/convenient storage space. Other administrative controls, which will be discussed in more depth elsewhere in this chapter, include algorithms to standardize decisions related to the type of equipment used and number of staff needed to perform tasks safely or use of lift teams as a care delivery option. Importantly, thorough staff training in use of patient-handling equipment, algorithms, identification of ergonomic risk factors, and other safe patient-handling techniques are administrative controls critical to reducing risk of injury in a direct patient care work environment.

Safe Patient-Handling Policies

A safe patient-handling policy is an evidence-based strategy used to reduce risk from patient handling and moving in health care. They are known by a variety of names, No Lift, Minimal Lift, or Zero Lift, but all emphasize the same critical element—avoiding manual patient handling and lifting except in emergency situations. Policies can be written at any level, including local facility, health care system (e.g., entire Veterans' Health Administration), state legislation (e.g., Texas, Washington, Rhode Island), or federal mandates (e.g., the United Kingdom, Australia, Canada, the Netherlands)

No-lift policies have been found to be an essential part of comprehensive programs (Nelson, 2003).[39] The American Nurses Association (ANA) issued a position statement that supports policies that eliminate the manual lifting of patients thus promoting a safe environment of care for nurses and patients.[40] In 2003, the Occupational Safety and Health Administration (OSHA) released ergonomics guidelines for nursing homes that serve as advisory recommendations rather than an enforceable standard.[41] Specific recommendations include

- Manual lifting of residents should be minimized in all cases and eliminated where feasible.
- If a manual lift is unavoidable, then the activity should be carefully assessed prior to completion to determine control measures for risk.
- Employers should implement an effective ergonomics process that provides management support, involves employees, identifies problems, implements solutions, addresses reports of injuries, provides training, and evaluates ergonomics efforts.

A facility-level policy template was developed by the VA from 2001 to 2002.[42] The template can be found in the VA/DoD *Patient Care Ergonomic Resource Guide* (http://www.visn8.med.va.gov/patientsafetycenter or Ref. [43]).

Clinical Decision-Making Tools

Historically, decisions related to patient handling were based on tradition and intuition, rather than evidence. This led to unnecessary variations in how tasks were performed. If a patient was dependent and could not assist moving from a bed to a chair, the caregiver would assess the situation and then make an equipment selection decision based on their nursing experience. Understandably, such decision making was not consistent, by the individual caregiver or within a group of caregivers. Often one caregiver might choose a certain type of equipment and another select a different type or none for the same patient and task. Not infrequently, the result of such decision making led to negative outcomes for the patient as well as the staff member. Fortunately, there is evidence that use of standardized patient-handling assessments and guidelines, along with effective communication systems to transfer this information, provide staff with equipment recommendations that facilitate evidence-based nursing practice (Nelson et al., 2004).[44,45] Such standardized tools include the "Assessment, Algorithms, and Care Plan for Safe Patient Handling" (Ref. [70], Chapters 5 and 6) and the Association of Operating

Room Nurses (AORN) guidance statement, Safe patient handling and movement in the perioperative setting[47] and will be mentioned below.

Nelson et al.[19] successfully implemented a Safe Patient-Handling Program for the VA that included the use of the "Assessment, Algorithms, and Care Plan for Safe Patient Handling", as an integral part of their comprehensive program. This program element was included in the OSHA's *Ergonomic Guidelines for Nursing Homes*[41] and has been updated.[42]

The intent of the assessment protocol, algorithms, and care plan was to provide a standardized method of giving direction to patient care providers on the most appropriate equipment to use and the number of staff needed for completion of high-risk patient-handling tasks based on cognitive, medical, and physical characteristics of each individual patient/resident. In using this program element, it was expected that the risk of injury for both staff and patients would decrease as well as the functional capabilities of patients improve as they would receive assistance appropriate for their functional level.

These tools have been shown to be an effective staff training tool, but unit implementation has sometimes been difficult. On the unit, use of the algorithms is sometimes considered time consuming and not practical, and the care plan difficult to update as recommended. However, when successfully implemented, there has been strong support from management, and the benefits have been great. Suggestions for facilitating successful implementation include thorough training on use and incorporating the algorithms into routine practice by prominently displaying them in locations easily accessible to staff.[48] But, no matter how the assessment and algorithms are utilized in a patient care setting, the information contained within is invaluable for safe patient handling, and staff should be trained on the algorithms initially and refreshed on an annual basis.

The latest trend is to customize the patient algorithms for various clinical care settings, including perioperative, rehabilitation, orthopedic, imaging, and many other clinical environments. Presently, the National American Orthopedic Nurses (NAON) Association is in the process of developing guidelines for orthopedic handling tasks, with expected publication late 2008. In 2007, the AORN released their guidance statement: Safe patient handling and movement in the perioperative setting.[47] It includes ergonomic tools to guide perioperative staff members in protecting themselves from ergonomic injury.

Use of Unit-Based Peer Leaders and Facility-Based Champions to Facilitate Program Implementation

Implementation of patient-handling programs in health care environments significantly changes the way patient care is performed. Caregivers are expected to give up manual patient-handling techniques they were taught in nursing school and have performed, often for years. Suddenly they are asked to take a little more time and insert a sling, or locate a piece of equipment, and even though they may have heard the science and rationale for moving patients with mechanical assistance, time pressures can jeopardize acceptance and adherence with new evidence-based practices. An effective strategy to foster acceptance and adherence to program elements includes the use of unit-based peer leaders,[42,49] and may contribute to sustaining the program over time.

Peer leaders are defined as credible informal leaders[50] able to foster positive transformations to improve circumstances for others and for their community.[51] Research has shown that such leader behavior can be shared by staff within a workplace.[52] Use of peer leaders showed higher correlations with positive performance than formal manager leadership[11] as well as an association with a high level of organizational performance.[52,53] Due to this finding, experts in management strategies support peer leaders as change agents that increase staff involvement in management issues.[54] Additionally, the OSHA *Ergonomic Guidelines for Nursing Homes*[41] notes that worker buy-in, and the consequential motivation for and acceptance of change, can be improved when employees are involved in the change process.[41]

Two evidence-based peer leader models, one in the United States and the other in the Netherlands, will be discussed. The Netherlands has a very strong peer leader program that is attributed for much of the success of their patient-handling initiatives. They are found throughout the Netherlands, with around 5000 registered ErgoCoaches. One or two are found on every unit of a hospital with this program. Their responsibility includes implementation and continuation of safe patient-handling programs, but more than that, they see themselves as the owner of their unit program and as the problem solver. They are nursing staff who have special training and this additional area of responsibility, so their role is considered collateral duty.[49]

EroCoaches are considered informal leaders and a critical function for them is to act as change agents and the facilitators of knowledge transfer. The Netherlands government established an ErgoCoach National Support Group that assists in sharing information between ErgoCoaches throughout the country. Through this venue, they learn about state-of-the-art technology and science and can communicate with their peers. As can be seen, ErgoCoaches are actively supported in their efforts at all levels.

Knibbe and Knibbe[49] conducted a cross-sectional pilot study to determine the effectiveness of ErgoCoaches and if they have a positive effect on the implementation of safe patient-handling guidelines. They found that the majority (81%) were specially trained (2.6 days on average on top of regular ergonomic training) and spent 2.1 h per week working on ErgoCoach activities: 76% felt supported by management; 89% regularly gave advice to colleagues and 47% were regularly asked for advice by colleagues; 88% considered their position to be crucial for long-term change and a significant 91% considered themselves change agents for their unit and their facility. The Knibbe's study found that facilities with ErgoCoaches were more compliant with safe patient-handling guidelines and were more successful in integrating their guidelines into patient care plans. They also found there was less sick leave taken and that equipment was better maintained.

As well, the VA's unit peer leader model focuses on the importance of knowledge transfer, or sharing information with and between coworkers, and in doing so, forging a direct connection between staff and safe patient-handling program goals.[19,55] By incorporating knowledge transfer strategies during the VA study, the peer leaders were successful in obtaining staff, management, and patient acceptance as well as maintaining the program over the time of the study.[19] The VA peer leaders shared information between themselves and management, staff, and other peer leaders. Management relied on these peer leaders to make them aware of the competence of their staff in use of new patient-handling techniques and in program acceptance. Weekly logs were used to relay the activity of each peer leader, the perceived acceptance of staff, management, and patients, and the perceived effectiveness of each program element.

The exchanges with coworkers were as simple as brief in-services and one-on-one interactions with coworkers. A critical knowledge exchange strategy used by the VA per leaders was the After Action Review (AAR) process (Matz, 2006a), sometimes known as a Safety Huddle. The benefit of the AAR was to give staff a venue for solving problems encountered during program implementation, as well as for discussing near miss and injury incidents and safety concerns. This 5–15 min facilitated and structured brainstorming session empowered staff by asking staff to raise and importantly, address unit and program issues. Peer leaders also incorporated social marketing strategies that allowed them to better promote and gain acceptance for the program.

Some peer leader programs were also responsible for staff training. This training incorporated the rationale for safe patient handling, their role as a patient-handling peer leader, information on the program, and equipment use. Usually, at the inception of a safe patient-handling program, soon after peer leader training, staff were given an awareness training to educate them on the problems inherent in lifting and moving patients and the technology available to reduce their risk. Staff training also might include information on unit injuries as well as plans for implementing a program and introduction of patient-handling equipment. Another tool used to educate staff is the use of a Perception of High-Risk Tasks. This can be found in the VA/DoD *Patient Care Ergonomic Resource Guide*, Chapter 3 (http://www.visn8.med.va.gov/patientsafetycenter or Ref. [56]). Although a survey, it brings to light the potential risk involved in each high-risk task staff are completing. Most importantly comprehensive and often repetitive staff training on equipment must be provided, often by the peer leader. And, this is just initial program training. Peer leaders must continue structured and unstructured training on the program elements, equipment, and other necessary information to not only the existing nurse population but to new staff too.

Regular communication between peer leaders was vital to successful programs. Face-to-face meetings were held on a biweekly basis within a facility, and conference calls were held on a monthly basis within a region in order to share best practices and challenges. Peer leaders were connected electronically as well.

Peer leaders need managerial support to be effective. Peer leaders must have time away from direct patient care to carry out their roles, and this translates into incorporating staffing strategies that allow for coverage. As well, evidence based on expert opinion shows that facility safe patient-handling coordinators with responsibility for "leading" the peer leaders is also a critical element in a successful program. In addition, these facility champions are responsible for coordinating equipment selection and purchase, tracking slings, equipment maintenance, tracking patient-handling injuries, and more. Essentially, they assume responsibility for the development, implementation, coordination, maintenance, and evaluation of the program, including integrated programs that cross service lines. As can be seen, unit peer leaders along with the facility coordinators are truly the forces of change for implementing safe work practices in direct patient care work environments.

Lift Teams

Lift teams have developed as a result of nursing staff not being able to perform transfers effectively. There are many research studies that have been conducted supporting the use of lifts teams.[57–65] The traditional definition of a lift team according to Meittunen[65] was two physically fit people, competent in lifting techniques, working together to accomplish high-risk client transfers. However, today, there are lift teams that consists of 10 or more members and there is more patient-handling equipment available for team members to safely perform transfers. The evidence shows that lift teams have benefits but also some disadvantages. Lift teams greatly assist where lifts are uncoordinated, personnel are unprotected, there are differences in caregiver heights, nurses may have a prior injury, fatigue, lack of knowledge of using mechanical devices or may be untrained lifters. All of these factors contribute to nursing back injuries and use of a lift team can be of great benefit. In a review study by Haiduven[64] use of lift teams were studied in nine programs and all reported a reduction in lost time work injuries or injuries related to lifting and transfer of client. There has been some recent work done on lift teams although not yet published.

Manon Short (RPT) injury prevention coordinator at Tampa General Hospital (TGH) has developed a successful lift team since 2002. TGH is a level 1 trauma center, 1000-bed facility. Since its inception, TGH has reduced employee patient-handling injuries by 62% through use of a lift team and has reduced cost by 96% from 1999 to 2006. In survey taken at TGH during 2007, 306 surveys were completed among staff from 27 departments. Results showed that 52% replied that the lift team response time was less than 15 min, 51% staff rated service as excellent, lift team coverage was best at days and evenings, and 71% staff stated that the lift team program was extremely important. Safer patient transfers were reported by 85% staff, followed by 79% of staff feeling less back discomfort, and 49% said there was increased patient satisfaction with transfers. Manon Short stated the following as key factors for a lift team.[66]

What makes a lift team program successful?

1. Adequate lifting equipment on each unit.
2. Supportive management/facility champion.
3. Ample coverage for number of beds: 1 lift team (2 people) for every 200 beds.
4. Acceptable response time (15 min).

5. Type of workers hired for lift team (mature, physically fit, able to communicate, educate staff on culture of safety).
6. Having the lift team involved in maintenance and inventory of lift equipment.
7. Support from frontline staff in working with lift team members.
8. Lift team members do not have other job duties. Primary job duty is to lift and transfer patients.

One important reason for a lift team member to be physically fit is that there is a manual component to the job. For example, lifting a leg of a bariatric patient to apply a sling involves a manual lift and the worker needs to be physically able to perform this task without risk of injury. With an increase in the bariatric population, use of lift teams will assist as these experts know the variety of technology available and how to use them quickly and efficiently to provide safe patient care.[67] Use of technology and proper training on equipment use has proven critical in the safe handling of patients and safety of all health care workers.

Conclusions

In summary, the science for safe patient handling has evolved. The challenge now is to apply this research to practice in support of safer working environments for patient care providers across settings of care. This means eliminating approaches where is strong evidence that they do not contribute to staff or patient safety, including manual patient handling, use of back belts, classes in body mechanics, or training programs that focus on techniques for manual patient handling. Rather, health care settings need to be more focused on evidence-based approaches, including use of patient-handling technologies, environment modifications, administrative controls (e.g., scheduling, assignments), no-lift policies, clinical decision-making tools (e.g., algorithms or patient care assessment protocols), staff training on safe use of patient-handling equipment, unit-based peer leaders, and/or lift teams.

Recommendations

While the past decade contributed science and new technologies toward the quest for safer working environments for patient care providers, much work is needed on how best to implement this evidence into practice. The purchase of equipment and designation of staff to perform as facility champion or unit-based peer leader can be costly. Early efforts to build a business case for patient care ergonomics have been successful[4,17,68] but more work is needed to move this important program to the forefront in health care, where there are many competing demands for time and resources.

Further, research is needed to address several high-risk tasks identified, for which there remains no evidence-based solutions that have been widely accepted. For example, more research is needed to address risks associated with standing for long periods in perioperative settings, or holding a limb for long periods while applying a cast in orthopedics. Of particular importance is finding safe, yet efficient technology solutions for repositioning a patient in bed (a high-risk and high-volume patient-handling task that covers all clinical areas).

In this era of an international nursing shortage, it is time for health care administrators, managers, frontline care providers, patient advocates, and policy makers to take action, finding creative ways to support evidence-based approaches across all clinical settings.

References

1. Roelofs PD, Bierma-Zeinstra SM, van Poppel MN, Jellema P, Willemsen SP, van Tulder MW et al. Lumbar supports to prevent recurrent low back pain among home care workers: A randomized trial. *Ann Int Med* 147(10):685–692, 2007.
2. Alexander A, Woolley SM, and Bisesi M. The effectiveness of back belts on occupational back injuries and worker perception. *Prof Safety* 10:22–26, 1995.

3. Wassell JT, Gardner LI, Landsittel DP, Johnston JJ, and Johnston JM. A prospective study of back belts for prevention of back pain and injury. *J Am Med Assoc* 284(21):2727–2732, 2000.
4. Nelson AL, Fragala G, and Menzel N. Myths and facts about back injuries in nursing. *Am J Nurs* 103:32–40, 2003.
5. Hemborg B and Moritz U. Intra-abdominal pressure and trunk muscle activity during lifting. *Scand J Rehabil Med* 17:5–13, 1985.
6. Marras WS, Lavender SA, Leurgans SE, Rajulu SL, Allread WG, Fathallah FA, and Ferguson SA. The role of the dynamic three-dimensional trunk motion in occupationally-related low back disorders. *Spine* 18(5):617–628, 1993.
7. Gilbertons LG, Krag MH, and Pope MH. Investigation of the effect of intra-abdominal pressure on the load bearing of the spine. *Trans Orthop Res Soc* 8:177, 1983.
8. Gracovetsky S, Farfan HF, and Lamy C. The mechanism of the lumbar spine. *Spine* 6:249–262, 1981.
9. Legg SJ. The effect of abdominal muscle fatigue and training on the intra-abdominal pressure developed during lifting. *Ergonomics* 24:191–195, 1981.
10. Marras WS, King AI, Joynt RL. Measurements of loads on the lumbar spine under isometric and isokinetic conditions. *Spine* 9(2):176–187, 1984.
11. McGill SM and Norman RW. Partitioning of the L4/L5 dynamic moment into disc, ligamentous and muscular components during lifting. *Spine* 11(7):666–678, 1986.
12. Pope MH, Andersson GB, Frymoyer JW, and Chaffin. *Occupational Low Back Pain: Assessment, Treatment and Prevention*. Mosby-Year Book, St. Louis, 1991.
13. Daynard D, Yassi A, Cooper JE, Tate R, Norman R, and Wells R. Biomechanical analysis of peak and cumulative spinal loads during patient handling activities: A sub-study of a randomized controlled trial to prevent lift and transfer injury health care workers. *Appl Ergon* 32:199–214, 2001.
14. Evanoff B, Wolf L, Aton E, Canos J, and Collins J. Reduction in injury rates in nursing personnel through introduction of mechanical lifts in the workplace. *Am J Indl Med* 44(4):451–457, 2003.
15. Garg A, Owen B, Beller D, and Banaag J. A biomechanical and ergonomic evaluation of patient transferring tasks: Wheelchair to shower chair and shower chair to wheelchair. *Ergonomics* 34:407–419, 1991.
16. Yassi A, Cooper JE, Tate RB, Gerlach S, Muir M, Trottier J, and Massey K. A randomized controlled trial to prevent patient lift and transfer injuries of healthcare workers. *Spine* 26:1739–1746, 2001.
17. Collins J, Nelson AL, and Fragala G. *Question/Answer Booklet for Nursing Home Administrators*. NIOSH, Washington, DC, 2004a.
18. Holliday PJ, Fernie GR, Plowman S. The impact of new lifting technology in long-term care. *AAOHN J* 42:582–589, 1994.
19. Nelson AL, Matz M, Chen F, Siddharthan K, Lloyd J, and Fragala G. Development and evaluation of a multifaceted ergonomics program to prevent injuries associated with patient Handling Tasks. *J Int Nurs Stud* 43:717–733, 2006.
20. Ronald LA, Yassi A, Spiegel J, Tate RB, Tait D, and Mozel MR. Effectiveness of installing overhead ceiling lifts. *AAOHN J* 50:120–126, 2002.
21. Villeneuve J. The ceiling lift: An efficient way to prevent injuries to nursing staff. *J Healthc Safety Compliance Infect Control* 2(1):19–23, 1998.
22. De Looze MP, Zinzen E, Caboor D, Heyblom P, Van Bree E, Van Roy P, Toussaint HM, and Clarijs JP. Effect of individually chosen bed-height adjustments on the low-back stress of nurses. *Scand J Work Environ Health* 20:427–434, 1994.
23. Knibbe JJ and Friele RD. Prevalence of back pain and characteristics of the physical workload of community nurses. *Ergonomics* 39:186–198, 1996.
24. Knibbe N and Knibbe JJ. Postural load of nurses during bathing and showering of patients. Internal Report, Locomotion Health Consultancy, the Netherlands, 1995.

25. Marras WS, Davis KG, Kirking BC, Bertsche PK. A comprehensive analysis of low-back disorder risk and spinal loading during the transferring and repositioning of patients using different techniques. *Ergonomics* 42(7):904–926, 1999.
26. Gagnon M, Akre F, Chehade A, Kemp F, and Lortie M. Mechanical work and energy transfers while turning patients in bed. *Ergonomics* 30:1515–1530, 1987a.
27. Gagnon M, Chehade A, Kemp F, and Lortie M. Lumbo-sacral loads and selected muscle activity while turning patients in bed. *Ergonomics* 30:1013–1032, 1987b.
28. Gagnon M, Roy D, Lortie M, and Roy R. Evolution of the execution parameters on a patient handling task. *La Travail Humain* 51:193–210, 1988.
29. Baptiste A, Boda SV, Nelson AL, Lloyd JD, and Lee WE III. Friction-reducing devices for lateral patient transfers: A clinical evaluation. *AAOHN* 54(4):173–180, 2006.
30. Bohannon R. Horizontal transfers between adjacent surfaces: Forces required using different methods. *Arch Phys Med Rehabil* 80:851–853, 1999.
31. Bohannon R and Grevelding P. Reduced push forces accompany device use during transfers of seated subjects. *J Rehabil Res Dev* 38:135–139, 2001.
32. Lloyd JD and Baptiste A. Friction-reducing devices for lateral patient transfers: A biomechanical evaluation. *AAOHN* 54(3):113–119, 2006.
33. Corlett EN, Lloyd PV, Tarling C, Troup JDG, and Wright B. *The Guide to Handling Patients*, 3rd edn. National Back Pain Association and the Royal College of Nursing, London, 1993.
34. Hignett S, Crumpton E, Ruszala S, Alexander P, Fray M, and Fletcher B. Evidence-based patient handling: Systematic review. *Nurs Stand* 17(33):33–36, 2003.
35. Owen BD. Decreasing the back injury problem in nursing personnel. *Surg Serv Manage* 5(7):15–21, 1999.
36. U.S. Department of Labor, Occupational Safety and Health Administration (OSHA). Informational Booklet on Industrial Hygiene, OSHA 3143 (revised), 1998. Retrieved February 7, 2008, from http://www.osha.gov/Publications/OSHA3143/OSHA3143.htm
37. Matz M. Understanding hazards and controls in health care. In: D Fell-Carlson (ed.), *Working Safely in Health Care: A Practical Guide*. Delmar Thomson Learning Publishing Company, New York, 2008.
38. Nelson A and Baptiste A. Evidence-based practices for safe patient handling and movement. *Nursing World/OJIN* 9(3):4, 2004.
39. Garg A. *Long-Term Effectiveness of Zero-Lift Program in Seven Nursing Homes and One Hospital.* NIOSH, Atlanta, GA, 1999.
40. American Nursing Association. Position statement on elimination of manual patient handling to prevent work-related musculoskeletal disorders, 2003. Access verified on March 12, 2004 at http://nursingworld.org/readroom/position/workplac/pathand.htm
41. U.S. Department of Labor, Occupational Safety and Health Administration. *Ergonomic Guidelines for Nursing Homes*, 2003. Retrieved February 7, 2008 from http://www.osha.gov/ergonomics/guidelines/nursinghome/final_nh_guidelines.html
42. Nelson AL (ed.). *Safe Patient Handling and Movement: A Practical Guide for Health Care Professionals.* Springer Publishing, New York, 2006.
43. Collins J. Safe lifting policies. In: AL Nelson (ed.), *Handle with Care: Safe Patient Handling and Movement*. Springer Publishing Company, New York, 2006.
44. Fragala G and Bailey LP. Addressing occupational strains and sprains: Musculoskeletal Injuries in Hospitals. *AAOHN* 51(6):252–259, 2003.
45. Health Care Occupational Health and Safety Association. *Transfer and Lifts for Caregivers.* HCOSHA Publications, Ontario, Canada, 1986.
46. Nelson A (ed.). *Patient Care Ergonomics Resource Guide: Safe Patient Handling and Movement.* Veterans Administration Patient Safety Center of Inquiry, Tampa, FL, 2001. Retrieved February 9, 2008 from http://www.patientsafetycenter.com/Safe%20Pt%20Handling%20Div.htm

47. AORN Workplace Safety Taskforce. *Safe Patient Handling and Movement in the Perioperative Setting.* Association of periOperative Registered Nurses (AORN), Denver, CO, 2007.
48. Matz M. Analysis of VA patient handling and movement injuries and preventive programs: Staff and management focus groups qualitative analyses report. Internal report to Chief Consultant, Occupational Health, Safety, and Prevention Strategic Healthcare Group, Veterans Health Administration, 2007.
49. Knibbe H. Ergonomic approach in the Netherlands: Experience. In *Presentation at the 5th Annual Safe Patient Handling and Movement Conference*, St. Pete Beach, FL, 2005.
50. House RJ. The social scientific study of leadership: Quo vadis? *J Manage* 23(3):409, 1997.
51. Astin HS and Astin AW. *A Social Change Model of Leadership Development Guidebook*, version 3. Higher Education Research Institute, University of California, Los Angeles, Los Angeles, CA, 1996.
52. Bowers DG and Seashore SE. Predicting organizational effectiveness with a four-factor theory of leadership. *Adm Sci Q* 11:238–263, 1966.
53. Eagleson G. *Notes for the Workshop on Leadership Development.* Unpublished manuscript, Australian Graduate School of Management, Sidney, Australia, 1996.
54. Hammer M and Champy J. *Reengineering the Corporation: A Manifesto for Business Revolution.* Harper Business, New York, 1993.
55. Matz M. Unit-based peer safety leaders to promote safe patient handling. In: AL Nelson (ed.), *Handle with Care: Safe Patient Handling and Movement.* Springer Publishing Company, New York, 2005.
56. Fragala G and Nelson A. Ergonomic workplace assessments of patient handling environments. In: AL Nelson (ed.), *Handle with Care: Safe Patient Handling and Movement.* Springer Publishing Company, New York, 2005.
57. Caska BA, Patnode RE, and Clickner D. Feasibility of nurse staffed lift team. *AAOHN J* 46(6):283–288, 1998.
58. Charney W. The lifting team method for reducing back injuries: A 10 hospital study. *AAOHN J* 45(6):300–304, 1997.
59. Charney W. The lifting team: Second year data reported (News). *AAOHN Journal* 40(10):503, 1992.
60. Charney W. Reducing back injury in nursing: A case study using mechanical equipment and a hospital transport team as a lift team. *J Healthc Safety Compliance Infect Control* 4(3):117–120, 2000.
61. Charney W, Zimmerman K, and Walara E. The lifting team: A design method to reduce lost time back injury in nursing. *AAOHN J* 39(5):231–234, 1991.
62. Davis A. Birth of a lift team: Experience and statistical analysis. *J Healthc Safety Compliance Infect Control* 5(1):15–18, 2001.
63. Donaldson AW. Lift team intervention: A six year picture. *J Healthc Safety Compliance Infect Control* 4(2):65–68, 2000.
64. Haiduven D. Lifting teams in health care facilities: A literature review. *AAOHN J* 51(5):210–218, 2003.
65. Meittunen EJ, Matzke K, McCormack H, and Sobczak SC. The effect of focusing ergonomic risk factors on a patient transfer team to reduce incidents among nurses associated with patient care. *J Healthc Safety Compliance Infect Control* 2(7):306–312, 1999.
66. Short M. Practical implementation of patient lift teams [plenary presentation]. In *Presented at the 7th Annual Safe Patient Handling and Movement Conference*, Disney's Contemporary Resort, Lake Buena Vista, FL. March 13, 2007.
67. Baptiste A, Leffard B, Rowen L, Ramos E, and Tyler R. Roundtable discussion on caregiver injury and safe patient handling. *Bariatr Nurs Surg Patient Care J* 2(1):7–16, 2007.
68. Collins JW, Wolf L, Bell J, and Evanoff B. An evaluation of a "best practices" musculoskeletal injury prevention program in nursing homes. *Inj Prev* 10:206–211, 2004b.

Supplemental Readings

Australian Nurses Foundation (Victoria Branch). *No Lifting Policy*. Australian Nurses Foundation, Melbourne, Australia, 1998.

Baptiste A. Safe client movement and handling. In: D Fell-Carlson (ed.), *Working Safely in Health Care: A Practical Guide*. Delmar Learning, Clifton Park, NY, 2008, pp. 94–145.

British Columbia Interior Health. *Musculoskeletal Injury Prevention (MSIP): A Practical Guide to Resident Handling*, 2004. Retrieved on February 8, 2008 from http://www.interiorhealth.ca/NR/rdonlyres/4AF034BF-78B6-48BF-A18D-2AA6B39F99FE/1911/Section2NoLiftPolicy.pdf

Centers for Disease Control and Prevention (CDC), Collins JW, Wolf L, Bell J, and Evanoff B. An evaluation of a best practices program in nursing homes (pg. 17). In *The State of the CDC, Fiscal Year 2003*. Author, Atlanta, GA, 2003. Retrieved February 13, 2008, from http://www.cdc.gov/about/stateofcdc/cdrom/SOCDC/SOCDC2003.pdf

Griffith R and Stevens M. Legal requirements for safe handling in community care. *Br J Community Nurs* 9(5):211–215, 2004.

Hayne C, personal correspondence, Cited in Collins M. *Occupational Back Pain in Nursing: Development, Implementation and Evaluation of a Comprehensive Prevention Program*. Worksafe Australia, National Occupational Health and Safety Commission, Australia, 1990.

Health and Safety Executive. *The Health and Safety (Miscellaneous Amendments) Regulations 2002*. The Stationery Office, London, 2002.

Hignett S. Work-related back pain in nurses. *J Adv Nurs* 23(6):1238–1246, 1996.

Kneafsey R. The effect of occupational socialization on nurses' patient handling practices. *J Clin Nurs* 9:585–593, 2000.

Manual Handling Operations Regulations (MHOR). *Guidance for Regulations*, L23 (2nd edition, 1998). HSE Books, London, 1992.

Matz M. After-action reviews. In: AL Nelson (ed.), *Handle with Care: Safe Patient Handling and Movement*. Springer Publishing Company, New York, 2005.

Manual Handling Operations Regulations (MHOR). The Stationery Office, London, 1992.

Monaghan H, Robinson L, and Steele Y. Implementing a no lift policy. *Nurs Stand* 12(50):35–37, 1998.

National Audit Office. A safer place to work: Improving the management of health and safety risks to staff in NHS trusts, 2003. Retrieved February 13, 2008, www.nao.gov.uk/publications/nao_reports/02-03/0203623.pdf

National Institute of Occupational Safety & Health (NIOSH). NIOSH facts: Back Belts, 2001. Retrieved December 10, 2007 from http://www.cdc.gov/niosh/backfs.html

Nelson AL, Owen B, Lloyd J, Fragala G, Matz M, Amato M, Bowers J, Moss-Cureton S, Ramsey G, and Lentz K. Safe patient handling & movement. *Am J Nurs* 103(3):32–43, 2003b. See also: patientsafetycenter.com

Owen BD, Keene K, and Olson S. An ergonomic approach to reducing back/shoulder stress in hospital nursing personnel: A five year follow Up. *Int J Nurs Stud* 39(3):295–302, 2002.

Owen BD, Keene K, Olson S, and Garg A. An ergonomic approach to reducing back stress while carrying out patient handling tasks with a hospitalized patient. In: M Hagberg, F Hofmann, U Stobel, and G Westlander (eds.), *Occupational Health for Health Care Workers*. ECOMED, Landsberg, Germany, 1995.

Owen B and Garg A. Patient handling tasks perceived to be most stressful by nursing assistants. In: A Mital (ed.), *Advances in Industrial Ergonomic and Safety*. Taylor & Francis, London, 1989, pp. 775–781.

Retsas A and Pinikahana J. Manual handling activities and injuries among nurses: An Australian hospital study. *J Adv Nurs* 31(4):875–883, 2000.

Royal College of Nursing. *Introducing a Safer Patient Handling Policy (Working Well Initiative) (Rep. No. RCN Code 000-603)*. Royal College of Nursing, London, 1996.

Wicker P. Manual handling in the perioperative environment. *Br J Perioper Nurs* 10(5):255–259, 2000.

An Ergonomic Approach to Reducing Back Stress in Nursing Personnel

Bernice D. Owen and Arun Garg

Back pain in health care personnel has been around for a long time! For at least the last 25 years, epidemiologic studies have documented a high prevalence of back pain in this work group.[21] Across occupations, the highest incidence and earliest appearance of work-related back problems have been found in heavy industry workers and nurses.[7,19] In fact, nursing personnel rank fifth nationally for filing workers' compensation claims; only heavy laborers such as miscellaneous workers, sanitation workers, warehouse workers, and mechanics surpass nursing.[17]

This problem may be even greater than published statistics indicate because through questionnaires, Owen[24] found that 38% of 503 nurses stated that they had episodes of occupationally related back problems but only one-third of those with back pain actually filed an incident report with their employer; this group also averaged 6.5 days of their own sick days for unreported back pain perceived to be occupationally related. The units where significantly more low-back pain (LBP) episodes occurred than were expected were intensive care, orthopedics, and rehabilitation. Twenty percent of the nurses who said they had back pain stated that they had made at least one transfer in order to decrease the amount of lifting/transferring of patients, e.g., they transferred to a different unit such as from surgical to obstetrics; changed employment settings such as from hospital to clinics or changed positions from staff nurses to educators. Another 12% indicated they were considering making a transfer and 12% stated they were thinking about leaving the profession of nursing due to occupationally related back pain. In England, Stubbs, Buckle, and Hudson[34] found 12% of all nurses intending to leave nursing permanently cited back pain as either a main or contributory factor.

Causes of Back Pain

Many authors believe the back pain problem is resultant from a combination of biomechanical and postural stressors. Variables such as the heavy load, the distance of the load (patient) from the lifter's center of gravity, the duration of the lift, awkward lifting positions, confined work space, unpredictable patient behaviors, and the amount of stooping and bending endured in the job, have an impact leading to excessive forces in the spinal area. Study into cause/effect relationships is important for this problem.

Lifting and transferring of patients have been perceived by nursing personnel to be the most frequent precipitating factors or causes of back problems.[15,16,24,32,36,40,41] For example, Owen[24] found that 89% of the back injury reports filed by hospital nursing personnel implicated a patient-handling task (PHT) as a precipitating factor. (A PHT was any task that involved lifting, moving, or handling the patient in some manner.) Stubbs et al.,[36] through questionnaires, reported 84% of the nurses perceived PHTs as important factors in their back pain. Jensen[16] (through workers' compensation records) found more than 73% of the back strain/sprain cases were reportedly triggered by these tasks.

Epidemiologic studies have reported that individuals who have more frequent back pain are more likely to report exposure to forward flexion, rotation, and lateral bending than those who have less back pain because it has been determined that these body movements produce large loads on the lumbar spine and can be harmful to the disk.[1,9]

While observing the body movements of nursing assistants (NAs), Nordin et al.[20] found the frequency and degree of trunk flexion to be high. A few studies have found high levels of biomechanical stress induced by patient lifting and transferring tasks.[10,12,37,38] In addition, high levels of postural stress (standing and stooping) are also cause for concern.[2]

Approaches to Prevention

Nursing personnel offered their ideas for prevention when asked through questionnaires and on incident reports.[22] Responses of 244 nurses indicated 89% stated good body mechanics during lifting and transferring of patients as the most important preventive measure. The second most frequent suggestion (n = 142, 58%) was that nurses be more willing to ask for help when they had judged they should not attempt the PHT alone. Only 29 nurses spoke of preventive approaches external to themselves, e.g., adequate staff, reduced work load, and increased patient participation with the task. Two nurses stated the side rails of the bed should have been lowered as well as the bed. The use of mechanical devices or "lifting aids" was mentioned but the only device specified was that of a draw sheet used for lifting or sliding the patient up in bed. Five nurses indicated efforts to prevent the LBP incident were futile because LBP is an "inevitable" part of nursing. Therefore, the vast majority of these nurses gave responses that focused on themselves such as body mechanics and their ability to ask for help.

Education and Training

Is training on body mechanics the answer? Most basic nursing textbooks have chapters pertaining to body mechanics. Unfortunately, some of the recommendations are confusing and even contradictory. For example, some authors suggest the nurse place at least 1 ft in the direction of the move when lifting or sliding a patient up in bed;[8,18] others say the nurse faces the side of the bed,[5,6] another gives no information about placement of feet.[30]

Snook et al.[31] studied the effectiveness of training as an approach to preventing occupationally related back injuries. They found training programs had no effect on reducing back injury rates; instead, those industries that had training programs had more back injuries than would be expected by chance. Some programs have experienced a decrease in back injuries after a training program, but with time the injury rates have returned to pretraining levels.[29]

Knowledge and application of body mechanics are important but not the full answer to prevention of back injuries in health care workers. Some research is even suggesting that the more body mechanics are taught, the higher the injury rate (H. Knibble, personal communication). Apparently when we focus so much on teaching nurses *how* to manually lift and transfer patients, a message is relayed that nurses can do anything with their bodies as long as they do it correctly. However, this message negates the impact of characteristics of the patient such as weight and combativeness, or elements of the environment such as confined work space and unevenness in lifting surfaces.

Pheasant (p. 295)[29] very aptly states "many people (both within the nursing profession and elsewhere) take the view that nurses have back problems because they are undertrained. The reality is that they are physically overloaded by their work activities. *In situations of this kind, training alone is necessary but not sufficient.* To make further progress, we need to identify the features of the working system which are responsible for the physical overload."

The Ergonomic Approach

Therefore, an approach to prevention must also incorporate job design, workplace design, and the impact that these factors have on patient care and the health and safety of health care workers. This ergonomic approach involves adjusting/changing the job to fit the capabilities and limitations of the worker rather than trying to change the worker in order to fit the job. The goals of this ergonomic approach are to identify those aspects of the job which are particularly hazardous and to redesign them so they are safer.[29] This may be done through such avenues as redesign of the task, the product, the work station, the environment, or the overall work organization.

Stubbs[33] suggests the ergonomic elements relate to the interaction between equipment, environment, task, and personnel. "Equipment" must be compatible with the strength and ability of nursing personnel. It must be compatible with other equipment such as the bathtub or bed and also with the environment such as with the floor/rug surfaces. Equipment should meet the needs of the user and a maintenance plan developed so it is available when needed. "Tasks" should not involve prolonged postural stress, there should be rest periods and high postural demands should be decreased through equipment that is comfortable for the nurse as well as the patient. Adequate staffing is also important to the task. The "environment" needs to be compatible with the equipment, the tasks, and the capabilities/needs of the nursing personnel. All of these elements should fit the capabilities of "personnel"; the nursing staff should not be expected to fit into a poorly designed system but should have available to them the support, equipment, etc. needed to do the job.

The tasks that nurses carry out and the work station where these tasks occur must be studied. Minimally, the manual handling actions that should be designed out of the work setting include lifting heavy loads, lifting away from the body, asymmetric lifting, and rotation of the torso. Lifting loads from below the knees or over the shoulder should also be avoided. In addition, there are factors about the load that contribute to the hazard of manual lifting; these include unpredictability and instability of the load, bulkiness of load and inability to grasp the load securely because of lack of handles.

Nursing personnel who have been involved in the lifting and transferring of patients certainly must realize that many of those tasks should be redesigned based on the above discussion.

Therefore, in order to decrease the back stress problem in nursing, nursing personnel must begin to look at their own capabilities, the tasks which they feel are stressful to the upper and lower back, the environment in which the tasks are carried out and equipment that may be helpful for themselves and the patients. This group of health professionals should play a vital role in delineating approaches to decreasing that stress. They must be encouraged to problem solve and work with management in striving for changes that could impact on this problem which is costly in relation to human suffering, staffing, and financial cost.

Example of Ergonomic Approach: Nursing Home

Through application of the ergonomic process the authors determined that biomechanical and perceived physical stresses could be reduced while lifting and transferring patients.[13] This process involved determining the most physically stressful patient care tasks, evaluating these tasks to determine the problem areas, testing out approaches to decrease the stress and evaluating the interventions to determine if stress was decreased. The following is a summary of the process these authors used which further resulted in a reduction of back injuries and lost work days due to back injuries.

This study is important to hospital personnel because: the lifting and transfer needs of hospital and nursing home residents are similar, and the ergonomic approaches used in the study can be used in the hospital setting.

Setting: Two floors of a county nursing home were selected for this study. There were 70 patients on each floor and most required help with many tasks.

Subjects: Thirty-eight of the 57 NAs employed at least part-time volunteered as subjects; 36 were females and two were males. They ranged in age from 19 to 61 years, with a mean of 32.8 years. Their average length of employment was 7.8 years (SD = 4.7 years). Seventy-five percent stated that within the last 3 years they suffered from back problems perceived to be related to work. However 60% stated they lost no work time within that period due to back problems, 15% missed 1 to 7 days and 25% lost 8 days or more.

Patients: The 140 patients ranged in age from 56 to 98 years with a mean of 84.7 years and about 64% were female. They had an average weight of 133 lb and height of 64 in. Most patients could not stand up or walk independently and their mental status indicated unpredictiveness for willingness to help with the task. In addition, the average ability for balance, body flexibility, and general physical ability was impaired. Patient characteristic data were collected at the beginning of the study and 6 months later to determine if the patient population had statistically changed; *t*-test findings indicated no significant differences between the first and second sets of data ($p > 0.05$).

Study design: The major steps (goals) in the study design were the following:

Goal 1

Determination of most stressful PHTs

Goal 2

Ergonomic evaluation of these tasks

Goal 3

Testing approaches to decrease stressfulness

(Locating assistive devices)

Goal 4

Laboratory study

Goal 5

Application of findings to clinical areas

Goal 1: Determining Stressful Tasks

The NAs listed the following 16 PHTs as most stressful in their patient care duties:[25]

Transferring patient from toilet to wheelchair (WC)
Transferring patient from WC to toilet
Transferring patient from WC to bed
Transferring patient from bed to WC
Transferring patient from bathtub to WC
Transferring patient from chairlift on bathtub to WC
Weighing patient (transferring patient from WC to scale chair)
Lifting patient up in bed
Repositioning patient within bed (e.g., side to side)
Repositioning patient in chair
Changing the absorbent pad worn by patient
Making bed with patient in it
Undressing patient
Tying "supports" to secure patient in WC
Feeding bed-ridden patient
Making bed when patient is not in it

The NAs then ranked these tasks according to stressfulness felt while performing the task. They also rated the amount of perceived physical exertion felt while carrying out each task; the Borg[4] scale was used for rating exertion to low back, upper back, shoulder, and whole body (this Likert-type scale progresses from 6 [very, very light] to 19 [very, very hard]).

The tasks ranked and rated as the most stressful were transfer on and off the toilet, in and out of bed, and transferring for the bathing and weighing processes (see Table 3.2). *Therefore, these tasks were selected for further study.*

TABLE 3.2 Ranking and Rating of Patient-Handling Tasks for Stressfulness

		Perceived Stress to Lower Back[b]	
Patient-Handling Task	Rank Order[a]	x	SD
Transferring patient from toilet to WC	1	14.3	2.7
Transferring patient from WC to toilet	2	14.1	2.8
Transferring patient from WC to bed	3	14.2	3.0
Transferring patient from bed to WC	4	14.1	2.9
Transferring patient from bathtub to WC	5	13.3	2.9
Transferring patient from chairlift to WC	6	13.4	3.2
Weighing patient	7	13.8	3.9

[a] Most stressful.

[b] Borg Scale; 6 = very, very light; 14 = very, very hard.

Summary

It was not surprising that these transfer tasks were selected as most stressful because the lifting postures for these tasks are biomechanically very stressful.

Also, these findings are in agreement with researchers who have found (through interview, questionnaire, and analysis of incident reports) that nursing personnel believe that much of their stress results from the lifting and transferring of patients.[15,22,32,41]

Goal 2: Ergonomic Evaluation of Stressful Tasks

Over a 6-month period the stressful tasks and the environment in which they were carried out were observed and videotaped.[14] Analysis of the data (including analysis through use of a static biomechanic model[11]) revealed the following findings:

Patient-transfer method. Patients who needed help with transfer were *manually lifted* 98% of the time. In executing this manual transfer, two nursing personnel stood facing the patient, each grasped the patient under the axilla with the upper arm, then lifted the patient up and carried him/her to a new location (Figure 3.7).

Assistive devices. There were gait belts and a mobile hydraulic hoist (Hoyer Lift) on each floor. The hoist was only used for two patients during these 6 months. A gait belt was observed being used with transfers of one patient. The reasons NAs gave for nonuse of assistive devices were devices were not available, they took too much time, they were unstable or tipped over, the patients did not like them, they were not safe, they bruised the frail skin of some patients, and some felt they did not have enough knowledge or skill to use them.

Six bathtubs had chairlifts (water pressure lifts shaped like a chair), which were used for the bathing process. The NAs transferred the patient from WC to chairlift and with water pressure guided the chairlift up over the edge of the bathtub and down into the water. To use the bathtubs without a chairlift, the NAs manually lifted the patient in and out of the water.

The patients were not showered in this nursing home because each shower had a drain guard that was 6 in. high which prevented a shower chair from being wheeled into the shower.

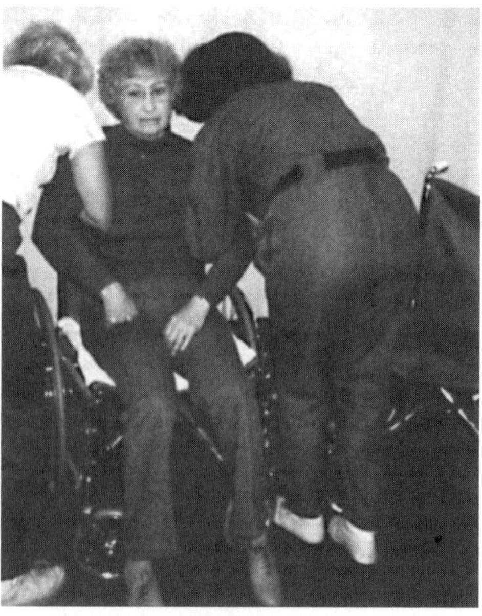

FIGURE 3.7 Manual transfer of patient using under-axilla technique.

Additional environmental factors were found to contribute to the difficulty of carrying out patient-lifting tasks. The floor of the bathrooms was 1/4 in. higher than the rest of the floor; this resulted in the NA having to push the WC or mechanical hoist several times against the elevation before rising over it. The beds were adjustable by manual crank only, so the NAs did not adjust the height of the bed to make it flush with the seat of the WC; the bed was 9 in. higher than the seat of the WC necessitating the NA to lift the patient up the 9 in. to get the patient into the bed. The toilet seat was 2 in. lower than the WC seat. The arm rests of the WCs were not adjustable; the NAs lifted the patients out of the WC rather than using a pull technique to transfer the patient. The toilets had side rails on them for safety; to transfer the patient, the NA lifted the patient onto the front of the toilet and then eased the patient to the back and down onto the toilet seat.

Frequency of patient lifting/transfer tasks. Each NA carried out an average of 24 most stressful lifting/transferring tasks per 8 h shift. Approximately half of these lifts were transferring patients on and off the toilet.

Postural stresses. AU the stressful tasks required multiple trunk flexions along with axial rotation of the spine and lateral bending. The mean trunk flexion for most tasks exceeded 30° and the average number of trunk flexions per occurrence of the task ranged from two to six (see Table 3.3).

Biomechanical stresses. The greatest amount of compressive force L_5S_1 and flexion moments occurred during the transfers of toilet to WC and WC to bed (see Table 3.4). However, the estimated compressive forces for all tasks exceeded the action limit of 3430 N recommended by U.S. Department of Health and Human Services.[39]

TABLE 3.3 Summary of Trunk Angles (in Degrees) per Task Obtained through Video Tapes

	Flexion (°)		Rotation (°)		Lateral Bending (°)	
Task	Mean	SD	Mean	SD	Mean	SD
Toilet to WC	51	14	33	8	25	8
WC to toilet	53	15	7	8	11	7
WC to bed	57	14	21	9	15	8
Bed to WC	44	13	18	8	17	7
WC to chairlift	48	16	15	9	13	8
Chairlift to WC	46	11	9	8	12	8
Weigh patients	10	3	6	3	6	2

TABLE 3.4 Summary of Biomechanical Analysis of Six Selected Patient-Handling Tasks for 50th Percentile Patient Weight

	Patient-Handling Task					
Variable	Toilet to WC	WC to Toilet	WC to Bed	Bed to WC	Chairlift to WC	WC to Chairlift
Applied hand force (N)	294	294	294	294	294	294
Compressive force (N)	4810	3680	4877	3991	4552	3680
Shear force (N)	788	886	805	801	792	699
Flexion moment (Nm)	202	153	217	161	196	139
Rotation moment (Nm)	85	61	519	82	64	92
Lateral bending moment (Nm)	49	65	46	54	49	45

Summary

This study showed that the NAs were subjected to high levels of biomechanical and postural stresses. Most of the transfers were done manually, few assistive devices were used and environmental factors influenced stressfulness. The high levels of biomechanical stresses found are in agreement with those reported by Stubbs et al.[35] based on intra-abdominal pressure, Gagnon et al.[10] based on compressive force and Torma-Krajewski[38] based on computations of action limits and maximum permissible limits recommended by the U.S. Department of Health and Human Services.[39]

Goal 3: Testing Approaches to Decrease Stressfulness (Locating Assistive Devices)

One ergonomic approach recommended for decreasing back stress is redesign of the task. This could be done through the use of assistive devices. Hence, the criteria for selection of assistive devices were established and devices located.[26]

Criteria for Selection of Devices

1. The device must be appropriate for the task to be accomplished. A device that could only be used for transfer of patient in a prone position could not be used for the stressful tasks of this study.
2. The device must be safe for both patient and nurse. It must be stable, strong enough to secure and hold the patient and permit the nurse to use safe biomechanics.
3. The device must be comfortable for the patient; this must also help to allay fears. It should not produce or intensify pain, bruising or tear the skin.
4. The device should be understood and used with relative ease. Bell[3] and Owen[23] found nursing personnel were reluctant to use assistive devices because they could not understand how to use them or lacked experience in their use.
5. The device must be efficient in the use of time. According to Bell[3] and Owen[23] the most frequent reason given for not using a device was the time needed for use.
6. Need for maintenance should be minimal. The above two authors found lack of proper functioning a major reason for nonuse.
7. The device must be maneuverable in a confined work space. Owen[23] and Valles-Pankratz[40] found space to be a problem.
8. The device should be versatile. It could be inferred from Bell's findings[3] that only a few assistive devices should be introduced at a time because the error rate and the need for time to execute the transfer increased when more than two devices were included in a teaching program.

Assistive devices. The following assistive devices were selected for further study: gait belt (Figure 3.8), MEDesign patient-handling sling (Figure 3.9), Posey walking belt with handles (Figure 3.10), Hoyer lift (Figure 3.11), Trans-Aid lift (Figure 3.12), and Ambulift C_3 hoist (Figure 3.13).

During the ergonomic study, it was found that several transfers could be eliminated for bathing if the patient was transferred from WC to a shower chair that could accommodate the patient for toileting and showering. Two shower/toileting chairs were selected for study. The heavy chair can be seen in Figure 3.14; the light weight chair with a plastic nonmovable seat and no foot rests is not pictured.

Summary

Criteria for selection of assistive devices were established. A number of devices such as slings, belts, and hoists were tested and the following were recommended for expanded study in the laboratory: gait belt, walking belt, MEDesign patient-handling sling, shower chair (Figure 3.14), Hoyer lift, Trans-Aid lift, and C_3 lift.

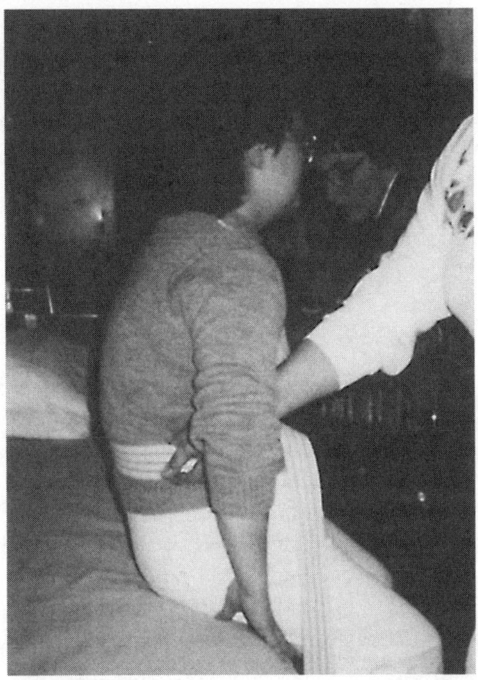

FIGURE 3.8 Gait belt is about 2 in. wide, of varying lengths with adjustable belt-like loop or buckle closure, has no handles and is made of cotton-canvas or nylon material. It should fit securely around patient's waist and is grasped with hand.

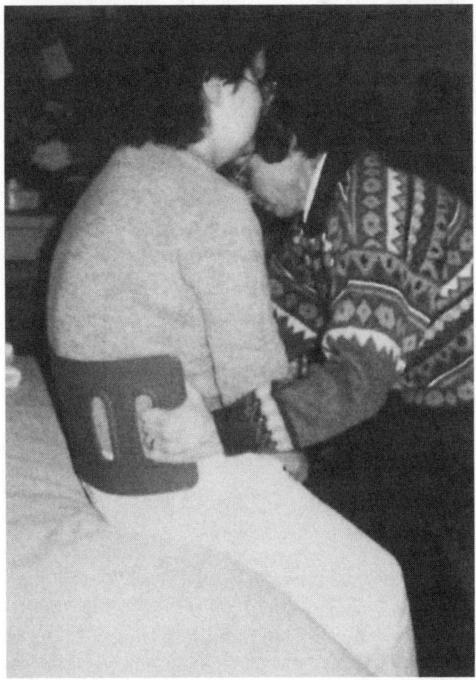

FIGURE 3.9 MEDesign patient-handling sling is 8 in. wide, 20 in. long, has a cut-out at each end allowing a hand grip, is made of flexible polymer material and is tucked securely around patient with bottom at buttock area.

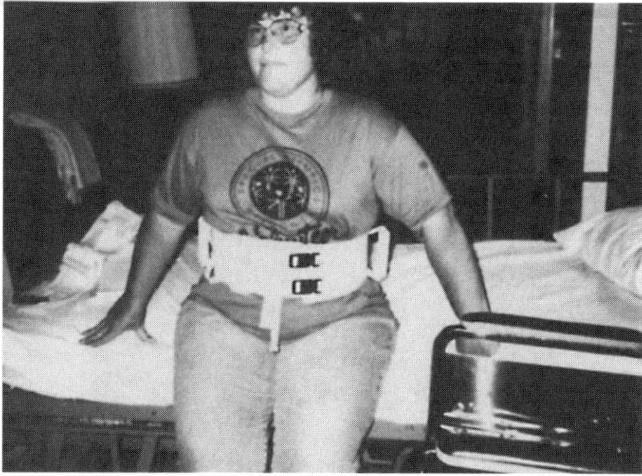

FIGURE 3.10 Posey walking belt is 5 in. wide, of varying lengths, has handles on each side, has velcro and two quick-release buckles for closure, is made of cotton-canvas-type material, fits snugly around lower abdomen and is grasped at handles.

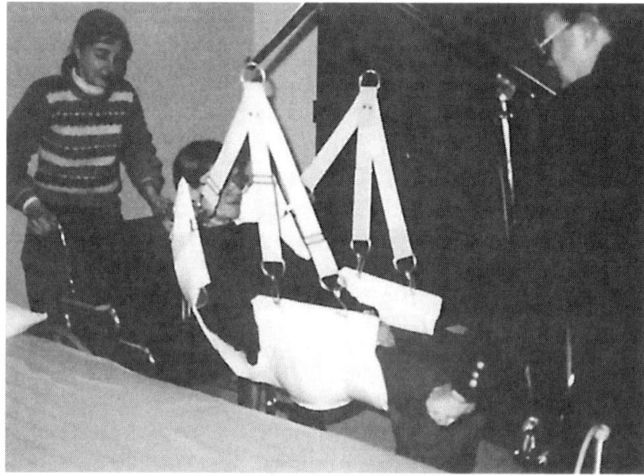

FIGURE 3.11 Hoyer lift is a hydraulic lift that has an adjustable base; a pump handle for raising and lowering the patient and a variety of slings that attach through hooks, chains, or web straps.

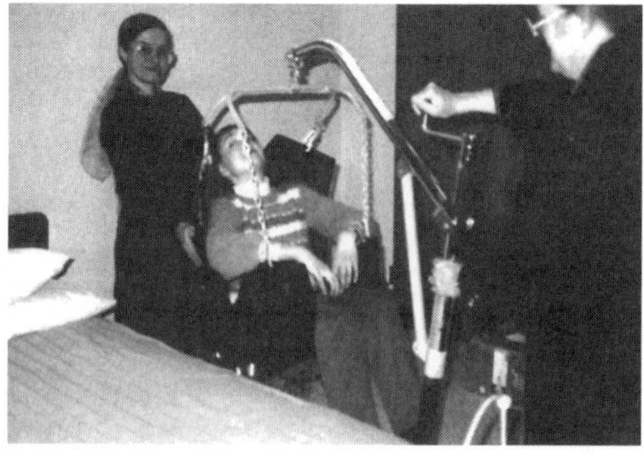

FIGURE 3.12 Trans-aid lift has a nonadjustable "C" base, a ball-bearing screw lifting mechanism with crank in horizontal plane and has a variety of slings that attach by hooks and dangling color-coded chains.

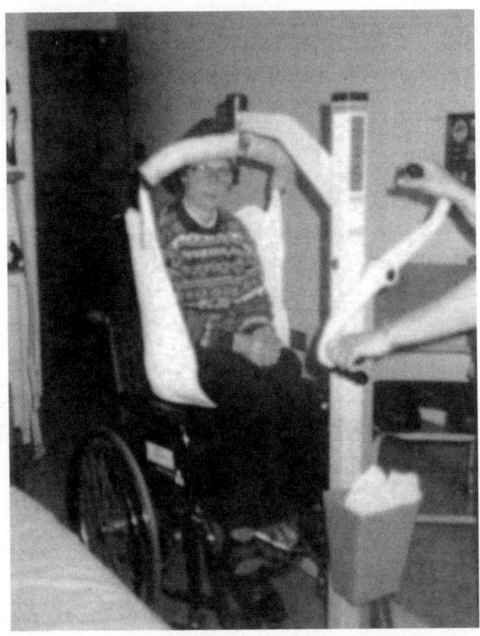

FIGURE 3.13 Ambulift C_3 has a semiadjustable base, a mechanical chain-winding mechanism for lifting/lowering with crank in vertical plane, and the sling attaches by loops and hooks.

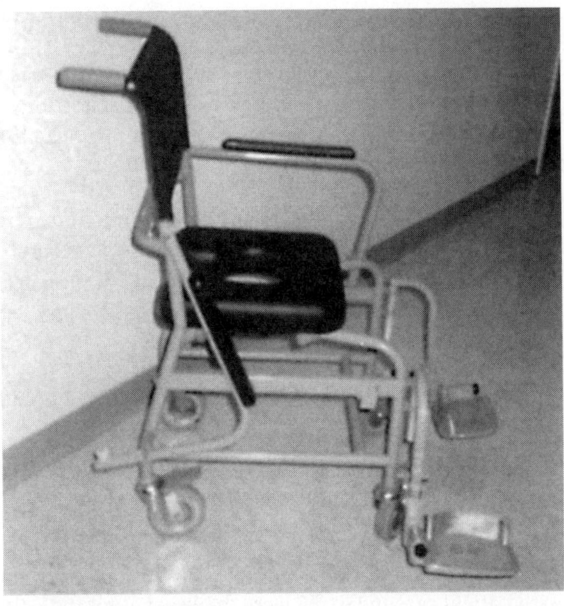

FIGURE 3.14 Shower/toileting chair is a heavy chair with padded removable seat and adjustable/removable foot rests and arm rests that can be lowered.

Goal 4: Laboratory Study

The purpose of this laboratory study was to evaluate eight different methods for carrying out the most stressful PHTs.[12,27] Methods found to reduce back stress would then be taken into the clinical area so nursing personnel could apply these to patient care (Goal 5).

Subjects. Six female senior nursing students were subjects and all participated both as nurse and "patient." They ranged in age from 21 to 23 years with an average weight of 139 lb and height of 65 in.. All stated they had no back problems. They were instructed not to support their own body weight while in a "patient" position.

Eight different methods. Five of the eight methods were manual transfers: the transfer method presently used in the nursing home (two NAs grasping the patient under the axilla area and lifting the patient to a new location); two NAs using a gait belt and with a gentle rocking movement pulling the patient to a new location; two NAs using the same rocking movement but pulling with the walking belt with handles; one NA using the walking belt with handles; one NA transferring via the MEDesign patient-handling sling.

Three of the methods involved transferring the "patient" via mechanical lifts; Hoyer (H), Trans-Aid (T), and Ambulift—C_3 (A).

Procedure. Each subject was studied using each of the eight methods while transferring a "patient" from WC to toilet, toilet to WC, bed to WC, WC to bed, shower chair (SC) to WC, and WC to SC.

The subjects were given time to learn the equipment and practice the methods. Immediately after each transfer, the subjects rated the physical stress felt for the shoulder, upper back, lower back, and whole body using a 10-point scale (0 = no stress, 9 = extreme stress). The "patients" rated their feelings of comfort and security using Likert scales of 0 = extremely comfortable and 7 = extremely uncomfortable; 0 = extremely secure and 7 = extremely insecure. After each transfer the subjects assumed their initial posture at the beginning of the transfer so body angles could be measured for biomechanical stresses and analyzed using the biomechanical model of Garg and Chaffin.[11] Tasks were also videotaped.

The following summarizes the combined data for all eight methods of transfer.[27]

Perceived physical stress. The traditional method of the manual axilla lift was the most stressful to all four body parts (Table 3.5). Following this, the gait belt was most stressful. Lift H was more stressful to all body parts than the manual techniques using the Posey Walking belt. The least stressful were the Posey walking belt and Lift A.

Biomechanical data. Compressive force at L_5S_1 was estimated to be about two times greater when transferring a patient using the traditional manual lifting method ($\bar{x}$= 4757 N), than when using any of the other methods for transfer (see Table 3.5). The least amount of compressive force was experienced while carrying out the two person transfers with gait belt ($\bar{x}$ = 2080 N) and with the walking belt ($\bar{x}$ = 2044 N).

"Patient" data. The traditional axilla lift and transfers using the gait belt and Lift H were uncomfortable and felt insecure (Table 3.6).

Summary

The present method of manually transferring the patient was the most stressful, least comfortable, most insecure, and provided the greatest amount of compressive force to L_5S_1. The methods taken into the clinical area to be used with patient care in Goal 5 were the Posey Walking Belt and Lift A.

TABLE 3.5 Summary of Stress Ratings and Biomechanical Data by Method of Transferring the "Patient"

	Method of Transfer															
	Manual Lifting (2)[b]		Gait Belt (2)		Walking Belt (2)		Walking Belt (1)		Patient Handling Sling (1)		Lift H (2)		Lift T (2)		Lift A (2)	
Stress[a] Body Part	$\bar{x}$	SD	$\bar{x}$	SD	$\bar{x}$	SD	$\bar{x}$	SD	$\bar{x}$	SD	$\bar{x}$	SD	$\bar{x}$	SD	$\bar{x}$	SD
Shoulder	6.5	1.2	5.1	1.3	3.4	1.7	3.2	0.9	4.1	1.5	5.2	1.8	4.1	1.5	2.5	1.1
Upper back	6.0	1.6	4.8	1.6	3.2	1.1	3.4	1.0	4.0	1.5	4.5	1.5	3.2	1.2	1.9	1.1
Lower back	6.0	1.5	4.9	1.8	3.1	1.4	3.3	1.6	4.4	1.9	3.7	1.9	2.8	1.7	1.3	1.0
Whole	6.2	2.3	4.9	1.4	3.1	1.0	3.3	1.1	4.1	1.5	3.5	1.6	3.1	1.6	1.2	0.8
Compressive force (N)	4757	244	2080	210	2044	176	2391	362	2446	430						

[a] Scale: 0 = no stress; 9 = extremely stressful.
[b] Indicates number of nurses making the transfer.

TABLE 3.6 Summary of "Patient" Data

	Method of Transfer															
	Manual Lifting (2)[a]		Gait Belt (2)		Walking Belt (2)		Walking Belt (1)		Patient Handling Sling (1)		Lift H (2)		Lift T (2)		Lift A (2)	
Variables	$\bar{x}$	SD	$\bar{x}$	SD	$\bar{x}$	SD	$\bar{x}$	SD	$\bar{x}$	SD	$\bar{x}$	SD	$\bar{x}$	SD	$\bar{x}$	SD
Comfort[b]	5.7	0.9	5.7	0.8	1.6	0.5	1.7	0.8	2.6	1.3	5.2	1.1	3.0	1.7	0.5	0.6
Security[c]	5.1	1.2	4.0	1.4	1.1	0.8	1.1	1.0	3.2	1.4	5.4	1.4	3.0	1.7	0.5	0.6

[a] Indicates number of nurses making the transfer.
[b] Scale: 0 = extremely comfortable; 7 = extremely uncomfortable.
[c] Scale: 0 = extremely secure; 7 = extremely insecure.

Goal 5: Application of Findings to Clinical Areas

The purpose of this part of the study was to take the positive findings from the laboratory and apply them to clinical patient care.[13] The focus was education of nursing personnel on use of the Posey walking belt and Ambulift C_3 when transferring patients in and out of bed, on and off the toilet and on and off the shower chairs. Postintervention data were collected for 8 months on floor 1 and 4 months on floor 2.

Subjects. All of the NAs were involved in this part of the study as were the 140 patients on floors 1 and 2 of the nursing home.

Procedure. The patients were categorized by amount of care needed. The environmental changes were made for worksite redesign, and all nursing staff on both floors received education and practice with transfer techniques. Two nurse observers were trained in the use of data collection instruments. The Borg[4] Rating of Perceived Exertion scale was used after completing a transfer. The two nurse observers randomly observed nursing personnel to collect the above data and to determine acceptability rates for the use of belts and Ambulifts. They also timed each PHT as it was being performed.

The Accident Investigative Reports and the OSHA 200 logs were reviewed to determine the number of back injuries that occurred on floors 1 and 2 for 4 years prior to intervention and 8 months postintervention.

Intervention. The 140 patients were grouped into three categories according to ability to assist with the transfer: independent ($n = 42$), dependent but weightbearing ($n = 49$) and dependent non-weightbearing ($n = 49$). The Ambulift C_3 was used with dependent nonweightbearing and heavy patients. The walking belt was recommended for dependent weightbearing patients weighing less than 150 lb. Each patient needing a walking belt was provided one at the bedside. An Ambulift C_3 was stationed on each wing (20 patients per wing).

Environmental changes were made such as toilet seat risings were placed on toilets so transfer surfaces were flush and showers were modified so shower chairs could be used.

All nursing personnel were trained in use of walking belt, Ambulift and shower chairs until they felt comfortable and demonstrated competency. The technique taught for using the walking belt was: two nursing personnel stood facing the patient, each had one hand gripping the belt handle; placement of feet was one foot facing the patient and other foot in the direction of the move; with flexed knees and backs straight, they used a synchronized, gentle rocking motion to create momentum and then pulled the patient toward themselves, shifted their weight to the foot facing the direction of the move, pivoted and transferred the patient. Instruction on use of the hoist followed the printed instructions provided by the manufacturer.

TABLE 3.7 Ratings of Perceived Exertion for Lower Back from Preintervention and Postintervention Phases for Selected Tasks

	Ratings of Perceived Exertion		
	Preintervention	Postintervention	
Task	Manual Lifting	Walking Belt	Ambulift
Bed to WC	14.1	8.6**	8.2**
WC to bed	14.2	9.3**	7.6**
WC toilet	14.1	9.5**	7.2**
Toilet to WC	14.3	8.8**	8.6**
Chairlift to WC	13.4	9.5**	6.0**
Weighing patients	13.8	—	6.3**

** Significant at $p \leq 0.01$.

The following summarizes the results of intervention.

Perceived exertion. In general, the average stressful ratings were about 14 (between "hard" and "somewhat hard") before intervention and 9 ("very light" for the walking belt) and 8 (between "very, very light" and "very light" for the Ambulift) after intervention (Table 3.7).

Acceptability rates. Most of the time the nursing personnel used the walking belt or hoist as directed according to the categorization of the patient as either dependent and weightbearing or nonweightbearing. Sometimes, however, the condition of the patient changed so the device used also changed. Ninety-five percent of the time the nursing personnel used either a hoist or walking belt when transferring a dependent, nonweightbearing patient from bed to WC or WC to bed. This was true 92% of the time for transferring patients on and off the toilet. It is probable that the hoist should have been used rather than the walking belt in some instances; however, the manual, under-the-axilla technique was not used!

Time. As expected it took less time to use the devices as personnel became more skilled in their use. During preintervention it took an average of 8–18 s to transfer a patient using the under-the-axilla manual method. It took an average of 25 s to transfer using the walking belt and 73 s for the hoist.

Injuries. There were 83 injuries per 200,000 work hours in the 4 years prior to the intervention stage of this study; these were back injuries that occurred on floors 1 and 2. During the 8 months postintervention there were 47 back injuries per 200,000 work hours. Most of the back injuries that occurred postintervention were during non-patient-transfer activities.

Summary

The results of this study indicate that an ergonomic intervention is appropriate for reducing physical and biomechanical stresses while transferring patients from one location to another. The most physically challenging tasks were redesigned or eliminated (e.g., through use of shower chairs for toileting and showering) and the acceptance rate of these new approaches was high.

Ergonomic Approach within a Hospital Setting

A similar study design has been implemented in the hospital setting. The PHTs ranked as most stressful by hospital personnel were very similar to those found in the nursing home study:

- Lift patient up from the floor
- Transfer patient from WC to bed

- Transfer patient from bed to cardiac chair
- Transfer patient from bed to commode
- Transfer patient from commode to bed
- Transfer patient from bed to WC
- Transfer patient from cardiac chair to bed
- Transfer patient from stretcher/cart to bed
- Transfer patient from bed to stretcher/cart

Criteria for the selection of *assistive devices* were applied and the following devices were studied:

Transfer on and off stretcher:

*Slipp Air Pal Dixie Smooth Mover

Transfer in and out of bed, on and off commode/toilet:

*MediMan Hoist *Sara Lift
*Posey Walking Belt Total Lift
Maxi Lift Dextra Lift
MediMaid

Lift up in bed:

Air Pal *Magic Sheet

Toileting in bed:

*Kimbro Pelvic Lift

Transfer from floor to WC or bed:

*Mediman Lift Maxi Lift

Those devices with an asterisk (*) were perceived to be least stressful, most comfortable, and secure, hand force was reduced and the estimated compressive force to L_5S_1 disk was lower than the NIOSH action limit;[39] these assistive devices were used within the intervention phase of the study (analysis of data is continuing).

Implementing an Ergonomic Program

A very important element in implementing any program is to have commitment from management and from the individuals who must carry out the program. Change from routine is difficult so definitive strategies must be planned to reinforce the driving forces and decrease the restraining forces.

Some driving forces may be the need to decrease the number of personnel with back problems in order to have a safer work environment and adequate staffing levels, a need to decrease workers' compensation insurance premiums, an opportunity to expend resources for primary prevention instead of at the tertiary or illness/injury level and a need to exhibit to employees personal concerns for their health and safety.

Some restraining forces may include time and financial resources. As shown in the nursing home study, the use of assistive devices does take more time and the equipment does demand financial investment.

A consultant with knowledge in ergonomics would be helpful in order to evaluate the interaction between equipment, environment, task, personnel, and overall work organization.

Someone on the staff should assume leadership within this back injury prevention program. Usually individuals in top-level management cannot take on the responsibilities because of ongoing commitments to other elements of the organization. The individual selected should be interested in the program, capable of providing leadership, have the respect of management and staff, be willing to test out various approaches to reducing back stresses, and have the ability to keep management informed of progress/problems in the program.

In some institutions it may be possible for a staff person from each unit to be assigned to this injury prevention team. Some advantages to this model are staff can relate to an individual who is already an integral part of the unit, her/his presence and work with this project will be a reminder of preventive efforts, this person can receive and funnel to the teams and the coordinator of the program ideas, concerns, etc. and one person can see that maintenance of equipment is up to date on that unit. This model of staff involvement incorporates important principles such as permitting the people who are at risk for the back problems to be involved in problem solving and even decision making in relation to this problem. This sharing in decision making also facilitates a feeling of ownership of and commitment to the program.

Tasks that are most stressful to carry out should be determined by those who normally carry out the work of the unit. The technique similar to the one described in the nursing home study can be used. Owen et al.[28] used four methods to determine stressfulness of tasks and found no statistically significant difference among them. The methods were asking the individuals to list stressful tasks and then to rank these tasks by stressfulness felt, asking them to rate the amount of stress felt to various parts of the body while carrying out those tasks cited as stressful (using Borg scale), estimating compressive force to L_5S_1 using the three-dimensional static biomechanical model of Garg and Chaffin, and estimating tensile force on the erector spinae muscles using a biomechanical model. Based on the results of the comparison of these four methods, it seems feasible that it is adequate to ask the individuals to rank and rate the stressfulness of those tasks perceived to be most stressful.

The task(s) found to be most stressful should be studied. Questions such as the following may help to crystallize the problem. How is the task performed? How much weight is lifted? Are good body mechanics being used? (Certain types of lifting carry a particularly high level of risk, e.g., lifting at a distance, lifting involving asymmetry and twisting.) Can manual lifting be eliminated? Are there environmental aspects that contribute to the stress such as unequal lifting surfaces or confined work space? Can a technique of pull or push be used instead of lift? Can the task be redesigned? Can it be eliminated? How can the workplace be changed?

Answers to some of these questions should be helpful in proposing alternative methods for carrying out the task. One alternative may be to redesign the task by introducing an assistive device for transfer of patients. It is helpful for staff who are involved in the transfer task to be involved in testing out various assistive devices and even to contribute to decision making for which devices to purchase. An adequate number of devices must be purchased so that they are readily available to staff.

Training in the alternative method is essential. Owen[23] and Bell[3] found personnel were reluctant to use assistive devices because they lacked knowledge and skill. Training to competency is important which includes demonstration and hands-on return demonstration in the classroom and at the bedside. A systematic schedule for retraining and also education of new staff should be instituted and maintained.

Management policies may be necessary to help assure/encourage staff to fulfill the goals of the project. It is important for personnel to see and experience management involvement and support throughout the program.

Maintenance of equipment is imperative also as this is often a reason for nonuse. Delegation of this responsibility to one person on the unit may be helpful.

A health management program including an element directed at back pain prevention should be built into every prevention program. This is often lacking in settings where health care professionals work such as hospitals and nursing homes. Personnel within the health management program should encourage the staff to participate in the health program, report symptoms early relating to back problems, conduct prompt follow-up, conduct worksite tours to study work practices and add their expertise to the ergonomic approach to prevention of occupationally related back problems.

Evaluation of outcomes of the project are important. Tools such as the Borg Rating of Perceived Exertion Scale or 7 to 10 point Likert scales are adequate to determine if there has been reduction in perceived physical stress. An on-going monitoring of the program is essential to assure staff are using the devices and using them correctly. A study of the incident reports and OSHA 200 logs also contribute to the evaluation process.

Summary

Many approaches to the prevention of occupationally related back pain have been attempted. However, nursing personnel continue to rank fifth nationally in workers' compensation claims filed for this problem. The ergonomic approach has been successful in decreasing the physical stress involved in the lifting and transferring patients.

References

1. Andersson GBJ. Epidemiological aspects on low-back pain in industry. *Spine* 6(1):53–60, 1981.
2. Baty D and Stubbs DA. Postural stress in geriatric nursing. *Int J Nurs Stud* 24:339–344, 1987.
3. Bell F. *Patient Lifting Devices in Hospitals*. Groom Helm, London, 1984.
4. Borg G. Physical performance and perceived exertion. Thesis, Gleerups, Lund, 1962.
5. Christensen B and Kockrow E. *Foundations of Nursing*. Mosby Year Book, St. Louis, MO, 1991.
6. Cole G. *Basic Nursing Skills and Concepts*. Mosby Year Book, St. Louis, MO, 1991.
7. Cust G, Pearson J, and Mair A. The prevalence of low back pain in nurses. *Int Nurs Rev* 19:169–178, 1972.
8. Earnest V. *Clinical Skills in Nursing Practice*, 2nd edn. J.B. Lippincott Co., Philadelphia, PA, 1993.
9. Frymoyer IW, Pope MH, and Costanza MC. Epidemiological studies of low-back pain. *Spine* 5: 419–423, 1980.
10. Gagnon M, Sicard C, and Sirois J. Evaluation of forces on the lumbo-sacral joint and assessment of work and energy transfers in nursing aides lifting patients. *Ergonomics* 29:407–421, 1986.
11. Garg A and Chaffin D. A biomechanical computerized simulation of human strength. *AIIE Trans* 7:1–15, 1975.
12. Garg A, Owen BD, Beller D, and Banaag J. A biomechanical and ergonomic evaluation of patient transferring tasks: Wheelchair to bed and bed to wheelchair. *Ergonomics* 34(3):289–312, 1991.
13. Garg A and Owen BD. Reducing back stress to nursing personnel: An ergonomic intervention in a nursing home. *Ergonomics* 35(11):1353–1375, 1992.
14. Garg A, Owen BD, and Carlson B. An ergonomic evaluation of nursing assistants' job in a nursing home. *Ergonomics* 35(9):979–995, 1992.
15. Harper P, Billet E, Gutowski M, Soo Hoo K, Lew M, and Roman A. Occupational low-back pain in hospital nurses. *J Occup Med* 27(7):518–524, 1985.
16. Jensen R. Events that trigger disabling back pain among nurses. In: *Proceedings of the 29th Annual Meeting of the Human Factor Society*. Human Factors Society, Santa Monica, CA, 1985.
17. Klein B, Jensen R, and Sanderson L. Assessment of workers' compensation claims for back strains/sprains. *J Occup Med* 26(6):443–448, 1984.
18. Kozier B, Erb G, Blais K, Johnson J, and Temple J. *Techniques in Clinical Nursing*, 4th edn. Addison-Wesley, Redwood City, CA, 1993, pp. 460–463.
19. Magora A. Investigation of the relation between low back pain and occupation. *Ind Med Surg* 39(11):31–37, 1970.
20. Nordin M, Ortengran R, and Andersson G. Measurements of trunk movements during work. *Spine* 5:465–469, 1984.

21. Owen BD and Damron C. Personal characteristics and back injury among hospital nursing personnel. *Res Nurs Health* 7:305–313, 1984.
22. Owen BD. The need for application of ergonomic principles in nursing. In: S Asfour (ed.), *Trends in Ergonomics/Human Factors IV*. Elsevier Science Publishers, North Holland, 1987, pp. 831–838.
23. Owen BD. Patient handling devices: An ergonomic approach to lifting patients. In: F Aghazadeh (ed.), *Trends in Ergonomics/Human Factors V*. Elsevier Science Publishers B.V., North-Holland, 1988.
24. Owen BD. The magnitude of low-back problems in nursing. *West J Nurs Res* 11(2):234–242, 1989.
25. Owen BD and Garg A. Patient handling tasks perceived to be most stressful by nursing assistants. In: A Mital (ed.), *Advances in Industrial Ergonomics and Safety I*. Taylor & Francis Ltd., Philadelphia, PA, 1989, pp. 775–781.
26. Owen BD and Garg A. Assistive devices for use with patient handling tasks. In: B Das (ed.), *Advances in Industrial Ergonomics and Safety II*. Taylor & Francis Ltd., London, 1990, pp. 585–592.
27. Owen BD and Garg A. Reducing risk for back pain in nursing personnel. *AAOHN J* 39(l):24–33, 1991.
28. Owen BD, Garg A, and Jensen RC. Four methods for identification of most back-stressing tasks performed by nursing assistants in nursing homes. *Int J Ind Ergon* 9:213–220, 1992.
29. Pheasant S. *Ergonomics, Work and Health*. Aspen Publishers, Inc., Gaithersburg, NO, 1991.
30. Smith S and Duell D. *Clinical Nursing Skills: Nursing Process Model Basic to Advanced Skills*, 3rd edn. Appleton & Lange, Norwalk, CT, 1992.
31. Snook S, Carnpanelli R, and Hart J. A study of three preventive approaches to low back injury. *J Occup Med* 20:478–481, 1978.
32. Stobbe T, Plummer R, Jensen R, and Attfleld M. Incidence of low back injuries among nursing personnel as a function of patient lifting frequency. *J Safety Res* 19(2):21–28, 1988.
33. Stubbs D. *Back Pain in Nurses: Summary and Recommendations*. University of Surrey, Guildford, U.K., 1986.
34. Stubbs D, Buckle P, Hudson M, Rivers P, and Baty D. Backing out: Nurse wastage associated with back pain. *Int J Nurs Stud* 23:325–336, 1986.
35. Stubbs D, Baty D, Buckle P, Fernandes H, Hudson M, Rivers P, Worringham C, and Barlow C. *Back Pain in Nurses: Summary and Recommendations*. University of Surrey, Guildford, U.K., 1986.
36. Stubbs D, Rivers P, Hudson M, and Worringham C. Back pain research. *Nursing Times* 77:857–858, 1981.
37. Stubbs D, Buckle P, Hudson M, Rivers P, and Worringham C. Back pain in the nursing profession. I. Epidemiology and pilot methodology. *Ergonomics* 26:755–765, 1983.
38. Torma-Krajewski J. *Occupational Hazards to Health Care Workers*. Northwest Center for Occupational Health and Safety, Seattle, WA, September 1986.
39. U.S. Department of Health and Human Services. *Work Practices Guide for Manual Lifting* (DHHS (NIOSH) Publication No. 81-122). Cincinnati, OH, 1981.
40. Valles-Pankratz S. What's in back of nursing home injuries. *Ohio Monit* 62(2):4–8, 1989.
41. Venning P, Walter S, and Stitt L. Personal and job-related factors as determinants of incidence of back injuries among nursing personnel. *J Occup Med* 28(10):820–825, 1987.

The Ceiling Lift: An Efficient Way of Preventing Injuries among Nursing Staff

Jocelyn Villeneuve

Introduction

The increase in the severity of patients' conditions, due to developments in medical technology and the aging of the population, has led directly to an increase in the number of employment accidents among

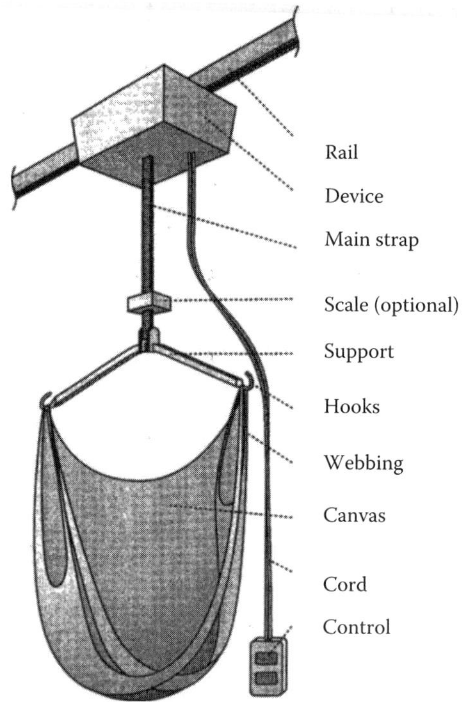

FIGURE 3.15 Ceiling lift.

nursing staff. State data on industrial injuries (Commission de la santé et de la sécurité de travail, Québec, 1992.) show that back injury is twice as prevalent in the health service sector as in any other industrial sector. Nearly half of all employment accidents in the health sector result in back injuries, accounting for 60% of lost work days. The situation is similar in the United States.[1,2]

Detailed analysis of the industrial injury data for 1991[3] has shown that such injuries are twice as common in institutions providing long-term physical care. Patients must be lifted more often, which requires a level of effort significantly in excess of the limits recommended by NIOSH.[4,5]

To be safe, all maneuvers in which patients are lifted must be performed by means of patient-lifting devices. Given the important role of such devices in the prevention of accidents during manipulation of patients, we decided to carry out a comparative study of the traditional free-standing mobile patient lift and the ceiling lift.[6,7] The latter has become increasingly popular in Quebec's health institutions since the beginning of the 1990s. Our study revealed that more than a hundred institutions had begun to use this type of lift to varying degrees (see Figures 3.15 and 3.16).

Methodology

The field study was carried out in five institutions, four of which were long-term care institutions and the other a rehabilitation center for young multiply handicapped people. In three of the five institutions both types of lifts were in constant use, and in all cases the staff had tried both types.

The goal of the study was to compare the overall performance of the ceiling lift and the traditional lift with respect to the following elements: level of satisfaction of direct users (nursing staff and patients) and management, impact on the staff posture and effort, and impact on operation time and workplace injury.

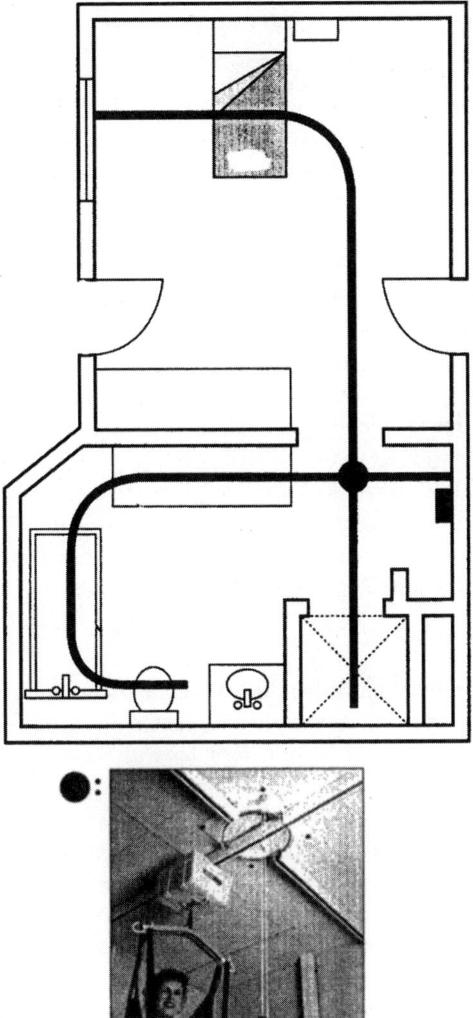

FIGURE 3.16 Example of rail running between the bedroom and bathroom with access to the bed, shower, toilet, and sink. The photo shows the overhead switching system at the junction of rails from different directions.

A variety of methods were used. A questionnaire was prepared on the basis of 21 performance criteria for patient lifts. It was distributed to 155 staff users in the five institutions. We received a total of 121 responses, for a response rate of 78.1%. Interviews were carried out with nursing staff (21), patients (13), and management (9 top managers and 10 middle managers). Observations and video recordings of various patient-transfer tasks were also made in the field.

In-depth analyses of state statistical reports from 1992 on employment injuries to nurses, nurses assistants, and attendants was used to estimate the cost of accidents caused by manipulation of patients in long-term care institutions. The study was completed by detailed examination of the employment injury registers of two of the long-term care institutions before and after introduction of the ceiling lifts.

Results

Questionnaires and Interviews with Users

When the "nursing staff" were asked which of the two types of lift they preferred, almost all (97.5% of respondents) selected the ceiling lift. The main reasons given were stability (100%), easy to use (99.1%), not bulky (97.5%), safe and comfortable for the patient (97.5%), safe and pleasant for the staff (97.4%), eliminates effort (96.5%), available (94.2%), and quick to use (95.3%). The preference was unequivocal, and the questionnaire responses were confirmed by the interviews.

The interviews with the "patients" produced similar results. Patients said they preferred the ceiling lift because it is comfortable and facilitates the work of the nursing staff. It made them feel less of a burden. Some patients were even able to operate the controls themselves, giving them more independence. In some cases, the ceiling lift enabled completely dependent patients to be moved from their beds.

We also observed the behavior of 10 other patients suffering from dementia, during successive transfers with the two devices. Three exhibited minor agitation during transfers with the traditional lift. No reaction was observed during transfers with the ceiling lift.

"Management" was equally satisfied with the choice of the ceiling lift, for similar reasons. They observed an overall reduction in the number of workplace injuries and an increase in satisfaction levels among the nursing staff and patients. A major advantage identified by the managers was that most transfers using the ceiling lift required only one staff member rather than two.

Posture and Effort

The most difficult task in manipulations using the traditional lift is handling the device itself. The most demanding operations are "pivoting" and "lateral movement of the device" when the patient is loaded. In both these operations, the operators exert quantifiable force with their arms, shoulders, and back, and the twisting movement required is exacerbated if the feet do not change position as the device is displaced.

Different movements with a traditional lift, when empty and when loaded with a 165 lb person, were measured using a dynamometer. Almost twice as much effort was required for the loaded movements (between 75 and 130 Nn). This level of effort, combined with postures in which the back is twisted, is likely to cause accidents. In addition, all the castors on the traditional lift tend to be of the pivoting kind, and the operator must exert force, through the back and arms, to steer the device in a particular direction. Also to be considered is the additional work involved in moving the lift to the patient's room or bathroom. The distances involved may be considerable if the lift is used with several different patients during a shift, as is generally the case in long-term care units. This step is completely eliminated by a fixed ceiling lift, and considerably reduced by a portable ceiling lift.

Statistical Summary of Two Institutions

Here are two examples of institutions that elected to introduce ceiling lifts throughout their long-term care units. Institution No. 1 covered all its requirements by installing 135 devices for 289 beds (most of the rooms held two beds). Institution No. 2 had already installed ceiling rails for the 200 long-term care beds (wing B) and, so far, had purchased 42 lifts.

One remarkable result obtained by both institutions is that no workplace injury related to patient transfers was recorded in rooms where ceiling lifts were available. Here, a patient transfer is defined as a transfer involving a change of surface, for example, from a bed to a chair, or vice-versa. It does not include maneuvers to reposition patients in the bed or chair, or to pick them up after falls.

Institution No. 1 achieved a perfect performance in 1996, with no employment accident related to patient-transfer maneuvers (Table 3.8). Also, all patient-transfer accidents since 1992 actually occurred in rooms with no ceiling lifts.

The figures provided by Institution No. 1 show that the $422,304 in workers compensation savings was almost sufficient to fund the entire cost of purchasing and installing the ceiling lifts ($450,000 for 105

TABLE 3.8 Accidents with Lost Time Involving Patient Transfers

Year	No. of Accidents	Days Lost	Workers' Comp. Contribution $ Can	No. of Beds	No. of Ceiling Lifts	Saving on Workers' Comp. ($ Can)
1988	24	278	22,032	237	—	—
1989	22	542	51,690	237	—	—
1990	18	342	36,704	237	—	—
1991	14	692	79,261	237	—	701
1992	4[a]	40	4,473	237	10	1,821
1993	10[a]	105	28,131	237	13	46,620
1994	6[a]	70	20,271	237	39	92,310
1995	2[a]	26	9,909	237	23	129,504
1996	0	0	0	289	50	151,348
Total					135	422,304

Note: Data is from Institution No. 1: CHSLD DRAPEAU DESCHAMBEAULT, contact was Ms. Francine Ouellette, Health and Safety Consultant. Data gathered from 1988 to 1996.

[a] No ceiling lift in room.

lifts with canvas and rails in 80% of the rooms and all the bathrooms in the main pavilion). Anticipated savings of $160,000 on workers compensation for 1997 should be sufficient to finance the installation of 56 rails and 30 additional devices in the new pavilion. Recurrent savings can thus be made in the operating budget for the next 10 years, since the devices have a life of approximately 15 years.

In 1991, Institution No. 2 introduced a prevention program for safe patient transfers, offered to all staff. The success of the program was reinforced by follow-up activities and corrective action. Table 3.9 shows a significant reduction in the days lost between 1992 and 1996. It shows all accidents related to patient moving transfers, including repositioning in the bed or chair and recovery after falls. However, no employment accidents occurred during patient transfers (change of surface) where ceiling lifts were available. Installation of the lifts began in 1995 and continued in 1996. In the next few years, enough lifts will be installed to cover all the institution's requirements.

The major savings, estimated at $638,250, on the cost of compensation for workplace injury, were sufficient to finance the cost of purchasing and installing the ceiling lifts, estimated at $247,000. (Balters S.

TABLE 3.9 Injuries with Loss of Time, Occurring While Helping Patients

Year	No. of Accidents	Days Lost	Estimated Workers' Comp. Contribution ($ Can)	No. of Beds (Wing B)	No. of Ceiling Lifts	Estimated Saving on Workers' Comp.[a] ($ Can)
1991	21	1571	589,125	200	—	—
1992	27	1039	389,625	200	—	199,500
1993	39	968	363,000	200	—	26,625
1994	18	354	132,750	200	—	230,250
1995	8	89	33,375	200	39	99,375
1996	5	45	16,875	200	3	82,500
Total					42	638,250

Note: Data is from Institution No. 2, CHRDL, contact person was Mr. Alain Saint-Pierre, Prevention Consultant, Health Service. Data was gathered from 1991 to 1996.

[a] Cost estimated at $125 per day, multiplied by a loading factor of 3.

Reclaiming lives. Team Rehab, November 1998.) The savings alone over the 2-year period (1995–1996) since the devices were installed amounted to $181,875.

Conclusion

The comparison of the overall performance of traditional patient lifts and ceiling lifts came down clearly in favor of the ceiling lifts, at the level of both user satisfaction and financial benefit. We recommend that the ceiling lift be given preference in long-term care institutions, because it provides a sure and effective way of preventing employment injury during manipulation of dependent patients. It would also be appropriate in acute care services such as orthopedics, neurology, radiology, and home-based care.

A new government policy now provides development project funding for the installation of ceiling rails in all the rooms of units where patient transfers are frequent. The devices and rails can be purchased out of the institution's own fixed and mobile equipment budget.

However, it is important to emphasize that ceiling lifts will not eliminate all accidents relating to patient assistance. An almost equally significant number of injuries occur during repositioning of patients in their bed or chair, and recovery of patients after falls, tasks for which the lift is unsuitable. Applications of the lift for repositioning patients could also be developed (sitting up in bed, changing sheets, arranging bedpans, etc.). It is to be hoped that technological developments will enable the device to be used to help patients walk and sit up in bed. The major investments required to install it will thus be all the more useful and profitable.

In addition to more suitable tools, the problem of accidents during patient manipulation requires a comprehensive approach involving training in safe handling methods, adequate working space, and an organization of work and time designed to avoid overwork.

Bibliography

1. Charney W, Zimmerman K, and Walara E. The lifting team. A design method to reduce lost time back injury in nursing. *AAOHN J* 39(5):231–234, 1991.
2. Charney W, Zimmerman K, and Walara E. A design method to reduce lost time back injury in nursing. In: *Essentials of Modern Hospital Safety*. Lewis Publishers, Ann Arbor, 1993, pp. 313–323.
3. Bigaouette M. Les lésions professionnelles dans le secteur de la santé et des services sociaux. *Objectif Prévention, ASSTSAS* 16(2):16–22, 1993.
4. NIOSH, Work practice guide for manual lifting, U.S. Department of Health and Human Services, Public Health Service, Cincinnati, OH, 1981, 183 p.
5. NIOSH, Documentation for the revised 1991 NIOSH lifting equation, National Technical Information Service—Scientific support documentation, U.S. Department of Commerce, Springfield, Document PB 91-226274, 1991.
6. Villeneuve J, Goumain P, and Elabidi D. Le lève-personne sur rail au plafond: Un outil de travail indispensable. *Objectif Prévention, ASSTSAS* 17(2):13–39, 1994.
7. Villeneuve J, Goumain P, and Elabidi D. Étude comparative de deux concepts de lève-personne pour les manutentions de bénéficiaires dans les soins de longue durée. *Actes du congrès de l'International Ergonomic Association (IEA)*, Toronto, 1994.
8. Bell F. *Patient-lifting Devices in Hospitals*. Croom Helt Ltd., London, 1984, p. 237.
9. Holliday P, Fernie G-R, and Plowman S. The impact of new lifting technology in long term care. *AAOHN J* 42(2):582–588, 1994.
10. Le Borgne D and Geoffrion L. *Maux de dos associés aux activités de déplacement des bénéficiaires dans un centre hospitalier à vocation gériatrique*. IRSST, Québec, 1989, 186 p.

Slings for Amputees

Justin Gilmore

Our facility currently is home to a double leg amputee resident. With much concern for the safety and also comfort of the resident, is there any other way to transfer a resident in this particular situation safely and comfortably other than using an amputee sling?

R.F. Jr., Country View Manor, Sibley, IA

Different suppliers have different design/safety/comfort objectives for their amputee slings. To give you just one example based on Liko's line of slings, the Model 70 Amputee Sling is designed specifically for patients with high amputations, thus a correctly fitted and applied Model 70 sling should be a perfectly safe and comfortable solution for the problem described, no matter what concerns the caregiver may have. If an alternative recommendation is required, a full-body sling such as the Liko Comfort Sling should be considered as a perfectly acceptable choice. This sling is somewhat similar to the sheet slings of the past and is normally applied in the bed by performing a bed roll to the patient, whereas the amputee sling can be applied in seated position.

The material choice of the sling is also affected by whether it is an amputee sling or a full-body sling. Amputee slings that are made of polyester or nylon are usually recommended when the sling is applied and removed in seated position. However, a full-body sling may require an airy or net-type material which usually allows for better airflow, or a cotton material for better comfort.

At the site of the bifurcation, your facility may also choose to provide a cushion on the seat of the chair or wheelchair for additional comfort and security.

General Sling Usage

My mother is in a special care home and I was wondering about the sling. Should it be left on at all times unless she is in her bed? Is this common practice?

C.H.
Concerned Sibling

Dear Cyril:

This is a fairly common question, and thanks for raising the issue with our general audience. Whether or not it is advisable to leave a sling under a patient for extended periods of time depends on a collective of patient-specific factors, which include the integrity of the patient's skin, type, and fabric of the sling, and application and removal.

Although we do not know each and every patient's skin integrity, we will always assume that we should be cautious in our approach. A patient's skin is important with the type of sling, a full-body or leg-support sling, because applying the full-body sling requires a bed roll whereas a leg-supported sling can be applied in seated position. The sling applied while seated takes less effort to apply and remove and requires less physical handling of the patient. When applying and removing a leg-supported sling, make sure that the material is smoothed out with no folds or wrinkles and that there are not any sling seams (outer sling edging) underneath the legs or against any bony prominence.

The sling fabric made of polyester/cotton mesh allows the skin to breathe and be more comfortable, whereas specific nylon or polyester material choices kept against a patient's body for a period of time could increase the ambient temperature near the patient's body.

While there is no commonly accepted definition of "extended periods," for convenience you can equate this time period to the amount of time when health care standards would normally move a patient to a different position—usually under 2 h. Please communicate this advice with your mother's special care provider to initiate further discussion regarding how best to help your mother with her care.

Removing Bariatric Patients from a Vehicle

I am looking for options for transferring bariatric patients out of a car when they arrive in emergency situations to the Emergency Room, particularly for those patients whose situation deteriorates during transport by a family member? Obviously vehicles vary greatly so I am interested in how others address this issue.

C.C.
Sequoyah Memorial Hospital Sallisaw, OK

In general, most mobile floor lifts are not able to be used for vehicle extractions due to the fact that their lift arms make contact with the vehicle's door opening or roof before the patient is able to be lifted off the seat. Liko's Golvo mobile lift, because it operates on a different principle (telescoping mast and retractable strap), would represent an acceptable alternative, but only for patients weighing up to 440 lb. Another alternative would be use of an overhead or ceiling-mounted system installed in an ambulance bay. A single-motor overhead system would be able to handle patients in the range up to 550 lb or more.

Safe Lifts Outside of an Emergency Department

We need ideas for safe lift assists outside our ED, especially out of family vehicles. A new ED is being planned and I would like to incorporate a safe lift system for this problem.

C.S
Maine Coast Memorial Hospital

In an existing facility you have the option of using a telescoping-mast mobile floor lift (see January '07 issue of the Lift Doctor) or installing an overhead or ceiling-mounted system. Your pending new construction project might offer an opportunity to "design in" an overhead system adjacent to the ED. In other words, your architect could be requested to design in an overhead system right in the ambulance bay. This overhead system should be able to assist in extracting patients weighing several hundred pounds. The patient could then be transferred to a wheelchair or gurney. Note that most mobile lifts are not able to be used for vehicle extractions.

Maneuvering Floor Lifts on Carpeted Surfaces

Our facility is carpeted and we find it difficult pushing hoists over this surface - do you have any solutions, maybe around wheels on hoists?

L.L., MercyAscot Hospitals, NZ

This is a wide ranging topic and I will attempt to provide as much information as possible in a limited amount of space. I am making an assumption that the wheels on your hoists are clean, move freely, and are the largest diameter available while still allowing access under beds and furniture. Further, even though this may not apply for your facility, readers need to remember that one of the major advantages of ceiling-mounted lifting systems is that they avoid complications arising from carpeting, furniture placement, medical equipment, and other complications.

Following are some general guidelines to employ when maneuvering a floor lift over a carpeted surface:

Pushing is most often easier on your back than pulling. When pushing you can lock your arms, maximize use of your legs, and put all of your weight into the lift.

When you push a patient in a lift, try to use your leg muscles more than your back or your arm muscles. Try the following technique to use your legs more: "lock" your arms in place by holding them into the starting push position. Try to keep your arms close to your trunk. Use your legs and shift your weight on them. Weight shifting requires that you have a wide base of support; feet should be at least a shoulder's width apart. Generally, forward to back weight shifting is preferable with your legs in a diagonal stance and weight shifted from front leg to back leg (think of it as a "lunge" type movement).

Changing direction. In order to change direction when stopped, lock one wheel and walk in a circular direction in order to reorient the lift rather than trying to push the lift sideways. Sideways pushing is a dangerous maneuver involving twisting of the back and high pressure on the lower spine, especially if the feet remain in a fixed position. And, remember, the most stressful time is when starting the lift in motion, so always be cautious when starting out. Once the lift is in motion, the effort decreases. Try to avoid twisting of the spine. Twisting puts a torque-type force on your low back. To avoid twisting point both your feet and arms in the same direction as the object you are trying to move. Instead of twisting your back, move your feet by taking small steps in the direction in which you are trying to move.

One final observation. The use of carpets in care facilities is diminishing over time due to modern facility hygiene standards and the difficulty in cleaning carpets versus cleaning hard surfaces. Carpets are also less desirable from an ergonomic point of view due to the difficulty in moving any equipment with wheels such as beds, patient lifts, and wheelchairs. Given a choice, the Lift Doctor would recommend against use of carpets in health care facilities for the above reasons.

Use of Ceiling Lifts with Hip Surgery Patients

Our Hospital has been working hard to decrease over exertion injuries through our Safe Lift Program. Orthopedics which is one of the heaviest in-patient units continues to have a high number of staff over exertion injuries. In an effort to reduce these injuries, we have recently installed overhead ceiling lifts but they are not being used. One of the barriers appears to be with the Orthopedic Surgeons who feel it is not safe to use lifts with patients who are recovering from hip surgery as hip alignment can be compromised. We have provided hammock slings which are designed to be used for these patients. Do you have any research or experience using lifts with this type of patient that may be helpful or are they correct in determining this can be unsafe? Do you have any suggestion for other assistive devices to help us reduce our injuries?

A.G., North York General Hospital

Obviously there are many different types of hip surgery. Hip replacement is one of them, and in the following comments we refer to the most common type of hip replacement. It is important to be informed regarding the type of hip surgery in order to choose the proper equipment to be used if the patient needs to be lifted.

One of the most common problems that might arise soon after hip replacement surgery is hip dislocation. Because the artificial ball and socket are smaller than the normal ones, the ball can become dislodged from the socket if the hip is placed in certain positions. The most dangerous position usually is pulling the knees up to the chest and letting the leg cross the midline of the patient's body. Therefore it is important that the sling you select provides a hip angle with no more than 90° flexion, for example a Comf0l1 or a Highback sling (Liko products). Note, however, that the Comfort sling does press the knees together when lifting, but the legs are not crossed over the midline. In order to separate the knees, a wedge pillow is probably advisable. Or, if there is a need to separate the knees even more, a sling with separate leg supports such as the Highback Original is a good choice.

Another possibility is to decrease the hip angle even more by using an Amputee sling in size medium together with side bars.

One final recommendation that often works: We recommend you invite the surgeon to attend a situation where the caregivers perform a lift using the proper sling. You may even wish to invite the surgeon to be lifted himself in order to demonstrate the safety and convenience of this solution which is used extensively throughout the world.

Sling Washing Instructions

> Washing of slings is a real issue for our LTACH. Washing Instructions state—wash in hot water with soap. I do not believe this will kill C. Dif and other pathogens. What do you suggest?
>
> **T.H., Transition Health Services**

Specific to C. difficile the literature states the following:

Clostridium difficile, C. difficile [klo-STRID-ee-um dif-uh-SEEL] is a bacterium. The bacteria are found in the feces. People can become infected if they touch items or surfaces that are contaminated with feces and then touch their mouth or mucous membranes. Health care workers can spread the bacteria to other patients or contaminate surfaces through hand contact. For safety precautions the following may be done to reduce the chance of spread to others: Wash hands with soap and water, especially after using the restroom and before eating. If your institution experiences an outbreak, consider using only soap and water for hand hygiene when caring for patients with *C. difficile*-associated disease as alcohol-based hand rubs may not be as effective against spore-forming bacteria. Clean surfaces in bathrooms, kitchens, and other areas on a regular basis with household detergent/disinfectants. Dedicate equipment whenever possible.

Hospital cleaning products can be used for routine cleaning. Hypochlorite-based disinfectants (e.g., household chlorine bleach) have been used with some success for environmental surface disinfection in those patient-care areas where surveillance and epidemiology indicate ongoing transmission of *C. difficile*. Note: EPA-registered hospital disinfectants are recommended for general use whenever possible in patient-care areas. At present there are no EPA-registered products with specific claims for inactivating *C. difficile* spores, but there are a number of registered products that contain hypochlorite. In the United States, liquid chemical germicides (disinfectants) are regulated by EPA and FDA. In health care settings, EPA regulates disinfectants that are used on environmental surfaces (housekeeping and clinical contact surfaces), and FDA regulates liquid chemical sterilants/high-level disinfectants (e.g., glutaraldehyde, hydrogen peroxide, and peracetic acid) used on critical and semicritical patient-care devices.

Laundering information from the CDC—June 2003 is as follows:

IV. Laundry process

- A. If hot-water laundry cycles arc used, wash with detergent in water > 160°F (>71°C) for >25 min (1270). Category IC (AIA: 7.3I.E3).
- B. No recommendation is offered regarding a hot-water temperature setting and cycle duration for items laundered in residence-style health care facilities. Unresolved issue.
- C. Follow fabric-care instructions and special laundering requirements for items used in the facility (364). Category II.
- D. Choose chemicals suitable for low-temperature washing at proper use concentration if low-temperature «160°F [<70°C]) laundry cycles are used (365–370). Category II.
- E. Package, transport, and store clean textiles and fabrics by methods that will ensure their cleanliness and protect them from dust and soil during interfacility loading, transport, and unloading (270). Category II.

Therefore, it appears that soap and water along with household and hospital detergents/disinfectants and cleaning products can be effectively used for routine cleaning for *C. difficile*. Chlorine bleach based disinfectants are recommended in patient-care areas where surveillance and epidemiology indicate ongoing transmission of *C. difficile*. Specific to slings, most manufacturers recommend washing with hospital cleaning products used for routine cleaning/disinfecting with the exception of chlorine bleach based.

As is indicated in italics above, it is recommended to dedicate equipment whenever possible. Reusable fabric slings are designed for single patient use, with cleaning/disinfecting between patients. The sling is meant to be kept with the specific patient using it, for the duration of that patient's hospital stay. It is sent to the laundry after the patient is discharged, or if it becomes grossly contaminated during the patient's stay. However, if laundering slings is an issue as you have stated, perhaps single patient use disposable slings would be an appropriate solution for your facility. Once the patient is discharged, or if the sling becomes grossly contaminated during the patient's stay, it is disposed of and a new disposable sling is obtained to fulfill the next need.

Once again, thank you for your question. Please do not hesitate to contact us with any questions or concerns regarding this information.

Mixing and Matching Equipment

I am inquiring about whether it is possible (or advisable) to use one of your Liko repositioning sheets with one of your competitors' ceiling lifts. I have two specific requirements: (1) The sheet must be less than 71 inches long; and (2) We would like to leave the sheet under the patient for a period of time.

Anonymous

1. Virtually all manufacturers discourage mixing their slings with lift equipment from other manufacturers. That's because, currently there is no third-party testing and certification agency with a valid and reliable protocol that has tested combinations of lifts and slings from different manufacturers. Further, there are numerous anecdotal reports of patients being injured when mix-and-match lifts/slings have been used, and in virtually all of those instances the facilities and patients have suffered due to lack of accountability between manufacturers.
2. That said, Liko does carry a short repositioning and transfer sheet (62 in. long) that can be left under the patient. It is made of a cotton/polyester blend; however, for safety, it is intended only for use with Liko lifting equipment and accessories.

Operating Room Transfers

What is the best way to transfer a patient from operating table to bed if the patient is anesthetized?

D.W., Huron Medical Center Bad Axe, Michigan

Transferring an anesthetized patient requires special care, and transfer of a patient following surgery requires yet another dimension in care and special preparation. The surgical patient's situation may become more complex as a result of the presence of IV tubing, intubation tubing, and possibly wound-related precautionary measures. Assuming you have taken those precautions into consideration and protected the patient against risk of additional injury, in general you will want to select a full-body sling or lift sheet that provides maximum support to the patient's entire body including the head. Log roll the patient into position to enable placement of the sling or transfer sheet per the manufacturer's instructions. When the lift begins, perform a "pre-lift" test by partially lifting the patient. Once the straps are taut, but while the patient's weight is still on the surface of the bed, cheek all connections as well as placement of the sling at all points of the patient's body. Complete the lift and move the patient into position over the bed as gently as possible. Lower the patient into position on the bed and log roll to allow removal of the sling.

One Person Transfers When Using Ceiling Lifts

> We have recently installed about 150 Liko ceiling lifts with plans to add a similar amount this year. Would you please give me your opinion on whether we can revise our lifts and transfers policy to allow one person transfers when using a ceiling lift? With all other lifting equipment (floor lifts, sit stand) we require two persons to be present.

While policy recommendations may differ from one manufacturer to another, Liko does recommend allowing single-caregiver assistance when using ceiling lifts assuming the caregiver has conducted an assessment and is fully confident that the lift-and-transfer operation can be conducted safely for both the patient and the caregiver! It is important to note that *not all situations are safe for single-person assistance.** For example, following are situations that might require two parsons to assist:

If medical equipment is attached to or accompanies the patient, two caregivers should be present during the transfer.

If the transfer is being made to a wheelchair or other unstable equipment, a second caregiver should be present to stabilize and properly position the chair.

Bariatric patients require a two person team.

Safe Lifting Policy Statements

> Do you have any written policies 1 can access? We are in the process of developing safe lifting policies for caregivers. A "no manual lift," or "minimum lift" policy would be helpful.

S.N., Kaiser LAMC

Thanks very much for raising an important issue that we believe is confronting health care institutions across the country. While policy statements are often created from scratch to meet the unique needs of an organization or culture, we believe there is a benefit to sharing those statements and policies across the entire health care industry. Sharing leads to consensus, and consensus will lead to universal adoption of safe lifting standards. If you need guidance immediately, here are a couple of options: (1) go to www.safeliftingportal.com, select the Safe Lifting Environment section, and click on "support" in the top navbar, then Policy Statement, (2) you might wish to review documents that are available online such as the "draft safe patient handling and movement policy."

Patient Assessment Tool

> Could you recommend an easy-to-use patient assessment tool or checklist to assist with determining a patient transferability?

C.B. Company: Beacon

* Please note that the patient's weight is not the only factor to consider when determining how many caregivers are needed. Even if the patient is not particularly heavy, he/she can be very difficult to turn to the side, or to keep on the side, when placing the sling into position. For example, the patient may be stiff, spastic, violent due to dementia, etc. Thus a risk assessment should always be performed and a decision made regarding whether the task requires one, two, or more caregivers, no matter whether the lift is a ceiling lift or a mobile lift. The objective of safe lifting should be to save caregivers, not to save on caregivers.

Making an assessment for using both proper lifts and slings is an important part of Safe Lifting. The patient needs to be assessed for their dependent to independent status, and then other factors that may impact the transfer. You must consider the task being performed to determine the correct lift equipment, such as: sit-to-sit transfer, lateral transfer, bathing and/or toileting. Here is a list of considerations when choosing a sling: need for head Support, back support, positioning, patient ability/disability, task to perform and patient size/weight. Factors that will influence THC choice of sling are design and construction, styles, materials, forces, infection control, and sling bar attachment. You will need to remember to take in to account the patient capabilities, sling application, hip angles, and whether the patient has any medical conditions that may influence THC transfer, such as amputees or chronic disease.

Each lift manufacturer should have their own safe lifting assessment tools, some of which may be specific to their equipment. We suggest your users contact their principal manufacturer to find out whether they provide tools and checklists for patient assessments.

Transport vs. Transfer

It is appropriate to use a lift to transfer a resident from the bed and from there to another room (i.e., bathroom). I have been told lifts are only to be used to transfer from a bed to a chair but not between rooms.

A.T.
Bethany Care Society

The issue of "Transport vs. Transfer" is a common concern. Patient safety and maintaining patient dignity arc our first concerns. The ISO 10535 Hoists for the Transfer of Disabled Persons standard does not prohibit the use of mobile lifts for transportation. Liko recommends that each facility take in to consideration their own circumstances to evaluate the appropriateness of mobile lifts and transport situations. Some of the important factors to bear in mind include: floor surfaces, thresholds, uneven surfaces, ramps, and privacy concerns for the patient. Transferring a patient from their bed or chair in to the bathroom is a common transport task. As I mentioned above, the floor surface needs to be taken in to consideration. However, eliminating a manual transfer task provides safety for both the patient and the caregiver. The use of a mobile lift as an "off unit" transfer device should be discouraged or at least substantially limited.

Leaving Lateral Transfer Aids under Patients during X-Ray

Are there any lateral transfer devices like the Slipp that can be left under a patient on an x-ray table.

M.l., Froedtert Hospital

One of your first considerations should be whether the x-ray is being used to examine the skeletal structure or whether soft tissue is the object for examination. The answer may affect whether the sling or transfer sheet can be left under the patient. Here's why.

Slings and lateral transferring devices can be radiolucent to varying degrees, depending on the specific material of construction used. The appearance of the fabric on a radiograph may be immaterial if skeletal structures are being examined (a common example would be shooting through a cast to observe a fracture site). However, when soft tissue is the object for evaluation, such as a chest x-ray, the appearance of fabric or a fabric weave pattern may have a negative effect (no pun intended). Of importance, the lower the density of the fabric, the less visible it is on the film. For example, cotton is less dense than polyester and would thus be less visible on the x-ray. You will also find that net polyester and plastic coated net will create a mesh pattern at certain energy and exposures and not others.

To summarize, if your sling or transfer sheet is cotton or a cotton/poly blend, you can probably leave it in place and not cause a problem, even when taking soft tissue x-rays. For example, it is not uncommon to shoot through an ER stretcher back support or several layers of cotton/poly bed linen. On the other hand, if your sling or sheet is a high-density poly material, it should be removed unless you are taking a skeletal series. To avoid the possibility of negative effects, you should always attempt to use a transfer or repositioning sheet made of a lower density fabric such.

Proning

We are struggling with safe "proning" in the OR. Any suggestions?

P.H.
Kaiser Permanente

"Proning," (i.e., prone repositioning) is a relatively new form of treatment used with ARDS patients who require high concentrations of inspired oxygen. Prone repositioning often improves oxygenation in patients who have ARDS by shifting blood flow to regions of the lung that are less severely injured and thus better aerated. Under close supervision of an attending physician, specially trained staff turn the patient face down from a supine position and may then be required to alternate between prone and supine repositioning as often as three times a day, or until the requirement for a high concentration of inspired oxygen is resolved.

Even with proper equipment, physically turning a critically ill patient has significant risks. For example, the patient may be connected to equipment with wires and tubes, or there may be a variety of clinical reasons to avoid the procedure. Patients whose heads cannot be supported in a face-down position, or very large patients, may not be recommended for the proning technique. It appears to the Lift Doctor that use of mechanical lifting equipment would be of limited value in most proning situations. However, it is technically feasible to use an overhead lift in combination with a repositioning sheet to transfer a patient from a supine to a prone position when repositioning from a gurney to a surgical table (or vice versa). Ideally there would have to be pillows placed over the edge between the two surfaces to allow the patient to be rolled over. Obviously you should practice this technique in advance of its use in an actual surgical environment. While this is a procedure I have seen work with a ceiling lift, it would likely be difficult to accomplish using a floor lift.

Establishing the Standard of Care™ with the Martin Chair-A-Table®

Willis Martin

Every once in a long while, a disruptive technology comes along that changes the rules. Martin Innovations, with their patented latching and patient transfer mechanism, has now introduced such a game changing technology.

In the last decade, the health care industry has recognized the impact of lifting injuries on medical professionals and has begun implementing policies and purchasing equipment designed to increase safety. Unfortunately, in the process of reducing risk, the costs have gone up. Safely transferring a patient from wheelchair to examination table, bed, or operating table either requires more people, more time, or both. In most cases, it also means a reduction in patient dignity as they move through the air suspended under a lift or strapped into an assistive device.

In a small town in North Carolina, there was a doctor who cared enough about both the patients and the staff to do something about it. He developed the concept of a system incorporating a latching wheelchair that would affect a patient transfer directly from the chair to the table or bed with a single medical professional and no external support equipment. After unsuccessfully trying to convince the larger players in the industry to bring such a device to market, he made the decision to retire early from practice and build a company that would focus on developing, building, and marketing a family of assistive devices.

Martin Innovations has now introduced a medical examination table that couples with a wheelchair; the Martin Chair-A-Table. As the only examination table in the world that couples with a wheelchair, it has garnered national and international attention and is the preferred power table of patients and practitioners who have used the product. The Chair-A-Table consists of two major parts; a wheelchair and an examination table capable of lifting 1000 lb. When operated together as a complete patient transfer system they offer man y advantages over existing equipment. Patients will no longer need to be lifted, a benefit clearly understood by the readers of this book. However, it can also be completely operated by a single staff member very quickly and efficiently, with a total patient transfer taking just a couple of minutes. Finally, because of the lack of rigging or suspending, the patient maintains their dignity and is not afraid of the experience. Practices can now examine more patients, cost effectively, while providing a higher degree of care without dangerous patient transfers or time consuming lift devices.

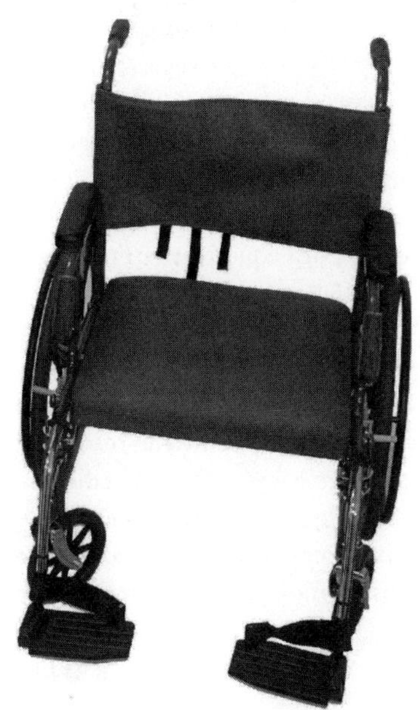

The patented wheelchair looks and functions the same as any other heavy duty wheelchair but has two special features. First, the seat has a fixture on the bottom that latches to the base of the table providing for safe and secure coupling. The wheelchair seat becomes the seat of the examination table without the patient ever having to lift their weight off the surface. Second, the wheels, sides, back, and foot rests of the wheelchair remove as a complete set with the press of a button, allowing for full access to the patient.

The patented table looks and operates like any other high capacity (bariatric) power examination table. It has powered articulated back and foot segments and raises and lowers the patient. The seat segment of the table can be removed to reveal a latch mechanism that attaches to the seat of the wheelchair and the exam table goes low enough (less than 16 in.) to couple with the underside seat height of the wheelchair.

The process of transferring a patient from the chair to the table and back to the chair is very simple, executable by one staff member, and requires absolutely "Zero Lift."® The value of the wheelchair and exam table becomes obvious when used together for automated patient transferring. The process is simple and requires virtually no training. Together these steps take a single person about one to two minutes, versus the 15–40 min it takes to use a lift.

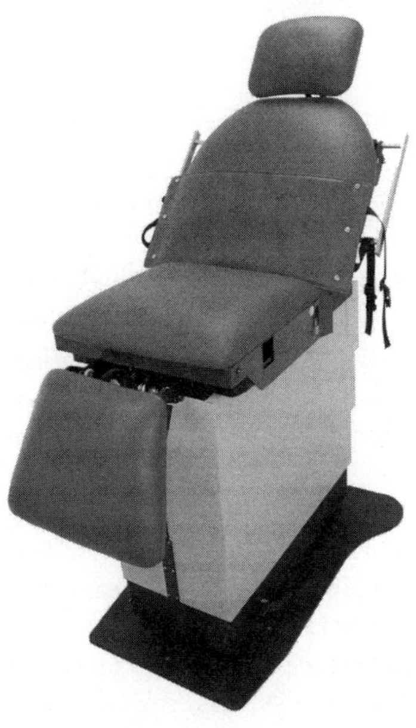

Step 1 The table functions as a normal power table with a very low minimum height when in use by ambulatory patients.

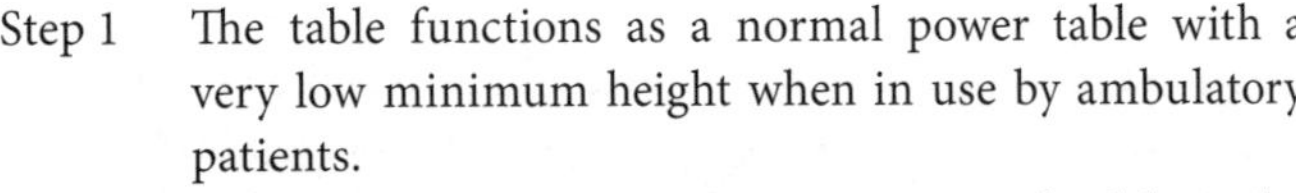

Step 2 When put into use to serve a patient in a wheelchair, the seat segment cushion is removed.

Step 3 The wheelchair, with the patient secured in it with the restraint strap, is wheeled backward so that the seat of the wheelchair takes the place of the removed seat cushion.

Step 4 The table and chair are raised with the patient aboard and the wheelchair sides are removed along with the wheels and armrests. The side rails on the examination table are then raised.

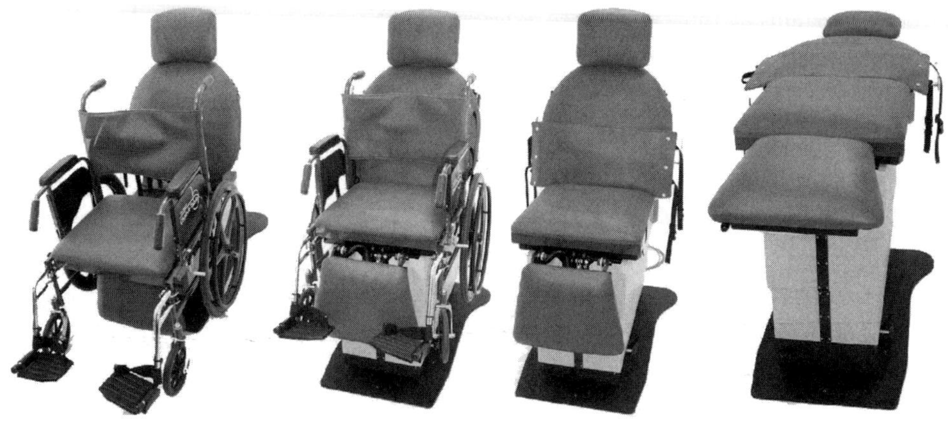

Step 5 The patient is reclined and/or feet raised as required by the doctor for examination.

Step 6 Upon completion of the examination, the process is reversed and the table is returned to its standard configuration for everyday use, eliminating the need for extra equipment in the exam room.

Benefits to Institutions

Since the equipment is dual use, for both ambulatory patients and users of wheelchairs, the return on investment of a Martin Chair-A-Table is quite short. Furthermore, the efficiency of the procedure of transfer means a significant cost reduction over traditional methods, allowing patients with handicaps to become profitable. By eliminating patient lifting altogether, workman's compensation claims will be dramatically reduced. Institutions will also decrease the amount of lost work days while providing a very useful recruiting tool for new staff and caregivers, as well as helping to retain those workers. In a recent study, 65% of nurses said that safer working conditions would solve the nursing shortage (Luke Snell Perry & Associates).

Benefits to Risk Managers and Insurers

The Martin Chair-A-Table dramatically decreases the number of injuries for staff and patients surrounding patient lifting and transportation. Having Chair-A-Tables will eliminate situations where injuries are common. Purchasing the equipment provides a solid, court defensible example of the institutions concern for staff and patient safety and establishes a higher Standard of Care. In 2006, there were over $2 billion in claims related to injuries suffered while lifting patients (U.S. Bureau of Labor & Statistics). The Chair-A-Table mitigates this threat and provides ADA compliance.

Benefits to Caregivers

The Martin Chair-A-Table creates a safer working environment for nurses and caregivers. Sixty percent (60%) of nurses cited a disabling back injury as their top health/safety concern, a higher rate than HIV/AIDS (ANA Health & Safety Survey, 2001). It allows caregivers to keep doing what they love and removes the risk of back injury. As the readers of this book have seen, the statistics vary, but somewhere between 12% and 18% of nurses leave their profession permanently each year due to back injury (Moses and Owen). The Chair-A-Table restores the feeling of being fully capable to perform the job every day and allows nurses with 25 lb lift limits to again contribute to patient care. By eliminating daily pressures, such as lifting 1.8 tons per shift (Tuohy-Main, 1997), caregivers will have a better quality of life without work related injury.

Martin Innovations is committed to continuing to bring an entire family of products to market in order to support those less fortunate and the medical professionals that try to make their lives better. We are constantly seeking ideas from the medical community on needs and product requirements. If you have any questions about our existing products or ideas for new ones, please visit www.MartinInnovations.com or email info@MartinInnovations.com

4

Emerging Infectious Diseases

William Charney
Healthcare Safety Consulting

J. H. Lange
Enviro-Safe Training and Consulting

G. Mastrangelo
University of Padua

Jacques Lavoie
Yves Cloutier
Jaime Lara
Geneviève Marchand
Institute of Research for Health and Safety

Jeanne Anderson
Andrew Geeslin
Andrew Streifel
Minnesota Department of Health

Emerging Infectious Diseases and Occupational and Public Health Unreadiness in the United States and Canada **4**-1
Introduction • Occupational Health Paradigms • Public Health Paradigms • National Agencies Response Paradigms • Effect of Globalization and Global Warming • Class
Are Hospitals or Public Health Systems Ready? A Conclusion **4**-8
References **4**-8
Personal Protective Equipment for Health Care Workers **4**-9
Summary • Introduction • Application and History of Personal Protective Equipment and Practices • Types of Common PPE • History • Protection against Emerging Infectious Diseases • Personal Protective Equipment • Respirators
Conclusion **4**-39
References **4**-39
Appendix 4.A.1 Guide on Respiratory Protection against Bioaerosols: Recommendations on Its Selection and Use **4**-47
Appendix 4.A.2 Airborne Infectious Disease Management: Methods for Temporary Negative Pressure Isolation **4**-85

Emerging Infectious Diseases and Occupational and Public Health Unreadiness in the United States and Canada

William Charney

Introduction

IF DISEASE IS AN EXPRESSION OF INDIVIDUAL LIFE UNDER UNFAVORABLE CONDITIONS, THEN EPIDEMICS MUST BE INDICATIVE OF MASS DISTURBANCES OF MASS LIFE

RUDOLF VERCHOW

Conditions ripe for flu disaster, *Seattle Times*, February 6, 2005.

Canada stockpiles drugs to combat global flu pandemic, *Vancouver Sun*, February 4, 2005.

Fatal plague outbreak feared in Congo, *Seattle Times*, February 19, 2005.

Stalking a Deadly Virus, Battling a Town's Fears, *New York Times*, April 17, 2005.

Bird flu virus mutation could spread worldwide, *Seattle Post Intelligencer*, February 22, 2005.

Lack of health insurance in the US will kill more people than Katrina, P. Krugman, (OP-Ed) *New York Times*, September 18, 2005.

Bird flu threat: Think globally, prepare locally, *Seattle Times*, April 15, 2005.

Naturally occurring emerging infectious diseases pose an immense threat to populations worldwide.[1] Almost every week, this topic of the threat of pandemics makes the headlines of major newspapers including but not limited to the *NY Times* as shown in the quotes above. Threat analysts are reporting almost weekly of the possibility and inevitability of a dangerous pathogen reaching our shores.[1] In North America, where we like to believe that we are protected by our science and technology, a dangerous ambivalence has somehow taken hold. Since 1993,[2] scientific texts have been warning and then urging health care facilities to step up their response capabilities for the potential of a virulent naturally occurring, airborne transmissible organism. Most of these warnings have been ignored. Secretary of Health and Human Services,[3] Tommy Thompson, said upon being purged from the Bush administration when asked what worried him most, Thompson cited the threat of a human flu pandemic. "This is really a huge bomb that could adversely impact on the healthcare of the world." And according to Davis in an article in *The Nation* magazine (July 18, 2005), despite this knowledge, the Department of Health and Human Services allocated more funds for "abstinence education" than to the development of an avian flu vaccine that might save millions of lives. This section on emerging infections concentrates on two vital themes: (1) the importance of a critical analysis of existing protocols and systems to protect the health care community during a naturally occurring infectious disease outbreak more appropriately called the *occupational health* outcome and (2) evaluating hospital readiness to respond to a naturally occurring infectious disease outbreak, which includes surge effectiveness, respiratory protection, and health care worker (HCW) training that is equal to risk.

One example of (1) is a quote from Robert Webster of St. Jude Hospital in Memphis, a respected influenza researcher, "If a pandemic happened today, hospital facilities would be overwhelmed and understaffed because many medical personnel would be afflicted with the disease."[3] And an example of (2), also cited by Davis, is that under the Democrats and the Republicans, Washington has looked the other way as the local health departments have lost funding and crucial "surge capacity" has been eroded in the wake of the HMO revolution.[3]

This section is designed to be a critical analysis and is part of a chapter on emerging infections disease and hospital preparedness that includes a section on respirators by Lange and an analysis of respirators by the Institute of Research on Health and Safety in Quebec (IRSST). One important proviso to keep in mind is the bioterror template does not necessarily bleed over to the naturally occurring infection paradigm either in training models or preparation. And despite some similarities being prepared for one does not mean we are prepared for the other. The billions that have been provided after 9/11 for the bioterror preparedness does not mean that it is money well spent for the naturally occurring pathogen response. Confusing the two can lead to dangerous myths that can leave us unprepared. According to Cohen[21] "the massive campaigns focusing on bio-terrorism preparedness have had adverse health consequences and have resulted in the diversion of essential public health personnel, facilities and other resources from urgent, real public health needs."

Occupational Health Paradigms

In the occupational health/protecting HCW arena, problems still seem to abound. Hospital designs do not provide for enough negative pressure isolation rooms to be able to isolate the numbers of patients during a pandemic. For example, there are no existing regulations either state, federal, or Joint Commission Accrediation Hospital Organization (JCAHO) that would require the needed amount of negative pressure isolation rooms during a pandemic either for patient care, triage, emergency trauma rooms, radiology, or for high-risk aerosolized procedures such as bronchoscopies. A limited amount of isolation capacity will not adequately defend health care systems against transmissions, especially in a patient "surge" situation. Hospitals in South Africa during the deadly tuberculosis (TB) epidemic in 2007 found that they could not isolate patients quickly enough, and cross-contaminations occurred (*NY Times* Science Section March 20, 2007). The problems and deficiencies in the training of HCW for respirator protection of airborne transmission or decontamination of surface removable contaminants are grossly underestimated in today's hospital climate. During the severe acute respiratory syndrome (SARS) outbreak in Canada, HCW were constantly cross-contaminating themselves during the change of gloves.[22] In certain studies, large percentages of HCW fail fit testings (see the section by Lange in this chapter). Triage area ventilation systems are not controlled for transmissions and at current design will often act as vectors of transmission. There is still controversy about the types of personnel protective equipment, especially types of respirators (see the section by Lange in this chapter) that can protect against an airborne virus. Nonclinical departments such diagnostic imaging or housekeeping have not been adequately prepared to deal with virulent infectious patients and health care facilities are still not cleaning adequately to defend against pathogens.[4] Health care systems are not being tested for their preparedness for naturally occurring pandemic scenarios. Regulatory agencies that set guidelines and rules that sometimes do not reflect current scientific literature on isolation and respiratory protection.

Avoiding the so-called Black Death syndrome, which was the fourteenth century's pandemic, is going to take putting the problems that exist today in our health care facilities on the radar screen and in many instances changing the "business as usual" criteria. Our health care systems are not set up to receive large populations of infectious patients either through design of the facilities or the way health care is administered. Codes for mechanical systems and pressure differentials would not apply. Mixing of infectious patients with noninfectious patients would not apply. Low level and inexpensive personnel protective equipment, now supplied, would in most cases not apply especially during clinical procedures that aerosolize. Current respirators now considered generally acceptable for protection against infectious agents would not apply. Training of HCW at present levels of readiness would not apply, as current training models would be inadequate to meet the severity of the toxicity. Community buildings may have to be used, and to date most communities have not scouted or prepared community buildings for large influxes of infectious patients.

Studies of cross-infection for contagious airborne diseases (e.g., influenza, measles, and TB) have found that placing patients in single rooms is safer than housing them in multi-bed spaces, which means current hospital designs might not apply.[5] SARS outbreaks in Asia and Canada dramatically highlighted the shortcomings of multi-bed rooms for controlling or preventing infections both for patients and HCW. SARS is transmitted by droplets that can be airborne over limited areas. Approximately 75% of SARS cases in Toronto resulted from exposure in hospital settings.[6] The pervasiveness in American and Canadian hospitals of multi-bed spaces in emergency departments and wards will severely impact infection control measures during an outbreak.[7] Quarantine models from state to state would have to be made more enforceable while implementation models for large-scale quarantines have yet to be tested.

SARS in Canada (HCW died and hundreds became infected) shines bright lights on the holes in the acute care responding systems and should be taken as messengers/harbingers of important information for the American health care community and the protection paradigms for occupational health outcomes. And despite the role that nosocomial infection transmissions have played in educating about airborne transmission to patients, protection of health care responders to potential infections has lagged. The classic studies of Riley[8] were very important to the comprehension of airborne transmission of TB in a health care setting. Charney's work developing a portable negative pressure unit to cheaply convert hospital rooms to negative pressure and air-scrubbing through high efficiency particulate air (HEPA) filtration is another example of an occupational health response to an emerging pathogen.[9] However, the totality of occupational protection to emerging infectious disease has not appeared on the radar screen with the intensity needed to protect this population of workers.

SARS in Canada accounted for an occupational transmission rate of 43% and in Hong Kong and China accounted for an occupational exposure rate of 20%.[10] With this particular coronavirus (CoV), mortality for HCW remained relatively low due to the lower toxicity and infectious virulence of the virus, not to excellent protection standards for HCW. In fact in Toronto, more money was spent hiring the Rolling Stones (1 million dollars) to promote tourism during the outbreak than was spent on protecting or training of HCW. The Canadian experience listed a number of factors that increased transmission to HCW: A brief list follows:[11]

1. Lack of HCW training in decontamination procedures
2. Protocols from the relevant regulatory agencies that changed almost on an hourly basis confusing HCW and their responses
3. Confusions as to the effectiveness of respirator selection and fit testing protocols
4. Lack of timely protocols for aerosol producing clinical procedures
5. Lack of training for first responders
6. Questions about isolation and negative pressure, especially in triage areas
7. Lack of timely protocols for airborne protection, especially in the early stages of the epidemic as it was labeled a "surface removal contaminant"
8. Following protocols but still seeing occupational transmission

Airflow in health care settings is ill-prepared for containing transmissions. A substantial number of viruses, bacteria, and fungi are capable of spread via the airborne route in hospitals.[12] Among the common exanthems, the evidence in support of airborne transmission is quite strong with respect to varicella-zoster and measles.[13] Rubella may also spread through the airborne route. A strong base of evidence of airborne transmission of respiratory syncytial virus and adenoviruses in pediatric wards also exists.[14] Hoffman and Dixon (1977) reported on the transmission of influenza viruses in hospital settings through airborne routes[15] and the strongest evidence of airborne transmission of influenza is a well-documented outbreak that occurred on a commercial aircraft.[16] All types of viruses can be spread throughout hospitals by airborne transmission. Even SARS, a coronavirus, which was mistakenly considered only a "surface removable" contamination was found to have an airborne transmission as well.[17] The Marburg virus, now occurring in parts of Africa has an airborne component exposure. There is evidence that certain enteric viruses may be transmitted through the air.[18] Sawyer[19] reported on a case of a viral-like gastroenteritis that occurred in a Toronto, Ontario hospital in 1985 where 635 hospital personnel were affected and the investigators found no common food or water source and believed contamination was through the airborne route. Most hospitals will rely on changing pressure differentials using their central HVAC systems. However, this is an unreliable methodology. It would mean decreasing supply to compensate, and the negative pressure realized using this method would in most cases not produce the needed pressure of a minimum of 0.01 in. of water gauge to maintain the negative pressure especially during the opening and closing of patient room doors. In a California[20] study of negative pressure rooms during the TB outbreak it was found that many hospitals though thinking they were providing negative pressure were in fact mistaken. In Toronto, Canada, during the SARS epidemic, Children's Hospital bought many portable HEPA negative pressure units to be able to provide adequate isolation. It was fortunate that this technology was available and could be provided in the numbers necessary. During a pandemic, lack of preparation on this scale will find hospitals trying to get equipment that may not be available at the last minute.

Centralization

Centralization of patients during a pandemic would be a wise and judicious use of resources. However, in the present climate, health care delivery systems see centralization as negative rather than positive. The positives are numerous, i.e., less training of personnel, better distribution of resources, less resources needed, less potential for cross-contamination, better control on variables and if mistakes are made better means of postcase analysis, better ventilation controls, less staff, and patients exposed. The negatives are described as patients often presenting with multiple symptoms, therefore a more flexible clinical resource is necessary. In Toronto, a hospital did centralize care during the SARS outbreak and found it very necessary. All the presenting flu-like cases were referred to one particular area of the facility that was trained and controlled for the specific type of clinical presentation. It is the belief of many pandemic experts that centralization is a necessary means to controlling adverse outcomes during a pandemic episode, especially of the type of virus is airborne and since the routes of transmission will not be known for at least months into the pandemic a proactive form of centralization should be high on the list of control priorities.

Surge and Other Effects

There is a looming sense that health care facilities would not be prepared for a surging population of victims or that cross-contamination and cross-transmissions could be prevented. SARS actually projected all the difficulties in protecting HCW from a natural emerging disease. From the perspectives of building design, patient flow, airflow parameters, disinfection principles for surface removable or air borne transmissions, personal protective equipment (PPE), and almost most importantly HCW training, the United States and Canada are underdeveloped and unprepared according to many experts. The occupational health dynamic is often the last item on the agenda when emerging disease is discussed. In an Op-ed piece in the *NY Times* written by Barack Obama and Richard Lugar entitled, "Grounding a Pandemic" there was not one word mentioned about how to protect transmission to health care workers (*NY Times* Op-Ed, June 8, 2005). This complacency is unsafe. We are taking for granted that our health care systems are going to be able to deal with thousands of sick and dying people, when in fact at the current level of preparedness they will be overwhelmed and chaos is quite predictable. Just from a standpoint of HCW protection technology, the national community and guideline agencies have not adopted a respiratory standard that seems acceptable to protect against airborne transmission, or provided health care facilities with enough acceptable respiratory protection equipment or models. Air scrubbers with HEPA or ultralow particulate air (ULPA) filters, that could scrub the air of viruses and bacteria and that would be an important ingredient to add protection factors in many health care rooms and spaces, are not currently required or used substantially.

HCW are substantially undertrained for emerging infections with regard to level of risk. This was apparent in Canada during the SARS outbreak. Constant cross-contamination for surface removable transmission was a problem, as well as a lack of knowledge about respirators.

Until the problems discussed in this section are admitted and addressed, the HCW is at increased risk, thereby putting community populations at greater risk. Risk assessment analysis stresses that all parts of the exposure whole be working intelligently together for positive outcomes. We are not there yet.

Public Health Paradigms

Emerging infectious diseases will endanger U.S. citizens at home and abroad.[1] "As Earth Warms Up, Virus from Tropics Moves to Italy" (*NY Times*, December 23, 2007, p. 21 on the outbreak of chikungunya virus).

Twenty well-known diseases, including super TB, malaria, and cholera have reemerged since 1973, often in more virulent and drug-resistant forms.

At least 30 previously unknown disease agents have been identified since 1973, including HIV, Ebola, hepatitis C, and Nipah virus for which no cures are available.[1]

- Newer diseases have emerged that are beginning to mutate and jump from animals to humans, H5N1 (bird flu), Moberg, etc.
- Annual infectious disease rates in the United States have nearly doubled to some 170,000 annually after reaching an historic low in 1980.[1]
- Influenza now kills some 30,000 Americans annually and epidemiologists generally agree that is not a question of whether, but when, the next killer pandemic will occur.

The American Public Health system is compromised by several "deficiencies" namely a shortage of personnel, communication problems, and time lags. Since the 1960s, the United States, like much of the rest of the world, has seen a decline in the ability of the public health system to address the threat of infectious disease. It has also been shown through the lens of analysis of Katrina that public health responses have been severely compromised, from the nonfunding of levies to protect the city of New Orleans to actual response and communications between federal, state, and city responders. At this point, 18% of public health laboratory positions are vacant and over 40% of public health epidemiologists lack training in the field.

Many of the agencies needed to respond during a natural public health disaster have suffered from cronyism to severe budget cuts in recent and past years. FEMA hired Mike Brown who had little or no disaster response experience. FEMA had become known as the "turkey farm" where high level positions were filled with political appointments. The Environmental Protection Agency (EPA), needed now more than ever in New Orleans, has been, crippled by cronyism. The agency has seen an exodus of experienced officials due to administrations', both Democrat and Republican, refusals to enforce environmental regulations. In an interview with the British journal, *The Independent*, on September 10, 2005, Hugh Kaufman, a senior policy analyst with the EPA, complained of severe budget cuts and inept political hacks in key positions. The Federal Drug Administration also has been accused of coziness with the drug companies and the agency's head of women's health issues resigned due to "politics" over health' in the delay of approving Plan B, the morning after pill.

The current Bush administrations' increasing focus on terrorism to the exclusion of natural disasters has been a concern for some time. A recent report by the Government Accountability Office showed that "almost 3 out of every 4 grant dollars appropriated to the Department of Homeland Security for first responders in fiscal year 2005 were for 3 primary programs that had explicit focus on terrorism." More than 2 billion dollars in grant money is available to local governments looking to improve the way they respond to terrorist attack but only 180 million is available under the grant program for natural disaster or pandemics. The Bush administration has even proposed cutting that to 170 million even though NEMA had identified a 264 million dollar national shortfall in natural disaster funding.

Katrina, like SARS, has put a spotlight on the flaws of the public health response systems that include response agencies, city, state, and federal and also health care delivery systems. "Confusion, desperation reigned at New Orleans city's hospitals" read the headlines in the *Seattle Post Intelligencer* on September 14, 2005. Evacuation of the infirm and sick did not take place in a timely life saving manner. Hospital back-up generators failed as electrical grids went off-line. Police communication systems failed. And even after 3 days, food and water supplies were not entering the city. Toxic waste issues will be evident and overbearing to underfunded agencies. And probably most importantly there was no plan to evacuate, feed, house 130,000 residents of New Orleans who live below the poverty line, drawing a class line in the sand of our public health readiness.

National Agencies Response Paradigms

I am somewhat perplexed at the Centers for Disease Control (CDC), Health Canada, and the World Health Organization's inability to be more cognizant and aware of the occupational health effects and protections necessary to assure HCW protection during the latest SARS outbreak. Their lack of preparedness on the occupational health front is not reassuring for the next potential pandemic. The CDC Guidelines for SARS became a questionable model of scientific inquiry. CDC protocols during the outbreak changed almost on a daily basis, confusing HCW and creating a climate of uncertainty especially on the issues of transmission and protection. John Lange (see section on respirators) shows that the respiratory requirements within the SARS Guidelines were a serious departure from the science of respiratory protection for the protection of HCW. The recommendation of a paper respirator, N95, with leakage factors of 10% at the face seal and 5% at the filter to protect HCW from an exposure that has a 15% mortality rate where airborne transmission was not ruled out (and later ruled in) does not follow any of the Occupational Safety and Health Administration (OSHA) rules for serious toxicity for respirator selection, especially when there was no dose/response relationship known. Dose/response relationships are the primary method to develop a "protection factor" which in turn drives the level of respiratory protection. This nonscientific attitude within the CDC continues in the position statement on plague where surgical masks would be allowed as the respirator of choice for HCW responding to plague and that the fit testing regulation could be waived; this despite the fact that the literature cites a 43% fatality rate using this method[19] and that the recommendation contradicts regulatory safeguards and contradicts peer-reviewed science.

Effect of Globalization and Global Warming

Globalization and the global economy have made it easier for diseases to spread from one country to the next. One can travel anywhere in the world now within 24h, transporting pathogens. The latest dengue epidemic in El Salvador was spread from Vietnam via Cuban workers, then to nearby islands in the Caribbean, on to the South American Continent, and into Central America. But developing countries are not the only ones affected. When East Nile virus appeared in the United States two years ago, health officials said 59 people in the New York City area were hospitalized. Since then federal researchers estimate about 1400 cases have been treated. C. Everett Koop has written that we have achieved the "globalization of disease."

Bird flu is only one of the six emerging global pandemics. They are super TB, H5N1 (bird flu), super-staph, SARS, super-malaria, and HIV. HIV alone has mutated and has gone from 2 to 400 strains in only 20 years. Influenza is justifiably feared. In 1918–1919, 40–50 million people (2%–3% of the world's population) died (ref. *MJA* 2004; 181(2) 62–63). Subsequent influenza pandemics occurred in 1957 and 1968.

Pneumonic plague, since last December, has resulted in 300 suspected cases and at least 26 deaths in eastern Congo. This is the largest plague outbreak since 1920. Hong Kong flu, which swept across the Pacific rim in 1968, reaching the United States in the same year, killed an estimated 34,000 Americans in 6 months. Asian flu claimed 70,000 American lives and a million worldwide. Spanish flu, which occurred in 1918, swept across the trenches in WW1, accounting for half the GI deaths. By some estimates, this flu infected at least a billion people worldwide killing 20 million. Most victims were healthy adults aged 20–50. There has been an outbreak of Marburg virus in Angola in 2005, and this dangerous virus is transmissible from person to person.

Global warming, as verified by 48 Nobel Prize winners, is another ingredient in the rapid growth of new bacteria and viruses. As global temperatures rise, conditions improve for pathogens to emerge. Global warming has also been associated with the intensity of hurricanes hitting the Gulf Coast this year, as the warming Gulf of Mexico feeds the ferocity of the hurricanes as they travel over water.

Class

The response to Katrina displayed the class bias that exists in the United States. Mexico has already warned that the next flu pandemic will affect the poorer countries disproportionately, and affect the global response capacity. It has long been argued that the first world riches were not creating a more level playing field for the third world, that the rising tides were not raising all ships. Warnings have been issued for years by epidemiologists, demographers, and political scientists that if more were not done to bridge the gap between rich and poor countries, the imbalances would affect the "global health." Even the report issued by the National Intelligence Council for the Central Intelligence Agency in January, 2000, warned. "New and reemerging infectious diseases will pose a rising global health threat and will complicate US global security over the next 20 years. These diseases will endanger US citizens at home and abroad and exacerbate social and political instability in key countries and regions in which the United States has significant interests." This report goes on to say, "development of an effective global surveillance and response system probably is at least a decade away owing to inadequate coordination and funding at the international level and lack of capacity and funds in many developing countries. The gap between the rich and poorer countries in the availability and quality of health care is widening.... compromising response."

Class is one of the most misunderstood and denied causes for worldwide infections and also response capabilities. Now with globalization and the rapid ways in which microbes can travel, the planet is shrinking. What will happen in Bangladesh could affect New York City within hours. A class approach to finding solutions to poverty in the third world is essential to protect the first world. Many diseases are poverty driven, like TB, etc. Antibiotic resistance in one country could lead to an epidemic in another. Poverty and class divergence is the petri dish for emerging diseases. A long-term strategy needs to be developed to combat poverty and class differences in the world and in nation states. This would be the best strategy to protect the citizens of the world against pandemics.

Are Hospitals or Public Health Systems Ready? A Conclusion

There seems to be accumulating evidence that our health care institutions will not be prepared for a pandemic scenario, and that our public health agencies, underfunded and understaffed, will have multiple problems responding to a pandemic scenario. In New Orleans, during Katrina, hundreds of patients died in hospitals and nursing homes despite some heroic stories of HCW sacrifices. There were failures in both health care and public health response on all levels. Katrina is surely the canary, as was SARS to our need to invest and solidify all aspects of preparedness, both for hospitals and public health, if we are to protect public health during a surge pandemic.

Necessary funding for public health has been diverted to terrorism, disproportionate to risk. This is the political disconnect that if anything Katrina has made us observe. As Frank points out, 3400 people died on September 11, but over 5200 people a day die from natural specific diseases that are preventable.

Unless we integrate this information into the body politique, insisting that the domestic pubic health problems get funded and repaired, the United States is at great risk. The breakdown of the levy systems in New Orleans was predicted based on integrated information, and the breakdown of the public health system was predictable based on the defunding of the systems. Tommy Thompson's Health and Human Resources budget for public health year after year was underfunded, and cut by large percentages by Congress during his administration. In America you get what you pay for. If we militarize space, 1/3 of children in this country go to bed hungry, we have made a disconnect. If we spend 2–7 billion dollars a month on a questionable foreign war and occupation when there are millions of Americans without health insurance, we have made a disconnect. If spending on abstinence training outpaces spending on public health, we have made another disconnect that leaves us underserved and quite vulnerable during times of either disasters or pandemics. Since 2001, the country has been waiting for a finished pandemic plan and at this writing it is still not completed, let alone tested and components verified for effectiveness.

Russian roulette belongs in the gambling casinos, not in the public health arena. There is too much at stake.

References

1. The global infectious disease threat and its implications for the U.S. U.S. Govt. in www.cia.gov/cia/reports/nie/report/nie99–17dhtml
2. Bierbaum P and Lippman M (eds.) In: *Proceedings of the Workshop on Engineering Controls for Preventing Airborne Infections in Workers in Healthcare*. U.S. Department of Health and Human Services, CDC, Cincinnati, OH, 1993.
3. Davis M. *The Nation Magazine*, July 18, 2005.

4. McCaughey B. *NY Times*, June 6, Op Ed 2005.
5. Gardner PS, Count SD, and Brokclebank JJ. Virus cross-infection in pediatric wards. *Br Med J* 2(5866):571–575, 1973.
6. Farquharson P and Baguley K. Responding to SARS. *J Emerg Nurs* 29(3):222–228, 2003.
7. Ulrich R and Zimring C. Role of the physical environment in the hospital of the 21st century. In: Report to the Center of Health Design, 2004.
8. Refer to the works of Richard Riley.
9. Charney W. HEPA surge tests for surrogate penetration for TB. In: W Charney (ed.), *Handbook of Modern Hospital Safety*. CRC Press, New York, 1999, pp. 221–246.
10. Ulrich R. et al. Role of physical environment in the hospital of 21st century. Report to the Center of Health Design, September 2004.
11. Bunja E. and McCaskell L. Presentation to the Commission to Investigate the Introduction and Spread of SARS. Ministry of Labour, Toronto, Ontario, Nov. 17, 2003. www.ona.org
12. Bierbaum P and Lippman M (eds.). In: *Proceedings of the Workshop on Engineering Controls for Preventing Airborne Infections in Workers in Healthcare and Related Facilities*. U.S. Department of Health and Human Services, CDC, Cincinnati, OH, July 1993.
13. Valenti W. Selected viruses of onsocomial importance. In: *Hospital Infections*, 3rd edn. Little and Brown, Boston, MA, 1992, pp. 789–821.
14. Chanock RW, Kim HW et al. Respiratory synctial virus. *JAMA* 176:647–653, 1971.
15. Hoffman PC and Dixon RE. Control of influenza in hospitals. *Ann Intern Med* 87:725–728, 1977.
16. Moser MR. An outbreak of influenza aboard a commercial airliner. *Am J Epidemiol* 110:1–6, 1979.
17. Yu I. TS, Li Y, Wong TW, Tam W, Chan AT, Lee JHW, Leung DYC, and Ho T, Evidence of airborne transmission of the severe acute respiratory syndrome virus. *N Engl J Med* 350: 1731–1739, Apr 22, 2004.
18. Kool J. Risk of person to person transmission with pneumonic plague. *Clin Infect Dis* 40:1166–1172, April 13, 2005.
19. Sawyer LA. 25–30 μm virus particles associated with a hospital outbreak of acute gastroenteritis with evidence of airborne transmission. *Am J Epidemiol* 127:1261–1271, 1988.
20. Sutton, P. California Department of Health Stud. In: *Handbook of Modern Hospital Safety*, 1st edn. CRC Press, New York, 1999.
21. Cohen H. *Influenza Biology*. In: Charney's *Emerging Infectious Diseases and the Threat to Occupational Health in the U.S. and Canada*. CRC Press, New York, 2006, pp. 267–305.
22. Bujna E. Presentation to the Commission to Investigate the Spread of SAR. In: Charney's *Emerging Infectious Disease and the Threat to Occupational Health in the U.S. and Canada*. CRC Press, New York, 2006, pp. 3–31.

Personal Protective Equipment for Health Care Workers

J. H. Lange and G. Mastrangelo

Summary

The risk of exposure by HCW to an emerging or historically highly infectious disease has increased over the past decade. This is best exemplified through the occurrence of the SARS in 2003, which caused a worldwide panic that a regional epidemic or global pandemic was on the verge of emergence. During this event, there became the realization of how important PPE, including respirators, are for HCW. However, it was soon realized that most had little experience or training using these devices and equipment as well as inadequate information on selection. When appropriate PPE and hygiene activities were implemented, infections from patients to HCW dramatically drop. The rate of infection from patient to patient and from HCW was also lowered. Certainly, the SARS event will not be the last, with numerous old and emerging diseases poised to initiate epidemics and, in some cases, pandemics. Concern associated with infectious microbes has become heightened through terrorist activities, as was seen with the anthrax event. It is now realized that HCW are on the frontline for these events, since they will be the first to see patients that have contracted these diseases and may be the second wave for exposure and subsequent infections. This chapter provides an overview of the hazards from infectious diseases to HCW and application of various forms of PPE, especially respirators. Implementation of PPE for HCW is presented and includes activities such as fit testing and selection. The importance of personal and institutional hygiene, including hand washing, is emphasized, along with a historical discussion of this topic.

Introduction

Hospital-related infectious disease has emerged as a major public health issue in the United States and throughout the world. As health care costs continue to rise along with an increasing incidence of nosocomial infections associated with the health care industry (HCI), there will be increased concern of disease transmission not only for HCW, but also the general community at large. Historically, diseases associated with health care settings were considered an inconvenience and part of the cost of care, but with a rise in antibiotic-resistant microbes, this has changed. Nosocomial diseases were traditionally considered to be a concern for patients, but slowly this scenario has changed, most notability after the SARS event in 2003. When few microbes had any degree of antibiotic resistance, little attention was paid to those that had became drug resistant. However, this has changed with drug-resistant microbes becoming a major concern in the HCI and clinical settings, and is now responsible for a considerable number of deaths (mortality), morbidity, and increased hospital costs. This is best represented by an event in May/June, 2007, of an individual who had reportedly extensively drug-resistant tuberculosis (XDR-TB) and traveled to Europe by plane potentially exposing other passengers (*USA Today*, June 11, 2007, p. 8D). Presently, there has been no other person(s) infected and it appears that this individual was not very contagious, but did warrant his quarantining, according to the Center for Disease Control and Prevention (CDC) for the treatment of TB. It is now reported that this person did not have XDR-TB, but multi-drug resistant (MDR) TB.

Most HCW are not aware that XDR-TB exists; although, many heard of MDR-TB and considered it a problem of third world countries, there are a large number of cases of TB in the United States.[1] For the most part this is true, but the overuse of antibiotics in both undeveloped and developed nations has increased the incidence of resistant organisms and this has spread drug resistance throughout the world. Presently, there remains an increase in drug resistance and it appears that even with dire warning there is little change in use of antibiotics. With the existence of rapid travel, spread of an infectious disease, as was seen with SARS, is a real and present hazard.[2,3] According to an article in *USA Today*, there have been 16,000 deaths worldwide from XDR-TB, including some in the United States and western European countries, indicating that this is not a third world problem, but one of global importance. The lesson from this TB event is how vulnerable today's society is for drug-resistant organisms and that a crisis relating to hospital-borne infections looms on the horizon, especially if appropriate action is not taken quickly. Thus, some of the old diseases considered to have disappeared, such as TB in the United States, will soon return and will likely be stronger and better adapted than ever before. What may be more alarming is that this is related to the diseases which are known and does not include those that are emerging, as was seen with SARS and now avian flu or avian influenza virus (AIV).

As such events increase, there will become a greater requirement to protect HCW from these infectious agents. Events like the one observed with XDR-TB demonstrate how vulnerable the population is to these microbes. To prevent or at least reduce the spread of these germs, both old and new practices will have to be implemented. The most basic and oldest practice to prevent the spread of infection is hand washing. However, hand washing can prevent fomite and related mucus membrane transmission (hand-to-mouth) of infectious agents. But this practice alone provides little protection against inhalation, which has been mostly ignored by HCW and for many microbes is an important route of transmission (e.g., SARS). This will require HCW to be informed on the use of respirators and related PPE. Use of these types of protective devices will be a big leap for the HCI. However, such applications of protective devices (PPE, which includes respirators) and related practices are not new and have been used in preventing spread of TB for decades.[4] Programs for protection of HCW against TB can serve as a starting point for a larger PPE program that is designed to protect against a wider range of infectious diseases.

As mentioned, there are also the hazards from emerging infectious diseases, which TB can be considered as well as others, and was recently exemplified through SARS. Thus, infectious diseases in this sense can be identified as either acute or chronic, with both becoming of greater concern. A recent report suggests that the United States and, most likely, all other countries are poorly prepared for such events.[5] The SARS event was an acute occurrence, while that associated with TB and antibiotic resistant microbes are chronic in nature and for the most part generated by man through inappropriate use of antibiotics. As was seen with SARS, these events occur unexpectedly and are difficult to manage and as a result can quickly become a pandemic event. Most consider that the next "emerging" infectious disease will be related to the influenza virus and most likely will be the AIV. Most of the cases from AIV are presently seen in Asia, although, cases are now been seen outside that region (e.g., Turkey).[6] It is agreed that it is not whether this flu will occur, but when and to what magnitude. However, it must also be noted that the flu is not the only emerging infectious disease and that there are many candidates, including some that cannot be named because they have yet to be identified. From this perspective alone, there has been a greater concern and awareness of the importance of PPE. However, as seen with the SARS event, most remain poorly prepared to implement PPE practices along with other related measures (e.g., engineering controls) at the time of an event. In general, emerging infectious disease can be defined as any infectious agent that is "unknown" or unrecognized or one that is known but has changed, such as through antibiotic resistance.

Application and History of Personal Protective Equipment and Practices

The use of PPE by HCW dates back to the time of Joseph Lister.[7] Here, gloves, one form of PPE, were used by surgeons to protect their hands from the effects of phenol and later implemented for preventing disease transmission and sterilization. In many ways, this was one of the first occupational practices that employed PPE in providing protection to HCW, although it was an indirect benefit. The use of respirators and other types of PPE, such as eye protection, have existed for many years in the HCI, although, until the occurrence of SARS was often given low priority[8,9] and in many ways still receives this categorization regarding protection of patients, visitors, and HCW. For the most part, in the western world, especially the United States, respirators were not considered important for HCW except when working with TB.[10] However, the emergence of unknown, and some could say, rare diseases, such as the Naphavirus, an increase in antibiotic-resistant microbes,[11] and a greater awareness of occupational nosocomial diseases (OND) (those contracted by HCW from infected patients in health care settings), and the importance and applicability of PPE have become a topical issue for those in the medical fields.[8,12] One of the issues relating to PPE is its applicability and appropriate implementation in the health care setting. It is well-known that PPE is used in other occupations (e.g., asbestos workers, construction, mining), but little has been published, at least in comparison, on its use for HCW.

PPE includes a wide variety of protective devices, the most common of these discussed in the literature being respirators.[8] However, since the route of transmission of organisms can vary, it is important to evaluate all forms of PPE when considering occupational disease prevention and practices and their frequency in use.[13,14] In addition, other basic practices must also be included as part of this protection, such as hand washing.[8] As more microbes become resistant to chemotherapeutic agents and new ones emerge, use of PPE will become more common and more important. However, even with PPE, the basic principles of exposure reduction must be considered for implementing the best and most appropriate protection. What must be remembered is that hand washing is one of the simplest and basic forms of prevention, with this basic practice needed as part of training for HCW.[15]

PPE including respirators have a long history in the industrial environment.[8] Here respirators have been used for decades and a great deal of information can be transferred from these practices to the HCI.[16] However, even here there are considerable gaps of information and misconceptions about PPE, especially when it comes to practical use by HCW. Many of the practical applications learned and used in the industrial environment can be extracted and adopted to the health care setting. Although industries do not commonly protect workers against infectious disease agents, there are many involved with toxic chemicals and substances. This can serve as a basic model for infectious disease agents in developing a protective program. However, it must be realized that for some infectious diseases, the number of viable units can be much smaller than that of an acute chemical that can cause disease.

The basic focus of this section will be on respirators; although, information relating to other forms of PPE (e.g., gloves) will be briefly presented. Until the SARS event, most commonly respirators were employed to protect HCW from TB and in some cases unusual viral outbreaks (e.g., Ebola).[3,10,17,18] A high infection rate, during the SARS outbreak, of HCW quickly resulted in the realization that protective measures were needed, especially in the form of respiratory protection.[19] It was soon realized that the SARS event would probably not be an isolated occurrence and many would follow, with the question becoming what organisms will be responsible for the next outbreak. This was seen on a local level by the occurrence of monkeypox in the United States. Today, one of the biggest concerns is avian flu and its potential to become a pandemic agent.[8] Many do not consider that there is a possibility of bird flu becoming pandemic,[20] but rather elude that it will eventually happen and from an OND prospective will put most HCW at great risk in contracting this virus.[21] It must also be realized that with 90,000 deaths a year in the United States from nosocomial infections, one can easily suggest that an outbreak is continuously occurring, but just not well recognized as such. One can term this a continuing pandemic. However, these numbers are estimates since there is currently no tracking system for infectious diseases in HCW in any country.[22]

Types of Common PPE

There are various forms of PPE that can be employed by HCW. Table 4.1 provides a list of commonly employed protective devices. Within each of these devices, there are many different kinds and forms in existence, along with variations among manufacturers. Each has its own complexities in use and form, as well as limitations in protection. Thus, selection for the application is necessary, with sometimes a compromise necessary to achieve protection for a range of possible events. This alone suggests that PPE is not 100% effective in preventing exposure to an agent, such as a type of bacteria, or for HCWs in the occurrence of disease, with the worker being the endpoint outcome for measuring effectiveness.

TABLE 4.1 List of Different Types of Commonly Used PPE

Type of Device	References[a]
Respirators	Lange[14]
Gloves	Mahler[125]
Eye glasses/goggles	Ho[95]
Face shield	Omokhodion[126]
Suit/coverings	Omokhodion[126]

[a] Identifies device in use by HCW.

History

Hospital acquired or associated infections (HAI) (nosocomial diseases/infections) have become an important topic/issue in health care. Historically, it was considered that patients in a facility were at risk from these diseases, which is true, and for the most part those with a deficient immune system or "open" wounds were at great risk. Today, this is no longer considered true, in that not only are patients at risk from infectious disease, but also HCW are as well.[8] Thus, nosocomial diseases not only include patients but the myriad all HCW, which include those in ambulatory settings along with some that were historically considered auxiliary to the HCI (e.g., maintenance and cleaning personnel). Thus, the population that can be influenced by these microbes has enlarged. The extent of drug resistance has also grown, whereas today cases now exist where there are no "useful" therapies against some strains of microbes. In the future, such occurrences will become the norm rather than the exception.

The word nosocomial is derived from the Greek term nosos, which means disease. Until the 1940–1950s, risk of acquiring disease in the hospital and clinic setting was unknown and probably considered a somewhat natural event that was unpreventable. This concept of infection control and prevention of disease transmission within a hospital changed through the revolutionary work of Ignaz Semmelweis who evaluated "infectious" disease at the Vienna General Hospital.[23,24] Here he observed the difference in infection rates involving *Streptococcus pyrogenes* (puerperal fever) in two different obstetric wards. This organism is now associated with what is commonly known as Strep A (Group A *Streptococcus*). One ward had a rate of about 10% that consisted of medical students and physicians treating and delivering babies and the other was about 3% and only had midwives providing these services.

Semmelweis observed that medical students and physicians often came from other services, including dissection, before treating patients on the ward (Best and Neuhauser, 2006).[25] He concluded that students and physicians carried infectious material, as we know it today, with them from these other services and this contamination resulted in the occurrence of puerperal fever in women.[26] Based on these observations, he instituted hygienic conditions, such as hand washing, and this resulted in the rate of disease dropping in the medical ward and was soon similar to that in the ward of midwives.[24] Although he is not given complete credit, he extended the hand washing to general disinfection and expanded the concept of disease prevention by also instituting cleaning of medical equipment.[26] In many ways, this set the stage for Louis Pasteur and Joseph Lister, with their contributions arising around the middle/end of Semmelweis's life.

Sadly, many of the lessons taught by Semmelweis have yet to be fully learned and appreciated by the medical community. Commonly, HCW do not accept or practice the advice they give to patients regarding prevention of infectious disease. For example, in a recent survey, the mean vaccination rate of patients in a care facility was about 82%, but for HCW it was 35%.[27] Many could say that the rate for those in a health care facility is low, but when examined in comparison with HCW is high. Not only are HCW at risk for influenza by not being vaccinated, they place the patients at risk when they become sick. Remember, the basic principle, "First do no harm."[27] Certainly, practices involving institutional sterilization and infection control have been well established and implemented, but many of the individual's have not adopted these practices. Hand washing, vaccination, and tuberculin skin testing can prevent a large number of nosocomial infections.[27] In the United States, it has been estimated that nosocomial infections result in 90,000 deaths a years, with about 5%–10% of all hospital patients contracting an infection during their stay. This adds about 5 billion dollars to heath care costs and increases the length of hospital stays.[11] By following the practices established by Semmelweis, the infection rate can be dramatically reduced as he observed when the practice of hand washing was first introduced, even when using nonsoap and water agents such as foam disinfectants.[28]

Emerging infectious disease have been known and recognized for centuries. Most of these diseases have disappeared in developed nations, but continue to exist in the underdeveloped world. Overall, infectious disease remains a major contributor to mortality and morbidity worldwide. Until recently, most in developed nations were not greatly concerned with these diseases or the observation of new ones arising. However, the occurrence of SARS and the emergence of AIV has changed much of this prospective. With the speed of travel, an endemic outbreak of a disease at a remote location in the world can quickly and rapidly spread to almost anywhere else and as such become a pandemic in the matter of a short time period. This was the fear related to SARS and is now the driving factor regarding evaluation of new flu strains.

Historically, infectious diseases were a major factor in the control and growth of the human population. For example, bubonic plaque that occurred in Europe in 1346 resulted in approximately one-third of this regional population succumbing to the disease. Some have estimated that this was between 20 and 30 million deaths.[29] Today bubonic plague is not a major hazard, as a result of adequate rodent control, but other diseases that existed and caused large numbers of deaths do remain viable and are important infectious diseases. Cholera and the Rota virus are major causes of death in developing countries and without sanitation controls would also be common in developed countries. Sanitation and public health measures in developed countries have not eliminated these and other diseases (e.g., TB), but have only established conditions that keep them in check and under control. Control of these diseases, which is mostly a result of improved hygiene, is primarily responsible for the rapid increase in life expectancy seen in developed nations.[30]

There has been numerous events in mans' history where man himself has initiated an infectious disease event as part of a military operation.[31] This can be said to be the start of bioterrorism. The use of biological agents for warfare is essentially a form of a poor man's nuclear weapon and can be said to be a potential replacement of nuclear activity by rogue nations and groups. However, the problem with releasing such an agent is that it cannot be controlled and its destruction may not be limited to only the enemy. This again has been clearly noted in the SARS event. Here, infected people in the prodromal stage were missed using rapid detection methods and their movement quickly spread the disease throughout the world. Thus, individuals can act as a carrier, in a suicide fashion, to transmit the disease. Such an occurrence can result in a disease emerging almost overnight and without warning. This would result in a horrific situation for HCW who would have to care for these sick individuals and would most likely have little information on the infectious agent. In many ways, this is what occurred in Canada relating to SARS where ultimately many of the cases ended up being HCW.

Many of the lessons presented by Semmelweis have not been fully realized and appear to occur over and over again. Semmelweis initially suggested that most of the hospital-derived infections arise from poor sanitary conditions or conditions that are ignored as contributing in the spreading disease, especially hand washing. As Semmelweis identified, poor "personal" hygiene by physicians, was the principle cause of puerperal fever, even though the existence of microbes had not yet formed in the scientific literature. Care and practice of disease control and prevention can eliminate many of these nosocomial diseases, and overall practice of sanitation in a community can eliminate most of the traditional infectious diseases. Even the disease associated with the medical symbol (Staff of Asclepius), *Dracunculiasis medinensis*, is on the verge of elimination through basic sanitation and public health practices.[32] Thus, use of simple control measures can prevent many of the historical infectious diseases that have been associated with man over the centuries, with these lessons highly applicable and valuable to today's hospital and health care environment.

Most infectious diseases that arise in health care environments are a result of poor or inadequate sanitary practices and over use of antimicrobial agents. When there is a good practice of hygiene, the spread of infections is often reduced, which includes both health care and community settings. Much of the current reduction in transmission-borne disease is a direct result of sanitation, especially related to diseases that were considered to be waterborne. Thus, treatment of the causation appears to be key in preventing these diseases. This was best illustrated by the investigation conducted by John Snow on cholera.[33] Cholera was a major infectious disease in England in the 1850s and is currently of great importance in many locations in the world. The studies by Snow in the 1850s have been referenced as the first true epidemiological investigations because he reported the results as a rate. After Snow's investigation, it was soon realized that water is not the only route of infection by "microbes," but they can occur due to man-to-man transmission. Within a clinical setting, other routes of transmission can occur as well, including those associated with fomites, aerolization, and those which are iatrogenic.[14]

Many diseases can be spread through aerolization and fomites, such as influenza and TB. Thus, adequate disinfection of surfaces is of great importance in disease prevention. However, today use of PPE has been a topical issue, especially related to HCW and the fact that this population in the Untied States is aging and becoming more susceptible to disease transmission.

The decline and in some cases the complete absence of historical infectious diseases has signaled to the public and HCW the end of the era of plaques and infectious disease outbreaks. This memory of plaques came to a quick end with the occurrence of SARS and related outbreaks (e.g., Giardia, West Nile virus). The emergence of bird flu and its discussion in the media has further raised the awareness of old and new infectious diseases that can result in endemic and epidemic events. Unfortunately, local events are often quickly forgotten (e.g., Giardia)[34] and are not long-lived in the memory of a modern population.

The occurrence of SARS in 2003 changed much of the way HCW and the public view emerging and in some ways past infectious diseases. SARS originated in China and then quickly spread to Hong Kong. If the disease remained in this region, it would have been noted as an interesting and intriguing epidemic in Southeast Asia. However, as it is well known, the disease quickly spread throughout the world and fears with past pandemics began to arise. As noted, this disease event, although tragic, did not emerge as true global pandemic, but fears of such did cause a global public health awaking, even in locations where there were no cases.[35] This disease demonstrated how fast an unknown agent can spread and how vulnerable those in the HCI are to such diseases.[36] Based on the distribution of cases, it appears that HCW were the most vulnerable to this disease.[14]

Many of the HCW that contracted SARS can be said to be a result of poor or nonexistent application or use of PPE.[37] When precautions were taken and PPE implemented there was still an infection rate of 2%–25%, with some of this attributed to inadequate use of PPE.[8] Initially it was thought that the virus was not spread by an airborne route. However, this was quickly shown not to be accurate. As with almost any agent, especially those that are viral, aerolization can occur, even if the agent is not a respiratory infection, suggesting that all infectious agents have to be taken into consideration as a requirement for respiratory protection. Understanding how an organism spreads is key for developing strategies for its prevention and protection of HCW.[38] The SARS event also illustrates how poorly trained and equipped HCW were in using PPE and serves as a warning for future events. A list of some microorganisms that can be spread or transmitted by an airborne route is shown in Table 4.2.

TABLE 4.2 List of Some Microorganisms That Can Be Spread by Inhalation and Related Routes

Organism Name	Type of Organism	Infection Site(s)
Acinetobacter baumannii	Bacteria	Blood
Aspergillus species	Fungi	Pulmonary
Candida species	Fungi	Gastrointestinal (GI)
Coronavirus (SARS)	Virus pulmonary	
Cryptococcus gatti	Fungi	Pulmonary, central nervous system (CNS)
Cryptococcus neoformans	Fungi	Pulmonary, CNS
Escherichia coli	Bacteria	Pulmonary, GI
Haemophilis influenza	Bacteria	Pulmonary
Influenza	Virus	Pulmonary
Legionella species	Bacteria	Pulmonary
Mycobacterium species	Bacteria	Pulmonary
Pseudomonas aeruginosa	Bacteria	Pulmonary, urinary, blood
Streptococcus pyogenes	Bacteria	Pulmonary, skin (Group A *Streptococcus*—GAS)
Staphylococcus aureus	Bacteria	Pulmonary, skin

As emerging diseases become a greater concern in the HCI, so will their association with HCW. SARS will not be the last emerging disease, but the beginning. Recently, other old, but also in some ways new, infectious agents have emerged (e.g., monkeypox, TB), with the greatest hazard on the horizon appearing to be avian flu. However, there must be a caveat to this, in that it appears as of now to be the next agent that may cause a pandemic. In the past century, there have been three to four pandemics related to influenza, depending on how you count an event. It appears based on world history that this is about the average occurrence in time for influenza pandemics. It should not be considered as whether this disease or a close relative will occur and result in a pandemic, but rather when will it occur[39] and what will be the consequences?

Emerging Infectious Diseases

There are a large number of identified emerging infectious diseases. Each of these pose a special hazard and concern for the HCW. However, in regard to protection, there are several basic tenants related to protecting workers from a disease entity. In order to understand protective practices for any specific disease, it is necessary to understand the organism's life history and when its most sensitive stage or stages exist. However, as mentioned, most if not all diseases are infectious pathogens, with some just opportunistic. A list of some emerging infectious diseases is shown in Table 4.3. Included in this table is TB, which is becoming commonly antibiotic resistant (e.g., XDR). Many of the historical diseases, like TB, can be considered emerging because they have changed their characteristics or have for the most part disappeared and are now reappearing or reemerging.

There are other diseases, which can be considered to be emerging, but have posed little hazard and may be more considered a novelty agent. This includes, for example, monkeypox, tanapox, and camelpox, most of which are zoonotic diseases. However, such agents may pose a hazard in the future if there is a large shift in its genetic information and ecological conditions supporting transfer to humans, as has been related to the Nipah virus in bats.[40] Such a change could result in the microbe being transferred among people, rather than being zoonotic.

One of the greatest hazards to HCW is that from nosocomial infections. A wide variety of organisms can be included in this group, including those that are considered emerging. Table 4.4 provides a list of some organisms that have been reported to be nosocomial. Certainly this list is not inclusive but provides a reference that one can start from.

A discussion of a few of these emerging diseases is presented. This presentation is not comprehensive, but for purposes of providing a basis to understand disease in relation to PPE.

SARS

The CoV is the agent that caused SARS in 2003. This infectious agent represents the scenario commonly discussed as an emerging potential pandemic. Occurrence of this outbreak demonstrated the potential hazard that Western counties can face from a new biological (infectious) agent. This RNA viral family (coronaviridae) was historically associated with common cold and disease in animals, especially dogs. Few studies before the 2003 outbreak of this disease were conducted, with most associated in animals, especially wildlife, related to CoVs.[41] Little was known about this viral family until SARS occurred, although, it was of importance in veterinary medicine and the agricultural industry.

Mortality for this disease (SARS) has been suggested to be about 3%, although, higher rates up to 15% have been reported.[42] These rates suggest that the disease is highly virulent. The wide variation in mortality is likely reflective of care and the amount of protection provided to HCW. In some locations, the largest number of people succumbing to this disease was associated with health care.

The disease was first observed in Guangdong Province (southern China) on November 16, 2002. Initially, some thought that the disease may be the avian (bird) flu or a variation of the influenza virus (AIV). However, it was soon determined that this outbreak of pneumonia was a result of a novel coronavirus (SARS-CoV). This virus has been shown to be fairly environmentally stable, at least for a few days, but can be easily inactivated with commonly used disinfectants (e.g., chlorine).

TABLE 4.3 Some Emerging Infectious Diseases

Name of Organism	Comment
Adenovirus	Has been associated in outbreaks in military trainees, children, and institutions and as a nosocomial disease. Vaccination has been suggested to be ineffective in preventing transmission and this virus may be responsible for more mortality than previously suspected.
Ebola	Has a high mortality rate. Named after the Ebola River Valley in the Congo where it was first observed in 1976. Appears to be transmitted by body fluids and direct contact (including mucus membrane); although, limited evidence exists that it can be transmitted by an airborne route.
Escherichia coli	Has been associated with food borne disease outbreaks and the agricultural industry may serve as a "reservoir" for this bacterium, especially in association with cattle farming and locations that are influenced by wildlife.
Influenza	A common causes of respiratory disease. Concern currently exists with the bird flu (H5N1 and H9N2) as a potential pandemic agent. Human cases associated with these viruses have mostly occurred in southeast Asia. Studies have suggested that some strains may develop resistance to antiviral agents. Often identified as a future pandemic agent.
Mycobacterium	Causes TB and leprosy. Occurrence of MDR has been shown to be transmitted through respiratory droplet. Transmission can be high, with one person infecting 10–15 others each year. The organism can become systematic, resulting in a form of TB called military TB. The occurrence of XDR and MDR forms are of particular importance in disease causation and carry a chance of becoming the common form in infections. These resistant forms are also very expensive and difficult to treat.
Nipah	First observed in Hendra, near Brisbane, Australia, as a respiratory and neurological disease associated with humans and horses. This virus is sometimes also called Hendra. The flying bat appears to be an important reservoir. Has a high fatality rate; although, unlikely easily transmitted among people. Has been reported outside Australia, including Singapore and Malaysia.
Prions	There are a number of prion-associated diseases. Considerable concern exists that these may be a jump across species, especially to humans, such as associated with chronic wasting disease (CWD) that is seen in deer. Has been reported as a nosocomial disease in a few rare cases, but the possibility of transmission by routes other than blood-borne or consumption exist; although, are not supported by experimental evidence. It is likely that many different types, which are currently unknown, of prions exit. Recent evidence suggests that CWD can be transmitted via a "soil" route, like being food borne in animals.
Rotavirus	Common cause of diarrhea in children. Is a highly contagious agent which is frequently spread by hand-to-mouth routes. Treatment to prevent dehydration can be effective, including oral rehydration therapy. A vaccine exists which appears to be effective in prevention of this disease.
Streptococcus	Group A (*Streptococcus pyrogenes*) and B (*Streptococcus agalactiae*) (Group A Streptococcus—GAS and Group B Streptococcus—GBS) are common causes of disease. These agents can result in a high mortality rate and can form biofilms. Recently, *Streptococcus suis*, a pathogen of swine, has been isolated and identified as an infection in people. Antibiotic resistance in these microbes can make them an emerging disease as is occurring with TB.

A close genetic isolate to the human SARS-CoV has been found in palm civets (*Paguma larvata*) which are sold in local southern Chinese markets and may be a reservoir or transmitting agent for this virus. It is not known if other animals, such as bats and ferrets, are also carriers of SARS-CoV, but there some speculation that they do act as reservoirs.

The disease appears to be spread by droplet transmission originating from the upper respiratory system. Historically, this agent was identified as a super-spreader; although, its association in many of the patients that became infected were HCW indicating that this agent was an important occupational hazard (disease). It has been suggested that if appropriate and effective PPE, especially respirators and hygienic conditions were implemented at the beginning of the outbreak, there would have been a much lower infection rate of HCW.[43] Others have suggested that the outbreak was eventually controlled, at least for HCW, through applicable implementation of PPE and infectious disease preventative

TABLE 4.4 List of Various Nosocomial Diseases

Aspergillus
Chicken pox (varicella-zoster virus), including shingles
Corynebacterium
Clostridium difficile
Cytomegalovirus
Ebola (Viral hemorrhagic fevers)
Enteroviruses
Hepatitis A, B, and C
Herpes Simplex
Influenza
Legionella
Mumps
Neisseria meningitides
Norovirus
Pertussis (whooping cough)
Human immunodeficiency virus (HIV)
Respiratory syncytial virus
Rotavirus or rota virus
SARS
Staphylococcus (MRSA and VRSA)
Streptococcus (GAS and GBS)
Rubella
TB

practices.[44] The hazard experienced by HCW is supported by Escudero et al.,[45] where the attack rate for HCW was higher than that of others in close contact with infected patients. Some of the infections in HCW can be attributed to treatment regimes given by HCW that acted as a spreading mechanism for the virus. For example, nebulization in providing breathing treatment of infected patients appears to be an ideal method for spread.[2] These activities, combined with inadequate ventilation in patient rooms, appear to be important in nosocomial transmission of the SARS-CoV in 2003 and provide a valuable learning tool for future outbreaks of infectious diseases.[46] It is likely that this will also occur for other organisms that can be "easily" transmitted through an airborne route, like avian influenza.

Other routes of transmission have also been suggested for this virus, besides aerolization.[47] These routes include urine and feces, with some suggesting that the sewage disposal system contributed to disease cases.[48] Since this was a novel disease and outbreaks of this nature were not, and still are not well understood, there was a sharp learning curve for implementing protective practices, especially as related to HCW. For the SARS event, concern also arose for what could be identified as auxiliary HCW. This population includes ambulance, x-ray (radiobiological), and maintenance personnel, as examples. Historically these groups were usually considered to be at least risk from hospital borne diseases, but after the SARS event this view has changed. However, lessons learned can be used in future outbreaks of disease and provide a practical case study in handling rapid pandemics of novel microbes. Differences among SARS rates in Chinese hospitals have been suggested to be a result of varying infectious disease preventative practices.[49] This suggests that preventive practices have value and benefit, although, not easily seen, especially before the SARS event. Those that had good nosocomial control measures appear to have had lower rates of infection among HCW.[50] This demonstrates the importance of infection control measures and use of PPE for all infectious diseases, including that which may be associated with influenza and other organisms that have been identified as super-spreaders. This disease event demonstrated the importance of training of all HCW on use of PPE.

Monkeypox

This disease was first identified in 1958 in a laboratory monkey, but was later shown to exist in wild populations in central and western Africa. There have been reports of monkeypox infections in other animals, including squirrels, rats, mice, and rabbits.[51,52] This microbe belongs to the viral family orthopoxviruses, which includes smallpox and cowpox. Although a few cases of human infection have occurred, most do not consider this organism to be an important disease. However, it has been identified as a potential emerging disease and may be considered like the CoVs (SARS). This alone makes the organism important for study, in that little is known and change can result in a pandemic, which has been recognized.[53] A recent local epidemic illustrates the importance of this organism as an emerging pathogen for humans.[54] The first human case related to this disease was in 1970.[55] In 2003 there was an outbreak of monkeypox in the United States, which was associated with exotic pets,[56] specifically prairie dogs.[57] Within a short period of time there were at least 71 reported cases or suspect cases of the disease, along with one suspected occupational case, which could be considered an OND.[56] This outbreak included a family cluster that had reported a variety of symptoms ranging from minor rash to a neurological infection (*Morbidity Mortality Weekly Report–MMWR*, 2003). The origin of the virus was traced to imported exotic animals from Ghana, West Africa, which through a pet store infected prairie dogs that were later purchased.[58] As a result of this outbreak, there was considerable concern that a wider epidemic would erupt, especially since there were reports in multiple states. What initially gave great concern is that the disease has some features that are similar to smallpox, except for monkeypox cases there is swelling of lymphnodes and this is a distinguishing factor between the two diseases.[51]

The disease does not appear to be easily transmitted from person-to-person; although, limited evidence exists that it may be transmitted through an aerosol route (Maskalyk, 2003). However, person-to-person transmission can occur and it has been suggesting that the organism does have the capability to evolve into a human-associated pathogen from one that is zoonotic (Maskalyk, 2003). Even with this limited transfer among people, the CDC has suggested that this virus is a risk for HCW. It appears that the smallpox vaccination provides protection against this pox as well and is warranted for HCW involved with the care of patients. Although the immunity provided is not absolute, it appears to be sufficient to allow the formation of herd immunity in a population.

Tuberculosis—Antibiotic Resistance

TB is caused by a variety of species of the genus *Mycobacterium* and is mostly a disease of the respiratory system. Species that are commonly associated with this disease include *M. tuberculosis*, *M. bovis*, *M. africanum*, *M. microti*, and *M. avium*. This organism is commonly spread through an airborne route and has been recognized as an occupational hazard for HCW for some time. Spread is usually a result of person-to-person transfer; although, it can be spread by other routes (e.g., milk). However, in most developed counties and with most milk and related products being pasteurized, food products have become an unlikely route of transmission. TB is an important disease in developing counties, with the WHO estimating that there are 2 million deaths as a result of this disease. One estimate suggests that approximately one-third of the human population is or has been infected with TB, with a higher percent associated with men than women. In many ways, this disease is an unrecognized plague and one that is gaining momentum through antibiotic resistance (e.g., MDR-TB and XDR-TB).[59]

The occurrence of TB that has drug resistance has become a major public health and occupational hazard. Traditionally, the drugs ioniazid and rifampin were the main choices in treating this disease. Normally drug-sensitive TB can be treated in a time period of about 6 month, if the therapy is continuous. Forms that are MDR can require 2 years or more of treatment with drugs having a large number of side effects and cost 7–22 times more than nonresistant forms to treat. It as been estimated that up to

50 million people may have MDR-TB worldwide and is on the rise due to improper therapy, inadequate dispensing of drugs, and failure to complete a treatment regime.[60]

The rise of XDR-TB is even of greater concern. XDR-TB is defined as being resistant to drugs ioniazid and rifampin, resistance to any fluoroquinolone and at least one second-line injectable drug (e.g., amikacin).[61] This form of TB is very difficult to treat and treatment is not practical in most developing countries further continuing its spread and is emerging as a global health problem.[62] In most cases, especially undeveloped countries, XDR-TB is almost always a fatal disease.

Resistance to TB arose from inadequate treatment and nonadherence to therapeutic regimes. Some populations, such as that associated with prisons have become locations for the development of antibiotic resistant TB. It has been reported that the case rate of TB in a prison can be 50 times greater than that seen in the general population and 50% greater in the number of MDR cases.[63] Employees and HCW in these institutions are at risk as well from infection. In a study of New York state prison employees, approximately one-third of the new cases of TB appeared to be a result of exposure from the person's employment.[63] This has even become more critical in other countries, like Russia, where antibiotic resistance for TB has become very common.[59] The rapid and continuous occurrence of antibiotic resistance in these locations has become a global health hazard.[64] This is also occurring in other locations where populations have inadequately controlled infectious disease health problems and poor sanitary/hygiene conditions exist (e.g., homeless).[65]

Approximately 4% of all new TB cases today are MDR in Eastern Europe, Latin America, Africa and Asia, and this trend is emerging in the United States as well. The emergence of such resistance is both a global health problem and one for HCW and is no longer restricted to third-world countries.[66] The impact from locations with high TB rates along with a high percent of MDR will have a strong impact on neighboring counties and make it difficult for those areas to control the disease.[67] Overall, these locations serve as a global reservoir for drug resistance and without control will allow continued spread to other parts of the world. It is suggested that the groups at greatest risk, at least initially, appears to be HCW, with such impact being seen in the locations where emergence is occurring, like Russia.[68]

Some have suggested that the Bacille Calmette-Guerin (BCG) vaccine should be administered to prevent TB.[69] This vaccine has been reported to provide protection in adults against the pulmonary form of the disease, and invasive complications in children, including the military form and meningitis. However, this vaccine is not without risk, where adverse effects result in about 5% of those vaccinated.[70] Senanayake and Collignon[70] suggested that when rates of this disease are low, control using isolation, negative pressure and PPE are more effective and efficient than that associated with vaccination.

For HCW, the risk of infection from TB greatly varies from location to location. For example, Australian HCW that were providing care overseas had an infection rate of 9.8 per 1000 persons-month (12% a year), which is much greater than that seen in Australia itself.[71] The rate in Australia is about 5 per 100,000. This becomes a greater concern in that many of the developed/undeveloped countries, like in the southeast Asia area, have a high rate of MDR-TB acting as a reservoir for these forms of TB. In Australia the rate of MDR TB is about 1%–2% of isolates, where as in some Chinese locations it is 2%–7%. A common question regarding this issue is whether HCW should receive the BCG vaccine for TB. It has been reported that vaccination can provide about a 40% protection from this disease in children that have household contacts.[71] For adults, the risk can be reduced to about 50%, with much of the reported variation a result of the differing strains used in vaccination, as well as age when vaccinated. It has been shown that the vaccine is fairly safe, with a fatality rate of 0.06–0.72 per million and most of these cases in immunocompromised persons. The major drawback given for the vaccine is that it interferes with testing. However, newer tests that are emerging and being developed are not affected by the vaccination possibly making this issue moot.

Influenza (Avian Flu)

Many consider influenza to be the most important emerging or exiting infectious disease agent that has the capability of causing a pandemic.[6,72] Influenza has existed since the beginning of man and appears to periodically cause worldwide pandemics and is in the viral family Orthomyxoviridae. Every year there is a new stain or set of stains of this virus that emerges. This results in millions of infectious cases and an unknown number of deaths. In most cases, the virus does not result in a true pandemic, but rather an "epidemic"; although, this event is by relative definition. Epidemics from influenza occur every year. However, it appears that about 3–4 times each century there is a major mutation in the virus that makes it highly virulent and in some ways more infectious. Pandemics that occurred in the twentieth century were in 1918, 1957, and 1968.

There are three major groups or types of influenza viruses, which are A, B, and C.[73] The type of greatest concern is A and is the causative agent of pandemics. This type can infect people as well as a wide variety of animals, including, horses, pigs, birds, camels, seals, mink, whales, dogs, and other animals.[74] Type B influenza on the other hand can cause epidemics; although, is more related to illness in children. Overall this virus generally causes mild illness and does not result in pandemics; although, can result in epidemics. Type C can infect people and animals, causes a mild form of the disease, but is not responsible for epidemics. The reason for the rapid and frequent changes associated with the influenza virus is the lack of proofreading mechanism(s) in its polymerases. This allows uncorrected changes in the RNA (genetic information) to be easily passed on, which permits rapid change in the organism. As a result there is a rapid antigenic drift occurring permitting a change in virulence and this leads to a functional antigenic shift. Each year there is a small change in the influenza virus, which is called antigenic drift. If the change is large, mostly resulting in a novel virus, this change is referred to as antigenic shift. When an antigenic shift occurs there is usually a resultant pandemic to follow, which is the concern associated with AIV.

It has been suggested that the next pandemic will be a result of the avian flu (AIV) (Influenza A/H5N1).[75] The nomenclature for the flu virus is the first letter is the strain, then followed by it origin, like Spain, then a strain number is given (the type strain), followed by the year it was isolated, and finally the viral subtype. So here, the one provided (A/H5N1) is type A, haemagglutinin 5 and neuraminidase 1. Haemagglutinin (HA) and neuraminidase (NA) are surface glycosylated proteins and are recognized by antibodies; thus, are used in serotyping the viruses. These receptors are also targets in drug therapy for this virus and are important for viral attachment and release. There are 16 HA and 9 NA types currently recognized. HA is the main site for attack by humeral immunity. Binding of HA to sialic acid residues is what accounts for much of the host's specificity to a specific influenza virus. These binding sites are found in the respiratory epithelial cells. NA helps in spreading this virus through cleaving linkages to sialic acid, which are glycosidic linkages, as well as on the virus. This is mechanism of many antivirals that are used in preventing influenza infections (e.g., tamiflu®).[76] The incubation period for influenza is around 2 days, with a range of 1–4; and can be infectious for about 5 days (adults, children about 10 days).[73]

The primary reservoir for the flu virus is birds, including chickens. In a few cases there has been transfer of the bird flu to man and this has resulted in limited disease.[6] This transfer of viruses between men allows the virus or reorganize its genetic information which can result in a new strain emerging, one that can be more effective in disease causation for man.[72] Historically, it has been considered that two viruses are needed to infect the host, which has been considered to be swine, with a resulting rearrangement of a new and novel virus. However, recent information suggests that this may occur through other means with a combination occurring in organisms other than swine, including man. Man can also be wholly infected by avian "specific" strains as well, thus, swine, as an intermediate, may not be necessary for the "formation" of a pandemic strain. Regardless, the occurrence of bird to man transfer and subsequent infection can result in the virus going from a low virulence in man to one that is high and easily transferred. Since birds migrate throughout the world, this movement can allow the virus to be spread via birds. The bird flu has occurred in many locations of the world resulting in the culling (killing) of millions of domestic fowl (e.g., chickens, ducks) with H5N1 appearing to be the dominant strain.[76] The most recent information suggests that this virus (H5N1) is epidemic in the southeast Asia area and exist in wildlife as well (e.g., wild birds). There have been a few outbreaks of this virus which involved people, and many of these resulted in high mortality rates, as was seen in the 1918 flu.[77]

In the 1918 flu some communities experienced morbidity rates of 25%–40%, with children, those under the age of 15 experiencing the highest rate.[72] The number of deaths in the United States from the

1918 flu was estimated to be about a half-million and worldwide between 20 and 50 million.[78] This may be a result of younger populations not having seen influenza viruses before and older individuals having some cross-reactive immunity with the emerging virus.[6] It is possible cross-reactivity may also exist for those that were vaccinated with a similar strain(s); thus, a long history of yearly vaccinations may have some benefit. Based on these values of 25%–40% morbidity, as an estimate, there would be more than 100 million deaths worldwide, with HCW likely having a disproportionately higher rate than the general population. At the present time, it is suggested that commonly used antiviral agents would not be effective against this strain.[77] It appears that the change needed for occurrence of a pandemic strain involves the HA and the pool of genetic information for such a change is large, especially when avian strains are examined.

Influenza has a good survival rate under normal conditions. For example, the virus can survive in water for at least 4 days with a temperature of 72°F and at 32°F this survival time can increase to 30 days. To effectively disinfect surfaces and hands, contact with the disinfecting agent needs to be at least 15 s, such as when washing hands; although, some suggest that the time period should be more in the range of 20 s for hand washing(with warm water).[79] Too effectively wash hands, cleaning of the wrists, fingers, and between fingers with soap and water must be undertaken for 20 s, with the suggestion of turning off the faucet using the towel which was previously employed for drying hands.[79–81] Hand washing can also be accomplished with nonsoap and water solutions such as alcohol-based solutions.[2]

There have been a number of cases of Avian flu transmission from birds to man[76] and now there have been a few reported cases of this virus going from man-to-man.[82,83] There were 13 cases and 9 deaths reported in Vietnam from the Avian virus suggesting that some strains are highly "pathogenic" and virulent to man.[84] Such occurrence suggests that jumps in the virus are beginning to occur and it is a matter of time until a strain emerges that is efficient in transmission and spread among people.[85] Parry[86] in a commentary reported that person-to-person transmission of the Avian flu was occurring in Cambodia. However, other outbreak events have not supported wide spread transmission, with some involving apparently no human transmission. These events do indicate that the species jump is being "attempted," as would be expected. As has been noted, when the event does occur, HCW will be on the frontline of exposure with vaccination and protective measures most paramount. Besides PPE and engineering controls, recent evidence suggests that cross-reactive immunity occurs with live flu vaccines against AIV's providing some protection for HCW that have been vaccinated with other strains. This alone makes a good case for yearly vaccination of HCW against the current flu.

Although vaccination is not a PPE-related activity, it can be considered an adjuvant. Vaccination of HCW to agents that they may be exposed is paramount and forms a basic frontline defense. Certainly vaccination is not perfect but it will provide some protection for most people. When this is included with PPE, the risk of becoming infected is greatly reduced.

Protection against Emerging Infectious Diseases

From the early beginnings of man, infectious disease has been a major health concern and historically was the major contributing factor for early death. Most of the diseases that are now identified as "emerging," which includes pox viruses, TB, and influenza, were common throughout the world, including developed nations. These diseases have been with man since antiquity and are well known in man's history. For example, TB was discussed in the early Greek literature (ca. 2400 BC) and was even considered, in that time period, that this disease could be an occupational hazard to physicians. Spread and occurrence of these diseases were mainly a result of inadequate knowledge of their life cycles. However, today, there are numerous poorly known and understood organisms (e.g., monkeypox, tanapox) that have the potential of becoming the next worldwide smallpox or SARS. What prevented the dissemination of many of these lesser known diseases was geographic isolation. Natural barriers kept many of the diseases in check and outbreaks local. Periodically, some have become regional or larger, as was seen with bubonic plaque in 1556–1559.[87] However, with the increased speed of travel and the world becoming a true community, spread of a new or old disease can occur within the matter of days and well before its "existence" can be realized. What is also not realized related to such outbreaks is the cost of carrying for the sick, which was evident with bubonic plaque,[87] and prevention can greatly reduce these costs. These economic impacts do not only occur for the HCI, but the community at large.

Epidemics occur on a frequent and regular basis, as seen for influenza. When they become global they are called pandemics. The group(s) on the frontline of these disease occurrences are HCW. When evaluating spread and transmission of disease, there are four primary ways of preventing spread, with three of these being of importance to HCW. These preventative methods are vaccination, barriers, and use of PPE, with the forth method being isolation. Isolation is commonly used for patients, like those with TB, but is not considered for HCW since they are the ones treating these diseases (e.g., pulmonary and military). Thus, HCW cannot be isolated from the infected patient. However, some practices used in isolation can be extended to HCW such as increasing ventilation rates; although, benefits of this nature are often greater for patients. Unfortunately there are many diseases which there is no vaccination or no effective vaccination. This was seen in the SARS outbreak and may occur with the Avian flu. Other diseases like monkeypox there is a cross-reactive vaccine, smallpox, which can provide protection. In such cases, vaccination is quickly needed for HCW since it can take 10 days to 2 weeks for immunity to be effective. It should also be realized that a vaccine is not usually 100% effective for all people. Thus, there will be some HCW for which no immunity is provided, even though they were vaccinated. However, this possibility cannot be a basis for not vaccinating HCW.

That leaves only two effective methods for prevention of disease in HCW: barriers and PPE. Barrier measures commonly include physical measures, such as tents, to prevent the spread and transmission of a disease from a patient. These methods have limited application here, since HCW need to be in close contract during the patient's treatment. Barrier methods are actually a form of engineering control; although, for some diseases, on a limited basis, may be of use. During the SARS event, increased air ventilation in rooms, which has been considered by some a barrier method, was shown to be effective in Vietnam.[128] There is always consideration for combinations of methods. Here, respirators were not readily available, and the increased number of air exchanges helped to prevent spread to the disease to others.[88] This does indicate that imaginative employment of barrier-type techniques can be useful in preventing and controlling, even in a limited way, the spread of some infectious diseases. Also, as has been known in the industrial setting, respirator is the least desirable method for protection with engineering control the most favorable in selection. Thus, expansion of these engineering control techniques along with other forms of protection can help to reduce the incidence of disease in HCW and most likely other patients as well.

There have been reports suggesting that patients who are infected with disease (respiratory) should use respirators. Koley[2] suggested that patients be issued eye protection and respirators. The problem with this is that the filtering mechanism of respirators is on the intake and not the exhalation port. Thus, any agents (infectious material) that are expelled by the patient will not be filtered; although, air being breathed in will be filtered. This would have little if any effect on hindering the expelled agent, such as a virus. There may also be difficulty in establishing a face seal in someone that is ill, especially for those that have pulmonary issues.[88] To increases the difficulty with the seal, it is likely that disposable respirators would be employed further complicating the situation. Physiological stress (e.g., cardiopulmonary) of breathing with a respirator would also occur for those using these devices.[89] For patient's that have compromised pulmonary systems, this may increase the risk of complications related to their disease and make breathing difficult. Thus, based on the information presented, it is inappropriate to have patient's employ respirators and likely most other forms of PPE for purposes of protecting HCW. Use of such mechanism may actually enhance the spread of disease and increase risks for patients.

Personal Protective Equipment

PPE includes a wide variety of equipment and in some cases materials or substances. However, in general, this is equipment or material that is employed by the individual for that person's protection against some specific or general hazard. Some of the most common forms of PPE are listed and identified in Table 4.1. It is common for HCW to use a half-mask respirator and studies have reported that these respirators are effective against airborne TB, especially when there is a good face seal.[90] This does not provide protection for the eyes; although, in some cases this site is not important in regard to protection against the agent, like in TB. Certainly TB can occur in the eye, but this is not common.[91] For other diseases, as was seen in SARS, eye protection is of great importance. What is commonly forgotten in regard to PPE is the importance of hand washing. It is usually suggested that hand washing be conducted for 15 s, vigorously, using soap and warm water. This will not kill all organisms and does not render the hands "sterile," but is for elimination of "pathogens" and by some is a cornerstone in infection disease control. However, this activity alone will not eliminate infectious agents. A common form of PPE for the hands is surgical gloves. These are routinely used in hospitals and are readily available. When these are applied with a gown, often the gloves can be pulled over the protective suit and creates a good seal. This may provide a "full" type body protection. When combined with a respirator, considerable protection can be afforded to the HCW. Again, if a half-mask respirator is employed, eye protection will be required as well. This can be in the form of goggles. In some cases a face shield may also be warranted.

In addition to these devices, other forms of PPE may be applicable such as shoe covers or foot protection. This is more for preventing the spread of microbes from the patient's room or location to other areas. Spills and leaks commonly occur and can be transmitted on shoes or related types of clothing and are not often thought of as a potential route of transmission. Although this covers the body, overall, it does not necessarily provide a complete seal against entrance of microbes through small openings or even tears may occur. It may under some circumstances require the HCW to shower after contact with a patient that is highly infectious; although, as in the SARS events, there was concern that spread of this virus was occurring through the sanitary sewage system. Fortunately, such extensive measures are not necessary for most microbes, even the ones identified as emerging. However, HCW should shower after work as a preventive measure. There are a few agents where such extensive activities may be warranted, such as Ebola-related viruses; here there is little known about the organism and mortality is nearly or approaching 100% for those infected. This may be applicable especially when the life history is unknown and the disease appears and then disappears as is seen with Ebola and Marburg agents making study and treatment difficult.[92] For some, the hazard of the agent cannot be underestimated, especially with such organisms like Marburg, where there is a 92% fatality rate and an amplification of the disease may occur in health care settings putting HCW at the greatest risk. Such an outbreak or release on a global scale would be disastrous and with the advent and occurrence of bioterrorism becomes a real concern. This alone raises the hazards associated with emerging diseases to a new level prompting the importance of PPE and its applicability in providing first line protection.[8]

Respirators

Respirators have been used by HCW for a considerable period of time. They are most commonly associated with care of patients having pulmonary TB. For the most part respirators are selected "off the shelf" handed out and used. In most cases this has been adequate and provided sufficient protection to HCW where there was a minimal spread of disease, but the occurrence of SARS changed much of the traditional thought on this concept. Emergence of the bird flu virus (influenza) has heightened the importance of respirators. The increased burden of antibiotic resistant microbes is also changing the landscape how infectious disease is viewed, especially as related to HCW. Much of the information related to respirator use comes from the industrial environment where various practices have been implemented for an effective and efficient program. Here, respirator fitting and testing is commonly performed, along with applicable training. Lessons from the industrial environment can provide a basis of practice for the HCI. Overall, it appears that a proper face seal is one of the critical components related to proper use of respirators.[93]

To achieve an effective respirator program it is necessary to have a training and selection program.[93] In many ways, each situation is a bit different; although, there are many characteristics in common which exist among programs regardless of the activity. One of the important aspects of a program is its maintenance and administration. Some of the programs will follow criteria established by the agency responsible for Occupational Safety and Health, here in the United States that is the U.S. OSHA. Most countries have established criteria that are similar or identical to that of OSHA; however, some of the European countries there are some differences. OSHA has a number of requirements for respirator use and activities and they would include those in health care. In the United States, requirements are described in the Code of Federal Regulations (CFR) and these can now be accessed through the internet. The construction, general, and maritime industries have their own codes although are similar. Thus, as indicated, there are three general categories of industry which OSHA applies regulations: construction, general, and maritime. Probably, the best to follow, generally, is the general industry, for health care settings. Here for example, the asbestos regulations are described in 29 CFR 1910.1001 and provide information on respirator use that are mostly related to HEPA filtration. These requirements have some information on use of respirators which are attempted to be more specific for that activity or industry. Unfortunately, there are no specific regulations under OSHA for the HCI; although, there is a regulation on blood borne pathogens. There are regulations on other forms of PPE, like for eye protection, head protection, and so forth. Most of these requirements have little applicability to HCW since these were designed for the construction and general industry.

Although these regulations for PPE are not designed for HCW they can be used in many ways as a general model in establishing a PPE program, especially for respirators. HCW do need to be aware that there are regulations for respirators. Overall, these regulations can be used to guide activities, especially in selection and fitting; although, are not designed for those being exposed to infectious disease agents. Since HCW may be exposed to agents that have not been previously known or recognized, there can be unique circumstances relating to PPE not addressed or considered by current regulations. In most cases, prevention of entry of the organism involves a physical barrier or mechanism of removal. For respirators, this will usually involve the use of a HEPA or a ULPA filters.

During the SARS outbreak, it was suggested that some cases of this disease in HCW was a result of inadequate or even improper use of respirators.[14] Selection of the appropriate respirator for SARS has been of considerable debate.[94] This confusion can be said to have resulted as a result of inadequate information on the life history of the organism. Initially little was known how the organism was spread with some suggesting that inhalation was not the most important route. However, since this was a pulmonary disease, it only makes sense that one of its likely routes of transmission would be inhalation. After information became available that it was transmitted by an aerosol, considerable attention was then given to respirators as a protective mechanism for HCW. In many ways this resulted in a panic and supplies of respirators often ran short. As with any emerging disease, including those that change characteristics, little will be initially known about the organism and considerable caution will be needed. Thus, the lesions from SARS must be taken for future events, especially when considering influenza. Here respirator protection was effective at minimizing exposure and subsequent disease. Other forms of PPE are also of great importance along with a myriad of other related activities.[95] This again supports the concept of engineering controls as being the most important method for preventing exposure, even for infectious diseases. In all cases, there must be a combination of practices implemented in protecting workers. It must be remembered, that the first line of protection is that associated with engineering controls and not respirators, but most consider respirator use to be more of a primary protective mechanism.

In any case, selection of the appropriate respirator is appropriate. Most do not consider surgical masks to be a form of a respirator.[12] These devices have little if any protection factor and were designed to prevent release of droplets and particles from the users mouth and nose to the environment in a "gross" way and themselves will not prevent release of viral particles or the like.[96] Thus, these devices will provide little protection against airborne microbes and are not a substitution for a functional respirator.

For the SARS event, which involved a virus (about 100 nm in size), most recommended N95 respirators.[97] Others[37,43] have suggested that this form of respiratory protection was not adequate for infectious agents of this nature. Initially Seto et al.[98] recommended the use of surgical masks, but not paper masks. Derrick and Gomersall[99] evaluated the use of multiple surgical masks and reported that they had a protection factor that was less than that reported for N95 respirators. Such effectiveness would even be lower for small viruses and other microbes. The N95 respirator has been suggested for HCW when working with TB patients and is the standard of protection against this organism.[100] They have been in use for TB patients since 1994 and are considered disposal respirators. Today, it is probably the most common respirator used in health care settings. However, it appears that there are differences in these respirators among manufacturers, at least in regard to filtration efficiency.[90] The most important factor in using these respirators appears to be associated with face seal (Lee et al., 2005).

Respirators are designed to provide protection for the respiratory tract. The respiratory system is often identified as an important organ of exposure; although, the largest number of occupational occurrences is associated with the skin. Usually, the most severe injuries, occupationally, are that in association with the lung and respiratory tract. Infectious disease agents can attack at any location in the respiratory tract. However, larger particles tend to be deposited more in the upper parts of the airway, like the nasopharyngeal area, where as particles that act more like a gas are deposited in the alveolar region. This large particle (say 30 μm) will be found in the nose, as an example. Particles which are about 5–30 μm would be in the nasopharyngeal location, those 1–5 μm in the tracho-bronchial, and about 1 μm and less the alveoli.[100] There are basically five mechanisms which particles are deposited in the respiratory tract: impaction, sedimentation, interception, electrostatic deposition, and diffusion. There have been three models established to describe deposition of particles, which are the American Conference of Governmental Industrial Hygienists (ACGIH), AEC (now NRC—Nuclear regulatory Commission, AEC—Atomic Energy Commission), and British Medical Research Council (BMRC). Although there are differences between these models, from a practical point of view they all follow the general concept of larger particles being removed by the upper part of the respiratory system and the smallest particles (say about 1 μm) being deposited in the alveolar region. Most consider the size and length of the microbe when evaluating its penetration into the filter. Some recent investigations have suggested length rather than aerodynamic diameter is most important in making this evaluation.[12]

Viruses are generally the smallest particles of concern; although, prions are smaller since they are infectious proteins. Viruses range in size from about 20–400 nm, which is 0.02–0.4 μm, and would be predominantly deposited in the lower part of the respiratory tract. However, they may also have a strong electronegative or positive charge, which influence the location of deposition. Bacteria are generally much larger in size and commonly range from about 0.4–10 μm in size. This would suggest a deposition in the upper regions of the lung, but again would be influenced by electrical charge. Mycobacterium has a size of about 0.9 μm, while mold, such as *Cladosporium cladosporioides* is around 2.1 μm.[17]

Routes of Exposure

The primary route of exposure to most airborne microbes is inhalation. The effect of filtration can have a large influence on the penetrability of microbes. Most respirator studies examine a constant flow rate of 20–85 L/m. However, in actual use there will be periodic high airflow rates during high work loads. Increased penetration of small particles has been suggested to occur at high flow rates; although, those that are small, 1 μm or smaller, there does not appear to be any effect.[12] When examining the size of microbes, this concern relating to penetration exists for the "smaller" viruses and prions. Although this is an important factor, the face seal is probably the most important especially since many HCW will not be likely fit tested before using a respirator.

An important issue often over looked is decontamination and disposal of respirators and filters. Decontamination can be an important aspect of any program, especially when an organism can survive for several days, like AIV. Maintenance and cleaning of respirators and parts must be included as part of any training program. This may be an unrecognized route of exposure to HCW. It should be noted that chlorine, especially prolonged exposure during cleaning can damage the respirator or components. Chlorine is a very electronegative element.

Types of Respirators

There are various types of respirators available and from a number of manufacturers. Respirators in the United States have to be National Institute of Occupational Safety and Health, Mine Safety and Health Administration (NIOSH/MSHA) approved. Each respirator has its application and limitation, but can be placed into six general groups. These are paper mask, quarter mask, half-mask, full-face mask, powered air-purifying respirator (PAPR), and supplied air respirator (SAR). Paper masks and surgical masks are not true respirators, however, have been used in HCI as respirators. All of these respirators are air-purifying respirators (APR), except SAR, in that they have a mechanic filter that removes the contaminant of interest, such as a bacteria. The most common respirator used by HCW for infectious diseases is the N95. Table 4.5 provides a summary of these types of respirators. In the United States, quarter masks are not used.

Most respirators used in the industrial environment are not disposable. In the HCI most respirators are disposable and for that reason the N95 is the most common used. Respirators that are APR in nature will not change the oxygen level in the surrounding environment. These respirators use a mechanical or chemical activity to remove the agent of concern from the air that is being breathed by the user. Most air contains about 21% oxygen and generally variation of oxygen is not a concern for HCW. The minimum oxygen level allowed by OSHA is 18.5%. Locations with oxygen lower than this is defined as oxygen deficient. Filters that are commonly used to remove microbes are HEPA and ULPA.

Surgical masks and paper mask have been designated by some as respirators. However, these devices do not allow an effective face seal to be established. During the SARS events, surgical masks were used by HCW for respiratory protection and some reported effectiveness.[98] Both surgical and paper masks are really barrier devices and have limited ability to prevent inhalation of microbes. Several studies have

TABLE 4.5 Common Types of Respirators and Some Characteristics

Respirator	Characteristic
Paper masks	Single use devices. Designed to protect the patient from droplets emitted from the wearer and does not protect user from airborne materials. Shown to be low in effectiveness in protecting against airborne infectious diseases. These devices are often not classified as a respirator.
Quarter mask	Not used in the United States. Does not cover the chin area as compared to half-masks. These respirators can be used with HEPA filters.
N95	Recommended respirator for TB protection. Common in health care and used by many HCW. Can provide some degree of protection, but does not provide full face protection. Disposable, easy to use and low in cost.
Half mask	Commonly used in the industrial environment. Considered effective in preventing inhalation of infectious diseases. Does not provide full face protection. Can be used with HEPA filters. A little more difficult to use, usually each person is assigned their own respirator, higher cost than N95.
Full face	Similar to half mask, but provides full face protection. Has limitation in that it can cause fear in patients and can fog up during use. Can be more difficult to maintain and more expensive than a half mask.
PAPR	A positive pressure respirator that employs a filter such as HEPA. Has a higher PF. Can be costly to maintain and more complex to use. When using with patients has limitations mentioned for the full face. Requires a battery for power and can add extra weight and difficulty with user. Highest cost among APR's.
SAR	Uses an air supply that is remote from the work area or patient location/setting. Has the highest level of protection. Complex to use and has limitations due to the difficulty in use. Not applicable for normal health care settings; although, used in research situations when working with highly dangerous organisms.

evaluated the effectiveness of these devices. A study by Nicas[18] reported that surgical masks allow the entry of 42% of droplets. This study does not indicate the PF afforded, but alone suggests a high amount of aerosolized material would enter the users' respiratory system. In another investigation on the filtration efficiency for surgical masks, Grinhpun[17] reported that they are about 20%. These results indicate that there is a large amount of penetration in surgical masks and that the efficiency of particle removal is poor to what some may even refer to as nonexistent. Based on these data, the paper masks, which have a poor fit, will likely have a low PF and be ineffective against airborne droplets and particles, such as viruses. Respirators, excluding paper masks, are commonly divided into three groups: quarter masks, half masks, and full-face masks. Quarter masks are not used in the United States and this type of device is not generally recommended as a respiratory device. These respirators do not go over or cover the chin. Half masks do go over the chin and are commonly used in the United States for respiratory protection. These respirators, however, do not provide protection to the eyes and can be used with various types of filters. Full-face masks do provide protection for the eyes and can also be used with a variety of filters as with half masks. All of these respirators are negative pressure air-purifying (APR), and it must be noted that they will have no change in the oxygen level for the users air; although, oxygen and gaseous environments are generally not a concern for most HCW. In addition, some respiratory systems have an air supply and these are self-contained breathing apparatus (SCBA) and SAR. SAR, including SCBA, are not practical under most conditions in health care setting. Such devices and systems can and are used in experimental situations, especially where the organism being evaluated is highly virulent, such as for some viruses (e.g., Marburg virus). Historically, the respirator used by HCW was the N95 and this respirator has been suggested for use in protection against TB.[102] However, it must be noted that one type of respirator will not be applicable for all situations.

For most diseases discussed in this chapter, their route of transmission is through the respiratory system; thus, the appropriate mechanism of transmission prevention is usually mechanical filtration. Mechanical systems trap particles, which in this case would be infectious disease agents. These devices collect particles by simple straining or depth filtration. Here simple straining prevents the particles from passing through due to its pore size and in depth filtration with the particles penetrating the filter becoming adhered inside the filter. Both of these devices are physical barrier (mechanical) systems that actually prevent the particle from entering and in this case the microbe acts as a particle. It should be noted that microbes often have a charge which aids in their filtration. Bacteria have a negative charge to them that will assist in remove from the air by a filter. Forces associated with the charge of the particle and mechanical filter or filtration device is called electrostatic interactions. In many cases this becomes important in filtration, especially for the smaller microbes. Other organisms, like bacteria responsible for TB (mycobacteria), are relatively large in size and are more likely physically captured. However, these organisms are also charged, and the stronger the charge attraction between the filter and particle the high the collection efficiency will be. Over time a filter can lose its charge and become primarily a mechanical filtration system; however, as filters collect particles they do become more efficient.

When evaluating efficiency of filters there are no filters for respirators that are 100% effective in removing all particles; although, as mentioned the degree of efficiency is highly dependent on a number of factors. These factors influencing capture efficiency include the particles size, its shape, charge, fit (seal), and the type of filter being employed (as related to these factors) with respirator. Grinshpun[17] suggested that the filter efficiency is related to the size of the microbe including its aspect ratio. This study also suggested that microbes do not grow on filters that are used with respirators; thus, not becoming a source of contamination. Filters on respirators were originally designed to protect against nonbiological agents, such as dust, mists, and fumes. There are two types of filters used for APR's: HEPA and ULPA. HEPA filters are commonly used in the industrial environment for dust, mists, and fumes are readily available from local suppliers, and has been shown to be effective against microbes.[8] These filters have the characteristics of being 99.97% efficient against monodispersed particles that are 0.3 μm (300 nm) in size and larger. When this efficiency and filtration are examined against the size of most bacteria it can be seen that there would be a high effectiveness in removal of the organisms from the air. In general, the size of bacteria ranges around 1–10 μm, with an organism like *Escherichia coli* being toward the lower end and *Bacillus megaterium* at the upper. However, there are some bacteria, such as Epulopiscium, which are rather large, at least on the "microscopic" scale and are visible with the naked eye; although, other organisms, like mycoplasma, are in a lower range of 0.1 μm (100 nm). Viruses, in general are around 10 to 200 nm in size. Here the Poliovirus is about 10 nm and the Poxviruses are around 200 nm. Thus, using the strict size characteristic of the HEPA filter, viruses will be less likely collected than bacteria due to size, but these measurements do not includes electrostatic factors, which are also of importance. For example, the SARS virus (SARS-CoV) has a size around 60–80 nm (0.060–0.080 μm), and is about 3 to 4 times smaller than HEPA's efficiency lower range of 0.3 μm (300 nm) for monodispersed particles. The size range for HEPA filters provided is that given by OSHA and many textbooks, but when these filters are evaluated against test agents, such as sodium chloride aerosol, there is a much lower range of filtration efficiency. According to one test report (http://www.cdc.gov/niosh/topics/respirators), a count median diameter (CMD) of 0.075 ± 0.020 μm with a geometric standard deviation (GSD) of 1.86 was observed. This suggests that the filtration range is much lower than that indicated by convention and that the range is within the size reported for the SARS-CoV virus, especially when electrical charge is considered. This also matches with observations during the SARS outbreaks that respirators with HEPA filters were effective in preventing infection of HCW.[98]

The size of the filter is not the only factor important in selection. The dose of the infectious agent is also a factor to be considered, which is how many infectious particles are need to induce the disease. This value has been estimated for some of the "common" infectious disease agents; however, the problem with an emerging disease is that there will be no information on the infectious dose and it may be one particle or organism. Under these conditions, a higher degree of concern is needed; thus, warranting a higher level of protection. This concern is not only associated with the filtration but also fit (seal) of the respirator as well. However, as noted, HEPA filters, based on standard reported filtration values, will be effective against most bacteria, even the ones at the lower size range. However, based on this criterion, these filters will be less effective against some viruses; although, when compared to the data provided in the sodium chloride aerosol test and that there will be some form of a charge, they will be effective. However, this size concern has resulted in a few suggesting that the ULPA be used instead of HEPA filters for emerging viruses, especially those that are less than 0.3 μm in size, such as for SARS-CoV.[43] But, as noted, evaluation with the SARS event demonstrated that HEPA filters were fairly effective in protecting HCW.[98] Thus, the standard value for these filters, as provided by OSHA, may be considered the upper range when evaluating efficiency.

ULPA filters have a greater efficiency in filtration as compared to HEPA filters. However, they are also more expensive and more difficult to obtain from local suppliers. These filters are commonly considered to have an efficiency of 99.999% for monodispersed particles that are 0.12 μm in diameter or larger. This filtration size is closer to that of most viruses (poxvirus 200 nm or 0.2 μm) and with electrostatic factors included will be highly efficient against these small particles. Even the smallest viruses, such as SARS-CoV would be efficiently collected. One of the main reasons for little use of these filtration devices is their cost and difficulty in obtaining them on short notice. These filters are usually not routinely stocked and upon occurrence of an outbreak would not be readily available and may not exist for all models of respirators. However, for HEPA filters, they are routinely used in the industrial environment and are a common item among safety suppliers and are easy to obtain. For those facilities that may have concern and be on the "front" line of emerging infectious disease, a supply of ULPA can be maintained. But again, as shown with the SARS event, proper use of respirators with HEPA filtration was effective and much of the "leakage" of respirators is a result on the face seal and not the filter.

Selecting the type of filter to be used can be as important as selection of the respirator. The filter should be selected based on the type of organism being encountered, with its size being of great importance. As discussed, organisms greatly vary in size, but in general, bacteria, fungi, and protozoons are larger than 1 μm; although, a few can be a bit smaller. Viruses are usually less than 1 μm and go into the range of 10 nm. The smallest organism is probably prions, which are infectious proteins. In general,

HEPA filters are highly effective in removing bacteria and larger microbes based on the standard classification of the size of filtration. When the sodium chloride test results are considered and electrostatic charge, efficiency of HEPA filters would even be greater and practical experience suggests that these filters are effective.

The same principle can be applied for viruses. Even though these agents, in general, are smaller than bacteria, their electrostatic charge allows them to be collected on the filter efficiently. This was demonstrated during the SARS event in 2003, especially when used properly. However, there may be little initially known of an emerging infectious disease causing the necessity to employ ULPA. Although, once the viral family is identified a general size range of the virus will be known. If PAPR are employed, they use a filtration system that is commonly HEPA and can suffer from the same limitation in filtration. However, these respirators do create a positive pressure in the mask and are "usually" full face. Alternatively, SCUBA and SAR systems can be employed, which provides air that is obtained from outside the contamination area. However, these respirators are difficult to use and for the most part impractical in a clinical setting. SARs have been used in research activities related to highly infectious agents (e.g., Ebola).

The most common type of respirator used by HCW is the N95. This respirator has been suggested for HCW to protect them from SARS and TB, and in general other emerging infectious diseases. N95 respirators were first recommended for use by HCW that were caring for TB patients. For SARS, some studies reported that the N95 respirator was effective in protecting HCW while others reported that it was not.[14] These respirators are single use and are not designed for reuse, although are often used over again. N95 respirators belong to the N class which also included N99 and N100. The value after the N refers to the efficiency of the respirator. A N100 respirator is actually not 100% efficient but rather 99.7%, while the N95 and N99 are 95% and 99%, respectively. The efficiency for these respirators is rated for an airflow rate of 85 L per minute with penetration of median aerodynamic particles of 0.3 μm in size. N designates that the respirator is not resistant for oil.[103] There are other respirators, which are designated as R and P for oil resistant and oil proof, respectively. However, since HCW are not likely to be exposed to conditions relating to oil they are of little importance in this discussion. However, there may be circumstances where repair to equipment is necessary in a patient's room and those undertaking this activity may need to use these types of devices. All of the N, R, and P "respirators" are disposable. These respirators can be fit tested; although, one study suggested that there is a high fail rate with these respirators and those passing approximately 20% will have a lower than expected rate of exposure protection.[104] N95 respirators have also been shown to be ineffective against small viruses (MS2, a bacteriophage for *Bacillus subtilis*) in experimental testing.[105] MS2 has been used as a simulating agent to evaluate effectiveness of respirators and is a small RNA virus, about 30 nm in size. Testing of surgical masks with this virus has shown that they are high variable and not effective.[105]

The "next" level in protection up from the N95 type of respirators is half-mask respirators. There are many different types and manufacturers of these respirators and the most common form of respiratory protection used in the industrial environment. These respirators are APR, light in weight, can be maintained by each individual, are reusable and usually are made in the sizes small, medium, and large. In some cases extra-large respirators can also be obtained. Most of the half-masks today are made from electrometric materials, plastics, or rubber-based and easily mold and conform to the face. These respirators can be used with a wide variety of filters including HEPA and ULPA. Half-masks have been used effectively by HCW in prevention of inhalation of infectious disease agents. Like the N95-type respirators, half-masks do not provide protection to the eyes and are negative pressure. All negative pressure respirators can cause difficulty in breathing and increase stress to the user's cardiopulmonary system resulting in confounding of existing health conditions.[106] These "preexisting" conditions may be of greatest importance for smokers. Although workers in the industrial environment frequently undergo medical examinations, including pulmonary function tests, this does not identify many of the potential underlying pathologies (e.g., cor pulmonale) that may exist in those using a negative pressure respirator. Some have suggested that this group of respirators should be selected over the N95 for HCW.[19] During the SARS event, half-masks with HEPA filters were shown to be effective in preventing disease among HCW, especially when they also employed eye protection.

Full-face respirators (FFR) are also APR, as are half-masks, but provide protection for the eyes and face, in general. These types of respirators are also negative pressure and have many of the same limitations as half-masks, although, due to the covering over the face also tend to fog up from humidity derived from exhalation. The major advantage of FFR is that they do offer eye protection. These respirators use the same filters as half-masks (e.g., HEPA). Facial protection with a respirator (e.g., face shields) has been shown to be important. Dentists using facial protection only, such as a shield, have resulted in the presence of antibodies against viruses, suggesting that a shield alone is not effective.[107] This demonstrates the importance of respiratory protection against microorganisms for HCI. Due to the bulkiness of these respirators, they are not commonly used in the HCI and they also create difficulties with communication limiting their usefulness in patient settings and care.

The next level of respiratory protection is PAPRs, which are positive pressure respirators. These respirators have a motor which supplies filtered air to the user. This air is filtered through some type of filter, like HEPA. These respirators are full-face. The main advantage of this type of respirator is that it is positive pressure and does not cause pulmonary stress on the user. The limitations for these are bulkiness; they use a battery for generating the positive pressure, and they are higher in cost. The time that they can be used is limited by the battery life. However, for all respirators that filter air, these do provide the highest PF and would exhibit the greatest degree of protection. These respirators have been given an assigned PF of 100. However, use of these devices around patients would not be practical and would certainly result in elevated levels of anxiety by patients seeing their providers wearing these respirators. For protection of HCW, this type of respirator would provide a high level of protection and would be economical especially when considering the potential cost associated with treating HCW that became infected; although, mentioned limitations, especially related to that from the prospective of the patient does restrict applicability. A PAPR would be applicable for HCWs that cannot use a negative pressure respirator or have pulmonary impairments. Thus, this may have some applicability to workers that are older and for those that may have underlying health issues. Koley[2] suggested that PAPRs be employed during the occurrence of emerging infectious diseases, which would include the bird flu. One report suggests that HCW do prefer PAPRs over N95s, generally, and over time use became acceptable.[108] This study indicates that HCW do adapt to various types of PPE. However, for any type of PPE there is a training time period required and in some ways for acceptance of the activities.

SARs are those that provide air from a source that is located outside the work area and the air provided must be at a minimum grade D. Requirements for air to be grade D it must be oxygen 19.5%–23.5%, hydrocarbons less than 5 mg/m^3, less than 1000 ppm carbon dioxide, less than 10 ppm carbon monoxide, and no pronounced odor. A SAR system is generally not applicable in a clinical setting due to the difficulty in operation and cost; however, has been used in research situations when working with dangerous microbes (e.g., Ebola). If such a system was to be employed, it would require establishment of special rooms and training for HCW.

Beside the respirator types mentioned, others have been evaluated for use in the HCI (surgical helmets). Derrick and Gomersall (2004) evaluated a helmet-type system during the SARS event and found that it did not provide adequate protection. This demonstrates that selection must be undertaken with care, especially when looking at systems that are not commonly used by HCW. The study of a helmet respirator, as a side note, did suggest that N100 respirators were effective. These respirators (N100) based on evaluation with a PortaCount® Plus quantitative test system had a PF of 100, which is above that commonly cited for half-masks, is 10. This study shows the effectiveness of this type of respirator and the importance of respirator evaluation. The Portacount system counts particles that are in the size range of 0.02 to 1 μm, which is similar to the size of many viruses that would be encountered by HCW (Derrick and Gomersall, 2004). This is a quantitative fit testing system and demonstrates the importance of fit testing. Most fit tests are not quantitative, but qualitative in nature.

Fit Testing and Fit Checks

One of the important aspects of any respirator program is fit testing and fit checks. OSHA requires that fit testing be conducted once a year and fit checks should be undertaken by the user each time they put their respirator on. Fit testing can assist in determining that there is an "appropriate" PF for the respirator and that it is fitting properly (face seal). To undertake this activity, each user needs some minimal training on use and care for the respirator that they will be employing. One study suggested that the actual method of training is not of great importance, but that instruction is provided.[109] HCW receiving what can be described as minimal training appear to have a similar "effectiveness" in passing fit tests as those undergoing what can be termed more exhaustive training.[109] Those having no training had a poorer outcome in ability to pass a fit. This can be attributed to a lack of ability to perform basic tasks in obtaining an appropriate fit. Based on these data, even minimal training will ensure that there is some level of protection being provided along with a good likelihood of being able to be properly fitted. The same is true for fit checks (positive and negative) and are an essential part of a respirator program and related training.[110] Fit testing/checks can only be accomplished with devices that have an adequate seal to the face. Fit testing can be either quantitative or qualitative in nature. However, it must also be emphasized that proper inspection of the respirator is needed to insure that the device is in working order.[110] Most testing conducted is qualitative and the PF is that assigned by OSHA (Table 4.6). Thus, paper and surgical masks, as examples, cannot be fit tested or be fit checked. A half-mask respirator and above can be tested, with this form of respirator having the lowest assigned PF. It must be noted that a PF is not an absolute number and it does take into account that some leakage will occur. Requirements for fit testing are described by OSHA and these can be found in conditions described for specific substances, like asbestos and benzene, and in the requirement for respirators (29 CFR 1910.134, Appendix A).

The respirator fit check is an activity the users perform each time they put their respirators on. This is also called the positive/negative pressure fit check because it involves evaluation of the exhalation and inhalation systems of the respirator (APR—e.g., half-mask). This type of testing is often considered less precise than fit testing but it does provide the user with information that they have established a seal of the respirator to the face. Most commonly there is leakage near the nasal part of the face, which is the upper area of the respirator. To conduct the positive fit check, the wearer must cover with the palm of their hands or a small plastic barrier, as an example, the exhalation portal (value) and gently exhale. A slight positive pressure should occur in the mask and if so the test can be considered to be successful. For a negative test, the same procedure is undertaken, except the inhalation value is covered. Here the wearer creates negative pressure by drawing air in and this should create a negative pressure inside the respirator. As noted, the most common area where leaks will occur is around the bridge of the nose location of the mask face seal. These tests can be conducted by the wearer and are rapid which give the procedure a good advantage as a quick test for adequate seal. Impairments that restrict or interfere with the face piece seal is an indicator that there is an inadequate fit and the wearer cannot employ the respirator. OSHA does not allow the user to have facial hair that interferes with the seal as well. Studies that evaluated face seal with facial hair suggest that there is a higher leakage and less fit.[111] Most people in the industrial environment who use respirators are aware of the issue of facial hair for respirator use, but those in health care may not.

TABLE 4.6 Types of Respirators and Published Protection Factors

Half-mask	10
Full-face mask	50
PAPR	100
SAR	1000

Thus, concern for fit as related to facial hair should be discussed in training. However, facial hair that does not directly affect the face to respirator seal, such as mustaches, are acceptable.

The requirements for fit test can vary among agencies, although, most follow the criteria established by OSHA.[101] Fit testing establishes that there is an adequate facial seal with the respirator and that the designated PF will be achieved. This provides the user with assurance that there is a good face-respirator seal. Employment of a respirator that does not have a good face seal can be dangerous to the user in that use of the respirator may create a false illusion of security against the environmental agent of concern. In some cases, this can be more hazardous than not using a respirator, especially when cardiopulmonary factors are considered with using a respirator.

Respirator use today is generally related to chronic toxicants and not acute agents such as an infectious agent. Historically, respirators were primarily for prevention of acute injuries. An estimate of the exposure dose cannot be easily obtained for most infectious diseases, especially those that are emerging. Due to the unknown exposure dose its degree of infectivity and the likely variation of dose from patient to patient (location to location) fit testing becomes of greater importance for the HCW. Although annual fit testing is required by OSHA, a more frequent testing at the beginning of a suspected outbreak may be appropriate as well. If there is a report of a global event, this may also trigger testing of key personnel in the event a case arises in the local area and such testing may also include first responders. However, such activities should be included in the written respiratory program so the procedure and process is understood by all personnel. During the SARS outbreak, it was soon realized that many of the HCW employing respirators were not adequately fit tested and this has been suggested to have contributed to some of the cases observed in this population. This demonstrated the importance of fit testing, especially for HCW that were dealing with highly infectious patients.[3] Although fit testing can be considered general in nature, training specific to hazards related to HCW should be incorporated into the training.[112] Thus, HCW should be given instructions on infectious diseases and preventative practices, beyond that which they routinely get in such courses as microbiology.

When conducting fit testing, the first part of this activity is the selection of a respirator. Manufacturers produce respirators in various sizes, which are usually small, medium, and large. However, the actual category can vary from manufacturer to manufacturer, so a small respirator provided by one manufacturer will not be the same as that from another. So when starting testing, each person needs to select a respirator that fits and is comfortable. OSHA provides guidance and criteria in conducting this selection. As with other forms of PPE, the same should be undertaken.

After each person selects a respirator, the type of fit testing that is to be conducted should be determined. There are two types of fit testing: quantitative or qualitative. The most common form of fit testing performed is qualitative and there are four agents generally discussed that can be employed for this type of testing (Table 4.7). Other agents may be used, but they must meet the criteria established by OSHA. For the most part, such an agent must be evaluated and this "evaluation" published in an industrial hygiene-type journal. A "standardized" PF is established for each type of respirator.

Based on the OSHA regulations, the PF for qualitative testing is 10. This means that when there is a concentration of 10 outside the mask there will be a concentration of 1 inside. In most cases the PF is

TABLE 4.7 OSHA Recognized Types of Qualitative Fit Testing Agents

Agent	Detection of Agent	References
Isoamyl acetate (banana oil)	Odor threshold	Gardner et al.[127]
Saccharin	Taste	Coffey et al.[104]
Irritant smoke	Respiratory irritation	Hannum et al.[109]
Bitrex™	Taste	Duling et al.[123]

adjusted for the type of respirator being used, like a full face as compared to a half-mask. The assigned PF for respirators is shown in Table 4.6. These values do not correspond well to experimental data, with most data suggesting that there is a greater amount of protection afforded. Published values for respirators are: surgical masks, N95, electrometric half-masks (with HEPA filter), and PAPR (with HEPA) 2.5, 17.5, 46.9, and 236, respectively.[8,88] Fit testing is most commonly conducted for half-mask respirators; although, can be performed for any type of respirator including those that are positive pressure (e.g., SAR). Quantitative testing can also be employed for these respirators.

There are some limitations regarding fit testing. For example, if the person has a change that alters their facial feature or interferes with the face-piece seal they will be required to be refit tested. These changes would include a "significant" gain or loss of weight (usually considered to be ±20 lb, (some use 10 lb) or 10% of the person's body weight),[113] cosmetic/plastic surgery that alters the seal or dental changes which also changes the "contour" of the face. Certainly, these issues and concerns which may require a "new" fit test should be discussed as a part of respirator training.

Regardless of the agent used for fit testing, each has advantages and disadvantages. However, if an agent is used properly, each can provide a determination as to the adequacy of fit. The most common agents used are saccharin, irritant smoke, and isoamyl acetate (banana oil). Recently, Bitrex™ (denatonium benzoate) has been introduced as a fit test agent and is commercially available for this purpose.[114] This agent has a bitter taste for the test subject when there is an inadequate fit.

The smoke test results in those being fitted that do not have a proper seal exhibit an involuntary cough. Many consider this as the ideal test for fit testing since it does not fully rely on the test subject advising that they do not detect the agent. Irritant smoke is generated through the production of hydrogen chloride gas and tin fumes. These "substances" react with moisture in the air resulting in a white colored smoke being generated with a pungent odor. The white smoke is actually a reaction of the stannic chloride with water (humidity). Thus, irritation results from the mixture of tin and HCl and these substances cause the irritation to the mucus membranes of the upper respiratory tract, particularity the nose and mouth. It should be noted, that these fumes can also cause irritation of the eyes. High concentrations of these substances can result in coughing, chest, pain, choking, and the eyes to water, ever after a short exposure. So there must be some care in limiting the amount of exposure. The OSHA PEL for HCl is 5 ppm and NIOSH identifies that the immediate danger to life and health concentration (IDLH) for this substance as being 100 ppm. However, stannic chloride has been reported to be much less toxic than HCl. Fit testing kits using tin and HCl can be purchased from many manufacturers, although, some do not recommend that these testing agents be employed for the reasons mentioned above. From a practical point of view the hazards exhibited from this testing, if used properly and reasonably, are low and can be considered nonexistent.

Fit testing agents, saccharin and banana oil (isoamyl acetate), involve the subject detecting the agent and reporting to the tester that exposure is occurring. Saccharin involves taste and banana oil odor, which smells like ripe bananas. The PEL for isoamyl acetate is 100 ppm and the IDLH level is 1000 ppm. There is no published PEL for saccharin. As mentioned, the disadvantage for these agents is that testing relies on detection by the person being tested, with some choosing not to report that they can detect the agent.

The most recently "approved" or recognized testing agent is Bitrex[114]. This agent has been tested against the other fit testing agents and has been suggested to be comparable or better in laboratory studies.[115] Historically, Bitrex has been used to denature alcohol and because of its bitter taste was added to prevent accidental consumption.[116] The effectiveness in preventing accidental poisoning of substances which it has been added has been recently questioned.[117] This highly bitter taste from Bitrex is often identified as its advantage over other agents such as banana oil and saccharin. Since irritant smoke has some toxicity, some consider Bitrex to be a better alternative due to its relatively low toxic properties. There is a report of this chemical causing utricaria and asthma, although this report is based on a single case event.[118] In a comparative study using N95 respirator and a PortaCount® testing system, a higher passing rate was observed when using this agent, especially when performed quantitatively.[115]

Regardless of which fit test agents is used, the user or one being tested must be able to detect the agent. This is usually performed before the person is fit tested, especially for odor (banana oil) and taste agents (saccharin). For others, cough at the end of the test, such as with irritant smoke, by the subject being tested can be sufficient. Most agents should not be tested on the person just before the test with a "low concentration of material" due to sensitization. This could artificially increase the chance (false positive) of passing the test by not being able to detect low concentrations of the agent. Here, the test subject is provided with a brief and small amount of the agent and will usually cough. Protocols for tests can be found in the OSHA regulations, like under asbestos. Here a detailed procedure is outlined. It should be noted that for some substances, not everyone can detect the agent and some may just report not detecting it to get the test over with. Generally, agents such as Bitrex and irritant smoke provide the best indicator for fit[104] and these factors may be considered when selecting a method, especially for HCW. However, substances like banana oil are commonly used and have been found to be effective and efficient fit test agents.

Fit testing using a quantitative procedure measures the actual amount of leakage for a respirator. Historically there were two types of tests employed for this kind of testing: sodium chloride and dioctylpthalate (DOP); although, testing procedures were time consuming, equipment was bulky and the procedure was difficult to perform. More recent testing agents have included: freon, ethylene gas, methylene blue, and paraffin oil. Today there are several commercial manufacturers of instruments for quantitative fit testing (e.g., PortaCount® Plus) and for the most part this has made such testing more readily available. These systems have replaced the "old versions" and practices for testing. This type of testing actually measures the concentration of "substance" inside and outside the persons mask to provide an individual PF. Measurements are performed by instrumentation associated with the fit testing system and this does not include the subjectivity associated with qualitative testing methods. The disadvantages of this type of testing are cost of instruments, need for a trained operator and it cannot be easily performed in the field. Quantitative fit testing systems are now computerized providing a readout (printout) of results.

Respirator fit testing must be conducted according to criteria established by OSHA. As mentioned, this is required to be performed once a year; although, some may consider that a more frequent testing be conducted. Certainly one of the limiting factors in undertaking this testing is cost (Kellerman et al., 1998). During fit testing personnel can also be instructed on proper use of the respirator as well as maintenance. It should be emphasized that since the agents of concern which is the reason for using a respirator in the first place can result in disease to the person (HCW) the wear factor should be 100%. This wear factor is the amount of time that the person actually uses the respirator in appropriate situations. Doffing (removing) the respirator during times of potential exposure can dramatically increase the risk of contracting the disease that these practices are trying to prevent.

Maintenance and Cleaning

One of the more important aspects related to respirators is maintenance and cleaning. Today, many of the respirators used are disposable and are designed for a single use by a single person. This can be an effective way to use and distribute respirators, but if they are used frequently it is not cost effective. There may also be issues with fit when using disposable respirators; although, one type should have the same fit, this may not be true in all cases. Variation observed in fit testing support that a person should be tested for the respirator they use.[104] In some larger institutions that use respirators frequently a central location for cleaning and maintenance can be established. However, for HCW, especially when dealing with infectious diseases, this system may not be easily instituted. Conversely having each person that uses a respirator clean and maintain their own can also create a hazard for that person as well as others in the area where it is being cleaned. People maintaining their own respirator may result in them being cleaned in an inappropriate location which can result in infectious agents being spread in previously uncontaminated areas. Hazards of spreading contamination and microbes through cleaning are one issue that can be discussed in training. Unlike most industrial environments, spread of the infectious agents can occur in the HCI by improperly maintaining and cleaning respirators as well as other forms of PPE. This situation is a bit different from most industrial environments.

Inspection of respirators is required each time they are used.[119] The person using a respirator needs to examine it to make sure it is in working order and functioning properly. Examination of value flaps inside the respirator needs to be included as well as evaluation for cracks or deformities. Worn or damaged parts can be replaced at the time of inspection. This requires that those using respirators have information on how to inspect and repair them. Respirators cannot be shared unless they are cleaned/disinfected. In most cases, it is more effective to provide each person with their own respirator.

Respirator cleaning can be divided into two general activities: cleaning and disinfection (sanitizing). To properly clean a respirator it must be disassembled before starting the process. At this time worn or damaged parts can be replaced. Cleaning can be undertaken in a clean bucket or similar "container" that can hold sufficient water (cleaning solution). Warm water should be used, generally less than 110°F. Do not use boiling water. The respirator can then be dissembled and the parts placed in the water. Some do not recommend removing the values because they can be easily lost, however, other do so they can be effectively cleaned. A neutral detergent should be used as the cleaning agent. Most dish detergent will work well for this purpose. Laundry soaps should not be used as well as those with lanolins and oils. A soft brush or similar should be used to clean the respirator and parts. This cleaning must be sufficient enough to remove "dirt" and grime. After cleaning, the parts should be rinsed with clean warm water. Rising must be through as to remove all soap residues. Disinfection can be performed with a number of agents. In some cases, commercial agents can be used for both cleaning and disinfection. Agents for disinfection include quaternary ammonia (usually one packet per 2 gal of water), Clorox (hypochlorite) (about 1 ounce per 2 gal of water), or an iodine solution (about 0.8 mL of tincture of iodine per liter of water). The water used should be 110°F. Disinfection time should be sufficient to kill all the microbes. Based on the food service industry, concentrations of hypochlorite should be 50 ppm, iodine 500 ppm and disinfection time for both being 2 min. For ammonium, a concentration of 200 ppm for 2 min has been suggested. Allowing a respirator to sit in a disinfecting solution too long can result in damage. The respirator and parts must then be thoroughly rinsed with clean water and air dried. There are some drying systems that can be purchased, but for most air drying in a clean location is sufficient. Respirators can then be put in storage. They can be stored in clean bags. However, they must be completely dry before being stored. Care must also be taken to reassemble the respirator properly.

Some have suggested using alcohol or similar wipes/pads for cleaning and disinfection of respirators. However, alcohol is not effective in killing microbes and for HCW may not be highly effective. Thus, this type of cleaning/disinfection, although common in the industrial environment, is not suggested for HCW who are dealing with infectious disease agents.

Proper Selection

The respirator that is selected by the person must properly fit that person. Lack of a fit can cause leaks and increase the exposure to the person as well as causing other problems for the user. One of the problems some workers have with respirators is occurrence of headaches.[120] These headaches are generally a result of the straps around the head (crown strap) being too tight.[121] The pressure on the head is usually a result of an inadequate fit (face seal) and the user attempts to obtain a better seal through tightening up the crown straps. In some cases this will allow the formation of a seal, although, this process causes pressure on the superficial nerves associated with the head resulting in a headache. These are for the most part tension headaches and can be prevented by using a properly fitting respirator. In some cases, this improper fit is a result of the person using the wrong-sized respirator.

Respirator Program

It is commonly suggested and reported that respirators are not properly employed.[122] One of the important aspects of a respirator program is training users on these PPE. Much of the poor use of respirators results from those employing these devices without any training or instruction and a lack of understanding of the importance of respirators by management. As with any activity in a health care setting, appropriate supervision and periodic refresher training on how to use a respirator is needed. During mold remediation after Hurricane Katrina, it was observed that only 24% of those donning respirators performed this task properly.[122] This emphasizes the importance of fit checks by the user along with proper inspections, maintenance, and cleaning.

As a general rule, OSHA requires that those using respirators, APR, be medically evaluated once a year and receive training on its use. It is also required that fit testing be conducted at least once a year and studies have reported that fit testing increases the effectiveness of respirator "use."[123] OSHA also requires that institutions have a written respiratory program. This program will provide information on how the program will be conducted, fit testing, respirator selection, and other activities. It should be designed toward the requirements of the institution and have as necessary any specific information related to activities performed. However, since HCW may encounter unknown agents, which are considered emerging, it will be difficult to provide specifics. The general OSHA requirements for respirators can be found in 29 CFR 1910.134. There are other OSHA sections and requirements for other forms of PPE; although, these conditions are mostly related to the industrial environment. However, this information can provide some guidance for those in the HCI. The Food and Drug Administration in 2007 provided guidance on non-APR that are used by HCW (Federal Register 72:36360-3).

One of the most important aspects of a respirator program as well as that for PPE is training. Most HCW do not normally receive training related to PPE, including respirators, as part of their formal education. This became evident during the SARS event where a rush for information on respirators occurred and it was suggested that the high rate of disease among HCW was a result of inadequate respirator use, some related to training. Some attributed this increased risk to HCW as a direct result of their lack of information on properly using respiratory devices.[124] Many HCW at the time of the SARS event did not have training on use and application of respirators. As a result of this event and that associated with the potential of bioterrorism, which in part was triggered by the anthrax episode in the United States, use of PPE is now included as part of training for HCW in some locations. Training today often includes fit testing, fit checks, cleaning, and maintenance of respirators and selection. Yassi et al.[88] suggested for long-term effectiveness of respirator use it is necessary to have frequent training. One of the effective ways that this can be accomplished is through peer-feed back along with periodic training. Most programs do not conduct such activities and this may reduce the effectiveness of even the initial instruction.[88] One of the reasons for lack of periodic training is the infrequency of outbreaks of emerging infectious diseases. However, inclusion of this information, in general, may make training more effective and provide a better understanding of its importance. Only providing respirator training will not make a program effective without a peer discussion of training. In many cases, the most highly trained persons in an organization are least trained on the basic practices related to preventative measures. Since disease does not select by categorization of occupation among HCW training and activities must include the entire gamet of personnel. In addition, peer comments and ideas can emerge from any person, regardless of their role and activity. Such activities can also emphasize some of the pit-falls that a program has and identify these deficiencies before an event.

For HCW, as well as others who use respirators, selection of an appropriate respirator is often a difficult and in some cases a time-consuming process. Table 4.8 provides some guidance on selecting the best respirator; although, other factors also need to be considered. One important issue that is commonly not discussed with HCW is the importance of the face seal (fit). This requires that the "face" be smooth and clean shaven. The table provides the general parameters of these respirators along with some advantages and disadvantages of each. When making any respirator selection there are a number of factors to consider. A list of some of these factors is provided in Table 4.6. Since HCW will be making a selection for an unknown disease there is an increased level of difficultly in this process. This process is also complicated by the issue of the infrequency of respirator use. However, information related to selection of respirators in regard to TB can be used as a primary guide. However, it has to be remembered that

TABLE 4.8 Selection of a Respirator for an Emerging Infectious Disease with a Discussion of Criteria

Type of Respirator	Criteria of Selection
Surgical mask/disposal paper mask	These forms of devices are generally not classified as Disposal paper respirators and are identified here as masks. Studies have suggested that their PF is around 1.2 with some possibly a bit higher (1.4). The OSHA recommended PF for a half-mask is 10, thus, these have a value less than five times or more the next "level" of respirator. These devices are not recommended as protection against an emerging infectious disease or other diseases/microbes (e.g., TB). Generally, there is no face seal with these respirators or masks and they cannot be fit tested nor can a respirator fit check be conducted. There have been studies which recommended use of these devices and reported lower rates of infection. This is likely due to the existence of low concentrations of microbes in the environment (infectious disease organisms). It is likely that use of such devices will create a false sense of security for users. These masks are commonly used due to low cost, simplicity in use and easy in availability.
N95	These respirators are probably the most common ones used by HCW. They have been designed for use with patients having TB. During the SARS event, there were reports that this respirator was not highly effective in preventing inhalation of the virus. Since these are half-masks, they do not provide protection to all the mucus membranes that may be exposed (e.g., conjunctiva). The respirator is also negative pressure. One advantage of these respirators is that they are low cost and disposable, but this can also be a disadvantage in that there may be a varying seal among the respirators. Seal has been suggested to be one of the limiting factors associated with these respirators with a suggested leakage of 10%. For many industrial compounds this may not be a major concern since most have a threshold dose. However, when dealing with infectious diseases, the threshold dose for infection may be small, such as one to two particles (infective units). Some studies have shown these this respirators is not applicable for procedures that involve aerosolization. Although very common in hospital settings, this respirator is not recommended for an emerging infectious disease, especially one that is difficult to treat (e.g., MDT) and those with high fatality rates (Ebola). However, they do provide a higher level of protection when compared to surgical or paper masks.
N100	These respirators have some of the same characteristics as N95s. They do have a higher level of protection; although, they are not recommended by some for emerging infectious disease. This is based on the same concept provided for N95s.
Half-masks	Half mask respirators can be fit tested and provide a designated PF. These respirators are elastomeric in form (or a soft rubber) and can easily mold to the face. However, they are negative pressure and are commonly called APR. Some studies have reported that the PF for these respirators are greater than 10, which is the set value by OSHA. The respirators can be used with HEPA and ULPA filter and have been reported to be effective in providing protection against emerging infectious diseases, however, do not provide protection to all the mucus membranes (e.g., conjunctiva) as indicated for N95s. Based on the SARS event, when combined with other types of PPE (e.g.,) and efficient engineering controls can be effective in preventing disease in HCW. OSHA recommended that fit testing be conducted annually for users of these respirators and a fit check be conducted before each application. These respirators are also some what low in cost and can be individually provided to workers. They can also be effectively cleaned.
Full-face	This type of respirator covers the entire face including the eyes. As with all APRs, they are negative pressure. Since they protect the eyes, they have been suggested to be the best protection for emerging infectious diseases. However, they suffer from the problem of being more difficult in application and can "fog-up" during use, especially in warmer and humid environments. Use of this type of respirator can also increase fear and anxiety of patients. Some users may also have claustrophobic issues when using these respirators. Respirator cartridges are the same as that for half-masks, HEPA, and ULPA. A higher protection factor is afforded to these respirators, according to OSHA, an assigned value of 50 is given. Disadvantages to these are cost and involve a higher degree of experience in use.
PAPR	PAPR are positive pressure respirators. They can use HEPA and ULPA for filtration. They have a higher level of protection than APRs. Due to the positive pressure system, there is additional weight and they are bulky. Use time is limited by the battery and each respirator may require several batteries for a day's operation. If aerolization procedures are being undertaken, they may be an appropriate selection for HCW, but suffer from the same issue associated with FFR in causing fear by the patient and possibly fogging up.

TB is a bacterium and many of the emerging diseases will be viral. The size difference and difference of infective dose will be an important issue regarding selection and use of respirators. Since these factors will be unknown for a "new" organism, the importance of fit will be critical. However, the importance of fit and face seal must be considered and may be the most important factor in selection and use.[93]

Conclusion

The primary control mechanisms for preventing exposure are through engineering controls. However, in most health care settings such controls are not highly feasible and secondary forms have to be considered, which commonly include PPE. Use of controls as a routine practice appears to provide a reduction in the overall rate of infectious diseases in hospital settings. One of the issues with emerging infectious diseases is that there is little warning as to occurrence and often there is little known about the microbe. This has increased reliance on PPE, such as respirators, but there appears to be a continuing lack of application in use, even after the SARS event. Most HCW have not received adequate training on use and maintenance of respirators, and this alone causes a reduction in the effectiveness that such devices provide. Since many of the emerging diseases are spread by an inhalation route, respiratory protection is paramount. A standardized training and practical program is needed to ensure the best protection, which includes related preventive measures associated with PPE. This includes fit checks, testing, maintenance, handing PPE that is contaminated, and disinfection practices. Overall, these activities are basic in nature, but commonly ignored or overlooked. Routine implementation of such programs will prepare an institution for the next pandemic or epidemic event as well as the current myriad of microbes.

A summary of characteristics of respirators are shown in Table 4.8. This provides a brief selection guide for the different types of "respiratory" devices that are available. Surgical masks and paper masks are presented in this table, although, most do not consider these devices to be actual respirators. As with all practices, it must be noted that this is only one form of protection and other routes of exposure occur (e.g., fomites). So any program must consider the range of exposure scenarios and not get lost in only implementing respiratory controls, even though, most consider this route to be paramount.

References

1. Baumann MH, Nolan R, Petrini M, Lee YC, Light RW, and Schneider E. Pleural tuberculosis in the United States: Incidence and drug resistance. *Hest* 131:1125–1132. 2007.
2. Koley TK. Severe acute respiratory syndrome: A preliminary review. *J Ind Med Assoc* 101:308–310, 2003.
3. Lange JH. SARS and respiratory protection. *Hong Kong Med J* 10:71–72, 2004.
4. Sepkowitz KA. Tuberculosis and the health care worker: Historical perspective. *Ann Intern Med* 120:71–79, 1994.
5. Harris G. Limited capacity is seen in flu defenses. *The New York Times*, Today's Headlines, NYTimes.com (accessed on July 18, 2007).
6. Chen J-M, Chen J-W, Dai J-J, and Sun YX. A survey of human cases of H5N1 avian influenza reported by the WHO before June 2006 for infection control. *Am J Infect Control* 35:351–353, 2007.
7. Toledo-Pereyra LH and Toledo MM. A critical study of Lister's work on antiseptic surgery. *Am J Surg* 131:736–744, 1976.
8. Lange JH and Mastrangelo G. Respirators and other personal protective equipment for health care workers. In: W Charney (ed.), *Emerging Infectious Diseases and the Threat to Occupational Health in the U.S. and Canada*. Taylor & Francis, Boca Raton, FL, 2006.
9. Chen WK, Wu HD, Lin CC, and Cheng YC. Emergency department response to SARS, Taiwan. *Emerg Infect Dis* 11:1067–1073, 2005.
10. Tablan OC, Anderson LJ, Besser R, Bridges C, and Hajjeh R. Guidelines for preventing health care associated pneumonia, 2003. *MMWR* 53(RR03):1–38, 2003.

11. Best M and Neuhauser D. Ignaz Semmelweis and the birth of infection control. *Qual Saf Health Care* 13:233–234, 2004.
12. Rengasamy A, Zhuang Z, and Berryann R. Respiratory protection against bioaerosols: Literature review and research needs. *Am J Infect Control* 32:345–354, 2004.
13. Whitby N and Whitby M. SARS: A new infectious disease for a new century. *Aust Family Physician* 32:779–783, 2003.
14. Lange JH. Respiratory protection and emerging infectious diseases: Lessons from severe acute respiratory syndrome (SARS). *Chin Med J* 118:62–68, 2005.
15. Sobotova L, Noskova T, Volskova J, and Aghova L. Practical training on nosocomial infections in a hospital environment. *Indoor Built Environ* 15:73–76, 2006.
16. Nicas M. Airborne pathogens: Selection of respiratory protection. In: W Charney (ed.), *Emerging Infectious Diseases and the Threat to Occupational Health in the U.S. and Canada*. Taylor & Francis, Boca Raton, FL, 2006, pp. 71–86.
17. Grinshpun S. Respirator performance with infectious agents (study with stimulant), http:www.cdc/gov/niosh/npptl/resources/pressrel/ (accessed April 26, 2005).
18. Nicas M. Respiratory protection and the risk of *Mycobacterium tuberculosis* infection. *Am J Ind Med* 27:317–333, 1995.
19. Lange JH. BIS/BTS SARS Guidelines. Correspondence/letter. *Thorax* 59:726–727, 2004a.
20. Normile D. Influenza: Is China coming clean on the bird flu. *Science* 314:905, 2006.
21. Hampson AW. Avian influenza: A pandemic waiting in the wings. *Emerg Med Aust* 18:420–429, 2006.
22. Sepkowitz KA and Eisenberg L. Occupational deaths among healthcare workers. *Emerg Infect Dis* 11:1003–1006, 2005.
23. Tan SY and Brown J. Ignac Philipp Semmelweis (1818–1865): Handwashing saves lives. *Singapore Med J* 47:6–7, 2006.
24. Nuland SB. *The Doctors' Plaque: Germs, Childbed Fever, and the Strange Story of Ignac Semmelweis*. W.W. Norton and Company, New York, 2003.
25. Dunn PM. Ignaz Semmelweis (1818–1865) of Budapest and the prevention of puerperal fever. *Arch Dis Child Fetal Neonatal Ed* 90:F345–F348, 2005.
26. Bencko V and Schejbalova M. From Ignaz Semmelweis to the present: Crucial problems of hospital hygiene. *Indoor Built Environ* 15:3–8, 2006.
27. Rea E and Upshur R. Semmelweis revisited: The ethics of infection prevention among health care workers. *Can Med Assoc J* 164:1447–1448, 2001.
28. Chlibek R, Hartmanova M, Severa J, Prymula R, and Splino M. The use of foam substances for disinfection. *Indoor Built Environ* 15:77–80, 2006.
29. Slack P. The black death past and present. *Trans R Soc Trop Med Hyg* 83:461–463, 1989.
30. De Flora S, Quaglia A, Bennicelli C, and Vercelli M. The epidemiological revolution of the 20th century. *FASEB J* 19:892–897, 2005.
31. Inglesby TV, Dennis DT, Henderson DA, Bartlett JG, Ascher MS, Eitzen E, Fine AD, Friedlander AM, Hauser J, Koerner JF, Layton M, McDade J, Osterholm MT, O'Toole T, Parker G, Perl TM, Russell PK, Schoch-Spana M, and Tonat K, for the Working Group on Civilian Biodefense. Plaque as a biological weapon. In: DA Henderson, TV Inglesby, and T O'Toole (eds), *Bioterrorims: Guidelines for Medical and Public Health Management*. AMA Press, Chicago, IL, 2002, pp. 121–140.
32. Barry M. The tail end of guinea worm—Global eradication without a drug or a vaccine. *N Engl J Med* 356:2561–2564, 2007.
33. Paneth N. Assessing the contributions of John Snow to epidemiology 150 years after removal of the Broad Street pump handle. *Epidemiology* 15:514–516, 2004.
34. Fong TT, Mansfield LS, Wilson DL, Schwab DJ, Molloy SL, and Rose JB. Massive microbiological groundwater contamination associated with a waterborne outbreak in Lake Erie, South Bass Island, Ohio. *Environ Health Perspect* 115:856–864, 2007.

35. Eichelberger L. SARS and New York's Chinatown: The politics of risk and blame during an epidemic of fear. *Soc Sci Med* 65(6):1284–1295 (Epub, May 31, 2007), 2007.
36. Varia M, Wilson S, Sarwal S, McGeer A, Gournis E, Galania E, Henry B, and Hospital Outbreak Investigation Team. Investigation of a nosocomial outbreak of severe acute respiratory syndrome (SARS) in Toronto, Canada. *Can Med Assoc J* 170:927, 2003.
37. Nicas M, Harrison R, Charney W, and Borwegan B. Respiratory protection and severe acute respiratory syndrome. *J Occup Environ Med* 46:196–197, 2004.
38. Aitken C and Jeffries DJ. Nosocomial spread of viral disease. *Clin Microbiol Rev* 14:528–546, 2001.
39. Webster RG and Govorkova EA. H5N1 influenza—Continuing evolution and spread. *NEJM* 355:2174–2177, 2006.
40. Maeda K, Hondo E, Terakawa J, Kiso Y, Nakaichi N, Endoh D, Sakai K, Morikawa S, and Mizutani T. Isolation of novel adenovirus from fruit bat (*Pteropus dasmallus yayeyamae*). *Emerg Infect Dis* 14:347–349, 2008.
41. Vijaykrishna D, Smith GJ, Zhang JX, Peiris JS, Chen H, and Guan Y. Evolutionary insights into the ecology of coronaviruses. *J Virol* 81:4012–4020, 2007.
42. World Health Organization. Update: Outbreak of severe acure respiratory syndrome—Worldwide. *J Am Med Assoc* 289:1918–1920, 2003.
43. Lange JH. The best protection. *Can Med Assoc J* 168:1524, 2003.
44. Low JGH and Wilder-Smith A. Infectious respiratory illnesses and their impact on healthcare workers: A review. *Ann Acad Med Singapore* 34:105–110, 2005.
45. Escudero IHG, Chen MI, and Leo YS. Surveillance of severe acute respiratory syndrome (SARS) in the post-outbreak period. *Singapore Med J* 46:165–171, 2005.
46. Li Y, Ching WH, Ching H, Yuen PL, Seto WH, Kwan JK, Leung JKC, Leung M, and Yu SCT. An evaluation of the ventilation performance of new SARS isolation wards in nine hospitals in Hong Kong. *Indoor Built Environ* 16:400–410, 2007.
47. Drosten C, Gunther S Preiser W, van der Werf S, Brodt HR, Becker S, Rabenau H, Panning M, Kolesnikova L, Fouchier RA, Berger A, Burguiere AM, Cinatl J, Eickmann M, Escriou N, Grywna K, Kramme S, Manugerra JC, Muller S, Rickerts V, Strumer M, Veith S, Klenk HD, Osterhaus AD, Schmitz H, and Doerr HW. Identification of a novel corona virus in patients with severe acute respiratory syndrome. *N Engl J Med* 348:1967–1976, 2003.
48. Chan-Yeung M and X Rh. SARS: epidemiology. *Respirology* 8(Suppl):S9–14, 2003.
49. Yu IT, Xie ZH, Tsoi KK, Chiu YL, Lok SW, Tang XP, Hui DS, Lee N, Li YM, Huang ZT, Liu T, Wong TW, Zhong NS, and Sung JJ. Why did outbreak of severe acute respiratory syndrome occur in some hospital wards but not in others? *Clin Infect Dis* 15:1017–1025, 2007.
50. Lai TS, Keung Nq T, Seto WH, Yam L, Law KI, and Chan J. Low prevalence of subclinical severe acute respiratory syndrome-associated coronovirus infection among hospital healthcare workers in Hong Kong. *Scand J Infect Dis* 37:500–503, 2005.
51. Tennessee department of Health Tennessee Epi-News. Monkeypox: An emerging infectious disease in North America. Communicable and Environmental Disease Services, Nashville, TN, July 2003.
52. Gordon SM and Longworth DL. SARS: here to stay? Monkeypox: Beware of exotic pets. *Cleve Clin J Med* 70:889–895, 2003.
53. Heeney JL. Zoonotic viral diseases and the frontier of early diagnosis, control and prevention. *J Intern Med* 260:399–408, 2006.
54. Rimoin AW, Kisalu N, Kehela-Ilunga B, Mukaba T, Wright LL, Formenty P, Wolfe ND, Shongo RL, Tshioko F, Okitolonda E, Muyembe JJ, Ryder RW, and Meyer H. Endemic human monkeypox, Democratic Republic of Congo, 2001–2004. *Emerg Infect Dis* 13:934–937, 2007.
55. Sbrana E, Xiao SY, Newman PC, and Tesh RB. Comparative pathology of North American and central African strains of monkeypox virus in a ground squirrel model of the disease. *Am J Trop Med Hyg* 76:155–164, 2007.

56. Cunha BE. Monkeypox in the United States: An occupational health look at the first cases. *AAOHN J* 52:164–168, 2004.
57. Fleischer AT, Kile JC, Davidson M, Fischer M, Karem KL, Tecalw R, Messersmith H, Pontones P, Beard BA, Branden ZH, Cono J, Seivae JJ, Khan AS, Damon I, and Kuehnet MJ. Evaluation of human-to-human transmission of monkeypox from infected patients to health care workers. *Clin Infect Dis* 40:689–694, 2005.
58. Ligon BL. Monkeypox: A review of the history and emergence in the Western hemisphere. *Semin Pediatr Infect Dis* 15:280–287, 2004.
59. Coninx R, Maher D, Reyes H, and Grzemska M. Tuberculosis in prisons in countries with high prevalence. *Br Med J* 320:440–442, 2000.
60. Rao NA. New drugs in resistant tuberculosis. *J Pak Med Assoc* 57:252–6, 2007.
61. Mortality Morbidity Weekly Report (MMWR). Extensively drug resistant tuberculosis—United States, 1993–2006. *MMWR* 56:250–253, 2007.
62. Dukes-Hamilton C, Sterling TR, Blumberg HM, Leonard M, McAuley J, Schlossberg D, Stout J, and Huitt G. Extensively drug-resistant tuberculosis: Are we learning from history or repeating it. *Clin Infect Dis* 45:338–342, 2007.
63. Laniado-Laborin R. Tuberculosis in correctional facilities: A nightmare without end in sight. *Chest* 119:681–683, 2001.
64. Kimerling ME, Slavuckii A, Chavers S, Persmtin GG, Tonkel T, Sirtkina O, Golubchikova V, and Baddeley A. The risk of MDR-TB and polyresistant tuberculosis among the civilian population of Tomsk City, Siberia. *Int J Tuberc Lung Dis* 7:866–872, 1999.
65. Mortality Morbidity Weekly Report (MMWR). Tuberculosis transmission in a homeless shelter population—New York, 2000–2003. *MMWR* 54:149–152, 2005.
66. Jones TF, Craig AS, Valway SE, Woodley CL, and Schaffner W. Transmission of tuberculosis in a jail. *Ann Intern Med* 131:557–563, 1999.
67. Samarina A, Zhemkov V, Zakharova O, and Hoffner S. Tuberculosis in St. Petersburg and the Baltic Sea region. *Scand J Infect Dis* 39:308–314, 2007.
68. Drobniewski F, Balabanova Y, Zakamova E, Nikolayevskyy V, and Fedorin I. Rates of latent tuberculosis in health care staff in Russia. *PLoS Med* 13:e55 (Epub), 2007.
69. Hussey G, Hawkridge T, and Hanekom W. Childhood tuberculosis: Old and new vaccines. *Pediatr Respir Rev* 8:148–154, 2007.
70. Senanayake SN and Collignon PJ. Should medical students be routinely offered BCG vaccination. *Med J Aust* 186:98, 2007.
71. Graham M, Howley TM, Pierce RJ, and Johnson PD. Should medical students be routinely offered BCG vaccination? *Med J Aust* 185:324–326, 2006.
72. Reid AH and Tauberberger JK. The origin of the 1918 pandemic influenza virus: A continuing enigma. *J Gen Virol* 84:2285–2292, 2003.
73. Hosbach P. Limiting the impact of seasonal influenza. *J Penn Acad Sci* 80:50–53, 2007.
74. Centers for Disease Control and Prevention (CDC). Influenza virus. http://www.cdc.gov/flu/avian (accessed February 24, 2005).
75. Monto AS. The threat of an avian influenza pandemic. *N Engl J Med* 352:323–325, 2005.
76. Webster RG, Guan Y, Poon L, Krauss S, Webby R, Govorkovai E, and Peiris M. The spread of the H5N1 bird flu epidemic in Asia in 2004. *Arch Virol Supp* 19:117–129, 2005.
77. Thomas JK and Noppenberger J. Avian influenza: A review. *Am J Health Syst Pharm* 64:149–165, 2007.
78. Erth HCJ. Influenza A virus pandemic. *J Penn Acad Sci* 80:45–49, 2007.
79. Food and Drug Administration (Department of Health and Human Services). How the FDA keeps produce safe. *FDA Consum* 41:12–19, 2007.
80. Massachusetts Medical Society. Good health is in your hands, Waltham, MA, May 11, 2001.

81. US Department of Labor. Avian flu guidance, http://www.osha.gov/dsg/guidance/avian-flu.html (accessed February 24, 2005).
82. Williams N. Alarms bells ring over bird flu threat. *Curr Biol* 15:R107–108, 2005.
83. Parry J. WHO confirms Avian flu outbreak in Hanoi. *Br Med J* 328:123, 2004.
84. Parry J. WHO investigates possible human to human transmission of Avian flu. *Br Med J* 328:308, 2004a.
85. Klempner MS and Shapiro DS. Crossing the species barrier—One small step to man, one giant leap to mankind. *N Engl J Med* 350:1171–1172, 2004.
86. Parry J. Mortality from avian flu is higher than in previous outbreak. *Br Med J* 328:368, 2005.
87. Dyer AD. The influence of bubonic plaque in England 1500–1667. *Med Hist* 22:308–326, 1978.
88. Yassi A, Moore D, FitzGerald JM, Bigelow P, Hon C-Y, and Bryce E, Research gaps in protecting healthcare workers from SARS and other respiratory pathogens: An interdisciplinary, multistakeholder evidence-based approach. *J Occup Environ Med* 46:613–622, 2005.
89. Lange JH. A questionnaire survey during asbestos abatement refresher training for frequency of respirator use, respirator fit testing and medical surveillance. *J Occup Med Toxicol* 2:65–74, 1993.
90. Qain Y, Willeke K, Grinshpan SA, Donnelly J, and Coffey CC. Performance of N95 respirators: Filtration efficiency for airborne inert particles. *Am Ind Hyg Assoc J* 59:128–32, 1998.
91. Jennings A, Bilous M, Asimakis P, and Maloof AJ. *Mycobacterium tuberculosis* presenting as chronic red eye. *Cornea* 25:1118–1120, 2006.
92. Ndayimirije N and Kindhauser MK. Marburg hemorrhagic fever in Angola—Fighting fear and a lethal pathogen. *N Engl J Med* 352:2155–2157, 2005.
93. Lee K, Slavcev A, and Nicas M. Respiratory protection against *Mycobacterium tuberculosis*: Quantitative fit test outcomes for five type of N95 filtering-facepiece respirators. *J Occup Environ Hyg* 1:22–28, 2004.
94. Thorne GD, Khozin S, and McDiamid MA. Using the hierarchy of control technologies to improve healthcare facility infection control: lessons fro severe acute respiratory syndrome. *J Occup Environ Med* 46:613–622, 2004.
95. Ho AS, Sung JJY, and Chan-Yeung M. An outbreak of severe acute respiratory syndrome among hospital workers in a community hospital in Hong Kong. *Ann Intern Med* 139:564–567, 2003.
96. Derrick JL and Gromerall CD. Surgical helmets and SARS infection; *Emerg Infect Dis* 10:277–279, 2004, http://www.cdc.gov/ncidod/EID/vol10no2/03-0764.htm (accessed May 30, 2005).
97. Fisher DA, Lim TK, Lim YT, Sing KS, and Tambyah PA. Atypical presentation of SARS. *Lancet* 361:1740, 2003.
98. Seto WH, Tsang D, Yung RWH, Ching TY, Ng TK, Ho LM, and Peiris JSM. Advisors of expert SARS group of hospital authority. Effectiveness of precautions against droplets and contact in prevention of nosocomial transmission of severe acute respiratory syndrome (SARS). *Lancet* 361:1519–1520, 2003.
99. Derrick JL and Gromerall CD. Protecting healthcare staff from severe acute respiratory syndrome: filtration capacity of multiple masks. *J Hosp Infect* 59:365–368, 2005.
100. Centers for Disease Control and Prevention (CDC). Guidelines for preventing the transmission of *Mycobacterium tuberculosis* in health care facilities. *MMWR* 43:1–27, 1994.
101. Rajhans DS. *Practical Guide to Respirator Usage in Industry*. Blackwell Publishing, Boston, MA, 1985.
102. Centers for Disease Control and Prevention (CDC). Public health guidance for community-level preparedness and response to severe acute respiratory syndrome (SARS), http://www.cdc.gov/ncidid/.sars/sarsprepplan.htm) (accessed October, 2003).
103. Minnesota Department of Health. N95 disposable respirators, St Paul, MN, http://www.health.state.mn.us/divs/idepc/dtopics/ (accessed April 26, 2005).
104. Coffey CC, Lawrence RB, Zhuang Z, Duling MG, and Campbell DL. Errors associated with three methods of assessing respirator fit. *J Occup Environ Hyg* 3:44–52, 2006.

105. Balazy A, Toivola M, Adhikari S, Sivasubramani SK, Reponen T, and Grinshpun SA. Do N95 respirators provide 95% protection level against airborne viruses, and how adequate are surgical masks? *Am J Infect Control* 34:51–57, 2006.
106. Lange JH. Health effects of respirator use at low airborne concentrations. *Med Hypotheses* 54:1005–1007, 2000.
107. Davis KJ, Herbert AM, Westmoreland D, and Bragg J. Seroepidemiological study of respiratory virus infections among dental surgeons. *Br Dent J* 176:262–265.
108. Khoo KL, Leng PH, Ibrahim IB, and Lim TK. The changing face of healthcare workers perceptions on powered air-purifying respirators during the SARS outbreak. *Respirology* 10:107–110, 2005.
109. Hannum D, Cycan K, Jones L, Stewart M, Morris S, Markowitz SM, and Wong ES. The effect of respirator training on the ability of healthcare workers to pass a qualitative fit test. *Infect Control Hosp Epidemiol* 17:633–635, 1996.
110. Delaney LJ, McKay RT, and Freeman A. Determination of known exhaust valve damage using a negative pressure user seal check method in full facepiece respirators. *Appl Environ Hyg* 18:237–243, 2003.
111. Stobbe TJ, daRoza RA, and Watkins MA. Facial hair and respirator fit: A review of the literature. *Am Ind Hyg Assoc J*. 49:199–204, 1988.
112. Bollinger NJ and Schutz RH. NIOSH guide to industrial respiratory protection. US Department of Health and Human Services, Washington, DC (Publication number 87–116), 1987.
113. Plog BA, Niland J, and Quinian PJ. *Fundamentals of Industrial Hygiene*. National Safety Council, Itasca, IL, 1996.
114. Mullins HE, Danisch SC, and Johnston AR. Development of new qualitative test for fit testing respirators. *Am Ind Hyg Assoc J* 56:1068–1073, 1995.
115. Clapham SJ. Comparison of N95 disposable filtering facepiece fits using Bitrex qualitative and TSI portacount quantitative fit testing. *Intl J Occupat Environ Health* 6:50–55, 2000.
116. Sibert JR and Frude N. Bittering agents in the prevention of accidental poisoning: Children's reactions to denatonium benzoate (Bitrex). *Arch Emerg Med* 8:1–7, 1991.
117. Mullins ME and Zane-Horowitz B. Was it necessary to add Bitrex (denatonium benzoate) to automotive products? *Vet Hum Toxicol* 46:150–152, 2004.
118. Bjorkner B. Contact urticaria and asthma from denatonium benzoate (Bitrex). *Contact Dermat* 6:466–471, 1980.
119. Brosseau LM and Traubel K. An evaluation of respirator maintenance requirements. *Am Ind Hyg Assoc J* 58:242–246, 1997.
120. Lim EC, Seet RC, Lee HK, Wilder-Smith EP, Chuak BY, and Ong BK. Headaches and the N95 facemask amongst health care providers. *Acta Neurol Scand* 113:199–202, 2006.
121. Lange JH, Priolo G, and Mastrangelo G. Respirators and headaches in industrial situations: Suggesting a preventative solution. *Acta Neurol Scand* 116:72, 2007.
122. Cummings KJ. Respirator donning in post-hurricane New Orleans. *Emerg Infect Dis* 13:700–707, 2007.
123. Duling MG, Lawrence RB, Slaven JE, and Coffey CC. Simulated workplace protection factors for half-facepiece respiratory protective devices. *J Occup Environ Hyg* 4:420–431, 2007.
124. Oh VMS and Linn TK. Singapore experience of SARS. *Clin Med* 3:448–451, 2003.
125. Mahler V. Skin protection in the healthcare setting. *Curr Probl Dermatol* 34:120–132, 2007.
126. Omokhodion FO. Health and safety in clinical laboratory practice in Ibadan, Nigeria. *Afr J Med Sci* 27:201–204, 1998.
127. Gardner PD, Hofacre KC, and Richardson AW. Comparison of stimulated respirator fit factors using aerosol and vapor challenges. *J Occup Environ Hyg* 1:29–38, 2004.

128. Ha L, Bloom SA, Nguyen QH, Maloney SA, Mai LQ, Leitmeyer KV, Anh BH, Reynolds MG, Montgomery JM, Comer JA, Horby PW, and Plant AJ. Lack of SARS transmission among public hospital workers, Vietnam. *Emerg Infect Dis* 10:265–268, 2004, http://www.cdc.gov/ncidod/EID/vol10no2/03-0707.htm (accessed May 30, 2005).
129. Jiang S, Huang L, Chen X, Wang J, Wu W, Xin S, Chen W, Zhan J, Yan L, Ma L, Li J, and Huang Z. Ventilation of wards and nosocomial outbreak of severe acute respiratory syndrome among health-care workers. *Chin Med J (Engl)* 116:1293–1297, 2003.
130. Nyka W. Studies on the infective particle in airborne tuberculosis. I. Observations in mice infected with bovine strain *M. tuberculosis*. *Am Rev Respir Dis* 85:33–39, 1962.
131. National Institute for Occupational Safety and Health. NIOSH pocket guide to chemical hazards. US Department of Health and Human Services (DHHS), DHHS (NIOSH), Washington, DC (Publication number 94–116), 1994.
132. North Safety Products. Material safety data sheet for denatonium benzoate (Bitrex), 2003, http://www.northsafety.com/usa/en/images/MSDS/ (accessed May 8, 2005).
133. Shen Z, Ning F, Zhou W, He X, Lin C, Chi DP, Zhu Z, and Schuchat A. Superspreading SARS events, Beijing, 2003. *Emerg Infect Dis* 10:256–260, 2004, http://www.cdc.gov/ncidod/EID/vol10no2/03–0732.htm (accessed May 30, 2005).
134. Tambyah PA. SRAS: Two years on. *Singapore Med J* 46:150–152, 2005.
135. Ungchusak K, Auewarakul P, Dowell SF, Kitphati R, Auwanit W, Puthavathana P, Uiprasertkul M, Boonnak K, Pittayawonganon C, Cox NJ, Zaki SR, Thawatsupha P, Chittaganpitch M, Khantang R, Simmerman JM, and Chunsutthiwat S. Probable person-to-person transmission of avian A (H5N1) *N Engl J Med* 352:333–340, 2005.
136. Wong SS and Yuen KY. Avian influenza virus infections in humans. *Chest* 129:156–68, 2006.
137. World Health Organization. Avian flu fact sheet, http://www.who.int/csr/don/2004_01_15/en/ (January 15, 2004).

Chemical Substances and Biological Agents

Studies and Research Projects

TECHNICAL GUIDE **RG-501**

Guide on respiratory protection against bioaerosols

Recommendations on its selection and use

Jacques Lavoie
Yves Cloutier
Jaime Lara
Geneviève Marchand

Established in Québec since 1980, the Institut de recherche Robert-Sauvé en santé et en sécurité du travail (IRSST) is a scientific research organization known for the quality of its work and the expertise of its personnel.

OUR RESEARCH
is *working* for you !

Mission

To contribute, through research, to the prevention of industrial accidents and occupational diseases as well as to the rehabilitation of affected workers.

To offer the laboratory services and expertise necessary for the activities of the public occupational health and safety prevention network.

To disseminate knowledge, and to act as scientific benchmark and expert.

Funded by the Commission de la santé et de la sécurité du travail, the IRSST has a board of directors made up of an equal number of employer and worker representatives.

To find out more

Visit our Web site for complete up-to-date information about the IRSST. All our publications can be downloaded at no charge.
www.irsst.qc.ca

To obtain the latest information on the research carried out or funded by the IRSST, subscribe to Prévention au travail, the free magazine published jointly by the IRSST and the CSST.
Subscription: 1-877-221-7046

Legal Deposit
Bibliothèque et Archives nationales
2007
ISBN: 978-2-89631-136-1 (print format)
ISBN: 978-2-89631-137-8 (PDF)
ISSN: 0820-8395

IRSST – Communications Division
505, De Maisonneuve Blvd West
Montréal (Québec)
H3A 3C2
Phone: 514 288-1551
Fax: 514 288-7636
publications@irsst.qc.ca
www.irsst.qc.ca

Chemical Substances and Biological Agents

Studies and Research Projects

 TECHNICAL GUIDE **RG-501**

Guide on respiratory protection against bioaerosols

Recommendations on its selection and use

Jacques Lavoie[1], Yves Cloutier[2], Jaime Lara[1] and Geneviève Marchand[3]

[1]Research Department, IRSST
[2]Strategic Watch and Quality Management Department, IRSST
[3]Research and Expertise Support Department, IRSST

Clic Research
www.irsst.qc.ca

This publication is available free of charge on the Web site.

This study was financed by the IRSST. The conclusions and recommendations are those of the authors.

MEMBERS OF THE FOLLOW-UP COMMITTEE

Daniel Boucher (CSST), Candide Fournier (CSST), Nicole Goyer (IRSST), Luc Ménard (CSST), Angélique Métra (Association paritaire pour la santé et la sécurité au travail, secteur affaires sociales, joint sector-based occupational health and safety association, social services sector) and Paule Pelletier (Agence de développement de réseaux des services de santé et des services sociaux de la Montérégie, Montérégie health services and social services network development agency).

ACKNOWLEDGMENTS

The authors thank Dr. Colette Lebâcle and Mrs. Isabelle Balty from INRS of France, Dr. Élizabeth Lajoie from the Agence de développement de réseaux des services de santé et des services sociaux de la Montérégie, Dr. Claude Ostiguy of the IRSST (Institut de recherche Robert-Sauvé en santé et en sécurité du travail du Québec, Occupational Health and Safety Research Institute Robert-Sauvé) and Mrs. Céline Lemieux du Ministère de la Santé et des services sociaux du Québec, Quebec Ministrys' of Health and Social Services. Additionally the authors thank Ms. Helen Fleischauer and Ms. Julie McCabe for the english translation.

Table of contents

PREFACE v

CONTEXT v

CHAPTER 1 Introduction 1

CHAPTER 2 Respirators 2

2.1 Air filtration mechanism 2
2.2 Assigned protection factor (APF) 4
2.3 Fit, seal and care of respirators 5
Fit tests 5
Seal tests 5
Care of respirators 6

CHAPTER 3 Respiratory protection against bioaerosols 7

3.1 Protection against infectious bioaerosols 7
3.1.1 Approach based on the significance of the risk and experts' recommendations 10
3.1.2. Criteria to be considered in selecting a respirator 14
3.1.3 Medical mask and respirator 15
3.2 Protection against non-infectious bioaerosols 16
3.2.1 Risk coefficient (RC) method 17
Tolerated background level for molds 18
Tolerated background level for bacteria 19
Tolerated background level for endotoxins 19
Minimum respiratory protection 20

CHAPTER 4 Limits and scope of this guide 21

CHAPTER 5 Examples of choices of a respirator 21

Personnel working in the room of a patient infected with the SARS virus 21
Operator cleaning a filter press in a wastewater treatment plant 21
Peat moss packager 22

GLOSSARY 23

BIBLIOGRAPHY 25

APPENDIX 1 Decision tree for selecting a respirator against bioaerosols 29

APPENDIX 2 Standards and regulations 30

LIST OF TABLES

1 Filter class designation according to NIOSH 42 CFR Part 84 approval 4
2 Infectious doses of microorganisms or diseases and inoculation routes 8
3 Behavior of bioaerosols in the air 10
4 Examples of choices of respirators for protection against some infectious bioaerosols based on the significance of the risk and experts' recommendations 12
5 Concentrations of bioaerosols normally found in various workplaces 17
6 Maximum use concentrations and corresponding protection factor 20

LIST OF FIGURES

1 Filtration efficiency based on particle diameter 3
2 Medical mask and filter respirator 16

PREFACE

The recommendations stated in this document apply to all people in charge of workers' respiratory protection against bioaerosols, whether infectious or not.

These recommendations summarize the state of knowledge of researchers at the Institut de recherche Robert-Sauvé en santé et en sécurité du travail (IRSST, Robert Sauvé Occupational Health and Safety Research Institute), in close collaboration with a follow-up committee made up of people working in occupational health and safety. This follow-up committee carried out the following role: to become familiar with the document's objectives; to ensure that the objectives meet the identified needs; and to decide on the feasibility of the selection process in order to ensure that the proposed protective measures are useful and relevant.

These recommendations in no way constitute a management plan for a possible pandemic. They are a complement to the Guide pratique de protection respiratoire (practical guidelines on respiratory protection) published by the IRSST (Lara and Vennes, 2003) to help in determining appropriate respiratory protection against bioaerosols, when the situation requires it. It is the responsibility of employers to ensure that risks involving respiratory protection are managed internally.

CONTEXT

All the recommendations specifically relate to respiratory protection against bioaerosols. For other types of exposure (direct or indirect contact with the body), other means of protection must be considered : wearing sealed goggles or visor, gloves and coveralls, hand washing, vaccination, etc.

The content of this document may eventually be updated to include the following aspects:

- the emergence of new infectious bioaerosols;
- updated recommendations of committees of experts;
- new knowledge on the toxicity of non-infectious bioaerosols, on fungal toxins (e.g., mycotoxins) and nanobacteria;
- new identification and counting methods used in microbiology such as genomic aerobiology;
- studies on the protection factors of respirators against bioaerosols;
- how to evaluate the mechanical filtration efficiency of filter materials against biological particles.

CHAPTER 1 **Introduction**

Microorganisms are present everywhere in our environment: water, soil, air, plants, animals and humans. When their presence in air is involved, they are called “airborne microorganisms” or “bioaerosols” (Goyer et al., 2001).

The American Conference of Governmental Industrial Hygienists (ACGIH) defines bioaerosols as being airborne particles consisting of living organisms, such as microorganisms (e.g., viruses, bacteria, molds, protozoa), or originating from living organisms (e.g., toxins, dead microorganisms or fragments of microorganisms) (ACGIH, 1999).

The European guideline 200/54/CE of September 18, 2000, on the risks of exposure to so-called “biological” agents in the workplace, includes microorganisms in this category, including those that are genetically modified, cell cultures, and human endoparasites likely to cause infection, allergy or poisoning (INRS, 2004).

There is growing interest in the risks of exposure to infectious bioaerosols for everyone who has a role to play in occupational health and safety (OHS). Choosing and using respiratory protection can be key decisions, among other things, in cases of exposure to severe acute respiratory syndrome (SARS), tuberculosis, avian or pig flu, anthrax, etc.

Since no general document on the respiratory protection of workers against bioaerosols existed, the members of the technical committee of the board of directors (3.33.1) of the CSST (Québec workers' compensation board), in the context of reviewing Schedule I of the Regulation respecting occupational health and safety (Order in council 885-2001), asked the IRSST to develop a guide on this subject. In Québec, this same concern had been addressed by the Comité ministériel sur les mesures de précaution contre le syndrome respiratoire aigu sévère (SRAS) (2004b) (Ministerial committee on precautions against SARS).

This is the framework for this document. Its objective is to guide in the selection of respirators against bioaerosols in hazardous situations for workers in different sectors: hospitals, household waste sorting centres, wastewater treatment centres, agriculture, food and beverage processing, etc.

This document first includes a brief description of respirators, air filtration mechanisms and the assigned protection factors for respirators, completed by information on their fit, seal and care. It then presents the respiratory protection required for infectious and non-infectious bioaerosols.

At the end of the document are a few examples on the choice and use of respirators for various work contexts. The appendices contain a decision tree for selecting a respirator against bioaerosols as well the current standards and regulations.

CHAPTER 2 **Respirators**

According to Lara and Vennes (2003), a respirator is a device used for protecting an individual against a risk of alteration of his/her health due to the inhalation of air contaminated by gases, vapors and aerosols (or bioaerosols) or to a lack of oxygen.

There are three categories of respirators: air-purifying (filtration), atmosphere-supplying (supplied-air or self-contained), and those combining supplied air and air purification. Their characteristics are described in the IRSST's Guide pratique de protection respiratoire by Lara and Vennes, published in 2003.

According to McCullough and Brosseau (1999), the choice of respirator is guided by the contaminant's physical nature and concentration. In the majority of situations, aerosols, including bioaerosols, require filtration. This is why the present guide essentially deals with air-purifying filter respirators, with this mechanism being sufficient to protect workers in most situations that involve contamination due to bioaerosols.

In the case of biological weapons, it is highly probable that the types of bioaerosols present, as well as their concentrations are unknown. It is then appropriate to use a self-contained breathing apparatus system (SCBA) (CDC, 2001b).

2.1 Air filtration mechanism

Bioaerosols that are living or dead, on solid or liquid substrates, behave in the air or on surfaces in the same way as other particles (Brousseau et al., 1997; Qian et al., 1997; CDC, 2001b; Lee et al., 2004). The fact that a particle is biologically active or not biological does not seem to affect either its retention by a filter or the deposition of the particle on the filter. As with inert particles, there may be re-aerosolization of a very small portion of the bacteria collected by the filter, following a cough or a violent sneeze by the carrier (Qian et al., 1997). Furthermore, particle filtration efficiency depends on the velocity, size, shape and the electrostatic and hygroscopic interactions of the particles (Hinds, 1982; Yassi and Bryce, 2004).

Figure 1 Filtration efficiency based on particle diameter

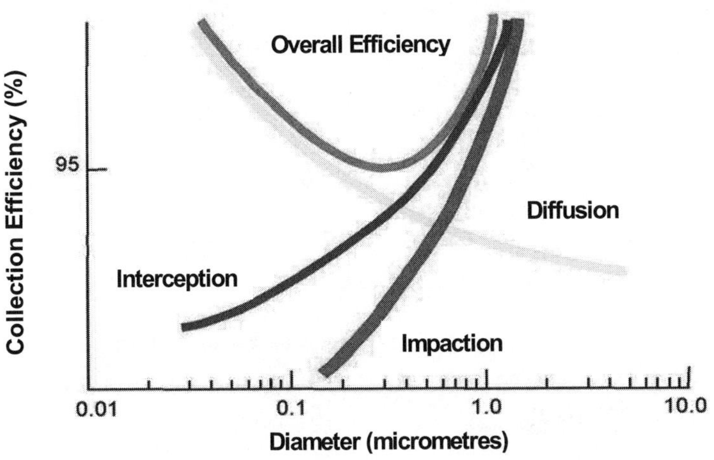

To properly understand filtration efficiency, it is important to know that several mechanisms affect the behavior of solid or liquid particles in the air: diffusion, sedimentation, impaction, interception and the electrostatic force. Three of them are critically important in how a particle is collected by a filter, namely interception, inertial impaction, and diffusion by Brownian motion (Yassi and Bryce, 2004; Willeke and Baron, 1993).

An increase in the size of a particle increases its capacity to be filtered by interception and inertial impaction, while a reduction in its size increases its collection by involving the diffusion mechanism (Willeke and Baron, 1993). As a result, there exists an intermediate particle size that brings two or more mechanisms into play, which operate simultaneously, without predominance. The total efficiency curve in Figure 1 indicates the range in which the particle's potential for penetration through the mechanical filter is at its maximum, and conversely, the one in which the filtration efficiency is at its minimum (Yassi and Bryce, 2004; Willeke and Baron, 1993).

For each type of filter, there is a particle diameter that minimizes the collection efficiency. Several well-known institutions consider the 0.3-micrometre (µm) particle as being the one that penetrates the deepest based on tests on respirator filters (Ruzer and Harley, 2005). Furthermore, it is the basic parameter for certification tests carried out on respirators in European standard EN149: 2001, NIOSH 42 CFR Part 84, and Australian Standard AS1716 (Yassi and Bryce, 2004).

The N95 filter respirator retains 95% of the most penetrating particles, meaning 0.3-micrometre (µm) particles, while the N99 retains 99%, and the N100, close to 100%. Filters in class N can be used in the presence of oil-free solid contaminants. There are also two other classes of filters, R (for one work shift) and P (for prolonged use), which are appropriate if there is oil present in the air (Lara and Vennes, 2003). Note that these designations, N95, N99 and others, must obtain NIOSH 42 CFR Part 84 approval. Table 1 groups them.

Table 1 Filter class designation according to NIOSH 42 CFR Part 84 approval

Minimum efficiency	Oil-free particles	Presence of oil in the particles (for one work shift - 8 hrs)	Presence of oil in the particles (for prolonged use)
95%	N95	R95	P95
99%	N99	R99	P99
99.97%	N100	R100	P100

The respirators must fit well. In addition, their seal must be checked before each use, except for some loose-fitting models (e.g., powered air-purifying respirator with loose-fitting facepiece/visor or supplied-air) (CDC, 2001b; Lara and Vennes, 2002, 2003).

2.2 Assigned protection factor (APF)

An assigned protection factor is given to each category of respirator in relation to its efficiency (Lara and Vennes, 2002, 2003). This factor is defined as being the ratio of the concentration of the contaminant measured outside (C_o) and inside (C_i) the respirator. The higher the assigned protection factor, the better the protection provided by the device.

$$APF = C_o / C_i$$

APFs were established by laboratory tests on well-shaved people trained in wearing respirators. For each class of filter, they determine the expected level of protection for a respirator under optimum use conditions. The various assigned protection factors are identified in two IRSST guides (Lara and Vennes, 2002, 2003).

To select a respirator with an appropriate APF, the risk coefficient (RC) must first be identified. It is the relationship between the concentration of the contaminant measured in the environment and the permissible exposure value (PEV) for chemical contaminants (Lara and Vennes, 2003). This same relationship is established from the tolerated background level, which can be considered equivalent to the PEV for non-infectious bioaerosols (Goyer et al., 2001). The concepts of risk coefficient and tolerated background level are defined in greater detail in section 3.2.1 of this document.

The important aspect to be remembered is the following: the APF must always be greater than the risk coefficient, based on the given situation in the workplace.

2.3 Fit, seal and care of respirators

Fit tests

Except for models with a loose-fitting facepiece/visor (such as some powered air-purifying or supplied-air respirators that do not require fitting), respirators must be properly adjusted to the user. They must be put on in such a way that they form a tight seal with the face to prevent contaminated air from entering the filtering facepiece around its periphery (Lara and Vennes, 2003).

Tests and devices help in selecting a respirator model and in verifying whether the fit is appropriate. Some types of tests "qualitatively" evaluate the fit by assessing the subjective perception of the odor or taste of a product (saccharine, bitrex or other) in an test chamber. Also, electronic instruments, such as the TSI Portacount and the Dynatech Fit Tester 3000, will fulfill the same objectives, quantitatively. The IRSST's Guide pratique de protection respiratoire (2003) explains in greater detail the difference between the qualitative and quantitative tests.

The equipment must also be chosen, fit, used and cared for according to CSA standard Z94.4-93 Selection, Use, and Care of Respirators. In addition, a respiratory protection program must be developed and applied by complying with this standard.

Seal checks

The seal of filtering facepieces should be checked before each use. These seal checks are of two types: positive pressure and negative pressure (CDC, 1994; Lara and Vennes, 2003; Comité ministériel sur les mesures de précaution contre le syndrome respiratoire aigu sévère (SRAS), 2004b).

In the case of a negative pressure test, the following guidelines must be followed:

- put on the respirator and tighten the straps (if adjustable) without it being uncomfortable;
- block the filter or the filtering surface with your hands without deforming it, for a short period of time;
- disconnect the hose or shut off the air supply if it is a powered or supplied-air respirator;
- inhale gently to create a vacuum and check whether the respirator flattens somewhat; if not, there is a leak in the face seal or in a component.

A positive pressure test, which must be carried out after the negative pressure seal checks, has only two steps:

- lightly cover the filter or the filtering surface with your hands without crushing it or deforming it;
- gently exhale into the facepiece.

If the seal is good, the facepiece will bulge slightly. If not, the respirator must be re-ajusted.

Care of respirators

A respirator's non-disposable components must be stored and cared for according to the manufacturer's recommendations, by considering the precautions pertaining to biological hazards. They must be stored and cared for, mainly in the case of an infectious bioaerosol, in such a way as to eliminate or minimize any risk of contamination of the users and transmission of infections.

In this same context, when a powered air-purifying respirator (PAPR) equipped with high efficiency filters could be used, a safe filter changing procedure must be implemented (Comité ministériel sur les mesures de précaution contre le syndrome respiratoire aigu sévère (SRAS), 2004b). Non-disposable respirators must be cleaned and disinfected according to the official procedures (OSHA, 2006). McCullough and Brosseau (1999) recommend that the filter of a powered air-purifying respirator be changed after each use in hantavirus-contaminated environments. The disposable components must be properly stored in compliance with certain guidelines. The advice of committees of experts must be referred to in establishing a safe procedure.

In the case of air-purifying respirators (or filters), it is preferable in the presence of bioaerosols to use disposable components and to discard them after use in order to avoid possible microbial proliferation in the filter as well as the immediate risks of contamination (Lavoie and Allard, 2004; Lavoie et al., 2004; Lavoie and Dunkerley, 2002; Lavoie and Guertin, 2001; CDC, 2001b; Lavoie, 2000).

When a respirator is removed, specific precautions must be taken, even if it is disposable, due to the risk of hand contamination by the infectious bioaerosols (Comité ministériel sur les mesures de précaution contre le syndrome respiratoire aigu sévère (SRAS), 2004b).

General information on fit and seal checks as well as information relating to respirator care can be found in the Guide pratique des appareils de protection respiratoire (Lara and Vennes, 2003).

It is important to note that any damaged, moistened or deformed non-disposable facepiece must be changed or repaired.

CHAPTER 3 **Respiratory protection against bioaerosols**

The sizes of bioaerosols are in the order of 0.02 to 0.25 µm for viruses, 0.3 to 15 µm for bacteria, and 1 to 50 µm for the majority of molds and yeasts (AIHA, 2000; Goyer et al., 2001).

According to Lee et al., (2005), most bacteria and molds are between 0.7 and 10 µm in size. The filtration efficiency of N95 filtering facepieces or half-facepieces should be greater than 99.5% for the majority of these sizes.

Evaluation of biological hazards and workers' respiratory protection must be based on the type of bioaerosol involved.

Two classes of biological hazards exist, based on the characteristics of the microorganisms making up a bioaerosol. They are:

1. infectious hazards (e.g., viruses, pathogenic bacteria);
2. non-infectious hazards (e.g., non-pathogenic bacteria and molds).

Infectious bioaerosols must be living to cause infections, which are defined as being the result of the penetration and development in a living being of microorganisms that can cause lesions by multiplying and eventually by secreting toxins or by spreading through the bloodstream (unofficial translation of definition in Le petit Larousse illustré, 2006).

The non-infectious bioaerosol category contains microorganisms found in the environment in general, which, even when dead, can produce immunological or toxic reactions when inhaled. Molds are a perfect example of these microorganisms.

3.1 Protection against infectious bioaerosols

Table 2 presents the infectious doses of some microorganisms (AIHA, 2000; Public Health Agency of Canada, 2005). Unless otherwise indicated, they apply to humans.

Table 2 Infectious doses of microorganisms or diseases and inoculation routes

Microorganisms or diseases	Infectious doses of microorganisms[1]	Inoculation routes
Histoplasma capsulatum	10 (mice)	Inhalation
Mycobacterium tuberculosis, Mycobacterium bovis	10	Inhalation
Coxsackie A21 virus (Enterovirus)	18 or fewer	Inhalation
Influenza	>790	Inhalation
Bacillus anthracis	8,000-50,000	Inhalation
Q fever (*Coxiella burnetii*)	10	Inhalation
Tularemia (*Francisella tularensis*)	5-10 10^6 - 10^8	Inhalation Ingestion
Adenovirus	>150	Intranasal
Respiratory syncytial virus	> 100-640	Intranasal
Syphilis	57	Intradermal
Malaria	10	Intravenous
Typhoid fever	10^5	Ingestion
E. coli	10^8	Ingestion
Enterohaemorrhagic *E. coli* 0157: H7	10	Ingestion
Bacillus cereus	10^6 or 10^5 per gram	Ingestion
Campylobacter jejuni	500 or fewer	Ingestion
Clostridium perfringens	10^5/gram of food	Ingestion
Hepatitis A virus	Estimated at 10-100 viruses	Ingestion, intravenous, others

[1]***When not specified, the dose is the number of organisms.***

Wearing a respirator does not protect us from all infectious bioaerosols. Some infections can be transmitted only by contact and not through the air. Other means of protection must then be used, such as vaccination, quarantine, etc.

For protection against diseases that can be contracted by bioaerosol inhalation, an appropriate and properly fitting respirator must be worn. In the hospital environment, among others, it is often the responsibility of local committees of experts such as the infection prevention and control committee, in close collaboration with occupational health and safety departments, to decide on the need for protection and the appropriate type of respirator. Decisions can also be made at higher levels should there be potential risks of epidemic or pandemic.

Molds and yeasts, generally classified as non-infectious bioaerosols, can travel without the help of a carrier. This is not the case for infectious bioaerosols, which, for the most part, are carried by larger diameter aerosols in the form of droplets (liquid particles) and dusts (solid particles).

During a sneeze, close to two million droplets can be expelled at a velocity of 100 m/sec (200 miles/hour), compared to fewer than 100,000 droplets from a cough . This significant difference is based on the origin of the secretions, which is deeper in the case of a cough (Yassi and Bryce, 2004). Several are large enough to contain thousands of microorganisms (ACGIH, 1999).

During the expulsion, the diameters of the droplets vary between 1 and 2,000 µm, 95% of which are in the order of 2 to 100 µm. However, they dry very rapidly. The drying times for 100 and 50 µm droplets in air at 50% relative humidity are 1.3 and 0.3 seconds, respectively (Lenhart et al., 2004b). This highlights the fact that relative humidity plays a role in the size and survival of infectious aerosols (Yassi and Bryce, 2004).

The size of infectious bioaerosols is probably between 0.1 and 10 µm (Yassi and Bryce, 2004; ACGIH, 1999). It even appears that the majority of viruses and bacteria that cause respiratory diseases in humans are usually inside bioaerosols with diameters greater than 5 µm.

All droplets and solid particles can contain microorganisms, proteins, mixtures of saliva, mucus and cell debris, which are carriers of respiratory infections (ACGIH, 1999).

Yassi and Bryce (2004) recommend specific precautions against measles, varicella and smallpox viruses due to their small size, which is close to 0.3 microns. In fact, in the case of viruses, the possibility will have to be considered that they can be carried by liquid particles (droplets) with initially large diameters but which, after drying, could shrink to the 0.3-micron size. Neglecting this aspect could lead to an inappropriate choice of respiratory protection for preventing the transmission of infections. This phenomenon and its impacts must be kept in mind in the prevention process.

Bioaerosols consisting of solid or liquid particles smaller than 10 µm in diameter remain suspended in the air for a sufficient time (a few hours) and are likely to be inhaled (Yassi and Bryce, 2004; ACGIH, 1999; Hinds, 1982). Table 3 indicates the time required for a bioaerosol to be deposited by sedimentation from a height of three metres.

Table 3 Behavior of bioaerosols in the air (Yassi and Bryce, 2004)

Diameter in µm	Time required for deposition from a height of 3 metres
100	10 sec.
40	1 min.
20	4 min.
10	17 min.
6 to 10	A few hours
0.06 to 6	Several hours

In fact, contrary to what is generally stated, where a distinction is made between micro-droplets (diameter ≤ 5 µm) and the largest droplets (diameter > 5 µm) which are assumed to be transported only over distances of less than 1 metre, it appears that solid or liquid particles between 6 and 10 µm can take a few hours before being deposited from a height of 3 metres (Lenhart et al., 2004b; Yassi and Bryce, 2004). The belief that droplets larger than 5 µm sediment before traveling a distance of one metre has no foundation (Lenhart et al., 2004b). The duration of suspension in the air must be taken into account, and a respirator chosen accordingly, in order to be well protected against an infectious bioaerosol.

3.1.1 Approach based on the significance of the risk and experts' recommendations

For infectious bioaerosols, the PEV and the tolerated background levels do not enter into the choice of a respirator because the infectious doses are unknown or inapplicable in most cases. In light of these facts, as recommended by CDC-NIOSH, the approach based on knowledge of the significance of the risk and experts' recommendations seems more appropriate (Lenhart et al., 2004a, 2004b).

This approach is a qualitative method that guides the choice of respiratory protection for an infectious inhalable bioaerosol and that requires a decision based on the judgment of several experts (Lenhart et al., 2004b). For example, different respirators will be chosen if the seasonal flu virus or the H5N1 avian influenza virus are involved. This method is used when the data are insufficient or unavailable for a quantitative approach to respirator selection.

The qualitative method has been used in recommending respirators for several types of exposure: to *Mycobacterium tuberculosis* (CDC 1994), *Histoplasma capsulatum* (Lenhart et al., 2004a), *Bacillus anthracis* (CDC, 2001a), to Hantavirus pulmonary syndrome (CDC, 2002), and to bioterrorism agents (CDC, 2001b).

Note that in cases where it is impossible to evaluate the significance of the risk involving an infectious inhalable bioaerosol, meaning when neither the agent nor its means of dissemination are known, the precautionary principle applies. The action rule allows a respirator to be selected whose APF is proportional to the perceived risk. For example, for protection against bioaerosols used as biological weapons, maximum respiratory protection must be favored (autonomous respirator with full-facepiece, positive pressure (SCBA)) (CDC, 2001b). Should the personnel in patient triage in a hospital admitting department be exposed to the SARS virus, the minimum protection required is a disposable N95 respirator (Comité ministériel sur les mesures de précaution contre le syndrome respiratoire aigu sévère (SRAS), 2004b).

Faced with the emergence of a virulent disease like SARS, the Comité ministériel sur les mesures de précaution contre le SRAS in Québec recommended the use of a NIOSH certified disposable N95 filtering half-facepiece respirator, that must be properly fit and used in conjunction with other necessary protective equipment (eye protection, gloves, coveralls, etc.), to prevent transmission, among other things, to the people assigned to ambulance transport and emergency department triage.

Situations can also arise that require respirators with efficiencies better than that of N95 (N99, N100, etc.) or of different classes (P or R) (OSHA, 2006). The Centers for Disease Control and Prevention therefore recommend the use of N100 filters for protection against the hantavirus (see *Table 4*) or a powered air-purifying respirator equipped with P100 filters adapted to certain personal characteristics (having a beard, etc.) (CDC, 2002).

Table 4 presents different choices of respirators made by committees of experts.

Table 4 Respirator choices for protection against some infectious bioaerosols based on the significance of the risk and experts' recommendations

Infectious bioaerosols	Significance of the risk of exposure[1]	Examples of work	Minimum respirator required according to the experts	Comments	Source
Mycobacterium tuberculosis	Low	Entry into an infected patient's room	N95 filtering half-facepiece	The size of this bacterium is >1 μm and can be found on micro-droplets < 5 μm.	CDC, 1994 (retrieved in 2006)
	Average or high	Bronchoscopy on infected patient, autopsy	Powered air-purifying (with P100 filter)[2]	Based on the characteristics of the tasks, a respirator with appropriate APFs is recommended.	
Bacillus anthracis	Low	Personnel doing mail sorting	N95 filtering half-facepiece	N95 filtering half -facepiece sufficient against this bacterium >1 μm.	CDC, 2001a (retrieved in 2006)
	Average or high	Personnel sampling *Bacillus anthracis* in a post office	Powered air-purifying with full-facepiece (P100 filter)	-The spore was possibly biologically modified. - The infectious dose and the exposure levels are unknown. - The type of work of the investigators is non-stationary. APF as high as possible for this situation	
Hantavirus	Low	Telephone installers, plumbers, electricians who can be in contact with rodents or rodents' nests.	Type of respirator to be determined by case	For low risks, the employer must inform the workers about the dangers of infection. For those who handle animals, an N100 filtering half-facepiece is recommended due to the size of the virus and its survival in the environment (approximately 1 week). The powered air-purifying respirator is recommended for those who cannot wear the N100 filtering half-facepiece.	CDC, 2002 (retrieved in 2004)

Infectious bioaerosols	Significance of the risk of exposure[1]	Examples of work	Minimum respirator required according to the experts	Comments	Source
	Average or high	People who frequently handle or are exposed to wild rodents (zoologists, exterminators, etc.)	Half-facepieceN100 or powered air-purifying half-facepiece (with P100 filter)		
Histoplasma capsulatum	Low	Inspection, collection of samples, etc.	N95 filtering hal-facepiece	*H. Capsulatum* is a mold > 1 µm.	Lenhart, 2004a
	Average	Cleaning or work outdoors	Powered air-purifying with disposable N95 half-facepiece	APF of 50	
	High	Chimney cleaning, work in attics and poultry houses	Full-facepiece with N95 filters	APF of 100	
SARS	Low	Personnel in charge of patient triage in a hospital emergency room in the presence of SARS	Disposable N95 filtering half-facepiece	By analogy with other committees of experts (Health Canada and WHO) and according to the reference framework in risk management for health in the Québec public health network	Comité ministériel sur les mesures de précaution contre le syndrome respiratoire aigu sévère (SRAS), 2004b
	Average or high	Personnel caring for an infected individual	Disposable N95 filtering half-facepiece alone or under a powered air-purifying respirator with complete disposable hood based on the HSC's[5] decision	Higher exposure level (higher APF) and protection of the worker when removing his protection.	
Influenza pandemic (recommendations solely for health institutions)	Low	When entering an infected patient's room	Surgical mask[3] or procedure mask	The route of transmission of the virus is by droplets (>5 µm)[4] (analogies with CDC and WHO)	CINQ, 2006
	Average or high	Personnel doing procedures that can generate bioaerosols	N95 filtering half-facepiece	Possibility of transmission in the form of aerosols (≤5 µm)[4]	

[1]: When the significance of the risk has not been evaluated, the precautionary principle applies.
[2]: Powered air-purifying respirators are always used with P100 filters (HEPA filters).
[3]: A surgical or procedure mask is not a respirator.
[4]: In occupational hygiene, droplets are aerosols; [5]: HSC -= Health and Safety Committee

These examples demonstrate that experts' recommendations can vary and that several criteria have to be considered to make a judicious selection of respirators against infectious bioaerosols.

3.1.2. Criteria to be considered in selecting a respirator

According to Lenhart et al., (2004b), a respirator for use against infectious bioaerosols is selected by considering several criteria:

- the possibility of inhalation;
- the state of knowledge;
- the transmission routes;
- the level of exposure;
- the microorganism;
    - its classification by infectious risk group (groups I, II, III, IV, Health Canada)
    - its virulence
    - its infectious dose
    - its size
    - its survival
    - its means of dispersion
    - its suspension time in the air
    - the drying time for the vector and the bioaerosol
- the characteristics of tasks known to involve potential or real exposure to infectious bioaerosols;
- knowledge relating to APFs;
- the advantages and disadvantages of wearing respirators.

In a respiratory risk management plan, the recommendations can be accompanied by directions to be followed (e.g., steps for removing the respirator, waste management, etc.) and/or means of prevention. In the case of SARS, for example, to avoid indirect contamination of the person and his environment, a well-established sequence must be followed when removing the different personal protective equipment (PPE), and hand washing is a vital measure (OSHA, 2006; Comité ministériel sur les mesures de précaution contre le syndrome respiratoire aigu sévère (SRAS), 2004b). Disposable PPEs contaminated by infectious bioaerosols must be handled as biomedical waste (CINQ, 2006; OSHA, 2006; INRS, 2003; CDC, 2002).

In its guide for protecting employees from avian influenza viruses (2006), OSHA proposes the following PPE removal sequence:

- remove protective clothing, except for gloves, before removing the respirator and goggles;
- then remove gloves and wash hands thoroughly with soap and water (avoid all contact of hands with mouth and face);
- remove eye protection and place in a designated receptacle for subsequent cleaning and disinfection;
- remove the disposable respirator and discard;
- wash hands a second time;
- discard the disposable PPE, considered as contaminated material;
- clean and disinfect non-disposable PPEs as specified in the official procedures.

This sequence is to be validated internally, based on actual work practices.

Also to be considered is the fact that some workers can be exposed simultaneously to harmful vapors and gases and to bioaerosols. Although prevention should be based in priority on measures of substitution, process modification, collection and ventilation at source, other respirators may be necessary. For example, in the case of simultaneous exposure to chemical substances and bioaerosols, two means must be combined: cartridges and filters.

3.1.3 Medical mask and respirator

Medical masks (procedure masks, surgical masks) are not respirators. They do not offer sufficient protection against the inhalation of bioaerosols because they cannot provide a close face seal (Yassi and Bryce, 2004; INRS, 2003; McCullough and Brosseau, 1999). In general, they are not made of sufficiently efficient filtering materials (Brosseau et al., 1997). The following figure (Figure 2) shows how a medical mask and a respirator function differently.

Despite the fact that a medical mask is a physical barrier against large droplets, secretions or excretions, its main function is to protect the patient against the aerosols expelled by the caregiver or a visitor (INRS, 2003; CDC, 1994; McCullough and Brosseau, 1999). In North America, all respirators must have NIOSH 42 CFR Part 84 approval (McCullough and Brosseau, 1999).

Figure 2 Medical mask and filtering facepiece respirator (source: INRS, 2003)

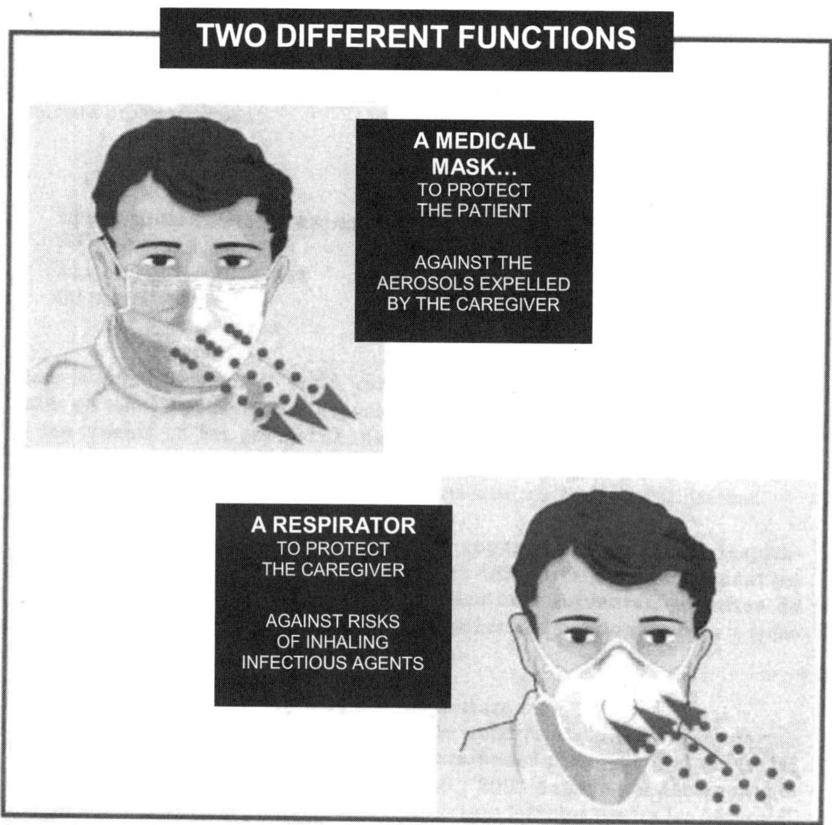

3.2 Protection against non-infectious bioaerosols

Although the inhalation dose-effect relationship has not been established for the majority of non-infectious bioaerosols, the scientific community agrees that some of these bioaerosols can cause health problems, particularly if they are present in sufficient concentration (Goyer et al., 2001; ACGIH, 1999).

As shown in Table 5, these concentrations can fluctuate greatly from one work environment to another.

Table 5 Maximum concentrations of bioaerosols found in scientific literature for various workplaces (Goyer et al., 2001)

Workplaces	Total bacteria (CFU/m³)[a]	Gram negative bacteria (CFU/m³)[c]	Molds (CFU/m³)
Outdoors	10^2	10^1	10^3
Agriculture (normal)	10^7	10^3	10^{3-4}
Agriculture (moldy hay)	10^9	10^3	10^9
Bakery	-[b]	-[b]	10^{2-3}
Composting plant	10^5	10^2	10^4
Wastewater treatment plant	10^4	10^4	10^3
Mushrooms (compost)	10^6	-[b]	10^4
Mushrooms (culture)	10^3	-[b]	10^2
Household waste (collection)	10^4	10^3	10^4
Office building	10^2	10^1	10^{2-3}
Paper mill effluents	10^4	10^3	10^4
Cutting fluid	10^6	10^4	10^5
Humidifier	10^3	10^3	10^{2-3}
Cotton mill	10^5	10^4	10^3
Paper mill	10^6	10^{23}	10^3
Swine confinement building	10^6	10^{3-4}	10^4
Sawmill	10^4	10^{3-4}	10^6
Peat bog	-[b]	-[b]	10^8
Sugar processing	10^5	10^3	10^3
Household waste sorting	10^4	10^3	10^4
Tobacco plant	10^3	10^2	10^4

[a]: CFU/m³ (colony forming units per cubic metre of air)

[b]: not documented

[c]: Gram negative bacteria are those that contain endotoxins in their exterior cell wall

As an example, concentrations of viable bacteria (that proliferate on gelose agar), in places where mushrooms are cultivated (compost), can reach 10^6 CFU/m³ of air (colony forming units per cubic metre of air). In a well-maintained office building, the same bacteria should not exceed a concentration of 10^2 CFU/m³ of air (Goyer et al., 2001). In some workplaces, when control at source is impossible and the concentrations of non-infectious bioaerosols are high, respirators may have to be used.

As with chemical contaminants, it is the concentrations ofr non-infectious bioaerosols present that allow the risk coefficient (RC) to be calculated, which, in turn, must dictate the choice of appropriate respirator, meaning one whose assigned protection factor (APF) is sufficient.

3.2.1 Risk coefficient (RC) method

Respirator selection based on the risk coefficient applies to non-infectious bioaerosols. This is because the scientific literature determines the tolerated background levels, for humans, that can be equivalent to the permissible exposure values (PEVs). These levels represent the thresholds below which the majority of individuals should not suffer from symptoms during exposure to non-infectious bioaerosols. As explained previously in section 2.2 for chemical contaminants, the RC is the relationship between the concentration of the contaminant in the ambient air (CC) and the PEV. For non-infectious bioaerosols, the tolerated background level replaces the PEV.

RC = CC/PEV or tolerated background level

As mentioned by Lara and Vennes (2003) in their Guide pratique de protection respiratoire, the RC must not exceed the APF value for the respirator. It is the RC that serves as a basis in choosing the type of respirator with the minimum appropriate APF.

Tolerated background level for molds

There are currently more than 100,000 species of molds in nature. Humans can easily be exposed to more than 200 species, several of which proliferate in a humid environment (Goyer et al., 2001).

Molds release their spores under the effect of significant air movement or as a reaction to unfavorable conditions such as a rapid increase or decrease in humidity or even in response to the need to reach a new source of food (Goyer et al., 2001).

Still according to Goyer et al. (2001), molds in our regions are active in the outdoor air from April to November and reach their peak of growth between July and the end of autumn. They persist despite the first frost, and although some can develop at temperatures below the freezing point, most are dormant. Snow cover considerably reduces the concentrations in the air, but does not kill molds. When the snow melts, they develop, mainly on dead vegetation. Also, temperature affects their rate of growth. Temperatures from 20 to 25°C correspond to the ideal zone of growth for the majority of them. The document by Goyer et al. (2001) provides additional information on molds.

In summer, the active proliferation period, outdoor mold concentrations can easily reach between 1,000 and 10,000 CFU/m^3 of air (Lavoie and Allard, 2004; OSHA, 2002; Hyvärinen et al., 2001). These concentrations compare with those measured in several studies carried out in Québec on the fungal content of outdoor air (Lavoie et al., 2004; Lavoie and Dunkerley, 2002; Goyer and Lavoie, 2001a, 2001b; Lavoie and Guertin, 2001; Lavoie, 2000; Lavoie and Alie, 1997; Lavoie et al., 1996; Lavoie and Comtois, 1993).

The National Allergy Bureau (NAB), a division of the American Academy of Allergy, Asthma and Immunology (AAAAI), states that the majority of individuals sensitized to mold spores could suffer from symptoms when the concentrations exceed 13,000 spores/m³ of air (IICRC, 2003). From this perspective, the Institute of Inspection, Cleaning and Restoration Certification (IICRC), in their document IICRC S520 entitled Standard and Reference Guide for Professional Mold Remediation, recommends using 10,000 spores/m³ of air as the background level for determining the necessity for wearing a respirator (IICRC, 2003).

It should be mentioned that the unit for this limit is spores, which, contrary to infectious agents, do not have to be alive or viable to retain their immunological properties (Burge, 1995). Clearly, the concentrations of spores/m³ of air are slightly higher than those reported in CFU/m³ because, in addition to the viable or cultivable fraction, the non-viable fraction is also counted. It should also be mentioned that, based on the current state of knowledge, this tolerated background level is also valid for the health effects of mycotoxins (included in the spores of some molds) (IICRC, 2003).

Tolerated background level for bacteria

For all cultivable bacteria, the Scandinavian countries as well as the IRSST, following their research on waste collection and treatment, are currently proposing a tolerated background level of 10,000 CFU/m³ of air (Lavoie and Allard, 2004; Goyer et al., 2001; Lavoie and Alie, 1997; Poulsen et al., 1995a, 1995b; Malmros et al., 1992; Malmros, 1990).

The level set for viable Gram negative bacteria is ten times less, or 1,000 CFU/m³ of air (Lavoie and Allard, 2004; Goyer and al., 2001; Lavoie and Alie, 1997; Poulsen et al., 1995a, 1995b; Malmros et al., 1992; Malmros, 1990).

Tolerated background level for endotoxins

There are also Relative Limit Values (RLVs) for endotoxins (ACGIH, 1999). These toxins present in the external cell wall of Gram negative bacteria can have non-infectious effects on workers' respiratory health (ACGIH, 1999). Based on the logic expressed in this document, RLVs can be equated to the tolerated background levels in calculating the risk level.

The RLVs proposed by the ACGIH are:

- 30 times the background concentration measured in the air at the reference site for healthy individuals;

- 10 times the background concentration in the air at the reference site for people with related symptoms.

The reference site for background sampling can be indoors or outdoors, depending on the season (ACGIH, 1999). For example, suppose that an average concentration of 1,000 endotoxin units (EU)/m³ of air was measured in a swine-confinement building, while the outdoor concentration (background) was 7 EU/m³ of air. This indicates a level 142 times greater than the background level. By wearing an N95 half-facepiece as described in Table 6, the concentration is reduced ten-fold (from 1,000 to 100 EU/m³ of air) and the RLV of 210 EU/m³ (30 times the background concentration of 7 EU/m³ of air) is respected for healthy individuals.

Minimum respiratory protection

Table 6, referring to the IICRC S520 guide, contains information on certain types of respirators and their assigned protection factors. This information is related to the concentrations of mold spores or viable bacteria that can be found in different work environments (IICRC, 2003).

Table 6 Maximum use concentrations and corresponding protection factors

Concentrations of molds in spores or total bacteria in CFU/m³⁽¹⁾	**Assigned protection factor**	**Minimum respirator required**
100,000	10	Half-facepiece with disposable N95 filters
250,000	25	Powered air-purifying or supplied-air with loose-fitting facepieces/visor
500,000	50	Powered air-purifying with half-facepiece
>500,000-1,000,000	100	Full-facepiece with disposable N95 filters
>1,000,000-10,000,000	1,000	Powered air-purifying with hood and hose and equipped with a high efficiency filter (P100 or HEPA)(2)
> 10,000,000	10,000	Self contained with full-facepiece (SCBA), positive pressure

(1): For a recommended background level of 10,000 spores/m³ of air for molds and 10,000 CFU/m³ of air for total bacteria.
(2): For the model with hood and high efficiency filter according to 3M. The hood and hose will have to be disposable for protection against infectious bioaerosols.

The various modes of respirator operation are described in the Guide pratique de protection respiratoire by Lara and Vennes (2003), and in the Guide des appareils de protection respiratoire utilisés au Québec, Lara and Vennes (2002).

As a complement to this information, Appendix 1 contains the selection tree for a respirator against infectious and non-infectious bioaerosols.

CHAPTER 4 **Limits and scope of this guide**

Experts' recommendations for respiratory protection may sometimes seem exaggerated. If any doubt exists, and the available data are insufficient to come to a definite conclusion, the precautionary principle applies, particularly regarding an infectious context.

As for future research needs, interest should now focus on ways to sterilize disposable N95 respirators so that they can be reused in the event of pandemics. With the help of colleagues in infection prevention or occupational health, knowledge about the behavior of microorganisms in the air should be developed further regarding the reduction of risks of disease transmission. The creation of a multidisciplinary network of specialists in respiratory protection against bioaerosols could meet these needs.

CHAPTER 5 **Examples of choices of a respirator**

Personnel working in the room of a patient infected with the SARS virus

In this scenario involving the presence of an infectious bioaerosol, the approach based on the significance of the risk and on experts' recommendations applies. In this regard, the ministerial committee recommended that personnel wear a disposable N95 filtering half-facepiece respirator, alone or under another respirator providing greater protection, such as a PAPR, equipped with a disposable hose and hood. This recommendation related to operations at high risk of producing aerosols (intubation, induced sputum, etc.) (Comité ministériel sur les mesures de précaution contre le syndrome respiratoire aigu sévère (SRAS), 2004b). Such a device is distinguished by a protection factor greater than that for the filtering half-facepiece; it therefore provides more protection. The half-facepiece respirator is used to protect the wearer from risks of contamination when removing the powered respirator before leaving the patient's room. Anyone having to work in this context should follow the procedure defined by the ministerial committee.

Operator cleaning a filter press at a wastewater treatment plant

In this situation, the risk coefficient (RC) method applies, since infectious bioaerosols are not involved and the tolerated background levels equivalent to the PEVs are available. The RC must be below the APF in choosing the respirator. For this type of operation, the literature describes total viable bacteria concentrations in the order of 50,000 CFU/m^3 of air (Goyer et al., 2001). The RC is therefore 5 (ambient air concentration of 50,000 CFU/m^3 of air/tolerated background level of 10,000 CFU/m^3 of air). Table 6 indicates that a disposable N95 filtering facepiece respirator (APF of 10) provides acceptable protection.

Peat moss packager

Since this work generally involves non-infectious bioaerosols, equipment selection is based on the risk coefficient method.

Evaluation of the air's fungal content, for a worker assigned to peat moss packaging, gives concentrations of 200,000,000 spores/m^3 of air (Duchaine et al., 2004). Using Table 6 as a basis, the worker should wear a respirator with a maximum APF, namely 10,000 (SCBA with full-facepiece at positive pressure). However, the characteristics of the tasks to be performed make this type of device unusable. As a result, after implementing control measures such as a local ventilation system and verifying its efficiency, the worker will be able to wear a PAPR (powered air-purifing respirator) with a hood and a P100 filter, if the new concentrations permit it. Because the bioaerosols present are not considered infectious, there's no need to use a disposable hood and hose.

GLOSSARY

Aerosol – Solid or liquid particle suspended in a gaseous medium such as air.

Assigned protection factor (APF) – Recommended value that provides an indication of the protection provided by a respirator. The higher this factor, the higher the protection provided the respirator.

Bioaerosols – Airborne particles consisting of living organisms such as microorganisms, or originating from living organisms, for example toxins, dead microorganisms or fragments of microorganisms (ACGIH, 1999).

CFU – Colony forming unit. It is the unit of measurement for viable microorganisms growing on gelose agars (culture media). One unit corresponds to one microorganism.

Diffusion – Mass transfer associated with the random movement of molecules or particles under the effect of a difference in concentration.

Droplet – Liquid particle.

Dust - Aerosol consisting of solid particles.

Endotoxins – Components of the exterior cell membrane of Gram negative bacteria, made up of lipopolysaccharides associated with proteins and lipids

Inertial impaction – Mechanism by which the particles that tend to travel in their original direction of movement will deviate from the lines of flow to impact on a surface. This tendency is even stronger when the particle is massive, its velocity is high, and the lines of flow diverge abruptly.

Infection – Result of the penetration and development in a living being of microorganisms that can cause lesions by multiplying and eventually secreting toxins or by spreading through the bloodstream.

Infectious dose – Quantity of microorganisms necessary to cause an infection in their host.

Interception – Phenomenon by which a particle is collected due to its physical size when it comes into contact with another particle or fiber.

Medical mask (for care, hygiene, anti-projection, procedure or surgical) – Their main function is to protect the patient against the aerosols expelled by a caregiver or visitor.

Mycotoxin – Secondary metabolite released by molds as defense against other microorganisms.

Particle – Small portion of solid or liquid material.

PEV (permissible exposure value) – According to the ROHS (Regulation respecting occupational health and safety, Schedule I), permissible limit of exposure for workers to an air contaminant.

Precautionary principle – Action rule that allows a respirator to be chosen with an APF proportional to the perceived risk

RLV – Relative limit Values for endotoxins. The proposed RLVs are:

- 30 times the background concentration in the air at the reference site for healthy individuals;
- 10 times the background concentration in the air at the reference site for people with related symptoms.

Respirator – Intended to protect an individual exposed to a risk of alteration of his health caused by the inhalation of air contaminated by gases, vapors, aerosols (including bioaerosols), or by a lack of oxygen.

Risk coefficient (RC) – Allows the minimum assigned protection factor (APF) for a respirator to be defined in relation to the contaminant's concentration. It is determined from the contaminant's concentration in the air divided by the tolerated background level for non-infectious bioaerosols. The APF must be higher than this coefficient.

Sedimentation – Phenomenon of particle displacement under the effect of gravity .

Virulence – Ability of a pathogenic organism to multiply in a living organism, resulting in sickness.

BIBLIOGRAPHY

American Conference of Governmental Industrial Hygienists (ACGIH). (1999). Bioaerosols: Assessment and control. American Conference of Governmental Industrial Hygienists, Cincinnati, OH, 322 p.

American Industrial Hygiene Association (AIHA). (2000). Biosafety Reference Manual. Second edition, American Industrial Hygiene Association publications, Fairfax, VA., 177 pages.

Brosseau, L.M., Vars McCullough, N., Vesley, D. (1997). Mycobacterial Aerosol Collection Efficiency of Respirator and Surgical Mask Filters Under Varying Conditions of Flow and Humidity. Applied Occupational and Environmental Hygiene 12(6):435-445.

Burge, H. (1995). Bioaerosols. Lewis Publishers, Ann Arbor, 318 pages.

Centers for Disease Control and prevention (CDC) (2002). Hantavirus pulmonary syndrome United States: Updated recommendations for risk reduction. Morbidity and Mortality Weekly Report, 51(RR09), 1-12. Retrieved January 1, 2004. http://www.cdc.gov/mmwr/PDF/rr/rr5109.pdf

Centers for Disease Control and Prevention (CDC) (2001a). Protecting investigators performing environmental sampling for *Bacillus anthracis:* Personal protective equipment. Retrieved July, 2006. http://www.cdc.gov/DocumentsApp/Anthrax/Protective/protective.asp.

Centers for Disease Control and Prevention (CDC) (2001b). Interim Recommendations for the Selection and Use of Protective Clothing and Respirators Against Biological Agents. Centers for Disease Control and Prevention and National Institute for Occupational Safety and Health, Cincinnati, OH, 2 pages. http://www.cdc.gov/niosh/unp-intrecppe.htm

Centers for Disease Control and Prevention (CDC) (1994). Guidelines for preventing the transmission of *Mycobacterium tuberculosis* in health-care facilities. Morbidity and Mortality Weekly Report, 43(RR13), 1-32. Retrieved January 1, 2006. http://www.cdc.gov/mmwr/preview/mmwrhtml/00035909.htm

Comité ministériel sur les mesures de précaution contre le syndrome respiratoire aigu sévère (SRAS) (2004a). Recommandations au directeur national de santé publique. Québec, mai, 2 pages.

Comité ministériel sur les mesures de précaution contre le syndrome respiratoire aigu sévère (SRAS) (2004b). Rapport final, Santé et services sociaux, Québec, 49 pages.

Comité sur les infections nosocomiales du Québec (CINQ) (2006). Mesures de prévention et contrôle de l'influenza pandémique pour les établissements de soins et les sites de soins non traditionnels. Direction des risques biologiques, environnementaux et occupationnels, Institut national de santé publique du Québec, cote INSPQ-2006-045, Québec, 63 pages.

Canadian Standards Association (CSA). (1993). Standard Z94.4-93, Selection, Use and Care of Respirators, Canadian Standards Association, 103 pages.

Duchaine, C., Cormier, Y., Mériaux, A., Pageau, P., Chabot, M., Israël-Assayag, E., Goyer, N., Cloutier, Y., Lazure, L. Santé respiratoire des travailleurs et qualité de l'air des tourbières du Québec possédant des sytèmes de dépoussiérage. Études et recherches, Rapport R-363, IRSST, 138 pages.

Goyer, N., Lavoie, J., Lazure, L., Marchand, G. (2001). Les bioaérosols en milieu de travail : guide d'évaluation, de contrôle et de prévention. Études et recherches, guide technique T-23, Institut de recherche Robert-Sauvé en santé et en sécurité du travail, 63 pages.

Goyer, N., Lavoie, J. (2001a). Emissions of chemical compounds and bioaerosols during the secondary treatment of paper mill effluents. American Industrial Hygiene Association Journal 62: 330-341.

Goyer, N., Lavoie, J. (2001b). Identification of sources of chemical and bioaerosols emissions in the work environment during secondary treatment of pulp mill effluents. TAPPI Journal 84 (2) :1-13.

Hinds, W.C. (1982). Aerosol Technology. Properties, Behavior, and Measurement of Airborne Particles. John Wiley and Sons, Toronto, 424 pages.

Hyvärinen, A., Vahteristo, M., Meklin, T., Jantunen, M., Nevalainen, A. (2001). Temporal and spatial variation of fungal concentrations in indoor air. Aerosol Science and Technology 35:688-695.

Institute of Inspection, Cleaning and Restoration Certification (ICRC). (2003). IICRC S520 Standard and Reference guide for professional mold remediation. Institute of Inspection, Cleaning and Restoration Certification, Vancouver, Washington, 176 pages.

Institut national de recherche et de sécurité (INRS). (2003). Appareils de protection respiratoire et métiers de la santé. Fiche pratique de sécurité ED 105, Institut national de recherche et de sécurité de France, 4 pages.

Institut national de recherche et de sécurité (INRS). (2004). Les agents biologiques. Fiche pratique de sécurité ED 117. Institut National de Recherche et de Sécurité, Paris, France, 4 pages.

Institut national de recherche et de sécurité (INRS). (2005). Risques infectieux en milieu de soins. Masques médicaux ou appareils de protection respiratoire jetables : quel matériel choisir ? Édition INRS ED 4136, Institut National de recherche et de Sécurité, Paris, France, 2 pages.

Lara, J. and M. Vennes. (2003). Guide pratique des appareils de protection respiratoire. 2e édition, Montréal, Commission de la santé et de la sécurité du travail du Québec and Institut de recherche Robert- Sauvé en santé et en sécurité du travail, 56 pages.

Lara, J. and M. Vennes. (2002). Guide des appareils de protection respiratoire utilisés au Québec. Montréal, Commission de la santé et de la sécurité du travail du Québec and Institut de recherche Robert- Sauvé en santé et en sécurité du travail, 86 pages.

Lavoie, J., Allard, R. (2004). Bioaérosols. Dans : Manuel d'hygiène du travail. Du diagnostic à la maîtrise des facteurs de risque. Chapitre 9, Modulo-Griffon éditeur, Québec, Canada, pp : 129-158.

Lavoie, J., Bourdouxhe, M., Guertin, S. (2004). Étude des agents biologiques et des contraintes ergonomiques lors de l'utilisation de camions avec bras assisté pour la collecte des ordures domestiques. PISTES 6(1) :1-26.

Lavoie, J., Dunkerley, C.J. (2002). Assessing waste collectors' exposure to bioaerosols. Aerobiologia 18 (3-4):277-285.

Lavoie, J., Guertin, S. (2001). Evaluation of health and safety risks in municipal solid waste recycling plants. Journal of the Air and Waste Management Association 51:352-360.

Lavoie, J. (2000). Évaluation de l'exposition aux bioaérosols dans les stations de traitement des eaux usées. Vecteur Environnement, 33(3) :43-50.

Lavoie, J., Alie, R. (1997). Determining the Characteristics to Be Considered from a Worker Health and Safety Standpoint in Household Waste Sorting and Composting Plants. Annals of Agricultural and Environmental Medicine, 4:123-128.

Lavoie, J., Pineau, S., Marchand, G. (1996). Aeromicrobial Analyses in a Wastewater Treatment Plant. In: Aerobiology, Muilenberg, M. and Burge, H. éditeurs, CRC Press/Lewis publisher, New York, New York, pp. 81-87.

Lavoie, J., Comtois, P. (1993). Microbial Decontamination of Ventilation Systems. Indoor Environment, 2:291-300.

Le petit Larousse illustré (2006). Larousse, Paris, France, 1856 pages.

Lee, S-A, Adhikari, A., Grinshpun, S.A., McKay, R., Shukla, R., Zeigler, H.L., Reponen, T. (2005). Respiratory Protection Provided by N95 Filtering Facepiece Respirators Against Airborne Dust and Microorganisms in Agricultural Farms. Journal of Occupational and Environmental Hygiene, 2(11):577-583.

Lee, K., Slavcev, A., Nicas, M. (2004). Respiratory protection against *Mycobacterium tuberculosis*: quantitative fit test outcomes for five type N95 filtering-facepiece respirators. Journal of Occupational and Environmental Hygiene, 1:22-28.

Lenhart, S.W., Schafer, M.P., Singal, M., Hajjeh, R.A. (2004a). Histoplasmosis. Protecting Worker at Risk. Revised edition, Centers for Disease Control and Prevention and National Institute for Occupational Safety and Health, Cincinnati, OH, Publication No. 2005-109, 32 pages.

Lenhart, S.W., Seitz, T., Trout, D., Bollinger, N. (2004b). Issues affecting respirator selection for workers exposed to infectious aerosols: emphasis on healthcare settings. Applied Biosafety, 9(1):2-36.

LSST (2003). Loi sur la santé et la sécurité du travail, L.R.Q., c.S-2.1, 76 pages. (Act respecting occupational health and safety (AOHS))

Malmros, P., Sigsgaard, T., Bach, B. (1992). Occupational health problems due to garbage sorting. Waste Management and research 10:227-234.

Malmros, P. (1990). Get wise on waste. A book about health and waste handling. Danish Working Environment Services. ISBN87-7534-400-9.

McCullough, N.V., Brosseau, L.M. (1999). Selecting Respirators for Control of Worker Exposure to Infectious Aerosols. Infection Control and Hospital Epidemiology 20(2):136-144.

National Institute for Occupational Safety and Health (NIOSH). (1999). TB Respiratory Protection Program in Health Care Facilities: Administrator's Guide. Cincinnati, U.S. Department of Health and Human Services, Public Health Service, Centers for Disease Control and Prevention, 116 p. http://www.cdc.gov/niosh/99-143.html.

Occupational Safety and Health Administration (OSHA). (2006). OSHA Guidance Update on Protecting Employees from Avian Flu (Avian Influenza) Viruses. U.S. Department of Labor, Occupational Safety and Health Administration, OSHA 3323-10N, 71 pages.

Occupational Safety and Health Administration (OSHA). (2002). Indoor air quality investigation. Washington, Occupational Safety and health Administration, U.S. Department of Labor, OSHA Technical Manual – Section III: chap. 2, 15 p.

Poulsen, O.M., Breum, N.O., Ebbehoj, N., Hansen, A.M., Ivens, U., van Lelevield, D., Malmros, P., Matthiasen, L., Nielsen, B.H., Nielsen, E.M., Schibye, B., Skov, T., Stenbaek, E.I., Wilkins, K.C. (1995a). Collection of domestic waste. Review of occupational health problems and their possible causes. The Science of the Total Environment, 168:1-19.

Poulsen, O.M., Breum, N.O., Ebbehoj, N., Hansen, A.M., Ivens, U., van Lelevield, D., Malmros, P., Matthiasen, L., Nielsen, B.H., Nielsen, E.M., Schibye, B., Skov, T., Stenbaek, E.I., Wilkins, K.C. (1995b). Sorting and recycling of domestic waste. Review of occupational health problems and their possible causes. The Science of the Total Environment, 168:33-56.

Public Health Agency of Canada. (2005). Material safety data sheet – infectious substances. www.phac-aspc.gc.ca/msds-ftss/ (agents infectieux)

Qian, Y., Willeke, K., Ulevicius, V., Grinshpun, S. (1997). Particle reentrainment from fibrous filters. Aerosol Science and Technology, 27(3): 394-404.

RSST (2007). Règlement sur la santé et la sécurité du travail, L.R.Q., c.S-2.1-r19.01. (Regulation respecting occupational health and safety (ROHS))

Ruzer, L.S., Harley, N.H. (2005). Aerosol Handbook. Measurement, Dosimetry and Health Effects. CRC Press, New York, 709 p..

Willeke, K., Baron, P.A. (1993). Aerosol measurement. Principles, techniques and applications. Van Nostrand Reinhold, New York, 876 pages.

Yassi, A., Bryce, E. (2004). Protecting the faces of health care workers: Knowledge gaps and research priorities for effective protection against occupationally-acquired respiratory infectious diseases. The Change Foundation, Ontario Hospital Association and Occupational Health and safety Agency for healthcare in BC, 103 pages.

APPENDIX 1 **Decision tree for selecting a respirator against bioaerosols**

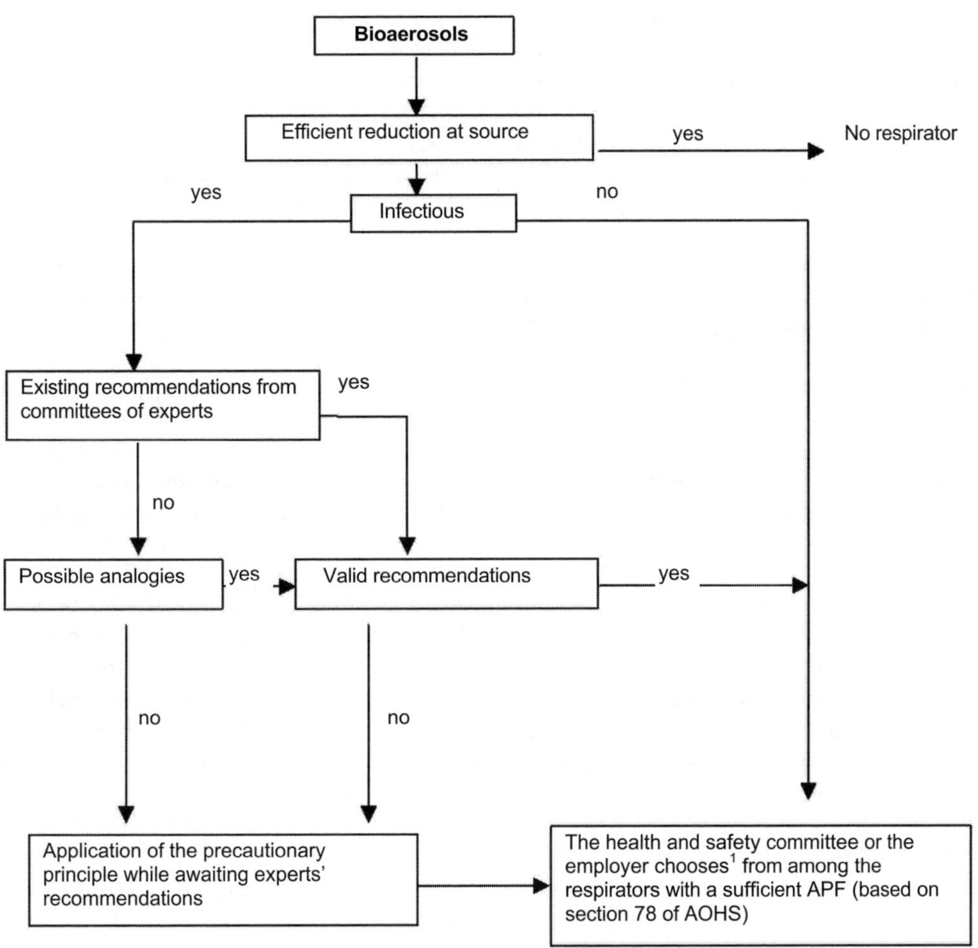

[1] The choice is based on the realities of the workplace.

APPENDIX 2 Standards and regulations

Section 1 of the Act respecting occupational health and safety of Québec (AOHS) defines a contaminant as solid, liquid or gaseous matter, a microorganism, a sound, a vibration, a radiation, heat or an odor, or any combination of these likely to alter in any way the health or safety of workers (AOHS, 2003). Employers and workers must know their obligations and their rights related to AOHS, as well as its pursuant Regulation respecting occupational health and safety (ROHS) (AOHS, 2003; ROHS, 2007). The rights and obligations concerning respiratory protection are particularly important in the presence of contaminants that are possible health hazards that cannot otherwise be controlled.

Although personal protective equipment (PPE) is vital for some workplaces or for performing certain tasks, it should however be remembered that the purpose of the AOHS is to eliminate, at source, hazards to the health, safety and physical integrity of workers (AOHS, 2003).

For respiratory protection against chemical contaminants, the ROHS prescribes more specific measures in sections 45 and 47. It states in particular that the employer must supply at no charge to the worker the respiratory protective equipment provided for in the *Guide des appareils de protection respiratoire utilisés au Québec* (Lara and Vennes, 2002) published by the IRSST (Québec occupational health and safety research institute) as it reads at the time that it is applied; the employer must ensure that the worker wears the equipment.

The equipment must be chosen, fit, used and cared for in compliance with CSA Standard Z94.4-93 Selection, Use and Care of Respirators. A respiratory protection program must be developed and applied in accordance with the standard. Furthermore, this program's specific requirements must also be followed for bioaerosols. In fact, other important organizations such as the Institut National de Recherche et de Sécurité (INRS) of France, the National Institute for Occupational Safety and Health (NIOSH), and the Centers for Disease Control and Prevention (CDC) recommend that a respiratory protection program be established when respirators are used, regardless of the contaminant involved (INRS, 2005; NIOSH, 1999).

In addition, according to section 78 of AOHS (2003), it is the responsibility of the establishment's health and safety committee to choose the respirators best adapted to the workers' needs.

Airborne Infectious Disease Management

Methods for Temporary Negative Pressure Isolation

Minnesota Department of Health

Office of Emergency Preparedness

Healthcare Systems Preparedness Program

Airborne Infectious Disease Management

Methods for Temporary Negative Pressure Isolation

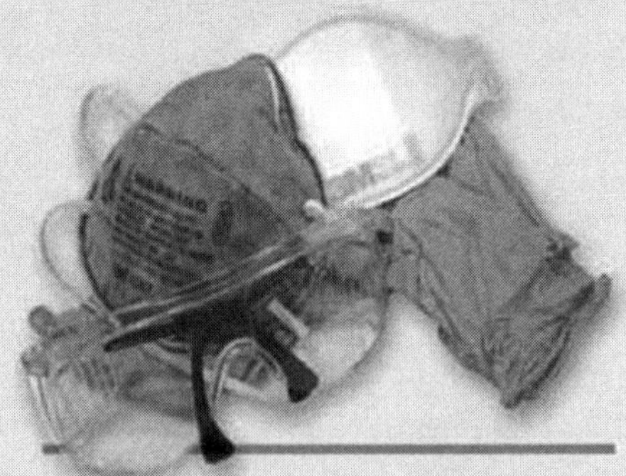

This guide has been produced by:

Office of Emergency Preparedness
Healthcare Systems Preparedness Program

Airborne Infectious Disease Management

Methods for Temporary Negative Pressure Isolation

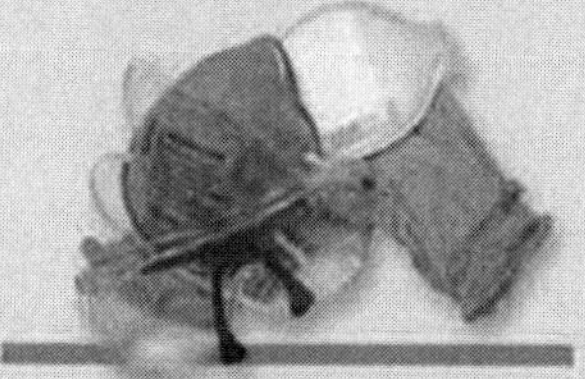

For further Information, please contact:

Office of Emergency Preparedness
Minnesota Department of Health
625 Robert Street North
P.O. Box 64975
St. Paul, MN 55164-0975

Phone: (651) 201-5701
Fax: (651) 201-5720

This user guide is available on the
Minnesota Department of Health Web site:
ttp://www.health.state.mn.us/oep/training/bhpp/isolation.html

Acknowledgements

This user guide has been written by the **Minnesota Department of Health** in conjunction with the **University of Minnesota** to assist hospital personnel in the management of airborne infection isolation.

AUTHORS:

Jeanne Anderson
Infection Control Practitioner
Office of Emergency Preparedness
Minnesota Department of Health

Andrew Geeslin
Engineering/Infection Control Intern
University of Minnesota

Andrew Streifel
Hospital Environmental Health Specialist
Environmental Health and Safety
University of Minnesota

Minnesota Department of Health does not endorse particular brands among competing products. Examples shown in these materials are for illustration only.

The authors gratefully acknowledge the **Minnesota Emergency Readiness Education and Training (MERET)** at the **University of Minnesota Centers for Public Health Education and Outreach** and the following individuals for their participation, assistance, and support of this project:

Keith Carlson
Director of Facilities Management
Mercy Hospital and Health Care Center
Moose Lake, MN

Gary Davis
Plant Engineer
LakeWood Health Center
Baudette, MN

Pete Swanson
Facility Services Manager
Pipestone County Medical Center
Pipestone, MN

The following individuals are gratefully acknowledged for their invaluable suggestions:

Judene Bartley
Vice President
Epidemiology Consulting Services Inc.
Beverly Hills, MI

Rick Hermans
Senior Project Manager
Center for Energy and Environment
Minneapolis, MN

Curtain TNPI photographs courtesy of Ken Meade, Research Mechanical Engineer, NIOSH/CDC, USPHS

MERET photos provided by Paul Bernhardt

References 37 | Appendix 20 | Surge capacity 16 | Portable anteroom 13 | TNPI Temporary Negative Pressure Isolation 5 | Environmental controls 3 | Principles of airborne infectious disease management 2 | **Introduction** 1

Introduction

Hospital preparedness for bioterrorism and other public health emergencies such as emerging airborne infectious diseases requires strategic planning to ensure that all components of respiratory protection programs, including environmental controls, are in place for **airborne infection isolation rooms (AIIRs).** Hospitals have insufficient facilities to provide airborne infection isolation for large numbers of patients with airborne infectious diseases presenting in a short time period.[1, 2] However, AIIRs have been increased recently, due to requirements of National Bioterrorism Hospital Preparedness Program.[3]

Without adequate environmental controls, patients with airborne infectious diseases will pose a risk to other patients and health care workers.[4] **Heating, ventilation, and air conditioning (HVAC)** expertise is essential for proper environmental management when planning control of airborne infectious disease outbreaks (natural or intentional). Design manuals and guidelines provide direction for infectious disease management.[5-11] *Refer to Appendix A, **"2006 AIA Criteria"** on page 21.*

This guide will assist health care facility plant maintenance and engineering staff, in coordination with infection control professionals, to prepare for a natural or terroristic event, involving an infectious agent transmitted by airborne droplet nuclei. Examples of such agents include measles, varicella, and tuberculosis.[5]

Audience for this Guide

The intended audience for this guideline includes health care:

- facility engineering and maintenance
- infection control
- environmental health and safety
- management personnel

Purpose of this Guide

- Provide **guidance** on environmental controls for airborne infectious disease management
- Provide a general guide for temporary **setup, installation, and operation** of portable HEPA machines when used to create negative pressure in a hospital room/area
- Provide **instruction on the use** of:
  Pressure gauges
  Particle counters
- Outline of **preventative maintenance** schedule for HVAC equipment related to AIIR

Goal of this Guide

A timely response is crucial for identification and containment of potentially infectious patients. The goal is for facilities to **develop a 12-hour response** to implement containment measures. Temporary negative pressure isolation methods are a safe alternative for hospitals that lack engineered AIIRs. These can be utilized in facilities to meet increased surge capacity for patient isolation. TNPI should also be used during hospital construction projects to reduce risks associated with airborne infectious diseases.

These temporary measures should be incorporated into the facility's infection control and emergency response plans. ▲

Principles of airborne infectious disease management

Airborne infection isolation is based on the following hierarchy of control measures.

- Administrative (work practice) controls
- Environmental controls
- Personal protective equipment (PPE)

These measures are intended to reduce the risk for exposure to airborne infectious disease agents by uninfected persons. AIIRs and hospital systems in general must be monitored to provide continual protective measures. *Refer to Appendixes B and C, **AIIR and HVAC System Maintenance Schedules,** on pages 22 and 23.*

Administrative (work practice) controls

- Managerial measures that reduce the risk for exposure to persons who might have an airborne infectious disease.
- Work practice controls include using infection control precautions while performing aerosol-generating procedures, closing doors to AIIRs, hand hygiene, and signage.

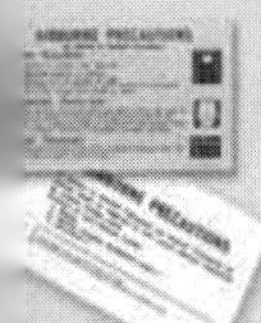

EXAMPLES

written policies and protocols to ensure the rapid identification, isolation, diagnostic evaluation, and treatment of persons likely to have an airborne infectious disease

Environmental controls

- Physical or mechanical measures (as opposed to administrative control measures) used to reduce the risk for transmission of airborne infectious diseases.

EXAMPLES

ventilation
filtration
ultraviolet germicidal irradiation
AIIRs
local exhaust ventilation devices

Personal protective equipment (PPE)

- Equipment worn by health care workers and others to reduce exposure to communicable diseases.

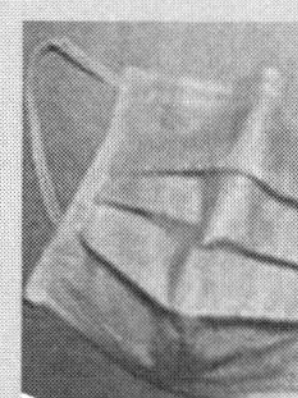

EXAMPLES

gowns
gloves
masks
respirators
eye protection

Environmental controls

This user guide will focus on the **environmental controls** necessary for airborne infection isolation.

The ventilation parameters essential for airborne infection isolation rooms/areas include:

- **Pressure management** for appropriate airflow direction;
- Room air changes for **dilution ventilation**; and
- **Filtration** to remove infectious particles.

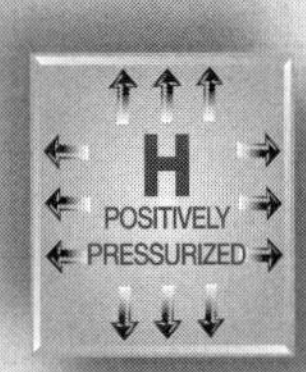

FIGURE 1
Illustrations used to identify ***Negatively*** *(top) and* ***Positively*** *(bottom) pressurized air space.*

Pressure management

For the purposes of this guide, **pressure** refers to the ***differential pressure between two spaces*** *(FIGURE 1)*.

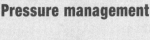

In health care settings, the two spaces are typically the isolation room and the corridor. For AIIR, the room **should be negatively pressurized** in relation to the corridor. This helps to prevent infectious particles from escaping the room envelope.

If an anteroom is present between the AIIR and the corridor, the AIIR may be negatively or positively pressurized to the anteroom. However, if the AIIR is positively pressurized to the anteroom, the anteroom must be negatively pressurized to the corridor.

A difference in pressure causes movement of air from areas at higher pressure to those at lower pressure. **The greater the pressure difference, the greater the resulting air velocity**. The movement of air is used to help provide containment of infectious particles by providing clean to dirty airflow. *Refer to Appendix D,* ***"Using a Pressure Gauge to Measure Relative Pressurization Between Two Spaces"*** *on page 24 for instructions on using a pressure gauge to determine differential pressure.*

The differential pressure or **pressure offset is established by mechanically adjusting the supply and exhaust air**. ***For a negative pressure room, the sum of the mechanically exhausted air must exceed the sum of the mechanically supplied air.*** This offset forces air to enter the room under the door and through other leakages and prevents infectious particles from escaping.[9]

In order to maintain consistent offset airflow, **the difference between exhaust and supply should create a pressure differential of about** ***0.01 inch water gauge (in. w.g.) or 2.5 Pascals (Pa)***.[9] Pressure in this application is used to induce airflow from adjacent spaces into the isolation room. ▲

References 37
Appendix 20
Surge capacity 16
Portable anteroom 13
TNPI Temporary Negative Pressure Isolation 5
Environmental controls **Dilation ventilation** **Filtration**
Principles of airborne infectious disease management 2
Introduction 1

Dilution ventilation

Mechanical ventilation is used to exchange the air in a space. The time required for removing a given percentage of airborne particles from a room or space depends on the number of ***air changes per hour (ACH)***, location of the ventilation inlet and outlet, and the physical configuration of the room or space (FIGURE 2).

*Refer to Appendix E, **"Using a HEPA Filter for Dilution Ventilation"** on page 26.* ▲

Filtration

For the purposes of this guide, **filtration** refers to the ***process of passing air through a filter***. Hospital buildings have some of the highest filtration requirements. Without filtration, particle concentrations accumulate in indoor environments. This can cause toxic effects even in healthy people.

Filtration reduces the risk for transmitting airborne infectious agents. Depending upon their size, particles may be deposited in the upper respiratory tract or the lower respiratory tract of humans. Particles can also be deposited in open wounds during dressing changes or invasive procedures. *See Appendix F, **"Microorganisms Associated with Airborne Transmission"** on page 28.*

When used correctly, portable HEPA filters prove to be an effective method for achieving an airborne isolation environment.[12] When properly installed and maintained, filters for clinical spaces should be able to ***remove at least 90% of particles (0.5 microns in size and larger)***[9] from outside and inside air.

*For evaluation of hospital HVAC systems and HEPA filters refer to Appendix G, **"Using a Particle Counter to Assess Indoor Air Quality and Filter Efficiency"** on page 29.*

*For information on filter selection and performance, see Appendix H, **"Data Interpretation"** on page 34. See Appendix I, **"Sample Log for Measuring Particle Counts"** on page 36.* ▲

FIGURE 2: *ACH AND TIME REQUIRED FOR REMOVAL EFFICIENCIES*

Time (minutes) required for removal of 90%, 99%, and 99.9% of airborne-contaminants.

ACH	90% EFFICIENCY	99% EFFICIENCY	99.9% EFFICIENCY
2	69	138	207
4	35	69	104
6	23	46	69
8	17	35	52
10	14	28	41
12	12	23	35
15	9	18	28
20	7	14	21
50	3	6	8

Modified from Table B.1, CDC Guidelines for Environmental Infection Control in Health-Care Facilities, 2003.[5]

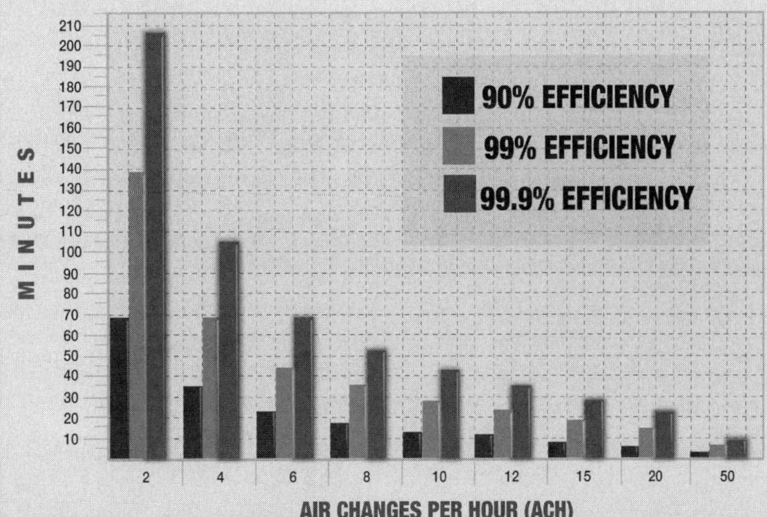

Perfect mixing of air is assumed. For rooms with stagnant air spaces, the time required may be much longer than shown. This is intended only as an approximation and is for ideal ventilation configurations.

References 37
Appendix 20
Surge capacity 16
Portable anteroom 13
TNPI Temporary Negative Pressure Isolation
Environmental controls 3
Principles of airborne infectious disease management 2
Introduction 1

Temporary Negative Pressure Isolation TNPI

TNPI Installations Must Observe Building & Fire Codes

All of the TNPI installations must observe state building and fire codes, and **National Fire Protection Association (NFPA) 101 (Life Safety Code)**. Being able to set up a TNPI requires planning and early contact with the authority that has jurisdiction *(your State Health Department)* to establish and verify compliance with applicable codes. In Minnesota, you can contact Fernando Nacionales, P.E., Engineering Services Section at (651) 201-3712 or Fernando.Nacionales@state.mn.us.

TNPI is considered when airborne infection isolation is needed and there are no available or insufficient AIIRs, such as can happen when there is an outbreak of an airborne infectious disease with large numbers of communicable patients. Temporary isolation is designed to protect patients and staff from contracting or transmitting highly infectious diseases.

Portable HEPA filters have been used in the past to isolate patients. These filters can also be used as an air scrubber. When used in this fashion, the filter simply cleans the air. **It does not provide pressure management for appropriate airflow direction.** The filter is placed in the room and turned on, without attached ductwork. **This is known as room recirculation and is not a preferred method for isolation.**

There are three types of temporary isolation. The ***two most effective methods*** of achieving temporary isolation are:

Discharging air to the outside

Discharging air to return air system

One less effective methods of achieving temporary isolation is:

Curtain TNPI

HEPA Filter Maintenance

If the HEPA filter is in place for an extended period of time, the pre-filter should be changed when lint buildup becomes visible (FIGURE 3). It is important to completely follow the manufacturer's directions for operation and maintenance of portable HEPA filter machines.

FIGURE 3
***Pre-filter** at inlet side of HEPA machine (is shown).*

1 Discharging air to the outside

Creates negative pressure room (airflow into patient room from the corridor or anteroom)

Steps for discharging air to the outside

1. Select a room
2. Set up pre-constructed window adapter
3. Set up HEPA machine and flex duct
4. Seal return air grille
5. Turn on HEPA machine and adjust flow

This option is one of the two preferred methods for achieving TNPI.

In this method, a HEPA filter is used to exhaust room air outside through the window. (Clearly, a window is required for this method.) *(FIGURE 4)*

- The two main purposes of the HEPA machine in this application are to **clean contaminated air** and **induce negative pressure** in the room.
- Because the discharged air is HEPA filtered, no extra consideration for air discharge location is required.

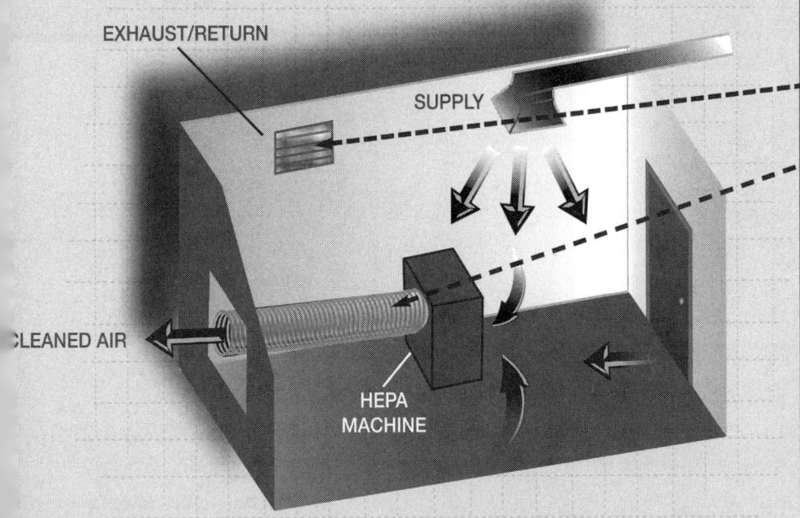

FIGURE 4

HEPA fan *exhausting clean air outside through the window.*

To prevent pulling air from return air system, the exhaust/return grilles should be sealed with tape.

Set up HEPA machine and flex duct.

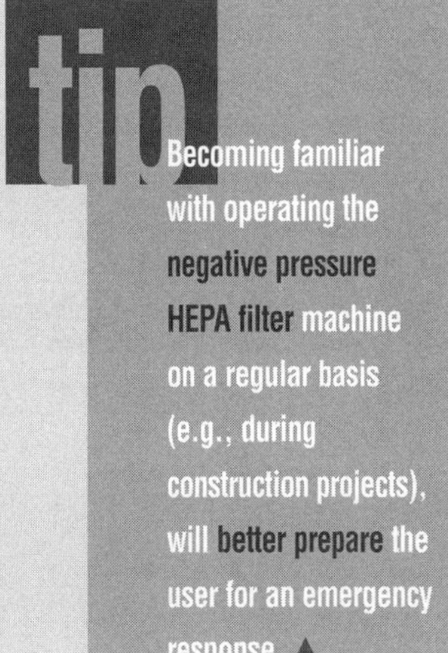

Becoming familiar with operating the negative pressure HEPA filter machine on a regular basis (e.g., during construction projects), will better prepare the user for an emergency response. ▲

Select a room

You should choose a room to set up TNPI. The HEPA machine should be **set up in advance to placing the patient** with a suspected airborne infectious disease in the room.

If possible, **select a room without transfer grilles.** If no such room exists, **completely seal the grilles** to promote negative pressure.

Set up pre-constructed window adapter

When using this option to establish TNPI, **a window adapter must be constructed.** This is used to provide a connection to the flex duct. *(FIGURE 5)*

The template should be constructed out of wood to fit into a standard window in your hospital's patient care rooms and should **provide an airtight fit.**

A piece of **sheet metal** *(a flange)* of the **same diameter as the flex duct should be fixed to the circular hole** in the wood to serve as an adapter between the wood template and the flex duct. The use of a window template, flex duct, and fan is common in hospital construction.

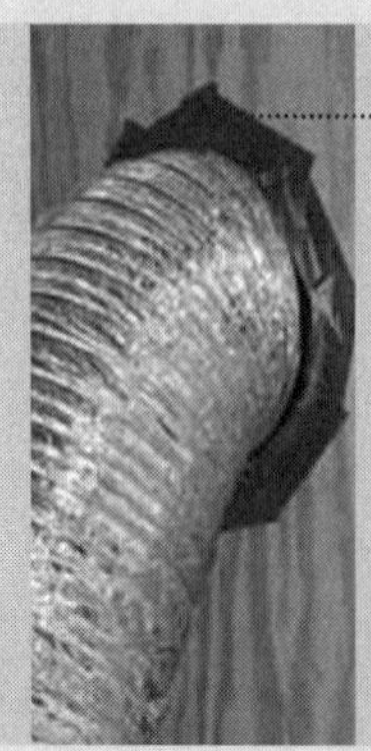

FIGURE 5
Window adapter *with flex duct attached.*

Set up HEPA machine and flex duct

Connect the flex duct to window adapter and HEPA machine. *(FIGURE 6)*

Seal return air grille

To prevent pulling air from return air system, the **return grilles should be sealed** with tape. A sheet of plastic or cardboard can be used provided that the edges are completely sealed with tape.

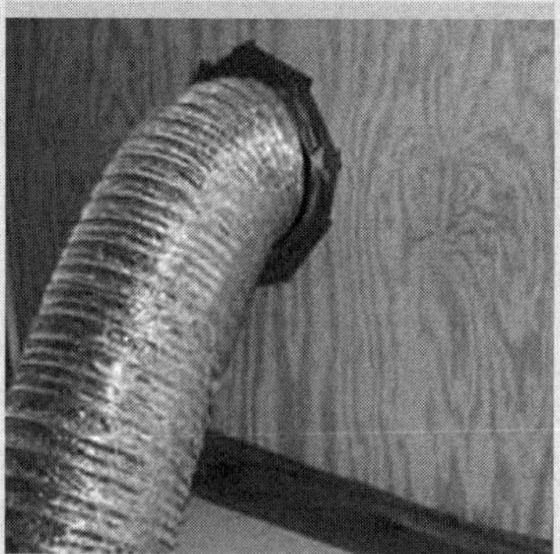

FIGURE 6
Flex duct *connected to window template. The other end of the flex duct should be connected to the HEPA machine.*

FIGURE 7
The flow rate *on this machine is adjusted with the dial on the right-hand side.*

Turn on HEPA machine and adjust flow

Once the flex duct is connected to the window template and the HEPA machine, the **machine should be turned on.** *(FIGURE 7)*

Adjust the flow on the output until the desired ***pressure differential of negative 2.5 Pa*** (pressure in corridor is greater than pressure in patient room) is reached. Increasing the flow will increase the pressure differential, and decreasing the flow will decrease the pressure differential.

The pressure differential should be measured with a hand-held digital pressure gauge. The **pressure should be monitored daily.**

Additional Notes for Discharging Air to the Outside

- In locations with seasonal cold (below-freezing) weather, it may be necessary to **install a damper or louver** on the outside of the window template. This will help to prevent airflow restriction due to condensation or ice/snow buildup.
- When no longer needed, **the HEPA filter and flex duct should be wiped down** with a hospital-approved disinfectant.
- HEPA filters can also be used to enhance dilution ventilation. *Refer to Appendix E,* ***"Using a HEPA Filter for Dilution Ventilation"*** *on page 26 for a discussion and example of dilution ventilation.* ▲

2 Discharging air to return air system

L NEGATIVELY PRESSURIZED

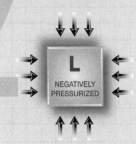

Creates negative pressure room (airflow into patient room from the corridor or anteroom)

Steps for discharging air to return air system

1. Select a room
2. Attach flex duct adapter to desired return grille
3. Set up HEPA filter machine and flex duct
4. Seal remaining return air grilles
5. Turn on HEPA filter machine and adjust flow

This option is one of the two preferred methods for achieving TNPI.

In this method, a HEPA filter machine is used to discharge room air into the return air system. *(FIGURE 8)*

- You must be careful when exhausting additional large volumes of air through the return air system.
- Because the air is HEPA filtered, it is okay to exhaust it through the return air system. If air was being exhausted by a **regular (non-HEPA filtered) large fan, it should not be discharged** through the return air system.

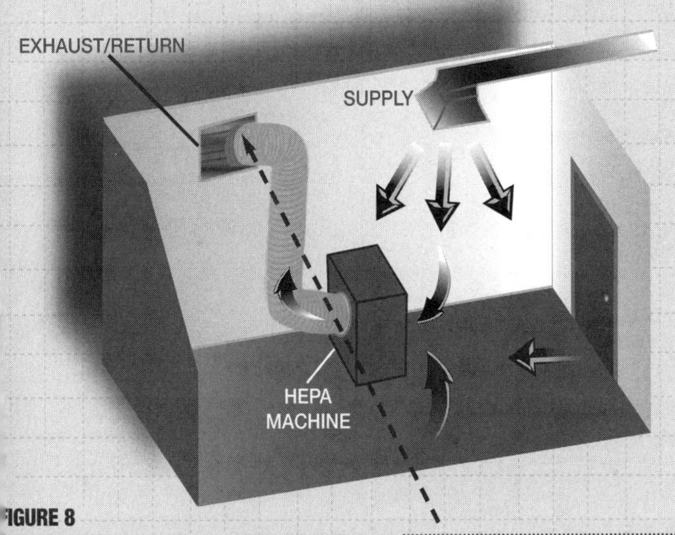

FIGURE 8
Discharging air *to return air system*

Attach flex duct adapter to desired return grille.

tip

Care must be taken when exhausting a large volume of air into the return air system, as it can over-pressurize the duct. Such changes can disturb the air-balancing of rooms on the same return/exhaust air system. ▲

Select a room

You should choose a room to set up TNPI. The HEPA machine should be **set up in advance to placing the patient** with a suspected airborne infectious disease in the room.

If possible, **select a room without transfer grilles.** If no such room exists, **completely seal the grilles** to promote negative pressure.

Attach flex duct adapter to desired return grille

Next, attach flex duct adapter to desired return grille. Ideally, if the return is low, connect the flex duct here. However, most returns are high on the wall or ceiling and these can be connected in the same way. It is important to provide a **tight seal between the flex duct and the adapter as well as between the adapter and the return grille.** *(FIGURE 9)*

Generally, it is easy to differentiate between a supply grille and return grille. If you are not sure, you can use a smoke stick to check airflow direction.

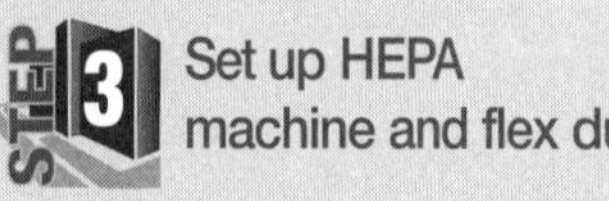

Set up HEPA machine and flex duct

Connect the flex duct to return grille adapter and HEPA machine. *(FIGURE 10)*

Seal remaining air grilles

To prevent pulling air from return air system, additional exhaust/return **grilles should be sealed with tape.** *(FIGURE 11)*

Most rooms have only one supply, a return grille, and one bathroom exhaust. ***100% of bathroom discharge is exhausted outside,*** so no extra consideration is required. If there is more than one return grille, which is uncommon, the additional grille(s) should be sealed. A sheet of plastic or cardboard can be used provided that the edges are completely sealed with tape.

FIGURE 9
A return grille ***adapter.***

FIGURE 10
A return grille adapter *with flex duct attached.*

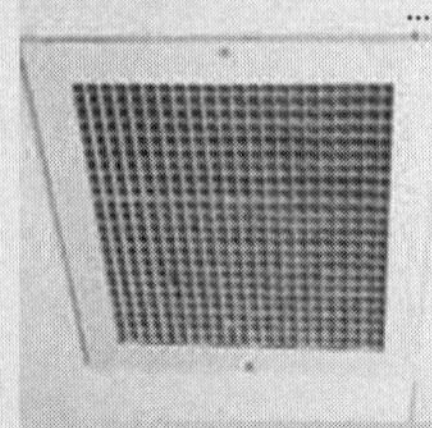

FIGURE 11
This return duct *must be sealed to prevent pulling air from return air system.*

Turn HEPA machine on and adjust flow

Once the flex duct is connected to the window template and the HEPA machine, the **machine should be turned on.**

Adjust the flow on the output until the desired ***pressure differential of negative 2.5 Pa*** (pressure in corridor is greater than pressure in patient room) is reached. Increasing the flow will increase the pressure differential, and decreasing the flow will decrease the pressure differential.

The pressure differential should be measured with a hand-held digital pressure gauge. The **pressure should be monitored daily.**

Additional Notes for Discharging Air to Return Air System

- You must be careful when exhausting additional large volumes of air through the return air system.
- When no longer needed, **the HEPA filter and flex duct should be wiped down** with a hospital-approved disinfectant.
- HEPA filters can also be used to enhance dilution ventilation. *Refer to Appendix E,* ***"Using a HEPA Filter for Dilution Ventilation"*** *on page 26 for a discussion and example of dilution ventilation.* ▲

3 Curtain TNPI

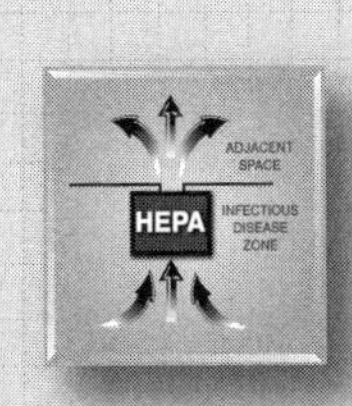

Curtain TNPI ***does not*** *provide negative pressure for the entire room, but induces airflow into the enclosed space from surrounding areas and discharges HEPA filtered air from the enclosure to the surrounding space.*

The TNPI methods ***1. Discharging Air to the Outside,*** and ***2. Discharging Air to Return Air System*** are preferred over ***3. Curtain TNPI.***

In this method, a HEPA filter is used to provide a type of relative negative pressure around a patient. *(FIGURE 12)*

- This method should be used for **non-ambulatory** patients.
- The area around the bed can be "sealed off" by **attaching fire-rated plastic sheeting** to the curtain track along with duct tape to attach the plastic to the floor.

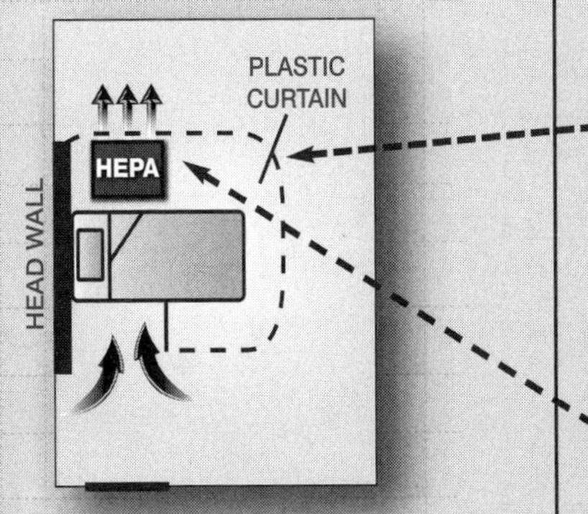

FIGURE 12
Curtain TNPI

The area *around the bed can be "sealed off" by attaching fire-rated plastic sheeting.*

The intake *must be within the plastic enclosure and the output must be outside the plastic enclosure.*

TNPI Installations Must Observe Building & Fire Codes

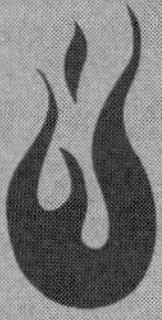

As stated earlier, Temporary Negative Pressure Isolation (TNPI) is considered when airborne infection isolation is needed and there is no available or insufficient AIIRs. This could happen when the number of infectious patients exceeds the number of available rooms.

All of the TNPI installations must observe state building and fire codes, and **National Fire Protection Association (NFPA) 101 (Life Safety Code)**. Being able to set up a TNPI requires planning and early contact with the authority that has jurisdiction *(your State Health Department)* to establish and verify compliance with applicable codes. In Minnesota, you can contact:
Fernando Nacionales, P.E.
Engineering Services Section
at (651) 201-3712 or
Fernando.Nacionales@state.mn.us.

Steps for setting up Curtain TNPI

1. Select a room
2. Attach fire-rated plastic to ceiling
3. Set up portable HEPA filter machine
4. Attach remaining surfaces with strong tape
5. Turn on portable HEPA filter

Select a room

If the patient is already in a hospital room, the room can be adapted for the Curtain TNPI method.

Due to the proximity of the machine to the patient, noise considerations and patient comfort must be addressed. *(FIGURE 13)*

As with the health care workers caring for the patient, the health care workers setting up Curtain TNPI **must also wear PPE.**

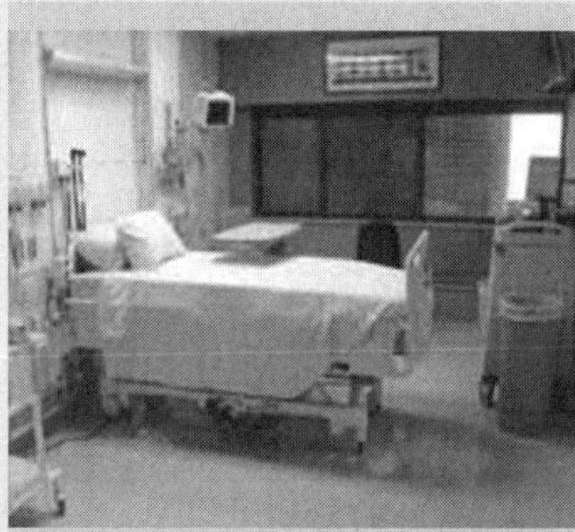

FIGURE 13
*Standard **patient room**.*

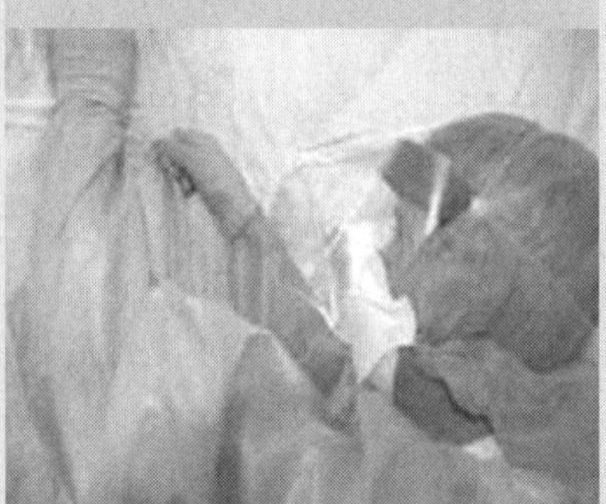

FIGURE 14
***Health care workers** setting up Curtain TNPI must wear PPE.*

Attach fire-rated plastic to ceiling

It is important to note that **the plastic must be able to fit nearly all the way around the bed.** *(FIGURE 14)* Also, ***the height should be at least six inches taller than the ceiling***, which allows for the plastic to be taped to the floor. For accessibility to the patient, the plastic should be ***at least three feet from the bed on all sides.***

Most patient care rooms have curtain tracks installed around the bed on the ceiling. These can be adapted for use with the plastic sheeting.

In addition, clips can be purchased that will allow the plastic sheeting to be attached to the ceiling grid (assuming the room has a suspended ceiling). *(FIGURE 15)*

FIGURE 15
***Plastic** can be attached to ceiling tile grid or curtain track with clips.*

Set up portable HEPA filter machine

The HEPA filter should be inserted into the plastic sheeting. *(FIGURE 16)* The **plastic needs to be cut so that the HEPA filter can be fit into the space.** The intake must be within the plastic enclosure and the output must be outside the plastic enclosure. This will draw the air from the enclosure, filter it, and exhaust it to the rest of the room. This will provide an airflow that will help to isolate the patient.

FIGURE 16
Portable HEPA filter *machine installed in plastic sheeting. The plastic should be taped around the HEPA machine to provide a seal.*

Attach remaining surfaces with strong tape

Once the HEPA filter is installed and sealed with tape, the remaining surfaces need to be sealed with tape. *(FIGURE 17)*

The hanging plastic curtain should be taped to the floor. This will help to minimize the "leakage area" of the enclosure and provide stronger airflow and containment.

The curtain near the HEPA filter must be taped to the head wall. This will also help to minimize the leakage area of the enclosure.

FIGURE 17
Plastic curtain *taped to wall.*

Turn on portable HEPA filter

The last step is to turn on the HEPA filter. If set up correctly, **the sides of the plastic will pull in** (as if under a vacuum). This will indicate that the enclosure is under negative pressure.

Additional Notes for Curtain TNPI

- This method **is not the preferred method** for isolating a patient. Health care workers will not be able to move the patient to another area in the building without completely disassembling the Curtain TNPI setup.

  In addition, the set up process can be time consuming and the majority of the set up takes place while the patient is within the enclosure which could increase transmission risk.

- When no longer needed, **the plastic curtain and HEPA filter should be wiped down** with a hospital-approved disinfectant.

- HEPA filters can also be used to enhance dilution ventilation. *Refer to Appendix E, **"Using a HEPA Filter for Dilution Ventilation"** on page 26 for a discussion and example of dilution ventilation.* ▲

Portable anteroom

Portable anterooms can be purchased with portable filters to create a space at the entrance to a standard room for TNPI. *(FIGURE 18)*

The advantage of a portable anteroom is the **relatively easy conversion of regular rooms** to AIIRs **without manipulation of the existing room ventilation.** Portable anterooms are easy to set up and convenient to use because any chosen room can be converted quickly.

A disadvantage of a portable anteroom is the inability to move beds through the door. Another disadvantage is that they do not depressurize the room unless the room door is open.

Steps for using a Portable Anteroom

1. Set up anteroom to manufacturer's recommendations
2. Attach the HEPA machine
3. Open the doors while HEPA machine is on
4. Close the doors after the patient and/or caregiver enters or leaves the room
5. Both doors should never be open at the same time
6. Clean and disinfect the portable anteroom cover

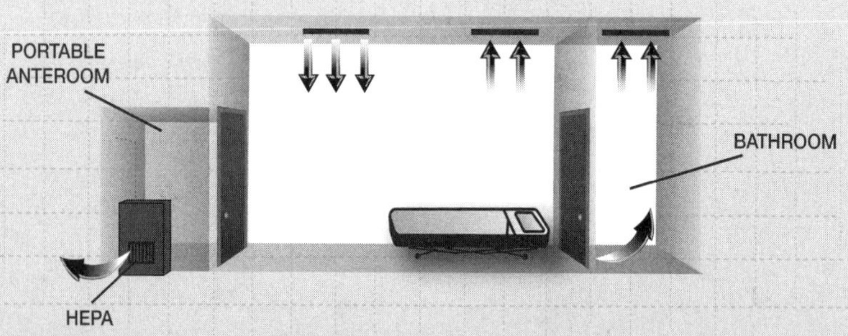

FIGURE 18
Portable Anteroom

Attach the portable HEPA filter to the appropriate sized portal on the side of the portable anteroom.

When turned on, the HEPA machine should cause the sides of the portable anteroom to pull inward.

The anteroom should be placed in front of the patient room door and taped to the wall to secure and cover the seams.

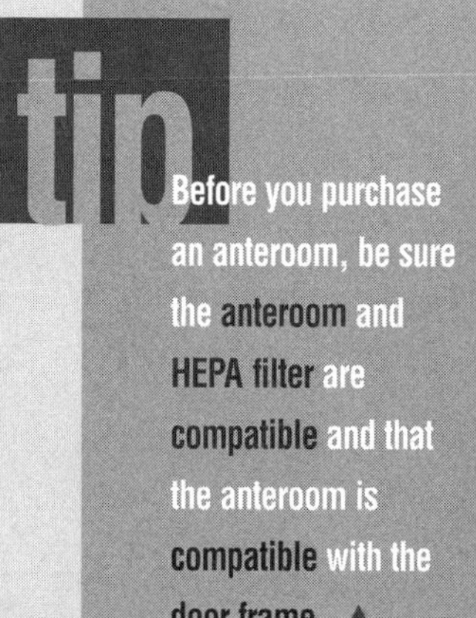

Before you purchase an anteroom, be sure the **anteroom** and **HEPA filter** are **compatible** and that the anteroom is **compatible** with the **door frame.** ▲

Set up anteroom

The anteroom should be set up according to manufacturer's recommendations. **The anteroom should be placed in front of the patient room door** and taped to the wall to secure and cover the seams.

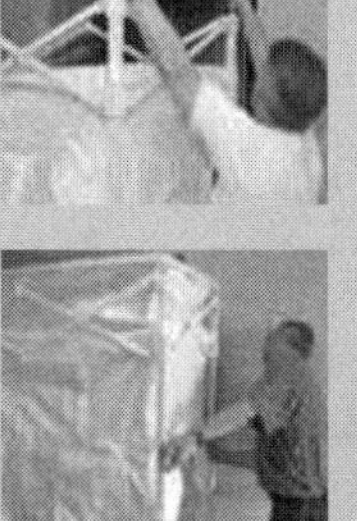

Attach the HEPA machine

Attach the portable HEPA filter to the appropriate sized portal on the side of the portable anteroom. When turned on, **the HEPA machine should cause the sides of the portable anteroom to pull inward,** indicating suction.

Open the doors

The door to the patient room and portable anteroom are **opened for patient transport** while the HEPA machine is on.

Close the doors

After the patient and/or caregiver enters or leaves the room, the room door and **portable anteroom door should be closed.**

Both doors not open at same time

If individual health care workers or visitors enter the room, one door should be closed before the other is opened. **Both doors should never be open at the same time.** This prevents short circuiting of air through an open doorway.

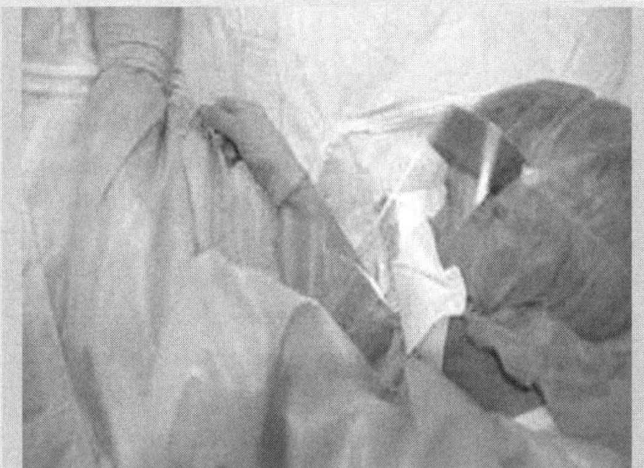

Clean portable anteroom cover

When no longer needed, **the portable anteroom cover should be cleaned and disinfected** with a hospital-approved disinfectant.

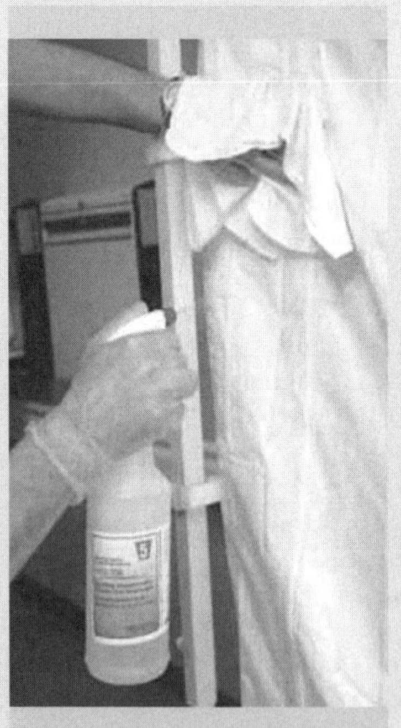

Additional Notes for Portable Anterooms

- The **door on the portable anteroom and patient room should be closed** except for entry and egress. In addition, **only one door should be opened** at one time.
- There are no standards on the magnitude of anteroom pressure. Therefore, there is no suggestion on airflow output or pressure differential for portable anterooms. It is only suggested that the **airflow is strong enough to pull in the sides** of the anteroom while **not so strong that it produces excess noise.** ▲

tip

Some permanently engineered AIIRs are designed with anterooms. ▲

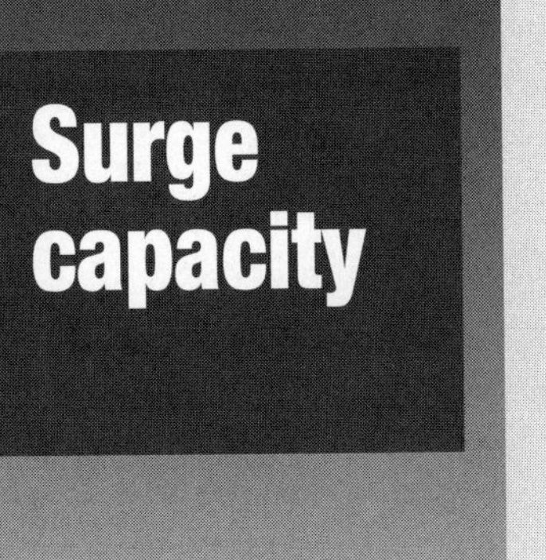

Surge capacity

Isolation surge capacity is the ability to manage high volumes of specialized patients.

Under extraordinary circumstances where the quantity of engineered airborne infection isolation rooms is insufficient to meet surge demand for patient isolation, hospitals can take various measures to protect patients and staff.[13] This section describes different methods of isolating large numbers of patients, including:

- **Smoke Zones**
- **Engineered System**
- **Temporary Surge Area**

The Goal of Isolation Surge Capacity

Housing large numbers of patients with airborne infectious diseases is challenging. The goal of surge capacity is to provide areas with safeguards to protect you and other patients from exposure to airborne infectious agents. *(FIGURE 19)*

An **infectious disease zone (IDZ)** is a *space used to isolate large numbers of patients*.

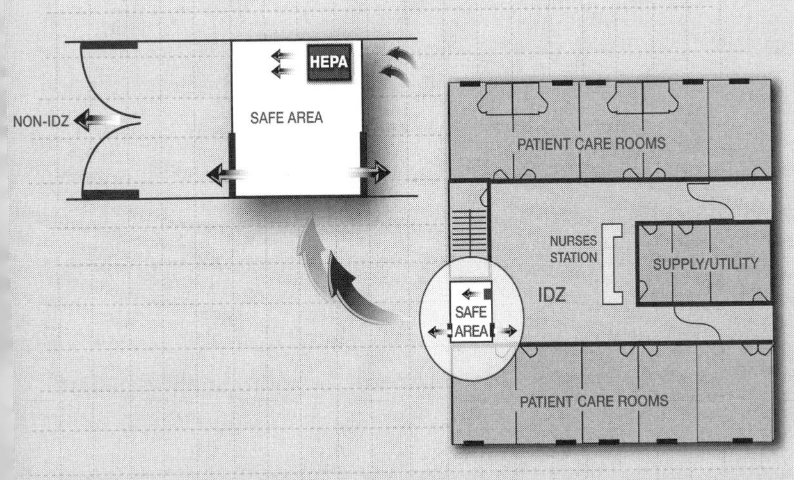

FIGURE 19

An infectious disease zone (IDZ) *is a space used to isolate large numbers of patients. A temporary constructed anteroom (safe area) for surge area is shown.*

Safe Areas

A safe area or anteroom *(FIGURE 19)* should be created between the IDZ and the rest of the hospital. **A portable HEPA filter is used to pressurize the anteroom.** This also provides clean air in the anteroom. Different models of HEPA filters have different configurations of the air intake. Regardless of the model, **the air intake should be drawing air into the HEPA filter from the IDZ.** The filtered exhaust should be discharged into the anteroom. An airtight seal should be made for all connections.

The anteroom should be large enough to accommodate function (for example, change clothes, hang PPE, dispose of waste, etc.). ▲

Smoke Zones

Hospitals have smoke zones that are designed to evacuate the smoke if a fire occurs. The zone has smoke-stopped barriers to prevent smoke movement between zones. **These airtight barriers would allow for potential isolation** of several hospital beds.

The patient rooms themselves may not be capable of being depressurized by the existing ventilation system. The smoke zone, however, is designed with fire rated doors that close as well as entire zone exhaust capability. Zone exhaust capability effectively isolates the wing, suite or space from other areas of the health care facility. These zones are routinely tested for fire management.

Smoke Zones as Perimeter of IDZ

The smoke zones can be used as the perimeter of an infectious disease zone (IDZ). *(FIGURE 20)* **This area can also be negatively pressurized in a method similar to TNPI #1 and #2.** That is, air can be discharged through the window, into the return air system, or discharged to an adjacent space using a portable HEPA filter machine.

As always, when discharging to the return air system, users must be cautious with the volume of air being discharged. ▲

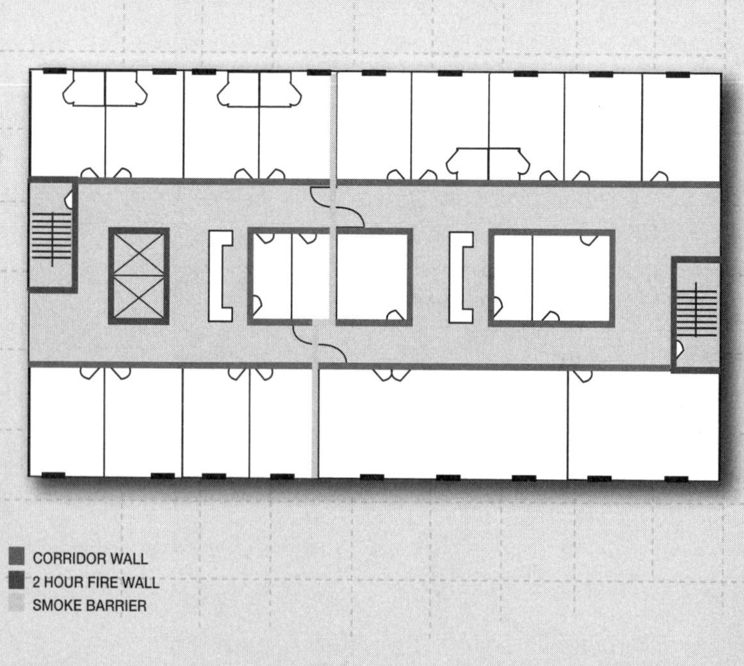

FIGURE 20

*Example of the **hospital smoke zones** that could be considered for potential isolation areas.*

References 37
Appendix 20
Surge capacity
Engineered System
Portable anteroom 13
TNPI Temporary Negative Pressure Isolation 5
Environmental controls 3
Principles of airborne infectious disease management 2
Introduction 1

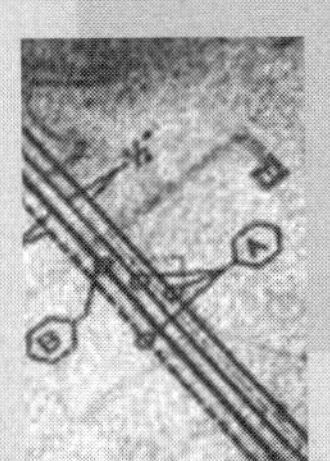

Engineered System

Ducted to the Outside

If the suite or wing is exclusively served by a single air handling unit, the facility engineer may be able to provide 100% exhaust from the area. *(FIGURE 21)*

In this method, the infectious disease zone **(IDZ) is depressurized using an exhaust diversion system** taking air from the IDZ and exhausting it to the outside ***25 feet away*** from public access and air intakes.

Damper Manipulation and Changes to the Ventilation System

This method will require damper manipulation and changes to the ventilation system that must be preformed by the facility's management. The exact steps required will vary from hospital to hospital. For example, some hospitals will need to change the settings on their Building Automation System (BAS). Others may be designed with pre-existing controls in the area that allow for this scenario.

PPE should be worn inside the IDZ. **The doors to the IDZ must be kept closed** except when entering or exiting.

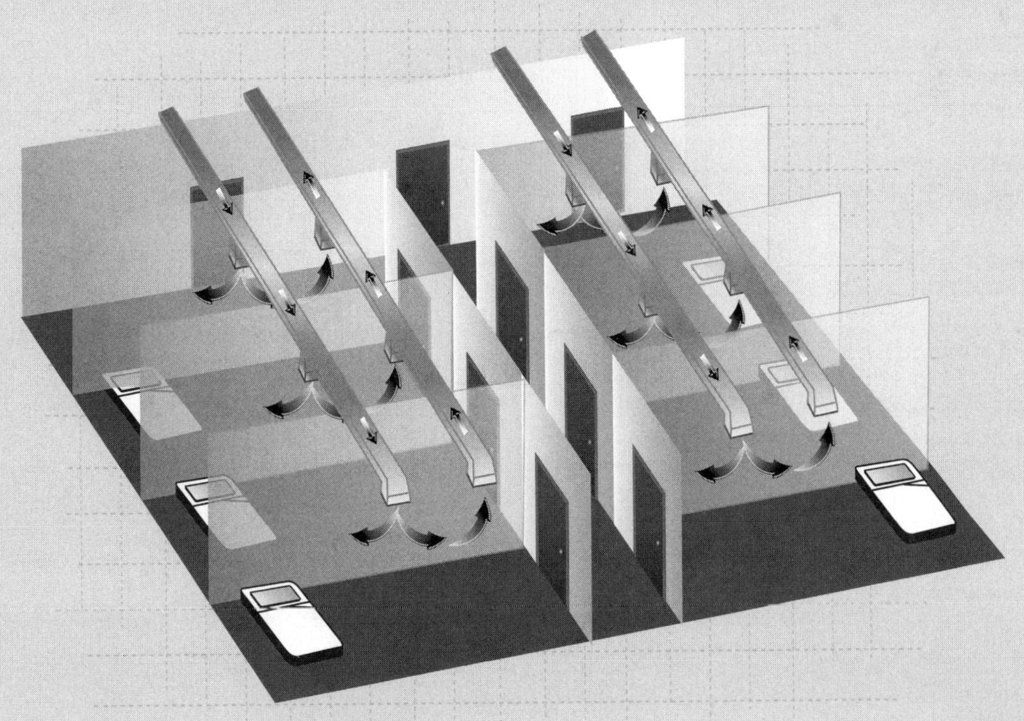

FIGURE 21
Engineered System

Temporary Surge Area

Using HEPA Filter

This method can be used to create a temporary surge area using a HEPA filter.

The three options for creating a temporary surge area using a HEPA filter are listed below. *(FIGURE 22)* They are similar to Types #1 and #2 of TNPI for a patient room. Each of these options should be used in conjunction with a safe area (anteroom).

- Discharging through a window
- Discharging to an adjacent space
- Discharging to a return air system

These three options may be used in combination to achieve desired pressure differential.

Discharging through a window

Some considerations to keep in mind when discharging air through a window are:

- A **window template can be created to exhaust** a suite of rooms through a flexible duct hookup from the HEPA filter to the window.
- Large **exhaust fans placed in the window** can be used instead of HEPA filters if the ***air is exhausted 25 feet away*** from public access and air intakes.
- **More than one machine** may be needed to establish appropriate pressure differential.

Discharging to an adjacent space

Some considerations to keep in mind when discharging air to an adjacent space are:

- The **air must be HEPA filtered.**
- The **air in the IDZ can be discharged to an adjacent area** outside the IDZ, within the building.

Discharging to a return air system

Some considerations to keep in mind when discharging air to a return air system are:

- The **air must be HEPA filtered.**
- A **return air grille adapter is used** to connect a flex duct (from the HEPA filter) to the return air grille.
- As always, when discharging to the return air system, users must be cautious with the volume of air being discharged.▲

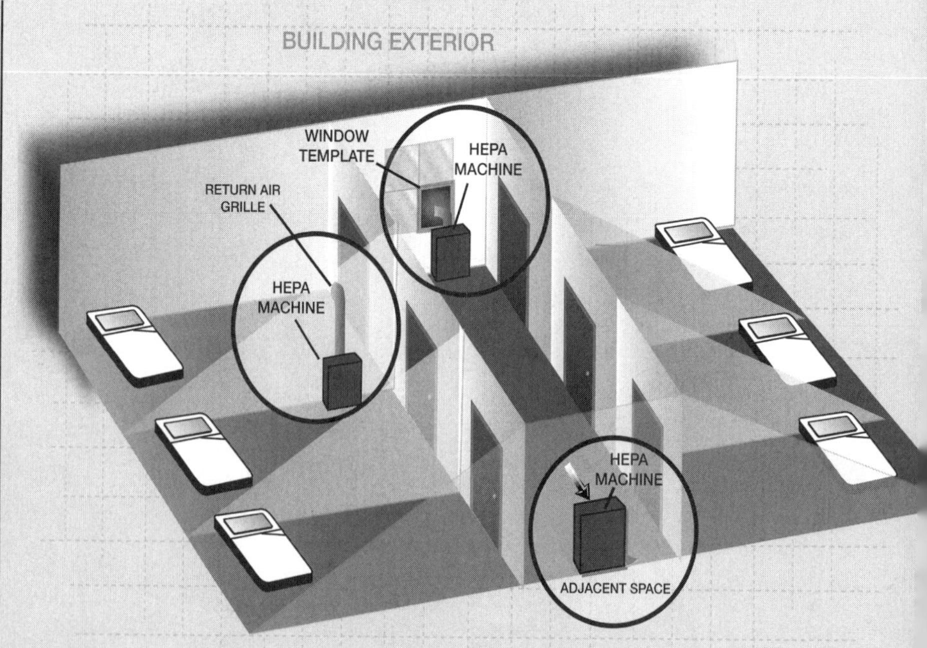

FIGURE 22
Engineered System *(shown with three different options).*

Appendix

A: 2006 AIA Criteria **21**

B: AIIR Maintenance Schedule **22**

C: HVAC System Maintenance Schedule **23**

D: Using a Pressure Gauge to Measure Relative Pressurization Between Two Spaces **24**

Using a Pressure Gauge **25**

E: Using a HEPA Filter for Dilution Ventilation **26**

F: Microorganisms Associated with Airborne Transmission **28**

G: Using a Particle Counter to Assess Indoor Air Quality and Filter Efficiency **29**

Portable HEPA Filters **30**

Testing Efficiency of Building Filtration System **31**

Building Filtration System: *Inside / Outside the Building* **32**

Building Filtration System: *Before / After the Filter* **33**

H: Data Interpretation **34**

I: Sample Log for Measuring Particle Counts **36**

2006 AIA Criteria

APPENDIX A

2006 American Institute of Architects (AIA) Guidelines
for Design and Construction of Hospital and Health Care Facilities

The following AIIR criteria are specified in the *2006 American Institute of Architects (AIA) Guidelines for Design and Construction of Hospital and Health Care Facilities*.[9]

Hospital AIIRs are required to meet criteria that was in place at the time of construction for new construction and major renovation. However, ***upgrading AIIRs to meet the criteria in the 2006 guidelines will better prepare hospitals*** to isolate patients with airborne infectious diseases. These criteria are consistent with *CDC Guidelines for Preventing the Transmission of Mycobacterium tuberculosis in Health-Care Settings, 2005.*

- ***Differential pressure shall be a minimum of 0.01-in. w.g. (2.5 Pascals).*** If alarms are installed, allowances shall be made to prevent nuisance alarms of monitoring devices. (AIA: Table 2.1-2)
- Provide ***ventilation to ensure ≥ 12 air changes/hour (ACH).*** (AIA: Table 2.1-2)
- Rooms with ***reversible airflow provisions*** for the purpose of switching between protective environment and AIIR functions ***are not acceptable.*** (AIA: Table 10.2.2.1 (3))
- Airborne infection isolation room perimeter walls, ceilings, and floors, including penetrations, shall be ***sealed tightly so that air does not infiltrate the environment*** from the outside or from other spaces. (AIA 3.2.2.4 (2a))
- Airborne infection isolation room(s) ***shall have self-closing devices on all room exit doors.*** (AIA: 3.2.2.4 (2b))
- ***Rooms shall have a permanently installed visual mechanism to constantly monitor the pressure status of the room when occupied*** by patients with an airborne infectious disease. The mechanism shall continuously monitor the direction of the airflow. (AIA: 3.2.2.4 (4))
- All areas for inpatient care, treatment and diagnosis, and those areas providing direct service or clean supplies such as sterile and clean processing, etc., shall have ***filter efficiency of 90% based on average dust spot efficiency*** per *American Society of Heating, Refrigeration and Air-conditioning Engineers (ASHRAE) 52.1-1992.* (AIA: Table 2.1-3) ▲

AIIR Maintenance Schedule

APPENDIX B

Sample Preventive Maintenance Schedule for AIIRs

For each item, place a "**X**" in the appropriate box.
"**Y**" indicates "Yes. Room is in compliance." "**N**" indicates "No. Room does not comply." "**NA**" indicates "Not Applicable to this room."

DATE	ROOM	WINDOWS CLOSED/SEALED	DOORS SELF-CLOSING OPERATIONAL	ALARMS ELECTRONICS FUNCTIONING	MECHANICAL DEVICE ZEROS WHEN DOOR OPENS	PRESSURE READING*		COMMENTS
						HAND-HELD	ELECTRONIC	
		Y N	Y N NA	Y N NA	Y N NA			
		Y N	Y N NA	Y N NA	Y N NA			
		Y N	Y N NA	Y N NA	Y N NA			
		Y N	Y N NA	Y N NA	Y N NA			
		Y N	Y N NA	Y N NA	Y N NA			
		Y N	Y N NA	Y N NA	Y N NA			
		Y N	Y N NA	Y N NA	Y N NA			
		Y N	Y N NA	Y N NA	Y N NA			
		Y N	Y N NA	Y N NA	Y N NA			
		Y N	Y N NA	Y N NA	Y N NA			
		Y N	Y N NA	Y N NA	Y N NA			
		Y N	Y N NA	Y N NA	Y N NA			
		Y N	Y N NA	Y N NA	Y N NA			
		Y N	Y N NA	Y N NA	Y N NA			
		Y N	Y N NA	Y N NA	Y N NA			
		Y N	Y N NA	Y N NA	Y N NA			
		Y N	Y N NA	Y N NA	Y N NA			
		Y N	Y N NA	Y N NA	Y N NA			
		Y N	Y N NA	Y N NA	Y N NA			
		Y N	Y N NA	Y N NA	Y N NA			

* Pressure should be checked daily when occupied, or monthly when unoccupied.

APPENDIX C

HVAC System Maintenance Schedule

Sample Preventive Maintenance Schedule for HVAC Systems

For each item, place a "**X**" in the appropriate box. "**Y**" indicates "Yes. Fan is in compliance." "**N**" indicates "No. Fan does not comply."

FAN ID/LOCATION: ______________________________

INSPECTION DATE: ______________________________

TASK	YES	NO	FOLLOW UP	COMMENTS
Inspect and clean exhaust grilles to prevent blockage & airflow retardation	Y	N	DATE DATE	
Visually inspect filter housing for holes and proper filter seal	Y	N	DATE DATE	
Clear outside air intake of debris	Y	N	DATE DATE	
Check return/exhaust dampers move freely	Y	N	DATE DATE	
Check filters for proper installation/spacers	Y	N	DATE DATE	
Check pressure set points	Y	N	DATE DATE	
Check steam/CW lines have no leaks	Y	N	DATE DATE	
Check return/exhaust belts are tight	Y	N	DATE DATE	
Check fan bearings/sheaves are lubricated	Y	N	DATE DATE	
Check humidifier controls are in working order	Y	N	DATE DATE	
Check fan lights are in working order/PSI	Y	N	DATE DATE	
Check fan cleanliness	Y	N	DATE DATE	

Using a Pressure Gauge to Measure Relative Pressurization Between Two Spaces

APPENDIX D

Airflow management requires monitoring of the ventilation system. Airborne infectious disease environments require airflow control to avoid potential infection. Although the airflow can be checked with a smoke stick for direction, it is important to know the magnitude of the flow. ***A pressure gauge is a quantitative method to measure the relative pressurization of two spaces.***

Steps for Using a Pressure Gauge to Measure Relative Pressurization Between Two Spaces

1. Turn on the digital pressure gauge (DPG).
2. Close the door of the area to be pressure tested.
3. Connect flexible rubber tubing to the DPG.
4. Place the tubing under the closed door.
5. Note the airflow direction.
6. Record the value and date.

Using a Pressure Gauge

Step 1: Turn on the Digital Pressure Gauge (DPG)

The first step is to turn on the Digital Pressure Gauge (DPG). The display should read zero when both hose connections are reading the same air pressure. Note the sensitivity of the machine (tapping one of the connections will cause a reading).

Step 2: Close the door of the area to be pressure tested

Make sure the door is closed.

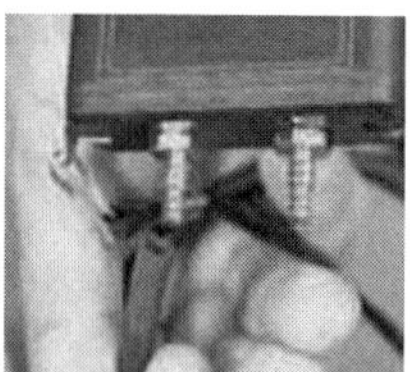

Step 3: Connect flexible rubber tubing to the DPG

Next, connect flexible rubber tubing to the DPG.

There is a set of connections on the DPG. One connection is for input pressure and the other is for the reference pressure. The DPG *(as shown in the graphic)* has two sets of connections.

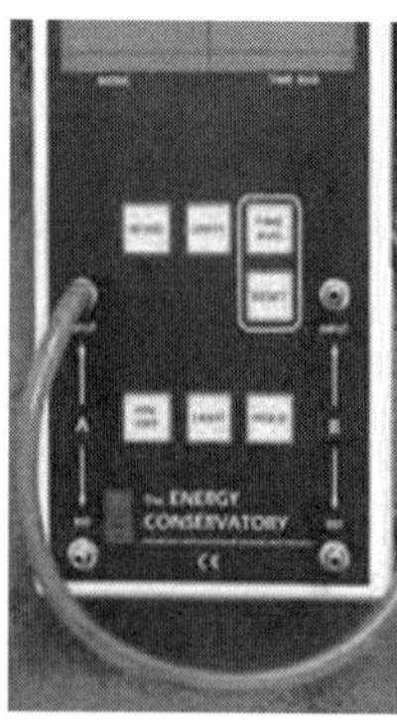

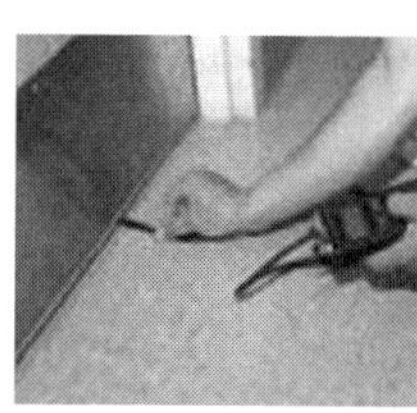

Step 4: Place the flexible rubber tubing under the door

To measure the relative pressure of a room from the corridor, connect the rubber tubing to the input pressure connection on the DPG. The reference connection is left open to corridor air.

With the door closed, ***place the tube at least four inches*** under the door.

Step 5: Note the airflow direction

If a negative sign (-) is displayed on the screen, the room is under "negative pressure." There will be no sign displayed on the screen if the room is positively pressurized. Record the sign of the reading on a log.

Step 6: Record the value and date

The final step in the process is to note the value on the screen and record the value and date in the log.

You should consult the pressure gauge user manual to change measurement units (in. w.g. or Pa) and to set the time interval over which the pressure will be averaged. A common time average interval is 1, 5, or 10 seconds.

Measurements should be recorded on a form similar to the ***"AIIR Maintenance Schedule"*** shown in Appendix B on page 22.

Using a HEPA Filter for Dilution Ventilation

APPENDIX E

Portable HEPA filter machines are used to provide temporary pressure management, dilution ventilation, and filtration (the three environmental controls required for airborne infection isolation). The ventilation systems that serve standard rooms are designed to provide dilution ventilation and filtration. For these two parameters, the HEPA filter acts as an enhancement. Because standard rooms are not designed to be negatively pressurized, the HEPA filter serves as the primary control for pressure management. As a result, the primary motivation for using a HEPA filter for TNPI is pressure (and airflow direction) management. Secondary to pressure management are dilution ventilation and filtration.

The ***recommended air exchange rate for AIIRs is 12 air changes per hour (ACH)***. As designed, most rooms provide some amount of air changes through mechanical ventilation. This is most likely less than the amount recommended for AIIRs. Although the primary purpose for HEPA filter usage is negative pressure, the HEPA filter output required to achieve 12 ACH can be determined by following the steps presented here.

Steps to Determine and Achieve Necessary Airflow

You are not required to follow these steps. They are given to simply provide an example of the required steps for achieving a given air exchange rate.

1. Calculate room volume.
2. Calculate necessary HEPA output airflow.
3. Measure airflow from HEPA filter with flow hood.
4. Adjust intensity control on HEPA filter.
5. Mark intensity level on machine.

Step 1: Calculate room volume

Determine the room volume by ***measuring the length, width, and height of the room.***

The floor area may be calculated by counting ceiling tiles (most are 2ft x 2ft or 2ft x 4ft) or floor tiles (many are 1 ft^2).
Note: A typical room has an 8 ft ceiling.

Example calculation given a room with the following dimensions:

Floor Area = 10 feet by 12 feet = 10' x 12' = 120 square feet
Ceiling Height = 8 feet
Volume = Floor Area x Ceiling Height = 120 sq. ft. x 8 ft. = 960 cu. ft.

Step 2: Calculate necessary HEPA output airflow

The recommended air change rate is 12 ACH. Using the volume and the recommended air change rate, ***calculate the necessary HEPA output airflow*** using the formula to the right:

ACH = Air changes per hour
Airflow = Mechanically exhausted airflow rate in cubic feet per minute (cfm)
Volume = Room air volume (length x width x height) in cubic feet (ft^3)

$$\text{Airflow} = \frac{\text{ACH x Volume}}{60}$$

Step 3: Measure airflow from HEPA filter with flow hood

As previously mentioned, the primary ***purpose for using a HEPA filter is pressure management.*** The minimum airflow required to achieve a pressure differential of negative 2.5 Pa was found by varying the HEPA filter intensity until the desired differential pressure was reached. When determining the airflow needed for dilution ventilation, the HEPA filter will already be operating at the minimum intensity required to induce the desired pressure differential.

A flow hood is used to measure airflow. Following the user manual for the flow hood, measure the airflow 'Q' (in cfm) of the HEPA filter. ***In order to provide 12 ACH in a room with a volume of 960 cubic feet, the required airflow is 192 cfm.***

Step 4: Adjust intensity control on HEPA filter

Once the airflow output has been determined, the intensity control on the HEPA filter can be adjusted and re-measured if the airflow is below the desired airflow. If the airflow is above that required for dilution ventilation, do not change the intensity of the HEPA filter.

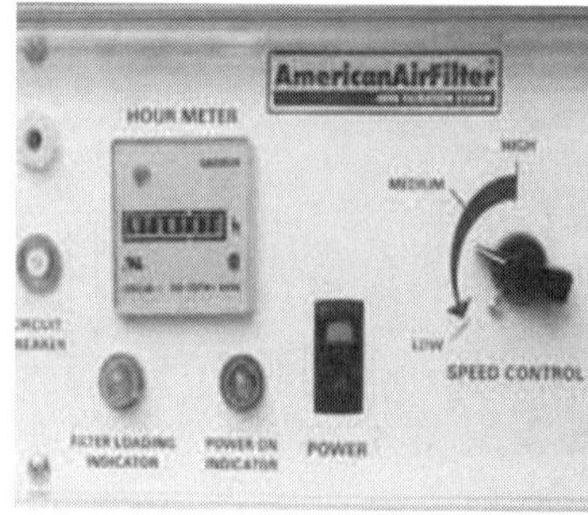

Step 5: Mark intensity level on machine

Mark the intensity level on the HEPA filter required to obtain 12 ACH. It is important to note that ***this airflow will only be constant for rooms of the same volume.*** ▲

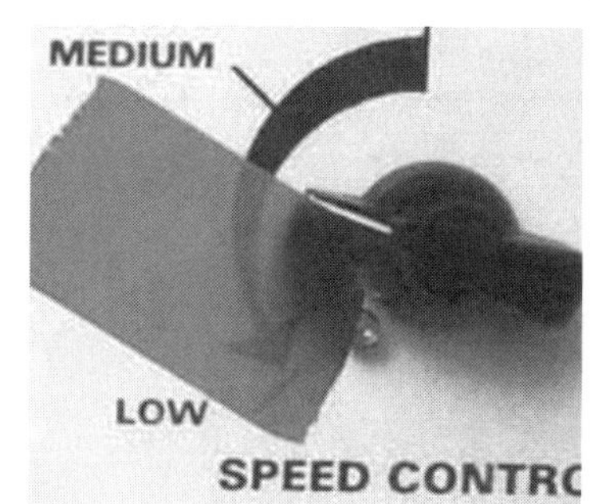

Microorganisms Associated with Airborne Transmission*

APPENDIX F

CDC Guidelines for Environmental Infection Control in Health-Care Facilities 2003 [5]

	FUNGUS	BACTERIA	VIRUS
Numerous reports in health-care facilities	Aspergillus spp.+ Mucorales (Rhizopus spp.)	Mycobacterium tuberculosis+	Measles (rubeola) virus Varicella-zoster virus
Atypical occasional reports	Acremonium spp. Fusarium spp. Pseudoallescheria boydii Scedosporium spp. Sporothrix cyanescens¶	Acinetobacter spp. Bacillus spp.¶ Brucella spp.** Staphylococcus aureus Group A Streptococcus	Smallpox virus (variola)§ Influenza viruses Respiratory syncytial virus Adenoviruses Norwalk-like virus
Airborne in nature airborne transmission in health care settings not described	Coccidioides immitis Cryptococcus spp. Histoplasma capsulatum	Coxiella burnetii (Q fever)	Hantaviruses Lassa virus Marburg virus Ebola virus Crimean-Congo virus
Under investigation	Pneumocystis carinii		

* This list excludes microorganisms transmitted from aerosols derived from water.

+ Refer to the text for references for these disease agents.

§ Airborne transmission of smallpox is infrequent. Potential for airborne transmission increases with patients who are effective disseminators present in facilities with low relative humidity in the air and faulty ventilation.

¶ Documentation of pseudoepidemic during construction.

** Airborne transmission documented in the laboratory but not in patient care areas

APPENDIX G

Using a Particle Counter to Assess Indoor Air Quality and Filter Efficiency

Particle counters measure the quantity of small particulate matter in the air.

Particle counters can also be used to determine the **efficacy of HEPA filters.** This can be done by ***comparing the particle count at the inlet*** to the HEPA filter ***with the particle count at the output*** of the HEPA filter when the HEPA filter is running.

The **efficacy of building filtration systems** can be monitored using a particle counter. This can be done by ***measuring the particle count before and after the final filters of the fans that serve those rooms***. Filtration can also be monitored by ***measuring the particle count outside the building and comparing it with the indoor particle count.***

Condensation and optical particle counters are two particle counters. Both types are acceptable to use when evaluating indoor air quality and filter efficiency. Directions for usage and examples of data interpretation are provided here.

Condensation particle counts are reported as *particles per cubic centimeter ranging from 0.02 to 1.0 µm in diameter.* **Optical particle counts** are reported as *particles per cubic foot ranging 0.5 µm in diameter.*

Steps to using particle counters to determine efficiency of different types of filters as well as examples of data interpretation are found in Appendix H, ***"Data Interpretation"*** on page 34.

Steps for Using a Particle Counter to Test the Efficiency of Portable HEPA Filters

1. Turn the particle counter on.
2. Measure the particle count at the air inlet of the HEPA filter.
3. Measure the particle count at the air output of the HEPA filter.
4. Log both measurements and note conditions at the time of the reading.

tip
It is important to verify filters are functioning properly before use and between usages. This can be done with a particle counter by comparing HEPA filter discharge air to inlet air. ▲

Portable HEPA Filters

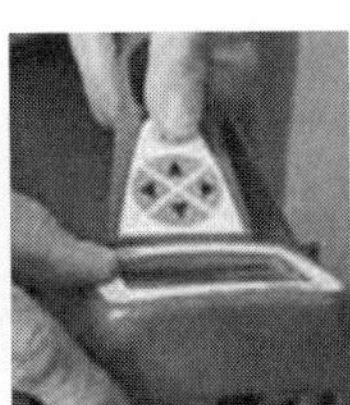

Step 1:
Turn on the particle counter

The first step is to turn on the particle counter.

Step 2:
Measure the particle count at the air inlet of the HEPA filter

In order to test the efficiency of a portable or permanently installed (wall or ceiling) HEPA filter, measure the particle count at the air inlet of the HEPA filter.

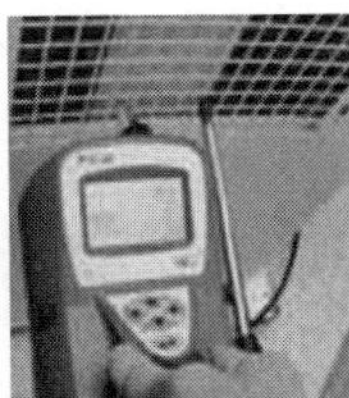

Step 3:
Measure the particle count air output of the HEPA filter

Measure the particle count at the air output of the HEPA filter output. This will be used to compare the air inlet and air outlet.

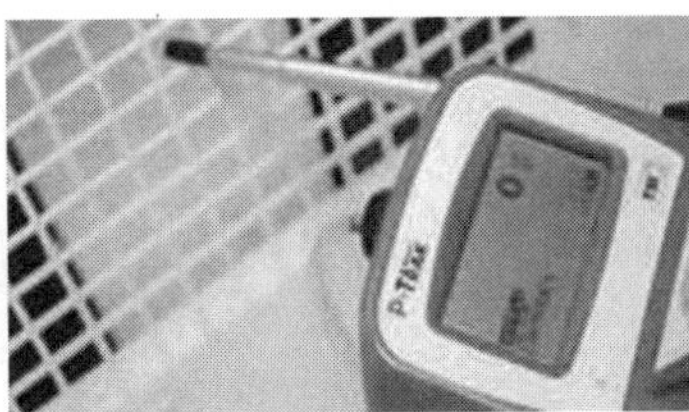

Step 4:
Log both measurements and note conditions at the time of the reading

These measurements should be recorded on a log similar to Appendix I, ***"Sample Log for Measuring Particle Counts"*** on page 36.

It is important to note that this isn't a scientific or laboratory test of the HEPA filter, but is useful in determining that the HEPA filter is working properly. Also, the test is only as accurate as the particle counter. Users should look to see a reduction of particles near the expected reduction for the machine.

Examples for data interpretation when using either optical or condensation particle counters can be found in Appendix H, ***"Data Interpretation"*** on page 34. ▲

Testing Efficiency of Building Filtration System

There are two options for using a particle counter to test the efficiency of a building filtration system.

Checking the efficacy of filters is important for a number of reasons. If the filters on the main air handling system are *working properly, there should be a significant (~90%) reduction in particles*.

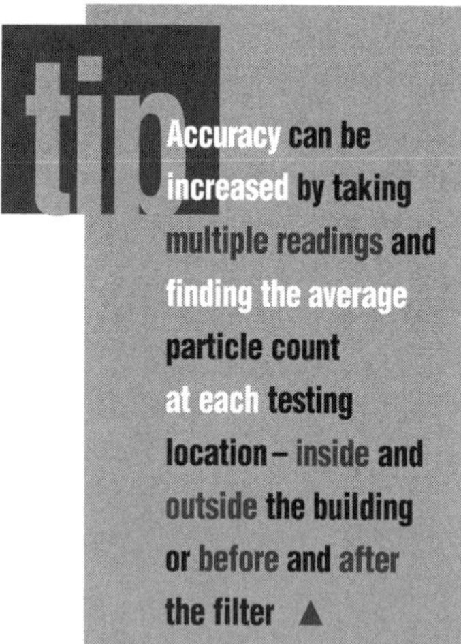

tip
Accuracy can be increased by taking multiple readings and finding the average particle count at each testing location – inside and outside the building or before and after the filter ▲

Steps for Using a Particle Counter to Test the Efficiency of a Building Filtration System

OPTION A: Inside / Outside the Building

1. Turn the particle counter on.
2. Measure the particle count outside the building.
3. Measure the particle count inside the building.
4. Record both measurements and note conditions at the time of the reading.

OPTION B: Before / After the Filter

If you have access to the air before and after the final filter, the filter efficiency can be checked here as well.

Hospitals often have mechanical spaces with fans and filter banks. The particle count can be tested before and after the final filter.

1. Turn the particle counter on.
2. Measure the particle count before the final filter.
3. Measure the particle count after the final filter.
4. Record both measurements and note conditions at the time of the reading.

Building Filtration System

OPTION A: INSIDE / OUTSIDE THE BUILDING

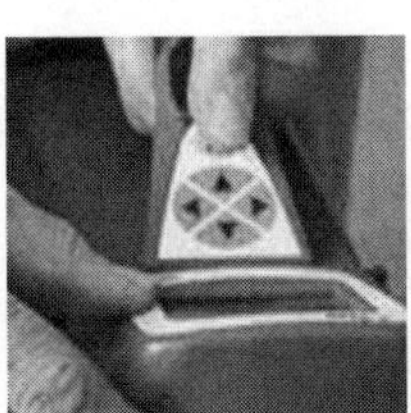

Step 1: Turn on the particle counter

The first step is to turn on the particle counter.

Step 2: Measure the particle count *outside* the building

Measure the particle count outside the building.

Accuracy can be increased by taking multiple readings outdoors and indoors.

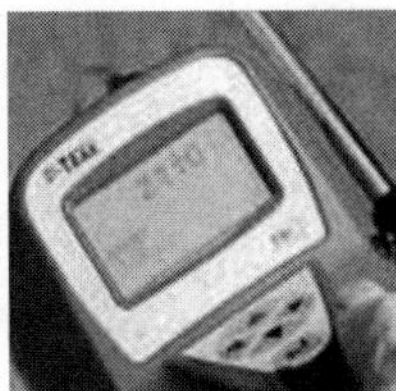

Step 3: Measure the particle count *inside* the building

Then measure the particle count inside the building (away from main doors).

Accuracy can be increased by taking multiple readings outdoors and indoors.

Step 4: Record both measurements and note conditions at the time of the reading

Since conditions can vary greatly from day to day and especially from season to season, it is important to record any abnormal readings and take notes on conditions at the time of the reading.

For example, it would be important to note if it is raining, if construction activities are going on inside or outside, if a medical nebulizer or housekeeping chemical is being used in the area, or if there are any other activities present that could possibly cause an abnormal reading.

Testing the building filtration system in this way can lead to unexpected results. For example, the outdoor particle count may be abnormally high due to vehicles nearby. Also, the indoor particle count may be abnormally high due to treatments or cleaning as described above. While this method is useful because it provides an overall view of the building, a more accurate test can be done to test the filters in particular.

Measurements should be recorded on a form similar to Appendix I, ***"Sample Log for Measuring Particle Counts"*** on page 36.

Building Filtration System

OPTION B: BEFORE / AFTER THE FILTER

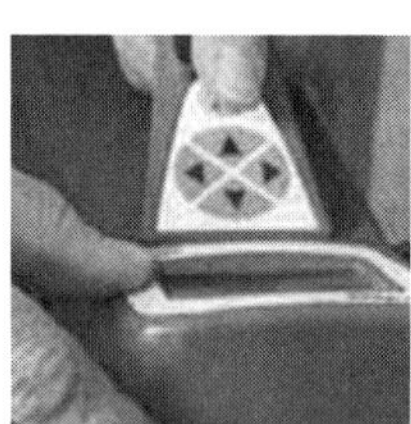

Step 1: Turn on the particle counter

The first step is to turn on the particle counter.

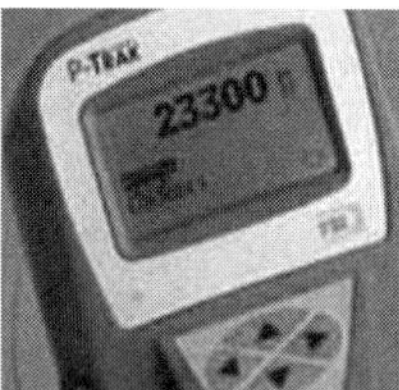

Step 2: Measure the particle count *before* the final filter

Measure the particle count **before** the final filter. If you have access to the air before and after the final filter, the filter efficiency can be checked.

Large hospitals have mechanical spaces with fans and filter banks, and the particle count can be tested before and after the final filter. Hospitals with smaller air-handling units can also test their filters in a similar manner.

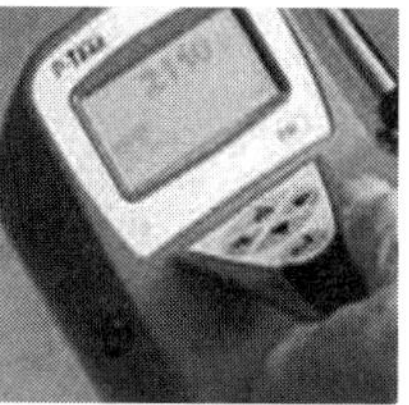

Step 3: Measure the particle count *after* the final filter

Measure the particle count **after** the final filter. If you have access to the air before and after the final filter, the filter efficiency can be checked.

Accuracy can be increased by taking a series of measurements and finding the average.

Step 4: Record both measurements and note conditions at the time of the reading

The final step is to record both measurements and note conditions at the time of the reading. This option provides a method to test the filters installed in the air-handling units. Although this isn't a scientific test, it is a reliable way to assess the performance of the filters.

A ***"Sample Log for Measuring Particle Counts"*** is shown in Appendix I on page 36. ▲

APPENDIX H

Data Interpretation

Filters are rated by industry standards on their ability to remove particulate matter from an airstream. Various rating systems exist. Efficiency can be measured by determining the concentration of material upstream and downstream of the filter. Particles should be reduced by the percentage efficiency of respective filtration systems.

You should consult the user manual of the particle counter for instructions on how to use it. Refer to the user manual for your particle counter to determine whether it is a **condensation particle counter** or an **optical particle counter.** If you're using a ***condensation particle counter, the size range should be 0.02 to 1.0 μm.*** The ***optical particle counter size range should be ≥ 0.5 μm.*** Some particle counters have data logging capabilities and directions for use vary from model to model.

Note: Percent reduction of particles is determined by ***subtracting the reading after the filter from the reading before the filter (outside air particles), dividing by the reading before the filter (outside air particles), and multiplying by 100.***

Examples of data interpretation for particle reduction in various filter efficiencies

TABLE A: CONDENSATION PC
PARTICLES REPORTED PER CC – RANGE 0.02 TO 1.0 μm

HOSPITAL FILTER RATING	OUTSIDE AIR PARTICLES	AFTER FILTER PC*	PERCENT REDUCTION
80	55000	11000	80%
90	55000	5500	90%
99.9	55000	17	99.97%

Condensation particle counts are reported as particles per cubic centimeter ranging from 0.02 to 1.0 μm.
*PC = particle counts

ASHRAE: American Society of Heating, Refrigeration and Air-conditioning Engineers

TABLE B: OPTICAL PARTICLE COUNTER
PARTICLES REPORTED PER CU.FT.- RANGE ≥ 0.5 μm

HOSPITAL FILTER RATING	OUTSIDE AIR PARTICLES	AFTER FILTER PC*	PERCENT REDUCTION
80	120000	24000	80%
90	120000	12000	90%
99.9	120000	36	99.97%

Optical particle counts are reported as particles per cubic foot ranging ≥0.5 μm.
*PC = particle counts

ASHRAE: American Society of Heating, Refrigeration and Air-conditioning Engineers

The numbers in Table A and Table B are an approximate reduction. These numbers will vary considerably from second to second as the environment is less controlled due to variations in ambient particle generation both indoors and outdoors. Because of this, the important reduction is in the order of magnitude reduction and not necessarily in the integer.

For example, with a 90% efficient filter and an outside particle count of 100000, it is expected that the particle count after the filter would be approximately 10000 (one tenth of the outdoor concentration). Because of the variability in conditions, an indoor particle of higher or lower than 10000 could be expected (e.g. 15000 or 5000, respectively).

If you don't see this type of reduction, you should ***check the filters for leakage***, ***worn out gaskets, worn out or broken clips, and proper installation of the spacers.*** ▲

Sample Log for Measuring Particle Counts

APPENDIX I

This log can be used for testing portable HEPA filters as well as whole building air filters.

"**PC INITIAL**" and "**PC FINAL**" refer to the initial and final particle counts, respectively.

For **testing a HEPA filter,** PC INITIAL refers to the particle count at the air intake, and PC FINAL refers to the particle count at the air output of the HEPA filter.

When testing a **whole building air filtration system**, PC INITIAL can refer to either the particle count outside or before the filter, and PC FINAL can refer to the particle count inside or after the filter.

DATE	FAN	PC INITIAL HEPA: *Filter Intake* Building Air Filter: *Outside / Before Filter*	PC FINAL HEPA: *Output of filter* Building Air Filter: *Inside / After Filter*	EXPECTED PERCENT REDUCTION	ACTUAL PERCENT REDUCTION $= \frac{\text{"PC I"} - \text{"PC F"}}{\text{"PC I"}} \times 100\%$	COMMENTS

References

1. Fraser VJ, Johnson K, Primack J, Jones M, Medoff G, Dunagan WC
   Evaluation of rooms with negative pressure ventilation used for respiratory isolation in seven midwestern hospitals
   Infect Control Hosp Epidemiol. Nov 1993; 14(11):623-628.

2. Francis J. Curry
   Isolation Rooms: Design, Assessment, and Upgrade
   National Tuberculosis Center; 1999.
   http://www.nationaltbcenter.edu/products/product_details.cfm?productID=WPT-04

3. U.S. Department of Health and Human Services
   Health Resources and Services Administration
   Health Care Systems Bureau, National Bioterrorism Hospital Preparedness Program
   Cooperative Agreement Continuation Guidance
   Federal Fiscal Year 2003, 2004, 2005.
   http://www.hrsa.gov/bioterrorism/

4. Nicas M, Sprinson JE, Royce SE, Harrison RJ, Macher JM
   Isolation rooms for tuberculosis control
   Infect Control Hosp Epidemiol. Nov 1993; 14(11):619-622.

5. CDC
   Guidelines for Environmental Infection Control in Health-Care Facilities
   Recommendations of CDC and the Healthcare Infection Control Practices Advisory Committee (HICPAC).
   MMWR Recomm Rep. Jun 6 2003; 52(RR-10):1-42.
   http://www.cdc.gov/mmwr/preview/mmwrhtml/rr5210a1.htm

6. **Guidelines for the classification and design of isolation rooms in health care facilities**
   Victoria Department of Human Services; *Standing Committee on Infection Control;* 1999.
   http://www.health.vic.gov.au/ideas/regulations/isolation.htm

7. **Isolation Rooms (Including Mechanically Ventilated Rooms): Best Practice Standards for Capital Planning**
   Belfast; Department of Health, Social Services and Public Safety
   Regional Advisory Committee on Communicable Disease Control; 2003.
   http://www.dhsspsni.gov.uk/hssmd41-04.pdf

8. ASHRAE
   HVAC Design Manual for Hospitals and Clinics. 2003: 27-45, 47-60, 87-113, 129-141.

9. AIA
   Guidelines for the Construction of Hospitals and Health Care Facilities. 2006.

10. CDC
    Guidelines for Preventing the Transmission of Mycobacterium tuberculosis in Health-Care Settings, 2005. *MMWR.* Dec 30 2005; 54(RR17).
    http://www.cdc.gov/mmwr/preview/mmwrhtml/rr5417a1.htm

11. Streifel AJ
    Design and maintenance of hospital ventilation systems and the prevention of airborne nosocomial infections.
    In: Mayhall CG, editor. *Hospital epidemiology and infection control.*
    Philadelphia: Lippincott, Williams & Wilkins; 2004. p.1577-1589.

12. Rutala W, Jones S, Worthington J, Reist P, Weber D
    Efficacy of portable filtration units in reducing aerosolized particles in the size range of mycobacterium tuberculosis
    Infect Control Hosp Epidemiol. July 1995; 16(7):391-398.

13. Mead K, Johnson D
    An evaluation of portable high-efficiency particulate air filtration for expedient patient isolation in epidemic and emergency response
    Annals of Emergency Medicine. Dec 2004; 44(6):635-645.

5

Disinfection and Infection Control

Balancing Infection Prevention with Safety and Health: Sterilization and Disinfection 5-1
Sterilization • Alternatives to Ethylene Oxide • High-Level Disinfection • Alternatives to Glutaraldehyde • Compatibility • Design for Health • Separation of Clean and Dirty • Train for Safety • Oversight • Continuous Quality Improvement
Glossary 5-11
Acknowledgment 5-12
References 5-12
Resources 5-13
Protecting Personnel: Integrated Model and Elements of an Health Care Facility Infection Prevention and Control (IPC) and Employee Health Services (EHS) Program 5-13
Introduction and Background
References 5-27
Health Care Construction and Infection Control 5-28
Pathogenic and Opportunistic Environmental Fungal Ecology • *Aspergillus*/Species • Patient Risk Factors • Pathobiology of Infection • Building Dynamics • HVAC System • Construction Infection Control Measures • ICRA and the ICRA Panel
References 5-39
Advances in Patient Safety: Contamination Studies Point to a New Strategy in the War against Hospital Infections 5-42
Background • Conclusion
References 5-46

Erica J. Stewart
Kaiser Permanente

Russell N. Olmsted
Saint Joseph Mercy Health System

Pier-George Zanoni
Michigan Department of Community Health

Laurence Lee
Pacific Industrial Hygiene

Jules Millogo
Retractable Technologies

Kathryn M. Duesman
Retractable Technologies

Balancing Infection Prevention with Safety and Health: Sterilization and Disinfection

*Erica J. Stewart**

Striving to provide quality care in a healing environment, health care facilities are taking a closer look at methods of sterilization and high-level disinfection for reusable medical equipment. Prevalent practices,

* Erica Stewart, CIH, HEM, is a Project Manager with Kaiser Permanente's National Environmental, Health and Safety Department. She has worked for Kaiser Permanente in various EH&S capacities for 17 years and worked in private EH&S consulting for seven years before joining Kaiser Permanente. She was the 2006 Chair of the AIHA Stewardship and Sustainability Committee and 2007 Chair of the AIHA Healthcare Working Group (erica.stewart@kp.org).

Device Classification	Devices (Examples)	Spaulding Process Classification	EPA Process Classification
Critical (enter sterile tissue or vascular system)	Implants, scalpels, needles, other surgical instruments, etc.	*Sterilization*: sporicidal chemical; prolonged contact	Sterilant/disinfectant
Semicritical (touches mucus membranes [except dental])	Flexible endoscopes, laryngoscopes, endotracheal tubes, and other similar instruments	*High-level disinfection*: sporicidal chemical; short contact	Sterilant/disinfectant
	Thermometers, hydrotherapy tanks	*Intermediate-level disinfection*	Hospital disinfectant with label claim for tuberculocidal activity
Noncritical (touches intact skin)	Stethoscopes, tabletops, bedpans, etc.	*Low-level disinfection*	Hospital disinfectant without label claim for tuberculocidal activity

FIGURE 5.1 Classification of devices, processes, and germicidal products. (From Rutala, W.A., *AJIC*, 24(4), 315, August 1996.)

including the use of ethylene oxide and glutaraldehyde, are not without safety and environmental impacts. Factoring in expensive and often incompatible instruments, the need for quick turnaround times and regulatory compliance, there is no simple solution to sterilization and high-level disinfection. How can we balance infection control requirements with a healing environment?

According to the Centers for Disease Control and Prevention (CDC), health care-acquired nosocomial infections account for 99,000 annual deaths.[1] While sterilization is done in a facility's Sterile Processing Department, high-level disinfection can occur in many places throughout a facility. The following is an overview of the available sterilization and high-level disinfection technologies, their advantages and disadvantages, health and safety hazards, and control recommendations.

Items that enter the sterile cavity of the body require destruction of all microorganisms, including bacterial spores. The process by which this is accomplished is sterilization. The Spaulding scheme (Figure 5.1) classifies the level of disinfection or sterilization needed for the type of medical device used in particular procedures.

Sterilization

Ethylene Oxide

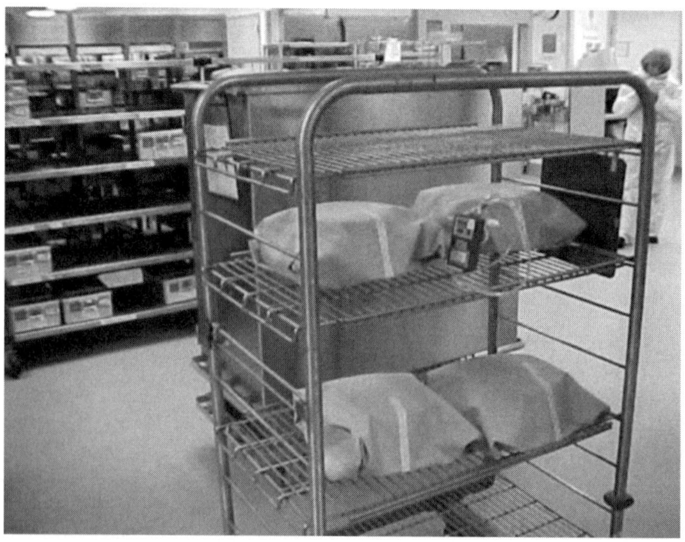

Steam sterilization is the gold standard. However, some medical devices, such as flexible endoscopes, cannot withstand the high temperature and pressure of steam autoclaves. For these devices, low-temperature ethylene oxide (EtO) gas sterilization has been the technology of choice for more than 30 years. Unfortunately, EtO is a known reproductive toxin, a suspected human carcinogen and a flammable gas. It will also cause frostbite on contact with bare skin and cause severe corneal burns.

EtO is highly regulated. It is one of the few chemicals for which the Occupational Safety and Health Administration (OSHA) has established a vertical standard, covering everything from occupational exposure assessments to work practices and equipment controls (29 CFR 1910.1047). Exposures are regulated to no more than an average of one part EtO to a million parts air (ppm) over an 8 h period (or time-weighted average, TWA). To control peak exposures, OSHA restricts exposures to no more than 5 ppm for any 15 min period. This is known as the short-term exposure limit (STEL).

EtO sterilizers must be housed in an equipment room provided with dedicated exhaust and the room must remain under negative pressure relative to adjacent spaces. Local exhaust ventilation is required above the sterilizer door and above floor drains and emergency relief valves. Special ventilated cabinets are required to house the compressed gas cylinders for sterilizers that use mixtures of EtO with hydro chlorofluorocarbon propellant. For sterilizers that use 100% EtO liquid in small canisters, the canisters must be kept in ventilated flammable storage cabinets.

To warn the staff of an emergency release, it is necessary to continuously monitor the work environment, the storage cabinet(s), and the sterilizer equipment room. Emergency response procedures must be developed and drilled for staff to follow in the event of an alarm. Responses to EtO releases require self-contained breathing apparatus (SCBA) due to the unknown concentration of the gas in air and due to the potential for an emergency release to result in atmospheres that are immediately dangerous to life and health. A combination of emergency eyewash and deluge shower is required to supply first aid measures.

The popularity of EtO has taken a plunge because of cost, risk, and hazards. Many facilities use multiple systems within one setting, striving to meet needs while reducing toxicity. Some facilities contract with third-party EtO sterilization services to sterilize heat- and moisture-sensitive instruments. This service can offer advantages (consolidation of a toxic material into one well-controlled environment; potentially reduced liability) but at a premium price. Consequently, some health care organizations either eliminate EtO sterilization (by looking into finding disposable alternatives and/or alternatives that can be reprocessed by other means) or consolidate EtO sterilization within regions, so that the few centers retaining EtO sterilizers use them at full capacity.

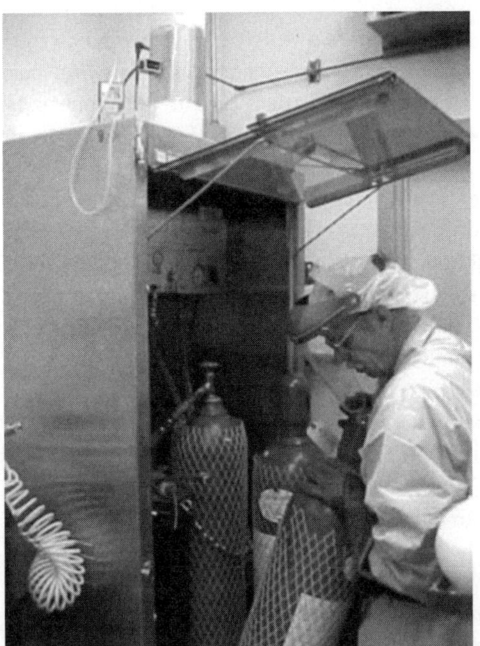

A few endoscope manufacturers are now developing or producing scopes that can be steam sterilized in response to these concerns. Exposure monitoring results typically do not detect any EtO, even when changing compressed gas cylinders, if engineering controls are well designed.[3] Nevertheless, because of the high construction and monitoring costs, as well as the relatively long turnaround time (20–24 h), and because of the problems outlined above, other alternatives have been sought.

Alternatives to Ethylene Oxide

Peracetic Acid

The first alternative to EtO was a just-in-time sterilizer developed specifically to quickly turnaround flexible and rigid endoscopes. The Steris System 1 (now called the P6000) was approved by the Food and Drug Administration (FDA) in 1999. It uses 35% peroxyacetic (or peracetic) acid in combination with 40% acetic acid, 6.5% hydrogen peroxide, and 1% sulfuric acid. Thirty-five percent peracetic acid is classified as a highly toxic material, and the S-20 Sterilant as a whole is classified as a Class II or III organic peroxide (by definition, <43% peroxyacetic acid), a Class 2 unstable (reactive) material (due to its hydrogen peroxide component), and a Class II combustible material (flash point = 115°F, above 100°F but below 140°F).

Precautions are necessary for working with this material. The packaging of its single use canisters is designed to "burp off" small amounts of chemical vapors, so canisters should be stored in well-ventilated areas. As many as six cases can be kept on open shelves in the same fire-rated areas if the area is sprinklered, but the room must not allow air to return to other occupied areas of the building. Reports of employees who have been burned when sprayed by incorrectly loaded reprocessors and by reprocessors opened while in cycle indicate that emergency eyewash facilities need to be available in the immediate work area.

While not as tightly regulated by OSHA as EtO, hydrogen peroxide, acetic acid, and sulfuric acid all have occupational exposure limits. Hydrogen peroxide has a permissible exposure limit (PEL) of 1 ppm as an 8 h TWA, but because the odor threshold is so high it has very poor warning properties. Fortunately acetic acid has excellent warning properties as its odor can be readily detected at levels that are well below those that cause harm. Because the sterilant mixture is irritating, staff who have a previous sensitivity to other respiratory irritants, like glutaraldehyde or formaldehyde, may not find relief from irritation when working with Steris S-20 Sterilant. Exposure monitoring during high volume processing using half a dozen processors at once detected acetic acid and hydrogen peroxide in the air but not at levels even approaching 10% of the OSHA limits.[4]

There is no established regulatory exposure limit or validated air monitoring method for peracetic acid, making it impossible to quantify workplace exposures. Additionally, little toxicological data is available on peracetic acid's effects on humans. Because of this, it is best to be prudent and to treat peracetic acid with the same care that you would other hazardous processes, such as by installing local exhaust ventilation over the processors and isolating processors from patient care areas. Rooms containing processors should be negative pressure, with nonrecirculating exhausts, and have at least 10 air changes per hour.

Once the sterilant is loaded into the processors and a cycle is started, it is immediately diluted with 10 L of water, which brings the working concentration of peracetic acid down to 0.2%. These solutions are nonhazardous and can be disposed of to drain without neutralization. Consequently, spill procedures need only follow general spill clean up guidelines. Partially dissolved canisters of sterilant may, however, be at a hazardous concentration which could harm employees or the environment if unexpectedly released so should be left in the sterilizer to run through another cycle to dilute it to safe levels.

It is worthwhile to note that because the process is wet, if the instrument tray is not covered and the instrument is not used within 2 h of reprocessing it is no longer considered a sterile instrument. For this reason, this reprocessor has been termed a "just-in-time" sterilization process. Indeed, many bacteria double their population in a wet environment within 20 min, so it is advisable to use the instrument immediately or dry it thoroughly using an alcohol purge followed by forced air drying. Once alcohol has been introduced to the instrument, however, it is no longer considered a sterile instrument, but only a high-level disinfected instrument.

In addition, the corrosivity of the sterilant appears to shorten the useful life of some flexible endoscopes, fogging lenses, and delaminating the glues that hold the plastic sheaths together, leading to fluid invasion and costly repairs. At least one gastroenterology endoscope manufacturer has published a letter of incompatibility with the Steris P6000 processor. Use of the processor will void its equipment warranties.

Hydrogen Peroxide Plasma Gas

The next alternative to EtO that was developed is a hydrogen peroxide plasma gas sterilizer that utilizes concentrated (50%) hydrogen peroxide in double-sealed ampoules arrayed in a cassette about the size of an ice cube tray. The liquid hydrogen peroxide is vaporized by radio frequency radiation into a highly reactive plasma gas inside the sealed evacuated chamber. Although hydrogen peroxide is a strong oxidizer and can cause burns on contact, this is not a problem because the ampoules in the multidose cassette are not pierced until the sterilizer is under vacuum. It is impossible for an employee to open the sterilizer during a cycle, so there is no chance of worker exposure to the hydrogen peroxide liquid or the plasma gas. The sterilizer requires no special ventilation or utilities to operate, and water vapor is its only waste product.

The single biggest drawback to this technology in regard to flexible endoscopes is that it is limited to processing only single channel flexible endoscopes with Teflon or polyethylene lumens of 850 mm length or less. Therefore it cannot reprocess multichannel endoscopes or the longer scopes used in gastroenterology. It can, however, process a variety of other heat- and moisture-sensitive instruments, including esophageal dilators, ophthalmic lenses, ultrasound probes and video cameras, and couplers, surgical power equipment, and their batteries.

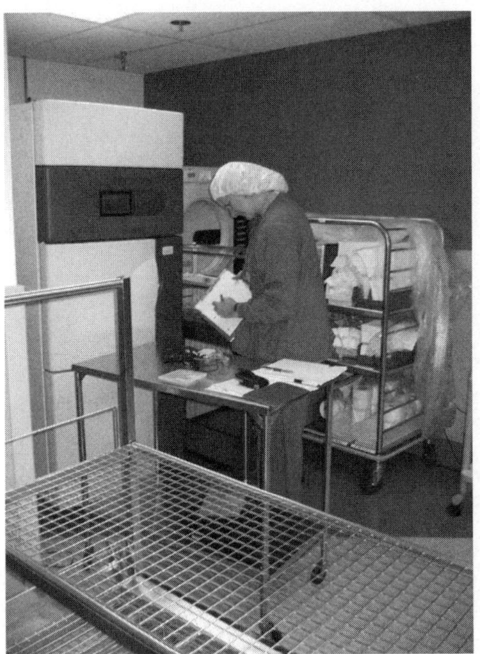

Ozone

The newest technology to offer an alternative to EtO is ozone. TSO_3 received FDA approval to market the 125L Ozone Sterilizer in 2007. The sterilizer generates ozone from on-site oxygen, water, and electricity. Ozone is an oxidative gas the odor of which is readily detected at levels far below those that cause adverse human health effects. Ozone is also a respiratory system irritant and a priority pollutant contributing to the formation of smog. It is therefore a public health concern. All the ozone in the sterilizer is theoretically converted to water vapor, requiring no special ventilation or utilities, although it would be prudent to place the sterilizer in an area of negative pressure nonrecirculating exhaust. Real-time air monitoring for ozone during an actual sterilizer cycle revealed very low levels present at the discharge outlet during the discharge phase.[5] Other concerns about the proper storage and monitoring of compressed oxygen cylinders may be relevant, depending on local code requirements.

This low-temperature sterilization process takes 4.5 h, a period comparable with steam sterilization, but is able to process many heat-labile materials (except for latex, Kraton®, and ether-based polyurethane; textile fabrics, copper and its alloys, zinc, or nickel are not recommended). The ozone system can process some lumened devices, such as arthroscopes, laryngoscopes, laparoscopes, resectoscopes, hysteroscopes, and bronchoscopes, as well as all types of urological scopes. It can also process ophthalmic lenses, cables and cords, power batteries, and Doppler probes. It cannot process flexible endoscopes, not only because these have very long, small lumens, but also because the ozone interacts with the plastics, making it tacky and vulnerable to delamination.

High-Level Disinfection

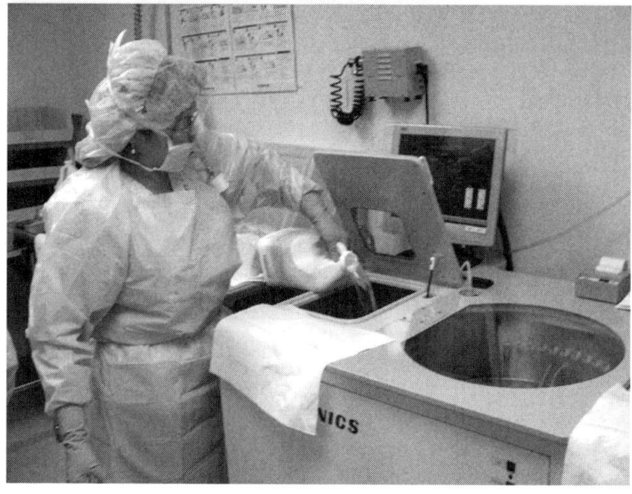

Many medical devices do not enter sterile tissues or vascular systems, including a variety of flexible endoscopes that are used in gastroenterology, head and neck surgery, urology, gynecology and anesthesiology, and the ultrasound probes used in cardiology, radiology, and obstetrics. According to the Spaulding scheme (see Figure 5.1), these can be high-level disinfected; they are defined as semicritical devices (they make contact with but do not penetrate mucous membranes and do not penetrate normally sterile areas of the body). Because of high medical device prices and the need for quick turnaround, high-level disinfection is often performed in the immediate treatment area instead of in the central Sterile Processing Department.

Glutaraldehyde

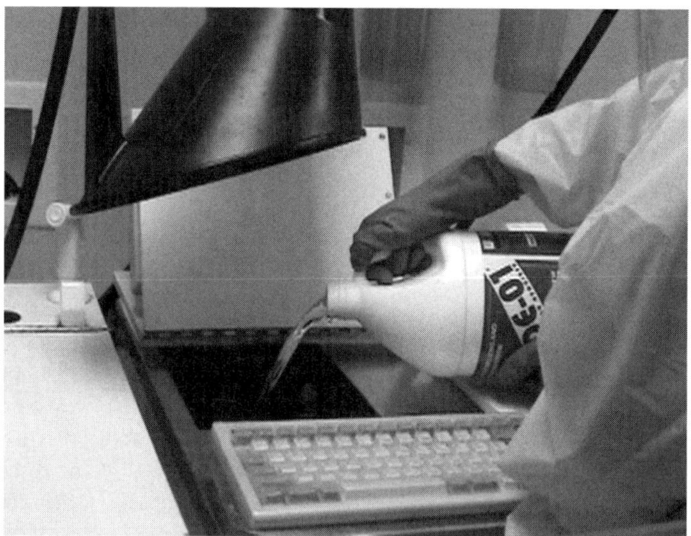

Glutaraldehyde has been the primary high-level disinfectant for the last 30 years, replacing formaldehyde as a disinfectant because of that chemical's carcinogenicity. While not a carcinogen or reproductive toxin, glutaraldehyde is an irritant, can induce asthma and respiratory sensitization, and can cause dermatitis. Workers complain of shortness of breath, irritation of mucous membranes, and skin rashes. Morale can be lowered when aerosolized vapors irritate staffers and their complaints go unresolved. Some nurses who share an acquired sensitivity to chemicals and fragrances attribute this sensitivity to workplace exposure to glutaraldehyde.

Federal OSHA no longer regulates exposure to glutaraldehyde, but many state programs have established ceiling limits to protect against the acute effects and respiratory sensitization potential. Notably, California recently lowered their ceiling limit to 0.05 ppm, a level that will take effect in July 2008.

Exposures can be dramatically reduced and symptoms virtually eliminated through the use of local exhaust ventilation and proper protective equipment, including goggles, face shields, aprons, and Nitrile or rubber gloves. While most latex surgical gloves provide adequate protection during prolonged contact from dermal exposure to glutaraldehyde, it is recommended that they not be used due to the risk of developing latex allergies. Surgical masks provide poor protection from disinfectant vapors and should also not be used. If respiratory protection is needed, selection of an appropriate respiratory should be based on the anticipated exposure levels.

Spills of glutaraldehyde are trickier to handle because of the moderate volatility of the chemical. Levels in an enclosed, poorly ventilated room can quickly exceed the ceiling limit, making it imperative that spills are quickly neutralized with sodium bicarbonate or glycine. If spills of even modest amounts (a cup or so) are not quickly addressed, an emergency response situation quickly develops.

Unfortunately, glutaraldehyde cannot be reliably monitored with real-time continuous meters. The structure of glutaraldehyde does not lend itself to monitoring by the traditional types of electrochemical sensors, infrared analyzers, or photoionization potential meters. The direct-reading instruments that are in the marketplace will react to glutaraldehyde but they lack a valid calibration standard and suffer interference from other chemicals, such as alcohol, which is likely to be present in the work area. The only validated monitoring method available is short-term integrated sampling, which requires laboratory analysis and therefore is of limited value in assessing peak concentrations, like which occur during spill situations. Analytical detection limits are getting lower and lower, allowing the industrial hygienist to take samples as short as 1–3 min and still achieve a satisfactory level of confidence in the result. This works well for characterizing short-duration tasks, such as activating and replenishing spent disinfectant, but does not truly estimate ceiling exposures.

Because of the eye irritation and skin sensitization potential, eyewashes should be provided in all areas where glutaraldehyde is used. In addition, if staff are required to mix and/or pour glutaraldehyde from containers of a gallon or more, an emergency deluge shower should be available in the work area to provide first aid in case of body splashes.

It should also be noted that some formulations of glutaraldehyde disinfectant at full strength may be considered a hazardous waste in some states, such as California. Testing end-of-use solutions for aquatic toxicity may be needed to ensure compliance with state regulations. Local Publicly Owned Treatment Works (POTW) may have even more stringent restrictions on aldehyde drain disposal. In general, neutralization of expired solutions with glycine will reduce free aldehydes to acceptable levels and such treatment may require approval by your local POTW or county hazardous waste permitting agency.

Alternatives to Glutaraldehyde

ortho-Phthalaldehyde

Not surprisingly, alternatives to glutaraldehyde have been sought to reduce human toxicity, while maintaining the biocidal activity of the disinfectant. In 1999, the FDA approved a 0.55% *ortho*-phthalaldehyde (OPA) disinfectant, and in 2007 approved a 0.6% OPA solution. These products achieve high-level disinfection in 12 min, versus glutaraldehyde's 45 min. In a heat-controlled automatic endoscope reprocessor, OPA is effective in 5 min, versus 20 min for 14 day glutaraldehyde. OPA is less irritating to the mucous membranes of staff and does not vaporize readily, although little human toxicological research is available.

The National Toxicology Program and The National Institute for Occupational Safety and Health (NIOSH) will be evaluating the chronic and acute health effects of OPA in their upcoming research agendas. To summarize what is known so far: In 2004, allergy literature reports of allergic reactions, in bladder cancer patients undergoing multiple cystoscopies using cystoscopes that had been reprocessed in OPA, prompted Johnson & Johnson to modify its 510(k) label and contraindicate OPA for reprocessing all urological instruments used to treat patients for bladder cancer. In addition, the label states that in rare instances health care workers (HCWs) have experienced irritation or possible allergic reaction that may be associated with exposure to Cidex OPA. According to the manufacturer, "in the majority of these instances, health care workers were not using it in a well-ventilated room or not wearing proper personal protective equipment."[6] Concentrated OPA is corrosive and while these effects were not demonstrated in the proprietary testing that Johnson & Johnson conducted for their FDA approval, it appears likely that the diluted product would be an eye and respiratory tract irritant or sensitizer.[7] Clinical reports in the literature, in addition to the cystoscopy reactions noted above, detail cases of asthma in a patient receiving repeat laryngoscopies, nurses disinfecting endoscopes with OPA in the absence of any glutaraldehyde exposure, and a nurse who switched from glutaraldehyde to OPA.[8] This last nurse also developed dermatitis on her lower legs. No further symptoms of asthma or dermatitis occurred once she was removed from the OPA workplace.

While NIOSH is investigating, and will hopefully publish validated monitoring methods and recommended exposure limits in 2009 or 2010, there is currently no way to assess occupational exposure to OPA.

Additionally, OPA is 3500 times more toxic to aquatic life than glutaraldehyde, requiring neutralization with glycine before drain disposal. As noted previously, state and county regulations for proper sewer disposal of high-level disinfectants will vary. Some counties may prohibit or limit the quantities of aldehydes disposed to their systems, regardless of free aldehyde concentration or acute aquatic toxicity test results.

The precautionary principle should be applied when dealing with OPA due to the lack of acute or chronic human health data available for this product. It is prudent to use the same engineering, administrative, and personal protective controls for OPA as for glutaraldehyde and to consolidate and isolate reprocessing rooms to areas of the building that have adequate general dilution ventilation (10–12 air changes per hour), that do not recirculate exhaust air to other areas of the building, that maintain a negative pressure relationship to adjacent spaces, and that are equipped with local exhaust ventilation or ductless fume hoods over soaking trays, tubes, and/or washers.

Spills of OPA should be neutralized with glycine before clean up (25 g of pure glycine, or approximately two level tablespoons, or one ounce, per gallon of solution). Because of OPA's superior biocidal activity, total aldehyde neutralization may take much longer than for typical glutaraldehyde disinfectants: 1 h versus 5 min at room temperature. Some commercial neutralization products use different amounts of glycine to achieve neutralization in shorter time frames. It is important to follow manufacturer recommendations for quantities and times. Given the unknown concentrations of OPA during a spill scenario and what little we know about the acute effects of OPA exposure, a similar spill clean up scenario should be used as for glutaraldehyde: Neutralize the spill within 1–5 min or call in the emergency response team.

Pouring of dilute OPA disinfectants from one gallon containers indicates the need for an emergency eyewash and shower, since OPA appears to be an even more potent skin sensitizer than glutaraldehyde is.

Concentrated OPA

In mid-2007, Advanced Sterilization Products (ASP) launched a fully automated washer and disinfector, a promising technology which purportedly eliminates the need for manually brushing the endoscope channels before reprocessing. The washer/disinfector uses a 5% concentrated OPA solution that is automatically diluted from a sealed bottle and neutralized before disposal to drain. Since the disinfectant is not reused, there is no need to manually test the solution's microbial effectiveness with a test strip. The unit is able to monitor how much fluid is flowing through each channel during the processor cycles and provides a printed record of all cycle parameters at the end. The unit is also equipped with an alcohol and compressed air cycle to facilitate drying.

This technology promises to relieve staff of the repetitive tasks of manually brushing channels and leak-testing endoscopes. While the promised benefits are certainly impressive, ASP is still in the process of answering concerns about the handling of alcohol waste, whether flammable atmospheres are created in the alcohol purge phase and whether processor lid gasketing eliminates the need for local exhaust ventilation. Any excess concentrated OPA left over from the sterilant supply is considered hazardous waste.

But again, the precautionary principle advises that solutions that are 10 times more concentrated than those in common use today require even more prudence when working around them, especially in spill situations. Emergency eyewash and shower stations should be required in all use areas.

Diluted Peracetic Acid

Steris also began marketing a new high-level disinfector in 2007, one that generates a dilute peracetic acid solution from a measured amount of bulk dry chemical mixed with water inside the processor and circulated throughout the processor and device lumens under pressure. The processor washing phase does not replace manual precleaning using a brush. The disinfector eliminates personal contact with chemical components by using a sealed bag for both the enzymatic detergent and disinfectant. The chemicals are not reused, so there is no need to manually test the solution's effectiveness. The waste from each cycle is purportedly nonhazardous, although independent laboratory tests of the waste effluent have not yet been verified. In addition, it is not possible to open the disinfector during a cycle, which eliminates skin exposure.

Although the actual ingredients and their concentrations are proprietary, from what has been published in the manufacturer's Material Safety Data Sheet it appears that spills could be handled similarly to the S-20 Sterilant used in the Steris P6000 once diluted; that is, only general spill clean up materials and precautions are needed. Because the components are completely sealed from operators no emergency first aid equipment is necessary.

Other

One more technology bears mention. Sterilox first received permission to market their electrolyzed saline generator in 2002; they reapplied in 2007 at a slightly different concentration and temperature range. While a promising technology because only saline, water, and electricity are needed to generate the single-use disinfectant, Sterilox is only currently available in the United Kingdom and European Union. Potential concerns with the technology are its large footprint and energy- and water-intensive resource use, and the fact that one flexible endoscope manufacturer will not honor warranties from scopes reprocessed in Sterilox.

Other liquid chemical disinfectants, using combinations of hydrogen peroxide and peracetic acid are also available in the marketplace. In general, these solutions allow lower temperatures to achieve high-level disinfection but have the disadvantage of the materials compatibility problems noted with the Steris P6000 processor. Therefore, these solutions do not have the market share that the other liquid chemical disinfectants listed above do and will not be discussed further.

Enzymatic Detergents

Of increasing interest and concern is the risk of health effects associated with exposure to proteolytic enzymes, (subtilisins) found in the majority of enzymatic detergents used to precleaning endoscopes. These products are recognized respiratory sensitizers. While historical health effects have been limited to the detergent manufacturing industry and associated with fine dry powders, there may be the potential for individuals to develop similar health effects, such as occupational asthma (as found with glutaraldehyde and OPA). Federal OSHA has established a regulatory limit of 0.0006 mg/m^3, a level approximately 100 times lower than for glutaraldehyde solutions. It is noteworthy that this concentration may be found in one droplet nuclei of aerosolized enzymatic detergent, so care must be taken to minimize splashing during pouring of concentrated detergents or when pushing detergents through lumens of endoscopes.

Compatibility

Regardless of what technology is in use, material compatibility is a must. The following reprocessing algorithm (Figure 5.2) is helpful in identifying the sterilization and disinfection options for various pieces of equipment. Manufacturers stipulate the sterilization processes approved for each device. It is imperative to consider reprocessing requirements and compatibility when purchasing new equipment.

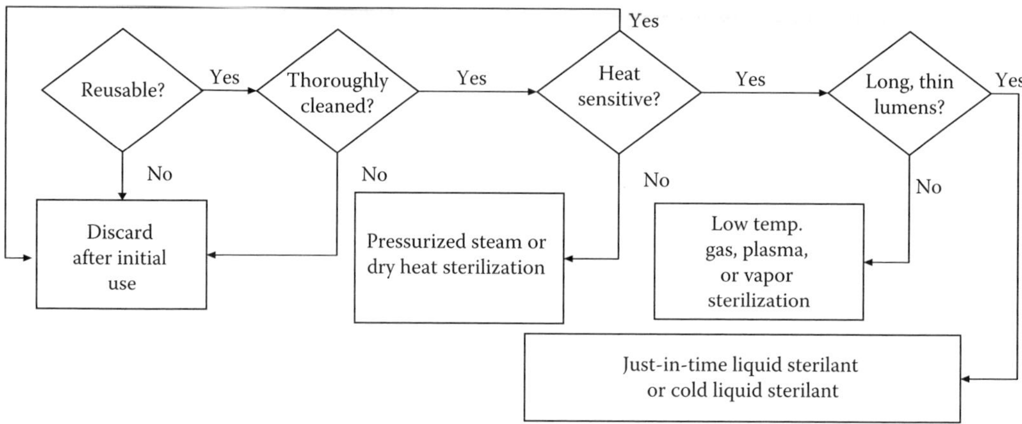

FIGURE 5.2 Reprocessing algorithm. (Reprinted from Muscarella, L.F., *Gastrointest. Endosc. Clin. North Am.*, 10(2), 245, April 2000. With permission.)

Design for Health

Given all of the factors considered above, how do we design workplaces to facilitate safe and environmentally healthy sterilization and high-level disinfection? The Joint Commission (JC) and federal and state OSHA programs require measurement of and set limits on levels of chemical vapors allowed in the air. The JC mandates plans and training for chemical spill response, and encourages standardized operations for high-level disinfection and sterilization practices. Industrial hygienists and infection control practitioners should be at the table for design and system development of all areas performing high-level disinfection and sterilization. Of equal importance, though, is reducing the number of areas where liquid or cold disinfectants and sterilants are used.

Separation of Clean and Dirty

CDC's Healthcare Infection Control Practices Advisory Committee (HICPAC) published the *Guidelines for Environmental Infection Control in Healthcare Facilities* in 2001.[10] It is a comprehensive summary of recommendations for the prevention and control of infectious diseases linked to health care environments. While it does not specifically call for physically separate clean and dirty utility rooms, it stresses the principle of segregating contaminated and clean equipment and supplies. In addition, the Association for Professionals in Infection Control and Epidemiology, Inc. (APIC) has guidelines for infection prevention and control in flexible endoscopy areas.[11] These guidelines recommend that besides separate hand washing and utility sink facilities, workflow should be designed to avoid the commingling of contaminated with clean equipment, and should promote good infection-control practices (Figure 5.3).

Train for Safety

The OSHA's Hazard Communication Standard requires documented staff training for those working with any chemical. The standard requires knowledge of the hazards associated with the chemical, proper use, methods of exposure, appropriate personal protective equipment, spill response, and access to material safety data sheets. The training should be given whenever anyone is introduced to the material, when and if there is a change of work practices, and annually. For departments that sterilize or high-level disinfect surgical instruments and medical devices, training should be specific to the hazards encountered with the chemicals in use. In addition, the Association for the Advancement of Medical Instrumentation (AAMI) requires annual assessments of staff competency to ensure staff follow all steps properly in the cleaning and disinfection of flexible endoscopes.

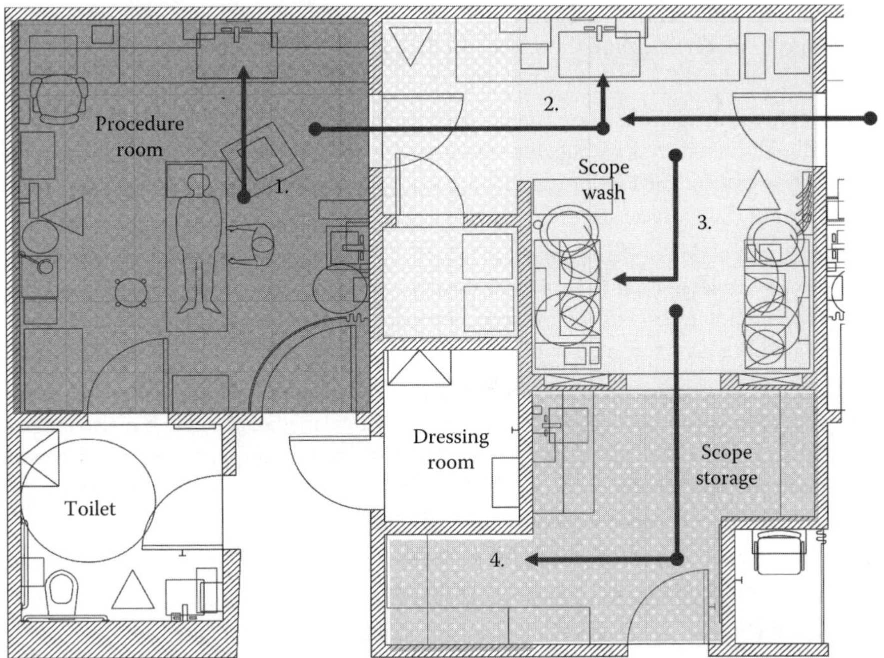

FIGURE 5.3 Typical endoscopy suite. After completion of the procedure the outside of the scope is wiped clean and flushed while still attached to the video tower (1). The scope is removed from the tower and transported to the scope wash room (2), where it is leak tested and manually cleaned with enzymatic cleaner and a brush, then rinsed and the outside dried. The cleaned and dried scope is placed in the automated endoscope reprocessor (3). At the completion of the cycle the scope is flushed with alcohol and dried with compressed air, then removed and stored (4) in a separate location. If functional spaces are not directly adjacent, then covered trays or carts should be used to transport scopes between functional rooms.

Oversight

Control of high-level disinfection and sterilization can be established through oversight by a Patient Safety or Environment of Care Hazardous Material and Waste Management Committee, as required by JC. Policies should be clearly and effectively posted. Worker training, validation of work process competency appropriate protective equipment, standardization and ongoing documented monitoring are integral to safety, compliance, quality control, and staff and environmental safety.

Continuous Quality Improvement

Taking a closer look at current practices provides an opportunity to standardize practices, and to improve safety protocols and the quality of the cleaning process. Safe and effective sterilization and high-level disinfection require oversight, but with controls in place, these practices can balance safety and healing.

Glossary[12]

Critical: The category of medical devices or instruments that are introduced directly into the human body, either into or in contact with the bloodstream or normally sterile areas of the body (e.g., surgical scalpel). These items are so called because of the substantial risk of acquiring infection if the item is contaminated with microorganisms at the time of use.

Decontamination: A process or treatment that renders a medical device, instrument, or environmental surface safe to handle. According to OSHA, "the use of physical or chemical means to remove, inactivate, or destroy bloodborne pathogens on a surface or item to the point where they are no longer capable of transmitting infectious particles and the surface or item is rendered safe for handling, use, or disposal" [29 CFR 1910.1030].

Disinfectant: A chemical agent used on inanimate objects (i.e., nonliving) (e.g., floors, walls, sinks) to destroy virtually all recognized pathogenic microorganisms, but not necessarily all microbial forms (e.g., bacterial endospores). The EPA groups disinfectants on whether the product label claims "limited," "general" or "hospital" disinfectant.

Disinfection: The destruction of pathogenic and other kinds of microorganisms by physical or chemical means. Disinfection is less lethal than sterilization, because it destroys most recognized pathogenic microorganisms, but not necessarily all microbial forms, such as bacterial spores. Disinfection does not ensure the margin of safety associated with sterilization processes.

Noncritical: The category of medical items or surfaces that carry the least risk of disease transmission. This category has been expanded to include not only noncritical medical devices but also environmental surfaces. Noncritical medical devices touch only unbroken (nonintact) skin (e.g., blood pressure cuff). Noncritical environmental surfaces can be further divided into clinical contact surfaces (e.g., light handle) and housekeeping surfaces (e.g., floors, countertops).

Semicritical: The category of medical devices or instruments (e.g., mouth mirror) that come into contact with mucous membranes and do not ordinarily penetrate body surfaces.

Sterilant: A liquid chemical germicide that destroys all forms of microbiological life, including high numbers of resistant bacterial spores.

Sterile/sterility: State of being free from all living microorganisms. In practice, usually described as a probability function, (e.g., the probability of a surviving microorganism being 1 in 1,000,000).

Sterilization: The use of a physical or chemical procedure to destroy all microorganisms including large numbers of resistant bacterial spores.

Acknowledgment

This section was expanded and adapted from an article written with Janet Brown of Hospitals for a Healthy Environment (www.h2e-online.org) and originally published by *Healthcare Design Magazine* in January 2008 (www.healthcaredesignmagazine.com).

References

1. Klevens RM, Edwards JR, Richards CL, et al. Estimating health care-associated infections and deaths in US hospitals, 2002. *Pub Health Rep* 122:160–166, 2007.
2. Rutala WA. APIC guideline for selection and use of disinfectants. *AJIC* 24(4):315, 1996.
3. Kaiser Permanente proprietary monitoring records, 1991–2007.
4. Kaiser Permanente confidential exposure monitoring records, 1998.
5. Kaiser Permanente proprietary monitoring records, 2007.
6. Johnson & Johnson's Advanced Sterilization Products Cidex OPA Label LC-20390-008 Rev. B (Rev. C or Rev. D).
7. Centers for Disease Control and Prevention (CDC). *Orthophthalaldehyde (OPA) Hazard Assessment Project Proposal*, p. 30, 2007.
8. Centers for Disease Control and Prevention (CDC). *Orthophthalaldehyde (OPA) Hazard Assessment Project Proposal*, pp. 30–31, 2007.
9. Muscarella LF. Automatic flexible endoscope reprocessors. *Gastrointest Endosc Clin North Am* 10(2): 245–257, 2000.
10. Centers for Disease Control and Prevention, Healthcare Infection Control Practices Advisory Committee (HICPAC). *Guideline for Environmental Infection Control in Healthcare Facilities*, Cincinnati, OH, 2001.

11. Alvarado CJ and Reichelderfer M. APIC guideline for infection prevention and control in flexible endoscopy. *AJIC* 28(2):138–155, 2000.
12. Division of Oral Health, National Center for Chronic Disease Prevention and Health Promotion, last reviewed: April 26, 2007.

Resources

- The Sustainable Hospitals Project—Alternatives, protective equipment. "Ten reasons to eliminate glutaraldehyde": http://sustainablehospitals.org/HTMLSrc/IP_Glutaraldehyde.html
- Hospitals for a Healthy Environment's Sterilants and Disinfectants Page: http://h2e-online.org/hazmat/steril.html
- APIC presentation on high level disinfection and sterilization by William Rutala, PhD, MPH, University of North Carolina, UNC Healthcare System, Chapel Hill, NC: http://www.apic.org/Content/NavigationMenu/Education/AnnualConference/2006AnnualConference/Program/Handouts/C2504.pdf
- Association for Professionals in Infection Control and Epidemiology: www.apic.org.
- CDC's Guidelines for Environmental Infection Control in Healthcare Facilities: http://www.cdc.gov/ncidod/dhqp/pdf/guidelines/Enviro_guide_03.pdf
- FDA list of approved cold liquid disinfectants: http://www.fda.gov/cdrh/ode/germlab.html
- OSHA's "What is ethylene oxide": http://www.osha.gov/OshDoc/data_General_Facts/ethylene-oxide-factsheet.pdf
- OSHA's "Best practices for safe use of glutaraldehyde in healthcare": http://www.osha.gov/Publications/glutaraldehyde.pdf
- NIOSH, "Glutaraldehyde—Occupational exposure in hospitals": http://www.cdc.gov/niosh/2001-115.html

Protecting Personnel: Integrated Model and Elements of an Health Care Facility Infection Prevention and Control (IPC) and Employee Health Services (EHS) Program

Russell N. Olmsted and Pier-George Zanoni

Introduction and Background

Contemporary infection prevention and control (IPC) programs have three overriding goals:[1,2]

- Protect the patient.
- Protect the HCW, visitors, and others in the health care environment.
- Accomplish the previous two goals in a cost-effective manner, whenever possible.

The first goal is the one shared with all direct care and support services (e.g., respiratory care, imaging, nutrition, environmental services, etc.) personnel and encompasses a driving desire to deliver care to those in need at the highest level of quality and as safe as possible. The second goal is one that infection control professionals (ICPs) collaboratively share with professionals in fields of employee/occupational health, facility safety, performance improvement, industrial hygiene, human resources, and organizational leadership.[3,4] Last, the intensity of resources dedicated to health care in the United States currently requires consideration and analysis of cost effectiveness of interventions and work practices however not at the expense of safety. These goals are applied across the range of care settings.

Emergence of Patient Safety

The focus on patient safety has emerged as a significant focus of health care delivery.[5] This phrase may suggest that it precludes or diminishes attention on safety of health care personnel (HCP) however the opposite is true. There is clear evidence that HCP who perceive their work environment as safe will provide a higher level of quality and safety of care to the patient.[6,7] Establishing, maintaining, and enhancing a culture of safety in an organization not only benefit the patient, but also personnel including mitigation of occupational exposure to infectious agents.[8–10] Moreover, one major benefit of the emergence of patient safety is that the goals for the IPC program outlined above have been extended beyond protection toward systematic improvements in care that continuously strive for a safer environment in the health care facility for all occupants.

Strategies for Creating a Culture of Safety

To create a culture of safety, organizations must address those factors known to influence employees' attitudes and behavior. Organizations must also direct measures to reduce hazards in the environment. Although many factors influence a culture of safety, the items listed below are key components of a safe environment and culture. These components are adapted from the CDC sharps injury workbook.[11]

Ensure organizational commitment. Organizations can use three important strategies to communicate their involvement in and commitment to safety:

- Include safety-related statements (e.g., zero tolerance for unsafe conditions and practices in the health care environment) in statements of the organization's mission, vision, values, goals, and objectives.
- Give high priority and visibility to safety committees, teams, and work groups (e.g., occupational health, infection control, quality assurance, pharmacy, and therapeutics), and ensure direct management involvement in the evaluation of committee processes and impact.
- Require action plans for safety in ongoing planning processes. (e.g., an action plan for improving the culture of safety for sharps injury prevention could be one element in an overall safety culture initiative).

Organizational leadership can also communicate a commitment to safety indirectly by modeling safe attitudes and practices. Health care professionals in positions of leadership send important messages to subordinates when they

- Handle sharp devices with care during procedures.
- Take steps to protect coworkers from injury.
- Properly dispose of sharps after use.

Similarly, supervisors should address sharps hazards in a nonpunitive manner as soon as they are observed and discuss safety concerns with their staff on a regular basis. This will positively reflect the organization's commitment to safety and build safety awareness among staff.

Involve personnel in the planning and implementation of activities that promote a safe health care environment. Involving personnel from various areas and disciplines while planning and implementing activities improves the culture of safety and is essential to the success of such an initiative. Those personnel who participate on committees or teams created to institutionalize safety serve as conduits of information from and to their various work sites. They also legitimize the importance of the initiative in the eyes of their peers.

Encourage reporting and removal of sharps injury hazards. Another strategy for institutionalizing a culture of safety is to create a blame-free environment for reporting sharps injuries and injury hazards. HCP who know that management will discuss problems in an open and blame-free manner are more likely to report hazards. Health care organizations can also actively look for sharps injury hazards by performing observational rounds and encouraging staff to report near misses and observed hazards in the work place. Once identified, hazards should be investigated as soon as possible to determine the contributing factors, and actions should be taken to remove or prevent the hazard from occurring in the future.

Develop feedback systems to increase safety awareness. A number of communication strategies can provide timely information and feedback on the status of sharps injury prevention in the organization.

One strategy incorporates findings from hazard investigations, ongoing problems with sharps injuries, and prevention improvements into articles in the organization's newsletter, staff memoranda, and/or electronic communication tools. It is important to communicate the value of safety by providing feedback when the problem is first observed and commending improvements. Another strategy is to create brochures and posters that enhance safety awareness. Such materials can reinforce prevention messages and highlight management's commitment to safety.

Promote individual accountability. Promoting individual accountability for safety communicates a strong message about the organization's commitment to a safe health care environment. In order for accountability to be an effective tool, all levels in the organization must comply. An organization can promote individual accountability for safe practices in general and sharps injury prevention in particular in many ways. One way is to incorporate an assessment of safety compliance practices in annual performance evaluations; for managers and supervisors, this might include evaluating methods used to communicate safety concerns to their subordinates. Organizations might also consider having staff sign a pledge to promote a safe health care environment. This could be incorporated into hiring procedures and/or as part of an organization-wide safety campaign.

Emerging models for preventing health care-associated infections (HAIs): The critical role of HCP

Morbidity and Mortality Associated with HAIs

Klevens et al. have recently estimated the total number of HAIs in U.S. hospitals, excluding those in intensive care units, to be 1.7 million in 2002 (see Figure 5.4).[12] Of these 98,987 subsequently died and these deaths were either caused or associated with the patient preceding HAI. This burden of mortality exceeds

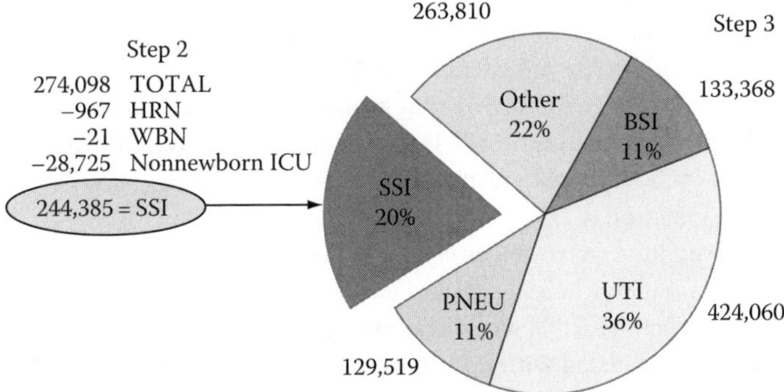

Notes: From the total number of surgical site infections (SSI) obtained from the National Hospital Discharge Dataset and the National Nosocomial Infections Surveillance (NNIS) system, we subtracted the number of SSI among newborns and adults and children in intensive care units. The remaining SSI were among adults and children outside of intensive care units. From hospital-wide surveillance in NNIS, we had the distribution of infections by major site and calculated the corresponding number of infections for pneumonias (PNEU), urinary tract infections (UTI), bloodstream infections (BSI), and other sites.

HRN = high-risk newborns

WBN = well-baby nurseries

ICU = intensive care unit

SSI = surgical site infections

BSI = bloodstream infections

UTI = urinary tract infections

PNEU = pneumonia

FIGURE 5.4 Calculation of the estimates of health care-associated infection in U.S. hospitals among adults and children outside of intensive care units, 2002. (From Klevens RM et al., *Public Health Rep* 122(2): 160–6, 2007.)

any other notifiable disease and places the impact from HAIs among the top 10 causes of death in the U.S. population. Figure 5.4 also illustrates that the frequency of the major sites of infection in descending order includes catheter-associated urinary tract infections (CA-UTIs), surgical site infection (SSI), pneumonia, and bloodstream infection (BSI). In terms of resources needed to treat these, however, there is a slight reordering. For example, ventilator-associated pneumonia (VAP), a subset of HA-pneumonia is estimated to cost $23,000/case on average and extends the length of stay by almost 10 days; for CA-BSI $18,432 and 12 extra days; $18,000 for an SSI after coronary bypass surgery and 26 days; and $1257 for a CA-UTI and less than 1 extra day.[13] While these estimates per infection are significant, much involve fixed costs or are affected by variation in reimbursement from payers. Of late, there is more attention on the impact of those who experience HAIs on efficiency of care delivery. For example, a patient with an HAI in and ICU will delay turn over of the patient room and at a certain frequency will hamper the ability of the hospital to accommodate new admissions appearing for care in the Emergency Department.[14] The delay in throughput has significant impact therefore on the operations of the hospital.

Consumers, Regulators, and Payers Response to the Problem of HAIs

Given the considerable burden that HAIs place on recipients of health care, there has been exponential growth in attention and concern involving these coincident with enhancing emphasis on patient safety. Consumer advocacy groups have joined with legislators to enact mandates for release of hospital-specific HAI data in over 24 states throughout the United States: http://www.apic.org/am/images/mandatory_reporting/mandrpt_map.gif

In addition, recent reports of the growing proportion of HAIs caused by multidrug-resistant organisms (MDROs) such as methicillin-resistant *Staphylococcus aureus* (MRSA) have launched legislative actions to mandate public reporting of isolates of MRSA in 6 states and another 10 have introduced study bills: http://www.apic.org/am/images/maps/mrsa_map.jpg

Other key influencing organizations such as the JC and the Centers for Medicare and Medicaid Services (CMS) have also paid increasing attention to the problem of HAIs. The JC Standards currently include a national patient safety goal (NPSG) that requires accredited members to ensure adherence with the CDC Hand Hygiene Guideline, 2002. They recently issued draft NPSGs for 2009 that address prevention of central line-associated bloodstream infections (CLABSI), SSIs, CA-UTIs, VAP, MRSA, and Clostridium difficle infection (CDI).

CMS a significant payer for care provided by most hospitals in the United States has launched a value-based purchasing initiative wherein changes to the inpatient prospective payment system (IPPS) scheduled for activation in October 2008 will elevate accountability by providers by no longer paying for certain adverse conditions associated with hospitalization that were not present on admission.[15] These include the following:

- Serious preventable events—object left in place during surgery, air embolism, or blood incompatibility
- CA-UTI
- Pressure ulcers (decubitus ulcers)
- Vascular CA infection
- Surgical site infection—mediastinitis after coronary artery bypass graft surgery
- Hospital-acquired injuries—fractures, dislocations, intracranial injury, crushing injury, burn, and other unspecified effects of external causes

Several additional conditions such as additional sites of HAI and certain pathogens like MRSA are under active consideration by CMS as additions to the value-based purchasing initiative in the near future.

HCP as the agent for prevention of HAIs—Models:
It is clear, given the intense focus on HAIs that HCP have and will continue to be the key element to drive prevention. Recently, there is a growing body of evidence indicating that a composite of various

processes of care aimed at prevention of HAIs can save lives. This composite or more typically referred to as "infection prevention bundles" involve a combination of elements that are enacted either in step-wise fashion and/or simultaneously. Each of the elements incorporated into the model are based on scientific evidence that have shown efficacy in preventing the unwanted outcome—a HAI. While none of the individual components are particularly novel, it is the bundling of these in a format that facilitates use at the bedside by HCP—using a tool such as a simple checklist—that has been extremely successful and captured widespread attention.[16] Some specific models from published studies using the bundle approach are as follows:

Central Line-Associated Bloodstream Infection Prevention

Pronovost et al. reported results of a statewide CLABSI prevention collaborative in 108 ICUs in Michigan hospitals which demonstrated a 66% reduction in incidence of CLABSI that was sustained for more than 1 year and saved over 1500 lives of patients cared for in participating hospitals.[17] The elements in the CLABSI prevention project included a checklist that was completed by a nurse who monitored the physician or other licensed health care professional inserting the central line in real time. The nurse was empowered to stop the insertion if there were any breeches in aseptic technique. These elements captured in a checklist format included the following:

Before the procedure, did the procedure provider:

Perform hand hygiene with antimicrobial soap or apply alcohol-based waterless hand cleaner immediately prior to procedure
Was hand hygiene directly observed?
Was skin antiseptic (2% chlorhexidine gluconate) applied to the insertion site?
Was a sterile full body drape put over the patient?

During the procedure, did the procedure provider:

Use sterile gloves
Wear head cover, mask and a sterile gown
Maintain a sterile field
Did all assistive personnel follow above precautions?
Did ancillary staff in the room follow appropriate precautions?

After the procedure:

Was a sterile dressing applied to the site?
Was a correction required to ensure compliance with infection prevention practices? If yes, explain.

The evidence-based practices that have been shown to be important in preventing CLABSI on which the checklist is based were

- Appropriate hand hygiene
- Use of chlorhexidine for skin preparation
- Use of full-barrier precautions during central line insertion
- Avoiding use of the femoral vein for insertion
- Removing unnecessary central lines
- Similar bundled interventions for prevention of CLABSI have been published (PRHI;[18] IHI Save 100,000 Lives Campaign[19])

Other collaboratives have demonstrated success with prevention of other sites of HAI such as VAP and SSI (Figure 5.4). All of these efforts, however, are dependent on engagement and support from HCP involved in direct care of the patient. Clearly, use of collaboratives will continue to grow but in addition to tools such as the checklist, the primary drivers for these involve champions at the level of the patient care unit who are willing to carry these evidence-based practices to the bedside. Organizational leadership also needs to ensure there are sufficient resources to support and sustain these critical prevention efforts.

Definition of Terms and Programmatic Guidance

The term HCP used throughout this chapter is one that encompasses all paid and unpaid persons working in health care settings who have the potential for occupational exposure infectious agents/microorganisms during the course of their work. Microorganisms can be present in and on patients and personnel, equipment, supplies, surfaces, and may be present in the air and water inside health care facilities. The types of personnel who work in health care facilities includes but is not limited to emergency medical service personnel, dental personnel, laboratory personnel, autopsy personnel, nurses, nursing assistants, physicians, technicians, therapists, pharmacists, students and trainees, contractual staff not employed by the health care facility, and persons not directly involved in patient care but potentially exposed to infectious agents (e.g., technicians associated with vendor equipment, clerical, dietary, housekeeping, maintenance, and volunteer personnel). The term HCP therefore more accurately reflects this broad range of job activities more so than the traditional "HCW." Risk of exposure to microorganisms includes a number of variables but those involved in direct or indirect (e.g., sterile processing department personnel who mechanically clean contaminated surgical instruments) contact with blood, body substances, and other potentially infectious materials (OPIM) are at higher risk than those with no or limited direct contact.

The U.S. CDC has published several guidelines that are a useful evidence-based foundation for the elements of an IPC program aimed at protection of personnel. Principal among these is the *Guideline for Infection Control in Healthcare Personnel* published in 1998. (CDC 1998) Subsequently, there have been several recent guidelines that while not covered in detail in this chapter are excellent resources upon which to draw. These include

2003 Environmental Infection Control (CDC EIC 03)[20]
2005 Tuberculosis (CDC TB 05)[21]
2007 Isolation (CDC Isol 07)

Mortality and Morbidity Associated with Occupational Exposure to Infectious Diseases

Sepkowitz and Eisenberg recently analyzed U.S. Department of Labor data and estimated that risk of mortality among HCP from occupational events, including infection, ranged between 17 and 57 per 1 million workers.[22] The rank order of incidence of occupational mortality was highest for emergency medical services personnel (64–95/million workers) followed by technologists/technicians (28), physicians (12–29), and nurses and aides/orderlies/attendants (8). Incidence of mortality for HCP fell just below the rate for the entire workforce (42.5/million) with certain occupations identified as highest, e.g., fisherman (1179/million), construction worker (1081–1452/million), and pilot (791–953/million). Subset analysis of mortality associated with specific infections found that between 9 and 42 HCWs per million die annually from occupational infection.[22]

Occupational sharps injury and exposure to bloodborne infectious diseases: The CDC has published an online workbook entitled, *Workbook for Designing, Implementing, and Evaluating a Sharps Injury Prevention Program* (Sharps Workbook 2004). This is a superb resource that places emphasis on prevention of sharps injuries. The topic of sharps injuries are reviewed in depth elsewhere in this text but some highlights from the CDC Workbook include the following.

There are over 20 pathogens in which transmission from a contaminated sharp object has been documented. These include some microbes such as *Mycobacterium tuberculosis* and *Neisseria gonorrhea*, which are more often transmitted via different routes. By far however hepatitis B virus (HBV), hepatitis C virus (HCV), and human immunodeficiency virus (HIV) are the pathogens most commonly transmitted by contaminated sharps. There has been a significant drop in annual number of cases of occupational HBV from a high of almost 12,000 in 1985 to 500 in 1997. Most of this is due to introduction of HB vaccine. Transmission risk for HBV after sharps injury remains, however, for those who are unimmunized and ranges between 6% and 30%. There is no vaccine against HCV, but the risk after exposure fortunately is much less, 1.8% (range: 0%–7%). Last, for HIV the average risk of HIV transmission after a percutaneous exposure is estimated to be approximately 0.3%; for a splash of blood/OPIM to mucous membrane the estimate is 0.09%.

Programmatic elements of a Personnel Health Program (PHP) with infection prevention focus

The elements of a PHP as described in the CDC 1998 Guideline (CDC workbook 2004) are

1. Coordination and collaborative network
2. Medical evaluations
3. Personnel health and safety education
4. Primary prevention: immunization programs
5. Management of job-related illnesses and exposures
6. Health counseling
7. Maintenance of records, data management, and confidentiality

Each of these is addressed in more detail in the following sections.

1. *Coordination and collaborative network*

A main link in the network of an effective PHP is open communication and collaboration between the architecture and staff in IPC and PHP departments. This is reflected in professional practice standards, the foundation of practice certification, and practice standards issued by professional organizations representing these two fields. The other key linkages in this network include human resources, safety, direct care, and support services.

2. *Preemployment and periodic health evaluations*

Preemployment and subsequent periodic health evaluations are an important component of preventing transmission of infection and enhancing safety of personnel. Preemployment evaluations should include an inventory of immunization status and obtaining histories of any conditions that might predispose personnel to acquiring or transmitting communicable diseases. This health inventory also is essential information for managing any subsequent exposures personnel might encounter during health care delivery. The inventory should focus on protection against vaccine-preventable diseases (VPDs) that include Hepatitis B, measles, mumps, rubella, varicella, and more recently pertussis. Recommendations for immunization against VPDs for HCP are updated frequently by the CDC's Advisory Committee on Immunization Practices (ACIP) and should be reviewed periodically for application by EHS and IPC (CDC ACIP recommendations). By illustration, personnel should be asked about prior history of infection with varicella-zoster virus (VZV) or documentation of receipt of varicella vaccine. For those with uncertain history of chicken pox, it is cost effective for EHS to serologically screen the employee for immunity against this virus. If susceptible by such testing, varicella vaccine can be offered early in the employee's work placement.

Another important component of preemployment evaluation is to determine if the employee has prior infection from *M. tuberculosis* (latent TB infection or LTBI) or disease. If there is no history, then a two-step tuberculin skin test (TST) is recommended at preemployment. Those found to have LTBI should be referred to local public health TB control programs for further evaluation and consideration of treatment. Personnel with history—ideally with documentation of LTBI—do not need any additional TST. Policy for provision of chest x-ray (CXR) for those with LTBI varies by facility but it is neither cost effective nor necessary to repeat CXRs thereafter. Instead, personnel with LTBI should be queried periodically for any symptoms suggestive of active TB disease. In a related matter, preemployment is a good time to fit test personnel at risk of exposure to *M. tuberculosis* as identified in the facility's TB risk assessment program (usually a collaborative effort between IPC and EHS programs). For most facilities, fit testing is a component of OSHA's required respiratory protection program and this testing must be repeated on an annual basis.[23] Other preemployment screenings may also include history of sensitivity to certain chemicals such as high level disinfectants (e.g., glutaraldehyde) and sensitivity to latex.

Periodic evaluations may be done as indicated for job reassignment, for ongoing programs (e.g., TB screening), or for evaluation of work-related problems or postexposure follow-up. Annual influenza vaccination of HCP plus consideration of other vaccines such as meningococcal and Tdap are additional types of periodic services that may occur after the preemployment evaluation.[24,25]

3. *Personnel health and safety education*

Education of personnel on prevention of infection is an important component of EHS. There are several requirements for specific topics such as avoiding exposure to bloodborne pathogens, TB, and safe work practices for working with chemicals. Content of such education should be appropriate to the type of work tasks, educational level, and literacy.

4. *Primary Prevention: Immunization programs*

Recommendations for immunization against select VPDs are provided in Table 5.1. Immunization of personnel against VPDs is the most effective elements of EHS. High levels of immunity among personnel protects not only personnel but patients, visitors, and others in the health care facility and their family members as well. This also avoids costly work restrictions and other interventions such as antimicrobial medications and additional laboratory testing after exposure of those susceptible to VPDs.

Several studies indicate that education and voluntary receipt of certain vaccines such as those that protect against influenza and hepatitis B work. For the latter, making receipt a work requirement has been shown to be the most effective policy and has resulted in very high proportion of protection among personnel against hepatitis B (CDC Workbook 2004). For the former, there is a recent JC standard that requires education for personnel on prevention of health care-associated influenza.[26] Making receipt of vaccination against seasonal influenza, a work requirement has been met with more resistance. However, there is considerable evidence that personnel are often the source of influenza that is transmitted to patients, other personnel, and visitors because virus is shed prior to onset of infection and personnel are not adherent with recommendations to stay away from work when experiencing influenza-like illness.[27] The CDC strongly recommends all personnel receive influenza vaccine each year and there are facilities that have made this a work requirement.[28]

5. *Management of job-related illnesses and exposures and intercurrent infections/colonization among personnel*

Exposure of personnel to communicable diseases calls on EHS to oversee prompt management of job-related illnesses and to provide appropriate postexposure prophylaxis in collaboration with IPC program.

Management can include diagnostic testing, immunization, provision of antimicrobial medications, and exclusion of personnel from work or patient contact. Decisions on work restrictions are based on the mode of transmission and the epidemiology of the disease. A summary of recommended work restrictions by select diseases is provided in Table 5.2. Additional background and discussion of risks and work restrictions are reviewed in CDC personnel health guidelines (CDC personnel health 1998).

Personnel can be a source of exposure to patients either due to active infection or colonization with certain microorganisms. Outbreaks or clusters of infection traced to personnel as a source include hepatitis B & C, methicillin-resistant *S. aureus*, Group A Streptococcus, TB, Salmonellosis, measles, and rubella.

6. *Health counseling*

Personnel should have access to adequate health counseling that is specific to the individual and may include information such as (a) the risk and prevention of occupationally acquired infections, (b) the risk of illness or other adverse outcome after exposures, (c) management of exposures, including the

TABLE 5.1 Immunobiologics and Schedules for HCP: Immunizing Agents Strongly Recommended for HCP

Generic Name	Primary Booster Dose Schedule	Indications	Major Precautions and Contraindications	Special Considerations
Hepatitis B recombinant vaccine	Two doses IM in the deltoid muscle 4 week apart; third dose 5 month after second; booster doses not necessary	HCP at risk of exposure to blood and body fluids	No apparent adverse effects to developing fetuses, not contraindicated in pregnancy; history of anaphylactic reaction to common baker's yeast	No therapeutic or adverse effects on HBV-infected persons; cost-effectiveness of prevaccination screening for susceptibility to HBV depends on costs of vaccination and antibody testing and prevalence of immunity in the group of potential vaccines; HCP who have ongoing contact with patients or blood should be tested 1–2 months after completing the vaccination series to determine serologic response
Influenza vaccine (inactivated whole or split virus)	Annual single-dose vaccination IM with current (either whole-or split-virus) vaccine	HCP with contact with high-risk patients or working in chronic care facilities; personnel with high-risk medical conditions and/or ≥65 years	History of anaphylactic hypersensitivity after egg ingestion	No evidence of maternal or fetal risk when vaccine was given to pregnant women with underlying conditions that render them at high risk for serious influenza complications
Measles live-virus vaccine	One dose SC; second dose at least 1 month later	HCP born in or after 1957 without documentation of (a) receipt of two doses of live vaccine on or after their first birthday, (b) physician-diagnosed measles, or (c) laboratory evidence of immunity; vaccine should be considered for all personnel, including those born before 1957, who have no proof of immunity	Pregnancy; immunocompromised[a] state; (including HIV-infected persons with severe immunosuppression) history of anaphylactic reactions after gelatin ingestion or receipt of neomycin; or recent receipt of immune globulin	MMR is the vaccine of choice if recipients are also likely to be susceptible to rubella and/or mumps; persons vaccinated between 1963 and 1967 with (a) a killed measles vaccine alone, (b) killed vaccine followed by live vaccine, or (c) a vaccine of unknown type should be revaccinated with two doses of live measles vaccine

(*continued*)

TABLE 5.1 (continued) Immunobiologics and Schedules for HCP: Immunizing Agents Strongly Recommended for HCP

Generic Name	Primary Booster Dose Schedule	Indications	Major Precautions and Contraindications	Special Considerations
Mumps live-virus vaccine	One dose SC; no booster	HCP believed to be susceptible can be vaccinated; adults born before 1957 can be considered immune	Pregnancy; immunocompromised[a] state; history of anaphylactic reaction after gelatin ingestion or receipt of neomycin	MMR is the vaccine of choice if recipients are also likely to be susceptible to measles and rubella
Rubella live-virus vaccine	One dose SC; no booster	HCP, both male and female, who lack documentation of receipt of live vaccine on or after their first birthday, or of laboratory evidence of immunity; adults born before 1957 can be considered immune, except women of childbearing age	Pregnancy; immunocompromised[a] state; history of anaphylactic reaction after receipt of neomycin	Women pregnant when vaccinated or who become pregnant within 3 months of vaccination should be counseled on the theoretic risks to the fetus, the risk of rubella vaccine-associated malformations in these women is negligible; MMR is the vaccine of choice if recipients are also likely to be susceptible to measles or mumps
Varicella-zoster live-virus vaccine	Two 0.5 mL doses SC, 4–8 weeks apart if ≥13 years	HCP without reliable history of varicella or laboratory evidence of varicella immunity	Pregnancy, immunocompromised[a] state, history of anaphylactic reaction after receipt of neomycin or gelatin; salicylate use should be avoided for 6 weeks after vaccination	Because 71%–93% of persons without a history of varicella are immune, serologic testing before vaccination may be cost effective

Source: Modified from ACIP Recommendations.

Note: IM, Intramuscularly; SC, subcutaneously.

[a] Persons immunocompromised because of immune deficiencies, HIV infection, leukemia, lymphoma, generalized malignancy, or immunosuppressive therapy with corticosteroids, alkylating drugs, antimetabolites, or radiation.

TABLE 5.2 Summary of Suggested Work Restrictions for HCP Exposed to or Infected with Infectious Diseases of Importance in Health Care Settings, in the Absence of State and Local Regulations

Disease/Problem	Work Restriction	Duration	Category
Conjunctivitis	Restrict from patient contact and contact with the patient's environment	Until discharge ceases	II
Cytomegalovirus infections	No restriction		II
Diarrheal diseases			
Acute stage (diarrhea with other symptoms)	Restrict from patient contact, contact with the patient's environment, or food handling	Until symptoms resolve	IB
Convalescent stage, *Salmonella* spp.	Restrict from care of high-risk patients	Until symptoms resolve; consult with local and state health authorities regarding need for negative stool cultures	IB
Diphtheria	Exclude from duty	Until antimicrobial therapy completed and two cultures obtained ≥24 h apart are negative	IB
Enteroviral infections	Restrict from care of infants, neonates, and immunocompromised patients and their environments	Until symptoms resolve	II
Hepatitis A	Restrict from patient contact, contact with patient's environment, and food handling	Until 7 days after onset of jaundice	IB
Hepatitis B			
Personnel with acute or chronic hepatitis B surface antigemia who do not perform exposure-prone procedures	No restriction[a]; refer to state regulations; standard precautions should always be observed		II
Personnel with acute or chronic hepatitis B e antigenemia who perform exposure-prone procedures	Do not perform exposure-prone invasive procedures until counsel from an expert review panel has been sought; panel should review and recommend procedures the worker can perform, taking into account specific procedure as well as skill and technique of worker; refer to state regulations	Until hepatitis B e antigen is negative	II
Hepatitis C	No recommendation		Unresolved issue
Herpes simplex			
Genital	No restriction		II

(continued)

TABLE 5.2 (continued) Summary of Suggested Work Restrictions for HCP Exposed to or Infected with Infectious Diseases of Importance in Health Care Settings, in the Absence of State and Local Regulations

Disease/Problem	Work Restriction	Duration	Category
Hands (herpetic whitlow)	Restrict from patient contact and contact with the patient's environment	Until lesions heal	IA
Orofacial	Evaluate for need to restrict from care of high-risk patients		II
HIV	Do not perform exposure-prone invasive procedures until counsel from an expert review panel has been sought; panel should review and recommend procedures the worker can perform, taking into account specific procedure as well as skill and technique of the worker; standard precautions should always be observed; refer to state regulations		II
Measles			
Active	Exclude from duty	Until 7 days after the rash appears	IA
Postexposure (susceptible personnel)	Exclude from duty	From 5th day after first exposure through 21st day after last exposure and/or 4 days after rash appears	IB
Meningococcal infections	Exclude from duty	Until 24 h after start of effective therapy	IA
Mumps			
Active	Exclude from duty	Until 9 days after onset of parotitis	IB
Postexposure (susceptible personnel)	Exclude from duty	From 12th day after first exposure through 26th day after last exposure or until 9 days after onset of parotitis	II
Pediculosis	Restrict from patient contact	Until treated and observed to be free of adult and immature lice	IB
Pertussis			
Active	Exclude from duty	From beginning of catarrhal stage through 3rd week after onset of paroxysms or until 5 days after start of effective antimicrobial therapy	IB
Postexposure (asymptomatic personnel)	No restriction, prophylaxis recommended		II
Postexposure (symptomatic personnel)	Exclude from duty	Until 5 days after start of effective antimicrobial therapy	IB
Rubella			
Active	Exclude from duty	Until 5 days after rash appears	IA
Postexposure (susceptible personnel)	Exclude from duty	From 7th day after first exposure through 21st day after last exposure	IB

Scabies			
S. aureus Infection	Restrict from patient contact	Until cleared by medical evaluation	IB
Active, draining skin lesions	Restrict from contact with patients and patient's environment or food handling	Until lesions have resolved	IB
Carrier state	No restriction, unless personnel are epidemiologically linked to transmission of the organism		IB
Streptococcal infection, group A	Restrict from patient care, contact with patient's environment, or food handling	Until 24 h after adequate treatment started	IB
Tuberculosis			
Active disease	Exclude from duty	Until proved noninfectious	IA
PPD converter	No restriction		IA
Varicella			
Active	Exclude from duty	Until all lesions dry and crust	IA
Postexposure (susceptible personnel)	Exclude from duty	From 10th day after first exposure through 21st day (28th day if VZIG given) after last exposure	IA
Zoster			
Localized, in healthy person	Cover lesions; restrict from care of high-risk patients[b]	Until all lesions dry and crust	II
Generalized or localized in immunosuppressed person	Restrict from patient contact	Until all lesions dry and crust	IB
Postexposure (susceptible personnel)	Restrict from patient contact	From 10th day after first exposure through 21st day (28th day if VZIG given) after last exposure or, if varicella occurs, until all lesions dry and crust	IA
Viral respiratory infections, acute febrile	Consider excluding from the care of high-risk patients[c] or contact with their environment during community outbreak of RSV and influenza	Until acute symptoms resolve	IB

Source: Modified from ACIP Recommendations.

[a] Unless epidemiologically linked to transmission of infection.

[b] Those susceptible to varicella and who are at increased risk of complications of varicella, such as neonates and immunocompromised persons of any age.

[c] High-risk patients as defined by the ACIP for complications of influenza.

TABLE 5.3 Pregnant HCP: Pertinent Facts to Guide Management of Occupational Exposures to Infectious Agents

Agent	Potential Effect on Fetus	Rate of Perinatal Transmission	Maternal Screening	Prevention
1. Cytomegalo-virus	Hearing loss; congenital syndrome[a]	15% after primary maternal infection; symptomatic 5%	Antibody provides some but not complete protection against clinical disease; routine screening not recommended	Standard precautions
2. Hepatitis B	Hepatitis; development of chronic infection in infant	HBeAg seropositive 90%; HBeAg negative 0%–25%	HBsAg routine screening recommended	Vaccine and HBIG to infant; standard precautions
3. Hepatitis C	Hepatitis	2%–5%	Anti-HCV; HCV RNA in reference labs; routine screening not recommended	Standard precautions
4. Herpes simplex	Mucocutaneous lesions, sepsis, encephalitis; congenital malformations (rare)	Unlikely from nosocomial exposure; primary 33%–50%, recurrent 4%	Antibody testing not useful; inspection for lesions at delivery	Standard precautions
5. HIV	AIDS by 2–3 year	8%–30%	Antibody by enzyme immunoassay, Western blot	Avoid high-risk behaviors; consider postexposure prophylaxis after high-risk needlestick exposure; intrapartum and postnatal zidovudine for HIV-seropositive mothers and their babies; standard precautions
6. Influenza	Inconsistent	Rare	None	Vaccine (safe during pregnancy); droplet precautions
7. Measles	Prematurity; abortion	Rare	History, antibody	Vaccine;[b] airborne precautions
8. Parvovirus B19	Hydrops, stillbirth	Rare, 3%–9% maximum adverse outcome	IgM and IgG antibody prepregnancy; antibody protective	Droplet precautions
9. Rubella	Congenital syndrome[a]	45%–50% overall; 90% in first 12 week	Antibody	Vaccine;[b] droplet precautions for acute infection; contact precautions for congenital rubella
10. Tuberculosis	Hepatomegaly, pulmonary, CNS	Rare	Skin test	Isoniazid ± ethambutol for disease; airborne precautions
11. Varicella-zoster	Malformations (skin, limb, CNS, eye); chicken pox	Total 25%; congenital syndrome (0%–4%)	Antibody	Vaccine;[b] VZIG within 96 h of exposure if susceptible; airborne and contact precautions

Source: Modified from Siegel, J.D., Risk and exposure to pregnant health-care worker, In Olmstead, R.N., editor, *APIC Infection Control and Applied Epidemiology: Principles and Practices*, Mosby, St Louis, MO 1996, pp. 22-2–22-3 (Table 22.1).

Note: HBeAg, Hepatitis B e antigen; CNS, central nervous system.

[a] Congenital syndrome: varying combinations of jaundice, hepatosplenomegaly, microcephaly, CNS abnormalities, thrombocytopenia, anemia, retinopathy, and skin and bone lesions.

[b] Live-virus vaccines are given routinely before pregnancy.

risks and benefits of postexposure prophylaxis regimens, and (d) the potential consequences of exposures or communicable diseases for family members, patients, or other personnel, both inside and outside the health care facility. One important example of this counseling involves the demographics of HCP. A majority of HCP are nurses who in turn are predominantly female—many of whom fall in the cohort of child-bearing years. Despite perception that pregnancy increases one's risk of exposure to infectious diseases, there is no scientific evidence that supports this perception. There are valid concerns by personnel of adverse effects of certain infections on the developing fetus. Table 5.3 summarizes risks and mitigation of these and are useful during health counseling sessions.

7. *Record keeping, data management, and confidentiality*

CDC guidelines recommend EHS establish and keep an updated record for all personnel and maintain the confidentiality of their records while ensuring that they receive appropriate management for occupational illnesses or exposures. Such records need to encompass other types of personnel such as volunteers, trainees, and contractual personnel. Specific requirements of such records including elements from OSHA are addressed elsewhere in this text.

References

1. Scheckler WA, et al. Requirements for infrastructure and essential activities of infection control and epidemiology in hospitals: A consensus panel report. *AJIC* 26:47–60, 1998.
2. Friedman C, et al. Requirements for infrastructure and essential activities of infection control and epidemiology in out-of-hospital settings: A consensus panel report. *Am J Infect Control* 27:418–430, 1999.
3. American Association of Occupational Health Nurses Inc. (AAOHN) Standards of Occupational and Environmental Health Nursing. Available at: http://www.aaohn.org/practice/standards.cfm
4. American College of Occupational & Environmental Medicine (ACOEM). Scope of occupational and environmental health programs and practice. Available at: http://www.acoem.org/guidelines.aspx?id = 736
5. Stelfox HT, et al. The "to err is human" report and the patient safety literature. *Qual Safety Health Care* 15(3):174–178, 2006.
6. Stone PW, et al. Nurse working conditions and patient safety outcomes. *Med Care* 45(6):571–578, 2007.
7. Warren N, et al. Employee working conditions and healthcare system performance: the Veterans Health Administration experience. *J Occupat Environ Med* 49(4):417–429, 2007.
8. Clarke SP, et al. Hospital work environments, nurse characteristics, and sharps injuries. *Am J Infect Control* 35(5):302–309, 2007.
9. Hooper J and Charney W. Creation of a safety culture: Reducing workplace injuries in a rural hospital setting. *AAOHN J* 53(9):394–398, 2005.
10. Lundstrom T, et al. Organizational and environmental factors that affect worker health and safety and patient outcomes. *Am J Infect Control* 30(2):93–106, 2002.
11. CDC. *Workbook for Designing, Implementing, and Evaluating a Sharps Injury Prevention Program* 2004. Available at: http://www.cdc.gov/sharpssafety/
12. Klevens RM, et al. Estimating health care-associated infections and deaths in U.S. hospitals, 2002. *Public Health Rep* 122:160–166, 2007.
13. Perencevich EN, et al. Raising standards while watching the bottom line: Making a business case for infection control. *Infect Control Hosp Epidemiol* 28:1121–1133, 2007.
14. Graves N, et al. Economics and preventing hospital-acquired infection: broadening the perspective. *Infect Control Hosp Epidemiol* 28:178–184, 2007.
15. Centers for Medicare & Medicaid Services (CMS). Medicare program: Changes to the hospital inpatient prospective payment systems and fiscal year 2008 rates. *Fed Regist* 72(162):47129–48175, 2007.

16. Gawande A. The checklist. If something so simple can transform intensive care, what else can it do? *The New Yorker*, December 10, 2007. Available at: http://www.newyorker.com/reporting/2007/12/10/071210fa_fact_gawande
17. Pronovost P, et al. An intervention to decrease catheter-related bloodstream infections in the ICU. *N Engl J Med* 355:2725–2732, 2006.
18. CDC. Reduction in central line-associated bloodstream infections among patients in intensive care units—Pennsylvania, April 2001–March 2005. *MMWR* 54(40):1013–1016, 2005.
19. IHI—go to their Web site, www.ihi.org
20. CDC. Guidelines for environmental infection control in health-care facilities: Recommendations of CDC and the Healthcare Infection Control Practices Advisory Committee (HICPAC). *MMWR* 52(RR-10):1–48, 2003.
21. CDC. Guidelines for preventing the transmission of mycobacterium tuberculosisin health-care settings. *MMWR* 54(RR17):1–141, 2005.
22. Sepkowitz KA and Eisenberg L. Occupational deaths among healthcare workers. *Emerg Infect Dis* [serial on the Internet]. 2005 Jul [date cited]. Available from http://www.cdc.gov/ncidod/EID/vol11no07/04-1038.htm
23. OSHA. Respiratory protection standard. Available at: http://www.osha.gov/SLTC/respiratoryprotection/standards.html
24. CDC. Prevention and control of meningococcal disease recommendations of the Advisory Committee on Immunization Practices (ACIP). *MMWR* 54(RR07):1–21, 2005.
25. CDC. Preventing tetanus, diphtheria, and pertussis among adults: Use of tetanus toxoid, reduced diphtheria toxoid and acellular pertussis vaccine. *MMWR* 55(RR17):1–33, 2006.
26. Joint Commission. Comprehensive accreditation manual for hospitals, 2007
27. Poland GA, et al. Requiring influenza vaccination for health care workers: Seven truths we must accept. *Vaccine* 23:2251–2255, 2005.
28. CDC. Influenza vaccination of health-care personnel. *MMWR* 55(RR02):1–16, 2006.
29. Siegel JD, Rhinehart E, Jackson M, Chiarello L, and the Healthcare Infection Control Practices Advisory Committee. 2007 Guideline for isolation precautions: Preventing transmission of infectious agents in healthcare settings, June 2007. Available at: http://www.cdc.gov/ncidod/dhqp/pdf/isolation2007.pdf
30. CDC. Prevention of varicella: Recommendations of the Advisory Committee on Immunization Practices (ACIP). *MMWR* 56(RR-4):1–56, 2007.
31. CDC. Immunization of health-care workers: Recommendations of the Advisory Committee on Immunization Practices (ACIP) and the Hospital Infection Control Practices Advisory Committee. *MMWR* 46(RR-18):1–42, 1997.
32. Boylard EA, Tablan OC, Williams WW, et al. Guideline for infection control in health care personnel. *Am J Infect Control* 26:289–354, 1998.

Health Care Construction and Infection Control

Laurence Lee

Health care-related construction costs in the United States were estimated at $23.7 billion in 2005 and are expected to continue to rise in the United States and abroad.[1] Construction activities vary widely in scope and include building new facilities, the renovation of existing facilities, seismic retrofitting, demolishing structures, building envelope repairs, tenant improvements, fiber-optic and cable installation, equipment (e.g., MRI installation) and computer facility upgrades, and mold removal and cleaning.

Health care construction guidelines have been issued as a result of reported outbreaks of hospital acquired (nosocomial) infections in immunosuppressed or otherwise compromised patients who were reported to be related to construction.[2,67] These infections were caused by environmental fungi such as

Aspergillus, Fusarium, Scedosporium, and other fungal species as well as bacteria such as *Legionella* and *Nocardia*.[2–5] Infection control teams are assembled and infection control risk assessments are made to prevent nosocomial infections related to construction.

Industrial hygienists and infection control practitioners frequently work together in assisting health care institutions during construction projects. It is important for each discipline to understand the ecology of opportunistic environmental fungi, patient risk factors, the pathobiology of infection, building dynamics, construction control options, and sampling from project inception to building commissioning to reduce the incidence of opportunistic fungal infection related to construction.

Pathogenic and Opportunistic Environmental Fungal Ecology

Fungi are ubiquitous in their distribution in buildings and outdoors in air, soil, and water.[6–8] There are approximately 100 environmental fungal genera that are capable of causing human infection.[9] There are four endemic, thermally dimorphic species in North America that are pathogenic and approximately 30 genera that are opportunistic.[9,10]

For the purpose of this chapter, pathogenic fungi are defined as those fungi capable of infecting an otherwise healthy individual. In North America, these include *Histoplasma capsulatum, Blastomyces dermatitidis, Cryptococcus neoformans*, and *Coccidiodes immitis*.[11] *H. capsulatum* is endemic to the Ohio and Mississippi River Valleys and typically associated with accumulations of bird and bat excrement.[12] *B. dermatitidis* is prevalent on the eastern seaboard and endemic in the Mississippi and Ohio River Valleys, the western shore of Lake Michigan in soil and wood associated with riverbanks, north into the Canadian provinces in soil[13] and pigeon manure.[14] *C. immitis* is endemic to desert soils of the southwestern United States and California.[15] *C. neoformans* infections occur worldwide and the organism has been isolated rarely in the environment. It has been found in soil and associated with pigeon droppings in North America.[16–18] Infection by *C. neoformans* rarely occurs in a healthy host and its role as a strictly pathogenic fungi has been questioned.[19]

All of the above pathogenic fungi are thermally dimorphic. That is, they have a hyphal (filamentous) form that produces spores (conidia) in the environment and have a yeast phase upon infection and growth in the human body at 37°C.[10,11]

Opportunistic fungi are capable of infecting a host individual that does not have normal defenses. That is, opportunistic fungi are typically harmless to hosts with normally functioning immune systems but have the "opportunity" to infect hosts without normally functioning immune defenses or the fungi are able to bypass the defenses such as the skin (e.g., implantation via surgery or trauma) and mucosal barriers. At-risk, immune compromised hosts generally have an underlying illness (e.g., leukemia, AIDS, tuberculosis, and COPD), are receiving medical treatments (e.g., corticosteroids, chemotherapy, and radiation), are undergoing organ transplantation or surgery, or other condition (e.g., burns) that impairs the ability of the body to defend against infection.[10]

Opportunistic fungi include the pathogenic fungi (molds and yeasts) that have environmental origins (e.g., *Aspergillus* and *Fusarium* species) as well as yeasts that are part of the normal human flora (e.g., *Candida* species). Nosocomial infections from environmental opportunistic fungi that include *Aspergillus, Fusarium, Acremonium, Scedosporium, Rhodotorula*, and *Paecilomyces* species; zygomycetes (i.e., *Mucor, Rhizopus*, and *Absidia* species); and others have been reported worldwide.[20,21]

The *Aspergillus* species will be the focus of this chapter as being representative of environmental fungi because they are distributed worldwide and grow on wet building materials,[7] have small spore size ranges (~2 to 6 microns in diameter),[9] and have high mortality rates ranging from approximately 25%–100% associated with infection depending on the site of infection and patient population.[22–24]

Aspergillus/Species

Aspergillus species are ubiquitous in the environment and are found in soil, air, and water.[7,25,26] Approximately 90% of all opportunistic fungal infections caused by *Aspergillus* species are caused by

Aspergillus fumigatus with *A. flavus*, *A. niger*, *A. terreus*, *A. nidulans*, *A. versicolor*, and *A. ustus* generally responsible for the remainder.[27,67] A newly identified drug-resistant species, *Aspergillus lentulus*, has also emerged.[28]

In hospital and other indoor environments, *Aspergillus* species have been found in air, dust, water, food, potted plants, clothing, building materials (e.g., wood, gypsum wallboard, fireproofing), HVAC systems, and wet or water-damaged building materials.[7,25,26,29–33]

Patient Risk Factors

Typical, healthy individuals inhale *Aspergillus* and other mold spores as a simple consequence of living on earth. Intact skin and mucosal barriers, and normally functioning innate (i.e., macrophage and neutrophils) and acquired (i.e., allergy mechanisms) immune systems prevent colonization and infection in the healthy host.[34,35]

Barrier defense function of the skin can be compromised by surgery, burns, catheterization and intravenous line insertions, and herpes virus lesions and resistance to colonization can be reduced due to treatment with broad-spectrum antibiotic that results in a reduction of normal microbiological flora.[35] Macrophage and neutrophil defenses can be impaired as a result of radiation and chemotherapy, high-dose corticosteroid, immune suppressive, and antirejection drug treatment regimens.[34–39]

Patients with underlying illnesses; or undergoing treatment and procedures affecting immune responses, such as leukemia; cancer; stem cell, bone marrow, and solid organ transplantation; cystic fibrosis; chronic granulomatous disease; chronic obstructive pulmonary disease (COPD); allergy; severe malnutrition; diabetes mellitus; late stage HIV/AIDS; TB and other cavitary illness; and premature infancy are considered to be at-risk for opportunistic fungal infection and colonization from *Aspergillus* and other environmentally acquired opportunistic fungi.[35,40–54]

Pathobiology of Infection

The most common route of exposure to *Aspergillus* spores is inhalation and as a consequence most disease initiates in the lungs or sinuses, however, direct inoculation of the damaged skin in burn, trauma or surgery patients has resulted in cutaneous infection.[22,40,55–57] The spores adhere to the tissue or mucous membrane at the point of contact and swell as part of the germination process.[22,40,55] The host defense directed against the spore is the macrophage and the neutrophil attacks hyphal growth, and tissue invasion occurs in the absence of a robust response.[22,40,55]

Building Dynamics

Construction, by its very nature, is disruptive. Construction can contribute to the infiltration of unfiltered air containing *Aspergillus* spores, disturb and aerosolize spores in dust and microbial reservoirs (e.g., in settle dust accumulations above suspended ceilings and growth from past/current water leaks hidden in wall cavities), and cause mold growth from water (e.g., leaks, pressure washing, and weather). Construction activities can release and distribute mold aerosolized mold spores via normal patterns of airflow in a facility. It is therefore important to recognize the effect that building dynamics can have on the distribution of construction dust in a health care facility.

HVAC System

Purpose

The heating, ventilation, and air conditioning system has several fundamental purposes that affect air movement in buildings:[59,60]

1. Pressurize the building positively with respect to the outside atmosphere to prevent the infiltration of unfiltered air to protect patients and occupants.
2. Provide proscribed internal room pressurization relationships (e.g., airborne isolation rooms are negatively pressurized to keep infectious agents such as TB in the room and procedure rooms are positively pressurized to prevent the infiltration of dust and infectious agents).
3. Filter and remove particulates (outdoor and indoor) from the air stream to protect the patients, occupants, and equipment.
4. Provide fresh outdoor air and sufficient air exchanges to dilute indoor "contaminants" (e.g., carbon dioxide in exhaled breath, body and chemical odors, etc.).
5. Provide for thermal comfort of the occupants.

Air Movement

Air movement in health care and other buildings is affected by overall building pressurization, internal room pressure relationships, and vertical building pathways. The aerosolization and distribution of dust containing mold spores from construction activity is affected by air movement. Dust generation in a discrete area of a hospital has the potential to travel throughout the building.

Overall building pressurization is positive in relation to the outside atmosphere. This means that excess air is pumped into the building in quantities over and above that is vented outside via the HVAC system, laboratory hoods, etc. Because buildings are rigid and cannot expand like a balloon, the excess air will "leak" out of the building via cracks/openings in the window frames and exterior walls and open doors. Loss of building pressurization to neutral or negative can result in the infiltration of unfiltered in via the same cracks/openings and doors.

Internal building pressure relationships are prescribed in the 2006 *AIA Guidelines for Design and Construction of Health Care Facilities*.[58] Potential airborne infectious are kept out of OR, delivery, procedure, clean, sterile, and procedure rooms, for example, by maintaining positive pressure in relationship to adjacent areas. Conversely, potential airborne infectious agents are contained within triage, ER waiting, bronchoscopy, radiology, laboratories, soil linen, ETO sterilizer, toilets, and soiled decontamination rooms, for example, by negative pressure. Air will move along internal pressure gradients in ways that are not always predictable.

Building air movement is also driven by means other than the HVAC system.[59] Air can move vertically based on the simple concept that warm air rises. This is commonly referred to as "the stack effect."[61] Vertical air movement is experienced in large open "shafts," such as stairwells, elevator shafts, and fire escapes; and through floor/ceiling pipe, conduit, telecom, laundry, and pneumatic tube penetrations. The stack effect tends to be more pronounced during the heating season and if the HVAC system is not up to task, the stack effect can result in the infiltration of outdoor air directly into the building as rising air needs to be "replaced."

Elevator shafts and dumbwaiters present a special case in air movement. At rest, air will move in accordance with the stack effect. However, the movement of the car in the shaft functions as a piston. When the cab is in motion, the air ahead of the direction of travel is positively pressurized and is driven out of the shaft via the elevator doors ahead. Simultaneously, the air is negatively pressurized on the following side of the cab and air is drawn by negative pressure into the shaft via the elevator doors behind. The effect of the piston is to "share" air between floors much like inhaling and exhaling.

It is important to recognize that aerosolized dust and mold spores generated during construction can be dispersed throughout a building by the HVAC system and natural processes and lead to unwanted patient exposure.

Legionellosis

Legionella pneumophila is a waterborne bacterium that is most commonly known as the disease agent of Legionnaires' disease, legionellosis, first came to widespread public attention in 1976 at an American

Legion Convention in Philadelphia.[69] *L. pneumophila* is a gram-negative bacterium that is ubiquitous in freshwater and is an intracellular protozoan parasite. The organism can proliferate in domestic and hospital water systems in biofilm and scale, particularly in hot water systems or inactive ("dead legs") sprinkler system or other piping with temperatures that range between 25°C and 42°C.[67,70,71] Hot water systems, evaporative cooling towers, and humidifiers are typically associated with the amplification of the organism and exposure by inhalation to aerosols, by ingestion and aspiration, via cooling towers, showers and drinking water faucets, respiratory therapy equipment, and room humidifiers has been documented.[67]

Legionellosis presents as two distinct clinical syndromes: the first is a mild influenza-like illness that typically resolves by itself and the second is a progressive pneumonia that can involve the cardiac, renal, and gastrointestinal systems.[67,70] Infection is largely opportunistic and patients at-risk from exposure and infection include immune-compromised patients similar to those at-risk for developing aspergillosis and include transplant and oncology patients, patients receiving corticosteroid treatment, surgical and dialysis patients, elderly patients, HIV, and smokers.[67,70] Legionellosis outbreaks have been reported to be associated with construction that involved the disruption of biofilm and scale due to changes in water pressure.[70]

Plumbing-related construction work that involves the potential disruption of biofilm and scale due to pressure changes, flushing, or other physical disturbance has the potential to increase the concentration of *Legionella* bacteria in the system. This may be less of a problem for facilities that routinely heat shock (>60°C), chemically treat (i.e., chlorine dioxide or monachloramine), and/or use copper-silver filters in their water systems to prevent potential outbreaks as a matter of course.[71–76]

To prevent the disruption and spread of biofilm and scale that potentially contains *Legionella* bacteria, it is important to identify the down stream flow and users and protect at-risk patients. Disconnecting and isolating the system for draining, cleaning, and maintenance purposes will prevent downstream disturbances. Heat shocking or chemically treating the system immediately prior to the work can reduce the overall number of viable organisms in the system that could be disturbed, although the effect of *Legionella* reduction lasts for a period of a few days and is transitory.[75,76] Heat shocking combined with the installation of copper-silver point of use filters for at-risk patients and critical services (i.e., food services), or the use of the filters alone can further reduce potential exposure.[72] Further, it has been suggested that the use of copper-silver filters is the superior method for treating water systems.[73,74]

Construction Infection Control Measures

Health care construction guidelines have been published in the United States, Canada, and Europe to prevent construction-related nosocomial infection and there is broad agreement regarding infection control measures.[2–5,58,67]

Barriers

Barriers are commonly constructed from floor to ceiling deck to physically isolate and enclose areas of construction. Barriers may be semipermanent for lengthy, large-scale projects and be constructed of gypsum wallboard on metal studs that is also insulated with fiberglass batting to control sound. Less substantial barriers constructed of fire-rated plastic sheeting and tape supported by telescoping pole extenders or metal-framed walls constructed of melamine panels are typically constructed for shorter, less involved projects. Plastic sheeting is more often used above suspended ceiling because it lends itself better to sealing around conduit, wiring, and ducting than a rigid barrier.

Regardless of the materials used, the purpose of the barriers is to prevent the physical movement of dust outside the construction area and prevent unauthorized entry by curious hospital staff and visitors. Seams between panels and plastic sheets, and the walls, floor, and ceiling are sealed with tape or caulk. Construction area entrances and exits can be double-flapped plastic or solid hinged doors. Hinged doors are commonly self-closing using spring hinges or rubber bungee cords. Vestibules/ante rooms are sometimes used as a staging area for clean clothing and changing space for large-scale, long-term projects.

Negative Pressure

Negative pressure is routinely established inside the construction area in an effort to prevent the movement of construction dust in air currents outside the construction area. HEPA-filtered negative air machines are typically used to depressurize the construction area in relation to the immediately adjacent rooms and hallways. Negative air machines, borrowed from the asbestos and lead abatement industries, range in capacity from 250 to 2000 cfm in the United States. Room is drawn through a prefilter and HEPA filter that removes particles down to a size of 0.3 microns with a 99.97% efficiency and exhausted outdoors through a window, door, or other suitable opening.

The exhaust and removal of air from the construction area requires a supply of "make-up" air to replace the air that is exhausted outdoors. Make up air from the "clean" surrounding areas of the building is drawn by negative pressure into the "dirty" construction area. The direction of air flow under these conditions is from "clean to dirty" which must be maintained throughout the construction project.[2] It is also important to consider matching the capacity of the negative air machines within the surplus supply air provided by the surrounding area to minimize changes or disruptions of established air flow patterns in the building. That is to say that if three 2000 cfm negative air machines were used in the construction area, and the surrounding area's supply air surplus was only 3000 cfm, an additional 3000 cfm would be drawn from beyond the immediate area by the negative air machines that could significantly change the local air flow patterns and perhaps draw unfiltered air into the building that would travel along this pathway and potential exposing a compromised patient traveling along or through this pathway.

Air flow needs to be monitored throughout the project and can be accomplished by using tell-tale flags the point in the direction of air flow, manometers, or smoke testing. It is important when using manometers on sites that have several construction areas to mark the desired pressure differential range on the manometer so that a lay person can determine visually if the work space is compliant. Pressure differentials can range from 0.001″ w.g. to 0.02″ w.g. depending on the preference of the institution. Smoke testing is routinely performed using ventilation smoke tubes that emit stannic oxychloride. The chemical reacts with the air and generates a visible plume of smoke. The smoke is released from the tube using a rubber bulb at the site and the direction of travel is observed. The chemical is an irritant and it is important to stay upwind and away from the plume. The plume can also trigger smoke detectors so it is important to identify their location and use minimal amounts of smoke.

Controlling Dust Movement

Dust can infiltrate and spread throughout a building as a result of moving and staging building materials, tools, and equipment in the building; workers tracking dust on their shoes and clothing; and conveying waste and debris. Designated construction pathways, stairways, and elevators are the preferred methods of separating construction from the rest of the hospital to prevent the transfer of dust and inadvertent exposure to compromised patients as a result of them leaving their rooms for tests of treatment. Tracking out dust outside the construction area is minimized by installing tacky, walk off mats immediately inside and outside the entrance to the construction area. It can be challenging training workers to step on to a mat that leaves a footprint because most people are trained since they were children not to leave footprints. In addition, HEPA-vacuuming coveralls and work clothing prior to exiting the construction area and/or wearing disposable coveralls and booties can also minimized the track out of dust on clothing. This also applies to tools and equipment that leaves the construction area and exits through the hospital. Routine, daily cleaning of the work area and adjacent areas throughout the day and at the end of the shift by mopping and HEPA-vacuuming is used to control dust generation at the source and control the spread of dust outside the work area.

The control of dust from the removal and conveyance of construction-related waste and debris is accomplished by using suitable, closed and covered waste containers when moving through the hospital. Waste containers of sealed, 6 mil or greater, plastic bags need to be wiped clean and be dust-free prior to leaving the construction area. Alternatively, on larger projects, it may be possible to convey the waste directly outdoors through a designated exit or temporary exterior elevator.

Vibration

Vibration is a fact of life in construction. Rotohammers, powder-actuated tools, bead blasting, wall and mechanical demolition, and heavy equipment (e.g., "bobcat" trucks, excavators, drill rigs, etc.) all contribute to vibration, and vibration can disturb and aerosolize dust. It is extremely important to consider vibration and recognize that vibration can affect surrounding areas in all directions and travel significant distances. For example, rotohammering duct hangers on the ceiling of the construction area will affect the floor immediately above and may travel along an adjacent steal beam, pipe, or duct affecting a space even further away from the point of attack. Bead blasting a floor can result in the delamination of fire proofing and concrete on the ceiling, as well as vibrate the suspended ceiling grid and tiles, below; resulting in the aerosolization of dust above the suspended ceiling, and the delamination of ceiling tile. Vibration and noise can also be upsetting to staff and patients as well as adversely effect the use of microscopes in microsurgery.

Because vibration is difficult, if not impossible, to control, coordinating high vibration activities such as rotohammering and bead blasting with all the surrounding occupants is critical. Scheduling high vibration activities off hours may be the only alternative is some cases. It is also important to anticipate and prepare for potential impacts such as dust aerosolization as best you can. For example, in a case where bead blasting a floor resulted in the delamination of concrete dust from the ceiling below and vibration of the ceiling grid and tiles caused the release of dust resulting in skin rashes, the ceiling tiles on the floor below were taped directly to the grid to, in effect, isolate and seal the above ceiling space and construction area from the floor below.

Protecting the Building

The wide-ranging physical impacts of construction have the potential to cause damage to the building and result in the aerosolization of dust and mold spores. It is critical that the building is protected from external, outdoor sources of dust; moldy building materials and dust/mold-contaminated equipment (i.e., HVAC system); and accidental water releases.

The construction and expansion of medical facilities on a given campus or neighboring construction projects can generate dust and odors that in turn can infiltrate an existing hospital building. It is important to close and seal windows, and protect HVAC system fresh air intakes. Windows can be locked, screwed shut, and sealed with caulk or tape to prevent the infiltration of dust. Fresh air intakes can be protected by placing air moving fans (i.e., 10,000 cfm or more) in a manner that disperses the on-coming dust cloud near the intakes, or locating them at the point of dust generation such as perimeter of an excavation or the dump truck loading station in order to direct the dust plume away from the intakes. Constructing and installing an extension to a fresh air intake is another option in the case of roofing work or other activities that take place in close proximity to a fresh air intake where fans are not practical.

Building materials such as gypsum wallboard (interior and exterior products), wood and composite wood products (i.e., desks, shelves, cabinets, etc.), insulating products, their packaging (e.g., cardboard boxes), and other products that contain wood or cellulose have the potential to support mold growth should they become wet. These products can become wet from weather exposure due to inadequate or outdoor storage at the factory, during shipment and delivery, during on-site storage and staging, or during installation. In order to prevent these mishaps, first, it is critical that the contractor and the hospital inspect building materials upon delivery and reject any materials with excess moisture and/or visible mold growth. Second, each must require that the materials be stored in a manner that protects them from the elements. Third, they must continue to inspect and monitor the condition of the materials during and postconstruction to insure that they do not become wet, moldy, or contaminated with dust.

HVAC system components and ducting are especially vulnerable to dust contamination. They need to be stored and staged in clean areas so that they are protected from the elements and dust, however, oftentimes during and postinstallation they can become contaminated by dust from other activities such as during the installation wall insulation and gypsum wallboard, and painting. Duct interiors, terminal reheat boxes, and VAV boxes can easily become contaminated with dust if they are not protected from other dust-generating construction activities that take place in close proximity. Fortunately, preventing dust contamination is simple. Many HVAC system components and ducting can be order with protective covering. Plastic bags and shrink wrap are used to prevent contamination when they leave the fabrication shop or distributor. The coverings are removed on a piece-by-piece basis as they are installed and may be wiped clean with a cloth and isopropyl alcohol or other suitable cleaning agent. At the end of the shift, any open ducts or components are covered and sealed with plastic.

Accidental water releases are a fact of life on construction projects. Sprinkler head breaks, freeze plug failures, pipe breaks, pressure washing, charging and testing drains, and plumbing systems, for example, can all release significant amounts of water in a facility. That water release can damage materials and furnishings as well as result in mold and microbial growth. All water releases may not be preventable but having a water release plan can be used to help respond to a release in a timely and effective manner.

Water release plans provide actions for prevention and minimization of water releases and contain steps for responding to a release. The most critical element of a water release plan is ensuring that all contractors, vendors, and in-house staff working on water systems determine exactly where the emergency shut off is located and the appropriate procedures to turn it off. Simply stated, the longer it takes to shut the water system off, the more water will be released and the greater the extent of the water damage will be.

The second step to take once the release is stopped is to determine the physical extent of the release. Visual observations of the immediate spill footprint are made, inspections of adjacent work areas and floor penetrations leading to floors below follow, and moisture meters are used identify "hidden" moisture in gypsum wallboard and wall cavities. Destructive means are often used to inspect wall cavity interiors for moisture in the metal sill plate or held in wall insulation. Moisture meters may not be able to "read" through ceramic tile, melamine panels, and multiple sheets of gypsum wallboard and additional destructive means will need to be used to access the wall or ceiling cavities for inspection.

Many institutions have adopted the criteria in the U.S. EPA Mold Remediation Guidelines for Schools and Commercial Buildings that recommends removing and discarding ceiling tile and insulation that has been wet for 24–48 h and removing and discarding gypsum wallboard wet for more than 48 h to prevent mold growth.[61] These guidelines may be overly conservative but they do emphasize the need to dry the wet building materials quickly.[62] Therefore, it is important for the water response plan to have already identified a qualified and experienced drying contractor or in-house (general contractor or hospital) who is ready to respond immediately. It is important that the drying contractor already be familiar with construction and infection control requirements and understand that they too must comply. It is not uncommon for a general contractor to engage in drying without the benefit of training or experience in structural drying in an effort to reduce cost. It might be rationalized as, "Drying, how hard can it be?" Attempts to dry a structure by untrained and inexperienced personnel can result in inadequate drying because the equipment was removed too soon and wet materials were not identified, fungal blooms because heating was used as a drying method and the percent relative humidity in the room approached 100% for several days, wall insulation remained wet because the wall cavity was not opened for air circulation, etc. There is an art and science to structural drying and it is best left to trained, experienced, and competent professionals. In the unfortunate event that building materials have not been dried quickly, removal of wet building materials in conformance with construction and infection control measures is the final option.

Infection Control Commissioning

Infection control commissioning is performed in varying degrees depending on the scope of construction and its potential impact upon patient safety. There are no specific guidelines published for infection control commissioning. The purpose of infection control commissioning is to ensure that problems such as moldy building materials, for example, are not "built into" the building, that building systems are functioning properly from an infection control standpoint, and that dust and fungal spore reservoirs are cleaned and removed before the building is approved for occupancy by staff and patients.

Infection control commissioning as described here has two distinct phases. The first phase consists of monitoring the contractor's infection control work practices and construction techniques in an on-going basis throughout construction. Infection control commissioning is thoroughly integrated in the construction process. The contractor is trained about construction and infection control and work practices controls and techniques to mitigate the spread of dust, prevent water events and mold growth occurring during construction, and routinely monitored by the hospital on a daily basis in order to identify problems and have them corrected immediately so that these problems are not "built into" the building. Checklists for compliance issues specific to the project, such as negative pressure, tacky mat usage and change out, water leaks, barrier integrity, building material protection, housekeeping, and overall cleanliness are developed and followed by the monitor. Environmental monitoring may be performed at various intervals throughout construction using laser particle counters and culturable fungal air sampling for opportunistic environmental fungi (e.g., *Aspergillus* species, *Fusarium* species, etc.) to identify dust and fungal spore releases and infiltration of unfiltered air associated with construction (see "Environmental monitoring," below).

The second phase is conducted at the end of construction following the completion of the punch list, HVAC system balancing and commissioning, and terminal cleaning by environmental services. It consists of a visual inspection to confirm that the work area is free of visible dust accumulations; particle counting to confirm cleanliness and HVAC system air filter efficiencies; a qualitative assessment of intrabuilding pressure relationships to confirm that they are appropriate (e.g., negative pressure rooms are negatively pressurized to the hallway); and culturable fungal air sampling for opportunistic environmental fungi to confirm cleaning and HVAC system operation.

Visual inspection for dust consists of observations above and below suspended ceilings. Contractors are typically required to leave the work area "broom clean" and above ceiling areas may reflect that. A terminally cleaned room or area is typically dust-free because every surface, horizontal and vertical, has been wiped clean. Wiping a finger (i.e., the "white glove test") along surfaces where dust typically accumulates such as the top of a door or door frame, the top of the floor cove base, drawer edges, and the like is conducted to confirm visual observations. The work area is required to be recleaned and reinspected if it fails inspection.

Visual observations using chemical smoke tubes are made to qualitatively confirm that appropriate intrabuilding pressure relationships have been established. The relationship is determined in advance per the AIA Guidelines as either positive or negative.[58] A puff of smoke is released from the tube at the entrance to the room. If the room is supposed to be positive, the smoke should be directed away and out of the room. If the room is negative the smoke should be directed into the room. This simple and direct testing method is extremely powerful when bringing improper room pressure relationships to the attention of the commissioning HVAC system engineer. Improper room pressure relationships need to be corrected and the infection control commissioning procedures repeated.

Particle counting consists of using a laser particle counter to take readings outdoors, in the supply air stream, and in the work area/room for the purpose of comparing between indoors and outdoors. This comparison is important in determining the actual reduction in indoor particle concentrations as a result of the filtration provided by the HVAC system. Both laser and condensate nuclei particle counters are appropriate for the task, however, the results produced by each differ because of the particle sizes they count and should not be used together. That is to say that either a laser particle counter or a condensate nuclei counter should be used exclusive of the other.

Most commercially available laser particle counters can provide cumulative particle counts down to a size of 0.3 microns (i.e., 0.3 to >10.0 microns range) and condensate nuclei providing cumulative particle counts down to a size of 0.1 microns. Cumulative particle counts above 0.3 microns for laser counters and above 0.1 microns for condensate nuclei counters are preferred because the smaller particles are typical of air pollutant sources and less likely to be generated by a process indoors. Outdoor measurements are made, followed by indoor measurements (i.e., supply air diffuser or 3′ to 5′ above the floor in a given room), and the percent reduction in indoor particle concentration is calculated:

[(Outdoor concentration – indoor concentration)/outdoor concentration] × 100 = % reduction

The actual reduction is compared to the estimated reduction based on filter efficiency. For example, a reduction of 85% is expected for an 85% efficient filter. The filters need to be inspected if they do not reach the desired efficiency. The filters may not be seated or installed properly, the filter gasket may not have been installed, filters may be missing, terminal cleaning may not be satisfactory, or there may be a reservoir of dust in the HVAC system, or in or adjacent the work area. If the filter efficiency is not reached the system needs to be checked, any deficiencies found corrected, the work area/room recleaned, reinspected, and retested.

Culturable air sampling for thermophilic opportunistic fungal pathogens is performed as a final check on cleanliness and building system operation after the successful conclusion of the visual inspection, particle counting, and the pressure relationship assessment. Culturable fungal air samples are preferred over nonculturable methods such as PCR because only pathogenic species, such as *Aspergillus fumigatus* for example, are capable of growing at 35°C–37°C.[55] For this reason, methods such as spore trapping, PCR, and culturable air sampling for mesophilic (i.e., incubated at 25°C) fungi are not appropriate. These methods can identify potentially opportunistic *Aspergillus* species, however, none of these methods are capable of determining if the species identified can grow at body temperature (35°C–37°C) and is therefore opportunistic.

Culturable air samples are typically collected using single-stage jet impactors (Andersen, Aerotech, and SAS) and slit to agar impactors (Mattson-Garvin and Casella).[63–65] High volume air sample volumes (~400 to 1400 L) are preferred in order to obtain low reporting limits (<2 CFU/m^3 to <1 CFU/m^3) in highly filtered and maintained health care environments. Several types of culture media have been used for *Aspergillus* investigations and include malt extract agar (MEA), inhibitory mold agar (MEA with 0.1% chloramphenicol), Sabouraud dextrose (with 0.1% chloramphenicol), and Czapek dox agar and the cultures are incubated for seven days at 37°C.[62,64,66]

There are no generally accepted guidelines regarding airborne opportunistic fungal concentrations for the purpose of infection control commissioning. The author proposes the following criteria for infection control commissioning purposes for a 484.5 L sample based on his experience:

0–2 CFU/m^3 *Aspergillus* species	Acceptable
>2–10 CFU/m^3 *Aspergillus* species	Reclean and retest
>10 CFU/m^3 *Aspergillus* species	Investigate potential sources, reclean and retest

The infection control commissioning process can be effective by ensuring that mold and water problems are not "built-in" during construction, confirm that critical building systems are working properly, and verify the effectiveness of cleaning prior to occupancy.

Environmental Monitoring

Environmental monitoring (particle counting and culturable air sampling) may be routinely conducted as a check on the efficacy of the controls used to control dust during construction. Measurements are made outside the construction area in neighboring, and typically occupied, spaces. Particle counting and culturable air sampling techniques are described above. These two techniques are most useful when baseline measurements are made prior to construction for comparison during construction, however determining baseline ranges is difficult. "Baseline" particle counts and airborne fungal concentrations fluctuate based on normal human activity (e.g., foot traffic, occupancy and use, and activity levels). It is expected that particle counts and air sample results will be markedly different at 9 AM and 9 PM simply due to the level of activity. The relatively short sample times associated with particle counting and fungal air sampling make them susceptible to bias due to the effect of changing local activity levels. This can be minimized in particle counting by programming the counter to take repeated samples over the course of a day or days in a single location; however, there is less flexibility with air sampling. Determining baseline ranges is no small endeavor and may involve the collection of hundreds or thousands of air samples over the course of 10 years.[65]

The author proposes a filter method air sampling method for culturable opportunistic fungi in order to address the limitations of conventional air sampling. Filter samples can be collected on a 24 h basis using 25 mm filter cassettes fitted with a 0.45 μm pore size PVC filter (i.e., TEM asbestos filter cassette). Air is drawn through the cassette and filter using a personal air sampling pump (Gillian GilAir 5) at flow rate of approximately 2 L/min. Each sampling pump requires an electrical outlet in order for the pump to operate on a 24 h basis and is placed on a shelf or cart to elevate the cassette approximately 3′ to 5′ above the floor. Approximately 10% unopened field blank samples are collected and submitted with the cassettes as a control and check on filter sterility.

The filter cassettes are delivered to the mycology laboratory for analysis. The filters are removed, and placed face down directly on a culture plate filled with malt extract agar (+0.01% chloramphenicol) and incubated at 37°C for a period of 7 days. The interior of the cassette nozzle is wiped with a sterile swab, and placed in a vial containing 1% TWEEN culture media solution. The solution (innoculum) is inoculated on MEA plates (as above) and incubated at 37°C for a period of 7 days. This method was compared to side by side 15 min (484.5 L) impactor samples in a highly filtered protected environment (i.e., 30% + 90% + HEPA point of use filters in series) and a nonprotected unit (i.e., 30% + 90% filters in series) and the filter method was found to have an increased sensitivity, lower reporting limits, and lower airborne fungal concentrations as compared to the impactor samples.

Particle counting and air sampling conducted during on a routine basis in areas adjacent construction can be extremely helpful in identifying dust releases provided accurate baseline ranges can be determined.

ICRA and the ICRA Panel

An infection control risk assessment (ICRA) is an evaluation of potential risk of infection to patients associated with construction.[58] The guidelines charge the health care organization and/or owner with convening a panel with expertise in infection control, direct patient care, risk management, facility design, construction and construction phasing, ventilation, safety, and epidemiology.

The ICRA panel is intended to be involved with all phases of construction including planning, design, and construction. The panel provides recommendations regarding design (i.e., the building features: isolation and protective environments, special ventilation and filtration requirements, water system, and finishes and furnishing; and the collection of construction, equipment, and mechanical and electrical drawings; equipment manuals; life safety plan; design data for future use); construction impacts (i.e., specific risk mitigation recommendations: disruption of essential services, control measures, patient safety, known hazards, etc. for indoor and outdoor construction); phasing; monitoring, evaluating, and updating/revising the infection control risk assessment recommendations as appropriate for construction and renovation projects; and commissioning (i.e., specifying and verifying HVAC system balance and filtration).

The primary focus of an ICRA panels is the anticipation, recognition, evaluation, and control of construction-related risks. The American Society for Healthcare Engineering (ASHE) has published an Infection Control Matrix to be used to assist the ICRA Panel with making infection control recommendations. The matrix assigns increasing levels of concern to different types of construction projects based on the invasiveness, scale and duration, and dust generating potential. It also assigns increasing levels of risk to potentially affected patient risk groups that range from lowest for offices to highest risk such as immune compromised patients. The construction type and affected patient risk groups are cross referenced in order to determine the appropriate level and type of precautions and engineering controls (Classes I through IV) necessary to protect patients. Class I represents the least level of precautions (e.g., dust minimization and cleaning) and Class IV represents the greatest level of precautions (e.g., negative pressure enclosures, other work practice and engineering controls, and terminal cleaning and inspection by infection control). The matrix is free from the ASHE Web site www.ashe.org and many institutions have modified the matrix to fit their individual needs and have also made it available on the Internet.

The ICRA process can be extremely effective in reducing infection. The University of Washington Medical Center (Seattle, WA) was engaged in a total renovation of four floors and the construction of a rooftop mechanical room penthouse of a high rise medical tower. The integration of the ICRA process into the quality management program resulted in the reduction of aspergillosis rates throughout the institution.[68]

References

1. Cauchon D and Appleby J. Hospital building booms in 'burbs,' *USA Today*, January 3, 2006.
2. Bartley JM. APIC state-of-the-art report: The role of infection control during construction in health care facilities. *Am J Infect Control* 28:156–169, 2000.
3. CDC. Guidelines for environmental control in health-care facilities. *MMWR* 52(RR-10), 2003, http://www.cdc.gov/mmwr/preview/mmwrhtml/rr5210a1.htm
4. National Disease Surveillance Centre. *National Guidelines for the Prevention of Nosocomial Invasive Aspergillosis during Construction/Renovation Activities*. Dublin, 2002.
5. Construction-related nosocomial infections for patients in health care facilities: Decreasing the risk of *Aspergillus*, *Legionella* and other infections. *Can Commun Dis Rep* 27(S2):1–55, 2001.
6. Carlile MJ and Watkinson SC. *The Fungi*. Academic Press Limited, San Diego, CA, 1994.
7. Levetin E. Fungi. In: HA Burge (ed.), *Bioaerosols*. CRC Press, Inc. Ann Arbor, MI, 1995.
8. Gregory PH. *The Microbiology of the Atmosphere*, 2nd edn. John Wiley & Sons, New York, NY, 1973.
9. Sutton DA, Fothergill AW, and Rinaldi MG. *Guide to Clinically Significant Fungi*. Williams and Wilkins, Baltimore, MD, 1998.
10. Fisher F and Cook N. *Fundamentals of Diagnostic Mycology*. W.B. Saunders & Company. Philadelphia, PA, 1998.
11. Sarosi GA and Davies SF. *Fungal Diseases of the Lung*. Lippincott Williams & Wilkins. Philadelphia, PA, 2000.
12. Wheat, JL. Histoplasmosis. In: GA Sarosi and SF Davies (eds.), *Fungal Diseases of the Lung*. Lippincott Williams & Wilkins, Philadelphia, PA, 2000, pp. 31–46.
13. Davies SF and Sarosi GA. Blastomycosis. In: GA Sarosi and SF Davies (eds.), *Fungal Diseases of the Lung*. Lippincott Williams & Wilkins, Philadelphia, PA, 2000, pp. 47–57.
14. Sarosi GA and Serstock DS. Isolation of *Blastomyces Dermatitidis* from pigeon manure. *Am Rev Respir Dis* 114(6):1179–1183, 1976.
15. Ampel NM. Coccidioidomycosis. In: GA Sarosi and SF Davies (eds.), *Fungal Diseases of the Lung*. Lippincott Williams & Wilkins, Philadelphia, PA, 2000.
16. Emmons C. Isolation of *Cryptococcus neoformans* from soil. *J Bacteriol* 62(6):685–690, 1951.
17. Emmons C. Saprophytic sources of *Cryptococcus neoformans* associated with pigeon droppings (*Columbia livia*). *Am J Hyg* 62:227–232, 1955.
18. Ruiz A, Fromtling RA, and Bulmer GS. Distribution of *Cryptococcus neoformans* in a natural site. *Infect Immun* 560–563, 1981.
19. Mitchell TG and Perfect JR. Cryptococcus in the era of AIDS—100 years after the discovery of *Cryptococcus neoformans*. *Clin Microbiol Rev* 515–548, 1995.
20. Fridkin SK and Jarvis WR. Epidemiology of nosocomial fungal infections. *Clin Microbiol Rev* 499–511, 1996.
21. Pfaller MA and Diekema DJ. Rare and emerging opportunistic fungal pathogens: Concern for resistance beyond *Candida albicans* and *Aspergillus fumigatus*. *J Clin Microbiol* 4419–4431, 2004.
22. Latge JP. *Aspergillus fumigatus* and Aspergillosis. *Clin Microbiol Rev* 310–350, 1999.

23. Lin SJ, Schranz J, and Teutsch SM. Aspergillosis case fatality rate: Systematic review of the literature. *Clin Infect Dis* 32:358–366, 2001.
24. Singh N and Paterson DL. *Aspergillus* infections in transplant recipients. *Clin Microbiol Rev* 44–69, 2005.
25. Labbe R and Garcia S (eds.). *Guide to Foodborne Pathogens*. John Wiley & Sons, New York, 2001.
26. Annaisse E, et al. Pathogenic molds (including Aspergillus species) in hospital water distribution systems: A 3-year prospective study and clinical implications for patients with hematologic malignancies. *Blood* 101(7):2542–2546, 2003.
27. Cohen J. Clinical manifestation and management of aspergillosis in the compromised patient. In: DW Warnock and MD Richardson (eds.), *Fungal Infection in the Compromised Patient*, 2nd edn. John Wiley & Sons, New York, 1991, pp. 85–116.
28. Balajee SA, et al. *Aspergillus lentulus* sp., a new sibling species of *A. fumigatus*. *Eukaryot Cell* 4: 625–632, 2005.
29. Loudon KW, et al. Kitchens as a source of *Aspergillus niger* infections. *J Hosp Infect* 32(3):191–198, March 1996.
30. Levetin E, et al. Effectiveness of germicidal UV radiation for reducing fungal contamination within air-handling units. *Appl Environ Microbiol* 3712–3715, 2001.
31. Lass-Florl C. *Aspergillus terreus* infections in haematological malignancies: Molecular epidemiology suggests association with in-hospital plants. *J Hosp Infect* 46(1):31–35, 2000.
32. Lacroix J, et al. Fungal contamination of hospital healthcare workers' overalls. *J Hosp Infect* 66(1): 88–90, 2007.
33. Carreras E. Preventing exposure to moulds. *Clin Microbiol Infect* 12(7):77–83, 2006.
34. Barnes PD and Marr KA. Aspergillosis: Spectrum of disease, diagnosis, and treatment. *Infect Dis Clin N Am* 20:545–561, 2006.
35. Bowden RA, et al. (eds.). Fungal infections in solid organ transplant recipients. In: *Transplant Infections*. Lipponcott-Raven, Philadelphia, PA, 1998.
36. Ogawa Y. A case of aspergillosis following radiation therapy. *Auris Nasus Larynx* 29(1):73–76, 2002.
37. Segal BH and Walsh TJ. Current approaches to diagnosis and treatment of invasive aspergillosis. *Am J Resp Crit Care Med* 73:703–717, 2006.
38. Kistemann T, et al. Role of increased environmental Aspergillus exposure for patients with chronic obstructive pulmonary disease (COPD) treated with corticosteroids in an intensive care unit. *Intl J Hyg Environ Health* 204(5–6):347–351, 2002.
39. Ader F, et al. Invasive pulmonary aspergillosis in chronic obstructive pulmonary disease: An emerging fungal pathogen. *Clin Microbiol Infect* 11(6):427–429, 2005.
40. Barnes PD and Marr KA. Aspergillosis: Pathogenesis, clinical manifestations and therapy. *Infect Dis Clin North Am* 16(4):875–894, 2002.
41. Morgan J, et al. Incidence of invasive aspergillosis in hematopoietic stem cell and solid organ transplantation: Interim results of a prospective multicenter surveillance program. *Med Mycol* 43(3):49–58, 2005.
42. Denning DW. Aspergillosis in "Nonimmunocompromised" critically ill patients. *Am J Resp Crit Care Med* 570:580–581, 2004.
43. Nunley DR, et al. Pulmonary aspergillosis in cystic fibrosis lung transplant recipients. *Chest* 114:1321–1329, 1998.
44. Geller DE, et al. Allergic bronchopulmonary aspergillosis in cystic fibrosis. *Chest* 116:639–646, 1998.
45. Meerssemann W, et al. Invasive aspergillosis in critically ill patients without malignancy. *Am J Resp Crit Care Med* 170:621–625, 2004.
46. Perfect JR, et al. The impact of culture isolation of Aspergillus species: A hospital-based survey of aspergillosis. *Clin Infect Dis* 33:1824–1833, 2001.
47. Rees JR, et al. The epidemiological features of invasive mycotic infections in the San Francisco Bay area, 1992–1993: Results of population-based laboratory active surveillance. *Clin Infect Dis* 27:1138–1147, 1998.
48. Vandewoude KH, et al. Aspergillosis in the ICU—The new 21st century proglem? *Med Mycol* 44:S71–S76, 2006.

49. Karas A, et al. Pulmonary aspergillosis: An analysis of 41 patients. *Ann Thorac Surg* 22:1–7, 1976.
50. Groll AH, et al. Invasive pulmonary aspergillosis in a critically ill neonate: A case report and review of invasive aspergillosis during the first 3 months of life. *Clin Infect Dis* 27(3):437–452, 1998.
51. Woodruff CA and Hebert AA.Neonatal primary cutaneous aspergillosis: Case report and review of the literature. *Pediatr Dermatol* 19(5):439–444, 2002.
52. Binder RE, et al. Chronic necrotizing pulmonary aspergillosis: A discrete clinical entity. *Medicine (Baltimore)* 61(2):109–124, 1982.
53. Cunha B. Infections in nonleukopenic compromised hosts (diabetes mellitus, SLE, steroids and asplenia) in critical care. *Crit Care Clin* 14(2):263–282, 1998.
54. Hartemink KJ, et al. Immunoparalysis as a cause of invasive aspergillosis? *Intens Care Med* 29(11):2068–2071, 2003.
55. Denning DW. Invasive aspergillosis. *Clin Infect Dis* (26):781–785, 1998.
56. Chakrabarti A, et al. Primary cutaneous aspergillosis: Our experience in 10 years. *J Infect* 37(1): 24–27, 1998.
57. Bretagne S. Fatal primary cutaneous aspergillosis in a bone marrow transplant recipient: Nosocomial acquisition in a laminar-air flow room. *J Infect* 36(3):235–239, 1997.
58. AIA, Facility Guidelines Institute. *Guidelines for Design and Construction of Health Care Facilities.* American Institute of Architects, Washington, DC, 2006.
59. Hays SM, et al. *Indoor Air Quality: Strategies and Solutions.* McGraw-Hill, New York, 1995.
60. Bearg D. *Indoor Air Quality and HVAC Systems.* CRC Lewis Publishers, Boca Raton, FL, 1993.
61. Bearg D. HVAC systems. In *Indoor Air Quality Handbook.* McGraw-Hill, New York, 2001.
62. U.S. Environmental Protection Agency. *Mold Remediation Guidelines for Schools and Commercial Buildings.* U.S. EPA, Washington DC, 2001.
63. Willeke K and Macher JM. Air sampling. In *Bioaerosols: Assessment and Control.* ACGIH, Cincinnati, OH, 1998.
64. Falvey D and Streifel A. Ten-year air sample analysis of *Aspergillus* prevalence in a university hospital. *J Hosp Infect* 67(1):35–41, 2007.
65. Streifel AJ, et al. *Aspergillus fumigatus* and other thermotolerant fungi generated by hospital building demolition. *Appl Environ Microbiol* 375–378, 1983.
66. Burge HA and Otten JA. Fungi. In *Bioaerosols: Assessment and Control.* ACGIH, Cincinnati, OH, 1998.
67. U.S. Department of Health and Human Services. *Guidelines for Environmental Infection Control in Health Care Facilities.* CDC, Atlanta, GA, 2003.
68. Larson EB. Measuring, monitoring, and reducing medical harm from a systems perspective: A medical director's personal reflections. *Acad Med* 77:993–1000, 2002.
69. Fields BS, et al. Legionella and legionnaires' disease: 25 years of investigation. *Clin Microbiol Rev* 506–526, 2002.
70. Stout JE, et al. Ecology of *Legionella pnuemophila* within water distribution systems. *Appl Environ Microbiol* 49(1):221–228, 1985.
71. Meintzner S, et al. Efficacy of thermal treatment and copper-silver ionization for controlling *Legionella pneumophila* in high-volume hot water plumbing systems. *Am J Infect Control* 25(6):452–457, 1997.
72. Stout JE, et al. Legionnaires' disease in a newly constructed long-term care facility. *J Am Geriatr Soc* 48(12):1589–1592, 2000.
73. Stout JE and Yu VL. Experiences of the first 16 hospitals using copper-silver ionization for Legionella control: Implications for the evaluation of other disinfection modalities. *Infect Control Hosp Epidemiol* 24(8):563–568, 2003.
74. Modol J, et al. Hospital-acquired legionnaires disease in a university hospital: Impact of the copper-silver ionization system. *Clin Infect Dis* 44(2):263–265, 2007.
75. Ragull S, et al. Superheat and flush effect on the control of hospital-acquired Legionella infection. *Med Clin (Barc)* 127(6):211–213, 2006.
76. Srinivasan A, et al. A 17-month evaluation of a chlorine dioxide water treatment system to control Legionella species in a hospital water supply. *Infect Control Hosp Epidemiol* 24(8):575–579, 2003.

Advances in Patient Safety: Contamination Studies Point to a New Strategy in the War against Hospital Infections

*Jules Millogo and Kathryn M. Duesman**

The luer tip of a standard syringe is exposed to the risk of transmission of pathogens that can lead to hospital-acquired infections. To address this problem, Retractable Technologies, Inc. designed the Patient Safe™ syringe. Two studies were conducted by an independent laboratory to determine whether the Patient Safe syringe reduces the risk of contamination, compared to the standard syringe.

Background

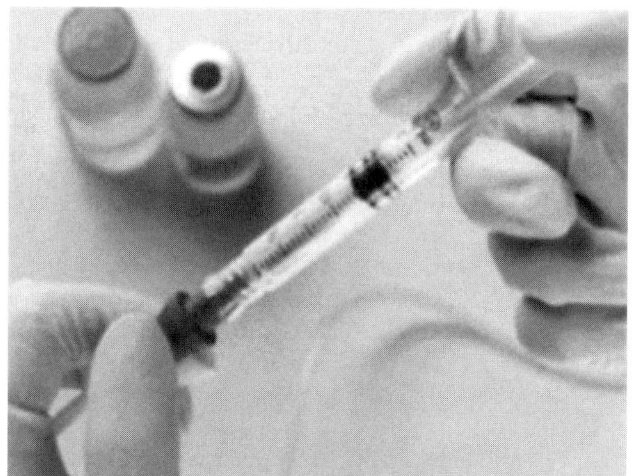

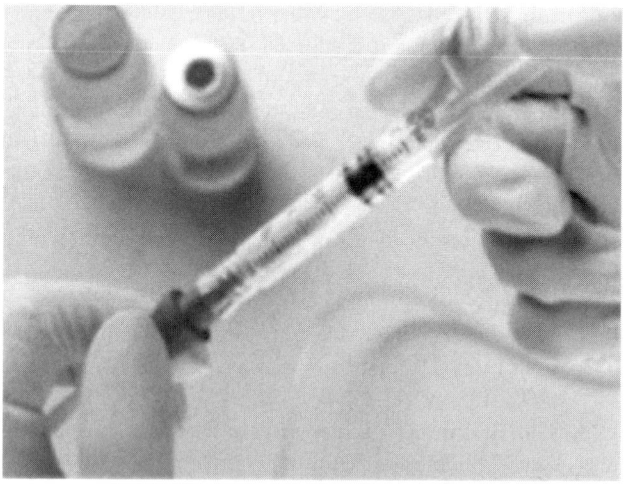

Each year there are about 1.7 million health care-associated infections in U.S. hospitals, and they result in approximately 99,000 deaths, according to the CDC.[1] MRSA infections and other antibiotic-resistant "super bugs" increasingly over the past few years have become a major concern in hospitals throughout the world. MRSA "is a major public health problem primarily related to health care."[2] About 14%, or 238,000 of the 1.7 million annual infections in U.S. hospitals, are bloodstream infections.[3] Worse yet, these numbers are growing at an alarming rate.

Besides the high toll in terms of lives lost and illnesses suffered, there is also the economic cost. The monetary cost in hospitals in the United States amounts to almost $5 billion annually. Beginning on October 1, 2008, CMS (Medicare) will no longer reimburse hospitals for eight hospital-acquired conditions, including vascular catheter-associated infections.[4] "Private insurers usually base their payments on Medicare's fee schedule."[5]

Hospitals have responded to the problem with a variety of approaches including hand washing, training, etc. However, until now there have been no

* Dr. Jules Millogo is medical director for Retractable Technologies, Inc. He holds a master of science degree in epidemiology of communicable diseases from the University College of London and an MD from the University of Ouagadougou, Burkina Faso.

Kathryn M. Duesman, BSN, RN, is executive director, global health, for Retractable Technologies, Inc. To contact the authors, email Ms. Duesman at RTIclinical@vanishpoint.com

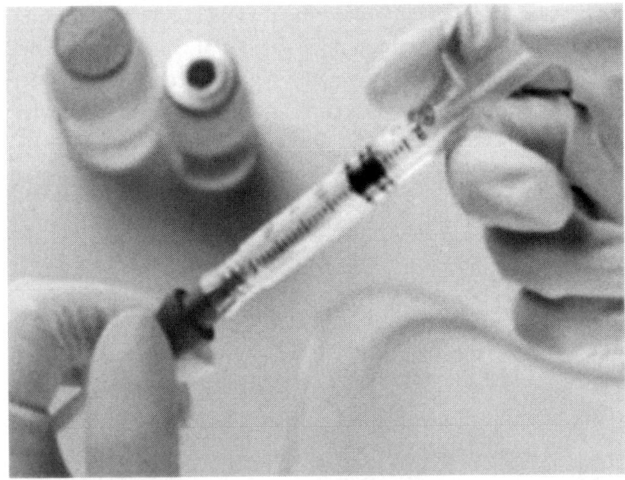

design or engineering controls available to prevent the risk of contact contamination associated with the luer end of a syringe. The luer tip of a standard syringe is exposed to the risk of contamination by pathogens that could be transmitted to patients via intravenous (IV) access valves and IV fluids, both of which can increase patients' risk for bloodstream infections.

Standard syringes with luer lock tips have a threaded collar surrounding the conical luer tip. The luer tip of standard syringes extends beyond the edge of the threaded collar. The exposure of the luer tip allows for contact contamination of the tip during medication preparation, transporting medication to patient areas, and handling prior to injection or infusion. Due to the proliferation of needleless IV systems in the United States, many intravenous injections are performed using luer tip syringes in conjunction with luer-activated valve (LAV) access devices.

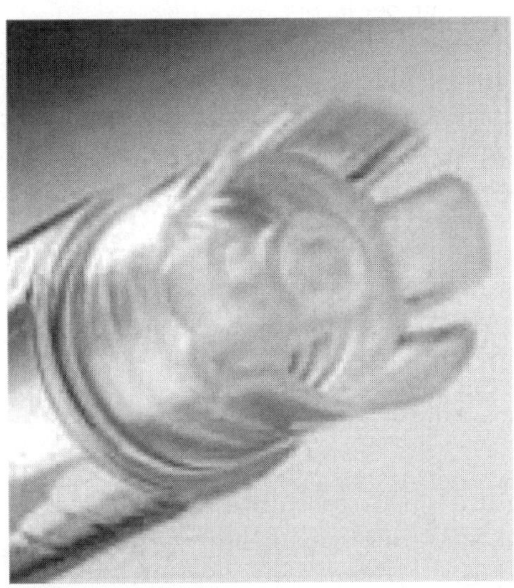

In an effort to avert contamination during these IV injections, it is common practice to swab the surface of LAV devices prior to access, and to swab the vial stoppers prior to medication preparation. However, the luer tip of a standard syringe is exposed to the risk of contact contamination by HCP, as well as through contact with multiple surfaces (e.g., countertops in medication and patient rooms, over-the-bed tables in patient rooms, bedding in patient rooms, and HCP's clothing). Since the luer tip directly interfaces with the access port, contamination of the luer tip has the potential for contaminating the access port and infusate, which increases the patient's risk of nosocomial infections, such as bloodstream infections.

Patient Safe syringes (a newly introduced product from Retractable Technologies, Inc.) are designed to connect to luer connection devices, such as standard hypodermic needles or vial access devices for the aspiration of fluids and medication, as well as LAV devices for the injection of fluids and medication.

The threaded collar surrounds and extends beyond the luer tip, protecting it from contact contamination. A unique "petal" design allows the luer guard to accommodate a variety of differently shaped interfacing devices, such as hypodermic needles with manually activated safety features. This extended petal design also serves as a guide to aid the user in aligning the syringe with the interfacing device, preventing inadvertent contamination.

Studies were conducted at an independent commercial laboratory to determine if Patient Safe syringes or currently marketed standard luer lock syringes prevent contact contamination of the syringe luer tip and subsequent contamination of liquids administered to patients either by direct injection or via an IV access system.

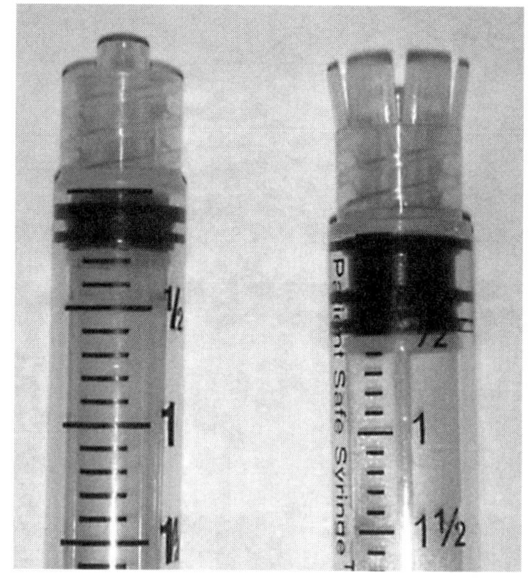

Two studies were conducted, with the following objectives:

1. To test the Patient Safe syringe and the standard luer lock syringe to quantify and compare the risk of syringe luer tip contamination transfer to the liquid drawn into the syringe.
2. To test the Patient Safe syringe and the standard luer lock syringe in order to quantify and compare the risk of syringe luer tip contamination transfer to LAV devices and infusate:
   a. The Patient Safe syringe's unique luer guard protects the luer tip against contact contamination, while the petal design allows the collar to expand, if needed.
   b. Patient Safe syringes are compatible with luer-activated valve access devices, such as the ICU Medical CLAVE® connector.
   c. The luer tip of a standard luer lock syringe is exposed, while the luer-locking collar of a Patient Safe syringe extends past the luer tip.
   d. Standard syringes transfer contaminants to luer-activated valves and intravenous fluid. Simulated contamination is visible under UV light.
   e. Patient Safe syringes do not transfer contaminants to luer-activated valves or patients. Simulated contamination is visible under UV light.

Study 1—Test of contamination transfer to liquid drawn into syringes
Test methods:

- Agar plates were inoculated with *Geobacillus stearothermophilus* and incubated for 24 h.
- The luer end of Patient Safe syringes and standard syringes were placed on the surface of the agar.
- Using aseptic technique, sterile hypodermic needles were attached to the syringes.
- Each syringe was filled with 2 mL of sterile saline taken from separate vials (one vial per syringe).
- The contents were expelled through the needle into a sterile container.
- The contents of each sample were incubated for at least than 48 h and the number of *G. stearothermophilus* colonies was counted.

Results: (Table 5.4)
The results demonstrated that the Patient Safe syringes protected the fluid pathway and subsequently prevented contamination of the fluid drawn into the syringe. The study demonstrated that the Patient Safe syringe showed no contamination of the syringe contents in any instance (0%), while the standard syringe demonstrated contamination of the syringe contents in every instance (100%).

TABLE 5.4 Colonies of *G. stearothermophilus* Counted

No.	Patient Safe with Valve A	Patient Safe with Valve B	Patient Safe with Valve C	Standard with Valve A	Standard with Valve B	Standard with Valve C
1	0	0	0	186	201	180
2	0	0	0	135	320	79
3	0	0	0	173	301	202
4	0	0	0	192	241	103
5	0	0	0	82	208	122
6	0	0	0	161	113	152
7	0	0	0	211	270	102
8	0	0	0	205	215	123
9	0	0	0	174	264	94
10	0	0	0	195	173	103
Neg. Control	0	0	0	0	0	0
Pos. Control	5.0 × 103	3.1 × 103	3.8 × 103	3.4 × 103	3.5 × 103	4.1 × 103

Note: Valves used were Valve A: BD Posiflow™; Valve B: ICU Medical CLAVE®; and Valve C: Baxter Clearlink.

Study 2—Test of contamination transfer to LAV and infusate
Test method:

- Agar plates were inoculated with *G. stearothermophilus* and incubated for 24 h.
- Using aseptic technique, sterile Patient Safe syringes and standard syringes were opened and attached to sterile 22G × 1″ standard hypodermic needles.
- Each syringe was filled with 2 mL sterile saline.
- The needles were removed.
- The luer ends of Patient Safe syringes and standard syringes were placed on the surface of the agar.
- The syringes were attached to separate valves (each opened using aseptic technique and swabbed with alcohol) and the contents expelled through the valve into a sterile container.
- The contents of each sample were incubated for at least 48 h and the number of *G. stearothermophilus* colonies was counted.

TABLE 5.5 Colonies of *G. stearothermophilus* Counted

Sample No.	Patient Safe Syringes	Standard Syringes[a]
1	0	307
2	0	147
3	0	298
4	0	310
5	0	190
6	0	238
7	0	174
8	0	180
9	0	357
10	0	309
Negative controls	0	0
Positive control one	3.3 × 103	3.6 × 103
Positive control two	2.4 × 103	3.1 × 103
Positive control mean	2.9 × 103	3.4 × 103

[a] Standard syringes used were BD syringes.

Results: (Table 5.5)
The results of Study 2 demonstrated that the Patient Safe syringe protects the fluid pathway, and subsequently the patient, from contamination when connecting the syringe to an access valve and pushing liquid through the valve. The study demonstrated that for each of the three brand of valves tested, the Patient Safe syringe showed no contamination beyond the valve in any instance (0%), while the standard syringe demonstrated contamination beyond the valve in every instance (100%).

Conclusion

These studies show that the Patient Safe syringe prevents contact contamination of the syringe luer tip and subsequent contamination of liquids administered to patients either by direct injection or via an intravenous access system, whereas currently marketed standard luer lock syringes do not.

References

1. Estimates of healthcare-associated infections, CDC document. Available online at: <http://www.cdc.gov/ncidod/dhqp/hai.html>
2. Klevens RM, et al. Invasive methicillin-resistant *Staphylococcus aureus* infections in the United States, *JAMA* 298(15):1763, 2007.
3. Klevens RM, Edwards JR, Richards CL Jr, Horan TC, Gaynes RP, Pollock DA, Cardo DM. Estimating health care-associated infections and deaths in U.S. hospitals, *Public Health Rep* 122(2):160–6, 2007.
4. Lubell J. CMS: Your mistake, your problem; eight hospital-acquired conditions won't be paid for, *Modern Healthcare* 37(33):10, 2007.
5. DoBias M. Nothing they don't know: Medicare report shows urgency of situation, *Modern Healthcare* 38(13):10, 2008.

6

Progress in Preventing Sharps Injuries in the United States

Jane Doe
San Francisco General Hospital

Jane Perry
University of Virginia Health System

Ahmed E. Gomaa
National Institute for Occupational Safety and Health

Janine Jagger
University of Virginia Health System

Emergency Care Research Institute

Introduction 6-1
Workers' Compensation? • Zero Risk? • Conclusion
Progress in Preventing Sharps Injuries in the United States 6-4
United States Legislation and Policy • OSHA Enforcement Actions • EPINet Sharps Injury Data • Impact of the NSPA: Market Data • Impact of the NSPA: Injury Rates • Efficacy of Safety Devices: Other Data • Issues in Implementing Safety Devices • Global Standard for Health Care Worker Protection
References 6-11
Occupational Needlestick Injuries 6-13
Needlestick-Prevention Devices • Risk and Prevention of Needlestick-Transmitted Infections • Risk Management • Product Reviews • Discussion and Recommendations: Selecting Needlestick-Prevention Devices for Four Clinical Applications • Analyzing Costs Associated with Needlesticks and Preventive Devices
References 6-51
Supplemental Reading 6-52

Introduction

Jane Doe

In September 1987, I learned that a recent blood exposure had rendered me human immunodeficiency virus (HIV)-positive. The only "dirty" needlestick of my career, this accident led me to become San Francisco General Hospital's first health care worker "known" to have acquired HIV occupationally. Nationally, I was the 13th. The years following would challenge me to attempt to comprehend the broader context in which such an accident could have occurred.

I came to San Francisco General Hospital a young but experienced nurse and received comprehensive training in universal precautions during a hospital orientation. My perception of occupational risk at that time was influenced by multiple factors: a still youthful sense of immortality, an uncritical assumption that the institution and the larger regulatory agencies overseeing it had employed every measure to provide a safe working environment, and a rudimentary understanding of health and safety tenets (my most salient memory of health and safety training during my blunt suture needles (BSN) education is being told not to bite my nails and thus creating open sores). My perceptions were politicized as well; problematically, the voices conveying alarm over occupational transmission were often tinged with homophobia, disdain for intravenous (IV) drug users, and proposals for withholding care. In the absence of a reasoned and comfortable forum for discussion of risk, a health care worker was left keeping concerns over occupational transmission to and from herself.

But I did not incur my injury because of an underestimation of risk or the inappropriate handling of a needle. I acquired HIV because I had no adequate means with which to discard a used, blood-filled, and unsheathed needle. At the moment of my needlestick, I found myself across a crowded room from the bathroom containing the sharps disposal container. In retrospect, I would recall a "gadget" I had used at another San Francisco hospital earlier that year. Marketed as the "Click Lock," this engineering device provided a cylindrical sheath over a needle that would have prevented my accident. Ironically, the Click Lock was being utilized in the other institution for the purposes of patient safety with central IV lines.

In December 1991, federal Occupational Safety and Health Administration (OSHA) issued its final standard on bloodborne pathogens. Effective in March 1992, the standard set a hierarchy of controls for bloodborne hazards in the workplace, emphasizing that engineering controls be used in preference to other control measures when possible. This emphasis was a victory for health care workers, who historically have had the onus for our safety placed on our own behavior. The standard called for each institution to develop an exposure control plan and required the employer to provide the hepatitis B vaccine to employees at risk free of cost.

It is known that prior to the development of the hepatitis B vaccine in 1982, 6000–8000 U.S. health care workers contracted the hepatitis B virus (HBV) occupationally each year, resulting in 200 deaths annually and leaving hundreds open to cirrhosis and liver cancer in the years to come. As late as 1992, statistics issued by Centers for Disease Control and Prevention (CDC) and federal OSHA indicated no significant reduction in the incidence of occupationally transmitted hepatitis B. Yet, it took federal OSHA a decade to mandate that the hepatitis B vaccine be made available to every health care worker at risk. As safety devices and needleless systems are implemented in our institutions at present, I am left to wonder where these devices have been as thousands of my coworkers have been dying of hepatitis B. Why has it taken a stigmatized, seemingly 100% lethal and ironically much less virulent disease to elicit a federal OSHA standard and to provoke an industry to design, manufacture, and market the devices Dr. Jagger tells us could prevent at least 85% of all needlestick? (Jagger, congressional testimony, February 7, 1992.)

In August 1990, the nation responded with shock and horror as the CDC reported a young woman's acquisition of HIV as a patient in a Florida dental practice. Later, it would be strongly suggested that inadequate sterilization techniques and infection control practices were responsible for the transmission of HIV to the young woman and four other of the dentist's patients. In July 1991, the U.S. Senate voted 81 to 18 for Jesse Helm's amendment requiring mandatory disclosure of HIV status by HIV-positive health care providers, who perform invasive procedures, with criminal penalties applying to those health care workers who do not comply. No definition of invasive procedures was provided by the amendment.

Though the amendment did not become law, the message of its overwhelming vote was clear. While infection control practices, the crux of the transmission issue, went unaddressed, the first legislative responses to this tragedy were punitive of health care workers.

At the time of this vote, 22 health care workers were "known" to have seroconverted occupationally, a statistic considered to be underreported. Health care workers still waited for a standard on bloodborne pathogens, still worked with unsafe devices, and in some areas continued to be blamed for their injuries. As we waited, federal funding for research on the efficacy of needlestick prevention technology remained inadequate. Clearly, the federal agencies charged by Congress through the OSH Act of 1970 had failed to demand that safety devices become an industry standard in a timely fashion. (The FDA had not developed performance standards for needle devices.) While the media inflamed public fear about contracting HIV from one's dentist, the risk to health care workers was dramatically underplayed.

Paradoxes exist as well in the responses of an institution renowned for its provision of acquired immunodeficiency syndrome (AIDS) care. Following my seroconversion, SanFrancisco general hospital

(SFGH) would create a leading needlestick response program within 2 years. Providing 24 h counseling, evaluation, Zidovudine therapy when appropriate, and blood exposure data collection, this program is notable for its management of occupational blood exposures. Prevention measures, however, would be implemented more gradually. The most obvious and facile prevention measure, the relocation of the sharps disposal container from the bathroom to the bedside (the point of use), would take 3 years. In 1991, when SFGH employed a safety IV stylet solely in its emergency department, workers in other units, together with SEIU, would launch a campaign to get the device distributed hospitalwide. It would take a class-action grievance to overcome institutional resistance and make the device available to all workers. Since 1991, a Needlestick Prevention Committee, with representation of occupational infectious disease personnel, management, and frontline workers, has reviewed safety products, evaluated data on needlesticks, and planned trainings. As of the end of 1993, the exposure rate has fallen to half the rate reported in 1989.

While any decrease in needlesticks is positive, the overall statistics are alarming and beg immediate action. With 1,000,000 needlesticks occurring nationally each year, approximately 20,000 or 2% are contaminated with HIV. If surveillance data indicating a 1/250–1/400 chance of contracting HIV from an HIV-positive needlestick are accurate, up to 50–80 health care workers a year are contracting HIV on the job (Jagger, Congressional Testimony, February 7, 1992) one health care worker each week. And thousands are plummeted into the shadows of worrying, deciding whether to embark on AZT therapy and undergoing testing over a period of 6 months. The cost of managing these needlesticks (counseling, blood testing, treatment) is estimated to be $750 million annually.

In the face of these compelling statistics made real to me by my own seroconversion, I attempt to understand what forces can account for the often fragmented, delayed, and obstructionist approaches to health and safety exhibited by our institutions and regulatory agencies. The health care industry is one of the most profitable industries in the nation. When profit is a motive, or resources are limited, institutions have not prioritized worker safety and have generally responded after tragedies occurred. This pattern is consistent with our health care delivery system, which is based on a disease management model rather than a disease prevention model. Our regulatory agencies are supported in their insufficient response by the political climate of antiregulatory sentiment and the resultant underfunding.

The vast majority of health care workers are women and people of color, and our well-being as workers has long been ignored. In addition, the socialization of care givers has taught us to prioritize the interests of the patient and the institution; a sense of self-advocacy is not cultivated or encouraged. Our sense of responsibility toward our patients is often exploited by an institution choosing to abdicate its responsibilities by placing the onus for safety on the worker rather than itself.

Workers' Compensation?

What happens to the health care worker after an occupational needlestick transmission? In the wake of my seroconversion, SFGH went on record as upholding my rights to remain employed and to have my confidentiality protected. In pursuing workers' compensation from the City and County of San Francisco, I requested a confidential means of processing my claim that would ensure minimal disclosure of my identity. The Workers' Compensation Division and City Attorney's Office did not agree to such a procedure for 19 months, and did so only after the involvement of my attorney, Service Employees International Union, ACT UP, numerous hospital administrators, the Board of Supervisors, the media, and finally the Mayor. That procedure (disclosure to two high-level personnel only) was later negotiated into the RNs' collective bargaining agreement, putting it into place for all other city RNs.

Workers' compensation will be my sole source of monetary compensation for becoming HIV-positive on the job. Its benefits include financing ongoing medical and mental health costs. In the event of disabling disease, it will provide no more than the cost of state disability as income maintenance. By law, my employer is protected from a suit for damages from my injury. In order to glean an additional 50% in workers' compensation benefits in the form of penalties, I would have to prove serious and willful misconduct on the part of my employer. One other option for the infected health care worker is to bring a product liability suit against the needle manufacturer for the absence of a safer device in the workplace. While the statute of limitations has passed for me, I know of a few health care workers who have won settlements in such cases. I am familiar as well with health care workers who have been obstructed by their employers or employer's insurers in the process of applying for job relocation, workers' compensation, and death benefits. Reimbursement of medical costs will be my only true compensation, and in the absence of a dependent at the time of my injury, my death benefit will go to the state of California rather than to my family or a partner.

Zero Risk?

Who determines what is an "acceptable" rate of needlesticks and at what pace an institution can afford to implement safer devices and needleless systems? Dr. Jagger's preliminary research tells us that at least 85% of needlesticks could be preventable; clearly a rate in excess of 15% of current needlesticks is unacceptable. Federal agencies and health care workers must intensify cooperative efforts to bring the rate to as close to zero as possible.

Health care workers, historically distanced from the health and safety infrastructures of our institutions, bring to the evaluation process the hands-on wisdom of how a device or work practice control may or may not decrease the risk of exposure and may or may not affect patient care. Active participation and leadership by frontline health care providers in the decision-making process about health and safety programs provide a sense of urgency and a more ethical time frame for needed changes. Effective participation will be contingent upon the development of a creative forum for discourse about hazards in the workplace and perceptions of risk, such as educational trainings based on interactive learning rather than brief, formal instruction. Workers and management must learn to talk freely about these issues without a sense of antagonism and cross-purposes. Health care worker involvement should be facilitated on work time or with compensation and at no risk of negative sanctions.

Conclusion

Six years after my needlestick, I understand more clearly the personal, institutional, and national politics as to how a needle intersected with my life, changing it irrevocably. I have witnessed the creative perseverance of health and safety advocates and union activists in demanding and shaping change from a recalcitrant industry. I have watched infected health care workers speak the truth with courage. I have heard, as well, administrative doublespeak in response to workplace hazards demanding urgent resolution. Sadly, this resistance exists despite the example of what has happened and continues to happen to human lives when health and safety is not prioritized.

What I have learned has provided me with valuable perspective, but it is with great sorrow that I realize that this understanding cannot restore to me and other infected health care workers the tremendous losses that result from this type of injury. My hope is that the lessons wrought from our experiences will be heeded so that further tragedies are prevented. Inherent in these lessons is the imperative to err on the side of caution and precaution, to harness strength and power from knowledge, and to righteously strive to save each others' lives.

Progress in Preventing Sharps Injuries in the United States

Jane Perry, Ahmed E. Gomaa, and Janine Jagger

In the last 10 years, significant progress has been made in the United States in preventing occupational exposures to bloodborne pathogens in the health care environment. Advances in sharps safety technology and other medical device equipment, coupled with federal legislation and stronger regulations, have combined to markedly reduce the rate of needlestick injuries to U.S. health care workers, and made the United States a leader in the market for safety-engineered needles and other sharp devices. We will review relevant legislation and policy actions, and discuss percutaneous injury (PI) data from the University of Virginia's exposure prevention information network (EPINet) sharps injury surveillance network, as well as from other sources, which document and reflect this progress.

United States Legislation and Policy

The Needlestick Safety and Prevention Act (NSPA) was signed into law by President Clinton in November 2000.[1] (photo of bill signing) The NSPA required that the OSHA revise the bloodborne pathogens standard (BPS) (29 CFR Part 1910.1030) in order to strengthen and clarify the language regarding the use of engineering controls to prevent sharps injuries. Specifically, it added a new term: "Sharps with engineered sharps injury protections," or SESIPs. SESIPs are defined as "a nonneedle sharp or a needle device used for withdrawing body fluids, accessing a vein or artery, or administering medications or other fluids, with a built-in safety feature or mechanism that effectively reduces the risk of an exposure incident."[2] In the revised BPS, OSHA stated that health care employers must "[d]ocument annually consideration and implementation of appropriate commercially available and effective safer medical devices designed to eliminate or minimize occupational exposure."[2] In the 2001 revision of the BPS compliance directive (a manual used by OSHA inspection officers to interpret and apply the standard), OSHA said further, "Where engineering controls will reduce employee exposure, either by removing, eliminating, or isolating the hazard, they must be used," thus making it clear that safety devices were not optional.[3]

OSHA Enforcement Actions

From April 2001, when the revised BPS became effective, through May 2002, OSHA issued four times the number of citations for failure to use safety devices and other engineering controls than it had issued over the previous decade for this specific section of the BPS.[4] "Special emphasis" programs were carried out by various federal and state OSHA offices across the country which focused on BPS compliance. In July 2003, federal OSHA fined a Pennsylvania nursing home $70,000—the maximum penalty OSHA can assess for a willful violation—for failure to use safety devices.[5] In September 2003, after a group of residents at Montefiore Medical Center submitted a complaint to OSHA and an inspection was conducted, the facility was fined $9000 for three serious violations of the BPS.[6] The citation listed 26 separate instances in which safety devices were not used. OSHA has continued to actively enforce the requirement to evaluate and implement the safety-engineered devices.

OSHA has shown a willingness, not just to issue citations and levy fines for noncompliance with the BPS, but also to litigate contested BPS citations. In December 2007, the federal Occupational Safety and Health Review Commission upheld a ruling from an OSHA administrative law judge affirming that an OSHA inspections officer correctly cited a Denver laboratory facility for removing needles from blood tube holders for the purpose of reusing the holders. The judge, and subsequently the Review Commission, found that section (d)(2)(vii) of the BPS plainly prohibited the removal of contaminated needles from reusable blood tube holders.[7] The case originated with a citation of a Denver phlebotomy facility operated by MetWest (a subsidiary of Quest Diagnostics, one of the largest laboratory companies nationwide). Quest company policy dictated the use of reusable blood tube holders in phlebotomy clinics; this, in turn, required removal of the double-ended phlebotomy needles, exposing users to potential injury from the tube-piercing end of the needle. Quest argued that, by using a push-button blood tube holder which allowed mechanical removal of the needle, risk to the worker was minimized. OSHA did not agree, however, and upheld its comprehensive ban on needle removal. This final decision rendered the practice of reusing blood tube holders all but obsolete in the United States.

EPINet Sharps Injury Data

In 2004, the International Healthcare Worker Safety Center (IHWSC) at the University of Virginia collected data on percutaneous injuries and blood and body fluid (BBF) exposures from 41 health care facilities in the United States that use the EPINet surveillance program to track exposure incidents. These facilities voluntarily participate in the collaborative EPINet network coordinated by the Center; their exposure data are combined into an aggregate database. Most of these facilities (38) are part of a state-wide network in South Carolina coordinated by Palmetto Hospital Trust Services; the other three facilities are located in Virginia, Pennsylvania, and Nebraska. Nine of the facilities are teaching hospitals, and 32 are nonteaching facilities.

In 2004, 22 facilities had an average daily census (ADC) of less than 100 occupied beds; 9 facilities had an ADC of 100–300; and 4 facilities had an ADC of greater than 300.[8] (ADC data were not available for six of the participating institutions, all of them are long-term care facilities.) Most of the facilities were acute-care or tertiary-care hospitals or medical centers; some had physicians' offices, home health agencies, and other outpatient settings affiliated with them. Participating facilities included an alcohol and drug abuse agency, a home hospice agency, a long-term acute-care facility, a skilled nursing facility, and a rehabilitation hospital.

In 2004, a total of 1155 percutaneous injuries were reported by network facilities (Figure 6.1). The overall PI rate for all network hospitals was 26.69 PIs per 100 occupied beds. The average PI rate for teaching hospitals was 33.19 injuries per 100 occupied beds and, for nonteaching facilities, 18.98 injuries per 100 occupied beds. For hospitals with an ADC of less than 100, the average rate was 21.9 per 100 occupied beds; for hospitals with an ADC of 100–300, 20.76 per 100 occupied beds; for hospitals with an ADC of greater than 300, 34.78 per 100 occupied beds.

EPINet data from 2004, as in previous years, revealed great variation among individual facilities in PI rates: two facilities had a zero injury rate, while four facilities had rates exceeding 50 PIs per 100 occupied beds. The reasons for such variation are not fully understood, but may include the mix of patients, injury underreporting rates, the extent to which a facility has converted to safety devices, and whether it is a teaching or nonteaching institution (teaching institutions tend to have higher injury rates).

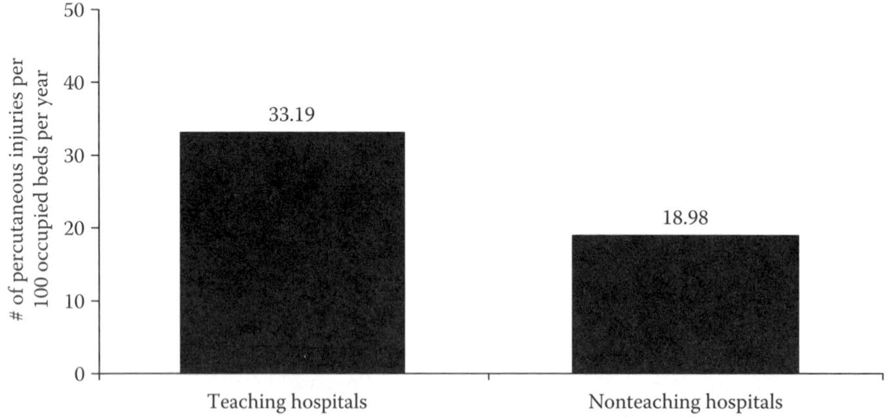

Data are for 41 healthcare facilities (9 teaching, 32 nonteaching). For teaching hospitals, the average daily, census was 2347, and total injuries were 779. For nonteaching hospitals, the average daily census was 1981 and total injuries were 376. Data are from the Exposure Prevention Information Network (EPINet), International Healthcare Worker Safety Center, University of Virginia. (*Note:* In 2004, a large teaching hospital with lower-than-average percutaneous injury rates dropped out of the EPINet network. This may have caused the PI rate for teaching hospitals to be higher in 2004 in comparison to 2003.)

FIGURE 6.1 The United States EPINet 2004: PI rates.

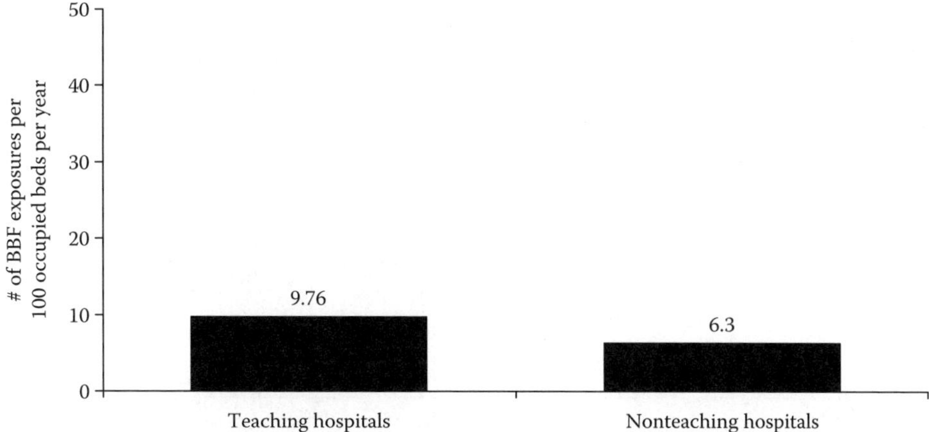

Data are for 41 healthcare facilities (9 teaching, 32 nonteaching). For teaching hospitals, the average daily census was 2347, and total exposures were 229. For nonteaching hospitals, the average daily census was 1981 and total exposures were 125.

FIGURE 6.2 The United States EPINet 2004: BBF exposure rates. (From International Health Care Worker Safety Center, University of Virginia Health System. U.S. EPINet Multihospital Sharps Injury and Blood Exposure Surveillance Network. Needlestick and Sharp-Object Injury Report, 2004 (1155 PLs, 41 hospitals contributing data). Report run February 2008.)

Because of these variables, it should not be assumed that a health care facility with a low PI rate necessarily has a superior safety record than a hospital with a higher rate. For example, a hospital with a higher-than-average PI rate may have more patients requiring invasive procedures than another facility with a lower rate. Thus, comparing rates among hospitals have limitations; it is more meaningful to track injury trends within a single institution over several years and make historical comparisons as prevention measures are implemented.

The EPINet surveillance system includes a separate report form for BBF exposures. The Center collects BBF exposure data from network hospitals as well as PI data. In 2004, a total of 354 BBF exposures were reported by network facilities (Figure 6.2).[9] The average BBF exposure rate overall was 8.18 per 100 occupied beds; for teaching hospitals, the rate was 9.76 per 100 occupied beds, and for nonteaching hospitals, 6.3 per 100 occupied beds.

Impact of the NSPA: Market Data

Market data provide one view of the impact of the NSPA on the implementation of safety-engineered devices in U.S. health care facilities. Between December 2000 and March 2003, the market share for safety-engineered IV catheters increased from 43% to 90% of the overall market for IV catheters (data are for hospital and alternate sites combined). Similarly, the market share for safety-engineered phlebotomy needles increased from 40% to 80% of the overall market during the same time period.[10] For safety-engineered syringes used for skin injection, the market share increased from 15% in 2000 to 50% in 2003.[10]

Of the major categories of sharp devices that cause injuries to HCWs, suture needles are the only one where there has been little or no growth in the market for safer alternatives since passage of the NSPA. BSNs, in particular, remain vastly underutilized by U.S. surgeons. BSNs are sharp enough to penetrate internal tissue, such as muscle or fascia, but not sharp enough, in most cases, to penetrate skin.[11] Most injuries occurring in the operating room (OR) are caused by suture needles, followed by scalpel blades and other surgical instruments;[12] studies have demonstrated the efficacy of BSNs in reducing injury risk to surgeons and other surgical personnel (Figure 6.3).[13,14] Although the American College of Surgeons

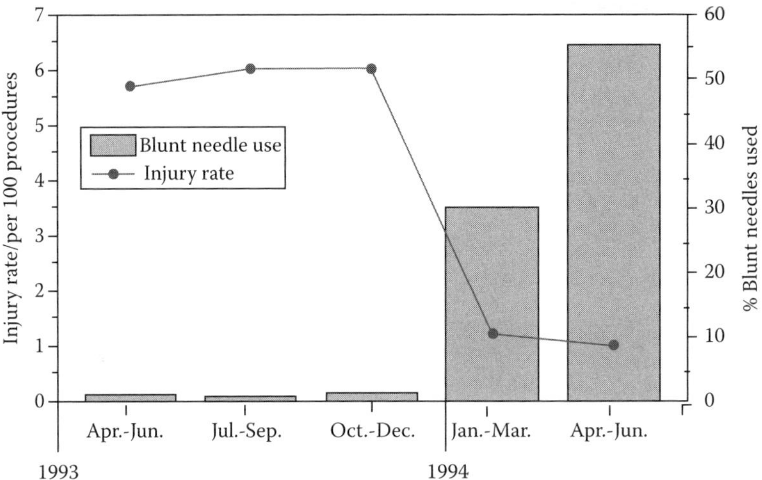

Source: U.S. Centers for disease control and prevention.

FIGURE 6.3 Rate of injury associated with use of curved suture needles during gynecologic surgical procedures and percentage of suture needles used that were blunt, by quarter—three hospitals, New York City, April 1993–June 1994. (From Centers for Disease Control and Prevention, *Morb. Mortal. Wkly. Rep.*, 46, 25, 1997; Berguer, R. and Heller, P.J., *J. Am. Coll. Surgeon*, 199, 462, 2004.)

issued a statement in 2005 supporting universal adoption of BSNs for suturing fascia[15] and OSHA issued a Safety and Health Information Bulletin recommending their use in 2007,[16] adoption rates for BSNs remain low. Market data from 2004 showed that blunt-tip suture needles had captured only 3% of the overall market for suture needles.[17] Adoption of safety-engineered scalpels, which provide protection from the scalpel blade during passing and after use, also remains low.[18]

Impact of the NSPA: Injury Rates

A comparison of EPINet surveillance data from 1993 to 2000 (before passage of the NSPA) and 2001 to 2004 (after passage of the law) revealed an overall decline of 36% in PI rates for hollow-bore needles (Figure 6.4).[19] PI rates dropped precipitously after the NSPA was enacted, and the lower rate has persisted. Injury rates for hollow-bore needles (i.e., injection, blood-drawing, and vascular access devices), which had shown little or no decline from 1993, when EPINet surveillance began, through 2000, dropped from 18.7 to 12.1 PIs per 100 occupied beds (rates are for conventional and safety devices combined). It should be noted that the injury rate for safety-engineered needles tripled after 2000; this may seem counter-intuitive but is actually to be expected, since safety-engineered devices reduce, but do not eliminate, needlestick injuries. The huge increase in the number of safety devices used in the health care workplace means, by necessity, that there will be more injuries caused by them. However, the sizable reduction in injuries from conventional needles has more than offset increased injuries from safety-engineered devices, resulting, overall, in a much lower PI rate.

Particularly encouraging is the fact that, while injury rates for hollow-bore needles declined by 36%, much higher declines were found for the device categories associated with the highest risk of blood-borne pathogen transmission. Injury rates for IV catheters declined by 53%; injury rates for blood collection (phlebotomy) needles declined by 60%.

These data point to a clinical area, where more rigorous enforcement and compliance efforts are needed, however. As noted above, the hospital setting least impacted by the law is the surgical environment

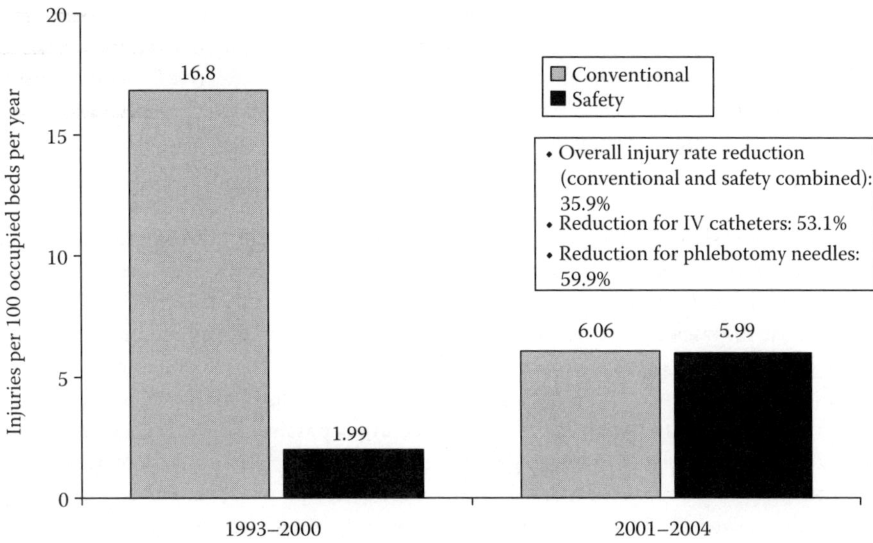

Data are for 87 hospitals and a total of 14,301 percutaneous injuries. Data from EPINet, International Healthcare Worker Safety Center, University of Virginia.

FIGURE 6.4 Percutaneous injury rates for all hollow-bore needles before and after passage of the NSPA. (Reprinted from Jagger, J., *Infect. Control. Hosp. Epidemiol.*, 28(1), 1, 2007.)

(where the use of hollow-bore needles is low compared to other clinical areas). Injury rates for the OR have remained virtually unchanged since passage of the NSPA: EPINet data from 1993 to 2000 revealed an OR sharps injury rate of 7.0 per 100 occupied beds; for 2001 to 2004, the rate was 6.99 per 100 occupied beds.[20]

Efficacy of Safety Devices: Other Data

Data from other studies confirm that requiring implementation of safety devices is an effective policy measure in reducing sharps injuries to health care workers. In 2004, researchers from the Memorial Sloan-Kettering Cancer Center in New York City published a study comparing PI data for the period 1998–2000, before safety devices were implemented hospitalwide (and before they were specifically required by the NSPA), to data from 2001 to 2002, after a near-total conversion to safety devices.[21] Because the conversion occurred within a very short-time frame rather than over a number of years (as is the case in most institutions), the study provided a clear "before-and-after" picture. The hospital's overall PI rate decreased by 58% after conversion to safety-engineered devices; injuries from hollow-bore needles (associated with the highest risk of bloodborne pathogen transmission) fell by 71%. Nurses had the largest decrease (75%) in PI rates of any occupational group. The authors concluded that "implementation of safety-engineered devices reduced PI rates across occupations, activities, times of injury, and devices."[21]

A 2006 review of 17 published studies of safety device implementation reached a similar conclusion: "In all studies reviewed, introduction of safety-engineered devices was followed by considerable reductions in PI rates (range: 22%–100%)."[22] Studies published in 2007 from France and Spain offered further evidence of the efficacy of safety devices in significantly reducing PI rates.[23,24]

Issues in Implementing Safety Devices

As facilities have worked to comply with the revised BPS and the requirement to evaluate and implement safety-engineered devices, a variety of issues have arisen. Facilities frequently ask whether

OSHA requires the use of specific types or brands of devices; however, OSHA has stated that "No one medical device is appropriate in all circumstances of use" and that "Employers must implement safer medical devices that are appropriate, commercially available, and effective."[25] On the other hand, OSHA does require that "employers evaluate the effectiveness of existing controls and review the feasibility of instituting more advanced engineering controls."[26]

Another frequently asked question in relation to safety devices is whether OSHA requires the use of "passive" devices. Active (or "clinician-controlled") safety devices require "a physical action by the user in order to activate the sharps safety feature, which is in addition to any actions needed to perform the primary function of the device."[27] A passive safety feature "automatically activates after use, i.e., does not require any additional action by the user to activate the sharps safety feature."[27] While OSHA does not require the use of passive safety devices, it states that "Passive features enhance the safety design and are more likely to have a greater impact on prevention."[26] On the other hand, the CDC's "Workbook for Sharps Injury Prevention" comments that "Many devices currently marketed as self-blunting, self-resheathing, or self-retracting imply that the safety feature is passive. However, devices that use these strategies generally require that the user engage the safety feature."[28] The Workbook further notes that "In certain situations, it is not practical or feasible for the device or...procedure to have a passive control. Therefore, whether a safety feature is active or passive should not take priority in deciding the merits of a particular device."[28]

One of the most significant challenges that facilities face regarding needlestick injuries and blood exposures is simply to remain vigilant and focused on the issue in the face of other pressing occupational safety concerns. A significant drop in PI rates was noted in EPINet data after passage of the NSPA in 2000, but rates have remained relatively stable since then. Data from the Sharps Injury Control Program of the California Department of Health Services reflect a similar phenomenon: a clear decline in hospital injury rates for the period 1998–2002, with a leveling-off of rates from 2002 to 2005.[29] While U.S. health care facilities have made great progress in reducing the risk of occupational blood exposures, these data underscore the need for ongoing efforts and a sustained focus if further risk reductions are to be achieved.

OSHA requires that facilities continue to evaluate, on an annual basis, safety-engineered sharps injury prevention technology; this is important because sharps safety technology continues to evolve in response to users' needs and regulatory trends. An example of this is seen with blood tube holders. OSHA clarified its position regarding reuse of blood tube holders in a Standard Interpretation letter issued in 2002[30]; this was followed by a Safety and Health Information Bulletin in 2003.[31] OSHA affirmatively stated that reuse of blood tube holders, which requires removal of the phlebotomy needle from the tube holder, was not permitted under the BPS. Since then, new designs for safety-engineered blood tube holders have emerged. Some offer protective mechanisms that shield both the front- and back-end needles after use; another integrates the needle and tube holder in one unit, thus eliminating the possibility of reuse altogether.

In order for a safety device to protect HCWs from injury after use, the protective feature must be activated. This is obvious, but gaining full compliance from users is not a simple or easy task, and inactivation of safety devices remains an ongoing issue. One key is to provide adequate training facility-wide when a new device is implemented, followed by annual in-servicing—particularly for devices that have a longer learning curve or require a change in technique. Some facilities hold an annual safety device "fair" to educate and update staff on available safety devices and give device manufacturers an opportunity to share new products. A hospital in North Carolina requires that each of its clinical departments conduct an annual competency review of high-risk procedures and devices; in addition, staff must attend a yearly "Sharps, Spills, and Spashes" session that includes hands-on demonstrations, assisted by product representatives, of the correct use of safety devices available in the institution.[32] The participation of product representatives in such meetings gives staff an opportunity to ask questions, voice concerns, and provide feedback about specific devices.

Global Standard for Health Care Worker Protection

In 2001, the CDC included the elimination of "occupational needlestick injuries among health care workers" on its list of seven "Healthcare Safety Challenges."[33] In order to meet this goal, continued vigilance will be necessary on the part of U.S. health care facilities to ensure that effective safety technology is both available and consistently used in hospitals and outpatient settings nationwide. In the United States, reducing the risk of needlestick injuries and blood exposures to HCWs has taken more than two decades of concerted effort, and the goal of full compliance with needle safety regulations has yet to be achieved. However, available data indicate that significant progress has been made, and the U.S. experience provides a model for other countries to follow. Since 2004, for example, Canadian provinces have passed needle safety regulations similar to those in the United States.[34]

In most industrialized countries, important advances have been made in reducing occupational infections from hepatitis B, hepatitis C, and HIV, through vaccination, postexposure medical intervention, antiviral therapies, and (in some countries) implementation of safety-engineered devices.[19] However, protecting health care workers from occupational infections in resource-poor countries—where HBV vaccine, HIV postexposure testing and treatments, and safety devices are less readily available—remains a significant challenge for the twenty-first century.[35] While the prevalence of bloodborne pathogens in many poor countries is high, documentation of infections caused by occupational blood exposures is scarce or nonexistent. One study found that although 70% of the world's HIV-infected population lives in sub-Saharan Africa, only 4% of worldwide cases of occupational HIV infection are reported from this region.[35] By contrast, 4% of the world's HIV-infected population lives in North America and western Europe, but 90% of documented occupational HIV infections are reported from these areas.[35,36]

Basic measures for protecting HCWs from life-threatening bloodborne diseases should be viewed as essential and included in the national health priorities of all nations. Such measures include universal hepatitis B vaccine for HCWs; provision of personal protective equipment, such as gloves and masks, for all high-risk exposure-prone procedures; safe containment and disposal of contaminated sharps and other medical waste; education and training in occupational exposure prevention, especially in medical and nursing schools; and adoption of safe work practices, such as hands-free passing of sharp instruments in surgical settings. Making such measures, a priority for major international health organizations is critical if greater protection for health care workers around the globe is to be achieved. The health care workforce in every country needs to be viewed as an invaluable, and increasingly scarce, resource.

References

1. Needlestick Safety and Prevention Act of 2000, Pub. L. No. 106–430, 114 Stat. 1901, November 6, 2000.
2. U.S. Department of Labor, Occupational Safety and Health Administration. Occupational exposure to bloodborne pathogens; needle-sticks and other sharps injuries; final rule (29 CFR Part 1910.1030). *Fed Regist* 66(12):5318–5325, 2001.
3. U.S. Department of Labor, Occupational Safety and Health Administration. *OSHA Instruction: Enforcement procedures for the Occupational Exposure to Bloodborne Pathogens. Directives Number CPL 2.2.69.* U.S. Department of Labor, Washington, D.C., November 27, 2001.
4. Perry J and Jagger J. Safer needles: Not optional. *Nursing* 32(10):20–22, 2002.
5. Perry J and Jagger J. OSHA cracks down on sharps safety violators. *Nursing* 34(3):68, 2004.
6. Perry J and Jagger J. Ground-breaking citations issued by OSHA for failure to use safety devices. *Adv Expo Prev* 6(6):61–64, 2003.
7. U.S. Occupational Safety and Health Review Commission. Secretary of Labor v Metwest, Inc. OSHRC Docket No. 04-0594. December 17, 2007. Decision available at: http://www.oshrc.gov/decisions/comm07.html

8. International Healthcare Worker Safety Center, University of Virginia Health System. U.S. EPINet Multihospital Sharps Injury and Blood Exposure Surveillance Network. Needlestick and Sharp-Object Injury Report, 2004 (1155 PIs, 41 hospitals contributing data). Report run February 2008.
9. International Healthcare Worker Safety Center, University of Virginia Health System. U.S. EPINet Multihospital Sharps Injury and Blood Exposure Surveillance Network. Blood and Body Fluid Exposure Report, 2004 (354 BBFs, 41 hospitals contributing data). Report run February 2008.
10. Perry J. Measuring progress: What market data show. *Adv Expo Prev* 6(5):58, 2003.
11. Jagger J, De Carli G, Perry J, Puro V, and Ippolito G. Occupational exposure to bloodborne pathogens: Epidemiology and prevention. In: RP Wenzel (ed.), *Prevention and Control of Nosocomial Infections*. Lippincott Williams & Wilkins, Baltimore, 2003, pp. 430–465. (See section on "Surgery: work practices and safety-engineered devices", pp. 451–452.)
12. Jagger J, Bentley M, and Tereskerz P. A study of patterns and prevention of blood exposures in OR personnel. *AORN J* 67:979–984, 986, 1998.
13. Centers for Disease Control and Prevention. Evaluation of blunt suture needles in preventing percutaneous injuries among health-care workers during gynecologic surgical procedures, New York City, March 1993–June 1994. *MMWR Morb Mortal Wkly Rep* 46:25–29, 1997.
14. Berguer R and Heller PJ. Preventing sharps injuries in the operating room. *J Am Coll Surg* 199(3):462–467, 2004.
15. American College of Surgeons (ACS). Statement on blunt suture needles. *Bull Am Coll Surg* 90(11): 24, 2005 (November). Available from http://www.facs.org/fellows_info/statements/st-52.html Accessed February 2008.
16. U.S. Department of Labor, Occupational Safety and Health Administration; Centers for Disease Control and Prevention, National Institute for Occupational Safety and Health. Safety and Health Information Bulletin: Use of blunt-tip suture needles to decrease percutaneous injuries to surgical personnel. SHIB 03-23-2007. DHHS (NIOSH) Publication No. 2008-101. Available from http://www.cdc.gov/niosh/docs/2008-101/
17. Perry JL, Pearson RD, and Jagger J. Infected healthcare workers and patient safety: A double standard. *Am J Infect Control* 34(5):314–319, 2006.
18. Anonymous. Helping your surgeons select safety scalpels. *Outpatient Surg Mag* 7(6):24, 2006 (June).
19. Jagger J. Caring for healthcare workers: A global perspective. *Infect Control Hosp Epidemiol* 28(1):1–4, 2007.
20. International Healthcare Worker Safety Center, University of Virginia Health System. U.S. EPINet Multihospital Sharps Injury and Blood Exposure Surveillance Network. Needlestick and Sharp-Object Injury Report; Operating Room; 1993–2000 (5,333 PIs, 87 hospitals contributing data) and 2000–2004 (1939 PIs, 67 hospitals contributing data). Excludes injuries occurring before use of device. Report run February 2008.
21. Sohn S, Eagan J, Sepkowitz KA, and Zuccotti G. Effect of implementing safety-engineered devices on percutaneous injury epidemiology. *Infect Control Hosp Epidemiol* 25(7):536–542, 2004.
22. Tuma S and Sepkowitz KA. Efficacy of safety-engineered device implementation in the prevention of percutaneous injuries: A review of published studies. *Clin Infect Dis* 42:1159–1170, 2006.
23. Lamontagne F, Abiteboul D, Lolom I, Pellissier G, Tarantola A, Descamps JM, and Bouvet E. Role of safety-engineered devices in preventing needlestick injuries in 32 French hospitals. *Infect Control Hosp Epidemiol* 28:18–23, 2007.
24. Valls V, Salud Lozano M, Yanez R, Martinez MJ, Pascual F, Lloret J, and Ruiz JA. Use of safety devices and the prevention of percutaneous injuries among healthcare workers. *Infect Control Hosp Epidemiol* 28:1352–1360, 2007.
25. U.S. Department of Labor, Occupational Safety and Health Administration. *OSHA Instruction: Enforcement Procedures for the Occupational Exposure to Bloodborne Pathogens. Directives Number CPL 2.2.69*. U.S. Department of Labor, Washington, D.C., November 27, 2001. Section D (Methods of Compliance), paragraph (d)(2)(i).

26. U.S. Department of Labor, Occupational Safety and Health Administration. How to prevent needlesticks: Answer to some important questions. Available from: http://www.osha.gov/Publications/OSHA3161/osha3161.html. Accessed February 2008.
27. U.S. Food and Drug Administration. Supplementary Guidance on Premarket Notifications for Medical Devices with Sharps Injury Prevention Features; Guidance for Industry and FDA. Dec. 31, 2002.
28. U.S. Centers for Disease Control and Prevention, Division of Healthcare Quality Promotion. Concept of 'active and passive' safety features. In: *Workbook for Designing, Implementing, and Evaluating a Sharps Injury Prevention Program (Appendix B-2)*. Published online February 12, 2004. Available from: http://www.cdc.gov/sharpssafety/index.html
29. Reinisch F, Gillen M, Kronlins E, and Harrison R. Sharps injury trends in California hospitals. Presentation at: Controlling Sharps Injuries in California: State of the Art Update for Occupational and Infection Control Professionals. Conference sponsored by the California Department of Health Services and the Center for Occupational and Environmental Health, University of California, Berkeley, Oakland, CA, September 15, 2006.
30. U.S. Department of Labor, Occupational Safety and Health Administration. Standard Interpretations: 6-12-2002 - Re-use of blood tube holders. Washington, D.C.; June 12, 2002. Available from: http://www.osha.gov/pls/oshaweb/owadisp.show_document?p_id = 24040&p_table = INTERPRETATIONS
31. U.S. Department of Labor, Occupational Safety and Health Administration. Safety and Health Information Bulletin: Disposal of contaminated needles and blood tube holders used for phlebotomy. SHIB 10-15-03. Washington, D.C.; October 15, 2003. Available from: http://www.osha.gov/dts/shib/shib101503.html
32. Perry J and Jagger J. Implementing safety devices: Two nurses share their experience. *Adv Expo Prev* 6(6):65–69, 2003.
33. Centers for Disease Control and Prevention. CDC's 7 Healthcare Safety Challenges. Released August 23, 2001. Available from: http://www.cdc.gov/ncidod/dhqp/about_challenges.html
34. Perry J and Jagger J. Needlestick prevention in Canada: The safety wave. *Adv Expo Prev* 7(4):40–41, 2006.
35. Sagoe-Moses C, Pearson RD, Perry J, et al. Risks to health care workers in developing countries. *New Engl J Med* 345:538–541, 2001.
36. Ippolito G, Puro V, Heptonstall J, Jagger J, De Carli G, and Petrosillo N. Occupational human immunodeficiency virus infection in health care workers: Worldwide cases through September 1997. *Clin Infect Dis* 28:365–383, 1999.

Occupational Needlestick Injuries*

Emergency Care Research Institute

Needlestick-Prevention Devices

According to the Centers for Disease Control (CDC),[1–4] needlestick injuries account for 80% of reported occupational HIV exposures, which could lead to AIDS. Although the fear of contracting AIDS has overshadowed the concern about acquiring the HBV through an accidental needlestick, the risk of acquiring—and dying from—HBV is actually much greater: CDC reports a 6%–30% chance of acquiring hepatitis B from a PI with an HBV-contaminated needle. Physicians, nurses, clinical laboratory technicians, pharmacy personnel, housekeeping staff, and waste handlers—all health care workers who may be exposed to patients' blood or body fluids—are at risk.

The October 19, 1987, Joint Advisory Note of the Departments of Labor and Health and Human Services states "Whenever possible, engineering controls should be used as the primary method to reduce worker exposure to harmful substances. The preferred approach…is to use, to the fullest extent feasible, intrinsically safe substances, procedures, or devices." Given this statement and the incidence of occupational AIDS and hepatitis B transmissions, the best way to reduce the risk of needlestick injury may be to use a needlestick-prevention device or to eliminate the needle from the procedure entirely.

Many manufacturers are marketing products that reduce the risk of needlestick injuries, and the number of different products available is rapidly increasing. In this Special Report and Product Review, we provide general background information and guidance on the need for and use of needlestick-prevention devices, as well as reviews of 26 such products from 22 manufacturers, divided into eight different Product Groups. Also, the Discussion and Recommendations section provides guidance on selecting products for four main clinical uses and includes a discussion of cost analysis, including a cost comparison model based on the Emergency Care Research Institute's (ECRI)[5,6] new Computer-Aided Health Devices™ (CAHD™) CAHDModel™ system.

Finally, we would like to thank CDC for its review of this report.

Risk and Prevention of Needlestick-Transmitted Infections

Risk Assessment

Accidental needlesticks occur frequently, posing a serious risk of transmitting fatal or chronic diseases to a wide range of health care workers. As one physician noted, "Rarely a day goes by in any large hospital where a needlestick incident is not reported."[7] Needlesticks have long been associated with transmitting both bacterial infections and viral infections, such as hepatitis B; now, they also serve as the agent of transmission of the HIV, which causes AIDS. This threat has spurred the development of the preventive devices examined and discussed in the pages that follow.

Sources of Risk

A number of clinical procedures and housekeeping activities carry an increased risk of needlestick injuries, including (1) disposing of or recapping used needles, (2) administering parenteral medications, (3) drawing blood, and (4) collecting linens and trash. In one study of 316 reported needlesticks (conducted before the use of needle disposal systems ["sharps" containers]), disposing of needles accounted for 24% of injuries; recapping needles, 12%; administering parenteral medications, 21%; drawing blood, 17%; and collecting linens and trash, 16%.[8] In another study of 286 reported needlesticks drawing blood constituted the highest risk, 20.6%; followed by recapping or corking needles, 18.2%; handling trash, 16.1%; and giving injections or infusions, 15.4%. Interestingly, injuries from needles poking out of overfilled needle disposal containers—which are designed "to protect" workers from accidental sticks—constituted 8.4% of the total injuries.[9]

According to another study of 1201 health care workers with blood exposures, 80% of which were due to needlesticks, 37% could have been prevented if recommended infection control precautions had been followed.[10] The exposures were caused by recapping needles by hand, 17%; improperly disposing of sharps, 14%; and exposing open wounds to sources of contamination, 6%. The remaining 63% of the exposures involved manipulating IV, phlebotomy, or arterial needles (36%) during an invasive procedure (8%), autopsy (2%), or other procedure (17%).

Accidental needlesticks have been associated with the transmission of AIDS, hepatitis (B and C), and many other viral, rickettsial, bacterial, fungal, and parasitic infections. In addition, personnel face a significant risk of injury from accidental sticks with needles contaminated by toxic antineoplastic drugs and immunotherapeutic agents. Although it is difficult to predict the exact degree of risk of acquiring a specific disease from an accidental needlestick because of pathogenicity dose, host susceptibility, and other factors, some information on the risks of acquiring AIDS and hepatitis B, the two diseases most commonly associated with needlestick injuries, is available.

Needlesticks and AIDS

Thus far, the available evidence indicates that health care workers face a small—yet nonetheless real—chance of acquiring HIV through accidental sticks with needles contaminated with the virus. According to two studies, the percentage of HIV seroconversions per known HIV exposures was 0.43% and 0.41%, respectively.[10,11] Prospective studies in the United Kingdom and Canada reported that 220 health care workers exposed to parenteral, mucous membrane, or cutaneous blood or body fluids in patients infected with HIV showed no evidence of HIV transmission.[12,13]

Two other studies assessed the risk of occupational transmission of HIV infection for health care workers. One study, conducted by the National Institutes of Health (NIH), tested 1344 health care workers with 170 documented needlestick exposures and 345 mucous membrane exposures to the blood or other body fluids of HIV-infected patients and reported only one HIV seroconversion as of December 1989.[14] The other study, done at the University of California, included 212 health care workers that had documented 625 needlestick injuries or mucous membrane exposures as of March 15, 1988, and again only one seroconversion following a needlestick was reported.[15] In addition, in the study reported above the 1201 health care workers who had been exposed to blood as of July 31, 1988, the serum of 963 health care workers (860 of whom had received a needlestick or cut from a sharp instrument) was tested for HIV at least 180 days after exposure, and only four tested positive for HIV after exposure to a needlestick injury.[10]

Based on the landmark studies of Marcus, Henderson, and Gerberding cited above, as well as other studies, Henderson concludes that "The risk of HIV-1 transmission with a percutaneous exposure to blood from an HIV-1-infected patient is approximately 0.3% per exposure." As of June 1990, only 19–24 HIV seroconversions were reported in the United States, but many more may not have been verified by or known to CDC.

According to Janine Jagger, MPH, PhD, of the University of Virginia, who gave a presentation at the Sixth International Conference on AIDS in San Francisco on June 22, 1990, current technology has the potential to prevent 85%–90% of needlesticks based on an estimated 800,000 needlesticks per year: "If 2% of hospital patients are HIV seropositive, and consequently 2% of needlesticks are HIV contaminated, then 64 health care workers will seroconvert per year, and 57 of those seroconversions would be preventable with technology that exists today."

Needlesticks and Hepatitis B

Although the opinions of health care experts vary about the risk of acquiring hepatitis B from an accidental stick with a needle contaminated by the virus, experts agree that the risk is "significantly" greater than that of acquiring HIV through a needlestick. According to the Hepatitis Branch of CDC, an estimated 12,000 health care workers exposed to blood become infected by HBV each year—500–600 are hospitalized, 700–1000 become HBV carriers, and approximately 250 die from acute or chronic consequences.[2] In contrast to the relatively small number of HIV transmissions, the annual incidence of infection with hepatitis B among laboratory staff, surgeons, physicians, and nurses is estimated at 37, 25, 11, and 4 per 100,000, respectively.[16] Chronic sequelae of HBV infection include both cirrhosis and at least 80% of all primary liver cancers. Hepatitis B has also been known to be transmitted from a mother to her fetus.

OSHA requires that hospitals offer hepatitis B vaccines free of charge to employees who are at substantial risk of direct contact with patients' BBFs. Available hepatitis vaccines provide over 90% protection against hepatitis B for seven or more years; however, many health care workers still do not receive the vaccine, taking action only when a needlestick injury actually occurs.

Risk Management

The risks associated with needlesticks underscore the importance of implementing effective risk management efforts that reduce exposure and the likelihood of transmission. First and foremost, health care workers at high or moderate risk should receive the hepatitis B vaccine, all health care workers should follow universal precautions (as outlined in NCCLS Document M29-T), and health care workers who sustain needlestick injuries should report them in accordance with hospital policies and procedures. In addition, hospitals should enhance worker safety by implementing policies and procedures, discussed in detail below, to prevent needlestick injuries. Failure to take protective measures, such as those recommended by CDC and enforced by OSHA,[17] could lead to increased employee injuries and the losses associated with such injuries. In addition, it could result in the imposition of citations and civil penalties by OSHA as a violation of its general duty clause. These recommendations (e.g., for sharps disposal) are also likely to be introduced via litigation as the standard of care for hospitals regarding prevention of AIDS or other transmissible diseases to health care workers.

Reporting

Just as health care workers often fail to be vaccinated, they often fail to report needlestick injuries. In a 1986 survey of 1473 nursing and medical personnel employed in two hospitals, 33.6% of the respondents had one or more needlesticks but did not report the incident.[18] A recent survey in a U.S. Air Force base hospital found that more than one-third of the 334 health care workers who had received a needlestick or other means of blood exposure did not report the injury through hospital reporting channels; the reasons given were lack of time or the feeling that it was not dangerous (Hospital Infection Control, August 1990).

As of June 1989, a 24 h needlestick hotline has been available at San Francisco General Hospital. The hotline provides immediate counseling and helps San Francisco General's health care workers determine the necessary prophylactic care. Also, if zidovudine (Retrovir®) is recommended immediately, it can be obtained through the hotline program. At the Sixth Annual International Conference on AIDS, it was reported that this service had prompted a 69% increase in the number of needlestick injuries reported at the hospital.

OSHA requires that "any needlestick requiring medical treatment (e.g., gamma globulin, hepatitis B immune globulin, hepatitis B vaccine, etc.) shall be recorded. In addition, since this type of treatment is considered absolutely necessary...such an injury cannot be considered minor."[17] Hospitals should strongly emphasize the importance of reporting a needlestick injury as soon as possible, to maximize the effects of postexposure follow-up.

Hospital Liability and Costs

In addition to their ethical concerns for employee well-being, hospitals face concerns about accidental needlesticks posing significant liability and costs. Hospital employees who contract an infection as the result of a needlestick are entitled to Worker's Compensation benefits. Because Worker's Compensation is a no-fault system, benefits are available regardless of whether the employee followed safety rules and preventive practices. Although Worker's Compensation is generally an exclusive remedy, some commentators have suggested that an employee might nevertheless be able, in some jurisdictions, to sue the hospital in tort on the theory that it intentionally created a hazardous work environment by disregarding proper safety precautions. And physicians and other nonemployees are not bound by Worker's Compensation and could bring a tort suit against the hospital. For example, in a highly publicized lawsuit, Veronica Prego, MD, contended that she acquired AIDS from an accidental needlestick at a New York City hospital where she had been an unpaid extern. The hospital settled the case for $1.3 million. The potential for hospital liability is likely to increase with safer product availability, especially if the hospital fails to provide such devices.

The combined costs of employee time lost, laboratory testing, case investigation, and, if necessary, treatment stemming from a needlestick injury can be significant. In one hospital study, conducted over a 47 months period from 1975 to 1979, the incidence of accidental needlestick injuries was as high as 81.8/1000 employees; the 316 needlesticks reported constituted one-third of all work-related injuries in the hospital during this period.[8] In addition, 1053 incidents of needlestick injuries were reported to the Worker's Compensation insurance of the University of Texas Medical Branch Hospitals in Galveston between November 1, 1984 and January 31, 1989; 61.6% of the incidents involved nurses, and 6.7% involved physicians.[19]

OSHA may impose fines up to $70,000 (for reported violations) on hospitals that fail to take measures to protect workers, giving hospitals a further incentive to abide by CDC recommendations. If hospital policies are at variance with CDC recommendations, documented—and defensible—reasons should be provided for the hospital's policies.

Methods of Prevention

Several methods of reducing the incidence of needlestick injuries are available. These include (1) increased education and training of all hospital personnel who come into contact with used—and thus potentially contaminated—needles, (2) a proposed OSHA ban on traditional recapping by the two-handed technique, (3) appropriate use of needle and syringe disposal containers, and (4) the use of needlestick-prevention devices.

Education and Training

Proper education of employees about the importance of following universal precautions; using approved disposal methods, including recommended impervious disposal containers; prompt emptying of disposal containers before they overflow; and using preventive devices are essential elements of a program to minimize needlestick occurrences. However, education, while indispensable, is not a panacea for needlestick prevention.

Several surveys of nursing and medical personnel have revealed that they do not perceive education as an effective means of reducing needlestick injuries. In one article, the authors note that "most respondents reported some knowledge of proper needle disposal techniques and perceived lack of knowledge as the least important reason for needlestick injuries."[18] Similarly, in another survey, nurses felt that "talks and information on preventing needle injuries and awards to individual nurses with good safety records" were the least effective solutions to needlestick injuries.[20] Instead, nurses viewed such practical measures as more frequent inspections and more frequent emptying of disposal containers as the best way to reduce injuries.

Despite these views, education is of documented benefit. The significantly higher rates of needlestick injuries for part-time employees, who "may be less familiar with the routines utilized for needle disposal and also less available for in-service education"[9] attest to the value of education. Educational efforts should be directed toward physicians, nursing personnel (including RNs, who in one study showed twice as many needlesticks as LPNs), clinical laboratory technicians, and housekeeping staff; special efforts should be made to educate part-time personnel and employees on all three shifts, as well as personnel with less than 1 year's experience, because these groups appear to be at greater risk for needlesticks.[9] Educating health care workers about the importance of receiving the hepatitis B vaccine, as discussed above, is also important.

An effective in-service educational program to prevent needlestick injuries should:

- Explain the hazards and risks associated with bloodborne pathogens, using the latest literature available.
- Stress hospital policies on needle use and disposal.
- Describe the steps for reporting and following up on a needlestick injury, should one occur.
- Provide the necessary training on any specific needlestick-prevention devices used.

The effectiveness of training should be assessed, for example, by questioning personnel on their understanding and knowledge, observing actual practice (handling and disposal), and monitoring the frequency of incidents. Training techniques, programs, and frequency should be modified as appropriate.

Needle-Recapping Controversy

CDC, Environmental Protection Agency, and Joint Commission on Accreditation of Healthcare Organizations all recommend that needles "not" be recapped by the traditional two-handed technique. In its latest proposed recommendations on prevention of HIV transmission in health care settings,

OSHA states: "Needles shall not be recapped (by the traditional two-handed technique), purposely bent or broken by hand, removed from disposable syringes, or otherwise manipulated by hand. Resheathing instruments, self-sheathing needles, or forceps shall be used to prevent recapping needles by hand."[17]

In support of this position, numerous studies show recapping to be the cause of a significant portion of all needlestick injuries.[21] According to a recent study performed at four large teaching hospitals, the percentage of injuries resulting from needle recapping was greater than 25%, and exceeded 50% in four instances.[22] The reasons for recapping were listed as inadequate knowledge (i.e., the misperception that recapping is a way to avoid needlesticks) concerns about personal risk, forgetfulness, and being too busy to follow universal precautions.

Some researchers and practitioners favor recapping with a wide-mouth needlecap, a one-handed scoop technique, or a recapping device. Proponents of recapping argue that while, in theory, disposing of uncapped needles into permanently sealable containers sounds like an ideal solution to the needlestick problem, in actuality this technique poses additional problems—not all needles are properly disposed of, and needles stuffed into overfilled containers may still be dangerous. Although the traditional two-handed recapping technique may be hazardous, handling an exposed contaminated needle is no safer.

As a result, recapping proponents claim, "over one-half of needle injuries, particularly those occurring during disposal, are not inflicted away from the bedside by uncapped needles, many of which probably could have been capped at the bedside."[23] One danger of these so-called downstream injuries, they argue, is that victims have no idea whether the needles they have been stuck with are contaminated, whereas those who recap needles at the bedside using safer recapping devices "are in a good position to permanently neutralize the needle and to seek appropriate treatment if they do suffer a stick."[23]

Problems with recapping underscore the need for proper training. In one recent study, the rate of needle recapping used with venipuncture and for percutaneous medication injections fell from 61% to 16%. In this 12 months period, an educational program was developed that reported the rate of needle recapping to employees.[24] In addition to such preventive measures as thorough education and training and the use of preventive devices, proper disposal techniques must be used.

Sharps Disposal

CDC stresses that disposal containers for needles should be located as close as possible to the point of use, presumably at the bedside in all patients' rooms.* However, such a plan poses significant operational problems.

To provide a box at the bedside, with a lock to affix it to the wall, requires personnel time and a regular maintenance schedule. This may be relatively easy in an intensive care unit (ICU), where boxes at each bedside are within a single confined area, the number of needles handled is great, and boxes often require emptying or changing on at least a daily basis. It is more difficult to implement in patient rooms that are widely dispersed throughout a multifloor building or several buildings, where needles are handled less frequently and at variable rates per room. These factors increase the difficulty in establishing a routine maintenance schedule because the need to empty or change boxes would vary from room to room and floor to floor.[18]

Similar problems are posed by affixing a needle disposal box to the nursing medication cart. Ideally, medication carts should be brought to the patient's bedside. In practice, however, many medication carts remain in the nursing station or hallway while nurses carry individual medications to patients by hand, a situation that "leaves the nurse at one end of the hall with a used needle and the dilemma of how to get it safely back to the disposal box in the nursing station."[18]

Hospitals purchasing a needle disposal system should make sure it is properly labeled and stands out as an infectious waste disposal container. Some disposal systems available are visually aesthetic and may be confused with a noninfectious unit (e.g., a towel dispenser).

---

* For further information on waste disposal containers, see ECRI's *Product Comparison Systems.*

While any needle disposal system is likely to have important limitations, it is an essential component of any needlestick-prevention program. In addition to ensuring that the system is easy to use and that containers are sturdy, it is important that the disposal container be located in all patients' rooms and other areas where needles are used and that a maintenance schedule that precludes overfills be established. Staff responsible for replacing containers should be clearly identified (e.g., nursing, housekeeping), and a mechanism for recognizing, reporting, and correcting any container hazards that may arise should be in place.

Preventive Devices on the Market

In the following section, we review 26 products from 22 manufacturers that represent some of the devices currently being marketed for needlestick prevention. We have divided the products into eight different Product Groups based on their intended use:

1. Needleless medication/vaccine injectors
2. Prefilled medication systems
3. IV starters with catheters
4. IV medication connectors
5. Blood collection systems
6. Disposable syringes
7. Needle guards
8. Needle-recapping devices

We reviewed products that are marketed as aids in reducing needlestick risks and that were provided to us for inclusion in this study; additional products may also be available. These devices were assessed for their ease of use and effectiveness in preventing needlesticks in various applications. Although we have included list prices for all devices and accessories, the actual selling prices may be substantially lower; thus, we have included guidance on performing a cost analysis in Analyzing Costs Associated with Needlesticks and Preventive Devices in the Discussion and Recommendations section. Also in this section, we address which of the Product Groups afford the greatest protection for four main applications in Selecting Needlestick-Prevention Devices for Four Clinical Applications.

Groups afford the greatest protection for four main applications in Selecting Needlestick-Prevention Devices for Four Clinical Applications.

Product Reviews

In this section, we examine 26 products from 22 manufacturers, categorized into eight Product Groups. Although we describe the general effectiveness of the Product Groups and comment on the individual devices in this section, readers should also refer to the Discussion and the Recommendations section, where we provide guidance on which Product Groups are appropriate for four main clinical applications, before making purchasing decisions. (Products are not shown to scale.)

Product Group 1: Needleless Medication/Vaccine Injectors

General Description

This group consists of gas-pressurized, needleless injectors that replace a syringe and needle. Current systems can deliver intramuscular (IM) or subcutaneous injections, primarily vaccines. Medication injectors intended for home use are not included in this review.

General Effectiveness

The single medication/vaccine injector we examined is effective in preventing needlesticks; however, it has limited applications. It may be useful in a large clinic that is administering large numbers of vaccinations.

BIOJECT BIOJECTOR®

Bioject Inc. [108133]
7620 S.W. Bridgeport Rd.
Portland, OR 97224
(503) 639-7221

Description and use: The Biojector is a reusable vaccination injector that uses pressurized CO_2. Users can fill disposable single-dose Bioject ampules (Figure 6.5).

According to the manufacturer, the Biojector is currently being used with DPT (diphtheria, pertussis, tetanus), influenza, tetanus, MMR (measles, mumps, rubella), yellow fever, and typhoid fever vaccines; it can also administer medications such as narcotics, analgesics, anticoagulators, vitamins, antibiotics, and hormones, but in only 0.5 mL doses.

List prices:

Cost/unit:
$475, Biojector

Additional cost/item:
$0.40, CO_2 cartridge (10–20 injections/cartridge)
$0.99, sterile single-dose disposable drug ampule

Comments: This product reduces cross contamination from needlesticks when administering IM or subcutaneous injections. A needle is still necessary to draw medication into the ampules, although it poses little risk of infection because it is not used for injections. The manufacturer states that prefilled variable-volume ampules will be available in the near future that will eliminate the need for a needle and allow the unit to be used for additional medications.

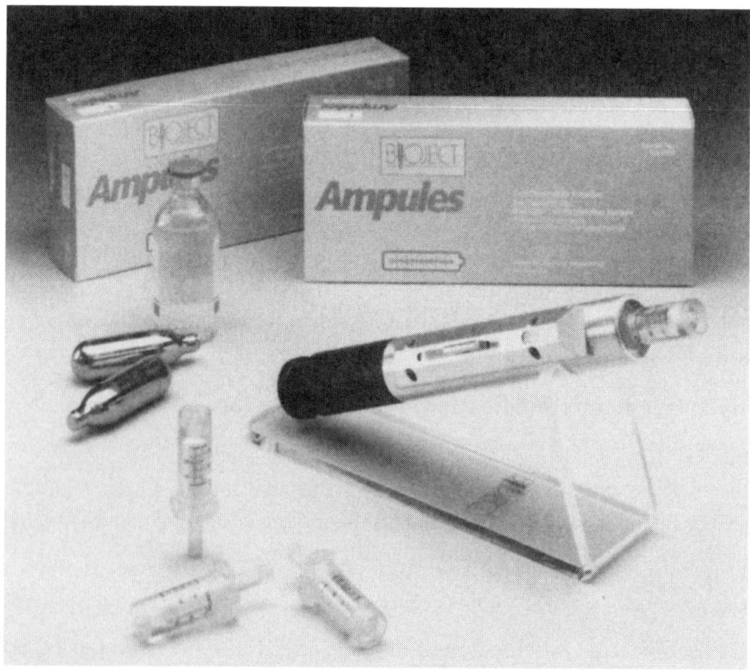

FIGURE 6.5 BiojectBiojector (with stand and supplies). (From BiojectBiojector, Portland, OR.)

Product Group 2: Prefilled Medication Systems

General Description

These systems, which are marketed as preventive devices by the manufacturers, were designed as convenient methods of administering medication and minimizing errors by supplying premeasured unit doses. Current products consist of a reusable cartridge holder and prefilled medication cartridge with a needle, and are intended to be dropped into a nearby needle disposal container after use.

General Effectiveness

None of the systems we examined effectively reduces the risk of needlesticks and none is recommended as a needlestick-prevention device. Prefilled medication systems are no safer than using a needle and syringe—the user is at risk from the exposed needle after use until it is disposed of. This especially affects those health care workers that do not have a disposal container nearby because of the risks of transporting the exposed needle and cartridge to a container. These products do eliminate the need for a needle for drawing medications into the syringe, but this procedure does not pose an infection risk.

WINTHROP PHARMACEUTICALS CARPUJECT®

Winthrop Pharmaceuticals
Div. Sterling Drug Inc. [104392]
90 Park Ave.
New York, NY 10016
(212) 907-2525; (800) 446-6267

Description and use: The Carpuject is available with a variety of medications and solutions. Users dispose of the cartridge by unscrewing the plunger, opening the blue cam lock, and releasing the cartridge directly into a disposal unit. No needle protection is provided after the cap is removed (Figure 6.6).

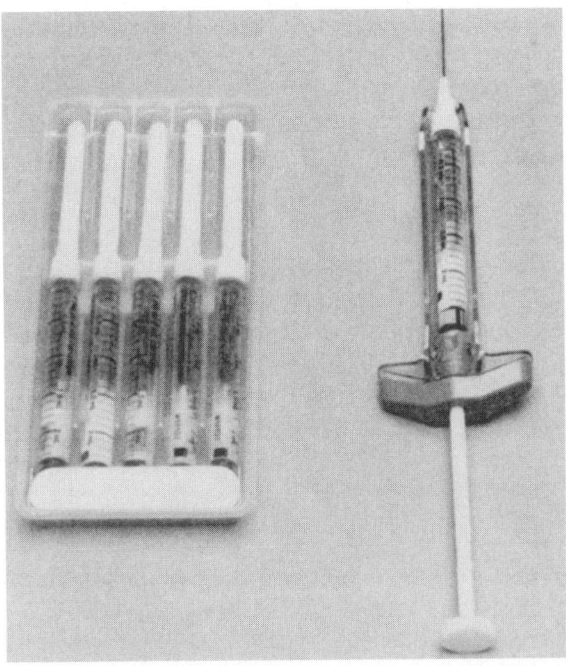

FIGURE 6.6 Winthrop Pharmaceuticals Carpuject (prefilled medication cartridges [left] and holder and cartridge set up for use). (From Winthrop Pharmaceuticals Carpuject, New York, NY.)

List prices:

Cost/unit:
No charge for Carpuject holder

Additional costs/item:
$0.40–$1.00 for the different cartridges

Comments: This system does not reduce the risk of needlesticks. If the disposal system is not located nearby, users may recap the needle or unscrew the cartridge and manually dispose of it, increasing the risk of needlesticks. Also, the reusable holders may be misplaced.

WYETH-AYERST TUBEX®

Wyeth-Ayerst Laboratories [101864]
P.O. Box 8299, Philadelphia, PA 19101
(215) 688-4400; (800) 424-8800

Description and use: The Tubex is also a needle cartridge system. Wyeth-Ayerst has disposable cartridges for a variety of medications and solutions. The design of the reusable holder is slightly different from that of the Winthrop Carpuject system. The cartridge is unscrewed from the holder and dropped vertically into a disposal container; the needle cannot be recapped. No needle protection is provided from the time the cap is removed to the time the cartridge is discarded (Figure 6.7).

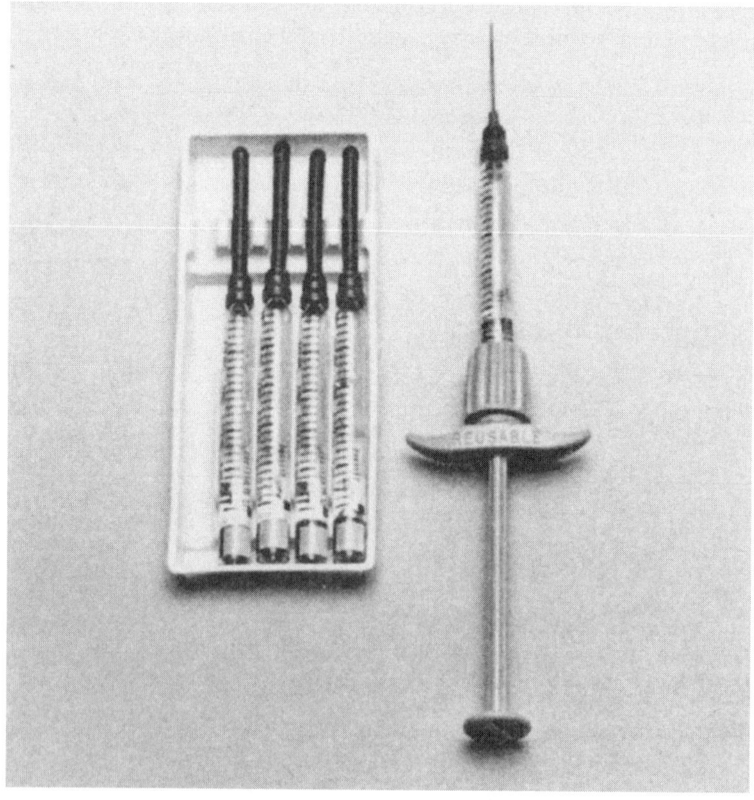

FIGURE 6.7 Wyeth-Ayerst Tubex (prefilled medication cartridges [left] and holder and cartridges set up for use). (From Wyeth-Ayerst Tubex, Philadelphia, PA.)

List prices:

Cost/unit:
No charge for Tubex holder

Additional costs/item (examples only; additional cartridges are available):
$0.52–$0.83, heparin lock flush solution
$0.55–$0.66, morphine
$0.63–$0.69, codeine
$0.73–$0.81, hydromorphone

Comments: This system does not reduce the risk of needlesticks. If the disposal system is mounted too high or is a horizontal-drop container, users may unscrew the cartridge and manually dispose of it, increasing the risk of needlesticks. Also, the reusable holders may be misplaced.

Product Group 3: IV Starters with Catheters

General Description

We examined three catheters and one needle infusion set used to administer IV therapy and draw or administer blood. These devices may incorporate a heparin lock for intermittent ("push") medication therapy.

General Effectiveness

The four devices we reviewed provide some safety when removing the introducer needle from the catheter or the infusion set needle from the arm. These products represent different needlestick-prevention designs, three of which appear to effectively reduce the risk of needlesticks: the Critikon ProtectIV can be used to replace many common catheters, the Menlo Care Landmark. Catheter is expensive and is primarily intended for intermediate-term and special applications, and the Ryan Medical Shamrock™ is appropriate only for short-term use and where its metal needle will not be a problem. The Deseret Intima reduces some of the risks of needlestick, but can still pose a significant hazard—for example, if misplaced in linens. Thus, hospitals will need to consider specific clinical needs and determine whether these or other products are appropriate (Figure 6.8).

CRITIKON PROTECTIV™

Critikon Inc.
A Johnson & Johnson Co. [101346]
4110 George Rd.
P.O. Box 31800
Tampa, FL 33631-3800
(813) 887-2000; (800) 237-2033

Description and use: The ProtectIV is and IV catheter with a built-in guard that covers and locks over the introducer needle as it is withdrawn from the vein. The catheter is available with different needle sizes (14 G–24 G).

List price:

Cost/unit:
$2.50, ProtectIV catheter

Comments: This product appears to be easy to use and to reduce the risk of needlesticks when IV therapy is started. However, like any such catheter, this device terminates in a Luer hub without a septum (or extension tubing), which exposes the user to the patient's blood. The manufacturer recommends using digital pressure above the catheter tip during the procedure, quickly connecting the IV set, and using gloves to minimize blood contact.

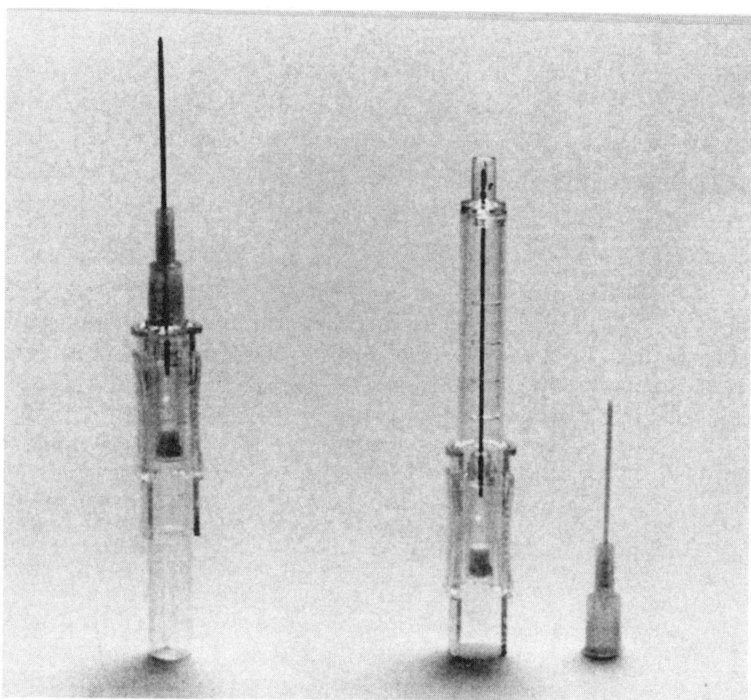

FIGURE 6.8 Critikon ProtectIV (before use [left] and after use showing protected needle for disposal [center] and catheter). (From Critikon ProtectIV, Southington, CT.)

DESERET MEDICAL INTIMA™

Deseret Medical Inc.
Becton Dickinson and Co. [101750]
9450 S. State St.
Sandy, UT 84070
(801) 255-6851; (800) 453-4538

Description and use: The Intima is an IV catheter with an introducer needle attached to a styles. After the catheter is in place, the needle-stylet assembly is removed through an injection adapter (which reduces blood leakage), exposing the needle. The Intima is available with or without a Y-site and in different needle sizes (Figure 6.9).

List prices:

Cost/unit:
$2.30, Intima IV catheter
$2.30, Intima IV catheter with Y-site
(PRN injection adapter included with catheter)

Comments: The flimsy styles on this product reduces the risk of needlesticks, although contact with the needle is possible while holding the styles. Of even greater concern, the styles could be left in the linen, and nursing or housekeeping staff could become injured by the needle end. Bright, visible color on the styles might help minimize, but not eliminate, this risk. Also, if the styles is removed improperly, blood could splash in the user's eye. The styles may also be more difficult to aim into a disposal container and may pop out of the top more easily.

FIGURE 6.9 Deseret Medical Intima (before use [left] and after use showing protected needle for disposal [center] and catheter). (From Deseret Medical Intima, Sandy, UT.)

Menlo Care Landmark®

Menlo Care Inc. [107575]
1350 Willow Rd.
Menlo Park, CA 94025
(415) 325-2500; (800) 752-8900

Description and use: The Landmark catheter is an over-the-needle device that is inserted in the antecubital fossa area and advanced up to 6 in. until the tip is in an upper-arm vessel. After the catheter is placed in the arm, the needle styles is removed and locked into a protective case (Figure 6.10).

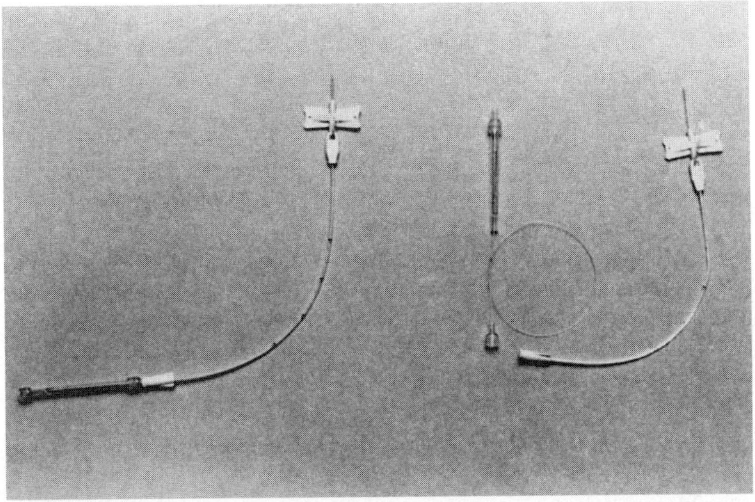

FIGURE 6.10 Menlo Care Landmark. (before use [left] and after use, showing protected needle for disposal [center] and catheter). (From Menlo Care Landmark, Menlo Park, CA.)

Applications include selected chemotherapies, hydration therapy, antibiotic therapies, blood delivery, and pain management. The catheter is made of Aquavene, which softens to reduce vein trauma and subsequent complications. This catheter is designed to be used for intermediate-term therapies (typically 10 days to several weeks) and is therefore introduced through a sterile procedure. It may be used as an alternative to a central line or multiple peripheral sticks for some applications. The manufacturer suggests that this catheter can be used by home infusion therapy patients (especially those with AIDS or cystic fibrosis); cardiac, pediatric, obstetric, and orthopedic patients; and some oncology patients.

List price:

Cost/unit:
$33 per Landmark catheter in case volume

Comments: This product reduces the risk of needlesticks, but is useful only for very specific applications.

RYAN MEDICAL SHAMROCK

Ryan Medical Inc. [108525]
Suite 201, 7106 Crossroads Blvd.
Brentwood, TN 37027
(615) 370-4242

Description and use: The Shamrock safety-winged, butterfly-needle infusion set provides a safety shield over the needle when it is removed from the vein. A visual indicator appears when the needle is locked inside the shield, which also produces an audible click (Figure 6.11).

List Price:

Cost/unit:
$0.58, (21, 23, or 25 G) Shamrock needle

Comments: This device reduces the risk of needlesticks. However, the nurses we spoke with were concerned that a steel needle is not as effective as a flexible catheter in minimizing trauma to the patient's vein. The manufacturer states that this device is intended for short-term IV therapy (for which venal trauma may not be as great a concern), or IV therapy for children. It is considerably less expensive than the catheters in this group.

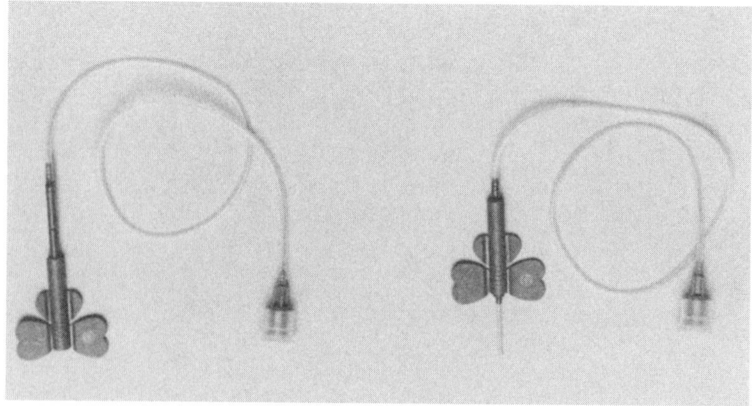

FIGURE 6.11 Ryan Medical Shamrock (before use [left] and after use). (From Ryan Medical Shamrock, Brentwood, TN.)

Product Group 4: IV Medication Connectors

General Description

These devices are intended to eliminate an exposed needle after administering medications through an IV delivery system or catheter. They may also be used for collecting blood through an IV line into a syringe or possibly a blood collection tube adapter (see Product Group 5). Several of the products we examined have a needle recessed in a plastic covering that can be inserted into a Y-site; the other end of the product has a Luer fitting for connecting a syringe or another infusion set. Three such products, the Baxter Needle*Lock™, IMS Stick-Gard™, and the Tri-State KleenNeedle®, provide a simple protective needle system for connecting to most standard septum-terminated Y-sites. Two other manufacturers, Burron Medical and ICU Medical, will soon introduce comparable products.

Other products in this group are similar, but are intended for use with their own specific injection adapters (e.g., those used as a heparin lock) that can be purchased separately or as part of the manufacturer's extension set; these additional components must be considered in the cost. Two manufacturers also have valves that can be used in place of a stopcock or injection adapter, and one manufacturer has a needle-free device that uses a plastic cannula.

These other systems are designed for special applications or may require special or additional products. Some of these devices used at a Y-site for administering drugs must be positioned relatively close to the catheter site; users should judge whether this will be inconvenient. None of these devices is designed for IM use, and, except as noted, all require a needle to draw up the medication. Users must carefully assess the cost, compatibility with existing systems, and benefits of these products.

General Effectiveness

All of the IV medication connectors reduce the risk of needlesticks when a standard needle would otherwise be used to give intermittent IV or piggyback medications; however, the clinical usefulness depends on the hospital's needs and existing products. Because a needle is used only for withdrawing medication from a vial and not for injecting the patient, there is little risk of infection; however, users may be tempted to use this needle for injection (Figure 6.12).

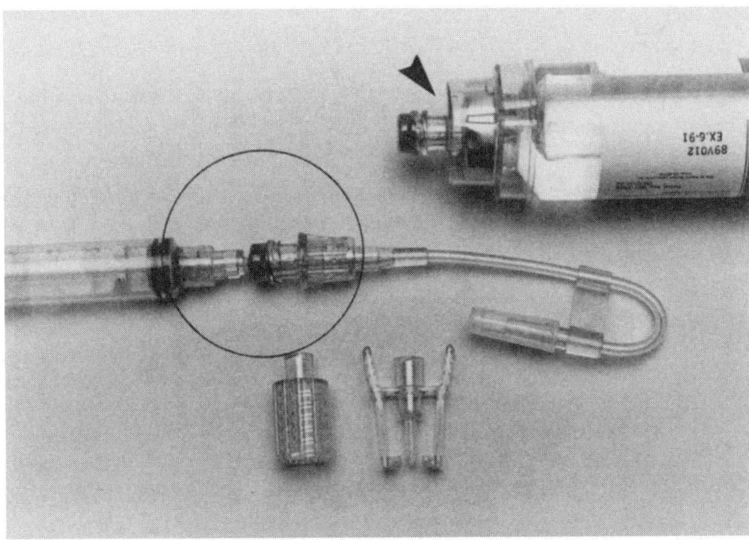

FIGURE 6.12 Baxter Healthcare InterLink IV Access System (multidose drug vial adapter [arrow], blunt cannula [in circle, left], injection adapter [in circle, right], and different types of cannulas [bottom]: threaded lock [left] and lever lock). (Baxter Healthcare InterLink, Deerfield, IL.)

A concern with these products is how often they need to be replaced to maintain asepsis. Depending on the way the product is designed and packaged, it may need to be replaced either every time a medication is given or possibly not until an IV set is changed. (This also affects cost per use.) Also, some concern may exist about the degree of asepsis protection as devices are connected and disconnected. For example, Luer connectors cannot be effectively wiped with alcohol, a common procedure with septum interfaces. Placing a cap over exposed Luer ports may help maintain an aseptic interface.

Most of these products have not had sufficient clinical trials to determine whether they pose any increased risk of patient infection. The Infection Control Committee and users should therefore review the use of these devices before implementing them and should monitor sepsis frequency. Again, the most effective and safe way of using these products should be determined and then monitored to confirm proper usage.

BAXTER HEALTHCARE INTERLINK™ IV ACCESS SYSTEM

Baxter Healthcare Corp., IV Systems Div. [106390]
1425 Lake Cook Rd.
Deerfield, IL 60015
(708) 940-5000

Description and use: The InterLink IV Access System, which is designed to replace the traditional needle and injection adapter, consists of two mating components: a special injection adapter connected to the Luer fitting of the catheter or extension set and a blunt plastic cannula that inserts into the special septum of the injection adapter. The cannula is available for syringes and IV sets (lever or threaded-lock connector) and prefilled syringes (heparin lock flush or sodium chloride). Also available is a multidose drug vial adapter to avoid having to use a needle when drawing the drug into the syringe.

List prices:

Cost/unit:

- $1.50, InterLink injection adapter (replaced with each catheter or IV set change)
- $0.25, blunt cannula
- $0.45, lever-lock cannula
- $0.50, threaded-lock cannula
- $2.50, drug vial adapter

Comments: This system eliminates the need for needles to draw up medication or solutions from multidose drug vials. The manufacturer states that it will offer a cannula for single-dose drug vial access sometime this year. The blunt cannula can be used only with the manufacturer's injection adapter. This adapter can be purchased separately or as part of the manufacturer's special extension sets with a Y-site. The full benefits of this product are achieved only when the entire system is used. Users should evaluate compatibility with existing IV products (e.g., extension sets) and carefully assess total system needs, costs, and benefits.

BAXTER HEALTHCARE NEEDLE*LOCK

Baxter Healthcare Corp.
IV Systems Div. [106390]
1425 Lake Cook Rd.
Deerfield, IL 60015
(708) 940-5000

Description and use: The Needle*Lock consists of a needle recessed in a plastic housing with a Luer connector on one end that can connect to any Luer connection (e.g., syringe, IV set). The needle housing at the other end of the device has a plastic hook that allows the Needle*Lock device to be secured on the Y-site. This device is available alone or as part of a secondary medication or IV set (Figure 6.13).

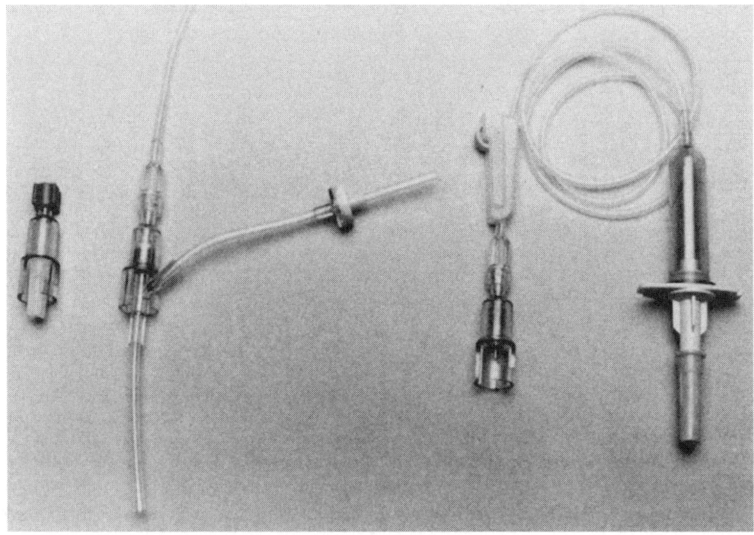

FIGURE 6.13 Baxter Healthcare Needle*Lock (device as packaged [left] and attached to a Y-site [center] and a secondary medication set). (From Baxter Healthcare Needle*Lock, Round Lake, IL.)

List prices:

Cost/unit:
$0.50, Needle*Lock (18 or 20 G)
$3.78, secondary medication set with Needle*Lock (18 G)
$4.25, IV set with Needle*Lock (20 G)

Comments: This device may be useful for piggybacking medications, because it provides a secure connection without taping. It cannot be secured directly to a catheter except at a Y-site. If this product is compatible with existing IV sets, it may be less expensive than some of the other devices in this group.

BURRON MEDICAL SAFSITE™

Burron Medical Inc.
Div. B. Braun of America Co. [101098]
824 12th Ave.
Bethlehem, PA 18018
(215) 691-5400; (800) 523-9695

Description and use: The Safsite is a reflex valve with a Luer connection at both ends. It opens when any standard Luer taper is connected and closes when the taper is disconnected. The valve is designed to replace an injection adapter and is compatible with standard IV sets. The Safsite can be purchased separately or with different extension sets. It is available with a Luer-lock replacement cap. A Luer adapter that can fit on a Winthrop Pharmaceuticals Carpuject cartridge is also available; this adapter replaces the Carpuject needle provided and fits onto the Safsite (Figure 6.14).

List prices:

Cost/unit:
$2.30, Safsite reflux valve with small-bore T-port extension set, male Luer-slip fitting
$2.20, Safsite reflex valve with 6″ small-bore extension set, male Luer-slip fitting
$1.07, Safsite reflux valve adapter
$0.25, Carpuject cartridge Luer adapter

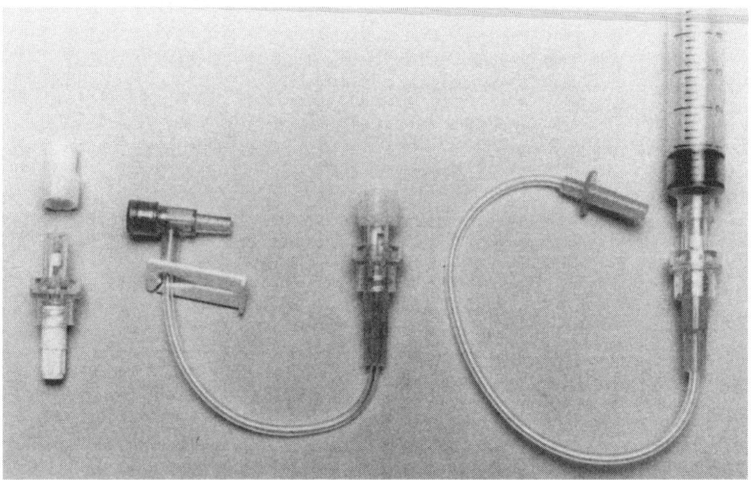

FIGURE 6.14 Burron Medical Safsite (reflex valves as packaged [left], T-port extension set with reflux valve attached to a T-port extension with a dead-end cap on valve [center], and reflux valve attached to an extension set as it might be used). (From Burron Medical Safsite, Bethlehem, PA.)

Comments: Some users may be concerned about environmental pathogens entering this system; however, the Safsite valve is available with a standard dead-end cap similar to port covers on a stopcock. This device can be used as an injection adapter, in place of a stopcock, or with the manufacturer's extension sets. It is not compatible with the injection adapters on existing IV sets.

Burron plans to offer its new Safsite Access Pin, which is designed to convert a Y-site on an administration set to a needle-free access site.

ICU MEDICAL CLICK LOCK™

ICU Medical Inc. [105658]
142 Technology Dr.
Irvine, CA 92718
(714) 753-1599; (800) 824-7890

Description and use: The Click Lock consists of a needle recessed in a plastic housing with a Luer connector on one end that can connect to any Luer connection (e.g., syringe, IV set) (Figure 6.15).

List prices:

Cost/unit:
- $1.60, Click Lock (with 18, 20, 21, or 23 G needle)
- $0.48, injection adapter
- $0.07, replacement needle (18, 20, 21, or 23 G)

Comments: This product is not compatible with some catheters unless the manufacturer's adapter is used, and it requires the manufacturer's adapter for piggybacking medications. A catheter or IV set with a Y-site or injection adapter may not be compatible with the Click Lock because the needle may not fit in the injection site or adapter. However, the manufacturer states that a new version will soon be available that ensures a secure and locked connection to injection ports on all standard IV administration sets with a Y-site. The manufacturer claims that the needles on the Click Lock can be replaced, which reduces costs. However, the risk of infection should be evaluated by the hospital before considering this procedure.

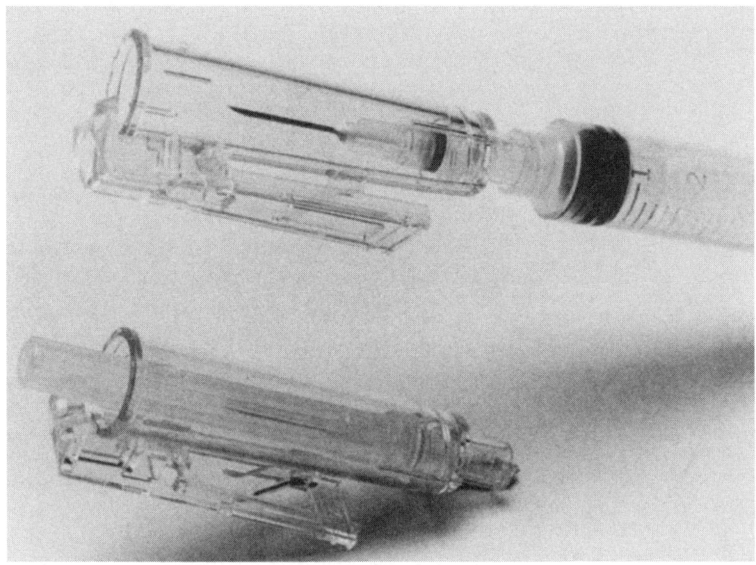

FIGURE 6.15 ICU Medical Click Lock (as packaged [bottom] and set up for use) (ICU Medical Click Lock, San Clemente, CA.)

IMS STICK-GARD

International Medication Systems (IMS) Ltd. [104648]
1886 Santa Anita Ave.
South El Monte, CA 91733
(818) 442-6757; (800) 423-4136

Description and use: The Stick-Gard consists of a needle recessed in a plastic housing to any Luer connector on one end that can connect to any Luer connection (e.g., syringe, IV set). This device can also be used with selected IMS MIN-I-JET prefilled syringes and is available in 18 or 20 G needles (Figure 6.16).

List price:

Cost/unit:
  $0.19, Stick-Gard

Comments: This product appears to work well for intermittent medications with IV sets or catheters and may be less expensive than some of the other devices in this group.

L & W TECHNOLOGY SAFEPORT™

L & W Technology Inc. [108520]
P.O. Box 69392
Los Angeles, CA 90069
(213) 275-7464; (800) 648-5920

Description and use: The Safeport injection site, which is used in place of an injection adapter, works like an automatic stopcock. This product is designed with an antibackflow and antiair embolism valve. It has Luer lock connections at each end, and a syringe or IV line can be locked onto the Safeport for continuous or intermittent use. The double- and triple-port manifolds have an in-line check valve in the main flow channel, allowing injections to be given without needing to manually pinch off the IV line (Figure 6.17).

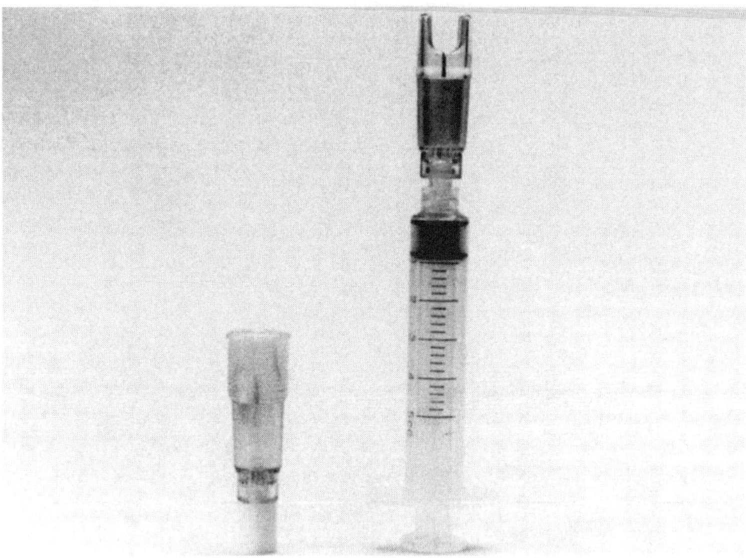

FIGURE 6.16 IMS Stick-Gard (as packaged [left] and set up for use). (From IMS Stick-Gard, South El Monte, CA.)

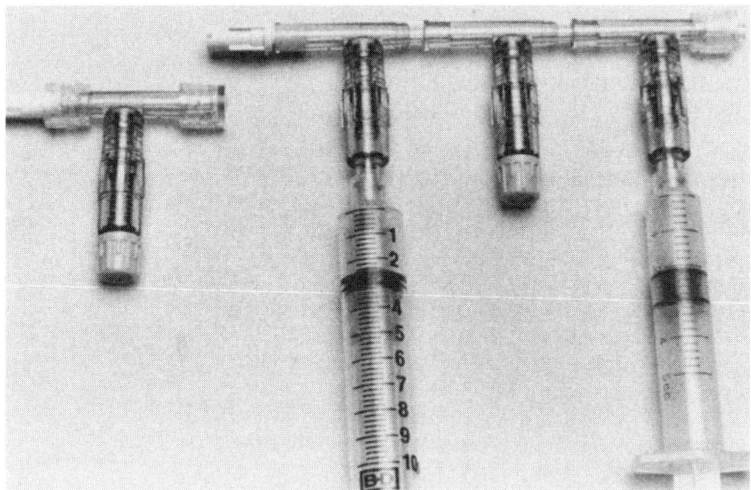

FIGURE 6.17 L&W Technology Safeport (single port as packaged [left] and triple port with syringes attached as it might be used). (From L&W Technology Safeport, Ottawa, ON.)

List prices:

Cost/unit:
  $2.35, single-port Safeport
  $4.70, double-port Safeport
  $6.95, triple-port Safeport

Comments: This device is designed to be used for ICU or anesthesia patients who may need to have a syringe left on the IV line while being closely supervised for intermittent medication doses (e.g., during slow administration of a drug such as Dilantin), for multiple simultaneous injections, or for other applications

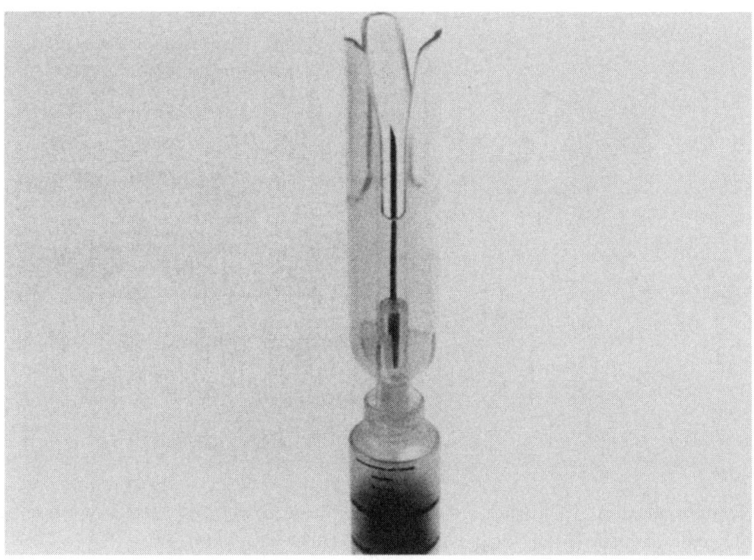

FIGURE 6.18 Pascall Medical SPIVE (attached to syringe). (From Pascall Medical SPIVE, Cocoa Beach, FL.)

in which a stopcock would normally be used. When this device is used in place of a stopcock, it does not reduce the risk of needlesticks, because a needle would not normally be used. (When a stopcock is used, it has a Luer connection so that a syringe would be hooked up directly; no needle is needed.) Users may be concerned about environmental pathogens entering the system; however, the Safeport is capped with a standard dead-end cap similar to port covers on a stopcock. This is a relatively expensive device and is intended for specialized use, and the user must ensure that Luer connections are present at the site of applications.

PASCALL MEDICAL SPIVE

Pascall Medical Corp. [108524]
Suite 618, Cape Royal Bldg.
1980 N. Atlantic Ave.
Cocoa Beach, FL 32931
(407) 784-4448; (800) 848-3036

Description and use: The SPIVE (Special Purpose IV Entering) device is a recessed needle device designed to be used only at a catheter with an injection adapter to administer intermittent medications. The device terminates in a Luer connector. This product is packaged as it is used, without a cap over the needle (Figure 6.18).

List price:

Cost/use:
$0.45, SPIVE

Comments: This product has just recently been introduced. It is not intended for use at a Y-site or for nursery applications.

RYAN MEDICAL SAF-T CLIK® IV IN-LINE CONNECTOR

Ryan Medical Inc. [108525]
Suite 201, 7106 Crossroads Blvd.
Brentwood, TN 37027
(615) 370-4242 (Figure 6.19)

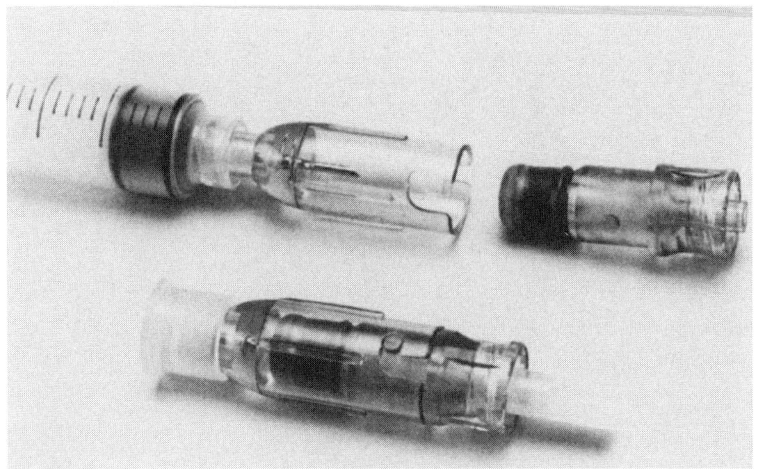

FIGURE 6.19 Ryan Medical Saf-T Clik (as packaged [bottom], attached to a syringe as it might be used [top left], and the protected needle adapter). (From Ryan Medical Saf-T Clik, Brentwood, TN.)

Description and use: The Saf-T Clik consists of two components that lock together; one component has a short needle with a protective case that can fit a Luer adapter, and the other is a special injection adapter that fits into the first component.

List prices:

Cost/unit:
- $0.98, Saf-T Clik
- $0.43, protected needle adapter
- $1.18–$1.30, connector with minibore tubing, extension set
- $0.56, heparin lock
- $0.90, extension set with minibore J-loop tubing
- $2.80, extension set with back-check valve
- $1.90, extenuation set with side-clamp

Comments: This product is compatible with only the manufacturer's extension sets or special injection adapters to the catheter. The short needle component is protected with a cap that allows reuse if multiple injections are necessary. However, there may be a risk of infection if a new protected needle component is not used each time. The manufacturer is developing an additional extension set.

TRI-STATE HOSPITAL SUPPLY KLEEN-NEEDLE®

Tri-State Hospital Supply Corp. [103919]
301 Catrell Dr.
P.O. Box 170
Howell, Ml 48843
(517) 546-5400; (800) 248-4058 (Figure 6.20)

Description and use: The Kleen-Needle system consists of a short needle recessed in a plastic housing with a Luer lock on one end; the other end screws into Tri-State's own heparin lock (injection adapter). However, the needle part of the system can be used alone on a standard Y-site.

List prices:

Cost/unit:
- $0.60, Kleen-Needle
- $0.80, heparin lock

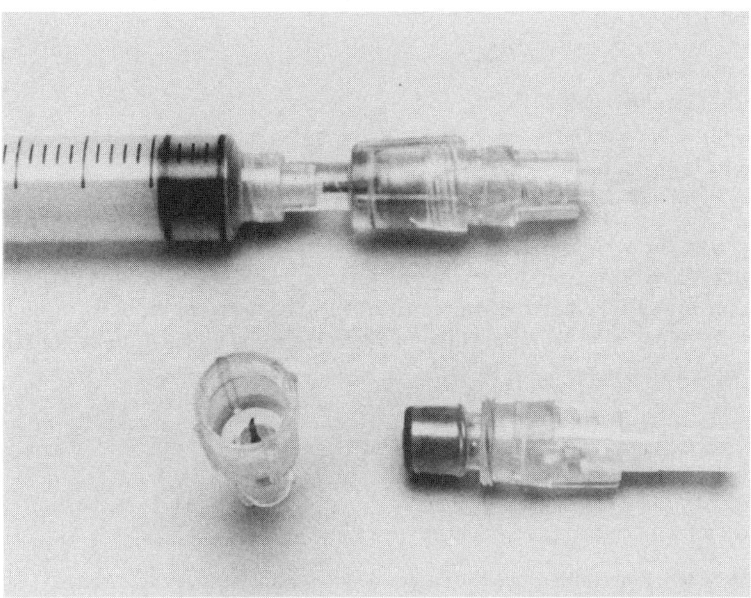

FIGURE 6.20 Tri-State Hospital Supply Kleen-Needle (attached to a syringe as it might be used [top] and Kleen-Needle [bottom left] and heparin lock as packaged). (From Tri-State Hospital Supply Kleen-Needle, MI.)

Comments: This product can be used for intermittent medications using existing IV sets or the manufacturer's heparin lock.

Product Group 5: Blood Collection Systems

General Description

This group is made up of two subgroups: blood collection tube adapters and an in-line collection system. Two of the devices are similar to conventional blood collection adapters that accept a needle and connect to the blood collection tube; the examined devices, however, have a protective sheath that can be positioned over the needle after use. The blood collection adapters are either reusable or disposable. (Standard blood collection devices are reusable; however, some hospitals may be disposing of them after each use.)

General Effectiveness

Both disposable and reusable systems have their respective advantages and disadvantages. A disposable system should be effective after its needle is retracted; however, it costs more than a reusable system and is quite bulky, and carrying several blood adapters in a tray can be inconvenient for a phlebotomist (the manufacturer does provide a pouch for carrying adapters). Also, a large disposal container is needed to dispose of the adapter and needle. Although reusable systems provide a relatively safe way of recapping, the risk of needlestick is still present because the inner needle is exposed after unscrewing the needle from the adapter.

The Baxter/Edwards VAMP (Venous/Arterial blood Management Protection) in-line system does not serve as a needlestick-prevention device, because a needle would not normally be used to draw blood from a transducer system; typically, a stopcock would be used for this purpose. However, this system does minimize blood leakage when drawing blood and may help minimize infection. Use of this device should also be monitored to determine any change in patient infection rate.

BAXTER/EDWARDS VAMP™

Baxter Healthcare Corp.
Edwards Critical-Care Div. [106431]
P.O. Box 11150
Santa Ana, CA 92711
(714) 250-2500; (800) 424-3278

Description and use: The VAMP system is a closed, needleless system for drawing blood from a blood pressure transducer line. Blood can be drawn into a syringe or blood collection tube. A blunt cannula adapter is attached to a syringe, and the sample site is designed to accept the blunt cannula. The invasive line has a reservoir to contain the initial blood in the line; this blood is reinfused into the patient after the sample is withdrawn (Figure 6.21).

List prices:

Cost/unit:
$10.00, VAMP kit
$33.00, VAMP kit with transducer system

Comments: This system does not reduce the risk of needlesticks because the standard system would use a stopcock, which does not require a needle. The manufacturer claims that this system may reduce the risk of infection associated with Luer connectors at a stopcock.

MEDICAL SAFETY PRODUCTS ACCI-GUARD®

Medical Safety Products Inc. (MSPI) [108521]
Suite 500
2696 S. Colorado Blvd.
Denver, CO 80222
(303) 756-5151; (800) 677-1115

Description and use: The Acci-Guard is a reusable blood collection holder. After withdrawing blood, the needle is withdrawn into a plastic case, where it can safely be recapped. The needle must then be unscrewed and disposed of. It is compatible with sampling needles up to 1½″ long (Figure 6.22).

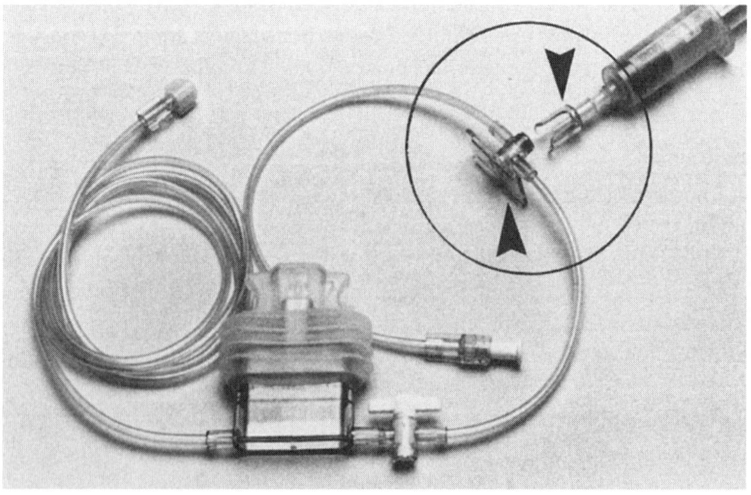

FIGURE 6.21 Baxter/Edwards VAMP (blunt cannula attached to a blood collection adapter [top arrow] and sample site [bottom arrow]). (From Baxter/Edwards VAMP, Irvine, CA.)

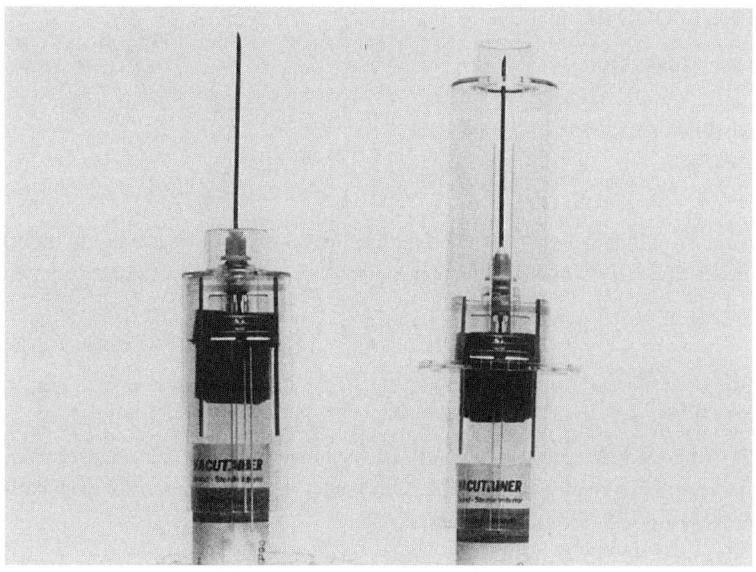

FIGURE 6.22 Medical Safety Products Acci-Guard (set up for use [left] and after use showing protected needle). (From Medical Safety Products Acci-Guard, Englewood, CO.)

List price:

Cost/unit:
$1.55–$1.95, Acci-Guard (estimated 200 uses)

Comments: This device helps reduce the risk of needlesticks. However, it may not be possible to use a screw-top disposal container. If the needle is manually unscrewed, the inner needle, which is covered only by soft rubber, presents the risk of needlestick (Figure 6.23).

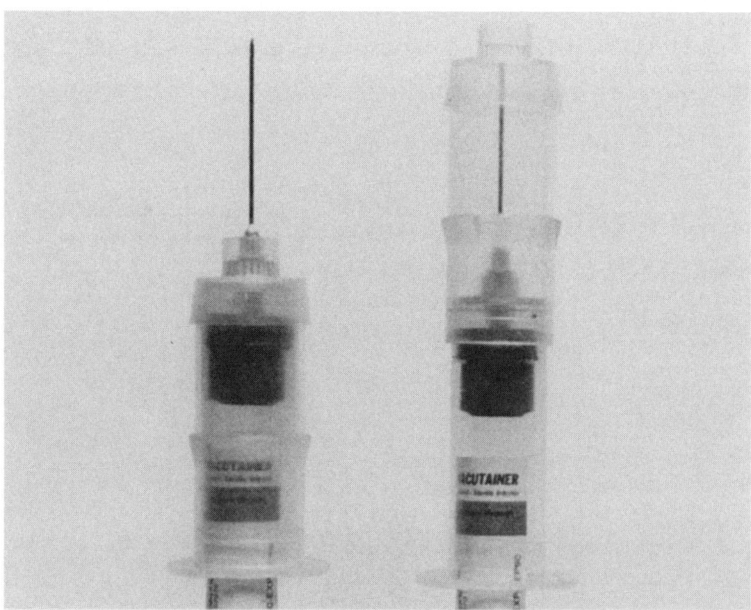

FIGURE 6.23 Ryan Medical Adapter (set up for use [left] and after use showing protected needle). (From Ryan Medical Adapter, Brentwood, TN.)

RYAN MEDICAL BLOOD ADAPTER

Ryan Medical Inc. [108525]
Suite 201
7106 Crossroads Blvd.
Brentwood, TN 37027
(615) 370-4242

Description and use: The Blood Adapter has a sheath that locks over the outer needle after use. It is compatible with sampling needles up to 1½″ long. The adapter is disposable and can be used only once.

List price:

Cost/unit:
$0.17, Blood Adapter

Comments: This device reduces the risk of needlesticks; however, the whole adapter must be discarded. This may be costlier and inconvenient to the user, who must carry an adapter for each patient (the manufacturer does provide a pouch for carrying adapters).

Product Group 6: Disposable Syringes

General Description

The products in this group have a syringe and needle with a plastic sheath that locks over the needle after use. These devices can be used for IV or IM therapy and can reduce the risk of needlesticks. However, they are available in only limited sizes. Two manufacturers have only a 3 cc syringe. One manufacturer has safety syringes in different sizes, but only up to 5 cc. Other disadvantages include the additional disposal bulk and higher cost.

General Effectiveness

These products are easy to use and offer excellent protection against needlesticks after the protective sheath is locked into position. However, available syringe sizes are extremely limited at present (Figure 6.24).

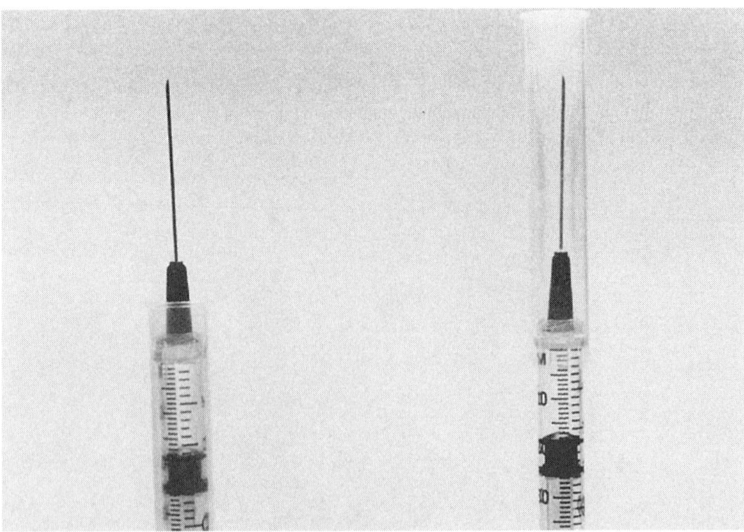

FIGURE 6.24 Becton Dickinson Safety-Lok Syringe (set up for use [left] and after use showing protected needle for disposal). (From Becton Dickinson safety-Lok Syringe, Rutherford, NJ.)

BECTON DICKINSON SAFETY-LOK™ SYRINGE

Becton Dickinson and Co.
Becton Dickinson Div. [101093]
One Stanley St.
Rutherford, NJ 07070
(201) 460-2000

Description and use: The Safety-Lok Syringe is a 3 cc syringe and needle combination used primarily to administer medication. It has a protective shield that the user locks in place over the needle after use and before disposal. This product is currently available in different needle sizes with only a 3 cc syringe.

List price:

Cost/unit:
$0.26, Safety-Lok

Comments: This product reduces the risk of needlesticks, but has limited applications.

NEEDLEPOINT GUARD SAFETY SYRINGE

NeedlePoint Guard Inc. [108060]
3969 Reuting Rd.
P.O. Box 1646
Grand Island, NE 68802-1646
(308) 384-3513; (800) 635-5878 (Figure 6.25)

Description and use: The Safety Syringe is a disposable syringe that has a protective shield that locks over the used needle. It is available in five different sizes—0.5 cc, 1 cc (insulin and tuberculin), 3 cc, and 5 cc—with various needle gauges and lengths. The 5 cc syringe may be an adequate size for blood samples. The needle and the protective shield can be separated from the syringe to allow recapping of the syringe (e.g., for obtaining blood gas samples).

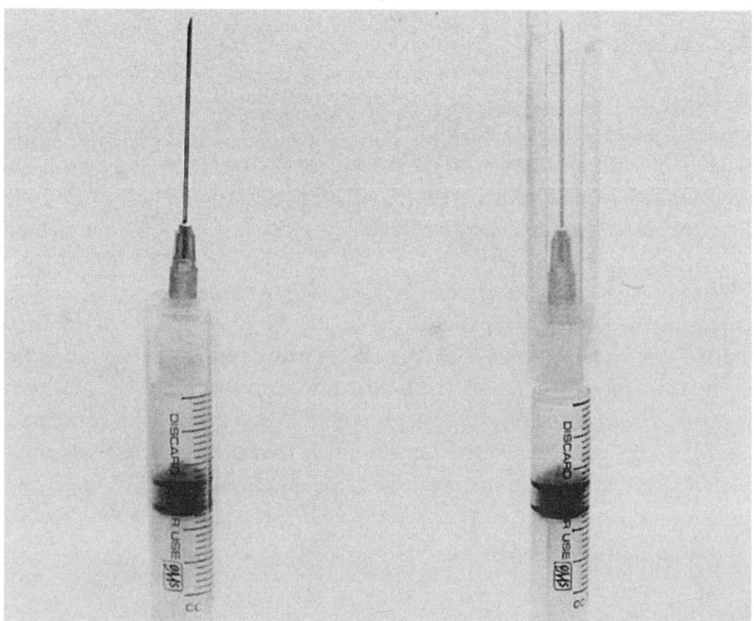

FIGURE 6.25 NeedlePoint Guard Safety Syringe (set up for use [left] and after use showing protected needle for disposal). (From NeedlePoint Guard Safety Syringe, Grand Island, NE.)

List prices:

Cost/unit:
$0.36, 0.5 and 1 cc Safety Syringe
$0.33, 3 and 5 cc Safety Syringe

Comments: This product reduces the risk of needlesticks and is the only product that is currently available in different sizes.

SHERWOOD MEDICAL MONOJECT®

Sherwood Medical Co.
Sub. American Home Products [101927]
1915 Olive St.
St. Louis, MO 63103
(314) 621-7788; (800) 325-7472

Description and use: The Monoject syringe has a protective sheath that extends over the needle and clicks into either of two positions. The first click position can be used to transport the medication to the patient's bedside without locking the sheath over the needle. The second click position is used to lock the sheath over the needle by turning the sheath in either direction. The Monoject is available with different gauge needles, but currently only in a 3 cc syringe.

List price:

Cost/unit:
$0.39, Monoject

Comments: This device reduces the risk of needlesticks. Protection against needlesticks during transport to the patient is an advantage, although the user may put the device at the first click position after the injection and assume it is permanently locked. Sherwood Medical also provides a safety feature in its standard syringes—a plastic holder for the needle cap that allows the user to perform one-handed recapping if properly trained, similar to a recapping device. This may be appropriate in some circumstances (Figure 6.26).

Product Group 7: Needle Guards

General Description

This group includes needles with a guard that can be positioned over them after use. These can be purchased separately or with syringes. One manufacturer also provides the guard without a needle. Products in this group can be used with a syringe (including some prefilled syringes) or (for one manufacturer's product) to replace a blood collection needle.

General Effectiveness

Needle guards provide one of the most adaptable designs for hospitalwide use of needlestick-prevention devices and reduce the risk of needlesticks after the protective sheath is locked into position. They also provide the hospital with one product that can be used to administer IV or IM medications, as well as to collect blood samples. However, the major limitations of this group include the cost increase over using a traditional needle and the possible awkwardness in using the device—the distance between the user's hand and the patient's vein is significantly greater because of the length of the needle guard; also, only limited needle gauges and lengths are currently available (Figure 6.27).

ICU MEDICAL HR NEEDLE™

ICU Medical Inc. [105658]
142 Technology Dr.
Irvine, CA 92718
(714) 753-1599; (800) 824-7890

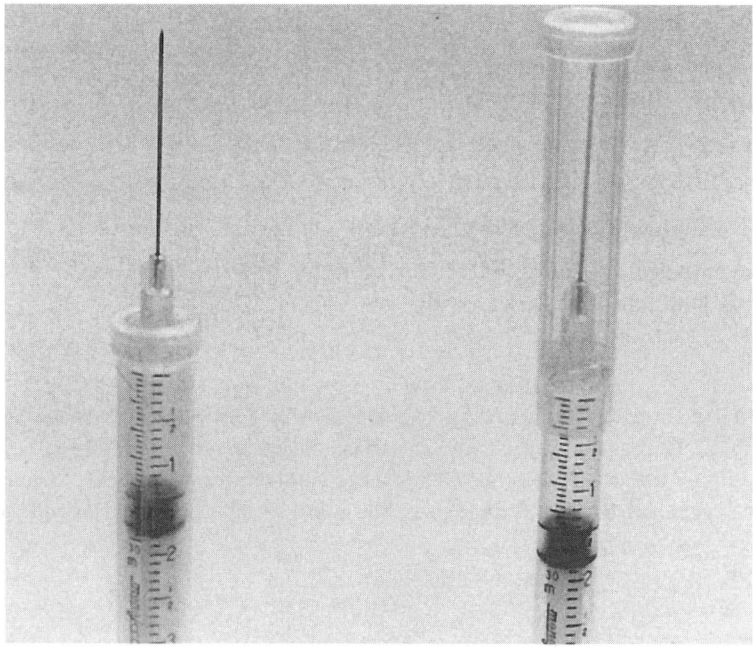

FIGURE 6.26 Sherwood Medical Monoject (set up for use [left] and after use, showing protected needle for disposal). (From Sherwood Medical Monoject, Sherwood.)

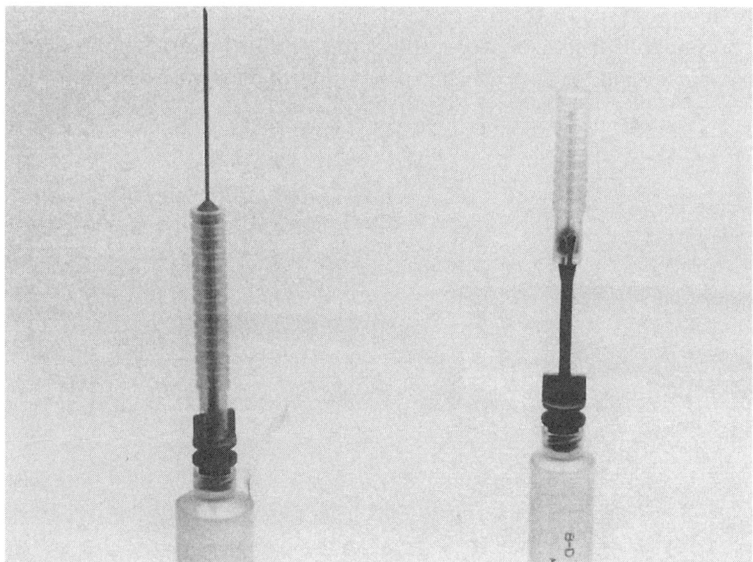

FIGURE 6.27 ICU Medical HR Needle (set up for use [left] and after use, showing protected needle for disposal). (From ICU Medical HR Needle, Irvine, CA.)

Description and use: The HR Needle has a plastic sheath that covers the needle after use. It is available in different sizes and in a blood collection adapter version for drawing blood.

List price:

Cost/unit:
$0.40, HR Needle

Comments: This device can reduce the risk of needlesticks. It is available for use with syringes or blood collection systems. The user could get stuck with the inner needle of a blood collection needle when disposing of it. However, some disposal containers allow the needle to be unscrewed directly into the container, avoiding this problem. Users may find the HR Needle awkward to use because the distance between the vein and the user's hand is doubled, which may make it difficult to control the needle during insertion.

NORTH AMERICAN MEDICAL PRODUCTS SAFE-SITE™

North American Medical Products (NAMP) Inc. [108522]
Bldg. 501, Rotterdam Industrial Park East Rd.
Schenectady, NY 12306
(518) 356-8110

Description and use: The Safe-Site is a protective guard that extends over the needle and locks in place. It is available with needle gauges 19, 20, 22, and 25 and needle lengths of 3/4″, 5/8″, and 1″. It is also available separately for use with a user-supplied needle. The manufacturer will provide customized packages of other needle gauges and different syringe sizes. The guard can also be moved down to help stabilize it on a Y-site for piggyback infusions.

List prices:

Cost/unit:
- $0.29, Safe-Site without needle and syringe
- $0.31, Safe-Site and needle
- $0.39, Safe-Site with needle and syringe

Comments: This device can reduce the risk of needlesticks. It can be used with a user-supplied needle of any gauge, but only up to 1″ length. However, this device doubles the distance between the user's hand and the patient's vein (or point of use), which may be awkward for some users (Figure 6.28).

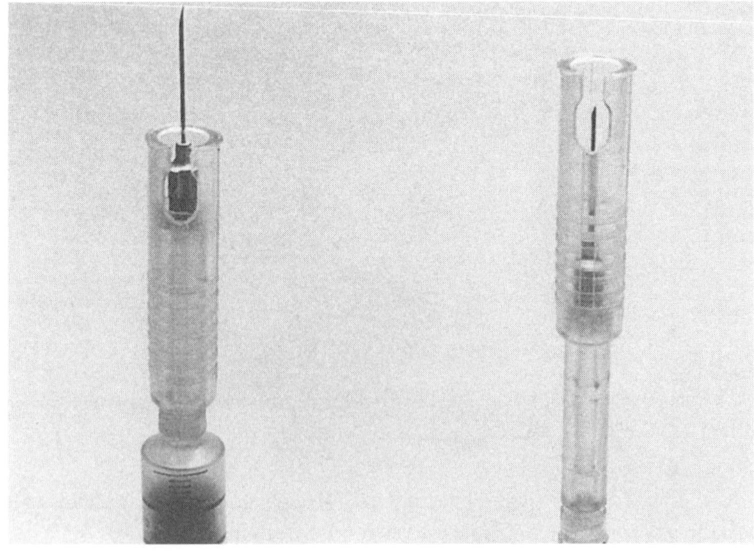

FIGURE 6.28 North American Medical Products Safe-Site (set up for use [left] and after use, showing protected needle for disposal). (From North American Medical Products Safe-Site, Schenectady, NY.)

Product Group 8: Needle-Recapping Devices

General Description

This group includes any device that can be used to recap a needle to help prevent a needlestick.

General Effectiveness

The two products examined have major limitations: (1) a recapping device is inconvenient because it will not always be readily available when it is needed, and users may avoid using it or forget to use it; and (2) it may not be compatible with some needle caps. An ideal needlestick-prevention device should be incorporated into the design of the needle so that it is readily available. A recapping device would need to be carried around by the user at all times and is not effective if the needle cap is misplaced or is not placed in the device initially. Because this device is separate from the needle, it is inconvenient, especially in emergency situations. In addition, the On-Gard Recapper™ is bulky, and needle caps sometimes fall into the device; the Terumo Safe-Guard shield appears too small to provide adequate protection.

However, there may be some applications for which a recapping device would be useful, and these devices are cost-effective. Some possible areas of use are the PACU, ICU/CCU, and dental clinics, where medications are administered serially from a single syringe and must be recapped to maintain sterility. There may also be a need to recap during surgical procedures and when obtaining blood gas samples (to prevent air contamination). Other methods of safe recapping, such as the one-handed scoop technique, require no additional devices, but do require a conveniently located flat surface.

ON-GARD SYSTEMS RECAPPER

On-Gard Systems Inc. [108523]
Suite 710, 1900 Grant St.
Denver, CO 80203
(303) 860-0723; (800) 878-8111

Description and use: The Recapper is used to recap needles after use. The cap is placed into the device before removing the needle. This device allows for the needle to then be safely recapped after use. Accessories available include a stand to allow one-handed recapping and a medical waste system that consists of two components: a sharps container and a container with disposable infectious waste bags (Figure 6.29).

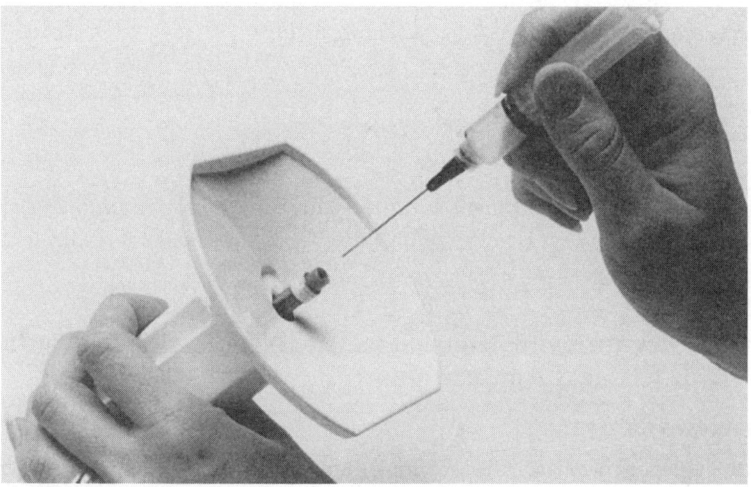

FIGURE 6.29 On-Gard Systems Recapper. (From On-Gard Systems Recapper, Denver, CO.)

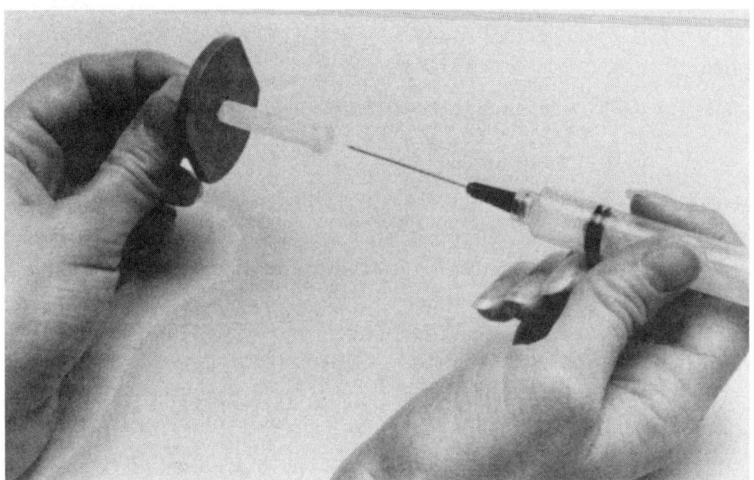

FIGURE 6.30 Terumo Safe-Guard Shield. (From Terumo Safe-Guard Shield, Somerset, NJ.)

List price:

Cost/unit:
$29.50, Recapper (reusable)

Comments: This product, like other recapping products, has major limitations as a needlestick-prevention device. It is bulky and inconvenient for the user to carry around, and during emergencies, the user may easily forget to use it. One hospital chained the device to each bed, but found that the health care practitioners were still not using it. We also found that some of the caps fall into the device and are then difficult to remove. However, the recapper may have a role in controlled environments, where recapping is necessary.

TERUMO SAFE-GUARD SHIELD

Terumo Corp. [101790]
2100 Cottontail Lane
Somerset, NJ 08873
(201) 463-1300; (800) 283-7866

Description and use: The Safe-Guard Shield is a plastic shield that holds the needle cap and allows recapping with the shield in front of the user's fingers (Figure 6.30).

List price:

Cost/unit:
$0.50, Safe-Guard Shield

Comments: This product appears too small to provide adequate protection from a needlestick during recapping of a needle. This product is not recommended because it would not readily be available to the user, especially during emergencies.

Discussion and Recommendations: Selecting Needlestick-Prevention Devices for Four Clinical Applications

General Recommendations

Although a great number of needlestick-prevention products are available, none of the devices or systems that we examined is without limitations with regard to cost, applicability, and effectiveness. Some devices marketed as risk-reducing products offer no protection from needlesticks. The needlestick-prevention market will likely change as new designs are introduced and needs are better defined. The assessments included in this report should help users evaluate new products as they become available.

When purchasing any needlestick-prevention devices, users should consider a number of factors. The most important is whether the product will protect users from sticks with contaminated needles. If the device is used for applications in which needles are not normally used, or if it leaves the needle exposed after use, it will not reduce the risk of needlesticks.

Another factor is whether the product is compatible with existing products and whether additional products are needed. For example, some of the IV medication connectors have secure locking mechanisms, but if the hospital is not using the manufacturer's extension sets, these devices would then have to be used at the catheter site, which may not be convenient.

Assessing the extent of user training required for a product and whether the product could be easily misused is another important factor. Generally, using a product with a well-designed, convenient, integral protective mechanism will be more effective than relying heavily on training. Ideally, the device should require no user action (e.g., should have a recessed needle); otherwise, the action should be minimal and convenient. Users will be tempted to ignore or bypass any protective mechanism that is inconvenient or requires extra steps.

We strongly recommend that a hospital perform a clinical evaluation of each product of interest to determine its effectiveness and utility. The physicians, nurses, and other health care workers who will be using the systems are critical participants in clinical trials. Their input is often over-looked, even though they are the day-to-day users of the devices and are likely to make valuable contributions to the selection process, both in reviewing demonstration products and during final decision making. Users who feel that a product is forced on them without their input are bound to find faults with the final selection and may fail to use it correctly. In addition, the Infection Control or Safety Committee should monitor these new devices for proper usage, for any increases in patient infection, and for any decrease in needlestick injuries, as well as the need for larger needle disposal containers or more frequent replacement of disposable materials (some prevention methods may add considerable bulk).

Because a single device will not meet all of a hospital's needs, users should carefully assess available products and attempt to select the minimum number of devices necessary. Using multiple devices increases the amount of training needed and the associated risk of confusion and misuse. The four main clinical applications we have identified in examining these devices are delivering IV medications, delivering IM/subcutaneous injections, introducing catheters, and collecting blood. In some cases, it may be possible to use the same product for more than one purpose (e.g., IV and IM injections).

Cost will also be a major concern. Hospitals should not pay an excessive amount to achieve only a minimal reduction in risk. In Analyzing Costs Associated with Needlesticks and Preventive Devices, we discuss cost in greater detail and provide a sample CAHDModel.

Product Group Recommendations by Application

Based on our observations and our discussions with clinical consultants, we have provided guidance on the Product Groups that are most suitable for specific clinical uses. We caution readers not to base purchasing decisions on this abbreviated listing alone, but to review this entire Special Report and Product Review to gain perspectives on the efficacy of the examined products and the clinical issues surrounding their use.

Several of the products we examined are not recommended as needlestick-prevention devices. Products currently available in Product Group 2, prefilled medication systems, are similar to using a needle and syringe—the user is at risk from the exposed needle after it is used and until it is disposed of. The in-line collection device in Product Group 5 does not require a needle to draw blood from a transducer system. And the needle recapping products in Product Group 8 are not recommended for most applications, because they are not incorporated into the design of the needle device and may not always be readily available.

TABLE 6.1 Recommended Product Groups by Application

	Application				
	Delivering IV Medications		Delivery IM/ Subcutaneous Injections	Introducing Catheters	Collecting Blood
Product Group	Intermittent	Secondary			
1. Needleless medication/vaccine injections	NA	NA	Yes	NA	NA
2. Prefilled medication systems	Not recommended as needlestick-prevention devices				
3. IV starters with catheters	NA	NA	NA	Yes	NA
4. IV medication connectors	Yes	Yes	NA	NA	NA
5. Blood collection system	NA	NA	NA	NA	Yes[a]
6. Disposable syringes	Yes	NA	Yes	NA	Yes
7. Needle guards	Yes	Yes	Yes	NA	Yes
8. Needle-recapping devices	Have limited applications[b]				

[a] In-line connectors are not recommended as needlestick-prevention devices.
[b] See Product Group 8 in the Product Reviews.

See Table 6.1 for a list of the eight Product Groups and the clinical applications for which they are suitable, and Table 6.2 for comments on individual products. Also, carefully review each Product Group before making a final purchasing decision. Below, we provide brief discussions on each application.

Applications 1 and 2: Delivering IV Medications and Delivering IM (and Subcutaneous) Injections

We have combined these two applications because of the overlap in the available alternatives. IV applications can be broken into two subcategories—administering intermittent (push) medications and delivering secondary medications (piggybacking). Some devices may be suitable for both IV subcategories, IM injections; other devices may be suitable for only one or two of these uses.

Also, some controversy exists among experts about the potential for cross-contamination when a needlestick injury occurs with a needle used for administering medication through the IV set. The risk of contact with the patient's blood appears to be less likely when administering drugs at a distance from the catheter site.

Product Group 1—Needleless Medication/Vaccine Injectors. The single product that we examined in Group 1, the Bioject Biojector, has some limitations for IM applications and is not designed for IV applications. Currently, it can deliver only a 0.5 mL dose and may be less convenient on a nursing floor, where several nurses may need the device at the same time. The Biojector is most useful where frequent vaccinations are being given, such as in a clinic.

Product Group 4—IV Medication Connectors. Most of the products in this group meet the needs of both IV applications (intermittent and secondary), although each product should be assessed for these applications individually. The simplest devices are those that have a recessed needle and can be used on most catheter and IV-set Y-sites. However, use of these special protective connectors may require the user to remove or discard a needle already provided with existing supplies (e.g., a needle used to draw solution into a syringe from a medication container or packaged inside a secondary set). The other systems in this group are useful for only specialized applications or require converting the

TABLE 6.2 Assessment of Manufacturers' Needlestick-Prevention Devices

Product Group	Manufacturer	Product	Prevents Needlestick	Comments
1. Needleless medication/ vaccine injectors	Bioject	Biojector	Yes	Limited IM applications; delivers only a 0.5 mL dose
2. Prefilled medication systems	Winthrop Pharmaceuticals	Carpuject	No	Requires sharps container, may help minimize medication errors
	Wyeth-Ayerst	Tubex	No	Requires sharps container, may help minimize medication errors
3. IV starters with catheters	Critikon	ProtectIV	Yes	
	Deseret Medical	Intima	Limited	Poses hazard if misplaced in linens or if splashed in eye
	Menlo Care	Landmark	Yes	Intended for intermediate-term use limited applications
	Ryan Medical	Shamrock	Yes	Uses a steel needle intended for short-term use
4. IV medication connectors	Baxter Healthcare	Interlink IV Access System	Yes	Need to use manufacturer's injection adapters
	Baxter Healthcare	Needle*Lock	Yes	Useful for piggyback medications
	Burron Medical	Safsite	Yes	Need to use manufacturer's injection adapters[a]
	ICU Medical	Click Lock	Yes	Not compatible with some IV sets and catheters[a]
	IMS	Stick-Gard	Yes	May be less expensive
	L&W Technology	Safeport	Yes	Limited applications
	Pascall Medical	SPIVE	Yes	Limited applications
	Ryan Medical	Saf-T Clik	Yes	Need to use manufacturer's injection adapters
	Tri-State Hospital Supply	Kleen-Needle	Yes	Comparable with existing equipment
5. Blood collection systems	Baxter/Edwards	VAMP	No	For use only with blood pressure lines
	Medical Safety Products	Acci-Guard	Yes	May pose risk of inner-needle exposure
	Ryan Medical	Blood Adapter	Yes	Effective, but may be inconvenient to user
6. Disposable syringes	Becton Dickinson	Safety-Lok Syringe	Yes	Available in only 3 cc syringes
	Needlepoint Guard	Safety Syringe	Yes	Available up to 5 cc size
	Sherwood Medical	Monoject	Yes	Available in only 3 cc size; has 2-position lock
7. Needle guards	ICU Medical	HR Needle	Yes	Used with syringes and blood adapters; for hospitalwide use, but is awkward
	North American Medical	Safe-Site	Yes	Available alone with needles/syringes; for hospitalwide use, but is awkward
8. Needle-recapping devices	On-Gard Systems	Recapper	See text	Not recommended for most locations; may not be available when needed
	Terumo	Safe-Guard Shield	See text	Not recommended for most locations; may not be available when needed

[a] Manufacturer plans to introduce a new product that will be compatible with other manufacturers' components.

entire system (e.g., using special extension sets and other components). Complete conversion will probably be more complex and costly, but may be more effective for some hospitals than using a simpler system.

If the device manufacturer offers a secondary medication set with the preventive device, the possibility of using a needle that is typically supplied with secondary sets would be eliminated. Some users reportedly use tape at the Y-site to secure a secondary-set needle in place; this is not a recommended procedure. Although additional support should not be required, some of the examined devices may provide more support or lock in place.

Product Group 6—Disposable Syringes. These syringes can be used for intermittent IV and IM injections, but not for secondary sets. At present, their overriding disadvantage is that they come in an inadequate range of sizes.

Product Group 7—Needle Guards. These devices are versatile enough to be used not only for intermittent and secondary IV applications, but also for IM applications. However, the extra length of the guard may make it awkward when using secondary sets and during IM injections. Hospital clinicians should investigate this before making a purchasing decision.

Application 3: Introducing Catheters

Product Group 3—IV Starters with Catheters. The four products that we examined represent different clinical applications and have different needlestick-prevention designs. Three of these products appear to effectively reduce the risk of needlesticks. The Critikon ProtectIV can be used to replace many common catheters. The Menlo Care Landmark catheter is expensive and is primarily intended for intermediate-term and special applications. The Ryan Medical Shamrock is appropriate only for short-term use and where its metal needle will not be a problem; it provides protection as long as the user remembers to properly cover the needle. The Deseret Intima reduces some of the risk of needlesticks, but can still pose a significant hazard—for example, if misplaced in linens. Thus, hospitals will need to consider specific clinical needs and determine whether these or other products are appropriate.

Application 4: Collecting Blood

Product Group 5—Blood Collection Systems. Both disposable and reusable systems have advantages and disadvantages. Although reusable systems do not eliminate the risk of needlesticks because the inner needle is exposed after unscrewing the needle from the adapter, they do provide a relatively safe way of recapping. The disposable adapters eliminate the need to recap (and also the exposure to the needle). However, disposable adapters are bulkier, making them more difficult to carry and creating greater disposal bulk, and they are costlier than reusables.

As mentioned above, in-line collection products are not needlestick-prevention devices.

Product Group 6—Disposable Syringes. Refer to the comments for Applications 1 and 2. Also, because of the syringe sizes, the amount of blood that can be collected would be limited.

Product Group 7—Needle Guards. Refer to the comments for Applications 1 and 2. Also, needle guards would be disposed of in a manner similar to reusable adapters (i.e., the inner needle would be exposed). Users should determine whether they would be awkward to use before purchasing these devices.

Analyzing Costs Associated with Needlesticks and Preventive Devices

Needlestick-prevention devices usually cost more than comparable nonpreventive devices currently being used in most hospitals. This cost increase, however, may be offset by savings that result from a decreased number of needlesticks. Because questions will likely be raised by studies published in the literature or presented by manufacturers (e.g., what factors must be considered in assessing cost?), we discuss below some of the costs associated with needlestick injuries and present the results of one published report, in which the authors compared the costs of using nonpreventive and preventive devices.

Rather than attempting to duplicate this study, and because needlestick-prevention devices are necessary for personal safety, hospitals should concentrate on conducting a cost comparison among preventive alternatives to determine the most cost-effective systems for their institution. The second half of this article provides guidance on how to perform such a comparison.

Cost of a Needlestick

The cost of occupational needlestick injuries will depend on the needlestick follow-up procedures taken; the procedures necessary will vary depending on the likelihood of infections. When the source of the needle is known and HIV or HBV contamination is unlikely, prophylactic treatment may be limited to testing the patient (source) for HIV and HBV, testing the person receiving the needlestick for adequate HBV and possibly HIV antibodies, and administering gamma globulin. When the source is unknown or is likely to have been HIV or HBV infected, additional measures may be appropriate, possibly including prophylactic zidovudine (Retrovir) treatments, which may help reduce the risk of AIDS. (The use of zidovudine in this context is still controversial.)

Assessing needlestick injury costs requires determining the number of needlesticks occurring in the hospital, classifying these incidents according to the extent (and cost) of treatment, and then calculating total post-needlestick prophylactic treatment. In addition, cost estimates associated with lost time, Worker's Compensation insurance, and possible legal costs and liability in the event of a lawsuit should be considered. Some of these data may be impossible to estimate accurately, especially liability costs; one successful liability claim could easily override all other considerations.

Cost estimates published in the literature or provided by manufacturers should be carefully examined. Determine whether these estimates are based on actual costs to the hospital or typical patient charges for those laboratory procedures. Personnel time costs (e.g., for the infection control nurse) should be included only if the amount of time is significant enough to justify staffing changes.

In a recent study,[25] the average cost for a needlestick (treatment, prophylaxis, and employee health department personnel time) was $405, with the largest portion of the total cost (60%) being hepatitis screens. Using the costs of six conventional needled devices currently in use, the authors calculated the cost of a needlestick as a percentage of the cost of the device that caused the injury. Because of the often large price differences between different devices, the percentages varied widely. For example, needlesticks caused by IV catheters (stylets) cost only 10% of the device's purchase price, whereas needlesticks caused by IV tubing needle assemblies cost 457%. (The other percentages were 23% for disposable syringes, 27% for butterfly needle IV sets, 57% for prefilled cartridge syringes, and 157% for vacuum tube phlebotomy sets.) On average, needlesticks cost an overall 36% above the purchase prices of the devices. Thus, a hospital could pay an additional 36% for preventive devices (assuming that they would totally eliminate the costs associated with needlesticks) without increasing total costs.

Cost Comparisons of Preventive Alternatives

Regardless of the cost of needlestick follow-up, hospitals are obligated to provide adequate preventive measures for their employees. Industry and government standards are likely to reinforce this obligation, as discussed in the Introduction to this report. Therefore, a primary concern for hospitals is selecting the most cost-effective preventive strategy. Needlestick-prevention devices will likely be a critical component in such a program; however, these products vary considerably not only in cost, but also in effectiveness, as discussed in this report.

The following discussion provides guidance on making a cost comparison among products under consideration for preventing needlesticks (effectiveness must be addressed using the guidance in the Product Review section and Selecting Needlestick-Prevention Devices for Four Clinical Applications earlier in this section). We also reemphasize the need for clinical trials. While we present some illustrative examples here, these should not be interpreted as necessarily typical or appropriate for an individual hospital. To enable each hospital to examine its own specific cost issues, we have provided a CAHDModel for one needlestick application that incorporates the principles below. Specific instructions for using the CAHDModel and entering institutional data are provided in the CAHDModel software.

Because most hospitals are concerned with the impact of needlestick-prevention devices on their budget, we suggest comparing the cost of using each new alternative to the cost of using current conventional (nonpreventive) devices, then using this difference for comparing alternative preventive products. This method will show how each new alternative under consideration compares with current costs and will also help organize institutional thinking in considering all the changes that will be required in converting from the current system to an alternative system.

Three major guidelines should be used in this cost assessment. First, calculate prices on an annual basis; using costs per use, cost per patient, or other similar comparisons can be misleading. Because the marketplace is likely to change rapidly, estimates should be made for no more than a 1 year period. Prices used should be those negotiated with the supplier, list prices can also be misleading.

Second, in performing a comparison, include all relevant costs in the analysis of each alternative. For example, if you will be switching the type of syringe used, include the cost of the old syringes in the cost of the current system and the cost of the new syringes in the alternative system. Do not include costs that are common to both alternatives (e.g., in the previous example, if the same syringe is used in both alternatives, exclude it from the calculation because it will have no impact on the cost difference).

TABLE 6.3 Sample CAHDModel Output

CAHDModel							
Application: Delivering IV Medications							
Alternative 1				Alternative 2			
Preventive Devices	Units/Year	$/Unit	$/Year	Preventive Devices	Units/Year	$/Unit	$/Year
Disposable syringes/ needles, 1 cc	20,000	$0.35	$7,000	Needle guards/needles for secondary med sets	20,000	$0.40	$8,000
Disposable syringes/ needles, 3 cc	50,000	$0.35	$17,500	Needle guards/ needles-syringes	110,000	$0.40	$44,000
Disposable syringes/ needles, 5 cc	30,000	$0.33	$9,900	Conventional syringes, 1 cc	20,000	$0.20	$4,000
Recessed needle for 10 cc syringes	10,000	$0.20	$2,000	Conventional syringes, 3 cc	50,000	$0.20	$10,000
Conventional syringes, 10 cc	10,000	$0.18	$1,800	Conventional syringes, 5 cc	30,000	$0.20	$6,000
Secondary med sets/ recessed needles	20,000	$3.50	$70,000	Conventional syringes, 10 cc	10,000	$0.18	$1,800
Total annual cost			$108,200	Total annual cost			$73,800
Currently Used Devices				Currently Used Devices			
Disposable syringes/ needles, 1 cc	20,000	$0.22	$4,400	Disposable syringes/ needles, 1 cc	20,000	$0.22	$4,400
Disposable syringes/ needles, 3 cc	50,000	$0.22	$11,000	Disposable syringes/ needles, 3 cc	50,000	$0.22	$11,000
Disposable syringes/ needles, 5 cc	30,000	$0.20	$6,000	Disposable syringes/ needles, 5 cc	30,000	$0.20	$6,000
Disposable syringes/ needles, 10 cc	10,000	$0.20	$2,000	Disposable syringes/ needles, 10 cc	10,000	$0.20	$2,000
Secondary med sets/ needles	20,000	$2.00	$40,000	Total annual cost			$23,400
Total annual cost			$63,400				
Total annual difference			$44,800	Total annual cost difference			$50,400

Third, the analysis should take into account all clinical applications. These will usually fall into the four categories identified in Selecting Needlestick-Prevention Devices for Four Clinical Applications: delivering IV medications, delivering IM (and subcutaneous) injections, introducing catheters, and collecting blood. The Recommended Product Groups by Application table (Table 6.1) shows the Product Groups discussed in our Product Reviews that might be appropriate for each application. Because of the overlap between applications and device types, the analysis should encompass all anticipated changes. For example, needle guards can be used for both IV and IM applications. An overall assessment allows alternatives to be compared and cost differences to be kept in perspective.

The sample CAHDModel output (Table 6.3) illustrates a cost analysis of products used for delivering IV medications. This model compared the costs of two proposed alternative systems using preventive devices with current systems using nonpreventive devices. Note the cost differences between Alternatives 1 and 2. (This sample is intended for illustrative purposes only; actual output may vary.)

Cost analysis following these principles will provide a useful guide in estimating the cost of preventive devices, compared both with current devices and with similar devices. CAHDModel will guide your analysis and perform the necessary calculations for you. Although cost is certainly important, product effectiveness, training requirements, and user acceptance are also key issues that must be considered before making a decision, as discussed in this report.

References

1. Centers for Disease Control. Recommendations for prevention of HIV transmission in health-care settings. *MMWR*, 36(2 S):3S–18S, 1987.
2. Centers for Disease Control: Guidelines for prevention of transmission of human immunodeficiency virus and hepatitis B virus to health-care and public-safety workers. *MMWR*, 38(5–6):1–37, 1989.
3. Centers for Disease Control: Protection against viral hepatitis. Recommendations of the immunization practices advisory committee (ACIP). *MMWR*, 9 39(52):1–25, 1990.
4. Centers for Disease Control: Public Health Service statement of occupational exposure to human immunodeficiency virus, including considerations regarding zidovudine postexposure use. *MMWR*, 39(RR-I):1–14, 1990.
5. ECRI: Accidental needlesticks. *Hosp Risk Control* 1987; *Infection Control* 8.
6. ECRI: AIDS:. Recommendations for preventing transmission. *Hosp Risk Control* 1987 *Infection Control* 7.
7. Roberts JR: Accidental needle stick. *EM & ACM*, May:6–7, 1987.
8. McCormick RD, and Maki DG: Epidemiology of needlestick injuries in hospital personnel. *Am J Med*, 70:928–932, 1981.
9. Neuberger JS, Harris J, Kundin WD, et al.: Incidence of needlestick injuries in hospital personnel: Implications for prevention. *Am J Infect Control*, 12(3):1716, 1984.
10. Marcus R: CDC Cooperative Needlestick Surveillance Group: Surveillance of healthcare workers exposed to blood from patients infected with the human immunodeficiency virus. *N Engl J Med*, 319(17):1118–1123, 1988.
11. Gerberding JL, Littell CG, Chambers HF, et al.: Risk of occupational HIV transmission in intensively exposed health-care workers: Follow-up. Abstract #343. *Presented at 1988 ICAAC Conference*, New Orleans. In: *Program and Abstracts of the 28th Interscience Conference on Antimicrobial Agents and Chemotherapy*, Los Angeles, Washington, D.C.: American Society for Microbiology. 1988, 169.
12. McEvoy M, Porter K, Mortimer P, et al.: Prospective study of clinical, laboratory, and ancillary staff with accidental exposures to blood or other body fluids from patients infected with HIV. *Br Med J*, 294:1595–1597, 1987.
13. Health and Welfare Canada: National surveillance program on occupational exposures to HIV among health-care workers in Canada. *Canada Dis Weekly Rep*, 13–37:163–166, 1987.

14. Henderson DK, Fahey BJ, Willy M, et al.: Risk of occupational transmission of human immunodeficiency virus type 1 (HIV-1) associated with clinical exposures. *Ann Intern Med*, 113:740–746, 1990.
15. Gerberding JL, Bryant-LeBlanc CE, Nelson K, et al.: Risk of transmitting the human immunodeficiency virus, cytomegalovirus, and hepatitis B virus to healthcare workers exposed to patients with AIDS and AIDS-related conditions. *J Infect Dis*, 156:1–8, 1987.
16. Finch RG: Time for action on hepatitis B immunization. *Br Med J*, 294:197–198, 1987.
17. OSHA Instruction CPL 2—2.44B: Enforcement procedures for occupational exposure to hepatitis B virus and human immunodeficiency virus, 1990.
18. Jackson MM, Dechairo DC, and Gardner DF: Perceptions and beliefs of nursing and medical personnel about needle-handling practices and needlestick injuries. *Am J Infect Control*, 14:1–10, 1986.
19. Mansour AM: Which physicians are at high risk for needlestick injuries? *Am J Infect Cont*, 18(3):208–210, 1990.
20. Feldman RHL: Hospital injuries. *Occup Health Saf*, 55(9):12–15, 1986.
21. Edmond M, Khakoo R, McTaggart B, et al.: Effect of bedside needle disposal units on needle recapping frequency and needlestick injury. *Infect Control Hosp Epidemiol*, 9(3):114–116, 1988.
22. Becker MH, Janz NK, Band J, et al.: Noncompliance with universal precautions policy: Why do physicians and nurses recap needles? *Am J Infect Control*, 18(4):232–239, 1990.
23. Sumner W:. Needlecaps to prevent needlestick injuries. *Infect Control*, 6(12):495–497, 1985.
24. Ribner BS, and Ribner BS: An effective educational program to reduce the frequency of needle recapping. *Infect Control Hosp Epidemiol*, 11:635–638, 1990.
25. Jagger J, Hunt E, and Pearson RD: Estimated cost of needlestick injuries for six major needled devices. *Infect Control Hosp Epidemiol*, 11(11):584–588, 1990.

Supplemental Reading

American Health Consultants: One-third of needlesticks go unreported at hospital. *Hosp Infect Control*, 17(8):107, 1990.

Beckmann SE, Fahey BJ, Gerberding JL, et al.: Risky business: Using necessarily imprecise casualty counts to estimate occupational risks for HIV-1 infection. *Infect Control Hosp Epidemiol*, 11:371–379, 1990.

Cooperative Needlestick Surveillance Group: Occupational risk of the acquired immunodeficiency syndrome among healthcare workers. *N Engl J Med*, 314(17):1127–1132, 1986.

Crow S: Disposable needle and syringe containers. *Infect Control*, 6(1):41–43, 1985.

Hamory BH: Underreporting of needlestick injuries in a university hospital. *Am J Infect Control*, 11(5):174–177, 1983.

Holthaus D: Suppliers heed call for protective products. *Hospitals*, 61:72–73, 1987.

Huber K and Sumner W: Recapping the accidental needlestick problem. *Am J Infect Control*, 15(3):127–130, 1987.

Jacobson JT, Burke JP, and Conti MT: Injuries of hospital employees from needles and sharp objects. *Infect Control*, 4:100–102, 1983.

Jagger J: Preventing HIV transmission in healthcare workers with safer needle devices. In: *Sixth International Conference on AIDS*, San Francisco, 1990.

Jagger J: Recapping used needles: Is it worse than the alternative? *J Infect Dis*, 162:784–785, 1990.

Jagger J, Hunt EH, Brand-Elnagger J, et al.: Rates of needlestick injury caused by various devices in a university hospital. *N Engl J Med*, 319(5):284–288, 1988.

Kirkman-Liff B, and Dandoy S: Hepatitis B—What price exposure? *Am J Nurs*, Aug:988–990, 1984.

McGuff J and Popovsky MA: Needlestick injuries in blood collection staff. *Transfusion*, 29:693–695, 1989.

National Committee for Clinical Laboratory Standards (NCCLS): *Protection of Laboratory Workers from Infectious Disease Transmitted by Blood, Body Fluids, and Tissue (Tentative Guidelines).* Villanova, PA: NCCLS, Villanova, PA, 1989 (NCCLS Document M29-T, Vol. 9, No. 1).

Neisson-Vernant C, Arfi S, Mathez D, et al.: Needlestick HIV seroconversion in a nurse. *Lancet* 2:814, 1986.

Oksenhendler E, Harzic M, Le Roux JM, et al.: HIV infection with seroconversion after a superficial needlestick injury to the finger. *N Engl J Med*, 315:582, 1986.

Randal J: Hepatitis-B epidemic. *Group Practice J*, July/Aug:10–17, 1989.

Reed JS, Anderson AC, and Hodges GR: Needlestick and puncture wounds: Definition of the problem. *Am J Infect Control*, 8:101–106, 1980.

Reuben FL, Norden CW, Rockwell K, et al.: Epidemiology of accidental needle puncture wounds in hospital workers. *Am J Med Sci*, 286(1):26–30, 1983.

Roberts S, and Scharf L: Appropriate needle disposal: Implementing change to reduce injury and lessen risk. *Am J Infect Control*, 14(5):32A–34A, 1986.

SF study finds minimal concern over AIDS and needle sticks. *Oncol Times*, Jan:18, 1986.

Swift C: Blood-borne diseases. *Group Practice J*, Nov/Dec:51–54, 1990.

Weiss SH, Saxinger WC, Rechtman D, et al.: HTLV-III infection among healthcare workers: Association with needle-stick injuries. *JAMA*, 254(15):2089–2093, 1985.

White K: "Why weren't you just more careful?" — What does it take to avoid occupational exposure to HIV? *AIDS Patient Care*, June:13–16, 1990.

UPDATES OF PRODUCT REVIEWS*

UPDATE TO GENERAL RECOMMENDATIONS

When you are conducting clinical evaluations of needlestick-prevention devices, consider special needs and situations that may arise, in addition to routine applications. For example, consider whether the device will hamper (slow) response during a code procedure or whether high pressures that occur during rapid infusion of a viscous fluid, such as dextrose, will cause disconnections.

Product Group 1: Needleless Medication/Vaccine Injectors

BIOJECT INC. [108133]

Bioject Biojector 50 (1.1 version)

As of November 1, 1991, a modified unit, described below, has superseded the unit that we assessed.

Bioject Biojector 50 (1.2 version)

New Product

The manufacturer states that the new Biojector 50 (1.2 version) has a wider operating temperature range of 50°F–100°F, is noiseless, and uses the same medication ampules as the 1.1 version; list price: $575.

Product Group 2: Prefilled Medication Systems

WINTHROP PHARMACEUTICALS
Div. Sterling Drug Inc. [104392]

Carpuject
No changes have been reported.

* ECRI Special Report and Product Review. *Health Devices* 20(5):154–180, 1991.

WYETH-AYERST LABORATORIES [101864]

Tubex
The Tubex injector was modified as of July 1991. The injector now has teeth inside the collar to grip the cartridges more securely. Also, as of September 1991, the modified injector can accommodate Dosette cartridges.

Product Group 3: IV Starters with Catheters

CRITIKON INC.
A Johnson & Johnson Co. [101346]

ProtectIV
No changes have been reported.

DESERET MEDICAL INC.
Becton Dickinson and Co. [101750]

Intima
No changes have been reported.

MENLO CARE INC. [107575]

Landmark
No changes have been reported.

RYAN MEDICAL INC. [108525]

Shamrock
No changes have been reported.

Product Group 4: IV Medication Connectors

Products from manufacturers included in the previous section of this chapter.

BAXTER HEALTHCARE CORP.
IV Systems Div. [106390]

InterLink IV Access System
NOTE: As of September 16, 1991, Becton Dickinson (Becton Dickinson Div. [101093], One Stanley St., Rutherford NJ 07070) has been manufacturing and marketing the InterLink cannula components of the InterLink IV Access System. Baxter will continue to manufacture and market only the injection site and any products that have the injection site (e.g., IV administration sets).

Baxter now has additional products available with the InterLink system:

- Vial access cannula—allows needleless access to single-dose vials and is compatible with all syringes.
- Three-way connector—compatible with any IV tubing to allow multiple needleless access to IV administration sets.
- IV administration sets and specialty extension sets—compatible with the InterLink cannula; a new Y-lock cannula is available to connect a secondary medication set to the primary IV administration set.
- The InterLink threaded-lock cannula with lipid-resistant material—suitable for administering lipid therapy.

COMMENT: At this time, the InterLink lever-lock cannula assessed in the May 1991 issue and the new Y-lock cannula should not be used for administering lipid therapy because of their construction. Certain plastics, sometimes used in devices of this type by various manufacturers, are not lipid resistant; users should verify that the product being used is suitable for lipid infusion. The packaging of the InterLink cannula is labeled with a warning that the product is not intended for this use (see *Health Devices Alerts* Action Items, Accession No. 01714, June 21, 1991).

Needle*Lock
Baxter now has four basic IV sets available for use with the 20 G Needle*Lock device.

BURRON MEDICAL INC.
Div. B. Braun of America Co. [101098]

Safsite
An access pin is now available to allow the Safsite valve to be used with a Y-site of any administration set. List prices: pin, $0.50; purchased with the Safsite valve, $1.80.

ICU MEDICAL INC. [105658]

Click Lock
No changes have been reported.

COMMENT: The manufacturer states that this product was designed for multiple-use on single patients to reduce the costs associated with intermittent drug administration. In the May 1991 issue, we recommended that hospitals review the infection control considerations of this use of the Click Lock.

Piggy Lock

New Product
Whereas the Click Lock is designed for connection to central and peripheral lines, the Piggy Lock is designed for piggybacking medications into any standard Y-site.

This new IV connection device is compatible with all Y-sites of standard administration sets. It has the same stainless-steel cannula as the Click Lock, but has a blue locking ring that locks around any Y-site (see Photo A). Like the Click Lock, the Piggy Lock is designed as a multiple-use, single-patient

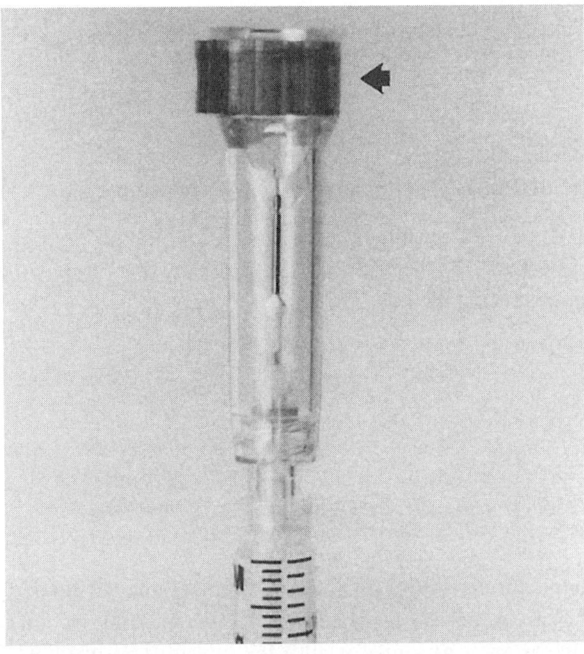

PHOTO A CU Medical Piggy Lock™ (the blue locking ring [arrow] locks around any Y-site). (From CU Medical Piggy Lock.)

product to reduce the costs associated with intermittent drug administration (see COMMENT above). List price: $1.60.

INTERNATIONAL MEDICATION SYSTEMS (IMS) LTD. [104648]

Stick-Gard
No changes to the product have been reported. However, Becton Dickinson is marketing the Stick-Gard as the Becton Dickinson Safety Gard IV needle.

L&W TECHNOLOGY INC. [108520]

Safeport
The new list prices for this product are $1.55 for the single-port Safeport (SIS-002) and $5.00 for the triple-port Safeport (SM-001).

L & W Technology now offers the following new extension sets, which allow needleless IV line access and aspiration of samples from the IV line because they do not contain the antibackflow valve found in other Safeport products; according to the manufacturer, all L & W Technology products can be used for lipid therapy.

- End-port access site (ES 100)—a 7-in. extension set with a Safeport valve on one end and a Luer connection on the other; list price: $2.10.
- Single-port extension set (ES 150)—an 11-in. extension set with a single needleless access port; list price: $2.40.
- Double-port extension set (ES 200)—a 20-in. extension set with two needleless access ports similar to the single-port extension set; list price: $3.70.
- Triple-port extension set (SM-001T)—a 48-in. extension set similar to the SM 001, but with the tubing included; list price: $5.25.

PASCALL MEDICAL CORP. [108524]

SPIVE
The list price is now $0.39.

RYAN MEDICAL INC. [108525]

Saf-T Clik
No changes have been reported.

TRI-STATE HOSPITAL SUPPLY CORP. [103919]

Kleen-Needle
No changes have been reported.

Products from manufacturers not included in previous section of this chapter.

BEECH MEDICAL PRODUCTS [108874]

P.O. Box 704
Washington Crossing, PA 18997
(215) 493-2785; (800) 235-5833

Beech Medical has a product line of different IV medication connectors.

Pro-Lok™
This shielded needle connector assembly has a recessed replaceable needle that can be used with any injection adapters or that can be locked onto the Pro-Lok intermittent injection cap ($0.50); four extension sets of different lengths are also available with the intermittent injection cap ($1.60–$1.80 with injection cap); list price: $1.65. (See Photo B.)

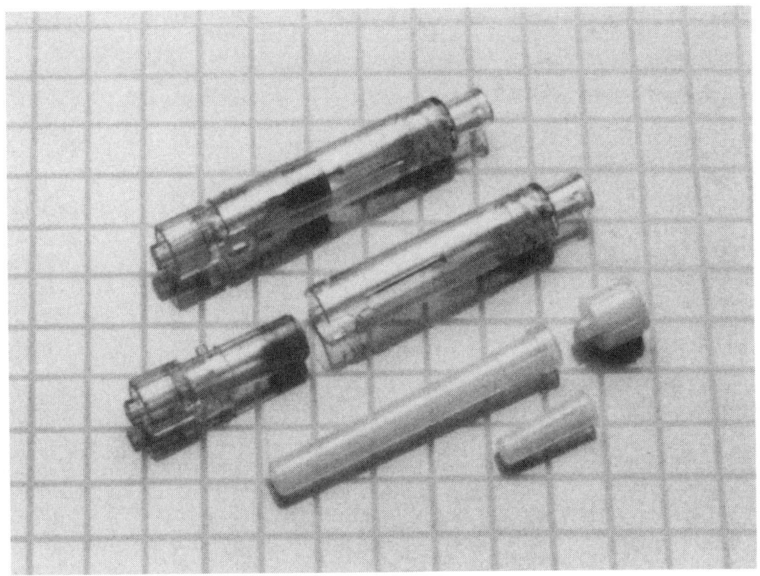

PHOTO B Beech Medical Pro-Lok (locked [top], unlocked [center], and needle packaging). (From Beech Medical Pro-Lok, Newtown, PA.)

Versa-Lok™

This shielded-needle Y-site connector has a recessed replaceable needle that locks to a secondary line at the Y-site (see Photo C); list price: $1.75.

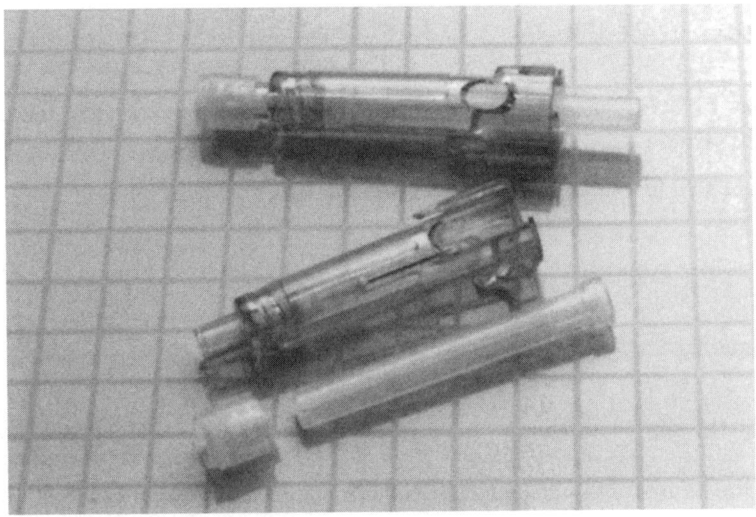

PHOTO C Beech Medical Versa-Lok (as it is packaged [top] and ready for use). (From Beech Medical Versa-Lok, Minneapolis, MN.)

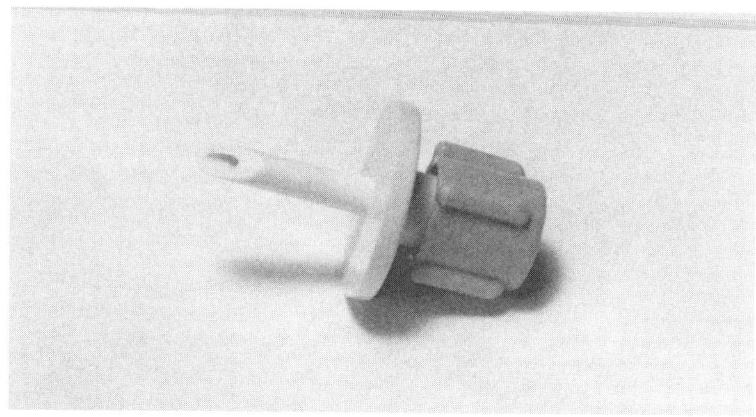

PHOTO D Edge Medical Safe-Draw multidose drug vial adapter. (From Edge Medical Safe-Draw, Boulder, CO.)

EDGE MEDICAL [108902]

1107 Fair Oaks Ave., Suite 106
South Pasadena, CA 91030
(213) 275-7654

Safe-Draw™
This multidose drug-vial adapter allows syringes to be filled from all sizes of drug vials without the use of a needle (see Photo D).

Product Group 5: Blood Collection Systems

Products from manufacturers included in previous section of this chapter.

BAXTER HEALTHCARE CORP.

Edwards Critical-Care Div. [106431]

VAMP
This product may reduce the risk of needlestick injuries where personnel are currently drawing blood samples from a catheter into a syringe and then transferring the blood to a blood collection tube with a needle.

MEDICAL SAFETY PRODUCTS INC. (MSPI) [108521]

Acci-Guard
No changes have been reported. MSPI will soon have available a biohazard container, described below, intended as a companion product to the Acci-Guard.

Biohazard Container

New Product
When used with the Acci-Guard, this container will provide a system for point-of-use needlestick protection and needle disposal without exposure to the inner or outer end of the blood collection needle. The container can also be used with standard blood collection holders. The list price is not yet available.

RYAN MEDICAL INC. [108525]

Blood Adapter
No changes have been reported.

Products from manufacturers not included in previous section of this chapter.

BECTON DICKINSON VACUTAINER SYSTEMS [103615]

One Stanley St.
Rutherford, NJ 07070
(201) 460-2000

Safety-Lok Needle Holder
Locking the safety sheath in place on this disposable safety blood collection tube holder requires two hands (see Photo E). List price: $0.49. A carrying pouch is also available (list price: $15.95, $5.25 for shoulder strap).

BIO-PLEXUS INC. [108876]

P.O. Box 826
Tolland, CT 06084
(203) 871-8601

Punctur-Guard™
This blood-collection needle has a blunt inner cannula that extends beyond the tip of the needle before it is removed from the vein. The Punctur-Guard is placed in a standard blood collection tube holder, and blood tubes are pressed into it until resistance is felt; the patient's blood can then be drawn. When the last blood tube has been filled, the user advances the tube until a click is heard, and the blunt inner cannula is released. When the user removes the needle, it has a blunt tip (see Photo F). List price: $0.25.

COMMENT: Hospitals should ensure that personnel are properly trained in using this product. The manufacturer provides complimentary in-service training. If the product is recapped and manually unscrewed from the blood adapter when disposing of the Punctur-Guard, the inner needle at the back end, which is covered by a latex boot, presents the risk of needlestick. The manufacturer recommends following CDC guidelines by not recapping the product with the traditional two-handed technique and

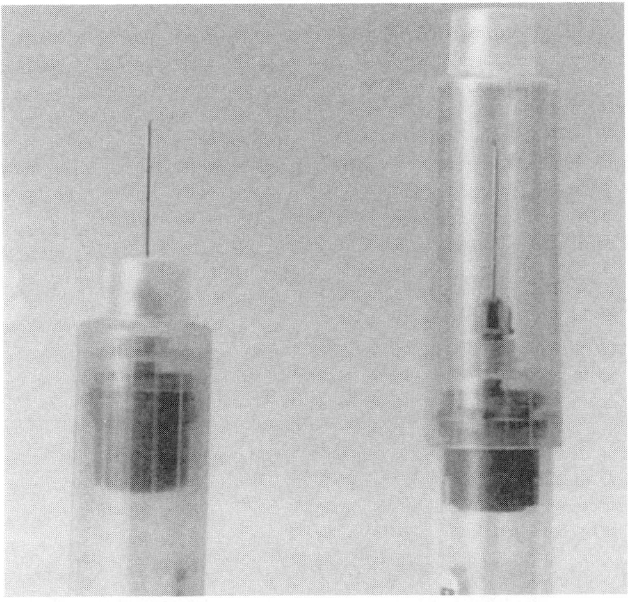

PHOTO E Becton Dickinson Vacutainer Systems Safety-Lok Needle Holder (set up for use [left] and after use, showing protected needle). (From Becton Dickinson Vacutainer Systems Safety-Lok, Rutherford, NJ.)

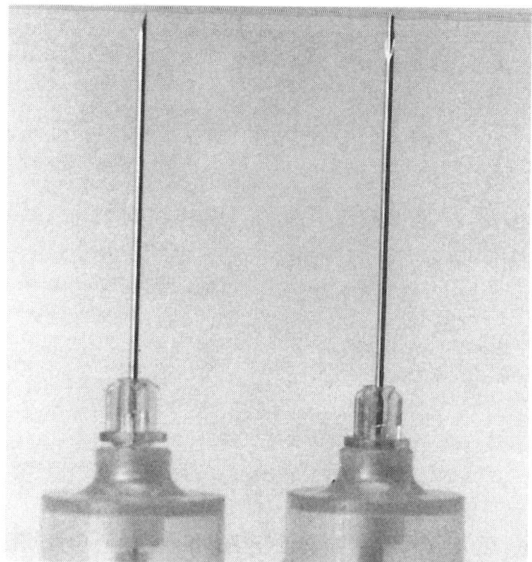

PHOTO F Bio-Plexus Punctur-Guard (set up for use [left] and with the blunt inner cannula extended beyond the tip of the needle after use). (From Bio-Plexus Punctur-Guard, Vernon, CT.)

by using it in conjunction with a screw-top waste-disposal unit. This method would eliminate the risk of needlestick injuries.

VIGGO-SPECTRAMED INC. [107830]

1900 Williams Dr.
Oxnard, CA 93030
(805) 983-1300; (800) 235-5945 (outside CA);
(800) 631-7015 (within CA)

Safedraw Closed-Loop Blood Sampling System
This blood pressure monitoring set allows the user to draw blood from arterial lines without using a needle. It is packaged with a pressure line, a volume-restricted syringe, and a blood-sampling septum with necessary connections (see Photo G). The Safedraw System is also available without a disposable transducer. It can be used with syringes and blood-collection tubes. List price: $14.95 for Safedraw System only.

Product Group 6: Disposable Syringes

BECTON DICKINSON AND CO.

Becton Dickinson Div. [101093]

Safety-Lok Syringe
No changes have been reported.

SHERWOOD MEDICAL CO.

SUB. AMERICAN HOME PRODUCTS [101927]

Monoject with Safety Shield
No changes have been reported.

Product Group 7: Needle Guards

Products from manufacturers included in previous section of this chapter.

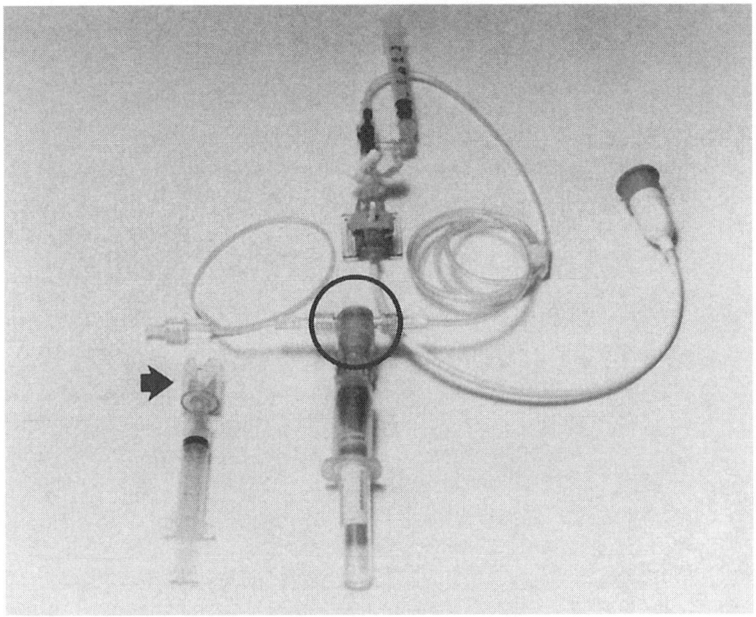

PHOTO G Viggo-Spectramed Safedraw™ Closed-Loop Blood Sampling System (note Safe Needle [needleless syringe adapter, arrow] and blood-sampling septum [in circle]). (From Viggo-Spectramed Safedraw, Oxnard, CA.)

ICU MEDICAL INC. [105658]

HR Needle
No changes have been reported.

NORTH AMERICAN MEDICAL PRODUCTS (NAMP) INC. [108522]

Safe-Site
The following list prices have changed: Safe-Site without needle and syringe, $0.34; Safe-Site with needle, $0.37; Safe-Site with needle and syringe, $0.46.

The Safe-Site is now available with 1¼-in. needles. A stopcock can be used with it to provide multiple drug administration through a Y-site. NAMP also provides vented or nonvented secondary medication sets with a Safe-Site needle guard. List prices: vented secondary medication set with Safe-Site, $1.80; nonvented secondary medication set with Safe-Site, $1.67.

Products from manufacturers not included in previous section of this chapter.

CONCORD/PORTEX [107113]

Kit St., P.O. Box 724
Keene, NH 03431-5911
(603) 352-3812; (800) 258-5464

Needle-Pro™
The Needle-Pro is a plastic sheath that fits on syringes before a needle is attached. After the needle is used, the plastic sheath can be pressed over it by leaning the needle against a hard surface (see Photo H). The Needle-Pro can be used with needles up to 1½-in. length. List price: $0.20.

COMMENT: This device will require appropriate training before use.

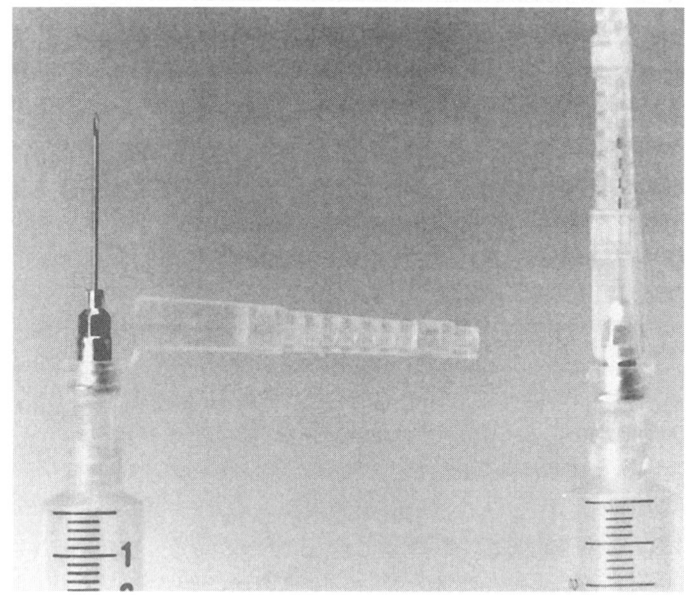

PHOTO H Concord/Portex Needle-Pro (set up for use [left] and after use, showing protected needle). (From Concord/Portex Needle-Pro, Keene, NH.)

Product Group 8: Needle-Recapping Devices

Products from manufacturers included in previous section of this chapter.

ON-GARD SYSTEMS INC. [108523]
New Address/Telephone Number:
1800 Fifteenth St., Suite 100
Denver, CO 80202
(303) 825-5210

Recapper
This device can be a cost-effective tool for reducing needlestick injuries in the applications listed in the May 1991 issue—that is, in controlled environments, such as PACUs, where recapping is necessary. However, this device is not recommended as a general, hospitalwide needlestick-prevention device because it is awkward to carry and therefore may not be available when needed.

TERUMO CORP. [101790]
New Telephone Number: (908) 302-4900

Safe-Guard Shield
No changes have been reported.

Products from manufacturers not included in previous section of this chapter.

J & T PRODUCTS LTD.
16B Regent St.
Newtownards, County Down,
BT23 4LH
Northern Ireland
Fax: 0232-243861
DisposiNeedle MK II™

Exclusive U.S. Distributor:
SEPTODONT INC. [108875]
P.O. Box 11926
Wilmington, DE 19850-1926
(302) 328-1102; (800) 872-8305

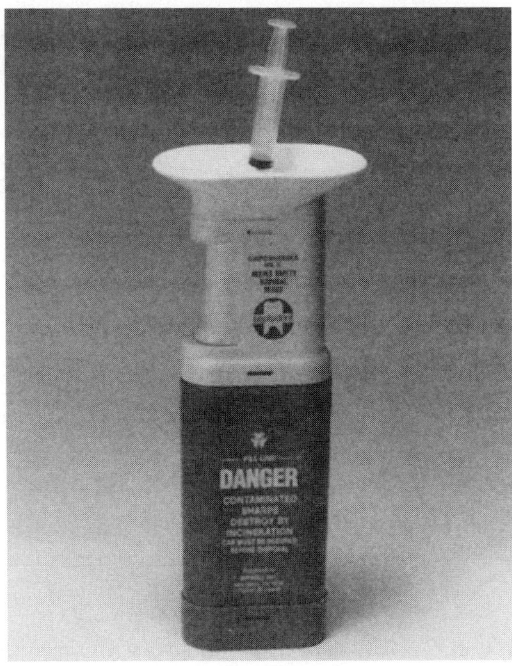

PHOTO I J & T Products (distributed by Septodont) DisposiNeedle MK II (with recapping device [white] attached to sharps container). (From J & T Products DisposiNeedle MK II, Lapeer, MI.)

This self-contained system is designed as a combination recapping device and sharps container; the container fits on the device to easily dispose of used needles (see Photo I). The recapping device can also be used with standard sharps containers. List price: $145.

Product Group 9: Sharps Pincushion Container

The following product is different from other needlestick-prevention devices and cannot be placed in any of the other eight product groups. Also, because it is not strictly a sharps waste disposal unit, despite its similarities with these devices, we are including it with this review of needlestick-prevention devices. (We will be evaluating sharps waste disposal units in a future issue of *Health Devices.*)

AIR SUPPORT MEDICAL [108873]

P.O. Box 99
199 Ramond St.
Hope, IN 47246
(812) 546-0050

Sharps Pin Cushion Container
This product, consisting of a plastic holder, a plastic container filled with foam material, and a plastic lid that locks the container when properly closed, is both a sharps container and a needlestick-prevention device. It is small and very easy to use and is intended to be used at the point of patient contact. Needles can be stuck into the foam portion of the unit immediately after use (see Photo J).

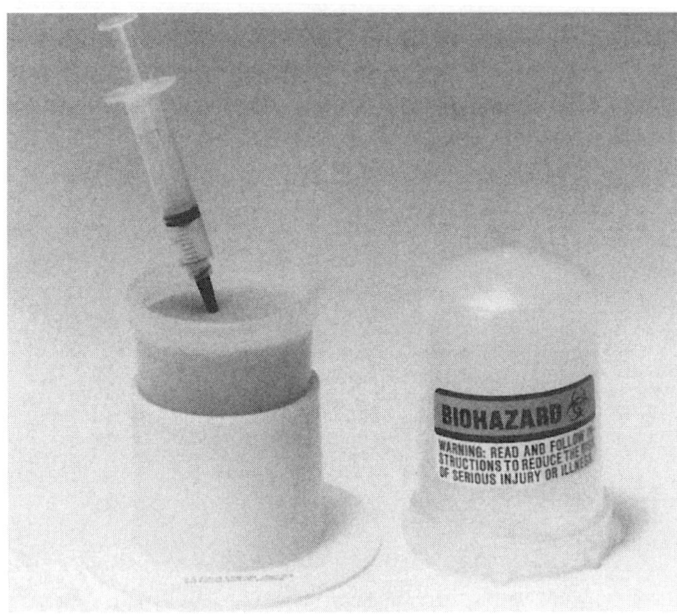

PHOTO J Air Support Medical Sharps Pin Cushion™ container. (From Air Support Medical Sharps Pin Cushion, Hope, IN.)

This product is designed for single-patient use and for no more than five sharps, except for butterfly needles, and it is primarily intended for use in an emergency room or OR. Users should evaluate whether this product is more effective than a small sharps container and whether it is cost-effective. List price: $1.50.

Source: ECRI Evaluation Updates. *Health Devices* 20(12):460–7, 1991.

7

Violence in the Health Care Industry

Introduction 7-2
Pressures That Cannot Be Overlooked 7-2
Preventing Workplace Violence in the Health Care Workplace 7-3
Typology • Magnitude and Severity of the Problem • Risk Factors • Types of Intervention Evaluated and Their Finding • What Needs to Be Done • Recommendations and Conclusions
Acknowledgments 7-16
References 7-16
Guidelines for Security and Safety of Health Care and Community Service Workers 7-18
Preface • Introduction: The Problem • Program Development • Program Elements • Medical Management • Recordkeeping • Training and Education • Evaluation of the Program
References 7-38
Further Readings 7-40
Glossary 7-41
Appendix 7.A.1 7-42
Appendix 7.A.2 7-44
Appendix 7.A.3 7-46
Appendix 7.A.4 7-47
Assault on the Job: We Can Do Something about It! 7-47
What Is Assault on the Job? • Posttraumatic Stress Disorder (PTSD) • Sample Contract Language to Protect Workers from on-the-Job Assault • Resource List • Organizing to Prevent Assault on the Job • Assault on the Job Survey—How Safe Is Your Work Environment? • Sample Assault Incident Report Form • Sample Grievance/Petition Form • Training Exercises
A Joint Labor/Management Experience in Implementing OSHA's Violence Prevention Guidelines in the NYS Office of Mental Health 7-68
Introduction • The Projects • Binghamton Psychiatric Center • Conclusion
Acknowledgments 7-102
References 7-102

Jane Lipscomb
University of California

Jonathan Rosen
New York State Public Employees Federation

Matthew London
Professional Employees Federation

Kathleen McPhaul
University of Maryland School of Nursing

Joyce A. Simonowitz
California Department of Industrial Relations

Service Employees International Union

Additional References 7-102
Appendix 7.A.5 7-103
Appendix 7.A.6 7-105
Appendix 7.A.7 7-108
Appendix 7.A.8 7-108

Introduction

A study by the International Association for Hospital Security reported that in 1990 virtually every hospital experienced on-site violence. One in four experienced an arson fire. One in five experienced an armed robbery, an on-site rape, or an on-site homicide.

Two disturbing future trends were noted. The first is an acceleration of lawsuits charging that health care facilities have failed to maintain safe premises. The second trend is toward increased kidnappings, extortion, and terrorism at health care facilities with prominent physicians and principals of hospitals becoming targets of threats and requiring protection.

Pressures That Cannot Be Overlooked

Even if the administrator of a health care facility were tempted to ignore their moral and ethical responsibilities to provide safe premises, there are three overriding sets of pressures that cannot be overlooked—lawsuits, staff moral, and bad publicity.

With hospitals no longer immune from suits by patients, visitors, staff, vendors, or even outsiders, the number of lawsuits for failure to provide adequate security, as well as the size of the awards or settlements, has become a matter of concern second only to malpractice. For example:

- Two lawsuits seeking $42.5 million in damages have been filed in federal court by families of two people shot to death at a Kansas Medical Center. The lawsuit stems from a fatal shotgun attack in the facility's emergency room. The victims were a hospital visitor and a doctor who was completing his residency training at the center.
- A $13 million negligence lawsuit has been filed against a hospital in Utah by lawyers for a 12-year-old mother who claimed she had been raped at 3:00 AM after giving birth at the hospital.
- A lawsuit asking $1 million in damages from a Tennessee Medical Center has been filed on behalf of two children whose mother was stabbed to death at the hospital while visiting her grandmother who was a patient there.
- A medical center in Springfield, MA, its chief of security, and others were named in a $44 million lawsuit for lack of security by the mother of a 7-year-old boy who was stabbed to death in the medical center's parking lot.
- In another parking lot incident, an anesthesiologist at a memorial hospital sued the hospital for $6 million for inadequate security following his being shot in the abdomen during a mugging.
- The director of a psychiatric center in Brooklyn, NY, was sued for $16 million by the brother of a nurse who was killed by a patient there.

Many hospitals have quietly settled a number of lawsuits, especially in connection with rape, other types of assaults, or infant kidnapping—for big dollars. Furthermore, the current trend is for juries to substantially increase the amount awarded to a plaintiff through punitive damages when they find a hospital negligent in providing adequate security.

Whether it is true or not, the perception by nurses, physicians, and other staff members, that security is inadequate at a hospital, frequently provokes protests and disputes. At a hospital in New Haven, CT, 700 employees signed a petition demanding tighter security following a mugging in the hospital's garage. At a medical center in Bakersfield, big headlines in local newspapers reported that employees were protesting the lack of security and indicated that some nurses were carrying revolvers to work. At a hospital in Brooklyn, interns and residents picketed the hospital demanding increased security at their residence hall. Of even greater consequence was the difficulty of a hospital in New Orleans, in attracting nurses to meet Joint Commission for Accreditation Healthcare (JCAH) staffing requirements. In a local newspaper, a former nurse charged that the lack of security hampered the recruitment of nurses. When JCAH denied accreditation to the hospital, the hospital, among other improvements, added security personnel.

A violent crime in a hospital often brings with it newspaper, radio, and TV publicity, sometimes national as well as local, and tends to distort the true picture of security at the institution. Sometimes the very nature of the crime, especially infant kidnappings, encourages the press to have a field day at the expense of the institution. Bad publicity of this kind can undo years of excellent public relations work.

In order to minimize the potential for violent crime, it is of paramount importance to establish a viable access control program. Institutional officers must be able to regulate the flow of visitors and patients, and everyone's identity must be easily recognized. At Children's Memorial Hospital in Chicago, the color of visitor passes has been set up on each floor. The hospital learned the hard way, after a baby was kidnapped there in 1980. At the Presbyterian Hospital in New York City, uniformed officers issue color-coded adhesive passes, which limit the visitor to a specific floor or area of the hospital. Visitors must state who they are visiting, and the name is checked on a computer printout. The two-visitors-at-a-time rule is also enforced. The result of such strict access controls? A 65% reduction in the number of reported crimes over an 18 month period.

The following, in three sections, outlines through epidemiology and design programs some of those steps that can be implemented to reduce violence to health care workers (HCWs).

Preventing Workplace Violence in the Health Care Workplace

Jane Lipscomb, Jonathan Rosen, Matthew London, and Kathleen McPhaul

Workplace violence is a leading cause of occupational injury for HCWs, especially staff who provide direct patient care in emergency rooms, psychiatric, geriatric, home care, and forensic and criminal justice settings. Although the main source of physical assaults is patients, perpetrators of violence against HCWs may include visitors and coworkers. Exposure to violence can also arise from strangers when HCWs conduct home visits in high crime areas. Domestic violence may also occur in the workplace.

Violence in any health care organization is a major deterrent to quality of care. Despite the obvious link between patient and staff safety, there is a prevailing bias that the risk of being assaulted by violent patients "is part of the job." All too frequently this dangerous bias is expressed by hospital administrators, criminal justice authorities, and even by HCWs themselves. Consequently, institutional acceptance of workplace violence is a major obstacle to developing effective violence prevention programs and protective regulations. Other barriers include lean staffing, nurse shortages, increases in more acutely ill patients, and an antiregulatory outlook among many hospital administrators and government officials.

In addition, government, certifying and/or licensing bodies, managers, and community organizations frequently prioritize patient safety over worker safety, without recognizing that they are inextricably linked. Staff working in a State hospital for the mentally retarded and developmentally disabled told the authors that their management interprets federal patient safety and accrediting criteria as requiring staff to "take the bullet" in the event that a patient is potentially at risk of assault by another patient.

There is a direct link between staff safety and patient safety. Violent patients/clients often assault and threaten other patients/clients in addition to their caretakers. Researchers have found that violence experienced by health care staff is associated with lower patient ratings of the quality of care.[32] Lean staffing has a major negative impact on timeliness and attentiveness to patient needs, frustrating patients, and caregivers alike. Staff shortages and injuries have led to excessive use of mandatory overtime and the resulting staff exhaustion and burnout. Additionally, replacement staff is often unfamiliar with patient's individual needs, triggers, and behaviors. As the quality of care declines, tension and conflict among caregivers and patients inevitably increases.

In this section we will review what is known about the scope of the problem and risk factors for workplace violence in a variety of health care settings. We will emphasize the key steps necessary for health care organizations to institute effective violence prevention programs. The discussion will include the community health care as well as the institutional setting, as some of the risk factors and prevention strategies differ. Although institution-based workers experience higher rates of physical assault, community health care poses a greater risk of fatal injury. We will describe two intervention research projects undertaken by the authors: (1) implementation and evaluation of a comprehensive violence prevention program in a State-operated mental health system; and (2) evaluating risk factors and establishing safety measures for community mental health workers providing care in client's homes in the context of State policy reform.

Typology

The California Division of Occupational Safety and Health (Cal/OSHA) developed a "typology" of workplace violence that describes the relationship between the perpetrator of violence and the victim (see Table 7.1). Although patient or client violence (type II) is the predominant concern; in the community setting there is the additional risk of being attacked by members of the public while traveling to clients' homes (type I). Coworker violence (type III) is highly prevalent in hierarchical health care organizations, while incidents of domestic violence (type IV) occur in health care work sites as in other industries.

TABLE 7.1 Typology of Workplace Violence

Type	Description
I: Criminal intent	The perpetrator has no legitimate relationship to the business or its employee, and is usually committing a crime in conjunction with the violence. These crimes can include robbery, rape, or other type of assault.
II: Customer/client	The perpetrator has a legitimate relationship with the business and becomes violent while being served by the business. This category includes customers, clients, patients, students, inmates, and any other group for which the business provides services. It is believed that a large portion of customer/client incidents occur in the health care industry, in settings such as nursing homes or psychiatric facilities; the victims are often patient caregivers.
III: Worker-on-worker	The perpetrator is an employee or past employee of the business or organization who attacks or threatens another employee(s) or past employee(s) in the workplace.
IV: Personal relationship	The perpetrator usually does not have a relationship with the workplace but has a personal relationship with the intended victim. This category includes victims of domestic violence assaulted or threatened while at work.

Sources: Adapted from Cal/OSHA 1995; Howard 1996; IPRC 2001.

In developing a workplace violence prevention program it is important for the planners to understand that mitigation of risks will vary depending on the type of violence that is being addressed. Effective programs require a site-specific risk evaluation and a targeted intervention program.

Magnitude and Severity of the Problem

The U.S. Department of Justice (DOJ) has estimated that annually between 1993 and 1999, 1.7 million "violent victimizations" defined as assaults, verbal threats, harassment, occur at work across all industries in the United States.[7] In addition, approximately 500–600 workplace homicides are reported annually to the U.S. Bureau of Labor Statistics (BLS 2007). Health care leads all other industries in the number of nonfatal assaults resulting in lost workdays in the United States; contributing 45% of all such assaults. According to the annual BLS Survey of Occupational Illness and Injuries, which captures only incidents defined as "OSHA reportable," the rate of nonfatal assaults to workers in the "nursing and personal care facilities" industry was 31.1 per 10,000 compared to 2.8 per 10,000 in the private sector as a whole.[8] By contrast, the more sensitive DOJ National Crime Victimization Survey (NCVS) report, estimated an overall average annual rate for nonfatal violent crimes at work of 12.5 per 1000 workers. In the health care field, the average annual rates between 1993 and 1999 were higher: 16.2 (physicians), 21.9 (nurses), and 68.4 (professional or custodial mental health workers).[7] In some state psychiatric hospitals and schools for the developmentally disabled, the annual rates are even higher, with several institutions reporting greater than 100 assaults per 100 staff.[9]

In December 2007, the Washington State Department of Labor and Industries issued a report titled "Violence in Washington State Workplaces 2000–2005" that presented data from the U.S. BLS' Survey of Occupational Injuries and Illnesses and from Washington State's workers' compensation claims related to assaults and violence.[25] The number of violence-related claims for the period 2000–2005 was approximately 12% lower than that for the period 1995–2000 (Figure 7.1).

Health- and social service-related occupations accounted for approximately one-half of all violence-related claims over the study period. Nurses' aides and orderlies, police officers, health aides, psychiatric aides, social workers, and private security guards had the highest number of violence-related

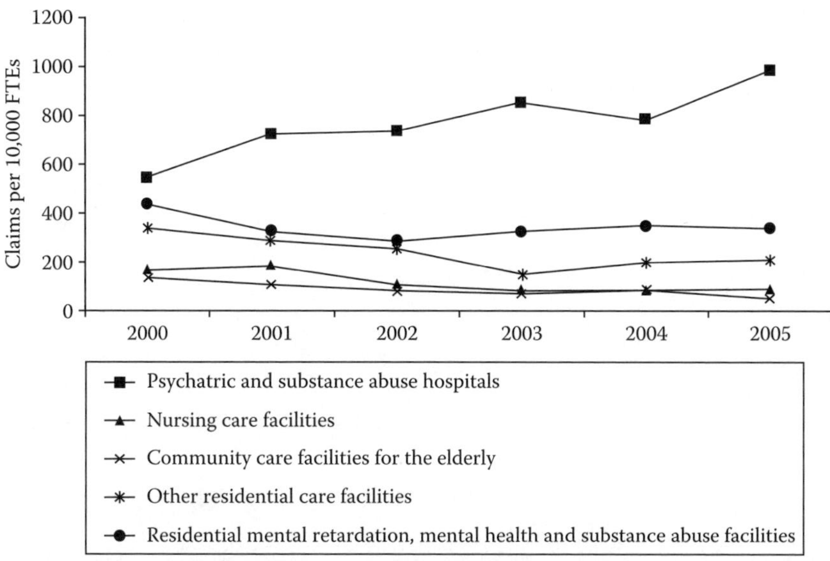

FIGURE 7.1 Violence-related claims rates for high risk state fund industries, 2000–2005.

claims. Health Care and Social Assistance is the highest risk major industry. Although violence-related claims rates fell by 35% during the study period, the claims rate in this industry is about five times higher than for all industries combined. Psychiatric and Substance Abuse Hospitals saw an increase of over 81% in their violence-related claims rates during the study period. In contrast, violence-related claims rates in other Health Care and Social Assistance industries fell throughout the study period (see chart below).

For nonfatal violence-related injuries, workers' compensation data ranked Social Services (142 per 10,000 workers) as the highest risk industry followed by Health Services (74.6 per 10,000 workers). Residential Care was ranked second among specific industry codes (301 per 10,000 workers); Individual Family and Social Services was ranked 10th (79 per 10,000 workers). Notably, these two specific industry sectors reported the second and third greatest percent increase in assault rates over the period 1995–2000, 46% and 33%, respectively.

Among Washington State workers across all industry codes, rates of compensable assault cases were substantially higher for state (30.4 per 10,000 workers) and local government workers (8.9 per 10,000 workers), compared with private sector workers (2.3 per 10,000 workers). Differences between public and private workers' risk can be explained in part by the nature of the patient client populations that public employees serve. The most difficult patients/clients who are indigent, criminal, dually diagnosed, or who are refused treatment in private or non-for-profit facilities are typically placed in state- and county-run institutions. This is in part the explanation for why the Crime Victims Survey finds that public employees are 16% of the workforce nationally, but 32% of the victims of workplace violence. While the incidence of workplace assaults has been trending sharply downward for private sector workers, both in Washington State and nationally, the trend has risen significantly for public sector employees.[26]

Risk Factors

In 1996, the National Institute for Occupational Safety and Health (NIOSH) issued a bulletin on workplace violence. Based on its review of the published scientific literature, NIOSH listed 10 risk factors for workplace violence, two of which included "working with unstable or volatile persons in health care, social service, or criminal justice settings" and "working in community-based settings".[10] See also Figure 7.2.

In guidelines published by the Occupational Safety and Health Administration (OSHA) that focused on health care and social service workers, the agency identified additional risk factors these workers face: the use of hospitals for the care of acutely disturbed and violent individuals; the increased number

Working directly with volatile people, especially, if they are under the influence of drugs or alcohol or have a history of violence or certain psychotic diagnoses
Working when understaffed—especially during meal times and visiting hours
Transporting patients
Long waits for service
Overcrowded, uncomfortable waiting rooms
Working alone
Poor environmental design
Inadequate security
Lack of staff training and policies for preventing and managing crises with potentially volatile patients
Drug and alcohol abuse
Access to firearms
Unrestricted movement of the public
Poorly lit corridors, rooms, parking lots, and other areas

FIGURE 7.2 Risk factors for violence in hospital settings. (From http://www.cdc.gov/niosh/2002-101.html)

of mentally ill patients who have been "deinstitutionalized" or released from psychiatric hospitals with inadequate follow-up care; isolated work with clients; lack of staff training; and inadequate staffing during off-shifts and at times of increased activities, such as meal time.[6]

Research

In 2001, a group of experts gathered for a Workplace Violence Intervention Research Workshop in Washington, DC at the invitation of NIOSH and the University of Iowa. The workshop's findings culminated in the publication of a report titled "Workplace Violence: A Report to the Nation" and also congressional action to dedicate a significant amount of NIOSH extramural research funding to workplace violence prevention research. Among its findings, the report noted the lack of systematic national data collection on workplace assaults, the paucity of data evaluating violence prevention strategies, and the methodological flaws in published intervention research to date.[12]

As background to this report, in 2000, researcher Carol Runyan and her colleagues reviewed the violence prevention intervention literature that was available at the time and found five studies that evaluated violence prevention training interventions, two that examined postincident psychological debriefing programs, and two that evaluated administrative controls to prevent violence.[13–23] Findings from these studies were mixed, with six reporting a positive impact and three reporting no or a negative impact (see Figure 7.4). All were quasiexperimental and lacked a formal control group. Runyan and her coauthors criticized the design of the published violence prevention interventions available at the time because they lacked systematic rigor.[13]

In the same year, Runyan et al. conducted their literature review, researchers Judith Arnetz and Bengt Arnetz reported on a randomized control trial of 47 health care workplaces. In the intervention facilities, there was "structured feedback" from supervisors following incidents.[13] They found significantly more (50%) violent incidents reported in the intervention facilities compared with the control group workplaces. The authors attributed this finding to an increased awareness of the workplace violence incidents on the part of staff and improved supervisory support at the intervention facilities.

Between 2000 and 2004, Lipscomb and her colleagues conducted an intervention effectiveness study to design and evaluate a comprehensive process for implementing the OSHA Violence Prevention Guidelines and to evaluate its impact in a large state operated mental health setting.[24] Program impact was evaluated by a combination of quantitative and qualitative assessments.

Recent work by Peek-Asa et al. examined a representative sample of hospitals in two states, one of which had enacted a law requiring a violence prevention program. The research included the use of onsite visits to assess architectural and design features as a component of workplace violence prevention programs. The most common environmental feature in the hospitals in both states was surveillance cameras. Eliminating areas where employees work alone or can become isolated was much less commonly achieved (8.8% and 0%). Controlled access was surprisingly low as well, 40% in the state with the workplace violence law and 22% in the comparison state.

Finding from this group of intervention studies is summarized in Figure 7.3.

Types of Intervention Evaluated and Their Finding

Institutional Settings: Risks and Preventive Strategies

Institutional health care settings where workers are especially vulnerable include mental health and acute psychiatric settings, emergency departments, geriatric and geropsychiatric units, and some intensive care areas such as neurotrauma.

For example, in a Washington State forensic hospital, a facility that treats the criminally mentally ill, staff completed a self-administered survey and reported an incidence rate of 415 assaults per 100 employees per year; on average, each employee suffered more than four assaults per year. Seventy-three percent of staff who completed the survey and reported that they had been assaulted by a patient during

Training
Lehman (1983) VA Hospital +
Infantino and Musingo (1985) +
Carmel and Hunter (1990) –
Parks (1996) –
Goodridge et al. (1997) +

Post incidence debriefing
Flannery et al. (1998) +
Matthews (1998) –

Other strategies
Drummond et al. (1989) – +
Hunter and Love (1996) +
Arnetz and Arnetz (2000) –

Comprehensive program
Lipscomb et al. (2006) +/–

Policy/regulation
Peek-Asa et al. (2007) +/–

FIGURE 7.3 Types of intervention evaluated and their finding.

the previous year, suffered at least a minor injury. Factors such as working in isolation, being a mental health technician, and working on the geriatric-medical hospital unit were associated with more severe employee injuries. Employees suffered less severe injuries if they received assault management training within the prior year.[28]

Many psychiatric settings now require that all patient care providers receive annual training in the management of aggressive patients. To date, few studies have examined the effectiveness of such training. Those that have done so have generally found improvement in nurses' knowledge, confidence, and sense of safety after being trained in how to manage aggressive behavior. While training can improve the individual skills of workers, it does not address the organizational improvements that can only be achieved through a comprehensive violence prevention program. These include environmental, administrative, and clinical approaches to violence prevention. Training is just one administrative approach.

The emergency department is another high risk setting for staff assaults. Weapons in emergency departments are a major factor in severe or fatal injuries. A National Institute of Justice report estimated that, in 1994, there were enough guns owned in the United States that every adult could have one.[29] The wide availability of firearms, as well as the gang violence that often spills into emergency departments, create a "perfect storm" for deadly incidents in our overwrought emergency rooms. A 14 year study conducted at a Los Angeles hospital found that between 1979 and 1993, 26% of major trauma patients were armed with deadly weapons. The hospital's screening process used either metal detectors or involved the removal and inspection of all clothing. This yielded an average of 5.4 weapons a day, of which 84% were guns and most of the remaining weapons were knives.[30] Focus groups of emergency department staff, conducted by the authors in a number of states, adds to the evidence that much of the emergency department violence is gang-related.

A recent study of a 770-bed acute care hospital, in Florida, surveyed nurses working in intensive care, the emergency department, and floor nurses. When asked about their experience with workplace violence during the prior year, every nurse in the emergency room reported being verbally abused and 82.1% reported being physically assaulted. Even among the floor nurses, 80.6% reported verbal abuse and 63.3% reported physical assault.[31]

Gates et al. examined workplace violence in five facilities with emergency departments in a midwestern city. Facilities included a Level 1 Trauma hospital with separate medical, psychiatric, and "medivac" care, and four facilities with a general emergency department. They found that 32% of surveyed staff ($n = 115$) worked in facilities where patient and triage areas are open to the public; 25% reported that weapons were easily brought into their facilities, and 22% noted a lack of metal detectors or alarms in their emergency department (ED). Sixty percent felt that long waiting times contributed to violence in their facilities.[62]

Although mental health and emergency departments have been the focus of research and even some legislation, no department within a health care setting is immune from workplace violence. Consequently, violence prevention programs should be developed for all departments and units.

Violence Prevention Policy and Programming

There is no federal standard that addresses how to develop a workplace violence prevention program. A number of states including California, Illinois, Kansas, Michigan, New York, and Washington have laws addressing some aspects of a violence prevention program in health care.

In 1993, after the murders of three emergency room physicians, Cal/OSHA published the first set of nonmandatory guidelines describing the components of a comprehensive workplace violence prevention program.[1,51] After several years of pressure from a multiunion task force on workplace violence, federal OSHA followed California's lead and, in 1996, issued a similar set of guidelines, entitled *Guidelines for Preventing Workplace Violence for Health Care and Social Service Workers*.[52]

The 1996 Federal OSHA guidelines provide a framework for addressing the problem of workplace violence and include the basic elements of any proactive health and safety program. We have found that establishing an ongoing process with appropriate management leadership and worker and union involvement is a critical first step. Also it is very important for participants to recognize that the goal of the program is organizational improvement and not strictly changing the practice of individual workers. Changing the cultural bias that "violence is part of the job" is a key starting point.

1. *Management commitment.* Management commitment must be evident in the form of high-level management involvement and support for a written workplace violence prevention policy and its implementation.
2. *Employee involvement.* Meaningful employee involvement in policy development, risk assessment, joint management–worker violence prevention committees, postassault counseling and debriefing, and follow-up are all critical program components. This must include frontline workers and, where a union exists, union representatives. The direct care staff has a wealth of knowledge about the risks and solutions and is key to implementing proposed interventions. Commitment of necessary time, resources, and personnel to violence prevention programs is the main determiner of their effectiveness.
3. *Worksite analysis.* A worksite analysis is the foundation on which an effective program exists. This analysis should use all available "data" sources and be repeated, at least in part, on a periodic basis. "Data" sources include: OSHA logs, unusual incident logs, overtime usage, patient incident records, and workers compensation data. This information can be invaluable in identifying trends and risk factors.[52]

   However, a number of researchers have identified significant barriers and disincentives to reporting workplace incidents and injuries.[53–55] In the Washington State forensic hospital survey, referenced earlier in the chapter, staff reported 415 assaults per 100 employees per year. By comparison, the hospital incident reporting system reflected a rate of only 35.3 assaults per 100 employees. Of those reporting a moderate, severe, or disabling injury related to an assault, only 43% filed for workers' compensation.[28]

   Thus, relying solely on existing data sources may severely underestimate the extent of the problem and will fail to take into account significant risk factors that have not been documented. The institutionally collected data sources should be complemented with information generated from staff surveys, focus groups, and other forms of direct communication with frontline workers.

Finally, regular walk-through surveys of all areas of the facility should be conducted, and should include staff from each area and from all shifts. Special attention should be paid to those areas where assaults have occurred. When a unit or facility is undergoing renovations or a "new design" it is very important to build safety into the design. It is almost always cheaper to building in safety than to retrofit.

4. *Hazard prevention and control.* Hazard prevention and control measures should be based on the factors identified during the risk assessment. A variety of controls should be used as follows: (1) Engineering controls that eliminate or reduce the risk through technology or design such as door locks, convex mirrors, bolting down furniture; (2) implementing administrative measures including policies and procedures, personal alarms, training, communication, teamwork, and appropriate staffing; and (3) clinical strategies such as patient programming, review of patient diagnoses and treatment plans, approaches to dealing with multiassaultive patients. It is also important for health care organizations to establish strong relations with criminal justice authorities so that crimes that are committed on hospital premises are appropriately prosecuted. Often criminal justice authorities and hospital administrators do not want to bother with violent patients who are already confined in a public institution. This fosters further violence in the affected health care setting and creates an unfair double standard of prosecution.

   Hazard prevention and control measures to consider include: modifying the layout of admissions areas, nurses' stations, medication rooms, lounges, patient rooms, or offices; limiting access to certain areas; evaluating all furnishings to ensure that they are not used as weapons; installing metal detectors; improving lighting, air quality, and noise levels; adding mirrors or cameras, removing trees or shrubbery from doorways and structures. Other options include hiring trained security personnel, increasing direct-care staff, or redeploying existing staff more effectively such as adding a second person for overnight duty. Developing and implementing appropriate policies and providing regular training are important administrative controls, for example issuing cell phones and personal alarm devices to workers.

   Additionally, programs need to be in place to provide support to assault victims and to their coworkers. Typically these are called Critical Incident Stress Management or Trauma Response Program. They feature trained teams that can debrief affected staff and refer them to easy access to medical and mental health services, assistance with the workers' compensation, and support in accessing the criminal justice system, when appropriate. The timing of debriefing can be very important in whether affected staff are willing and ready to participate. Further, many of these programs suffer because they do not include follow-up actions such as individual referrals for counseling or group discussions about lessons learned to prevent similar incidents. A major impact of workplace violence is posttraumatic stress disorder (PTSD) among staff, which often goes untreated (FLANNERY R date). Finally, to the extent possible exposure to the potential violence should be eliminated. An example would be not accepting certain Axis II personality disorder patients in a facility that is not prepared to address their behaviors and needs. It is also important to intervene early with patients who are multiassaultive by trying a variety of clinical strategies, transfer, to a more secure facility, or prosecution, as appropriate.

5. *Training and education.* At the time of hiring and annually thereafter, worksite- and job-specific training should be provided covering the risk factors, prevention measures, and relevant policies and procedures. This training should not be generic. For direct care staff, training should include skills in preventing and managing aggressive behavior.

6. *Recordkeeping and program evaluation.* Recordkeeping and program evaluation are linked and should include incidents of physical and verbal assaults. Early intervention at early signs of agitation may prevent violence from escalating into physical assault. It is a mistake to rely solely on workers' compensation claims or OSHA logs. Staff should be actively encourages to report all incidents. There should be systems for communicating about agitated patients within the treatment team, across shifts, and among nonunit staff such as recreation, program, vocational, or

occupational therapists. Incidents should be investigated promptly, with the focus on prevention, and any results or recommendations relayed to the individual and team who made the report. Retaliation of any kind against employees for filling out an incident report or filing a workers' compensation claim shall not be tolerated. Reporting, investigation, evaluation, and communication of incidents are a critical means of preventing workplace violence. It is also key to providing historical data for evaluating the effectiveness of the workplace violence prevention program and evaluating effectiveness of control measures.

The OSHA Guidelines merely provide a framework for developing a violence prevention program. Because these guidelines are "performance-based" rather than prescriptive, the responsibility lies with each employer to develop a specific and effective process for tailoring the guidelines to the specific needs and conditions in their facilities. As stated earlier, experience has proven that input from staff and unions (where present) is key. It should be noted that a number of international professional and governmental agencies have also issued guidance on violence prevention in health care settings.[56–61]

Case Example: Institutional Psychiatric Facility. Implementation of the OSHA Violence Prevention guidelines was the basis of a 4 year research collaboration (2000–2004) that included an academic research team, a state mental health system, and a joint-labor management health and safety committee.[24]

The agency and labor unions had already been involved for years in collaborative efforts to reduce workplace violence by implementing programs including training, a written policy entitled "Safe & Therapeutic Environment Plan", data collection, and postassault care. Preliminary work by the joint health and safety committee in two hospitals demonstrated the feasibility of the intervention, using a participatory approach, to facility and ward-specific risk assessments and interventions.[33]

The NIOSH funded research project involved three intervention facilities and three comparison or "usual care" facilities. The intervention facilities included two adult and one children's in-patient psychiatric hospital. What follows is a brief description of the project's activities.

Each intervention hospital formed a facility level advisory group comprised of direct care staff, management, and union representatives. These labor-management committees met regularly and provided input during each stage of the project and directed every aspect of the intervention at their facilities. The project included four hazard analysis activities: data analysis, staff focus groups, an environmental audit, and staff surveys.

The focus groups were conducted by a trained facilitator and were recorded and reported. Participants were direct care staff. Supervisors were not included so as to avoid staff discomfort or inhibition among focus group attendees. Using a standard set of questions, the facilitator engaged participants in discussion about their experiences, perception of risk factors, and ideas for solutions to workplace violence on their wards.

An architect with decades of experience in designing secure institutional settings conducted the environmental audits in each of the intervention facilities; facility advisory group members participated in these audits. The architect's approach was unique in that he made a strong effort to merge his knowledge of materials, hardware, and design with the clinical design and objectives of each ward. He did this by talking to the project advisory groups (PAGs) prior to conducting the audits and informally interviewing direct care staff during the inspections. For example, the design of the nurses' station has a significant impact on the ability of ward staff to observe and interact with patients. Each facility received a report that included short- and long-term recommendations for improvements.

(*continued*)

(continued)

The data analysis looked at number of incidents and severity (lost time) of incidents by ward/unit, time of day, activity, and location. These data helped the facility advisory groups select target wards/units for the intervention study and also focus efforts to higher risk activities and time frames. Additionally, some facilities analyzed the number of assaults that derived from patients who were repeat assaulters. In the adult facilities there was a startling trend that small number of patients who were repeat assaulters often was responsible for more than 50% of the assaults annual.

A staff survey, informed by the focus group findings, was developed and administered to all staff during work time. The survey included questions about verbal threats of assaults, as well as physical assaults with mild, moderate, severe, and disabling injuries. Other questions assessed procedures, communication, and work organization, as well as staff perceptions regarding the quality of the OSHA Violence Prevention elements at their facility and their perceptions of the quality of violence prevention training. The survey also included a section on staff satisfaction. This instrument was administered early in the project and repeated two years later to evaluate the impact of the program.

The agency had already institutionalized training for direct care staff that consisted of a 2.5 day initial and 1 day annual Prevention and Management of Crisis Situations (PMCS) training program. This curriculum was developed by internal training experts and had been recently revised and updated. Therefore the emphasis of the project's training component focused on communicating ongoing findings from the risk analysis phase of the project and consolidating efforts to implement solutions. This strategy was operationalized midway through the intervention when direct care staff were brought together for a series of "solutions-mapping" sessions where the results of the risk assessments were presented and then small groups worked collectively prioritize the actionable items with facility managers.

Each intervention facility developed their own hazard control plan based on the findings of their data analysis, focus groups, environmental audit, and staff survey. Control strategies included any engineering, administrative, or clinical activities designed to protect workers and patients from violence. Examples included:

- Improved change of work shift communication and teamwork, especially between psychiatrists, psychologists, social workers, and direct care staff
- Installation of a new personal alarm system as a statewide agency initiative
- Elimination of long waits in food lines
- Addition of a telephone in the day room
- Change in locks and other hardware
- Replacement of portable furnishings with secure items that cannot be used as weapons

A comparison of pre- and postintervention survey data indicated an improvement in staff perception of the quality of the facility's violence prevention program as defined by the OSHA elements in both intervention and comparison facilities over the course of the project. Many health care organizations lack adequate quantitative data systems for tracking incidents, lost time, and severity. This combined with the dynamic environments in which intervention programs are implemented makes traditional approach to evaluating effectiveness difficult. The impact on the rates of staff-reported physical assaults was equivocal, however. Qualitative findings included reports from intervention sites of a number of project successes, including a violence prevention training coach at one study site and the adoption of one facility's written violence prevention program into the facility's overall strategic plan. Intervention sites generated a list of violence prevention best practices that were then shared with the Directors of their sister facilities within the state system.

The agency had implemented multiple interventions system-wide that made it particularly difficult to isolate the impact of the study interventions in any quantitative way. Furthermore, the national trend toward elimination of the use of restraint and seclusion was in full effect at the time of the study. In some facilities its introduction was done without adequate communication and

(continued)

> preparation of direct care staff. In these instances there was often confusion about direct care staff roles when a patient became violent. The goal of this effort has been beneficial in eliminating the unnecessary use of restraint and seclusion and focusing on prevention efforts. However, the policy and its implementation have yet to clarify what steps should be taken when prevention fails.

Community Settings: Risks and Preventive Strategies

Advocates of the developmentally disabled have successfully pushed for reforms resulting in a transferring of care from restrictive institutional environments to group homes, family care, and day programs. While this transformation in public policy has been largely positive and beneficial, it has not included the commensurate provision of staffing and resources necessary to provide for client and staff safety. The largely nonprofit sector pays very low wages to direct care staff and minimal benefits. For these reasons, and due to poor working conditions, one of our society's most vulnerable patient populations is being served by a transient and unstable workforce. In general, these facilities are regulated for the protection of the clients and subject to frequent audits and inspections by public agencies. Considerably less attention is given to the safety of the work environment, despite substantial workers' compensation costs and high staff turnover. Social service workers, particularly those employed by public and nonprofit agencies that provide outpatient and residential services to the mentally retarded, severely developmentally disabled, or mentally ill, often care for clients whose behavior is not sufficiently stable for these non-institutional settings. This poses a risk to fellow clients and staff.

The risks to visiting social service and HCWs have been documented by a number of researchers.[34–38] Homicides of visiting social workers and nurses have been reported in Texas, Maryland, Michigan, Kansas, New York, Washington, Ohio, and Kentucky. In response to these deaths, legislation has been introduced in several of these states (Michigan, Kansas, New York, and Washington) to strengthen and/or require specific safety measures for these at-risk workers.[27,39–41] These measures include community-specific violence prevention training, the use of cell phones or pagers, having a coworker come along when visiting high risk clients,[27] and maximum caseloads. Important risk factors emerging from these homicides include: (1) the client's perception, that the visit would result in removing children from the home (New York, Texas, and Michigan); (2) care provider unknown to the patient (Washington and Maryland); (3) client in mental health crisis requiring possible involuntary commitment (Washington, Kansas); (4) relevant criminal history not known to case worker (Washington, Kansas, and New York); and (5) worker visiting alone (all).[42–49]

Protecting home visiting mental health workers present substantial and complex challenges. Each home entered by the employee is a unique and unpredictable work environment. Unlike some institutional settings, there are no security guards screening for weapons and contraband, no alarm systems or panic buttons, no video monitors, mirrors, special lights, quick release locks, or drop phones, and no additional staff to assist if a patient becomes agitated during the course of the visit.

In addition, home care workers often conduct visits without information about a patient's history of violence, criminal justice background, or recent stability and behavior. In community and home-based settings, unlike institutional settings, patient behavior is not monitored by qualified clinicians on an around-the-clock basis. Case workers are often the first to discover during a home visit when patients are decompensating, refusing medications, or having social problems that affect their potential dangerousness. The tools available in an inpatient setting such as chemical restraints, one to one observation, or other clinical interventions are not applicable to the community setting.[24,50]

Fortunately, the majority of developmentally disabled and/or mentally ill clients who live at home do so because their condition is stable. Home visiting services enable these clients to remain in the community, where they can be close to family and friends. Yet, clients may decompensate and become upset, agitated, or out of control posing a "high risk" to a home visiting worker. Home visits which are identified as being "high risk" should always be conducted by pairs of workers, in the company of a police officer or sheriff, or at a safe neutral setting. The assessment of risk should be based on the patient's history, clinical recommendations, and the input of family and friends.

Case Example: Community Setting. Following the murder of a visiting social worker in a northwestern state, the authors of this chapter were invited to conduct a consultative field study. The victim's union arranged for access to multiple administrators, agencies, and groups of community-based field workers caring for the mentally ill. Factors reported to be associated with violence included high caseloads, sicker clients, paperwork burdens, management emphasis on productivity rather than quality or safety, and other concerns such as lack of access to criminal history and patients' inability to obtain needed psychiatric medications. Ironically, a number of states have passed laws requiring criminal background checks on staff who do not have a legal right to access criminal history information on potentially violent clients. The federal reimbursement for home visits is based on a "per visit" system. This creates a financial disincentive for more than one staff to participate in a home visit.

The frequency of violent encounters varied by mental health setting. By far, Community Emergency and Treatment Centers (E and Ts) reported having the most frequent experience with combative and assaultive clients. The very purpose of E and Ts is short-term institutionalization for the acutely and dangerously mentally ill. Staff make the final judgment for involuntary commitment, often in settings like emergency rooms, clinics, and client homes. Generally, outpatient and home visiting settings report less overt violence, but more verbal aggression and fear associated with the potential for violence. They also report a lack of safety and security programs, including training. Unpredictable behavior, a symptom of some mental illnesses, increases the risk of assault to workers. Many outpatient case managers and mental health workers are responsible for transporting clients, frequently without a coworker present. Workers describe instances of unpredictable behavior on the part of their passenger/client, such as changing the gearshift or trying to exit the vehicle while it is in motion.

Staff in the community mental health system reported the full spectrum of violence, from sexual harassment, verbal threats of harm, spitting, profane language, and unwanted touching and physical contact, to major assaults with physical injuries or even death. Generally, physical assaults are the primary cause of concern, as they result in lost work time, medical costs, pain and suffering, psychological trauma, and even permanent disability or death. Much of the verbal and sexual aggression and minor assaults are not reported and, thus, not officially recognized. The relationships between lower level violence and patient safety, quality of care, absenteeism, staff retention, and job satisfaction have not been researched, but merit additional investigation.

What Needs to Be Done

Establishing a task force, workgroup, or using an existing health and safety committee or forum is the first step in developing a comprehensive violence prevention program. As stated earlier this process must include appropriate management decision makers, union representatives, direct care staff, and discipline chiefs. Once the team is established, existing and new risk assessment measures should be planned, reviewed, and analyzed. The focus should be implementing practical improvements whenever possible. The team's efforts must be communicated through a variety of ways so that all hospital stake holders are a part of the program. It is especially important to improve systems of providing risk information to direct care staff. The program should work at changing organizational culture so that violence whether verbal or physical is never considered "part of the job." Strong messages, written and oral, from top management, union leaders, and supervisors should encourage the reporting of verbal and physical violence. Staff will begin to "believe" when they see that their reports do not lead to recrimination and do cause protective interventions. In our experience, real or perceived patient privacy concerns and a "blame the victim" mentality discourage the sharing of important information. The following barriers deserve specific discussion:

Incomplete access to client history of violence and criminal background: In the aftermath of the tragic death of a mental health worker in Washington State, administrators began to investigate the barriers to obtaining relevant information, such as a client's criminal history and evidence of past violent behavior. In most states there are legal and regulatory barriers to accessing criminal background records that may be relevant to client and staff safety. In the Washington State and New York fatality cases, relevant information about several past episodes of violent behavior involving law enforcement were not known to the case workers when they visited the client alone. This type of information is critical for all psychiatric care workers to have, but is of the utmost importance to community-based mental health workers. Some solutions to this dilemma include increased interagency cooperation about patient histories; using a specific patient assessment tool for history of violence; and new legislation and regulations that allow increased access to criminal background information to case workers.

More acutely ill and dually diagnosed individuals: Deinstitutionalization of the mentally ill has led to a change in the patient demographics among those who remain in the inpatient facilities. They tend to be the more chronic patients, dually diagnosed with addiction and psychiatric illness, and criminal justice backgrounds. Further, public policy has led to the transfer of large numbers of psychiatric patients into the state and county prison systems that are ill equipped to provide care for such individuals. In 1996, the cost of inpatient psychiatric care in New York State (NYS) was $225,000 per patient compared to $28,000 per state prison inmate. The pressures to reduce state budgets, cut taxes, and reduce "big government" are clearly factors in this policy shift. These public policy shifts have also led to limits on public funding for social services, lean staffing in hospitals, and cutbacks in public health programs.

NYS has just passed legislation to increase mental health treatment and staffing in the prison system, especially directed toward prisoners in Special Housing Units, discipline units where inmates had previously been kept in isolation for 23 h per day, with only 1 day for recreation and little to no mental health programming. However, it is clear that a real policy shift will require a sea change in society's priorities to increase resources for the mentally ill and disabled.

Support for filing criminal charges: A significant barrier to violence prevention efforts is the practice of organizations discouraging or interfering with workers who want to press criminal charges against patients who have assaulted them Workers are often led to believe that they check their legal rights at the door upon arriving at work. Health care and social services workers are often told that they should expect to be assaulted due to the nature of their work. This is a double standard as an assault in a bar or on the street would not be tolerated. Health care organizations should meet with district attorneys, police chiefs, and judges to develop understanding, and in some cases, written agreements, on how they will work together to prosecute appropriate cases. It is especially important that programs include an evaluation component, so that progress can be measured, various interventions graded and modified as indicated, and that programs can be sustained.

Recommendations and Conclusions

In order to curtail violence in health care and social service agencies, we recommend a federal OSHA standard should be established that that mandates comprehensive violence prevention programs for all health care and social service workplaces in the public, private, and non-for-profit sectors. Until that time, states should promulgate legislation and regulations that require performance-based programs that are tailored to the actual needs and conditions of each regulated organization. They should mandate staff and union participation in all aspects of the program. They should be reviewed and revised periodically, especially following a serious incident. They should include all elements of the OSHA Guidelines. These standards should be incorporated into current institutional licensure and accreditation criteria. This has been successfully done with respect to bloodborne pathogen exposure in health care, where both OSHA and the Joint Commission (formerly the Joint Commission on Accreditation of Health Care Organizations) have provided specific requirements for health care facilities. There must be an effective complaint and enforcement mechanism so that staff who are victimized have recourse. We have found that without legislation, regulation, and enforcement most employers will not commit the resources necessary to establish effective programs.

Additionally, severe workplace assaults must be evaluated for their possible criminality; and, where appropriate, referrals to the criminal justice system must be made.

Policies to safeguard the rights of the mentally ill and developmentally disabled, must be balanced with the right of workers to return home safely to their family at the end of each workday. Ignoring the dangers to the workforce has already contributed to reduced morale, an exodus of experienced staff, and staff shortages—all of which negatively impact the quality of care and increase the likelihood of violence.

In this section we have not addressed staff on staff violence or domestic violence in the health care workplace in a comprehensive way. In part this is due to the need to focus on Type II violence. It is also because there is a paucity of data, research, or industry experience in developing effective programs addressing Types III and IV violence. Clearly, it is time to put resources into developing and evaluating programs to curtail all types of violence in health care.

Being assaulted at work should never be part of anyone's job.

Acknowledgments

The authors would like to thank Kelly Flannery, University of Maryland doctoral student, for all of her research and assistance. Most importantly, we would like to thank the thousands of nurses, aides, social workers, and others who provide compassionate care and service to the most vulnerable members of our society. Their wisdom and concern has been communicated to us through surveys, focus groups, and personal conversations. It is to them that we dedicate this chapter.

References

1. Centers for Disease Control and Prevention/National Institute for Occupational Safety and Health. *The National Occupational Research Agenda*. National Institute for Occupational Safety and Health, Cincinnati, OH, 2001.
2. Lipscomb J and Love C. Violence toward health care workers: An emerging occupational hazard. *Am Assoc Occupat Health Nurses*, 40:219–228, 1992.
3. Rippon TJ. Aggression and violence in health care professions. *J Adv Nurs*, 31:452–460, 2000.
4. Toscano G and Weber W. *Violence in the Workplace*. U.S. Department of Labor, Bureau of Labor Statistics, Washington, 1995.
5. Warchol G. *Workplace Violence, 1992–96*. U.S. Department of Justice, Office of Justice Programs, Washington, DC, July 1998, NCJ 168634.
6. Occupational Safety and Health Administration. *Guidelines for Preventing Workplace Violence for Health Care & Social Service Workers*. U.S. Department of Labor, Washington, DC, Publication 3148, 2004.
7. Duhart DT. *Violence in the Workplace, 1993–99*. U.S. Department of Justice, Office of Justice Programs, Washington, DC, NJC 190076, 2001.
8. Bureau of Labor Statistics, *Survey of Occupational Inquiries and Illnesses*. U.S. Department of Labor, Washington, DC, USDL 96-11, 1996.
9. Manning S. Occupational hazards, injury among direct health care staff, an unheralded occupational safety and health crisis. *Professional Safety*, 31–40, 2005.
10. National Institute for Occupational Safety and Health. Violence in the workplace: Risk Factors and Prevention Strategies. CIB #57 1996.
11. http://www.cdc.gov/niosh/2002-101.html
12. University of Iowa Injury Prevention Research Center. *Workplace Violence—A Report to the Nation*. University of Iowa, Iowa City, IA, 2001.
13. Runyan CW, Zakocs RC, and Zwerling C. Administrative and behavioral interventions for workplace violence prevention. *Am J Prevent Medic*, 18:116–127, 2000.

14. Carmel H and Hunter M. Compliance with training in managing assaultive behavior and injuries from inpatient violence. *Hosp Commun Psychiatry*, 41:558–560, 1990.
15. Goodridge D, Johnston P, and Thomson M. Impact of a nursing assistant training program on job performance, attitudes, and relationships with residents. *Educ Gerontol*, 23:37, 1997.
16. Infantino JA Jr and Musingo SY. Assaults and injuries among staff with and without training in aggression control techniques. *Hosp Commun Psychiatry*, 36:1312–1314, 1985.
17. Lehmann L, Padilla M, Clark S, and Loucks S. Training personnel in the prevention and management of violent behavior. *Hosp Commun Psychiatry*, 34:40–43, 1983.
18. Parkes J. Control and restraint training: A study of its effectiveness in a medium secure psychiatric unit. *J Foren Psychiatry*, 7:525–534, 1996.
19. Flannery M, Rosen H, and Turner K. Victims no more: Preventing home care aide abuse. *Caring*, 17:48–50, 52–43, 1998.
20. Matthews LR. Effects of staff debriefing on post-traumatic stress symptoms after assaults by community housing residents. *Psychiatr Serv*, 49:207–212, 1998.
21. Arnetz JE and Arnetz BB. Implementation and evaluation of a practical intervention programme for dealing with violence towards health care workers. *J Adv Nurs*, 31:668–680, 2000.
22. Drummond D, Sparr LF, and Gordon GH. Hospital violence reduction among high-risk patients. *J Am Med Assoc*, 261:2531–2534, 1989.
23. Hunter ME and Love CC. Total quality management and the reduction of inpatient violence and costs in a forensic psychiatric hospital. *Psychiatr Serv*, 47:751–754, 1996.
24. Lipscomb J. et al. Violence prevention in the mental health setting: The New York State experience. *CJNR*, 38:96–117, 2006.
25. Washington State Department of Labor and Industries. Violence in Washington workplaces, 1995–2000. 39-4-2002, 2002.
26. Foley M. *Violence in Washington Workplaces, 1995–2000*. Washington State Department of Labor and Industries, Olympia, Washington. Technical Report Number 39-4-2002, 2002.
27. http://www.leg.wa.gov/pub/billinfo/2007-08/Pdf/Bills/House%20Passed%20Legislature/1456-S.PL.pdf.
28. Bensley L. et al. Injuries due to assaults on psychiatric hospital employees in Washington State. *Am J Ind Med*, 31:92–99, 1997.
29. Travis J. *Guns in America: National Survey on Private Ownership and Use of Firearms*. National Institute of Justice, Washington, DC, NCJ 165476, 1997.
30. Ordog GJ, Wasserberger J, Ordog C, Ackroyd G, and Atluri S. Weapon carriage among major trauma victims in the emergency department. *Acad Emerg Med*, 2:109–114, 1995.
31. May DD and Grubbs LM. The extent, nature, and precipitating factors of nursing assault among three groups of registered nurses in a regional medical center. *J Emerg Nurs*, 28:11–17, 2002.
32. Arnetz JE and Arnetz BB. Violence towards health care staff and possible effects on the quality of patient care. *Soc Sci Med*, 52:417–427, 2001.
33. Rosen J. A joint labor/management experience in implementing OSHA's violence prevention guidelines in the NYS Office of Mental Health. In: W Charney (ed.), *Handbook of Modern Hospital Safety*. CRC Press, Lewis Publishers, Boca Raton, FL, 1999.
34. Fitzwater E and Gates D. Violence and home care. A focus group study. *Home Health Nurse*, 18:596–605, 2000.
35. Fazzone P, Barloon LF, McConnell SJ, and Chitty JA. Personal safety, violence, and home health. *Pub Health Nurs*, 17:43–52, 2000.
36. Barling J, Rogers AG, and Kelloway EK. Behind closed doors: In-home workers' experience of sexual harassment and workplace violence. *J Occup Health Psychol*, 6:255–269, 2001.
37. Bussing A and Hoge T. Aggression and violence against home care workers. *J Occup Health Psychol*, 9:206–219, 2004.
38. Schulte SM, Nolt BJ, Williams RL, Spinks CL, and Hellsten JJ. Violence and threats of violence experienced by public health field-workers. *JAMA*, 5:439–442, 1998.
39. http://www.pef.org/stopworkplaceviolence/files/judi_scanlon_bill.pdf

40. http://www.moore.house.gov/nr.asp?nr_id=495
41. http://www.michigan.gov/dhs/0,1607,7-124-5459_7341-15406–,00.html
42. Public Employee Safety and Health. Notice of violation and order to comply. General Duty Citation to the New York State Office of Mental, 1999.
43. Gillespie C. Ohio children's service worker stabbed to death, The Nando Times Nation, 2001.
44. Ly P. Slaying raises mental care questions, 2002, at www.washingtonpost.com
45. Newhill CE. *Client Violence in Social Work Practice*, The Guilford Press, New York, 2003.
46. Sedensky M. Kansas: Social worker killed August 17 & her 17-yr-old client is charged with murder, 2004, at http://www.fightcps.com/2004_08_29_archive.html
47. Austin L. Social workers rethink job safety a Texas slaying brings new steps, 2006, at http://workforce.socialworkers.org/whatsnew/WFC-MediaWatch-Web.pdf
48. http://www.foxnews.com/story/0,2933,225422,00.html
49. http://www.komw.net/artman/publish/printer_1811.shtml
50. Occupational Safety and Health Administration. OSHA fact sheet: Workplace violence, 2002.
51. Simonowitz J. Violence in the workplace: You're entitled to protection. *RN*, 57:61–63, 1994.
52. Occupational Safety and Health Administration. *Guidelines for Preventing Workplace Violence for Health Care and Social Service Workers.*, Washington, DC, OSHA # 31481996.
53. Rosenman KD. et al. How much work-related injury and illness is missing by the current national surveillance system? *JOEM*, 48:357–365, 2006.
54. Fan ZJ, Bonauto DK, Foley MP, and Silverstein BA. Under reporting of work-related injury or illness to workers' compensation: Individual and industry factors. *JOEM*, 9:914–922, 2006.
55. Azaroff LS, Levenstein C, and Wegman DH. Occupational injury and illness surveillance: Conceptual filters explain underreporting. *Am J Pub Health*, 9:1421–1429, 2002.
56. www.aacn.org/AACN/pubpolcy.nsf
57. www.nursingworld.org/dlwa/osh/wp6.htm
58. www.cnanurses.ca/CNA/documents/pdf/publications/FS22_Violence_Workplace_e.pdf
59. www.cna-aiic.ca/CNA/documents/pdf/publications/PS57_Violence_March_2002_e.pdf
60. www.ilo.org/public/english/bureau/inf/magazine/26/violence.htm
61. www.who.int/violence_injury_prevention/violence/activites/workplace/en
62. Gates D, Ross C, and McQueen L. Violence against emergency department workers. *J Emerg Med* 31(3):331–337, 2006.

Ordog, 1995 a quarter of patients admitted to an inner city ED were carrying weapons Peek Asa (2002) overall improvement in hospital security programs after implementation of state law (Hospital Security Act).

Guidelines for Security and Safety of Health Care and Community Service Workers

Joyce A. Simonowitz

Preface

Violence is an escalating problem in workplaces, and health care settings are not exempt from this trend. In the past, health care professionals have generally regarded themselves as immune to harm arising from their work. When workplace violence resulted in injuries, administrators and supervisors often expressed sentiments that the HCW might have been at fault or that these incidents were "part of the job." These guidelines were developed in response to the increasing number of severe injuries, some resulting in death, experienced by health care and community service workers. A variety of individuals, organizations, unions, and state and local government agencies have requested assistance from the Cal/OSHA to control this serious occupational health hazard.

These guidelines are designed to assist all health care, community workers, and support staff who may be exposed to violent behavior from patients, clients or the public. Because every work situation that could present a threat to worker safety cannot be covered, the focus is on major health care and community service locations, such as public and private medical services (acute-care hospitals, emergency rooms, long-term care facilities, public health and other clinics, home health services and prehospital care); psychiatric service (inpatient, clinic, residential and home visiting); alcohol and drug treatment facilities and social welfare agencies including unemployment and welfare eligibility offices, homeless shelters, parole and public defender services and child welfare services. The recommended courses of action, however, may also be applied to prevent violence in any facility.

Measures to prevent assaults should include engineered systems and administrative measures as well as training. Alarm systems are one of the most important protective measures for hospitals and clinics. Of course, back-up security staff to respond to the alarm must accompany any alarm system. Many of the basic problems leading to violence may be traced to inadequate staffing levels, therefore an effective administrative measure is to ensure that appropriate staffing is maintained at all times. Training of all personnel is a necessary preventive as well as protective measure. Although the temptation to place the responsibility for controlling violence on the trained employee is great, it will not be sufficient to prevent serious injuries. A concerned administration that implements and maintains a well-developed program should be able to succeed in reducing the incidence of assaults and injuries in the workplace.

This section does not cover public safety work, such as in police or corrections departments where exposure to violence is well recognized and already addressed by department guidelines or regulations. These guidelines do not specifically address conditions of service industry workers in the private sector, such as sales personnel or restaurant workers. Some of the guidelines, however, may be adapted to protect these workers from violent acts that may occur in their contacts with the general public.

I am grateful to all who have contributed to this section. I am aware that updates will be needed as needs are identified and technology is refined.

I want to specifically acknowledge Melody M. Kawamoto, MD, for her assistance in editing this chapter.

Introduction: The Problem

During the past two decades, we have seen a sharp increase in violence in our cities, country, and society. Estimates show that nearly one-third of all Americans are victimized by crime each year (Poster and Ryan, 1989). Violence in the workplace is a manifestation of this problem, with homicide being the third leading cause of occupational death among all workers in the United States from 1980 to 1988 (Jenkins et al., 1992) and the leading cause of fatal occupational injuries among women from 1980 to 1985 (Levin et al., 1992).

Higher rates of occupational homicides were found in the retail and service industries, especially among sales workers (Jenkins et al., 1992). This increased risk may be explained by contact with the public and the handling of money (Kraus, 1987). Research into the causes of the increasing incidence of death and serious injury to HCWs has led to the theory that exposure to the public may be an important risk (Lipscomb and Love, 1992; Lavoie et al., 1988). The risk is increased particularly in emotionally charged situations with mentally disturbed persons or when workers appear to be unprotected.

During the past few years, violence resulting in the death of California health care and community workers occurred in emergency rooms, psychiatric hospitals, community mental health clinics, and social service offices. Assaults, hostage taking, rapes, robbery, and other violent actions are also reported at these and other health care and community settings. In a study by Conn and Lion (1983), assaults by patients in a general hospital occurred in a variety of locations. Although 41% of assaults occurred in the psychiatric units, they also occurred in emergency rooms (18%), medical units (13%), surgical units (80%), and even pediatric units (7%).

Carmel and Hunter (1989) found that the psychiatric nursing staff of a maximum security forensic hospital in California sustained 16 assault injuries per 100 employees per year. This investigation used the OSHA definition for occupational industry: an injury that results in death, lost work days, loss of consciousness, restriction of work or motion, termination of employment, transfer to another job, or medical treatment other than first aid (Bureau of Labor Statistics, 1986). Work-related injuries reported on OSHA forms and reported to the BLS for 1989 occurred at a rate of 8.3 per 100 full-time workers in all industries combined. The highest rate, 14.2 per 100 full-time workers, was seen in the construction industry (Bureau of Labor Statistics, 1991). In comparison, data collected by Carmel and Hunter suggest that some psychiatric workers may be at a higher risk for injuries from all causes in the country's most hazardous industry (Lipscomb and Love, 1992).

Madden et al. (1976), Lanza (1983), and Poster and Ryan (1989) have reported that 46%–100% of nurses, psychiatrists, and other therapists in psychiatric facilities experienced at least one assault during their career. Research on the causes and methods of prevention of violence in psychiatric facilities was funded by the California Department of Industrial Relations after the death of a psychiatric hospital worker in 1989. This investigation is in progress at the forensic hospital at the present time.

Lavoie et al. (1988), investigated 127 large, university-based hospital emergency departments and reported that 43% (55) had at least one physical attack on a medical staff member per month. Of the reported acts of violence in the last 5 years, 7% (9) resulted in death. Emergency room personnel face a significant risk of injury from assaults by patients, but in addition, may be abused by relatives or other persons associated with the patient. Further, the violence that occurs in the emergency room is often shifted into the hospital when the patient is transferred to the receiving unit.

Bernstein (1981) reported that 26% of reported assaultive behaviors in a study of California psychotherapists occurred in the outpatient setting. The death of an outpatient psychiatric worker in 1989 in California at the hands of a homeless client underscores the risk that exists in this setting. Investigations by OSHA officials in two California counties identified a nearly complete lack of security measures in outpatient facilities, leaving workers unprotected and vulnerable to abuse and assaults.

Community service workers are at risk of hostile behavior from the public when they visit clients at hotels, apartments, or homes in unfamiliar or dangerous locations, especially at night. Child welfare workers have reported that parents of children who are being taken to foster homes or other types of court action have become violent and assaulted workers with knives and fists. Sexual assaults with serious injury, other physical assaults, and robberies have been reported by workers in the hospital and community. In addition, clients or their relatives and friends may direct their anger, which can be extreme or violent, at community workers. In Canada, in community settings, physical attacks by patients were reported by 1.1%–14.1% of nurses surveyed by the Manitoba Association of Registered Nurses (Liss, 1993).

Few research investigations have focused on the incidence of violence to community workers, but reports have been received from many sources such as union workers or parking enforcement workers who have suffered abusive and at times violent behavior from hostile motorists. Hotel housekeepers are currently being studied after complaining of sexual abuse and threats in hotels in which they work. Such research is needed to identify the scope of violence in the medical field and the community as a whole.

Risk Factors

Risk factors may be viewed from the standpoint of (1) the environment, (2) work practices, and (3) victim and perpetrator profile.

Environmental Factors

Health care and community service workers are at increased risk of assaults because of increased violence in our society. This increase in violence is thought to be a result of such factors as the easy availability of guns and weapons; the use of violence by many in the population as a means of solving problems; the increase in unemployment poverty and homelessness; the decrease in social services to the poor and mentally ill; the increase in gang-related activity and drug and alcohol use; violence depicted in television and movies and the increasing use of hospitals by police and criminal justice systems for acutely disturbed patients. These may be thought of as a partial listing, which may have a direct contribution to the safety and security of workers.

An important risk factor at hospital and psychiatric facilities is the carrying of weapons by patients and their family or friends. Wasserberger et al. (1989) reported that 25% of major trauma patients treated in the emergency room carried weapons. Attacks on emergency rooms in gang-related shootings have been documented in two Los Angeles hospitals (Long Beach Press Telegram, 1990). Goetz et al. (1991) found that 17.3% of psychiatric patients searched were carrying weapons.

Other risk factors include the early release from hospitals of the acute and chronically mentally ill, the right of patients to refuse psychotropic treatment, inability to involuntarily hospitalize mentally ill persons unless they pose an immediate threat to themselves or others and the use of hospitalization in lieu of incarceration of criminals. McNeil et al. (1991) found that police referrals were significantly more likely to have displayed violent behavior such as physical attacks and fear-inducing behavior during the two weeks before coming to the psychiatric emergency service and during the initial 24 h of evaluation and treatment.

Work Practices

Many studies have implicated staffing patterns as contributors to violence. Both Jones (1985) and Fineberg et al. (1988) found that shortage of staff and the reduction of trained, regular staff increased the incidence of violence. Assaults were associated with meal times, visiting times, and times of increased staff responsibilities. This suggests that staffing evaluations do not take into account the potential hazards associated with increased activity in the units or for times when transportation of clients is needed. Assaults were also noted at night when staffing is usually reduced. Frequency of exposure to and interaction with patients or clients are known factors that increase a health care or community worker's vulnerability. Work in high crime areas, at an isolated work station or working alone without systems for emergency assistance may increase the risk of assaults. In addition, typical work activities may arouse anger or fear in some patients and result in acts of violence. Long waits in emergency rooms and inability to obtain needed services are seen as contributors to the problem of violence. This was evidenced in the emergency department shooting in Los Angeles where three doctors were shot by an angry, dissatisfied, and disturbed client.

Perpetrator and Victim Profile

It is difficult to predict when or which patients/clients will become violent, since the majority of assaults are perpetrated by a minority of persons. More acute and untreated mentally ill persons are being admitted to and quickly released from psychiatric hospitals and are in need of intensive outpatient treatment and services. These services are often lacking due to funding cuts. Further, clearly only a small percentage of violence is perpetrated by the mentally ill. Gang members, distraught relatives, drug users, social deviants, or threatened individuals are often aggressive or violent.

A history of violent behavior is one of the best indicators of future violence by an individual. This information, however, may not be available, especially for new patients or clients. Even if this information were available, workers not directly involved with the individual client would not have access to it. At times violence is not aimed at the actual care giver. Keep et al. (1992) reported on the gunshot death of a nurse and an emergency medical technician student who were targets of a disturbed family member of a patient who died in surgery the previous day.

Workers who make home visits or community work cannot control the conditions in the community and have little control over the individuals they may encounter in their work. Dillon (1992) reported the shooting death of four county workers in upstate New York and the beating death of a case worker who removed a 7-year-old child from a violent home. The victim of assault is often untrained and unprepared to evaluate escalating behavior and to know and practice methods of defusing hostility or protecting themselves from violence. Training, when provided, is often not required as part of the job and may be offered infrequently. However, using training as the sole safety program element creates an impossible burden on the employee for safety and security for him or herself, coworkers or other clients. Personal protective measures may be needed and communication devices are often lacking.

Cost of Violence

Little has been done to study the cost to employers and employees of work-related injuries and illnesses, including assaults. A few studies have shown an increase in assaults over the past two decades. Adler et al. (1983) reported 42 work days lost over a 2 year period due to violence to 28 workers, an increase from the previous 2 years in which 11 workers lost 62 work days. Carmel and Hunter (1989) reported that of 121 workers sustaining 134 injuries, 43% involved lost time from work with 13% of those injured missing more than 21 days from work. In this same investigation, an estimate of the costs of assault was that the 134 injuries from patient violence cost $766,000 and resulted in 4291 days lost and 1445 days of restricted duty. Lanza and Milner (1989) reported 78 assaults during a 4 month period. If this pattern were repeated for the remainder of the year, 312 assaults could be expected with a staggering cost per year from medical treatment and lost time. Additional costs may result from security or response team time, employee assistance program or other counseling services, facility repairs, training and support services for the unit involved, modified duty and reduction of effectiveness of work productivity in all staff due to a heightened awareness of the potential for violence.

True rates of violence at health care and community service facilities, however, must be assumed to be higher than documented rates. Episodes of violence are often unreported. If reported, records are not necessarily maintained. Nurses and other health care professionals are reluctant to report assaults or threatening behavior when the prevailing attitude of administrators and supervisors and sometimes other staff members, is that violence "comes with the territory" or "health professionals accept the risk when they enter the field." Administrators, peers and even the victims themselves, may initially assume that the violent act resulted from a failure to deal effectively or therapeutically with the client or patient and thus attribute the incident to professional incompetence. Lanza and Carifio (1991), in a study to determine causal attributions made to nurses who are victims of assault, found that women are blamed more than men, and that if injured, "the nurse must have done something wrong."

In addition to the blame and potential for improper evaluation of the worker's skills, physical and emotional injury may have occurred. Poster and Ryan (1989a) report that cognitive emotional and physical sequelae may be present long after the victim has returned to work. Davidson and Jackson (1985), Lanza (1983, 1985b), and Poster and Ryan (1989a) reported that assaulted workers experience feelings of self-doubt, depression, fear, posttraumatic stress syndrome, loss of sleep, irritability, disturbed relationships with family and peers, decreased ability to function effectively at the workplace, increased absenteeism and flight from the health care profession. The mental costs to the victim of violence should be recognized and even if physical injury did not occur, professional counseling services may be required to aid in an employee's recovery. The articles referenced all describe the need for and the conduct of counseling programs. Ryan and Poster (1989b) document the benefits of counseling for rapid recovery after assault. The costs to the employee are often unrecognized and thus are not included in any cost accounting of the problem.

White and Hatcher (1988) discuss costs to the organization and the victim of violence pointing to the increased costs due to the "second injury" phenomenon of perceived rejection of the victim by the agency, coworkers and even family, resulting in filing of lawsuits. These suits may cause substantial long-term costs to the agency.

Prevention

Although it is difficult to pinpoint specific causes and solutions for the increase in violence in the workplace and in particular health care settings, recognition of the problem is a beginning. Some solutions to the overall reduction of violence in this country may be found in actions such as eliminating violence in television programs, implementing effective programs of gun control, and reducing drug and alcohol abuse. All companies should investigate programs recently instituted by several convenience store chains of robbery deterrence strategies such as increased lighting, closed circuit TV monitors, visible money handling locations, if sales are involved, limiting access and egress and providing security staff.

Other methods of preventing assault may be in expanding the national data base with standardized reporting and information collection systems. It may also be necessary to fund and conduct research on postassault outcomes, the need for rehabilitation for returning to work, the length of employment after assault and on techniques of preventing injury and death from occupational violence.

In a San Francisco hospital, methods have been developed to attempt to deal with violence issues with the formation of two focus groups. One group, the Violence Task Force, functions to advise the administration regarding modification of hospital policy toward reducing incidents of violence. The second group, San Francisco Emergency Workers Critical Incident Stress Debriefing Team, counsels victims of physical, sexual or verbal assault. This group also provides needed support to staff who may be exposed to bloody and brutal scenes in their work environment.

White and Hatcher (1988) have outlined management and medical objectives and responses to violence-induced trauma as well as decision trees and checklists to aid in assessing and constructing a response plan. Although not necessarily incident preventing, a response plan should be incorporated into an overall plan of prevention.

Training employees in management of assaultive behavior (MAB) or professional assault response has been shown by Carmel and Hunter (1990) to reduce the incidence of assaults to hospital staff. Infantino and Musingo (1985) and Blair and New (1991) also found that new and untrained staff were at highest risk for injury.

Keep and Gilbert (1992) report that legislation is being proposed in California to make violence to emergency personnel reportable to local police and criminal charges pressed if there is sufficient evidence. This action is also recommended by Morrison and Herzog (1992), especially in relation to emergency department staff. Other staff of facilities such as psychiatric units should be advised and policies established to assist in the decision of the appropriateness and effectiveness of such action.

Administrative controls and mechanical devices are being recommended and gradually implemented, but the problems appear to be escalating. Although long ignored by hospital and other administrators and professionals, the problem of workplace violence is being recognized. Increasing numbers of health care and community service workers as well as OSHA professionals have come to the conclusion that injuries related to workplace violence should no longer be tolerated. In the past, little was done to protect workers from violence. Currently, as discussed, a variety of health care, community service facilities, unions and researchers are seeking solutions to the problem. Managers and administrators are being advised to make the provision of adequate measures to prevent violence a high priority. Some safety measures may seem expensive or difficult to implement but are needed to adequately protect the health and well being of health care and community service workers. It is also important to recognize that the belief that certain risks are "part of the job" contributes to the continuation of violence and possibly the shortage of trained health care and community service workers.

Cal/OSHA recognizes its obligation to develop standards and guidelines to provide safe workplaces for health care and community service workers. These workplaces should be free from health and safety hazards, including fear and the threat of assaults. The Injury and Illness Prevention Program, as defined under the General Industry Safety Order, Section 3203, requires all employers to develop an Injury and Illness Prevention Program for hazards unique to their place of employment. This Injury and Illness Prevention Program should provide the framework for each employer's program of preventing assaults—one of the major hazards of work in health care and community service and perhaps in the community as a whole.

These Cal/OSHA guidelines are designed to assist managers and administrators in the development and implementation of programs to protect their workers. Although not exhaustive, these guidelines include philosophical approaches as well as practical methods to prevent and control assaults. The potential for violence may always exist for health care and community service workers, whether at large medical centers, community-based drug treatment programs, mental health clinics, or for workers making home visits in the community. Because of the potential for injury to workers, health care and community service organizations must comply with Title 8 of the CCR, Section 3203. This regulation requires an Injury and Illness Prevention Program which stipulates that responsible persons perform worksite analyses, identify sentinel events, and establish controls and training programs to reduce or eliminate hazards to worker health and safety. A copy of the State of New Jersey OSHA Guidelines (Appendix 7.A.3) on measures and safeguards in dealing with violent or aggressive behavior in public sector health care facilities is provided as an example of one of the first state OSHA recommendations addressing this serious issue. We anticipate more states and federal OSHA will eventually follow suit.

Many health care providers, researchers, educators, unions, and OSHA enforcement professionals contributed to the development of these guidelines. The cooperation and commitment of employers is necessary, however, to translate these guidelines into an effective program for the occupational health and safety of health care and community service workers.

Program Development

The guidelines are divided into two major divisions: (1) general provisions and program development and (2) specific work setting requirements. General provisions and program development include provisions that must be adopted by all high risk industries to assess risk and to develop needed programs.

Within the specific work setting, guidelines will be subdivided into (a) engineering controls, (b) work practices, (c) personal protective measures, and (d) individualized training measures by major worksite category, i.e., inpatient psychiatric hospitals and psychiatric units, hospital and emergency rooms, outpatient facilities, and community workers.

General Program Essentials

Management Commitment and Employee Involvement

Commitment and involvement are essential elements in any safety and health program. Management provides the organizational resources and motivating forces necessary to deal effectively with safety and security hazards. Employee involvement, both individually and collectively, is achieved by encouraging participation in the worksite assessment, developing clear effective procedures, and identifying existing and potential hazards. Employee knowledge and skills should be incorporated into any plan to abate and prevent safety and security hazards.

Commitment by Top Management

The implementation of an effective safety and security program includes a commitment by the employer to provide the visible involvement of administrators of hospitals, clinics, and agencies, so that all employees, from managers to line workers, fully understand that management has a serious commitment to the program. An effective program should have a team approach with top management as the team leader and should include the following:

1. The demonstration of management's concern for employee emotional and physical safety and health by placing a high priority on eliminating safety and security hazards.
2. A policy which places employee safety and health on the same level of importance as patient/client safety. The responsible implementation of this policy requires management to integrate issues of employee safety and security with restorative therapeutic services to assure that this protection is part of the daily hospital/clinic or agency activity.
3. Employer commitment to security through the philosophical refusal to tolerate violence in the institution and to employees and the assurance that every effort will be made to prevent its occurrence.
4. Employer commitment to assign and communicate the responsibility for various aspects of safety and security to supervisors, physicians, social workers, nursing staff, and other employees involved

so that they know what is expected of them. Also to ensure that recordkeeping is accomplished and utilized using good principles of epidemiology to aid in meeting program goals.

5. Employer commitment to provide adequate authority and resources to all responsible parties so that assigned responsibilities can be met.
6. Employer commitment to ensure that each manager, supervisor, professional, and employee responsible for the security and safety program in the workplace is accountable for carrying out those responsibilities.
7. Employer develops and maintains a program of medical and emotional health care for employees who are assaulted or suffer abusive behavior.
8. Development of a safety committee in keeping with requirements of GISO 3203 and which evaluates all reports and records of assaults and incidents of aggression. When this committee makes recommendations for correction, the employer reports back to the committee in a timely manner on actions taken on the recommendation.

Employee Involvement

An effective program includes a commitment by the employer to provide for, and encourage employee involvement in the safety and security program and in the decisions that affect worker safety and health as well as client well-being. Involvement may include the following:

1. An employee suggestion/complaint procedure which allows workers to bring their concerns to management and receive feedback without fear of reprisal or criticism of ability.
2. Employees follow a procedure which requires prompt and accurate reporting of incidents with or without injury. If injury has occurred, prompt first aid or medical aid must be sought and treatment provided or offered.
3. Employees participate in a safety and health committee that receives information and reports on security problems, makes facility inspections, analyzes reports and data, and makes recommendations for corrections.
4. Employees participate in case conference meetings, and present patient information and problems which may help employees to identify potentially violent patients and discuss safe methods of managing difficult clients. (Identification of potential perpetrators.)
5. Employees participate in security response teams that are trained and possess required professional assault response skills.
6. Employees participate in training and refresher courses in professional assault response training such as PART® to learn techniques of recognizing escalating agitation, deflecting, or controlling the undesirable behavior and, if necessary, of controlling assaultive behavior, protecting clients and other staff members.
7. Participation in training as needed in nonhospital work settings, such as "dealing with the hostile client" or even the police department program of "personal safety" should be provided and, required to be attended by all involved employees.

Written Program

Effective implementation requires a written program for job safety, health, and security that is endorsed and advocated by the highest level of management and professional practitioners or medical board. This program should outline the employer's goals and objectives. The written program should be suitable for the size, type, and complexity of the facility and its operations and should permit these guidelines to be applied to the specific hazardous situation of each health care unit or operation.

The written program should be communicated to all personnel regardless of number of staff or work shift. The program should establish clear goals and objectives that are understood by all members of the organization. The communication needs to be extended to physicians, psychiatrists, etc. and all levels of staff, including housekeeping, dietary, and clerical.

Regular Program Review and Evaluation

Procedures and mechanisms should be developed to evaluate the implementation of the security program and to monitor progress. This evaluation and recordkeeping program should be reviewed regularly by top management and the medical management team. At least semiannual reviews are recommended to evaluate success in meeting goals and objectives. This will be discussed further as part of the recordkeeping and evaluation.

Program Elements

An effective occupational safety and health program of security and safety in medical care facilities and community service includes the following major program elements: (A) worksite analysis, (B) hazard prevention and control, (C) engineering controls, (D) administrative controls, (E) personal protective devices, (F) medical management and counseling, (G) education and training, and (H) recordkeeping and evaluation.

Worksite Analysis

Worksite analysis identifies existing hazards and conditions, operations and situations that create or contribute to hazards, and areas where hazards may develop. This includes close scrutiny and tracking of injury/illness and incident records to identify patterns that may indicate causes of aggressive behavior and assaults.

The objectives of worksite analyses are to recognize, identify, and to plan to correct security hazards. Analysis utilizes existing records and worksite evaluations including:

Record Review

1. Analyze medical, safety, and insurance records, including the OSHA 200 log and information compiled for incidents or near incidents of assaultive behavior from clients or visitors. This process should involve health care providers to ensure confidentiality of records of patients and employees. This information should be used to identify incidence, severity, and establish a baseline for identifying change.
2. Identify and analyze any apparent trends in injuries relating to particular departments, units, job titles, unit activities or work stations, activity, or time of day. It may include identification of sentinel events such as threatening of providers of care or identification and classification of clients anticipated to be aggressive.

Identification of Security Hazards

Worksite analysis should use a systematic method to identify those areas needing in-depth scrutiny of security hazards. This analysis should do the following:

1. Identify those work positions in which staff is at risk of assaultive behavior.
2. Use a checklist for identifying high risk factors that includes components such as type of client, physical risk factors of the building, isolated locations/job activities, lighting problems, high risk activities or situations, problem clients, uncontrolled access, and areas of previous security problems.
3. Identify low risk positions for light or relief duty or restricted activity work positions when injuries do occur.
4. Determine if risk factors have been reduced or eliminated to the extent feasible. Identify existing programs in place and analyze effectiveness of those programs, including engineering control measures and their effectiveness.
5. Apply analysis to all newly planned and modified facilities, or any public services program to ensure that hazards are reduced or eliminated before involving patients/clients or employees.
6. Conduct periodic surveys at least annually or whenever there are operation changes, to identify new or previously unnoticed risks and deficiencies and to assess the effects of changes in the

building design, work processes, patient services, and security practices. Evaluation and analysis of information gathered and incorporation of all this information into a plan of correction and ongoing surveillance should be the result of the worksite analysis.

Hazard Prevention and Control

Selected work settings have been utilized for discussion of methods of reducing hazards. Each of the selected work situations—psychiatric hospitals and psychiatric wards, hospitals and emergency rooms, outpatient facilities, and community work settings—will be addressed with general engineering concepts, specific engineering and administrative controls, work practice controls, and personal protective equipment as appropriate to control hazards. These methods are contained in (B) through (F).

Engineering Administrative and Work Practice Controls for All Settings

General Building, Workstation and Area Designs

Hospital, clinic, emergency room, and nurse's station designs are appropriate when they provide secure, well-lit protected areas which do not facilitate assaults or other uncontrolled activity.

1. Design of facilities should ensure uncrowded conditions for staff and clients. Rooms for privacy and protection, avoiding isolation are needed. For example, doors must be fitted with windows. Interview rooms for new patients or known assaultive patients should utilize a system which provides privacy but which may also permit other staff to see activity. In psychiatric units "time out" or seclusion rooms are needed. In emergency departments, rooms are needed in which agitated patients may be confined safely to protect themselves, other clients, and staff.
2. Patient care rooms and counseling rooms should be designed and furniture arranged to prevent entrapment of the staff and/or reduce anxiety in clients. Light switches in patient rooms should be located outside the room. Furniture may be fixed to the floor, soft or with rounded edges and colors restful and light.
3. Nurse stations should be protected by enclosures which prevent patients from molesting, throwing objects, reaching into the station otherwise creating a hazard or nuisance to staff—such barriers should not restrict communication but should be protective.
4. Lockable and secure bathroom facilities and other amenities must be provided for staff members separate from client rest rooms.
5. Client access to staff counseling rooms and other facility areas must be controlled; that is, doors from client waiting rooms must be locked and all outside doors locked from the outside to prevent unauthorized entry, but permit exit in cases of emergency or fire.
6. Metal bars or protective decorative grating on outside ground level windows should be installed (in accordance with fire department codes) to prevent unauthorized entry.
7. Bright and effective lighting systems must be provided for all indoor building areas as well as grounds around the facility and especially in the parking areas.
8. Curved mirrors should be installed at intersections of halls or in areas where an individual may conceal his or her presence.
9. All permanent and temporary employees who work in secured areas should be provided with keys to gain access to work areas whenever on duty.
10. Metal detectors should be installed to screen patients and visitors in psychiatric facilities. Emergency rooms should have available handheld metal detectors to use in identifying weapons.

Maintenance

1. Maintenance must be an integral part of any safety and security system. Prompt repair and replacement programs are needed to ensure the safety of staff and clients. Replacement of burned out lights, broken windows, etc. is essential to maintain the system in safe operating conditions.
2. If an alarm system is to be effective, it must by used, tested, and maintained according to strict policy. Any personal alarm devices should be carried and tested as required by the manufacturer

and facility policy. Maintenance on personal and other alarm systems must take place monthly. Batteries and operation of the alarm devices must be checked by a security officer to ensure the function and safety of the system as prescribed by provisions of GISO 6184.

3. Any mechanical device utilized for security and safety must be routinely tested for effectiveness and maintained on a scheduled basis.

Psychiatric Hospital/In-Patient Facilities

Engineering Control

Alarm systems are imperative for use in psychiatric units, hospitals, mental health clinics, emergency rooms, or where drugs are stored. Whereas alarm systems are not necessarily preventive, they may reduce serious injury when a client is escalating in abusive behavior or threatening with or without a weapon.

1. Alarm systems which rely on the use of telephones, whistles, or screams are ineffective and dangerous. A proper system consists of an electronic device which activates an alert to a dangerous situation in two ways, visually and audibly. Such a system identifies the location of the room or location of the worker by means of an alarm sound and a lighted indicator which visually identifies the location. In addition, the alarm should be sounded in a security area or other response team areas which will summon aid. This type of alarm system typically utilizes a pen-like device which is carried by the employee and can be triggered easily in an emergency situation. This system should be in accordance with provisions of California Title 8, GISO Section 6184, Emergency Alarm Systems (State of California, Department of Industrial Relations GISO). Back-up security personnel must be available to respond to the alarm.
2. "Panic buttons" are needed in medicine rooms, nurses stations, stairwells, and activity rooms. Any such alarm system may incorporate a telephone paging system in order to direct others to the location of the disturbance but alarm systems must not depend on the use of a telephone to summon assistance.
3. Video screening of high risk areas or activities may be of value and permits one security guard to visualize a number of high risk areas, both inside and outside the building.
4. Metal detection systems such as handheld devices or other systems to identify persons with hidden weapons should be considered. These systems are in use in courts, boards of supervisors, some Departments of Public Social Service, schools, and emergency rooms. Although controversial, the fact remains that many people, including homeless and mentally ill persons do or are forced to carry weapons for defense while living on the streets. Some system of identifying persons who are carrying guns, knives, ice picks, screw drivers, etc. may be useful and should be considered. In psychiatric facilities, patients who have been on leave or pass should be screened upon return for concealed weapons.

Administrative Controls

A sound overall security program includes administrative controls that reduce hazards from inadequate staffing, insufficient security measures, and poor work practices.

1. In order to enable staff members to identify and deal effectively with clients who behave in a violent manner, the administrator must insist on plans for patient treatment regimens and management of clients which include a gradual progression of measures given to staff to prevent violent behavior from escalating. These measures should not encourage inappropriate use of medication/restraints or isolation. However, the least restrictive yet appropriate and effective plan for preventing a client from injuring staff, other clients and self must be developed and be part of every unit and care plan. This enables a staff member to take primary prevention steps to stop escalating aggressive behavior. These procedures should cover verbal or physical threats

or acting out of disturbed clients to help both the client and staff to feel a sense of control within the unit.

2. Security guards must be provided. These security guards should be trained in principles of human behavior and aggression. They should be assigned to areas where there may be psychologically stressed clients such as emergency rooms or psychiatric services.
3. In order to staff safely, a written acuity system should be established that evaluates the level of staff coverage vis-à-vis patient acuity and activity level. Staffing of units where aggressive behavior may be expected should be such that there is always an adequate, safe staff/patient ratio. The provision of reserve or emergency teams should be utilized to prevent staff members being left with inadequate support (regardless of staffing quotas) overwhelmed by circumstances of case load that would prevent adequate assessment of severity of illness. This also requires administrators to analyze and to identify times or areas where hostilities take place and provide a back-up team or staff at levels which are safe, such as in admission units, crisis or acute units or during the 9 h or meal times or any other time, or activity identified as high risk.

   Provision of sufficient staff for interaction and clinical activity is important because patients/clients need access to medical assistance from staff. Possibility of violence often threatens staff when the structure of the patient/nurse relationship is weak. Therefore, sufficient staff members are essential to allow formation of therapeutic relationships and a safe environment.
4. It is necessary to establish on-call teams, reserve or emergency teams of staff who may provide services in hospitals such as, responding to emergencies, transportation or escort services, dining room assistance, or many of the other activities which tend to reduce available staff where assigned.
5. All oncoming staff or employees should be provided with a census report which indicates precautions for every client. Methods must be developed and enforced to inform float staff, new staff members, or oncoming staff at change of shifts of any potential assaultive behavior problems with clients. These methods of identification should include chart tags, log books, census reports, and/or other information system within the facility. Other sources of information may include mandatory provision of probation reports of clients who may have had a history of violent behavior. However, the need for a program of "Universal Precautions for Violence" must be recognized and integrated in any patient care setting.
6. Staff members should be instructed to limit physical intervention in altercations between patients whenever possible unless there are adequate numbers of staff or emergency response teams, and security called. In the case where serious injury is to be prevented, emergency alarm systems should always be activated. Administrators need to give clear messages to clients that violence is not permitted. Legal charges may be pressed against clients who assault other clients or staff members. Administrators should provide information to staff who wish to press charges against assaulting clients.
7. Policies must be provided with regard to safety and security of staff when making rounds for patient checks, key and door opening policy, open vs. locked seclusion policies, evacuation policy in emergencies, and for patients in restraints. Monitoring high risk patients at night and whenever behavior indicates escalating aggression, needs to be addressed in policy as well as medical management protocols.
8. Escort services by security should be arranged so that staff members do not have to walk alone in parking lots or other parking areas in the evening or late hours.
9. Visitors and maintenance persons or crews should be escorted and observed while in any locked facility. Often they have tools or possessions which could be inadvertently left and inappropriately used by clients.
10. Administrators need to work with local police to establish liaison and response mechanisms for police assistance when calls are made for help by a clinic or facility, and conversely to facilitate the hospital's provisions of assistance to local police in handling emergency cases.

11. Assaultive clients may need to be considered for placement in more acute units or hospitals where greater security may be provided. It is not wise to force staff members to confront a continuously threatening client, nor is it appropriate to allow aggressive behavior to go unchecked. Some programs may have the option of transferring clients to acute units, criminal units, or to other more restrictive settings.

Work Practice Controls

1. Clothing should be worn which may prevent injury, such as low-heeled shoes, use of conservative earrings or jewelry, and clothing which is not provocative.
2. Keys should be inconspicuous and worn in such a manner to avoid incidents yet be readily available when needed.
3. Personal alarm systems described under engineering controls must be utilized by staff members and tested as scheduled.
4. No employee should be permitted to work alone in a unit or facility unless backup is immediately available.

Clinics and Outpatient Facilities

Engineering Controls

1. An emergency personal alarm system is of the highest priority. An alarm system may be of two types: the personal alarm device as identified under hospitals and inpatient facilities or the type which is triggered at the desk of the counselor of medical staff. This desk system may be silent in the counseling room, but audible in a central assistance area and must clearly identify the room in which the problem is occurring. "Panic buttons" are needed in medicine rooms, bathrooms, and other remote areas such as stairwells, nurses stations, activity rooms, etc.

   Such systems may use a backup paging or public address system on the telephone in order to direct others to the location for assistance but alarm systems must not depend on the use of telephone to summon assistance.
2. Maintenance is required for alarm systems as outlined in the Appendices, GISO, Section 6184.
3. Reception areas should be designed so that receptionists and staff may be protected by safety glass and locked doors to the clinic treatment areas.
4. Furniture in crises treatment areas and quiet rooms should be kept to a minimum and be fixed to the floor. These rooms should have all equipment secured in locked cupboards.
5. First-aid kits shall be available as required in GISO Section 3400.

All requirements of the Bloodborne Pathogen Standard, GISO Section 5193, apply to clinics where blood exposure is possible.

Work Practice and Administrative Controls

1. Psychiatric clients/patients should be escorted to and from waiting rooms and not permitted to move about unsupervised in clinic areas. Access to clinic facilities other than waiting rooms should be strictly controlled with security provisions in effect.
2. Security guards trained in principles of human behavior and aggression should be provided during clinic hours. Guards should be provided where there may be psychologically stressed clients or persons who have taken hostile actions, such as in emergency facilities, hospitals where there are acute or dangerous patients or areas where drug or other criminal activity is commonplace.
3. Staff members should be given the greatest possible assistance in obtaining information to evaluate the history of, or potential for, violent behavior in patients. They should be required to treat and/or interview aggressive or agitated clients in open areas where other staff may observe interactions but still provide privacy and confidentiality.

4. Assistance and advice should be sought in case management conference with coworkers and supervisors to aid in identifying treatment of potentially violent clients. Whenever an agitated client or visitor is encountered, treatment or intervention should be provided when possible to defuse the situation. However, security or assistance should be requested to assist in avoiding violence.
5. No employee should be permitted to work or stay in a facility or isolated unit when they are the only staff member present in the facility, if the location is so isolated that they are unable to obtain assistance if needed, or in the evening or at night if the clinic is closed.
6. Employees must report all incidents of aggressive behavior such as pushing, threatening, etc. with or without injury, and logs maintained recording all incidents or near incidents.
7. Records, logs, or flagging charts must be updated whenever information is obtained regarding assaultive behavior or previous criminal behavior.
8. Administrators should work with local police to establish liaison and response mechanisms for police assistance when calls are made for help by a clinic. Likewise, this will also facilitate the clinics provision of assistance to local police in handling emergency cases.
9. Referral systems and pathways to psychiatric facilities need to be developed to facilitate prompt and safe hospitalization of clients who demonstrate violent or suicidal behavior. These methods may include: direct phone link to the local police, exchange of training and communication with local psychiatric services, and written guidelines outlining commitment procedures.
10. Clothing and apparel should be worn which will not contribute to injury such as low-heeled shoes, use of conservative earrings or jewelry, and clothing which is not provocative.
11. Keys should be kept covered and worn in such a manner to avoid incidents, yet be available.
12. All protective devices and procedures should be required to be used by all staff.

Emergency Rooms and General Hospitals

Engineering Controls

1. Alarm systems or "panic buttons" should be installed at nurses stations, triage stations, registration areas, hallways, and in nurse lounge areas. These alarm systems must be relayed to security police or locations where assistance is available 24 h per day. A telephone link to the local police department should be established in addition to other systems.
2. Metal detection systems installed at emergency room entrances may be used to identify guns, knives, or other weapons. Lockers can be used to store weapons and belongings or the weapons may be transferred to the local police department for processing if the weapons are not registered. Handheld metal detection devices are needed to identify concealed weapons if there is no larger system. Signs posted at the entrance will notify patients and visitors that screening will be performed.
3. Seclusion or security rooms are required for containing confused or aggressive clients. Although privacy may be needed both for the agitated patient and other patients, security, and the ability to monitor the patient and staff is also required in any secluded or quiet room.
4. Bullet-resistant glass should be used to provide protection for triage, admitting, or other reception areas where employees may greet or interact with the public.
5. Strictly enforced limited access to emergency treatment areas are needed to eliminate unwanted or dangerous persons in the emergency room. Doors may be locked or key-coded.
6. Closed circuit TV monitors may be used to survey concealed areas or areas where problems may occur.

Work Practices and Administrative Controls

1. Security guards trained in principles of human behavior and aggression must be provided in all emergency rooms. Death and serious injury have been documented in emergency areas in hospitals, but the presence of security persons often reduces the threatening or aggressive behavior

demonstrated by patients, relatives, friends, or those seeking drugs. Armed guards must be considered in any risk assessment in high volume emergency rooms.

2. No staff person should be assigned alone in an emergency area or walk-in clinic.
3. After dark, all unnecessary doors are locked, access into the hospital is limited and patrolled by security.
4. A regularly updated policy should be in place directing hostile patient management, use of restraints, or other methods of management. This policy should be detailed and provide guidelines for progressively restrictive action as the situation calls for.
5. Any verbally threatening, aggressive, or assaultive incident must be reported and logged.
6. Name tags need to be worn at all times in the hospital and emergency room. Hospital policy must demand that persons, including staff, who enter into the treatment area of the emergency room have or seek permission to enter the area to reduce the volume of unauthorized individuals.

   When transferring a hostile or agitated patient (or one who may have relatives, friends, or enemies who pose a security problem) to a unit within the hospital, security is required during transport and transfer to the unit. This security presence may be required until the patient is stabilized or controlled to protect staff who are providing care.
7. Emergency or hospital staff who have been assaulted should be permitted and/or assisted to request police assistance or file charges of assault against any patient or relative who injures, just as a private citizen has the right to do so. Being in the helping profession does not reduce the right of pressing charges or damages.

General Hospitals

1. Information must be clearly transmitted to the receiving unit of security problems with the patient. Charts must be flagged clearly noting and identifying the security risks involved with this patient.
2. If patients with any disorder or illness have a known history of violent acts, it is incumbent upon the administration to demand health care providers or physicians to disclose that information to hospital staff at the onset of hospitalization.
3. Whenever patients display aggressive or hostile behavior to hospital staff members, it must be made part of the care plan that supervisors and managers are notified and protective measures and action are initiated.
4. Prompt medical or emotional evaluation treatment must be made available to any staff who has been subjected to abusive behavior from a client/patient, whether in emergency rooms, psychiatric units, or general hospital settings.
5. Visitors should sign in and have an issued pass particularly in newborn nursery, pediatric departments, or any other risk departments.
6. Any patient who may be deemed at risk should be placed on a "restricted visitor list." Restricted visitor lists must be maintained by security, nurses station, and visitor sign-in areas.
7. Social service/worker staff should be utilized to defuse situations. In-house social workers are an important part of the hospital staff as are employee heath staff.

Home/Field Operations—Community Service Workers

Engineering Controls

1. In order to provide some measure of safety and to keep the employee in contact with headquarters or a source of assistance, cellular car phones should be installed/provided for official use when staff are assigned to duties which take them into private homes and the community. The workers may include (to name a few) parking enforcers, union business agents, psychiatric evaluators, public social service workers, children's service workers, visiting nurses, and home health aides.

2. Handheld alarm or noise devices or other effective alarm devices are highly recommended to be provided for all field personnel.
3. Beepers or alarm systems which alert a central office of problems should be investigated and provided.
4. Other protective devices should be investigated and provided such as pepper spray.

Work Practice and Administrative Controls

1. Employees are to be instructed not to enter any location where they feel threatened or unsafe. This decision must be the judgment of the employee. Procedures should be developed to assist the employee to evaluate the relative hazard in a given situation. In hazardous cases, the managers must facilitate and establish a "buddy system." This "buddy system" should be required whenever an employee feels insecure regarding the time of activity, the location of work, the nature of the client's health problem, and history of aggressive or assaultive behavior or potential for aggressive acts.
2. Employers must provide for the field staff a program of personal safety education. This program should be at the minimum, one provided by local police departments or other agencies which include training on awareness, avoidance, and action to take to prevent mugging, robbery, rapes, and other assaults.
3. Procedures should be established to assist employees to reduce the likelihood of assaults and robbery from those seeking drugs or money, as well as procedures to follow in the case of threatening behavior and provision for a fail-safe backup in administrative offices.
4. A fail-safe backup system is provided in the administrative office at all times of operation for employees in the field who may need assistance.
5. All incidents of threats or other aggression must be reported and logged. Records must be maintained and utilized to prevent future security and safety problems.
6. Police assistance and escorts should be required in dangerous or hostile situations or at night. Procedures for evaluating and arranging for such police accompaniment must be developed and training provided.

Medical Management

A medical program which provides knowledgeable medical and emotional treatment should be established. This program shall assure that victimized employees are provided with the same concern that is often shown to the abuse client. Violence is a major safety hazard in psychiatric and acute care facility emergency rooms, homeless shelters, and other health care settings and workplaces. Medical and emotional evaluation and treatment are frequently needed but often difficult to obtain.

The consequences to employees who are abused by clients may include death and severe and life-threatening injuries, in addition to short- and long-term psychological trauma, posttraumatic stress, anger, anxiety, irritability, depression, shock, disbelief, self-blame, fear of returning to work, disturbed sleep patterns, headache, and change in relationships with coworkers and family. All have been reported by HCWs after assaults, particularly if the attack has come without warning. They may also fear criticism by managers, increase use of alcohol and medication to cope with stress, suffer from feelings of professional incompetence, physical illness, powerlessness, increase in absenteeism, and experience performance difficulties.

Administrators and supervisors have often ignored the needs of the physically or psychologically abused or assaulted staff, requiring them to continue working, obtain medical care from private medical doctors or blame the individual for irresponsible behavior. Injured staff must have immediate physical evaluations, be removed from the unit and treated for acute injuries. Referral should be made for appropriate evaluation, treatment, counseling, and assistance at the time of the incident and for any required follow-up treatment.

Medical Services

1. This should include provision of prompt medical evaluation and treatment whenever an assault takes place regardless of severity. A system of immediate treatment is required regardless of time of day or night. Injured employees should be removed from the unit until order has been restored. Transportation of the injured to medical care must be provided if it is not available on-site or in an employee health service. Follow-up treatment provided at no cost to employees must also be provided.

Counseling Services

1. A trauma-crisis counseling or critical incident debriefing program must be established and provided on an ongoing basis whenever staff are victims of assaults. This "counseling program" may be developed and provided by in-house staff as part of an employee health service, by a trained psychologist, psychiatrist, or other clinical staff member such as a clinical nurse specialist, a social worker or referral may be made to an outside specialist. In addition, peer counseling or support groups may be provided. Any counseling provided should be by well-trained psychosocial counselors whether through employee assistance program (EAP) programs, in-house programs, or by other professionals away from the facility who must understand the issues of assault and its consequences.
2. Reassignment of staff should be considered when assaults have taken place. At times it is very difficult for staff to return to the same unit to face the assailant. Assailants often repeat threats and aggressive behavior and actions need to be taken to prevent this from occurring. Staff development programs should be provided to teach staff and supervisors to be more sensitive to the feelings and trauma experienced by victims of assaults. Some professionals advocate joint counseling sessions including the assaultive client and staff member to attempt to identify the motive when it occurs in inpatient facilities and to defuse situations which may lead to continued problems.
3. Unit staff should also receive counseling to prevent "blaming the victim syndrome" and to assist them with any stress problems they may be experiencing as a result of the assault. Violence often leaves staff fearful and concerned. They need to have the opportunity to discuss these fears and to know that administration is concerned and will take measures to correct deficiencies. This may be called a defusing or debriefing secession and unit staff members may need this activity immediately after an incident to enable them to continue working. First-aid kits or materials must be provided on each unit or facility.
4. The replacement and transportation of the injured staff member must be provided for at the earliest time. Do not leave a unit short staffed in the event of an assault. The development of an employee health service staffed by a trained occupational health specialist, may be an important addition to the hospital team. Such employee health staff can provide treatment, arrange for counseling, refer to a specialist and should have procedures in place for all shifts. Employee health nurses should be trained in posttraumatic counseling and may be utilized for group counseling programs or other assistance programs.
5. Legal advice regarding pressing charges should be available, as well as information regarding workers' compensation benefits, and other employee rights must be provided regardless of apparent injury. If assignment to light duty is needed or disability is incurred, these services are to be provided without hesitation. Reporting to the appropriate local law enforcement agency and assistance in making this report is to be provided. Employees may not be discouraged or coerced when making reports or workers' compensation claims.
6. All assaults must be investigated, reports made and needed corrective action determined. However, methods of investigation must be such that the individual does not perceive blame or criticism for assaultive action taken by clients. The circumstances of the incident or other information which will help to prevent further problems, needs to be identified, not to blame the worker for incompetence and compound the psychological injury which is most commonly experienced.

Recordkeeping

Within the major program elements, recordkeeping is the heart of the program, providing information for analysis, evaluation of methods of control, severity determination, identifying training needs, and overall program evaluations.

Records shall be kept of the following:

1. OSHA 200 log. OSHA regulations require entry in the Injury and Illness Log 200, of any injury which requires more than first aid, is a lost-time injury, requires modified duty or causes loss of consciousness. Assaults should be entered in the log. Doctors' reports of work injury and supervisors' reports shall be kept of each recorded assault.
2. Incidents of abuse, verbal attacks, or aggressive behavior which may be threatening to the worker but not resulting in injury, such as pushing, shouting, or an act of aggression toward other clients requiring action by staff should be recorded. This record may be an assaultive incident report or documented in some manner which can be evaluated on a monthly basis by department safety committee.
3. A system of recording and communicating should be developed so that all staff who may provide care for an escalating or potentially aggressive, abusive violent client will be aware of the status of the client and of any problems experienced in the past. This information regarding history of past violence should be noted on the patient's chart, communicated in shift change report and noted in an incident log.
4. An information-gathering system should be in place which will enable incorporation of past history of violent behavior, incarceration, probation reports, or any other information which will assist health care staff to assess violence status. Employees are to be encouraged to seek and obtain information regarding history of violence whenever possible.
5. Emergency room staff should be encouraged to obtain a record from police and relatives, information regarding drug abuse, criminal activity, or other information to adequately assist in assessing a patient. This would enable them to appropriately house, treat, and refer potentially violent cases. They should document the frequency of admission of violent clients or hostile encounters with relatives and friends.
6. Records need to be kept concerning assaults, including the type of activity, i.e., unprovoked sudden attack, patient-to-patient altercation, and MAB actions. Information needed includes who was assaulted and circumstances of the incident without focusing on any alleged wrongdoing of staff persons. These records also need to include a description of the environment, location, or any contributing factors, corrective measures identified, including building design, or other measures needed. Determination must be made of the nature of the injuries sustained: severe, minor or the cause of long-term disability, and the potential or actual cost to the facility and employee. Records of any lost time or other factors which may result from the incident should be maintained.
7. Minutes of the safety meetings and inspections shall be kept in accordance with requirements of Title 8, Section 3203. Collective actions recommended as a result of reviewing reports or investigating accidents or inspections need to be documented with the administration's response and completion dates of those actions should be included in the minutes and records.
8. Records of training program contents and sign-in sheets of all attendees should be kept. Attendance records at all "PART" or "MAB" training should be retained. Qualifications of trainers shall be maintained along with records of training.

Training and Education

General

A major program element in an effective safety and security program is training and education. The purpose of training and education is to ensure that employees are sufficiently informed about the safety and security hazards to which they may be exposed and thus, are able to participate actively in their own and coworkers protection. All employees should be periodically trained in the employer's safety and security program.

Training and education are critical components of a safety and security program for employees who are potential victims of assaults. Training allows managers, supervisors, and employees to understand security and other hazards associated with a job or location within the facility, the prevention and control of these hazards, and the medical and psychological consequences of assault.

1. A training program should include the following individuals:
   a. All affected employees including doctors, dentists, nurses, teachers, counselors, psychiatric technicians, social workers, dietary, and housekeeping, in short all health care and community service staff and all other staff members who may encounter or be subject to abuse or assaults from clients/patients
   b. Engineers, security officers, maintenance personnel
   c. Supervisors and managers
   d. Health care providers and counselors for employees and employee health personnel
2. The program should be designed and implemented by qualified persons. Appropriate special training should be provided for personnel responsible for administering the training program.
3. Several types of programs are available and have been utilized, such as MAB, PART, Police Department Assault Avoidance Programs, or Personal Safety training. A combination of such training may be incorporated depending on the severity of the risk and assessed risk. These management programs must be provided and attendance required at least yearly. Updates may be provided monthly/quarterly.
4. The program should be presented in the language and at a level of understanding appropriate for the individuals being trained. It should provide an overview of the potential risk of illness and injuries from assault and the cause and early recognition of escalating behavior or recognition of situations which may lead to assaults. The means of preventing or defusing volatile situations, safe methods of restraint or escape, or use of other corrective measures or safety devices which may be necessary to reduce injury and control behavior are critical areas of training. Methods of self-protection and protection of coworkers, the proper treatment of staff and patient procedures, recordkeeping, and employee rights need to be emphasized.
5. The training program should also include the means for adequately evaluating its effectiveness. The adequacy of the frequency of training should be reviewed. The whole program evaluation may be achieved by using employee interviews, testing and observing, and/or reviewing reports of behavior of individuals in situations that are reported to be threatening in nature.
6. Employees who are potentially exposed to safety and security hazards should be given formal instruction on the hazards associated with the unit or job and facility. This includes information on the types of injuries or problems identified in the facility, the policy and procedures contained in the overall safety program of the facility, those hazards unique to the unit or program and the methods used by the facility to control the specific hazards. The information should discuss the risk factors that cause or contribute to assaults, etiology of violence and general characteristics of violent people, methods of controlling aberrant behavior, methods of protection and reporting procedures, and methods to obtain corrective action.

   Training for affected employees should consist of both general and specific job training. "Specific job training" is contained in the following section or may be found in administrative controls in the specific work location section.

Job-Specific Training

New employees and reassigned workers or registry staff should receive an initial orientation and hands-on training prior to being placed in a treatment unit or job. Each new employee should receive a demonstration of alarm systems and protective devices and the required maintenance schedules and procedures. The training should also contain the use of administrative or work practice controls to reduce injury.

1. The initial training program should include
   a. Care, use, and maintenance of alarm tools and other protection devices
   b. Location and operation of alarm systems
   c. MAB, PART, or other training
   d. Communication systems and treatment plans
   e. Policies and procedures for reporting incidents and obtaining medical care and counseling
   f. Injury and Illness Prevention Program (8 CCR 3203)
   g. Hazard Communication Program (8 CCR 5194)
   h. Bloodborne Pathogen Program if applicable (8 CCR 5193)
   i. Rights of employees, treatment of injury and counseling programs.
2. On-the-job training should emphasize development and use of safe and efficient methods of de-escalating aggressive behavior, self-protection techniques, methods of communicating information which will help other staff to protect themselves and discussions of rights of employees vis-à-vis patient rights.
3. Specific measures at each location, such as protective equipment, location and use of alarm systems, determination of when to use the buddy system, and so on as needed for safety, must be part of the specific training.
4. Training unit coworkers from the same unit and shift may facilitate team work in the work setting.

Training for Supervisors and Managers Maintenance and Security Personnel

1. Supervisors and managers are responsible for ensuring that employees are not placed in assignments that compromise safety and that employees feel comfortable in reporting incidents. They must be trained in methods and procedures which will reduce the security hazards and train employees to behave compassionately with coworkers when an incident does occur. They need to ensure that employees follow safe work practices and receive appropriate training to enable them to do this. Supervisors and managers, therefore, should undergo training as comparable to that of the employee and such additional training as will enable them to recognize a potentially hazardous situation, make changes in the physical plant, patient care treatment program, staffing policy, and procedures or other such situations which are contributing to hazardous conditions. They should be able to reinforce the employer's program of safety and security, assist security guards when needed and train employees as the need arises.
2. Training for engineers and maintenance should consist of an explanation or a discussion of the general hazards of violence, the prevention and correction of security problems, and personal protection devices and techniques. They need to be acutely aware of how to avoid creating hazards in the process of their work.
3. Security personnel need to be recruited and trained whenever possible for the specific job and facility. Security companies usually provide general training on guard or security issues. However, specific training by the hospital or clinic should include psychological components of handling aggressive and abusive clients, types of disorders, and the psychology of handling aggression and defusing hostile situations. If weapons are utilized by security staff, special training and procedures need to be developed to prevent inappropriate use of weapons and the creation of additional hazards.

Evaluation of the Program

Procedures and mechanisms should be developed to evaluate the implementation of the safety and security programs and to monitor progress and accomplishments. Top administrators and medical directors should review the program regularly. Semiannual reviews are recommended to evaluate success in meeting goals and objectives. Evaluation techniques include some of the following:

1. Establishment of a uniform reporting system and regular review of reports.
2. Review of reports and minutes of Safety and Security Committee.
3. Analyses of trends and rates in illness/injury or incident reports.
4. Survey employees.
5. Before and after surveys/evaluations of job or worksite changes or new systems.
6. Up-to-date records of job improvements or programs implemented.
7. Evaluation of employee experiences with hostile situations and results of medical treatment programs provided. Follow up should be repeated several weeks and several months after an incident.

Results of management's review of the program should be a written progress report and program update which should be shared with all responsible parties and communicated to employees. New or revised goals arising from the review identifying jobs, activities, procedures, and departments should be shared with all employees. Any deficiencies should be identified and corrective action taken. Safety of employees should not be given a lesser priority than client safety as they are often dependent on one another. If it is unsafe for employees, the same problem will be the source of risk to other clients or patients.

Managers, administrators, supervisors, and medical and nursing directors should review the program frequently to reevaluate goals and objectives and discuss changes. Regular meetings with all involved including the Safety Committee, union representatives, and employee groups at risk should be held to discuss changes in the program.

If we are to provide a safe work environment, it must be evident from administrators, supervisors, and peer groups that hazards from violence will be controlled. Employees in psychiatric facilities, drug treatment programs, emergency rooms, convalescent homes, community clinics, or community settings are to be provided with a safe and secure work environment and injury from assault is not to be accepted or tolerated and is no longer "part of the job."

References

Bell C. Female homicides in United States workplaces, 1980–1985. *Am J Publ Health* 81(6):729–732, 1991.

California Department of Industrial Relations, California Code of Regulations, Title 8, General Industry Safety Orders. Sections 3203, 6184, and 3400.

Centers for Disease Control (CDC). Occupational homicides among women—United States, 1980–1985. *MMWR* 39:543–544, 551–552, 1990.

Cohen S, Kamarck T, and Mermelstein R. A global measure of perceived stress. *J Health Soc Behav* 24:385–396, 1983.

Craig TJ. An epidemiological study of problems associated with violence among psychiatric inpatients. *Am J Psychiatry* 139(10):1262–1266, 1982.

Cronin M. New law aims to reduce kidnappings. *Nurse Week* 5(3):1 and 24, 1991.

Edelman SE. Managing the violent patient in a community mental center. *Hosp Commun Psychiatry* 29(7):460–462, 1978.

Eichelman E. A behavioral emergency plan. *Hosp Commun Psychiatry* 35(10):1678, 1984.

Engle F and Marsh S. Helping the employee victim of violence in hospitals. *Hosp Commun Psychiatry* 37(2):159–162, 1986.

Goetz RR, Bloom JD, Chenell SL, and Moorhead JC. Weapons possessed by patients in a university emergency department. *Ann Emerg Med* 20(1):8–10, 1981.

Gosnold DK. The violent patient in the accident and emergency department. *R Soc Health J* 98(4):189–190, 1978.

Haffke EA and Reid WH. Violence against mental health personnel in Nebraska. In: JR Lion and WH Reid (eds.), *Assaults within Psychiatric Facilities*. Grune & Stratton, Inc., Orlando, FL, 1983, pp. 91–102.

Hatti S, Dubin WR, and Weiss KJ. A study of circumstances surrounding patient assaults on psychiatrists. *Hosp Commun Psychiatry* 33(8):660–661, 1982.

Hodgkinson P, Hillis T, and Russell D. Assaults on staff in psychiatric hospitals. *Nurs Times* 80:44–46, 1984.

Ionno JA. A prospective study of assaultive behavior in female psychiatric inpatients. In: JR Lion and WH Reid (eds.), *Assaults within Psychiatric Facilities*. Grune & Stratton, Inc., Orlando, FL, 1983, pp. 71–80.

Kurlowitcz L. Violence in the emergency department. *Am J Nurs* 90(9):34–37, 1990.

Kuzmits FE. When employees kill other employees: The case of Joseph T. Wesbecker. *J Occupat Med* 32(10):1014–1020, 1990.

La Brash L and Cain J. A near-fatal assault on a psychiatric unit. *Hosp Commun Psychiatry* 35(2):168–169, 1984.

Lanza ML. Factors affecting blame placement for patient assault upon nurses. *Issues Ment Health Nurs* 6(1–2):143–161, 1984a.

Lanza ML. A follow-up study of nurses' reactions to physical assault. *Hosp Commun Psychiatry* 35(5): 492–494, 1984b.

Lanza ML. Victim assault support team for staff. *Hosp Commun Psychiatry* 35(5):414–417, 1984c.

Lanza ML. Counseling services for staff victims of patient assault. *Admin Ment Health* 12(3):205–207, 1985a.

Levy P and Hartocollis P. Nursing aides and patient violence. *Am J Psychiatry* 133(4):429–431, 1976.

Lion JR and Pasternak SA. Counter transference reactions to violent patients. *Am J Psychiatry* 130(2):207–210, 1973.

Lion JR and Reid WH (eds.). *Assaults within Psychiatric Facilities*. Grune & Stratton, Inc., Orlando, FL, 1983.

Lion JR, Snyder W, and Merrill GL. Underreporting of assaults on staff in a state hospital. *Hosp Commun Psychiatry* 32(7):497–498, 1981.

Lusk SL. Violence experienced by nurses aides in nursing homes. *AAOHS J* 40(5):237–241, 1992.

Mantell M. The crises response team reports on Edmond, Oklahoma, Massacre. *Nova Newsletter* 11, 1987.

Meddis SV. 7 cities lead violence epidemic. *USA Today*, April 29, 1991.

Monahan J and Shah SA. Dangerousness and commitment of the mentally disordered in the United States. *Schizophr Bull* 15(4):541–553, 1989.

Morrison EF and Herzog EA. What therapeutic and protective measures, as well as legal actions, can staff take when they are attacked by patients. *Journal of Psychosocial Nursing* 30(7):41–44, 1992.

Navis ES. Controlling violent patients before they control you. *Nursing* 87(17):52–54, 1987.

Ochitill HN. Violence in a general hospital. In: JR Lion and WH Reid (eds.), *Assaults within Psychiatric Facilities*. Grune & Stratton, Inc., Orlando, FL, 1983, pp. 103–118.

Phelan LA, Mills MJ, and Ryan JA. Prosecuting psychiatric patients for assaults. *Hosp Commun Psychiatry* 36(6):581–582, 1985.

Rossi AM, Jacobs M, Monteleone M, Olson R, Surber RW, Winkler E, and Wommack A. Violent or fear-inducing behavior associated with hospital admission. *Hosp Commun Psychiatry* 36(6):643–647, 1985.

Ruben I, Wolkon G, and Yamamoto J. Physical attacks on psychiatric residents by patients. *J Nerv Ment Dis* 168(4):243–245, 1980.

Ryan JA and Poster EC. When a patient hits you. *Can Nurse* 87(8):23–25, 1991.

Schwartz CJ and Greenfield GP. Charging a patient with assault of a nurse on psychiatric unit. *Can Psychiatr Assoc J* 23(4):197–200, 1978.

Scott JR and Whitehead JJ. An administrative approach to the problem of violence. *J Ment Health Admin* 8(2):36–40, 1981.

Sosowsky L. Explaining the increased arrest rate among mental patients: A cautionary note. *Am J Psychiatry* 137(12):1602–1605, 1980.

State of California/Internal Memorandum Employee lost workday injuries from client violence, 1973–1980, 1980.

Tardiff K. A survey of assault by chronic patients in a state hospital system. In: JR Lion and WH Ried (eds.), *Assaults within Psychiatric Facilities*. Grune & Stratton, Inc., Orlando, FL, 1983, pp. 3–20.
Tardiff K and Koenigsberg HW. Assaultive behavior among psychiatric outpatients. *Am J Psychiatry* 142(8):960–963, 1985.
Tardiff K and Sweillam A. Assault, suicide and mental illness. *Arch Gen Psychiatry* 37(2):164–169, 1989.
Tardiff K and Sweillam A. Assaultive behavior among chronic inpatients. *Am J Psychiatry* 139(2):212–215, 1982.
Teplin L. The prevalence of severe mental disorder among male urban jail detainees: Comparison with the epidemiologic catchment area program. *Am J Pub Health* 80(6):663–669, 1990.
Wasserberger J, Ordog GJ, Harden E, Kolodny M, and Allen K. Violence in the emergency department. *Top Emerg Med* 14(2):71–78, 1992.
Whitman RM, Armao BB, and Dent OB. Assault on the therapist. *Am J Psychiatry* 133(4):426–429, 1976.
Wilkinson T. Drifter judged sane in killing of mental health therapist. *Los Angeles Times*, December 11, 1990, pp. Bl–B4.
Winterbottom S. Coping with the violent patient in accident and emergency. *J Med Ethics* 5(3):124–127, 1979.
Yesavage JA, Werner PD, Becker J, et al. Inpatient evaluation of aggression in psychiatric patients. *J Nerv Ment Dis* 169(5):299–302, 1981.
Zitrin A, Herdesty AS, Burdock EL, and Drossman AK. Crime and violence among mental patients. *Am J Psychiatry* 133(2):142–149, 1976.

Further Readings

Adler WN, Kreeger C, and Ziegler P. Patient violence in a psychiatric hospital. In: JR Lion and WH Reid (eds.), *Assaults within Psychiatric Facilities*. Grune & Stratton, Inc., Orlando, FL, 1983, pp. 81–90.
Bernstein HA. Survey of threats and assaults directed toward psychotherapists. *Am J Psychother* 35(4): 542–549, 1981.
Blair T and New SA. Assaultive behavior. *J Psychos Nurs* 29(11):25–29, 1991.
Carmel H and Hunter M. Staff injuries from inpatient violence. *Hosp Commun Psychiatry* 40(l):41–46, 1989.
Carmel H and Hunter M. Compliance with training in managing assaultive behavior and injuries from in-patient violence. *Hosp Commun Psychiatry* 41(5):558–560, 1990.
Conn LM and Lion JR. Assaults in a university hospital. In: JR Lion and WH Reid (eds.), *Assaults within Psychiatric Facilities*. W.B. Saunders & Co., Philadelphia, PA, 1983, pp. 61–69.
Davidson P and Jackson C. The nurse as a survivor: Delayed post-traumatic stress reaction and cumulative trauma in nursing. *Intl J Nurs Stud* 22(1):1–13, 1985.
Dillon S. Social workers: Targets in a violent society. *New York Times*, November 18, 1992, pp. A1 and A18.
Fineberg NA, James DV, and Shah AK. Agency nurses and violence in psychiatric ward. *Lancet* 1:474, 1988.
Infantino AJ and Musingo S. Assaults and injuries among staff with and without training in aggression control techniques. *Hosp Commun Psychiatry* 36:1312–1314, 1983.
Jenkins LE, Layne L, and Kesner S. Homicides in the workplace. *J Am Assoc Occupat Health Nurs* 40(5):215–218, 1992.
Jones MK. Patient violence report of 200 incidents. *J Psychos Nurs Ment Health Serv* 23(6):12–17, 1985.
Keep N, Gilbert P, et al. California Emergency Nurses Association's informal survey of violence in California emergency departments. *J Emerg Nurs* 18(5):433–442, 1992.
Kraus JF. Homicide while at work: Persons, industries and occupations at high risk. *Am J Publ Health* 77:1285–1289, 1987.
Lanza ML. The reactions of nursing staff to physical assault by a patient. *Hosp Commun Psychiatry* 34(1):44–47, 1983.
Lanza ML. How nurses react to patient assault. *J Psychos Nurs* 23(6):6–11, 1985b.

Lanza ML and Carifio J. Blaming the victim: Complex (non-linear) patterns of causal attribution by nurses in response to vignettes of a patient assaulting a nurse. *J Emerg Nurs* 17(5):299–309, 1991.
Lanza ML and Miller J. The dollar cost of patient assaults. *Hosp Commun Psychiatry* 40(12):1227–1229, 1989.
Lavoie F, Carter GL, Denzel DF, and Berg RL. Emergency department violence in United States teaching hospitals. *Ann Emerg Med* 17(11):1227–1233, 1988.
Levin PF, Hewitt J, and Misner S. Female workplace homicides. *AAOHN J* 40(8):229–236, 1992.
Lipscomb JA and Love C. Violence toward health care workers. *AAOHN J* 40(5):219–228, 1992.
Liss GM. Examination of workers' compensation claims among nurses in Ontario for injuries due to violence. Unpublished report, Health & Safety Studies Unit—Ministry of Labor, 1993.
Long Beach (CA) Press Telegram (1990), April 15, 1.
Madden DJ, Lion JR, and Penna MW. Assaults on psychiatrists by patients. *Am J Psychiatry* 133(4):422–425, 1976.
McNeil DE, et al. Characteristics of persons referred by police to psychiatric emergency room. *Hosp Commun Psychiatry* 42(4):425–427, 1991.
Poster EC and Ryan JA. Nurses' attitudes toward physical assaults by patients. *Archives of Psychiatric Nursing*, 3(6):315–322, 1989.
Ryan JA and Poster EC. The assaulted nurse: Short-term and long-term responses. *Archives of Psychiatric Nursing*, 3(6):323–331, 1989a.
Ryan JA and Poster EC. Supporting your staff after a patient assault. *Nursing* 89(12):32k, 32n, 32p., 1989b.
U.S. Department of Labor, Bureau of Labor Statistics, *Occupational Injuries and Illnesses in the United States by Industry*, 1989. Bulletin 2379, 1991.
U.S. Department of Labor, Bureau of Labor Statistics, *A Brief Guide to Recordkeeping Requirements for Occupational Injuries and Illness*, 29 CFR 1904, 1986.
Wasserberger J, Ordog GJ, Kolodny M and Allen K. Violence in a community emergency room. *Arch: Emerg Med*, 6:266–269, 1989.
White SG and Hatcher C. Violence and trauma response. In: RC Larsen and JS Felton (eds.), Psychiatric injury in the workplace. *Occupat Med State Art Rev* 3(4):677–694, 1988. Hanley & Belfus, Inc., Philadelphia.

Glossary

Abusive behavior: Actions which result in injury such as slapping, pinching, pulling hair or other actions such as pulling clothing, spitting, threats, or other fear-producing actions such as racial slurs, posturing, damage to property, and throwing food or objects.

Assault: Any aggressive act of hitting, kicking, pushing, biting, scratching, sexual attack, or any other such physical or verbal attacks directed to the worker by a patient/client, relative, or associated individual which arises during or as a result of the performance of duties and which results in death, physical injury, or mental harm.

Assaultive incident: An aggressive act or threat by a patient/client, relative, or associated individual which may cause physical or mental injury, even of a minor nature, requiring first aid or reporting.

Community worker: All employed workers who provide service to the community in private homes, places of business, or other locations which may present an unsafe or hostile environment. Examples of such workers include, but are not limited to, parking enforcement officers, psychiatric social workers, home health workers, union representatives, visiting or public health nurses, social service workers, and home health aids. The location of the workplace may be mobile or fixed.

Inpatient facility: A hospital, convalescent hospital, nursing home, board and care facility, homeless shelter, developmentally disabled facility, correction facility, or any facility which provides 24 h staffing and health care, supervision, and protection.

Injury: Physical or emotional harm to an individual resulting in broken bones, lacerations, bruises and contusions, scratches, bites, breaks in the skin, strains and sprains, or other pain and discomfort immediate or delayed, caused by an interaction with a patient/client or in the performance of the job.

MAB: A training program which trains staff to prevent assaultive incidents and to implement emergency measures when prevention fails.

Mental harm: Anxiety, fear, depression, inability to perform job functions, posttraumatic stress syndrome, inability to sleep, or other manifestations of emotional reactions to an assault or abusive incident.

Outpatient facility: Any health care facility or clinic, emergency room, community mental health clinic, drug treatment clinic, or other facility which provides drop-in or other "as needed care" or service to the community in fixed locations.

PART: A training program designed to provide a systematic approach to recognition and control of escalating aggressive and assaultive behavior in a patient/client or of other hostile situations.

Psychiatric inpatient facility: Public or private psychiatric inpatient treatment facilities.

Threat: A serious declaration of intent to harm at the time or in the future.

Threat or verbal attack: Any words, racial slurs, gestures or display of weapons which are perceived by the worker as a clear and real threat to their safety and which may cause fear, anxiety, or inability to perform job functions.

Appendix 7.A.1

General Industry Safety Orders (Section 3203): Section 3203. Injury and Illness Prevention Program

(a) Effective July 1, 1991, every employer shall establish, implement, and maintain an effective Injury and Illness Prevention Program (Program). The Program shall be in writing and, shall, at a minimum:

(1) Identify the person or persons with authority and responsibility for implementing the Program.

(2) Include a system for ensuring that employees comply with safe and healthy work practices. Substantial compliance with this provision includes recognition of employees who follow safe and healthful work practices, training and retraining programs, disciplinary actions, or any other such means that ensure employee compliance with safe and healthful work practices.

(3) Include a system for communicating with employees in a form readily understandable by all affected employees on matters relating to occupational safety and health, including provision designed to encourage employees to inform the employer of hazards at the worksite without fear of reprisal. Substantial compliance with this provision includes meetings, training programs, posting, written communications, a system of anonymous notification by employees about hazards, labor/management safety and health committees, or any other means that ensure communication with employees.

Exception: Employers having fewer than 10 employees shall be permitted to communicate to and instruct employees orally in general safe work practices with specific instructions with respect to hazards unique to the employees' job assignments as compliance with subsection (a)(3).

(4) Include procedures for identifying the evaluating work place hazards including scheduled periodic inspections to identify unsafe conditions and work practices. Inspections shall be made to identify and evaluate hazards.

(A) When the Program is first established.

Exception: Those employers having in place on July 1, 1991, a written Injury and Illness Prevention Program complying with previously existing section 3203.

(B) Whenever new substances, processes, procedures, or equipment are introduced to the workplace that represent a new occupational safety and health hazard; and
(C) Whenever the employer is made aware of a new or previously unrecognized hazard.
(5) Include a procedure to investigate occupational injury or occupational illness.
(6) Include methods and/or procedures for correcting unsafe or unhealthy conditions, work practices, and work procedures in a timely manner based on the severity of the hazard:
(A) When observed or discovered; and
(B) When an imminent hazard exists which cannot be immediately abated without endangering employee(s) and/or property, remove all exposed personnel from the area except those necessary to correct the existing condition. Employees necessary to correct the hazardous condition shall be provided the necessary safeguards.
(7) Provide training and instruction:
(A) When the program is first established,

Exception: Employers having in place on July 1, 1991, a written Injury and Illness Prevention Program complying with the previously existing Accident Prevention Program in section 3203.

(B) To all new employees;
(C) To all employees given new job assignments for which training has not previously been received;
(D) Whenever new substances, processed procedures or equipment are introduced to the workplace and represent a new hazard;
(E) Whenever the employer is made aware of a new or previously unrecognized hazard; and
(F) For supervisors to familiarize them with the safety and health hazards to which employees under their immediate direction and control may be exposed.

(b) Records of the steps taken to implement and maintain the Program shall include:
(1) Records of scheduled and periodic inspections required by subsection (a)(4) to identify unsafe conditions and work practices, including person(s) conducting the inspection, the unsafe conditions and work practices that have been identified and action taken to correct the identified unsafe conditions and work practices. These records shall be maintained for 3 years; and

Exception: Employers with fewer than 10 employees may elect to maintain its inspection records only until the hazard is corrected.

(2) Documentation of safety and health training required by subsection (a)(7) for each employee, including employee name or other identified training dates, (a)(s) of training, and training providers. This documentation shall be maintained for 3 years.

Exception no. 1: Employers with fewer than 10 employees can substantially comply with the documentation provision by maintaining a log of instruction provided to the employee with respect to the hazards unique to the employees work assignment when he is first hired or assigned new duties.

Exception no. 2: Training records of employees who have worked for less than 1 year for the employer need not be retained beyond the term of employment if they are provided to the employee upon termination of employment.

(c) Employers who elect to use a labor/management safety and healthy committee to comply with the communication requirements of subsection (a)(3) of this section shall be presumed to be in substantial compliance with subsection (a)(3) if the committee:

(1) Meets regularly, but not less than quarterly;
(2) Prepares and makes available to the affected employees, written records of the safety and health issues discussed at the committee meetings and, maintained for review by the Division upon request;
(3) Reviews results of the periodic, scheduled worksite inspections;
(4) Reviews investigations of occupational accidents and causes of incidents resulting in occupational injury, occupational illness, or exposure to hazardous substances and, where appropriate, submits suggestions to management for the prevention of future incidents;
(5) Reviews investigations of alleged hazardous conditions brought to the attention of any committee member. When determined necessary by the committee, the committee may conduct its own inspection and investigation to assist in remedial solutions;
(6) Submits recommendations to assist in the evaluation of employee safety suggestions; and
(7) Upon request from the Division, verifies abatement action taken by the employer to abate citations issued by the Division.

Note: Authority cited Sections 142.3 and 6401.7, Labor Code Reference. Sections 142.3 and 6401.7 Labor Code.

History

1. New section filed 4-1-77, effective 30th day thereafter (Register 77, No. 14). For former history, see Register 74, No. 43.
2. Editorial correction of subsection (a)(1)(Register 77, No. 41).
3. Amendment of subsection (a)(2) filed 4-12-83, effective 30th day thereafter (Register 83, No. 16).
4. Amendment filed 1-16-91, operative 2-15-91 (Register 91, No. 8).
5. Editorial correction of subsections (a), (a)(2), (a)(4)(A), and (a)(7)(Register 91, No. 31).

Appendix 7.A.2

General Industry Safety Order (Section 6184) Article 165. Employee Alarm Systems

(a) Scope and application
(1) This section applies to all emergency employee alarms. This section does not apply to those discharge or supervisory alarms required on various fixed extinguishing systems or to supervisory alarms on fire suppression, alarm or detection systems unless they are intended to be employee alarm systems.
(2) The requirements in this section that pertain to maintenance, testing and inspection shall apply to all local fire alarm signaling systems used for alerting employees regardless of the other functions of the system.
(3) All predischarge employee alarms shall meet the requirements of subsection (b)(1) through (b)(4), (c), and (d)(1) of this section.
(4) The employee alarm shall be distinctive and recognizable as a signal to evacuate the work area or to perform actions designated under the emergency action plan.
(5) All employees shall be made aware of means and methods of reporting emergencies. These methods may be but not limited to manual pull box alarms, public address systems, radios or telephones. When telephones are used as a means of reporting an emergency, telephone numbers shall be conspicuously posted nearby.
(6) The employer shall establish procedures for sounding emergency alarms in the workplace. For those employers with 10 or fewer employees in a particular workplace, direct voice

communication is an acceptable procedure for sounding the alarm provided all employees can hear the alarm. Such workplaces need not have a back-up system.

(b) General requirements

(1) Where local fire alarm signaling systems are required by these orders, they shall meet the design requirements of the National Fire Protection Association's "Standard for the Installation, Maintenance, and Use of Local Protective Signaling Systems for Watchman, Fire Alarm and Supervisory Service," NFPA No. 72A–1975 and the requirements of this section.

(2) The employee alarm system shall provide warning for necessary emergency action as called for in the emergency action plan, or for reaction time for safe escape of employees from the workplace or the immediate work area, or both.

(3) The employee alarm shall be capable of being perceived above ambient noise or light levels by all employees in the affected portions of the workplace. Tactile devices may be used to alert those employees who would not otherwise be able to recognize the audible or visual alarm.

(c) Installation and restoration

(1) The employer shall assure that all devices, components, combinations of devices or systems constructed and installed to comply with this standard shall be approved. Steam whistles, air horns, strobe lights or similar lighting devices, or tactile devices meeting the requirements of this section are considered to meet this requirement for approval.

(2) All employee alarm systems shall be restored to normal operating condition as promptly as possible after each test or alarm.

(d) Maintenance and testing

(1) All employee alarm systems shall be maintained in operating condition except when undergoing repairs or maintenance.

(2) A test of the reliability and adequacy of nonsupervised employee alarm systems shall be made every two months. A different actuation device shall be used in each test of a multiactuation device system so that no individual device is used for two consecutive tests.

(3) The employer shall maintain or replace power supplies as often as is necessary to assure a fully operational condition. Back-up means of alarm, such as employee runners or telephones, shall be provided when systems are out of service.

(4) Employee alarm circuitry installed after July 1, 1981, shall be supervised and provide positive notification to assigned personnel whenever a deficiency exists in the system. All supervised employee alarm systems shall be tested at least annually for reliability and adequacy.

(5) Servicing, maintenance, and testing of employee alarms shall be performed by persons trained in the designed operation and functions necessary for reliable and safe operations of the system.

(e) Manual operation

(1) Manually operated actuation devices for use in conjunction with employee alarms shall be unobstructed, conspicuous and readily accessible.

Note: Authority and reference cited: Section 142.5. Labor Code.

History

1. New Article 165 (Section 6184) filed 9-8-81; effective 30th day thereafter (Register 81, No. 37).
2. Editorial correction of subsections (b)(1) and (e)(1) filed 11-9-81; effective 30th day thereafter (Register 81, No. 45).
3. Editorial correction of subsections (b) and (d) filed 6-30-82 (Register 82, No. 27).
4. Change without regulatory effect deleting Title 24 reference (Register 87, No. 49).
5. Editorial correction of subsection (e)(1) deleting obsolete Title 24 reference (Register 88, No. 9).

Appendix 7.A.3

The New Jersey Department of Labor Bulletin. Guidelines on Measures and Safeguards in Dealing with Violent or Aggressive Behavior in Public Sector Health Care Facilities

Introduction

The State of New Jersey and its local government agencies are concerned with the management of clients with violent or aggressive behavior toward employees. The level of awareness and sensitivity to effective and available safeguards must be continuously emphasized.

These guidelines are being issued by the New Jersey Department of Labor (NJDOL) under the Public Employees Occupational Safety and Health (PEOSH) Program to assist public employers in health care facilities in adopting measures and procedures which will protect the safety of their employees. This is consistent with the legislative intent of the New Jersey PEOSH Act, N.J.S.A. 34: 6A-25 et seq.

Authority

Pursuant to N.J.S.A. 34:6A-31, and the Safety and Health Standards for Public Employees, N.J.A.C. 12:100, Commissioner of Labor is authorized to promulgate safety and health standards for employees working within public facilities in the State of New Jersey.

Scope

Health care facility workers employed in jobs with patients/clients assessed as a safety risk because of violent or aggressive behavior.

1. Safety measures and procedures shall include:
    a. A system for patient/client assessment with respect to client/patient behaviors.
    b. A system for communicating such information to employees assigned to work with the assessed patient/client.
    c. A system for summoning assistance in a cottage or ward if a patient/client behaves in a violent or aggressive manner.
    d. Provision or creation of an area where violent or aggressive patients/clients can be contained if necessary to ensure the safety of others.
    e. A procedure for patient/client restraint to be used when necessary to ensure the safety of others.
2. Instruction and training in the management of violent and aggressive patients/clients shall be provided to all patient/client care providers during orientation and a review program shall be offered periodically.
3. First-aid supplies and personnel trained in first aid shall be available in all cottages and wards.
4. A system of supportive intervention shall be made available to any employee involved in an incident with a violent or aggressive patient/client.
5. Safe staffing levels as determined by facility or division policy or mandated by regulation shall be maintained.
6. In instances of violent or aggressive patient/client activity, systems shall be in place to initiate appropriate interventions. Concerns about the possibility of a violent patient/client incident shall be reported to an immediate supervisor who shall immediately investigate and take whatever action is necessary to manage the situation in a manner that prevents or minimizes harm.

PEOSH Reporting Requirements

In compliance with the New Jersey PEOSH Act N.J.S.A. 34:6A-25 et seq., all public employers shall maintain records and file reports on occupational injuries and illness occurring to their workers. Such documentation shall be maintained on the PEOSH Program Log and Summary of Occupational Injuries and Illness (NJOSH No. 200).

Appendix 7.A.4

Sources of Assistance in Training and Program Development

Note: These programs and groups have not been evaluated, not recommended but are provided as an available source.

1. PART—Professional Assault Response Training
   Paul A. Smith, PhD
   Professional Growth Facilitators
   Post Office Box 5981
   San Clemente, CA 92674-5981
   Telephone & Fax #(714)498-3529
   PART—Basic Course
   PART—Advance Consultation
   PART —Training for Trainers
   PART—Trainer Recertification
2. Nonviolent Crises Intervention
   A two-tape video series
   National Crises Prevention Institute
   3 315-K N. 124th Street
   Brookfield, WI 53005
   1(800)558-8976
3. Service Employees International Union (SEIU)
   Maggie Robbins & John Mehring
   Western Health & Safety Coordinator
   3055 Wilshire Boulevard, Suite 1050
   Los Angeles, CA 90010

Assault on the Job: We Can Do Something about It!

Service Employees International Union

This material is by the SEIU, representing 1 million workers employed in the public and private sector. The SEIU is also the nation's largest HCWs union representing over 450,000 HCWs.

On any given day, in any given place, assaults are a reality of our working lives.

Joe, a mental health aide, must often work alone due to cutbacks in staffing. In the middle of the night he is attacked by a patient suffering cuts, bruises, and a back injury. This is his second assault in a year.

As a receptionist in a social service agency, Donna is on the front line, dealing with clients who are frustrated and impatient with the system. Yet she has little power to do anything to help them. The abuse, hostility, and threats of violence take their toll on Donna's health. She fears the day when someone will attack her with a weapon.

The workers are repairing gas lines when they get caught in the cross fire of a gang shootout.

Maria works in a nursing home. One of her patients continually hassles her with insults and sexual advances. When her attempts to transfer fail, she begins to suffer frequent headaches and has trouble sleeping.

What Is Assault on the Job?

As the above cases show, assault is not just physical violence. It also includes near misses, verbal abuse, unwanted sexual advances, or the threat of any of these. Even if a worker is never physically injured, the stress from the fear of assault may lead to serious health problems.

Assault on the Job Can Also Mean Death

The NIOSH estimates that during the early 1980s, 13% of the 7000 annual work-related deaths were no accidents—they were preventable homicides. According to the Centers for Disease Control and Prevention, murder is the leading cause of death for women in the workplace.

How Big a Problem Is It?

State workers' compensation agencies do not keep figures on job-related assaults so it is not known how many work-related injuries are caused by on-the-job violence. And workers often do not report assaults. Because so many attacks go unreported, employers, the government and often workers as well, do not always recognize violence as a workplace health and safety issue. Still, common sense should tell us that society's problems with violence does not stop outside the walls of our workplace.

Why Workplace Assault Is Usually Not Reported

The reasons are many:

- "Part of the job" syndrome: In certain jobs workers are expected to put up with attacks, threats and verbal abuse.
- A violent society is to blame: The assault is considered a consequence of living in a violent society, rather than working in an unsafe workplace.
- Fear of blame or reprisal: Workers may be afraid that they will be held responsible for any violent act that involved a patient or client.
- Lack of management support: Workers are often discouraged from reporting problems.
- No serious injuries: When physical injuries are minor, or when a worker does not miss a day of work, the injury is not reported to the worker's compensation board.
- Not worth the effort: If workers think nothing will be done, they feel there is no reason to file a report.

Who Is Affected?

Workers who are most likely to be assaulted are

- Those who work with the public
- Those who must work alone
- Those who handle money
- Those who come in contact with patients or clients who may be violent
- New employees

Many SEIU members are affected, especially those who work in hospitals, nursing homes, mental health institutions, and facilities for the developmentally disabled, shelters, prisons, social service departments and social/family service agencies, or educational institutions, as well as visiting nurses, home care providers, and utility workers.

Why Are Workers Assaulted?

Each incident of violence has its own set of causes. Working with clients or patients who may be frustrated, anxious, impatient, angry, in shock, mentally disturbed, or under the influence of drugs or alcohol inevitably carries with it the potential for violence. These people may lash out against whoever is closest to them—often an employee. Assaults are often unprovoked but also occur when workers are performing their duties in restraint and seclusion procedures.

Some specific factors which commonly play a role are

- Understaffing, which forces people to work alone or without enough staff to provide good coverage, thus allowing tensions to rise among patients or clients
- Deinstitutionalization, which leaves institutions with patients and clients who need greater attention and care and, in the absence of proper staffing and safeguards, can become dangerous
- Lack of training for workers in recognizing and defusing potentially violent situations
- Failure to alert workers to which patients or clients have a history of violence
- Failure to design safe workplaces and emergency procedures
- Failure to identify hazardous conditions and develop proper controls, policies, and education programs

Many of these factors result from budget cuts that are all too common in health care institutions and government agencies these days. These cutbacks make it hard for workers to give the care and service patients and clients deserve. Everyone suffers as a result. Increased violence is one very real effect of the budget axe.

Is Sexual Harassment a Form of Assault on the Job?

Sexual harassment is unwanted, repeated sexual attention at work. It may be expressed in the following ways:

- Unwelcome touching or patting
- Suggestive remarks or other verbal abuse
- Staring or leering
- Requests for sexual favors
- Compromising invitations
- Physical assault
- Offensive work environment (pinups/pornography)

Sexual harassment is another form of assault. Even if a worker is never physically injured, the stress of a repeated verbal abuse or fear of impending violence can result in serious health problems. Those who have experienced severe harassment cite a long list of physical symptoms including headaches, backaches, nausea, stomach ailments, fatigue, and sleep and eating disorders.

Sexual harassment is a form of sex discrimination and it is illegal under Title VII of the Civil Rights Act of 1964.

Specifically, sexual harassment is illegal if

- Your job or promotion depends on your saying yes
- The harassment creates an intimidating, hostile, or offensive workplace

A Widespread Problem

Sexual harassment happens in every kind of work environment, at all levels. It is carried out by superiors and subordinates, coworkers, and clients. Every race, gender, or age group can become a target. Men are also sexually harassed, though on a much smaller scale.

The victim does not have to be of the opposite sex. In fact, according to Equal Employment Opportunity Commission (EEOC) guidelines, the victim does not even have to be the person harassed but could be anyone affected by the offensive conduct.

Nancy and Susan work together. Susan is being harassed by a male coworker who uses foul language and tells off-color jokes to embarrass Susan. He also leaves pornographic pictures on Susan's desk. Nancy overhears these remarks and sometimes sees the pictures. In this instance, Susan is not the only person who is being sexually harassed!

- A survey of federal employees in 1980 showed that 42% of females and 15% of males said they had been harassed on the job. A follow-up survey in 1987 yielded nearly identical results.
- Since 1980, more than 38,500 charges of sexual harassment have been filed with the federal government but this figure is just the tip of the iceberg because many workers never file a complaint.

What Is Your Employer's Responsibility?

A significant number of workplaces—especially small- and medium-sized firms—have no policy regarding sexual harassment, but they should. While there is no one model for a good sexual harassment policy, all policies should send a clear message: *sexual harassment will not be tolerated.*

Elements of a Good Sexual Harassment Policy

- Union representatives are involved in policy development.
- Employees are involved at all stages of policy development.
- The policy is in writing and widely distributed through an employee handbook and orientation materials.
- The policy is publicized to both staff and clients.
- Top management is seen actively supporting the policy.
- Training is ongoing and occurs on work time.
- Procedures for how complaints are to be reported, recorded and investigated are clear and concise.
- Managers and supervisors are clearly instructed to begin investigations within seven days after a formal complaint is made.
- Discipline and counseling procedures are clearly stated. Discipline can range from verbal and written warnings to formal reprimands, suspension, transfer, probation, demotion, or dismissal. Counseling and/or sensitivity training may be appropriate.
- Confidentiality of an incident is maintained throughout the process.

Sexual Harassment: Fighting Back!

Dealing with sexual harassment may be difficult, but ignoring sexual harassment does not make it go away. Use the following guidelines to establish a strong case and fight back!

1. *Say no clearly.* State frankly that you find the harasser's behavior offensive. Firmly refuse all invitations. If harassment persists, write a memo asking the harasser to stop; keep a copy.
2. *Document the harassment.* Detail what, when, and where it happened, and include your response. This information is vital when a pattern of offensive conduct must be proven.
3. *Get emotional support from friends and family.* Do not try to fight this alone.
4. *Keep records of your job performance.* Your harasser may question your job performance in order to justify his/her behavior.
5. *Look for witnesses and other victims.* Two accusations are harder to ignore.
6. *Grieve it.* Get your steward and union involved right away.
7. *File a complaint.* Contact your state antidiscrimination agency or the federal EEOC if you decide to pursue a legal solution. You do not need an attorney to file a claim, but it may help to speak with a lawyer who specializes in employment discrimination.
8. *Act promptly.* There are state and federal time limits on how long after an act of harassment a complaint can be filed.
9. *Don't be intimidated.* It is unlawful for employers to retaliate after a complaint is filed. File another complaint based on the retaliation.

10. *Negotiate* a strong clause in your union contract that protects workers from sexual harassment.
11. *Educate and agitate*. Organize discussions on sexual harassment. Find out if others are experiencing the problem. Use posters, buttons, and flyers to send a strong message to management that workers will not accept such hostile working conditions.

Sexual Harassment Can Be Stopped

Break the silence. Use all means to demonstrate that sexual harassment will not be tolerated in the workplace!

Posttraumatic Stress Disorder (PTSD)

Assault victims have something in common with combat veterans and victims of terrorism, natural and man-made disasters, street crime, rape, and incest—an increased risk of PTSD.

What Is PTSD?

PTSD is the way a person reacts to emotional stress or physical injury, assault, or other forms of extreme stress outside everyday experience. It includes physical pain from the assault, as well as anger, anxiety, depression, fatigue, and preoccupation with the event. Other common symptoms are depression, flashbacks, and nightmares. PTSD can also do serious damage to family relations and social life.

Should PTSD Be Treated?

Yes. Voluntary individual counseling is the best form of treatment for an assaulted worker. Often, however, assault victims fail to seek help and blame themselves for the incident. They may also seek a psychiatric referral for treatment as an indication of mental illness, rather than as simply continuing treatment for their injuries.

What Are the Employers' Obligations?

Employers are responsible for providing employees a safe, healthy work environment free from recognized hazards. The location of the assault, which is likely to be the everyday worksite, can provoke anxiety and flashbacks (sometimes for weeks or months) and make it difficult for an employee to return to work. Therefore, injured workers need far more than a few minutes of counseling. They need ongoing support. Most assaulted workers depend on the informal support of family and friends which may not be enough.

Workers will also feel less victimized if they feel they have some control of their safety on the job. A staff's ability to make changes in policies and procedures that lead to a safer workplace will help foster a sense of empowerment and control.

What Can the Union Do?

A system for providing ongoing support, counseling, and assistance for assaulted members is a necessary part of any policy dealing with on-the-job assault. As trade unionists, we must educate our bosses and our members to the reality of posttraumatic stress. We must push to have PTSD recognized as a consequence of violence in the workplace.

Sample Contract Language to Protect Workers from on-the-Job Assault

One of the most effective ways to provide a safe and healthy workplace is to negotiate specific contract language. Here are some sample clauses relating to on-the-job assault.

General Clause

The employer is responsible for taking all necessary steps to protect employees from assault on the job.

Employer's Policy for Dealing with Assault on the Job

The Employer shall develop written policies and procedures to deal with on-the-job assault. Such policies must address the prevention of assault on-the-job, the management of situations of assault and the provision of legal counsel and posttraumatic support to employees who have been assaulted on the job by clients or the public.

This policy shall be part of the Employer's comprehensive health and safety policy. A written copy of the policy shall be given to every employee. The Employer must also establish a procedure for the documentation of all incidents, and shall take immediate and appropriate action, as outlined in the written policy, to deal with each incident.

Policy/Plan for Health Care/Social Service Facilities

The Employer shall conduct an ongoing security and safety assessment, and develop a security plan with measures to protect personnel, patients, and visitors from aggressive or violent behavior. The security and safety assessment shall examine trends of aggressive or violent behavior. A security plan shall include, but not be limited to, security considerations relating to all of the following:

- Physical layout
- Staffing
- Security personnel availability
- Policy and training related to appropriate responses to aggressive or violent acts

Joint Labor/Management Health and Safety Committee

The purpose of the committee is to identify and investigate health and safety hazards and make recommendations on preventive measures.

The committee shall be composed of an equal number of representatives from the Union and the Employer, and the Union shall have the sole power to appoint its representatives to the committee. The committee shall make recommendations on policies to prevent on-the-job assault, on the management of violent situations and on how to provide support to workers who have experienced or face on-the-job assault.

All incidents of assault will be brought to the attention of the Health and Safety Committee. The parties agree that the Health and Safety Committee shall:

- Assist in the development of policies and workplace design changes that will reduce the risk of assault on the job.
- Regularly review all reports of incidents of assault.
- Assist in the development and implementation of training programs that will reduce the risk of on-the-job assault.

Staffing Levels

The employer agrees to provide an adequate level of trained staff. Employees will not be required to work alone in potentially violent situations.

Employees will be notified as to potentially violent or aggressive patients, residents or clients, and will work/travel in pairs when required to work under such circumstances.

Workplace Design

Where appropriate, the Employer shall institute additional security measures including, but not limited to

- Installation of metal detectors
- Installation of surveillance cameras
- Limiting public access to the facility and specific departments or units
- Installation of bullet-proof glass
- Installation of emergency "panic" buttons to alert security personnel

Two-Way Radios, Alarms, and Paging Systems

The Employer shall provide two-way radios, alarms and/or paging systems, or other electronic warning devices or means of summoning immediate aid where employees ascertain a need. All equipment shall be maintained and periodically tested, and employees will receive training in the operation of the equipment.

Training

The Employer shall provide training to all employees at risk of assault on how to defuse potentially violent situations and verbal confrontation. Employees shall also be trained in self-protection. Training should include, but not be limited to: discussion of how to recognize warning signs and possible triggers to violence; how to resist attack and avoid escalation of the situation; how to control and defuse aggressive situations; and a full review of the Employer's Written policy for dealing with assault on the job.

All employees at risk of assault by patients, clients, or the public shall receive security education and training relating to the following topics:

- General safety measures
- Personal safety measures
- The assault cycle
- Aggressive and violence-predicting behavior
- Obtaining patient history from a patient with aggressive or violent behavior
- Characteristics of aggressive and violent patients, and victims
- Verbal and physical maneuvers to diffuse and avoid violent behavior
- Strategies to avoid physical harm
- Restraining techniques
- Appropriate use of medications as chemical restraints
- Critical incident stress briefing
- Available employee assistance programs

Posttraumatic Stress/Referral Services

The Employer shall, in the event of an incident of assault, provide counseling and support for the affected employee(s). Employees are to be compensated for lost days of work, counseling sessions, hospitalization, and other relevant expenses.

The Employer shall offer referral information and assistance to any employee who is assaulted by a patient, visitor or member of the public. Such information shall include, but not be limited to, the employee's legal right to press charges in a court of law.

Prosecuting Offenders

The Employer shall assist the assaulted employee in any legal actions that she/he undertakes against the offender. If the employee decides not to press charges, the Employer must provide a written explanation to the Union and to the employee of its decision.

No-Retaliation Clause

No employee shall be discharged, penalized, or disciplined for his/her victimization in an incident of assault. The Employer agrees not to retaliate or discriminate against that employee.

Union Nonliability

The Employer has the sole responsibility to provide a safe workplace and to correct health and safety hazards, and that nothing in this Agreement shall imply that the Union has undertaken or assumed any portion of that responsibility.

Other Union Rights

The Employer agrees that the Union has the right to bring into the workplace any union staff or other union representatives to assist investigating health and safety conditions.

Nothing in this chapter shall be deemed to wave any statutory rights that the Union may have.

Resource List

California Occupational Safety and Health Administration (Cal/OSHA) has *Guidelines for Security and Safety of Health Care and Community Service Workers*

Cal/OSHA
Department of Industrial Relations
455 Golden Gate Avenue, Suite 5202
San Francisco, CA 94102
(415)703-4341

NJDOL/Public Employee Occupational Safety and Health Administration has *Guidelines on Measures and Safeguards in Dealing with Violent or Aggressive Behavior in Public Sector Health Care Facilities*

NJDOL/Office of Public Employees Safety
CN 386
Trenton, NJ 08625
(609)633-3796

Organizations

Service Employees International Union, Health & Safety Staff

International Office
1313 L Street, NW
Washington, DC 20005
(202)898-3200

Eastern Regional Office
145 Tremont Street, Suite 202
Boston, MA 02111
(617)482-4471

New York Regional Office
330 West 42nd Street, Suite 1905
New York, NY 10036
(212)947-1944

New England Office
14 Quentin Street
Waterbury, CT 06706
(203)574-7966

West Coast Office
3055 Wilshire Blvd., Suite 1050
Los Angeles, CA 90010

Western Region Field Office
150 Denny Way
P.O. Box 19360
Seattle, WA 98109
(206)448-7348

Central States Regional Office
228 S. Wabash, Suite 300
Chicago, IL 60604
(312)427-7637

Michigan State Council Office
419 S. Washington Street
Lansing, MI 48933
(517)372-0903

SEIU Canadian Office
75 The Donway West, Suite 1410
Don Mills, ONT M3C2E9
Canada
(416)447-2311

9 to 5 National Association of Working Women
1224 Huron Road
Cleveland, OH 44115
(216)566-9308

National Institute for Occupational
Safety and Health
(NIOSH is the research arm of the Occupational Safety and Health Administration, and will conduct a health hazard evaluation at your workplace on request.)
1600 Clifton Road, NE
Atlanta, GA 30333
(800)356-4673

Occupational Safety and Health Administration (OSHA)
(OSHA develops and enforces workplace health and safety standards.)

OSHA Offices

National Office
200 Constitution Avenue, NW
Washington, DC 20210
(202)523-8091

Region I (Connecticut, Maine, Massachusetts, New Hampshire, Rhode Island, Vermont)
133 Portland Street, 1st Floor
Boston, MA 02114
(617)565-7164

Region II (New Jersey, New York, Puerto Rico, Virgin Islands)
201 Varick Street, Room 670
New York, NY 10014
(212)337-2378

Region III (Delaware, District of Columbia, Maryland, Pennsylvania, Virginia, West Virginia)
Gateway Building, Suite 2100
3535 Market Street
Philadelphia, PA 19104
(215)596-1201

Region IV (Alabama, Florida, Georgia, Kentucky, Mississippi, North Carolina, South Carolina, Tennessee)
Suite 587
1375 Peachtree Street, NE
Atlanta, GA 30367
(404)347-3573

Region V (Illinois, Indiana, Michigan, Minnesota, Ohio, Wisconsin)
32nd Floor, Room 3244
230 S. Dearborn Street
Chicago, IL 60604
(312)353-2220

Region VI (Arkansas, Louisiana, New Mexico, Oklahoma, Texas)
525 Griffin Street, Room 602
Dallas, TX 75202
(214)767-4731

Region VII (Iowa, Kansas, Missouri, Nebraska)
911 Walnut Street, Room 406
Kansas City, MO 64106
(816)844-3061

Region VIII (Colorado, Montana, North Dakota, South Dakota, Utah, Wyoming)
Federal Building, Room 1576
1961 Stout Street
Denver, CO 80294
(303)844-3061

Region IX (Arizona, California, Hawaii, Nevada, American Samoa, Guam, Trust Territory of the Pacific Islands)
71 Stevenson Street, 4th Floor
San Francisco, CA 94105
(415)744-6670

Region X (Alaska, Idaho, Oregon, Washington)
Federal Office Building
Room 6003
909 1st Avenue
Seattle, WA 98174
(206)442-5930

OSHA Federally Approved State Plan Offices

Alaska
Alaska Department of Labor
P.O. Box 1149
Juneau, AK 99802
(907)465-2700

Arizona
Division of Occupational Safety and Health
Industrial Commission of Arizona
800 West Washington
Phoenix, AZ 85007
(602)255-5795

California
Department of Industrial Relations
525 Golden Gate Avenue
San Francisco, CA 94102
(415)557-3356

Connecticut
(Public Employees Only)
Connecticut Department of Labor
200 Folly Brook Boulevard
Wethersfield, CT 06109
(203)566-5123

Delaware
Department of Labor
820 North French Street
Wilmington, DE 19801

Hawaii
Department of Labor and Industrial Relations
830 Punchbowl Street
Honolulu, HI 93813
(808)548-3150

Indiana
Division of Labor
1013 State Office Building
100 North Senate Avenue
Indianapolis, IN 46204
(317)232-2665

Iowa
Division of Labor Services
1000 East Grand Avenue
Des Moines, IA 50319

Kentucky
Kentucky Labor Cabinet
U.I. Highway 127 South
Frankfort, KY 40601
(502)564-3070

Maryland
Division of Labor and Industry
Department of Licensing and Regulations
502 St. Paul Place
Baltimore, MD 21202
(301)333-4176

Michigan
Department of Public Health
3423 North Logan Street
P.O. Box 30195
Lansing, MI 48909
(517)335-8022

Michigan
Department of Labor
309 N. Washington
P.O. Box 30015
Lansing, MI 48909
(517)373-9600

Minnesota
Department of Labor and Industry
443 Lafayette Road
St. Paul, MN 55101
(612)296-2342

Nevada
Department of Industrial Relations
Division of Occupational Safety and Health
Capitol Complex
1370 South Curry Street
Carson City, NV 89710
(702)885-5240

New Mexico
Environmental Improvement Division
Health and Environment Department
1190 St. Francis Drive, N2200
Santa Fe, NM 87503
(505)827-2850

New York
(Public Employees Only)
New York Department of Labor
Division of Safety & Health State Campus
Bldg. 12, Suite 159
Albany, NY 12240
(518)457-5508

North Carolina
Department of Labor
4 West Edenton Street
Raleigh, NC 27603
(919)733-7166

Oregon
Department of Insurance and Finance
21 Labor and Industries Building
Salem, OR 97310
(503)378-3304

Puerto Rico
Department of Labor and Human Resources
Prudencio Rivera Martinez Building
505 Munoz Rivera Avenue

Hata Rey, Puerto Rico 00918
(809)754-2119/2122

South Carolina
Department of Labor
3600 Forest Drive
P.O. Box 11329
Columbia, SC 29211
(803)734-9594

Tennessee
Department of Labor
501 Union Building
Suite "A," 2nd Floor
Nashville, TN 37219
(615)741-2582

Utah
Utah Occupational Safety and Health
160 East 300 South
P.O. Box 5800
Salt Lake City, UT 84110-5800
(801)530-6900

Vermont
Department of Labor and Industry
120 State Street
Montpelier, VT 05602
(802)828-2765

Virgin Islands
Department of Labor
P.O. Box 890, Christiansted
St. Croix, Virgin Islands 00820
(809)773-1994

Virginia
Department of Labor and Industry
P.O. Box 12064
Richmond, VA 23241-0064
(804)786-2376

Washington
Department of Labor and Industry
General Administration Building
Room 344-AX31
Olympia, WA 98504
(206)753-6307

Wyoming
Occupational Health and Safety Department
604 East 25th Street
Cheyenne, WY 82002
(307)777-7786/7787

Committees on Occupational Safety and Health (COSH Groups)

Alaska
Alaska Health Project
420 W. 7th Avenue, Suite 101
Anchorage, AK 99501
(907)276-2864

California
BACOSH (San Francisco Bay Area COSH)
c/o Mr. Glenn Shor
Labor Occupational Health Program
Institute of Industrial Relations
2521 Channing Way
Berkeley, CA 94720
(415)642-5507

LACOSH (Los Angeles COSH)
2501 South Hill Street
Los Angeles, CA 90007
(213)749-6161

Sacramento COSH
c/o Fire Fighters Local 522
3101 Stockton Boulevard
Sacramento, CA 95820
(916)444-8134 or (916)924-8060

SCCOSH (Santa Clara Center for Occupational Safety and Health)
Occupational Safety and Health
760 N. 1st Street
San Jose, CA 95112
(408)998-4050

Connecticut
ConnectiCOSH
P.O. Box 3117
Hartford, CT 06103
(203)549-1877

District of Columbia
Alice Hamilton Center for Occupational
Safety and Health
410 Seventh Street, SE
Washington, DC 20003
(202)543-0005

Illinois
CACOSH (Chicago COSH)
37 South Ashland
Chicago, IL 60607
(312)666-1611

Maine
Maine Labor Group on Health, Inc.
Box V
Augusta, ME 04332
(207)622-7823

Massachusetts
MassCOSH
555 Amory Street
Boston, Ma 02130
(617)524-6686

Michigan
SEMCOSH (Southeast Michigan COSH)
2727 Second Street
Detroit, MI 48201
(303)961-3345

New York
ALCOSH (Allegheny Council on Occupational Safety and Health)
100 East Second Street
Jamestown, NY 14701
(716)488-0720

CYNCOSH (Central New York COSH)
615 West Genessee Street
Syracuse, NY 13204
(315)471-6187

NYCOSH (New York COSH)
275 Seventh Avenue
New York, NY 10001
(212)627-3900

ROCOSH (Rochester COSH)
797 Elmwood Avenue
Rochester, NY 14620
(716)244-0420

WYNCOSH (Western New York COSH)
2495 Main Street, Suite 438
Buffalo, NY 14214
(716)833-5416

North Carolina
NCOSH
P.O. Box 2514
Durham, NC 27715
(919)286-9249

Pennsylvania
PHILAPOSH (Philadelphia Project on Occupational Safety and Health)
3001 Walnut Street, 5th Floor
Philadelphia, PA 19104
(215)925-SAFE (7233)

Rhode Island
RICOSH
340 Lockwood Street
Providence, RI 02907
(401)751-2015

Tennessee
TNCOSH
1514 E. Magnolia, Suite 406
Knoxville, TN 37917
(615)5252-3147

Texas
5735 Regina
Beaumont, TX 77706
(409)898-1427

Wisconsin
WISCOSH (Wisconsin COSH)
1334 South 11th Street
Milwaukee, WI 53204
(414)643-0928

Canada
VanCOSH (Vancouver COSH)
616 East 10th Avenue
Vancouver, British Columbia V5T2A5

WOSH (Windsor Occupational Safety and Health Project)
1109 Tecumseh Road East
Windsor, Ontario N8WlB3
(519)254-4192

Organizing to Prevent Assault on the Job

Unions have long led the fight for safe and healthy working conditions. That fight includes keeping our workplaces free of assaults. Union members in many different jobs have identified assault as a serious safety and health issue. Safe workplaces, free from assault, should be a goal of every union member. Like every union goal, our success depends on how well we organize.

1. How can you determine whether or not assault on the job is a problem at your workplace? *Talk, listen and encourage.*

Talk to your coworkers and find out if they share your concerns about safety. Develop a short survey (like the one in this chapter) to distribute in your facility. The SEIU Health and Safety Department can send you samples or help you design your own. Make sure you involve as many people in this activity as possible. Compile the results of your survey, consult with stewards and other union leaders, and then decide a group the most effective way to use these results.

One of the most important activities to get people involved in is documenting the problem. Urge members to document all assault incidents, and all the near incidents, using the Incident Report Form in this section. Review this data on a regular basis and update members every time there is an incident of assault.

Hold lunch-time meetings or form committees to research specific problems and develop solutions once the hazards have been identified. Keep members informed through the local union newsletter and on bulletin boards. Always encourage members to document incidents.

2. How do you solve the problem? *Develop a plan of action.*

A work environment in which people fear attack is NOT a healthy or safe place. Employers must recognize that it is their legal duty to protect workers from assault just as much as any other health and safety hazard. Assault is not something that "just happens." Taking the right steps, we can control and prevent violence in the workplace.

Use your Health and Safety Committee or form one to deal with on-the-job assault. Keep members involved in developing the solution. Think through what steps you want management to take. When you have gathered your documentation and survey results, bring a group of workers to meet with management. Present your complaint, and ask them to correct it. If your demands are met, make sure you publicize your success.

3. What if management refuses to take action? *Grieve, Negotiate, Agitate, and Organize.*

If management refuses to respond to your demands, be prepared to take the following steps:

- File a grievance.
- Refuse to work alone or under certain conditions, but check with your local union leadership first.
- Contact OSHA and demand action.
- Publicize the problem using the media.
- Create a slogan campaign like "Understaffing Kills." Wear buttons, post notices, and hand out flyers.
- Negotiate health and safety language in your next contract.
- Build coalitions and lobby for laws that provide proper protection for workers from being assaulted.

SEIU members and locals throughout the country that have grieved, negotiated, agitated, and organized have succeeded in making their workplaces safer.

SEIU Success Stories

SEIU locals have fought for and won important protection for their members.

Staffing and a New Felony Law

When Georgia State Employees Union, Local 1985, organized a campaign to protest the 20 assaults in a 3 month period at a youth development center, their actions resulted in a 10% staff increase, the allocation of $5 million to build a new facility for violent offenders, plus a new law which made assault on a staff member a felony.

Mandatory Training

After Local 1199NE in Connecticut discovered that 8% of all workers' compensation injuries were patient related-with most due to assaults—the union's actions led to the creation of a 21 h mandatory training program for 9000 members across the state. At one mental health hospital, injuries due to assaults dropped 60% after the training and the implementation of an employee wellness program.

Better Patient to Staff Ratios

After several serious assault incidents at their mental health facility in West Virginia, Local 1199WO members filed a mass grievance. The outcome was an increase in staffing which changed the ratio of patient to provider from 11 to 1 to < 6 to 1.

Assault on the Job Survey—How Safe Is Your Work Environment?

Employee Commitment/Workplace Policy

Does your employer…

1. Place a high priority on eliminating hazards associated with assault on the job?
   ☐ Yes ☐ No
2. Have a policy that places employee safety on the same level of importance as patient/client safety?
   ☐ Yes ☐ No
3. Discuss assault openly at Health and Safety Committee meetings?
   ☐ Yes ☐ No
4. Investigate and document all instances of assault and/or harassment?
   ☐ Yes ☐ No
5. Have a written policy concerning assault on-the-job?
   ☐ Yes ☐ No
6. Involve employees in developing the policy?
   ☐ Yes ☐ No
7. Have a program to provide support for victims of assault?
   ☐ Yes ☐No
8. Support employees who have been involved in an assault incident, rather than discipline them?
   ☐ Yes ☐ No
9. Provide legal counsel for assault victims?
   ☐ Yes ☐ No
10. Encourage reporting assault incidents to the police and support prosecution of offenders?
    ☐ Yes ☐ No

Staffing

1. Is staffing adequate?
   ☐ Yes ☐ No
2. Does your employer make sure you don't work alone?
   ☐ Yes ☐ No
3. Is there an adequate number of security staff?
   ☐ Yes ☐ No
4. Is backup staff always scheduled?
   ☐Yes ☐ No

Workplace Design

1. Are all work areas well-lit?
   ☐ Yes ☐ No
2. Are private washrooms provided for staff?
   ☐ Yes ☐ No
3. Is access to office areas/employees' work stations restricted to only authorized staff and clients?
   ☐ Yes ☐ No
4. Are there electronic alarm systems, closed-circuit TV or two-way radios?
   ☐ Yes ☐ No
5. Is furniture well placed so employees cannot get trapped in a room with a client?
   ☐Yes ☐ No
6. Are employees who do field work provided with personal alarm systems or beepers?
   ☐ Yes ☐ No
7. Are parking lots, garages, and other areas that employees need to walk through, secure and well-lit?
   ☐Yes ☐ No

Training

1. Do all employees receive adequate training on how to protect themselves from being assaulted on the job?
   ☐ Yes ☐No
2. Were employees involved in developing the training programs?
   ☐ Yes ☐ No
3. Do all new staff receive training upon hire?
   ☐ Yes ☐ No
4. Do backup staff receive training?
   ☐ Yes ☐ No

Sample Assault Incident Report Form

A sample of the form for you to complete and send it to your union representative so that we can (1) accurately record the incidents of assault and/or harassment that occur, (2) notify and/or follow up with your employer regarding the problem, and (3) plan strategies to prevent these problems from recurring:

1. Date of incident: mo ___ day ___ yr ________________
2. Name, or pseudonym, of member(s): ________________
3. Work location: ________________
4. Local union: ________________
5. Please describe the incident: ________________
6. Where did the incident occur? ________________
7. Did the incident involve a weapon? ________________
8. Were you injured? ________________
9. To what extent? ________________
10. Did you lose any work days? How many? ________________
11. Have you applied for worker's compensation? ________________
12. Was the person who assaulted you a coworker, supervisor, patient, client (or someone else)? ________________
13. Were you singled out, or was the assault directed at more than one individual?________________
14. Were you alone when the incident occurred? If yes, why? ________________
15. Did you have any reason to believe an incident might occur? If yes, why?________________
16. Did you report the incident to your supervisor? ________________
17. Have you filed a police report? ________________
18. If so, was the attacker charged? ________________
19. To your knowledge, has this type of incident ever happened before to your coworkers?
    ____
20. Have you had any counseling or support since the incident? ________________
21. Do you have any thoughts on how this incident could have been avoided? What should your employer do to avoid similar incidents in the future? ________________

Remember, the more statistics we have on violent encounters in the workplace, the more compelling our arguments are for demanding improved workplace design, adequate staffing levels, better security and ongoing training.

Action Taken by Local
Talked with management (date): ________________
Grievance filed (date): ________________
OSHA complaint filed: ________________
Other actions: ________________
Comments: ________________

Sample Grievance/Petition Form

Filing a group grievance or presenting management with a petition of protest is a good way to bring the issue of on-the-job assault to your employer's attention. If you have negotiated contract language that protects members from on-the-job assault, then use the grievance process to get some action. If you do not have contract language, or you want to include employees who may not be in your union (they may be managers or members of another union), then the group petition can be an effective action.

Here is a sample mass grievance/petition. You can adapt this to meet the specific needs of your campaign.

SEIU Local 9999
Group Grievance/Petition

To: Director(s) or person(s) in charge

From: The undersigned members of Local 9999

Date: January 1, 1994

Statement of problem: The Employer is failing to provide a safe and healthy workplace, free from recognized hazards associated with assault on the job. (If you have contract language, cite the article.)

Relief sought: Provide a safe and healthy work environment (as per Article of the contract). Install alarms and other appropriate monitoring equipment in areas where the public has access, or where employees may be working alone. Develop a comprehensive policy to protect workers from being assaulted on the job, including the provision of adequate staffing, incident reporting procedures, the provision of appropriate and adequate training for employees who are at risk of assault and support services for victims of assault.

Name____________________	Title ____________________
____________________	____________________
____________________	____________________

Training Exercises

Case Studies

In each of the case studies below, you play the role of union activists. Work together in small groups to discuss what actions you would take to address these problems concerning assault on the job.

1. Joe, a mental health aide, must often work alone due to cutbacks in staffing. Recently, he was attacked by a patient, and suffered cuts, bruises, and serious back injury. Joe is the fourth member of your union who has been assaulted by a patient in the last 2 months.
2. Donna works at the front desk in your social service agency. The clients, who are often frustrated and impatient with the system, often take their anger out on Donna. The abuse, hostility, and threats of assault have taken their toll on her, as she continually talks about her fear of someone attacking her with a weapon.
3. You and your coworkers do field work, and most of the time you work alone. Years ago that was fine, but is seems the streets have gotten more dangerous, and you all feel very uncomfortable traveling alone. No one has gotten hurt yet, but there have been several near-misses, and everyone wants the union to do something about it.
4. Maria works in the nursing home where you work. One of her patients continually hassles her with insults and sexual advances. She has also been harassed by the son of another patient on that same floor. She recently asked to be transferred so she could avoid these situations, but her request was denied. She is now suffering frequent headaches and has trouble sleeping.
5. You and your coworkers drive to work, and you park your car in the building's parking lot. There have been several incidents in the last few months, and one coworker was recently mugged as she approached her car.

6. The emergency room in your hospital has recently been the sight of two serious assaults on workers. Staff wants something done about it, but the employer keeps saying that they cannot predict when violence is going to break out.

Response to Assault on the Job

The purpose of this exercise is

- To discuss *what is currently being done* to protect workers from on-the-job assault
- To discuss *what should be done* to protect workers from being assaulted on the job
- For you and your coworkers to *develop an action plan* to deal with the issue of on-the-job assault

How to Complete the Exercise

1. In small groups, participants should relate incidents of assault that have occurred on the job. AR incidents should be listed on a chalkboard or chart paper. (Participants may want to look at "Case Studies" in this section for other examples of on-the-job assault.)
2. After all incidents have been listed, participants should complete the "Assault Response Chart" by answering the two questions for each group of people listed.
3. Based on your responses, have the group set some priorities for actions to take to address the issue of assault on the job.

Step 1: Discussing the incidents of assault at your workplace

Example:
Incident
Joan, a social service worker, was assaulted by one of her clients. She missed a week of work, and she is afraid to come back, because she fears the client will assault her again.

Discuss and list the incidents that have occurred among your members.

Step 2: What are these people doing about it?

Assault responses

How do they respond, and what are they doing about it?	How should they respond, and what should they be doing about it?
Workers:	Management:
____________________	____________________
____________________	____________________
The Union:	
____________________	____________________
____________________	____________________
OSHA/Other Government Agencies:	
____________________	____________________
____________________	____________________
Researchers/academics:	
____________________	____________________
____________________	____________________

Step 3: Setting priorities

Based on your discussion, list the actions that you and your coworkers will take to reduce the risk of assault on the job.

____

____

____

A Joint Labor/Management Experience in Implementing OSHA's Violence Prevention Guidelines in the NYS Office of Mental Health

Jonathan Rosen

Introduction

In 1996, a joint labor/management initiative was launched in the NYS Office of Mental Health (OMH) to implement the OSHA's violence prevention guidelines on two units in each of two state mental-health facilities. This section will provide an overview of the problem of violence in the health care industry and then describe the case studies within OMH.

The key stake holders involved in initiating the projects included representatives from the OMH multiunion Health & Safety Committee. This agency-level committee has had a productive history due to the commitment of its members to a collaborative consensus process that features sharing of resources, expertise, and focusing on key hazards. The committee is comprised of management personnel from OMH Bureau of Employees Relations and several high-level facility managers as well as elected Union officials and staff.

The state worker unions in New York have extensive contact language on health and safety that include requirements for joint health and safety committees at the agency and worksite levels. The unions represented in this forum include the NYS Public Employees Federation (PEF), the Civil Service Employees Association (CSEA), and Corrections and Law Enforcement Council 82. Council 82 represents Corrections Officers, Secure Hospital Treatment Aides, and Safety and Security Officers. CSEA represents Mental Health Therapy Aides (MHTA), and Licensed Practical Nurses, Social Service, Maintenance, Food Service, and Administrative workers. PEF represents professional, scientific, and technical workers including 15,000 HCWs employed by state agencies such as Corrections, Mental Health, Division for Youth, and the State University Health Science Centers. Council 82 and CSEA are affiliated with the American Federation of State, County, and Municipal Employees International Union. PEF is affiliated with the SEIU and the American Federation of Teachers.

Assaults on staff by inmates, patients, and other institutional service recipients have been such a significant problem that PEF has had an Attack Insurance plan from Lloyd's of London since 1989. The benefit pays $2000 dollars to any worker who is assaulted, files a police report, receives medical attention, and is off from work for 5 days or more. The plan was the first of its kind and since its inception, 430 out of 700 claims have involved staff of the OMH which currently operate 29 psychiatric hospitals. CSEA has a similar insurance program.

PEF and CSEA have extensive Occupational Safety & Health Departments that are funded by the unions, joint funds with the employer, and grants. These resources allow PEF and CSEA to develop innovative prevention programs to address key hazards such as workplace violence.

Background

Many researchers have described various aspects of the epidemic of workplace violence in the health care industry. Although this is not a new phenomenon to veteran HCWs and administrators, it is beginning to be recognized within the industry as an occupational hazard due to efforts of regulatory agencies, health and safety professionals, labor unions, and academics.[1]

A 1993, nationwide analysis by the NIOSH[2] revealed that workplace violence is the leading cause of death for female workers and the second leading cause of death for males. While workplace homicide mainly affects males, taxi cab drivers, convenience store workers, and other retail employment, nonfatal violence mainly affects females, HCWs, and especially nursing home workers (see Table 7.2).

TABLE 7.2

Characteristic	Fatal		Nonfatal	
Number of Cases		1,063		22,396
Percent		100%		100%
Sex injured more frequently	Men	82%	Women	56%
Most frequent violent act	Shooting	82%	Hitting, kicking, beating	47%
Primary perpetrator	Robber	75%	Health care patient	45%
Occupations with largest share	Taxi driver	9%	Nurse's aides	30%
	Cashier	9%		
Industries with largest share	Grocery stores	17%	Nursing homes	27%
	Restaurants	14%		

Source: BLS Report 891, June 1995.
Note: Data for fatalities is for 1993: data on nonfatal assaults is for 1992.

Violence in Health Care Facilities

In a meta-analysis, Jane Lipscomb and Colleen Love[1] described various studies of violence in health care facilities that analyzed the scope of the problem, types of health care facilities affected, and the main causal factors. Rates of violence, affected job titles, and causal factors varied depending on the type of health care facility being studied. For example, a 1989 report by Poster found that 74% of psychiatric nurses surveyed agreed that staff members working with mentally ill patients can expect to be physically assaulted during their careers. Other settings that are considered high risk include emergency departments, pediatric units, facilities for mentally retarded, medical–surgical units, and nursing homes.

Weapon carrying in hospitals has been an increasing concern in emergency departments, especially in large urban areas. In a 1988 study at a Los Angeles Level I trauma center, Wassenberger found that 4796 weapons were confiscated from 21,456 patients over a 9 year period and at least 25% of the major trauma patients were carrying lethal weapons.

A counterproductive trend has been an overemphasis on the issue of worker on worker violence by some hospital administrators, many management consultants, and some union leaders. According to 1994 BLS data only 4% of the fatalities and 20% of the nonfatal assaults relate to this category of violence. Sensational headlines about berserk postal workers have created an environment where this issue is getting more attention than it deserves. Certainly, this type of conflict in the workplace is important to address. However, the first priority in health care institutions is usually dealing with prevention of attacks on staff by people in their care. A specific hospital analysis of injury trends will help each organization to establish its priorities.

OSHA Guidelines

OSHA published, "Guidelines for Preventing Workplace Violence for Health Care and Social Service Workers" in 1996.[3] These guidelines were developed in response to NIOSH's research, petitioning by HCW unions, and a growing awareness of the impact of the problem. OSHA's guidelines were largely based on previous work developed by California OSHA.[4] The guidelines provide an overview of the problem and a framework for addressing it. The list below contains the key elements of a health care violence-prevention program according to OSHA'S guidelines.

- Management commitment and employee involvement
- Written program
- Worksite analysis

    - Records analysis and tracking
    - Monitoring trends and analyzing incidents
    - Screening surveys
    - Workplace security analysis
- Hazard prevention and control
    - Engineering controls and workplace adaptation
    - Administrative and work practice controls
- Medical management and postincident response
- Training and education
    - All employees
    - Supervisors, managers, and security personnel
- Recordkeeping and evaluation of the program

The health care industry has had an increasing interest in workplace violence since the distribution of the OSHA guidelines and due to advocacy from Unions, patient advocacy groups, pressure from the press, and in efforts to drive down compensation costs.

Although the OSHA guidelines provide an outline for developing a violence prevention program, it is up to the stakeholders within the industry to do the painstaking work of implementing them in a manner that will yield results. The devil is in the details and it is particularly difficult to launch innovative employee-based programs in the current environment of restructuring, and all too often, cutbacks.

Organizational Culture and Political Will

As in many of our society's institutions, the health care industry is dominated by hierarchical organizational structures. Empowering frontline workers to address workplace violence is key to meaningful implementation of OSHA's guidelines. An obstacle in most institutions is that HCWs are expected to be subservient to the chain of command in the hospital-management and clinical-care structures. Often, physicians and other health care specialists buy into the notion that their years of sacrifice at the academic altar should place them in a superior position in their working lives. There are numerous problems presented by these types of relationships:

- Direct care workers, especially nurses and aides have the greatest exposure and are most often assaulted. They are closest to the problems and often have rich ideas on how to solve them.
- A purely academic approach often alienates direct care staff.
- Hierarchical structures are usually resistant to changing policies, environments, and treatment programs identified as problematic in a workplace violence-prevention analysis.

HCWs are taught to sacrifice their own well-being for their patients. This notion contradicts the basic idea of a workplace health and safety program: "workers should be able to come home from their jobs in relatively the same condition as when they left for work." The dominant culture tends to blame workers for their injuries rather than analyzing equipment, environments, and policies. In the field of industrial hygiene a basic principle is to follow a hierarchy of control measures in addressing workplace hazards as follows:

- The first choice is to use *engineering controls* which remove the hazard at the source.
- The second choice is *administrative controls* which reduce the hazard by altering work practices or training workers.
- The least desirable control measure third choice, is *personal protective equipment* which does nothing to reduce the hazard, is usually uncomfortable for workers, and may interfere with work performance and patient care.

Implementation of OSHA's violence prevention guidelines is often negatively affected by the above-described cultural norms. Instead of blaming workers for their injuries, an enlightened approach assures they are the motor force for change. Successful programs require management commitment of time and

TABLE 7.3

Title	Number	Percent of Staff
Registered Nurses	78,588	13.4
Other mental health workers (less than BA)	126,878	21.7
Total	205,466	35.1

resources and an openness to new ideas. Commitment is also essential from union leaders and other key stakeholders in developing a consensus, participatory process to confront workplace violence in any health care setting.

National Profile

In 1992 there were 5498 organizations providing mental health services including state, county, and private mental hospitals, general hospitals with separate psychiatric services, VA medical centers, residential treatment centers, and outpatient clinics.[5] Overall the Mental Health Industry employed 585,972 workers nationwide during 1992. Direct-care staff are the workers who are most at risk, and their number nationwide in 1992 (Table 7.3).

While there is no national profile on violence in the industry, NYS's experience is informative.

New York State

The NYS OMH operated 29 psychiatric centers, employing 23,500 workers, and providing treatment and care for 7,500, patients in 1996. Of the hospitals, 18 are primarily for adults, 3 are forensic, and 6 are children's institutions.[6] The agency is undergoing dramatic changes. The workforce has been cut in half over an 8 year period and the inpatient census has declined by about two-thirds during the same time frame. Facilities are being closed and/or consolidated and patients are being moved to community residences and outpatient clinics. Many former patients are ending up in the prison system.[7] Estimates range from 10% to 30% of the current inmate population in NYS Department of Corrections are former mental health patients or are receiving mental-health care (Table 7.4).

In addition to changes in staffing, there has been an overall tightening of resources, and a change in patient demographics. The current patient population is more often male, younger, dually diagnosed with mental illness and chemical abuse. A larger proportion of patients have had involvement or are coming directly out of the criminal justice system. This has created a climate where an aging workforce is being forced to adapt to a more violent and manipulative clientele.

OMH maintains an occupational injury and illness reporting (OIRS) database which was used to analyze injury trends. The 29 hospitals employed a total of 23,552 people during 1996 and experienced 2751 reported injuries related to restraint or seclusion or assault by patients. This is an average rate of 11.7 injuries per 100 workers statewide. An analysis of higher risk units and job titles in the case studies revealed much greater rates. This microanalysis revealed the limitation of calculating hospital-wide injury rates compared to unit and job title specific rates (see the injury analysis data from Binghamton PC for more detail). A hospital-wide analysis includes many people who have no exposure to the hazard of violence. On the other hand, a microanalysis, while helping to focus on key units and job titles, produces very unstable rates due to the small number of workers in the sample. These limitations should be considered when analyzing injury data.

Over the past several years the agency has been making a concerted effort to reduce the use of restraint and seclusion. In part, this was due to the revelation that 111 patient deaths that occurred during the 10 year period of 1984–1993 associated with restraint and seclusion.[8] An interagency task force was formed including oversight, patient advocacy, and other interested groups to study the problem. They found

TABLE 7.4 Changes in Adult Inpatient Population, Fiscal Year 1986–1996, NYS OMH

	1986	1996	Change(%)	Change in Proportion of Census/ Admissions(%)
Admissions	23,790	8,091	−66	—
Census	20,249	7,588	−63	—
Geriatric admissions	1,403	323	−77	−32
Geriatric census	8,306	1,431	−83	−53
Census aged 35–49	3,746	2,944	−21	+109
Males	10,311	4,777	−54	+24
Males aged 35–49	2,382	2,065	−13	+130
MICA admissions	2,640	1,702	−36	+91
MICA census	544	940	+73	+360
Multicultural census	5,467	3,451	−38	+64
Multicultural male census	3,321	2,405	−28	+93
Legal status of admissions				
Involuntary	14,764	6,024	−59	+19
Civil	13,939	5,112	−63	+8
Forensic	825	912	+11	+223
Voluntary	8,963	2,046	−77	+33

Source: *OMH Statewide Comprehensive Plan for Mental Health Services, 1997–2001.*
Note: MICA stands for mentally ill chemical abuser.

that high rates of usage of restraint and seclusion did not correlate with staff/patient safety. This led to a change in philosophy and policy causing a drastic reduction in use.

The graph in Figure 7.4 illustrates the differences in rates caused by assaults vs. restraint or seclusion and between hospitals. There were significantly higher rates of injury at the three forensic hospitals—Kirby, Mid-Hudson, and Central New York—as well as the children's facilities (e.g., Children's Psychiatric Center [CPC]). In general, there are greater numbers of injuries due to assaults than to restraint or seclusion. However, in six institutions the reverse was true and in four there were equal numbers of cases.

The Projects

Shortly after the OSHA guidelines were published in 1996 the OMH Multi-Union Health and Safety Committee distributed them to all hospital directors and union leaders. A cover letter was attached encouraging each hospital to review policies and compare them to OSHA's recommendations. Prior to this action, the Multi-Union committee had proposed development of an agency policy on "Trauma Response." The resulting agency-wide directive required each psychiatric hospital to establish a policy on providing emotional support and counseling to injured staff and other affected employees. The agency, with union consultation, recently completed a comprehensive training program for people assigned to trauma-response counseling and postincident debriefing. During 1994, the unions and management also were awarded a state labor grant which allowed for the development of a train-the-trainer program entitled *Safety Training for Community Mental Health Workers.*

During this era, when the PEF Health & Safety Department learned of a significant assault on a member, arrangements were made for a site visit. The visits were conducted by the author of this chapter and personnel from OMH Bureau of Employee Relations, who served as management reps

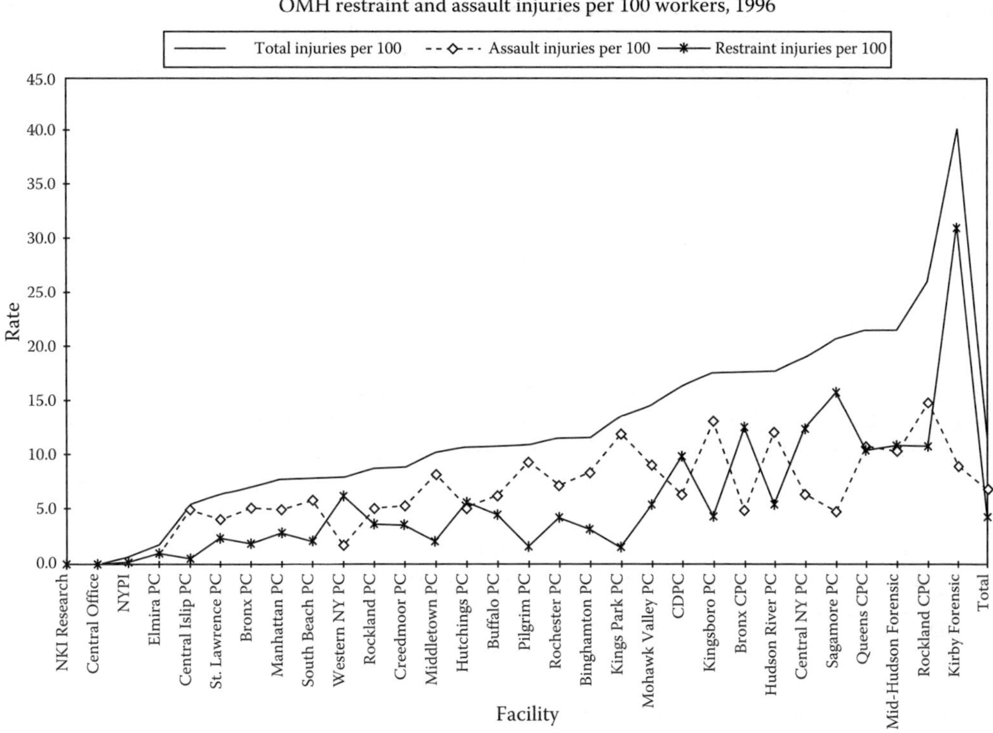

FIGURE 7.4 Illustration of the differences in injury rates as caused by assaults vs. restraints.

on the Multi-Union Health & Safety Committee. Typically, these tours included a meeting with the Hospital Director, Department Heads, Union Representatives, and, in some cases, the injured worker. The facility management would outline their programs to prevent patient violence and specific actions taken in response to the case involved. The Union would introduce the NIOSH data, the OSHA guidelines, and discuss the feasibility of implementing elements of the OSHA program that were not in place. The meetings were followed by a tour of the facility and a closing session.

While these interactions were educational for all parties involved, they did not lead to significant actions being taken. Part of the scene in the institutions is frequent inspections by central agency, Joint Commission for Accreditation Healthcare Organization (JCAHO), and other authorities. Most facilities are very experienced in dealing with outside oversight and the labor or management visits were handled accordingly. These experiences illustrated that if new approaches were going to be attempted, local management and labor leaders needed to commit time and resources with support and guidance from the central agency and union leadership. This recognition, together with the initiative from two facilities, led to the pilot projects below described.

Traditional approaches to controlling patient violence have focused on changing worker behavior through training and clinical or behavioral strategies. Rarely have physical and organizational environments been considered important in addressing the problem. In part this is due to a perception that assaults on mental HCWs are part of the job and are not predictable or preventable. Since psychiatric hospitals are supposed to be "clinical" environments, outbursts of patient violence have typically been seen as clinical issues. The OMH experience shows that the emergence of occupational health guidelines for violence prevention presents an important opportunity to blend the best from the expertise of psychiatric and occupational health and safety disciplines. This strategy, as well as commitment of time and resources, were key to the pilot projects. There were some significant differences in the two projects due to different levels of management commitment and skill in fostering a consensus-based process.

Buffalo Psychiatric Center

In 1996, the Buffalo Psychiatric Center (BPC) was awarded a state labor department grant to reduce assault-related job injuries, targeting staff from the Intensive Treatment Units 75 and 76 which are chronic intensive single sex wards with about 30 patients each, many of whom are assaultive.

Through a competitive bidding process, the NYS PEF Health & Safety Department was selected as the contractor. PEF also contributed, in kind, resources it had obtained through the collective bargaining agreement with NYS and its Department of Labor Grant.

At the time of the study, 1996, BPC employed 870 workers. Units 75 and 76 employed 6 nurses and 11 MHTAs across three shifts. The downsizing of BPC had led to a patient population that tended to be more chronic and more severely disturbed, leading to an increase in staff injuries. Each unit had 28 patient beds. During 1993 and 1994, BPC averaged slightly more than 20 staff per month out on occupational injury leave. The Intensive Treatment Unit had 101 staff injuries in 1994 with 29 resulting in lost-time. Of the 101 injuries, 71 were related to patient assaults or restraint and seclusion of patients. Of the 71, 19 resulted in lost-time.

The goal of the project was to reduce the risk of injury to workers from patient assaults. The objectives were to identify factors that could predict or prevent assaultive situations and to provide training.

Methods

Phase I—needs assessment

1. Formation of project advisory committee
2. Records review and rate analysis
3. Environmental assessment and recommendations—Review environmental factors and control methods
   a. Ward design
   b. Security measures
4. Policies and procedure assessment and recommendations
   a. Behavioral assessment
      - Predicting violence
      - Early recognition of escalating behavior
      - Worker based risk reduction planning
   b. Intervention techniques
      - Assessing the crisis
      - Intervention teamwork
      - Defusing escalating behavior
      - Personal safety and self-protection
   c. Awareness of relative risk and security control plan
   d. Assault incident procedures, medical management, and postincident protocols

Phase II—planning interventions and curriculum writing

1. Proposed interventions will be reviewed by the PAG and management leadership team.
2. After appropriate interventions based on the needs assessment have been approved, PEF and the Director of Staff Growth and Development will develop new materials to be combined with portions of existing programs including
   a. OMH's "Safety Training for Mental Health Workers in the Community"
   b. OMH's "Crisis Management"
   c. OMH PC 701 "Patient Care—Restraint and Seclusion"
   d. OMH's OM 410 Policy on "Response to Traumatic Events"

Results

The needs assessment included analysis of environmental, organizational, and clinical factors relating to patient assaults on staff. These analyses were used to develop recommendations to reduce these injuries and conduct a 2 day team-training program.

A PAG was organized including appropriate management, departmental, union, and patient representatives to guide the project. Although rich input was obtained from line staff through focus groups, surveys, and direct observation, participation of direct care workers in the project advisory committee meetings was inconsistent. The result was that the PAG was not well represented by the staff who are most highly affected by patient assaults. This reduced the effectiveness of the PAG in reviewing, prioritizing, and implementing appropriate interventions.

Focus Group Report

Focus groups were conducted by a Union Health & Safety Specialist and an Injury Epidemiologist and involved 20 staff from the two units.

The purpose of the focus groups was to stimulate discussion among staff members to identify their views on factors relating to assaults on staff. This was not meant to obtain objective data. Rather, the purpose was to obtain rich feedback from affected workers. The responses were used to design a formal survey:

1. Clinical or treatment concerns
   a. Consider infection control during assaults
   b. Patient characteristics that lead to assault
      - Self-abuse
      - Verbal abuse, cursing, gruesome stories
      - Near misses, e.g., "I'm going to get me a blonde," resulted in patient-on-patient assault and could have been blonde staff member
      - Displaced anger—"Take everything out on you because you were there"
      - Threats/fist up
      - Leaving without permission/inappropriate behaviors when LOWC
      - Physiology of drugs such as nicotine and caffeine
      - Harassing phone calls
      - Patients who refuse discharge
      - Wheeling and dealing in contraband
      - History of assaults
      - Sexual preoccupation or recall of past abuse
   c. Medication problems
      - Blood drawings for medication surveillance may be refused or difficult
      - Periods of medication change may lead to hostile or wild behavior*
      - Caffeine and nicotine are drugs and may change behavior
      - Patients in later shifts may be undermedicated, or overstimulated from contraband caffeine
   d. Smoking problems
      - Building wide restrictions—why are they in place?
      - Cigarettes become contraband or are stolen or sold
      - Illegal smoking in rooms and bathrooms have caused fires
      - Panhandling of cigarettes occurs
      - Difficult to monitor outdoor smoking privilege

* In one of the PAG meetings, it was revealed that a federal court ordered a reduction in the use of medication. Members of the PAG felt that this was leading to instances of under use of medications.

e. Environmental concerns
    - Centralized nursing stations become targets
    - Back of ward is cul de sac, staff could be isolated
    - Lockable bathrooms during days that can be wedged shut by patient's body
    - Removable ceiling tiles, ideal spot for contraband and weapons
    - Room decor items (pictures, balloon valences) ideal spot for contraband and weapons
    - Day room has no lockable TV control or TV plexiglass shield

f. Administrative concerns
    - Problem patients sent to one ward
    - Change of shift duties
    - Off-shift scheduling
    - Retired workers used for vacation fills
    - Female staff on male ward
    - Need more staffing in day room and smoking excursions outside
    - Pressing of charges against assailant difficult and inconvenient
    - Incentives for behavior change require money to prevent panhandling, etc.

2. Suggestions for prevention of assault
    a. Identify past experience
    b. Identify potentially assaultive patients and communicate the information to other workers, other shifts
    c. Spread problems out or decrease census on problem wards
    d. Programming on ward: smaller groups on program; go out on excursions
    e. Restriction of privileges
    f. Use of wrist alarms: better ways to retain wrist alarms
    g. Code Greens successful in limiting assault, but arrive too late to prevent injury
    h. Increase level of knowledge: train retired staff and those who fill in for vacations
    i. Survey workers: use payday sign-up, use nurse administrator to encourage surveys being completed
    j. That assaultive behavior be reviewed for changes
    k. Assaults and damage to property (punching holes in mirrors, etc.) has consequences
    l. Map risk areas
    m. Training needs
        - Repeat behavior management training
        - Verbal skills: talking down violence
        - Crisis management
        - Returning or weekend workers
        - Teamwork: building teams or liaison with doctors
        - Predicting violent behavior
        - Personal self-protection
        - Communication

Observation of the Units

Contract staff visited the units on nine occasions for 1–3 h to observe patient–staff interaction and to informally interview staff and patients regarding their views on the problem. It was very valuable for the project staff to observe the climate on the wards first hand, including several outbursts and a few violent episodes. These visitations revealed that staff perceptions regarding the causes of workplace violence were often quite different than those of administrators. Staff focused on issues such as appropriateness of patient programs to their cultural experience, unit rules, and denial of privileges to patients caught smoking in the units. The latter would result in smoking addicted patients, many with a violent history, being denied smoking privileges for 24–48 h. Administrators, on the other hand, were more concerned with teamwork, communication, and individual staff skills.

Survey

A survey of staff on Wards 75 and 76 was conducted in November 1996 after the PAG approved the survey instrument (Appendix 7.A.4). Questions were gleaned from the focus groups and observations and from using an instrument designed in the Washington State study.[9] The surveys and a cover letter describing the project were given to staff by the nurse administrator on each ward who also collected them. Names were omitted to maintain staff confidentiality and the evaluation of responses was done by an independent consultant. Of the staff, 30 were eligible for the survey and 27 were actually returned, a 90% response rate.

Participation on the two wards was evenly divided, with 13 responses from Ward 75 and 12 from 76: 11 respondents worked on the day shift, 6 on evenings, and 8 on nights. Of those surveyed, 15 were MHTAs, 8 were RNs, 2 had other job titles, and 2 did not list their titles.

Of the staff, 25 reported that they had been assaulted at work by patients: 93% of all respondents. Of those 25 staff, 6 reported being assaulted once, 8 were assaulted twice, and 2 reported being assaulted three or more times in the past year, 10 reported no assaults during that time period, and 1 listed an "other" response. The majority of staff reported wearing a wrist alarm either at all times (10) or occasionally (10). However, 6 reported that they never wore a wrist alarm, and 2 did not answer the question.

When asked to rate the promptness of an emergency response, 11 reported that the response was always or usually prompt, and 14 reported that the response was sometimes delayed. An additional question was asked about previous training in Managing Crisis Situations (MCS). This question was eliminated from the analysis since it was determined that a number of the respondents had confused this type of training with the annual right-to-know training.

There were three additional components to the survey. The first part was a list of factors that might contribute to work-related assaults. Respondents were asked to choose their top five concerns and to prioritize them (see Figure 7.5). The scores for the top five concerns were weighted to give the highest priorities the greatest weight. The top five concerns were

1. Adequate numbers and deployment of personnel
2. Identify patient with potentially assaultive (e.g., agitated) behavior
3. Staff training in self-defense and/or restraint procedures
4. Staff clinical and interpersonal skills
5. Legal penalties for competent assaultive patients

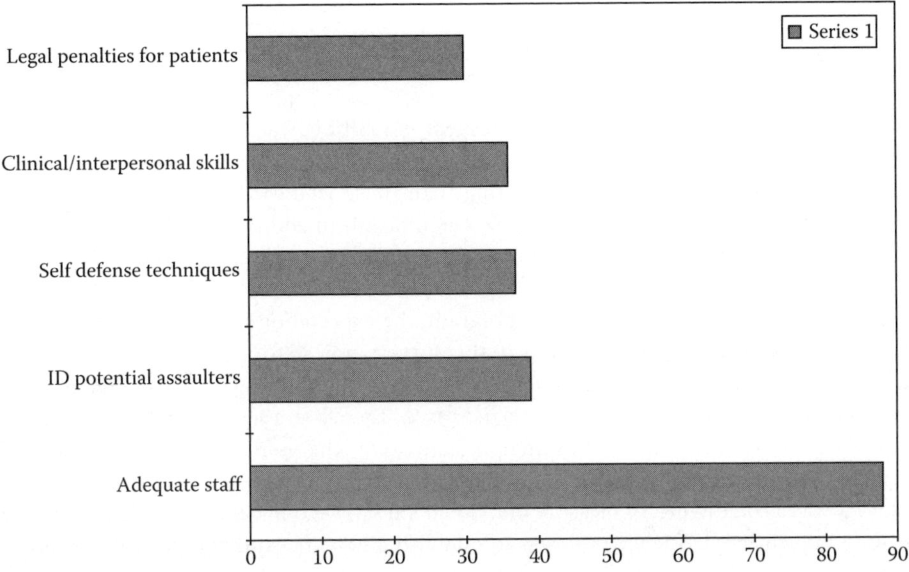

FIGURE 7.5 Top five concerns of workers.

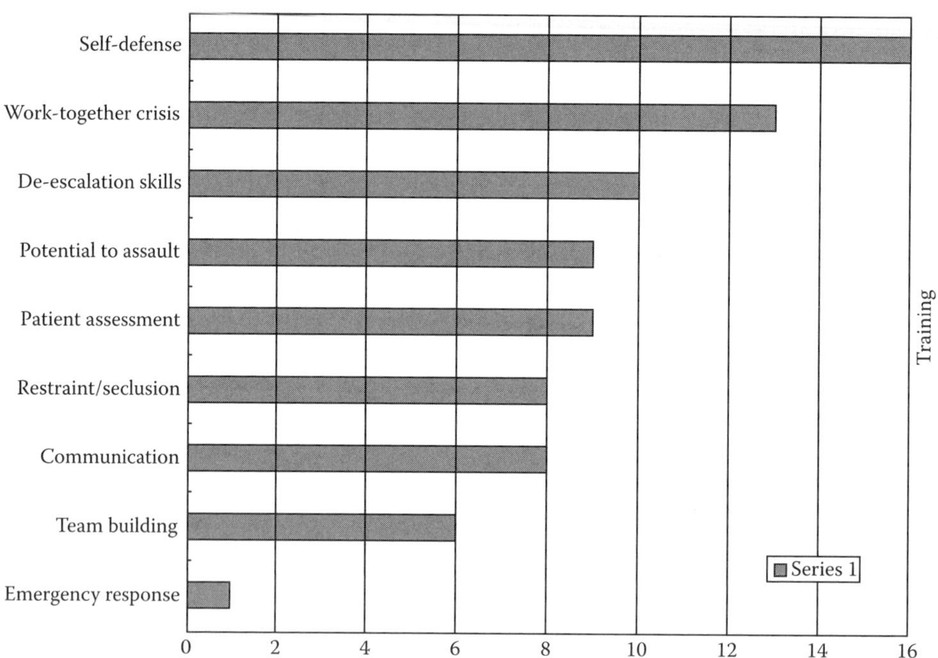

FIGURE 7.6 Training staffers would like to have.

The results of this component were consistent with the feedback from focus groups that took place on both wards.

The second portion of the survey related to the types of training programs that staff would like to see at BPC. A variety of types of training were listed and staff were asked to select as many as they felt would be appropriate (see Figure 7.6). This information was gathered to assist in tailoring the grant-related training sessions. The top five topics selected for training programs were

1. Self-defense
2. Working together in crisis
3. Verbal de-escalation
4. Policies on patient assessment for violence
5. Assessing potential assaultiveness

Data in this section were also analyzed with respect to job title, ward, and shift. Training interests were similar throughout, except for two instances. Those who work on the evening and night shifts were more likely to request self-defense training than those who work on the day shift. While the overall sample size was small, this difference was significant. In addition, staff on Ward 75 were significantly more likely to request training on working together in crisis than those individuals who work on Ward 76.

When evaluating a new health and safety program, the expectation is that the number of episodes will decrease. Research, as well as the results of this survey, indicates that reporting of assaults is usually well below their actual occurrence. When staff participate in training sessions to reduce assaults, a frequent result of their heightened awareness is increased. Therefore, a successful program may initially see increased cases reported. In the Washington State Study, the numbers of cases did not decrease. However, the amount of lost-time was reduced significantly.

The survey was also designed to measure individual job satisfaction before and after the recommendations were implemented. The purpose was to establish a baseline so that a postintervention survey could measure the difference in worker satisfaction between the two time periods. In this section of the survey, staff were asked, for example, to rate how satisfied they were with supervisors and administrators, and how satisfied they were with their job in general. In addition, staff were asked to rate their present level of health. Analysis by shift, ward, and job title showed similar responses between groups. Of those surveyed, 96% reported that their supervisor was somewhat willing or very willing to listen to work-related problems. Additional results indicated that between 70% and 80% of the staff reported being either somewhat or very satisfied in the following areas:

- The way supervisors and administrators treat workers
- The competence of supervisors
- Being able to turn to fellow workers for help when something is troubling
- The responses of fellow workers to displays of emotion such as anger, sorrow, or laughter
- Acceptance and support of new ideas and thoughts by fellow workers
- Overall job satisfaction

The level of satisfaction with the amount of praise one got for doing a good job was slightly lower, with 60% of respondents somewhat or very satisfied. The level of satisfaction with the way policies we implemented was even lower, with only 45% of the staff feeling either somewhat or very satisfied in this area.

While 57% of respondents would not recommend their job to someone else to some degree, 77% of those surveyed would be either somewhat likely or very likely to take their job again. There was a significant difference between staff working on different shifts for this question. While the responses of day shift workers were distributed fairly evenly between not at all likely, not too likely, somewhat likely and very likely to take this job again, evening- and night-shift workers almost exclusively reported that they were somewhat likely to take their job again.

Individuals were then asked to rate their level of exhaustion following a day of work. While 36% reported being seldom exhausted, 64% stated that they were often very tired after work. Of respondents, 60% self-reported either good or very good levels of health and 70% stated that their health was either good, very good, or excellent in comparison to the health of others. In the area of self-reported health, however, MHTAs were more likely to report either fair or very good health, while RNs reported evenly distributed levels of fair, good, and very good health.

Environmental Assessment

Background

The environmental assessment of Unit 76 was conducted by a Union Industrial Hygienist, the author of this chapter. Because the layout of the two units is identical, the recommendations were applied to both units.

An additional component of the environmental assessment involved Kevin Murrett, AIA, Principal of Architectural Resources, a firm with a history of working with OMH. He conducted an inspection on Unit 75 on October 9, 1996 and his report and recommendations were consistent with the original findings.

The methods used in the inspections were visual observation and worker interviews. At the time of the initial visit there were three MHTAs and one nurse on the ward.

Observations and Worker Comments

Doors

1. Many of the doors to the units were opened and locked with skeleton-type keys (Figure 7.7). This system required an individual to manually turn the key to lock the door behind them. Historically, skeleton keys were used so that, in case of fire, a worker could quickly identify the proper key to quickly allow people to escape harm.

   Recommendation: Security could be increased by replacing skeleton key locks with "passive" mortise door locks, where doors lock automatically when closed. The issue of fire safety could be accommodated by using color-coded keys.

FIGURE 7.7 Skeleton key locks require workers to "remember" to manually lock them.

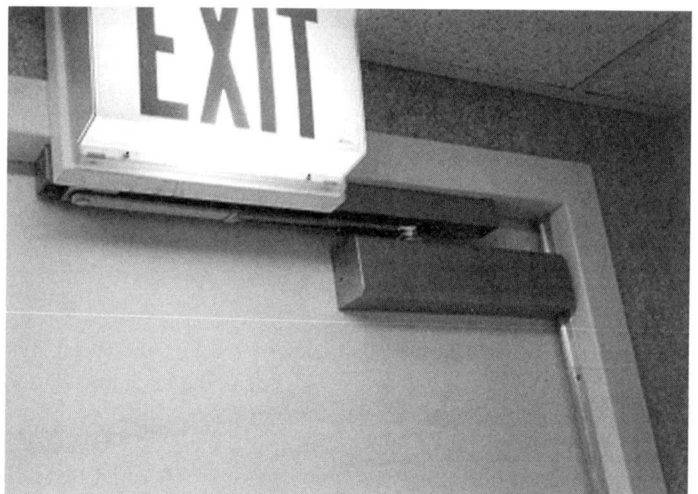

FIGURE 7.8 Typical door closer.

2. A worker reported that the closers on the doors in the unit were sometimes very slow to close (Figure 7.8). It was also mentioned that, in one instance, an agitated patient attempted to force his way into a utility closet, where a nurse was attempting to push the door closed so that it could be locked.

   Recommendation: Institute a preventive maintenance program for door closers.

Ceiling Tiles

3. Ceiling tiles in the corridors are easily removed and patients have put contraband above them (Figure 7.9). Staff related that (a) underwear soiled in feces, pieces of metal, a suitcase, coffee, and cigarettes have been found above ceiling tiles; (b) in one instance a patient tried to commit suicide

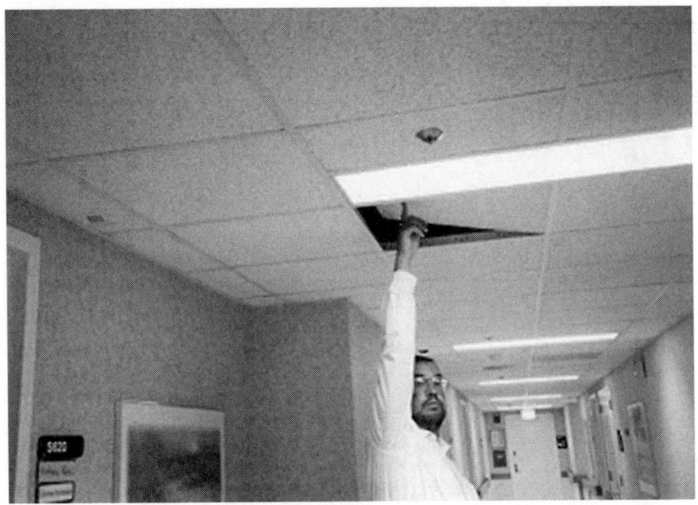

FIGURE 7.9 Tiles are easily removed for hiding contraband.

by hanging from the aluminum suspension grid; and (c) clips used to hold the ceiling tiles in place have been used as weapons.

Recommendation: An engineering/architectural evaluation should be conducted to design a method to secure the ceiling tiles.

Personal Safety & Security

4. Staff reported that when personal alarms were activated security announced "Code Green" emergencies over the PA system which initiated support from outside the unit. However, many staff members do not wear personal alarms routinely.

   Recommendation: Staff should be educated about the importance of wearing personal alarms and required to use them.

5. The back hallway in-between the corridors next to the mechanical equipment room provided a potential hiding place.

   Recommendation: A plastic convex ceiling mirror could be strategically placed to increase visibility and eliminate blind spots in that area (Figure 7.10) (see Kevin Murrett's report for an engineering solution to this problem).

6. The store room in the back corridor was an isolated area and needed to be controlled (Figure 7.11).

   Recommendation: BPC should consider implementing a policy requiring a minimum of two staff to enter this area and other team members should be informed of coworkers whereabouts.

Smoking

7a. Staff reported that many of the incidents on the ward involve cigarettes and smoking. Since there was no smoking in the Strozzi, a staff member would have to escort patients who smoke out of the building every hour. This resulted in fewer staff on the unit to deal with patient needs or to respond to a crisis.

   Recommendation: Use of the specially ventilated day room for patient smoking should be reconsidered.

7b. Staff reported that when patients were caught smoking in bedrooms or elsewhere on the unit, hospital policy was to deny them smoking privileges for a period of time. This often caused

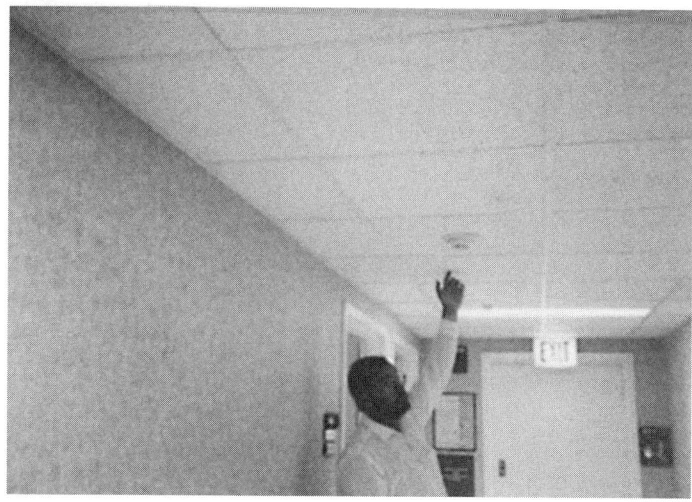

FIGURE 7.10 Approximate location of proposed convex mirror ceiling tile.

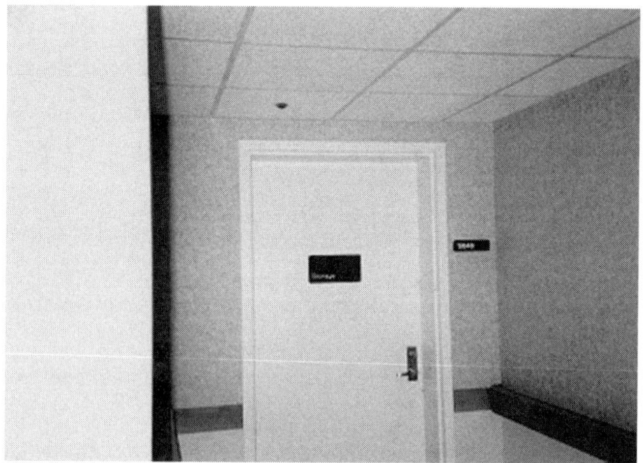

FIGURE 7.11 Door to isolated back store room.

smoking-addicted patients to become agitated and even assaultive. One staff member reported that a patient threatened to kill a nurse who caught him smoking in his bedroom.

Recommendation: Hospital management should consider alternative consequences for patients caught smoking which will not lead to increased patient agitation and potential assaultive behavior.

Decorations and/or Worksite Design

8. The decorative pictures on the corridor walls could easily be removed and used as a weapons. In fact, we observed that a picture had been removed (Figure 7.12).

   Recommendation: A more secure method of hanging pictures is to use toggle bolts and Plexiglas.

FIGURE 7.12 Pictures can be pulled from the walls.

9. The design of the nurses' station (Figure 7.13) sometimes contributed to patient/staff tension in that patients could easily access the space by "jumping" over the structure or reaching over to grab things. Also staff could not observe activities in the day room when seated at this station.

   Recommendation: Consideration should be given to redesigning the nurses stations or modifying them to provide greater security and increase visibility of patients in the day room (also see recommendations in Kevin Murrett's report).

10. The telephone in the nurses station was not secured and had been a source of contention between staff and patients.

    Recommendation: The phone should be installed on the wall or under the desk to eliminate this source of patient/staff discord.

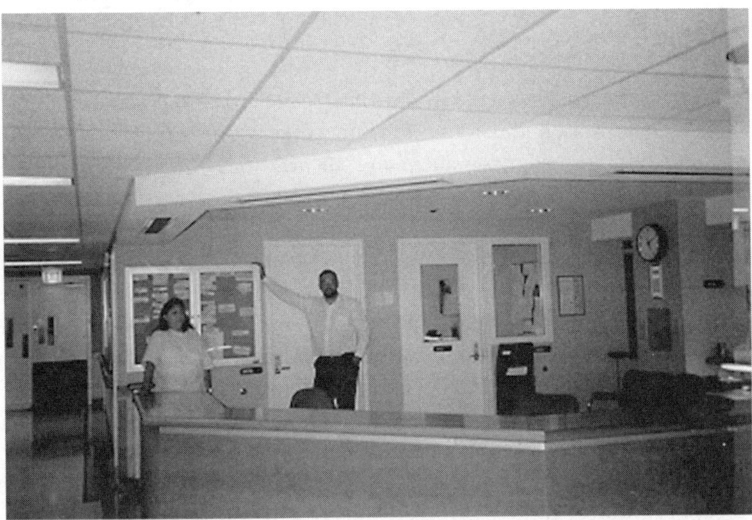

FIGURE 7.13 Nurses station.

Managing Crisis Situations

11. Regarding MCS Training, one staff member reported never having been trained during 9 years of employment. Others reported viewing a short video a couple of years ago. The sister facility, Binghamton PC, had mandatory MCS training in place and experienced considerably less lost-time injuries in 1995.

    Recommendation: BPC should institute mandatory MCS training with annual updates, especially for workers in units that are experiencing a high number of incidents.

12. A staff member reported that a "code green" which involved a particular patient occurred on day shift and the staff on evening shift never were informed what happened. Consideration should have been given to improving cross shift-communication, especially regarding escalating patient aggressiveness.

    Recommendations: (a) An incident log to transfer information between shifts regarding patient agitation and/or assaultiveness should be tried on a pilot basis; (b) any "code greens" or assaults should be reported to oncoming staff during change-of-shift meetings; and (c) an attempt should be made to hold more treatment team meetings at times when representatives from across the three shifts may attend.

Program

13. Staff reported that forcing patients to go to programs sometimes caused them to become aggressive. Also, many of the programs that were available frustrated patients because they would jibe with their interests and cultural backgrounds. When asked what kind of programs would be more suited to patients on that unit, staff suggested auto mechanics or wood shop.

    Recommendation: Consideration should be given to providing programs that match patient interests and backgrounds.

14. During the visit we observed a hand broom that had been left in the corridor and a can of disinfectant spray in a patient bathroom. These objects could be used as weapons and should not have been left around the unit.

    Recommendation: Include housekeeping staff in training programs and emphasize that they can contribute to preventing patient/staff injuries. Have oncoming staff conduct a 5 min visual inspection of the unit to check the ward environment at the beginning of each shift.

Training

Staff (Figure 7.14) were trained on two separate days in December 1996 and January 1997. To emphasize the goal of promoting teamwork and communication, workers from all three shifts trained together. The agenda was tailored to the needs assessment and the priorities identified through worker surveys. An expert trainer with over 20 years of experience in mental health and the criminal justice system helped conduct the sessions together with Union Health & Safety and facility Staff Development personnel. A risk-mapping exercise allowed workers to illustrate, on a schematic of the unit, where assaults had taken place.

The training involved teaching cognitive knowledge about pathways to violence, the findings of the project team (focus groups, surveys, and environmental assessments), and nonverbal, verbal, and physical skills in preventing violence. The program began with a strong statement of management commitment to the process from the deputy of administration. The training was concluded when the participants were divided into three groups in which they outlined their top five concerns that were presented to the facility director and the top cabinet officers. Although there was a considerable amount of venting, this method allowed workers to directly identify their priorities for addressing the violence problem with their facility leadership.

FIGURE 7.14 Staff workers.

Training Evaluation

Results of written evaluations by attendees of Day 1 are shown in Tables 7.5 and 7.6.

Buffalo Project Conclusion

The Project Team advised that the recommendations should be fully considered and prioritized by BPC's leadership with appropriate input from all stakeholders. To that end, the PEF Health and Safety Department offered its support in the following ways:

- Providing a presentation on the project and its results to BPC's leadership
- Assisting with follow-up surveys and training sessions
- Helping to facilitate expansion of the project to other units as deemed appropriate

Summary of Recommendations and Best Practices

The following recommendations and best practices were based on the assessment phase of the project. Some of the recommendations were based on innovative approaches that were implemented at sister OMH facilities. The project team acknowledged that the interventions needed to be considered in terms of their economic and technical feasibility. Although, some of the steps could be taken right away, others

TABLE 7.5 Unit 75

Evaluation Criteria	Score	Possible Score
Training useful to me	3.3	4
Materials relevant	3.2	4
Presentation well-organized, understandable	3.4	4
Sufficient discussions/ interacting time	3.3	4
Trainer(s) evaluation	4.58	5

Note: Attending, 19; Evaluating, 19 (100% response rate).

TABLE 7.6 Unit 76

Evaluation Criteria	Score	Possible Score
Training useful to me	3.5	4
Materials relevant	3.5	4
Presentation well-organized, understandable	3.6	4
Sufficient discussions/ interacting time	3.6	4
Trainer(s) evaluation	4.43	5

Note: Attending, 14; Evaluating, 14 (100% response rate).

would require further study and experimentation. The intent of the recommendations were to reduce the number and severity of violent incidents by creating a safer and more therapeutic hospital environment and by better preparing staff to recognize and anticipate patient needs.

Summary of Recommendations

1. *MCS training should be mandatory for all staff who have patient interaction, especially in units that experience significant numbers of assaults.* Training staff on methods of preventing crisis through nonverbal, verbal, and physical interventions is essential to reducing the number of incidents and the severity of incidents. Interestingly, Buffalo reports fewer incidents than its sister facility Binghamton PC. However, Binghamton, which requires MCS training, has fewer lost-work days.
2. *Annual updates on MCS training should be mandatory for all staff who have patient interaction, especially in units that experience significant numbers of assaults.* Annual updates will help to ensure that staff enact the methods and practice them. This will help staff to deal with patients needs in the most sensitive manner and to avoid incidents.
3. *Environmental recommendations are listed in the environmental assessment report (attached) and should be prioritized and fully considered.* It appears that many incidents involve smoking, therefore, these concerns warrant special attention.
4. *Emergency responses (code greens) may be improved by using a system of dedicated teams, trained in this function.* Rockland Children's PC has established such a system. Currently, BPC uses a system where all available staff are supposed to assist during a code green. Problems include response time and people from outside the unit who are unfamiliar with the patients. The dedicated-team approach involves a weekly assignment of an emergency-response-team leader and an adequate number of responders. Rockland also uses an emergency-response kit which includes medication and gloves.
5. *All staff should be required to wear personal alarms.* The survey and observations indicate that a significant number of staff do not wear their personal alarms. There were problems with insufficient alarms being available. Rockland Children's PC has installed a state-of-the-art system which lights up a board in the safety office showing the exact location of an incident. It appears that the BPC system needs upgrading regarding its use of alarms, response time, and its ability to pinpoint incidents.
6. *Communication may be improved by holding some team meetings between shifts.* Clinicians should also meet with direct care staff to discuss particular patients treatment plans and needs, especially those who are assaultive and display a tendency toward a manipulative cognitive path towards violence.
7. *Assessment of patient violence should be routine, especially on high-risk units.* Intervening in Violence in the Psychiatric Setting should be considered for use, especially in high-risk units. (A copy of this instrument is attached as Appendix 7.A.8).

8. *Consider using a violence calendar and violence log.* Middletown PC has instituted these systems facility-wide under the direction of the Hospital Director. The interventions were developed by a QAI team and the program is administered by the staff development specialist. The program includes a "Credo" for patient and staff to establish baseline values on violence, respect, trust, caring, responsibility, and fairness. The calendar and log are simple, not time consuming, and the information they provide is used in team, community, and staff meetings. Call the staff development specialist to request a presentation on this innovative program.
9. *Develop an on-the-job system of mentoring direct care staff on the skills and techniques learned in the classroom.* Mentoring on the unit by team leaders, other professionals, and peers who are expert at diffusing crises is essential to fully implementing the methods taught in training sessions. Implementation of a mentoring system should be considered.
10. *Institute a postincident support program for staff.* OMH has issued a policy directive requiring all facilities to develop such a policy. The purpose is to provide emotional support and counseling to staff who have been assaulted or seriously injured. Prompt medical management of staff injuries and appropriate counseling and debriefing of injured workers, coworkers, and patients should be in a written policy and implemented hospital-wide. To assure confidentiality, counseling and debriefing should only be conducted once it has been determined that no disciplinary charges will be made against an affected staff member.
11. *The methods and training developed in this project should be extended to other units at BPC, especially those with high assault rates.*
12. *Follow-up evaluation of this project should be conducted to track changes in incident rates, lost-time, and worker satisfaction.* PEF will assist in conducting a follow-up survey of staff on units 75 and 76 and also reviewing accident and injury data to help measure the program's effectiveness.

Binghamton Psychiatric Center

In response to the distribution of the OSHA Violence Guidelines by the Multi-Union Health & Safety Committee, the Binghamton PC Health & Safety Committee established a subcommittee to compare its policies with OSHA's recommendations. Additionally, the committee, at the request of the Local PEF leader invited this author and central agency management to visit the facility. The visit began with a meeting which included the director of the facility, her top staff, union representatives, and direct-care workers. When the director inquired about the scope of the visit, the union proposed establishing a joint project team that would analyze violence on two units and develop appropriate interventions. The facility director and local union leaders quickly embraced the proposal.

The, "Safe Unit Environment Project" focused on the secure Unit 95 and the Admissions Unit 97. Methods used in the Buffalo project were repeated at Binghamton. The committee established a goal:

To enhance the work environment by substantially reducing the causes of serious injury to both patients and staff.

Injury Data Analysis

In contrast to Buffalo, more detailed injury data was provided in the Binghamton project that allowed for a more thorough analysis of injury trends.

A first step in any injury prevention project is to collect and analyze available data to see if distinct injury patterns exist. Once a pattern is recognized, targeted interventions can be developed. Data sources used in this analysis include Binghamton PC assault and restraint injury reports, Occupational Injury Reporting System (OIRS)* statewide reports, and staffing information reported by the facility

* OIRS captures all reported injuries including those with and without lost-time. However, it is well accepted in the research literature that most occupational injury and illness data systems tend to underestimate incidence due to inconsistent reporting. Threats, for example, are not routinely reported and/or recorded in OIRS.

management. The statewide rates were calculated using OIRS data from April 1, 1995 to December 31, 1995. The rates for units 94 and 97 were calculated using Binghamton PC data from June 1, 1995 to September 30, 1996. Rates are used to allow for a comparison between facilities employing different numbers of workers.

Comparison with Other Facilities

A look at statewide injury rates revealed how Binghamton compared to other facilities. Figure 7.15 displays the estimates for each of the 29 facility's rates during the 1995 calendar year. It was based on *all reported injuries,* including those with and without lost-time. Binghamton PC ranked 10th highest of 29 facilities statewide, with a 40.6 per 100-reported injury rate.

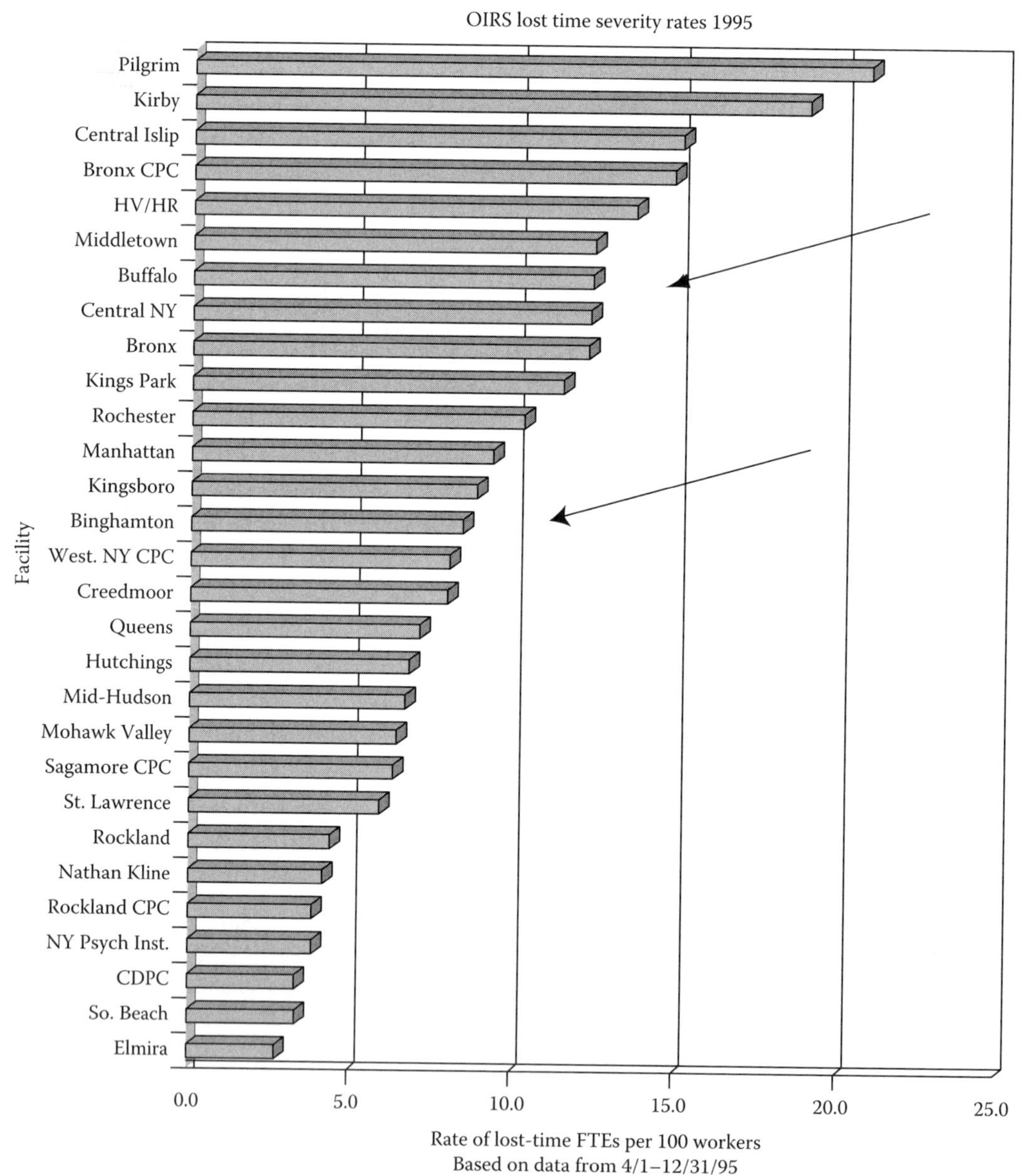

FIGURE 7.15 Statewide injury rate estimates for 1995.

These rates represent the number of cases per 100 workers during 1995.

Figure 7.16 shows the rates of *lost-time injury cases* for the 29 facilities. Binghamton ranked 22nd with a 6.5 per 100 lost-work-time injury rate. The final comparison shown in Figure 7.17 shows *lost-time severity* rates and Binghamton ranked 14th with an 8.5 per 100 rate. The last two measures are most significant in that lost-time injuries represent the greatest pain and suffering, worker's compensation costs, and disruption to staffing and operations. The *lost-time case rate* is calculated by taking the number of lost-time injury cases divided by the number of full-time workers per year. The *lost-time severity rate* showed that the number of workdays lost to injury at Binghamton in 1995 was equivalent to the time of 55 full-time employees. This rate is used to estimate the equivalent number of workers lost to injury per year and is calculated by dividing the number of lost work days by the number of full-time workers.

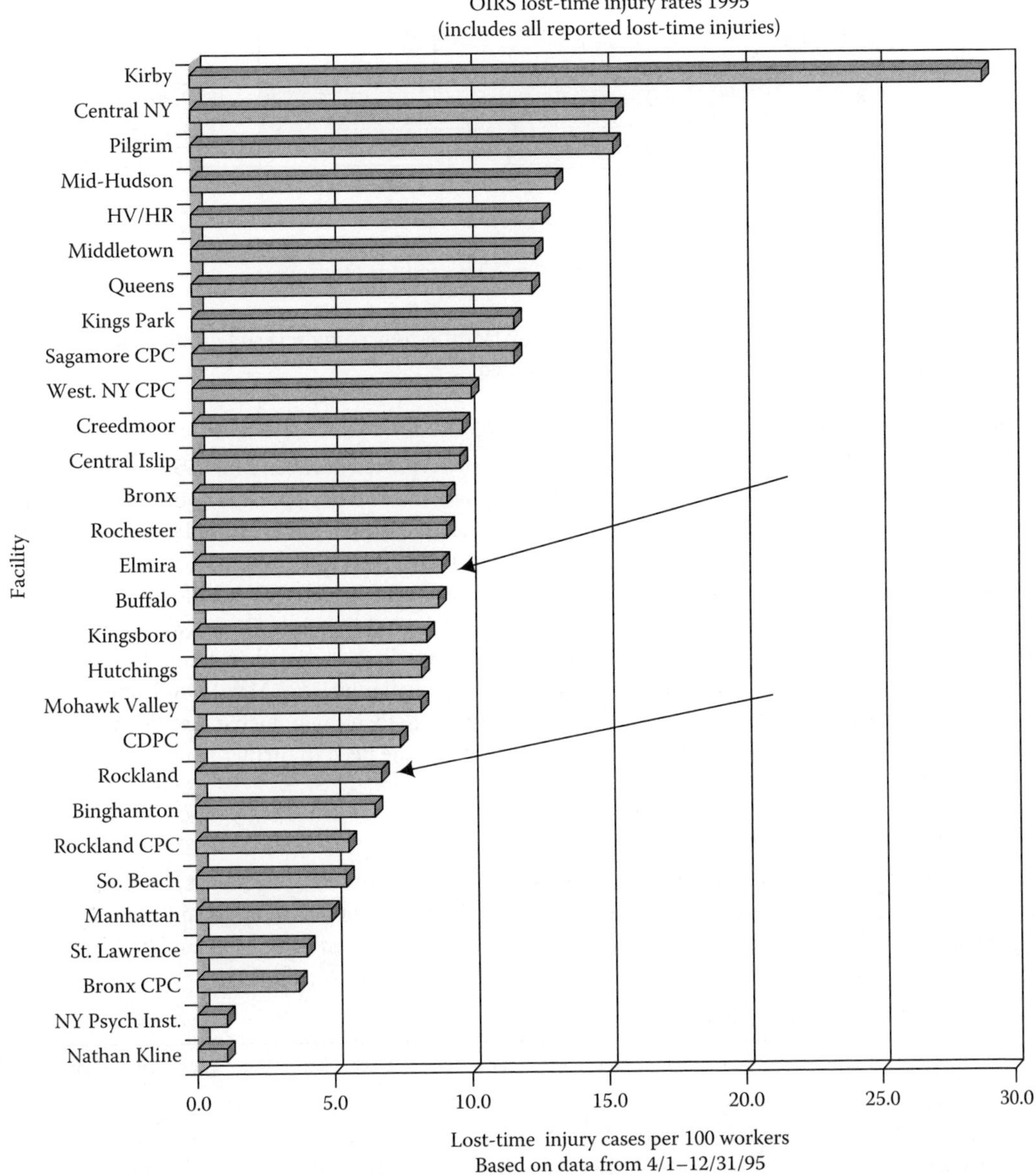

FIGURE 7.16 Lost-time injury rates by OIRS for 1995.

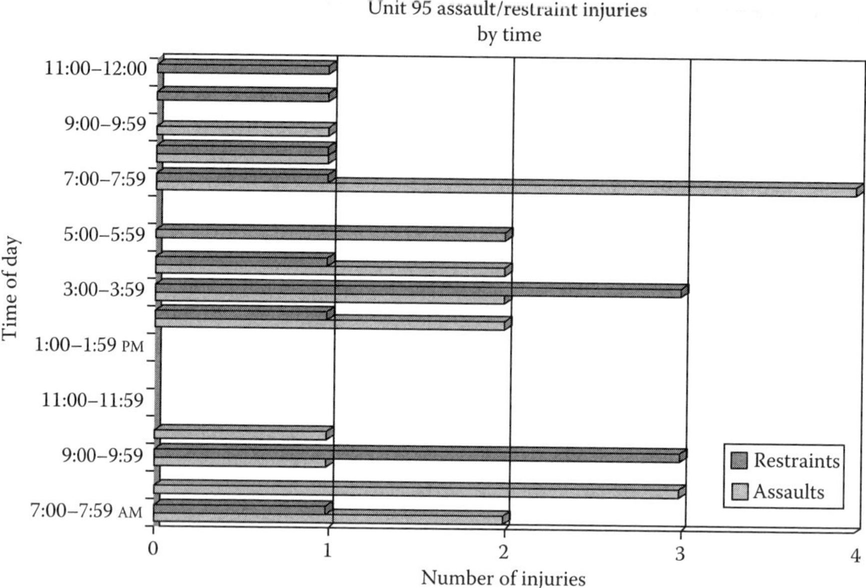

FIGURE 7.17 Lost-time severity rates by OIRS for 1995. *FTE = full-time employees.

While Binghamton is doing much better than many facilities regarding severity of injuries, continued efforts to further reduce injuries were clearly warranted.

Assault and Restraint Injuries

According to the Binghamton data, there were 290 injuries reported between May 1995 and May 1996. Of these injuries, 33% (95) were caused by restraining patients or from patient assaults. The monthly number of restraint and assault injuries is displayed in Table 7.7. The numbers varied greatly by month and although assault injuries increased between December 1995 and April 1996, it does not appear that the difference represents a trend or seasonal variation. The average number of assault injuries over the year is 4.77/month, and the average number of restraint injuries is 2.5/month. These two causes contribute more to injury in the facility than any other cause. The contribution varied on a monthly basis from a high of 46% of all injuries in February 1996, to a low of 6% of all injuries in May 1996. These data supported the decision by management and labor at Binghamton to focus on reducing assault and restraint injuries. Furthermore, the randomness of these data enforced the need to further study the increase in assaults between December 1995 and April 1996.

Assault and Restraint Injury Rates in Units 95 and 97

There are eight patient units in the Binghamton PC. Units 81, 85, and 89 are geriatric, units 2, 3, and 4 are adult, 97 is admissions, and 95 is a secure unit. Over the 15 month period between June 1, 1995 and September 1, 1996, there were 44 assault and restraint injuries reported in units 95 and 97 involving a

TABLE 7.7 Reported Injuries by Cause and Month

1995–1996	May	June	July	Aug.	Sept.	Oct.	Nov.	Dec.	Jan.	Feb.	Mar.	Apr.	May
Restraint (R)	6	5	2	2	1	2	4	2	2	2	2	3	0
Assault (A)	4	5	2	4	4	3	4	6	7	11	5	6	1
Others	19	16	17	11	11	16	15	11	17	15	17	14	16
% R & A	34	38	19	35	31	24	35	42	35	46	29	39	6

staff of 35. Of these injuries (23 per 100), 10 occurred in Unit 97 which had 18 employees, and 34 of the injuries (77 per 100) occurred in Unit 95 which had 17 people. When the number of assault and restraint injuries in each unit was divided by the number of staff employed by that unit and annualized, an annual assault and/or restraint injury rate was determined. The annual injury rates for units 97 and 95 were 44 and 160 per 100, respectively.

An annual injury rate of 100 represents that, on average, every employee will experience an injury during the course of the year. Therefore, a rate greater than 100 represents that more than one injury per employee is occurring annually.

These data indicated that the injury rate on Unit 95 was significantly greater than on 97. To properly focus preventative efforts, it was recommended to conduct a comparative analysis of all Binghamton PC units assault and restraint injury experience.

Assault and Restraint Lost-Time Injuries

Of the 44 assault and restraint injuries reported from units 95 and 97, 5 (11%) resulted in lost-time: 1 from Unit 97 and the other 4 from Unit 95. Three were caused by assaults and two from restraining.

The lost-time restraint/assault injury rates for units 95 and 97 were 24 and 6 per 100, respectively. This compares to an overall rate of 6.5 per 100 facility wide.

Two lost-time assault injuries occurred in February 1996, the month with the most injuries. Further investigation revealed that one problematic patient was responsible for multiple assaults on staff during this period.

Units 95 and 97 Assault and Restraint Injury by Shift

When analyzing each of the units by shift over the same 15 month period, more injuries were reported during the day shift and none occurred during nights (Tables 7.8 and 7.9):

Assault and Restraint Injuries by Times of the Day

The times of the day when these injuries occurred have been separated by unit in Figures 7.18 and 7.19. The time periods which appeared to have increased incidents were

Unit 95	Unit 97
7:00–9:00 AM	10:00–11:00 AM
2:00–5:00 PM	1:00–2:00 PM
7:00–8:00 PM	

TABLE 7.8 Unit 97—Admission

Shift of Injury	Assault Injuries	Restraint Injuries	Total Injuries
Day	4	4	8
Evening	1	1	2
Night	0	0	0
Total	5	5	10

TABLE 7.9 Unit 95—Secure Unit

Shift of Injury	Assault Injuries	Restraint Injuries	Total Injuries
Day	11	8	19
Evening	8	7	15
Night	0	0	0
Total	19	15	34

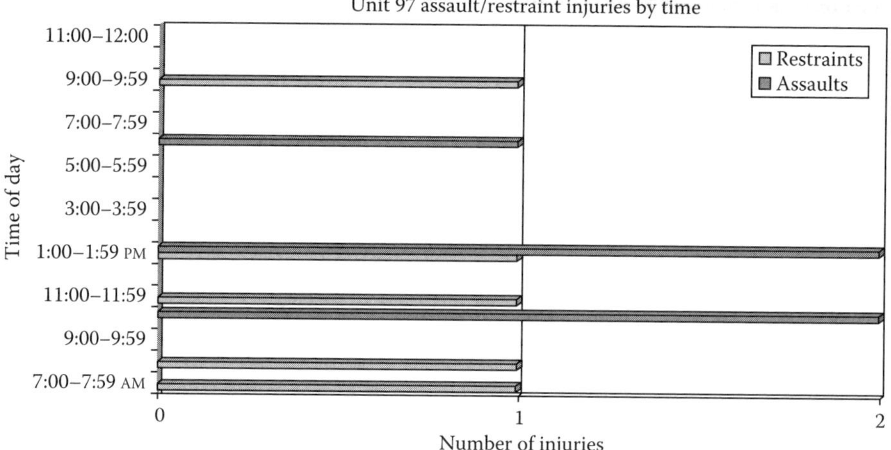

FIGURE 7.18 Unit 97—Assault and restraint injuries by time.

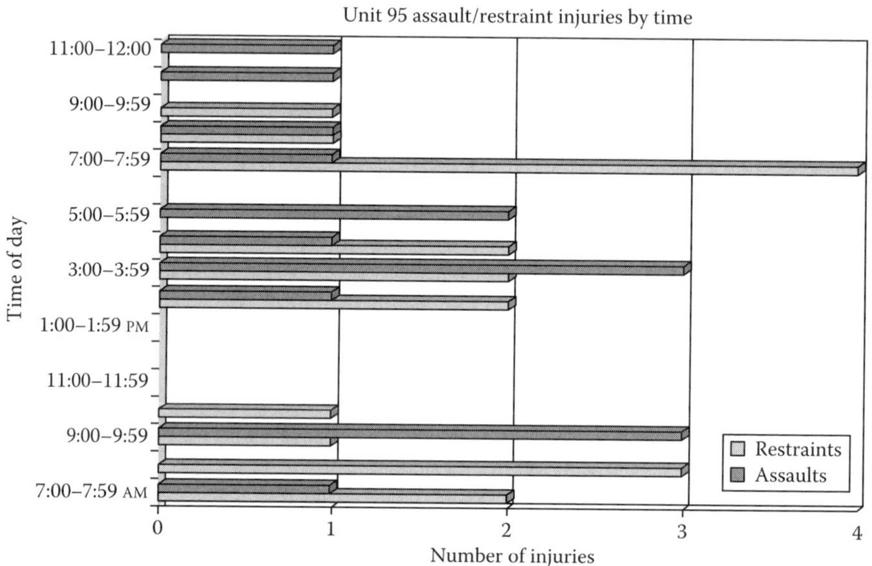

FIGURE 7.19 Unit 95—Assault and restraint injuries by time.

Further investigation may reveal if there is any correlation between activities and incidents during these periods, especially with respect to activities, communication, and coordination at shift change. The investigation should include mapping where the incidents take place and characterizing predictive factors, if any exist.

Assault and Restraint Injury Rates by Job Title

Injury rates were also determined by job title by dividing the number of injuries in each title by the number of employees holding that title. Using these rates, those job titles presenting the greatest risk can be clearly identified.

The injury rate in the Secure Unit 95 is five times the facility average for RNs and LPNs at 220 and 160 per 100, respectively. The SCTAs have a rate that is more than twice the facility average at 88 per 100. In Admissions Unit 97 the nurses have a rate that is double the facility average, 80 per 100. However, the MHTAs have a rate that is lower than the facility average, 27 per hundred.

Facility Average Injury Rate, All Units: 40.6 per 100
Facility Average Lost-Time Injury Rate, All Units: 8.5 per 100

The lost-time rates reveal a similar pattern. On Secure Unit 95, nurses have a lost-time rate four times the facility average at 32 per 100 and SCTAs have a rate that is twice the facility average at 16 per 100. On Admissions Unit 97, the nurses have a lost-time rate that is somewhat greater than the facility average, 13 per 100. However, the MHTAs had no lost-time injuries during the 15 month period.

This analysis indicates that nurses are experiencing a much greater rate of injury than MHTAs and SCTAs. Further inquiry into why this condition exists is warranted. Also, efforts to understand the high rates of injury to SCTAs in the secure unit are clearly justified. An analysis of injury rates by job title, facility wide should be made to further reveal units and jobs that are at high risk and targets for intervention.

Tables 7.10 and 7.11 show the type of injury (restraint and assault) by job title. This comparison reveals that nurses are more frequently assaulted, while SCTAs and MHTAs are more frequently injured during restraining activities. This difference should be considered as we develop a prevention strategy (Tables 7.12 and 7.13).

Number of Repeat Cases

Repeat cases are illustrated in Figures 7.20 and 7.21. Of the population of 35 workers, there were 26 who reported injuries in units 97 and 95 in this 15 month, time frame. Of those who reported injuries, 11 (46 per 100) reported an injury more than once and 5 (19 per 100) at least three times. Further investigation regarding the reasons for the repeat cases may help reveal factors such as:

- Activities or tasks that are related to injuries
- Environmental conditions

TABLE 7.10 Unit 95 Secure Unit: Assault and Restraint Rates by Title

	Assault Injuries	Restraint Injuries	Totals	Number of Employees	Annualized Injury Rates per 100 Workers
RN	8	3	11	5	220
LPN	2	2	4	2	160
SCTA	4	7	11	10	88
Totals	14	12	26	17	127

TABLE 7.11 Unit 97 Admissions: Assault and Restraint Rates by Title

	Assault Injuries	Restraint Injuries	Totals	Number of Employees	Annualized Injury Rates per 100 Workers
RN	4	2	6	6	80
MHTA	1	3	4	12	27
Totals	5	5	10	18	44

TABLE 7.12 Unit 95 Secure Unit: Lost-Time Assault and Restraint Rates by Title

	Lost-Time Assault Injuries	Lost-Time Restraint Injuries	Total Lost-Time	Number of Employees	Annualized Injury Rates per 100 Workers
RN	2	0	2	5	32
LPN	0	0	0	2	0
SCTA	1	1	2	10	16
Totals	3	1	4	17	19

TABLE 7.13 Unit 97 Admissions: Lost-Time Assault and Restraint Rates by Title

	Lost-Time Assault Injuries	Lost-Time Restraint Injuries	Total Lost-Time	Number of Employees	Annualized Injury Rates per 100 Workers
RN	0	1	0	6	13
MHTA	0	0	0	12	0
Totals	0	1	1	18	4

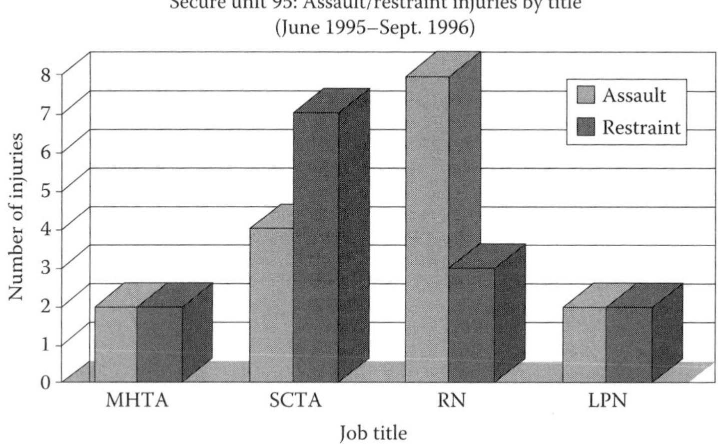

FIGURE 7.20 Secure unit 95: Assault/restraint injuries by title (June 1995–September 1996).

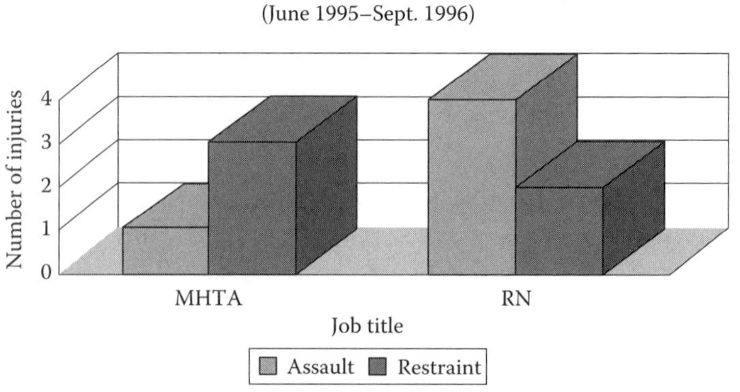

FIGURE 7.21 Admissions unit 97: Assault/restraint injuries by title (June 1995–September 1996).

- Randomness or predictability of assaults
- Effectiveness of emergency response to crisis
- Adequacy of staffing and staff deployment
- Relative levels of staff training and skills in diffusing agitated patients

Conclusion

The above analysis illustrates the scope of the worker injury problem at Binghamton PC and on units 95 and 97 in particular. The information enforces the decision by labor and management to initiate a project regarding ward safety. Combined with worker input and further study, this analysis can be used by a committee to develop a prevention strategy. The following is a summary of the recommendations made throughout the report.

Recommendations

1. Continued efforts to further reduce injuries are clearly warranted. The decision by management and labor at Binghamton to focus on reducing assault and restraint injuries is supported by the data and should continue.
2. To properly focus preventative efforts, a comparative analysis of all Binghamton PC units' assault and restraint injury experience should be undertaken. This review should include an analysis of injury rates by job title, facility wide to further reveal units and jobs that are at high risk and targets for intervention.
3. The reasons for the increase in assaults between December 1995 and April 1996 merits further study to determine if there was a trend based on seasonal attributes or particular patient and/or staff factors. Further investigation may reveal if there is any correlation between activities and incidents during these periods, especially with respect to activities, communication, and coordination at shift change. The investigation should include mapping where the incidents took place and characterizing predictive factors, if any exist. It would also be worth checking to see if specific factors contributed to a heightened level of stress on the secure unit in February 1996 when two lost-time injuries occurred; such as
    - Activities or tasks that are related to injuries
    - Environmental conditions
    - Randomness or predictability of assaults
    - Effectiveness of emergency response to crisis
    - Adequacy of staffing and staff deployment
    - Relative levels of staff training and skills in recognizing and diffusing agitated patients
4. This analysis indicates that nurses are experiencing a much greater rate of injury than MHTAs and SCTAs. The nurses are more frequently injured due to assaults, while MHTAs and SCTAs are more frequently injured during restraining activities. Therefore, it makes sense to focus on these occupations and recognize these differences while developing an intervention. Calculation of how often restraining patients leads to staff injury will also be helpful in addressing this problem.
5. Further investigation regarding the reasons for the repeat cases will also help in identifying possible solutions (Table 7.14).

Conclusion

The projects at Binghamton and BPCs are continuing. Over time an evaluation of the impact will help to assess the effectiveness of the methods used in the pilot studies. There are a number of confounding factors that will make it difficult to fully evaluate the impact of these projects.

TABLE 7.14 Units 95 and 97 Focus Group Interviews January 27 and 28, 1997

Worker Concerns	Number of Comments	Comments
Training	10	MCS mandatory for clinical and voluntary for nonclinical workers; has high rate of attendees
		MICA training is mandated
		Outflow of information from patients at discharge leads to more worker re-training
		Training generated by ex-clients that are well motivated but not typical of usual patients
		Physical safety training good but verbal skills need to be emphasized; former verbal bluffs now considered abusive
		MCS good for handling patients
		MCS training adequate—used often—more verbal for geriatric; more physical for others; mandatory for professionals
		Have MICA training
		Potentially hazardous to bring in staff from other units to take over during training due to patients decompensating—difficult to return to unit—half staff should attend classes and use regular staff left on all three shifts
		Annual training not enough
Type of patients/clients	26	Sociopathic; noisy when going to dining room
		Self-abusive
		Street persons/MICA
		Dangerous patients
		Patients are "bottom of the barrel," deinstitutionalization, and jammed facilities and gridlock in the community result in lengthier stays for those ready to be released
		Many alcohol and drug issues—patients not really mentally ill
		System is bogged
		ACLU and patient-rights issues result in patients not taking medications after leaving and entering community; arduous to get on court-ordered medications
		Single patient in 4- and 5-point restraints 45 times—generates a lot of overtime
		Patients have no support system outside the hospital
		CPLs sent for long-term care with no ultimate control other than forensic committees—have 5 now
		Taking away of choices causes aggression
		Most first-time patients, need institutionalized living orientation—should be 90 days
		Who is really psychotic, who is not
		HIV/AIDS patients have no incentive to control activities
		Right to refuse medication when actively psychotic—delay in getting court-ordered medications, may be bad for patients health; should go right to court and intramuscular and per-oral should be ordered or stat dose if assaultive
		Infection control issues with HIV-positive patients
		System is manipulated by MICA clients who pretend suicide, get SSI, then return to system after partying
		Medicating for immediate behavior is "band aid"
		Discharged patients may have assaultive potential—follow them

TABLE 7.14 (continued) Units 95 and 97 Focus Group Interviews January 27 and 28, 1997

Worker Concerns	Number of Comments	Comments
		Patients know how to verbally and physically abuse staff because they know how to use the system—too much leniency given
		Less physical activity for patients due to lock down and staffing
		No forcible medication unless imminent danger; push to get out in community where if they act out, police in flak jackets and 9 mm gms presses charges; patient sent back to psychiatric center
		A psychiatric diagnosis now equates to money—people know such a diagnosis will get you money
		Secure care unit accepting from other catchment areas/CPLs not always admitted to secure care—should come to secure care due to skills of manipulation learned in corrections
		New younger male MICA clients; police do not want to deal with mental health issues—make occasional sweeps
Environmental	13	Telephone should be boxed
		Telephone in day room is problem
		Weapons are available, e.g., chairs not bolted down, cigarettes
		Plexiglas kicked out of smoking room
		Air conditioning installation helped situation
		Objects (boots, shoes, walkman cords, shoelaces) need to be restricted
		Plugs put in holes of window levers in some locations, not in others
		Windows are breakable and can be used as weapons
		Plexiglas on pictures can also be broken and used as weapon; pictures are bolted to walls but have sharp edges
Psychosocial		Secure care unit is threatened with loss, not operating as per 7/95 policy Appendix III
		New behavior program with 10 beds
		Requirement for secure care patients diagnosed as self-abusive, etc. to be stabilized in 60 days, limited patients, may destroy secure care
		Crowded with 25 (present census was 24)
Communication	11	Lack of control and continuity from shift to shift
		Worker should check with regular staff when on someone else's unit floating
		Statistics for worker injury are unknown
		Histories of assaults should come with patient; history is not accessed or patient may be new to system due to overcrowding elsewhere
		Communication is by (1) day book, (2) word of mouth, and (3) patient report of aggressive behavior, also meds in change of shift transfer; good communication on this unit
		Communication levels between services vary in quality; paperwork and notes lacking
		Did not get cited for chart deficiencies, etc.; decisions made and followed through
		Reports generated tend to criticize workers
		Administration expects employees to know location of all other employees
		Incident documentation needed: who sees?

(*continued*)

TABLE 7.14 (continued) Units 95 and 97 Focus Group Interviews January 27 and 28, 1997

Worker Concerns	Number of Comments	Comments
Policy issues	34	No therapy aides in on treatment planning
		Lack of consistency—need policy change or consistent compliance; more consistency, less confusion among patients
		Policies can be rescinded by doctors orders; policies should be revised from input from all three shifts and posted for clients to read
		Privileges (e.g., separate rooms in 95) are sometimes sought by acting out, then difficult to return
		CPLs should first come to Unit 95
		Between a rock and a hard place—clients qualify for secure care, but central office policy is to get them out the door
		People will remain who need controlled environment
		Those put out are not getting support upon discharge or are released before stabilization, or they do not comply with treatment
		What will happen regarding the money being spent on changes?
		Patients report abuse directly to the top instead of to charge nurse, etc. who would understand the issues
		Patients who should be on secure care are on admissions and vice versa, e.g., secure care patients were escorted on a bowling trip, or suicidal patients transfer to admissions because of space problems in community
		Revolving admissions: MICA client from jail; "dumping" from the community
		Programs added onto weekends, leaving clients no fresh air exposure, no free time, no sleep-in time, locked dorms during programs, changed visitor hours; should revisit mandated programs, on weekends especially
		With ratio of 6:1, can take patients for walks, etc.
		Program nonparticipators disruptive, e.g., sleep in halls
		Upon new admissions, there should be drug withdrawal for 3 days before interventions occur
		Difficult to get bodies in or out to programs and downtown for appointments; time requirements require program attendance; separate staff for programs needed
		One-on-ones with patient in restraint severely restricts available staff
		No management policy on general outside hospital duty; should be "eyes and ears" only, no assistance given during hospital visit by employee; "sitters" required by hospital policies if patient is overactive
		No scheduled overtime as in past occurred with increased census—now upon assignment; voluntarily when one-on-ones occur, mandatory overtime required if no volunteers

TABLE 7.14 (continued) Units 95 and 97 Focus Group Interviews January 27 and 28, 1997

Worker Concerns	Number of Comments	Comments
		What is cabinet level responsibility (upon documentation of an incident)? Are we at "brick wall" with care for patient not improving, is alternative treatment needed and can therapeutic relationship be maintained?
		Help is available from administration—go down and get interrogated
		(Nurses) have no confidence in improvements to system—more being asked, e.g., medical clinic being set up, charge nurse asked to man treatment mall and new clinic as well as admissions; "semantic" MICAs have constant calls for evaluations, discharge surveys that are negative lead to complaints from administration of staff negativity
		Staffing requests languish unfilled, or fill skilled slot with overtime; transferred workers or notice sent to charge nurse who assigns
		Need acknowledgment of job worth—no positive statements are received
Smoking policy		Cigarette restrictions
		Smoking is workable now: smoke in designated area in unit with observer outside small unit window, going smoke-free could cause problems
		Due to lack of money, cigarettes not provided to new patient until patient obtained job; staff provided cigarettes in interim
		It is inconsistent that treatment plan is individualized, but smoking privilege is blanket policy
		It would be a war zone if the building went smoke-free
		Cigarette restrictions due to lack of money leads to agitation in patient
		Building going smoke-free in 1999
		Cigarettes big issue—no cigarettes when they first enter unit
		Want facility smoke-free—patients cannot smoke on unit due to "short-term" care that keeps getting extended; one employee escorts 19 patients to small enclosed smoking area; patients are referred because Binghampton BPC known to have smoking areas
Posttrauma response	5	Lack of criminal involvement with police for assaulters—only restraint—resistance by police and due to in-house hope to keep it quiet
		Experience in legal system is poor, "part of your job" per police. Basketted cases by safety are now turned in as assaults but police not cooperative. District attorney does not appear to listen; we want issues recorded in proper way; liaison with police concept a good one
		After event, safety did not help; workers' comp was a mess
		Employee-assistance-program crisis intervention team is under way with union involvement

(*continued*)

TABLE 7.14 (continued) Units 95 and 97 Focus Group Interviews January 27 and 28, 1997

Worker Concerns	Number of Comments	Comments
		Verbal assaults, name calling, ethnic slurs lead to no retaliation except possible follow-up by nurse, which can cause complaints and interrogations against employee—causes employee to be pulled
Worker characteristics	19	Old school techniques
		RIFed so many times that there are only 2 social workers and 30 admissions last month
		Staff capable, not under trained
		RIFs cause increased age in workers
		Lack of staffing most important thing—patients sense it; is money used in cost-benefit analysis of staffing vs. patient assaults?
		Staffing gender nonspecific unless for emergency room visits or "sensitive" care units; need clarification for how gender staffing decisions are made, as in incontinence or rape cases
		Juggling staff around—nights "raided"
		Differences in privileges granted—no follow-through, no sitting and talking about it, no decision making
		Patients have rights, but what happened to us?
		Retirees returning but some situations not appropriate, workers feel protective of them if older
		5-Point restraints may require 6 people, preferably male
		Secure unit is threatened with loss, not operating as per 7/95 policy Appendix III
		Complacency among staff
		All jobs important
		Used to working together in teams on shift, problems in working together cross-shifts—only sporadic
		Hard to learn to keep frustrations down, especially when dealing with pedophiles, etc.; burnout
		Nurses are more frequently assaulted because they are who patient identifies as the person who communicates denial of privilege
		Morale plummets with tired, discouraged workers
		Therapy aides on Unit 97 bidding out—everyone wants off the unit

Binghamton Psychiatric Center Training Initiative on Reducing Workplace Injury for Units 95 and 97

Time	Topic
8:00 AM	Introduction and welcome
	Management commitment and employee involvement
	Margaret Dugan, Chief Executive Officer, BPC
8:05	Project Background
	Patricia Walsh, R.D., Co-Chair Health & Safety Committee
8:10	Union Support
	Laura Kittredge, Public Employees Federation Council Leader, Co-Chair Health & Safety Committee
	Mark Mandyck, CSEA Local President

TABLE 7.14 (continued) Units 95 and 97 Focus Group Interviews January 27 and 28, 1997

Time	Topic
8:20	Goals and objectives of the training:
	Violence injuries in OMH and Binghamton PC: Data analysis
	Deb Wagoner, OMH Employee Relations and
	Jonathan Rosen, Director Health & Safety, PEF
8:30	Results of the BPC focus groups
	Janet Foley Health & Safety Project Developer, CSEA
8:40	Defining and predicting violence (Safety Training, pp. 11 and 15)
	Ken Duszynski, Intensive Case Manager
9:25	Risk reduction planning (Safety Training, p. 24)
	Ken Duszynski
10:10	Break
10:25	Consumer focus
	Nona McQuay, Health and Safety Specialist, PEF
11:00	Role play/role play video
	Critical incident debriefing (Safety Training, p. 53)
	Ken Duszynski
12:00 PM	Lunch
1:00	Critical incident debriefing on actual cases; Participants
1:45	Posttrauma responses—agency policy
2:00	Break
2:15	Partnerships with law enforcement agencies
	Chief Safety Officer David Jones, Wayne Gove, Ken Duszynski
2:30	Presentation of environmental study/risk mapping
	Nona McQuay, Jonathan Rosen
2:40	Participant recommendations: environmental, organizational, and clinical priorities; All
3:15	Wrap-up and evaluation
3:30	Dismissal

Note: Agenda for March 20, 1997 and repeated March 21, 1997; located at Garvin Building, 3rd Floor, Unit 91, Staff Development, Classroom 112.

- A number of facility and agency initiatives are under way to address the same problem. These include programs for improving patient assessment, mandatory training, and changes in clinical treatment and rehabilitation programs.
- There are significant differences in the skills, commitment to the problem, and relationships among managers and union leaders in the 28 hospitals operated by OMH. These differences may have a significant effect on the ability of a particular hospital organization to effectively implement the process described in this chapter.
- The use of a microanalysis of job titles and units, while very helpful in identifying problem areas, produces unstable rates which can be sharply affected by small changes in incidence.

Despite these limitations, the pilot projects in OMH illustrate that when there are strong labor/management relationships and organizational will, the implementation of OSHA's guidelines for prevention of violence in health care facilities is feasible.

Acknowledgments

A special word of thanks to the project staff which included Nona McQuay and Marian Arbesman. Also I would like to thank: Janet Foley of CSEA; Deborah Wagoner and Charles Thompson of OMH; Dan Brown, the project team, and labor and management leaders at Buffalo PC; and Margret Dugan, Laura Kittredge-Copp, Mark Mandyk, the project team members, and the labor and management leaders at Binghamton PC.

References

1. Lipscomb JA and Love CC. Violence toward health care workers, an emerging occupational hazard. *AAOHN J* 40:5, 1992.
2. Violence in the workplace, risk factors and prevention strategies. *NIOSH Current Intelligence Bulletin 57*, June 1996.
3. *Guidelines for Preventing Workplace Violence for Health Care and Social Service Workers*, US Department of Labor, Washington, DC, OSHA 3148, 1996.
4. Joyce A. Simonowitz. *CAL/OSHA Guidelines for Workplace Security and Safety of Health Care and Community Service Workers*, Medical Unit, Division of Occupational Safety & Health, Department of Industrial Relations, 1993.
5. Center for Mental Health Services, Mental Health, United States, 1996, US Department of Health and Human Services, Public Health Service, Substance Abuse and Mental Health Services Administration, Rockville, MD, DHHS Publication No. (SMA)96-3098, 1996.
6. *Statewide Comprehensive Plan for Mental Health Services, 1997–2001*, OMH, 1997.
7. Berger BR. *J Am Acad Psychiat Law* 25:1, 1997.
8. *Restraint and Seclusion Practices in New York State Psychiatric Facilities*, NYS Commission on Quality of Care for the Mentally Disabled, 1994.
9. Bensley L, Nelson N, Kaufman J, Silverstein B, and Kalat J. *Study of Assaults on staff in Washington State Psychiatric Hospitals.* State of Washington Department of Labor and Industries, Olympia, WA, 1993.

Additional References

Askari E. Tools for developing a workplace violence prevention program. *Occupat Med State Art Rev* 11(2):336–374, 1996.

Bensley, L, Nancy N, et al. *Studies of Assaults on Staff in Washington State Psychiatric Hospitals: Final Report*, State of Washington, Department of Labor and Industries, December 1993.

Dvoskin J, Massaro J, Nerney M, and Howie the Harp. *Safety Training for Mental Health Workers in the Community: Trainer's Manual*, New York State Office of Mental Health, Albany, NY, The Information Exchange, New York, June 1995.

Fisher A. Intervening in violence in the psychiatric setting, Chapter 34. In: HS Wilson (ed.), *Psychiatric Nursing*, 5th edn. Addison-Wesley, New York, 1996, pp. 816–838.

Martin K. Improving staff safety through an aggression management program. *Arch Psychiat Nurs* 9(4):211–215, 1995.

Managing Crisis Situations, New York State Office of Mental Health.

New York State Employee Assistance Program, *Traumatic Incidents at the Workplace: The Role of EAP in Trauma Response*, NYS EAP, Albany, NY, September 26, 1994.

Simonowitz JA. Health care workers and workplace violence. *Occupat Med State Art Rev* 11(2), 1996.

U.S. Dept. of Health and Human Services, Public Health Service, National Institute for Occupational Safety and Health. *Violence in the Workplace: Risk Factors and Prevention Strategies*, NIOSH Current Intelligence Bulletin 57, Cincinnati, OH, June 1996.

U.S. Department of Labor, OSHA. *Guidelines for Preventing Workplace Violence for Health Care and Social Service Workers*, OSHA 3148, 1996.

Appendix 7.A.5

REDUCING ASSAULT RELATED INJURY ➡ STAFF SURVEY

ALL RESPONSES WILL BE KEPT CONFIDENTIAL, AND NO NAMES WILL BE USED ON ANY PART OF THE SURVEY.

What shift do you regularly work on? ❑ Days ❑ Evenings ❑ Nights

What ward are you regularly assigned to? ❑ 75 ❑ 76

What is your job title? ❑ MHTA ❑ RN ❑ Other

What do you think are the most important factors contributing to assaults on staff on your ward?

Please indicate only your top **5** concerns. Place a "1" next to the issue that you think is the top priority, and a "2" next to the issue that you think is the next highest priority and so forth. If you have no opinion or don't know, please check "Don't know".

_______ a. Staff training in self-defense/restraint procedures
_______ b. Staff clinical and interpersonal skills
_______ c. Staff physical fitness
_______ d. An effective security alarm system
_______ e. An effective emergency response program
_______ f. Adequate numbers/deployment of personnel
_______ g. Adequate locks on doors
_______ h. Hospital practices (e.g. handling patients' money or cigarettes)
_______ i. Physical environment (e.g. noise, ceiling tiles, lighting)
_______ j. Identifying patients with a history of assaults
_______ k. Identifying patients with potentially assaultive (e.g. agitated) behavior
_______ l. Transfer of information at shift change about the potentially assaultive patients
_______ m. Adequate control of patient movement and whereabouts
_______ n. Procedures for reporting assaults to administrators
_______ o. Procedures for reporting assaults to police
_______ p. Legal penalties for competent assaultive patients
_______ q. Structured psychological support for assaulted staff
_______ r. Other _______________________
_______ s. Don't know

Please check the types of training you would like to see at your Psychiatric Center (You may check more than one).

_______ Interpersonal communication
_______ Verbal de-escalation
_______ Restraint/seclusion procedures
_______ Training on emergency response systems
_______ Policies on patient assessment for violence
_______ Assessing potential assaultiveness
_______ Self-defense
_______ Team/consensus building
_______ Working together in crisis

Other _______________________

Have you ever been assaulted (hit, bite, kicked, choked, shoved) while at work by one of the patients?

❑ Yes ❑ No

Have you been assaulted within the past year by one of the patients?

❑ No ❑ Twice
❑ Once ❑ 3 times or more

over →

ALL RESPONSES WILL BE KEPT CONFIDENTIAL, AND NO NAMES WILL BE USED ON ANY PART OF THE SURVEY.

Have you ever been trained in managing crisis situations?
- ❑ Never
- ❑ At time of hire
- ❑ Annually
- ❑ Within the past year

Do you wear your personal wrist alarm:
- ❑ Always
- ❑ Occasionally
- ❑ Never

Do you find emergency response to crisis to be:
- ❑ Always prompt
- ❑ Usually prompt
- ❑ Sometimes delayed
- ❑ Always too late

The remaining part of the survey deals with how satisfied you are with various aspects of your work

How satisfied are you with:

	Not at all satisfied	Not too satisfied	Somewhat satisfied	Very satisfied
The way supervisors and administrators treat workers?	❑	❑	❑	❑
The way policies are put into practice?	❑	❑	❑	❑
The competence of your supervisors?	❑	❑	❑	❑
The praise you get for doing a good job?	❑	❑	❑	❑
How satisfied are you that you can turn to fellow workers for help when something is troubling you?	❑	❑	❑	❑
How satisfied are you with the way your fellow workers respond to your emotions such as anger, sorrow or laughter?	❑	❑	❑	❑
How satisfied are you that your fellow workers accept and support you new ideas and thoughts?	❑	❑	❑	❑
All in all, how satisfied are you with your job?	❑	❑	❑	❑

How strongly would you recommend your job to someone else?
- ❑ Not at all strongly
- ❑ Somewhat strongly
- ❑ Not too strongly
- ❑ Very strongly

If you were looking for a job now, how likely is it that you would decide to take this job again?
- ❑ Not at all likely
- ❑ Somewhat likely
- ❑ Not too likely
- ❑ Very likely

To what extent is your supervisor willing to listen to your work-related problems?
- ❑ Not at all willing
- ❑ Somewhat willing
- ❑ Not too willing
- ❑ Very willing

How often are you physically exhausted after work?
- ❑ Never
- ❑ Often
- ❑ Seldom
- ❑ Always

Overall, how would you rate your health at the present time?
- ❑ Poor
- ❑ Very good
- ❑ Fair
- ❑ Excellent
- ❑ Good

How would you rate your health compared to other persons your age?
- ❑ Poor
- ❑ Fair
- ❑ Good
- ❑ Very good
- ❑ Excellent

PREPARED BY BPC/LABOR MANAGEMENT COMMITTEE: PROJECT TO REDUCE WORKFORCE VIOLENCE

PEF CSEA

Appendix 7.A.6

Monday, November 4, 1996
Mr. Jonathan Rosen
New York State Public Employees Federation
1168-70 Troy-Schenectady Road
PO Box 12414
Albany, NY 12212-2414

Dear Jonathan:

Workplace Study
AR 146.01

I appreciate you contacting me in the course of preparing this study of inpatient facility workplace safety. Architectural Resources has developed a strong resume of health care experience including survey, programming, design, and planning work with the NYS OMH.

In response to our telephone conversation I arranged to meet Ms. Nona McQuay and Mr. Dan Brown at Buffalo PC on Wednesday, October 9, 1996. Ms. McQuay was unable to attend but was represented by Ms. Marian Arbesman.

Dr. Brown and Ms. Arbesman described the conditions on Ward 75 (sixth floor south, acute men) and the comments they had received from a survey of on-ward staff. We walked through the ward and program areas on the second floor. As there were very few residents on the ward during our tour, I expressed interest in returning later in the day to observe behavior and staff response. Late afternoons were described as a high-stress time because clients are returning from day activities and the staff shift changes.

This letter describes comments made by Brown and Arbesman and my response. It also includes notes made during my observation and tour of the inpatient ward and reviews comments from the OMH 1994 renovation project manager. In each case, I have tried to balance between what appears to be a very pleasant therapeutic environment and a more safe work environment. None of this should be taken as criticism of the original design, which appears to be well developed, or suggestions to create a highly secure, more institutional setting (Figure 7.A.1).

Reported Staff Comments

Nurse station counter: Concern about low height, clients leap over top. Staff suggested partial Plexiglas shield.

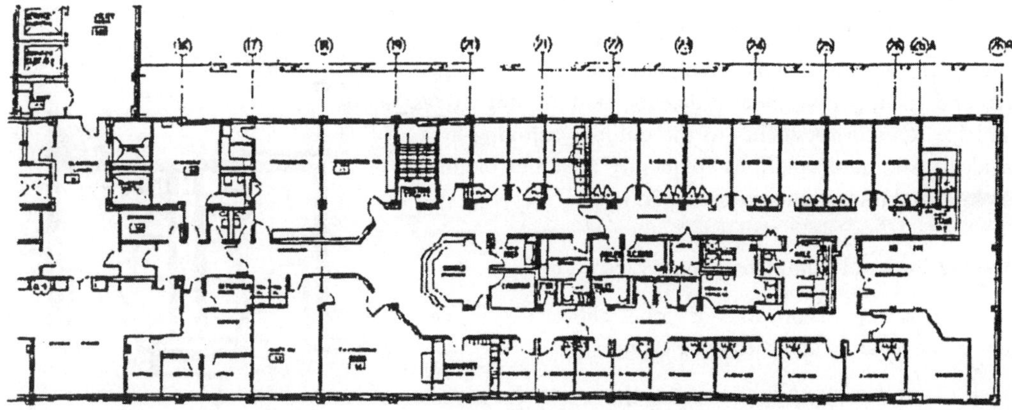

FIGURE 7.A.1 Ward floor plan.

Nurse station half doors: Swing gates on either side of the nurse station are too easy to open. Locks should be concealed in some way.

Ceiling tiles: Easily removed and used for stashing contraband. (Corridor ceilings are suspended acoustic tiles, two foot square, set into an aluminum suspension grid. All room ceilings are solid drywall construction.)

Curtin balloon valances: *Used for hiding contraband. (Ward staff with whom I spoke did not feel this was a concern. If so suggested a simple straight valance.)*

Nurse station visibility: Staff are reluctant to let residents use "Quiet Room" and "Program Room" because they are not visible from nurse station.

Cross corridor at end of ward: Staff feel residents can hide in the connecting corridor next to the Mechanical Room. Maintenance staff report finding scores of cigarette butts above the ceiling in this area. I agree, this is the kind of space that needs to be controlled. The two-corridor design really creates two separate areas. In fact, this works fine on coed wards. I would suggest closing off this area with an additional door and installing a motion detector. Staff would be able to leave doors open at night or leave the area completely locked off (Figure 7.A.2).

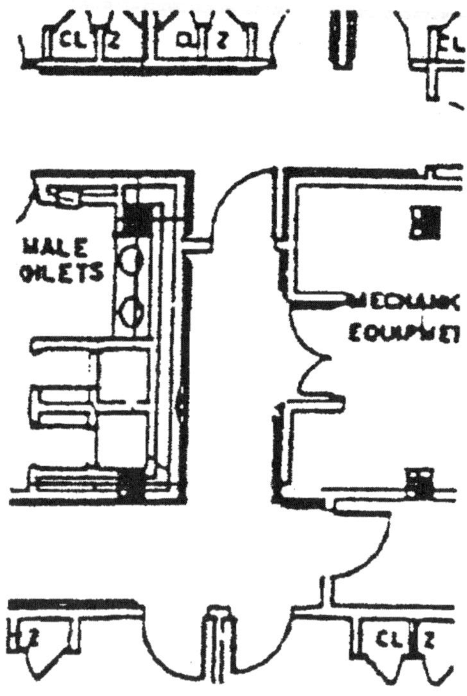

FIGURE 7.A.2

Day toilets: Staff noted residents can wedge themselves into these rooms and block staff from opening the door. Building codes will not permit reversing the swing of these doors (it would block the corridor). Another solution is the old door within a door although I am not aware of any fire-rated models. The better solution would be a removable hinge door developed by OMH (Figure 7.A.3).

Entry vestibule: Staff are somewhat concerned about being locked in sally port entry vestibule while escorting residents to and from the unit. I agree. While not without exception, very few residents escape by rushing a single door as it is opened. With the implementation of the building security plan a "bolter" would get no further than the main lobby. I suggest removing the interior set of doors completely or perhaps replacing with a half door.

Smoking policy: The "smoke-free" building policy presents staff and client management problems. Staff are not always available to escort residents to the outdoor smoking area. Incidents and confrontations frequently arise out of illicit smoking and the inability to smoke on ward.

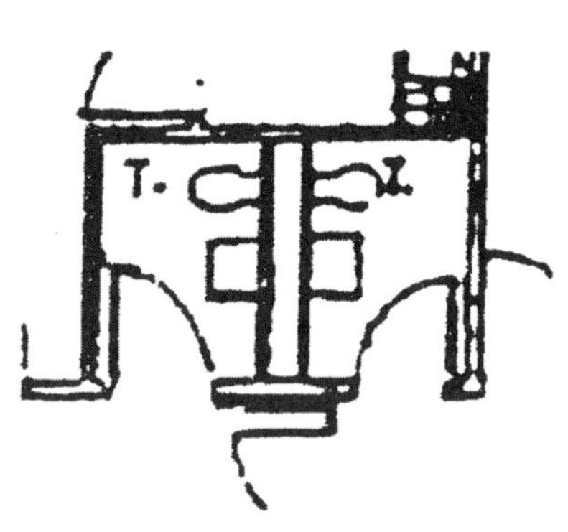

FIGURE 7.A.3

Observations and Comments from Ward Staff

Suspended ceilings in corridors (described above): Cross members can be removed and used as a weapon.

Wood chairs: Heavy upright wood chairs in lounges can be picked up and used in fights. Light metal framed dining room (Lowenstien) chairs are actually preferred.

Television in Day Room: Staff concerned a resident could remove TV and throw it. Also feel many arguments start over which program to watch and volume. Staff would like to enclose TV in Plexiglas and use a remote control.

Door hardware: ADA-compliant lever handles are of the open-end type. I would suggest the closed-end style. Staff report no injuries, just catching and sometimes tearing clothing.

Corridor handrail: Plastic covered product suffers from physical abuse, especially where used as a door stop (adjacent to seclusion room). Access doors and fire extinguisher cabinets also show physical damage.

Bulletin board frames: Several bulletin boards are covered by aluminum-framed Plexiglas enclosures. The corners are very sharp and could easily cause injuries. Ideally these should be recessed. As a minimum I recommend adding a wood perimeter trim with radiused corners.
The following are observations I made during my afternoon visit.

Picture frames: Corridor artwork consists of posters set in aluminum sectional frames screwed to the wall. These have sharp corners and could be pried loose and used as a weapon. Artwork significantly contributes to the ambiance and should be retained. An alternative on the more difficult wards would be to laminate art to wood panels with eased edges.

Nurse station counter: Opinions vary on the height of the nurse station. Most seem content with the present design, feeling it is either adequate or that nothing short of a full height Lexan wall will deter aggressive clients. I suggested a raised but open counter to raise the height without reducing the openness.

The Nurse Station itself seems excessive, particularly with fewer staff and staff who are encouraged to be with clients as much as possible. However simply reducing the counter size is meaningless unless this new found space is somehow put to use. Therefore I do not feel such an extensive redesign would serve any beneficial purpose (Figure 7.A.4).

Acoustics: With few absorptive surfaces the noise level is very high and surely affects behavior of residents and staff. With every room closed off by fire doors hardware is constantly latching and unlatching. Conservations are very loud. PA announcements add to the generally high-ambient noise level. I would suggest installing a high performance carpet in the corridors that is specifically designed for health care environments. This will reduce noise levels and injuries from accidental falls.

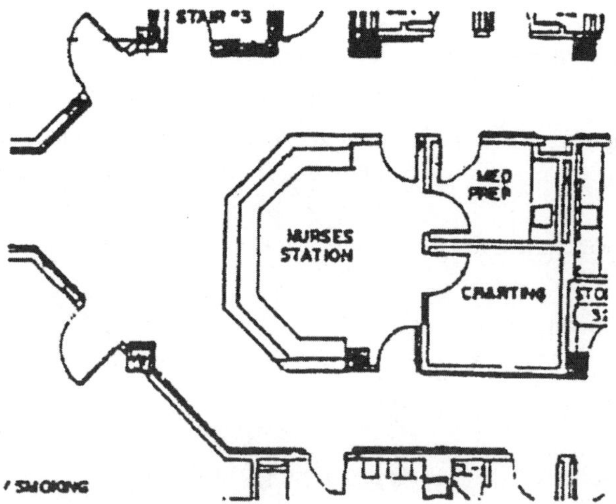

FIGURE 7.A.4

Environment controls: Staff commented on the lack of fresh air, built-up odors and generally stagnant air. I could not form an opinion on this (I have smelled much worse) and noted a fully calibrated evaluation would be necessary before making such a statement.

Smoke detection: Bathrooms are used by residents for smoking, which of course is not permitted. Staff suggested installing sensitive "local" detectors and alarms. Technically this could be unwise. Discerning between local/general alarms and real/false alarms may not be possible. Further, patients may be able to easily defeat such a system by covering the detector or exhaling directly into the exhaust vent.

Program areas: With limited staff and the need for better visibility it may be useful on such active wards to combine. Recreation and Program Rooms on each side of the center corridor. This would create larger, more easily supervised spaces which can be broken down if desired by furniture arrangements, lighting, and other interior landscape elements.

I hope you find these comments useful in your study. It has been a pleasure working with you and the staff at Buffalo PC in the preparation of these materials.

Sincerely,
Kevin D. Murrett, AIA

Appendix 7.A.7

Memo

To: Mr. Jonathan Rosen
NYS PEF
Health and Safety Department
1168-70 Troy-Schenectady Road
PO Box 12414
Albany, NY 12212-2414

From: Kevin D. Murrett, AIA

Date: November 19, 1996

Subject: Environment Assessment Report
Buffalo Psychiatric Center
AR 146.01

Johnathan:

I received your draft report and have only a few comments.

Page 2 Item 1: In addition to the keys, door locks should be color coded as well.

Page 3 Item 5a: The smoking problem can also be alleviated by providing additional secure outdoor recreation areas. The limited area east of Strozzi Building is inadequate for the number and clinical profile of most residents.

Good luck with your work, please feel free to call me if you have any questions.

Kevin

Appendix 7.A.8

I. Clinical history
   A. Diagnosis at discharge
      Axis I: ____________________
      ____

F. Age at onset:

G. Psychotropic medications:
   ______Taking prior to admission
   ______Not taking prior to admission

Axis II: ______________ Medications:
______________ ______________________
______________ ______________________

B. Age:__________ ______________________

C. Sex:__________ ______________________

D. Admitting status

___72-HR hold___Voluntary Previous criminal history

___14-DAY cert.___Other ___Yes ___No

___Temp conservatorship ______________________

E. Previous experience in seclusion/restraint ______________________

___Yes ___No I. Use of alcohol/street drugs

Reaction to seclusion/restraint ___Yes ___No

______________________ ______________________

______________________ ______________________

II. Violence history

A. Previous institutional violence___Yes___No

Type of institution:__________ Date(s):_____ ___

Number of incidents:__________ _____ ___

Type of violence: Against person ___Yes___No Date________________

Family ___Yes___No Date________________

Stranger ___Yes___No Date________________

Inmate/clinet ___Yes___No Date________________

RN/LPT/MD ___Yes___No Date________________

Other ___Yes___No Date________________

Who________________

Weapon used ___Yes___No Date________________

Against property ___Yes___No Date________________

Weapon used ___Yes___No Date________________

Type ______________________

Verbal threat (only) ___Yes___No Date________________

Situational factors: Time of day ________________

Location ________________

Engaged in therapeutic activity ___Yes___No

Type of activity _______

Other Factors______________________

____

____

Interactional factors: Engaged in interaction with victim ___Yes ___No

Type of interaction ______________________

____

With whom ______________________

Content of conversation, request:

____

____

Response to violence: Medications ___Yes___No

Type and close: ________________________________

____

____

Seclusion only ___Yes___No

Seclusion/restraint ___Yes___No

Milieu management ___Yes___No

Combination ___Yes___No

(list) ________________________________

____

____

Client's response to intervention(s): ________________________________

____

____

B. Community violence

Previous violence:___Yes___No

Number of incidents:___Dates(s):__________ _________

__________ _________

__________ _________

Type of violence: Against person _____Yes_____No Date_________

Family _____Yes_____No Date_________

Stranger _____Yes_____No Date_________

Inmate/client _____Yes_____No Date_________

RN/LPT/MD _____Yes_____No Date_________

Other _____Yes_____No Date_________

Who_________

Weapon used _____Yes_____No Date_________

Against property _____Yes_____No Date_________

Type ______________________________________

Verbal threat (only) _____Yes_____No Date______________

Situational factors: Alcohol _____Yes_____No Amount___________

Street drugs _____Yes_____No Date______________

Type _____________________________________

____

____

Time of Day _________ Activity ________________

____

Location _____________________________________

Other factors ________________________________

____

International factors: Engaged in interaction with victim ___Yes___No

Type of interaction ________________________________

____

____

Others present: ________________________________

____

Content of conversation, request, argument, or dispute:

____

____

8

Prevention of Slip, Trip, and Fall Hazards for Workers in Hospital Settings

Studies of Work-Related Injury in Hospitals Attributed to Slips, Trips, and Falls 8-2
Distribution of Hospital STF Incidents by Occupation 8-2
Body Part and Nature of Injury 8-2
Age Group, Length of Employment, and Gender 8-3
Circumstances of STFs 8-3
Friction Characteristics of Footwear and Flooring 8-4
Laboratory Testing of Shoe–Floor Slip Resistance • Flooring Testing and Classification
Evaluation of the Effectiveness of a Comprehensive STF Prevention Program for Hospital Employees 8-6
Identifying STF Hazards 8-6
Review Past Injury Records and Conduct Investigations of Future STF Incidents
Hazard Assessments 8-7
Strategies for Prevention 8-7
Written Housekeeping Programs • Keep Floors Clean and Dry • Prevent Pedestrian Access to Wet Floors • Slip-Resistant Shoes
Minimize Tripping Hazards 8-9
Operating Rooms 8-9
Ice and Snow Removal 8-10
Lighting 8-10
Stairs and Handrails 8-11
Conclusion 8-11
References 8-11

James W. Collins
National Institute for Occupational Safety and Health

Jennifer L. Bell
National Institute for Occupational Safety and Health

Slip, trip, and fall (STF) incidents are a significant source of workers' compensation claims for staff in hospitals. In 2006, 13,750 STFs accounted for 23% of all work-related injuries in hospitals requiring at least 1 day away from work (Bureau of Labor Statistics [BLS], 2007a). BLS (2007b) also reported that the

* The findings and conclusions in this report are those of the author(s) and do not necessarily represent the views of the National Institute for Occupational Safety and Health.

incidence rate of lost-workday injuries from same-level STFs in hospitals was 35.2 per 10,000 full-time equivalent (FTE), which is almost 60% greater than the average for all other private industries combined (20.2 per 10,000 FTE). Although extensive work has been initiated to prevent patient falls in health care settings, there has been minimal research or systematic efforts to develop evidence-based guidelines for preventing STFs for workers in health care settings.

Ensuring safe, effective, and quality health care is a nationwide public health priority (Institute of Medicine, 1999). One of the key factors contributing to the delivery of quality of health care is maintaining the health and safety of health care workers. The huge negative impact of STF injuries on health care workers led a multidisciplinary team to conduct research to identify STF risk factors (Courtney et al., 2001; Lombardi et al., 2007) and to conduct laboratory and field researches to evaluate the effectiveness of interventions (Collins et al., 2006) for reducing work-related STF injuries among hospital employees. Historically, STF incidents have been considered largely nonpreventable and the blame has been placed on the carelessness of the fall victim (Lacroix and Dejoy, 1989; Sotter, 2000; Lehane and Stubbs, 2001). Contrary to the popular belief that STFs are random events that cannot be prevented, recent research (Collins et al., 2006) has shown that a comprehensive STF-prevention program can significantly reduce workers' compensation claims for same-level falls involving hospital staff.

Studies of Work-Related Injury in Hospitals Attributed to Slips, Trips, and Falls

Detailed information on STF risk factors is available from three studies in hospital settings (Courtney et al., 2001; Lombardi et al., 2007; Bell et al., 2008). The Bell et al. (2008) study, conducted by the National Institute for Occupational Safety and Health (NIOSH), analyzed 472 STF incidents involving staff from three hospitals from January 1, 1996 through December 31, 2005. During the 10 year study period, a dynamic cohort of 16,900 individual employees worked 80,506,017 h, representing 40,253 worker-years. A total of 2263 workers' compensation claims were filed, 21% ($n = 472$) of which involved STFs, resulting in 1.2 workers' compensation injury claims attributed to STF incidents per 100 FTE workers.

As part of a collaborative research effort with NIOSH, a case-crossover study of employees in seven hospitals was conducted to identify risk factors for STFs (Courtney et al., 2001; Lombardi et al., 2007). A total of 153 hospital employees who slipped, tripped, or fell at work (both indoors and outdoors) were interviewed and asked to describe STF circumstances and risk factors in the hospital environment. Respondents had a mean age of 46 (range = 19–67) years, and had worked for the hospital for an average of 9.3 years. Eighty-six percent of the respondents were women.

Distribution of Hospital STF Incidents by Occupation

In the Bell et al. (2008) study, food service workers suffered the highest rate of STF workers' compensation claims, with 4.0 claims per 100 FTE (Table 8.1). Most hospitals have a food service department on the premises that provides meals around the clock for patients, visitors, and staff. Nursing staff incurred the most STF claims ($n = 141$), but because they comprised the largest proportion of the total work hours (33.6%) they had a much lower claim rate (1.0 STF workers' compensation claim per 100 FTE). In the Courtney et al. (2001) and Lombardi et al. (2007) studies, 50% of the participants worked in an occupation that provided direct care to hospital patients with a distribution similar to those in the Bell et al.'s (2008) study (Table 8.1).

Body Part and Nature of Injury

The most commonly injured body part from STF events was a lower extremity (44.9%), followed by an upper extremity (17.7%), multiple body parts (16.7%), back/trunk (16.2%), and head/neck (4.5%). The

TABLE 8.1 STF Workers' Compensation Claim Rates by Job Group

Job Group	# STF Claims	Hours Worked	% of Total Hours	Rate per 100 FTE
Food services	57	2,872,015	3.6	4.0
Parking, valet, transport	10	618,607	0.8	3.2
EMS	11	997,357	1.2	2.4
Custodial, housekeeping	41	4,000,007	5.0	2.1
Maintenance, groundskeeping	11	1,504,635	1.9	1.6
Teachers, including childcare	14	2,064,253	2.6	1.4
Unknown	1	174,391	0.2	1.1
Nursing and nursing-related	141	27,055,196	33.6	1.0
Other health professions	26	5,169,605	6.4	1.0
Security	3	798,905	1.0	1.0
Physical/occupational therapy	10	1,997,623	2.5	1.0
Medical, laboratory, and other technicians	44	8,864,009	11.0	1.0
Office/administrative	95	20,378,625	25.3	0.9
Physicians	2	1,724,854	2.1	0.2
Total	472	80,506,017	100	1.2

nature of injury was most often sprains, strains, dislocations, and tears (48.1%). STF injuries were significantly more likely to result in fractures, multiple injuries, and bruises, contusions, and concussions than non-STF injuries and were less likely to result in cuts, lacerations, punctures, and abrasions than non-STF injuries ($\chi^2 = 213.1$, $p < 0.0001$). Lower extremities were much more likely to be injured by a STF incident than with a non-STF incident, and upper extremities were less likely to be injured after an STF injury than with a non-STF injury ($\chi^2 = 404.0$, $p < 0.0001$).

Age Group, Length of Employment, and Gender

In the Bell et al.'s (2008) study, STF injury rates were nearly twice as high for females, older workers, and those employed for less than 6 months. Bell et al. (2008) showed that 88% of the total STF claims ($n = 412$) occurred to females and the STF claims rate for females (1.27 per 100 workers) was significantly higher than for males (0.77 per 100 workers). Workers employed for less than 6 months experienced the highest claims rate (2.0 STFs per 100 workers), followed by workers employed for greater than 6 months and less than 1 year (1.7 STFs per 100 workers); both of these groups had significantly greater STF claim rates than workers employed greater than 1 year (1.1 STFs per 100 workers). STF claim rates were significantly greater for employees greater than 45 years of age (1.6 STFs per 100 workers) than for employees less than 45 years (1.0 STFs per 100 workers). Older employees of both genders had higher STF claims rates compared to younger employees (Figure 8.1), and no interaction between age and length of employment was found.

Circumstances of STFs

Of the 472 STF incidents, 85% ($n = 405$) were same-level STFs and 15% ($n = 70$) were falls from elevation that primarily occurred on stairs, from stepstools, ladders, or from hospital shuttle buses. STFs due to liquid contamination (water, grease, ice, soapy detergent, floor stripper, and wax) were the most common cause of STF incidents. In the food services department, the most common slippery conditions

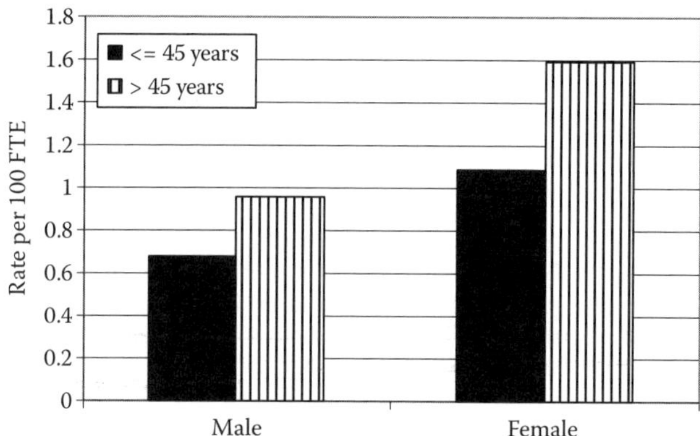

FIGURE 8.1 STF claim rates by age group and gender.

consisted of food and grease on the floor in food preparation and cooking areas, spilled drinks and ice in food serving areas, and soapy detergent on the floor in dishwashing areas.

In the Courtney et al.'s (2001) study and the Lombardi et al. (2007) study, workers reported slipping (55%) more frequently than tripping (33%); 41% of workers indicated they fell forward, 23% fell to the side, and 21% fell backward. Workers who fell most often cited the hands, knees, or buttocks as the primary points of impact with the floor or ground and the back, knees, and ankles/feet were the most frequently injured body parts. Forty-four percent of workers described the walking surface where they experienced their STF as clean and dry while 53% reported the presence of some type of contaminant including water, ice, body fluids, etc. Sixty-four percent of STFs occurred at a transitional area—wet to dry/dry to wet (32%), from one floor type to another (20%), or involving uneven surfaces (15%). Overall, 114 workers reported 228 injuries as a result of their STF event, an average of two injuries per event; 7% ($n = 8$) of study participants were not injured in the fall incidents. These results highlight the importance of managing surface contamination and surface transitions in hospitals.

Friction Characteristics of Footwear and Flooring

Laboratory Testing of Shoe–Floor Slip Resistance

One of the significant considerations regarding fall prevention is the slip-resistance characteristics of footwear and flooring (Collins et al., 2006; Thorpe et al., 2007). The coefficient of friction between the footwear and the floor is affected by the footwear material, floor, and contamination condition (Chang and Matz, 2001; Li et al., 2004; Li and Chen, 2004). While the slip-resistance characteristics of safety shoes and general footwear have been studied (Grönqvist, 1995; Leclercq et al., 1995), footwear commonly worn in hospitals settings has not been systematically evaluated.

In conjunction with the NIOSH study to prevent falls among hospital workers, laboratory studies were conducted to evaluate the slipperiness of shoes most commonly worn in hospitals and promising slip-resistant shoes, and hospital flooring (existing and slip-resistant) tested with "soapy" and "oily" contaminants (Collins et al., 2006) to simulate common contaminant conditions in health care settings.

The apparatus for testing the slip-resistance of shoes consisted of a movable artificial foot controlled by a computer and activated by three hydraulic cylinders. This slipmeter apparatus closely simulated the movements of a human foot and the forces applied between the shoe sole and the floor during heel strike in normal gait (Grönqvist et al., 1989). For this study, the following test parameters were used: normal force 500 N, sliding velocity 0.4 m/s, and heel contact angle 5°C. The dynamic friction coefficient (DCOF) was computed during the time interval 100–150 ms from heel strike, which represents a critical moment for a slip and fall under conditions of level walking (Grönqvist et al., 1989).

Seven shoe types were first pretested on a stainless steel surface (roughness, Rz 1.6 μm) under test conditions "new" (intact heel and sole) and "abraded"—after abrasion, as stipulated in the draft standard (European Standard EN 13287, 2004) section 9 (preparation of the sole). Since most shoes perform well on clean and dry surfaces, this test examined the slip-resistance of shoes under slippery flooring conditions ("oil/grease") commonly encountered in food service areas in hospitals.

Combinations of "slippery" flooring conditions were then tested, using glycerol 85 wt.% to simulate the "oily" contaminant condition and natrium lauryl sulfate 0.5 wt.% in water to represent the "soapy" contaminant condition. Ten different flooring surfaces were tested using two shoe types that were determined to have significantly different slip-resistance characteristics, as determined by the pretest procedure; separate trials were conducted for "new" conditions and "abraded" conditions. Paired *t*-tests compared the differences between initial (intact) and abraded shoe heels and soles, and among the different flooring and contaminant test conditions. The following slip-resistance classification for dynamic coefficient of friction was used:

DCOF	Slip-Resistance Class
>0.30	Slip-resistant
≥0.20–0.30	Moderately slip-resistant
<0.20	Slippery

The laboratory study identified the combination of slip-resistant shoes and flooring that performed optimally under soapy and oily conditions. The results for the shoes are presented in Table 8.2, and for the floorings with oily and soapy conditions in Table 8.3. Results confirmed previous results (Grönqvist, 1995) showing that heel and sole abrasion significantly improved slip-resistance.

Flooring Testing and Classification

Slip-resistance evaluations of 10 hospital floorings were conducted with both good and poor performing shoes (nos. 2 and 3, Table 8.2) (Collins et al., 2006). Shoe no. 2 was classified as "slip-resistant" (DCOF > 0.30) and shoe no. 3 as "slippery" (DCOF < 0.20) on stainless steel ("oily") test condition.

TABLE 8.2 Shoes Pretested on the Reference Stainless Steel Surface—"Oily" Condition

	DCOF (SD)		
Shoe No. and Type	Intact Heel/Sole	Abraded Heel/Sole	*p*-Value*
1 Nursing shoe with laces	0.159 (0.001)	0.198 (0.022)	<0.01
2 Slip-resistant shoe with laces—a	0.328 (0.026)	0.375 (0.028)	<0.001
3 Athletic shoe with laces commonly worn by hospital staff—b	0.149 (0.015)	0.173 (0.013)	<0.001
4 Commonly worn nursing clog	0.073 (0.012)	0.141 (0.008)	<0.001
5 Shoe with open heel	0.084 (0.011)	0.142 (0.012)	<0.001
6 Walking shoe with laces commonly worn by hospital staff—c	0.113 (0.009)	0.138 (0.019)	<0.01
7 Slip-resistant shoe with laces	0.140 (0.009)	0.153 (0.010)	<0.05

* Statistically significant difference between new and abraded soles (*t*-test paired two-tailed).

TABLE 8.3 Floorings Tested in the "Oily" and "Soapy" Conditions with Two Shoes

Flooring No. and Type	DCOF (SD) Shoe No. 2[a] "Oily" Condition	"Soapy" Condition	Shoe No. 3[b] "Oily" Condition	"Soapy" Condition
1 Waxed vinyl tile	0.356 (0.044)	0.263 (0.029)	0.133 (0.015)	0.163 (0.008)
2 Slip-resistant—a	0.325 (0.027)	0.452 (0.034)	0.129 (0.012)	0.254 (0.015)
3 Slip-resistant—b	0.369 (0.027)	0.353 (0.021)	0.155 (0.023)	0.283 (0.012)
4 Slip-resistant—c	0.367 (0.034)	0.378 (0.021)	0.144 (0.017)	0.290 (0.011)
5 Slip-resistant—d	0.335 (0.023)	0.277 (0.022)	0.131 (0.013)	0.202 (0.007)
6 Quarry tile	0.580 (0.026)	0.753 (0.021)	0.288 (0.021)	0.539 (0.018)
7 Safety—a	0.352 (0.023)	0.405 (0.016)	0.163 (0.008)	0.243 (0.034)
8 Safety—b	0.311 (0.018)	0.319 (0.012)	0.146 (0.008)	0.242 (0.023)
9 Safety—c	0.351 (0.022)	0.483 (0.019)	0.154 (0.011)	0.248 (0.032)
10 Safety—d	0.365 (0.020)	0.437 (0.014)	0.168 (0.008)	0.267 (0.025)

[a] Shoe No. 2—Slip-resistant shoe with laces.
[b] Shoe No. 3—Common tennis shoe with laces.

All DCOF differences between the two contaminant conditions "oily" vs. "soapy" were statistically significant ($p < 0.01$) for all floorings, except flooring no. 4 and no. 8 tested with shoe no. 2 (Table 8.3). All DCOF were significantly different (p-values < 0.01) between the two test shoes (no. 2 and no. 3) (Table 8.3). Quarry tile was the only tested flooring that was slip-resistant with both test shoes under all contaminant conditions (Table 8.3).

Evaluation of the Effectiveness of a Comprehensive STF Prevention Program for Hospital Employees

Bell et al. (2008) conducted a 10-year intervention trial to evaluate the effectiveness of a comprehensive STF prevention program in three acute-care hospitals (Bell et al., 2008). The prevention program included (1) analysis of injury records to identify common causes of STFs, (2) on-site hazard assessments, (3) environmental changes, (4) changes to housekeeping procedures and products, (5) changes to ice and snow removal procedures, (6) campaigns to highlight the importance of STF prevention among hospital workers, (7) flooring changes, and (8) slip-resistant footwear for certain employee subgroups. The hospital's total STF workers' compensation claims rate declined by 58% from a preintervention rate of 1.66 claims per 100 FTE workers to a postintervention STF injury rate of 0.76 claims per 100 FTEs.

Identifying STF Hazards

A review of historical injury records and on-site hazard assessments can identify conditions, circumstances, locations, and patterns of work-related STF incidents that can be targeted for prevention.

Review Past Injury Records and Conduct Investigations of Future STF Incidents

Contrary to the frequently held societal belief that falls are inevitable and not likely to be prevented, the examination of the detailed circumstances of STF incidents among hospital employees revealed that many STF injuries are preventable (Collins et al., 2006). The review of historical STF-related workers' compensation claims and incident reports used in conjunction with a process to investigate falls when they occur can be an effective way to identify the most common STF circumstances and identify job groups at highest risk. Incident investigations should include detailed interviews with workers who have fallen. It is possible to identify potential "hot spots" by reviewing the incident descriptions to identify locations where multiple STF incidents have occurred.

Hazard Assessments

Hazard assessments are useful for identifying conditions that might increase the risk of STF incidents. Specific conditions that can pose hazards include the condition of walkway surfaces, walkway levelness, objects and contaminants on the floor, projecting objects, cords, lighting, handrails, and drains (watch for standing water where drainage is not functioning properly). Areas inside the hospital that should be inspected include the hospital's entrances, stairs, ramps, operating rooms (ORs), the emergency room, scrub sink areas, nursing stations, the pharmacy, the histology laboratory, hallways, the kitchen (including dishwashing areas and the cafeteria), patient rooms including bathrooms, surgical instrument decontamination areas, laundry rooms, engineering and carpenter shops, and the morgue. Areas outside the hospital that should be examined include parking areas, streets, handicap ramps, and sidewalks.

Strategies for Prevention

The following prevention strategies are based on the 10 year NIOSH study in three acute-care hospitals that demonstrated a comprehensive STF prevention program can be highly effective for reducing STF workers' compensation claims incurred by hospital staff (Bell et al., 2008). Hospital safety and health staff, housekeepers, and all hospital staff can use this information to take preventive action to mitigate hazardous environmental conditions in and around their hospitals to minimize hospital-specific STF hazards. The following sections present a variety of strategies to reduce the risk of STF injuries in hospital settings.

Written Housekeeping Programs

A written housekeeping program can help to ensure the quality and consistency of housekeeping procedures. The program should describe

- Procedures for routine floor care and procedures for promptly cleaning spills or other unexpected contaminants on floor surfaces
- Materials and housekeeping products
- Use of "wet floor," "caution" signs, and barriers
- Cleaning schedules
- Cleaning methods

Keep Floors Clean and Dry

Contaminants such as water, body fluids, spilled drinks, and grease are the most common hazards that make walking surfaces slippery and lead to STF incidents for hospital employees. These hazards primarily occur at building entrances where rain and snow are tracked in and in-food service areas such as the kitchen, cafeteria, serving line, freezers, dishwashing areas, and near sinks, ice machines, soap dispensers, and water fountains. Floors can also be wet in areas where surgical instruments are decontaminated. Most walking surfaces are slip-resistant when they are clean and dry. The following prevention strategies can help keep floors clean and dry:

- Encourage workers to cover, clean, or report spills promptly.
- Advertise telephone/pager numbers through e-mails, posters, and general awareness campaigns so that all hospital staff are familiar with the number to call housekeeping to quickly clean up spills.

- Conveniently locate wall-mounted paper towels or other clean-up materials throughout the hospital (near elevators, nursing stations, outside the cafeteria, water fountains) so that all employees have easy access to clean-up materials.
- Conveniently locate pop-up warning signs so that staff can quickly place them over a spill while waiting. For large spills, it is important to block off the area rather than just place unconnected cones that may be ignored by pedestrians.
- Optimal floor cleaning procedures can prevent slips and falls. Research by Quirion (2004) found that damp mopping alone is not the best floor cleaning method. A two-step immersion mopping process was found to be superior to damp-mopping. In the two-step process, (1) cleaning solution is applied on a section of the floor with a dripping mop, and (2) after a few minutes, the cleaning solution is removed with a wrung mop, before the solution has had a chance to dry (see the recommended reading section for more details).
- Not all floor cleaning products are equally effective. Make sure the product used is suited for the environmental contamination conditions.
- Make sure cleaning products are mixed according to manufacturer's recommendations.
- Effective procedures to degrease floors should be implemented in areas where food is prepared, cooked, and served. The appropriate degreaser/cleaner should be used and the manufacturer's instructions should be carefully followed. Common mistakes include not letting the cleaner stay on the floor for the proper length of time, not using a stiff deck brush to loosen contaminants, and not providing a thorough clean rinse (for cleaning products that require it).
- Place umbrella bags by building entrances to trap water that would otherwise drip on the hospital floor.
- Provide a sufficient number of water-absorbent walk-off mats with beveled edges at hospital entrances. The mats should be large enough for multiple steps to fall on the mat and wide enough to cover the entire doorway. As a general rule, when a person steps off the last mat, the soles of their shoes should not leave tracks on the floor. During inclement weather, it may be necessary to add additional mats or replace mats that have become saturated.

Prevent Pedestrian Access to Wet Floors

Mopping, disinfecting, stripping, and waxing floor surfaces in hallways, patient rooms, med rooms, bathrooms, kitchens, and cafeterias create slipping hazards for hospital staff. Simple steps to reduce this risk include

- Barrier signs (tension rod across bathroom doorways, cones with chains, hallway barriers) can be effective for blocking access to public bathrooms in the hospital.
- Use high visibility, taller "wet floor" caution signs/cones that can be joined by plastic chains or warning ribbons/tape to warn pedestrians of slippery conditions. Cordon off slippery areas and direct pedestrian traffic on a clear dry lane. Cones alone are not effective for keeping pedestrians off wet floors.
- Wet floor signs should be promptly removed after the floor is clean and dry to prevent staff from becoming complacent about the sign's intended warning.
- Completely block off pedestrian access when stripping or applying wax.

Slip-Resistant Shoes

Slip-resistant shoes are an important component of a comprehensive STF prevention program and staff who work on walking surfaces that are continually wet, greasy, or slippery may benefit from slip-resistant shoes. Job classifications at highest risk of an STF injury that may benefit from slip-resistant shoes are food service workers, housekeepers, custodians, maintenance workers, dishwashers, and instrument decontamination workers. Because nursing personnel suffer the highest total number of STF claims in hospital settings (Bell et al., 2008), they should also be included in a slip-resistant footwear program. Specialized shoes, designed to reduce the risk of slipping while stripping or applying wax, should be provided to housekeeping staff.

Anecdotal evidence suggests that use of slip-resistant shoes by employees is enhanced when the employer either provides the shoes or a payroll deduction for approved shoe purchases. Additionally, some shoe vendors will make periodic site visits so that employees may try on shoes at their workplace to ensure an appropriate fit. Shoe fit, comfort, and style are important factors that determine whether employees will wear slip-resistant shoes. It may be useful for employees to have the opportunity to try on shoes to obtain the proper fit before purchasing. Slip-resistant overshoes are a low-cost way to provide slip-resistance to staff.

Minimize Tripping Hazards

Exposed cords stretched across walkways and cords under workstations can catch an employee's foot and lead to a trip and fall incident. Clutter in walkways, storage areas, and hallways can potentially lead to a trip and fall incident. The following should be considered to minimize tripping hazards inside the hospital:

- Keep hallways, work areas, and walkways clear of objects and clutter.
- Use cord organizers to bundle and secure loose cords and wires under nursing stations, computer workstations, and patient rooms.
- Reroute cords so that they do not cross walking paths.
- Organize ORs to minimize equipment cords across walkways.
- Consider retractable cord holders on phones in patient rooms and nursing stations.
- Replace or restretch loose or buckled carpet.
- Replace mats with curled or ripped edges; secure edges with carpet tape.
- Remove, patch underneath, and replace indented or blistered floor tiles.
- Patch or fill cracks in walkways greater than 1/4″.
- Create visual cues for pedestrians by highlighting changes in walkway elevation with yellow warning paint or tape.
- Consider replacing smooth flooring materials with rougher-surfaced flooring with a higher coefficient of friction when renovating or replacing hospital flooring.

To minimize tripping hazards outside hospitals, the following strategies should be considered:

- Patch and repair holes, deep grooves, and cracks greater than 1/2″ in cement, asphalt, or other surfaces in parking areas and sidewalks.
- For adjoining walkway surfaces with changes in walkway level greater than 1/4″, bevel the surface by providing a ramp or provide a visual cue by painting uneven floor surfaces a bright contrasting color (i.e., yellow).

Operating Rooms

Although STFs occur throughout a hospital, the OR is of special interest because a fall in the OR can cause direct patient injury, disrupt the surgical procedure, contribute to surgical errors, or delay future surgeries (Brogmus et al., 2007). Brogmus et al. (2007) recommends that well-designed ORs include the following key features to reduce STFs in ORs:

- Slip-resistant flooring
- Procedures to control contaminants
- Proper floor-cleaning methods

- Policies and procedures to investigate fall incidents
- Minimized tripping hazards through securing and routing cords/cables out of walking paths
- Preplanned placement of low-profile equipment and supplies so that equipment and supplies remain off the floor and are accessible from mobile utility booms
- Use of slip-resistant absorptive mats to control contaminants, and removal of mats when they become saturated
- Maintenance of an unobstructed view of the walking pathways in the OR
- Lighting requirements are critical and enhanced OR lights should be used so that general lighting corresponds to the lighting level needed at the surgical site
- Efficient placement of equipment and supplies to minimize the walking distance to obtain instruments and supplies and to access waste containers
- Ample waste receptacle systems and contaminant clean-up materials placed in strategic locations
- Mandated planning briefings to discuss clean-up duty assignments and equipment and tube arrangement and routing to ensure OR team efficiency
- Minimize fatigue by ensuring all OR personnel are well rested for each procedure and receive scheduled time off and breaks during the work shift
- Participative architectural design; have architects, engineers, builders, and hospital administrators collaborate on OR-design decisions with end users such as surgeons, circulating nurses, and scrub technicians

Ice and Snow Removal

The most important aspect to controlling risks during winter weather is to remove snow and ice as soon as possible after it has accumulated. Hospital administrators should work with their snow removal staff or vendors to ensure frequent removal when needed. In addition,

- Encourage employees to report icy conditions; prominently display phone or pager numbers for staff to report icy conditions.
- Provide ice cleats or slip-resistant shoe covers for home health workers, maintenance workers, and other workers who work outdoors.
- Distribute winter weather warnings by e-mail when ice and snow storms are predicted.
- Conveniently place bins containing ice-melting chemicals near the top and bottom of outdoor stairways, parking garage exits, and heavily traveled walkways that are prone to refreezing so that any employee can apply ice melting chemicals when they notice icy patches.
- When renovating exterior entrances, build well-lit, good-draining, covered walkways leading to entrances to provide walk-off areas that allow for water, snow, and other contaminants to be removed from footwear before entering the building.

Lighting

Inadequate lighting impairs vision and ability to see hazards. The hazard can occur anywhere, but particular attention should be paid to lighting levels in parking structures, storage rooms, hallways, and stairwells. Adequate lighting helps to illuminate areas which makes walking safer and easier and allows employees to see their surroundings. Prevention strategies include

- Installing more light fixtures in poorly lit areas
- Verifying that light bulbs have an appropriate brightness
- Installing light fixtures that emit light from all sides

Stairs and Handrails

Uneven and poorly marked stairs can lead to missteps and can cause employees to trip and fall. Handrails that are not of the appropriate height or poorly maintained can also lead to a fall. Prevention strategies should

- Confirm all handrails are up to code (34–38″ from flooring).
- Ensure that discontinuous handrails are of a consistent height.
- Paint the edge (nosing) of each step, including the top and bottom, to provide a visual cue of a change in elevation.
- Ensure that stairwells and steps have adequate lighting.

Conclusion

Injuries related to STF events can cause serious injuries to hospital staff and are one of the leading causes of workers' compensation claims in hospital settings. Little emphasis has been placed on fall prevention among hospital and nursing home staff because of the widespread perception that these incidents are not preventable. However, examination of the details surrounding STF incidents among hospital staff indicates that many of these incidents are preventable. Research provides evidence that implementation of a comprehensive STF prevention program can significantly reduce STF injury claims involving hospital staff. Because STFs result from a wide variety of circumstances, a coordinated effort is required by the safety department, the housekeeping staff, and essentially every hospital staff member to have an impact on the prevention of STF incidents. Shared responsibility among hospital staff for maintaining safe floor conditions should be emphasized. In addition to all of the products and procedures that can be implemented to promote clean and dry floors, one of the key components of a successful STF prevention program is to raise awareness regarding the importance of STF prevention and to empower every employee to share in the responsibility of eliminating STF hazards. Whether this involves cleaning spills, applying ice melting chemicals to icy patches in parking areas or sidewalks, or cordoning off an area to alert fellow employees while waiting for housekeeping staff to arrive, a successful STF program requires that all hospital staff share the responsibility for prevention.

References

Bell JL, Collins JW, Wolf L, Grönqvist R, Chiou S, Chang WR, Courtney TK, Sorock GS, Lombardi D, and Evanoff B. Evaluation of a comprehensive STF prevention program for hospital employees. *Ergonomics*, 51(12):1906–1925, 2008.

Brogmus G, Leone W, Butler L, and Hernandez E. Best practices in OR suite layout and equipment choices to reduce slips, trips, and falls. *AORN J* 86(3):384–398, 2007.

Bureau of Labor Statistics. Occupational injuries and illnesses and fatal injuries profiles tool: Case and demographic numbers. U.S. Department of Labor, Washington DC, 2005. http://data.bls.gov/GQT/servlet/InitialPage (accessed January 7, 2008).

Bureau of Labor Statistics (BLS). Table R4. Number of nonfatal occupational injuries and illnesses involving days away from work by industry and selected events or exposures leading to injury or illness, 2006. Bureau of Labor Statistics, U.S. Department of Labor, Washington, DC, Survey of occupational injuries and illnesses in cooperation with participating State agencies, 2007a.

Bureau of Labor Statistics (BLS). Table R8. Incidence rates for nonfatal occupational injuries and illnesses involving days away from work per 10,000 full-time workers by industry and selected events or exposures leading to injury or illness, 2006. Bureau of Labor Statistics, U.S. Department of Labor, Washington, DC, Survey of occupational injuries and illnesses in cooperation with participating State agencies, 2007b.

Chang WR and Matz S. The slip resistance of common footwear materials measured with two slipmeters. *Appl Ergon* 32:540–558, 2001.

Collins J, Bell JL, Gronqvist R, Courtney TK, Sorock GS, Chang WR, Wolf L, Chiou S, and Evanoff B. Slip, trip, and fall (STF) prevention in health care workers. *International Ergonomics Association Triennial Congress Proceedings*, Maastricht, the Netherlands, 2006.

Courtney TK, Sorock GS, Manning DP, Collins JW, and Holbein-Jenny MA. Occupational slip, trip, and fall-related injuries—Can the contribution of slipperiness be isolated? *Ergonomics* 44:1118–1137, 2001.

European Standard EN 13287. Personal protective equipment—Footwear—Test method for slip resistance, March 2004.

Grönqvist R. Mechanisms of friction and assessment of slip resistance of new and used footwear soles on contaminated floors. *Ergonomics* 38(2):224–241, 1995.

Grönqvist R, Roine J, Järvinen E, and Korhonen E. An apparatus and method for determining the slip resistance of shoes and floors by simulation of human foot motions. *Ergonomics* 32:979–995, 1989.

Institute of Medicine. *To Err Is Human: Building a Safer Health System*. Institute of Medicine of the National Academies, Washington, DC, 1999.

Lacroix D and Dejoy D. Causal attribution to effort and supervisory response to workplace accidents. *J Occup Accid* 11:97–109, 1989.

Leclercq S, Tisserand M, and Saulnier H. Assessment of slipping resistance of footwear and floor surface. Influence of manufacture and utilization of the products. *Ergonomics* 38:209–219, 1995.

Lehane P and Stubbs D. The perceptions of managers and accident subjects in the service industries towards slip and trip accidents. *Appl Ergon* 32:119–126, 2001.

Li KW, Chang WR, Leamon TB, and Chen CJ. Floor slipperiness measurement friction coefficient, roughness of floors, and subjective perception under spillage conditions. *Safety Sci* 42(6):547–565, 2004.

Li KW and Chen CJ. The effect of shoe soling tread groove width on the coefficient of friction with different sole materials, floors, and contaminants. *Appl Ergon* 35(6):499–507, 2004.

Lombardi DA, Courtney TK, Verma SK, Brennan MJ, Wellman HM, Sorock, GS, Bell, JL, and Collins, JW. Risk factors for slips, trips, and falls: A case-crossover study of U.S. health care workers. *Oral Presentation at the Annual Meeting of the American Public Health Association*, November 3–7, 2007, Washington, DC, 2007.

Quirion F. Optimal cleaning for safer floors. In PT McCabe (ed.), *Contemporary Ergonomics*. CRC Press, Warrington, U.K., 2004, pp. 28–32.

Sotter G. *Stop Slip and Fall Accidents*. Sotter Engineering Corporation, Mission Viejo, CA, 2000.

Thorpe S, Loo-Morrey M, Houlihan R, and Lemon P. Slip and fall accidents in workplace environments—The role of footwear. *Proceedings of the International Conference on Slips, Trips, and Falls 2007—From Research to Practice*, August 23–24, 2007, IEA Press, Hopkinton, MA, 2007, pp. 188–192.

9

Ergonomics

Ergonomic Challenges in Hospital Ancillary Departments........9-1
Introduction • Environmental Services • Diagnostic Imaging • Clinical Laboratories • Pharmacy • Food Service • Summary
Acknowledgment........9-22
References........9-22
Ergonomic Design in the Workplace in Health Care Facilities........9-23
Introduction • Problem Situation • Levels of Functional Building Analysis • Dynamic Simulation of Future Activities • Conclusion
References........9-41
Reducing Occupational Back Pain Disability among Health Care Workers through Ergonomics........9-42
Back Pain and Health Care Workers • What Is Acceptable to Lift? • What Research Studies Say • Manual Lifting beyond Reasonable Capabilities • Previous Approaches to the Prevention of Back Injuries • Ergonomics as the Answer • Ergonomic Systems Approach • Summary
References........9-51

Margaret Wan
Kaiser Permanente

Jocelyn Villeneuve
Association for Health and Safety Sector Social Affairs

Guy Fragala
James A. Haley Veterans' Hospital

Ergonomic Challenges in Hospital Ancillary Departments

Margaret Wan

Introduction

Discussions of health care ergonomics have been focused on injuries of nursing staff as a result of patient transfers. The challenges faced by ancillary departments that provide supporting services have not received as much attention in the literature. Yet, as early as 1987, it was noted that upper extremity musculoskeletal disorders (MSDs) in hospital workers were prevalent not only among nursing staff, but also among those engaged in manual work such as food preparation and laundry.[1] More recently, Goldman et al.[2] identified high-risk areas for back injury in a large teaching hospital. The authors found that non-nursing areas, which demonstrated increased rates for back injury, included environmental services, pharmacy, radiology, and dietary work.

This chapter reviews some ergonomic issues encountered in various hospital ancillary departments and provides examples of successful interventions. These issues and interventions are by no means exhaustive. Since the scope of ancillary services varies and is often dependent on the size of the hospital, only the most common services are discussed. They are environmental services, diagnostic imaging, clinical laboratories, pharmacy, and food service.

Environmental Services

The environmental services department is responsible for general cleaning and processing linen, trash, and biohazardous waste. The major tasks performed may be divided into cleaning and materials handling. Ergonomic risk factors in environmental services work are well known. In a study of 941 unionized hotel room cleaners, only 5% of participants reported no bodily pain during the past 4 weeks.[3] Most participants experienced severe back or neck pain, which was associated with physical workload, work intensification, and ergonomic problems related to equipment and supplies. In the hospital setting, the risks may be higher. For instance, infection control concerns may demand more frequent removal of trash from patient rooms, compared with hotel guest rooms. According to a study of injury rates of various jobs at two private hospitals, janitors and building cleaners had the highest injury rate of 21.3 per 100 full-time employees.[4] Another hospital with 140 employees in the environmental services department had 80 injuries over a 2 years period. Fifty-three percent of these injuries were caused by lifting, pushing or pulling, slips, and needlesticks. About 70% of the injuries occurred during cleaning and 30% of the injuries occurred during linen, trash, and biohazardous waste processing. These percentages align with the distribution of employees between the two broad categories of tasks in this department. To determine the solutions, we must first consider the possible causes of the injuries in each category.

Cleaning

The majority of hospital cleaners are women. Women generally experience greater relative spine compression loads than men and are more at risk of injury under the same physical work demands.[5] Nordander et al.[6] compared muscle activities of hospital cleaners and office workers. They found that cleaning was associated with a much higher static load and cleaners had much less muscular rest, measured as a percentage of total registered time in one working day.

The cleaning method applied may make a difference in the physical and perceived load. Hagner and Hagberg[7] evaluated 11 healthy female cleaners when they mopped the floor using the "push" method or the "figure-of-eight" method. There was less local muscle loading and perceived exertion when the former method was used. Nevertheless, even the "push" method seemed to result in levels of static loading and oxygen uptake that might be harmful.

New technology has produced equipment with improved ergonomic designs. Conventional mops have wooden handles and are heavy, especially when they are wet. The new microfiber mops are lightweight and easily maneuverable. Their use requires minimal physical exertion. Ergonomic mop handles and extendible shafts promote neutral body postures. Backpack vacuum cleaners have been found to have a biomechanical design superior to upright cleaners.[8] Backpack vacuum cleaners are also more efficient. Experienced workers using backpack vacuum cleaners could clean much faster than using the upright design at similar levels of energy expenditure and perceived effort.[9]

Implementing a participatory ergonomics approach has been found effective in reducing the risk and severity of injuries among 137 hospital cleaners.[10] Such a program may include purchasing floor coverings that are easier to clean, modifying cleaner's trolleys so they are easier to move, and increasing task rotation to prevent overexertion. Wearing nonskid shoes and drying wet surfaces promptly will reduce the risk of slips, trips, and falls. Selection of flooring based on coefficients of friction must be balanced between the ease of moving trolleys and the probability of slips and falls.

Materials Handling

The handling of linen and trash is a major cause of ergonomic injuries in the environmental services department. Typically, linen is placed in bags which may weigh up to 30 lb each. The bags are transported in large, tall linen carts (Figure 9.1). One load averages about 240 lb. The height of the trash carts is less—the top edge and the handle are approximately 43 and 37 in., respectively, from the floor. However, in order to reduce the number of runs, employees tend to overload the carts (Figure 9.2). A load of trash may weigh 200–380 lb.

FIGURE 9.1 Tall linen carts.

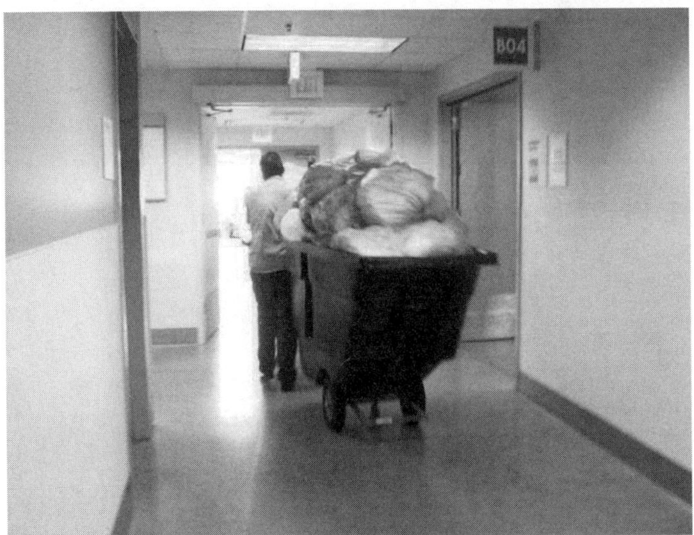

FIGURE 9.2 Overloaded trash cart.

Repetitive heavy lifting is required in the process of filling these linen and trash carts. As the carts are filled up, lifting above shoulder height occurs. The problem is aggravated for persons of short stature. In spite of materials handling guidelines to push rather than pull, employees pull the carts because their line of sight would be obstructed by the load if they walk behind the cart. Hospitals use linen carts similar to those used in the hospitality industry. In a study of hotel room cleaners, 84% of participants considered the linen carts too heavy.[3]

When the linen carts are filled, they are transported to the loading dock for pick up by a contractor or to the hospital's in-house laundry, where unloading takes place. Similarly, when the trash carts are filled, the contents must be unloaded into a dumpster or compactor. If this task is performed manually, an employee must reach into the cart to grasp and remove the bags of trash and load them into the dumpster or compactor, most likely at a height above shoulder level. The task involves extended horizontal and vertical reach and repetitive non-neutral postures.

Issues related to the handling of biohazardous waste are similar to those related to handling other trash, except that usually a contractor for biohazardous waste disposal will unload the biohazardous waste from the containers after environmental services employees have collected the waste. In the collection process, the biohazardous waste containers weighing up to 55 lb are lifted. They have inadequate handles and no caster. Nevertheless, in one hospital, sprains and strains accounted for only 25% of injuries related to the processing of biohazardous waste. The other 75% of injuries were caused by needlesticks.

In addition to scheduled rest breaks and job rotation, interventions have been devised to reduce the physiological workload and increase the efficiency of handling linen and trash:

- Linen or trash chute systems are designed to transport laundry or trash to lower levels of the building (Figures 9.3 and 9.4). Their advantage is that manual materials handling and ergonomic stressors are minimized. Their disadvantage is that it will be extremely costly and difficult, if not impossible, to install these systems in existing facilities. Some systems may create problems with noise or odor but manufacturers have introduced designs to overcome such issues. Infection control concerns may require that chutes not be used for biohazardous waste. The chutes should be designed properly to avoid items being stuck inside, requiring an employee to adopt an awkward posture to retrieve them.
- Soiled linen collection systems such as the one shown in Figure 9.5 eliminate many musculoskeletal stressors in handling linen. They also improve the containment of airborne bacteria and odor and help prevent incidents caused by misplaced sharps. One disadvantage of some systems is that the lids of the containers must be lifted manually as there is no foot pedal. Furthermore, existing facilities must determine if the ceiling height provides sufficient clearance for the system to operate. Other space constraint may apply.

FIGURE 9.3 Trash chute intake.

FIGURE 9.4 Trash chute discharge and receiving cart.

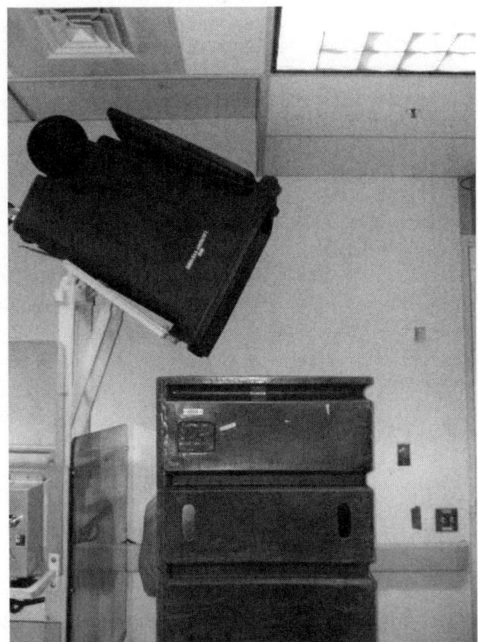

FIGURE 9.5 Soiled linen collection system.

- Motorized equipment is available to assist in the transfer of linen and trash carts. Figure 9.6 shows one example. When motors are used, it is possible to join linen carts together with a hitch to optimize working time (Figure 9.7). As in the case of patient lift equipment, employees do not always take advantage of the available equipment because they consider it cumbersome or time consuming to use. Employee education is essential to ensure acceptance and proper use of the equipment.

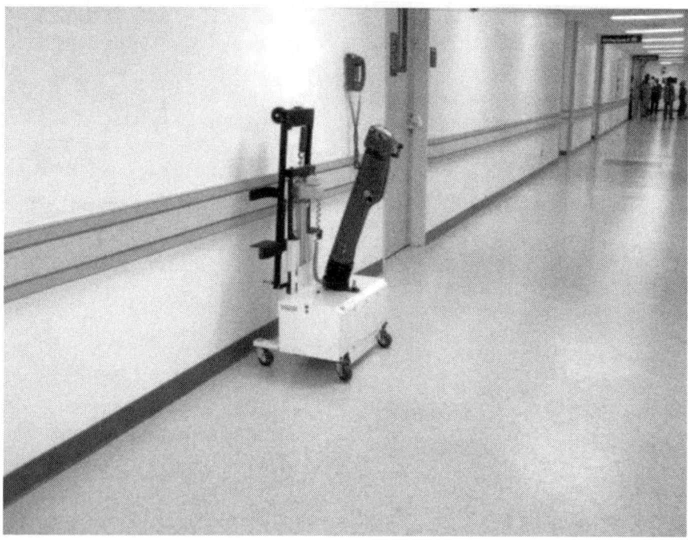

FIGURE 9.6 Motor for transporting carts.

FIGURE 9.7 Hitch for joining linen carts.

- Automated "no-touch" cart dumpers can be employed to eliminate manual transfer of trash from the trash carts to the dumpster or compactor.
- Using carts with a spring platform can optimize the hand working heights when employees unload linen from the carts.
- Regular preventive maintenance of the equipment, whether manual or motorized, improves efficiency and reduces the body force required to handle the equipment.
- Reducing the height of the linen carts is effective in reducing the load and the height of lifting (Figure 9.8). The smaller capacity of each cart may mean longer collection times due to more runs, which can be offset by using the motorized equipment and hitch. If a hospital contracts out its linen service, the linen carts used by employees may be the property of the contractor and the hospital will have to work with the contractor to make modifications.

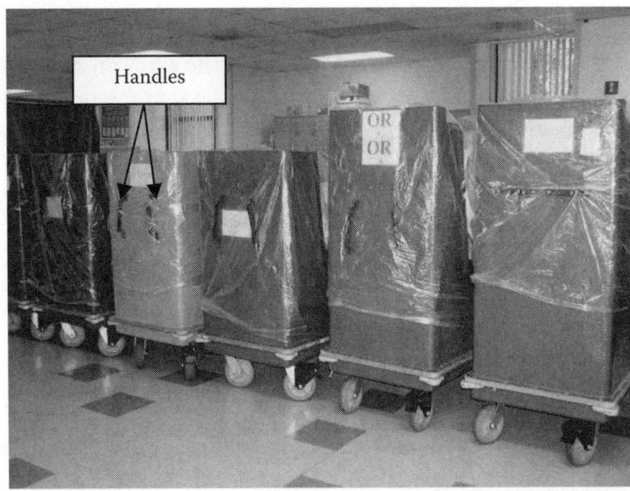

FIGURE 9.8 Comparison of the shorter and taller linen carts. Handles have been added for better grip.

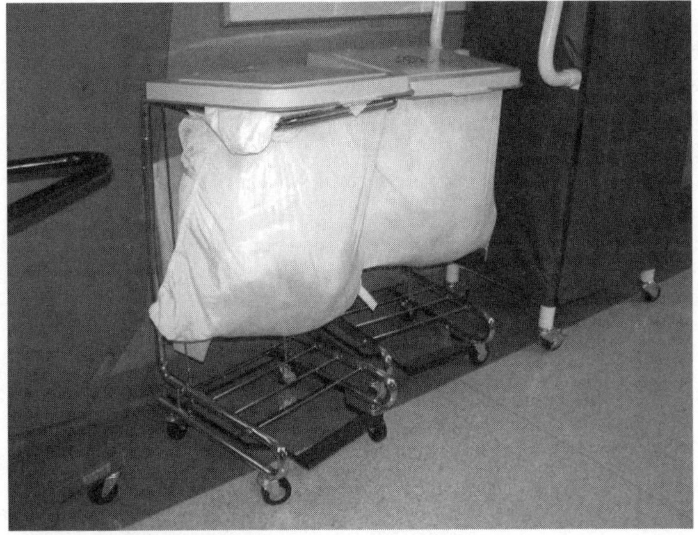

FIGURE 9.9 Smaller linen bags reduce weight and force exertion.

Handles have been added for better grip.

- Adding handles to the carts provides a better grip and reduces the force required to move the carts (see Figure 9.8).
- Decreasing the size of the linen bags helps reduce the weight and the force exertion during pick-up and transfer of the bags (Figure 9.9).
- Using larger wheels (6–8 in.) or casters with tapered roller bearings for the carts increases mobility and decreases pushing force.

Diagnostic Imaging

A range of technology is used in diagnostic imaging, including x-ray, ultrasound, computed tomography, and magnetic resonance imaging. In diagnostic imaging, ergonomic risk factors exist in both the scanning process and the interpretation process.

Scanning Process

The scanning process is usually performed by a technician. The exact procedures vary by the specific technology. In general, the employee positions the patient in the appropriate place relative to the equipment and operates the equipment to acquire images of the targeted body parts. Positioning the patient can be a biomechanically demanding task if the patient is not ambulatory. Safe patient handling is covered elsewhere in this book. Manipulation of the equipment, data entry into the computer, viewing of the monitor, and handling of supplies and materials may pose ergonomic risks associated with forceful pinch grips, prolonged awkward postures, or extended reach.

X-Ray Technologists

In a study of x-ray technologists, some of the tasks were found to be stressful due to repetition.[11] An example was the handling and positioning of cassettes. Other activities might be performed at awkward joint angles, such as pushing and pulling x-ray tubes.

In one hospital, two technologists sustained injuries and other employees complained of wrist, hand, arm, and finger pain and discomfort due to handling of heavy cassettes. A job analysis revealed the following:

- Each employee might take care of 20 patients a day.
- Four cassettes per patient were needed for most patients.
- Each cassette could weigh up to 20 lb and was moved three to four times, between the storage rack, the x-ray table, and the x-ray film processor. In other words, one technologist might move cassettes up to 320 times in a day. The average travel distance was 75 ft each time.
- The technologists had to twist the wrist three to four times while using a cassette, so the total wrist extensions, flexions, and deviations were about 1280 times a day.
- The cassettes had no handles for a good grip (Figure 9.10).
- Placing a cassette into the x-ray film processor required lifting the cassette at shoulder level.
- One x-ray tube was heavy and moving it required a 40 lb force.

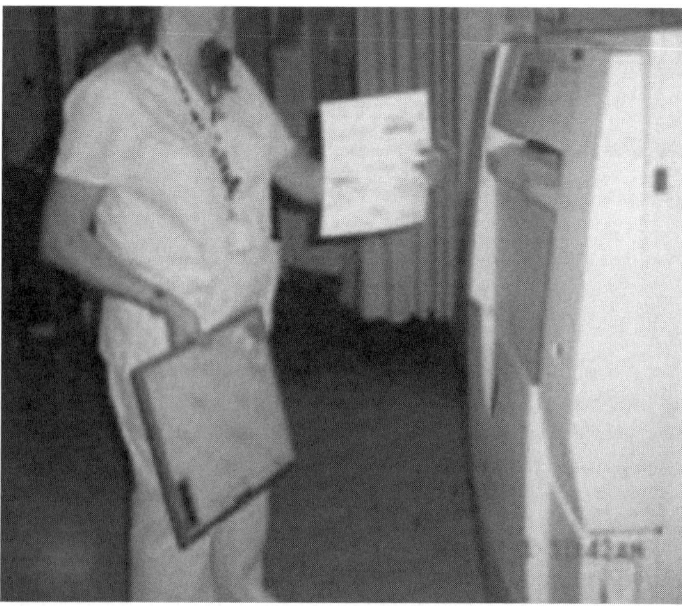

FIGURE 9.10 Holding cassette without handle.

Some of the above factors could not be modified without changing the x-ray equipment or the room layout, which would incur considerable expenses. Other factors could be modified:

- Change the configuration of the storage rack so that the heaviest cassettes could be placed horizontally on the upper portion of the rack to facilitate access.
- Transport the cassettes with a cart rather than by hand.
- Train the technologists in body mechanics and stretching; advise them to lift cassettes with both hands and close to the body, to maintain neutral postures as much as possible, and to take frequent stretch breaks.

If the x-ray equipment can be changed, conversion to digital radiology will eliminate the insertion or removal of cassettes, which requires a forceful pinch grip. The filmless technology is also a preferable method of archiving since the life of films in general radiology is 7 years. Going digital will eliminate the pinch grip used by employees when carrying the large processed films.

Sonographers

The National Institute for Occupational Safety and Health (NIOSH) has found that sonographers are at risk of developing work-related MSDs in the upper extremities, neck, and back.[12] Figure 9.11 demonstrates the scanning task performed by a sonographer at a standard workstation. Positioning of the equipment leads to awkward postures—twisting of the neck, flexion and abduction of the shoulder, and extension of the elbow. Holding the transducer requires a pinch grip.

Measures to minimize these ergonomic risk factors may include solutions often employed in office ergonomics. Adjustable workstations, monitors, and chairs can reduce awkward postures and eyestrain. Furniture and equipment readily available in the market can be effective in alleviating employee complaints about shoulder, arm, wrist, and back pain or soreness. Figure 9.12 shows a sonographer's workstation, where an armrest is attached to the patient's bed and the sonographer sits on a saddle chair. In one hospital, implementing this combination of armrest and saddle chair has reduced arm and shoulder discomfort by 70%–90%.

Interpretation Process

Radiologists view and interpret medical images in the office setting. However, the workstation design is more complex than commonly found at office workstations. The picture, archiving, and collection (PAC) system which the radiologists use have multiple monitors that are viewed side by side. It may

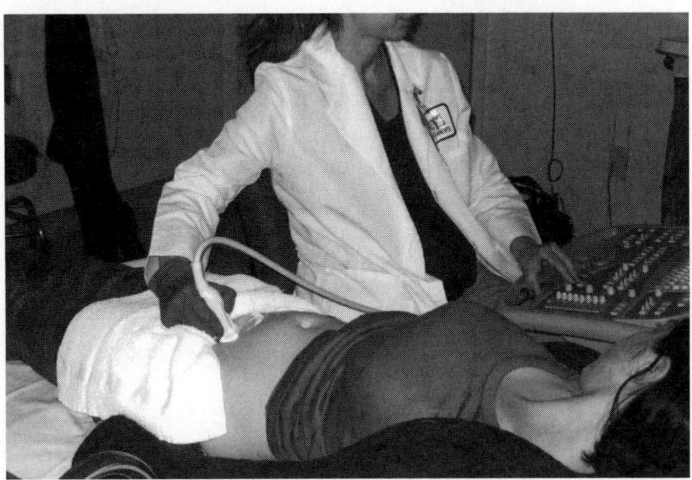

FIGURE 9.11 Sonographer performing scanning task at standard workstation.

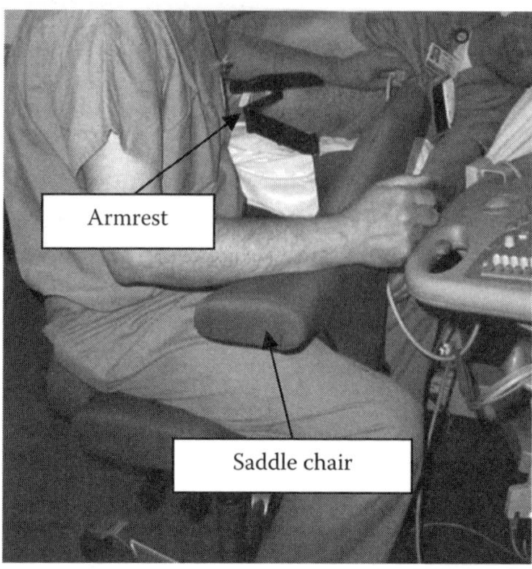

FIGURE 9.12 Sonographer performing scanning task at modified workstation.

FIGURE 9.13 Radiology reading station with chair that adjusts for spine postures.

appear that this setup would reduce neck strain, since the head of the radiologist is no longer fixed in one position but moves to slightly different positions to view different monitors. However, potential working-memory strain exists, particularly among new or infrequent users.[13] Hospitals are testing innovative designs of radiology workstations and reading rooms to reduce ergonomic stressors as much as possible. The environmental design of the radiology department in one hospital takes into account the intensity and quality of the ambient light, temperature, relative humidity, as well as layout of the workstation and other factors. Figure 9.13 presents an example of one of their workstations.

Clinical Laboratories

Hospital laboratory employees perform a variety of tasks that range from drawing blood to analyzing specimens. Many of these activities are associated with ergonomic risk factors. Ramadan and Ferreira[14] evaluated predominant work activities of workers at a laboratory of clinical pathology and found that a higher percentage of MSD symptoms were reported by workers performing operational activities, compared to workers performing nonoperational activities (72.1% vs. 27.9%). Operational activities included tasks most commonly performed at this type of laboratories, whereas nonoperational activities involved administrative and management tasks. Kilroy and Dockrell[15] found that the 3 months prevalence of musculoskeletal symptoms among female biomedical scientists was 79%, although this percentage decreased to 54% after instituting ergonomic interventions that comprised workplace changes and education on risk factors.

The following sections discuss some of the operational activities in clinical laboratories that pose ergonomic challenges.

Phlebotomy

Smith[16] described a case of lateral epicondylitis in a phlebotomist, who venesected up to 150 patients per shift. The injury was thought to be caused by the forceful gripping and repetitive twisting needed to break the seals on green vacutainer needles. Subsequently, the hospital works department created a device to break the seal. The hospital also enforced rest breaks and increased the number of phlebotomist to reduce work pressure. Eventually, the manufacturer of the vacutainer needles changed the seals so that they were smaller and easier to break.

Phlebotomists who draw blood in a standing position may experience pain in the back and lower extremities due to the static load. Antifatigue mats are desirable in reducing the load. If the staff switch between standing and sitting positions throughout the day, antifatigue runners or flooring is preferable to avoid having to roll stools across the beveled mat edges.

Blood draw stations can have adjustable tables and chairs to accommodate clinicians and patients of various heights and to promote neutral body postures (Figure 9.14). The benefits of the adjustable blood draw station may not always be fully realized due to ignorance or disregard of ergonomic principles. As seen in Figure 9.15, the patient's adjustable chair had been replaced by a non-adjustable chair. As a consequence, the top surface of the table was not adjusted to the proper height, so the phlebotomist had

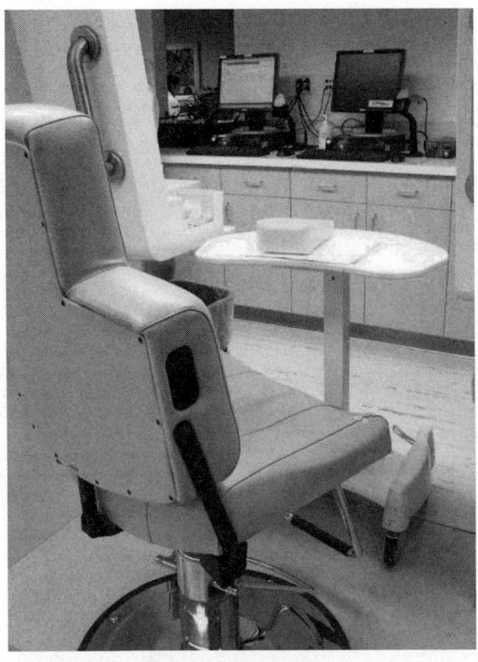

FIGURE 9.14 Blood draw station with adjustable table and chair.

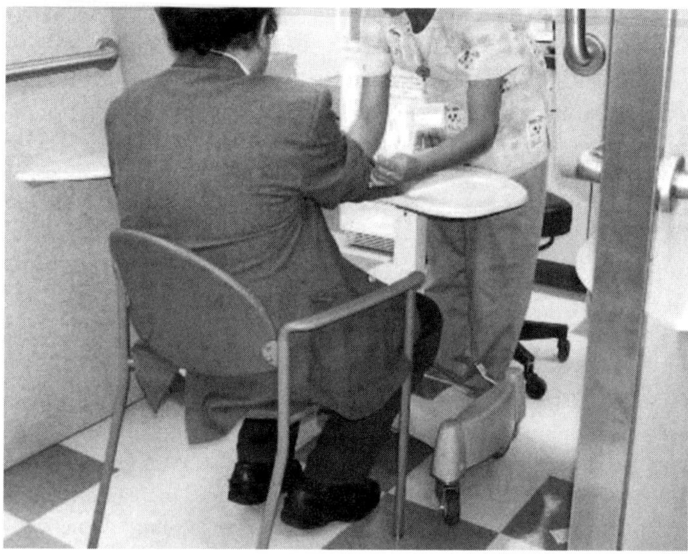

FIGURE 9.15 Blood draw station without proper chair and table adjustments. The phlebotomist had to bend the trunk to draw blood.

to bend forward while drawing blood. Phlebotomists should be trained on the importance of following ergonomic guidelines and using ergonomically designed furniture and equipment.

The phlebotomist had to bend the trunk to draw blood.

Pipetting

Laboratory personnel often perform the task of pipetting. Repeated use of pipettes can cause MSDs in the upper extremities such as carpal tunnel syndrome, lateral epicondylitis, and thumb tenosynovitis. That is because in pipetting, tip insertion, plunger manipulation, and tip ejection require relatively high levels of hand forces. Prolonged static postures reduce blood flow in the upper extremities. Repetitions lead to wear and tear on the tendons and joints. Some pipettes are designed so that the wrist must be bent to operate the pipette. Minuk et al.[17] reported on pipetter's thumb injuries. Fredriksson[18] found that the strain on the thumb for a woman with weak muscular structures was unacceptably high and that symptoms increased with the amount of time spent with pipetting. Bjorksten et al.[19] determined that the prevalence of hand ailments among laboratory assistants using plunger-operated pipettes was twice that among female state employees in general. An increased risk of hand and shoulder ailments was associated with more than 300 h per year of pipetting. David and Buckle[20] discovered an association between the use of pipettes and upper-limb disorders. They observed a dose–response relationship in that the percentage of those reporting hand complaints increased as the duration of continuous use increased.

During pipetting, the user must hold the pipette firmly and aim the tip of the pipette precisely into the receiving vessel. Some pipettes require a 4-finger grasp. Lee and Jiang[21] designed a pipette that required a power grip, which improved the precision and comfort and shortened the time required to perform the task. Another way to reduce ergonomic risk factors while pipetting is to select pipettes with minimal plunger travel and low tip-ejection and plunger forces. Pipettes that offer a comfortable grip, promote a neutral position of the wrist, and minimize thumb extension are desirable. An electronic pipette can reduce force and repetition since a motor controls the plunger and the pipette is programmable for mixing, diluting, and dispensing multiple samples. Good work practices to lessen the loading on the tendons include taking rest breaks, keeping supplies within comfortable reach, alternating pipetting with other activities and between right and left hands if possible, using only the minimal amount of force to press the plunger, varying the way the pipette is held, and maintaining neutral postures.

Specimen Processing, Analyses, and Storage

In the pathology laboratory, employees use a microtome to cut specimens into thin segments for microscopic examination. Conventional microtomes are mechanical devices that require repetitive hand-cranking and bending at the wrist (Figure 9.16). The introduction of the laser microtome removes these ergonomic risk factors while improving the quality of tissue-processing. Until laser microtomes are widely used, administrative controls such as scheduled rest breaks and job rotation are needed to prevent MSDs among microtome users.

Examining specimens under a microscope for prolonged periods exposes employees to continuous static muscular load and an increased risk of MSDs in the neck, shoulder, upper extremities, and back. According to Sillanpaa et al.,[22] 75% of microscope workers in a research center suffered from pain in the shoulder region, 57% suffered from pain in the neck region, 49% suffered from pain in the lower back, and 39% suffered from pain in the upper back. The authors designed an adjustable microscope table that improved the neck position, allowed the user to keep the head in an upright position, and supported the forearms with less flexion of the upper arm. The design was effective in reducing the static load on the neck and other muscles. Other ergonomic solutions may include the use of adjustable eyepieces or eyepiece extensions to reduce awkward neck and back postures, addition of rubber eyecups to minimize soft tissue compression, or installation of a video system or eyepieceless microscopes to allow greater head and body movement while viewing slides.

Some pathology laboratories hire employees to transcribe notes dictated by the pathologists or pathologist assistants as they analyzed specimens. The transcriptionist spends long hours listening to the recording and typing the notes on the computer. Applying basic office ergonomics principles in the workstation design and encouraging regular stretch breaks will minimize eyestrain and other problems associated with prolonged use of computers. Speech recognition software designed for the field of medicine offers an alternative to manual transcription and facilitates accurate and timely documentation.

Another area of concern in the pathology laboratory is manual materials handling, in this case the materials being heavy slide drawers. It is common for medical assistants to transfer drawers that are

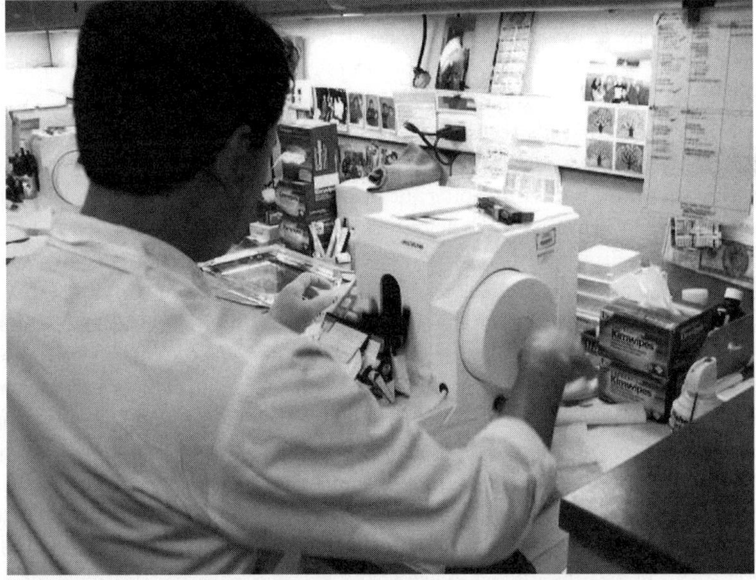

FIGURE 9.16 Employee operating microtome. The right wrist shows extreme flexion.

FIGURE 9.17 Handling slide drawers in an awkward posture.

loaded with slides and weigh about 5 lb. This weight may not seem a lot, but the drawers are long and narrow, making it hard to balance them in the hands. The employee must be very careful and must hold the drawers tightly. In Figure 9.17, an employee was seen in an awkward posture trying to balance the weight of the drawer and place it into the lower basket of a cart. Furthermore, the manner in which slide drawers are stacked requires the employee to bend over to reach the bottom drawers and to extend the arms above shoulder height to reach the upper levels. Performing these job tasks is prone to cause back, shoulder, and hand injuries.

Pharmacy

A hospital pharmacy department may be divided into inpatient, outpatient, and infusion center. These operations have similar ergonomic challenges. For example, in both the inpatient and outpatient pharmacies, pharmacists must count pills when filling prescriptions. Automatic pill counters can reduce the time the pharmacist's neck is positioned in a non-neutral posture. There are also differences among these pharmacy operations, as explained in the following sections.

Inpatient Pharmacy

In inpatient pharmacy, pharmacists receive the prescriptions and prepare the documentation on the computer (Figure 9.18). Unless a robotic system is used, dispensing requires that the drugs be manually picked up from inventory bins. These bins are stacked on shelves against the wall, almost to ceiling height (Figure 9.19). Employees must reach below knee level or above shoulder level when they perform the tasks of stocking and dispensing. They may need to climb on a ladder or stool to reach the top shelves. A well-constructed, stable step ladder with handrails will prevent trips and falls (Figure 9.20).

Employees preparing the drugs to fill the orders may be working in a standing position much of the time. Antifatigue runners can help relieve the static load on the back and lower extremities.

Automated robotic systems can be used to dispense medications and avoid repetitive motion injuries. They offer other advantages such as reduced medication errors and improved inventory management.

Medications and supplies are transported to patient rooms in pharmacy carts. One hospital measured the push–pull force associated with the use of a pharmacy cart carrying a typical load. On a smooth surface with tile flooring, the peak push force and peak pull force were approximately 20 lb. The peak push force and peak pull force increased to 35 and 28 lb, respectively, on carpeted floor. Smooth floor surfaces make it easier to roll the carts and are also easier to clean, an advantage for environmental services personnel.

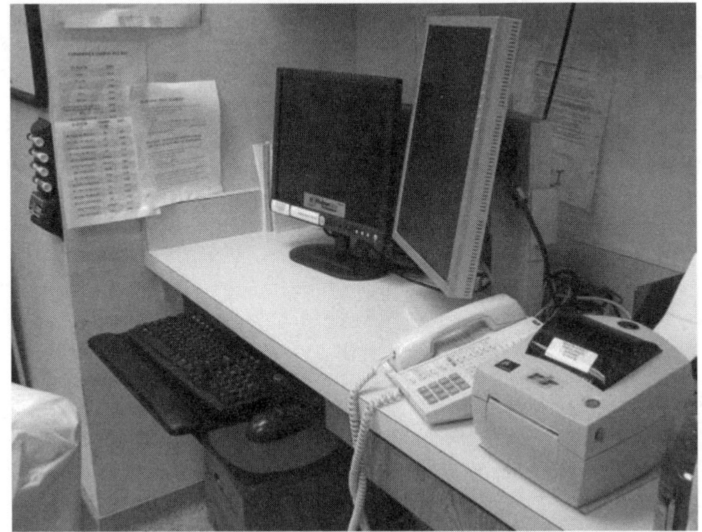

FIGURE 9.18 Pharmacist's workstation in inpatient pharmacy.

FIGURE 9.19 Inventory bins stacked on shelves against the wall.

Outpatient Pharmacy

In the outpatient pharmacy, prescriptions are frequently filled while the patient or a family member is waiting. The outpatient pharmacy has workstations set up for interaction with customers and for cash or credit card point-of-sale transactions (Figure 9.21). Other workstations are located close to the inventory bins. Sometimes, these workstations are set up on the top of the counter and they are not adjustable. The monitor and keyboard are placed at a height that is not suitable for either standing or sitting work, causing much elbow flexion (Figure 9.22).

FIGURE 9.20 Step ladder with handrails in inpatient pharmacy.

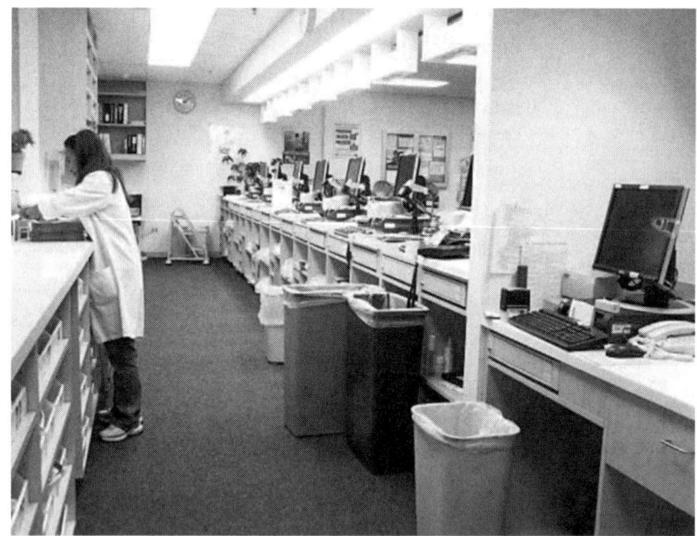

FIGURE 9.21 Outpatient pharmacy point-of-sale workstations.

Other ergonomic challenges in the outpatient pharmacy are similar to those in the inpatient pharmacy, such as stocking and retrieving inventory from bins on the upper and lower shelves and moving heavy carts that carry medications and supplies.

Infusion Center

Infusion centers may serve outpatients who come to the hospital for treatments or who need the medications for use at home. In one hospital, employees were ripping open the seals of medicine containers by

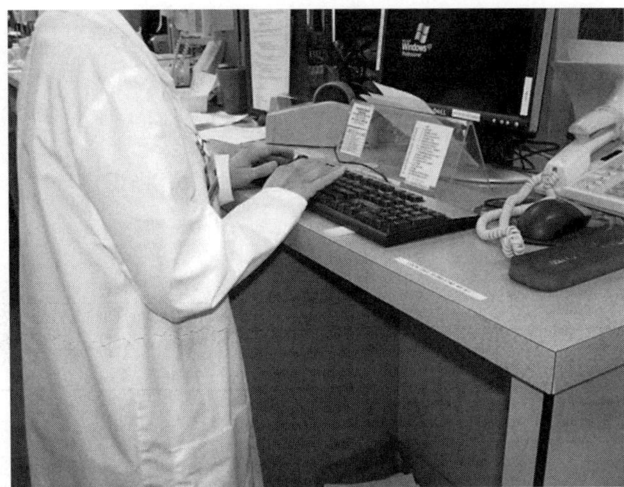

FIGURE 9.22 Outpatient pharmacy workstations on counters.

FIGURE 9.23 Overuse of fingers to open seals.

hand, as frequently as 28 times a shift, causing MSD symptoms (Figure 9.23). The in-house ergonomist recommended the use of a simple and inexpensive tool—a pair of pliers—to open the seals and that eliminated the problem (Figure 9.24).

In the same infusion center, employees short in stature had difficulty hanging bags of intravenous medication onto the horizontal bar inside the laboratory hood. They were able to use a stepping platform (Figure 9.25). Later on a pair of brackets was added to the sides of the hood so that the horizontal bar could be positioned at a lower level (Figure 9.26). These examples show that effective ergonomic solutions can be found in readily available and inexpensive materials.

Food Service

In addition to heat stress, burns, slips, trips, and falls, employees working in the kitchen face a combination of risk factors for the development of MSDs. For example, dishwashing required frequent bending

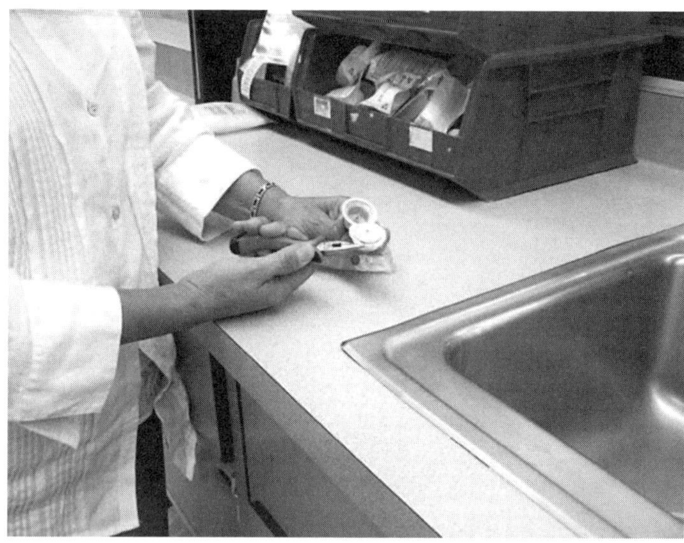

FIGURE 9.24 Using pliers to open seals.

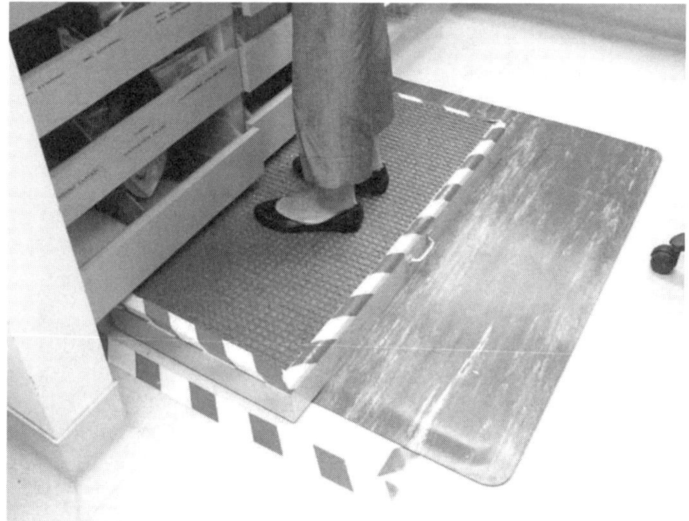

FIGURE 9.25 Stepping platform.

and twisting of the trunk, lifting, repetitive movements of the hands, non-neutral wrist and shoulder postures, and use of hand force.[23] Most kitchen work is done when standing or walking. It was not surprising, therefore, that only 13% of 495 female kitchen workers reported no musculoskeletal pain symptoms during the past 3 months.[23] Neck pain was the most predominant symptom reported but pain in the back and upper extremities was also common. Aminoff et al.[24] studied six female and three male hospital kitchen workers while working on a conveyor belt to collect and sort dirty plates, glasses, and cutlery for cleaning. The mean oxygen uptake corresponded to 41% of the individual's peak oxygen uptake. Such high work intensity has the potential of creating excessive physiological strain.

One hospital conducted an ergonomics analysis of its food service department with 50 employees working at food preparation and cafeteria server workstations. The employees complained about pain in

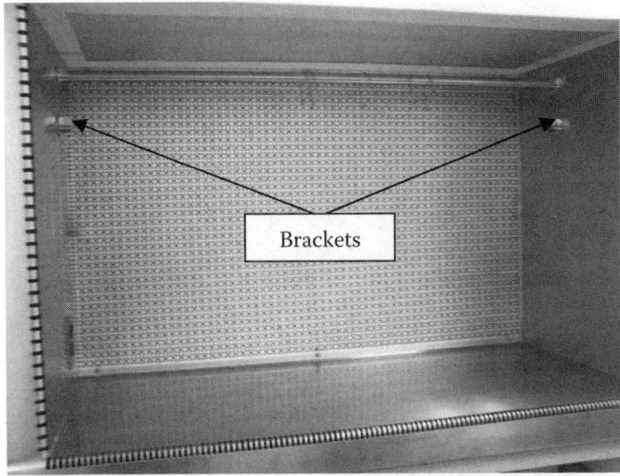

FIGURE 9.26 Brackets added to laboratory hood to allow additional position of horizontal bar.

the upper extremities, neck, and shoulder. The number of injuries increased from 13 to 16 between the years 2005 and 2006. The following risk factors were identified:

- Repetitive motion of the hands, wrists, and fingers was used to cut fruits and vegetables (Figure 9.27).
- Food bags and boxes weighing more than 30 lb each were lifted and placed at multiple levels that might be lower than knee height or higher than shoulder height (Figure 9.28).
- Food preparation workstations were not height-adjustable, resulting in continuous bending of the neck for many employees (Figure 9.29).
- Server workstations were poorly designed, requiring 180° of body twisting and neck rotation in order for a server to prepare the food and communicate with the waiting customers (Figure 9.30).

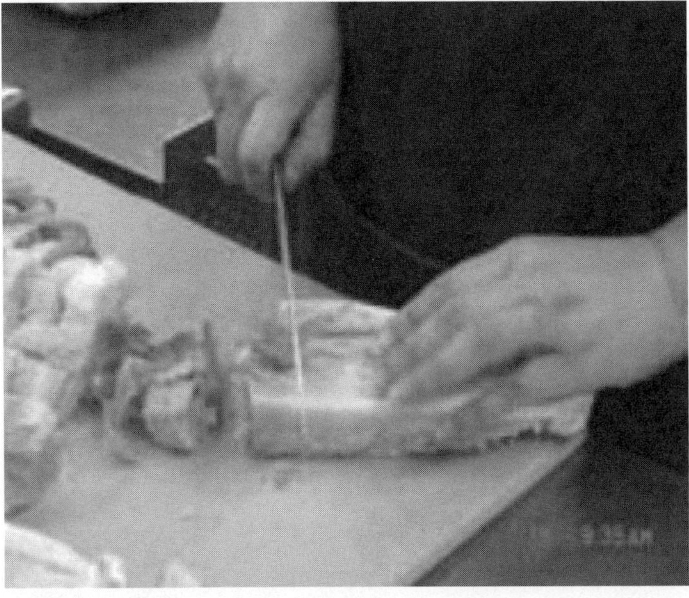

FIGURE 9.27 Cutting vegetables requires repetitive motion of the hands.

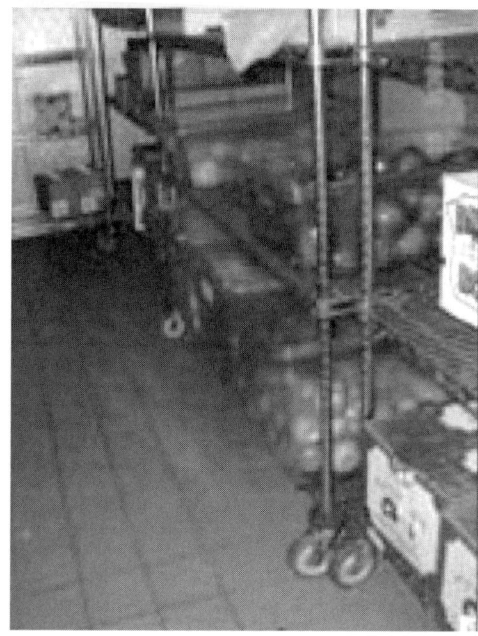

FIGURE 9.28 Multilevel racks with bags and boxes of food.

FIGURE 9.29 Food preparation workstation.

FIGURE 9.30 Cafeteria server workstation.

- Forceful grip of tongs and repetitive hand and wrist motions were observed in the employees at the server workstations. Hamburger was the most popular order. On the average, more than 150 hamburgers were sold in a day. Approximately, 20 movements of the hands, wrists, and fingers were needed to prepare one hamburger, amounting to more than 3000 movements per day.
- Practically, all work was done standing or walking.

Recommendations to overcome these challenges included the following:

- Purchase a lettuce cutter or order lettuce precut to reduce the amount and time of cutting food.
- Buy food in smaller, lighter packages to reduce the risk of lifting injuries.
- Use an adjustable cutting board platform to reduce neck strain.
- Create a self-service system for some food items or eliminate serving hot food altogether. Having customers serve themselves will cut down the repetitive motion required of employees.
- Implement scheduled rest breaks and job rotation.
- Train employees in body mechanics and the importance of maintaining neutral postures.

Another ergonomic risk factor among food service employees arises from the need to transport prepared food to patient rooms. Motorized food service trams are preferable to manual carts as pushing and pulling forces are minimized. Several trams can be joined in a train, so this method is safer and more efficient than manual handling.

Summary

This chapter has provided examples of ergonomic challenges and solutions in five hospital ancillary departments. Whereas some solutions may be specific to one department's unique situation, the following recommendations have universal applications:

- Work with manufacturers and contractors to create or improve a product so that it has a better ergonomic design.
- Better ergonomic design promotes not only workplace safety, but also patient safety.

- New technology can incorporate better ergonomic design while improving the efficiency and outcomes of clinical procedures.
- Scheduled rest breaks, job rotation, and training employees in body mechanics, lifting techniques, and stretch exercises are effective in reducing ergonomic injuries.

Acknowledgment

I thank Dr. Khan Siddiqui of the VA Maryland Health Care System, Mr. Dan Williams of Christiana Care Health System, Ms. Nancy Terrebonne and Mr. Soleiman Djavaheri, both of Kaiser Permanente, for providing some of the information and photographs included in this section. The contents of this section are solely the responsibility of the author and do not necessarily represent the official views of Kaiser Permanente.

References

1. Punnett L. Upper extremity musculoskeletal disorders in hospital workers. *J Hand Surg* 12(5 Pt. 2): 858–862, 1987.
2. Goldman RH, Jarrard MR, Kim R, Loomis S, and Atkins EH. Prioritizing back injury risk in hospital employees: Application and comparison of different injury rates. *J Occup Environ Med* 42(6): 645–652, 2000.
3. Krause N, Scherzer T, and Rugulies R. Physical workload, work intensification, and prevalence of pain in low wage workers: Results from a participatory research project with hotel room cleaners in Las Vegas. *Am J Ind Med* 48(5):326–337, 2005.
4. d'Errico A, Punnett L, Cifuentes M, Boyer J, Tessler J, Gore R, et al. Hospital injury rates in relation to socioeconomic status and working conditions. *Occup Environ Med* 64(5):325–333, 2007.
5. Marras WS, Davis KG, and Jorgensen M. Spine loading as a function of gender. *Spine* 27(22): 2514–2520, 2002.
6. Nordander C, Hansson GA, Rylander L, Asterland P, Byström JU, Ohlsson K, et al. Muscular rest and gap frequency as EMG measures of physical exposure: The impact of work tasks and individual related factors. *Ergonomics* 43(11):1904–1904, 2000.
7. Hagner IM and Hagberg M. Evaluation of two floor-mopping work methods by measurement of load. *Ergonomics* 32(4):401–408, 1989.
8. Standley S, Murray C, and Woellner RA. Back talk: Key considerations for backpack vacuum ergonomics. *Health Facility Manage* 17(12):33–36, 2004.
9. Mengelkoch LJ and Clark K. Comparison of work rates, energy expenditure, and perceived exertion during a 1-h vacuuming task with a backpack vacuum cleaner and an upright vacuum cleaner. *Appl Ergon* 37(2):159–165, 2006.
10. Carrivick PJW, Lee AH, Yau KKW, and Stevenson MR. Evaluating the effectiveness of a participatory ergonomics approach in reducing the risk and severity of injuries from manual handling. *Ergonomics* 48(8):907–914, 2005.
11. Kumar S, Moro L, and Narayan Y. A biomechanical analysis of loads on x-ray technologists: A field study. *Ergonomics* 46(5):502–517, 2003.
12. National Institute for Occupational Safety and Health. *Hazard evaluation and technical assistance report: University of Medicine and Dentistry of New Jersey, St. Peter's University Hospital, Piscataway, New Jersey* (NIOSH Health Hazard Evaluation Report No. 99-0093). Cincinnati, OH, 1999.
13. Beard DV, Hemminger BM, Pisano ED, Denelsbeck KM, Warshauer DM, Mauro MA, et al. Computed tomography interpretations with a low-cost workstation: A timing study. *J Digit Imaging* 7(3): 133–139, 1994.
14. Ramadan PA and Ferreira MJ. Risk factors associated with the reporting of musculoskeletal symptoms in workers at a laboratory of clinical pathology. *Ann Occup Hyg* 50(3):297–303, 2006.

15. Kilroy N and Dockrell S. Ergonomic intervention: Its effect on working posture and musculoskeletal symptoms in female biomedical scientists. *Brit J Biomed Sci* 57(3):199–206, 2000.
16. Smith NAL. Lateral epicondylitis in a hospital phlebotomist—An ergonomic solution. *Occup Med (London)* 51(8):513–515, 2001.
17. Minuk GY, Waggoner JG, Hoofnagle JH, Hanson RG, and Pappas SC. Pipetter's thumb. Case report. *New Engl J Med* 306(12):751, 1982.
18. Fredriksson K. Laboratory work with automatic pipettes: A study on how pipetting affects the thumb. *Ergonomics* 38(5):1067–1073, 1995.
19. Bjorksten MG, Almby B, and Jansson ES. Hand and shoulder ailments among laboratory technicians using modern plunger-operated pipettes. *Appl Ergon* 25(2):88–94, 1994.
20. David G and Buckle P. A questionnaire survey of the ergonomic problems associated with pipettes and their usage with specific reference to work-related upper limb disorders. *Appl Ergon* 28(4): 257–262, 1997.
21. Lee Y-H and Jiang M-S. An ergonomic design and performance evaluation of pipettes. *Appl Ergon* 30(6):487–493, 1999.
22. Sillanpaa J, Nyberg M, and Laippala P. A new table for work with a microscope, a solution to ergonomic problems. *Appl Ergon* 34(6):621–628, 2003.
23. Haukka E, Leino-Arjas P, Solovieva S, Ranta R, Viikari-Juntura E, and Riihimäki H. Co-occurrence of musculoskeletal pain among female kitchen workers. *Int Arch Occup Env Hea* 80(1):141–148, 2006.
24. Aminoff T, Smolander J, Korhonen O, and Louhevaara V. Physiological strain during kitchen work in relation to maximal and task-specific peak values. *Ergonomics* 42(4):584–592, 1999.

Ergonomic Design in the Workplace in Health Care Facilities

Jocelyn Villeneuve

Introduction

Following observation in a number of health care facilities, certain dysfunctions in workplace design were identified, both in the patient-care units themselves and in other related areas such as laboratories, food services, administrative offices, and so on. Ergonomic interventions were required to correct these situations, when in fact the problems could easily have been avoided at the planning stage. Nearly three quarters of industrial accidents involving nursing staff occur in the rooms of long-term care patients. A key factor is lack of space in the rooms for equipment such as patient lifts, stretchers, and wheelchairs. Bathrooms and toilets are often poorly designed, and accidents occur when nursing staff handle patients. Many similar examples exist in all hospital services.

This chapter describes an ergonomic procedure aimed at ensuring that workplace designs provide the best possible health and safety conditions and satisfy the requirements of nursing staff in terms of comfort, quality, and efficiency.

The procedure in question is a support procedure for design projects and is based on a specific "approach" and specific "methods" in "design ergonomy." It is derived from a trend developed recently in France.[1] Although focused on employees, it also has a positive impact for clients and other users. In fact, the methods employed can be applied to all categories of users.

Approach

The approach is "participatory" in that, to succeed, it requires the direct participation of and structured consultation between the management, employees, and clients in the health care institution.

The approach is "prospective," and concentrates on anticipation of future activities, to ensure that the design concept will provide suitable conditions for the performance of those activities.

The approach also gives "decision-making support" to designers and institutional players, by helping them to understand the demands of the work and thus to make appropriate design choices. It therefore provides an excellent complement to the work of the architect, who is concerned more with the construction of the building itself.

Methods

The methods used to apply the approach basically comprise "simulations" of predetermined future activity scenarios. The simulations are generally performed on a full-scale or enlarged plan. Their goal is to test the suitability of the design concepts for the future activity scenarios considered most important. A method of analyzing the functionality of existing and future buildings is also proposed.

The approach presented is adapted to the context of major renovation or construction projects. The methods would have to be modified for smaller projects. Nevertheless, the objective remains unchanged to ensure that the facilities resulting from the project provide an environment in which the work can be performed safely, comfortably, and efficiently.

We have tested and perfected this approach over the last 4 years in more than 200 institutions in Quebec's hospital sector. Given the scope of the testing, we believe the approach described here is both useful and effective.

Problem Situation

Many design projects are completed without sufficient organized dialog between the direct users (management, personnel, and clients). All too often, the fundamental decisions regarding design choices are made in a vacuum by a committee of technical professionals who do not give enough consideration to overall working conditions and the conditions in which the users of the building—employees, clients, visitors, and suppliers—will have to conduct their activities. Major structural problems exist in project realization procedures, at the level of dialog mechanisms between the various parties and the management of relevant information.

Generally speaking, construction standards do not consider ergonomic aspects or health and safety at work. As a professional discipline, ergonomy is a fairly recent phenomenon, and perceptions of its role are limited. The term "ergonomy" tends to be associated with workstation design and the selection of furniture (chairs, desks, etc.). In the context of the approach described here, however, it is defined as a discipline specialized in the design of the user/environment interface, and thus covers a much broader field of application.

All these factors together mean that serious design faults sometimes occur when, in fact, they could have been avoided by a more rigorous analysis of user activities. Situations such as this generate discontent and can affect operational efficiency. Sometimes, new buildings have to be altered immediately after construction. This is both costly and unproductive.

Our goal is to help designers, as far as possible, to eliminate design errors from the working and living environment, by applying a new ergonomic approach based on the analysis of user activities.

Levels of Functional Building Analysis

In the functional analysis of existing or future buildings, it is useful to distinguish three levels, each involving different types of activities (see Figure 9.31):

- A *macroscopic level*, involving the building as a whole in its relation with site and the location of the main activity zones within the building
- A *mesoscopic level*, involving the functional relationships between the different facilities within each activity zone, i.e., a department or related departments
- A *microscopic level*, involving workstations and the performance of activities on the premises

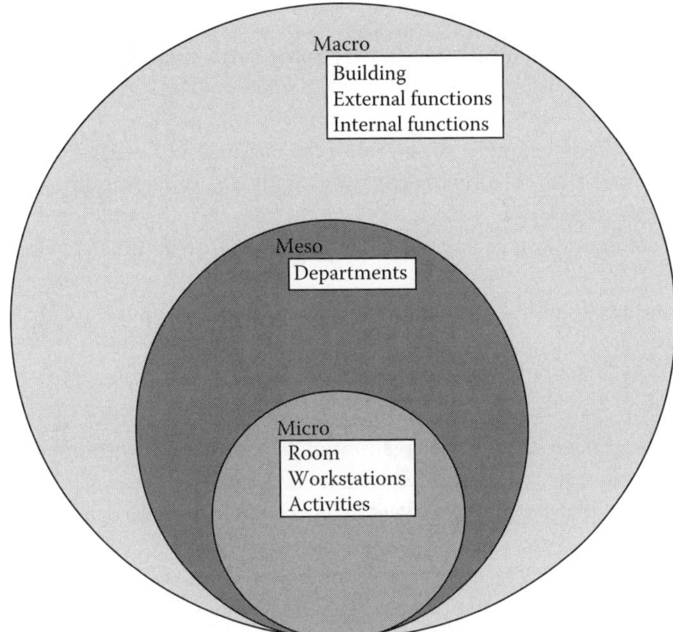

FIGURE 9.31 Levels of project analysis. Overall, the building is viewed as a broad system comprising a number of subsystems (departments), which themselves contain smaller subsystems (workstations). All are open systems, which maintain close functional relationships between the various levels.

From this standpoint, the building as a whole is seen as a broad system comprising a set of subsystems (the departments), which themselves comprise smaller subsystems (the workstations). All the systems are open. Close functional relationships exist between the different levels.

Macroscopic Level

This level is concerned with the relationship between the position of the building and the characteristics of the site: its size, position in relation to the cardinal points and neighboring streets, future easements, and the type of soil and vegetation.

The limits and possibilities of the site with respect to the "external functions" of the building and required surface areas significantly affect the design concept in terms of the volume of the building—basically, its shape and the number of floors. The external functions of a building include pedestrian and vehicular entrances and exits, service access and delivery of goods, terraces, gardens, and parking. The access points must be positioned so as to comply with safety and entrance/exit surveillance requirements. An enclosed outdoor space—a garden or terrace—will be needed, for example, in facilities housing patients who are likely to wander.

The other major aspect to be considered concerns the position of the main activity zones in relation to the "internal functions" of the building: reception, administration, care units, dietary service, specialized facilities, community facilities, staff facilities, and technical facilities. These zones must be positioned so as to obtain the best possible proxemic relationship among them and with the outside of the building to facilitate internal circulation of people and materials.

The proximity of the various zones to one another will depend on their respective functions and the compatibility or incompatibility of their activities. For example, contact between soiled and clean materials should be avoided.

Analysis of horizontal (floor by floor) and vertical (floor to floor) circulation of people and materials will enable the activity zones, stairwells, lifts, laundry chutes, and technical channels to be located in the best possible position in the building.

Mesoscopic Level

The mesoscopic level is concerned with the activity zones themselves. Each zone may contain one or more related services. Here, it is important to assess the functional relationship of each facility with the others.

If we take a care unit as an example, the layout of the facilities will depend on the exterior shape of the building and the internal circulation of people and materials. Whatever the shape, the indoor space will almost always be arranged on either side of a long corridor or at the junction of two or three shorter corridors. Different configurations permitting loop circulation are recommended for elderly people who are likely to wander.

At this level, the layout should be such that it minimizes staff movement and facilitates communication between the facilities in the activity zone.

Microscopic Level

The microscopic level is concerned with the organization of workstations within the facilities. A workstation is always related to a greater or lesser extent to a set of other workstations. It is important to identify these relations and ensure that the physical layout takes account of chronological work sequences so that the physical and informational links are as efficient as possible. The layout of workstations in a laboratory, for example, should be designed to avoid unnecessary handling of samples and should reflect the chronology of the operations performed on the most common types of samples. This obviously requires prior analysis of operational sequences and the paths taken by the different samples during handling by laboratory technicians. The specific constraints of each workstation must also be identified, especially if employment injuries or significant incidents have been declared in the past.

Environmental considerations such as the location of windows according to the orientation of the building, potential sources of noise, and ventilation requirements must be considered when designing workstations according to the type of activity to be performed. A computer-based activity in an administrative office does not involve the same design constraints as a workshop activity requiring the use of noisy mechanical tools and toxic products.

Equipment and tools are generally selected at the microscopic level, on the basis of ergonomic criteria that take account of safety and comfort objectives for users and also production objectives. The choice of a specific production system sometimes has an impact on the design as a whole. Such choices are made at the macroscopic level because they affect the initial design concept. The decision to use a tunnel washer rather than washers and dryers in a laundry, for example, will condition the design of the laundry area and also the upstream areas, e.g., dirty laundry reception area, weighing area, and tunnel feed area, and the downstream areas, e.g., sorting, drying, and ironing.

Dynamic Simulation of Future Activities

The primary requirement of a design project is to produce a design that is consistent with the future activities of users. Consequently, the activities to be performed in the new facilities must be anticipated. Simulation is the best way of doing this.

We therefore propose a "dynamic simulation of future activities" in order to test the proposed design concepts in light of user needs. The simulation procedure can be applied to all steps of the design phase—programming, design, preliminary plans and specifications, and detailed plans and specifications. It can also be reproduced at the three functional analysis levels—in other words, at the level of the building itself (macroscopic), the various departments (mesoscopic), and the individual workstations (microscopic).

The general logic of the simulation process is to develop priority future activity scenarios on the basis of the data used to define the design project. The scenarios are simulated by placing real users in a layout representing the proposed design concept. The suitability of the concept for the predetermined future

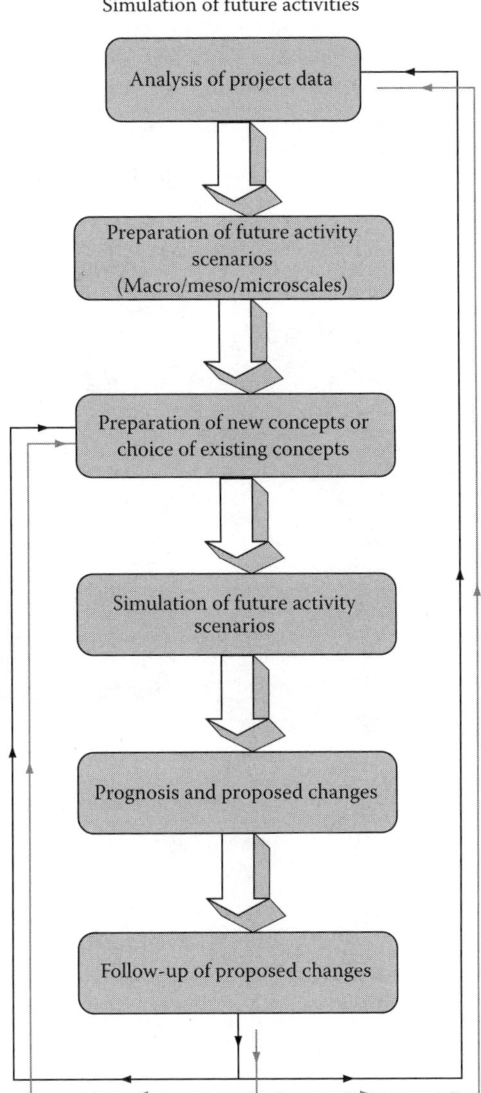

FIGURE 9.32 Simulation of future activities.

activity scenarios is then diagnosed. This gives rise to proposals for changes to the concept or the scenarios and, finally, to a design that fully satisfies user expectations.

Figure 9.32 shows the simulation process in diagram form. The process is described step by step in the following paragraphs. To illustrate the text, we have used the example of recent major renovation work at a long-term care center in Quebec.

Analysis of Project Data

To obtain a clear vision of the project, a set of knowledge, in writing or not, must be collected and analyzed. In addition to the official documents that usually accompany design projects (evaluation assessment reports, architectural program, plans and specifications), other data are needed to ensure that the future activities are properly understood (see Figure 9.33).

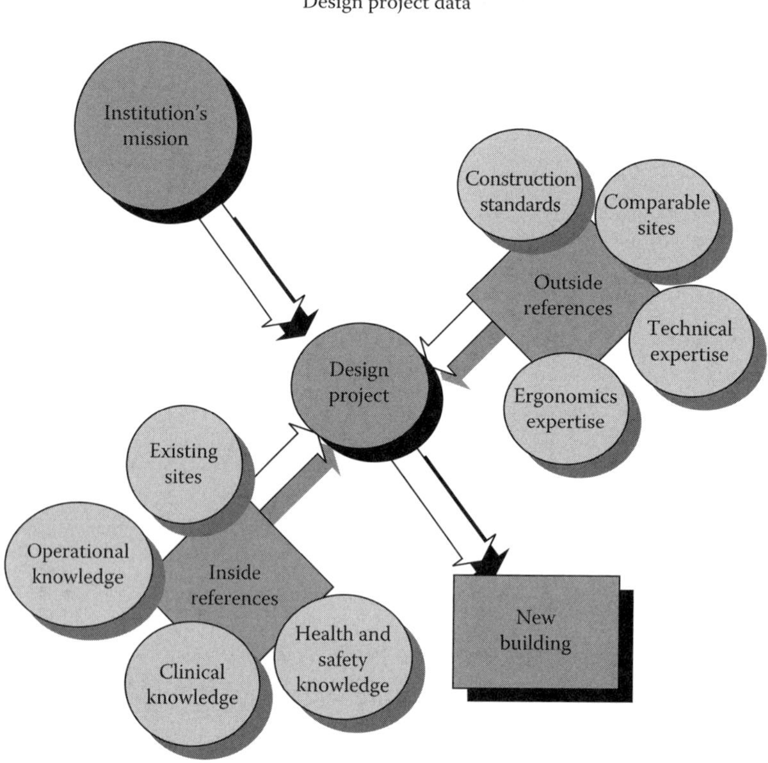

FIGURE 9.33 Design project data. Set of data to be considered in defining a design project to obtain a clear view of the future activities of users.

These data are concerned, first, with the institution's existing and planned operating methods with respect to its new mission, and second, with the health, safety, and comfort of users.

Potentially vulnerable situations must be identified at the outset, so that they can be eliminated or controlled in the new facilities. To do this, the circumstances causing these situations must be diagnosed and solutions must be prepared to prevent repetition in the future.

Example 9.1

At B long-term care center, three diagnostic studies were carried out on the institution's operation and orientation prior to drawing the architectural program. First, a back-injury prevention program provided a detailed portrait of the main causes of workplace injuries and proposed a number of solutions. As an extension of this report, an environmental ergonomic study was ordered to obtain a complete diagnosis of the center's dysfunctional aspects.[5] Later in the same year, the management filed a strategic report recommending a major reorientation of the institution's mission, in which the institution would be transformed into a specialized psychogeriatric center.

The center was built in 1977 and was designed for independent elderly people. However, over time it began to accept increasingly severe cases and the average number of care hours per client increased by 68% in just 5 years. Nearly half of its clients suffered from cognitive disorders.

The existing building, composed of five separate one-story pavilions constructed around a central zone with a cafeteria, had become a source of enormous functional problems and, clearly, no longer

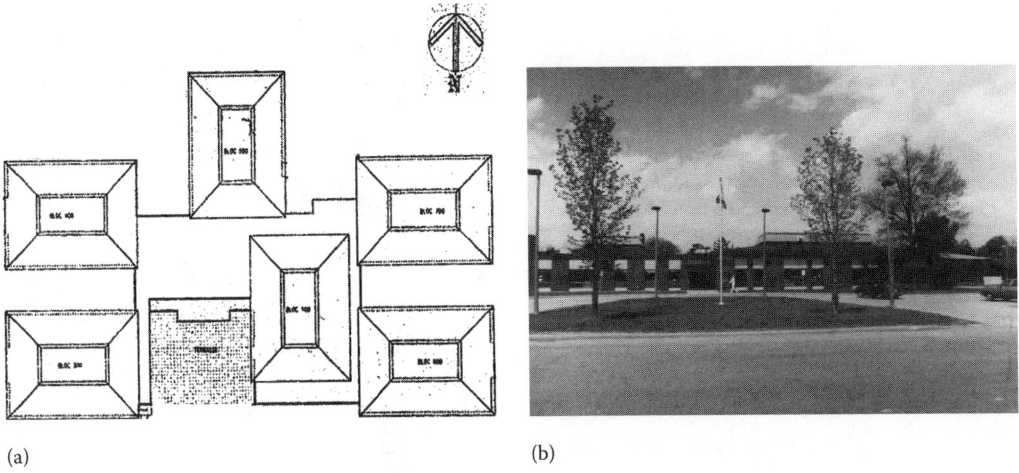

(a) (b)

FIGURE 9.34 (a) Building implementation plan before renovation. (b) Photo of center before renovation.

satisfied the needs of its existing and future clients (see Figure 9.34). The architectural scheme, based on client independence, had a number of detrimental effects on operations, client services, and staff health.

- The separation of the building meant that the staff often had to travel long distances which considerably increased both the workload and the amount of unproductive working time (see Figure 9.35).
- The maze of corridors produced an overall "labyrinth effect," which added to the wandering problems of confused clients (see Figure 9.36).
- Staff supervisory duties were made more difficult because there was only one nursing station located at the main entrance. A secondary station had been added, but did not have a direct view of the pavilions (see Figure 9.37).
- The work teams were broken up in the center and experienced communication and supply problems due to the distances involved.
- The original rooms, toilets, and bathrooms were much too small and were designed for independent users. The most likely employment accident scenarios were in fact related to these critical sites for patient transfer operations (see Figures 9.38 through 9.40).

The new design concept was therefore tailored to solve these major functional problems in the existing building.

Preparation of Future Activity Scenarios

A scenario is defined as "a series or sequence of actions by an individual or a group in a work-related situation occurring in hypothetical conditions and a hypothetical organizational context."[7] A scenario is therefore an operational proposal aimed at providing a realistic image of the future activity, which is consistent with the work dynamics. Many scenarios are built on common representations known to experienced health sector players. This facilitates the task in that, generally speaking, these mental constructs are easily crystallized. It is sufficient to explain them in further detail and write them down.

The description of the scenarios at all three levels (macro, meso, and micro) must provide a clear and realistic view of the course of operations over time and in space. It must give priority to potential prejudices to the health, safety, and comfort of the people involved in the situations described.

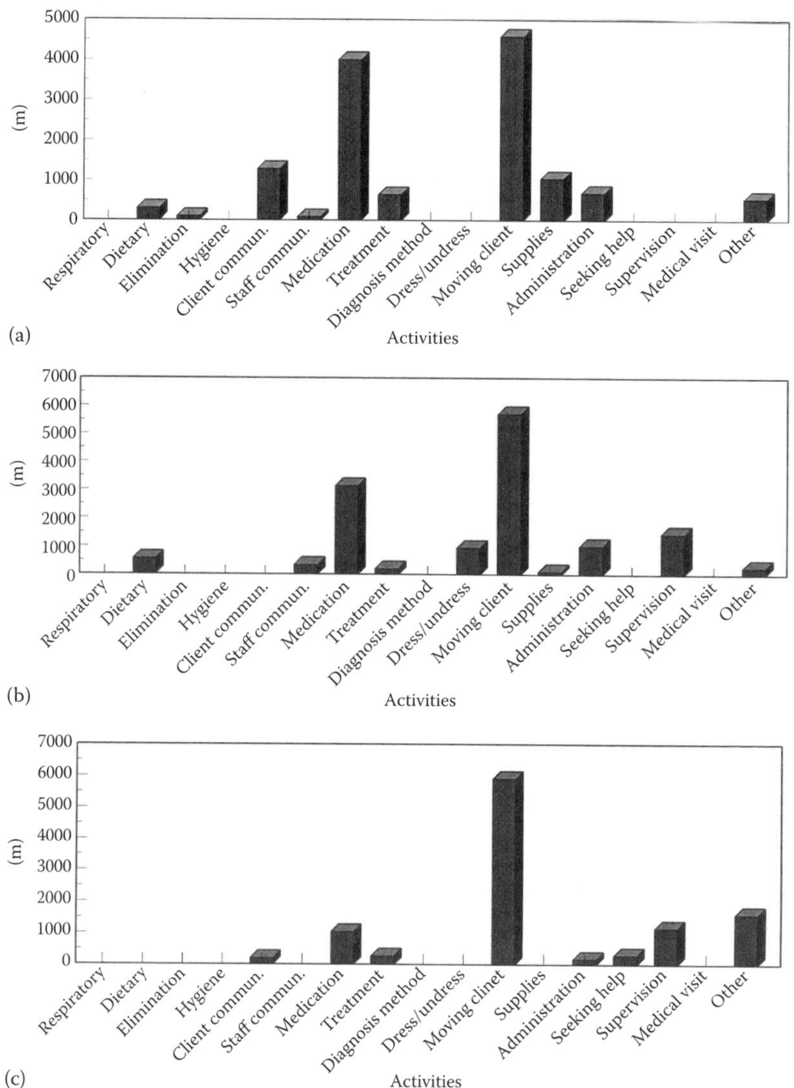

FIGURE 9.35 Staff traveling distances—unfavorable cases. Distances traveled by a nurses' assistant (a) on the day shift totaled 4,029 m or 13,220 ft and (b) on the night shift totaled 4,153 m or 13,625 ft, and (c) the distance traveled by a nurse on the night shift, which totaled 6,289 m or 20,634 ft.

In developing a future activity scenario, the following are required:

- Detailed chronological step-by-step description of the activities to be simulated
- Description of the characteristics of the people involved (managers, employees, and clients, as the case may be)
- Description of the organizational framework
- Description of the working methods used
- Description of fixed and mobile equipment and any tools used
- Brief description of the desired physical framework

The scenarios can take many different forms: a descriptive text accompanied by a chronological table of activities, an organization flow chart, a series of photographs, a process diagram, and so on. It may be

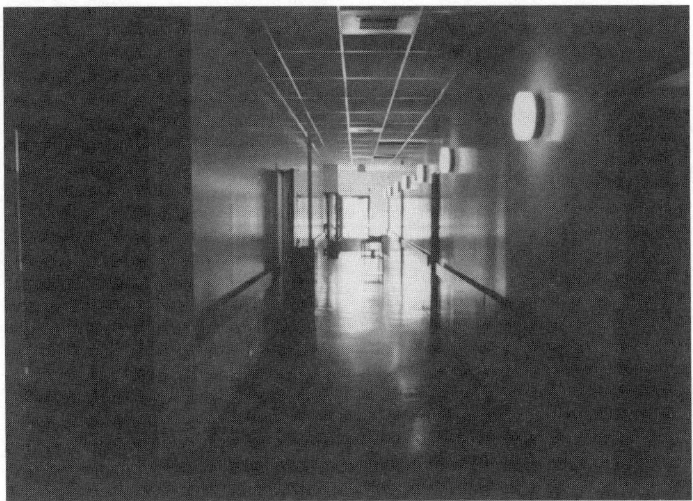

FIGURE 9.36 Labyrinth effect produced by the existing building and the similarity of the corridors.

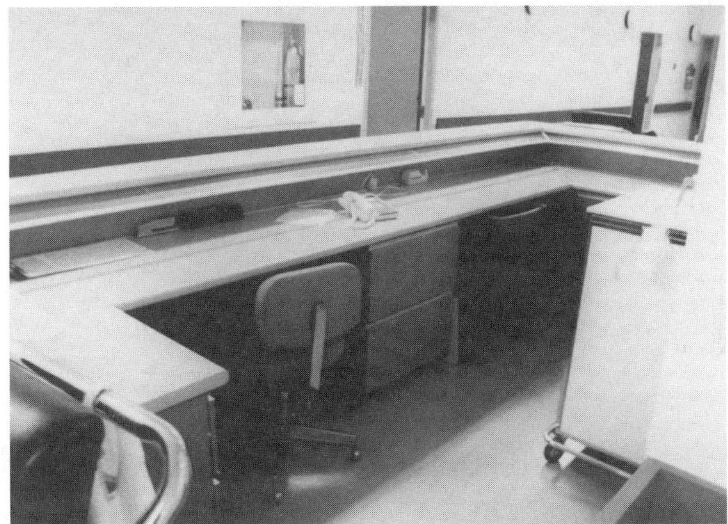

FIGURE 9.37 Nursing station located opposite a wall.

interesting to combine several different forms, for example, a short descriptive text accompanied by a chronological table of operations and a layout plan.

Example 9.2

At B long-term care center, a number of priority scenarios were developed. The following paragraphs contain three sample scenarios used as a basis for the concept definition for the renovated building.

At the macroscopic level, the general aim in developing the scenarios was to obtain maximum grouping of homogeneous or related operations in areas that were better defined geographically to solve the

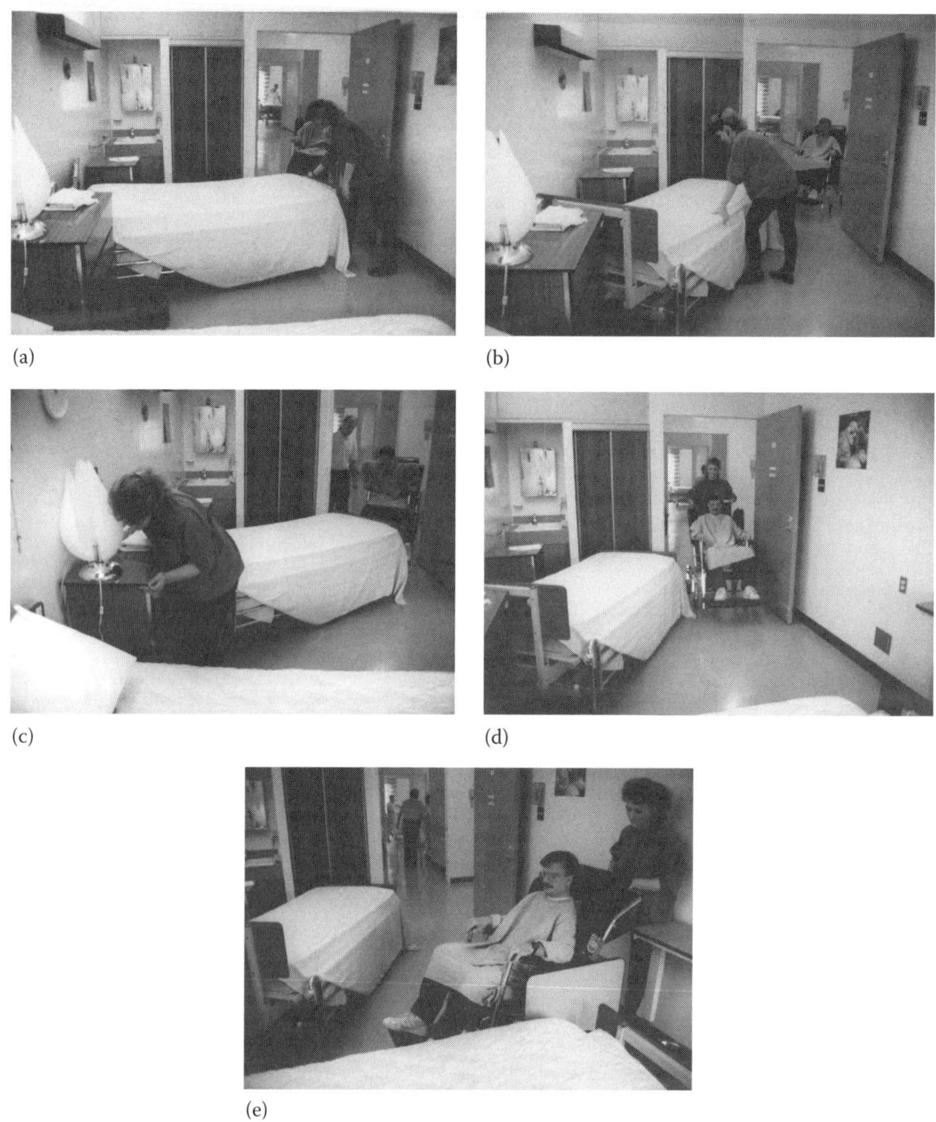

FIGURE 9.38 Difficult access to the second bed in a semiprivate room (simulation). (a) Move the bed, (b) clear the entrance, (c) move the bedside cabinets, (d) bring the patient, and finally (e) replace the bed and the furniture.

general problem of activity dispersal throughout the existing building. One of the main requirements was to control the circulation of confused patients by breaking up the "labyrinth effect" created by the maze of similar corridors. The major scenario therefore took shape around a new room layout in two separate 40-bed care units, one with door controls, specifically for confused patients, and the other for lucid patients. Movement would thus be reduced to a minimum and task distribution would be improved by reorganizing the staff into two independent teams.

At the mesoscopic level, the common service areas (dining room, bathrooms, recreation room overlooking the garden, and clean/soiled utility rooms) were relocated in the central zone near the nursing station, thus enabling services dispensed outside the unit to be brought into the unit. The nursing station was relocated to facilitate supervision, especially in the unit housing patients suffering from mental confusion. In the reorganized unit, communications and local supplies would be much easier.

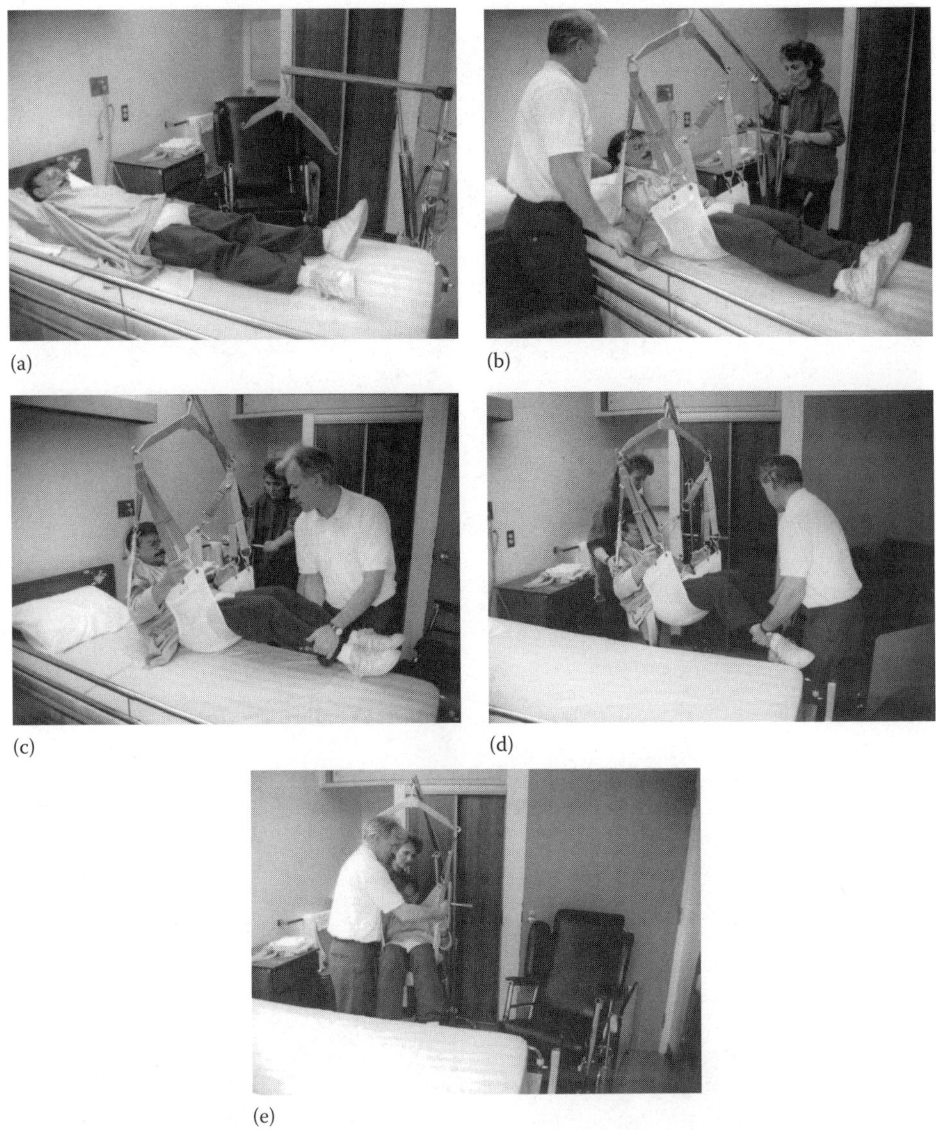

FIGURE 9.39 Single rooms that are too small to transfer patients from the bed to a geriatric chair using a traditional patient lift (simulation). The worker must (a) bring the patient lift into the room, (b) lift the patient from the bed, (c) free the patient's legs, (d) move the lift and swing the patient toward the chair, and (e) failure of the maneuver due to lack of space.

At the microscopic level, a scenario was developed to test the plans for the renovated and newly constructed rooms. Spatial requirements were dictated by the use of equipment such as mobile free-standing patient lifts, geriatric chairs, and wheelchairs. Rooms had to be spacious enough to allow the staff and semi-independent patients to move around the bed safely.

A detailed description of the scenarios was produced for the purposes of a full-scale or enlarged simulation.

Development of a Design Concept

The scenario describes a way of performing a set of activities according to predefined requirements. To be acted out, the projections require a design proposal within which the operating hypothesis is

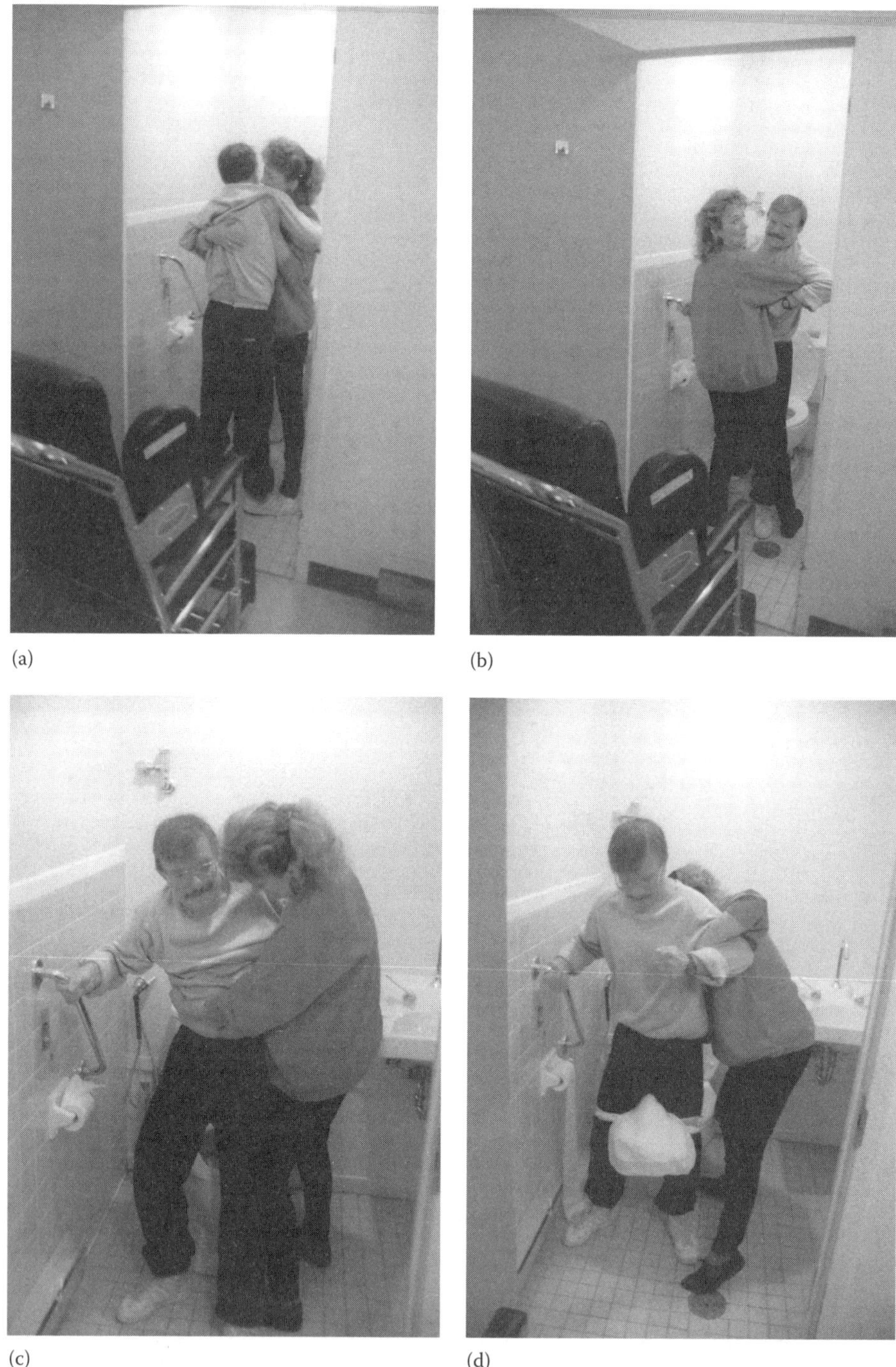

(a) (b) (c) (d)

FIGURE 9.40 Toilets and bathroom space too small to help dependent patients (simulation). (a) Transfer the patient from the chair to the toilet. (b) Swing the patient. (c) Remove the patient's garments while supporting him/her with one arm and avoiding intrusive structures. (d) Clean the patient while supporting him/her with one arm and blocked access that leads to defective posture with effort.

possible. This is known as the design concept. It may be useful to submit more than one concept to offer users a choice. The concepts may already have been defined by standards, specimen plans or models tested elsewhere or they may have to be designed specifically for the scenario. The design concept provides a physical framework for the scenario. It is generally presented in the form of sketches, which may or may not be related to a scale plan.

Example 9.3

At the B long-term care center, the reassembled unit concept was designed on the basis of the objectives set by the scenario. Two new extensions were also planned at the eastern and western extremities of the building, to enable the 80 beds to be grouped into two separate units of 40 beds each (see Figure 9.41). Each new extension would enable eight new rooms to be added, and would house all the unit's central services: the nursing station and staff room, the sanitary complex, the dining room, the recreation room adjoining the garden, and the utility and local supply rooms (see Figure 9.42). In the existing pavilions, the central island housing the services originally designed for independent patients would be demolished to make way for larger rooms with adjacent toilets (see Figure 9.43). For the newly constructed rooms, the model room proposed in the government program was modified by the addition of a ceiling rail between the bed and the adjacent toilet. These rooms would be reserved for patients who were entirely dependent on staff assistance for transfers (see Figure 9.44).

Dynamic Simulation of Future Activity Scenarios

Simulation is a prospective exercise that enables the functional viability of a design concept to be assessed more accurately on the basis of a predetermined future activity scenario. In some ways, it is a dynamic "staging" of anticipated activities within a selected design concept. It should reproduce, as faithfully as possible, an anticipated reality that exists only in the representation of the design players or with reference to a comparable situation.

It also provides an opportunity to confront different viewpoints in a positive way. The work will be viewed very differently by an architect, engineer, department head, employee, or ergonomist. This diversity of viewpoints is not an obstacle to project development. On the contrary, it provides an overall vision of the projected situation without which the design exercise may be defective. Simulation is thus an excellent way of confronting viewpoints and reaching a creative compromise.

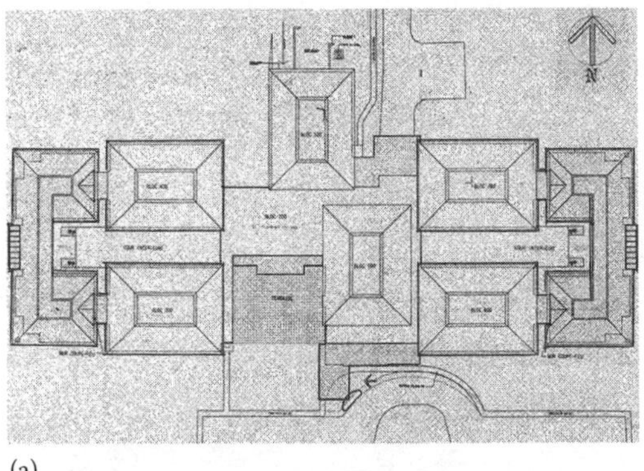
(a)

(b)

FIGURE 9.41 (a) Implementation plan for the renovated center. The boxed sections on the left- and right-hand sides of the plan are the two 40-bed care units. (b) Photo of center after renovation.

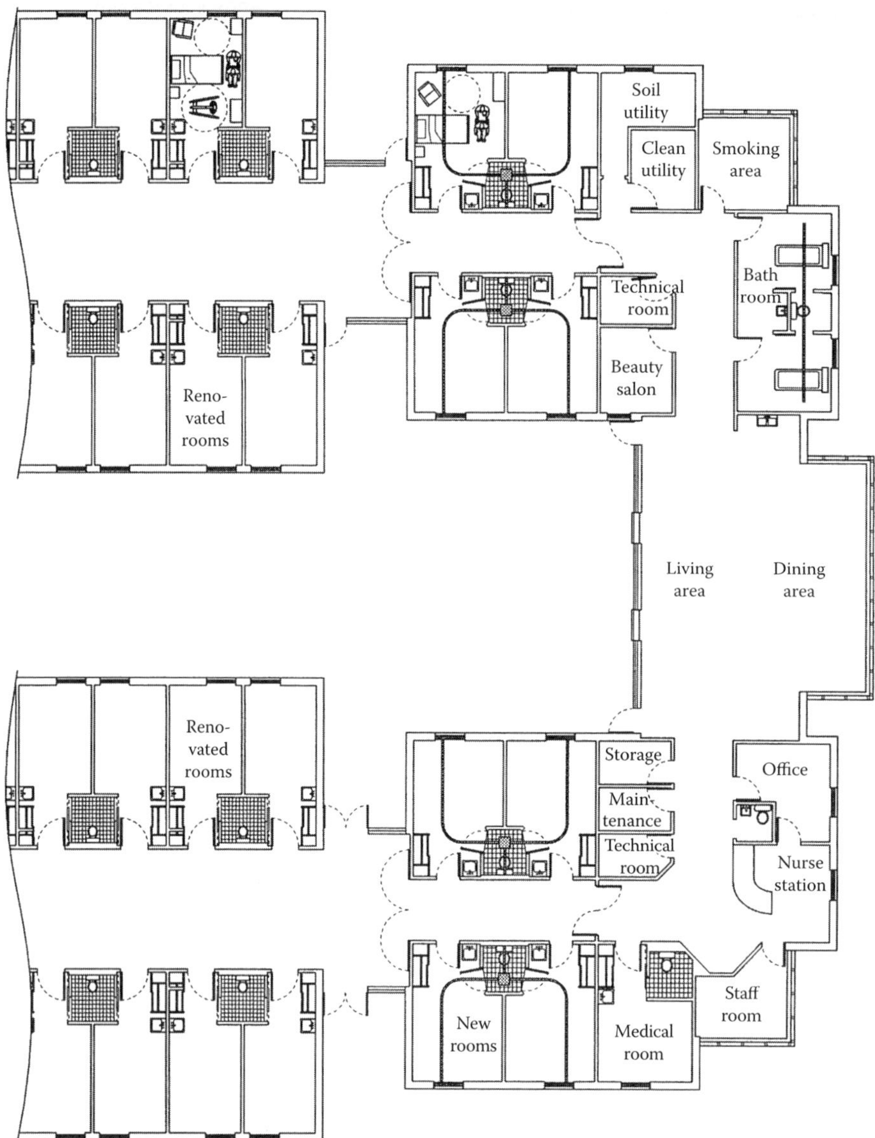

FIGURE 9.42 Reduced plan of a renovated care unit showing part of the former building and the new construction housing the common service rooms.

Simulation Props

The props used to create dynamic future activity simulations are "the enlarged plan, full-scale simulation, and the prototype."

Enlarged Plan

Users are not skilled at reading plans and some basic instruction will be needed to enable them to make a useful contribution to group discussions.

The plans should be enlarged to a minimum scale of 1:20 or 1:50. All equipment and circulation must be shown. Elevations of some areas and perspective drawings may be required to facilitate the representation of volumes.

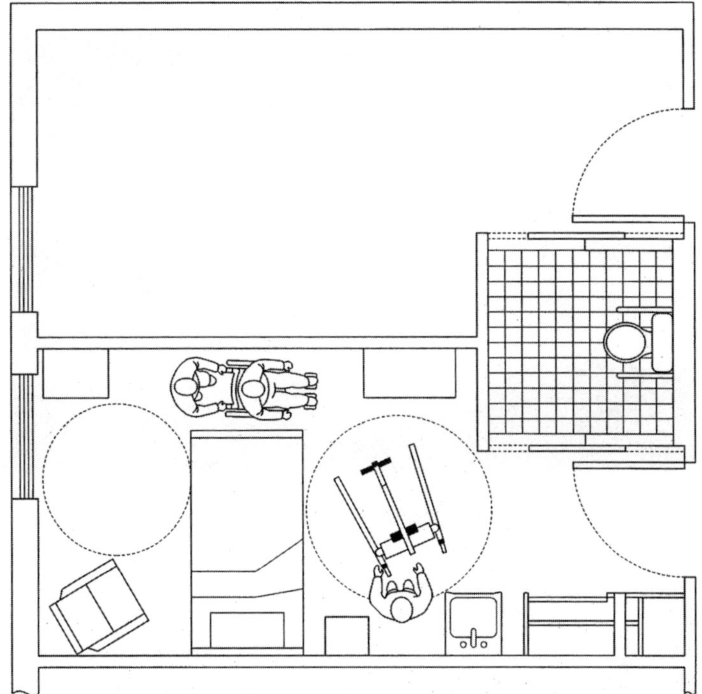

FIGURE 9.43 Model plan of renovated room. The space around the bed and in the bathroom is sufficient for safe use of transfer equipment such as patient lifts and wheelchairs.

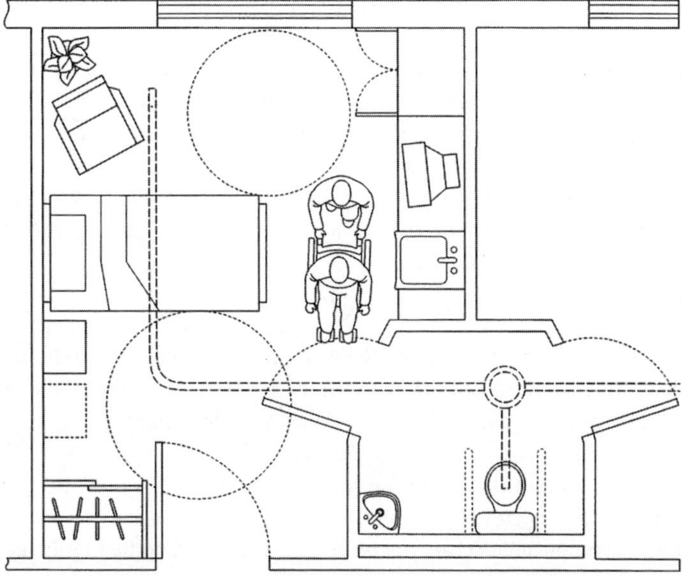

FIGURE 9.44 Model plan of new room. The new rooms are equipped with patient lifts on ceiling rails running between the bed and the bathroom. A comfortable space has been left around the bed and in the bathroom.

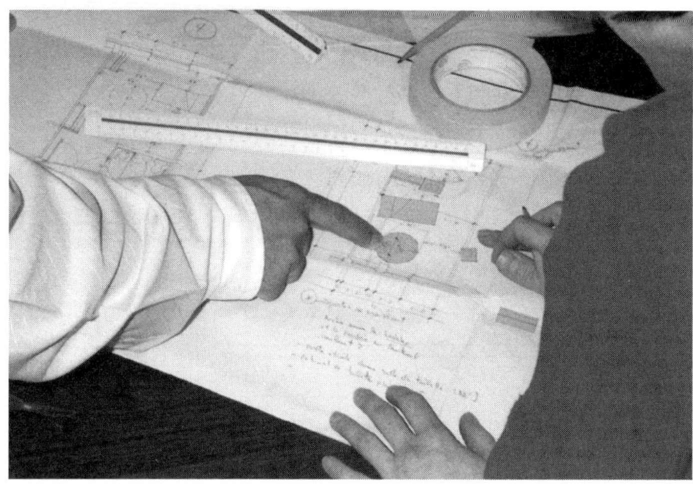

FIGURE 9.45 Simulation of future activities (enlarged plan).

A simple first-level method of simulating activities on an enlarged plan is to make scale cut-outs of fixed and movable objects in colored cards and to act out activities in the target area. This exercise is often useful for testing facilities, which do not present major operating problems (see Figure 9.45). However, the enlarged plan has certain limits which require the use of other props to provide better ways of reproducing anticipated situations, where doubts as to their functionality persist.

Full-Scale Simulation

Full-scale simulation is a sure way of testing design concepts. It requires a full-scale model of surface areas and volumes, and has the advantage of involving real users. It is relatively simple to organize, either in comparable existing facilities or in a room that is large enough to reproduce the situation with all fixed and movable equipment.

Simulations such as this are appropriate when copies of the same layout model will be reproduced a number of times in the building. This is often the case in hospitals, where the different floors all follow the same model. They are also appropriate for testing operations of a critical nature, where errors may have serious human or financial consequences. Experience has shown that full-scale simulations are never superfluous even when preliminary testing of the plan has been performed. They always culminate in modifications—sometimes major ones—to the initial concept, derived mostly from direct observation of the players as they carry out their activities.

Prototype

The prototype can be used not only to simulate the work, but also to test different forms of procedures and physical installations. Ultimately, it is designed in the same way as a laboratory. Its implementation is therefore more sophisticated, because the designer must be able to move equipment (furniture, tools, etc.) and even room divisions without difficulty. Because prototypes require special implementation, they are generally used only in unusual working situations and usually for a specific workstation or a model room that will be reproduced several times. It is commonly used in office building development.

Although it is a more complicated setup, the prototype may prove to be highly useful in large-scale projects or during implementation of new concepts for which no comparable and tested reference sites exist. Prototypes are particularly appropriate for accurate testing of high-risk work situations, for example, laboratory analysis of highly infectious samples.

Example 9.4

A set of simulations was conducted at the project definition stage of the B long-term care center. It was decided to test the functionality of the model renovated room, which had been the site of most past employment accidents. A full-scale simulation was performed, because there were potential space problems at the foot of the bed and on the transfer (door) side if a free-standing mobile patient lift was used.

The entire room layout was reproduced in the corner of a larger room. It was furnished with real furniture and equipment. Fixtures were indicated by means of adhesive tape on the floor, and volumes (wardrobe, washbasin, etc.) were replaced by cardboard dummies. Two activities were identified as critical (Figure 9.46):

- Transfer of the patient from the bed to the geriatric chair by means of the patient lift operated by two employees, one attendant, and one nurses' assistant
- Self-propulsion by wheelchair-bound patients past the foot of the bed

The members of the pilot committee—an extended version of the health and safety committee—and the architect were all called to the simulation. Employees were relieved of their usual duties to take part in the process. The session was led by the general manager, who was also in charge of the project. The goals of the exercise and the methods used were clearly stated at the beginning of the simulation.

When the process was complete a prognosis was made. The space allowed was insufficient for the transfer of the patient by means of a patient lift operated by two nurses' staff. There was also a risk that the patient would hit an elbow against the bed or catch a hand on the crank during self-propulsion of the wheelchair along the foot of the bed.

Following this prognosis, it was decided to extend the room toward the center of the pavilion, so as to leave enough room for two nurses to handle the patient lift comfortably. Unfortunately, it was impossible to increase the space at the foot of the bed because of the location of the windows, which prevented the partition walls from being moved. A palliative procedure was therefore introduced. If patients wished to travel past the foot of the bed in a wheelchair, they would always be assisted by an employee. In any case, very few patients were sufficiently independent to perform this maneuver without help.

Prognosis and Follow-Up

The simulation is aimed at testing the viability of a design concept in light of the predetermined future-activity scenario. When it is complete, the people involved are in a position to decide whether or not the

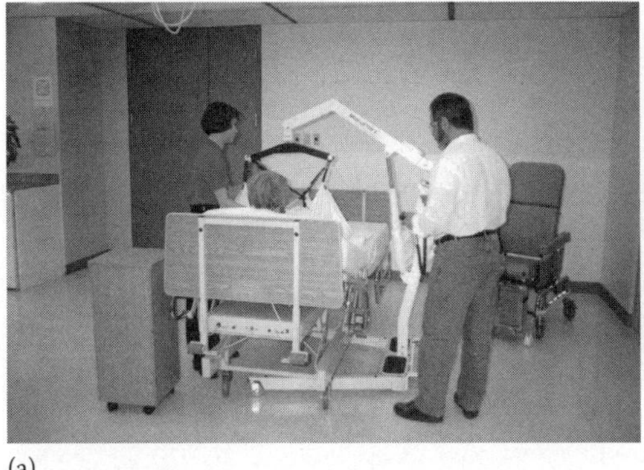
(a)

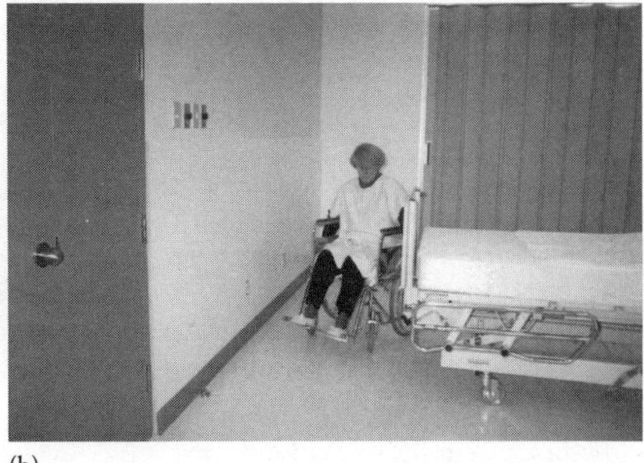
(b)

FIGURE 9.46 Simulation of future activities (full-scale). (a) Simulation of transfer from bed to geriatric chair. (b) Simulation of passage of wheelchair past foot of bed.

concept is appropriate for the work context as defined. This is known as the prognosis. The prognosis can take three directions:

- Concept works well in the context defined by the scenario.
- Concept does not work well and has to be reviewed in whole or in part (modified concept).
- Concept would work in a different context (modified scenario).

Example 9.5

In Example 9.4, the proposed concept had to be reviewed for the renovated rooms, by significantly increasing the surface area on the transfer side, i.e., the door side. The simulation also led to a change in the initial scenario, by the introduction of the requirement that patients wishing to travel past the foot of the bed in a wheelchair must seek assistance from a staff member (see Figure 9.47).

The design concept must be adapted to the future activity scenario and not vice versa. In design projects involving substantial sums of money, users have a right to expect that the proposed design will meet their expectations. The designer must satisfy users' needs instead of trying to change working methods to suit the proposed design. This is not always easy and creative compromise is often required, especially in renovation projects where additional constraints are imposed by the existing building.

The results of the simulation must be written up, and rigorous follow-up is required to define new measures until a satisfactory solution is found, in terms of operational functioning and the health, safety, and comfort of users.

(a)

(b)

FIGURE 9.47 Model renovated room. (a) Sufficient space around the bed and (b) adapted toilet with access at both sides—the arm supports retract into the wall.

Conclusion

This ergonomic procedure is not intended to be carried out in parallel to the normal procedure but should form an integral part of it and be conducted in close cooperation with everyone involved: the project leader, management, employees and their representatives, professionals, architects, engineers, and government representatives. Its aim is to add an essential aspect to project realization: the consideration of working and living activities in design concepts.

To do this, it is vital that internal project mechanisms be well structured and that provision be made for true cooperation between the social players in the institution and, in particular, the direct users of the future premises, i.e., management and employees, as well as clients and other users. They should be involved from the very beginning of the project definition process, because their expertise and knowledge are essential to the success of the project.

The concepts proposed at the different design stages should be tested through appropriate simulations of future activities. Obviously, simulations will not be required in every situation. The functionality of many facilities can be established by professional expertise or reference to construction standards and comparable sites. The situations in which more elaborate simulations are required must be properly defined and carefully prepared.

There is a tendency to think that a design will be functional and need not be tested either because it is the norm, things have always been done like that, there is no time, etc. Reference to users is unfortunately too evasive, insufficient, and even totally absent in some projects. This can lead to major design faults that are only discovered after construction.

Construction professionals focus their attention on the building and technical aspects and it is absolutely vital that they should do this. However, the approach proposed here focuses on users and, therefore, provides an excellent complement to the work of the architect. The time invested is profitable for a number of reasons:

- It leads to the creation of facilities that are better suited to the needs of their users.
- The designer is able to draw an optimal version of the plans and specifications in a shorter time frame.
- Postconstruction modifications and related expenses are reduced significantly.

The challenge of this user-focused approach is to obtain more success with design projects at a lower cost.

References

1. Daniellou F. Le statut de la pratique et des connaissances dans l'intervention ergonomique de conception, Université de Toulouse, Le Mirail, 1992.
2. Bertrand G and Morissette L. La chambre: milieu de vie et lieu de travail, Collection Parc, No. 2, ASSTSAS, 1996.
3. Estry-Béhart M. Ergonomie hospitalière: théorie et pratique, ESTEM, 1996.
4. Garrigou A, Bellemare M, and Richard J-G. La simulation dynamique des activités futures, Intervention IRSST/SECAL, Final Report, IRSST, 1995.
5. Goumain P and Villeneuve J. Étude d'ergonomie architecturale dans un centre d'hébergement pour personnes âgées, *Proceedings of the Canadian Ergonomics Association*, 1991.
6. Ledoux E. Ergonomie et conception des espaces de travail. *Travail et Santé,* 12(2):34–37, 1995.
7. Maline J. Simuler le travail. ANACT, Collection outils et méthodes, 1994.
8. MSSS, Direction de la construction, Répertoire des normes et procédures, Vols. 1–7, 1994.
9. Villeneuve J, et al. Objectif prévention, *Dossier Thématique PARC,* 17(5):13–33, 1994.
10. Villeneuve J. Le programme PARC: aide à la conduite des projets architecturaux, Collection PARC No. 1, ASSTSAS, 1996.

Reducing Occupational Back Pain Disability among Health Care Workers through Ergonomics

Guy Fragala

Back Pain and Health Care Workers

In a review of injury statistics as presented by the Bureau of Labor Statistics, one might be surprised to find that health care workers are among the leading occupations at risk when considering occupational injuries, and back injuries are a large part of the problem. The problem of back injuries suffered by workers in the health care industry is well documented.[1–8] In fact, the problem has been increasing over recent years. When considering occupational injury rates, the long-term care industry traditionally is one of the top industries regarding injury experience. From 1980 through 1992, for the long-term care industry, the overall occupational injury incidence rate for total cases demonstrates a steady rise from 10.7 in 1980 to 18.6 in 1992. For the same time lost work day cases rise from 5.6 in 1980 to 9.3 in 1992. When considering a lost work day which resulted from injury disability, the rise again has been quite steady with an incidence rate of 83.5 in 1980 rising 155%–212.5% in 1992. Acute care hospitals not only demonstrate somewhat better injury experience, but also have high rates and demonstrate similar trends of increase. The 1993 edition of *Accident Facts* discusses the percent of total injury claims filed, which were for back injuries for some industries. For convalescent or nursing homes, the number is 52.5, therefore, over 50% of the high injury incidence experience in the long-term care industry is related to back injury claims.*

What Is Acceptable to Lift?

Some researchers have begun to try and assess the acceptability of tasks involving lifting of patients and residents to determine if these tasks are acceptable for the worker. When considering patient or resident handling tasks, two basic questions arise. First, are there optimum techniques to be used for the many complex movements and transfers required in the health care industry, and second, are the loads which workers are required to move beyond reasonable capabilities? In considering these two basic questions, we may find that many of the manual transfers of patients or residents required in the health care industry may in themselves be intrinsically unsafe.

In spite of the lack of scientifically justified proof, various so-called correct lifting techniques have been recommended. However, there is a lack of consensus on proper lifting techniques and experts are not in complete agreement concerning which procedures are best for specific manual transfers.[9,10] Considering the complexity of manual lifting tasks in the health care profession, one might assume that there is not one single correct lifting technique, which is right for all those involved in patient-care. It is dependent upon the current situation, the individual worker, and the patient involved. How then can the worker be expected to constantly try to minimize the ever-present risk factors associated with manual transfers by applying body mechanics?

In addition to the consideration of what might be the proper lifting technique, there are the loads involved in lifting in these manual transfers. Assuming that there is a limit beyond which it is no longer safe to lift a relative load, the question arises, "What is the limit?" This is not an easy question to answer. There have been some investigations, but the matter is complex. Many variables play a part and they are often interrelated. The acceptable load differs per individual by daily variation, such as fatigue, training conditions, or motivation. Also, the load cannot simply be expressed in a number of

* All injury incidence rate data taken from Accident Facts, National Safety Council, Itasca, IL, 1981 through 1993 editions.

pounds. The spacing between the load and the spine determines the final load on the vertebrae that are situated at a lower level of the lumbar section of the spine and on the other body parts involved in the lifting process.

What Research Studies Say

Some investigations have attempted to recommend acceptable loads for lifting tasks related to patient or resident handling. Although different criteria have been used and there is variation in the final results, generally, the loads involved are found to be beyond what is considered to be a safe limit. Some information from these representative studies will be presented.

A quantitative method which has received much attention is the *Work Practices Guide for Manual Lifting* developed by the NIOSH.[11] These guidelines were revised in 1991 to provide for more realistic lifting conditions.[12] Using the original NIOSH Formula for Nursing Care Tasks, a number of the standard conditions do not apply. For example, no rotation of the trunk must take place and spatial restrictions often present in the health care setting are not permitted. Further, the fact that a living load may oppose or make unexpected movements is not taken into account. This means that in itself this useful method is not totally applicable to the lifting and moving of people in the health care setting. However, some investigations have tried to make appropriate considerations to apply the original formula to health care patient or resident handling tasks.

When considering stressful patient-handling tasks, some estimates were made to relate high-risk tasks to the NIOSH Lifting Formula by Owen and Garg.[13] In this study, the biomechanical results are estimated because the confined space in the laboratory made it difficult for all of the joint angles to be viewed via video tape. The estimates of compressive force to the L5/S1 disk according to percentile of patient weight were all above the "action limit" permitted as safe by the *NIOSH Work Practices Guide*. Results obtained closely approached the "maximum permissible levels." According to the *Work Practices Guide*, administrative controls are needed when conditions exceed the action limit. Administrative controls might include preventive measures such as worker selection and training of workers. Engineering controls are suggested when action limits are exceeded and needed at or above the maximum permissible level. These engineering controls are related to redesign of the job such as transferring patients with mechanical lifts.

In a study by Knibbe and Knibbe,[14] the load on the lumbar spine at the L5/S1 area was calculated through a simple static biomechanical method and partly simulated by a computer. The possible amount of assistance from the patient was varied and also followed by computer simulation. Modeling was done in a series of static postures. Results were compared to the NIOSH formula recommendations. Although the method was highly simplified, some conclusions were drawn as indicators for practice and perhaps, direction for future research. The task evaluated was the work load on the back during the transfer from wheelchair to toilet. The average NIOSH action limit was calculated to be 19.8 lb. The maximum permissible limit was 59.4 lb. This indicates that unless the patient can assist the health care worker so that no more than 19.8 lb is lifted for each nurse involved in the lift, the lift can be considered unsafe and place the health care worker at risk of injury.

In a study by Stubbs et al.,[15] interabdominal pressure was measured for various lifting tasks. It was concluded that the techniques utilized had little effect on reducing the risks associated with back injury. In the study, an interabdominal pressure of 50 mmHg or higher was considered to be unacceptable. This is based on data gathered by Davis in a study entitled, "Acceptable Magnitude of Loads for Handling by Females," HSE Report, 1/MS/126/637/78. In this study conducted by Stubbs, some of the lifting tasks conducted exceeded this 50 mmHg. The patients used in these studies were only approximately 117 lb. This is generally well below the average loads nurses may be required to move. It was hypothesized that with the general population, many of the lifting tasks required of nurses would exceed this acceptable limit. The study further recommends that the approach that should be considered is one of establishing safer systems of work within an ergonomics framework.

The lifting burden of a nursing aide in a geriatric ward was determined by using a force place by Dehlin.[16] Three lifting operations were performed for all patients needing lifting assistance. At the same time as the forces obtained from a force plate were recorded, the lifts were photographed. The lifting burden during these nursing transfers often equaled or exceeded the recommendations of various authors concerning permissible maximum weight loads during different types of lifts. The lifts were often performed under unfavorable conditions and seldom with an ideal lifting technique. Patients used in the study were also considered to be lighter than the general population.

Zuidema et al.[17,18] have compared various proposals for lifting standards and have arrived at some recommendations. An example constructed from their recommendations further highlights the acceptability of loads in health care lifting tasks. Considering the maximum load to be lifted under ideal circumstances for women who are involved in occasional lifting during the day, the average recommended limit for all age groups is 21.25 lb. This means that, under these circumstances, a patient of 132 lb has to be lifted by a minimum of six women. For frequent lifting and difficult circumstances, recommended lifting loads according to this survey of standards are even less. This is unrealistic in practice which implies that one should look for other solutions. When considering the male population, real expectations during work situations also exceed these standards.

Manual Lifting beyond Reasonable Capabilities

One thing that many will agree upon is that heavy manual lifting is a prime contributor to back pain and for those who have back pain, it is quite difficult to do a job which involves manual lifting and this results in people being away from their jobs—occupational disability. As stated earlier, there have been reports that the health care industry is the riskiest in the nation because of the high incidence of occupational injuries occurring (and remember many of these injuries involve occupational back disorders). The numbers are climbing, and preventive action is required. We must accept the fact that manual lifts and transfers of patients and residents are beyond what we can expect from the workforce.

Previous Approaches to the Prevention of Back Injuries

Over the years, there have been many attempts to try and prevent occupational back disorders and reduce the resulting disability. However, in spite of these efforts, the problem has grown worse. Why have traditional programs failed? A major focus of many back injury prevention-training programs in health care has been on training workers to use proper body mechanics when lifting. These sessions usually focus on proper ways to conduct a lift followed by demonstration and practice. Researchers have studied the effectiveness of training in reducing the impact of occupational back injuries and back pain and are finding that traditional training methods have not had a significant impact in reducing the problem.[19–28]

Why is it that these programs are not improving the problem? When trying to teach someone to use a certain set of techniques, there is an attempt to modify behavior. It is very difficult to modify behavior. If some success is achieved by behavior modification, it is usually short lived. Once workers have left the controlled setting of a training classroom and they are out doing their jobs, many times they do not receive any good follow-up monitoring, and new behaviors can deteriorate quickly. The principles they are taught are theoretical optimum principles and when workers get into the real work environment it may be difficult to apply these principles. As a final point, one must ask is there a best way for all people to conduct a certain type of lift—probably not. Different instructors teach different methods and differences among individual workers may require different methods to conduct a lift or transfer. As we discussed earlier, if the loads being lifted are excessive and beyond what is safe for a health care worker to lift no matter how the lift is conducted, it will be intrinsically unsafe for the worker. We should not abandon our training efforts, however, we must redirect our training to be more effective to prevent and reduce the impact of this back injury problem.

Ergonomics as the Answer

The science or discipline of ergonomics whereby we match job tasks to the capabilities of workers is receiving much attention today. The Occupational Safety and Health Administration (OSHA) has made ergonomics an emphasis for the 1990s. Ideas presented for a proposed standard on ergonomics requires employers to study their workplace for the presence of signal risk factors, one of which is frequent or forceful manual lifting. The tasks involving patient or resident transfers and handling definitely do include this signal risk factor. If a job contains a signal risk factor, it must be studied closer and if determined to be a problem job, a method to improve the job must be determined. Through the principles of ergonomics, jobs must be redesigned and improved to be within reasonable limits. The basic principles of ergonomics seem to offer the best hope in improving the problems associated with occupational back disorders; however, ergonomics is not a magical solution and to be effective, a well-thought-out system of implementation or an ergonomics management program must be developed. The remainder of this chapter will present an overview of a five-step methodology to design and implement an ergonomics management program within a health care facility.

Ergonomic Systems Approach

Before beginning the actual implementation of an ergonomics systems approach, an appropriate foundation must be laid in order for the program to have a chance to succeed. As with any program within an organizational structure, top management must be committed to the implementation of an ergonomics-based systems approach aimed at the prevention of back injuries. Without this support from top management, chances for success will be diminished and limited. Some managers may be very well aware of the back injury problem within their organizations, and others may not be aware of the magnitude of the problem or may have the issue low on their list of priorities. In order to secure this commitment from top management, some ground work might be necessary to establish the need for a back injury prevention program. A review of injury statistics and costs are probably the two most important key factors in establishing this need. As with any program, goals and objectives should be established. Examples of possible goals and objectives might include reduction in the incidence of back injuries, improved working conditions for employees, increased efficiency related to patient handling, improved quality of care, and standardization of lifting procedures.

Next, the personnel who will work on this problem within the organization must be identified. In a large organization, it may be assigned to an appropriate operational unit. In a smaller organization, a committee or task force may be organized to work on the problem. With this groundwork in place, the organization is now prepared to embark on the implementation of an ergonomics-based system. The method being suggested here involves a five-step methodology. The five steps include

1. Risk identification and assessment
2. Risk analysis
3. Formulation of recommendations
4. Implementation of recommendations
5. Monitoring and evaluation system to measure impact of the program

Step 1: Risk Identification and Assessment

In the risk identification and assessment phase, an organization should develop a system to determine what they perceive to be the high-risk jobs and high-risk tasks. For example, some studies have indicated that bathing tasks, toileting tasks, and transfers from beds to chairs are high stress tasks for patient handlers. Through job observation, questionnaires to employees or brainstorming sessions with patient handlers, individual sites should determine the high-risk activities within their organization. Some specific data-gathering tools may be designed and utilized. In constructing data-gathering tools, capabilities and limitations of the organization should be considered.

KEY POINTS

- Identify the perceived high-risk situations.
- Determine the methods of assessment.
- Develop the specific data collection tools.

Step 2: Risk Analysis

Once a determination has been made of what are perceived as high-risk areas and high-risk jobs, a risk analysis should be conducted whereby actual experience will be correlated with what is perceived to be a high-risk situation. That is, are the perceived high-risk jobs actually resulting in losses associated with occupational injuries? In order to conduct this analysis, an organization will need a well-maintained database of occupational injuries. OSHA Form 200 Injury Logs should be available to review summaries of occupational injury and illness experience. If there is not a correlation between perceived high-risk situations and actual experience, why should reasons be determined? It might be that some preventive activities are effective in minimizing the problems in some areas, or characteristics of the work force can have an impact on a unit's experience. Whatever the reason, high-risk and high loss areas need to be identified with some understanding of why problems are present or why there is potential for problems, and true high-risk or high loss areas need to be selected. Through risk analysis, a prioritization will be developed as to which tasks, jobs, or areas should actually be placed on the top of priority lists for implementation of prevention programs. In addition, techniques such as job safety analysis may be employed to better define high-risk components of certain jobs in attempts to determine what optimum methods for improvement might be.

KEY POINTS

- Examine the injury experience.
- Determine the correlations.
- Match with perceived high-risk activities.
- Set the priorities.
- Analyze the jobs for risk factors.
- Prioritization.

Step 3: Formulation of Recommendations

As a result of risk analysis, high-risk situations or job tasks were identified. In the health care industry, these might include reaching and lifting with loads far from the body, lifting of heavy loads, twisting while lifting, unexpected changes in load demand during the lift, reaching low or high to begin a lift or moving a load a significant distance—that is carrying a load. For progress to result, significant changes in job design must occur. Some examples of ergonomic recommendations might include matching of height-adjustable stretchers and beds, which will result in slide transfers rather than lifts. The use of sliding sheets or boards may ease these sliding transfers further. Wheelchairs and other chairs can be better designed to facilitate lifting tasks, which will reduce the risk factors. Probably, the most effective ergonomic intervention related to the risk associated with patient handling would be to eliminate the lift where possible using effectively designed patient-handling equipment. Lifting equipment has been available for a number of years in the health care industry; however, it does not appear as though it has been effectively used to minimize the risks associated with patient handling. If patient-handling equipment is to be effectively utilized, it must be properly designed and integrated into the overall system of patient care as part of the caretaking process of patients. This is why an effective system of implementation is required.

KEY POINTS

- Where is change needed?
- Consider the engineering controls available.
- Administrative controls are another option.

Recommendation Strategies

"Engineering design solutions" and "administrative solutions" are the two strategic areas. The engineering design solutions" usually involve a physical change to the way a job task is conducted or physical modification to the workplace. The changes can be observed as workers conduct the job task in a new way. "Examples" might include the introduction of lifting aid devices such as lateral transfer aids or mechanical lifting aids. The introduction of height adjustable beds is to match with stretcher heights or the use of wheelchairs, which can be converted into stretchers. The "advantages" and "disadvantages" of engineering design solutions result because these are usually more permanent solutions to problems. They may have a higher initial cost but may have a lower cost over the long term as a result of cost reductions realized from the implementation of the changes.

Administrative solutions usually only involve the workers in the way the work is done and do not involve a physical change to the workplace. Changes are apparent by watching how the work is conducted or how workers actually do their job. "Examples" might include job redesign such as changes in scheduling, minimizing the amount of times a patient or resident must be transferred, job rotation where more people are involved in the process of transfers, worker preselection (where appropriate), and the introduction of lifting teams to increase the number of staff involved in a transfer. There are "advantages" and "disadvantages" of the administrative solutions. These recommendations are usually relatively fast and easy to implement, and may have a low initial cost. However, implementation requires continual enforcement and reinforcement and although short-term successes may be realized, it is difficult to achieve long-term change and improvement.

Redesign of Patient/Resident Transfer Tasks

Through engineering controls, changes are made in job design to minimize or eliminate the risk factors. Let us consider some high-risk patient-handling activities with the idea of changing the high-risk components of the job. Tasks involving a bed-to-chair or chair-to-bed transfer can be very difficult. First, let us consider moving someone out of a bed and into a chair. The difficulty of the task will vary depending upon the dependency level of the person to be moved. Considering a totally dependent person, staff must reach across an obstacle (the bed) to have access to the person they need to assist. This involves reaching and it is usually not possible to position oneself with bent knees since the worker is usually leaning against a bed. The patient needs to be physically lifted and considering weight, the loads involved in the lift are unacceptable. Movement into a chair involves moving the person being assisted to a different height level and there is usually some carrying involved. The unacceptable risk factors of this job involve reaching, lifting a heavy load, suboptimal lifting postures, and carrying a load a significant distance. In order to redesign this task effectively, the optimum situation would be to eliminate these high-risk activities. Lifting aid devices are applicable to this situation. These lifting aid devices include full-body slings, which are very useful for the totally dependent patient or resident. In addition, the bed-to-chair transfer can be converted into a bed-to-stretcher transfer. Through the use of convertible wheelchairs which bend back and convert into stretchers and with height adjustment capabilities, a slide transfer rather than a lift may result.

If the patient is not totally dependent, a transfer such as bed to chair may be done by first getting the patient to a sitting posture. Again the amount of assistance required will depend upon the patient's status. Once in a sitting posture, a stand and pivot transfer can be conducted. Some health care workers are highly skilled in this transfer technique and have done it many times without suffering any occupational injuries. However, loads involved are heavy and if the patient does something unexpectedly such as collapses from a weakness in the legs, the health care worker must react and often times these unexpected occurrences result in occupational injuries. Again through application of some lifting aid devices, the risk associated with this type of transfer can be minimized. A device which could be considered in this situation would be a standing and repositioning lift, which is a lifting device with a simpler sling for patients or residents with weight-bearing capabilities.

Lifting Aid Equipment It has been established that patient/resident handling activities need to be redesigned. Physical changes need to be made in the way workers lift and handle patients. Engineering controls are considered by thinking of what new and additional tools might be added to the way this work is accomplished. In past years, there have been some engineering controls available, but presently there is much activity regarding the development of equipment which can assist in patient-transferring tasks. As with other types of work tools, these new devices or tools have been developed to make work easier. We are now seeing many tools being developed to make the work involved with patient-handling easier. The following is a general introductory discussion of some of the types of equipment, which can be very useful in redesigning manual patient-handling tasks.

Sliding Boards For bed-to-stretcher type transfers, low-cost sliding boards are available. Sliding boards are usually made of a smooth rigid material with a low coefficient to friction. The lower coefficient of friction allows for an easier sliding process. These boards assist when lying slide transfers are done. Force is still required to move the patient. Sliding boards, however, do offer some improvement at a minimal cost. Sliding boards offer a starting point, with a low initial investment to begin to improve the way patient transfers are conducted.

Air-Assisted Lateral-Sliding Aids These are devices where a flexible mattress is placed under a patient in the same manner as a sliding board. There is a portable air supply attached to the mattress which inflates the mattress. Air flows through perforations in the mattress and the patient is moved on a cushioned film of air allowing staff to perform the task with much less effort.

Friction-Reducing Lateral-Sliding Aids These devices are positioned under the patient or resident similar to a sliding board but rather than moving with the patient they provide a surface for the patient to be slid over more easily due to the friction-reducing properties of the device. These are simple, low-cost devices which are very easy to store.

Flat Stretchers with Transfer Aids Stretchers are now available, which are height adjustable and have a mechanical means of transferring a patient onto and off the stretcher. Some are motorized and some use a hand crank mechanical device. These devices eliminate the need to manually slide the patient.

Convertible Wheelchairs Since bed-to-chair transfers are difficult because lifts are involved, some new wheelchairs can convert into stretchers, where the back of the wheelchair pulls down and the leg supports come up to form a flat stretcher. These devices eliminate the need to do transfers into and out of wheelchairs. There are wheelchair devices that convert to stretchers which also have a mechanical transfer aid built in for a bed-to-stretcher or stretcher-to-bed type transfer.

Gait Belts An object with handles improves the grasp opportunity for the worker and reduces the risk. Gait belts are belts which go on patients or residents usually around the area of the waist which provide handles for a worker to grasp when assisting or transferring a dependent patient or resident.

Full-Body Sling Lifts Probably the most common lifting aid device in use is a full-body sling lift. There are a number of models and configurations available. The majority of sling lifts are mounted on a portable base. However, use of ceiling mounted sling lifts is growing. Portable base and the ceiling mounted devices have their advantages. With a ceiling mounted device, there is no need to maneuver over floors and around furniture with bases. These units are quite easy to use. However, transfers are limited to where overhead tracks have been installed. As time goes on we may see more overhead tracks in use. Where overhead tracks are not available or practical, portable bases can be used to suspend full-body sling lifts. At this time portable, base units are much more prevalent than overhead track mounts. Sling lifts are usually used for highly dependent patients or residents. They can be used to move patients out of beds, into and out of chairs, for toileting tasks, bathing tasks, and for any type of lift where a patient or resident is highly dependent. These lifts are available with all types of features, and there is a wide variation in the types of slings available. Newer sling design is much easier to put on the patient or resident and in some cases does not even require the patient or resident to be lifted or rolled to place the sling under them. Improvements are constantly being developed with these devices.

Standing Assist and Repositioning Lift These lifts are a relatively new concept, where a simple sling is used which passes under the arms of the patient or resident similar to the type of sling used for helicopter rescues at sea. The sling is of simple design and very easy to place on the patient. These types of lifts are very useful when patients are somewhat dependent and have some weight-bearing capabilities. They are excellent to move patients into and out of chairs and for toileting tasks. They approach toilets from the front and can maneuver in small bathrooms with restricted areas. There are some variations in the sling design, but the basic concept is to put it around the patient's back and under their arms.

Standing Assist and Repositioning Aids Some patients or residents may only need a little support to stand. In this case, they can lift themselves if they have a support to grasp. Various types of devices can be provided to assist a patient from a seated to standing position by allowing them to hold onto a secure device and pull themselves upright.

Bathing Lifts There are a wide variety of bathing lifts available, some are integrated with the tub unit or may operate independently. A lift which can be used to lift patients or residents from beds in residential areas, then used as a transport device and further used as a bathing lift, will minimize the number of transfers required. Bathing lifts can be used in conjunction with ergonomically designed bathing systems, such as height-adjustable tubs which allow for easy transfer of the patient or resident from the tub then bringing the patient or resident to the appropriate height to reduce static bent over postures by the worker.

Other Ergonomic Bathing Devices New and innovative bathing devices are constantly being developed. Some new devices can be brought right to the patient or resident's bed minimizing the transfers and transports required. These units have lifting and turning capabilities.

Administrative Controls for Patient/Resident Handling Tasks

Administrative controls may be applied to patient-handling tasks. For example, the number of patient transfers may be reduced by effectively scheduling procedures the patients may require over the day. Rather than transferring the patients from a bed to a wheelchair or transport device for a particular procedure or diagnostic test then bringing them back to their room, putting them back to bed and redoing the transfer for a number of other procedures during the day, scheduling can be planned better. Scheduling might be done so that the patient will be transferred out of bed, brought from place to place for various necessary procedures, and then returned to their room.

Here is an example of how administrative controls can be used involving rescheduling to minimize a high concentration of lifting activities for direct patient-care staff. It takes place at a state department of mental retardation involving facilities housing highly dependent residents who are in need of much assistance to be moved. One of the most demanding times for resident transfers involved the part of the day when staff were preparing residents to be picked up in buses and transported to their daily activities. Because of the way activities were scheduled and how the buses ran, staff were rushing and highly stressed in order to prepare residents for transport in a short time. Lifting aid equipment was considered and did improve the situation; however, the short window of time to get residents out of bed and prepared for transport was creating the problem. This was not an issue that the direct patient-care staff caring for the residents in the residential area could solve themselves; it involved many people throughout the entire facility including those responsible for scheduling resident activity programs, meals, and the organization, which had been contracted to provide transport services. These groups other than the direct resident-care staff were unaware of the problems encountered with the short time window provided to prepare residents for transport. After an initial meeting was held with these other operational groups at the facility, they understood the problem and were more than willing to consider options to improve the situation. Scheduled activities were adjusted and methods of transport pickups were also changed. This resulted in distributing the number of required transfers over a larger period of the workday and allowed for better use of lifting aid equipment. The implementation of this administrative control took some careful planning and presentation of the problem and cooperation from a wide segment of many operational groups within the facility. The end result, however, was positive to all involved including the residents who received better care because the direct resident-care staff had more time in preparation for the transport process and could give more individual attention to residents.

Administrative controls also might work in conjunction with engineering controls through the use of lift teams. The concept of lift teams had received attention in a specific study done at San Francisco General Hospital,[29] where back injuries related to patient-handling tasks were reduced significantly as a result of the use of lift teams. With lift teams, the lifting activity is done by a specially trained group of workers. However, the workers in this particular study at San Francisco General were also provided with lifting aid equipment. Each organization must decide what will work best in their setting. Whether only specifically trained workers will do transfers with lifting aid devices or all patient-care workers will be trained in the use of lifting aid equipment. Another example of administrative controls would include increase in staff required to do necessary transfers to reduce the amount of force on an individual worker. However, as cost containment becomes more of an issue, increases in staff will be difficult to achieve.

Step 4: Implementation of Recommendations

Implementation of recommendations will involve changes to the workplace. In order to enhance chances for success, a well thought out process should be developed. If engineering solutions such as patient lifting aid equipment are to be introduced, programs for educational awareness and training are necessary. The implementation team must formulate a plan and each player on the team must understand their role in the plan. The objective of this ergonomic-based injury prevention program is to redesign high-risk job tasks in manual patient transfers. Through educational awareness sessions, this message should be delivered throughout the organization. Through hands-on training sessions, staff must be taught new techniques to be applied in transferring patients. During this implementation phase, it must be remembered that changes are being made in the way work is done. To achieve success, staff must feel like they are part of the program development. In this implementation phase, through new directions of training, patient-care-handling staff should learn to assess risk factors in their job and be encouraged to minimize these risk factors with the assistance of management and the implementation team. Workers need to understand what the limits of lifting are and prepare for the unexpected. Remember, the goal is to engineer solutions which will modify patient lifting tasks to reduce risk factors. It is important that everyone understands and excepts their role in the overall ergonomic injury management program.

When back pain does occur, case management will continue to be an important part of the overall program. Through effective medical management, lost time can be reduced and workers can return to work without lengthy disability. Systems for early intervention to detect any early worker's symptoms should be set up for injury prevention and when injuries do occur mechanisms for close follow-up to monitor workers' progress should be developed. The objective should be to get workers back in the workplace as soon as possible after an injury does occur. To accommodate any temporary occupational disability a worker might experience, modified duties should be designed to allow workers to return to the workplace. Recognizing the worker is not at 100% capacity, but they can still be a productive contributor.

KEY POINTS

- Establishing implementation team
- Your education and training efforts
- Specific mechanisms for implementing the system
- Establishing buy-in and creating enthusiasm
- The medical management system

Step 5: Measurement and Evaluation

A system to measure the impact of the program after the program has been put in place and has been operating for a period is the final component. A system for monitoring and evaluation should be developed to determine what successes and failures have occurred so appropriate adjustments can be considered. The monitoring and evaluation system is also important in maintaining an adequate level of interest and attention to the program. The monitoring function also requires a system for data collection as did risk assessment. It must be determined what information will be useful in the evaluation process. Relevant information might include: injury experience, lost work days resulting from occupational injuries, costs associated with occupational injuries, attitudes of staff, how effectively new equipment is utilized or comparisons between units. A set of predetermined indicators should be developed for the purpose of consistent data collection. An indicator is a measurable variable, such as the number of back injuries related to patient-handling tasks. The data collected must be converted to useful information and communicated to appropriate personnel within the organization. As a result of this monitoring function, proper evaluation can be accomplished and modifications made as necessary for continuous improvement.

KEY POINTS

- Selecting the indicators
- Collecting and organizing the data
- Presenting the data as useful information
- Modifying the system as needed
- Remembering the continuous improvement cycle

Summary

The problem of back injuries in the health care industry has been with us for a long time and there is not a quick and easy solution. However, the implementation of an ergonomic-based prevention program offers hope for improvement. Whether workers compensation costs or pressures from OSHA drive the need, ergonomics will become more important to injury prevention in the health care industry. Not only will an ergonomic-based system help reduce losses associated with workers compensation, but also it will help you comply with future OSHA ergonomic regulations. The concepts of ergonomics make sense and can help save money and improve the quality of work life for workers and the quality of patient care.

References

1. Dehlin O and Lindberg B. Lifting burden for a nursing aide during patient care in a geriatric ward. *Scand J Rehab* 7:65–72, 1975.
2. Stellman JM. Safety in the health care industry. *Occupat Health Nursing* 30:17–21, 1982.
3. Stubbs DA, et al. Back pain in the nursing profession. II. Epidemiology and pilot methodology. *Ergonomics* 26:755–765, 1983.
4. Stubbs DA, et al. Back pain in the nursing profession. II. The effectiveness of training. *Ergonomics* 26:767–779, 1983.

5. Klein BP, et al. Assessment of worker's compensation claims for back strains/sprains. *J Occupat Med* 26:443–448, 1984.
6. Wachs JE and Parker-Conrad JE. Predictors of registered nurses' lifting behavior. *AAOHN J* 37: 131–140, 1989.
7. Jensen RC. Back injuries among nursing personnel related to exposure. *Appl Occup Environ Hyg* 5:38–45, 1990.
8. Owen B and Garg A. Reducing the risk for back pain in nursing personnel. *AAOHN J* 39:24–33, 1991.
9. Garg A. Basis for guide: Biomechanical approach, *Revised Work Practices Guide for Manual Lifting*, NIOSH, Cincinnati, OH, 1990.
10. Venning PJ. Back injury prevention among nursing personnel: The role of education. *AAOHN J* 36:327–333, 1988.
11. *Work Practices Guide for Manual Lifting*, U.S. Department of Health and Human Services, NIOSH, Cincinnati, OH, March, 1981.
12. National Technical Information Service, *Scientific Support Documentation for the Revised 1991 NIOSH Lifting Formula*, PB91-226274, U.S. Department of Commerce, Springfield, VA, 1991.
13. Owen B and Garg A. Reducing the risk for back pain in nursing personnel. *AAOHN J* 39:24–33, 1991.
14. Knibbe JJ and Knibbe NE. The workload on the back during the transfer from the wheelchair to the toilet. *Locomotion*, Zwaag, the Netherlands, 1990, pp. 1–10.
15. Stubbs DA, et al. Back pain in the nursing profession—The effectiveness of training. *Ergonomics* 26:767–779, 1983.
16. Dehlin O and Lindberg B. Lifting burden for a nursing aide during patient care in a geriatric ward. *Scand J Rehab Med* 7:65–72, 1975.
17. Zuidema H, Vam Akkerveeken PF, et al. Report of preliminary directives for maximum acceptable weights which may be moved by hand. *Foundation Industrial Health Services for the Building Industry*, Amsterdam, the Netherlands 1983.
18. Zuidema H. Back loading factors, a practical approach. *Mag Public Health Care* 5:182–184, 1985.
19. Brown JR. *Manual Lifting and Related Fields, An Annotated Bibliography*, Labor Safety Council of Ontario, Ontario, Canada, 1972.
20. Dehlin O, et al. Back symptoms in nursing aides in a geriatric hospital. *Scand J Rehab Med* 8:47–53, 1976.
21. Snook SH. A study of three preventive approaches to low back injury. *J Occup Med* 20:478–481, 1978.
22. Anderson JAD. Back pain and occupation. In: MIV Jayson (ed.), *The Lumbar Spine and Back Pain*, 2nd edn. Pitman Medical, London, 1980, pp. 57–82.
23. Stubbs DA, et al. Back pain in the nursing profession. I. Epidemiology and pilot methodology. *Ergonomics* 26:755–765, 1983.
24. Daws J. Lifting and moving patients, a revision training programme. *Nursing Times* 77(48):2067–2069, 1981.
25. Buckle P. A multidisciplinary investigation of factors associated with low back pain. PhD thesis, Cranfield Institute of Technology, Cranfield, England 1981.
26. Stubbs DA, et al. Back pain in the nursing profession. II. The effectiveness of training. *Ergonomics* 26:767–779, 1983.
27. Owen BD and Garg A. Reducing risk for back pain in nursing personnel. *AAOHN J* 39:24–33, 1991.
28. Harber P, Pena L, et al. Personal history, training and worksite as predictors of back pain of nurses. *Am J Ind Med* 25:519–526, 1994.
29. Charney W, Zimmerman K, and Walara E. The lifting team. *AAOHN J* 39:231–234, 1991.

10

Chemodrugs

Occupational Exposure to Hazardous Drugs: Background and Health Effects **10**-1
Environmental Monitoring **10**-4
Occupational Exposure Pathways **10**-6
Dermal Pathways • Other Worker Exposure Pathways • Biological Monitoring
Control of Hazardous Drugs in the Workplace **10**-8
Hazard Assessment • Policies, Programs, and Procedures • Receiving and Storage • Preparation Areas • Hazardous Drug Transportation • Administering Hazardous Drugs • Spill Management • Engineering Controls • Ventilated Cabinets • Closed System Devices • PPE • Administrative Controls • Training • Medical Surveillance • Safe Levels of Personnel Exposure—Guidelines, Standards, Regulations, and the Development of Occupational Exposure Limits (OEL)
Emerging and Other Issues **10**-18
Chemotherapy in Interventional Radiography and the OR • Aerosol Administrations and Nanotechnology • Gene Therapy
Conclusions **10**-20
References **10**-21

Thomas P. Fuller
Illinois State University

Occupational Exposure to Hazardous Drugs: Background and Health Effects

Medicinal mixtures have been used for the successful treatment of illness and injury for over a century. The pharmaceutical industry has progressed from the alchemist's trial and error ministrations to a sophisticated worldwide enterprise employing millions of workers and utilizing modern research and manufacturing techniques.

As the pharmaceutical industry has expanded, so too has the number of workers potentially exposed to hazardous agents, both inside and outside of the health care sector. Today, workers in research, manufacturing, distribution, health care, veterinary services, and waste disposal or treatment are all potentially exposed. Within the health care sector, the workers most likely to be exposed include pharmacy staff, nursing staff, physicians, operating room (OR) personnel, environmental services staff, the shipping and receiving department, and hazardous waste handlers. Exposures are not only limited to hospitals, but can also occur in extended care facilities, outpatient clinics, physician offices where chemotherapy and biotherapy are administered, retail pharmacies, and home health care.

As the pharmaceutical industry expands with hundreds of new products each year, many employees enter the workplace with little knowledge of the chemical constituents, possible exposure pathways, or potential health effects of the pharmaceutical products they encounter. More drugs are being used, by more workers, in increasingly varied types of applications and clinical settings. At the same time, the drugs are becoming more potent, and are often being administered in combination with other agents, for which little is known about possible synergistic effects. Often, little is known about the metabolic pathways of specific drugs in patients, or possible exposure routes to workers.

There is a pervasive misconception among many workers in the health care industry that only antineoplastic or chemotherapeutic drugs can be hazardous. Some confusion is understandable, since definitions of the term "hazardous" have changed many times over the past 40 years as new information becomes available. Hazardous drugs may include antineoplastic or cytotoxic agents, biologic agents, antiviral materials, or immunosuppressive agents. In addition, many investigational and new drugs should be considered potentially hazardous until definitive information regarding their safety becomes available.

The carcinogenic potential of anticancer drugs was reported as early as the 1970s.[1–5] It is clear that the general nature of most anticancer drugs, particularly with regard to their intended use at high dosages, means that they would be cancer-causing in other normal tissues. In 1981, the International Agency for Research on Cancer (IARC) published Monographs on the Evaluation of the Carcinogenic Risk of Chemicals in Humans and included chemotherapeutic agents in the list.[6]

In 1990, the American Society of Health-System Pharmacists (ASHP) defined hazardous drugs as those that are

1. Genotoxic (i.e., mutagenic and clastogenic in short-term test systems)
2. Carcinogenic in animal models, in the patient population, or both, as reported by the IARC
3. Teratogenic or fertility impairing in animal studies or treated patients
4. Toxic at low doses in animal models or treated patients[7]

In 2004, the National Institute of Occupational Safety and Health (NIOSH) provided a new definition of hazardous drugs as those that are

1. Carcinogenic
2. Teratogenic or developmentally toxic
3. Toxic to reproductive systems
4. Toxic to organs at low doses
5. Genotoxic
6. New drugs in which the structure and toxicity profiles mimic existing drugs that have already been determined to be hazardous by the above criteria

In 2004 publication, NIOSH provided a sample list of 136 drugs that met the definition of hazardous drugs. Many were antineoplastic agents, but the list also included a large number of drugs that are not used for chemotherapy, such as vaccines, gonadotropins, estrogens, contraceptives, antivirals, and progestins.[8] Despite the NIOSH report and many other listings of hazardous drugs, many workers still do not realize that the agents they handle and administer on a daily basis are dangerous.[9]

The NIOSH list was not meant to be a complete and singular list of hazardous drugs. It is expected that as new agents become available, each medication would be evaluated by the available information on the product and added to the list, if necessary. In accordance with the Occupational Safety and Health Administration standard for hazard communication (29CFR1910.1200), each employer is expected to develop a hazardous communication program appropriate for their unique workplace.[10] In terms of hazardous drugs, this requires health care organizations to review the drugs used at their facilities to determine whether they are "hazardous," and if so, to include them in their hazardous communications and safety programs.

Since the initial associations were made between cancer and antineoplastic doses in cancer patients in the 1970s, there have been numerous studies on the health effects from exposure to pharmaceuticals. In the 1980s, research focused on the mutagenicity of biological indicators such as blood and urine in workers handling cytotoxic agents.[11–15] The results primarily demonstrated that oncology and pharmacy department workers who handled chemotherapeutic agents tended to have increased incidence of a variety of biological markers that indicated exposure and a possible relation to carcinogenesis, including urinary thioether excretion, urine mutagenic activity, and lymphocytic chromosomal damage.

Early health studies of reproductive health effects were inconsistent. A study by Rogers in 1987 found that exposure to antineoplastic agents was associated with a statistically significant increase in spontaneous abortions for workers handling those materials during pregnancy.[15] Those results were supported in a study in 1985 by Seleven that observed a significant association between fetal loss and occupational exposure to antineoplastic agents during the first trimester of pregnancy.[16] Spontaneous abortion was also associated with women involved in cancer chemotherapy perfusion preparation in later studies.[17–19] Adversely, a study by Hemminki in 1985 found that cytostatic drugs did not affect the frequency of spontaneous abortion in occupationally exposed nurses, but they were associated with malformations in the offspring.[20]

A 2005 paper by Meirow reported increased abortion and malformation risk for women who became pregnant shortly after chemotherapy treatment.[21] It was noted that women receiving chemotherapy treatment during the first trimester experienced increased spontaneous abortion and malformation rates in the offspring, while women receiving second and third trimester treatments experienced an increased risk of stillbirth, fetal growth restriction, and premature birth. A 2007 paper by Fransman concluded through an epidemiological investigation that antineoplastic drugs may reduce fertility and increase poor neonatal outcomes among occupationally exposed oncology nurses.[22]

Throughout the 1990s, numerous studies were conducted analyzing DNA damage to workers handling antineoplastic agents. More sophisticated analytical methods were developed to analyze such measures as DNA strand breaks, other chromosomal aberrations, alkali-labile sites in blood cells indicating chemical damage, and even cyclophosphamide (CP) in urine. In several studies, an association between analytical indicators and occupational exposure was evident; as more accurate measures of exposure became available, the associations between specific worker exposure and activities became more apparent. This research resulted in improved work practices that began to control exposures.[23–27]

Despite the importance of more recent and numerous studies that have demonstrated associations between worker exposures to antineoplastic agents and chromosomal aberrations, the "gold standard" that relates worker exposures to increased incidence of cancer at lower chronic occupational-type exposures remains elusive.[28–31] Despite the assumption and understanding that DNA breaks, chromosomal aberrations, or antineoplastic agents in worker urine could be associated with increased incidence in worker cancers, it has not yet been demonstrated. The worker exposures occur over extended periods and demonstrate a potential chronic exposure to a broad group of workers, but cancer risk has not yet been proven.

A 1994 paper described the results of an occupational risk assessment to health care workers handling cytotoxic agents. Data on worker metabolism of CP were correlated with other animal study and cancer risk data to conclude that the exposed workers faced a cancer risk of 1.4–10 per million.[32] Another study of cancer morbidity among Danish female pharmacy technicians indicated slightly increased incidence of nonmelanoma skin cancer; however, without adequate assessment of numerous exposures to chemicals, including workplace solvents, it is not possible to strictly associate the results with drug exposures.[33]

A meta-analysis of health care providers who work with cancer drugs conducted by Dranitsaris in 2005 provides a comprehensive review and discussion of a large number of earlier papers.[34] Again, due to limited data, the Dranitsaris analysis had difficulty in addressing cancer risk in health care workers. This study, using modern meta-analytical techniques and based on the composite results of numerous prior papers, was able to conclude that an association existed between occupational drug exposure and spontaneous abortions. Perhaps, the most pertinent conclusion of this late research is that there remains a great need for continued health effects research in the area of occupational exposure to drugs.

Environmental Monitoring

Despite the wealth of knowledge about the documented and postulated health effects associated with occupational exposure to hazardous drugs, environmental monitoring in health care settings has been nearly nonexistent, except in research settings. Without such monitoring, the accurate assessment and control of occupational exposure are very difficult, if not impossible.

Through the observation of activities of occupationally exposed workers, it has been noted that methods of handling and preparation vary widely between staff, departments, and facilities. Some workers wear protective clothing such as gloves and gowns, and some do not.[35] Through observation of handling of small spills and cross contamination, variations in possible exposure routes can be postulated.

Early monitoring studies demonstrated good correlation between concentrations of CP in environmental air samples and the levels of CP observed in nursing and pharmacy worker urine.[36] Methods have been developed and published for CP measurement and analysis using both high-performance liquid chromatography and mass spectrometry (MS) from samples collected on glass fiber filters with limits of detection (LOD) of 1.0 and 0.05 $\mu g/m^3$, respectively. Measurable levels in CP manufacturing were easily documented at numerous locations in the production facility.[37] Airborne levels of fluorouracil were documented in the air of a pharmaceutical plant, and methotrexate and CP were documented in the air in pharmacies.[38–40]

More recent air monitoring techniques for CP have also been investigated and reported. Larson reviewed the use of glass fiber filters for collection and found that solid sorbent tubes Anasorb 708, a methacrylic acid polymer, offered improved collection and quantification efficiencies, not only for CP but for fluorouracil, doxorubicin, and paclitaxel as well.[41] LOD were reported as low as 0.7 $\mu g/m^3$. Another report by Hedmer describes air sampling and analytical techniques with liquid chromatography and tandem mass spectroscopy.[42]

In most cases, the percentage of air samples containing measurable airborne concentrations of hazardous drugs was low, and the actual concentration of the drugs was quite low, when present. However, due to the physical and chemical characteristics of most antineoplastic drugs in use, surface contamination sampling is a more useful environmental monitoring method to indicate potential occupational exposure.

Most hazardous drugs have low vapor pressures, high molecular weights, and relatively high densities. This makes them slower to evaporate or even less likely to become airborne from routine handling procedures. It also makes them more environmentally stable and capable of remaining in the workplace on materials, equipment, and surfaces without evaporating readily. It may also make them somewhat more difficult to clean up with cleaners commonly and historically used in health care and pharmacy settings, such as alcohols.

Over the past 20 years, numerous published articles have described surface sampling methods and analytical laboratory procedures.[39,40,42–62] To varying degrees, hazardous drugs have been shown to persist in measurable levels in nursing stations, patient care and treatment areas, pharmacy preparation workbenches, patient laundry, toilets and bathrooms, protective masks, gloves, gowns, floors, keyboards, telephones, nursing stations and administrative areas, shoes, carts, storage bins, waste containers, isolators, workers hands, window panes, door handles, furniture, biosafety cabinets, and infusion bags. Several recent studies indicated surface contamination on packaging and the outside of drug vials received from the manufacturer.[53,54,58,63,64]

Most surface contamination environmental monitoring has been performed with common antineoplastic drugs, such as CP, fluorouracil, and methotrexate.[65] Additional studies have included the drugs ifosamide, etoposide, cytarabine, gemcitabine, cisplatin, platinum, carboplatin, and chlorambucil.[22,56,60,66] Various sampling and analytical collection techniques have been reported in detail. With more than 2000 pharmaceuticals in use or development, however, the capacity to environmentally monitor fewer than 10 is a shortcoming. The few that can be sampled remain as markers of potential environmental contamination and demonstrate the "likelihood" of similar levels of contamination in other agents that cannot be directly monitored.

The correlation of worker observation with environmental surface monitoring has demonstrated contamination resulting from a wide spectrum of health care activities. Significant levels of area and worker contamination have resulted from pharmaceutical preparation by both pharmacists and nurses. Some specific conditions of exposure include

Reconstituting powdered or lypholized drugs and further diluting either the reconstituted powder or concentrated liquid forms of hazardous drugs
Expelling air from syringes filled with hazardous drugs
Administering hazardous drugs by intramuscular, subcutaneous, or intravenous (IV) routes
Counting out individual, uncoated oral doses, and tablets from multidose bottles
Unit-dosing uncoated tablets in a unit-dose machine
Compounding potent powders into custom-dosage capsules
Generating aerosols during the administration of drugs, either by direct IV push or by IV infusion
Priming the IV set with a drug-containing solution at the patient bedside
Handling body fluids or body-fluid-contaminated clothing, dressings, linens, bedding, and other materials
Handling contaminated wastes generated at any step of the preparation or administration process
Performing certain specialized procedures (such as intraoperative intraperitoneal chemotherapy) in the OR
Handling unused hazardous drugs or hazardous-drug-contaminated waste
Decontaminating and cleaning drug preparation or clinical areas
Transporting infectious, chemical or hazardous waste containers
Removing and disposing of personnel protective equipment (PPE) after handling hazardous drugs or waste.[8]

Some observations that have been made about worker activities as they relate to environmental monitoring indicate that surface contamination is closely related to the work practices, but with some interesting twists. It has been shown that different pharmacies in different hospitals, and even those within the same hospital, may have very different safety policies in place that affect the levels of surface contamination measured. Hospitals and departments with written policies and programs are observed to have less surface contamination. Workers who use protective clothing and equipment regularly have been shown to have less area surface contamination associated with their activities. Of course, larger departments that prepare more products, administer more patient doses, and handle more concentrated drugs also have higher surface contamination events and concentrations.

Studies have shown that there is also a fair amount of variation in contamination levels, even between the same worker on different days and for different activities. Some workers tend to have better work practices and lower levels of contamination, but even those same workers may still periodically have a spill or small measurable contamination event. This tends to lead to the conclusion that these small drip or splash events are unavoidable with existing control technologies.

Despite the inconsistencies in environmental monitoring for hazardous drug contamination in the workplace, many papers have been published reporting measurement methods with adequate LOD. The contamination levels vary significantly and are sometimes widespread. The presently limited availability of commercial laboratories available to perform analyses and limitations in the number of hazardous drugs that can be measured are shortcomings that should be overcome in order to better assess individual workplaces, document safe work practices, and alert occupational health and safety professionals to site-specific problems and issues. Regulatory agencies should begin to consider putting standards in place and developing regulations to protect workers. OSHA should begin to enforce the implementation of internationally recognized accepted safety practices, including environmental monitoring that documents adequacy of safety programs.

Occupational Exposure Pathways

Dermal Pathways

Due to the chemical nature of most antineoplastic agents and the differences in the levels monitored in the air and on surfaces, it has been suggested that dermal exposure is the primary route of worker exposure.[65] This has also been demonstrated in measures of surface contamination and drug levels in workers urine that concluded inhalation is of minor importance for internal exposure, compared with dermal routes.[67]

A study by Fransman demonstrated that hands, forearms, and forehead accounted for 87% of cyclophosphamide total body exposure in pharmacy technicians and oncology nurses. This study also showed that dermal contamination of the workers was common.[68]

In addition, a study by Kromhout showed widespread environmental contamination with hazardous drugs associated with spillage of patient urine, and concluded that elevated levels of antineoplastic drugs in worker urine were the result of dermal contamination and improper handling and control techniques. The Kromhout paper also described a contamination scoring method using a black light to observe, measure, and quantify the surface contamination in a mock workplace setup. Significant contamination was found to be associated with the handling of patient contaminated urine from urinals, patient toilets, and utility rooms. Contamination was found on the work surfaces, floors, shoes, nurses' skin, and most frequently on the skin of patients themselves. The fluorescent measurement methods demonstrated contamination on counters, urinals, and bathroom tiles. Contamination of nurses and environmental service cleaning staff were directly related to contact with contaminated urinals, floors, and bed linens.[69]

The importance of the dermal pathway in occupational exposure to hazardous drugs requires the inclusion and further development of qualitative and quantitative measures of dermal exposure. Several methods for dermal assessment have been introduced and expanded upon in recent years.[70–73] Models can predict median potential dermal exposure rates for hands and the rest of the body from the values of relevant exposure determinants. By comparing the dermal exposure data from a broad range of activities and different industries, models can be developed to predict the likelihood of dermal exposures for various jobs and tasks, which can be used in quantitative risk assessment. These methods could be easily adapted to monitor occupational exposure to hazardous drugs in the clinical setting.

The dermal exposure assessment and pathway papers are important because they begin to put quantifiable measures to the likelihood of dermal exposure. The amount of suspect ingredients reaching the skin for given workplace factors, work patterns, worker tasks, techniques, production processes, protective controls in place, and even the equipment used in the work processes can all be included in the likelihood of exposure assessment. These methods can then be used to estimate and compare exposures, and evaluate the effectiveness of engineering, administrative, or PPE controls. These assessments could also be used to support later epidemiological studies of worker groups. In a paper by van Wendel de Joode, another observational semiquantitative assessment method (DREAM) was developed to assess dermal exposures by systematically evaluating exposure determinants using preassigned default values based on a conceptual model of dermal exposure. The results, reported for a variety of tasks and work scenarios, offer good insight into the likely dermal exposure pathways for workers handling and preparing mixtures of hazardous drugs in health care.[71]

Other Worker Exposure Pathways

The increased use of aerosol pathways of drug administration to patients poses a respiratory exposure risk to workers also. Little information is available regarding the environmental concentrations of medications being administered by inhalation. Conservative controls to limit possible worker inhalation of the drugs should be implemented to minimize exposures. These would include the use of hoods and administration tents under negative pressure to adjacent workspaces, sufficient ventilation in the dose administration rooms, and continued use of these controls over the period in which the patient may be exhaling the drug or other hazardous metabolites in their breath.

Ingestion of hazardous drugs is a likely scenario; however, little information is available on this pathway. It is common in the United States to see food and beverages being consumed at the oncology clinic nursing stations. Despite the cross contamination of drugs on surfaces and vials, workers may not always be vigilant about wearing gloves or washing their hands before eating. Strict enforcement of contamination control requirements and elimination of food and beverages is encouraged.

Percutaneous exposures and needlesticks are unfortunately still common exposure pathways in health care. Workers exposed to biohazards from sharps and needlesticks have an additional risk when the needle was also used to administer a hazardous drug. When these exposures are reported, they should be identified as a special circumstance. Workers exposed by this pathway should be seen by a physician for evaluation.

Biological Monitoring

Biomonitoring is the general term used to describe the use of biomarkers to estimate exposure to environmental agents. Occupational biomonitoring focuses on the measurement of these biomarkers in occupationally exposed workers. Biological monitoring of workers is defined as the measurement of chemical markers of exposure to physical, chemical, and biological agents. Biological monitoring reflects exposure from all possible routes, and is especially useful when dermal pathways are the primary routes of exposure, or when other monitoring options or indicators of potential exposure are not available.

Biological monitoring is used to look for contaminants in the workers themselves as key indicators of the adequacy of safety controls. Routine evaluation of constituents and metabolites in urine or blood is performed in several industries. Urine is routinely evaluated for tritium in workers who work in heavy water nuclear reactors or handle tritium while performing research. Blood lead levels are evaluated in 120 industries, including smelting, to periodically evaluate the levels to which workers are being exposed. Workers who use radioactive materials or work at nuclear facilities periodically undergo whole-body radiation scans to determine whether they have inhaled, ingested, or dermally absorbed a variety of radioactive materials as a result of their work activities. Workers handing certain levels of radioactive iodine routinely receive thyroid scans to see if the radionuclide is collecting in that organ at unacceptable levels.

Monitoring biological determinants of exposure is a highly developed and broadly used strategy in the field of occupational health and safety. OSHA currently requires routine biological monitoring for approximately 20 different chemicals used in numerous industries. For example, OSHA requires that workers in an industry using cadmium be included in an extensive medical surveillance program that includes various activities, such as monitoring urine or blood. Depending on measured levels, the employer may be required to take extra actions to protect the worker from additional exposures, including removal from the job.

Biomarkers can range from measurements of a chemical, its metabolite, or specific genes that may be affected by exposure. The benefits of biomonitoring include assessing current exposures, linking exposures with disease, identifying unknown or unsuspected exposures like dermal or oral routes, following trends of exposure over time and evaluating the mechanisms of action of a particular agent in the workplace. Drawbacks exist as well, in that taking biological samples (such as drawing blood) can be somewhat invasive, and in new industries biomarker monitoring can take a long time to be accepted as a measure of exposure. Biological measurements tend to have greater variability than other exposure assessment tools, due to confounding variables such as non-workplace exposures, diet, genetic makeup, previous exposures at work, and lifestyle factors. Workers may fear biological monitoring and perceive it as an infringement on their privacy. They may fear that positive results are a negative reflection on their work performance. Employers may not favor biological monitoring because of the added expense and limitations in data interpretation and relationships to the exposing agents in the workplace. Employers also fear that positive results indicate inadequate controls and possible liabilities. Until accurate risk associations between the biological indicators and the potential health affects can be quantified, the information obtained in biological monitoring is at best only an indicator of working conditions, not an indicator that workers are "safe" or "not safe."

The American Conference of Governmental Industrial Hygienists (ACGIH) publishes an annual list of biological exposure indices (BEIs). These BEIs are used to assess overall worker exposure to chemicals that are present in the workplace, through measurement of the appropriate determinants in biological specimens collected from worker-exhaled air, urine, blood saliva, hair, finger nails, sputum, or other biological specimens. The BEIs are reference values intended as guidelines for the evaluation of potential health hazards in the workplace. Recommended limits of BEIs are based upon accepted understanding of health effects and are generally compared to similar exposure limits for respirator exposures. The primary usefulness of BEIs is to assess the working conditions and potential for worker exposures in lieu of other accurate workplace environmental monitoring techniques, as a means to determine the adequacy of existing safety controls.

Biological monitoring of workers handling hazardous drugs has been performed for over 20 years. The first studies were performed as a means to determine whether nurses, pharmacists, and other hospital personnel were being exposed to hazardous drugs that they handled as part of their jobs. A study by Hirst in 1984 identified cisplatin in urine of workers via gas chromatography and MS.[74] Another method demonstrated that same year also analyzed and found cisplatin in urine via atomic-absorption spectroscopy.[75] Later studies continued to find cisplatin in exposed workers, but these new studies also expanded to other drugs, including CP, platinum, fluorouracil, ifosamide, and methotrexate.[26,76–78]

Other significant potential pathways of exposure include ingestion, injection, and inhalation through drug vaporization. The identified spread of hazardous drugs in the health care workplace described in previous sections indicates a significant potential for worker contamination. Food and beverages should be prohibited in and around locations where these medications and treatments are administered.

The use of safe needle devices is doubly important in units, where patients receive injections. Not only are blood-borne pathogens a concern, but the needles have hazardous drugs inside them, enhancing worker safety concerns. Needlesticks of employees working with hazardous drugs should be reported to Occupational Health Specialists and Safety Professionals immediately whenever they occur.

Although environmental monitoring of hazardous drugs in the air of health care environments has demonstrated fairly low levels, recent research indicates that hazardous drugs can evaporate into the air. Open drug containers, uncovered waste or disposal containers, spills that are not cleaned up immediately, and small, undetected spills throughout the work day all contribute to potentially elevated air levels. Other evidence shows that hazardous drugs initially collected by ventilation cabinet high efficiency particulate air (HEPA) filters can reevaporate off the filter into the hood exhaust stream. Depending on the type of cabinet, the drugs could recirculate in the cabinet or be exhausted to the pharmacy room air.[79]

Control of Hazardous Drugs in the Workplace

Hazard Assessment

The first step in controlling occupational exposure to hazardous drugs in the workplace is to perform an assessment of the hazards that workers will be exposed to in the work setting. The physical layout of the work area should be noted, in addition to an overview of the entire work environment.

A list of the types and quantities of drugs being handled should be obtained and reviewed. The methods used to prepare and administer drugs should be considered in addition to the dose volumes, frequency, and form (tablets, coated vs. uncoated, powder, and liquid). The inventory of drugs and methods should be maintained and updated periodically.

The available equipment, including ventilated cabinets, closed-system drug transfer devices, glovebags, and needleless systems, should be evaluated. The area general ventilation systems should be evaluated in regard to supply and exhaust volume, air changes per hour, and differential pressures that indicate containment air directionality.

Conditions for exposure have been identified in a variety of operations. Those activities using the greatest volumes and concentrations are of greatest concern for exposure potential. The Oncology Nursing Society provides detailed instructions for nursing activities involving the safe administration of drugs in such activities as IV infusions, injections, intramuscular or sub-muscular injections, oral and topical agents, and intra-cavity administrations. The use of aerosolized drug administrations has also gained in popularity in recent years. Other new modalities include the increased use of combination chemotherapy with interventional radiology procedures or in the OR.

Policies, Programs, and Procedures

General policies should be written that clarify the various roles and responsibilities for drug handling and safety within a given health care organization. The directives should be implemented and approved by top levels of hospital management to be the most effective. The main objectives of the policy are to describe (1) how workers are to be protected from health hazards associated with hazardous drugs and (2) how occupational exposures will be minimized.

Written programs that outline the overall attributes of a hazardous drug safety program should be developed. These programs should address the general requirements for the control of hazardous drugs, including labeling, storage, personnel issues, spill control, and waste handling. A hazardous drug safety plan should be developed and be readily accessible to all employees, contractors, and trainees.

The Plan should require detailed Standard Operating Procedures (SOPs) to be developed on a departmental or local area basis, establishing discrete actions and responses to be taken to ensure the safe use of hazardous drugs with regard to specific medical applications. These procedures should address specific medical applications for the use of the drugs, but they should also include specific safety steps or requirements within the context of each individual procedure. The Hazardous Drug Plan should specify criteria that the employer uses to determine and implement control measures that reduce employee exposures to hazardous drugs, including engineering controls, PPE, and administrative policies such as training and record keeping.

Other administrative controls that should be described in the Plan include circumstances that require specific approvals for use of particularly hazardous drugs or procedures, provisions for medical surveillance of employees, and occupational or environmental monitoring of drug exposure or contamination. The Plan should be maintained as current, and should be reviewed and updated whenever new drugs or methods are introduced to the program and at least once per year.

Receiving and Storage

Occupational exposure control must begin where hazardous drugs enter the facility. Workers who receive hazardous drug packages are at most risk from contamination from damaged and leaking containers. These workers should be able to consult written procedures and training materials on hazardous drug handling. Both the procedures and training should include discussion of how to respond to leaking or damaged packages, whom to contact, and what to do if they are personally contaminated. Detailed instructions should be provided that identify the receipt of the package, inventory, security, and proper methods to transfer the package to the pharmacy or nursing areas where they will be used. These workers must be informed about the potential hazards of exposure to the drugs and possible health effects.

Purchasing and distribution staff should receive training on the labeling of hazardous drug packaging and the practice of separating hazardous materials from nonhazardous materials. Any personnel responsible for opening hazardous drug packaging must be provided the appropriate PPE and be trained on the proper methods for donning and doffing.

Workers who receive hazardous drug packages and are expected to open them should be provided, and should wear an impermeable apron, laboratory coat, eye protection, and gloves. All hazardous drugs should be stored and transported in closed containers that minimize the likelihood of breakage, and be contained in a secondary container. All internal package materials such as vials or containers should be considered contaminated.

The room used to receive and temporarily store hazardous drug shipments should be large enough and have adequate ventilation to ensure sufficient exhaust to dilute and remove any airborne contaminates that may occur from spills or damaged packages.

Drug receiving areas should always be included in program assessments. Access to these receipt and temporary storage areas should be restricted to essential administratively approved staff. Measures should be considered and made available to ensure minimization of exposures to staff in these ancillary areas.

Preparation Areas

The sophistication of drug preparation areas varies widely; from small hospitals, outpatient treatment clinics, hospital patient areas, doctor's offices, and hospital specialty treatment centers, to large state-of-the-art hospital pharmacies. Similarly, the types of drugs in use, state of the facilities, safety programs, and medical treatments performed vary drastically. The fact that a clinic or doctors office is small and does not treat a large number of patients does not negate potential for significant exposure from accidental contamination incidents, improper procedures, or inadequate facility design. There is no clear evidence that the drugs used in smaller facilities are either less potent or less toxic.

Hazardous drugs may be transported directly to the point of use, or may proceed to another prior location to be reconstituted, transferred to different containers, or otherwise manipulated before being administered to patients. Even when the utmost care is taken, there are many opportunities for occupational exposures.

Common manipulations that have been associated with splattering, spraying, and aerosolization of drugs include withdrawal of needles form drug vials, drug transfer using syringes and needles of filter straws, breaking open ampules, and expulsion of air from a drug-filled syringe.[35] In addition, many opportunities for surface contamination exist that are likely to lead to worker or area contamination.

Full protective clothing, including labcoats, protective eyewear, and double gloves must be worn. Internal packages and vials should be handled as if contaminated and first removed and wiped down with solutions of either isopropyl alcohol or sodium hypochlorite (bleach) to ensure that they are clean. The gauze used to perform the decontamination should be disposed of as a hazardous waste.

Before work begins at a bench or laminar flow hood, all surfaces should be cleaned with gauze and a preparation of either isopropyl alcohol or sodium hypochlorite. This is a precaution to ensure the bench is clean and that no residual hazardous drugs remain behind from previous users.

When drug preparation is complete, the final product should be sealed in a leakproof plastic bag or other sealable container for transport before removing the material from the ventilation cabinet. Pharmacy staff should prime all IV tubing with nondrug containing fluid and syringes inside the cabinet before removing. All waste containers in the cabinet should be sealed and wiped before they are removed from the cabinet. Before leaving the ventilated cabinet work area, workers should remove and dispose of the outer layers of gloves and sleeve covers as hazardous waste.

The use of closed-system transfer devices, glove boxes, and needleless systems when transferring hazardous drugs from primary packaging (such as vials) to dosing equipment (such as infusion bags, bottles, or pumps) has been shown to reduce local contamination of work areas and materials. Closed systems also reduce the potential for generating aerosols.[27,57,80] Closed-system devices must always be used inside the ventilated cabinet; the same PPE requirements apply even when using these systems and devices.

Hazardous Drug Transportation

Once drugs are prepared within the institution, and are properly labeled, they should be transported to the place of use in leakproof zippered bags within protective containers. Containers should be labeled as "hazardous." Personnel who transport the drugs should be aware of the hazards and emergency procedures in case of a spill, including the location and use of spill kits.

Administering Hazardous Drugs

The Oncology Nursing Society has published various guides for the administration of hazardous and chemotherapeutic drugs.[81,82] It should be kept in mind, however, that not all hazardous drugs are antineoplastic or only administered in an oncologic setting. Despite this, it is recommended that the same practices for administration be applied whatever the patient setting, be it clinic, doctor's office, nursing home, or home care. Other safe handling guides also exist outside the field of oncology.[8,83,84]

Administration of drugs to patients is generally performed by nurses or physicians. Drug injection into the IV line, clearing of air from the syringe or infusion line, and leakage at the tubing, syringe, or stopcock connection present opportunities for skin contact and aerosol generation, and area contamination. Subsequent removal and disposal of syringes and materials can also lead to environmental exposures.

Excreta in urine, feces, sweat, mucus, and from the respiration from patients who have received hazardous drug administrations have been shown to sometimes contain high concentrations of the hazardous drugs or other hazardous metabolites.[75] Any workers that may come in contact with patient excreta should be made aware of potential hazards and health effects, proper PPE, and handling methods. This includes housekeeping staff that may be handling contaminated linens and/or cleaning patient toilet facilities.

Spill Management

All spills and leaks of hazardous drugs should be considered serious and should receive immediate attention. If spills are not cleaned up quickly, the spread of contamination becomes more likely and more difficult to control.

Personnel who notice the spill should warn others in the area and post a warning sign, if necessary, to keep others from entering the area while the cleanup proceeds. The spill should be isolated from other workers and patients. The person cleaning up the spill should don protective clothing, including an impermeable gown, shoe covers, two pairs of protective gloves, splash goggles, and a face shield. If the spill is in dry powdered form, or if a dry powder is being used to absorb the spill, it may be appropriate to wear an N95 respirator to prevent inhalation of small particles. These types of respirators, however, provide no protection from evaporated drugs in the breathing zone. Only workers who are specifically trained to wear an N95 respirator and who have been fit-tested within the past year should do so.

Cleanup efforts should proceed by wiping up liquids with disposable absorbent towels or spill-control pillows. Absorbent powders such as vermiculite may also be appropriate for larger spills. The materials that have absorbed the drugs should then be scooped up using a plastic scraper or dustpan to place the materials in heavy-duty waste disposal bags and then a puncture-proof hazardous waste drum. Cleanup of the area should continue a minimum of three repetitive times working inward from the least-contaminated area towards the most-contaminated area. At the end of the process, detergent followed by a water rinse should be used to clean the drug from the surfaces. Once the area is clean, the PPE can be removed and the waste containers should be labeled and moved to hazardous waste storage.

In the event of an accidental exposure, great care should be taken not to cross-contaminate greater areas of the worker's body, other workers, areas, or patients. Other workers in the area, also wearing PPE, should assist the contaminated worker in taking off the PPE without spreading the contamination.

The PPE should be disposed of as hazardous waste. The contaminated area should be washed for a minimum of 15 min with soap and water or more specific cleaning regimens recommended by the material safety data sheet (MSDS), if available. Medical attention should be sought immediately either through occupational medicine/health department in the organization or at the local emergency department.

In the event of an eye exposure to hazardous drugs, the eye should immediately be flushed for 15 min in water or another material indicated by the MSDS or department-specific emergency response procedures. In some cases, the flushing material recommended may be specific to the drug. Any information available on the hazardous drug being administered should be provided to the medical or emergency department personnel attending to the exposed worker. All workplace settings should maintain continually stocked spill kits, including outpatient clinics, nursing homes, and doctors offices. Nurses providing home care treatments with hazardous drugs should carry portable spill kits to respond to uncontrolled releases and contamination events.

Engineering Controls

The engineering controls and ventilation required to ensure that hazardous drugs can be prepared safely depend on the types of drugs and quantities being handled, and vary between different facilities and departments. It is important that qualified engineers are consulted for system design and are clearly aware of industry design standards and the hazards of the drugs that are proposed for use in the facility. The design of hazardous drug handling facilities has been raised to levels of sophistication that only experienced and qualified engineers can address. Too often, problems with ventilation systems become evident after a facility is constructed and it is too late to make expensive corrections. In addition to meeting specific design specifications, a building commissioning process should always be conducted before occupancy and use of hazardous drugs begins.

Knowledge about ventilation airflow direction objectives has changed in recent years and is a complex factor in facility design. Contrary viewpoints remain. Some of the primary objectives of the facility are to (1) keep hazardous drugs within the ventilated cabinet workspaces to prevent exposures and to minimize exposure of workers handling the agents, (2) keep the hazardous drugs within the room, so in the event of a spill or airborne release the chemicals do not reach the public and are not spread throughout the facility, and (3) protect the medications from biological contamination.

In order to achieve these objectives, a fine balance of airflow is necessary, and the monitoring systems that maintain the airflow, pressure differentials, and airflow directions must be very sophisticated. These systems must be able to react quickly to the changes in outdoor weather conditions and the activities of workers in the area as they perform activities and open doors to enter and exit the workrooms.

In the past, pharmacies were typically designed to be kept under positive pressure in relation to adjacent areas, such as hallways, in order to ensure that air flowed from the clean area to dirtier areas in order to keep the pharmacy clean and germs out of the preparations. This may still be the case in many facilities.

The latest guidelines of the USP 797 recommend that oncology drug preparations take place in an ISO class 5 ventilated cabinet or room, within an ISO class 7 room that is negative in relation to an anteroom, which is also at either ISO class 7 or 8 levels, and also under negative pressure in relation to adjacent external hallways. The rationale for this change is the need to keep hazardous drugs inside the interior preparation room in the event of an accidental release or spill.[85] Each room should have a minimum of 12 air changes per hour.

There are some pitfalls in the USP 797 criteria, however, in that infectious agents in the adjacent hallway or uncontrolled areas will be directed toward the pharmacy anteroom, and then the prep-room, by the flow from more positive areas to the most negative internal room. An alternative to this design that offers both product protection from infectious agents and the control of accidental drug releases within the pharmacy is to maintain the innermost prep-room under positive pressure in relation to the anteroom. This design actually allows the internal drug preparation and handling room to be under positive pressure to keep microbial contamination out, while the anteroom is kept under negative pressure in relation to both the hallway and the prep-room. This design ensures that if there is an accidental release of hazardous materials in the preparation room, they cannot reach public hallways, while at the same time microbial contaminates from the hallway would be stopped in the anteroom and cannot make it to the clean prep-room.

A fine balance must be struck between all three rooms to ensure that the pressures remain constant and can perform their duties even when workers enter and leave the rooms, and to ensure that the ventilation safety cabinets continue to function at their design specifications at all times. It may take fine-tuning of room supply and exhaust rates, and close coordination with the monitoring systems, but these designs can be most efficient at meeting the objectives stated above. To help ventilation systems perform at maximum efficiency and to reduce the possibility of cross-contamination of both infectious agents and hazardous drugs, it is recommended that workers minimize the number of times per day that they enter and leave the internal drug handling room. It should also be required that workers should not enter or exit the room when another worker is handling hazardous drugs in the ventilated cabinet. The use of pass-boxes or pass windows is encouraged to reduce the disturbance and turbulence created each time the door to the anteroom is opened.

Ventilated Cabinets

Ventilated cabinets have a dual purpose when working with hazardous drugs. Firstly, they are used to eliminate or reduce occupational exposures to the drugs or infectious agents. Secondly, they may be used to protect the product from being contaminated by the external environment and the worker in order to maintain sterility.

When asepsis is not required for product preparation, a Biological Safety Cabinet (BSC) Class I may be acceptable for handling hazardous drugs. When hazardous drugs are being prepared that require sterile techniques, however, the work should be performed in either a Class II or Class III BSC. These types of cabinets protect the worker from the agents and the materials being handled from contamination.

There are four main types of Class II BSCs. They all have downward airflow and HEPA filtration. They are differentiated by the amount of air recirculated within the cabinet, whether this air is vented to the room or to the outdoors, and whether the ducts are under positive or negative pressure.

Type A cabinets recirculate approximately 70% of the cabinet air through the HEPA filters and back into the cabinet. The rest of the cabinet air is released into the preparation room. Contaminated ducts are under positive pressure.

Type B1 cabinets have higher velocity air inflow, and recirculate 30% of the cabinet air, and exhaust the rest to the outdoors through HEPA filters. Ducts and plenums operate under negative pressure.

Type B2 systems are similar to Type B1, except that no air is recirculated within the cabinet.

Type B3 cabinets are similar to Type A. They recirculate approximately 70% of the cabinet air, with the remaining 30% discharged to the outdoors under negative system pressure.

Class III cabinets are totally enclosed with gas tight construction. The entire cabinet is under negative pressure and the operations are performed through gloves attached to the device. The exhaust air is released outdoors through HEPA filters.

The ideal cabinets to be used for preparation of hazardous drugs that require sterile techniques are the Class II Type B2 or the Class III cabinets. This is because each design offers maximum protection to workers and the product at the same time. Cabinets that recirculate the hood air back through the workers' workspace subjects them to possible additional exposures. As noted earlier, it has also recently been demonstrated that some of the chemicals captured by HEPA filters can rerelease hazardous agents when they evaporate off the filter and subject workers in the room to hazardous exposures.[79]

BSC exhaust fans should remain on at all times to ensure minimal contamination of the work surfaces, unless the unit is being mechanically repaired or moved. If the blower is turned off, it should be decontaminated before use. Each BSC should be equipped with a continuous monitoring device to allow confirmation of adequate airflow and cabinet performance. Cabinets should be located in areas with minimal turbulence, in order to minimize back-drafts and escape of contaminants to the room.

Cabinet maintenance activities should only be performed by qualified professionals with understanding of not only the mechanical systems, but the types of hazardous agents being used in the cabinet. These workers should be familiar with the appropriate types of PPE to be worn when working on the systems and other controls to be used to prevent the spread of contamination in the preparation room and adjacent facility areas. The cabinet should be emptied of all agents and decontaminated prior to initiation of maintenance activities. Cabinet materials and parts should be decontaminated and placed in bags before being removed from the area. Filters and filtration media should be placed in plastic bags and disposed of as hazardous and/or biological infectious waste.

The hood should be certified by these technicians prior to use, and every 6 months thereafter. HEPA filters should be replaced periodically in accordance with manufacturer specifications and activities and workloads of the facility. Certification and maintenance stickers should be prominently displayed on the front of the cabinet.

In order to minimize the possibility of worker or product contamination, hazardous drug handling cabinets should be cleaned often. It is recommended that surfaces are cleaned at the start of each shift, at least once in the middle of an 8 h shift, and again at the end of the workday. Whenever a spill takes place in the hood, decontamination should also be completed before work continues.

Decontamination should consist of surface cleaning with water and detergent followed by thorough rinsing. The use of detergents or bleach is recommended to ensure thorough deactivation of the chemical agents that were used in the hood. The use of 70% isopropyl alcohol or ethyl alcohol is recommended for disinfection of the work surfaces.

Closed System Devices

Closed system protective devices have been developed that are effective in reducing the spread of contamination during preparation and administration of anticancer drugs. One popular brand-name device called PhaSeal utilizes a double membrane system to ensure leak-free transfers and connections to prevent the release of drug aerosols and droplets during preparation. Studies have shown that the use of such devices greatly reduces the amount of air or surface contamination that results from handling procedures. They also reduce the levels of contamination found on workers.[51,57,80,86,87] Since there is still a possibility of an aerosolization or contamination incident when using these types of devices, it is recommended that preparations be conducted with appropriate PPE, and in ventilated cabinets, if possible.

PPE

Protective gloves should be specifically selected for the drugs being used as indicated on the MSDS or other information from the manufacturer or other reputable source. Studies have shown that different glove materials have drastically different permeability ratings for different drugs. Latex gloves should not be used for drug preparation, due to the toxic properties associated with latex and hypersensitivity reactions. In addition, one study found that latex gloves were completely permeable to carmustine, thiotepa, and CP.[88] In another study of 13 different gloves to 13 cytotoxic agents by Wallemacq, it was shown that neoprene and nitrile gloves provided good protection to all of the agents tested.[89]

Two layers of gloves should be worn. Gloves should be inspected prior to donning and they should be changed every 30 min, whenever they are torn or punctured, or are contaminated. When gloves are taken off, they should be discarded as hazardous waste and hands should be washed with soap and water before new gloves are donned.

Disposable gowns with closed fronts, long sleeves, and elastic or knit cuffs made of a nonabsorbent material should be worn when working with hazardous drugs. Disposable sleeve covers may be worn as additional protection to wrists close to the work area, and disposed of periodically. Each gown type and material should be selected by carefully reviewing the capabilities and drugs to be handled. In a study of six disposable protective gown materials for splash protection from 15 antineoplastic drugs, only one gown provided complete protection from all 15 agents. Two of the gowns tested let all 15 antineoplastic agents through.[90]

Administrative Controls

In addition to policies, programs, and procedures described in prior sections, other requirements for administrative controls should specifically be considered. Record keeping may be of prime importance with regard to worker exposures and the quantities and types of materials that they handled or administered. Controls over worker exposures should be established for those that may demonstrate health effects believed to be associated with drug exposure, or women who are pregnant or trying to become pregnant. The responsibilities to minimize exposures to pregnant women should be delineated at employment by both the employee and employer. In lieu of exposure limits, prudent practice would deem that exposures be minimized for the entire workforce wherever possible, through engineering and PPE controls. For sensitive populations, it may be more protective to remove workers from duties that require handling or administering hazardous drugs. This all must be done without constraining the employees' rights to work.

Other administrative controls include using signage to identify the location of hazardous drugs as notifications to both employees and the public. The minimization of the public and elimination of food from patient dose administration areas are also part of a thorough administrative program.

Training

Even though most of the workers who handle and administer hazardous drugs in health care settings are highly skilled and educated, it is important that they receive specific safety training on the individual facility Hazardous Drug Plan and the associated safety practices and systems in place. The engineering controls at each institution are different, as are programmatic roles and responsibilities, not to mention site-specific SOPs. Highly skilled personnel also need to know the specific hazards associated with the drugs in use at a given facility, and the emergency and spill response protocols to be followed when necessary. These workers should be made aware of, and enrolled in, the medical surveillance program and should be told when and how often to report for evaluation. Safety training should be conducted annually as a review and to discuss any program, control, system, or regulatory changes that have occurred over the past year.

Training should also be conducted for all other workers, such as housekeepers and aides, in accordance with the Hazard Communication Standard. Related information should be provided at the time of the workers' assignment to work with or near hazardous drugs or treated patients. Records of all training attendance should be maintained for the duration of the workers' employment plus 30 years. Training records should include the dates of the training sessions, summary of the training content, names, and qualifications of the person conducting the training, and names and job titles of persons attending the sessions.

Hazardous drug safety training must include the following elements:

- Methods and observations to be used to detect the presence or release of hazardous drugs into the work area
- Use of PPE
- Associated physical or health hazards resulting from exposure to the drugs including the carcinogenic potential of exposure
- Proper disposal methods
- Measures employees can take to protect themselves from exposures

- Management of spills
- Management of accidental exposures
- The details of the Hazard Communication Program, including the labeling and signage systems, location of MSDSs, and sources of other related information

Medical Surveillance

Medical surveillance involves the collection and interpretation of data regarding the workforce in order to detect changes in the health status of a working population potentially exposed to hazardous drugs. The medical surveillance program first establishes a baseline of a worker's health and then monitors their ongoing health as it relates to their potential exposure to the hazardous agents. Medical surveillance is one part of a comprehensive approach in evaluating and minimizing worker exposures to toxic or hazardous agents. It is a key part of the overall health and safety program and should be closely coordinated with other aspects such as industrial hygiene, environmental monitoring, safe work practices, training, and the use of PPE.

The primary purpose of surveillance is to identify the earliest reversible biologic effects so that exposure can be reduced or eliminated before the employee sustains irreversible damage. The occurrence of exposure-related diseases or other adverse health effects should prompt immediate reevaluation of primary preventive measures.

Employers should ensure that health care workers who are exposed to hazardous drugs are routinely monitored as part of their medical surveillance program.[7,81,84] This should include workers who directly handle hazardous drugs such as nurses, pharmacists, and physicians. Other workers who come in contact with patients or patient wastes within 48 h of administration of hazardous drugs should be included in the medical surveillance program.

Postexposure examinations should take place in the event of needlesticks, spills, or personnel contamination incidents. The examination should focus on the involved exposure route as well as potentially affected organ systems. Treatment and further laboratory studies may follow as indicated and guided by emergency response protocols.

Medical evaluation should include the reproductive status of the worker. The reproductive toxicity of many hazardous drugs should be described to workers, along with the discussion of other possible health effects from exposures.

Despite the comprehensive database of information on the biological monitoring of hazardous drugs and related metabolites, or genetic outcomes in the urine and blood of workers over the past 25 years, and the potential benefits to the administration of a more successful health and safety program that could minimize occupational exposure to the drugs, few organizations actually implement such programs. This is due to a variety of obstacles.

First, it is not common practice in health care for workers to submit urine or blood samples, particularly for drug analyses. Second, it might be assumed that neither the workers nor hospital managements want to know if (and to what extent) workers are being exposed to hazardous drugs. Third, there is a lack of direction from the federal government on the topic. Although Biological Monitoring requirements and limits have been developed by OSHA and the ACGIH for other chemicals in other industries, there are no regulations or guidelines for these measurements to be done on workers exposed to hazardous drugs. Finally, since there is little impetus or interest in pursuing this type of monitoring, commercial sources of monitoring and measurement for these drugs or metabolites are not available and have not been standardized, except in specialty literature and research publications. The ability for any one individual health care system to develop and implement a biological monitoring program is limited by resources, staffing, technical capabilities, and interest in the results. Unfortunately, this status is unlikely to change without external motivation by the government, despite the benefits seen by such programs in other industries.

Safe Levels of Personnel Exposure—Guidelines, Standards, Regulations, and the Development of Occupational Exposure Limits (OEL)

The 1970 OSHA Act requires employers to provide employees safe working conditions and empowers OSHA to prescribe mandatory occupational safety and health standards. Some of OSHA's most controversial projects have been in setting regulatory limits and enforcing occupational exposures to toxic chemicals.

While manufacturers of food additives, drugs, and pesticides must demonstrate the safety of their products to consumers prior to marketing them, no employer need obtain advance approval of new processes or materials or conduct tests to ensure that its operations will not jeopardize worker health. OSHA must first discover that a material already in use threatens worker health before it can attempt to control exposure. Standards for toxic chemicals typically set maximum limits on employee exposure and prescribe changes in employer procedures or equipment to achieve this level.

The OSHA Act specifies that in regulating toxic chemicals it shall adopt the standard "which most adequately assures, to the extent feasible, on the basis of the best available evidence, that no employee will suffer material impairment of health or physical capacity." The meaning of these contradictory phrases was a source of controversy for many years. Court decisions made it clear that the "best available evidence" did not require proof of causation or non-positive epidemiological studies; animal data alone could support regulation of a toxic substance. The debate focused on OSHA's obligation to weigh the economic costs of its standards. The agency acknowledged that it was required to consider technological achievability and viability for industry, but it denied that it was obliged to balance the health benefits and economic costs.

Legal challenges to OSHA standards have clarified OSHA's responsibilities. The Supreme Court overturned OSHA's benzene standard in 1980 because the agency had not shown that prevailing worker exposure levels posed a "significant" health risk. This prerequisite proved difficult when OSHA attempted to establish standards for 428 air contaminants in a single proceeding in 1989. OSHA failed to show that each individual contaminant posed a "significant" risk at current levels. However, the Supreme Court earlier upheld OSHA's cotton dust standard, rejecting arguments that the agency was obligated to weigh the costs of individual standards for concededly hazardous substances.[91]

Today, it continues to be difficult to pass occupational health limits of exposure through OSHA. In the last 25 years, only about two dozen additional Permissible Exposure Limits (PELs) have been published. Despite the fact that the IARC lists cisplatin as "probably causes cancer in humans" and CP as "causes cancer in humans," neither one has an OSHA PEL requiring employers to protect their employees.

A limited study recently completed by the California Environmental Protection Agency found that 44 chemicals known to cause cancer do not have OELs set by OSHA. Other organizations, such as the IARC and the National Toxicology Program, also list numerous chemicals, including hazardous drugs, among carcinogens that are not regulated by OSHA.

To date, there are published regulatory limits for only a handful of chemicals considered "hazardous drugs" used in health care. Some drugs, such as cisplatin, have PELs for various constituents that make up the compound, such as hydrogen chloride and *cis*-diaminedichloroplatinum. But others, like methotrexate, which is a known teratogen and mutagen, have no OELs.

Several papers have identified issues and methods for the development of OELs for pharmaceuticals.[92,93] Due to the lack of regulation and research in the area, however, OELs have been mostly developed by private industries and enterprises. This is done primarily to ensure the health and safety of pharmaceutical manufacturing staff.

Risk assessment and control of occupational exposures in high potency pharmaceutical manufacturing includes some core elements common to most companies. Hazard evaluation and review includes documentation of toxicological and clinical data, physical and chemical properties, and basic manufacturing processing safety criteria such as ventilation and PPE availability. Methods to estimate and measure occupational exposure levels are developed including exposure indices, environmental sampling, and laboratory analyses. A performance-based exposure control limit strategy for pharmaceutical active ingredients was developed by Naumann, which provides useful criteria for manufacturers assessing safe levels of exposure and corresponding control criteria.[94]

Unfortunately, the OELs, analytical methods, and control criteria developed by manufacturers for internal use are not generally published or available to the public or even the consumer. The lack of transparency or availability of all relevant information regarding new pharmaceuticals, such as toxicity or long-term health effects, along with a lack of regulatory oversight, are just a few shortcomings of the manufacturer-based methods of creating OELs.

The ability to measure hazardous drugs in the environment and in bioassays of workers is an indication of exposure. In cases where the drugs are known or probable carcinogens, teratogens, or mutagens and there are no OELs or PELs, it should be considered that there is no safe level of occupational exposure. Health care employers must assume the responsibility for analyzing their own inventory of hazardous drugs and without given OELs design programs and controls to eliminate or minimize all occupational exposures.

Emerging and Other Issues

Chemotherapy in Interventional Radiography and the OR

The use of chemotherapy drugs in both radiology and surgery has been expanding in recent years, and the technologies and drugs being administered have expanded accordingly. Regional infusion of the chemotherapy drugs in international radiography (IR) or the OR allows direct exposure to the neoplasm and reduces other systemic toxicities. Infusion of intra-arterial chemotherapy began in the 1980s, but numerous expanded regimens have regained popularity. New methods in IR and the OR using chemotherapy perfusion, intra-arterial melphalan percutaneous hepatic perfusion, and systemic liposomal dioxorubicin administration with concomitant radiofrequency ablation have been reported.

Along with previous IR procedures that administer large doses of doxorubicin, cisplatin, or mitomycin C, the new procedures introduce the expanded potential for additional occupational exposures to pharmacists, nurses, physicians, and technicians. Strategies for worker protection and controls need to be reevaluated to minimize the hazards from these new procedures. Each health care organization must take responsibility for performing workplace assessments and implementing controls. In the case of new methods of administration in new departments, training of staff who may not have worked with hazardous drugs in the past is a crucial part of the upgraded safety program. New work locations will need to be evaluated and different types of injection equipment must be understood before hazardous drugs can be used in them.

Aerosol Administrations and Nanotechnology

Some institutes have reported the expanded and expanding use of nebulized aerosol administrations of drugs to patients. The creation of aerosols may provide added therapeutic benefits to patients, but at the same time, by creating airborne pathways, they greatly increase the likelihood of airborne occupational exposures. In some cases, it could be expected that these pathways of exposure could be much more efficient in getting into the workers with more toxic or hazardous impacts.

A non-published study by A. Streifel at the University of Minnesota Medical Center found peaks of aerosolized drugs as high as $3.5\,\mu g/m^3$, with an average of about 1.0 in the work area around the patient, for periods of up to an hour.[95] Depending on the agent used and the likelihood that a nurse or other worker would remain in the work area for the treatment period, exposures through inhalation could be significant. The hazard potential would vary depending on the toxicity of the agent. As these types of treatments expand in popularity, it will be important to implement adequate controls and monitor work places for airborne levels and worker exposures.

Nanotechnology involves the manipulation of matter on the atomic or molecular scale, normally 1–100 nm. Nanomedicine will potentially revolutionize the future delivery of medications, diagnostics, and treatments in a targeted approach. The unusually small sizes of nanomedicines confer special capabilities useful in human physiology and medical applications. These same characteristics, however, present new questions regarding the toxic effects to both the patients and the workers handling the nanomedicines. The hazards associated with nanotechnology are not yet fully understood. Methods to accurately measure nanomaterials in the environment and workplace are just beginning to be developed.

Gene Therapy

Gene therapy is an experimental treatment that involves introducing genetic material (DNA or RNA) into a person's cells to fight disease. Typically, a "normal" gene is inserted into the genome to replace an "abnormal," disease-causing gene. A carrier molecule called a vector must be used to deliver the therapeutic gene to the patient's target cells. The most common vector is a virus that has been genetically altered to carry normal human DNA. Viruses encapsulate and deliver their genes to human cells in a pathogenic manner, and scientists are trying to use this capacity to manipulate the virus genome to remove the disease-causing genes and replace them with the therapeutic genes.

In the most common methods, target cells in the patient, such as cancer cells in the lung, are infected with the viral vector. The vector then unloads its genetic material containing the therapeutic human gene into the target cell. The generation of a functional protein product from the therapeutic gene restores the target cell to a normal state. Other possible therapies include swapping abnormal genes with normal genes through homologous recombination, repairing abnormal genes through selective reverse mutation, or regulation of a gene's activities to turn it off or on.

Other vectors used in gene therapy may include retroviruses to be integrated into the chromosomes of host cells, adenoviruses, adeno-associated viruses, and herpes simplex viruses. Besides viruses, gene therapy can be performed by direct insertion of pure DNA into target cells, or the use of artificial lipid spheres called liposomes to carry DNA through the cells membrane to the nucleus, among others.

Sixty-five percent of gene therapies are conducted for cancer diseases. According to the *Journal of Gene Medicine*, there have been 1180 gene therapy clinical trials approved worldwide since 1989.[96] These are only the trials and research activities that are funded by either the National Institutes of Health or government agencies of the European Union, and have therefore been reported.

Although there is much hope in the development of gene therapy as a viable and useful treatment, some significant shortcomings exist in the outcomes of the clinical trial patients. One of the most significant negative patient outcomes was the organ failure and death from severe immune responses to the adenovirus carrier in 1999. Other, later problems included difficulties in long-term benefits due to the short-lived nature of the genes in the body, normal patient immune responses that sometimes try to fight off the injected agents, control of the inserted genes once inside the body to only affect the disease cells and leave other normal cells and systems alone, and many diseases that are caused by disorders in more than one gene.

The negative outcomes in the patients tend to also point toward the potential hazards to the workers who prepare and handle these agents. Potential health effects to healthy workers handling these agents have not been fully investigated or reported. But based on the observed results of clinical trials, some expected outcomes of occupational exposure would include cancer or any number of impacts of organ function or physiological processes that might be impacted by an accidental or chronic exposure. The gene therapy agent's potential to affect normal health tissues or cellular function makes it difficult to determine where it might go in a healthy exposed worker, and difficult to determine what long-term health effects might be. Since there is no known threshold for the harmful effects of these agents, even a small exposure may have significant consequences, such as the irreversible creation of a cancer cell. Potential routes of exposure to workers include inhalation, injection, ingestion, and dermal. The environmental viability of retroviral and viral vectors is probably only a few minutes;

however, adenoviral vectors may be viable as long as a couple of hours depending on the delivery agent. Either way, workers preparing and administering these agents can be exposed to viable agents in the process.

In clinical trials, it is common for patients to receive numerous and large doses of treatment agents. A dosage of 40 intramuscular injections of 0.5 mL each over a period of days or weeks would not be unusual. Each dosage might contain as many as a billion DNA virons per milliliter. Typically, doses are in liquid form, although other forms are being evaluated. Other modes of administration, such as IV and subcutaneous methods, are also in use in some trials.

Gene therapy agents currently follow product trials similar to those of other pharmaceuticals. At this point, agents are made by pharmaceutical companies and provided to hospitals or research facilities for the clinical trials. The agents tend to be delivered to the hospital pharmacies in small sealed containers of vials. Once at the pharmacy, the pharmacists must redistribute the agents to the delivery devices such as syringes or IV bags. These materials are then transported to the clinical or patient areas to be administered by physicians or nurses.

The gene therapy mixtures should be handled in a similar fashion to any other hazardous drug. Impermeable gowns, face shields, protective eyewear, and two pair of protective gloves should be worn whenever preparing, handling, or administering these agents. And preparation should always be done in a hazardous drug preparation area in a Class II Type B2 ventilated cabinet. Materials should be transported in a gasket sealed shatterproof container.

Work areas should be periodically decontaminated with 70% ethanol or other standard hospital disinfectants to ensure product protection from biological contaminates. Spills should be cleaned up immediately and absorbed in towels or vermiculite and deactivated with the addition of 10% bleach. All waste materials, including needles and contaminated administration materials should be disposed of as infectious waste and ultimately incinerated.

There is not a lot of information available regarding the metabolism of gene therapy agents in patients. It is expected that there is little likelihood of the administered vectors leaving the body in any excreta. Patient excreta and contaminated materials such as bed linens and environmental contamination should be handled with caution, however, until more information becomes available.

Conclusions

The pace of pharmaceutical development continues to dramatically increase every year. Medicine rushes to take advantage of new advances, first in clinical trials, and soon afterward, via implementation across the broad health care industry. The potential risks and hazards to workers developing, handling and administering the agents are often undocumented, but potentially severe. The methods to monitor the agents in the workplace or in exposed workers are undeveloped. Safe methods of control must be implemented as if the agents are highly toxic until proven otherwise. All currently recommended methods of control and worker protection must be thoroughly implemented.

Unfortunately, worker awareness and the effective, consistent implementation of safety controls lag behind the development of new drugs. Many workers are unaware of the hazards associated with the drugs they prepare and administer. Employers do not institute safety programs and associated worker training to the levels comparable to the use and handling of other toxic agents. These shortcomings are to a large extent due to the lack of regulation in the field and little regulatory oversight of important safety activities necessary for adequate exposure control and worker protection.

Occupational safety and health professionals must push to develop and implement advances in commercially available and standardized environmental monitoring techniques, commercially available and standardized biological monitoring methods, and regulations over allowable exposure levels, BEIs, and safety programs. They must keep constant vigilance and work closely with the research community to keep pace with the latest innovations and potential occupational hazards.

References

1. Harris C. The carcinogenicity of anticancer drugs: A hazard in man. *Cancer* 37:1014–1023, 1976.
2. Weisburger J, Griswold D, Prejean C, Wood H, and Weisburger E. The carcinogenic properties of some of the principal drugs used in clinical cancer chemothera. *Recent Res Cancer* 52:1–17, 1975.
3. Benedict W, Baker M, and Haroun L. Mutagenicity of cancer chemotherapeutic agents in the Salmonella/microsome test. *Cancer Res* 37:2209–2213, 1977.
4. Harris C. Immunosuppressive anticancer drugs in men: Their oncogenic potential. *Radiology* 114(1): 163–166, 1975.
5. Falck K, Grohn P, Sorsa M, Vainio H, Heinonen E, and Holsti L. Mutagenicity in urine of nurses handling cytostatic drugs. *Lancet* 313(9):1250–1251, 1979.
6. International, A.f.R.o.C. *Monograph—Evaluation of the Carcinogenic Risk of Chemicals in Humans.* IARC, Geneva, 1981.
7. ASHP, C.o.P.A. ASHP guidelines on handling hazardous drugs. *Am J Health Syst Pharm* 63:1172–1191, 2006.
8. NIOSH. *Alert—Preventing Occupational Exposures to Antineoplastic and Other Hazardous Drugs in Health Care Settings.* National Institute of Occupational Safety and Health, Cincinnati, OH, 2004.
9. Fuller T, Bain E, and Sperrazza K. A survey of the status of hazardous drug awareness and control in a sample of Massachusetts nursing population. *J Occup Env Hyg* 4:D113–119, 2007.
10. OSHA, O.S.a.H.A. Hazard Communication. Code of Federal Regulations, 2008. 29CFR1910.1200.
11. Bayhan A, Burgaz S, and Karakaya A. Urinary thioether excretion in nurses at an oncologic department. *J Clin Pharm Ther* 12(5):303–306, 1987.
12. Benhamou S, Callais F, Sancho-Garnier H, Min S, Courtois YA, and Festy B. Mutagenicity in urine from nurses handling cytostatic agents. *Eur J Cancer Clin Oncol* 22(12):1489–1493, 1986.
13. Chrysostomou A, Seshadri R, and Morley A. Mutation frequency in nurses and pharmacists working with cytotoxic drugs. *Aust New Zealand J Med* 14(6):831–834, 1984.
14. Nikula E, Kiviniitty K, Leisti J, and Taskinen P. Chromosome aberrations in lymphocytes of nurses handling cytostatic agents. *Scand J Work Environ Health* 10(2):71–74, 1984.
15. Rogers B and Emmett E. Handling antineoplastic agents: Urine mutagenicity in nurses. *J Nurs Scholarship* 19(3):108–113, 1987.
16. Seleven S, Lindbohm M, Polsci M, Hornung R, and Hemminki K. A study of occupational exposure to antineoplastic drugs and fetal loss in nurses. *New Engl J Med* 313:1173–1178, 1985.
17. Stucker I, Caillard J, Collin R, Gout M, Poyen D, and Hemon D. Risk of spontaneous abortion among nurses handling antineoplastic drugs. *Scand J Work Environ Health* 16(2):102–107, 1990.
18. Skov T, Maarup B, Olsen J, Rorth M, Winthereik H, and Lynge E. Leukaemia and reproductive outcome among nurses handling antineoplastic drugs. *Br J Ind Med* 49(12):855–861, 1992.
19. Valanis B, Vollmer W, and Steele P. Occupational exposure to antineoplastic agents: Self-reported miscarriages and stillbirths among nurses and pharmacists. *J Occup Env Med* 41(8):632–638, 1999.
20. Hemminki K, Kyyronen P, and Lindbohm M. Spontaneous abortions and malformations in the offspring of nurses exposed to anaesthetic gases, cytostatic drugs, and other potential hazards in hospitals, based on registered information of outcome. *J Epi Com Health* 39:141–147, 1985.
21. Meirow D and Schiff E. Appraisal of chemotherapy effects on reproductive outcome according to animal studies and clinical data. *J Natl Cancer Inst Monogr* 34:21–25, 2005.
22. Fransman W, Roeleveld N, Peelen S, de Kort W, Kromhout H, and Heederik D. Nurses with dermal exposure to antineoplastic drugs: Reproductive outcomes. *Epidemiology* 18(1):112–119, 2007.
23. Fuchs J, Hengstler J, Jung D, Hiltl G, Konietzko J, and Oesch F. DNA damage in nurses handling antineoplastic agents. *Mutat Res* 342:17–23, 1995.
24. Undeger U, Basaran N, Kars A, and Guc D. Assessment of DNA damage in nurses handling antineoplastic drugs by the alkaline COMET assay. *Mutat Res* 439:277–285, 1999.

25. Maluf S and Erdtmann B. Follow-up study of the genetic damage in lymphocytes of pharmacists and nurses handling antineoplastic drugs evaluated by cytokinesis-block micronuclei analysis and single cell gel electrophoresis assay. *Mutat Res* 471:21–27, 2000.
26. Sessink P, Cerna M, Rossner P, Pastorkova A, Bavarova H, Frankova K, Anzion R, and Bos R. Urinary cyclophosphamide excretion and chromosomal aberrations in peripheral blood lymphocytes after occupational exposure to antineoplastic agents. *Mutat Res* 309(2):193–199, 1994.
27. Sessink P and Bos R. Drugs hazardous to healthcare workers: evaluation of methods for monitoring occupational exposure to cytostatic drugs. *Drug Safety* 20(4):347–359, 1999.
28. Anwar W, Salama S, Serafy M, Hemida S, and Hafez A. Chromosomal aberrations and micronucleus frequency in nurses occupationally exposed to cytotoxic drugs. *Mutagenesis* 9(4):317–317, 1994.
29. Burgaz S, Karahalil B, Canli Z, Terzioglu F, Ancel G, Anzion RBM, Bos RP, and Hünttner E. Assessment of genotoxic damage in nurses occupationally exposed to antineoplastics by the analysis of chromosomal aberrations. *Hum Exp Toxicol* 21(3):129–135, 2002.
30. Milkovic-Kraus S and Horvat D. Chromosomal abnormalities among nurses occupationally exposed to antineoplastic drugs. *Am J Ind Med* 19(6):771–774, 2007.
31. Sarto F, Trevisan A, Tomanin R, Canova A, and Fiorentino M. Chromosomal aberrations, sister chromatid exchanges, and urinary thioethers in nurses handling antineoplastic drugs. *Am J Ind Med* 18(6):689–695, 2007.
32. Sessink P, Kroese E, van Kranen H, and Bos R. Cancer risk assessment for health care workers occupationally exposed to cyclophosphamide. *Int Arch Occup Environ Health* 67(5):317–323, 1995.
33. Hansen J and Olsen J. Cancer morbidity among Danish female pharmacy technicians. *Scand J Work Environ Health* 20(1):22–26, 1994.
34. Dranitsaris G, Johnston M, Poirier S, Schueller T, Milliken D, Green E, and Zanke B. Are health care providers who work with cancer drugs at an increased risk for toxic events? A systematic review and meta-analysis of the literature. *J Oncol Pharm Pract* 11:69–78, 2005.
35. Stellman J, Aufiero B, and Taub R. Assessment of potential exposure to antineoplastic agents in the health care setting. *Prev Med* 13(3):245–255, 1984.
36. Sorsa M, Pyy L, Salomaa S, Nylund L, and Yager J. Biological and environmental monitoring of occupational exposure to cyclophosphamide in industry and hospitals. *Mutat Res* 204(3):465–479, 1988.
37. Pyy L, Sorsa M, and Hakala E. Ambient monitoring of cyclophosphamide in manufacture and hospitals. *Am Ind Hyg Assoc J* 49(6):314–317, 1988.
38. Sessink P, Timmersmans J, Anzion R, and Bos R. Assessment of occupational exposure of pharmaceutical plant workers to 5-fluorouracil. Determination of alpha-fluro-beta-alanine in urine. *J Occup Med* 36(1):79–83, 1994.
39. Sessink P, Anzion R, Van den Broek P, and Bos R. Detection of contamination with antineoplastic agents in a hospital pharmacy department. *Pharm Weekbl Sci* 14(1):16–22, 1992.
40. Sessink P, Wittenhorst B, Anzion R, Bos R, and Rob P. Exposure of pharmacy technicians to antineoplastic agents: Reevaluation after additional protective measures. *Arch Environ Health* 52(3): 240–244, 1997.
41. Larson R, Khazaeli M, and Dillon H. A new monitoring method using solid sorbent media for evaluation of airborne cyclophasphamide and other antineoplastic agents. *Appl Occup Environ Hyg* 18(2):120–131, 2003.
42. Hedmer M, Jonsson B, and Nygren O. Development and validation of methods for environmental monitoring of cyclophosphamide in workplaces. *J Environ Monit* 6(12):979–984, 2004.
43. McDevitt J, Lees P, and McDiarmid M. Exposure of hospital pharmacists and nurses to antineoplastic agents. *J Occup Med* 35(1):57–60, 1993.
44. Labuhn K, Valanis B, Schoeny R, Loveday K, and Vollmer W. Nurses' and pharmacists' exposure to antineoplastic drugs: Findings from industrial hygiene scans and urine mutagenicity tests. *Cancer Nurs* 21(2):79–89, 1998.

45. Floridia L, Pietropaolo A, Tavazzani M, Rubino F, and Columbi A. High-performance liquid chromatography of methotrexate for environmental monitoring of surface contamination in hospital departments and assessment of occupational exposure. *J Chromatogr* 726:95–103, 1999.
46. Floridia L, Pietropaolo A, Tavazzani M, Rubino F, and Columbi A. Measurement of surface contamination from nucleoside analogue antineoplastic drugs by high-performance liquid chromatography in occupational hygiene studies of oncologic hospital departments. *J Chromatogr* 724:325–334, 1999.
47. Rubino F, Floridia L, Pietropaolo A, Tavazzani M, and Colombi A. Measurement of surface contamination by certain antineoplastic drugs using high-performance liquid chromatography: Applications in occupational hygiene investigations in hospital environments. *Meicina del Lavoro* 90(4):572–583, 1999.
48. Connor T, Anderson R, Sessink P, Broadfield L, and Power L. Surface contamination with antineoplastic agents in six cancer treatment centers in Canada and the United States. *Am J Health Syst Pharm* 56:1427–1432, 1999.
49. Micoli G, Turci R, Arpellini M, and Minoia C. Determination of 5-fluorouracil in environmental samples by solid-phase extraction and high-performance liquid chromatography with ultraviolet detection. *J Chromatogr B Biomed Sci Appl* 750(1):25–32, 2001.
50. Nygren O and Gustavsson B, and Eriksson R. A test method for assessment of spill and leakage from drug preparation systems. *Ann Occup Hyg* 49(8):711–718, 2005.
51. Crauste-Manciet S, Sessnik P, Ferrari S, Jomier J, and Brossard D. Environmental contamination with cytotoxic drugs in healthcare using positive air pressure isolators. *Ann Occup Hyg* 49(7):619–628, 2005.
52. Acampora A, Castiglia L, Miraglia N, Pieri M, Soave C, Liotti F, and Sannolo N. A case study: Surface contamination of cyclophosphamide due to working practices and cleaning procedures in two Italian hospitals. *Ann Occup Hyg* 49(7):611–618, 2005.
53. Hedmer M, Georgiadi A, Bremberg R, Jonsson A, and Eksborg S. Surface contamination of cyclophosphamide packaging and surface contamination with antineoplastic drugs in a hospital pharmacy in Sweden. *Ann Occup Hyg* 49(7):629–637, 2005.
54. Connor T. Hazardous anticancer drugs in healthcare: Environmental exposure assessment. *Framing the Future in Light of the Past: Living in a Chemical World, 3rd International Scientific Conference*, Bologna, Italy, 2005.
55. Connor T, Sessink P, Harrison B, Pretty J, Peters B, Alfaro R, Bilos A, Beckmann G, Bing M, Anderson L, and DeChristoforo R. Surface contamination of chemotherapy drug vials and evaluation of new vial-cleaning techniques: Results of three studies. *Am J Health Syst Pharm* 62:475–484, 2005.
56. Schmaus G, Schieri R, and Funck S. Monitoring surface contamination by antineoplastic drugs using gas chromatography-mass spectrometry and voltammetry. *Am J Health Syst Pharm* 59:956–961, 2002.
57. Nygren O, Gustavsson B, Strom L, Eriksson R, Jarneborn L, and Friberg A. Exposure to anticancer drugs during preparation and administration. Investigations of an open and a closed system. *J Environ Monit* 4(5):739–742, 2002.
58. Mason H, Morton J, Garfit S, Iqbal S, and Jones K. Cytotoxic drug contamination on the outside of vials delivered to a hospital pharmacy. *Ann Occup Hyg* 47(8):681–685, 2003.
59. Fransman W, Vermeulen R, and Kromhout H. Dermal exposure to cyclophosphamide in hospitals during preparation, nursing and cleaning activities. *Int Arch Occup Environ Health* 78(5):403–412, 2005.
60. Harrison B, Peters B, and Bing M. Comparison of surface contamination with cyclophosphamide and fluorouracil using a closed-system drug transfer device versus standard preparation techniques. *Am J Health Syst Pharm* 63(18):1736–1744, 2006.
61. Nygren O. Wipe sampling as a tool for monitoring aerosol deposition in workplaces. *J Environ Monit* 8(1):49–52, 2006.

62. Meijster T, Fransman W, Veldhof R, and Kromhout H. Exposure to antineoplastic drugs outside the hospital environment. *Ann Occup Hyg* 50(7):657–664, 2006.
63. Connor T. Hazardous anticancer drugs in health care—Environmental exposure assessment. *Ann NY Acad Sci* 1076:615–623, 2006.
64. Nygren O, Gustavsson B, Strom L, and Friberg A. Cisplatin contamination observed on the outside of drug vials. *Ann Occup Hyg* 46(6):555–557, 2002.
65. Baker E and Connor T. Monitoring occupational exposure to cancer chemotherapy drugs. *Am J Health Syst Pharm* 53:2713–2723, 1996.
66. Fransman W, Huized D, Tuerk J, and Kromhout H. Inhalation and dermal exposure to eight antineoplastic drugs in an industrial laundry facility. *Int Arch Occup Environ Health* 80(5):396–403, 2007.
67. Sessink P, Van de Kerkhof M, Anzion R, Noordhoek J, and Bos R. Environmental contamination and assessment of exposure to antineoplastic agents by determination of cyclophosphamide in urine of exposed pharmacy technicians: Is skin absorption and important exposure route? *Arch Environ Health* 49(3):396–397, 1995.
68. Fransman W, Vermeulen R, and Kromhout, H. Occupational dermal exposure to cyclophosphamide in Dutch hospitals: A pilot study. *Ann Occup Hyg* 48(3):237–244, 2004.
69. Kromhout H, Hoek F, Uitterhoeve R, Huijbers R, Overmars R, Anzion R, and Vermeulen R. Postulating a dermal pathway for exposure to anti-neoplastic drugs among hospital workers. Applying a conceptual model to the results of three workplace surveys. *Ann Occup Hyg* 44(7):551–560, 2000.
70. Rajan-Sithamparanadarajah R, Roff M, Delgado P, Eriksson K, Fransman W, Gijsbers J, Hughson G, Makinen M, and van Hemmen J. Patterns of dermal exposure to hazardous substances in European Union workplaces. *Ann Occup Hyg* 48(3):285–297, 2004.
71. van Wendel de Joode B, Vermeulen R, van Hemmen J, Fransman W, and Kromhout H. Accuracy of a semiquantitative method for dermal exposure assessment (DREAM). *Occup Environ Med* 62:623–632, 2005.
72. Schneider T, Cherrie JW, Vermeulen R, and Kromhout H. Dermal exposure assessment. *Ann Occup Hyg* 44(7):493–499, 2000.
73. Warren N, Marquart H, Christopher Y, Laitinen J, and van Hemmen J. Task-based dermal exposure models for regulatory risk assessment. *Ann Occup Hyg* 50(5):491–503, 2006.
74. Hirst M, Mills D, Tse S, Levin L, and White D. Occupational exposure to cyclophosphamide. *Lancet* 323(8370):186–188, 1984.
75. Venitt S, Crofton-Sleigh C, Hunt J, Speechley V, and Briggs K. Monitoring exposure of nursing and pharmacy personnel to cytotoxic drugs: Urinary mutation assays and urinary platinum as markers of absorption. *Lancet* 323(8368):74–77, 1984.
76. Sessink P, Friemel N, Anzion R, and Bos R. Biological and environmental monitoring of occupational exposure of pharmaceutical plant workers to methotrexate. *Int Arch Occup Environ Health* 65(6):401–403, 1984.
77. Ensslin A, Pethran A, Schierl A, and Fruhmann G. Urinary platinum in hospital personnel occupationally exposed to platinum-containing antineoplastic drugs. *Int Arch Occup Environ Health* 65(5):339–342, 1984.
78. Ensslin A, Huber R, Pethran A, Rommelt H, Schierl R, Kulka W, and Fruhmann G. Biological monitoring of hospital pharmacy personnel occupationally exposed to cytostatic drugs: Urinary excretion and cytogenetics studies. *Int Arch Occup Environ Health* 70(3):205–208, 1997.
79. Opiolka I, Schmidt K, Kiffmeyer K, and Schoppe G. Determination of the vapour pressure of cytotoxic drugs and its effects on occupational safety. *J Oncol Pharm Pract* 6(1):9–35, 2000.
80. Connor T, Anderson R, Sessink P, and Spivey S. Effectiveness of a closed-system device in containing surface contamination with cyclophosphamide and ifosfamide in an i.v. admixture area. *Am J Health Syst Pharm* 59(1):68–72, 2002.
81. Polovich M. Safe handling of hazardous drugs. *Online J Issues Nurs* 9(3), 2004.

82. Polovich M. *Safe Handling of Hazardous Drugs*. Oncology Nursing Society, Pittsburgh, PA, 2003, pp. 1–45.
83. NIOSH. *Antineoplastic Agents—Occupational Hazards in Hospitals*. NIOSH, Cincinnati, OH, 2004, pp. 1–13.
84. OSHA. *Controlling Occupational Exposure to Hazardous Drugs*. Occupational Safety and Health Administration, Washington DC, 1999, pp. 1–35.
85. USP, 797.*Pharmaceutical Compounding—Sterile Preparations*. The United States Pharmaceutical Compounding Convention, Rockville, MD, 2008, pp. 1–61.
86. Wick C, Slawson M, Jorgenson J, and Tyler L. Using a closed-system protective device to reduce personnel exposure to antineoplastic agents. *Am J Health Syst Pharm* 60(22):2314–2320, 2003.
87. Sessink P. How to work safely outside the biological safety cabinet. *J Oncol Pharm Pract* 6(1):9–35, 2000.
88. Laidlaw J, Connor T, and Theiss J. Permeability of latex and polyvinyl chloride gloves to 20 antineoplastic drugs. *Am J Hosp Pharm* 41(12):2618–2623, 1984.
89. Wallemacq P, Capron A, Vanbinst R, Boeckmanns E, Gillard J, and Favier B. Permeability of 13 different gloves to 13 cytotoxic agents under controlled dynamic conditions. *Am J Health Syst Pharm* 63(6):547–556, 2006.
90. Harrison B and Kloos M. Penetration and splash protection of six disposable gown materials against fifteen antineoplastic drugs. *J Oncol Pharm Pract* 5(2):61–66, 1999.
91. Merrill R. Regulatory toxicology. In: CA Doull (ed.), *Toxicology—The Basic Science of Poisons*, Chapter 34. McGraw-Hill, New York, 1996.
92. Naumann B and Weideman P. Scientific basis for uncertainty factors used to establish occupational exposure limits for pharmaceutical active ingredients. *Hum Ecol Risk Assess* 1(5):590–613, 1995.
93. Naumann E and Sargent B. Setting occupational exposure limits for pharmaceuticals. *Occup Med* 12:67–80, 1997.
94. Naumann B, Sargent E, Starkman B, Fraser W, Becker G, and Kirk G. Performance-based exposure control limits for pharmaceutical active ingredients. *Am Ind Hyg Assoc J* 57:33–42, 1996.
95. Streifel A. Telephone conversation. In: T Fuller (ed.), Minneapolis, MN, 2008.
96. *Journal of Gene Medicine*, Gene therapy clinical trials worldwide, 2008.

11

Tuberculosis Engineering Controls

Lorraine M. Conroy
University of Illinois at Chicago

John E. Franke
University of Illinois at Chicago

Byron S. Tepper
Johns Hopkins University and Johns Hopkins Hospital

John Mehring
Service Employees International Union

Derrick Hodge
FOJP Service Corporation

Daniel Kass
Hunter College

Olmstead Environmental Services

William Charney
Healthcare Safety Consulting

June Fisher
University of California, San Francisco

Patrice M. Sutton
Public Health Institute

Mark Nicas
University of California, Berkeley

Robert J. Harrison
California Department of Health Services

Introduction 11-2
An Industrial Hygiene Approach to Tuberculosis Control 11-3
Introduction • Hazard Characterization • Regulations and Guidelines • Source Control • Research Needs
References 11-26
Portable HEPA Filtration for TB Isolation in Hospitals and Clinics 11-30
Introduction • Mechanics of Filtration • HEPA Filtration in Health Care Facilities • Respiratory Isolation for Tuberculosis (TB) • Portable In-Room HEPA Filtration • Respiratory Protection • Source Control • Discussion
References 11-40
Preventing TB in the Workplace: What Did We Learn from HIV? Policies Regarding the HIV-Infected Health Care Worker 11-41
Reducing the Spread of Tuberculosis in Your Workplace 11-47
Introduction • Checking Your Building's Ventilation • Air Quality Survey • How to Interpret the Results of the Air Quality Survey • Low-Cost Ways to Improve Natural Ventilation to Reduce the Spread of TB • Change the Way You Interact with People • Move Desks, Chairs, and People • Summary and Follow-Up • Acknowledgments
Appendix 11.A.1 11-57
Appendix 11.A.2 11-64
Appendix 11.A.3 11-77
Appendix 11.A.4 11-79
Engineering Control Options for Tuberculosis in Health Care Settings 11-84
Introduction • Discussion • Conclusion
References 11-88
The Inefficiency of Surgical Masks for Protection against Droplet Nuclei TB 11-89
References 11-90
Appendix 11.A.5 11-91
The Economics of Implementing an Engineering Tuberculosis Control Plan in a County Hospital 11-102

Introduction • Background • Emergency Department • Centralization of Sputum Induction • Negative-Pressure Isolation Rooms • Pressure Differential Monitoring of Negative Pressure • Bronchoscopy • Outpatient Clinics • Economics

References **11**-108

Implementing a Quality Assurance Program for Tuberculosis Control **11**-109

Background • Guidelines for Implementing a Quality Assurance Project • Survey Results for Adherence to CDC Guidelines for TB Patient Isolation • Conclusions

References **11**-115

Introduction

Since the recent increases in health care worker conversions to tuberculosis (TB) and in some cases death, the need to understand and control the transmigration of droplet nuclei has become of paramount importance. Engineering controls, air changes per hour (ach), negative pressure, high-efficiency particulate air (HEPA) filtration, ultraviolet germicidal irradiation (UVGI), capture at the source, enclosure devices are now industry vocabularies that need to be understood by a variety of scientific disciplines in the health care setting. Selection and implementation of control options has become labor intensive and in some cases controversial. Preparing a health care facility to meet the standards for a TB Control Program is multidimensional as there needs to be scientific understanding of the aerosolization of TB, control procedures, budgetary considerations, air volume measurements of areas in the hospitals where suspected or identified cases or high-risk medical procedures are to be done, types of engineering controls selected for the various areas of concern, maintenance of engineering controls selected for the various areas of concern, etc. Different scientific disciplines within each medical facility must work together harmoniously in order to achieve optimum results, for example, Infection Control, Engineers, Industrial Hygienists, Nursing, and Administrators. In some states, regulatory codes are in conflict with TB control guidelines and over time these conflicts must be resolved to mutual satisfaction.

The purpose of this chapter is to provide some additional information and design characteristics for engineering controls for TB. The reader is also referred to the CDC Guidelines, Part II, Department of Health and Human Services, Draft Guidelines for Preventing the Transmission of TB in Health Care Facilities for recommendations and additional references for engineering controls. This chapter contains eight sections namely “An industrial hygiene approach to tuberculosis control”; “Portable HEPA filtration for TB isolation in hospitals and clinics”; “Preventing TB in the workplace: What did we learn from HIV? Policies regarding the HIV-infected health care worker”: *Note*: At the time of writing the CDC Guidelines, there was no challenge data confirming the clearance efficiency of HEPA filters. The data presented here, though not yet in peer review but tested by independent laboratories funded by manufacturers, nonetheless give some positive indications as to the functionality and clearance rates of these systems; “Reducing the spread of tuberculosis in your workplace”; “Engineering control options for tuberculosis in health care settings”; “The inefficiency of surgical masks for protection against droplet nuclei TB”; “The economics of implementing and engineering TB control plan in a county hospital”; and “Implementing a quality assurance program for tuberculosis control.”

An Industrial Hygiene Approach to Tuberculosis Control

Lorraine M. Conroy and John E. Franke

Introduction

Industrial hygienists address occupational health problems using a three-step approach: recognition, evaluation, and control. The approach involves answering three questions: (1) What are the potential health hazards? (2) Are the potential hazards truly hazardous? and (3) How can the hazard be eliminated or minimized? The toxic hazard of a material is defined as the likelihood of injury of a person by other than mechanical means.[1] The evaluation of the potential hazard is usually done by conducting air sampling or exposure monitoring and comparing the exposure with some established guideline or standard.

In the case of *Mycobacterium tuberculosis* (MTb), there are no established exposure limits. There is also no practical method for evaluating the MTb exposure with air sampling. Several studies indicate that a single bacteria particle can cause disease.[2–5] The evaluation of the hazard potential is influenced by the background incidence of disease in the population served by the institution, the number of TB cases treated by the institution, the tuberculin skin test (TST) conversion rate in the workforce of the institution, types of treatments and other aerosol-generating processes performed in the institution, and engineering and administrative controls in place at the institution.

Source identification and isolation are the key steps in controlling transmission of TB in health care facilities.[6] The central theme of this section is the application of basic industrial hygiene principles to the control of the TB transmission hazard. Industrial hygiene methods could enhance infection control efforts to identify TB sources in a facility. The application of ventilation to isolate known or suspected TB sources would also be improved by following industrial hygiene principles.

Hazard Characterization

The number of reported cases of TB in the United States in 1991 was 26,283, 2% higher than in 1990.[7,8] There had been an annual decline in the number of cases of approximately 5% since the 1950s.[7] In 1958, Dublin reported the death rate in the United States from TB as less than 8 per 100,000, a 96% reduction from 1900. This led him to conclude “We cannot say exactly when control will be complete, but there is every indication that it will be some time in the course of the next 20 years.”[9]

There was a 6%–7% annual decline in TB cases from 1981 to 1984. Using the trend from 1981 to 1984 to estimate the expected number of cases for 1985–1991, it is estimated that more than 39,000 excess cases of TB occurred between 1985 and 1991.[7]

Persons living in the same household, those who travel in the same vehicles, and those who share air with an infectious person through a common ventilation system for a prolonged time[10,11] are at risk of acquiring TB infection. Ventilation systems have contributed to the transmission of TB.[2] Twenty-seven of 67 susceptible office workers became infected following 160 h of exposure to air shared by an infectious office worker in the same building.[12]

Several studies have documented higher than expected TST conversion rates in hospital personnel.[13–22] At least one case resulted in occupational transmission of active disease.[21] Procedures such as bronchoscopy, endotracheal intubation and auctioning with mechanical ventilation, open abscess irrigation, and autopsy have been implicated in nosocomial transmission.[13,16,20]

The Centers for Disease Control and Prevention (CDC) and others have investigated several outbreaks of multiple drug-resistant TB (MDR-TB) in New York and Florida.[22–25] One of these investigations indicated that two categories of factors contributed to the outbreaks.[22] The first category included delays in diagnosing TB in HIV-infected persons and delays in recognizing drug resistance. The second category included delays in acid fast bacilli (AFB) isolation on admission or readmission, maintaining AFB isolation for an inadequate amount of time, lapses in AFB isolation, such as open doors and patients leaving isolation rooms, AFB isolation rooms with inadequate negative pressure, and inadequate numbers of AFB isolation rooms. The outcome of these factors can be improved using industrial hygiene principles outlined in this section.

Regulations and Guidelines

In the United States, the Occupational Safety and Health Administration promulgates and enforces standards for protecting the health of American workers. The authority for standard setting was established with the passage of the Occupational Safety and Health Act of 1970 (OSH Act). The Act requires employers to "provide a safe and healthful workplace free from recognized hazards."[26] This is often referred to as the "general duty" clause of the OSH Act. The OSH Act also prescribes a formal standard setting procedure. Health and safety standards established under the OSH Act generally have two forms. Many chemical agents are regulated under the Air Contaminants Standard.[27] This standard sets maximum exposure limits called Permissible Exposure Limits (PEL) for several hundred chemical agents. Several substance-specific standards have also been promulgated. These standards set a PEL, but also establish other requirements of the employer such as training, medical surveillance, and recordkeeping.

The PEL must be met through a combination of engineering and work practice controls. If engineering controls do not reduce the exposure below the PEL or while engineering controls are being implemented, respiratory protection may be used to achieve compliance with the standard. The second type of standard does not establish a maximum exposure limit but outlines requirements of the employer necessary to reduce the exposure to the lowest feasible level. An example of this type of standard is the Bloodborne Pathogens Standard.[28] The Bloodborne Pathogens Standard requires employers to implement a written exposure control plan; prepare an exposure determination; use a combination of universal precautions, engineering and work practice controls, and personal protective equipment to minimize or eliminate exposure; implement proper disposal and labeling procedures; provide hepatitis B vaccinations; and provide information and training on the risks of exposure to bloodborne pathogens, procedures and practices necessary to prevent exposure, and the requirements of the standard.[28]

At the present time, no OSHA standard specifically regulates exposure to MTb. However, OSHA Region 2 has published enforcement guidelines for occupational exposure to TB.[29] The guidelines describe when the general duty clause, Section 5(a)(1) of the OSH Act, may be cited. Citations can only be issued to employers whose employees work on a regular basis in health care settings, correctional institutions, homeless shelters, long-term care facilities, and drug treatment centers. The document lists examples of feasible and useful abatement methods for TB control and states that "the non-use of any of these methods is likely to result in the continued existence of a serious hazard and may, therefore, allow citation under 5(a)(1)." The methods are (1) medical screening of employees; (2) work removal of employees who have current pulmonary or laryngeal TB until adequate treatment is instituted, their cough is resolved, or until a physician certifies that the person is no longer infectious; (3) training and education of employees about the hazards and control of TB; (4) respiratory isolation of infectious TB patients in negative pressure rooms exhausted to the outside; and (5) respiratory protection use by the patient, if possible, and the employee during patient transport.

The document also describes other OSHA standards that may apply. These include recordkeeping[30] and respiratory protection.[31] Additionally, OSHA Region 2 considers TB infections (positive skin test) and TB disease as recordable on the OSHA 200 log where TB has been identified as a hazard.[29]

In 1993 two hospitals in Madison, Wisconsin were inspected and received citations for failing to adequately protect workers from TB.[32] The first hospital was cited under Section 5(a)(1) of the OSH Act for failing to provide medical surveillance and failing to ensure that an isolation room was properly ventilated to operate under negative pressure. The hospital was also cited for four violations of 1910.134 (respiratory protection) including using respirators that were not NIOSH approved, allowing workers with beards, sideburns, and skullcaps to wear respirators and not having or maintaining a written respiratory protection program.[32]

The second hospital was a Veterans Hospital. OSHA has the authority to inspect federal facilities and to issue citations but cannot assess penalties. The hospital was cited under Section 1-201(a) of Executive Order 12196, which is equivalent to the general duty clause of the OSH Act. The violations in this case were due to deficiencies in medical surveillance and failing to record an active case of TB on the injury and illness log.[32]

CDC has published several sets of guidelines for preventing TB in health care settings,[33,34] migrant farm workers,[35] long-term care facilities,[11] and correctional institutions.[10] The most recent draft of the guidelines for health care settings was published in the Federal Register in October 1993.[36]

The current CDC guidelines have the following steps: (1) implement control measures that follow an established hierarchy of administrative and engineering controls and personal respiratory protection; (2) perform a risk assessment at each health care facility and develop a written TB control plan; (3) provide early identification and management of persons with TB; (4) implement a medical surveillance program for employees using purified protein derivative (PPD) skin testing; and (5) educate, train, and communicate with health care workers about the risks of TB and the measures used to prevent MTb exposure.[36]

Regulations and guidelines have also been issued by state agencies and professional associations. The state of California, for example, has proposed regulations for occupational TB control that specify an exposure control plan, TB surveillance and employee notification, medical evaluation and preventive therapy, methods of exposure control, training, and recordkeeping.[37] The American Society for Hospital Engineering has guidelines that discuss risk assessment and management for TB control in hospitals.[38] The American Society for Heating, Ventilating and Air Conditioning Engineers has general guidelines for health facilities that give design criteria and mention TB in the discussion.[39]

In addition to occupational health requirements, state and local building codes must be considered when designing or implementing control strategies for MTb. The recommendation of keeping TB patients in isolation rooms under negative pressure relative to the corridor violates state and local codes in many areas. A variance may be needed or the use of an anteroom may be required to meet both the CDC guidelines and building codes.

Source Characterization

Mycobacterium tuberculosis is a rod-shaped bacteria which varies in width from 0.2 to 0.6 μm and from 0.5 to 4.0 μm in length.[40,41] The bacteria are expelled from infected persons through coughing, sneezing, talking, and singing[2,3] and become aerosolized as droplets.[7,33] The largest droplets (e.g., exceeding 100 μm) settle onto surfaces and are removed from the air.[42] Droplets less than 100 μm evaporate to form stable, nearly spherical, droplet nuclei in the 1–4 μm size range.[42] A study by Loudon and Roberts[43] indicated that 30% of droplet nuclei, by number, resulting from coughing were less than 3 μm in diameter.

The droplets are small enough that room air currents keep them airborne and spread them throughout a room or building.[44] Their small mass means that the droplet nuclei have negligible inertia and are unable to travel through air on their own. Instead they must follow the burst of air released by a cough, for example, and then follow room air currents when the air burst slows down. Air currents always exist in rooms and random room air velocities are typically in the range of 20–40 ft/min (fpm).

The fact that infectious particles must follow the air currents that they encounter has important implications for control. Isolation design criteria aimed at controlling the contaminated air currents have a good chance to control the airborne infection hazard. Characterization of the room environment is, therefore, as important as characterizing the droplet nuclei source when specifying controls.

Because room air is not quiescent, droplet nuclei can remain airborne for prolonged periods of time (hours, at least).[45] Anyone who breathes air that contains these droplet nuclei can become infected with TB.[45] After inhalation, droplet nuclei can penetrate to the alveolar region of the lung, where infection is initiated.[40,41]

A study by Kent[46] indicated that the TB bacterium is highly resistant to environmental stresses, probably survives for extended time in the environment and could be resuspended in an infectious state from settled dust. The viability of the resuspended particles depends, in part, on the moisture retained in the dust and the temperature and relative humidity of the room environment.[47]

It appears that a single *Mycobacterium tuberculosis* cell is adequate for infection.[2–5] It also appears that everyone who converts to a positive TB skin test acts as a low-level source of infection, at least temporarily.[2,48]

Several studies indicate that the risk of infection is a function of several factors including concentration of droplet nuclei, cumulative time that air containing droplet nuclei is breathed, and the worker's pulmonary ventilation rate.[49] Harris and McClement[50] describe the risk of airborne transmission to be a function of several factors including rate and concentration of expelled organisms, the physical state of the airborne discharge and the volume and rate of exchange of the air in the space where the bacilli are ejected. They state, however, that the most important risk factor is the length of time an individual shares a volume of air with an infectious case of TB. CDC lists the following environmental factors that enhance transmission: contact between susceptible persons and an infectious patient in relatively small, enclosed spaces; inadequate ventilation that results in insufficient dilution or removal of infectious droplet nuclei; and recirculation of air containing infectious droplet nuclei.[34]

Droplet nuclei concentration is higher near the source, especially after a burst release such as a cough. The droplets diffuse throughout the space with time, reducing this concentration gradient until another burst is released. Therefore, the risk of exposure to infectious droplets is highest and most variable near the source.[51] Controlling the length of time a worker shares this near-field air volume with a patient is crucial. Isolating this air space should also be the starting point for source control and ventilation design.

Since droplet nuclei follow the local air currents, it is possible to evaluate contaminated airflow patterns in existing buildings with smoke tube and tracer gas methods. The smoke tube can visually simulate the travel of a burst of droplets, for example, from the head position of a patient bed. The test can be used to specify work positions and local exhaust hoods. Smoke tubes release a momentarily irritating acid aerosol and caution must be used to avoid eye contact with and inhalation of the smoke.

Tracer gases are nonreactive, nontoxic gases that do not normally exist in the test space and are measurable over a wide range of low concentrations.[52] Tracer gases can be released in ways that simulate a source emission. Monitoring the air at critical exposure locations gives information about the transport of droplet nuclei in rooms and buildings (e.g., reception desks in clinic waiting rooms). Tracer gas testing is usually performed by specialists.

Source Control

The hierarchy of control strategies for any hazardous substance is (1) control at the source; (2) control between the source and the worker; and (3) control at the worker. Examples of control at the source are substitution with a less hazardous substance or process, source isolation, and local exhaust ventilation. Examples of control between the source and worker are dilution ventilation, shielding, and use of UV lights. An example of control at the worker is the use of personal respiratory protection.

Ventilation

Ventilation is one of the most important engineering techniques available for maintaining and improving workplace environments. Dilution or general ventilation refers to the dilution of contaminated air with uncontaminated air in a general area, room, or building. Local exhaust ventilation refers to the capture of pollutants at the source. Local exhaust ventilation is preferred for controlling atmospheric concentrations of airborne hazards because the capture and control of contaminants can be complete and workers' exposure can be prevented. With general ventilation the contaminant concentration is diluted, but exposure is never completely prevented. Since one infectious droplet can transmit the disease, local exhaust controls should be considered before relying on dilution ventilation methods.

Other advantages of local exhaust ventilation are (1) less exhaust air is required for equivalent control, making possible lower operating costs and (2) the contaminant is controlled by a smaller air volume, thus reducing costs of associated air cleaning.

A local exhaust ventilation system consists of hoods or enclosures, ductwork leading to an exhaust fan, often an air cleaning device and a discharge point. There are several general guidelines for design and operation of local exhaust ventilation systems.

1. Hood should physically enclose the source as completely as practicable.
2. Contaminated air should be captured with sufficient velocity so that it is always directed into the enclosure.
3. System should be designed to direct the contaminated air away from workers' breathing zones.
4. System should be operated so that workers are not placed in the airflow path between the source and the hood.
5. Sufficient make-up air must be supplied to replace the exhausted air at the design pressure differential between the room and the adjoining spaces.
6. Contaminated exhaust air should be discharged away from building air inlets and occupied areas.

There are three general categories of hood types: enclosures, receiving hoods, and exterior hoods. Enclosures surround the point of emission or contaminant generation, either completely or partially. Complete enclosures require the lowest exhaust rate of the three hood types.

Receiving hoods are those hoods which use some characteristic of the process to help air contaminants flow into the hood. Two examples of receiving hoods are grinding wheel hoods and canopy hoods for heated processes. Grinding wheels release particles with a high velocity. Placing the hood directly in the path of the high-velocity particles aids in capture. With hot sources, the contaminant is released, upward from the source, with a high velocity due to the buoyancy of heated air. Placing the canopy hood directly above the source also aids in contaminant capture.

Exterior hoods differ from enclosures in that they must capture contaminants being generated at a point outside the hood. They differ from receiving hoods in that they must capture contaminants without the aid of supplemental forces. Exterior hoods are sensitive to external conditions, especially crossdrafts, which may interfere with their "reach." Even slight drafts can cause some exterior hoods to become ineffective. Exterior hoods are used when processes require easy access to the source and enclosures obstruct performance of the job. For a given contaminant capture efficiency, exterior hoods require the most airflow.

The capture efficiency of local exhaust hoods depends on many factors including: physical state of the contaminant (gas/vapor or particle); temperature of source; direction and velocity of contaminant release; distance of the hood from the source (especially important with exterior hoods); drafts (significantly affect exterior hood performance but can also affect booths and partial enclosures); and worker activity, such as reaching into the zone of influence of the hood.

The current design method for local exhaust systems is given in the *Industrial Ventilation Manual.*[53] The first step is to eliminate or minimize air movements in the area of the hood and locate the hood around or as close as practicable to the source. The capture velocity is then determined. The manual[53] defines capture velocity as "the air velocity at any point in front of the hood or at the hood opening necessary to overcome opposing air currents and to capture the contaminated air at that point by causing it to flow into the hood." The value of capture velocity depends on the velocity of the contaminant at release and the magnitude of room air currents. The CDC recommends a capture velocity of 200 fpm for exterior-type hoods. The recommendations also include having the patient face directly into the hood opening so that coughing or sneezing will be directed into the hood.[36]

Several researchers[54–63] have developed empirical expressions for centerline velocity as a function of hood shape, airflow into the hood, distance of the source from the hood, and hood area. The airflow necessary to obtain the desired capture velocity is calculated using one of these empirical expressions. The most commonly used expressions are those of DallaValle[54] and Silverman[55–57] and are given by

Plain round or rectangular hood:

$$Q = v_c(10x^2 + A) \tag{11.1}$$

Flanged round or rectangular hood:

$$Q = 0.75v_c(10x^2 + A) \tag{11.2}$$

Plain slot hood:

$$Q = 3.7v_c Lx \tag{11.3}$$

Flanged slot hood:

$$Q = 2.8v_c Lx \tag{11.4}$$

where

- Q is the hood airflow (volume/time)
- v_c is the desired capture velocity (length/time)
- x is the distance between source and hood (length)
- A is the hood area (length squared)
- L is the slot length (length)

and a slot hood is defined as having a length to width ratio ≥5.

Flanges are flat plates attached to the hood, usually parallel to the hood face, which limit the flow of air from behind the hood. Flanging improves the efficiency of exterior hoods by forcing air to flow from the zone directly in front of the hood, where the air is contaminated and drawing less air from behind the hood, where the air is not contaminated.

For booth type hoods, the required hood flow is calculated by multiplying the capture velocity by the area of the booth opening:

$$Q = vA \tag{11.5}$$

where

- Q is the hood airflow (volume/time)
- v is the velocity at the booth opening (length/time)
- A is the area of booth opening (length squared)

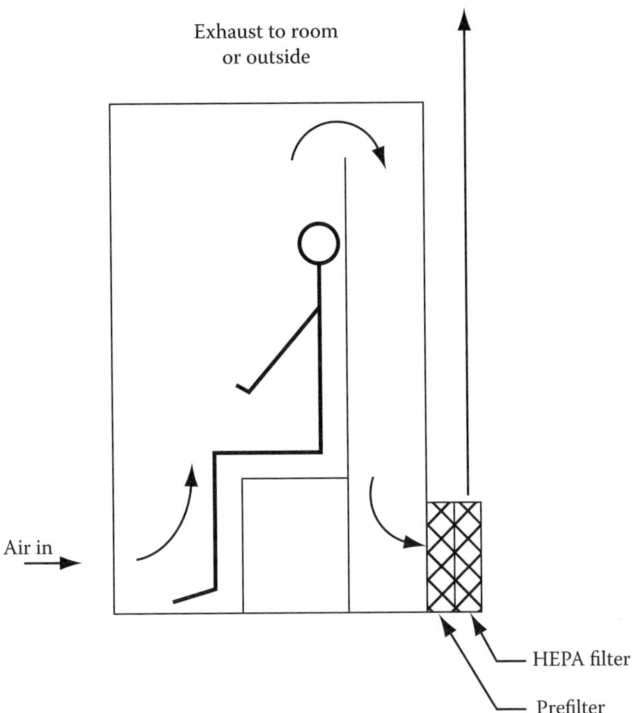

FIGURE 11.1 Sputum induction booth. (From Centers for Disease Control and Prevention, *Federal Register* 58(195), 52809, October 12, 1993.)

Figure 11.1 shows a booth used for sputum induction used for MTb control. Other examples of local exhaust ventilation systems used in hospitals are waste anesthetic gas scavenging systems and laser plume extraction systems.

In addition to the design procedure outlined above, the *Industrial Ventilation Manual*[53] gives design plates for hoods for many different industrial operations. The design plates are taken from designs used in actual installations of local exhaust ventilation systems. The manual cautions that modifications to the design plates may be necessary for special conditions, such as drafts. The manual also cautions against the design data being indiscriminately applied to highly toxic materials.

Many of the design plates given in the manual[53] could be used as a starting point for design of hoods for MTb control. The efficacy of the designs for MTb control would have to be validated before the designs could be widely used.

Some examples of design plates that might be considered for MTb control are a welding hood, for cough-inducing procedures such as bronchoscopy or intubation, and a table slot hood for autopsy procedures. Laboratories handling infectious samples can refer to biological safety cabinet designs.[64]

Capture efficiency is a more quantitative index of hood performance than centerline capture velocity. Capture efficiency is defined as the fraction of contaminant generated that is captured directly by the hood. Roach[65] recognized that velocity is not the only determinant of exhaust effectiveness. Using dimensional analysis, capture efficiency, being dimensionless, must be equated with some other nondimensional expression. Some physical quantity or quantities other than air velocity must be involved. Ellenbecker et al.[66] have shown capture efficiency to be a function of hood airflow, hood area, distance from the hood, crossdraft velocity, and source temperature.

Several investigators have studied capture efficiency of exhaust hoods.[67–77] Many of these studies have been laboratory studies under controlled conditions. Conroy et al.[76] and Prodans et al.[77] have conducted field-based studies of local exhaust hood performance for hoods used for control of vapor degreasing solvents. Further development and validation of local exhaust hood designs for MTb control is needed using a capture efficiency approach.

All local exhaust systems need to be tested initially and periodically to ensure that they are operating as designed. One design criterion is the hood flow rate. Air velocity at some point in the system is measured and the hood airflow is calculated using Equation 11.5. The velocity can be measured in the duct downstream of the hood using a Pitot tube and manometer or at the face of the hood using a thermoanemometer, swinging vane anemometer, or rotating vane anemometer. The cross-sectional area of the measurement location is also needed to calculate hood airflow.

Another design criterion is capture efficiency. Capture efficiency can be measured using a tracer gas. A tracer gas is released at a known rate at the source location and the concentration of the tracer gas is measured in the duct downstream of the hood. The hood airflow is also measured. The capture efficiency is calculated using

$$\eta = \frac{C_d Q}{S} \tag{11.6}$$

where

η is the capture efficiency (dimensionless)
C_d is the concentration of tracer gas in duct (mass/volume)
Q is the hood airflow (volume/time)
S is the tracer gas release rate (mass/time)

A description of these instruments and procedures used for testing local exhaust systems is given in Chapter 9 of the *Industrial Ventilation Manual*[53] and Chapters 3, 5, and 13 of Burgess et al.[78]

Example 11.1:

The booth shown in Figure 11.1 is used for sputum induction from suspected or known TB patients. The booth opening is 84 × 32 in. with the door open. With the door closed there is a 3/4 × 32 in. gap below the door. (a) With the door closed, what hood airflow is needed to maintain a capture velocity of 200 fpm through the booth opening? (b) What hood airflow would be necessary if the door is left open?

Use Equation 11.5:

$$Q = vA \tag{11.7}$$

Door closed:

$$A = (3/4 \times 32\ \text{in.})\left(\frac{1\,\text{ft}^2}{12^2\,\text{in.}^2}\right) = 0.1667\,\text{ft}^2$$

$$v = 200\,\text{fpm}$$

$$Q = vA = 200\,\text{fpm} \times 0.1667\,\text{ft}^2 = 33\ \text{cfm}$$

Door opened:

$$A = (84 \times 32\,\text{in.})\left(\frac{1\,\text{ft}^2}{12^2\,\text{in.}^2}\right) = 18.67\,\text{ft}^2$$

$$v = 200\,\text{fpm}$$

$$Q = vA = 200\,\text{fpm} \times 18.67\,\text{ft}^2 = 3733\ \text{cfm}$$

Designing the booth for the closed door position would be the more economical choice. (c) If the closed door design is installed, what capture velocity would result if a staff member left the door open during sputum induction?

Rearranging Equation 11.5 to solve for v gives

$$v = \frac{Q}{A} = \frac{33\ \text{cfm}}{18.67\,\text{ft}^2} \approx 2\ \text{fpm}$$

Isolation of the patient in the booth would be lost because the air velocity through the opening is less than disturbing room air currents typically found in rooms.

Directional Airflow

Directional airflow uses directed air movement without benefit of enclosures to isolate workers from MTb sources. This concept has many potential applications in health care facilities. However, it is in the second tier of ventilation control designs after local exhaust.

Directional airflow systems use velocity and bulk air movement to direct contaminated air away from health care workers. The most studied application of directional airflow is the clean room. The technology has been applied successfully in critical health care settings such as operating rooms. Although federal guidelines regulate cleanroom performance, worker protection is not considered in cleanroom design standards.[53] The purpose of the room is protection of a product or patient.

Nevertheless, some of the cleanroom features can be applied to droplet nuclei control. The unidirectional flow system moves filtered, recirculated air in either a vertical or horizontal direction.[64] Because they distribute large air volumes across opposing room surfaces, these designs provide constant velocity, bulk air flows through the rooms in predictable paths that small particles must follow. The health care worker can be protected if their movements do not place them downstream of the patient or source.

Cleanrooms have several other features which need consideration when applying the designs to TB control:

1. Excess supply air keeps out dust from adjoining spaces.
2. A series of prefilters and high-efficiency filters reduces the number concentration of dust particles to very low levels.
3. The life of the filters is extended by maximizing the recirculated airflow and minimizing the fresh outdoor airflow.
4. A unidirectional control air velocity of about 100 fpm is usually attained across the entire room.
5. Preventive maintenance time is high.
6. Cleanroom initial costs are high.

For application to MTb control, the first feature, excess supply air, should generally be avoided. The workers and other patients need protection from the TB patient. The system can be balanced to provide an excess of exhaust air to isolate the room from the adjoining ones. If the room has a positive pressure to protect an immune-compromised patient, for example, a negative pressure anteroom can be added to isolate the room from the adjoining spaces. However, workers entering the anteroom will need to be protected just as if they were in the patient room.

The second feature, filtration, is not needed for one-pass airflow rooms. However, the airflow needed for a 100 fpm control velocity is large and the energy cost is high. For example, an isolation room that is 10 ft wide and 10 ft high needs a 10,000 cfm (cubic feet per minute) capacity system (Q = [10 ft × 10 ft × 100 fpm] using Equation 11.5). If this were a one-pass system, the loss of conditioned air would be significant.

The third feature, recirculated air with filtration, would have a higher equipment expense but lower operating expense associated with energy savings. The example above would still need a 10,000 cfm fan. However, only a small percentage of this flow would need to be exhausted with proper fresh airflow for comfort.

Cleanroom systems are not practical for most TB control situations. However, there may be some critical applications in health care facilities where the high exposure hazard justifies the extra expense. For example, emergency rooms in hospitals that serve patient populations with high risks of TB infection may need rooms or curtained areas with reasonably good droplet nuclei control. Modified, unidirectional flow cleanroom designs could provide that control in ways that do not obstruct emergency room activities.

Multidirectional flow rooms are more commonly used for TB control. If the clean supply air is introduced at a position where the health care workers normally stand, the exposure time and risk in a patient isolation room, for example, would be reduced. Although the short-circuited airflow impairs the mixing needed for good dilution of air contaminants, good positioning of the worker (near the supply diffuser) and the patient (near the exhaust grill) could improve the exposure risk.

Figure 11.2 shows the concept proposed in the CDC guidelines.[36] Directional flow only exists very close to the supply and exhaust points in the room because the control airflows and velocities are much less than ones used in cleanrooms. This setup hopes that directional airflow patterns are established to transport droplet nuclei away from the worker. However, thermal air currents and airflows through window cracks can easily upset these patterns. The bulk airflow patterns over long averaging times may indeed be away from workers in these rooms. Whether or not the airflow patterns provide adequate protection from short-term emission activities like coughing has not been validated.

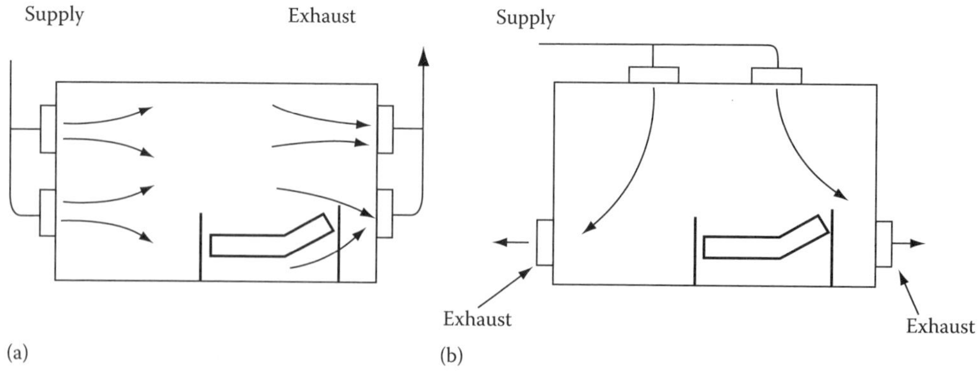

FIGURE 11.2 Directional flow patterns. (From Centers for Disease Control and Prevention, *Federal Register* 58(195), 52809, October 12, 1993.)

Example 11.2:

A patient bed in a 2000 ft^3 room is located near the exhaust grill (Figure 11.2a). The 18 in. (1.5 ft) square grill is 3 ft from the center of the bed. (a) What airflow is needed to attain a 50 fpm control velocity at this point? (b) If the room cross-sectional area is 100 ft^2 and the supply air comes from the opposite wall, what is the average air velocity in the room?

a. Use Equation 11.2 for a flanged hood (the wall acts as a flange).

$$Q = 0.75v_c(10x^2 + A) = 0.75(50\,\text{fpm})(10(3^2\,\text{ft}^2) + 1.5^2\,\text{ft}^2) = 3459\,\text{cfm}$$

b. Use Equation 11.5

$$v = \frac{Q}{A} = \frac{3459\,\text{cfm}}{100\,\text{ft}^2} = 35\,\text{fpm}$$

Although this system could provide reasonably good directional airflow control, the expense associated with the fan and filters would be much more than normally invested in an isolation room.

Another directional airflow strategy is the air shower. This concept places an air supply hood over a fixed work location. The hood should have side curtains for better directional control and a thermostat for thermal comfort. Typical air velocities are in the range of 100–700 fpm. The hood provides a supplied air "island"[79] that workers can use to isolate themselves from contaminated air. Good communication about the use of the air shower is needed for workers to properly use this and any other directional airflow control.

A disadvantage of cleanrooms, and of directional airflow systems in general, is discomfort associated with drafts. Thermal comfort is mainly a function of room temperature, air velocity, air turbulence, the temperature difference between the room and the drafts, and the part of the body in contact with the draft.[80]

Another factor to consider is the effect of drafts on the eyes. Occupant complaints about indoor air quality are sometimes associated with exposure of the eyes to draft velocities greater than 100 fpm.[81,82] The effects include eye irritation, contact lens problems, dry eyes, and headache.

Control with Concentration Reduction

Dilution Ventilation

The use of dilution ventilation is most effective for low generation rates of nontoxic or low-toxicity gaseous or vaporous contaminants where there is source–worker separation. Rather than isolate the worker from the patient, dilution ventilation reduces the residence time of droplet nuclei in the worker's air space. Prohibitively high airflow rates are needed for high generation rates and/or high toxicity contaminants. The worker will be exposed before the contaminant is diluted if the worker is very close to the source.

The current design method for general ventilation uses a completely mixed space mass balance approach. The model assumes perfect and instantaneous mixing of contaminants, no concentration of contaminant in the supply air, and a constant generation rate. A safety/mixing factor is introduced to account for deviations from these assumptions.[53] The design calculation uses

$$t_2 - t_1 = \frac{KV}{Q}\ln\frac{G - \dfrac{C_1Q}{K}}{G - \dfrac{C_2Q}{K}} \tag{11.8}$$

where

t_1 is the initial time
t_2 is the time of interest
C_1 is the contaminant concentration at time t
C_2 is the contaminant concentration at time t_2
Q is the airflow through space
V is the volume of space
G is the contaminant generation rate
K is the safety/mixing factor

At steady state, Equation 11.8 reduces to

$$C_{ss} = \frac{GK}{Q} \tag{11.9}$$

where C_{ss} is the steady-state contaminant concentration.

In order to calculate the required airflow (Q), an estimate of the contaminant generation rate (G) and the acceptable contaminant concentration (C_{ss}) is needed. In the case of MTb, both of these estimates are unavailable. Generation rates are difficult to measure and will vary from person to person, with various activities (e.g., talking, coughing, sneezing will all result in different generation rates), and with various procedures (e.g., sputum induction, bronchoscopy, etc.). An acceptable exposure concentration has not been determined for MTb. Again there will be high variability in this parameter due to the infectivity of the droplet and the susceptibility of the exposed individual.

The use of Equation 11.9 for MTb control is limited because steady-state concentrations are rarely, if ever, achieved. The most common generation source is a person coughing. In that case, there will be a sudden large increase in the concentration of droplet nuclei with subsequent decay when the person stops coughing. Conceptually, an average concentration (the integral of the concentration during the sudden increase and subsequent decay) would be the appropriate concentration for use in designing the ventilation system.

In addition to estimates of the generation rate and acceptable concentration, the mixing factor must be determined or estimated. The mixing factor describes the effectiveness of the supplied ventilation in reducing contaminant concentrations. The *Industrial Ventilation Manual*[53] gives estimates of mixing factors for several room configurations. The mixing factor can also be estimated through experimental measurement of tracer gas concentration decay and mechanical ventilation rates. A tracer gas is released throughout the space of interest and the concentration is measured at a location representative of the concentration in the space, e.g., an exhaust grill. The decay of the tracer gas concentration with time and measured ventilation rate and space volume are used in Equation 11.8 to calculate the mixing factor, K.

Example 11.3:

A patient room has a volume of 1500 ft^3. The mechanical exhaust ventilation is 45 cfm. The mechanical supply ventilation is 30 cfm and the airflow under the door to the room is 15 cfm. Tracer gas concentration as a function of time was as follows:

Time	Elapsed Time (min)	Concentration (Parts per Billion [ppb])
1518	0	8.12
1532	14	7.49
1538	20	5.76
1552	34	4.27

What is the mixing factor for this room? Assume that there is no other air infiltration or exfiltration.

Following the release of the tracer gas, the generation rate is 0 and Equation 11.8 reduces to

$$t_2 - t_1 = \frac{KV}{Q} \ln \frac{C_1}{C_2}$$

which can be rearranged to

$$\ln C_2 = \ln C_1 - \frac{Q}{KV} \Delta t$$

Plotting $\ln(C_2)$ versus the change in time (Δt) results in a line with an intercept equal to $\ln(C_1)$ and a slope equal to $(-Q/(KV))$. This is shown in Figure 11.3. The slope of the "best-fit" line through the data is $-0.0085\,\text{min}^{-1}$. The mixing factor, K, is calculated as follows:

$$K = \frac{1}{\text{slope}} \frac{Q}{V} = -\frac{1}{-0.0085\,\text{min}^{-1}} \times \frac{45\,\text{cfm}}{1500\,\text{ft}^3} = 3.5$$

The current recommendations for ventilation rates are given in Table 11.1.[39] The recommendations in Table 11.2 are based on comfort criteria and experience and will not necessarily provide adequate protection from MTb. The recommendations should be used as minimum design criteria and higher ventilation rates will result in lower contaminant concentrations. Using Equation 11.9 shows that an infinite airflow is necessary to maintain the concentration of MTb at zero if there is any generation of particles. Without an estimate of generation rate and acceptable concentration, it is impossible to calculate the required airflow.

One calculation that is possible is the amount of time needed to reduce the concentration to some specified fraction of the initial concentration once generation has ceased, for example, when an infected person is no longer in the space. Equation 11.8 can be rearranged to give

$$t_2 = -\frac{VK}{Q} \ln \frac{C_2}{C_1} \tag{11.10}$$

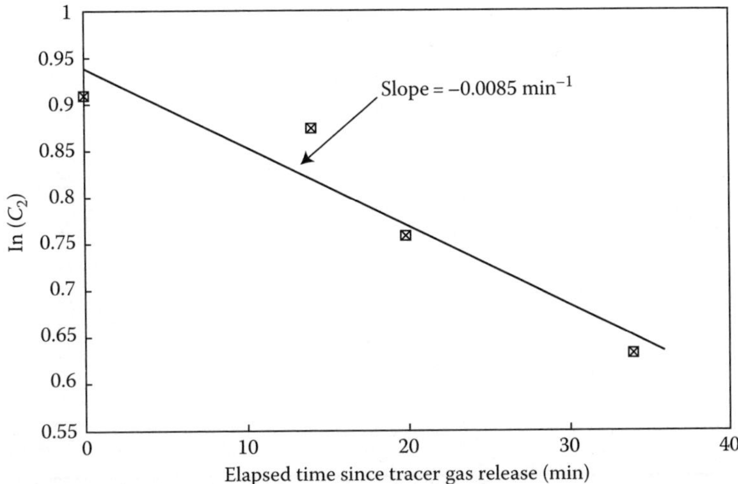

FIGURE 11.3 Tracer gas decay curve.

TABLE 11.1 Recommended Ventilation Rates

Function	Pressure	Air Changes per Hour	
		Outside Air	Total
Operating	Positive	15	15–25
Recovery	Equal	2	6
Trauma	Positive	5	12
Intensive care unit	Positive	2	6
Isolation	Negative, positive	2	6
Isolation anteroom	Negative, positive	2	10
Bacteriology laboratory	Negative	2	6
Treatment	Negative, positive	2	6

Source: American Society of Heating, Refrigerating and Air-Conditioning Engineers, Inc. *Heating, Ventilating, and Air-Conditioning Applications*. American Society of Heating, Refrigerating and Air-Conditioning Engineers, Inc., Atlanta, GA, 1991.

TABLE 11.2 Air Changes per Hour and Time in Minutes for Removal Efficiencies of 90%, 99%, and 99.9% of Airborne Contaminants

Air Changes per Hour	Minutes Required for a Removal Efficiency of		
	90%	99%	99.9%
1	138	276	414
2	69	138	207
3	46	92	138
4	35	69	104
5	28	55	83
6	23	46	69
7	20	39	59
8	17	35	52
9	15	31	46
10	14	28	41
11	13	25	38
12	12	23	35
13	11	21	32
14	10	20	30
15	9	18	28
16	9	17	26
17	8	16	24
18	8	15	23
19	7	15	22
20	7	14	21
25	6	11	17
30	5	9	14
35	4	8	12
40	3	7	10
45	3	6	9
50	3	6	8

Source: Centers for Disease Control and Prevention, *Federal Register* 58(195), 52809, October 12, 1993.

Table 11.2[36] gives the number of minutes necessary to reduce the concentration by 90%, 99%, and 99.9% for various air exchange rates. The time given in Table 11.2 must be multiplied by the appropriate mixing factor for the space.

Example 11.4:

A sputum-induction booth is shown in Figure 11.4. (a) What value of K (mixing factor) would you assign for this space? How can you determine K? (b) If the airflow through the space Q is 41 cfm, what is the air change rate? (c) For the K value assigned in step (a) and the air change rate in step (b), how long would be needed between users for a 90% reduction in concentration? (d) How long for a 99% reduction?

a. Using Ref. [53, p. 2–4], the situation that most resembles this situation has $K = 2.5$ as a minimum. For MTb control you might want to add in a safety factor for toxicity and increase K to 3–5. For this example, pick $K = 4$. You could also determine K experimentally by releasing a tracer gas and measuring its decay.
b. The air change rate is the airflow through space divided by the space volume.

$$V = 4.33\,\text{ft} \times 7.67\,\text{ft} \times 9.33\,\text{ft} = 310\,\text{ft}^3$$

$$\text{Air change rate} = \frac{Q}{V} = \frac{41\ \text{cfm}}{310\ \text{ft}^3} \times \frac{60\ \text{min}}{\text{h}} \approx 8\,\text{ach}$$

c. For 90% reduction and 8 ach, Table 11.2 gives $t = 17\,\text{min}$. The total time necessary would be $t \times K = 17\,\text{min} \times 4 = 68\,\text{min}$. For 99.9% reduction and 8 ach, Table 11.2 gives $t = 52\,\text{min}$. The total time would be $52\,\text{min} \times 4 = 208\,\text{min}$ or 3.5 h. For a busy emergency room or clinic, 3.5 h or even 68 min might be too long. To reduce the amount of time, the airflow could be increased, the volume of the space reduced, or the mixing characteristics of the space improved.

Air Filtration

Another method of concentration reduction is the use of filters to remove droplet nuclei from the air. HEPA filters are filters that are 99.97% efficient for particulates 0.3 μm in diameter. Ultra-low penetration

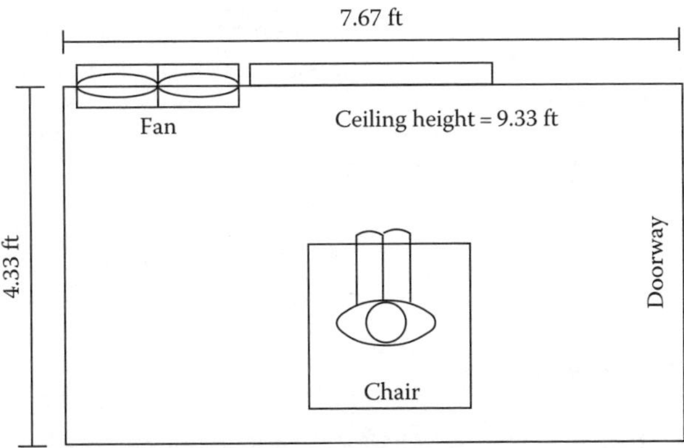

FIGURE 11.4 Sputum-induction room.

air (ULPA) filters are 99.999% efficient for particulates 0.3 μm in diameter. The filter efficiency increases with increase in particle size. Since droplet nuclei which carry MTb are approximately 1–4 μm, a HEPA filter can be expected to be at least 99.97% efficient. However, the efficiency of HEPA filters for MTb has not been validated.

HEPA filters can be applied to TB control in two ways: (1) duct filtration and (2) room air filtration. Duct filtration places the filter in the exhaust duct of the space. All the air being exhausted from the space passes through the filter before discharge to the outside or recirculation to occupied spaces. In-room filtration involves the use of a self-contained unit to reduce the concentration in the room. The unit will have a fan which pulls room air through the filter before discharging the air back into the room. The success of both filtration methods depends on all of the contaminated air in the space passing through the filter, which in turn depends on the mixing characteristics of the room.

The use of portable filtration units, for reducing particle concentrations, is limited by the number of air changes that can be generated by the fan. Larger flow rates through a filter with constant cross-sectional area will result in higher filter face velocities, which may reduce the filter efficiency. Larger flow rates also require larger fans which may make the units too large for the space and/or unacceptably noisy. Additionally, the inlet and outlet locations on portable units are relatively close to each other. This may mean that the units are repeatedly cleaning the same air, i.e., the relative location of the inlet and discharge may result in air "short-circuiting" the space. More research in actual settings is needed on the effects of these units on room particle concentrations and mixing characteristics. Limited research by the manufacturers suggests that these units may be beneficial in reducing particle concentrations (unpublished data).

The CDC guidelines specify recirculation of HEPA filtered air as a supplemental control, with local and one-pass dilution ventilation used as the primary control. Air that cannot be discharged to the outside away from air intakes or occupied spaces should be exhausted through a HEPA filter. The guidelines further state that "in any application, HEPA filters need to be carefully installed and meticulously maintained to ensure adequate function."[36]

Example 11.5:

This example is meant to illustrate the effect of a HEPA filtration unit on the contaminant concentration reduction in an existing isolation room. In order to completely solve the example, an estimate of the generation rate is needed. For this example, assume $G = 18{,}000$ particles/min while the patient is coughing and $G = 90$ particles/min while the patient is not coughing.[43] The true generation rates for a patient coughing and not coughing are unknown and highly variable.

Case 1: Isolation room without HEPA filtration unit.

Room volume = 1350 ft^3, mechanical exhaust ventilation rate = 90 cfm, mechanical supply ventilation rate = 90 cfm, air leakage under door from pressurized anteroom = 7.7 cfm, approximate mixing factor = 4, supply air is 50% outside air.

a. What is the air change rate?

$$\text{Exhaust airflow} = \frac{90\ \text{cfm}}{1350\ \text{ft}^3} \times \frac{60\ \text{min}}{\text{h}} = 4\ \text{ach}$$

b. What is the outdoor air change rate?

$$\text{Outdoor airflow} = \frac{(0.5)(97.7\,\text{cfm})}{1350\,\text{ft}^3} \times \frac{60\,\text{min}}{\text{h}} = 2.2\,\text{ach}$$

Case 2: Isolation room with HEPA filtration unit.

Room volume = 1350 ft^3, mechanical exhaust ventilation rate = 90 cfm, mechanical supply ventilation rate = 90 cfm, air leakage under door from pressurized anteroom = 7.7 cfm, approximate mixing factor = 4, supply air is 50% outside air, HEPA filtration flow = 400 cfm, HEPA filter efficiency = 99.97%.

c. What is the new air change rate?
With the HEPA unit the total airflow rate is increased according to

$$Q = Q_E + FQ_H$$

where
Q_E is the mechanical exhaust rate
Q_H is the HEPA unit exhaust rate
F is the filter efficiency

$$\text{Exhaust airflow} = \frac{90\,\text{cfm} + (0.9997)(400\,\text{cfm})}{1350\,\text{ft}^3} \times \frac{60\,\text{min}}{\text{h}} = 22\,\text{ach}$$

d. What is the outdoor air change rate with the unit operating?

$$\text{Outdoor airflow} = \frac{(0.5)(97.7\,\text{cfm})}{1350\,\text{ft}^3} \times \frac{60\,\text{min}}{\text{h}} = 2.2\,\text{ach}$$

e. How would the room concentration change with the introduction of the HEPA filtration unit?
Rearranging Equation 11.8 to include the recirculation of air and to solve for concentration results in[83]

$$C_2 = \frac{KG}{Q_E + FQ_H}\left(1 - \exp\left(\frac{(Q_E + FQ_H)\,t}{KV}\right)\right) + C_1 \exp\left(\frac{(Q_E + FQ_H)\,t}{KV}\right) \tag{11.11}$$

For simplicity, assume the patient coughs for the first 5 min of each hour. Figure 11.5 shows the concentration curve as a function of time, using Equation 11.11, for each case. The example assumes that the mixing factor is not affected by the HEPA filtration unit. If the HEPA filtration unit is affecting only a pocket of air in its general vicinity and not cleaning the rest of the room then the mixing factor would be different and the curve would have to be adjusted.

This example illustrates the theoretical effect of the addition of a HEPA filtration unit. These units have not been thoroughly tested in actual workplace settings and more research is needed to understand their performance.

Ultraviolet Irradiation

The third method of reducing the concentration of infectious MTb droplet nuclei is with the use of UVGI. Ultraviolet (UV) radiation is in the 100–400 nm wavelength range of the electromagnetic radiation spectrum. The UV spectrum is divided into three regions. The UV-A range is from 320–400 nm. The UV-B range is from 290 to 320 nm. The UV-C range is from 100 to 290 nm. Commercially available germicidal lamps are low-pressure mercury-vapor lamps which are operated in the UV-C range with a predominate wavelength of 254 nm.

UVGI used in exhaust ducts has been shown to be effective in disinfecting air of TB bacilli in experiments with guinea pigs.[84] Other studies have shown the effectiveness of UVGI in reducing transmission

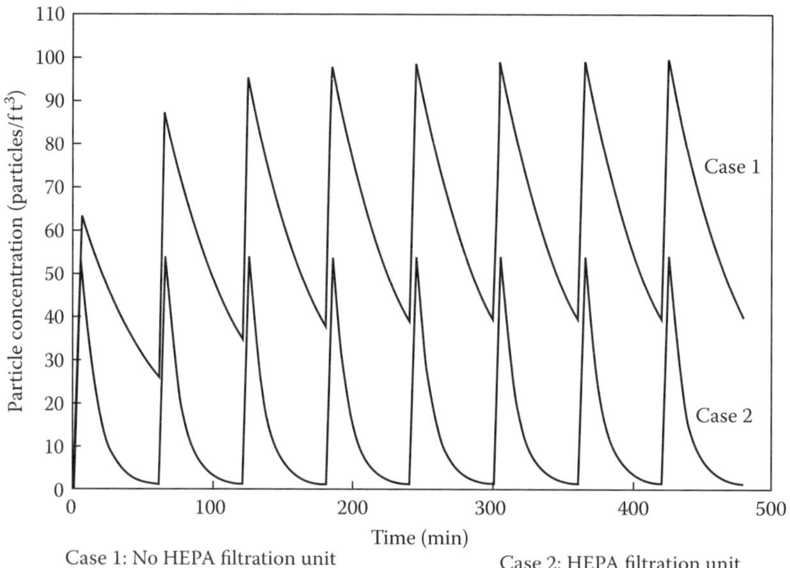

FIGURE 11.5 Concentration vs. time with and without the use of a HEPA filtration unit.

of other infections in hospitals,[85] classrooms,[86–88] and military housing.[89] The use of UVGI has been suggested by CDC as a supplemental control measure. A recent study where UVGI was evaluated in an outpatient waiting room showed a 14%–19% reduction in culturable airborne bacteria.[90] The researchers in that study concluded that environmental factors such as open doors and windows and the mechanical ventilation in the space may limit the effectiveness of UVGI as a control measure for airborne bacteria.

As with HEPA filtration, two methods of UVGI may be used: (1) duct irradiation and (2) upper air irradiation. Unlike HEPA filtration, the UV lamps cannot be located in the occupied zone of the room because of the health hazards associated with UV exposure.

Duct irradiation involves placing the UV lamps in the exhaust duct to disinfect the air before it is recirculated. If properly designed and maintained, high UV intensities can be used in the duct without exposure to humans, except during maintenance activities. To prevent exposure to maintenance personnel, the system should be designed to prevent access to the duct until the UV lamps have been turned off.

The CDC guidelines[36] outline the situations where duct irradiation may be used and situations where it should not be used. Duct irradiation may be used for isolation and treatment rooms to recirculate the air back into the isolation or treatment room and for other patient rooms and general use areas, such as waiting areas and emergency rooms, where there may be unrecognized TB. Duct irradiation is not recommended as a substitute for HEPA filtration for recirculation from isolation rooms to other areas of the facility or when discharge to the outside is not possible.

Upper room air irradiation involves suspending UV lamps from the ceiling or mounting them on the walls. The lamp must be high enough to prevent eye or skin exposure to persons in the room. The bottom of the lamp must be shielded and the ceiling and walls should be nonreflective surfaces. Upper room air irradiation depends on contaminated air in the lower part of the room moving to the upper part and remaining there long enough to kill the bacteria. This in turn, depends on the mixing characteristics of the room. Unlike dilution ventilation and HEPA filtration, there is a trade-off in effectiveness

TABLE 11.3 Maximum Permissible Exposure Times for UV Radiation

Duration of Exposure	Effective Irradiance (μW/cm^2)
8 h	0.1
4 h	0.2
2 h	0.4
1 h	0.8
30 min	1.7
15 min	3.3
10 min	5.0
5 min	10.0
1 min	50.0
30 s	100.0

Source: Centers for Disease Control and Prevention, *Federal Register* 58(195), 52809, October 12, 1993.

with increasing air change rates and mixing. Good mixing improves the effectiveness of UVGI if there is adequate residence time of the air in the upper portion of the room. High air change rates may not allow for adequate residence time in the upper portion of the room.

UV radiation has been shown to cause keratoconjunctivitis and skin erythema.[91] Broad spectrum UV has been associated with increased risk of skin cancer[92,93] and the International Agency for Research on Cancer (IARC) has listed UV-C as a Group 2A (probable human) carcinogen.[93] NIOSH and the American Conference of Governmental Industrial Hygienists (ACGIH) have published exposure guidelines for UV radiation. The NIOSH Recommended Exposure Limit (REL)[91] and the ACGIH Threshold Limit Value (TLV)[94] are essentially the same. The guidelines are intended to protect nearly all workers from the acute effects of UV exposure (keratoconjunctivitis and skin erythema). The exposure guidelines are not intended to protect against long-term effects such as skin cancer or to protect individuals who may be photosensitized due to disease or exposure to photosensitizing chemicals.

The REL and TLV limit the amount of time an individual may be exposed to a certain UV intensity or conversely the UV intensity for a given time period. Table 11.3 shows the allowable time for various UV intensities. The values given are for UV at a wavelength of 270 nm, considered to be the most hazardous. For other wavelengths, the intensity values in Table 11.3 should be divided by $S\lambda$, given in Table 11.4. For UV irradiation at a wavelength of 254 nm, the allowable 8 h exposure is 0.2 μW/cm^2.

In addition to limiting the exposure to 0.2 μW/cm^2 for an 8 h average, CDC[36] recommends that when UVGI is used in a facility (1) employees should be trained in the general principles of UVGI, the potential hazardous effects of UVGI, the potential for photosensitivity from certain medical conditions or pharmaceuticals, and general maintenance procedures for UVGI fixtures; (2) warning signs (Figure 11.6) should be posted on UV lamps and at accesses to ducts where UV lamps are used; (3) a preventive maintenance program should be implemented; and (4) a regularly scheduled evaluation of UV exposure to hospital personnel and patients should be conducted. There are direct reading instruments that can be used for measuring UV intensities in a room.

Patient Isolation Rooms

Negative pressure isolation rooms protect the occupants of adjoining rooms and spaces. People who enter the isolation room are not isolated from the TB patient unless the room has local exhaust or directional airflow controls. Instead, the exposure hazard is reduced by diluting the droplet nuclei concentration in the room.

TABLE 11.4 Relative Special Effectiveness for Several Wavelengths of UV Radiation

Wavelength (nm)	Relative Spectral Effectiveness ($S\lambda$)
240	0.300
250	0.430
254	0.500
255	0.520
260	0.650
265	0.810
270	1.000
275	0.960
280	0.880
285	0.770
290	0.640

Source: Centers for Disease Control, National Institute for Occupational Safety and Health. *Criteria for a Recommended Standard-Occupational Exposure to Ultraviolet Radiation.* U.S. Department of Health, Education, and Welfare, NIOSH (Publication no. (HSM) 73-110009), Washington, DC, 1973.

CAUTION
ULTRAVIOLET ENERGY
PROTECT EYES AND SKIN

CAUTION
ULTRAVIOLET ENERGY
TURN OFF LAMPS BEFORE
ENTERING ROOM

FIGURE 11.6 Example warning signs for UV radiation. (From Centers for Disease Control and Prevention, *Federal Register*, 58(195), 52809, October 12, 1993.)

Existing health care facilities serving high-risk populations have difficulty providing the number of isolation rooms needed to care for suspected and confirmed cases of TB. Many facilities do not have isolation rooms that meet the current requirements for dilution airflow (Table 11.1), airflow into the room at all points (negative pressure) and airflow distribution within the room (Figure 11.2). Some issues that need to be considered when specifying improvements or new construction of the rooms are

1. How many rooms are needed for acute care patients with suspected TB?
2. What control performance criteria will be used to design the acute care isolation rooms?
3. Does the emergency room need isolation controls?
4. What are the performance criteria for the emergency room?
5. How many rooms are needed for long-term care TB patients and what are the performance criteria?
6. Where should the isolation rooms be located within the building?
7. What are the intended room layouts, equipment locations, room penetrations, and window openings in the rooms?
8. What interim measures will be used and how will these controls be evaluated?
9. Are anterooms necessary for good isolation in these locations?
10. Are there local building codes or fire safety regulations that conflict with the isolation room specifications?
11. Is there a detrimental effect of the room air balances on the central ventilation system in the building?

12. How will the isolation room ventilation controls be monitored?
13. Are the room isolation components accessible for maintenance?
14. Who will maintain the systems?

The proposed CDC guidelines provide the minimum design criteria for isolation rooms.[36] They are based on American Institute of Architects (AIA) guidelines.[95] The following example presents these criteria as applied to a review of architectural drawings for a proposed renovation project.

Example 11.6:

A dental clinic serving clients with high risk of TB infection decided to convert its two dental operatories into negative pressure, isolation rooms. The patient waiting room also needed to be isolated from the rest of the facility. A review of the architectural drawings gave the following information.

The existing floor area for each room is 90 ft² and the ceiling height is 9 ft There are 1/2 in. openings along the bottoms of the 7 × 3 ft, sealed doors. The rooms will use existing supply air ducts connected to the central ventilation system. Each room will have one 175 cfm supply air diffuser near the entry.

Existing return air grills and ducts will be removed. Two 115 cfm exhaust grills will be installed in each room at a high and low location on the wall opposite from the door. A new exhaust fan will be installed on the roof 50 ft from air intakes. A UV light will be used in the exhaust duct to disinfect the discharged air. The design uses rectangular duct with several branches and elbows up to the exterior wall. Circular duct is specified for the vertical run up to the fan. The specified fan delivers 675 cfm at 0.625 in. of static pressure and 1165 rpm.

Evaluate the plans with respect to the proposed CDC guidelines.[36]

1. Guidelines: Maintain 10% more exhaust than supply airflow or at least a 50 cfm difference. Each room design specifies 175 cfm supply and 230 cfm exhaust. The difference in the balance is −30% or 55 cfm more exhaust.
2. Guidelines: Maintain an air velocity of at least 100 fpm through openings around the door in its normal position. If the door is normally shut, the air velocity through the bottom is 440 fpm 55 cfm/(0.125 ft²). If the door is ajar with a 1/2 in. opening along all sides, the air velocity is 66 fpm. If the door is wide open, the air velocity through the opening is only 3 fpm.
3. Guidelines: Provide dilution airflow volumes of at least 6 ach, 2 of which must be outdoor air. The total air change rate, based on the exhaust airflow, is

$$\frac{230\,\text{cfm}}{(90\,\text{ft}^2)(9\,\text{ft})} \times \frac{60\,\text{min}}{\text{h}} = 17\,\text{ach}$$

   This meets the 6 ach guideline if the room air mixing factor is at least 2.8 ($K = 17/6$). The locations of the supply air grill at the ceiling and high/low exhaust air grills across the rooms may indeed provide the necessary mixing.

   The outdoor air change rate, based on the supply airflow and an assumed outdoor air percentage of 20%, is

$$\frac{(0.2)(175\,\text{cfm})}{(90\,\text{ft}^2)(9\,\text{ft})} \times \frac{60\,\text{min}}{\text{h}} = 2.5\text{ ach}$$

4. Guidelines: Discharge the exhaust airflow outdoors away from walkways and air intakes or filter it before recirculating.

The exhaust fan location on the roof is stated to be more than 30 ft away from intakes. The UV light in the duct will help clean the discharged air. However, the light may accumulate dust and lint and need frequent cleaning.

Maintaining the design exhaust airflows in the rooms (as well as the proper balance with supply air) is critical to the success of this design. The accuracy of the static pressure estimate for this system should be checked to verify the selected fan size. The fittings and use of rectangular duct suggest that the air leakage into the duct could be as high as 15% of the total airflow. Since the specified air balance is set at the minimum recommended difference of 50 cfm, there is no room for error. The designers should offer some assurance that the selected fan will deliver more than the designed airflow rate. Sealing the duct seams and fittings could reduce the leakage. Choosing a belt-driven fan would offer some adjustability of airflow.

Measuring airflows in existing isolation rooms should be part of the preventive maintenance program at the facility. Airflows from supply and exhaust air grills can be measured with a swinging vane anemometer attached to an airflow hood.[53] Flow from an obstructed diffuser should be checked with a smoke tube if a reliable measure of airflow cannot be made.

Airflow under the isolation room door can be measured with a tape measure and an air velocity device such as a thermoanemometer. The average of six or more velocity readings along the bottom are multiplied by the area of the opening to get airflow (Equation 11.5). Airflow direction should be checked with a smoke tube at the bottom and around the door perimeter with the door in its normal position. The airflow direction between the anteroom and corridor should also be checked with smoke.

The room pressurization can be checked with an electronic manometer if the resolution is very sensitive (0.001 in. water) and the meter is calibrated. The meter senses pressure in the room and the hall with two probes and reports the differential pressure between the two spaces. The suggested measurement points are at the center of the door in the corridor and at the kick plate near the door bottom inside the room.

Example 11.7:

The following measurements were made in a patient isolation room. Compare them with the CDC guidelines.

The room layout has two beds, a bathroom and an anteroom. The room pressurization is switchable from positive to negative with a control in the anteroom. The negative pressure mode is achieved by increasing the supply airflow in the anteroom. Airflows in the isolation room are not changed by the pressure control. The facility ventilation system uses a minimum of 50% outdoor air.

The floor area is 150 ft^2 and the ceiling is 9 ft high. A large, hinged window is on the outdoor wall. There are three access panels to the false ceiling space above the beds and bathroom.

The room has three supply air diffusers above the beds and at the window. A curtain rail blocks one diffuser. There are exhaust grills above the door and in the bathroom. The 7 × 3.5 ft door is gasketed and has a 0.25 in. opening at the bottom.

Airflow Measurement				
Location	Method	Velocity	Direction	Airflow
Bed 1 diffuser	Airflow hood		Into room	30 cfm
Bed 2 diffuser	Airflow hood		Into room	30 cfm
Window diffuser (by curtain rail)	Smoke tube		Into room	? (Assume 30 cfm)
Bath grill	Airflow hood		Out of room	50 cfm
Grill by door	Airflow hood		Out of room	40 cfm
Room door at bottom	Thermoanemometer (6 point traverse)	105 fpm	Into room	
Anteroom diffuser	Airflow hood		Into anteroom	80 cfm
Anteroom grill	Airflow hood		Out of anteroom	30 cfm
Hall door at bottom	Smoke tube		Out of anteroom	
Both sides of room door	Differential pressure gauge		Into room	−0.002 in. H_2O

1. Guidelines: Patients with known or suspected TB should be placed in a private room. The isolation room in this example has two beds, which would not meet the guideline.
2. Guidelines: Maintain 10% more exhaust than supply airflow or at least a 50 cfm difference.

$$\text{Exhaust airflow} = 50\,\text{cfm} + 40\,\text{cfm} = 90\,\text{cfm}$$

$$\text{Area under door} = \frac{0.25\,\text{in.}}{12\,\text{in./ft}}3.5\,\text{ft} = 0.073\,\text{ft}^2$$

$$\text{Airflow under door} = 105\,\text{fpm} \times 0.073\,\text{ft}^2 = 7.7\,\text{cfm}$$

$$\text{Mechanical supply airflow} = 30\,\text{cfm} + 30\,\text{cfm} + (30\,\text{cfm}) = 90\,\text{cfm}$$

$$\text{Total supply airflow} = 90\,\text{cfm} + 7.7\,\text{cfm} = 97.7\,\text{cfm}$$

The mechanical exhaust and supply airflows in the room are equal. Although the 50 cfm excess exhaust flow guideline is not met, the smoke tube test and differential pressure measurement at the door indicate that the isolation room is in a negative pressure mode compared to the anteroom. This is because the anteroom is pressurized by excess supply airflow. The exhaust grill near the door will promote the flow of contaminated air toward it. Opening either door disrupts the pressure difference with the room so that a slug of contaminated air can escape into the corridor.

Closing the air balance in existing rooms is difficult because of unknown infiltration airflows through the window cracks and room wall penetrations. Airflows around closed windows and penetrations can be significant. They cannot be measured accurately with conventional instruments. Smoke testing the ceiling access panels is needed in this room to detect contamination of the false ceiling space.

3. Guidelines: Maintain an air velocity of at least 100 fpm through openings around the door in its normal position.
   The closed door air velocity is 105 fpm into the room. Although this meets the guideline, a better guideline is 120 fpm because of the measurement method error (about 20%).
4. Guidelines: Provide dilution airflow volumes of at least 6 ach, 2 of which must be outdoor air.

$$\text{Room volume} = 150\,\text{ft}^2 \times 9\,\text{ft} = 1350\,\text{ft}^3$$

$$\text{Exhaust airflow} = \frac{90\,\text{cfm}}{1350\,\text{ft}^3} \times \frac{60\,\text{min}}{\text{h}} = 4\,\text{ach}$$

$$\text{Outdoor airflow} = \frac{(0.5)(97.7\,\text{cfm})}{1350\,\text{ft}^3} \times \frac{60\,\text{min}}{\text{h}} = 2.2\,\text{ach}$$

The total airflow is less than 6 ach. The outdoor airflow is more than the 2 ach guideline. Increasing the exhaust airflow in the room to 140 cfm so that it exceeds the supply airflow by 50 cfm would increase the total air change rate to

$$\text{Exhaust airflow} = \frac{90\,\text{cfm}}{1350\,\text{ft}^3} \times \frac{60\,\text{min}}{\text{h}} = 6.2\,\text{ach}$$

The balance between supply and exhaust air in the anteroom should also be adjusted to provide an excess of 50 cfm supply air. This will promote airflow into the isolation room without promoting airflow into the corridor.

Upgrading the system in this way (increasing exhaust flow to 140 cfm) would help meet guidelines 1 and 2. Guideline 3 would be met only if the room had near perfect mixing of the air. The location of supply air diffusers over the beds and exhaust grills near the door is the opposite of the recommended mixing pattern (Figure 11.2).

Research Needs

This section described the application of basic industrial hygiene principles to the control of the TB transmission hazard. Several areas needing research were also identified. The first was the need to develop state and local codes which are consistent with infection control methods while still meeting fire safety and other considerations.

In order to design effective control systems, the droplet nuclei generation rate for various activities must be known. Droplet nuclei generation rates and emission factors need to be developed and validated for representative particles or surrogate microorganisms. Additionally, it is important to determine if there is an acceptable exposure concentration that can be used for design.

The application of directional airflow to MTb control should be investigated and validated in real situations. There is also a need for experimental determination and validation of the "near field" of the source. At what distance from the source does the droplet nuclei concentration equal the background concentration?

The design and validation of local exhaust hoods for MTb control needs to undertaken. The research should evaluate design factors such as capture velocity and efficiency. Acceptance of the hoods by the workers using them should also be optimized.

With the design of isolation rooms, several questions need to be addressed. How many isolation rooms are necessary in a facility? Where should clinic and isolation rooms be located in the facility to minimize the hazard to other occupants of the hospital? What is the minimum airflow difference necessary to achieve effective isolation? What is an adequate negative pressure? What is an adequate control velocity at the openings to the room? What factors (room size, furniture placement, etc.) affect the necessary airflow difference, negative pressure, and control velocity?

The performance of portable HEPA filtration units and UVGI in actual facilities needs to be validated. The hazard reduction and associated costs need to be quantified for these controls. Using them to reduce reentry hazard of air discharged outdoors is another application that needs validation.

This is only a partial list of research needs. Scientists are once again investigating some of these items. Development and maintenance of engineering controls for infectious diseases is a continuing challenge. Hopefully, motivation and resources needed by the health care community to control the renewed TB hazard will be realized.

References

1. Peterson JE. *Industrial Health*. American Conference of Governmental Industrial Hygienists, Inc., Cincinnati, OH, 1991.
2. Houk VN. Spread of tuberculosis via recirculated air in a naval vessel: The Byrd study. In: RB Kundsin (ed.), *Airborne Contagion. Annals of the New York Academy of Sciences*, Vol. 353. New York Academy of Sciences, New York, 1980.
3. Riley RL. Indoor airborne infection. *Environ Int* 8:317–320, 1982.
4. Nardell EA. Dodging droplet nuclei, reducing the probability of nosocomial tuberculosis transmission in the AIDS era (editorial). *Am Rev Respir Dis* 142:501–503, 1990.
5. Bloom BR and Murray CJL. Tuberculosis: Commentary on a reemergent killer. *Science* 257:1055–1064, 1992.
6. Melius J. Source characterization and control. In: *Workshop on Engineering Controls for Preventing Airborne Infections in Workers in Health Care and Related Facilities*, CDC, NIOSH, July 14, 1993.
7. American Thoracic Society. Control of tuberculosis in the United States. *Am Rev Respir Dis* 146: 1623–1633, 1992.
8. Centers for Disease Control. Summary of notifiable diseases, United States. *MMWR* 40(53): 57, 1992.

9. Dublin LI. The course of tuberculosis mortality and morbidity in the United States. *Am J Pub Health* 48(11):1439–1448, 1958.
10. Centers for Disease Control. Prevention and control of tuberculosis in correctional institutions: Recommendations of the Advisory Committee for the Elimination of Tuberculosis. *MMWR* 38: 313–320, 325, 1989.
11. Centers for Disease Control. Prevention and control of tuberculosis in facilities providing long-term care to the elderly. *MMWR* 39(RR-10), 1990.
12. Nardell EA, Keegan J, Cheney SA, and Etkind SC. Airborne infection-theoretical emits of protection achievable by building ventilation. *Am Rev Respir Dis* 144:302–306, 1991.
13. Catanzaro A. Nosocomial tuberculosis. *Am Rev Respir Dis* 125:559–562, 1982.
14. Brennan C, Muder RR, and Muraca PW. Occult endemic tuberculosis in a chronic care facility. *Infect Control Hosp Epidemiol* 9(12):548–552, 1988.
15. Malaskey C, Jordan T, Potulski F, and Reichman LB. Occupational tuberculous infections among pulmonary physicians in training. *Am Rev Respir Dis* 142:505–507, 1990.
16. Centers for Disease Control. *Mycobacterium tuberculosis* transmission in a health clinic Florida, 1988. *MMWR* 38(15):256–258, 263–264, 1989.
17. Craven RB, Wenzel RP, and Atuk NO. Minimizing tuberculosis risk to hospital personnel and students exposed to unsuspected disease. *Ann Int Med* 82:628–632, 1975.
18. Dooley SW, Villarino ME, Lawrence M, Salinas L, Amil S, Rullan JV, Jarvis WR, Bloch AB, and Cauthen GM. Nosocomial transmission of tuberculosis in a hospital unit for FUV-infected patients. *JAMA* 267(19):2632–2634, 1992.
19. Ehrenkranz NJ and Kicklighter JL. Tuberculosis outbreak in a general hospital: Evidence for airborne spread of infection. *Ann Int Med* 77(3):377–382, 1972.
20. Frampton MW. An outbreak of tuberculosis among hospital personnel caring for a patient with a skin ulcer. *Ann Int Med* 117(4):312–313, 1992.
21. Haley CE, McDonald RC, Rossi L, Jones WD, Haley RW, and Luby JP. Tuberculosis epidemic among hospital personnel. *Infect Control Hosp Epidemiol* 10(5):204–210, 1989.
22. Centers for Disease Control. Nosocomial transmission of multidrug-resistant tuberculosis among IUV-infected persons—Florida and New York, 1988–1991. *MMWR* 40(34):585–591, 1991.
23. Edlin BR, Tokars JI, Griego MH, Crawford JT, Williams J, Sordillo EM, Ong KR, et al. An outbreak of multidrug-resistant tuberculosis among hospitalized patients with the acquired immunodeficiency syndrome. *N Engl J Med* 326(23):1514–1521, 1992.
24. Fishl MA, Uttamchandani RB, Daikos GL, Poblete RB, Moreno JN, Reyes RR, Boota AM, Thompson LM, Cleary TJ, and Lai S. An outbreak of tuberculosis caused by multiple-drug-resistant tubercle bacilli among patients with HIV infection. *Ann Int Med* 117(3):177–183, 1992.
25. Pearson ML, Jereb JA, Frieden TR, Crawford JT, Davis BJ, Dooley SW, and Jarvis WR. Nosocomial transmission of multidrug-resistant *Mycobacterium tuberculosis. Ann Int Med* 117(3):191–196, 1992.
26. Public Law 91–596. 91st Congress, S.2193, December 29, 1970.
27. Air contaminants, *Code of Federal Regulations* Title 29, Pt. 1910.1000.
28. Bloodborne pathogens final rule, *Federal Register* 56:235 (December 6, 1991), pp. 64003–64182.
29. Occupational Safety and Health Administration Region 2. Enforcement Guidelines for Occupational Exposure to Tuberculosis, May 1992.
30. Access to employee exposure and medical records, *Code of Federal Regulations* Title 29, Pt. 1910.20.
31. Respiratory protection, *Code of Federal Regulations* Title 29, Pt. 1910.134.
32. TB control plans focus of settlements between OSHA, two Wisconsin hospitals. *Occupational Safety and Health Reporter*, 23(18), The Bureau of National Affairs, Inc., Washington, DC, September 29, 1993.
33. Centers for Disease Control. Prevention of TB Transmission in Hospitals. U.S. Department of Health and Human Services, Public Health Service, CDC, DHHS Publ. No. (CDC) 82–8371, Atlanta GA, October 1982.

34. Centers for Disease Control. Updated Guidelines for Preventing Transmission of Tuberculosis in Health Care Settings with Special Emphasis on HIV Related Issues. Draft 5, January 15, 1993.
35. Centers for Disease Control. Prevention and control of tuberculosis in migrant farm workers. *MMWR* 41(RR-10), 1992.
36. Centers for Disease Control and Prevention. Draft guidelines for preventing the transmission of tuberculosis in health-care facilities, 2nd edn: Notice of Comment Period, *Federal Register* 58(195):52809–52854, October 12, 1993.
37. *Title 8, California Code of Regulations*. Div. 1, Ch. 4, Subchapter 7, Group 16 Control of Hazardous Substances. Article 109, Section 5197, Occupational Tuberculosis Control, Draft III (March, 1993).
38. Gershom RM, McArthur BR, Thompson E, and Grimes MJ. TB control in the hospital environment. *Healthcare Facilities Management Series*. American Society for Hospital Engineering of the American Hospital Association, Chicago, 1993.
39. American Society of Heating, Refrigerating and Air-Conditioning Engineers, Inc. *Heating, Ventilating, and Air-Conditioning Applications*. American Society of Heating, Refrigerating and Air-Conditioning Engineers, Inc., Atlanta, GA, 1991.
40. Smith D. *Mycobacterium tuberculosis* and tuberculosis. In: WK Joklik and DT Smith (eds.), *Zinsser Microbiology*, 15th edn. Ashton-Centary-Crofts, New York, 1972.
41. Breed RS, Murray EGD, and Smith NR. *Bergey's Manual of Determinative Bacteriology*, 7th edn. The Williams & Wilkins Company, Baltimore, ND, 1957.
42. Riley RL and O'Grady F. *Airborne Infection—Transmission and Control*. Macmillan, New York, 1961.
43. Loudon RG and Roberts RM. Droplet expulsion from the respiratory tract. *Am Rev Respir Dis* 95:435–442, 1967.
44. Wells WF. Aerodynamics of droplet nuclei. *Airborne Contagion and Air Hygiene*. Harvard University Press, Cambridge, MA, 1955.
45. Centers for Disease Control. National action plan to combat multidrug-resistant tuberculosis: Recommendations of the CDC TB Task Force. *MMWR* 41(RR-11), 1992.
46. Kent DC. Tuberculosis as a military epidemic disease and its control by the Navy Tuberculosis Control Program. *Dis Chest* 52:588–594, 1967.
47. Cole EC. Aerosol characterization. In *Workshop on Engineering Controls for Preventing Airborne Infections in Workers in Health Care and Related Facilities*, CDC, NIOSH, July 14, 1993.
48. Kent DC, Reid D, Sokolowski JW, and Houk VN. Tuberculin conversion: The iceberg of tuberculous pathogenesis. *Arch Environ Health* 14:580–584, 1967.
49. Centers for Disease Control, National Institute for Occupational Safety and Health. *NIOSH Recommended Guidelines for Personal Respiratory Protection of Workers in Health-Care Facilities Potentially Exposed to Tuberculosis*. U.S. Department of Health and Human Services, Public Health Service, CDC, NIOSH, Atlanta, GA, September 14, 1992.
50. Harris HW and McClement JH. Pulmonary tuberculosis. In: PD Hoeprich (ed.), *Infectious Diseases—A Modern Treatise of Infectious Processes*, 3rd edn. Harper & Row, Philadelphia, PA, 1983.
51. Woods JE, Braymen DT, Rasmussen RW, Reynolds GL, and Montag GM. Ventilation requirement in hospital operating rooms, Parts I and II. *ASHRAE Trans* 92(Part 2), 1986.
52. Grimsrud DT, Sherman MH, Janssen JE, Pearman AN, and Harrje DT. An intercomparison of tracer gases used for air infiltration measurements. *ASHRAE Trans* 86(Part 1), 258–267, 1980.
53. American Conference of Governmental Industrial Hygienists. *Industrial Ventilation—A Manual of Recommended Practice*, 21st edn. American Conference of Governmental Industrial Hygienists, Committee on Industrial Ventilation, Lansing, NE, 1992.
54. DallaValle JM. Studies in the design of local exhaust hoods. Doctoral thesis, Harvard University, Boston, MA, 1930.
55. Silverman L. Fundamental factors in the design of lateral exhaust hoods for industrial tanks. *J Ind Hyg Tox* 23(5):187–266, 1941.
56. Silverman L. Centerline velocity characteristics of round openings under suction. *J Ind Hyg Tox* 24(9):259–266, 1942.

57. Silverman L. Velocity characteristics of narrow exhaust slots. *J Ind Hyg Tox* 24(9):267–276, 1942.
58. Garrison RP. Nozzle performance and design for high velocity/low volume exhaust ventilation. Doctoral thesis, University of Michigan, Ann Anbor, MI, 1977.
59. Garrison RP. Centerline velocity gradients for plain and flanged local exhaust inlets. *Am Ind Hyg Assoc J* 42(10):739–746, 1981.
60. Garrison RP. Velocity calculation for local exhaust inlets-empirical design equations. *Am Ind Hyg Assoc J* 44(12):937–940, 1983.
61. Fletcher B. Centerline velocity characteristics of rectangular unflanged hoods and slots under suction. *Ann Occup Hyg* 20:141–146, 1977.
62. Fletcher B. Effect of flanges on the velocity in front of exhaust hoods. *Ann Occup Hyg* 21:265–269, 1978.
63. Fletcher B and Johnson AE. Velocity profiles around hoods and slots and the effects of the adjacent plane. *Ann Occup Hyg* 25:365–372, 1982.
64. *Procedural Standards for Certified Testing of Clean Rooms.* National Environmental Balancing Bureau, Vienna, VA, pp. 4.14, 1988.
65. Roach SA. On the role of turbulent diffusion in ventilation. *Ann Occup Hyg* 24:105–132, 1981.
66. Ellenbecker MJ, Gempel RF, and Burgess WA. Capture efficiency of local exhaust ventilation systems. *Am Ind Hyg Assoc J* 44(10):752–755, 1983.
67. NIOSH Research Report. *Ventilation Requirements for Grinding, Buffing, and Polishing Operations.* National Institute for Occupational Safety and Health, Pub. No.: (NIOSH) 75–107, 1974.
68. Jansson A. Capture efficiencies of local exhausts for hand grinding, drilling, and welding. *Staub-Reinhalt* 40:111–113, 1980.
69. Rake BW. Influence of crossdrafts on the performance of a biological safety cabinet. *Appl Environ Microbiol* 36(2):278–283, 1978.
70. Fuller FH and Etchells AW. The rating of laboratory hood performance. *ASHRAE J* 49–53, 1979.
71. Fletcher B and Johnson AE. The capture efficiency of local exhaust ventilation hoods and the role of capture velocity. In: HD Goodfellow (ed.), *Ventilation '85*. Elsevier Science, Amsterdam, the Netherlands,1986.
72. Flynn MR and Ellenbecker MJ. Capture efficiency of flanged circular exhaust hoods. *Ann Occup Hyg* 30(4):497–513, 1986.
73. Flynn MR and Ellenbecker MJ. The potential flow solution for air flow into a flanged circular hood. *Am Ind Hyg Assoc J* 46(6):318–322, 1985.
74. Conroy LM and Ellenbecker MJ. Capture efficiency of flanged slot hoods under the influence of a uniform crossdraft: Model development and validation. *Appl Ind Hyg* 4:135–142, 1989.
75. Conroy LM and Ellenbecker MJ. Capture efficiency of flanged slot hoods and area sources under the influence of a uniform crossdraft. In: JH Vincent (ed.), *Ventilation '88*. Pergamon Press, Oxford, England, 1989, pp. 41–46.
76. Conroy LM, Prodans RS, Fergon SM, Lachman M, Yu X, Franke JE, and Barbiaux M. Field study of vapor degreaser local exhaust hood performance. In: *Presented at Ventilation '91*, Cincinnati, OH, September 19, 1991.
77. Prodans RS, Conroy LM, Fergon SM, Lachman M, Franke JE, and Barbiaux M. Field study of local exhaust performance for hoods exhausting vapor degreasers. In: *Presented at the American Industrial Hygiene Conference*, Salt Lake City, UT, May 21, 1991.
78. Burgess WA, Ellenbecker MJ, and Treitinan RD. *Ventilation for Control of the Work Environment.* John Wiley & Sons, New York, 1989.
79. Burton DJ. *Industrial Ventilation Work Book*, 2nd ed. IVE Inc., Bountiful, UT, 1992.
80. American Society of Heating, Refrigerating and Air-Conditioning Engineers Inc. *Fundamentals* (Chapter 8). American Society of Heating, Refrigerating and Air-Conditioning Engineers Inc., Atlanta, GA, 1993.
81. Franck C. Eye symptoms and signs in buildings with indoor climate problems (office eye syndrome). *ACTA Ophthalmol* 64:306–311, 1986.

82. Wyon NM and Wyon DP. Measurement of acute response to draught in the eye. *ACTA Ophihalmol* 65:385–392, 1987.
83. Wadden RA and Scheff PA. *Indoor Air Pollution*. John Wiley & Sons, New York, 1983.
84. Riley RL, Wells WK, Mills CC, Nyka W, and McLean RL. Air hygiene in tuberculosis: quantitative studies of infectivity and control in a pilot ward. *Am Rev Tuberc* 75:420–431, 1957.
85. McLean RL. General discussion: the mechanism of spread of Asian influenza. *Am Rev Respir Dis* 83:36–38, 1961.
86. Wells WF, Wells MW, and Wilder TS. The environmental control of epidemic contagion. I. An epidemiologic study of radiant disinfection of air in day schools. *Am J Hyg* 35:97–121, 1942.
87. Wells WF and Holla WA. Ventilation in the flow of measles and chicken pox through a community. Progress report, January 1, 1946 to June 15, 1949. Airborne Infection Study, Westchester County Department of Health. *JAMA* 142:1337–1344, 1950.
88. Perkins JE, Bahlke AM, and Silverman HF. Effect of ultra-violet irradiation of classrooms on spread of measles in large rural central schools. *Am J Pub Health* 37:529–537, 1947.
89. Willmon TL, Hollaender A, and Langmuir AD. Studies of the control of acute respiratory diseases among naval recruits. I. A review of a four-year experience with ultraviolet irradiation and dust suppressive measures, 1943–1947. *Am J Hyg* 48:227–232, 1948.
90. Macher JM, Alevantis LE, Chang L-L, and Liu K-S. Effect of ultraviolet germicidal lamps on airborne microorganisms in an outpatient waiting room. *App Occup Environ Hyg* 7(8):505–513, 1992.
91. Centers for Disease Control, National Institute for Occupational Safety and Health. *Criteria for a Recommended Standard-Occupational Exposure to Ultraviolet Radiation*. U.S. Department of Health, Education, and Welfare, NIOSH (Publication No. (HSM) 73-110009), Washington, DC, 1973.
92. Urbach F. The biological effects of ultraviolet radiation (with emphasis on the skin). In: F Urbach (ed.), *Proceedings of the 1st International Conference Sponsored Jointly by the Skin and Cancer Hospital, Temple University Health Sciences Center and the International Society of Biometeorology*. Pergamon Press, Oxford, England, 1969.
93. World Health Organization, International Agency for Research on Cancer. *IARC Monographs on the Evaluation of Carcinogenic Risks to Humans: Solar and Ultraviolet Radiation*, Vol. 55. World Health Organization, International Agency for Research on Cancer, Lyon, France, 1992.
94. American Conference of Governmental Industrial Hygienists. *Threshold Limit Values for Chemical Substances and Physical Agents*. American Conference of Governmental Industrial Hygienists, Inc., Cincinnati, OH, 1993.
95. AIA Committee on Architecture for Health. *Guidelines for Construction and Equipment of Hospital and Medical Facilities*. American Institute of Architects Press, Washington, DC, 1993.

Portable HEPA Filtration for TB Isolation in Hospitals and Clinics

Byron S. Tepper

Introduction

High-efficiency particulate air (HEPA) filters, also referred to as ultra-high efficiency air filters or absolute filters, have been generally accepted as capable of removing viable and nonviable particulates from airstreams, producing ultraclean, microbiologically sterile air. HEPA filters were developed during the 1940s and 1950s by the U.S. Army Chemical Corps and Naval Research Laboratories and the Atomic Energy Commission to provide respiratory protection from biological warfare agents for military personnel and to contain radioactive dust and other airborne radioactive particulates generated in the developing nuclear materials industry. Since then, HEPA filters have been used in a large number of medical, research, electronic, pharmaceutical, and industrial applications where clean air is essential to the work or where emissions of toxic particulates or hazardous biologic agents must be controlled.

A knowledge of the construction, performance, and mechanics of HEPA filtration is essential to the successful application of this filter in contamination control. Familiarity with the mechanisms of filtration and the methods used to test their efficiency are also necessary to dispel the erroneous assumption that HEPA filters only remove airborne particles down to a diameter of 0.3 μm.

Mechanics of Filtration

HEPA filters are composed of continuous sheets of borosilicate glass filter paper. Although glass may provide the best mesh of fibers of circular cross-sections and a small diameter, cellulose, asbestos, and other mineral fibers have been used, and today, many HEPA filters use fibers of plastic materials. The mesh of fibers, often referred to as a "depth filter," offers little resistance to airflow since it consists principally of empty spaces. The fibers, however, present an enormous surface area of the fibers within the matrix for adsorption of particles by mechanisms which will be described below. HEPA filters are not sieve or screen type filters as many erroneously assume. Sieve filters contain pores or holes which are smaller than the smallest particle to be eliminated; the larger particles are entrapped in the pore, the smaller pass through the pore. Particles are almost entirely trapped on the surface of sieve filters. Resistance to airflow is very high, increasing as the pores become occluded until the pores are totally occluded. The sieve filter obviously has little application in the bulk cleaning of air.

By definition, a HEPA filter has an efficiency of 99.97% for particles 0.3 μm in diameter. The 0.3 μm particle was selected for filter challenge because theoretical studies[1] have shown that filtration efficiency should be at a minimum for that size particle and that efficiency increases for particles smaller or larger than 0.3 μm (Figure 11.7). Concern has often been expressed that viruses, which are much smaller than 0.3 μm, will readily pass through a HEPA filter. Figure 11.7 indicates that the filtration efficiency should be greater than 0.03% penetration (99.97% efficiency) for the test particles; actual tests with Tl coliphage, which has a diameter of 0.1 μm, have recorded penetrations in the order of 0.0002%–0.002% (99.9998%–99.998% efficiency). It is unlikely that aerosols are generated that are pure monodispersates of viruses. In nature, viruses are usually associated with body fluids (saliva, sputum, blood, urine) which, when dried,

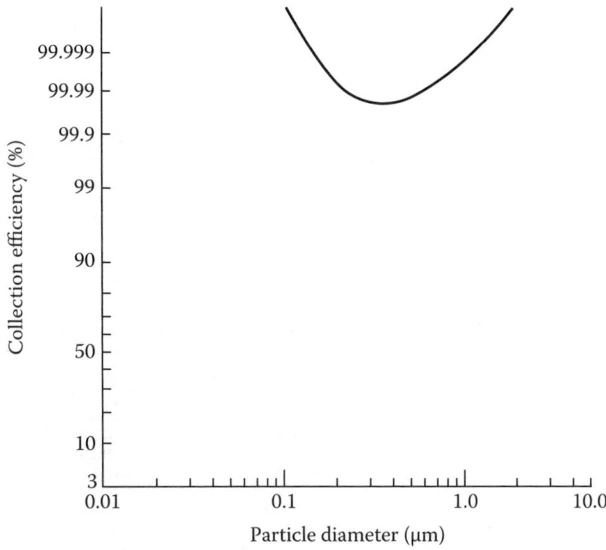

FIGURE 11.7 Theoretical HEPA filter collection efficiency. (From National Cancer Institute. *A Workshop for Certification of Biological Safety Cabinets*. Office of Biohazards and Environmental Control, Publication No. BH 74-01-11, Rockville Bio-Engineering Services, Dow Chemical U.S.A., 8–15, 1974.)

contribute to the size of the particle. Until there is evidence to the contrary, it is assumed that virus aerosols will produce droplet nuclei similar in size and nature to bacterial aerosols.

Airborne particles can be divided into two classes, those that are large enough to settle rapidly near the source and those in the form of droplet nuclei which remain suspended in air and behave as a gas. In HEPA filtration, the large particles may actually be sieved and remain loosely attached to the surface of the filter rather than firmly attached to the fibers. There are at least five mechanisms by which the small particles are collected on a single fiber[3]: (a) inertial impaction, (b) diffusion, (c) direct interception, (d) sedimentation, and (e) electrostatic attraction.

The air flowing to and around a single fiber is shown in Figure 11.8a. Note that the airstream bends as it goes past the fiber. Inertial impaction occurs when a relatively large particle (0.5 μm or above) carried in the air fails to follow the airstream around the fiber. Such particles, having a specific gravity greater than air, follow a trajectory in the direction of the obstruction and are impacted and adhere to the fiber by van der Waals forces. The role of inertial impaction in filtration increases markedly as the air velocity increases.

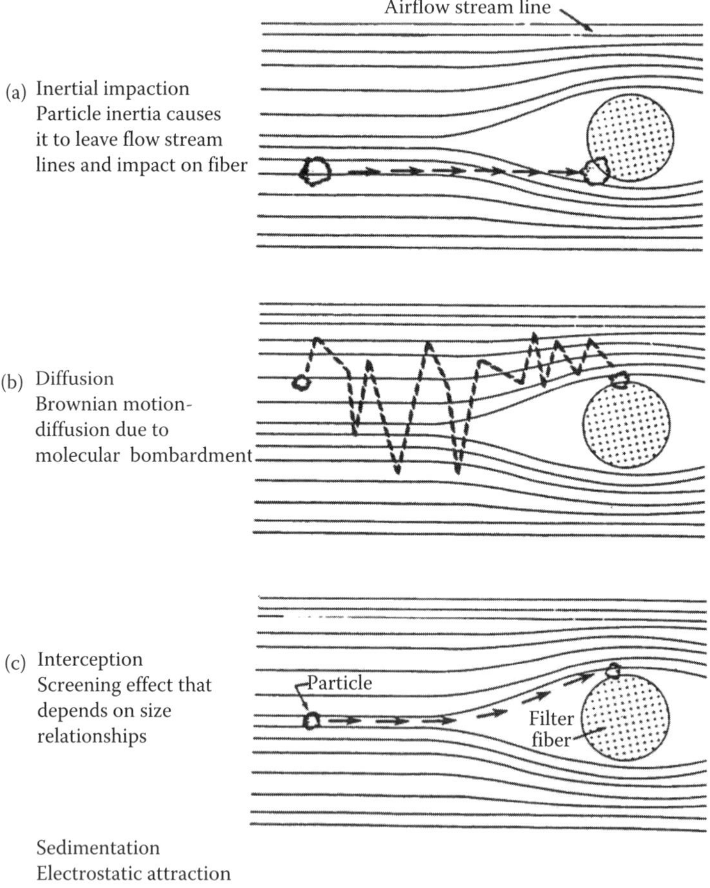

FIGURE 11.8 Air filtration theory particle collection mechanisms. (From National Cancer Institute. *A Workshop for Certification of Biological Safety Cabinets*. Office of Biohazards and Environmental Control, Publication No. BH 74-01-11, Rockville Bio-Engineering Services, Dow Chemical U.S.A., 8–15, 1974.)

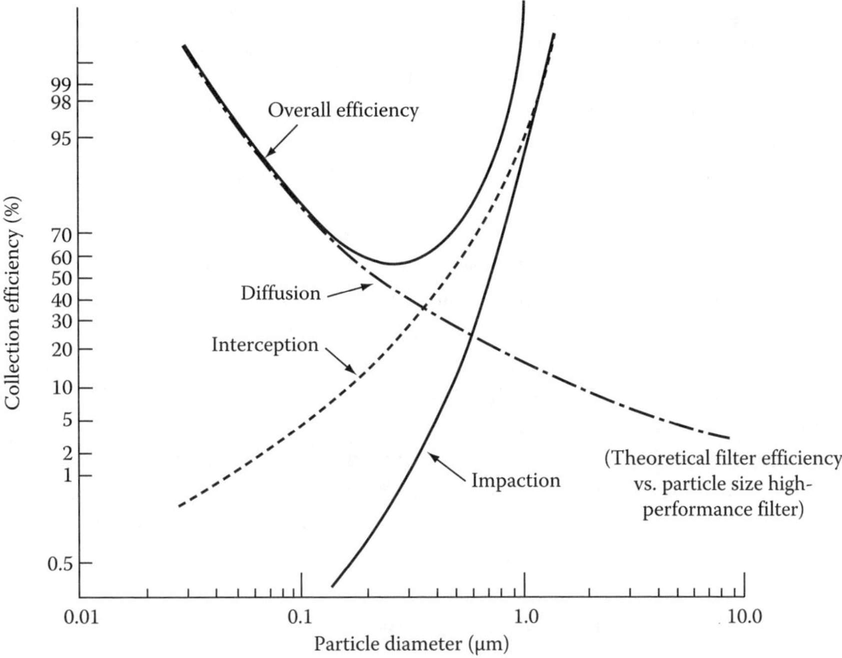

FIGURE 11.9 Relative effect of particle collection mechanisms.

Particles too small to be influenced by the inertial effect but small enough to exhibit Brownian motion may diffuse across airstream lines with "a high probability" of impaction on a filter fiber (Figure 11.8b). This diffusion mechanism should be most efficient for the removal of particles the size of nondispersed viruses. Increasing air velocity can smooth out Brownian motion and shorten the course of the particle through the filter bed, reducing the chance of collision.

Direct interception operates in the case of particles too large to exhibit Brownian motion and too small to be captured by the inertial effect (Figure 11.8c). Such particles tend to remain in the stream lines and are the particle size that most readily penetrates the filter bed. Test methods involving 0.2–0.3 μm particles rely on this principle and, hence, are a useful indication of the maximum penetration (lowest efficiency) to be expected for the working air velocity of a given filter.

Sedimentation in accordance with Stokes law is of relatively little importance under normal conditions as it will only apply to heavy particles. However, as flow decreases sedimentation may become increasingly important for particles as small as 0.5 μm in diameter. Electrostatic attraction is of relatively minor importance, since current filter materials fail to develop a significant charge. The role of electrostatic charge as newer plastic fibers are used in HEPA filters.

The contribution of each mechanism of filtration to the overall efficiency of HEPA filter is shown in Figure 11.9.

HEPA Filtration in Health Care Facilities

It is not surprising that techniques using HEPA-filtered, sterile air and laminar airflow were tested in the early 1960s for the control of airborne infections and for protective isolation of patients in the health care environment.

No area of a hospital requires more careful control of environmental contaminants than does the surgical suite. The laminar airflow concept developed for industrial clean room use has been successful in reducing airborne contaminants generated by the activities of the surgical team in the vicinity of the opening table. Laminar airflow in surgical operating rooms is defined as ultraclean (HEPA-filtered) airflow that is predominantly unidirectional, either vertical or horizontal, when not obstructed. The unidirectional laminar airflow pattern is commonly attained at a velocity of 90 ± 20 fpm. In a commonly used alternative to laminar airflow in the operating room, HEPA-filtered supply air is delivered from the ceiling, with downward movement to several exhaust inlets located low on opposite walls. This is probably the most effective air movement pattern for maintaining the concentration of contamination in the operating room at acceptable levels.

HEPA-filtered supply air systems have also been used and are recommended by ASHRAE[4] for rooms used for clinical treatment of patients with a high susceptibility to infection, such as leukemia, burns, bone marrow transplant, organ transplant or acquired immunodeficiency syndrome (AIDS). Laminar airflow systems have been used to protect patients during and after treatments which suppress immunity. Some physicians prefer to house patients in rooms with air changes of 15 ach (air changes per hour) or greater in which sterile air (HEPA-filtered), supplied by nonaspirating diffusers, is drawn over the patient and returned near the floor at the door to the room.

Stand-alone, bench-type laminar airflow units routinely provide protective sterile air environments in hospital pharmacies, tissue banks, and blood banks. Biological safety cabinets, which combine HEPA filtration and controlled airflow, have found a place in hospital laboratories not only to protect the laboratory workers from exposure to the potential pathogens in the diagnostic specimens they handle daily.

Respiratory Isolation for Tuberculosis (TB)

The resurgence of TB, compounded by nosocomial outbreaks of multiple drug-resistant tuberculosis (MDR-TB), has posed an immediate challenge to hospital engineers. Whereas hospital wound infection and, possibly, infection of patients undergoing chemotherapy may be the result of exposures to relatively large and heavy particulates averaging 13 μm, TB is transmitted primarily by airborne droplet nucleic 1–5 μm in diameter produced when infected individuals cough, sneeze, or speak. Such particles are respirable and are retained in deep pulmonary spaces. For TB, a single organism deposited in the lungs may be all that is needed to cause infection. There is no doubt that the risk of acquiring TB in a medical environment is almost exclusively a function of the concentration of the infectious particles in the air. Air control is of obvious importance for eliminating or reducing the airborne contaminants; this is accomplished by dilution ventilation and local exhaust ventilation.

In general ventilation systems rely on dilution ventilation to control airborne contaminants, the contaminated air is continually exhausted while room air is replaced and mixed with uncontaminated air. The resultant gradient reduction of airborne contaminants is related to the number of air changes per hour; the greater the number of air changes the more rapid the reduction in airborne contaminants. In dealing with particulate contaminants, i.e., TB droplet nuclei, the uncontaminated air can be fresh (outdoor) air or air that has been "cleaned" by passage through an appropriate filter.

Current CDC guidelines for the prevention of transmission of TB in health care facilities[5] recommend that any patient suspected or known to have infectious TB should be placed in TB isolation in a private room with a minimum of 6 ach. The recent draft revision of the guidelines[6] suggests that airflows greater than 6 ach, up to 37 ach, would be expected to result in a greater dilution of droplet nuclei. In addition to the high ventilation rates, other specifications for isolation include (a) maintenance of the room under negative pressure (airflow into the room) to prevent airborne contaminants from escaping the isolation room, (b) airflow patterns in the room designed to assure proper mixing to prevent stagnation or "short circuiting" of air, (c) air from the isolation room should be exhausted to the outside (single pass air) and dispersed so that it is not entrained in the building or neighboring building fresh air supplies; if reentrainment cannot be avoided the exhaust must be HEPA filtered before discharge, and (d) if the hospital building recirculates air, HEPA filters should be used to remove contaminants before the air is returned to the general ventilation system. Similar isolation ventilation is recommended for areas where high-risk patients are examined and treated. Such areas include triage rooms, waiting rooms, examining rooms in the emergency department, and ambulatory care areas including radiology suites. Isolation ventilation is most important where high-risk procedures, i.e., sputum induction, bronchoscopy, and pentamidine or other aerosol therapy are performed.

A recent evaluation of isolation facilities in seven hospitals in a midwestern metropolitan[7] area showed that (a) there were very few rooms designed to have suitable air change and negative pressure ventilation suitable for respiratory isolation, (b) only three hospitals had intensive care respiratory isolation rooms, (c) none had isolation rooms in the emergency department, and most important, and (d) only 55% of the isolation rooms demonstrated negative pressure airflow with the doors closed. The survey documented that not only was there an inadequate number of respiratory isolation rooms, but those that were designed for the purpose were poorly maintained. The term "adequate" is defined by the CDC[5] as "enough TB isolation rooms to appropriately isolate all patients with suspected or confirmed active TB. This number should be derived from the risk assessment of the health care facility." For some metropolitan hospitals with AIDS treatment programs this could mean ten or more TB isolation rooms; all acute care hospitals should have at least one. The cost of providing isolation rooms with single pass air exhausted directly to the outside may be prohibitive in this era of health care cost containment, especially since the risk/benefit analyses have not been done. Retrofitting of existing buildings could even be more difficult considering that most HVAC systems in older buildings are not adequately sized to provide and condition (heat, cool, dehumidify) the increased volumes of air required. Exhausting air from a room is only a minor engineering problem; providing the conditioned supply or make-up air, without disturbing the air balance in the rest of the facility, is a major problem.

Recirculation of HEPA-filtered air to other areas of the facility or recirculation of HEPA-filtered air back into the isolation room are safe alternatives which could conserve energy and increase the air changes of uncontaminated air without disturbing air balance. Fixed systems are not without problems. Filter housings must be located as close as possible to the isolation room to minimize the amount of potentially contaminated exhaust or return ductwork. They must be accessible for decontamination, maintenance, and filter changes. Fans may have to provide the static pressure required to overcome the resistance to airflow caused by the filter media; fixed fans are often noisy and may require sound suppression to avoid disturbing patients. Location of supply air diffusers and returns must be located to avoid dead air spaces and short circuiting to exhaust and return systems. Systems should be tested to assure that proper air mixing and dilution ventilation occur after installation and after furnishings which may obstruct airflow are provided.

Portable In-Room HEPA Filtration

Portable in-room HEPA filtration systems are available which can remove particulates and recirculate uncontaminated air into the isolation rooms. Portable filtration units are available that can recirculate 30,000–48,000 cfh, which can supplement the general ventilation air changes in a 1600 ft^3 isolation room by as much as 30 ach. Although not all portable filter units are quiet, at least one portable HEPA filtration system can deliver 800 cfm of filtered air at the extremely low noise level of 55 dBA. The portable HEPA filtration systems have an advantage over the fixed systems in that they can be removed from the patient room and moved to a remote location for maintenance, decontamination, and filter changes. Portable HEPA filtration systems are available which can quickly convert a standard patient room into a TB isolation room by increasing the number of air changes per hour and, by using inexpensive duct work connections, can create a negative pressure (inward airflow) condition by delivering a fraction of the HEPA-filtered air to the building exhaust system, to air recirculation return ducts, and/or out the room window.

The efficiency of portable HEPA filters to remove particulates under actual conditions of use can be evaluated and, if not acceptable, the filter unit can be readily located to optimize the dilution ventilation. Lastly, portable HEPA filtration units are inexpensive; multiple units can be acquired for the cost of designing and installing fixed filtration systems.

The efficiency of a portable HEPA filtration unit, designed for use at the Johns Hopkins Hospital, has been tested in a typical patient room. The room measured 11 ft W × 18 ft D × 8 ft H (1584 ft^3) and had a ventilation rate of 8 ach. The room was deliberately chosen because it was 20 cfm positive to the corridor. The filter unit, 3 ft W × 2 ft D × 6 ft H, had an inlet at floor level and the supply at the top; the unit was designed to deliver 550 cfm or 21 ach at a sound level of 55 dBA. The room was challenged with bis-(2-ethylhexyl) sebacate (dioctyl sebacate, DOS), average particle size 0.3 μm, at 250,000 particles per cubic foot. Particle counts were taken with and without the HEPA filtration. Figure 11.10 shows the decay curve, average of three runs, for the normal ventilation and the additional dilution ventilation provided by the HEPA filtration unit; the time to reduce particulate levels to ambient background was reduced from 27 to 7 min. Within the next 7 min, the ambient particle level was reduced by 60% and then a steady state was achieved between the dilution ventilation and the particles in the supply air. The two 7 min segments represent two aspects of contamination control. The first segment represents the burst contamination, the cough or sneeze that adds a relatively large number of droplet nuclei to the environment. The decay curve shows rapid clearance. The second segment models the continuous generation of infectious droplet nuclei from a TB patient. The cough frequency at the time of admission to the hospital may average as many as 15 coughs per hour before treatment; the decline in cough frequency during treatment is rapid with most patients, reducing their count to half the initial value within 2 weeks of treatment with antituberculous drugs.[8] The decay curves show that the counts are reduced but are not cleared; a steady state is achieved.

Figures 11.11 and 11.12 demonstrate the air curtain effect of the filtration unit placed between the patient bed and the room door. When particles of DOS are generated at the bed headboard with the filter running, only a small percentage, approximately 1%, escape to the door side of the filter; the unit appears to establish an anteroom or airback effect in the patient room even in a room positive to the corridor. The air curtain theoretically will capture any droplet nuclei that would, without the air curtain, be entrained in the wake following attending staff who are leaving the room. In standard isolation rooms,

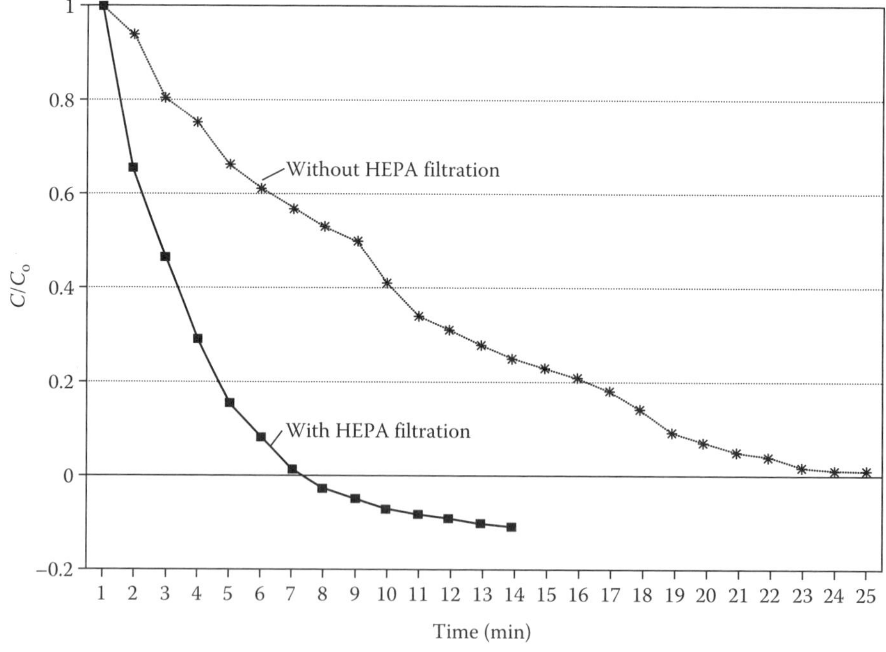

FIGURE 11.10 Clearance of particles generated in a patient room with the portable HEPA filtration unit off and with the filtration unit on.

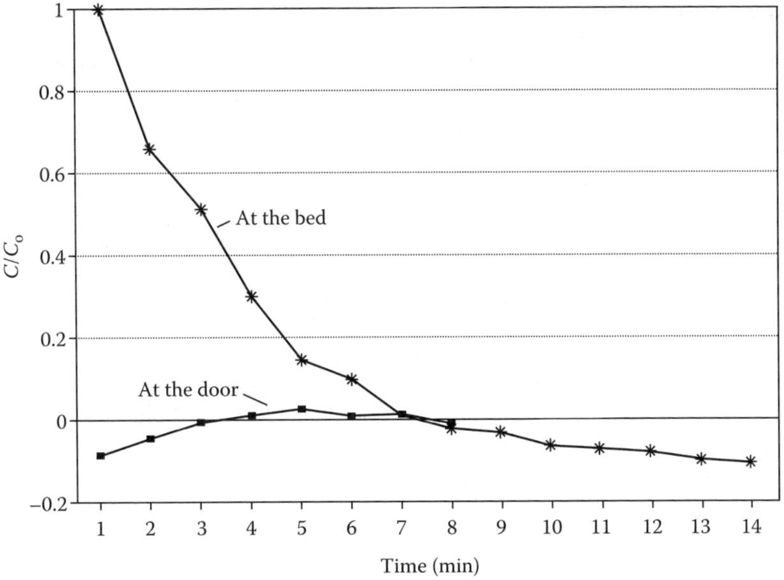

FIGURE 11.11 Clearance of particles generated in a patient room with the portable HEPA filtration unit on.

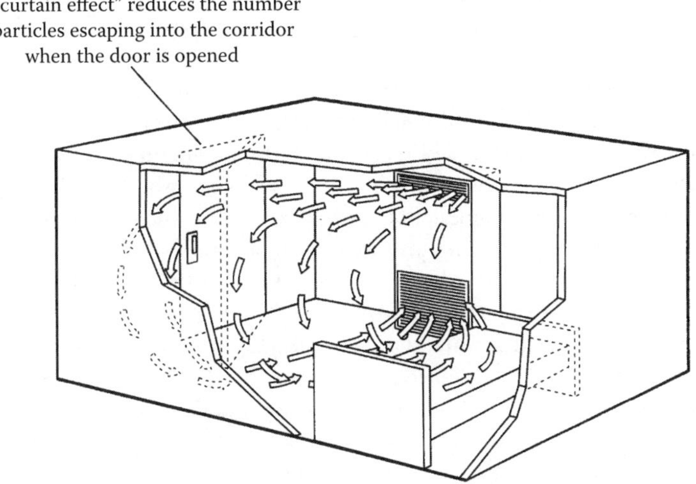

FIGURE 11.12 Placement of portable HEPA filtration unit in patient room which produces "air curtain effect." See Figure 11.11 for data.

only relatively small airflows are needed to control the direction of flow through the cracks around closed doors. The opening of a door instantaneously reduces any existing pressure between the separated areas to such a degree as to nullify the effectiveness of the pressure.[9] The air supply rates needed to control airflow direction through an open door are considerable and impractical. The air curtain produced by the portable HEPA filtration unit located near the door provides the added protection with the door open as well as closed, reducing transfer on entering or leaving the room.

The data in this model system show that HEPA filtration can rapidly reduce airborne contaminants, thus reducing the exposure of health care workers to infectious droplet nuclei.

Respiratory Protection

It should be apparent that dilution ventilation only reduces the risk of infection but never entirely eliminates it. The isolation room cannot protect the health care worker from infectious droplet nuclei generated while they are attending the patient. Respiratory protection is obviously required for all health care workers and visitors entering rooms where patients with known or suspected infectious TB are isolated. The type of respirator to be worn is controversial. Without entering and documenting the arguments, it appears that the minimal protection required will be a NIOSH-approved HEPA particulate air respirator. NIOSH-approved respirators providing greater protection will also be acceptable. This is the standard outlined in the October 8, 1993, OSHA directive on "Enforcement Policy and Procedures for Occupational Exposure to Tuberculosis."[10] Along with the requirement to use the HEPA respirators is the responsibility of health care facilities to be in full compliance with the OSHA standard 29 CFR 1910.134: Respiratory Protection, which includes medical examinations, training, and qualitative or quantitative fit testing of all employees required to wear respiratory protection.

Source Control

The most efficient control of airborne contaminants is source control, the capture of the contaminant at its source and its removal without exposing persons in the area.

Historically, studies on the transmission of TB have shown that providing patients with tissues to cover their mouths and noses when coughing or sneezing significantly reduces the number of droplet nuclei released to the environment. Similarly, providing patients surgical masks to be worn in common areas and during transport has been shown to reduce infectious particles. Since both of these controls provide only partial protection and since they require the cooperation of the patient, they must be supplemented with other control methods.

Local exhaust is a source-control technique which removes airborne contaminants at or near the patient who is the source of infectious droplet nuclei. This technique is especially important during the performance of medical procedures likely to generate aerosols containing infectious particles or during procedures likely to induce coughing or sneezing. These high-risk procedures include endotracheal auctioning, bronchoscopy, and sputum induction; aerosol treatments such as pentamidine therapy; cough-inducing; and other aerosol-generating procedures. Hoods, booths, or tents provided with exhaust systems can be used to capture and exhaust the droplet nuclei directly to the outside of the building. If reentrainment is possible or when the exhaust cannot be directed to the outside, the exhaust containing the droplet nuclei and the medication should be discharged only after HEPA filtration.

At Johns Hopkins, the specially designed free-standing hood (reverse flow clean air bench) shown in Figure 11.13 has been used since 1974 in sputum induction to protect health care workers from exposure to potentially infectious aerosols generated during the procedure. More recently, the same hoods have been used for the control of health care worker exposure to fugitive pentamidine during aerosol administration.[11] Sampling of the area and in the breathing zone of health care personnel as positive for pentamidine during administration of the drug without the engineering control; samples were all below the analytical limits of detection when the engineering control was used with the hood velocity at 100 LFM at the exterior edge (Table 11.5). The exhaust from the hood which had its own HEPA filter was captured by a thimble connection and exhausted to the outside. The use of the thimble exhaust connection provided directional airflow from the treatment room door to the hood and then to the outside. The data show the effectiveness of the hood in capturing fugitive pentamidine and provide indirect evidence that the hood had been providing protection from infectious agents, including TB, since its use was initiated.

Booths and tents which totally enclose patients have been used during sputum induction and pentamidine administration procedures to capture droplet nuclei from infectious patients. These devices, usually connected to local exhaust, undoubtedly are effective; however, the effectiveness of such units has not been adequately evaluated.

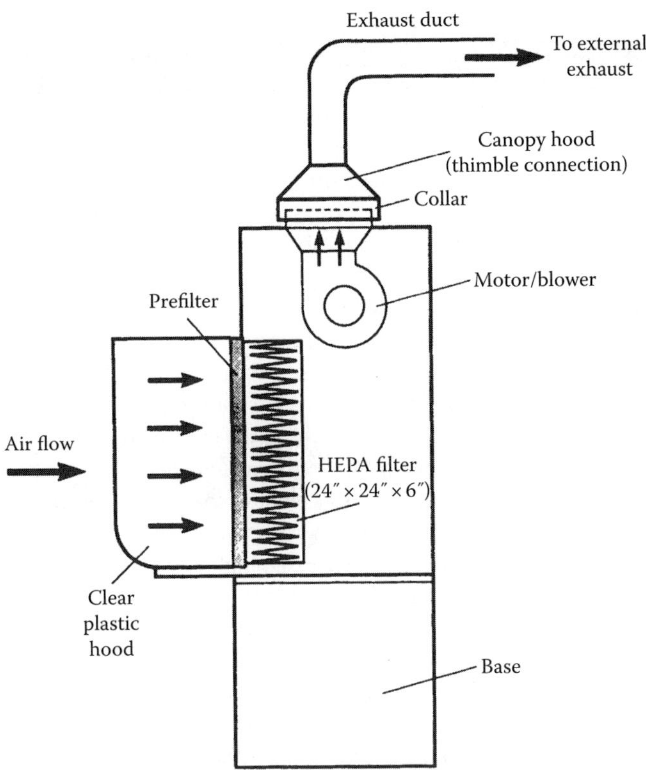

FIGURE 11.13 Design of a reversed flow, horizontal flow containment hood. The diagram shows the location of filters and the direction of airflow. (From McDiarmid, M.A., Schaefer, J., Richard, C.L., Chaisson, R.E., and Tepper, B.S., *Chest*, 102, 1764, 1992.)

TABLE 11.5 Air Sampling Results for Pentamidine

	Equipment Controls	
Sample Type	Without Hood (No. of Samples >LD[a])	With Hood (No. of Samples >LD)
Area	11/13 (85%)	0/7 (0%)
Personal	5/8 (63%)	0/7 (0%)

Source: Adapted from McDiarmid, M.A., Schaefer, J., Richard, C.L., Chaisson, R.E., and Tepper, B.S., *Chest*, 102, 1764, 1992.

[a] LD = limit of detection ~ 0.00033 mg/m^3.

Discussion

Since 1988, outbreaks of TB, including MDR-TB involving over 200 patients, have been reported to the CDC. In outbreaks in seven hospitals and one prison at least 16 health care workers have developed MDR-TB infection; at least five health care workers have died. Nationwide, 700 health care employees have become infected after workplace exposure and have required medical treatment. These events have sensitized the health care industry to the resurgence of TB. In 1990, in response to concerns about TB transmission, the CDC issued guidelines for preventing the transmission of TB in health care settings. The engineering controls, specifically the dilution ventilation and "negative pressure" isolation rooms specified in the guidelines have presented an engineering and financial challenge to many hospital administrations.

As recently as 1992, surveys have shown that 27% of acute care hospitals do not have isolation rooms that were designed to meet the CDC specifications, and the evidence suggests that if all the existing isolation rooms were tested, most would not meet their design specifications. It can be assumed further that the number and location of existing isolation rooms would not be appropriate or adequate for the preventive measures recommended in the CDC guidelines.

A survey of our isolation rooms showed results similar to the Fraser et al.[7] study. The isolation rooms in our newer buildings were designed for protective isolation (positive pressure). Over the years clinical services have been relocated so that many of the isolation facilities with anterooms and directional airflow were not in hospital spaces which patients with TB could occupy. Most of the isolation rooms in areas admitting patients at high risk for TB did not achieve the optimal combination of directional airflow, air changes, and external exhaust. In addition, there were no isolation rooms in the ER, ICU, radiology, or other areas at high risk for nosocomial transmission of TB. The renovation of existing isolation rooms and the addition of new rooms in appropriate locations was given high priority; however, design and construction will be extensive because of the age of existing buildings and inadequate HVAC systems. Adding isolation rooms to the already limited space in intensive care and emergency rooms has challenged our engineers and facility planners. It was obvious that we needed an interim environmental control strategy while optimizing our isolation room capacity.

The interim solution was to use portable, recirculating, high flow rate HEPA filter units in the existing isolation rooms and in patient rooms used for isolation. Operationally, each patient with possible pulmonary TB was placed in AFB isolation, which carries the requirement that a HEPA filtration unit be placed in the isolation room. The filters were placed in the rooms as shown in Figure 11.12. The requirement was also established that all staff wear a high-filtration (dust/mist) mask when entering the rooms of patients on AFB isolation.

In 1 year, 284 patients were placed on AFB isolation. Of these, 10 patients were culture positive for *Mycobacterium tuberculosis;* 4 of the 10 were AFB smear positive and presumed to have transmissible TB. To date, we have not observed PPD conversions in any staff attending these patients.

In our hospital, the portable HEPA filtration units have also been used in bronchoscopy rooms, bronchoscopy recovery rooms, intensive care cubicles, renal dialysis cubicles, in the ER triage room, and in a special examination/treatment room in the ER. As described earlier, we have used a containment hood for sputum indication and pentamidine administration for several years without associated occupational TB infection. Probably the longest interim use of the portable HEPA filtration units will be in outpatient waiting rooms in high-risk areas such as the ER, AIDS clinics, and substance abuse clinics. These areas which usually are large, open, and poorly ventilated have been supplied with multiple portable recirculating HEPA filtration units until such time as the areas are enclosed and the ventilation is upgraded.

Review of our data on AFB isolation reveals that over the course of the past year, there have been at least 5 days in which 10 or more patients have been on AFB isolation with the portable HEPA filtration units in place. The total number of AFB isolation rooms in areas which could be occupied by patients with suspected or confirmed TB has not yet been established, but undoubtedly there will be occasions where the number of patients with suspected TB exceeds the number of permanent isolation rooms. It appears that even after the upgrading and additions there still may be a niche for the portable HEPA filtration units.

References

1. Langmuir I. *Filtration of Aerosols and Development of Filter Materials.* O.S.R.D. Report 865. Office of Technical Services, Washington, DC, 1943.
2. National Cancer Institute. A Workshop for Certification of Biological Safety Cabinets. Office of Biohazards and Environmental Control, Publication No. BH 74-01-11, Rockville Bio-Engineering Services, Dow Chemical U.S.A., 1974, pp. 8–15.

3. Darlow HM. Air filters for recirculating systems: Minimum efficiency for bacteria and viruses. In: JF Ph Hers and KC Winkler (eds.), *Airborne Transmission and Airborne Infection*. Oosthoek Publishing Co., Utrecht, the Netherlands, 1973, pp. 516–519.
4. ASHRAE. Health facilities. In: *HVAC Applications* (Chapter 7). ASHRAE Handbook, Atlanta, GA. 1991.
5. Centers for Disease Control and Prevention. Guidelines for preventing the transmission of tuberculosis in health-care settings with special focus on FUV related issues. *MMWR* 39(RR-17):1–29, 1990.
6. Centers for Disease Control and Prevention. Draft guidelines for preventing the transmission of tuberculosis in health-care facilities, Second edition: Notice of Comment Period. *Federal Register* 58(195):52810–52854, 1993.
7. Fraser VJ, Johnson, Johnson K, Primack J, Jones M, Medoff G, and Dunagan WC. Evaluation of rooms with negative pressure ventilation used for respiratory isolation in seven Midwestern hospitals. *Infect Control Hasp Epidemiol* 14:623–628, 1993.
8. Loudon RG and Spohn SK. Cough frequency and infectivity in patients with pulmonary tuberculosis. *Am Rev Resp Dis* 99:109–111, 1969.
9. Caplan KL. Ventilation and air conditioning. In: RG Bond, GS Michaelsen, and RL DeRoos (eds.), *Environmental Health and Safety in Health-Care Facilities*. Macmillan Publishing Co., New York, 1973, pp. 66–121.
10. Occupational Safety and Health Administration. Memorandum for Regional Administrators: Enforcement Policy and Procedures for Occupational Exposure to Tuberculosis. October 8, 1993.
11. McDiarmid MA, Schaefer J, Richard CL, Chaisson RE, and Tepper BS. Efficacy of engineering controls in reducing occupational exposure to aerosolized pentamidine. *Chest* 102:1764–1766, 1992.

Preventing TB in the Workplace: What Did We Learn from HIV? Policies Regarding the HIV-Infected Health Care Worker

John Mehring

The HIV/AIDS epidemic brought about a revolutionary change in our practice of infection control in health care and other settings. With the adoption of universal precautions as a control methodology, the risk of exposure to infected blood and other body fluids or materials is acknowledged to exist potentially anywhere, rather than linked exclusively to a confirmed diagnosis.

However, we know that some tasks and procedures place workers at greater risk than others, and now efforts are being made to evaluate these tasks and procedures to modify them, to eliminate them, if feasible, or to institute safer devices or barriers with the goal of further decreasing the risk of exposure to potentially infectious materials.

The concept of applying universal precautions to patients and clients and other members of the public is on firm ground despite efforts to implement mandatory screening programs to determine HIV infection in individual patients and clients.

However, the concept of applying universal precautions to workers is on less firm ground and efforts have been made by various sectors of our society, including the Centers for Disease Control and Prevention (CDC), to discuss and implement mandatory, or less coercive, screening programs to determine HIV infection in individual health care workers and restrict infected health care workers' job duties and responsibilities.

Organizations in the field, such as labor unions, which have argued for increased infection control vis-à-vis patients and clients, agree that there is a concomitant need for increased infection control vis-à-vis workers, although there is significantly less risk of bloodborne infection from workers as opposed to patients and clients. We believe, however, that the same principles that apply in decreasing the risk to workers should apply to decreasing the risk to patients and clients: that is, new, more aggressive infection control methods must be researched and implemented in order to make the workplace safer for all parties. Again, these methods include evaluating the tasks and procedures which place these parties at greater risk of infection, and modifying them, or eliminating them, if feasible, or instituting safer devices or barriers with the goal of decreasing the risk of exposure to all potentially infectious materials.

While there is general support for a more aggressive approach to infection control to provide protection against bloodborne pathogens, this support has not been implemented in concrete ways. All three relevant federal government agencies, the CDC, the Food and Drug Administration (FDA), and the Occupational Safety and Health Administration (OSHA) have been extremely conservative in recommending, supporting, or requiring infection control efforts beyond elementary universal precautions. The CDC, which has been preoccupied over the past few years regarding the risk of infection posed by invasive procedures, has never recommended, ironically, the use of new procedures or devices to reduce this risk.

A dichotomy has thus developed: there appears to be a pronounced lack of will in these federal agencies and other quarters to go beyond what has already been achieved, while parties on the front lines shift attention away from an individual's infection status, and try to refocus attention on a more advanced program of infection control that protects society across-the-board.

Regardless of our society's inability to move more quickly to a more effective infection control program, the fact remains that our progress in this area has been substantial. Our mileposts include adoption of universal precautions as a recommendation by the CDC in 1987, and the promulgation of the Bloodborne Pathogens Standard by federal OSHA in 1992. These actions by government have not only prevented exposure to potentially infected body fluids more effectively than previous infection control programs, but they also allowed health care workers and health care patients and clients to resist successfully repeated attempts to discriminate against them. In addition, the passage and implementation of the Americans with Disabilities Act of 1990 provided solid legal ground to protect HIV-infected health care workers and health care patients and clients from discrimination. This is a model that must be carried over and applied to the new struggle to control and prevent tuberculosis (TB) infection and disease, and protect HIV-infected health care workers and health care patients and clients from discrimination.

For a decade, HIV-infected health care workers were the subject of concerted efforts to remove them from the workplace to protect patients and clients from exposure to HIV infection. Now HIV-infected health care workers are being pressured to leave the health care workplace because of the risk they face from TB infection.

HIV-infected health care workers, and other immunocompromised workers, are at greater risk for acquiring TB infection when coming into contact with the airborne bacillus and developing active TB once they are infected. This risk is exacerbated by the emergence of multidrug resistant TB, which has an extremely poor prognosis. HIV-infected health care workers are being counseled by some individuals and organizations to leave the health care workplace in order to decrease or eliminate this risk of TB infection and disease.

Labor unions which have grappled with this issue have decided against the medical removal of a class of individuals who can be medically protected in the workplace if an effective infection control program to prevent the transmission of airborne TB is instituted. Of course, we know these programs do not exist in most workplaces. Institutions have been slow to adopt the CDC's recommendations. Many employers are reluctant to commit financial resources to TB prevention or are opposed philosophically to an industrial hygiene approach to TB prevention. Federal OSHA has dragged its feet in moving toward an enforceable Prevention of Occupational Tuberculosis Standard. In the meantime, workers are fending for themselves. In this vacuum, confusion and inaction predominate, allowing advocates of the medical removal of HIV-infected health care workers to appear sensible and humanitarian.

What can HIV-infected health care workers do to protect their health and their employment rights? HIV-infected health care workers should assess their risk for TB infection, and they must be allowed, in principle and in practice, to reduce the risk of TB infection by asking for, and receiving, "reasonable accommodation," a right given them by the Americans with Disabilities Act of 1990. Nevertheless, most HIV-infected health care workers will remain on the job where they are currently, because that is what they want to do, or because they are ignorant of their infection status. That is, and should remain, their right. And just as we know that we could test everyone for HIV infection and still not know accurately what everyone's infection status is, we also know that the occupational transmission of TB will continue even when pressure is exerted to make the health care workforce "HIV-free." In terms of infection control, a medical removal program for HIV-infected workers is as productive an outcome as a mandatory HIV screening program.

If employers are committed to a safe work environment, they will give their employees the education and training to make their individual risk assessment, and they will provide an occupational environment that protects all workers and the public, across-the-board. The general application of industrial hygiene principles will protect these populations at risk for TB infection, just as the specific application of another industrial hygiene principle—universal precautions—lays the foundation for protecting these populations at risk for bloodborne infections.

For HIV-infected workers this means getting critical information and services in a way that guarantees their confidentiality. It is very important that immunocompromised workers screen themselves carefully for TB infection and disease. Workers who may be immunocompromised include those who are HIV-infected, as well as those with leukemia, lymphoma or those who are using various drug or radiation treatments. Accurate screening for TB in immunocompromised persons is difficult to accomplish because the standard TB test, the purified protein derivative (PPD) skin test, requires a healthy immune system to work correctly. A compromised immune system may not be able to mount an immune response, which means that the person is anergic, and therefore the PPD test result may not be accurate. A negative PPD skin test may actually be a false negative. The person may be infected with TB, but the test may not reveal that.

HIV-infected health care workers should test themselves for anergy. Testing for anergy involves using at least two other skin test antigens such as mumps antigen, tetanus toxoid, and candida antigen. If a person does not react to these antigens, the person can be considered anergic. This person should be medically monitored for TB symptoms.

Practically speaking, an effective occupational TB prevention program can only be built when all workers are informed about the connection between immunocompromised status and the risk of TB infection, so that the target population of immunocompromised workers is reached successfully, and their privacy rights are respected. For example, employers should provide anergy testing in tandem with PPD testing or after a negative result has been obtained, to all employees who elect it, with no explanation required, as well as offering an outside referral.

All workers need to make informed decisions about their risk and be given the appropriate equipment to protect themselves. In the debate over appropriate respirator use, we must agree upon a minimum level of protection, while an additional level of protection must be provided to those workers who choose it, a choice which will be made based on workers' education and training and individual risk assessment.

HIV-infected health care workers will only be as protected to the degree all other health care workers are protected. We must learn the lessons from the HIV/AIDS epidemic and apply those lessons to the TB epidemic: (1) an exposure control program which casts a broad net of infection protection, incorporating an industrial hygiene hierarchy that goes beyond diagnosis, or infection status alone and (2) policies and procedures that protect every health care worker and patient from discrimination.

Organizations with TB recommendations specific to immunocompromised health care workers

1. American Association of Physicians for Human Rights, TB: Recommendations for the HIV-Infected Health Care Provider, 1992.
2. American Medical Association, Multiple-Drug Resistant TB: A Multifaceted Problem, 1992; Update on TB, 1993.

3. American Nurses Association, Position Statement on Tuberculosis and HIV, 1993.
4. Centers for Disease Control and Prevention, Guidelines for Preventing the Transmission of TB in Health-Care Facilities, 2nd edn. (Draft), 1993.
5. Labor Coalition to Fight TB in the Workplace: AFGE, AFSCME, AFT, District 1199, SEIU, Fed-OSHA Petition for TB Prevention Standard, 1993.
6. State of New York Department of Health, Recommendations for HIV-Infected Health-care Workers Regarding Tuberculosis, (Draft), 1994.

Organizations with no TB recommendations specific to immunocompromised health care workers

American Hospital Association
Association for Practitioners in Infection Control
Federal OSHA

Organization	Document	Year	Recommendations Specific to Immunocompromised Health Care Workers
American Association of Physicians for Human Rights	TB: Recommendations for the HIV-Infected Health Care Provider	1992	(1) Know HIV status, and receive regular medical care. (2) Perform PPD and anergy testing every 6 months. (3) Be well acquainted with signs and symptoms of TB. (4) HEPA respirators should be worn with known or suspected TB. (5) Must weigh personal risk of TB infection. A change in job setting or career may be appropriate. (6) HIV-infected providers who risk exposure to MDR-TB should strongly consider a change in job setting or career.
American Hospital Association	Tuberculosis Control in Hospitals	1992	None
American Medical Association	Multiple-Drug Resistant TB: A Multifaceted Problem	1992	(1) HIV-infected health care workers should be carefully apprised of their risk of clinical TB. Risks and benefits of caring for persons with active or suspected TB should be carefully considered.
	Update on TB	1993	(1) Powered air-purification respirators (PAPRs) may be useful for the protection of immunocompromised health care workers who care for patients with infectious TB.
American Nurses Association	Position Statement on Tuberculosis and HIV	1993	(1) HIV-positive nurse to know their TB status. (2) Self-limit their nursing practice based on a case-by-case assessment of their TB status. (3) Self-restrict their contact with patients, co-workers, and visitors if symptoms associated with TB are present. (4) Adhere to prescribed medication regime for TB to decrease the opportunity for transmission of the disease.
Association for Practitioners in Infection Control	Position Statement on TB Prevention and Control	1992	None

(continued)

Organization	Document	Year	Recommendations Specific to Immunocompromised Health Care Workers
Centers for Disease Control and Prevention	Guidelines for Preventing the Transmission of Tuberculosis in Health-Care Facilities, 2nd Edition (Draft)	1993	(1) All health care workers should know if they have a medical condition or are receiving a medical treatment that may lead to severely impaired cell-mediated immunity. (2) All health care workers should be counseled about potential risks, in severely immunocompromised persons, associated with taking care of persons with some infectious diseases, including TB. (3) Severely immunosuppressed health care workers should avoid exposure to *M. tuberculosis*. Health care workers with severely impaired cell-mediated immunity (due to HIV infection or other causes) who may be exposed to *M. tuberculosis* should consider a change in job setting. (4) Employers should make reasonable attempts to offer alternative job assignments to an employee with a documented condition compromising cell-mediated immunity who works in a high-risk setting for TB. The facility should offer, but not compel, a work setting which the health care worker would have the lowest possible risk of occupational exposure to *M. tuberculosis*. (5) All health care workers should be informed that immunosuppressed health care workers need to have appropriate follow up and screening for infectious diseases, including TB. Health care workers who are known to be HIV-infected or otherwise severely immunosuppressed should be tested for cutaneous anergy at the time of PPD testing. Consideration should be given to retesting immunocompromised health care workers with PPD and anergy tests at least every 6 months because of the high risk of rapid progression to active TB should infection occur. (6) Information provided by health care workers regarding their immune status should be treated confidentially.
Federal OSHA	Enforcement Policy and Procedures for Occupational Exposure to Tuberculosis	1993	None
Labor Coalition to Fight TB in the Workplace: AFGE, AFSCME, AFT, District 1199, SEIU	Fed-OSHA Petition for TB Prevention Standard	1993	(1) Prior to PPD test, all employees will be counseled regarding increased risk if immunocompromised. Upon the employee's request, the employees will be tested with at least two delayed-type skin-test antigens, in addition to PPD. Those employees found to be anergic will receive a follow-up evaluation by an appropriate health care professional and monitored for development of TB-related symptoms. The PPD and anergy tests will be offered at least every 6 months for immunocompromised workers.

(continued)

(continued)

Organization	Document	Year	Recommendations Specific to Immunocompromised Health Care Workers
			(2) The employer must provide a confidential referral at the request of immunocompromised health care workers to an employee on an individual basis regarding his/her risk of TB. (3) Employers must inform all health care workers with severely cell-mediated immunity who may be exposed to *M. tuberculosis* about their rights to reasonable accommodations under the Americans with Disabilities Act (ADA), including voluntary transfers to work areas and activities in which there is the lowest possible risk of *M. tuberculosis*. (4) Employer must treat confidentially all information provided by a health care worker regarding their immune status. The employer must have written procedures on confidential handling of such information.
State of New York Department of Health	Recommendations for HIV-Infected Health Care Workers Regarding TB (Draft)	1994	(1) Employers should educate all workers at risk for occupational exposure to TB about the risks associated with TB and the precautions they should take to avoid exposure. That TB presents an especially serious risk to people who are immunocompromised for any reason, including HIV infection, must be a part of the education message. Discussions about relative risk of continued employment in a specific setting should be included. (2) All employees at risk for occupational exposure should be routinely assessed for TB infection. Immunosuppressed individuals with TB infection will not develop a positive PPD when cutaneous anergy is present. If there is a high likelihood that an anergic worker has been exposed to TB, then the worker and his or her health care provider should consider further diagnostic evaluations (such as chest x-ray) and possible chemoprophylaxis. (3) Employers should not mandate reassignment of immunosuppressed workers for worksites with a high prevalence of TB solely because of the risks to the worker. If an immunosuppressed worker requests reassignment or modification of duties to reduce risk, the employer should consider the situation carefully, and accommodate the worker's request whenever possible. (4) Employers must ensure that workers with infectious TB be removed from the workplace. (5) All individuals who may be exposed to HIV through personal behavior, blood products, or occupational exposure should be counseled to seek HIV testing so that they may benefit from medical management. Recent outbreaks of TB, including drug-resistant strains, make the reasons for knowing one's HIV status even more compelling.

(continued)

Organization	Document	Year	Recommendations Specific to Immunocompromised Health Care Workers
			(6) Individuals who are infected with TB should be encouraged by their health care providers to undergo HIV testing...since their HIV status may influence the value and interpretation of the tuberculin skin (PPD) test as well as recommendations for preventive therapy and treatment of TB disease.
			(7) Individuals with HIV who are at risk for TB infection should be under the care of a physician. Decisions regarding diagnosis, prophylaxis, and treatment should be made by the individual and his or her physician.

Reducing the Spread of Tuberculosis in Your Workplace

Derrick Hodge and Daniel Kass

Introduction

In the last few years, rates of tuberculosis (TB) infection have climbed. In New York City, poverty, homelessness, conditions in homeless shelters, high rates of HIV and AIDS, and overcrowded jails and prisons have all contributed to the epidemic. Some get TB because they have no home and are forced to spend night after night in crowded shelters. Others are sick with AIDS or weakened by HIV, making them more susceptible to active TB. Even housed and healthy people have become infected with TB from continued contact with others who are sick. Each of these people has something in common—they got TB after breathing in the air coughed out by someone who was sick with TB. If your work puts you in contact with people who may have active TB, then you might also be at risk for infection.

When people with active TB cough or sneeze, they may release small particles into the air called "droplet nuclei." People with active TB are infectious to others. People with TB infection, or those who do not have symptoms, are not infectious to others. Indoors, incoming sunlight, and fresh air reduce the amount of particles that remain in the air and infect others. Since you usually cannot control the amount of sunlight at work, you need to focus on making sure there is a lot of fresh air to get rid of the particles in the air.

This fact sheet will help you to

- Look at your workplace for problems with the ventilation and quality of air which increase the risk of TB transmission.
- Try some low-cost, easy-to-do solutions to improve the problems you identify in your survey.

This fact sheet recommends low- and no-cost changes to your workplace. Alone, these cannot eliminate the risk of TB transmission. To significantly reduce this risk, a combination of approaches is needed, which may include training your staff about TB prevention and control, improving or adding ventilation systems, installing ultraviolet lights, and selecting the proper personal protective equipment, such as respirators. Most important, your agency needs to identify those with active TB promptly and help them get treatment.

The first part of this fact sheet tells you how to evaluate the ventilation and flow of air in your workplace.

Checking Your Building's Ventilation

What Is Ventilation?

The purpose of a ventilation system is to move stale or contaminated air out of a space, and bring in fresh air. When the air is moved by a motor and fan through vents and ducts, it is called forced air ventilation. Even without forced air, air will naturally move in and out of doorways, windows, and other openings. This is called natural ventilation (Figure 11.14). Natural ventilation can be improved by using portable fans and air conditioners to help move the air and add fresh air. This fact sheet will help you evaluate natural ventilation in your building and improve it, if necessary.

Good air quality depends on good ventilation. When there is close and continuous contact with a person who has active TB, poor ventilation can lead to TB transmission if

1. There is little or no fresh air moving into the building.
2. Contaminated air is removed very slowly or not at all.

To prevent TB transmission, and for general comfort, your natural ventilation must bring in an adequate amount of fresh air. The best way to do this, of course, is to leave an outdoor window open wide. If your office does not have a window, try to increase the amount of fresh, outside air moving through your work space.

Air Quality Survey

It is difficult to test physically the amount of fresh air moving into a space from natural ventilation. One way to check is to see if you are experiencing any symptoms of working in a stuffy environmental. Frequent headaches, drowsiness, and fatigue are common among occupants of stuffy environments. In these environments, colds and flu may spread usually quickly, widely, and linger longer. Allergies may develop or be aggravated, leading to frequent throat, eye, and sinus irritation. Though each of us experience these symptoms at one time or another, stuffy, unhealthy environments may cause them to occur more frequently.

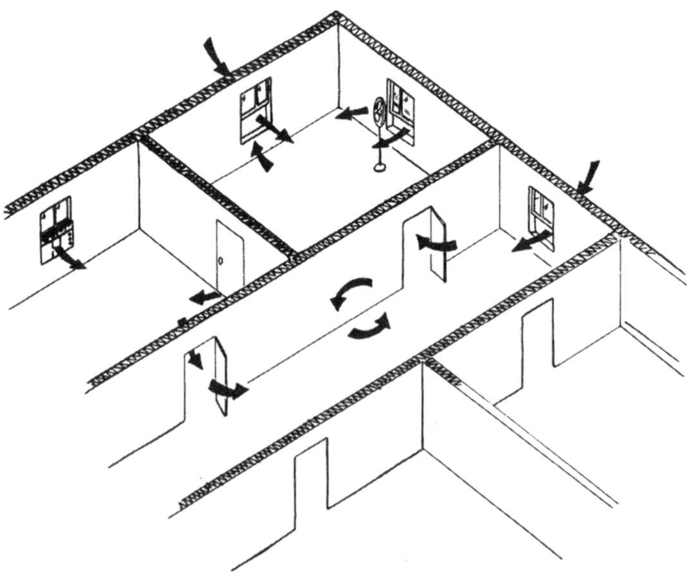

FIGURE 11.14 Air naturally moves through windows, doors, and other openings. Some rooms also have window air conditioners and/or desktop, window, or floor-standing fans.

While stuffy air alone does not mean that there is a risk of TB transmission or infection, it may indicate that air movement needs to be improved.

Another way to assess and improve the air quality in your building is to look at how portable fans and air conditioners are used in your area. The following sections will help to determine the amount of fresh air and the proper movement of air in your environment.

Answer the following questions by checking Yes, No, or Not Applicable (N/A).

1. If there are windows in your office area, are they easy to open and close?
   ☐ Yes ☐ No ☐ N/A
2. Do odors or smells go away fairly quickly?
   ☐ Yes ☐ No ☐ N/A
3. If you have floor fans, window fans or tabletop fans installed are they clean and do all of their features work (such as speeds and position adjustments)?
   ☐ Yes ☐ No ☐ N/A
4. If there is a window air conditioner, is its filter clean?
   ☐ Yes ☐ No ☐ N/A
5. If there is a window air conditioner, is the fresh air knob or vent open and working?
   ☐ Yes ☐ No ☐ N/A
6. How does the air move naturally in the area? To find out, you need to buy several incense sticks, and draw a simple floor-plan of the work area in question. The floor plan should show the location of doors, windows and where people are located or seated in the area. The following test involves making smoke and looking at the direction it moves in your area. Since incense gives off an odor, the time it takes for the odor to travel and go away will help you figure out how the air is traveling throughout the building (see Figure 11.15).

Why is it useful to know the direction of air movement?

Knowing how the air flows in your work area can help you decide where to place fans, how to position desks, and how to arrange seating. If someone is coughing, you want the air to carry the droplets from the cough away from other people, not toward them.

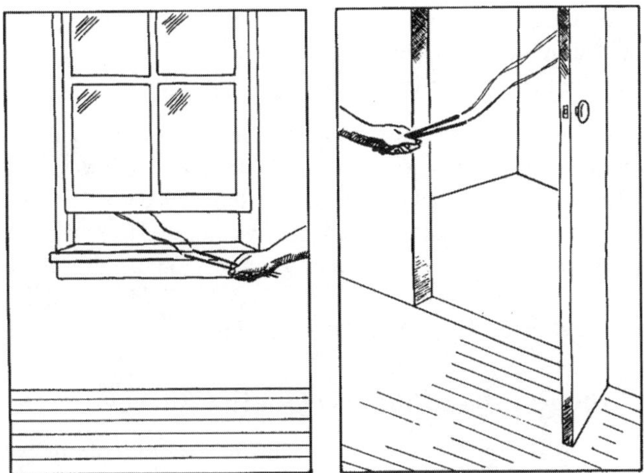

FIGURE 11.15 Light two incense sticks. After they start to burn, blow them out so that smoke rises. Hold them together horizontally near the base of each window and doorway in the room (if your window or doorway is usually wide-open or closed, keep it that way). Look to see which direction the smoke is blowing. Ask people in adjacent rooms to tell you when they begin to smell the incense.

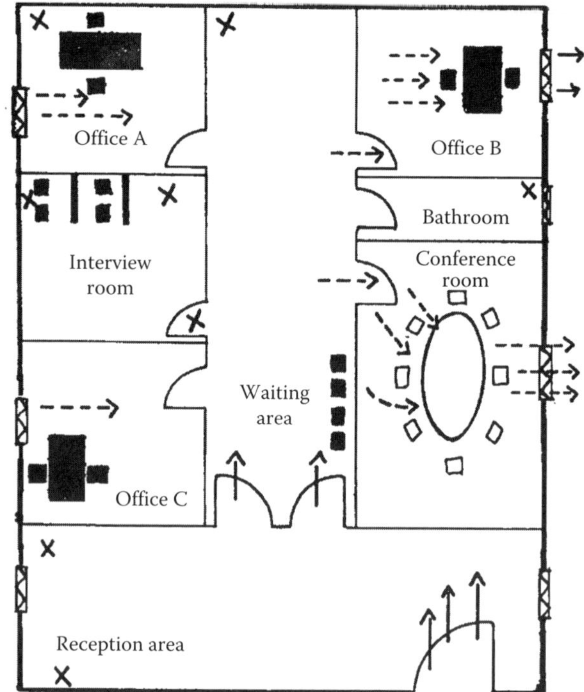

FIGURE 11.16 This completed floor plan shows airflow through doorways and windows, and across room and hallways. This facility has no forced air ventilation system and relies upon open windows and doorways for fresh air.

Knowing how the air is moving helps to

1. Find areas where there is little or no air movement. Knowing this can help you make decisions about where and how to place fans and air conditioners to create an airflow.
2. Position fans so that they are not blowing against the natural flow or air, but helping it. It is easier to add to the natural flow than to try to reverse it. When you do want to change the direction of the air, you will need more powerful fans.
3. Decide where to place furniture and people.

Using the floor plan, draw arrows indicating the direction smoke is blowing from windows, doorways, hallways, and in the middle of the room. You can also make a note of how strongly the air is flowing. (Use a single line for strong airflow, dotted line for light airflow. See Figure 11.16.) Adjacent areas where the incense smell travels first or is strongest probably receive more air from the testing area than others. Since windows usually stay open in summer and closed in winter, make sure you repeat this test in both seasons.

Figure 11.16 is one example of a completed airflow diagram.

How to Interpret the Results of the Air Quality Survey

If you and your coworkers answer no to one or more of the questions in the air quality survey, then the low-cost improvements on this page can help you find ways to improve air quality. If your airflow floor plan from question 6 shows that you have poor air movement or that the airflow could carry droplet nuclei from one person directly to another, then use the next section to help you to position people and fans to make improvements.

Low-Cost Ways to Improve Natural Ventilation to Reduce the Spread of TB

There are many ways to improve the ventilation of a building: an office can be remodeled, new openings can be cut into doors and walls, a forced air system can even be installed, but each of these strategies requires expert help and can be very expensive. Because of the cost, many building owners refuse to make these changes. Fortunately, there are some small, inexpensive improvements which will increase the amount of fresh air in the area where you work. This section offers some ideas about how to make these changes. Alone, these suggestions will not be enough to eliminate the risk of TB transmission, but they may help to reduce the risk.

Most of the suggestions below can be done by staff without the help of maintenance workers. You should consider all of the suggestions in the order they are listed. Together, they can help to improve air quality of your building.

Open Your Windows

Windows should be opened as much as possible. Opening a window is the simplest way to bring fresh air into a building. Some windows can be opened at the top and the bottom. If this is the case in your area, open both as much as possible. If the windows are hard to open past a certain point, use a lubricant so they can be opened more easily. For wooden windows, try rubbing candle wax where the window slides against the frame. Metal parts can be lubricated with household oil, WD-40, or similar products. If these suggestions do not work, ask your maintenance staff to make the repairs.

In winter months, keep your windows as open as possible. If you cannot get enough heat to keep the windows open, there are two things you can do.

Make Your Heating System Work Better

If radiators are your main source of heat, ask your maintenance workers to service them. They may need to have water or air removed. This will help them produce more heat.

Add Space Heaters

Check with your building maintenance staff first to make sure the electrical system can handle the added load otherwise, they could be fire hazards. Since space heaters dry the air, you might consider putting in a humidifier, or keeping a pan of water on top of a hot radiator. If you use a humidifier or pan of water, make sure you change the water frequently.

Install and Properly Position Fans

Fans can do three things:

1. Move the fresh air in.
2. Blow the contaminated air out of a work area.
3. Circulate the existing air to thin out the amount of TB droplets in an area.

All three, in combination, may reduce the risk of TB transmission. Without adding fresh air or removing contaminated air, a fan will only keep circuiting bad air and could do more harm than good.

There are three kinds of fans that can be added to your work area that can help improve the quality of air: window fans, floor-standing fans, and portable tabletop fans (see Figure 11.17).

Fans blow air in one direction. Do not use fans just to create wind. Without a supply of fresh or clean air, fans will just recirculate any TB droplets which are present, possibly doing more harm than good. This section of the fact sheet describes ways to control the proper movement of air to reduce exposure to TB. Which direction do you want the fans to blow in your own area? Here is what you can do to help decide.

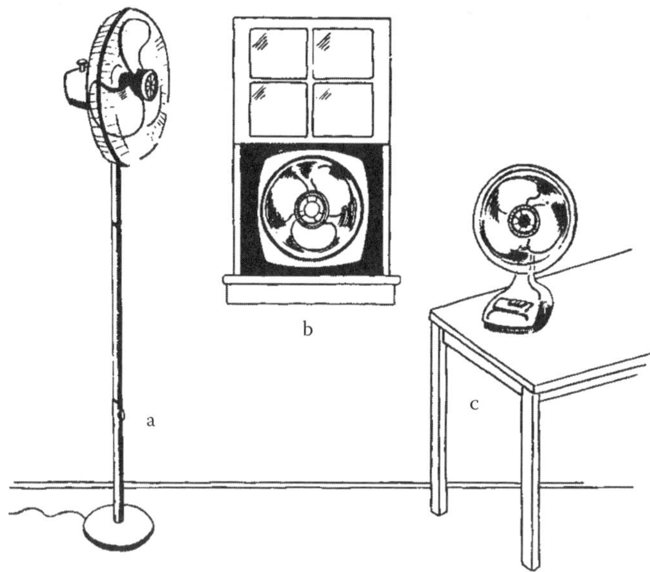

FIGURE 11.17 (a) Floor-standing fan, (b) window fan, and (c) portable tabletop fan.

Window Fans and Floor-Standing Fans

Window fans and floor fans can either bring fresh outside air in, or blow "dirty" indoor air out. If there are people coughing in an area next to your office, then have fresh air come in from your window. If people are coughing directly in your work area, then have the window fan blow the air out. You always want to move dirty air away from other areas and directly to the outdoors.

Sometimes a window fan's air direction can be changed just by turning it around in the window. If you are going to buy a window fan, buy one that can easily reverse the direction of the air. A floor-standing fan can also be put near a window to blow air in or suck air out. Use the same rules as those for window fans, above.

Figures 11.18 through 11.21 show ways of using fans to control airflow when a waiting room is located in an office and when a waiting area is located in a hallway.

Portable Tabletop Fans

Tabletop fans can also help control the direction of air. They can be useful if you are sitting with someone who may be coughing. If the person is facing you, have a fan positioned behind you, blowing air over your shoulder. The fan should blow the air diagonally and away from you. If this is uncomfortable for the person coughing, point the fan so it blows air between you. Make sure that the fan is not pointed at anyone else in the work area. Also, have the air flowing toward a window where air exits the building or toward an air purifier.

Clean and Maintain Your Air Conditioner and Fans

Window (see Figure 11.22) and wall-mounted air conditioners need to be checked to make sure they are blowing as much fresh air as they are able to.

You should regularly clean or replace the filter on the air conditioner. Remove the cover or face plate to get to the filter. Most can be gently rinsed. Also, make sure that the dial or sliding knob which adjusts the fresh air vent works. If the knob does not work, have someone fix it. Always keep the air conditioner set to provide the maximum amount of fresh air. If the air conditioner does not work after trying these suggestions, it may need professional servicing.

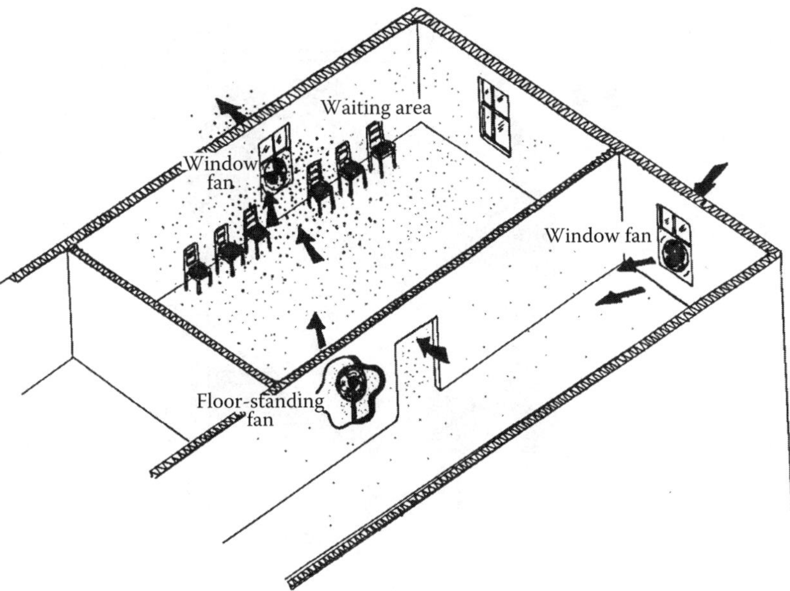

FIGURE 11.18 In this diagram, the dots represent air contaminated with TB droplets. To minimize contamination and exposure, a window fan draws air out of the waiting area, helped by a floor-standing fan. Another window fan brings fresh air into the hallway. Whenever a fan is added to force air out, make sure that you bring in more fresh air.

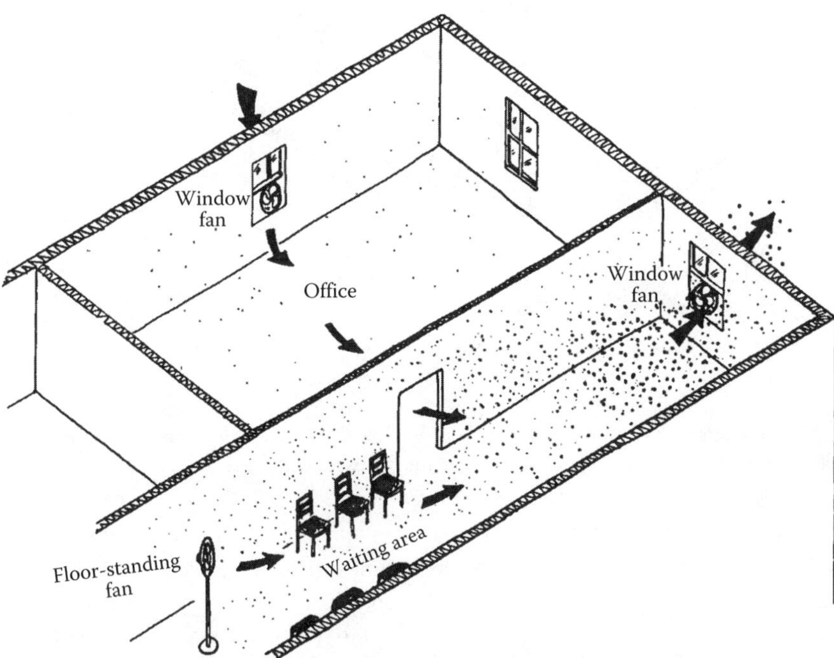

FIGURE 11.19 Here, a floor-standing fan and window fan move air out of the hallway. Fresh air is added through a window fan in the office.

FIGURE 11.20 When two people sit across from one another, a fan can be positioned so that it blows air between the two people, directly toward an open window. If your fan oscillates back and forth, turn off that feature so that the fan always blows in the same direction. If you are buying a fan, make sure it can stay in the same position.

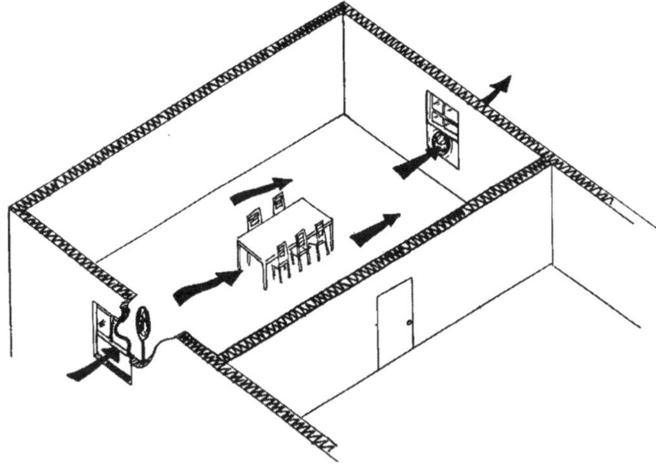

FIGURE 11.21 In rooms with windows on opposite sides of the room, fans can be used to create a good flow of air across the whole room. For small spaces, window fans could be used. For larger rooms, use floor-standing fans.

Use a Portable Room Air Cleaner

The most reliable ventilation strategy to lower the risk of getting TB at your workplace is to have as much fresh air circulating as possible at all times. In rooms without windows, or in winter months, an air cleaner may help.

Air cleaners filter out tiny particles in the air like smoke, pollen, and bacteria. Though no one is yet sure, it is thought that some portable air cleaners can help filter out the "droplet nuclei" which spread TB. Air cleaners which may help eliminate TB from the air contain filters which are called "high-efficiency" or "HEPA" filters.

FIGURE 11.22 Typical window air conditioner.

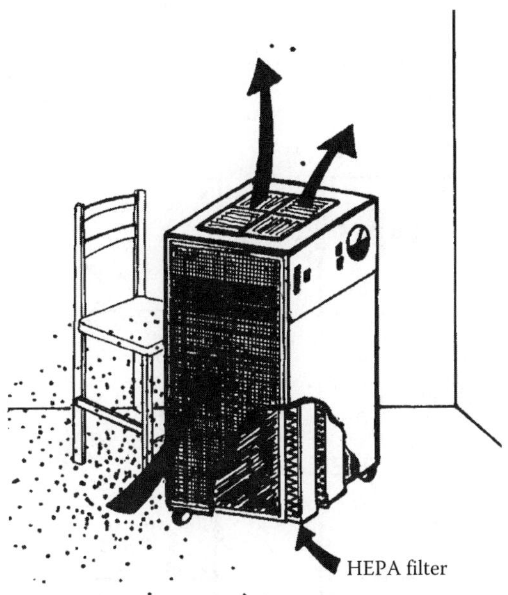

FIGURE 11.23 Typical portable room air conditioner.

Room air cleaners (see Figure 11.23) can be placed in conference rooms, office rooms, or any area where a large number of people gather at the same time. Larger models generally filter more air, at a faster rate, than smaller ones. Expect to pay $500 or more for a good one.

Here are some tips for shopping for an air purifier:

1. Buy the largest you can afford.
2. Buy the newest model—they tend to filter more air and smaller particles.
3. You may need more than one for large rooms.
4. Make sure the cleaner has filters which are labeled "high efficiency" or "HEPA".
5. Have a supply of extra filters and change them often (follow manufacturer's direction).

Change the Way You Interact with People

Make Sure People Cover Their Mouths when Coughing or Sneezing

Have a good supply of tissue in your work areas at all times. Do not be shy about asking people to use them. Ask people to throw these tissues in a covered garbage can and make sure to empty it frequently.

Leave the Room Momentarily

If a person starts to sneeze repeatedly or cough heavily, you can hold your breath and calmly leave the area for a few minutes until the attack subsides and the air has a chance to clear.

Decrease Crowding as Much as Possible

Spread staff and clients out evenly throughout the building so fewer people are around one another at one time. If there are many people coming for some service, try to limit the number coming at any given time. Serve fewer people over a longer period of time rather than all at once.

Move Desks, Chairs, and People

Put people where the airflow is best. Move people away from areas where there is little or no ventilation.

If you are meeting one on one with someone who may be coughing, have them positioned so that they are not facing you directly. Limit close contact with anyone who shows signs of TB. As noted before, getting prompt treatment for people with active TB is the best strategy for protecting clients and staff.

The floor plan in Figure 11.24a shows the results of an airflow survey using incense sticks in one facility. Notice the following problems:

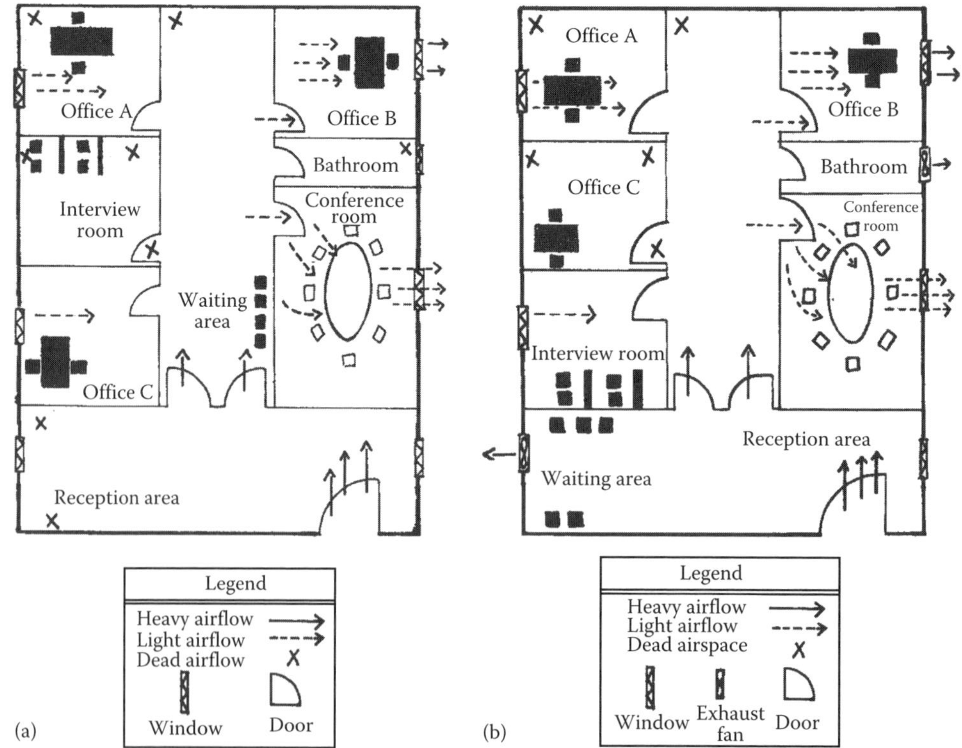

FIGURE 11.24 An example of workplace improvements which help to reduce the space of tuberculosis: (a) before improvements and (b) after improvements.

1. Interview room has no window and no detectable airflow, potentially resulting in a concentration of airborne particles.
2. In Office A, there is a dead air space (an area where no air flows) where people sit at a table across from one another.
3. When people are sitting in the chairs in Office B, air flows from one to the other. If one coughs, the particles may go toward the other.
4. In the bathroom, there is a dead air space.
5. Waiting area is in a hallway. The airflow pattern may carry any airborne TB droplets into the conference room and other areas.

The floor plan in Figure 11.24b shows improvements. Using the suggestions offered in the previous section, several changes have been made:

1. Interview area has been moved to a room with better airflow.
2. Tables and chairs in Office A and Office B have been repositioned so that air flows between people.
3. Waiting area has been moved from the hallway to the front of the facility.
4. Window exhaust fans have been added in the bathroom and in the new waiting area to draw air out of the room.

Summary and Follow-Up

Get Additional Help

Alone, the suggestions offered above will not solve complex problems in facilities where people with active TB reside, use, or visit. Once you have completed the steps suggested here, or if you feel they do not apply to your facility, you should seek additional technical advice and assistance. Organize staff training on TB and help your agency establish TB protocols. Regulations may be forthcoming which require your facility to do more. Stay in touch with health professionals and TB advocates to keep up to date with recommendations and guidelines for airflow in facilities like yours.

For help in finding these resources, call the Hunter College Center for Occupational and Environmental Health or the Center on AIDS, Drugs and Community Health.

Acknowledgments

This fact sheet was jointly prepared by the Hunter College Center for Occupational and Environmental Health ((212)481-8790) and the Hunter College Center on AIDS, Drugs and Community Health ((212)481-7672).

Source: The Center for Occupational Health and the Center on AIDS, Drugs and Community Health at Hunter College, 425 East 25th Street, New York, NY 10010. Reproduced with permission.

Appendix 11.A.1

Test Method

a. Perform rudimentary supply air volume measurements in 3Dl5 to determine nascent room air exchange rate.
b. Measure room/hallway positive pressure with doors and windows closed.
c. Conduct FS209E room qualification to establish background.
d. Elevate particulate load in room with Bolus and record concentration versus time data to create clearance curve with machine off.
e. Turn air cleaning device on at T.B.D. air volume rate and record concentration versus time decay after Bolus AB. The room is quiescent to the hall today, positive pressure <0.005 w.c. smoke reveals room slightly positive. Heating, ventilating, and air conditioning (HVAC) air estimated to be approximately 150 cfm (cubic feet per minute).

1. Test: 6/9/94, room 3Dl5 HVAC recovery test, no air scrubbing. [a]Count = 6 s @ 1.0 cfm = 0.1 ft^3, Channel = >l.0 μm.
2.

Time (m)	Count[a]	Comment
0.0	1,983	Background
0.5	1,777	Background
1.0	1,686	Background
1.5	1,643	Background
2.0	2,029	Inj. Bolus
2.5	46,128	Rising
3.0	110,128	Rising
3.5	192,962	Peak
4.0	181,394	Decay
4.5	188,361	Decay
5.0	191,927	Decay
5.5	175,957	Decay
6.0	174,710	Decay
6.5	176,260	Decay
7.0	167,960	Decay
7.5	165,877	Decay
8.0	150,716	Decay
8.5	146,112	Decay
9.0	139,033	Decay
9.5	119,913	Decay
10.0	120,664	Decay
10.5	103,928	Decay
11.0	97,353	Decay
11.5	91,656	Decay
12.0	88,046	Decay
12.5	74,029	Decay
13.0	61,180	Decay
13.5	51,444	Decay
14.0	49,709	Decay
14.5	51,665	Decay
15.0	47,204	Decay
15.5	38,419	Decay
16.0	36,334	Decay
16.5	31,325	Decay
17.0	32,356	Decay
17.5	30,267	Decay
18.0	26,881	Decay
18.5	21,615	Decay
19.0	20,875	Decay
19.5	19,097	Decay
20.0	18,142	Decay
20.5	16,515	Decay
21.0	14,022	Decay
21.5	13,451	Decay
22.0	13,000	Decay
22.5	11,055	Decay

(continued)

Time (m)	Count[a]	Comment
23.0	10,010	Decay
23.5	9,925	Decay
24.0	8,579	Decay
24.5	8,549	Decay
25.0	8,170	Decay
25.5	7,980	Decay
26.0	7,493	Decay
26.5	7,804	Decay
27.0	7,155	Decay
36.0	3,498	♦ Note time change, this is start of next test run
36.5	3,563	
37.0	3,348	Decay
37.5	3,498	Decay

1. Test: 6/9/94, room 3Dl5 HVAC, machine on at approximately 400 cfm throughout. [a]Count = 6 s @ 1.0 cfm = 0.1 ft^3, Channel = >l.0 μm.
2.

Time	Count[a]	Comment
0.0	164	Background
0.5	155	Background
1.0	139	Background
1.5	85	Background
2.0	146	Background
2.5	100	Background
3.0	85	Background
3.5	3,032	Add DOP; peak decay
4.0	49,296	Add DOP; peak decay
4.5	46,944	Add DOP; peak decay
5.0	33,173	Add DOP; peak decay
5.5	31,736	Add DOP; peak decay
6.0	19,985	Add DOP; peak decay
6.5	14,639	Add DOP; peak decay
7.0	10,868	Add DOP; peak decay
7.5	9,165	Add DOP; peak decay
8.0	8,719	Add DOP; peak decay
8.5	7,368	Add DOP; peak decay
9.0	5,991	Add DOP; peak decay
9.5	4,983	Add DOP; peak decay
10.0	4,198	Add DOP; peak decay
10.5	3,469	Add DOP; peak decay
11.0	3,036	Add DOP; peak decay
11.5	2,356	Add DOP; peak decay
12.0	2,113	Add DOP; peak decay
12.5	1,714	Add DOP; peak decay
13.0	1,229	Add DOP; peak decay

(*continued*)

(continued)

Time	Count[a]	Comment
13.5	1,078	Add DOP; peak decay
14.0	865	Add DOP; peak decay
14.5	742	Add DOP; peak decay
15.0	577	Add DOP; peak decay
15.5	471	Add DOP; peak decay
16.0	424	Add DOP; peak decay

1. Test: 6/9/94, room 3Dl5, repeat test with machine on at 400 cfm continuously. [a]Count = 6 s @ 1.0 cfm = 0.1 ft^3, Channel = >l.0 μm.
2.

Time	Count[a]	Comment
0.0	577	Background
0.5	471	Background
1.0	424	Background
1.5	5,370	Add Bolus
2.0	35,539	Rising peak decay
2.5	124,092	Rising peak decay
3.0	100,655	Rising peak decay
3.5	83,286	Rising peak decay
4.0	56,860	Rising peak decay
4.5	45,435	Rising peak decay
5.0	32,057	Rising peak decay
5.5	21,427	Rising peak decay
6.0	16,271	Rising peak decay
6.5	12,485	Rising peak decay
7.0	11,821	Rising peak decay
7.5	8,976	Rising peak decay
8.0	7,674	Rising peak decay
8.5	6,502	Rising peak decay
9.0	5,464	Rising peak decay
10.0	3,963	Rising peak decay
11.0	2,589	Rising peak decay
11.5	1,985	Rising peak decay
12.0	1,639	Rising peak decay
12.5	1,296	Rising peak decay
13.0	1,052	Rising peak decay
13.5	878	Rising peak decay
14.0	676	Rising peak decay
14.5	644	Rising peak decay

HEPA Test Mode 880 cfm

Time	Concentration
0.0	612,300
0.5	596,440
1.0	577,300
1.5	545,800

(continued)

Time	Concentration
2.0	519,320
2.5	458,350
3.0	417,870
3.5	378,800
4.0	294,170
4.5	235,600
5.0	190,900
5.5	146,890
6.0	106,150
6.5	75,710
7.0	61,530
7.5	43,310
8.0	23,640
8.5	29,140
9.0	23,570
10.0	19,670
10.5	17,650
11.0	16,260
11.5	16,160
12.0	14,380
12.5	16,710

RS 1000 HEPA Challenge Test

San Francisco General Hospital Medical Center

Technical Safety Services of San Ramon, CA was retained by San Francisco General Hospital to test the capacity of the portable RS 1000 HEPA recirculating negative air machine utilized in 10 TB isolation rooms. Clearance rates/decay rates were determined by challenging the machine with DOP, fogging the room with specific particle counts per cubic foot of air, and timing the decay curve to background counts. The curve slope is considered an activity of time versus a reduction in quantified particle count in air (Figure 11.A.1).

Test equipment: Discreet particle counter: manufactured by particle measurement systems Mode 1# Lasaire 310: factory calibration date 4/94

Sample flow rate: 1 cfm

Room size: 1300 ft^3

HVAC: Supply 150 cfm

Probe site: Exhaust grille, ceiling supply diffuser

Machine speeds:

1. Trial test without machine operating
2. Machine at 400 cfm setting actively used in isolation rooms as a balance between noise and clearance values
3. Machine set at HEPA test mode 880 cfm
4. A setting providing micron range: 1 μm setting

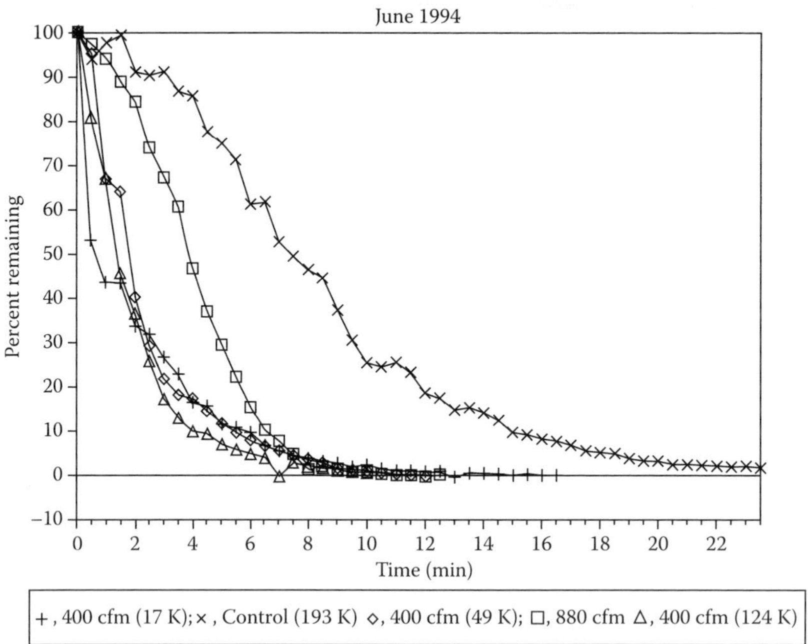

FIGURE 11.A.1 RS 1000 challenge test.

Discussion of Results

The RS 1000 functioned as an air scrubber in the two basic modes tested, 400 and 880 cfm (HEPA Test Mode). The unit was tested at these volumes with three challenge trials each and all decay slopes corresponded.

The first challenge test without the machine running indicated a 37.5 clearance rate or 16 min clearance time to 90% of background. This corresponds to approximately 9 air changes per hour (ach) (see table, Federal Register, Air Changes Per Hour and Time, Vol. 58, No. 195), if one assumes a mixing factor of 1 (perfect mixing). However, the formula: Average cfm × 60 room volume of 400 cfm × 60, 1300 which equals 18 ach).

Three trials at 400 cfm showed a decay time of 5.5, 6.5, and 4 min to decay to 90% of background. This range corresponds to approximately 20–25 ach when consulting Table S3 of the Federal Register, Vol. 58, No. 195. This in turn is slightly greater than the results with the air change formula due to better mixing rate.

Three trials at 880 cfm (HEPA Test Mode) showed the average recovery time to be 4.5 min. Adding the 880 cfm to the 150 cfm supplied by the house HVAC yields 46 ach. Applying this value to Table S3-1 suggests that the 90% recovery time should be 3 min, therefore the overall room performance does not compare well with the expected value. This can be explained by a less than 1 mixing value.

One other explanation is that the turbulence created by a high volume current creates a delay in decay, or that at the higher rate of air exchange, a homeostasis is reduced with performance. (See Table S3-1 Federal Register, Vol. 58, No. 195). Between 30 and 50 air changes, little increase in performance is noted.

The RS 1000 which brings air in at the bottom through a 17 ≤ × 17 ≤ grille and exhausts air through the top, showed in all test modes to clear the air and mix the air extremely well in a 1300 ft^3 room. It proved that it can remove particles in an order of magnitude much greater than the central ventilation system due to a higher volume efficiency and better mixing factor.

Time	0	0.5	1	1.5	2	2.5	3	3.5	4	4.5	5	5.5	6
880 cfm	612,300	596,440	596,440	545,800	519,320	458,350	417,870	378,800	294,170	235,600	190,900	148,890	106,150
Count	596,300	580,440	580,440	529,800	503,320	442,350	401,870	362,800	208,170	219,600	174,900	130,890	90,150
Background % remain	100	97.34	97.34	88.85	84.41	74.18	67.39	60.84	46.65	36.83	29.33	21.95	15.12
400 cfm (112 K)	17,651	9,416	7,700	7,672	5,964	5,634	4,728	4,029	2,933	2,779	2,061	1,919	1,751
Count	17,566	9,331	7,615	7,587	5,879	5,549	4,638	3,944	2,848	2,694	1,976	1,834	1,666
Background % remain	100	55.12	43.38	43.19	33.47	31.59	26.40	22.45	16.21	15.34	11.25	10.44	9.48
400 cfm (49 K)	49,296	46,944	33,173	31,736	19,985	14,639	10,868	9,165	8,719	7,368	5,991	4,983	4,198
Count	48,872	46,520	32,749	31,312	19,561	14,215	10,444	9,741	8,295	6,944	5,567	4,559	3,774
Background % remain	100	95.19	67.01	64.07	40.02	29.09	21.37	17.89	16.97	14.21	11.39	9.33	7.72
400 cfm (124 K)	124,092	100,655	83,286	56,860	45,435	32,057	21,427	16,271	12,485	11,821	8,976	7,674	6,502
Count	123,448	100,011	82,642	56,216	44,791	31,413	20,783	15,627	11,841	11,177	8,332	7,030	5,058
Background % remain	100	81.02	66.96	45.56	36.30	25.47	16.86	12.69	9.62	9.08	6.78	5.73	4.78
400 cfm (193 K)	192,962	181,394	188,361	191,927	175,957	174,710	176,260	167,960	165,877	150,716	146,112	139,033	119,913
Count	189,462	177,894	184,861	188,427	172,457	171,210	172,760	164,460	162,377	147,216	142,612	135,533	116,413
Background % remain	100	93.69	97.57	99.45	91.02	90.37	91.18	86.80	85.70	77.70	75.27	71.54	61.44

Ben Gonzales, PE, CSP, Technical Safety Services.
For: William Charney, DOH, Director Environmental Health, San Francisco General Hospital.

Appendix 11.A.2

Olmstead Environmental Services

Challenge Data for Portable HEPA Filter Systems for TB

Introduction

Hazardous Material Technologies Corporation (HMTC) was retained by Biological Controls to test the effectiveness of the MICROCON® as an auxiliary ventilation system in filtering simulated airborne pathogenic organisms such as TB bacillus and the AIDS virus. HMTC was asked to test the MICROCON's ability to control an airborne concentration of respirable particulate of 1–3 μm in diameter (see Appendix 11.A.2.2 for detailed information of particulate characteristics).

The MICROCON is a portable, self-contained freestanding air purification device. It is designed to function as an internal ventilation system capable of both filtration (99.97% of particulate ≥0.3 μm in diameter) and disinfection of pathogenic species. The unit can be operated at any one of three flow rates.

The test method used consisted of generating an airborne particulate concentration within the test rooms, in the range of 10–20 mg/M^3. The goal of the test was to determine the ability of the MICROCON to remove airborne particulate from a typical hospital environment under simulated operating conditions. AU testing was performed in a soon to be renovated section of Horton Hospital, Middletown, New York, on December 22, 1992. The two test rooms used in the test were a special isolation room and a patient residence room.

Scope of Evaluation

The intent of this study was to evaluate by impartial objective means the ability of biological control's MICROCON to filter a simulated human generated bioaerosol intended to mimic the type of aerosol produced by a human cough or sneeze.

The particulate medium chosen to simulate the bioaerosol was Arizona Road Dust. The test dust had a mean diameter of 0.76 μm. This medium was chosen because its aerodynamic properties are within the human respirable range. This medium also permits sufficient residency for consistent and reproducible measurement of airborne concentrations.

In general, the test consisted of generating an airborne dust concentration in the test room and measuring the change in the particulate concentration. The internal volume of the test isolation room and regular patient room were 1090 and 1612 ft^3, respectively.

Prior to the start of each test, the particulate monitoring equipment was adjusted and a consistent airborne dust concentration generated. After an appropriate period of time to allow for stabilization of the dust concentration, the MICROCON was placed in central location (as possible) within the test room prior to the start of the test.

Once a stable airborne dust concentration was achieved and a baseline established the MICROCON was then activated. The concentration of airborne dust within the test rooms was monitored using four direct-reading portable aerosol monitors which measured airborne particulate levels at two levels in each of the two locations. The monitoring equipment chosen for monitoring particulate levels were battery powered, handheld Haz-Dust™ particulate monitors manufactured by Environmental Devices Corp. The Haz-Dust monitor has a digital readout as well as a DC voltage output for data recording.

Four of these monitors were used in each test. Two test stands were used in each test cell and each contained two monitors mounted at different heights. Monitors 1 and 3 were mounted on test stand number 1 with monitors 2 and 4 mounted on test stand number 2. Monitors 1 and 2 were mounted at a height of 66 in. with monitors 3 and 4 mounted at a height of 32 in.. The height of 66 in. was chosen as representative of the breathing zone height of the average standing adult. The height of 32 in. was chosen as an approximation of the height of a supine patient's breathing zone. The test stands were located as follows:

1. Test stand #1—(holding monitors 1 and 3) was placed near the front side of the bed approximately 36 in. from the wall near the head of the bed.
2. Test stand #2—(holding monitors 2 and 4) was placed in the corner, at the foot of the bed, near the window approximately 24 in. from the corner of the rooms.

The test protocol was designed to address possible increased settling attributable to the greater density of the test dust as compared to that of a bioaerosol. Initially, in an effort to compensate for the lack of air currents necessary for uniform particle dispersion, one 24 in. diameter fan was employed to distribute the test dust and to produce a more uniform concentration of the test dust within the test room. The primary reasons for the diminution of the test dust concentrations appeared to be impact losses on fan blades, associated surfaces, and gravity-induced settling or drop-out. Another source of loss was the scavenging effect produced by static charges on the room's interior surfaces.

As a result of scavenging, drop-out, and impact losses the quantity of test dust required to produce a stable concentration of 10–20 mg/M^3 in each test room was somewhat greater than calculations indicated. A handheld air-powered nebulizer was used to disperse the test dust in each test cell. Once the dust loading requirements were met, it took 5 to 7 min to achieve a stable and acceptably uniform test dust concentration within the desired range.

Discussion of Test

Unless otherwise noted, all tests were performed with the MICROCON flow setting selector in the medium position. The tests began after a 4–5 min settling period to allow the dust concentration to stabilize. After this settling period the MICROCON unit was activated. During this stabilization period no additional mixing was provided other than that produced by the room's own dynamics.

Chart 1—Background Dust Levels—Room 336

This test represents the background or residual airborne dust present in the test room prior to the introduction of the test dust. This dust is undoubtedly the result of lack of use because this section of the hospital was not being used.

Chart 2—Settling Profile—Room 336

This test represents normal settling characteristics of the dust used in the tests, as well as any inherent flow dynamics of the test room. An airborne dust concentration was generated, mixed, and then allowed to settle naturally.

Chart 3—Test Number 1—Room 336

This test shows the performance of the MICROCON (air machine) in the test room with an elevated level of dust present. After the initial dust application it was noted that additional openings through which air could flow into the room needed to be sealed if a true test of the MICROCON was to be obtained. After these openings were sealed a second dust application was performed (see Chart 3).

Chart 4—Test Number 2—Room 336

In this test all room HVAC vents and openings were opened (this condition is more typical of normal usage conditions).

Chart 5—Test Number 3—Room 336

The test conditions were the same as those indicated for Test Number 2, but the MICROCON was operated with the flow setting on low.

Chart 6—Test Number 4—Room 335

The test conditions were the same as those indicated for Test Number 2.

Chart 7—Test Number 5—Room 335

The test conditions were the same as those indicated for Test Number 2, but the MICROCON was operated with the flow setting on low.

Discussion of Findings/Results

An analysis of the purge test data indicates that the MICROCON is effective in capturing and removing significant quantities of respirable airborne particulate from the immediate surrounding area. The effectiveness of the disinfection option of the unit was not evaluated as part of the purge testing reported herein.

The test data collected consisted of airborne particulate concentrations measured as a function of location and elevation (within the test room) and time. These data are presented in graph form in Appendix 11.A.1. A total of five purge tests were conducted. These included operating the MICROCON alone and in tandem with the test room's HVAC system. The MICROCON appeared to function most efficiently when operated with a room's own HVAC system but is also capable of operating well even without such aid. This was most likely due to the added turbulence and air movement provided by the HVAC system. It would also appear that the MICROCON benefits when it is placed in a location where it can take best advantage of the induced air currents produced by the HVAC system when these currents do not comprise the design effectiveness of the unit.

It should be noted when reviewing the data continued in the charted test data in accordance with the Haz-Dust manufacturer's directions all monitors and subsequent readings are subject to a background correction of approximately 1%–2% of the highest dust concentrations encountered. This correction is a function of the optics used in the monitors and *is noncorrectable by zero correction*. Hence, each final reading should be viewed accordingly.

Because we were asked to perform the evaluation tests under the most stringent conditions possible, the reported data were not adjusted in any way for background buildup. When the background buildup is taken into account and the final readings so corrected *the actual levels or percentage of total dust removed by the MICROCON (Air Machine) during the test is even greater than indicated*.

The lack of zero correction and the high levels of airborne dust used during these tests were of particular concern with regard to the data depicted by sensors number 4, 3, 3, and 3 on Charts Number 4 (Test No. 2), Chart Number 5 (Test No. 3), Chart Number 6 (Test No. 4), and Chart Number 7 (Test No. 5), respectively. These sensor readouts show clear evidence of this phenomena. If the background corrections recommended by the manufacturer are performed on all data the end point or final readings at these locations would be more consistent with those observed for the other sensor readings obtained during the respective test.

Conclusions

The test data indicate that the MICROCON performs its intended task "removal of respirable size airborne particulate from the air within its effective area of influence." The size of the effective area of influence is to a large extent dependent upon a number of factors including drafts, obstructions, and machine operating speed. All these factors can influence the aerodynamic behavior of airborne particulate and should be considered in any proposed applications. If these factors are adequately addressed and the manufacturer's instructions followed then the MICROCON should perform as indicated.

Purge Test Charts 1–7

Background Dust Levels—Room 336

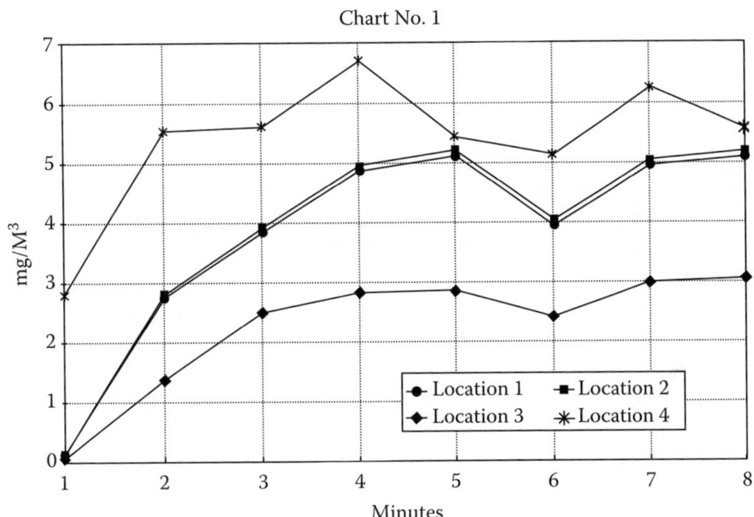

Settling Profile—Room 336

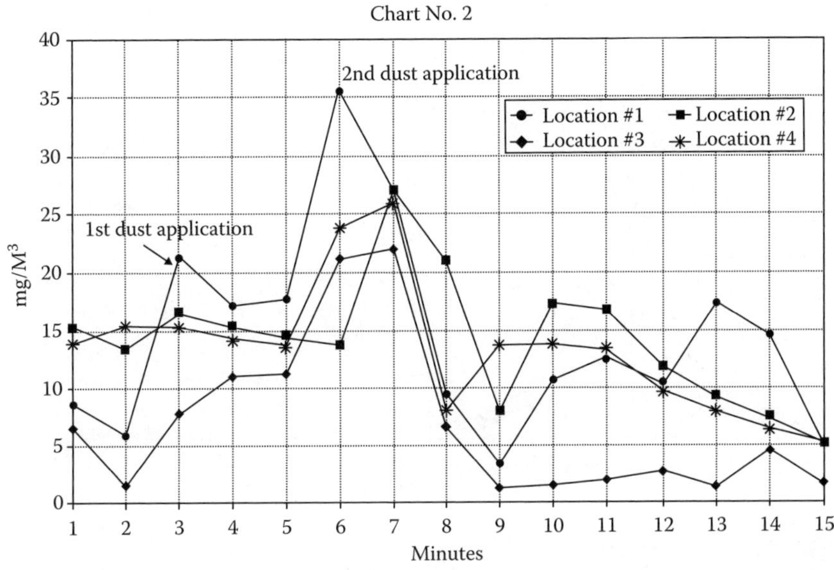

Test No. 1—Room 336

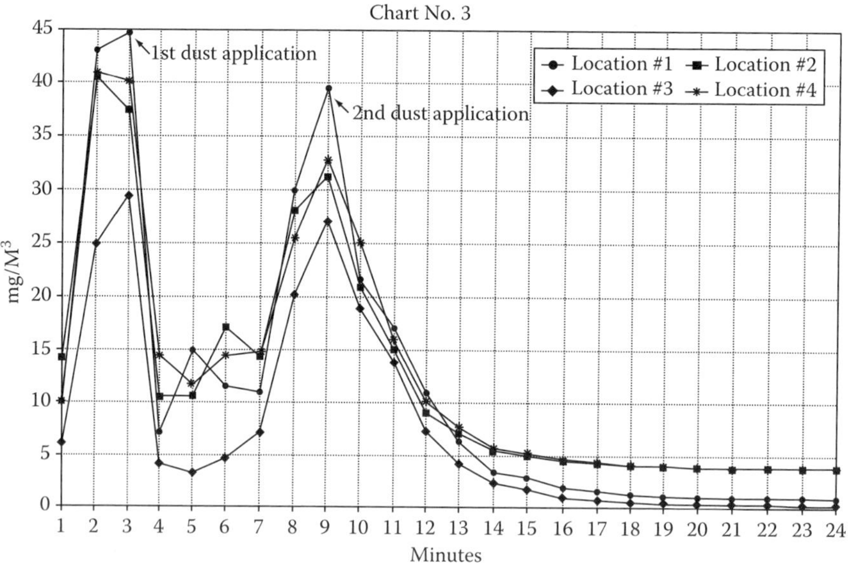

Test No. 2—Room 336

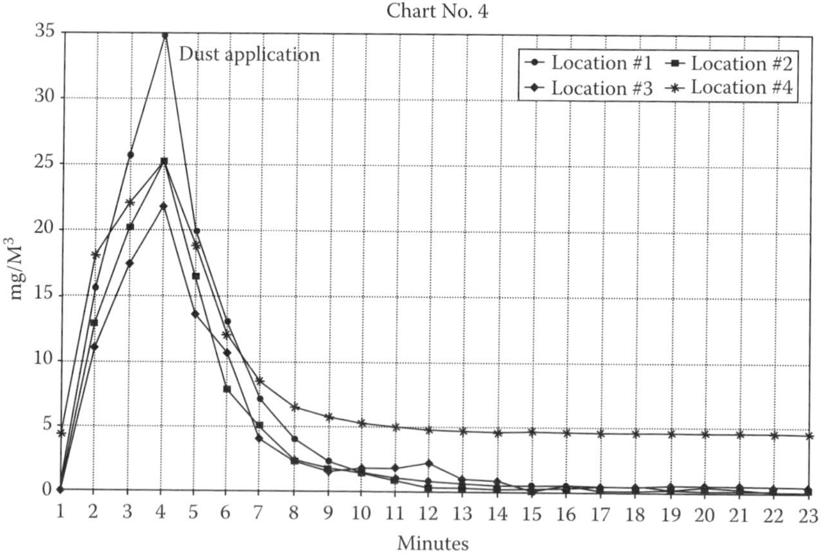

Test No. 3—Room 336

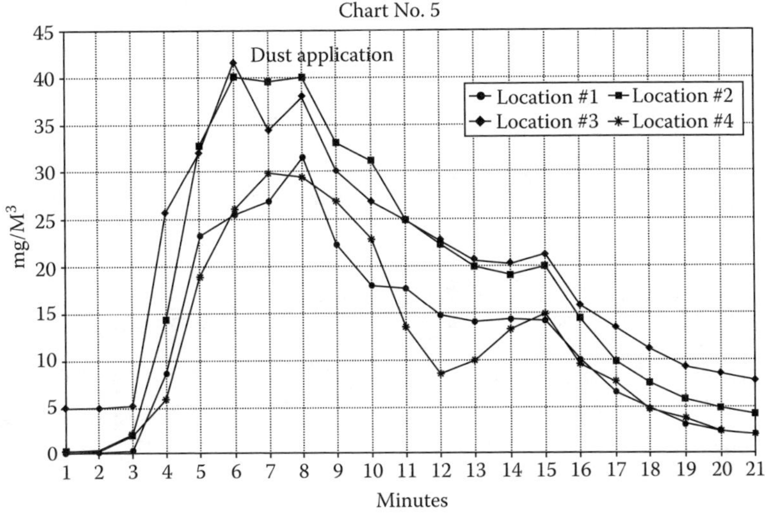

Test No. 4—Room 335

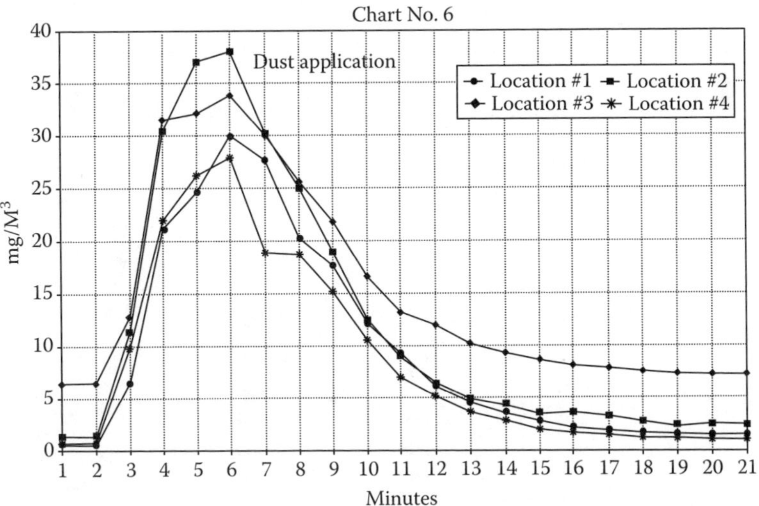

Test No. 5—Room 335

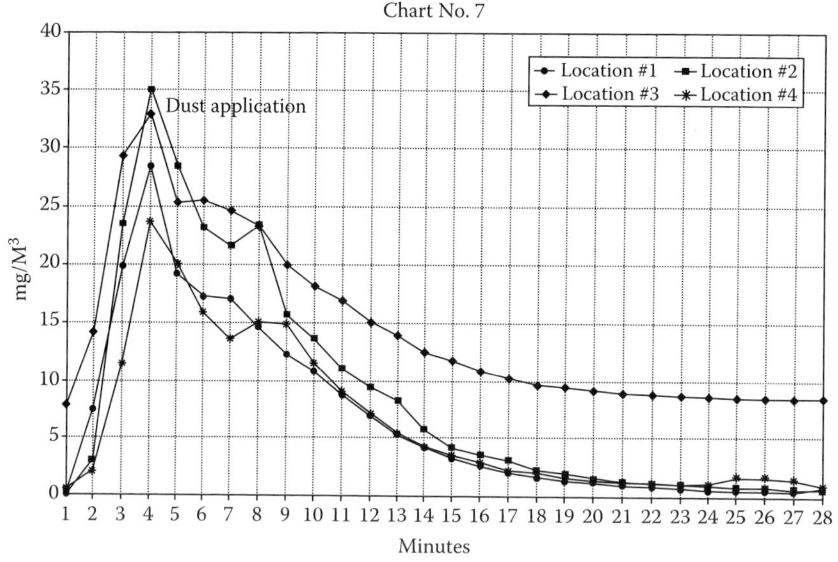

Particulate (Dust) Size Data

Coulter Counter Model TAII Particle Size Analysis	
Customer: General	Material: Arizona Road Dust
Date: 10-5-89	No. of runs: 2
Sample No: 4281H	Aperture: 30
PTI Run No: 4281 F.F.	Cust. Lot No: 60029J
Operator: TAF	Mean volume: 0.79 µm

Total Data	Diff %	Cum %>	Diameter (µm)	Active Channels
0	0.00	100.00	0.198	
0	0.00	100.00	0.250	
0	0.00	100.00	0.315	
0	0.00	100.00	0.397	0–2.5 µm
483	23.47	100.00	0.500	99.13
567	27.55	76.53	0.630	
513	24.93	48.98	0.794	
277	13.46	24.05	1.000	0–5 µm
135	6.56	10.59	1.260	100.00
48	2.33	4.03	1.590	
17	0.83	1.70	2.000	
8	0.39	0.87	2.520	5–10 µm
6	0.29	0.49	3.170	0.00
4	0.19	0.19	4.000	
0	0.00	0.00	5.040	
0	0.00	0.00	6.350	10–20 µm
0	0.00	0.00	8.000	0.00
0	0.00	0.00	10.080	
0	0.00	0.00	12.700	
0	0.00	0.00	16.000	20–40 µm

(continued)

Total Data	Diff %	Cum %>	Diameter (μm)	Active Channels
0	0.00	0.00	20.200	0.00
0	0.00	0.00	25.400	
0	0.00	0.00	32.000	
0	0.00	0.00	40.300	40–80 μm
0	0.00	0.00	50.800	0.00
0	0.00	0.00	64.000	
0	0.00	0.00	80.600	
0	0.00	0.00	101.600	80 + μm =
0	0.00	0.00	128.000	0.00
0	0.00	0.00	161.000	
0	0.00	0.00	203.000	
0	0.00	0.00	256.000	
0	0.00	0.00	322.000	
0	0.00	0.00	406.000	
0	0.00	0.00	512.000	
0	0.00	0.00	645.000	

Powder Technology Inc. (612) 894–8737.

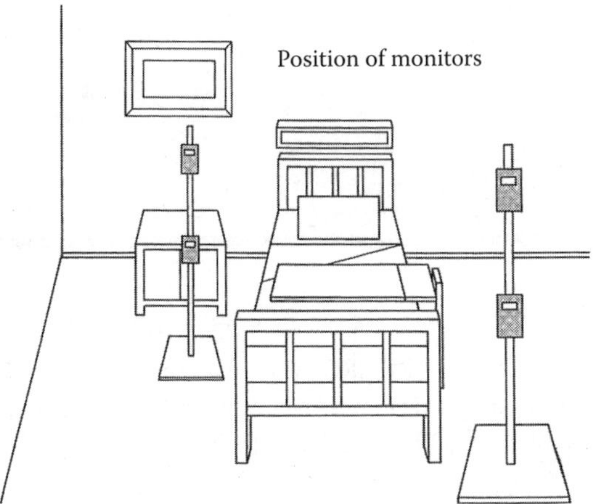

Position of monitors.

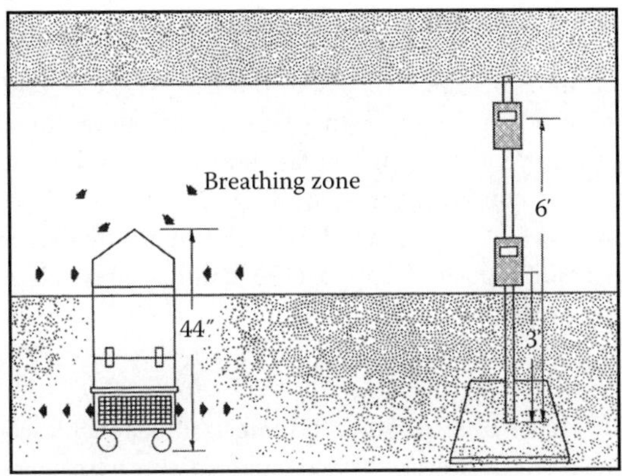

Breathing zone.

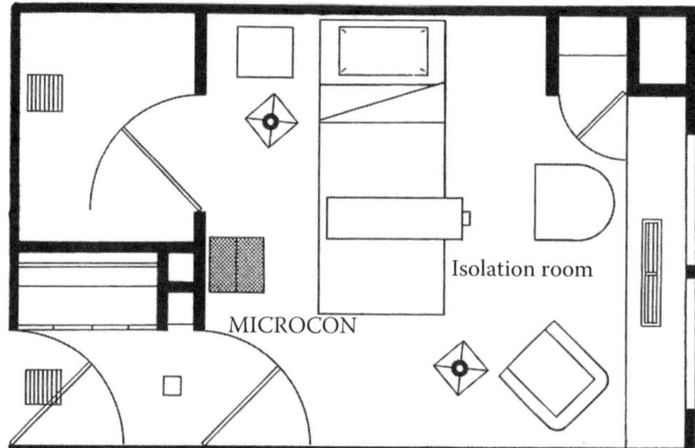

Isolation room.

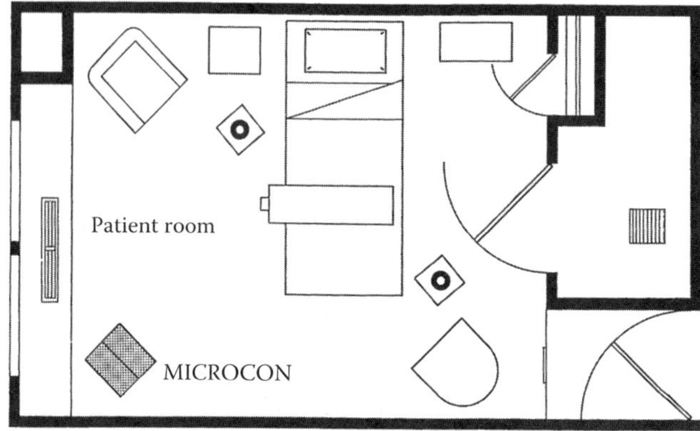

Patient room.

Source: Evaluation report purge test of MICROCON in a typical hospital environment using a simulated human bioaerosol. Prepared by Hazardous Materials Technologies Corporation, Mount Laurel, NJ. Prepared for Mr. Gary Messina, Biological Controls, Eatontown, NJ. Report date: January 27, 1993.

Investigation of Isolation Room with the MICROCON HEPA Air Filtration System

An investigation was made into the effectiveness of the MICROCON HEPA Air Filtration System. A description of the equipment is included in this report. The survey involved sampling the air for microbiological viable organisms and total particulate levels. The samples were taken before running the MICROCON and while the equipment was running. The results indicate that the MICROCON HEPA Air Filtration System was effective at removing airborne particulate material and viable microbiological organisms. These results appear to support the manufacturer's claim that the filtration system will remove airborne droplet nuclei produced by persons actively infected with TB.

Background

Over the past decade the incidence of respiratory TB infection has increased steadily in the New York Metropolitan area. Many hospitals in Newark have treated patients with actively infectious TB. Patients with active cases of TB can release infectious droplet nuclei containing TB into the air thereby exposing health care personnel in proximal areas. The risk of infection to the exposed persons is related to the quantities of TB bacillus in the air, the duration of exposure and the general health of the individual. The Centers for Disease Control and Prevention (CDC) has a criteria document that suggests methods that may reduce the risk of TB exposure. These methods include

1. Using isolation rooms for infectious persons. These rooms should be under negative pressure to the surrounding areas and should receive a minimum of 6 ach.
2. Providing adequate ventilation to reduce concentrations of TB in the air.
3. Using high-efficiency particulate in air (HEPA) filtration systems for recirculated air.
4. Using ultraviolet lamps to kill airborne TB.
5. Provide respiratory protection to exposed individuals in high-risk areas such as isolation rooms.

This survey involved the evaluation of the MICROCON Air Filtration System. This unit consists of a fan that pulls air into the top of the equipment through a HEPA filter and discharges the air back into the room through the base of the equipment. The unit operates at three fan settings:

1. 725 cfm
2. 675 cfm
3. 400 cfm

For a patient room that is 12 ft^2 and with an 8 ft ceiling this unit would provide over 20 ach at the lowest setting.

Methods

Air sampling was conducted following NIOSH method 500 for total particulate in air. The 5.0 μm poly-vinyl chloride (PVC) filters were preweighted and analyzed by the ITT Hartford Environmental Laboratory in Hartford, CT. This is an AIHA accredited laboratory. See Appendix 2 in this section for the laboratory report from the Hartford laboratory.

Fungi samples were collected on 3% malt extract agar using an Anderson N6 impaction sampler. Sample petri dishes were prepared and analyzed by P&K Microbiology Services of Cherry Hill, NJ.

Results

Both the dust samples and the fungi samples indicate a reduction of airborne particulate levels. Table 11.A.1 summarizes the dust level sampling results. Levels prior to operating the air cleaner were 180 micrograms per cubic meter (mcg/m^3) of air. After running the air cleaner levels were below the limit of detection of 10 mcg/m^3.

Table 11.A.2 summarizes the bioaerosol monitoring the fungi levels in air. Levels were reduced from 40 cfu/m^3 prior to running the air cleaner to between 5 and 7 cfu/m^3 after running at low speed for approximately 1 h.

TABLE 11.A.1 Airborne Particulate Sampling Room Tested

Sample No.	Description	Result (μg/m^3)[a]
14024	Pretest—in room B521 prior to operating the MICROCON Air Filtration System	180
14023	Pretest—In the hallway outside room prior to operating the MICROCON Air Filtration System	20
14021	After operating the MICROCON Air Filtration System Inside the room	Below 10
14016	Field blank	<25 μg per filter

[a] μg/m^3 indicates micrograms per cubic meter.

TABLE 11.A.2 Microbial Evaluation Room Tested

Sample No.	Location—Description	Species Identified	Result (cfu/m^3)
9306281	Inside room, pretest; before running the MICROCON air cleaner 141.5 L	Cladosporium, penicillium, yeasts	42.4
9306282	Outside room, pretest; before running the 141.5 L	Cladosporium	14.1
9306283	Field blank	No growth	
9306284	Inside room after turning on the MICROCON air cleaner 141.5 L	No growth	<5
9306285	Inside room after turning on the MICROCON air cleaner 198.1 L	Cladosporium	5
9306286	Inside room after turning on the MICROCON air cleaner 424.5 L	Cladosporium, sterile fungi	7.1

Discussion

The room tested is designed to operate as an isolation for TB patients. The room is designed to have 6 ach and is under negative pressure to the hallway. This design is adequate for preventing the spread of airborne TB to the corridor, but it does not appreciably reduce the level of airborne contaminants inside the room. Operating the air cleaner in the room reduced levels of particulate and bioaerosols.

The results of this monitoring survey indicate the ability for HEPA filtration devices to reduce overall bioaerosol levels inside the isolation room. These results are consistent with our findings in a similar test conducted in May of 1993. Although we have not used the system on TB, we believe the results support the contention that these air filtration devices are effective at reducing airborne droplet nuclei.

There is no single way to protect persons from TB transmission and it is recommended that a combination of methods be used. These sample results indicate that HEPA filtration in combination with negative pressure in isolation rooms will provide an improved level of protection to the medical staff.

It should be noted that the MICROCON HEPA Air Filtration System is an effective means of lowering the level of bioaerosols in the air. There is no engineering control that will eliminate TB exposure in the isolation room.

Source: Edward Olmsted, CIH, CSP, Olmsted Environmental Services, Inc., RR I Box 480, Garrison, NY 10524.

Evaluation of MICROCON HEPA Air Filtration System in an Emergency Room for Controlling Bioaerosols

An investigation was made into the effectiveness of the MICROCON HEPA Air Filtration System. The survey was conducted on October 13, 1993, by Edward Olmsted, CIH, CSP. A description of the air filtration equipment is included in this report. The survey involved sampling the air for microbiological viable organisms and total particulate levels. The samples were taken before running the MICROCON and while the equipment was running. The results indicate that the MICROCON HEPA Air Filtration System was effective at reducing viable microbiological organisms. The particulate levels were not appreciably reduced by the MICROCON HEPA Air Filtration System. These results appear to support the manufacturer's claim that the filtration system will remove airborne droplet nuclei produced by persons actively infected with TB; and thereby, reduce the number of airborne infectious organisms.

Background

Over the past decade, the incidence of respiratory TB infection has increased steadily in the New York Metropolitan area. Many hospitals in Newark have treated patients with actively infectious TB. Patients with active cases of TB can release infectious droplet nuclei containing TB into the air, thereby, exposing health care personnel in proximal areas. The risk of infection to the exposed persons is related to the quantities of TB bacillus in the air, the duration of exposure, and the general health of the individual. The CDC has a criteria document that suggests methods that may reduce the risk of TB exposure. These methods include

1. Using isolation rooms for infectious persons. These rooms should be under negative pressure to the surrounding areas and should receive a minimum of 6 ach.
2. Providing adequate ventilation to reduce concentrations of TB in the air.
3. Using high-efficiency particulate in air (HEPA) filtration systems for recirculated air.
4. Using ultraviolet lamps to kill airborne TB.
5. Provide respiratory protection to exposed individuals in high risk areas; such as isolation rooms.

This survey involved the evaluation of the MICROCON Air Filtration System. This unit consists of a fan that pulls air into the top of the equipment through a HEPA filter and discharges the air back into the room through the base of the equipment. The unit operates at three fan settings:

1. 725 cfm
2. 675 cfm
3. 400 cfm

Previous tests of the MICROCON have indicated that the system is effective at reducing airborne particulate and viable organisms in an AFB isolation room and in a general use patient room. In this case, the system was tested in an emergency room (ER) waiting area. The ER was under normal use by the hospital during all sampling. Figure 11.A.2 gives a layout of the ER.

ERs provide unique problems for hospitals attempting to develop infection control programs for TB. Large numbers of persons visit the ER each day; and in urban areas such as Newark, NJ, there is an increased potential for actively TB-infected parsons to come to the ER. Controlling TB in ERs has mostly consisted of the following:

1. Isolating persons that show signs and symptoms of TB
2. Installing UV lamps
3. Installing HEPA filtration systems

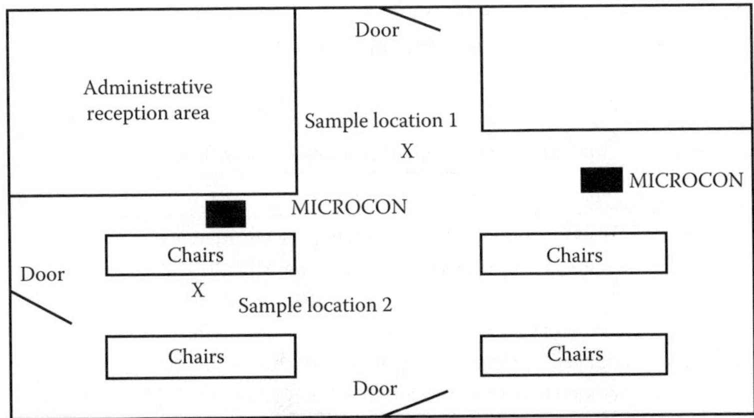

FIGURE 11.A.2 Emergency room.

Very little is known about the effectiveness of these methods at controlling TB. The results of this survey indicate that the MICROCON HEPA Air Filtration System does reduce airborne bioaerosols and may reduce infectious TB as well.

Methods

Air sampling was conducted following NIOSH method 500 for total particulate in air. The 5.0 μm PVC filters were preweighted and analyzed by the ITT Hartford Environmental Laboratory in Hartford, CT. This is an AIHA accredited laboratory. The samples were collected using a volume flow rate of 15 L per minute calibrated using a rotameter. The laboratory report from the Hartford Laboratory is attached to this report.

Bacteria in air samples were collected using Tryptose soy agar and were collected using the Anderson N6 single stage impaction sampler operated at a flow rate of 28.3 lpm. The sampler was calibrated using an in-line rotameter. Sample petri dishes were prepared and analyzed by P&K Microbiology Services of Cherry Hill, NJ.

Samples were collected before operating any air cleaning equipment in the waiting area. After the pretest, two MICROCONs were operated at the medium speed. Figure 11.A.2 indicates the locations of the MICROCON air cleaners. Samples of the air were taken after 90 min of air cleaning.

Results

Bioaerosol Sampling

Attached to this report are the analytical results from P&K Microbiological Services. Table 11.A.3 gives a summary of the results.

Particulate Sampling

Attached to this report is the analytical results for particulate sampling. Table 11.A.4 summarizes the results.

Discussion

These results indicate that the MICROCON HEPA Air Filtration System is effective at reducing airborne levels of bacteria. The results for particulate sampling did not show a reduction in airborne particulate levels. We believe that the level of particulate in the air was too low for statistically significant results. Nonetheless, the bioaerosol sample results clearly support the contention by the manufacturer that airborne viable bacteria are reduced by the system.

It should be noted that these units cannot be expected to eliminate exposure to TB in any setting. Other factors that are important include the rate of generation of TB in the room. This is dependent upon the degree of infectiousness of the patient. Our sampling results indicate that there is a fourfold

TABLE 11.A.3 Bacteria in Air Sampling in Emergency Room Waiting Room

Sample No.	Description	Result (cfu/m^3)
1	Location #1—Pretest; before operating the MICROCON	388.7
2	Location #2—Pretest; before operating the MICROCON	388.7
3	Location #3—After running the MICROCON 90 min	88.3
4	Location #2—After running the MICROCON 90 min	150.2
5	Location #1—After running the MICROCON 3 h	91.9
6	Location #2—After running the MICROCON 3 h	134.3

Note: After Bioaerosol Measurements (cfu/m^3): Before the MICROCON was run: 388.7; After 90 min of air cleaning via MICROCON: 119.2; After 180 min of air cleaning via MICROCON: 113.1.

TABLE 11.A.4 Particulate Sampling in Emergency Room Waiting Area

Sample No.	Location	Result μg/m³
13969	Location #1—Pretest; before operating the MICROCON	30
13965	Location #2—Pretest; before operating the MICROCON	20
13966	Location #1—After running the MICROCON 90 min	<20
13968	Location #2—After running the MICROCON 90 min	50

Note: Average levels: Before running the MICROCON HEPA Air Filtration System: 25 μg/m³; After 90 min of running the MICROCON: 35 μg/m³.

reduction in airborne bacteria after running the equipment for 90 min, but no additional reduction is achieved with continual operation of the MICROCON HEPA Air Filtration System.

Source: Edward Olmsted, CIH, CSP, Olmsted Environmental Services, Inc., RR 1 Box 480, Garrison, NY 10524.

Appendix 11.A.3

ClestraClean Room Air Sterilizer Performance Testing

On March 12, 1993 and April 12, 1993 several performance tests were conducted on the new ClestraClean Room Air Sterilizer. Listed below are the tests that were performed and the results of those tests.

1. Airflow measurements

   Description of testing method:
   - All airflow measurements were accomplished by taking duct traverses either at the temporary inlet duct or at the 4 ≤ flex duct attached to the side of the unit.

   Equipment used:
   - Alnor Velometer, Model 6000 P

   Results:
   - Mode: High-speed in full recirculation, 525 cfm
   - Mode: Low-speed in full recirculation, 230 cfm
   - Mode: High-speed in portion of total (4 ≤ round discharge) to ambient, 430 cfm discharge to room, 135 cfm (4 ≤ round) to ambient
   - Mode: Low-speed in portion of total (4 ≤ round discharge) to ambient, 140 cfm discharge to room, 70 cfm (4 ≤ round) to ambient
   - Mode: High-speed in portion of total (4 ≤ round suction) from ambient, 470 cfm from room, 75 cfm (4 ≤ round) from ambient
   - Mode: Low-speed in portion of total (4 ≤ round suction) from ambient, 200 cfm from room, 40 cfm (4 ≤ round) from ambient

2. Volts and amp draw measurements

   Equipment used:
   - AMP Probe, Model RS-3

   Results:
   - High speed in full recirculation mode; 110 V and 3.0 amps
   - Low speed in full recirculation mode, 110 V and 6.0 amps

3. Sound level measurements

   Equipment used:
   - Extech, Model 407703

   Results:
   - High speed in full recirculation mode; 62 dBA (background with unit off was 48 dBA)
   - Low speed in full recirculation mode; 51 dBA (background with unit off was 48 dBA)

4. Pressurization measurements

   Description of testing method:
   - In a reasonably tight room (approximately 1355 CU FT) the pressurization testing was per formed. Negative pressurization was obtained by discharging a portion of the supply air to the ambient through the 4 ≤ ø flex duct connected to the 4 ≤ ø discharge port. Positive pressurization was obtained by drawing a portion of the return air from the ambient through the 4 ≤ ø flex duct connected to the 4 ≤ ø suction port.

   Equipment used:
   - Shortridge Air Data Meter, Model ADM 860

   Results:
   - Room was −0.03 ≤ W.C. to ambient with unit on high speed.
   - Duct discharging a portion of supply air to ambient.
   - Room was −0.01 ≤ W.C. to ambient with unit on low speed and 4 ≤ ø duct discharging a portion of supply air to ambient.
   - Room was +0.002 ≤ W.C. with unit on high speed and 4 ≤ ø duct drawing a portion of return air from ambient.
   - Room was +0.001 ≤ W.C. with unit on low speed and 4 ≤ ø duct drawing a portion of return air from ambient.

5. PARTICLE COUNT MEASUREMENTS

   Description and testing method:
   - Individual particle count measurements were sampled for 1 min with a volume of 1 CU FT of air.
   - Baseline particle counts were measured with the unit off. With the unit on, the particle counter sensing probe was placed approximately 8′ from unit at 30≤ above the floor. Several measurements were taken with the unit on high speed, low speed, and high speed with the 4 ≤ ø duct discharging a portion of the supply to simulate a room under negative pressurization.

   Equipment used:
   - Met One Particle Counter, Model 200

   Results:
   - Baseline particle counts (μm)

0.3	0.5	1.0	2.0	5.0	10.0
256,252	149,066	19,341	4,548	372	86

With the unit on high speed, full recirculation, particle counts were taken again after 15 min:
Results:

0.3	0.5	1.0	2.0	5.0	10.0
10,965	4,670	481	98	11	3

With the unit on low speed, full recirculation:
Results:

0.3	0.5	1.0	2.0	5.0	10.0
11,403	5,099	657	195	21	8

With the unit on high speed, 4≤ ø duct discharging a portion of supply to maintain a slight negative room pressurization:
Results:

0.3	0.5	1.0	2.0	5.0	10.0
30,325	13,488	1,641	548	124	57

Appendix 11.A.4

The following is a partial list of commercially available HEPA filters. There are ceiling mount designs, roll-in floor designs. There are designs that offer air scrubbing and negative pressure in one module. Prices range from $100 to $20,000 depending on type, power, and add-ons.

Each facility needs to make individual assessments of needed capacity and function accordingly.

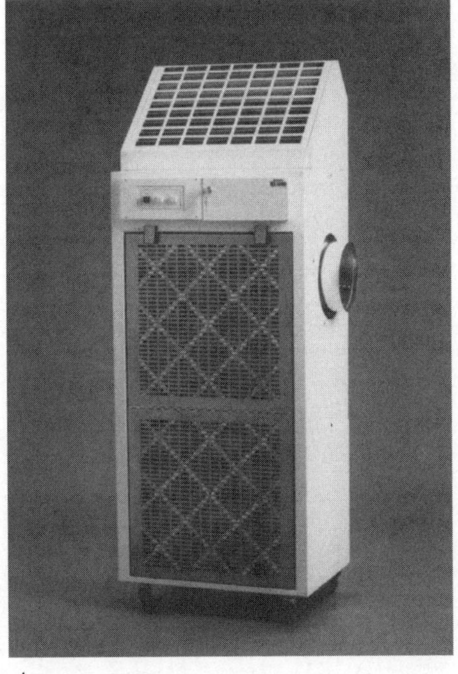

A

Model RS 1000 Portable Air Scrubber/Negative Pressure Unit. Available from BioSafety Systems, Inc., San Diego, CA.

B

MICROCON. Available from Biological Controls. One Industrial Way, Building D, Unit F, Eatontown, NJ 07724-3319, Telephone: 908-389-3319, Fax: 908-389-8821.

MICROCON WallMAP™. Available from Biological Controls. One Industrial Way, Building D, Unit F, Eatontown, NJ 07724-3319, Telephone: 908-389-3319, Fax: 908-389-8821.

Enviro™ Airlock from Medical Safety Products, Inc., Marietta, GA.

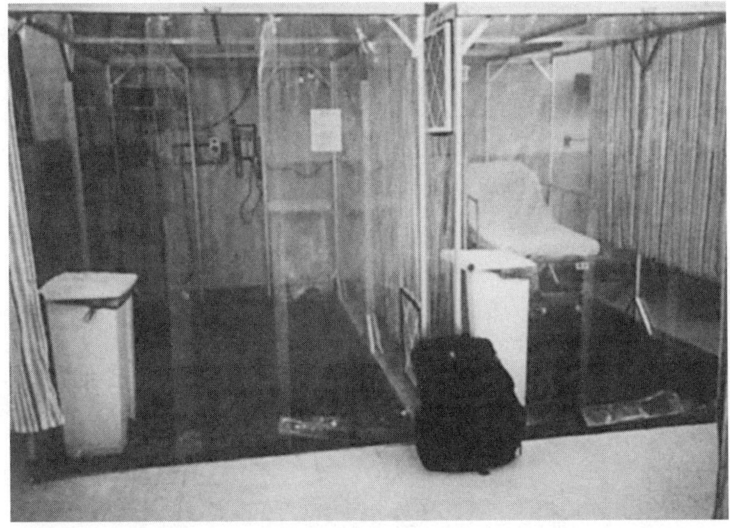

Enviro™ Airlock Isolation Chamber in Emergency Room. Available from Medical Safety Products, Inc., Marietta, GA.

Enviro™ Sputum Induction Chamber in Emergency Room. Available from Medical Safety Products, Inc., Marietta, GA.

HEPAPORT. HEPA Fan/Exhaust. To create a negative pressure room as recommended by the Center for Disease Control (CDC), a filtered room exhaust may be installed in a window or wall powerful enough to overcome under-door or under-curtain leaks. Available from Modern Medical Systems Company, A Division of A. Kingsbury Co., Inc., 1655 Jericho Turnpike, New Hyde Park, NY 11040, 1-800-426-5304.

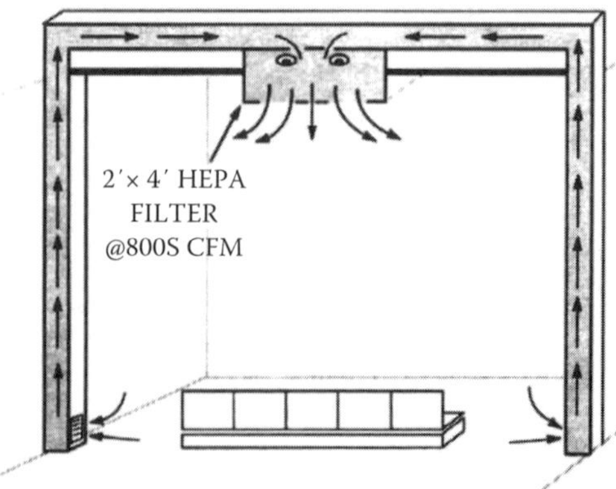

THE CRAB, Ceiling Recirculating Air Blower. For HEPA Air Filtration of waiting rooms, patient rooms, and treatment rooms. Available from Modern Medical Systems Company, A Division of A. Kingsbury Co., Inc., 1655 Jericho Turnpike, New Hyde Park, NY 11040, 1-800-426-5304.

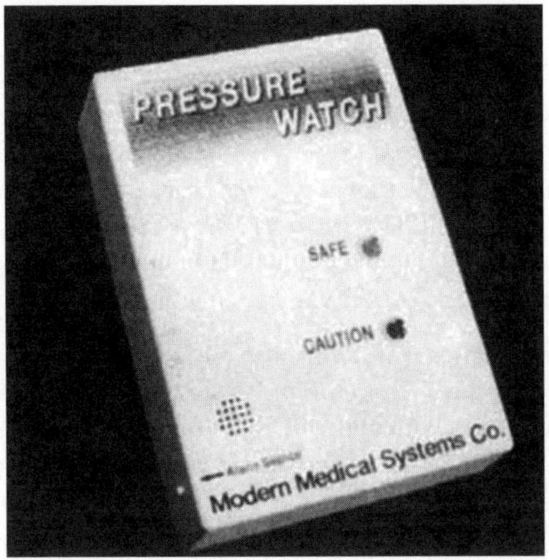

PRESSURE WATCH. For continuous monitoring of negative or positive pressure rooms. Available from Modern Medical Systems Company, A Division of A. Kingsbury Co., Inc., 1655 Jericho Turnpike, New Hyde Park, NY 11040, 1-800-426-5304.

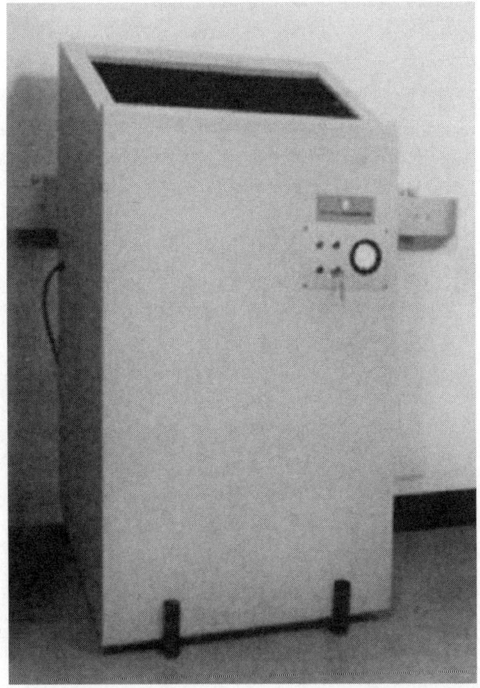

BioShield®. High-efficiency hospital and medical grade air filtration unit. Available from Airo Clean, Inc., Pickering Creek Industrial Park, 212 Philips Road, Exton, PA 19341, Telephone: 610-524-8100. Fax: 610-524-8135.

Engineering Control Options for Tuberculosis in Health Care Settings

William Charney

Introduction

The transmission of tuberculosis (TB) from patient to health care worker (HCW) is again on the rise.[1] Hospitals need engineering methodologies to help prevent the transmission of droplet nuclei TB. Administrative controls include screening and rapid identification of active disease cases. This section will not cover the clinical or administrative controls, but rather the engineering control options available to health care settings.

The engineering control options available include the following:

1. High-efficiency particulate air (HEPA) filter portable units placed in waiting rooms and patient rooms to scrub droplet nuclei from the air (see Figure 11.25).
2. Isolation of patient using a bed canopy HEPA filter.
3. Upper room UV and UV lights in return air duct in a recirculating HVAC system (see the section "Upper Room UV").
4. Centralization and utilization of a sputum induction hood for all sputum inductions.
5. Negative pressure isolation rooms as recommended by the CDC.[2]
6. Isolation of procedures within Plexiglas™ barriers.
7. Upgrading of respiratory protection from surgical mask to a respirator National Institute of Safety and Health (NIOSH) certified for micron range of 1–5 mm (see Respiratory Controls, page 11–38).

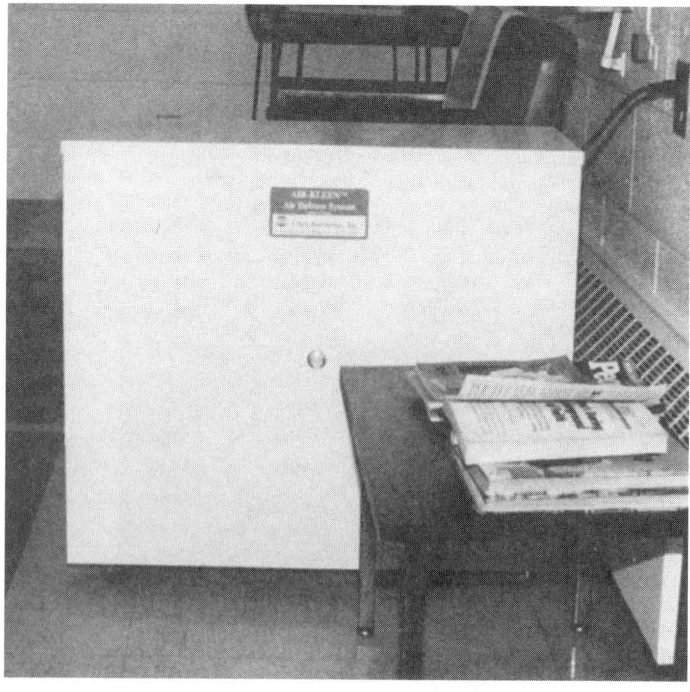

FIGURE 11.25 HEPA filter portable unit.

Discussion

Droplet Nuclei TB

Droplet nuclei TB is still somewhat of an enigma. It is still not known how many droplet nuclei, when dispersed in the ambient air, are required to produce infection in the human host.[3] It is also understood that the longer one is exposed to a given quantity of droplet nuclei TB, the greater the potential for infection.

Duguid[4] showed that 97% of these droplet nuclei were between 0.5 and 12 μm in diameter, the majority of these being 1–2 μm in diameter. These small droplets can remain suspended in a room for more than 90 min.[5]

These droplet nuclei follow air currents within buildings, potentially causing infection well beyond the range of direct person to person contact.

Ventilation

Recirculated air within buildings promotes TB transmission.[6] However, just providing adequate outside air may not solve the transmission problem.

Ventilation systems, however, can lower the probability of the spread of TB in three ways:

1. Quantity of airflow
2. Quality of airflow or scrubbing
3. Direction of airflow

HEPA Filtration

It has been shown that HEPA filters will effectively remove droplet nuclei TB.[7,8]

The first design type HEPA unit is the standing, portable HEPA unit. These units will assist in removing droplet nuclei from the ambient air of a patient's room, waiting room, bronchoscopy suite, etc. These machines are sized and fitted to rooms to create additional room air changes, providing a scrubbing of room air of droplet nuclei, as HEPA filters remove 99.97% of particulate matter down to 0.3 μm. The price range is from $250 to $2000, depending on portability, size, and power of the units and the types of additional filters desired.

HEPA filters can also be designed to fit an existing HVAC unit. However, space is necessary within the mechanical room and these filters create significant additional resistance to airflow beyond the capacity of most residential and light commercial air handling equipment.

HEPA Bed Tent

A second engineering control consideration is the HEPA Bed Tent (see Figure 11.26). This unit fits over the bed, provides a negative pressure environment within the tent, and is scrubbed by a HEPA filter unit. This unit is indicated for relatively nonambulatory patients who are cooperative. It has ports for access and the whole side can be lifted if total patient access is necessary; approximate cost $1500.

Upper Room UV

Both upper room UV and fresh air ventilation decrease the concentration of airborne organisms in logarithmic fashion.[9] Riley et al. described lamps installed in exhaust ducts from the rooms of patients with infectious TB were shown to prevent infection of guinea pigs which are highly susceptible to TB.[10] CDC continues to recommend UV lamps as a supplement to ventilation in settings where the risk of TB infection is high.[11]

Sputum Induction

A fourth method of control is the utilization of a sputum induction hood when performing sputum induction. Sputum induction is a high-risk activity. CDC recommends an individual room or booth with negative pressure. A sputum induction booth (see Figure 11.27) provides negative pressure at the face (face velocity >150 fpm), with a prefilter and a HEPA filter. The unit shown (cost approximately $1000) is

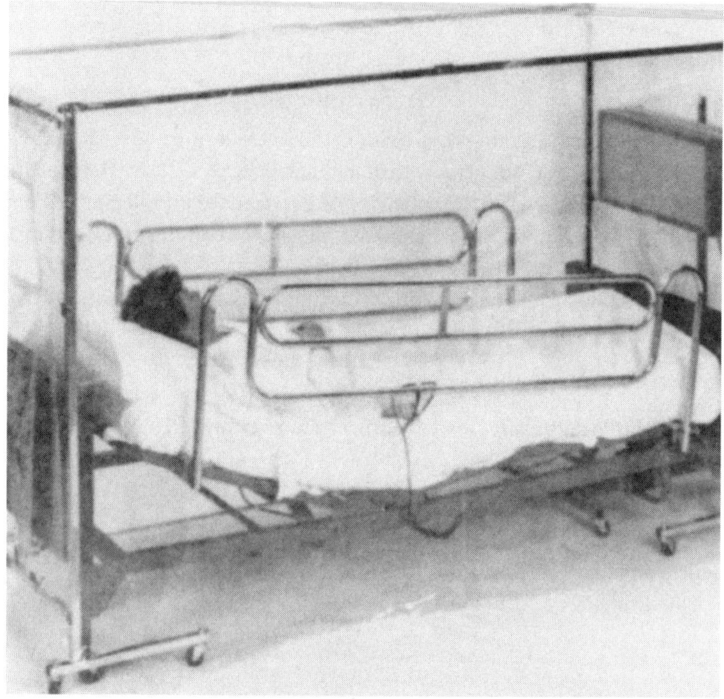

FIGURE 11.26 HEPA bed tent. (Available from Peace Medical.)

FIGURE 11.27 Sputum induction booth.

mobile and can be centralized in one area or moved to a nonambulatory patient room. A sputum induction booth is designed to contain the actual droplet nuclei containing tubercle bacilli. (These areas should not have recirculated air. The benefit of isolation barriers is dependent on nonrecirculated air.)

Isolation Barriers

Plexiglas barriers can be used as separation controls in areas as Emergency Admitting, clinical interviewing, etc. These barriers protect the HCW from potential transmission by providing a physical barrier. This prevents aerosolized coughs, sneezes, etc., from having direct contact with the HCW. These areas should be kept under positive pressure relative to their external surroundings.

HEPA-Assisted Recovery Rate Tests

The purpose of this report is to present the results of HEPA-assisted recovery rate testing in Room 5R1 of the San Francisco General Hospital Medical Center. This testing was performed as a part of the qualification of the benefits of adding an in-room HEPA filtration unit to assist in the reduction in airborne particulates.

Summary

1. TSS measured the airborne particulate recovery rate in ICU Room 5R1 on a single day, May 05, 1998.
2. The in-room HEPA filtration unit under test was an RxAIR Model 3000 (Figure 11.28).
3. Although there were no specific acceptance criteria for this test tour, the overall results agree with natural expectations. As summarized in the following table, using the RxAIR Model 3000 was readily seen to increase the effective ventilation rate in the room:

Case	Effective Ventilation Rate (ach)
HEPA unit off	4.6
HEPA unit on "high"	23
HEPA unit on "low"	19.7

4. This test battery was limited in scope to the execution of TSS SOP 9501, "Airborne Particle Recovery Test for Mechanically Ventilated Spaces" (v1, 4/11/95) in room 5R1 under three cases: (1) with the RxAIR unit off; (2) with the RxAIR unit on "high"; and (3) with the RxAIR unit on "low."
5. Test results are discussed in section 6 of this report.
6. Pertinent additional documentation, including diagrams, instrument calibration records and copies of the field data are in section 7.

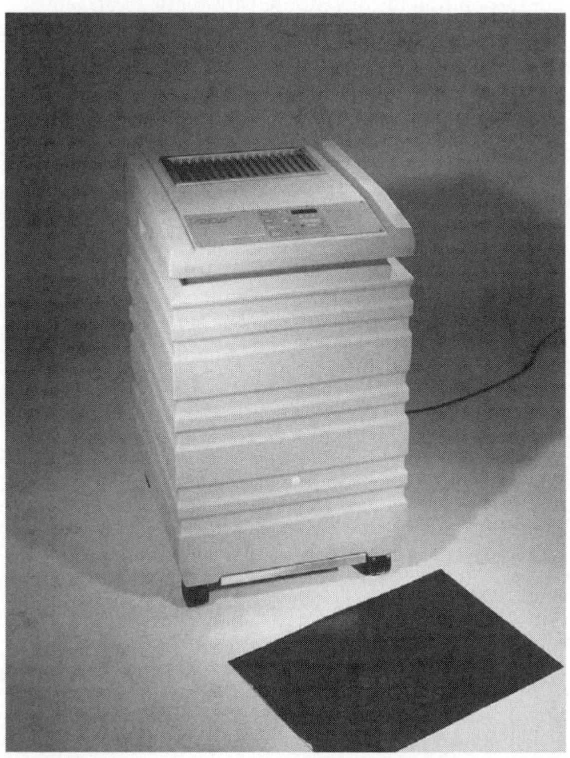

FIGURE 11.28 RxAIR Model 3000.

Test Results

Airborne particle recovery test for mechanically ventilated spaces:

1. In the manner of IES-RP-CC006.2 Section 6.7, as codified by TSS SOP 9501, TSS briefly injected aerosol at the supply register on Room 5R1 while simultaneously and continuously measuring the concentration of airborne particulates > = 0.5 μm. In this manner, TSS recorded the increase and decay of the particulate tracer.
2. Three cases were considered for testing: (1) with the RxAIR unit off; (2) with the RxAIR unit on "high"; and (3) with the RxAIR unit on "low."
3. Four trials were conducted according to the following schedule:

Trial	Case	Result, ach_{EFF}	Comment
1	RxAIR off	4.6	Anticipated low ach_{EFF}
2	RxAIR on high	19.7	
3	RxAIR on low	19.7	
4	RxAIR on high	23	Best ach_{EFF}, as expected

4. Although trials #2 and #3 demonstrated that the RxAIR unit could clean the air in the room, the test method did not resolve its "high" performance from its "low." Accordingly, the RxAIR unit was set to "high" and subjected to trial #4 to better resolve ach_{EFF}, with the result that ach_{EFF} rose solidly to 23.

Conclusion

TB transmission is problematic for hospitals, HCWs, homeless shelters, health clinics, and HIV shelters, and with the fourth and fifth generation multiple resistant strains being seen, it is an even greater problem as these strains are not responding to conventional antibiotic treatments.

On the positive side, there are engineering control methodologies that can help provide a safer environment for HCWs who are exposed to suspected or known cases of TB. These include standing HEPA filter units, ultraviolet lighting in rooms or placed in ventilation ducts, and physical Plexiglas barriers. Though these methods have not been adequately tested, they theoretically are scientifically sound, and until absolute data are available, provide options to help prevent the spread of droplet nuclei TB.

Engineering controls, upgrading respiratory protection, and good administrative controls (rapid screening, etc.), will contribute to containing the spread of TB in the health care setting.

References

1. Centers for Disease Control. Guidelines for preventing the transmission of tuberculosis in health care settings, with special focus on HIV related issues. *MMWR* 39:2, 1990.
2. Centers for Disease Control. Guidelines for preventing the transmission of tuberculosis in health care settings. *MMWR* 39:RR17, 1990.
3. Riley R and Nardell E. Clearing the air. *Am Rev Respir Dis* 139:1266–1294, 1989.
4. Duguid JB. Source significance and control of indoor microbiological aerosols. *Publ Health Rep* 3:229, 1983.
5. Wells WF. *Aerodynamics of Droplet Nuclei*. Harvard University Press, Cambridge, MA, 1955, pp. 13–19.
6. Centers for Disease Control. Mycobacterium tuberculosis transmission in a health clinic, Florida. *MMWR* 38:256–64, 1989.
7. Wickramanayake GB. Decontamination technologies for release of bioprocessing facilities. *Environ Rev* 9(5):447, 1989.
8. Gibraltar Biological Laboratory. Report #54837—Capacity of peace HEPA filter to retain certain pathogenic organisms. Fairfield, 1990.

9. Riley R and Nardell E. Clearing the air, the theory and application of UV air disinfection. *Am Rev Respir Dis* 139:1257–1294, 1989.
10. Riley RL, Mills CC, et al. Infectiousness of air from a tuberculosis ward. *Am Rev Respir Dis* 88: 511–525, 1962.
11. U.S. Department of Health and Human Services (Public Health Service). Guidelines for preventing tuberculosis transmission in health-care settings, with special focus on HIV related issues. *MMWR* 39:RR17, 1990.

The Inefficiency of Surgical Masks for Protection against Droplet Nuclei TB

William Charney and June Fisher

The general perception of the average health care worker is that the generic surgical mask protects the wearer against airborne droplet nuclei *Mycobacterium tuberculosis*. However, a review of the literature shows that droplet nuclei tuberculosis (TB) with a diameter of 1–5 µm will not be stopped by the standard surgical mask. Surgical masks do not have a tight seal around the face; therefore, air seeking the path of least resistance will transmigrate around the unsealed edges into the respiratory zone of the wearer.

Pippin[1] tested the efficiency of surgical masks and found that the passage of inspired air around the periphery of the two types of masks tested compromised the masks' capacity to screen airborne contaminants. This study used tracer particles of 22.4 µm, which are 4×–20× larger than the droplet nuclei TB, 1–5 µm.[2] Tuomi[3] of the Department of Industrial Hygiene and Toxicology, Helsinki, Finland tested two types of surgical masks, the Surgene and Aseptek. The respirators were tested on a test head connected to a breathing simulator. The mass median diameter (MMD) of the polydispersed corn oil aerosol used in this study ranged from 0.52 to 3.2 µm. The overall efficiency for the surgical masks was 68% for Surgene and 33% for Aseptek.

Riley and Nardell[4] also state that ordinary clinical masks are ineffective as a precaution against airborne infection, rendering them ineffective.

The Centers for Disease Control[5] states that standard surgical masks are not effective in preventing inhalation of droplet nuclei because they are not designed to provide a tight face seal or to filter out particulates in the droplet nucleus range 1–5 µm.

Wentzell[6] also commented on the questionable protective effect of surgical masks with regard to potentially infectious particulates, since the average surgical mask is designed and tested to prevent contamination from the health care provider to the patient and not to protect the health care worker against inspired particulates.

The principle of respiratory protection is clear, that respirators shall be fitted properly and shall be tested for facepiecetoface seal: 29 CFR 1910.134 (e)(5). The federal regulation also states that respirators shall not be worn when conditions prevent a good face seal: 29 CFR 1910 (e)(5)(i).

The standard surgical mask does not meet this condition for qualitative fit testing. (Qualitative fit test helps insure maximum efficiency between facepiece and wearer.) The standard surgical masks were also, as stated by Wentzell, designed to protect the patient, not the wearer. TB infection occurs when droplet nuclei are inhaled by a susceptible person, as the TB particles are small enough that they remain suspended and follow general air currents.[7] And since there is no definitive dose–response relationship between the number of droplet nuclei one needs to inhale in order to provoke disease,[8] the best possible respirator fit is recommended.

Therefore, a NIOSH/MSHA certified respirator (see Figure 11.29) that can prevent penetration of aerosolized particles in the range of 1–5 µm is indicated for health care workers exposed to known or suspected TB patients, and the dependency on surgical masks is neither quantitatively nor qualitatively compatible with protecting health care workers from droplet nuclei TB for short duration or long duration exposure.

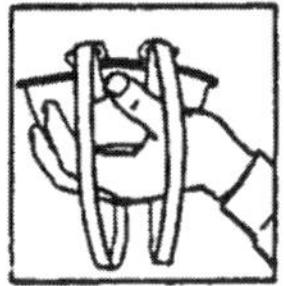

1. Untwist straps. Cup mask in hand with nose contour (narrow end) at fingertips, allowing headstraps to fall below hand.

2. Place mask under chin with nose contour (narrow end) up.

3. Pull bottom strap over head and below ears, to neck. Raise top strap to crown of head. Untwist straps.

4. If mask does not fit snugly over face, remove mask from head. To adjust tension on straps, gently pull straps to correct position by pulling ends of straps.

5. Put mask back on face. Do not adjust straps with mask on face!

FIGURE 11.29 MOLDEX 2200 Fume, Dust Mist Respirator. Fitting instructions.

The cost of the upgrade to a NIOSH/MSHA approved mask is approximately 67 cents as compared to 23 cents for the standard surgical mask. The 44 cents' difference, however, can be offset by HCWs using the NIOSH respirator more than one time, which is permissible.

References

1. Pippin J. Efficiency of face masks in preventing inhalation of airborne contaminants. *Oral Maxillofac Surg* 45:319–23, 1987.
2. U.S. Department of Health and Human Services (Public Health Service). Guidelines for preventing tuberculosis transmission in healthcare settings, with special focus on HIV related issues. *MMWR* 39:RR17, 1990.
3. Tuomi T. Face seal leakage of half masks and surgical masks, *Am Ind Hyg Assoc J* 45(6):308–12, 1985.
4. Riley R and Nardell E. Clearing the air, the theory and application of ultraviolet air disinfection. *Am Rev Respir Dis* 139:1290, 1989.
5. Dooley SW Jr, Castro KG, Hutton MD, Mullan RJ, Polder JA, and Snider DE Jr. Guidelines for preventing tuberculosis transmission in health care settings. *MMWR Recomm Rep* 39, 1–29, 1990.

6. Wentzell JM. Physical properties of aerosols produced by dermabrasion, *Dermatol* 125:1637–43, 1989.
7. Wells WF. *Aerodynamics of Droplet Nuclei*. Cambridge University Press, New York, 1955, pp. 13–19.
8. Riley R and Nardell E. Clearing the air, the theory and application of ultraviolet air disinfection. *Am Rev Respir Dis* 139:1286, 1989.

Appendix 11.A.5

Using Ultraviolet Radiation and Ventilation to Control Tuberculosis

Introduction

Tuberculosis (TB) infection and disease remains a worldwide burden despite efforts to identify and treat cases. After many years of slow but steady decrease in the incidence of TB in the United States, this disease is making a comeback. It is a serious public health problem among the homeless, some groups of immigrants, inmates of correctional institutions, and people infected with or at high risk for infection with the human immunodeficiency virus (HIV). Concerns about the continued spread of TB infection have led to an increased interest in using environmental control measures to protect the abovementioned groups of people, health care personnel, and the general public from TB infection and disease.

How TB Spreads

TB is spread (transmitted) from person to person by tiny airborne particles containing tubercle bacilli. Particles, called "droplet nuclei," containing these bacteria are coughed up by persons with untreated or inadequately treated, clinically active, pulmonary or laryngeal TB. Droplet nuclei are carried on air currents and disperse rapidly throughout a room. The longer that a susceptible (uninfected) person shares the same air space with someone who has active pulmonary or laryngeal TB, the greater the chance that the susceptible person will inhale airborne TB bacilli. Therefore, the goal of environmental infection control is to prevent the spread of TB by protecting susceptible people from inhaling airborne particles generated by infectious individuals.

Environmental Control Measures

Airborne infections such as TB can be prevented by killing the infectious microorganisms in the air with ultraviolet (UV) radiation. UV lamps are inexpensive, easy to install and maintain, and are effective at killing airborne microorganisms. Therefore, some health authorities recommend UV lamps for certain high-risk environments.

Although environmental control measures such as air disinfection and ventilation can decrease the transmission (i.e., airborne spread) of TB, they are only supplements to the usual control measures. Environmental controls cannot replace conventional interventions such as the prompt detection and treatment of cases, and tracing of contacts. However, if the identification and follow-up of TB cases is difficult, UV air disinfection and exhaust ventilation can provide an extra measure of protection.

UV Radiation

Questions and Answers on the Use of UV Radiation for TB Control

How are UV lamps used to disinfect the air? Germicidal UV lamps (low-pressure mercury vapor lamps which emit radiation near 254 nm) can be used in ceiling fixtures suspended above the people in a room or within air ducts of recirculating systems. The first method is called overhead or upper air irradiation. The fixtures are shielded on the bottoms so that the radiation is directed only up toward the ceiling and out to the sides (see Figure 11.A.3). The bottoms of the fixtures should be at least 7 ft.

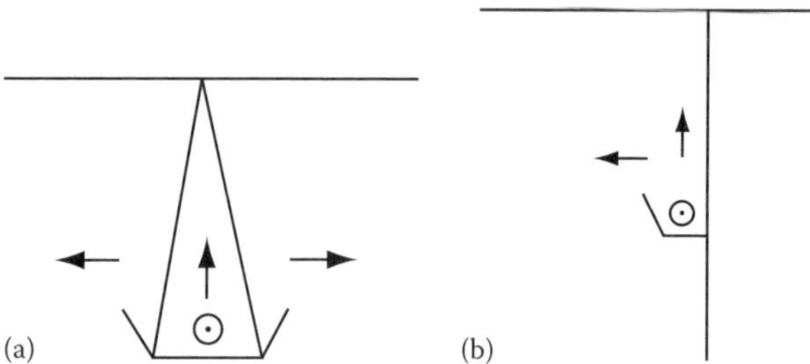

FIGURE 11.A.3 Examples of shielded UV fixtures: (a) ceiling-mounted fixture and (b) wall-mounted fixture (the arrows indicate the direction of the radiation).

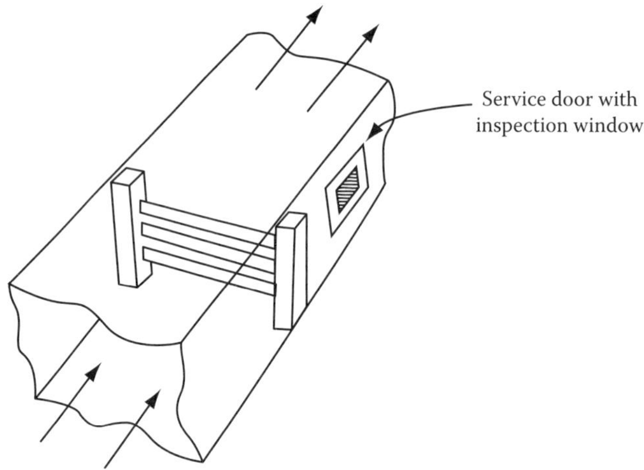

FIGURE 11.A.4 UV lamps located inside ventilation system duct.

above the floor, so that people will not bump into them or look directly at the bare tubes.[1,2] Overhead UV lamps are most effective in rooms in which the ceilings are at least 9 ft high, because this allows a significant portion of the air to be irradiated.

Fixtures with adjustable louvers which control the direction of the UV rays may be used (with caution) in areas with lower ceilings.

The air entering or leaving a room can be disinfected with UV lamps placed inside the ventilation system ducts, as illustrated in Figure 11.A.4. This method of air disinfection could be used, for example, in a TB clinic from which air is recirculated to other parts of the building, but in which overhead disinfection cannot be used because the ceiling is too low. UV lamps could be placed instead inside the exhaust duct to kill any microorganisms in the air leaving the clinic. Because people are not exposed to the UV radiation, very high levels can be used inside ducts.

How effective is UV radiation at killing TB bacteria?

The following factors, which are discussed in detail below, play a part in determining whether or not exposure to UV radiation will kill a microorganism:

1. Type of microorganism
2. Dose of radiation to which it is exposed
3. Amount of moisture in the air

Type of microorganism: UV radiation rapidly kills airborne bacteria and viruses, but it is less effective against airborne fungal and bacterial spores. The bacteria that cause TB are more resistant to UV than some others, but they are not as difficult to kill as spores. UV radiation does not penetrate well through matter, therefore, bacteria carried in large particles of dried sputum, for example, might be shielded from the disinfection radiation. However, these larger particles do not remain suspended in the air for very long nor do they reach the lower lung if inhaled, and therefore do not pose as great a risk of infection as do smaller, more UV-sensitive particles.

Radiation dose: For air disinfection to be effective, the TB bacteria must receive a sufficient dose of UV radiation. This can be achieved by using lamps of the correct wavelength and intensity (wattage) and by exposing the bacteria for sufficiently long periods of time. There must be good mixing between the irradiated upper room air and the air lower in a room where people are. In this way, contaminated droplets are moved into the irradiated zone and the disinfected air dilutes the contaminated lower air. Portable or ceiling fans can be used to increase air circulation.

Moisture in the air: At typical indoor temperatures, relative humidities above 70% reduce the bactericidal efficiency of UV radiation. This level of moisture is not common in air conditioned buildings (unless the air conditioning equipment is not operating properly) or during the colder months of the year when airborne transmission is more likely. However, if excess humidity is a problem, air disinfection must exclusively rely on other forms of environmental control, such as increased ventilation or duct irradiation.

For what types of facilities should overhead UV radiation be considered as a means to reduce the transmission of infection?

Overhead UV lamps are useful in crowded and poorly ventilated buildings where the conventional control methods are inadequate. Examples of such areas are

Shelters for the homeless[3]
Correctional institutions[4]
Nursing homes
Hospitals: emergency rooms, operating rooms, intensive care areas, some laboratories
TB clinics: sputum collection rooms, aerosol treatment areas, bronchoscopy rooms
AIDS clinics: aerosol pentamidine treatment areas[5]

In such places, there occasionally are unidentified TB cases. UV lamps also might be appropriate in pediatricians' offices, where the disinfecting radiation would help control the transmission of other airborne infections such as measles.

Is the airborne spread of TB in high-risk settings best controlled by mechanical ventilation, air cleaners, or UV air disinfection?

The American Society of Heating, Refrigerating and Air Conditioning Engineers (ASHRAE) recommends an outdoor air supply of 15–20 cubic feet of air per minute for each person in a room (15–20 cfm/person) for comfort purposes.[6] However, this amount of ventilation would not significantly reduce the number of airborne TB bacteria that a highly infectious person could produce. It would take more than 100 cfm/person of outdoor air to reduce the number of airborne bacteria to the same level that a UV lamp could.[1,2] Supplying that much outdoor air would be impractical in many situations, because it is very expensive to heat or cool such large quantities of outdoor air and to move it through a ventilation system.

Indoor air that might contain infectious particles cannot safely be recirculated back into a room unless the airborne microorganisms have been removed or inactivated. Filtering the air, installing UV lamps inside the air ducts, and other forms of air cleaning can effectively remove or kill bacteria and are suitable, or even necessary, in certain facilities. However, high-efficiency air cleaners are more expensive to install and maintain than overhead UV lamps, and generally are not practical for treating the air in large open rooms. See the section beginning on page 11–47 for more details on the use of general and local exhaust ventilation for the control of TB.

It is often best to design a ventilation system to provide at least 15 cfm/person for comfort, to use exhaust hoods and special booths to contain airborne particles produced by high-risk procedures, and to consider using overhead UV radiation to further reduce the airborne spread of infectious particles.

Checklist of Features to Consider before Recommending Overhead UV Air Disinfection

	Yes	No
A potentially high risk of TB transmission in the population that cannot be controlled adequately by conventional interventions	()	()
Sufficiently high ceilings (at least 9 ft) that are relatively free of obstructions	()	()
Relative humidity (RH) below 70%	()	()
The capability of maintaining and operating the UV lamps effectively and safely	()	()
A means of measuring the reduction in airborne infections among the occupants	()	()
If the answer to any of the above questions is "no," overhead UV disinfection might not be suitable for the proposed situation.		

How are UV lamps used inside heating, ventilating, and air conditioning (HVAC) system ducts?
When duct irradiation is used, the lamps should be placed at right angles to the direction of airflow and installed at the center of the longest available run, as illustrated in Figure 11.A.4. Because the effectiveness of germicidal UV radiation is dependent on both the level of radiation and the exposure time, it is best, whenever possible, to install UV lamps where the velocity of the air is lowest.

The access door for servicing the lamps should have an inspection window (ordinary glass is sufficient to filter out the UV rays) through which the lamps periodically are checked, and a sign in appropriate languages alerting maintenance personnel to the health hazard of looking directly at bare tubes. The lock for this door should have an electric switch so that the lamps turn off automatically whenever the door is opened.

How do I decide how many overhead UV lamps an area needs?
The number of fixtures needed depends on the size of the room the height of the ceilings, and the location of the supply air diffusers. The higher the UV radiation intensity, the more rapidly microorganisms are inactivated. Therefore, the more lamps the better, so long as human exposure to UV is kept at safe levels. As a general rule of thumb, install the equivalent of one 30 W fixture for every 200 ft^2 of floor area or for every seven people in the room, whichever is greater. Using two 15 W lamps instead of one 30 W lamp might provide better irradiation to the upper room air and reduce the risk of exposing the occupants to unsafe levels of UV radiation. This amount of air disinfection is considered the equivalent of 20 room air changes per hour (ach) or approximately 100 cfm/person.[1,2] Ceiling fixtures should be oriented so that the radiation is directed into the longest part of a room (see Figure 11.A.5). If multiple fixtures are installed, they should be spaced evenly so that there is minimal overlap of irradiation.

Can upper air UV irradiation cause any problems?
UV radiation is divided into three regions. UV-A (long UV 400–320 nm), and UV-B (midrange UV 320–290 nm) produce sunburn and tanning. Prolonged exposure can cause skin cancer and cataracts.[7] Germicidal UV radiation is in the UV-C region of the spectrum (short UV 290–100 nm). UV-C may cause reddening of the skin and conjunctivitis (a feeling of sand in the eyes), but not skin cancer or cataracts in humans.[8] Because UV-C is absorbed in the outer layers of the skin and eyes, the irritation produced by overexposure is superficial. Although serious burns can be incapacitating at the time, the effects usually disappear within 24 h without lasting effects.[8]

Intense UV radiation can fade colored paints and fabrics, speed up the deterioration of plastics, and cause plants to wilt. Because prolonged exposure to UV radiation can damage skin and eyes, one should never look directly at the lighted tubes for more than a few seconds unless wearing appropriate safety goggles or a face shield. Also, one should wear a long-sleeved shirt and gloves if it is necessary to work near lighted lamps. Maintenance personnel should exercise appropriate precautions against breakage when replacing and disposing of these tubes. Modern UV lamps do not produce ozone.

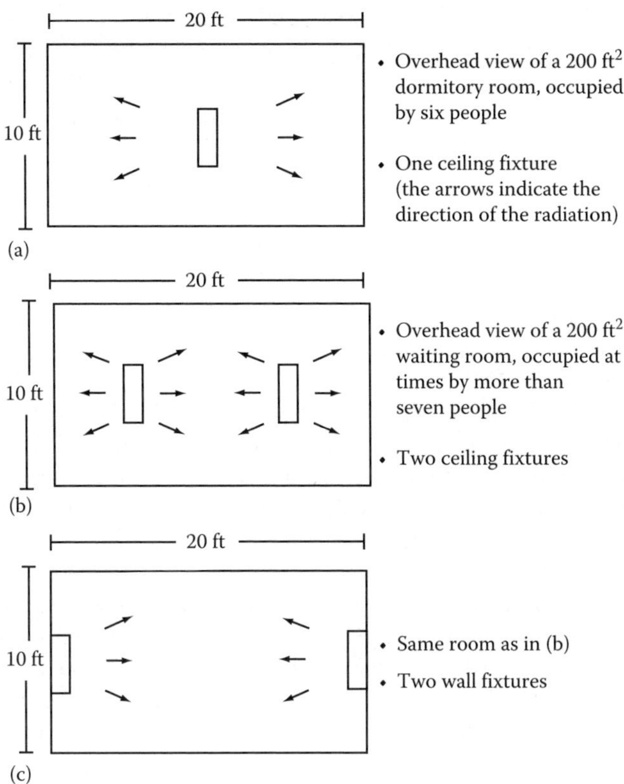

FIGURE 11.A.5 Placement of UV ceiling fixtures.

It also must be remembered that UV lamps produce some visible violet light. While the color is scarcely noticeable during the day, at night it might disturb people who are trying to sleep. Nonetheless, the lamps should be operated whenever people occupy the controlled rooms. In buildings that are occupied all day, the lamps should be left on continuously. Frequent on–off cycling tends to shorten the life of some types of lamps.

A person should be assigned responsibility to maintain the UV fixtures, to train personnel in safety procedures, and to keep a maintenance log. A maintenance and safety log should include information on tube cleaning, baseline and periodic meter readings for irradiation areas (between lamp and ceiling), and for employee/client exposure areas, training, and complaints of eye or skin irritation. There is no substitute for on-site meter readings to ensure that germicidal levels are attained between tubes and ceilings or walls, while at the same time ensuring that room occupants are not exposed to injurious levels. Employees should be trained in the hazards of exposure to UV radiation. This training should include

1. Information as to the proper eye protection clothing to be used.
2. Instructions on how to recognize the symptoms of eye and skin damage due to UV radiation.

How can I be sure that new UV lamps have been installed safely?

When the lamps are installed, an experienced person should measure the level of UV radiation at the locations where people will be exposed. All readings must be below the American Conference of Government Industrial Hygienists' (ACGIH) limit, which for germicidal lamps is 0.2 μW/cm^2 for an 8 h period.[8] If you plan on purchasing a UV radiometer, make certain that it can measure the necessary range of wave lengths and is sufficiently sensitive. (The exposure limit refers to 254 nm UV radiation, which is the predominant wavelength produced by germicidal lamps.) The permitted exposure limit

CAUTION	CUIDADO
High intensity Ultraviolet energy	Alto Intensidad Energía Ultravioleta
Protect eyes & skin	Protege Ojos y Piel

FIGURE 11.A.6 Warning labels.

is 6 mJ/cm^2 or 6000 μW s/cm^2. For an 8 h day, the average exposure level limit is 0.2 μW/cm^2. For 24 h exposure, the limit would be 0.07 μW/cm^2.

To reduce UV exposure in the occupied portions of a room, install baffles to prevent people from looking directly at the tubes. If necessary, use paints containing titanium dioxide (TiO_2) to reduce reflection from ceilings and walls, or otherwise treat them to reduce reflection. Once a room has been checked to ensure that all occupied areas are below the NIOSH exposure limit, these measurements need not be repeated every time the lamps are changed.

If the UV lamp manufacturer or installer does not supply warning labels, make or purchase your own to identify the fixtures as dangerous for direct eye and prolonged skin exposure. Figure 11.A.6 is a sample warning label recommended by NIOSH.[9] The NIOSH document also states that for this type of lamp, warning labels are required on the lamp and housing, but not in the work area. Make certain that the switches to turn the UV fixtures off for inspection, cleaning, and tube changing are accessible when needed but are out of the reach of unauthorized people so that the lamps cannot be fumed on or off accidentally.

How often must the tubes be changed?
Unless a facility has an appropriate radiometer and a trained staff member who can check lamp output periodically, we recommend that the tubes be replaced each year or at the end of their rated life if that is less than twelve months.

How often must the lamps be cleaned?
TURN THE LAMPS OFF BEFORE INSPECTING THEM. The tubes and the fixture trough should be checked periodically (perhaps every three months) and cleaned when needed. We do not recommend frequent lamp reading or cleaning, unless performed by a properly trained and equipped person, because of the risk of accidents.

How much does it cost to use overhead UV air disinfection?
New UV fixtures cost between \$100 and \$500, depending on their design, and new lamps cost between \$50 and \$100 per tube. You can estimate how much it will cost to operate a lamp continuously for one year by multiplying the input wattage by the number of hours of use (8760 h per one year) and the local cost of electricity (see the formula below). UV lamps should be purchased as they are needed, because prolonged storage results in a loss of UV intensity.

$$\left(\frac{\text{Input wattage}}{1000\,\text{W/kW}}\right) \times (\text{hours of use}) \times \left(\frac{\text{cost}}{\text{kW-h}}\right) = \text{operation cost}$$

Where can I find someone to install germicidal UV fixtures?
You may be able to locate suppliers by contacting your local health department, a hospital, or microbiology laboratory, since these entities may already have experience of their own with UV fixtures. Prior to purchase and installation of tubes and/or fixtures, you should review the information in this

booklet with your supplier and/or installer to ensure that appropriate fixtures are provided and properly installed. If your supplier/installer has not recently been involved in these types of installations, familiarization with the material in this booklet, especially as it pertains to safety, is all the more important.

Recommendations

Suggestions on the proper use of upper air UV irradiation

1. Use germicidal lamps that emit short wave (254 nm) UV radiation.
2. Install the equivalent of one 30 W ceiling or wall fixture for every 200 ft^2 of floor area or for every seven people in a room, whichever is greater. Always check that your installation does not exceed safe exposure levels for workers or clients.
3. Position the lamps to irradiate the greatest area.
4. Place labels on the fixtures to warn people to avoid direct eye and prolonged skin exposure.
5. Check that the level of UV radiation where people will be exposed does not exceed 0.2 μW/cm^2 (assumes an 8 h exposure). Recheck whenever reflective surfaces or room contents are changed.
6. Change the tubes annually or when meter readings indicate tube failure.
7. Inform all employees of the potential hazards of overexposure to UV radiation and provide adequate personal protection to anyone who needs it.
8. Delegate a person to maintain the lamps and fixtures to see that safety measures are followed and to keep a log of when the lamps are replaced.

Technical Details about UV Radiation

As mentioned previously, UV radiation is divided into three regions by wavelength: long, middle, and short waves. UV radiation for the purpose of killing microorganisms is produced by lamps that emit radiation in the short wave region near 254 nm (there are 25 million nanometers in an inch), which is near the peak of bactericidal effectiveness. All UV wavelengths are shorter than visible light; UV rays are therefore invisible to the human eye and should be referred to as UV radiation or UV energy, not as UV light. However, UV lamps also emit a small amount of visible violet-blue light.

Two important principles (the reciprocity and inverse square laws), which are helpful in understanding how UV lamps can be used for air disinfection, are discussed below.

Reciprocity law: The dose of radiation to which microorganisms are exposed is equal to the intensity of the radiation times the duration of exposure. Equal doses of UV radiation have the same disinfecting action. Therefore exposure to 100 μW/cm^2 for 10 s has the same killing power as exposure to 50 μW/cm for 20 s.

Inverse square law: As with visible light, the intensity of UV radiation diminishes as the square of the distance from a source. For example, two feet from a point source the intensity of the radiation would be only $(1/2)^2$ or 1/4 of what it was at one foot, and at four feet it would be $(1/4)^2$ or 1/16 of the level at 1 ft.

Distance from UV Source (ft)	Measured UV Intensity (μW/cm^2)
1	800
2	200
4	50

Because UV lamps are long, narrow sources of radiation rather than point sources, it is difficult to predict the UV intensity at a given distance from a lamp. Therefore, it is more reliable to measure the actual exposure level with a sufficiently sensitive UV meter.

Ventilation

The concentrations of contaminants in a room can be controlled with mechanical ventilation in two ways. The first is called local exhaust ventilation. The second is referred to as dilution ventilation. These methods are described below.

Local Exhaust Ventilation

A local exhaust ventilation system removes airborne contaminants at or near their sources and can contain infectious aerosols very effectively. An example would be the safety hoods that laboratory workers use when handling specimens. Similar enclosures often are used for aerosol-generating activities such as sputum collection and aerosol therapy. Detailed information on the design of local exhaust systems is available in *Industrial Ventilation: A Manual of Recommended Practice.*[10]

The major components of a local exhaust ventilation system are (1) the hood or enclosure, (2) the ductwork through which the air moves, (3) an air cleaner, and (4) the fan (see Figure 11.A.7). An air cleaner, such as a high-efficiency air filter, a UV lamp, or an air incinerator, is needed when the contaminated air cannot be exhausted safely or is recirculated to a room.

The booth illustrated in Figure 11.A.8 is designed for use during sputum induction or aerosolized pentamidine treatment. Patients enter the booth through a door in the side and either sit or lie on a bench. Air is drawn into the chamber through a prefilter (1) located in the front, then moves up past the patient and over the top of a back panel (2). The air is filtered through a HEPA (high efficiency particulate air) filter (3) before it is exhausted back into a room (4).

Also on the market are vinyl chambers that isolate the patient in an enclosure similar to an oxygen tent canopy. The canopy falls to within 2 in. of the floor. The air is drawn upward from the open area

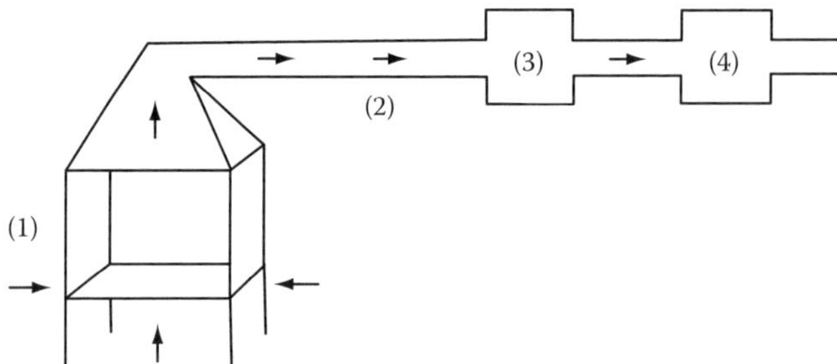

FIGURE 11.A.7 Sputum collection/aerosol therapy cabinet (the arrows indicate the direction of air movements).

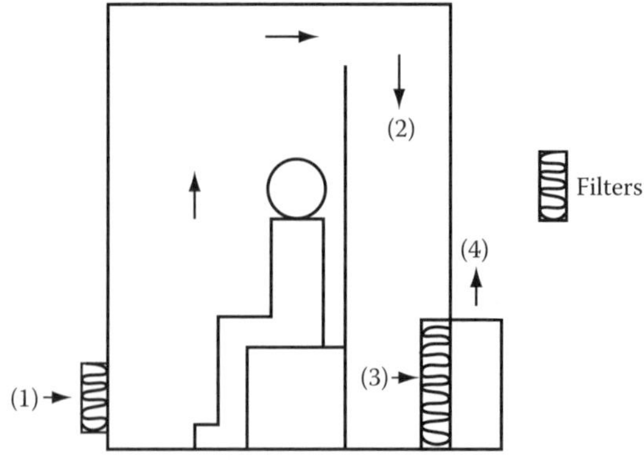

FIGURE 11.A.8 Side view of a sputum collection/aerosol therapy booth.

around the skirt and exhausted from the enclosure by a fan through a HEPA filter. The chamber is collapsible, conveniently sized so that it takes little space and is on wheels for increased mobility.

Dilution Ventilation

Dilution ventilation is a process of supplying pollutant-free air into a space to reduce the level of contamination. Some ventilation systems provide 100% outdoor air and exhaust all return air. Other systems, in the interest of energy conservation, recirculate part of the return air.

If a ventilation system is used to protect people from infectious particles, the system must draw the contaminated air away from the people in the room and carry it safely outside. The American Institute of Architects (AIA) has developed construction and operation guidelines for hospitals and medical facilities.[11] A design that they recommend for reducing exposure to air contaminants generated by people is to place the air supply diffusers in the ceiling and the exhaust air grilles (also called registers) near the floor (see Figure 11.A.9). The idea behind this arrangement is that clean air enters above the occupants and the contaminated air is removed at floor level. This design would not be appropriate if overhead UV lamps were being used because the contaminated air would be directed away from the lamps. When overhead UV lamps are used, the supply air diffusers should be near the floor, and the exhaust air grilles near or in the ceiling.

Contaminated air removed from a room should not be recirculated back into the room or to other parts of a building unless it passes through a high efficiency filter, or an air incinerator, past a high intensity UV lamp, or is otherwise disinfected. Untreated exhaust air should be discharged well away (at least 25 ft.) and downwind from the air intakes of any buildings.[10] Air often is exhausted at the roof of a building so that any contaminants it contains are well diluted before they reach street level. Detailed guidelines for designing ventilation systems for infection control can be obtained from the AIA[11] and from the ASHRAE.[6]

Room Air Distribution

Short circuiting of air. If a supply air diffuser is only a few feet away from a return air grille, part of the supply air will be exhausted before it mixes with the air in the room. This is called "short circuiting."

Air stagnation. Given average room air movement and the nonuniformity of natural convection currents, stagnant regions are formed in a room near the ceiling during cooling and near the floor during heating. The air in a stagnant region does not mix well with the air in the rest of the room.

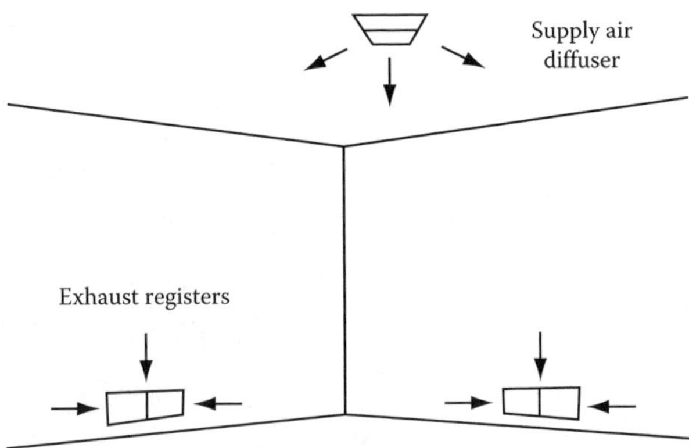

FIGURE 11.A.9 Typical ventilation system design for a hospital (the arrows indicate the direction of air movement).

Air mixing. The mixing of supply air with the air in a room is affected by (1) the configuration of the supply diffusers and the return air grilles, (2) the volume of air supplied to the room, and (3) the operating mode of the ventilation system, that is, whether it is heating or cooling the air.

Calculating the Number of Room Air Changes per Hour

Engineers sometimes express dilution ventilation in terms of the air change rate, which is measured as the number of room air changes per hour (ach). An air change rate of 1 ach means that an amount of air equal to the volume of the room is removed and replaced once every hour. CDC recommends that rooms for respiratory isolation have (1) separate exhaust ventilation to the outside, (2) negative air pressure in relation to adjacent areas, and (3) provide at least 4–6 ach of outdoor air.[4,12] To prevent the transmission of TB in high risk areas, the CDC and the American Thoracic Society (ATS) recommend 20 ach.[13] Reference 17 cites the American Society of Heating, Refrigerating and Air Conditioning Engineers and the Health Resources and Services Administration (our Reference 6) as recommending that acid fast bacilli isolation rooms: “should have at least six total air changes per hour including at least two outside air changes per hour…” A facility could meet this recommendation by supplying 4 ACH of air recirculated from other parts of the building and 2 ach of outdoor air. Reference 17 also gives ventilation recommendations for intensive care units and emergency and autopsy rooms, and discusses the use of personal respirators for patients while they are being transported, and for health care providers while they are treating patients with suspected or confirmed infectious TB. Although an air change rate as high as 20 ach may be desirable in high risk areas and has been recommended in earlier documents, we realize that this level of ventilation is impractical for many facilities and that the resultant air movement may cause drafts.

For a given air change rate and room size, the required amount of supply air in cubic feet per minute (cfm) can be calculated, as follows:

$$\frac{(\text{Room volume in cubic feet}) \times (\text{number of ach})}{60\,\text{min/h}}$$

For example, a 20 ft × 10 ft room with a 10 ft ceiling would need 667 cfm of supply air to have an air change rate of 20 ach:

$$\frac{(20\,\text{ft} \times 10\,\text{ft} \times 10\,\text{ft}) \times (20\,\text{ach})}{60\,\text{min/h}}$$

You can find out what the design airflow rates are for a given area by asking the building manager to check the original plans. The operator of your heating, ventilating, and air-conditioning (HVAC SYSTEM DUCTS) system or a contractor can determine the actual supply airflow rate for a particular room by measuring it with special instruments.

As mentioned, ASHRAE recommends a ventilation rate of at least 15 cubic feet of outdoor air per minute per person (15 cfm/person) for comfort.[6] The ASHRAE standard assumes that there are 10 person/1000 ft^2 in hospital treatment rooms, and 30 persons/1000 ft^2 in lobby areas. In a well-designed facility, 15 cfm/person will sufficiently dilute the odors that people produce, but it should not be relied on to reduce exposure to highly infectious air contaminants.

Referring again to the 2000 ft^3 room used as an example above, when occupied by seven people, this room would need a supply of at least 105 (cfm) or outdoor air (7 people × 15 cfm/person = 105 cfm) to conform with the ASHRAE standard. However, this would supply only three air changes an hour [(105 cfm × 60 min/h)/2000 ft^3 = 3 ach]. This amount of dilution air would not be sufficient to control the spread of TB.

For comparison, if the 20 ach that the CDC/ATS[13] recommends was supplied with outdoor air, the ventilation rate would have to be 95 cfm/person (667 cfm/7 people = 95 cfm/person), an impractically high ventilation rate for most facilities. However, the equivalent of this amount of TB-free air could be provided with upper air UV disinfection, as discussed in Section II.

Supplying Make-Up Air

When an exhaust system is used, air must be supplied to the room to makeup for the volume removed through the ventilation system or the hood. If the ventilation system is balanced to provide less than the required amount of makeup air, the room will be under negative pressure. This vacuum will result in the movement of air from cleaner adjacent rooms into the contaminated room and not the reverse. The U.S. Public Health Service suggests an infiltration rate of 50 cfm for containment laboratories[14] or approximately 10% of the return air volumetric flowrate.[15]

It is important that a building's air supply and exhaust system be designed to accommodate the pressure changes caused by the opening and closing of doors, and by the movement of staff and patients. An instrument to measure air pressure (a manometer) with an audible alarm can be installed to warn people if room pressure changes.

Proper ventilation is an important determinant of a comfortable indoor environment. It also can provide a measure of protection against airborne infection. Periodic monitoring and maintenance of air supply and exhaust systems is necessary for effective operation. The main recommendations of the section beginning on page 11–47 are summarized below.

Suggestions for Good Ventilation

1. Use local exhaust ventilation (e.g., cabinets or booths) to contain activities that generate infectious particles. Use general exhaust ventilation to remove contaminated air from a room, and to keep the room under negative pressure relative to other areas. Supply at least the minimum recommended amount of outdoor air.
2. Install supply air diffusers and general exhaust air grilles so that contaminated air moves away from the breathing zones of the people in a room. Make certain that short-circuiting and stagnation of air are minimized.
3. Discharge the contaminated air outdoors at least 25 ft away and downwind from any outdoor air intakes. Install high-efficiency air filters, UV lamps, or other disinfection units if the air must be exhausted closer than 25 ft from an air intake or if a large percentage of the potentially contaminated air will be recirculated to a building.

In order to prevent the transmission of TB and other respiratory infections in certain high-risk settings, health care workers are paying increasing attention to environmental control measures. The references listed below provide additional details on the subject. For further information, contact your local health department or TB control officer; the Indoor Air Quality Program, Air & Industrial Hygiene Laboratory, California Department of Health Services, Berkeley CA 94704-1011, or the Tuberculosis Control and Refugee Health Programs Units, Infectious Disease Branch, California Department of Health Services, 714/744 P Street, P.O. Box 942732, Sacramento, California 94234-7320.

References

1. Riley RL. Ultraviolet air disinfection for control of respiratory contagion. In: R Kundsin (ed.), *Architectural Design and Indoor Microbial Pollution.* Oxford University Press, New York, 1988, pp. 174–197.
2. Riley RL and Nardell EA. Clearing the air: The theory and application of ultraviolet air disinfection. *Am Rev Respir Dis* 139:1286–1294, 1989.

3. Nardell EA. Ultraviolet air disinfection to control tuberculosis in a shelter for the homeless. In: R Kundsin (Ed.), *Architectural Design and Indoor Microbial Pollution*. Oxford University Press, New York, 1988, pp. 296–308.
4. Centers for Disease Control. Prevention and control of tuberculosis in correctional institutions: Recommendations of the advisory committee for the elimination of tuberculosis. *MMWR* 38:313–9, 1989.
5. U.S. Department of Health and Human Services, Public Health Service, Centers for Disease Control. *TB/HIV: The Connection: What Health Care Workers Should Know*. Atlanta, GA, 1993.
6. American Society for Heating, Refrigerating and Air Conditioning Engineers (ASHRAE). *Standard 62–89. Ventilation for Acceptable Indoor Air Quality*. ASHRAE, Atlanta, 1989.
7. Tayler HR, West SK, Rosenthal FS, et al. Effect of ultraviolet radiation on cataract formation. *N Engl J Med* 319:1429–1433, 1988.
8. Murray WE. Ultraviolet radiation exposure in a mycobacteriology laboratory. *Health Phys* 58:507–510, 1990.
9. National Institute for Occupational Safety & Health (NIOSH). *Occupational Exposure to Ultraviolet Radiation. DHHS (NIOSH)* Publication No. 73-11009, Cincinnati, OH, 1972.
10. American Conference of Governmental Industrial Hygienists (ACGIH). *Industrial Ventilation: A Manual of Recommended Practice*. Lansing, MI, 1991.
11. American Institute of Architects, Committee on Architecture for Health. *Guidelines for Construction and Equipment of Hospital and Medical Facilities*. American Institute of Architects, Waldorf, MD, 1987.
12. Centers for Disease Control. *Guidelines for Prevention of TB Transmission in Hospitals.*U.S. Department of Health and Human Services, Public Health Service, DHHS (CDC), Atlanta, GA, 1982, 82-8371.
13. Centers for Disease Control/American Thoracic Society. *Diagnostic Standards and Classification of Tuberculosis and Other Mycobacterial Diseases*, 14th edn. (Reprinted from *Am Rev Respir Dis* 123(3):343–358, 1981.)
14. U.S. Public Health Service. *National Institutes of Health Laboratory Safety Monograph. Supplement to the NIH Guidelines for Recombinant DNA Research*. U.S. Department of Health, Education and Welfare, National Institutes of Health, Bethesda, MD, 1978.
15. Woods JE and Rask DR. Heating, ventilation, airconditioning systems: The engineering approach to methods of control. In: Kundsin R (ed.), *Architectural Design and Indoor Microbial Pollution*. Oxford University Press, New York, 1988, p .142.

Source: California Indoor Air Quality Program, Air and Industrial Hygiene Laboratory, and the Tuberculosis Control and Refugee Health Programs Unit, Infectious Disease Branch, Kenneth W. Kizer, M.D., M.P.H., Director, Department of Health Services, 1990.

The Economics of Implementing an Engineering Tuberculosis Control Plan in a County Hospital

William Charney

Introduction

Tuberculosis (TB) has reemerged in recent years as a public health problem.[1] California reported 5382 cases of TB in 1992, a 54% increase in reported cases since 1985.[1] Subsequently, health care workers are at increasing risk of occupationally acquired TB and, in some sectors, occupationally acquired TB has resulted in mortality.[2] Although the California Health and Safety Code (Title 22, Section 70723) requires hospitals and health care facilities to perform baseline preemployment and yearly skin testing for TB, there is no state-wide data on the incidence of TB skin test conversions among an estimated 270,000 at-risk health care workers in California.

San Francisco General Hospital (SFGH) is an inner-city county health facility serving a diverse patient population. The patient population is ethnically diverse: 31% Hispanic, 27% white, 23% African-American, 15% Asian, 4% other/unknown. The hospital provides care to about one-half of Medi-Cal and low-income patients in San Francisco and is virtually the only provider for indigent adults not eligible for Medi-Cal in the city. SFGH is the only hospital in San Francisco serving prisoners: 384 inpatients and 868 outpatients in 1993. SFGH has an estimated 4000 employees and in 1995 treated 275 cases of active TB. On any given day, SFGH will have between 4 and 10 patients hospitalized with pulmonary TB and the hospital performs over 500 sputum inductions per month to rule out active TB. The hospital has already had patients in the facility who are resistant to two and three of the more effective antituberculosis drugs. SFGH has a patient population that is almost 25% human immunodeficiency virus HIV, and many health care providers who are HIV, and consequently more susceptible to TB.

Background

In order to comply with the Federal and State regulations as well as the Center for Disease Control (CDC) recommendations for *Mycobacterium tuberculosis* (MTb) for isolation rooms and air changes per hour (ach) in designated TB areas, high-efficiency particulate air (HEPA) filtration was the engineering control option chosen. This technology was selected in order to economically increase a low air volume output by the central heating and ventilation system (HVAC) which could neither provide negative pressure or a minimum of 6 ach in designated areas.

HEPA filters have been shown to capture MTb[3] and are listed as a control option by both the CDC[4] and California (Cal) Occupational Safety and Health Administration (Cal-OSHA).[5] Areas of the hospital needing retrofitting for engineering controls were chosen using (l) preliminary conversion data; (2) a plan for directionality of TB patient flow; and (3) definitions of high-risk medical procedure.

Emergency Department

The emergency department (ED) is the first area of contact for unscreened patients seeking medical attention. Waiting rooms and observation wards (where highly suspected patients may wait for up to 24h before being admitted to a ward room) were fitted with 800 cfm ceiling-mounted HEPA units (Figure 11.30) that provide additional air scrubbing and air changes. Isolation barriers were provided

FIGURE 11.30 800 cfm ceiling-mounted HEPA unit.

for the clerical staff who perform the initial "sign-in" of incoming patients. The barrier consists of a plexiglass window shield with communication slots. The barrier is designed to protect clerical staff from the initial aerosol contact with patients and the space is under positive pressure. Two triage rooms were retrofitted with booster fans to provide negative pressure. Coughing patients are triaged to these rooms. The booster fans allow for a negative pressure of 0.002 in. of water gauge. A negative-pressure room was added to the Observation Ward to allow a coughing patient to remain until a ward room was found.

Centralization of Sputum Induction

The centralization of sputum induction using a HEPA filter engineering control is an important part of the TB Control Plan. There are different types of HEPA engineering control devices that can be purchased. Ambulatory patients can be transported to a specified area with the engineering control device (Figure 11.31). Nonambulatory patients can receive their sputum induction in their room with the mobile unit (see Figure 11.32).

Negative-Pressure Isolation Rooms

Regulatory language stipulates that hospitals provide negative-pressure isolation rooms with a minimum of 6 ach. Negative pressure, as defined by the American Society of Heating, Refrigeration and Air Conditioning (ASHRAE) and adopted by the CDC, is a minimum of 10% greater exhaust over supply air or a minimum of 50 cfm increased exhaust over supply. These are virtually minimum values and greater directionality of airflow and volumes should be attempted to obtain higher protection factors. In order to provide negative pressure, SFGH installed a HEPA portable recirculation/negative-pressure unit, an RS 1000, (Figure 11.33). This unit provides a top-end air volume of 800 cfm on a variable speed fan system that can be adjusted as needed. It has a built-in damper system that allows for 40%–60% of the air to be directed out of a window creating negative pressure. RS 1000 challenge test results are shown in Figure 11.34. SFGH purchased 12 negative-pressure units in order to create 10 negative-pressure ward rooms with two back-up units in storage to replace a failed unit. Additional negative-pressure rooms were installed due to an expanding need for isolation rooms. The 10 initial rooms have expanded to 17 isolation rooms using the same portable system to create the negative pressure.

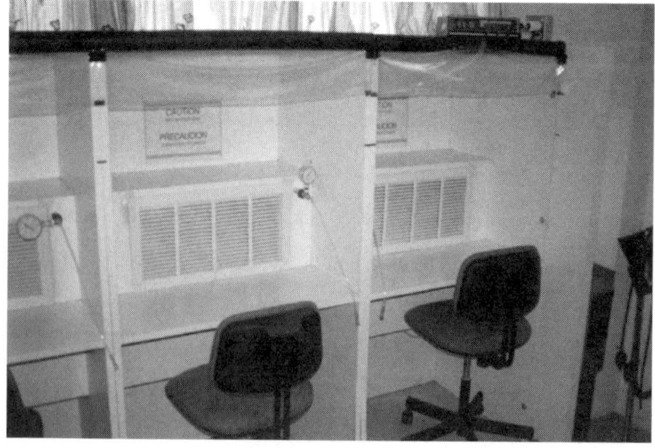

FIGURE 11.31 Built-in sputum-induction exhaust ventilation.

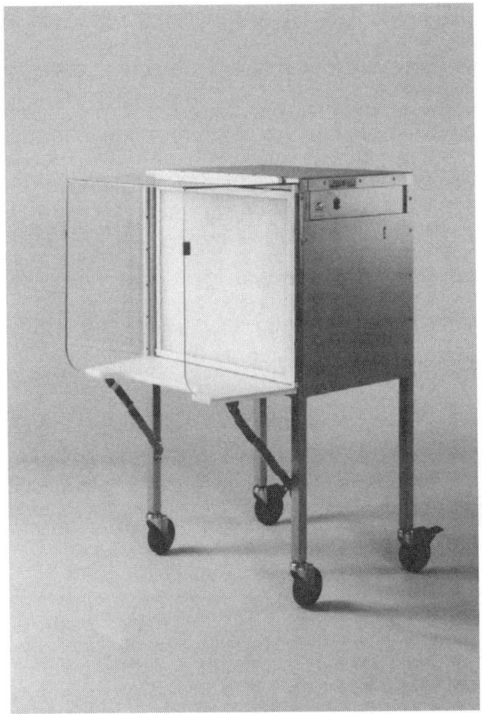

FIGURE 11.32 Portable or mobile HEPA sputum-induction unit.

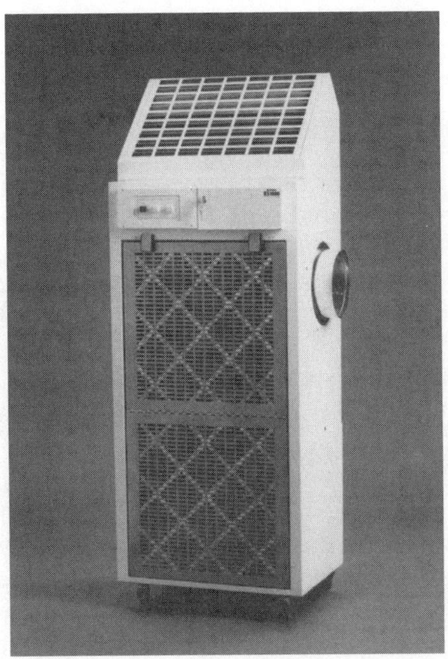

FIGURE 11.33 The RS 1000, a HEPA portable negative-pressure recirculation unit.

Pressure Differential Monitoring of Negative Pressure

Pressure differential monitors were selected for installation on all negative-pressure rooms in order to have a constant display module for alerting personnel to potential changes in pressure. It is important to have constant readings of negative pressure for the following reasons:

1. External atmospheric forces on the building may change pressure in the rooms.
2. Contractors working on HVAC systems in the hospital may cause changes in the system thereby de-balancing pressure.
3. Doors ajar, or negative pressure machine malfunctioning.

The monitors on the wall (Figure 11.35), in front of each room communicates to ward staff the pressure conditions within the room at the moment of interest.

Bronchoscopy

Bronchoscopy is defined as a "high-risk medical procedure." The room needs to be under negative pressure with a high air exchange rate. SFGH fitted the bronchoscopy suite with an RS 1000 unit that provides negative pressure and 10 ach.

Outpatient Clinics

SFGH has several outpatient clinics serving high-risk populations for TB. These clinics are mostly located in older brick buildings with nonexistent or underpowered ventilation systems. Portable recirculating HEPA units were installed in treatment rooms, phlebotomy rooms, and waiting areas to increase air scrubbing (Figure 11.36).

An outpatient waiting room was fitted with two active Viotec UV ceiling fan systems, manufactured by JJI Lighting. These systems provided another 500 cfm of air scrubbing.

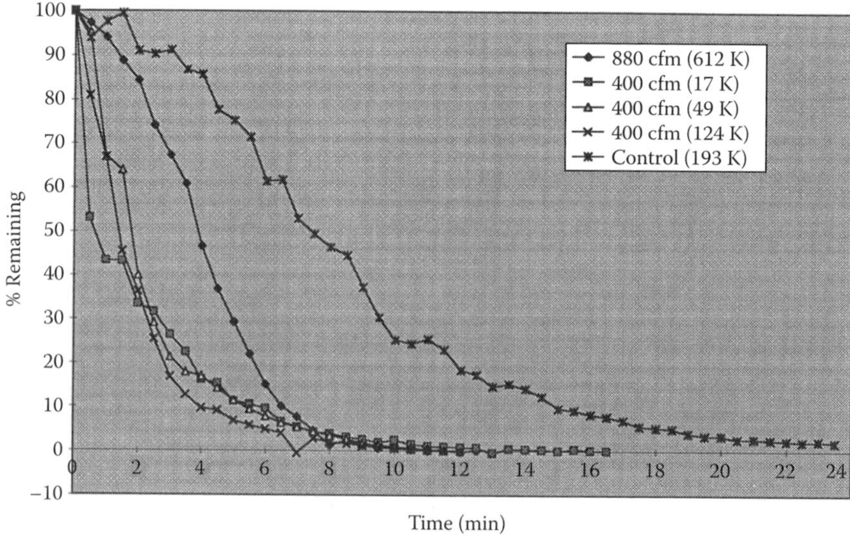

FIGURE 11.34 RS 1000 challenge test.

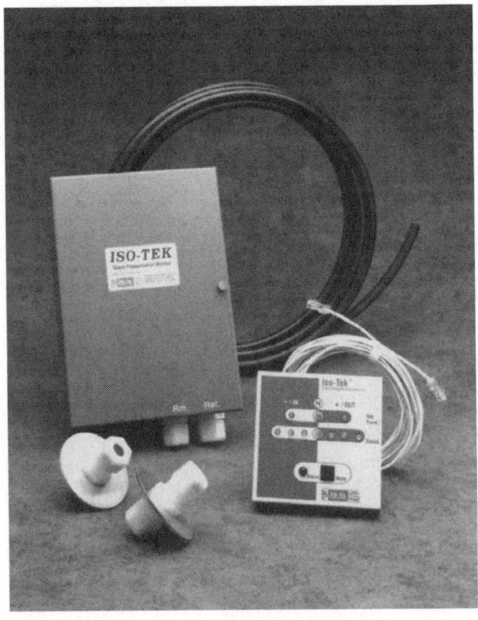

FIGURE 11.35 Room-pressure monitors that are displayed on the wall in front of the room.

FIGURE 11.36 Portable recirculating HEPA filter.

TABLE 11.6 Economics of Engineering Controls

Area	Unit Type	No. of Units	Unit Cost	Total Unit Cost	Installation Cost	Total Cost
Emergency room	Ceiling HEPA 800 CFM	5	$1,600.00	$8,000.00	$15,000.00	$23,000.00
Ward rooms	RS 1000	12	$3,000.00	$36,000.00 (approx)	$2,000.00	$38,000.00
Outpatient areas	Portable EV 35	15	$300.00	$4,500.00	N/A	$4,500.00
	Portable EV 15	10	$200.00	$2,000.00	N/A	$2,000.00
HVAC system—rebalance to produce 10 negative-pressure rooms						$2,000.00
	5-Station sputum-induction booth	1	$15,000.00			$15,00.00
	Portable sputum-induction booth	1	$1,500.00			$1,500.00
All negative-pressure rooms	Pressure monitor	22	675.00		N/A	14,850.00

Note: The total cost for TB Engineering Control Program in approximate dollars was $100,850.00. This covers 18 negative-pressure isolation rooms, 6 sputum-induction booths, 25 portable HEPA units used in outpatient settings, and rebalancing the central HVAC system in 10 rooms.

Economics

Table 11.6 shows a breakdown of the approximate costs of implementing an engineering control plan for TB. One hundred thousand dollars is the estimated cost; at the time of publication, however, another $25,000 was incurred due to an addition of seven isolation rooms using RS 1000 technology, bringing the total approximate cost to $125,000 for a 350-bed facility. SFGH is meeting the challenge to comply with the TB regulations and safe-practice guidelines for engineering controls. These controls do stress existing capital equipment budgets and at times equipment needs had to wait for budget allocations.

References

1. California Tuberculosis Elimination Task Force, Stuart Egie, Plan for TB Control and Elimination, CA Department Health Services, Berkeley, CA, January 1994.
2. Valway S, Pearson ML, Ikeda R, and Edlin BR. HIV-infected (HIV+) health care workers with multidrug-resistant tuberculosis (MDR-TB), 1990–92. In: *Program and Abstracts of the 33rd Interscience Conference on Antimicrobial Agents and Chemotherapy*, New Orleans, October 17–20, 1993. American Society for Microbiology, Washington, DC, Abstract, 1993, p. 231.
3. Wickramanayake GB. Decontamination technologies for release of bioprocessing facilities. *Environ Rev* 9:447, 1989.
4. Guidelines for Preventing Tuberculosis Transmission in Health Care Settings, Centers for Disease Control, U.S. Department of Health and Human Services, July 1990.
5. Hazardous substances, Article 109, Hazardous Substances and Processes, California OSHA, Title 8, California Code of Regulations, Division I, Chapter 4, Subchapter 7, Group 16, October 1993.
6. Tepper B. Portable HEPA filtration for TB isolation in hospitals and clinics. In *Essentials of Modern Hospital Safety*, 3. Lewis Press, Ann Arbor, MI, 1994, pp. 143–157.
7. Smith D. Mycobacterium tuberculosis and tuberculosis. In *Zinsser Microbiology*, 15 ed., Eds., Joklik, WK and Smith DT, Acc, New York, 1972.
8. Riley H and O'Grady F. *Airborne Infection—Transmission and Control.* Macmillan, New York, 1961.

9. Wells WF. *Aerodynamics of Droplet Nuclei in Airborne Contagion and Air Hygiene*. Harvard University Press, Cambridge, MA, 1955.
10. Center for Disease Control. Maternal action plan to control multi-drug resistant tuberculosis: Recommendations of the TB task force. *MMWR* 41:11, 1992.

Implementing a Quality Assurance Program for Tuberculosis Control

Patrice M. Sutton, Mark Nicas, and Robert J. Harrison

Background

Health care workers have historically been a population at risk for tuberculosis (TB). Sepkowitz has shown that despite significant epidemiologic evidence of the risk for TB among health care workers, the view of TB as an occupational hazard did not emerge until the 1950s when TB among the general population rapidly declined.[1]

The increased incidence of active pulmonary TB in the U.S. population in the 1980s has led to renewed appreciation of the infection risk posed to hospital workers.[2] In response to the resurgence of TB, the U.S. Centers for Disease Control and Prevention (CDC) published guidelines to prevent the transmission of TB in health care facilities.[3] Between 1994 and 1995, as part of a multiyear Cooperative Agreement with the National Institute of Occupational Safety and Health (NIOSH), the California Department of Health Services (CDHS) evaluated adherence to CDC Guidelines at three high-risk acute care facilities.[3]

The CDHS study found a lack of full adherence to CDC guidelines at all three hospitals.[4] Reports about work practices varied among hospital staff, and documentation to verify that hospital policy had been implemented was often absent. For example, the TB control plan included a list of "designated TB isolation rooms," but the ventilation performance of these rooms was not documented. At one hospital, ventilation measurements from the previous year showed that the isolation room currently in use for those TB patients considered most infectious was actually under positive pressure.

CDHS' findings suggest there can be a discrepancy between the written TB control policy and actual hospital practice. We therefore undertook a quality assurance project to determine the extent to which a hospital's TB control plan was implemented, and to identify obstacles to implementation. This chapter presents guidelines for implementing a quality assurance project and reports findings from a 582-bed hospital that admitted 72 patients with active pulmonary TB and had a staff of 5500 in 1996.

Guidelines for Implementing a Quality Assurance Project

A quality assurance project that is not regulatory in nature or linked to disciplinary action is a useful tool to improve TB control. Such an approach is a relatively simple way to gather and quantify data prospectively, is nonintrusive to patients, and is minimally intrusive to staff. The information collected tends to be reliable because the method incorporates the experience and knowledge of TB control from a wide range of health care workers with a minimum of recall bias. As the data reflect the actual experience of many (although not all) health care workers responsible for implementation of the TB control plan, it can provide a common basis for decision making. This is particularly important in the absence of common agreement of what constitutes acceptable risk for TB infection among health care workers.

Propose the Project to a Wide Range of Health Care Workers

Initiation of a quality assurance project involves soliciting the wide participation of hospital staff. Participation of line staff is needed to identify hospital locations and groups of workers at risk, to determine which TB control measures should be assessed, to develop meaningful evaluation criteria, and to ensure their cooperation with the project. Feedback on the proposal should be elicited from all staff with responsibility to oversee hospital TB control efforts. Because workers in "hands on" jobs are often in the best position to observe developing or threatening problems,[5] it is essential to have the participation of representatives of employees responsible for implementing TB control measures on a daily basis, and from supervisory and nonsupervisory staff from affected hospital locations.

The proposal should include the following:

- A clear statement of the project goals and objectives
- The nature of the information to be collected
- Who will be collecting, analyzing, and reporting the data
- How the data collected will be used and disseminated
- The time and effort individuals will need to contribute to the project, if any
- The voluntary nature of the study

Providing this information to health care workers at the start of the project allows them to: (1) decide if they support the goals of the project; (2) evaluate if the type of information to be collected will be helpful to them (rather than critical of their efforts); (3) understand what trigger events will lead to specific actions or interventions; and (4) assess their ability to provide the time (if any) needed to implement the project. Cooperation with the project should be voluntary, and an opportunity to decline participation should be provided.

In the CDHS study, regularly scheduled meetings of the Infection Control, Environmental Health and Safety, and Quality Assurance Committees served as forums to discuss the project at the onset and throughout its course. A written one-page project description was also provided to participants.

Develop a Protocol

Discussions with hospital staff provide the basis for developing a written protocol that describes the following:

- TB control measures that will be assessed—Consideration should be given to each of the three categories of control measures: administrative, engineering, and personal respiratory protection. Select control measures that are considered fundamental to TB control by all parties and that are simple to quantify.
- Assessment criteria—Objective measures must be developed to assess whether or not the control measure is being implemented.
- Method used to measure if the criteria have been met—The methodology should be quantitative, prospective, routine, nonintrusive to patients, and minimally intrusive to staff. The methodology should include precise definitions of all criteria and terminology, e.g., negative pressure, suspect TB patient, NIOSH approved respirator, hospital locations included in the survey.

CDHS evaluated hospital adherence to the CDC guidelines for the isolation of suspect and known TB patients. Once a week CDHS project staff went to all hospital locations designated *a priori* by hospital infection control staff as areas where TB patients might be located. At each location nursing supervisory staff were asked, "Do you have any TB patients here today?" If they responded yes, we asked, "Are they a suspect or known case?" and "What room are they in?" Therefore, the unit of observation was the "TB patient room," and the clinical status of each TB patient was categorized by the nursing supervisor (i.e., a "suspect TB patient" and "known TB patient" were defined as any patients considered to be a suspect or known case by the nursing supervisor). If the same individual was found on more than one occasion, this was considered a separate observation.

Each TB patient room was observed for the following:

- Is the room a "designated" TB isolation room?
- Is there a sign on the door?
- Is the door closed?

- If a designated TB isolation room, is the room under negative pressure?
- Is there a supply of NIOSH approved respirators at the room entry?

In the course of making the TB patient room observations, the use of respiratory protection by any health care workers entering the isolation rooms was also noted. The protocol used by CDHS is summarized in Table 11.7.

Perform the Audits

The frequency of data collection, the length of time of the project overall, and the person(s) responsible for collecting the data must be determined. It may be preferable to perform the assessment at various times of the day and week to reduce the chance that the observations are in some way biased by when they are made.

A notification plan must be agreed upon *before data collection begins.* Key factors to consider when developing this component of the project are

- Who needs to get the information?
- What type of information do they need?
- The timeliness of notification.
- That the project is not regulatory in nature.
- That the project is not linked to disciplinary action.

Developing a notification plan that accounts for all of these issues is a difficult but critical undertaking. A well defined project goal and the full participation of health care workers from the outset are valuable tools for weighing the many issues that must be incorporated into these decisions.

Timely notification can greatly enhance the value of a quality assurance project, because problems can be addressed as they are recognized and health care worker exposures to *Mycobacterium tuberculosis* (MTb) aerosol can be prevented. Nursing supervisors all chose the option of receiving the results of the survey in their respective areas immediately after the investigators made the observations. Findings were reported to the director of environmental health and/or the infection control coordinator at the end of each weekly audit. Summary data were disseminated to the physician and nursing staff through presentations to the Infection Control and the Quality Assurance Committees, and to employee representatives at the Environmental Health and Safety Committee.

The potential for any individual to be "blamed" for problems that are identified should be minimized. One approach is to not collect any data that clearly documents the actions of an identifiable health care worker. For example, the names of health care workers observed entering TB isolation rooms without or improperly using respiratory protection were not collected, and only summary data without location information were disseminated. Although this approach minimizes the opportunity for the data to be used to discipline a health care worker, the opportunity to prevent a potential exposure is temporarily lost, because a potentially hazardous work practice is not corrected immediately.

Evaluate the Project and Disseminate the Results

Results of the project should be compiled, summarized, and reported in writing to all parties. An opportunity to review a draft of the written report should be provided to ensure that the final report is accurate and clear. When evaluating the project, consider the following:

- Were the project findings linked to change?
- Is there a need to continue the project?
- How could the project be improved?

A summary of the results of CDHS' assessment of adherence to CDC TB isolation Guidelines shows that CDHS staff observed 1–10 TB patient rooms each week over the 14-week survey period, for a total of 91 TB patient room observations. The patient was designated by nursing staff as "suspect TB" in 75

TABLE 11.7 CDHS Approved TB Patient Isolation Data Collection Form

TB PATIENT ISOLATION DATA COLLECTION FORM				**Location**___________		**Date** ___________		**Time**___________	
Nursing supervisor reporting:_______________				**Observations by:**_______________					
Reported By Nursing Supervisor				**Observed By CDHS HCW Project Staff**					
Patient Name	Patient status	Patient room #	Room door closed ?	Sign on door ?	Supply of NIOSH approved HCW respirators available at entrance to iso. room ?	Supply of HCW masks available at entrance to isolation room?	Is patient in a designated isolation room ?	Pressure differential under room door (in. H_2O)	Room under negative pressure ?
Comments:	RO TB		Y N	Y N	Y N	Y N	Y N		Y N
Comments:	RO TB		Y N	Y N	Y N	Y N	Y N		Y N
Comments:	RO TB		Y N	Y N	Y N	Y N	Y N		Y N
Comments:	RO TB		Y N	Y N	Y N	Y N	Y N		Y N
Comments:	RO TB		Y N	Y N	Y N	Y N	Y N		Y N

rooms (82%) and as "known TB" in 16 rooms (18%). Rooms with suspect and known TB patients housed 73 and 10 individuals, respectively.

Survey Results for Adherence to CDC Guidelines for TB Patient Isolation

Place TB Patients in Designated TB Isolation Rooms

Of the known and of the suspect were placed in designated isolation rooms. The fact that only a small fraction of suspect TB patients, 81% will be infectious, and the erroneous perception by some staff that placing a portable high-efficiency particulate air (HEPA) unit in a patient's room (i.e., the unit is only cleaning, not exhausting room air) changes the room into an isolation room, may have reduced concerns among infection control staff about the practice of placing known and suspect TB patients in nonnegative pressure isolation rooms.

The response most often given as to why TB patients were not placed in designated TB isolation rooms was that such a room was not available. The need for specialized medical care also competed with TB control in making patient placement decisions. For example, one suspect TB patient needed monitoring and care available only in the intensive care unit. In another instance, it was reported that a psychiatric patient with active TB was not placed in a designated isolation room because nursing staff wanted the patient in a room with a window in the door so that the patient could be observed from outside the room.

Place a Warning Sign on the Isolation Room Door

A "respiratory precautions" sign was placed on the outside of the door in 98% of the 91 TB patient rooms observed. Nursing staff did not place a sign on one room because their attention to this task was diverted by the need to attend to another patient who had fallen. We were unable to identify the reason the second room did not have a sign.

Close the Isolation Room Door

Of the 91 isolation room doors, 90% were closed. Reasons for isolation room doors being open were

1. The door was not self-closing and had not been pulled closed ($n = 4$).
2. The patient had left the room and did not close the door behind him or her. On one of these room doors the patient had taped the note "went to smoke" ($n = 2$).

   It appears to be a common practice that TB patients leave their rooms. If they are wearing a mask, it is not a breach of isolation. However, some health care workers believed that if the TB patient exited the isolation room, there would be no hazard with the door open. Assuming that the isolation room meets CDC's minimum recommended 6 ach (air changes per hour) and that room air is perfectly mixed, it would take 46 min for the room air to be considered 99% disinfected. Prior to this time, if patients leave their room door open, MTb aerosol in the room may contaminate adjacent areas. In addition, housekeeping and other staff who routinely enter "empty" TB isolation rooms may be at unrecognized risk if sufficient time has not passed to disinfect the air, or if they are not wearing appropriate respiratory protection.

3. On the jail ward, safety concerns competed with TB control. Here, the door was intentionally held open by health care workers when they entered the room to prevent them from being automatically locked inside ($n = 2$).
4. Nursing staff did not close the door because they were diverted by the need to attend another patient who had fallen ($n = 1$).
5. For one room, we were unable to identify the reason that the door was open.

Maintain the Isolation Room under Negative Pressure

Negative pressure was only measured in rooms designated for TB isolation. Of the 72 designated isolation rooms tested 90% were under negative pressure. In five of the seven isolation rooms under positive pressure, administrative lapses were the cause. HEPA units are used to effect negative pressure in some isolation rooms by exhausting room air through a duct directly outside through the room window. In three instances the isolation room was under positive pressure because the HEPA unit was either not turned on, or was operating at a setting too low to create sufficient exhaust. Positive pressure resulted from an open door in one room and from an open window in another.

The temporary disruption of negative pressure in one isolation room appears to have resulted from strong winds outside the hospital that caused a change in the pressure relationship between the interior and the exterior of the building. This in turn resulted in a change in the pressure differential between the isolation room and the hallway. One room was under positive pressure because the ventilation system did not have the power to exhaust enough room air to create negative pressure.

NIOSH-Approved Respirators Available and Used Properly

NIOSH-approved N95 respirators were available for use at the entrance of 90% of the TB isolation rooms. Health care workers used type N95 respirators during 92% of the 37 entries into TB isolation rooms that we observed. No respiratory protection was used by one physician. One dietary worker and one health care provider donned surgical masks intended for patient use that were also available at the entrance of 55% of the rooms observed. Nursing staff suggested that the location of patient masks at the room entry may encourage misuse of these masks by health care workers.

The respirator was used properly in only 29% of the 34 entries observed. Incorrect usage included using only one strap ($n = 19$), positioning both straps worn on the neck ($n = 3$), holding the respirator to the face without using straps ($n = 1$), and use by a health care worker with a full beard ($n = 1$). A lack of adherence to the fit testing components of OSHA's respiratory protection requirements may have contributed to reduced awareness by health care workers regarding the importance of respirator fit.

Conclusions

This quality assurance study documents that at this hospital TB control program implementation deviated from the written plan. TB patients were not always placed in negative pressure isolation rooms, isolation room doors were not always closed, isolation rooms were not always under negative pressure, and respirators were occasionally not used, and/or frequently not used properly.

The inadequacy of and/or lack of adherence to the hospital's TB control plan can lead to inaccurate conclusions regarding the risk of health care worker exposure to MTb aerosol. An incomplete understanding regarding implementation of TB control measures may confound efforts to evaluate their efficacy. The quality assurance project serves as a reminder of the factors that will influence assumptions about risk, i.e., what you ask, who you ask, where, how, and how often you look for hazards. The data also suggest a potential for health care workers throughout the hospital to be exposed to MTb aerosol due to imperfect containment in isolation rooms; this widespread exposure makes it difficult to identify a truly "unexposed" group of health care workers.

Although many meetings were held with medical and employee committees to discuss the project protocol, the study could have been improved by further increasing the participation of health care workers. For example, our efforts to assess patient masking practices in the emergency department waiting room entailed 17 min over 14 weeks observing 119 individuals. However, none of these patients were observed to be coughing. Increased participation of health care workers in the emergency department may have resulted in a more effective assessment tool.

In another example, our a priori assumptions of locations where TB patients "might be" proved to be incomplete. The identification of an active TB patient who had been on a psychiatric ward with unrecognized TB prompted us to expand our weekly audits to seven hospital locations that had not been designated a priori as having TB patients. Although none of these locations had any designated TB isolation rooms, nursing supervisors in all of these areas reported the potential for caring for TB patients. In the course of eight weekly audits we found eight patients identified as suspect TB in these "undesignated TB locations": five in the postsurgical unit, two in a psychiatric unit, and one in the step-down intensive care unit.

A quality assurance project is an early warning system. However, to be effective the results must be linked to a mechanism for timely change. Our measurements of adherence to the plan after 8 weeks were virtually the same as after 14 weeks of observations. However, the recommendation to improve TB control efforts by increasing the number and distribution of isolation rooms throughout the hospital was implemented at the completion of this project. Six months after completion of the project, additional recommendations to provide better training to health care workers and to install self-closing doors on isolation rooms have not yet been addressed. This underscores that a quality assurance project is an important adjunct to, not a substitute for, interventions that must be undertaken if TB among health care workers is to be prevented.

References

1. Sepkowitz KA. Tuberculosis and the health care worker: A historical perspective. *Ann Intern Med* 120:71–79, 1994.
2. Centers for Disease Control and Prevention. Guidelines for preventing the transmission of *Mycobacterium tuberculosis* in health care facilities. *MMWR* 43:13, 1994.
3. National Institute for Occupational Safety and Health Cooperative Agreement #U50/CCU910074-04. Occupationally Related Tuberculosis in Health Care Workers, Robert Harrison, MD, MPH, Principal Investigator.
4. Sutton PM, Nicas M, Reinisch F, and Harrison R. Evaluating the control of tuberculosis among health care workers: Adherence of three high-risk hospitals to CDC guidelines.
5. Sawyer HJ. Occupational health concerns in the health care field. *Patty's Industrial Hygiene and Toxicology: General Principles*, 4th edn. John Wiley & Sons, New York, 1991, pp. 371–372.

12

Electrocautery Smoke: Reasons for Scavenging

Stackhouse, Incorporated

Jacob D. Paz
J&L Environmental Services, Incorporated

Charles J. Bryant
Richard Gorman
John Stewart
Wen-Zong Whong
National Institute for Occupational Safety and Health

Lindsey V. Kayman
Department of Environmental and Occupational Health Support Services

Association of Operating Room Nurses

William Charney
Healthcare Safety Consulting

Scott Miller
Shawn Campbell
Bob Reynolds
San Francisco General Hospital

Importance of Understanding Surgical Smoke 12-2
Known Content of Surgical Smoke • Current Research on Surgical Smoke • Standards and Recommendations
References 12-3
Supplemental Reading 12-4
Electrocautery Health and Safety 12-4
History of Electrocautery • Types of Electrocautery • Electrocautery Health Hazards • Control of ESU Smoke and RF/MW Radiation • Conclusion
References 12-9
Health Hazard Evaluation Report 12-10
Preface • Summary • Introduction • Background Methods and Materials • Evaluation Criteria • Results • Conclusions • Recommendations
References 12-23
Authorship and Acknowledgments • Distribution and Availability of Report
Laser Hazards in the Health Care Industry 12-24
Background Information on Lasers • Laser Hazards Eye Hazards • Skin Damage • Laser Plumes—Just a Bad Smell? • Tips on Using Smoke Evacuators • Fires • Electrical Hazards • Laser Safety Policy • Laser Safety Program • Laser Safety Personnel • Laser Maintenance • Training and Approval of Laser Operators • General Operational Guidelines • Personnel Protection • Operating Precautions • Engineering Control Measures • Fire Safety • Laser Safety Training • Government Regulation of Lasers • Laser Resources
Proposed Recommended Practices: Electrosurgery 12-41
Purpose • Recommended Practice I • Recommended Practice II • Recommended Practice III • Recommended Practice IV • Recommended Practice V
Glossary 12-48
References 12-48
Suggested Readings 12-49
Laser Plume Quantification 12-50
Sampling Instrumentation • Method • Discussion • Recommendations

Laser Plume: Case Study.......... 12-52
Introduction • Background • Hypothesis • Investigation • Discussion • Clinical Trial of Modified Unit • Conclusion
Reference.......... 12-55
Appendix 12.A.1.......... 12-56

This chapter concentrates on the health and safety criteria for plume control during laser and electrocautery surgery. Many articles recommend scavenging during these procedures.[1–5] Presented in this chapter are (1) the importance of understanding surgical smoke, (2) electrocautery health and safety, (3) health hazard evaluation report, (4) laser hazards in the health care industry, (5) proposed recommended practices, and (6) laser plume quantification. For recommended standards and regulations, the reader is referred to the American National Standards Institute (ANSI Z136.3) and OSHA, respectively.

Importance of Understanding Surgical Smoke*

Stackhouse, Incorporated

The advent of laser technology in surgery raised concerns over the impact of smoke on the health of patients and the surgical staff.

In the early days of laser surgery, the primary motivation for evacuating smoke was the acrid odor that was present during most procedures.[1]

Over time, other concerns emerged regarding the inhalation of vaporized human tissue. These concerns were shared by leading safety organizations and health care workers alike and resulted in a great deal of research into the hazards of all forms of surgical smoke, including what is generated during electrosurgery.

Understanding the dangers of surgical smoke begins with an overview of the gases and particulate matter that are generated when human tissue comes in contact with laser and electrosurgery heat sources.

Known Content of Surgical Smoke

New evidence suggests that there is little difference between the smoke generated from electrosurgery and smoke from lasers. Each contains carbonized tissue, blood, and virus.[1–10] In addition, surgical smoke contains the gases benzene, toluene, formaldehyde, and polycyclic aromatic hydrocarbons; these are known carcinogens.[11–14] These gases create the acrid smell, but the real danger from smoke comes from the particle content of the smoke.

Knowledge that each form of particulate matter and gases present in surgical smoke are known to pose health risks, a number or researchers set out to gain a better understanding of the specific health risks to surgical teams.

Current Research on Surgical Smoke

One of the first and most comprehensive animal studies on the physiological effects of surgical smoke was conducted by Dr. Michael Baggish in 1992 who was the president of the American Society of Lasers in Medicine and Surgery and currently the chairman of the Obstetrics and Gynecology Department at Ravenswood Hospital in Chicago, Illinois.[1] His objective was twofold:

* Reproduced from: Hazards of surgical smoke (video study guide), Stackhouse, Inc., Riverside, CA, 1992.

- Identify any physiologic risks associated with the inhalation of surgical smoke.
- Determine the effectiveness of commercially available evacuation equipment in protecting rats from these effects.

Dr. Baggish vaporized pig skin with a CO_2 laser to create surgical smoke. He found that accumulated particulate matter on the rat's lung tissue caused interstitial pneumonia, bronchitis, and emphysema. The severity of disease increased in proportion to increased exposure to smoke.

Dr. Baggish verified the value of protection from smoke 1 year later by studying the value of smoke evacuation.[5] Again, vaporizing pig skin, this time Dr. Baggish divided the rats into two groups. The control group experienced the same accumulated particulate matter on their lungs as in the first study. However, rats receiving air filtered through a smoke evacuator equipped with an ultralow penetration air (ULPA) filter experienced no pathological changes over the course of the study.

In 1987, Garden and O'Banion found living virus (human papilloma) in a CO_2 laser smoke.[3] In 1989, Sawchuk and colleagues found viral DNA in the vapor of warts.[10] These findings have led to current concerns that the thermal effects of laser and electrosurgery may not be counted on to destroy all viable DNA from viruses.

In 1988, Dr. Barry Wenig found the conclusions of Baggish regarding the damaging effects of smoke on lung tissue applied also to the use of electrosurgical units (ESU) and the ND:YAG laser. Wenig also recommended the use of an efficient smoke evacuator in all procedures that generate surgical smoke.

Further studies by Baggish have shown the presence of HIV DNA captive in the suction tubing of the evacuator used to suction the surgical smoke while lasing 10 mL of HIV-infected cells.[7] This further substantiates the importance of proper smoke evacuation.

There have been no documented cases of health care workers being infected with these viruses by means of surgical smoke. Considerably, more research needs to be done in this area clearly defining the risk of airborne pathogens. However, preliminary indications have captured the attention of regulatory agencies charged with setting forth policies that assure health work environments.

Standards and Recommendations

The American National Standards Institute (ANSI) is responsible for setting safety standards for health care workers while on the job. ANSI has compiled evidence that airborne contaminants during class 4 laser surgery (laser procedures that use a power setting of 0.5 W or greater) can cause lacrimation, nausea, vomiting, and abdominal cramping. As a result, ANSI has recommended that all class 4 laser procedures producing smoke must have it removed by localized exhaust ventilation.[16]

The National Institute of Occupational Safety and Health (NIOSH) has made recommendations at specific hospitals employing either electrosurgical equipment or lasers. Based on the mutagenicity of compounds NIOSH collected at these facilities, they recommended that smoke evacuation units be used to reduce the potential for chronic health effects.[13,14]

The Emergency Care Research Institute (ECRI) and independent medical device test organization have issued statements that the treatment of electrosurgery smoke possesses the same danger as laser smoke.[15] The Food and Drug Administration (FDA) has actively challenged current acceptance of smoke in the surgical suite during any procedure.

The position of these regulatory agencies makes it clear that surgical smoke should be evacuated, not only for the safety of the surgical staff, but also for the safety of the patient as well.

References

1. Baggish MS and Elbakry M. The effects of laser smoke on the lungs of rats. *Am J Ob/Gyn* 156(5):1260–1265, 1987.
2. Voien D. Intact viruses in CO_2 laser plumes spur safety concern. *Clin Laser Mon* 5(9):101–103, 1987.

3. Garden JM, O'Banion MK, Shelnitz LS, Pinski KS, Bakus AD, Reichmann ME, and Sundberg JP. Papillomavirus in the vapor of carbon dioxide laser-treated verrucae. *JAMA* 259(8):1199–1202, 1988.
4. Byrne PO, Sisson PR, Oliver PD, and Inghan HR. Carbon dioxide laser irradiation of bacterial targets *in vitro*. *J Hosp Infect* 6:84–86, 1986.
5. Baggish MS, Baltoyannis P, and Sze E. Protection of the rat lung from the harmful effects of laser smoke. *Laser Surg Med* 8:248–253, 1988.
6. Carlson SE, Schwartz DE, and Wentzel JM. Properties of aerosolized blood produced during a dermabrasion medical procedure. *American Industrial Hygiene Conference*, St. Louis, MO, 5/19/89.
7. Baggish MS, Poiesz BJ, Joret D, Wiliamson P, and Refai A. Presence of human immunodeficiency virus DNA in laser smoke. *Laser Surg Med* 11:197–203, 1991.
8. Research confirms earlier study on plume hazard: Viral contaminants. *Clin Laser Mon* 1989.
9. Some problems about condensate induced by CO_2 laser irradiation. *Laser Safety* 1985.
10. Sawchuk WS, Wedber PJ, Lowy DR, and Dzubow LM. Infectious papillomavirus in the vapor of warts treated with carbon dioxide laser or electrocoagulation: Detection and protection. *J Amer Acad Derm* 21(1):44–49, 1989.
11. Kokosa JM and Doyle DJ. Chemical composition of laser-tissue interaction smoke plume. *J Laser Appl* 1(3):59–63, 1989.
12. Kokosa JM and Doyle DJ. Chemical by-products produced by CO_2 and Nd:YAG laser interaction with tissue. *SPIE* 908:51–53, 1988.
13. Bryant C, Gorman R, Stewart J, and Whong WZ. Health Hazard Report HETA 85–126–1932, Bryn Mawr Hospital, Bryn Mawr, PA, 1988.
14. Moss CE, Bryant C, Stewart J, Whong WZ, Fleeger A, and Gunter BJ. Health Hazard Report HETA 88–101–2008, University of Utah Health Sciences Center, Salt Lake City, UT, 1990.
15. ECRI, *Health Devices* 19, 1, 1990.
16. American National Standard for the safe use of lasers in health care facilities; ANSI Z136.3–1988.

Supplemental Reading

AORN Standards and Recommended Practices for Perioperative Nursing, 1990.

Baumann N. The plume hazard: ICALEO panel calls for more suction. *Laser Med Surg News Adv* 16–18, 1989.

Neufeldt V and Guralnik DB. *Webster's New World Dictionary of American English*, 3rd College edn. 1988.

Newman Dorland WA. *The American Illustrated Medical Dictionary*, 21st edn. 1947.

OSHA cites surgicenter for "serious" safety violations. *Clin Laser Mon* 69–70, 1990.

Electrocautery Health and Safety

Jacob D. Paz

History of Electrocautery

Bovie[1] in 1928 discovered that high-frequency alternating current in the range 250,000–2,000,000 Hz could be used to incise coagulated tissue to obtain homeostasis. This technique was first popularized by Cushing[2] in neurosurgery and was subsequently used in other types of surgery. Heat achieves homeostasis by denaturation of protein, which results in coagulation of large areas of tissue. With actual cautery, heat is transmitted from the instrument by conduction directly to the tissue; with electrocautery, heating occurs by induction from an alternating current source.

When electrocautery is employed, the amplitude setting should be high enough to produce prompt coagulation, but not so high as to set up an arc between the tissue and the cautery tip. Strict control prevents burns outside the operative field and the exit of current through electrocardiographic leads and other monitoring devices.

A negative plate should be placed beneath the patient whenever cautery is used so that severe skin burns do not occur. The advantage of cautery is that it saves time; its disadvantage is that more tissue is killed than with precise ligature. In the past, certain anesthetic agents, such as cyclopropane, could not be used with electrocautery because of the hazard of explosion.[3]

Types of Electrocautery

Currently, two types of electrocautery devices are in use, unipolar and bipolar.

The unipolar ESU is used both for surgical dissection and homeostasis. When undampened high-frequency electrical current is passed through tissue, the active electrode functions as a bloodless knife, and the cells at the edges of the wound disintegrate. A mild thermal injury occurs away from the plane of cutting, and blood vessels thromboses. When the oscillations are dampened, homeostasis is accomplished without cutting. During this procedure, the cells rapidly dehydrate, the vessels within the tissue coagulate and damage to adjacent tissue may be extensive. The precise tip of the divided vessel is all that requires coagulation, however, and the power of the unit should be set at the lowest level possible.

The bipolar electrocautery unit confines the damage to tissues between the tips of the cauterizing forceps. Notably, the bipolar instrument can be used in a wet environment, and it is indicated to control bleeding in microvascular and microneural surgery.

Electrocautery Health Hazards

Explosion, burns, cardiac effects, and muscle excitation are the main health hazards often associated with ESUs are well documented in the literature and are discussed elsewhere.[4] To this list, we can now add electromagnetic (EM) radiation, smoke, and bioaerosol hazards, which are discussed here.

EM Radiation

Microwave (MW) and radiofrequency (RF) radiation are EM energies in the form of waves that travel at the speed of light; RF ranges from 100 KHz to 300 MHz and MW from 300 MHz to 300 GHz. They are classified as nonionizing radiation because the energy of these photons is relatively low. Only limited information on exposure to electric or magnetic fields generated by ESU has been reported in the literature.

Adverse biological effects may occur from the heating of deep body tissues by MW and RF radiation. Such heating may produce damaging cell alterations, and other non-dermal effects, such as neurological, behavioral, and immunological changes resulting in leukemia, cataract, mood swings, and dizziness.[5,6]

Effects due to heating are well documented in animals, but the evidence is incomplete and disputable for those effects occurring without an accompanying increase in tissue temperature. Thermal effects occur in direct proportion to the field strength or power density. When the amount of heat generated from the absorbed energy is too great to be released into the surrounding environment, the temperature of the body gradually increases and can lead to heat stress.

Reports of human effects cover a series of clinical and epidemiological investigations into the association between RF radiation and damage to the eyes, central nervous system, and reproductive capability. Only a few studies have assessed RF/MW exposure levels to medical personnel. The Occupational Safety and Health Administration (OSHA) recommended standard for exposure to MW s is 10 mW/cm^2. Both ANSI and American Conference of Governmental Industrial Hygienists (ACGIH) have published guidelines for occupational exposure to EM radiation.[7,8]

Any area where RF/MW radiation exposure exceeds permissible levels should be considered potentially hazardous. The area should be clearly identified, and warning signs posted. Interlocks may be used to prevent unauthorized entry. Basic protective measures include the provision of shields or absorbing enclosures for equipment. Personal protective equipment may be used (e.g., gonad shields, protective suits, and wire-netting helmets). Although protective goggles have been developed, they may not provide sufficient protection. Implementing such precautions, however, would be difficult in the operating room (OR).

Carpenter,[9] Birenbaum,[10] and Guy et al.[11] have studied the relationships of MW to subsequent developments of ocular cataracts. The severity of ocular damage induced depends upon MW intensity, wavelength, and duration of exposure. Cataractogenesis required a power density above 100 mW/cm^2 and localized ocular hypothermia of 41°C. In addition to power density, time and frequency are important factors with respect to lens changes. Not only the penetration depth of energy, but also the specific absorption rate is distribute frequency dependent. It appears that a critical range of frequency extended from 0.8 to 10 GHz. No lens changes were detected at frequencies of 386,486 MHz or at 35 and 107 GHz.

Fox et al.[12] were among the first to measure the frequency spectrum and power density of ESU. They reported that the spark gap ESU emission extended up to 1 GHz and maximum energies concentrated below 100 MHz with a peak of 2.4 MHz. They reported that the ESU power density was 150 mW/cm^2 (which exceeds the present OSHA standard of 10 mW/cm^2 for 6 min of exposure), and stated that "...the surgeon and patient are most exposed because the active electrode is manipulated by the surgeon and the radiating lead is draped over the patient." Urologists doing transurethral resections are heavily exposed since the radiating active lead enters near the eye. We do not have sufficient information about the new ESU.

A literature review indicates that ESUs are operated at a frequency of 650 Hz.[23] However, in many cases spectrum energy of ESU could not be found in the ESU manual as well as in the professional literature. Electric field (E) strength measurements of ESU by Ruggera and Segerson[13] in 1977 found that the H field was 1000 mW/cm^2 at 16 cm from the ESU source.

Paz et al.,[5] in 1987, conducted a simulation study, to evaluate electric and magnetic field strength generated by a monocular ESU. Test results are listed in Tables 12.1 and 12.2. Experimental data indicated that the surgeons may expose themselves to high levels of electric and magnetic fields exceeding ACGIH threshold values. Ocular exposures were especially high: 20 cm from the active lead at the eye/forehead position the E field ranged from 4.0×10^4 to 9.0×10^6 V^2/m^2 and the magnetic field was 3.5 A^2/m^2 in the coagulation mode, compared to 0.01 A^2/m^2 in the cutting mode.

A real-time survey of E fields was conducted during surgery by Paz.[14] Experimental data illustrated in Table 12.3 showed that the electric field strength in the cutting mode ranged from 1.0×10^3 to 1.0×10^5 V^2/m^2. By comparison, a higher electric field value was noticed in the coagulation mode to the

TABLE 12.1 Electric Field Strength V^2/m^2: Bipolar ESU Simulation Studies

Body Organ	Coagulation Mode (V^2/m^2)	Cutting Mode (V^2/m^2) ($\times 10^4$)
Eye/forehead	9.0×10^6	2.0
Neck	1.0×10^6	3.0
Chest	1.0×10^5	3.0
Upper arms	1.0×10^5	5.0
Lower arms	5.0×10^4	4.0
Waist	1.0×10^4	4.0
Gonads	1.0×10^4	4.0

TABLE 12.2 Magnetic Field Strength A^2/m^2 Bipolar ESU Simulation Studies

Body Organ	Coagulation Mode (A^2/m^2)	Cutting Mode (A^2/m^2)
Eye/forehead	3.50	0.01
Neck	0.50	0.01
Chest	0.04	0.06
Upper arms	0.05	0.04
Lower arms	0.05	0.05
Waist	0.04	0.03
Gonads	0.01	0.02

TABLE 12.3 Real-Time Monitoring of Electric Field Strength V^2/m^2 Bipolar ESU

Operation Type	Low Value (V^2/m^2)	Peak Value (V^2/m^2)	Mean Value (A^2/m^2)
Cutting mode	1.0×10^5	7.0×10^7	1.0×10^6
Cutting mode	1.0×10^3	5.0×10^5	2.9×10^5

extent of 1.0×10^3 to 7.0×10^7 with a peak of $>10 \times 10^7$ V^2/m^2. Electric field strength (mean values) measurements both in cutting mode and coagulation mode exceeded both ACGIH and ANSI standards.

Electrocautery Smoke

In 1976, Goldstein and Paz[15] reported increasing levels of NO_2 where high energy devices were used. The average NO_2 level in the OR during 11 procedures in which ESU and x-ray were used was 0.114 ppm, and in the control area was 0.57 ppm. Area monitoring for NO_2 by a chemiluminescent probe located about 3 m on a perpendicular line from the middle of the OR table showed an NO concentration of about 0.6 ppm and NO_2 levels of 0.14 ppm. High levels of >5.0 ppm NO_2, NO, and NO_x were recorded sparking ESU in a system of air, and N_2O was measured using a chemiluminescence monitor. They hypothesized that high energy devices may modify OR atmosphere and may lead to the formation of other new toxic byproducts.

Paz and Milliken[17] studied the effect of ESU sparking on N_2O, air, and halothane in an environmental chamber. Experimental results showed an increase in values which reached levels of about 40 ppm NO_2, measured by use of Miran 1A. Sparking halothane mixture in air caused an increase in 8.8 μ infrared (IR) wavelength, indicating that new material was being produced. One explanation is that the new IP, peak on the spectrogram, indicated the formation of halothane degradation products.

The experimental data of Bosterling and Trudell[18] using gas chromatography-mass spectrometry (GC-MS) demonstrated that ultraviolet (UV) irradiation of halothane in air is capable of producing free radicals to form new toxic byproducts. This work confirmed the earlier hypothesis of Paz et al.[17]

A recent study by Gatti et al.[19] revealed electrocautery smoke produced during breast surgery contained unidentified organic compounds undetected by current analytical techniques. Using the Ames test, Gatti reported that these compounds, found in air samples from an OR, were mutagens.

Electrocautery and Bioaerosol Hazards

The AIDS epidemic has focused attention on the routes by which HIV virus may be transmitted. One potential exposure route is inhalation of blood-containing aerosols infected with the virus in the OR. Early studies reported that common surgical power tools produced blood-containing aerosols composed of particles in the <5 mm size range.[20] Johnson and Robinson[21] reported that some of the aerosols found to be infected by HIV-infected blood, which also were generated by a surgical tool, had the ability to infect T-cell tissue culture.

In more recent studies, Jewett et al.[22] studied the potential hazard of blood aerosol generated by a variety of power tools, such as the Hall drill, Shea drill, and ESU operating in cutting and coagulation modes during surgery. The experimental data showed that surgical tools capable of generating a wide distribution of particle sizes produced blood-containing particles in the respirable range. Surgical masks do not provide adequate respiratory protection against these aerosols. The surgical mask does not provide effective protection for removal of bioaerosol particles, the authors speculated that the heat produced by ESU may inactivate the virus.

Control of ESU Smoke and RF/MW Radiation

RF and MW

Any area where RF/MW radiation exposure exceeds permissible levels should be considered potentially hazardous. The area should be clearly identified and warning signs posted. Interlocks may be used to prevent authorized entry. Basic protective measures include the provision of shields or absorbing enclosures for equipment. Personal protective equipment may be used (e.g., gonad shields, protective suits, and wire-netting helmets). Implementing such precautions, however, would be difficult in the OR.

Engineering Control of ESU Smoke

Both the National Institutes of Health and the Laser Institute have recommended smoke evacuators and filtration be used during laser and ESU procedures to eliminate odor and reduce the health risks associated with generation of toxic byproducts during surgery. An earlier version of smoke evacuator was designed with 0.5 μ particle filtration with charcoal to evacuate noxious odors. Today, as more and more questions arise over the presence of airborne chemicals, microorganisms, and viable DNA in smoke, an evacuator is considered more than an instrument of convenience; it is an important part of protective equipment.

A typical smoke evacuator consists of a Vacuum pump, operated at 50 CFM, which provides high static suction; and three filters completely enclosed. First, the filter stage draws air throughout a pre-filter, to capture large particles, collecting up to 80 cm^3 of fluid. At a second stage, ultraefficiency filters capture potentially viable microorganisms and carbonized tissue as small as 0.01 μ with the efficiency of 99.9999%. At the third stage, activated carbon adsorbs organic compounds and odors by products. As air is drawn into the final stage, flow velocity reduced by filtration allows the gases more time to be absorbed on activated carbon and thereby increasing adsorption efficiency. Figure 12.1 illustrates a smoke filtration system.

Conclusion

The author recommends that immediate research is needed to (1) identify ESU toxic byproducts, mutagen, and potential carcinogen generated during ESU surgery; (2) perform real-time exposure and dosimetry to RF/MW on OR personnel; (3) test and analyze all ESU for spectrum energy; (4) conduct

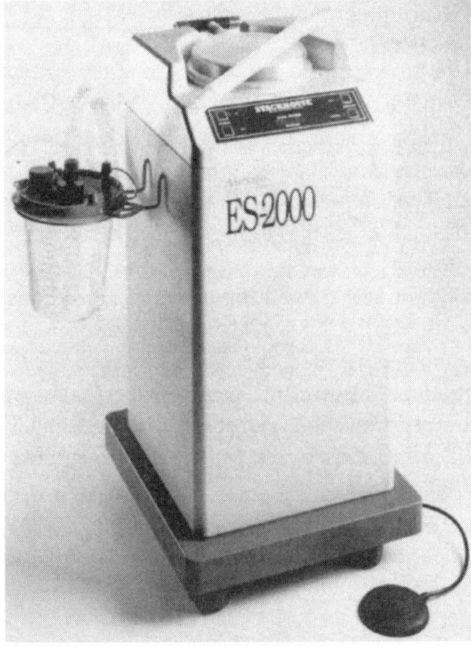

FIGURE 12.1 Stackhouse AirSafe® ES-2000 Electrosurgical Smoke Filtration System (Riverside, CA).

epidemiological studies on the potential health effects of ESU smoke; (5) and educate OR personnel on health hazards associated with the use of ESU.

References

1. Sabiston CD. *Textbook of Surgery: The Biological Basis of Modern Surgical Practice*, pp. 135–136.
2. Sabiston CD. *Textbook of Surgery: The Biological Basis of Modern Surgical Practice*, pp. 214–215.
3. Bailey MK, Bromley HR, Allison JM, Conroy JM, and Krzyzaniak W. Electrocautery-induced airway fire during tracheostomy. *Anesth Analg* 84:2376–2382, 1991.
4. Schellhammer FP. Electrocautery: Principle hazards and precaution. *Urology* 3:261–268, 1991.
5. Paz JD, Milliken R, Ingram WT, and Frank A. Potential ocular damage from microwave exposure during electrosurgery: Dosimetric survey. *J Occup Med* 29:580–583, 1987.
6. Peterson RC. Bioeffects of microwaves: A review of knowledge. *J Occup Med* 25:103–111, 1993.
7. American Conference Governmental Industrial Hygiene, Threshold Limit Value for Chemical Substances and Physical Agent and Biological Exposure Indicts, 1993–1994.
8. ANSI C95. 1–1992/IEEEC95. 1–1991.
9. Carpenter RL and Van Ummersen CA. The action of microwave on the eye. *J Microwave* 3:3–19, 1968.
10. Birenbaum L, Grosof GM, Rosental SW, and Zaret MM. Effects of microwave on the eye. *IEEE Trans Bio-Med Eng* BME-161:7–14, 1969.
11. Guy AW, Lin JD, Kramer PO, and Emary AF. Effects of 2450 MHz radiation on rabbit eye. *IEEE Trans Microwave Techq* MTTT-23 6:492–498, 1975.
12. Fox J, Kendel RT, and Brook HR. Radiofrequency in the operating room theater. *Lancet* 31:962, 1976.
13. Ruggera PS and Segerson DA. Quantitative measurements near electrocautery unit. *AAMI 12th Annual Meeting*, Arlington, VA, March, 1977.
14. Paz JD. Real time non-ionized radiation survey in the operating room. *Paper Presented at the Annual Meeting AIHA*, San Francisco, May, 1988.
15. Goldstein B, Paz JD, Giufreida JJ, Palmes ED, and Ferrand FD. Atmospheric derivatives of anesthesia gases as a possible hazard to operating room personnel. *Lancet* 31:235–237, 1976.
16. Tomita A, Shigenobu M, Kazgata N, Setsuo U, Masakazu F, Minoru H, and Tomio H. Mutagenicity of smoke condensate included by CO_2 laser irradiation and electrocauterization. *Mut Res* 89:145–149, 1981.
17. Paz JD and Milliken RA. Radiofrequency degradation of anesthetic gases as a possible health hazard. *Paper Presented at the Annual Meeting of the American Chemical Society*, Kansas City, September, 1982.
18. Bosterling B and Trudell JR. Production of 5- and 15-hydroperoxyeicosatetaenoic acid from arachidonic acid by halothane-free radicals generated by UV-irradiation. *Anesthesiology* 60:209–213, 1984.
19. Gatti JE, Bryant CJ, Barrett M, and Murphy JB. The mutagenicity of electrocautery smoke. *Plast Reconst Surg* 89:781–785, 1992.
20. Heinsohn P, Jewett DL, Balzer L, Bennett CH, Seipel P, and Rosen A. Aerosol created by some surgical power tool: Particles size distribution, and qualitative hemoglobin content. *Appl Occup Env Hyg* 6:773–776, 1991.
21. Johnson GK and Robinson WS. Human immunodeficiency virus type 1 in the vapor of surgical masks. *Am Ind Hyg Assoc J* 46:308–312, 1985.
22. Jewett DL, Heinsohn P, Bennett C, Rosen A, and Neuilly C. Blood-containing aerosols generated by surgical techniques: A possible infectious hazard. *Am Ind Hyg Assoc J* 53:229–231, 1993.
23. Kopencky KK, Steidle CP, Eble JN, Birhrle R, Dreesen RG, Sutton GP, and Becker GJ. Endoluminal radio-frequency electrocautery for permanent urethral occlusion in swine. *Radiology* 170:1043–1046, 1989.

Health Hazard Evaluation Report

Charles J. Bryant, Richard Gorman, John Stewart, and Wen-Zong Whong

Preface

The Hazard Evaluations and Technical Assistance Branch of NIOSH conducts field investigations of possible health hazards in the workplace. These investigations are conducted under the authority of Section 20(a)(6) of the Occupational Safety and Health Act of 1970, 29 U.S.C. 669(a)(6), which authorizes the Secretary of Health and Human Services, following a written request from any employer or authorized representative of employees, to determine whether any substance normally found in the place of employment has potentially toxic effects in such concentrations as used or found.

The Hazard Evaluations and Technical Assistance Branch also provides, upon request, medical, nursing, and industrial hygiene technical and consultative assistance (TA) to federal, state, and local agencies; labor, industry, and other groups or individuals to control occupational health hazards and to prevent related trauma and disease.

Summary

On January 7, 1985, the National Institute for Occupational Safety and Health (NIOSH) received a request from a group of plastic surgeons at the Bryn Mawr Hospital in Bryn Mawr, PA, to evaluate exposure to emissions generated by the use of electrocautery knives during reduction mammoplasty surgical procedures. Numerous health effects (headache, nausea, upper respiratory, and eye irritation) reported by OR personnel were cited in the request.

An initial on-site survey was conducted on February 14, 1985, with follow-up surveys performed on December 12, 1985 (Pennsylvania Hospital), April 28, 1987, and August 26, 1987. Industrial hygiene sampling was conducted to evaluate exposure to hydrocarbons, nitrosamines, total particulates, benzene soluble fraction, polynuclear aromatic compounds (PNAs) and airborne mutagens. In addition, since very little data have previously been collected for this exposure situation, sampling was performed to obtain qualitative exposure data utilizing a variety of solid sorbent tubes (high volume sampling), Fourier transform IR spectroscopy (FTIR) and aldehyde screening sorbent tubes (Orbo-23).

Personal (breathing zone) and area samples collected for hydrocarbons contained isopropanol at concentrations well below all relevant criteria.

None of the 7 nitrosamines or 16 PNAs that were evaluated were found in detectable quantities. Since several nitrosamines and PNAs are carcinogenic, any detectable levels would have been considered potentially significant.

Concentrations of airborne particulates ranged from 0.4 to 9.4 mg/m^3 of air with a mean of 2.75 mg/m^3. Although these levels all were below the OSHA PEL (15 mg/m^3) and ACGIH TLV (10 mg/m^3) for total nuisance particulates, it is not known at this time whether this particulate is biologically inert; comparison with the nuisance dust evaluation criteria may not be appropriate.

The benzene-soluble fraction of the particulate samples ranged from 0.5 to 7.4 mg/m^3, averaging 2.4 mg/m^3. Seven of the 11 samples exceeded the NIOSH recommended exposure limit of 0.1 mg/m^3 and OSHA PEL of 0.2 mg/m^3. The purposes of these exposure criteria are to minimize worker exposure to carcinogenic PNA compounds. However, this is based on industrial settings (coke ovens, asphalt, petroleum coke) and may not apply to a nonindustrial hospital environment.

Sorbent tubes (charcoal, silica gel, Tenax-TA) that were utilized at high sampling volumes, qualitatively revealed a trace (between the limit of detection and quantitation) amount of hydrocarbons. All of the concentrations were far below evaluation criteria and would not be expected to cause any health effects. FTIR analysis identified a component of the smoke as a compound or compounds related to fatty acid esters. None of the aldehydes (C_1–C_8 aldehydes) evaluated were detected.

Solvent extracts of airborne particles were mutagenic (with microsomal (S9) activation, and slightly mutagenic without activation) to the *Salmonella typhimurium* TA 98 strain, clearly indicating OR personnel exposures to potentially genotoxic agents. However, whether exposure of OR personnel to agents that are mutagenic to bacteria or the level of these agents to which workers are exposed poses any genotoxic hazards is not known.

> On the basis of the mutagenicity of the airborne compounds collected during this evaluation and the acute health effects reported by OR personnel, NIOSH investigators determined that there is a potential hazard from exposure to smoke generated by electrocautery knives during reduction mammoplasty surgical procedures.

Key words: SIC 8062 (General Medical and Surgical Hospitals), ORs, electrocautery knives, electrocautery smoke, reduction mammoplasty, mutagenicity assessment, polynuclear aromatics (PNAs), nitrosamines, and benzene soluble fraction.

Introduction

On January 7, 1985, NIOSH received a request for a health hazard evaluation at the Bryn Mawr Hospital, Bryn Mawr, PA. The request was submitted by a group of surgeons, who were concerned about exposure to emissions generated by electrocautery knives when performing reduction mammoplasties. NIOSH investigators conducted environmental surveys at the Bryn Mawr and Pennsylvania Hospitals on February 14 and December 12, 1985, April 28, 1987, and August 26, 1987.

Background

The surgical procedure known as breast reduction is one of the most common procedures, where considerable smoke is produced. The plastic surgeons at Bryn Mawr Hospital became concerned about the chemical composition and toxicity of this smoke, after noticing that several OR personnel were experiencing acute health effects during this procedure. Reported health effects included upper respiratory and eye irritation, headache, and nausea (obnoxious odors).

The electrosurgical knife (ESK) is presently used for a wide variety of surgical procedures in many health care facilities throughout the United States. Currently, there may be as many as 30–40 U.S. manufacturers of ESK devices. These devices cut or coagulate body tissues utilizing an EM field that is focused onto the body site. The presence of this EM field requires the use of a grounding pad to be placed on the opposite side of the body being cut in order to collect all fields produced by the ESK devices. The ESK units used in this evaluation were a Valley Laboratory (model SSE2L) and a Neo-Med (model 3000). The operating parameters used on these systems during surgical procedures were the same (i.e., mid-range cut and coagulate settings estimated to be 120 W delivered to the cutting area).

On February 14, 1985, NIOSH personnel conducted an initial environmental survey at the Bryn Mawr Hospital. Personal (breathing zone) and area air samples were taken for hydrocarbons, nitrosamines, total particulates, benzene soluble fraction, and PNAs. Findings from this visit were presented in a letter dated May 7, 1985.

A follow-up visit was made on December 12, 1985, at the Pennsylvania Hospital in Philadelphia, PA. Environmental samples were taken (at the suggestion of NIOSH chemists) for PNAs, total particulates, benzene soluble fraction, qualitative organic sorbent tube sampling (charcoal, silica gel, Tenax-TA) and FTIR for qualitative organic analysis. Results were reported on February 14, 1986.

On April 28, 1987, NIOSH conducted an additional follow-up study at the Bryn Mawr Hospital. Environmental air samples for qualitative aldehyde scans were obtained. Monitoring for airborne mutagens was performed on April 28 and August 26, 1987. Results were reported in a letter dated December 1, 1987.

Although the purpose of this evaluation was to determine the nature of the emissions produced by ESK devices in the OR, it should be realized that there are other potential occupational health issues in addition to the chemical and environmental concerns. One issue is the production of EM radiation. Previous NIOSH research work on such systems has indicated that RF radiation at 0.5 MHz is produced by these systems.[1] This finding has also been confirmed in another report.[2]

Methods and Materials

Total Particulates, Benzene Solubles, and PNAs

Personal and area air samples were collected utilizing a sampling train consisting of a Zefluor 2 μ filter (Membrana Co.) and a cellulose acetate O-ring in a cassette, followed by a 7 mm O.D. glass tube containing two sections of prewashed XAD-2 resin (100 mg/50 mg) connected to a battery-operated sampling pump calibrated at a flowrate of 2.0 L/min.

Total particulate weights were determined by weighing the samples plus the filters on an electrobalance and subtracting the previously determined tare weight of the filters. The instrumental precision is 0.01 mg.

The benzene soluble fraction of the filter samples was determined by placing the filters in screw-cap vials with 5 mL of benzene and sonifying for 15 min. The extract was filtered through a Millex-HV 0.45 μm filter and collected in a screw-cap vial. Each sample was transferred into a tared Teflon® cup and evaporated to dryness in a vacuum oven at 40°C. The Teflon cups were again weighed and the difference recorded. The analytical limit of detection is 0.05 mg/sample.

The filter and tube samples were analyzed for PNAs following NIOSH Technical Bulletin TB-001 issued December 1, 1982. The filters and tubes were desorbed in 5 mL of benzene and sonicated for 30 min. The resulting solution was filtered through a 0.45 μm nylon filter. The samples and standards desorbed in benzene were solvent exchanged to acetonitrile by alternate and multiple additions of acetonitrile and evaporation. The samples and standards were not allowed to go to dryness at any time during the exchange. Analysis was then performed by high performance liquid chromatography (HPLC) with a fluorescence/UV detector. The retention times of the analytes in the standards were compared to the retention times in the sample chromatograms for analyte identification. The standard analytes and their associated analytical limits of detection (LOD) are listed below:

Analyte	LOD Nanograms/Sample
Acenaphthene	100
Acenaphthylene	500
Anthracene	250
Benz(a)anthracene	25
Benzo(a)pyrene	25
Benzo(b)fluorathene	25
Benzo(e)pyrene	50
Benzo(k)fluoranthene	25
Benzo(g,h,l)perylene	50
Chrysene	25
Dibenz(a,h)anthracene	25
Fluoranthene	50
Fluorene	100
Indeno(1,2,3,c,d)pyrene	50
Phenanthrene	100
Pyrene	50

Nitrosamines

Personal and area air samples for nitrosamines were collected on Thermosorb/N tubes attached to battery-operated sampling pumps operating at a flowrate of 2.0/min. The tubes were desorbed with 2.0 mL of a solution of 75% methylene chloride and 25% methyl alcohol. The samples were then analyzed by GC with a thermal energy analyzer in the nitrosamine mode. The analytical LOD for this method ranged from 10 to 100 ng per sample (depending upon the particular nitrosamine that was to be identified).

Hydrocarbons

The air samples for hydrocarbons were collected by drawing air through a glass tube containing 150 mg of activated charcoal at a flowrate of 1.0/min (qualitative samples) and 0.21 pm (quantitative samples) using calibrated, battery-operated sampling pumps. The samples were desorbed with 1 mL of carbon disulfide and analyzed by GC with a flame ionization detector (FID). Some of the samples were concentrated and analyzed by GC using a mass spectrometer for major compound identification.

Sorbent Tubes—Qualitative Organic Analysis

Personal and area air samples for qualitative organics analyses were collected on charcoal, silica gel, and Tenax-TA tubes attached to battery-operated sampling pumps operating at a flowrate of 1.0/min (Tenax-TA, 0.51 pm).

The high volume Tenax tubes were analyzed first by thermal desorption. A Telanar model 4000 dynamic head space concentrator equipped with a heated sampler module and capillary cryo focusing interface was used for this procedure. The concentrator unit was interfaced directly to a GC/MS system. Front 100 mg sections of the Tenax sample tubes were put into the sampler module heated to 200°C. The head space was continually purged during this time and the effluent trapped on an internal Tenax trap. The trap was then thermally desorbed onto the front end of a 30 m DB-1 capillary column, flash heated, and injected into the gas chromatograph and mass spectrometer for analysis.

Both the charcoal and silica gel sorbent tubes were screened by GC (FID) and GC/MS. Charcoal tubes were desorbed with 1 mL carbon disulfide and the silica gel tubes were desorbed with 1 mL ethanol. Both front sections and the front glass wool plugs were desorbed together. All analyses were performed using 30 m DB-1 fused silica capillary columns (splitless mode).

FTIR Qualitative Organic Analysis

Personal and area samples for FTIR analysis were collected by drawing air through a Zefluor filter at a flowrate of 1.0/min using calibrated, battery-operated sampling pumps.

Of the six samples submitted, one area and one personal sample were selected for analysis after visually inspecting the filters. The initial analysis involved the analysis of the filters using attenuated total reflectance spectroscopy (ATR). The area sample filter was removed from its cassette and placed in a Barnes Model 305 Horizontal ATR cell. The crystal in the cell was KRS-5 (thallous bromide). Spectra were collected with a Nicolet 60SX Fourier Transform IR Spectrometer using a combined Indium Actinimide/Mercury Cadmium Telluride detector at 0.5 cm^{-1} spectral resolution. After correcting the recorded sample spectra for the background effects of the filter, there appeared to be some compound present. When the filter was removed from the ATR cell, an oily residue was noted to remain on the ATR cell crystal. A spectra of this material were recorded and corrected for the background of the ATR cell. The corrected spectra were compared with the Aldrich FTIR spectral search library using Nicolet searching software. This filter was then desorbed with 1,1,2-trichlorotrifluoroethane. This solution was then evaporated onto the ATR crystal. The spectra were similar to that obtained from the residue of the filter. This desorption procedure was used for the personal sample.

Qualitative Aldehyde Screen

Samples for airborne aldehydes were collected by drawing air through Orbo-23 tubes at a flowrate of 0.08/min. Samples were desorbed with 1 mL of toluene in an ultrasonic bath for 60 min. Aliquots of the sample extracts were then screened by GC (FID) twice; first with a 30 m DB-1 GC column, and second with a 30 m DB was fused silica capillary column (splitless mode).

Airborne Mutagens

Airborne particles were collected on glass–fiber filters (type A/E, 4″ diameter) using Hi-Vol pumps (General Metal Works) at flow rates between 17 and 24 CFM. Filters were changed if the flow rate dropped below 17 CFM.

Samples from the first survey (April 28, 1987) were first extracted with 150 mL of methylene chloride (DCM) then with 150 mL of acetone plus methanol (A + M). Samples from the second survey (August 25, 1987) were divided, because of the quantity of particles. One half was extracted as that in the first survey, the other half was extracted with an XAD-2 resin column. Each extract was filtered and concentrated to a final volume of 0.45 and 0.3 mL in dimethyl sulfoxide for the first and second surveys, respectively.

The same sampling sites were used for both surveys. In the OR, air was sampled 3 ft directly above the operation. As a control (CR), samplers were placed 1/2 ft above the floor in the anteroom.

All extracts were tested for the mutagenic activity in both tester strains TA98 and TA100 of *S. typhimurium* using the Salmonella/microsomal microsuspension test.[3] The system is characterized by adding increased numbers of bacterial cells (approx. 10^9), which are exposed to airborne particle extracts with or without S9 in a concentrated treatment mixture. For metabolic activation, 0.065 mL of S9 mix (10% S9) was also added to each treatment tube. The S9 was prepared from the livers of male Fischer rats pretreated with Aroclor 1254 (500 mg/kg body wt). The micro-suspension test is a suitable assay system for limited quantities of test materials. After 90 min pre-incubation at 37°C, the mixture is processed according to the standard Ames test protocol.[4] The mutagenic activity was scored in tester cells from histidine-dependence to histidine-independence.

In the *in situ* assay, samples were taken at intervals of 2, 4, and 6 h postoperation from the trapping media and were plated on the appropriate agar plates to determine survival and mutation frequencies.[5] Plates were scored after incubation at 37°C for 2 days.

Evaluation Criteria

As a guide to the evaluation of the hazards posed by workplace exposures, NIOSH field staff employ environmental evaluation criteria for assessment of a number of chemical and physical agents. These criteria are intended to suggest levels of exposure to which most workers may be exposed up to 10 h/day, 40 h/week, for a working lifetime, without experiencing adverse health effects. It is, however, important that not all exposures are maintained below these levels. A small percentage may experience adverse health effects because of individual susceptibility, a preexisting medical condition, and/or hypersensitivity (allergy).

In addition, some hazardous substances may act in combination with other workplace exposures, the general environment or with medications or personal habits of the worker to produce health effects, even if the occupational exposures are controlled at the level set by the evaluation criteria. Also, some substances are absorbed by direct contact with the skin and mucous membranes, and thus, potentially increasing the overall exposure. Finally, evaluation criteria may change over the years as new information on the toxic effects of an agent becomes available.

The primary sources of environmental evaluation criteria for the workplace are (1) NIOSH criteria documents and recommendations, (2) the ACGIH TLVS, and (3) the U.S. Department of Labor (OSHA) occupational health standards. Often, the NIOSH recommendations and ACGIH TLVs are lower than the corresponding OSHA standards. Both NIOSH recommendations and ACGIH TLVs usually are based on more recent information than are the OSHA standards. The OSHA standards also may be required to take into account the feasibility of controlling exposures in various industries, where the agents are used; the NIOSH recommended standards, by contrast, are based primarily on concerns relating to the prevention of occupational disease. In evaluating the exposure levels and the recommendations for reducing these levels found in the report, it should be noted that industry is legally required to meet those levels specified by an OSHA standard.

A time-weighted average (TWA) exposure refers to the average airborne concentration of a substance during a normal 8–10 workdays. Some substances have recommended short-term exposure limits or ceiling values, which are intended to supplement the TWA, where there are recognized toxic effects from high short-term exposures.

Isopropanol

Isopropyl alcohol causes mild irritation of the eyes, nose, and throat. High vapor concentrations may cause drowsiness, dizziness, and headache. Repeated skin exposure may cause drying and cracking. NIOSH recommends an exposure limit of 980 mg/m^3.[6] The OSHA standard and ACGIH TLV are the same.[7,8]

Nitrosamines

Nitrosamines are a class of compounds, which are readily formed by the interaction of secondary amines and nitrites or oxides of nitrogen. Because these precursors are ubiquitous, nitrosamines have been found in air, water, tobacco smoke, cured meats, cosmetics and in many industrial processes, including leather tanneries, pesticide formulations, and tire and rubber manufacture facilities.[9]

Nitrosamines are considered to be among the most potent of animal carcinogens. Of more than 150 nitrosamine compounds tested approximately 80% have been found to be carcinogenic in at least one species of animal. To date, there are no standards for employee exposure to airborne nitrosamines. OSHA has a regulation regarding work practices and handling of liquid and solid *N*-nitrosodimethylamine in concentrations greater than 1%.[8] In addition, the FDA has limited the amount of nitrosamines allowed in beer to 5 ppb and the United States Department of Agriculture has limited nitrosamine concentration in cooked bacon to 10 ppb. The International Agency for Research on Cancer recommends that *N*-nitrosodimethylamine, *N*-nitrosodiethylamine, *N*-nitrosodibutylamine, and *N*-nitrosomorpholine be regarded for practical purposes as if they were carcinogenic to humans.[10] NIOSH policy on human exposure to known or suspected carcinogens is to reduce exposure to the lowest feasible level (LFL).[11]

PNAs and Benzene of Cyclohexane Solubles

PNAs are condensed ring aromatic hydrocarbons normally arising from the combustion of organic matter. They are commonly emitted into the air when coal tar, coal tar pitch, or their products are heated, but can result from burning the heavy petroleum fraction used in petroleum coke.[12] A number of PNAs, including benzo(a)pyrene and anthracene are carcinogenic (lung and skin). There are no federal standards pertaining to airborne concentrations of individual PNAs. In 1967, the ACGIH adopted a TLV of 0.2 mg/m^3 for coal tar pitch volatiles (CTPV), described as a "benzene-soluble" fraction, and listed certain carcinogenic components of CTPV. The TLV was established to minimize exposure to the listed substances believed to be carcinogens, viz, anthracene, BaP [benzo(a)pyrent], phenanthrene, acridine, chrysene, and pyrene. CTPVs are among the seven substances listed as "Human Carcinogens" in Appendix A of the current ACGIH TLVs. This group consists of "a substance, or substances, associated with industrial processes, recognized to have carcinogenic or cocarcinogenic potential with an assigned TLV." The TLV was promulgated as a federal standard under the Occupational Safety and Health Act of 1970 (29 CFR 1910.1000).[13] In 1972, the Federal Register (37:24749, November 21, 1972) contained an interpretative rule of the term "...coal tar pitch volatiles include the fused polycyclic hydrocarbons which volatilize from the distillation residues of coal, petroleum, wood, and other organic matter." This has been reprinted as 29 CFR 1910.1002. The general philosophy behind this interpretation was that "all of these volatiles have the same basic composition and...present the same dangers to a person's health."[14]

In the development of the NIOSH recommended standard, it was concluded that CTPVs are carcinogenic and can increase the risk of lung and skin cancer in workers. Since no absolutely safe concentration can be established for a carcinogen, NIOSH recommended the exposure limit be the lowest concentration that can be reliably detected by the recommended method of environmental monitoring. At that time (September, 1977), the lowest detectable concentration for CPTVs was 0.1 mg/m^3 for the recommended sampling method.

Although the benzene or cyclohexane extractable fraction offers an easier, less expensive method of analysis than PNA quantitation, there is no certainty that there is a correlation between the two. The analytical method for measuring the benzene soluble fraction is not limited to PNAs but will include all other organic compounds collected on the filter and soluble in benzene.[15]

Mutagenicity Assay

All the overlayed plates were scored for *his* + revertants after 2 days of incubations. An extract was considered mutagenic if the number of revertants in any of the four concentrations tested (undiluted, 1–2, 1–4, 1–8) was twofold or greater than the control, and showed a dose-related response.

Results

Hydrocarbons

Table 12.4 presents the results of the air samples taken for hydrocarbons. Four of the five samples taken contained isopropanol in concentrations ranging from 1.4 to 4.9 mg/m³. All of the samples were well below the evaluation criteria for 980 mg/m³. The likely source was not the emissions from surgery, but the isopropanol used as a sanitizing agent in the OR.

Nitrosamines

The results of the air samples taken for nitrosamines are presented in Table 12.5. None of the seven nitrosamines evaluated were detected. The specific compounds evaluated included the nitrosamines of dimethyl, diethyl, dipropyl, and dibutylamine, plus those of pyrrolidine, piperidine, and morpholine. The limit of detection for these compounds ranged from 25 to 150 ng/sample. Since several nitrosamines are carcinogenic, any detectable levels would have been considered significant.

Particulates, Benzene Solubles, and PNAs

Tables 12.6, 12.8, and 12.9 contain the data from the analysis of the Zenfluor filters and the Orbo-43 tubes. Each filter and tube sampling train provided the following three types of data:

1. Total Particulates—represent the total weight of the smoke per cubic meter of sampled air. Five personal breathing-zone air samples for total particulate ranged from 0.4 to 2.0 mg/m³ with a mean of 1.0 mg/m³. Six area samples ranged from 0.7 to 9.4 mg/m, with a mean of 4.2 mg/m³.
2. Benzene Soluble Fraction—represents the total weight of the smoke that is benzene soluble per cubic meter of sampled air. No benzene soluble fraction was detected in any of the particular samples

TABLE 12.4 Isopropanol, Bryn Mawr Hospital, Bryn Mawr, PA, HETA 85–126, February 14, 1985

Sample Location/Job	Sample Type	Sampling	Isopropanol (mg/m³)[a]
Assistant surgeon	Personal	13:30–15:20	1.4
Surgeon	Personal	13:30–15:20	2.1
Anesthesia area or OR lights	Area	13:30–15:20	1.4
Surgical nurse	Personal	13:30–15:20	4.9
Laser used on breast tissues	Area	15:30–15:41	ND[b]
Evaluation criteria:		NIOSH	980
(8 h TWA)		ACGIH	980
		OSHA	980

[a] All air concentrations are reported as TWA s for the time sampled.
[b] Non-detectable. Limit of detection is 0.01 mg/sample, which would correspond to an atmospheric concentration of 0.48 mg/m3 when the average sample air volume (21 L) is considered.

TABLE 12.5 Nitrosamines, Bryn Mawr Hospital, Bryn Mawr, PA, HETA 85-126, February 14, 1985

Sample Location/Job	Sample Type	Sampling	Nitrosamines[a]
Assistant surgeon	Personal	13:30–15:20	ND[b]
Surgeon	Personal	13:30–15:20	ND[b]
Anesthesia area or OR lights	Area	13:30–15:20	ND[b]
Surgical nurse	Personal	13:30–15:20	ND[b]
Laser used on breast tissues	Area	15:30–15:41	ND[b]
Evaluation criteria			LFL[c]

[a] NIOSH currently uses a seven standard mixture to calibrate and identify specific nitrosamines. The mixture contains the nitrosamines of dimethyl, diethyl, dipropyl, and dibutylamine plus those of pyrrolidine, piperidine, and morpholine.

[b] Non-detectable. LOD range from 25 to 150 ng/sample (aid adjusted concentrations would range from 114 to 682 μg/m^3).

[c] No evaluation criteria have been established for nitrosamines. Exposure should be reduced to LFL.

TABLE 12.6 Particulates, Benzene Solubles, and PNAs, Bryn Mawr Hospital, Bryn Mawr, PA, HETA 85-126, February 14, 1985

			Exposure Concentrations[a]		
Sample Location/Job	Sample Type	Sampling Period	Total Particulates (mg/m^3)	Benzene Soluble (mg/m^3)	PNAs[b]
Assistant surgeon	Personal	13:30–15:20	1.6	ND	ND
Surgeon	Personal	13:30–15:20	ND[c]	ND	ND
Anesthesia area or OR lights	Area	13:30–15:20	1.2	ND	ND
Surgical nurse	Personal	13:30–15:20	0.4	ND	ND
Laser used on breast tissue	Area	15:39–15:41	9.4	7.4	ND
Evaluation criteria		NIOSH	—	0.1	[d]
(8 h TWA)		ACGIH	10	0.2	[d]
		OSHA	15	0.2	[d]

[a] All air samples are reported as TWA s for the time period sampled.

[b] Represents the following EPA priority PNAs: acenapthene, acenaphthylene, anthracene, benz(a)anthracene, benzo(a)pyrene, benzo(a)fluorathene, benzo(e)pyrene, benzo(k)fluoranthene, benzo(g,h,i)perylene, chrysene, dibenz(a,h)anthracene, fluoranthene, fluorene, ideno(1,2,3,c,d)pyrene, phenanthrene, pyrene.

[c] Non-detectable. LOD are 0.01 mg/sample for total particulates, 0.05 mg/sample for benzene soluble fraction, and 25–500 ng/sample for the various PNAs analyzed.

[d] No criteria currently exist for total PNAs; however, a number of individual PNAs are carcinogenic (benzo(a)pyrene, anthracene, chrysene) and exposures should be controlled to the LFL.

taken on February 14, 1985. All of the samples taken on December 12, 1985 had a benzene soluble fraction ranging from 0.7 to 6.7 mg/m^3, well above the NIOSH evaluation criteria of 0.3 mg/m^3.

3. PNAs—represent the analysis for 17 polynuclear aromatic hydrocarbons. None of the 17 PNAs, which are monitored in the NIOSH standard method, were detected in any of the samples.

Sorbent Tubes—Qualitative Organic Analysis (Table 12.7)

Table 12.7 lists the substances that were identified by sorbent tube sampling during the survey on December 12, 1985. The air samples indicate that trace amounts (less than 10 μg) of hydrocarbons were

TABLE 12.7 Sorbent Tube Sampling, Pennsylvania Hospital, Philadelphia, PA, HETA 85-126, December 12, 1985

Sorbent Tube	Substances Identified
Tenax-TA[a]	Isoflurane (anesthetic gas)
	Halothane (anesthetic gas)
Charcoal[b]	Isoflurane
	Isopropanol
	1,1,1-Trichloroethane
	Trichloroethylene
	Toluene
	Perchloroethylene
	Xylene
	Several aliphatic hydrocarbons
Silica gel	None detected

[a] Thermally desorbed tubes cannot be quantified.

[b] Substances were present in trace quantities, between the limit of detection (1–5 μg/sample) and limit of quantitation (5–10 μg/sample).

TABLE 12.8 Zefluor Filter Sample/FTIR Analysis, Pennsylvania Hospital, Philadelphia, PA, HETA 85-126, December 12, 1985

Sample Type	Substances Identified
Zefluor (area)	Compound or compounds related to fatty acid esters
Zefluor (personal)	Compound or compounds related to fatty acid esters

TABLE 12.9 Particulates, Benzene Solubles and PNAs, Pennsylvania Hospital, Philadelphia, PA, HETA 85-126, December 12, 1985

			Exposure Concentrations[a]		
Sample Location/Job	Sample Type	Sampling Period	Total Particulates (mg/m^3)	Benzene Soluble (mg/m^3)	PNAs[b]
Surgeon	Personal	11:30–12:40	2.0	1.4	ND
Assistant surgeon	Personal	11:30–12:40	0.9	0.7	ND
Handheld/operative site	Area	11:35–12:35	0.7	1.3	ND
Handheld/operative site	Area	11:35–12:35	0.7	1.7	ND
Handheld/breast tissue/ post-surgery	Area	13:00–13:15	8.7	6.7	ND
Handheld/breast tissue	Area	13:00–13.15	4.7	6.7	ND
Evaluation criteria	NIOSH		—	0.1	c
(8 h TWA)	ACGIH		10	0.2	c
	OSHA		15	0.2	c

Note: ND, non-detectable; LOD are 0.1 mg/sample for total particulates, 0.05 mg/sample for benzene soluble fraction, and 25–500 ng/sample for the various PNAs analyzed.

[a] All air samples are reported as TWA s for the time sampled.

[b] Represents the following EPA priority PNAs: acenapthene, acenaphthylene, anthracene, benz(a)anthracene, benzo(a)pyrene, benzo(a)fluorathene, benzo(e)pyrene, benzo(k)fluoranthene, benzo(g,h,i)perylene, chrysene, dibenz(a,h)anthracene, fluoranthene, fluorene, ideno(1,2,3,c,d)pyrene, phenanthrene, pyrene.

[c] No criteria currently exist for total PNAs; however, a number of individual PNAs are carcinogenic (benzo(a) pyrene, anthracene, chrysene) and exposures should be controlled to the LFL.

present within the OR. The substances identified would not be expected to cause ill-health effects in most people at the levels detected.

FTIR Qualitative Organic Analysis

The search area of the spectra indicated that the compounds found on both the area and personal filters were related to fatty acid esters, based on spectral similarities. A search of the library using the absolute derivative search algorithm indicated that olive oil, cottonseed oil, methyl stearate, and castor oil were all close matches to the sample spectra for the bulk sample. The goodness-of-fit values associated with each of these matches were all equivalent and higher than what would be expected for an exact match. A low goodness-of-fit value indicates a good match of library spectra. For the qualitative sample, matches included ethyl stearate, castor oil, and other straight chain hydrocarbon compounds. Based on the source of the sample, the identification of the sample as fatty acid esters is logical.

Attempts made to analyze these samples by GC were unsuccessful. The samples were not volatile enough to allow chromatography.

As a final attempt at characterization of the samples, the two desorbed filter solutions were submitted for direct probe mass spectral analysis. The solutions were evaporated onto the direct probe of the mass spectrometer and heated under vacuum. Results from this analysis indicated that the samples gave mass spectral detail related to straight chain hydrocarbons, i.e., methylene group (CH_2) fragments. Since fatty acid esters contain long chain alkyl groups, this fragmentation pattern was not unexpected. The mass spectra were compared to methyl stearate and stearyl palmitate and were found to be similar to the sample spectra but were not an exact match, indicating that there was probably a mixture of fatty acid esters in the samples.

Qualitative Aldehyde Screen

Table 12.10 presents the results of the area and personal samples taken for qualitative aldehyde scan analysis. No aldehydes were detected. It should be noted that the aldehyde scan has only been tried on the low molecular weight (C_1–C_8) aliphatic aldehydes.

Airborne Mutagens

Airborne particles collected on glass–fiber filters from both surveys were found to be mutagenic (TA98). Samples from the first survey on April 28, 1987 (Table 12.11) showed a positive response only with S9, but the second survey (August 26, 1987) samples also showed a slight response without activation. The mutagenic response of extracts from the organic solvent extraction of the second survey was higher than those from the XAD-2 column extraction (Table 12.12). No significant mutagenic response was found with the *in situ* assay system in either survey (Table 12.13). The population of TA98W for the *in situ* testing was too low and gave sporadic results. This was not reported.

TABLE 12.10 Qualitative Aldehyde Scan, Bryn Mawr Hospital, Bryn Mawr, PA, HETA 85-126, April 28, 1987

Sample Location	Type Sample	Aldehydes Identified[a]
Scrub nurse	Personal	None
Surgeon	Personal	None
Assistant surgeon	Personal	None
Anesthetist	Area	None
OR	Area	None

[a] Aldehyde scan has only been tried on the low molecular weight (C_1–C_8) aliphatic aldehydes.

TABLE 12.11 Mutagenicity of Airborne Particle Extracts, Bryn Mawr Hospital, Bryn Mawr, PA, HETA 85-126, April 28, 1987

			His. Rev./Plate			
			TA98		TA100	
Sample Location	Particles μg/Plate	Air Vol. m³/Plate	−S9	+S9	−S9	+S9
DCM extraction						
OR	78	0.25	4	6	54	55
	155	0.49	6	8	64	52
	310	0.98	10	19	62	50
	620	1.95	10	31	62	55
Control room	9	0.25	4	8	51	43
	18	0.49	7	9	60	47
	37	0.98	8	8	48	45
	73	1.95	8	11	53	48
Filter control			5	8	54	45
Negative control			7	5	44	46
Positive control[a]			1608			1926

Note: His. Rev., histidine revertants.
[a] 2-Aminoanthracine: 2.5 μg/plate.

TABLE 12.12 Mutagenicity of Airborne Particle Extracts, Bryn Mawr Hospital, Bryn Mawr, PA, HETA 85-126, August 26, 1987

			His. Rev./Plate		
			TA98		TA100
Sample Location	Particles μg/Plate	Air Vol. m³/Plate	−S9	+S9[a]	+S9
DCM extraction					
OR	265	0.35	5	37	54
	530	0.70	6	69	67
	1060	1.41	12	92	79
Control room	17	0.73	4	8	52
	33	1.46	2	7	73
Filter control			3	6	64
XAD2 column extract					
OR	265	0.35	4	24	57
	530	0.70	6	45	45
	1060	1.41	10	57	57
Control room	17	0.73	2	8	55
	33	1.46	5	8	49
Filter control			4	7	59
Negative control			4	7	68
Positive control[b]				1551	1610

Note: His. Rev., histidine revertants.
[a] Average of two experiments.
[b] 2-Aminoanthracine; 2.5 μg/plate.

TABLE 12.13 *In Situ* Testing of Airborne Particle Extracts, Bryn Mawr Hospital, Bryn Mawr, PA, HETA 85-126

	Control Room				OR			
	C		T		C		T	
Hour	% Sur.	Rev./10^7	% Sur.	Rev./10^7	% Sur.	Rev./10^7	% Sur.	Rev./10^7
April 28, 1987								
2	100	8	104	7	100	7	97	8
4	100	5	111	7	110	6	100	7
6	104	8	114	7	124	6	107	8
2 S	100	7	79	10	100	11	123	10
4 S	74	11	58	12	85	14	108	10
6 S	40	17	33	21	52	20	92	13
August 25, 1987								
2	100	14	164	12	100	12	131	12
4	118	13	182	10	108	8	123	11
6	109	12	200	13	108	10	131	12
2 S	100	21	127	16	100	26	112	23
4 S	97	21	116	21	102	24	138	19
6 S	64	33	80	30	57	37	75	35

Note: TA100 only; S, with S-9 activation; C, recirculating closed system (control); T, ambient room air; % Sur, percent survival; Rev./107, number of revertants/107 living cells.

An extract was considered mutagenic if the number of revertants in any of the four concentrations tested (undiluted, 1–2, 1–4, 1–8) was twofold or greater than the control and showed a dose-related response.

Conclusions

A. No specific organic vapors, other than isopropanol, were quantitatively identified during the surgical procedures.
B. There were no PNAs or nitrosamines detected during the procedures.
C. The exposures to particulates, which ranged from 0.4 to 2.0 mg/m^3, confirm that the visible emissions are more than just water vapor. However, there are no exposure criteria with which to compare this exposure. Exposure criteria of 10–15 mg/m^3 have been established for nuisance dust; however, to apply this criteria, the particulate would need to be biologically inert, which in this case, is not known. Airborne particulate can contribute to the eye irritation, which has been reported during these procedures. We would have expected the samples from the surgeons and some of the area samples (hand-held approximating breathing-zone) to be similar. The "not-detected" on one surgeon suggests that we encountered a flow problem with that sample train even though the flow looked fine at the end of the sampling period.
D. The sample (Table 12.6) taken during a brief demonstration of the laser-cutting technique measured 9.4 mg/m^3 of total particulate of which 7.4 mg/m^3 was found to be benzene soluble. It is difficult to make conclusions based on one sample, but these data suggest that there is a tendency for the laser method to produce more particulate that is soluble in benzene. Whether this produces more PNAs is unknown.
E. All of the particulate samples taken on December 12, 1985 contained benzene soluble fractions above the evaluation criteria. However, it should be noted that the following three issues complicate the interpretation of the benzene soluble fraction data:

1. Three of the four area samples contained benzene soluble fractions that were higher than the corresponding total particulate values. This is improbable; the benzene soluble fraction can be equal to, but never exceeds the total particulate value. A review of the blank values revealed no discrepancies in the analytical procedures. No explanation for this anomaly can be offered at this time.
2. None of the electrocautery samples taken during the February 14 survey contained a benzene soluble fraction, while all the samples taken on December 12 had a benzene soluble fraction above the NIOSH evaluation criteria. Sampling conditions were almost identical. Again, no explanation for the difference in concentrations can be given.
3. PNAs are benzene soluble and would therefore be contained in the benzene soluble fraction of the total particulate samples. It follows that the higher the value for benzene solubles, the more potential for PNAs and, therefore, the greater the risk from the exposure. However, it should be noted that this concept is based on industrial exposures (CTPV, petroleum coke, asphalt fumes, etc.). Although the benzene extractable fraction offers an easier, less expensive method of analysis than PNA quantitation, there is no certainty that there is a correlation between the two (especially in a nonindustrial setting).

F. Trace amounts of hydrocarbons were identified (utilizing high volume sorbent tube sampling) and would not be expected to cause ill-health effects in most people at the levels detected.

G. Based on the FRIT (qualitative organic) analysis, the major component of the samples is a compound or compounds related to fatty acid esters.

H. Aldehydes (C_1–C_8) aliphatic were not present in quantifiable levels.

I. The results of the studies for airborne mutagens indicate that the solvent extracts of airborne particles collected from the hospital OR using cauterization were mutagenic. The mutagenic activity varies from patient to patient: age, fat content, and size. The patient in the first survey was older with more fat in the tissue than the patient in the second survey. By comparison, samples from the second survey showed at least double the mutagenic activity than those of the first. Whether exposure of OR personnel to agents that are mutagenic to bacteria or the level and condition of these agents to which workers are exposed pose any genotoxic hazards is not known. Limited information suggests that there is a correlation between the bacterial mutagenicity level of airborne particles and lung cancer incidence.[16,17] Index of the mutagenicity of air particles has been considered to be a more powerful measure of the human health hazard of air pollution than the traditional indices of particulate concentration.[17] This information is yet to be validated by further epidemiological studies, where the mutagenic activity of collected air samples is known. In the meantime, it may be prudent to monitor OR personnel for any adverse health effects and to reduce mutagenically active contaminants whenever possible in the OR.

J. OR staff experience acute health effects (upper respiratory and eye irritation, headache, nausea [obnoxious odors]) during this type of surgery, where electrocautery techniques are used for a substantial part of the total operative procedure.

Recommendations

1. Engineering ventilation controls (smoke evacuation units) should be utilized to minimize the acute health effects and further reduce the potential for any chronic health effects. The smoke evacuation units will also eliminate the emissions that can impair the surgeon's vision.
2. Any further acute or chronic health effects experienced by the OR staff should be evaluated and documented.
3. Exposure to electrocute smoke should be reevaluated if other techniques for identifying and quantitating the smoke emissions can be found or developed.

References

1. Moss CE. Evaluation of body currents from exposure to radiofrequency fields. *Paper Presented at the American Industrial Hygiene Conference*, San Francisco, CA, May 15–20, 1988.
2. Paz J, Ingram TW, Milliken R, and Hartstein G. Real time dosimetric survey during electrocautery surgery. *Paper Presented at the American Industrial Hygiene Conference*, San Francisco, CA, May 15–20, 1988.
3. Kado NY et al. A simple modification of the Salmonella liquid-incubative assay. *Mutat Res* 121: 25–32, 1983.
4. Ames BN et al. Methods for detecting carcinogens and mutagens with Salmonella/mammalian microsome mutagenicity test. *Mutat Res* 31:347–363, 1976.
5. Whong W-Z et al. Development of an *in situ* microbial mutagenicity test system for airborne workplace mutagens: Laboratory evaluation. *Mutat Res* 130:45–51, 1984.
6. National Institute for Occupational Safety and Health. *Current Intelligence Bulletin 48: Organic Solvent Neurotoxicity*. National Institute for Occupational Safety and Health, Cincinnati, OH, 1987 (DHHS Publication No. (NIOSH) 87-104).
7. American Conference of Governmental Industrial Hygienists. *Threshold Limit Values and Biological Exposure Indices for 1987–1988*. Cincinnati, OH: ACGIH, 1987.
8. Occupational Safety and Health Administration. OSHA safety and health standards. 29 CFR 1910.1000. Occupational Safety and Health Administration, revised 1987.
9. Frank CW and Berry CM. *N*-nitrosamines. In: *Patty's Industrial Hygiene and Toxicology*, Vol. IIB, Chapter 43, 3rd edn. John Wiley & Sons, New York, 1981.
10. Bogovski R, Preussman EA, Walker EA, and Davis W. *Evaluation of Carcinogenic Risk of Chemicals to Man. IARC Monographs*, Vol. 1. International Agency for Research on Cancer, World Health Organization, Lyon, France, 1972.
11. National Institute for Occupational Safety and Health. *Working with Carcinogens*. National Institute for Occupational Safety and Health, Cincinnati, OH, 1977 (DHEW publication no. (NIOSH) 77-206).
12. Scala RA. Toxicology of PPOM. *J Occup Med* 17:784–788, 1985.
13. National Institute for Occupational Safety and Health. *Criteria for a Recommended Standard: Occupational Exposure to Coal Tar Products*. National Institute for Occupational Safety and Health, Cincinnati, OH, 1978 (DHEW publication no. (NIOSH) 78-107).
14. National Institute for Occupational Safety and Health. *Criteria for a Recommended Standard: Occupational Exposure to Asphalt Fumes*. National Institute for Occupational Safety and Health, Cincinnati, OH, 1978 (DHEW publication no. (NIOSH) 78-106).
15. National Institute for Occupational Safety and Health. *Petroleum Refinery Workers Exposure to PAHs at Fluid Catalytic Cracker, Coker, and Asphalt Processing Units*. National Institute for Occupational Safety and Health, Cincinnati, OH, 1981 (NIOSH Contract No. 210-78-0082).
16. Kaiser C, Keer A, McCalla DR, Lockington JN, and Gibson ES. Use of bacterial mutagenicity assays to probe steel foundry lung cancer hazard. *Polynuclear Aromatic Hydrocarbons. 5th International Symposium Chemical and Biological Effects*, Ohio, 583–592, 1981.
17. Walker RD, Connor TH, MacDonald EJ, Trieff NM, Legator MS, MacKenzie KW, and Dobbins JG. Correlation of mutagenic assessment of Houston air particulate extracts in relation to lung cancer mortality rates. *Environ Res* 28:303–312, 1982.

Authorship and Acknowledgments

Report prepared by Charles J. Bryant, C.I.H., Industrial Hygienist, Industrial Hygiene Section; Richard Gorman, C.I.H., Industrial Hygiene Engineer, Industrial Hygiene Section; John Stewart, B.S., Research Microbiologist, NIOSH, Morgantown, West Virginia; Wen-Zong Whong, Ph.D. Geneticist, NIOSH, Morgantown, West Virginia.

Field Assistance: James Boiano, C.I.H., Industrial Hygienist, Industrial Hygiene Section.

Originating Office: Hazard Evaluations and Technical Assistance Branch, Division of Surveillance, Hazard Evaluations, and Field Studies.

Report typed by Kathy Conway, Clerk-typist, Industrial Hygiene Section.

Distribution and Availability of Report

Copies of this report are currently available upon request from NIOSH, Division of Standards Development and Technology Transfer, Publications Dissemination Section, 4676 Columbia Parkway, Cincinnati, OH 45226. After 90 days, the report will be available through the National Technical Information Service (NTIS), 5285 Port Royal, Springfield, VA 22161. Information regarding its availability through NTIS can be obtained from NIOSH Publications Office at the Cincinnati address. Copies of this report have been sent to 1. John E. Gatti, M.D., J. Brien Murphy, M.D., and R. Barrett Noone, M.D.; 2. NIOSH, Boston Region; 3. OSHA, Region III.

For the purpose of informing affected employees, copies of this report shall be posted by the employer in a prominent place accessible to the employees for a period of 30 calendar days.

Source: HETA 85-126-1932, September 1988, Bryn Mawr Hospital, Bryn Mawr, PA. NIOSH Investigators: Charles Bryant, M.S., C.I.H., Richard Gorman, M.S., C.I.H., John Stewart, B.S., and Wen-Zong Whong, Ph.D.

(Mention of company names of products does not constitute endorsement by the NIOSH.)

Laser Hazards in the Health Care Industry

Lindsey V. Kayman

Lasers have opened new vistas in many branches of health care. As with any technology, lasers are not without risks. Therefore, special precautions must be taken to prevent serious potential health hazards, which can pose a threat to both medical staff and patents. This guide will summarize major laser safety issues, discuss regulation of lasers, and provide resources for further information.

Background Information on Lasers

The word "laser" means light amplification by the stimulated emission of radiation. Lasers are now used in many industries, including building construction and telecommunications, and in consumer products such as compact disk players and office printers. In health care settings, lasers are widely used for microscopic surgery and for measuring immunoglobulins and other elements in the blood.

Lasers create biological effects in tissue because they focus large amounts of light energy on a small surface area. The precise effect depends on the length of exposure time. When tissue is exposed to very short laser pulses, it becomes extremely hot and promotes a micro-explosion with a shock wave. The laser energy causes coagulation, cutting, or vaporization of the tissue. At longer exposure times, the absorbed energy is spread out over a larger area, and the effect is smaller.

Commonly used lasers in health care settings are the carbon dioxide laser and argon laser which use gas, and the Nd:Yag laser which uses a crystal medium. Continuous wave lasers deliver energy as long as the laser is activated; pulsed lasers deliver short bursts of energy. Each laser system has its own unique set of optical properties, controls, and output characteristics. The chart below lists some lasers used in health care and their application.

Lasers are being used in health care settings precisely for their ability to heat and cut biological tissue. Yet, this ability of lasers to damage biological tissue is also the reason why lasers are potentially dangerous to users and patients, unless lasers are properly operated and maintained.

Types of Lasers Used in Health Care

Type	Application
Carbon dioxide	Incision and excision by vaporization in neurosurgery, otolaryngology, gynecology, podiatry, and general surgery
Argon	Coagulates tissue, sealing blood vessels in retina. Also used in plastic surgery, gastroenterology, and dermatology
KTP 532 (Green)	Cutting, coagulation, and vaporization of tissue. Used in otolaryngology, gynecology, neurosurgery, urology, podiatry, general surgery, etc.
Nd:YAG (continuous wave)	Thermal effect and deep penetration used in gastrointestinal, urological, gynecological, and general surgery
Nd:YAG (Q switched)	Breaking apart a target tissue. Used primarily in ophthalmology
Helium-neon	For aiming the invisible beams of carbon dioxide and Nd:YAG lasers
Ruby	Destroying tissues in dermatology and plastic surgery
Tunable dye	Treating malignant tissues

Source: Abstract from ECRI, Health Care Environmental Management Systems, "Special Issues in Healthcare Safety," Volume 3, April 1991.

Various professional organizations and government agencies have recognized the potential dangers of laser use. They have categorized lasers into four major hazard classes according to the wavelength and the power density of the laser beam. Most lasers used in surgery are considered Class 4 high-risk laser products.

Every employer that utilizes Class 3B or Class 4 lasers must have a trained Laser Safety Officer (LSO), who has the knowledge and authority to monitor and enforce the control of laser hazards. The LSO must

- Confirm the classic of lasers used at the workplace.
- Approve the standard operating procedures (SOPs).
- Ensure that workers exposed to lasers have received proper training.
- Limit the access of nonessential personnel to laser work areas.
- Maintain the laser equipment properly.
- Ensure that laser equipment is properly installed.
- Reduce or eliminate other risks in the work area that could make the use of lasers more dangerous to users.
- Recommend the appropriate protective equipment such as eyewear and protective clothing.
- Specify the appropriate warning signs.

Laser Hazards

Several hazards are associated with lasers in the health care industry: damage to the eyes, skin burns, inhalation of toxic chemicals and pathogens, fires, and electrical shock. Each hazard and how it can be controlled are discussed below.

Eye Hazards

The eye is especially vulnerable to injury from a laser beam; it is generally considered to be the organ at greatest risk. Exposure of unprotected eyes for a fraction of a second as well as chronic lower-power exposures from scattered, diffused, and reflected laser beams (such as from surgical instruments and tissue) can cause serious, irreversible damage. The damage will be located in the part of the eye, where the laser energy is absorbed. The invisible beam of the laser which operates in the IR region of the light spectrum is absorbed in the cornea and may cause corneal scarring and loss of vision. Lasers which transmit light in the visible or near IR regions are focused by the lens of the eye to produce an intense concentration of light energy on the retina. The energy is converted to heat and causes a burn and loss of vision, especially if the burn is located in the macula, the center of visual acuity of the retina. Reflected laser beams can also injure the eye.

Eye Protection Devices

Engineering controls, which are built into laser equipment by the manufacturer, are an important means of controlling laser hazards. Examples of engineering controls are protective housing, fail-safe interlocks, master switch controls and beam stops or attenuators to reduce output emissions. However, these measures may not completely control the beam stray reflections in many situations. Therefore, OSHA requires eyewear in addition to engineering controls whenever accessible emission levels exceed maximum permissible exposure levels.

Unfortunately, eyewear creates many problems for users, which explains why many workers are reluctant to wear it: eyewear is often uncomfortable. It also reduces visibility—it can fog up; it can cause tunnel vision, making it difficult to see the laser beam and the patient. For these and other reasons, the LSO must select eyewear that is effective and also "user-friendly"; i.e., eyewear that has good visibility and is comfortable to wear.

Several complex factors must be taken into consideration when selecting eyewear. For example, appropriate eyewear is dependent on the wavelength and power density of the laser in use: eyewear for the carbon dioxide beam may not protect against the Nd:YAG laser. Many other considerations affect the quality and appropriateness of eyewear. These are discussed below.

Optical Density

The filtering ability of laser eyewear is rated by a factor called optical density (OD). The laser manufacturer or LSO should determine the OD needed for protective eyewear. A high OD at a given wavelength indicates greater laser beam absorbency. The OD and the wavelength the eyewear is designed for must be imprinted on the lens or frame. Note that the OD does not take into account the mechanical strength of the eyewear.

Although it is more protective from a safety viewpoint, eyewear with a high OD causes reduced visibility, difficulty seeing a beam which normally may be visible, and eye fatigue.

The visibility allowed by protective eyewear is rated by its visible luminous transmission (LT) expressed as the percentage of light seen through the glasses; the higher the LT, the better the visibility. Eyewear used during procedures should have the highest LT at the safest OD.

Goggles vs. Glasses

Goggles rather than glasses are generally recommended to protect against back reflection or side entrance of a stray beam. With side shields, properly fitted, lightweight laser protective spectacles provide a good alternative to goggles if discomfort might result in their not being worn. Side shields made of the same material as the lenses allow for adequate peripheral vision as well.

Reflective vs. Absorptive Laser Protective Eyewear

The lenses of protective eyewear are either reflectors or absorbers. Both can be made out of plastic or glass. Reflective lens filters have a thin surface coating, which is designed to reflect a laser beam away from the eye. An advantage of reflective coatings is good visibility. They can be designed to selectively reflect a given wavelength while transmitting as much of the remaining visible spectrum as possible.

A surface scratch is a serious problem in a reflective filter because it could allow penetration of the laser beam, causing possible injury to the eye. In addition, the angle at which the laser beam hits the lens could affect the protection afforded by the eyewear. Another problem with reflective lenses is that the laser beam remains a safety hazard when it is reflected. Reflected beams are the most common type of safety hazard from lasers. Coworkers may be exposed to the reflected beam. There is a danger of fire if the beam hits a combustible or flammable item.

Rather than reflecting the beam, absorptive filters convert the incoming laser energy to a heat, which is harmlessly diffused through the lens. Absorptive lenses have many benefits compared to reflective lenses:

The protection afforded by absorptive filters is not affected by surface scratches; they are less likely to create hazardous beam reflections; and the protection they afford is not affected by the angle at which the beam hits the lens. However, many absorbing glass filters cannot be easily annealed (thermally hardened). Consequently, they do not provide adequate impact resistance. In some goggle designs, impact-resistant polycarbonate filters are used together with non-hardened glass filters to provide good impact resistance. Another problem with absorbing plastic filter materials is that the organic dyes which are used as absorbers are affected by heat and/or UV radiation, which can cause the filter to darken or decrease its ability to absorb laser energy.

Plastic vs. Glass Lenses

There are pluses and minuses of both glass and plastic (polymeric) protective lenses. Unless thermally hardened, glass does not resist physical impact as well as plastic. Glass is more easily scratched than many polymers. Although glass can usually withstand higher laser exposure levels than plastic, plastic boasts a high heat-deflection temperature, which enables them to withstand laser beams with high energy densities.

Plastic is more lightweight than glass, and it may be molded into comfortable shapes. Plastic materials generally also display a lower threshold for laser beam penetration. According to OSHA Guidelines, plastic eyewear is only appropriate when the wearer is more than an arm's length from the target area where the beam is focused (typically 0.5 m or about 18 in.); when the direct "raw beam" exposure cannot exceed 20 W level.

Plastic eyewear should be adequate for support staff standing at a distance from the laser, but plastic filters are not considered protective enough for technicians servicing the laser. A 20 W "raw beam" exposure would be more likely to occur to workers during servicing of the laser or to the operator of a laser while working at a close distance.

Other Considerations

- Some eyewear frames cannot withstand the same exposures that the lenses are designed to tolerate. Certain frames are available which are coated with a laser-absorbent material to correct for this.
- Dyes used in eyewear to absorb laser radiation can be bleached or darkened by long-term exposure to light and heat or can simply deteriorate over time.
- Buildup of humidity within tight fitting eyewear is a problem. OSHA has cited a hospital for use of eyewear with side shields that had air circulation holes since open holes can increase the risk of beam exposure. Covered vents are more acceptable. Fogging may be reduced by the use of antifogging cleaning solutions.
- If protective eyewear which provides multiwavelength eye protection is to be used, each wavelength and corresponding OD should be confirmed in advance and understood by each wearer.
- Some brands of eyewear are designed to improve visibility by selectively altering and reflecting the hazardous wavelengths while transmitting a great deal of the remaining visible spectrum.
- Certain eyewear filters make it difficult to see certain colors found in the beam, blood and tissue, colored warning lights, laser emission indicators, and other important instrument displays. Alignment eyewear may be useful in these situations. It is used for low power visible laser beams which align the high power or invisible beam. Wearing alignment eyewear with a low power visible beam allows beams, blood and instrument displays to be seen while providing some protection from diffuse radiation. Alignment eyewear should never be worn during the operation; it is also not meant for use with power or invisible beam lasers.

Controlling Eye Hazards

The employer's LSO should develop a written safety program including a protocol for eye hazards. The protocol should include the following:

- *The determination of a nominal hazard zone (NHZ).* The NHZ is the space within which the level of direct, reflected, or scattered radiation during normal operation exceeds the applicable maximum permissible exposure; it is the area around the laser, where eye protection is required when the laser is activated. The NHZ may be determined by using the tables in the ANSI's laser standard, Z136.1 Section 8 and Appendix B. (See Resources Section for information about ANSI.) The NHZ can also be derived from information supplied by the laser manufacturer.
- *Periodic eye examinations.* The frequency and content of eye examinations are found in the eye examination chart.
- *Selection of eyewear.* A variety of high quality models that provide good visibility and comfort should be available. Joint labor/management safety committees or product evaluation committees can be helpful to the LSO in evaluating and selecting appropriate eyewear.
- *Eyewear inspection and maintenance program.* Eyewear should be properly stored when not in use. It should be periodically cleaned and inspected and should be replaced if discolored, pitted, or cracked.
- *Using additional controls to reduce exposure.* Eyewear is not the only protection against vision hazards. The LSO should also make sure that additional protective measures are also in place to reduce potential exposures. Here are a couple of examples:

A thorough examination of all operating room materials and instruments must be conducted to, reflecting surfaces. Those that could be in the path of the laser must be replaced, modified, or covered. Sometimes, instruments that appear to be nonreflective may in fact be reflective to infrared radiation. Only special nonreflective laser instruments should be used for laser surgery.

Laser absorbent filters are available to shield OR observation windows and other areas. Use of filters allows observers to view laser procedures through a window without wearing protective eyewear.

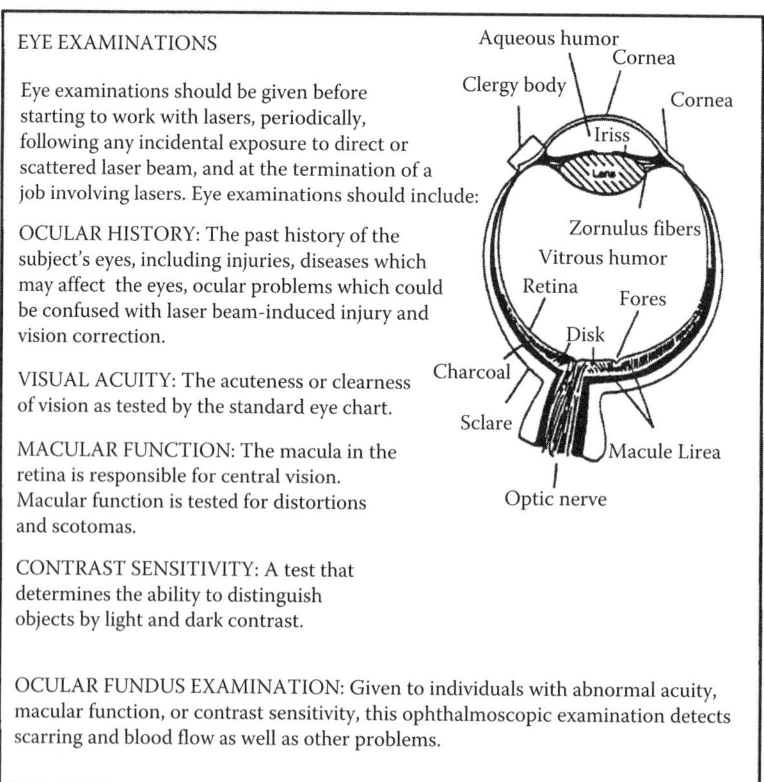
EYE EXAMINATIONS

Eye examinations should be given before starting to work with lasers, periodically, following any incidental exposure to direct or scattered laser beam, and at the termination of a job involving lasers. Eye examinations should include:

OCULAR HISTORY: The past history of the subject's eyes, including injuries, diseases which may affect the eyes, ocular problems which could be confused with laser beam-induced injury and vision correction.

VISUAL ACUITY: The acuteness or clearness of vision as tested by the standard eye chart.

MACULAR FUNCTION: The macula in the retina is responsible for central vision. Macular function is tested for distortions and scotomas.

CONTRAST SENSITIVITY: A test that determines the ability to distinguish objects by light and dark contrast.

OCULAR FUNDUS EXAMINATION: Given to individuals with abnormal acuity, macular function, or contrast sensitivity, this ophthalmoscopic examination detects scarring and blood flow as well as other problems.

Source: From ECRI, Health Devices, 18(1), 373, 1989; Abstracted from ANSI Z136.3, Sections 6.2, 6.3, and 6.4.

Skin Damage

Damage to the skin from laser beam exposure may range from localized reddening to charring and deep incision. The amount of damage is largely dependent upon wavelength and energy density of the beam and duration of exposure.

Skin effects for exposure to various wavelength radiation are detailed in Table 12.14. Surgical gloves, gown, cap, mask, and laser safety eyewear are considered adequate attire when working near lasers. The LSO should determine whether special fire-resistant clothing is necessary.

Laser Plumes—Just a Bad Smell?

Plume Hazards

Medical lasers work by vaporizing, coagulating, etc. human tissues. The resulting vapors, smoke, and particle debris, called "laser plume," are composed of both gaseous and particulate pollutants. The amount and type of plume is dependent on the type of surgery being done, the type of laser, the surgeons technique, and other factors.

Laser surgery team members may suffer acute (short term) health effects from laser plume exposure including: eye, nose; and throat irritation; tearing of the eyes; abdominal cramping; nausea; vomiting; nasal congestion; poor inspiratory effort; chest tightness; flu-like symptoms; and fatigue. Symptoms may persist for 24–48 h after exposure.

There are no studies at this time documenting chronic health effects caused by long-term exposure to plume. However, carcinogens, mutagens, irritants, and fine dusts have been found in laser plumes, as well as viable bacteria spores, cancer cells, and viral DNA. A discussion of these follows here.

Plume Contents

Lung Damaging Dust

In one study, smoke was collected from nearby patients' abdomens during carbon dioxide laser laparoscopic treatment of endometriosis. Smoke particles were found to be spheres with the size range 0.1–.8 μm. When particles of this size are inhaled they can penetrate to the alveoli, the deepest regions of the lung. There is concern that high exposure to these particles in the laser plume over time can cause lung problems similar to that found with other similar-sized dust, including coal, cotton and grain dusts and cigarette smoke particles. Surgical masks are not able to filter out such small particles. Studies of the effects on laser smoke in the lungs of rats found identifiable lesions as a result of chronic exposure.

Toxic Chemicals

Some of the chemicals which have been documented from laser beam contact with human and animal tissue are listed below. The number and type of chemicals depend on the type of surgery and other factors.

- Benzene
- Formaldehyde
- Acrolein
- Aldehydes
- Polycyclic hydrocarbons
- Methane
- Hydrogen cyanide and cyanide compounds
- Water

Benzene and formaldehyde are known carcinogens. Acrolein produces a bad odor and is an irritant at extremely low concentration. Hydrogen cyanide and cyanide compounds are irritants.

TABLE 12.14 Effects of Laser Exposure to Eyes and Skin

Type of Laser	Wave Length (μm)	Affected Part of Eye	Acute Effects of Eye Exposure	Chronic Effects of Eye Exposure	Acute Effects of Skin Exposure	Chronic Effects of Skin Exposure
CO_2	Far IR invisible beam 10.6	Cornea, lens	Burns; ulcers; intense pain	Cataracts; blindness	Skin burns	—
Nd:YAG	Near IR invisible beam 1.064	Retina	Because retina has no pain sensing nerve, exposure to invisible beam may go unnoticed until considerable damage is done. Q-switched Nd: YAG eye exposure causes audible pop. Retinal coagulation, blind spots, ulcers	Cumulative retinal injuries: (at first, difficulty detecting blue–green); progressive damage; blindness	Skin burns; photosensitive reactions	—
Argon	Blue–green 0.488 and 0.514 visible	Retina	Bright flash of laser beam color followed by after-image of complementary color. Retinal coagulation; blind spots; blindness	Cumulative retinal injuries; progressive damage; blindness	Skin burns; photosensitive reactions	Retinal changes, particularly to color vision and small angle acuity
Helium–neon	Red 0.632 visible	Retina	Same as above	Same as above	Same as above	—
Gold vapor	Red 0.632 visible	Retina	Same as above	Same as above	—	
KTP/532	Green 0.532 visible	Retina	Same as above	Same as above	Same as above	—
Argon fluoride	0.193 UVC	Cornea conjunctiva	Photo keratitis: inflammation of outer layer of cornea; redness, tearing, discharge from conjunctiva; corneal surface cell layer splitting; stromal haze	—	Skin burns; skin tanning	Skin cancer
Krypton chloride	0.222 UVC	Cornea conjunctiva	Same as above	—	Same as above	Same as above
Krypton fluoride	0.248 UVC	Cornea conjunctiva	Same as above	—	Same as above	Same as above
Xenon chloride	0.308 UVB	Cornea	Photokeratitis	—	Increased accelerated pigmentation; skin burns	Skin aging; skin cancer
Xenon fluoride	0.351 UVA	Cornea, lens	Redness and tanning of eyelid skin	Cataracts; blindness	Tanning; skin burns	—

If the laser beam contacts material other than tissue, additional chemicals will be given off. Using the laser to remove methyl methacrylate bone cement generates formic acid formaldehyde, acrolein, and methyl methacrylate monomer.

Teflon-coated products should never be used in the vicinity of the laser. Lethal hydrofluoric acid can be given off if the laser beam touches the material or if the material was involved in a fire involving an oxygen-enriched atmosphere.

Biological Agents

The possibility of transmitting pathogens through the particles in the laser smoke is still being researched. However, in various studies, intact cells, identifiable cell parts, bacterial spores, and intact viral DNA have been collected from the plume. Here are some examples:

- Human papillomavirus (HPV) is associated with warts, other lesions as well as benign and malignant skin tumors. Lasers are commonly used to vaporize warts and other lesions caused by HPV. Intact DNA from HPV has been found in the plume. However, a study found that most doctors who contracted warts got them on their hands, probably from direct contact, rather than from the laser plume.
- A CO_2 laser was used to vaporize HIV culture medium in a laboratory. HIV DNA was found in the smoke evacuator hose used to collect the laser plume. The HIV particles were found to be infective, although long-term replication appeared to be impaired.
- At low irradiance levels, viable biological spores were found in the plume from CO_2 lasing of bacteria-treated skin.

Further studies are certainly needed for a better understanding of the potential infectivity of laser plumes with varying characteristics.

Controlling Plume Exposures

If smoke is not adequately scavenged, contaminants will build up in the air. Laser plume buildup not only contains materials which can be hazardous to health; the plume can also obstruct workers' field of vision during surgical procedures. If there is inadequate room ventilation, contaminants may remain in the air for significant periods even after the surgery is completed, which could result in exposure to employees and to subsequent patients. Therefore, it is essential to reduce the plume as much as possible, if not to eliminate it altogether.

Ventilation

In-House vs. Dedicated Exhaust Systems The most important control measure for gaseous and particulate emissions is local exhaust ventilation. Local exhaust ventilation eliminates the plume at its source. Many hospitals use the wall suction system for this purpose. In-line filters must be used, or the suction lines will become quickly clogged. However, the high efficiency filters which are necessary to capture submicron-sized particles may offer too much resistance for the system and may result in too little suction capability. In addition, in-house vacuum lines may terminate in a machine room in some very old units. In such situations, maintenance employees may become inadvertently exposed to plume contaminants.

A suction hose which is attached to a dedicated exhaust (a duct which exhausts air directly out of the building rather than recirculating it) is preferable to the in-house vacuum system. The blower should be located on top of the roof. This will keep the ductwork under negative pressure, preventing leakage of toxic, or malodorous contaminants. Finally, various nozzles should be available for improving collection efficiency of plume contaminants for different procedures.

Smoke Evacuation Systems If a dedicated exhaust system is unavailable, plume buildup and exposures can be adequately controlled by the proper and diligent use of a smoke evacuation system. Smoke evacuators are used to suction the smoke generated by laser procedures. Activated charcoal beds are used to reduce odors and remove certain organic vapors. High efficiency particulate air (HEPA) or ULPA filters are used to remove airborne particles. HEPA filters generally remove 99.97% of 0.3 μ particles. ULPA filters generally remove 99.99% of 0.12 μ particles.

Studies have shown that smoke evacuator performance can be affected by

- *Angle placement of the nozzle.* The nozzle should be placed in the same direction as external air flow of plume production.
- *How close the nozzle is to the treatment site.* An orifice 1 cm from the treatment site is 98.6% efficient. If the laser collection device is moved only 2 cm away from the exposed tissue, the collection efficiency is reduced to 50% and drops further as the distance is increased.
- High power vs. low power laser procedures. High power laser procedures emit smoke in many directions at a greater velocity compared to low power procedures. Higher smoke evacuator air flow speeds are necessary for high power procedures.

Respiratory Protection

Respirators should never replace air cleaning. Respirators should be used as additional protection, where needed. Like other protective equipment, such as eyewear discussed above, respirators come with a host of problems. These problems are discussed here.

Standard surgical masks are designed to protect patients from the germs of their health care provider, but surgical masks do not provide adequate employee protection against plume contaminants including:

- Chemical vapors such as formaldehyde, acrolein, cyanide which may be found in the plume.
- Very tiny dust particles, including viruses, which once inhaled, can reach the lung's alveoli and possibly cause damage.
- Cellular debris and bacteria. Surgical masks may filter much of these larger particles; however, particles can enter the breathing zone at the nose piece and other loose-fitting areas.

More research is needed on which respirators should be worn to protect employees from inhaling laser plume. One possible alternative to surgical masks in use at some institutions is industrial-type disposable respirators, which are available from industrial respirator suppliers (e.g., 3 M or Moldex). Disposable respirators contain materials which are more effective at filtering out small particles than the paper used for surgical masks. Disposable respirators also provide a better fit than surgical masks, and they are comfortable to wear. However, at this time, it is not known if their use would offer significantly more protection than standard surgical scrub masks.

Rubber or silicone half mask respirators with disposable cartridges are another alternative. Such respirators are even more protective than disposable respirators, but they are considered impractical for the surgical setting. They are uncomfortable, difficult to wear with eye projection, and they cannot be worn by persons with mustaches, beards, or sideburns. They may also be too distracting for the intense concentration needed during laser surgery. Half-mask respirators should be considered if there is the likelihood of exposure to biohazardous aerosols. The cartridges should contain a combination of HEPA filters and organic vapor adsorbent materials to prevent exposure to both particles and chemical vapors.

If disposable respirators, half-mask respirators, or any other type of respirator are to be used by employees, OSHA requires that a comprehensive respirator program must be put in place. OSHAs respirator program includes fit-testing training and proper maintenance. (See OSHA Standard 29 CFR 1910.134 for more details.)

Tips on Using Smoke Evacuators

- Alternate nozzles may be useful for different surgical procedures.
- The smoke evacuation system should be left on for a short time after lasing is completed in order to clear remaining pollutants from the air. The smoke evacuator must be regularly maintained. Charcoal and particulate filters must be changed when they have reached capacity.
- Whenever possible, the air leaving the smoke evacuator should be exhausted directly outdoors to prevent contaminants which have not been adsorbed or filtered from reentry into the building.
- Smoke evacuator filters must be changed once they have reached their capacity. This is best determined by a pressure drop across the filter. Charcoal filters are used to remove gaseous hydrocarbons. Higher grade filters are able to adsorb more carbon tetrachloride. A CTC-60 grade charcoal is capable of adsorbing 60% carbon tetrachloride by weight, whereas a CTC-80 grade charcoal can adsorb 80% carbon tetrachloride by weight.
- All interconnecting parts of the smoke evacuator system must be kept free of kinks and clogs, and they must be maintained at optimal efficiency at all times. Collection efficiencies of filters may vary with the manufacturer. Leakage around the filter can occur if the filter is not properly fitted in its housing, or if the filter has been damaged. Ability to suck will be reduced when the filter has reached its capacity.

Skin Protection

Surgical gloves, gown, and cap are considered adequate attire when working near lasers. These garments will protect skin from exposure to laser plume and tissue debris.

Most frequently reported laser accidents are

- Unanticipated eye exposure during alignment
- Misaligned optics
- Available eye protection not used
- Equipment malfunction
- Improper methods of handling high voltage
- Intentional exposure of unprotected personnel
- Use of unfamiliar equipment
- Lack of protection from ancillary hazards
- Improper restoration of equipment following service

Source: Abstracted from D. Terrible, *Preview of the 21st Century. Ohio Monitor,* February, 1986, p. 6.

Fires

Causes of Laser Fires

Fires caused by lasers are uncommon, but they do happen. A fire can be started by inadvertently misdirecting the beam or the reflection of the beam onto a combustible material. Reflective surfaces that are concave in shape can focus the beam, making it potentially more harmful.

If oxygen or nitrous oxide is being used, there will be extra oxygen in the air. The oxygen-enriched atmosphere will allow many materials to ignite that are not normally flammable—even those marketed as fire resistant. In an oxygen-enriched environment, the flames will be especially intense.

Conventional fires can also be a problem. When the laser is in the standby mode, the hot tip can start a fire if it touches surgical drapes or other flammable materials. The toxic fumes produced when certain plastics and fabrics burn can be life-threatening for patients and staff.

Items which may be ignited by a laser beam include:

- Intestinal gas (flatus)
- Hair (facial hair is particularly vulnerable due to its proximity to anesthetic gases)

- Skin that has been treated with acetone- or alcohol-based skin preparation solutions
- Paper products
- Surgical drapes (even if they are marketed as fire resistant)
- Rubber and plastic instruments including nonmetallic endotracheal tubes: PVC, silicone red rubber
- Human tissue, miscellaneous materials including plastic adhesive tapes, oil-based lubricants, ointments, gloves, dry gauze, cotton

Preventing Laser Fires

- All personnel should be trained in the causes and prevention of fires as well as appropriate responses to a fire involving a laser.
- Precise control of the laser beam must be maintained at all times.
- Rooms, instruments, and equipment should be inspected to detect surfaces, which can cause unwanted reflection of the beam.
- During surgery, the laser should be left in the stand-by mode at all times except when the hand-piece is in the hands of the surgeon.
- Skin preparation solution vapors should fully vaporize before covering the area with surgical drapes.
- When in the stand-by mode, the hot tip of the laser should not be allowed to touch combustible items.
- During surgery in or near the bowel, proper bowel preparation is necessary to avoid ignition of rectal gas.
- During laryngotracheal surgery an endotracheal tube is sometimes used. Special precautions must be taken to reduce the risk of combustion while keeping the patient well-oxygenated. Silicone, PVC, red rubber, and even specially designed laser-resistant endotracheal tubes may be ignited in oxygen-rich atmospheres. Stainless steel endotracheal tubes or non-endotracheal methods to ventilate the patient should be used.

Controlling Laser Fires

- A basin of sterile saline and a syringe should be kept on hand to douse a small fire and to keep protective dressings wet.
- Portable fire extinguishers should be conveniently located. Personnel should be trained in how to use them.
- National Fire Protection Association Standard 99 (1990) Appendix C-12.4 "Suggested Procedures in the Event of a Fire or Explosion-Anesthetizing Locations" should be in effect.
- A Halon or CO_2 fire extinguisher should be used rather than water if power cannot be disconnected during a fire.

Electrical Hazards

Electrical hazards are the largest cause of fatalities in accidents involving lasers. Many lasers use high voltage and high amperage currents. Electrical hazards are usually minimized by enclosures around high voltage devices within the laser cabinet. Electrocutions have primarily involved technicians who opened the protective covering of the laser. The high voltage DC capacitors of some lasers can remain energized for an extended period after the laser has been unplugged from the wall outlet. Danger arises mainly when an untrained or unauthorized person attempts internal laser maintenance.

The ANSI recommends special precautions for equipment servicing within 24 h after the presence of high voltage within the unit. A special grounding rod is used to ensure discharge of high voltage capacitors. Additional details for servicing equipment are included in ANSI Z 136.3 Section 7.1.

ANSI Z 136.3 Section 7.1 contains additional electrical safely requirements:

- Grounding of metallic parts of laser equipment
- Short circuit tests for combustible components in electrical circuits
- Prevention of shock hazard
- Prevention of electrical hazards from gas-laser tubes and flash lamps
- Labeling lasers with electrical rating, frequency, and watts
- Preventing EM interference
- Preventing explosions in high pressure arc lamps and filament lamps

Laser Safety Policy

Every employer utilizing a laser should have a written policy. The elements of a policy are outlined here.

Laser Safety Program

The employer has a written laser safety program which includes:

- Formation of a laser safety committee (LSC)
- Credentialing of medical personnel
- Education and training
- Safety precautions including general precautions, eye protection, eye examinations, preoperative setup, operating precautions, and anesthesia guidelines
- Maintenance and service
- Scavenging of plume contaminants
- New laser acquisitions
- Appointing an LSO

Laser Safety Personnel

1. An LSC has been established which
   - Reviews and approves new laser technology and the design of laser facilities.
   - Reviews and approves protocols and safety measures for lasers used in medical treatment.
   - Appoints an LSO.
   - Investigates unusual occurrences related to laser uses and reports the results to the Quality Assurance Program.
2. The LSO is trained in laser operation, clinical applications, and safety measures. The LSO has the authority to monitor and enforce the control of laser hazards and effect the knowledgeable evaluation and control of laser hazards. The LSO acts as the laser education coordinator.
3. Laser technicians report to the LSO on all technical and clinical issues. Laser technicians are responsible for the day-to-day operation of surgical lasers.
4. The laser user is responsible for performing a daily visual inspection of the laser system to check for apparent abnormalities.

Laser Maintenance

1. All persons who install maintain and/or service lasers have been properly trained and approved by the LSC.
2. Enclosed lasers have built-in access panel interlocks and automatic shuttering to protect maintenance persons.

3. Maintenance on a laser is only done when another person is present to render emergency medical aid or to call for assistance in the event of an injury.
4. Initial installation and subsequent maintenance or servicing of lasers that may affect their performance are followed by performance and safety testing prior to the laser's use in the OR.

Training and Approval of Laser Operators

1. Only personnel with specific credentials are authorized by the LSC to operate lasers.
2. All personnel who are present during laser operation or maintenance have received appropriate training on the principles of operating lasers, their applications, attendant risks to patient and staff, safety control measures, and equipment care.

Authorization is given to specified physicians (and specified other persons to use nonsurgical lasers) for specific types of laser delivery systems and procedures. Authorization is dependent on at least one of the following criteria

- Documented attendance and completion of a formal laser training course offered by a recognized authority
- Documented completion of a formal training program offered by the manufacturer of the laser
- Residency or on-the-job training with proven proficiency while assisting a certified laser physician

General Operational Guidelines

1. An NHZ has been established by the LSO. Outside the NHZ, the level of direct, reflected, or scattered radiation is not expected to exceed the applicable maximum permissible exposure limit.
2. The laser is locked when not in use or unattended to prevent unintentional or unauthorized activation. The key is stored in a secure location.
3. The laser technician ensures that all necessary equipment is ready for operation, and that all required safety precautions have been implemented, including visual checks of the surgical laser and its control settings.
4. All reflective surfaces which are likely to be contacted by the laser beam are removed from the path of a fixed laser or from the OR prior to laser surgery.
5. All windows shall be covered as necessary.
6. Where necessary, safety latches or interlocks are used to prevent unexpected entry into laser controlled areas.
7. All access doors to laser ORs are posted with a laser warning system, which includes a warning light and sign to prevent unexpected interruption of the laser operator.

Personnel Protection

1. Laser users and support staff have pre-placement and periodic eye examinations for adverse effects.
2. Appropriate laser eye protection is worn by all persons present prior to powering the laser system and throughout the entire procedure. This includes all staff, bystanders, and the patient. The eye protection is labeled with its OD and the wavelength it protects against.
3. Eyewear is appropriately cleaned after each use.
4. Respirators which are effective at filtering out particles as small as 0.3 μm with 99% efficiency are used in conjunction with scavenging of the laser smoke when there is a possibility of generating biohazardous aerosols.

Operating Precautions

1. The laser is not switched to the operating mode until the procedure is ready to begin and the laser has been aligned and positioned.
2. The laser beam is never aimed at a person, except for the therapeutic purposes.
3. The laser is positioned to avoid placing the beam at eye level whenever possible.

Engineering Control Measures

1. Local exhaust ventilation is used to control airborne contaminants. Whenever possible, the exhaust is vented directly out of the building. Otherwise, smoke evacuators employing HEPA and charcoal filtration are used. Filters are changed regularly in accordance with the manufacturer's directions.
2. Microscopes have a fail-safe method to project the users' eyes against laser beam reflections (i.e., built-in filters, separate optical paths for intermittent viewing and firing, shutters that automatically close prior to firing).

Fire Safety

1. Special precautions are taken if laser energy is to be used near an endotracheal tube.
2. Patients are ventilated with nonoxygen-enriched room air when possible. If oxygen therapy is necessary, it is administered at the minimum concentration necessary to properly support the patient.
3. Staff are knowledgeable about procedures to follow in the event of various types of fires.
4. Flammable prepping solutions are not used.
5. Liquids are never placed on top of the laser to avoid short circuits.
6. A basin of saline or water is on hand for patient-related fires.
7. A UL-approved fire extinguisher is readily available in the event of equipment or material fire.
8. Combustible materials such as OR gowns, drapes, and towels are kept out of the laser path to avoid combustion.

Laser Safety Training

The LSO must ensure that an appropriate training program is in place for each group of employees who operate lasers or who are exposed to lasers. Here is an outline of topics to be included in a training program for health care support staff as recommended by OSHA and ANSI. (For information about OSHA and ANSI, see "Laser Resources" section.)

Laser Beam

a. What it is and what it can do
b. The hazards of lasers
c. Eye and skin hazards
d. Other laser hazards

Safety Measures in Laser Surgery

a. Eye protection
b. Reflected beam hazards
c. Explosions

d. Smoke evacuation
e. Fire hazards
f. Details of SOPs for the OR

Methods and Procedures to Assure Safety

a. Boundary of NHZ
b. Laser area warning signs
c. Entry way controls
d. Availability of personal protective equipment: eyewear, nonflammable gowns, etc.
e. Control of unauthorized personnel to prevent access to the laser
f. Techniques for safety
g. Use of surgical drapes in laser surgery procedures
h. Proper laser system controls (e.g., foot switch)

Laser Safety Checklist

Date: ______________ **Location:** ________________
Laser status: (A) not in use (B) after use (C) in use

Requirement:	YES	NO	N/A
Room:			
Appropriate signs posted outside procedure room when laser is in use (sign includes wavelength, class, "Warning," CW or pulsed)			
All windows and doors covered with a nontransparent, nonreflecting material			
Appropriate fire extinguisher present			
Gas tanks stored properly			
Laser keys locked in designated place when not in laser			
Auxiliary accessories stored properly			
Laser Users:			
All persons including patient, wearing appropriate eye wear when laser in use			
Laser:			
Electrical cords intact			
Outlets intact			
Water pressure adequate for cooling			
No leakage from laser hoses			
Operating procedures on laser			
Maintenance up-to-date			
Properly stored			
Laser log complete			
Laser log accurate			
Smoke Evacuator:			
Used with laser			
Time recorded on fliter			
Filter changed as necessary			
Valid safety sticker			
Properly stored			

COMMENTS ____

____

Name of Inspector ____________________________________

Government Regulation of Lasers

1. **OSHA**. (Look in the blue pages of your telephone book under U.S. Government, Department of Labor for OSHA office nearest you.)

OSHA is a regulatory agency within the U.S. Department of Labor that is responsible for overseeing the health and safety of America's private sector workers. (Public employees in many states are covered by state OSHA plans that are administered by their state Department of Labor.) OSHA promulgates worker safety regulations and has the authority to enforce them. Employers that do not comply with OSHA regulations must correct the problem and pay fines.

Although OSHA has not yet set a formal standard for medical lasers, OSHA does have very detailed guidelines on lasers, which its inspectors use. These guidelines, which were issued on August 5, 1991, are called Guidelines for Laser Safety and Hazard Assessment (OSHA Instruction PUB 8–1.7, Directorate of Technical Support). The guidelines incorporate many of the recommendations for laser safety written by a professional organization called ANSI (see Resources below). The guidelines include detailed information in the following areas:

- The principles of laser operation and use
- Effects of laser light on the eye and skin
- Standards for laser safety
- Hazard evaluation
- Control measures
- Personal protective equipment
- Laser training requirements for various classes of lasers

In addition to the guidelines, OSHA has cited employers for unsafe conditions under a variety of its existing standards. For example:

- Eye & Face Protection Standard (29 CFR 1910.133). OSHA can issue citations to health care employers.
- If the wrong type of eyewear is provided for the surgery being done.
- If employees fail to use eyewear.
- If the eyewear is not protective enough. For example, OSHA has issued a citation against a hospital for providing eyewear without side shields and for providing eyewear with vented side shields (because vents can allow the laser beam to penetrate through the opening).
- Respiratory Protection Standard (29 CFR 1910.134). OSHA can issue citations to health care employers.
- If the employer is using respirators without determining that wearers are medically able to wear them.
- If workers have not been trained how to use respirators.
- If respirators have not been properly fit on the wearer.
- If respirators are not properly stored.
- If reusable respirators are not cleaned and disinfected.
- Fire Protection (29 CFR 1910.155–165). There are several standards for fire protection, including alarms, detection systems, and extinguishing systems.
- General-Duty Clause, Section 5(a)(1) of the OSHA Act of 1970. The General Duty Clause states that it is the employers' duty to provide a safe workplace. The clause is a catch-all for the enforcement of health and safety measures that are not specifically included in its Regulations. OSHA has used the General Duty Clause to issue citations against hospitals that do not take proper precautions to prevent the laser from being inadvertently activated.
- Air Contaminants (29 CFR 1910.1000 Subpart Z). Despite the fact that the plume may contain ingredients that pose a real hazard to health, it is unlikely that OSHA will find violations of any of its chemical standards. In most cases, the specific chemical contaminants in the plume will be present in concentrations which are allowed by OSHA.

Furthermore, OSHA does not currently regulate exposure to biological agents in laser plumes. However, the ANSI standard, which is incorporated into the OSHA Guidelines, includes a recommendation that effective smoke evacuators be used during laser surgery.

A more appropriate agency to investigate both chemical and biological hazards of laser plumes is the NIOSH.

2. **NIOSH.** U.S. Department of Health and Human Services, U.S. Public Health Service, Centers for Disease Control, 4676 Columbia Parkway, Cincinnati, OH 45226, (513)841–4382.

NIOSH was created by the OSHAct to be the research arm of OSHA. NIOSH does research on occupational hazards, evaluates control measures, and makes recommendations to OSHA for standards. Although its recommendations do not have the force of law, health care workers can call in NIOSH to perform a health hazard evaluation of the workplace, including the hazards of laser plume.

3. **Center for Devices and Radiological Health (CDRH).** Office of compliance (HFZ-300), 8757 Georgia Ave., Silver Spring, MD 20910, (301)443–4190.

The CDRH is a regulatory bureau within the Federal FDA of the U.S. Department of Health and Human Services. All laser products manufactured since 1976 must comply with CDRH specifications. Manufacturers obtain premarket approval or clearance of their laser surgical devices through CDRH. The CDRH issues "The Compliance Guide for Laser Products," which summarizes the requirements of the U.S. Federal Laser Product Performance Standard (21 CFR Part 1000, 1040.10, 1040.11). Manufacturers should use this performance standard in order to comply with CDRH requirements for labeling and classifying lasers.

4. **State Regulations.** Laser regulations vary considerably from state to state. Such regulations are generally concerned with the registration of lasers and the licensing of operators and institutions.

New York laser regulations, Code Rule 50, are enforced by the Department of Labor. Massachusetts regulations, 105 CMR 21, are administered by the Department of Health. New Jersey and Connecticut do not have state regulations on lasers at this time. "Suggested State Regulations for Lasers" has recently been promulgated by the Conference of Radiation Control Program Directors. This may lead to changes in state regulations.

Laser Resources

1. Government Agencies OSHA, NIOSH, FDA. See previous section on Government Regulation of Lasers.
2. American National Standards Institute (ANSI), 11 West 42nd St., 13th floor, New York, NY 10036, (212)642–4900.

   ANSI is a professional organization of engineers from many fields which has issued the leading consensus standard on the safe use of lasers (ANSI Z 136.1) and on safe use of lasers in medicine (ANSI Z 136.3). OSHA relies in large part on ANSI's research for its policies regarding laser safety.
3. American Conference of Governmental Industrial Hygienists (ACGIH), 6500 Glenway Avenue, Building D-7 Cincinnati, OH 45211, (513)661–7881.

   ACGIH is a professional organization of industrial hygienists and safety experts. The ACGIH has established maximum exposure limits (MELs) known as threshold limit values for employee eye and skin exposure to laser radiation. These MELs are used to select protective eyewear, determine NHZ s, and other safety precautions for laser use.
4. Joint Commission of Accreditation of Health Care Organizations (JCAHO), One Renaissance Blvd., Oakbrook Terrace, IL 60181, (708)474–7028.

   The JCAHO is a private, professional organization that accredits health care facilities. JCAHO is in the process of adopting the ANSI standard for medical lasers—ANSI Z-136.3. An institution's accreditation can be held up if it is not in compliance with ANSI.

5. The American Society for Laser Medicine and Surgery, Inc. 2404 Stewart Square, Wausau, WI 54401, (715)845–9283.

   This organization is composed of scientists, physicians, nurses, and paramedical personnel. It holds annual meetings and publishes a bimonthly journal, *Lasers in Surgery and Medicine.*
6. Laser Institute of America, 12424 Research Parkway, Suite 130, Orlando, FL 32826, (407)380–1553.

   The Laser Institute of America is a nonprofit educational society, which conducts continuing education courses, holds technical symposia, offers educational materials, and publishes the peer-reviewed journal, *Laser Topics.* It promotes the advancement of laser technology and applications.
7. Association of Operating Room Nurses (AORN), 2170 S. Parker Road, Denver, CO 80231, (303)755–6300.

   This professional organization publishes general information of interest to operating room personnel. See Proposed recommended practices: Laser safety in the operating room. *AORN J* 49(l):284–91, 1989.
8. The Emergency Care Research Institute (ECRI), 5200 Butler Pike, Plymouth Meeting, PA 19462, (215)825–6000.

   ECRI is an independent, nonprofit research, and consulting organization which, for a fee, provides training seminars and publishes information on health technology, including surgical lasers and accessories, laser fires, credentialing and training recommendations, and safety programs. It also performs comparative evaluations of surgical lasers, laser resistant endotracheal tubes, smoke evacuators, and laser protective eyewear.

Source: This document is reprinted courtesy of Communications Workers of America District 1.

Proposed Recommended Practices: Electrosurgery

Association of Operating Room Nurses

The following draft is being published for review and comment by AORN members. The AORN Recommended Practices Committee (RPC) is interested in receiving comments on this proposal from members and others.

These recommended practices are intended as achievable recommendations representing what is believed to be an optimal level of practice. Policies and procedures will reflect variations in practice settings and/or clinical situations that determine the degree to which the recommended practices can be fulfilled.

AORN recognizes the numerous different settings in which perioperative nurses practice. The recommended practices are intended as guidelines adaptable to various practice settings. These practice settings include traditional ORs, ambulatory surgery units, physicians' offices, cardiac catheterization laboratories, endoscopy rooms, radiology departments, emergency departments, and all other areas where surgery may be performed.

Although nonmembers may submit comments, the intent of the committee is to reach a consensus among AORN members. All comments will be acknowledged and considered by RPC before final approval of these recommendations by the committee and the AORN Board of Directors. Comments should be sent to Recommended Practices Committee, AORN, Inc., 2170 S. Parker Road, Suite 300, Denver, CO 80231–5711, Attention: Mary O'Neale, RN, BS, CNOR.

Purpose

These recommended practices provide guidelines to assist perioperative personnel in the use of electrosurgical equipment in their practice settings. Proper care and handling of electrosurgical equipment is essential to patient and personnel safety. Electrosurgery is used routinely to cut and coagulate body tissue with high RF electrical current. These recommended practices do not endorse any specific product. Biomedical services in practice settings should develop detailed, routine safety and preventive maintenance inspections and maintain records.

Recommended Practice I

The ESU, dispersive electrode, and active electrode selected for use should meet performance and safety criteria established by the practice setting.

Interpretive Statement 1

Information regarding adequate safety margins, in-factory testing methods, warranties and a manual for maintenance, and inspections should be obtained from the manufacturer.

Rationale

Equipment manuals assist in developing operational, safety, and maintenance guidelines.[1] The ESU should be used according to the manufacturer's written instructions.[2]

Interpretive Statement 2

The ESU should be designed to minimize unintentional activation.

Rationale

Unintentional activation may result in patient and personnel injury.[3]

Interpretive Statement 3

The ESU cord should be of adequate length and flexibility to reach the outlet without stress or use of an extension cord. Kinks, knots, and curls should be removed from the ESU cord before it is plugged into the wall outlet.

Rationale

Tension increases the risk that the cord will become disconnected or frayed, which may result in injury to patients and personnel.[4] Use of extension cords may result in macroshock or microshock.[5] Cords that do not lie flat on the floor produce a potential for tripping and/or accidental unplugging.[6]

Interpretive Statement 4

The ESU plug, not the cord, should be held when it is inserted into or removed from an electrical outlet.

Rationale

Pulling on the ESU cord may cause it to break at the point, where the wire is attached to the plug.[7] Cord breakage is dangerous to patients and personnel and is inconvenient, and replacements are costly.[8]

Interpretive Statement 5

The ESU should be inspected before each use. An ESU that is not working properly or is damaged should be reported, labeled, and removed immediately to be checked by the biomedical department.

Rationale

Equipment is checked to ensure it is in good working order.[9] The manufacturer's written safety precautions are followed for the well-being of the patient and personnel involved with the procedure.[10]

Interpretive Statement 6

The ESU should be grounded properly.

Rationale

Proper grounding reduces the risk of electrical shock to the patient and perioperative personnel.[11]

Interpretive Statement 7

The ESU should be mounted on a movable stand that will not tip.

Rationale

Safety measures for perioperative personnel and patients to prevent injury and damage to the ESU require the stand to be tip resistant and moved carefully.[12]

Interpretive Statement 8

The ESU and all reusable parts are cleaned with care following use according to the manufacturer's written instructions.

Rationale

The ESU surface should not be saturated or have fluid poured over it because this could permit chemical germicide into the generator and cause malfunction.[13]

Interpretive Statement 9

When the ESU foot switch is used, perioperative personnel should cover it with a clear, impervious cover if recommended by the manufacturer.

Rationale

Placement in a clear, impervious cover protects the foot switch from fluid spillage.[14]

Interpretive Statement 10

During the procedure, perioperative personnel should check the entire circuit if higher than normal power settings are requested by the operator.

Rationale

The dispersive electrode, generator, or connecting cords may be at fault and should be checked for any possible malfunction or hazard. Shock to those touching the patient may result. The patient and/or perioperative personnel may be burned.[15]

Interpretive Statement 11

Each ESU should be assigned an identification number/serial number.

Rationale

An identification number/serial number allows for documentation of inspections, routine preventive maintenance, and tracking of equipment function and problems.[16]

Recommended Practice II

Perioperative personnel should demonstrate competency in the use of the ESU in the practice setting.

Interpretive Statement 1

Perioperative personnel should be instructed in the proper operation, care, and handling of the ESU before use.

Rationale

Instruction and return demonstration in proper usage help prevent injury and extend the life of the ESU.[17]

Interpretive Statement 2

A detailed manual of operating instructions should be obtained from the manufacturer and be available in the practice setting. A brief set of operational directions should be on or attached to the ESU.

Rationale

Each type of ESU has specific manufacturer's written operating instructions that should be followed for the safe operation of the unit.[18]

Recommended Practice III

The FSU, active electrode, and dispersive electrode should be used in a manner that reduces the potential for injury.

Interpretive Statement 1

The ESU should

- Not be used in the presence of flammable agents (e.g., alcohol, tincture-based agents).
- Have safety features (e.g., lights, activation sound) and be tested before each use.
- Have the cord, plug and foot switch cord checked for exposed wires or frays in the insulation.
- Have power settings confirmed orally with the operator before activation and determined in conjunction with the manufacturer's recommendations.
- Be protected from spills.
- Be operated at the lowest effective power settings for coagulation and/or cutting.

Rationale

Inspections of the ESU and all safety features should be performed before each use because of potential hazards.[19] The volume of the activation indicator should be adjusted to an audible level to alert perioperative personnel immediately when an ESU is activated inadvertently.[20] Ignition of flammable agents by the active electrode has resulted in patient and perioperative personnel injury.[21] Fluids should not be placed on top of the ESU, because unintentional activation or device failure may occur if liquids enter the ESU generator.[22]

Interpretive Statement 2

The active electrode should

- Fasten directly into the ESU in a labeled, stress-resistant receptacle (if an adapter is used, it should be one that is approved by the manufacturer and does not compromise the generator's safety features).
- Be inspected at the field for damage before use.
- Be placed in a clean, dry, well-insulated safety holster (i.e., recommended by the manufacturer for use with the ESU) in a highly visible area when not in use during a procedure.
- Be impervious to fluids.
- Be disconnected from the ESU if allowed to drop below the sterile field.
- Have a tip that is secure and easy to clean of charred tissue.

Rationale

Incomplete circuitry, unintentional activation, and incompatibility of the active electrode with the generator may result in patient injury.[23]

Interpretive Statement 3

The dispersive electrode should

- Be inspected before each use for wire breakage or fraying.
- Be the appropriate size for the patient (i.e., neonate/infant, pediatric, adult) and never be cut to reduce size.
- Be placed on the positioned patient on a clean, dry skin surface, over a large muscle mass and as close to the operative site as possible (i.e., bony prominence, scar tissue, skin over an implanted metal prosthesis, hairy surfaces, pressure points should be avoided).
- Fasten directly into the ESU in a labeled stress-resistant receptacle if an adapter, which is approved by the manufacturer and does not compromise the generator's safety features, is used.
- Have connections that are intact, clean, and make effective contact.
- Maintain uniform body contact (potential problems include tenting, gaping, and liquids that interfere with adhesion).

Perioperative personnel should check the status of the dispersive electrode and connection of the cable if any tension is applied to the cord or if the surgical team repositions the patient.

If reusable, the electrode should have periodic inspections by the biomedical service for electrical integrity and as recommended by the manufacturer.

Rationale

Wire breakage and frays can deviate current flow.[24] Incomplete circuitry may lead to patient injury.[25] Adequate tissue perfusion promotes electrical conductivity in the area and dissipates heat at the electrode contact surface.[26] Hair should be removed before applying the dispersive electrode according to the manufacturer's written instructions. Hairy surfaces have poor adhesion and tend to insulate.[27]

There is potential for superheating if a dispersive electrode is placed on the skin over an implanted metal prosthesis. The important factor in the dispersive electrode is the actual surface area in contact with the patient. The amount of surface area affects heat buildup at the dispersive site.[28]

Discussion

During some surgical procedures, it may be desirable to use two ESUs simultaneously on the same patient. Perioperative personnel should place each dispersive electrode as close as possible to the respective surgical sites and ensure that there is no possibility of the two dispersive electrodes touching. The two ESUs must be of the same technology (e.g., both grounded ESUs, both isolated ESUs). The biomedical service should test ESUs to ensure that simultaneous operation will not create any microshock hazards.

Interpretive Statement 4

The bipolar ESU should be used with its foot switch or a hand switching forceps according to the manufacturer's written instructions.

Discussion

In bipolar electrosurgery, a forcep is used for the coagulation of body tissue. One side of the forceps is the active electrode and the other side is the inactive electrode or ground. A dispersive electrode is not needed because current flows between the tips of the forceps rather than through the patient. The operator uses a foot switch to control the bipolar unit to provide precise hemostasis without stimulation or current spread to nearby structures.[29]

Interpretive Statement 5

Patients with pacemakers should have continuous electrocardiogram (ECG) monitoring when an ESU is being used.

Rationale

Use of the ESU may interfere with pacemaker circuitry. The bipolar unit may be used when operating on a patient with a pacemaker.[30]

Discussion

Modern pacemakers are subject to interference; most are designed to be shielded from radio frequency current during ESU use. Perioperative personnel should implement additional actions for the pacemaker patient that include, but are not limited to

- Making the distance between the active and dispersive electrodes as close as possible and placing both as far from the pacemaker as possible.
- Ensuring that the current path from the surgical site to the dispersive electrode does not pass through the vicinity of the heart.
- Keeping all ESU cords and cables away from the pacemaker and the leads. Having a defibrillator available in the room.
- Checking with the pacemaker's manufacturer regarding its function during use of ESUs.
- Evaluating the pacemaker postoperatively for proper function.[31]

Interpretive Statement 6

A patient with an automatic implantable cardioverter defibrillator (AICD) should have the device deactivated before the procedure and have his or her ECG monitored continuously if an ESU will be used.

Rationale

Electrosurgery must not be used on a patient with an activated AICD because it may trigger the device to shock.[32]

Interpretive Statement 7

The patient's skin integrity should be evaluated and documented before and after ESU use. Particular areas to observe are under the dispersive electrode, under ECG leads and at temperature probe entry sites.

Rationale

Assessment will allow evaluation of skin condition for possible injury. Alternate pathway burns have been reported at ECG electrode sites and temperature probe entry sites.[33]

Interpretive Statement 8

If an adverse skin reaction or injury occurs the ESU and active and dispersive electrodes should be sent with their packages to the biomedical service for a full investigation. Device identification, maintenance/service information, and event information should be included in the report from the practice setting.

Rationale

Retaining the ESU and electrodes allows for a complete systems check to determine system integrity.[34]

Recommended Practice IV

Patients and perioperative personnel should be protected from inhaling the smoke generated during electrosurgery.

Interpretive Statement 1

An evacuation system should be used to remove surgical smoke.

Discussion

There may be a potential hazard from exposure to smoke generated during electrosurgery.[35] Further research must be performed to determine the actual magnitude of smoke exposure under practical electrosurgical conditions.[36]

Interpretive Statement

Smoke evacuation systems should be used according to manufacturer's written instructions.

Discussion

Health care facilities may use the AORN "Recommended Practices for Product Evaluation and Selection for Patient Care in the Practice Setting" and AORN "Recommended Practices for Laser Safety in the Practice Setting" to assist in selecting a smoke evacuation system.

When the evacuation system is used for the filtration of electrosurgical smoke, placement of the evacuator suction tubing should be as close to the source of the smoke as possible. This will maximize smoke capture and enhance visibility at the surgical site.

Research findings suggest that there is little difference between the smoke generated from electrosurgery and from lasers. There is an undefined potential for bacterial and viral contamination of smoke. Toxicity and mutagenicity of the gaseous byproducts exist.[37] High filtration surgical masks may be worn by perioperative personnel during procedures that generate surgical smoke.

Recommended Practice V

Policies and procedures for electrosurgery should be developed, reviewed annually, and available within the practice setting.

Discussion

These policies and procedures should include, but are not limited to

- Equipment maintenance programs
- Reporting of injuries
- Sanitation of ESU
- Documentation of the ESU brand name, ESU identification number/serial number, settings used, dispersive electrode and ECG pad placement, patient skin condition before and after electrosurgery, and other electrical devices used

These recommended practices should be used as guidelines for the development of policies and procedures in the practice setting. Policies and procedures establish authority, responsibility, and accountability. They also serve as operational guidelines.

An introduction and review of policies and procedures should be included in orientation and ongoing education of personnel to assist in the development of knowledge, skills, and attitudes that affect patient outcomes. Policies and procedures also assist in the development of quality assessment and improvement activities.

Glossary

Active electrode: The accessory that directs current flow to the operative site. Examples include pencils with various tips, resectoscopes, and fulguration tips.

Current: A movement of electricity analogous to the flow of a stream of water.

Dispersive electrode: The accessory that directs current flow from the patient back to the generator (often called the patient plate, return electrode, inactive electrode, or grounding plate/pad).

Electrosurgery: The cutting and coagulation of body tissue with a high RF current.

ESU: For the purposes of this document, the ESU is defined as the generator, the foot switch and cord (if applicable), and the electrical plug, cord, and connections.

Generator: The machine that produces RF waves (often called a cautery unit, power unit, or Bovie).

Grounded electrosurgery: The dispersive electrode is grounded to the metal chassis of the generator. Current with flow from the active electrode when it touches any grounded object in the room.

Isolated electrosurgery: No reference to ground. For current to flow, there must be a complete circuit path from the active terminal to the patient terminal.

Macroshock: Occurs when current flows through a large skin surface, as during inadvertent contact with moderately high voltage sources, such as electrical wiring failures, that allow skin contact with a live wire or surface at full voltage.

Microshock: Occurs when current is applied to a small area of skin, as when current from an exterior source flows through the cardiac catheter or conductor.

References

1. Shaffer MJ and Gordon MR. Clinical engineering standards, obligations, and accountability. *Med Instrum* 13:209–215, July/August, 1979.
2. Atkinson LJ. *Berry and Kohn's Operating Room Technique*, 7th edn. Mosby-Year Book, Inc., St Louis, 1992, p. 253.
3. Emergency Care Research Institute. Electrosurgical units. *Technol Surg* 8:3, November 1987; Moak E. Electrosurgical unit safety: The role of the perioperative nurse. *AORN J* 53:745, March 1991.
4. Groah LK. *Operating Room Nursing: Perioperative Practice*, 2nd edn. Appleton & Lange, Norwalk, CT, 1990, p. 299.
5. Moser ME. Electrical shock: An orientation study guide. *Point of View* 23:4–5, May 1, 1986.
6. Groah LK. *Operating Room Nursing, Perioperative Practice*, 2nd edn. Appleton & Lange, Norwalk, CT, 1990, p. 299.
7. Meeker MH and Rothrock JC. *Alexander's Care of the Patient in Surgery*, 9th edn. Mosby-Year Book Inc., St. Louis, 1991, p. 43; Atkinson LJ. *Berry and Kohn's Operating Room Technique*, 7th edn. Mosby-Year Book, Inc., St Louis, MN, 1992, p. 253.
8. Meeker MH and Rothrock JC. *Alexander's Care of the Patient in Surgery*, 9th edn. Mosby-Year Book, Inc., St Louis, MN, 1991, p. 43.
9. Kneedler JA and Dodge GH. *Perioperative Patient Care: The Nursing Perspective,* 2nd edn. Jones and Bartlett Publishers, Boston, 1991, p. 382.
10. *Ibid.*
11. Schellhammer PF. Electrosurgery: Principles, hazards, and precautions. *Urology* 3:261–267, March 1974.
12. *Ibid,* p. 170.
13. *Ibid,* p. 232; *Ibid,* p. 300.
14. Moak E. Electrosurgical unit safety: The role of the perioperative nurse. *AORN J* 53:746, March 1991.
15. *Ibid,* p. 172; *Ibid,* 746; *Ibid,* p. 300.

16. Shaffer MJ and Gordon MR. Clinical engineering standards, obligations, and accountability. *Med Instrum* 13:170–171, July/August 1979; *Ibid,* pp. 170–171.
17. Skreenock JJ. Electrosurgical quality assurance: The view from the OR table. *Med Instrum* 14:261–263, September/October 1980; *Ibid*, p. 43.
18. Skreenock JJ. Electrosurgical quality assurance: The view from the OR table. *Med Instrum* 14:261–263, September/October 1980; *Ibid*, p. 170.
19. *Ibid.*
20. Emergency Care Research Institute. Update: Controlling the risks of electrosurgery. *Health Devices* 18:431, December 1989.
21. Bowdle TA, et al. Fire following use of electrocautery during emergency percutaneous transtracheal ventilation. *Anesthesiology* 66:697–698, May 1987; Freund PR and Radke HM. Intraoperative explosion: Methane gas and diet. *Anesthesiology* 55:700–701, December 1981.
22. *Ibid*, 262; *Ibid*, 746; *Ibid*, p. 300.
23. Reeter AK. Bipolar forceps misconnection hazardous. *OR Manager* 6:13, February 1990; *Ibid*, p. 172.
24. Emergency Care Research Institute. Update: Controlling the risks of electrosurgery. *Health Dev* 18:430–431, December 1989.
25. *Ibid.*
26. Neufeld GR and Foster KR. Electrical impedance properties of the body and the problem of alternate site burns during electrosurgery. *Med Instrum* 19:83–87, March/April 1985.
27. *Ibid*, p. 300; *Ibid*, p. 171.
28. *Ibid.*
29. *Ibid*, p. 301.
30. *Ibid*, 748.
31. *Ibid*, p. 301.
32. Lee BL and Mirabal G. Automatic implantable cardioverter defibrillator: Interpreting, treating vehicular fibrillation. *AORN J* 50:1226, December 1989; Moser SA, Crawford D, and Thomas A. Updated care guidelines for patients with automatic implantable cardioverter defibrillators. *Crit Care Nurse* 13:70, April 1993.
33. Schneider AJL, Apple HP, and Braun RT. Electrical burns at skin temperature probes. *Anesthesiology* 47:72–74, July 1977; Finley B et al. Electrosurgical burns resulting from use of miniature ECG electrodes. *Anesthesiology* 41:263–269, September 1974; *Ibid*, 748–749; *Ibid*, p. 299.
34. Gendron F. Burns occurring during lengthy surgical procedures. *J Clin Eng* 5:19–26, January–March 1980; *Ibid*, 749, 752.
35. Emergency Care Research Institute. ESU smoke—Should it be evacuated? *Health Devices* 19:12, January 1990.
36. *Ibid.*
37. Baggish MS, et al. Presence of human immunodeficiency virus DNA in laser smoke. *Laser Surg Med* 11:202–203, 1991; Y. Tomita et al. Mutagenicity of smoke condensates induced by CO_2 laser irradiation and electrocauterization. *Mut Res* 89:145, 1981; Sawchuk WS, et al. Infectious papillomavirus in the vapor of warts treated with carbon dioxide laser or electrocoagulation: Detection and protection. *J Am Acad Dermatol* 21:41, July 1989.

Suggested Readings

Becker CM, Malhotra IV, and Hedley-Whyte J. The distribution of radio-frequency current and burns. *Anesthesiology* 38:106–122, February 1973.

Buczko GB and McKay WPS. Electrical safety in the operating room. *Can J Anesth* 34(3):315–322, 1987.

Emergency Care Research Institute. ESU monitoring systems. *Technol Surg* 8:1–3, December 1987.

Gatti JE, et al. The mutagenicity of electrocautery smoke. *Plast Reconstr Surg* 89:781–784, May 1992.
National Institute for Occupational Safety and Health. *Health Hazard Evaluation Report*. Publ No. HETA 85-126-1932. U.S. Department of Health and Human Services, Washington, DC, 1988.
Pearce J. Current electrosurgical practice hazards. *J Med Eng Technol* 9:107–111, May/June 1985.
Soderstrom RM. Electrosurgery's advantages and disadvantages. *Compr OB/GYN* 35:35–47, October 15, 1990.
Tucker RD and Ferguson S. Do surgical gloves protect staff during electrosurgical procedures? *Surgery* 110:892–895, November 1991.
Voyles CR and Tucker RD. Education and engineering solutions for potential problems with laparoscopic monopolar electrosurgery. *Am J Surg* 164:57–62, July 1992.

Laser Plume Quantification

William Charney

Quantification data of laser plume to test the efficiency of a laser scavenging system are presented. These data were compiled during a simulated case and during a real case of removal of a condyloma. The scavenging device tested was a Baxter Class 1 Smoke Evacuator equipped with both HEPA and ULPA filters with an efficiency to 0.01 μ.

Sampling Instrumentation

A Miniram PDM 3, a light screening aerosol monitor of the nephelometric type that continually senses the combined scattering from the population of particles present, was used. The Miniram uses a GaAlAs light-emitting source, which generates a narrow-band emission. The radiation scattered by airborne particles is sensed over an angular range of approximately 45°–90° by a silicon-photovoltaic hybrid detector.

- Measurement ranges: 0.01–10 mg/m^3 and 0.01–100 mg/m^3
- Particle size range: 0.01–10 μ

Method

During simulation and real case, the hose of the scavenging device was held approximately 2 cm from the burn site. The wattage of the laser power was set at 10 and the burn times were normally 10 s with a 20 s burn. The scavenger was turned on simultaneously with the burn by the use of a foot activator switch.

Discussion

During both the simulation and the real case, all the smoke was evacuated and zero level exposure was quantified. During simulation when the scavenging device was turned off at a median level of 12 mg/m^3 particulate, staff without proper respirators complained of symptoms.

Recommendations

Use of the scavenging system reduces quantified levels of particulate to zero. The scavenger hose when held within the range of 2 cm from the burn site has good capture and it is assumed that capture capacity decreases as the distance from the burn site increases. Therefore, the scavenging system should be used for all laser cases that produce a plume to protect the staff and the patient from the content of the plume. (See Tables 12.15 and 12.16.)

TABLE 12.15 Laser Plume Quantifications: Simulation; April 29, 1992

Sl. No.	Type	Positions in Feet	Laser (W)	No. of Seconds	mg/m^3
1	Background (9:29)	5	Off	Off	0.00
2	Background (9:30)	5	Off	Off	0.00
3	Background (9:31)	5	Off	Off	0.00
4	Background (9:34)	5	Off	Off	0.00
5	With scavenging	5 (SBZ)	10	10	0.00
6	With scavenging	5 (SBZ)	10	10	0.00
7	With scavenging	5 (SBZ)	10	10	0.00
8	With scavenging	5 (SBZ)	20	10	0.00
9	With scavenging	5 (SBZ)	20	10	0.00
10	With scavenging	5 (SBZ)	20	20	0.00
11	With scavenging	5 (SBZ)	20	20	0.00
12	With scavenging	5 (SBZ)	10	10	0.00
13	With scavenging	5 (SBZ)	10	10	0.00
14	With scavenging	5 (SBZ)	10	10	0.00
15	With scavenging	5 (SBZ)	10	10	0.00
16	With scavenging	5 (SBZ)	10	10	0.00
17	With scavenging	5 (SBZ)	10	10	0.00
18	With scavenging	5 (SBZ)	10	10	0.00
19	With scavenging	5 (SBZ)	10	10	0.00
20	With scavenging	5 (SBZ)	10	10	0.00
21	No scavenging	5 (SBZ)	10	10	12.7
22	No scavenging	5 (SBZ)	10	10	13.8
23	No scavenging	5 (SBZ)	10	10	14.2
24	No scavenging	5 (SBZ)	10	10	13.7
25	No scavenging	5 (SBZ)	10	10	12.7
26	No scavenging	5 (SBZ)	10	10	15.8
27	No scavenging	5 (SBZ)	10	10	12.7
28	No scavenging	5 (SBZ)	10	10	12.7
29	No scavenging	5 (SBZ)	10	10	12.7
30	No scavenging	5 (SBZ)	10	10	12.7
31	With scavenging	5 (NBZ)	10	10	0.00
32	With scavenging	5 (NBZ)	10	10	0.00
33	With scavenging	5 (NBZ)	10	10	0.00
34	With scavenging	5 (NBZ)	10	10	0.00
35	With scavenging	5 (NBZ)	10	10	0.00
36	With scavenging	5 (NBZ)	10	10	0.00
37	With scavenging	5 (NBZ)	10	10	0.00
38	With scavenging	5 (NBZ)	10	10	0.00
39	With scavenging	5 (NBZ)	10	10	0.00
40	With scavenging	5 (NBZ)	10	10	0.00

Note: SBZ, surgeon breathing zone; NBZ, nurse breathing zone.

TABLE 12.16 Laser Quantifications: Real Case

Sl. No.	Type	Position	Laser (W)	No. of Seconds	mg/m^3
1	Background	SBZ	Off	Off	0.00
2	Background	SBZ	Off	Off	0.00
3	Background	SBZ	Off	Off	0.00
4	Background	SBZ	Off	Off	0.00
5	Laser on	SBZ	10	10	0.00
6	Laser on	SBZ	10	10	0.00
7	Laser on	SBZ	10	10	0.00
8	Laser on	SBZ	10	10	0.00
9	Laser on	SBZ	10	10	0.00
10	Laser on	SBZ	10	10	0.00
11	Laser on	SBZ	10	10	0.00
12	Laser on	SBZ	10	10	0.00
13	Laser on	SBZ	10	10	0.00
14	Laser on	SBZ	10	10	0.00
15	Laser on	SBZ	10	10	0.00
16	Laser on	SBZ	10	10	0.00
17	Laser on	SBZ	10	10	0.00
18	Laser on	SBZ	10	10	0.00
19	Laser on	SBZ	10	10	0.00
20	Laser on	SBZ	10	10	0.00
21	Laser on	SBZ	10	10	0.00
22	Laser on	SBZ	10	10	0.00
23	Laser on	SBZ	10	10	0.00
24	Laser on	SBZ	10	10	0.00
25	Laser on	SBZ	10	10	0.00

Note: SBZ, surgeon breathing zone.

Laser Plume: Case Study

William Charney, Scott Miller, Shawn Campbell, and Bob Reynolds

Introduction

Medical laser procedures are becoming more and more common as laser technology develops. They are used in the following medical specialties: gynecology, neurosurgery, general surgery, dermatology, urology, ophthalmology, gastroenterology, otolaryngology, podiatry, and dentistry, as well as in experimental laser surgery. The following case study is not written as a comprehensive guide to laser safety (the reader is referred to the American National Standard for the Safe Use of Lasers in Healthcare Facilities, ANSI 2 136.3., 1988); however, it is an experiential clinical case that provides interesting symptomatologies, hypotheses, and interventions.

Each hospital using lasers should develop uniform laser safety policies and procedures based on ANSI 2 136.3., The ECRI Model Policy, or from the Laser Institute of America guidelines. Generally, these should include

1. Formation of a laser committee
2. Credentialing of medical personnel
3. Education and training

4. Safety precautions, including general precautions, eye protection and baseline examinations, preoperative setup, operating precautions, and anesthesia guidelines
5. Maintenance and service
6. Scavenging systems
7. New laser acquisitions

Included as an appendix to this case study is a draft laser policy.

Background

Two registered nurses assisting on procedures using a carbon dioxide laser for treatment of patients with cervical dysplasia, warts, or condylomata caused by HPV complained of the following symptoms:

1. Nasal congestion
2. Poor inspiratory effort
3. Throat irritation
4. Upper respiratory chest tightness
5. Flu-like symptoms
6. Fatigue

Symptoms persisted for 24–48 h after exposure to the laser procedure. The procedure in question is performed in an outpatient clinic room. The laser has a smoke plume evacuation system that consists of the following parts:

1. Pickup orifice attached to the vaginal speculum
2. Half-inch Tygon tubing that connects the pickup orifice to the filter housing box
3. Filter box housing filter
4. Capture box

Air is then circulated back to ambient room air.

Hypothesis

Laser plume transmigrating into ambient air in respirable (lodging in alveoli) particle size was being inspired by the nurses, provoking symptoms. The laser plume was escaping through a dysfunctioning scavenging system.

Investigation

Preliminary investigation of the nurses' complaints began with engineering assessment of the laser plume speculum scavenging system, with individual assessment of

1. Pickup orifice
2. Capture box
3. Efficiency of air flow
4. Potentiality of blood/particulate matter blocking air velocity flow (AVF) within the pickup orifice or capture bore

Assessment of the vaginal speculum revealed that the relatively small diameters of the speculum's smoke aspiration tubing and orifices had the potential of becoming plugged with bodily material, either by tissue or blood particulates. Three or four speculums tested were completely plugged, thereby decreasing AVF to nearly zero. With the tubing blocked, the laser plume was not being evacuated but rather was being aerosolized into the ambient air of the room. It could not be determined how many surgical cases were required for the speculums to become blocked.

The Biomedical Engineering Department was consulted. Consensus opinion and assessment of the speculum were

1. Diameter of the orifices on the speculum aspiration tubing was of marginal size (2 mm) for the capture capacity of the laser plume.
2. Inside diameter of the speculum aspiration was too small (3 mm).
3. The inside of the original pickup orifice was found to have flashing from the drilling of the holes that contributed to the decrease in flow rate of air.
4. On one orifice, there was spot welding penetration into the tubing which obstructed air flow within the tubing.
5. Because of the residual flashing inside the pickup orifices, it was difficult to clean the tubing with a cleaning rod to empty particulate debris.

Biomedical engineering was assigned the task of redesigning and modifying the speculums using the following criteria:

1. The aspiration tube diameter was increased from 3 to 6.35 mm using stainless steel tubing (see Figure 12.2a). The assessment of this modification was that by doubling the size, potential blockage would be decreased and unrestricted flow would be increased.
2. Larger pickup orifices were machined not only to increase the capture face potential, but also to reduce the eddies.

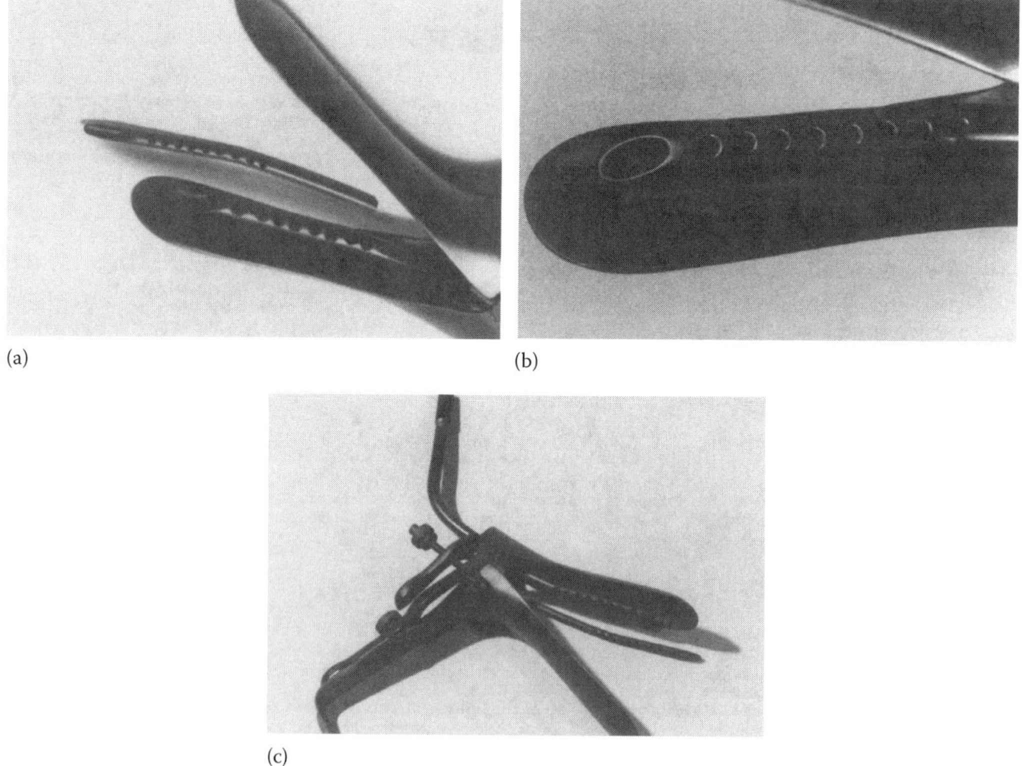

FIGURE 12.2 Vaginal speculum used in the laser procedure. (a) Middle pickup orifice (aspiration tube) is of original size (3 mm); attached orifice is of the new 6.35 mm. (b) Re-ebonized 6.35 mm pickup orifice (aspiration tube). (c) Complete speculum with smaller aspiration tube in the middle for contrast.

3. The tube was then welded into place with a stainless steel filler rod, keeping the entire speculum assembly stainless steel.
4. Extra steps were taken to correct the cleaning rod problem (orifice needs to be cleaned after each procedure to remove blood, tissue, etc., to maintain air flow). After drilling the orifices, the aspiration tube was again rebored, then the orifices chamfered to remove all sharp edges, thus allowing for easy access for cleaning rods.
5. To complete the modification, the refabricated speculum was re-ebonized with a thin, flat finish (see Figure 12.2b). The complete speculum is shown in Figure 12.2c.
6. *Filtration System.* The original Laser Smoke Filtration System was Stockhouse LFA-100. It is a 0.5 μ filtration at 99.07% efficiency. It was decided to upgrade the filtration capacity. A Stockhouse PT-1000 Point One 0.01 μ filtration at 99.9999% efficiency was installed. Though Brownian motion plays a role in particle size distribution with filtration, the decision to upgrade the filter was based on viral particle size (smallest pure 18–26 nm) and the fact that there are no published data about the particle size of virus in laser plumes.

Discussion

The respiratory symptoms experienced by the attending nurses could have either been caused by the "number" and "size" of unscavenged particles or the "type" of particles being respired. Gordon et al.[1] showed that vapor produced by the carbon dioxide laser during the vaporization of papillomavirus-infected verrucae had viral DNA present and that this viral DNA was liberated into the air with the vapor of laser-treated verrucae. Also, the CDRH (*Bulletin,* Vol. XXII, No. 10) cited their recent study demonstrating that viable bacterial viruses can be present in laser plume. It is not in the scope of this case study to determine the exact etiologic agent causing the symptoms. However, it is felt that the escaping plume caused the symptoms of the nurses.

Clinical Trial of Modified Unit

A clinical trial was set up to evaluate the modified speculum unit. The procedure involved was similar to the procedures done before the modification of the unit. The doctor in charge verified that this modified unit worked very well to eliminate the plume and that there were no interferences from the larger bore sizes. The staff attending had no respiratory symptoms after the procedure or for the days following.

Conclusion

The elimination of the plume at the source is the design criterion of the scavenging system and must be achievable. All the different interconnecting parts of the scavenging system speculum, pickup orifice, bore, tubing, and filtration system must be kept at optimal efficiency at all times. If plume escapes into the ambient air, symptoms can potentially be expected either due to size and number of respirable particles or viral contamination. Not all medical laser treatments will produce the same type of plume or need the same type of evacuation system. Each laser does need to be assessed by qualified personnel and attending staff.

Reference

1. Garden JM, O'Banion MK, Shelnitz LS, Pinski KS, Bakus AD, Reichmann ME, and Sundberg JP. Papillomavirus in the vapor of carbon dioxide laser-treated verrucae. *J Am Med Assoc* 259(8):1199–1202, 1988.

Appendix 12.A.1

Sample Hospital Policy Statement: Laser Safety Procedures

Subject: Laser Committee

Policy: It is the policy of San Francisco General Hospital Medical Center to establish and maintain a Laser Committee reporting to the Hospital Safety Committee.

Purpose: The goals of the Committee are

1. To develop the Hospital's physician credentialing criteria for laser use.
2. To establish rules and practices to ensure laser safety.
3. To assist in the training and education of physicians, nurses, and support personnel.
4. To make recommendations to the Capital Equipment Committee for laser modality acquisition.
5. To investigate unusual occurrences related to laser use.

Procedures:

1. The Chief of Service of each user department shall designate a representative to the Committee. At minimum, the Committee shall also include the Director of OR Nursing, the Director of Environmental Health and Safety, and the Director of Biomedical Engineering.
2. The Committee shall appoint an LSO to monitor safe laser practice. The LSO will report Committee activities semiannually to the Hospital Safety Committee.

Effective: DRAFT 11/21/88

Laser Committee

I. **Function.** The Laser Committee, as designated by the attached Hospital Policy, is responsible for the Laser Safety functions at SFGHMC.

II. **Membership**

1. The current physician membership of the Laser Committee consists of users/physicians from the following services:
   (a) Ophthalmology
   (b) Gastroenterology
   (c) Urology
   (d) Gynecology
   (e) Anesthesia
2. Nursing and other departments represented are
   (a) Head nurse, gastroenterology
   (b) Director, OR nursing
   (c) Director, biomedical engineering
   (d) Electronic technician supervisor, biomedical engineering
   (e) Director, environmental health and safety
   (f) Director, plant services

III. **Duties**

1. **LSO.** The Laser Committee shall appoint an LSO from among the users who will act as the laser advocate and laser education coordinator. The LSO shall be fully trained in laser operation and applications, clinical operation and use, and safety. Specific duties include:

(a) Convene the Laser Committee at least semiannually, or as necessary to investigate unusual occurrences or to formulate recommendations to the Capital Equipment Committee for laser-related equipment requests.

(b) Report Laser Committee activities semiannually to the Hospital Safety Committee.

2. **Members.** In addition to participating actively in Committee agenda, members shall be responsible for laser education and training of personnel for their respective departments.

Credentialing

I. Criteria. The Laser Committee shall authorize or certify individual physicians to perform laser procedures based on at least one of the following criteria:

1. Documented completion of an approved formal laser training program offered by a recognized authority (e.g., the American Society of Lasers in Medicine and Surgery), which includes a minimum of 4 h each in didactic and clinical hands-on educational workshops.
2. Documented completion of a manufacturer's formal training program that includes a demonstration of clinical competence under the management of a previously certified laser physician.
3. Preceptorship or residency training and proven proficiency while assisting an experienced, certified laser physician. In this case, the certified physician or chairman of the residency program shall provide a formal letter stating the number of hours dedicated to such training, equipment used, procedures undertaken, and the increasing levels of responsibility assigned to the physician-in-training and monitored by the certifying physician.

II. Documentation. In addition to the originals of certification or approval letters that the physician should retain, copies of such documents should be filed with the Laser Committee.

III. Exceptions. A physician shall not use a laser for any procedures other than those for which he has the Laser Committee's aforementioned approval. In emergencies, laser use without prior approval may be unavoidable, but the physician must then notify the Laser Committee in writing after such use, noting the circumstances that necessitated emergency use and postoperative plan for the patient.

Education and Training

I. Initial Training. All OR and clinical personnel who are likely to be present during laser operation or maintenance are required to attend a training seminar that describes the principles of operating lasers, their clinical applications, attendant risks to patient and staff, safety procedures, and the care of equipment.

Safety Precautions

I. General Precautions

1. All access doors to laser ORs shall be posted with warning signs to prevent unexpected interruption of the laser operator.
2. A UL-approved fire extinguisher should be readily available in the event of equipment or material fire.
3. All reflective surfaces likely to be contacted by the laser beam shall be removed from the OR. Matte or anodized finished instruments shall be used when possible.
4. All windows in the laser OR shall be covered.

II. Eye Protection

1. Protective eyewear of the appropriate OD for the type of laser being used shall be worn by all persons present prior to powering the laser system and during the entire procedure. This includes all staff, bystanders, and the patient.
2. Microscopes shall have a failsafe method to protect the user's eyes against laser beam reflections (i.e., builtin filters, separate optical paths for intermittent viewing and firing, shutters that automatically close prior to firing).
3. Users and support staff who are normally in the OR during laser activation shall have eye examinations at least every 3 years designed to detect any adverse effects that may have been caused by inadvertent laser radiation. Eye examinations must also be performed during laser orientation to establish a baseline.

III. Preoperative Setup

1. The key to the laser system must be removed and kept in a secure location when the laser is not in use.
2. The physician shall test the focusing and the limits of the power range intended for the procedure before using the laser.
3. When possible, and except for ophthalmic lasers, the physician should avoid placing the laser beam path at eye level (sitting or standing).
4. The laser shall never be switched to the operating mode until the procedure is ready to begin and the laser is in the proper position.
5. The laser shall never be aimed at a person except for therapeutic purposes.
6. When a laser plume will be generated, it must be evacuated by a smoke evacuator or by wall suction. The vacuum tubing or probe used to evacuate the plume should not be used to suction blood or fluids unless it is specifically designed to perform both functions simultaneously.

IV. Operating Precautions

1. Combustible material, such as OR gowns, drapes, and towels, should be kept out of the laser beam path to avoid combustion.
2. When possible, skin and tissue adjacent to the lasing site should be draped with moist towels to avoid inadvertent damage from direct or reflected irradiation. Towels must be remoistened as necessary to avoid combustion.
3. Alcohol based or other flammable prepping solutions should not be used. Betadine or sterile water can be used if the area is then dried thoroughly.
4. When procedures are performed in the perianal area, the flammability of the intestinal gas must be considered. Cotton balls with a water-based jelly applied may be inserted into the rectum.
5. An alternative means of controlling bleeding should be readily available to the surgical team during surgery if the laser is to be used in a procedure where profuse bleeding is a recognized risk.

V. Anesthesia Guidelines

1. Flammable anesthetics must not be used in conjunction with laser surgery.
2. If laser energy is to be used near an endotracheal tube, special precautions must be taken. Red rubber or silicone elastomer tubes wrapped in reflective tape, or laser-resistant tubes should be utilized. Vinyl tubes must be avoided, as toxic hydrochloric acid fumes will be emitted if the tube is burned.
3. If laser energy is to be delivered through an endoscope and used near an endotrachael tube, a rigid endoscope with a nonreflective surface or a flexible fiberoptic bronchoscope with a nonflammable sheath should be used. In addition, laser energy should be applied slowly and intermittently to avoid scattering flaming carbonized particles that, in the presence of high oxygen concentrations, could promote endotracheal tube fires.

4. The anesthesiologist should ventilate the patient with nonoxygen-enriched (room) air when possible. If oxygen therapy is necessary, it should be administered at the minimum concentration necessary to properly support the patient.
5. The anesthesiologist shall be knowledgeable of the procedures to follow in the event of an endotracheal tube fire. (The endotracheal tube should be immediately disconnected from the gas source and then removed. Light taping of the tube to the patient to facilitate rapid extubation is recommended.)
6. Oil-based endotracheal tube lubricants shall not be used. (Water soluble solutions of local anesthetics are acceptable alternatives.)
7. The anesthesiologist and the surgical team shall agree, prior to the induction of anesthesia, on a management plan in the event of total airway obstruction.
8. If possible, patients with compromised airways shall not be premedicated with narcotics, diazepam, or barbiturates. Standard inhalation or intravenous techniques can be employed in patients without compromising airways.
9. A muscle relaxant to immobilize the surgical field should be considered.
10. Blind intubation shall be avoided in patients with airway lesions. All necessary equipment (e.g., laryngoscope blades, various size endotracheal tubes, tracheostomy set, bronchoscopes) shall be present prior to beginning intubation. A surgeon qualified to perform a tracheostomy shall be accessible.

Unusual Occurrences

I. The Laser Committee shall be responsible for investigating all unusual occurrences related to laser use and reporting the results to the Quality Assurance Program.

Maintenance and Service

I. **Routine Inspection.** The laser user is responsible for performing a daily visual inspection of the laser system to check for any apparent abnormalities. Equipment abnormalities should be reported to Biomedical Engineering or to the service contract vendor.

II. **Cleaning**

1. **Laser System.** It is recommended that each laser system be wiped off with alcohol between each laser case.
2. **Carbon Dioxide Accessories**
   (a) Handpieces should be steam sterilized or flash sterilized for three minutes prior to use.
   (b) Lens should be washed carefully with soap and water, dried, and carefully stored after each use.
3. **Nd:Yd Laser Accessories**
   (a) Endoscopes should be cleaned as described in the Hospital's Infection Control Manual.
4. **Argon Accessories**
   (a) The dermatological hand piece shall be soaked in Cidex for 10 min prior to use with each patient. After each use, it shall be washed with soap and water and rinsed well.
   (b) The contact lens used with Ophthalmology Argon Lasers is carefully cleaned with soap and water, dried, and wiped with lens paper after each patient use.

III. **Maintenance.** Only qualified manufacturer's representatives or in-house technicians properly trained and approved by the Laser Committee shall install, maintain, and/or service the laser. Documentation of laser maintenance training shall be submitted to the Laser Committee.

Initial installation of a laser and any periodic maintenance or repair that could affect laser performance or safety shall be followed by performance/safety tests prior to laser use.

New Laser Acquisition

I. **Equipment.** The Laser Committee shall review and make recommendations to the Hospital Capital Equipment Committee for all laser technology acquisitions.

II. **Space.** The Laser Committee shall plan and coordinate all laser facility design and modifications with the Plant Services Department.

Source: San Francisco General Hospital Medical Center, 1988.

13

Safe Use of Ethylene Oxide in the Hospital Environment

Use of Ethylene Oxide in Medical Care 13-1
Health Effects 13-2
Chemical Properties 13-2
Sterilization Process 13-4
Regulatory Environment 13-7
Occupational Safety and Health Administration
Evaluating the EtO Hazard 13-9
Exposure Control 13-10
Equipment Design • Local Exhaust Ventilation System • Dilution Ventilation Systems • Substitution • Changes in the Work Process • Isolation • ETO-Related Programs • Maintenance Programs • Training Programs • Emergency Action Plan • Respiratory Protection Equipment
Community Exposure 13-16
Air Emissions • Sewage Emissions • Potential Control Technologies • Adsorption/Reaction
References 13-17
Suggested Reading 13-18
Appendix 13.A.1 13-18
Appendix 13.A.2 13-19
Appendix 13.A.3 13-19
Appendix 13.A.4 13-21
Appendix 13.A.5 13-21
Appendix 13.A.6 13-22

Arthur R. Reich
University of California, San Francisco

Use of Ethylene Oxide in Medical Care

In 1985, almost 3 million tons of ethylene oxide (EtO) was produced in the United States. Hospitals and other health care facilities routinely use ethylene oxide as an agent to sterilize medical devices and equipment. Its use is especially important in the sterilization of heat- and moisture-sensitive items that cannot be sterilized by steam. There is no suitable substitute at the present time for EtO sterilization within hospitals, and its continued use is essential for the control of nosocomial infections. Less than

0.5% of all EtO produced in the United States is used as a sterilant; however, this use poses possibly the most significant occupational hazard owing to the greater potential for employee exposure and the number of workers involved.[1] In 1977, the National Institute for Occupational Safety and Health (NIOSH) estimated that approximately 75,000 health care workers were potentially exposed to EtO.[2] Perhaps due to increased awareness of the potential health hazards of EtO, recent estimates of the number of people exposed have been reduced. In 1985, NIOSH estimated that 22,230 hospital workers were potentially exposed to EtO.[3] Approximately 60%–70% of industrially sterilized medical devices use EtO as a sterilant. It is estimated that the health industry sterilizes 10–12 billion items per year. As a conservative estimate, hospitals, clinics, and doctors sterilize an additional 200 million items per year.[4]

Health Effects

In vivo EtO is rapidly distributed throughout the body. Acute and chronic exposure leads to respiratory tract irritation and central nervous system depression as well as other pathological changes. At high doses, EtO can induce dominant lethal mutations and cause embryotoxicity in rodents. One epidemiologic study suggested an increase in spontaneous abortions due to EtO exposure.

EtO, presumably due to its ability to alkylate DNA, causes gene mutations in both prokaryotic and eukaryotic cells and leads to sister chromatid exchanges and chromosomal damage, including the formation of aberrations and micronuclei.[5,6] Several types of tumors have been induced in rats by EtO. Epidemiologic studies of people occupationally exposed to EtO strongly suggest increased incidences of stomach cancer and leukemia.[7] The International Agency for Research on Cancer (IARC) concluded that there is sufficient evidence that carcinogenicity is limited. Overall, based on both the animal and human data, IARC considers that EtO is probably carcinogenic in people.[8] Typical EtO exposure patterns are given in Table 13.1.

The primary effects of acute, subchronic, or chronic exposure to EtO are similar to humans and animals: central nervous system depression and respiratory tract irritation. EtO can irritate the eyes, nose, and throat of most people when air concentrations reach about 200 parts per million (ppm). Levels above 1000 ppm can cause coughing, lung irritation, breathing difficulties, and chest pain. Short-term high level exposure to EtO has a mild depressant effect on the brain, somewhat like alcohol. Skin or eye contact with liquid EtO causes burning. Skin effects may not appear until 1–5 h later. Sensitization and cataract formation are also associated with repeated exposure in humans.

Repeated inhalational exposure to high concentrations of EtO is associated with neuropathy. Observations of neurotoxicity in humans have been confirmed in some, but not all, studies by histopathologic observation. In laboratory animals, pathologic changes have been observed in the lungs, kidney, liver, testicles, and blood. Epidemiologic studies have not revealed any toxic, nonmalignant effects after long-term exposure to levels of EtO below 10 ppm.

In 1977, NIOSH recommended a level for occupational exposure to EtO that was designed to protect workers against the acute and chronic nonmalignant health effects of EtO. They recommended continued observation of the then-current occupational standard of 50 ppm as a time-weighted average (TWA) for an 8 h shift. No noncarcinogenic effects were expected below this exposure level. However, in 1984, an occupational standard of 1.0 ppm as a permissible exposure limit (PEL) for an 8 h TWA exposure was promulgated based on evidence of EtO's carcinogenicity (OSHA 1984).[9,10] OSHA has recently added a short-term exposure limit of 5.0 ppm for a 15 min TWA.

Chemical Properties

EtO is a colorless gas at room temperature and a liquid when compressed. The chemical structure of EtO is

$$H_2C\overset{}{-}CH_2 \text{ (bridged by O)} \qquad C_2H_4O$$

TABLE 13.1 Typical EtO Exposure Pattern: IR Survey Results

	Uncontrolled		Controlled		
Exposure Pattern	Concentration (ppm)	Time (h)	Concentration (ppm)	Time (h)	Exposure Control Approach
Pre-cycle room levels	3	0.25	1	0.25	At least 15 air changes per hour. Local exhaust system-aerators.
Cylinder storage room	10	0.25	1	0.25	Leak test pressure lines. Check valves on tank lines. Local exhaust system.
Sterilizer door leak	75	2.00	0	2.00	Leak test door, maintenance. Local exhaust system.
Purge line discharge	300	0.25	10	0.25	Local exhaust system. Remote location.
Opening door	200	0.25	1	0.25	Local exhaust system. Crack door and move away for a time period determined by IR monitoring.
Packing aerator	10	0.25	2	0.25	Load in baskets. Protective gloves. Arm's length handling.
Other duties	2	4.75	0	4.75	Isolate sterilizer location.

$$\text{TWA} = \frac{C_1T_1 + C_2T_2 + C_3T_3 + \cdots + C_nT_n}{8}$$

$$\text{Uncontrolled TWA} = \frac{(3\times 0.25)+(10\times 0.25)+(75\times 2.00)+(300\times 0.25)+(200\times 0.25)+(10\times 0.25)+(2\times 4.75)}{8}$$
$$= 36\,\text{ppm}$$

$$\text{Partially controlled TWA} = \frac{(3\times 0.25)+(10\times 0.25)+(75\times 2.00)+(10\times 0.25)+(1\times 0.25)+(10\times 0.25)+(10\times 0.25)+(2\times 4.75)}{8}$$
$$= 21\,\text{ppm}$$

$$\text{Controlled TWA} = \frac{(1\times 0.25)+(1\times 0.25)+(0\times 2.00)+(10\times 0.25)+(1\times 0.25)+(2\times 0.25)+(0\times 4.75)}{8}$$
$$= 0.47\,\text{ppm}$$

EtO has a molecular weight of 44.05, a boiling point of 51°F, a vapor pressure of 1.091 mm Hg at 20°C, a specific gravity of 0.87 (water = 1), and a vapor density of 1.5 (air = 1). One hundred percent EtO has explosive limits of 3%–100% and a flash point of −6°C (tag open cup). EtO is an alkylating agent that reacts directly with -COOH, -NH2, -SH, and -OH groups. It also reacts with the ring nitrogen of purine and pyrimidine bases and the amino groups of amino acids and proteins. It is relatively noncorrosive to most materials except some rubbers.

EtO is provided to most hospitals as 100% pure, undiluted EtO in small cartridges in up to 134 g containers that contain less than 5 fluid ounces of EtO or as a nonflammable gas mixture known as 88/12 (12% EtO in 88% Freon 12 by weight). Cylinders used are usually 140 lb. The cartridges are used to operate sterilizers with chamber volumes of less than 4 ft^3. These cartridges contain pressurized liquid that will vaporize when released. It is assumed that as long as the sterilizer is in a well-ventilated environment, the lower flammable limit (3% or 30,000 ppm) will never be reached. The EtO–Freon mixture is preferred because of its nonflammability, its low operating pressure in the sterilizer (usually about 8 psig), and the ease in finding leaks with an ordinary Freon detector.[11] Cold temperatures do no harm to the cylinder contents, but cylinders should be placed at room temperature for at least 24 h before the cylinder is placed into service as a working cylinder. Pressure in a cylinder will almost double when moved from a 40°F environment to room temperature (about 65 psig).

Sterilization Process

The advantage of EtO sterilization lies in its effectiveness against microorganisms and spores in many types of materials with little physical damage to the material itself. In addition, the ability of EtO to readily diffuse and penetrate allows for the effective sterilization of prepackaged, preassembled items. In hospitals, EtO is used to sterilize delicate instruments that would otherwise be damaged or destroyed by the heat and steam produced by a conventional autoclave. Various optical instruments, for example, would be ruined if water vapor entered the optic elements, and high temperatures can damage or destroy the numerous plastic and rubber goods used in hospitals today. EtO sterilization methods have allowed for the proliferation of disposable medical products that can be sterilized after packaging in paper or plastic shelf packs, assuring sterility of the product until use. Although the first application of EtO as a sterilant occurred in the food industry, the Environmental Protection Agency (EPA) now approves its use only to fumigate specific items such as spices and black walnut meats.

The sterilization mechanism of EtO is based on a reaction between EtO and nucleic acids within the target cell. Alkylation of nucleic acids by EtO disrupts protein synthesis and, ultimately, cell reproduction. The nature of the alkylation reaction requires an ionic medium; therefore, humidification of the product is often necessary for effective sterilization. As in most chemical reactions, gas concentration, temperature, and exposure time are also important rate-limiting factors that must be considered in the sterilization process.[12]

Sterilization chambers are designed to control gas concentration, temperature, humidity, and exposure time to obtain optimum results. EtO gas concentration determines the number of EtO molecules available to react with the microorganisms and ultimately affects the speed and efficacy of the sterilization. For most processes, precise gas concentrations are not critical and are effectively controlled by injecting a known volume or weight of gas and closely regulating the chamber vacuum and/or pressure. Although an increase in temperature speeds the sterilization process, only moderately elevated temperatures are employed to avoid damage to the product. Depending on the items to be sterilized, and possibly the product packaging, sterilization temperatures range from ambient to 60°C (140°F). Experience has shown that items to be sterilized should be humidified to 30% relative humidity or higher, depending on the selected sterilizing temperature. Again, care must be taken not to over-humidify because items may be damaged by water vapor. The duration of exposure to EtO required for complete sterilization will depend necessarily on the aforementioned factors and the ability of EtO to permeate the specific material.

Although a wide variety of equipment designs have been used for EtO sterilization, they basically fall into two categories: one uses pure (100%) EtO; the other, a mixture of EtO and a dilutant gas such as chlorofluorocarbons or carbon dioxide. Pure EtO is quite flammable. Well-designed sterilizers using 100% EtO operate at subatmospheric conditions to eliminate potential gas leakage problems that may lead to fire or explosion. Sterilizers designed for use with nonflammable EtO mixtures can incorporate both subatmospheric conditions and pressurized states into the sterilization cycle.

Regardless of the type of gas used, the sterilization cycle has four phases. The sterilizing equipment regulates three of these phases: the conditioning phase, where the items to be sterilized are brought to a specified temperature and humidity; the EtO exposure phase; and the evacuation phase, when the chamber is flushed with filtered air to remove excess EtO from the chamber. At this point, the items are sterile but not ready for immediate use. Residual EtO remains dissolved in freshly sterilized items and can produce chemical burns to handlers or users of these items. Hence, during the fourth phase, an aeration phase, EtO is allowed to dissipate from the product. To hasten this process and control fugitive

EtO, sterilized items may be placed in an aeration chamber where elevated temperatures and increased air flow adequately aerate most items within 12 h. New sterilization equipment has recently combined both the sterilization and aeration processes so that one does not have to transfer materials to the aerator until the aeration cycle is completed.

Occupational exposure to EtO is dependent in certain respects on the design of the sterilizer, the integrity of the chamber and gas supply and exhaust lines, and the efficiency of the purge cycle. Small (2–4 ft^3), tabletop sterilization chambers (used primarily in health care facilities) are self-contained units that use individual cartridges of 100% EtO to supply gas for each load. Because flammable, pure EtO is used, the entire sterilization cycle proceeds under negative pressure (Figure 13.1). When the chamber has achieved the proper vacuum, temperature, and humidity, the EtO cartridge is automatically punctured,

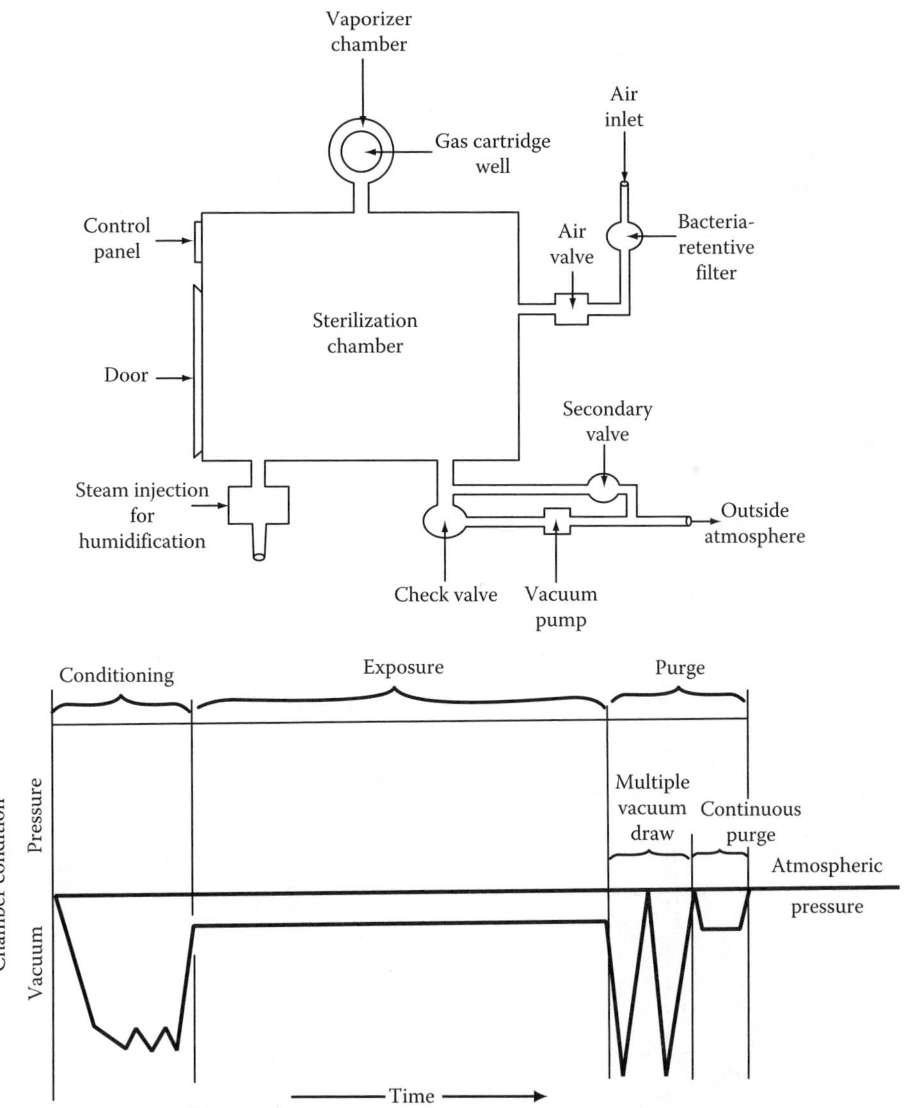

FIGURE 13.1 Sterilizer for use with 100% EtO.

releasing gas into the chamber. At the end of the gas exposure phase, a final vacuum is drawn—approximately ~7 in. of mercury (7 ≤ Hg)—to remove excess EtO, and the chamber is purged by flushing with air. A slight negative pressure is maintained until the cycle is complete. In these small units, purged EtO leaves the unit via copper tubing to the outside atmosphere. Intermediate-sized (9–30 ft^3), built-in sterilizers generally use an 88% Freon and 12% EtO (88/12) gas mixture supplied in gas cylinders. Although items are preconditioned under a vacuum, the actual sterilization phase is conducted above atmospheric pressure (Figure 13.2). The post-sterilization purge consists of one or more deep vacuum draws (in some cases 30 ≤ Hg) and an air flush at slightly negative pressure. Large (over 200 ft^3), freestanding sterilizers

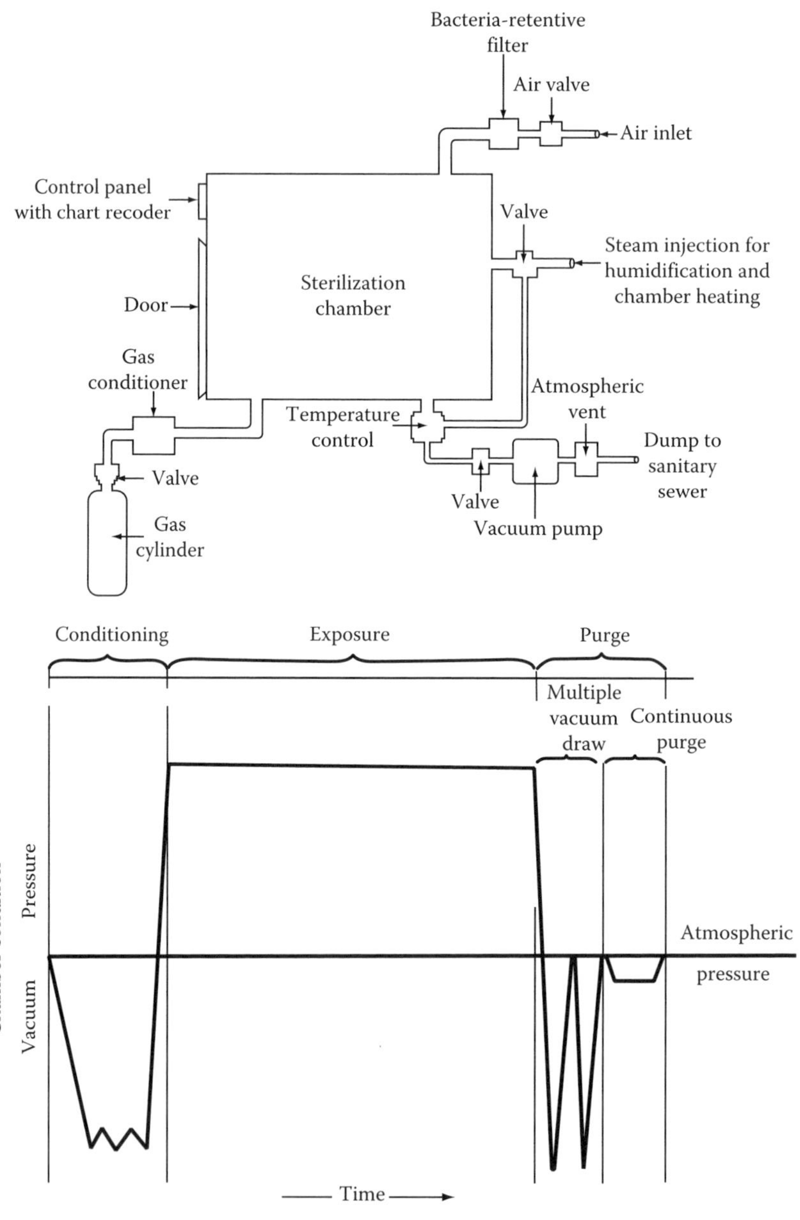

FIGURE 13.2 Sterilizer for use with EtO gas mixtures

are found in commercial sterilization operations. Such sterilizers are designed to use either 100% EtO or nonflammable gas mixtures and are generally built to the user's specifications.

Depending on the efficiency of the chamber purge cycle and the size and type of load, a percentage of the original EtO charge remains in the sterilization chamber and presents the main EtO exposure source. Some of this residual EtO may be released into the surrounding work area when the door is opened at the end of the cycle. The rest is dissolved, absorbed, or trapped within the sterilized items and is slowly released over time, a process referred to as off-gassing.

The potential for EtO exposure exists during the purge cycle of medium- to large-sized sterilizers if the vacuum outlet is plumbed to an open drain. In these sterilizers, EtO is purged from the chamber by a water-sealed vacuum pump. The water outlet line empties either locally or remotely into an open drain because hard piping into a sewer line without an air gap is not approved by most plumbing codes. It is often believed the EtO dissolves in the water or reacts to form ethylene glycol; this is not the case. It has been estimated that ~85% of the EtO used in sterilization can be released at the drain opening, 10% remains in the water entering the sewer system, and the remaining 5% stays in the chamber.

At all times, whether or not the sterilizer is in operation, leaks from supply tanks, valves, and fitting may continuously emit EtO into the workplace. During the sterilization cycle, EtO exposure can result from a leak of the door gasket or a failure of the door-locking mechanism. In addition, sterilizers using nonflammable EtO gas mixtures are equipped with an emergency relief valve that will trip to release pressure before the pressure rating of the chamber is exceeded. If the relief valve outlet is not vented to the outside, a trip of the valve can result in a very large release of EtO into the work environment.

Regulatory Environment

Occupational Safety and Health Administration

On June 22, 1984, the Occupational Safety and Health Administration (OSHA) released its standard on occupational exposure to EtO. The majority of workers exposed to EtO are employed in hospitals and medical products firms. The cost of compliance for this segment of the industry is estimated to be $70 million. The PEL to EtO is 1.0 ppm, determined as an 8 h TWA air concentration. The standard also established an action level (AL) of 0.5 ppm determined as an 8 h TWA. Users of EtO should keep in mind that these standards are performance standards; that is, as greater and greater concentrations are exceeded, regulatory compliance such as environmental and medical monitoring increases (Table 13.2).

In 1988, OSHA amended its standard to provide for a short-term employee exposure limit of 5.0 ppm determined as a 15 min TWA. The basis for the standards is a determination by OSHA, based on animal and human data, that exposures to EtO present a carcinogenic, mutagenic, reproductive, neurologic, and sensitization hazard to workers. The general sections of the federal OSHA standard for EtO and related OSHA standards can be found below:

Specific Sections of OSHA EtO Standard

29CFR1910.1047 (D)	Exposure Monitoring
29CFR1910.1047 (E)	Regulated Areas
29CFR1910.1047 (F)	Methods of Compliance
29CFR1910.1047 (G)	Respiratory Protection and Personal Protective Equipment
29CFR1910.1047 (H)	Emergency Situation
29CFR1910.1047 (J)	Communication of Hazards to Employees
29CFR1910.1047 (K)	Recordkeeping
29CFR1910.1047 (L)	Observation of Monitoring
29CFR1910.1047	Ventilation

Related OSHA Standards

29CFR1910.134	Respiratory Protection
29CFR1910.20	Access to Employee Exposure and Medical Records

TABLE 13.2 EtO Standards (Vary with Concentration)

No.	Exceeds 8 h TWA (>1.0 ppm)	Exceeds AL (0.5–1.0 ppm)	Less than AL (<0.5 ppm)
1.	Monitor every 3 months. May reschedule to every 6 months if two consecutive measurements taken 7 days apart are below 1.0 ppm. May discontinue if consecutive measurements are below 0.5 ppm.	Monitor every 6 months. May discontinue if two consecutive measurements taken 7 days apart are below 0.5 ppm.	Required initial monitoring and when process or engineering change. May discontinue if results are less than 0.5 ppm.
2.	Report or post all monitoring results within 15 days after receipt. Written notification to employees of corrective action being taken.	Report or post all monitoring results within 15 days after receipt.	Report or post all monitoring results within 15 days after receipt.
3.	Environmental measurement records must be kept for 30 years.	Environmental measurement records must be kept for 30 years.	Environmental measurement records must be kept for 30 years.
4.	Regulated area must be established and demarcated.	Regulated area not necessary.	Regulated area not necessary.
5.	Establish and implement a written control program of engineering, administrative, and personal protective equipment measures.	No control program required.	No control program required.
6.	Schedule periodic leak detection surveys.	No leak detection surveys required.	No leak detection surveys required.
7.	Written plan for emergency situations.	Written plan for emergency situations.	Written plan for emergency situations.
8.	Means for alerting employees.	Means for alerting employees.	Means for alerting employees.
9.	Medical surveillance required.	Medical surveillance required for employees exposed at or above the AL for at least 30 days per year. Rotation of employees is prohibited.	Medical surveillance not required.
10.	Employee information and training required.	Employee information and training required.	Employee information and training required.
11.	Material safety data sheet available.	Material safety data sheet available.	Material safety data sheet available.

29CFR1910.132 Personal Protective Equipment
29CFR1910.141 Sanitation
29CFR1910.151 Medical Services and First Aid
29CFR1910.133 Eye and Face Protection
29CFR1910.101 Compressed Gases (General Requirements)
29CFR1910.1200 Hazard Communication—Requires chemical manufacturers and importers to assess the hazards of chemicals that they produce or import, and all employers (SIC codes 20 through 39) to provide information to their employees concerning hazardous chemicals by means of hazard communication programs including labels, material safety data sheets, training, and access to written records. Note: Individual state regulations differ substantially in coverage and implementation.

Some states have adopted an occupational exposure limit for EtO, while others have environmental standards for EtO that may be more stringent than the current federal standard. States that do not have their own approved occupational safety and health plans must follow the federal OSHA standard. State agencies should be contacted for more information.

The EPA registers sterilants such as EtO as a pesticide. The EPA's jurisdiction is mandated through the Federal Insecticide, Fungicide, and Rodenticide Act passed by Congress in 1972. This act provides that no pesticide can be registered by EPA unless it is shown to be safe and effective when used as directed. It requires users to show that they know the correct way to use and handle pesticides, including antimicrobial agents. EtO is also registered with the EPA as a fungicide for fumigation of books; dental, pharmaceutical, medical, and scientific equipment and supplies; drugs; paper; soil; clothing; spices; and transportation vehicles, such as jet aircraft and railroad passenger cars. Under other laws and regulations that it administers, the EPA presumably has the authority to control EtO effluents and emissions that could pollute the environment. The EPA is investigating EtO as a potential toxic air pollutant under the Clean Air Act.

Evaluating the EtO Hazard

Each hospital must determine for itself the strategy it will use in developing a control program that will reduce EtO exposure to acceptable limits. To do so, a hospital must be able to separate and analyze the component parts of EtO exposure. It must then determine how to respond to its analysis, implement a plan that improves its particular situation, if necessary, and, on a periodic basis, monitor the results of the changes.

Exposure and relative exposure probability of EtO are the result of three independent components: personnel practices, equipment conditions, and ventilation characteristics. A hospital will be able to develop effective and efficient ways to reduce exposure only when it understands the degree to which each of these elements contributes to overall personnel exposure. As these elements are independent in their relationship to one another, if any one of them could be completely controlled and eliminated as a factor, no exposure would then exist. For example, if the sterilization cycle could be completed before personnel handle the materials, no exposure would be possible regardless of the equipment's condition or ventilation characteristics. Or, if the equipment conditions were such that EtO could be contained through the sterilization process, which includes the aeration cycle, there would be no reason to be concerned about personnel practice. This is now a reality in some new equipment. As in all discussions of the ideal, however, these examples are not representative of real-life situations, where more than one exposure element needs to be addressed. It is never possible to eliminate the exposure potential of any one element. The examples demonstrate that not all components need to be addressed completely for exposure and exposure potential to be near zero. (Total exposure potential is a product of the individual exposure potential elements.) If one of these percentages is near zero, its multiplication by any other percentage will keep the total exposure potential near zero.

Once the components of exposure are understood, the hospital should review its current operation with these elements in mind. The methodology may be as simple as (1) observing personnel practices while having a passive monitor attached to key personnel responsible for EtO sterilization, (2) testing for leaks with a refrigeration leak detector (88/12 sterilizers), and (3) reviewing plans for characteristics of the ventilation system. More sophisticated data may be sought by contracting for a real-time EtO analysis and for air-balance studies. Whichever method is used, the studies should result in a list of control options for each exposure component (see Table 13.1).

As all options need not be implemented for an effective control program, only those that are most effective from a benefit-cost perspective should be installed. Once installed, an evaluation of the program's effectiveness should be conducted, and a monitoring program should be started to confirm periodically the program's continued performance.

An air sampling program should comply with the following requirements:

1. The monitoring method shall be accurate to a confidence level of 95%, to within ±25% at the 1.0 ppm PEL and to within ±35% at the 0.5 ppm AL. It is prudent to use the same personal

monitoring methods that OSHA compliance or NIOSH industrial hygienists use. For example, the U.S. Department of Labor OSHA, Directorate of Technical Support has validated two methods for use by OSHA compliance officers. The primary method used for OSHA compliance purposes uses a hydrogen bromide coated charcoal tube as a sampling medium across which 24 L of air is drawn at a flow rate of 0.1 L/min. Analysis is by gas chromatography with electron capture detection. This method has a sampling and analytical error of 0.10. A secondary method requires drawing 1 L of air at a flow rate of 0.05 L/min across two standard activated charcoal tubes arranged in series. This method has a sampling and analytical error of 0.11.

2. All monitoring results must be kept for 30 years.
3. Employees are to be notified of all monitoring results within 15 working days of receipt of results. This requirement can be met by posting all results in the work area.
4. Whenever the PEL (1.0 ppm) is exceeded, the employer must notify the employees of the corrective action that will be taken.
5. OSHA requires breathing zone samples (preferably a sampler worn by an employee) to determine compliance.
6. Monitoring records must have the following information:
   a. Date of measurement
   b. Operation
   c. Sampling and analytical methods used and their accuracy
   d. Name, social security number, and exposure of the employees whose exposures are represented
   e. Protective equipment worn
   f. Number, duration, and results of samples taken

The letter of the law exempts employers from additional monitoring if initial monitoring results are below the AL. However, if engineering changes occur that would be expected to change exposure patterns, a reevaluation of employee exposures must be performed. The sensitivity of direct reading instruments to achieve less than 0.5 ppm is not necessarily required. A measurement system that provides reliable real-time monitoring below 1.0 ppm is invaluable in characterizing emissions. Be sure to remember that legal compliance should be done with personal sampling. The use of continuous monitoring systems and direct-reading instruments may be used; however, there is difficulty in characterizing the true employee exposure because hospital central supply employees typically change their work patterns and often leave the area for other parts of the hospital for long periods of time. Infrared (IR) monitoring continues to provide the most useful information for evaluating exposure patterns, engineering controls, and equipment function. It is recommended that IR monitoring and personal sampling be performed every 6 months when environmental conditions are stabilized.

Exposure Control

Equipment Design

EtO sterilization chambers without basic safety features such as an automatic door-locking mechanism, a controlled release of EtO into the chamber, and an EtO purge and air-flush system leading to a remote or controlled location should not be used. A few facilities had such equipment, but claimed they were never used. These antiquated systems not only have no means of controlling employee exposures to the entire EtO charge, but also have a higher potential for incomplete sterilization.

An effective purge of the sterilization chamber before the chamber door is opened can limit the exposures of operators when they open the sterilizer door. Several companies in the medical products industry and some sterilizer manufacturers are experimenting with variations in the purge cycles for optimal control of residual EtO. Monitoring results gave indications that variables such as the depth of the vacuum draw, the number of draws, and continuous purge systems can decrease the amount of residual EtO in the chamber at the end of the cycle (see Figures 13.1 and 13.2). Another factor that may affect the efficiency of the purge is the quality of the air used to flush the chamber. If the air intake is located in an area contaminated with EtO (such as a maintenance access room without adequate ventilation at the exhaust drain), EtO is circulated back into the chamber.

Implementing controls that reduce contact with freshly sterilized items can be very effective in reducing exposures. A few manufacturers of the smaller sterilizers are beginning to produce equipment that both sterilizes and aerates, eliminating the need for a worker to open the sterilizer and handle freshly sterilized items. Automatic door-opening systems, as seen on some of the larger sterilizers, can eliminate exposures received when manually opening sterilization chambers. One facility had coupled this with a mechanism for automatically unloading the sterilized products into an aeration area.

Local Exhaust Ventilation System

The purpose of local exhaust systems is to capture EtO emissions at their source before the EtO can contaminate air in the employee's breathing zone. The most common uses are at the sterilizer door and exhaust drain (Figure 13.3), although provisions to control large releases of EtO from chamber emergency relief valves on medium-to-large sterilizers should also be considered (Figure 13.4). Many of the observed local exhaust systems were ineffective owing to poor design. Such systems should be designed by an engineer or industrial hygienist familiar with the principles of industrial ventilation to assure that the air volume capacity of the fan and the proportions of the control hood(s) and duct work create sufficient air velocities for emission control. To capture EtO before it reaches the operator's breathing zone, hoods over sterilizer doors must be located as near to the door opening as possible, and systems installed at the exhaust drains and pressure relief valves should enclose the source as much as possible.

Large, walk-in type units often have chamber exhaust systems pulling from the rear of the chamber to flush residual and off-gassing EtO away from the operator when the door is opened. These systems can quickly reduce levels within the chamber and also control the escape of the EtO into the workplace.

Dilution Ventilation Systems

Dilution systems are designed to reduce EtO levels in the general work area. However, they alone cannot limit the operator's exposure because EtO is not removed at the source of emission. Where

FIGURE 13.3 EtO drain with an exhaust connected to a dedicated exhaust system in the mechanical room. (Photo by Frank Maniscalco.)

FIGURE 13.4 EtO sterilizer. Note the exhaust grill above sterilizer. This sterilizer also has a built-in Enviroguard system. (Photo by Frank Maniscalco.)

dilution ventilation is installed in lieu of local exhaust, as observed in many hospitals, operators are potentially exposed to EtO from the emission source before the dilution system can produce any effect. Dilution ventilation can be used successfully, however, to augment local exhaust systems where complete control of the emission source cannot be achieved owing to the size of the equipment or the amount of EtO released. At medical product and spice plants, dilution systems minimize ambient EtO level and produce air flow patterns away from general work areas. The design of such systems should be reviewed by an industrial hygiene ventilation professional.

Substitution

Because of the potential health hazards associated with EtO, only those items that must be sterile and cannot be sterilized by any other method should be treated with EtO. It was noted in a study conducted by California OSHA that many items are being gas sterilized which do not require this treatment. At several hospitals, specific guidelines determining those items that will be gas sterilized have reduced the number of gas sterilization loads and thereby the potential exposures of employees. These hospitals employ a more efficient use of autoclaves and pasteurization methods. The medical products industry is substituting gamma radiation sterilization techniques for some of its products; however, certain plastics will still require sterilization with EtO.

Changes in the Work Process

In most facilities, operators crack the door of the sterilizer and leave the area for 10–20 min prior to unloading to allow EtO remaining in the chamber to dissipate. This may be particularly effective in reducing exposures where the purge cycle of the sterilizer is less than adequate or where chamber exhaust systems flush the open chamber with uncontaminated air prior to unloading, as observed in large, walk-in units.

Procedures that reduce the time required to transfer the product from the sterilizer to the aerator can reduce exposures. Such procedures include the use of baskets, carts, and pallets and locating the aeration area as close to the sterilizer as possible.

Isolation

Isolating the sterilization and aeration processes to limited-access work areas (Figure 13.5) may reduce the number of employees potentially exposed to EtO, but without additional controls (such as shown in Figure 13.6) this has little effect on operator exposure. For small sterilization operations, physical isolation may not be necessary, but consideration should be given to locating equipment away from highly

FIGURE 13.5 Separate EtO sterilizer room with 20 air changes per hour. (Photo by Frank Maniscalco.)

FIGURE 13.6 Emergency shutoff valve connected to EtO cylinder. (Photo by Frank Maniscalco.)

populated work areas. In large operations where control of EtO emissions is difficult, physical isolation of sterilization and aeration areas becomes more important.

ETO-Related Programs

The appropriate selection and implementation of the above EtO control measures should be combined with the development of programs to assure their continued effectiveness. In several cases, unnecessary employee exposures could be related to deficiencies in the administration of otherwise effective EtO controls.

Maintenance Programs

Maintenance programs are necessary not only for the effective operation of the sterilizers, but also to ensure control of employee exposures. A written program complying with the manufacturers' specifications should establish equipment maintenance schedules and checklists. An annual inspection of ventilation systems designed for contaminant control must include the measurement of air flow rates and the inspection of fan blades and power drives. Direct-reading instruments should be available to make periodic leak checks around door seals and gas lines. Written procedures for tank change operations should define safe practices such as use of one-way valves, gloves and face shield, leak testing, and emergency and first-aid procedures.

Training Programs

Although sterilizer operators were generally well trained in the hazards of EtO exposure, proper work practices, and emergency procedures, maintenance employees were rarely included in these programs. Even though maintenance workers may not be exposed to EtO on a daily basis, duties that require them to repair sterilizers, perform leak checks, change EtO cylinders, or respond to EtO emergencies necessitate that they receive appropriate training. In addition, such programs must be scheduled regularly to present any new information, instruct new employees, and provide a refresher as necessary. OSHA regulations address employee training programs and require employers to provide training in respect to the hazards and safe work practices associated with EtO.

Emergency Action Plan

OSHA and many hazardous material regulations now require written emergency action plans that adequately address procedures for an EtO emergency such as a major leak or spill. Each employee who may be affected by an EtO emergency must be aware of his or her responsibilities and the specific actions to be taken. The plan should specify when employees must evacuate the area; which employees, if any, should respond to the emergency; when and how outside assistance such as the fire department should be contacted; and testing procedures for determining safe reentry.

EtO is used as a gas in a pressure system. It is obvious that, from time to time, leaks will occur from the various components of the system or mechanical failure can occur. There are six elements necessary for an emergency action plan:

1. Selection of an early warning system.
2. What constitutes an emergency condition.
3. What emergency action will be taken.
4. What first aid and medical procedures are to be followed.
5. What administration actions will be taken.
6. Protective equipment.

The plan should be written, be reviewed by all departments involved, and become part of employee training.

Early Warning Sensor System

OSHA requires that a means be provided to promptly alert employees during an EtO emergency. The implication is that an EtO gas sensor be installed in areas where gas leaks, equipment breakdowns, or ventilation malfunctions might occur. The most important consideration is the location of the sensor. Typically, a sensor should be located in the workroom near the front of the sterilizer (this sensor will monitor for door seal leaks and door ventilation controls). A second location is the sterilizer equipment room (this sensor monitors equipment leaks and ventilation control efficiency at the gas discharge point). A third location might be the compressed gas tank area. Additional locations may be selected depending upon the system purchased and specific need (i.e., adjacent areas). It is not necessary for the sensor to achieve a sensitivity of the 0.5 ppm AL. Sensors are early warning devices, not occupational health monitoring devices. Most sensor systems are hydrocarbon detectors that are nonspecific, unless one is willing to pay thousands of dollars for specific gas analyzer systems. OSHA has not suggested or required this action. Economical solid-state sensors with a sensitivity of 10–20 ppm are available. Be aware that nonspecific sensors may respond to other gases and vapors in the area such as natural gas and alcohols. A two-threshold system set at 10–15 and 100 ppm (required protective clothing) is preferable.

Emergency Conditions

An emergency can be caused by (1) sensor alarm, (2) ventilation failure during sterilization cycle, (3) equipment damage or malfunction, and (4) gas cylinder leaks.

Emergency Action

The emergency action element of the plan should tell employees when to evacuate and where to relocate. It should have a notification (call list) procedure.

First Aid and Medical Procedures

All employees should know the first-aid procedure for EtO exposures and how medical treatment decisions are to be made. It is wise to preplan a potential event with the hospital emergency room.

Administration Actions

Decisions have to be made with respect to various responsibilities and notifications during an episode (e.g., regulatory agency notification, vendor notification, surgery notification, employee reentry, etc.).

Protective Equipment

1. Respiratory protection (assumes in-house response). The only acceptable respirator is a positive-pressure self-contained breathing apparatus.
2. Protective clothing, gloves (assumes in-house response). No engineering studies are published that evaluate EtO permeability of protective clothing. As a rule, response people should spend no more than 3 min at any one time in an area with greater than 100 ppm EtO. It should not take longer than 3 min to turn off the gas. Aerate all exposed equipment.
3. An eye wash/drench hose should be located within 30 ft of the tank changing area for immediate flushing of the eyes and body.

Respiratory Protection Equipment

The use of ineffective, unapproved respirators was prevalent throughout the CAL-OSHA study.[11] Air-purifying respirators, such as surgical or dust masks and canister or cartridge respirators, are not effective for protection from EtO exposure. The ability of EtO to diffuse and penetrate renders surgical and dust masks totally useless. In fact, their use may present a hazard by promoting a false sense of security to employees in emergency situations. Although gas masks with canisters designed for EtO are available, their use is not currently approved because of the high odor threshold and poor warning properties of EtO. A worker wearing a gas mask may be unknowingly exposed to EtO when the service life of the canister has been exceeded and EtO breaks through the canister. Self-contained breathing apparatus (SCBA) or airline respirators are the only respirators acceptable for EtO. In emergency situations, SCBA must be worn.

Approved respirators must be worn when concentrations of EtO are not known, such as when entering large, walk-in chambers for the unloading process. Once EtO levels are tested and shown to be within acceptable levels, respirators need not be worn.

Community Exposure

Air Emissions

EtO has been identified as a major toxic air pollutant. The Office of Air Quality Planning and Standards of the EPA designated EtO as an intent-to-list compound in the *Federal Register* in early 1986.

Estimations of community contamination from hospital emissions of EtO are made with dispersion modeling techniques. This method is used because the highest levels of EtO anticipated to be found in ambient air are about 20,000 parts per trillion (ppt), and current detection limits for ambient air monitoring analysis are greater than 100,000 ppt. Using this approach, the California State Air Resources Board[13] modeled an area in Los Angeles County that included EtO emissions from 55 hospitals, one large industrial sterilization facility, and two compressed gas repackaging facilities. The 1985 modeling study indicated that the approximately 7 million people in the exposure area were exposed to a population-weighted annual mean EtO concentration of about 50 ppt. About 350,000 (5%) of these people were exposed to an average annual concentration of greater than 160 ppt. The report pointed out that by excluding the three industrial sources, the annual average EtO concentration for the exposure area due only to hospitals was estimated to be 9 ppt. Risk assessment calculation made in the report predicted an upper 95% confidence limit of 55 excess lifetime cancer cases for the exposure area population exposed to 50 ppt EtO.

Sewage Emissions

Most EtO discharged to sewers enters the wastewater through the water-sealed vacuum pumps used to evacuate sterilizer chambers after sterilization is complete. EtO volatilizes from natural waters within a period of hours, while hydrolysis occurs over a period of days. A half-life of 12–14 days may be expected at normal water temperature and greater than 20 days for bio-oxidation in the presence of microorganisms. As a result, in a typical sewer system, most of the dissolved EtO would be expected to evaporate within 2 mi of its influent point.

Potential Control Technologies

A recent EPA report[13] discusses the development of EtO control technology for hospital sterilizers. The chemical reactions of EtO and dichlorodifluoromethane were reviewed to evaluate sterilizer emission control options. Catalytic oxidation and acid hydrolysis were shown to be especially suitable. The following selection criteria were developed for EtO control technology options (in order of priority): (1) cost, (2) effectiveness and environmental safety, (3) state of development, (4) complexity, (5) space requirements, and (6) safety.

Nine potential control options were examined. Six options were eliminated for the following reasons: carbon adsorption (high operating cost), thermal incineration (toxic by-product), condensation (explosion hazard), ozonation (high cost), corona discharge (toxic by-product), and ultraviolet photolysis (toxic by-product).

Catalytic Oxidation

A control system has been developed in which relatively dilute mixtures of air and EtO (12/88) are passed through a catalyst bed at 300°F–350°F. The EtO is oxidized to carbon dioxide and water; the Freon passes through unchanged. The system is characterized by relatively high flow rates (500–1000 cfm) and relatively dilute concentrations of EtO (5–500 ppm). The system is designed to treat ETO emissions in the sterilizer exhaust and those in the ventilation air from the aeration cabinets and other areas. The system has had 2 years of apparently trouble-free operation at a hospital in Philadelphia. The unit is claimed to be 99.9% efficient in controlling EtO.

Acid Hydrolysis

Another control system has been developed that consists of a countercurrent-packed column in which EtO (in 88/12) is hydrolyzed to ethylene glycol using sulfuric acid at pH 1 (the Freon is unaffected). The system is characterized by relatively high concentrations of EtO (250,000 ppm) and very low and highly variable flow rates (0.1–5 cfm). Many industrial-sized units have been installed, and test data on these units show that they are 99+% efficient. A hospital system was installed in March 1987. Another type of acid hydrolysis system has been developed in which EtO is bubbled through diffusers into aqueous sulfuric acid. A unit of this kind has been designed for hospitals and is claimed to be 99.2% efficient.

Adsorption/Reaction

Some exploratory work has been done on a proprietary process that uses a combination of adsorption and reaction. The process is in the developmental stage and is not ready for full-scale application.

References

1. American Conference of Governmental Industrial Hygienists. *Documentation of Threshold Limit Values*. ACGIH, Cincinnati, OH, 1988.
2. National Institute for Occupational Safety and Health. *Current Intelligence Bulletin No. 35—Ethylene Oxide*, May 22, 1981. National Institute for Occupational Safety and Health, U.S. Department of Health and Human Services.
3. Gun RT, Mullan RJ, Brown D, et al. *Report of the Division of Surveillance Hazard Evaluations and Field Studies (DSHEFS) Task Force on Hospital Worker Health*, December 1985. National Institute for Occupational Safety and Health, U.S. Department of Health and Human Services.
4. Health Industry Manufacturer's Association. *Ethylene Oxide*, Technical Report No. 78-3, 1978.
5. Hogstedt B, Gullberg B, Hedner K, Konig A, Mitelman F, Skerfving S, and Widegren B. Chromosome aberrations and micronuclei in bone marrow cells and peripheral blood lymphocytes in humans exposed to ethylene oxide. *Hereditas* 98:105–113, 1983.
6. Stolley PD, Soper KA, Galloway SM, et al. Sister-chromatid exchanges in association with occupational exposure to ethylene oxide. *Mutat Res* 129:89–102, 1984.
7. Hogstedt C, Aringer L, and Gustavsson A. Epidemiologic support for ethylene oxide as a cancer causing agent. *J Am Med Assoc* 255:1575–1578, 1986.
8. International Agency for Research on Cancer. Monographs on the evaluation of carcinogenic risk of chemicals to man. *Int Agency Res Cancer* 11:61–163, 1976.

9. Health Evaluation System and Information Service. *Ethylene Oxide*, Fact Sheet No. 9, Berkeley, CA, August 1987.
10. California Department of Health Services—Air Toxics Unit. *Health Effects of Ethylene Oxide*, Part B, Berkeley, CA, September 1986.
11. CAL-OSHA Special Studies Unit. *Occupational Exposures to Ethylene Oxide in Hospitals. Medical Products Industries and Spice Plants*, Berkeley, CA, April 1984.
12. American Society for Hospital Central Service Personnel. *Ethylene Oxide Use in Hospitals*, 2nd edn. American Hospital Association, Chicago, IL, 1986.
13. Environmental Protection Agency. *Ethylene Oxide Control Technology Development for Hospital Sterilizers*, EPA/600/2–88/028. National Technical Information Service, May 1988.

Suggested Reading

Association for the Advancement of Medical Instrumentation. *AAMI Recommended Practice. Good Hospital Practice: Ethylene Oxide Gas—Ventilation Recommendations and Safe Use*, EO-VKSU. AAMI, Arlington, VA, 1981.

Appendix 13.A.1

Local Exhaust System Survey Log

LOCATION ______________________ ADDRESS ______________

DEPT.________ BLDG. ____ RM # ______ HOOD #______ FAN # ______________

HOOD USE ______________________ OPERATION TIME ______________

SASHES REQUIRED ______________ FACE AREA ____ (H) × ____ (W) = ____

OTHER HOODS ON SYSTEM ______________________________

OTHER INFORMATION ______________________________

____

APPLICABLE STANDARDS:

FLOW MEASUREMENT POSITIONS:

DATE	BY	FLOW MEASUREMENTS RESULTS (fpm)													COMMENTS
		1	2	3	4	5	6	7	8	9	10	11	12	AVE	

Appendix 13.A.2

EtO Freon Leak Testing Record

Hospital:
Location:
Equipment:

Date	Sur-veyed by	Test Equip	Cylinder Valve	Flex Connec-tions	Filters Manifold	Supply Line	Pressure Relief Valve	Equip-ment Sources	Sterilizer Door(s)	Com-ments

Mark OK if Satisfactory
Report Leaks to Department Manager

Appendix 13.A.3

Sample Emergency Plan

Note: The following is a sample plan for use as a draft to generate a well-developed plan for a hospital. Resources vary from hospital to hospital. It assumes fire department response.

_____Hospital
Emergency Plan

I. The Following Events Constitute Emergency Conditions:

A. Ventilation system shutdown during gas sterilizer gassing and evacuation cycles
B. Gas sensor alarms
C. Obvious equipment damage or malfunction (i.e., pressure buildup, gas odor, leaks that can be heard, etc.)

II. Emergency Action Procedures:

A. All employees should evacuate the department and relocate at _____. Close all doors to the department.
B. The supervisor shall alert the hospital switchboard operator:
   1. Describe the condition or cause, if possible
   2. Describe the gas (i.e., 88/12 or 100% EtO)
   3. Describe and request medical assistance, if injury occurs
C. The supervisor shall ensure that all entrances to the department are locked or secured, so that persons outside the department cannot inadvertently enter the area. Close all doors to the department. Remain in area to assist fire department.
D. The hospital switchboard operator shall
   1. Notify hospital security
   2. Notify plant engineering, who will
   3. Alert the local fire department of the event

III. First Aid Procedures:

A. Exposed individuals should be moved into a fresh air environment as soon as possible. If breathing has stopped, give artificial respiration. Call "Code Blue."
B. Liquid EtO can cause a severe skin irritation or chemical burn depending upon the length of exposure. Remove all contaminated clothing and flush the contact area for at least 15 min.
C. Eye contact will require immediate flushing of the eye for at least 15 min.
D. All exposed individuals shall be seen at the hospital emergency department for evaluation.

IV. Administrative Actions:

A. The fire department will be expected to turn off the gas, if possible.
B. The equipment service vendor should be contacted for advice and repair.
C. The hospital operating room shall be notified if surgery schedules are expected to be disrupted.
D. It is prudent to reenter the department 1 h after the gas is turned off, unless the problem is readily recognizable and correctable or monitoring equipment is available.
E. _____ must be notified of the incident by phone within 24 h followed by a written report within 15 days.
F. An industrial hygiene consultation may be necessary if the problem is not readily recognizable and resolved.

V. Outside Resources:

A. Industrial hygiene consultation
B. Medical consultation, hospital occupational medicine department EtO vendor, and poison control center
C. Equipment vendor
D. Gas vendor

Appendix 13.A.4

Ethylene Oxide Exposure Measurements
(Sample Form)

Department: ____________________

Employee Name: ____________________ Social Security #: ____________________

Date Work Period	Work Description (%)	Measurement System Sample ID	Sample Time (Total Time)	Sample TWA (ppm)	8-Hour TWA (ppm)
5/13/36	OS Dept—20%	Personal	1500–1800	0.00	1.1
1500–2300	Issue—10% Decon—30%	AMSCO scan #213	(180 min)		
	Isolat—35% Transfer	Personal 3M Badge Y 4028	1830–2300 (270 min)	2.0	
5/13/86	During gas Ster cycle	Personal AMSCO Scan # 214	1700–1900	0.1	Short-term
5/13/86	During Evac Cycle	Personal 3M Badge Y 6273	1830–1900	<0.03	Short-term
5/13/86	Unloaded EtO	BZ	1900–1910	2.0 ave	Short-term
1500–2300	Sterilizer & Aerator Transfer	Infrared		20.0 high <1.0 low	
5/14/86	Changed Tanks	BZ	0700–0705	5.0 ave	Short-term
0700–1500		Infrared		10.0 high <1.0 low	

Sample 8 Hr TWA Calculation: $\frac{180(0.00) + 270(2.0) + 30(0.00)}{480} = 1.1$ ppm

Common Work Descriptions: Equip Room - Sterilizer Equipment Room; Decon - Decontamination; Isolat - Isolation Preparation; Issue - Material Issue; OS Dept - Outside Department; Tank Chge - Cylinder Tank Change; Transfer - Transferred Sterilized Materials; Tray Prep - Tray Preparation; Wrapping - Wrapping Table

Appendix 13.A.5

Work Practices to Prevent Ethylene Oxide Exposure

The following work practices and procedures are suggested for employees that work directly with gas sterilization activities:

I. **Personal Protective Equipment.** Protective clothing manufacturer's report that materials made of polyvinyl alcohol (PVA) show resistance to EtO. No good engineering studies exist that evaluate EtO protective gloves and clothing. It is prudent to wear heavy rubber gloves when handling damp materials. Disposable surgical gloves under cloth or terry-cloth gloves have been the rule in most hospitals. Heavy leather gloves or PVA gloves are recommended for tank changing. A face shield should be worn when tanks are changed. Long-sleeved shirts or gowns should be worn by employees to prevent skin exposure when unloading the sterilizer or changing tanks.

II. **Administrative Controls.** Exposure to employees can be minimized by
   A. Reducing unnecessary traffic in the sterilization and equipment areas.
   B. Ensuring that materials are placed in baskets. This practice results in minimizing exposure time when transferring materials to aerators.
   C. Locating aerators next to sterilizers to shorten the transfer route.
   D. Keeping gas sterilized materials at arm's length and as far away from the breathing zone. Concentration levels are related to distance from emission.
   E. A leak testing program for gas sterilization equipment and cylinder and supply lines can give another order of protection. Be sure to use the appropriate leak detector (i.e., a Freon detector will not detect 100% EtO).

III. **Employee Information and Training.** Employee training on EtO is required for those employees who are potentially exposed at or above the AL. Training should be performed at the time of initial assignment and at least annually thereafter. It is recommended that all hospital sterilization staff be provided training on the safe handling of EtO. Employees should be informed of the following:
   A. Available safety information and information in appendices A and B of the standard. The labeling system should be explained. An explanation of the material safety data sheet.
   B. Explanation of the operation involving the use of EtO.
   C. Location of the OSHA standard.
   D. Medical surveillance program, if instituted.
   E. Work practices instituted to protect employees.
   F. Explanation of the environmental monitoring program and ventilation system(s).
   G. Details of the hazard communication program developed by the employer.
   H. Emergency procedures.

   It is important to document health and safety training when it is performed.

IV. **Signs.** Entrances to areas where there is potential for exposure above the PEL (1.0 ppm) shall have signs bearing the following legend:

   CAUTION
   CONTAINS ETHYLENE OXIDE
   CANCER AND REPRODUCTIVE HAZARD
   AUTHORIZED PERSONNEL ONLY
   RESPIRATORS AND PROTECTIVE CLOTHING MAY
   BE REQUIRED TO BE WORN IN THIS AREA

   It is prudent to put the sign on doors to rooms that house EtO equipment and compressed cylinders.

Appendix 13.A.6

Review of Engineering Controls and Physical Design Requirements for Ethylene Oxide

The following recommendations are made on the basis of controlling EtO during normal operations and during leaks and episodes. It is recognized that all hospitals cannot meet them because of architectural and physical facility constraints. One should think of them as least preferable to most preferable, keeping in mind that the OSHA standard is a performance standard.

I. **Facility Location:**
   A. **Gas cylinders.** EtO gas cylinders should be located exterior to the building (most preferable) or in a dedicated room (next preferable) or equipment room (next preferable). Locating gas cylinders in the work area of employees provides a potential for employee exposure.
   B. **Sterilizer equipment (including aerators).** Sterilizer equipment should have its own room (most preferable) or the equipment body and gas discharge point should be within an equipment room (next preferable). Isolating EtO activities limits the number of employees that can be potentially exposed. It is preferable to have an EtO-regulated area.

II. **Equipment Modifications:**
   A. Sterilizer safety relief valves should be hard piped to an exhaust system.
   B. Vendors should be contacted to see if evacuation (purge) cycles efficiency can be improved to minimize gas residuals on sterilized materials.
   C. Sterilizer equipment manufacturers provide equipment such as liquid gas separators, ventilation capture boxes, etc., to aid in ventilation control.
   D. Tee valves and a purge line that discharges to the exhaust system should be installed on the gas supply line system. The tee valve protects employees when changing tanks. The purge system protects employees when working on supply line components such as the gas filters.

III. **Ventilation:**
   A. **Room ventilation.** Room ventilation should be part of EtO local exhaust control system (preferable). It is not advisable to connect any other rooms or operations on the same system as the EtO removal system. All rooms that house EtO equipment should be negative pressure with respect to adjacent areas. The room should have at least 10 air changes per hour.
   B. **Exhaust removal ducts/fan locations.** The most preferable exhaust duct will be dedicated to the exterior of the building. It should exit 50 ft from any openable windows or air supply intake. The fan should be at the end of the system and preferably on emergency power. The next preferable system would have a dedicated exhaust to a large building exhaust shaft. The least preferable system would use the building exhaust system. Using building exhaust systems generally results in concern with dampers, potential cross-contamination problems, and competition with other activities for ventilation needs.
   C. **Location of local exhaust hoods.** The purpose of local exhaust hoods is to control EtO at the source of emission. These sources are the door (major source of routine employee exposure), the gas cylinder area, transfer operation to aeration cabinet, aeration cabinet discharge, and the sterilizer chamber discharge point (usually the floor drain in the equipment room).

It should be kept in mind that a short-term exposure standard has been added to the current EtO OSHA standard. One should also not forget that the above emission points are precisely where OSHA would require placement of early warning sensors for leak detection.

1. The use of booster fans to achieve control velocities and volumetric flow rates is less preferable than a sized exhaust system that is always under negative pressure. In this situation, it is necessary to have two fans. The failure of the building removal system affects the function of the positive pressure system.
2. EtO ventilation exhaust systems should have flow-monitoring devices installed in the system that signal at the location of the sterilizer equipment operation panel.
3. Exhaust hoods should be installed over gas cylinders stored inside buildings (preferable). The reason for this is that this is a common area for gas leaks. The area should have an early warning gas sensor to alert employees. The hood should cover the entire manifold and cylinder valve apparatus. Flip-up Plexiglas doors are useful to see controls. As a general rule, each cylinder requires 100 cfm exhaust. Gas cylinder cabinets are the next preferable choice. The number of air changes in the cabinet is limited because of the size of the makeup air grill.

4. All aerators should be connected to the dedicated exhaust system. Usually it requires 50–100 cfm. Provide a damper to adjust flow rates. Some aerator temperature gradients are affected when connected to excessive exhaust. Too little exhaust can result in excessively pressurized cabinets, which can result in leaks.
5. The discharge point of the sterilizer is usually the floor drain. Sterilizer equipment companies sell liquid gas separators, in-line scavenging devices, etc. It is important to follow their installation instructions when using such devices. For example, floor drains may have to be completely sealed, a minimum building exhaust removal system may be required, etc. It should be understood that air concentrations measured as below limits of detection are achievable during the evacuation cycles. This should be the standard of control, particularly in a freestanding or pass-through unit or an equipment room that houses other sterilization equipment. If a sterilizer receives the air for the evacuation cycle from an uncontrolled equipment room, the gas can be introduced back into the sterilizer.
6. The door exposure pattern varies from sterilizer to sterilizer. A sterilizer that uses cart unloading and transfer should have an anteroom in front of the door (most preferable). The anteroom should be exhausted through ceiling registers into the dedicated system. This situation provides a regulated area, encloses the door (most containment), and provides an area to room-aerate. It should be understood that even after a 15 min waiting period, when the cart is pulled out, a bolus of EtO develops, which results in employee exposure. The next preferable system would be a hood over the door. It should be located as close to the swing of the door as possible. Use Plexiglas access panel for access to controls. Install side panels to each side of the door to maximize enclosure. Slot exhaust systems are least preferable because they provide minimum enclosure and can have their control air movement affected by drafts of external origin. This type of hood generally moves the least volume of air per unit of time. A hood that provides a volumetric flow rate of 300 cfm or higher (depending on size of door) is necessary.

14

Monitoring Aldehydes

Monitoring Aldehydes in the Hospital **14**-1
Introduction and Background • Characteristics of Any Gas Monitor • Aldehyde Analytical Problem in the Hospital
References **14**-11
Appendix 14.A.1 .... **14**-12
Appendix 14.A.2 .... **14**-12
Appendix 14.A.3 .... **14**-13
Control of Formaldehyde Exposure in Hospitals **14**-14
Introduction • Variables of Exposure and Population Exposed • Monitoring for Formaldehyde • Methods of Compliance
References **14**-20
Appendix 14.A.4 .... **14**-21
Hidden Toxicities of Glutaraldehyde **14**-24
Introduction • Population Exposed • Glutaraldehyde Investigations • Case Studies • Methods of Compliance
References **14**-28
Appendix 14.A.5 .... **14**-29
Appendix 14.A.6 .... **14**-29
Appendix 14.A.7 .... **14**-30

Edward W. Finucane
High Technology Enterprises

William Charney
Healthcare Safety Consulting

Monitoring Aldehydes in the Hospital

Edward W. Finucane

Introduction and Background

Among the more irritating and difficult to quantify volatile organic compounds that are routinely found in the hospital are the two aldehydes, formaldehyde, and glutaraldehyde. Table 14.1 shows the currently established and acceptable Occupational Safety and Health Administration (OSHA) permissible exposure limits (PELs), American Conference of Governmental Industrial Hygienists (ACGIH) threshold limit values (TLVs), National Institute for Occupational Safety and Health (NIOSH) recommended exposure limits (RELs), and Deutsche Forschungsgemeinschaft's (DFG, German Research Foundation) maximum allowable concentrations (MAKs) for each of these two organic compounds.

TABLE 14.1 Occupational Exposure Values

Regulatory/Agency's Occupational Exposure Value	For Formaldehyde	For Glutaraldehyde
U.S. Dept. of Labor, OSHA, 8 h TWA permissible exposure limit (PEL-TWA)	1.0 ppm	None established
U.S. Dept. of Labor, OSHA, 15 min TWA permissible exposure limit (PEL-STEL)	2.0 ppm	None established
U.S. Dept. of Labor, OSHA, ceiling value permissible exposure level (PEL-C)	None established	0.20 ppm
ACGIH 8 h TWA Threshold Limit Value [TLV-TWA]	1.0 ppm	None established
ACGIH 15 min TWA threshold limit value (TLV-STEL)	2.0 ppm	None established
ACGIH ceiling value threshold limit value (TLV-C)	None established	0.20 ppm
NIOSH 8 -h TWA recommended exposure limit (REL-TWA)	0.016 ppm	None established
NIOSH 15 min TWA recommended exposure limit (REL-STEL)	None established	None established
NIOSH ceiling value recommended exposure limit (REL-C)	0.10 ppm	0.20 ppm
German D.F.G.* 8 h TWA maximum concentration value (MAK-TWA)	0.50 ppm	0.20 ppm
German D.F.G.* 15 min TWA maximum concentration value (MAK-STEL)	None established	None established
German D.F.G.* peak value maximum concentration value (MAK-peak)	1.0 ppm	0.40 ppm

Source: *Guide to Occupational Exposure Values—1991*, compiled by the American Conference of Governmental Industrial Hygienists, 1991.

* D.F.G., Deutsche Forschungsgemeinschaft.

It is understood, at the present time, that an overall reevaluation of new information and data on the toxicity and irritability of formaldehyde is underway at both NIOSH and the U.S. Department of Labor's OSHA. It appears very likely that the outcome of this effort will be a reduction of the PEL-TWA from its current 1.0 ppm level to a new value of 0.75 ppm. This reevaluation will almost certainly also affect the PEL-STEL.

Characteristics of Any Gas Monitor

In undertaking the monitoring of any material, one must consider and weigh several different, and usually competing, factors that relate to the analytical problem. Of greatest importance among the factors that must be examined is the following group of seven:

- *Sensitivity* to the material being analyzed—this is usually expressed in terms of the analytical method's minimum detection limit, or MDL. Typically, one would like to see the analytical method's MDL be less than 10% of the "significant concentration" for the material to be monitored; the significant concentration is most commonly understood to be an appropriately selected occupational exposure value for that material—usually its PEL-TLV.
- *Selectivity* for the material being analyzed—this is a measure of the analytical method's ability to distinguish the material to be monitored from *anything else* that might also be in the ambient matrix that is to be monitored (i.e., some other volatile vapor that might potentially interfere with the measurement). Interferences can be of two types—namely, positive or negative. Positive interferences will tend to cause the analytical reading to be greater than the true concentration value. Negative interferences, on the other hand, will cause a diminishment or "quenching" of the analytical reading to values less than the true concentration.
- *Reproducibility* of the analytical approach—this is a measure of the method's ability to provide consistent readings for the same concentration of the material to be measured: either at different times or under different ambient conditions, or by different analysts. It is also a measure of the method's ability to remain in calibration over extended periods of time.
- *Purchase costs* of the analytical instrumentation—this is an obvious factor.
- *Operating costs* for the analytical method—this, too, is an obvious factor.
- *Timeliness* of the analytical method—this is a measure of the time interval required for the analytical method to provide its concentration level answer for the material being analyzed. This time interval

can be very broad, ranging from instantaneous (or real-time) to very extended periods—as when, for example, a passive dosimeter must be sent to some distant external laboratory for evaluation.

- *Continuity* of the analytical method—this is a measure of whether the analytical method is capable of providing continuous concentration readouts for the material being monitored; or, alternatively, can only give periodic discrete measures of that concentration.

Aldehyde Analytical Problem in the Hospital

Identifying exactly how to test for either formaldehyde or glutaraldehyde in the various hospital locations where these two materials are routinely found and used is the most difficult problem. What follows are separate discussions of each of these two problems, as well as specific case studies. To assist the reader in selecting an analytical method, a matrix based upon the relationship of the previously listed seven factors (sensitivity, selectivity, etc.) to the possible analytical methods will be developed for each of the two aldehydes of interest. In these tabular listings (one each in the Formaldehyde (Table 14.2) and the Glutaraldehyde (Table 14.3) sections of this chapter), all the currently recognized analytical methods that may be used for the determination of the ambient concentrations of either of these two aldehydes will be examined. Each method will be rated in the seven previously listed categories, with ratings ranging from 1 to 5. These ratings will represent the following qualitative judgments:

Rating #	Sensitivity [I] Selectivity [II] Reproductivity [III]	Purchase Cost [IV] Oper. Cost [V]	Timeliness [VI]	Continuity [VII]
5	Excellent	Very high cost	Real-time	Continuous
4	Good	High cost	<2 min delay	—
3	Fair	Average cost	<10 min delay	Discrete pts.
2	Poor	Low cost	>10 min delay	—
1	Unacceptable	Very low cost	>1 day delay	Average dose

The mnemonics that will be used to designate the analytical methods that are to be evaluated for their capabilities in quantifying concentrations of these two aldehydes are listed below:

AIR Infrared spectrophotometric absorbance-based analyzers
PIR Photoacoustic infrared spectrophotometric-based analyzers
NDIR Nondispersive infrared-based analyzers

TABLE 14.2

Analytical Method	I Sensitivity	II Selectivity	III Reproducibility	IV Purchase Costs	V Operating Costs	VI Timeliness	VII Continuity
AIR	3	2	4	4	1	5	5
PIR	4	5	5	5	1	4	3
NDIR	1	2	4	3	1	5	5
EC-A	1	1	2	2	3	5	5
EC-P	1	1	2	1	2	4	5
FID	3	1	3	4	3	5	5
PID	4	1	3	4	3	5	5
OID	2	1	3	4	3	5	5
WC	5	5	5	3	3	3	5
GC	5	5	3	4	4	3	3
PD	5	3	4	4	—	1	1
AD	4	4	3	4	5	1	1

TABLE 14.3

Analytical Method	I Sensitivity	II Selectivity	III Reproducibility	IV Purchase Costs	V Operating Costs	VI Timeliness	VII Continuity
AIR	1	2	2	4	1	5	5
PIR	1?	4?	2?	5	1	4	3
NDIR	1?	1	?	3	1	5	5
EC-A	1?	1	?	2	3	5	5
EC-P	1?	1	?	1	2	4	5
FID	2?	1	?	4	3	5	5
PID	3?	1	?	4	3	5	5
OID	1?	1	?	4	3	5	5
WC	5	4	5	3	3	3	5
GC	3?	5	?	4	4	3	3
PD	5	3	4	4	—	1	1
AD	4	4	3	4	5	1	1

EC-A Pumped electrochemical-based analyzers
EC-P Passive electrochemical-based analyzers
FID Flame ionization detector-based survey analyzers
PID Photoionization detector-based survey analyzers
OID Other ionizing detector-based survey analyzers
WC Wet chemistry calorimetric-based analyzers
GC Gas chromatography-based analyzers
PD Passive dosimetric-based systems
AD Active dosimetric-based systems

Formaldehyde

Formaldehyde is most frequently encountered in the hospital as a water solution (Formalin) that has been stabilized with up to 15%, by weight, of methanol. Formalin, and consequently both formaldehyde and methanol vapors, can be encountered in the Pathology Department, in the Morgue, and in locations where activities such as kidney dialysis bundle regeneration occur. The vapors of formaldehyde are both pungent and intensely irritating, particularly to the mucous membranes. Monitoring for formaldehyde vapors in the hospital is a very difficult task. There are several different types of analytical methods that have been, and are now being used in this area; however, as can be seen from Table 14.2, none of them is completely attractive.

It will be instructive to discuss next each of the foregoing analytical methods, in the order listed in Table 14.2.

AIR spectrophotometric analyzers suffer from severe sensitivity and selectivity problems. The typical formaldehyde MDL for this category of analyzer falls in the range of 0.4–0.5 ppm, which, though less than both the recent past 1.0 ppm, and the new 0.75 ppm PEL-TWA, can be considered, at best, borderline acceptable. In addition, this method's lack of selectivity, particularly because of the positive interferences that are caused by methanol vapors, also comes close to disqualifying it for this application.

PIR spectrophotometric analyzers have a slightly lower MDL (in the range of 0.2–0.3 ppm for formaldehyde) and are therefore slightly better suited for this application. It must be noted, however, that these units' MDL is still higher than the current target level of 0.1 ppm, and if the predicted downward shift of the formaldehyde PEL-TLV is eventually instituted, then this class of analyzer, too, will become similarly disqualified. The principal limitation of this method is the fact that it is not continuous; it is capable of providing only a set of discrete formaldehyde concentration values, each one of which could be separated from its nearest neighbor, in time units, by up to 90 s.

Nondispersive infrared analyzers (NDIR) are basically disqualified from being used to measure formaldehyde vapors because of their even poorer and, therefore, totally unacceptable MDL, which is in the 3–5 ppm range. This class of analyzers also suffers from a severe lack of selectivity.

Electrochemical analyzers, either pumped or passive (ED-A & EC-P), are even less sensitive than any of the previously listed units (MDLs in the range of 5–10 ppm), and this factor alone disqualifies them from making useful formaldehyde measurements. In addition, they also suffer from unacceptably poor levels of selectivity.

All of the flame, photo, and other ionizing detector-based analyzers (FID, PID, and OID) suffer from an absolute lack of selectivity (they all function as "universal detectors"); none can distinguish among the possible different vapors that would almost certainly be present in any environment that might contain formaldehyde; and they are all, therefore, not applicable to making measurements for these aldehydes.

Unlike any of the previously listed analytical methods, the wet chemistry (WC) method is both more than adequately sensitive (its MDL for formaldehyde is in the range of 0.006–0.01 ppm), and completely selective in *any* ambient matrix. The preferred approach here makes use of the unique and characteristic color change of an acid-bleached pararosaniline solution, brought about by its exposure to ambient formaldehyde vapors. The color change of the pararosaniline solution can be readily measured colorimetrically at 550 nm, and this color change will be proportional to the ambient formaldehyde concentration that caused it. The most important limitation of this method is the fact that the chemical reaction kinetics involved requires 7–9 min to proceed to an adequately complete color change endpoint to be able to be quantified at the low concentration levels of formaldehyde that must be monitored and documented. Thus, although this method does give continuous results, it provides those results approximately 8 min after the fact. There are no known interferences for this method. Setting up and operating a wet chemistry of the type described here is both a difficult and a time-consuming process; it does, however, provide excellent and incontrovertible results.[2]

The gas chromatographic (GC) method also shows great promise—both from the perspective of its sensitivity and its selectivity—in monitoring ambient formaldehyde vapors. The MDL for a gas chromatograph that has been properly set up to monitor formaldehyde should be approximately 0.05 ppm, or possibly even better. Its principal limitation is in the area of its continuity and its timeliness. Each individual gas chromatographic analysis would likely require a minimum of 3 min; this method would thus only be able to provide a set of discrete ambient formaldehyde concentrations, each separated in time from the next by the duration of the analytical period.

The final methodology to be considered is the dosimetric approach—both passive and active (PD & AD). Both of these methods provide very adequate sensitivity and selectivity, and this will remain kite even if the current PEL-TLV for formaldehyde were to decrease in the future to the 0.75 ppm level. There are numerous commercially available passive dosimetry badges and/or tags on the market, and virtually all will perform adequately in quantifying formaldehyde dose. Active dosimetry can also be performed using personal sampling pumps and appropriately selected sorbent tubes. Such methods develop very adequate data by analyzing the material desorbed from these tubes, using either reverse phase high pressure liquid chromatography or standard gas chromatography. Active dosimetry suffers from the same inadequacy as does the passive approach: neither is capable of providing continuous ambient concentration data. Each can provide only *average* concentration or dose data over some specified exposure period.

On balance, the combination of a wet chemistry approach—to obtain the continuous ambient formaldehyde concentration values—and some appropriate dosimetry method—to obtain individual dose data—probably provides the best possible combination of capabilities for this application.

Formaldehyde Monitoring Case Study

A 3 h continuous evaluation of the ambient levels of formaldehyde in the Pathology Department of a major hospital was undertaken in mid-1989. The primary purpose of this effort was to determine the adequacy of the local exhaust hood that was in operation at the Grossing Pathologist's work station. Passive dosimetry evaluations with this individual had earlier provided ambient formaldehyde dose values that ranged from "trace levels" (<0.005 ppm) to 0.60 ppm for periods that ranged from 3 to 4 h. It was decided that a profile of the ambient levels of formaldehyde would be useful, and the study that will be described in this section was, therefore, undertaken.

Two different analytical methods were employed simultaneously. The first used a pararosaniline-based Wet Chemistry Analyzer that had been set up in the manner described above. The second employed an Infrared Spectrophotometer set up to monitor formaldehyde. Both analyzers had been calibration-checked on the previous day with pure formaldehyde standards that had been developed from a stable permeation device source. These calibration checks had, in both cases, been very successful—both analyzers had responded with a precision within ±5% of the analytical reading, or better, when compared to the challenge calibration standards. The Wet Chemistry Analyzer was calibration-checked in the range 0–1.0 ppm (±0.01 ppm), whereas the Infrared Analyzer was checked in the range 0–10 ppm (±0.1 ppm).

At the Grossing Pathologist's station, each analyzer was set up to sample the air at a point 3 in. above the work bench, and at points immediately to the left (the Wet Chemistry unit), and to the right (the Infrared unit) of the position where the pathologist worked. This location was approximately 18 in. below and in front of the pathologist's breathing zone (BZ). The slot exhaust hood at the back of the work bench was also at a height of 3 in. above this surface, and was located approximately 12–14 in. back from the position where the grossing pathology and, therefore, the sources of formaldehyde vapor were located. It is unlikely that there was much, if any, difference in the ambient concentration levels of formaldehyde that each of these analyzers would be measuring during this evaluation period. The actual sample points were 17 in. apart.

The results of these two analyzer evaluations of the formaldehyde levels are shown graphically in Figure 14.1 (for the Wet Chemistry unit) and Figure 14.2 (for the Infrared unit). Note that the two graphs show formaldehyde concentrations, on the vertical axis, plotted against time, on the horizontal axis. Note also that the two plots have been adjusted to the same time basis—that is to say, the graph of the Wet Chemistry unit's results has been shifted to the left by approximately 7.5 min to compensate for the relatively slow reaction kinetics of this method, while the graph of the Infrared unit's results has been presented exactly as it was determined. Finally, note that the full-scale concentrations represented on

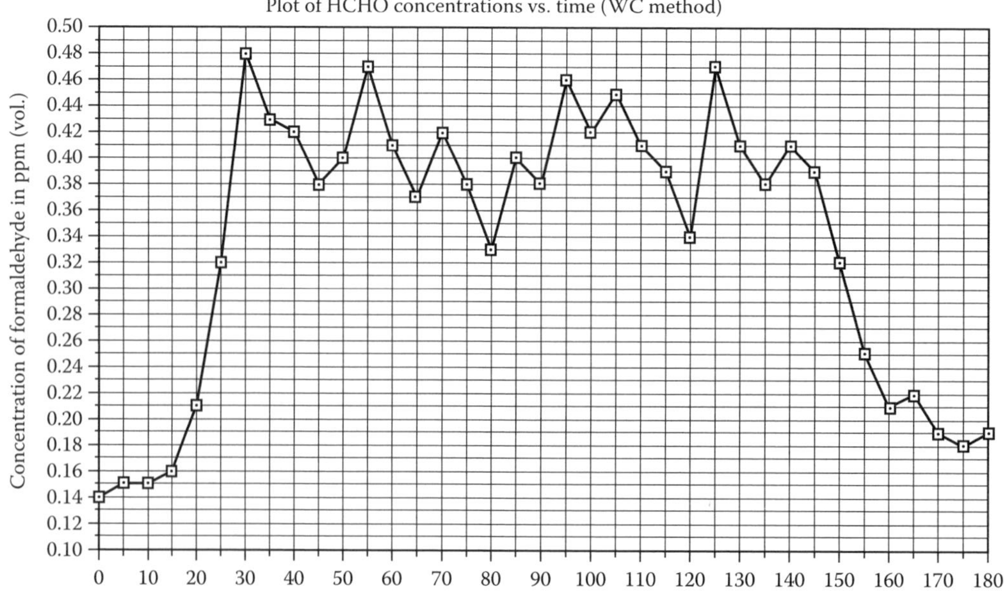

FIGURE 14.1 Time from the start of monitoring, in minutes.

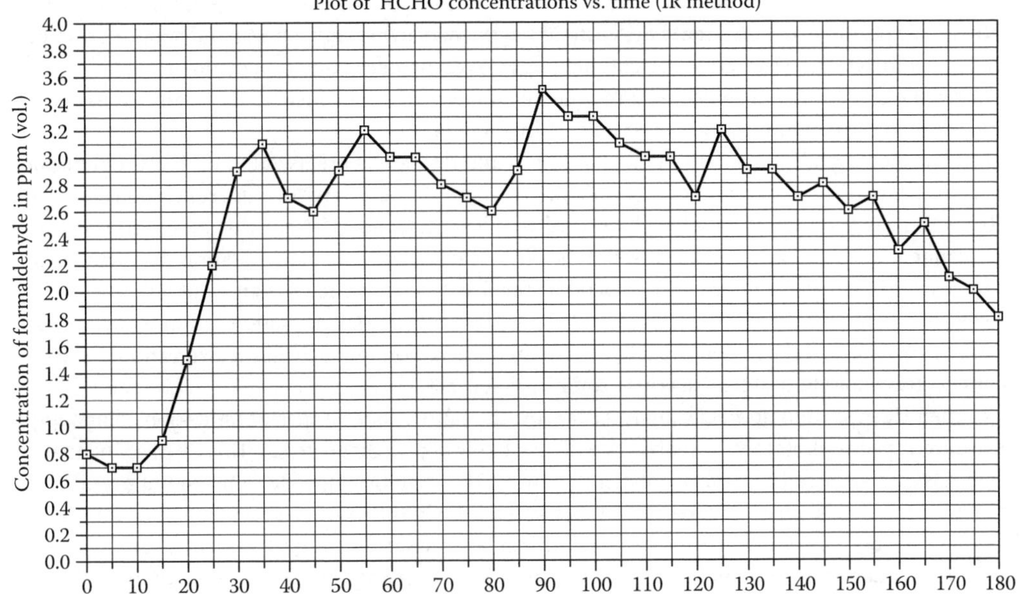

FIGURE 14.2 Time from the start of monitoring, in minutes.

these two plots are different; for the Wet Chemistry unit, full scale was set at 0.5 ppm, while the Infrared Spectrophotometric unit, full-scale value was 5 ppm. The Wet Chemistry unit readings, as represented by Figure 14.1, must be regarded as the more accurate and representative ones, since this method—unlike its Infrared counterpart, as shown in Figure 14.2—has no known interferences.

Monitoring in this situation was initiated as soon as the grossing pathologist arrived and set up for work. The actual pathology characterizations started about 30 min into the overall monitoring process, and were completed at approximately the 138 min point. Monitoring continued until the 180 min point, while the pathologist cleaned up and checked recorded comments.

It can be seen from these two plots that there was a significant variation in the results reported by these two different methods. There was a very good general formaldehyde concentration trend correlation, as represented by the similarity of these two plots. The problem was the dramatic difference in the magnitude of the concentration results reported by the two different methods. On the average, the Infrared unit gave answers that were almost one order of magnitude greater than those provided by the Wet Chemistry unit—actually an average of ~7.5 times larger. The most likely explanation for this discrepancy would be the impact of the interference of methanol vapors on the Infrared unit's indicated concentration.

Glutaraldehyde

Glutaraldehyde is commonly encountered in the hospital in any location where a cold sterilization or disinfection process would be performed. The locations of such processes in any hospital will depend largely on how that specific hospital's infection control policies and procedures are carried out. Both because of the unusually low OSHA established Ceiling PEL level for glutaraldehyde (0.2 ppm), and the potential adverse effects that individuals who are exposed to this material can experience, it is extremely important to be able to monitor employee exposure levels.

Monitoring of the ambient levels of glutaraldehyde vapors is presently possible with only one or two different methods, as shown in Table 14.3. This tabulation will use the same analytical method mnemonics and the same rating system (for the same seven categories) that was used in Table 14.2. Note that several of the numerical ratings in Table 14.3 are followed by "?s" and several other spaces are listed simply as "?s"; the former was done in cases for which it was important to make an estimate of what value a rating in a specific situation should be, without having had any specific, first-hand supportive experience upon which to rely; the latter was chosen when no reasonable estimation could be made.

As was the case for the formaldehyde-monitoring section, it will be instructive to discuss each of the foregoing analytical methods, in the order listed in Table 14.3.

Infrared spectrophotometric analyzers have been and frequently are used to monitor for glutaraldehyde. An infrared spectrophotometer is capable of providing relatively specific, continuous, real-time ambient concentration level data for glutaraldehyde; however, the MDL for this material in this type of instrument is in the range of 0.8–1.0 ppm. For this reason, the only concentration data that such an analyzer would be capable of providing would be for situations in which the ambient levels of glutaraldehyde were *far* above the existing PEL-C of 0.2 ppm. That these analyzers are used in this application cannot be denied; that they offer virtually no useful information for such a task also cannot be denied. Their selectivity for glutaraldehyde is also somewhat suspect.

For the PIR spectrophotometric analyzer family, the same general types of limitations that apply to the infrared spectrophotometric analyzers will also likely apply. That is to say, this category of analyzer will almost certainly fail to have a sufficiently low MDL for glutaraldehyde to make it a useful choice for monitoring this material—its sensitivity would likely be in the 0.5–0.6 ppm range. It is unlikely that there has *ever* been a successful calibration of this type of analyzer for glutaraldehyde at any concentration level; thus all the comments and conclusions listed here are purely speculative. In the event a photoacoustic infrared analyzer were to be successfully calibrated for glutaraldehyde, it would doubtless be able to provide a very acceptable selectivity for this material.

NDIR glutaraldehyde analyzers do not exist at this time; thus no specific comments can be listed for them either. It is unlikely that this type of analyzer could ever be made sufficiently glutaraldehyde-sensitive to justify manufacturing one for this application.

Electrochemical analyzers for glutaraldehyde, whether pumped or passive (EC-A and EC-P), also do not appear to exist at this time. It is quite possible that some sort of a relatively specific electrochemical analyzer could be produced to monitor glutaraldehyde levels; however, such a unit would almost certainly lack adequate sensitivity to be very useful in this application.

As was true in the case of monitoring formaldehyde levels, ionizing detector-based analyzers (FID, PID, and OID), would also function, in this case, as "universal detectors." Thus, such an analyzer would lack the selectivity required to function as a useful glutaraldehyde monitor. It is possible that analyzers in this category might be made sufficiently sensitive—particularly, the photoionization units and, to a lesser extent, their flame ionization counterparts—to have some potential for monitoring glutaraldehyde in an ambient matrix where there were *no other* volatile organic compounds present; however, such an interference-free situation is unlikely ever to occur in the hospital.

The most effective method for determining ambient concentration levels of glutaraldehyde employs a WC analyzer. The analytical process employed by a glutaraldehyde wet chemistry analyzer differs from that described for analyzing formaldehyde. A glutaraldehyde analyzer makes use of the chemical reaction between *any* aldehyde and MBTH, or 3-methyl-2-benzothiazolone hydrazone hydrochloride. The reaction between the aldehyde vapor and MBTH produces an azine that can be easily oxidized to form a distinctive blue cationic die. The intensity of the blue color formed–measured calorimetrically at a wavelength of 550 nm—will be proportional to the original ambient glutaraldehyde concentration. Both the sensitivity (MDL = 0.004 ppm), and the selectivity (the only interferences are from *other* aldehydes, none of which is present in *any* of the commercially available cold sterilizing solutions) of this method make it more than adequate for this task. In fact, its only limitation is that it is not real-time. As with its formaldehyde counterpart, the reaction kinetics of this wet chemistry-based procedure require the passage of 6–8 min before an adequate color change endpoint has been achieved. Setting up and operating a wet chemistry analyzer for glutaraldehyde is a difficult and time-consuming process; once it has been set up, however, such a system would provide excellent in controversial results.[3,4]

Sufficiently selective and sensitive analyses of ambient glutaraldehyde concentration levels using GC analyzers are theoretically very possible; however, difficulties in calibrating such analyzers have limited their use in this application. In addition, gas chromatograph-based ambient concentration data for glutaraldehyde would be neither real time nor continuous.

PD for glutaraldehyde are widely available from several different manufacturers, and virtually all such units can, and do, provide useful and specific dosimetry data. The principal problem with the passive dosimetry approach lies in the definition of the OSHA Ceiling PEL. A Ceiling PEL is defined as "a concentration level that should not be exceeded during any part of the working exposure." As such, a Ceiling PEL should be regarded as a parameter that *must* be evaluated on a continuous and/or instantaneous basis. There is a provision in the OSHA definition, however, that stipulates that "if instantaneous monitoring is not feasible, then this factor can be evaluated as a 15 min time-weighted average (TWA), EXCEPT when the substance vapors can cause *immediate irritation* with *exceedingly short exposures.*"[5]

Because it is virtually impossible to evaluate ambient glutaraldehyde concentrations on a continuous or instantaneous basis (see Table 14.3), the exception to the definition would appear to apply. On the other hand, this is a vapor that can produce *immediate irritation* with *exceedingly short exposures*; thus the exception to the OSHA Ceiling PEL definition should *not* apply! The reality of the situation, however, dictates that we overlook the latter factor (immediate irritation), and perform personal dosimetry monitoring. It must be noted that such monitoring *must* be performed over 15 min periods, and that the specific monitoring periods be selected so that they constitute "*the worst possible 15 min periods*"—i.e., those periods when the cold sterilizing solutions are either being poured into their sterilizing pans or being poured out of these pans down the drain in the disposal process, or when the sterilized items are being "harvested" from these pans.

AD can also be performed using personal sampling pumps and approximately selected sorbent tubes. Such methods develop very adequate data by desorbing the tubes, and then analyzing the desorbed material using reverse phase high pressure liquid chromatography and/or standard gas chromatography. Active dosimetry suffers from the same inadequacy as does its passive badge counterpart, namely, it cannot provide instantaneous ambient concentration data; it is, therefore, correspondingly at odds with the OSHA Ceiling PEL definition. In addition, active dosimetry procedures will usually require sampling periods greater than 15 min in order to be able to absorb sufficient quantities of glutaraldehyde from the air to provide for high-confidence analyses.

On balance, for glutaraldehyde, as was the case for formaldehyde, the combination of a wet chemistry approach—to obtain the *continuous* ambient glutaraldehyde concentration values—and some appropriate dosimetry method — to obtain *individual dose* data—will probably provide the best possible combination of capabilities for this application.

Glutaraldehyde Monitoring Case Study

For a moderately extended period in mid-1990, a Sterile Processing Department employee at a large metropolitan hospital had been experiencing significant throat and eye irritation problems. Although the magnitude of these problems varied between moderate and severe, the general trend had been negative. It seemed that these problems were caused by exposure to one of the commonly used proprietary cold sterilization products required to be used several times each shift as part of the regular job. This employee usually had to complete at least four—and sometimes as many as six—separate cold sterilization runs each day. Each "sterilization run" required a 45–60 min process, which consisted of loading the sterilization pan with the implements to be sterilized, adding the cold sterilizing solution, waiting for an adequate soak, harvesting and rinsing the sterile implements, and finally, pouring out the "spent" sterilizing solution. All of these procedures were carried out in what appeared to be a reasonably well ventilated room. The Cold Sterilization Work Station itself did not have any special or unique exhaust ventilation arrangements.

In response to these difficulties, this hospital's Engineering Department had chosen to enclose the Cold Sterilization Work Station completely. This work station, which consisted of both the cold sterilization pan and the sink that was used both for rinsing the sterile implement, and disposing of "spent" sterilizing solution, was, in effect, transferred into a well-designed and carefully sealed exhaust hood. The hospital wanted to obtain an objective answer regarding the efficacy of their improvements. Although the employee who had made the original complaints appeared to be experiencing less and less difficulty, the department manager wanted a still more objective answer. The Engineering Department wanted both:

1. An assessment of the ambient glutaraldehyde levels immediately in front of the hood access doors of their "new" Cold Sterilization Work Station, during the period when a batch of implements was being cold sterilized
2. An additional second assessment of the glutaraldehyde levels above the sterilization pan and the sink, inside the hood, also during a cold sterilization process

The two 1 h duration sets of data that are shown in Figure 14.3 (the ambient glutaraldehyde levels *outside* of the new hood during a procedure) and Figure 14.4 (the levels *inside* this same hood during a subsequent procedure) clearly document this monitoring effort.

The results of the two evaluations, expressed as 1 h time-weighted averages of the two different glutaraldehyde concentration levels, one outside and the other inside the hood were as follows:

From Figure 14.3: $C_{\text{outside}} = 0.0083\,\text{ppm} = 8.3\,\text{ppb}$
TWA for a 1.0 h period
From Figure 14.4: $C_{\text{inside}} = 0.5142\,\text{ppm} = 514.2\,\text{ppb}$
TWA for a 1.0 h period

To the extent that these calculated TWAs for the two monitoring runs were truly representative—and it was believed that they were—then the "objective and quantitative" improvement factor provided by the new installation (i.e., the engineering modifications) would simply be the ratio of these two TWAs, with the "after" being divided by the "before," thus:

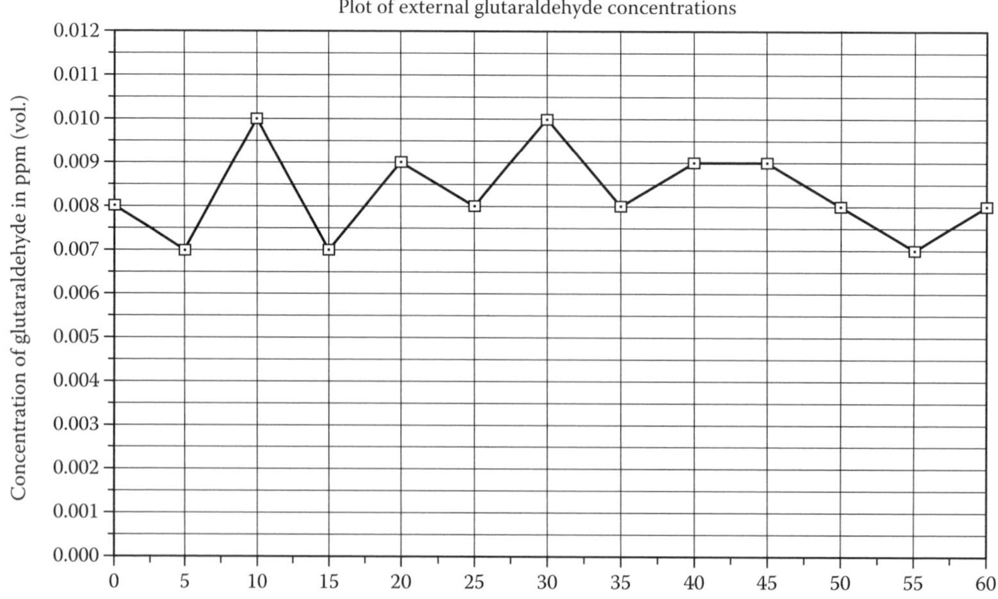

FIGURE 14.3 Time from the start of monitoring, in minutes.

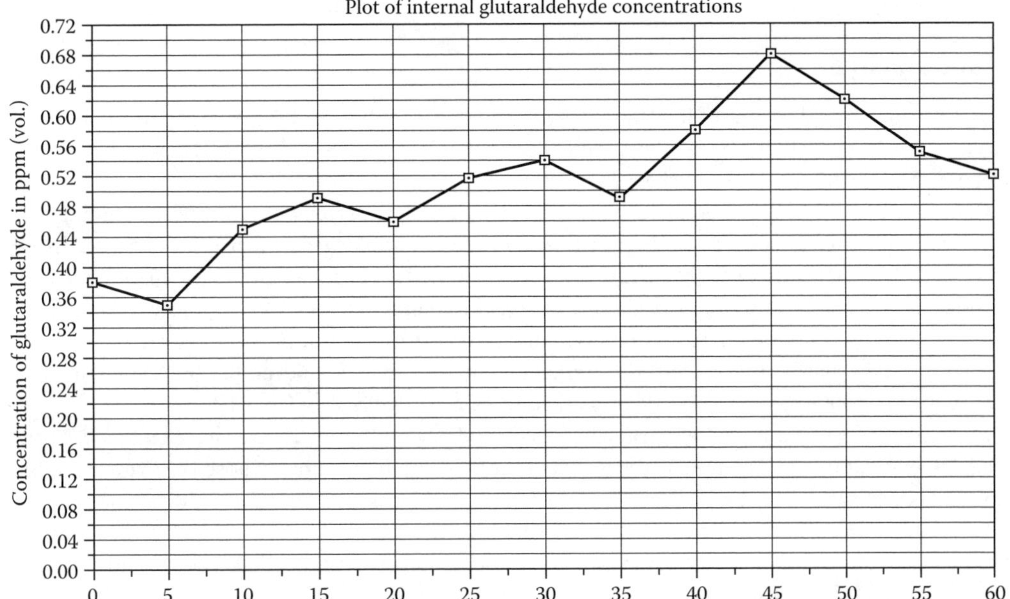

FIGURE 14.4 Time from the start of monitoring, in minutes.

$$\text{Improvement factor} = \frac{C_{\text{outside}}}{C_{\text{inside}}} = \frac{0.0083}{0.5142} = 0.0162$$

The inverse of this factor represents the decrease in experienced concentrations of glutaraldehyde, thus:

$$\text{Decrease} = \frac{1}{0.0162} = 61.70$$

It can be seen, therefore, that the engineering modifications resulted in a 61.70-fold decrease in the levels of glutaraldehyde vapors to which the employee would be exposed in the future. In essence, the external glutaraldehyde concentrations—to which the employee had been exposed prior to the modifications, and would, in the future, continue to be exposed—was reduced from an unacceptable level that was approximately 257% of the existing PEL-C, to a new, lower level that was only 4.2% of this value. The engineering modifications were, therefore, judged to have been very successful.

References

1. *Guide to Occupational Exposure Values—1991*, compiled by the American Conference of Governmental Industrial Hygienists, 1991.
2. Lyles, Dowling, and Blanchard. A quantitative method for monitoring formaldehyde. *J Air Pollut Cont Assoc* 15:106, 1965.
3. Sawicki E, et al. *J Analyt Chem* 33:93, 1961.
4. Lodge JP Jr. *Methods of Air Sampling and Analysis*. Lewis Publishers, Boca Raton, FL, 1989, pp. 279–284.
5. OSHA: *Final Rule Air Contaminants—Permissible Exposure Limits, Title 29 Code of Federal Regulations Part 1910.1000*, Federal Register, 1989.

Appendix 14.A.1

Glutaraldehyde

OSHA previously had no limit for glutaraldehyde and proposed establishing a ceiling limit of 0.2 ppm, based on the ACGIH (1986/Ex. 1–3) recommendation. NIOSH (Ex. 8–47, Table N1) concurred with this proposal, and the final rule established this limit.

Glutaraldehyde is an aliphatic dialdehyde that forms colorless crystals. It is strongly irritating to the nose, eyes, and skin (*Human Sensory Irritation Threshold of Glutaraldehyde Vapor* 1976, as cited in ACGIH 1986/Ex. 1–3, p. 285) and can cause allergic contact dermatitis from occasional or incidental occupational exposure (Jordan, Dahl, and Albert 1972/Ex. 1–1056). The rat oral LD_{50} has been variously reported as 250, 820, and 2380 mg/kg (Stonehill, Krop, and Borick 1963/Ex. 1–1066; Smyth 1963 and NIOSH 1975f. both as cited in ACGIH 1986/Ex. 1–3, p. 285). The dermal LD_{50} in the rabbit is 2560 mg/kg, and the 4 h inhalation LD_{50} in the rat is 5000 ppm (NIOSH 1975f, as cited in ACGIH 1986/Ex. 1–3, p. 285).

Mice exposed to alkalinized glutaraldehyde at 8 and 33 ppm for 24 h have shown marked nervous behavior with panting and compulsive washing of the face and limbs; those exposed to 33 ppm exhibited signs of toxic hepatitis at autopsy (Varpela, Otterstrom, and Hackman 1971/Ex. 1–1072).

In a study of a cold-sterilizing operation in which the operator was exposed for 12 min to an activated 2% aqueous solution, a measurement of 0.38 ppm glutaraldehyde was taken in the operator's BZ; the operator and the investigators experienced severe eye, nose, and throat irritation, as well as sudden headache at the end of this procedure (Schneider and Blejer 1973, as cited in ACGIH 1986/Ex. 1–3, p. 285). Another study employing very precise methods of airborne concentration measurement reported the irritation response level for glutaraldehyde to be 0.3 ppm and the odor recognition threshold to be 0.04 ppm (Colwell 1976, as cited in ACGIH 1986/Ex. 1–3, p. 285).

Other than the NIOSH submission, OSHA received no comments on its proposal to establish a ceiling level of 0.2 ppm for glutaraldehyde. The agency found that the human evidence cited above clearly demonstrated a significant risk of irritation to the eyes, nose, and throat associated with short-term exposures to glutaraldehyde at concentrations of 0.3 ppm or above. OSHA considered the irritation effects associated with exposure to glutaraldehyde to be material impairments of health. Therefore, OSHA established a 0.2 ppm ceiling limit for this substance in the final rule.

Source: Federal Register/Vol. 54, No. 12/Thursday, January 19, 1989/Rules and Regulations, pp. 2464. Glutaraldehyde: CAS:111-30-8; Chemical Formula: OCH $(CH_2)_3$CHO, H.S. No. 1187.

Appendix 14.A.2

Acrolein and/or Formaldehyde

Method no.:	50
Matrix:	Air
Target concentrations:	Acrolein—0.1 ppm (0.23 mg/m^3)
(OSHA PELs)	Formaldehyde—3 ppm (3.7 mg/m^3)
Procedure:	Air samples are collected by drawing known volumes of air through sampling tubes containing XAD-2 adsorbent, and which have been coated with 2-(hydroxymethyl) piperidine. The samples are desorbed with toluene and then analyzed by gas chromatography using a nitrogen-selective detector.

Recommended sampling rate and air volumes:	0.1 L/min and Acrolein–		48 L
		Formaldehyde–	24 L
		Acrolein and Formaldehyde–	24 L

	Acrolein	Formaldehyde
Reliable quantitation limit: (based on recommended air volumes)	2.7 ppb (6.1 μg/m³)	16 ppb (20 μg/m³)
Standard error of estimate at the target concentration: (See Section 4.7.)	7.1%	7.3%

Special requirements: The sampling tubes should be obtained at the OSHA laboratory.

Status of method: A sampling and analytical method that has been subjected to the established evaluation procedures of the Organic Methods Evaluation Branch.

Source: Warren Hendricks, Organic Methods Evaluation Branch, OSHA Analytical Laboratory, Salt Lake City, Utah, March, 1985.

Appendix 14.A.3

Glutaraldehyde

Method no.:	64
Matrix:	Air
Target concentration:	0.2 ppm (0.8 mg/m³) (ACGIH TLV-Ceiling)
Procedure:	An air sample is collected by drawing a known volume of air through an open-face air monitoring cassette containing glass-fiber filters, each of which is coated with 2,4-dinitrophenylhydrazine and phosphoric acid. The sample is extracted with acetonitrile and analyzed by high-pressure liquid chromatography (HPLC) using a UV detector.
Recommended air volume and sampling rate:	15 L at 1 L/min
Reliable quantitation limit:	4.4 ppb (18 μg/m³)
Standard error of estimate at the target concentration:	6.2%
(Section 4.7.) Status of method:	Evaluated method. This method has been subjected to the established evaluation procedures of the Organic Methods Evaluation Branch.

Source: Warren Hendricks, Organic Methods Evaluation Branch, OSHA Analytical Laboratory, Salt Lake City, Utah, June, 1987.

Control of Formaldehyde Exposure in Hospitals

William Charney

Introduction

Health Effects

Formaldehyde gas is an irritant to the eyes and respiratory tract.[1] As a liquid in solution, it can cause both primary irritation and sensitization dermatitis. Formaldehyde exposure has also been linked to occupational asthma in the hospital setting,[2] and in other work environments.[3] Formaldehyde has also been found by the U.S. Environmental Protection Agency (EPA) to be a probable human carcinogen (*New York Times,* April 17, 1987, p. 8). The ACGIH lists formaldehyde as a suspected carcinogen.[4] Therefore, exposure must be controlled to maintain the lowest levels possible.

Exposure Limits

The ACGIH has set a 1 ppm threshold limit value for an 8 h time-weighted average with a ceiling level of 2 ppm that is not to be exceeded even for short durations.[4] The National Institute for Occupational Safety and Health (NIOSH) recommends as a guide to peak exposures that no worker be exposed to formaldehyde concentrations greater than 0. 1 ppm for any 15 min sampling period. NIOSH also recommends that no worker should be exposed to concentrations in excess of 0.016 ppm as an 8 h time-weighted average concentration.[5] The odor threshold has been reported to be 1 ppm.[6]

Formaldehyde in the Hospital Environment

As of December 1985, NIOSH had conducted seven Health Hazard Evaluations (HHEs) which focused on formaldehyde in hospital environments. Five of these seven HHEs found air concentrations which exceeded the NIOSH recommended limit at the time of 0.8 ppm. Air concentrations of formaldehyde detected in these studies ranged from 0.04 to 1.28 ppm in a hemodialysis unit, from 0.08 to 2.58 ppm in a histology laboratory, and from 0.01 to 2.2 ppm in a gross anatomy laboratory.[7] Since these health hazard evaluations were completed, NIOSH has reduced the recommended maximum exposure concentration to 0.1 ppm for a 15 min ceiling level and 0.016 as an 8 h time-weighted average.[5]

While the recent NIOSH exposure recommendations pose a serious challenge for exposure control strategies, the experience of this author suggests that the ACGIH-recommended levels are easily achievable, and, in fact, an action level of 0.5 ppm can be maintained in all areas using formaldehyde or formalin solutions. This section will not discuss at length formaldehyde's toxicity, biological effects, or other environmental data, but rather will focus on the methods to evaluate and control exposure in the hospital setting.

Variables of Exposure and Population Exposed

Formaldehyde is used in hospitals to fix and preserve tissue. A 10% buffered formalin solution is made from 37% formaldehyde solution. Formaldehyde is also used for reticulin staining. This is usually 20% formalin solution (concentrations may differ).

The population in hospitals exposed to formaldehyde includes:

1. *Operating room*: doctors and nurses in the transferring of excised tissues to formalin. These tissues are then sent to Pathology for gross description and selective sectioning.
2. *Pathology*: doctors, technicians', and autopsy assistants
   a. During gross description and sectioning. A doctor or technician may be exposed to formalin vapors for greater or lesser time spans depending upon the size of the specimen to be described. The times may vary anywhere from 30 s for small biopsies to 1.5 h for larger specimens such as colon or larynx.

b. During preparation of formalin and formalin-containing solutions. Technicians may be exposed to formaldehyde in the preparation of solutions. Ten percent buffered formalin, Brazil solution, cell block fixative, and Bouin solution are all solutions that require formalin. They are fixatives (maintain cell integrity, stop autolysis, and putrefaction) and contain formaldehyde in varied concentrations. Formalin may also be used in some staining procedures, mainly reticulin staining (some Trichrome procedures use Bouin solution), in very minimal concentrations.
c. In transferring organs to formalin for storing and/or further sectioning (autopsy assistants only).
d. During setup of processing machines. The technicians who set up the processing machine (e.g., Technicon, Histomatic) are exposed to formalin when changing the solution in the containers or when placing the tissues to be processed.

3. *Research*: Exposures similar to Pathology.
4. *Clinics*: Many doctors have their own supply of formalin so patient biopsies specimens can be placed in formalin immediately and then sent to Pathology.
5. *Other laboratories*: Some older laboratory methods, such as the isolation of worms and parasites in feces, may be used in microbiology. However, these methods are becoming obsolete and may be found in smaller hospitals or research laboratories.
6. *Hemodialysis*: Disinfection of machines (internal and surface).

Differences in quantified exposure for all the personnel listed above are explained by several factors:

1. Individual technique: Did the doctor wash the specimen with water before analysis?
2. Availability of local exhaust ventilation.
3. Design and maintenance of local exhaust ventilation to guarantee 100 LFPM face velocity.
4. Number of air changes per hour of dilution ventilation in the area or work.
5. Use of personnel protective equipment.

Monitoring for Formaldehyde

Industrial hygiene monitoring is essential to quantify formaldehyde concentrations for safety and legal compliance in hospitals. A variety of methods have been documented for accuracy both for short-term measuring and long-term monitoring. This section will allude to several methods among many. A critical review of formaldehyde monitoring methods can be found in Balmont.[8] Monitoring should be done in all areas where formalin (formaldehyde) is manipulated: autopsy rooms, pathology laboratories, and specimen storage rooms. Manipulations, such as cutting of tissue, dilution of 37% HCHO to 10% HCHO, waste disposal transfers, and pouring, should be quantified by reliable monitoring technique to determine exposure values for time-weighted average (TWA), compliance with short-term exposure limit, and ceiling units.

Air quantification is used to establish the level of ambient air exposure and individual exposure and to monitor on a regular basis the efficiency of whatever plan has been adopted to reduce level to legal compliance. The following method for air monitoring has been used successfully at the Jewish General Hospital, Montreal, Quebec. First is evaluation of any area according to National Institute for Occupational Safety and Health method 3500. Two impingers are connected in series using 20 mL of distilled water. Minimum sample time is 40 min. The limit of detection of this method is generally 0.04, but detection limits can be increased with increased volumes. However, given the ceiling, short-term exposure limit and time-weighted average compliance of 1 ppm or greater, this volume seems adequate for compliance. The interferences for this method are phenol and other organics. Quantified results of midget impingement done at the Jewish General Hospital in the pathology cutting laboratory are shown in Table 14.4. Figure 14.5 shows the locations of sampling sites, work and storage areas, and exhaust hoods in the laboratory.

TABLE 14.4 Quantified Results of Midget Impingement of Formaldehyde, Pathology Laboratory, Jewish General Hospital, Montreal, on Two Testing Dates in 1984

Post	Time	Minutes	Number of Samples	Concentration (ppm)	Operation
February 21					
3	9:50	40	1651	0.35	No work being done in laboratory at this moment
1	10:00	40	1647	0:10	No work at this station
1	11:15	40	1646	0.14	No work at this station
2	11:30	20	1650	0.20	Preparing solution
4	12:50	20	1654	0.06	Just before start of work
1	13:30	40	1645	0.24	Beginning to handle the specimens
4	13:40	20	1649	0.36	Beginning to work on large specimens
4	14:00	20	1653	0.29	Pathologist cutting
1	14:15	40	1648	0.39	Pathologist continues to cut
4	14:30	20	1644	1.20	Work of pathologist cutting
March 19					
3	9:40	40	1714	0.08	No cutting
1	9:40	20	1710	0.57	No cutting
1	10:20	40	1706	0.16	No cutting
1	12:25	40	1707	0.33	Beginning of work
4	15:10	20	1715	1.98	Cutting 6δ from BZ of pathologist
4	15:10	20	1716	0.49	12δ from pathologist BZ during cutting
4	15:40	20	1708	1.30	6δ BZ of pathologist during cutting

Notes: No local exhaust ventilation hoods at these times.

Post #1. Corresponds to respiratory zone of two workers—local exhaust provided—situated side by side while working on specimens.

Post #2. Air analysis during preparation of formaldehyde solution.

Post #3. Situated in open shelving area when specimens are stored.

Post #4. (BZ) of pathologist cutting table. No local exhaust ventilation.

Once all areas using formaldehyde have been quantified (personal BZ and ambient air) using midget impingement, short-term and ceiling-level grab sampling can be performed using a Miran infrared spectrophotometer. The Miran is not very reliable below 1 ppm and interferences can still occur at 3.58 wavelength (not alcohol). As seen in Table 14.5, during pouring of HCHO, levels rise above the 1-ppm level.

Monitoring programs for formaldehyde are facilitated by the use of passive dosimetry, especially for time-weighted average compliance. There are many different types of passive dosimeters on the market, some of which have been evaluated in the literature.[9,10] Choice depends on performance data, detection limits, availability, user preferences, and price of analysis.

The three different types of monitoring—impingement, infrared spectrophotometry, and passive dosimetry—can cover the major inhalation exposure possibilities. The infrared quantifications shown in Table 14.5 reveal the necessity for protective measures during the pouring, dilution, and mixing of formalin solution. The use of a fume hood for these applications would be advisable.

The quantifications shown in Table 14.4, pathologist BZ, show a degree of exposure approaching 50% of the TLV of 1 ppm. Although these levels are below legal compliance, pathologists and technicians still had irritative symptoms. Formaldehyde can cause symptoms at below the 1 ppm guideline and individual susceptibility still plays an important role in a prevention program. It is for these reasons that pathology cutting tables should be fitted with local exhaust ventilation.

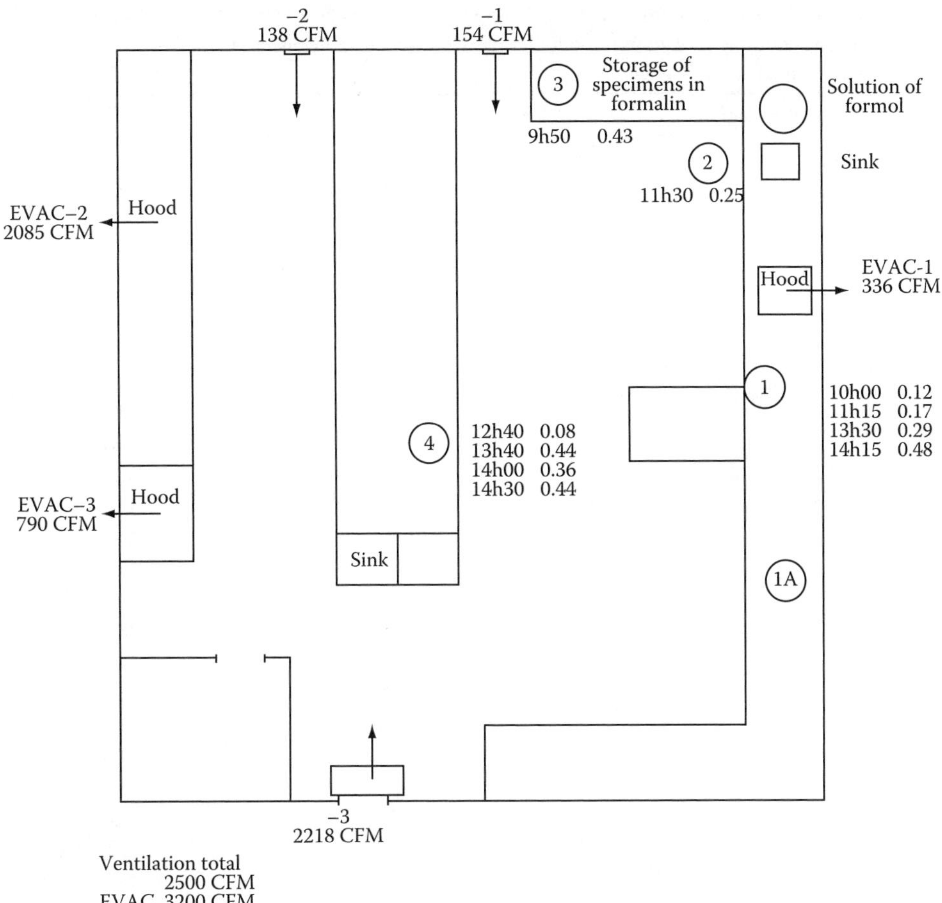

FIGURE 14.5 Pathology laboratory, Jewish General Hospital, Montreal, February–March 1984. No local exhaust hoods.

TABLE 14.5 Formaldehyde Infrared Monitoring Data, Miran 1B Infrared Spectrophotometer, Autopsy Room, Veterans Administration Hospital, Madison, Wisconsin, February 1987

Time of Sample (PM)	Activity	Location of Sample	Result (ppm)
3:01	Pouring		3.3
			0.2
		Floor level	5.8
3:02	Filling		2.1
		BZ	1.9
		BZ	8.7
		BZ	3.4
		Floor	1.4

(*continued*)

TABLE 14.5 (continued) Formaldehyde Infrared Monitoring Data, Miran 1B Infrared Spectrophotometer, Autopsy Room, Veterans Administration Hospital, Madison, Wisconsin, February 1987

Time of Sample (PM)	Activity	Location of Sample	Result (ppm)
3:04	Pouring	BZ	2.0
		BZ	1.2
		BZ	0.8
		BZ	1.6
3:05	Adding H_2O	BZ	4.1
		BZ	2.2
		BZ	3.8
		BZ	4.0
		BZ	4.1
		BZ	6.4
		BZ	7.7
		BZ	16
		BZ	21
		BZ	11
3:06		BZ	23
		BZ	18
		BZ	14
		BZ	6.5
	Closed	BZ	5.3
3:08		BZ	2.6
			2.7
			2.4
		BZ	1.6
		Over container	26
		Over BZ	32
3:30	While pouring	BZ	3.0
		BZ	5.5
		BZ	15.0
		BZ	22.0
		BZ	7.5
	Mixing	BZ	1.5
3:30	Ambient air	DG-215	1.5
			0.8
			0.7
3:30	Dumping tissue	BZ	3.0
		BZ	3.8
		BZ	4.3
		BZ	3.6
		BZ	4.1
		BZ	8.2
		BZ	8.3
		BZ	9.7
		BZ	7.7
		BZ	12
		BZ	13

Methods of Compliance

Design of Local Ventilation

Pathology cutting tables need to be fitted with local exhaust ventilation either of the slot type (Figure 14.6) or the Plexiglas hood type (Figure 14.7). Minimum face velocity should not be less than 100 FPM. Slot or hood design should pull air back and away from the BZ of the cutter. The exhausts need to be dedicated exhausts, and elimination of eddies is an important design feature. These hoods need a good maintenance program, and face velocity tests need to be done routinely. The laboratory area should be under negative pressure relative to the corridor with a minimum of 10 air changes an hour. Specimen storage shelves can be exhausted if convenient specimens are kept in well-sealed containers. Storage rooms containing specimens preserved in formaldehyde need to have good dilution and exhaust ventilation (Figure 14.8).

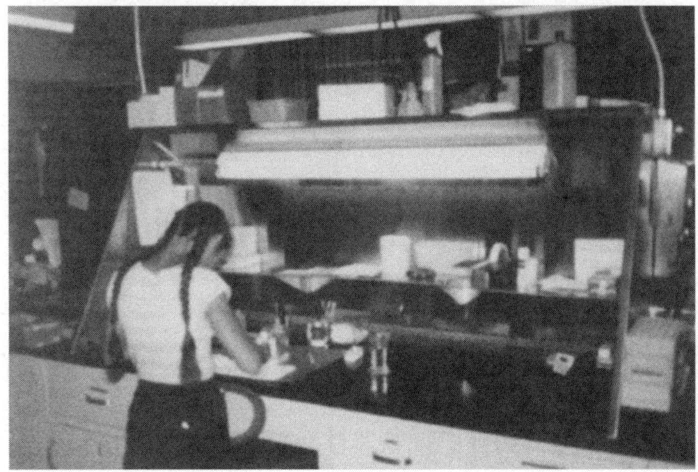

FIGURE 14.6 Slot-type hood for tissue mounting and specimen cutting. Notice the slot at the back of the hood. (Photo by Frank Maniscalco.)

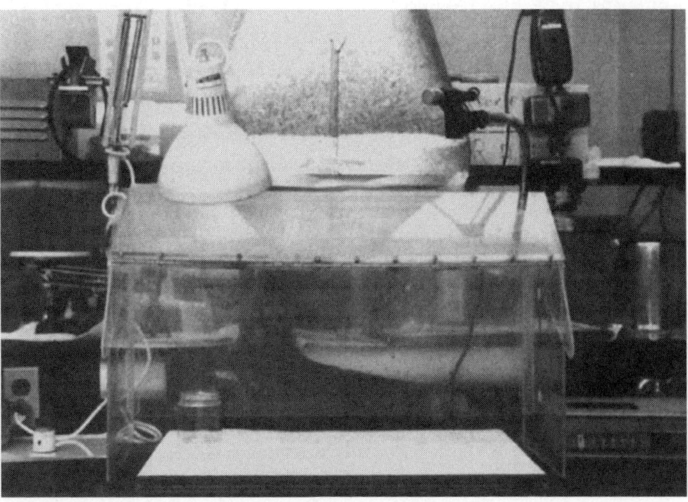

FIGURE 14.7 Plexiglas hood for cutting table. (Photo by Frank Maniscalco.)

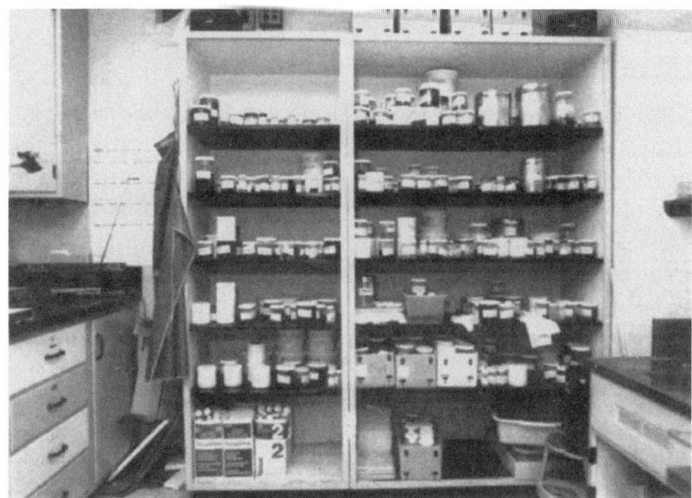

FIGURE 14.8 Specimen storage without enclosure or ventilation of formaldehyde. (Photo by Frank Maniscalco.)

Personal Protective Equipment

Working with formaldehyde requires the use of personal protective equipment:

1. Gloves made of rubber will protect against hand or skin contact.
2. Aprons can be worn to protect against splashes to the torso.
3. Framed safety goggles should be worn by all employees manipulating liquid solutions of formaldehyde to protect eyes. Goggles will also protect the eyes from contact with formaldehyde vapor.
4. Face shielding or plastic shielding to protect the face is required when the work process involves risk of splashes. A combination of safety goggles and plastic full-face shield can be used if the work process involves the risk of formaldehyde solution penetrating face shield.

Medical Surveillance Program

A medical surveillance program should be initiated for all employees with repeated contact with formaldehyde. The program should take into consideration skin, eyes, and the respiratory tract. Baseline pulmonary function tests can be a useful tool in annual assessment.

Formaldehyde is a sensitizing agent. Therefore, employees who have been exposed to legal compliance exposures over the long term may still suddenly develop symptoms. Patch testing employees who are suspected of reacting to formaldehyde is one method to determine sensitization.

Hazard communication (United States) and the Work Health Material Information System (WHMIS) both regulate the need to inform employees of the toxicity hazards of the chemicals in the workplace. As part of the medical surveillance program, employees should be given the Material Safety Data Sheet on formaldehyde.

References

1. *NIOSH Criteria for Recommended Standard, Subtext Occupational Exposure to Formaldehyde*, U.S. Department of Health, Education and Welfare, December 1976, p. 4.
2. Hendrick D and Lane D. Formalin asthma in hospital staff. *Br Med J* 1(5958):607–608, 1975.
3. Haavisto P. *Asthma Investigated Using Exposure Chamber*. Work Health Safety, Finnish Institute of Occupational Health, 1988, pp. 34–35.
4. American Conference of Government Industrial Hygienists: *Threshold Limit Values and Biological Exposure Indices for 1987–88*. American Conference of Governmental Industrial Hygienists, Cincinnati, OH, 1987, p. 42.

5. National Institute for Occupational Safety and Health: *NIOSH Pocket Guide to Chemical Hazards*, NIOSH Publication No. 85–114. National Institute for Occupational Safety and Health, February 1987.
6. Leonardos G, Kendall D, and Barnard N. Odor threshold determinations of 53 odorant chemicals. *J Air Pollut Control Assoc* 19(2):91–95, 1969.
7. Gun RT, Mullan RJ, and Brown D, et al. *Report of the Division of Surveillance, Hazard Evaluations, and Field Studies (D SHEFS) Task Force on Hospital Worker Health.* U.S. Department of Health and Human Services, National Institute for Occupational Safety and Health, 1985.
8. Balmont JL. *Formaldehyde Methods Manual.* Formaldehyde Institute, Scarsdale, NY, 1983.
9. Coyne LB, Cook RE, Mann JR, Bouyowcos S, McDonald OF, and Baldwin CL. Formaldehyde: a comparative evaluation of four monitoring methods. *Am Ind Hyg Assoc J* 46(10):609–619, 1985.
10. 1984 Report on the Consensus Workshop on Formaldehyde. *Environ Health Pers* 58:323–381, 1984.

Appendix 14.A.4

Summary of the Proposed OSHA Formaldehyde Standard

I. Scope and application. Applies to all occupational exposures to formaldehyde.

II. Definitions. Formaldehyde is defined as gaseous formaldehyde, solutions containing greater than 0.1% formaldehyde, and solids capable of releasing formaldehyde. *Action level* is defined as 50% of the proposed PEL (see below).

III. Permissible exposure limit (PEL). "The employer shall ensure that no employee is exposed to an airborne concentration of formaldehyde which exceeds 1.0 part formaldehyde per million parts of air (1 ppm) or 1.5 parts of formaldehyde per million of air (1.5 ppm) as an 8 h TWA."

IV. Exposure monitoring. Except where employer documents via objective data that exposure at or above Action Level (0.5 or 0.75 ppm) could not be generated, each employer shall monitor employees to determine their exposure to formaldehyde.

V. Initial monitoring. Employer shall monitor any employees who may be exposed and accurately determine their exposure. If all employees who may be exposed are not monitored, a representative sampling strategy must ensure that employees monitored are representative of all employees, based on investigation of work sites: (1) for formaldehyde-releasing potential; (2) where complaints or symptoms have been reported; and (3) for analysis of exposure patterns.

VI. Repeated monitoring. The above monitoring must be repeated "each time there is a change in production, equipment, process, personnel, or control measures which may result in new or additional exposure to formaldehyde."

VII. Periodic monitoring. Employer shall periodically monitor all employees shown by initial monitoring to be exposed at or above the Action Level (0.5 or 0.75 ppm).

VIII. Termination of monitoring. Employer may discontinue monitoring if employee exposure is below Action Level for two consecutive sampling periods.

IX. Accuracy of monitoring. Monitoring shall be accurate to within +25% at PEL and +35% at Action Level.

X. Employee notification of monitoring results. Employees shall be notified in writing of monitoring results within 15 days of employer's receiving results. If an employee's exposure is above the PEL, employer shall distribute and implement a written plan for corrective action.

XI. Observation of monitoring. Affected employees or their representatives shall be permitted to observe monitoring.

XII. Regulated areas. Post the following legend at access points to areas which may exceed PEL, limit access to authorized persons, and communicate the restriction to all affected personnel.

DANGER—FORMALDEHYDE—POTENTIAL CANCER HAZARD—RESPIRATORS REQUIRED—ADDITIONAL PROTECTIVE EQUIPMENT MAY BE REQUIRED

XIII. Methods of compliance. Institute Engineering Controls and Work Practices to maintain exposure below the PEL, where feasible, and institute Respiratory Protection until these modifications can be made, in emergencies, and in areas where these measures will not be sufficient to achieve compliance with the PEL.

XIV. Respirators. Select respirators approved by MSHA and NIOSH under 30 CFR 11 as specified in the regulation according to the expected severity of exposure. Where respirators are used, employers shall comply with 29 CFR 1910.134 (b), (d), (e), (f), and with Medical Surveillance requirements of this Standard. Employer shall perform documented and appropriate Quantitative Fit tests and appropriate respirators shall be selected. Air-purifying cartridges, if used, shall be replaced each work shift. Canisters, if used, shall be replaced after two workshifts unless canister displays an end-of-life indicator. Employer shall permit employee to wash face and respirator mouthpiece as needed.

XV. Protective equipment and clothing. Where skin contact with liquid or solid formaldehyde is possible, employer shall supply appropriate protective equipment and clothing in accordance with 29 CFR 1910.132.132/.133 at no cost and assure its use. Gas-proof goggles shall be supplied at no cost for eye irritation, if required. Contaminated clothing shall be stored to minimize exposure until cleaned and shall be labeled as described under Regulated Areas except for the respirator and protective equipment statements. No employee shall be permitted to take home contaminated equipment or clothing. Employer shall decontaminate any contaminated equipment or clothing, making certain that such materials are handled by trained persons and that persons receiving contaminated materials for cleaning are properly notified.

XVI. Hygiene protection. Employer shall provide change rooms and showers (29 CFR 191.141) for employees who need protective clothing. If employees may be splashed with formaldehyde, employer shall provide conveniently located quick-drench showers and eyewash fountains and assure their use.

XVII. Housekeeping. Employer shall conduct a program to detect leaks and spills, including visual inspections, equipment maintenance, and leak surveys, and shall assure containment, decontamination, and disposal of spills promptly by employees having proper equipment and training. Formaldehyde-containing waste shall be sealed in containers labeled "CAUTION—CONTAINS FORMALDEHYDE—AVOID INHALATION AND SKIN CONTACT."

XVIII. Emergencies. Employer shall develop a written plan, in accordance with 29 CFR 1910.38 (a), for each workplace where there is exposure to formaldehyde, including provision for adequate training, equipment, evacuation, medical assistance, and cleanup.

XIX. Medical surveillance. Employer shall institute Medical Surveillance programs for any employees required to wear a respirator to comply with the PEL and for any employees exposed in emergencies, to consist of examination by a physician, initially (as soon as possible in the case of an emergency), and at least annually thereafter, including (a) medical and work history with emphasis on eye, nose, or throat; (b) physical exam with emphasis on respiratory system; (c) pulmonary function test including FVC and FEV; (d) any other test the physician deems necessary; and (e) counseling of employees having conditions aggravated by formaldehyde exposure or respirator use.

Employer shall supply this Standard to examining physicians along with a description, including (a) employee's duties as they relate to exposure; (b) the typical exposure level for that job; (c) any personal protective equipment or respiratory protection associated with that job; and (d) previous available medical history.

Examining physician shall supply written opinion containing findings related to occupational exposure, including (a) whether employee would be "at an increased risk of material impairment of health" from further formaldehyde exposure or respirator use; (b) whether respirator should be reevaluated; (c) recommended limitations on employee's further exposure or use of protective equipment; (d) a statement that the employee has been informed by physician of any formaldehyde-aggravatable conditions, whether these may have resulted from exposure and whether there is a need for further explanation or treatment. The employer shall obtain results of examination and tests and shall supply the physician's written opinion to the employee within 15 days.

XX. Hazard communication. Employer shall comply with labeling requirements of 29 CFR 1910.1200 with respect to containers of formaldehyde or formaldehyde-containing products and in-process materials (exception: garments, bed clothes, draperies, furniture need not be labeled). Labels shall include the legends: CAUTION—CONTAINS FORMALDEHYDE—EYE AND RESPIRATORY SYSTEM IRRITANT—POTENTIAL CANCER HAZARD—AVOID INHALATION. Employers who manufacture or import formaldehyde-releasing materials shall develop and update Material Safety Data Sheets as required in 29 CFR 1910.1200 (g) and provide these to employers purchasing such materials.

XXI. Employee information and training. Employees assigned to workplaces covered by this standard shall be trained as of their initial assignment when any new hazard is introduced and receive ongoing training, at minimum, on an annual basis. The training program shall include (a) a copy of this Standard; (b) explanation of the Material Safety Data Sheet for formaldehyde; (c) purpose for and description of the Medical Surveillance Program, including health effects, signs, and symptoms of exposure; (d) description of safe procedures for each job in each area with formaldehyde exposure; (e) purpose and limitations of any personal protective equipment and clothing; (f) handling of spills emergency and cleanup; (g) explanation of engineering and work practice controls; and (h) review of emergency procedures.

Employer shall make available all training materials to employees and to the Assistant Secretary and Director (OSHA) on request.

XXII. Recordkeeping. Documentation of Personal Exposure Measurements shall include (a) date; (b) operation monitored; (c) methods used and evidence of accuracy; (d) number, duration, and results of monitoring; (e) protective devices worn; and (f) name, job, social security number, and exposure of each employee monitored.

Documentation of any objective data supporting a decision that Monitoring is not required.

Documentation of Medical Surveillance shall include (a) name and social security number; (b) physician's written opinions; (c) health complaints which may be related to formaldehyde; and (d) examination results, including medical history, questionnaire, and any medical tests.

Documentation of Respirator Fit Tests shall include (a) protocol of test; (b) results; (c) size and manufacturer of respirators; and (d) date of test, name, and social security number of employee, respirator type, and facepiece selected.

Employer shall retain Personal Exposure records for 30 years and Medical Surveillance records for the duration of employment plus 30 years. Only the most recent Respirator Fit record need be retained. Employer shall make records available, on request, to affected employee, former employee, or valid representatives and transfer records, as required by 29 CFR 1910.20. If employer ceases to do business, records shall be transferred to successor employer or to the Director (OSHA).

XXIII. Effective date. This Standard becomes effective 60 days after publication of the Final Rule except that Exposure Measurements, Medical Surveillance, and the Emergency Plan required by the rule must be completed by 6 months after its effective date. Engineering Controls and Work Practice Controls required by the Standard shall be implemented within 1 year of the effective date. Businesses with fewer than 20 employees have 2 years to implement these modifications.

Hidden Toxicities of Glutaraldehyde

William Charney

Introduction

Glutaraldehyde is a saturated dialdehyde with the formula CHO–CH_2–CH_2CH_2–CHO. It has a molecular weight of 100.12. Glutaraldehyde has a pungent odor, with a threshold recognition level of 0.04 ppm. Eye and respiratory irritation are noted at a level of 0.3 ppm, but individual sensitivity reaction can occur at lower quantified levels. The ACGIH recommends a 0.2 ppm ceiling level. Neither federal- nor state-PEL has been established. At present, there are neither NIOSH criteria nor an OSHA standard.

Glutaraldehyde is the active ingredient in Cidex, Sporicidin, Sonocide, Metracide, and other commercially available products and is used extensively throughout hospitals as a cold disinfectant. It has been implicated in the literature as a cause of asthma and rhinitis (in hospital personnel in endoscopy units);[1] contact dermititis;[2–4] burning eyes, nose, and throat, lung irritation, and chest tightness;[5] and breathlessness and reduced lung function.[6] Glutaraldehyde can be found in almost any area of the hospital. It can be used to clean and disinfect ear, nose, and throat instruments, dialysis instruments, surgical instruments, suction bottles, Vitalograph tubing, bronchoscopes, endoscopes, medical sputum mouth pieces, corrugated tubing, transducers, and more.

Normal submersion time in activated glutaraldehyde solution is 10 min; however, the tuberculosis bacillus and other stronger bacteria and viruses take a 10 h soak time. The buffered, activated solution usually sits in an open 1 L container (see Figure 14.9). Most often the general-dilution ventilation rates in these units are unpredictable. The only personal protection used by medical staff is sterile latex gloves, not considered adequate skin protection against glutaraldehyde, which is percutaneous.

Population Exposed

The population exposed can vary from hospital to hospital depending on infection control criteria and policy. The list that follows is based on personal experience:

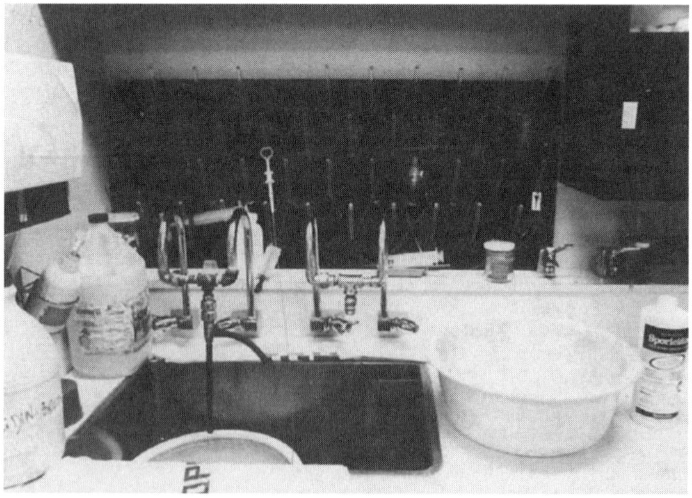

FIGURE 14.9 Sporicidin or 2% glutaraldehyde sitting in an open container on a tabletop and in a sink in a gastroenterology laboratory, allowing for glutaraldehyde vapors to accumulate in the ambient air. (Photo by Frank Maniscalco.)

- Nurses on units with a cold sterilizing procedure that uses glutaraldehyde; dialysis nurses, intensive care nurses, endoscopy nurses, emergency room nurses, and operating room nurses if Cidex or Sonocide solutions are used in infection control procedures
- Doctors who work in operating theaters, dialysis departments, or endoscopy units
- Central service employees who use glutaraldehyde as a sterilant
- Research technicians, researchers, and pharmacy personnel who activate solution
- Laboratory technicians who sterilize benchtops

General symptoms reported by hospital staff exposed to glutaraldehyde at the National Jewish Hospital in Denver, Colorado; the Jewish General Hospital, Montreal, Quebec; and University of Wisconsin Hospital, Madison, Wisconsin are as follows: (1) burning eyes, (2) headaches, (3) rhinitis, (4) skin sensitization, (5) chest tightness, (6) asthma and asthma-like symptoms, (7) staining of the hands (brownish/tan), (8) throat and lung irritation, (9) flu-like symptoms, and (10) hives.

Types of exposure are varied. Manual cleaning of instruments can expose staff to high levels of glutaraldehyde vapor. While the activated, buffered solution sits in containers, staff can be exposed to residual vapor. Mixing and activating the solution is also a route of inhalation exposure. Retrieval of instruments soaking in solution with unprotected or poorly protected hands is a means of percutaneous exposure.

Glutaraldehyde Investigations

Case studies implicating glutaraldehyde in clinical pathology with accompanying quantitative data are rare. NIOSH[5] in a Health Hazard Evaluation Report, Paul Pryor, NIOSH investigator, investigated complaints at the National Jewish Hospital, Denver, Colorado. Eight personal BZ and 13 area air samples were taken. Sampling times ranged from 30 to 80 min using sorbent tubes at a rate of 0.2 L/min. Samples analyzed using reverse phase high liquid chromatography; glutaraldehyde concentrations in the eight personal BZ samples ranged from N.D. to 1.5 mg/m^3 with six (46%) exceeding the TLV. In the units where 82% (9 of 11 exposed workers) reported irritative eye and throat symptoms, evaluation of the ventilation system showed it to be inadequate to remove glutaraldehyde vapors.

In 1985 at Mercy Medical Center, Paul Pryor, NIOSH investigator, quantified glutaraldehyde exposures in the range of N.D. to 1.98 mg/m^3 for eight personal samples and N.D. to 0.74 mg/m^3 for area samples, using sorbent tubes at 0.2 L/min for 30–45 min. On the basis of the environmental results it was concluded a health hazard did exist from glutaraldehyde exposure.

Case Studies

Two case studies will now be discussed involving health problems associated with glutaraldehyde exposure. The first concerns the University of Wisconsin Hospital; the second concerns Jewish General Hospital.

Case Study I: University of Wisconsin Hospital

Nineteen health care employees in module K4/2 at the Center for Health Sciences, University of Wisconsin Hospital complained of hives, chest tightness, and watery eyes. Glutaraldehyde, stored in 1 L baths on countertops, was being used in this area to disinfect bronchoscopes. The ventilation system was an independent recirculating type with 10% fresh air provided. No personal protection was used.

A calibrated Miran 1A infrared spectrophotometer was used to determine the glutaraldehyde concentration. The minimum reliable detectable limit for glutaraldehyde concentration for the Miran is 0.1 ppm. No chemical vapors capable of causing interference at the wavelength were noticed in any sampling locations. Monitoring results are given in Table 14.6.

TABLE 14.6 BZ and Ambient Room Air (ARA) Glutaraldehyde Concentration (ppm) in CSC Endoscopy (K4/2)

Location	Date	Sample Type	Number of 1 min Samples	Mean (+STD Error)	Range
K4/261	1/6/87	BZ	10	0.126 + 0.013	0.1–0.18
	1/6/87	ARA	8	0.07 + 0.007	0.05–0.1
	1/8/87	BZ[a]	10	0.076 + 0.005	0.05–0.1
	1/8/87	ARA[a]	8	0.053 + 0.003	0.04–0.06
K4/259	1/6/87	BZ	8	0.049 + 0.003	0.04–0.06
Olympus EW-10 washers	1/6/87	BZ[b]	6	0.195 + 0.013	0.16–0.25
Machines (2)	1/6/87	ARA[c]	10	0.087 + 0.007	0.06–0.13

Source: Richard, A. and Johnson, R.S., Environmental Health Specialist, University of Wisconsin-Madison.
[a] Sampling conducted during refilling of instrument disinfection containers with fresh CIDEX.
[b] Sampling conducted during the 3 min disinfection cycle.
[c] Ambient air sampled for 5 min following disinfection cycle.

Considering the poor sensitivity of the infrared spectrophotometer to glutaraldehyde at concentrations less than 0.1 ppm, it is possible that much of the absorbance below 0.001 AU (at 3.7 μm) is due to instrument noise. In an effort to error on the side of safety, any possible instrument noise was considered to be absorbance due to the presence of glutaraldehyde. As the monitoring data indicate, even when this safety margin is factored in, personnel exposure levels do not exceed the ACGIH recommended ceiling limit of 0.2 ppm in air. This does not necessarily mean that predisposed individuals (i.e., asthmatic sensitive) will not experience discomfort or respiratory stress or that personnel will not find the odor objectionable.

One month after application of the following recommendations, the nurses' symptoms subsided:

1. Reprogram the air handling system to allow for nonrecirculating for 100% fresh air.
2. Change all glutaraldehyde containers to air-tight models.
3. Use neoprene gloves.
4. Install local ventilation hoods for glutaraldehyde stations.
5. Check staff for sensitivity. Skin provocation tests were performed on five staff, with two showing signs of reaction to glutaraldehyde. Transfer requests were granted to these staff.

Case Study II: Jewish General Hospital

Twenty-two dialysis nurses complained of upper respiratory irritation, rhinitis, lacrimation, and sinusitis.[1] Formaldehyde (HCHO) was being used by this unit to disinfect the dialysis machines. Glutaraldehyde was being used in open 1 L baths for cold sterilization of instruments and tubing. General dilution air ventilation system provided approximately 7–10 air changes per hour. Formaldehyde testing using midget impingement at 200 cc/min with chromotropic acid analysis was performed, showing a range of N.D. to 0.4 mg/mL during the course of the workday. Formaldehyde was removed from this unit for a period of 2 weeks. All instruments were sent to Central Supply for gas sterilization. Symptoms persisted. It was then noted that all the nurses immersed their unprotected hands into the glutaraldehyde baths to retrieve the instruments. This skin contact was then suspected as the cause of their symptoms. Glutaraldehyde was removed from this unit on a trial basis. Within 1 week all symptoms of the dialysis personnel were eliminated.

As glutaraldehyde use in hospital settings increases, the need for industrial hygiene surveillance increases. Glutaraldehyde should be considered an etiologic agent for allergic respiratory reaction, asthma, upper respiratory irritations, dermatitis, hives, and other more esoteric individual responses, if other etiologic causes are eliminated through scientific method. All of the irritative symptoms can occur at quantified levels below the ceiling threshold of 0.2 ppm. Glutaraldehyde should be used with the same type of caution as formaldehyde. Industrial hygiene intervention can include the following: substitution; local exhaust ventilation; personal protection equipment.

Substitution is still a controversial issue. The nonglutaraldehyde-based disinfectants are usually not recommended by the surgical equipment manufacturer; however, gas sterilization by ethylene oxide is an acceptable alternative to the cold sterilization technique. Due to lack of funds for enough equipment to affect the turnaround time for EtO gas sterilization, hospitals use the cold sterilization methods. At the Jewish General Hospital, substitutions were successfully used for glutaraldehyde-based products in the following areas: dialysis, intensive care unit, operating theaters, and recovery room.

- *Dialysis unit.* All aldehydes were removed, including formaldehyde and glutaraldehyde, based on the sensitization of the dialysis staff to aldehydes. All instruments that were being cold sterilized with glutaraldehyde (clamps and scissors) are gas sterilized in the Central Service area. Internal disinfection of the COBE dialysis machines is accomplished using 0.25% hypochlorite (Javex); external disinfection of the COBE dialysis machine is accomplished with another chlorine-buffered solution.
- *Intensive care unit.* By substitution of a type of transducer, Savlon soap could be used as the disinfection agent. The Head Nurse of this unit had developed a glutaraldehyde sensitivity.
- *Operating theaters.* Cidamatic washing machines (glutaraldehyde-based) were replaced by a pasteurization unit. There is now another commercially available substitute for glutaraldehyde manufactured by Steris Corporation that uses peracetic (peroxyacetic) acid in a closed system. As San Francisco General Hospital we have successfully eliminated glutaraldehyde in our operating theaters and endoscopy unit.

The Centers for Disease Control and Prevention (CDC) and EPA Classification Schemes for Sterilants, Disinfectants, and Sanitizers can be found in Appendix 14.A.5.

Methods of Compliance

Engineering

Regarding local ventilation: General dilution ventilation is not adequate to protect workers from glutaraldehyde exposure unless 20 air changes an hour are maintained; local ventilation is more acceptable. Glutaraldehyde baths should be kept under a fume hood. Despite its low vapor pressure, this compound is found readily in air and even at low concentration can cause respiratory reaction (see Figure 14.10).

The areas using glutaraldehyde should be kept under negative pressure to avoid transmigration of chemical vapor and its odor.

Personal Protection

All skin contact with glutaraldehyde must be avoided. Heavy neoprene or nitrile gloves must be worn. This chemical is percutaneous and will pass readily through unprotected skin.

Industrial Hygiene

Air monitoring, using acceptable methodologies with the proper detection limits should be used in areas using glutaraldehyde. Results should be extrapolated and responsible recommendations made based on quantitative data. Industrial hygiene methods include:

FIGURE 14.10 Local exhaust hood designed using existing exhaust of 300 FPM. Plexiglas was used to contain all the glutaraldehyde-based disinfectants. (Photo by Frank Maniscalco. With permission.)

- Impingement method: deionized water (H_2O) 1 L/min for at least 30–60 L. Sensitivity of method: 0.01 ppm.
- The Miran infrared can be used to determine both area and BZ quantifications. However, reliability decreases below 1 ppm. Each Miran must be calibrated specifically for glutaraldehyde for reliable results.

Medical Surveillance

All medical personnel who regularly have contact with glutaraldehyde-based compounds should be followed by a medical surveillance program. In periodic physical examinations, the employee's general health should be evaluated. Special attention can be given to skin and respiratory symptoms. Hives, flu-like recurring complaints, and chest tightness can be indications of sensitization. Baseline, then annual, pulmonary function tests should be performed to determine decrease in function. Patch testing to determine sensitization can be part of an overall medical surveillance program.

References

1. Corrado OJ. Asthma and rhinitis after exposure to glutaraldehyde in endoscopy units. *Hum Toxicol* S 325–327, 1986.
2. Hansen R. Glutaraldehyde occupational dermatitis. *Contact Dermatitis* 9(l):81–82, 1983.
3. Jardon W. Contact dermatitis from glutaraldehyde. *Arch Dermatol* 103: January 1972.
4. Bardazzi F. Glutaraldehyde dermatitis in nurses. *Contact Dermat* 14(5):319–320, 1986.
5. *NIOSH: Health Hazard Evaluation Report*, HETA, 83, 074–1525. Paul Pryor, Investigator.
6. Benson WG. Case report: Exposure to glutaraldehyde. *J Soc Occ Med* 34:63–64, 1984.

Appendix 14.A.5

CDC and EPA Classification Schemes for Sterilants, Disinfectants, and Sanitizers

EPA Product Classifications	CDC Process Classifications
Sterilant/disinfectant	*Sterilization* (sporicidal chemical—prolonged contact time)
	High level disinfection (sporicidal chemical—short contact time)
Hospital disinfectant (with label claim for TB activity)	*Intermediate level disinfection*
Hospital disinfectant	*Low level disinfection*
Sanitizer Examples of common disinfectants used in endoscopy settings High level disinfectants (used on instruments, *not* on environmental surfaces)	
	• Glutaraldehyde or • Hydrogen peroxide-based chemicals
Intermediate level disinfectants (*not* used on instruments)	
	• Idophors • Chlorine compounds • Phenolics
Low level disinfectants (*not* used on instruments)	
	• Quarternary ammonium compounds

Appendix 14.A.6

Veterans Administration Memorandum

Date: January 29, 1987

From: Mark Reicheiderfer, M.D.

Subject: Endoscope Disinfection in Fiberoptics Laboratory

To: Bill Charney, Industrial Hygienist

This memo is a letter of appreciation for your involvement in the endoscope disinfection issue and, in particular, your decision to switch from glutaraldehyde to the new, less toxic disinfectant. All of us in the laboratory are appreciative, as none of us are happy about the current use of glutaraldehyde for disinfecting endoscopes in this hospital. There is no question that glutaraldehyde is toxic to the personnel who work in this area. I myself have an allergy with eye burning, nasal stuffiness, sneezing, and upper airway irritation upon exposure to glutaraldehyde. I am aware of other personnel in other institutions having asthma and allergic rashes as a result of exposure. You may be interested that such symptoms are being reported at the University (where glutaraldehyde is used) and there is now an investigation of the prevalence of these symptoms in the gastroenterology clinic.

The second reason we are happy with the new disinfectant is that I am convinced the glutaraldehyde is toxic to the instruments and will shorten their half-life. It seems likely to me that any increase in cost related to using the less toxic disinfectant will be outweighed by extending the lives of the endoscopes, thereby saving the VA money overall. This to me is a secondary consideration, however, to the safety of the laboratory personnel.

Once again, thanks for your efforts in this area. We are excited about getting glutaraldehyde out of the laboratory finally and hope that the new stuff arrives soon.

Appendix 14.A.7

San Francisco General Hospital Memoranda

What follows are two memoranda concerning the elimination of glutaraldehyde from the Operating Theaters at San Francisco General Hospital. This action was taken for two primary reasons: first, the complaint of some Operating Room staff who developed hypersensitivity to glutaraldehyde and second, the toxicity data mentioned earlier in this chapter.

The SFGHMC is now using a Steris unit that is a closed system that applies peracetic acid. The units are capable of disinfecting 98% of all instruments used in an Operating Theater. The units are installed under local exhaust ventilation with a dedicated exhaust to the outside.

Date: September 22, 1989

To: Cedric Bainton, M.D., Chief of Anesthesia

From: Ada Tarkington, A.D.N., Director, Operating Room Nursing Services

Enclosed is a copy of my memo to the staff regarding the glutaraldehyde solution. As we discussed the goal of the actions I am taking is to eliminate the use of glutaraldehyde from the Operating Room to the fullest possible extent. The Arthroscope camera is, to my knowledge, the only item frequently used which cannot be immersed in the Steris system and will continue to require glutaraldehyde soaking for appropriate disinfection. This soaking will have to be done in the actual Operating Room so that the camera can be placed directly onto a sterile field after disinfection.

We have agreed that whenever glutaraldehyde must be used for any reason the Head Nurse/Charge Nurse will discuss the situation and the need for use of this chemical with the anesthesia attending physician to determine if Mr. X is present or expected to be on duty. The goal is to provide Mr. X with an environment which will not compromise his health to the maximum extent possible or to provide an opportunity to discuss the situation with him so that he can be adequately informed regarding the potential exposure to glutaraldehyde.

I believe this summarizes the major points of our discussions. If I have missed anything or misquoted any of our agreements please let me know immediately.

Date: September 22, 1989

To: All O.R. Nursing Personnel

From: Ada Tarkington, Divisional Director, Operating Room Services

EFFECTIVE IMMEDIATELY THE STANDING GLUTARALDEHYDE COLD STERILIZATION FROM CYSTO & DECONTAMINATION WILL BE PERMANENTLY REMOVED.

The Steris System should be used for cold sterilization of all endoscopes. Any flexible endoscope which cannot be leak tested will be processed and soaked in glutaraldehyde by the anesthesia technician.

If the Steris System is down for any reason the only available backup will be the use of glutaraldehyde. If it becomes necessary to use glutaraldehyde, authorization must be given by either Ada Tarkington, Terry Bertrand, HN, Cedric Bainton, MD or HN/CN on shift or WE.

When approval has been given, a fresh solution of glutaraldehyde should be prepared, the instrument soaked appropriately. This solution will be discarded immediately after use. Under no circumstances shall active glutaraldehyde solution be left standing when not immediately needed for use.

For the present the Arthroscope camera will continue to require soaking in glutaraldehyde in the operating room where the procedure is scheduled to be done. The need for use of this camera should be discussed with the HN/CN and anesthesia attending before the glutaraldehyde solution is prepared.

15

Pentamidine

Occupational Risks of Pentamidine: A Review of Current Data 15-1
Health Concerns • Engineering Controls • Conclusions
References 15-3
Appendices 15-4
Appendix 15.A.1 • Appendix 15.A.2 • Appendix 15.A.3
Inhalation Hazard Control for Health Care Workers during the Administration of Aerosolized Pentamidine 15-7
Background • Cal/OSHA Compliance Investigations • Medical Interviews and Review of Medical Records • Discussion
References 15-11
Aerosolized Pentamidine Exposures in Health Care Workers 15-12
Introduction • Background • UCSF Health Effects Study • Results • Medical Results • Discussion • Recommendations
References 15-17
Verification of the Capture Efficiency of HEPA Filtration for Aerosolized Pentamidine 15-18
Study Rationale • Sample Strategy • Analyzer Parameters and Method • Results • Discussion of Results
References 15-22
Engineering Controls 15-22

William Charney
Healthcare Safety Consulting

Lawrence Rose
California Department of Industrial Relations

Patricia Quinlan
University of California, Berkeley

Occupational Risks of Pentamidine: A Review of Current Data

William Charney

Health Concerns

Two major occupational health concerns arise from the administration of aerosolized pentamidine (AP): long-term chronic effects to respiratory therapists from AP and the potential for *Mycobacteria tuberculosis* transmission.[1]

Chronic Effects

AP is used to treat *Pneumocystis carinii* pneumonia (PCP) infections in immunosuppressed patients. In general, the procedure for aerosolization calls for 600 mg of lympholyzed pentamidine to be reconstituted with 6 mL of sterile water. The pentamidine solution is then placed in the Respiragard 11 nebulizing system. The solution is nebulized for approximately 20 min. The Respiragard 11 system is specifically designed for the AP. It consists of a Wright nebulizer and a series of one-way valves that act as both a baffle to trap large particles and a direct exhalation to a bacterial filter. The system is designed to prevent AP from being dispersed into the room.

Exposure time of staff administering AP can fluctuate depending on number of patients and dosage; however, exposure time normally varies from 4 to 8 h.

No published studies, to date, have been conducted on the effect of AP on the normal human lung. However, in an animal study,[2] rats receiving high-dose AP showed no ill effects and no histopathologic evidence of airway inflammation or damage.

An air study to quantify pentamidine exposure was conducted by Montgomery et al.[3] An Anderson sampler was used with a sampling rate of 28.3 L/min for a 4 h period. Three sets of samples were collected on separate days. The study states that assuming a 40 min exposure at V_e 61/m and a deposition fraction of 0.5, the mean pulmonary dose a health care worker (HCW) could receive is 4.2×10^{-6} mg and the estimated dose the patient receives is 15 mg for a 150 mg dose. The Montgomery study extrapolates, therefore, that HCWs receive one-millionth as much as the patient. Montgomery also states that high patient absorptions during therapeutic exposures have not shown pulmonary toxicity (Montgomery, unpublished data). Another study by Montgomery et al.[4] shows that of eight AP patients' serum samples only one AP patient had a measurable amount of pentamidine in serum (13 ng/mL). The particle mass median aerodynamic diameter (MMAD) of the aerosol particle, as measured in the Montgomery study, is 8 μ. As a result of his published studies and studies to be published, Montgomery believes the risk of occupational exposure to AP to be low.

Tuberculosis Transmission

The second major concern to HCWs administering AP is the risk of tuberculosis (TB) transmission. The *MMWR* reported 42% of staff members at a western Palm Beach County, Florida clinic were identified as having positive tuberculin skin test reactions. Ventilation studies at the clinic revealed that >90% of the air of the building was recirculated and that 0.48 fresh air changes occurred per hour. Considering that AP is administered to a population of patients with human immunodeficiency virus (HIV) infection who may also have TB, strict precautions must be taken to avoid transmission.

A tight screening process for patients undergoing pentamidine treatment is essential to identify possible TB. A chest radiograph and sputum smears for acid-fast bacilli (AFB) is suggested by the *California Morbidity Weekly Report*.[5] "If the chest radiograph is not suggestive of active TB, and two out of three sputum smears are negative for AFB, AP treatments can be initiated." However, the report continues to say that any patient suspected of having an infection of TB should begin anti-TB therapy before starting AP. Screening for TB should be ongoing in case a patient develops TB during the course of AP therapy (see Appendix 15.A.1).

Engineering Controls

Ventilation control design is crucial in controlling potential TB transmission and in controlling symptoms related to AP (symptoms reported have included scratchy throats and burning eyes, and reduced lung function was reported in a respiratory therapist who had delivered AP for months in an inadequately ventilated room).[6,7]

Engineering ventilation control recommendations vary in scope. The Massachusetts Department of Health has issued guidelines on AP. Its first choice is to place patients in negative chamber booths with high efficiency particulate air (HEPA) filters. These booths are about the size of telephone booths, and cost about $2500 (see Appendix 15.A.2). Another design method is local ventilation designed into open cubicles where the AP patient sits. The exhausted air can be filtered, then exhausted to the outside. Both the booth design and the cubicle design allow for total environmental control, thus eliminating the need for the ongoing TB screening.

Escape of the AP from the device used to administer it can be eliminated or minimized by modifying the device and or the way it is used, for example, changing the device so it releases the aerosolized drug on demand only (i.e., only at the time of patient inhalation), rather than continuously as is done at present. Alternatively, patients receiving the drug should be instructed to shut off the nebulizer before they remove the mouthpieces from their mouths.

Conclusions

1. According to the Montgomery studies, there appears to be low risk from administering AP, due to the low quantifiable dose, the size of the median particle, and the low system absorption following aerosol administration. However, continued studies need to be conducted on toxicity to rule out any variables.
2. *Mycobacteria tuberculosis* transmission is a risk, and strict procedures must be implemented. Special ventilation principles must be followed to eliminate the risk of transmission.
3. The type of ventilation engineering control is up to the facility Infection Control Department.
4. NIOSH/MSHA masks can be worn by therapists in contact with the exposure as a means of control for AP, but engineering controls are the preferred means of protection. 3M 9970 respirator is an alternative, but fit testing must be accompanied with respirator program.
5. Because existing data on the teratogenic effects of pentamidine are limited, female respiratory therapists should be made aware of the limited nature of the information and should be allowed to make an informed consent regarding administering AP. Hospitals should oblige with a pregnancy transfer policy to these personnel.
6. Due to the scarcity of toxicity data to long-term effects to low-level exposure to pentamidine, every effort needs to be made to isolate employees from patients and to protect the environment from the potential for contamination. This is best accomplished by (1) engineering controls, and (2) local exhaust ventilation and proper ventilation guidelines in rooms where AP is administered.

References

1. CDC, *Mycobacterium tuberculosis* transmission in a health clinic, Florida, 1988. *Morb Mortal Wkly Rep* 38:256, April 21, 1989.
2. Debs RJ, Blumenfeld W, Brunette EN, Staubinger RM, Montgomery AB, Lin N, and Papahajopoulos D. Successful treatment with aerosolized pentamidine of *Pneumocystis carinii* pneumonia in rats. *Anti Microb Agent Chem* 31:37–41, 1987.
3. Montgomery AB, Corkery K, Terrell B, Brunette EN, and Debs RJ. Second hand exposure to aerosolized pentamidine. *Am Rev Respir Dis* 139(4):Al49, 1989.
4. Montgomery AB, Debs RJ, Luce J, Corkery KJ, Turner J, Brunette E, Lin E, and Hopewell P. Selective delivery of pentamidine to the lung by aerosol. *Rev Respir Dis* 137:477–478, 1988.
5. Infectious Disease Branch. *California Morbidity Weekly Report* 14, April 14, 1989.
6. *AIDS Alert*, August 1989, pp. 134–135.
7. Gude JK. Correspondence. *Am Rev Respir Dis* 139(4):1060, April 1989.

Appendices*

Appendix 15.A.1

Aerosolized Pentamidine Standards: Tuberculosis Screening

Statement of the problem: AP has proven efficacy in preventing PCP, the leading cause of death in AIDS patients. TB occurs at some time in the course of 4%–5% of AIDS patients nationwide, and about 2% of AIDS patients in San Francisco. This poses a risk of TB infection and disease to staff and other patients at AP sites. In order to minimize this risk, three factors must be addressed:

1. Screening of clients for TB prior to initiating AP prophylaxis.
2. Monitoring clients during the time they are in an AP prophylaxis program.
3. Decreasing the amount of infectious TB droplet nuclei that could persist in the AP facility should a patient with open TB be unknowingly treated at the facility. Ventilation and UV light are the modalities to consider.

Initial screening shall consist of the following:

1. All prospective AP clients should have a chest x-ray done within 1 month of the initiation of AP.
2. All prospective AP clients should be free of new pulmonary symptoms prior to initiation of AP.
3. All AP clients should have a PPD skin test done within 1 year and, if it is positive to ≥5 mm, should be offered INH prophylaxis.
4. If the screening x-ray is abnormal or new pulmonary symptoms compatible with TB occur, sputums for AFB smears × 2 should be done and negative prior to initiation of AP.

Ongoing monitoring shall consist of the following:

Prior to each treatment, the client will be interviewed by the respiratory therapist with a short questionnaire. This questionnaire will include the following:

1. Have you developed a new cough since your last treatment?
2. Have you developed sputum production or has your sputum increased?
3. Have you been diagnosed with possible or definite TB in the past month?

If the questionnaire suggests a potential TB problem, the client should be referred back to his/her primary provider for assessment and clearance.

Appendix 15.A.2

Aerosol Eater

The aerosol eater is a chamber designed to remove infectious aerosols produced by patients during sputum induction or other cough-producing procedures and to remove airborne drug particles produced when patients are treated with aerosolized drugs, such as pentamidine (Figure 15.A.1).

The patient enters through a door in the side of the chamber and sits facing forward. When the door is closed and the switch turned on, room air enters through a pre-filter located in the top panel. The air passes down in front of the patient and out through an HEPA filter located at the bottom of the back panel. A blower behind the filter pulls air through the system and discharges upward into the room.

The front and side panels are of Plexiglas or glass. The back panel is solid except for an opening at the bottom for the HEPA filter. The patient compartment is 72 in. high, 39 in. front to back, and 26 in. wide. The pre-filter is 20 × 20 × 1 in., and the HEPA filter is 24 × 24 × 6 in. The chamber accommodates a folding porch chair.

* Edward A. Nardell and Sue Etkind. With permission.

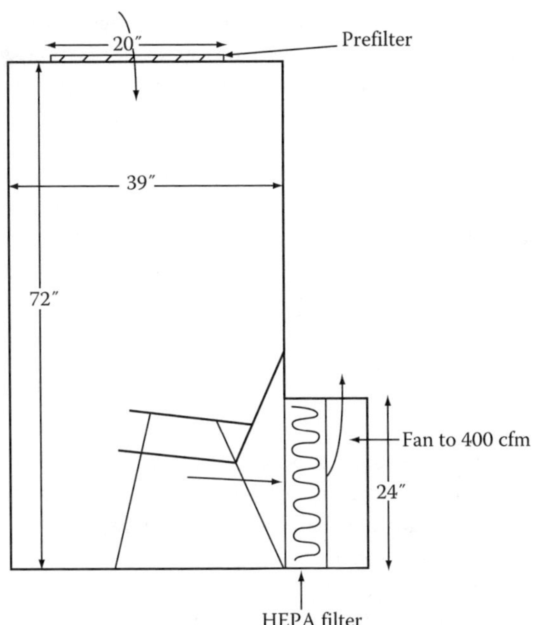

FIGURE 15.A.1 Aerosol eater chamber for control of airborne drug particles produced when patients are treated with aerosolized drugs. Patient enters through Plexiglas door and sits on a folding porch chair. Legend: 1 in. equals 1 ft. (Designed by Richard L. Riley, MD, Professor Emeritus, Johns Hopkins School of Medicine, and Edward A. Nardell, MD, Professor of Medicine, Harvard University School of Medicine.)

The pre-filter has 25% efficiency according to ASHRAE standard number 52–76. The HEPA filter (Boston Filter Co., Athol, MA) takes out 99.96% of particles down to 0.3 μ in size. A Dwyer gauge measures the pressure drop across the HEPA filter and indicates when the filter should be changed. The squirrel cage blower is rated at 400 cfm with a 0.6 in. pressure gradient. A rheostat controls the speed of the blower. Wire mesh protects the outlet.

The blower should be run at about half speed. The pre-filter should be changed when visibly dirty, and the HEPA filter when the Dwyer gauge so indicates.

Appendix 15.A.3

Guidelines to Reduce the Chance of Airborne Tuberculosis Transmission from HIV-Infected Patients

TO: Massachusetts Health Care Providers Involved in Pentamidine Aerosol Treatments

FROM: Edward A. Nardell, M.D., Tuberculosis Control Officer, MDPH

Sue Etkind, R.N., M.S., Director, Tuberculosis Program, MDPH

RE: Guidelines to reduce the risk of airborne tuberculosis (TB) transmission

Extensive TB transmission associated with pentamidine aerosol treatments has been reported,[1] and several other similar occurrences are now being investigated by the CDC. HIV-infected patients are at increased risk of reactivating TB. The risk of transmitting unsuspected TB to other HIV-infected patients and to staff appears to be enhanced by the coughing induced by pentamidine aerosol treatments. The aerosols produced by sputum induction, bronchoscopy, and by suctioning patients with artificial airways may present similar risks to persons breathing the same air. To detect occult TB transmission, regular Mantoux PPD skin testing of staff involved in these activities is strongly recommended. While inadequate fresh air ventilation in rooms where these procedures occur may greatly increase the probability of transmission, adequate ventilation alone cannot eliminate the risk. The enclosed guidelines are designed to reduce the risks of transmission.

I. **TB Transmission Associated with Aerosolized Pentamidine Treatment.** AP may induce coughing and thereby facilitate transmission of undiagnosed pulmonary TB disease from HIV-infected patients to persons sharing the same breathing space during or after treatment.[1] Other HIV-infected persons are especially at risk; so are healthy staff in treatment centers and household contacts if treatments are given at home. The potential for TB transmission depends on the prevalence of TB infection in the HIV-infected population being served, and on other factors such as (1) fresh air room ventilation, (2) the number of infectious droplet nuclei generated by the patient, and (3) duration of exposure.

A. **Tuberculosis screening:**

1. All HIV-infected patients should be screened for TB infection or disease before starting pentamidine aerosol therapy. Screening should include a history and physical examination, Mantoux skin test, including control antigens, chest x-rays, sputum examination, and other diagnostic tests, where appropriate.
2. Patients with a history of previous PPD positivity, or with current PPD reactions >5 mm, with no evidence of TB disease, should receive INH preventive therapy, regardless of age, as per current CDC/ATS guidelines.[2]
3. Regardless of PPD status, patients receiving aerosol treatment should be reevaluated for TB if they develop persistent cough, fever, or symptoms compatible with TB disease. Aerosol therapy should be suspended pending the result of diagnostic tests.
4. Treatment centers serving patients in high risk groups for TB infection (intravenous drug users, Haitians, and others from endemic areas for TB infection, prisoners, homeless, etc.) should rescreen patients and staff for TB infection and disease every 6 months. However, if TB infection is known to be unusual among the patient population served, rescreening of patients and staff may be less frequent. However, all patients and staff must be screened initially and at least annually thereafter.

B. **Environmental precautions:**

1. *Minimum precautions.* Fresh air ventilation in the treatment area should be at least 20 room changes per hour (about four times normal) to reduce the chance of person-to-person transmission. Air pressure should be negative with relation to adjacent rooms and hallways. Air from the treatment area should not be recirculated within the building, but exhausted to the outside, with precautions that the discharge is not adjacent to the air intakes of other buildings. Where applicable, ultraviolet (UV) air disinfection is a less expensive alternative to high-level fresh air ventilation to reduce person-to-person TB transmission. However, UV fixtures must be planned, manufactured, installed, and maintained in accordance with published guidelines.[3] For effective UV air disinfection, room humidity must be <60%, which may preclude application in some treatment centers due to aerosol treatment itself, or climatic conditions.
2. *Optimal precautions.* Neither high-level fresh air ventilation nor UV air disinfection can provide complete protection against person-to-person TB transmission if aerosols are administered in rooms occupied by both TB cases and susceptible hosts. Optimal protection would be provided by treatment administered within a booth occupied by only the patient. The booth should receive high levels of fresh air ventilation, the exhaust of which must either be vented safely outside or disinfected by UV light, HEPA filters, or a combination of both methods. Detailed plans for such a booth are available upon request.

C. **Treatment precautions.** Current AP treatment protocols are designed to minimize coughing, and treat it when it occurs. Variations in equipment or technique may result in increased cough, with consequent greater risk of transmission if coexistent pulmonary TB disease exists.

II. Sputum Induction by Aerosol. Hypertonic saline aerosol is widely used to induce coughing in patients unable to generate satisfactory sputum samples for diagnostic purposes, and therefore may be used for AIDS patients with undiagnosed pulmonary problems. There is potential for transmission of undiagnosed TB pulmonary disease to others, depending on the prevalence of infection in the population. Sputum induction should be performed under conditions that assume that TB may be present.

A. Minimum precautions. Patients should be screened for TB by chest x-ray. As few people as possible should be present during the induction. Ventilation or UV air disinfection should be as outlined above.

B. Optimal precautions. The same type of booth used for aerosol therapy should be effective protection for induction. It would be unwise to mix symptomatic patients awaiting induction with otherwise well patients on prophylaxis.

III. TB Transmission Associated with Bronchoscopy, Endotracheal Intubation, and Tracheostomy. Bronchoscopy is extensively used to diagnose infections in HIV-infected patients. Many patients presenting respiratory failure and requiring intubation have HIV-related infections, including TB. Both fiber-optic and rigid bronchoscopy have the potential to induce coughing and thereby facilitate transmission of TB. Artificial airways require periodic auctioning, which also induces coughing. The highest rate of TB infectivity yet recorded occurred during bronchoscopy and endotracheal intubation in an intensive care unit.[4]

Precautions. Brochoscopy should be performed in a room equipped to handle contaminated air by either high-level ventilation (20 room changes per hour exhausted safely to the outside, with negative room air pressure relative to halls and adjacent rooms) or properly installed and maintained UV light. Staff should be tested regularly for TB infection.

Intensive care units should have isolation rooms equipped to handle contaminated air for critically ill patients with abnormal chest x-rays, until TB can be ruled out by several sputum examinations. Closed systems for auctioning endotracheal tubes should be evaluated for possible use to reduce the production of infectious aerosols. Staff should be screened regularly for TB.

References

1. CDC. *Mycobacterium tuberculosis* transmission in a health clinic—Florida, 1988. *Morb Mortal Wkly Rep* 38:256–264, April 21, 1989.
2. CDC. Tuberculosis and human immunodeficiency virus infection: Recommendations of the advisory committee for the elimination of tuberculosis (ACET). *Morb Mortal Wkly Rep* 38:236–250, 1989.
3. Riley RL and Nardell EA. Clearing the air, the theory and application of ultraviolet air disinfection. *Am Rev Respir Dis* 139:1286–1294, 1989.
4. Catanzaro A. Nosocomial tuberculosis. *Am Rev Respir Dis* 125:559–562, 1982.

Inhalation Hazard Control for Health Care Workers during the Administration of Aerosolized Pentamidine

Lawrence Rose

Background

AP has proven to be effective in preventing PCP, the leading cause of death in AIDS patients, and has been approved for preventive therapy by the Food and Drug Administration.[1]

The most frequent use of AP is for prophylaxis of PCP in HIV-positive patients. Each pentamidine therapy session lasts about 30 min: 300 mg of pentamidine mist is delivered to the patient's mouth through a Respirgard II nebulizing system specifically designed for administration of AP. Thus, the optimum mist droplet size (1–3 μ) is delivered to the alveoli.

The administering HCW or respiratory therapist must remain in the room in close attendance due to frequent and sudden patient respiratory distress (coughing paroxysms, wheezing, hypoxia, etc.) and the need, at times, for rapid intervention (bronchodilators and oxygen administration).

There is a visible aerosol released into the room when the mouthpiece of the nebulizer is removed; frequent coughing spasms also release ambient persistent AP droplets into the room. The adverse health effects of systemically administered pentamidine are well known.[2,3] Evidence of adverse pulmonary effects in workers has been confirmed in a report of a decrease in diffusion capacity in a respiratory technician administering pentamidine.[4] There is another recent report of asthma in respiratory therapists[5] exposed to AP. AP-induced conjunctivitis has been reported.[6] The mutagenicity, carcinogenicity, and teratogenicity data are incomplete. Further testing is required to characterize these areas of health concern fully.

Pentamidine does inhibit dihydrofolate reductase and is therefore a potential reproductive hazard.[7] Adverse reproductive outcomes with other aerosolized drugs have been reported in HCWs, which prompted the adoption of safe handling guidelines and the use of well-ventilated safety hoods for drug mixing.[8–10] Pentamidine has been detected in the ambient air in AP therapy rooms during AP administration.[11]

The transmission of TB has occurred in a facility where AP was administered to patients with HIV;[12] CDC now recommends control guidelines.[13] In addition, the cost effectiveness of UV radiation for transmission control in AP therapy rooms is now being evaluated.[14]

Cal/OSHA Compliance Investigations

Outpatient AP has been administered at several major medical centers in the Bay Area for the past 3–4 years. The number of patients receiving this treatment is rapidly increasing. Cal/OSHA has investigated five medical centers with AP clinics. Twenty-seven AP clinic HCWs have been interviewed during on-site inspections.

Medical Interviews and Review of Medical Records

Administering HCWs frequently experienced moderate to severe chest tightness, wheezing, coughing, and shortness of breath with exertion; at times, self-treatment with bronchodilator inhalants was necessary. These reactions occasionally lasted up to 5 days. One worker was hospitalized on two occasions with severe acute asthmatic bronchitis exacerbation.

There were also mild to moderate eye, nose, and throat irritative reactions, with increased mucus discharge, stuffy nose, hoarseness, tearing, burning eyes, and sneezing complaints. In addition, there were complaints of mild fatigue, headaches, light headedness, and bitter metallic taste.

Associated Factors

The aforementioned HCW's health reactions were directly related to the number of patients, number of treatment chairs per room, and daily duration of exposure secondary to the following factors:

1. Inadequate room ventilation.
2. Side by side patient chairs—up to eight in one room.
3. The volume of patients per working day (2–60 per day).
4. AP mist coming out of mouthpieces when removed because of coughing paroxysms (or improper patient handling of the pressurized aerosolized mouthpiece end, and pronounced visible release of a fine AP mist that would suspend in the room air).
5. The necessity for close HCW-patient observation and proximity during treatments.
6. No effective HCW eye protection, respirators, or personal protection available.
7. Inadequate HCW preemployment and periodic medical screening.

Cal/OSHA Program Requirements to Reduce AP HCW Risks

1. An HCW medical screening and surveillance program with particular emphasis on workers with respiratory problems (e.g., a history of asthma), pregnancy or planning a family within 8 weeks, upper respiratory–external eye reaction histories, or those on certain medications.
2. A TB screening-surveillance control programs should be established by the Hospital Infectious Disease Control Committee and/or Employee Health, including patient screening.
3. Effective control of the AP ambient air mist, requiring exhaust ventilation with sufficient negative pressure to allow for complete control of the environment surrounding the patient. And an exhaust HEPA filter (that can filter out droplets 2 μμ in diameter) if the air is to be recirculated back into the room.
4. As a temporary measure, prior to the installation of appropriate hood stalls or booths, high-efficiency organic mist respirators, gloves, eye protection with side shields, and disposable gowns should be worn. This personal protection should be worn when the HCW must enter a booth to assist the patient.
5. To eliminate or minimize AP mist escape, a more effective automatic shutoff system should be engineered in the AP delivery system tubing.
6. A reproductive policy for HCWs should be adopted that includes alternate work assignments for pregnant workers or workers planning to be pregnant within 8 weeks.
7. HCWs with any acute or prolonged upper respiratory problems or other AP health reactions should be offered alternate work assignments after medical evaluation.

Notes on Table 15.1: Summary of Symptoms

Column A: Each number represents one HCW administering AP in one hospital setting. Each letter (A through E) represents a hospital.

Column B: Symptomatic bronchospasm is graded 1 to 4+ according to the intensity of chest tightness, wheezing, shortness of breath, cough, and medical treatment (bronchodilators, etc.).

Summary: There were a total of 27 HCWs interviewed at five different Bay Area hospitals.

- Sixteen had bronchospastic reactions.
- Nine could taste the AP.
- Thirteen noticed throat reactions, some with hoarseness.
- Thirteen experienced stuffy noses, some with mucus discharge.
- Eight developed ocular tearing, burning, or redness.
- Nine felt mild to severe headaches.
- Ten noticed varying degrees of fatigue, worsening toward the end of the day.

There were seven HCWs that terminated work in AP hospital clinics; the individual worker reactions were

1. Severe, incapacitating headaches
2. Severe bronchospasm, treated by physician
3. Severe shortness of breath, with bronchospasm treated by physician
4. Severe coughing, wheezing, and bronchospasm treated by physician
5. Mild to moderate symptomatic wheezing and chest tightness
6. Moderate cough, with symptomatic wheezing and chest tightness
7. Concern for possible adverse reproductive effects (in the first trimester of pregnancy)

Column J: Exposure categories are graded mild to severe, from 1 to 4+, based upon the exposure severity factors listed below:

1. Room ventilation factors and the estimated air exchanges per hour
   a. Volume of room
   b. Ventilation system airflow

TABLE 15.1 Summary of Symptoms

A	B	C	D	E	F	G	H	I	J
HCW	Symptomatic Bronchospasm	Bitter Metallic Taste	Throat Irritation	Nasal Discharge	Eye Irritative Reaction	Headache	Fatigue	Comments	Exposure Category
1a	4+	Yes	Yes, sore throat	Yes	Yes	No	No	Self-treated with bronchodilators	4
2a	4+	No	Yes, hoarseness	No	No	Yes	No	Self-treated with bronchodilators	4
3a	4+	No	Yes, sore throat	Yes, severe postnasal drip	Yes	Yes, severe	Yes	Physician treated with bronchodilators	4
4a	3+	No	Yes	Yes	Yes	No	No	Self treated with bronchodilators	4
5a	0	No	Yes, postnasal/ sore throat	No	No	No	Yes		4
6a	2+	No	No	No	No	No	No		
7a	0	No	No	No	No	No	No	Supervisor (minimal exposure)	1
8a	3+	No	No	No	No	No	Yes	Pregnant 4 months, then removed	4
9b	4+	No	No	No	No	No	Yes	Severe asthmatic, physician treated-systemic steroids	4
10b	4+	No	Yes	Yes	No	No	No		4
11b	3+	Yes	Yes	Yes	Yes	Yes	Yes	Self-treated with bronchodilators	4
12b	4+	Yes	No	Yes	No	No	No	Hospitalized 1 week, recurred with one exposure	4
13b	+	Yes	Yes	No	No	No	Yes		4
14b	+	No	Yes	Yes	No	No	Yes	Physician treated with bronchodilators	4
15c	0	No	No	No	No	No	No		1
16c	0	Yes	No	Yes	No	No	No	Physician treated with bronchodilators	2
17d	4+	No	Yes	Yes	Yes	Yes	No		4
18d	4+	No	Yes	Yes	Yes	No	No	To supervisor (minimal exposure)	4
19d	0	No	No	Yes	No	No	No	Physician treated with bronchodilators	4
20e	0	No	No	No	No	Yes	No	Contact hand dermatitis, self-treated with bronchodilators	2
21e	4+	Yes	Yes	Yes	Yes	No	No		2
22e	0	No	No	No	No	No	Yes	Asymptomatic	2
23e	0	No	No	No	No	Yes	No	Self-treated with bronchodilators	2
24e	0	Yes	No	No	Yes	No	No	Severe headaches	2
25e	0	Yes	Yes	No	No	No	Yes	Asymptomatic	2
26e	2+	No	No	No	No	No	Yes		2
27f	0	Yes	No	Yes	No		No	Contact hand dermatitis	1

   c. Open windows and prevailing winds
   d. Open doors
   e. Position of treatment chairs and air flow
2. Treatment frequency per 8 h working shift
   a. Number of patients treated in an 8 h working day
   b. More than one patient treatment chair in operation
   c. Number of patient chairs in the treatment room
3. Personal protection, adequacy, and compliance
   a. Training and education program adequacy
   b. Respirator program
   c. Availability of personal protective gear
4. AP health reaction recognition and reporting in Employee Health Systems
5. Patient responses
   a. Effectiveness of patient instructions on how to minimize escape when the mouthpiece is removed from the mouth.
   b. Frequency of paroxysmal patient pulmonary reactions with coughing, dyspneal wheezing, and hypoxia, necessitating rapid close HCW patient observation and intervention (oxygen and bronchodilators).

Discussion

The HCW acute irritative health reactions and the degree of exposure correlated well. Many of these HCWs had administered AP for over 3 years, and were associated with the initial AP clinical research trials. They are highly motivated, dedicated, well trained, and experienced in administering AP. Their observations regarding their own acute upper respiratory reactions are reliable and quite understated. There was a reluctance to participate in anything that might interfere with the efficiency and accessibility of this lifesaving AP program.

In view of the obvious acute upper respiratory ocular HCW reactions, Cal/OSHA regulatory action was mandatory. This action was a carefully thought out balance of effective engineering controls for HCW personal protection vs. lifesaving health care patient needs. Past Cal/OSHA regulatory actions to protect HCWs have been necessary for chemotherapeutic drugs, ribavirin, and anesthetic gas exposures.

This AP experience once again points up the need for appropriate hospital committees responsible to set policies to prevent HCW exposures to aerosolized drugs. The makeup of these committees would best be a mix of several professional disciplines (e.g., physicians, nurses, respiratory therapists, industrial hygienists, ventilation engineers, etc.).

References

1. Nightingale SC. From the Food and Drug Administration. *J Am Med Assoc* 262:184, 1989.
2. Pearson RD and Hewlett EL. Pentamidine for the treatment of pneumocystis carinii pneumonia and other protozoan diseases. *Ann Intern Med* 103:782–786, 1985.
3. Sands M, Kron MA, and Brown BB. Pentamidine: A review. *Rev Infect Dis* 7:625–634, 1985.
4. Gude JK. Selective delivery of pentamidine to the lung by aerosol. *Am Rev Respir Dis* 139:1060, 1989.
5. Kern DG and Franklin H. Asthma in respiratory therapists. *Ann Intern Med* 110:767–773, 1989.
6. Lindley DA and Schleupner CJ. Aerosolized pentamidine and conjunctivitis. *Ann Intern Med* 109:988, 1988.
7. Waalkes TP and Makula D. Pharmacologic aspects of pentamidine. *Natl Cancer Inst, Monogr* 43:171–176, 1976.

8. Rogers B and Emmett EA. Handling antineoplastic agents: Urine mutagenicity in nurses. *J Nurs Scholarsh* 19:108–113, 1987.
9. Hemminki K, Kyyrgronen, and Lindbohn, ML. Spontaneous abortions and malformations in the offspring of nurses to anesthetic gases and other hazards. *J Epidemiol Community Health* 39: 141–147, 1985.
10. Selevan S, Lindbohm ML, et al. A study of occupational exposure to antineoplastic drugs and fetal loss in nurses. *N Engl J Med* 313:1173–1178, 1985.
11. Montgomery AB, Debs RJ, Luce JM, et al. Aerosolized pentamidine as second line therapy in patients with AIDS and pneumocystis carinii pneumonia. *Chest* 95:747–750, 1989.
12. CDC. *Mycobacterium tuberculosis* transmission in a health clinic—Florida, *Morb Mortal Wkly Rep* 38(15):257, 1988.
13. CDC. Tuberculosis and human immunodeficiency virus infection: Recommendations of the advisory committee for the elimination of tuberculosis (ACET). *Morb Mortal Wkly Rep* 38(5-5):1–9, 1989.
14. Kizer WK. DHS, Using ultraviolet radiation and ventilation to control tuberculosis, 1989 California Indoor Air Quality Program, Air and Industrial Hygiene Laboratory and the Tuberculosis Control and Refugee Health Program Unit, Infectious Disease Branch, Berkeley, CA, 1989.

Aerosolized Pentamidine Exposures in Health Care Workers

Patricia Quinlan

Introduction

Pentamidine, an aromatic diamidine derivative anti-protozoal agent, is used as a treatment for patients with *Pneumocystis carinii* (PC).[1] Pentamidine use has increased in recent years in preventive therapy and treatment of PC pneumonia, as there is a high frequency of the occurrence of this organism in HIV-infected individuals. To get the most effective treatment to the lung tissue, where the infection occurs, an aerosolized inhalation delivery method has been evolved for this drug. For example, bronchoalveolar ravage fluid concentrations of the drug are at least 5–10 times higher following oral inhalation of AP than after intravenous pentamidine administration. Pentamidine absorption from the lungs to the systemic circulation is limited, so that delivering the maximum dose concentration to the lungs through an aerosol is an efficient method of treatment.

Because of the sense of urgency that accompanied the development of effective treatments for HIV-infected patients, the potential consequences to HCWs who administer the drug in aerosolized form did not receive critical attention until recently. Delivering the drug as an aerosol presents the risk that workers as well as patients will become exposed. Pentamidine has been detected in the ambient air of therapy rooms, indicating that secondary exposure to medical personnel does occur.[2–4] The consequences of these exposures are still under study. At least one respiratory therapist has been reported to have developed a transient decrease in diffusing capacity, which resolved after discontinuation of AP administration.[5]

In addition to the direct health effects from pentamidine exposure, there is concern that those who administer pentamidine may be at risk for inhalation exposure to TB. This concern is based on findings that as many as 5%–15% of AIDS patients may be infected with TB.[6] Moreover, there have been reports of positive TB skin test conversions among HCWs administering this agent.

More information is needed. We need to better understand HCWs' exposure to AP during administration of the drug, and the effects of occupational exposure to AP on pulmonary function and the development of respiratory symptoms and any other chronic effects. As this information becomes available, we will be better able to design and evaluate appropriate controls to protect the HCWs who administer this agent.

Background

Pentamidine isothionate is commercially available as a sterile lyophilized powder. It has an aqueous solubility of approximately 100 mg/mL at 25°C.

For aerosolization, the lyophilized pentamidine isethionate (usually 300 mg) is reconstituted with 6 mL of sterile water to make a solution. The pentamidine solution is then placed in a nebulizer. The solution is nebulized for approximately 20–30 min. One nebulizer commonly used in the United States is the Respirgard II nebulizing system, which is specifically designed for administration of AP. It consists of a nebulizer and a series of one-way valves that trap particles, with direct exhalation to a bacterial filter. The system is designed to prevent release of AP to the surrounding environment. It is disposable and designed for single-treatment use. In Canada, FISONeb ultrasonic nebulizers are frequently used. With these nebulizers, 60 mg of pentamidine isethionate are dissolved in 3 mL of distilled water.

The most frequent use of AP is for prophylaxis of PCP in HIV-positive patients. The administration of AP is usually done in a therapy room, where one to eight patients receive pentamidine therapy at the same time. Each pentamidine therapy session lasts approximately 30 min. When therapy is in progress, the respiratory therapist or other HCW typically remains in attendance in the therapy room or in an adjoining room. In some facilities, the patients are in booths or in separate rooms equipped with exhaust filtration systems, while in other facilities, treatment is administered in clinic rooms that have only dilution ventilation. The HCW often does not wear any respiratory protection while working in the AP therapy room. The patients are scheduled to come in at 10–15 min intervals. Thus, there is a constant turnover of patients starting and ending therapy. Pretreatment bronchodilator inhalation may be given for those patients who have developed breathing difficulties with AP administration in the past. If a new patient develops respiratory symptoms, AP should be stopped and nebulized bronchodilator given before continuation of AP therapy. When the patient stops inhaling AP and comes off the mouthpiece of the nebulizer, there is visible aerosol released to the surrounding environment, because of the continuous flow of air through the nebulizer. There are some shutoff valves available for the nebulizer, but these have not met with wide acceptance because of difficulty of use. Another source of ambient pentamidine is patient coughing during inhalation of the drug.

Adverse respiratory effects in patients are common with AP. Beta-agonist inhaler therapy or pretreatment is required for some patients.[6] Reversible bronchospasm has been demonstrated in some patients receiving AP. In a series of patients treated prophylactically with AP, peak expiratory flow rates transiently decreased after pentamidine administration, to an average of 74.5% of pretreatment values.[7] Occasionally, a cough may be severe enough to require discontinuation of therapy.[8] The long-term effects of AP on pulmonary tissue and function are not known.

Pentamidine has been detected in the ambient air in AP therapy rooms.[2–4,9] Although the doses HCWs receive may be relatively small compared to the doses patients receive, the degree of exposure may be dependent on AP therapy room environmental conditions, number of nebulizers in simultaneous use, or different patient behaviors.

Symptoms of scratchy throat, burning eyes, and reduced lung function were reported by one respiratory therapist who delivered AP in an inadequately ventilated room. After 14 months of delivering nebulized pentamidine for the prophylaxis of PCP, another respiratory therapist showed a reversible moderate decrease of carbon monoxide diffusing capacity (DL_{co}).[5] A survey of health complaints in five facilities in northern California was undertaken in 1990. Symptoms reported by the investigator included shortness of breath, wheezing, cough, and chest tightness.[10] Except for these anecdotal reports, there is no other information available on changes in pulmonary function, airway responsiveness, or prevalence of respiratory symptoms among respiratory therapists delivering AP.

Following oral inhalation of AP, bronchoalveolar ravage fluid concentrations of the drug are substantially higher (at least 5–10 times) than those attained following IV administration.[11] Pentamidine appears to undergo limited absorption from the respiratory tract into the systemic circulation; peak plasma concentrations appear to occur at completion of the administration and appear to be 5% or less of the levels attained following IV administration.[11]

To date, very little is known about the mutagenic, teratogenic, or carcinogenic effects of AP. One report stated that pentamidine was negative in the Ames assays.[12] The report did state, however, that the drug does have the ability to inhibit dihydrofolate reductase. Other similar antimetabolites are known reproductive toxic agents.[9] The LD_{50} in mice is 100 mg/kg intraperitoneal. The MSDS from the manufacturer states that chronic information on the toxicity of the drug is not known. Other symptoms reported in patients receiving IV pentamidine include cardiac arrthymias, acute renal failure, hypotension, hypoglycemia, nausea, and a metallic taste in the mouth (Lyphomed MSDS). The manufacturer cautions against ingestion, inhalation, and skin absorption. Information from the American Society of Hospital Pharmacists cautions female workers about handling pentamidine, given the current lack of information of the reproductive toxicity of the drug.[12] One *in vitro* study showed that pentamidine was transported across the isolated placenta.[13] Other researchers reported a small increase in postimplantation loss in a study of pregnant rats.[14]

Exposure Assessment

There have been several reports to date of exposure to AP among HCWs. In 1989, a study was conducted at San Francisco General Hospital to assess exposure to AP.[2] Four hour area samples were collected on three separate sampling days. The researchers used an Anderson cascade impactor with paraffin on each of the sampling stages. They collected the samples at 92 L/min. The plates were eluted with 10 mL of distilled water and the eluent was analyzed by high pressure liquid chromatography (HPLC), following a method previously developed for the analysis of pentamidine isethionate in blood and urine.[15] They reported an average concentration of $4.5 \pm 3.6 \times 10^{-2}$ μg/m^3. The MMAD for the sampled AP was reported to be 8.0 μ ± 2.7.

In 1990, researchers at Johns Hopkins collected area and breathing zone AP samples before and after instituting engineering controls. Samples were collected on glass fiber (GF) filters and extracted with sterile water. The eluent was analyzed by UV spectroscopy. The researchers reported obtaining AP levels ranging from non-detectable to 45 μg/m^3 in area samples and from non-detectable to 18 μg/m^3 on personal samples in treatment rooms where no controls were in place. The number of treatments per day ranged from 8 to 17. The researchers measured again after installing engineering controls. All of these later samples were below the limit of detection (LD < 0.33 μg/m^3).[16]

A third study was conducted at a local pentamidine clinic by researchers from McGill University in Montreal in May 1990. Personal samples were collected with Marple personal cascade impactors, using a 34 mm (0.8 μ pore size) polyvinyl chloride (PVC) substrate. Area samples were collected on 37 mm cassettes containing PVC filters (0.8 μ pore size). Flow rates were 3 L/min for the cascade impactor and 3.4 L/min for the filter cassettes. Samples were collected on three different occasions. The samples were eluted with HPLC-grade distilled water and analyzed by HPLC, following techniques developed by Lin et al.[15] The nebulizer in use was the FISONeb ultrasonic nebulizer, which uses a 60 mg treatment (as compared with the 300 mg treatment used with the Respirgard II nebulizer). When the sampling was conducted the treatment room was under negative pressure and had approximately 14 air changes per hour (ACHs). Between 11 and 15 patients received treatment on the days of sampling. The HCW spent most of the day in the treatment room supervising patients. Average pentamidine concentrations of 0.231 μg/m^3 ± 0.081 were reported for the personal samples. The average for the area samples in the treatment room were 0.499 μg/m^3 ±0.1. No pentamidine was detected in the hallway outside of the treatment room. The detection limit for the method was <0.5 ng/mL.[3]

A fourth study by McIvor et al.[4] also found measurable levels of pentamidine in one AP treatment facility in Toronto. The clinic also used the FISONeb ultrasonic nebulizer. Ambient air samples were collected in various locations in this facility: the treatment rooms, the waiting room for the AP clinic, and in the corridor outside the clinic. Samples were collected in impingers filled with 15 mL of distilled water at a flow rate of 1 L/min. The samples were collected over an 8 h period. The samples were analyzed by HPLC, with a detection limit of 20 ng/mL. The AP treatment rooms are separate from each other and each contains a high-volume air filtration unit, which is equipped with HEPA filters. The doors to the three treatment rooms are closed during the AP administration. On the day of the study, 38 patients were treated with AP. The level detected in the treatment room itself was 17 μg/m^3, while a level of 13 μg/m^3 was detected in the waiting room, and 1.6 μg/m^3 was detected in the corridor outside the clinic. Between 1.3 and 4.5 μg/m^3 were detected in other locations in the building. No personal samples were collected.[4]

UCSF Health Effects Study

In 1990, researchers from University of California, San Francisco were requested to study potential adverse health effects from exposure to AP. The overall objective of this study[18] was to assess the extent of occupational risks to HCWs associated with AP administration. The study was designed to measure personal and general room area concentrations of pentamidine during nebulizer administration to patients, determine urine concentrations of pentamidine among HCWs administering AP, and assess the potential for respiratory effects among HCWs from secondary exposure to AP.

A total of 16 workers from 9 facilities in Northern California participated in the study. The amount of time spent administering pentamidine ranged from 3 to 40 h per week. There also was a wide range in the average number of patients treated per day at each facility. The larger facilities averaged more than 25 per day, while the smaller facilities averaged as few as 2 per week.

The respirator outcomes measured included nonspecific airway responsiveness (measured by methacholine inhalation challenge), cross-shift forced expiratory volume in 1 s (FEV_1) and forced vital capacity (FVC), carbon monoxide diffusing capacity (DL_{co}), and peak expiratory flow rates (PEFR). The subjects also kept a log book for 3 weeks of the study to record symptoms on work and nonwork days. Urine samples were collected at the beginning and end of the workweek for pentamidine testing.

Personal breathing zone and area samples were collected for each subject during the entire shift while administering pentamidine. The time of collection ranged from 2 to 12 h, depending on the length of the shift. Concurrent with the collection of the environmental data, assessment of the ventilation and measurements of the temperature and relative humidity were undertaken. The number of treatments administered on the sampling days was also recorded.

As there were no NIOSH sampling and analytical methods for pentamidine, the researchers worked with NIOSH on the development of a method.[17] Personal and area samples were collected on Gelman Type AE GF filters at 2 L/min, for the entire workshift. (In the later part of the study, samples were collected on both PVC and GF substrates.) Opaque cassettes were used because it was thought at the time that pentamidine might degrade if exposed to light. To determine the size of the aerosol, a Marple personal cascade impactor (model 298, Andersen) was used for personal sampling concurrent with the GF cassette. Mylar substrate was placed on 4 stages of the 8-stage impactor. The stages used were 1, 2, 3, 6, and the backup filter with corresponding cutoff of 20, 15, 10, 2, and <2 μ, respectively.

The samples were analyzed by an HPLC process, which was a modification of an existing method developed to measure pentamidine in plasma.[15] The extraction solute consisted of 3 mL of methanol or ethanol mixture, with small amounts of sodium heptane sulfonate, phosphoric acid, and tetramethyl ammonium chloride. The beakers were sonicated, the filters were removed, and the extract was analyzed by HPLC with fluorescence detection. The recently completed NIOSH method for collection of AP calls for the use of PVC filters and the analysis is similar to that described above.

Results

Environmental Sampling

Personal breathing zone samples from the facilities ranged from non-detectable to 132 μg/m^3. Area samples obtained in the treatment rooms at these facilities ranged from non-detectable to 95.6 μg/m^3.

TABLE 15.2 Airborne Exposure Levels and Conditions

Site	Average Number of Treatments	ACH	Average of Personal Samples (μg/m³)[a]	Average of Area Samples (μg/m³)[a]	LEV	HCW in Room During Treatment
A (pre)	25	2.4	9.59	5.71	No	Yes
A (post)	22	2.4	ND (below LOQ)[b]	0.102	Yes, stationary booths	Yes
B	16	15	36.2	30.93	No	Yes
C	8	28	0.84[b]	ND[b]	Yes, portable enclosed booth	No
D	7	9	0.675	0.39	No	No
E	24	6.8	23.82	38.77	No	Yes

[a] All the results reported are for pentamidine isethionate; there is 1 mg of pentamidine in 1.74 mg of pentamide isethionate.

[b] This result represents only one sample.

The limit of detection (LOD) was 8 ng/sample and the limit of quantification (LOQ) was 50 ng/sample. (These results are reported as pentamidine isethionate; there is 1 mg of pentamidine in 1.74 mg of pentamidine isethionate.) For the Marple cascade impactor samples, the majority of the sample was collected on the backup filter, with the largest portion of the remainder on stage 6 (cutoff of 2 μ) indicating a probable MMAD of <2 μ. The number of patients treated on the day of sampling ranged from a low of 2 to a high of 37. The room ventilation rates ranged from 2.4 to 62 ACH. Personal protective equipment (PPE) was worn at only one institution.

Airborne exposure levels and conditions on the day of sampling for five of the institutions are shown in Table 15.2. As can be seen from the table, two of the sites had local exhaust ventilation (Sites A and C). The average AP levels, both area and personal, were lower at Site A after the installation of the stationary local exhaust system. A commercial, portable booth was used at Site C. The area sample collected in the room outside the booth was non-detectable, and the personal sample was below the levels measured at most of the other institutions. For three of the institutions (Sites A, B, and E), the HCWs remained in the room during the treatments, as their desks were in the treatment rooms. In the other two institutions, the HCW would enter the room to connect or disconnect the nebulizer from the patient; otherwise they remained outside of the treatment room, usually in an adjoining room.

Medical Results

Pentamidine was not detected in the urine samples analyzed by the USCF analytical laboratory (LOD of 229 ng/mL). Results of pulmonary function testing showed a mean decrease in cross-shift FEV_1 on days that the HCWs were administering pentamidine. This decrease was significant when compared to the mean cross-shift change in FEV_1 on days when the HCWs did not administer pentamidine. There also was greater diurnal variation in the PEFR on the days they administered AP, as compared to the days when they did not. The clinical significance of these findings remains to be determined. There were no significant changes detected in DL_{co}, methacholine responsiveness, or FVC that could be related to exposure to AP.

Discussion

This study detected measurable levels of pentamidine in the ambient air of the treatment room as well as on the personal breathing zone samples, as did other studies. The results from this study appear to be higher for many of the facilities investigated than the results obtained in either the Montreal or the Toronto study. There were several differences between the Canadian clinics and the clinics that participated in the UCSF study, which may have contributed to the differences in results. As mentioned earlier, a smaller dose of pentamidine is administered using the FISONeb ultrasonic nebulizer than when the Respirgard II nebulizer is used (60 vs. 300 mg/treatment, respectively). Additionally, the differences among the ventilation systems in these facilities may also have contributed to the differences in the results.

The results varied widely among the facilities studied by UCSF. There was no direct correlation between number of treatments of ACH with either breathing zone or area exposure levels. For example, Site B (see Table 15.2) administered fewer treatments than did Site A (16 vs. 25) and had more ACH (pre-LEV [local exhaust vent]), yet both the breathing zone and area samples were higher than those at Site A (pre-LEV). From this author's observations of the work practices and patient practices at each facility, it is unlikely there were significant differences between the facilities in these practices.

As reported above, the results of the pulmonary function testing did show a relationship between AP exposure and several significant pulmonary endpoints (FEV_1 and PEFR). There was, however, no clear-cut dose–response relationship between exposure and these changes in pulmonary function.

There is no recommended exposure level for AP. However, given the uncertainty concerning the possible reproductive, carcinogenic, and other chronic health effects, coupled with the pulmonary function changes reported in the UCSF study, and given the fact that without controls in place, pentamidine has been detected in the ambient air in AP treatment rooms, it is best to minimize exposure. Additionally, and perhaps more significantly, controls should be put in place to reduce the potential for transmission of TB.

Recommendations

Nebulizer Valve Controls

Automatic shutoff valves for the nebulizer may help to reduce ambient AP levels in the treatment room. There are some valves available that are difficult for the patients to use and thus have frequently been removed from the devices. Other refinements of the nebulizer system, such as nebulization only during inspiration, should be investigated.

Administrative Controls

Prompt recognition and treatment of TB in HIV-positive patients should help in reducing transmission of TB. Screening and surveillance of workers should place emphasis on their respiratory complaints. Workers with symptoms should be more extensively evaluated and may need to be offered alternative assignments. Reproductive policies, such as allowing pregnant workers the right to transfer to other jobs, may need to be available until such time as it is established that exposure does not pose a reproductive risk. Finally, all workers administering pentamidine as well as those handling pentamidine waste should be instructed in proper work procedures.

Personal Protective Equipment

If proper engineering and administrative controls are instituted, there should be no need for PPE. However, there may be situations where exposure cannot be prevented. In these instances, it may be necessary to wear PPE to prevent exposure to AP and TB. Current recommendations include the use of a HEPA mask, such as the 3M 9970.

References

1. Nightingale SL. From the Food and Drug Administration. *J Am Med Assoc* 262:184, 1989.
2. Montgomery AB, Corkey K, et al. Occupational exposure to aerosolized pentamidine. *Chest* 98: 386–388, 1990.

3. Boulard M. Evaluation of exposure of health care personnel to aerosolized pentamidine isethionate (an antiprozoal agent) at hospital X. Master's thesis, School of Occupational Health, McGill University, Montreal, Quebec.
4. McIvor RA, et al. Risk of second hand exposure to aerosolized pentamidine. Report from the Toronto Central Aerosol Pentamidine Clinic, Toronto, Canada, 1991.
5. Gude JK. Correspondence: Selective delivery of pentamidine to the lung by aerosol. *Am Rev Respir Dis* 139:1060, 1989.
6. Montgomery, et al. Aerosolized pentamidine as second line therapy in patients with AIDS and pneumocystis carinii pneumonia. *Chest* 95:747–750, 1989.
7. Smith DE. Reversible bronchoconstruction with nebulized pentamidine. *Lancet* 11:905, 1988.
8. Girard PM, et al. Prevention of *Pneumocyctis carinii* pneumonia relapse by pentamidine aerosol in zudovudine treated AIDS patients. *Lancet* I:1348–1352, 1989.
9. McDiarmid M, et al. Aerosolized pentamidine in public health. *Lancet* I:863–864, 1989.
10. Rose L. Inhalation hazard control for health care workers during the administration of aerosolized pentamidine. Cal/OSHA Report, San Francisco, CA, 1991.
11. Conte JE, et al. Concentration of AP in bronchoalveolar ravage, systemic absorption and excretion. *Antimicrob Agents Chemother* 32:1490–1493,1988.
12. Anonymous. Pentamidine isethionate, In: GK McEvoy (ed.), *American Hospital Formulatory Service Drug Information*. American Society of Hospital Pharmacists, Bethesda, MD, 1990, pp. 458–466.
13. Fortunato SJ, et al. Determination of pentamidine transfer in the in vitro perfused human cotyledon with high-performance liquid chromatography. *Am J Obstet Gynecol* 160:759–761, 1989.
14. Harstaad TW, et al. Embryofetal effects of pentamidine isethionate administration to pregnant Sprague-Dawley rats. *Am J Obstet Gynecol* 163:912–916, 1990.
15. Lin, et al. High performance liquid chromatography determination of pentamidine in plasma. *J Liq Chromatogr* 9:2035–2046, 1986.
16. McDiarmid MA, personal communication.
17. NIOSH. Analytical Method 5032, *NIOSH Manual of Analytical Methods*, Cincinnati, OH, 1994.
18. Centers for Disease Control. *Mycobacterium tuberculosis* transmission in a health clinic—Florida, 1988. *Morb Mortal Wkly Rep* 38:256–264, 1989.

Verification of the Capture Efficiency of HEPA Filtration for Aerosolized Pentamidine

William Charney

Aerosolized medications, now being used for medical therapies, can pose clinical health risks to hospital personnel administering the therapies. Ribavirin, used in the treatment of RSV, is a teratogen,[1] and pentamidine, used in the treatment of PCP, has had serious enough occupational side effects that California OSHA adopted special guidelines for the administration of the drug.[2] The generic engineering design for both ribavirin and pentamidine incorporates HEPA filtration. This study tested the capacity of HEPA filtration to block aerosolized pentamidine isethionate.

Pentamidine isethionate, as AP, has been approved by the Food and Drug Administration[3] for prophylaxis of PCP in patients infected with the HIV. Pentamidine is an aromatic diamidine derivative antiprotozoan agent. It is normally aerosolized at $0.5\,\mu^3$ and in air studies the MMAD was shown to be around $2.0\,\mu$ (see "Medical Results" section on page 15-16).

Due to the aerosolization of the drug, there is greater potential for exposure. For aerosolization, the lyophilized pentamidine isethionate, usually 300 mg, is reconstituted with 6 mL of sterile water to create the solution. The solution is then placed in a nebulizer. Pentamidine has been detected in the ambient air in AP therapy rooms.[4–7]

Study Rationale

Local ventilation designs to decrease ambient air levels of pentamidine have utilized HEPA filters. However, there is no documentation that HEPA filters will in fact capture the AP. This study's objective was to determine the effectiveness of HEPA filtration to capture pentamidine when aerosolized in the clinical setting.

Sample Strategy

Pentamidine isethionate ($C_{23}H_{36}N_4O_{12}S_2$) was nebulized upstream from the HEPA filter. The pentamidine aerosol was drawn through the HEPA filter by negative pressure created by a fan (Figure 15.1). The fan system is the same type that would be found in a local exhaust ventilation system. An aerosol particulate analyzer, QCM, model PC24 probe was placed at the exhaust grill of the engineering system in order to confirm breakthrough of pentamidine. This is a direct readout instrument by cascade impact system that reads particulates as frequency changes.

Analyzer Parameters and Method

The cascade impactor of the PC-2 is a series of 10 aerodynamic inertial impactors designed in accordance with well-established principles in aerosol technology. The 10 impactor stages are arranged in a cascade, with jets that segregate the larger aerosol particles on top. The jet diameters of the impactor stages become progressively smaller in the lower stages. Unlike conventional cascade impactors, in which aerosol samples are collected over periods of minutes or hours on filters or other mechanical collection plates and weighed externally, the PC-2 impactor system utilizes "active" piezoelectric crystal sensors in each stage for sample collection within seconds, providing real-time signals proportional to the mass of the aerosol samples collected.

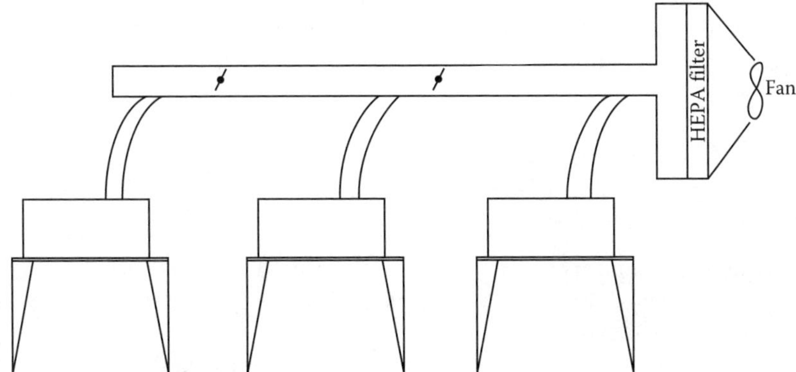

FIGURE 15.1 Pentamidine delivery system design. Specifications: (1) negative pressure, (2) HEPA filtration, (3) 1/4 HP fan for 400 cfm, (4) flexible ducting 4 in., (5) main duct 6 in., (6) air recirculates into the room after HEPA filtration, and (7) approximate cost for three stations is $5000.

The aerosol stream entering the instrument encounters the largest nozzle first, and the nozzles become progressively smaller in subsequent stages. After passing through the 10th (lowest) stage, the air flows through a flow meter to a pump, where it is exhausted to the atmosphere.

The capture efficiency vs. particle size for the standard 10-stage instrument is shown below. It is customary to designate a stage by the size at which there is a 50% probability of capture of particles of a specified mass density.

Stage	Dp50 (μ)	Stage	Dp50 (μ)
1	25.0	6	0.80
2	12.5	7	0.40
3	6.3	8	0.20
4	3.2	9	0.10
5	1.6	10	0.50

n Frequency signals, one from each stage, emanate from the sensing stack and are fed to *o*, the control unit, where they are stored and processed by a microprocessor. By monitoring and recording the frequency from each stage before and after aerosol sampling, the range in frequency in each stage caused by mass loading is calculated by the microprocessor automatically. The sampling time is also fed to the microprocessor through the setting of the thumbwheel switch on the control unit's front panel. The ratio of frequency change over time is used by the microprocessor to automatically calculate the mass concentration, in units of $\mu g/m^3$, in each stage, using the following relationships:[8]

$$\Delta f / t = \sigma V C$$

where

Δf is the frequency change (Hz)
σ is the sensitivity factor
t is the sample time (min)
C is the concentration (mg/m^3)
V is the sample flow rate (L/min)

Background samples were taken with the aerosol particulate analyzer to confirm room background particulate. Any frequency counts larger than these background counts were considered to be pentamidine. The sensitivity of the aerosol particulate analyzer to the pentamidine particulate was confirmed in a closed test chamber. Pentamidine was nebulized. The analyzer reacted at all 10 impactor stages.

Results

As shown by the computer readouts of the aerosol particulate generator, there was no breakthrough of pentamidine through the HEPA filter.

Readout #1 is the background of room air with the exhaust of the system not running. The readings in the left-hand column are frequency readings (*F*) that change in reaction to mass particles.
Readout #2 is a confirmation of background.
Readout #3 is reading at exhaust port with exhaust motor running as background.
Readout #4 is a confirmation of #3.
Readout #5 is nebulizing pentamidine for 10 s and sampling at exhaust port for 5 s.
Readout #6 is nebulizing pentamidine for 20 s and sampling at exhaust port for 10 s.
Readout #7 nebulizes for 90 s and samples for 30 s.
Readout #8 nebulizes for 75 s and collects sample 60 s.

#1		#2		#3		#4		#5		#6		#7		#8	
CH1		CH1		CH1		CH1		CH1		CH1		CH1		CH1	
$\Delta T = 030$		$\Delta T = 030$		$\Delta T = 030$		$\Delta T = 030$		$\Delta T = 030$		$\Delta T = 010$		$\Delta T = 030$		$\Delta T = 060$	
Total	0.02050	Total	0.02146	Total	0.00348	Total	0.00618	Total	0.00066	Total	0.19590	Total	0.01126	Total	0.00335
CH ΔF	mg/m^3	CH ΔF	mg/m^3	CH ΔF	mg/m^3	CH ΔF	mg/m^3	CH ΔF	mg/m^3	CH ΔF	mg/m^3	CH ΔF	mg/m^3	CH ΔF	mg/m^3
1-0001-	0.00140	1 003	0.00420	1-0005-	0.00700	1 0000	0.00000	1-0006-	0.00840	1 0001	0.00420	1 0005	0.00700	1 0001	0.00070
2 0000	0.00000	2 0000	0.00000	2 0000	0.00000	2 0001	0.00134	2 0000	0.00000	2 0000	0.00000	2 0001	0.00134	2 0000	0.00000
3 0001	0.00088	3.0000	0.00000	3 0000	0.00000	3 0000	0.00000	3–0001-	0.00088	3 0000	0.00000	3 0000	0.00000	3 0000	0.00000
4 0003	0.00198	4 0002	0.00132	4 0000	0.00000	4 0001	0.00066	4 00001	0.00066	4 0000	0.00000	4 0001	0.00066	4-0001-	0.00033
5 0005	0.00290	5 0004	0.00232	5 0001	0.00058	5 0002	0.00116	5-0001-	0.00058	5 0000	0.00000	5 0000	0.00000	5-0011-	0.00319
6 0002	0.00112	6 0000	0.00000	6-0002-	0.00112	6 0001	0.00056	6-0003-	0.00168	6-0001-	0.00168	6-0001-	0.00056	6 0002	0.00056
7 0011	0.00594	7 0011	0.00594	7 0003	0.00162	7 0001	0.00054	7 0000	0.00000	7 0001	0.00162	7 0003	0.00162	7 0003	0.00081
8 0005	0.00320	8 0006	0.00384	8 0000	0.00000	8 0001	0.00064	8-0001-	0.00064	8 0099	0.19008	8 0000	0.00000	8 0003	0.00096
9 0005	0.00320	0 0004	0.00256	9 0000	0.00000	9 0002	0.00128	9 0000	0.00000	9-0001-	0.00192	9 0001	0.00064	9 0001	0,00032
10 0002	0.00128	10 0002	0.00128	10 0002	0.00128	10 0000	0.00000	10-0002	0.00128	10 0000	0.00000	10 0000	0.00000	10-0002-	0.00064

Discussion of Results

Comparing room background levels with those at the exhaust port of the system shows no elevation. The interpretation of these results is that the HEPA filter is capturing all the pentamidine at the 10 different micron range sizes of the cascade impactor (14.00–0.07 m).

References

1. Ferm VH. Teratogenic effects of ribavirin on hamsters and rat embryos. *Teratology* 17:93–102, 1978.
2. Rose L. Inhalation hazard control for health care workers during administration of aerosolized pentamidine. Cal/OSHA Report, San Francisco, CA, 1991.
3. Nightingale SC. Food and Drug Administration. *J Am Med Assoc* 2621–2814, 1989.
4. Montgomery AB. Occupational exposure to AP. *Chest* 98:386–388, 1990.
5. Bouland M. Evaluation of exposure of health care personnel to AP at Hospital X. Master's thesis, School of Occupational Health, McGill University, Montreal, Quebec.
6. McIvor RA. Risk of secondhand exposure to aerosolized pentamidine. Report from Toronto Central Aerosol Pentamidine Clinic, Toronto, Canada, 1991.
7. McDiarmid MA. Aerosolized pentamidine and public health. *Lancet* 1:863–864, 1989.
8. Cascade Impactor Information Design Bulletin.

Engineering Controls

William Charney

Appropriately designed ventilation systems are essential to the reduction of AP exposure. As minimums, the CDC guidelines on reducing TB transmission recommend that the treatment room be under negative pressure, have six ACHs, that the room be exhausted away from the building intakes, and that filters be placed in the exhaust. Additionally, they recommend that HCWs in the treatment room wear appropriate protective equipment, including respirators and protective clothing. These recommendations will help to prevent AP and TB bacillus from leaving the AP treatment room. However, these recommendations will not prevent transmission between patients in the treatment room, or to the staff present in the treatment room who are not wearing appropriate protective equipment.

Properly designed local exhaust ventilation systems are the preferred method of control for reduction of exposure to both AP and TB. There are several commercially available booths, which are under negative pressure and which recirculate the exhaust air through a HEPA filter prior to releasing the air back into the room. One such booth is the Demistifier™ isolation chamber (Figure 15.2), manufactured by Peace Medical. It exhausts approximately 220 cfm through a pre-filter, HEPA filter, and charcoal filter. It comes with disposable canopy, which they recommend changing frequently. It is easily moved from room to room. Another fairly portable device is the Emerson aerosol treatment chamber, model 7AT (Figure 15.3). It too consists of a pre-filter and HEPA filter and exhausts 150–270 cfm through these filters. San Francisco General Hospital and Medical Center has designed their own booths approved by Cal/OSHA. Such a booth is shown in Figure 15.4. As with the other two booths, the air is drawn through a pre-filter and a HEPA filter before being exhausted into the room. The difference with this design is that the patient is not completely enclosed. Figure 15.5 is another design. Finally, Figure 15.6 shows a portable aerosol treatment station. It too exhausts the air through a pre-filter and HEPA filter and recirculates the air back into the room. Again, with this and any recirculating system, it is important to test the HEPA filters on a regular basis to ensure their effectiveness.

FIGURE 15.2 Demistifier isolation chamber from Peace Medical, \$1200.

FIGURE 15.3 Emerson model 7-AT aerosol treatment and sputum induction chamber, \$4000.

FIGURE 15.4 AP ventilation control system designed by SFGHMC.

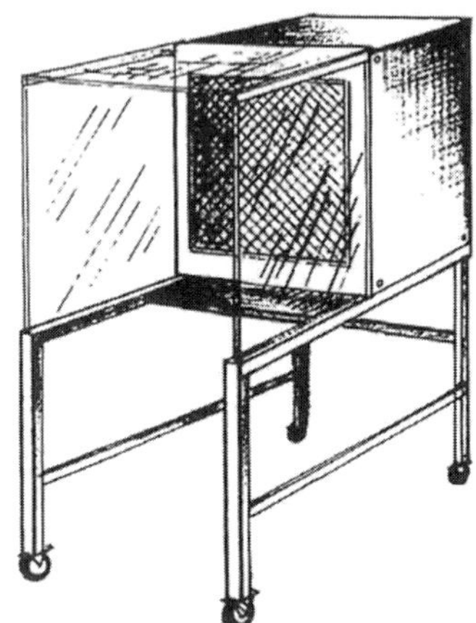

FIGURE 15.5 Aerosol treatment guard from HR, Inc., $2650.

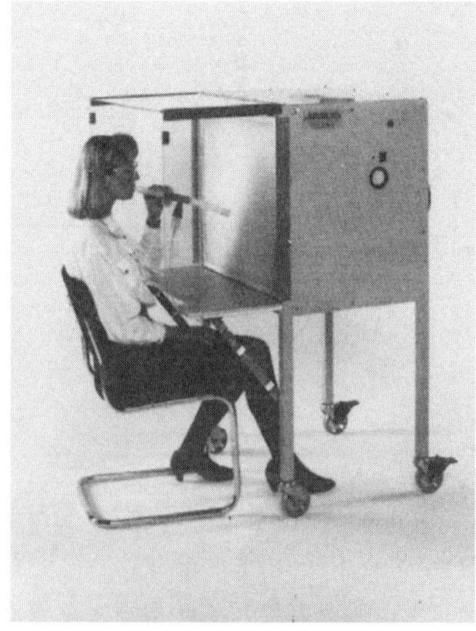

FIGURE 15.6 Portable aerosol treatment station.

16

Ribavirin

Ribavirin Aerosol: Does a New Method of Drug Delivery Place Workers at Risk? **16**-1
Summary • Introduction • Toxicology • Acceptable Exposure Levels • Exposure Assessment Methods • Exposure Levels • Reported Symptoms • Discussion and Conclusions • Recommendations • Policies in Place
References **16**-8
Appendix 16.A.1 .... **16**-9
Appendix 16.A.2 .... **16**-10
Ribavirin Aerosol Engineering and Administrative Controls to Contain Aerosolized Ribavirin: Results of Simulation and Application to One Patient **16**-13
Introduction • Materials and Methods • Results • Discussion • Conclusion
Acknowledgments **16**-18
References **16**-19
Validation of Control System for Ribavirin Aerosol **16**-20
Reference **16**-21
Appendix 16.A.3 .... **16**-21
Appendix 16.A.4 .... **16**-40

Jim Bellows
California Department of Health Services

Robert J. Harrison
California Department of Health Services

William Charney
Healthcare Safety Consulting

Kevin J. Corkery
San Francisco General Hospital

Lee Wugofski
San Francisco General Hospital

Ribavirin Aerosol: Does a New Method of Drug Delivery Place Workers at Risk?

Jim Bellows and Robert Harrison

Summary

In 1987, the California Occupational Health Program (COHP), California Department of Health Services (CDHS), conducted a study of health care workers' (HCW) exposures to the antiviral drug ribavirin. HCWs at several hospitals had raised concern about the safety of ribavirin, based on three factors: the drug is a potent animal teratogen; it is usually administered as a respirable aerosol, with the excess discharged into patient care areas; and the staff in those areas often include pregnant women.

Air monitoring and biological monitoring were used to assess the exposures of 10 nurses and two respiratory therapists (RTs) at four hospitals. The highest exposures (mean 161 μg/m^3) occurred when the drug was administered via oxygen tent. For the nurse with the highest exposure (316 μg/m^3), the estimated daily absorbed dose was 30 μg/kg, about 30% of the dose levels that have produced fetal resorptions and malformations in animals. Ribavirin was not detected in any urine or serum samples, but 0.44 μg/mL was present in the red blood cells of the nurse who had the highest measured airborne exposure.

Exposures were lower during delivery via mist mask and much lower during delivery via mechanical ventilator (mean 4 μg/m^3). No symptoms were reported by any of the workers monitored, but other HCWs handling ribavirin have reportedly experienced a variety of symptoms, including skin and eye irritation and difficulty breathing.

Because little margin of safety is maintained, worker exposure to ribavirin during administration via oxygen tent may pose a reproductive risk to HCWs. This should be considered one of the most serious reproductive hazards yet identified in hospitals. Workers should be advised of the reproductive risk, and comprehensive policies should be implemented to protect those who are pregnant or are actively attempting to conceive. Several specific control measures are recommended at the end of this section.

Introduction

Ribavirin (1-b-D-ribofuranosyl-1,1,4-triazole-3-carboximide, Virazoleä™) is a synthetic nucleoside analog which inhibits a broad spectrum of viral infections. The drug was approved in 1986 by the U.S. Food and Drug Administration (FDA) for aerosol treatment of infants and young children with severe respiratory syncytial virus (RSV) infection. RSV is rarely serious in adults and healthy children, but hospital outbreaks can have substantial fatality rates. Most cases occur during the winter.

An aerosol generator (SPAG IIä™) supplies respirable particles (mass median diameter 1.3 μm) at 12.5 L/min,[1] approximately two to five times the minute volume of a typical pediatric patient. During administration by oxygen tent or mist mask, the excess is exhausted directly into the room. When the drug is delivered via ventilator (in the United States, currently approved only on an investigational basis), the excess aerosol can be filtered to limit the release of ribavirin.[2] Duration of treatment is generally 3–5 days for 12–20 h each day, although the drug may be administered for longer periods. Patients are typically treated in pediatric intensive care units (PICUs).

Despite some controversy about the efficacy of the drug, ribavirin treatment for RSV is apparently increasing, both in the number of hospitals using the drug and in the number of patients treated. Clinical trials are also in progress for oral or intravenous ribavirin treatment of other viral diseases, including AIDS.

Ribavirin has been found to cause fetal resorptions or malformations in many rodent species.[3,4] Based on the animal data, use in pregnant women is contraindicated.[5] Ribavirin also causes tubular atrophy in the testes of adult rats.[3]

In one previous study of occupational exposure, no ribavirin was detected in the red blood cells, plasma, or urine of nurses caring for patients receiving ribavirin aerosol; air samples were not collected.[6] Absorbed ribavirin is cleared rapidly from human plasma but accumulates in red blood cells and other tissues, reaching a plateau 3 days after administration of a single dose.[7] From red cells, it is eliminated with a half-life of approximately 40 days.[1]

During 1986, the CDHS/COHP received several inquiries about ribavirin. Two principal questions were asked: Are pregnant workers who handle the drug at risk for adverse pregnancy outcomes? and What precautionary measures should be taken? Insufficient data were available to provide a solid basis for answering either question. A study to evaluate the occupational health risks was requested by the Associate Director of Pharmaceutical Services of a San Francisco medical center and was conducted in the winter of 1986–1987.[8] Three other area hospitals were eventually added to the study to include a diversity of exposure conditions.

Toxicology

In experimental animals, the most sensitive known toxicity endpoints are reproductive: embryolethality, teratogenesis, and testicular atrophy. In rabbits, daily oral doses of 1.0 mg/kg during gestation increased fetal resorptions, but 0.3 mg/kg did not.[3] In hamsters, single intraperitoneal doses of 1.2 mg/kg (the lowest tested) induced both fetal resorptions and gross structural malformations.[4] Similar effects have been observed in other animals at comparable or slightly higher dosages. Reproductive effects in all species occurred in the absence of maternal toxicity.

In baboons given 60–120 mg/kg daily for 4 days during various stages of gestation (including some *after* the most sensitive period of organogenesis), one of seven fetuses aborted spontaneously and no major malformations were observed.[3] However, this study cannot be considered to demonstrate absence of reproductive toxicity in primates, because of the small number of animals and the timing of the doses.

Daily administration of 16 mg/kg (lowest dose tested) produced tubular atrophy in the testicle of adult rats.[3] Carcinogenesis bioassays are pending.

Disposition of ribavirin in humans and nonhuman primates is significantly different from that in rodents. Rats eliminate most ribavirin in urine within 24 h.[9] By contrast, human volunteers eliminated only 33% after 24 h.[7] The drug has been shown to accumulate in the red blood cells of humans and monkeys.[7,9]

Effects of ribavirin on pregnancy or testicular function have not been studied in humans. In clinical trials, ribavirin has been administered to adult humans orally, by inhalation, and by intraperitoneal injection. It is well absorbed by all routes, and has been shown to cause skin rash, eye irritation, anemia, and (in adults with underlying pulmonary disease) decreased pulmonary function.[1]

Acceptable Exposure Levels

No government agencies or professional associations have established regulations for occupational exposure to ribavirin aerosol. In order to evaluate the worker exposures measured in its study, COHP derived a provisional exposure limit of 2.7 μg/m³.[10] This exposure limit is not an enforceable standard, but is meant to be used as a working guideline, representing a concentration unlikely to produce adverse reproductive effects.

In the absence of human data regarding reproductive toxicity, COHP's provisional exposure limit for ribavirin aerosol was based on the application of a safety factor to the experimental no-observed-effect level (NOEL) for the most sensitive endpoint (embryolethality) in the most sensitive species tested (rabbits). A safety factor of 1000 was selected, based on guidelines issued by the U.S. Environmental Protection Agency (EPA).[11,12]

Although the risks are most obvious for pregnant women, a single airborne ribavirin concentration was selected as the exposure limit for all employees. This was based on the considerations that other women could be subject to reproductive toxicity during as-yet-undetected pregnancies, and that testicular atrophy has been observed in experimental animals at the lowest dose tested.

No limits were proposed for ribavirin concentrations in blood or urine because pharmacokinetic parameters, which could relate absorbed dose to ribavirin concentrations in blood and urine, are not known for low doses. Distribution and elimination have been investigated at much higher therapeutic doses, but they are known to be dose-dependent. Even using the available high-dose pharmacokinetic values, which tend to overestimate the ribavirin concentrations that would be present in blood and urine, full-shift exposure at the provisional exposure limit (2.7 μ/m³) would not be expected to produce detectable concentrations in biological fluids.[10]

Exposure Assessment Methods

Air monitoring and biological monitoring were conducted by COHP in four PICUs. Exposures of 10 nurses and 2 RTs were evaluated. Each cared for a ribavirin-treated patient for two consecutive days during his or her participation in the study, but had cared for no such patients during the previous 30 days; none were pregnant. Participants were instructed to perform their patient care duties according to standard hospital policies, without any change in work practices. At the end of the work shift, each participant recorded the amount of time spent at the patient's bedside, the number of times the drug delivery system was disturbed, and any use of gloves or face masks. Decisions about route of administration and duration of therapy were made by the treating physicians.

To measure airborne exposures, full-shift air samples were collected in participants' breathing zones and at the patients' bedsides. Ribavirin was analyzed by high performance liquid chromatography (HPLC), with a quantitative limit of detection of 1 μg/m^3. In four cases, paired samples were collected; these were analyzed by two independent methods, HPLC and radioimmunoassay (RIA), to validate the analytical techniques. To evaluate ribavirin absorption, plasma, red blood cell, and urine samples were obtained before and immediately following the first work shift, and 3–7 days later; all were analyzed by RIA, with a quantitative limit of detection of 0.02 μg/mL (0.1 μM).[13] Daily absorbed doses were estimated as

$$D = c \times t \times V \times W^{-1} \times 70\%$$

where

D is the estimated absorbed dose (μg/kg)
c is the measured average air concentration (μ/m^3)
t is the exposure time (min)
V is the standard minute volume (0.019 m^3/min for women, 0.020 m^3/min for men)[13]
W is the standard body weight[14]
70% is the fraction of the inhaled dose that is believed to be absorbed[15]

For participants who wore surgical masks, a mask penetration of 50% was assumed, although the fraction of respirable particles penetrating the masks might be greater.[16]

Exposure Levels

Participants' work shifts lasted 7–12 h, of which an average of 50% (range 20%–80%) was spent at the bedsides of the treated patients. Surgical masks were worn by four of eight subjects caring for patients receiving ribavirin by oxygen tent and two of three nurses caring for ventilated patients. Oxygen tents were lifted to perform patient care duties 5–17 times per shift. All ventilators were equipped with filters to remove ribavirin from exhausted air, according to an established method.[2]

Airborne exposures of the participants are summarized in Table 16.1. The highest exposures occurred during administration via oxygen tent (mean 161 μ/m^3). The highest individual exposure (316 μg/m^3) was measured for a nurse whose work practices seemed similar to others' during this mode of administration. Exposures were lowest for three nurses caring for ventilated patients (mean 4 μg/m^3). One nurse caring for a patient receiving ribavirin by mist mask was exposed to a concentration of 62 μg/m^3. Overall, the nurses' exposures were comparable to the RTs'.

General area samples showed ribavirin concentrations generally higher than the corresponding personal samples, averaging 317 μg/m^3 during administration via oxygen tent.

For the participants who cared for patients receiving ribavirin via oxygen tent, some factors that could influence workplace air concentrations in personal and area samples are shown in Table 16.2. For this small group of samples, no statistically significantly associations were identified between these factors and the measured exposures.

TABLE 16.1 Ribavirin Exposure of Study Participants, Measured by Personal and Area Air Samples

			Air Concentrations (μg/m³)	
Participant	Job[a]	Mask Use	Personal	Area
Oxygen tent				
3	RN	+	95	263
4	RT	+	134	263
5	RT	–	69	86
6	RN	–	168	86
7	RN	–	100	277
8	RN	–	316	1048
9	RN	+	112	69
10	RN	+	159	127
Ventilator				
1	RN	+	6	4
2	RN	+	6	4
13			<1	<1
Face mask				
11	RN		65	283

[a] RN, registered nurse; RT, respiratory therapist.

TABLE 16.2 Ribavirin Exposures of Health Care Personnel Using Oxygen Tent Administration, and Some Parameters That Could Affect Exposure Levels

					Exposures μg/m³	
Participant	Work Shift (h)	Time at Bedside (%)	Times Lifted Oxygen Tent	Ventilation, Rate (h¹)	Personal	Area
RNs						
3	7	—	—	6.1	95	263
6	8	60	10	6.1	168	86
7	8	20	12	—	100	277
8	7	40	10	—	316	1048
9	11	80	12	3.4	112	69
10	12	75	17	3.4	159	127
RTs						
4	7	—	—	6.1	134	263
5	10	23	5	6.1	69	86

Estimated absorbed doses are shown in Table 16.3. The most highly exposed group, nurses administering ribavirin via oxygen tent without wearing masks, had a mean estimated daily absorbed dose of 13.5 μg/kg. This dose is 13% of the dose that caused embryolethality in rabbits and 11% of the dose that caused malformations in hamsters. For the most highly exposed individual, the estimated absorbed dose was 30 μg/kg, approximately 30% of the dose causing reproductive toxicity in animals.

Ribavirin was not detected in any urine or serum samples, or in any of the preexposure red blood cell samples. In the red cell sample collected 5 days after first exposure from one participant, 0.44 μg/mL of ribavirin was measured. The airborne exposure level for this nurse was the highest observed.

TABLE 16.3 Estimated Absorbed Doses for Participants Administering Ribavirin Aerosol

Participant	Exposure (μg/m³)	Dose (μg/kg)
Oxygen tent		
3	95	4
4	134	5
5	69	7
6	168	17
7	100	10
8	316	30
9	112	7
10	159	12
Ventilator		
1	6	0.4
2	6	0.4
13	1	<0.1
Mist mask		
11	65	10

During the study, investigators observed that substantial amounts of ribavirin condensed inside oxygen tents, on bedsheets, and at the joints of the ventilator tubing.

Reported Symptoms

The COHP study was not designed to produce substantial information about health effects experienced by workers caring for ribavirin-treated patients. No symptoms were reported by any of the participants. However, the investigators have received accounts of employees experiencing symptoms associated with ribavirin administration. Eye irritation, skin rashes, cough, and difficulty in breathing have been reported by both nurses and RTs. At this time, no definitive data are available to indicate whether these symptoms occur more frequently during ribavirin administration than at other times.

Discussion and Conclusions

A previous report had concluded, after detecting no ribavirin in workers' blood or urine, that occupational exposure to ribavirin was unlikely to pose a reproductive risk.[6] However, these data should be interpreted with caution in view of the fact that detectable ribavirin levels are not expected in biological fluids after exposure at the provisional limit of 2.7 μg/m³.[10]

The airborne exposures measured by COHP, and the estimated absorbed doses, indicate that during oxygen-tent delivery, HCWs may receive doses approaching the levels that cause embryolethality and teratogenicity in animals. The average safety factor during this mode of administration was <100. Therefore, ribavirin may pose a reproductive risk to these workers. The concentrations in air samples collected at patients' bedsides indicate that visitors who spend many hours in the vicinity may also be at risk. The single personal air measurement during mist-mask administration suggests this may also produce hazardous exposures; however, the single case limits any further conclusions.

During administration by ventilator, with exhaust filters in place, ribavirin exposures were much lower and an appropriate margin of safety was maintained. However, the lower exposures with this mode of administration do not justify its recommendation as a hazard-control measure, because mechanical ventilation poses substantial risks to the patient and can be ordered only when clinically necessary.

Exposure levels higher than those presented above are likely to occur, for several reasons. First, because of the study's small sample size, it is unlikely to have captured the most extreme conditions. Second, RSV cases (and ribavirin therapy) were relatively infrequent during the year of the study. At times, ribavirin is administered to several patients per room, and treatment periods for various patients often overlap, requiring almost continuous ribavirin delivery. These circumstances probably produce worker exposures that exceed those measured in this study.

During periods of heavy ribavirin use, some HCWs may be exposed repeatedly. Accumulation of ribavirin in red blood cells could then produce concentrations far higher than those that occurred in study participants after just two days of exposure.

The risk assessment presented above has some limitations that should be recognized. It is based largely on absorbed doses that were estimated using assumptions (rather than actual determinations) of participants' ventilation rates and their absorption of the drug once inhaled, and about the protective value of surgical masks. If different values were used in the model, then the estimated absorbed doses would be higher or lower, with corresponding changes in the margin of safety. Additionally, the use of a safety factor is a rather simplistic method for establishing exposure limits, but no better approach is currently available for reproductive toxicants.

The presence of detectable ribavirin in the red blood cells of one participant indicates that absorption may indeed occur following occupational exposure to the aerosol. This finding must be interpreted with caution because it was measured in only a single sample. However, the fact that it occurred in the most highly exposed participant suggests that the observation is not spurious. The lack of low-dose pharmacokinetic data prevents direct comparison of the airborne and biological concentrations.

Aerosolized drug administration has been found to be uniquely efficient in some applications. This mode of administration is being investigated for a variety of other drugs and may be more common in the future. For this reason, development of effective recovery systems is essential.

Recommendations

Overall, data from the COHP study indicate that ribavirin poses a substantial reproductive risk to exposed HCWs. Hospitals using ribavirin should make provisions for preventing overexposures to the drug, especially among women who may be pregnant and among workers (men or women) actively trying to begin a pregnancy. To prevent reproductive toxicity, the following steps are recommended:

1. HCWs who are pregnant or are actively attempting to conceive should be advised of the reproductive risk associated with ribavirin exposure during oxygen tent administration, and should be offered alternative job responsibilities with no loss of pay or other employment benefits.
2. Available methods of minimizing exposure should be used whenever feasible. Included among these are
   a. Turning off the aerosol generator when an oxygen tent must be lifted for more than a few seconds
   b. Minimizing the number of times the oxygen tent must be lifted
   c. Using exhaust filters during administration via ventilator
   d. Arranging mechanical ventilators to cycle the aerosol through the ventilator (instead of discharging it into the room) during periods when it is disconnected from the patient
   e. Increasing overall airflow rates in treatment areas
   f. Segregating ribavirin-treated patients into designated rooms, to reduce the exposures of workers caring for other patients
3. Common surgical masks or dust masks are unlikely to provide adequate protection, and should not be used as the primary means of limiting exposure.[15] Other types of respirators are not practical in patient care settings.
4. Devices specifically designed to capture excess ribavirin before it enters the workplace should be used, if such devices become available and are shown through field testing to be effective,

practical, and reliable. One such device, developed by the drug's manufacturer, has been reviewed by the U.S. FDA and released for marketing. However, few data are available on the drug-scavenging effectiveness of this device. Limited information, shown in Appendix 16.A.2, suggests that the device produces exposures lower than those that occur with standard oxygen tents but still exceeding the provisional exposure limit. Other devices are currently under development at several academic medical centers.

5. For hospital workers such as laundry or janitorial workers who may handle ribavirin-contaminated equipment, exposures should be limited by a control program which includes:
   a. Labeling of all ribavirin-contaminated materials
   b. Prompt and thorough cleaning of contaminated equipment before it is reused
   c. Special provisions for handling contaminated materials (e.g., bedsheets of patients treated via oxygen tent should not be shaken as they are prepared for washing)
   d. Training in the hazards of ribavirin and in safe work practices, for all staff who might have contact with the drug
6. Family members and visitors of patients treated with ribavirin should be informed of the reproductive risk associated with ribavirin exposure during oxygen tent administration. Individuals who are pregnant or attempting to conceive may wish to limit exposure to the aerosol.

To implement these recommendations, hospitals may choose to incorporate control of ribavirin exposures into a comprehensive reproductive protection program. For hospitals that have no such program, the risks associated with ribavirin administration may be among the best reasons to institute one.

Policies in Place

Hospitals in the COHP study, and certain many others as well, have established policies to implement the above recommendations. As an example of the control measures that can be successfully incorporated into routine hospital functions, Appendix 16.A.1 shows the ribavirin-control policies of a major teaching center. Some of the COHP recommendations above are not spelled out in these policies, and in some cases the policies go beyond the COHP recommendations to meet the needs of the hospital. Hospitals in any setting should be able to implement protection programs as protective as that shown in Appendix 16.A.1. As an alternative, some hospitals have chosen to sharply restrict ribavirin use until more information about the drug's efficacy and safety become available.[17]

References

1. ICN Pharmaceuticals, Inc. Virazole™ (Ribavirin) Prescribing information, 1986.
2. Demers RR, Parker J, Frankel LR, and Smith DW. Administration of ribavirin to neonatal and pediatric patients during mechanical ventilation. *Respir Care* 31:1188–1195, 1986.
3. Hillyard IW. The preclinical toxicology and safety of ribavirin. In: RA Smith and W Kirkpatrick (eds.), *Ribavirin: A Broad Spectrum Antiviral Agent*. Academic Press, New York, 1980.
4. Kilham L and Ferm VH. Congenital anomalies induced in hamster embryos with ribavirin. *Science* 195:413–414, 1977.
5. *FDA Drug Bulletin* 16:7, 1986.
6. Rodriguez WJ, Dang Bui RH, Connor JD, et al. Environmental exposure of primary care personnel to ribavirin aerosol when supervising treatment of infants with respiratory syncytial virus infections. *Antimicrob Agents Chemother* 31:1143–1146, 1987.
7. Catlin DH, Smith RA, and Samuels AI. 14-C Ribavirin: Distribution and pharmacokinetic studies in rats, baboons and man. In: RA Smith and W Kirkpatrick (eds.), *Ribavirin: A Broad Spectrum Antiviral Agent*. Academic Press, New York, 1980.

8. Harrison R, Bellows J, Rempel D, Rudolph L, Kizer KW, Jin A, Guglielmo J, and Bernard BB. Assessing exposures of health-care personnel to aerosols of ribavirin—California. *Morbid Mortal Week Rep* 37:560–563, 1988.
9. Ferrera EA, Oishi JS, Wannemacher RW, and Stephen EL. Plasma disappearance, urine excretion, and tissue distribution of ribavirin in rats and rhesus monkeys. *Antimicrob Agents Chemother* 19:1042–1049, 1981.
10. Occupational Health Surveillance and Evaluation Program: *Health Care Worker Exposure to Ribavirin Aerosol*, Report no. FI-86–009. Berkeley, CA, California Department of Health Services, 1988.
11. *Federal Register* 50(219):46946, 1985.
12. U.S. EPA. Guidelines for the health assessment of suspect developmental toxicants. *Fed Regis* 51(185):34028–34040, 1986.
13. Austin RK, Trefts PE, Hintz M, Connors JD, and Kagnoff MF. Sensitive radioimmunosassay for the broad-spectrum antiviral agent ribavirin. *Antimicrob Agents Chemother* 24:696–701, 1983.
14. International Commission on Radiological Protection. *Report of the Task Group on Reference Man*. ICRP no. 23, Pergamon Press, New York, 1975.
15. Connor JD, Hintz M, and Van Dyke R. Ribavirin pharmacokinetics in children and adults during therapeutic trials. In: RA Smith, V Knight, and JAD Smith (eds.), *Clinical Applications of Ribavirin*. Academic Press, New York, 1984.
16. Tuomi T. Face seal leakage of face masks and surgical masks. *Am Indust Hyg Assoc J* 46:308–312, 1985.
17. Guglielmo BJ, Jacobs RA, and Locksley RM. The exposure of health care workers to ribavirin aerosol. *J Am Med Assoc* 261:1880–1881, 1989.

Appendix 16.A.1

Ribavirin Protection Policies of a Major Teaching Hospital

General Guidelines for All Patients

1. Patients receiving ribavirin are to be placed individually in single rooms or cohorted in common rooms. The door is to remain closed with the precautions clearly posted outside.
2. No pregnant individual is to enter the room.
3. No male or female individual who is actively trying to conceive within 6 weeks is to enter the room.
4. All other individuals can enter the room including lactating women (the reproductive risk posed by ribavirin is limited to effects on the fetus).
5. Gowns and gloves are to be worn by all individuals having contact with the ribavirin crystals. Handwashing is recommended after ungloving.
6. Masks are optional since they do not ensure adequate protection against the reproductive risk. Both blue masks and white 3 M masks are to be available outside the room for those individuals not at risk who choose to wear a mask.
7. Soft contacts should be removed before entering the room to prevent any damage based on the findings of one clinician. Hard contacts or glasses do not seem to be affected adversely.
8. Individuals with asthma and/or other pulmonary conditions have experienced episodes of respiratory difficulty; therefore, a HCW with a history of respiratory disease should be provided an alternate assignment when possible.

Ventilated Patients

Ribavirin will be delivered via a filtered recirculating system which essentially limits dispersion of crystals into the general environment.

1. The set-up maintenance and treatment of this system is covered in the Respiratory Therapy guidelines
2. Ventilator tubing is to remain capped during patient suctions to provide an intact system

Nonventilated Patients

The amount of ribavirin crystals dispersed into the general environment depends on the particular method of delivery. The method chosen should expose the smallest surface area to potential contamination while allowing for the optimal delivery of nursing care, patient treatment, and comfort. The following additional precautions apply to all nonventilated patients since they all have ribavirin crystals deposited on all room surfaces.

1. Only the necessary amount of supplies are to be taken into the room and these are to be stored as far as possible from the bedside to decrease the deposit of ribavirin crystals on the surfaces. The same guidelines apply to the patient's belongings.
2. All equipment is to be thoroughly cleaned before being taken out of the room. Any cleaning solution is usually adequate as ribavirin is water-soluble.
3. Environmental Services is to ensure that items used to clean the room are disposed of or thoroughly rinsed of ribavirin before use in another area. It is recommended that these rooms be cleaned last.
4. Meals are to be on the disposable isolation trays.
5. Trash is to be double-bagged.
6. Linen is to be treated with a minimum of shaking to decrease further dispersing of ribavirin crystals. The linen bag is to be clearly marked "Ribavirin" before being sent to the laundry to alert the staff there to the special handling required.
7. The room is to be high-cleaned at the conclusion of the course of ribavirin or at the time of the patient discharge.

Appendix 16.A.2

Industrial Hygiene Survey Report

Purpose and Scope

Industrial hygiene monitoring was conducted in an isolation pediatric intensive care room at a Northern California hospital (not included in the study described in the body of this chapter). The purpose of this monitoring was to investigate the extent to which ICN Pharmaceuticals Aerosol Delivery Hood minimized the escape of aerosolized ribavirin (AR) during a May 1989 administration of the drug to a patient.

During the administration of the drug to a patient with his (her) head enclosed by the ICN Aerosol Delivery Hood, personal samples were taken to evaluate the exposure of two registered nurses (RNs) providing care to that patient. In addition, two general area samples were taken at the locations specified in Table 16.A.1.

Volumetric airflow rate measurements were made at ventilation system ceiling exhaust and supply openings located in the room (see Table 16.A.2).

Equipment and Survey Procedures

Samples were collected on glass–fiber filters and were analyzed by HPLC by Alpha Chemical and Biomedical Laboratories (Petaluma, CA) using a modification of National Institute for Occupational Safety and Health Method Number 5027.

Supply and exhaust air ventilation flow rates were measured using a Shortridge Model CFM-83 Flowhood equipped with a 2′ × 2′ top.

TABLE 16.A.1 Atmospheric Ribavirin Concentrations in Employees Breathing Zones and Patient Care Areas, Northern California Hospital's Isolation PICU, May 1989

Sample Number	Description	Time Start/ Stop	Concentration (μg/m³)[a]
California DHS Recommended Limit[b]			2.5
First RN			
RIB08	Provided primary care for 4-month-old patient with RSV. Spent approx. 70 min of sampling period in the isolation room and disturbed ICN aerosol hood seven times. Disturbances varied from removal of hood from over patient to slightly raising and shifting the hood. Drug nebulizer remained in operation during these disturbances. Because patient was stable, the number of hood disturbances needed was minimal.	1644/2352	8.29
Second RN			
RIB14	Assisted first RN with patient. Spent only approximately 12 min of sampling period inside the isolation room.	1820/2358	<2.09
General area samples			
RIB11	Sampler suspended on post (at breathing zone height) at end of bed away from wall and farthest from the nebulizer and aerosol hood (right-rear bed post).	1644/2352	6.77
RIB10	Sampler suspended 6 in. above (at breathing zone height) the SPAG unit and vacuum/filter portion of ICN aerosol hood assembly which were located on right-hand side of bed at its head.	1644/2352	12.55

Note: Irregularly shaped room (11.5 ft by 12 ft—dimensions at greatest width and length). "<" means less than the detection limit for the sampling and analytical method used.

[a] Concentrations are expressed in micrograms per cubic meter of air.

[b] The CDHS has recommended an acceptable occupational airborne concentration for an 8 h workshift in order to limit the ribavirin uptake via inhalation to one one-thousandth of the lowest NOEL. Maintaining workplace exposure below this occupational exposure limit is recommended to minimize the risk among employees of adverse reproductive effects (teratogenic effects and testicular atrophy) observed in laboratory animals.

TABLE 16.A.2 Ventilation Measurements, April 1989, a Northern California Hospital's Isolation PICU

Location/Description	Volumetric Airflow (cfm)[a]
Isolation room used for ribavirin administration	
Ceiling exhaust opening in corner above nebulizer	60
Ceiling supply diffuser beyond end of bed near room entrance	245

[a] Cubic feet per minute.

Results and Conclusions

The personal samples for the two nurses taken during the May 1989 administration of the drug to a patient showed their time-weighted average (TWA) exposures to be <10 μg/mL. The measured TWA exposure of the nurse who spent the least amount of time in the isolation room was below the detection limit for the air sample taken in her breathing zone (see Table 16.A.1).

The personal and general area sampling results obtained indicate that the ICN Pharmaceuticals Aerosol Delivery Hood can provide an effective mechanism for reducing airborne room contamination from ribavirin during its administration to small pediatric patients. Good work practices should be used to maximize the effectiveness of the hood in minimizing occupational ribavirin exposure (see "Discussion").

Smoke tube testing indicated that purging of an aerosol from the food's interior requires several minutes. Of course, hood clearance time depends upon the volumetric flow rate of the purging gas and how well the incoming gas mixes with the aerosol.

Ventilation system volumetric flow rates for the isolation room are presented in Table 16.A.2.

Recommendations

1. Use the ICN Pharmaceuticals Aerosol Delivery Hood as an alternative to oxygen tents for the administration of ribavirin aerosol.
2. Develop the written work practices for aerosol delivery hood use that will enable respiratory therapy, nursing staff, and other health care providers to minimize their exposures to ribavirin.
3. Offer the alternative work to personnel who are pregnant or attempting to conceive.
4. Discuss the results of environmental monitoring and its implications with all affected employees and make the data available to them.

Discussion

In its report *Health Care Worker Exposure to Ribavirin Aerosol* dated February 1, 1988, the CDHS recommended the development of a control system specifically designed to reduce ribavirin aerosol emissions from oxygen tents. The environmental monitoring data collected during this survey indicate that the ICN Pharmaceuticals Aerosol Delivery Hood is such a control system. However, exposures measured while the device was in use still slightly exceeded the Department's recommended exposure limit of 2.7 $\mu g/m^3$.

The hood substitutes for an oxygen tent and has mechanisms attached to it that scavenge much of the ribavirin aerosol escaping from it. According to ICN Pharmaceuticals, the vacuum unit that collects aerosolized drug includes a high efficiency filter with a collection efficiency of 99.97% for an aerosol with a mass median diameter of 0.5 micron. This collection efficiency is adequate for removing the ribavirin aerosol from the vacuum device's exhaust air prior to its release back into the patient care room.

It must be understood that the effectiveness of the hood in minimizing employee ribavirin exposure depends to a large extent of adherence to work practices designed to reduce the amount of ribavirin released form the hood when patient care procedures require its periodic removal. Work practices expected to be successful in this regard include:

1. Shutting off the nebulizer*
2. Switching to a medical air/oxygen gas stream that does not pass through the nebulizer
3. Waiting several minutes to allow the alternate medical air/oxygen gas stream to purge residual ribavirin form the interior of the hood prior to its removal

If an emergency requires immediate removal of the delivery hood, the nebulizer can simply be shut off prior to the hood's removal.

Source: Adapted from a report prepared by Stephen W. Hemperly, MS, CIH, Industrial Hygienist, Kaiser Foundation Hospitals, Berkeley, CA.

* A shut-off valve can be installed between the medical air/oxygen supply and the nebulizer to permit staff to turn off the nebulizer without having to adjust flow rate controls on the nebulizer.

Ribavirin Aerosol Engineering and Administrative Controls to Contain Aerosolized Ribavirin: Results of Simulation and Application to One Patient

William Charney, Kevin J. Corkery, Roger Kraemer, and Lee Wugofski

Introduction

Ribavirin is an antiviral agent introduced in the 1980s for the treatment of RSV infections in infants and children.[1] Although a number of clinically useful drugs are routinely administered by aerosolization, particularly for the treatment of bronchospastic disorders, ribavirin differs because the particle size (mass median diameter 1.3 micron)[2] allows deep penetration into the lung; administration usually occurs 12–18 h daily for three to seven consecutive days.[1] The drug has been shown to be teratogenic in some animal species,[3–7] and, consequently, concern over the occupational exposure of HCWs has prompted medical centers to reevaluate their policies on indications for and use of ribavirin.[8] Some have decided to provide respirators with protection factors (PFs) of 10 or greater.[9] (The PF is the ratio of the concentration of a particular substance in ambient air to concentration inside the mask.)

Because personal protective equipment or barrier protection (e.g., a respirator) is considered the least appropriate control measure for protecting people from potentially hazardous substances, and because the components of a respiratory protection program can be difficult to comply with,[10] a system that reduces occupational exposure to a level that eliminates the need for personal respiratory protection is desirable. A respiratory protection program involves semiannual fit testing of the respiratory mask, annual medical surveillance of all personnel involved, employee instruction in the use of the mask and about the hazardous substances used, recordkeeping, safe storage of respirators not in use, and routine respirator inspection.[10] In our experience, the communication necessary among personnel in the Respiratory Care, Nursing, and Occupation Health and Safety Services to schedule and implement the many facets of a personal protection program is time consuming and difficult to maintain.

Despite numerous investigations of HCW exposure to ribavirin aerosols,[2,9,11,12] no consensus has been reached concerning the precautions necessary for safe administration of this agent. The CDHS[13] has determined that an acceptable airborne ribavirin concentration can be calculated by applying an EPA safety factor of 1000 to the experimental NOEL for the most sensitive endpoint (the concentration at which an effect is seen) in the most sensitive species tested.[13,14] We adopted the NOEL observed by Harrison et al.[11] for the rabbit (0.3 mg/kg) to estimate an acceptable airborne concentration for an 8 h work shift as a TWA of 2.7 µg/m^3, assuming that the employee weighs 58 kg, has a 1 min ventilation of 19 L, and retains 70% of the inspired ribavirin. TWA is the average concentration of the substance in question to which nearly all workers may be exposed in an 8 h workday without adverse effect. TWA assumes that workers are usually not exposed to a substance for their entire shift and is calculated by the equation:

$$\mathrm{TWA} = \frac{(C_1^3 \cdot T_1) + (C_2^3 \cdot T_2) + (C_N^3 \cdot T_N)}{T_{\mathrm{TOT}}}$$

where

C is the concentration of substance being measured
T is the time of exposure to substance being measured
T_{TOT} is the total number of minutes worked on shift
N is the number of times exposed

We determined the room concentration of ribavirin to be 456 µg during a 12 h treatment when ribavirin was being administered directly into a mist tent (unpublished data, San Francisco General Hospital,

1989). In a preliminary study, we evaluated the Aerosol Delivery Hood System by ICN Pharmaceutical and found the highest concentration of ribavirin to be 36.2 μg/m³.[15] We consider these two concentrations to be unacceptable because ribavirin exposure was above 2.7 μg/m³. The Aerosol Delivery Hood System may also have limited usefulness due to the size of infants that it will accommodate. It has been a goal of our Health and Safety Service to work in conjunction with the Respiratory Care and Nursing Services to devise a method of administering ribavirin in a safe and efficient one that would obviate the need for personal protective equipment. In the pilot study, we sought to determine whether the scavenging system that we constructed could decrease HCW exposure to a ribavirin TWA of 2.7 μg/m³ or less.

Materials and Methods

We assembled and evaluated a double containment system, which consists of a 25″ × 30″ × 28″ oxygen tent canopy over a smaller 12″ × 12″ × 12″ oxygen hood. A Small Particle Aerosol Generator (SPAG-2) delivers AR into one port of the hood at a flow rate of 12 L/min (nebulizer flow rate of 7 L/min plus a drying flow of 5 L/min). A heated and humidified gas mixture is supplied through the other port at a flow rate of 12 L/min to provide consistent oxygenation to the hood (and thus to the patient) when the SPAG-2 nebulizer is turned off. A single blender provides gas at a controlled concentration to the SPAG-2 and to the supplemental delivery system, which remains in continuous operation. The flow rate is continuously monitored with an IL408 oxygen analyzer. The space between the tent canopy and the oxygen hood is evacuated continuously by two vacuum units equipped with high-efficiency particulate air (HEPA) filters. Two pieces of smooth-bore aerosol tubing, one 40″ and one 24″ in length, are attached at one end to the vacuum units and to aerosol T-adapters taped to the oxygen hood at the other end. The T-adapters decrease the likelihood of accidental occlusion. The SPAG-2 unit is equipped with a manual shut-off valve for easy interruption of aerosol delivery. The free edges of the tent canopy are tucked beneath the mattress to help eliminate leakage. The HEPA filters are replaced daily.

At the start of the tests, evacuation flow was measured at the T-piece and of the aerosol tubing. A Jaeger pneumotachograph (calibrated with a Fisher-Porter flowtube and an I-L syringe) was used to measure flow. In order to determine whether the flow output of the vacuum units was reduced by the resistance offered by the pneumotachograph, we estimated flow indirectly by the following gas-dilution technique: A 40″, 22 mm ID smooth-bore tube was attached to a vacuum unit. A bleed flow of 10 L/min oxygen was delivered by a Porter flowtube to the distal end of the tube. The unit was turned on and the concentration of the evacuation flow was allowed to stabilize. The concentration of oxygen (% O_2) in the evacuation flow was measured by mass spectrometer connected to a gas sampling port at the proximal end of the tube. The flow was calculated by the equations:

$$F_{\dot{V}evacO_2} = \frac{(\dot{V}_{bleed})(F_{\dot{V}bleedO_2}) + (\dot{V}_{amb})(F_{\dot{V}ambO_2})}{\dot{V}_{bleed} + \dot{V}_{amb}}$$

and

$$\dot{V}_{evac} = \dot{V}_{bleed} + \dot{V}_{amb}$$

where

$F_{\dot{V}evacO_2}$ is the fractional oxygen concentration of evacuation flow
$\dot{V}_{bleed}$ is the bleed flow
$F_{\dot{V}bleedO_2}$ is the fractional oxygen concentration of bleed flow
$\dot{V}_{amb}$ is the ambient flow
$F_{\dot{V}ambO_2}$ is the fractional oxygen concentration ambient flow
$\dot{V}_{evac}$ is the evacuation flow

$\dot{V}_{bleed}$, $F_{\dot{V}bleedO_2}$, $F_{\dot{V}ambO_2}$, and $F_{\dot{V}evacO_2}$ are known; the first equation is solved for $\dot{V}_{amb}$, and then the second equation is solved for $\dot{V}_{evac}$.

Personal and area-monitoring samples were taken first during ribavirin administration to a simulated patient and later to a 2-month-old infant with RSV infection in the Medical-Surgical Intensive Care Unit. (Our institution does not have a separate PICU.) For the simulation, the test system was set up and three 5 h area samples and one 100 min personal sample were taken. For the patient application, ribavirin administration was begun at 21:00 on one evening and discontinued at 15:00 the following day. This was the patient's fifth and final dose of ribavirin. The first four doses had been administered through the patient's endotracheal tube via mechanical ventilator. The patient continued to improve and was discharged from the ICU the morning following the fifth ribavirin treatment. The patient was discharged from the hospital 2 days later.

Three HCWs were equipped with sample pumps with filters positioned within their breathing zone (i.e., filters were placed on their collars) and were monitored over two shifts. In addition, area sampling was performed at the head of the bed, 6 ft above floor level—the location considered to be in the breathing zone of persons who might be in the room but not directly attending the patient. The HCWs were instructed to shut off the nebulizer with the manual cutoff valve 5 min before breaking into the system for patient care activities. The log indicated that compliance was universal, and the 5 min delay posed no problems during the course of treatment. Interruptions in aerosol administration were logged for the 18 h period (Table 16.4). All bedside personnel (nurses and therapists) were required to wear a fit-tested disposable respirator with a personal protection factor (PPF) of 10. Fit testing was performed according to Occupational Safety and Health Administration (OSHA) Standard 1910.134(c)(5).[10] No other persons were allowed to enter the patient's room during the time that ribavirin was being administered.

Airborne ribavirin samples were collected on Gelman 37 mm glass fiber filters with a cellulose ester backup pad in a 2-piece cassette using an ALPHA-1 Air Sampler. Pump samples were collected at an

TABLE 16.4 Aerosol Interruptions and Total Exposure Time during 18 h Treatment

Time Off–On	Total Minutes Off
Night shift	
22:18–22:38	20
22:57–23:00	3
24:00–00:20	20
01:25–01:40	15
03:35–04:05	30
05:25–05:40	15
06:20–06:45	*25*
Number of interruptions = 7	128
Day shift	
08:42–08:59	17
09:20–09:45	25
10:00–10:30	30
13:00–13:30	*30*
Number of interruptions = 4	102
Replacement nurse day shift	
10:24–10:42	*18*
Number of interruptions = 1	18

average flow rate of 2 L/min for all samples. Pumps were calibrated using a Gilibrator Standard System (a calibration method that utilizes a primary flow calibrator and a bubble generator) and were checked periodically in the field with a rotameter calibrated against the Gilibrator standard.

The rooms in the ICU are designed for "positive pressure" ventilation, with approximately 6.5 air exchanges/hour (350 cfm supply, 300 cfm exhaust). Ventilation is by dispersion airflow through the perforated ceiling in a room with a volume of approximately 3200 ft^3. The positive pressure design is intended to prevent the transmigration of microorganisms into the ICU rooms. Although some might consider it more desirable to administer ribavirin in a room with "negative pressure" ventilation, such a room was not available to the patient at the time of this pilot study.

Samples were analyzed by HPLC according to National Institute for Occupational Safety and Health (NIOSH) Method 5027. As part of the laboratory's procedure, a spike sample (a known amount of ribavirin) and a bland sample (containing no ribavirin) are analyzed concurrently with the test samples, to verify test results. Detection limits for this analysis are variable due to difference in total volume collected for each sample. Detection limit for sample volume collected appears in Table 16.5. The volume collected varied according to the worker's exposure. The greater the volume of air drawn through the sample pump, the lower the detection limit. The lowest level of detection (LOD) for this method is 0.7 μg/sample.

Results

The mean (SD) evacuation flow rate for each vacuum unit was 91 (1.1) L/min when measured by the Jaeger pneumotachograph and 97 (1.1) L/min when measured by gas dilution.

No other personnel entered the room during ribavirin administration. The oxygen concentration in the hood (35%) was unaffected by the evacuation flow.

Ribavirin was not detected in area or personal breathing zone (PBZ) samples. Table 16.5 provides the exposure time, sample pump flow rate, volume of air passed through the pump filters, and the detection limit of the analytical method based on the volume collected.

No special patient surveillance measures—other than the continuous electrocardiographic, oxygen-concentration, and pulse-oximetry monitoring routinely performed on ICU patients—was deemed necessary.

TABLE 16.5 Personal and Area Sampling Results from Bench Testing and Patient Application

Filter #	Sample	Exposure Time (min)	Pump Flow Rate (L)	Volume Collected (L)	Detection Limit (μg/m³)	Results
Bench data						
1	Top of oxygen hood, center	300	2	600	3.93	ND[a]
2	Top of oxygen hood, right	300	2	600	3.93	ND
3	Top of oxygen hood, left	300	2	600	3.93	ND
4	Personal sample, center	100	2	200	11.8	ND
Patient data						
8199 GF1	Personal—night shift/day	218	2	436	5.41	ND
8200 GF3	Personal—day shift	440	2	880	2.68	ND
8201 GF2	Area sample night shift/day	630	2	1260	1.87	ND
8202 GF4	Area sample day shift	440	2	880	2.68	ND
8203 GF5	Personal replacement	300	2	600	3.93	ND

[a] ND = none detected.

Discussion

Environmental levels measured in a California State Department of Health's Field Investigation[13] have demonstrated that the total weighted average (TWA) of 2.7 μg/m^3 (or lower) can be achieved during patient administration via a mechanical ventilator equipped with appropriate filters. For patients who are not intubated, some success in reducing environmental levels has been reported[16] utilizing the Aerosol Delivery Hood System developed by the manufacturers of ribavirin, but application of this system is limited to small infants (less than about 6 kg in weight). In addition, environmental levels measured and reported in a preliminary study suggest that personal protection equipment may still be needed.[15] The highest airborne concentrations (area samples of 557 μg/m^3 reported by Gladu and Ecobichon[12] and 317 μg/m^3 reported by the California Department of Health[13]) have been seen when the drug has been administered by simple oxygen tent. Because patients with RSV may be treated with ribavirin in an attempt to avoid intubation, institutions have had to provide personal respiratory protection to employees to prevent excessive exposure.

The numerous components of a personal respiratory protection program must be adhered to if the employer supplies respirators to personnel. In addition to the necessary training and medical evaluation, each employee must be fit-tested semiannually so that adequate protection can be empirically demonstrated and inappropriate use and a false sense of security can be avoided.

Since late 1989, our institution has been fit-testing employees involved in ribavirin administration, and has found that five different brands of respirators (nonpowered, air-purifying type, equipped with HEPA filters) of various sizes are required to fit the diverse work force properly. The OSHA proscription of most types of facial hair because of hair's interference with adequate respirator seals has further complicated fitting and staffing. Powered air-purifying respirators may address some of these concerns, but their substantial cost and noisiness may limit their usefulness. Use of disposable respirators such as the 3 M 9970 has certain advantages in that they may be less offensive to the wearer and cause less concern to the patient's family, but they too require following the OSHA Standard 1910.134(e)(5) for fit testing.[10]

The system evaluated in this study differs from those previously reported[12,15,16] in incorporating two separate evacuation components, each utilizing high efficiency filters (99.97% efficient with aerosol of mass median diameter 0.3 micron). Our experience with the simulated patient encouraged us to try the system on patients requiring ribavirin. Unfortunately for our experimental purposes and fortunately for our patient population, only one patient reported at this location required ribavirin administration between January 1 and June 1, 1990.

The fact that no ribavirin was detected, in our opinion, is due to exhaust volume and the integrity of the HEPA filters. We had used the same sampling technique for measuring and the same laboratory for analyzing ribavirin during two previous studies (unpublished data) that did not utilize the same engineering and administrative controls. Ribavirin was detected during those studies. Samples collected by the same sampling technique and analyzed by the same laboratory in studies by another large medical center in the San Francisco Bay Area have yielded ribavirin concentrations (personal communication, Steven Henperly, Industrial Hygienist, Kaiser Medical Center–San Francisco, 1990). We believe that the combination of the high evacuation flow in the double-tent system and the 5 min interval of no aerosolization before the tent is opened allows the ribavirin to be scavenged through the HEPA filters. It may be difficult to adhere the 5 min shut-off stipulation if a number of ribavirin treatments are occurring simultaneously and an emergency situation arises. Further evaluation of this technique is necessary before the use of personal respiratory protection can be eliminated.

The evacuation flow measurements suggest that the vacuum unit is sensitive to the resistance of the evacuation tubing. Flow measured by the pneumotachograph was lower than flow estimated by gas dilution. Total evacuation flow from the two units estimated by gas dilution dropped from 194 to 144 L/min when the units were bench tested with a 6 ft corrugated aerosol tube in place of the smooth-bore tube. This suggests that any change in circuit resistance (increasing the length of the evacuation circuit; using corrugated tubing, which causes turbulent flow; partially blocking the tubing with bedsheets; or crimping tubing with the bedrail) could lower evacuation flow and increase the risk of ribavirin exposure. This emphasizes the importance of maintaining a high evacuation flow and of measuring the flow before initiating ribavirin aerosolization.

Conclusion

Our system appears to maintain environmental levels of ribavirin at or below the TWA recommended by the California State Department of Health. With airborne levels kept at the range detected in this evaluation, rooms with negative air pressure relative to outside halls may not be necessary.

Our preliminary results suggest that personal protective devices (i.e., respirators) may not be needed if the double-tent system and the administrative measure of shutting off the aerosol generator for 5 min prior to opening the tent for patient care are followed. The efficacy and importance of interrupting aerosolization to allow proper evacuation and setting time cannot be overemphasized. We still require personnel to use respirators when administering ribavirin to nonintubated patients. Until more results with this system have been reported from our institution or others, we will require the wearing of the respirator.

Finally, the risk of ribavirin exposure to the developing fetus has not been established. We believe that employees who are pregnant or who are actively attempting to conceive should be assigned to areas of the hospital in which ribavirin exposure will not occur.

Acknowledgments

We thank Rita Smith, RN, Robert Levin, PharmD, Arthur Glickman, RN, Moses Grossman, MD, and Romy Rafanan, MS, RN for their technical assistance, advice, and support. The cost of analyzing the samples was underwritten by San Francisco General Hospital's Occupational Health and Safety Service.

Product and Service Sources

Hoods and tents:

- Oxygen tent canopy, THS 523, Tri-Anim Health Services, Inc., Glendale, CA
- Oxygen hood, Tot Hut #1001, Peace Medical, West Orange, NJ
- Aerosol delivery system, #6005, ICN Pharmaceuticals, Inc., Costa Mesa, CA

Evacuation components:

- Vacuum units, #4031, ICN Pharmaceuticals, Inc., Costa Mesa, CA
- 22 mm smooth-bore tubing, Corr A Tube II Part 1413, Hudson Oxygen, Temecula, CA

Aerosol generator:

- Small Particle Aerosol Generator, SPAG-2, ICN Pharmaceuticals, Inc., Costa Mesa, CA

Calibration devices:

- Jaeger pneumotachograph, Jaeger Medical Instruments GmbH & Co. KG, Hoechberg, Germany
- I-I, syringe, Jaeger Medical Instruments GmbH & Co. KG, Hoechberg, Germany
- Flowtube, FPI/2GS-27-G-10 Float 1/2GSVT45A, Fisher-Porter Co., Warminster, PA
- Flowtube, R-104, Porter Instrument Co., Hatfield, PA
- Precision Rotameter, BGI, Inc., Waltham, MA
- Gilibrator Standard System, P/D 800-26 and P/D 800-2868N, Gilian Instrument Corp., Wayne, NJ

Analytical equipment:
  Gelman 37-mm glass fiber filters, Type A/E#61652 micron nominal pore size, Alpha Biomedical, Petaluma, CA
  Alpha-1 Air Sampler, DuPont Instrument Systems, Kennett Square, PA
  Mass spectrometer, Perkin-Elmer, Pomona, CA

Liquid chromatography:
  Alpha Chemical and Biomedical Laboratory, Petaluma, CA

Respirators:
  Disposable respirator, 3 M 9970, 3 M, St. Paul, MN
  Permanent respirator, Wilson Half Mask Respirator #6140, Pedley-Knowles, San Francisco, CA

References

1. American Academy of Pediatrics. Ribavirin therapy of respiratory syncytial virus. *Pediatrics* 79:475–478, 1987.
2. Rodriguez WJ, Dand Bui RH, Connor JD, et al. Environmental exposure of primary care personnel to ribavirin aerosol when supervising treatment of infants with respiratory syncytial virus infections. *Anitmicrob Agents Chemother* 31:1143–1146, 1987.
3. Ferm VH, Willhite C, and Kilham L. Teratogenic effects of ribavirin on hamster and rat embryos. *Teratology* 17:93–102, 1978.
4. Kochar DM, Penner JD, and Knudsen TB. Embryotoxic, teratogenic and metabolic effects of ribavirin in mice. *Toxic Appl Pharmacol* 52:100–112, 1980.
5. Fernandez H, Banks G, and Smith R. Ribavirin: A clinical overview. *Eur J Epidemiol* 2:1–14, 1986.
6. Hillyard IW. The preclinical toxicology and safety of ribavirin. In: RA Smith and W Kirkpatrick (eds.), *Ribavirin—A Wide Spectrum Agent*. Academic Press, New York, pp. 59–71.
7. Kilham L and Ferm VH. Congenital anomalies induced in hamster embryos with ribavirin. *Science* 195:413–414, 1977.
8. Guglielmo GJ, Jacobs RA, and Locksley RM. The exposure of health care workers to ribavirin aerosol (letter). *JAMA* 261(13):1880–1881, 1989.
9. Fackler JC, Flannery K, Zipkin M, and Mclutosh K. Precautions in the use of ribavirin at the Children's Hospital (letter). *N Engl J Med* 332:634, 1990.
10. Occupational Safety and Health Administration. *Respiratory Protection*. Department of Labor, Washington DC, 29CFR 1910:134.
11. Harrison R, Bellow J, Rempel D, et al. Assessing exposures of health-care personnel to aerosols of ribavirin. *MMWR* 37:560–562, 1988.
12. Gladu JM and Ecobichon DJ. Evaluation of exposure of health care personnel to ribavirin. *J Toxicol Environ Health* 28:1–12, 1989.
13. California Department of Health. Health care workers' exposure to ribavirin aerosol. Field Investigation FI-86-009, Berkeley, CA, 1988.
14. United States Environmental Protection Agency. Guidelines for the health assessment of suspected developmental toxicants. *Fed Reg* 51(185):34028–34040, 1986.
15. Corkery K, Eckman D, and Charney W. Environmental exposure of aerosolized ribavirin (abstract). *Respir Care*, 34:1027, 1989.
16. Krilov LR, Jacobson JM, and Karol MS. Containment of environmental escape of aerosolized ribavirin using a double hood system (abstract). Abstracts of the 90th Annual Meeting of the American Society of Microbiology:A115.

Source: *Respir Care* 35:11, 1990.

Validation of Control System for Ribavirin Aerosol

William Charney

In November 1990, my colleagues and I reported on the results of the application of engineering and administrative controls to contain AR.[1] I present here the results of further testing (Table 16.6). The reader is referred to our earlier paper for the details of the engineering and administrative controls imposed and the sampling equipment and analytical method employed, which were duplicated during the monitoring reported here.

The child under treatment while this sampling was done was 10 months of age, extremely active, and in an agitated state. Medical personnel from the Pediatrics service and the ICU decided to sedate him in order to contain him within the area of the engineering control. With a more subdued child, results might have been even more favorable, but that is conjecture until additional monitoring has been done.

The area sample measurements shown in Table 16.6 indicate that for casual entry into the room, no personal protective equipment (PPE) is necessary if the engineering and administrative controls are being applied. It also indicates that no transmigration of aerosol to the corridor or adjacent areas occurs.

The personal samples reflect a slight excursion above the recommended 2.7 μg/m³ (4.05, ND, 4.54, 4.73). The nondetectable result occurred during a period of fewer break-ins (3) and less caregiver time in tent (68 min) compared to periods of more break-ins and greater total caregiver time in tent (125, 185, and 197 min). With the PPE all results were well below the recommended standard of 2.7 μg/m³ with a PF of 10×.

We conclude and recommend that (1) these results show the integrity of the engineering system in reducing levels of ribavirin aerosol to safe levels for those casually entering the room. Caregivers who must break into the tent should use a personal, fit-tested respirator with a tenfold PF; (2) the child's

TABLE 16.6 Results of Personal and Area Sampling with Engineering and Administrative Controls in Force

Shift	Number of Break-Ins	Total Caregiver Time in Tent (min)	Caregiver Time in Room (h)	Volume Collected (L)	Detectable Limit (μg/m³)	PPE PF10	Result without PPE (μg/m³)	Result with PPE (μg/m³)
Personal sample								
9:00 PM–6:30 AM								
Day 1	8	125	9.5	1140	2.07	9970	4.05	0.4
Area sample (foot of bed)								
9:00 PM–6:30 AM								
Day 1	NA	NA	9.5	1140	2.07	NA	ND	NA
Personal sample								
7:30 AM–3:30 PM								
Day 2	13	185	8.0	960	2.46	Wilson HF	4.54	0.45
Personal sample								
9:00 PM–6:30 AM								
Day 2	3	68	9.0	1080	2.19	Wilson HF	ND	ND
Personal sample								
7:30 AM–3:30 PM								
Day 3	7	197	8.5	1020	2.31	Wilson HF	4.73	0.47

Note: PPE, personal protective equipment; PF, protection factor; NA, not available; ND, none detectable; HF, half-face air purifier.

activity may have affected the levels of ribavirin detected but this remains to be demonstrated; and (3) the engineering system should be used when ribavirin therapy is delivered.

Reference

1. Charney W, Corkery KJ, Kraemer R, and Wugofski L. Engineering and administrative controls to contain aerosolized ribavirin: Results of simulation and application to one patient. *Respir Care* 35:1042–1048, 1990.

Appendix 16.A.3

Hazard Evaluation and Technical Assistance
Interim Report HETA 91-104
Florida Hospital
Orlando, Florida
January 1992

Hazard Evaluation and Technical Assistance Branch Division of Surveillance,
Hazard Evaluations and Field Studies National Institute for Occupational Safety and Health
4676 Columbia Parkway
Cincinnati, Ohio

NIOSH Investigators:
John A. Decker, M.S., R.Ph.
Ruth A. Shults, R.N., M.P.H.

Source: *Respir Care* 36(6):626–629, 1991.

Summary

In response to a Florida Hospital management request, the NIOSH conducted an industrial hygiene and medical evaluation of HCWs' exposure to AR. The evaluation was conducted during three visits to Florida Hospital: February 2–4, 1991, April 3–5, 1991, and October 18–20, 1991. This interim report contains air sampling data for AR collected during all three visits and urinary ribavirin sample results collected during the first two visits.

Forty-six full-shift and short-term personal air samples for AR were collected from nurses and RTs. Fifty area air samples were collected. Sixty-one urine samples have been analyzed for ribavirin to date.

Analysis of variance (ANOVA) and Tukey's honestly significant difference (HSD) tests were used to compare the full-shift mean PBZ concentrations among nurses. There were statistically significant differences in mean PBZ concentrations among the evaluated methods of ribavirin administration. The mean concentration associated with administration through the Aerosol Delivery Hood® (ADH®) was significantly greater than the ADH enclosed by the Demistifier® scavenging tent or ventilator. The mean concentration associated with croup tent was significantly greater than the ADH/Demistifier combination or ventilator administration.

Among nurses, full-shift samples ranged from 18.7 to 31.0 μg per cubic meter air (μg/m^3) for the ADH, 12.0–28.2 μg/m^3 for the croup tent, nondetected to 13.2 μg/m^3 for the ADH/Demistifier combination, and <3.3–4.8 μg/m^3 for the ventilator. Short-term samples among RTs ranged from <12.1 for the ADH, 33.3–83.8 μg/m^3 for the croup tent, 8.3–55.5 μg/m^3 for the ADH/Demistifier combination, and nondetected for the ventilator.

Detectable levels of ribavirin were consistently found in the urine of nurses following occupational exposure, despite the use of engineering controls and high-efficiency disposable respirators. Port-shift urinary ribavirin levels were significantly higher in nurses than in RTs. A statistically significant linear relationship was present between full-shift PBZ ribavirin concentrations and postshift urinary ribavirin.

Introduction

On February 2–4, 1991, April 3–5, 1991, and October 18–20, 1991, representatives of the NIOSH visited Florida Hospital in response to a request by hospital management to evaluate employee exposures to AR. The objectives of the NIOSH investigation were to characterize workers' exposures and evaluate the effectiveness of engineering controls, work practices, and personal protective equipment.

This interim report includes information previously reported to Florida Hospital in letters dated May 16, 1991 and November 1, 1991. Participants in the biological monitoring from the first two visits were informed of their individual results via letter during May–August 1991.

Background

The administration of pharmaceutical aerosols is rapidly expanding in medicine. Asthma, chronic obstructive pulmonary disease, and pulmonary infections are frequently treated with aerosols of sympathomimetics, beta-agonists, corticosteroids, and antimicrobials. The advantages to the patient include rapid onset of therapeutic action, optimized delivery of the drug to the site of action, and reduction of unwanted systemic side-effects. Aerosol delivery, however, results in increased exposure to the HCW, compared with other administration routes. The difficulty in controlling the spread of aerosols, along with the small particle size, contributes to the risk of occupational exposure. Much of the concern about occupational exposure to pharmaceutical aerosols has centered around the use of ribavirin. The adverse reproductive effects of ribavirin exposure in animal studies have raised concerns among HCWs who administer ribavirin; many of these workers are in their reproductive years.

Florida Hospital is an 801-bed medical center and teaching hospital. The investigation was conducted in the PICU, where AR is administered to infants and children.

Uses of Ribavirin

Ribavirin is a synthetic nucleoside that is licensed by the U.S. FDA for the short-term treatment of severe RSV infections.[1] Its antiviral activity is thought to result from inhibition of RNA and DNA synthesis, which subsequently inhibits protein synthesis and viral replication.[2] AR has also been used to treat both influenza B pneumonia and RSV pneumonia in immunocompromised adults.[3,4] Clinical trials have studied the use of ribavirin in the treatment of influenza in otherwise healthy adults.[5,6]

Commercially available ribavirin is available as a sterile, lyophilized powder, which is initially reconstituted by adding 50–100 mL additive-free sterile water to a 6 g vial. The initial solution is transferred to a sterile wide-mouthed flask, which serves as the reservoir for the aerosol generator and is further diluted to a final volume of 300 mL with sterile water.

Ribavirin aerosol is generated by a Small Particle Aerosol Generator® (Model SPAG-2® nebulizer) marketed by the drug manufacturer, ICN Pharmaceuticals. The SPAG-2 nebulizer delivers AR at a rate of approximately 14 L/min. When the recommended starting solution of 20 mg of ribavirin per milliliter (mg/mL) sterile water is used, the average concentration of aerosol generated by the unit is expected to be 190 mg per cubic meter (mg/m^3), according to the manufacturer.[7] The small particle size (1.0–1.3 μm mass median diameter) of the ribavirin aerosol permits deep penetration of the drug into the patient's lungs.[8]

The aerosol can be delivered to the patient by a variety of methods, including face mask, head hood (i.e., Aerosol Delivery Hood), croup or mist tent, oxygen hood, or direct coupling to tracheostomy. During these applications aerosol may escape into the environment and be inhaled by hospital staff caring for the patient or working nearby.

Shift Assignments and Hospital Policies

The Pediatric Intensive Care nurses work 12 h shifts, while RTs work 8 h shifts; RTs occasionally work a double shift of 16 h. Nurses care for one or two patients receiving ribavirin, depending on the patient's condition and time required for care and feeding. Nurses spend about 20%–40% of the shift, or 2.5–5 h, giving bedside care. When not providing care, they sit at a makeshift desk directly outside the patient's room. As the primary health care providers, nurses generally have the highest potential for exposure to AR. RTs are generally assigned to one patient receiving ribavirin, and they spend approximately 1–1.5 h per shift in the patient's room. During the remainder of the shift, the RTs rotate to different patients throughout the hospital.

Setting-up and dismantling the ribavirin delivery system is the responsibility of the RTs. This entails transferring the liquid ribavirin solution to the SPAG-2 unit's reservoir, securing the reservoir in the unit, turning on the unit, checking/adjusting the airflow settings to the manufacturer's specifications, and ensuring that the delivery equipment is secure and functioning properly. The child is then placed into the administration device. AR is delivered from the SPAG-2 to the ADH or tent through tubing.

Every 3–5 h, the RTs checked the patient's vital signs, solution volume, and nebulizer function. Bronchodilator medications were also administered at this time, if ordered by the physician. Each visit usually lasted 15–45 min.

The hospital's written policy states that employees involved with ribavirin administration are required to wear isolation gowns, shoe covers, latex gloves, and 3 M 9970® high efficiency disposable respirators while in the treatment rooms. The hospital policy also states that the aerosol generator (SPAG-2) must be turned off at least 5 min before the administration hood or tent is opened.

Engineering Controls

The hospital has recently implemented several engineering controls related to ribavirin administration. Newly constructed ventilation systems in ribavirin treatment rooms were designed to provide 22 air changes per hour while maintaining negative pressure with respect to the adjacent hallway. The return air is vented to the outside.

The ADH (supplied by the drug manufacturer, ICN Pharmaceuticals) is equipped with an evacuation system, which is intended to remove ribavirin from around the area where the child's body enters the hood. The supply of AR from the SPAG-2, however, is not dependent on the operational status of the evacuation system.

A Demistifier isolation tent (Peace Medical), which scavenges ribavirin aerosol escaping from the ADH, is currently being used on a trial basis by the hospital. During administration of ribavirin, the plastic isolation tent is placed over the ADH. Air within the tent is exhausted into the room through a HEPA filter system.[9]

Evaluation Criteria

Toxicology of Ribavirin

In animal studies, ribavirin has been shown to be teratogenic and embryolethal in rats, mice, and hamsters, and embryolethal in rabbits.[10–13] The single primate study did not show teratogenic effects; however, due to a limited number of test animals, this study did not provide adequate evidence to evaluate reproductive outcome.[14] Three studies in rats showed degenerative or histopathologic testicular effects. Eight other studies in rats, mice, dogs, and monkeys induced no testicular effects.[15] Ribavirin was found to be toxic to lactating animals and their offspring.[16]

The adverse reproductive effects seen in animal studies have raised concerns among HCWs who administer ribavirin; many of these workers are in their reproductive years. Ribavirin has not been linked to fetal abnormalities in humans; however, given the wide spectrum of teratogenic potential in several animal species, avoidance of ribavirin prior to pregnancy, during pregnancy, and during lactation has been recommended.[17] At present, the potential reproductive health effects of occupational exposure to ribavirin are unknown.

Adverse effects occur infrequently in patients receiving AR; the more commonly reported effects include respiratory and cardiovascular disturbances, rash, and skin irritations.[16,18] Hemolytic anemia and suppression of erythropoiesis can occur when the drug is given orally or parenterally.[16,18]

Acute effects due to environmental exposure to ribavirin aerosol include rhinitis, headache, and eye irritation.[16,18] The drug has been reported to precipitate on contact lenses, and eye irritation has been reported in employees wearing contact lenses.[16,19]

Pharmacokinetics of Ribavirin

Following inhalation, ribavirin is deposited in the respiratory tract. It is then redistributed from the respiratory tract into the circulation with eventual accumulation in erythrocytes. The extent of accumulation following inhalation has not been established, but following oral administration of a single dose of ribavirin, plasma and erythrocyte levels initially increased in parallel. Within 2 h after administration, the plasma levels began to fall while erythrocyte levels continued to rise.

Erythrocyte levels rose to a plateau at about 4 days and then declined, with an apparent half-life of 40 days.[20]

Ribavirin is believed to be metabolized in the liver. The major route of elimination of ribavirin and its metabolites appears to be renal. In healthy adults with normal renal function, excretion of ribavirin administered orally indicates that approximately 53% of a single dose is excreted within 72–80 h.[16] An additional 15% is excreted in the feces.[16] No data are available regarding cutaneous or mucocutaneous absorption.

Exposure Recommendations—Ribavirin

The toxicological data currently available for ribavirin are insufficient to determine the health risks to HCWs. No occupational exposure criteria for ribavirin have been published by the Occupational and Health Administration (OSHA), NIOSH, or the American Conference of Governmental Industrial Hygienists (ACGIH).

The CDHS has suggested that an occupational exposure limit, based on a risk assessment model, can be calculated by applying a safety factor of 1000 to the NOEL in the most sensitive animal species.[21,22] Using this procedure, the CDHS has proposed a limit of 2.7 $\mu g/m^3$ as an 8 h TWA. This calculation was based on a minute ventilation of 19 L, an employee weight of 58 kg, and a pulmonary ribavirin retention rate of 70%. The model was based on pharmacokinetic data collected after administration of therapeutic doses, which may not be a reliable indicator of lower-dose occupational exposure. Additionally, the minute ventilation rate was probably in excess of normal conditions.

Room Ventilation Recommendations

The American Institute of Architects (AIA) Committee on Architecture for Health has published ventilation recommendations for hospitals. Isolation rooms are recommended to have a minimum of six total air changes per hour and should be under negative pressure. Regular patient rooms are required to have a minimum of two total air changes per hour.[23]

Methods

Background

NIOSH investigators collected PBZ samples and urine specimens for ribavirin analysis from nurses and RTs who administered the drug. Each patient had a private room. The monitoring was conducted in conjunction with the following administration methods: ADH with ICN evacuation device in operation, ADH and Pup® tent enclosed by the Demistifier scavenging tent, croup tent,

TABLE 16.A.3 Sampling Summary, Florida Hospital, Orlando, Florida. HETA 91–104

	February 2–4, 1991	April 3–5, 1991	October 18–20, 1991
Patients	3 infants	1 infant	2 infants
		2 older children	1 older child
Administration unit	Croup tent	ADH/Demistifier	ADH/Demistifier
	ADH alone	Ventilator	Pup/Demistifier
	ADH/Demistifier		ADH alone
Number of full-shift air samples	9	3	7
Number of short-term air samples	12	3	1
Number of area air samples	32	4	12
Number of urine samples	47	41	60

and direct coupling to a ventilator. The ICN scavenging system was not in operation when the Demistifier was used. Ventilation measurements were made to characterize the effect of room ventilation on AR concentrations. A summary of the number and types of air and urine samples can be found in Table 16.A.3.

February 2–4, 1991

During the February visit, three children received ribavirin for treatment of RSV infection. One was treated with the ADH enclosed by the Demistifier scavenging tent, the second was treated within a croup tent, in which the ribavirin aerosol was supplied directly into the tent, and the remaining child was treated with the ADH alone.

The nurses' short-term PBZ samples were collected while full-shift sampling was in progress. Area samples were collected within the treatment rooms and the nurses' station, which is located across the hall from the ribavirin patient rooms. Three bulk samples of ribavirin solution were collected from the SPAG-2 for analysis.

April 3–5, 1991

During the April visit, three children were treated with ribavirin; one via an ADH enclosed by a Demistifier scavenging tent, and the remaining two via ventilators through tracheostomies.

No scavenging devices, such as the Demistifier, were used with the ventilator administration.

In addition to the samples collected during ventilator administration (Table 16.A.3), five PBZ samples for AR were collected from three nurses and two RTs assigned to the infant receiving ribavirin aerosol in the ADH enclosed by the Demistifier scavenging tent. Eleven area samples were collected. Because of an equipment malfunction in the SPAG-2, the results of these air samples will not be reported; they are not included in Table 16.A.3.

Five sets of samples from within the ADH and the Demistifier tent were also collected. These five sets of samples were collected on 37 mm glass fiber filters at a flow rate of 1.0 L/min for 10 min within the ADH and 15 min within the Demistifier tent.

October 18–20, 1991

During the October visit, three children were treated with ribavirin. One was treated via an ADH enclosed by the Demistifier scavenging tent. The second was treated within a Pup tent (Peace Medical), in which the ribavirin aerosol was supplied directly into the tent. The Demistifier scavenging tent was placed over the Pup tent. The third patient was treated with an ADH equipped with the ICN evacuation device (no Demistifier).

Three samples from inside the ADH, and two from inside the Demistifier were collected at a flow rate of 1 L/min for 5 and 10 min, respectively.

Air Sampling Methodology and Laboratory Analysis

Participants and Sample Types

Personal samples were collected in the workers' breathing zone. Full-shift samples were generally collected from nurses, who provided care continually throughout their shift. Short-term samples were generally collected from RTs, who provided care approximately four times per shift. Exposure monitoring was conducted only on employees. No in-mask sampling from the respirators was conducted.

Sampling Methodology

Air sampling for AR was conducted according to NIOSH method 5027, utilizing 37 mm diameter, 1.0 μm glass fiber filters.[24] AR was collected on the filters at a flow rate of 2.0 L/min for full-shift personal and area samples. A flow rate of 3.0 L/min was utilized for the short-term samples.

Laboratory Analysis

The glass fiber filters containing ribavirin were extracted with 3 mL sulfuric acid solution (pH = 2.5) in an ultrasonic bath and analyzed by HPLC using a cation exchange resin column. The HPLC was equipped with an ultraviolet detector set at the 210 nm wavelength. The reported limits of detection were 0.3 and 1 μg/sample for the February and October visits, respectively. The reported limits of quantitation were 0.8 and 3.1 μg/sample, for the February and October visits, respectively.

Statistical Analysis of Results

ANOVA and Tukey's HSD[25] were used to compare simultaneously group means of the different administration methods (ADH, ADH/Demistifier, croup tent, ventilator). Full-shift PBZ concentrations among nurses working with different administration methods were compared.

Biological Monitoring Methodology

Background

Previous studies of ribavirin exposure in HCWs have attempted to measure ribavirin and/or its metabolites in urine, plasma, and erythrocytes. Harrison et al. collected 30 urine, 30 erythrocyte, and 30 plasma samples from 10 ribavirin-exposed HCWs. Ribavirin was detected in only one erythrocyte sample (at a concentration of 0.44 μg/mL) collected from a nurse 5 days after exposure.[26] Environmental samples collected during the work shift of this nurse detected a PBZ level of 316 μg/m^3 and an area concentration of 1048 μg/m^3. Bradley et al. collected erythrocytes and plasma from 7 ribavirin-exposed HCWs who were not wearing respirators, but no ribavirin was detected.[27] PBZ levels ranged from 1.0 μg/m^3 to 1328 μg/m^3.

In a previous study, NIOSH investigators found a measurable level of ribavirin in the urine of one of three ribavirin-exposed HCWs, but not in plasma or erythrocytes.[28] With this knowledge, NIOSH investigators decided to collect only urine from ribavirin-exposed HCWs. Based on the pharmacokinetics seen in clinical trials, NIOSH investigators chose to collect three urine specimens: prior to exposure, immediately following exposure, and 24 to 48 h postexposure.

Participants

All nurses and RTs providing direct care to patients receiving ribavirin were invited to participate in the study; 45 of 48 (94%) HCWs submitted urine samples. HCWs received their patient assignments prior to being recruited for the study and without regard to their willingness to participate in the study. Each participant completed a brief questionnaire to document job title, work area, and recent history of ribavirin exposure.

Sample Collection

In order to avoid contamination of the urine stemming from ribavirin, potentially present on their hands, HCWs were instructed to wash their hands prior to contributing a urine specimen. Samples were collected in disposable paper cups and placed in clean glass or plastic transport tubes without preservative. The samples were frozen and shipped by overnight express to the contract laboratory.

Laboratory Analytical Method

HPLC and RIA were used to measure the urinary ribavirin levels.[29,30] The laboratory reported the lowest value at which the amount of ribavirin in the urine can be accurately quantified, or the limit of quantification (LOQ), to be 0.02 micromoles per liter of urine (μmol/L).[31] However, using data from replicate control and field samples, NIOSH chemists estimated the LOQ to be 0.01 μmol/L.[32] At levels below the LOQ the precision with which the laboratory method measures ribavirin is reduced. All of the urine ribavirin values were creatinine-adjusted, and those values were used in statistical analyses regardless of whether the unadjusted ribavirin value was above or below the LOQ.

Creatinine Correction of Urine Samples

Urine samples collected over a 24 h period would have provided the most accurate measurement of HCWs' excretion of ribavirin. However, it was impractical to collect 24 h urine samples (an urine over a 24 h period), so "spot" urine samples were used. To "standardize" the concentration of substances (i.e., to make the results comparable from one time to another and from one person to another), it is common practice to correct the results for the dilution of the urine, and creatinine correction is the preferred method for very concentrated and very dilute urine samples.[33,34] Creatinine is a normal metabolic product that is excreted by the kidney at a daily rate that is constant for an individual.

The creatinine-corrected urinary ribavirin result was obtained by dividing the urinary ribavirin result by the creatinine concentration. The unit of measure used in reporting the creatinine-corrected urinary ribavirin level is micromoles of ribavirin per gram of creatinine (μmol/g).

Data Analysis

The results of the "next day" urine samples were not included in the analyses because not all participants provided these samples and the intervals between collection of postshift samples and next day samples varied widely. Some of the HCWs had cared for patients who were receiving ribavirin therapy before the NIOSH team arrived to begin biological monitoring. Because some participants had prior exposure and others did not, it was necessary to control for the preshift urine ribavirin levels. This was accomplished using analysis of covariance (ANCOVA), with the preshift ribavirin level and job title as independent variables. The null hypothesis was that, after adjusting for differences due to preshift urine ribavirin levels, the mean postshift urinary ribavirin level of nurses did not significantly differ from that of RTs. Logarithmic transformation was done on the postshift urinary ribavirin levels to obtain a normal distribution.

Linear regression was performed to test for an association between the full-shift PBZ ribavirin levels of nurses and their corresponding postshift urinary ribavirin levels. The preshift urine ribavirin levels and the logarithmic transformation of the PBZ levels were the independent variables, and the logarithmic transformation of the postshift ribavirin levels was the dependent variable.

Ventilation Evaluation

February 1991

Smoke tubes were used to visualize the direction of airflow between the treatment rooms and the adjacent hallway, to determine if ribavirin could potentially migrate out of the treatment rooms.

April 1991

Room ventilation flow rates were measured in the ADH/Demistifier treatment room (#6322) using an Airdata® flow meter. Four sets of measurements (supplies and exhausts) were made between 1305 h on April 4 and 0650 h on April 5, 1991. Measurements were recorded with the front door closed and the bathroom door open. Hospital personnel adjusted the room ventilation in the morning on April 4, 1991.

The patients receiving ribavirin by ventilator were not in the specially designed ribavirin treatment rooms. No measurements were made in these rooms.

October 1991

Ventilation flow rates (supplies and exhausts) were measured in each of the treatment rooms (room 6321, 6322, and 6330-PICU bed #7) with a Shortridge Instruments Air Data Flow Meter, CFM-88, Series 8400. Measurements were made with the front door closed and the bathroom door open.

Results and Discussion

Personal Air Sampling Results—Summary

Since most of the employees wore NIOSH/Mine Safety and Health Administration (MSHA) approved 3M 9970 disposable high efficiency respirators while working in the treatment rooms, the actual exposures were probably lower than the breathing zone concentrations that are reported.

Table 16.A.4 summarizes all the personal air monitoring results from the February and October visits. Tables 16.A.5 and 16.A.6 present a detailed list of personal full-shift and short-term sample results. The ADH/Demistifier data from the April visit are not included because the volume of ribavirin solution used by the SPAG was later found to be much less than recommended by the manufacturer. Air samples collected inside the ADH were much less than the concentration recommended by the drug manufacturer.

The use of the scavenging tent, which enclosed the ADH, lowered personal exposures for both nurses and RTs. The mean (direct average of the TWAs), full-shift, TWA breathing zone concentration for nurses was 4.4 μg/m^3 (range <3.3–4.8 μg/m^3) with the use of the scavenging tent, versus 24.9 μg/m^3

TABLE 16.A.4 Summary Data, Aersolized Ribavirin Personal Exposure Concentrations, Florida Hospital, Orlando, Florida, HETA 91-104

Administration Method	Job Category	Number of Samples	Sample Type	Range of Conc. (μg/m^3)[a]	Mean of Conc. (μg/m^3)[b]
ADH + tent[c]	Nurses	8	Full-shift	ND–13.2[d]	4.4
ADH + tent	Nurses	2	Short-term	11.9–13.9	12.9
ADH + tent	RTs[e]	1	Full-shift	ND	ND
ADH + tent	RTs	6	Short-term	8.3–55.5	22.0
Croup tent	Nurses	4	Full-shift	12.0–28.2	22.9
Croup tent	Nurses	2	Short-term	58.8–95.2	77.0
Croup tent	RTs	4	Short-term	33.3–83.3	58.1
ADH alone and ADH + tent	Nurses	1	Full-shift	78.0	78.0
ADH alone	Nurses	2	Full-shift	18.7–31.0	24.9
ADH alone	RTs	2	Short-term	<7.5–<12.1	<12.1
ADH alone	RTs	1	Full-shift	5.9	5.9
Bear Cub® Ventilator	Nurses	3	Full-shift	<3.3–4.8	4.3
Bear Cub® Ventilator	RTs	3	Short-term	ND	ND

[a] "Range of conc." refers to the range in concentrations of the individual samples, expressed in micrograms per cubic meter (μg/m^3).

[b] "Mean of conc." refers to the mean of the individual samples.

[c] "ADH + tent" refers to an Aerosol Delivery Hood enclosed by the Peace Medical Demistifier isolation tent. This scavenging tent was placed over the ADH. The ADH did not have the ICN evacuation apparatus connected.

[d] "ND" means non-detected. ND concentrations were treated as zero for calculating the "Mean of conc."

[e] "RTs" signifies respiratory therapists.

TABLE 16.A.5 Personal Samples of Ribavirin, Full-Shift Samples, Florida Hospital, Orlando, Florida. HETA 91–104

Job Title/Unit	Date	Sample Period	Sample Time (min)	Time in Room (min)	Percent Time in Room	Conc. (μg/m³)[a]	TWA Conc.[b]
Nurse[c]	2/2/91	1904–2246	222	80	36	11.3	
Demistifier		2246–0645	479	160	33	12.8	12.4
Nurse[c]	2/3/91	0739–1337	358	170	47	9.8	
Demistifier		1445–1839	236	65	28	8.5	9.3
Nurse[c]	2/3/91	1915–2320	185	175	95	22.4	
Demistifier		2320–0629	369	85	23	7.9	13.2
Nurse[d] Demistifier	10/18/91	1640–2255	375	30	8	ND	ND
RT[d] Demistifier	10/18/91	1724–2305	341	50	14	ND	ND
Nurse[d] Demistifier	10/18/91	0756–1451	415	75	18	ND	ND
Nurse[d]	10/19/91	0729–0955	146	60	41	ND	ND
Demistifier		1245–1646	242	115	47	ND	ND
RT[e] ADH alone	10/18/91	0750–1447	417	60	14	5.9	5.9
Nurse[e] ADH alone	10/18/91	1532–1807	157	58	36	18.7	18.7
Nurse ADH alone	10/18/91	0734–1228	294	50	17	31.0	31.0
Nurse[d]	2/2/91	1906–2245	221	55	25	33.9	
Croup tent		2245–0645	480	150	31	25.6	28.2
Nurse[d]	2/3/91	0708–1315	367	130	35	36.8	
Croup tent		1315–1600	165	44	27	9.1	28.2
Nurse[d]	2/3/91	1915–2319	244	135	55	38.9	
Croup tent		2319–0629	430	112	26	14.0	23.0
Nurse[d] Croup test	2/4/91	0730–1510	460	178	39	12.0	12.0
Nurse Ventilator	4/3/91	1403–1903	300	150	50	<3.3	<3.3
Nurse Ventilator	4/3/91	1944–0620	636	192	30	4.7	4.7
Nurse Ventilator	4/4/91	1951–0620	630	480	76	4.8	4.8
Nurse[c,d] Croup and Demistifier	2/2/91	1525–1856	211	61	28	43.7	43.7
Nurse[c,f] ADH and Demistifier[g]	2/4/91	0730–1505	455	255	56	78.0	78.0

[a] Conc. (μg/m³) = ribavirin concentration in micrograms per cubic meter air, computed over the sampling period.

[b] TWA Conc. = time-weighted average concentration of ribavirin in micrograms per cubic meter air, computed over the entire work shift.

[c] Administration in room 6322.

[d] Administration in room 6321.

[e] Administration in room 6330.

[f] Administration in room 6320.

[g] ADH = Aerosol Delivery Hood, ICN Pharmaceuticals.

TABLE 16.A.6 Personal Samples for Ribavirin, Short-term Samples, Florida Hospital, Orlando. HETA 91–104

Job Title	Administration Unit	Sampling Date	Sampling Time (minutes)	Concentration ($\mu g/m^3$)[a]
RT[b]	Croup Test	2/2/91	8	83.8
RT	Croup Tent	2/2/91	14	71.4
RT	Croup Tent	2/2/91	15	44.4
RT	Croup Tent	2/3/91	32	33.3
Nurse	Croup Tent	2/2/91	88	95.2
Nurse	Croup Tent	2/3/91	47	58.8
RT	Demistifier[c]	2/2/91	30	33.3
RT	Demistifier	2/2/91	36	55.5
RT	Demistifier	2/3/91	32	9.4
RT	Demistifier	2/3/91	28	11.9
RT	Demistifier	2/3/91	10	8.3
RT	Demistifier	2/4/91	20	13.3
RT	Demistifier	10/19/91	38	<40.7
Nurse	Demistifier	2/2/91	22	11.9
Nurse	Demistifier	2/3/91	48	13.9
RT	ADH[d]	2/4/91	36	<7.4[e]
RT	ADH	2/4/91	22	<12.1[g]
RT	Ventilator[f]	4/3/91	30	ND[g]
RT	Ventilator	4/4/91	25	ND
RT	Ventilator	4/4/91	40	ND

[a] $\mu g/m^3$ = micrograms per cubic meter air over the specified sample time.
[b] RT = respiratory therapist.
[c] Demistifier = Scavenging tent (Peace Medical, Inc.) placed over the Aerosol Delivery Hood.
[d] ADH = Aerosol Delivery Hood (ICN Pharmaceuticals).
[e] "<" = indicated value below the LOQ of 3.1 μg/sample.
[f] Bear Cub pediatric ventilator, administered through tracheotomy.
[g] ND = Below the LOD of 1 μg/sample.

(18.7–31.0 $\mu g/m^3$) without the scavenging tent. The ribavirin concentrations in the full-shift RT sample was below the limit of detection (LOD) with the scavenging tent, versus 5.9 $\mu g/m^3$ without the scavenging tent. The croup tent administration resulted in a full-shift mean breathing zone concentration of 22.9 $\mu g/m^3$ (12.0–28.2 $\mu g/m^3$).

Comparatively low exposures (full-shift mean of 4.3 $\mu g/m^3$ for nurses) occurred with the ventilator administration. None of the three short-term air samples from the RTs had detectable ribavirin. This finding was not unexpected, since the pediatric ventilator was essentially a closed system with a filter on the exhalation circuit.

The highest full-shift personal exposure (78.0 $\mu g/m^3$) was collected from a nurse caring for two children, one child treated with the ADH enclosed by the Demistifier scavenging tent. The nurse did not always turn the aerosol generator off 5 min before opening the administration device. The highest short-term exposures (Table 16.A.3—means of 58.1 and 77.0 $\mu g/m^3$) occurred with the croup tent, which was reasonable to expect since a substantial amount of ribavirin remained inside the relatively large tent when it was opened by the HCW.

Short-term samples collected from one RT while using the ADH alone had ribavirin concentrations below the LOQ (<7.4–<12.1 $\mu g/m^3$). The RT was in the habit of fuming off the aerosol generator and leaving the room for 10 to 15 min before starting his work. During this interim period, a large percentage of AR was probably removed by the room ventilation system, which provided a measured 18–19 air changes per hour (ACH). It should be noted that about 7 min are required for 90% removal efficiency of airborne contaminant, assuming 19 ACH and perfect mixing.[35] In addition to the flushing effect of the ventilation system, the level of ribavirin exposure is probably related to the proximity of the employee to the administration hood during the shut-off period before the hood is opened.

ANOVA and Tukey's HSD were used to simultaneously compare group means of the different administration methods (ADH, ADH/Demistifier, croup tent, and ventilator). Full-shift PBZ concentrations among nurses were compared. Administration with the ADH alone resulted in statistically significant greater exposures than the ADH/Demistifier combination. The croup tent and ADH resulted in exposures that were significantly greater than the ventilator administration. The croup tent was significantly greater than the ADH/Demistifier. No other pair-wise comparisons (ADH versus croup tent, etc.) were statistically significant.

The overall ANOVA had an F-value of 10.09 with an associated p-value of 0.0017 (degrees of freedom were 3, 11). For the purpose of statistical data analysis, values below the LOD were assigned the LOD value divided by the square root of 2. Several other methods were used to adjust the values below the LOD (such as counting these values at the LOD and at zero). Regardless of the value used, the result was the same. The results, however, should be interpreted with a note of caution, due to the limited sample size.

February 1991 Area Sample Results

Table 16.A.7 lists the results of area air samples within the treatment rooms and the nurses' station. Measurable levels of ribavirin (mean 9.7 μg/m^3) were found at the nurses' station, indicating that the level of negative pressure within the treatment rooms relative to the hallway was insufficient. Smoke tube tests also indicated that the pressure in the treatment rooms was neutral or under marginally negative pressure.

An area sampler was placed on each side of the croup and the Demistifier tents. As expected, the average ribavirin concentrations were highest on the side of the tent that was opened when the children required attention. The TWAs on either side of the croup tent were 54.0 and 40.0 μg/m^3; for the Demistifier, the values were 17.0 and 9.8 μg/m^3. Area concentrations were generally lower for the Demistifier than for the croup tent. In all but one measurement, area concentrations at the sides of the Demistifier tent were lower than for the ADH alone.

Two bulk samples of ribavirin solution, taken before placement into the SPAG unit, were analyzed to determine the ribavirin concentration. The sample taken from room 6322 contained 20 mg/mL. The sample taken from room 6321 was damaged during shipment. A sample collected from SPAG in room 6322 at the end of the shift contained 32 mg/mL. This finding confirms the increase in ribavirin concentration in the reservoir, which occurs during treatment because the SPAG discharges water vapor in addition to aerosol particles containing the drug.[36] The mechanism is probably the dilution air blowing continuously through the reservoir, and causing evaporation of water.

Ventilation observations. Rooms 6320 and 6322 were under marginally negative pressure with respect to the adjacent hallway (air was moving into treatment room). Room 6321 was under weakly positive pressure with respect to the hallway (air movement out of the treatment room). All supply and return vents were functioning.

April 1991 Area Sample Results

Ribavirin concentrations in all personal and area samples collected in association with the child receiving ribavirin aerosol in the ADH enclosed with the Demistifier were below the LOD.

Short-term air sampling for ribavirin was conducted inside the ADH to determine if there was a relationship between occupational exposures and hood concentrations. The results of five 5 min samples inside the ICN hood ranged from 4.9 to 25 mg/m^3, well under the concentration of 190 mg/m^3 recommended by the manufacturer. The amount of ribavirin solution used in the aerosol generator was later found to be much less than expected, according to a telephone conversation with the Director of Respiratory

TABLE 16.A.7 Ribavirin Area Air Samples, Florida Hospital, Orlando, Florida, February 2–4, 1991. HETA 91–104

Sampling Location or Room Number	Date	Sample Period	Conc. (μg/m³)[a]	Average Conc.[b]
Nurses' station, across hall from treatment rooms	2/2/91	1515–1907	4.3	9.7
		1915–2247	2.2	
		2247–0647	8.6	
	2/3/91	0710–1242	9.0	
		1242–1845	13.9	
		1845–0629	15.9	
	2/4/91	0756–1504	4.7	
Room 6321, croup tent. 3 ft from tent and floor	2/2/91	1517–1911	47.2	
		1916–2250	52.9	
Employees worked from this side of the tent	2/3/91	2250–0649	92.3	
		0715–1241	32.2	
		1241–1849	28.5	54.0
Room 6321, croup, tent. Sampler set near SPAG. 2 ft from tent, 3 ft from floor	2/2/91	1919–2249	38.1	
	2/3/91	2249–0645	52.6	
		0712–1243	46.8	
		1243–1849	36.9	
		1858–0630	33.3	
	2/4/91	0759–1506	35.1	40.0
Room 6321, croup. Above tent on light fixture	2/2/91	1529–1910	84.7	
Room 6322, Demistifier. Near SPAG. 3 ft from floor, 2 ft from tent	2/2/91	1525–1856	33.5	
		1921–2253	21.2	
		2253–0650	16.6	
Employees worked from this side of tent	2/3/91	0719–1245	16.9	
		1245–1850	15.1	
	2/4/91	0758–1507	8.4	17.0
Room 6322, Demistifier. 3 ft from tent and floor	2/2/91	1922–2254	11.8	9.8
		2254–0650	13.6	
	2/3/91	0716–1245	10.6	
		1245–1850	11.0	
		1900–0630	7.2	
Room 6322, Demistifier. Above tent on light fixture	2/2/91	1520–1911	82.3	
Room 6320, ICN hood alone. Sampler 4 ft off floor, 3 ft from hood	2/4/91	0800–1507	32.8	32.8

[a] μg/m³ = micrograms per cubic meter air over the actual sampling period.
[b] Average Conc. = TWA concentration of ribavirin in micrograms per cubic meter air.

Care in May 1991. Therefore, the PBZ and area air monitoring results were lower than normal and probably not representative of typical exposures.

Administration of ribavirin to a patient through a Bear Cub® ventilator was also conducted during the April visit. A scavenging unit was not used during this administration. The results of area samples in the treatment room ranged from nondetected to 12.9 μg/m³. Concentrations at the nurses' station were from nondetected to <2.4 μg/m³. Specific results can be found in Table 16.A.8.

Ventilation observation measurements were made in the treatment room with the Demistifier (room 6322). Using an estimated room volume of 1330 ft³, ACH were calculated using the exhaust measurements for the room. The room air was reportedly exhausted to the outside. The results, which ranged

TABLE 16.A.8 Ribavirin Area Air Samples, Florida Hospital, Orlando, Florida, April 3–5 and October 18–19, 1991. HETA 91–104

Sampling Location or Room Number	Date	Sample Period	Conc. (μg/m³)[a]
Nurses' station, across hall from ICU #6	4/4/91	0740–1920	ND[b]
	4/4/91	1946–0621	<2.4
Head of bed, Bear Cub Ventilator, Room ICU #6	4/3/91	1305–1902	11.2
	4/3/91	1925–0622	12.9
	4/4/91	0654–1921	ND
	4/4/91	1950–0622	11.1
Nurses' station, across hall from 6320 and 6321	10/18/91	0707–1746	ND
	10/18/91	1804–0650	ND
	10/19/91	0720–0955	ND
Nurses' station, across hall from 6330	10/18/91	0746–1746	ND
	10/18/91	1805–0100	ND
Room 6321, Demistifier enclosing ADH, 4 ft from tent, 3 ft off floor	10/18/91	0712–1746	ND
	10/18/91	1804–0650	ND
	10/19/91	0720–0955	37.3
	10/19/91	1249–1647	ND
Room 6322, Demistifier enclosing Pup tent	10/18/91	1710–1500	16.0
Room 6330, ADH alone, 4 ft away, 3 ft off floor	10/18/91	0735–1747	161.8
	10/18/91	1806–0653	ND
Room 6321, inside ADH	10/18/91	1622–1627	64,000
Inside the Demistifier	10/18/91	1622–1632	1,500
Room 6321, inside ADH	10/19/91	1020–1025	30,000
Inside the Demistifier	10/19/91	1020–1030	2.400
Room 6322, inside ADH	10/19/91	0730–0735	78,000

[a] Conc. (μg/m³) = ribavirin concentration in micrograms per cubic meter, air, computed over the sampling period.

[b] ND = below the limit of detection of 0.8 μg/m³/sample.

from 18 to 19 ACH, were well in excess of the minimum rates (6 ACH) recommended by the AIA. The results of ventilation measurements can be found in Table 16.A.9.

Smoke tube tests indicated that room 6322 was under significant negative pressure with respect to the adjacent hallway, and that the rooms used for the ventilator patients (PICU #3 and #6) were under very slight negative pressure. The patients receiving ribavirin by ventilator were not in the specially ribavirin treatment rooms. Room ventilation flow rates were not measured in these rooms, since the location and design of the return and supply diffusers did not permit the use of airflow equipment.

October Area Air Sample Results

Some samples collected inside the ADH were less than the expected concentration of 190 mg/m³, specified by the drug manufacturer.[37] Five minute sample results ranged from 30 to 78 mg/m³ ribavirin with the nebulizer airflow set at 7 L/min (see Table 16.A.8). However, the amount of ribavirin solution used by the SPAG-2 was normal. Other investigators have found that AR concentrations within the administration hood vary as a function of time and nebulizer airflow.[8] AR concentrations within the treatment hood might also vary depending on the sampling methodology and location within the administration hood. Two samples collected from inside the Demistifier unit had ribavirin concentrations of 1500 and 2400 μg/m³.

TABLE 16.A.9 Room Ventilation Measurements, PICU, Florida Hospital, Orlando, Florida. HETA 91–104

Room Number and Date	Supply (cfm)	Return (cfm)	Bathroom Exhaust (cfm)	Air Changes per Hour
6322 4/3/91	414	−357	−74	19
6322 4/4/91	234	−331	−71	18
6322 4/4/91	228	−342	−72	19
6321 10/18/91	271	−358	−93	20
6321 10/19/91	257	−371	−97	21
6322 10/18/91	443	−367	−124	22
6322 10/19/91	369	−295	−95	18
6330 PICU #7 10/18/91	171	−229	None	10
6330 PICU #7 10/19/91	167	−243	None	11

Ventilation flow rates were measured in each of the treatment rooms (6321, 6322, and 6330-PICU bed #7). Measurements were recorded with the front door closed and the bathroom door open. Table 16.A.9 lists the results, which ranged from 18 to 22 ACH in rooms 6321 and 6322 and 10 to 11 ACH in room 6330 (PICU #7). The room ventilation rates were well in excess of the minimum rate (6 ACH) recommended by the AIA for isolation rooms.[34] Using tissue paper, it was visually determined that all of the treatment rooms were under significant negative pressure at the doorway.

Biological Monitoring Results

As previously described, urine samples were collected immediately prior to the start of the shift, at the end of the shift, and 24–48 h postshift. Urine samples were collected from 18 HCWs during the February and April NIOSH visits; 61 samples were analyzed for ribavirin and creatinine: 21 preshift, 23 postshift, and 17 next day. Nine RTs, eight male and one female, and nine nurses, all females, participated. Their ages ranged from 22 years to 47 years, with a mean age of 33 years. Results from eight additional HCWs who cared for a patient whose malfunctioning SPAG prevented the patient from receiving a therapeutic dosage were excluded from analysis. No air sample taken in the room had detectable levels of ribavirin.

Urine ribavirin levels prior to creatinine adjustment ranged from nondetectable (<0.01 μmol/L) to 0.24 μmol/L of urine. The creatinine-adjusted ribavirin levels, reported in micromoles of ribavirin per gram (μmol/g) of creatinine, ranged from nondetectable to 0.140 μmol/g. Correlation between unadjusted postshift ribavirin levels and adjusted postshift ribavirin levels was 0.77 (Pearson correlation coefficient, $p = 0.001$) and 0.90 (Spearman correlation coefficient, $p = 0.001$).

For statistical analyses, multiple sets of urine results from an individual were not considered to be independent of each other. Some HCWs worked two shifts during the 3 day sampling period, and

TABLE 16.A.10 Urinary Ribavirin Levels, Florida Hospital, Orlando, Florida. HETA 91–104

Job Title	Preshift Unadjusted[a] (μmol/L)	Preshift Adjusted[b] (μmol/g)	Postshift Unadjusted (μmol/L)	Postshift Adjusted (μmol/g)
RT	0.0006	0.000	0.0025	0.002
RT	0.0118	0.005	0.0069	0.002
RT	0.0025	0.001	0.0075	0.004
RT	0.0054	0.002	0.0105[e]	0.004
RT	0.0097	0.003	0.0097	0.004
RT	0.0087	0.010	0.0056	0.006
RT	0.0090	0.017	0.0110[e]	0.009
RT	0.0054	0.002	0.0999[e]	0.036
Nurse[c]	0.0000	0.000	0.0020	0.002
Nurse[c]	0.0000	0.000	0.0139[e]	0.009
Nurse[d]	0.0042	0.000	0.0943[e]	0.054
Nurse	0.0037	0.002	0.0357[e]	0.019
Nurse	0.0010	0.000	0.0228[e]	0.023
Nurse	0.0055	0.012	0.0166[e]	0.035
Nurse[d]	0.0130[e]	0.015	0.0933[e]	0.042
Nurse	0.0229[e]	0.013	0.0443[e]	0.092
Nurse	0.0598[e]	0.085	0.2411[e]	0.140

[a] Micromoles of ribavirin per liter of urine.
[b] Micromoles of ribavirin per gram of creatinine.
[c] No preshift urine sample was obtained: a value of "0.000" was assigned.
[d] Each value represents the mean of two samples; one was collected during the February visit and the other was collected during the April visit.
[e] Unadjusted ribavirin value above the 0.01 μmol/L limit of quantification.

therefore they contributed two sets of urine samples during one visit; the urine results from the second shift worked were excluded from analysis in all but one case; in this case, no preshift sample was obtained for the first shift the HCW worked, so those results were excluded. Other HCWs contributed one set of samples during each of two visits; their results were averaged by person.

Two additional sets of results did not include preshift urines. Neither HCW had been exposed to ribavirin in the 14 days prior to the shift during which the sampling was conducted; therefore, each preshift ribavirin was assigned a value of "0.000" and included in the sets in the analysis.

ANCOVA of 17 urine sample sets showed that nurses had significantly higher postshift urinary ribavirin levels compared to RTs ($F = 7.76$, d.f. = 2,14, $p = 0.01$). Table 16.A.10 lists the preshift and postshift values before and after creatinine adjustment. Eight of the nine nurses, but only three of the eight RTs had postshift urinary ribavirin labels, before adjusting for creatinine, greater than the 0.01 μmol/L limit of quantification.

TWAs of PBZ air samples and urine ribavirin labels were available for seven nurses over nine work shifts. Linear regression showed that the overall model was significant ($F = 9.61$, d.f. = 2,6, $p = 0.01$). Further, after adjusting for preshift urine ribavirin labels, a significant linear relationship remained between the PBZ levels of ribavirin and the HCWs' postshift urine ribavirin levels ($T = 3.32$, $p = 0.01$).

Study Limitations

NIOSH investigators requested that work practices remain as usual during the study period. While this policy increases the generalizability of the results, the differing shift lengths, the working of double shifts, having exposure to ribavirin in the 24 h prior to the start of the study, and the working of more than one shift during the study period by some HCWs place limitations on the interpretation of the biological monitoring data.

Most employees complied with the hospital's respiratory protection policy during the NIOSH visits. Employees who wore respirators were fit-tested with the 3 M® saccharin aerosol qualitative system. Some HCWs were observed wearing surgical masks instead of respirators while in patients' rooms. Nurses reported that it was sometimes necessary for them to immediately enter a patient's room, and in these instances respirators were not always worn.

Six of the 17 urine sample sets had postshift urinary ribavirin values, before adjusting for creatinine, far less than the 0.01 μmol/L limit of quantification. The laboratory method used to measure ribavirin is less precise at levels below the limit of quantification.

The use of respirators did not affect the PBZ air concentrations. However, respirators probably decreased urinary ribavirin levels by decreasing the actual inhaled dose. Disposable respirators, such as the 3 M 9970 respirator, have an assigned PF of 5 (the factor is 10 when the respirators are assigned to employees based on the results of quantitative fit testing).[38] The assigned PF is the minimum anticipated protection provided by a properly functioning respirator to a given percentage of properly fit-tested and trained users. A respirator with an assigned PF of 5 will presumably reduce the inhaled dose fivefold. Florida Hospital's respiratory protection policy specifies a qualitative saccharine fit-test; therefore, the assigned PF is 5.

Since the level of protection afforded by respirators is sometimes variable because of face-seal leakage, the use of respirators presumably weakens the PBZ urine ribavirin relationship. The level of protection afforded by surgical masks is unknown. Thus, the effect of surgical masks on urine ribavirin concentration is unknown.

HCWs did not change clothes prior to providing postshift urine samples, so contamination from their uniforms could have occurred. However, HCWs wore isolation gowns over their uniforms while in patients' rooms, so the potential contamination of urine from ribavirin on the uniform was presumably reduced.

Conclusions

Detectable levels of ribavirin were consistently found in the postshift urine of nurses despite the aggressive use of control measures. Postshift urinary ribavirin levels were significantly higher in nurses than in RTs. A statistically significant linear relationship was present between PBZ ribavirin levels and postshift urinary ribavirin readings for nurses. These findings suggest the presence of a dose–response relationship between occupational exposure to AR and excretion of ribavirin in the urine.

Variables that can affect HCWs' exposure to AR include the method of administration, use of scavenging devices, and implementation of certain work practices, such as turning off the aerosol generator before opening the administration device. Other factors that may affect exposure but were not fully evaluated include the concentration of AR produced by the aerosol generator, room ventilation rates, and effectiveness of the respirators.

Although patient care considerations typically determine the route of ribavirin administration, hospital staff should be aware that in this study, PBZ exposures were greatest when ribavirin was administered by croup tent or ADH, and least with the ventilator or ADH/Demistifier combination.

Florida Hospital is utilizing current engineering controls to reduce ribavirin exposure among HCWs. Work policies, including turning off the aerosol generator prior to providing care to the patient and permitting alternative job assignments for individuals who are actively trying to conceive or are lactating are practiced with some inconsistency. The use of personal protective equipment, including disposable respirators, is required by hospital policy, and is practiced by most HCWs.

Recommendations

The following recommendations are offered to minimize exposure of HCWs and other individuals who may enter rooms where ribavirin is administered. Many of these recommendations have already been implemented at Florida Hospital.

1. Training programs should be developed to educate HCWs about potential risks of ribavirin exposure. Education should not be limited to direct care personnel, but should include ancillary personnel such as phlebotomists, housekeepers, maintenance staff, and others who enter the room during treatment or must clean contaminated rooms, waste, and bedding. The staff should be educated to recognize situations that could result in increased occupational exposure. Female HCWs who are pregnant or lactating, or who may become pregnant, and male HCWs whose sexual partners are not actively avoiding pregnancy should be counseled about risk reduction strategies, such as alternate job assignments. Family members and visitors, who may stay in the room for long periods of time during treatment, should be notified of potential health effects from ribavirin exposure.
2. Various ribavirin administration and scavenging systems result in different levels of environmental contamination. All administration systems should include a mechanism to reduce environmental exposures to ribavirin. It is the responsibility of hospital management to implement more effective control measures as they become available. Administration and scavenging equipment should be inspected by respiratory therapy staff on a regular basis.
3. Rooms where ribavirin is administered should conform to the AIA recommendations for isolation rooms.[30] Rooms should provide a minimum of six total ACH, and should be under negative pressure. Room air should be exhausted to the outside rather than recirculated to other areas of the hospital. At Florida Hospital the air from the specially designed isolation rooms is reportedly exhausted to the outside.
4. Air pressure in the ribavirin treatment rooms should be evaluated before therapy begins and daily thereafter. Ideally, ribavirin treatment should begin only if room air pressure is negative with respect to the hallway. This can be accomplished by observing the direction of airflow at the doorway by holding a piece of tissue paper at the cracked doorway.
5. The aerosol generator should be fumed off for a minimum of 5 min prior to the HCW entering the room to provide routine care. This could be accomplished by placement of a remote switch outside the room.
6. During aerosol therapy, ribavirin precipitate is deposited on the patient and on the surrounding area. To prevent the dust from becoming airborne, care should be taken when ribavirin-contaminated clothing, bedding, or equipment is handled.[39] Although dermal absorption is not thought to be significant, dermal exposure should be avoided to prevent unintentional oral ingestion or ocular contact. The use of personal protective equipment, including gloves, gowns, and air-tight goggles should be considered.
7. Ribavirin has been found to deposit on contact lenses, so HCWs should be discouraged from wearing lenses when working with ribavirin. If contacts are worn, air-tight goggles should be used.
8. Individual hospitals may choose to use respirators to further reduce HCW exposure to ribavirin. NIOSH/MSHA-approved high efficiency disposable particulate respirators, assigned to employees based on the results of quantitative fit tests, have been found by in-mask sampling to reduce exposure to AR to the analytical limit of detection.[40] OSHA standard (29 CFR 1910.134) requires that all occupational respirator use must take place within the context of a respiratory protection program that includes evaluation of worker fitness to use a respirator, training, fit testing, and maintenance. Surgical masks should not be relied upon to provide personal protection from occupational exposure to ribavirin.[41]

9. In order to help reduce exposure of HCWs to ribavirin, medically unnecessary use of it should be avoided. Accordingly, medical staff should remain mindful of the American Academy of Pediatrics recommendations[42] and other current knowledge regarding ribavirin therapy.

References

1. Food and Drug Administration. Ribavirin aerosol approved for severe cases of RSV in infants and young children. *FDA Drug Bull* 16(1):7, 1986.
2. Smith R. Background and mechanisms of ribavirin action. In: R Smith, V Knight, and J Smith (eds.), *Clinical Application of Ribavirin*. Academic Press, Orlando, FL, 1984, pp. 1–18.
3. Bell M, Hunter J, and Mostafa S. Nebulized ribavirin for influenza B viral pneumonia in a ventilated immunocompromised adult. *Lancet* 2(8169):1084–1085, 1988.
4. Murphy D and Rose C. Respiratory syncytial virus pneumonia in human immunodeficiency virus-infected man (letter). *JAMA* 261:1147, 1989.
5. Smith C and Fox J. Double-blind evaluation of ribavirin in naturally occurring influenza. In: R Smith, and W Kirkpatrick (eds.), *Ribavirin: A Broad Spectrum Antiviral Agent*. Academic Press, New York, 1980, pp. 147–164.
6. Gilbert B, Wilson S, Knight V, et al. Ribavirin small particle aerosol treatment of influenza in college students 1981–1983. In: R Smith, V Knight, and J Smith (eds.), *Clinical Applications of Ribavirin*. Academic Press, Orlando, FL, 1984, pp. 125–143.
7. ICN Pharmaceuticals, Inc. *Small Particle Aerosol Generator Model SPAG-2 6000 Series Instructions for Use*. ICN Pharmaceuticals, Inc., Costa Mesa, CA, 1987.
8. Arnold S and Buchan R. Exposure to ribavirin aerosol. *Appl Occup Environ Hyg* 6(4):271–279, 1991.
9. Peace Medical. *Demistifier® Isolation Chamber Manufacturer's Literature*. Peace Medical Company, Orange, NJ, 1990.
10. Kilham L and Ferm V. Congenital abnormalities induced in hamster embryos with ribavirin. *Science* 195:413–414, 1977.
11. Hillyard I. The preclinical toxicology and safety of ribavirin. In: RA Smith, and W Kirkpatrick (eds.), *Ribavirin: A Broad Spectrum Antiviral Agent*. Academic Press, New York, 1980, pp. 55–71.
12. Ferm V, Willhite C, and Kilham L. Teratogenic effects of ribavirin on hamster and rat embryos. *Teratology* 17:93–102, 1978.
13. Kochhar D, Penner J, and Knudsen T. Embryotoxic, teratogenic, and metabolic effects of ribavirin in mice. *Teratol Appl Pharmacol* 52:99–112, 1980.
14. Gallagher R, Givan N, Hazelden K, McGregor D, Ridgway P, and de S. Wichramaratne G. Teratogenicity testing in baboons of virazole. Unpublished report from ICN Pharmaceuticals, 1977.
15. Marks M. Adverse drug reactions: United States experience II. *Pediatr Infect Dis J* 9:S117–S118, 1990.
16. American Hospital Formulary Service. *AHFS Drug Information*. American Society of Hospital Pharmacists, Inc., Bethesda, MD, 1990, pp. 372–379.
17. Waskin H. Toxicology of antimicrobial aerosols: A review of aerosolized ribavirin and pentamidine. *Respir Care* 36:1026–1036, 1991.
18. Janai H. Adverse drug reactions: United States experience I. *Pediatr Infect Dis J* 9:S115–S116, 1990.
19. Diamond S and Dupuis L. Contact lens damage due to ribavirin exposure (letter). *Drug Intell Clin Pharm*, 23:428–429, 1989.
20. Catlin D, Smith R, and Samuels A. ^{14}C-ribavirin: Distribution and pharmacokinetics studies in rats, baboons, and man. In: R Smith, and W Kirkpatrick (eds.), *Ribavirin: A Broad Spectrum Antiviral Agent*. Academic Press, New York, 1980, pp. 83–98.

21. California Department of Health Services. *Health Care Worker Exposure to Ribavirin Aerosol: Field Investigation FI-86-009.* California Department of Health Services, Occupational Health Surveillance and Evaluation Program, Berkeley, CA, 1988.
22. Environmental Protection Agency. Guidelines for health assessment of suspected development toxicants. *Federal Register* 51(185):34028–34040, 1986, United States Environmental Protection Agency, Washington, DC.
23. American Institute of Architects. *Guidelines for Construction and Equipment of Hospital and Medical Facilities.* American Institute of Architects Press, Washington, DC. 1987.
24. National Institute of Occupational Safety and Health. Ribavirin: Method 5027 (issued 5/15/89). In Eller P (ed.): *NIOSH Manual of Analytical Methods*, 3rd edn. U.S. Department of Health and Human Services, Public Health Service, Centers for Disease Control, National Institute for Occupational Safety and Health, Cincinnati, OH, 1984.
25. Neter WW, Wasserman W, and Kutner M. *Applied Linear Statistic Models*, 3rd edn. Irwin Publishers, Homewood, IL, 1990, 580–581.
26. Centers for Disease Control. Assessing exposures of health-care personnel to aerosols of ribavirin–California. *MMWR* 37(36):560–563, 1988, U.S. Department of Health and Human Services, Public Health Service, Centers for Disease Control, Atlanta, GA.
27. Bradley J, Connor J, Compogiannis L, and Eiger L. Exposure of health care workers to ribavirin during therapy for respiratory syncytial virus infections. *Antimicrob Agents Chemother* 34(4):668–670, 1990.
28. Deitchman S. Letter, April 6, 1990, to Michael Brady, Columbus Children's Hospital, Columbus, OH,U.S. Department of Health and Human Services, Public Health Service, Centers for Disease Control, National Institute for Occupational Safety and Health, Hazard Evaluation and Technical Assistance Branch, Cincinnati, OH,1990, pp. 89–145.
29. Austin RK, Trefts PE, Hintz, Connor JD, and Kagnoff MF. Sensitive radioimmunoassay for the broad-spectrum antiviral agent ribavirin. *Antimicrob Agents Chemother* 24(5):696–701, 1983.
30. Granich GG, Krogstag DJ, Connor JD, Desrochers KL, and Sherwood C. High-performance liquid chromatography (HPLC) assay for ribavirin and comparison of the HPLC assay with radioimmunoassay. *Antimicrob Agents Chemother* 33(3):311–315, 1989.
31. Sherwood J. Report of June 26, 1991, regarding urinary ribavirin analyses. Antiviral Assay Lab, La Jolla, CA, University of California, San Diego, 1991, pp. 91–104.
32. Taess A. Memorandum, October 15, 1991, to R. Shults, NIOSH, regarding urinalysis for ribavirin for HETA 91–104. U.S. Department of Health and Human Services, Public Health Service, Centers for Disease Control, National Institute for Occupational Safety and Health, Division of Biomedical and Behavioral Science, Cincinnati, OH, 1991.
33. Elkins H, Pagnotto L, and Smith H. Concentration adjustments in urinalysis. *Am Ind Hyg Assoc J* 35:559–565, 1974.
34. Lauwerys R. *Industrial Chemical Exposure: Guidelines for Biological Monitoring.* Biomedical Publications, Davis, CA, 6–7, 1983.
35. Centers for Disease Control: Guidelines for preventing the transmission of tuberculosis in health-care settings, with special focus on HIV-related issues. *MMWR* 39 RR-17, 1990.
36. McClung HW, Knight V, Gilbert BE, Wilson SZ, Quaries JM, and Divine GW. Ribavirin aerosol treatment of influenza B virus infection. *JAMA* 249(19):2671–2674, 1983.
37. Schumacher M and Dowd A. (eds.) *Physician's Desk Reference.* Medical Economics Data, Montvale, NJ, 1992.
38. National Institute for Occupational Safety and Health. *NIOSH Respirator Decision Logic.* U.S. Department of Health and Human Services, Public Health Service, Centers for Disease Control, National Institute for Occupational Safety and Health, Cincinnati, OH, 1987.
39. Prows CA. Ribavirin's risk in reproduction—How great are they? *Matern Child Nurs J* 14:400–404, 1989.

40. National Institute for Occupational Safety and Health. HCA Wesley Medical Center, HETA 91–155. U.S. Department of Health and Human Services, Public Health Service, Centers for Disease Control National Institute for Occupational Safety and Health, Cincinnati, OH, 1991.
41. Tuomi T. Face seal leakage of half masks and surgical masks. *Am Ind Hyg Assoc J* 46:308–312, 1985.
42. American Academy of Pediatrics. Ribavirin therapy of respiratory syncytial ViNS. In *Report of the Committee on Infectious Diseases*. American Academy of Pediatrics, Elk Grove, IL, 1991, 581–587.

Appendix 16.A.4

Health Hazard Evaluation Report
HETA 90–155–2169
HCA Wesley Medical Center
Wichita, Kansas

NIOSH Investigators

Steven W. Lenhart and John A. Decker

Summary

A management request was received from the RT educator of HCA Wesley Medical Center in Wichita, Kansas, for a Health Hazard Evaluation of the effectiveness of procedures used at the hospital to control exposures of RTs and nurses to aerosolized ribavirin. Ribavirin (l-β-D-ribofuranosyl-1,2,4-triazole-3-carboxamide) is a synthetic nucleoside analogue which is licensed in the United States for the short-term treatment of RSV infection.[1] Occupational exposure criteria have not been established for ribavirin. Because the drug has been found to be teratogenic and/or embryolethal in most animal species in which it has been tested,[1–3] there is concern about its potential reproductive effects in humans.

Twelve-inch cubical "Care Cube" disposable oxygen-delivery hoods and Viratek Small Particle Aerosol Generators are used for administration of Virazole® aerosol (ribavirin) at HCA Wesley Medical Center. In addition, each "Care Cube" hood is adapted with a scavenging system in an attempt to limit the amount of ribavirin released from the hood into a patient's room. As an extra precautionary measure, hospital management requires that a 3 M 8710 dust/mist respirator (NIOSH/MSHA approved number TC-21C-132) be worn by every person who enters a room where ribavirin is being administered.

Three site visits were made in association with this Health Hazard Evaluation. The purposes of the first site visit were to evaluate exposures of RTs and nurses to ribavirin during aerosol administration with infant mannequins used to simulate patients, and to conduct a pilot study to develop a sampling technique for evaluating in-mask exposures to ribavirin. Area samples were collected within the "Care Cube" disposable hoods used for administration of ribavirin, beside beds upon which the hoods were located, and at a location where no exposure to ribavirin was expected. Four RTs and two nurses agreed to participate in personal exposure monitoring, which consisted of the simultaneous collection of lapel and in-mask samples for ribavirin. While hospital management requires that a 3 M 8710 respirator be worn by everyone entering a room where ribavirin is administered, 3 M 9920 dust/fume/mist respirators (NIOSH/MSHA approval number TC-21C-202) were used during this site visit because they can better support a sampling probe and cassette without affecting the fit of the respirator. Quantitative fit testing was not conducted.

During the second site visit, area air samples were collected within "Care Cube" disposable hoods that were not adapted with scavenging systems and beside beds upon which the hoods were located. Quantitative fit testing using a Portacount&permile; Respirator Fit Tester was also conducted for respirator assignment to RTs and nurses. The purpose of the third site visit was to collect simultaneous in-mask and lapel samples to evaluate the level of protection received by RTs and nurses during ribavirin administration to a patient.

The results of simultaneous lapel and in-mask sampling during the first site visit produced 8 h TWA lapel concentrations of ribavirin ranging from 87 to 323 μg/m^3. Fifty-nine percent (10/17) of the in-mask samples had no detectable ribavirin, four others contained trace quantities, and the three remaining samples contained quantifiable amounts of ribavirin. Respirators which were assigned based on the results of quantitative fit test conducted during the second site visit were worn by four RTs and three nurses during the third site visit. All of the respirators were high efficiency half-masks jointly approved by the NIOSH/MSHA. The results of lapel sampling produced 8 h TWA lapel concentrations ranging from 40 to 120 μg/m^3. Seven of the eight in-mask samples obtained simultaneously with the lapel samples had no detectable ribavirin, and one sample was reported at the limit of detection.

The sampling results from the Health Hazard Evaluation suggest that notable concentrations of AR were present in the rooms where the drug was administered despite the addition of a scavenging system to the "Care Cube" disposable hood. High efficiency air-purifying respirators approved by NIOSH/MSHA and assigned to employees based on the results of quantitative fit tests were found by in-mask sampling to reduce exposures to AR to the limit of detection of the analytical method. Therefore, a recommendation was made to continue the use of respirators and initiate a complete respiratory protection program that would remain until technically feasible devices and/or procedures for the administration of ribavirin are developed and implemented that would alone reduce exposures of health care providers at HCA Wesley Medical Center to AR.

Introduction

A management request was received from the RT educator of HCA Wesley Medical Center in Wichita, Kansas, for a Health Hazard Evaluation of the effectiveness of procedures used at the hospital to control exposures of RTs and nurses to AR. Ribavirin (l-β-D-ribofuranosyl-1,2,4-triazole-3-carboxamide) is a synthetic nucleoside analogue which is licensed in the United States for the short-term treatment of RSV infection.[1] Ribavirin aerosol is indicated in the treatment of carefully selected hospitalized infants and young children with severe lower respiratory tract infections due to RSV.[2]

Occupational exposure criteria have not been established for ribavirin. Because the drug has been found to be teratogenic and/or embryolethal in most animal species in which it has been tested,[1–3] there is concern about its potential reproductive effects in humans. Health hazard assessment data available for ribavirin aerosol are currently insufficient to assess accurately the health risk to exposed HCWs.[3] Ribavirin has not been linked to fetal abnormalities in humans; however, given the wide spectrum of teratogenic potential in most animal species, avoidance of ribavirin prior to pregnancy, during pregnancy, and during lactation has been recommended by the author of a review of the toxicology of antimicrobial aerosols.[1]

HCA Wesley Medical Center is licensed for 760 beds, employs approximately 3000 full-time and part-time employees, and has a 640-member medical staff. The hospital's PICU has 18 beds. To address the concerns of the RTs and nurses who worked on this unit regarding their potential exposures to AR, the management of the hospital implemented procedures in March 1987, in an attempt to limit environmental exposures to ribavirin for the health care providers.

The HCA Wesley Medical Center program for limiting environmental exposure to ribavirin consists of the following procedures:[4]

- A patient who will receive ribavirin is to be admitted to a private room. The door to the room is to be closed during ribavirin treatment.
- When entering a ribavirin patient's room, all employees and visitors are to wear a respirator with a capability of filtering particles with aerodynamic diameters ranging from 0.4 to 0.6 μm (3 M model 8710).

- When ribavirin is administered via ventilator, a filter is to be used in the ventilator circuit to prevent the release of ribavirin aerosol into the room. When the ventilator is disconnected, a second filter is to be attached to the patient's connector to prevent release.
- When ribavirin is administered via an oxygen-delivery hood, a "Care Cube" is to be used. An air-entrainment adapter with filter is to be used to scavenge excess ribavirin from the outflow port of the "Care Cube."
- After the administration of ribavirin is completed, employees and visitors are to continue to wear their respirators for 30 min. After 30 min, the room's ventilation exchange system is expected to have cleared any escaped ribavirin aerosol. (The air exchange rate is estimated by the hospital's engineering staff to be 12 ACH.)

Three site visits were made in association with this Health Hazard Evaluation: (1) July 17–18, 1990, (2) January 15–16, 1991, and (3) February 28–March 1, 1991. The purposes of the first site visit were to evaluate exposures of RTs and nurses to ribavirin during aerosol administration with infant mannequins used to simulate patients, and to conduct a pilot study to develop a sampling technique for evaluating in-mask exposures to ribavirin. After the first site visit, an interim letter dated December 5, 1990, was mailed to the requester. During the second site visit, samples were collected both inside and outside the hoods used to administer ribavirin to patients, and respirator quantitative fit tests were conducted using different air-purifying respirators. The purpose of the third site visit was to collect simultaneous in-mask and lapel samples to evaluate the level of protection received by RTs and nurses during ribavirin administration to a patient.

Background

Twelve inch cubical "Care Cube" disposable hoods and Viratek Small Particle Aerosol Generators (model SPAG-2, 6000 series, ICN Pharmaceuticals, Inc., ICN Plaza, 3300 Hyland Avenue, Costa Mesa, California 92626) are used for administration of Virazole aerosol (ribavirin) at HCA Wesley Medical Center. The drug manufacturer's recommended concentration of 20 mg of ribavirin per milliliter of sterile USP water (mg/mL) is used as the starting solution for the drug reservoir of the SPAG-2. This drug concentration is expected to produce an aerosol concentration of 190 mg/m^3 inside a ribavirin administration hood for a 12 h period.[2]

The operating parameters of the SPAG-2, as used at HCA Wesley Medical Center during this study, were a regulator pressure of 26 lb per square inch gauge (psig), a nebulizer airflow rate of 6.5 L/min, and a drying airflow rate of 6.5 L/min. These values are within the ranges of the operating parameters recommended by the manufacturer.[2] (The aerosol delivery rate of 13 L/min is at least twice the 1 min volume of a typical patient.[5]) In addition, each "Care Cube" hood is adapted with a scavenging system in an attempt to limit the amount of ribavirin released from the hood into a patient's room. The scavenging system consists of a 24% venturi oxygen mask adapter operated at 2 L/min, corrugated plastic tubing, and a model BB-50 T Pall breathing circuit filter (Pall Biomedical Products Corporation, East Hills, New York) placed on top of the hood. Air is exhausted from the filter into the room environment.

As an extra precautionary measure, a 3 M 8710 dust/mist respirator (NIOSH/MSHA approval number TC-21C-132) is required by hospital management to be worn by every person who enters a room where ribavirin is being administered. An exposure limit has not been established or recommended for ribavirin, and industrial hygiene sampling had not been conducted previously at the PICU of HCA Wesley Medical Center to evaluate exposures of RTs and nurses to ribavirin. Therefore, it was not possible to calculate the minimum level of protection that a respirator would need to achieve in order to be selected for this application. The 3 M 8710 respirator was selected by hospital management based upon a knowledge of the filter efficiency of the classes of respirators to which the 3 M 8710 belongs and of the reported particle size of the ribavirin aerosol generated by the Viratek SPAG-2.

Single-use dust and mist respirators must demonstrate 99% efficiency against silica dust particles with a geometric mean diameter of 0.4–0.6 µm and a geometric standard deviation not greater than 2 to be certified by the NIOSH.[6] Health professionals at HCA Wesley Medical Center believed that the 3 M 8710 would adequately protect a user against the ribavirin aerosol with a reported mass median diameter of approximately 1.3 µm.[2] Consideration was not given regarding the contribution of face-to-facepiece seal leakage to the overall performance of a respirator. In addition, the measures generally incorporated in an acceptable respiratory protection program (e.g., medical examinations, fit testing, and training) required by the Occupational Safety and Health Administration (described in 29 Code of Federal Regulations Part 1910.134), were not present at HCA Wesley Medical Center.

Methods

Site Visit #1 (July 17 and 18, 1990)

Twenty milligrams per mililiter is the recommended concentration of ribavirin prepared as the starting solution for the drug reservoir of the Viratek small particle aerosol generator (SPAG) model SPAG-2 used at HCA Wesley Medical Center. Bulk samples of ribavirin solutions were collected on both days of sampling and were analyzed to confirm that the solutions contained the recommended concentration of ribavirin.

Area air samples were collected within the "Care Cube" disposable hoods used for administration of ribavirin, beside beds upon which the hoods were located, and at a location where no exposure to ribavirin was expected. RTs and nurses were asked to participate in personal exposure monitoring, which consisted of the simultaneous collection of lapel and in-mask samples for ribavirin. The RTs and nurses simulated the activities that would normally have been necessary if a patient was actually receiving care, but did so with infant mannequins for the purposes of this site visit.

While a 3 M 8710 respirator is required by hospital management to be worn by everyone entering a room where ribavirin is administered, 3 M 9920 dust/fume/mist respirators (NIOSH/MSHA approval number TC-21C-202) were used for this evaluation because they can better support a sampling probe and cassette without affecting the fit of the respirator. Both respirators belong to the class of respirators described as disposable air-purifying half-mask respirators and are selected for protection against particulate exposures. This class of respiratory protection has an assigned PF of 5, but a PF of 10 can be used if they have been properly fitted using a quantitative fit test.[7]

A plastic Liu probe for in-mask sampling was positioned at a location directly below the exhalation valve of each 3 M 9920 respirator used during this evaluation.[8] Quantitative fit tests were not conducted during this site visit. Respirators were always donned and removed at a location designated as the "IH office," where no exposure to ribavirin was expected. Area air sampling was conducted at this location to determine whether ribavirin was present at concentrations that would influence the overall results of the pilot study. Sampling pumps were always started after respirators were in place and were stopped before respirators were removed to ensure the integrity of each in-mask sample. All study participants were observed to ensure that they wore their respirators properly during all periods of potential exposure to ribavirin. Each sampling probe was washed at the conclusion of each full-shift sampling period with 10 mL of purified water, and the resulting liquid was submitted for ribavirin analysis. This was done to ensure that all ribavirin entering a sampling probe was collected within a sampling cassette.

Equipment for each area, personal, or in-mask sample consisted of a three-piece, closed-face 37 mm cassette containing a glass fiber filter (type A/E, Product Number 61652, Gelman Sciences, Inc., Ann Arbor, M1 48106–9990) and a cellulose backup pad. Each cassette was connected by flexible tubing to a personal sampling pump operated at 1.0 L/min for samples collected inside the "Care Cube" hoods and at 2.0 L/min for all other samples. Field blanks were prepared and submitted for analysis along with the sample cassettes.

Samples were analyzed in accordance with NIOSH analytical method 5027 Issued May 15, 1989.[9] Each filter sample was removed from its cassette, folded in half, and inserted into a culture tube for extraction with 3 mL of deionized water with sulfuric acid added (90 mL/L, pH = 2.5) in an ultrasonic bath. Each sample was agitated for 15 min. After each filter sample solution or each bulk sample of ribavirin solution was filtered through a syringe filter, an injection volume of 30 μL was analyzed using a HPLC equipped with an ultraviolet detector.

Site Visit #2 (January 15 and 16, 1991)

Area air samples were collected within "Care Cube" disposable hoods that were not adapted with scavenging systems, and beside beds upon which the hoods were located. Samples were collected in the same manner as for the first visit, and each sample was likewise analyzed in the same manner, according to NIOSH analytical method 5027. Bulk samples of ribavirin solutions were collected both before and after administration and analyzed for ribavirin concentration.

Quantitative fit testing was also conducted for respirator assignment using a Portacount % Respirator Fit Tester (TSI, Inc., P.O. Box 64394, St. Paul, MN 55164). Quantitative fit factors of respirator wearers measured with the Portacount % have been reported on a group basis as having a high degree of correlation to those obtained by a recognized photometer quantitative fit test system.[10] A Portacount % was used during this study because it is less cumbersome to transport to a study site than are conventional quantitative fit test systems. The group of respirators used for fit testing consisted of Moldex 2300 dust/mist disposable respirators (NIOSH/MSHA approval number TC-21C-350), Moldex 3400 dust/fume/mist disposable respirators (NIOSH/MSHA approval number TC-21C-418; Moldex-Metric, Inc., 4671 Leahy Street, Culver City, CA 90232), 3 M 8710 dust/mist disposable respirators (NIOSH/MSHA approval number TC-21C-132), 3 M 9920 dust/fume/mist disposable respirators (NIOSH/MSHA approval number TC-21C-202), medium and large 3 M 9970 high efficiency dust/fume/mist disposable respirators (NIOSH/MSHA approval number TC-21C-437; 3 M Occupational Health and Environmental Safety Division), and small, medium, and large MSA half-mask respirators with high efficiency dust/fume/mist cartridges (NIOSH/MSHA approval number TC-21C-135; Mine Safety Appliances Company, P.O. Box 439, Pittsburgh, PA 15230).

Each employee's fit test started with a series of screening tests consisting generally of one of the Moldex respirators, one of the 3 M low efficiency respirators, one of the 3 M high efficiency respirators, and one of the MSA half-mask, high efficiency respirators. Each screening fit test was conducted while the employee breathed normally. The respirator with the highest screening fit factor was selected for a complete fit test during which the employee performed the following six exercises: normal breathing (NB1), deep breathing (DB), moving head side to side (SS), moving head up and down (UD), talking (TK), and normal breathing (NB2). An overall fit factor (Fl7) was then calculated using the following equation:[10]

$$\text{Overall FF} = 6/[(1/NB1_{FF}) + (1/DB_{FF}) + (1/SS_{FF}) + (1/UD_{FF}) + (1/TK_{FF}) + (I/NB2_{FF})]$$

Site Visit #3 (February 28 and March 1, 1991)

Area air samples were collected within a "Care Cube" disposable hood used for administration of ribavirin to an infant patient, beside the bed upon which the hood was located, and at locations where no exposure to ribavirin was expected. Bulk samples of ribavirin solutions were collected before administration at the beginning of each day of testing. Study participants were selected for participation in the personal exposure monitoring phase of this site visit from among the nurses and RTs available during the work shifts when ribavirin was being administered.

Based upon the results of the quantitative fit tests conducted during the second visit, the RTs or nurses scheduled to work on one of the days of sampling who had the highest overall fit factors were asked to participate in the personal exposure monitoring phase of the study, which consisted of the simultaneous collection of lapel and in-mask samples for ribavirin. Samples were collected in the same manner as for the first and second site visits, and each sample was likewise analyzed in the same manner, according to NIOSH analytical method 5027.

Results and Discussion

Site Visit #1 (July 17 and 18, 1990)

Two bulk samples collected on July 17, 1990, from different flasks containing starting solutions of ribavirin, were analyzed, and both were reported to contain 30 mg/mL, which is greater than the recommended concentration of 20 mg/mL. A sample collected on July 18, 1990, was reported to contain 21 mg/mL.

The results of sampling with cassettes placed inside "Care Cube" hoods are presented in Table 16.A.11. The results of the 21 short-term samples (10–18 min in duration) range from 1.3 to 123 mg/m^3. The five mean concentrations range from 3.6 to 93.2 mg/m^3. These concentrations are lower than the expected concentration of 190 mg/m^3, possibly because of decreased delivery pressures, obstructed delivery

TABLE 16.A.11 Concentrations of Ribavirin Inside "Care Cube" Disposable Hoods, HCA Wesley Medical Center, Wichita, Kansas, Site Visit #1 (July 17–18, 1990). HETA 90–155

Sampling Location	Sample Number	Ribavirin Sampling Period	Concentration (mg/m^3)	Mean (SD) Concentration (mg/m^3)
July 17, 1990				
Room 512, Bed #3	7	0810–0820	4.1	62.4 (32.5)
	15	0936–0948	82.5	
	18	1204–1215	45.5	
	20	1235–1246	74.5	
	24	1356–1406	76.0	
	26	1440–1450	92.0	
Intensive Care	13	0921–0931	48.0	41.2 (26.1)
	6	1006–1016	64.0	
	5	1102–1112	47.0	
	19	1209–1219	69.0	
	22	1316–1326	6.4	
	25	1422–1433	12.7	
July 18, 1990				
Room 512, Bed #1	40	0938–0951	76.9	71.6 (54.2)
	47	1056–1106	15.0	
	53	1346–1359	123.0	
Room 512, Bed #3	41	0938–0951	76.9	93.2 (16.6)
	46	1056–1106	110.0	
	54	1346–1400	92.8	
Intensive Care	43	0954–1012	6.1	3.6 (2.4)
	48	1059–1112	1.3	
	55	1348–1401	3.5	

Note: LOD: 1.0 µg/sample; LOQ, 4.0 µg/sample; SD, standard deviation; TWA, time-weighted average.

TABLE 16.A.12 Results of Area Air Sampling for Ribavirin, HCA Wesley Medical Center, Wichita, Kansas, Site Visit #1 (July 17–18, 1990). HETA 90–155

Sampling Location	Sample Number	Ribavirin Sampling Period	8 h TWA Ribavirin Concentration (μg/m³)	Concentration (μg/m³)
July 17, 1990				
Room 512, beside bed #3	8	0816–1508	11	9
Room 512, on desk	9	0819–1458	71	59
PICU, beside bed	14	0923–1511	330	239
"IH office"	12	0848–1514	ND	—
July 18, 1990				
Room 512, beside bed #1	38	0802–1505	378	333
Room 512, beside bed #3	37	0802–1504	652	573
PICU, beside bed	39	0809–1501	243	209
"IH office"	30	0729–1509	(3)[a]	(3)[a]

Note: PICU, pediatric intensive care unit; ND, none detected; TWA, time-weighted average.
[a] Values in () represent a quantity of ribavirin between the LOD [1.0 μg/sample] and the LOQ [4.0 μg/sample], and should be considered trace concentrations with limited confidence in their accuracy.

nozzles, or removal of excessive amounts of ribavirin by the scavenging system. Research has been conducted which demonstrates that fluctuations in delivery concentrations can occur as a function of nebulizer airflow rate.[5] Decreasing the nebulizer airflow rate resulted in a significant reduction of aerosol concentration. A concentration of 190 mg/m³ at a nebulizer airflow rate of 7 L/min was reduced to only 7 mg/m³ at a nebulizer airflow rate of 4 L/min. Based upon the results of this study,[5] a concentration of approximately 140 mg/m³ would be predicted from the nebulizer airflow rate of 6.5 L/min used at HCA Wesley Medical Center.

The results of area air sampling conducted in room 512 and the intensive care unit of the pediatric ward are presented in Table 16.A.12. There is no obvious or apparent explanation for the discrepancies between the ribavirin concentration estimate of 9 μg/m³ recorded beside Bed #3 in room 512 and the concentration estimate of 59 μg/m³ recorded approximately 25 ft away at the nurse's desk in room 512. Similarly, there is no explanation for the difference between the concentration estimates recorded beside Bed #3 on the first day of sampling (9 μg/m³) and the second day (573 μg/m³). Sampling at these locations was initiated on both days after ribavirin administration was started, and sampling was stopped after ribavirin administration was completed. Therefore, the differences are not associated with variabilities related to sampling methods, or to sampling during only certain phases of a ribavirin administration procedure. It is also unlikely that the administration techniques of the RTs and nurses were contributing factors, since the mean concentrations inside the delivery hoods used at Bed #3 were similar (62.4 and 93.2 mg/m³). It is plausible that differences in the orientation of the sampling cassettes with respect to the delivery hoods might have been a contributing factor in the observed differences.

The sampling location identified in Table 16.A.12 as "IH office" was an area of the pediatric ward where respirators and personal and in-mask sampling cassettes were donned and removed. As anticipated, there was essentially no ribavirin exposure at this location, and it is unlikely that the trace concentration present on July 18, 1990 affected the overall results of the pilot study.

The results of simultaneous lapel and in-mask sampling are presented in Table 16.A.13. The 8 h TWA lapel concentrations range from 87 to 323 μg/m³. Fifty-nine percent (10/17) of the in-mask sampling results were reported as nondetected; the analytical limit of detection was 1 μg/sample. An additional four samples contained quantities of ribavirin between the limit of detection (1.0 μg/sample) and the limit of quantitation (4.0 μg/sample); they should be considered trace amounts and limited confidence should be placed in their accuracy. Three samples contained quantities of ribavirin which exceeded the LOQ.

TABLE 16.A.13 Results of Lapel and In-Mask Sampling for Ribavirin. HCA Wesley Medical Center, Wichita, Kansas. Site Visit #3 (February 28 and March 1, 1991). HETA 90–155

Job/Emp./Mask	Lapel Sample Number	Sampling Duration (min)	Actual/8 h TWA Ribavirin Concentration (μg/m³)	In-Mask Sample Number	Sampling Duration (min)	Ribavirin Concentration (μg/m³)
February 28, 1991						
RN/E/MSA M (7 AM–7 PM)	3	88	438/120	1	88	ND
RT/U/MSA M (7 AM–3 PM)	4	89	236/44	2	89	ND
RT/S/MSA L (3 PM–11 PM)	15	46	413/40	14	46	ND
March 1, 1991						
RN/F/MSA L (7 AM–3 PM)	31	92	408/78	32[a]	92	(4)
RN/H/MSA M (3 PM–11 PM)	43	78	494/80	44	78	ND
RT/Q/3 M 9970 L (7 AM–3 PM)	29	102	279/59	30	15	ND
				37	87	ND
RT/Y/3M 9970 L (3 pm–11 pm)	45	45	733/69	46	45	ND

Note: The value in () represents a quantity of ribavirin between the LOD [0.7 μg/sample] and the LOQ [2 μg/sample], and should be considered a trace concentration with limited confidence in its accuracy.

Emp, employee identification letter; L, large; M, medium; ND, none detected; RN, registered nurse; RT, respiratory therapist; TWA, time-weighted average.

[a] The cassette of sample 32 was disconnected accidentally from the respirator facepiece for approximately 2 s during sampling

To conservatively approximate the workplace PFs achieved during this phase of the study, in-mask samples that were reported as nondetected were replaced with the limit of detection (1 μg/sample), and 8 h TWA concentrations were determined. Also ribavirin found from the probe wash of the one sample (RT/PICU on July 17, 1990) was included in the calculation of its associated in-mask concentration. Probe wash concentrations that were reported as nondetected were assumed to contain no ribavirin. The workplace PFs achieved are presented in Table 16.A.14. A workplace PF was not calculated for the RN in room 512 on July 17, 1990, because the quantity of ribavirin reported for the probe wash associated with this set of samples was twice the amount reported for this individual's lapel sample. This situation suggests that the probe, or probe wash solution, was contaminated and that using this value would not reflect a true indication of respirator performance.

Disposable respirators like the 3 M 8710 and the 3 M 9920 have an assigned PF of 5.[7] While the respirators used at Wesley Medical Center were not quantitatively fit tested, the five calculated workplace PFs exceed 5, and range from 15 to 29. Although the sample size is small, an assigned PF of 12 results from these five values, using a calculation method described elsewhere.[11,12]

Site Visit #2 (January 15 and 16, 1991)

The results of analysis of bulk samples taken from preadministration and postadministration ribavirin solutions used during the 2 days of testing are presented in Table 16.A.15. The solutions ranged from 22 to 38 mg/mL, and all exceeded the recommended concentration of 20 mg/mL.

Because the concentrations within the "Care Cube" hoods evaluated during the first visit were all less than the expected concentration of 190 mg/m³, testing was conducted during the second visit to

TABLE 16.A.14 Workplace PFs for RNs and RTs during Ribavirin Administration, HCA Wesley Medical Center, Wichita, Kansas. Site Visit #1 (July 17–18, 1990). HETA 90–155

	(A)	8 h TWA	(B) Approximate 8-h TWA	
Job Location	Ribavirin Probe Wash Number	Lapel Concentration (μg/m²)	In-mask Concentration (μg/m³)	Workplace PF (A/B)
July 17, 1990				
RN/Room 512	160	87	—	—
RT/Room 512	ND	87	3	29
RT/PICU	(10)[a]	219	14	16
July 18, 1990				
RN/Room 512	ND	323	21	15
RT/Room 512	ND	281	11	26
RT/PICU	ND	125	8	16

Note: PICU, pediatric intensive care unit; ND, none detected; TWA, time-weighted average.
[a] The value in () represents a quantity of ribavirin between the LOD [6 μg/10 mL] and the LOQ [19 μg/10 mL], and should be considered a trace amount with limited confidence in its accuracy.

TABLE 16.A.15 Concentrations of Ribavirin Solutions Prepared for Reservoirs of SPAGs. HCA Wesley Medical Center, Wichita, Kansas. Site Visit #2 (January 15–16, 1991). HETA 90–155

SPAG Location	Sample Number	Ribavirin Pre-Administration	Concentration (mg/mL) Post-Administration
July 15, 1991			
Room 512, Bed #1	63	—	37
Room 512, Bed #3	64	—	38
Room 512, Bed #3	65 and 67	31	38
Room 511, Bed #8	66 and 68	22	36
Purified Water	69	None detected	None detected

Note: LOD, 0.0003 mg/mL; LOQ, 0.00009 mg/mL.

evaluate the concentrations of ribavirin inside hoods without scavenging systems. The results of testing are presented in Table 16.A.16. The 32 short-term samples (10–11 min in duration) range from 0.3 to 180 mg/m³. The four mean concentrations range from 91.0 to 124 mg/m³. While nurses and RTs simulated their normal activities during the first visit, which included frequent opening of the hood, this was not done during the second visit. Although all of the short-term sampling results were still <190 mg/m³, seven equaled or exceeded the predicted concentration of 140 mg/m³.

The results of area air sampling conducted in Room 512 and the intensive care unit of the pediatric ward are presented in Table 16.A.17. Overall, the 8 h TWA ribavirin concentrations are slightly higher than those recorded during the first visit.

The results of quantitative fit testing for 32 nurses and RTs are presented in Tables 16.A.18 and 16.A.19. The 35 fit factors achieved during normal breathing with low efficiency disposable respirators (Moldex 2300 and 3400, and 3 M 8710 and 9920) range from 2 to 92 and have a geometric mean of 15 with a geometric standard deviation of 2.54. Generally, quantitative fit tests of negative pressure respirators are conducted using high-efficiency filters, achieving the test's primary purpose of evaluating face-to-facepiece seal leakage only. The fit factors reported here for low efficiency disposable respirators might

TABLE 16.A.16 Concentrations of Ribavirin Inside "Care Cube" Disposable Hoods. HCA Wesley Medical Center, Wichita, Kansas. Site Visit #2 (January 15–16, 1991). HETA 90–155

Sampling Location	Sample Number	Sampling Period	Ribavirin Concentration (mg/m^3)	Mean (SD) Concentration (mg/m^3)
January 15, 1991				
Room 512, Bed #1	1	0945–0955	0.3	124 (65)
	9	1025–1035	48.0	
	11	1110–1120	120	
	13	1235–1245	150	
	15	1325–1335	160	
	17	1420–1431	164	
	19	1525–1535	180	
	21	1650–1700	170	
Room 512, Bed #3	2	0945–0955	0.4	91.0 (44)
	8	1025–1035	87.0	
	10	1110–1120	110	
	12	1235–1245	110	
	14	1325–1335	110	
	16	1420–1431	118	
	18	1525–1535	140	
	20	1650–1700	53.0	
January 16, 1991				
Room 512, Bed #3	31	0910–0920	58.0	92.2 (19)
	33	1007–1017	77.0	
	35	1105–1115	95.0	
	37	1205–1215	84.0	
	39	1305–1315	98.0	
	41	1407–1417	110	
	43	1507–1517	120	
	45	1542–1552	96.0	
Room 511, Bed #8	32	0855–0905	110	109 (27)
(Intensive care)	34	0950–1000	110	
	36	1050–1100	99.0	
	38	1152–1202	110	
	40	1250–1300	110	
	42	1355–1405	130	
	44	1455–1505	150	
	46	1558–1608	55.0	

Note: LOD, 0.8 μg/sample; LOQ, 2.0 μg/sample; SD: standard deviation.

be underestimates, since in-mask concentrations could have represented not only leakage of the challenge aerosol between the face-to-facepiece seal, but also an unknown amount of penetration of the aerosol through the filter material.

The 40 fit factors achieved with the high efficiency disposable respirator (3 M 9970 medium and large) range from 2 to 25,000 and have a geometric mean of 142 with a geometric standard deviation of 17.9. The 34 fit factors achieved with the MSA half-mask with high efficiency cartridges range from 22 to 13,000 and have a geometric mean of 1120 with a geometric standard deviation of 4.60. The high degree of variability associated with the fit factors achieved with the 3 M 9970 respirator serves to emphasize the importance of conducting quantitative fit tests prior to providing respirators to employees.

TABLE 16.A.17 Results of Area Air Sampling for Ribavirin. HCA Wesley Medical Center, Wichita, Kansas. Site Visit #2 (January 15–16, 1991). HETA 90–155

Sampling Location	Sample Number	Sampling Period	Ribavirin Concentration (μg/m³)	8 h TWA Ribavirin Concentration (μg/m³)
Room 512 (January 15, 1991)				
Bed #1, beside SPAG	4	0919–1249	381	646
	24	1257–1722	868	
Bed #1, opposite SPAG	3	0919–1251	424	833
	23	1257–1725	1157	
Bed #3, beside SPAG	7	0920–1252	472	677
	22	1257–1721	852	
Bed #3, opposite SPAG	5	0920–1252	448	792
	25	1257–1721	1080	
on nurse's desk	6	0919–1251	354	666
	26	1257–1722	924	
Rooms 512 and 511 (January 16, 1991)				
Bed #3, beside SPAG (Room 512)	50	0848–1247	875	667
	56	1247–1557	579	
Bed #3, opposite SPAG (Room 512)	51	0847–1247	1062	760
	57	1247–1557	579	
On nurse's desk	49	0847–1247	312	271
	55	1247–1557	290	
Bed #8, beside SPAG (Room 511)	47	0835–1235	292	313
	53	1236–1610	374	
Bed #8, opposite SPAG (Room 511)	48	0835–1235	542	584
	54	1236–1611	698	
Outside of Room 511	52	0853–1253	(4)[a]	(3)[a]
	58	1253–1613	(2)[a]	

Note: TWA: time-weighted average.
[a] Values in () represent a quantity of ribavirin between the LOD [0.8 μg/sample], and should be considered trace concentrations with limited confidence in their accuracy.

TABLE 16.A.18 Screening Quantitative Fit Factors for Nurses and RTs. HCA Wesley Medical Center, Wichita, Kansas. Site Visit #2 (January 15–16, 1991). HETA 90–155

Employee	2300	Molde x 3400	8710	9920	3M 9970 M	9970L	Small	MSA Medium	Large
Nurses									
A	5	—	—	—	3700	10	—	9100	—
B	—	72	—	—	—				
C	—	4	—	—	2				
D	6	—	—	—	480				
E	—	—	—	31	—				
F	—	6	—	—	—				
G	—	—	—	43	7				
H	—	—	16	—	2400				
I	—	30	—	—					
J	—	4	—	—					
K	—	—	—	36					
L	—	—	12	—					
M	6	—	—	—					

TABLE 16.A.19 Overall Quantitative Fit Factors for Nurses and RTs. HCA Wesley Medical Center, Wichita, Kansas. Site Visit #2 (January 15–16, 1991). HETA 90–155

Nurses			RTs		
Employee	Respirator/Size	Overall Fit Factor	Employee	Respirator/Size	Overall Fit Factor
First shift					
A	MSA/Medium	5735	N	3M 9970/medium	7088
B	MSA/medium	3106	O	MSA/medium	6210
C	MSA/medium	2226	P	MSA/large	1948
D	MSA/medium	1594	Q	3M 9970/medium	1282
E	MSA/medium	1335	R	MSA/medium	636
F	MSA/large	226	S	MSA/large	467
G	MSA/small	88	T	MSA/large	374
			U	MSA/medium	210
			V	3M 9970 large	149
			W	MSA/medium	58
			X	MSA/medium	29
Second shift					
H	MSA/medium	8070	Y	3M 9970/large	9339
I	3M 9970/large	739	Z	MSA/medium	1476
Third shift					
J	MSA/medium	8393	AA	MSA/medium	4298
K	3M 9970/medium	2244	BB	3M 9970/large	3881
L	MSA/large	610	CC	MSA/medium	3422
M	MSA/medium	366	DD	3M 9970/large	1900
			EE	MSA/large	981
			FF	3M 9970/medium	361

As stated in Appendix 16.A.3, the majority of assigned PFs for the various classes of respirators have been based upon the results of efficiency disposable respirators, 1.2 for the 3 M 9970 high efficiency disposable respirator, and 92 for the MSA half-mask respirator with high efficiency cartridges.

Overall fit factors ranked from highest to lowest according to job and work shift are presented in Table 16.A.19 for the respirators with the highest screening fit factors achieved by the nurses and RTs who were tested. Assignment of a respirator to each nurse and RT based upon the highest overall fit factor achieved during the fit tests resulted in 72% of the employees (23/32) being assigned a half-mask respirator with high efficiency cartridges (MSA), and the remainder of the employees (9/32) being assigned a high efficiency disposable respirator (3 M 9970) quantitative fit testing. While the fit factors presented in Table 16.A.18 represent the values achieved during only the normal breathing exercise, it is of interest that the assigned PFs achieved with these values are 3.2 for the low

Site Visit #3 (February 28 and March 1, 1991)

Testing during the administration of ribavirin to an infant patient occurred on the second and third days of a 3 day treatment period. The analysis of bulk samples of other preadministration ribavirin solutions were reported as 21 mg/mL.

The results of sampling with cassettes placed inside the “Care Cube” hood with a scavenging system used for ribavirin administration are presented in Table 16.A.20. The results of 16 short-term samples (10–21 min in duration) range from 52.7 to 180 mg/m^3. The two mean concentrations are 85.2 and 148.6 mg/m^3. Interestingly, the mean concentration for the treatment given on February 28 is less than

TABLE 16.A.20 Concentrations of Ribavirin Inside "Care Cube" Disposable Hood. HCA Wesley Medical Center, Wichita, Kansas. Site Visit #3 (February 28 and March 1, 1991). HETA 90-155

Sampling Location	Sample Number	Sampling Period	Ribavirin Concentration (mg/m³)	Mean (SD) Concentration (mg/m³)
February 28, 1991				
Room 511, Bed #7	5	1110–1121	77.3	85.2 (21.8)
	9	1150–1200	70.0	(9 samples)
	10	1241–1252	52.7	
	11	1405–1220	62.7	
	13	1517–1530	107.7	
	16	1635–1646	100.0	
	19	1809–1821	108.3	
	21	2046–2104	77.8	
	23	2216–2226	110.0	
March 1, 1991				
Room 511, Bed #7	33	1032–1045	138.5	148.6 (22.0)
	38	1136–1157	157.1	(7 samples)
	39	1235–1245	130.0	
	41	1424–1434	130.0	
	42[a]	1434–1636	ND	
	50	1656–1706	130.0	
	51	2041–2051	180.0	
	53	2240–2252	175.0	

Note: LOD, 2 μg/sample; LOQ, 5 μg/sample; SD; standard deviation.

[a] Air was not drawn through sample 42 in order to evaluate the potential for migration.

the predicted concentration of 140 mg/m³. Two of the three short-term concentrations which exceeded 140 mg/m³ occurred at the end of the study on March 1, 1991. Because the patient's activity within the "Care Cube" had made it difficult to keep the opening of the hood closed securely, a nurse used two of the "gator" clips from the NIOSH sampling equipment to effect a better seal. While the infant's movements were not restricted by this innovation, ribavirin concentrations inside the hood were apparently increased.

The results of area air sampling are presented in Table 16.A.21. The patient's parents were unable to visit their child on the days of testing, and therefore only area samples were collected on a window curtain near a chair for visitors in the patient's room. The sampling locations identified as Room 511 (Bed #8) and Room 512 were areas where respirators, and personal and in-mask sampling cassettes were donned and removed. As anticipated, there was essentially no ribavirin exposure at these locations, and it is unlikely that the trace concentrations present on February 28, 1991, in Room 511 (Bed #8) affected the overall results of this phase of the study.

The results of simultaneous lapel and in-mask sampling are presented in Table 16.A.22. The 8-h TWA lapel concentrations range from 40 to 120 μg/m³. Seven of the 8 in-mask sampling results were reported as nondetected and the remaining sampling result was reported at the limit of detection of the analytical method (0.7 μg/sample). As with the personal sampling data from the first visit, approximate workplace PFs were calculated and are presented in Table 16.A.23. They range from 50 to 133.

Conclusions/Recommendations

The sampling results of this study demonstrate that notable concentrations of AR were present in the rooms where the drug was administered despite the addition of a scavenging system to the "Care Cube" disposable hood.

TABLE 16.A.21 Results of Area Sampling for Ribavirin. HCA Wesley Medical Center, Wichita, Kansas. Site Visit #3 (February 28 and March 1, 1991). HETA 90-155

Sampling Location	Sample Number	Sampling Period	Ribavirin Concentration (μg/m³)	8h TWA Ribavirin Concentration (μg/m³)
February 28, 1991				
Room 511, Bed #7 (on window curtain)	8	0914–1317	181	170
	12	1317–1717	229	
	20	1717–2049	170	
	22	2049–2255	218	
Room 511, Bed #8	7	0916–1542	(1.3)[a]	(1.6)
(on privacy curtain)	17	1542–2248	(2.4)	
Room 512	6	0916–1542	ND	ND
	18	1542–2248	ND	
March 1, 1991				
Room 511, Bed #7 (on window curtain)	36	0855–1220	366	279
	40	1220–1520	306	
	48	1522–1918	275	
	52	1918–2250	330	
Room 511, Bed #8	35	0856–1514	(2.6)	(3.7)
(on privacy curtain)	47	1514–2240	5.6	
Room 512	34	0847–1525	ND	ND
	49	1525–2233	ND	

Note: ND, none detected; TWA, time-weighted average.

[a] Values in () represent quantities of ribavirin between the LOD [0.7 μg/sample] and the LOQ [2 μg/sample], and should be considered trace concentrations with limited confidence in their accuracy.

Therefore, the policy established in 1987 by the management of HCA Wesley Medical Center regarding the wearing of respirators by HCWs in rooms where ribavirin is administered should remain in force. However, the use of respirators should involve all of the components of a complete respiratory protection program and should be developed in accordance with the requirements for an acceptable program established by the OSHA as described in 29 Code of Federal Regulations Part 1910.134. The respiratory protection program should be considered an interim measure that should remain in place until technically feasible devices and/or procedures for the administration of ribavirin aerosol are developed and implemented that will, by themselves, reduce exposures of health care providers at HCA Wesley Medical Center to AR.

The respiratory protection program should include the performance of quantitative fit tests to assign the particular respirator that each individual should wear from a group of respirators that are approved jointly by the National Institute for Occupational Safety and Health and the Mine Safety and Health Administration, and that represent a variety of manufacturers and facepiece sizes. Additionally, a required minimum overall quantitative fit factor should be selected before a specific respirator is assigned to an individual. While no data have been reported to demonstrate a relationship between quantitative fit factors and the workplace performance of a respirator, safety factors have been used to establish minimum acceptable quantitative fit factors. A safety factor of 10 has been applied to half-mask respirators,[13–15] but other factors (e.g., 20[16] and 25[11]) have also been used. The minimum overall quantitative fit factor selected for use at HCA Wesley Medical Center should be determined with caution and with recognition of the uncertainty of its effectiveness.[7]

A sampling surveillance program should be developed to routinely monitor the effectiveness of the respiratory protection program. Such a program can consist of lapel sampling alone by determining whether or not 8 h TWA concentrations experienced by nurses and RTs potentially exposed to AR remain similar to those levels reported here. However, if it is feasible to conduct simultaneous lapel and

TABLE 16.A.22 Results of Lapel and In-Mask Sampling for Ribavirin. HCA Wesley Medical Center, Wichita, Kansas. Site Visit #3 (February 28 and March 1, 1991). HETA 90–155

Job/Emp./Mask	Lapel Sample Number	Sampling Duration (min)	Actual/8 h TWA Ribavirin Concentration (μg/m³)	In-Mask Sample Number	Sampling Duration (min)	Ribavirin Concentration (μg/m³)
February 28, 1991						
RN/E/MSA M (7 AM–7 PM)	3	88	438/120	1	88	ND
RT/U/MSA M (7 AM–3 PM)	4	89	236/44	2	89	ND
RT/S/MSA L (3 PM–11 PM)	15	46	413/40	14	46	ND
March 1, 1991						
RN/F/MSA L (7 AM–3 PM)	31	92	408/78	32[a]	92	(4)
RN/H/MSA M (3 PM–11 PM)	43	78	494/80	44	78	ND
RT/Q/3 M 9970 L (7 AM–3 PM)	29	102	279/59	30	15	ND
				37	87	ND
RT/Y/3M 9970 L (3 PM–11 PM)	45	45	733/69	46	45	ND

Notes: The value in () represents a quantity of ribavirin between the limit of detection [0.7 μg/sample] and the limit of quantitation [2 μg/sample], and should be considered a trace concentration with limited confidence in its accuracy; Emp: employee identification letter; L: large; M: medium; ND: none detected; RN: Registered Nurse; RT: respiratory therapist; TWA: time-weighted average.

[a] The cassette of sample 32 was disconnected accidentally from the respirator facepiece for approximately 2 seconds during sampling.

TABLE 16.A.23 Workplace PFs for RNs and Respiratory Therapists (RTs) during Ribavirin Administration. HCA Wesley Medical Center, Wichita, Kansas. Site Visit #3 (February 28 and March 1, 1991). HETA 90–155

Job/Emp./ Mask Shift	Ribavirin in Probe Wash (μg/10 mL)	(A) 8 h TWA Lapel Concentration (μg/m³)	(B) Approximate 8 h TWA In-Mask Concentration (μg/m³)	Workplace PF (A/B)
February 28, 1991				
RN/E/MSA M (7 AM–7 PM)	ND	120	1.1	109
RT/U/MS M (7 AM–3 PM)	ND	44	0.7	63
RT/S/MSA L (3 PM–11 PM)	ND	40	0.8	50
March 1, 1991				
RN/F/MSA L (7 AM–3 PM)	ND	78	0.8	98
RN/H/MSA M (3 PM–11 PM)	ND	80	0.6	133
RT/Q/3M 9970 L (7 AM–3 PM)	ND	59	0.6	98
RT/Y/3M 9970 L (3 PM–11 PM)	ND	69	0.8	86

Note: Emp, employee identification letter; L, large; M, medium; ND, none detected; TWA, time-weighted average.

in-mask sampling for ribavirin, this approach is preferred for ensuring the continued effectiveness of the respiratory protection program.[7]

As mentioned previously, the engineering staff of HCA Wesley Medical Center estimated that the rooms used for ribavirin administration had 12 ACH. The 1991 Aerosol Consensus Statement of the American Association for Respiratory Care contains guidance that patient rooms where ribavirin is administered should have a minimum of 6 ACH.[17] The ventilation system of each room where ribavirin is administered at HCA Wesley Medical Center should continue to be monitored routinely to ensure that optimal operation is maintained. The ventilation system of each room should also be maintained at a slight negative pressure to prevent AR from entering other occupied areas of the PICU.

Addendum A

Protection Factors for Respirators

Because differences exist among the various classes of respirators with regard to their protective capabilities, respirators are assigned PFs as guidance for their selection. A PF is the ratio of the concentration of a contaminant in the environment surrounding a respirator wearer to the concentration of the contaminant inside the respirator wearer's facepiece. The majority of assigned PFs are based on quantitative fit factors rather than workplace PFs. *Quantitative fit factors* are determined from tests in which a group of respirator wearers perform a specific regimen of head and body movements for a short period of time while in a laboratory test chamber containing a challenge aerosol. A *workplace PF* is a measure of the protection provided in a workplace under the actual conditions of that workplace by a properly functioning respirator which is correctly worn and used.[18] An *assigned PF* is the minimum expected workplace level of respiratory protection that would be provided by a properly functioning respirator, or class of respirators, to a stated percentage of properly fitted and trained users.[18,19] This proportion has usually been specified as 95% for test data derived from both quantitative fit factors[20] and workplace PFs.[11,12]

The maximum use concentration for a respirator is generally determined by multiplying the assigned PF of the respirator by a contaminant's lowest occupational exposure limit (i.e., the lowest value among a contaminant's PEL of the OSHA, REL of NIOSH, and threshold limit value of the American Conference of Governmental Industrial Hygienists). Alternatively, the minimum level of protection necessary for a specific occupational application can be calculated after exposure estimates have been determined for environmental contaminants. This is usually done by dividing the highest 8 h TWA exposure estimate of an airborne contaminant by the contaminant's lowest occupational exposure limit. Then, a class of respiratory protection is selected with an assigned PF equal to or exceeding the needed level of protection. For example, if a set of industrial hygiene samples collected during a particular operation produced 8 h TWA exposure estimates ranging from 8 to 50 mg/m^3 for a contaminant with an occupational exposure limit of 10 mg/m^3, then a respirator with an assigned PF of at least 5 (50/10 = 5) should be selected. Such a respirator would reduce the highest exposure concentration to an in-mask concentration equal to, or less than, the contaminant's exposure limit for the majority of respirator wearers.

After implementation of a respiratory protection program, simultaneous lapel and in-mask sampling should be performed on a sample set of respirator wearers to ensure that the respirator selected is indeed sufficient to protect its user during all conditions of use. Such sampling should be conducted periodically to further ensure that there have been no significant changes in the conditions of respirator usage that might reduce the effectiveness of the particular respirator in service.

References

1. Waskin H. Toxicology of antimicrobial aerosols: A review of aerosolized ribavirin and pentamidine. *Respir Care* 36:1026–1036, 1991.
2. Barnhart ER. *Physicians' Desk Reference*, 45th edn. Medical Economics Company, Oradell, NJ, 1991, pp. 1096–1098.

3. Matlock D, Buchan RM, and Tillery M. A local exhaust ventilation system to reduce airborne ribavirin concentrations. *Am Ind Hyg Assoc J* 52:428–432, 1991.
4. HCA Wesley Medical Center. Procedures for ribavirin administration. In *Respiratory Therapy Procedure Manual*. Wichita, KS.
5. Arnold SD and Buchan RM. Exposure to ribavirin aerosol. *Appl Occup Environ Hyg* 6:271–279, 1991.
6. National Institute for Occupational Safety and Health. *NIOSH Guide to Industrial Respiratory Protection*. U.S. Department of Health and Human Services, Public Health Service, Centers for Disease Control, National Institute for Occupational Safety and Health, Cincinnati, OH, 1987.
7. National Institute for Occupational Safety and Health. *NIOSH Respirator Decision Logic*. U.S. Department of Health and Human Services, Public Health Service, Centers for Disease Control, National Institute for Occupational Safety and Health, Cincinnati, OH, 1987.
8. Liu BYH, Sega K, Rubow KL, Lenhart SW, and Myers WR. In-mask aerosol sampling for powered air purifying respirators. *Am Ind Hyg Assoc J* 45:278–283, 1984.
9. National Institute for Occupational Safety and Health. Ribavirin. Method 5027 (issued 5/15/89), In: PM Eller (ed.), *NIOSH Manual of Analytical Methods*, 3rd edn. U.S. Department of Health and Human Services, Public Health Service, Centers for Disease Control, National Institute for Occupational Safety and Health, Cincinnati, OH, 1984.
10. Rose JC, Oestenstad RK, and Rose VA. A comparison of respirator fit factors determined by portable condensation nuclei counting and forward light-scattering photometric methods. *Appl Occup Environ Hyg* 5:792–797.
11. Lenhart SW and Campbell DL. Assigned protection factors for two respirator types based upon workplace performance testing. *Ann Occup Hyg* 28:173–182, 1984.
12. Reed LD, Lenhart SW, Stephenson RL, and Allender RL. Workplace evaluation of a disposable respirator in a dusty environment. *Appl Occup Environ Hyg* 2:53–56, 1987.
13. Birkner LR. Respiratory Protection: *A Manual and Guideline*. (BJ Held and BL Held, eds.), American Industrial Hygiene Association, Akron, OH.
14. Code of Federal Regulations. Appendix C to 29 CFR 1910.1001 (Asbestos, tremolite, anthophyllite, and actinolite). U.S. Government Printing Office, Federal Register, Washington, DC, 1989.
15. Code of Federal Regulations. Appendix D to 29 CFR 1926.58 (Asbestos, tremolite, anthophyllite, and actinolite). U.S. Government Printing Office, Federal Register, Washington, DC, 1989.
16. Williams FT. An analytical method for respirator performance prediction utilizing the quantitation fit test (QNFT). Unpublished paper presented at the First International Respirator Research Workshop, Morgantown, WV, 1980.
17. American Association for Respiratory Care. Aerosol consensus statement—1991. *Respir Care* 36:916–921, 1983.
18. Myers WR, Lenhart SW, Campbell D, and Provost G. Letter to the editor, Respirator performance terminology. *Am Ind Hyg Assoc J* 44:B25–26, 1985.
19. Guy HP. Respirator performance terminology. *Am Ind Hyg Assoc J* 46:B-22 and B.24.
20. Hyatt EC. Respiratory protection factors. Informal report, LA-6084-MS, Los Alamos, NM, 1976.

Source: Hazard Evaluations and Technical Assistance Branch, Division of Surveillance, Hazard Evaluations and Field Studies. Report prepared by Steven W. Lenhart, CIH, Industrial Hygienist, Industrial Hygiene Section. Field assistance by John A. Decker, MS, Industrial Hygienist, Industrial Hygiene Section, Hazard Evaluations and Technical Assistance Branch, and Debra Fox, RRT, HCA Wesley Medical Center, 550 North Hillside, Wichita, KS 67214.

17

Trace Anesthetic Gas

Edward W. Finucane
High Technology Enterprises

Anthony Schapera
San Francisco General Hospital

William Charney
Healthcare Safety Consulting

Patricia A. Heinsohn
San Francisco Department of Public Health

D.L. Jewett
University of California, San Francisco

Inhalation Anesthesia Agents 17-1
Introduction and Background • Existing Exposure Standards • Currently Established Limits and Standards • Anesthesia Agent Toxicology • Nitrous Oxide • Halogenated Anesthesia Agents—Generalized • Halothane • Methoxyflurane • Sources of Waste or Fugitive Anesthesia Agents • Waste Anesthetic Gases • Fugitive Anesthetic Gases • Monitoring Anesthesia Agents • Basic Analytical Parameters • Analytical Methods • Halogenated Anesthetic Gas Analyses
References 17-15
Anesthesia Mask Gas-Scavenging System 17-16
Introduction • Methods • Results • Discussion • Conclusions
References 17-20
Personal Exposure to Blood Aerosols among Operating Room Personnel 17-20
Introduction • Methods • Results • Discussion • Conclusion
Acknowledgments 17-25
References 17-25

Inhalation Anesthesia Agents

Edward W. Finucane

Introduction and Background

Considerable valid concern exists for the potential harm—particularly to certain categories of exposed individuals—that could occur as a result of an exposure to volatile inhalation anesthesia agents. Included in this general broad category of chemicals are the following seven, tabulated below, along with some of their more important physical properties.

Nitrous Oxide

Synonyms:	Laughing gas, hyponitrous acid anhydride, factitious air, and/or hippie crack
Name and formula:	Dinitrogen monoxide or nitrous oxide: N_2O
Structure:	N ∫ O = N
Physical properties:	Boiling point = −88.5°C (gas at room temperature)
	Molecular weight = 44.01 amu

Gaseous nitrous oxide specific gravity = 1.53 (the specific gravity of air = 1.00); Colorless gas with a slightly sweetish odor and taste
Normally supplied in metal cylinders containing a vapor phase above a liquid phase, at an approximate internal pressure of 800 psig.

Halothane

Synonyms: Fluothane and/or rhodialothan

Name and formula: 2-Bromo-2-chloro-1,1,1-trifluoroethane: $C_2HBrClF_3$

Structure:

```
  F  Br
  |  |
F-C--C-F
  |  |
  F  Cl
```

Physical properties:
Boiling point = 50.2°C (liquid at room temperature)
Molecular weight = 197.39 amu
Liquid density = 1.871 g/cm^3
Nonflammable and highly volatile liquid, with a sweetish but not wholly unpleasant odor
Normally supplied in specially keyed glass bottles.

Enflurane

Synonyms: Ethrane, efrane, alyrane, NSC-115944, compound 347, and/or methylflurether

Name and formula: 2-Chloro-l-(difluoromethoxy)-1,1,2-trifluoroethane
2-chloro-1,1,2-trifluoroethyl difluoromethyl ether
$C_3H_2ClF_5O$

Structure:

```
  F     F  F
  |     |  |
H-C--O--C--C-H
  |     |  |
  F     F  Cl
```

Physical properties:
Boiling point = 56.5°C (liquid at room temperature)
Molecular weight = 184.50 amu
Liquid density = 1.517 g/cm^3
Stable, volatile, and nonflammable colorless liquid, with a faint but characteristic ethereal odor
Normally supplied in specially keyed glass bottles.

Isoflurane

Synonyms: Forane, forene, aerrane, and/or compound 469

Name and formula: 2-Chloro-2-(difluoromethoxy)-1,1,1-trifluoroethane
or
1-chloro-2,2,2-trifluoroethyl difluoromethyl ether
$C_3H_2ClF_5O$

Structure:

```
  F  H     F
  |  |     |
F-C--C--O--C-H
  |  |     |
  F  Cl    F
```

Physical properties:
Boiling point = 48.5°C (liquid at room temperature)
Molecular weight = 184.50 amu
Liquid density = 1.450 g/cm^3
Stable, highly volatile, and nonflammable colorless liquid, with a mild but characteristic ethereal odor
Normally supplied in specially keyed glass bottles.

Methoxyflurane

Synonyms: Metofane, penthrane, pentrane, and/or DA 759

Name and formula: 2,2-Dichloro-1,1-difluoro-l-methoxyethane
or
2,2-dichloro-1,1-difluoroethyl methyl ether
$C_3H_4Cl_2F_2O$

Structure:

```
  H     F Cl
  |     |  |
H-C-O-C-C-H
  |     |  |
  H     F Cl
```

Physical properties: Boiling point = 105°C (liquid at room temperature)
Molecular weight = 164.97 amu
Liquid density = 1.425 g/cm^3
Stable, moderately volatile and nonflammable liquid with a moderately strong and pungent ethereal odor
Normally supplied unkeyed glass bottles.

Comments: Limited to veterinary anesthesia.

Suprane

Synonym: Desflurane

Name and formula: 1-(Difluoromethoxy)-1,2,2,2-tetrafluoroethane
or
1,2,2,2-tetrafluoroethyl difluoromethyl ether
$C_3H_2F_6O$

Structure:

```
  F     F F
  |     | |
H-C-O-C-C-F
  |     | |
  F     H F
```

Physical properties: Boiling point = 22.8°C (liquid, when in a check valve sealed bottle at room temperature)
Molecular weight = 168.04 amu
Liquid density = 1.467 g/cm^3
Stable, extremely volatile and nonflammable colorless liquid with a moderately strong and pungent ethereal odor
Normally supplied in unkeyed glass bottles equipped with a pressure check valve

Comments: Approved by the U.S. Food and Drug Administration for human inhalation anesthesia use in the United States in mid-1993.

Sevoflurane

Synonyms: Sevofrane

Name and formula: 2-Fluoromethoxy-l,1,1,3,3,3-hexafluoropropane
or
1,1,1,3,3,3-hexafluoro-2-propyl fluoromethyl ether
$C_4H_3F_7O$

Structure:

```
      F
      |
 FF-C-F  H
  |   |    |
F-C-C-O-C-F
  |   |    |
  F   H    H
```

Physical properties: Boiling point = 58.6°C (liquid at room temperature)
Molecular weight = 200.06 amu

	Liquid density = 1.525 g/cm^3
	Stable, volatile, and nonflammable colorless liquid, with a fairly pungent ethereal odor
	Normally supplied in unkeyed glass bottles
Comments:	Not yet approved by the U.S. Food and Drug Administration for human inhalation anesthesia use in the United States.

Existing Exposure Standards[1]

The currently established Government and Professional Society Exposure Limits and Standards for these seven volatile agents are tabulated below; these listings will use appropriate portions of the following sets of abbreviations:

Applications applicable to all areas, organizations, and agencies:

TWA: Time-weighted average (usually evaluated over 8 h).
STEL: Short-term exposure limit (usually evaluated over a 15 min period, as a time-weighted average).
C: Ceiling value (a "never-to-be-exceeded" concentration).

Abbreviations applicable to the American Conference of Government Industrial Hygienists (ACGIH):

TLV: Threshold limit value (the concentration level above which an exposed individual might reasonably anticipate the onset of adverse effects—as determined by the ACGIH in specific evaluation efforts covering these effects).

Abbreviations applicable to the U.S. Department of Labor, Occupational Safety and Health Administration (OSHA):

PEL: Permissible exposure limit (the concentration level above which an employer can be cited for failure to maintain a "safe" workplace).

Abbreviations applicable to the National Institute for Occupational Safety and Health (NIOSH):

REL: Recommended exposure limit (the concentration level above which NIOSH data and experience have indicated the possible onset of adverse effects in exposed individuals).

Abbreviations applicable to the Deutsche Forschungsgemeinschaft Agency of Germany:

MAK: Maximum arbeitsplatz konzentration (Maximum Concentration Value for the Workplace) (the concentration level above which German employers may be cited for failure to maintain a "safe" workplace).

Currently Established Limits and Standards

Nitrous oxide:	ACGIH 8 h TLV-TWA = 50 ppm
	NIOSH 8 h REL-TWA = 25 ppm
Halothane:	ACGIH 8 h TLV-TVA = 50 ppm
	NIOSH 60 min REL-C = 2 ppm
	DFG 8 h MAK-TWA = 5 ppm
Enflurane:	ACGIH 8 h TLV-TWA = 75 ppm
	NIOSH 60 min REL-C = 2 ppm
Isoflurane:	No established exposure limits or standards of any type from any agency or organization
Methoxyflurane:	NIOSH 15 min REL-C = 2 ppm

Suprane:	No established exposure limits or standards of any type from any agency or organization
Sevoflurane:	No established exposure limits or standards of any type from any agency or organization

Anesthesia Agent Toxicology

Much of the data on the toxicology of the seven agents have been developed as a result of observing and cataloging the complaints, illnesses, and various other complications that have been reported by individuals who have worked in locations, where there are nonzero ambient concentration levels of these same agents. In such cases, it is invariably difficult to establish direct, irrefutable "cause-and-effect" relationships between the reported ailment and the suspect causal exposure to the particular volatile anesthesia vapor. Comparative statistical studies on exposed versus unexposed populations have tended to confirm several of these hypothetical "cause-and-effect" relationships; however, several important unanswered questions still remain.

Much of the actual systematic work on the toxicology of these chemicals has involved the use of laboratory animals; and much of this work has involved exposing these animals to concentration levels that "are far higher" than currently recognized exposure standards—as tabulated above. The evident primary goal of most of these laboratory investigations has been to identify and document the actual health risks of relatively large scale exposures to these agents. This fact notwithstanding, it is largely these data—and particularly, the extrapolation of these data back to lower and lower concentration ranges—that have provided the most reliable basis for identifying the health hazards associated with volatile anesthesia agents.

The data in this area, thus far as it is currently known and recognized, will next be listed (1) by each specific volatile agent and (2) by the various health difficulties for which that agent is thought to be responsible.

Nitrous Oxide

There is a steadily growing body of data that supports the contention that exposures to nitrous oxide (at 8 h TWA concentration levels in excess of 25 ppm), by second and third trimester pregnant women can produce significant increases in their rates of (1) spontaneous abortion, (2) stillbirth, and (3) congenital abnormalities among their children.[2–7] Although there is not yet total agreement on these "cause-and-effect" relationships, it is safe to say that the general attitude of health professionals is that nitrous oxide exposures—particularly to second and third trimester pregnant women—must be carefully monitored and controlled.

In addition to the foregoing, there have been several epidemiological studies and a lesser number of laboratory investigations, the results of which have suggested that exposures to trace levels of nitrous oxide might also be responsible for increases in hepatic as well as renal disease (excluding renal pyelonephritis and cystitis).[2–4,7] These studies have indicated greater adverse impacts on exposed women than on men. Considerably more investigative work needs to be done in these areas before specific toxicological conclusions can be definitively reached.

There have also been scattered reports of ambient nitrous oxide exposures causing certain genetic abnormalities—as measured by Sister-Chromatid Exchange Analyses.[8] As it is true in the case of the previously mentioned situation involving hepatic and renal disease, considerably more investigative work needs to be done in these areas before specific genetic toxicological conclusions can be confirmed.

Halogenated Anesthesia Agents—Generalized

Exposures to the halogenated inhalation anesthesia agents will be treated, first, by identifying any of the similar broadly based negative effects that can be or have been caused by "any" of these six compounds (not an unreasonable approach, since these six are relatively similar, chemically), and second, by identifying health problems that appear to be uniquely caused by any single member of this group—to the extent that data specific to any of these individual agents have been developed.

The most commonly reported health risk associated with the halogenated anesthesia agents is similar to those reported and documented for many of the other halogenated organic compounds (i.e., refrigerant, vapor degreasing solvents, etc.). These include various chronic difficulties in the areas of psychomotor, hematopoietic, central nervous, hepatic and renal system diseases, and dysfunctions.[3,4,7] Clearly, all six of these agents function as acute central nervous system depressants. The chronic ailment category most commonly reported among individuals exposed to these volatile agents involves the liver, and includes a number of actual cases of chemical hepatitis hypothetically projected to have been caused by exposures to these vapors. Considerable experimental effort has been devoted to evaluating the relative liver toxicity of four of these anesthesia agents, as determined by monitoring the intracellular K^+ content (this, as an indicator of the time, concentration and oxygen-dependent cytotoxicity of the biotransformation processing of these agents in the liver). The result of these undertakings has been a ranking—from the most to the least harmful—as potential hepatotoxins—this ranking is: halothane, isoflurane, enflurane, and sevoflurane.[9]

A small number of reports indicate that repeated low level exposures to these agents may also be responsible for increased susceptibility to infections and neoplastic disease.[4]

Finally, there have been a few reports of ambient-halogenated anesthesia agent exposures causing certain genetic abnormalities—again, as measured by Sister Chromatid Exchanges—however, the clear determination that one, or even any, of these agents was the principal causal factor in these cases, as contrasted to nitrous oxide being the causal agent, has not been conclusively made.[8]

Halothane

Halothane is the member of this group that (1) is not an ether and (2) contains bromine. As such, it would not be surprising if there were at least some differences between it and the other halogenated agents, from the perspective of their effects on exposed individuals, and indeed there are some differences. Chemical (or clinical) hepatitis is far more closely associated with halothane exposures than is the case with any other of these agents.[10–13] The most likely mechanism for this chemical induced hepatitis is the hypersensitivity reaction to liver neo-antigens that are produced by the halothane metabolite, 2-chloro-1,1,1-trifluoroethane.[10]

Various other studies have implied that halothane may be teratogenic and/or embryotoxic. These conclusions have been developed through studies of the resultant ultrastructural effects in rats exposed to halothane at 10 ppm (vol.) or more versus the same effects observed in unexposed rats.[13] Finally, ambient occupational halothane exposures have been linked to increased and potentially unhealthy plasma bromide concentrations.[14] In contrast, when enflurane is the inhalation anesthesia agent being used in surgical procedure, this plasma bromide concentration buildup in exposed individuals is completely absent.[15]

Methoxyflurane

Methoxyflurane—administered to male Fisher rats, at a concentration of 50 ppm (vol.), over a 14 weeks period—has been found to be both growth depressing and hepatotoxic. At the end of the exposure period, the livers of all of the exposed rats were examined, and all showed focal hepatocellular degeneration and necrosis, as well as evidence of liver cell regeneration.[16]

Sources of Waste or Fugitive Anesthesia Agents

In the overall consideration of the important and/or significant sources of potentially harmful ambient occupational concentration levels of these anesthetic agents, we must consider two principal sources;

namely: (1) waste and/or (2) fugitive sources. A "waste anesthetic gas" is any gas or vapor that has been, or is being, exhaled by an anesthetized person, either during or after a procedure that has involved the use of inhalation anesthesia agents. A "fugitive anesthetic gas" is any gas or vapor that has escaped from any system that is functionally upstream of the patient (i.e., the anesthetic gas vaporizer/ventilator, the high pressure nitrous oxide supply system outlets, etc.). By definition, fugitive anesthetic gases exist only in surgical suites, operating rooms, and/or the piping systems that extend between the master gas supply manifold and the outlet fittings wherever they may be located. These shall each be considered in order.

Waste Anesthetic Gases

Typically, in the modern hospital operating room, there will be two separate and distinct systems that have been engineered, in whole or in part, to mitigate problems associated with the potential buildup of waste anesthetic gases. The first of these, a vacuum scavenging system, is designed to function solely as a local exhaust hood, situated directly at the anesthesiologist's (or anesthetist's) station. Its purpose is simply to capture and remove the patient's exhaled breath during the surgical procedure. The second of these systems is the overall ventilation system, the main function of which is to circulate and replenish the room's air.

Vacuum scavenging systems should be (and are) used as the principal mechanism for removing waste anesthetic gases, whether the anesthesia mixture is administered to the patient by open mask or endotracheal tube induction. When such a system is of sufficient design capacity (exhaust flow rate volume ≥25 L/min, typically), and is utilized properly, it will be effective—or even very effective—in removing the waste anesthetic gases from the surgical suite, thereby minimizing the exposure of the surgical team.[17] Alternatively, for a scavenging system that has less than this level of exhaust capacity, or—even worse—for an operating room that is not equipped with any type of waste gas scavenging system, the buildup of potentially unsatisfactorily high concentration levels of waste anesthetic gases will be almost a certainty.

In the most general sense, the scavenging of waste anesthetic gases will always be more difficult in the cases of either open-circuit or mask induction—either in marked contrast to induction via endotracheal tube. In these difficult types of situations, the exhaust connection between the anesthesia induction circuit and the waste gas scavenging system will be both more tenuous and much more difficult to maintain; thus the potential for unscavenged anesthetic gases collecting in unsatisfactorily high concentration levels will be much greater.

As a fairly typical example, in 1990, ambient air measurements were made in seven different hospitals in Vienna, Austria. In one of these measurement situations, during an otolaryngological surgical procedure that necessitated anesthesia induction by open-circuit, and in a facility without a waste gas scavenging system (a textbook worst possible case scenario), peak concentration levels of nitrous oxide and halothane were found to be 2600+ ppm and 150+ ppm, respectively. These astonishingly high concentration levels contrasted markedly to their counterparts in operating rooms at other Viennese hospitals, all of which had properly designed and operating waste gas scavenging systems. For this second group of operating rooms, and in situations where the anesthetic gases were administered to the patient via endotracheal tube, the concentrations, expressed as 8 h TWAs, were found to be in the range 8–15 ppm (mean = 11 ± 3 ppm) for nitrous oxide; and 0.1–0.6 ppm (mean = 0.3 ± 0.2 ppm) for Halothane. For the same or similar operating rooms, when the anesthetic gases were mask induced, these concentrations, also expressed as 8 h TWAs, were found to be in the range 24–211 ppm (mean = 83 ± 49 ppm) for nitrous oxide, and 0.3–1.3 ppm (mean = 0.8 ± 0.3 ppm) for halothane.[17]

Clearly, in the operating room, the occupational waste anesthetic gas exposure of every individual involved can be significantly mitigated through the correct use of an adequately designed and properly functioning exhaust scavenging system.[18–25]

General operating room ventilation systems, too, contribute to the removal of waste anesthetic gases; however, the primary function of these systems is to circulate and replenish the air in the room. This replenishment function must always be emphasized, since this aspect of the operation of a general ventilation system is always its prime function. In accomplishing this, however, the ventilation system will clearly supplement the vacuum scavenging system in the removal of waste anesthetic gases. Replenishment here should be understood to imply that "all" of the air that is removed from the operating room should ultimately be exhausted from the building by that operating room's ventilation system. None of the air should ever be recirculated.

In addition, for any operating room, the ventilation system's capacity expressed in terms of its overall room air exchange rate should be at least 15 room volumes/h, and preferably 18–20 room volumes/h.[18,19,26] An operating room with a ventilation system having this sort of overall air handling capacity (assuming also that the overall ventilation flow patterns in the room include no short circuits) will generally be an occupationally safe place in which to work. Such a ventilation system can be regarded as sufficient, both in and of itself, and as a functional backup to the room's waste anesthetic gas-scavenging system. For reference, a short-circuiting air flow pattern is one in which inlet air will flow directly to an exhaust register, without having swept out every volume or space within the room itself—such upswept spaces are called "dead volumes." For an operating room that has "dead volumes," the potential for harmful ambient occupational anesthetic gas buildups may be very great, and this is true even if the overall room air exchange rate was high enough to appear to guarantee against such situations.

General surgical recovery areas, too, are potential sources for dangerously high concentration levels of waste anesthetic gases. Recovering surgery patients will tend to exhale exponentially decreasing concentrations of those gases they have inhaled during their surgery. For any patient whose surgical procedure has required an extended period (i.e., more than 4 h), the initial concentration of waste anesthetic gases in their exhaled breath can be very high—nitrous oxide concentrations in excess of 2000 ppm have been measured in the breathing zone of patients immediately after their arrival in the recovery area. Additionally, the time period during which these patients must be monitored—in order to fully complete their "offgasing" process—can be quite prolonged. Nurses, and any other medical practitioner, who must tend to these patients or who must work in areas where they are recovering are, themselves, very good candidates for receiving an unsatisfactorily high level exposure to waste anesthetic gases.

In general, the ventilation requirements for a surgical recovery area will be the same in terms of the necessary minimum air exchange rate as was the case for an operating room; and the same comments with respect to short circuiting and "dead volumes" also apply. An even more effective way to ensure against the "dead volume," that may, or actually does, contains high concentration levels of waste anesthetic gases, involves the use of individual slot exhaust hoods at each patient location. Such an approach will always provide the best possible engineering solution, insofar as worker exposures to waste anesthetic gases are concerned.[27]

A brief comment with respect to veterinary surgical suites is in order at this point. The risks of waste anesthetic gas exposures in these operatories are probably considerably greater than those that would be expected to occur in their human operating room counterparts. The reasons for this are several and include all of the following: (1) veterinary clinics frequently employ portable gas delivery carts that are not designed to capture waste anesthetic gases; (2) the costs of specific waste gas scavenging systems may be prohibitively high, and therefore beyond the reach of the relatively smaller veterinary clinic; and (3) the overall effectiveness of such scavenging systems has yet to be fully verified in the veterinary area, and there are, therefore, neither current OSHA recommendations nor standards for veterinary scavenging systems (to say nothing of the fact that there are no OSHA PELs for any of the anesthetic agents).[28]

In a typical survey of 14 different veterinary operatories, each performing surgeries in which anesthetic mixtures of nitrous oxide and halothane were used, the following geometric mean TWA concentrations (as well as the overall concentration ranges measured) were as follows: 100 ppm (range 14–1700 ppm) and 2.6 ppm (range 0.5–119 ppm), respectively, for nitrous oxide and halothane. Since a typical veterinary surgery rarely lasts more than 4 h, the same geometric mean, TWA concentrations, now extrapolated out to cover an 8 h period, and again for nitrous oxide and halothane were as follows: 34 ppm (range 5–530 ppm) and 1.3 ppm (range 0.5–34 ppm), respectively.[29]

Fugitive Anesthetic Gases

Anesthetic gas leaks, as implied earlier, will typically originate from one of the following three sources: (1) from the high pressure piping systems that deliver nitrous oxide to the operating suites (including the low-pressure connections to the anesthesia circuit); (2) from leakage on or around the anesthetic gas vaporizer/ventilator machine—in particular, from any and all of the low-pressure connections whose many seals and joints should always be suspect (these machines must have components of this type in order to facilitate the required frequent cleanings and replacements); or (3) from the poor work practices and/or habits of the anesthesiologist/anesthetist who utilizes this equipment. In the event that the source of the unsatisfactorily high ambient concentration levels of anesthetic gases turns out to have been attributable to the poor work practices of the anesthesiologist/anesthetist, then the ambient anesthetic gas vapors present can be in either the waste or the fugitive category. Again, as stipulated earlier, an overall air exchange ventilation rate of 18–20 room volumes/h can help mitigate these types of problems; however, elimination of the specific leaks would be the preferred solution.[30]

Monitoring Anesthesia Agents

Monitoring ambient and occupationally significant concentration levels of the various anesthesia agents can be accomplished in several ways; the choice of the method or approach involved will usually be dictated by the type of information required (i.e., real-time concentrations, individual exposure dosimetry, etc.); and/or the nature, type, and number of potential interferants that might be present.

Basic Analytical Parameters

These functional requirements of "any" gas analysis can be understood, most beneficially, in terms of the four characteristic factors that apply to "every" ambient analytical application, namely: (1) the sensitivity, (2) the selectivity, (3) the reproducibility, and (4) the timeliness of the method. The following four descriptions will hopefully adequately define and describe these functionality parameters:

1. A method's "sensitivity" is usually expressed in terms of its minimum detection limit—or MDL—to the material being analyzed. Ideally, a method's MDL should be equal to, or less than, 10% of some "significant concentration" (usually, the PEL-TLV) of the material to be monitored.
2. A method's "selectivity" is the measure of its ability to distinguish the material to be monitored from "everything else" that might be in the matrix to be monitored (i.e., interfering chemicals). Interferences can be either positive or negative. Positive ones increase the analytical reading; negative ones, on the other hand, decrease or "quench" it.
3. A method's "reproducibility" is a measure of its ability to provide consistent readings for the same concentration of the material to be measured: (a) at different times, (b) under different ambient conditions, or (c) by different analysis. It is also a measure of the method's ability to remain in calibration for extended period.
4. A method's "timeliness" is a measure of the time interval required for it to provide its assessment of the unknown concentration answer for the material being analyzed. This time interval can be quite broad, ranging from instantaneous (real-time) to very extended periods—as, for example, when a passive dosimeter must be sent to some distant external location for evaluation.

Analytical Methods

There are a number of viable analytical methods that purport to monitor ambient concentration levels of the anesthetic gases. In general, analyzers in each of the method categories listed below—in Tables 17.1 and 17.2—have been advertised as being both well suited to and available for these tasks. Certain units employing these analytical methods can be provided as portables, others are available in fixed or

TABLE 17.1 Nitrous Oxide

IR-n	Infrared spectrophotometric absorbance analyzers
PAS-n	Photoacoustic infrared spectrophotometric analyzers
GC-n	Gas chromatographic analyzers
PD-n	Passive dosimetric systems
AD-n	Active dosimetric systems

TABLE 17.2 Halogenated Anesthesia Agents

IR-n	Infrared spectrophotometric absorbance analyzers
PAS-h	Photoacoustic infrared spectrophotometric analyzers
AEC/SS-h	Pumped or active electrochemical/solid state analyzers
PEC/SS-h	Passive electrochemical/solid state analyzers
GC-h	Gas chromatographic analyzers
PD-h	Passive dosimetric systems
AD-h	Active dosimetric systems

installed arrangements. In addition, some of the fixed units can be provided in single or multiple point monitors. In a few, systems are available in both of these configurations.

The following two listings cover those analytical methods that appear to the author to offer the greatest potential for successfully analyzing the anesthetic gases. Table 17.1 focuses on the analysis of nitrous oxide, while Table 17.2 addresses the six halogenated anesthesia agents.

This following Functional Score Sheet represents the structure within which the author's assessment of the relative merits and/or capabilities of each of the several different analytical methods (analyzer types) listed above will be presented, with separate and distinct assessments for each of the two different analyses—namely, the one for nitrous oxide and the one for the halogenated anesthetic gases.

Functional Score Sheet

Rating No.	(1) Sensitivity	(2) Selectivity	(3) Reproducibility	(4) Timeliness
5	<0.2 ppm	Excellent	Excellent	Real time
4	<0.5 ppm	Very good	Very good	<2 min delay
3	<1.0 ppm	Good	Good	≤10 min delay
2	≤5.0 ppm	Fair	Fair	>10 min delay
1	>5.0 ppm	Poor	Poor	>1 day delay

From Table 17.1, it can be seen that there are five potentially viable analytical methods available for this task. That there will be additional methods available in the future is a virtual certainty. That there may be even additional ones available today cannot be denied; however, the author is unaware of any other current candidates. There are, to be sure, well-established methods that analyze nitrous oxide at concentrations in the percent range; however, for ambient analyses focused on occupational health risks, the five-member listing is quite complete.

The following two tabulations list the author's evaluation of the relative merits of each analytical method in successfully completing ambient, occupationally significant nitrous oxide analyses. Table 17.3 focuses on "portable" systems, while Table 17.4 provides the same assessment for their installed or fixed counterparts.

Clearly, if the real-time or near real-time concentration data must be obtained, the best currently available methods are either (1) long pathlength infrared spectrophotometry (IR-n) or (2) photoacoustic infrared spectrophotometry (PAS-n).

TABLE 17.3 Ratings of the Various Portable Nitrous Oxide Analyzers and/or Systems

Analytical Method	1 Sensitivity	2 Selectivity	3 Reproducibility	4 Timeliness
IR-n	5	4	4	5
Pas-n	5	5	5	4
PD-n	3	4	4	1–2
AD-n	4	3	3	1–2

TABLE 17.4 Ratings of Various Fixed Nitrous Oxide Analyzers and/or Systems

Analytical Method	1 Sensitivity	2 Selectivity	3 Reproducibility	4 Timeliness
IR-n	5	5	5	4
PAS-n	5	5	5	4
GC-n	5	5	4	2

If it is necessary to actually monitor real-time for ambient levels of nitrous oxide, it should be noted that there is an existing NIOSH Standard (Method No. 6600) that specifies the use of long pathlength infrared for this task. This approach, which accomplishes N_2O analyses extremely well, calls for making absorbance measurements at a mid-infrared wavelength of 4.48 μm (a wave number of 2232 cm^{-1}), and an appropriately long pathlength, usually—for ambient situations in which the N_2O concentrations are less than 200 ppm—at a pathlength of 20.25 m.[31]

There are both portable and fixed long pathlength infrared analyzers that are very well suited to analyzing ambient levels of nitrous oxide. Fixed systems have the capability of being multiplexed so that a single analyzer can be used to monitor as many as 24 separate locations, simultaneously data logging and summarizing the analysis at each location on any time or sequence basis that may be required. Multiplexing a single unit to analyze at many separate points will offer both advantages and disadvantages. Among the former are the following two: (1) lower purchase costs, since only a single analyzer must be obtained and (2) lower operating costs that arise from having to calibrate and service only a single analyzer. On the negative side, this approach will always extend the time interval between subsequent analyses at each point. Finally, fixed systems can also be set up to compensate for potential interferants, thereby improving the selectivity of the system.

Near real-time measurements (i.e., delays of less than 2 min for each discrete N_2O concentration readout) can be very adequately accomplished by photoacoustic infrared spectrophotometric analyzers. Although NIOSH Method No. 6600 neither specifies nor identifies photoacoustic infrared methods as suitable for nitrous oxide analyses, the author feels that the methods are sufficiently similar that this technology could be regarded as having been included in this method. A photoacoustic infrared spectrophotometric analyzer will employ the same mid-infrared wavelength in its analytical setup as does its long pathlength infrared cousin, namely, 4.48 μm (or a wave number of 2232 cm^{-1}).

As was the case with the long pathlength infrared systems, there are both portable and fixed photoacoustic infrared analyzers that are very well suited to analyzing ambient levels of nitrous oxide. Fixed systems of this type also have the capability of being multiplexed, with the same simultaneous data logging and summarizing capability as was the case for the long pathlength infrared systems. In addition, multiplexing a single photoacoustic infrared analyzer to work at many points will offer the same generic advantages as was the case for its long pathlength infrared system cousin.

An active dosimetric system (AD-n) is one that utilizes a battery-powered personal air sampling pump—which the pumping rate can be very accurately set and maintained for a specific time period. Connected in series with this pump will be a sorbent filled tube. This combination system will then be clipped to an individual for whom nitrous oxide exposure data are being sought. This active dosimeter, with its pump operating, will be worn by this person for a prescribed time period, at the end of which, it will be retrieved and forwarded to an analytical laboratory. In this laboratory, the sorbent material in the tube will be desorbed and its contents analyzed chromatographically.

A passive dosimeter (PD-n) is simply a card or tag, on which will be mounted a porous pad impregnated with some material that will react chemically with nitrous oxide. An initially unexposed card or tag will be clipped to the collar or the shirt pocket of the individual for whom nitrous oxide exposure data are being sought. During the time period when this tag is being worn, nitrous oxide will diffuse into the pad and undergo the desired chemical reaction. This chemical reaction will cause a change of some type (color, reflectivity, etc.) that can be measured accurately so as to quantify the nitrous oxide exposure.

Finally, both active and passive dosimetry systems are available for nitrous oxide measurements. In general, these two approaches are clearly the very best methods for documenting a single individual's exposure to nitrous oxide, over either a full 8 h workday, or any shorter, but specific and well-documented time interval. Dosimetry systems that will permit the user to obtain an individual's dose directly on site and at the conclusion of any identified time interval are currently under development. Such systems will eliminate the requirement for sending exposed dosimeters to some remote location or laboratory for evaluation.

Fixed gas chromatographic systems (GC-n) are also capable of completing the analysis for nitrous oxide. The simplest and least expensive of these systems—one that would employ a thermal conductivity detector (TCD)—will be limited to ambient concentrations equal to or greater than 20 ppm. More sophisticated systems that use a discharge ionization detector (DID) can be set up to work at ambient concentrations down to less than 1.0 ppm. For this application, one would choose a chromatographic column packed with one of the readily available, common, rigid structured, uniform pore-sized, porous polymers such as Poropak or Chromosorb. Capillary columns coated with dimethyl polysiloxane will likely also be effective for this separation. Successful analytical results could be obtained by operating the chromatograph's oven either isothermally or under temperature programming. In all of the cases listed above, one would probably choose a helium carrier gas for ambient nitrous oxide analyses. Although this analytical method will provide excellent selectivity, gas chromatographic systems will provide excellent selectivity, gas chromatographic systems will almost always be more difficult and expensive to calibrate, maintain, and operate.

Halogenated Anesthetic Gas Analyses

From Table 17.2, it can be seen that there are seven potentially viable analytical methods available for this task. Again, it is certain that there will be additional methods available in the future, and there may even be other current ones available to perform these analyses. For the ambient analysis of the halogenated anesthetic gases, however, and with a focus on occupational health risk significant concentration levels, the seven-member listing of Table 17.2 will be both very satisfactory and relatively complete. The following two tabulations list the author's evaluation of the relative merits of each of these halogenated anesthetic gas analytical methods. Table 17.5 focuses on portable analyzers, while Table 17.6 provides the same assessment for installed or fixed systems.

Although there are no current NIOSH (or other) Standard Methods for measuring any of the halogenated anesthesia agents, clearly both the long pathlength infrared (IR-h) and the photoacoustic infrared (PAS-h) approaches offer fine solutions to obtaining these types of data. For reference, the following tabulation, Table 17.7, identifies both the "preferred" and a "second choice" wavelength (wave number) applicable to either of the infrared analytical approaches listed above, for any of the six halogenated anesthesia agents. For the long pathlength system, assuming occupationally important ambient concentration levels of any of these six chemicals (i.e., in the range of 0–25 ppm), the pathlength should be set at 20.25 m. There is no pathlength consideration for the photoacoustic infrared system.

TABLE 17.5 Ratings of the Various Portable Halogenated Anesthetic Gas Analyzers and/or Systems

Analytical Method	1 Sensitivity	2 Selectivity	3 Reproducibility	4 Timeliness
IR-h	5	3	4	5
PAS-h	5	4	5	4
AEC/SS-h	2	2	3	4
PEC/SS-h	2	2	3	3
PD-h	3	2	3	1–2
AD-h	3	5	3	1–2

TABLE 17.6 Ratings of the Various Fixed Halogenated Anesthetic Gas Analyzers and/or Systems

Analytical Method	1 Sensitivity	2 Selectivity	3 Reproducibility	4 Timeliness
IR-n	5	4	5	2–4
PAS-n	5	5	5	2–4
AEC/SS-n	2	2	3	3–4
PEC/SS-n	1	2	3	4
GC-n	5	5	4	1–3

TABLE 17.7 Analytical Wavelength Choices for the Halogenated Anesthetic Gases

	Preferred		Second Choice	
Anesthesia Agent	Wavelength (μm)	Wave Number (cm^{-1})	Wavelength (μm)	Wave Number (cm^{-1})
Halothane	7.85	1274	8.83	1132
Enflurane	8.67	1153	12.05	830
Isoflurane	8.57	1167	8.17	1224
Methoxyflurane	9.20	1087	7.61	1314
Suprane	8.19	1222	8.46	1182
Sevoflurane	8.09	1236	9.73	1027

There are portable and fixed long pathlength infrared spectrophotometric analyzers (IR-h) that are very well suited to analyzing ambient levels of the six halogenated anesthetic gases. The fixed systems all have the capability of being multiplexed so that a single analyzer can be used to monitor many locations, and these systems can even do this for up to four of the six different halogenated anesthetic gases, simultaneously. These systems will also simultaneously datalog as well as generate periodic data reports that summarize the measurements at each location. As was the case for the nitrous oxide analytical situation, multiplexing a single unit to analyze at many separate points will offer both advantages and disadvantages. There appear to be two advantages, namely: (1) lower purchase costs, since only a single analyzer must be obtained and (2) lower operating costs that arise from having to calibrate and service only a single analyzer. Since fixed systems can be set up to compensate for potential interferants, they can monitor for up to four of these halogenated anesthetic gases simultaneously. They accomplish this by adjusting their response to each of these six different gases by the effect that each of the others has on it.

Because of this, the significantly improved however, the broad similarity in the infrared absorptive fingerprints of these six halogenated anesthetic agents means that the analytical selectivity will never be perfect. On the negative side, this multipoint and/or multicomponent analytical approach will always extend the time interval between subsequent analyses at each point.

Near real-time measurements (i.e., delays of less than 2 min for each discrete halogenated anesthetic gas concentration readout) can be very adequately accomplished by photoacoustic infrared spectrophotometric analyzers (PAS-h). This type of analyzer will employ the same preferred and second choice mid-infrared wavelengths (wave numbers), as listed in Table 17.7, as will its long pathlength infrared spectrophotometric cousin.

As was the case with the long pathlength infrared systems, there are both portable and fixed photoacoustic infrared analyzers that are very well suited to analyzing ambient levels of the six halogenated anesthetic gases. Fixed systems of this type, also, can be multiplexed to many different points, while providing highly useful and functional datalogging and summarizing capabilities. These analyzers can also simultaneously analyze for up to four different halogenated anesthetic gases, and in doing so, achieve the same general benefits (and disadvantages) as do their long pathlength infrared cousins.

The pumped or active electrochemical/solid state analyzer (AEC/SS-h) is a unit that uses either fairly analyte-specific electrochemical cells or solid state detectors to provide their concentration readouts. These units also employ internal sampling pumps to aspirate air through their respective detector sections. Their passive electrochemical/solid state analyzer counterparts (PEC/SS-h) are identical in every respect "except" that they do not have internal sampling pumps—rather, they rely on the ambient diffusion of the target analyte to, and ultimately into, their detectors. This type of analyzer will usually also be equipped with some sort of a selective membrane that is positioned in front of its detector. This membrane will have been selected, because it is permeable to only a limited number of chemicals, and its placement in the analytical path will have the effect of improving the analyzer's selectivity.

Active and passive electrochemical/solid state analyzers are both available in virtually any desired configuration—portable or fixed, single or multipoint. These units will not, however, be able to analyze simultaneously for different halogenated anesthetic gases. The two principal advantages of these types of analyzers over any of their alternatives are their very low original cost and their previously mentioned wide configurational flexibility. They are particularly well suited to determining the existence and/or the location of any significant leak of any of the halogenated anesthetic gases; however, their relatively poor selectivity and, particularly, their very poor sensitivity usually disqualify them from active consideration for any monitor task at ambient and occupationally important concentration levels.

Both active (AD-h) and passive dosimetry systems (PD-h) are available for the measurement of the halogenated anesthetic gases. As was the case with nitrous oxide dosimetry, a dosimeter of either type will clearly be the very best method available for documenting a single individual's exposure to these materials. Dosimetry systems that will permit the use to obtain an individual's dose directly on site and at the conclusion of any identified time interval are currently under development. Such systems will eliminate the requirement for sending exposed dosimeters to some remote location or laboratory for evaluation.

Fixed gas chromatographic systems (GC-h) are also very capable of completing the analysis for any, or all, of these halogenated anesthesia agents. Gas chromatographs designed to analyze for these materials could do so successfully using either a flame ionization detector (FID) or a photoionization detector (PID). MDLs for any of these gases would be well below 1.0 ppm. For this application, one would choose a chromatographic column packed with a low polarity, porous polymer such as Graphitized Carbon, Gas Chrom 220, Poropak Q, or Chromosorb 102. Successful analytical results could be obtained by operating the chromatograph's oven either isothermally or under temperature programming. Although this analytical method will almost certainly provide the very highest levels of analytical sensitivity and selectivity, gas chromatographic systems will always be more difficult and expensive to calibrate, maintain, and operate.

References

1. *Guide to Occupational Exposure Values—1993*, compiled by the American Conference of Government Industrial Hygienists, 1993.
2. Foley K. AANA journal course: Update for nurse anesthetists—occupational exposure to trace anesthesias: Quantifying the risk. *J Am Ass Nurs Anesth* 61(4):405–412, August, 1993.
3. Buring JE, Hennekens CH, Mayrent SL, Rosner B, Greenberg ER, and Colton T. Health experiences of operating room personnel. *Anesthesiology* 62(3):325–330, March, 1985.
4. Green CJ. Anesthetic gases and health risks to laboratory personnel: A review. *Lab Anim* 15(4): 397–403, October, 1981.
5. Gardner RJ. Inhalation anesthetics—An update. *Tox Subst Bull* 9:11, June, 1988.
6. Gardner RJ. Inhalation anesthetics—An update. *Tox Subst Bull* 17:3, December, 1991.
7. Rogers B. A review of the toxic effects of waste anesthetic gases. *J Am Ass Occup Health Nurs* 34(12):574–579, December, 1986.
8. Sardas S, Cuhruk H, Karakaya AE, and Atakurt Y. Sister-chromatid exchanges in operating room personnel. *Mutat Res* 279(2):117–120, May, 1992.
9. Ghantous HW, Fernando J, Gandolfi AJ, and Brendel K. The toxicity of the halogenated anesthetic gases in guinea pig liver slices. *Toxicology* 62(l):59–69, May, 1990.
10. Sutherland DE and Smith WA. Chemical hepatitis associated with occupational exposures to halothane in a research laboratory. *J Vet Hum Toxicol* 34(5):423–424, October, 1992.
11. Lind RC, Gandolfi AJ, and Hall PD. Hepatotoxicity of subanesthesia halothane in the guinea pig. *J Anesth Analg* 74(4):559–563, April, 1992.
12. Lings S. Halothane related liver affections in an anaesthetist. *Brit J Ind Med* 45(10):716–717, 1988.
13. Toker K, Ozer NK, Yalcin AS, Tuzuner S, Gogus FY, and Emerk K. The effect of chronic halothane exposure on lipid peroxidation, osmotic fragility and morphology of rat erythrocytes. *J Appl Toxicol* 10(6):407–409, December, 1990.
14. Baeder C and Albrecht M. The embryotoxic/teratogenic potential of halothane. *Int Arch Occup Environ Health* 62(4):263–271, June, 1990.
15. Carlson P, Ekstrand J, and Hallen B. Plasma fluoride and bromide concentrations during occupational exposure to enflurane and halothane. *Acta Anaesthesiol Scand* 29(7):669–673, 1985.
16. Plummer JL, Hall PD, Jenner MA, Ilsley AH, and Cousins MJ. Hepatic and renal effects of prolonged exposure to rats to 50 ppm methoxyflurane. *Acta Pharmacol Toxicol* 57(3):176–183, September, 1985.
17. Gilly H, Lex C, and Steinbereithner K. Anesthetic gas contamination in the operating room—An unsolved problem? *Anaesthesist* 40(11):629–637, November, 1991.
18. Girompaire D, Landais A, and Allegrini D. Pollution of operating rooms. *Cahiers D'Anesthesiologie* 39(4):253–256, 1991.
19. Gardner RJ. Inhalation anaesthetics—An update. *Tox Subst Bull* 17:3, December, 1991.
20. Snef L, Stapf F, Mueller W, and Schimmel G. Investigations on operating theatre staff with long-term exposure to halothane. *Zeitschrift fiir de Gesamte Hygiene und ihre Grenzgebeit* 35(8):480–483, September, 1989.
21. Breum NO and Kann T. Elimination of waste anaesthetic gases from operating theatres. *Acta Anaesthesiol Scand* 32(5):388–390, July, 1988.
22. Sonander H, Stenqvist O, and Nilsson K. Nitrous oxide exposure during routine anaesthetic work. *Acta Anaesthesiol Scand* 29(2):203–208, February, 1985.
23. Carlsson P, Ljungqvist B, and Hallen B. The effect of local scavenging on occupational exposure to nitrous oxide. *Acta Anaesthesiol Scand* 27(6):470–475, December, 1983.
24. Dula DJ, Skiendzielewski JJ, and Snover SW. The scavenger device for nitrous oxide administration. *Ann Emerg Med* 12(12):759–761, December, 1983.

25. Thompson JM, Sithamparanadarajah R, Hutton P, Robinson JS, and Stephen WI. Evaluation of the efficacy of an active scavenger for controlling air contamination in an operating theatre. *Brit J Anaesth* 53(3):235–240, March, 1981.
26. Witner CG, Hamm G, Hueck U, Apel G, and Lamprecht E. Toxicity risk in the exposure of surgical personnel to gaseous anaesthetics. *Zeitschrift fiir de Gesamte Hygiene und ihre Grenzgebeit* 33(12):622–627, December, 1987.
27. Gardner RJ. Inhalation anaesthetics—Exposure and control. *Ann Occup Hyg* 33(2):159–173, 1989.
28. Burkhart JE and Stobbe TJ. Real-time measurement and control of waste anesthetic gases during veterinary surgeries. *Am Ind Hyg Assoc J* 51(12):640–645, December 1990.
29. Gardner RJ, Hampton J, and Causton JS. Inhalation anaesthetics—Exposure and control during veterinary surgery. *Ann Occup Hyg* 35(4):377–388, 1991.
30. Pothmann W, Shimada K, Goerig M, Fuhlrott M, and Schulte am Esch J. Pollution of the workplace by anesthetic gases, causes and prevention. *Anaesthesist* 40(6):339–346, June, 1991.
31. Burroughs GE and Woebkenberg ML. NIOSH/DPSE. Nitrous oxide analyses—Criteria for a recommended standard, occupational exposure to waste anesthetic gases and vapors. *NIOSH Method No. 6600*, February 15, 1984.

Anesthesia Mask Gas-Scavenging System

Anthony Schapera and William Charney

Introduction

The significance of contamination of the operating room by anesthesia gas is controversial. Several studies and reviews have described an increased risk to the health of operating room staff in association with exposure to waste anesthetic gases.[1–4] It is reasonable to assume that reducing environmental contamination will lessen the health risk to operating room staff. The National Institute of Occupational Safety and Health (NIOSH)[5] recommends that levels of N_2O in the operating room do not exceed 25 ppm over the course of a general anesthetic. In our operating rooms, levels of N_2O measured in the immediate area or "breathing zone" of anesthesiologists often ranged from 80 ppm to greater than 200 ppm during delivery of inhalation anesthesia by an anesthesia mask. We therefore sought a way in which to reduce this level of contamination.

Leaks from a semi-closed breathing circuit will reduce the effectiveness of any anesthesia gas-scavenging system. Common sources of leaks are at the CO_2 absorber gaskets, in the anesthesia ventilator bellows, at the pressure release valve ("pop-off valve"), at the reservoir bag or at the high pressure gas connections at the central gas manifold. These leaks can be eliminated by routine maintenance checks and preanesthesia equipment checks. An additional major leak, however, occurs from beneath the edge of an anesthesia mask during administration of inhalation anesthesia. Both the contour of the patient's face and the design of the mask may contribute to an imperfect seal between the mask and face, allowing leakage of gas from beneath the edge of the anesthesia mask. Anesthesia masks have a soft inflatable rim which improves the seal but the fit is often imperfect, especially in patients with thick beards or those with loss of subcutaneous tissue around the cheek area.

The purpose of this study was to test the effectiveness of a device designed in order to prevent anesthesia gases that had escaped from beneath the edge of the anesthesia mask from contaminating the operating room environment.

Methods

The scavenging device (Figures 17.1 and 17.2) consists of a conventional anesthesia mask with a soft inflatable rim (King Systems, Model #1065, size 6, large adult mask). A 5 mm internal diameter 6 ft

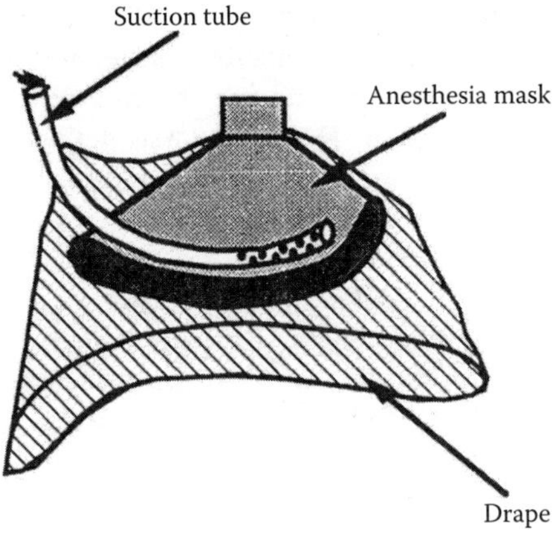

FIGURE 17.1 Anesthesia mask with scavenging attachment.

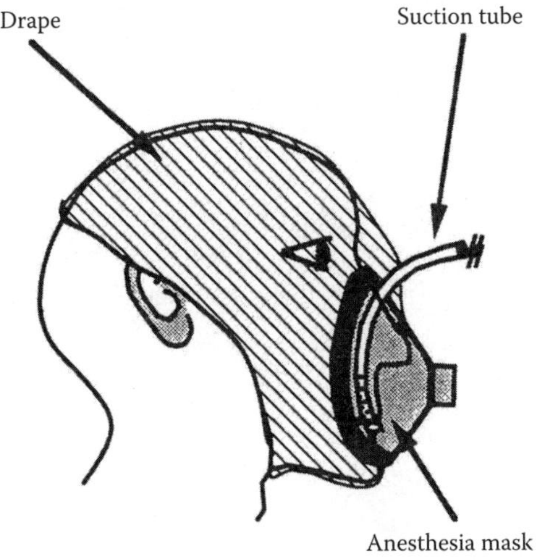

FIGURE 17.2 Method of applying scavenging drape over patient's face.

length of suction tubing is attached with one open end adherent to the right side of the anesthesia mask. A distal inch of the suction tubing has several side-holes in order to allow continued suction flow if the tip were to be occluded. A clear polyethylene plastic drape of approximately 12 in. diameter and 8 in. tall is attached to the anesthetic mask over the suction tubing. The free rim of this drape is tailored to conform to the contours of a normal adult face, and the edges are elasticized to provide a more effective seal where it contacts the skin. The suction tube exits through an opening in the plastic drape and is connected to a source for vacuum (300–500 mmHg). The anesthesia face mask is applied in the usual fashion during induction and maintenance of anesthesia. The suction tubing is connected to the source of suction, and as anesthesia is induced, the clean plastic drape is brought over the patient's face and adjusted so as to provide the best negative pressure beneath the edge of the face mask will be scavenged. In the event that the anesthesiologist needs to gain rapid access to the patient's face, the drape may be easily removed.

We studied the effectiveness of the mask gas-scavenging system in five healthy patients undergoing elective surgery with inhalation anesthesia that included N_2O. The study was approved by the University of California San Francisco Committee on Human Research and informed consent was obtained from all study subjects.

The operating rooms used ranged in size from 8,000 to 10,000 ft^3 and all rooms underwent 10 complete air changes per hour.

A Miran 1A infrared spectrophotometer (calibrated to manufacturer's specifications) was used to measure N_2O levels. This has a detection limit of 1 ppm. We first sampled the operating room air for background levels of N_2O and next tested the breathing circuit for leaks by pressurizing the anesthesia circle system with N_2O. We sampled for N_2O at various locations in the breathing circuit and at scavenger reservoir bag. If obvious leaks were found, we corrected the leak or, if not correctable, we did not conduct the study in that operating room.

Anesthesia was then induced in the test patients with a combination of intravenous agents and volatile inhaled agents (halothane or isoflurane as determined by the anesthesiologist) in oxygen. The test period began when a steady state of anesthesia had been achieved and the anesthesiologist was satisfied that the airway was secured. An oral or nasal airway was used at the discretion of the anesthesiologist, but if used was left in place throughout the study period.

For the study period, patients breathed spontaneously while anesthesia was maintained with either halothane or isoflurane and an N_2O (3 L/min)/O_2 (2 L/min) gas mixture. For the first half hour of the study, the anesthesia mask gas-scavenging system was used to administer anesthesia.

The anesthesiologist held the anesthesia mask with the left hand and applied the mask to the patient's face in the manner preferred by the anesthesiologist. We did not advise the anesthesiologist in the method of application of the mask itself, but did ensure that the plastic shroud was draped over the patient's face in the most effective way to collect escaping anesthetic gasses. Head straps were not used. In no cases were oral or nasal esophageal stethoscopes or temperature probes used. The breathing circuit safety-release ("pop-off") valve was adjusted so that the reservoir bag on the circuit completely refilled during exhalation and partially emptied during inhalation.

Samples of air were obtained continuously from the environment adjacent to the anesthesia mask (~2 ft from the anesthesia mask), and the highest N_2O-specific light absorbance in each minute time frame was recorded for 30 min. The anesthesiologist was unaware of the level of N_2O being measured. After 30 min, we disconnected the scavenging device and carefully removed it from the anesthesia mask while the anesthesiologist continued to apply the mask to the patient's face as before. We continued to sample room air for N_2O over the next 30 min.

Data were described as the mean values ±1 SD for N_2O concentration over each study period.

Results

Patient characteristics are shown in Table 17.8. The scavenging system markedly reduced the average N_2O concentration over the course of 30 min of inhalation anesthesia in each case (Figure 17.3). The N_2O concentration was maintained at consistently low levels throughout the test period as illustrated in the data from patient number 5 (Figure 17.4).

TABLE 17.8 Patient Characteristics

Patient	Age	Sex	Bearded	Dentition
1	33	Male	No	Full
2	63	Female	No	Full
3	49	Male	No	Full
4	38	Female	No	Edentulous
5	48	Male	No	Edentulous

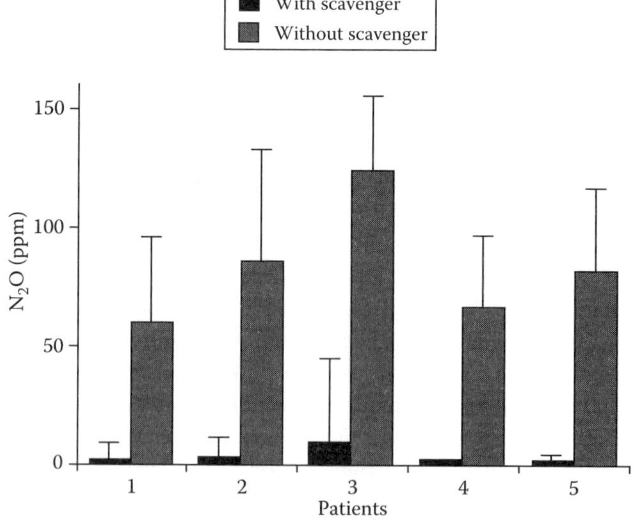

FIGURE 17.3 N_2O concentrations (mean ± SD) in breathing zone of anesthesiologist while delivering inhalation anesthesia by mask to five patients.

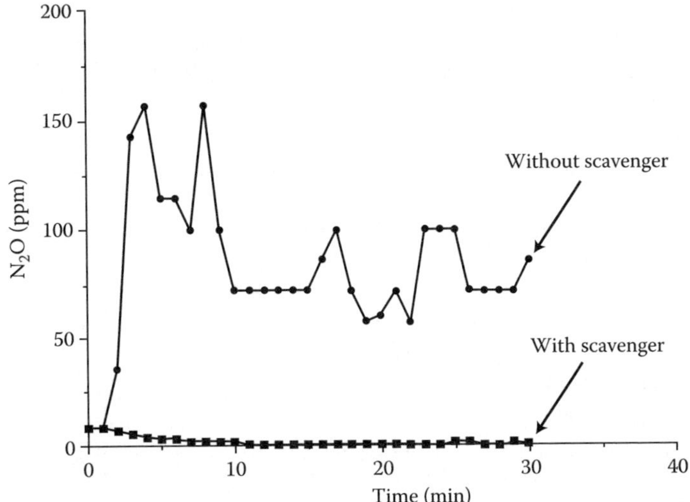

FIGURE 17.4 Plot of N_2O concentration vs. time in the breathing zone of anesthesiologist delivering inhalation anesthesia by mask to patient number five. Values shown are the highest value sampled during each minute of the test period.

Discussion

The most significant result of this study is the demonstration that the newly designed anesthesia mask-scavenging attachment reduces contamination with N_2O with the anesthesiologist's breathing zone to a concentration well below the 25 ppm level recommended by NIOSH. Without the attachment, levels of N_2O greatly exceeded the recommended limit. Using a double mask scavenging device (essentially an inner mask surrounded by an outer rigid shell into which escaping gas is evacuated), N_2O levels were reduced from 145 to 15 ppm in one study.[6] Use of this device results in a further sevenfold reduction of N_2O in the anesthesiologist's breathing zone.

The scavenging attachment has several design points that deserve emphasis. We used the main-line hospital vacuum system as a source of suction for the evacuation of escaping gases. The static suction pressure in this system is not constant, but varies between 300 and 500 mmHg and results in an air flow through 5 mm ID tubing of about 60 L/min. This rate of flow should be more than adequate to evacuate gas escaping from beneath an anesthesia mask (fresh gas flow in most anesthesia systems seldom exceeds 10 L/min).

Because most anesthetists hold the anesthesia mask with the left hand, the right side of the mask is usually less effectively sealed against the patient's face. We attached the suction tubing to the right hand side of the anesthesia mask in order to more effectively contain and evacuate gas escaping from the right edge of the mask. The distal inch of the suction tubing has several side-holes in place in order to minimize occlusion of the suction by the plastic drape being sucked up against a single orifice.

The plastic drape itself is of clear polyethylene plastic. The contour is designed so that the edges fit well over the occiput and chin reducing the chance of slippage of the edges during use. A loose elasticized edge assists the fit of the drape. Although the drape fits well during use, its overall fit is fairly loose so that it can be instantly removed if necessary. The clear plastic enables the anesthetist to see the patients face clearly during the procedure. The patient's eyes should be taped shut during use of the device in order to prevent corneal abrasion by the plastic. In contrast to the double mask device,[6] the attachment of the drape does not add to the bulk of the anesthesia mask, an important factor in minimizing hand fatigue when delivering inhalation anesthesia by mask.

Conclusions

We have demonstrated a new scavenging device that effectively reduces environmental contamination with anesthetic gases during inhalation anesthesia delivered by a standard anesthesia mask.

References

1. Cohen EN, Bellville JW, and Brown BW. Anesthesia, pregnancy and miscarriage: A study of operating room nurses and anesthetists. *Anesthesiology* 35:343–347, 1991.
2. Report of an ad hoc committee of the effect of trace anesthetics of the health of operating room personnel. American Society of Anesthesiologists. Occupational disease among operating room personnel: A national study. *Anesthesiology* 41:321–340, 1974.
3. Corbett TH, Cornell RG, Endres JL, and Lieding K. Birth defects among children of nurse anesthetists. *Anesthesiology* 41:341–344, 1974.
4. Brodsky JB, Cohen EN, Brown BW, Wu ML, and Whitcher CE. Exposure to nitrous oxide and neurologic disease among dental professionals. *Anesth Analg* 60:297–301, 1981.
5. National Institute for Occupational Safety and Health. Criteria for a Recommended Standard—Occupational Exposure to Waste Anesthetic Gases and Vapors. *DHEW Pub. No (NIOSH)* 77–140. Cleveland, US Department of Health, Education and Welfare, National Institute for Occupational Safety and Health; page 3, 1977.
6. Reiz S, Gustavsson A-S, Häggmark S, et al. The double mask—A new local scavenging system for anaesthetic gases and volatile agents. *Acta Anaesthesiol Scand* 30:260–265,1986.

Personal Exposure to Blood Aerosols among Operating Room Personnel

Patricia A. Heinsohn and D. L. Jewett

Introduction

An exposure assessment was performed based on the same underlying assumption as that of universal precautions, a clinical infection control technique. Universal precautions involve using a prescribed set of operating procedures in patient care based on the assumption that all blood and body fluids are potentially bloodborne pathogen infective.[1] Since all blood and body fluids are presumed to be infectious, the presence of the pathogen does not have to be demonstrated. Exposure to blood constitutes exposure to any bloodborne pathogen present. Exposure to hemoglobin, present in the red blood cells suspended in whole blood, constitutes exposure to blood.

Universal precautions are intended to prevent exposure via skin splash, parenteral inoculation, and exposure to the mucous membranes of the eye, nose, and mouth. The barrier precautions and techniques used in universal precautions are intended to control these exposures, because they involve known or suspected routes of transmission of human immunodeficiency virus (HIV) or other bloodborne pathogens; needlestick is a common example. Even so, the adequacy of universal precautions has been questioned. For example, the use of gloves and other protective devices does not appear to offer substantial protection against hepatitis B virus (HBV) exposure in oral surgeons.[2]

The use of personal protective equipment prescribed under universal precautions does not, and was never intended to, eliminate blood aerosol exposure. Surgical masks[3] are generally ineffective at filtering particles below 5 μm diameter, and do not seal to the face. Unfortunately, they are perceived by many health care workers as respiratory protective devices.

The inhalation route of exposure to bloodborne pathogens via blood-containing aerosol has not been extensively studied. Of the 10 bloodborne pathogens listed in the Federal Register,[4] the three of most significance to health care workers include *Mycobacterium tuberculosis,* HBV, and HIV. Of these, *M. tuberculosis* is the only pathogen shown unequivocally to be transmitted by inhalation. It is quite infectious; one tubercle bacillus is capable of initiating infection in a susceptible host. Respiratory transmission of HBV has been suspected and investigated by Petersen, who sampled for hepatitis B surface antigen and hemoglobin in renal dialysis centers. Failure to detect either one in the air samples[5] indicates a failure to collect a blood aerosol. Failure to collect any blood aerosol does not necessarily mean that none was generated or that personnel were not exposed to blood aerosol. The data may be explained by an inability to meet the air sampling limit of quantitation given the sampling and analytical methods described.

It remains troubling that HBV oropharyngeal transmission was reported in personnel wearing surgical mask protection in the preparation of human blood products.[6] More recently, Reingold demonstrated a strong correlation between years in practice and seropositivity among oral surgeons; this correlation was unaffected by the use of gloves, face masks, or eye protection. Oral surgeons, like other surgeons, use powered instruments capable of aerosol generation. The significant association with the number of years in practice could be explained by a continuous low dose aerosol exposure.

There are no epidemiologic data suggesting that aerosol transmission of HIV occurs. Further, it may be impossible to demonstrate aerosol exposure as a risk factor by present epidemiological study design, given that exposed personnel are also exposed to known risk factors. The household contact study[7] suggests that transmission involving naturally produced aerosols (coughs and sneezes) does not result in disease transmission of HIV. However, such aerosols do not generally contain blood. Coughs and sneezes by nature are random events, unlike the intentional mechanical aerosol generation performed in the operating room, which involves exposure on a routine basis.

A theoretical basis of respiratory infection with HIV exists. The alveolar macrophages carry the glycoprotein CD4 receptor site, by which HIV glycoprotein 120 is known to infect cells. HIV can infect alveolar macrophages *in vitro.*[8] HIV is known to replicate in alveolar macrophages, which serve as a reservoir of infection in that they do not succumb to the cytotoxic effect of HIV infection.[8] In addition, a laboratory strain of HIV-1 was shown to survive aerosolization and remain T-cell culture infective.[9] Therefore, theoretically, an occupational exposure to blood aerosol small enough to penetrate to the alveoli could result in binding of the HIV glycoprotein 120 of viable HIV to the CD4 receptor sites on the surface of alveolar macrophages, with subsequent macrophage infection.

The mucous membrane lining of the upper respiratory tract may also be a target site of infection. CDC/NIH cautions that, at least in the laboratory setting, "the skin…and mucous membranes of the eye, nose, mouth, and possibly the respiratory tract should be considered as potential pathways for entry of the virus."[10] Although the existence of a pathway certainly does not guarantee infection, it would also be prudent to prevent aerosol exposure to the upper respiratory tract.

An obvious and fundamental epidemiologic principle requires that exposure data be collected before any association between exposure and disease can be made. Where blood splash and needlestick are readily recognized, personal blood aerosol exposure must be determined by sampling from the breathing zone. Thus, there is an obvious need to identify personnel at risk of exposure and quantify their exposure.

Methods

The sampling method used was chosen to yield size-selective hemoglobin results. It was undesirable to characterize only the respirable mass fraction, since deposition in the upper respiratory tract may be hazardous. Breathing zone samples were taken during surgery with Marple personal cascade impactors (Andersen Instruments, Inc., Atlanta, GA) and analyzed with Hemastix (Miles, Elkhart, IN). Both the sampling and analytical methods are discussed in detail elsewhere.[11] The surgeries monitored included orthopaedic, urologic, vascular, and cardiac surgery. Personnel monitored consisted of surgeons and primarily first assistants.

TABLE 17.9 Breathing Zone Hemoglobin Concentrations by Specialty (μg/m^3)

Specialty/Task	Mean	SD	Range	*N*
Orthopedic				
Surgeon	2.15	2.69	0.17–7.32	6
Assistant	2.30	1.63	0.51–4.90	8
Urologic				
Surgeon			0.26–0.51	3
Assistant			0.276–0.35	2

Results

The breathing zone blood aerosol exposures ranged from 0.17 to 7.32 μg hemoglobin/m^3, with an arithmetic mean and SD of 1.39 and 2.09 μg/m^3, respectively, among all surgeons ($n = 11$) monitored. The exposure concentrations among all assistants ranged from 0.26 to 4.90 μg/m^3, with an arithmetic mean and SD of 1.78 and 1.54 μg/m^3, respectively, among all assistants ($n = 12$) monitored.

Table 17.9 shows the summary statistics among orthopedic personnel and urology personnel. With respect to the size fraction and hemoglobin relationship, stage 2 (effective cut-off aerodynamic diameter or ECAD = 15 μm) of every sample was positive at some hemoglobin concentration.

Seventy-five percent of all samples for stage 5 (ECAD = 3.5 μm) were positive and 40% of all samples for stage 8 (ECAD = 0.8 μm) were positive. Breathing zone samples from all personnel monitored contained inhalable blood aerosols.

Discussion

The overall mean exposure is actually higher for the assistant than the surgeon, but the range indicates that the surgeon experienced a higher individual exposure within this data set. When the results are broken down according to the two most frequently monitored specialties, the surgical assistant in orthopedic surgery has a higher mean exposure than the surgeon, but the range is greater for the surgeon. It cannot be concluded, however, that higher exposures occur in orthopedic surgery than in urology, because of the small sample size.

Other operating room employees not sampled may be at risk of exposure, e.g., the sterile nurse, circulating nurse, and anesthesiologist. The surgeon and first assistant were chosen as the most likely to be at greatest risk, because they are closest to the site of aerosol generated as a result of their job duties.

Sampling times ranged from 105 to 390 min. As such, they reflect only the total exposure concentration incurred over the length of the one operation. Thus, these results do not reflect the total exposure for personnel who participate in more than one operation per shift.

The data document an exposure to blood aerosol of a size distribution capable of being inhaled and deposited throughout the respiratory tract, including the gas exchange region. Control methods must be implemented if the exposure is unacceptable. Although engineering controls are preferable, it is not readily apparent that an effective local exhaust system could be designed that could be used in all operative situations. It has also been shown that drawing in and away from the operative site (as local exhaust ventilation would do) increases the risk of nosocomial infection. Research involving specific surgical tools needs to be done to determine effective means of aerosol reduction without compromising performance. Administrative controls, such as rotating personnel, are not practical or may be impossible. The remaining control, use of respiratory protective equipment (RPE), although the least desirable, is the most feasible until the aforementioned research is done.

Unfortunately, the selection of the appropriate RPE is not as simple as it may seem. The use of most RPE in the sterile field of the operating room necessarily requires that it be sterile for each use to protect against any particles that might be shed into the sterile field. Steam sterilization and/or ethylene oxide sterilization may bring about the deterioration of the elastomeric parts of nondisposable industrial RPE. The hospital infection control committee may not accept common decontamination procedures (e.g., surface decontamination with hypochlorite, iodophor, or quaternary ammonium solutions) used by industrial hygienists to clean respirators.

Sterilization of non-disposable RPE is also impractical. The ideal product would include a disposable sterile mask, helmet, or hood. A sterilizable mask, helmet, or hood would also be satisfactory but could result in higher costs; the industrial hygienist administering the respiratory protection program must have adequate time and resources for inspection of equipment after each sterilization and before each use.

The two air-purifying devices pictured in Figures 17.5 and 17.6 are manufactured in the United States. They are marketed as infection control devices rather than RPE. Such devices were originally made for the operating room to offer maximum protection for the particularly compromised or susceptible patient from the flora exhaled by the wearers, and ambient microorganisms. They accomplish that protection by high efficiency particulate air (HEPA) filtering of the air exhaled by the wearer and the excess air supplied by the blower. The manufacturers have added a HEPA filter to the supply air to provide some measure of protection for the wearer.

Unfortunately, neither of these devices is a NIOSH/MSHA-approved respirator. Such approval would demonstrate that the device met a minimum level of performance as a respiratory protection device. Although no NIOSH/MSHA approvals exist for respirators used specifically to control infectious aerosol exposures, there are approvals given for RPE used to control other particulate hazards.

One such approved respirator is a disposable negative pressure air-purifying half mask respirator capable of HEPA filtration. It is not sterile but need not be in that it covers approximately the same area as the surgical mask. This half mask would have to be worn with a surgical mask over it, thus covering the exhalation valve, to offer the standard level of protection to the patient. Filtering exhausted air would be required of any RPE used in the operating room. This is obviously a limitation of all industrial RPE presently available.

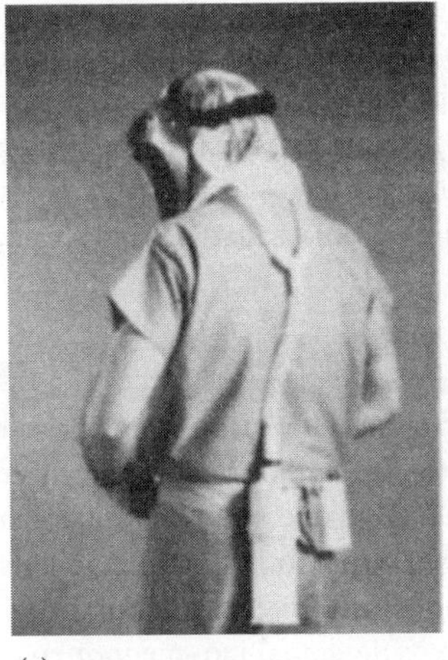

(a)

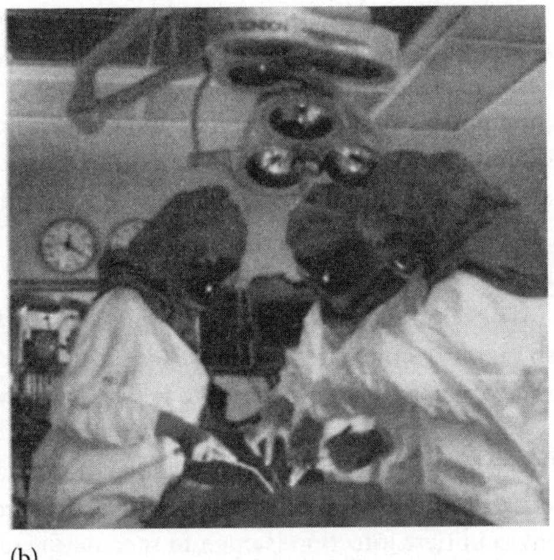

(b)

FIGURE 17.5 DePuy Sterile™ View System: (a) rear view of the device as worn prior to donning sterile hood and (b) the device in use as worn with sterile hood covering. (From DePuy Sterile™ View System, Warsaw, IN.)

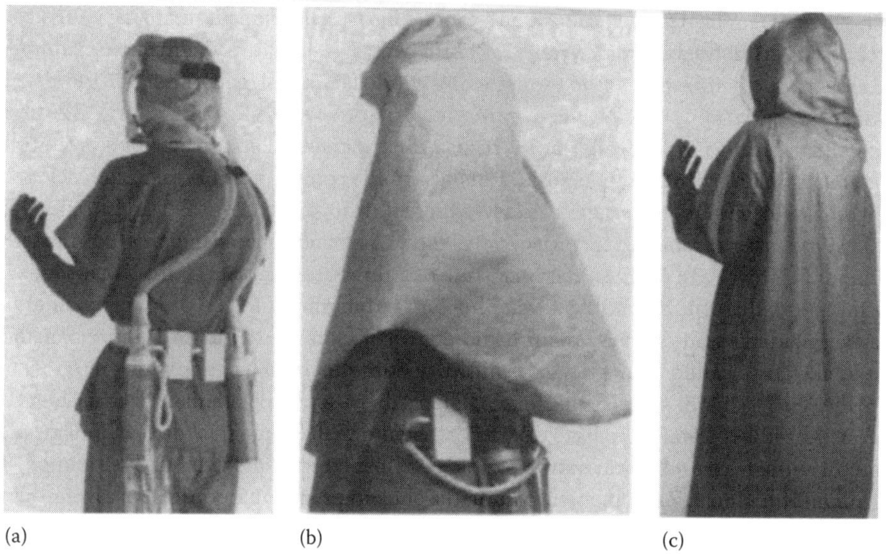

FIGURE 17.6 Stackhouse Freedom Mark III Surgical Helmet System: (a) rear view of the device as worn prior to donning sterile hood or gown, (b) the device with sterile hood donned, and (c) the device as worn with gown. (From Stackhouse Freedom Mark III Surgical Helmet System, Riverside, CA.)

Infectious or potentially infectious aerosols are also particulate by definition. It is believed that the present testing and certification procedures for other particulate approvals are applicable to infectious aerosols. This is one of the issues being examined by a new subcommittee of the ANSI Z88.2 respiratory protection committee. Another issue is the apparent failure of HEPA filters when wet.[12]

Another dilemma in proper RPE selection is deciding what level of protection the present data warrant. Further, direct application of the assigned protection factors may not be appropriate. Assigned protection factors are based on mass penetration data as opposed to the number of particles penetrating. Yet, it is the number of viable organisms inhaled and deposited, which is critical.

In the absence of a prescribed decision logic for RPE selection, consistent approvals, and a defined safe level of exposure, user compliance issues and costs may dominate the decision. Generally speaking, wearers tend to dislike the increased resistance to breathing when using negative pressure air-purifying RPE. The sterility requirement is a limitation of industrial RPE presently available.

Use of RPE must be done as part of a respiratory protection program[13] and requires oversight by a respiratory protection officer.

Conclusion

Although the data reported here do not constitute a complete exposure assessment, they do demonstrate that surgeons and assistants are at risk of exposure to blood aerosols, which may contain bloodborne pathogens for which surgical masks offer inadequate protection. Additional sampling should be done across all specialties and employees. Interpretation of these data beyond documenting exposure is difficult due to the lack of information on the infectivity of HIV, HBV, and other bloodborne pathogens via the inhalation route. Whether or not blood aerosol exposure could constitute a deposited and absorbed dose sufficient to initiate infection is open to speculation. Yet professional judgment often dictates that industrial hygienists assess occupational exposure to suspected airborne hazards and/or recommend control measures to minimize or prevent exposures before any hazard is proven.

The personal protective equipment and techniques used as part of universal precautions are inadequate as control measures for blood aerosol exposure. Engineering and administrative controls are not feasible control methods at the present time. The only viable control remaining is the use of RPE, which is an entirely acceptable practice prior to implementation of effective engineering controls. Careful consideration must be given to the proper selection of appropriate RPE.

Acknowledgments

This work was supported in part by a grant funded through a NIOSH and ERC, Inc. cooperative agreement.

References

1. Centers for Disease Control. Recommendations for preventing transmission of infection with human T-lymphotropic virus type III/lymphadenopathy-associated virus in the workplace. *MMWR* 34:681–686, 691–695, 1985.
2. Reingold AL, Kane MA, and Hightower AW. Failure of gloves and other protective devices to prevent transmission of hepatitis B virus to oral surgeons. *JAMA* 259:2558–2560, 1988.
3. Tuomi T. Face seal leakage of half masks and surgical masks. *Am Ind Hyg Assoc J* 46(6):308–312, 1985.
4. 29 CFR 1910 Occupational Exposure to Bloodborne Pathogens; Proposed Rule and Notice of Hearing, Tuesday 30, 1989.
5. Petersen NJ. An assessment of the airborne route in hepatitis B transmission. *Ann NY Acad Sci* 353:157, 1980.
6. Kuh C and Ward WE. Occupational virus hepatitis: An apparent hazard for medical personnel. *JAMA* 143(7):631–635, 1950.
7. 29 CFR 1910 Occupational Exposure to Bloodborne Pathogens; Proposed Rule and Notice of Hearing, Tuesday 30, 1989.
8. Salahuddin SZ, Rose RM, Groopman JE, Markham PD, and Gallo RC. Human T lymphotropic virus type III infection of human alveolar macrophages. *Blood* 68(1):281–284, 1986.
9. Johnson GK and Robinson WS. Human immunodeficiency virus-1 (HIV-1) in the vapors of surgical power instruments. *J Med Vir* 33:47–50, 1991.
10. HIV Agent Summary Statement in Biosafety in Microbiological and Biomedical Laboratories. USDHHS PHS CDC and NIH. HHS Publication No. 88–8395.
11. Heinsohn PA and Jewett DL. Exposure to blood containing aerosols in the operating room: A preliminary study. Submitted. *J Am Ind Hyg Assoc* 54:446–453, 1993.
12. Vincent ME and Glorioso JC. Evaluation of vacuum/suction safety devices in preventing transmission of human virus pathogens. *Am Clin Lab* 8:26–29, 1989.
13. 29 CFR 1910.134 Part 11, Volume 37, Number 202. Respiratory protection.

18

Cost–Benefit

William Charney
Healthcare Safety Consulting

Donald Maynes
Equitable Safety Solutions

Cost–Benefit Analysis in Health Care **18**-1
Introduction
Politics of Cost–Benefit **18**-2
Formulas for Calculating Cost–Benefit Ratios **18**-2
Culture of Safety
References **18**-3
Making the Business Case for Injury Prevention in Health Care—In Control? **18**-3
Modeling Control • Taking Control • Process
References **18**-10

Cost–Benefit Analysis in Health Care

William Charney

Introduction

Workers on the job injury protection in health care are seldom discussed as an ethical issue. Rather the real discussions are about money, return on investment, making a profit, etc. If a prevention program is expensive and budgets for the program are not easily achievable, the program is seldom implemented and workers' health and safety are sacrificed. Another denial apparatus used by health care systems to refuse to adequately fund health and safety for its employees is that it will take needed dollars away from patient safety programs. This attitude in itself is a self-defeating formula as published science now concludes employee safety is critical to controlling negative patient outcomes.[1]

There are many citations in the literature that prove health and safety programs are cost–benefit positive.[2–5] Despite this, many health care delivery systems claim that they have no money to invest in protecting their employees. Prevention is always cheaper than paying for the injury. This truism becomes lost in an administrative inability to understand and incorporate the real cost of the injury into a cost–benefit analysis. This is happening for several reasons:

- Lack of study and understanding of the science of cost–benefit by industry CEOs and CFOs
- Lack of study and an inability to translate the dollar costs due to decrease in productivity because of an injury
- A lack of understanding of the monetary impact of indirect costs for an injury, which can often be considered 4× the direct costs
- A lack of understanding of the impact on patient care when health care workers are injured

It is striking that despite the hundreds of millions of dollars of incurred costs in the hospital industry due to occupational injury, hospital officials are not reacting boldly. Hospitals and nursing homes continue to incur high rates of injury, higher rates in most states, according to Bureau of Labor Statistics

(BLS) than mining, agriculture, manufacturing, construction. Almost 1 in 10 health care workers reports a lost day injury per year (10%) if the two NAIC codes are added (e.g., hospital rates in 2003 8.5 per 100FTE + Nursing Homes 13.7 per 100FTE yields 23.2 per 200FTE or 10%).

The real loss in dollars is in billions when all the hard and soft costs are added. The loss of productivity, morale, and replacement costs is extremely high. The uninsured costs commonly referred to as "indirect costs" amount to a loss for each specific facility to 4× the direct costs. These include

- Productivity losses due to an injured workers absence and reduction in coworkers productivity
- Lost workdays: (wages × hours lost)
- Lost time for managers filling out paperwork
- 21% lost productivity for back injured worker following injury
- Overtime paid to others during lost workdays
- Personnel and training time to hire replacements
- Cost of emergency equipment
- Cost for light duty program management
- Claims processing costs
- Recruitment cost

A number of researchers have attempted to estimate the magnitude of indirect costs to employers. Estimates of the size of indirect costs are in a wide range from 10% to 2000% of direct costs.

Politics of Cost–Benefit

The United States is 27th on the list for providing per capita occupational safety for its workers at $3.97 per worker. In this age of the nursing shortage with thousands of unfilled positions, it is of utmost importance to integrate the concept of a safe working environment into the equation of the nursing shortage. In an OSHA testimony hearing on the Federal Ergonomics Bill, it was stated that the United States loses 5% of its nurses annually because of disabling ergonomic injury. In a survey conducted by the American Federation of Teachers, 56% of nurses surveyed said that they would leave the profession due to the excess of stress and physical demands. Turnover rates in some nursing homes are at 100% and the national turnover rate of RNs in acute care is 14%. Vacancy rates, absenteeism due to job dissatisfaction, inability to recruit new employees are all testimonies that the status quo ante needs to change. To accomplish this, a scientific cost–benefit approach must be taken.

Formulas for Calculating Cost–Benefit Ratios

Some formulas in use give compelling arguments as to what a competent safety program can save.

Profit ratio formula: Estimated cost saving of the program divided by the profit ratio = dollars not having to be billed. Hospital A spends $170,000 on a back prevention program and calculates the 1 year saving of workers compensation costs to be $54,000. The $54,000 is divided by the profit ratio of 2.8% (mean of profit ratio in health care) equals 2 million dollars the system does not have to bill out to cover the cost of the injuries.

Turnover rate formula: Since turnovers are extremely expensive, it is important to have a calculation for determining the success at lowering turnover after implementation of a program:

Formula: Number of terminations or loss of employees divided by average # of FTEs × 100

Vacancy rate formula: Open FTEs/current FTEs + Open FTEs

Calculating for absence rate: # of days absent in a month/# of average employees during the month × # of workdays

Workers compensation cost per employee: Total WC costs for year/average # of employees

Cost rate formula: Cost rate = dollars actually spent/100FTE. To calculate cost rate numbers, use cost rate = ($$ spent in an area per year) × 200,000 h worked/100FTE then divided by number of hours worked in an area per year

This formula actually lets you compare dollars saved after implementation of a program.

These above formulas will assist in calculating the progress and success of any given occupational health program implemented. Despite the copious number of peer review science (refer to Nelson, Fragala, Collins, Charney, et al.) on the positive cost–benefit of implementing prevention programs, the difficulty in convincing systems to invest is high. Nelson cites that purchasing equipment to prevent back injury will pay for itself within 12–18 months. A typical savings in workers compensation are between $300 and $1000 per year per bed with a pay back period of about 2 years. Savings will depend on injury experience (SEIU data 1199, New Jersey).

Culture of Safety

Building a true culture of safety means investing on both sides of the health care equation equally, staff and patients. In an anecdotal survey done in the state of Washington safety budgets amounted to one-tenth of 1% of the total operational budget (0.0001%). If we accept the philosophy that "you get what you pay for" this amount buys very little safety. By not protecting staff, an increase and direct relationship exists for downstream negative patient outcome. So the loop is complete, not investing in employee protection programs yields a higher rate of patient dissatisfactions. By not understanding the science of cost–benefit, by not be able to mathematically evaluate safety programs, health care is left in a fog.

References

1. Charney W. and Schirmer J. Nursing injury rates and negative patient outcomes: Connecting the dots. *AAOHN J* 55(11):470–475, 2007.
2. Villeneuve J. the ceiling lift: An efficient way to prevent injuries to nursing staff. *J Health Safety Compliance Infect Control* 2(1):19–23, 1998.
3. Santoro M. Lifting teams can help hospitals eliminate costly back injuries to nurses. *Hosp Employee Health* 13(7):81–87, 1994.
4. Fragala G. Injuries cut with lift use in ergonomics demonstration project. *Provider* 19(10):39–40, 1993.
5. Garg A. Long term effectiveness of "zero lift programs" in seven nursing homes and one hospital. Contract report no. U60/CCU512089-02, Univeristy of Wisconsin-Milwaukee, Milwaukee, WI, 1999.

Making the Business Case for Injury Prevention in Health Care—In Control?

Donald Maynes

Making the business case for safe patient handling and injury prevention efforts has long been the challenge for those who are intimately familiar with the results when such an effort is left to happenstance. Unfortunately, that represents the majority of acute health care settings in the United States today.

Management or happenstance? When it comes to safe patient handling and injury prevention techniques, those are your only two choices. It is highly interesting, at least to me, that those who choose management actually get something accomplished.[1] But of course, that's everybody in health care, isn't it?

This is not to say that such efforts are not underway—they are. Nonetheless, the odds still favor the fact that you will one day walk into a hospital that is less prepared to deal with these issues, than one that is totally capable. The injury rates of caregivers[1] continue to reflect the factual actions of those who participate in and manage this sector of the economy.

Over the course of the past 15 years, we have experienced a series of changes in the way that health care is delivered in this country. Reimbursement models are changed, and then changed again; managed care programs continue to go through metamorphoses; forms of treatment have advanced significantly in some areas; but the techniques used to actually move, pick up, slide over, and transport patients still (in the majority) dates back decades.

These all seem so elementary. The majority of hospitals are currently dependent upon outdated technology to provide safe patient handling services (Figure 18.1), caregivers are being injured at high rates, risk-related charges continue to grow overall, patients and their families are becoming increasing disgruntled,

Loss Analysis Report

For Anytown Hospital
Valued as of 10/01/07

Actual Historical
Losses by Period

Causative Agent	Claim Code	Claim Count	Incurred Values	Development Factors	Developed Incurred	Percentage Total (%)
Period: 06/01/2004–5/31/2005						
Patient handling	A	33	34,627	1,273	44,080	6
Repositioning	B	14	139,624		177,741	23
Cumulative trauma	C	6	143,994		183,304	24
Slip and falls	D	46	94,327		120,079	16
Not otherwise classified	E	172	182,876		232,802	31
Total by period		271	595,448		758,005	100
Period: 06/01/2005–5/31/2006						
Patient handling	A	20	13,465	1,514	20,386	2
Repositioning	B	19	60,940		92,262	10
Cumulative trauma	C	8	64,174		97,159	11
Slip and falls	D	39	170,963		258,839	29
Not otherwise classified	E	134	280,959		425,372	48
Total by period		220	590,500		894,018	100
Period: 06/01/2006–5/31/2007						
Patient handling	A	32	140,941	2,148	302,742	15
Repositioning	B	20	171,243		367,830	18
Cumulative trauma	C	3	199,726		429,011	21
Slip and falls	D	46	205,917		442,310	22
Not otherwise classified	E	209	235,962		506,846	25
Total by period		310	953,789		2,048,739	100
Grand totals:		801	$2,139,737			

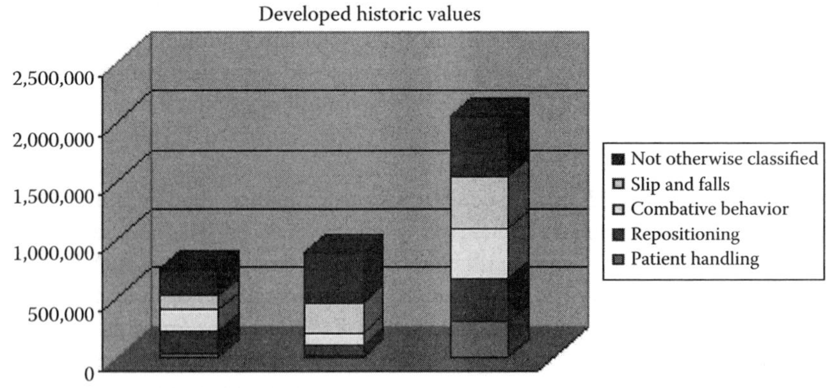

Developed historic values

2,500,000
2,000,000
1,500,000
1,000,000
500,000
0

Not otherwise classified
Slip and falls
Combative behavior
Repositioning
Patient handling

FIGURE 18.1 Caregiver injuries related to patient handling have increased over time.

and there is a nursing shortage. Hmmm! Imagine that. Seems the old cliché, "An ounce of prevention is worth a pound of cure" no longer applies. Or has it simply been forgotten or ignored for so long that we have accepted the "lack of prevention" as practice. May happenstance abound!

Nursing injury rates have been directly connected to

- The national nursing shortage
- The reduction of nursing hours at bedside
- The vast majority of reasoning behind nurses leaving the occupation
- In combination, increased negative patient outcomes.[2]

Happenstance should not be confused with "risk." The concept of risk is simple enough to understand; at least on the surface. Arguably, the word risk is generally understood to mean that there is an element of uncertainty about the outcome of any given situation. More succinctly put "Risk is a condition in which there is a possibility of an adverse deviation from a desired outcome that is expected or hoped for."[3] Too often, management thinks of the two terms as nearly synonymous in nature; they are not.

Happenstance is created by a state of uncertainty. Uncertainty can only exists in someone's mind; not as a physical attribute of a building, process, or required task. Hence, it is only logical that uncertainty leads to happenstance, which (in turn) acquiesces to resident risk, and can only have one outcome…and that one's not so good.

Management, on the other hand, can use the definition of risk to confront the "certainties" that are easily and readily identifiable. Those certainties include the requirements of care for any patient or resident of a health care facility anywhere in the world. People are in those places because they

- Are under care
- Require assistance in getting into and out of bed
- Need help to get to a chair or a bathroom
- Must gain access to specialized equipment to assist in their diagnosis or treatment

The basic tenet of nursing is centered on "care," but that has become ill-defined to include serving as the pack mule to accomplish the latter of the three above shown points; and the statistics proves it. Add "timeliness" to each of the above reflected needs, and without management, the place could deteriorate completely into chaos.

Developing a manageable alternative becomes a matter of economics. "To an economist, cost refers to foregone opportunities or alternatives—sacrificing something to get something else."[4] In the case of safety programming in general, those making the sacrifices initially, however, are those who are trying to perform the very services required. In the longer term, "implicit costs"[5] of the nursing shortage, cost of recruitment, and time spent in managing coworkers' injuries reduces the income that could be generated.

In an additional economic sense, and when arbitrary management practices lag behind obviously prudent actions, "industrial justice" steps in on behalf of groups who are being challenged in this case caregivers. Currently, there are seven states that have enacted safe patient handling laws, and at least 10 more are considering them. At the Federal Level, Representative John Conyers has presented legislation that would establish a Federal Safe Patient Handling Standard to prevent musculoskeletal disorders for direct-care registered nurses and other health care providers. On September 26, 2006, Congressman Conyers introduced the "Nurse and Patient Safety & Protection Act" (H.R. 6182). This would require OSHA to develop and implement a standard that will eliminate manual lifting of patients by direct-care registered nurses through the use of mechanical devices, except during a declared state of emergency. The legislation also requires hospitals to develop a plan to comply with the standard (with input from RNs), provides protection for RNs through refusal of assignment and whistleblower provisions, and requires the Secretary to perform audits.

These actions follow the enactment of "PL 109-41," July 29, 2005, 119 Stat 424, which is simply entitled, "Patient Safety and Quality Act of 2005." That law, in part, is what has empowered Medicare to change its position of the reimbursement for care for injuries that can be attributed to "in-house"-developed injuries.

There is an obviously direct relationship between employee injury and the increased costs of workers' compensation claims. The story does not stop there, however, because where there are high workers' compensation claims costs, typically professional liability exposure and its costs are also high. If that is not enough, there is a further relationship between professional liability costs and the overall image of the facility. That image drives Net Patient Satisfaction Scores and that, in turn, drive revenues and profits.[6]

The American Nurses Association has also studied the etymology of nurse injury, and has come to several conclusions. Injuries are being driven by

- Changing United States demographics

(Population growth, aging of the population, and the prevalence of obesity)

- Increased demands on the health care system

(Advances in technology, growing need for disease management, and disease issues associated with elderly/Bariatric patients).[7]

At the risk of sounding overly negative, the climate today has been defined as one that is fraught with problems that continue to weaken the staff of caregivers, which then creates an environment prone to needless costs, weakens the health care system generally, and simply wastes huge sums of money. It sounds like happenstance is winning.

Modeling Control

The ultimate goal of any injury prevention program consists of two objectives:

1. To enhance safety
2. To create a more efficient working environment

This combined focus leads to a more profitable operation. Safety is enhanced through the identification and mitigation of the risks resident in any given location. The workflow design is an outgrowth of the retooling of the environment required to mitigate the initial risks that cause caregiver injury. "Across the board, "integration" of programs—with core goals, with the organization's culture, and with each other—is found to be the key to success."[8]

Developing the proficiencies required to successfully operate an injury prevention program starts with a basic understanding of the ultimate desired outcome. As Yogi Berra puts it, "If you don't know where you're going, you might end up someplace else." In order to avoid that happenstance, and to better define the safe patient handling "destination," then it is imperative that we first understand the nature of the conundrum.

Over the course of the last 14 years, we have studied hospitals throughout Canada, the United Kingdom, and the United States. While the economic system that supports health care is dramatically different, the risks associated with lifting, transporting, and repositioning patients are identical. Much like those laws recently passed or currently under consideration in the United States, the United Kingdom (through their Safety Executive) has had safe patient systems mandated for nearly the past 10 years, and started 5 years before that.

"The 1992 regulations call for risky "manual" handling to be avoided "so far as is reasonably practicable". This means weighing up the risk of injury against the cost or effort required to introduce new measures. Doing nothing can only be justified if the cost of measures greatly outweighs the risk. Given the high risk of injury associated with lifting patients, the absence of a safer handling policy is almost certain to amount to a breach of the regulations. The cost of handling equipment is not only "reasonable" in legal terms (cost versus risk), but also in real terms too.

A large health employer implementing a safer handling policy might initially expect to spend 0.3% of the annual budget on new equipment. After that, the annual maintenance and replacement bill should be no more than between 0.03% and 0.04% of the budget (RCN 1999). One large civil claim from an injured nurse would amount to several times the cost of the equipment needed. There are also huge indirect costs in both financial and human terms that must also be taken into account. Good employers have little option but to work toward implementing a safer handling policy as soon as possible.[9]

While the law in the United Kingdom preceded those in the United States, that is not to say that efforts were not underway to develop strategic controls for safe patient handling, and caregiver safety. As early as 1992, and again in 1993, Doctors Arun Garg and Berniece Owens were conducting ergonomic studies on long-term care facilities. These studies looked at the requirements of not only having equipment present to mitigate the risk, but also actually putting that equipment into practice.

One of the earliest acute care settings to adopt safe patient handling efforts was the Jim Thorpe Rehabilitation Hospital, a part of Integris Health, located in Oklahoma City. "To get a better handle on preventing these common injuries, Integris launched its No-Lift policy in 1997 to shift the culture from just moving patients to moving the patients safely by using new lifting equipment."[10] In 2006, Bill Wandel, Vice President of Risk and Finance for Integris, was presented the Theodore Roosevelt Award for workplace safety; an annual award made by Risk & Insurance Magazine. In previous years, Integris also received the Award for Excellence from the Oklahoma Safety Council (2003), and the Crest Award from the Department of Labor (2004).

At the time, Integris implemented their safe patient handling program, workers' compensation losses alone were running $4.8 million annually. By 2000, those costs had been reduced to less than 50%, and today they are only a fraction of what once was a huge problem. In total, this approach has saved Integris Health over $15.0 million in the past 5 years alone.

Integris Health represents excellence in melding together the two objectives associated with safe patient handling. First, they went about equipping the hospitals with some equipment (subsequently adding to that mix) to enhance the safety of their caregivers, and then created work processes that are much more efficient than those in place prior to program implementation.

Unfortunately, Integris Health remains the exception and not the rule. Most of the hospitals we have analyzed to date (over 350) in the United States continue to experience the majority of workers' compensation claims from causes directly attributable to patient lift, transfer, and transport.

We have identified several examples of hospitals that commenced safe patient handling programs (some only in part, others comprehensive), and then later abandoned those efforts. Typically that management decision is made when the business case is not clear to new leadership. One such case recently resurfaced in our offices, and the results are really quite telling.

This particular hospital must remain nameless for obvious reasons. After all, nobody would want to lay claim to the increased costs that have been borne both by the employer and the caregivers who experienced traumatic injuries. This particular operation had claims numbers that were proximate to those at Integris, and had similar demographics. Upon the investment of over $800,000 into a safe patient handling program, they experienced claims savings of over $3.3 million in their first 3 years of operations. Following that, it appears that the claims savings began to level somewhat, but still remained at only a fraction of the comparative pre-programmed values. Then the operation was abandoned altogether when new management took control of the operation, and was seeking methods to cut FTE's. It should be only too logical to know where their claim values stand today. Yet, right back up to the pre-programmed levels. Worse yet, that particular organization experienced huge turnover rates, and had to cover those additional costs as well.

Over the course of time, programs have expanded into more locations, and the results are astonishingly similar. Earlier this year, St Mary's Hospital, Amsterdam, New York (and part of Ascension Health) was awarded the Best Practice Award for the Northeast Region of NYONE. Because of similar efforts throughout their system, Ascension Health is scheduled to be the winner of the Best Practice Award system-wide at the upcoming Safe Patient Handling Conference sponsored by the Veteran's Administration VISN 8 Program.

Still, there is work to be done. In a recent analysis of over 100 hospitals, we determined that nearly one-third of all workers' compensation claims can be attributed to patient handling tasks. So how much could be saved by that group alone? Conservatively over $30 million.

Taking Control

To exercise management of safe patient handling and its benefits, you must start by first understanding what the current situation is at your facility. That is a relatively simple task, but can take some time depending on the record keeping methods that you have in place.

Safe patient handling programs consist of

- Investigative phase
- Implementation phase
- Operations phase

The investigative phase is further broken down into two separate and distinct analyses. The first one centers on the historical workers' compensation experience, and the latter concentrates on the risks and exposures that drive those costs. We use your own experience to tip us off as to where to look for those hidden tasks and efforts that ultimately injure your most valued clinical staff.

Process

1. Hospital "Loss Runs" are obtained for the proceeding 3 years. A loss run is a detail of worker injury claims that includes date, description of the injury, what caused the injury, and the current value of the claim.
2. Each claim listed on the Loss Run is grouped into one of five categories:
   A. Patient lifting claims
   B. Patient repositioning claims
   C. Cumulative trauma claims
   D. Slip and fall claims
   E. Not otherwise classified

Experience has proven that Categories A and B generally represent the largest cost of claims. These are also the categories that can achieve the greatest reduction through the introduction of patient handling equipment and policies that support their use.

3. Once the classification of claims is complete, the expected reductions from the prevention program are then factored for each classification.

Classification	Expected Reduction (%)
Patient lifting claims	64.0
Patient repositioning claims	64.0
Cumulative trauma claims	8.0
Slip and fall claims	5.4
Not otherwise classified	2.7

The Expected Reduction percentages are based on experience as defined and approved by London Market underwriters.

4. Insurance companies use 3 years of historic data to project what the most likely outcome will be in the future. We use the same methods by determining what the average amount of claims in each category has been over the past 3 years. In other words, this value is what has happened historically, and is likely to be the case again if an injury prevention program is not put into place.
5. Since claim values change over time, we multiply the number shown on the loss run in each category by a factor that reflects the expected future change in these values. This factor, known

Future Trend

Projected Losses without Intervention

Average Values	Causative Agent	Claim Code	Number of Claims	Average Value	Developed Claim Values	Percent Of Total (%)	Projected 3 Years Total	Percent of Projected (%)
	Patient Handling	A	28	63,011	122,402	10	367,207	10
	Repositioning	B	18	123,936	212,611	17	637,834	17
	Cumulative Trauma	C	6	135,964	236,491	19	709,474	19
	Slip & Falls	D	44	157,069	273,743	22	821,228	22
	Not Otherwise Classified	E	172	233,266	388,340	31	1,165,019	31
	Total By Period		267 $	713,246	$ 1,233,587	100	$ 3,700,762	100

Additional Projected Savings

Expected 3 Years Savings	Causative Agent	Claim Code	Reduction Percentage (%)	Annual $ Reduction	Total 3 Years Savings	Percent of Projected (%)
	Patient Handling	A	64	78,338	235,013	6
	Repositioning	B	64	136,071	408,214	11
	Cumulative Trauma	C	8	18,919	56,758	2
	Slip & Falls	D	5	14,782	44,346	1
	Not Otherwise Classified	E	3	10,485	31,456	1
	Total By Period		N/A	$ 258,595	$ 775,786	21

FIGURE 18.2 Analysis of losses and projected savings over a three year period.

as a Development Factor, is supplied by either the facilities insurer, actuaries, or through various resources organizations that accumulate such information.

6. Once we have determined the “developed average,” we then determine the cost all equipment and services needed to operate an effective injury prevention program through the site and task analyses conducted on site at the facility.
7. Next, we subtract the cost of the program from the average value of projected claims.
8. The result is what claims are most likely to be in the future, even with an effective injury prevention program (Figure 18.2).

The concepts we use to help clients attain a viable financial plan are really quite simple. If the facility does nothing, as they have in the past, then claims costs will continue at the historic rate. By implementing and operating an injury prevention program, these claims will go away. In most facilities, the amounts of money that they have spent on claims historically are more than sufficient to cover the cost of prevention.

It is important to remember that the key to any injury prevention program is found within its operation; not through simply the purchase of equipment. To attain the results expected in our Loss Analytics, it is assumed that the facility will enact policies and procedures that we recommend.

Once policies are put into place, ongoing measurement of the success of the program must be maintained, monitored, and reported. It is essential that this be done by one person, and Internal Coordinator, to assure that the measurements are being consistently reviewed, and remedial actions taken at the earliest of opportunities should they become necessary.

It is an established fact that health care facilities today are challenged to find the capital that is needed to venture into anything that will ultimately assist them in more profitable operations. By using loss analytics as the guideline, proof can be shown as to the effectiveness of the application of these valuable resources.

References

1. United States Bureau of Labor Statistics, 2002.
2. Charney W and Schirmer J. Nursery injury rates and negative patient outcomes—Connecting the dots. *AAOHN J* 55(11):470–475, November, 2007.
3. Vaughn EJ and Vaughan T. The problem of risk. *Fundamentals of Risk and Insurance*. Wiley & Sons, Inc., New York, 2003.
4. Albrecht WP Jr. *Economics*, 4th edn. Prentice-Hall, Englewood Cliffs, NJ, 1986.
5. Implicit costs are opportunity costs that are not reflected in a firm's accounting statements.
6. Kim H. In his remarks. *Ascension Health Risk Management Conference*, July, 2007, Chicago, IL.
7. Hughes NL. *Providing the Evidence: Safe Patient Handling* copyright ANA, 2007.
8. Institute of Medicine, Improving Health, as presented by the National Business Group on Health, Executive Summary: Mission Control, 2007.
9. RCN Code of Practice for Patient Handling, Royal College of Nursing, Updated, October, 2007.
10. Gurevitz S. Tales of uplift, *Risk & Insurance Magazine*, November 2006.

19

Clinical Approach to Glove Dermatitis and Latex Allergy

Introduction 19-1
Illustrative Case 19-2
Diagnostic Considerations • Resolution • Summary
Epidemiology of Latex Allergy 19-5
Diagnostic Approach to Glove Dermatitis and Latex Allergy ... 19-7
Background: Pathophysiology • History and Physical Examination • Allergy Testing • Workup of Respiratory Symptoms
Prevention and Treatment 19-11
Primary Prevention • Secondary Prevention (Screening) • Tertiary Prevention (Treatment) • Related Issues
Conclusions 19-12
Acknowledgments 19-13
References 19-13

Dennis Shusterman
University of California, San Francisco

Introduction

Prevention of occupationally transmitted infectious diseases is of major concern to health care workers. In this context, increasing emphasis upon the so-called universal precautions translates into increased use of protective gloves. With the increased use of gloves has come an enhanced recognition of two general entities: glove dermatitis and latex allergy. These two diagnostic categories are overlapping, but not synonymous. Glove dermatitis is a subcategory of hand dermatitis, in this case resulting from glove-wearing. Latex allergy, on the other hand, is generally an immediate (Type I) allergic reaction to antigens from the rubber tree, *Hevea brasiliensis*.* A variety of preexisting skin diseases (and susceptibility states) may underlie a clinical state of glove dermatitis, and, at times, must be distinguished from allergy as a causal factor. Further, although skin symptoms are the most frequently recognized manifestation of latex allergy, the respiratory tract (and cardiovascular system) can also be affected, at times in the absence of skin lesions. These complex relationships are schematized in Figure 19.1, and alluded to in the following case study and ensuing discussion.

* Less commonly, allergic contact dermatitis (Type IV hypersensitivity) typically manifests itself as a reaction to antioxidants and accelerators added during the processing of gloves.

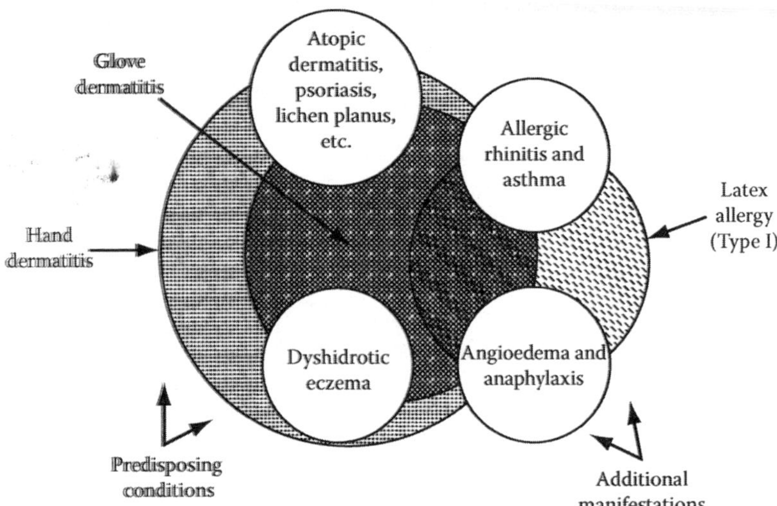

FIGURE 19.1 Schematized relationship between hand dermatitis, glove dermatitis, and latex allergy.

Illustrative Case

A 35-year-old registered nurse was seen at the employee health clinic of a university hospital for evaluation of hand dermatitis. She had a history of atopic dermatitis since adolescence but her hand symptoms (itching, peeling, cracking, and redness) had only been significant for the prior 4 years. During her 8 years in nursing, the first 4 years were spent in inpatient ward duties, and the most recent four in an outpatient endoscopy service. At the time of initial evaluation, she wore unpowdered latex examination gloves approximately 6h a day, and washed her hands as many as 20 times a day. When using disinfectants (glutaraldehyde and paracetic acid), she normally wore gloves. She had tried to use glove liners, but found that with double-gloving, the liners interfere excessively with her manual dexterity. She had not tried wearing nonlatex gloves for any extended period, but generally noted improvement of her hands when away from work for extended periods. She had used a variety of over-the-counter hand creams (e.g., Neutragena®), but without significant improvement.

The patient gave an additional history of acute itching and burning of her hands when she handled selected foods, including tomatoes, pineapple, strawberries, spinach, and raw chicken. During the course of her evaluation, she reported onset of similar acute symptoms when donning latex gloves. The patient reported hay fever symptoms for the first time the previous spring; however, she denied work-related respiratory symptoms, including rhinorrhea, nasal congestion, sneezing, cough, sputum production, or wheeze. She had no history of childhood asthma and was a lifelong nonsmoker.

On examination, the patient had diffuse erythema of the palms and palmar aspects of the fingers bilaterally; the dorsa of several fingers were also involved. In addition to erythema, the affected areas were lichenified, with scattered fissuring (Figures 19.2 through 19.4).

Diagnostic Considerations

Skin conditions rendering individuals susceptible to nonspecific occlusive effects from gloves include atopic dermatitis, dyshidrotic eczema, psoriasis, and other, less common disorders. Although in individuals with intact skin, allergic contact urticaria is the *sine qua non* of latex-specific skin diseases,

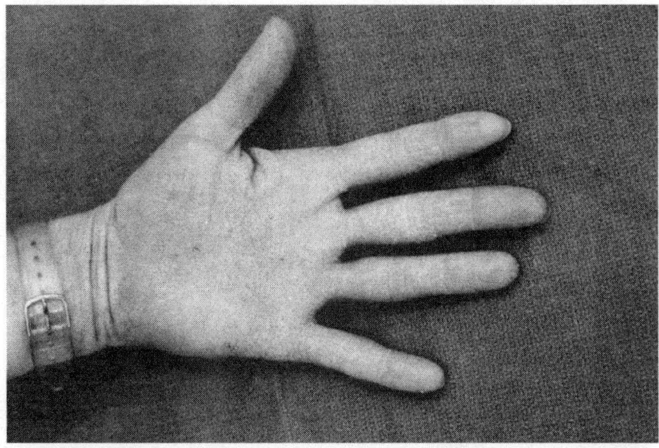

FIGURE 19.2 Volar aspect of the left hand (note areas of fissuring in palm).

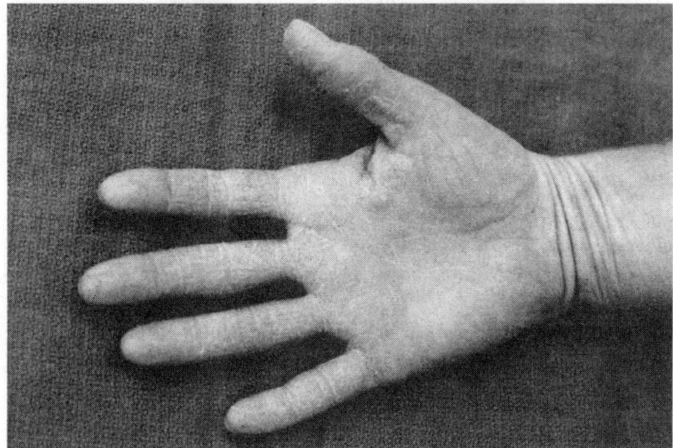

FIGURE 19. 3 Volar aspect of the right hand.

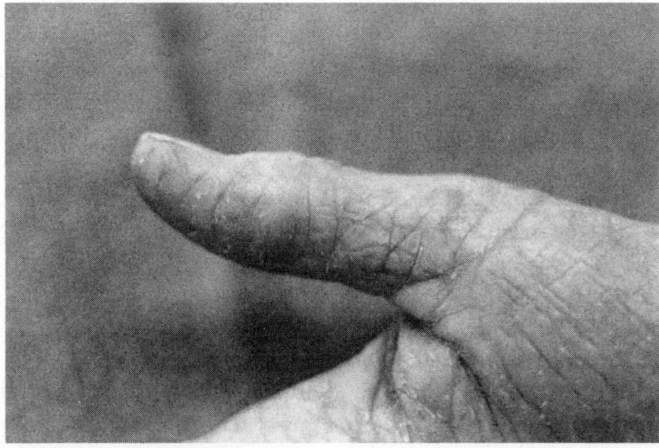

FIGURE 19.4 Close-up of the volar aspect of the right thumb showing hyperkeratosis and fissuring.

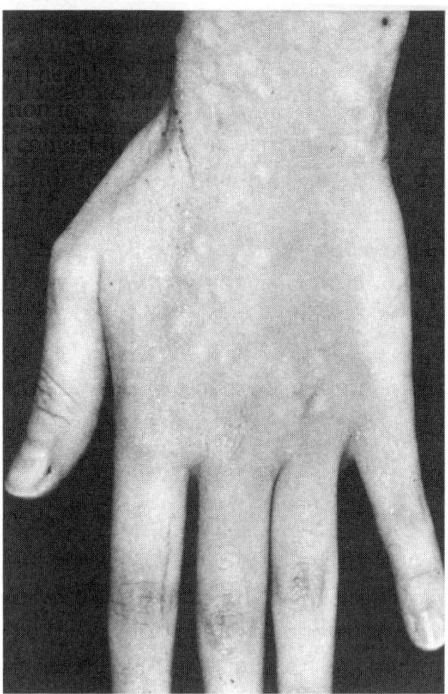

FIGURE 19.5 Contact urticaria in a glove distribution. (Photo courtesy of Kristiina Turjanmaa.)

patients who already have thickened and fissured skin may experience contact urticaria as burning and/or itching (acutely), or as worsening of their skin condition (chronically). Thus, the classic appearance of this condition as an isolated (and rapidly reversing) wheal- and-flare reaction (Figure 19.5) is one that is frequently masked by preexisting skin conditions.

In this case, the patient's history of preexisting atopic dermatitis gave the author an initial impression that her skin changes were a nonspecific effect of glove occlusion. However, the patient failed to respond to topical steroids alone (triamcinolone acetonide, 0.1% cream), and was unable to tolerate cloth glove liners. More importantly, she began to report acute burning and itching after skin contact with latex gloves, similar to the reaction she had with selected foods. This history was consistent with contact urticaria, initially to foods and later to latex exposure. Because of her chronic skin changes, the patient would not necessarily exhibit the characteristic appearance of contact urticaria and ancillary testing was necessary to establish the diagnosis.

Resolution

The patient was referred for (48 h) patch testing. This was conducted using latex glove material on intact skin, and was read as negative. Concurrently, an *in vitro* (ELISA) test for latex-specific IgE (ALA-Stat, Diagnostic Products Corporation, Los Angeles, California) was reported as 0.38 IU/mL, a weakly positive reaction. This pattern was interpreted as being consistent with early latex allergy. The patient was advised that skin contact with latex-containing products (and other known contact urticants) should be avoided in both the workplace and home. A potent topical steroid (clobetasol propionate, 0.05% ointment) was prescribed for 4 weeks, followed by a medium-potency steroid (triamcinolone acetonide, 0.1% cream). With the above regimen, the patient reported significant improvement of symptoms, including reduced redness, itching, and fissuring; physical examination confirmed the improvement.

Summary

A 35-year-old nurse with preexisting atopic dermatitis developed worsening skin involvement associated with frequent hand-washing and extensive use of unpowdered latex examination gloves in an endoscopy unit. Over-the-counter emollient creams were not helpful and the patient began to experience acute-onset hand itching and stinging after donning gloves, as she had previously experienced when her hands contacted certain foodstuffs. She denied having respiratory symptoms. Although patch testing with latex glove material was negative on intact skin, an *in vitro* test for latex-specific IgE was weakly positive. The patient was diagnosed with early latex allergy (contact urticaria) superimposed on preexisting atopic dermatitis, was treated with topical steroids and latex avoidance, and experienced significant improvement. She continued with her usual and customary employment without any lost work time.

Epidemiology of Latex Allergy

Although isolated observations of urticarial reactions to latex products date back to the 1920s, current awareness of immediate-onset latex-associated skin reactions derives from two publications. In 1979 Nutter in the United Kingdom[1] and in 1980 Forstrom in Finland[2] described wheal-and-flare reactions among latex glove-wearers, a reaction they each labeled as latex-induced contact urticaria. Since then, numerous cases of rhinoconjunctivitis, asthma, angioedema, and even anaphylaxis have also been ascribed to latex allergy.[3] The Food and Drug Administration (FDA) did much to increase awareness of this issue when, in 1991, it issued a bulletin describing anaphylactic reactions to rubber-tipped barium enema catheters.[4] This information encouraged anesthesiologists to investigate hitherto unexplained intraoperative hypotensive events and to conclude that, in some cases, anaphylactic reactions to surgeon's gloves may have been responsible.[5–8]

As awareness of the diagnosis has spread, several at-risk groups were identified. Health care workers, because of their daily potential for exposure to latex in gloves and other medical devices, are an obvious group (see below). Children with spina bifida have also emerged as being at risk, as are other patients who require indwelling urinary catheters and/or who have undergone multiple operative procedures.[8–12] Since the use of latex gloves is also common among housekeeping personnel and food service and greenhouse workers, increasing attention has been paid to allergic sensitization among these groups as well.[13,14]

Several surveys have examined either skin test reactivity or latex-specific IgE among unselected patients in allergy practices, at times excluding patients referred specifically to rule out latex allergy. Fuchs added latex-glove extract to the skin-prick testing panel for unselected patients referred for allergy testing. Among 1288 patients tested, 4.2% reacted to latex, 30% of whom had no clinical history suggestive of latex allergy.[15] Hadjiliadis and coworkers[16] likewise included latex extract in the skin test panel of a hospital-based allergy and asthma referral clinic. In that study, 4.5% of 224 subjects tested exhibited an unambiguous positive response to latex, a history of current or prior latex glove use was elicited from all 10 responders. Moneret-Vautrin and colleagues[17] performed a similar study on 569 allergy-clinic patients referred for various indications, and found that 6.9% reacted to latex extract. Of those reacting, 85% had other documented allergies, one-half had a history of latex exposure, and about 40% gave a history of reacting to latex-containing products.

Finally, two studies looked for latex-specific IgE in blood, one among allergy patients and the other among volunteer blood donors. Reinheimer and Ownby[18] performed a latex-specific ELISA test on 200 consecutive serum samples submitted for a total IgE determination. (Samples were excluded from this study if a latex antibody titer had also been requested.) Of this sample, 12% were positive for latex-specific IgE. A disproportionate number of the positives were among children (none of whom had meningomyelocele), and only two positive patients had a history of recurrent latex exposure. Ownby et al.[19] applied the same ELISA assay to sera from volunteer blood donors and found that 6.4% exhibited

TABLE 19.1 Surveys of Latex Allergy/Glove Dermatitis among Health Care Workers

Target Group	Number	Symptoms (%)[a]	Positive RAST (%)	Positive ELISA (%)	Positive Skin Test (%)	Author/Year
Physicians	101				10	Arellano et al. (1992)
Dentists	1628	13.7[b]				Berky et al. (1992)
	131	17[b]			10	Tarlo et al. (1997)
Nurses	140	76			22	Douglas et al. (1997)
	741			8.9		Grzybowski et al. (1996)
Nurses (O.R.)	268	41			10.7	Lagier et al. (1992)
Anesthesiologists and nurse anesthetists	101		3		15.8	Konrad et al. (1997)
Lab technicians	230	53	2			Salkie (1993)
Various	534	44				Kujala and Reijula (1995)
	134	23[b]		6.7	8.2	Kibby and Akl (1997)
	1472	30.9			6.8 (of 139)	Leung et al. (1997)
	1351	[c]			12.1	Liss et al. (1997)
	512				4.5	Turjanmaa (1987)
	273				5	Vandenplas et al. (1995)
	202		2		2	Wrangsjo et al. (1994)
	224				17	Yassin et al. (1994)

[a] Any glove-related dermal or respiratory symptoms, unless otherwise indicated.

[b] Studies in which subjects were screened for symptoms consistent with latex allergy (i.e., self-reported pruritis, uritcaria, rhinoconjunctivitis, or wheezing within minutes of latex exposure).

[c] Prevalence estimates generated for individual symptoms only.

latex-specific IgE, with 2.3% being strongly positive. This study is important in terms of population-wide prevalence estimates because, unlike the others cited, it did not select for the presence of allergies among study subjects.

A range of prevalence estimates have been offered for latex sensitization among health care workers (Table 19.1). Arellano and coworkers[20] reported positive skin-prick test results to latex extract in 10% of 101 hospital-based anesthesiologists, radiologists, and surgeons compared to 3% in a non-glove-exposed control group. Physicians with a prior history of atopy were at higher risk than nonatopics. Salkie[21] reported that 53% of medical laboratory technologists responding to a survey reported symptoms related to use of latex gloves; however, only 3/123 (2% of subjects) were positive for antilatex IgE on RAST testing.[21] Vandenplas et al.[22] found that 5% of a sample of 273 nurses, laboratory technologists, and cleaning staff at one hospital were skin-prick test positive to latex extract. All positive workers reported glove-related urticaria; 12 of 13 reported rhinoconjunctivitis symptoms, and 5 of 13 reported asthma symptoms. Berky and associates[23] reported that 13.7% of army dentists responding to a questionnaire survey gave clinical histories compatible with latex glove allergy. Lagier et al.[24] found a 41% prevalence of self-reported glove-related health problems among 268 operating room nurses surveyed; of 197 receiving skin-prick testing with latex extract, on the other hand, 10.7% were reactive. By comparison, Kujala and Reijula[25] surveyed 534 physicians, nurses, and housekeeping personnel in a large Finnish hospital and found that 44% reported at least one dermal symptom associated with wearing latex gloves. Yassin and coworkers[26] performed latex skin-prick testing on 224 nurses, laboratory technicians, dental staff, physicians, and respiratory therapists, and found that 17% had positive results. Grzybowski and colleagues[27] found latex-specific IgE in sera from 8.9% of 741 registered nurses tested at a large metropolitan hospital. Together with other similar surveys,[28–35] these studies suggest that approximately 10% of potentially exposed health care workers will manifest laboratory or skin test results consistent with some degree of latex sensitization.

Although there are no dependable estimates of trends in the prevalence of latex allergy, there is a widespread perception that latex allergy is becoming more common over time. This may be a real phenomenon or may constitute a diagnostic bias (i.e., the increasing attention paid to latex allergy in the medical literature may make it more likely that the condition will be correctly diagnosed when present). The institution of universal precautions has required glove-wearing by a greater number of health care personnel, and for a larger portion of their work day. At the same time, the demand for more low-cost gloves has led to changes in the manufacturing processes (e.g., shortened wash times), resulting in higher levels of residual antigen in some glove types. As noted in the above epidemiologic reviews, preexisting atopy appears to predispose individuals to latex sensitization. Thus, potential diagnostic bias notwithstanding, a combination of more widespread and intense exposure of medical workers to latex antigens, along with a subpopulation of health care workers at increased risk for developing atopic disorders, make the possibility of increasing rates of latex sensitization a plausible one.

Diagnostic Approach to Glove Dermatitis and Latex Allergy

Background: Pathophysiology

From the standpoint of disease causation, gloves can be as important for what they keep "in" as for what they keep "out." In the health care setting, gloves are employed as an impermeable and pliable barrier to prevent microorganisms (and caustic cleaning agents) from contacting the worker's skin. The worker's hands, in turn, are a source of heat and moisture, the dissipation of which is inhibited by glove-wearing. For the average worker with intact skin, the accumulation of perspiration and heat within gloves constitutes a minor nuisance. For the patient with preexisting atopic dermatitis or dyshidrotic eczema, however, any occlusive glove, regardless of its chemical constituents, may produce an exacerbation of the underlying condition. Thus, an early task in evaluating patients with glove dermatitis is to determine the relative likelihood of specific allergy vs. nonspecific irritation or exacerbation of preexisting skin disease. That is to say, the clinician must estimate the relative contributions of allergic contact urticaria, allergic contact dermatitis, irritant contact dermatitis, and/or other processes to the presenting symptoms.

Beyond diagnosis, the clinician must evaluate the importance of what are usually multiple occupational exposures in the genesis of the disease process. These exposures may be as exotic as anthraquinone antioxidants in gloves, and as mundane as frequent hand-washing with soap and water (Table 19.2). Thus, a hand dermatitis may represent an irritant process from frequent hand-washing, a contact urticarial

TABLE 19.2 Etiologic Agents in Hand Dermatitis

Irritant Contact Dermatitis
Nonspecific occlusive effect (gloves)
Frequent hand-washing
Disinfectants and cleaning agents
Contact Urticaria (Type I)
Latex rubber (gloves)
Food antigens
Disinfectants (e.g., chlorhexidine)
Allergic Contact Dermatitis (Type IV)
Antioxidants and accelerators in gloves
Disinfectants (e.g., glutaraldehyde, chlorhexidine, etc.)
Adhesives (e.g., cyanoacrylate—may penetrate gloves)

Three distinct types of skin testing may be indicated in the workup of glove dermatitis and latex allergy.[39] Skin testing for contact urticaria is performed as a so-called use test. In the use test, the patient dons a glove finger (or full glove, if there is no initial reaction), and the clinician observes for contact urticaria after 15 min. A nonlatex glove is typically worn on the opposite hand to rule out physical urticaria as a cause for reaction. Also, some clinicians apply the glove to a wetted hand. Skin-prick testing (for a wheal-and-flare reaction), although considered the gold standard for latex allergy diagnosis, does have its limitations. Specifically, standardized commercial skin-test antigens are currently pending approval by the FDA and practitioners generally find it necessary to elute antigens from commercial gloves prior to testing. This creates an obvious potential for variability of results.[40]

Conventional (48 h) patch testing involves either cutting patches from commercial gloves (including both the customary glove worn by the affected patient and a nonlatex control) or applying patches with known concentrations of glove allergens known to produce delayed-type hypersensitivity. This type of testing is most appropriate when the history and dermatologic findings suggest a Type IV allergic contact dermatitis (as may be seen with sensitization to rubber accelerators and antioxidants). Although patch testing with a glove sample also raises the possibility of provoking a Type I reaction to latex (contact urticaria), the patch test may be negative when the use test is positive. This apparent paradox occurs because the patch test is normally applied to the back of the torso (normally intact skin) and the use test involves the hands (i.e., often nonintact skin).

Workup of Respiratory Symptoms

If respiratory symptoms are present, additional diagnostic efforts are warranted (see algorithm in Figure 19.8). Rhinitis/conjunctivitis symptoms typically include itchy eyes and nose, sneezing, rhinorrhea, and nasal congestion. Asthma may present as wheezing, chest tightness, cough, or some combination thereof. Patients with suspected latex-induced respiratory symptoms may benefit from keeping symptom logs. If bronchospasm is suspected, serial peak flow measurements are useful, but keep the following guidelines in mind:

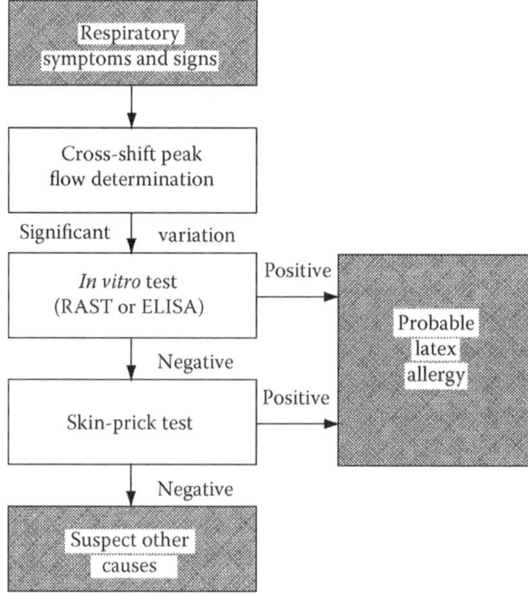

FIGURE 19.8 Diagnostic algorithm in suspected respiratory tract latex allergy.

- Obtain the readings over 1–2 weeks period.
- Include the early morning, midday, afternoon, and evening readings.
- Include the weekend data.

Early- and/or late-phase asthmatic reactions, as well as the so-called ratcheting of symptoms (and of peak flow) over the work week, may be apparent.[41,42] Of note, nasal peak (inspiratory) flow can also be documented on a cross-shift basis using a specially modified peak-flow meter and may provide ancillary documentation of latex-induced rhinitis.[43]

Prevention and Treatment

Primary Prevention

Primary prevention is intervention aimed at preventing the initial development of a disease state. In the case of latex allergy, such intervention takes the form of employee screening and counseling, exposure control, or both. An employee screening and counseling program identifies individuals with preexisting skin conditions and/or an allergy history, and counsels them in both their personal work practices and medical utilization patterns. For such workers (indeed, for all workers), it is important to view employee health services as an accessible place in which a priority is placed on developing solutions to workplace problems. Exposure control (reduction and avoidance of latex-antigen exposure) can target sensitized individuals (see secondary and tertiary prevention), individuals with preexisting conditions putting them at risk, or all workers in a given facility. As noted below, the approach to exposure control is often governed, at least in part, by economics.

Latex gloves vary considerably in their extractable protein antigen content, and some so-called hypoallergenic gloves may have a decreased sensitization potential.[44] (It is important to distinguish between sensitization and exacerbation, since there are no current data to suggest that low-antigen gloves are safe for individuals with preexisting latex allergy.) In addition, several studies have shown that cornstarch from powdered gloves is the most important carrier for airborne latex allergen in medical facilities.[45–47] Thus, health care workers with respiratory allergies may have their symptoms triggered via passive exposure to antigen from others' powdered gloves, regardless of the type of glove that they, themselves, are wearing. To minimize this problem, synthetic, nonlatex gloves—or nonpowdered latex gloves—can be substituted for powdered latex gloves for routine use. In the case of nonsterile (examination) gloves, this substitution carries little in the way of cost differential. However, the situation is quite different for sterile (surgeon's) gloves, with a five- to sixfold cost differential between powdered latex and its alternatives at the time of writing. Thus, although considerable potential exists for reducing latex antigen exposure within health care facilities, cost remains a major perceived barrier (from the standpoint of hospital administration). The overall cost-benefit analysis, however, should factor in the cost of potential adverse health events (among both health care workers and patients), as well as the cost of disability and of retraining of skilled health care personnel. These factors, as well as potential product liability issues, may dramatically realign the economics of glove manufacturing, distribution, and purchasing in the not-too-distant future.

Secondary Prevention (Screening)

Secondary prevention of latex allergy consists of early recognition of clinical and subclinical states of sensitization and intervention before full-blown sensitization occurs. One approach to secondary prevention is an extension of the initial (pre-placement) screening and involves periodically asking a series of focused questions regarding any symptoms experienced while wearing gloves (or when near coworkers wearing gloves). Individuals can then be identified with early or impending latex sensitization. Another possibility for secondary prevention would involve routine screening (with consent) of exposed workers for subclinical latex sensitization, for example, using an *in vitro* (RAST or ELISA) test for latex-specific IgE. Follow-up would consist of counseling and intervention (latex avoidance) among those showing evidence of early sensitization, even in the absence of symptoms. Potential drawbacks of such an approach, however, include cost, the invasiveness of obtaining venous blood, and an inevitable yield of false-positive results. For example, assuming a baseline prevalence of latex sensitization of 10%—and a test sensitivity and specificity of 87% and 85%, respectively—the positive predictive value of this screening exercise would only be 39% (i.e., only two out of five individuals with a positive test would actually have early latex sensitization). Perhaps the most cost-effective approach to early diagnosis is maintaining an atmosphere of open communication between health care and occupational health workers about glove health issues. Helpful in this regard is a policy of open access to employee health services, and a willingness of supervisors to attempt to provide work practice and environmental accommodations.[48] Without such an atmosphere, it is not unusual for health care personnel to conceal or deny that a condition exists, at least in part out of concern for job security.

Tertiary Prevention (Treatment)

Tertiary prevention consists of intervention to arrest the progression of disease and to limit associated disability. For patients with fixed skin lesions (i.e., skin changes beyond the reversible wheal-and-flare of contact urticaria), topical steroid preparations are generally indicated. In addition, a brief respite from any glove wearing may occasionally be useful. Once diagnosed with glove dermatitis and/or latex allergy, the affected individual must also be supplied with appropriate gloves. For the individual with atopic dermatitis and/or dyshidrotic eczema alone, the use of cloth glove liners is far more important that substituting glove materials. For the individual with true latex skin allergy, substitute gloves are available fabricated from a variety of materials, including vinyl (polyvinyl chloride), nitrile (polynitrile), polychloroprene, and substituted polystyrene polymers.[49,50] Allergic reactions to synthetic glove materials are unusual, but not unheard-of.[51] The most challenging situation is posed by individuals with significant respiratory symptoms. Cornstarch on powdered latex gloves acts as an airborne antigen carrier, and can therefore passively trigger respiratory symptoms in individuals not wearing such gloves. Thus, in order to allow an individual with respiratory manifestations of latex allergy to safely return to work, it may be necessary to replace all powdered latex gloves with unpowdered gloves in part or all of a facility. As noted above, such a measure may have profound economic consequences. On the other hand, excluding a health professional from their customary employment may also be a costly proposition. In the words of the Americans with Disabilities Act of 1990, a qualified individual with a disability has a right to reasonable accommodation with respect to working conditions, while at the same time, essential functions of the job cannot pose an imminent risk of harm. Thus, prudent occupational health practitioners carefully document any decision-making processes that affect the initial or continuing employment status of latex-sensitized workers.

Related Issues

In household settings, latex can be encountered in a variety of products, including chewing gum, condoms, balloons, clothing, and recreational equipment.[52] Problems may also occur with the consumption of fruits known to cross-react with latex, including avocado, banana, chestnut, kiwi fruit, papaya, and possibly others.[53,54] These (potentially hidden) hazards should be discussed during counseling of employees with latex allergy.

Conclusions

Latex allergy and glove dermatitis are two overlapping sets of conditions with serious implications for patients, health care workers, and hospital policy makers. Successful environmental control strategies dealing with both skin and respiratory manifestations of latex allergy may require looking beyond the issue of individual glove-wearing practices, and taking into account airborne allergen from the gloves of nonaffected individuals. Intervention, particularly primary prevention, may raise serious cost-benefit issues for those in charge of allocating scarce resources. These competing priorities, in turn, can only be appreciated in their full complexity by informed health and safety professionals and hospital administrators.

Acknowledgments

The author wishes to acknowledge the assistance of Kim Sullivan and Dr. Curt Hamann of SmartPractice, Phoenix, Arizona, for supplying the photographs of contact urticaria and allergic contact dermatitis.

References

1. Nutter AF. Contact urticaria to rubber. *Br J Dermatol* 101:597–598, 1979.
2. Forstrom L. Contact urticaria from latex surgical gloves. *Contact Dermatitis* 6:33–34, 1980.
3. Sussman GL, Tarlo S, and Dolovich J. The spectrum of IgE-mediated responses to latex *J Am Med Assoc* 265:2844–2847, 1991.
4. Food and Drug Administration, U.S. Department of Health and Human Services. Medical alert: Allergic reactions to latex-containing medical devices, FDA Medical Device Bulletin, April, 1991.
5. Binkley K, Cheema A, Sussman G, Moudgil G, O'Connor M, Evans S, et al. Generalized allergic reactions during anesthesia. *J Allergy Clin Immunol* 89:768–774, 1992.
6. McKinstry LJ, Fenton WJ, and Barrett P. Anaesthesia and the patient with latex allergy. *Can J Anaesth* 39:587–589, 1992.
7. Spears FD, Littlewood KE, and Liu DW. Anaesthesia for the patient with allergy to latex. *Anaesth Intensive Care* 23:623–635, 1995.
8. Pollard RJ and Layon AJ. Latex allergy in the operating room: Case report and a brief review of the literature. *J Clin Anesth* 8:161–167, 1996.
9. Kelly KJ, Pearson ML, Kurup VP, Havens PL, Byrd RS, Setlock MA, et al. A cluster of anaphylactic reactions in children with spina bifida during general anesthesia: Epidemiologic features, risk factors, and latex hypersensitivity. *J Allergy Clin Immunol* 94:53–61, 1994.
10. Ellsworth PI, Merguerian PA, Klein RB, and Rozycki AA. Evaluation and risk factors of latex allergy in spina bifida patients: Is it preventable? *J Urol* 150:691–693, 1993.
11. Nieto A, Estornell F, Mazon A, Reig C, Nieto A, and Garcia-Ibarra F. Allergy to latex in spina bifida: A multivariate study of associated factors in 100 consecutive patients. *J Allergy Clin Immunol* 98:501–507, 1996.
12. Pittman T, Kiburz J, Gabriel K, Steinhardt G, Williams D, and Slater J. Latex allergy in children with spina bifida. *Pediatr Neurosurg* 22:96–100, 1995.
13. Carillo T, Blanco C, Quiralte J, Castillo R, Cuevas M, and Rodriquez de Castro F. Prevalence of latex allergy among greenhouse workers. *J Allergy Clin Immunol* 96:699–701, 1995.
14. Sussman GL, Lem D, Liss G, and Beezhold D. Latex allergy in housekeeping personnel. *Ann Allergy Asthma Immunol* 74:415–418, 1995.
15. Fuchs T. Latex allergy [letter]. *J Allergy Clin Immunol* 93:951–952, 1994.
16. Hadjiliadis D, Khan K, and Tarlo SM. Skin test responses to latex in an allergy and asthma clinic. *J Allergy Clin Immunol* 96:431–432, 1995.
17. Moneret-Vautrin DA, Beaudouin E, Widmer S, Mouton C, Kanny G, Prestat F, et al. Prospective study of risk factors in natural rubber latex hypersensitivity. *J Allergy Clin Immunol* 92:668–677, 1993.
18. Reinheimer G and Ownby DR. Prevalence of latex-specific IgE antibodies in patients being evaluated for allergy. *Ann Allergy Asthma Immunol* 74:184–187, 1995.

19. Ownby DR, Ownby HE, McCullough J, and Shafer AW. The prevalence of anti-latex IgE antibodies in 1000 volunteer blood donors. *J Allergy Clin Immunol* 97:1188–1192, 1996.
20. Arellano R, Bradley J, and Sussman G. Prevalence of latex sensitization among hospital physicians occupationally exposed to latex gloves. *Anesthesiology* 77:905–908, 1992.
21. Salkie ML. The prevalence of atopy and hypersensitivity to latex in medical laboratory technologists. *Arch Pathol Lab Med* 117:897–899, 1993.
22. Vandenplas O, Delwiche JP, Evrard G, Aimont P, van der Brempt X, Jamart J, et al. Prevalence of occupational asthma due to latex among hospital personnel. *Am J Respir Crit Care Med* 151:54–60, 1995.
23. Berky ZT, Luciano WJ, and James WD. Latex glove allergy. A survey of the U.S. Army Dental Corps. *J Am Med Assoc* 268:2695–2697, 1992.
24. Lagier F, Vervloet D, Lhermet I, Poyen D, and Charpin D. Prevalence of latex allergy in operating room nurses. *J Allergy Clin Immunol* 90:319–322, 1992.
25. Kujala VM and Reijula KE. Glove-induced dermal and respiratory symptoms among health care workers in one Finnish hospital. *Am J Ind Med* 28:89–98, 1995.
26. Yassin MS, Lierl MB, Fischer TJ, O'Brien K, Cross J, and Steinmetz C. Latex allergy in hospital employees. *Ann Allergy* 72:245–249, 1994.
27. Grzybowski M, Ownby DR, Peyser PA, Johnson CC, and Schork MA. The prevalence of anti-latex IgE antibodies among registered nurses. *J Allergy Clin Immunol* 98:535–544, 1996.
28. Douglas R, Morton J, Czarny D, and O'Hehir RE. Prevalence of IgE-mediated allergy to latex in hospital nursing staff. *Aust NZ J Med* 27:165–169, 1997.
29. Kibby T and Akl M. Prevalence of latex sensitization in a hospital employee population. *Ann Allergy Asthma Immunol* 78:41–44, 1997.
30. Konrad C, Fieber T, Gerber H, Schuepfer G, and Muellner G. The prevalence of latex sensitivity among anesthesiology staff. *Anesth Analg* 84:629–633, 1997.
31. Leung R, Ho A, Chan J, Choy D, and Lai CK. Prevalence of latex allergy in hospital staff in Hong Kong. *Clin Exp Allergy* 27(2):167–174, 1997.
32. Liss GM, Sussman GL, Deal K, Brown S, Cividino M, Siu S, et al. Latex allergy: Epidemiological study of 1351 hospital workers. *Occup Environ Med* 54:335–342, 1997.
33. Tarlo SM, Sussman GL, and Holness DL. Latex sensitivity in dental students and staff: A cross-sectional study. *J Allergy Clin Immunol* 99:396–401, 1997.
34. Turjanmaa K. Incidence of immediate allergy to latex gloves in hospital personnel. *Contact Dermatitis* 17(5):270–275, 1987.
35. Wrangsjo K, Osterman K, and van Hage-Hamsten M. Glove-related skin symptoms among operating theatre and dental care unit personnel. II. Clinical examination, tests and laboratory findings indicating latex allergy. *Contact Dermatitis* 30(3):139–143, 1994.
36. Kelly KJ, Kurup VP, Reijula KE, and Fink JN. The diagnosis of natural rubber latex allergy. *J Allergy Clin Immunol* 93:813–816, 1994.
37. Kelly KJ, Kurup V, Zacharisen M, Resnick A, and Fink JN. Skin and serologic testing in the diagnosis of latex allergy. *J Allergy Clin Immunol* 91:1140–1145, 1993.
38. Kurup VP, Kelly KJ, Turjanmaa K, Alenius H, Reunala T, Palosuo T, et al. Immunoglobulin E reactivity to latex antigens in the sera of patients from Finland and the United States. *J Allergy Clin Immunol* 91:1128–1134, 1993.
39. Turjanmaa K, Makinen-Kuljunen S, Reunala T, Alenius H, and Palosuo T. Natural rubber latex allergy: The European experience. *Immunol Allergy Clin N Am* 15:71–88, 1995.
40. Fink JN, Kelly KJ, Elms N, and Kurup VP. Comparative studies of latex extracts used in skin testing. *Ann Allergy Asthma Immunol* 76:149–152, 1996.
41. Brugnami G, Marabini A, Siracusa A, and Abbritti G. Work-related late asthmatic response induced by latex allergy. *J Allergy Clin Immunol* 96:457–464, 1995.

42. Valentino M, Pizzichini MA, Monaco F, and Governa M. Latex-induced asthma in four healthcare workers in a regional hospital. *Occup Med (Oxf)* 44:161–164, 1994.
43. Ahman M and Wrangsjo K. Nasal peak-flow-rate recording is useful in detecting allergic nasal reactions—A case report. *Allergy* 49:785–787, 1994.
44. Yunginger JW, Jones RT, Fransway AF, Kelso JM, Warner MA, and Hunt LW. Extractable latex allergens and proteins in disposable medical gloves and other rubber products. *J Allergy Clin Immunol* 93:836–842, 1994.
45. Heilman DK, Jones RT, Swanson MC, and Yunginger JW. A prospective, controlled study showing that rubber gloves are the major contributor to latex aeroallergen levels in the operating room. *J Allergy Clin Immunol* 98:325–330, 1996.
46. Tarlo SM, Sussman G, Contala A, and Swanson MC. Control of airborne latex by use of powder-free latex gloves. *J Allergy Clin Immunol* 93:985–989, 1994.
47. Tomazic VJ, Shampaine EL, Lamanna A, Withrow TJ, Adkinson F, and Hamilton RG. Cornstarch powder on latex products is an allergen carrier. *J Allergy Clin Immunol* 93:751–758, 1994.
48. Thompson RL. Educational challenges of latex protein allergy. *Immunol Allergy Clin N Am* 15:1159–1174, 1995.
49. Mellstrom GA and Boman AS. Gloves: Types, materials, and manufacturing. In: GA Mellstrom, JE Wahlberg, and HI Maibach (eds.), *Protective Gloves for Occupational Use*. CRC Press, Boca Raton, FL, 1994, pp. 21–35.
50. Hamann CP and Kick SA. Diagnosis-driven management of natural rubber latex glove sensitivity. In: GA Mellstrom, JE Wahlberg, and HI Maibach (eds.), *Protective Gloves for Occupational Use*. CRC Press, Boca Raton, FL, 1994, pp. 131–156.
51. Truscott W. The industry perspective on latex. *Immunol Allergy Clin N Am* 15:89–121, 1995.
52. Kelly KJ. Management of the latex-allergic patient. *Immunol Allergy Clin N Am* 15:139–157, 1995.
53. Fisher AA. Association of latex and food allergy. *Cutis* 52:70–71, 1993.
54. Kurup VP, Kelly T, Elms N, Kelly K, and Fink J. Cross-reactivity of food allergens in latex allergy. *Allergy Proc* 15:211–216, 1994.

20

Latex Glove Use: Essentials in Modern Hospital Safety

Wava M. Truscott
Safeskin Corporation

Introduction **20**-1
Barrier Protection **20**-2
Bloodborne Pathogens Final Ruling • Barrier Protection Profile of a Glove
Glove-Associated Reactions **20**-7
Irritation • Delayed-Type Hypersensitivity (Type IV) • Immediate-Type Hypersensitivity (Type I) • Respiratory Complications • Hospital Protocols
Powder vs. Powderfree Gloves **20**-16
Absorption • Protein • Chemicals • Microorganisms • Erroneous Laboratory Results
Proper Glove Removal and Disposal **20**-18
Summary **20**-18
Conclusion **20**-20
References **20**-20

Introduction

Hands have long been recognized as one of the major vectors for the spread of disease. In the mid-1800s, Ignaz Semmelweis was able to reduce the death rate from puerperal fever from <18% to 2.5% by the simple act of having attendants wash their hands in chlorine water.[1] Gloves were first used to protect the wearer during postmortem examinations in 1847, but they were not used in surgery until 1879, when they were donned to protect the hands of the scrub nurse from the harsh effects of disinfectants.

The importance of the glove as a primary protective barrier against the spread of disease was again emphasized with the publication of the series, universal precautions, promulgated by the Centers for Disease Control and Prevention (CDC) in 1987.[2] These admonitions were followed by mandatory regulations, issued by the Occupational Safety Hazards Act (OSHA) in 1991.[3]

The subsequent increase in glove usage and potential risks associated with their failure have focused attention on, and research into, numerous glove-related issues. These include the performance and preservation of barrier protection, glove-associated reactions, infection control practices, critical powder issues, and glove selection criteria.

Barrier Protection

Since the advent of the human immunodeficiency virus (HIV) and increased concerns over the rising number of hepatitis B virus (HBV)-infected individuals, gloves have been officially identified in CDC Universal Precautions as personal protective equipment (PPE) worn to provide protection from occupational exposure. It is estimated that 1% of all hospital patients are carriers of HBV and that many of them are asymptomatic.[4] New hepatitis strains continue to emerge, including virulent hepatitis C virus. Add to this exposure, the fact that 1 in every 250 patients hospitalized in America is reported to be infected with HIV (many of whom are also asymptomatic) and the importance of barrier protection becomes paramount.[5]

Studies have shown that the use of gloves is associated with decreased rates of HBV infection, as shown in a study performed by Gonzales and Naleway[6] (Table 20.1), where health care workers performing the same procedures with three different gloving practices were tested for HBV.

This table illustrates the importance of gloves in reducing the risk of occupational exposure to the bloodborne pathogens. Gloved hands are our primary contact with patients and, as such, are frequently exposed to pathogenic organisms. Hands are prone to surface injury, often marred with cuts, scrapes, abrasions, cracks, hangnails, and splinters, providing easy access for microorganisms when hands are unprotected. Gloves provide a barrier to prevent direct exposure.

Another occupational hazard is the risk of needlestick injury. Although gloves cannot completely prevent these incidents, they have been shown to provide some protection. One study demonstrated up to a 50% reduction in the amount of blood on a needle as it passed through the glove, due to a wipe-off effect of the material.[7] Natural rubber latex tends to close around penetrating instruments, providing greater wiping action than most synthetic materials, which rupture without recoil.

Several studies have shown that the use of double latex surgical gloves more than doubles the protection for the wearers by reducing the actual penetration of sharps through the inner glove from 60% to 80%.[8] The number of organisms remaining on the needles that do breach the inner glove should be further reduced by the wipe-off from the second glove. The give and elastic properties of latex are also thought, to some degree, to deflect the penetration angle of invading sharps. Two layers would increase the deflection.

Bloodborne Pathogens Final Ruling

Recognizing the importance of gloves in reducing the risk of infection when employees are exposed to bloodborne pathogens, OSHA mandated the use of gloves where potential contact with these infectious agents was increased. Utilizing the Blood Borne Pathogens Final Ruling, OSHA enforces regulations with regard to gloving requirements whenever there is an anticipated hand contact with potentially infectious materials, including patient mucous membranes, non-intact skin, invasive procedures, or contact with potentially infectious materials such as

TABLE 20.1 Hospital Study: Glove Use vs. HBV

Health Care Workers Gloving Frequency	HBV Infected (%)
Never	15.3
Intermittently	9.5
Routinely	6.3

Note: In this 1988 study, health care workers were divided into three groups, those who *never* wore gloves; wore them *intermittently*; and wore gloves *routinely*. Each of the three groups were tested for HBV infection. It was determined that individuals who elected to wear gloves routinely had a lower risk of acquiring HBV.

- Amniotic fluid
- Blood
- Cerebral spinal fluid
- Pericardial fluid
- Peritoneal fluid
- Pleural fluid
- Saliva (during dental procedures)
- Semen
- Serum
- Synovial fluid
- Any unfixed tissue or organ
- Any body fluid visibly contaminated with blood
- Any body fluid where it is difficult or impossible to determine whether or not blood is present
- Any HIV cell or tissue culture
- Tissue from an animal infected with HIV or HBV

The wearing of gloves, however, is not limited by the regulation. They should be worn whenever a biological or chemical hazard exposure is possible. Although perhaps obvious to all, it never hurts to emphasize the fact that non-bloodborne pathogens, such as infectious diarrhea or tuberculosis, can be just as deadly and require similar isolation precautions.[9]

The selection of gloves includes disposable surgical and examination gloves in addition to reusable utility gloves. For individuals allergic to latex, alternatives must be made available. The regulations identify hypoallergenic gloves, glove liners, and powderfree gloves that must be made readily accessible to employees who cannot use the standard equipment.[10,*] Gloves must be changed as soon as is reasonably possible when they become torn or punctured, and hands must be washed after glove removal. Although not specified in the OSHA standard, gloves should be changed between patients in accordance with standard infection control practices.

Barrier Protection Profile of a Glove

It is an FDA requirement that all gloves meet specific manufacturing specifications. Limits for pinholes and major defects (e.g., tears or foreign objects) are set at 1.5% for surgeons and 4.0% for examination gloves. This is known as the acceptable quality assurance level for unused gloves. The evaluation is conducted by filling a specified number of gloves with 1000 mL of water and observing for leaks over a 2 min period. Although this method is adequate for detecting gross manufacturing defects, it does not indicate if protection will be maintained during use, creating a potentially inappropriate assumption of security.

Gloves vary in performance characteristics under different circumstances. The material of construction, fit, physical characteristics, storage conditions, and work practices all influence in-use glove barrier protection.

Glove Composition

The molecular structure of the material from which a glove is made is one of the primary determinants of barrier performance. For example, there is a great deal of difference in material performance between natural rubber latex and vinyl.

Natural rubber latex still maintains its reputation for superior performance, due primarily to a combination of structural attributes. These include molecular coiling that provides the stretch and recoil

* Since the bloodborne pathogens ruling was written, we have gained a clearer understanding of the various types of reactions associated with gloves and can recommend more appropriate alternatives to use when necessary. These are discussed in the section "Glove-Associated Reactions".

(elasticity) characteristics for which latex is well known. The elastic coils are held together with sufficient double-bonded side chains to provide strength. Sulfur vulcanization provides natural rubber latex with thermal stability, the ability to maintain flexibility and strength over a wide range of temperatures, and extend shelf life.

Vinyl, often selected as an alternative for latex sensitive individuals, is very rigid at the molecular structure level. Molecules of this polyvinyl chloride material are linked extensively with a dense network of inflexible branches. Because vinyl is an inherently weaker material than latex, it was necessary to develop separate, less stringent requirements for vinyl glove performance. Inequities in physical requirements are specified by the American Standards and Test Materials (ASTM) as shown in Table 20.2.

The differences in these ASTM requirements are quite significant when one realizes that in-use barrier performance or durability is a result of the combined interaction of these characteristics. When vinyl is snagged by instruments, repeatedly jabbed by fingernail tips, or is pulled apart where stretch is required, molecular branch attachments are disrupted, creating structural breaks. This type of breakdown is often not visually apparent, but can be demonstrated in performance studies. A summary of some of the more recent research is provided in Table 20.3. There are, however, many circumstances where vinyl gloves are appropriate. These include situations where gloves are only minimally stressed for short periods of time, and where biohazard risks are minimal.

Nitrile gloves, another synthetic alternative for latex-sensitive individuals, utilize acrylonitrile butadiene as a replacement for the natural rubber latex. The chemical constituents and molecular structure are basic to latex formulations, which is why the material has more structural strength and stretch than vinyl (Figure 20.1).

One will, however, still need to select nitrile gloves noted to have low chemical residuals to prevent irritant and contact dermatitis. A recent study by Rego and Dufresne[11] demonstrates that nitrile

TABLE 20.2 ASTM Requirement Inequities

	Vinyl	Latex
ASTM number	D5250-92	D3578-95
Tensile strength[a]	9 MPa	14 MPa
Elongation[b]	300%	700%
Finger thickness	0.05 mm	0.08 mm

[a] Tensile strength, measured in megapascals is a demonstration of material strength.

[b] Elongation represents the capacity of the glove to stretch, rather than break.

TABLE 20.3 Vinyl vs. Latex Barrier Performance Studies

	Percent Failure			
Date	Vinyl	Latex	Method of Detection	Author
1989	53	3	Dye	Korniewicz
1990	63	7	Virus: ØX174	Korniewicz
	22	<1	Virus: Lambda	Klein
	56[a]	<1[a]	Lambda	
1993	43	9	Bacteria gram (−)	Olsen

Note: Several studies have evaluated the in-use barrier durability of gloves by conducting different clinical- and/or simulated-use studies. Barrier integrity of the used gloves were assessed using dyes, different virus strains, or bacteria.

[a] Gloves were first exposed to disinfectant (70% ethanol).

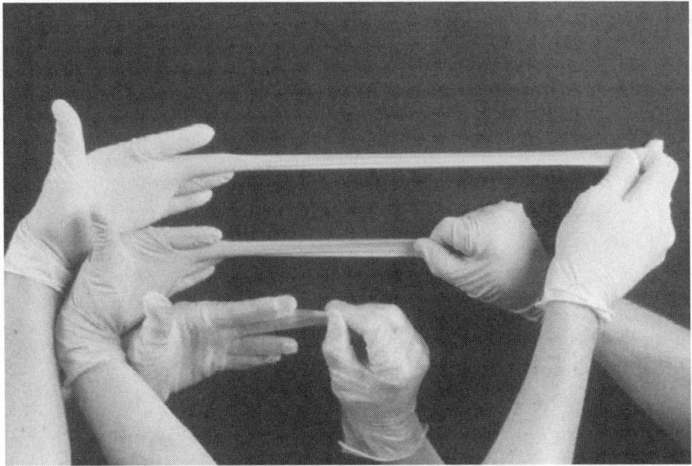

FIGURE 20.1 The stretchability of a glove is dramatically dependent on the material of manufacture. Shown here from top to bottom are natural rubber latex, nitrile, and vinyl.

maintained its barrier protection comparable to latex during rigorous simulated use studies (1% and 2% failures, respectively), while vinyl had significantly more failures (30%).

Other synthetic gloves currently available for latex-sensitive individuals are frequently more expensive and used primarily as surgical rather than examination gloves. Barrier performance characteristics of these products should be requested from the manufacturer or evaluated by the hospital (Figure 20.2). Glove manufacturers continue to search for new synthetic materials that will provide excellent barrier protection and comfort without being cost prohibitive.

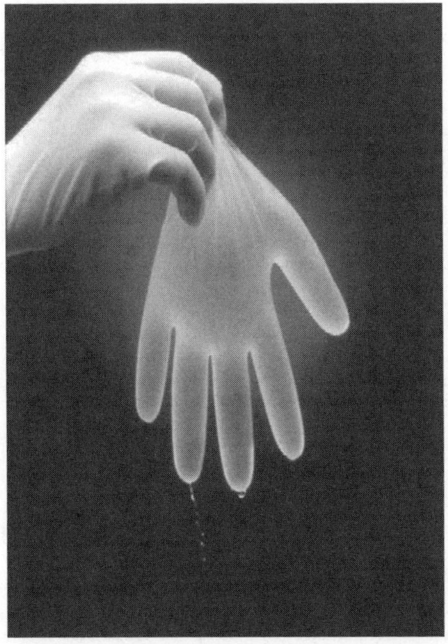

FIGURE 20.2 Following clinical- or simulated-use studies, fill the glove with water and observe for leaks.

General Chemicals

Gloved-hand contact with potentially sensitizing, toxic, carcinogenic, or teratogenic chemicals occurs daily in almost every area of the hospital, including patient care, diagnostic and research laboratories, and even housekeeping. In each of these situations, whether or not chemicals will permeate or physically degrade a glove is critical information for the wearer. Generally, chemical compatibility depends upon the type of material from which the glove is made (e.g., natural rubber, neoprene, vinyl, butyl rubber, nitrile, etc.) and the specific chemical to which it is exposed (e.g., organic solvent, acid, base, etc.). However, the glove formulation, its thickness, and the presence and composition of any coatings can make a major difference in a glove's potential to maintain its barrier protection. Other important factors in barrier maintenance would be concentration, temperature, duration, and extent of chemical contact (splash or submersion).

Questions regarding specific chemical compatibility should be addressed with the glove manufacturer. Even if the chemical is relatively innocuous (e.g., ethanol or isopropanol), if it breaks down the barrier integrity of a glove, subsequent contact with infectious substances can result in employee exposure.

There are chemical gloves made for use when handling specific chemical types. They are usually thick, coated, reusable gloves. If the instructions for use of substances with which employees are working or if in-house procedures specify the use of a chemical glove, examination or surgical gloves are not acceptable substitutes.

Chemotherapeutic and Cytotoxic Drugs

Another area in which the compatibility and proper use of gloves is a critical safety concern is in the preparation and administration of chemotherapeutic and cytotoxic drugs. Whether handling drugs in the laboratory or pharmacy, during patient care in the hospital or at alternate care sites, the selection of appropriate gloves and adherence to recommended practices is critical. Because many of these agents can affect the body in very low doses over extended periods of time, even seemingly minor breaches in protocol that happen repeatedly over the employee's career, mean unnecessary risk exposure. A brief outline of recommended practices is presented in Table 20.4.

Fit, Comfort, and Grip

Additionally, there are partially material-related factors that can put employees and procedures at risk.

Fit. If a glove is too tight or stiff (high modulus), fatigue can develop, affecting fine motor skills and potentially aggravating carpel tunnel syndrome. If the glove is too loose, tactile sensitivity may be reduced and the baggy material may interfere with procedures, or snag in equipment.

Grip. Gloves should be appropriate to the task. For example, in handling of moist glass specimen containers, a glove should have excellent wet grip. For surgical procedures, gloves should have excellent wet grip on metal to minimize instrument slippage. Because of their use in multiple work situations, examination gloves are usually challenged with the need for excellent wet and dry grip on a multitude of surfaces.

TABLE 20.4 Recommendations for Handling of Cytotoxic Drugs and Chemotherapy

Physical	Procedural
Latex	Double glove
Powderfree	Change every 30–60 min[a]
Thickness = ≥0.009 in. at the fingertips (9 mL)	Remove outer glove immediately if it is contaminated, torn, or punctured
Disposable	Always wash and rinse hands immediately upon glove removal
	Foam disinfectants do not remove cytotoxic chemicals

[a] OSHA and American Society of Hospital Pharmacists specify 60 min, and Oncology Nursing Society (ONS) states 30–60 min.

Gloves with antislip characteristics may be textured or formulated with an added tack surface, free of slippery lubricants such as silicone. In some departments, such as phlebotomy, individual preference may call for a smooth, non-tack glove for procedures such as palpation.

In-Use Practices

Lotions and barrier creams containing high concentrations of oils (including mineral, lanolin, coconut, palm, or jojoba), Vaseline®, and other petroleum-based products should not be worn under gloves. These break down chemical bonds in latex and many synthetic materials, weakening the gloves. Water and glycerin-based lotions are acceptable. Ask the manufacturer for test data demonstrating that their lotion is compatible with the glove material.

Long fingernails and jewelry can tear and rip gloves. Moderation in fingernail length and the removal of jewelry should be standard practice while wearing gloves in order to avoid related punctures and to facilitate adequate hand washing and rinsing procedures.

Poor donning techniques of sterile surgical gloves can result in rips and tears. Perioperative personnel should position the glove correctly and avoid excessive stretching of the cuff (Figure 20.3). Surgeons must thoroughly dry hands to allow them to slide into the gloves. This is particularly true for powder-free gloves.

Tapes, labels, Tegaderm, and electrocardiogram leads can cause ruptures, tears, and pinholes, compromising barrier integrity and decreasing performance efficiency. This may necessitate glove removal, increasing exposure risks and glove replacement waste. Some gloves resist tape adhesion better than others. If this is an issue in specific departments, this characteristic should be evaluated.

Storage conditions affect gloves. Heat, light, moisture, and ozone can all degrade latex and synthetic gloves. Ozone is created by electrical equipment such as generators, UV or fluorescent light, and x-ray equipment. Signs of ozone degradation include white creases that have small holes reminiscent of a perforation line, especially where a glove is folded on itself. Stored in the original packaging, they should be in a cool, dry place (optimally <90°F and 40% humidity) away from direct light and electrical generating equipment. Once boxes are opened, they are especially vulnerable to attack. Gloves should not be dumped into large bins as this increases their oxygen exposure, and thus oxidative degradation. Reuse of disposable gloves is not an acceptable practice. Washing does not adequately remove microbial contamination nor is the glove constructed to withstand resterilization.

Reusable gloves, often used in central processing, x-ray laboratories, and housekeeping, may be disinfected and reused if barrier integrity is still intact and there are no signs of deterioration (e.g., brittleness, tackiness, elongation, or growth at the fingertips). Obtain information from the manufacturer regarding resterilization if such practices are anticipated.

Glove-Associated Reactions

Glove-associated reactions can range in severity from a simple annoyance to anaphylactic shock. Although referenced by several terms, there are three basic types of reactions associated with the use of gloves: irritation, allergic contact dermatitis, and immediate-type hypersensitivity (Figure 20.4). These three reactions are often referred to by other terms, confusing those unfamiliar with these aliases. Several of these cross references are delineated in Table 20.5.

It is important to be able to identify symptoms, isolate causative agents, manage recovery, and institute preventative programs for glove-related reactions. You, or your employees, may unnecessarily increase the risk of infection due to microbial transmission through fissures and lesions of damaged epidermis, or of sensitization to latex proteins or chemicals found in many gloves. It is critical that staff members understand the differences between these conditions enabling them to readily aid in their management. Each reaction will be discussed separately. Clinical examples and additional information are provided in Chapter 19.

Proper glove donning technique: A surgical perspective

In compliance with AORN recommended standards

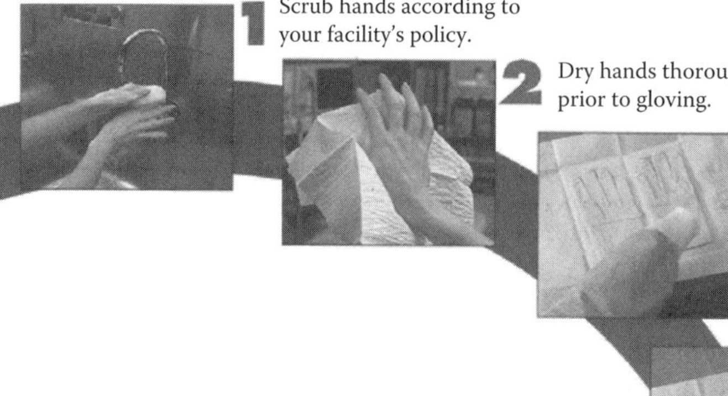

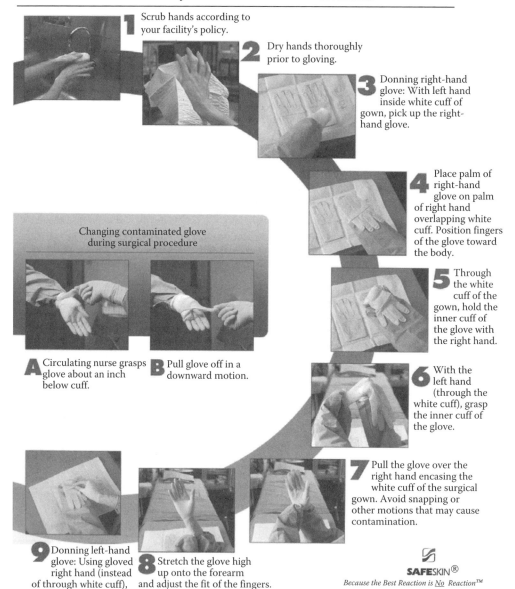

1 Scrub hands according to your facility's policy.

2 Dry hands thoroughly prior to gloving.

3 Donning right-hand glove: With left hand inside white cuff of gown, pick up the right-hand glove.

4 Place palm of right-hand glove on palm of right hand overlapping white cuff. Position fingers of the glove toward the body.

5 Through the white cuff of the gown, hold the inner cuff of the glove with the right hand.

6 With the left hand (through the white cuff), grasp the inner cuff of the glove.

Changing contaminated glove during surgical procedure

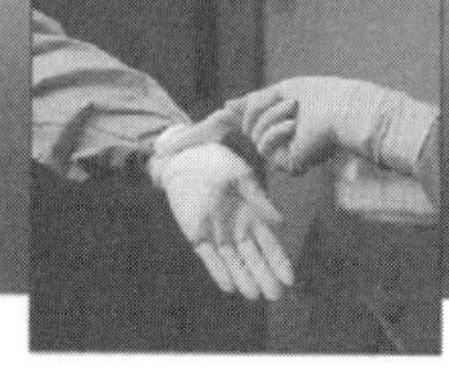

A Circulating nurse grasps glove about an inch below cuff.

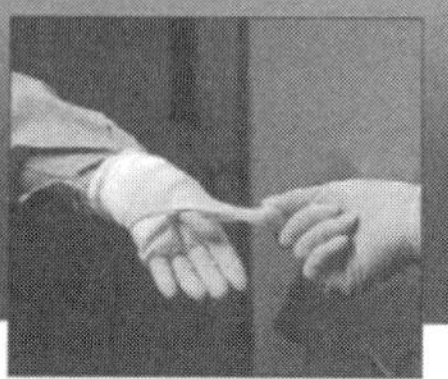

B Pull glove off in a downward motion.

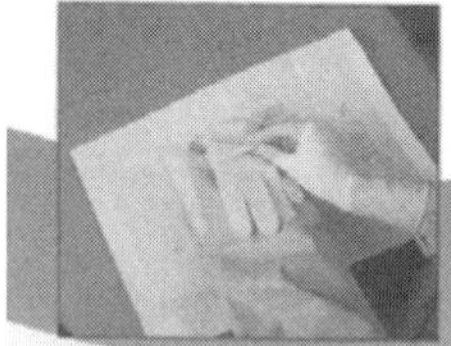

9 Donning left-hand glove: Using gloved right hand (instead of through white cuff), repeat steps 3 through 7.

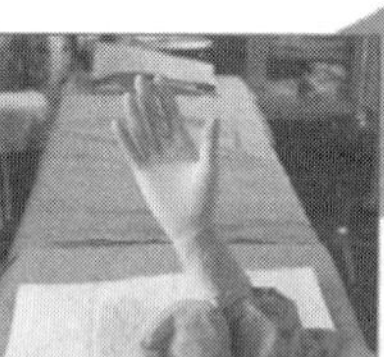

8 Stretch the glove high up onto the forearm and adjust the fit of the fingers.

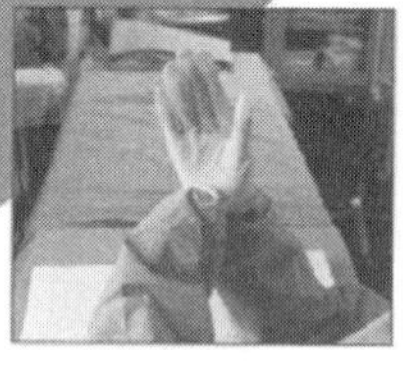

7 Pull the glove over the right hand encasing the white cuff of the surgical gown. Avoid snapping or other motions that may cause contamination.

Technical Advisors: Joanne Selva, BS, CIC; Maria Ninivaggi, RN, MSN, CIC; from Infection Control Consultants, Inc.

FIGURE 20.3 Proper donning techniques.

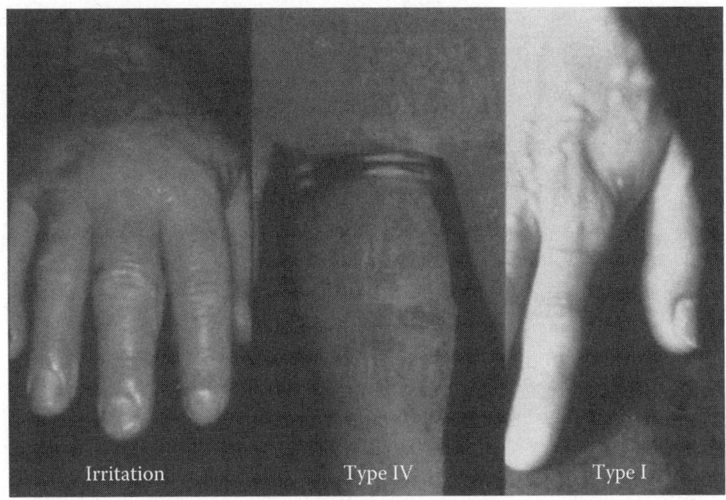

FIGURE 20.4 There are three types of glove-associated reactions.

TABLE 20.5 Glove-Associated Reactions

Reaction	Other Terms
Irritation	Irritant contact dermatitis
Allergic contact dermatitis	Delayed-type hypersensitivity
	Type IV hypersensitivity
	Chemical allergy
Immediate-type hypersensitivity	Type I hypersensitivity
	Protein allergy

Note: Glove-associated reactions are often referred to by a variety of terms.

Irritation

Everyone is susceptible to irritation, a nonallergenic condition. Irritation occurs as a result of either direct or indirect cell injury. The first symptom is usually redness with associated burning or itching. Repeated or continued contact with the irritant results in chronic inflammation, characterized by dry thickened skin, fissures, and papules (hard bumps). This state of an irritation is generally referred to as irritant contact dermatitis. More intense symptoms often appear where gloves may be especially tight, such as on knuckles, on the back of the hand, or on the wrist. Irritant reactions are limited to the area covered by the glove and do not extend beyond the cuff, even after long-term exposure.

While investigating the cause of an irritation, non-glove-related irritants should not be overlooked. Non-glove-related agents of irritant dermatitis include placing the hands (even occasionally) into disinfectants, surface-cleaning agents, anesthetics, or compounding agents. Insufficient rinsing after washing hands may lead to irritation due to soap residuals. This can be most apparent when jewelry is worn, as it may interfere with effective rinsing. Soap residue can cause pH alterations and can remove protective lipids between epidermal cells, thereby eliminating the epidermal buffering capacity and damaging the protective mantle.

Epidermal breakdown can also occur when gloves are worn for prolonged periods of time. Trapped perspiration is absorbed into the cells, which become hyperhydrated and fragile, increasing vulnerability to maceration. The problem is augmented when protective lipids, surrounding and insulating the epidermal cells, are removed by gloving powder, exposing the cells to increased injury. Abrasion caused by the powder particles further aggravates irritant reactions. Endotoxin that accumulates in the powder during the manufacture of powdered gloves can also cause irritant dermatitis.[12] Endotoxins are the pyrogenic lipopolysaccharide cell walls of dead gram-negative bacteria. Excessive moisture on warm occluded skin increases the multiplication rate of normal flora, whose metabolic by-products can cause further irritation.

Chemicals utilized in the manufacture of both synthetic and natural rubber latex gloves can also cause irritation. Excessive use of processing chemicals, biocides, and preservatives, as well as inadequate leaching or washing can produce gloves with high irritant potential. Because they are formulated and processed to remove excess chemical additives, hypoallergenic gloves are very appropriate for individuals prone to irritation.

To reduce the incidence of irritant contact dermatitis and the risk of microbial transmission through the disrupted integument,

- Remove the jewelry.
- Wash the hands with a mild soap each time gloves are removed.
- Avoid the use of hot water whenever possible.
- Rinse thoroughly.
- Dry thoroughly.
- Select the gloves processed with minimal chemicals.
- Wear the powderfree gloves.
- Choose the gloves low in overall chemical content.
- Change the gloves frequently.

A larger glove may help increase air circulation and minimize pressure aggravation, but caution must be taken to avoid excessive bagginess that may interfere with job performance.

Delayed-Type Hypersensitivity (Type IV)

Delayed-type hypersensitivity or allergic contact dermatitis is a cell-mediated allergic reaction initiated by sensitizing chemicals called contact sensitizers. Only individuals genetically programmed to be able to develop an immunological response to those specific chemicals used in manufacturing the glove will ever be capable of experiencing this type of glove-associated reaction. Repeated contact with these chemicals increases the number of sensitized T cells specifically recognizing that antigen (contact sensitizer). The sensitization period, which may take years to develop, is asymptomatic.

When an individual's sensitization threshold level is reached, the clinical symptoms that follow subsequent exposures often begin with redness and itching, progressing to small, clustered vesicles that elicit pain when scratched. After prolonged, repeated exposure to the chemical, a chronic condition may develop with dry thickened skin, cracking, crusting, peeling, scaling, lesions, and papules. The appearance of allergic contact dermatitis may easily be confused with that of irritant dermatitis. However, it differs in that the most intense level of symptoms appear from 6 to 48 h after exposure (hence the term "delayed-type hypersensitivity"). In chronic cases, symptoms may eventually extend up the arm, beyond the glove cuff.[13]

Contact sensitizers are found in both synthetic and natural rubber latex gloves. The additives most frequently associated with reactions are the chemicals known as accelerators that function as catalysts. Fortunately, there are only four basic categories of accelerators: thiurams, thiazoles, carbamates, and thioureas. They are, however, used in varying amounts and combinations by different manufacturers. Approximately 15%–25% of glove-related cases of allergic contact dermatitis are caused by chemicals other than accelerators. These potential contact sensitizers include antioxidants, colorants, preservatives, resins, plasticizers, biocides, and soaps.[14] Contact sensitizers utilized in the manufacture of natural rubber latex and synthetic gloves are listed in Table 20.6.

TABLE 20.6 Latex and Synthetic Glove-Associated Contact Sensitizers and Irritants

Accelerators/Curing Agents	Antioxidants/Antiozonates	Processing Agents
Aldehyde–amine reaction products	Amines	Surfactants
Benzothiazoles	Phenols	Retarders
Dithiocarbamates	Sulfides	*N*-nitrosodiphenylamine
Dithiophosphates	Donning agents	Phthalic anhydride
Guanidines	Powders	Sulfonamide derivatives
Thirourea	Plasticizers	Stabilizers
Thiurams	Paratoluene sulfonamide	Dibutyl tin dilaurate
Thiocabamyl sulfonamides	Phthalates	Dibutyl tin maleate
Alkylphenol disulfides	Naphthylamines	Epoxy resins
Paraphenylenediamine derivatives		

Note: The manufacture of synthetic and natural rubber latex gloves utilizes a number of similar chemicals. Their selection and the amount used together with the thoroughness of their cleansing process dictates a glove's potential for eliciting irritant or allergic contact dermatitis.

Determination of the specific chemicals responsible for an individual's allergic contact dermatitis may be made by a dermatologist through patch testing. The North American patch test kit, for example, has a selection (#16—Rubber Chemicals) that contains the major sensitizers in latex and synthetic gloves. Alternatively, information supplied by the glove manufacturer may be sufficient to identify the causative agent. Once the responsible contact sensitizer has been identified, the employee can be provided with gloves containing a different accelerator mix or those having lower overall chemical levels (hypoallergenic). A general misconception is that switching to a synthetic glove will alleviate dermal reactions. Because many of the same chemical sensitizers utilized in the manufacture of latex are used in the manufacture of synthetic gloves, switching to a synthetic glove may not be helpful.

Gloves labeled "hypoallergenic" were developed to address the needs of individuals who experience either irritant reactions or allergic contact dermatitis. These gloves are specifically formulated and processed in order to minimize the level of residual chemicals in general and contact sensitizers in particular. For a company to use the "hypoallergenic" label, their gloves must undergo repeated challenges with a modified Draize test on 200 individuals, under the supervision of a dermatologist.[15] The results of the study, along with a description of the manufacturing processes used to obtain the hypoallergenic quality, must be submitted to the FDA in order to obtain clearance for market distribution. Complications with continued use of the term "hypoallergenic" have arisen due to the recent emergence of allergic reactions to protein allergens in natural rubber latex. The label designation "hypoallergenic" has been misinterpreted to mean low in proteins as well as chemicals. Consequently, this term will be changed by the FDA in the near future to more clearly identify gloves with minimal chemical content for those individuals prone to irritant or allergic contact dermatitis, without giving a false sense of security to those individuals allergic to latex protein.

The development of allergic contact dermatitis is dose and rate dependent. Individuals who are predisposed to react to specific sensitizers found in gloves, but always wear gloves with minimal extractable chemicals, may never have enough exposure to reach their symptom expression threshold and may remain symptom-free throughout their careers. Many of those already sensitized to one or more of the chemicals have found that wearing hypoallergenic gloves allows them to work symptom-free. One such study showed a reduction of irritation and allergic contact dermatitis in 80%–85% of the participants.[16]

Some individuals have found relief from irritant and allergic contact dermatitis by using glove liners. Liners provide a buffer between the skin and the glove. They also absorb sweat, thereby decreasing the incidence of maceration resulting from the combined conditions of hyperhydration and friction. Liners must, however, be changed each time the gloves are changed or they can readily become fomites for the spread of infection. Most liners can be washed for reuse.

Although the term "barrier" is not appropriate for the lotions and creams currently being marketed, many provide a degree of buffering or resistance to chemical penetration that some individuals find helpful. Care must be taken to ensure compatibility of the lotion with the type of glove worn (latex, vinyl, nitrile, neoprene, etc.). To prevent barrier degradation, care should be taken to follow the manufacturers' instructions for use. For example, excessive amounts of special lotions, salves, or creams may occlude the skin sufficiently to prevent air exchange and produce another form of adverse skin condition.

> Caution: Though a potential aid for individuals with irritant and allergic contact dermatitis, neither glove liners nor "barrier" lotions are appropriate for individuals who have a Type I, immediate-type hypersensitivity to the proteins in natural rubber latex.

Immediate-Type Hypersensitivity (Type I)

Immediate-type hypersensitivity is an IgE-mediated allergic response. This Type I hypersensitivity has the potential to elicit severe reactions. It is seen in individuals allergic to non-glove allergens, such as penicillin, sulfa drugs, peanuts, and bee stings. In gloves, the allergens are the proteins in natural rubber latex.

Protein allergens come from the source of latex, *Hevea braziliensis*, the rubber tree. The two reactions already discussed (irritation and delayed-type hypersensitivity) have been reported since the 1930s.[17] Immediate-type hypersensitivity, however, was only recently reported. The first documented case of this reaction was to household gloves in Great Britain in 1979.[18] Since the early 1990s, more than 250 individuals have experienced anaphylactic reactions, including some fatalities, to the small natural rubber balloon (cuff) on silicone enema catheters.[19]

Genetically predisposed individuals who are repeatedly exposed to high enough levels of latex protein allergens, may become sensitized. To be genetically predisposed means that the individual has the DNA map to construct IgE antibodies that fit or recognize latex proteins specifically. For example, individuals allergic to cottonwood pollen have the DNA sequences that direct the creation of the cottonwood pollen IgE antibodies, but are not necessarily capable of creating IgE antibodies to the proteins in natural rubber latex. When allergen specific IgE antibodies are made, they attach to basophil and mast cells, proliferating these sensitized cells with each exposure. No symptoms are experienced during this period of sensitization. If, however, a threshold level is reached where there are a sufficient number of these sensitized cells to produce a perceptible response, subsequent exposure may elicit Type I symptoms as the activated cells release histamine and other vasoactive mediators.[20] Symptoms of Type I reactions include urticaria (hives, which may remain local or spread systemically), itching of the eyes, facial swelling, rhinitis, asthma, abdominal cramping, diarrhea, nausea, pharyngeal swelling, headaches, disorientation, tachycardia, hypotension, dizziness anaphylactic shock, and, rarely, death.

Individuals who are already expressing any of the symptoms described should contact their occupational or employee health department and see an allergist. Confirmation of latex sensitivity (Type I) is performed by skin-prick testing or serum IgE studies (ELISA, RAST inhibition, etc.). Latex protein-sensitive individuals, if so advised by their allergist, should

- Wear an alert bracelet or necklace
- Carry an EpiPen® or other source of epinephrine
- Wear only synthetic gloves
- Work in environments free of powdered latex gloves
- Avoid contact with natural rubber products as much as possible
- Be under the care of an allergist or physician familiar with latex protein allergy (Type I)

Individuals who are at increased risk for a Type I reaction to latex are those who have occupational exposure to latex, are allergy prone (atopic) in general, or have specific allergic reactions to the fruits, vegetables, and plants listed in Table 20.7.

TABLE 20.7 Cross-Reactive Allergens

Avocados	Grapes	Passion fruit
Bananas	Kiwis	Potatoes
Celery	Mangos	Ragweed
Cherries	Melon	Strawberries
Chestnuts	Peaches	Tomatoes
Ficus		

Note: Individuals allergic to the proteins in natural rubber latex (Type I) are at increased risk of developing an allergy to the items in this table. The reverse is also true.

A factor that dramatically increases the risk of developing this type of hypersensitivity is very early surgery (first few hours to days of life) followed by multiple surgeries and procedures. This is usually accompanied by extensive natural rubber latex exposure through both invasive and mucosal routes. Spina bifida patients appear to be at highest risk with sensitivity rates reported between 27% and 68%.[21] Since we cannot yet look at a genetic map and tell whether a health care worker can create NRL protein specific antibodies, it is important to decrease the risk of employee sensitization by purchasing only low protein gloves. Does this mean that everyone should switch to synthetic gloves? No. As discussed under "Barrier protection profile of a glove," many synthetics cannot withstand the rigors of use. They break down or rupture, increasing the risk of pathogen or hazardous agent exposure.

It is, however, important to have low protein gloves. To say gloves vary in their level of total protein or protein allergen content is an understatement. Dr. Yunginger of the Mayo Clinic reported differences of over 3000-fold in gloves from different manufacturers.[22] Wearing just one high protein allergen glove can expose an individual to as much allergen as in wearing 3000 low allergen gloves. Taking into account a potential of 20 glove changes a day, the use of high allergen gloves would be equivalent to exposure to 60,000 low-allergen gloves! Another way to visualize the potential impact of high allergen vs. low allergen gloves is the rate with which a susceptible (predisposed) individual reaches their threshold for symptom expression. Figure 20.5 is a theoretical illustration of twins, with the same genetic makeup and the same basic environmental exposure prior to their entrance into their health

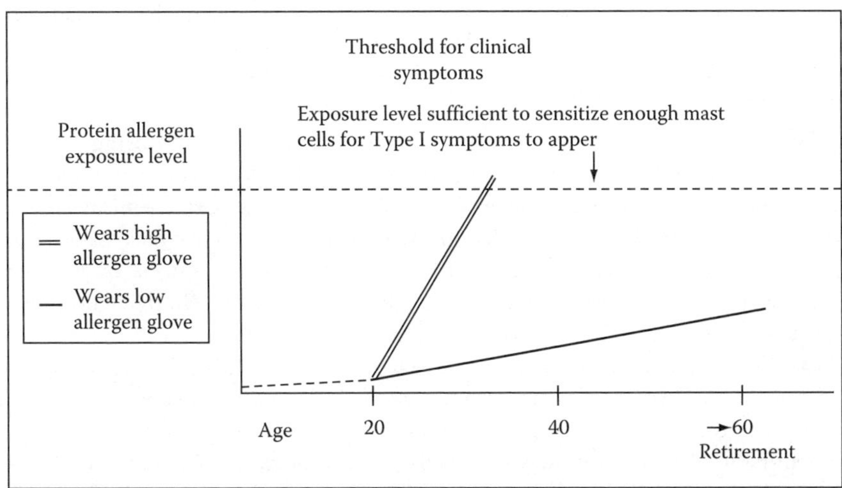

FIGURE 20.5 Theoretical illustration of genetically susceptible twins with the same latex protein allergen level of exposure until they begin careers at different health care institutions.

care careers at different hospitals. One hospital supplies high allergen gloves and the other uses only low allergen gloves. The twin depicted by the double line, at the high allergen facility, has reached his symptom threshold before the age of 30. The other twin, depicted by the single line, retired at age 65, long before reaching that threshold and never experiencing symptoms. Although everyone's threshold and exposure levels will vary, the principles remain the same.

Presently, there are many test methods for evaluating protein in latex products. However, the FDA recognizes ASTM D5712-95, Standard Test Method for Analysis of Protein in Natural Rubber and Its Products, as the standard for measuring total protein levels in latex gloves. This method determines the amount of total water-extractable protein in a glove. While not every protein is an allergen, the total protein content of a glove is a good indicator of potential allergens. The thought process is that if all proteins are kept at a minimal level, then potential allergens will also be low.

This testing method is the basis for the FDA clearance for marketing their product with a protein content statement. While there is no official definition of "low protein," the FDA recognizes levels of ≤50 μg/g as the lowest level confidently detected utilizing this test method. The protein claim will appear as

> This glove contains equal to or less than 50 μg/g of total water soluble protein.

Unfortunately, this statement is not yet a requirement and is utilized by only a few manufacturers. Reasons for not having a protein content statement on the label include the following:

- Several brand names actually represent distributors, packaging gloves purchased from a number of different sources, making control of protein levels very difficult.
- Protein content level might be high and the manufacturer may want to avoid announcing the fact.
- Company's government submission may not yet have been completed.

Ask the manufacturer for the protein content of their product, as determined using the ASTM D5712-95, Modified Lowry method. Test results from different methodologies should not be compared.

Because low-protein latex gloves are intended to prevent individuals from becoming sensitized, not to prevent reactions in sensitized individuals, the following FDA recommended cautionary statement (or one similar in nature) is required:

> Caution: Safe use of this glove by or on latex-sensitized individuals has not been established.

Respiratory Complications

Asthma and other pulmonary restrictive diseases, such as pneumonitis, are being experienced by a growing number of hospital personnel. It is important to identify glove-related and non-glove-related causes of these reactions. Inhalation of glove powder alone, or coated with chemicals and endotoxins from the glove, may cause inflammatory reactions. Disinfectants, sterilants, and various other chemicals ubiquitous in the hospital are also often irritants and/or contact sensitizers. These chemicals can adhere to the powder particles, become aerosolized, and readily inhaled by individuals in the area. This can also initiate nasal, throat, and respiratory symptoms.

Protein from the glove surface can, similarly, attach to powder particles. Latex-sensitive individuals (Type I) exposed to latex allergens may or may not have respiratory symptoms. Symptoms for those who do experience respiratory involvement range from restrictive airways to full-blown asthma. A diagnosis for latex-protein-induced asthma includes a latex positive skin-prick test or serum IgE assay.

Close monitoring and detailed cause/effect histories are necessary for proper diagnosis of personnel experiencing respiratory symptoms in the workplace. Note that it may be possible that individuals experience a delayed reaction when contact sensitizers are involved.

Hospital Protocols

Hospital protocols must be established to address the various aspects of glove-associated reactions.

Employees

Protocols must be developed outlining the program that will use to identify the cause of employee reactions and the appropriate actions to be implemented for the individuals and their work environments. A brief diagram to facilitate the differentiation process is provided in Figure 20.6. Actual protocols, however, should be reviewed by staff allergists and immunologists prior to approval.

Prevention

Protocols should be established for the selection and proper use of gloves to minimize employee risks of developing glove-associated adverse conditions. A brief outline of important issues to be addressed is included in Table 20.8.

Although the purchasing department may find the initial cost of low-chemical, low-protein, powder-free gloves to be more expensive, careful cost-of-reactions records kept by the occupational or employee health department may well change this assessment. Meticulous tracking of the cost of attending to employee medications, physician visits, sick leave, temporary replacement salary, retraining, and, potentially, workers' compensations claims has yielded interesting results. Some health care workers have had to discontinue their vocation of choice due to severe latex allergies. The cost can be up to $215,000, with as much as 90% of it paid by the employer.[23] Another report listed costs of $194,000.[24] At Hamilton Civic Hospitals where there are more than 3000 full-time workers and 750 beds, over 300 of these employees have been diagnosed as latex allergic (Type I).[24] Prevention, up front, more than pays for itself.

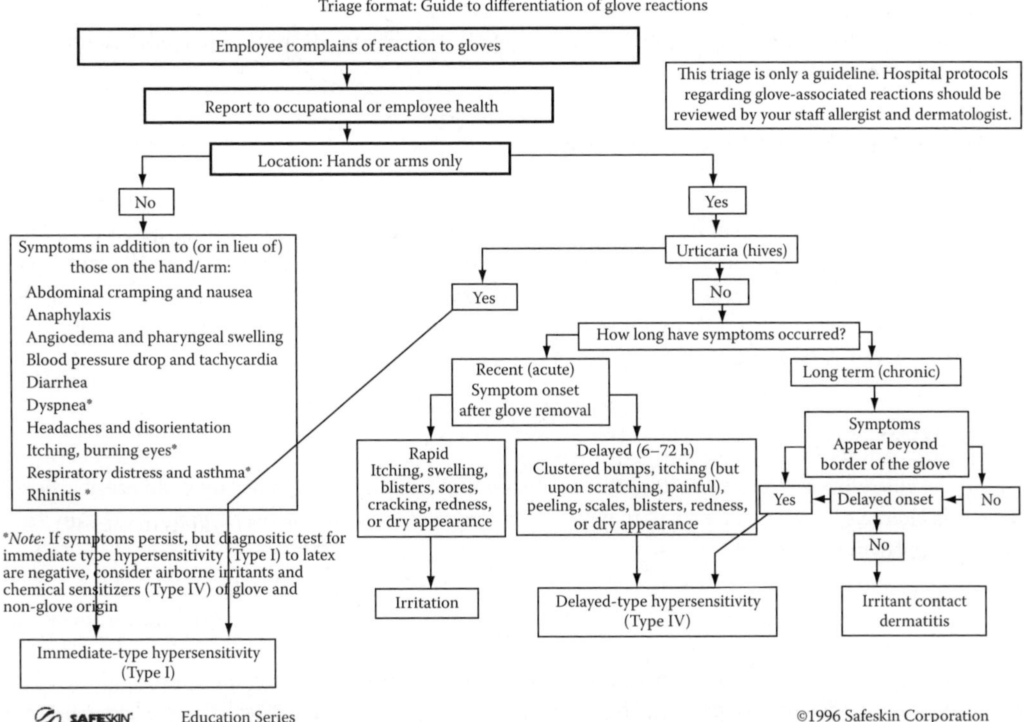

FIGURE 20.6 Sample employee protocol to facilitate differentiation of glove reactions.

TABLE 20.8 Prevent Development of Glove-Associated Adverse Conditions

Reaction	Glove Characteristic for Prevention
Irritation (irritant dermatitis)	Fit properly, avoid friction
	Powderfree to prevent absorption of protective lipids and abrasive action
	Low in chemicals (referred to as hypoallergenic until an alternate terminology is selected)
	Changed frequently
	Glove liners may be helpful
Delayed-type hypersensitivity (Type IV, allergic contact dermatitis)	Low in chemicals (referred to as hypoallergenic until an alternate terminology is selected)
	Powderfree to prevent transfer of adherent chemical antigens
	Glove liners may be helpful
Immediate-type hypersensitivity (Type I)	Latex low in proteins (allergens)[a] or an appropriate synthetic
	Powderfree to prevent aerosolization and transfer of adherent protein allergens
Respiratory complications	Powderfree gloves low in proteins and chemicals

[a] Individuals who have already become latex-protein sensitized (Type I) must be in synthetic gloves with those around them in powderfree.

Patients

Although the focus of this chapter is on the employee, it is imperative that hospital-wide (floor, operating rooms, pharmacy, clinics, etc.) protocols be established for the identification and care of latex-sensitive (Type I) patients. A number of individuals, hospitals, and organizations who have established such guidelines are noted in the reference section.[25] As a safety precaution, all spina bifida patients should be considered latex positive.

Powder vs. Powderfree Gloves

Powder serves two purposes. First, it functions as a manufacturing aid by preventing the glove from sticking to itself or other gloves during processing and packaging. Second, it facilitates donning. Most glove powder is composed primarily of USP absorbable dusting powder (chemically treated cornstarch). To produce a powderfree glove the powder must be removed and the glove surface treated to reduce tack (stickiness) and ease donning. Despite the added costs required for powder removal, the demand for powderfree gloves is escalating due to a number of increasing concerns.

Absorption

One of the advantages of powder is its ability to absorb moisture, as is evidenced by rapid sweat absorption. This characteristic can also result in negative consequences, including overdrying of the skin leading to chapped, cracked hands.

Protein

Powder absorbs a number of substances other than moisture. Several articles have been published on the adsorption (adherence) of proteins, from the surface of latex gloves to powder particles. When the protein-laden particles are aerosolized, the protein allergens they carry increase allergen exposure. Levels will depend on a myriad of factors, including

- *Protein content.* In general, the higher the protein content of the glove, the higher the concentration on each particle.
- *Size of the room.* The smaller the room, the more rapidly the airborne allergen will increase in concentration.
- *Number of individuals.* The more people in, or in and out of the room, the higher the allergen level can rise due to the powder/protein contribution from their individual gloves and the re-aerosolization of powder that had already settled.
- *Efficiency of the air filtration system.* If the system is extremely efficient, it may keep allergen levels lower than if it is not. Poor systems encourage stagnation and may contribute to build up over time. If systems are poorly constructed, the powder may actually be transported to remote locations.

Studies have demonstrated that there is no significant difference in aeroallergen (latex protein) levels when using powderfree latex gloves or when no gloves at all are worn.[26] It should be emphasized again that latex-sensitive individuals (Type I) should wear only synthetic gloves and work in a low-protein, powderfree environment.

Chemicals

Just as proteins can be adsorbed from the gloves to the powder particle surface, so can chemicals. Non-glove chemicals such as glutaraldehyde, formaldehyde-based disinfectants, chemotherapeutic agents, and medications can also be absorbed by the powder and aerosolized.

Microorganisms

The practice of wearing gloves and changing them between patients is based on a principal demonstrated by Semmilweis that states eliminating the vehicle of disease transport between patients, as well as between patient and health care professional, prevents cross contamination. However, because they have reduced the incidence of disease transfer so dramatically, we have neglected to note some critical areas of concern where gloves have actually increased exposure to infectious agents. As gloves contact contaminated body substances, microorganisms adhere readily to the hundreds of thousands of particles. These efficient fomites await transport to unsuspecting new hosts via direct contact, indirect transfer (e.g., computer terminals, charts, water pitchers, pens, etc.), or aerosolization (Figure 20.7). Although not specifically citing the role of powder in microbial aerosolization, Pike[27] stated "Inhalation of infected airborne particles (aerosols) released during many common laboratory manipulations have probably caused the largest number of laboratory associated diseases."

Studies have shown that the presence of powder in wounds can significantly decrease the resistance to infection.[28] Add to this its potential role as a fomite, and it is readily apparent that there are a number of immune-compromised and infectious-disease areas where powder may increase the incidence and spread of nosocomial infections.

Erroneous Laboratory Results

Powder has been reported to have caused the following:

- PCR amplification interference[29]
- False negative HIV tests[30]
- Erroneously low cyclosporin concentration determinations[31]
- False pregnancy results[31]
- Interference with assays dependent on optical scanning (plate readers, spectrophotometers, electron microscopy, etc.)[31]

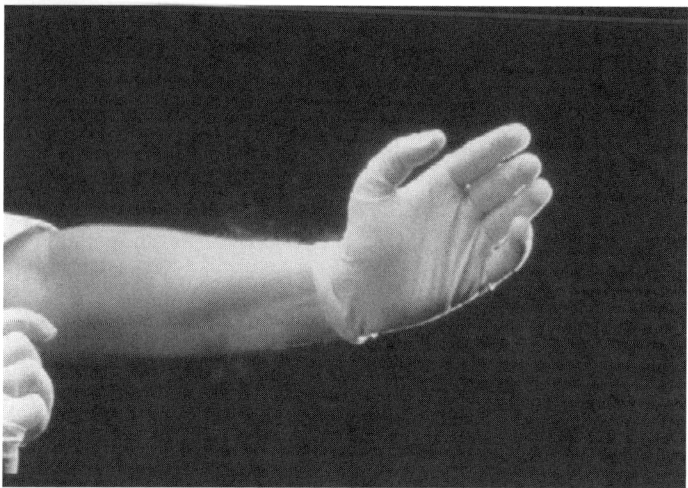

FIGURE 20.7 When donning, manipulating, or removing gloves, powder is readily aerosolized. Proteins and chemicals from the gloves, as well as pathogens that may have contaminated the glove during patient care or specimen handling, can attach to the powder and similarly be disseminated.

Proper Glove Removal and Disposal

Contaminated gloves must be removed with consideration for the potential spread of infection to yourself or individuals in the vicinity. The pictures in Figure 20.8 illustrate correct removal technique. Care must be given to the avoidance of skin or apparel contact with the exterior or dirty side of the glove. Avoid snapping gloves and the sling-shot method of disposal.

Summary

Gloves reduce the daily risks of cross-infection in the hospital environment. As with all PPE, careful consideration must be given for appropriate selection. Selection criteria should include the following:

- *Durability.* Should have adequate in-use barrier protection for tasks to be performed. Prevention is cost effective.
- *Low in proteins.* Gloves should have a low level of protein allergens or be made of a suitable synthetic material to reduce the risk of developing immediate-type hypersensitivity (Type I). Once an individual is sensitized to latex proteins, they must be in a synthetic glove in a powderfree environment. Prevention is cost effective.
- *Low in chemicals (hypoallergenic).* Latex and synthetic gloves should be low in chemicals to reduce the risk of developing irritant or Type IV allergic contact dermatitis. Prevention is cost effective.
- *Powderfree.* Gloves should be powderfree to eliminate the role of powder particle contact transfer and aerosolization of adherent protein allergens, chemicals, and microorganisms. Powderfree gloves reduce laboratory assay interference. Prevention is cost effective.
- *Snug but comfortable fit.* Gloves should fit snugly enough to eliminate baggy interference but should not be so tight that they impede movement.
- *Wet and dry grip.* Should be appropriate for the task performed.

Once the appropriate glove has been selected it should be stored, worn, and removed properly.

Storage recommendations:

- Cool (<90°F)
- Dry (<40% humidity)

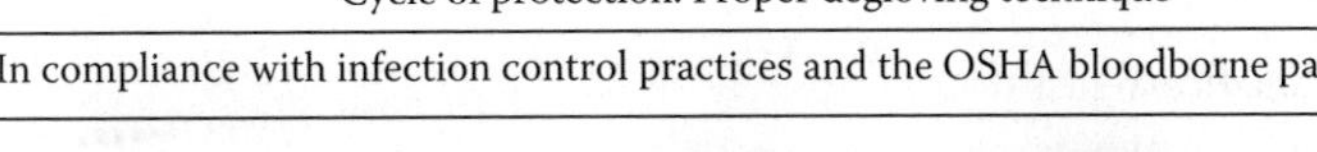

Cycle of protection: Proper degloving technique
In compliance with infection control practices and the OSHA bloodborne pathogen rule

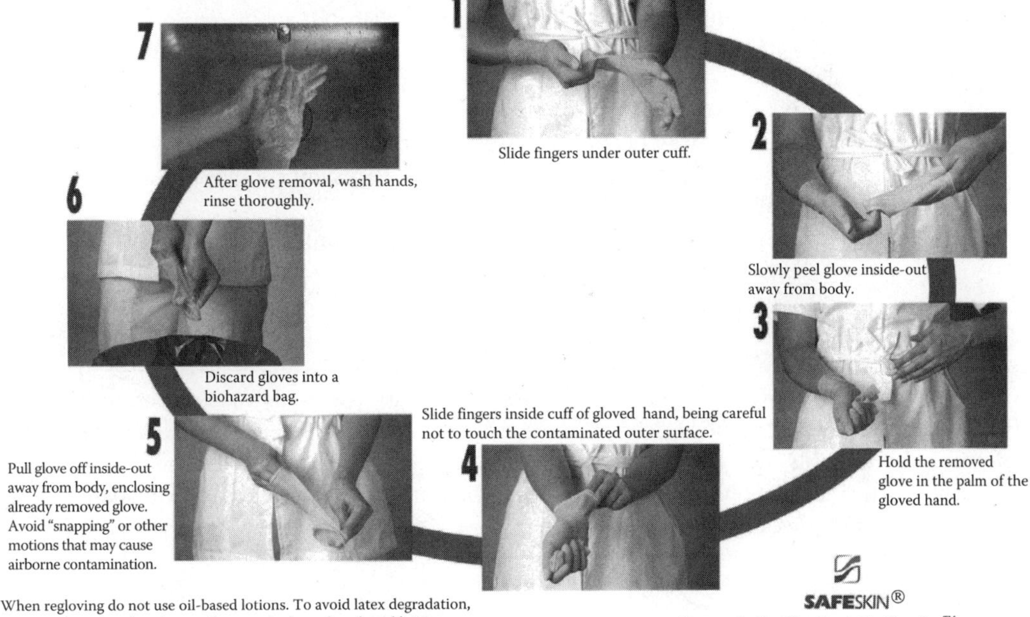

1
Slide fingers under outer cuff.
2
Slowly peel glove inside-out away from body.
3
Hold the removed glove in the palm of the gloved hand.
4
Slide fingers inside cuff of gloved hand, being careful not to touch the contaminated outer surface.
5
Pull glove off inside-out away from body, enclosing already removed glove. Avoid "snapping" or other motions that may cause airborne contamination.
6
Discard gloves into a biohazard bag.
7
After glove removal, wash hands, rinse thoroughly.

When regloving do not use oil-based lotions. To avoid latex degradation, use only water, glycerin, or other non-hydrocarbon-based lotions.

Technical Advisors: Maria Ninivaggi, RN, MSN, CIC and Joanne Selva, RN, BS, CIC from Infection Control Consultants, Inc.

SAFESKIN®
Because the Best Reaction is No Reaction™
For an introduction to Safeskin's Gloving Management Program call 1-800-462-9989.

FIGURE 20.8 Proper glove removal. Glove removal should always be followed by proper hand washing and extensive rinsing.

- Away from direct sunlight, ultraviolet or fluorescent light, x-ray machines, or any other electrical generating sources that produce ozone
- Maintain in primary packaging
- First-in first-out rotation

In-use guidelines:

- Trim the long and sharp fingernails.
- Remove the jewelry.
- Make certain that any lotion worn under gloves will not degrade the glove material.
- Don properly.
- Observe the glove while donning for any holes, tears, or rips and discard if noted.
- Double the glove when the potential for cross-infection and the use of sharps deems the practice appropriate.

Glove removal recommendations:

- Remove the gloves between patients (consider changing between procedures on the same patient if there is a potential for cross contamination).
- Remove the gloves as soon as possible when a tear or perforation is suspected.
- Glove removal should be performed with consideration for contamination and aerosolization of surface microorganisms or particles.
- Wash and rinse the hands thoroughly each time gloves are removed.

Observe for glove-associated reactions:

- First report to occupational health
- Determine if dermal reaction is
    - An irritation (irritant contact dermatitis), dermatologist
    - Allergic contact dermatitis (chemical allergy, Type IV, or delayed-type hypersensitivity), dermatologist
    - Immediate-type hypersensitivity (protein allergy and Type I hypersensitivity), allergist
- Determine the cause of any occupationally associated respiratory problems:
    - Asthma (if glove, may be NRL protein related), must wear synthetic gloves in a powderfree environment, allergist
    - Pneumonitis (irritant or chemical-associated inflammation), consider powderfree environment, allergist or respirator specialist

Conclusion

Glove protection is no longer defined solely by physical integrity. A glove must also protect the wearer from developing glove-associated reactions, whether dermal, respiratory, or systemic. Prevention is cost effective. Although this chapter addresses occupational concerns, readers should extend the issues discussed here to their patients, being aware of the possibilities not only from dermal and inhalant contact, but also through invasive exposure.

References

1. Wyklicky H and Skopec M. Ignaz Philipp Semmelweis, the prophet of bacteriology. *Infect Control* 4:5, 1983.
2. Centers for Disease Control. Recommendations for prevention of HIV transmission in health-care settings. *Morb Mortal Wkly Rep* 36(Suppl. no. 2S), 1987.
3. Occupational Safety and Health Administration. Occupational exposure to bloodborne pathogens: Final rule, 29 CFR Part 1910: 1030. *Federal Register* 56, December 6, 1991.
4. Occupational exposure to bloodborne pathogens: Final rule, 64014.
5. Ibid., 64011.
6. Gonzales E and Naleway C. Assessment of the effectiveness of glove use as a barrier technique in the dental operatory. *J Am Dent Assoc* 117:467–469, 1988.
7. Pugliese G. *OSHA's final bloodborne pathogen standard: A special briefing*. Chicago: American Hospital Association (AHA), 1992; AHA Item No. 155904.
8. Gerberding JL, Littell C, Tarkington A, Brown A, and Schecter W. Risk of exposure of surgical personnel to patients' blood during surgery at San Francisco General Hospital. *N Engl J Med* 322:1788–1793, 1990.
9. Garner JS and Simmons BP. Guideline for isolation precautions in hospitals. *Infect Control* 4, 1983.
10. Occupational exposure to bloodborne pathogens: Final rule, 64177.
11. Rego A and Roley L. In-use barrier integrity of gloves: Latex and nitrile superior to vinyl. *Am J Inf Control* 27(5):405–410, 1999.
12. Shmunes E and Darby T. Contact dermatitis due to endotoxin in irradiated latex gloves. *Contact Derm* 10:240–244, 1984.
13. Truscott WT and Roley L. Glove-associated reactions: Addressing an increasing concern. *Dermatol Nurs* 7:283–292, 1995.
14. Truscott WT. The industry perspective on latex. *Immunol Allergy Clin Nort Am* 15(1), 1995.
15. Food and Drug Administration. Hypoallergenic cosmetics. Proposed definition, 21 CFR Part 1. *Federal Register* 7288–7292, February 25, 1974.

16. Boyer EM. The effectiveness of a low-chemical, low-protein medical glove to prevent or reduce dermatological problems. *J Dent Hyg* 69:2, 1995.
17. Downing JG. Dermatitis from rubber gloves. *N Engl J Med* 208:196, 1933.
18. Nutter AF. Contact urticaria to rubber. *Br J Dermatol* 101:5497, 1979.
19. Tomazic VJ, et al. Short analytical review—Latex-associated allergies and anaphylactic reactions. *Clin Immunol Immunopathol* 64(2):89–97, 1992.
20. Gonzalez E. Latex hypersensitivity: A new and unexpected problem. *Hosp Pract* 27:2, 1992.
21. Sussman G. Allergy to latex rubber. *Ann Int Med* 122:43–46, 1995.
22. Yunginger JW, Jones RT, Fransway AF, et al. Extractable latex allergens and proteins in disposable medical gloves and other rubber products. *J Allergy Clin Immunol* 93:837–842, 1996.
23. Sussman G and Beezhold D. Safe use of natural rubber latex. *Allergy Asthma Proc* 17:2, 1996.
24. Sussman G, Ownby D, and Keith P. The implications of latex allergy. Presented at the American Academy of Allergy, Asthma and Immunology (AAAAI), Latex Allergy Section, 1996.
25. Children's Hospital, 300 Longwood Avenue, Boston MA 02115; American Association of Nurse Anesthetists (AANA), 222 South Prospect, Park Ridge IL 60068-4001; Spina Bifida Association of America, 4590 MacArthur Blvd. NW, Suite 250, Washington DC 20007-4226.
26. Swanson M, Bubak M, Hunt L, Yunginger J, et al. Quantification of occupational latex aeroallergen in a medical center. *J Allergy Clin Immunol* 94(3):445–451, 1994.
27. Pike MR. Laboratory-associated infections: Incidences, fatalities, causes and prevention. *Annu Rev Microbiol* 33:41, 1979.
28. Jaffray DC and Nade S. Does surgical glove powder decrease the inoculum of bacteria required to produce an abscess? *Surg Glove Powder* 28:4, 1983.
29. deLomas JG, et al. False-negative results by polymerase chain reaction due to contamination by glove powder. *Transfusion* 32:83, 1992.
30. Lampe AS, Pieterse-Bruins HJ, and Egter Van Wissekerke JCR. Wearing gloves as cause of false-negative NIV tests. *Lancet* 2(8620):1140–1141, 1988.
31. Hamlin CR, Black AL, and Opalek JT. Assay interference caused by powder from prepowdered latex gloves. *Clin Chem* 37:8, 1991.

21

Health and Safety Hazards of Shiftwork: Implications for Health Care Workers and Strategies for Prevention

Joseph Schirmer
American Federation of Labor and Congress of Industrial Organizations

Introduction 21-1
Hidden Costs of Shiftwork: Public Safety 21-2
Night Work in the Health Care Industry • Effects of Shiftwork on Workers • Nurses • Accidents • Psychological Effects of Shiftwork • Shiftwork and Reproductive Function • Shiftwork and Gastrointestinal Function • Shiftwork and Heart Disease
Strategies 21-12
Control of Light in the Environment • Direction of Shift Rotation • Naps and Shiftwork • Personal Medical Factors • Physical Fitness • Controlling the Frequency of Shift Rotation • Limiting the Hours Worked at Night • Education
Conclusion 21-14
Checklist for Reducing Hazards of Night Work and Shiftwork
References 21-15

Introduction

Many physiological functions follow a daily rhythm consistent with the rising and setting of the sun. During daytime, body temperature, heart rate, blood pressure, and adrenalin secretion achieve peak output as physical and mental work capacity rise.[1] This daytime rise in work capacity coincides with the period of peak activity for most people. Conversely, body temperature, heart rate, adrenalin release, and mental alertness drop to their lowest values during night. These rhythmic drops in physical function and mental alertness coincide with periods of decreasing activity, rest, and sleep for most people. It is important for health and well-being to maintain this rhythm by obtaining full rest and sleep. The nighttime low-energy period is also a time of some risk, especially for those who are ill. The most common time of death, based on a composite study of 437,511 deaths, was between 4 A.M. and 6 A.M.[2] Is there also some risk to healthy workers who try to reverse the normal sequence of day work and night rest?

Since the introduction of electric power in the late nineteenth century, it has become possible to provide sufficient power and lighting to enable work to continue through the night. In heavily capitalized manufacturing industries, night work enables employers to achieve more rapid returns on their capital investment and greater efficiency through continuous production. In service industries such as transportation, utilities, communications, health care and hospitals, or police and fire protection, night work is increasingly necessary. Indeed, there is a common expectation that 24 h service will be available.

Among all U.S. workers in 1980, approximately 26% of men and 18% of women worked a variable shift schedule.[3] Shiftwork is particularly common in young families. In the United States, approximately 50% of young couples with children under 5 years old include one spouse who works nonstandard hours.[4] Most shiftworkers (72%) do so involuntarily; only a minority (28%) report voluntary selections.[5]

Hidden Costs of Shiftwork: Public Safety

What do Chernobyl, Valdez, Three Mile Island, and Bhopal have in common? They were all disastrous accidents resulting in environmental contamination which occurred during night shifts, between 11 P.M. and 7 A.M.

The accident at Bhopal began at 12:56 a.m.; *Exxon Valdez* ran aground approximately at midnight. The Chernobyl nuclear power facility incident began at 1:23 A.M. The Three Mile Island nuclear power facility incident began at 4 A.M., and then between 4 A.M. and 6 A.M., as shiftworkers did not recognize the loss of core coolant resulting from a broken valve, the radioactive core overheated and radioactive material escaped into the environment. The radiation releases at Chernobyl and Three Mile Island have had national and international political repercussions. Public confidence in the safety of nuclear power has declined and electric utilities have shifted investment toward other sources of power. The Chernobyl accident near Kiev may also have contributed to regional distrust of the central authority in Moscow and to a sharper appreciation of the value of political independence for the Ukraine.

The methyl isocyanate release at a chemical facility in Bhopal occurred during the night shift at 12:56 A.M. After the fatalities occurred in Bhopal, public opinion in the United States supported a major federal policy change. Employers are now required to report to the federal and local governments comprehensive information about the identities and amounts of chemicals used, stored, and released into the environment.

It is appropriate that public demands for safe sources of power and for access to information about toxic materials should guide our society. Unfortunately, it has taken well-publicized disasters for these needs to be recognized. In a sense, the disasters act as catalysts to speed up decisions which would otherwise occur more slowly. Disasters focus public opinion into effective action. The demands for safety and increased public access to information which followed these disasters may well have been too narrowly focused. For example, the timing of these accidents suggests a need to critically evaluate current patterns of shiftwork. Are current shift schedules compatible with public safety or responsible patient care?

There is a growing understanding of the importance of shiftwork. The official report on the Three Mile Island incident recognized that the weekly rotating shift schedule may have been a contributing factor to the operator errors.[6] Other incidents in the U.S. nuclear power industry have also involved human errors during the early morning hours.

The oil tanker *Exxon Valdez* ran aground in Prince William Sound, Alaska around midnight, spilling 258,000 barrels of crude oil and causing terrible environmental contamination. The cost of a partial cleanup exceeded two billion dollars. An investigation by the National Transportation Safety Board determined that the shift schedule provided inadequate rest and the resultant third mate's fatigue was one of the five major factors responsible for the accident.[7] Insufficient staff were another of the major factors cited.

The official report on the *Challenger* accident in 1986 cited the fact that key managers had had less than 2 h sleep and were under considerable time pressure.[8] Operator fatigue as a result of shiftwork was also recognized as a major factor; it nearly caused a disaster to the shuttle *Columbia* 3 weeks before the

Challenger accident. The operators who drained 9 ton of oxygen from the shuttle 5 min before the scheduled launch were in the eleventh hour of a 12 h shift after working three 12 h night shifts in a row.

Our traditional attitudes about shift and staffing patterns have minimized and individualized the problems of night work. As scientific studies of night work continue and as public awareness of these issues increases, our attitudes will also evolve. As we come to appreciate the costs of night workers' mistakes, we can weigh the potential costs of future mistakes against the costs of altering current staff schedule patterns. A recent report by the Office of Technology Assessment on shiftwork, "Biological Rhythms," shrewdly focused on nuclear power plant workers, doctors, nurses, and military personnel—groups whose work and fatigue-driven mistakes have the potential to affect everyone.[9] This report raises the question for national consideration: "Are current shift schedules compatible with public safety or responsible patient care?"

Health care workers must often work at night and they routinely make decisions which have potential life and death consequences. What are the impacts of staffing and scheduling patterns?

Night Work in the Health Care Industry

A greater percentage of health service workers work at night (10.3%) than is the case in any other occupational employment sector. Hospitals, the biggest employers in the health care field, employ more night shiftworkers (6.6%) than any other industry.[5] Nurses are the largest group of health care workers and the largest single group affected by night work. There are more than two million registered nurses (RNs) in the United States.[9] Approximately 71% of all hospital-employed RNs work outside the day shift at least part of the time; only 29% work straight day shifts. Of all RNs in hospitals, 21% work only evenings; 17% work nights permanently, and 33% rotate shifts. Of those who rotate, some rotate to all shifts and some rotate only to two shifts. In total, 42% of all hospital-employed RNs work nights: 17% work full-time nights and 25% rotate to nights.[10]

There are three quarters of a million health technologists or technicians and 30% work nonstandard hours: evenings, nights, or rotating shifts.[5]

Physicians also work nights and long hours when completing postgraduate clinical training in specialties such as internal medicine, pediatrics, or family practice. There are approximately 85,000 physicians in U.S. postgraduate "residency" programs. Postgraduate programs which required physicians to live at the hospital (hence the term "resident") began at Johns Hopkins University approximately 100 years ago. Residents were expected to be on call 24 h a day and they were discouraged from marrying.[9] Today "residents" no longer live in the hospitals full time and marriage is a matter of personal choice, but the hours still include 24 h shifts. The most inexperienced doctors, first year residents, are the most likely to be required to work 24 h shifts and other excessive hours. Until residents in New York went on strike in 1975, they were on-call every second night, a duty often involving 24 h of continuous work. Now, on-call shift schedules in New York and elsewhere are usually limited to every 3 days. Still, the hours for residents would be considered excessive by most standards; many residents routinely work more than 80 h a week and 24 h shifts are commonplace.[9]

What are the implications and results of this apparent contradiction between normal physiology and night work for health care workers? What are the effects of working when our circadian rhythms suggest sleep? Do night workers experience more stress, fatigue, disease, or injury as a result of working when their bodies would normally rest? Are patients jeopardized by these shift schedules? Are there individual differences in tolerance for shiftwork? Are some people more suited for night work? Can adaptation occur? What strategies maximize adaptation and minimize health hazards? In the last decade, several reviewers have evaluated the adverse health effects of shiftwork and set forth ideas for prevention.[9,11–15]

The goals of this essay are to review the effects of shiftwork, particularly for health care workers, and to review what strategies may be useful in dealing with these issues. Because of the significance of good medical care and because so many health care workers work at night, shiftwork information obtained from other industries should be studied because of its implications for the health care field.

Effects of Shiftwork on Workers

Sleep Disorders

For more than 7 million U.S. workers who must work at night,[5] the most obvious problem they experience is sleep disruption and fatigue. Although workers attempt to ignore or overpower their normal circadian rhythms, most cannot. The normal internal biological rhythms persist. External factors such as light, noise, family, and social demands also interfere with sleep. Stress and fatigue result.

Sleep after night work is shorter compared to normal sleep. Night workers on permanent shift have been found to get about 1 h less sleep than do afternoon or day workers, and night workers who rotate shifts get 2 h less sleep than those who rotate to other shifts. Night workers who rotate get about 5 h of sleep.[16] The problem gets worse with age. Among permanent night shiftworkers, those 18–29 years old sleep 40–50 min longer than those aged 40–49.[17]

The disruption in sleep pattern is more severe the later the onset of sleep is postponed toward noon. Night shiftworkers' subjective feelings of sleepiness have been confirmed by electroencephalography (EEG). Despite staying up all night, many night workers experience insomnia during the day.[18,19]

Fourteen studies have examined nurses' sleep disorders as function of shiftwork.[23] In general, the larger studies report that nurses who work nights or rotating shifts get less sleep than other nurses.[9] Hospital nurses who rotate shifts have been found to have unique sleep patterns, distinct from those of regular sleepers, shiftworkers, other night workers, or other known sleep disorder patterns.[20] Nurses who permanently work nights report getting the least sleep of any group. Part-time night nurses report less sleep than full-time night nurses,[18] probably because they revert to a normal schedule while not working. Lack of sleep is a greater problem for those with children; the more children, the greater the problem.[21]

Adaptation seems to occur to some extent. Part-timers and those who rotate shifts experience greater difficulty than those who permanently work nights. Nurses who rotate shifts experience more fatigue than do fixed shiftworkers, even when the study controlled for the potentially confusing effects of age and marital status.[22]

A 6268-person survey in Finland examined how sleep disorders varied with occupation. Those occupations with most sleep problems tended to be lower in social class and more likely to be required to work nights. For example, among female staff nurses, 14.4% reported difficulty falling asleep, almost twice as many as among head nurses (7.8%) and nearly three times more than male doctors (4.9%). More than one-fourth (26.4%) of female hospital aides reported usually waking up three times or more at night, while other health care occupations reported disturbed sleep less often: female ward nurses = 9.4%, female head nurses = 8.9%, and male physicians = 1.6%.[23]

Sick Time

A 30 month study of shiftwork sponsored by the National Institute for Occupational Safety and Health reviewed health and accident files and surveyed by questionnaire 1200 nurses and a similar number of food processors to evaluate the health consequences of shiftwork.[22] They found that nurses on rotating shifts tended to take more sick time than did nurses on fixed shifts. Also, the rotating shift nurses tended to have more severe reasons for absence from work than did fixed shift nurses. Rotating shiftworkers (nurses and food processors) tended to use worksite clinics more than did fixed shiftworkers. When the reasons given for visiting the clinics were examined, it became clear that severity was a problem as well as frequency of illness. Rotators seemed to suffer from more serious problems, and used sick time only for the most serious problems. They went to worksite clinics for less severe problems, problems that might have caused fixed shiftworkers to miss work.

Job Performance under Night Work and Shiftwork Conditions

Working during normal sleep hours results in poorer job performance and increased frequency of fatigue-related accidents at night.

Measures of job performance over a 24 h period indicate that reaction times are increased at night and ability to do mental arithmetic is diminished; the observed decreases in performance are similar to those caused by moderate alcohol consumption.[24] These trends have been observed in many occupations. A British study found that residents who were sleep-deprived had less ability to do math.[25] The frequency of minor accidents to hospital staff is highest during third shift, with observable peaks in frequency occurring around changes of shift. Operator speed in answering switchboards and textile workers' speed of joining threads is lowest during the night shift.[26] The distribution of 75,000 Swedish meter reader errors over a 20 year period showed a major peak between 2 A.M. and 4 A.M.[27] A study of over 2200 locomotive train driver errors resulting in emergency braking situations indicated that peak error frequencies occur in the early morning hours between 3 A.M. and 6 A.M. and again in the early afternoon between 1 P.M. and 3 P.M.[28]

Nurses

Nurses who rotate shifts are perceived to be less efficient by their supervisors. Nursing supervisors ranked nurses who work rotating shifts lower in terms of job performance, motivation, and patient care than fixed shiftworkers.[29] The psychological effects of shiftwork may result in a more negative or depressed attitude toward work. In a study of circadian rhythms in nurses, 12% reported episodes of temporary gross motor paralysis, lasting an average of 4 min per incident, while working nights. Typically these incidents occurred in the latter part of the shift, during early morning, when nurses were sitting and writing in patients' charts or reading.[30]

One might expect that the effect of sleep deprivation on physicians' and nurses' work performance would have been well studied. There is actually a great need for more research on the effects of night work, extended hours, and sleep deprivation on nurses' and physicians' performance. Deaconson et al. cite only 10 studies on physicians.[31] Most including Wilkinson,[32] Poulton,[33] Friedman,[34] and Hawkins[35] report that physicians function less reliably during extended periods of work, including night work. Others found little or no effect from sleep deprivation on residents.[31,36,37]

Friedman[42] evaluated the effects of sleep deprivation on a group of 14 interns in New York City. Subjects were asked to read a 20 min electrocardiogram strip and to characterize their emotional and psychological state both when rested and when fatigued. In the fatigued condition, interns had slept an average of 1.8 h in the preceding 32. In the rested condition, interns had slept an average of 7 h in the preceding 32. When sleep-deprived, interns made an average of 85% more errors in electrocardiograph interpretation. In addition, 3 of 14 interns (21%) took 7.3 min longer to interpret the electrocardiogram when fatigued than when rested. Fatigued interns also felt sadder, less socially affectionate, and less vigorous than rested interns. They judged themselves to have perceptual, cognitive, and psychological problems when tired; 12 (86%) reported difficulty thinking, 10 reported depression, 9 hyperirritability, 6 depersonalization and inappropriate effect, and 5 recent memory problems. A minor flaw in this study is that the time of day when testing occurred was not reported or controlled. This is, however, the *only* study to examine the effects of sleep loss on a typical clinical task.

Since performance varies with time of day in a circadian fashion, independently of sleep loss, the time when the test is taken may affect the outcome. Thus, for example, in a test of grammatical reasoning, a 4% reduction in speed and a 5% increase in errors occurred at 11 P.M., compared to 8 P.M.[38]

Beatty examined the effect of no more than 2 h sleep in the preceding 24 h on anesthesiology residents' ability to perform simulated surgical tasks.[39] They found slight deficits in vigilance, but a major slowdown (23%) in grammatical reasoning after a single night period of sleep deprivation.

Poulton[41] evaluated how sleep loss impaired the abilities of 30 junior residents to perform card sorting and laboratory data interpretation. Sleep loss was calculated by comparing hours of sleep in the 24 h period preceding the tests (exposure) to the average hours of sleep during the previous 4 weeks (control period). These doctors had worked an average of 65 h per week during the month under study and therefore exposure to sleep loss occurred both in the 4 week period and in the 24 h preceding the testing. This study design has an intrinsic tendency to underestimate the effect of the recent sleep loss, since sleep loss occurred during both periods. The subjects' performance was measured with 3 min tests eight times during 1 month. The study found that a 3 min card sorting test was negatively affected by sleep deficits of 5 h, and score during 3 min of interpreting laboratory tests was negatively affected by sleep deficits of 8 h. The authors note that shorter sleep deficits, such as 3 h, would have produced negative effects if performance had been tested over a longer period, for example, a 20 min electrocardiogram rather than a 3 min test.

Wilkinson[40] surveyed all 6500 junior doctors in England, Scotland, and Wales in 1973 with the question "Do you think that your hours of duty are so long as to impair your ability to work with adequate efficiency?" The 2452 responding checked one of five choices; the results were as follows: always = 3.3%, often = 34%, occasionally = 47.6, rarely = 12.2%, never = 2.9%. Thus over one-third considered that their efficiency was inadequate due to overwork. Specialties with more than one-third responding "always" or "often" to the question included obstetrics, orthopedics, surgery, pediatrics, general medicine, and emergency. Respondents reported that in the previous 24 h, they had worked an average of 12.76 h and been on call 8.24 h, with 3 h of free time; they had an average of 6 h of sleep. The majority of respondents volunteered comments; 105 (4%) thought that patients seen at the end of a period of extended duty received noticeably worse treatment; 141 (6%) mentioned actual errors of varying seriousness. This study, while relying on physicians' subjective judgments about the effect of extended hours, is the largest, most ambitious, and most thorough study of the subject.

Reznik[45] considered the effect of sleep deprivation on 12 surgical residents in a rural community. Sleep deprivation was defined as less than 3 h of sleep in a 24 h period. The parameters tested were factual recall, interpreting laboratory test data, and manual dexterity. No significant differences in performance were found, although the authors note that longer periods of sleep debt would be more likely to result in measurable effects.

Deaconson[44] tested 26 surgical residents daily for 19 days, and found that sleep loss (as defined by lack of 4 h continuous sleep during the preceding 24 h) did not affect their performance on a battery of psychometric tests. The tests were all conducted between 6 A.M. and 8 A.M., at a time when normal circadian rhythms tend to produce peak performance. The authors suggest that these tests are a reasonable stand-in for the delivery of patient care. It is not clear, however, how the test indices, such as placing small objects into a pegboard in a prescribed sequence, relate to medical competence to diagnose or treat patients.

According to Asken and Raham's review, behavioral and psychomotor tests show a consistent decline with sleep deprivation, while emotional and psychological parameters are more even more severely affected.[40] They ask, given the weight of the evidence showing negative effects, where is the evidence showing a positive justification for long hours for residents either in terms of education or medical care? They warn that the depersonalization learned during long hours of residency may carry over and become a permanent feature of the physician's practice.

In 1975, interns in New York City hospitals went on strike over hours and conditions. In response to this strike, many hospitals revised their schedules, allowing doctors in residency who worked every day to be on call every third night instead of every other night. By 1989, this limitation to on-call duty to one night out of every three was adopted as policy for all residents in internal medicine throughout the United States.[41]

In 1984, an 18-year-old woman died 8 h after entering the emergency room of New York Hospital, Cornell University, for treatment. She was treated by resident doctors who had been working for 18 h. Her father charged that she had received substandard care. A grand jury investigating her death found "that the number of hours that interns and residents are required to work is counterproductive to providing quality medical care."[42] The grand jury recognized that the long hours worked by junior doctors provide cheap labor for hospitals, but that this results in a decreased quality of care. When inexperienced physicians without supervision work long hours, mistakes involving patient care are more likely to occur. Although the grand jury did not indict anyone, it did recommend that the New York State Department of Health limit the hours that residents work.

Subsequently, in 1988, the New York Commissioner of Health issued regulations limiting residents' hours in emergency rooms to shifts no longer than 12 h. Also, residents' hours in other acute care specialties were limited to no more than 24 consecutive hours and no more than an average of 80 h per week.[9] These changes have increased costs of medical care by 3%, because more supervision is needed and work that was formerly done by unpaid residents is now done by paid staff.

During 1988 and 1989, the Accreditation Council for Graduate Medical Education developed new standards for working hours and supervision of residents. Through the Residency Review Committees, 16 out of 24 specialty boards have adopted policies on residents' work hours. In emergency medicine, residents may work no more than 60 h per week, in 12 h shifts. Internal medicine residents can work no more than 80 h per week, with continuous duty in the emergency room limited to 12 h shifts.

There has been little published research on the relationship between quality of patient care and the shift schedules worked by health care providers, although some studies have compared the effect of 8 h and 12 h shifts on patient care. Shift schedules can affect patient care either directly or indirectly. Shift schedules which are more consistent with circadian rhythms are likely to contribute to more alert staff, who can then provide more attentive care. Shift schedules can affect patient care indirectly: schedules that are more subjectively satisfying to staff could result in less staff turnover, more continuity of care, and better staff cohesiveness.[9]

Measures that compare job performance between shifts are complicated by the fact that job tasks may vary with different shifts; working conditions, such as supervision and lighting, also vary. Nonetheless, shiftwork, particularly night shiftwork, affects job performance negatively. Continued night work has an additive negative effect on performance, because of chronic sleep loss and fatigue.

Accidents

The published data concerning accident rates in industrial settings by shift and time of day are inconsistent and insufficient to draw firm conclusions,[9] but data from the transportation industries clearly indicate that accidents occur more frequently at night. The probability of truck drivers being involved in collisions is higher at night.[43,44] A study of 338 single-vehicle accidents in January 1976 found that 46% of the accidents occurred between midnight and 8 A.M., although only 19% of the total traffic was on the road at that time. U.S. interstate trucking data indicate that peak time for accidents occurs between 4 A.M. and 6 A.M.[45] Mitler et al. compiled data from New York, Israel, and Texas, concerning single-vehicle accidents. They found that accidents as a result of the driver falling asleep at the wheel peaked between 1 A.M. and 4 A.M.[2] Another U.S. study of 493 single-vehicle truck accidents showed a peak time of occurrence between 1 A.M. and 7 A.M.[44]

Working extended shifts also contributes to risk. Truck drivers who drive for longer than 8 h are almost twice as likely to be involved in a crash.[46] A study of accidents involving Navy aircraft found Naval pilots who worked more than 10 h were more likely to be at fault than those who worked less than 10 h before the mishaps.[47]

A review of accident data for 1200 nurses indicated that rotating shift nurses had significantly more accidents of all kinds than did fixed shiftworkers. Nurses on fixed night shifts had significantly more finger injuries than fixed day or evening shift nurses, but rotating shift nurses had the most finger injuries of any group.[22]

Neuberger et al. examined 3 years of workers' compensation reports of needlestick injuries in a teaching hospital and found that 1 needlestick injury per year occurred for every 100 RNs, but that the rate of needlestick injuries to nursing staff was higher on night shift than during evenings or days.[48] The increase was attributed to fewer staff to help with difficult patients, increased fatigue, poor lighting, and less opportunity for night shift staff to attend in-service education.

Psychological Effects of Shiftwork

There is considerable evidence that shiftworkers report lower subjective health and well-being than those working days, but among shiftworkers there is considerable variation in the magnitude of the effect. It seems that some workers are predictably less able to adapt to shiftwork. Those who are more alert in morning ("morningness"), and those with more rigid sleep habits and less ability to overcome drowsiness are generally less likely to adapt to shiftwork.[49] There is some evidence that introverted and neurotic personalities are less able to adapt to shiftwork than are extroverts, but these cross-sectional studies do not clearly distinguish which is the causal and which is the dependent variable in the relationship between personality type and adaptation to shiftwork.[50] A survey of 1200 nurses found that night shift and rotating shift nurses experienced more physiological and psychological problems than fixed shift day or evening workers, and that the greater the departure from the conventional day shift schedule, the greater the effects.[22]

A cross-sectional questionnaire survey of 3446 male blue collar workers in Germany found that shiftworkers and former shiftworkers who left shiftwork on medical advice reported more psychosomatic complaints, such as headaches, than did day workers or those who left shiftwork for nonmedical reasons. Potentially confusing factors such as skill level and stress were examined and their effects were controlled.[51] The authors, Frese and Semmer, point out that social context is a hidden factor in many shiftwork studies. Thus, for example, a survey conducted by an employer, asking if employees experience medical problems, is likely to be answered negatively, while a similar survey conducted by a labor union is likely to be answered positively. In the study by Frese and Semmer, for example, if a doctor has already recommended leaving shiftwork, employees may feel more free to report problems associated with shiftwork.

First-year physicians experience more psychiatric problems, including irritability and hostility, than other professionals, and sleep deprivation is the major cause, overshadowing the effects of increased responsibility, anxiety, and loss of free time.[11] It has been reported that physicians in general suffer from problems such as alcoholism, drug dependence, depression, and suicidal tendencies, at a rate greater than that of the rest of the population. In addition to the intrinsic stress and great responsibility of the work, some of these problems have been attributed to a syndrome of overwork, which may include fatigue, irritability, and sleep disturbance.[40]

A mail-back survey of 463 nurses found that night work did interfere with sleep, but, unexpectedly, shiftwork was not associated with increased depression in nurses.[52] This study, by a management team, raises the possibility that many of those on shiftwork may elect to do so voluntarily and thus do not experience depression because they experience shiftwork as a controllable aspect of their life.

A study of 60 student nurses during their first 15 months on shiftwork found that shiftwork had a significant negative effect on their psychological well-being as measured by questionnaire. Psychological symptoms were determined, with 12 questions including ability to concentrate, lost sleep due to worry, usefulness, decision-making ability, depression, happiness, confidence, self-worth, etc. The questions were asked before beginning shiftwork, after 6 months, and again after 15 months of shiftwork.

This study also found that those who perceived their supervisors to offer a high degree of social support experienced fewer psychological symptoms. The nurses' intrinsic personality traits did not influence the long-term effects of shiftwork, but external forces, such as supervisors and other organizational, behavioral, and physiological factors could moderate the effects of shiftwork.[49] Similarly, in a British survey of residents, many spontaneously reported their supervisors and administrators to be unsympathetic to the problems of extended hours.[32]

Shiftwork and Reproductive Function

There have been few studies published to date, if any, concerning the effect of shiftwork on male reproductive function.

A study of 146 female nurses in Cincinnati found no differences in menstruation by shift.[53] A study of 1200 nurses by the National Institute for Occupational Safety and Health found differences in menstruation by shift. Nurses working fixed day shifts reported more irregular periods, while rotating shift nurses reported fewer cramps but more tension, weakness, and sickness during menstruation, and longer menstrual periods, than did fixed shift nurses.[22]

Studies of women who work nonstandard hours and rotating shifts indicate that they are at slightly higher risk for miscarriage, premature delivery,[54] and delivering infants with low birth weight[55] and smaller size.[56] A survey of 2264 working women in Japan concerning their reproductive and menstruation histories and their working hours found that women who worked at night reported higher rates of menstrual irregularity and more painful menstruation than women who worked days. Pregnancy and childbirth rates were lower among night workers, and abortion and stillborn delivery rates were higher.[57]

Shiftwork and Gastrointestinal Function

Nurses and others who work rotating shifts have reported more symptoms of digestive system disorder than have day workers.[22,58,59] Rutenfranz reviewed studies of several patterns of shiftwork involving over 8000 workers, and concluded that night shift is particularly well correlated with both sleep disturbances and gastrointestinal complaints. Day work and days with rotation to evenings were far less disruptive than either permanent nights or rotations which included nights.[60] One problem is that shiftwork alters routine eating schedules. More than 70% of coal miners reported that night work interfered with their eating habits, while less than 10% said that day work interfered.[61] Shiftwork may also alter the type of food consumed. It may be difficult to get a good hot meal at 3 A.M., for example. Night workers are more likely to increase their consumption of readily available junk food or convenience foods high in sugar, fat, and caffeine to fight off fatigue.

Other objective physiological data support an association between shiftwork and gastrointestinal disease.[62] The physiological mechanisms which link shiftwork exposure to gastrointestinal dysfunction and increased risk of disease are described by Vener et al.[63] They reviewed human and animal physiological data and developed a thesis that shiftwork changes adrenal secretions, pineal rhythmicity, digestive enzyme secretion, feeding patterns, and gastrointestinal motility. These changes cause desynchronization, thus creating stress-like responses and preulcerative conditions. The physiological effects of shiftwork mimic the effects of stress-induced ulcers.

Rutenfranz reviewed 14 studies and found a greater incidence of ulcers in former shiftworkers and in current shiftworkers whose work schedules included nights than among dayworkers or shiftworkers whose rotation excluded nights.[60] Segawa et al.[64] studied 11,657 Japanese workers over a 4 year period and found shiftworkers' ulcer incidence rates to be double that of day workers. Shiftworkers had a 2.38% incidence of gastric ulcers and 1.36% incidence of duodenal ulcers. Ex-shiftworkers had an incidence of 1.52% for gastric ulcer and 0.62% for duodenal ulcers. The incidence rate in day workers was only 1.03% for gastric ulcers and 0.69% for duodenal ulcers; these rates for day workers are half the ulcer incidence rates experienced by shiftworkers. Shiftwork doubles the risk of gastric ulcers.

Shiftwork and Heart Disease

Shiftworkers have greater cardiovascular risk factors than do day workers. Shiftworkers suffer from lack of sleep and fatigue. Their normal circadian rhythms are disrupted and full adjustment does not occur. The shape of the curve showing normal daily variation in blood pressure and heart rate (higher in day, lower in night) is flattened by shiftwork, indicating that cardiac output is affected. Greater cigarette consumption has been reported among shiftworkers.[65] Perhaps the stimulating effect of nicotine helps to counteract sleepiness. The greater rate of smoking among night workers may be associated with the observations that less educated workers are more likely to be required to work night shifts than professional or white-collar workers, and that increasing education is associated with lower rates of smoking.

However, differences in smoking behavior between shiftworkers and day workers do not explain all the cardiovascular health effects observed in shiftworkers.

Several studies have found higher serum cholesterol levels, triglycerides, and serum lipids in shiftworkers.[65–67] There may be several factors contributing to these findings, including circadian changes in metabolism and differences in activity, diet, smoking, alcohol consumption. Shiftworkers experience changes in mealtime and diet.[68] It may be that shiftworkers eat different kinds of food; it may also be that the way food is metabolized is different among shiftworkers. A study by DeBacker et al. indicates that the changes in serum lipids result from changes in metabolism rather than changes in diet, or other factors. Independent of smoking, obesity, diet, and leisure time activity, workers with the most irregular hours have higher total and LDL cholesterol and lower HDL cholesterol.[69]

Akerstedt et al. have written an outstanding review of the literature on heart disease and shiftwork.[70] They present data that clearly support an association between shiftwork and increased risk of cardiovascular disease. They argue that many epidemiology studies have not properly evaluated exposure to shiftwork. Cross-sectional studies that contrast health outcomes between current day workers and current shiftworkers often omit from analysis those workers who worked shiftwork in the past, but who now work day shifts. Including the health outcome of these workers can help in the calculation of a dose-response relationship between the time exposed to shiftwork (dose) and the health outcome under study (response).

Workers may enter shiftwork either involuntarily, because of lack of seniority, for example, or voluntarily, perhaps for extra pay. It may be that those who elect to do shiftwork may be more physically fit than others. In every available study of shift schedules at work, some group of people switch from shiftwork to day work. A study of 2137 workers with at least 1 year of exposure to shiftwork found that one-fourth transferred to day work within 10 years. Of these, 42% transferred for medical reasons (gastrointestinal, nervous, or circulatory disorders).[71] Therefore, those who stay in shiftwork will generally include workers in better health. This is similar to the "healthy worker effect," in which epidemiology studies find working people to be healthier than the general population, because those who are ill, injured, and elderly are generally not working.

A study by Aanonsen that found shiftworkers had a lower incidence of myocardial infarction and angina pectoris than two shift day workers[72] has been soundly criticized on methodological grounds.[70] A major problem with the Aanonsen study is that any worker who did not stay with the employer throughout the 5 year period was excluded from analysis. Thus the shiftwork exposures and disease experiences of those who left employment because of retirement, death, disease, or for other reasons such as inability to adjust to shiftwork were filtered out of the results. Even with these limitations, an analysis of this study does provide clues to the effects of shiftwork. Those who stayed with the employer but transferred out of shiftwork for "nonmedical" reasons had a higher incidence of myocardial infarction and angina than either of the over two groups and they were absent from work due to cardiovascular disease more than five times as often as day workers or shiftworkers. It may be that the nonmedical transfers were reluctant to seek medical attention for their symptoms. Those who transferred to day work for "medical" reasons clearly benefited; their absentee rate due to cardiovascular disease was half that of the day or shiftworkers and 10 times less than that of those who transferred for nonmedical reasons.

Koller et al. examined 270 refinery workers over a 5 year period to compare day workers to three shiftworkers.[73,74] For comparison, the workers exposed to shiftwork were matched with nonexposed workers by age and experience. Shiftworkers had the highest incidence (20%) of diseases of the circulatory system and day workers were lowest (7%), while ex-shiftworkers were intermediate (15%). The triple risk seen in shiftworkers was greater for those with more years of shiftwork exposure. A dose–response relationship was observed.

Taylor and Pocock also found ex-shiftworkers to be at high risk; they looked at the age-corrected incidence of death from cardiovascular disease among 8700 workers from 10 companies over a 12 year period, compared to national averages.[75] Death from cardiovascular disease was less than expected in day workers, average for shiftworkers, but more than expected among ex-shiftworkers; the standard mortality ratio (SMR) was 0.92 for day workers, 1.02 for shiftworkers, and 1.33 for ex-shiftworkers. An SMR of 1 is the expected death rate for workers of a given age. For death from atherosclerotic heart disease, the ratios were 0.94 for day shift, 1.03 for shiftworkers, and 1.25 for ex-shiftworkers. For hypertension as a cause of death, the ratio of deaths to expected values was 0.74 for day workers, 1.30 for shiftworkers, and 2.08 for ex-shiftworkers.

In perhaps the best designed study to date, Knutsson et al. followed 504 papermill workers for 15 years and found that the incidence of ischemic heart disease was higher among shiftworkers than among day workers.[76] The observed effect was independent of sex difference, age, or smoking history. An important dose–response relationship was observed. Workers with more years' exposure to shiftwork had a progressively increasing risk for ischemic heart disease. This trend was observed in five categories of exposure from 0 to 20 years. Those with 2–5 years' rotating shift experience had a relative risk of 1.5 times the expected ischemic heart disease rate. Those with 6–10 years' shiftwork had a relative risk of 2.0. Those with 11–15 years' shiftwork experience had a risk of 2.2. Those with 1–20 years had a risk of 2.8 times the expected rate for their age. The dose–response relationship was not observed for those with 21 or more years' shiftwork exposure. This group of experienced shiftworkers actually had a decreased risk, suggesting that those who remained after 20 years were particularly fit. Only the increased relative risks among workers with 11–15 and 16–20 years' experience were high enough to satisfy traditional measures of statistical significance.

Interestingly, had the authors not separated and ranked the shiftworkers by years of shiftwork exposure, the effects of shiftwork would have been diluted and lost. The only flaw in this study is that the shiftworkers were process operators and they were compared to day workers whose occupation was maintenance. In U.S. papermills, the jobs are quite different in content and relative autonomy. Process operators tend machines that are in continuous production. Maintenance workers fix machines when they break down; they thus have relatively greater autonomy than process operators. Because job stress is inversely associated with autonomy, process operators may be at higher risk of heart disease than maintenance workers, regardless of shift schedule. If this same distinction is true for Scandinavian papermill workers, then this factor may have influenced the results. Even with this limitation, however, the Knutsson examination stands out as an excellent study, linking duration of exposure to shiftwork with increased risk of heart disease.

This study of papermill workers indicates a clear association between increased years of shiftwork exposure and ischemic heart disease. Those with 11–20 years of shiftwork experienced from 2.2 to 2.8 times the incidence of ischemic heart disease than did day workers with blue collar jobs in the same industry, when controlling for age and smoking history. These results have greater significance because they are consistent with findings of increased heart disease risk factors in shiftworkers, such as increased serum lipids.

Akerstedt et al. (1984) reported on two major population-based studies.[70] Alfredsson et al. ranked 271 occupations for various risk factors, including shiftwork, using the results of 14,500 interviews. All male cases of myocardial infarction in Stockholm over a 2 year period were selected and age-matched with controls from the same area. When occupations were compared, only shiftwork and monotony were found to be risk factors. The risk from shiftwork was similar to the risk observed from smoking. Controlling for smoking history reduced, but did not eliminate, the risk from shiftwork.

Another study of 958,096 persons, among them 2530 with diagnoses of ischemic heart disease, found a strong correlation between occupations with exposure to shiftwork and death from ischemic heart diseased.[70] For males, shiftwork was the best predictor of ischemic heart disease of any occupational variable studied. For females, shiftwork ranked second after "no influence on work breaks." This finding is particularly relevant for nurses who report that they often ignore their scheduled breaks to provide consistent and timely patient care. Shiftwork was also associated with greater risk of accidents and low-back pain. Low points in circadian rhythms, decreased alertness, and staff shortages on night shifts may contribute to the observed excess in accidents and low-back pain. Decreased staffing levels on night shifts may also contribute to more back strain. In this population-based study, shiftwork was not associated with other health conditions, including cerebrovascular, gastrointestinal, or psychiatric disease, alcoholism, suicide, cancer, or diabetes. Unfortunately, such large population-based studies can only crudely estimate exposure to shiftwork, but their findings of increased risk of heart disease are consistent with the smaller, more well-designed studies on oil refinery and papermill workers and with the findings of increased serum lipids in shiftworkers. In summary, the data available indicate that shiftwork exposure is correlated with an increased risk of heart disease, the risk increasing with the duration of exposure.

Strategies

Given that the human body cannot undergo a complete adaptation to shiftwork,[77] what strategies are available to help individuals and organizations to minimize the harmful effects of shiftwork?

Control of Light in the Environment

There are data to suggest that the human physiological clock, left to itself, would operate on a cycle longer than 24 h, but that it is synchronized to a 24 h schedule by light and by social norms. In comparing the shiftworker during transition to the jet traveler, both may have to adjust to schedules that may be up to 12 h divergent from their normal schedule. The jet traveler expects to adapt to the work and social schedule in the new location. The dominant visual and social cues help the jet traveler to adapt. Thus jet travelers adapt more rapidly if they stay outside in natural light in their new location rather than in their hotel rooms.[78] But for the shiftworker who has been up all night, the pattern of day and night and the social norms of increased daylight activity are disruptive and certainly at variance with their needs for rest during the day. Studies of air travelers indicate that it takes about 1 day to adjust to a 1 h shift.[11] If it then takes 8 days to adjust to an 8 h schedule change, then the minimum period for a shift rotation should be 8 days.

The influence of light and dark cycles on circadian rhythms is significant. Czeisler et al. have found that exposure to bright light during the night shift and exposure to nearly complete darkness during the day helped achieve a shift in the circadian rhythms as indicated by sleep patterns, temperature, plasma cortisol concentration, urinary excretion rate, subjective assessment of alertness, and measured cognitive performance.[77] They caution that the success of this strategy is dependent on the timing, intensity, and duration of light exposure. It is important that those who work at night try to sleep as soon as they get home, preferably in a very dark environment, rather than staying up until midday and then trying to sleep.

Direction of Shift Rotation

There is clear scientific consensus that shift rotation should be clockwise. The human body would naturally follow a 25 h cycle if it were not influenced by daylight as a result of the earth's rotation, so rotation to a later shift is easier than rotation to an earlier shift. Similarly, adaptation to jet travel in a westerly direction is easier than is true with travel to the east. Travel to the east means that dawn arrives earlier, while travel to the west means that you get to sleep late. Shift rotation should move from morning to afternoon to nights.[79]

Naps and Shiftwork

Napping strategies have also been studied. Apparently, if the next main sleep period will be at night, it is better to forego a nap so that the next sleep will be of good quality.[80] Otherwise, naps can be helpful.

Personal Medical Factors

There is good evidence that people differ in their biological ability to tolerate working a rotating shift pattern.[81] Medical attention for shiftworkers should include both preplacement and periodic examinations.

Certain preexisting medical problems are risk factors indicating likely intolerance to shiftwork, especially digestive system problems such as ulcers, and other risk factors for ischemic heart disease. Those with diabetes and epilepsy should also probably avoid shiftwork.[60] There are also differences among people in terms of their internal circadian rhythms; some are more alert and productive early in the day ("morningness") while others are more alert in the evening and at night. Morningness influences sleep quality and quantity, and fatigue among those who work different shifts.

Physical Fitness

Nurses who are in good physical condition will tolerate shiftwork better.[82] Exercise-induced variations in maximal oxygen consumption and muscle strength are powerful positive factors that can lower fatigue, improve sleep quality, and reduce muscular skeletal symptoms, according to a study of a group of 128 nurses and nursing aides.[83] Physical exercise and increased fitness also increase efficiency in tasks requiring memory, and they improve subjective alertness. The effects of exercise also help shiftworkers to adapt to shift changes more rapidly.[11]

Controlling the Frequency of Shift Rotation

The frequency of rotation is an important issue in shiftwork and there is not yet clear consensus in the scientific community on what rotation frequency is best. The most common shift pattern in U.S. industry is weekly variation.[22] It seems clear that this pattern is the least optimal and that it causes the greatest disruption of circadian rhythms. Either shorter durations, such as 1 or 2 nights, or longer periods, such as 3–4 weeks, are preferable. Rapid rotation (changing everyday or every over day) may often be preferred when compared to schedules which change on a weekly basis. Air traffic controllers prefer rapid rotation, for example.[84] Many workers find it easier to recover from the disruption caused by a short stint of nights than from the more severe disruption caused by longer periods, such as 1 week. One advantage is that the worker can more easily attempt to keep in sync with others in the family or the rest of society. Rapid rotation shifts are particularly popular in Europe. However, rapid rotation means that the body never really gets adjusted to nights, but rather suffers some turmoil during the short period of night work.

A longer period of rotation may offer some advantage, because there is more chance for some adjustment in the internal circadian rhythms to occur. In 1943, Teleky recommended that shifts should not change more often than once a month.[85] A study of a shift schedule which rotated to a later shift every 21 days found that this schedule offered significant advantages in worker productivity, subjective health estimates, schedule satisfaction, and decreased worker turnover when compared to schedules which changed weekly in a counterclockwise direction.[86]

Limiting the Hours Worked at Night

Limiting the hours worked at night is an important concept which has not yet received enough attention. Nurses who work weekends often get 40 h pay for 24 h work. This type of thinking and reimbursement should also be applied to night work. However, there is real danger regarding night work and rotating shiftwork as conditions for which extra pay can provide compensation. The physiological effects of night work and shiftwork cannot be lessened with money. Rest and time to recover and readjust to normal day schedules are more important from a physiological perspective. Extra compensation without a reduction in hours may help to outweigh the dissatisfaction and motivate employees, but without solving the specific negative effects of night work and shiftwork. The danger in this "hazard pay" approach is that people will begin to believe that the extra compensation does balance the effects of shiftwork. Money cannot buy health if hazardous work schedules are maintained. This argument has been well developed by Thierry[87] and it deserves wider attention.

Rutenfranz suggests that night workers should receive extra time off to allow their bodies to adjust to the transition.[60] Night work and rotations including nights (11 P.M. to 7 A.M.) are more disruptive than evening (3 P.M. to 11 P.M.) shifts. Mitler et al. (1988), in their discussion of catastrophes and public policy, recommend limiting active duty to ensure that adequate time for sleep is available between shifts.[2]

In Japan, night work is illegal for women except in certain occupations, including telephone operators and health care providers.[57] Laws governing shiftwork in Belgium recognize implicitly the difficulty of night work by establishing the right to transfer to day work for workers who have worked nights for 20 years. This can be done at age 55 for any reason and at age 50 for medical reasons. Also, weekend workers are provided additional compensation. Weekly working time is reduced to 28 h, but wages are paid for a full work week (40 h).[88]

Collective agreements in Denmark set a norm of 37 weekly hours for day work and 35 for second and third shifts.[89]

The International Labor Organization in its 77th Session in June, 1990 developed recommendations concerning shiftwork, including:

Recommendation #178

1. Normal hours of work for night workers should not exceed 8 in any 24 h period in which they perform night work, except in the case of work which includes substantial periods of mere attendance or stand-by.
2. The normal hours of work of night workers should generally be less on average than, and, in any case, not exceed on average those of workers performing the same work to the same requirements by day in the branch of activity or the undertaking concerned.[90]

Adler and Roll suggest that time-off schedules may be more significant than work schedules for those on shiftwork.[91] For many workers, it is important to schedule their time off at convenient times so that they can attend to the other aspects of their lives, such as child care, time with spouse or family, or school schedules. The time that they work is of less significance than obtaining the time off that fits their needs. These choices and desires have an effect on the wages paid to various shifts. Although night work is socially and physiologically difficult, weekend work is considered even more invasive of social norms. U.S. hospitals in 1989 paid an average of $1.60 per hour differential over base wages to RNs for weekend work, but only $1.36 per hour differential for night work.[92] Hospitals are willing to pay more for weekend work than for night work. Perhaps RNs value time off on weekends more highly than time off at night for social and family reasons.

Education

Workers involved in shiftwork should be provided with education about accident frequency, and gastrointestinal, cardiovascular, psychological, and over risk factors and successful prevention strategies, so that they can take advantage of the lessons that have been learned through experience. They can then reduce their personal risk factors for ulcers, heart disease, and psychological stress with changes in their individual diet, smoking practices, sleep behavior patterns, home sleep environment, and physical fitness. As a group, workers can also seek to change dangerous shiftwork practices through collective bargaining, education, and legislation.

Conclusion

RNs are the largest single group of night workers in the United States. Nurses earned extra wages of $1.36 per hour on average in 1988 for working at night. If these workers realized that they were at twice the risk of ulcers and that they are at increased risk of heart disease as well as losing sleep and suffering disruption of normal social, family, and biological rhythms, they might not be willing to settle for $10.88 per shift as an incentive. In fact, many are not willing to do so. Permanent night shift nursing positions are running at a 30% vacancy rate nationally,[92] while for acute care nursing positions generally, the vacancy rate is 11%.[93] Only as pay rates for night work are increased and hours per week are decreased to compensate for the health risks and the stress implicit in night work will these night shift vacancies be filled. From the perspective of reducing the health hazards of night work, reducing the hours is more important than increasing the pay. Realistically, however, both increases in pay and decreases in hours will probably be necessary to fill existing night shift vacancies.

Checklist for Reducing Hazards of Night Work and Shiftwork

1. Do rotating shiftworkers rotate in a clockwise pattern?
2. Is the frequency of rotation either very rapid (one or two nights at a time) or very slow (3 or 4 weeks of consecutive nights)?
3. Are the hours of work reduced for night workers regardless of the compensation offered?
4. Are night and shiftworkers provided with encouragement and opportunities to develop and maintain physical fitness?
5. Are facilities available for night workers to prepare or be served good meals, so that their dependence on food from vending machines can be reduced, and so that their risks of stomach upset, gastric ulcers, and heart disease can also be reduced?
6. Are night shiftworkers provided with extra illumination at work and encouraged to completely darken their bedrooms during the daytime when they sleep? Such a policy will take advantage of the circadian rhythm dependence on cues of light and dark and help to achieve temporary shifts in circadian rhythms. This could be especially useful for employees who work long periods of nights. For those on rapid rotation it would be less useful, since they are not trying to change their baseline circadian rhythms during night work.
7. Are employees and employers provided with education about the increased risks for gastric ulcers, accidents, cardiovascular disease, and risk reduction strategies?

References

1. Wotjtczak-Jaroszowa J. Physiological and psychological aspects of night and shiftwork. U.S. Department of Health, Education and Welfare, National Institute for Occupational Safety and Health, Washington, DC, 1977.
2. Mitler MM, Carskadon MA, Czeisler CA, Dement WC, Dinges DF, and Graeber RC. Catastrophes, sleep, and public policy: Consensus report. *Sleep* 11:100–109, 1988.
3. Gordon NP, Cleary PD, Parker CE, and Czeisler CA. The prevalence and health impact of shiftwork. *Am J Public Health* 76:1225–1228, 1986.
4. Presser HB. Shiftwork and child care among young dual-earner couples. *J Marriage Family* 50:133–148, 1988.
5. Mellor EF. Shiftwork and flextime: How prevalent are they? *Monthly Labor Rev* 109:14–21, 1986.
6. U.S. President's Commission on the Accident at Three Mile Island. U.S. Government Printing Office, Washington, DC, 1979.
7. National Transportation Safety Board. Marine accident report, Grounding of the Exxon Valdez on Bligh Reef, Prince William Sound, AK, March 24, 1989, NTSB/MAR-90/04. National Technical Information Service, Springfield, VA, 1990.
8. Report of the Presidential Commission on the Space Shuttle Challenger Accident. U.S. Government Printing Office, Washington D.C., 1986.
9. U.S. Congress, Office of Technology Assessment. *Biological Rhythms: Implications for the Worker*, OTA-BA-463. U.S. Government Printing Office, Washington, DC, 1991.
10. Adams C. *Report of the Hospital Nursing Personnel Surveys*. American Hospital Association, Chicago, 1989.
11. Fossey E. Shiftwork can seriously damage your health! *Professional Nurse* June:476–480, 1990.

12. Jung F. Shiftwork, its effect on health performance and well-being. *Am Assoc Occup Health Nurs J* 34:161–164, 1986.
13. Haider M, Koller M, and Cervinka R. *Night and Shiftwork: Long-Term Effects and Their Prevention*. Verlag Peter Lang, Frankfurt, Germany, 1986.
14. Monk TH and Folkard S (eds.). *Hours of Work*. Wiley, Chichester, England, 1985.
15. Reinberg A, Vieux N, and Andlaver P. *Night and Shiftwork: Biological and Social Aspects*. Pergamon Press, Oxford, England, 1981.
16. Colligan M and Tepas D. The stress of hours of work. *Am Ind Hyg J* 47:686–695, 1986.
17. Tepas D. Department of Psychology, University of Connecticut, 1990. Cited in U.S. Congress, Office of Technology Assessment. *Biological Rhythms: Implications for the Worker*, OTA-BA-463. U.S. Government Printing Office, Washington, DC, 1991.
18. Folkard S, Monk TH, and Lobban MC. Short and long-term adjustment of circadian rhythms in "permanent" night nurses. *Ergonomics* 21:785–799, 1978.
19. Akerstedt T. Sleepiness as a consequence of shiftwork. *Sleep* 11:17–34, 1988.
20. Torii S, Okudaira N, Fukuda H, Kanamoto H, Yamashiro Y, Akiya M, Nomoto K, Katayama N, Hasegawa M, Sato M, Hatano M, and Nemoto H. Effects of night shift on sleep patterns of nurses. *J Human Ergol* 11(Suppl.):233–244, 1982.
21. Gadbois C. Women on night shift: Interdependence of sleep and off-the-job activities. Night and shiftwork: Biological and social aspects. In: *Proceedings of the 5th International Symposium on Night and Shiftwork*. Pergamon Press, Oxford, 1981.
22. Tasto DL, Colligan MJ, Skjei EW, and Polly SJ. Health consequences of shiftwork. U.S. Department of Health and Human Services, National Institute of Occupational Safety and Health, Washington, DC, 1978.
23. Partinen M, Eskelinen L, and Tuomi K. Complaints of insomnia in different occupations. *Scand J Work Environ Health* 10:467–469, 1984.
24. Akerstedt T, Torsvall, and Gillberg M. Sleepiness in shiftwork. A review with emphasis on continuous monitoring of EEG and EOG. *Chron Intern* 4:129–140, 1987.
25. Leighton K and Livingston M. Fatigue in doctors. *Lancet* 1:1280, 1983.
26. Monk TH. Shiftworker performance. In: AJ Scott (ed.), *Shiftwork, Occupational Medicine State of the Art Reviews*. Hanley and Belfus, Philadelphia, PA, 1990.
27. Bjerner B, Holm A, and Swensson A. Diurnal variation in mental performance: A study of three-shiftworkers. *Brit J Ind Med* 12:103–110, 1955.
28. Hildebrandt G, Rehmert W, and Rutenfranz J. Twelve and 24 hour rhythms in error frequency of locomotive drivers and the influence of tiredness. *Int J Chron* 2:175–180, 1974.
29. Coffey LC, Skipper JK, and Jung FD. Nurses and shiftwork: Effects on job performance and job related stress. *J Adv Nurs* 13:245–254, 1988.
30. Folkard S, Condon R, and Herbert M. Night shift paralysis. *Experientia* 40:510–512, 1984.
31. Deaconson TF, O'Hair DP, and Levy MF et al. Sleep deprivation and resident performance. *J Am Med Assoc* 260:1721–1727, 1988.
32. Wilkinson RT, Tyler PD, and Varey C. Duty hours of young hospital doctors: Effects on the quality of work. *J Occup Psychol* 48:219–229, 1975.
33. Poulton EC, Hunt GM, Carpenter A, and Edwards RS. The performance of junior hospital doctors following reduced sleep and long hours of work. *Ergonomics* 21:279–295, 1978.
34. Friedman RC, Bigger JT, and Kornfield DS. The intern and sleep loss. *New Eng J Med* 285:201–203, 1971.
35. Hawkins MR, Vichick DA, Silsby HB et al. Sleep deprivation and performance of house officers. *J Med Ed* 60:530–535, 1985.
36. Reznick RK and Folse JR. Effect of sleep deprivation on the performance of surgical residents. *Am J Surg* 154:520–525, 1987.

37. Klose JK, Wallace-Barnhill GL, and Craythorne MWB. Performance test results for anesthesia residents over a 5-day week including on-call duty, abstracted. *Anesthesiology* 63:A485, 1985.
38. Folkard S. Diurnal variation in logical reasoning. *Br J Psychol* 66:1–8, 1975.
39. Beatty J, Ahern S, and Katz K. Sleep deprivation and the vigilance of anesthesiologists during simulated surgery. In: R Mackie (ed.), *Vigilance, Theory, Operational Performance and Physiological Correlates*. Plenum, New York, 1977, pp. 1–18.
40. Asken MJ and Raham DC Resident performance and sleep deprivation: A review. *J Med Educ* 58:382–388, 1983.
41. Page L. ACGME review panel sets up hour limits for internist residents. *Am Med News* 18:2, 1988.
42. Report of the Fourth Grand Jury for the April/May term of 1986 concerning the care and treatment of a patient and the supervision of interns and junior residents at a hospital in New York County, New York Supreme Court, New York County, 1986.
43. Hamelin P. Lorry driver's time habits in work and their involvement in traffic accidents. *Ergonomics* 30:1323–1333, 1987.
44. Mackie RR and Miller JC. Effects of hours of service, regularity of schedules and cargo loading on truck and bus driver fatigue. U.S. Department of Transportation, Bureau of Motor Carrier Safety and National Highway Traffic Safety Administration, 1978.
45. Harris W. Fatigue, circadian rhythm and truck accidents. In: RR Mackie (ed.), *Vigilance, Theory, Operational Performance, and Physiological Correlates*, Vol. 3 of *Nato Conference Series*. Human Factors. Plenum Press, New York, 1977, pp. 133–146.
46. Jones IS and Stein HS. *Effect of Driver Hours of Service on Tractor-Trailer Crash Involvement*. Insurance Institute for Highway Safety, Arlington Virginia, 1987.
47. Borowsky MS and Wall R. Naval aviation mishaps and fatigue. *Aviat Space Environ Med* 54:241–249, 1983.
48. Neuberger JS, Harris JA, Kundin WD, Bischone A, and Chin TDY. Incidence of needle-stick injuries in hospital personnel: Implications for prevention. *Am J Inf Cont* 12:171–176, 1984.
49. Bohle P and Tilley AJ. The impact of night work on psychological well-being. *Ergonomics* 32:1089–1099, 1989.
50. Colquhoun WP and Condon R. Introversion-extraversion and the adjustment of the body-temperature rhythm to night work. In: A Reinberg, N Vieux, and P Andlauer (eds.), *Night and Shiftwork: Biological and Social Aspects*. Pergamon Press, Oxford, 1981, pp. 223–228.
51. Frese M and Semmer N. Shiftwork, stress and psychosomatic complaints: A comparison between workers in different shiftwork schedules, non-shiftworkers and former shiftworkers. *Ergonomics* 29:99–114, 1986.
52. Skipper JK, Jung FD, and Coffey LC. Nurses and shiftwork: Effects on physical health and mental depression. *J Adv Nurs* 15:835–842, 1990.
53. Kuchinski BB. The effects of shiftwork on the menstrual characteristics of nurses. Doctoral dissertation, Johns Hopkins University, Baltimore, 1989.
54. Axelsson G, Rylander R, and Molin I. Outcome of pregnancy in relation to irregular and inconvenient work schedules. *Br J Ind Med* 46:393–398, 1989.
55. Armstrong BG, Nolin AD, and McDonald AD. Work in pregnancy and birth weight for gestational age. *Br J Ind Med* 46:196–199, 1989.
56. Nurminen T. Shiftwork, fetal development, and the course of pregnancy. *Scand J Work Environ Health* 15:395–403, 1989.
57. Uehata T and Sasakawa N. The fatigue and maternity disturbances of night workwomen. *J Human Ergol* 11(Suppl.): 465–474, 1982.
58. Angerspach D, Knauth P, Loskant H, Karvonen M, Undeutch K, and Rutenfranz J. A retrospective study comparing complaints and diseases in day and shiftworkers. *Int Arch Occup Environ Health* 45:127–140, 1980.

59. Frese M and Okonek K. Reasons to leave shiftwork and psychological and psychosomatic complaints of former shiftworkers. *J App Phys* 69:509–514, 1984.
60. Rutenfranz J. Occupational health measures for night and shiftworkers. *J Human Ergol* 11(Suppl.): 67–86, 1982.
61. Duchon JC and Keran CM. Relationship among shiftworker eating habits, eating satisfaction and self-reported health in a population of miners. *J Work Stress* 4:111–120, 1990.
62. Moore-Ede MC and Richardson GS. Medical implications of shiftwork. *Ann Rev Med* 36:607–617, 1985.
63. Vener KJ, Szabo S, and Moore JG. The effect of shiftwork on gastrointestinal function: A review. *Chronobiologia* 16:421–439, 1989.
64. Segawa K, Nakazawa S, Tsukamato Y, Kurita Y, Goto H, Fukui A, and Takano K. Peptic ulcer is prevalent among shiftworkers. *Digest Dis Sci* 32:449–453, 1987.
65. Knutsson A and Zamore K (1982) cited in Knutsson A, Akerstedt T, Jonsson BG, Orth-Gomer K. Increased risk of ischemic heart disease in shiftworkers. *Lancet* 12:89–92, 1986.
66. Thelle DS, Forde OH, Try K, and Lehmann EH. The Tromso heart study. *Acta Med Scand* 200: 107–118, 1976.
67. Theorell T and Akerstedt E. Day and night work. Changes in cholesterol, uric acid, glucose and potassium in serum and circadian patterns of urinary catecholamine excretion. *Acta Med Scand* 200:47–53, 1976.
68. Rutenfranz J, Knauth P, and Angersbach D. Shiftwork research issues: In: LC Johnson, DI Tepas, WP Colquhoun, and MJ Colligan (eds.), *Biological Rhythms, Sleep and Shiftwork, Advances in Sleep Research*, Vol. 7. Spectrum, New York, 1981, pp. 165–196.
69. DeBacker G, Kornitzer M, Peters H, and Dramaix M. Relation between work rhythm and coronary risk factors. *Eur Heart J* 5(Suppl. 1):307, 1984.
70. Akerstedt T, Knutsson A, Alfredsson L, and Theorell T. Shiftwork and cardiovascular disease. *Scand J Work Environ Health* 10:409–414, 1984.
71. Thiis-Evensen E (1949) cited in Akerstedt T, Knutsson A, Alfredsson L, and Theorell T. Shiftwork and cardiovascular disease. *Scand J Work Environ Health* 10:409–414, 1984.
72. Aanonsen A. Shiftwork and health. Universitetsforlaget, Oslo 1964.
73. Koller M. Health risks related to shiftwork. *Int Arch Occup Environ Health* 53:59–75, 1983.
74. Koller M, Kundi M, and Cervinka R. Field studies of shiftwork at an Austrian oil refinery: I. Health and psychosocial wellbeing of workers who drop out of shiftwork. *Ergonomics* 21:835–847, 1978.
75. Taylor PJ and Pocock SJ. Mortality of shift and day workers 1956–68. *Br J Ind Med* 29:201–207, 1972.
76. Knutsson A, Akerstedt T, Jonsson BG, and Orth-Gomer K. Increased risk of ischemic heart disease in shiftworkers. *Lancet* 12:89–92, 1986.
77. Czeisler CA, Johnson MP, Duffy JF, Brown EN, Ronda JM, and Kronauer RE. Exposure to bright light and darkness to treat physiologic maladaption to night work. *New Eng J Med* 322:1253–1259, 1990.
78. Klein KE and Wegmann HM. The resynchronization of human circadian rhythms after transmeridian flights as a result of flight direction and mode of activity. In: LE Scheving, F Halberg, and JE Pauly (eds.), *Chronobiology*. Igaku Shoin, Tokyo, 1974, pp. 564–570.
79. Rose M. Shiftwork: How does it affect you? *Am J Nurs* 4:442–447, 1984.
80. Akerstedt T, Torsvall I, and Gillberg M. *Sleep and Alertness, Chronobiological, Behavioral and Medical Aspects of Napping*. Raven Press, New York, 1989.
81. Smolensky MH and Reinberg A. Clinical chronobiology: Relevance and applications to the practice of occupational medicine. *Occup Med* 5:239–272, 1990.
82. Harma MI, Ilmarinen J, Knauth P, Rutenfranz J, and Hanninen O. Physical training intervention in female shiftworkers: The effects of intervention on fitness, fatigue, sleep and psychosomatic symptoms. *Ergonomics* 31:39–50, 1988.

83. Harma M, Ilmarinen J, and Knauth P. Physical fitness and other individual factors relating to the shiftwork tolerance of women. *Chronobiol Int* 4:417–424, 1988.
84. Smith RC. Job attitudes of air traffic controllers: A comparison of three air traffic controller specialties. Industrial Medicine and Surgery, FAA Office of Aviation Report, #73-2, 1973.
85. Teleky L. Problems of night work. *Ind Med* 12:758–779, 1943.
86. Czeisler CA, Moore-Ede MC, and Coleman RM. Rotating shiftwork schedules that disrupt sleep are improved by applying circadian principles. *Science* 217:460–463, 1982.
87. Thierry H. Compensation for shiftwork: A model and some results. In: LC Johnson, DI Tepas, WP Colquhoun, and MJ Colligan (eds.), *The Twenty-Four Hour Workday: Proceedings of a Symposium on Variations in Work-Sleep Schedules*. U.S. Department of Health and Human Services, PHS, CDC, NIOSH, Cincinnati, 1981.
88. International Labor Office. *Conditions of Work Digest*. International Labor Office, Geneva, Switzerland, 1990, pp. 233–234.
89. International Labor Office. *Conditions of Work Digest*. International Labor Office, Geneva, Switzerland, 1990, p. 240.
90. International Labor Office. The hours we work. *Conditions of Work Digest* 9(2):224, 1990.
91. Adler A and Roll Y. Characterizing shiftwork patterns by the off-time distribution. *J Human Ergol* 11(Suppl.):409–415, 1982.
92. Hospital Nursing Personnel Survey. American Hospital Association. Chicago, IL, 1989.
93. American Hospital Association cited in *Am J Nurs* 86, 1991.

22

Radiation Protection in Hospitals

Introduction 22-1
Physical Properties of Radiation 22-2
Nonionizing Radiation and Ionizing Radiation • General Properties of Radioactive Material Used in Hospitals • Half-Life • Energy of Ionizing Radiation • Half-Value Layer • Inverse Square Law
Biological Effects of Ionizing Radiation 22-5
Dose-Equivalent • Types of Exposure • Major Sources of Radiation Exposure
Dose Limits 22-7
Nonstochastic Effects • Stochastic Effects • Occupational Exposure of Women of Reproductive Capacity • Occupational Exposure of Pregnant Women • Internal Exposure • Licenses and Authorization
Environmental Surveillance and Dosimetry 22-9
Environmental Surveillance • Personnel Dosimetry
Diagnostic Radiology 22-10
Fluoroscopic Procedures • Mobile Units • Dental Radiology
Nuclear Medicine 22-11
Clinical Laboratories and Research Facilities 22-12
Radiation Therapy 22-12
Teletherapy • Intracavitary Brachytherapy • Interstitial Brachytherapy • Radiation Therapy with Open Sources
Bibliography 22-14
Appendix 22.A.1 22-15

Jean-Pierre Gauvin
Montreal Joint Hospital Institute

Introduction

X-ray-emitting equipment is used in radiology, in operating rooms, and in in-patient units. Radioactive solutions are prepared and administered to patients in nuclear medicine. Clinical laboratories make use of radioactive preparations as part of their analytical protocol. Research laboratories are very often housed in hospital institutions and require the use of various radioelements. In radiation therapy, patients are treated with radiation.

An extensive program of radiation protection must accompany the use of radiation in hospitals. Such a program should aim at protecting patients from excessive exposure to radiation during diagnosis or treatment, and the public and personnel against radiation leakage emitted by radiation-emitting equipment, radioactive sources, and patients who have undergone an isotopic investigation or treatment.

Physical Properties of Radiation

One of the basic laws of thermodynamics explains that within the universe, all systems in which levels of organization rise require an outside supply of energy. If the Earth were a closed system cut off from the rest of the universe, its transformation from a body of inert matter billions of years ago into a planet where highly organized biological life exists would never have happened.

What is responsible for the continuation of life on Earth? The primary cause, the phenomenon that is essential to preservation of life on Earth, is the supply of energy provided by the sun. Without this energy supply, neither animal nor vegetable would be capable of survival.

The spectral distribution of sunlight that reaches the Earth includes ultraviolet rays, visible light, infrared light, and even radio waves. Although the visible portion of the solar spectrum is the most apparent, it constitutes less than 40% of the whole spectrum.

In addition to the sun, the cosmos provides numerous sources of radiation known as cosmic rays. Cosmic rays are constituted of electromagnetic radiation (light having a wavelength or "color" undetectable by the human eye) and particles (small bits of matter) projected to the Earth at very high speeds.

Nonionizing Radiation and Ionizing Radiation

Nonionizing Radiation

Nonionizing radiation, similar to visible light, has the ability to increase the temperature of a target material. Different types of nonionizing radiation include

- Radio waves
- Microwaves
- Infrared light
- Visible light
- Ultraviolet light

Health effects of nonionizing radiation are related to an increase in the temperature of exposed tissues (e.g., thermal burns).

Ionizing Radiation

Ionizing radiation has the same properties as nonionizing radiation plus the ability to create ions in exposed materials. Such a production of ions could result in direct damage to the genetic material of the cells (the cell is the basic constituent of biological material) and/or in the production of cellular poison (e.g., peroxide).

The different types of ionizing radiation are

- Alpha particles
- Beta particles
- Neutrons
- X-rays
- Gamma rays

General Properties of Radioactive Material Used in Hospitals

Radioactive isotopes are used extensively in hospitals for diagnosis, treatment, and research purposes. Radioactive materials are elements with chemical properties identical to their nonradioactive counterparts, having, in addition, the ability to emit several types of ionizing radiation. A radioactive element can emit any of the types of ionizing radiation previously mentioned.

Alpha Particles

This type of radiation, an emission from the nucleus of a radioactive atom, consists of a particle with two protons and two neutrons (a nucleus of helium) traveling at very high speeds (approximately one-twentieth the speed of light). Alpha rays do not penetrate very deeply. Airborne alpha particles might travel only a few centimeters. A sheet of paper would suffice to stop the alpha particles completely.

Due to their low penetration ability, alpha particles do not constitute any risk of external exposure. Alpha particles are completely attenuated by the surface of the skin, which is composed of dead cells. Any intake of radioactive material-emitting alpha particles would result in the emission of alpha particles inside the body. This would then result in intense exposure to those organs where radioactive material is concentrated. Alpha emitters are generally not used in biomedical laboratories. An important source of alpha emissions is radon gas, which is naturally present in the environment in low concentrations.

Beta Particles

Beta radiation consists of the emission of electrons by the nucleus of a radioactive atom. Beta particles have a low mass (one-seventy-three-hundredth of the mass of an alpha particle). They are ejected from the nucleus of radioactive atoms at high speeds (approaching the speed of light). The penetration power of beta particles depends on their energy level (or speed). It is generally very low (less than 1 mm of water for C-14), but could reach 1 cm of water in the case of high-energy radiation (such as P-32).

When high-energy beta particles are slowed down or stopped by a heavy object, x-rays are produced by a process known as "Bremsstrahlung." The creation of Bremsstrahlung radiation by beta particles tends to increase in importance when the atomic number of the absorbent material increases. Plexiglas (as well as water) is very unlikely to produce Bremsstrahlung radiation, while lead and other heavy materials (having an atomic number greater than 82) are likely to do so. When using high atomic number material to shield a source of beta particles, the total thickness of the shielding must be sufficient to (1) stop the beta rays and (2) attenuate sufficiently the Bremsstrahlung radiation (which has a much higher penetration level than the original beta particles and thus is considerably more difficult to shield). High-energy beta particles (e.g., P-32) could constitute a risk of external exposure.

Beta emitters such as tritium (H-3), sulfur-35 (S-35), and carbon-14 (C-14) are not energetic enough to produce Bremsstrahlung radiation and usually do not constitute a risk of external exposure. Only in the case of an accident involving ingestion, contamination of the skin, injection, or inhalation could a significant dose be received from beta exposure.

Positron Particles

Positrons are composed of electrons with a positive electrical charge. Upon emission, these particles behave in the same manner as beta particles. When a collision occurs between a positron and an electron, the positron–electron group is destroyed and replaced by two gamma rays (two sparks of light) with an energy equal in mass and speed to those particles that were destroyed in the collision. Sodium-22 (Na-22), a positron emitter, is currently used in research laboratories. Due to the production of such annihilation radiation (gamma rays), when positron emitters and an electron collide, they can constitute a risk of external exposure.

Gamma Rays

Gamma rays consist of electromagnetic waves (light). These pulses of light are spontaneously emitted by the nucleus of a radioactive atom during its transformation.

Unlike alpha and beta radiation, gamma rays are characterized by having a very high penetration (they can penetrate several centimeters of concrete). Due to such a high penetration, gamma rays can constitute a risk of external exposure. Current medical uses in hospitals include chromium-51 (Cr-51), iodine-125 (I-125), cobalt-57 (Co-57), technetium 99-m (Tc-99 m), gallium-67 (Ga-67), and iodine-131 (I-131).

X-Rays

X-rays consist of electromagnetic rays emitted from the extranuclear part of an atom. The properties of x-rays are, in essence, the same as gamma rays. The only difference between the two is their origin in the atom. Generally, x-rays emitted by radioactive isotopes have an energy level far lower than gamma rays. Iron-55 (Fe-55) is an x-ray emitter currently used in hospital laboratories.

The "activity" of a source of radioactive isotopes is the average number of nuclear transformations produced in a 1 s period. Each transformation might result in the emission of one or more photons or ionizing particles (alpha, beta, gamma, etc.). The "becquerel" is the unit specially reserved for the activity. The activity is the unit that relates to the number of radioactive isotopes present in a given sample.

1 becquerel (Bq) = 1 nuclear transformation per second
1 kilobecquerel (kBq) = 1,000 nuclear transformations per second
1 megabecquerel (MBq) = 1,000,000 nuclear transformations per second

The concept of activity of a source is the most important principle in radiation protection. The evaluation of the risks associated with the use of radioactive substances requires good information regarding the activity and the nature of the isotope involved. For all radioactive isotopes, there is an activity below which no significant risk of exposure can occur.

Half-Life

Radioactive material is not radioactive for eternity. Similar to a wood fire, radioactive material is consumed at a rate that depends on the nature of the isotope.

The "half-life" of a radioactive isotope is the time required to reduce by half the initial activity of any amount of this radioisotope. At the end of each period corresponding to one half-life, the initial activity is reduced by half. Depending on the isotope, the half-life of a radioactive isotope is measured in time frames spanning anywhere from a fraction of one-millionth of a second up to several millions of years. Meanwhile, the half-life is rigorously maintained at a constant level for the same isotope.

Energy of Ionizing Radiation

All forms of radiation can be characterized by their energy. In the case of ionizing particles, energy signifies speed. The unit of measurement that designates energy is the electron volt (eV) (1 eV corresponds to the gain of energy of one electron accelerated by a potential of 1 V). Since 1 eV is a relatively small energy value, the energy is often expressed in terms of kiloelectron Volts (1 keV = 1000 eV) or millions of electron Volts (1 MeV = 1,000,000 eV).

In the case of ionizing radiation in the form of "electromagnetic waves" (synonyms: photons, light, gamma rays, x-rays), the energy of radiation signifies "wavelengths" (or color). The shorter the wavelength of photon radiation, the higher the energy level and the greater the power of penetration.

Half-Value Layer

Regarding radiation in the form of photons (x-rays, gamma rays), the penetration depends on the energy of the photons emitted. The unit of measure for the penetration of photon radiation is called the "half-value layer" (HVL), which is the thickness of absorbent material required to reduce by half the intensity of the radiation (2 HVL reduces intensity four times, 4 HVL reduces intensity eight times, etc.). The HVL varies depending on the energy of the radiation and the nature of the absorbent material

TABLE 22.1 Half-Value Layers According to Various Energy Levels

Photon Energy (keV)	HVL Water (mm)	HVL Aluminum (mm)	HVL Lead (mm)
120 keV (ex: Cobalt-57)	45	18	1
6 keV (ex: Iron-55)	15	—	—
30 keV (ex: Iodine-125)	20	0.8	0.2
320 keV (ex: Chromium-51)	60	30	1.7
511 keV (ex: annihilation of B + (Sodium-22))	70	32	4
1 100 keV (ex: Iron-59)	100	40	9

TABLE 22.2 Maximum Penetrations of Beta Emitters Currently in Use

Energy	Air (mm)	Water (mm)	Plexiglas (mm)	Aluminum (mm)
18 keV (Tritium, H-3)	10	0.05	—	—
140 keV (Carbon-14, C-14)	250	0.25	0.2	0.15
140 keV (Sulfur-35, S-35)	250	0.25	0.2	0.15
1700 keV (Phosphorus-32, P-32)	5000	7.5	5	2.5

used. Table 22.1 shows the HVLs for gamma rays in water, aluminum, and lead for a variety of different isotopes.

The beta rays of those isotopes currently in use in medical laboratories have a penetration ability that is generally too low to constitute a risk of exposure unless there is contamination of the skin, injection, inhalation, or ingestion of the radioactive substance. The maximum penetrations of beta emitters currently in use are given in Table 22.2.

Inverse Square Law

A concept that is common to all types of radiation is that their intensity is inversely proportional to the square of the distance between the "detector" and the "radiation source." In practice, this means that if the distance between a source and a detector is double, then the exposure will be reduced by four; at triple the distance, the intensity of the radiation would be nine times less and at "*n*" times the distance, the intensity would be *n* squared times less. A worker who is aware of the law of the inverse square of distance would handle highly active solutions of gamma emitters with tongs rather than with bare hands.

Biological Effects of Ionizing Radiation

Dose-Equivalent

Defined for the purpose of radiation protection, "dose-equivalent" expresses, on a standardized scale for all types of ionizing radiation, the deleterious effects of exposure to radiation. The special name for dose-equivalent is the "sievert" (Sv); one sievert equals dose expressed in grays (Gy) times *Q*. *Q* is a quality factor that depends on the nature of the radiation involved.

Any assessment of the dose-equivalent received by a person is incomplete unless the organs or tissues that were exposed are specified. Since the health effects associated with an exposure to radiation are closely related to the organs involved, the notion of dose-equivalent should always be differentiated by "whole body," "hands," "thyroid," "gonads," etc.

Types of Exposure

Two main types of radiation exposure should be distinguished:

- *Acute exposure.* A one-time, major accidental exposure received in a short period of time is called an acute exposure. The effects of this particular type of exposure will be experienced in the hours and days following the accident. Such effects are said to have a nonstochastic nature because the magnitude of the effects produced by the exposure is proportional to the radiation dose received.
- *Chronic exposure.* A long-term, low-level exposure is called a chronic exposure. The effects of this type of exposure cannot be predicted with certainty. Such effects are said to be of a stochastic nature, which means that the risk that the exposure has produced a health effect is proportional to the radiation dose received. The severity of the effect is, in such cases, assumed to be constant, and it is only the risk of having an effect that varies with the radiation dose received.

Acute Exposure

The effects of an acute exposure vary according to the radiation dose received and depend on the organs that were exposed. For doses of less than 0.2 Sv, there are no clinically detectable effects. When doses exceed 1 Sv, the effects can vary from a simple skin erythema (redness of the skin) at 2 Sv, baldness at 3 Sv, destruction of the hematopoietic system at 6 Sv, and local destruction of exposed tissues at 10 Sv. Whole-body exposure to doses higher than 6 Sv leads to death in 80% of the cases in the 2 months following exposure. Hospital laboratories, other than radiation therapy facilities, do not generally use quantities of radioactive material that are large enough to have accidents likely to lead to acute exposure to radiation.

Chronic Exposure

The effects of low-level exposure to radiation are divided into two categories: somatic and genetic. Somatic effects pertain only to those individuals exposed to radiation, whereas genetic damage could be transmitted to future generations.

The somatic effects of a low-level exposure to radiation consist of an increase in the risk of cancer to the exposed tissues. In the case of an exposure of a pregnant woman at the abdominal level, the exposure of the fetus could lead to congenital malformation as well as an increase in the risk of early cancer for the unborn child. These health effects, however, are not specific to radiation exposure and, consequently, should an exposed person develop cancer, it would be impossible to demonstrate that such effects have effectively been produced by the exposure. Epidemiological studies conducted among the survivors of the first detonations of nuclear weapons and among patients treated with radiation therapy have shown an increase in the risk of developing a fatal cancer of the order of 1:100,000/mSv received uniformly over the whole body. According to our state of knowledge, there is no minimum dose of radiation below which a risk equal to zero could be associated. Even small doses of radiation are considered to correspond to some risk of developing a disease. The higher the radiation dose, the higher the risk. It is in this sense that radiation doses are said to be cumulative.

An important corollary of this principle is that radiation doses should be maintained as low as reasonably achievable (ALARA) taking into account economic and social factors.

TABLE 22.3 Annual Whole-Body Dose-Equivalent Received on Average by the North American Population

Source	Annual Dose (mSv)
Natural sources	
Cosmic rays	0.28
Land sources	0.26
Isotopes present in the human body	0.28
(C-14, Ra-226, Pm-222, K-40)	0.82
Medical sources	
Medical radiology	0.77
Dental radiology	0.01
Nuclear medicine	0.14
	0.92
Other	
Nuclear fallout	0.05
Nuclear power station	0.01
Construction material	0.05
TV sets	0.005
Plane rides (at high altitudes)	0.005
	0.12
Total	1.86 mSv/year

Major Sources of Radiation Exposure

Since first appearing on the Earth, mankind has always been exposed to radiation. Natural radiation sources fall into three categories: cosmic rays, radioactive material present in trace concentrations, and radioactive elements naturally present in the human body. In addition to natural radiation, medical and dental radiological procedures constitute an important source of exposure for the population. Added to medical and natural sources of radiation are artificial sources. Table 22.3 gives the annual whole-body dose-equivalent received on average by the North American population.

Dose Limits

The principle according to which the effects of radiation vary proportionally with the dose received, without any threshold, signifies that any exposure to radiation involves certain risk. A system of dose limits for workers therefore necessitates some knowledge of the level of risk currently accepted by society. Industries generally considered to be adequately safe show fatal accident rates in the order of one death per 10,000 workers per year. Based on this information and based on the health risk associated with radiation exposure, the International Commission on Radiological Protection (ICRP publication No. 26) has recommended the following dose-equivalent limit: 50 mSv to the whole body for workers occupationally exposed to ionizing radiation. In case of partial body exposure, it recommends the following dose limits.

Nonstochastic Effects

The ICRP's recommendations are intended to prevent nonstochastic effects and to limit the occurrence of stochastic effects to acceptable levels. The ICRP believes that nonstochastic effects can be prevented by applying a dose-equivalent limit of 0.5 Sv in a year to all tissues except the lens, for which the ICRP recommends a limit of 0.5 Sv in a year. These limits apply irrespective of whether the tissues are exposed singly or together with other organs, and they are intended to constrain any exposure that fulfills the limitation of stochastic effects.

Stochastic Effects

For stochastic effects, the ICRP's recommended dose limitation is based on the principle that the risk should be equal whether the whole body is irradiated uniformly or whether there is nonuniform irradiation. This condition will be met if

$$(W_{\text{gonads}} \times \text{dose to gonads}) + (W_{\text{breast}} \times \text{dose to breasts}) + \cdots < H_{\text{wb,L}}$$

or

$$\Sigma w_t H_t < H_{\text{wb,L}}$$

where

w_t is a weighting factor representing the proportion of the stochastic risk resulting from tissue (t) to the total risk, when the whole body is irradiated uniformly

H_t is the annual dose-equivalent in tissue (t)

$H_{\text{wb,L}}$ is the recommended annual dose-equivalent limit for uniform irradiation of the whole body, namely, 50 mSv

The values of w_t, recommended by the ICRP are

Tissue	w_t
Gonads	0.25
Breast	0.15
Red bone marrow	0.12
Lung	0.12
Thyroid	0.03
Bone surfaces	0.03
Remainder	0.30

The values of w_t for the remaining tissues requires further clarification. The ICRP recommends that a value of $w_t = 0.06$ is applicable to each of the five organs or tissues, with the remainder receiving the highest dose-equivalents, and that the exposure of all other remaining tissues can be neglected. (When the gastrointestinal tract is irradiated, the stomach, small intestine, upper large intestine, and lower large intestine are treated as four separate organs).

Occupational Exposure of Women of Reproductive Capacity

When women of reproductive capacity are occupationally exposed under the limits recommended, and when this exposure is received at an approximately regular rate, it is unlikely that any embryo could receive more than 5 mSv during the first 2 months of pregnancy. Having regard to the circumstances in which such exposures could occur, the ICRP believes that this procedure will provide appropriate protection during the essential period of organogenesis.

Occupational Exposure of Pregnant Women

It is likely that any pregnancy of more than 2 months' duration would have been recognized by the woman herself or the physician. The ICRP recommends that when pregnancy has been diagnosed, arrangements should be made to ensure that the woman can continue to work only where the annual exposure will be unlikely to exceed three-tenths of the dose-equivalent limit.

Internal Exposure

Internal exposure takes place as a result of an intake of radioactive material (following an ingestion or inhalation of radioisotopes). The "annual limit on intake" (ALI) developed by the ICRP corresponds to the magnitude of an intake by inhalation or ingestion that would result in a radiation dose-equivalent to the dose limits. Internal exposure is likely to take place in a contaminated environment; therefore, it is essential that any laboratory making use of open sources of radioactive material conduct a good "environmental surveillance program" (i.e., regular evaluations of levels of contamination on all working surfaces). In addition to a good environmental surveillance program, "bioassay procedures" (i.e., evaluation of internal doses through an assessment of the concentration of radioisotopes present in the body) should be carried out under the guidance of an expert in radiation protection, in the case of an accident or when significant radiation doses are expected. An example of such procedures are the thyroid scans performed on workers using volatile solutions of radioactive iodine (I-125, I-131).

Licenses and Authorization

In most countries of the world, the principles of radiation protection established by the ICRP are well incorporated in various regulations, and licenses and authorization from governmental agencies are usually required for the purchase and use of radioactive material.

Environmental Surveillance and Dosimetry

A radiation protection program is generally composed of two elements: (1) environmental surveillance and (2) personnel dosimetry.

Environmental Surveillance

Environmental controls consist of all the measurements of radiation exposure rate and levels of contamination. The environmental controls aim at assessing and controlling the levels of radiation exposure to which workers, visitors, and the public are subjected.

Personnel Dosimetry

Personnel dosimetry consists of the assessment of external exposure to radiation with the use of a radiation detector placed directly on the body. The radiation detector consists of a film badge, thermoluminescent dosimeters (TLDs), or electronic dosimeters and integrates the radiation dose received over an extended period of time. A personnel dosimetry program using TLD monitors or film badges is required for workers subjected to working conditions involving a significant risk of external exposure to radiation. Examples of such tasks are the regular use in a laboratory of more than 40 MBq of radioisotopes emitting x-rays or gamma rays or very energetic beta particles (excluding H-3, C-14, and S-35, which are low-energy beta emitters) or work in radiation therapy, nuclear medicine, or radiology departments.

Most of the tasks involving the use of radioactive material in a medical laboratory, with the exception of a nuclear medicine or radiation therapy laboratory, do not involve any significant risk of external exposure to radiation. In these cases, a well-conducted environmental surveillance program will be sufficient to confirm that the personnel are not exposed externally and internally. Since the cost of a personnel dosimetry program is minimal, personnel dosimetry is very often used to supplement the environmental controls.

Diagnostic Radiology

Diagnostic radiology facilities must be designed in such a way that all unnecessary doses to the public or the personnel are avoided. Except for special procedures, the equipment must be energized from a control booth that is properly shielded against radiation. The room must be shielded with full consideration of the occupancy in adjacent rooms. The shielding plans must have been reviewed by a health physicist at the planning stage, and the facilities must be inspected initially and every 2 years thereafter. This is to ensure that the operation of the equipment does not constitute a risk for the patients and the personnel and to confirm that the general shielding, based on the actual use of the equipment, is still adequate. The following guidelines should be applied to all radiology procedures:

- An x-ray room must not be used for more than one radiological examination simultaneously.
- Except for those persons whose presence is essential for the examination, no person must be in the x-ray room when the exposure is carried out.
- Operators should remain inside the control booth or behind protective screens during an x-ray exposure. In cases where this is impractical, protective clothing must be worn.
- When there is a need to support children or weak patients, holding devices should be used. If parents, escorts, or other personnel are called to assist, they must be provided with protective aprons and, if possible, protective gloves and positioning so as to avoid the primary beam.
- All entrance doors to an x-ray room must be kept closed while a patient is in the room.
- The entrance doors of x-ray rooms must be labeled with a warning sign (CAUTION: X-RAYS).
- All personnel working in radiology must wear a dosimeter (film badge or TLD monitor). The dosimeter must be positioned so that it gives a good indication of the radiation dose uniformly received on the body. When a lead apron is worn, the dosimeter must be positioned inside.

Fluoroscopic Procedures

Some fluoroscopic procedures require the presence of the personnel in the room during the exposure. When this is the case, protective aprons (equivalent to 0.25 mm of lead) must be worn. The lead shields or curtains mounted on the fluoroscopic unit must not be considered a sufficient substitute for the wearing of protective clothing. Protective gauntlets should be worn by the radiologist during each fluoroscopic examination. Depending on the exact procedure performed, eye protection could also be required when personnel are exposed to a high radiation field at eye level.

Mobile Units

Mobile units are used on in-patient units when the condition of the patient is such that it is inadvisable for the examination to be performed in the main x-ray department. During the operation of a mobile x-ray unit, the primary beam should be directed away from occupied areas and every effort must be made to ensure that the beam does not irradiate any other individuals in the vicinity of the patient. A distance of 3 m from the patient undergoing radiology with a mobile x-ray unit is generally considered adequate when exposure to the primary beam is prevented. The operator should use a lead apron or a protective screen when exposures are made. Mobile x-ray units should never be left unattended when they are ready to be energized.

Dental Radiology

In dental radiology, the operator should always energize the x-ray tube from a location where he is adequately protected. The general safety procedures that apply to general radiology also apply to dental radiology. The risk of radiation exposure in dental radiology is, however, much lower because of the lower output of the x-ray tube.

Nuclear Medicine

In nuclear medicine, high activities of radioactive material are prepared and administered to patients. The radioactive material used consists mainly of the following isotopes:

- Technetium-99 m (half-life—6 h)
- Gallium-67 (half-life—78 h)
- Iodine-131 (half-life—8 days)
- Xenon-133 (a gas having a half-life of 5 days)

The radioactive material stored in the department requires a shielding designed in function of the nature and total activity of radioisotopes. The total activity of the radioactive substance injected into a patient for a radioisotopic investigation is generally low enough so that the patient does not constitute a risk of external exposure for the public or the personnel. In the advent that a nuclear medicine patient gives a blood or urine sample for further analysis in biochemistry, the activity of such a sample is low enough not to require any special precautions. Due to the large activity of radioactive material handled in a nuclear medicine department, personnel must monitor their external exposure to radiation—a film badge or TLD dosimeter must be worn—and the following rules should be applied when handling radioactive solutions or radioactive gas:

- The radioactive preparations shall be used only in rooms specially designed for that effect and be provided with an adequate ventilation system, easily washable bench surfaces, and a fume hood when the procedures are likely to generate radioactive aerosols, vapors, or gas.
- Smoking, eating, drinking, and storage of food or drink are prohibited in any area where there is a possibility of contamination by radioactive material.
- All procedures involving radioactive materials should be carried out on trays or on benches lined with disposable absorbent material.
- Laboratories must be kept locked when not in use.
- When hand or clothing contamination is possible, protective gloves and clothing must be worn. In order to minimize the spread of contamination, such clothing should not be worn outside the laboratory.
- Wipe tests must be performed on surfaces and equipment likely to become contaminated with radioactive material. These tests should be made either at regular intervals or after each significant workload, and, if necessary, decontamination should be carried out.
- No removable contamination should be tolerated.
- Records of the tests must be properly documented and filed.
- Documentation on the purchase and use of all radioisotopes in the form of a log book must be maintained in all departments and laboratories handling radioactive materials.

Clinical Laboratories and Research Facilities

In clinical laboratories, radioactive solutions are used for trace studies and radioimmunological procedures. The total activity used for each preparation is generally low and does not constitute any risk of external exposure. Personnel dosimeters (film badge or TLD badge) are generally not required.

All steps should be taken in order to minimize the spread of contamination on laboratory surfaces or equipment, because this would increase the likelihood of skin contamination and internal exposure. The procedures outlined in the section "Nuclear Medicine" regarding the use of radioactive solutions should be implemented to the fullest extent.

In research laboratories, the total activity of radioactive material handled, as well as the nature of the isotope involved, very often constitute a risk of external exposure, and, in addition to the control of contamination mentioned previously, external exposure should also be monitored and personnel dosimeters might have to be worn. Handling of high activities of strong beta emitters such as P-32 might require the monitoring of the radiation dose received on extremities—in which case, wrist dosimeters or finger dosimeters must be worn.

When handling activities of iodine-125 or iodine-131 higher than 5 megabecquerels (MBq), bioassay procedures (thyroid scans) should be instituted in order to ensure that the internal dose received by the personnel is not excessive. All new experiments involving the use of radionuclides should be well planned, and their radiation safety should be assessed at the planning stage by a radiation protection expert.

Radiation Therapy

Radiation therapy consists of the use of ionizing radiation for the treatment of illnesses. In the year following the discovery of x-rays by Roentgen in 1895, x-rays were already being used in the treatment of certain kinds of skin cancer. Although the initial results were debatable for lack of truly scientific methodology, radiation therapy has evolved to the point where all major hospitals now have radiation therapy departments.

The equipment used in radiation therapy has the potential, when not properly maintained or operated, for extremely high accidental exposures and thus requires an extensive program of radiation protection.

Teletherapy

Teletherapy is performed with a device producing a high radiation field on the anatomical regions under treatment. The following devices are used for teletherapy:

- *Equipment housing radioactive sources* (cesium-137 or cobalt-60). Such equipment consists of a radioactive source housed in an appropriately shielded container provided with a shutter system. It produces, during a preset length of time, a high exposure on the area under treatment. When turned off, these units have an exposure rate in their vicinity of less than 0.025 mSv/h and do not constitute a risk of radiation exposure. When operated, their dose rate is in the range of 10,000 mSv/h (a chest x-ray gives a radiation dose in the range of 0.2 mSv).
- *X-ray irradiators and accelerators*. Such equipment produces a radiation field of the same magnitude as the equipment previously mentioned. However, they differ in the fact that they need electricity to produce radiation. When turned off, they do not produce any leakage of ionizing radiation.

During a treatment, only the patient should be present in the radiation therapy room. The equipment should always be operated from a control panel located outside the therapy room. For equipment housing radioactive material, a leak test should be performed every 6 months in order to ensure that the radioactive source is still perfectly sealed. All therapy installations must be provided with warning lights showing when the radiation beam is on. Such warning lights must be present on the control panel at the entrance of the treatment room as well as inside the treatment room. The opening of the door of the treatment room must automatically shut off the exposure.

Intracavitary Brachytherapy

In brachytherapy procedures, a sealed source is applied directly on the surface of the body or inside the patient's body. Mainly used in the treatment of cancer of the cervix, brachytherapy generally consists of the application of a very active source inside the vagina (cobalt-60, cesium-137, or radium-192).

Prior to the 1960s, the source was always inserted manually by the medical personnel. During such insertions, the personnel were likely to receive significant radiation exposure. In most hospital centers, this method has been replaced by "after loading methods" during which a tube instead of the source itself is inserted into the patient's vagina. When the tube is properly positioned and the patient is isolated in a properly shielded room, the source is transferred from its shielded container through the tube into the patient with no exposure to the personnel.

Interstitial Brachytherapy

Interstitial brachytherapy consists of the insertion directly in a tumor of small sources of radioactive material. The insertion could be temporary (this requires removal at the end of the treatment) or permanent, in which case the radioisotopes chosen should have relatively short half-lives. The radiation protection program used in all kinds of brachytherapy procedures must be designed by a health physicist or a radiation oncology specialist. The following guidelines must be enforced:

- The location of the treatment room must be sufficiently remote or adequately shielded in order to maintain at a low level the radiation exposure to the persons present in adjacent rooms.
- The nursing personnel must have received enough information to ensure their own protection, the protection of the patients, and the protection of visitors.
- When required, warning signs must be posted at the entrance of the patient's room.
- The personnel likely to receive a radiation dose higher than 5 mSv/h must be informed of this fact and personnel dosimetry must be performed with a TLD badge or film badge.
- At the end of the treatment, controls should be enforced in order to ensure that the patient does not constitute any risk for the public and that the room is free of contamination before it is once again occupied by another patient.

Radiation Therapy with Open Sources

Some radiation therapy treatments are performed with radioactive substances administered orally. The most common example is radioiodine treatment (iodine-131) used in the case of thyroid cancer or hyperthyroidism. Depending on the total activity of the radionuclides administered, the patients under treatment could constitute a risk of occupational exposure for personnel. The secretions of these patients could contain high concentrations of radioactive material, and, therefore, such patients constitute a risk of contamination to the environment. The patients for whom an activity of less than 1.2 GBq has been administered generally produce an exposure rate such that in normal practice the radiation dose likely to be received by personnel is low and does not necessitate any special requirements regarding the nature and the duration of nursing care. However, in the case of incontinent patients, protective gloves should be worn in order to prevent skin contact with the patient's biological fluids. When the activity administered is higher than 1.2 GBq, the same precautions as applied in brachytherapy should be implemented, with, in addition, all controls related to the handling of radioactive solutions (see the "Biological Effects of Ionizing Radiation" section).

Bibliography

Association des physicians et ingénieurs biomedicaux du Québec. *Manuel de radioprotection—Usage des radiations ionisantes en milieu hospitalier*. Association des physicians et ingénieurs biomédicaux du Québec, Médecine nucléaire et radiobiologie, C.H. Universitaire de Sherbrooke, 3001, 12e Avenue Nord, Sherbrooke, Québec, JIH SN4, 1983.

Cember, H. *Introduction to Health Physics*, 2nd edn. Pergamon Press, New York, 1983.

Health and Welfare Canada. *X-Ray Equipment in Medical Diagnosis: Recommended Safety Procedures for Installation and Use*, Safety Code No. 20A.

Health and Welfare Canada. *Radiation Protection in Dental Practice*, Safety Code No. 22.

Health and Welfare Canada. *Laboratory Facilities for Handling Radioisotopes*, Safety Code No. APB-SC-12.

Institut conjoint hospitalier de Montréal. *Manuel de radioprotection*. Institut conjoint hospitalier de Montréal, 2155 rue Guy, Montréal, QC, H3H 2R9, 1982.

International Commission on Radiological Protection. *Limits for Intakes of Radionuclides by Workers*, ICRP publication No. 20.

International Commission on Radiological Protection. *The Handling, Storage, Use and Disposal of Unsealed Radionuclides in Hospitals and Medical Research Establishments*, ICRP publication No. 25.

International Commission on Radiological Protection. *Recommendations of the International Commission on Radiological Protection*, ICRP publication No. 26.

International Commission on Radiological Protection. *Problems Involved in Developing an Index of Harm*, ICRP publication No. 27.

International Commission on Radiological Protection. *Statement from the 1978 Stockholm Meeting of the International Commission on Radiological Protection—The Principles and General Procedures for Handling Emergency and Accidental Exposures of Workers*, ICRP publication No. 28.

National Council on Radiation Protection and Measurements. *Radiation Protection in Educational Institutions*, NCRP Report No. 32. NCRP, Washington, DC, 1966.

National Council on Radiation Protection and Measurements. *Dental X-Ray Protection*, NCRP Report No. 35. NCRP, Washington, DC, 1970.

National Council on Radiation Protection and Measurements. *Precautions in the Management of Patients Who Have Received Therapeutic Amounts of Radionuclides*, NCRP Report No. 37. NCRP, Washington, DC, 1970.

National Council on Radiation Protection and Measurements. *Review of the Current State of Radiation Protection Philosophy*, NCRP Report No. 43. NCRP, Washington, DC, 1975.

National Council on Radiation Protection and Measurements. *Radiation Protection for Medical and Allied Health Personnel*, NCRP Report No. 48. NCRP, Washington, DC, 1976.

National Council on Radiation Protection and Measurements. *Structural Shielding Design and Evaluation for Medical Use of X-Rays and Gamma Rays of Energies up to 10 MeV*, NCRP Report No. 49. NCRP, Washington, DC, 1976.

National Council on Radiation Protection and Measurements. *Review of NCRP Radiation Dose Limit for Embryo and Fetus in Occupationally Exposed Women*, NCRP Report No. 53. NCRP, Washington, DC, 1977.

National Council on Radiation Protection and Measurements. *Medical Radiation Exposure of Pregnant and Potentially Pregnant Women*, NCRP Report No. 54. NCRP, Washington, DC, 1977.

National Council on Radiation Protection and Measurements. *Radiation Exposure from Consumer Products and Miscellaneous Sources*, NCRP Report No. 56. NCRP, Washington, DC.

National Council on Radiation Protection and Measurements. *Instrumentation and Monitoring Methods for Radiation Protection*, NCRP Report No. 57. NCRP, Washington, DC, 1978.

National Council on Radiation Protection and Measurements. *A Handbook of Radioactivity Measurements Procedures*, NCRP Report No. 58. NCRP, Washington, DC, 1978.

National Council on Radiation Protection and Measurements. *Operational Radiation Safety Program*, NCRP Report No. 59. NCRP, Washington, DC, 1978.

National Council on Radiation Protection and Measurements. *Management of Persons Accidentally Contaminated with Radionuclides*, NCRP Report No. 65. NCRP, Washington, DC, 1980.

National Council on Radiation Protection and Measurements. *Dosimetry of X-Ray and Gamma-Ray Beams for Radiation Therapy in the Energy Range 10 keV to 50 MeV*, NCRP Report No. 69. NCRP, Washington, DC, 1981.

National Council on Radiation Protection and Measurements. *Radiation Protection in Pediatric Radiology*. NCRP, Washington, DC, 1981.

US Department of Health, Education and Welfare. *Radiological Health Handbook*. US HEW, Bureau of Radiological Health, Public Health Service, Rockville, MD, 1970.

Appendix 22.A.1

San Francisco General Hospital Radiation Safety Handbook for Hospital Personnel

Introduction

A. **General.** During the course of your duties, you might be assigned to care for patients who have received radioactive material, or you might work near an x-ray machine. This handbook is prepared to provide you with the general principles of radiation protection for yourself while caring for the patient.

B. **Types of radiation.** All living and nonliving things are made up of atoms. Some of these atoms may be unstable and undergo nuclear changes which result in the emission of particles or rays. The radioactive atoms are called radionuclides. The emission process is called radioactivity. In the hospital, you will find three types of radioactivity (alpha, beta, gamma) and x-ray.

   1. Alpha radiation is seldom encountered in the usual medical application of radiation. Alphas are easily shielded and do not generally constitute an external exposure hazard to nurses or other staff members in the hospital. The range of alphas in air is less than about 2 in. The sealed metal capsule of radium sources used in radiotherapy will absorb all the alphas emitted.
   2. Beta radiation will be stopped by about one-half inch of wood, plastic, water, tissue, etc. Therefore, patients who have received radioactive material which gives off only beta radiation do not become an external radiation hazard to nurses or others. Problems may arise, however, due to contamination of bedding, dressings, etc. An example of a beta emitter is radioactive phosphorus (P-32).
   3. Gamma radiation can penetrate many inches of iron, concrete, wood, plastic, water, etc. Patients who have received large doses of radioactive material that emit gamma rays (for example, for some therapy procedures) may be a source of exposure to nurses and other personnel. Many gamma emitters are used in medicine, such as radioactive iodine (I-131), radium (Ra-226), gallium (Ga-67), thallium (T1–201), and technetium (Tc-99 m).
   4. X-ray machines are commonly found in the hospital. X-rays are a type of energy similar to light, and, like gamma rays, they pass easily through fairly thick materials. X-ray machines have built-in shielding. The useful beam is restricted by a cone or an adjustable collimator. All permanent installations of diagnostic and therapy x-ray machines are housed in well-shielded rooms.

C. **Radioactive decay.** All radionuclides undergo radioactive decay at a specific rate. When half of the original radioactive atoms have decayed, the material is said to have gone through one half-life. Some elements such as radium have a very long half-life, (1640 years). Others, such as thallium-201 (Tl-201) and iodine-131 (I-131), have fairly short half-lives, approximately 3 and 8 days, respectively. Therefore, the level of radioactivity diminishes relatively rapidly. Phosphorus-32 (P-32) has a half-life of 14 days. Radionuclides which are used for diagnostic purposes, scans and images have short half-lives. For example, the commonly used radionuclide technetium (Tc-99 m) has a half-life of 6 h. Thus, in 42 h (7 half-lives) 99.3% of the initial activity will have decayed.

D. **Units of radioactivity.** The activity of a radionuclide is specified in terms of the number of atoms disintegrating per second. A millicurie (mCi) is 37 million disintegrations per second; a microcurie (mCi) or 1/1000th of a mCi, is 37,000 disintegrations per second.

The activity of radionuclides used in most diagnostic procedures is in the microcurie and low millicurie range, while almost all therapeutic procedures use millicurie amounts.

E. **Hazards of ionizing radiation exposure.** Like other useful tools, radiation may create a potential hazard unless used with strict adherence to safety procedure guides.

The safety rules which govern all uses of ionizing radiation are concerned with preventing genetic damage as well as with protecting the health of the exposed individual. Therefore, a basic radiation protection principle is "to keep such exposures as low as reasonably achievable."

At San Francisco General Hospital and on the UCSF campus, there is a continuous program of evaluation and control of radiation hazards carried out by the Office of Environmental Health and Safety and the Committee on the Use of Radioisotope and Radiation Safety. All uses of radionuclides and radiation sources are subject to review by this committee.

F. **Exposure limits.** Limits which are established by the California Administrative code for persons occupationally exposed to radiation are as follows:
   1. 1250 mrem per quarter (3 months) to the whole body or major portion thereof, head and trunk (5000 mrem/year).
   2. 7500 mrem per quarter to the skin only (from non-penetrating radiation, e.g., very soft x-rays and beta particles) (30,000 mrem/year or 6 times the whole-body dose).
   3. 18,750 mrem per quarter to hands and forearms, feet and ankles (75,000 mrem/year or 15 times the whole-body dose).

The mrem (millirem) is a unit with which we estimate radiation dose. A person will receive about 100 mrem in San Francisco each year from natural background radiation.

Limits recommended by the National Council on Radiation Protection (NCRP) for special situations are as follows:

4. 15,000 mrem/year (not to exceed 5000 mrem/quarter) to other organ tissue or organ systems (including skin).
5. 500 mrem to a fetus (due to exposure of the expectant mother).

The University of California EH&S Office has established an administrative guideline for whole-body exposure for persons exposed to ionizing radiation while working on this campus. This guideline is 500 mrem/year for fertile women and 1250 mrem/year for other radiation workers. This administrative guideline determines when exposure numbers from film badges are investigated. Numbers in excess of these guidelines are not violations or overexposures. The intent of the program is to keep exposures to a minimum.

G. **Film badge.** Film badges are used to measure the radiation dose that you receive while attending patients undergoing radionuclide treatment. Film badges are to be worn only while on duty. Never leave the badge near a radiation source, wear a badge assigned to another person, or wear it during any medical or dental x-ray procedure that you may undergo. The badge must be exchanged at the end of each month for reading.

H. **The basic principles of radiation protection.** The basic methods of reducing your exposure to radiation, regardless of the specific source of radiation, are as follows:
   1. Keep the length of time of exposure to a minimum.
   2. Maintain a safe distance from the source.

3. Place a shield between yourself and the source.
4. Protect yourself against radioactive contamination.

I. Application of the basic principles of radiation safety to care of patient

1. Always provide adequate care for the patient—but work efficiently. Observe the recommended time limits listed on the "Physician's Order" form in the patient's chart. Whenever practical, without harm or discomfort to patients, encourage the patients to provide for themselves. In all cases care should be taken to avoid excessive hurrying, assuming awkward positions that might hinder your efficiency in performing a task, or causing undue alarm to the patient.
2. Maintain distance from patient, except when necessary for patient's care. You should note that if you double your distance from the source of radiation, the dose rate will be one-fourth, triple the distance one-ninth, and so forth. This is known as the inverse square law of dose vs. distance.
3. Use of shielding may be a more difficult problem, because most of the gamma ray emitting nuclides which are used in therapy require very thick lead to reduce the radiation levels to any significant degree.
4. When the "Physician's Orders" stipulate that contamination control measures are required, wear disposable rubber gloves when handling potentially contaminated items such as dressings, urine containers, etc.

Diagnostic Uses of Radionuclides

Frequently patients may undergo diagnostic tests involving radioactive materials such as bone scans, liver scans, lung scans, thyroid studies, radioimmunoassays, etc. Some of these procedures do not involve the administration of radionuclides to the patient but are tests performed on blood samples only. Even when the patient receives radioactivity, the levels are low, so no precautions are required.

The radiation dose levels in the vicinity of these patients are low and do not require use of film badges or restrictions on visitors or nursing time. The radioactive materials used in diagnosis generally have short half-lives, so urine spilled a day or two after the study does not have significant radioactivity.

Therapy Uses of I-131 and P-32

This type of therapy involves the administration of soluble radioactive material to a patient orally or by IV injection. The nuclides used include I-131 and P-32 (soluble phosphate). Because the radioactive materials circulate in the blood, all body fluids may contain radioactivity. Special precautions will be required to control radioactive contamination.

A. Contamination control. If P-32 is given, about 5%–10% of the dose administered is excreted in the first 12 h. Saliva is also contaminated. If I-131 is administered, up to 50% of the dose is excreted in the first 48 h. Most of the excretion is in the urine; however, significant levels of contamination may be present in the perspiration. Since the dose is usually given orally, the contents of the stomach are also radioactive for the first few hours after ingestion of the I-131. Thus, anything handled by the patient may be contaminated. Disposable gloves should be worn to handle the patient or items that come in direct contact with patient. No material, including trash, shall be removed from patient's room without the authorization of Nuclear Medicine.

Whenever a patient receives such therapy, an appropriate "Physician's Order Form" will be completed and placed in the patient's chart and a "Caution Radioactive Materials" label affixed to the chart cover. Also, a "Caution Radioactive Materials" sign will be posted on the patient's door and personnel instructed not to remove any material from the patient's room.

B. Nursing care. Nursing care is to be restricted for the term of treatment to those activities essential to the well-being of the patient. Disposable gloves should be worn to perform routine patient care. If special nursing care is required, the problem will be worked out by Nuclear Medicine, and the head of the Nursing Unit will be informed. Where visiting and nursing times are restricted, no pregnant women (including staff) or visitors under 18 years old are permitted in the room.

C. **Dishes.** Disposable dishes and utensils should be used. The leftover food, whenever possible, should be flushed down the toilet to avoid creating an unpleasant odor. The dishes together with other waste should be bagged and held in the room for disposal by Nuclear Medicine.

D. **Telephone.** The telephone in the patient's room should be covered with a small plastic bag to prevent contamination.

E. **Linens.** It is not necessary to change bed linens daily if they are not soiled. Linens, towels, and launderable items should be held in the room until cleared by Nuclear Medicine.

F. **Toilet instructions.** The patient should use the toilet facilities in his/her room. Instruct the patient to flush the toilet three times to clear the waste lines and dilute the radioactivity. The patient should also be instructed to take special care to avoid splashing urine when voiding and to wipe the toilet seat with tissue after using the toilet, or toilet seat covers can be used to avoid contamination. The sink should be rinsed after use, especially after brushing teeth. If the urine is to be collected for assay, a special, unbreakable labeled container is provided by Nuclear Medicine. The bathroom floor will be covered with plastic-backed absorbent paper. If the urine must be collected by attending personnel, or when bed pans are required, gloves should be worn to prevent contamination. The gloves should be disposed of in the waste container and hands should be washed thoroughly. Nuclear Medicine will instruct the attending personnel to take safety precautions for handling and storing specimens.

G. **Baths.** Unless it is necessary, the bath should be postponed for the first 48 h for I-131 or 12 h for P-32 patients. If possible, the patient should bathe himself and should rinse the shower or tub thoroughly afterward.

H. **Housekeeping.** Housekeeping should be postponed until a Nuclear Medicine survey has been made at the conclusion of use of the room. Bag paper wastes, tissues, etc., and hold in the room to be checked by Nuclear Medicine.

I. **Film badges.** When iodine-131 is administered in doses exceeding 10 mCi, film badges are to be worn by attending staff. Phosphorus (P-32) emits beta particles which are totally absorbed by the patient's body. The small activity levels present as contamination on the body are not sufficient to require the use of badges.

J. **Restriction of visiting time.** Visitors to patients receiving more than 30 mCi of radioactive iodine-131, should be limited to no more than 1 h/day. They must avoid direct contact such as kissing the patient and stay 2 m away from the patient. All visitors and staff personnel should stop at the nurse's desk for special instructions and follow the instructions posted on the door. Children, and particularly fetuses, are more sensitive to radiation effects; for this reason, exposure of unborn infants should not be allowed. Nurses should consult Table 22.A1 on iodine-131 for the maximum permissible working times.

Since P-32 emits beta particles that are totally absorbed by the patient's body, nursing staff and visitors should be aware there are nevertheless some low radioactivity levels present as contamination on the body. Wearing disposable gloves and washing hands are sufficient for caring for the patient.

TABLE 22.A1 Recommended Maximum Daily Working and Visiting Times for I-131

Activity in Patient (mCi)	Maximum Hours for Nursing Personnel Near Bedside	Maximum Hours for Visitor at Average Distance of 6 ft
0–9	Unlimited	Unlimited
10–19	8	Unlimited
20–29	4	Unlimited
30–39	2	8
60–89	1	4
90–150	3/4	3

K. **Room survey prior to release of room.** At the conclusion of the use of a room for this type of therapy, prior to having the room cleaned, call Nuclear Medicine to arrange for a radiation survey of the room.

L. **Spills.** If a patient who has received I-131 should vomit or be incontinent during the first 48 hours, call Nuclear Medicine immediately. Do not attempt to clean up the spill. Take interim steps to check the spread of contamination as follows:
   1. Restrict the area—allow no one to enter except for urgent treatment of the patient.
   2. Keep people 2 m from the spill.
   3. Persons who have been in the contaminated area should stay there until surveyed and cleared by Nuclear Medicine.
   4. Remove contaminated clothing. Place in a plastic bag brought to the area.
   5. Cleanse contaminated skin using facilities in the room.
   6. If there is appreciable liquid spilled, absorb with paper towels.
   7. Retain all contaminated or suspected materials in area until cleared by Nuclear Medicine.

If a patient who has received soluble P-32 vomits or is incontinent during the first 12 hours, the same procedure should be followed.

M. **Emergencies.** Nonradiation: Seizures, cardiac arrest, trauma, etc. Follow normal emergency procedure. Call the Nuclear Medicine or a Nuclear Medicine physician. Hold potentially contaminated items and personnel until they can be checked.

N. **Death of patient.** Notify the attending physician and the Nuclear Medicine and Pathology departments of the death of a patient who contains radioactive material. Make sure the pathologist has a copy of the "Physician's Order Form" and that the chart records the type and amount of radioactive material.

Doctor's Orders for Patients Who Have Received Radionuclide Therapy

Patient received _____ millicuries of I-131 orally at _____ am/pm on _____ 20_____. Initial exposure rate at 1 meter from patient _____ mr/hr measured by _____ (Nuclear Medicine).
(signature)

STANDARD INSTRUCTIONS (Nuclear Medicine Physician should check off appropriate items or line them out and initial to countermand)

ALL PATIENTS:

_____ 1. Interview patient and discuss pages 5 thru 8 (THERAPY) of the Radiation Safety Handbook for Hospital Staff with patients and family members.

_____ 2. Special instructions for patient and members of patient's family (especially children and pregnant women).

_____ 3. If patient is readmitted to the hospital or dies before (date), notify Nuclear Medicine. After normal working hours, call the Nuclear Medicine physician who is on-call.

Inpatients Only:

_____ 1. Patient must be in a private room unless prior approval has been obtained from Nuclear Medicine.

_____ 2. Patient must remain in room.

_____ 3. Patient to use disposable dishes and utensils. Dispose in a plastic bag provided by Nuclear Medicine.

_____ 4. Patient may have baths (shower). Instruct patient to bathe himself and then to rinse shower thoroughly. The shower will be checked by Nuclear Medicine before release to nonradioactive patient use.

_____ 5. Instruct patient to flush toilet three times after each use.

_____ 6. No visitor under 18 years of age or who is (or may be) pregnant. Staff and visitors are to follow the radiation exposure guidelines.

_____ 7. Post "CAUTION RADIOACTIVE MATERIALS" and "INSTRUCTIONS TO WARD PERSONNEL" on door. Affix "CAUTION RADIOACTIVE" on patient's wristband.

_____ 8. In the event of spill, follow the instructions in the "Radiation Safety Handbook for Hospital Staff" on page 7.

_____ 9. Hold all linens and disposable waste in room until cleared by Nuclear Medicine.

_____ 10. When a patient is discharged, request Nuclear Medicine to survey the room prior to admitting housekeeping to the room.

SPECIAL ORDERS

____

____

In the event of any difficulty, call:	days	nights
Dr. ______________________________	______________	______________
Dr. ______________________________	______________	______________

If he/she cannot be reached promptly, call Nuclear Medicine Dept. during the day or the Nuclear Medicine physician on-call.

PATIENT'S DEATH: If a patient dies, immediately notify Nuclear Medicine Physician.

FOR FURTHER INFORMATION: Consult "Radiation Safety Handbook for Hospital

Personnel."

______________________________M.D.

______________________________(Date)

Safe Handling of Cadavers Containing Radioactive Isotopes

A. Procedure after death of patient

1. If a patient containing less than 5 mCi radioactive material dies in the hospital, precautions should simply be taken to avoid contamination.
2. If a patient dies in the hospital and contains more than 5 mCi, the responsible physician signing the death certificate should inform the pathologist and the Radiation Safety Officer of this fact. The Funeral Director's form should be completed. (NCRP Report #37 Appendix V).
3. If there is an autopsy, it may be necessary for the pathologist to take the precautions detailed in step B while performing the autopsy.

B. Conduct of autopsy

1. When a cadaver suspected of containing any radioactive materials is to be autopsied, the Radiation Safety Officer should be notified.
2. The amount of activity remaining in the body should be estimated by reference to the time since the administration of the isotope and its biological fate.
3. If the remaining amount is less than 5 mCi, no special precautions are necessary other than the usual wearing of gloves, except in cases of I-131 therapy or therapy with insoluble P-32, where the body cavity contains most of the activity and should be handled as rapidly as possible.
4. Where the residual activity exceeds 5 mCi, the following procedures should be followed:
   a. Monitor the body with a survey meter before it is opened to establish the maximum working time if necessary. Film badges or ring badges may be required.
   b. Drain carefully all body fluids and save for assay. In cases of I-131 therapy, the blood and particularly the urine will be radioactive.

c. After the body is opened, a second survey should be made to estimate the level of beta dose from P-32 or other pure beta-emitting radionuclides.
d. In cases of I-131, the thyroid gland will produce a gamma dose of 0.5 R/min near its surface for each 10 mCi in it and, consequently, should not be touched by hand directly. Its removal, depending on the activity level, should be accomplished using a long instrument.
e. Highly radioactive fluids should be stored behind a shield. Consult Radiation Safety or Nuclear Medicine for disposal.
f. All instruments and clothing involved in the autopsy should be monitored after the procedure and stored or decontaminated before being returned to general use or dispatch to a laundry. The autopsy room should also be monitored and decontaminated if necessary.

Emergency Admission of Patient Involved with Radiation Accident Procedures

If a radiation accident patient is brought to the emergency room, the following procedures are to be followed:

1. Notify the Nuclear Medicine physician.
2. If there is radionuclide contamination, all exposed individuals should be monitored, isolated, and decontaminated before being examined and treated unless life saving procedures are necessary. Personnel caring for patients should wear surgical clothing.
3. If patient is ambulatory: Confine to a small area. Spread sheet or paper for patient to stand on. Patient should disrobe; put clothing in a plastic bag for later monitoring. Save all samples of clothing, blood, urine, stool, vomitus. Label with patient name, date, time. Patient should take a shower in the nearest available facility and be resurveyed, rewashing repeatedly if necessary until decontaminated.
4. If patient is nonambulatory: Place patient on sheet and cut off clothing. Clothing should be saved as above for monitoring. Patients should be washed with repeated monitoring until decontaminated.
5. For treatment of radiation injury, call the Nuclear Medicine physician. Off-duty hours, contact the Nuclear Medicine physician who is on-call.

Radioisotope Policies: Conditions of Use

1. **General safety precautions**. Safety is a result of following careful procedures in the laboratory. The safety of each operation or manipulation must be considered both separately and in relation to the overall experimental design. Periodic self-evaluation, modified for the facility or need, is suggested for users of ionizing radiation. The following precautions should be followed regardless of the amount or type of isotope involved:
   a. Protective clothing, including shoes and laboratory coats, must be worn. Coats should never be worn out of the work area.
   b. Rubber or disposable gloves must be worn during all manipulations that could result in contamination.
   c. Mouth pipetting is prohibited.
   d. Smoking, eating, drinking, and applying makeup in areas with radioisotopes, or in areas that may be contaminated, are prohibited.
   e. Refrigerators containing, or having contained, radioisotopes, may not be used for food storage.
   f. Unbreakable containers must be used whenever possible. If glass is used, secondary containers must also be used.
   g. Widespread use of absorbent paper with impervious backing is recommended for covering work areas.
   h. Radioactive work should be confined to small areas. This is to simplify containment, shielding, and cleanup in cases of contamination.

   i. Glove boxes must be used if appreciable amounts of radioisotopes are being manipulated or if there is potential for contamination, volatilization, or aerosol formation.
   j. Use of protective equipment (masks, coats, gloves, shoe covers, etc.) must never substitute for the minimization of hazardous conditions.
   k. Standard labeled and shielded waste storage containers should be used. Eleven quart step cans are available from the Storehouse.
   l. Work areas and clothing should be monitored daily for radioactive contamination.
   m. Individuals with cuts on their hands or arms should refrain from work with unsealed sources.
2. **Users and locations.** Only personnel having valid training numbers and approved forms on file with the Radiation Safety Officer may use radioisotopes.

Only those facilities listed on the application form and approved by the Radiation Safety Officer may be used for radioisotope work.

3. **Administration of radioisotopes to animals.** Before administering any isotopes to animals, a plan for handling the animals must be filed with the Director of Animal Care. Items to be addressed include:
   a. Animals given radioactive materials should be caged separately from other animals. Supplement B must be submitted with the basic radioisotope application form, if these animals are to be housed in the Animal Care Facility.

Cages must be labeled with appropriate radiation warning signs. The name of the isotope, the quantity, date(s) of administration, and the name of the person responsible for the experiment must all be on the label.

   b. The handling and disposal of radioactive excretions, animal carcasses, or tissues must be done as specified by the Radiation Safety Officer. If the excrete is potentially radioactive, arrangements must be made for its handling and disposal, to minimize contamination of cages, workers, and surrounding areas.
   c. If experimental conditions are such that significant quantities of radioisotope could be released during animal respiration, metabolic cages fitted with suitable filters or scrubbers may be required by the Radiation Safety Officer or the Director of Animal Care.
   d. Users are responsible for assuring that caretakers and custodians are aware of the potential hazards of handling radioactive animals, and, if animals are not housed in the Animal Care Facility, that personnel caring for the animals are suitably trained and supervised in the necessary precautions.
   e. Administration of radioisotopes to animals that are not the property of the University requires special consideration and must receive prior approval by the Radiation Safety Officer.
4. **Administration of radioisotopes to humans.** Whenever humans are to receive radioisotopes in a research context, Supplement C must be completed and submitted to the Radiation Safety Committee, along with a basic radioisotope application. This category includes the use of those standard clinical procedures requiring approval of the Committee on Human Research, as described in their guidelines. The Committee on Human Research requires a copy of the approved Radiation Safety Committee application before giving its approval.
5. **Exposure**
   a. Maximum level. The maximum permissible level of investigators working with radiation and radionuclides is 100 mRem, whole-body exposure, in one month. The exposure of personnel not directly involved with the use of radiation on campus should not be greater than 500 mRem per year.

Periodic surveys shall be provided by the Radiation Safety Officer for all areas in which radioisotopes are used or stored.

b. Pregnancy. For pregnant workers, the maximum limit of exposure is 500 mRem over the 9-month gestation period. Pregnant employees are strongly encouraged to inform their supervisors immediately, so appraisal of exposure can be made. Appropriate monitoring is recommended. The Radiation Safety Officer is available for consultation and advice to employees and supervisors.

Source: San Francisco General Hospital, San Francisco, California, 1989.

23

Reproductive Hazards in Hospitals

Introduction—What Is a Reproductive Toxin? 23-1
Background 23-2
Attributes of an Acceptable Policy 23-3
Nondiscrimination • Employee Privacy and Confidentiality
Elements of a Comprehensive Program to Confront Reproductive Hazards 23-4
Integration into Hazard Communication Program: Information and Training about Reproductive Toxins • Reproductive Hazards Committee • Control Measures and Policies for Reproductive Toxins
Hierarchy of Control Measures to Achieve ALARA Exposures 23-7
Substitution • Engineering Controls • Work Practices Monitoring • Personal Protective Equipment • Medical Removal Protection
Compensation for Employees Receiving Medical Removal Protection 23-9
Conclusion 23-10
References 23-11

Lindsey V. Kayman
University of Medicine and Dentistry of New Jersey

Introduction—What Is a Reproductive Toxin?

Reproductive toxins are materials that can interfere with reproductive functions or can cause damage to an exposed adult's ova, sperm, embryo, fetus, or child. Examples of reproductive effects include the following:

- Effects on the reproductive organs (e.g., enlarged breasts, atrophied testicles, damaged ova)
- Effects on adult sexual functions (e.g., ovulation, libido fertility, menstruation)
- Effects on the offspring of males or females who were exposed, by causing structural abnormalities, functional deficiencies, diseases or altered growth, or death of the conceptus
- Effects on the health of the neonate by concentrating in breast milk
- Increased risk of cancer early in life or in adulthood from transplacental carcinogens crossing the placenta

"Mutagens" affect offspring through changes in the DNA of paternal spermatogonia or maternal oocytes prior to conception. "Teratogens" affect the developing embryo or fetus via exposures in the womb. Effects may be apparent at birth or may be detectable only years after birth. An infant may be exposed to "environmental chemicals" during early postnatal life by drinking contaminated breast milk, exposure to toxins brought home on the work clothes of parents, or exposure to toxins in its environment.

Background

Occupational reproductive hazards have always been present in the hospital workplace. Hospitals have had many years' experience dealing with some of these. For example, ionizing radiation may cause a variety of reproductive effects in males and females and may result in injury to their offspring. Certain biological agents found in hospitals are teratogens.

Despite the potential long-term impact on society and the extreme human suffering which may be caused by reproductive dysfunction or birth defects, attention is only recently being paid to identifying and preventing occupational exposure to reproductive toxins. Of the 60,000–70,000 chemicals currently used commercially, only ionizing radiation and three chemicals (ethylene oxide, dibromochloropropane, and lead) have been specifically regulated by Occupational Safety and Health (OSHA) to prevent reproductive effects.

In a 1985 report, National Institute for Occupational Safety and Health (NIOSH) listed reproductive impairment as sixth of 10 leading work-related diseases based on the number of workers occupationally exposed to reproductive toxins.[1] Part 2 of this report, concerning national strategies for the prevention of leading work-related diseases, is scheduled to be released in the near future. It will address disorders of reproduction.

OSHA's lack of regulatory guidance concerning occupational reproductive toxins is due to a number of factors, including scientific, legal, political, and philosophical issues. OSHA does not have the authority to promulgate regulations unless there is a preponderance of scientific evidence which demonstrates the existence of a significant health risk. However, there has been very little research done on a number of reproductive issues, including effects on adult sexual function such as libido, menstruation, lactation, and ovulation. There is a dearth of information on risk of chemically induced genetic mutations in workers who may parent children in the future.

OSHA has not regulated any chemicals solely on the basis of animal studies, but has published an intent to regulate certain glycol ethers based on reproductive effects in animals. Epidemiologic studies are undertaken infrequently due to a variety of technical considerations, such as confounding factors and difficulties in detecting the reproductive outcomes. Most epidemiologic studies that have been performed have focused on teratogenic effects. Many of these studies addressed only the mother's exposures, ignoring the potential contributions of paternal exposures.

The interpretation of animal studies to predict similar reproductive effects in humans remains controversial. The EPA has published a method of determining safe exposure levels in the general population based on animal data.[2,3] The maximum dose which produces "no observable adverse effect level" (NOAEL) in the most sensitive animal tested is determined. The airborne concentration which would cause a similar dosage in humans is determined and then a safety factor is applied. A 10-fold factor is utilized to account for variation in sensitivities among the human population. Another 10-fold factor accounts for uncertainties in extrapolating from data in lab animal species to humans. An additional 10-fold safety factor is used when a "lowest observed adverse effect level" (LOAEL) is used rather than an NOAEL. Additional modifying factors may be utilized to account for professional judgment of the entire database of the chemical.

The 1000-fold safety factor, which is often employed for exposures to the public rather than for occupational exposures, remains controversial. Usually standards designed for the general public are more stringent than standards designed for workers, on the premise that in order to work, one must be relatively healthy. The general public contains members who are elderly, infirm, or hypersusceptible.

However, for reproductive toxins, workers are in an age group that is potentially at greater risk than is the general public. In addition, female workers may actually exhibit an "unhealthy worker effect" whereby women who are infertile or who have suffered miscarriage(s) remain in the work force rather than leave the work force to care for children.

Attributes of an Acceptable Policy

Nondiscrimination

Protection of workers' reproductive health and the health of their offspring must not have a discriminatory impact on any group of employees' wages, benefits, seniority, or right to employment.

To date, most policies addressing reproductive hazards on the job are aimed at protecting the fetus, which in some circumstances is more susceptible to injury from certain chemicals on the job than are adult workers. These policies, known as "fetal protection policies" usually result in the involuntary exclusion of female or pregnant employees from certain jobs.

In general, fetal protection policies do not address preventing exposures by making the job as safe as possible using engineering and other control measures, but instead rely on excluding "sensitive workers." However, nonteratogenic reproductive effects often occur at exposure levels lower than that required to cause teratogenesis.

Fetal protection policies are often administered arbitrarily without a careful study to evaluate the hazards of jobs which are made off-limits or the positive effects expected as a result of the policy.

Fetal protection policies which exclude all women (or all fertile women) have a negative impact on women's employment rights. In many cases they are in violation of Title VII of the Civil Rights Act of 1964 or its amendment, the Pregnancy Discrimination Act of 1978. This amendment prohibits discrimination based on pregnancy, childbirth, or related medical conditions.

Under Title VII, fetal protection policies are assumed to be discriminatory unless there is evidence that exposures at levels encountered in the workplace may cause a significant risk of harm to the unborn children of women employees but will not cause harm to the unborn children of male employees. There must also be evidence that the policy will significantly reduce this risk of harm.

Even if the required evidence is obtained, fetal protection policies may still be in violation of Title VII if it can be proved that an equally effective alternative policy would have a less adverse impact on one sex.

A comprehensive discussion of the legality of fetal protection policies is covered in the OTA publication, *Reproductive Health Hazards in the Workplace.*[4] Fetal protection policies are usually liability-avoidance measures. Employees who are injured on the job are barred from tort suits against their employers in most cases, and may instead only receive workers' compensation. Live-born children, injured as a result of their parents' exposure, may sue for damages.[5] Ironically, despite the focus on reproductive hazards to female employees, case law to date involves male employees rather than females. Suits have been filed on behalf of children injured as a result of their father's exposure to Agent Orange, radiation, dibromochloropropane, pesticides, herbicides, and others.

Employee Privacy and Confidentiality

Reproductive function is one of the most sensitive and intimate areas of life. For this reason, it is recommended that policies that have special procedures for employees who are trying to parent a child or who are pregnant should rely on voluntary reporting of reproductive status by the employee.

It is a violation of the Pregnancy Discrimination Act of 1978 to refuse to hire or promote a woman because she is or may become pregnant. Questions about reproductive status should not be listed on any employment form.

Employee cooperation can be improved by an informed work force which does not fear losing jobs or receiving lower pay as a result of informing the company of future plans.

Elements of a Comprehensive Program to Confront Reproductive Hazards

A program to confront reproductive hazards must provide feasible solutions both financially and in the daily running of the departments in order to be effective. The following elements of an overall program will help accomplish this goal:

- Integration into hazard communication program: Information and training about reproductive toxins
- Reproductive hazards committee
- Control measures and policies for reproductive hazards

Integration into Hazard Communication Program: Information and Training about Reproductive Toxins

Hospitals have been regulated under the OSHA Hazard Communication Standard since August 1988. This standard requires that employers inventory the chemicals used throughout the facility and also provide health and safety information to employees. Required health and safety information includes chemical information sheets, known as Material Safety Data Sheets, and labeling of chemical containers. New employees must receive health and safety training before beginning work. Current employees must receive training before working with new chemicals and to review health and safety information as appropriate.

Identification of Reproductive Toxins

Material Safety Data Sheets, required under the OSHA Hazard Communication Standard, should contain information about reproductive effects associated with each chemical. Employees should be encouraged to provide copies of Material Safety Data Sheets to their physicians when they are planning to attempt to conceive.

Unfortunately, many older Material Safety Data Sheets are deficient in providing information on reproductive effects. However, up-to-date Material Safety Data Sheets include this information when it is available. The rule stipulates that up-to-date sheets be utilized in compliance with the standard.

Chemical companies and software companies offer computerized Materials Safety Data Sheet subscription services where the Materials Safety Data Sheets are updated quarterly. These services are available for personal computer or mainframe applications. A benefit of many of these types of services is that a search can be conducted to identify chemicals with specific reproductive hazard data, such as information on teratogenicity or mutagenicity.

Technology is available to make personal computer databases on CD-Rom, available on a mainframe very economically.

The hospital community should be encouraged to utilize Material Safety Data Sheets to factor health and safety information into their choice of products for all applications in the hospital.

Additional Information on Reproductive Toxins

A wide variety of additional resources are available to identify reproductive toxins. Computerized literature searches can provide a variety of information. A few of the computer-based data retrieval systems which can provide information on occupational or environmental reproductive hazards are as follows:

- *REPROTOX*: an information system developed by the Reproductive Toxicology Center in Washington, DC.[6] It provides affordable computerized information on potential hazards in the environment to human pregnancy and reproduction to members who dial a toll-free phone number.
- *Registry Toxic Effects Chemical Substances (RTECS)*: provides study results on toxicity, animal studies, specific paternal- and maternal-mediated effects, and fertility. In addition, dose levels for positive studies are supplied.

- *Medline*: an information system developed by the National Library of Medicine, Bethesda, MD. Contains references from 3000 medical journals.
- *Toxline*: developed by the National Library of Medicine, Bethesda, MD. Contains references from 400,000 toxicology and epidemiology studies.
- *On-line Catalog of Teratogenic Agents*: Central Laboratory for Human Embryology (Seattle, WA), database on teratogenic effects of approximately 2000 substances.

Many states have reproductive hazard teratogen phone information lines. Local March of Dimes chapters may be able to provide assistance in locating additional services in states which are not listed. Most information lines only service callers from the state in which the information lines are located. Examples of these phone information lines are as follows:

- Arkansas Genetics Program Teratogen Screening (501) 686-5994
- Connecticut Pregnancy Exposure Information Service (800) 325-5391, (203) 674-1465
- Colorado Teratogen Information Program (800) 322-2082, (303) 861-6395
- The State of Florida Teratogen Information Service (904) 392-4104
- Illinois Teratogen Information Service (800) 252-4847, (312) 883-7095
- Massachusetts Teratogen Information Service (800) 322-5014, (617) 787-4957
- Nebraska Teratogen Project (402) 559-5071
- New Jersey Teratology Information Network (800) 441-0025
- New York Teratogen Information Service (716) 831-2018, (716) 833-4300
- Pregnancy Healthline (Philadelphia, PA) (215) 829-KIDS
- Pregnancy Safety Hotline (Pittsburgh, PA) (412) 687-SAFE
- Pregnancy Riskline (Utah/Montana) (800) 822-BABY, (801) 583-2229
- Vermont Pregnancy Risk Information Service (800) 531-9800, (802) 658-4310
- Washington State Poison Control Network (800) 732-6985, (206) 526-2121
- Wisconsin Teratogen Project (800) 352-3020

Training

The initial training required under the Hazard Communication standard should contain a discussion of hazards and controls employees will encounter on the job. Some institutions have expanded their Hazard Communication training session to include nonchemical hazards and controls.

The Hazard Communication training program should be standardized to include a discussion of the lack of conclusive studies on reproductive effects for most chemicals and the resulting need to minimize all exposures as much as possible. Known or suspected reproductive toxins should be identified, as well as any protocols that have been developed to control exposures. Both male and female reproductive toxins should be discussed.

Training may also include details of any services provided at the hospital such as pregnancy testing, counseling about reproductive hazards on the job and pregnancy planning, air sampling or health, and safety evaluations of work areas. During the training, employees should be notified who to call to evaluate any potential hazardous working conditions.

Reproductive Hazards Committee

Development of a comprehensive reproductive hazards policy involves extensive knowledge of the institution and technical expertise in many areas. A committee can be specifically designed to have expertise in necessary areas. Committee members can include representatives from personnel, academic and research departments, the safety office, representatives of each union who may be affected, and members who are proficient in toxicology and reproductive biology.

The committee will participate in the development of a policy and should ensure that it is feasible and ethical and does not contain any illegal provisions.

Once a policy has been adopted, the committee should continue to function as a formal committee. If the committee has enough expertise it can classify materials as reproductive toxins. The committee can also be involved in making decisions concerning control measures to be implemented. As much flexibility as possible, with the participation of affected employees, will be of more aid in achieving reasonable solutions than will deciding the solution to all circumstances in advance.

Another benefit of decision making by a committee is to allow decisions to be made using more information and in a more reasoned manner than if the decisions were made informally by a Safety Officer or an Employee Health Services Manager. However, contingency plans should be in the policy for decisions that have to be made immediately, i.e., to prevent imminent exposures to a pregnant woman or potential parent.

Control Measures and Policies for Reproductive Toxins

Designation of Current Potential Parents

Current potential parents include male and female employees who are able to parent a child and are having relations without using contraceptives. Pregnant women are a separate category of current potential parents. Special precautions may be necessary for current potential parents in certain circumstances.

The embryo/fetus is most susceptible to the harmful effects of workplace chemicals in the first trimester of gestation while the organs are forming. However, pregnancy is often not confirmed till well into the first trimester. Continued exposure may result during this time if exposures are only addressed once pregnancy is recognized.

Males who are attempting to parent a child may also need special precautions. Reproductive toxins on the job may affect male reproductive organs, which is often manifested as reduced fertility.

Spermatogonia are continuously developed during the course of the lives of adult males. Spermatogonia continuously develop and mature over a 73–86 day period in adult males and are susceptible to mutations or other damage during this time. Birth defects or death of the conceptus and heritable changes in chromosomes are possible if conception occurs from a sperm which has been damaged or mutated.

Breast feeding employees may also need to take special precautions to prevent exposures to chemicals which could concentrate in breast milk resulting in exposure to their babies.

As Low as Reasonably Achievable Exposures to Protect All Employees from Reproductive Hazard

A policy which focuses only on preventing exposure to pregnant women alone does not ensure that damage is not done to the reproductive system or future offspring of male employees, employees who do not know that they are pregnant, or employees who are not ready to conceive.

At this time there are few chemicals for which a NOAEL has been identified for reproductive effects. Exposures to chemicals which have not been adequately studied for reproductive effects should be minimized until they are proven safe. For this reason a prudent course of action is to use substitution, engineering, and administrative control to keep exposures as low as reasonably achievable (ALARA).

For example, in its criteria standard for anesthetic gases, NIOSH recognized adverse reproductive effects in exposed workers and their offspring as the main health effect to be prevented in developing exposure guidelines.[7] Since a safe level of exposure was not possible to define, NIOSH recommended that exposures to anesthetic gases be kept below air concentrations which can be detected by approved air sampling and analysis methods.

These provisions will prevent damage to the reproductive health of employees and the health of their future offspring. In addition, exposures to employees during unplanned pregnancies or while sperm is developing prior to conception will be minimized.

Hierarchy of Control Measures to Achieve ALARA Exposures

Substitution

As with other toxic materials, substitution with safer materials is the first control measure to consider. Examples of safe substitutes for reproductive toxins are as follows:

- Housekeeping supplies and pesticides are available that do not contain carcinogens or reproductive toxins.
- Digital thermometers and pressure-sensing devices may be substituted for those containing mercury.
- Nonchromium-based glass cleaning materials are available to replace chromic sulfuric acid.
- Aqueous-based scintillation cocktail is available to replace toluene-based preparations.
- Lead-free solder is available for general use, replacing lead-tin solders.
- Laboratory procedures generally employ the choice of a number of solvents. Procedures which commonly use reproductive toxins should be examined to determine if a safer substitute is available.
- Premixed or prediluted materials containing toxins should be purchased rather than diluting or preparing the mixture on-site.

Hospitals should offer positive incentives to all departments to routinely research available substitutes. Incentives may include awards, positive publicity, and commendations. Many reproductive toxins are also serious environmental pollutants. Significant cost savings may be realized in the hazardous waste budget if substitutes are used rather than toxic materials.

Engineering Controls

As mentioned earlier, all exposures to potential reproductive toxins should be kept as low as reasonably achievable by the use of engineering controls. An industrial hygienist should investigate reports of toxic chemicals used without local exhaust ventilation. Some examples of engineering controls which should be considered to minimize exposures include:

- *Laboratories.* Automated, ventilated autotechnicons in the histology department are available to prevent the release of xylene, formaldehyde, and ethanol vapors into the lab. Leakage often occurs around the blower in old-style laboratory fume hoods which have the blower located directly on top of the hood. These fume hoods should be modified so that the blower is near the roof. Alarms should be installed on laboratory chemical fume hoods which will alert users if the hood is not functioning properly.
- *Operating rooms.* Active scavenging systems are generally more effective than passive scavenging systems to remove waste anesthetic gases in operating rooms. Because postoperative patients may be exhaling anesthetic gases, recovery rooms must have sufficient ventilation.
- *Portable processes.* Special tasks that are performed in a variety of locations may generate contaminants (such as occasional pouring of solvents) in an area that does not have fixed local exhaust ventilation can be controlled using flexible ducting attached to a dedicated exhaust grill placed nearby the contaminant source. However, it must be confirmed that the exhaust grill does not recirculate the air to other locations in the hospital, but instead leads directly to the roof of the building. Flexible ducting which has a blower on one side can be used to remove contaminants away from employees in temporary operations (such as portable welding operations).
- *Pharmacy and nursing areas.* Chemotherapeutic drugs should be mixed in a biological safety cabinet which has been certified to have correct airflow and an intact filter. The cabinet must be recertified yearly and every time it is moved from one location to another.

- *Animal research areas.* Animals that are administered chemotherapeutics or other toxins during research will excrete these materials in their urine and feces. Animal handlers can be exposed to these materials via skin contact or by breathing in aerosolized materials while moving bedding. High-efficiency filter-top cages and meticulous work practices should be used in these circumstances.

Work Practices Monitoring

Employee work practices for use of reproductive hazards and other toxins should be monitored. Training sessions covering safe work practices could include simulation of work utilizing a fluorescent dye to disclose the presence of contamination on work surfaces and skin. The fluorescent dye can also be used as a disclosing medium to evaluate the effectiveness of work practices at minimizing exposures during simulation of work with toxic materials. Tubing and connections on anesthesia machines can be easily leak-tested prior to use with common anesthetics and an inexpensive Freon detector.

Personal Protective Equipment

Respirators

In some cases, respirators may be a useful backup for employees who are trying to conceive as long as appropriate measures are taken to ensure that exposures do not occur as a result of skin absorption. Disposable dust masks do not provide adequate protection against toxic materials. If used properly, half-face air purifying respirators will lower exposures by a factor of 10. Full-face air purifying respirator can lower exposures by a factor of 100. Under less than optimal conditions, the actual protection offered by these respirators will be reduced. In addition, respirators are uncomfortable and are generally not feasible for extended time periods. Detailed standard operating procedures, training, and good work practices are necessary to utilize respirators effectively.

Gloves

Surgical latex gloves are not designed to protect against chemicals. Glove manufacturers provide information about matching gloves to specific chemicals to be used.

Medical Removal Protection

Medical removal protection is the removal of pregnant employees or potential parents from work involving reproductive toxins. Substitution, engineering controls, good work practices, and job redesign as well as elimination of specific tasks will be sufficient to address the great majority of concerns related to preventing exposures in current potential parents due to a reproductive hazard. Examples of circumstances where employees can remain in their job but avoid potentially hazardous exposures include:

- In Central Processing, current potential parents can work in areas where ethylene oxide is not used. This area should be under positive pressure to prevent potentially contaminated air from leaking in. As a double insurance, a Freon detector can be used to detect the presence of ethylene oxide when it is used in a 12% ethylene oxide, 88% Freon mixture.
- In the operating rooms, current potential parents can work in pre-op or if possible in areas where anesthetics are not used.
- In patient-care areas current potential parents can refrain from caring for patients receiving chemotherapy drugs. Housekeeping should also refrain from handling soiled linens, etc., of these patients.
- Female current potential parents can refrain from caring for patients who have teratogenic viruses. (A good discussion of this problem is included in the chapter by Nelson and Sullivan-Bolyai.[8])

There are very few jobs where total job medical removal protection is necessary. Usually, informal arrangements are made in each department for an employee-at-risk to refrain from certain tasks. Currently, these arrangements are common only for pregnant women, are rare for women attempting to conceive, and are almost nonexistent for men attempting to parent a child.

Mandatory Medical Removal Policies

Most litigation has found mandatory medical removal policies to be in violation of Title VII of the Civil Rights Act. Many employers who institute mandatory medical removal policy will provide alternative work in attempt to reduce the incidence of discrimination suits. Many union contractors and some employer policies guarantee no loss in wages, seniority, or other employment benefits to workers who are reassigned due to a mandatory policy. Even with these provisions, a mandatory medical removal policy may incur Title VII liability. California and Connecticut have regulations prohibiting the requirement that any employee be sterilized as a condition of employment.[9,10]

Employee Requests for Medical Removal Protection

Employees who request partial or full medical removal protection to avoid exposures during pregnancy or attempts at conception are generally in a difficult position, unless the employer agrees that the medical removal protection is necessary. With the exception of California, the employer is under no legal obligation to find the employee alternative work; to provide paid or unpaid leave; or to retain salary, benefits, or seniority if the employee refuses to work at a job because of fear of injury to an embryo or fetus.

The California Fair Employment and Housing Act Section 12945 (c)(1) states:

> It shall be unlawful employment practice unless based upon a bona fide occupational qualification: (c)(1) for an employer who has a policy, practice, or collective bargaining agreement requiring or authorizing the transfer of temporarily disabled employees to less strenuous or hazardous positions for the duration of the disability to refuse to transfer a pregnant female employee who so requests.
>
> (c)(2) for any employer to temporarily transfer a pregnant female employee to a less strenuous or hazardous position for the duration of her pregnancy if she so requests with the advice of her physician where such transfer can be reasonably accommodated. Provided, however, that no employer shall be required by this section to create additional employment which the employer would not otherwise created, nor shall employer be required to discharge any employee, transfer any employee with more seniority or promote any employee who is not qualified to perform the job.

Many hospitals will attempt to provide an alternative job with a comparable salary and will retain benefits and seniority when there is sufficient staff to perform tasks which the pregnant employee is refraining from. Usually a letter from the employee's private physician is required. Employees who are potential fathers or mothers rarely ask for or receive medical removal protection.

Compensation for Employees Receiving Medical Removal Protection

As mentioned earlier, medical removal protection should be an issue only when alternative measures are not feasible or adequate enough to protect employees during a period of temporary increased risk.

Policies which require employees who are current potential parents to remain in a job which cannot be made adequately safe for them to conceive and bear healthy children, or else suffer loss of job, pay and benefits puts workers in an untenable position. Providing temporary pay and benefits for workers who must be removed from their jobs due to the employers inability to make the job safe enough allows the hospital to communicate its commitment to the health and safety of employees and their families, to continue the employment of valuable experienced employees and provides improved employee morale and loyalty in a more humane atmosphere.

Many hospitals have light duty programs or temporary labor pools which can provide a source of alternative work for employees who are temporarily disabled. These can be a source of alternate work for employees on medical removal protection if comparable work is not available. If no alternative work is available to the employee, a layoff or leave of absence may be necessary.

Workers' compensation is not available as a source of compensation for workers who have been temporarily removed from a job to avoid a toxic exposure.

Many hospitals offer disability benefit plans, which vary in duration and extent of coverage. When current potential parents are removed from a job to prevent exposure to reproductive toxins, they are also barred from receiving disability payments. Disability insurance is only designed to compensate workers who are disabled from working. The type of situations which could be covered depends on the hospital's specific insurance plan. Examples of situations which could be covered under hospital disability plans include

1. Disability related to pregnancy, labor, and delivery
2. Disability caused by pregnancy complications
3. Aggravation of preexisting medical conditions due to pregnancy (e.g., diabetes, hypertension)
4. Physiologic changes as a result of pregnancy which make the employee more at risk from job conditions (e.g., susceptibility to heat stress, backache)
5. Physiologic changes as a result of pregnancy which make the employee (not fetus) more susceptible to hazardous exposures on the job (e.g., some odors might make a pregnant employee nauseated)

Many hospital disability plans only cover the first three situations listed. When an employee is disabled as a result of pregnancy, the derivative guidelines of the Pregnancy Discrimination Act stipulate that employers must treat the disabled pregnant employee in the same manner as employees with other disabilities. The reasonableness of job modification, transfer, or leave must be viewed in the context of the hospital's past practices regarding the accommodation of employees temporarily disabled for other health reasons. Generally medical leave is considered only if the employee's job cannot be reasonably modified.

When workers are provided with a temporary alternate job due to disability or to avoid a reproductive hazard, employers are not required to make up the difference if the salary of the temporary job is not as high as the employee's permanent salary.

Because workers' compensation and disability benefits are not available for workers who are temporarily removed from their jobs to avoid exposure to a reproductive toxin, some companies lay off employees in this situation, allowing them to retain their benefits and collect unemployment compensation for a specified time period. They are then reinstated when they no longer need special protection from hazards on the job.

In certain cases, unions have negotiated that employees who are transferred or on leave as a result of medical removal protection should remain at full salary, benefits, and seniority until they can continue working.

Conclusion

The development of an equitable reproductive hazards policy is often complicated by a lack of information on dose–response relationships and on male-mediated reproductive effects. A systematic evaluation of the materials used in each department, as required under the OSHA Hazard Communication Standard will allow the hazards and controls as well as the need for any special precautionary measures to be evaluated for reproductive toxins. A combination of eliminating the use of reproductive toxins where feasible, keeping exposures as low as reasonably achievable at all times, and providing selective medical removal protection where necessary offers a conservative approach to the problem until more information is available.

References

1. NIOSH recommendations for Occupational Safety and Health Standards. *MMWR Suppl* 34:15, July 19, 1985.
2. U.S. Environmental Protection Agency. EPA proposed guidelines for assessing female reproductive risk: Notice. *Fed Reg* 53(126):24834–24847, June 30, 1988.
3. U.S. Environmental Protection Agency. Proposed guidelines for assessing male reproductive risk and request for comments. *Fed Reg* 53(126):24850–24869, June 30, 1988.
4. US Congress, Office of Technology Assessment. *Reproductive Health Hazards in the Workplace,* OTA-BA-266. US Government Printing Office, Washington, DC, December 1985.
5. Bertin J. Reproduction, women, and the workplace: Legal issues. In: ZA Stein and MC Hatch (eds.), *Occupational Medicine State of the Art Reviews: Reproductive Problems in the Workplace*. Hanley and Belfus, Inc., Philadelphia, PA, July–September 1986, vol. 1, No. 3.
6. REPROTOX is a subscription service available from the Reproductive Toxicology Center, Columbia Hospital for Women Medical Center, Washington, DC.
7. NIOSH. *Criteria for a Recommended Standard... Occupational Exposure to Waste Anesthetic Gases and Vapors.* US Department of Health, Education and Welfare, Cincinnati, OH, Publication No. 77–140, March 1977.
8. Nelson KE and Sullivan-Bolyai JZ. In: EA Emmett (ed.), Preventing teratogenic viral infections in hospital employees: The cases of rubella, cytomegalovirus, and varicela–zoster virus. *Occupational Medicine, State of the Art Reviews: Health Care Workers.* Hanley and Belfus, Inc., Philadelphia, PA, 471–498, July–September 1987, vol. 2, No. 3, 471–498.
9. California Fair Employment and Housing Act, Ch. 619, L. 1980, Section 12945.5, effective January 1, 1981.
10. Public Act 81–281, SB 1185, effective October 1, 1981.

24

Medical Waste

Michael L. Garvin
University of Iowa Hospitals and Clinics

Herb B. Kuhn
American Hospital Association

Elaine Peters
KPMG Peat Marwick

Medical Waste Management: The Problem and Solutions 24-1
Introduction • Overview of Medical Waste • Lack of Standard Terminology Hinders the Discussion of the Subject • Warning Symbol • Defining PIW • Disposal Technology • Regulations That Affect PIW Management • Summary of Recommendations
Appendix 24.A.1 24-11
Emerging Trends in Infectious Waste Management 24-12
Introduction • Historical Activity in Regulating Infectious and Other Medical Waste • Conclusion
Bibliography 24-20
Appendix 24.A.2 24-21
Appendix 24.A.3 24-21

Medical Waste Management: The Problem and Solutions

Michael L. Garvin

Introduction

During the 1980s, the task of waste management was made increasingly more difficult for hospitals. Potentially infectious waste (PIW), generated by every hospital, clinic, and doctor's office in the country, received intense scrutiny by the media and regulators. The main reason for this attention can be traced to the increase of AIDS cases in the early 1980s and the general lack of understanding of how the HIV virus could be transferred. The public became concerned that the HIV virus could be spread through medical waste placed in landfills. Legislation was passed that banned PIW from the landfill. At the same time, states rewrote definitions thereby classifying a significantly larger percentage of hospital waste as potentially infectious.

Just as the 1980s ended with the passage of the overly comprehensive and unnecessarily costly Medical Waste Tracking Act (MWTA), the 1990s opened with the voice of reason in the form of a congressional mandated report from the Agency for Toxic Substance and Disease Registry (ATSDR). This report[1] concluded that PIW presents no more risk to the environment and the public than general household waste. It went on to say that a hospital is better advised to focus its resources on reducing occupational exposures to the waste. The Occupational Safety and Health Administration (OSHA) has picked up that theme in writing a new Bloodborne Pathogens Standard.

Hospitals and health care associations would be wise to seize this opportunity and assess waste management options in the light of scientific data and reason. The objective of this report is to further a rational discussion on the future of PIW management.

Overview of Medical Waste

In the course of daily activities, a hospital produces waste. Some of that waste is discharged into the sanitary sewer system and some of it is released in gaseous form through laboratory hood vent ducts, but most of it is "solid waste." Technically, solid waste comprises the largest percentage of hospital-generated waste and includes such waste types as general office trash, food service waste, and even the fastest growing waste type, recycled waste. In addition, solid waste includes three types of waste which fall under federal or state regulation: radioactive, chemical, and potentially infectious. These three types of waste comprise "regulated medical waste." Simply put, they are medically generated waste, which are governed by regulation. In the MWTA of 1988, the term "regulated medical waste" was used loosely to apply to those items identified by the federal Environmental Protection Agency (EPA) as being potentially infectious. While those EPA identified items fall under the general term of regulated medical waste, they are by no means the only medical wastes regulated at the present time. Figure 24.1 shows the seven different medical waste types of combinations, which are governed by regulations. The combination waste types will become the hot topic of future regulations.

The focus of this report is PIW. A review of the literature also found a number of measurement units for PIW generated. The most useful unit of measurement is "pounds per patient day" as it considers utilization as well as bed capacity. It is difficult to provide exact statistics on how much of this waste is generated. The difficulty arises due to the wide range of definitions hospitals use for PIW. The American Hospital Association (AHA) estimates that an average hospital will produce approximately 20 lb of solid

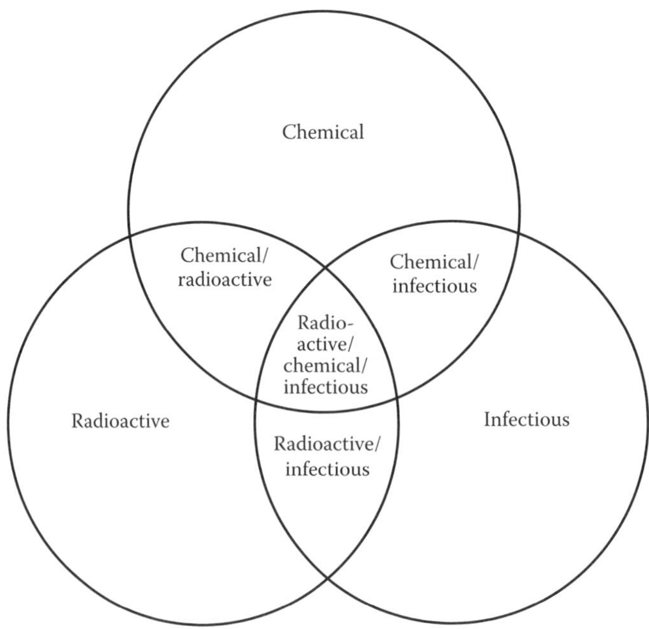

FIGURE 24.1 Regulated medical waste.

TABLE 24.1 Comparison of Categories Included in Definitions of Infectious Waste

Types of Waste	AHA	CDC	IOWA	EPA	MWTA
1. Microbiological (e.g., stocks and cultures)	Yes	Yes	Yes	Yes	Yes
2. Blood and blood products	Yes	Yes	Yes	Yes	Yes
3. Pathological (e.g., tissues)	Yes	Yes	Yes	Yes	Yes
4. Sharps (used)	Yes	Yes	Yes	Yes	Yes
5. (Communicable disease isolation CDCC Class 4)	No	Yes	Yes	Yes	Yes
6. Contaminated animal carcasses	No	No	Yes	Yes	Yes
7. Other isolation	No	No	No	Yes	Yes
8. Contaminated laboratory waste	No	No	No	Optional (1)	Yes (2)
9. Surgery and autopsy waste	No	No	No	Optional (1)	Yes (2)
10. Dialysis waste	No	No	No	Optional (1)	Yes (2)
11. Contaminated equipment	No	No	No	Optional (1)	Yes (2)
12. Sharps (unused)	No	No	No	Optional (1)	Yes (2)

waste for every patient day. Of that solid waste produced, approximately 10%–12% is considered potentially infectious. So on the average, a hospital would generate about 2.0–2.4 lb of PIW for every patient day if the facility includes only those categories of waste identified by the AHA as potentially infectious. Table 24.1 provides a comparison of which categories of waste are included in definitions supported by different regulatory agencies. The statistic most often quoted in the literature is that PIW comprises 15% of all hospital waste.

Not strongly recommended but EPA suggests that these items be considered saturated with blood as well as intravenous.

A study[2] conducted by the University of North Carolina of 441 randomly selected hospitals reported the following:

- Hospitals with fewer than 100 beds generated 1.5 lb per patient day
- Hospitals with 100–300 beds generated 2.4 lb per patient day
- Hospitals with 300–500 beds generated 2.7 lb per patient day
- Hospitals with more than 500 beds generated 2.5 lb per patient day

The range of this data is probably due to a combination of different definitions and levels of medical care. One recent study of a large university tertiary care facility found that 14.8% of all waste is considered potentially infectious. For every patient day, 3.9 lb of PIW is generated. This study more accurately assesses the amount of PIW generated at University of Iowa Hospitals and Clinics (UIHC) relative to total waste. Depending on changes in PIW definition, procedures and products purchased, the statistic may continue to be dynamic.

In a literature search, no data could be found on the impact that clinic activity has on the generation of PIW. With clinics playing a larger role in health care delivery, such data would be very helpful. Another issue is the effect that intensity of care has on PIW. A tertiary care hospital will probably produce more of this waste than a primary or secondary care facility.

Lack of Standard Terminology Hinders the Discussion of the Subject

In reviewing the literature in preparation for this report, one major observation was the lack of standard terminology used when discussing this topic. That portion of regulated medical waste which presents a potential infectious risk is called medical waste, regulated medical waste, biohazard waste, isolation waste, infectious waste, and PIW. Since it is hospital-generated waste which, if not contained and managed properly, "could" result in the transfer of infection, the phrase "PIW" is the most accurate term.

"Regulated medical waste" has become the common phrase used by the federal EPA. If the MWTA is expanded this year, we may all soon be using that term. Chemical waste was called just that until the federal regulation started using the term "hazardous waste."

The most accurate term for this waste is "potentially infectious." Realistically, a great stride forward would be achieved if all parties, including regulators, would agree on either "potentially infectious" or simply "infectious."

A glossary of other terms is included in Appendix 24.A.1.

Warning Symbol

For decades, the biohazard symbol had served as a strong warning that a serious biological hazard existed with a labeled substance. Extremely concentrated strains of biological agents as well as biological weapons warranted the symbol. During the waste crisis, which followed the increase in AIDS cases, the biohazard label was adopted for infectious waste. Now that same symbol is placarded on doors leading to biological weapons as well as those leading to holding areas for PIW. There is absolutely no comparison in the degree of risk involved with the two different items. The adoption of this symbol has overstated the risk involved with handling PIW while diluting the effectiveness of its original purpose, yet a suitable replacement has not been designed. If a warning symbol is to be used, a new symbol needs to be developed.

Defining PIW

In the early 1980s, the problem was that no one had properly defined PIW. In the early 1990s, the main problem was that there were too many definitions. This waste has now been defined by the Centers for Disease Control and Prevention (CDC), the AHA, most state department of natural resources, the EPA, and by Congress through the MWTAMWTA. These definitions can include any or all of the following hospital-generated waste items:

1. Microbiological waste (stocks and cultures)
2. Blood and blood products
3. Pathological waste (tissues, etc.)
4. Sharps (used)
5. Communicable disease isolation waste (exotic diseases known as class 4 by CDC)
6. Other isolation waste
7. Contaminated animal carcasses
8. Contaminated laboratory waste
9. Surgery and autopsy waste
10. Dialysis waste
11. Contaminated equipment
12. Sharps (unused)

Table 24.2 shows a comparison of the different PIW definitions. The manner in which this waste is defined directly impacts the volume generated by a facility, and volume directly relates to cost. Reports[3] have indicated that the program has increased waste disposal costs dramatically. With a solid majority of states now having PIW regulations, the MWTA is seen as a costly and needless regulation which hospitals, facing enough financial pressures, should not be required to bear.

As is the case in states such as Iowa, hospital associations and state regulatory agencies have come to agreement on a reasonable definition for this waste. This definition parallels closely the CDC recommendation definition.

TABLE 24.2 Waste Management Summary

Waste Product Annually	Amount (lb)
Landfill	4,995,360
Incinerator (PIW)	946,782
Recycled cardboard[a]	293,350
Recycled white paper	96,000
Chemical waste	30,000
Radioactive waste	4,000
Total	6,365,492

Notes: Total amount of waste generated per patient day: 26.3 lb.

Percentage of PIW to total waste: 14.8%.

Pounds of PIW per patient day (based on 1989–1990 data): 3.9 lb per patient day.

[a] $13.75 per ton for recycled cardboard and avoids a charge of $39 per ton to landfill it.

1. Microbiological waste
2. Blood and blood products
3. Pathological waste
4. Used sharps
5. Class 4 communicable disease isolation waste
6. Contaminated animal carcasses

Disposal Technology

The activities of the last 10 years have placed hospital waste disposal programs in a precarious position. As mentioned previously, the concern surrounding the increase in AIDS cases resulted in banning PIW from landfills. With most hospitals having easy access to an incinerator, they simply burned the waste. Studies[2] have shown nearly 80% of all hospitals depend on incineration for disposal of at least part of their PIW. With the dawn of the 1990s, clean air regulations are being written on both the state and federal levels. Some researchers[4] believe that unless these regulations are reasonable and consider the financial impact on hospitals, on-site incineration will be financially beyond the reach of most facilities. Some rural hospitals may suffer the most. One study[5] indicates that air emission standards can increase incineration purchase and operation cost by a factor of 10.

So where will that leave facilities? The full impact of new air emission regulations will be felt by hospitals in the mid-part of this decade. Their options will include

1. Developing a hospital consortium-owned regional disposal facility
2. Hospital or associations contracting with a commercial disposal firm
3. Paying for an incinerator upgrade
4. Purchasing an alternative disposal technology

Depending on the severity of the new regulations and the particular characteristics of the hospital's situation, any of these options could be the best option.

Option 1: *Developing a hospital consortium-owned regional disposal facility.* Madison, Wisconsin area hospitals have run a consortium-owned incinerator for 5 years. At the present time, there are very few of these types of facility in the country. The Volunteer Hospital Association of Iowa is forming a group, which will run a regional unit for about 20 hospitals in eastern Iowa.

Option 2: *Hospital or associations contracting with a commercial disposal firm.* At least for the short term, hospitals and hospital associations are choosing this option. When dealing with reputable firms, hospitals receive reasonable rates and favorable contract conditions. Dealing with less than reputable firms or companies having a monopoly on such service can cost hospitals dearly. There is a need for research on the commercial disposal rates in different sections of the country. The research could consider the cost impact on rural versus urban hospitals.

The Iroquois Health Care Consortium based in Albany, New York is supporting the construction of a steam sterilization/grinding facility by Browning-Ferris, Inc. (BFI). The Maryland Hospital Association has encouraged a commercial firm to site a large incinerator in the Baltimore area. The Milwaukee Area Hospital Council is in the process of doing the same in eastern Wisconsin.

A study[6] showed that 35% of Iowa hospitals are choosing (or are forced to choose) to contract waste disposal to a commercial firm for at least part of their PIW.

Option 3: *Purchasing an alternative disposal technology.* Hospitals which want to control advantages of an on-site technology but can no longer afford incineration will most likely choose an alternative disposal technology. This has been the case in California, where air emission regulation has forced most hospitals away from incineration.

Table 24.3 shows a comparison of on-site waste disposal technologies, which provide an alternative to incineration.

The following is a discussion of alternative disposal technologies.

Steam Sterilization

Except for incineration, steam sterilization is the most common form of infectious waste treatment used by hospitals. This technology requires that the waste be subjected to steam, in some cases under pressure, for a certain amount of time. Liquids produced in the process can be drained and the decontaminated solid waste can be placed directly in a landfill or it can be compacted or shredded prior to final disposal.

Advantages

The major advantages for steam sterilization include the fact that hospitals are familiar with the technology. It has been used to treat infectious waste in hospitals for many years and has well developed quality control protocols. The technology requires little floor space and is simple to operate. It has one of the lowest "purchase" and "operating" costs of all infectious waste technologies. Steam sterilization produces no air pollution emission which allows for easy siting of a unit, and, when coupled with either compaction or shredding, the technology can reduce waste volume by as much as 80% and render the waste unrecognizable.

Disadvantages

The disadvantages of steam sterilization include the need for special waste packaging. In most cases, the technology requires that the waste be placed in bags, which are opened to allow steam penetration. In other cases, bags "melt" when exposed to the heat steam thus requiring secondary containment. Steam sterilization does not reduce the consideration if the local landfill charges by weight. Steam sterilization is also not recommended for wastes such as pathology wastes. Some studies indicate that staff who work around medical waste autoclaves may be exposed to volatile organic compounds in excess of safe levels.

It is suggested that a hospital which chooses this technology clearly explains its plans to the condition of the waste when it reaches the landfill.

Purchase costs for a steam sterilization/compaction unit to service hospitals ranging from

50 to 500 beds—$50,000 to $150,000
Operating costs—2.5 to 3.5 cents per pound.

TABLE 24.3 Comparison of Potentially Infectious Alternative Waste Treatment Technologies

Technology	Advantages	Disadvantages	Investment	Operating Cost	Volume Reduction
Steam sterilization and compaction	Low cost Small space requirement Easy to install Simple to operate Some volume reduction	Not suitable for pathological waste Requires special containment bags May create odor problem Waste appearance is not altered Weight is not reduced	50–500 bed facility \$50,000–\$150,000	2.5–3.5 cents per pound	70%
Grinding and chemical disinfection	Medium cost Small space requirement Easy to install Simple to operate Some volume reduction Alters appearance Simple to permit	Not suitable for pathological waste Weight is not reduced Liquid discharge may require special permit	50–500 bed facility \$50,000–\$400,000	6–10 cents per pound	80%
Grinding and microwave treatment	Low cost Small space requirement Easy to install Simple to operate Volume reduction Alters appearance Simple to permit Reliable performance	Not suitable for pathological waste Weight is not reduced	400–1000 bed facility \$400,000–\$500,000	7–14 cents per pound	80%

Chemical Disinfecting/Shredding

The disinfecting/shredding technology grinds the waste in the presence of a hypochloride solution. The hypochloride solution disinfects the waste while the shredding process reduces total volume and renders the waste unrecognizable. The disinfectant solution is then drained from the waste and, in most cases, disposed of to a sanitary sewer. In some cases, pretreatment may be necessary. The remaining solid waste can be placed in a landfill.

Advantages

The disinfecting/shredding technology can reduce the volume of infectious waste by as much as 80% as well as make the waste unrecognizable. The technology requires little floor space and is simple to operate and produces no toxic air emissions.

Disadvantages

In some cases, wastewater discharged from the chemical disinfecting/shredding unit may have to be permitted by local water treatment authorities. Depending on state emission standards, the cost of a large unit may be as high as a small incinerator. This technology does not reduce the weight of the waste.

Purchase costs for a chemical disinfecting/shredding unit:

Smaller unit—$45,000
Larger unit (1500 lb per hour)—$390,000
Operating costs—6 to 10 cents per pound

Grinding/Microwaving

The grinding/microwave technology has been imported from Europe. It was developed by a company in Germany, has had over 3 years of experience in that country and is now in operation in France, Switzerland, Italy, and the United States. The technology shreds the waste in a high-efficiency particulate air (HEPA)-filtered chamber and then subjects that waste to microwaves. Steam is ejected into the treatment chamber to aid in disinfection. This process differs from steam sterilization in that the heat is generated from the center of the waste mass.

Advantages

The grinding process reduces the waste volume by as much as 80% and renders the waste unrecognizable. The siting process for a grinding/microwave unit is greatly simplified by the fact that there are virtually no air emissions for the process. The equipment has demonstrated the ability to be very reliable.

Disadvantages

Like the steam sterilization option, this technology will not reduce the weight of the waste and hospitals will have to provide evidence to the landfill that the waste has been disinfected. This technology is not recommended for pathology waste.

Purchase costs for a grinding/microwave unit to service hospitals ranging from

400 to 1000 beds—$375,00 to $650,000
Operating costs—smaller unit—9 to 14 cents per pound
Larger unit—5 to 7 cents per pound

Incineration: (For Comparison with Alternative Technologies)

Incineration technology uses carefully controlled high temperature combustion to destroy infectious waste. There are a number of designs available within the incineration technology including multiple-chamber, rotary kiln, and controlled air systems. Developments in pollution abatement hardware have brought incinerators in compliance with the strictest air emission standards, and, although not as cost effective as 10 years ago, waste heat recovery systems can produce energy in the form of steam.

Advantages

Hospitals are comfortable with incinerators. The technology has been the preferred choice for infectious waste treatment in the past. Incineration reduces both volume and weight by approximately 95%. The technology can treat all infectious waste items including pathology wastes.

Disadvantages

Because of air emission concerns, it is becoming increasingly difficult to site an incinerator, especially in urban areas. New sophisticated designs and pollution control equipment required by many states make this technology have the most expensive "purchase" and "operating" costs of all technologies. New legislation on ash management and required operator training is driving operating costs still higher. Public opposition seems to be highest against incineration in comparison to the other treatment alternatives.

Purchase costs of an incinerator which would service hospitals ranging from

200 to 1000 beds—$100,000 to $3,000,000 (depending on pollution control equipment required)
Operating costs—12 to 70 cents per pound

Table 24.4 shows the recommended disposal or treatment method for different types of PIW.

A number of new technologies are currently under development. Gamma Wave Irradiation and Electrothermal Disinfection uses a sealed cobalt source and low frequency radio waves, respectively, to treat PIW. Experiments in burning the waste with lasers and melting it with a welding-type arc also are underway. None of these technologies are available for on-site installation.

Much of this information comes from three technology assessment reports.[7–9]

Regulations That Affect PIW Management

Both federal and state regulators have taken a more aggressive approach to overseeing how hospitals manage their PIW. Attachment 7 shows the major regulation generated by the EPA for both air quality and solid waste management. It also details the federal and state OSHA regulation affecting waste management. Even though both the EPA and OSHA have had authority to produce regulations involving PIW management for over 20 years, there had not been any reason to develop rules until concern arose over the increase in AIDS cases.

TABLE 24.4 Recommended Methods for Disposal of PIW

Type of Waste	Recommended Disposal Technology
1. Microbiological	Incineration
	Disinfection and landfilling (disinfection technology might involve steam, microwave, electrothermal, or chemical treatment)
	Grinding is required by the MWTA and some state regulation
2. Blood and blood products	Sewer disposal
	Disinfection/solidification/landfilling (powder is added to liquid waste—waste is solidified and disinfected)
3. Pathological waste and contaminated animal carcasses	Incineration
4. Sharps	Incineration
	Disinfection/grinding/landfilling

Summary of Recommendations

The public concern over the management of PIW continues to grow. Unfortunately, a rational discussion is hindered by a lack of consensus on such critical issues as terminology and unit of measurement. The implementation of the following recommendations would assist in addressing these needs:

1. *Standard Terminology*—The phrase "PIW" most accurately describes the portion of medical facility-generated waste, which has the "potential" of conveying disease. Further discussion and regulation should avoid the misnomer "regulated medical waste."
2. *Standardize Unit of Measurement*—The amount of PIW generated is best measured by "pounds per patient day." This unit of measurement incorporates both bed capacity and utilization. Such measurement tools as "pounds per bed" are misleading without a utilization factor.
3. *Standardize Definition*—Because it has a foundation in scientific assessment, the definition recommended by the CDC should be accepted as standard. Such a definition would include:
   a. Microbiological waste
   b. Blood and blood products
   c. Pathological waste
   d. Used sharps
   e. Class 4 communicable disease isolation waste
   f. Contaminated animal carcasses

   Further state and federal regulation should acknowledge this standard and not needlessly add financial burden to health care facilities.
4. *Need for Further Research*—More research needs to be conducted concerning who is most affected by PIW regulations. One study[9] indicates that when regulation increases waste volumes and/or limits disposal technology, smaller rural hospitals are most negatively affected. A survey conducted by the Iowa Hospital Association found that of the 37 hospitals, which have contracts with disposal firms, seven state that they had only one firm to provide such a service. More research needs to be done on the impact of including doctors' offices under regulation as well as the impact of health care's shift from inpatient to outpatient services.

   Armed with the data such research would provide, hospitals and hospital associations could strengthen their communication with state and federal legislators. Should there not be air emission in standard exemptions for small or rural facilities? Should a 1 ton-a-day incinerator in rural Iowa be required to have the same pollution equipment as a 200 ton-a-day unit in Los Angeles?

   Knowing now that landfilling PIW does not pose an environmental risk, would it not be prudent to allow small generators, such as doctors' offices, clinics and small hospitals, to once again landfill that waste as long as occupational risks while transporting are addressed?
5. *Review of Management Practices*—Hospitals, in agreement with OSHA and ATSDR assertions that PIW constitutes an occupational hazard, need to fully address that hazard. Hospitals should conduct comprehensive reviews of their waste management programs in light of the new safety standards. Complete assessments of waste separation, packaging, handling, and disposal procedures need to be made. Hospital product evaluation and safety committees need to design cost/benefit analyses on new safety-enhancing products such as sheathable syringes, sharps disposal boxes, and powders that disinfect and solidify potentially infectious liquid waste. The present cost of treatment for sharps injuries and other PIW exposures needs to be assessed.
6. *Implementation of Waste Reduction Program*—Hospitals can reduce the amount of waste generated. Medical supply manufacturers should be requested to provide products and packaging, which can be recycled or which lend themselves to "low polluting" incineration or biodegradation. In addition, hospitals can encourage practices, which shift waste from the disposal to the recycle bin. The state of Iowa requires that 50% of all waste generated be recycled trash by the year 2000. This will not happen without waste reduction.

The standardization of PIW terminology, units of measurements, and definitions can only hasten the resolution of this waste management problem as it assists facilities in evaluating current programs and deciding on cost effective options for the future.

Appendix 24.A.1

Recommendations for Standardized Medical Waste Terminology

Airborne Waste—particles and gases discharged to the air by way of an incinerator stack or a facility exhaust flue.

Biohazard Waste (Biological Waste)—often used synonymously with infectious waste. Traditionally, the term was applied to waste with very high concentrations of infectious agents. Such waste might be found in research or clinical laboratory settings as well as biological defense installations.

Chemical Waste—any waste which includes toxic substances as defined by the EPA.

Disinfection—a reduction of populations of disease-producing microorganisms. PIW regulation requires that the waste be disinfected but not sterilized.

General Waste—waste categories such as office and food services trash, which present no potential hazard and require no special handling procedures.

Hazardous Waste—term used by the EPA to refer to any waste, which may pose a hazard to humans or the environment. The term is used synonymously with "chemical waste."

Infectious Waste—a slightly looser term than "PIW" but is used synonymously. The term "infectious" incorrectly implies that the waste has been positively identified as having the capability of transmitting infection.

Isolation Waste—all waste generated by a patient, who has been placed on isolation precautions.

Medical Waste—often incorrectly used as a term for "PIW," medical waste is simply all waste generated by a medical facility.

Pathological Waste—waste generated by pathological services usually containing tissue, gross specimen, and limbs.

PIW—the most accurate term for waste, which has the "potential" for transmitting infectious. The term is used synonymously with "infectious waste."

Radioactive Waste—waste with radioactive properties greater than normal background radiation level.

Regulated Medical Waste—waste generated by a medical facility, which are subject to regulation. The waste includes radioactive, chemical, and potentially infectious. The term has been used by the EPA to refer to PIW.

Sewage—general term used to refer to all waste discharged into a sanitary sewer.

Solid Waste—defined by state and federal regulators as all wastes generated by a facility, which are disposed of by any means except discharge to the air. The term includes wastes placed in landfills and waste incinerated but not waste discharged from an exhaust flue or an incinerator stack.

Sterilization—the complete destruction of microbial life.

References

1. Rettig PC. ATSDR: Medical waste poses no threat to public. *Health Facil Manage,* 60–66, June, 1990.
2. Rutala WA, Odette RL, and Samsa GP. Management of infectious waste by US hospitals. *JAMA,* 262:635–1640, 1989.

3. Sedor P. Costs soar under EPAs waste tracking program. *Health Facil Manage*, 3:24, 26–8, 30, 1990.
4. Brodsky R. Strict incinerator regulations may make it too expensive for hospitals to go it alone. *Mod Healthc* March 18, 1991.
5. Garvin M. Waste disposal costs show wide variation from state to state. *Mod Healthc* March 18, 1991.
6. Iowa Hospital Association. *Survey on Infectious Waste Management in the State of Iowa,* 1990.
7. Doucet L. *Infectious Waste Treatment and Disposal Alternatives* (AHA Technical Document Series), AHA, Chicago, 1989.
8. Cross F. *Evaluation of Alternative Techniques for the Treatment of Solid Medical Waste.*
9. U.S. Congress Office of Technology Assessment. *Medical Waste Treatment Technologies. OTA N3-2045.0.* U.S. Government Printing Office, Washington, DC, March, 1990.

Emerging Trends in Infectious Waste Management

Herb B. Kuhn and Elaine Peters

Introduction

The summers of 1987 and 1988 produced some graphic and disheartening signals of the solid waste problems plaguing the United States. The floating garbage barge and the Eastern shore washups of commercial and residential waste were, according to many, the symptoms of a society grappling with the inevitable effects of its throwaway habits. In an August 30, 1988 editorial, *The Washington Post* stated that the problems of improper disposal of solid waste are not confined to "just a few violators. It is a country of 246 million people who have never been very tidy in throwing out their trash."

Unfortunately, some of these washups contained medical waste. Pictures of syringes, test tubes, and other medical waste played nightly on the national news, and coupled with the public's growing fear of AIDS, these events brought new demands for stringent regulation of the health care waste stream. While many of the incidents were later linked to small clinics typically exempt from regulation, most of the attention focused on hospitals as the largest generators of medical waste.

Although it is clear that some improvements in the waste management system are indeed needed, it has also become increasingly evident that the policy process is being driven by the public's perception that medical waste poses a significant health risk. These concerns, however, have not been borne out by scientific and epidemiologic evidence. In addition, there appears to be new attention to the esthetic character of waste disposal—whether or not it poses a risk. The threatening appearance of medical materials has prompted calls for special handling of much of the health care waste stream.

Most policy makers agree that the medical waste issue will not be resolved until the larger problem of solid waste management is adequately addressed. However, moved by reports of poor handling of the medical waste stream, various political subdivisions—from localities to states to the federal government—have begun the process of requiring special handling of not only what was traditionally considered infectious, but also other disposable materials found in the medical setting. While several states have successfully addressed this issue in an atmosphere that has led to reasonable and appropriate regulation, others have enacted unnecessarily burdensome and costly standards. Although the implications for federal regulatory activity are still being played out in Congress and the executive branch, it is clear that hospitals may face significant changes in waste management requirements over the next few years.

Protecting the public is the preeminent goal of the health care industry. A hospital's mission is to care not only for its patients, but also for the community at large. However, policy makers and health care managers need to find a balance—a balance between protecting the public health and the environment while ensuring the wise use of scarce health care resources. In an era of cost containment, it is critical that any new standards be grounded in scientific evidence of need and that costly regulation of benign waste be avoided.

This section examines the political and regulatory environment in which states and the federal government are regulating medical waste. It also looks at problems involved in building a consensus on waste definitions and disposal methods and examines the differences between occupational and community risks. It discusses why hospitals should have a strong and meaningful infectious waste program in place and looks at some of the trends the health care field may encounter in the future.

Historical Activity in Regulating Infectious and Other Medical Waste

To understand the current political and regulatory environment, it is helpful to examine how management of medical waste has evolved over the last two decades. In the mid-1970s, growing concerns about the environment led Congress to enact a broad blueprint for managing solid waste, known as the Resource Conservation and Recovery Act (RCRA), P.L. 94–580. RCRA gave the EPA broad authority to regulate "cradle-to-grave" management of those wastes it identified as hazardous. Included in the list of materials that might be deemed hazardous was waste with infectious characteristics. However, EPA never issued final regulations for the management of infectious wastes, citing the need for and apparent lack of "considerable evidence that these wastes cause harm to human health and the environment."

Instead, EPA opted to develop guidance materials in response to the numerous requests for information on infectious waste management. In 1982, the agency published a *Draft Manual for Infectious Waste Management,* which was subsequently revised and updated in 1986 as the *EPA Guide for Infectious Waste Management.* The *EPA Guide* provided technical advice on waste segregation, packaging, storage, transport, treatment, and disposal. Most important, it recommended that six basic categories of waste be designated as infectious:

- Cultures and stocks of infectious agents and associated biologicals
- Contaminated sharps
- Human blood and blood products
- Pathological waste
- Contaminated animal carcasses and bedding
- Isolation waste

In addition, EPA identified a broad "optional" category of miscellaneous contaminated wastes, reflecting a lack of consensus on the need for special handling of a number of other materials. The agency recommended that each facility make its own determination of whether these miscellaneous wastes should be designated as infectious, including, for example, surgery and autopsy wastes, laboratory and dialysis wastes, and contaminated equipment. Thus, while the recommendation allowed for flexibility in designating a waste management program, it also left some doubts about the types of waste that should truly be considered infectious.

The picture was further complicated by the parallel development of waste handling guidelines by the federal CDC. Issued in 1985, the CDC guidelines defined infectious waste somewhat more narrowly than EPA. The CDC called for four basic infectious waste categories: cultures and stocks of infectious agents; contaminated sharps; blood and blood products; and pathological waste. Isolation wastes were not included in the CDC guidelines nor were the EPA's optional groupings. Rather, the guidelines suggested that "prudent handling" of these wastes was adequate and that individual institutions should devise their own handling procedures.

Thus, federal guidance in the early 1980s provided hospitals with a fair amount of flexibility in devising waste management strategies—flexibility that most experts and health care managers found desirable, particularly in light of the lack of scientific evidence implicating hospital waste in community disease. However, it is important to note that because neither the EPA document nor the CDC guidelines represented formal regulatory requirements and because there was not complete agreement between the two agencies on waste definitions, hospitals and state governments were left with an unclear message about the appropriate approach to take with infectious waste.

In this uncertain environment, state governments took divergent paths in regulating hospital waste practices. Some chose to do nothing at all or to develop minimal guidance, while others promulgated extensive regulations that vastly expanded the definition of infectious waste. As a result, although most institutions were very conscientious about following the basic CDC guidelines, the treatment of infectious waste varied considerably across the country. For example, using the CDC guidelines a hospital could expect to classify from 5% to 10% of its waste stream as infectious, while some state laws meant that up to 70% of waste required special handling. Without clear guidance from federal agencies and often lacking adequate local expertise, many state regulators chose to err on the side of "safety," and in the process, established new and sometimes unnecessary standards for waste management.

Interestingly enough, state activity in infectious waste regulation seemed to run counter to the prevailing trend in environmental regulation. In other areas, states seemed content to mimic or enforce federal environmental programs. However, a recent study by the Council of State Governments (CSG) (1988) notes that states have created a set of standards for infectious waste management "in the process of meeting the public's demand for protection."

Demand for Protection

This "demand for protection" came to light in 1987, when scattered reports of improper medical waste disposal first surfaced in the media. Particularly frightening was an incident in Marion, Indiana, where children were found playing with vials of blood left in the dumpster of a nearby clinic. When testing showed that some of the samples were positive for the AIDS virus, public concern escalated. Meanwhile, less dramatic but equally troublesome pressures were prompting states to reexamine their waste regulations. Local landfill space was shrinking rapidly in many regions, leaving fewer options for disposal and forcing hospitals to opt for expensive transport of waste out-of-state. Those states receiving waste for land filling began to express concern over the lack of controls regarding treatment and transport. Moreover, many landfills that had previously accepted health care waste now shunned hospital business, fearing that workers could contract AIDS from contaminated patient care materials. Even treated waste was suddenly rejected by some fearful disposal workers.

These developments accelerated the pace of state regulatory activity. A 1988 survey conducted by the CSG reported that 88% (39) of all states were regulating or were planning to regulate infectious waste, compared to only 57% (27) in 1986. A survey conducted by the AHA's State Issues Forum in October 1988 found that 23 states anticipated additional legislative or regulatory action on infectious waste by the end of 1989.

Some distinct patterns have emerged in state regulation, including a tendency to remove infectious waste from the list of hazardous wastes and create either a special waste category or to regulate it as nonhazardous. In addition, a majority of states are now requiring, or indicate an intent to require (56%), infectious waste to be treated prior to disposal, either by steam sterilization or incineration. Seventy-two percent of these states indicated in the CSG study that incineration is recommended as the preferred method.

Another interesting development is the growing number of states turning their attention to small waste generators. Studies by CSG and the National Solid Waste Management Association (1986) found that approximately half of all states list doctors, dentists, and veterinary clinics as generators, while three-fourths include health care clinics. However, the dominant focus of regulation continues to be hospitals as the largest generators of infectious waste.

Federal Reaction

The events of 1987 also prompted reaction in Congress and at the EPA, although initially at a somewhat slower pace than at the state level. Although RCRA had been reauthorized twice since 1976, no real action on infectious waste was taken by Congress until the fall of 1987. Following media reports, numerous bills were introduced calling for federal regulation of infectious and medical waste and in October, 27 Senate members sent a letter to EPA Administrator Lee Thomas urging him to implement a manifest system for the tracking of medical waste. Although subsequent congressional hearings were held on the issue, no new initiatives were approved before the close of the year.

Meanwhile, EPA responded to these new concerns by convening a meeting of experts in November 1987 to again wrestle with the issues of definitions, proper management techniques, and the risks posed by infectious waste. In a report made available to the public, the expert panel concluded that community exposures to infectious waste were isolated incidents and that the primary risk associated with the health care waste stream was occupational. The panel recommended that EPA direct its efforts toward the development of guidance documents and public education.

However, the summer of 1988 transformed the political environment surrounding the health care waste issue. New washups of residential and commercial waste—some containing medical materials—forced closings of Eastern shore beaches at the height of the tourist season, creating severe economic hardships for many beach communities. The nation's solid waste dilemma reached the cover of several national news magazines as reports of washups filtered in from other parts of the country, including the Great Lakes and the Gulf of Mexico. While many of these incidents appeared to be linked to a few small clinics, hospitals—because of the volume of waste they generate—became the central focus of renewed political activity.

The Senate mobilized support first, approving a measure on August 3, 1988 calling for the creation of a demonstration program for the tracking of medical waste in the states of New Jersey, New York, and Connecticut, where the most serious problems have arisen. Spearheaded by Sen. Frank Lautenberg (D-NJ) and endorsed by EPA, the legislation defined medical waste quite broadly and gave EPA broad discretion on the types of waste to be included in the tracking system. The bill also sought to exempt small quantity generators from tracking requirements, over the objections of many experts.

Meanwhile, the House moved a similar measure on a more deliberate track. In early September, the Transportation, Tourism, and Hazardous Materials Subcommittee began consideration of legislation originally introduced in October 1987 by the subcommittee's chairman, Rep. Thomas Luken (D-OH). This measure went through several iterations, moving from a very prescriptive, national program for the management of infectious and other medical waste to a limited demonstration program for tracking medical waste.

The House demonstration program became the initiative that ultimately was signed into law on November 1, 1988 (P.L. 100–582). P.L. 100–582 established 2 years demonstration program for tracking and handling medical waste in 10 states—New Jersey, New York, Connecticut, and the states bordering the Great Lakes. The law identifies 10 categories of medical waste that must be tracked (both the recommended and optional categories designed in the *EPA Guide for Infectious Waste Management*). It permits EPA to exclude the "optional" category wastes from the list if a finding is made that they do not pose a hazard to human health or the environment. However, by explicitly listing categories, Congress created substantial pressure for EPA to include all the groupings in the demonstration program, raising concerns that inappropriate waste designations would be established in regulations and practice.

The program must begin 9 months after enactment, and states have the option of withdrawing from the program (although coastal states must have a state program that is no less stringent than the federal program in order to exercise this option). Likewise, states not listed to participate in the program can petition to be included in the demonstration. The law also gives EPA the authority to exempt generators of 50 lb or less of listed waste per month. States have the authority to bring enforcement action against anyone who imports medical waste into a state that does not comply with the provisions of the bill.

This new law contains two additional points that are worth noting because of the standard they may set for the future. First, generators who incinerate on site are exempted from tracking. In other words, if medical waste is rendered unrecognizable at the point of generation, special handling in terms of tracking is not required. Second, the legislation calls for three studies—two interim and one final report. These interim reports are particularly critical because they will provide Congress with early information on the success of the program and data on which to make future legislative recommendations. Accordingly, the health care industry should not assume Congress will wait for the conclusion of the demonstration program before moving to address this issue again.

Such movement seemed likely in late 1988, inasmuch as the 101st Congress was preparing to consider reauthorization of RCRA. The Senate Environment and Public Works Subcommittee on Hazardous Wastes and Toxic Substances, under the chairmanship of Sen. Max Baucus (D-MT), held hearings in late September 1988 on this effort. On September 9, Sen. Baucus introduced his RCRA reauthorization bill (S.2773, the Waste Minimization and Control Act of 1988), which contained one section on infectious waste management, requiring the segregation, packaging, treatment, and transportation (including manifesting) of this waste stream. The timing of RCRA reauthorization seemed especially significant because of its potential as a vehicle for requiring a nationwide program for medical and infectious waste management. Given congressional dissatisfaction with the patchwork approach of state regulation and federal guidance, such a program appears imminent.

Although much of EPA's response to the events of 1988 seemed destined to be driven by congressional action, the agency also appears to be stepping up its own efforts to reduce the confusion around the infectious/medical waste issue. In June 1988, EPA published a request for public comment in the *Federal Register,* seeking input on five major issues:

- Appropriate definition of infectious waste
- Nature of the problem posed by such waste
- EPA's role in infectious waste management
- Need for a tracking system for infectious waste
- Exemptions from an infectious waste control program

Of particular interest in the responses EPA received was the lack of any evidence presented by the commenters suggesting that properly handled infectious waste poses a public health problem. However, many of the respondents went on to call for a reconciliation of waste definitions between EPA and CDC, and some suggested that a model regulatory program was needed to avert unnecessary and confusing regulation at the local level. The majority of commenters agreed that the greatest concern posed by infectious waste was the potential for occupational exposure of health care workers and waste handlers. The response appeared mixed on whether small-quantity generators should be exempted from regulation, and a number of commenters appeared to feel that hazardous waste manifesting systems would be too complicated for infectious waste, although a simpler state-administered program might be acceptable.

In late summer of 1988, EPA created an agency-based Medical Waste Task Force and charged it with coordinating activities on infectious/medical waste. Thus, the agency seemed poised to assume a federal leadership role in the infectious waste controversy. However, a number of issues continue to loom as obstacles to a more rational approach to waste management.

Solving the Problem: The Struggle over Definitions and Risks

In looking toward the development of a coherent national program on infectious waste management, the biggest obstacle facing hospitals and regulators has been the lack of a consistent and rational definition of infectious waste. Historical disagreements have recently been complicated by congressional interest in requiring special handling of a broader category of "medical waste"—items that may not pose a risk of infection but are viewed as special because of their threatening or distasteful appearance. While all health care waste should be handled prudently, the high cost of infectious waste disposal (exceeding that of municipal waste disposal by as much as 50-fold) suggests chat this additional expense should be justified by evidence of the need for greater disease prevention and environmental protection.

In its 1988 request for public comment on the infectious waste issue, EPA cited its definition of infectious waste as "waste *capable* of causing infectious disease" (emphasis added). The agency correctly went on to identify five factors that contribute to the risk of introduction of a disease. These include

- Presence of a pathogen
- Sufficient virulence

- Dose of the pathogen
- Portal of entry
- Resistance of the host

Unfortunately, this requisite chain of infection sometimes has been obscured by the tendency of most individuals to equate the presence of a pathogen with infectious waste. However, the presence of a pathogen alone is not sufficient to represent a risk of infection. All environments (except those maintained under sterile conditions) harbor some population of pathogenic microorganisms. What has been difficult for the public and legislators to grasp is that all the factors cited by EPA must be present simultaneously for disease transmission to occur.

In practical terms, individuals seeking a workable definition of infectious waste generally look to specific categories or types of waste. Although this approach is more simplistic, it can result in a tendency to de-emphasize all five factors in the chain of reaction. Therefore, it is essential that care be given in designing programs based on waste source or type and that waste designations be based on a reasonable assessment of risk.

A review of scientific literature shows that there has been no epidemiologic evidence of disease transmission in the community from improper handling or disposal of hospital waste, excluding the occupational risk associated with sharps (W. Rutala, oral communication, 1988). This finding is supported by the experience reported in a 1988 AHA survey of a random sample of U.S. hospitals. However, there are valid reasons for requiring special handling of the four basic waste categories traditionally endorsed by CDC: contaminated sharps, cultures and stocks of infectious agents, blood and blood products, and pathological waste.

The AHA has consistency supported special handling for these categories, although the rationale for doing so varies according to the degree of risk posed by each grouping. Two of these categories—contaminated sharps and cultures and stocks of infectious agents—represent the greatest risk of injury or transmission of infection from mismanagement or improper disposal, and although the risk is primarily occupational, strict adherence to disposal guidelines should be ensured. Sharps (e.g., needles and scalpels) have been associated with injury and disease transmission in occupational settings (e.g., hepatitis B and human immunodeficiency virus infection). However, the risk is believed to be related to their recent contamination with sufficient quantities of pathogenic material (usually blood) and the provision of a portal of entry into a host through punctures or cuts.

Cultures and stocks of infectious agents pose a potential risk due to their heavy concentration of biological materials and the fact that they usually are in glass containers that, if broken, become contaminated sharps. However, while sharps present a real occupational hazard, the environmental risk that they pose, if disposed of properly, is negligible.

The two remaining categories, blood and blood products and pathological waste, are not associated, epidemiologically or microbiologically, with infectious disease transmission but are still perceived by the public to pose a substantial risk of harm. To remain responsive to community concerns and to acknowledge the "learning curve" that must be realized through public education, AHA has supported the continuing inclusion of these categories in the definition of infectious waste. It should be stressed, however, that adherence to prudent management practices at the point of generation should be sufficient to prevent disease transmission from these two waste categories. Although blood and blood products can present an "occupational" hazard, particularly in the presence of sharps (a portal of entry), they do not in themselves present an environmental risk, because of the fragility of viruses and other organisms once away from the body. Pathological wastes generally present a hazard to the public only from an emotional or esthetic point of view, and thus their prudent handling and disposal are adequate.

As alluded to above, there are different risks found in the occupational setting and in the environment. These differences are sometimes difficult to understand but are central to identifying appropriate waste management practices. The recent emphasis in health care settings on the use of "universal precautions" for contact with blood and body fluids of all patients should not be allowed to confuse the issue. Although the concept of universal precautions regards blood and certain body fluids from all patients as potentially infectious, these precautions are designed to prevent occupational exposure of mucous membranes and non-intact skin to bloodborne diseases. As such, barrier precautions are employed in the health care setting, where the risk of such exposure is intense and frequent. Once waste is properly contained, and in the absence of gross negligence by the waste handler, two essential factors for disease transmission are no longer present—mode of transmission and portal of entry. In addition, once blood and other fluids are away from the body, a suitable environment is no longer available for most pathogens to thrive.

For these reasons, many experts feel that isolation wastes and any other waste potentially contaminated with blood and body fluids (e.g., surgery or dialysis waste) should not be classified as infectious. Studies quantifying the microbial load associated with waste from different sources, such as surgical suites or intensive care units, have found that this waste is no more contaminated with microorganisms than household waste. Indeed, a study by Kalnowski et al. (1983) found that hospital waste was anywhere from 10 to 10,000 times "less" microbially contaminated than household waste. However, in the current political environment, hospitals may be faced with demands to follow special procedures for a greater percentage of their waste stream.

Treating Health Care Waste

In addition to ongoing concerns about appropriate waste designation, many hospitals increasingly have expressed concern about the diminishing range of disposal options. Availability of sanitary landfill space has decreased dramatically in the last few years, forcing many hospitals to ship their waste hundreds or thousands of miles across the country. Moreover, trends in state and federal activity suggest that landfilling, particularly of untreated waste, is falling out of favor. Because the risks of infectious wastes are primarily occupational, if properly packaged, disposal in a sanitary landfill has been considered acceptable in a number of states. Increasingly, however, treatment of all waste by autoclaving or incineration is being recommended or required to reduce or eliminate pathogens.

Despite the growing emphasis on waste treatment, many hospitals have encountered regulatory restrictions on the use of on-site incineration. Conflicting and excessive permit requirements at the local level have threatened hospital incineration capability in some areas—a troubling development given the efficacy of incineration in destroying pathogens and rendering waste unrecognizable. Consequently, there is a significant need for guidance on appropriate standards for treatment methods that can ensure the safe and effective operation of technology without further constraining disposal options.

The growing demand for special handling cannot be met without attention to improved technology and new disposal alternatives. Unfortunately, little legislative energy to date has focused on the development of long-term management strategies. Hospitals and other generators must press for investment in technology research and testing if economical options are to be available over the next several years. Renewed interest in waste minimization techniques and recycling may also offer possibilities for improved management of general municipal and health care waste streams.

Tracking Infectious Waste

Although greater use of on-site treatment may lessen the need for off-site transportation and thus lowering the potential for mishandling of waste, considerable attention is currently being given to manifesting and tracking systems for infectious waste. State investigators have been repeatedly frustrated by their inability to track the violators responsible for ocean dumping. In an attempt to prevent future breaches of waste disposal, regulators have focused on the manifesting model used for following hazardous chemical and radioactive waste.

The value of applying this model to infectious waste seems questionable. Unlike hazardous waste, infectious waste poses little risk to the environment, and any associated risk rapidly diminishes with time. While radioactive waste may be of concern for several decades, the primary concern with infectious materials is in the occupational setting. Moreover, questions have been raised about whether tracking is likely to prevent the infrequent occurrence of gross negligence by unscrupulous waste haulers and whether its use is likely to further diminish the already small risk of disease transmission. Indeed, it is possible that implementation of a complex tracking system may cause significant new problems for generators, handlers, and regulators without recognizable benefit and at considerable increased expense.

Nevertheless, Congress and some states have shown strong interest in manifesting as a means of addressing the infectious waste issue. Hospitals should anticipate the likely implementation of some sort of waste tracking for treated and untreated infectious waste as well as some "medical" wastes.

Beyond Hospital Doors: Small Generators of Infectious Waste

Although improvements may be made in hospital waste management, it seems unlikely that the overall issue of infectious waste can be resolved without attention to small quantity generators. While Congress has been reluctant to regulate these generators, the explosion of out-of-hospital care has made the actions of these facilities more significant than ever. Given that some of the widely publicized incidents have been linked to doctors' offices and clinics, it is even more imperative that these sites follow appropriate waste-handling practices.

The volume of waste from small quantity generators is difficult to estimate, but the scope of the problem may be projected by even a cursory look at the number of facilities under consideration. Drawing from 1985 data, 13,200 freestanding laboratories, 180,000 private physicians' offices, and 98,400 private dentists' offices should observe proper infectious waste management guidelines. By 1986 estimates, approximately 2900 outpatient clinics and 16,400 nursing homes generate medical waste. And in 1987, there were approximately 650 ambulatory surgery centers and 860 freestanding dialysis centers. These figures do not capture all of the health care facilities and home care settings that generate medical waste—a number that is certainly growing with the shift in emphasis to outpatient care. These facilities together represent a significant population of generators.

The small generator issue is important if a consistent and rational approach to waste management is sought by the health care industry and the public. If the determination of potential infection is truly based on scientific, epidemiologic evidence, presence of risk alone should be sufficient to require consistent management controls, regardless of the setting. the realization that infectious waste should be handled consistently should create greater incentives to develop cost-effective alternatives for waste disposal and monitoring that can ensure wise use of scarce resources.

Conclusion

Trends in the management of infectious waste are coming into sharper focus. Despite the lack of evidence documenting a real community and environmental risk, and a call by waste management professionals for additional education rather than regulation, Congress is certain to continue asserting its authority into the picture. As the surveys reviewed earlier indicate, states will continue to move to address this issue regardless of federal activity.

But will these new and intensified efforts—particularly efforts aimed at requiring the tracking of medical waste—provide solutions to the problems of improper dumping of medical wastes? A high-ranking EPA official has stated that tracking may keep a lot of medical waste out of dumpsters, but it will not halt the washing ashore of such waste. This same official felt that tracking provides no guarantees that medical waste will not show up on beaches in the future. This could particularly be true if Congress, as with many states, exempts certain small-quantity generators from prudent management practices.

Further compounding hospital managers' efforts to deal with infectious waste management is the push for broader designation of waste categories for special handling. No longer are regulators, lawmakers, and the public talking about "infectious waste" and the risks posed by those wastes. "Medical waste" is the operative term—a term that could ultimately become synonymous with a health care facility's entire waste stream. As pointed out at the beginning of this section, such definition could lead to costly regulation of benign waste.

Finally, the possibility that new and additional regulation could lead to dramatic constraints on (or even obsolescence of) technology is a very real, serious trend. The irony is that such regulation also could lead to new problems. For example, the Office Technology Assessment, an analytical agency that serves Congress, contemplated in an October 1988 report that new air pollution controls would lead to more off-site treatment of waste. This, in turn, could mean more ground transportation of waste, thus increasing the possibility of illegal contractor dumping or spills. While the complete environmental picture must be considered, a balance must be found.

Policy makers need to understand the risks posed by infectious waste and fully understand the problem before any rational policy can be adopted. Unfortunately, the effort to ensure such an objective view is and will continue to be clouded by news headlines and emotional reactions.

One report observed that of the 2000 items of medical waste found in 1988 on beaches in the New York area, all could fit into six shopping bags. This compares to New Jersey, where the state removes 26,000 tons a year of solid waste discarded by beach users. This comparison obviously provides little comfort to beach users or others who may come in contact with medical waste, but nevertheless represents an important illustration of the extent of the problem and the need for a reasonable, rational response.

Hospitals have a legal and ethical responsibility to ensure safe handling of infectious materials. In addition, the hospital role in community health underscores the need for a strong commitment to appropriate waste management practices. Failure to adhere to proper procedures undermines public confidence in the hospital's mission. However, in an era of constrained reimbursement systems and competing demands, hospital managers must help forge a rational policy that protects the community while conserving diminishing resources. This challenge argues for hospital leadership in working with Congress and federal and state agencies as well as the general public. Moreover, addressing the infectious waste dilemma will require a commitment to seeking long-term solutions to the broader problems of waste disposal confronting the country today.

Bibliography

Denit JD. Testimony of the U.S. Environmental Protection Agency before the House Small Business Subcommittee on Regulation and Business Opportunities, Washington, DC, August 9, 1988.

Hazardous and Infectious Waste State-by-State Report, State Issues Forum, Washington, DC, October 1988.

Jemison T. EPA moving on limited trash-tracking rules. U.S. *Medicine,* November 1988, 34.

Kalnowski G, Wiegand H, and Rüden H. The microbial contamination of hospital waste. *Zentralbl. Bakteriol Mikrobiol Hyg [B].* 178:364–379, 1983.

Kurtz H. Vacation land to wasteland—Beach pollution rooted in old problems. *Washington Post,* August 7, 1988, 6–7.

McVicar JW. Testimony of the CDC before the House Energy and Commerce Subcommittee on Transportation, Tourism, and Hazardous Materials, Washington, DC, October 21, 1987.

Office of Technology Assessment: *Issues in Medical Waste Management.* Washington, DC, Congress of the United States, Office of Technology Assessment, October 1988.

Porter JW. Testimony of the U.S. Environmental Protection Agency before the House Energy and Commerce Subcommittee on Transportation, Tourism and Hazardous Materials, Washington, DC, October 21, 1987.

Prindiville S. Testimony of the National Solid Waste Management Association before the House Energy and Commerce Subcommittee on Transportation, Tourism and Hazardous Materials, Washington, DC, October 21, 1987.

State Infectious Waste Regulatory Programs, The Council of State Governments, Washington, DC, February 1988.

US Environmental Protection Agency: *EPA Guide for Infectious Waste Management,* Report No. EPA/530-SW-86-014. Washington, DC, Office Solid Waste, U.S. Environmental Protection Agency, May 1986.

US Environmental Protection Agency: *Report on the Proceedings of the EPA Infectious Waste Management Meeting.* Washington, DC, November 1987.

Appendix 24.A.2

EPA Assessment Plan for Management of Medical Waste

1. To develop a universally accepted definition of "medical waste" to facilitate appropriate control and/or regulation of this waste.
2. To evaluate effectiveness of existing state programs in controlling the medical waste problem, and, in particular, to identify those components of state programs that are successful.
3. To quantify the extent of the problem by determining the amount and types of medical waste that are generated, current treatment and disposal practices, the relative contribution of each source (doctors' offices, clinics, hospitals), and compliance costs.
4. To develop the most effective means of tracking and reporting the handling of medical waste and to ensure the proper management and destruction of the waste.
5. To determine which transportation, treatment, storage, and disposal methods are most effective in minimizing environmental release by each waste type, and what resource requirements are required for each method.
6. To determine the environmental, economic, and health risks of improper disposal of infectious waste.
7. To determine whether regulations are adequate for hospital incinerators. The agency is completing an examination of available control technologies by the end of the year and will result in
   - Hospital waste combustion study report
   - Hospital incinerator operator training manual
   - Hospital incinerator inspection manual
8. To ensure that the general public and interest groups are provided with the information necessary to understand the nature of this problem, and kept fully informed of all program developments, including program implementation requirements. To provide educational information for the affected industry, regulators, and home medical product users.

Source: Reproduced from *Environmental Backgrounder—Medical Waste,* Office of Public Affairs, U.S. Environmental Protection Agency, Washington, DC, August 1988.

Appendix 24.A.3

Chronology of EPA Action on Infectious and Medical Waste

1982— Published draft guidance for states and health care community on infectious waste management.

1982— Began ongoing educational program by providing instructors and speakers for continuing education programs for health care workers and trade and professional association meetings and symposia.

1986— Published *EPA Guide for Infectious Waste Management,* which finalized the 1982 draft guidance.

1986— EPA announces Near Coastal Waters Strategy to protect our overstressed coastlines.

10/87—Issued a draft study of hospital-waste combustion.

11/87—Called together a group of experts that included representatives from the CDC, the National Institutes for Health, the American Medical Association, states, and the Environmental Defense Fund to discuss infectious waste management. The panel agreed risks were primarily occupational and that public exposures were isolated. The group also agreed that EPA's initial efforts should be through guidance and education.

1988—Each region designated a Medical Waste contract person and two consulting firms were contracted to develop educational materials and guidance, and to conduct studies of state programs.

4/88—EPA, the National Oceanic and Atmospheric Administration (NOAA), and the Department of Transportation (DOT) begin to implement Section 2204 of the Marine Plastic Pollution Research and Control Act of 1987 to educate the public on the prevention of plastics pollution. EPA is developing a series of fact sheets and will share in the distribution of NOAA's materials.

5/88—EPA began to assist New York and New Jersey with developing a bistate tracking system on medical waste, which was completed in August 1988.

6/88—Published a *Federal Register* Notice in which EPA requested comments on issues related to medical wastes. The comment period closed on Monday, August 1, 1988. Began to develop an inspection manual for hospital incinerators and a training manual for the operators of hospital incinerators. Began preparing brochures, posters, and a bibliography of abstracts of infectious waste studies to supplement the agency's guidance document on infectious waste management. Guidance materials to assist states in implementing infectious waste management programs under existing municipal solid waste authorities are also under development.

7/88—Creation of an Office Pollution Prevention, to maximize and focus the agency's attention on reducing waste and pollution before it becomes a disposal or cleanup problem.

8/88—Appointment of EPA Medical Waste Task Force, chaired by Dr. John Moore, to coordinate Agency activities on medical waste. Creation of a task force on solid waste in EPA to work specifically on the national municipal solid waste dilemma.

9/88—Region 2 is hosting a Medical Waste Conference in mid-to-late September with state health and environmental commissioners from New York City, New York, New Jersey Rhode Island, Connecticut, Pennsylvania, and Maryland.

—Federal officials from EPA, the Department of Defense, Department of Transportation, Department of Commerce, National Science Foundation, State Department, and the Department to Interior will meet to discuss improved federal sharing of enforcement and resources to combat improper disposal of medical waste.

10/88—EPA is sponsoring a two-day Ocean Pollution Enforcement Conference in Point Judith, Rhode Island during the week of October 3. Representatives of the National Association of Attorneys General, various state and federal agencies (including the Coast Guard and the FBI), and the Attorneys General of the 14 East Coast states are expected to attend. The discussions will cover a broad range of state and federal ocean pollution enforcement issues, including medical waste disposal.

—Oceans '88 Conference in Baltimore [MD] with medical waste panel included on agenda (10/31–11/2).

Source: Reproduced from *Environmental Backgrounder—Medical Waste,* Office of Public Affairs, U.S. Environmental Protection Agency, Washington, DC, August 1988.

25

The Occupational Hazards of Home Health Care

Introduction 25-1
Trends in Home Health Care 25-1
Job Injury and Illness in Homecare 25-3
Occupational Hazards 25-4
Setting 25-4
Personal Safety 25-4
Ergonomics and Back Injuries 25-5
Infectious Diseases 25-7
Bloodborne Pathogens • Universal Precautions • Tuberculosis
Conclusion 25-9
References 25-10

Elaine Askari
University of California, Berkeley

Barbara DeBaun
California Pacific Medical Center

Introduction

Health care workers who provide services in the home face a substantial risk of job-related injury and illness. These providers may be challenged by numerous environmental factors in the home, including poor sanitation, lack of ventilation, pets, rodent or insect infestations, infectious diseases, limited supplies, and unpredictable conditions and events. They are susceptible to musculoskeletal injuries due to heavy or awkward lifting and may even encounter the threat of violence when in the home or nearby.

This chapter is a survey of some of the occupational hazards in home health care work. It will examine the explosive growth of the homecare field in recent years and review available statistics about occupational injury and illness rates. Finally, it will focus on several key health and safety issues for homecare providers—personal safety, ergonomics, back injuries, bloodborne pathogens, and tuberculosis.

It should be noted that, to date, relatively little research has been done on home health care hazards. Many of the available injury and illness statistics are incomplete. This is clearly a field in need of further study.

Trends in Home Health Care

Health care is one of the largest industries in the United States. Home health care is the fastest growing segment of the industry. According to the Bureau of Labor Statistics (BLS), 555,400 U.S. employees worked for private-home health care agencies in 1994.[1] This figure does not include employees of public

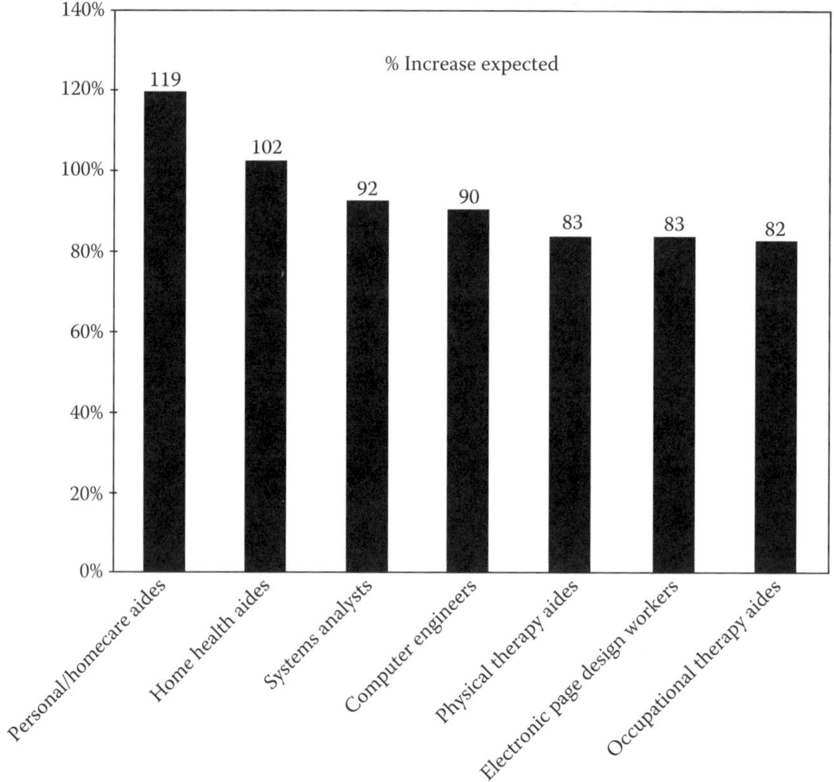

FIGURE 25.1 Fastest growing occupations in the United States, 1994–2005. (From Bureau of Labor Statistics, Office of Employment Projections, November 1995.)

agencies or hospital-based agencies. Marion Merrell Dow, a private marketing firm, conducted a survey in 1994 and found that 900,000 U.S. workers were employed in home health care overall.[2]

From 1988 to 1993, home health care employment had an annual growth rate of 16.4%. This compared with 4.3% for total health services and 2.8% for hospitals.[3] In 1994, home health care became the second fastest growing industry segment in the entire U.S. economy. The BLS projects an increase of more than 500,000 jobs in home health care between 1994 and 2005[4] (see Figure 25.1).

The number of U.S. clients for homecare services increased from 1.5 million in 1988 to 3.5 million in 1993. In part, this growth resulted from the expansion of Medicare coverage due to a federal district court decision (*Duggan v. Bowen*)[5] in 1988. Other contributing factors include an increasing elderly population and the lower costs for homecare compared to hospital care (primarily due to the lower wages paid in homecare). In addition, technological changes have made it easier to provide complex medical services in the home.

The U.S. Office of Management and Budget defines home health care services as "establishments primarily engaged in providing skilled nursing or medical care in the home, under the supervision of a physician."[6] Home health aides make up the largest proportion of home health care workers (31%). Registered nurses constitute 20%, homemaker aides 13%, and licensed practical nurses 7%.[3] Other homecare occupations include therapists (occupational, respiratory, physical, and speech), social workers, pathologists, and physicians. Services in the home range from basic personal care like feeding, cleaning, and bathing to more complex tasks like home infusion therapy. The skill level of home health care workers must equal or even surpass that of providers in an acute-care setting. Many homecare workers must be able to perform an array of advanced medical procedures in a less than ideal setting and they often lack the resources they need.

The majority of the homecare clients are elderly. Many of these patients suffer from heart disease, cancer, stroke, or orthopedic problems. In recent years, many people with HIV/AIDS have also been cared for in the home. A 1986 study found that approximately 17%–22% of AIDS patients used home health care services at any given time.[7] A recent study of homecare agencies in the San Francisco Bay Area showed that approximately 95% of them provide care for people with AIDS.[8]

Job Injury and Illness in Homecare

According to the BLS, there were 6588 occupational fatalities in the United States in 1994.[9] Of these, 86 fatalities occurred in health care services, including 19 in home health care. Transportation accidents caused 79% of the home health care fatalities. This can be explained by the amount of driving required between patient visits.

In the same year, there were 217,817 nonfatal occupational injuries in health care services. There were 98,196 injuries in hospitals, 83,450 in nursing homes, and 18,812 in home health care.[10] Of the home health care injuries, 59% involved strains and sprains and 49% affected the trunk (mostly the back). Common sources of injury included overexertion (39%) and interaction with the patient (36%) (see Figure 25.2). The homecare figures probably reflect under-reporting. It is often difficult for people who

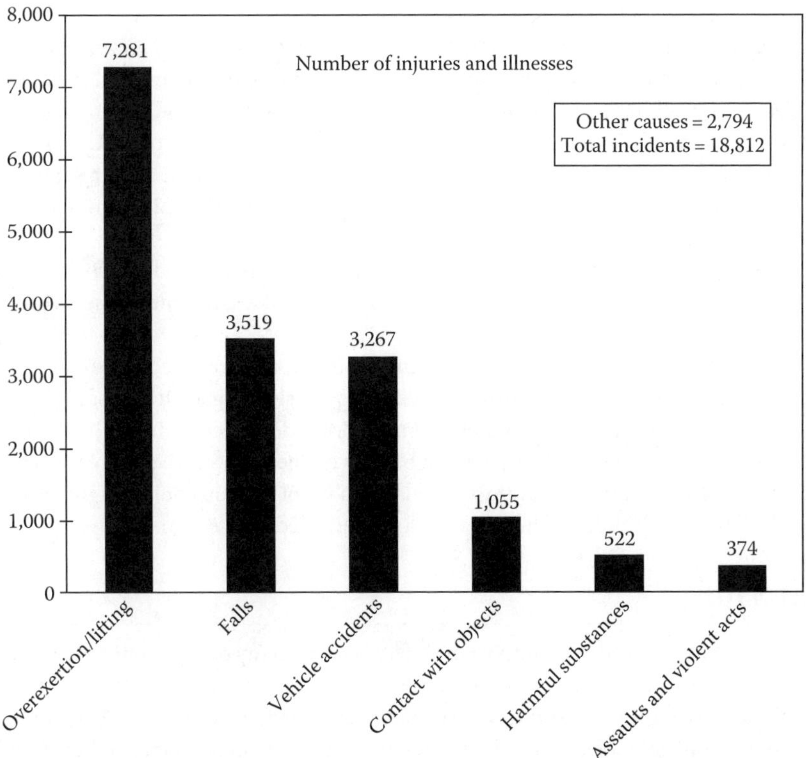

FIGURE 25.2 Leading causes of injury and illness in homecare, 1994. (From Bureau of Labor Statistics.)

work in the field to report their injuries. Homecare agencies usually do not have a trained employee health nurse or other occupational health staff. Therefore home health care workers may not have access to immediate consultation, assistance, or clear reporting procedures when they sustain an injury in the home. Many of these workers are also on tight schedules, which may deter them from driving to an office or clinic out of their work area to report an injury.

Occupational Hazards

In hospitals, nursing homes, and other health care industry workplaces, employees may be exposed to numerous job hazards, including communicable diseases, chemical and biological agents, carcinogens, ionizing and nonionizing radiation, ergonomic hazards, and psychosocial risks.[11] Workers in home health care are exposed to many of these same hazards. But in addition, they work in a unique environment where they can also be faced with a range of other problems. This chapters will address a few typical homecare hazards. However, the authors stress that there is a scarcity of research on occupational hazards in home health care. Therefore, we will describe relevant information from the literature and add examples from our own experience with home health care employees.

Setting

To understand the hazards faced by homecare workers, one must take into account the unique conditions found in the homecare setting. Smith identified three distinct types of homecare work environments: rural, suburban, and urban inner city.[12] According to Smith, common physical risk factors in all types of environments include pets and other animals; driving and parking; home structure, maintenance, and hygiene; personal safety/vulnerability; weather; and time of day. These and other factors result in a lack of worker control over the environment.

Smith conducted interviews with 29 home health care workers, and observed them as they went about their work. Workers in rural areas related several stories about the unusual risks they faced. For example, a nurse was walking up to a house when a pig "ran towards me and, wham, it bit me. I was just so stunned. It was hard to live that one down with my fellow workers and the ER workers who treated me." Another nurse was confronted by a gaggle of geese "hissing and flapping their wings." Other workers described the hazards of driving on unpaved country roads full of pot holes in bad weather. Some had experienced car trouble with no one nearby to help.

In a suburban area, one nurse described working in homes where people let their pets "do their business" all over the house. Another worker reported coming out of a suburban home "with flea bites and covered in dog or cat hair."

Many of the problems described by workers in the inner city involved personal safety. Workers described coming in contact with pit bulls (extremely aggressive dogs). Others said they had to work around loaded guns and other kinds of weapons. One worker described some of the hazards associated with housing projects: "The high rise projects can be very dangerous. Stairwells are especially bad spots (with a lot of drug use and dealing). Elevators can also be a problem. You could call the elevator and walk in on a drug deal. In some buildings the elevators are controlled by the dealers."

Personal Safety

Violence on the job has emerged as an important safety issue in many occupations today. The BLS reports that in 1994 there were approximately 20,000 nonfatal violent incidents in the nation's workplaces (counting only those that resulted in lost workdays). Women were the victims in nearly 60% of these incidents. The most common nonfatal violent acts (about 40%) involved "hitting, kicking, and beating."[13]

Service workers are at especially high risk of nonfatal assault. Of the service workers who are assaulted, nursing aides, and orderlies account for more than half. These and other workers in health and residential care occupations (e.g., social workers) are assaulted primarily by clients who resist their help. There have been no studies to date that examined the risk of assault in homecare specifically.

Federal OSHA issued its "Guidelines for Preventing Workplace Violence among Health Care and Social Service Workers" in March, 1996.[14] OSHA identified homecare workers as a group at risk for workplace violence because they "work alone...may have to work late night or early morning hours... often work in high-crime areas...and work in homes, where there is extensive contact with the public."[15]

The risks to personal safety in home health care have not been adequately studied, and statistics on the prevalence of the problem may be unreliable. This is another area where under-reporting may be significant. Some researchers have suggested that homecare workers (like other health care workers) often accept assaultive behavior as part of the job. In addition, formal incident reports may be filed only for assaults that require medical treatment. Often no reports are filed for verbal threats, sexual harassment, or various other types of assault.[16]

One recent study sought to determine whether the quality of care is affected when home visits present threats to personal safety.[17] Nearly 100 homecare employers and administrators were asked about their perception of risk and response to it. Among the factors that respondents associated with risk were geographic location, high incidence of crime, inappropriate patient or caregiver behavior, and evening assignments. Respondents emphasized that visits are now made to the home 24 h a day. Approximately 66% of respondents said that they leave a situation "as soon as possible" if they perceive it to involve high risk. A larger study is now in progress by the same authors.

Under U.S. health and safety law, the employer is responsible for maintaining a safe and healthful workplace. OSHA has outlined several steps that employers can take to prevent violence in the homecare setting (Figure 25.3):

- Provide the safety education for employees.
- Establish a communication system (such as cellular phones).
- Utilize a "buddy system" or security escorts.
- Establish the policies to discourage robbery (e.g., do not carry a purse).
- Assure the proper maintenance of vehicles.
- Provide the handheld alarms to employees.
- Establish the procedures for employees and the employer to follow when violence does occurs.[15]

Ergonomics and Back Injuries

According to BLS figures for 1994, overexertion was the most common cause of lost-time occupational injury in the private sector U.S. workforce.[13] Nursing homes and scheduled airline/air-courier services led all other industries in overexertion incidents. The same year, the BLS rated the 10 occupations with the most lost-time injuries and illnesses of all types and found that nursing aides/orderlies ranked third (truck drivers were first, and nonconstruction laborers were second).

The BLS found that sprains and strains were, by far, the leading type of work-related injury in every major U.S. industry. The trunk, including the back, was the body part most affected. Overexertion while maneuvering objects led all other disabling events and was cited in 16%–33% of the cases in every industry.

One study examined low back injuries in health care during 1984–1986.[18] Incident reports were collected for nursing assistants in a large hospital and home health aides in two large homecare agencies. The researchers found the rate of back injury to be much higher among the home health aides (15.4 per 100 full-time equivalent workers) than among the hospital nursing assistants (5.9 per 100 full-time equivalent workers).

The study showed that 40% of the back injuries in both groups involved activities at the patient's bedside. About 66% of the back injuries occurred during planned patient care activities such as transferring

✔

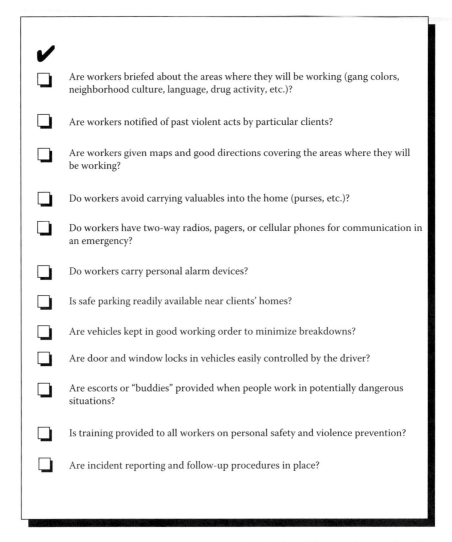

- ☐ Are workers briefed about the areas where they will be working (gang colors, neighborhood culture, language, drug activity, etc.)?
- ☐ Are workers notified of past violent acts by particular clients?
- ☐ Are workers given maps and good directions covering the areas where they will be working?
- ☐ Do workers avoid carrying valuables into the home (purses, etc.)?
- ☐ Do workers have two-way radios, pagers, or cellular phones for communication in an emergency?
- ☐ Do workers carry personal alarm devices?
- ☐ Is safe parking readily available near clients' homes?
- ☐ Are vehicles kept in good working order to minimize breakdowns?
- ☐ Are door and window locks in vehicles easily controlled by the driver?
- ☐ Are escorts or "buddies" provided when people work in potentially dangerous situations?
- ☐ Is training provided to all workers on personal safety and violence prevention?
- ☐ Are incident reporting and follow-up procedures in place?

FIGURE 25.3 Personal safety tips for home health care workers. (Adapted from *Violence on the Job: A Guidebook for Labor and Management*, LOHP, Berkeley, CA, 1997.)

a patient to a wheelchair. For homecare, other activities at the time of injury included helping a patient in or out of a bed, chair, or tub; catching a patient during a fall; turning a patient; stooping over a patient in bed; and helping a patient on or off a toilet. No lifting equipment was used in 75% of the incidents involving nursing assistants and 80% of the incidents among home health aides. Home health aides were working alone in 88% of the incidents and nursing assistants were working alone in 39% of the incidents.

One of the problems in the homecare setting is that the worker has to adapt to an unpredictable environment, which differs from one home to the next. There is no opportunity to engineer safety problems out of the home as can be done with fixed workplaces. Adaptation is even more difficult because it must typically be done by a single employee working alone, for many clients in many different homes. For example, the "lifting team" concept developed for the acute-care setting cannot be applied to homecare. Charney describes the lifting team concept as a method that incorporates proper body mechanics, personal protective equipment (such as transfer belts), mechanical lifting devices, and a two-person team.[19] Candidates for the lifting team are screened for flexibility and strength. Those selected for the team are trained on proper lifting techniques, how to use mechanical lifting devices, and how to coordinate the lift. Although this full concept is not applicable to homecare because most employees work by themselves, it is at least important to provide mechanical lifting devices and to train homecare workers in how to use them.

Owen recommends such an approach to reduce back injuries in the homecare setting.[20] The patient's needs should be assessed, and the proper techniques selected to handle, lift, and move the patient safely. These techniques should then be clearly communicated to the staff who will be doing the work. This communication should be included in new employee orientation programs, and there should be frequent in-service training. Owen stresses that "assistive devices or mechanical lifts" should be used to lift or transfer patients and to move patients in bed. Devices used in the home will have to be different from the heavy equipment often used in hospitals and other institutions. Devices should be easy to maneuver and carry into the home. Owen suggests transfer belts, transfer boards, portable commodes, chairs and cushions that lift, and lift poles that a patient can use to stand independently.

Infectious Diseases

Home health care workers must be protected against communicable diseases at all times. A range of personal protective equipment should be readily available, including gloves, face shields, impervious gowns, and cardiopulmonary resuscitation (CPR) masks. Respirators may be needed when working with tuberculosis (TB) patients. Equipment should also be available for protection against bloodborne diseases, including safe needle devices that protect against accidental needlestick injury, good sharps disposal containers, and access to immediate treatment when there is a significant blood exposure.

Bloodborne Pathogens

Occupational exposure to bloodborne pathogens continues to be a threat to health care workers both in institutions and in the home. Today there is a great deal of concern about exposure to the hepatitis B virus (HBV) and the Human Immunodeficiency Virus (HIV) both of which are transmitted through blood contact. Many more health care workers become infected with HBV than with HIV but HIV is still considered a significant risk.

The Centers for Disease Control and Prevention (CDC) recently reported on 51 documented cases of occupational transmission of HIV to health care workers.[21] (There are 106 additional cases where occupational transmission is considered a possibility but has not been documented.) Percutaneous exposure was the source in the majority (44) of the 51 documented cases. (It is generally believed that CDC figures represent considerably less than the actual number of cases, due to under-reporting.)

The occupational risk of exposure to bloodborne pathogens in the homecare setting has not been the focus of a great deal of investigation or research. For example, the number of needlestick injuries in home health care is unknown, although it has been estimated that 800,000 needlesticks occur each year in acute-care institutions in the Unite States.[22] Activities which pose a risk of blood exposure to home health care workers include intravenous infusions, vascular access, intravenous nutritional support, wound care, and respiratory therapy. The risk of a needlestick injury in this setting may be increased due to the less controlled environment (poor lighting, clutter, cramped quarters, pets, etc.). One home health care nurse reported getting a needlestick because a client's cat jumped on the bed while she was drawing blood. Also, proper equipment may not be available in homecare.

In 1994, the CDC conducted an epidemiological survey involving bloodborne pathogens in the home. CDC summarized eight reported cases of home caregivers who became infected with HIV. In four of these cases, the caregivers had significant direct contact with an infected person's body secretions or excretions.[23]

One of the present authors was involved in a 1994 study that compared needlestick injury rates between homecare nurses and hospital nurses.[24] An anonymous questionnaire, completed by 76 Northern California homecare nurses, sought to determine the number of reported and unreported needlestick injuries in the group. Results showed that a substantial number of needlestick injuries occur to homecare nurses and that the majority of them go unreported. The rate of reported needlestick injuries among homecare nurses was 50% of the rate among the acute-care hospital nurses. However, the homecare rate may be much greater than these figures indicate because of under-reporting. The survey found that 87% of the needlestick incidents in homecare were unreported.

Universal Precautions

Most home health care agencies are required to comply with federal OSHA's Bloodborne Pathogens standard [29 CFR §1910.1030]. The standard mandates a set of protective measures termed "Universal Precautions." This approach is based on treating all blood and body fluids as if they are infectious. Employers must provide, at no cost, equipment such as gloves, gowns, masks, mouthpieces, resuscitation bags, and sharps disposal containers, as well as employee training.

One of the present authors conducted a 1995 study to assess compliance with Universal Precautions in home health care.[25] A questionnaire was completed by 214 home health care nurses in Northern California. Results showed that compliance rates were highest for glove use and proper sharps disposal and lowest for wearing fluid resistant gowns, face shields, and eye protection.

When asked about the obstacles encountered to use of Universal Precautions in the home, 66% of the respondents said they had no face shields available, 38% had no gowns, and 28% had no eye protection. The respondents identified availability of equipment as the major factor determining compliance with Universal Precautions. Approximately 47% of the respondents also indicated that "feeling comfortable using protective equipment" affected their decision to use it.

Respondents said that they had problems with sharps containers, on the average, about 25% of the time. Problems included needle apparatus got caught in the container (37%); no container nearby (32%); container full (28%); and being forced to use a bottle or can in place of a real sharps container (27%).

In response to this identified need, National Institute of Occuptional Safety and Health (NIOSH) recently awarded new funding to the Training for Development of Innovative Control Technology Project (TDICT) in San Francisco and the Labor Occupational Health Program (LOHP) at U.C. Berkeley to design a prototype safety device specifically to reduce exposure to bloodborne pathogens in homecare. Homecare workers will be fully involved throughout the design process.

In 1993–1994, LOHP received a grant from federal OSHA to train California homecare workers on the Bloodborne Pathogens standard. The training was conducted with the Service Employees International Union (SEIU). LOHP and SEIU trained 285 homecare workers. Over 75% were aides and attendants (not nurses). Approximately 81% were women and 67% were people of color. Anecdotal remarks made by participants during the training revealed that most of these workers had never received any information from their employers about bloodborne pathogens or how to protect themselves on the job. In addition, the majority had to purchase their own protective equipment (gloves, aprons, etc.). They were among the lowest paid workers in the entire health care industry and were clearly not aware of protections mandated by the OSHA standard.

Tuberculosis

Since 1984, the annual number of new TB cases in the United States has increased by 18%—from 22,255 cases in 1984 to 26,283 cases in 1991.[26] A disease which was once thought to be under control has reemerged. Since 1993, however, there has been a slight decline in the annual number of new cases. The CDC reports that 24% of TB cases in the United States occur in people aged 65 and older. This has important implications for homecare workers who see many elderly patients.

Studies have shown that health care workers have a significant risk of occupationally acquired tuberculosis. Surveillance studies conducted recently show that some groups of workers who care for TB patients become infected at an annual rate of 5%–10% (PPD conversion rate).[27] However, it must be stressed that none of these studies have surveyed workers in homecare.

Less than one page of the CDC's 132-page *Guidelines for Preventing the Transmission of Mycobacterium Tuberculosis in Health-Care Facilities* is devoted to home health care.[28] It is critical that more study be done and that home health care agencies establish comprehensive TB training, education, counseling, and screening programs. There should be provisions for identifying workers and clients who have active TB, baseline two-step PPD skin testing, and follow-up skin testing at intervals appropriate to the degree of risk.

Conclusion

Homecare is one of the fastest growing industries in the United States, and homecare aides are projected to be the fastest growing occupation in the industry. These aides are mainly middle-aged women who have not graduated from high school.[29] In the San Francisco Bay Area, they are primarily women of color. They are some of the lowest paid workers in health care and have the least knowledge about the risks they face on the job. They receive little training. Although they are the fastest growing sector of the workforce, they have very limited power to make changes in their work environment.

Turnover rates are high. Although their interests might best be served by joining a union that can bargain for better working conditions, only about 10% of U.S. homecare workers are organized to date.

Worker safety, client safety, quality of care, and client autonomy are intertwined very tightly in the homecare setting. There is little documentation or research on the hazards faced by homecare workers. It is clear, however, that they do face unique health and safety challenges, which are increased due to an unpredictable work environment as well as poor training and equipment. The home environment is often beyond the control of both the worker and the client. Crime in the neighborhood, unsafe or poorly maintained buildings, inadequate lighting, and poor sanitation can impact them both.

Following are some problems that have been addressed in this chapter:

- Many homecare workers are not given adequate training on how to safely lift, transfer, or reposition a patient. Many are not provided assistive devices that can help them perform these tasks in a proper ergonomic manner. The result can be potential worker and client injuries.
- Few steps are being taken to analyze assaults and violent acts against homecare workers. Also there is often a lack of education and training, safety procedures, and security devices to protect these workers.
- Many homecare workers are not provided adequate training on infection control policies and procedures (especially regarding bloodborne pathogens and TB). This puts both the patient and worker at risk of infectious disease.
- Many homecare nurses are performing technologically advanced medical procedures that may entail high risk. For example, intravenous catheter insertion and care, wound care, and tracheotomy care may now be done in the home. The hazards of performing such procedures in the unique environment of homecare have not been adequately studied. Safety devices designed specifically for the homecare setting may need to be developed.
- OSHA does not inspect home workplaces. Thus there is no objective way to assess compliance with its regulations, such as the Hazard Communication and the Bloodborne Pathogens standards.
- Sometimes no personal protective equipment is supplied to homecare workers or they are expected to supply it themselves. Equipment that is provided may be inadequate.

As employers, homecare agencies should establish procedures and policies to address the hazards faced by their workers. Workers' input should be solicited. A safer work environment should serve to foster a better working relationship between clients and those who care for them.

References

1. *Occupational Outlook Quarterly*, U.S. Department of Labor, Bureau of Labor Statistics. Fall 1995.
2. The home care industry, Service Employees International Union, Research Department, 1995.
3. Freeman L. Home-sweet-home health care, *Monthly Labor Rev* March 3–11, 1995.
4. Fastest growing occupations 1994–2005, U.S. Department of Labor, Bureau of Labor Statistics, Office of Employment Projections, November 1995.
5. *Duggan v. Bowen*. 691 F. Supp. 1487 (D.D.C. 1988).
6. *Standard Industrial Classification Manual*, U.S. Office of Management and Budget, 1987.
7. Lusby G, Martin JP, and Schietinger H. Infection control at home: a guideline for caregivers to follow, *Am J Hospice Care*, March/April, 1986.
8. White M and Smith W. Home health care: Infection control issues, *Am J Infect Control*, 21(3): 146–150, 1993.
9. Fatal occupational injuries by industry and event or exposure, U.S. Department of Labor, Bureau of Labor Statistics, 1994.
10. Survey of occupational injuries and illnesses, U.S. Department of Labor, Bureau of Labor Statistics, 1994.
11. Sterling D. Overview of health and safety in the health care environment, *Essentials of Modern Hospital Safety*, Vol. 3, chap. 1, Boca Raton, FL, 1994.
12. Smith W. Occupational Risk Perception in Home Health Care Workers. Doctoral dissertation, UMI# 9542205, 1995.
13. Characteristics of injuries and illnesses resulting in absences from work, U.S. Department of Labor, Bureau of Labor Statistics, 1994.
14. Guidelines for preventing workplace violence among health care and social service workers, U.S. Department of Labor, Occupational Safety and Health Administration, 1996.
15. Protecting community workers against violence, Fact Sheet No. OSHA 96-53, U.S. Department of Labor, Program Highlights, 1996.
16. Lipscomb J and Love C. Violence toward health care workers—An emerging occupational hazard, *AAOHN J* 40, 219–228, 1992.
17. Kendra MA, Weiker A, Simon S, Graut A, and Shullick D. Safety concerns affecting delivery of home health care, *Public Health Nursing*, 13, 83–89, April 1996.
18. Myers A, Jensen RC, Nestor D, and Rattiner J. Low back injuries among home health aides compared with hospital nursing aides, *Home Health Care Serv Q* 14, 149–155, 1993.
19. Charney W, Zimmerman K, and Walara M. The lifting team, *AAOHN J* 39, 231–234, 1991.
20. Owen B. Back injuries in the home health care setting, *Home Health Care Consultant*, 3, 25–39, 1996.
21. HIV/AIDS Surveillance Report, U.S. Centers for Disease Control and Prevention. June 1996.
22. Jagger F, Pearson R.D. Universal precautions: still missing the point on needlesticks, *Infect Control Hosp Epidemiol* 12(4):211–213, 1991.
23. U.S. Centers for Disease Control and Prevention. Epidemiologic notes and reports, human immunodeficiency virus transmission in household settings—United States. *MMWR*, 19, 347–356, 1994.
24. Askari E. and Lipscomb J. Bloodborne pathogens—Risks in the home health care setting, *AAOHN J* submitted for publication, 9(4):589–608, 1997.
25. Askari E. Compliance with universal precautions in home care, Unpublished study, 1995.
26. U.S. Centers for Disease Control and Prevention.
27. Markowitz S. Epidemiology of tuberculosis among health care workers, *Occup Med State Art Rev* 9, 1994.
28. Guidelines for preventing the transmission of mycobacterium tuberculosis in health-care facilities. U.S. Centers for Disease Control and Prevention, 1996.
29. Current population survey, U.S. Bureau of the Census, 1994.

26

Caring Until It Hurts: How Nursing Work Is Becoming the Most Dangerous Job in America

Executive Summary **26**-2
Back Injuries Are Epidemic • Understaffing Is a Key Cause of Injuries • Employers Do Not Take Safety Concerns Seriously • Unsafe Working Conditions Exact a High Cost from Taxpayers • Recommendations
Introduction **26**-4
Dangers of Nursing Home Work **26**-6
Back Injuries Predominate **26**-7
Why Is Nursing Home Work So Hazardous? **26**-8
Risks to Nursing Home Workers Have Been Increasing **26**-8
Understaffing Puts Residents and Workers at Risk • Employers Favor Gimmicks Rather than Real Solutions
Safety Saves Money **26**-10
Taxpayers Ultimately Pay the Cost of Unsafe Working Conditions • Safety Can Save Money
Recommendations **26**-12
Improve Staffing Standards • Establish an Ergonomics Standard • Change OSHA's Enforcement Focus
References **26**-13

Service Employees International Union

This study was prepared by the Service Employees International Union (SEIU). SEIU is the largest union of health care workers in the United States, with 475,000 health care members working in nursing homes, hospitals, HMOs, home care agencies, and other facilities. With a total of 1.1 million members in the United States and Canada, SEIU is the third largest and fastest growing union in the AFL-CIO.

Executive Summary

What is the most dangerous job in America? Mining? Construction? Working in a steel mill? Trucking?

No. More dangerous than all of these—and fast becoming the most dangerous job in the United States—is work in a health care facility, or nursing home work.

Ironically, nursing home work is also one of the fastest-growing jobs in America.

While working conditions in many other industries have gradually improved over the last decade, nursing homes have become far more dangerous places to work, as caregivers themselves suffer an epidemic of crippling workplace injuries.

These injuries, devastating as they are to a growing workforce of committed caregivers—and costly as they are to taxpayers—also point to a broader, equally dangerous problem. That problem is a developing crisis in staffing and conditions in nursing homes nationwide that threatens the quality of care for millions of America's most vulnerable citizens. It is a crisis that is occurring just as the $85 billion mostly taxpayer-financed industry is experiencing its most radical transformation in a quarter century.

The SEIU has found that

- Occupational illness and injury rates for nursing home workers are higher than for workers in other industries with well-documented hazards, such as mining and construction.
- Nursing home workers are injured at an alarming rate of just under 17% a year—more than twice the rate of private sector workers generally.
- Occupational illness and injury rates for nursing home workers increased by 53% between 1983 and 1994—with more than 200,000 injuries reported in the industry every year.
- Of those injured nursing home workers who must take time off, more than a quarter require more than two work-weeks to recover. Less than a third are able to return within a day or two.
- The nursing home industry is the most dangerous fast-growing industry in the United States. Of the 20 fastest-growing industries in the United States, nursing homes have the highest rate of occupational illness and injury.

Back Injuries Are Epidemic

Back injuries, widely agreed to be among the most serious and costly of injuries, are the most common type of injury suffered by nursing home workers.

- While back injuries account for 25% of all injuries reported in the private sector, they account for 43% of all injuries in nursing homes.
- Nurse aides, who provide most of the patient care in nursing homes, are particularly at risk. Injuries to the back and trunk account for more than half of all injuries to nurse aides working in nursing homes.

Understaffing Is a Key Cause of Injuries

The shift to prospective payment in both private and public health insurances has led to earlier hospital discharges and an overall increase in the acuity levels—of nursing home patients. Unfortunately, staffing levels have not increased to match the increased workload.

- The National Academy of Sciences' Institute of Medicine has released data showing that the number of nurse aide hours per resident day only increased from 2.0 in 1991 to 2.1 in 1993, a statistically insignificant increase.
- A 1993 survey of nursing home workers by the SEIU found that 86% reported current staffing levels in their facilities to be inadequate and 91% reported that their workloads had increased over the past year.

Employers Do Not Take Safety Concerns Seriously

Rather than hiring additional staff or redesigning work to eliminate hazards, nursing home employers have tended to favor gimmicks that do little to prevent injuries. Much of the focus has been on teaching proper lifting techniques, which are of limited use in patient care settings. In addition, SEIU has also found that

- Inspections of nursing homes by the Occupational Safety and Health Administration (OSHA) have often found that safety training programs for nurse aides are not effective. Nurse aides generally do not use the techniques outlined in the program, either because there are not enough staff on hand or because they have not received adequate instruction in the techniques.
- Back belts are a favorite intervention of nursing home operators. But there is no evidence that they reduce the hazards of repeated lifting, pushing, twisting, and bending.
- Many employers sponsor programs such as "safety bingo," that encourage workers not to report injuries in exchange for the chance to win a television set or other merchandise.

Unsafe Working Conditions Exact a High Cost from Taxpayers

Nursing home industry executives often resist hiring additional staff or purchasing safety equipment because of the cost. The truth is that running an unsafe workplace increases costs in a variety of hidden as well as more obvious ways.

- The staggering workers' compensation insurance rates paid by nursing home employers are a reflection of the magnitude and severity of injuries and illnesses in nursing homes. The nursing home industry as a whole paid close to $1 billion in workers' compensation insurance payments in 1995.
- The high rate of injury in nursing homes has contributed to high rates of turnover—as high as 100% at some facilities. Constant turnover forces employers to invest resources in recruitment and training of new workers.
- Conditions that contribute to worker injuries, such as understaffing, also add to patient injuries. The nursing home industry spends $1.2 billion to heal preventable decubitus ulcers (bedsores) caused by lack of nutritional hydration, mobility, and cleanliness and $4.3 billion on incontinent care because residents are not toileted frequently enough.

Because public programs pay for most nursing home care, it is taxpayers who ultimately bear the burden of unsafe working conditions in the nursing home industry. Taxpayers pay for 75% of the nursing home care provided in the United States, primarily through the Medicaid program.

A number of academic studies have shown that there are cost-effective interventions that can reduce the number of injuries in nursing homes and save money for employers, patients, and taxpayers.

Recommendations

Based on the information presented in this study, SEIU makes the following recommendations:

Improve Staffing Standards

The cornerstone of worker safety and quality resident care is adequate staffing. SEIU recommends the establishment of acuity-based staffing ratios, with specific minimum ratios for nurse aides. The U.S. Health Care Financing Administration (HCFA) should monitor the relationship of staffing to acuity levels, resident outcomes, and worker and patient injury rates on an ongoing basis.

Establish an Ergonomics Standard

OSHA is in the process of developing an ergonomics standard for private sector employers. Such a standard would require worksite risk analysis, the evaluation and implementation of feasible methods of preventing and reducing those risks, a program of prompt treatment for workers who report first signs of an injury and education and training in how to prevent injuries.

Change OSHA's Enforcement Focus

Since its inception, the focus of OSHA's enforcement efforts has been in industries with widely recognized hazards, especially those in the manufacturing sector. With over 70% of workers now employed in the service sector, this focus is out of date. Stepping up enforcement in the nursing home industry (by far the most hazardous service sector industry) would allow OSHA to improve its ability to effectively enforce workplace safety laws in the service sector.

Without a bold, concerted effort to address staffing issues and related conditions in nursing homes nationwide, not only will the explosive growth in nursing home worker injuries continue, but also the quality of care provided to residents can only be expected to deteriorate.

Introduction

The nursing home industry employs roughly 1.7 million workers who provide around-the-clock patient care and assistance in the activities of daily living to chronically ill and disabled individuals. The nursing home workforce is disproportionately female (87%), minority (28%), and low-wage earners. The average nursing home worker earns just over $17,000 a year, and many work under stressful and hazardous conditions that put their own health, as well as the health of their patients, at risk.

Unlike hospitals, where registered nurses provide most of the hands-on patient care, the bulk of patient care in nursing homes is provided by nursing aides. Nursing aides constitute 42% of the nursing home workforce (see Figure 26.1).

Many of the workers employed in nursing homes earn poverty-level wages. As Figure 26.2 makes clear, wages for nurse aides and a range of other nursing home job classifications are below the poverty line for a family of four.

Nursing home workers not only suffer the insult of below-poverty-level wages, but working in a nursing home is rapidly becoming one of the most dangerous jobs in America. While most people believe that nursing homes are safe and clean—they are health care facilities, after all—compared to other kinds of workplaces, working in a nursing home is actually more dangerous than working in a coal mine, a steel mill, a warehouse, or a paper mill.

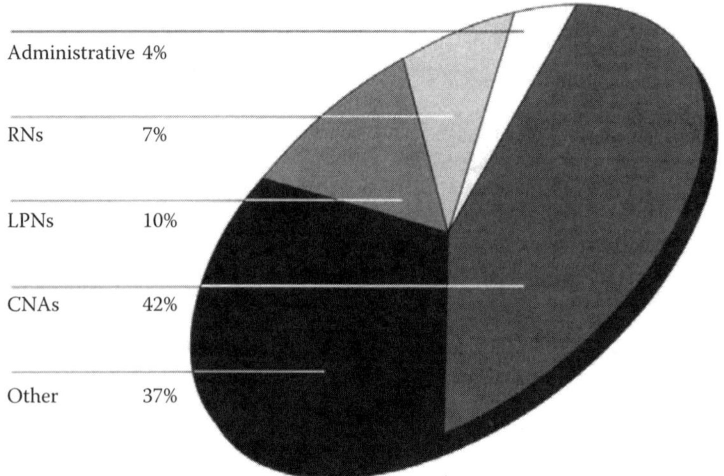

FIGURE 26.1 Nurse aides are the largest part of nursing home workforce. (From Bureau of Labor Statistics.)

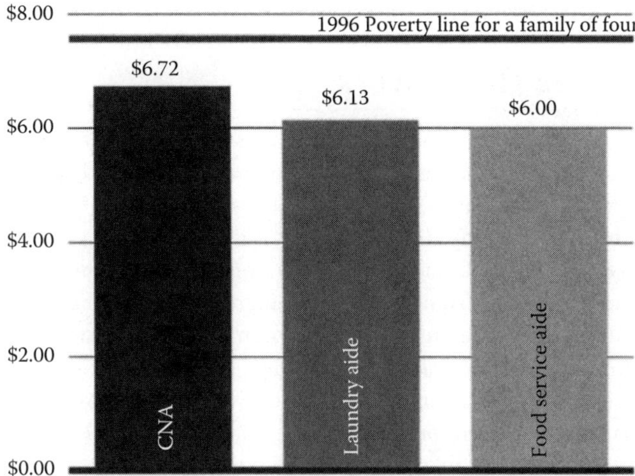

FIGURE 26.2 Wages for most nursing home jobs are below the poverty line. (From Nursing Home Compensation Report, 1996–97, Zabka and associates.)

While working conditions in many other industries have gradually improved over the past decade, nursing homes have become more dangerous to workers. Between 1983 and 1993, the injury rate for nursing home workers increased by 50%. Today, nursing home workers are injured at more than twice the rate of private sector workers generally.

Nursing homes also hold the distinction of being the most dangerous fast-growing industry in America. Of the 20 fastest-growing industries in the United States, nursing homes have the highest illness and injury rate and the largest number of lost workdays. The nursing home industry is expected to add three-quarters of a million new jobs between now and the year 2005. The industry is undergoing its most radical transformation in a quarter century.

While back injuries—widely agreed to be among the most serious and costly of injuries—account for a quarter of all injuries reported in the private sector, they account for almost half the injuries in nursing homes. The consequences of this epidemic of workplace injuries extend to all Americans: to workers, who suffer debilitating and disabling injuries; to their families, whose standard of living may be severely reduced by a breadwinner's disability; to nursing home employers, who pay ever-increasing costs for workers' compensation, for retraining and recruitment, and for decreased productivity; and to taxpayers, whose tax dollars finance Medicaid and Medicare, which finance most of the nursing home care in the United States. For too long, back injuries and other strains have been viewed as an inevitable part of nursing home work: "just part of the job." We can no longer afford to take that view. It is time for the nursing home industry, federal and state legislatures, government agencies, and nursing home workers to join together to defeat the terrible epidemic of crippling workplace injuries in nursing homes.

If we fail to act, we risk losing the committed, skilled workers who provide long-term care to millions of Americans. These workers know that the longer they continue to work in a nursing home, the more likely they will suffer a permanently disabling injury. Not only will the industry be unable to retain experienced workers, but it will also become increasingly difficult to recruit new staff members who cannot accept the high risk of injury.

This report, prepared by the SEIU reviews the extent and the character of injuries in nursing homes, describes their major causes, and illustrates their effects on workers and patients. It also provides estimates of the cost that this new epidemic of workplace injuries imposes on the public.

The bottom line is that working in a nursing home is so demanding that workers' bodies are literally breaking down; and worker injuries are a portent of an even broader crisis in nursing home care that affects residents as well as workers.

Dangers of Nursing Home Work

Contrary to popular belief, working in a nursing home is more dangerous than working in a coal mine, a steel mill, a warehouse, or a paper mill. In 1993, the injury and illness incidence rate[1] for nursing homes was an alarming 17.3 per 100 full-time workers, greater than coal mining (10.3), blast furnaces and steel mills (14.0), warehousing and trucking (11.9), and paper mills (8.7).

Nursing homes have become much more dangerous over the past decade. Between 1983 and 1993, the illness and injury rate increased from 11 to 17.3, an increase of 55%. The bitter irony, as Figure 26.3 makes clear, is that injury rates for nursing homes have been rising while rates for other high-risk industries such as manufacturing and mining have been falling.

These injuries are not minor. In the majority of cases, workers must take time away from work to recuperate. Over the last decade, the number of lost work days per 100 full-time nursing home workers (see Figure 26.4) has almost doubled, increasing from 98.2 to 186.9.

In 1993, nursing homes reported 220,800 injuries on the job. The nursing home industry ranked third in total injuries and illness, only exceeded by meat-products processing plants and manufacturing plants for motor vehicles and related equipment.

Another dubious distinction for the nursing home industry is that it is the most dangerous fast-growing industry in the United States. Of the 20 fastest-growing industries in the United States, nursing homes rank 18th if ranked by rate of growth. But if those 20 industries are ranked by injury incidence rates, nursing homes rank first.

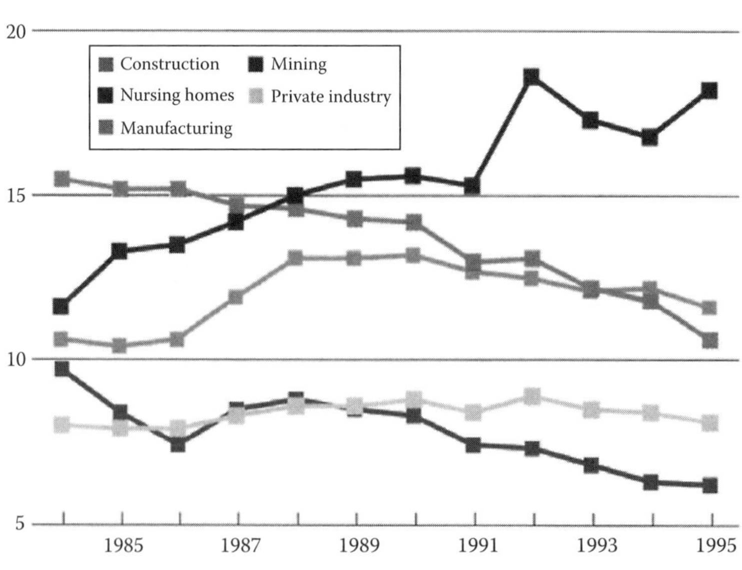

FIGURE 26.3 Nursing home injury rates have soared since 1983. Rates given are per 100 full-time workers in selected industries. (From Bureau of Labor Statistics.)

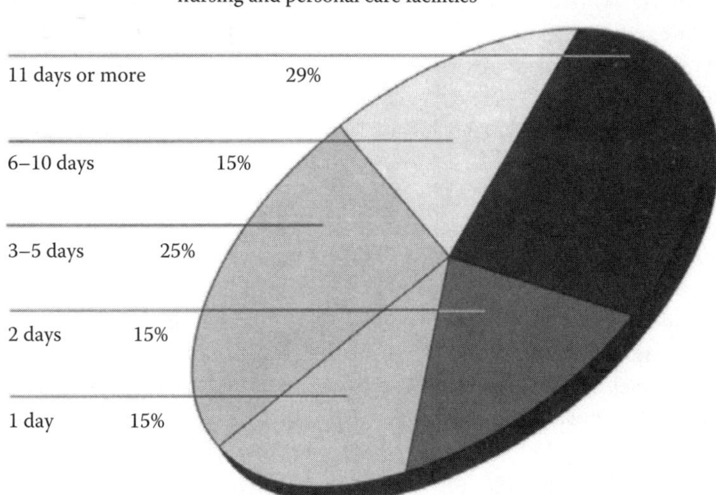

FIGURE 26.4 The rise of lost work days. Lost work days is calculated as per 100 full-time workers in selected industries. (From Bureau of Labor Statistics, 1994.)

Back Injuries Predominate

Nursing home workers are more likely to be injured on the job than other workers, and their injuries more often involve the back, widely agreed to be among the most serious and costly of injuries. A 1990 study by the U.S. Bureau of Labor Statistics (BLS) found that back injuries make up 25% of injuries reported in private industry, but account for 43% of all injuries in nursing homes.[2] The risk of back injury is even higher if only nurse aides, who provide the bulk of patient care, are considered. Data from the 1992 BLS Survey of Occupational Injuries and Illnesses found that injuries to the back and trunk accounted for over half of injuries to nurse aides (see Figure 26.5).[3]

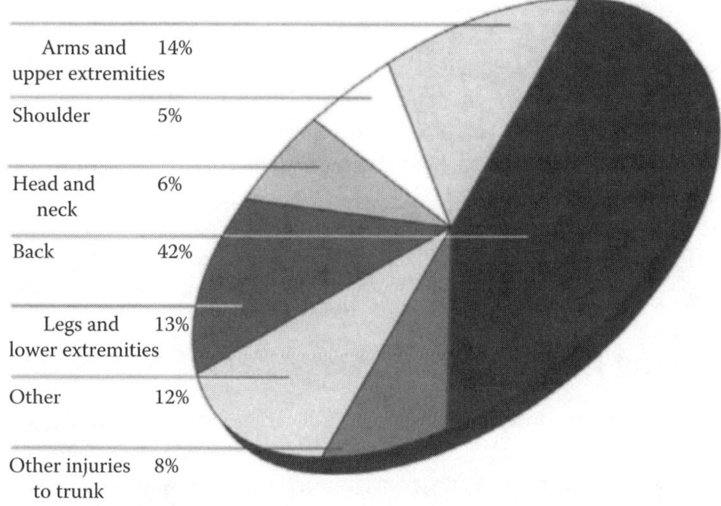

FIGURE 26.5 Back injuries are the most common injuries for nursing aides. (From Location of Injury for All, Nursing Home Workers, Bureau of Labor Statistics, 1994.)

Data from the 1992 survey for nurse aides found that nursing home workers are particularly at risk for strain and sprain injuries, with 80% of shoulder injuries and 85% of back injuries to nurse aides being categorized as strains and sprains.

Nursing home workers are also at greater risk for repetitive strain disorders, one of the most rapidly growing categories of occupational illness. A recent study in Washington state found that, statewide, the nursing home industry was responsible for the largest number of cases of carpal tunnel syndrome, a painful and crippling disorder of the hands and wrists. Nurse aides were more than twice as likely as the general public to suffer this disorder. Unlike acute injuries, repetitive strain disabilities are rarely reported as work related, so the actual rate is likely to be much higher.

Reporting patterns may also lead to underestimating the true risk. Strains are usually reported as work related only if they are acute. In nursing homes, stressful activities that do not necessarily cause acute lead to cumulative trauma that can end in debilitating injury or disease. Unless they can point to a single causal incident, workers are discouraged from reporting the injury or making workers' compensation claims. In many facilities, employer safety programs such as "safety bingo" give workers financial or other incentives not to report their injuries.

Why Is Nursing Home Work So Hazardous?

As far back as 1980, the injury and illness incidence rate for nursing homes was 10.7, higher than the rate of 8.7 for all private industry.

Much of the reason for this is the physically demanding nature of nursing home work. Nurse aides lift and turn patients, help them in and out of baths, make beds, and take residents to and from the toilet. Housekeepers vacuum, dust, mop, clean walls and windows, and collect trash. Dietary workers carry heavy boxes and pans, clean their work areas, and cut and prepare large amounts of food. Laundry workers lift heavy linens, both wet and dry. They load and unload large washing machines and dryers. All nursing home workers use large and small muscles over and over again, often to move bodies or objects that weigh more than they do.

The more patient handling a worker is required to do, the greater the risk. Unpublished data from OSHA show that, for nurse aides in nursing homes, 81% of shoulder injuries and 78% of back injuries were caused by "health care patient or resident of health care facility." Over half of all back injuries (and just under half of shoulder injuries) to nurse aides in nursing homes result from "overexertion in lifting," which often occurs while handling a patient.[4]

Nurse aides, who provide most of the direct patient care in nursing homes, are particularly at risk. OSHA inspections of a number of Beverly Enterprises nursing homes found that in 1992 short-staffing was forcing nurse aides to perform many patient lifts and transfers alone. On many shifts, a nursing assistant would perform up to 40 resident lifts and transfers of residents weighing up to 260 lb without the help of another employee or mechanical hoist. In short, many nursing aides were lifting over *10,000* lb *per shift*.[5]

These amounts far exceed recommended safety levels. The International Labor Organization and the National Institute for Occupational Safety and Health (NIOSH) recommend that manual lifting be limited to about 50 lb twice an hour, for objects with a center of gravity of about 15 in. from the lifter.[6]

Risks to Nursing Home Workers Have Been Increasing

While nursing home work has always been hazardous, there is evidence that conditions for nursing home workers have become much more hazardous in recent years. A number of factors have converged to make nursing home work much more dangerous to workers than it was 10 or 15 years ago.

The most important factor has been a reduction in the length of hospital stays, especially for the Medicare population. Many analysts trace the decline in length of stay to the federal government's introduction of prospective payment into Medicare in 1983, although other factors, such as improvements in medical technology, have also played a role. The average hospital stay for an elderly patient in 1991 was about 1.8 days shorter than in 1981, an 18% reduction. Average stay for the adult population as a whole fell by 10% over the same period.

With patients leaving the hospital earlier, the burden of providing "subacute" and specialty care is increasingly falling on nursing home workers. According to the American Health Care Association (the nursing home industry's trade association), there are currently between 10,000 and 15,000 subacute beds and more than 60,000 beds in specialty care units nationally.[7] Modern Healthcare's 1993 survey of nursing homes found that 56% report subacute units.[8] Clinically, these changes have meant that residents tend to need more assistance with activities of daily living. An increasing number of residents also need rehabilitative services, oncology, wound care, and infusion services. Nursing home workers also have to treat patients with more complex medical conditions such as Alzheimer's disease and AIDS.

Understaffing Puts Residents and Workers at Risk

Nursing home operators have failed to address either the problem of unsafe working condition or the dramatic worsening of those conditions over the past few years. Staffing levels, for example, have not kept up with the changes in patient mix.

An analysis of Medicare and Medicaid cost report data by Health Care Investment Analysis (HCIA) shows full-time employees per average daily census barely increasing from 0.75 in 1989 to 0.76 in 1991.[9] Marion Merrell Dow, in its annual survey of 200 nursing homes, showed that over a 3 year period from 1990 to 1993, the number of nursing home employees per setup bed only increased by 5%. For-profit institutions showed an even smaller rate of increase 2.5% from 1991 to 1993.[10]

In 1992, SEIU surveyed 10,000 nurses across the United States. Of those employed in nursing homes, 86% reported that current staffing levels were inadequate and 91% reported that their workloads had increased over the past year. Nursing home nurses felt that they should be spending twice as much time with their patients as they were able under current conditions.[11]

When working in a chronically understaffed facility, the pressure to pull one's own weight is enormous, even when workers are not physically up to the task because of injury or fatigue. Workers often are forced to lift and turn patients by themselves because no help is available and many are forced to make difficult choices between remaining with patients who need assistance or leaving to help their coworkers when called. Many workers end up injuring themselves further by doing additional lifts in order to avoid overburdening injured coworkers. More experienced workers are sometimes injured when trying to help orient new employees.

Employers Favor Gimmicks Rather than Real Solutions

Rather than hiring additional staff or redesigning work to eliminate hazards, nursing home employers have tended to favor gimmicks that do little to prevent injuries. In 1987, for example, Beverly Enterprises distributed a training program to all its facilities entitled Lift With Care, designed to teach "proper lifting techniques" to nursing home staff.[12] While the concept of good body mechanics is the battle cry of trainers, such techniques—back straight, legs bent, and objects between legs or close to body—are not easily applied to nursing home work.

In many cases, for example, a single lift may require leaning over a bed to lift a resident, holding him while turning 90° and lowering him into a chair. Patients may have to be lifted over chair arms, or from a low chair to a higher bed, or up from the floor. Transfers often involve contortion of the trunk, uneven weight distribution, and exertion from awkward postures. The additional stress of sustaining awkward positions, as when stooping to feed or pulling a wheelchair while ambulating a resident, further stretches and weakens muscles.

OSHA inspections of nursing homes have found that safety training programs for nurse aides are not effective. Although the Lift with Care program has several component parts, in many cases only the

25 min video is used. Nursing aides generally do not use the lifting techniques outlined in the program, either because they require more staff than are on hand for a given lift or because they have not received adequate instruction in the techniques.[13]

Another recent innovation is to require workers to wear girdle-like devices called "back support systems" or "back belts" designed to keep workers' backs rigid. Although they have not been comprehensively tested in nursing homes, there is no evidence to date that such devices are effective in preventing injuries. A 2 year study by the Back Belt Working Group of NIOSH concluded that "back belts do not mitigate the hazards to workers posed by repeated lifting, pushing, pulling, twisting or bending." The Working Group "does not recommend the use of back belts to prevent injuries among uninjured workers, and does not consider back belts to be personal protective equipment."[14]

More significant than what nursing homes do is what they do not do. Nursing aides report that the nursing care plans for each patient usually fail to provide written directions so as to the number of staff needed to transfer the patient or the safest way to do so, and supervisors rarely provide this information.[15]

Nursing home operators have also resisted investing in mechanical lifting equipment that could reduce the hazard to nursing home workers. Although advanced equipment is available, most nursing homes continue to use outmoded lifting equipment. In many cases, the lifts are used for weighing residents and are unavailable for use in transfers. Many nurses' aides report, however, that they usually do not use lifting equipment, even when it is available, because it takes too much time, it is not safe, and they do not feel comfortable using it.[16] These problems have been confirmed by a number of academic studies.[17]

The bottom line is that the nursing home industry has not committed itself to a comprehensive program that would prevent workplace injuries. Many employers continue to deny the extent of the problem, contesting OSHA citations and refusing to negotiate legitimate solutions. External pressure from organized labor and government agencies will undoubtedly be needed to force the industry to make significant changes in the way it treats workplace hazards and injuries.

> One of the most egregious employer initiatives is so-called "safety bingo." "The company constantly claims their compensation insurance is too high. But instead of putting on more staff, they introduced "safety bingo." The grand prize was a 19 in. color remote-controlled television set. They drew a bingo number whenever a week passed without a reported injury. If an injury was reported, everyone would have to throw their bingo card away and start again. Some people were so committed to winning the television that they wouldn't report injuries."[18]
>
> **—Beverly Enterprises Nursing Home Worker, Pennsylvania**

Safety Saves Money

Employers in the nursing home industry often resist hiring additional staff or purchasing safety equipment because of the cost. But running an unsafe workplace does not save money in the long run. It actually increases costs in a variety of hidden as well as more obvious ways.

- *Workers' compensation*—The staggering workers' compensation insurance rates paid by nursing home employers are a reflection of the magnitude and severity of injuries and illnesses in nursing homes. There are a number of states where the insurance rate for nursing homes is twice or even three times the rate for an average private sector employer. The nursing home industry as a whole paid close to $1 billion in workers' compensation insurance costs in 1994.
- *Injuries drive workers from the profession*—Turnover in the nursing home industry is very high—in excess of 100% at some facilities. The injury rates, low pay, and stress due to understaffing discourage workers from remaining in the industry for very long. New workers are shocked and dismayed at the frenetic pace they are expected to maintain. More experienced workers burn out as they find themselves unable to provide the kind of care they believe nursing home residents need. Constant turnover forces employers to invest resources in recruitment and training of new workers.

- *Reduced tax base*—People on workers' compensation do not pay income taxes. They have less money available to spend on taxable goods. They may require vocational retraining through state-funded programs as well as other forms of assistance.
- *Health insurance redlining shifts costs to other industries and workers*—Because nursing homes are such hazardous places to work, many insurance companies will not sell health insurance to nursing home operators. A recent University of Michigan survey of insurance companies found that two-thirds of those surveyed listed nursing homes among a group of employers ineligible for coverage under any circumstances and an additional quarter of those surveyed would only offer coverage with certain restrictions.[19] When nursing home workers are unable to obtain health insurance at work, the cost of their medical care is shifted to other employers and workers.
- *Injuries to patients drive up the cost of care*—Workers are not the only ones put at risk by unsafe working conditions. Patients also suffer. The industry spends $1.2 billion to heal preventable decubitus ulcers (bedsores) caused by lack of nutritional hydration, mobility, and cleanliness, and $4.3 billion on incontinent care because residents are not toileted frequently enough.[20]

Taxpayers Ultimately Pay the Cost of Unsafe Working Conditions

Because public programs pay for most nursing home care, it is taxpayers who ultimately bear the burden of unsafe working conditions in the nursing home industry. Taxpayers pay for over 75% of the nursing home care provided in the United States, primarily through the Medicaid program (see Figure 26.6).

Of the $1 billion paid out by nursing home operators in workers' compensation insurance premiums, for example, well over three-quarters of that amount was paid by taxpayers. Similarly, the industry was able to shift to taxpayers over 75% of the $1.2 billion spent on healing preventable bedsores, the $2.6 billion spent on treating residents injured in falls, and the $4.3 billion spent on treating conditions associated with incontinence.[21]

Safety Can Save Money

The above data suggest that employers have significantly underestimated the costs associated with operating an unsafe workplace. The irony is that they also overestimate the cost of having a safe workplace.

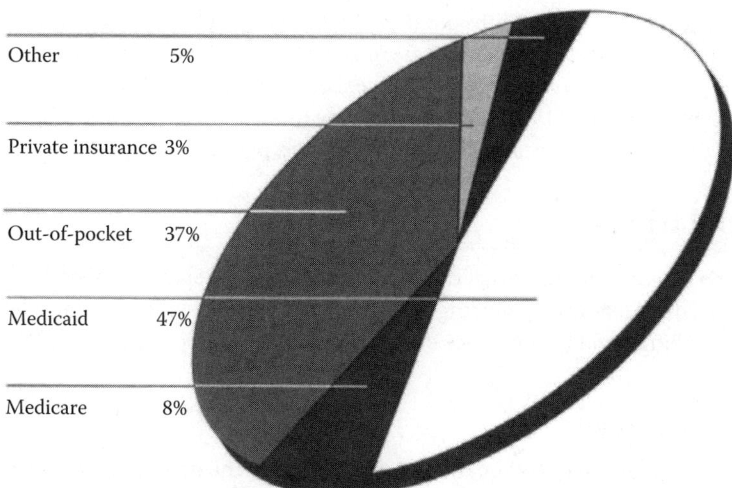

FIGURE 26.6 Workers' compensation costs (as percent of payroll). (From Health Care Financing Administration, 1996.)

A typical nursing home with 100 workers pays between $50,000 and $100,000 a year in workers' compensation insurance payments. This money could pay for one or two extra nursing assistants per shift, which could go a long way toward reducing injuries due to understaffing. These funds could also be used to invest in technology that would make lifting and transferring patients less hazardous. One year's worth of workers' compensation payments for a typical nursing home would pay for 10–15 mechanical devices, which would allow staff to lift and transfer patients without having to lift too much weight.[22] A wide range of affordable examples are available:

- Chairs are available that can double as a toilet and a shower stool to eliminate unnecessary transfers. Currently, patients are often lifted from bed to wheelchair, wheelchair to toilet, toilet to wheelchair, wheelchair to shower, shower to wheelchair, and wheelchair back to bed. Instead, patients could be lifted from their bed to a shower chair (where toileting and showering are done) and from there back to bed, eliminating a number of lifts.
- Portable or ceiling-mounted patient hoists can substitute a mechanical device for a human lifter. Most homes possess only antiquated lifting equipment; many safer and more versatile varieties are now available.
- Hoist belts or lifting belts can allow workers to lift patients up with less strain on the worker's back by placing a belt around the hips of the patient.
- Ambulation belts are available which help stabilize patients while they are walking, reducing the risk of falls.
- Friction-reducing sliding boards can be placed beneath patients so they can slide instead of being lifted into a wheelchair.
- Other patient-handling devices are available, including roller boards, handles on sheets, turntables for pivot transfers, bath boards, and transfer devices for many different situations.

A number of academic studies have shown that these types of interventions have the potential to reduce significantly the number of injuries to nursing home workers. A study by Bernice D. Owen and Arun Garg, for example, examined the impact of a number of ergonomic interventions on injury rates in two nursing home units with a total of 140 beds and 57 nursing assistants. The program involved determining which patient care tasks involved the highest risk of injury, modifying the jobs, and introducing new patient transferring devices in order to reduce the risk of injury. In the 12 month period following the intervention, the back-injury rate was cut nearly in half. Of the remaining back injuries, over 80% were not related to the stressful tasks identified in the study. Lost and restricted workdays also declined, dropping to zero during the last 4 months of the post-intervention phase of the study.[23]

For nursing home operators to claim that these interventions are "too expensive" is absurd. While the industry continues to plead empty pockets, its balance sheets have never looked better. From 1990 to 1992, the combined rate of profit for all U.S. nursing homes rose from 28% to 37.2%, a 30% increase. The largest nursing home chain in the United States, Beverly Enterprises, posted a 1994 profit of $74 million.[24]

Recommendations

Our nation can no longer afford the staggering costs imposed by the epidemic of back injuries and other cumulative trauma disorders among nursing home workers. The following steps must be taken if we are to reverse the disturbing trend in workplace injuries in nursing homes.

Improve Staffing Standards

The cornerstone of worker safety and quality resident care is adequate staffing. While many states and the federal government have established staffing standards, employers are exploiting loopholes in the regulations. In some states, employers are able to include administrative staff when calculating whether they met the required staffing levels. SEIU recommends

- Enforceable staffing ratios linked to the acuity of residents. Such ratios should be enforced on a floor or unit level to prevent facilities from understaffing in particular units while still meeting a facility-wide target.
- Specific minimum staffing ratios for nurse aides. Since nurse aides provide most patient care, it is important that specific standards be set for them, independent of case mix. An expert panel of the National Committee to Preserve Social Security and Medicare has proposed nurse aid-to-patient ratios of 1:8 (days), 1:10 (evenings), and 1:15 (nights). While SEIU remains concerned that the ratio of workers to residents may not be sufficient, they are lower (i.e., better) than we observe in many facilities.
- HCFA should study and monitor the relationship of staffing to acuity levels, resident outcomes, and worker and patient injury rates on an ongoing basis.

Establish an Ergonomics Standard

OSHA is in the process of developing an ergonomics standard for private sector employers. An ergonomics standard would provide clear guidance to nursing home employers on how to prevent disabling strain and sprain injuries. Such a standard should require that employers develop an injury prevention program that includes the following elements:

- A worksite analysis to identify risk factors that cause injuries (e.g., certain types of patient care tasks, such as lifts and transfers, that pose a particularly high risk).
- The evaluation and implementation of feasible methods to prevent and reduce those risk factors, including the use of mechanical lifting equipment, improved staffing levels, and better communication to staff of a resident's need for assistance in activities of daily living.
- A program of early treatment and ongoing medical management for workers who are injured or who report early warning signs of strains and sprain injuries.
- Education and training on how to identify and prevent injuries caused by lifting and repetitive tasks.

Change OSHA's Enforcement Focus

Since its inception, the focus of OSHA's enforcement activity has been on industries with widely recognized hazards, especially in the manufacturing sector. With well over 70% of workers now employed in the service sector, there is a need for OSHA to focus on the hazards that are endemic to this sector. This will require new education and training for OSHA inspectors and the investment of additional resources in service sector enforcement. Stepping up enforcement in the nursing home industry (by far the most hazardous service sector industry) would allow OSHA to improve its ability to effectively enforce workplace safety laws in the service sector.

References

1. In most cases, incidence rates are the number of illnesses or injuries (or both) per 100 full-time equivalent workers.
2. Personick ME. Nursing home aides experience increase in serious injuries. *Monthly Labor Rev.*, 113:30–37, 1990.
3. Bureau of Labor Statistics. Unpublished data from the 1992 Survey of Occupational Safety and Health Administration.
4. Unpublished data. Occupational Safety and Health Administration.
5. Occupational Safety and Health Administration. Pretrial Brief, Robert B. Reich, Secretary of Labor, U.S. Department of Labor v. Beverly Enterprises Inc., d/b/a Blue Ridge Haven, Convalescent Center West, Beverly Manor of Monroeville, Meyersdale Manor, Richard Manor, Carpenter Care Center and its successors.

6. DeClercq NG and Lund J. *NIOSH Lifting Formula Changes Scope to Calculate Maximum Weight Limits*, Occupational Health and Safety, February, 1993.
7. American Health Care Association, *Facts and Trends: The Nursing Facility Sourcebook*. AHCA, Washington, 1994.
8. O'Connor J. Subacute care thrust into the limelight. *McKnight's Long Term Care News*, June 1994.
9. Health Care Investment Analysts, and Arthur Andersen and Company. *The Guide to the Nursing Home Industry*, 1993.
10. *Managed Care Digest: Long Term Care Edition*. Marion Merrell Dow, Kansas City, MO, 1992 and 1994 editions.
11. The National Nurse Survey. Service Employees International Union, 1992.
12. Reich v. Beverly Enterprises, op. cit.
13. These examples are taken from inspection reports from several Pennsylvania nursing homes submitted by Dr. Bernice Owen to the Occupational Safety and Health Administration in 1991.
14. Workplace use of back belts, U.S. Department of Health and Human Services, National Institute for Occupational Safety and Health, July 1994.
15. These examples are taken from inspection reports from several Pennsylvania nursing homes submitted by Dr. Bernice Owen to the Occupational Safety and Health Administration in 1991.
16. Ibid.
17. See particularly Bell F. *Lifting Devices in Hospitals*, Groom Helm, London, 1984; Owen BD, Patient handling devices: An ergonomic approach to lifting patients. In: VF Aghazadeh (ed.), *Trends in Ergonomics/ Human Factors*. Elsevier, North-Holland, 1988.
18. Testimony of Aimee Miller. Certified Nurse Aide, Member of Local 1199P, Service Employees International Union before the Pennsylvania House of Representatives, Committee on Labor Relations, November 19, 1992.
19. Small business and health care reform, University of Michigan School of Public Health, 1994.
20. Nursing home resident rights: Has the administration set a land mine for the landmark OBA 1987 nursing home reform law, A Staff Report by the Subcommittee on Aging, Senate Committee on Labor and Human Resources, June 13, 1991.
21. Ibid.
22. Cost of abatements from Reich v. Beverly Enterprises, op. cit.
23. Garg A and Owen B. Reducing back stress to nursing personnel: An ergonomic intervention in a nursing home. *Ergonomics*, 35:11, 1992. See also Owen B and Garg A. Assistive devices for use with patient handling tasks. In: *Advances in Industrial Ergonomics and Safety II*. Das B (ed.), Taylor & Francis, London, U.K., 1990; Jensen RC and Hodous TK. *A Guide for Preventing Back Injuries among Nursing Assistants Working in Nursing Homes*, U.S. Department of Health and Human Services, National Institute for Occupational Safety and Health, 1991.
24. Health Care Investment Analysts. *Guide to the Nursing Home Industry*, 1994.

27

Laboratory Safety

Introduction 27-1
Organization for Laboratory Safety 27-2
Safety Policy • Responsibilities for Safety
Laboratory Safety Information Systems 27-3
Labeling • Material Safety Data Sheets • Laboratory Safety Training Programs
Controlling Chemical Hazards 27-5
Storage of Chemicals • Peroxidizable Chemicals • Explosive Chemicals • Incompatible Chemicals • Management of Hazardous Research Chemicals
Controlling Biological Hazards 27-7
Design of Biohazard Facilities • Biological Safety Cabinets • Animal Care and Handling
Controlling Radiation Hazards 27-8
Controlling Mechanical Hazards 27-9
Compressed Gases • Glassware Safety • Centrifuges
Personal Protective Equipment and Safety Devices 27-10
Protective Clothing • Protective Gloves • Eye Protection • Respiratory Protection • Emergency Wash Devices
Laboratory Design 27-12
Laboratory Ventilation
Spill Response Contingencies 27-14
Monitoring Hazards in the Laboratory 27-15
Environmental Monitoring • Monitoring Processes • Medical Surveillance
Suggested Readings 27-16

Wayne Wood
McGill University

Introduction

Assessment of risks in hospital laboratories is particularly difficult when one considers the range of possible hazards: fires; explosions; inhalation of toxic gases, aerosols, and vapors; splashes of corrosive chemicals on the skin or in the eyes; thermal burns; cryogenic burns; and accidental injections, falls, and cuts. Of these, the most difficult to assess are exposures to chemicals, radiation, or infectious agents. Exposures in the laboratory are typically short in duration, intermittent, and involve small quantities (relative to an industrial setting) of mixtures of agents. Little is known about the health effects of such an exposure profile.

While it may be relatively easy to record laboratory accidents resulting in bodily harm such as burns, chemical splashes, and cuts, little is known about the incidence of occupational diseases. One of the reasons for this is the difficulty in correlating diseases with occupational exposures, especially in the laboratory setting. For many agents used in the laboratory, there is a lack of toxicological data and no established permissible exposure levels on which one may base a risk assessment. To complicate matters further, available toxicological information is usually based on exposures to one agent and does not take multiple exposures into consideration, but the combined health effects may surpass the total effects of the individual agents (synergy).

Due to these difficulties in quantifying risks, an effective approach to laboratory safety involves development of universal control measures. The term "universal control" refers to the use of measures such as ventilation, personal protective equipment, substitution, and storage, handling, and disposal procedures in order to minimize or even eliminate exposures, irrespective of the agents involved. Undoubtedly, there are exceptionally dangerous agents that dictate the most rigid of controls, but for the most part, universal controls can be applied to almost any laboratory in a health care facility. The main thrust of this chapter is on controls, with minor emphasis on principles of recognition and evaluation of risks. Parts have been written in a format similar to a manual and may be used as a base for establishing a safety manual for most clinical or research laboratories.

This chapter is not intended to cover all facets of laboratory safety—the focus is on aspects that, in the opinion of this author, are characterized by neglect and pose avoidable dangers to those working in hospital laboratories. For topics that are not included in this chapter, readers are referred to the list of suggested readings at the end.

In keeping with the international flavor of this text, minimal reference is made to specific occupational health and safety regulations. For this type of information, readers are referred to their local government occupational health and safety regulatory agencies. Instead of citing specific regulations, references are made to advisory bodies whose guidelines are recognized internationally and are frequently incorporated into health and safety legislation.

Organization for Laboratory Safety

Safety Policy

The first requirement for an effective laboratory safety program is a clear statement, signed by the chief executive officer, stating the institution's mission, baseline standards, and roles and responsibilities at all levels within the organization. Individuals involved in the safety program need clearly defined mandates and must be given full support of the institution in carrying out their functions.

Responsibilities for Safety

Administration

Administration's key responsibilities lie in the provision of a safe physical work environment, that is, ensuring that the facilities accommodate the nature of the work being performed and in assuming institutional financial and legal responsibilities for the general pattern of safety practices.

Department and Unit Heads

Department and unit heads are responsible for overseeing the application of health and safety programs and ensuring that supervisory personnel reporting to them assume their responsibilities for adhering to safety policy. They are also responsible for reporting any problems that cannot be solved at the departmental level.

Supervisors and Principal Investigators

Supervisors and principal investigators are responsible for establishing safety and emergency procedures for all areas under their direction and for ensuring that employees are well informed of workplace hazards, precautions necessary for the safe performance of work, use of safety and emergency equipment, and emergency procedures. They are also responsible for advising their department heads of problems that cannot be solved without further intervention.

Laboratory Workers

Laboratory workers are responsible for adhering to all safety procedures and for conducting their work in such a manner so as not to jeopardize their own or others' safety. They are responsible for correcting hazards or reporting to their supervisors situations that they are unable to remedy.

Safety Committee

An active safety committee is the backbone of a successful safety program. Mandates vary from institution to institution, and in some jurisdictions the mandate and the committee makeup are specified by occupational health and safety regulations. Benefits of having such committees include making the best use of internal expertise, maximizing participation at all levels in the organization, providing a forum for open discussions of health and safety matters, promotion of health and safety, education of the committee and by the committee, and development of policy and procedures. Key functions of the laboratory safety committee in the area of laboratory safety are in ensuring that inspections are performed on a regular basis, overseeing the training program, and establishing baseline standards and policies for laboratory safety.

Industrial Hygienists and Safety Managers

Mandates of industrial hygienists and safety managers may vary from institution to institution, but one role most have in common is as internal advisor. In addition to acting as educators and internal consultants, they should be actively involved in identifying and monitoring hazards; formulating remedial measures; specifying facilities requirements; formulating institutional safety policies and procedures; receiving and investigating reports of accidents, dangerous incidents, and occupational diseases; liaising with external resources and regulatory agencies; and keeping the institution abreast of developments in the area of health and safety.

Traditionally, many industrial hygienists and safety managers have played the role of enforcers; however, the notion that they can be everywhere at once to act as watchdogs is unrealistic. Today's trends are toward their adopting the roles of advisors, which is more consistent with the belief that responsibility for workplace safety is to be shared with workers and management. Industrial hygienists and safety managers should be expected to intervene primarily when this system shows signs of failure.

Laboratory Safety Information Systems

Labeling

In most locales, vendors of hazardous agents are obliged by law to display identification and precautionary information on products' containers. These legal requirements usually specify the format of the precautionary information, which may be in the form of symbols and written statements. It should not be assumed that all workers are capable of understanding this information, because some may be faced with fluency or literacy problems. Therefore, instruction in how to interpret precautionary information must be an integral part of the laboratory safety training program. Workers with fluency or literacy problems need help to develop an understanding of the meanings of the various precautionary symbols.

When materials are transferred from vendors' containers, identification and precautionary information must be displayed on the new containers. Failure to do so may not only be in violation of labeling regulations, but may also give rise to waste disposal problems if the material cannot be properly identified.

After all labeling requirements are met, the real challenge begins—getting people to read the labels. Grave errors can occur from failure to read the labels. Many compounds have names that are similar, but differ dramatically in physical, chemical, and toxicological properties. Users of laboratory reagents should be trained to read the label once when taking a container from its storage shelf or compartment, a second time before opening the container, and a third time before removing any of the contents.

Material Safety Data Sheets

As with precautionary labeling in most jurisdictions, there are legal requirements for the provision, format, and use of material safety data sheets (MSDSs). Simply making them available is not enough—it is imperative that safety training programs include instruction in the meaning of terms and symbols displayed in MSDSs. It is also crucial that collections of MSDSs be kept up-to-date at all times and that they be made readily accessible.

Laboratory Safety Training Programs

All too often laboratory accidents are the result, albeit direct or indirect, of lack of awareness of workplace hazards and their appropriate controls. For this reason, safety training is the cornerstone of a successful safety program. The safety committee and industrial hygienist or safety manager should play an integral role in establishing institutional standards for safety training.

Safety training is necessary at two levels: departmental and institutional. At the departmental level, all employees must be given safety training specific to their work. Instructions must include information on the hazardous properties of the agents being used; necessary precautions to be observed; and procedures to be followed for spills of hazardous agents, fires, and medical emergencies. Employees should be shown the location of all safety equipment (such as eyewashes, emergency showers, first-aid kits, spill control kits, and fire fighting equipment) and instructed in their operation.

Individual supervisors must assume responsibility for ensuring all their employees are provided with adequate safety training. This implies that supervisors are themselves adequately trained to assume this responsibility but, more often than not, they have never been provided formal training in laboratory safety. Therefore, supervisors must be provided supplementary training in the principles of laboratory safety and be made aware of applicable regulations and codes of practice to enable them to oversee the training of their employees.

Training to be provided institution-wide should cover the universal controls applicable to most laboratories, for example, principles of fire safety, radiation protection, safe handling and storage of chemical agents, biosafety, and other aspects of laboratory safety. Key players in safety training should be the safety manager, industrial hygienist, health physicist, fire marshal, and safety committees, but the success of the program does not have to depend entirely on these people. Much of this training may be accomplished through the use of audiovisual aids, which may take the form of videos, interactive videos, films, slide shows, computer programs, and posters.

Safety training is a dynamic process, not be considered as a one-time affair. The program must be designed to keep up with turnover of employees, changes in procedures, new legislation, and changes in knowledge about workplace hazards. These factors are to be used when determining the frequency of training sessions and the target audience. New employees must be given an orientation session in safety before commencing any work in the laboratory. All employees (including supervisors) require regular training sessions in universal controls and institutional and legislative regulations regarding laboratory safety. At the departmental level, supervisors must inform those under their jurisdiction of changes in procedures and associated precautionary implications. To prevent employees from missing out on essential training, up-to-date records must be maintained. These records should indicate the content of training sessions, names of the participants, dates of training sessions, and projected dates for refresher courses.

Written communiqués are also invaluable sources of safety information. Laboratory safety manuals, newsletters, and memoranda help maintain a steady flow of safety information. It is important to avoid creating a situation of information overload whereby the volumes of reading material become tiresome. Written communiqués should be kept as a brief as possible without simplifying the messages to the point where the meanings are altered. To accommodate the enthusiasts who desire more information, a collection of more detailed safety references should be available.

Controlling Chemical Hazards

Storage of Chemicals

Temperature control, ignition control, ventilation, segregation, identification, and isolation are universal controls that will serve to greatly reduce the risk of fire, explosions, or other accidents associated with storage of laboratory reagents. These risks can be minimized by observing the following guidelines:

- Minimize quantities of chemicals kept on hand in the laboratory. Table 27.1 lists the permissible container sizes for flammable liquids.
- Isolate flammable liquids from heat sources such as direct sunlight, ovens, burners, and hotplates.
- Refrigerate flammable liquids when possible. By doing so, the evaporation of vapors will be minimized, thus reducing the risk of fire or explosion. Refrigeration devices (including freezers) used for storage of flammable liquids must be specially constructed to eliminate spark sources (from thermostats or light fixtures) within the storage compartments. Such devices should meet National Fire Protection Association (NFPA) requirements as specified in codes 45 and 56C.
- When transferring flammable liquids to or from metal drums, attach a grounding wire to the drum and all accessories used for the transfer procedure.
- Do not store flammable solvents out in the open such as on bench tops or floors.
- Use NFPA-approved flammable liquid storage cabinets and containers.
- Store incompatible reagents in separate compartments. For example, store flammable liquids separate from oxidizing agents such as nitric acid, organic peroxides, chromic acid, and permanganates.
- Transfer flammable liquids in well-ventilated areas to avoid having the vapors build up to an explosive concentration.
- Store large containers on lower shelves. Do not place any at or above eye level. Whenever possible, place bottles in a tray to contain leaks.

TABLE 27.1 Maximum Allowable Container Sizes for Flammable and Combustible Liquids in Laboratories

	Flammable Liquids						Combustible Liquids			
	Class IA[a]		Class IB[b]		Class IC[c]		Class II[d]		Class IIIA[e]	
Type of Container	Liters	Gal	Liters	Gal	Liters	Gal	Liters	Gal	Liters	Gal
Glass	0.5	0.12	2	0.25	4	1	4	1	20	5
Metal (other than DOT approved drums) or approved plastic	4	1	20	5	20	5	20	5	20	5
Safety cans	7.5	2	20	5	20	5	20	5	20	5

Source: Modified from National Fire Prevention Association Code 30.

[a] Liquids having flash points below 73°F (22.8°C) and boiling points below 100°F (37.8°C).

[b] Liquids having flash points below 73°F (22.8°C) and boiling points at or above 100°F (37.8°C).

[c] Liquids having flash points at or above 73°F (22.8°C) and boiling points below 100°F (37.8°C).

[d] Liquids having flash points at or above 73°F (37.8°C) and boiling points below 140°F (60°C).

[e] Liquids having flash points at or above 140°F (60°C) and below 200°F (93.4°C).

Peroxidizable Chemicals

Many chemicals, most notably ethers, are susceptible to decomposition, resulting in explosive products. Ethers, liquid paraffins, and olefins form peroxides on exposure to air and light. Because most of these products have been packaged in an air atmosphere, peroxides can form even if the containers have not been opened. Though there are tests available to detect the presence of peroxides, it is most prudent to adhere to the following guidelines:

- Unopened containers of ethers should be discarded after 1 year.
- Containers of ethers should be discarded within 6 months of opening.
- Ethers beyond their expiry date should not be handled. Removal should be conducted by hazardous materials technicians trained to stabilize and remove the material.
- The following are just a few examples of compounds prone to peroxide formation:
  - Diethyl ether (ether)
  - Dioxane
  - Isopropyl ether
  - Tetrahydrofuran
  - The label and MSDS will also indicate whether a chemical is unstable.

Explosive Chemicals

Many chemicals are susceptible to rapid decomposition or explosion when subject to forces such as being struck, vibrated, agitated, or heated. Some become increasingly shock-sensitive with age. A typical example of this is picric acid, which if allowed to become dry, is extremely explosive. Typical atomic groupings that are associated with the possibility of explosion are nitrates, azides, nitrites, chlorates, nitros, diazos, nitrosos, diazoniums, fulminates, perchlorates, peroxides, hydroperoxides, and picrates. For a more complete list of explosive compounds and atomic groupings, readers are advised to refer to texts on hazardous chemical reactions. Whenever working with compounds that are potentially explosive, the following guidelines should be observed:

- Refer to the label and MSDS to determine if a chemical is explosive.
- Write the date received and opened on all containers of explosive or shock-sensitive chemicals.
- Discard the opened containers after 6 months and closed containers after 1 year, unless inhibitors have been added by the manufacturer.
- Work with small quantities.
- Perform the experiments behind a face shield.

Some materials are not intrinsically explosive but, if used under certain conditions, may give rise to explosive products. A classic example of this is perchloric acid, a powerful oxidizing agent that may react explosively with reducing agents and organic matter. Perchloric acid vapors tend to condense on the inside of fume hoods and the inner linings of ducts, eventually forming perchlorate crystals, which are shock-sensitive explosives. Many accidents, some of them fatal, involving the use of perchloric acid have been recorded. Perchloric acid should only be used in a water washdown hood of noncombustible construction, and when wet digestions are performed, organic matter should first be treated with nitric acid to destroy easily oxidizable matter.

Incompatible Chemicals

Those trained in the basic principles of chemistry should be able to predict some of the more obvious incompatible combinations of chemicals, such as oxidizing reagents with reducing reagents or strong oxidizers with combustible or flammable materials; however, certain hazardous combinations can occur even between chemicals of the same classifications. An example of this is the reaction between acetic acid and chromic acid. Therefore, it is not sufficient to rely solely on one's knowledge of chemistry when looking at chemical compatibility. Readers are advised to refer to reagents' labels, MSDSs, and lists of hazardous chemical reactions to consider the possibility of incompatible combinations.

Management of Hazardous Research Chemicals

It is not unusual for researchers to be using chemical agents for which there are little or no toxicological data, for example, cytotoxic or antineoplastic drugs. Some of these agents will dictate safety measures beyond the universal controls outlined in this chapter. Determination of the handling and storage procedures should be determined jointly by the principal investigator and the industrial hygienist and be based on the philosophy that if little is known of the agent's toxicological properties, it should be handled as if it were highly toxic (see also Chapter 5).

Proper management of hazardous chemicals must involve inventory controls and simple housekeeping. Unfortunately, it has been observed that these areas are frequency neglected, resulting in the presence of unnecessary hazards. For example, if labels are not checked for expiry dates, unstable chemicals may be present in the laboratory unbeknownst to its occupants. Inventories of laboratory reagents should be conducted at least annually, in an effort to discard all reagents beyond their expiry dates, along with unwanted, contaminated, and surplus chemicals. This serves to minimize quantities kept on hand, provide the opportunity to evaluate the storage practices, and ensure that chemicals are properly identified, segregated, and contained.

A hazardous waste disposal service is an essential element of management of hazardous research chemicals for reasons of safety and compliance with hazardous waste regulations. Practices that were acceptable in the past, such as using drains as a means of disposal, are no longer tolerated by environmental protection agencies or by the public. Aside from environmental considerations, pouring chemical agents down drains can result in adverse reactions if incompatible mixtures are created. This can occur as a result of the discharge of chemicals into the drains of one laboratory or into common drains and traps shared by different laboratories. Another negative consequence of improper disposal is the release of vapors from volatile materials that may migrate along drain pipes and appear in different locations of the building. These vapors can give rise to transient odors and also can pose an explosion hazard in the case of flammable solvents that are immiscible with water. Although in certain circumstances disposal via the sewer system is acceptable, it is prudent to let waste management personnel determine disposal methods.

Another useful measure for controlling chemical hazards is substitution, which involves replacing chemicals with ones that have less hazardous properties and which can produce the same results. An example of a sensible substitution is the replacement of chromic acid cleaning solutions with commercially available cleansers. Another example is the use of aqueous-based scintillation fluids instead of organic-solvent-based fluids. These fluids are nonflammable, biodegradable (thereby reducing the costs of waste disposal), and less toxic. Many institutions have gone beyond just suggesting these substitution measures and have made them policy.

Controlling Biological Hazards

Design of Biohazard Facilities

The first step in designing a facility for handling biohazardous agents is to determine the level of containment required. It should be based on the nature of the species to be used and the experimental protocol. This protocol should be developed jointly by the principal investigator and a biohazards committee using the guidelines established by external agencies dealing with biosafety and disease control. Once the experimental protocol and level of containment have been established, facility design and layout should be performed according to standard specifications and, once complete, should be inspected and certified regularly. Most jurisdictions require that this be done a minimum of once per year.

Biological Safety Cabinets

Biological safety cabinets should be used for any experiments with a risk of generating pathogenic aerosols. Devices that typically generate aerosols include blenders, pipettes, syringes, stirrers, vortexers, and sonicators.

- Biological safety cabinets should be certified at the time of installation, annually, and whenever they are relocated.
- Nothing should be placed over the front intake or rear exhaust grilles. Equipment should be placed at least 4 in. inside the cabinet window.
- Transfers of viable materials should be performed as deeply in the cabinet as possible.
- Interior surfaces of the cabinet should be disinfected regularly, and all experimental apparatus should be disinfected before removal.
- The cabinet's fan should be allowed to run 2–3 min before beginning an experiment in order to allow sufficient time to purge airborne contamination. The fan should also be allowed to run an additional 2–3 min after completion of an experiment.
- Air turbulence should be minimized both outside the cabinet (which may be caused by nearby pedestrian movement) and inside the cabinet (such as that produced by flames). If a flame must be employed, a burner with a pilot light should be used.
- Only biological safety cabinets attached to exhaust systems can be used for procedures involving toxic vapors or gases.

Animal Care and Handling

Use of animals in the laboratory dictates special control measures in order to protect workers from zoonoses (diseases transmissible from animal to man). Zoonoses may be contracted via a number of vectors, such as animal bites, scratches, contact with animal blood, tissues, and excrete, or exposure to aerosols generated by animal care activities. In order to determine the necessary control measures, the type of species involved and the nature of contact should be taken into consideration. The nature of contact should not be limited to that with the animal, as it should also include contact with animal cages, accessories, body fluids, excrete, and carcasses. Two important elements of an animal care and handling program are veterinary surveillance for the animals and medical surveillance for the employees. Veterinary surveillance is especially valuable to ensure that incoming animals are in good health and are not carrying any transmissible diseases. Unfortunately, this is not always possible, because some conditions may not be clinically evident, for example, Q-fever. Medical surveillance of the employees is important both before employment and during employment after they have been exposed to animals. Before an employee is able to work with animals, it should be determined whether he/she has any predispositions to asthma, allergies, or other diseases that could be aggravated by contact with animals. For more information on medical and veterinary surveillance programs, readers are advised to consult guides for the care and use of laboratory animals.

Controlling Radiation Hazards

Radiation protection is treated as a separate chapter in this book (Chapter 10), but it should be stressed here that radioisotopes are chemicals and should be considered not only for their radiation-emitting properties, but also for their chemical, physical, and toxicological properties when their hazards are assessed. It is important to not lose sight of this fact and to incorporate the guidelines established for nonradioactive chemicals into experimental protocol.

Controlling Mechanical Hazards

Compressed Gases

Cylinders of compressed gases present a multitude of chemical and mechanical hazards to their users and to the occupants of laboratories. The type of gas involved should be considered for the following properties: flammability, reactivity, explosivity, corrosivity, and asphyxiation potential. It should be noted that many gases are not necessarily chemical asphyxiants but may nonetheless result in asphyxiation via displacement of air.

Most gases, when vaporized, cause a cooling effect that may be sufficient to cause freezing of the skin and tissues. Gases also pose the risk of causing embolism if the gas stream is powerful enough to penetrate the skin and arteries.

Cylinders have been referred to as "sleeping giants." The high pressures in compressed gas cylinders make cylinders potential bombs. If the gas is suddenly released, the resulting energy is sufficient to propel a cylinder through concrete walls, wreaking havoc along its path.

The following guidelines should be observed for the storage and handling of gas cylinders:

- All gas cylinders, full or empty, should be securely supported using racks, chains, or stands.
- When cylinders are not in use or when they are being transported, remove the regulator and attach the protective cap.
- When transporting cylinders, chain or strap the cylinder to the cylinder cart.
- Verify that the regulator is appropriate for the gas being used and the pressure being delivered. Do not rely upon the pressure gauge to indicate the maximum pressure ratings; check the regulator's specifications.
- Do not use adapters or Teflon®* tape to attach regulators to gas cylinders.
- Never bleed a cylinder completely empty, leave some residual pressure.
- Do not lubricate the high pressure side of an oxygen cylinder.
- Avoid subjecting cylinders to temperature extremes.

Glassware Safety

Handling glass apparatus poses the risk of cuts in any laboratory, but, particularly in hospital laboratories, cuts can result in grave consequences due to the handling of infectious agents.

When handling glass rods, tubes, or pipettes

- Avoid the sharp tips; fire polish the ends if any chips or sharp edges are evident.
- Lubricate with water or glycerine.
- Ensure the openings in stoppers or pipette fillers are properly sized, not too small.
- Insert carefully, with a slight twisting motion, keeping hands close together.
- Use the gloves or cloth towel to protect hands.

For all types of glassware

- Broken glass should be picked up with a dust pan and brush, not with hands.
- Discard the broken glass in a rigid container separate from regular garbage and label appropriately. Contaminated items should be sterilized prior to disposal.
- Glass is weakened by all types of stresses (e.g., heating, bumping, etc.). Used glassware merits extra caution when it is handled. Discard or repair all glassware that is chipped, cracked, or star-cracked because these vessels cannot tolerate normal stresses.

---

* Registered trademark of E.I. duPont de Nemours and Company, Inc., Wilmington, Delaware.

Centrifuges

Centrifuges are potential sources of contamination of laboratory equipment, producers of aerosols, and mechanical hazards. The following are guidelines for the use of centrifuges:

- Select the centrifuges with interlock devices that prevent access to the inside until the rotor has come to a complete halt. Do not disconnect or tamper with these devices.
- Before putting a centrifuge into operation, ensure that it is properly installed by a qualified service person. The centrifuge should be well anchored, properly balanced, electrically grounded, and all moving parts should be inspected to verify that none became loose during shipping.
- Instruct all users in the importance of checking the balancing prior to each use.
- When centrifuging any pathogenic or toxic liquids, use capped centrifuge tubes to prevent the release of aerosols.
- Use the tubes of unbreakable materials whenever possible.
- Clean the spills promptly. Thoroughly decontaminate centrifuges regularly and prior to servicing.

Personal Protective Equipment and Safety Devices

Laboratory workers should be made aware of the types, uses, limitations, and availability of personal protective equipment. The laboratory supervisor is responsible for ensuring workers are provided adequate quantities and appropriate types of personal protective equipment. The types of protection used should reflect the degree of risk for a given procedure and location.

Protective Clothing

The minimum protection needed is a lab coat for protection from splashes and contamination. In certain circumstances, such as when working with infectious agents, gowns with closures at the back are preferable. Other types of protective clothing, such as shoe covers, head covers, and sleeve bands, are commonly used in laboratories such as autopsy rooms, where they are useful to protect workers from contamination as well as to contain contamination within the laboratory. Disposable clothing has the greatest advantages in terms of protecting workers because there is no need for laundering; the disadvantages are cost and waste disposal requirements. If reusable clothing is employed, a laundry service must be provided to the workers, who should not be required to take work clothing home and possibly expose family members to workplace contaminants.

Protective clothing should be discarded when workers leave the laboratory area so as not to contaminate other areas such as cafeterias, other workers, or patients. Failure to do so defeats much of the purpose of using protective clothing.

Protective Gloves

There is no one type of protective glove that can be used in all situations and meet all the requirements for dexterity, impermeability, and resistivity. Careful selection of the correct glove must be considered for the type of work being performed. Information on gloves' resistivity and impermeability is available from suppliers for a limited number of agents and is sometimes inconsistent. It is advisable to refer to more than one source before selecting the type of glove material. In cases where there is no information pertaining to the agents being used, it may be necessary to perform tests involving soaking the glove material with the agents in question and observing for chemical degradation and breakthrough of the material.

Gloves should be removed whenever contaminated and must not be worn when materials not involved in a given experiment are touched so as to prevent contamination of telephones, door handles, and other objects touched by other persons.

Eye Protection

Eye protection should be required for everyone in laboratories where chemicals, biohazardous agents, and radioisotopes are handled. For light work, standard glasses with unbreakable lenses may suffice. For light to moderate work, glasses should be equipped with side shields. For work with a significant risk of splashes or projectiles, safety goggles are advisable, and in situations where there is a danger of damage to the eyes and face from projectiles, splashes, or explosion, a full face shield should be used.

Respiratory Protection

Although engineering controls (local exhausts, etc.) are the primary means of protection from aerosols, gases, and vapors, respiratory protection devices may be needed as a supplementary safety measure. The type of device required should be determined only by someone thoroughly trained in their selection, care, and use, such as an industrial hygienist. One should not rely solely on suppliers to specify needs.

There are two basic categories of respiratory protective devices:

1. Air-purifying: (a) particulate-removing that use mechanical filtration (b) gas- and vapor-removing that use chemical media, and (c) a combination of (a) and (b).
2. Air-supplying: (a) self-contained breathing apparatus and (b) air-line supplied respirators.

Air-purifying respirators are most commonly used in the laboratory, but in situations where an oxygen-deficient atmosphere is likely or when toxic gases or vapors with poor warning properties (odor, taste, irritation, etc.) are being used, air-supplying respirators are needed. For emergency response (e.g., fire, hazardous spills, etc.), positive-pressure air-supplying respirators are the most appropriate.

A successful respiratory protection program dictates the implementation of good quality control and must contain the following elements:

- Written standard operating procedures
- Respirator selection based on the hazard to which workers may be exposed
- Instruction on the care and use of respirators
- Assignment of respirators to individuals for their exclusive use, wherever practicable
- Regular cleaning and disinfection of respirators
- Storage of devices in convenient and sanitary locations
- Inspections of respirators during cleaning, replacement of worn or deteriorated parts, inspection of emergency respirators monthly assignment of tasks requiring respiratory protection to only those medically fit to perform work under the stress of added breathing resistance, and medical surveillance to determine the health status
- Regular inspections and evaluations to determine the effectiveness of respiratory protection program

Emergency Wash Devices

Readily available supplies of water are needed for situations where the eyes or body of personnel may be exposed to injurious corrosive, toxic, or irritating agents. Such devices must be conveniently located in the workplace for immediate emergency use. Usefulness of laboratory emergency wash devices may be limited if they are not given the necessary attention. It is imperative that all laboratory workers are thoroughly familiar with the operation of these devices and learn to use them without hesitation whenever the situation dictates. Unfortunately, workers have in certain instances hesitated to use them for reasons such as lack of confidence that the device is functional, embarrassment about being involved in an accident or making a mess, or fear of repercussions from causing a scene.

Regardless of the types of emergency wash devices in place, training and maintenance will be the factors that determine their effectiveness. Training, preferably hands-on in the operation of these devices, must be included in the laboratory safety training program. Regular inspection and testing are required to ensure they will be functional when the need arises. The American National Standards Institute (ANSI) adopted a standard for emergency eyewash and shower equipment in 1981 (Standard No. Z358.1–1981), which specifies that this equipment should be activated weekly to flush the lines and to verify proper operation. This standard has yet to be translated into reality, especially in the area of showers, which are not convenient to test without the services of a plumber and a means of controlling the spread of water if the shower is located in an open area. Even in the more avant-garde institutions, semiannual testing is considered a high standard. It is certainly more feasible to meet the ANSI standard with respect to eye washes, because they can easily be activated frequently by the laboratory workers.

Eye/Face Washes

Some devices are designed to direct water at the eyes only, whereas a growing number of devices are being designed to wash both the eyes and face. This is a sensible design, as it is likely that when something is splashed into the eyes, the face will be affected as well. The difference lies in the dispersion of the water stream and the flow rate. It is recommended that the flow rate be a minimum of 1.5 L/min for eyewashes and 11.4 L/min for eye/face washes, and they should be able to deliver these flows for a minimum of 15 min. The ANSI standard calls for location of these devices to be within 10 s and 100 ft of the work area. The control valve should be simple to operate and be able to turn on in 1 s or less and remain on without the use of the operator's hands. The standard does not address the issue of temperature control, which can be a serious problem in cold winter climates. Few wash devices have been designed to control temperature reliably, with the exception of portable devices that have reservoirs. These portable devices present problems of their own related to fungal and bacterial contamination and are not recommended for use except as a temporary measure. As a rule, all emergency washes should have plumbed supplies of freshwater.

Emergency Showers

The ANSI standard for emergency showers also calls for location within 10 s and 100 ft of the work area. The minimum flow rate is 113.6 L/min to be connected to a 1 in. water feed. Although not covered by the standard, floor drains are advisable for showers located in the laboratory. This may give rise to odors emanating from the plumbing system if the drain traps are allowed to dry out. Regular flushing of the drains should be considered as part of the maintenance program for emergency wash devices. The valve actuator for the shower should be low enough to allow access for all laboratory workers. This may require special adaptation for workers in wheelchairs.

Drench Hoses

Handheld drench hoses cannot be considered as replacements for eyewashes, since the water stream can only irrigate one eye at a time. However, they can be invaluable for splashes on localized areas of the skin in situations when a full body dousing with a shower could be considered overkill. For typical splashes that one may encounter in a hospital laboratory, a drench hose may be all that is needed for proper irrigation. Drench hoses have the advantage that they can be installed close to a sink where they are likely to be used for other purposes, resulting in the employees being more familiar with and confident in their operation. The recommended flow rate for hand-held drench hoses is 11.4 L per minute.

Laboratory Design

Many laboratory safety problems stem from inadequate design, which may occur during the construction stage or as a result of changes in procedures for which the facility was not designed. Clearly, it is preferable to design a laboratory properly from the onset, but in some instances changes in procedures may necessitate retrofitting laboratories to accommodate safety requirements. The key areas to consider with respect to design of laboratories are

- Fire prevention, detection, and suppression
- Ventilation systems
- Chemical storage
- Safety and emergency response equipment
- Laboratory water and potable water supply
- Ergonomics

Specifications for most of these design considerations are not within the scope of this chapter and are well covered in standard laboratory reference manuals. One area that is frequently neglected and merits special mention is ergonomics, the science which deals with the interactions between man and his working environment. In the laboratory, the relationship between workers and their tools of the trade is very intimate. A productive laboratory is one designed to enhance the harmony of this relationship by means of good illumination, well-designed benches and work space, noise control, proper tools, and an overall pleasant environment. These factors serve to reduce all types of stress experienced by the workers, contributing to an overall productive environment. The connection with the subject at hand lies in the belief that a well-designed laboratory is conducive both to productivity and safety.

Laboratory Ventilation

General Ventilation

The primary purposes of general ventilation are to maintain comfort levels (i.e., temperature and humidity) for human occupancy and to provide supply air to exhaust ventilation devices. In some circumstances, general ventilation may be required to also maintain comfort levels for animals or to provide the conditions needed for particular experiments. Although general ventilation provides a degree of dilution, thereby reducing the concentration of all airborne contaminants as well as minimizing the possibility of creating an explosive concentration of flammable contaminants, it should not be relied upon as a primary means of protection. In the laboratory, airborne pollutants should be captured at the source using local exhausts.

There are two main components in a general ventilation system: a supply and an exhaust. Regardless of the arrangement, no air exhausted from the laboratory should be recirculated through the building. The system should provide a minimum fresh air exchange rate as specified by regulatory agencies. The American Society of Heating, Refrigerating, and Air-Conditioning Engineers (ASHRAE) recommends a rate of 20 ft^3/min per person based on an occupancy rate of 30 persons per 1000 ft^2 of floor area (Standard No. ANSI/ASHRAE 62–1981R). The optimum operative temperatures are 71°F (21.7°C) in the winter and 76°F (27.2°C) in the summer.

The final requirement of the general ventilation system is that it be balanced such that the laboratory is at a negative pressure with respect to adjacent zones. This will control the migration of contaminants to offices, corridors, patient rooms, and other areas of the building.

Exhaust Ventilation

All procedures involving the liberation of toxic or flammable vapors should be carried out in a fume hood or other form of local exhaust device. In conjunction with personal protection and good practices, exhaust ventilation can provide a situation of zero exposure, a concept almost unheard of industrial sectors.

The velocity at the face of a fume hood must be sufficient to prevent the escape of contaminants from the hood into the laboratory. The American Conference of Governmental Industrial Hygienists (ACGIH) recommends a linear velocity of 100 ft/min at the face of the hood. Wind turbulence caused by pedestrian traffic, the movement of doors, or excessive clutter inside the hood may compromise its ability to provide containment. Therefore, it is recommended that hoods be placed in areas of low traffic and as far away from doors as possible. Fume hoods should not be used for storage of chemicals due to these turbulence problems.

Fume hoods should be equipped with some form of meter or visual indicator which verifies for the user that it is functioning properly. For very hazardous procedures, audible indicators are recommended to alert the user that operating conditions have changed. The hoods should be regularly inspected and their performance monitored with an anemometer that has been accurately calibrated. The ACGIH recommends inspections monthly for critical or new installations, and quarterly or semiannually for all others.

Face velocities on most fume hoods, with the exception of those designed for variable flow, are maximized by minimizing the hood opening. Users should be instructed in the importance of keeping the hood doors closed when not in use, and to minimize the opening when performing experiments.

Local exhausts should be designed to enclose the process as much as possible. Canopy hoods, which are commonly used to exhaust certain processes such as atomic absorption spectrophotometry, are not designed as enclosures and therefore have severe limitations. If such devices are used, they should be placed as close as possible to the sources of emissions and be arranged so that the flow of contaminants does not pass through the breathing zone of the operator. For most applications, canopy hoods are not suitable for hospital laboratories.

Spill Response Contingencies

Immediately after a spill or other accidental discharge of hazardous agents is not the time to begin planning how to respond. Predetermining spill response procedures will serve to greatly reduce the extent of damage to personnel and property in the event of such an incident. The first step in preparing a spill response plan is to evaluate the types of agents used in the laboratory in order to determine the materials and information needed for spill response. Although the institution can provide the support and backup systems, individual laboratory supervisors must be held responsible for determining the procedures specific to their areas. In some instances, only the laboratory specialists may be able to determine the remedial measures if they are working with highly specialized substances. The factors that must be considered when establishing contingency plans are as follows:

- Categories of agents and their chemical, physical, and toxicological properties
- Quantities that may be released
- Possible locations of release
- Personal protective equipment needed
- Types and quantities of neutralizing, disinfecting, or absorbing material needed

Establishment of lines of communication is a necessary component of contingency plans. Procedures must be established for summoning assistance in dealing with spills and for alerting support personnel of the incident. This is especially crucial for after-hours incidents, when minimal support may be available and it may become necessary to contact persons at home.

The general guidelines for initially responding to a spill situation are as follows:

- If a spill is minor and of known limited danger, it should be cleaned up immediately.
- If a spill is of unknown composition or potentially dangerous (e.g., explosive, toxic, etc.), all persons present should be first alerted and instructed to leave the room.
- Cleanups should be performed only by those fully equipped with all necessary protective equipment and materials necessary for decontamination or cleanup and trained in their use.
- If a spill cannot be safely handled utilizing the available materials and personnel, the area should be secured so as to ensure personnel safety and then additional help must be summoned.

Monitoring Hazards in the Laboratory

There are three complementary approaches to monitoring hazards in the laboratory. One can monitor the environment, the processes, and the workers.

Environmental Monitoring

Environmental monitoring is generally for estimating or determining workers' exposures to hazardous agents by measuring airborne concentrations or levels of surface contamination. Methods for most common agents are generally well established, but difficulties arise when evaluating agents for which there are no established sampling and analytical procedures. It is not unusual to come across this dilemma in a hospital laboratory environment where new agents are frequently introduced. This may at times dictate development of special techniques by the industrial hygienist with a chemist for monitoring chemical agents, a microbiologist for biological agents, and a health physicist for radioactive isotopes.

Interpretation of monitoring data also poses peculiar challenges, due the absence of permissible exposure levels for many agents and also due to the difficulty in determining the effects of exposure to more than one agent. In using monitoring data to formulate control measures, it is advisable to lean toward the conservative side and to adopt the philosophy of health physicists, which is to maintain exposures to levels as low as reasonably achievable.

Whenever possible, air monitoring should utilize personal samples to estimate exposures to workers. This is preferred over simply using stationary samplers that merely provide environmental data and that may not directly relate to the exposures experienced by the workers. Sampling protocol should strive to determine the dose by sampling for typical work periods, such as a workday. Short-term samples may also be required.

Monitoring Processes

Laboratory inspections must be conducted to adequately monitor laboratory processes and facilities. Such evaluations should examine not only the inherent properties of hazardous agents being used, but also the ways they are being used and the control measures in place. Successful inspection programs require participation not only from the safety audit teams, but also from the laboratory workers who should be given the opportunity to solicit advice and to submit their suggestions regarding laboratory safety.

There are direct connections between monitoring processes and preventive maintenance. The performance of emergency wash devices, fire prevention devices, ventilation systems, and other safety and emergency equipment must be part of this program, and responsibilities for these evaluations must be clearly defined.

Medical Surveillance

Medical surveillance for workers exposed to occupational hazards should be performed by physicians trained in occupational health practices. Physicians without such training may not be able to recognize links between diseases and occupational causes.

Medical surveillance programs must be carefully designed to suit the needs of the workers and the types of agents to which they may be exposed. The key elements of such a program are

- *Preemployment or preassignment screening.* Assessments of health status should be designed to determine if employees have predispositions to certain medical problems that may be aggravated by certain types of work. For example, someone with severe allergies may be advised not to perform work with animals or certain agents. These assessments may consist of a physical examination, specific tests, or simply a medical questionnaire.

- *Periodic health appraisals.* Periodic follow-up evaluations should be performed to assess whether employees' medical statuses remain compatible with their work assignments or to detect occupationally related illnesses. For certain types of work, specific tests may be conducted to look for the presence of agents that may have entered the body or for their metabolites This is referred to as biological monitoring. For certain conditions that can be monitored only by the presence of blood antibodies, serum banking may also be advisable.
- *Diagnosis and treatment.* Ideally, these measures should be aimed at rehabilitation and be prompt in delivery. Emergency treatment for sudden illnesses or injuries should be readily available and not be dependent on the periodic appraisals.
- *Immunization.* Work with certain human pathogens, infectious agents, and some species of animals will dictate immunization prophylaxis, for example, hepatitis B vaccinations. Due to human rights questions, many of these programs involve voluntary participation; therefore, the success of such programs will require ongoing education of employees in the benefits of participation.
- *Medical recordkeeping.* As with all medical records, the confidentiality must be maintained by the program physician. The records must be complete and accurate. Statistical analyses must be performed to examine possible epidemiologic phenomena.

Suggested Readings

American Conference of Governmental Industrial Hygienists. *Industrial Ventilation—A Manual of Recommended Practices*, 19th edn. ACGIH, Cincinnati, OH, 1986.

American Society of Heating, Refrigerating, and Air-Conditioning Engineers. *Thermal Environmental Conditions for Human Occupancy*, ANSI/ASHRAE Standard No. 55–1981. ASHRAE, Atlanta, GA, 1981.

American Society of Heating, Refrigerating, and Air-Conditioning Engineers. *Ventilation for Acceptable Indoor Air Quality*, ANSI/ASHRAE Standard No. 62–1981R. ASHRAE, Atlanta, GA, 1981.

Centers for Disease Control and National Institutes of Health. *Biosafety in Microbiological and Biomedical Laboratories*, 3rd ed. US Government Printing Office, Washington, DC, 1993.

Matheson Gas Products. *Guide to Safe Handling of Compressed Gases*. Matheson Gas Products, Secaucus, NJ, 1983.

National Fire Prevention Association. *Fire Prevention Handbook*, 16th edn. NFPA, Quincy, MA, 1986.

National Research Council. *Prudent Practices for Disposal of Chemicals from Laboratories.* National Academy Press, Washington, DC, 1983.

Pipitone DA. *Safe Storage of Laboratory Chemicals*, 2nd ed. John Wiley & Sons, New York, 1991.

Steere N. *CRC Handbook of Laboratory Safety*, 2nd edn. CRC Press Inc., Boca Raton, FL, 1971.

Young JA. *Improving Safety in the Chemical Laboratory: A Practical Guide,* 2nd ed. John Wiley & Sons, New York, 1991.

28

Biological Exposure Index Testing: Two Case Studies

William Charney
Healthcare Safety Consulting

Martin Lipnowski
Jewish General Hospital

Lee Wugofski
San Francisco General Hospital

Toluene 28-1
Discussion • Summary and Conclusions
Xylene 28-4

Toluene

William Charney and Martin Lipnowski

Technicians in a histology laboratory at Jewish General Hospital, Montreal, Quebec perform a cover slipping process during which skin contact with toluene ($C_6H_5CH_3$) is inevitable. Air exposure quantifications were done showing low ambient air concentration (Table 28.1) due to an effective local exhaust ventilation hood. However, because of the percutaneous nature of toluene and the lack of hand protection, it was decided to quantify the amount of absorbed toluene by venous blood methodology.

Cover slipping or mounting is a procedure in which slides are taken from a 500 mL container of toluene (Figure 28.1a through e). Slides are permanently sealed by placing 1 mL of Eukitt (glue) on a cover slip. Skin contact with toluene occurs when the slide is retrieved from the toluene. Gloves are not worn by the technicians due to the loss of finger dexterity. Forceps are sometimes used to avoid initial skin contact; however, skin contact still occurs when the technician manipulates the slide. All cover slipping is done under a local exhaust ventilation hood that provides >100 linear feet per minute (LFPM) of negative pressure at the face (Figure 28.2).

Air sampling was done during the period of cover slipping using two methodologies: (1) a Miran infrared spectrophotometer with a path length of 6.75 m and wave length of 13.7 mm to determine breathing zone concentration and (2) 3M 3550 organic vapor passive dosimetry to cross-correlate the personal breathing zone concentrations. The badge was placed as close to the breathing zone as possible. The analytical method was NIOSH 127, Perkin-Elmer Sigma 2000 gas chromatograph. The two methodologies corresponded 10% (Table 28.1).

Toluene in venous blood was measured by purge and trap gas chromatography mass spectrometry. Samples were taken immediately after workshift rotation. Samples were collected following the American Conference of Governmental Industrial Hygienists (ACGIH) recommendations using vacuum tubes with heparin anticoagulant. Samples were stored and transported upright at 2°C–5°C in dry ice.

Toluene is not usually present in blood in humans; however, levels up to 0.015 mg/L have been reported in controls. The ACGIH recommends a method to determine toluene in venous blood. The biological

TABLE 28.1 Toluene Absorbed by Histology Laboratory Technicians during Cover Slipping Process, Jewish General Hospital

Technician #	Time (min)	Air Level: Infrared (ppm)	Air Level: Passive (mg/m)	Venous Blood Level (mg/L)
1	62	6	11.5	0.01
2	41	6	13.4	0.01
3	35	6	13.5	0.03
4	195	7	15.5	0.002

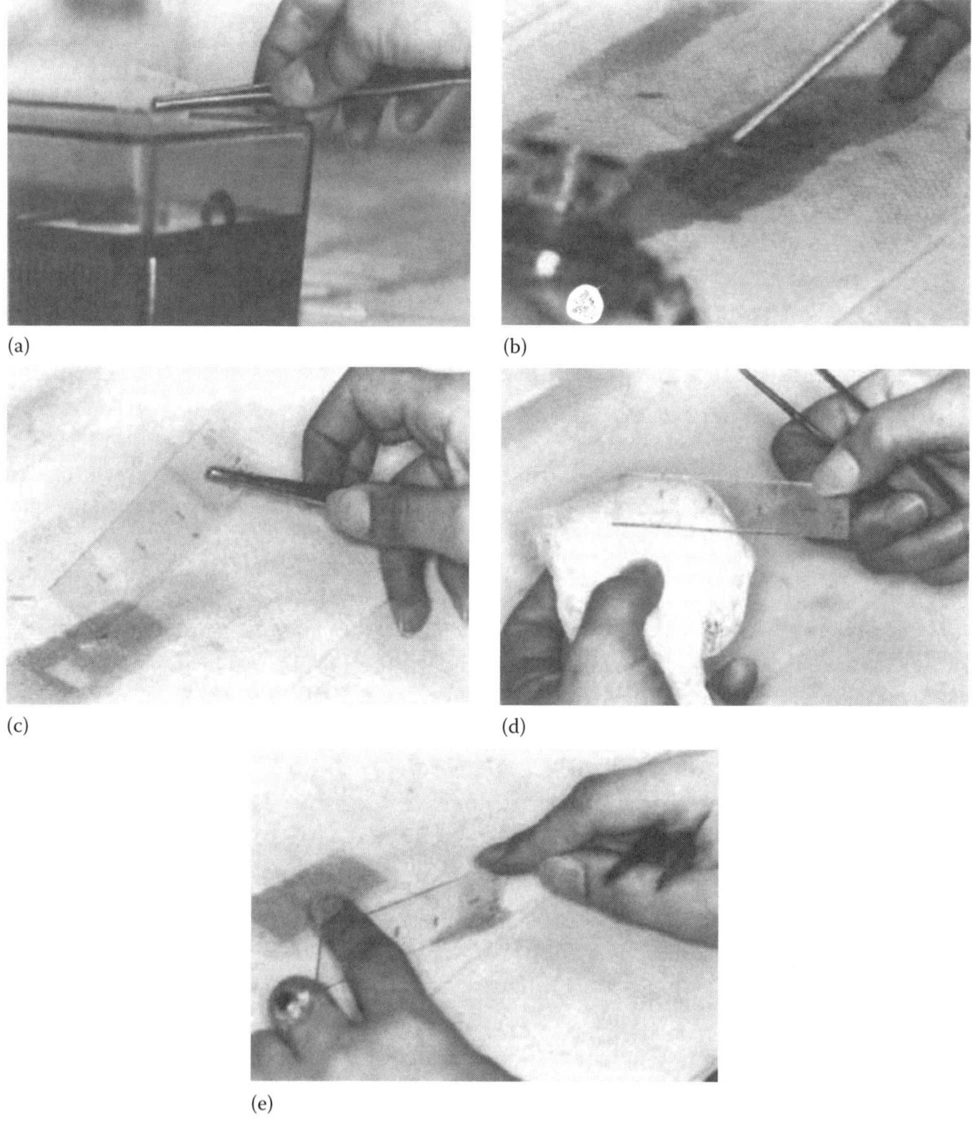

(a) (b) (c) (d) (e)

FIGURE 28.1 Cover slipping or mounting procedure at Jewish General Hospital in which slides are taken from a 500 mL container of toluene: (a) removing slide with forceps, (b and c) sealing the cover slip with glue, and (d and e) manipulation of the slide.

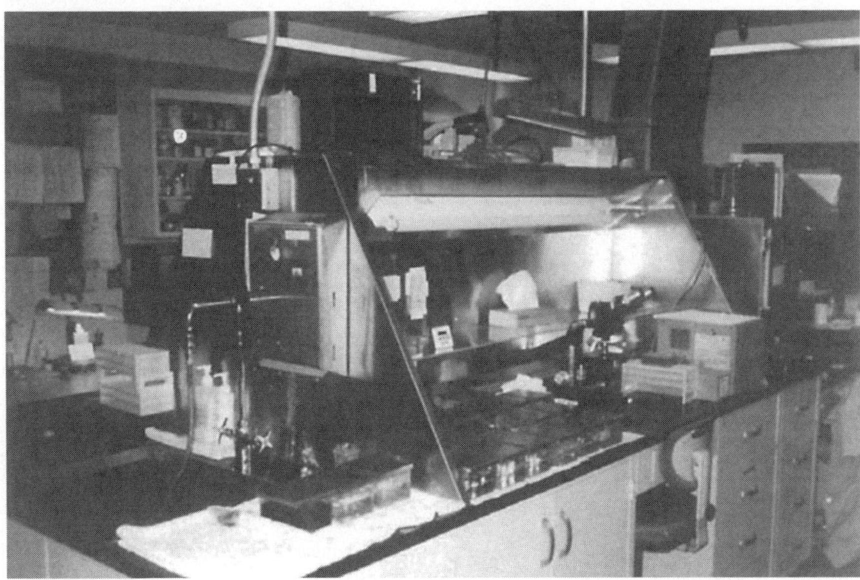

FIGURE 28.2 Workstation for toluene and alcohol procedures and storage with slot-type local exhaust ventilation at Jewish General Hospital. Slot air exhaust is located at the back of the hood.

exposure index (ACGIH BEI 19 [84]) is 1.0 mg/L. Toluene in venous blood is the recommended method when air quantification levels are below 50 ppm.

Discussion

The levels found in this study correspond to background levels found in controls as determined by other studies. Technician #4 had the highest exposure time of 195 min, but showed the lowest absorbed dose. This could be due to technique. The major route of exposure to toluene is inhalation of the vapor, and although there is no significant absorption occupationally from skin absorption of the vapor, percutaneous absorption should be taken into consideration for assessment when skin contact with toluene liquid occurs. Venous blood was chosen as the technique for this study because the air quantifications were well below 50 ppm.

Summary and Conclusions

Toluene in venous blood was used as a method to correlate absorbed skin dose of toluene from four hospital laboratory histology technicians with the ACGIH biological exposure index (BEI) for toluene. All four technicians showed only background levels of toluene in venous blood. A dose–response relationship was not evident in this study, because the technician with the highest exposure time had the lowest absorbed dose.

Given these results, we concluded

1. Technique in manipulating the cover slipping process can reduce skin contact with liquid toluene.
2. There does not appear to be any significant absorbed dose in the mounting process.
3. Further studies are recommended. See Table 28.2 for processes in which hospital personnel are exposed to toluene.
4. Venous blood should be used as a technique of monitoring absorbed dose in hospital histology and cytology technicians with skin contact with toluene as part of a global residual surveillance program. The ACGIH BEI should be used as the safety threshold.

TABLE 28.2 Toluene Exposure in Hospitals

Where	Who	Why
Pathology (histology cytology)	Technicians	Processing tissues Toluene is used as a cleaning agent Toluene is poured into recipient containers of the processing machine (histomatic) Used toluene is poured from machine containers into disposal apparatus Staining of slides Toluene is used for deparaffining slides at the onset of staining Toluene is used as a cleaning agent toward the end of staining Mounting of slides Mounting media (e.g., Eukitt & Permount) are toluene based General cleaning and degreasing of equipment Cleaning microtome knives, forceps, and machines countertops
	Pathologists	Staining of frozen-section slides Cleaning dirty or sticky slides due to oil or mounting media
Hermatology and microbiology	Technicians	Mounting and cleaning slides In microbiology, may use toluene in one or two tests, but these tests are very rare (once in 3 months) and do not use much toluene (≈5/mL)
Biochemistry	Technicians	A toluene-based solution (scintillation cocktail) is used on a regular basis for certain radioimmunoassay (RIA) tests Another test, that for hydroxyproline, used toluene for its analysis. Quantity to toluene used is not great; however, exposure time to it may be relatively long (tubes containing toluene should be stoppered)
Research	Doctors and technicians	Certain tests requiring the separation or extraction of elements from the specimen may use toluene to do so
Maintenance	Painters, cleaners, and electricians	Many paints, cleaning solutions, and other solvents may contain toluene (or other benzenoid hydrocarbons)

Xylene

William Charney and Lee Wugofski

Technicians in a histology laboratory at San Francisco General Hospital perform a cover slipping process where skin contact with xylene is inevitable. Cover slipping or mounting is a procedure where slides are taken from a 500 mL container of xylene. Skin contact with xylene occurs when the technician manipulates the slide.

Two air-sampling technologies were employed to quantify inhalation exposure. A Miran infrared spectrophotometer was used to detect grab samples during the course of the cover slipping. Passive dosimetry, organic vapor badges, and 3M 3550 were used to determine time-weighted average (TWA) and to cross-correlate the Miran results.

Xylene in urine was determined by methylhippuric acid in urine. The BEI as determined by the ACGIH is 1.5/g creatinine. (It has been recommended that urinary methylhippuric acid results be expressed as grams methylhippuric acid per gram urinary creatinine. The isomers of methylhippuric acid are not normally found in the urine of humans not exposed to xylenes.)

Sample collection was done on three technicians. Technician #1 samples were collected prior to exposure, immediately after exposure, and at the end of the week. Technician #2 had a sample drawn immediately postexposure and one sample drawn end of shift postexposure. For technician #3, one end of shift sample was drawn postexposure and one immediate sample was drawn after exposure. High-performance thin layer chromatography was the analytical method with a detection limit of 0.1 g/L. Exposure to 100 ppm xylene produces urine concentration of 3 g/L. All results, shown in Tables 28.3 through 28.5, were non-detectable.

TABLE 28.3 BEI Testing Results for Xylene, San Francisco General Hospital Histology Technicians

LNAME	FNAME	ID #	Date	Time	SK ID #	Methylhippuric Acid Result
Technician #1	Rosemarie	H-9030	8/14/89	8:00	U291038	Non-detectable
	Rosemarie	H-9030	8/17/89	13:30	U2910390	Non-detectable
	Rosemarie	H-9030	8/22/89	16:30	U2910841	Non-detectable
Technician #2	Norma	H-7693	9/25/89	11:00	U291082	Non-detectable
	Norma	H-7693	10/2/89	15:00	U2910819	Non-detectable
Technician #3	Martha	H-8108	8/29/89	15:00	U2910830	Non-detectable
	Martha	H-8108	9/8/89		U2910852	Non-detectable

TABLE 28.4 Xylene Sampling Data, Miran Infrared Spectrophotometer, Histology/Pathology Department, San Francisco General Hospital

Time	Absorbance (ABs)	Sample Location or Description	ppm	Analyze for (Give Specific Substances)
10:45	04	Staining trays	20	
	08		39	
	3		147	
10:48	05	Cover slipping (no hood)	24	
	17		83	
	14		69	
10:49	2		98	
	25		123	
10:50	04	Cover slipping with hood	20	
	03		15	
10:55	02		10	
	01		5	
	02		10	
11:00	01		5	
	01		5	

(*continued*)

TABLE 28.4 (continued) Xylene Sampling Data, Miran Infrared Spectrophotometer, Histology/Pathology Department, San Francisco General Hospital

Time	Absorbance (ABs)	Sample Location or Description	ppm	Analyze for (Give Specific Substances)
11:01	01	Cover slipping	5	
	01		5	
	02		5	
11:03	01		5	
	01		5	
11:07	01		5	
	01		5	
11:09	01		5	
	01		5	
11:15	01		5	
	01		5	
	01		5	
11:20	01		5	
	01		5	
	01		5	
11:22	03	Cover slipping	15	
	02		10	
	02		10	
	02		10	
11:25	01	Changed underpadding	5	
	02		10	
	01		5	
11:28	01		5	
	01		5	
	04		20	
11:30	04		20	
	03		15	
	07		34	
	03		15	
	02		10	
11:30	01	Cover slipping	5	
	03		15	
	02		10	
	02		10	
	02		10	
	02		10	
	01		5	
	01		5	
	01		5	
	01		5	
	04		20	
	05		30	
	02		10	
	02		10	
	01		5	

Note: TWA: xylene, 100 ppm; Stel, 150 ppm.

TABLE 28.5 Laboratory Results for Xylene—Passive Dosimetry, San Francisco General Hospital

LABNO	SMPLNO-ID	Air (LT)	Front (mg)	Back (mg)	Total (mg)	mg/m^3	ppm
9463	1078-CR	9.83			2.10	21.4	9.3
9466	0962-MG	11.57			0.0590	5.14	1.18
9468	0913-NS	14.74			0.0779	5.28	1.22
9469	0822-IP	11.47			0.0334	2.91	0.670
9470	0995-RN	11.47			0.167	14.6	3.36
9471	0942-IP	13.10			0.0408	3.11	0.716
9472	0663-MG	11.47			0.285	24.8	5.71
9473	0946-IP	13.10			0.0444	3.39	0.781

Note: Assay, Aromatic Hydrocarbons (NIOSH 1501), Matrix, B3M.
LT = Lifetime

29

Functions and Staffing of a Hospital Safety Office

Introduction 29-1
Assessment of Needs 29-2
Fire/Life Safety • Electrical Safety • Hospital Waste Program • Safety Committee/Accident/Incident Follow-Up • Hazard Communication/Industrial Hygiene/ Chemical Health and Safety • Industrial Hygiene Monitoring of Chemical Use • Chemical Spill Cleanups • Liaison with Other Departments • Liaison with Regulatory and Other Agencies • Committee Membership • Renovation/ Planning • Product Recalls • Technical Service/Educational Services • Meeting with Vendors and Consultants • Professional Maintenance • Universal Precautions • Administration of the Safety Office
Environmental Regulations Affecting Hospitals 29-10
Appendix 29.A.1 29-11

Lindsey V. Kayman
University of Medicine and Dentistry of New Jersey

John F. Clemons
Boston University Medical Center

Introduction

The safety director of a large urban medical center recently remarked, "I have trouble sleeping at night. I am living on the jagged edge of fear. I am so severely understaffed that I can't possibly inspect the entire facility let alone make sure any problems I find are corrected!" Many hospitals are drastically understaffed in light of the need for compliance with a wide variety of regulations. The number and severity of potential hazards to employees, visitors, and patients are comparable to some of the most hazardous industries.

This chapter is designed to serve as a tool to assess the staffing needs for a comprehensive hospital safety and health program. It can also be used as an audit tool to assure that all potential health and safety hazards are addressed. The Hospital Health and Safety Office is generally staffed by professionals with expertise in industrial hygiene, safety, occupational health, or environmental health. The analysis in this chapter is based on a general hospital with the following characteristics: (1) 300 beds, (2) approximately 500,000 sq ft of floor space, and (3) three buildings for patient care and two ancillary buildings. A 300-bed teaching hospital with adjacent research facilities under hospital control was also considered, based on the following assumptions: (1) approximately 200 laboratories and (2) approximately 60,000 sq ft of floor space.

Similar assessments may be performed on other institutions by revising the allotted times for listed tasks, depending on the size and needs of each facility. To determine the number of full- or part-time employees needed, add together the number of hours needed per month to implement each function. Multiply the total hours per month by 12 months/year. Then divide by 2080 h per year per person. This will yield the required number of employees to staff the office.

The analysis presented in this chapter demonstrates the need for two professional employees for the hospital alone. An additional 1.5 full-time employees are necessary if a research facility is included. This brings the staffing level up to 3.5 full-time employees. At least one secretarial/administrative employee per two professionals should be added to complete the staffing. A listing of reference books is offered in the appendix at the end of the chapter.

Assessment of Needs

Fire/Life Safety

- 55 h/month, hospital only
- 80 h/month, including research facility

The Safety Office staff should lead the training and response effort in all areas of life safety. This effort includes inspection of all egress corridors, stairs, exit doors, exits, and emergency lighting at least semi-annually. Walk-throughs of heavily trafficked areas on a daily or weekly basis will add essential depth to the program.

A Safety Office representative should be present at all drills, whether a fire or smoke or system malfunction or practice drill. Participation in the follow-up critiques is also essential to assure correction of any actions taken (or not taken), that is, that systems have been repaired or that any weak points noticed in practice drills are addressed.

The safety director often has primary responsibility during a fire at the hospital. The safety director's knowledge of special hazards throughout the institution will enable the fire department to use the most appropriate fire fighting techniques and minimize risk to patients, staff, visitors, and firefighters.

It is essential that all new personnel be given fire safety orientation on their first day of employment. This will include review of drill and evacuation rules, and basic instructions on use of fire extinguishers, fire alarms, and telephone communications in fire emergencies.

Hands-on instruction for fire extinguisher use should be made available at least annually to all personnel.

Patient care employees should receive additional orientation on horizontal evacuation, how to move patients in an emergency, and emergency life support for patients with special needs (e.g., oxygen-dependence, pacemaker, respirator, etc.).

Although not strictly "life safety," the following should be addressed as part of the inspection program:

- Inspection of children's play areas for safe toys and climbing equipment
- Evaluation of adequacy of security for violent patients in the psychiatric ward and training for staff on emergency protocols
- Inspection of special hazards, such as electrical hazards, in newborn nurseries

Electrical Safety

- 4 h/month, hospital only
- 6 h/month, including research facility

All patient care equipment and electrical receptacles must be tested semiannually. The Safety Office should monitor the safety standards of electrical systems, receptacles, and especially patient care equipment during its routine inspections. Because maintenance of this equipment requires specialized training as an electrician or electronics technician, the management usually falls on the electrical maintenance shop, an electronics shop, or biomedical equipment shop.

Hospital Waste Program

- 30 h/month, hospital only
- 80 h/month, including research facility

While housekeeping will have the primary responsibility for waste collection and disposal of general hospital trash, maintenance may operate an incinerator and infection control will write the rules and do the training. A continuing audit by the Safety Office is essential to assure continued high standards. This will include follow up of any reports of incorrectly segregated waste and sharps or other items improperly disposed of.

Additionally, the Safety Office should have the responsibility for managing the chemical waste program. Hazardous chemical wastes to be collected and disposed of include photographic developers, chemotherapeutic agents, laboratory reagents, and others.

The Safety Office will generally pick up chemical wastes and transport them to the chemical waste room or will establish office hours, during which chemical waste may be dropped off by generators. The benefits of the former method are that the Safety Office staff can check chemical waste for proper packaging and labeling prior to arrival at the chemical waste room and can confirm that proper transport vehicles are used and that incompatible chemicals will be kept segregated during transport. Unlike service departments such as housekeeping or materials transport, Safety Office personnel must have the expertise to know what to do in the event of a spill (see *Chemical Spill Cleanups*).

Safety Office staff will be responsible for the appropriate segregation of chemicals once they reach the waste room. The Safety Office will maintain all required records, inventories, manifests, and the budget.

The Safety Office may be responsible for bulking liquids from 1 gal containers into 55 gal drums, provided that proper ventilation and safety controls are in place in the chemical waste room. One hazardous waste company charges approximately $60 to dispose of a 1 gal waste solvent container as opposed to $4/gal for liquid consolidated in 55 gal drums. This could represent a significant cost savings to the institution.

The Safety Office will call in the hazardous waste removal company as needed and monitor packing and shipping of chemical waste. The Safety Office will sign applicable manifests and maintain records to ensure that all Resource Conservation and Recovery Act (RCRA) requirements are met.

Disposal of cytotoxic waste (chemotherapeutic agents) deserves special mention. Most state hazardous waste regulatory agencies require that cytotoxic waste be incinerated at 2000°F. Some hospitals have on-site incineration that can guarantee these temperatures. Other hospitals have to contract the waste hauling. In each case, a separate waste stream must be initiated for cytotoxic waste. In selecting a private contractor for disposal of cytotoxics, the hospital should make certain that the disposal company is an insured carrier and that the hospital receives a "total destruction" manifest from the waste hauler.

Cytotoxic waste may be stored in plastic, covered containers in a certified storage area within the hospital while awaiting the every 45- or 90-day pickup by the waste hauler. Under normal storage circumstances, the cytotoxic waste residue is not considered volatile, given the molecular structures of the chemicals. However, both for storage on inpatient units and storage in the pickup area, plastic containers should be lined with double plastic bags. Each container should have a list of exactly what drug-contaminated waste is inside the container. Whoever does the internal transportation of the cytotoxic waste within the hospital needs to be trained as to the toxicity of antineoplastic drugs (Hazard Communication Act 29 CFR 1910.1200).

Safety Committee/Accident/Incident Follow-Up

- 16 h/month, hospital only
- 20 h/month, including research facility

The Safety Committee, which meets at least quarterly, is responsible for developing, implementing, and maintaining a comprehensive hospital-wide safety program. The Safety Committee is an important vehicle by which safety considerations are taken into account in institutional planning, goal setting, and budgeting. The safety director and other members should prepare reports for and make recommendations to the Safety Committee. The Safety Committee should then make formal recommendations to higher administration. Once changes have been decided upon, the safety director has administrative responsibility for carrying them out.

The chair of the Safety Committee varies from institution to institution. One format is to have a senior administrator chair the committee while the safety director acts as the operative arm of the committee. Another format is to have the safety director chair the committee, which allows for more control of committee activities and direction of the overall program. Safety Office administrative support may record and disseminate the minutes of the meeting.

The Safety Office should conduct a monthly review on accidents and incidents. (Risk management may perform this function for incidents involving patients.) Causes and resolutions of incidents and accidents should be reported to the Safety Committee. Unusual accidents or incidents may require extra attention by the Safety Office.

Hazard Communication/Industrial Hygiene/Chemical Health and Safety

- 50 h/month, hospital only
- 100 h/month, including research facility

Although the Occupational Safety and Health Administration (OSHA) Hazard Communication Standard preempted state right-to-know laws involving workers, many states still have community right-to-know laws. These state laws have requirements in addition to the federal community right-to-know law, which is Title III of the federal Superfund Amendment and Reauthorization Act (SARA). In certain instances, hospital reporting exemptions are permitted for the federal regulation. The state regulations usually require lists of chemicals or collections of Material Safety Data Sheets, (MSDSs) to be provided to fire departments and other local agencies.

The Hazard Communication Standard requires a written plan, chemical inventory, MSDSs for each chemical, employee training, and chemical labeling.

Material Safety Data Sheets

MSDSs must be available for each hazardous product in use at the hospital for employees on all work shifts. The Safety Office will maintain a central computerized or paper file of these sheets. Upon receiving an inventory of chemicals from each department, the Safety Office may provide an MSDS for each chemical listed. These can be kept in notebook or the departmental safety manual of each department. However, some larger hospitals that have a network computer system are making MSDSs available by computer. Each department can then access its own MSDSs from the computer as needed. The American Society of Safety Engineers publishes a catalog of commercial safety and health software programs which includes MSDSs.

Employee Training

One format is for the Safety Office to conduct all required training. Another format, depending on the size of the institution, is for the Safety Office to train designated personnel from each department so that they, in turn, may conduct training for current and new employees. Trained staff would discuss each question with new employees. Written responses from employees would reinforce learning and could be used later for reference purposes. A record of this training should be placed in the personnel files or maintained by the Safety Office. For this type of training, the Safety Office would be responsible for developing health and safety educational materials to supplement the training.

Chemical Labeling

Monitoring of compliance with the labeling provisions of the Hazard Communication Standard can be included in routine safety inspections.

Industrial Hygiene Monitoring of Chemical Use

- 55 h/month, hospital only
- 85 h/month, including research facility

Specific attention must be given to ensure that toxic or hazardous materials are used in such a manner as to reduce employee exposures to lowest possible levels, assuring compliance with all OSHA standards. If the Safety Office does not have an industrial hygienist as a staff member, then the budget should include an allocation for industrial hygiene consultants. The industrial hygienist will review work practices, personal protective equipment, and ventilation to determine the need for employee training, air sampling, and additional ventilation controls. Air sampling may be performed to determine employee exposures or room concentrations, demonstrate compliance with OSHA regulations, and evaluate effectiveness of engineering controls or work practices.

Examples of areas needing specific chemical monitoring are as follows:

- Oncology/pharmacy—antineoplastic drugs
- Central processing, various departments—ethylene oxide
- Operating room, respiratory therapy—glutaraldehyde, anesthetics
- Recovery room, emergency room, patient areas—anesthetic gases
- Pathology—formaldehyde, solvents, picric acid
- Hospital wide—pesticides, solvents used in construction, asbestos
- Mechanical areas—asbestos, polychlorinated biphenyls (PCBs) in transformers
- Neurological laboratories—collodion (ether)/acetone
- Histology laboratories—xylene, alcohols, stains
- Dialysis—formaldehyde
- Laboratories—chromic sulfuric acid, various chemicals

Depending on results of air monitoring, training programs may need to be developed. A respirator program may be instituted as an interim measure before engineering controls are installed.

For toxic materials that may cause exposures due to skin contamination, innovative training methods have been employed. For example, some hospitals utilize a fluorescent dye during training to illustrate possible contamination from mishandling of antineoplastics or other materials.

Other nonchemical industrial hygiene surveys may include microwave ovens, electrocautery surgery, nonionizing radiation, lasers, noise-generating devices, hot work areas (laundry, incinerator room, boiler room), or biological agents in nonpatient care areas. Recombinant DNA, animal dander, and excreta (which may contain drugs) may be an issue in research institutions.

Chemical Spill Cleanups

- 8 h/month, hospital only
- 12 h/month, including research facility

Hospitals are required by federal law to develop emergency contingency plans for chemical spills and other potential emergencies. This function is usually performed by the Safety Office.

Common chemical spills include the following:

- Mercury
- Formaldehyde
- Chromic sulfuric acid
- Organic solvents
- Acids
- Antineoplastic-contaminated body fluids

Chemical spills generally occur in the laboratories. Occasionally a spill will happen in the receiving area, or during delivery or transport of chemical waste. A chemical spill will frequently occur when the transport cart wheel becomes stuck in an elevator door track. These situations can lead to serious fire or exposure emergencies.

After making a basic assessment of possible spills, the Safety Office needs to develop the following:

- *Chemical spill cleanup materials.* Chemical spill cart equipped with solvent, acid, and alkaline cleanup sorbents, chemical-resistant gloves, respirators appropriate for a variety of chemical hazards, eye protection, gloves, aprons, and shoe covers. Provisions must be made to replenish the cart as items are used.
- *Selection of an emergency response team.* Provisions should be made so that an emergency responder is available 24 h/day. If the Safety Office has enough personnel, one should always be on call. Additional technically trained individuals may be recruited from laboratory staff. Only personnel with appropriate expertise and training should clean up the spills. Security and other departments may perform ancillary duties, such as keeping the area clear of unauthorized persons and shutting off utility systems.
- *Training of emergency response team members.* Training for emergency response team members will include hazard assessment, personal protective equipment selection, emergency communication network, first-aid procedures, chemical neutralizing techniques, proper cleanup, and packaging for disposal and critiquing of incidents. All members should receive training prior to cleaning up a chemical spill. Training updates should be conducted at least annually. Self-contained breathing apparatus training should be repeated at least every 6 months. *Note:* The OSHA Respirator Standard (29 CFR 1910.134) requires training and other standard operating procedures for anyone who may use a respirator in the course of his/her job. A medical evaluation is required for personnel who are assigned to use respirators.
- *Patient area spills involving antineoplastic drugs.* Spills of patient body fluids contaminated with antineoplastics may occur in the patient units. These spills should be cleaned up by nursing staff using spill kits designed for this purpose. The kit consists of labeled waste-holding bags, absorbent materials, gloves, masks, and other protective equipment. Commercial cytotoxic spill kits sometimes come incomplete. They often will not have the required amount of absorbent pads or the correct type of broken glass sweep. Each hospital's appropriate committee should evaluate their cytotoxic spill kits to make certain the kits are capable of handling larger spills (100 mL or more). The Safety Office may train nurses on how to use the spill kits. Sometimes the Safety Office will perform this function.

Liaison with Other Departments

- 12 h/month, hospital only
- 20 h/month, including research facility

Continuing contact with interior designers, engineers, and architects is a must to assure proper specifications and furnishings (fire retardant) and space layouts that do not violate egress requirements. The Safety Office should ensure that adequate ventilation controls are installed considering the type of work that will be done.

Reviews with nursing, housekeeping, maintenance, and other departments should be performed to assure that training in lifting, waste handling, and so on is appropriate. Laboratory managers may also need review of chemical handling and disposal procedure and proper control of airborne contaminants.

The Safety Office may work with the legal office to determine the institution's response to new regulatory requirements as well as inquiries from regulatory agencies. Employee Health Services should contact the Safety Office when employees report safety incidents or health effects associated with their jobs. The Safety Office can monitor the work environment and make recommendations concerning the need for administrative, ventilation, and work process controls. Safety and employee health services should work together to develop institutional respirator standard operating procedures and plans for medical surveillance of employees working with specific substances. Safety and employee health should review all departmental policies on hazardous materials.

The Safety Office will generally be responsible for informing appropriate departments about changes in OSHA, EPA, and other environmental and health and safety regulations. This may be accomplished in private meetings, manager's meetings, or through publications.

When making recommendations that will affect patients, the Safety Office will interact with the appropriate medical and administrative departments.

The Safety Office will assist all departments in updating all safety-related standard operating procedures and other records in preparation for surveys by the Joint Commission on Accreditation of Healthcare Organizations (JCAHO).

Administration will need periodic updates from safety and engineering on required upgrading of fire protection systems, asbestos removal, ventilation, and other code conformity requirements. In many situations, the Budget Office will be involved in these discussions.

Because there are many departments in the hospital concerned with safety, the safety department will often play the role of ombudsman and fill in wherever needed. For example, the Safety Office may audit interdisciplinary systems in the hospital which can affect patient safety, such as assuring that

- Patient mobile equipment is adequately cleaned
- Patient equipment may be obtained quickly by nursing staff
- Incoming laundry remains clean and uncontaminated by biological agents

The Safety Office will have contact with departments not listed above when specific needs arise or during environmental rounds or during the semiannual safety survey.

Liaison with Regulatory and Other Agencies

- 10 h/month, hospital only
- 18 h/month, including research facility

The Safety Office can perform a vital role for the hospital by acting as primary contact for the local fire department, the building inspector, the public health inspector, OSHA, EPA, and other agencies that regularly survey the hospital for code compliance. The Safety Office, in this role, can be a useful buffer for the administration or coordinator, where two agencies may be in conflict. The Safety Office may also act as an internal coordinator where several departments may be affected by imposed requirements. The safety director usually accompanies the hospital team during JCAHO inspections and is the primary hospital spokesperson during the safety records review. The Safety Office, along with Engineering and Risk Management, reviews deficiency reports and other recommendations made by the insurance carrier and develops a plan for corrective action.

Committee Membership

- 8 h/month, hospital only
- 16 h/month, including research facility

In addition to the Safety Committee, Safety Office representation is usually necessary on the following committees:

- Disaster
- Infection control
- Radiation safety
- Risk management/quality assurance
- Renovation planning
- Laser safety

Many institutions are forming a reproductive hazards committee (see Chapter 7). A labor management health and safety committee may be an important tool for addressing the concerns of workers in unionized or nonunionized institutions. If the institution includes a research facility, there is often Safety Office representation on the following additional committees:

- Institutional biohazards
- Animal handling (the safety officer would address risks to employees)

In addition, the safety director will often set up ad hoc committees to address specific problems.

Renovation/Planning

- 12 h/month, hospital only
- 20 h/month, including research facility

It is important for the Safety Office to participate in the review of new construction or renovation needs with in-house staff and consultants, architects, and construction companies for compliance with safety requirements, noise, and other annoyances to personnel during construction. The Safety Office will also review plans for compliance with codes, placement of safety apparatus, adequate storage space, and fire egress.

In addition, the Safety Office must monitor construction programs to ensure that construction workers and hospital employees are not exposed to toxic materials or other safety hazards during the work. Major construction programs necessitate a safety manager on the job. The Safety Office should meet with construction personnel to ensure that all necessary safety controls are in place during construction. Egress routes for adjacent areas must be maintained unobstructed at all times.

All asbestos-related projects require detailed coordination of hospital departments and close supervision by the Safety Office. The Safety Office should have input as to which contractor is selected and may hire outside industrial hygiene consultants to write specifications and monitor the jobs. The Safety Office should review air sampling and other reports regularly during ongoing work.

Product Recalls

- 4 h/month, hospital only
- 8 h/month, including research facility

In many institutions, Purchasing will handle product recalls because they have records of equipment purchases and would most likely receive the recall notice from the vendor. The Safety Office's role is to assure that all users of recalled equipment get the notices and respond appropriately. The Safety Office may choose to subscribe to the "FDA Enforcement Reports" and the National Recall Alert Center as a backup to ensure that no recall notices are missed. In this case, the Safety Office would coordinate the system of distributing recall information to the appropriate departments and may retain documentation of action taken.

Technical Service/Educational Services

- 35 h/month, hospital only
- 60 h/month, including research facility

The Safety Office can provide a wide variety of direct services to the hospital community. The following illustrates the types of technical services the Safety Office commonly provides:

- Determining the cause of and then correcting odor complaints
- Helping researchers choose the least toxic materials to use
- Helping to identify the cause of medical symptoms of hospital personnel
- Designing engineering controls and changes in procedures or work practices to promote safety
- Providing advice on proper handling techniques for highly toxic or reactive materials
- Conducting health and safety literature searches on various topics

A safety newsletter is an excellent way to provide regulatory information, technical information, and other safety information of interest to employees. A newsletter that is well written and attractive will increase the visibility and credibility of the Safety Office. Researchers may be attracted by a section on safety trivia. Questionnaires and surveys of employee safety concerns are important means of obtaining information, involving employees, and keeping in touch with the hospital community.

Meeting with Vendors and Consultants

- 8 h/month, hospital only
- 16 h/month, including research facility

Meetings with vendors of safety devices and publications, and also of general hospital equipment (preferably by invitation!) are essential to keep the safety staff up to date on state-of-the-art devices and costs thereof.

In addition to traditional safety equipment, the Safety Office may recommend the purchase of a variety of equipment that is used in the hospital. Examples are solvent-recovery stills, laboratory refrigerators, hazardous waste services, chemical-resistant gloves, down-draft tables for the morgue, and other items. For these purposes, it is useful for the Safety Office to have supply catalogues of equipment used by various disciplines in the hospital. This will prove an important means of finding out about safety devices used for specific applications.

The Safety Office will periodically require the services of specialists whose expertise is beyond that of the staff, or who can supplement the limited availability of staff. Examples are

- Industrial hygiene asbestos consultants
- Water sampling for possible chemical contamination
- Ventilation engineers for special projects
- Industrial hygienists
- 24 h emergency spill consultants
- Trainers for special programs
- Experts in integrated pest management

Professional Maintenance

- 20 h/month, hospital only
- 30 h/month, including research facility

To maintain professional competence, safety staff members must devote a reasonable amount of time for reading current safety, environmental, and industrial hygiene journals, regulations, newsletters, and relevant studies. They must also attend professional seminars (at least in their field of specialization; however, attendance at a range of seminar topics will increase the attendees' breadth of knowledge and value to the institution).

Attendance at professional seminars is mandatory for the professional to maintain certified status. Certification in safety or industrial hygiene is important for assuring that professional staff are qualified in their respective fields, know how to write a detailed and accurate report, and will be considered "expert" in the event of a regulatory or court challenge.

Universal Precautions

- 3 h/month, hospital only
- 5 h/month, including research facility

The OSHA Universal Precautions Standard is usually implemented by the infection control staff. There may be a need for the Safety Office to periodically audit proper handling of trash.

Administration of the Safety Office

- 18 h/month, hospital only
- 33 h/month, including research facility

The safety director will work with the staff so that time allocation to various projects is appropriate to meet the hospital's needs and objectives. The director will need to assure that all inquiries and correspondence are answered in a timely fashion and that project reports are also promptly issued. Any hazardous situations must be immediately addressed. Logs should be maintained of complaint calls, chemical pickup requests, air sampling results, and other items of importance. Technical advice given over the phone should be confined in writing (at least for the office file). Other administrative functions include maintaining required records, ordering equipment and supplies, budgeting, keeping track of accounts, writing and disseminating minutes of meetings, and maintenance of MSDS files and other technical information, including training records. Regular meetings with administration are necessary to ensure high level support for Safety Office goals, concerns, activities, and needs. Ongoing communication with the person whom the safety director reports to is essential to assure that the program is consistent with the hospital's goals and for maintaining the required authority of the office.

Throughout this chapter comments are made about the need for adequate records. For any hospital accredited by JCAHO, we must reemphasize the need for complete recordkeeping. The JCAHO *Plant Technology & Safety Management* handbook spells out very clearly the recordkeeping requirements for each part of the safety program.

Environmental Regulations Affecting Hospitals

- *OSHA General Industry Standards*: Asbestos; Ethylene Oxide; Formaldehyde; Hazard Communication; Universal Precautions; Lab Standard (coming soon).
- *NFPA Codes*: Health Care Facilities—99; Life Safety Code—101; National Electric Code—70; if research laboratories are present, Laboratories Using Chemicals—45
- *EPA*: Superfund Amendment and Reauthorization Act; Clean Drinking Water Act; Resource Conservation and Recovery Act; Clean Air Act
- *Local and State Regulations*: Building Codes; Fire Department Regulations; Department of Public Health
- *Nuclear Regulatory Commission*: isotopes used in radiation therapy; isotopes used in research
- *Industry Standards*: Joint Commission on Accreditation of Healthcare Organizations; American Hospital Association

Appendix 29.A.1

Basic Library for the Health Care Safety Professional

Accreditation Manual for Hospitals, Joint Commission on Accreditation of Hospitals & Health Care Facilities, Chicago, IL.

Management & Compliance Series Vol. I - Hazardous Waste Management, American Society of Hospital Engineers, Chicago, IL.

Universal Precautions Guidelines—OSHA—CDC.

OSHA General Industry Standards—29 CFR 1900–1910.

OSHA Construction Industry Standards—29 CFR 1926.

OSHA Hazard Communication Standard—29 CFR 1910–1928 (August 1987).

State and Local "Right-to-Know" Laws.

State and Local Building Codes (or BOCA).

State and Local Fire Codes.

State and Local Department of Public Health Regulations.

State and ANSI Regulations on use of Lasers.

National Fire Protection Association—National Fire Codes. (A full set is preferred because of many cross-references and references in OSHA, state and local fire codes.)

If a full set of Fire Codes is not affordable, the following codes are bare minimum:

NFPA 101—Life Safety Code—Handbook, if possible.

NFPA 99—Health Care Facilities—Handbook, if possible.

NFPA 70—National Electric Code.

Other helpful codes are
NFPA 10—Portable Fire Extinguishers.

NFPA 30—Flammable & Combustible Liquids Code.

NFPA 50—Bulk Oxygen Systems.

NFPA 704—Fire Hazards of Materials.

If research is being done in laboratories on-site:

NFPA 45—Laboratories Using Chemicals

Other research lab references:
Prudent Practices for Handling Hazardous Chemicals in Laboratories—National Academy Press, Washington, DC.

Prudent Practices for Disposal of Chemicals in Laboratories—National Academy Press, Washington, DC.

Guidelines for Laboratory Design—Wiley-Interscience, New York.

Source: Compiled by John F. Clemens, PE/CSP, Secretary/Treasurer to Assistant Administrator.

30

Education for Action: An Innovative Approach to Training Hospital Employees

Merri Weinger
California Department of Health Services

Nina Wallerstein
University of New Mexico

Purpose of Training **30**-2
Preparation for Training **30**-4
Training Session **30**-5
Content of Training • Training Objectives • Training Methods • Training Evaluation
Conclusion **30**-7
Appendix 30.A.1 .... **30**-7

You've just been handed the job of training your hospital's 3000 employees. Your task is to bring the hospital into compliance with OSHA and JCAHO requirements for health and safety training. The list of requirements is long, your JCAHO inspection is quickly approaching, but when you attempt to enlist managers and supervisors, they are concerned about staff time: "Can you do it in an hour? We have patients to take care of." You're also faced with the challenge of training employees on all shifts. "What about a 15-minute training video which would be shown to all new employees?" supervisors ask.

As you grapple with the quantity of information to include, you also worry how best to present the material to get the point across. You remember last year's program on hazardous materials for the Housekeeping Department. The day after the training, a porter threw a carton of hazardous chemicals into the dumpster and almost created an explosion. You were asked skeptically, "So what did the training accomplish?" As you ponder this dilemma, you realize that staff turnover means starting over as soon as you have made it through one round of training.

Does this situation or the frustrations that emerge sound familiar? This chapter examines the reality of training in the hospital setting and provides tools and methods to develop a comprehensive training program. Section one starts with the purpose of training, the broad educational approach, and the constraints within hospitals. Section two outlines how to prepare for training sessions. Section three covers the training session: its content, checklist for planning training, a more detailed description of methods, and a sample format for training curricula.

Purpose of Training

Before becoming overwhelmed by the specifics of planning a training session, it is important to step back and think about the question, Why train employees? One answer is that JCAHO(PL1.8.) requires safety orientation for all new personnel and staff participation in continuing safety education and training. Beyond legal requirements, however, training can be one of the most important elements of a health and safety program.

Hospital workers face a multitude of hazardous conditions and substances. They need to know how to protect themselves on a daily basis. Failure to train can be disastrous for everyone. Workers get hurt. Patient care is compromised. Hospitals can become decertified.

Unfortunately, hospitals may only give lip service to the importance of training. Management often provides inadequate release time for vital training sessions. Trainers and employees need to demand the time necessary. Training provides people with skills and support to adopt safer work practices, to prevent accidents, to use their rights on the job, and to identify strategies with their coworkers to improve the work environment.

Training fundamentally involves the people exposed to hazards as active participants in creating and maintaining a healthy and safe work environment. Successful training encourages employees to participate in health and safety issues and problem-solving actions long after the training sessions end.

There are many types of trainings used in hospital settings, but not all are effective in promoting this goal of active participation.

- Training is not effective when people passively receive information, such as through a lecture or video presentation. Adult education theory tells us that people retain information better when they are actively involved in the learning process.
- Training is not effective when only information is presented. Learning theory tells us that people need to observe and practice skills to change their behaviors.
- Training is not effective if the focus becomes narrowed to teaching behavioral changes or specific competencies. Training also needs to include the context of behaviors and the analysis of obstacles and actions to improve health and safety.

In one hospital, for example, the trainer was continually frustrated that maintenance staff did not regularly use respirators even with extensive training. She discovered later that respirators were kept in one locked central office, which made access extremely difficult. If the training had considered obstacles, participants could have informed her of the dilemma and steps could have been taken to change policy.

The perspective presented in this chapter is that effective training means "Education For Action." This is an approach which (1) is based on real-life experiences, (2) incorporates dialog between teacher and learners, (3) is participatory, and (4) has the goal of empowerment.

To motivate participants, the content of training sessions should be based on the reality of people's jobs. With careful advance preparation (described in the next section), trainers can learn about their trainee's needs and hazards. To most effectively promote learning, however, the training should cover more than people's job experiences. Training should be based on the concerns, fears, or other emotions that motivate people to want to learn and, more importantly, act to change situations. If people's concerns remain hidden, they can block learning and block the use of safety information. Conversely, these concerns or "hidden voices" have the power to inspire learning and action.

After an extensive hospital training in San Francisco, for example, the emergency room technicians were still not observing universal precautions when working with severely injured patients. After investigation, it was discovered that the technicians felt they could not compromise patient care to take the time to suit up. These feelings were never discussed in the training. As a result, the technicians ignored the policy and continued to potentially expose themselves to AIDS. If the feelings had been discussed, people together could have developed a better way to protect themselves and their patients.

Methods which encourage dialog enable people in a safe group environment to share their concerns and their job experiences and to analyze the hazards in their environment. Dialog should start with the feelings and work issues that motivate people to care about health and safety. By sharing their feelings, people learn that they are not alone, that together they can analyze the barriers to a safe and healthy environment and can identify their role in protecting themselves and their coworkers.

Participatory education suggests that training sessions are microcosms of the level of involvement you as the trainer would like participants to assume. If you want people to become users of information and problem solvers on the job, they need to adopt those roles during the training session itself. Participatory methods encourage workers to listen to and learn from each other and respect each other's opinion. In addition to interactive methods, a participatory approach stresses the role of employees throughout all stages of the training process: in the planning (by identifying problems and content areas to address); in the training session; and in the follow-up actions after the training.

The goal of Education for Action is to empower people to act to improve health and safety for themselves, for their coworkers, and for their departments. One trainer (along with safety staff) in a hospital cannot ensure that health and safety regulations are being followed. It becomes important for all employees to become active participants in health and safety on a daily basis. Empowerment therefore starts with employees learning how to individually protect themselves. But it extends to people confronting and seeking to change unsafe conditions, to adopting preventive practices, and to working together to identify and solve health and safety problems. This could be through staff meetings, department health and safety committees, or union activities.

Education for Action is based on the writings of Brazilian educator Paulo Freire. In Freire's work, the purpose of education is to promote participation of people in gaining control over their lives and having a voice in improving their work environment. A Freire process requires three stages: active listening for the issues in people's lives (or conducting an ongoing needs assessment); dialog (or incorporating the issues into the training session for discussion); and actions (or identifying strategies to address the unhealthy or unsafe conditions that are discussed in the training).

But at this point you are probably wondering, "How will it work in the hospital setting?" You're confronted with management's double message: "Train, but there's no time." Additional constraints include few staff with skills in providing training and a high rate of hospital staff turnover, which requires ongoing training resources. This chapter suggests that Education for Action is a long-term process. You can begin with small adjustments to your existing programs that lead in this direction. Initial steps would be to consider many of the participatory methods presented in the next sections.

As discussed earlier, however, training is vital to protecting workers on the job. If this is our goal, management's mixed message is unacceptable. In the long run, all of us as trainers need to demand the time it takes for an effective training program: a program which assesses people's needs and concerns, includes interactive sessions, and promotes worker-initiated actions in each department and becomes a regular and continuing part of the work process.

In addition to more time and resources for training sessions, supervisors need to feel confident and competent in providing ongoing training to their staff. As interpreted by law, supervisors are responsible for implementing training. Since the training role and the use of participatory methods may be unfamiliar, it will be important to develop a train-the-trainer program for supervisors and shop stewards. This train-the-trainer program would review specific training responsibilities (i.e., MSDSs, infection control procedures) curricula, methods, and how to use them. More trainers ensure small interactive sessions and a decentralized approach, which promotes better follow-up and ongoing surveillance throughout the hospital.

Preparation for Training

As the new trainer, the hospital management has told you to immediately develop a Right-to-Know Training. The hospital's Hazard Communication Program, including training, should have been in place several years ago. You decide to begin with the nursing staff and find yourself bombarded with questions. Which chemicals in our department are hazardous? How do we find out? Where are the binders of Material Safety Data Sheets? Unfortunately, you don't have the answers. The inventory of hazardous materials is incomplete, MSDSs are sketchy and you are facing a frustrated group of trainees.

Advance preparation might have prevented this scenario. A broad range of information about each hospital department must be collected: the hazards and potential exposures faced by employees; the composition of the group to be trained, their job categories, backgrounds and experience; and your trainees' needs and concerns. This information will enable you to tailor your educational program to the particular group's needs, to anticipate potential obstacles to achieving your training objectives, and to ensure that the department can effectively implement the training messages.

The listening starts with an objective assessment of the work environment. Conduct a walk-through inspection of the hospital department and observe working conditions and practices. Collect information on the hazardous materials and processes being used. Environmental monitoring, if available, could quantify potential exposures. Learn the group's health and safety history. Have there been any accidents, injuries, or illnesses? Any outstanding citations for violation of health and safety regulations? Collect the appropriate forms: monitoring records; OSHA Log 200s; citations.

Identify whether health and safety procedures and protocols are in place. Training may lead to employee frustration when policies have not been previously established and departments cannot provide adequate follow-up.

If a department, for example, is scheduled for a training on Emergency Response, check the department's procedures. Are there written protocols for spills? Appropriate spill kits and equipment? Have evacuation routes been designated? If not, work with the department to develop procedures before the training session.

After collecting information on the work environment, assess the group you plan to train by listening to their needs and concerns. During the walk-through, you can talk to individual employees to gain an understanding of their attitudes: which safety practices work for them and which don't. Written surveys are useful tools for gathering anonymous and aggregate information about the group's needs. Try to assess employee experience with the subject matter, their level of literacy, and any special language needs.

To promote Education for Action and the goal of empowerment, look for issues that will motivate people to act. Are there problems that are making employees angry, afraid, or excited? Do people hold attitudes which become barriers to working safely?

> At a San Francisco hospital, for example, a training was planned for radiation technicians about the hazards of ionizing radiation. During the needs assessment, the trainer discovered that X-ray technicians were angry about the multiple and conflicting demands by physicians, about the broken gurneys, and about inadequate staffing which they felt jeopardized their safety. Discussion during the training led to proposed actions to address these safety issues.

The issues that emerge during the listening stage of the needs assessment form the basis of the training session: its objectives, the concerns that require dialog, and the methods to meet the objectives. Listening, however, does not stop with the formal needs assessment. It continues during the training session (Did the topics meet people's needs? Did other issues surface for future sessions?) and after the session (What follow-up would be helpful?).

In addition to the needs assessment, preparation includes the logistics of time; space (a comfortable meeting room with movable chairs); and materials (audiovisual aids, handouts). A major planning task involves attendance; how to assure adequate notification beforehand and staff coverage during training sessions. A written notice from hospital management which reminds administrators and supervisors that training is mandatory is often helpful in encouraging full attendance. Documentation of participant attendance and the content taught is also required by JCAHO. (A checklist of all preparation activities is included in Appendix 30.A.1.)

Training Session

Content of Training

Given the variety and complexity of hazards in the hospital work environment, the universe of potential training topics is extremely broad. Many hospitals include an orientation to basic health and safety precautions and hospital resources. Legal mandates require training on the life cycle of hazardous materials: their purchase, storage, use, and disposal. The Hazard Communication Standard requires specific training on the safe use of particular hazardous materials, such as ethylene oxide, asbestos, and anesthetic gases for employees with potential exposures. Required safety training includes topics such as fire and electrical safety, back care, disaster preparedness, and radiation safety. Infection control is a standard component of the training agenda. Other health and safety topics may range from working safely on video display terminals to occupational stress and personal safety or self-defense.

Training curricula should have certain common elements:

- *How to recognize the hazard.* Information should cover activities or processes that might lead to potential exposures; who might be exposed; routes of exposure; and how much exposure is too much (legal standard and dose–response relationship).
- *Health effects.* They should include acute and chronic effects, symptoms, diagnosis, and treatment (if applicable).
- *Prevention.* It includes engineering controls (substitution, administrative controls, and rotating workers); safe work practices; and the proper use of protective equipment.
- *Obstacles to working safely and strategies for addressing them.* In this part of the program, the training moves from the ideal to the reality of how to implement health and safety standards and better work practices. Even with the best of intentions, it is often difficult to make health and safety changes. Budgets are tight. Management may be reluctant to make the necessary financial commitment. Hospital staff face conflicting demands of patient care and self-protection. Employees may also deny their exposure to risks. Training sessions, therefore, must include current and potential problems, the barriers to change, and short- and long-term action steps to resolving the problem.
- *Legal rights in health and safety and resources.* They are available to employees, both in-house and area wide. This section reinforces employees' shared responsibility for protecting their health and safety on the job. The safety officer or industrial hygienist cannot be everywhere and emergency situations can occur without warning. Employees must be aware of whom to call for help, and of their rights to take appropriate actions to ensure their personal protection. While the discussion of rights will not be all-inclusive, employees should be aware of their "basic four" rights to protection under the law:
- The "right to know" about hazardous substances they may be exposed to on the job and how to work with them safely (as well as their right to training on these substances, monitoring, and medical examination records).
- The "right to file an anonymous complaint" with the Occupational Safety and Health Administration and request a workplace inspection. Of course, internal resources and the in-house chain of command should be utilized first and reviewed in the training session.
- The "right to refuse unsafe work" if the employee believes and can prove that doing the job would be a threat to life and limb.

- The "right to protection against discrimination" for exercising their right to a safe and healthy workplace.
- *Evaluation*. This section should assess whether learning objectives were met and invite subjective feedback from participants on the training session.

Training Objectives

After defining the content or hazards to be covered (such as cytotoxic drugs), the next step is to define objectives or anticipated outcomes of the training. These outcomes fall into four categories: changes in level of information, attitudes/emotions, behaviors, and action/problem-solving skills. It is important to include objectives from each category to promote the goals of active participation and empowerment. Once objectives are determined, methods can be selected which best achieve the desired objective.

Information objectives include the knowledge you hope participants will acquire during the session. Examples might be the hazards of exposure to asbestos or ethylene oxide, or the phone number to call if a hazardous situation is identified.

Attitudes and emotions may influence how people implement health and safety standards on the job. You may have heard comments like: "Most accidents at work are caused by careless or accident-prone workers," "I've been working with this stuff for years and I feel fine," or, "You want me to do MSDS training? That's not my job. I already have enough to do."

Attitudes in the same workplace may vary enormously and, if they are not identified, may present serious barriers to learning and positive actions. Supervisors who feel that most accidents happen at work because employees are careless may not perform an objective accident investigation but immediately blame the employee instead. Employees who feel invulnerable to hazardous materials or infection because of working for years without becoming sick may fail to take appropriate precautions. For empowerment education, attitudes and emotions need to be the starting point for discussion and incorporated throughout the training.

Behavioral skills are the desired competencies that employees will acquire as a result of training. Examples are the ability to respond to a mercury spill, read, and interpret a Material Safety Data Sheet, or safely lift a patient. Training often focuses on behaviors to the exclusion of the other three objectives.

Action/problem-solving skills include the ability to analyze a particular problem, identify its causes, potential solutions, and action steps for resolving it. Action objectives may include those steps that employees can take in their jobs to protect their health and safety, such as identifying and reporting potential hazards, requesting information on new materials introduced into the workplace, and working with fellow staff members and their union to ensure the work environment is safe. Often, as a result of a training, people identify multiple actions to change a situation. The problem cannot be easily solved and requires long-term commitment. In these instances, the efforts of the trainees are not merely as problem solvers, but also problem posers (in Freire's terms). Through dialog, they are "posing" for discussion the complexity of actions necessary to address the problem over an extended period.

Training Methods

There are many different ways to conduct a training session. The criteria for selecting methods include both an understanding of how adults learn most effectively and the training goals and objectives you hope to achieve. It is important to remember that adults expect to learn information based on their real experiences which can help solve problems. They learn best when they are actively engaged in participatory activities, dialog about issues, or hands-on practice.

To empower people to demand their health and safety rights, all training sessions should include action/problem-solving objectives in addition to the more common information, attitudinal, and behavior objectives.

In Appendix 30.A.1, a detailed description of methods is presented. The methods are grouped under the category of objective which they best achieve.

Training Evaluation

Evaluation is an essential part of the ongoing listening process: How effective was your training? Did you accomplish your classroom objectives? Did the training measure up to participants' expectations? Is your program having the desired impact on work practices back on the job? Did your training uncover new issues and therefore new objectives to incorporate in future training sessions?

Your evaluation will attempt to answer these questions and should be developed before you train. Not unlike training methods, evaluation techniques are based on your objectives. A common approach for measuring information objectives is to administer a post test at the conclusion of training. Observation by the trainer in the classroom and health and safety personnel and supervisors back on the job is the most effective way to assess how well behavioral skills have been learned.

Action skill objectives must also be evaluated over time. To assist you in this process, a list of desired indicators of activity can be generated and assessed at proscribed intervals. Sample indicators are

- Requests for information or equipment
- Participation in health and safety committee meetings
- Reports of potential hazards or violations

You also want subjective feedback from participants about what they liked most and least about the training, the instructor's performance, and recommendations for future training. Questions which elicit this information can be added to the post test. When distributing the evaluation form to participants, it is helpful to remind them that their feedback will be used to shape the program. You can cite examples from the past where employee input has been incorporated.

Conclusion

Implementing the Education for Action approach in the hospital environment is a challenge, given the multiple demands on limited training staff, time, and resources. However, it is also a responsibility. Training is required and essential to the success of any health and safety program. Hospital workers are on the front line exposed to hazards and the stresses (and rewards) of patient care. They need to become active participants in protecting their health and safety on the job. Management needs to recognize the value and time commitment required for action-oriented training. Education for Action, in the context of a comprehensive health and safety program, can pave the way toward legal compliance, an empowered work force and a safer workplace.

Appendix 30.A.1

Training Methods

Checklist for Planning a Training Session

1. **Group to be trained**
   Number of employees
   Work shifts
   Job categories

Prior health and safety training
Special training needs (language, literacy)
Health and safety concerns or problems

2. **Work History**
   Accidents and injuries, OSHA log 200
   OSHA, JCAHO citations or violations
3. **Work Environment**
   Hazardous materials used
   Work processes, practices, and potential exposures
   Personal protective equipment
   Policies and procedures
   - Ex: Hazard Communication (inventory, MSDS binder, labels, written plan, etc.)
   - Emergency response (procedure, spill kits, and equipment)

   Training records
4. **Logistics**
   Comfortable room (movable seats)
   Refreshments
   Equipment and materials (handouts, AV equipment, blackboard, etc.)

Methods for Information Objectives

The presentation of factual information is an important component in most training sessions. Our challenge as trainers is to present this information in the most interesting and interactive way possible. The methods which follow proceed from the most passive (lectures and certain types of audiovisuals) to more active approaches. Teaching methods are summarized in Table 30.A.1.

Lecture

While the lecture may be appropriate for reaching large audiences, an effort should be made to involve the audience. Interesting anecdotes can enliven the session and provide examples of ideas. For example, lectures on the right-to-know law can begin with the retelling of the incident which led to the passage of the law. Workers in a pesticide-formulating plant in California discovered that many in the plant had been trying to have children without success. Their suspicions about a common cause led to an investigation which revealed that their infertility was the result of a chemical in the pesticide they produced. Participants' interest is maintained as they hear an example of worker-initiated investigation of the work environment.

In addition to telling stories which involve the audience, you can encourage active participation by posing questions which require students to apply information being presented. For example, "If you thought that a product on your job was making you sick, what would you do?"

In the interest of maintaining audience interest, lecture time should be kept to a limit. Some people say that 20 min is the longest time anyone can assimilate lecture information.

Audiovisuals

The most basic use of visuals is posting a colorful, written agenda in front of the room which lists training topics to be covered and approximate time estimates.

Prepared videos and slide shows can be an entertaining way of teaching factual information which both maintains audience attention and are effective for large groups. The audience, however, remains passive. Interaction with the participants can be enhanced by preparing questions to discuss after the show.

There is a current trend toward the use of interactive video with individuals to encourage audience involvement. In this method, the participant watches a video presentation which requests answers to questions or written input from the learner at certain intervals. While this approach is effective for ensuring that the learner is involved, the interaction is with a machine rather than with

TABLE 30.A.1 Education for Action: Teaching Methods Chart

Teaching Methods	Strengths	Limitations	Objectives Achieved
Lecture	Presents factual material in direct/logical manner	Experts may not always be good teachers	Information skills
	Contains experiences which inspire	Audience is passive Learning difficult to gauge	
	Stimulates thinking to open a discussion	Needs clear introduction and summary	
	For large audiences	Needs time and content limits to be effective	
Worksheets and questionnaires	Allows people to think for themselves without being influenced by others in discussion	Can be used only for short period Handout requires preparation time	Information skills Attitudes/emotions
	Individual thoughts can then be shared in small or large groups		
Brainstorm	Listening exercise that allows creative thinking for new ideas	Can become unfocused	Information skills
	Encourages full participation because all ideas equally recorded	Needs to be limited to 10–15 min	Attitudes/emotions
	Ideas equally recorded		
	Can be used to quickly catalog information		
Planning deck	Allows students to learn a procedure by ordering its component parts	Requires planning and creating of multiple planning decks	Information skills
	Provides experience in planning as a group		
Audiovisual materials (films slideshows, etc.)	Entertaining way of teaching content and raising issues	Too many issues often presented at one time to have a focused discussion	Information skills
	Keeps audience's attention Effective for large groups	Discussion will not have full participation	

(*continued*)

TABLE 30.A.1 (continued) Education for Action: Teaching Methods Chart

Teaching Methods	Strengths	Limitations	Objectives Achieved
Audiovisuals as triggers	Develops analytic skills Allows for exploration of solutions	Discussion may not have full participation	Action/problem-solving skills
Case studies (trigger)	Develops analytic and problem-solving skills Allows for exploration of solutions Allows students to apply new knowledge and skills	People may not see relevance to own situation Case and tasks for small group must be clearly defined to be effective	Action/problem-solving skills
Role-play session (trigger)	Introduces problem-situation dramatically Develops analytic skills Provides opportunity for people to assume roles of others Allows for exploration of solutions	People may be too self-conscious Not appropriate for large groups	Action/problem-solving skills
Report-back session	Allows for large group discussion of role plays, case studies, and small group exercise Gives people a chance to reflect on experience	Can be repetitive if each small group says the same thing Instructors should prepare questions to focus discussion so not repetitive	Action/problem-solving skills
Prioritizing/planning activity	Ensures participation by students Provides experience in analyzing and prioritizing problems Allows for active discussion and debate	Requires a large wall or blackboard for posting Posting activity should proceed at a lively pace to be effective	Action/problem-solving skills
Hands-on practice	Provides classroom practice of learned behavior	Requires sufficient time, appropriate physical space, and equipment	Behavioral skills

Source: Adapted from *Labor Educator's Health and Safety Manual*, Labor Occupational Health Program, University of California, Berkeley.

an instructor who can respond to questions or with fellow learners who raise provocative issues and questions.

Worksheet Questionnaires

Questionnaires can serve as excellent discussion starters to assess how much participants already know and to provide information that would ordinarily be given in a lecture. Even if people are guessing the answers to many of the questions, they enjoy the process and become interested in knowing the information when the instructor reviews the answers. Questionnaires can also be used at the end of a session for review and to evaluate how much information has been retained. They can be administered in a variety of ways. You can invite participants to complete the questionnaire independently. You then ask for a show of hands as you read each potential response. Volunteers with different responses are encouraged to justify their response, leading to a lively discussion. Then you give the correct answer. Worksheets can also be completed by participants in small groups, encouraging consensus.

Sample information questions are

1. Which of the following has the highest exposure to anesthetic gases?
   a. Scrub nurse
   b. Surgeon
   c. Anesthesiologist
2. Asbestos is dangerous in all forms (pot holders, tiles, wallboard, etc.).
   a. True
   b. False

Planning Deck

The planning deck is an activity which involves participants in identifying and ordering the components of a task or procedure to be learned. You prepare for the activity by listing each step in the procedure (such as responding to a mercury spill or implementing the right-to-know law) on a large card. Steps for spill response might include: inform your supervisor, call the industrial hygienist, isolate the area, locate the spill equipment, etc. Multiple decks then need to be created to allow for working in teams. The task of each team is to reach a consensus on the order for the steps. The first team to complete the task reports and explains their procedure to the group. Groups with a different order can justify their positions, followed by discussion and confirmation of the desired order for the procedure.

Methods for Attitude and Emotion Objectives

Since attitudes about health and safety can effectively block learning and implementation, an effort to identify and explore group attitudes should be incorporated early in the training session. By bringing attitudes into the training session directly, people's feelings are given recognition and learning is facilitated. Emotions can also emerge through the methods that promote action/problem-solving objectives. Two of the more straightforward approaches to exploring attitudes are described below.

Worksheet Questionnaires

The same questionnaire described above which is used to provide factual information can include questions which elicit attitudes. Sample attitudinal questions are

1. Most accidents are caused by careless workers.
   a. True
   b. False
2. If your job was making you sick, you would know it.
   a. True
   b. False

A discussion of attitudes is encouraged by inviting participants with conflicting viewpoints to present and justify their opinions. Your role as instructor is to ensure that participants consider all points of view, rather than making a definitive statement about the "correct" attitude. For example, rather than concluding the discussion about accidents with the assertion that there is no such thing as a careless worker, support the need for an accident investigation, acknowledging that accidents are caused by a variety of factors.

Brainstorm

A brainstorm is a listing exercise which can be utilized to quickly catalog a body of information, as well as to explore attitudes. It encourages full participation because all ideas are equally recorded. In a training session on infection control, you can ask participants for their opinions on why needlesticks continue to occur in the hospital or why universal precautions are not being observed. You record the items on a large sheet of paper. When all ideas have been recorded, you can help group the items. New approaches for dealing with the issue can also be elicited by asking participants how they think the problem should be addressed.

Brainstorms are also an excellent tool for presenting factual information, building on the group's preexisting knowledge. In a training session on hazardous waste, participants can be asked to name the hazardous wastes generated by their department. You record the items, add and correct the list, and arrive at the complete list for the department. While brainstorms are a lively way to accomplish a variety of objectives, they can become unfocused. You should be sure to limit the discussion to 10–15 min.

Methods for Action/Problem-Solving Skill Objectives

Action/problem-solving methods are essential to promoting Education for Action. These methods not only facilitate active participation, but also promote dialog leading to actions and strategies to improve the work environment. The tool for these methods (from Freire) is the use of a "discussion trigger," followed by an inductive questioning method that promotes dialog leading to action. A discussion trigger can be any form: a role play, a case study, a slide or series of slides, a song, or a trigger video.

The content for the trigger emerges from the listening, needs assessment stage: the critical issues or emotions that people care about related to health and safety. A trigger brings together these issues into a physical form that allows trainees to project their emotional and social responses in a focused fashion. An effective trigger has the following characteristics:

- It is familiar, a situation immediately recognized by the group.
- It is presented as a problem that has both individual and environmental components (i.e., respirators may be used infrequently because of individual attitudes and inaccessibility).
- It is presented as an open-ended problem with no solution; the solutions will come from the group discussion.
- Problem should not be overwhelming, but should allow for people to come up with small actions for change.

To lead a dialog on a trigger, there is a five-step questioning process guided by the acronym SHOWED that enables people to share their personal reactions, analyze their environment, and come up with actions. People are asked to describe what they "See" in the trigger. What are people expressing on the surface? The next step is to define what's really "Happening." This deeper level often emerges with the question, What do different people in the trigger (i.e., role play or case study) feel about their situation? The third step is to ask how this trigger relates to "Our" lives. Is it the same or different from our situations on the job? By allowing people to share their experiences and their own feelings, they become motivated to act to change the problems raised. The fourth step is to ask "Why is" this a problem in our department or in the hospital. With this step, people perceive the context of the problem and identify barriers to changing hazardous conditions. The final question is, what can we as individuals or as members or a group "Do" about the problem? This is the step that leads to action and to promoting the goal of empowerment and long-term involvement in health and safety. Examples of triggers in their different forms and with sample follow-up discussion questions are presented below.

Role Plays

Role plays introduce problem situations dramatically and allow for analysis and discussion of solutions. They can also provide a forum for exploring group attitudes on a particular subject. The content for the role play is drawn from listening for concerns of the trainees. Rather than using a case study or a slide to portray the problem, you as instructor translate the trainees' concerns into a dialog to be read by participants during the class.

For example, in preparing for a training session on AIDS for a group of custodians, you may hear a range of attitudes and perceptions: denial of personal risk, incorrect information, skepticism about information being presented by authorities, hysteria, and perceived problems of lack of equipment. The following role play contains these emotions and attitudes and identifies a character to represent each one.

JOE: With that new AIDS ward in the hospital, I'm sure we'll all have AIDS in no time.
SAM: What are you worried about? You're not gay.
JOE: They say I can still get it, but who knows what's true anymore because there's a new study every day.
SARA: All I know is that I want to have gloves on before I touch *anything, anywhere* in this hospital… if I could ever find any!
SAM: You two are making a big deal out of nothing. I've got work to do.

During the training, three people can become actors and read the role play. Following the reading, you lead a discussion, eliciting the group's responses to the issues being portrayed: What's happening here? How does each character feel? Have you ever felt this way? Why is there a problem getting gloves? What can be done about it? You may also get some important information during the session, such as employee suggestions for equipment access and distribution. Role plays can also be created spontaneously by participants. You distribute roles to volunteers, such as the skeptical employee who practices self-denial or the fearful, somewhat hysterical employee, to be played using their own words. While spontaneous role plays foster creativity, participants are often self-conscious and difficult to recruit. A written dialog is often less intimidating.

Audiovisuals

Some of the liveliest additions to a training program are slides including the individuals who will be trained and their work environment. Photography should be a regular component in the pretraining walk-through at the hospital. Subjects would include safe and unsafe work practices, locations of MSDS binders and spill equipment with some candid shots of department employees. It is best to include people in the photos.

While slides selectively organized can present information, such as the steps in implementing the right-to-know law, they can also trigger analysis of particular problems which were identified in the needs assessment and listening phase. If back injuries from improper lifting are a problem, for example, a picture of an employee lifting a patient by himself could be used to explore the causes of the problem. The same inductive questioning process outlined in the beginning of action objectives could be utilized to explore the source of the problem: lack of information, time, staff, and equipment. The discussion will eventually arrive at action steps, both short- and long term.

Video can similarly be used to trigger discussions of barriers to working safely. A videotape on reducing needlestick incidents was produced at San Francisco General Hospital. Using volunteer health care workers as actresses, part of the tape portrayed a nurse who, while at lunch with her coworker, admits that she stuck herself that morning and did not seek treatment at Employee Health. Responding to the concern of her coworkers, she explains that the patient was a child who could not have had HIV disease. In addition to eliciting a strong emotional reaction, the video provides a focus for analyzing the problem and exploring solutions.

Case Studies

The cases should be drawn from real and familiar situations that have occurred and problems that were identified during the planning process. Participants read and respond to questions about the case in small groups, followed by a report-back session in the large group. When used at the end of a session, case studies allow participants to apply new knowledge and skills.

The following case, based on a real hospital incident, was utilized in a hazardous materials training for nurses.

You are the charge nurse when a coworker in the dialysis unit spills a glass bottle of liquid while sterilizing the equipment. She calls the Housekeeping Department to clean up the spill. The porters arrive and begin to soak up the liquid with rags. They are soon overcome by the fumes.
Questions for discussion:

1. What is the problem here?
2. How did each person perceive the problem?
3. Do you know of other similar incidents?
4. Why do these incidents occur?
5. Could this situation have been avoided? If so, how? What recommendations would you make to prevent problems like this from occurring in the future?

Prioritizing/Planning

Identifying potential problems in the workplace can be very overwhelming. Not all problems are equally hazardous, nor will they be dealt with immediately. During the training session, employees can be encouraged to participate in the planning process by ranking problems of concern and breaking down an action plan into its component steps. In this activity, the instructor asks a general question, such as "What is one obstacle that you will face back on the job when trying to put into practice what you have learned in training?" or, to be more specific, "What is one obstacle to preventing back injuries in the hospital?"

Each participant writes one obstacle in large print on a piece of paper. The instructor then asks for a volunteer to share his obstacle and pass the piece of paper to the front of the room for posting. The instructor then calls for obstacles with a similar theme, posting each piece of paper under the former to create a vertical column. A new column is created for each new theme. Proceeding in this manner, a visual representation of the most pressing problems is created—with the longest list usually reflecting the problem of greatest concern. Following the identification phase, the instructor can initiate a discussion of each obstacle, barriers to resolving it and positive action steps that can be taken.

For construction of a quick plan for immediate action, the same process can be used by asking the question, "What is one step you can take back on the job to help create a safer work environment?" The steps generated by the group can then be evaluated and prioritized.

This activity helps the instructor get a better understanding of the problems encountered by the learners and provides practice in planning and problem solving skills for participants.

Methods for Behavioral Skill Objectives

The most effective way to achieve objectives for behavioral skill development is to provide the learners with opportunities to practice in the class.

Hands-On Practice

Since all employees need to learn how to read and interpret Material Safety Data Sheets (MSDSs) as part of the training component of the Hazard Communication Standard, they should experience reading an MSDS during the training. The initial step in the training process is to help employees understand the utility of the MSDS by analyzing one that is well written with the instructor. The instructor asks the employees what they would want to know about this product before they started to work with it. Based on the employees responses, the class looks for the information on the data sheet—and in the process learns what the sheet contains, its strengths and limitations. With more time, the class is divided into small groups and given the task of reading different MSDSs as a group. The group assignment is to make a list of the most important things a new employee should know about the product. Back in the large group, spokespeople for the small groups deliver brief "new employee orientations" on each product.

Other examples of learning behavioral skills by doing include using spill kits to clean up mock spills, drafting hazardous waste labels in small groups, and practicing evacuation procedures. You can evaluate competency by circulating and observing the small group performances. You can also ask the small groups to demonstrate and explain a certain procedure for evaluation by the large group.

Demonstrations

Volunteer demonstrations of the proper use of protective equipment, with additions and corrections from the large group, can be very effective. In a training session on hazardous materials for the Housekeeping Department, the porter's cart of cleaning agents and protective equipment was brought to the classroom. The instructor posed hypothetical situations such as "You need to remove graffiti from the elevator walls. What product and protective equipment would you use? What about disinfecting the nursery?"

A Sample 1 H Agenda

The participatory training methods which form the core of Education for Action are more time consuming. Although you may not have the time or staffing to initiate the ideal, comprehensive training program right now, you can begin by making small, action-oriented adjustments to your current program.

In one example from San Francisco, trainers were faced with the task of providing hazardous materials training to several thousand municipal hospital employees. To maximize participation, the session was delivered on all shifts to groups of no more than 30 people. The trainers adapted some of the methods described above to create a 1 h curriculum.

The sessions began with a brainstorm, asking the group, "What are some hazardous materials you work with?" Using this list, they continued the dialog with the question, "What is it about these products that make them hazardous?" A worksheet questionnaire provided the basis for discussion of symptoms, dose–response, and legal rights. Proceeding to the requirements of hazardous materials regulations, the trainer introduced the discussion by telling the story about pesticide formulators in California. Hazardous materials regulations were reviewed by using slides of familiar people and hazardous locations to trace the path of chemicals in the hospital from purchase to disposal.

Although the ideal training would use participatory methods to review MSDS information for all potentially hazardous products that employees work with, this shorter training focused on the data sheet of a common product (isopropyl alcohol). As alcohol users, participants were asked what they wanted to know about the product. Together, the instructor and the class looked for answers to these questions on the MSDS.

To clarify responsibility for cleanup of different types of spills, the trainer used brief vignettes on spill response, such as, "For a toxic (or body substance? or unknown substance?) spill, who would you call?"

A portion of this sample 1 h agenda is included in a format, which can be used for all health and safety training. It outlines the time allotment, objectives, activities, and materials needed for each topic, providing a useful guide for all the trainers who will teach this course.

Was 1 h an adequate time allotment to achieve the training objectives? No, but the path toward a comprehensive Education for Action program is a process. While lobbying for more time and resources, the trainers used participatory methods to reach hundreds of workers with much needed information. A sample format for a 1 h training session is shown in Table 30.A.2.

TABLE 30.A.2 Sample Format: Part of 1 h Hospital Training

Time	Topic	Objectives	Activities	Materials
2 min	Introduction	1. Introduce training	1. Trainers introduce themselves 2. Review today's agenda	1. Agenda paper
10 min	Hazardous materials and health safety	1. List hazardous materials in their workplace 2. Define hazardous materials	1. Ask: "What are some hazardous materials you work with?" (List on board) 2. Ask: "What is it about these products that make them hazardous?" (List on board)	1. Board & markers 2. Hand out: "How to ID hazardous materials" 3. Hand out: "Definition of hazard classes" 4. Questionnaire—Part 1
7 min		1. Define how hazardous materials affect the body	1. Use "Intro to Occupational Health and Safety Part 1" to provide basis for discussion on symptoms close-response, and protection	
8 min	Laws	1. List requirements of Hazardous Materials Regulations	1. Introduce with DBCP story 2. Use slide show to provide basis for discussion on a. Chemical life cycle b. Right-to-know: Inventory, MSDS, MSDS accessibility, training, labeling requirements, secondary container labeling, labeling problems including unknowns, written plan, and personal protective equipment. c. Permit and disclosure ordinance d. Emergency release response e. Hazardous waste	1. Series of slides 2. List of right-to-know steps on paper

31

Stress Factors

Union Perspective on Employee Education 31-1
Introduction • Limits of Enforcement • HCWs and Their Own Health • Rodney Dangerfield Is Not Alone • Some Successful Ideas I Didn't Invent • Examples of Worker Education Success Stories • Some Resources for Worker Health and Safety Training • Conclusion
Occupational and Environmental Health Training for Hospital House Staff 31-7
Training Issues • Needs Assessment • Hazards in the Health Care Environment • How Is This Information Provided? • Summary
Appendix 31.A.1 31-10
Appendix 31.A.2 31-11
Appendix 31.A.3 31-12
Appendix 31.A.4 31-13
Appendix 31.A.5 31-15
Nursing Student Health and Safety Training in the Hospital Workplace: Risks and Responsibilities 31-17
Introduction • Hospital Workplace Hazards Faced by Nursing Students • Special Risks and Problems Facing Nursing Students in Hospital Settings • Special Issues and Problems Facing Hospitals • What Nursing Students Are Currently Taught about Hospital Workplace Hazards • Key Components of an Effective Approach to Educating Nursing Students about Hospital Workplace Hazards • Conclusion
Acknowledgments 31-24
References 31-24

Laura A. Job
New York State Department of Labor

Wendy E. Shearn
Kaiser Permanente Medical Center

Kathleen Kahler
Kaiser Foundation Hospital

Jolie Pearl
San Francisco General Hospital

Marian McDonald
Tulane University

Union Perspective on Employee Education

Laura A. Job

Introduction

I worked in a five-and-ten, a supermarket, and a cafeteria before I worked as a laboratory technician in the blood bank of a major medical center in New York City. I worked in that blood bank for close to 9 years. The second half of those 9 years differed from the first in two ways: I worked part-time in another hospital and I had become a union activist.

I left my blood bank job to help develop, coordinate, and eventually direct the occupational safety and health program for the National Union of Hospital and Health Care Employees. This gave me the opportunity to work with health and safety specialists; union activists; and federal, state, and local officials. Of much more importance, I had the opportunity to work with all kinds of health care workers (HCW) throughout the United States and occasionally with their administrators or safety directors as well.

So, for close to 20 years, I have been involved in the health care field, and for more than a dozen I have paid particularly close attention to the injuries and illnesses of others. I have experienced a lot, learned a lot, and thought a lot about the whole experience. And I have come to one main conclusion the best way to ensure that hospital workers or any other working people are not sickened, injured, disabled, or killed by their jobs is to give them the information and ability to protect themselves.

Limits of Enforcement

I think of government occupational safety and health standards as "minimum wage health and safety." A standard is important because it says very clearly what unacceptable working conditions are. A standard, whether wage or health, has some token penalties for violators and some modest system to enforce compliance. But even if government and business had a total commitment to a maximum enforcement effort, which they do not, there could not be an industrial hygienist and safety inspector in every workplace around the clock. We also could not expect that enforcement staff would be experts on the specifics of each workplace.

There are three things I would like you to think about: minimum wage does not cover all workers; plenty of employers pay workers who are covered by the standard less than the minimum wage anyway; and minimum wage provides a standard of living below the poverty level. That is, you can't live on it.

When you apply the analogy to government health and safety standards, there is quite a resemblance: not all workers are covered by government standards and many known hazards are unregulated; plenty of employers knowingly or unknowingly violate job safety standards; and workers can be harmed at legal levels of exposure. That is, you can't live on it.

Take it even further. Does this analogy hold up in the health care setting? Do the health and safety standards set by the hospital effectively cover all workers? (The word "effectively" is important here. We have all seen those dusty policy books that are never opened until after the injury.) Do all department heads and supervisors enforce the safety and health standards set by the administration? And can workers live on it—that is, are injuries and illnesses still occurring in hospitals with safety and health standards? You bet they are.

Clearly, we need to get comprehensive coverage, stronger standards, and stricter enforcement to improve the survival of health care employees, but just as clearly, legal remedies and a conscientious health care management cannot do the whole job either.

HCWs and Their Own Health

I am far from the first person to think of worker education as the solution to this problem. But because others did, as a worker I experienced the benefits of worker education, and later as a labor educator and activist I saw it benefit many others. Education works and works well.

Training workers to recognize hazards and know how to get them abated is a successful protection strategy for many reasons, some of them obvious. Train 10 employees or supervisors and there are now 10 more people in the hospital on the lookout for potential hazards. Train 20 and you have doubled your safety patrol. If you train 100 people from each shift and department, you have round-the-clock hospital wide coverage 365 days a year. Your safety patrol has expert knowledge about the tasks they perform and conditions they work under, for they experience them every working day. And your safety patrol is highly motivated, for they are saving their own lives and protecting their unborn.

Both compliance with health and safety regulations and morale will improve when employees, including managers, know the good reasons for the regulations. This strikes at the heart of the issue of respect for the workforce, whether they are working in entry-level positions or as highly educated professionals.

Rodney Dangerfield Is Not Alone

I can say with confidence that Mr. Dangerfield has plenty of company among HCWs, who "don't get no respect" either. The most consistent and persistent theme that has been raised by the HCWs I have taught and worked with was that their concern for and interest in their own health and safety was not taken seriously. "Don't worry, it's safe" was a commonly reported answer to questions about potentially hazardous situations, even though it often turned out that things were not as safe as they should have been. Often the response was even more patronizing and paternalistic, especially if the worker was female, of color, and/or in a "less-skilled" position.

Of course, HCWs should be worried about their job exposures. Every day they work with pathogens, toxic drugs and reagents, and radioactive materials in environments that are rarely well controlled. Unless HCWs are trained and educated at work, they will bring to the job enough information to be concerned, but not enough to differentiate between the truly dangerous situation and one that is safe.

When I worked in the lab, I used to pass a door that had one of those purple propeller radiation signs on it. I recognized the symbol from the old fallout shelter days of the 1950s, but had no idea what I was (or was not) supposed to do. Stay away from the door? Hold my breath if I had to go inside? I never knew and always had an eerie feeling when I went past. On the other hand, working in the blood bank and at high risk of contracting hepatitis B infection from the exposure to blood and blood products, I only received anecdotal information about my risk and protective measures from one of the old-timers, who had herself been trained in calmer, stricter days. Well, as it turns out now, I should have worried less about radioactivity, and more about the hepatitis B virus. The point is that there are a lot of working people worried when they don't need to be and not at all worried when they should be, because the people with the information won't share it.

I have heard the arguments—and I don't buy them—that information about potentially dangerous substances such as asbestos or cancer drugs will just alarm employees; or that pathogens, toxics, and radioactive materials are too technical for the average person to understand and therefore again, employees will be confused and upset, not enlightened. It has been my experience that when presenting information causes concern, it is more likely that the fault is in the presentation, not in the goal of educating people about self-protection. Certainly, microbiology, toxicology, radiology, and industrial hygiene can be extremely technical, complicated, dry, and remote subjects, but they don't need to be, and the basic information that workers need to protect themselves can be understood by anyone capable of securing a position in the health care industry. Successful and effective educational programs don't happen easily or by chance; they take thought and commitment to the task. If you don't think education will work—it won't.

Some Successful Ideas I Didn't Invent

There is a large community of labor educators and health and safety activists who attend each other's sessions, distribute their manuals, fact sheets, and training materials, and adapt and modify each other's work to suit the task at hand. I am therefore indebted to others, known and unknown, for all of the following. I pass these ideas on because they worked for me as an educator; and the only credit I can take is knowing a good thing when I see it!

Be Clear about What You Want to Do

You would be surprised how many times I got myself out of a mental muddle by reestablishing my goal. I have found this to be a useful idea in situations other than worker education, but it is very important in an educational program. After all, how can you get your message across if you aren't too sure exactly what the message is? Once you have clarified your goal, you can set priorities, emphasize, simplify, and make your material useful and relevant to your target group. Otherwise, you are just throwing out information, not educating.

Have the People You Want to Educate Help Plan the Session

This is important in several ways. On the most obvious level, it makes sure your program covers all their concerns. It also will keep you from trying to educate a group about safe handling of the toner for the photocopier when there is a rumor racing through the hospital that the patient who hemorrhaged all over the emergency room yesterday is infected with the AIDS virus. But most important, as in the session itself, you will learn a lot and the session will be better for it. And better received, because you will be responding to the needs of the group, not telling them what you think they need to hear.

Respect the Knowledge of Your Target Group

Working people have a lot of knowledge about the work they do and about the potential hazards they face at work. The role of the health and safety educator is to help people discover how much they know and to aid them in putting this information in a context that gives them confidence about safe and unsafe conditions and procedures. There are teaching techniques that are useful for doing this—lecture is not one of them. Limit your lecture time to the very small body of information that only you have, and that the class can't figure out.

Techniques that do work include asking each member of the class to list two or three examples of "whatever" from their own experience. When the "instructor" compiles a list of all the examples from the entire class, that list is close to being comprehensive. Not only is the class not bored to tears by the "teacher" intoning a long list of all the whatevers, but each person will also remember the whatevers better, especially those whatevers that apply to themselves.

Another technique is the use of problem-solving exercises. Pose a situation similar to one you wish the class to be able to handle on their own and give them the information they need to address it. Several people work on the same problem, so no one person is on the spot. Afterward, the class can point up the strengths and weaknesses of each other's solutions. By "practicing" in a class setting, participants see in advance where they need help and are better prepared when it counts.

Keep It Simple

Don't clutter your message with knowledge for the sake of knowledge, or use technical or lofty language where the vernacular will do. A good rule that was passed on to me when I was first starting out is to say "Workers need to know this because…" before including a topic in a presentation. If you don't have a good ending to that sentence, save the topic for an accredited course.

Make It Relevant

Class exercises, fact sheets, examples, etc. should clearly relate situations and information that the target group can recognize and identify with. Your planning committee can be very helpful here.

Want to teach about Material Safety Data Sheets? Have the class get one on something they work with and are worried about instead of showing them one on a chemical they have never heard of! AIDS and herpes were the best-researched diseases in my sessions on infection and isolation techniques.

Use Humor

Sure, this is serious stuff. After all, you are literally saving lives. Nevertheless, a session that is occasionally amusing is occasionally entertaining and therefore well remembered. Funny examples can be pulled from real life, and made funnier if exaggerated and the characters are given descriptive names and personalities to match. Most of my problem-solving exercises took place at Makeumwell Hospital. You get the picture.

Plan for Your Follow-Up

The expression "Use it or lose it" applies to more than just sex and algebra. If you are talking to working people about their occupational health, your goal is to encourage or discourage specific behaviors on the job. You should plan now how to make that as easy as possible after your session is over. Usually this can be done by way of giving the class a "tool" to break down what you want into small, easy steps. Do you want the class to spot and report certain hazardous conditions? Design a simple form or checklist the class can use easily to do this. Do you want the class to change its procedure to a safer one? After discussing the reason for the change and going over the new procedure in class, preferably with a hands-on practice session, you may have the class draw up a list of necessary supplies, etc. and develop a step-by-step instruction plan to post at the work site.

Examples of Worker Education Success Stories

The stories are true; the names are false. I am not including them to brag, but to testify that occupational safety and health education do work, sometimes wonderfully! (OK, I am bragging just a little. But "just" a little.)

Eve Beattie

Eve Beattie took the first class I ran for the union on hazards in health care institutions. Eve was a tough, active, outspoken union delegate from the dietary department of a large medical center in New York City. We had assigned a class project for each participant to identify a hazard in his or her workplace and take the steps necessary to get that hazard corrected. In class, we discussed the different kinds of hazards and the different means of getting them abated. One week Eve came to class clearly angry and frustrated. She volunteered (Eve was not shy!) that she was "not" going to be able to hand in her project. "Why, what's the problem?" I asked, wondering if she had selected something too hard, or was having trouble with her supervisor, or 10 or 12 other impediments to the completion of her assignment. I just couldn't wait to hear the problem so I could try out my new role as counselor. "I can't do my assignment, because every time I point a hazard out to my boss, they go and fix it right away." Eve thought she failed because she did not get the Occupational Safety and Health Administration (OSHA) inspector in. Getting five hazards corrected in 4 weeks sounded like success to me!

Ellen Dunston

Ellen Dunston is a nurse at one of the major hospitals in Connecticut. She attended a session on radioactive substances in hospitals. In the session, I recommended two reference books that I had found to be very helpful. Ellen bought the books; Ellen read the books; Ellen studied the books. Ellen hunted down and critiqued the hospital's policy manual on radioactive materials. Ellen then critiqued the way radioactive materials were really handled day to day in the hospital. When she was finished, Ellen had a list of 11 ways to improve the protection of HCWs and patients from unnecessary exposure to ionizing radiation. Ellen took this document to management, which at first tried to pooh-pooh her concerns (the old "Don't worry, it's safe" routine), and then tried to stonewall ("You are wrong. We are right."). Ellen persisted. She knew her stuff. Then management revised their manual, tightened their procedures, and set up a union management committee to see that the ALARA principle (exposures to radiation kept As Low As Reasonably Achievable) was adhered to by the hospital.

Anonymous Hospital Safety Committee, Anonymous, PA

Following the principles outlined above as "Keep it simple" and "Make it easy," I had spent weeks in front of the word processor taking the OSHA standard on ethylene oxide (which is written in a language I call "federal register") and turning it into a checklist written in English. So I was ready when I got a call from Susie Helper of the Anonymous Hospital for information on ethylene oxide. Susie called on behalf of the union safety committee members, who passed out the checklist to the workers in central supply where the gas is used. When the checklist was completed they had found 14 violations of the standard. This information was formally presented to the hospital management at the hospital's safety committee meeting. The hospital was clearly wrong and began to correct the violations immediately. Some of the improvements took months to achieve, but Anonymous Hospital is now in compliance with the OSHA standard on ethylene oxide.

And then there was the AIDS checklist passed out at each workshop...but you get the idea.

Some Resources for Worker Health and Safety Training

Hopefully, you are now sold on the idea that the best way to protect workers is to educate them to protect themselves. There are numerous sources of information, education, training, and technical assistance in the United States to help you do it. Labor unions, trade and other employer organizations, educational institutions, and nonprofit organizations including "committees for occupational safety and health" (COSH groups) now supplement the traditional safety councils in offering programs, training materials, and assistance on dealing with job hazards.

This was not always the case.

In the 1970s, the OSHA created the "New Directions" grant program. Under this program over $1 million is awarded each year to organizations to develop an occupational health and safety program to serve their constituency. The program consists of a 1 year planning grant followed by a 5-year developmental grant. The level of funding is set each year when the grantees submit a new proposal. The goal of the New Directions program is to create competent, self-sufficient health, and safety resource centers.

By and large, the program succeeded. Perhaps only a small number of granted programs were truly self-sufficient after 6 years, but the New Directions program spawned not only a network of resource centers and tons of training material, but also hundreds of thousands of educated people, workers, and employers aware of the issue of occupational health.

Once a person becomes aware, it is permanent. No one can undo it. There developed a demand for more information as everyone involved learned enough to realize how big the problem is and how effective worker education can be to address it. This demand forced the Reagan administration to keep some semblance of New Directions in place, although its effectiveness was curtailed.

In 1985, this demand led to the passage of legislation in New York State which created the Occupational Safety and Health Training and Education Grant Program. This program, which is expected to pioneer other state-funded training programs, is premised on the concept that it is better to spend money to educate workers to prevent occupational injury and illness than it is to pay compensation after a worker is harmed. Therefore, New York increased employers' assessment for Workers' Compensation to fund $4.3 million in grants for worker education programs and other activities designed to prevent occupational illness and injury. This grant program trains over 50,000 people each year.

If your organization isn't quite ready to launch a full-blown occupational safety and health program, but you are ready for a modest training session, one of the best ways to get either direct help or referred to an appropriate resource in your community is to contact the "COSH group" in your area. COSHs are associations of labor unions, safety and health professionals, and individual members. COSHs provide training, technical assistance, and serve as a communication network on issues affecting occupational safety and health. COSHs also have committees that undertake projects such as producing a pamphlet on reproductive hazards found in workplaces. COSHs are usually named after the area they serve, for example: New York Committee for Occupational Safety and Health, NYCOSH. But, if you can't find a COSH in your area, why not start one?

Conclusion

Worker education prevents occupational injuries, illnesses, and deaths. I know because I have seen it work. There are resources available in most areas to help new worker education programs get started, and there are successful models to establish resources for those areas where help is scarce. There are thousands of people like myself ready to help. I think we are ready to make a successful assault against unsafe workplaces. I think we can do it!

Occupational and Environmental Health Training for Hospital House Staff

Wendy E. Shearn and Kathleen Kahler

Why does the occupational and environmental health training for house staff need to be any different from any other HCWs' training? This is the question we were asked by many occupational and environmental health professionals. Clearly, we must establish a database of occupational and environmental exposure issues for this group of health care providers. By not providing hospital-specific occupational and environmental health training, a significant risk of financial and criminal penalties is created for medical center administrators and the corporations which own such institutions.

Given traditional extreme hours of work often in excess of 70 h per week, there are increased risks for health care professionals due to sleep deprivation, including depression, substance abuse, suicide, and accidents. They may be exposed to chemicals and physical hazards.

Chemical hazards which place the house staff at risk include antineoplastic drugs, formalin, anesthetic gases, ribavirin, ethylene oxide, nitrous oxide, and lasers. Use, possible routes of exposure, health effects, engineering controls, and protective equipment need to be addressed in the circumstances in which they are encountered by house officers.

Historically, the house staff have been identified as students, and as such house staff have been underpaid for the number of hours actually worked. This situation has resulted in confusion about the house staff's status as employees. OSHA has assisted medical centers in clarifying the status of house staff by defining physicians as employees. If a physician is employed by a health maintenance organization or a corporation, all OSHA standards apply to them, and compliance with regulations, including training, is required.

Training Issues

Ideally, occupational and environmental health and safety information should be provided to residents at their orientation. Compliance with OSHA's Bloodborne Pathogen Standard and the Hazard Communication Standard should be stressed. The initial orientation may be the only opportunity to provide this type of information to the residents, because it may be the only time they are together and available.

Some programs rotate the house staff through several departments, while others concentrate on only one. Some residents rotate through different hospitals. This variation in programs makes it difficult to identify who has the responsibility for the house staff's occupational and environmental health and safety training. Interdepartmental and interfacility policies and practices often vary.

Frequently, types of equipment are worn to protect against body fluid exposures. Safer needle devices require special training. Systems for handling biohazardous and hazardous waste should be reviewed by the department or a facility.

House staff should be trained on proper disposal of specimens, needles, bodily fluids, and common chemicals such as formalin and alcohol. Violations of environmental and occupational health regulations carry fines, penalties, and potential criminal liability.

Clinical decisions and practices of the house staff impact the health and safety of other staff members. For example, the house staff need to know potential health effects of hospital-administered aerosolized pentamidine on other staff such as nurses and respiratory therapists and what precautions are needed. If they were informed about this, they might order alternate therapy.

Procedures such as chest drainage may create "mixed waste" when body fluids and mercury are combined. The resulting waste requires special decontamination to avoid exposing support staff to bodily fluids and mercury vapors.

Needles left on patient tables, in bed linens, waste baskets, sleep rooms, or floors, or on top of lockers or in pockets of laboratory coats can be fatal since transmission of hepatitis B or AIDS to HCWs by needlesticks has been documented.

Needs Assessment

Facility-specific information on common problems should be obtained through a survey of various hospital departments including nursing, pharmacy, and environmental services.

Statistics could be reviewed to determine the type of injuries or illnesses incurred by residents in the recent past. Records of health and safety complaints filed by residents should be reviewed.

Based on experience and needs assessment, a hospital house staff orientation program should be developed.

Hazards in the Health Care Environment

Cytotoxic drugs are sometimes corrosive and present a reproductive hazard. Because house staff may at times be required to administer them, they need to be educated regarding protective equipment, spill cleanup, and management of contaminated materials.

When handling specimens and at autopsies, formalin exposure may occur. Formalin must be disposed of properly according to environmental regulations.

Glutaraldehydes used for sterilization of instruments have been left in open containers in procedure rooms, emergency departments and obstetrics, and gynecology for quick turnaround of instruments. Handling of glutaraldehydes in uncovered containers could result in exposure or sensitization.

Nitrous oxide is used as a cryogen in ophthalmology and obstetrics and gynecology. Exposure of the house staff, patients, and other staff can occur when the nitrous oxide is vented into the room or into a recirculating ventilation system.

Mercury poses occupational health and environmental hazards. Many facilities still use mercury to weight gastrointestinal tubes. In this circumstance, the risk of a mercury spill is significant. In addition, mercury should not be disposed of in the sewer because this method violates environmental regulations. A list of common equipment and locations where mercury is found in a medical center, along with alternative products or work practices is in Appendix 31.A.2.

Alcohols and xylenes, which are found in pathology laboratories, should be disposed of in a specific manner and not via the drain.

"Prevention Point," a videotape designed for physicians in training (Altschul Group Corporation, Evanston, IL), has been used to demonstrate safe needle management. The importance of reporting an exposure and receiving appropriate follow-up care is stressed in the video tape.

New needle devices or "safer" needles should be addressed as part of orientation. Many new needleless systems and "safer" devices are being introduced at medical centers. Some require new skills or techniques. House staff require training using these devices.

Ergonomics and physical hazards must be addressed. Back injuries may result or be aggravated by improper height of an examination table. Other hazards include stress and sleep deprivation.

Emergency preparedness needs to be included in orientation. Local hazards such as earthquake or flood should be addressed.

Respirators are now required for house staff working with suspected or diagnosed tuberculosis (TB) patients under certain circumstances. Annual training and fit-testing for respirators are a requirement.

House staff should be reminded of their right to refuse to perform hazardous work and to ask for training in a procedure for their own protection as well as for that of the patients.

How Is This Information Provided?

Many of the new regulations affecting health care require the physical presence of a trainer to ensure that employee questions are answered. To ensure accuracy and consistency of practices and procedures related to occupational and environmental health and safety, utilizing local experts from the departments within the medical center is important. Orientation for house staff should be presented in accordance with their educational background.

Managers are traditionally used in orientation to discuss procedures for providing or accessing services from their clinical departments. Managers must communicate to those departments that information on health and safety is essential.

Key questions when planning the training should include: (1) topics relevant to the medical center; (2) time required and time available; and (3) who provides the training. See *Essentials of Modern Hospital Safety,* Volume 1, Chapter 20, "Education for action: An innovative approach to training hospital employees" is an excellent resource for developing this type of training.

Monitoring compliance with policies and procedures is the key to ensuring their practice. In addition, mechanisms for addressing noncompliance with occupational and environmental health and safety procedures should also be established.

All training records, including attendance records, and the goals and the objectives of the training should be maintained.

Summary

Occupational and environmental health and safety training should be required for house staff orientation. As new risks arise and present risks are illuminated, additional training will be needed.

We anticipate further development of resources through continued educational research concerning house staff. Education is a critical part of an effective prevention plan.

Appendix 31.A.1

"Safer" Medical Devices: Preventing Needlesticks and Sharps Exposures Fact Sheet

Increased awareness of occupational exposures to bloodborne pathogens, specifically hepatitis B and HIV, has increased interest in the development of safer medical devices to prevent needlesticks and sharps exposures. Additionally, the Federal OSHA recently introduced the Bloodborne Pathogen Standard. In this standard, OSHA concludes that exposures can be minimized or eliminated by using provisions that include engineering controls (e.g., use of self-sheathing needles), work practices (e.g., universal precautions), and personal protective clothing and equipment.

Selecting new devices which utilize engineering controls is a difficult process for all health care facilities because of the following issues:

1. Design
   a. Many designs are in their initial stage and therefore may not be the "best" design.
   b. Many initial designs were directed at high-use products, not at high-risk products.
2. Protection
   a. Advertised "safety" devices do not always consider exposures to ancillary staff.
   b. New products do not have statistical data to support their use.
3. Availability
   a. Manufacturers have been unable to supply quantities needed for larger facilities.
   b. Products may not be directly exchanged with nonsafety devices.
4. Education
   a. Staff may need new skills to use product correctly.
5. Regulation
   a. CDC and OSHA encourage engineering controls, yet no regulating agency has specifications for new safety designs.

The following safety devices, which utilize engineering controls to prevent needlesticks/sharps injuries, have been selected for the Medical Center with the assistance of the Needlestick Prevention Subcommittee and the Product Selection and Evaluations Subcommittee:

1. Lancets
   a. Ames Glucolet 2™ Automatic Lancing Device
   b. Tenderlett™
2. IV Tubing/Medication Connections
   a. Baxter Protective Needle Lock™
   b. Baxter Needle Less IV Access System—INTERLINK™ (In-service to begin May/June)
3. IV Catheters (Stylets)
   a. Critikon PROTECTIV™ IV
4. Winged Infusion (Butterfly)
   a. Ryan Medical SHAMROCK™ (Currently being piloted)
5. Safety Syringes
   a. Becton Dickenson SAFETY-LOK™ Syringe (Currently being piloted)
6. Vacuum Collection Systems
   a. Ryan Medical SAF-T CLICK® (Currently being reevaluated)

Future evaluations of safety devices which provide engineering controls will be targeted at high-risk procedures for which work practices alone have made it difficult to control exposures (e.g., blood culture collection, blood gas kits, scalpels, and suture needles).

Source: Kaiser Permanente Medical Center, San Francisco, Safety Office, May, 1992.

Appendix 31.A.2

Wendy Shearn, Kathleen Kahler, and Jennifer McNary

Mercury in Health Care Settings at Kaiser Permanente

Sources of Mercury	Locations Used	Alternative Work Practices	Alternative Products
1 Bougie/red maloney esophageal dilator (up to 3 lb)	Operating rooms	Inspect tubes for damage or deterioration before use	Use stainless steel balls instead of mercury to weight
	Gastrointestinal laboratories	Remove from service if damage is noted.	Other suggestions are tungsten or barium
	Endoscopy procedures		
2 Cantor gastrointestinal tube (6–9 g of mercury)	Medical surgical units	Process for handling mercury and waste disposal should be specific.	No alternative product available.
3 Chest drainage unit (i.e., Pleur-Evac)	CVICU/ICU/medical surgical units	Use water and portable suction units to achieve greater suction.	Many products can achieve adequate suction by utilizing water or by using a vent plug with wall suction.
4 Feeding tubes	Medical surgical units	Substitute/replace with tungsten weighted tube.	New products are weighted with tungsten instead of mercury.
5 Thermometers		Substitute/replace with electronic thermometers.	Electronic thermometers.
Room	All locations		
Older incubator hood	Nursery/ICN		
Body 1.5 g	All locations		
Laboratory 3–4 g	Laboratory		
6 Sphygmomanometers (200 g of mercury)	All locations	Substitute/replace with aneroid gauges for mercury sphygmomanometers.	Aneroid gauges.
7 Hematology (some analyzers [hydraulic systems used] to draw mercury samples through counting chamber)	Laboratory	Replace with newer model.	Newer models do not use mercury in the hydraulic system.
8 Strip chart recorder used in cardiac ultrasound dry silver type (7 mg of mercuric bromide)	CV/ICU, ICU	Switch to thermal head recording system.	Thermally sensitive paper.
9 Red mercuric oxide	Laboratory	Substitute	Not required for laboratory use.
10 Mercuric chloride	Pathology (preparing tissue specimens)	Substitute reliable analytical methods.	Zinc can be used, but many pathologists state that results lack clarity.
11 Fluorescent tubes (40 mg of mercury)	All locations	Correct handling in the disposal process. No rushing.	No alternative product available at this time. Designers are working to develop tubes with less mercury.

(*continued*)

(continued)

Sources of Mercury	Locations Used	Alternative Work Practices	Alternative Products
12 Barometers	Laboratories	Substitute electronic barometers for mercury containing devices.	Electronic barometer.
13 Alkaline batteries and mercury batteries	All locations	Consideration should be given to disposal process. Recycle products.	Low-mercury batteries.

Note: All products containing mercury need to be disposed of as either hazardous or extremely hazardous wastes. New products contaminated with biohazardous material need to be decontaminated before disposing of material as hazardous waste.

Appendix 31.A.3

TB, Exposures, and HCWs

TB is on the rise in the United States today. This fact sheet will address the concerns of HCWs with an explanation of what TB is and how it is spread. It will also discuss the detection and prevention of TB.

How Is TB Spread?

TB is a disease caused by bacteria and is spread by coughing. You do "not" "catch" TB by sharing a glass or by touching something handled by a person with active TB. (Active TB generally means TB in the lungs.) Coughing people with active TB expel bacteria into the air in very small numbers. When a patient with active TB coughs directly in the face of a HCW, that worker can become infected if a closed area is shared for a prolonged period.

What Kinds of Precautions Are Taken at Kaiser Medical Center to Prevent the Spread of TB?

The Kaiser Foundation Hospital uses 100% fresh air in its ventilation system. This means that the air from a patient's room or any room is vented to the outside. Air is not reused or recycled in the hospital. Therefore, HCWs not working in an enclosed area with a patient with TB are at minimal risk of contracting TB.

Respiratory precautions are another method of protecting HCWs from exposure to TB. A sign stating "STOP—See nurse before entering" may be posted outside the door of a patient with TB. Anyone entering the room of a patient with TB should wear appropriate personal respiratory protection. In actuality, the "patient" should wear a mask, although compliance with this seems difficult.

Respiratory precautions are recommended even though the air in the hospital rooms "gets changed" by the local exhaust ventilation two to three times an hour. Here's why: it still takes from 2 to 3-1/2 h to remove 99.9% of the air, including the airborne bacteria.

What Is a TB Exposure and How Can It Happen?

Usually, a TB exposure occurs when someone has prolonged contact with a patient who has TB but who has not yet been diagnosed with it. The lack of a diagnosis is usually not due to an oversight or error: TB can be hard to diagnose. The physicians have to make an evaluation of all the illnesses (including TB) that might cause the same symptoms that TB causes before they can make the diagnosis. Although a chest x-ray can be helpful, the preliminary diagnosis of TB is made from a positive sputum smear. Depending on the patient's condition, it may be difficult to get a good sputum smear. All of these factors may delay the correct diagnosis of TB.

How Will I Know If I Have Been Exposed?

You and your coworkers will be notified of the TB exposure by the normal means of communication in your department (e.g., posted memo, communication log, staff meeting, etc.).

What Should I Do If I Have Been Exposed?

The Infection Control Committee recommends that routine annual TB skin testing (or semiannual testing for certain designated areas) is sufficient for most exposures. Exposed employees requesting testing/evaluation before their routine testing date should contact Employee Health. Employee Health recommends that employees who had significant prolonged contact and/or who are immunocompromised (e.g., HIV+, taking anticancer drugs or steroids) should contact their health care practitioner or the Employee Health Department to discuss their exposure. Depending on the circumstances of your exposure and your state of health, you may not need follow up or you may need to get a TB skin test 12 weeks after exposure to see if your test becomes positive.

What Will Happen If I Develop a Positive TB Skin Test?

A positive skin test probably means that you, like 10 million other people living in the United States, are infected with TB, but are not contagious or sick with it. To see if you have active TB, you will need to get a chest x-ray. If your skin test was positive but your chest x-ray does not show TB in your lungs and if you are not coughing, having drenching night sweats, losing weight or spitting up blood, then you have dormant or latent, not active, TB. You cannot infect anyone else.

You may need to get medication for the latent infection to reduce your chances of developing active TB in the future. You will need to discuss the results of your test with your health care practitioner or with the health care practitioner in Employee Health to decide whether or not you will need medication.

Prevention

To prevent or minimize exposures, the Infection Control Committee recommends that you

1. Teach all patients to cover their mouths when coughing
2. Follow the universal precautions
3. Observe the postings on patient doors

Prepared by: G. Denton, NP, Employee Health, K. Kahler, MPH, Safety Office, and W. Shearn, MD, Chief of Occupational Medicine and Employee Health; Kaiser Permanente Medical Center: San Francisco, Employee Health Center.

Appendix 31.A.4

Scabies Fact Sheet

What Is Scabies?

Scabies is an infectious disease caused by a human parasitic mite, *Sarcoptes scabiei*.

What Are the Symptoms of Scabies?

Symptoms of "classic" or "typical" scabies include:

Itching: more intense at night

Skin lesions: particularly around finger Webs, sides of toes and fingers, wrists, elbows, knees, ankles, abdomen, and thighs.

In people with depressed immune systems, scabies may not cause itching or lesions. Rather, the skin may appear scaled and crusted, most often on the hands, feet, elbows, and knees. These symptoms are characteristic of "crusted" or *Norwegian scabies.*

What Is the Incubation Period?

In a previously unexposed individual, 2–6 weeks may elapse between exposure and onset of itching. In people who have had scabies in the past and are sensitized to the mite, reexposure may produce itching as soon as 48 h after exposure.

How Is Scabies Transmitted?

Scabies is transmitted by skin-to-skin contact with "infested persons" or, less commonly, by contact with their freshly infested clothing or other personal objects.

What Causes the Itching and the Bumps?

The mite and its secretions, feces, and eggs (all foreign protein material) stimulate the body's immune responses, causing the symptoms of scabies.

How Is Scabies Diagnosed?

The diagnosis of scabies is made by history taking and physical examination. Sometimes, a microscopic examination of a skin scraping can demonstrate the presence of the mite, its eggs or its fecal pellets.

How Is Scabies Treated?

There are various creams or lotions that may be prescribed for scabies: Permethrin (Elimite), Lindane (Kwell) or Crotamiton 10% (Eurax). In general, all medication should be applied after a brief shower or bath. The skin should be dry and the medication applied to the body from the neck down with special attention to the areas around the finger Webs, wrists, elbows, axillae, breasts, buttocks, and genitalia. It should be left on for 12 h. If the medication is removed—such as by hand washing—it should be reapplied. After 12 h, a cleansing bath or shower should be taken.

Sheets, pajamas, towels, and any clothes worn during the previous week should be washed on the hot cycle and dried in a clothes dryer. Non-washable items should be sealed in plastic bags and not opened for 2 weeks. It is not necessary to wash your carpets, furniture and all the clothes in the closet. For asymptomatic individuals, a single treatment is recommended. For those who are symptomatic, treatment should be repeated in 1 week.

Can I Still Work?

For asymptomatic personnel, no work restrictions are necessary. If an exposed employee is symptomatic, he or she may return to work after one treatment with medication has been completed. Family members or close contacts generally do not need to be treated unless the employee is symptomatic was exposed to a patient with Norwegian or crusted scabies.

Prevention

Hand washing before and after patient contact may prevent infestation. The mite roams freely on the skin for up to 4 h before burrowing, so washing with soap and water can prevent it from making your body its home. When caring for patients with Norwegian scabies who have not yet been treated with Elimite, keep the ends of your gown sleeves tucked in the ends of your gloves. Complaints of itching and/or even a minimal rash should be reported to your supervisor so that a physician can examine the patient to make the diagnosis.

Source: Kaiser Permanente Medical Center, San Francisco.

Appendix 31.A.5

Hepatitis and HCWs

Introduction

Hepatitis is an inflammation of the liver that can be caused by a number of agents, including viruses, toxic chemicals (e.g., alcohol), certain drugs, and parasites. This fact sheet, which was written for HCWs, describes hepatitis B and hepatitis C, the most likely forms of "viral" hepatitis to which HCWs could be exposed. This fact sheet briefly addresses hepatitis A as well.

Transmission of Hepatitis A

Viral hepatitis A is spread primarily through the "fecal–oral" route, which means, for example, by eating contaminated food, particularly shellfish. Being an HCW does "not" make a person at increased risk of contracting this illness.

Transmission of Hepatitis B

Viral hepatitis B is not spread like hepatitis A (by food or feces), nor is it transmitted through everyday contact as with a cold. It is transmitted through contact with blood and other body fluids (saliva, semen, vaginal fluids, etc.). Hepatitis B is most commonly spread through contaminated needles, sexual contact, and perinatal transmission (from mother to infant around the time of delivery). HCWs are potentially exposed when they work around blood and body fluids.

Symptoms

A person who contracts hepatitis B may not have any symptoms or he/she may have any or all of the following: loss of appetite, loss of taste for cigarettes, fatigue, headache, stiff or aching joints, low grade fever, nausea, vomiting, abdominal pain, jaundice (yellowing of the skin and eyes), dark urine, and light colored stools. Similarly, a person with hepatitis A may or may not have the same symptoms. Blood tests can determine if someone currently has or has had either hepatitis A or B.

Incubation and Contagious Periods

The hepatitis B incubation period (the time from exposure to the onset of symptoms) varies from 6 weeks to 6 months. A person infected with hepatitis B is contagious as long as the virus remains in the blood: 1–6 months for most people but indefinitely for some.

Complications of Hepatitis

For hepatitis B, 90% of the patients recover spontaneously and develop lifelong immunity to the virus. However, 5%–10% continue to have symptoms for more than 6 months and may develop some degree of chronic hepatitis, which can cause cirrhosis or liver cancer. Mothers can pass the virus on to their babies at delivery.

Work-Related Exposures to Hepatitis B

Every year, 200–300 HCWs die from hepatitis B or related illnesses. HCWs can be infected by hepatitis B if they are stuck by a needle or another sharp instrument that is contaminated with infected blood. Workers can also be infected through a splash of blood or other infectious body fluids to the eyes, nose, mouth, or open skin (cuts, sores, rashes, etc.). Bites that penetrate the skin can also transmit hepatitis B. Statistically, about 6%–30% of HCWs who are exposed to hepatitis B infected blood or body fluids will contract the disease. After an exposure, HCWs may need gamma globulin or a special hyperimmune globulin.

Prevention

The most important ways to protect yourself from hepatitis B are the following:

1. Get the hepatitis B vaccine. *Take all three doses.*
2. Always follow universal precautions. Treat all blood and body fluids as potentially infectious and protect yourself.
3. Wash your hands before and after physical contact with any body fluids. Even if you were wearing gloves, you should wash your hands after removing the gloves.
4. Wear gloves and other protective equipment (goggles, masks, face shields) when you expect to come in contact with blood and other body fluids.
5. Dispose of needles and other sharps safely in the sharps container. Never bend, break, or cut needles before disposing of them. If you need to recap them, use the one-handed scoop technique.

Transmission of Hepatitis C

Hepatitis C is thought to be transmitted in a manner similar to that for hepatitis B: by contact with blood and body fluids. Intravenous drug use and sexual contact with someone who is hepatitis C positive may place a person at risk of contracting hepatitis C. At the present time, there are conflicting reports about perinatal transmission of hepatitis C.

Transfusions also place the people receiving them at risk of contracting the hepatitis C virus. Transfusion blood is tested for both hepatitis B and C, but hepatitis B is much easier to detect and the test for it is currently more accurate than the test for hepatitis C.

Symptoms and Incubation

The symptoms of hepatitis C are similar to those of hepatitis A and B. Hepatitis C's incubation period of approximately 60 days is longer than hepatitis A's but shorter than hepatitis B's.

Chronic Hepatitis C

Like hepatitis B patients, many hepatitis C patients develop chronic infection; in fact, half the people infected with hepatitis C develop chronic hepatitis.

Work-Related Exposures to Hepatitis C

HCWs may become infected if they are stuck by a needle or another sharp, which is contaminated with infected blood. Splashes of body fluids to the eyes, nose or mouth or open cuts or rashes seem less likely to transmit hepatitis C than hepatitis B. About 3%–1% out of every 27 HCWs exposed to hepatitis C positive blood will contract the virus.

Prevention

Although no vaccine is currently available for hepatitis C, gamma globulin may be given to HCWs after blood or body fluid exposure to protect them from contracting hepatitis C. In addition, the same prevention methods listed for hepatitis B should be followed by HCWs to protect themselves from work related exposures to hepatitis C.

If you think you have been exposed to any kind of hepatitis, you should notify your supervisor and the Employee Health Center. If you have any questions regarding hepatitis, please call the Employee Health Center.

Nursing Student Health and Safety Training in the Hospital Workplace: Risks and Responsibilities

Jolie Pearl and Marian McDonald

Introduction

How well prepared are nursing students to protect themselves from occupational hazards when they enter the hospital workplace? How prepared are hospitals to provide a safe environment for nursing students? What are the potential consequences of not providing a safe workplace for nursing students—for students and institutions alike?

This chapter discusses hazards encountered by nurses in hospital settings and the special risks and problems faced by nursing students during clinical practicums, as well as issues faced by hospitals that have nursing students on site. The concerns raised are also relevant for new nursing graduates, many of whom may not be prepared to avoid hazardous exposures in the hospital environment. Current nursing school approaches and curriculum content regarding occupational hazards faced by nursing students are described and deficiencies addressed. This section then offers an overview of what an approach to educating nursing students about workplace hazards should include.

Hospital-based nurses continue to be at high risk for exposure to occupational hazards and workplace injuries and illnesses. Nurses-in-training (e.g., nursing students) are at "especially" high risk for exposure to workplace hazards on the job, because of the following factors:

- Widespread lack of appropriate education about avoiding exposure to workplace hazards
- Inexperience
- Their role as patient providers
- Lack of or limited protection under existing regulations because of student status

Nursing students are not the only ones at risk, however. Hospitals are vulnerable as well, both because of their legal responsibility to protect patients and the potential legal ramifications of the provision of a clinical environment for student practicums with access to patients and exposure to hazards. Hospitals' vulnerability to litigation is exacerbated by the fact that most nursing students receive little or no formal health and safety education or training, and that which they do receive is rarely monitored or evaluated by the hospitals in which students do their practicums.

Hospital departments of nursing, as well as other departments with responsibility for staff education and/or compliance with health and safety training, have a critical role and responsibility in the prevention of occupational illnesses and injuries in hospitals and in the training, education, and protection of nursing staff and students. Hospital nursing departments are also responsible for coordination and oversight of on-site nursing student clinical practicums. It is to these hospital departments, as well as educators interested in the health and safety training of nursing students and future nursing staff, that this chapter is addressed.

Hospital Workplace Hazards Faced by Nursing Students

Nurses are exposed to a wide variety of occupational hazards in the hospital work environment and suffer high rates of injuries and illnesses on the job. Nurses encounter virtually every type of hazard present in the hospital environment. These include

- Infectious hazards, such as TB, hepatitis B, HIV, and CMV
- Chemical hazards, such as ethylene oxide, formaldehyde, and anesthetic gases

- Pharmaceutical hazards, such as antineoplastic agents, ribavirin, and pentamidine
- Physical hazards, such as lifting, radiation, and assault
- Psychological hazards, such as stress and shiftwork
- Reproductive hazards, such as ethylene oxide, infectious agents, and ribavirin

There is no doubt that nursing is a hazardous profession. Unfortunately, the health care industry has often dragged its feet in responding to health threats to its employees and been slow to act upon its responsibility to provide a safe and healthy work environment.

A case in point is occupational exposure to hepatitis B. Hepatitis B has been a serious and well-documented occupational health hazard to HCWs for many years.[1] As many as 200 HCWs die each year from occupational exposure to hepatitis B, with many others suffering other serious health effects.[2] HCWs with both frequent and infrequent blood contact were known to be at risk for acquiring hepatitis B infection.

Despite the availability and proven effectiveness of a vaccine against the virus, many health care institutions did not offer it to staff at risk, or they required staff to cover the cost of the vaccine themselves. It was not until the promulgation of the OSHA Bloodborne Pathogens Standard (BBPS) in 1991 that provision of the vaccine at no cost to health care employees was required by law.[3]

For nursing students, workplace hazards include those faced by fully trained and practicing nurses. The likelihood of workplace exposure may be even greater among nursing students, however, due to lack of training and experience, and the medical and financial implications more dire.

Little data are available on injuries and illnesses sustained by nursing students during hospital practicums. In addition, few studies of occupational injuries and illnesses among nursing or other health care staff include nursing students in their study cohorts. Consequently, the true magnitude of occupational injuries and illnesses among nursing students remains unknown.

One study of blood and body fluid exposures among HCWs in a large teaching hospital did include nursing students in its cohort. The study found that these students sustained the third highest overall incidence rate of needlestick and blood and body fluid exposures among all staff categories in the institution. Nurses in the "General" and "Respiratory/Rehabilitation" areas sustained the first and second highest incidence rates, respectively. Nurses and nursing students sustained 78.8% of all reported exposures in this study.[4]

Nursing students are at high risk for acquiring hepatitis B. However, hepatitis B vaccination rates among nursing students are low. The percentage of nursing schools requiring students to be vaccinated was as low as 4.2% in 1988,[5] despite Centers for Disease Control and Prevention recommendations that students in the health care professions be vaccinated during their training period, prior to workplace contact with blood.[6]

The impact of the BBPS on hepatitis B vaccination rates among students in schools of nursing is not known. Under the standard, hospitals are not required to provide the vaccine to nursing students.

Special Risks and Problems Facing Nursing Students in Hospital Settings

Nursing students today face a number of challenges. Nursing students perform many of the same job functions and provide almost as many patient care activities as do nurse employees. They face the same workplace hazards as staff nurses. Yet nursing students may be at even greater risk than staff nurses, for the following reasons:

- As students, they do not benefit from required health and safety training that staff nurses are entitled to receive.
- They lack the protection of regulatory bodies such as the OSHA and, in many states, Workers' Compensation.
- Occupational surveillance and screening data on nursing students are not being gathered on a systematic basis.

Historical Background

The special risks and problems facing nursing students during clinical practicums are not new. Historically, nursing students were seen as low cost or free labor. Nursing students were expected to function fully as staff, facing widespread and dramatic rates of occupational illness and injury.

For example, in Minnesota in the 1920s, very high rates of TB were detected among nursing and medical students. They were particularly high among those who worked in TB sanitoria, where rates of active disease among nursing students ranged from 5% to 19%, and TB infection rates reached 100% in some instances.[7]

In that period, nursing students were required to train in TB wards. It was not until the institution of the "contagious disease technique" in 1933 that a "serious effort was made to set up a barrier between patients and students for the protection of the latter against tubercle bacilli" in the Minnesota University Hospitals.[8] Advocates of this approach also understood the importance of education: "It has been shown that no matter how cognizant the student nurse is of the presence of TB among her (sic) patients she cannot protect herself against tubercle bacilli unless she is provided with the necessary equipment, encouraged to use it, and taught every necessary step in carrying out contagious disease technic."[9]

Exemption from Regulation

The OSHAct (1970) covers only those people who are employees of a specified employer.[10] Nursing students are not employed and therefore are not covered. Because of this, nursing students are not covered under OSHA. This means nursing students do not benefit from a number of the OSHAct's provisions.

Under OSHA's Hazard Communication Standard (1983) and the BBPS (1991), hospitals are required to provide training to employees on the various occupational hazards they will encounter in the workplace. Nursing students are likely to be excluded from these trainings.

When nursing students do receive education about hazards in the hospital work setting, generally it is at the discretion of the nursing school and individual nursing instructors. Hospitals that provide student practicums on site may also request or require student education on specific workplace hazards prior to coming to their clinical practicums. More routinely, the focal point of hospital concern is on prevention of disease transmission from nursing students to patients, with documentation of freedom from childhood diseases and active TB often required of students prior to patient contact.

Since OSHA standards do not cover unpaid students, the likelihood of nursing students being assured of routine and ongoing in-service training on health and safety is not high. Without adequate training, these students are at risk of inappropriate use of equipment, which could lead to injury to themselves and/or others. Devices which might seem routine to staff in a hospital can be potentially dangerous to the untrained.

A recent episode in a Bay Area hospital provides an example. A nursing student who had not been properly trained in the correct disposal of contaminated needles sustained a needlestick injury (and exposure to HIV-infected blood) while disposing of a needle in a "sharps" container.[11] While it could be argued that the cause of the event was the design of the container, the student was at a disadvantage. She had not received training on the recommended usage of the device, nor had she practiced needle disposal techniques with a less hazardous substance.

Traditionally, hospitals and nursing schools have provided a certain amount of site-specific training to nursing students on workplace practices for infection control and, more recently, on practices designed to prevent infection transmission to employees, in the form of universal precautions. This site-specific training is often provided by the nurse preceptor in conjunction with a nursing instructor from the nursing school, who is on site with the students. Often during clinical practicums, a nursing student will be assigned to work with a staff nurse from the institution, who will be responsible for demonstrating to the student techniques and practices of patient care. The quantity and quality of those demonstrations will be reflective of the abilities and knowledge of the individual nurse. The likelihood is not high that a staff nurse will be fully versed in the theoretical or conceptual underpinnings of all potential health hazards and the correct ways to work safely with known hazards in the work environment; and such an expectation places an undue burden on the individual nurse. Nurses are not routinely trained as teachers in this area and are often overburdened with patient care duties as a result of understaffing.

Nursing students are exempted from OSHA protections such as provision of the hepatitis B vaccine, which hospitals must now provide at no cost to employees at risk. As mentioned above, only 4.2% of nursing schools required hepatitis B vaccination of their students in 1988. Of these 26 schools, 77% required students to cover the cost of the vaccine themselves.[12] The high cost of the vaccine may be prohibitive to some students.

Hospitals do not routinely include student nurses in ongoing staff screening programs. Hospitals may require that student nurses be free from active TB and have various vaccinations. Students are expected to get these services from their own private practitioners or from a student health service. If a student sustains an injury during a clinical practicum, she/he generally must seek consultation and treatment outside the facility. The injury may or may not be reported to the health care facility; and since follow-up treatment for the injury or illness is generally not done on site, it is unlikely student illnesses and injuries will be included in statistical data gathering on occupational injures or illnesses for any given worksite.

Exemption from Workers' Compensation

The definition of who is covered by workers' compensation varies from state to state. In California, for example, nursing students are considered to be volunteers under state labor law, and they are not covered by the disability provisions of workers' compensation laws. If a student nurse were to sustain and appropriately document an injury or illness during the clinical practicum, he or she could, in theory, be eligible for only medical coverage under Workers' Compensation.[13] Nursing students may, on the other hand, exercise their tort law rights to sue a hospital which has been negligent and therefore contributed to a nursing student's illness or injury, as will be discussed below.

Exclusion from Surveillance

It is worth noting that interns and residents (physicians-in-training) are much more likely to be included in hospital surveillance cohorts than are nursing students. The more frequent inclusion of physicians-in-training may be due to the extended period of time which they spend in clinical settings, or to the fact that during training, they are paid and therefore afforded employee status. Inequities in status are certainly a factor as well.[14] In the hospital hierarchy, physicians are afforded greater status than nurses; nursing students are on yet a lower rung of the prestige ladder. The marginalization of nursing students is perpetuated by their exclusion from these study cohorts.

For more insidious problems such as exposure to TB, the exclusion of nursing students from routine hospital staff screening ensures that new conversions among students are never linked to a specific worksite or patient exposure. With certain types of exposures, such as HIV, immediate evaluation and prophylactic treatment are warranted. Yet because the BBPS does not apply to students, they are not guaranteed either appropriate follow-up evaluation or prophylaxis. Additionally, without the immediate establishment of a baseline negative HIV test, students might be ineligible for medical coverage under workers' compensation and in danger of losing evidence for future legal recourse if they were to seroconvert.

Schools of nursing or student health centers may compile data on injuries and illnesses among nursing students, but these data are not easily accessible, nor are aggregate data available for the full range of workplace hazards encountered by student nurses.

These problems are compounded by lack of communication and clear protocols between health and safety programs and on-site student nursing programs. For example, recently a nursing student at a San Francisco Bay Area hospital did an extensive (3 h) intake interview with a newly admitted patient who had an undiagnosed cough. The patient was later placed on respiratory precautions to rule out active TB. The nursing student was not informed, as staff caregivers were, of potential exposure to TB, but discovered the information inadvertently.[15] This situation illustrates two critical points. First, the student probably had not been adequately educated about potential symptoms of active TB, nor about the protocol for initiating respiratory isolation for the patient. Second, she was not notified of her potential exposure, and so was not alerted to the potential need for appropriate follow-up evaluation.

Special Issues and Problems Facing Hospitals

Hospitals that allow nursing students on-site for clinical practicums need to be concerned about providing appropriate health and safety education of these students for several reasons:

- Possible legal liability if students have occupational exposures on site
- Hospital concern about compliance with Joint Commission on the Accreditation of Healthcare Organizations (JCAHO)
- The legal obligation to provide a safe work environment
- The accurate collection of screening and surveillance data on occupational illnesses and injuries on site

Hospital Liability for Occupational Injuries and Illnesses Sustained by Nursing Students

In the fall of 1992, a California nursing student sustained a needlestick injury while disposing of a needle contaminated with HIV-infected blood. As a result of that incident, the student has brought a civil lawsuit against the hospital where the event occurred, as well as against the nursing school, the nursing instructor, and the manufacturer of the needle disposal box.[16] The arguments against the hospital are compelling and raise the question of whether other hospitals are vulnerable to similar lawsuits if they do not change their practice regarding health and safety training and protection for nursing students.

The arguments being used against the hospital address three important issues. The first is the hospital's liability, because it provided the clinical environment in which the student was functioning and in which the injury occurred. The second is the hospital's failure to properly ascertain the competency of the nursing instructor who had oversight of the nursing student at the institution. The third issue concerns the lack of guidelines from the hospital (or from the nursing school) to address how students are to be protected from exposure to communicable diseases while they are learning new skills, such as the proper handling and disposal of needles and other hazardous equipment.[17]

The first two arguments are straightforward and are similar to the liability a hospital has for the safety and appropriate care of its patients. For example, if a patient was harmed by a nursing student or instructor, the hospital would be held liable for the harm done to the patient (despite the fact that neither students nor instructors are employees of the institution), because the institution allowed them access to a patient who was under its care.

The third argument, regarding policies about student access to patients with communicable diseases, raises critical issues. If guidelines were developed which allow student choice regarding working with patients with communicable diseases, implicit in such a policy would be the identification of patients who have specific diseases—namely AIDS or infection with HIV. Such a policy could easily entail the revelation of confidential information about patients, and could be a breach of state law, ethical guidelines (including providers' duty to treat), existing hospital policy, and patient trust. Because of the "window period" in which the AIDS virus may be present without antibodies being detected, there is no way to ensure that any given patient is not infected with HIV and therefore free of the infection at the time a student may work with that patient. In short, such a policy would be unethical, if not illegal, and essentially unworkable.

This case demonstrates a range of complex issues regarding hospital liability, which can arise when a nursing student is injured during a practicum. Failure to address these issues can leave a hospital open to lawsuits.

Compliance with JCAHO

Despite student nurses' exclusion from worker health and safety regulations, guidelines exist which may apply to students, for which hospitals could be responsible. The guidelines for JCAHO apply to all hospitals accredited by that organization and may eventually be modified to address students specifically.[18]

Under the 1994 JCAHO guidelines on "Orientation, Training, and Education of Staff," the term "individual" (not "employee") is used to describe those for whom the institution is responsible for providing appropriate "orientation."[19] "Orientation" may include discussion of specific safety hazards and related policies applicable to the individual's assigned duties. The hospital's responsibility for provision of this orientation includes individuals from off-site agencies who provide nursing care to patients. "Whether done by the hospital or the off-site agency, the hospital is responsible for assuring that each individual from an off-site agency has completed an adequate and timely orientation to the hospital."[20] It is possible that "off-site agency" could be interpreted to include nursing schools.

Providing a Safe Work Environment

A hospital could be in violation of OSHA's requirements for providing a safe and healthy workplace, if an illness or injury sustained by an employee could be shown to have resulted from the actions of a student nurse. For example, a student nurse who had not been properly trained in the safe handling of contaminated needles could endanger hospital staff.

Accurate Surveillance

Finally, hospital-based occupational exposure screening and surveillance data will be incomplete and misleading if they do not include nursing and other students who have patient contact and are at risk for exposure to communicable diseases. It behooves hospitals to develop systems for accurately gathering exposure data on students. Such data would be extremely valuable to students and institutions alike.

What Nursing Students Are Currently Taught about Hospital Workplace Hazards

"Material for this section was developed through interviews with faculty and students in San Francisco Bay Area and East Coast nursing schools."

With the advent of HIV/AIDS, a new awareness has developed about occupational hazards found in the health care setting. Nursing practice and education have begun to change to reflect an increased understanding of hospitals as potentially hazardous environments, and new standards of practice have been developed which are designed to better protect nurses on the job.

Despite important advances in teaching about bloodborne hazards, the education nursing students receive is still inadequate and does not address the full range of hazards they will face during their clinical practicums. The current overall approach to this aspect of nursing education does not ensure that students will know how to adequately protect themselves in the clinical environment. No specific standards for health and safety curriculum in nursing schools exist; the content, methodology, and evaluation of the health and safety education nursing students receive varies considerably. Criteria are lacking to evaluate existing curricula on the attitudes, knowledge, and behavior of nursing students regarding occupational hazards.

With the growth in the theory and practice of universal precautions, nursing students have benefited from a greater emphasis in their education on avoiding needlestick injuries and exposure to patients' bodily fluids. Universal precautions have become a primary focus of health and safety education in nursing schools. However, education about other workplace hazards and other health and safety concerns are generally not taught as a separate or specific content area.

When nursing students are taught about the hazards of an infectious agent, it is in the context of the education they receive about a particular disease, and the related prevention of nosocomial infections, rather than as part of a health and safety curriculum. For example, when students are taught about the pathophysiology of TB, they may also receive information about the importance of preventing its spread to other patients and themselves. Relatively, few links may be made between the study of the disease and the appropriate protective measures students should take. To compound this problem, many nursing schools have students begin clinical rotations during their first semester, before having studied disease pathophysiology and associated protective measures.

Because of current general concern about HIV/AIDS and TB, students benefit from relatively greater attention to the prevention of these diseases. Additionally, prevention of back injuries is usually covered in nursing school curricula. Education is minimal about most other occupational hazards encountered in the hospital. Protection from exposure to chemical and reproductive hazards may be absent altogether.

Key Components of an Effective Approach to Educating Nursing Students about Hospital Workplace Hazards

Responsibilities regarding nursing student health and safety education and training are different for nursing schools and hospitals. The general parameters of what each institution should address are outlined below.

Responsibilities of Nursing Schools

The focus of nursing school education about hazards should be the general and applied principles of occupational health and safety in the health care workplace. It should delineate the following:

- The full range of hazards, including where and under what circumstances and work procedures they will be encountering
- Health risks of exposures
- Protective measures, including the roles and limitations of engineering and administrative controls and personal protective equipment
- Medical and reporting postexposure protocols
- Legal rights and protections, as well as resources, including clarification of what protections are "not" extended to nursing students

Such training can be based upon health and safety curricula and materials developed by nursing professionals, occupational health professionals, and unions. A number of such articles, training materials, and handbooks are available.[21–26]

Responsibilities of Hospitals

Hospital health and safety education and training of nursing students should address the specifics of the workplace and should be as comprehensive as that provided to all hospital staff nurses. It should accomplish the following:

- Address the institution's policies and procedures for infection control and prevention of infectious disease transmission to staff and patients.
- Provide specific instruction on both existing and new procedures or devices utilized or practiced by nursing staff in all patient care areas where nursing students rotate.
- Review the protective measures in use and/or recommended or required by the institution.
- Delineate post-hazard exposure protocols of the institution, including screening and reporting requirements.
- Include all workplace hazards nurses may encounter, and review as well fire and electrical safety, and personal and hospital security.
- Evaluate the effectiveness of the education and training done on site, as well as determine the baseline health and safety knowledge students have received from their school.

Hospitals and nursing schools will need to work together to ensure that nursing students are prepared to enter the hospital workplace with sufficient knowledge and skill to protect themselves and others.

Conclusion

Nursing students are tomorrow's healers. Yet, current health and safety education and training practices of both nursing schools and hospitals leave nursing students in the dark. As a result, nursing students are uninformed and unprotected during hospital practicums.

Because of these practices, nursing students are being forced to make unacceptable sacrifices which can lead to problems for them, their nursing schools and for hospitals. Nursing students deprived of health and safety training and education are especially vulnerable to occupational illnesses and injuries. Their patients and coworkers may be disadvantaged as well. Hospitals offering practicums may face liability issues.

These problems can be addressed if the key institutions involved—nursing schools and hospitals—take heed and act upon their distinct and interconnected responsibilities for providing nursing students a safe learning environment. This can be done by adding workplace health and safety to the nursing school curriculum, by including nursing students in hospital health and safety training, screening, and surveillance and by strengthening interinstitution communication and evaluation.

Such policy and training changes would benefit all parties involved. These changes would serve to strengthen the health, quality, and commitment of future generations of nurses while enhancing patient care and the overall safety of the hospital environment.

Acknowledgments

The authors would like to extend heartfelt thanks to the many people throughout the country who granted time and helped think through the different aspects of this chapter. Without their expertise and interest this chapter would not have been possible.

References

1. Nelson KE. Prevention of hepatitis in health care workers. In: EA Emmett (ed.), Health Problems of Health Care Workers. *Occup Med State Art Rev* 2(3), July–September, 1987.
2. Centers for Disease Control. Guidelines for prevention of transmission of human immunodeficiency virus and hepatitis B virus to health-care and public-safety workers. *MMWR* 38(S-6):5, 1989.
3. OSHA. Final rule, occupational exposure to bloodborne pathogens. *Fed Regist* 56(235):64145–64182, 1991.
4. Yassi A and McGill M. Determinants of blood and body fluid exposure in a large teaching hospital: Hazards of the intermittent intravenous procedure. *Am J Infect Control* 19(3):129–130, June, 1991.
5. Goetz A and Yu V. Hepatitis B and Hepatitis B vaccine requirements in schools of nursing in the United States: A national survey. *Am J Infect Control* 18(4):243, August 1990.
6. Centers for Disease Control. Protection against viral hepatitis: Recommendations of the immunization practices advisory committee (ACIP). *MMWR* 39(RR-2):14, 1990.
7. Myers AJ. *Invited and Conquered—Historical Sketch of Tuberculosis in Minnesota.* Minnesota Public Health Association, St. Paul, MN, 1949, pp. 602–603.
8. *Ibid*, p. 604.
9. *Ibid*, pp. 612–613.
10. 29 USCS §652, no. 9.
11. Nursing Student Sues USF Over Tainted Needle. *San Francisco Chronicle,* September 25, 1993.
12. *Ibid,* Goetz, p. 243.
13. Berg G. *The Noetics Group*. Personal communication, San Francisco, CA, December, 1993.

14. Butter I, et al. *Sex and Status: Hierarchies in the Health Workforce.* American Public Health Association, Washington, DC, 1985.
15. Burgel B. Associate Clinical Professor, School of Nursing, University of California, San Francisco. Personal communication, November, 1993.
16. *San Francisco Chronicle, Ibid.*
17. Silver M. Attorney for University of San Francisco nursing student. Personal communication, November, 1993.
18. Dubor G. Nursing Instructor, Samuel Merritt College School of Nursing. Oakland, CA. Personal Communication, November, 1993.
19. Joint Commission on the Accreditation of Healthcare Organizations. *1994 Accreditation Manual for Hospitals—Volume II Scoring Guidelines. Section 4: Orientation, Training and Education of Staff.* p. 2. J.C.A.H.O.: Oakbrook Terrace, IL, 1993.
20. *Ibid,* p. 2.
21. California Nurses Association. *Health and Safety Issues for Nurses.* San Francisco, CNA, 1986.
22. California Nurses Association. *If It Happens to You. What You Need to Know About Occupational Exposure to HIV, HBV and Other Blood-Borne Pathogens.* San Francisco, CNA, 1990.
23. Service Employees International Union (SEIU). *Health and Safety Manual.* SEIU, Washington, DC, 1987.
24. Labor Occupational Health Program (LOHP). *Hospital Hazards: A Union Guide to Inspecting the Workplace.* Labor Occupational Health Program, Berkeley, CA, 1991.
25. Weinger M and Wallerstein N. Education for action: An innovative approach to training hospital employees. In: W Charney and J Shirmer (eds.), *Essentials of Modern Hospital Safety,* Vol. 1. Lewis Publishers, Boca Raton, FL, 1990.
26. Coleman L and Dickinson C. The risks of healing: The hazards of the nursing profession. In: W Chavkin (ed.), *Double Exposure Women's Health Hazards on the Job and at Home.* Monthly Review Press, New York, 1984.

32

Case Studies of Health Care Workers in the Compensation System

Lenora S. Colbert
National Health & Human Service Employees Union

Background 32-2
Health Care Industry and Workers' Compensation: Cost-Containment through Preventing Injuries and Illness vs. Cost-Containment by Preventing Compensation of Claims • Politics of Workers' Compensation • Health Care Work Environment Is among the Most Hazardous • Workers' Compensation and the Health Care Worker with an Occupational Disease • Controversions and Delay • Workers' Compensation for Occupationally Contracted Tuberculosis • Case of Sharon B • Case of Ethel T • Toward a Reform Agenda Linking Workers' Compensation to Prevention of Workplace Injuries and Illness • New York's Training and Education Grant Program Funded by Workers' Compensation Premiums • Occupational Health Diagnostic Clinical Centers • 1199 National Benefit Fund • Case Study #1 • Case Study #2 • Integrating the Workers' Compensation Medical Component in National Health Care System Reforms
References 32-9

My education about workers' compensation began in 1986 when I started working for 1199, National Health and Human Service Employees' Union. As Occupational Safety and Health Director, I suddenly became aware of issues I had previously overlooked as a worker. I could now relate the dangerous incidents I had personally witnessed to issues of occupational health and safety.

I have worked in the health care field for 26 years in several different capacities including a nurses' aide, central supply packaging and sterilizing equipment worker, file clerk, messenger and driver, and also a radiology technologist. I have experienced many work-related injuries that went unreported.

My real education about workers' compensation began when I came on staff at 1199. I could connect the incidents I had personally experienced to these issues. At 1199, the need for strong advocacy in the prevention of injuries, illness, and diseases attributed to the work place are understood.

This chapter examines the experience of health care workers with the state-based workers' compensation systems by focusing on certain salient and fairly typical problems encountered by workers sustaining workplace injuries and, especially, work-related illnesses. The scenarios described are based on case studies selected from the files of the New York workers' compensation law firm of Pasternack, Popish, and Reiff, which for more than two decades has represented the majority of workers' compensation claimants from the 1199 National Health and Human Service Employees' Union. Other scenarios are case histories from the 1199 National Benefit Fund Disability Department.

While these case histories are from New York State, I want to provide a generic treatment of how the state systems too often fail to adequately and promptly compensate health care workers for job-related ailments, particularly in relation to occupational illness rather than workplace accidents.

Background

The workers' compensation system is the country's earliest forms of social insurance, introduced in most states over three quarters of a century ago. Since no federal minimum standards exist, the laws, their administration, and the ability of workers to receive adequate compensation vary from state to state. Generally, state systems impose "no-fault" liability on employers for job-related disabilities. Workers relinquish their common-law rights to legal action. In return, employers promise workers prompt and adequate compensation, complete medical care, and rehabilitation services for work-related injuries and illness. In most states workers are compensated at a rate pegged to two thirds of their gross average weekly wages up to a maximum generally at or near the state average weekly wage for the period in which they are temporarily totally disabled and some proportionate amount thereof if they are classified as "partially disabled."

Medical evaluations of the degree of workers' disabilities are the basis for both the duration and amount of wage replacement benefits. These are sometimes referred to as the "indemnity" portion of the benefits and provide entitlement to medical and rehab care at the employer's expense. For claims of occupational disease, medical evidence of the work-relatedness of illness is the basis for determining whether or not the illness will be treated as compensable under workers' compensation.

Health Care Industry and Workers' Compensation: Cost-Containment through Preventing Injuries and Illness vs. Cost-Containment by Preventing Compensation of Claims

Since employers' liability for the cost of both medical care and wage replacement benefits are at stake, as we shall see below, employers or their insurers relentlessly pursue cost-containment policies which make medical expenses and wage replacement benefits the subject of litigation. They challenge (controvert) the work-relatedness of occupational disease claims or, in respect to both accident and illness claims, seek to cut off or reduce wage replacement benefits as soon as possible, while workers who may be too disabled to return to work must struggle in the system to maintain wage replacement weekly payments at a level which can sustain themselves and their families.

Similarly, medical costs are supposed to be paid in full under workers' compensation systems, but as medical costs have increased in recent years they too are the focus of employers' efforts to cut or shift such costs. Workers in need of diagnostic tests, surgical procedures or palliative treatment are often forced to litigate to obtain needed medical care or face having such costs borne by the union's health benefit plan with workers picking up the costs of copayments and deductibles. This economic conflict is manifested on the medico/legal terrain in the form of disputes about the work-relatedness of occupational illness claims, medical treatment and the degree and duration of workers' disabilities.

While health care workers, particularly those employed at hospitals with employee health services and emergency rooms close at hand, have less difficulty accessing medical care than many other workers, it is by no means assured that the work-relatedness of their ailments will be recognized, diagnosed, and recorded so that their entitlements under workers' compensation systems will be established without hardship or impediments. Doctors employed by health care institutions are often the first to see health care workers presenting symptoms of illness, either in employee health service offices or emergency rooms. Like their counterparts in other practice settings they usually lack training in occupational health, fail to take thorough occupational histories and are primarily interested in diagnosis and treatment rather than in determining whether the work or work environment has been the cause of, or a significant contributory factor in the origins of the illness or injury. Such shortcomings play a role not only in the failure to recognize occupational diseases attributable to health care workers' chemical exposures and infectious diseases but also in failing to connect certain chronic diseases of the musculoskeletal system to previous incidents of workplace acute trauma or cumulative trauma due to repetitive tasks involving heavy lifting, grasping, and the like.

In the case of claims for occupational disease, employers—and health care institutions are not exceptions—have controverted a high percentage of claims on the issue of compensability, frequently by exploiting medical and scientific uncertainty surrounding the etiology of occupational disease. While the claim is contested and awaiting adjudication or settlement, sick and disabled health care workers do not receive workers' compensation wage benefits and might not receive medical benefits since health insurers usually do not cover work-related illness.

These problems are compounded by the marked trend in recent years toward the corporatization of health care.[1] The "new" medical–industrial complex is characterized by the advent of profit-making health care institutions and the impact they have had on so-called nonprofit health care providers which emulate their profit maximizing orientation and methods. Hospital administrators are not interested in establishing systems or structures to ensure that occupational injuries and illness are properly identified and recorded since costs associated with such injuries and illnesses will have to be borne by the institutions. Indeed, the lack of such systems will facilitate the shifting of such costs onto union health insurance funds, state disability systems partly funded by workers, social security disability systems partly funded by workers, social security disability and public assistance systems funded by taxpayers or onto the injured workers and their families. And the ability of the health care institutions to externalize such costs will tend to undermine efforts toward prevention since less incentive will exist to reduce such costs and health and safety measures cost money, at least in the short term. So, health care institutions can be seen as reflecting contradictory missions. Within the institutions are professional caregivers who are dedicated to treating their own health care workers when they are injured and ill, as well as administrators who are dedicated to minimizing costs, including labor costs like those associated with compensation and treatment of the institutions' employees.

Politics of Workers' Compensation

It should be noted, moreover, that the state-based nature of workers' compensation has worked to the disadvantage of workers and to the advantage of employers. Unlike the social security system which covers everyone in the nation and makes its entitlement structure virtually immune from the political depredations of those politicians who might pursue cost-cutting in the name of deficit reduction, when it comes to the state-based workers' compensation systems, business forces often threaten to move out of state, pointing to cheaper workers' compensation costs in other states or regions to induce nervous state legislators to enact changes in the law which have the effect of eroding workers' entitlements. Thus, workers' political power to defend their interests is fragmented state by state rather than concentrated as it is in a uniform federal system. Accordingly, most of the problems outlined below are likely to get worse rather than better, just as declining real wages for most segments of workers in the face of rising worker productivity may be attributable, in part, to declines in unionization, historically the means by which workers exercised some power in defense of their interests and secured a rising portion, during the postwar expansion, of their contribution to production.

Health Care Work Environment Is among the Most Hazardous

According to the industry risk index (IRI) which ranks industries by degree of hazard, hospitals, and health care facilities (SIC codes 8062 and 8099) are among the 50 most hazardous.[2] Health care workers are exposed to a wide array of illness-inducing hazardous substances. Autopsy workers are exposed to formaldehyde, x-ray technicians to radiation hazards, and respiratory therapists to pentamidine. Instruments are sterilized with glutaraldehyde, contaminated needles may transmit bloodborne infectious diseases like hepatitis B or HIV, housekeepers may be exposed to biohazardous medical waste. Approximately one-fourth of all workers' compensation indemnity expenditures in eight states were for back injuries. Nurses and patient aides appear to be at high risk for work-related back injury.[11,12] Multistate analyses of back injury risk among worker groups revealed that nursing aides and garbage collectors ranked first or second in each state. Nurses' aides, porters, and laundry workers suffer low back injuries and chronic diseases of the musculoskeletal system after long-term employment involving heavy lifting, bending, and transporting tasks, while clerical workers at ergonomically unsound computer keyboard workstations are turning up with repetitive stress injuries like carpal tunnel syndrome, tendinitis, de Quervains' syndrome, tenosynovitis, and other upper extremity disorders.

Workers' Compensation and the Health Care Worker with an Occupational Disease

While health care workers may contract a variety of occupationally induced diseases, there is a significant gap between the incidence of such illness and the number of claims filed and compensated. The recognition that the system, historically designed around workplace accidents and "no-fault" principles, has broken down in relation to occupational diseases is widely shared by diverse observers who have examined it.[3,4]

No discussion of health care workers' experience in the workers' compensation system would be complete without mention of the fact that many workers who contract occupational diseases do not file claims for compensation and therefore never enter the system. A variety of factors are responsible:

1. Relatively little is known about the potential health effects of most synthetic chemicals. No information is available regarding the toxicity of many chemicals in the health care setting.
2. Doctors are not trained to consider work as a cause of disease. Surveys show that adequate work histories are reflected on less than ten percent of hospital charges.[5] Accordingly, many illnesses of occupational origin are mistakenly diagnosed and attributed to their causes, such as smoking, aging, or lifestyle factors.
3. Doctors receive little or no training in occupational medicine in medical school—an average of only four hours in four years of medical school.[6]
4. Health care workers are typically exposed to more than one hazardous substance and are often not informed that they have been exposed to hazardous chemicals or substances. Symptoms may become manifest only after many years from the onset of exposure.

Controversions and Delay

Those health care workers who know or suspect that their illness is work related are confronted with a high incidence of controversions and delay. "Individual employer liability appears to provide a strong incentive for employers (or their agents) to adopt a defensive litigation strategy which results in extensive litigation within a no-fault system."[7]

Workers' Compensation for Occupationally Contracted Tuberculosis

Health care workers who contract tuberculosis through workplace exposure are often the victims of dual system failures—a failure of occupational health policy to prevent disease transmission—often because of inadequate infection control procedures and lack of proper ventilation in treatment facilities, and a failure of the compensation system to identify and compensate claims for work-related tuberculosis.

The failure of the workers' compensation system should be understood not only in terms of the legal barriers which may in the end defeat a worker's claim for compensation, but also as a failure of the system to educate workers and require health care institutions responsible for diagnosing and treating those who contract the disease to take occupational histories and determine whether the disease was "to a reasonable degree of medical certainty" the outcome of a workplace exposure.

Case of Sharon B

Sharon B was, at the time of her illness in 1992, a 42-year-old laboratory technician who assisted in the performance of autopsies, including cutting open cadavers, removing, weighing, and bagging body organs where the cause of death was determined to have been tuberculosis. She carried out such work without any protection other than a dust mask. After becoming ill and missing work intermittently for some time, she was laid-off for excessive absences. About 6 months after she was discharged from work, she was admitted to a hospital with symptoms diagnosed in June, 1992 as multidrug-resistant tuberculosis.

She was treated in a hospital for 3 months before being discharged and underwent an extended period of convalescence. At no time during her stay in the hospital was any attention given to whether she might have contracted the disease through her occupation. Doctors are focused principally on the diagnosis and treatment of disease not on which insurers have liability for the payment of the bills. All bills were presented to Medicaid. In a hospital, understandably, the main objective was to restore the patient to health.

But from the perspective of the worker who will not necessarily be able to return to work upon discharge from the hospital, and who will not necessarily have been paid while sick and unable to work, the issue of income maintenance is critical. Sharon B was reduced to being a recipient of public assistance.

When her plight came to the attention of the union that had formerly represented her during her employment, she was assisted in obtaining legal representation and filed a claim for compensation in January 1993. Her case was controverted by her former employer. The doctor who diagnosed and treated her in the hospital provided a report which indicated the diagnosis but expressed no opinion regarding whether her disease was likely caused by her work and work environment.

When the doctor's opinion regarding the work-relatedness of Sharon B's illness was solicited by her attorneys, the physician communicated to the firm his view that she could not have contracted tuberculosis by assisting in autopsies in as much as tuberculosis is an airborne infectious disease which could not be transmitted even by infected cadavers since they "don't breathe."

The doctor was clearly ignorant of the literature regarding nosocomial transmission of tuberculosis during procedures such as autopsy.[8–10]

The case of Sharon B will have to be litigated at a hearing in which an expert physician trained in occupational medicine will be required to overcome the uninformed opinion of the doctor who diagnosed and treated Sharon B. Her illness has left her deaf in one ear and with permanent nerve damage. In the past 6 months, all 21 of her coworkers have been tested, and 10 have tested positive for the first time.

Case of Ethel T

Ethel T is a hospital coding clerk who has performed data entry work for over 20 years. Over the course of that period of time she has worked on several computer keyboards and work stations. Recently, she was diagnosed with bilateral carpal tunnel syndrome after seeing a hospital-affiliated doctor to whom she was referred by the hospital's employee health service to which she had reported with complaints of persistent, severe pain in her hands, wrists, and arms.

As in the case of Sharon B., the doctor who diagnosed Ethel failed to render an opinion regarding whether or not the disease was the result of Ethel's long years of keyboard work. Under the New York workers' compensation law, Ethel's work need not be the sole cause or even the main cause of the disease. The law requires that her work be a significant contributing factor in the cause, or aggravation of her condition. Nonetheless, her claim was controverted. This means that she is not eligible for wage replacement benefits and has been refused authorization for a surgical procedure known as a carpal ligament release to relieve the pain caused by compression of the median nerve in each wrist by inflamed tendons. Her claim may take 1 or 2 years to resolve before the Workers' Compensation Board.

If doctors wish to treat such patients conservatively in the hope that a period of several weeks off the keyboard together with physical therapy might reverse the progress of the disease and prevent the need for surgery and permanent impairment, the employer's controversion of the claim usually makes such a course impossible. First, the worker usually cannot stop working without a prospect of wage replacement benefits. Second, physical therapists will not treat the worker because they are not assured of payment by the hospital or its carrier or agent. Even if they might be paid when the claim is ultimately established, they are usually not willing to wait a year or 18 months until the claim is established. It is ironic that an employee of a health care institution is having compensation and medical treatment blocked by the hospital.

Toward a Reform Agenda Linking Workers' Compensation to Prevention of Workplace Injuries and Illness

Barth and Hunt recommended the dissemination of information on workplace hazards to both workers and employers as a means to contribute to the prevention of occupational disease. In New York, the state legislature, responding to an initiative of the New York Committee for Occupational Safety and Health (NYCOSH), established a $5 million grant program, funded by an assessment on workers' compensation premiums, to train workers concerning occupational hazards, right to know and workers' compensation in 1985. NYCOSH led a coalition effort in which labor union, COSH groups, and occupational health professionals strongly lobbied for the program.

New York's Training and Education Grant Program Funded by Workers' Compensation Premiums

This program elicited worker-training proposals totaling more than $11 million in its first year, attesting to both the need for such training and the strong interest on the part of unions, COSH groups, employers, and health professionals in the academic sphere.

Training courses and curricula, films, and other resource materials specifically targeted to worker populations at risk of occupational disease have been produced and disseminated. Such programs constitute an approach to the prevention of occupational disease that follows up on the right-to-know movement of the early 1980s.

A mere legal "right-to-know" can hardly make a substantial contribution to preventing workplace exposures and illnesses, without the training that educates workers to the hazards they face and the means to protect themselves. This program is likely to have an important educational impact on employers, particularly smaller employers, who as Barth points out, "[i]n some instances...know, little more than their employees about the hazards to which they are exposed."

Occupational Health Diagnostic Clinical Centers

Under New York law, the workers' compensation system does not pay the medical costs of occupational disease screenings unless such screening results in a positive diagnosis of an occupational illness. Even if an individual or group of workers has clearly been exposed to dangerous levels of toxic substances in the workplace and a screening is medically indicated, no compensation for the costs of such evaluations will be made.

The statewide Occupational Health Diagnostic Clinical Center network is a promising effort toward the prevention of occupational illness, and it's reasonable that the law should be amended to require that the workers' compensation system contribute to the cost of such efforts through payment for occupational disease screenings for workplace exposures associated with disease.

An ancillary and closely related element in New York's approach to preventing occupational illness will be the development of a statewide data collection system incorporating workers' compensation claims information to identify hazardous occupational exposures and diseases and thereby target enforcement and preventive efforts toward hazardous industries with worker populations at high risk.

As the Interim Report put it, "individual employer liability appears to provide a strong incentive for employers (or their agents), to adopt a defensive litigation strategy which results in extensive litigation within a no-fault system." That report made four important recommendations addressed to the state workers' compensation system's treatment of occupational disease claims (pp. 99, 100).

1. Establish the legal presumptions to reduce the difficulty of proving the cause of occupational disease.
2. Establish an employer— and/or producer—financed trust fund to pay benefits.
3. Eliminate the artificial barriers to occupational disease claims in the law.
4. Establish a neutral administrative body to administer the compensation of occupational disease claims.

To these I would add the following provision: insurer-mandated payment of medical expenses for claims. To achieve a more equitable and expeditious payment of medical expenses incurred by individuals filing for workers' compensation benefits, a system should be developed in which health insurers are required initially to pay the cost of medical care for individuals seeking care for a suspected work-related illness. If an individual's claim is sustained by the workers' compensation system, the health care insurer would be reimbursed at the time of settlement or award. If no award is made and the worker is not financially responsible, the insurer could be reimbursed through a special fund. This system would remedy the present situation in which often neither medical insurance nor workers' compensation coverage is available to pay the cost of health care for individuals. As a result of this lack of coverage, individuals often defer a much needed medical evaluation, and secondary prevention interventions are often foregone.

1199 National Benefit Fund

The 1199 National Benefit Fund for hospital and health care employees grew out of a small medical plan whose coverage was limited to basic surgery and hospitalization. This plan which initially covered pharmacists was expanded to include hospital workers in 1962. This plan, moreover, which began in 1945, today covers over 78,000 members employed in health care and human services.

The fund provides disability benefits for its members. Included in the Disability Benefits Plan is an Intervention Program to assist members during the initial stages of their disability and to help them successfully return to work at the earliest possible date. This is best accomplished through early intervention.

Early intervention in its broadest sense may be viewed as that initial contact with the patient and medical provider when the first disability claim is submitted. This intervention is essential in assisting the patient to return to employment at the earliest and most feasible date possible. The benefits of early intervention are implicit in the proven belief that disabled workers overwhelmingly want to return to the workplace forthwith for both financial security and for their own sense of esteem and well being.

An important component of the Intervention Program is identification of workers' compensation claims to ensure that the costs associated with such claims are not shifted into our fund.

The following are two case histories[13,14] which can be used as examples of what members sometimes experience when engaged in the process of obtaining workers' compensation benefits. Although the data that was collected were from actual claims that were handled in the Disability Benefits Department, the names have been changed to maintain confidentiality. In both instances, the role that the Fund played was as a result of our Intervention Program which is part of the initial processing, upon receipt of a disability claim or notice of a potential disability claim. The only exception would be a claim of very short duration, particularly where the member has already returned to work at the point the claim is being processed.

Case Study #1

A. Jones, a 55-year-old respiratory therapist, submitted a claim to the fund on January 23, 1992. A. Jones is a male hospital worker employed at a hospital in New York City that treats AIDS patients. The initial diagnosis on the claim form was pulmonary tuberculosis. The member had last worked on October 16, 1991 and shortly thereafter filed a claim for workers' compensation with the employer's plan. When no benefits were forthcoming after approximately two months the member filed for disability benefits. Upon review of the claim form at the fund office it was noted that the member, the doctor and the employer failed to indicate that the illness may have been job related. It appears the member did not acknowledge the possibility of workers' compensation due to his experience with the compensation carrier up to that point. The employer failed to indicate the possibility of a job-related illness for obvious reasons. The doctor failed to make a connection since there may have been a relationship between the doctor and the hospital.

As part of our intervention program the claims examiner having connected the member's illness with the type of job performed contacted the member to verify the type of employment and to raise the question of workers' compensation. As a result, the claims examiner was able to convince the member that to pursue the workers' compensation claim was in his best interest. The member filed a claim with the carrier and the board. After several months a hearing was scheduled.

In the meantime, the Benefit Fund paid 26 weeks in disability benefits and filed an appropriate notice with the carrier and the board. The member received $5,590.00 in disability benefits from the fund. However, it took more than six months before the member's case was finally adjudicated by the board. The Workers' Compensation Board rendered a decision that the member's illness was job related and made an award. The fund was subsequently reimbursed by the carrier and the member received workers' compensation benefits beyond the 26 week period which the fund had paid.

Case Study #2

B. Roberts, a 36-year-old female nursing assistant employed at a hospital in Brooklyn, New York was diagnosed with carpal tunnel syndrome with chief complaints of pain in the right wrist radiating to the elbow and numbness in the right hand involving all fingers. The member filed a claim for disability benefits on May 26, 1992. An initial payment was made by the fund on June 10, 1992 at which time an intake was completed by the claims examiner, which includes pursuing the issue of workers' compensation.

It was determined, with the assistance of a workers' compensation attorney that this member was a good candidate for the Mount Sinai Occupational Health Clinic to determine if the member's condition was job related. The member was very reluctant at first due to her perception of how she would be treated by the workers' compensation carrier and the possibility of being left without funds for a substantial period of time. After being assured that the fund would continue the payment of disability benefits while her case was being adjudicated by the board, the member agreed to be examined by the Mount Sinai Clinic. Accordingly, a claim was filed with the workers' compensation board on or about September 10, 1992. It was subsequently determined that the member's condition was job related and evidence of this was submitted to the board and the fund was reimbursed for 26 weeks of disability benefits paid in the amount of $5,590.00. The member subsequently returned to work in November, 1992. The period of disability paid by the fund was April 22, 1992 to October 20, 1992. It should be noted that the hearing took place almost one year after the member filed a claim for workers' compensation benefits.

Had the member not returned to work after receiving the maximum of 26 weeks in disability payments, the claims examiner would have referred the member to the fund's Members' Assistance Department. The Members' Assistant Department is staffed with certified social workers who among other duties, provide counseling and assistance to members who have received maximum benefits from the fund.

Integrating the Workers' Compensation Medical Component in National Health Care System Reforms

The following comments are regarding Title X-Coordination of the Medical Portion of Workers' Compensation and Automobile Insurance, from the Health Security Act, 1993.

The most significant areas of concern we have about the legislation in its current form are as follows:

An individual with a work-related occupational disease or injury should be guaranteed the right to choose the provider for both the diagnosis and treatment of his/her occupational condition.

Since both the diagnosis and the disability determination are used to decide on the benefit level for an affected worker, it is not appropriate for workers to be evaluated by physicians hired either by an employer or by any other health care provider who may not be able to provide an independent assessment of the diagnosis, work-relatedness and extent of disability.

The proposed legislative language speaks only in terms of the treatment of work-related conditions, failing to recognize that the diagnosis and determination of work-relatedness frequently require evaluation by specialized occupational medicine physicians.

Case members should also be responsible for coordination of return to work with work place modification and exposure cessation, as this is the mainstay of treatment of occupational diseases. Finally, the case manager should coordinate treatment via referrals to appropriate providers. However, it is not appropriate that this be done in consultation with the workers' compensation carrier; rather, it should be done in consultation with the occupational medicine specialist responsible for the affected workers care.

Experience rating of employers' and workers' compensation experience needs to be added to provide incentives toward the reduction of these preventable illnesses and injuries. Therefore, we urge that incorporation of wording and recognizing the importance of experience rating in assessing workers' compensation premiums and the mechanisms by which this can be linked to disease prevention.

References

1. Salmon JW. *The Corporate Transformation of Health Care: Issues and Directions*. Baywood Publishing Co., Amityville, NY, 1990.
2. Pederson D, Young R, and Sundin D. A *Model for the Identification of High Risk Occupational Groups Using RTECS and NIOSH Data*. NIOSH Technical Report 83-117. US Government Printing Office, Washington, DC, 1983.
3. An Interim Report to Congress on Occupational Diseases, US Department of Labor, 1980, pp. 54–78; Barth P and Hunt A. *Workers' Compensation and Work-Related Illnesses and Diseases*. MIT Press, Cambridge, MA, pp. 163–178, 1980.
4. APHA Policy Statement No. 8329 (PP): Compensation for and prevention of occupational disease. APHA Policy Statements 1948–present, cumulative. APHA, Washington, DC. Reprinted in *Am J Pub Health*, 74(3):292.
5. Rosenstock L. Occupational medicine: Too long neglected. *Ann Intern Med*, 95:774–776, 1981.
6. Levy BS. The teaching of occupational health in US medical schools: Five-year follow-up of an initial survey. *Am J Pub Health*, 75:79–80, 1985.
7. Interim Report, An Interim Report to Congress on Occupational Diseases, US Department of Labor, 1980, pp. 98–99.
8. Lundgren R. Tuberculosis infection transmitted at autopsy. *Tubercle*, 68:147–150, 1987.

9. Kantor HS, Poblete R, and Pusateri SL. Nosocomial transmission of tuberculosis from unsuspect disease. *Am J Med*, 84:833–838, 1988.
10. Federal Register. "Nosocomial transmission of TB has been associated with close contact with infectious patients or health care workers, and during procedures such as…autopsy." Tuesday October 12, 1993, p. 52812.
11. Harber P, Billet E, Gutowski M, Soottook, Law M, and Roman A. Occupational low back pain in hospital nurses. *J Occup Med*, 27:518–524, 1985.
12. Cust G, Pearson JDG, and Mair A. The prevalence of low back pain in nurses. *Nurs Rev*, 19:169–179, 1972.
13. Spratley, S, Assistant Director, 1199 National Benefit Fund, 1199 National Health and Human Service Employees Union, 310 West 43rd Street, New York, NY, 10036.
14 Tuminaro, Dominick, Pasternack, Popish & Reiff, PC., 111 Livingston Street, 22nd Floor, Brooklyn, New York 11201.

Index

A

Accidents, in shiftworkers, 21-7; *see also* Health care orker (HCW); Shiftwork, health and safety hazards
Accreditation Council for Graduate Medical Education, 21-7
ACGIH, *see* American Conference of Governmental Industrial Hygienists
ACH, *see* Air changes per hour
Active dosimetric system (AD-n), 17-11–17-12
Active electrochemical/solid state analyzer (AEC/SS-h), 17-14
ADH®, *see* Aerosol Delivery Hood®
Administrative actions, EtO exposure, 13-20; *see also* Ethylene oxide (EtO)
AD-n, *see* Active dosimetric system
AEC/SS-h, *see* Active electrochemical/solid state analyzer
Aerosol Delivery Hood® (ADH®), 16-17, 16-21
Aerosol eater chamber, 15-4–15-5
Aerosolized pentamidine (AP)
  airborne tuberculosis transmission and treatment, 15-5–15-7
  engineering ventilation control, 15-2–15-3, 15-22–15-24
  health care workers
    exposures, 15-12–15-17
    health effects, 15-9–15-10
  HEPA filtration efficiency, 15-18–15-22
  occupational risks, 15-1–15-2
  inhalation hazard control HCW in administration, 15-7–15-11
  tuberculosis screening, 15-4
Aerosolized ribavirin (AR), HCW's health effects; *see also* Ethylene oxide (EtO)
  acceptable exposure levels, 16-3
  control system validation, 16-20–16-21
  engineering and administrative controls, 16-13–16-14
    materials and methodologies, 16-14–16-16
    outcomes, 16-17–16-18
  engineering controls, 16-23
  exposure levels and assessment methods, 16-4–16-6
  health effects, 16-2–16-3
  health hazard assessment data, 16-41
  hospital's written policy and shift assignments, 16-23
  industrial hygiene survey report, 16-10–16-12
  limitations, 16-35–16-36
  methodologies, 16-24–16-28
  outcomes, 16-28–16-35
  pharmacokinetics and exposure recommendations, 16-24
  policies, recommendations, 16-8
  protection policies, 16-9–16-10
  recommendations, 16-7–16-8, 16-37–16-38
  recommended concentration, 16-43–16-55
  room ventilation, 16-24
  symptoms, 16-6
  toxicology, 16-3, 16-23–16-24 uses of, 16-22
Aerosol therapy cabinet, 11-98
Agency for Toxic Substance and Disease Registry (ATSDR), 24-1
AHA, *see* American Hospital Association
AIA, *see* American Institute of Architects
AICD, *see* Automatic implantable cardioverter defibrillator
Airborne aldehyde screening; *see also* Health Hazard Evaluation Report, NIOSH
  methodology, 12-13
  outcomes, 12-19
Airborne infectious disease management
  administrative controls, 4-89

AIIR maintenance schedule, 4-109
airborne infection isolation rooms (AIIRs), 4-88
American Institute of Architects (AIA), 4-108
assess indoor air quality and filter efficiency
building filtration system, 4-118–4-120
particle counters measure, 4-116
portable HEPA filters, 4-116–4-117
test efficiency, 4-118
data interpretation, 4-121–4-122
determine and achieve necessary airflow, 4-113–4-114
discharging air
flex duct and flow rate, 4-94
grille adapter, 4-96
HEPA filter machine, 4-93
return air system, steps, 4-95–4-96
steps, 4-93–4-94
window adapter, 4-94
environmental controls
air changes per hour (ACH), 4-91
dilution ventilation and filtration, 4-91
physical/mechanical measures, 4-89
pressure management, 4-90
guidelines, 4-88
HEPA filter maintenance, 4-92
HVAC system maintenance schedule, 4-110
measure relative pressurization, 4-111–4-112
measuring particle counts, 4-123
microorganisms, 4-115
personal protective equipment (PPE), 4-89
portable anteroom
HEPA machine, 4-101
steps, 4-100–4-102
portable HEPA filter machines, 4-113
pressure gauge steps, 4-112
principles, 4-89
surge capacity
anteroom, 4-103
engineered system, 4-105
goals, 4-103
infectious disease zone (IDZ), 4-103
smoke zones, 4-104
temporary surge area, 4-106
ventilation system, 4-105
temporary negative pressure isolation (TNPI)
curtain, steps, 4-97–4-99
health care workers, 4-98
National Fire Protection Association (NFPA), 4-92, 4-97
plastic curtain, 4-99
portable HEPA filter machine, 4-99
Airborne particles screening; *see also* Health Hazard Evaluation Report, NIOSH
methods, 12-14
outcomes, 12-19–12-21
Airborne tuberculosis transmission, 15-5–15-7
Air changes per hour (ACH), 16-31
Air monitoring and biological monitoring, PICUs, 16-4
Air quantification, 14-15
Air sampling methodology and laboratory analysis, ribavirin, 16-26
AIR spectrophotometric analyzers, 14-4
ALARA exposures, control measures, 23-7–23-9; *see also* Hospital, reproductive hazards
Aldehydes monitoring, hospitals; *see also* Health care worker (HCW)
analytical problem, 14-3–14-4
formaldehyde, 14-4–14-7
glutaraldehyde, 14-7–14-11
historical perspective, 14-1–14-2
Allergic contact dermatitis, *see* Delayed-type hypersensitivity
Allergy testing, glove dermatitis, 19-9–19-10; *see also* Glove dermatitis; Latex allergy
ALPHA-1 Air Sampler, 16-15–16-16
Alpha particles, properties of, 22-3; *see also* Radiation safety procedures, hospitals
American Conference of Governmental Industrial Hygienists (ACGIH), 12-5, 14-1, 16-24, 27-13, 28-1
American Health Care Association, 26-9
American Hospital Association (AHA), 24-2
American Institute of Architects (AIA), 11-23, 11-30, 11-90, 16-24
American National Standards Institute (ANSI), 12-3, 12-40, 27-12
American Nurses Association (ANA), 3-19–3-20
American Society of Heating, Refrigerating, and Air-Conditioning Engineers (ASHRAE), 11-34, 11-93, 11-100, 11-104, 27-13
American Standards and Test Materials (ASTM), 20-4
Americans with Disabilities Act (ADA) of 1990, 19-12
Amputee slings/full-body sling
bariatric patients, 3-115–3-116
ceiling lifts, 3-120
hip surgery patients, 3-117–3-118
lateral transfer aids, 3-121–3-122
maneuvering floor lifts, 3-116–3-117
matching equipment, 3-119
operating table, 3-119–3-120
patient assessment tool, 3-120–3-121
proning, 3-122
safe lift assists outside, 3-116
safe lifting policy statements, 3-120
transport *vs.* transfer, 3-121
washing instructions, 3-118–3-119
Analysis of variance (ANOVA), 16-21
Anesthesia mask gas-scavenging system
methodologies, 17-16–17-18
outcomes, 17-18–17-19
Animal studies, physiological effects of surgical smoke, 12-2–12-3
Annual limit on intake (ALI), 22-9

ANOVA, *see* Analysis of variance
ANSI, *see* American National Standards Institute
AORN, *see* Association of Operating Room Nurses
Argon laser, 12-24
ASHRAE, *see* American Society of Heating, Refrigerating, and Air-Conditioning Engineers
Assigned PF, definition, 16-55
Association of Operating Room Nurses (AORN), 12-41
ASTM, *see* American Standards and Test Materials
ATR, *see* Attenuated total reflectance spectroscopy
ATSDR, *see* Agency for Toxic Substance and Disease Registry
Attenuated total reflectance spectroscopy (ATR), 12-13
Audiovisuals training, hospital employees, 30-8–30-11, 30-13–30-14; *see also* Hospital employees, education
Automatic implantable cardioverter defibrillator (AICD), 12-46

B

Back injuries; *see also* Home health care; Nursing home workers
- amputee slings/full-body sling
  - bariatric patients, 3-115–3-116
  - ceiling lifts, 3-120
  - hip surgery patients, 3-117–3-118
  - lateral transfer aids, 3-121–3-122
  - maneuvering floor lifts, 3-116–3-117
  - matching equipment, 3-119
  - operating table, 3-119–3-120
  - patient assessment tool, 3-120–3-121
  - proning, 3-122
  - safe lift assists outside, 3-116
  - safe lifting policy statements, 3-120
  - transport *vs.* transfer, 3-121
  - washing instructions, 3-118–3-119
- Australia, patient-handling policy, 3-17–3-18
- back pain
  - additional environmental factors, 3-97
  - assistive devices, 3-96
  - biomechanical stresses, 3-97–3-98
  - causes, 3-92
  - clinical areas, 3-104–3-105
  - ergonomic approach, 3-105–3-108
  - laboratory studies, 3-102–3-104
  - lifting/transferring tasks, 3-97
  - low-back pain (LBP) episodes, 3-92
  - patient-handling task (PHT), 3-93–3-96
  - patient-transfer method, 3-96
  - postural stresses, 3-97
  - prevention, 3-92–3-95
  - testing approaches, 3-98–3-101
- bariatric task force, 3-59–3-60
- Care™, 3-122–3-125
- ceiling lift
  - industrial sector, 3-109–3-110
  - institutions, 3-112–3-114
  - methodology, 3-110–3-111
  - nursing staff, 3-112
  - posture and effort, 3-112
- clinical management
  - atypical pressure ulcers, 3-53
  - continuous positive airway pressure (CPAP), 3-54
  - discharge planning, 3-55–3-56
  - EMS services, 3-50–3-51
  - goals, 3-55
  - MRI scanner, 3-51–3-52
  - operating room (OR) table, 3-52–3-53
  - peripherally inserted central catheter (PICC), 3-51
  - WLS patients, 3-50
  - wound healing, 3-54
  - x-ray beams, 3-52
- home health care, 25-5–25-7
- Hudson, Anne
  - bed frame, 3-40–3-41
  - degenerative disk disease, 3-39–3-40
  - drag, cradle and hug lift, 3-36
  - essential functions, 3-42
  - herniated lumbar disks, 3-37
  - lifting patients, 3-35–3-36
  - lumbar fusion surgery, 3-42–3-43
  - lumbar strain injury, 3-39
  - medical care units, 3-35
  - monkey pole, 3-41
  - neurologist, 3-38–3-39
  - nursing skills, 3-34
  - physical therapy, 3-37
  - spine surgery, 3-43
  - Team Spirit Award, 3-40
  - voc rehab evaluation, 3-41–3-42
  - workers' compensation system, 3-38
- interdisciplinary team, 3-58–3-59
- lift teams
  - advantages, 3-10–3-11
  - anecdotal evidence, 3-8–3-10
  - biomechanics, 3-5
  - causes, 3-3–3-4
  - historical background, 3-2–3-3
  - implementation steps, 3-11–3-12
  - manual lifting, 3-5
  - mechanization, 3-4
  - nursing schools, 3-5–3-6
  - strategies, 3-6–3-7
  - successes, 3-7–3-8
- Martin Chair-A-Table®
  - caregivers, 3-124–3-125
  - game changing technology, 3-122
  - institutions, 3-124
  - risk managers and insurers, 3-124

steps, 3-123–3-124
Nurse and Patient Safety and Protection Act of 2007
D-MI, 3-29
HR 378, 3-30–3-31
nursing homes, 3-18–3-19
nursing home workers, 26-2, 26-7–26-8
obesity
bariatrics, 3-43–3-44
body mass index (BMI), 3-44–3-45
cardiac and metabolic treament, 3-45
clinical care, 3-46
definition, 3-45–3-46
historical studies, 3-46–3-47
vs. overweight, 3-44
waist-to-hip ratio, 3-45
weight loss surgery (WLS), 3-43
patient-handling tasks (PHTs)
administrative controls, 3-82–3-83
back belts, 3-76–3-78
bed mover, 3-81
clinical decision-making tools, 3-81–3-82
equipment/devices, 3-78–3-81
lift teams, 3-86–3-87
manual lifting, 3-82
policies, 3-83
powered stretcher, 3-82
transport, 3-81–3-82
unit-based peer leaders, 3-84–3-86
patient-handling technologies
ambulatory and mobility aids, 3-73–3-74
bathing equipment, 3-74
intellidrive system, 3-74–3-75
lateral transfer aids, 3-70–3-73
mattresses and beds, 3-74
patient-transfer equipment design, 3-75
powered full-body sling lifts, 3-65–3-70
patient lifting
manual handling, 3-13–3-15
musculoskeletal disorders (MSDs), 3-14
non-manual handling, 3-15
safe work environment
bariatric task force, 3-56–3-57
criteria-based protocol, 3-56
ergonomic assessment, 3-57–3-58
United Kingdom, patient-handling policy, 3-16–3-17
United States, patient-handling policy
American Nurses Association (ANA), 3-19–3-20
California, 3-21–3-22
causes, solutions, and impacts, 3-20
evidence-based policy, 3-21
Florida, 3-23
Hawaii, 3-25
hazard assessments, 3-24–3-25
Illinois, 3-23
Iowa and Nevada, 3-22
Maryland, 3-26–3-27
Massachusetts, 3-21
Michigan and Ohio, 3-23
Minnesota, 3-27–3-28
Missouri and New York, 3-23
nursing homes, 3-18–3-19
Rhode Island, 3-25–3-26
safety legislation, 3-20–3-21
Texas, 3-23–3-24
Vermont, 3-23
Washington, 3-24
weight loss surgery (WLS)
deep venous thrombosis (DVT), 3-48–3-49
Minnesota Multiphasic Personality Inventory-2 (MMMP-2), 3-48
psychosocial implications, 3-49–3-50
Roux-en-Y gastric bypass (RYGB), 3-48
wound care, 3-49
Back pain
additional environmental factors, 3-97
assistive devices, 3-96
biomechanical stresses, 3-97–3-98
causes, 3-92
clinical areas, 3-104–3-105
ergonomic approach
assistive devices, 3-106
program implementation, 3-106–3-108
study design, 3-105–3-106
laboratory studies, 3-102–3-104
lifting/transferring tasks, 3-97
low-back pain (LBP) episodes, 3-92
patient-handling task (PHT), 3-95–3-96
patient-transfer method, 3-96
postural stresses, 3-97
prevention
education and training, 3-93
ergonomic approach, 3-93–3-95
nursing personnel, 3-92–3-93
testing approaches
Ambulift C_3, 3-101
criteria, 3-98
gait belt, 3-99
Hoyer lift, 3-100
MEDesign patient-handling sling, 3-99
Posey walking belt, 3-100
shower/toileting chair, 3-101
Trans-Aid lift, 3-100
BBPS, *see* Bloodborne Pathogens Standard
BEI, *see* Biological exposure index
Benzene solubles fraction analysis; *see also* Anesthesia mask gas-scavenging system; Health Hazard Evaluation Report, NIOSH
complications, 12-21–12-22
determination methods, 12-12
outcomes, 12-16–12-17

Beta particles, properties of, 22-3; *see also* Radiation safety procedures, hospitals
BFI, *see* Browning-Ferris, Inc.
Binghamton Psychiatric Center, joint labor/management experience
  assault and restraint injuries
    cause and month, 7-90–7-91
    job title, 7-92–7-93
    lost-time injuries, 7-91
    repeat cases, 7-93–7-95
    shift, 7-91
    times of the day, 7-91–7-92
  injury data analysis, 7-87–7-88
  lost-time injury rates, 7-89
  lost-time severity rates, 7-89–7-90
  recommendations, 7-95
  reducing workplace injury, 7-100
  safe unit environment project, 7-87
  statewide injury rate comparison, 7-88
Bioaerosols
  approach, 4-64–4-65
  avian influenza viruses (AIV), 4-69
  bacteria, 4-73
  biological hazards, 4-61
  concentrations, 4-70–4-71
  criteria, 4-68
  decision tree, 4-83
  endotoxins, 4-73–4-74
  hazards and electrocautery, 12-7
  infectious doses, 4-61–4-62
  measles, varicella and smallpox viruses, 4-63
  medical mask, 4-69–4-70
  molds, 4-72–4-73
  non-infectious, 4-61–4-62, 4-70–4-71
  peat moss packager, 4-76
  protection factors, 4-74
  research guidlines, 4-75
  risk and experts' recommendations, 4-64–4-68
  risk coefficient (RC) method, 4-72
  SARS virus, 4-75
  wastewater treatment plant, 4-75
Biological exposure index (BEI), 28-3
Biological hazards control, 27-7–27-8; *see also* Laboratory safety
Biological monitoring methodology, ribavirin, 16-26–16-27
Bipolar electrocautery unit, 12-5
Blood aerosols exposure determination, operating room personnel, 17-20–17-24
Bloodborne pathogens; *see also* Home health care; Latex glove
  gloves, importance, 20-2–20-3
  health care workers, 25-7–25-8
Bloodborne Pathogens Standard (BBPS), 31-18
Brainstorm exercise, in hospital employees, 30-12; *see also* Hospital employees, education
Breast reduction technique, 12-11
Bremsstrahlung radiation, production, 22-3; *see also* Radiation safety procedures, hospitals
Bronchoscopy, 15-7
Browning-Ferris, Inc. (BFI), 24-6
Buffalo Psychiatric Center (BPC), joint labor/management experience
  approaches, 7-76
  clinical/treatment concerns, 7-75
  competitive bidding process, 7-74
  decorations and/or worksite design, 7-82–7-83
  doors and ceiling tiles, 7-79–7-82
  environmental assessment, 7-79
  managing crisis situations, 7-84
  nurses station, 7-83
  personal safety and security, 7-81
  phase I and II, 7-74
  project advisory groups (PAG), 7-75
  recommendations, 7-86–7-87
  satisfaction levels, 7-79
  smoking, 7-81–7-82
  training evaluation, 7-85–7-86
  training programs, 7-78
  unit observation, 7-76
  workers concern, 7-77
Bureau of Labor Statistics (BLS), 2-14–2-15, 18-1–18-2, 25-1

C

Cadavers, safe handling of, 22-20–22-21; *see also* Radiation safety procedures, hospitals
CAHD™, *see* Computer-Aided Health Devices™
California Department of Health Services (CDHS), 11-109–11-112, 16-1
California Occupational Health Program (COHP), 16-1
Cal/OSHA regulatory action, AP HCW risk, 15-8–15-9
Carbon dioxide laser, 12-24
Carcinogenic effects, of EtO, 13-2
Cardiopulmonary resuscitation (CPR), 25-7
Care™, 3-122–3-125
Care Cube disposable hoods, ribavirin administration, 16-40
CDC, *see* Centers for Disease Control and Prevention
CDHS, *see* California Department of Health Services
CDRH, *see* Center for Devices and Radiological Health
Center for Devices and Radiological Health (CDRH), 12-40
Centers for Disease Control and Prevention (CDC), 14-27, 20-1, 24-4
  directional flow patterns, 11-12
  epidemiological survey, 25-7
  guidelines, 11-6
  health care profession, 1-3
  infectious waste, 24-13

sputum induction booth, 11-9
tuberculosis control, 11-109
U.S. model, 3-18
UV radiation, 11-22
Chemical disinfecting/shredding technology, hospital waste treatment, 24-8; *see also* Medical waste
Chemical hazards control, laboratory safety, 27-5–27-7; *see also* Laboratory safety
Chemical spill cleanups, hospitals, 29-5–29-6; *see also* Hospital safety office, need assessment
Chemodrugs
American Society of Health-System Pharmacists (ASHP), 10-2
emerging issues
aerosol administrations and nanotechnology, 10-18–10-19
gene therapy, 10-19–10-20
interventional radiography, chemotherapy, 10-18
environmental monitoring
antineoplastic drugs, 10-4
limits of detection (LOD), 10-4–10-5
hazardous drugs control, workplace
administering hazardous drugs, 10-11
closed system devices, 10-14
control, administration, 10-15
drug transportation, 10-11
engineering controls, 10-12–10-13
hazard assessment, 10-8–10-9
medical surveillance, 10-16
personnel exposure, safe levels, 10-17
polices, programs and procedures, 10-9
PPE, 10-14–10-15
preparation areas, 10-10
receiving and storage, 10-9–10-10
spill management, 10-11–10-12
training, 10-15–10-16
ventilated cabinets, 10-13–10-14
hazardous drugs exposure, effects, 10-1–10-3
dranitsaris analysis, 10-3
International Agency for Research on Cancer (IARC), 10-2
Cigarette consumption, shiftworkers, 21-9–21-10; *see also* Shiftwork, health and safety hazards
Civil Rights Act of 1964, 23-3
ClestraClean room air sterilizer, 11-77–11-79
Coal tar pitch volatiles (CTPV), 12-15
COHP, *see* California Occupational Health Program
Committees for Occupational Safety and Health (COSH), 31-6
Computer-Aided Health Devices™ (CAHD™), 6-14
Contact sensitizers, definition of, 20-10–20-11; *see also* Gloves in hospital
Continuous wave lasers, 12-24
Control technologies for EtO, selection criteria, 14-16–14-17
COSH, *see* Committees for Occupational Safety and Health
Cost benefit analysis, in health care; *see also* Health care worker (HCW)
cost-benefit ratios, calculating formulas, 18-2–18-3
politics, 18-2
safety culture, 18-3
Council of State Governments (CSG), 24-14
County hospital, TB engineering controls
bronchoscopy, 11-106
emergency department (ED), 11-103–11-104
negative-pressure isolation room, 11-104–11-105
outpatient clinics, 11-106–11-108
pressure differential monitoring, 11-106
San Francisco General Hospital (SFGH), 11-103–11-104, 11-108
sputum induction, centralization, 11-104
CP, *see* Cyclophosphamide
CPR, *see* Cardiopulmonary resuscitation
Cross-reactive allergens, immediate-type hypersensitivity, 20-12–20-13; *see also* Immediate-type hypersensitivity
CSG, *see* Council of State Governments
CTPV, *see* Coal tar pitch volatiles
Cyclophosphamide (CP), 10-3–10-4
Cytotoxic drugs and gloves usage, 20-6; *see also* Latex glove

D

Delayed-type hypersensitivity, 20-10–20-12; *see also* Gloves in hospital
Demistifier™ isolation chamber, 15-22
Deutsche Forschungsgemeinschaft (DFG), 14-1
Diagnostic imaging, ergonomic challenges
interpretation process
modified workstation, 9-10
picture, archiving, and collection (PAC) system, 9-9
scanning process
sonograhers, 9-9
x-ray technologists, 9-8–9-9
DID, *see* Discharge ionization detector
Dilution ventilation systems, EtO exposure control, 13-11–13-12
Discharge ionization detector (DID), 17-12
Disinfection, *see* Infection prevention and control (IPC) program
Droplet nuclei generation rates, 11-6, 11-21, 11-26
Dynamic friction coefficient (DCOF), 8-5–8-6
Dynamic simulation, health care ergonomics
design concept development
enlarged plan, 9-36
full-scale simulation, 9-38
future activity scenarios, 9-35
operating hypothesis, 9-33

prototype, 9-38
simulation props, 9-36
enlarged plan, 9-36–9-38
full-scale simulation, 9-38
functional analysis levels, 9-26
prognosis and follow-up
model renovated room, 9-40
viability testing, 9-39
project data analysis
building implementation, 9-29
design, 9-28
prototype, 9-38
scenario activity
chronological table, 9-30
geriatric chair, 9-33
labyrinth effect, 9-31

E

ECRI, *see* Emergency Care Research Institute
Electrical hazards, lasers, 12-34–12-35; *see also* Health care
Electrical safety, hospital, 29-2–29-3; *see also* Hospital safety office, need assessment
Electrocautery smoke, 12-7
Electrochemical analyzers
formaldehyde monitoring, 14-5
glutaraldehyde monitoring, 14-8
Electroencephalography (EEG), 21-4
Electrosurgery, recommended practices, 12-41
recommended practice I, 12-42–12-43
recommended practice II, 12-43–12-44
recommended practice III, 12-44–12-46
recommended practice IV and V, 12-47
Electrosurgical knife (ESK), 12-11
Electrosurgical units (ESU), 12-3
future research, 12-8
health hazards, 12-5–12-7
history, 12-4–12-5
smoke control and RF/MW radiation, 12-7–12-8
types, 12-5
ELISA test, latex-specific IgE, 19-4; *see also* Glove dermatitis
Emergency action plan for EtO exposure, procedures, 13-14–13-15, 13-19; *see also* Ethylene oxide (EtO)
Emergency Care Research Institute (ECRI), 12-3, 12-41
Emerging infectious diseases, 4-1–4-2
exposure route, 4-26–4-27
fit testing and checks
chronic toxicants, 4-33
dioctylpthalate (DOP), 4-35
limitations, 4-34
Occupational Safety and Health Administration (OSHA), 4-32
PortaCount®, 4-34–4-35
quantitative/qualitative, 4-31–4-35
types, 4-32
maintenance and cleaning, 4-35–4-36
personal protective equipment (PPE)
avian influenza virus (AIV), 4-10–4-11
cholera and rota virus, 4-14
hazards, 4-16
HCW dates, 4-12–4-13
health care industry (HCI), 4-10–4-11
influenza (avian flu), 4-20–4-22
microorganisms, 4-15
monkeypox, 4-19
multi-drug resistant (MDR) TB, 4-10
nosocomial disease, 4-13
occupational nosocomial diseases (OND), 4-11–4-12
pathogens, 4-24–4-25
SARS, 4-16–4-19
Streptococcus pyrogenes, 4-13
transmission-borne disease, 4-14–4-15
tuberculosis–antibiotic resistance, 4-19–4-20
types, 4-12
proper selection, 4-36
protection, 4-22–4-23
respirators
aerodynamic diameter, 4-26
airborne microorganisms, 4-55
air filtration mechanism, 4-56–4-58
air-purifying respirators (APR), 4-27
assigned protection factor (APF), 4-58
bioaerosols, 4-61–4-76, 4-83–4-84
biological weapons, 4-56
categories, 4-56
characteristics, 4-27
filtration efficiency, 4-57
fit tests, 4-59
full-face respirators (FFR), 4-31
HCW, 4-30
helmet-type system, 4-31
HEPA filters, 4-29–4-30
mechanical systems, 4-28
N95 type, 4-30–4-31
PAPRs, 4-31
powered air-purifying respirator (PAPR), 4-60
poxviruses, 4-29
program, 4-36–4-39
pulmonary TB, 4-24
requirements, 4-25
SARS event, 4-25–4-26
SAR system, 4-31
SCUBA and SAR systems, 4-30
seal checks, 4-59
self-contained breathing apparatus (SCBA), 4-28
standards and regulations, 4-84
surgical and paper masks, 4-27–4-28
ULPA filters, 4-25

United States and Canada
class, 4-7–4-8
globalization and global warming effects, 4-7
national agencies response paradigms, 4-6–4-7
occupational health paradigms, 4-2–4-5
public health paradigms, 4-5–4-6
Employee education, union perspective on, 31-1–31-7
Employee health services (EHS) program
coordination and collaborative network, 5-19
immunization programs, 5-20
immunobiologics and schedules, 5-20–5-22
job-related illnesses and exposures, 5-20, 5-27
occupational sharps injury and exposure, 5-18–5-19
personnel health and safety education, 5-20
preemployment and periodic health evaluations, 5-19–5-20
pregnant HCP, 5-26–5-27
record keeping, data management, and confidentiality, 5-27
work restrictions, 5-20, 5-23–5-25
Endotoxins and powdered gloves, 20-10; *see also* Latex glove
Engineering controls and physical design, EtO exposure control, 13-22–13-24
Environmental and occupational health training, for hospital house staff, 31-7–31-9
health care environment, hazards, 31-8–31-9
hepatitis and HCWs, 31-15–31-16
mercury, 31-11–31-12
needs assessment, 31-8
nursing student health and safety training, 31-17–31-18
education, of nursing students, 31-23–31-24
hospital workplace hazards, 31-17–31-18, 31-22–31-23
issues and problems facing hospitals, 31-21–31-22
problems and risks in, 31-19–31-21
safer medical devices, 31-10
scabies fact sheet, 31-13–31-14
TB exposures and HCWs, 31-12–31-13
training issues, 31-8
Environmental monitoring, laboratory, 27-15; *see also* Laboratory safety
Environmental Protection Agency (EPA), 13-4, 24-2
federal reaction by, 24-14–24-16 (*see also* Medical waste)
infectious waste, definition of, 24-16–24-17
Environmental regulations affecting hospitals, 29-10; *see also* Hospital safety office, need assessment
Environmental services in hospital, ergonomics
cleaning, 9-2
materials handling
biohazardous waste handling, 9-4
overloaded trash cart, 9-3
soiled linen collection system, 9-5
tall linen carts, 9-3
trash chute systems, 9-4
EPINet, *see* Exposure prevention information network
Ergonomics
hospital ancillary departments
clinical laboratories, 9-10–9-14
diagnostic imaging, 9-7–9-10
environmental services, 9-2–9-7
food service, 9-17–9-21
pharmacy, 9-14–9-17
reducing occupational back pain disability
health care workers, back pain, 9-42
research study, 9-43–9-44
systems approach, 9-45–9-51
workplace design, health care facilities
dynamic simulation, 9-26–9-41
functional building analysis, 9-24–9-26
problem situation, 9-24
ESK, *see* Electrosurgical knife
ESU, *see* Electrosurgical units
Ethylene oxide (EtO); *see also* Aerosolized pentamidine (AP), occupational risks; Health care worker (HCW)
administrative actions, 13-20
chemical structure, 13-2–13-4
community exposure, 13-16–13-17
engineering controls and physical design, 13-22–13-24
exposure control
emergency action plan for, 13-14–13-15
equipment designing for, 13-10–13-11
isolation and EtO control measures, 13-13–13-14
local exhaust and dilution ventilation systems, 13-11–13-12
maintenance and training programs, 13-14
respiratory protection equipment, 13-15–13-16
substitution and changes in work process, 13-12–13-13
first aid procedures for exposure of, 13-20
hazard evaluation of, 13-9–13-10
health effects of, 13-2
in medical care, 13-1–13-2
OSHA EtO standard, 13-7–13-9
peracetic acid, 5-4
short-term exposure limit (STEL), 5-3
sterilization, 5-2–5-3
sterilization mechanism of, 13-4–13-7
work practices and procedure for exposure prevention of, 13-21–13-22
EtO–Freon mixture, 13-3
Explosive chemicals, in laboratory safety, 27-6; *see also* Laboratory safety
Exposure prevention information network (EPINet), 6-5

Eye hazards; *see also* Ethylene oxide (EtO)
EtO in, 13-2
laser beam, 12-25
controlling of, 12-27–12-28
eye protection devices, 12-26–12-27
Eye protection, laboratories, 27-11; *see also* Laboratory safety
Eyewears, eye hazard protection, 12-26–12-27

F

FDA, *see* Food and Drug Administration
Fetal protection policies, 23-3
FEV_1, *see* Forced expiratory volume in 1 s
FID, *see* Flame ionization detector
Film badge, role of, 22-16; *see also* Hospitals, radiation safety procedures
Fire safety, in hospital, 29-2; *see also* Hospital safety office, need assessment
First aid procedures, EtO exposure, 13-19; *see also* Ethylene oxide (EtO)
FISONeb ultrasonic nebulizers, 15-13–15-14
Flame ionization detector (FID), 12-13, 17-14
Food and Drug Administration (FDA), 12-3, 16-3, 19-5
ASTM D5712-95, 20-14
gloves manufacturing specifications, 20-3–20-7
Forced expiratory volume in 1 s (FEV_1), 15-15
Formaldehyde; *see also* Aldehydes monitoring, hospitals
exposure control
methods, 14-19–14-20
monitoring, 14-15–14-18
variables and population exposed, 14-14–14-15
monitoring, 14-4–14-7
Fourier transform IR spectroscopy (FTIR), 12-10
FTIR, *see* Fourier transform IR spectroscopy
FTIR qualitative organic analysis, 12-13, 12-19; *see also* Health Hazard Evaluation Report, NIOSH
Fugitive anesthetic gases, sources of, 17-9

G

Gamma rays, properties of, 22-3–22-4; *see also* Radiation safety procedures, hospitals
Gas chromatographic (GC) method, 14-5
Gas chromatography-mass spectrometry (GC-MS), 12-7
Gas monitor, characteristics of, 14-2–14-3
Gastrointestinal function and shiftwork, 21-9; *see also* Shiftwork, health and safety hazards
GC-MS, *see* Gas chromatography-mass spectrometry
Gene mutations and EtO exposure, 13-2
Geobacillus stearothermophilus, 5-44–5-46
Glove dermatitis
diagnostic considerations, 19-2–19-4
diagnostic methods
allergy testing, 19-9–19-10
history and physical examination, 19-8–19-9
pathophysiology, 19-7–19-8
respiratory symptoms, 19-10–19-11
prevention and treatment
primary prevention, 19-11
secondary prevention, 19-11–19-12
tertiary prevention, 19-12
relationship between hand and latex dermatitis and, 19-1–19-2
resolution, 19-4
Gloves in hospital, 20-1
barrier protection
bloodborne pathogens, 20-2–20-3
glove profile, 20-3–20-7
glove-associated reactions, 20-7–20-9
delayed-type hypersensitivity, 20-10–20-12
hospital protocols, 20-15–20-16
immediate-type hypersensitivity, 20-12–20-14
irritation, 20-9–20-10
respiratory complications, 20-14
in laboratory safety, 27-10
powder *vs.* powderfree gloves, 20-16–20-18
proper glove removal and disposal, 20-18
Glutaraldehyde; *see also* Aldehydes monitoring, hospitals
monitoring, 14-7–14-11
toxicities
case studies, 14-25–14-27
investigations, 14-25
methods of compliance, 14-27–14-28
population exposer, 14-24–14-25
Goggles, usage of, 12-26; *see also* Health care
Government regulation, of lasers, 12-39–12-40
Grinding/microwave technology, in hospital waste management, 24-8; *see also* Medical waste

H

Half-value layer (HVL), 22-4–22-5
Halogenated inhalation anesthesia agents; *see also* Inhalation anesthesia agents
health effects of, 17-5–17-6
monitoring of, 17-9–17-14
Halothane, health effects of, 17-6; *see also* Inhalation anesthesia agents
Hand dermatitis
case study, 19-2
etiologic agents, 19-7
Hazard Communication Standard
hospitals, 29-4–29-5
training, 30-5–30-6
Hazard evaluations and technical assistance branch, NIOSH, 12-10

Hazardous drugs, workplace
  administering hazardous drugs, 10-11
  closed system devices, 10-14
  control, administration, 10-15
  drug transportation, 10-11
  engineering controls, 10-12–10-13
  hazard assessment, 10-8–10-9
  medical surveillance, 10-16
  occupational exposure limits, 10-17–10-18
  permissible exposure limits, 10-17
  personnel exposure, safe levels, 10-17
  polices, programs and procedures, 10-9
  PPE, 10-14–10-15
  preparation areas, 10-10
  receiving and storage, 10-9–10-10
  spill management, 10-11–10-12
  training, 10-15–10-16
  ventilated cabinets, 10-13–10-14
Hazardous research chemicals, management, 27-7
HBV, *see* Hepatitis B virus
HCA Wesley Medical Center program, ribavirin exposure limitation, 16-41–16-42
HCFA, *see* Health Care Financing Administration
HCIA, *see* Health Care Investment Analysis
HCW, *see* Health care worker
Health care
  business case for injury prevention in, 18-3–18-6
    modeling control, 18-6–18-7
    process for, 18-8–18-10
    safe patient handling and benefits, management, 18-8
  cost benefit analysis, 18-1–18-2
    cost–benefit ratios, 18-2–18-3
    politics of, 18-2
    safety culture, 18-3
  industry, laser hazards
    electrical hazards, 12-34–12-35
    eye hazards, 12-25–12-28
    laser fires, 12-33–12-34, 12-37
    laser plumes, 12-29, 12-31–12-32
    safety policy, 12-35–12-40
    skin damage, 12-29
  lasers types, 12-25
Health Care and Social Assistance (HCSA) sector
  injury, illness and surveillance systems
    Bureau of Labor Statistics (BLS), 2-14–2-15
    data source limitations, 2-44
    disease and cause-specific mortality data, 2-38–2-44
    exposures and sources, 2-24–2-29
    fatal occupational injuries, 2-31–2-32
    infectious disease data, 2-34–2-38
    key facts, 2-33
    nonfatal occupation, 2-18–2-24
    occupational injuries, 2-16–2-18
    OSHA, 2-14
    populations, 2-29–2-31
    Survey of Occupational Injuries and Illnesses (SOII), 2-14
    total nonfatal occupational injuries and illnesses, 2-16–2-18
    underestimate, 2-33–2-34
  subsectors and industry
    employer and self-employed establishments, 2-1–2-2
    employment and forecast, 2-4
    employment growth, 2-4–2-5
    occupations and forecast, 2-4–2-9
    types, 2-1
    wage and salary workers, 2-2–2-3
  workers, demographic characteristics
    occupation, 2-12–2-13
    workplace, 2-10–2-12
Health care-associated infections (HAIs)
  catheter-associated urinary tract infections (CA-UTIs), 5-15–5-16
  health care personnel (HCP), 5-16–5-17
  multidrug-resistant organisms (MDROs), 5-16
Health care environment, hazards, 31-8–31-9
Health Care Financing Administration (HCFA), 26-3
Health care industry, night work, 21-3; *see also* Shiftwork, health and safety hazards
Health Care Investment Analysis (HCIA), 26-9
Health care professions, 1-1–1-3
  antineoplastics, 1-4
  associations, 1-5–1-6
  governmental agencies, 1-4–1-5
  growth, 1-6–1-7
  hazards and engineering methods, 1-2–1-3
  hospital accreditation and JCAHO, 1-5
  magnetic resonance imaging (MRI), 1-4
  musculoskeletal injuries, 1-3
  OSHA standards, 1-2, 1-4
  unions, 1-6
  worker compensation acts, 1-3
Health care waste, treating, 24-18; *see also* Medical waste
Health care worker (HCW), 11-84, 11-86, 11-88, 11-90, 16-1, 31-2, 32-1–32-2, 32-5–32-6; *see also* Shiftwork, health and safety hazards
  aerosolized pentamidine risks, 15-1–15-2
    exposures of, 15-12–15-17
    health effects of, 15-9–15-10
    inhalation hazard control, 15-7–15-11
    tuberculosis transmission in, 15-6
  ANSI safety standards for, 12-3
  case studies of, 32-8–32-9
  compensation
    and health care industry, 32-2–32-3
    politics of, 32-3
  health care environment, 32-3–32-4
  job injury protection and cost benefit analysis, 18-1–18-2
  1199 National Benefit Fund, 32-7–32-8

New York's training and education grant program, 32-6
occupational disease and health care environment, 32-4
occupational health diagnostic clinical centers, 32-6–32-7
personal protective equipment (PPE)
avian influenza virus (AIV), 4-10–4-11
cholera and rota virus, 4-14
hazards, 4-16
HCW dates, 4-12–4-13
health care industry (HCI), 4-10–4-11
influenza (avian flu), 4-20–4-22
microorganisms, 4-15
monkeypox, 4-19
multi-drug resistant (MDR) TB, 4-10
nosocomial disease, 4-13
occupational nosocomial diseases (OND), 4-11–4-12
pathogens, 4-24–4-25
severe acute respiratory syndrome (SARS), 4-16–4-19
Streptococcus pyrogenes, 4-13
transmission-borne disease, 4-14–4-15
tuberculosis–antibiotic resistance, 4-19–4-20
types, 4-12
respirators, 4-30
ribavirin aerosol health effects, 16-2–16-3
acceptable exposure levels and exposure assessment methods, 16-3–16-4
control system validation for, 16-20–16-21
engineering and administrative controls of, 16-13–16-18
engineering controls, 16-23
exposure levels for, 16-4–16-6
health hazard assessment data, 16-41
hospital's written policy and shift assignments, 16-23
industrial hygiene survey report, 16-10–16-12
limitations of, 16-35–16-36
methodologies for, 16-24–16-28
outcomes of, 16-28–16-35
pharmacokinetics and exposure recommendations, 16-24
policies for recommendations implementation, 16-8
protection policies of major teaching hospital, 16-9–16-10
recommendations for, 16-7–16-8, 16-37–16-38, 16-52–16-55
recommended concentration of, 16-43–16-55
room ventilation recommendations, 16-24
symptoms for, 16-6
toxicology of, 16-3, 16-23–16-24 uses of, 16-22
TB, workers compensation for, 32-4–32-5
United States and Canada, 4-1–4-8
workplace injuries and illness, prevention of, 32-6
Health Hazard Evaluation Report, NIOSH; *see also* Electrosurgical units; Health care
evaluation criteria for workplace in, 12-14–12-16
historical perspectives of, 12-11–12-12
materials and methodologies for, 12-12–12-14
outcomes of, 12-16–12-21
recommendations for, 12-22
Health Hazard Evaluations (HHEs), 14-14
Health Security Act, 1993, 32-9
Heart disease and shiftwork, 21-9–21-12; *see also* Shiftwork, health and safety hazards
Heating, ventilating, and air conditioning (HVAC) system duct, 11-35, 11-40, 11-57–11-59, 11-103
HEPA, *see* High-efficiency particulate air
HEPA bed tent, 11-85–11-86
Hepatitis A virus, transmission of, 31-15
Hepatitis B virus (HBV), 20-2, 25-7
cost analysis, 6-48–6-49
needlestick-prevention devices, 6-13–6-14
occupational infections, 6-2
respiratory transmission, determination of, 17-21
risk source, 6-15
transmission of, 31-15
Hepatitis C virus, transmission of, 31-16
Hevea brasiliensis, 19-1, 20-12
HHEs, *see* Health Hazard Evaluations
High-efficiency particulate air (HEPA), 11-18–11-20, 11-30–11-40, 11-61–11-64, 12-32, 15-18–15-22, 16-14, 24-8
High performance liquid chromatography (HPLC), 12-12, 15-14, 16-4
Homecare work environment, types of, 25-4
Home health care
ergonomics and back injuries, 25-5–25-7
homecare setting, 25-4
infectious diseases in, 25-7–25-9
job injury and illness in, 25-3–25-4
occupational hazards in, 25-4
personal safety, 25-4–25-5
trends in, 25-1–25-3
Honestly significant difference (HSD), 16-21
Hospital ancillary department, ergonomic challenges
clinical laboratories
phlebotomy, 9-11–9-12
pipetting, 9-12
specimen processing, analyses and storage, 9-13–9-14
diagnostic imaging
interpretation process, 9-9
scanning process, 9-8
environmental services
cleaning, 9-2
materials handling, 9-2–9-7

food service
MSDs, 9-17
multilevel racks, 9-20
neck pain, 9-18
recommendations, 9-21
risk factors, 9-19
pharmacy
infusion center, 9-16–9-17
inpatient, 9-14–9-15
outpatient, 9-15–9-16
Hospital employees, education, 30-1
preparation for training, 30-4–30-5
sample 1 H agenda, 30-15–30-16
training methods, 30-8–30-11
action solving skill objectives, 30-12–30-14
attitude and emotion objectives, 30-11–30-12
behavioral skill objectives, 30-14
training session, checklist for planning, 30-7–30-8
training, purpose of, 30-2–30-3
training session, 30-5–30-7
Hospital environment, EtO
administrative actions for exposure of, 13-20
chemical structure of, 13-2–13-4
community exposure of, 13-16–13-17
engineering controls and physical design, 13-22–13-24
exposure control
emergency action plan for, 13-14–13-15
equipment designing for, 13-10–13-11
isolation and EtO control measures, 13-13–13-14
local exhaust and dilution ventilation systems, 13-11–13-12
maintenance and training programs, 13-14
respiratory protection equipment, 13-15–3-16
substitution and changes in work process, 13-12–13-13
first aid procedures for exposure of, 13-20
hazard evaluation of, 13-9–13-10
health effects of, 13-2
in medical care, 13-1–13-2
OSHA EtO standard, 13-7–13-9
sterilization mechanism of, 13-4–13-7
work practices and procedure, exposure prevention, 13-21–13-22
Hospital house staff, occupational and environmental health training, 31-7–31-9
health care environment, hazards, 31-8–31-9
hepatitis and HCWs, 31-15–31-16
mercury, 31-11–31-12
needs assessment, 31-8
nursing student health and safety training, 31-17–31-18
education, of nursing students, 31-23–31-24
hospital workplace hazards, 31-17–31-18, 31-22–31-23
issues and problems facing hospitals, 31-21–31-22
problems and risks in, 31-19–31-21
safer medical devices, 31-10–31-12
scabies fact sheet, 31-13–31-14
TB exposures and HCWs, 31-12–31-13
training issues, 31-8
Hospital loss runs, categories, 18-8
Hospital personnel, radiation safety, 22-15–22-17
Hospital policy statement, laser safety procedures, 12-56–12-60
Hospital protocols and glove-associated reactions, 20-15–20-16; *see also* Latex glove
Hospital, reproductive hazards, 23-1–23-2
ALARA exposures, 23-7–23-9
background of, 23-2–23-3
medical removal protection, 23-9–23-10
policy in, 23-3
program to confront reproductive hazards, 23-4–23-6
Hospital safety, latex glove role in, 20-1; *see also* Latex glove
barrier protection
bloodborne pathogens, 20-2–20-3
glove profile, 20-3–20-7
glove-associated reactions
delayed-type hypersensitivity, 20-10–20-12
hospital protocols, 20-15–20-16
immediate-type hypersensitivity, 20-12–20-14
irritation, 20-9–20-10
respiratory complications, 20-14
powder *vs.* powderfree gloves, 20-16–20-18
proper glove removal and disposal, 20-18
Hospital safety office, need assessment, 29-1–29-2
chemical spill cleanups, 29-5–29-6
chemical use, industrial hygiene monitoring of, 29-5
committee membership, 29-7–29-8
electrical safety, 29-2–29-3
environmental regulations affecting hospitals, 29-10
fire/life safety, 29-2
hazard communication standard, 29-4–29-5
hospital waste program, 29-3
liaison, 29-6–29-7
library in, 29-11
planning and product recalls, 29-8
professional maintenance, 29-9–29-10
safety committee/accident, 29-4
safety office, administration of, 29-10
technical service, 29-9
universal precautions, 29-10
vendors and consultants, meeting with, 29-9
Hospitals, radiation safety procedures, 22-2
cadavers, safe handling of, 22-20–22-21
clinical laboratories and research facilities, 22-12
diagnostic radiology, 22-10–22-11

dose limits, 22-7–22-9
environmental surveillance and dosimetry, 22-9–22-10
I-131 and P-32, therapy uses of, 22-17–22-19
ionizing radiation, biological effects of, 22-5–22-7
nuclear medicine, 22-11
radiation accident patient, in emergency room, 22-21
radiation, physical properties
half life, 22-4
half-value layer, 22-4–22-5
inverse square law, 22-5
ionizing radiation, energy of, 22-4
nonionizing radiation and ionizing radiation, 22-2
radioactive materials in hospitals, 22-2–22-4
radiation therapy, 22-12–22-13
radioisotope policies, 22-21–22-23
radionuclides, diagnostic uses of, 22-17
radionuclide therapy of patients, 22-19
Hospital waste, 24-2–24-3
disposal technology, 24-5–24-9
EPA assessment plan, 24-21–24-22
historical activity in regulating infectious, 24-13–24-19
infectious waste management, 24-12–24-13
management of, 24-1–24-2
PIW, 24-4–24-5
PIW management, 24-9–24-11
program, 29-3 (*see also* Hospital safety office, need assessment)
standard terminology, 24-3–24-4
warning symbol, 24-4
HSD, *see* Honestly significant difference
Hudson, Anne
bed frame, 3-40–3-41
degenerative disk disease, 3-39–3-40
drag, cradle and hug lift, 3-36
essential functions, 3-42
herniated lumbar disks, 3-37
lifting patients, 3-35–3-36
lumbar fusion surgery, 3-42–3-43
lumbar strain injury, 3-39
medical care units, 3-35
monkey pole, 3-41
neurologist, 3-38–3-39
nursing skills, 3-34
physical therapy, 3-37
spine surgery, 3-43
Team Spirit Award, 3-40
voc rehab evaluation, 3-41–3-42
workers' compensation system, 3-38
Human Immunodeficiency Virus (HIV), 20-2, 25-7
HVL, *see* Half-value layer
Hydrocarbons screening; *see also* Health Hazard Evaluation Report, NIOSH
determination methods, 12-13
outcomes, 12-16

I

I-131 and P-32, therapy, 22-17–22-19; *see also* Radiation safety procedures, hospitals
IARC, *see* International Agency for Research on Cancer
ICRP, *see* International Commission on Radiological Protection
ICU Medical CLAVE®, 5-44
Immediate-type hypersensitivity, 20-12–20-14; *see also* Gloves in hospital
Incineration technology, in hospital waste management, 24-8–24-9; *see also* Medical waste
Industrial hygiene; *see also* Health care worker (HCW)
monitoring, importance of, 14-15
sampling, importance of, 12-10
survey report, for ribavirin, 16-10–16-12
Industrial hygiene approach, TB control
air filtration
HEPA filtration unit, 11-20
ultra-low penetration air (ULPA) filters, 11-17
dilution ventilation, concentration reduction control
contaminant concentration (*C*ss), 11-14
generation rates, 11-14
directional airflow
design, TB control, 11-12
MTb sources, 11-11
multidirectional flow rooms, 11-12
droplet nuclei generation rates, 11-26
hazard characterization, 11-3–11-4
industrial ventilation manual, 11-7, 11-9–11-10, 11-14
Mycobacterium tuberculosis (MTb), 11-3, 11-103, 11-111
patient isolation rooms, 11-21–11-25
American Institute of Architects (AIA) guidelines, 11-23
droplet nuclei concentration, 11-21
warning signs, UV radiation, 11-22
regulations and guidelines, 11-4–11-6
droplet nuclei concentration, 11-7
Mycobacterium tuberculosis, 11-6
Occupational Safety and Health Act (OSH Act), 11-4–11-5
permissible exposure limits (PEL), 11-4
source characterization, 11-5–11-6
research needs, 11-26
source characterization, 11-5–11-6
source control, 11-6–11-11
ultraviolet irradiation

duct irradiation, 11-20
HEPA filtration unit, 11-19–11-20
ultraviolet germicidal irradiation (UVGI), 11-19–11-21
ventilation
capture efficiency, 11-10
empirical expressions, 11-8
flanges, 11-8
industrial ventilation manual, 11-10
local exhaust systems, 11-7
slot hood, 11-8
sputum induction booth, 11-9
tracer gas decay curve, 11-15
Industrial hygiene approach, tuberculosis control
Industrial hygiene monitoring, 29-5; *see also* Hospital safety office, need assessment
Industry risk index (IRI), 32-3
Infection control risk assessment (ICRA) panels, 5-38–5-39
Infection prevention and control (IPC) program
construction guidelines
barriers, 5-32
building protection, 5-34–5-35
commission, 5-35–5-37
dust movement, 5-33
environmental monitoring, 5-37–5-38
infection control risk assessment (ICRA) panels, 5-38–5-39
negative pressure, 5-33
vibration, 5-34
health care construction
Aspergillus species, 5-29–5-30
building dynamics, 5-30
construction guidelines, 5-32–5-38
HVAC system, 5-30–5-32
pathobiology, 5-30
pathogenic and opportunistic environmental fungal ecology, 5-29
patient risk factors, 5-30
HVAC system
air movement, 5-31
Legionnaires' disease, 5-31–5-32
purposes, 5-30–5-31
mortality and morbidity
coordination and collaborative network, 5-19
immunization programs, 5-20
immunobiologics and schedules, 5-20–5-22
job-related illnesses and exposures, 5-20, 5-27
occupational sharps injury and exposure, 5-18–5-19
personnel health and safety education, 5-20
preemployment and periodic health evaluations, 5-19–5-20
pregnant HCP, 5-26–5-27
record keeping, data management and confidentiality, 5-27
work restrictions, 5-20, 5-23–5-25
patient safety, 5-42–5-43
Geobacillus stearothermophilus, 5-44–5-46
liquid drawn, 5-44–5-45
luer-activated valve (LAV), 5-43
objectives, 5-44
syringes, 5-43–5-44
protecting personnel
central line-associated bloodstream infection (CLABSI) prevention, 5-17–5-18
consumers, regulators and payers response, 5-16–5-17
health care-associated infections (HAIs), 5-15–5-16
mortality and morbidity, 5-18–5-27
patient safety, 5-13–5-15
terms and programmatic guidance, 5-18
sterilization and disinfection
classification, 5-1–5-2
clean and dirty, 5-10
concentrated OPA, 5-8–5-9
diluted peracetic acid, 5-9
ethylene oxide (EtO), 5-2–5-3
glutaraldehyde, 5-6–5-7
health, 5-10
hydrogen peroxide plasma gas, 5-5
material compatibility, 5-9–5-10
ortho-phthalaldehyde, 5-7–5-8
oversight, 5-11
ozone, 5-5
peracetic acid, 5-4
proteolytic enzymes, 5-9
quality improvement, 5-11
reprocessing algorithm, 5-10
safety, 5-10–5-11
Sterilox, 5-9
Infectious diseases, in home health care, 25-7–25-9; *see also* Home health care
Infectious hospital waste; *see also* Medical waste
historical activity in, 24-13–24-14, 24-19
demand for protection, 24-14
federal reaction, 24-14–24-16
health care waste, treating, 24-18
problem solving, 24-16–24-18
tracking infectious waste, 24-18–24-19
management, trends in, 24-12–24-13
Infrared spectrophotometer analyzer (IR-h), 17-13
formaldehyde monitoring, 14-6–14-7
glutaraldehyde monitoring, 14-8
Inhalation anesthesia agents
anesthesia mask gas-scavenging system, 17-16–17-19
exposure standards, 17-4–17-5
monitoring of, 17-9–17-14
physical properties of, 17-1–17-4
sources of waste/fugitive, 17-6–17-9
toxicology of
halothane and methoxyflurane, 17-6

nitrous oxide and halogenated inhalation anesthesia, 17-5–17-6
International Agency for Research on Cancer (IARC), 10-2, 10-17, 12-15, 13-2
International Commission on Radiological Protection (ICRP), 22-7
International Labor Organization, shiftwork recommendations, 21-14
Interstitial brachytherapy, 22-13
Intracavitary brachytherapy, 22-13
Ionizing detector-based analyzers
 formaldehyde monitoring, 14-5
 glutaraldehyde monitoring, 14-8
Ionizing radiation; *see also* Hospitals, radiation safety procedures; Radiation safety procedures, hospitals
 biological effects of, 22-5–22-7
 energy of, 22-4
 in radiation therapy, 22-12–22-13
IR-h, *see* Infrared Spectrophotometer analyzer
IRI, *see* Industry risk index
Isopropyl alcohol, biological effects of, 12-15

J

Jaeger pneumotachograph, usage of, 16-14
JCAHO, *see* Joint Commission on Accreditation of Health Care Organizations
Jewish General Hospital, glutaraldehyde study, 14-26–14-27
Job assault, violence
 employer's responsibility, 7-50
 factors, 7-49
 guidelines, 7-50–7-51
 posttraumatic stress disorder (PTSD), 7-51
 prevention organization, 7-62–7-63
 reasons and victims, 7-48
 resource list, 7-54–7-62
 sample assault incident report form, 7-65
 sample contract language
  employer's comprehensive health and safety policy, 7-52
  two-way radios, alarms and/or paging systems, 7-53
  union rights, 7-54
  workplace design, 7-52
 sample grievance/petition form, 7-66
 sexual harassment, 7-49
 survey, 7-64–7-65
 work environment, 7-49–7-50
 work-related deaths, 7-48
Job injury and illness, home health care, 25-3–25-4; *see also* Home health care
Joint Commission on Accreditation of Health Care Organizations (JCAHO), 12-40, 29-7, 31-21
Joint labor/management experience
 assaults *vs.* restraints, 7-73
 Binghamton Psychiatric Center
  assault and restraint injuries, 7-90–7-95
  injury data analysis, 7-87–7-88
  lost-time injury rates, 7-89
  lost-time severity rates, 7-89–7-90
  recommendations, 7-95
  reducing workplace injury, 7-100
  safe unit environment project, 7-87
  statewide injury rate comparison, 7-88
 Buffalo Psychiatric Center (BPC)
  clinical or treatment concerns, 7-75
  competitive bidding process, 7-74
  decorations and/or worksite design, 7-82–7-83
  doors and ceiling tiles, 7-79–7-82
  environmental assessment, 7-79
  managing crisis situations, 7-84
  nurses station, 7-83
  personal safety & security, 7-81
  phase I and II, 7-74
  project advisory groups (PAG), 7-75
  recommendations, 7-86–7-87
  satisfaction level, 7-79
  smoking, 7-81–7-82
  suggestions, 7-76
  training evaluation, 7-85–7-86
  training programs, 7-78
  unit observation, 7-76
  workers concern, 7-77
 focus group interviews, 7-96–7-101
 national profile, 7-71
 New York State (NYS)
  adult inpatient population changes, 7-72
  microanalysis, 7-71
 organizational culture and political will, 7-70–7-71
 OSHA guidelines, 7-69–7-70
 violence, 7-69

K

Kaiser Foundation Hospital
 mercury, 31-11–31-12
 TB prevention, 31-12

L

Laboratory safety, 27-1–27-2
 biological hazards, 27-7–27-8
 chemical hazards, 27-5–27-7
 controlling mechanical hazards, 27-9–27-10
 controlling radiation hazards, 27-8
 hazards monitoring in, 27-15–27-16
 information systems, 27-3–27-5
 laboratory design, 27-12–27-14
 organization for, 27-2–27-3
 safety devices and personal protective equipment, 27-10–27-12
 spill response procedures, 27-14

Labor Occupational Health Program, 25-8
Laminar airflow systems, 11-33–11-34
Laser
  absorbent filters, usage of, 12-28
  hazards, health care industry
    electrical hazards, 12-34–12-35
    eye hazards, 12-25–12-28, 12-30
    laser fires, 12-33–12-34
    laser plumes, 12-29, 12-31–12-32
    safety policy for, 12-35–12-40
    skin damage, 12-29–12-30
  historical information of, 12-24–12-25
  maintenance, processes in, 12-35–12-36
  operational guidelines, 12-36
  smoke and electrosurgery smoke, difference of, 12-2
  technology in surgery, disadvantages of, 12-2
Laser operators, training and approval of, 12-36
Laser plume
  case study for, 12-52–12-55
  quantification data of, 12-50–12-52
Laser Safety Committee (LSC), 12-35, 12-56–12-57
Laser Safety Officer (LSO), 12-25
Laser Safety Personnel, 12-35
Laser safety program, 12-35
Laser safety training, 12-37–12-38
Laser users and support staff, personnel protection, 12-36
Latex allergy, 19-1
  diagnostic methods
    allergy testing, 19-9–19-10
    history and physical examination, 19-8–19-9
    pathophysiology of, 19-7–19-8
    respiratory symptoms, 19-10–19-11
  epidemiology of, 19-5–19-7
  prevention and treatment
    primary prevention, 19-11
    secondary prevention, 19-11–19-12
    tertiary prevention, 19-12
  relationship between hand and latex dermatitis and, 19-2
Latex glove, 20-1
  barrier protection
    bloodborne pathogens, 20-2–20-3
    glove profile, 20-3–20-7
  glove-associated reactions, 20-7–20-9
    delayed-type hypersensitivity, 20-10–20-12
    hospital protocols, 20-15–20-16
    immediate-type hypersensitivity, 20-12–20-14
    irritation, 20-9–20-10
    respiratory complications, 20-14
  powder *vs.* powderfree gloves, 20-16–20-18
  proper glove removal and disposal, 20-18
LFL, *see* Lowest feasible level
LFPM, *see* Linear feet per minute
Life safety, hospital, 29-2; *see also* Hospital safety office, need assessment
Lift aid equipment, ergonomics, 9-48
Limits of detection (LOD), 12-12
Linear feet per minute (LFPM), 28-1
LOAEL, *see* Lowest observed adverse effect level
Local exhaust ventilation technique; *see also* Infrared Spectrophotometer analyzer
  EtO exposure control, 13-10–13-11
  for formaldehyde monitoring, 12-19–12-20
  for plume control, 12-31
LOD, *see* Limits of detection; Lowest level of detection
LOHP, *see* Labor Occupational Health Program
Lowest feasible level (LFL), 12-15
Lowest level of detection (LOD), 16-16
Lowest observed adverse effect level (LOAEL), 23-2
LSC, *see* Laser Safety Committee
LSO, *see* Laser Safety Officer

M

Magnetic resonance imaging (MRI), 1-4
MAKs, *see* Maximum allowable concentrations
Management of assaultive behavior (MAB), 7-23
Martin Chair-A-Table®
  caregivers, 3-124–3-125
  game changing technology, 3-122
  institutions, 3-124
  risk managers and insurers, 3-124
  steps, 3-123–3-124
Material Safety Data Sheets (MSDSs), 27-4, 29-4, 30-15
Maximum allowable concentrations (MAKs), 14-1
Maximum exposure limits (MELs), 12-40
MBTH, *see* 3-methyl-2-benzothiazolone hydrazone hydrochloride
MDL, *see* Minimum detection limit
Mechanical hazards control, laboratory safety, 27-9–27-10; *see also* Laboratory safety
Medical laser procedures, usage of, 52
Medical removal protection, 23-8–23-10
Medical waste, 24-2–24-3
  disposal technology, 24-5–24-9
  EPA assessment plan, 24-21–24-22
  historical activity, regulating infectious, 24-13–24-19
  infectious waste management, 24-12–24-13
  management of, 24-1–24-2
  PIW management, 24-9–24-11
  potentially infectious waste (PIW), 24-4–24-5
  standard terminology, lack of, 24-3–24-4
  warning symbol, 24-4
Medical Waste Tracking Act (MWTA), 24-1
MELs, *see* Maximum exposure limits
Menstruation and shiftwork, 21-9; *see also* Health care worker (HCW); Shiftwork, health and safety hazards
Methoxyflurane, health effects of, 17-6
3-Methyl-2-benzothiazolone hydrazone hydrochloride (MBTH), 14-8

MICROCON HEPA air filtration system
controlling bioaerosols, emergency room, 11-74–11-77
isolation room investigation, 11-72–11-74
Microwave (MW) and radiofrequency (RF) radiation
biological effects, 12-5–12-6
and control of ESU smoke, 12-7–12-8
Mine Safety and Health Administration (MSHA), 16-28
Minimum detection limit (MDL), 14-2
Miniram PDM 3, 12-50
Miran infrared spectrophotometer, 14-16, 17-17
MSDs, *see* Musculoskeletal disorders
MSDSs, *see* Material Safety Data Sheets
MSHA, *see* Mine Safety and Health Administration
Musculoskeletal disorders (MSDs), 9-1, 9-9, 9-12, 9-13, 9-17
MWTA, *see* Medical Waste Tracking Act
Mycobacterium tuberculosis (MTb), 11-3, 11-103, 11-111

N

1199 National Benefit Fund, hospital employees, 32-7–32-8
National Council on Radiation Protection (NCRP), 22-16
National Fire Protection Association (NFPA), 27-5
National Institute for Occupational Safety and Health (NIOSH), 9-9, 9-43, 10-2, 11-89–11-90, 12-3, 23-2, 25-8, 26-8
National Institutes of Health and the Laser Institute, role of, 12-8
NCRP, *see* National Council on Radiation Protection
NDIR, *see* Nondispersive infrared analyzers
Nd:Yag laser, 12-24
Needlestick Safety and Prevention Act (NSPA), 6-5
Neurotoxicity in humans, EtO in, 13-2
New York Commissioner of Health, in shiftwork regulations, 21-7
New York Committee for Occupational Safety and Health (NYCOSH), 32-6
New York State (NYS)
acutely ill and dually diagnosed individuals, 7-15
competitive bidding process, 7-74
occupational injury and illness reporting (OIRS) database, 7-71
Office of Mental Health (OMH), 7-68
Public Employees Federation (PEF), 7-68
New York's training and education grant program, 32-6
NFPA, *see* National Fire Protection Association
NHZ, *see* Nominal hazard zone
Nightwork, 21-1–21-2; *see also* Health care worker (HCW); Shiftwork, health and safety hazards
checklist for reducing hazards of, 21-15
hidden costs of shiftwork, 21-2–21-3
accidents, 21-7
effect on workers, 21-4–21-5
gastrointestinal function and, 21-9
health care industry, night work in, 21-3
heart disease and, 21-9–21-12
nurses, 21-5–21-7
psychological effects of, 21-8
reproductive function, 21-8–21-9
strategies of
education, 21-14
hours limitation at night, 21-13–21-14
light control in the environment, 21-12
personal medical factors, 21-12–21-13
physical fitness and shift rotation frequency, 21-13
shift rotation and napping strategies, 21-12
NIOSH, *see* National Institute for Occupational Safety and Health (NIOSH)
Nitrile gloves, role of, 20-4
Nitrosamines; *see also* Health Hazard Evaluation Report, NIOSH
air samples, results of, 12-16
biological effects of, 12-15
determination of, 12-12–12-13
Nitrous oxide; *see also* Inhalation anesthesia agents
health effects of, 17-5
monitoring of, 17-9–17-12
role in ophthalmology, 31-8
NOAEL, *see* No observable adverse effect level
NOEL, *see* No-observed effect level
Nominal hazard zone (NHZ), 12-28
Nondispersive infrared analyzers (NDIR), 14-5, 14-8
Nonfatal occupational injuries and illness, HCSA sector
fatal occupational injuries
distribution, 2-31–2-32
incidence rates, 2-32
sex, race and ethnic origin, 2-31
incidence rates
assaults and violent acts, 2-27
back injuries, 2-26
body parts, 2-20, 2-22
events, 2-20, 2-23
floor and ground surfaces, 2-28
health care patient, 2-27
home health care services, 2-24
hospitals, 2-19, 2-21
nature, 2-20, 2-22
nursing and residential care facilities, 2-16–2-18
overexertion and repetitive motion, 2-25
private industry employers, 2-19–2-20
skin diseases/disorders, 2-18–2-19
slips, trips and falls, 2-26
sources, 2-20, 2-23

sprains and strains, 2-24–2-25
subsector, 2-20–2-21
worker motion, 2-28–2-29
industry level data
age category, 2-30
occupational data, 2-31
sex, race and ethnic origin, 2-29–2-30
nursing and residential care facilities, 2-16–2-18
private and service-providing industries, 2-18
No observable adverse effect level (NOAEL), 23-2
No-observed effect level (NOEL), 16-3
NSPA, *see* Needlestick Safety and Prevention Act
Nuclear medicine and radioactive materials, 22-11; *see also* Hospitals, radiation safety procedures in
Nurse and Patient Safety and Protection Act of 2007, 3-29–3-31
Nurses, in shiftwork, 21-5–21-7; *see also* Shiftwork, health and safety hazards
Nursing home industry, 26-4–26-6, 26-10–26-12
Nursing home workers
back injuries in, 26-2, 26-7–26-8
dangers of, 26-6–26-7
injury and illness incidence rate of, 26-8
recommendations for, 26-3–26-4, 26-12–26-13
risks to, 26-8–26-10
understaffing, 26-2
unsafe working conditions, 26-3
Nursing student health and safety training, hospital workplace, 31-17–31-18
education, of nursing students, 31-23–31-24
hospital workplace hazards, 31-17–31-18, 31-22–31-23
issues and problems facing hospitals, 31-21–31-22
problems and risks in, 31-19–31-21
NYCOSH, *see* New York Committee for Occupational Safety and Health
NYS, *see* New York State

O

Occupational back pain disability reduction
back injuries prevention, 9-44
manual lifting, 9-44
research study, 9-43–9-44
system approach
formulation recommendations, 9-46–9-50
implementation recommendations, 9-50–9-51
measurement and evaluation, 9-51
risk analysis, 9-46
risk identification and assessment, 9-45–9-46
Occupational exposure pathways
biological monitoring, 10-7–10-8
dermal pathways, 10-6
worker exposure pathways, 10-6–10-7
Occupational health hazards, home health care, 25-4; *see also* Home health care
Occupational injury reporting system (OIRS), 7-87
Occupational needlestick injuries
needlestick-prevention devices, 6-13–6-14
analyzing costs, 6-48–6-51
CAHDModel output, 6-50–6-51
general recommendations, 6-44–6-45
product group applications, 6-45–6-48
prevention methods, needlestick-transmitted infections
education and training, 6-17
needle-recapping controversy, 6-17–6-18
preventive devices, 6-19
sharps disposal, 6-18–6-19
risk, needlestick-transmitted infections
assessment, 6-14
management, 6-15–6-17
sources, 6-14–6-15
workers' compensation, 6-3–6-4
zero risk, 6-4
Occupational reproductive hazards, 23-2–23-3
Occupational risks, of aerosolized pentamidine, 15-1–15-2; *see also* Health care worker (HCW)
airborne tuberculosis transmission and treatment of, 15-5–15-7
engineering ventilation control for, 15-2–15-3, 15-22–15-24
in health care workers, 15-9–15-10, 15-12–15-17
HEPA filtration efficiency for, 15-18–15-22
inhalation hazard control for HCW in administration of, 15-7–15-11
tuberculosis screening, 15-4
Occupational Safety and Health Act (OSH Act), 11-4–11-5
Occupational Safety and Health Administration (OSHA), 1-2, 1-4, 6-2, 12-5, 23-2, 24-2; *see also* Health care
bloodborne pathogens standard, 25-8
enforcement efforts, in nursing home industry, 26-4
ergonomics standard, for private sector employers, 26-13
for ETO exposure, 13-7–13-10
formaldehyde standard for, 14-21–14-23
homecare workers, definition of, 25-5
nursing workers, inspections of, 26-3
Occupational Safety Hazards Act in 1991, 20-1
Office of Mental Health (OMH)
environmental assessment, 7-79
New York State, 7-71
occupational health guidelines, 7-73
occupational injury and illness reporting (OIRS) database, 7-71
pilot projects, 7-101
planning interventions and curriculum writing, 7-74
postincident support program, 7-87

OIRS, *see* Occupational injury reporting system
Operating room personnel, blood aerosols exposure determination, 17-20–17-24
OSHA, *see* Occupational Safety and Health Administration
OSH Act, *see* Occupational Safety and Health Act

P

PAG, *see* Project advisory groups
Papermill workers and heart disease risk, 21-11; *see also* Shiftwork, health and safety hazards
PAS-h, *see* Photoacoustic infrared spectrophotometric analyzers
Passive electrochemical/solid state analyzer counterparts (PEC/SS-h), 17-14
Patch testing, in glove dermatitis, 19-10; *see also* Glove dermatitis
Patient-handling policy
  Australia, 3-17–3-18
  United Kingdom, 3-16–3-17
  United States
    American Nurses Association (ANA), 3-19–3-20
    California, 3-21–3-22
    causes, solutions, and impacts, 3-20
    evidence-based policy, 3-21
    Florida, 3-23
    Hawaii, 3-25
    hazard assessments, 3-24–3-25
    Illinois, 3-23
    Iowa, 3-22
    Maryland, 3-26–3-27
    Massachusetts, 3-21
    Michigan, 3-23
    Minnesota, 3-27–3-28
    Missouri, 3-23
    Nevada, 3-22
    New York, 3-23
    nursing homes, 3-18–3-19
    Ohio, 3-23
    Rhode Island, 3-25–3-26
    safety legislation, 3-20–3-21
    Texas, 3-23–3-24
    Vermont, 3-23
    Washington, 3-24
Patient handling programs, features of, 18-8
Patient-handling tasks (PHTs), back injuries
  administrative controls, 3-82–3-83
  back belts, 3-76–3-78
  bed mover, 3-81
  clinical decision-making tools, 3-83–3-84
  equipment/devices
    ceiling lifts, 3-78–3-79
    clinical trials, 3-80–3-81
    floor-based lift, 3-78–3-79
    lateral transfer solutions, 3-81
    PRS system, 3-79–3-80
    repositioning aids, 3-79
  lift teams, 3-86–3-87
  manual lifting, 3-82
  policies, 3-83
  powered stretcher, 3-82
  transport, 3-81–3-82
  unit-based peer leaders
    definitions, 3-84
    ErgoCoaches, 3-85
    program implementation, 3-86
    VA peer leaders, 3-85
Patient-handling technologies; *see also* Back injuries
  ambulatory and mobility aids, 3-73–3-74
  bathing equipment, 3-74
  intellidrive system, 3-74–3-75
  lateral transfer aids
    concepts, 3-70
    friction-reducing properties, 3-70–3-72
    repositioning devices, 3-72
    sliding boards, 3-71
    transfer chairs, 3-72–3-73
  mattresses and beds, 3-74
  patient-transfer equipment design, 3-75
  powered full-body sling lifts
    advantages, 3-67
    ceiling-mounted lifting system, 3-65–3-66
    floor-based lifts, 3-67–3-68
    gait/transfer belt, 3-69–3-70
    powered standing lifts, 3-68
    spinal cord injury/disorders (SCI/D), 3-66
    standing assist and repositioning aids, 3-69
Patient injury prevention program, goal of, 18-6
Patient Safe™, 5-42
PBZ, *see* Personal breathing zone
PCP, *see Pneumocystis carinii* pneumonia
Peak expiratory flow rates (PEFR), 15-15
PEC/SS-h, *see* Passive electrochemical/solid state analyzer counterparts
Pediatric intensive care units (PICUs), 16-2
PEF, *see* Public Employees Federation
PEFR, *see* Peak expiratory flow rates
PEL, *see* Permissible exposure limits
Pentamidine, occupational risks of, 15-1–15-3
  airborne tuberculosis, 15-5–15-7
  engineering ventilation control for, 15-2–15-3, 15-22–15-24
  health care workers
    exposures of, 15-12–15-17
    health effects of, 15-9–15-10
  HEPA filtration efficiency for, 15-18–15-22
  inhalation hazard control, HCW, 15-7–15-11
  tuberculosis screening, 15-4
PEOSH, *see* Public Employees Occupational Safety and Health
Peracetic acid, 5-4
Percutaneous injury (PI), 6-5

Permissible exposure limits (PEL), 11-4, 13-2, 14-1
Peroxidizable chemicals, laboratory safety, 27-6; *see also* Laboratory safety
Personal breathing zone (PBZ), 16-16
Personal protection factor (PPF), 16-15
Personal protective equipment, 20-2; *see also* Health care worker (HCW)
  for ethylene oxide exposure, 13-21–13-22
  for formaldehyde exposure, 14-20
  for RF/MW radiation exposure, 12-5
  and techniques, for blood aerosol exposure, 17-22–17-24 (*see also* Anesthesia mask gas-scavenging system)
Personal protective equipment (PPE)
  avian influenza virus (AIV), 4-10–4-11
  cholera and rota virus, 4-14
  hazards, 4-16
  HCW dates, 4-12–4-13
  health care industry (HCI), 4-10–4-11
  influenza (avian flu), 4-20–4-22
  microorganisms, 4-15
  monkeypox, 4-19
  multi-drug resistant (MDR) TB, 4-10
  nosocomial disease, 4-13
  occupational nosocomial diseases (OND), 4-11–4-12
  pathogens, 4-24–4-25
  severe acute respiratory syndrome (SARS)
    emerging infectious diseases, 4-16–4-17
    HCW, 4-17–4-18
    nosocomial diseases, 4-18
    novel coronavirus (SARS-CoV), 4-16–4-17
    urine and feces, 4-18–4-19
  Streptococcus pyrogenes, 4-13
  transmission-borne disease, 4-14–4-15
  tuberculosis–antibiotic resistance, 4-19–4-20
  types, 4-12
Personnel health program (PHP), 5-19
Pharmacy, ergonomic challenges
  infusion center
    pair of pliers, 9-17
    ripping open seals, 9-16
  inpatient
    antifatigue runners, 9-14
    workstation, 9-15
  outpatient
    inventory bins, 9-15
    step ladder, 9-16
Phlebotomy, ergonomics, 9-11–9-12
Photoacoustic infrared spectrophotometric analyzers (PAS-h), 17-14
Photoionization detector (PID), 17-14
PI, *see* Percutaneous injury
PICUs, *see* Pediatric intensive care units PID, *see* Photoionization detector
PIR spectrophotometric analyzers, limitation of, 14-4
PIW, *see* Potentially infectious waste
Plastic eyewear, 12-27; *see also* Health care
Plume exposures, controlling of, 12-31–12-32
PNAs, *see* Polynuclear aromatic compounds
Pneumocystis carinii pneumonia (PCP), 15-2
Polynuclear aromatic compounds (PNAs), 12-10, 12-12, 12-15–12-16
Portable HEPA filtration, TB isolation
  challenge data
    MICROCON®, 11-64–11-67
    purge test data, 11-66
    stabilization period, 11-65
  filtration, mechanics
    depth filter, 11-31
    relative effect, 11-33
    single fiber, 11-32
  health care facilities, 11-33–11-34
  in-room HEPA filtration
    air curtain effect, 11-37
    antituberculous drugs, 11-36
    negative pressure, 11-35
    particle clearance, 11-36
  interim solution, 11-40
  mechanics of filtration, 11-31–11-33
  respiratory isolation, TB
    isolation ventilation, 11-35
    multiple drug-resistant tuberculosis (MDR-TB), 11-34
  respiratory protection, 11-38
  source control
    containment hood, 11-39
    fugitive pentamidine, 11-38
Positron particles, properties of, 22-3; *see also* Radiation safety procedures, hospitals
Posttraumatic stress disorder (PTSD), 7-51
Potentially infectious waste (PIW), 24-1, 24-4–24-5, 24-10–24-11
PPE, *see* Personal protective equipment
PPF, *see* Personal protection factor
Pregnancy Discrimination Act of 1978, 23-3
Pregnant women, occupational exposure, 22-9; *see also* Hospitals, radiation safety procedures
Project advisory groups (PAG), 7-11
Proper glove donning techniques, 20-8; *see also* Latex glove
Protein level, in latex gloves, 20-14; *see also* Latex glove
PTSD, *see* Posttraumatic stress disorder
Public Employees Federation (PEF)
  attack insurance plan, 7-68
  Binghamton Psychiatric Center, 7-87
  Buffalo project conclusion, 7-85
  competitive bidding process, 7-74
  health & safety department, 7-72
  innovative prevention programs, 7-68
  planning interventions and curriculum writing, 7-74
Public Employees Occupational Safety and Health (PEOSH), 7-43–7-44

Public safety; *see also* Shiftwork, health and safety hazards
checklist for reducing hazards, 21-15
shiftwork, hidden costs of, 21-2–21-3
accidents, 21-7
effect on workers, 21-4–21-5
gastrointestinal function and, 21-9
health care industry, night work in, 21-3
heart disease and, 21-9–21-12
nurses, 21-5–21-7
psychological effects of, 21-8
reproductive function, 21-8–21-9
strategies of
education, 21-14
hours limitation at night, 21-13–21-14
light control in the environment, 21-12
personal medical factors, 21-12–21-13
physical fitness and shift rotation frequency, 21-13
shift rotation and napping strategies, 21-12
Pulsed lasers, 12-24

Q

Quantitative fit testing, ribavirin exposure, 16-43

R

Radiation hazards control, laboratory safety, 27-8; *see also* Laboratory safety
Radiation safety procedures, hospitals, 22-2; *see also* Hospitals, radiation safety procedures
cadavers, safe handling of, 22-20–22-21
clinical laboratories and research facilities, 22-12
diagnostic radiology, 22-10–22-11
dose limits, 22-7–22-9
environmental surveillance and dosimetry, 22-9–22-10
I-131 and P-32, therapy uses of, 22-17–22-19
ionizing radiation, biological effects of, 22-5–22-7
nuclear medicine, 22-11
radiation accident patient, 22-21
radiation, physical properties
half life, 22-4
half-value layer, 22-4–22-5
inverse square law, 22-5
ionizing radiation, energy of, 22-4
nonionizing radiation and ionizing radiation, 22-2
radioactive materials, 22-2–22-4
radiation therapy, 22-12–22-13
radioisotope policies, 22-21–22-23
radionuclide therapy, 22-17, 22-19
Radioimmunoassay (RIA), 16-4
Radioisotope policies, usage of, 22-21–22-23; *see also* Hospitals, radiation safety procedures
RCRA, *see* Resource Conservation and Recovery Act
Recommended exposure limits (RELs), 14-1
Recommended Practices Committee (RPC), 12-41
Reducing occupational back pain disability, ergonomics
back injury prevention, 9-44–9-45
health care workers, 9-42
lifting techniques, 9-42
manual lifting, 9-44
NIOSH formula, 9-43
research study, 9-43–9-44
systems approach
assessment, 9-45–9-46
evaluation and measurement, 9-51
implementation, 9-50–9-51
recommendations, formulation, 9-46–9-50
risk analysis, 9-46
Reducing tuberculosis spread, workplace
air quality survey
air movement direction, 11-49–11-50
natural ventilation, 11-48
checking building ventilation, 11-48
interacting way, 11-56
move desks, chairs, and people
floor plan, 11-57
workplace improvements, 11-56
natural ventilation, low-cost ways
installing properly position fans, 11-51–11-54
opening windows, 11-51
portable room air cleaner, 11-54–11-55
Reflective laser protective eyewear, 12-26–12-27; *see also* Health care
Registered nurses (RNs), 21-3
Registry toxic effects chemical substances (RTECS), 23-4
Regulated medical waste, definition of, 24-2; *see also* Medical waste
RELs, *see* Recommended exposure limits
Reproductive hazards committee, 23-5–23-6
Reproductive risk, 23-1–23-2; *see also* Hospital, reproductive hazards
ALARA exposures, 23-7–23-9
background of, 23-2–23-3
medical removal protection, 23-9–23-10
policy, 23-3
program to confront reproductive hazards, 23-4–23-6
Reproductive toxin
effects of, 23-1–23-2 (*see also* Hospital, reproductive hazards)
identification, 23-4
policies for, 23-6
Resident transfer tasks, 9-47–9-48
Resource Conservation and Recovery Act (RCRA), 24-13, 29-3
Respirators
aerodynamic diameter, 4-26

airborne microorganisms, 4-55
air filtration mechanism, 4-56–4-58
air-purifying respirators (APR), 4-27
assigned protection factor (APF), 4-58
bioaerosols
approach, 4-64–4-65
avian influenza viruses (AIV), 4-69
bacteria, 4-73
biological hazards, 4-61
concentrations, 4-70–4-71
criteria, 4-68
decision tree, 4-83
endotoxins, 4-73–4-74
infectious doses, 4-61–4-62
measles, varicella and smallpox viruses, 4-63
medical mask, 4-69–4-70
molds, 4-72–4-73
non-infectious, 4-61–4-62, 4-70
peat moss packager, 4-76
protection factors, 4-74
research guidlines, 4-75
risk and experts' recommendations, 4-64–4-68
risk coefficient (RC) method, 4-72
SARS virus, 4-75
wastewater treatment plant, 4-75
biological weapons, 4-56
categories, 4-56
characteristics, 4-27
filtration efficiency, 4-57
fit tests, 4-59
full-face respirators (FFR), 4-31
HCW, 4-30
helmet-type system, 4-31
HEPA filters, 4-29–4-30
mechanical systems, 4-28
N95 type, 4-30–4-31
powered air-purifying respirator (PAPR), 4-31, 4-60
poxviruses, 4-29
programs
criteria and types, 4-37–4-39
HCW, 4-37
OSHA requirements, 4-36–4-37
plume control, 12-32
pulmonary TB, 4-24
requirements, 4-25
SARS event, 4-25–4-26
SAR system, 4-31
SCUBA and SAR systems, 4-30
seal checks, 4-59
self-contained breathing apparatus (SCBA), 4-28
standards and regulations, 4-84
surgical and paper masks, 4-27–4-28
ULPA filters, 4-25
Respiratory protection, laboratories, 27-11; *see also* Laboratory safety
Respiratory protective equipment (RPE); *see also* Health care worker (HCW); Personal protective equipment
blood aerosols exposure, 17-22–17-24
EtO exposure, 14-15
Respiratory symptoms, glove dermatitis, 19-10–19-11; *see also* Glove dermatitis; Latex allergy
Respiratory syncytial virus (RSV), 16-2
Ribavirin aerosol, HCWs' health effects of, 16-3; *see also* Aerosolized pentamidine (AP), occupational risks
acceptable exposure levels, 16-3
control system validation for, 16-20–16-21
engineering and administrative controls, 16-13–16-14
materials and methodologies, 16-14–16-16
outcomes, 16-17–16-18
engineering controls, 16-23
exposure assessment methods, 16-4
exposure levels for, 16-4–16-6
health hazard assessment data, 16-41
hospital's written policy and shift assignments, 16-23
industrial hygiene survey report, 16-10–16-12
limitations, 16-35–16-36
methodologies, 16-24–16-28
outcomes, 16-28–16-35
pharmacokinetics and exposure, 16-24
policies, recommendations, 16-8
protection policies, 16-9–16-10
recommendations for, 16-7–16-8, 16-37–16-38, 16-52–16-55
recommended concentration, 16-43–16-55
room ventilation, 16-24
symptoms, 16-6
toxicology of, 16-3, 16-23–16-24
uses of, 16-22
RPC, *see* Recommended Practices Committee
RPE, *see* Respiratory protective equipment
RSV, *see* Respiratory syncytial virus
RTECS, *see* Registry toxic effects chemical substances

S

Safety Committee, hospitals, 29-4; *see also* Hospital safety office, need assessment
Safety Office, hospital, 29-1–29-2; *see also* Hospital safety office, need assessment
chemical spill cleanups, 29-5–29-6
chemical use, industrial hygiene monitoring of, 29-5
committee membership, 29-7–29-8
electrical safety, 29-2–29-3
environmental regulations affecting hospitals, 29-10
fire/life safety, 29-2
hazard communication standard, 29-4–29-5

hospital waste program, 29-3
liaison, 29-6–29-7
library in, 29-11
planning and product recalls, 29-8
professional maintenance, 29-9–29-10
safety committee/accident, 29-4
safety office, administration of, 29-10
technical service, 29-9
universal precautions, 29-10
vendors and consultants, meeting with, 29-9
Safety policy, laboratories, 27-2; *see also* Laboratory safety
San Francisco General Hospital (SFGH), 11-103, 11-104, 11-106, 11-108
San Francisco General Hospital memoranda, glutaraldehyde elimination, 14-30–14-31
SARA, *see* Superfund Amendment and Reauthorization Act
Sarcoptes scabiei, 31-13
Scabies, 31-13
prevention of, 31-14
symptoms of, 31-13–31-14
transmission and treatment of, 31-14
SCBA, *see* Self-contained breathing apparatus
Security and safety, health care
clinics and outpatient facilities
engineering control, 7-30
work practice and administrative controls, 7-30–7-31
emergency rooms and general hospitals
engineering control, 7-31
general hospitals, 7-32
hazard prevention and control
general building, workstation and area designs, 7-27
maintenance, 7-27–7-28
home/field operations
engineering controls, 7-32–7-33
work practice and administrative controls, 7-33
medical management, 7-33–7-34
occupational homicides, 7-19
prevention, 7-22–7-24
program development
commitment by top management, 7-24–7-25
employee involvement, 7-25
management commitment and employee involvement, 7-24
regular program review and evaluation, 7-26
written program, 7-25
program evaluation, 7-37–7-38
psychiatric hospital/in-patient facilities
administrative controls, 7-28–7-30
engineering control, 7-28
work practice controls, 7-30
recordkeeping, 7-35
risk factors
environmental factors and work practices, 7-20–7-21
perpetrator and victim profile, 7-21–7-22
sexual assaults with serious injury, 7-20
training and education
general, 7-35–7-36
job-specific, 7-36–7-37
supervisors and managers maintenance and security personnel, 7-37
violence cost, 7-22
worksite analysis
record review, 7-26
security hazards identification, 7-26–7-27
SEIU, *see* Service Employees International Union
Self-contained breathing apparatus (SCBA), 13-15–13-16
Service Employees International Union (SEIU), 7-47–7-48, 25-8, 26-1
recommendations on nursing home workers, 26-3–26-4
research on nursing home workers, 26-2
staffing standards improvement, recommendations on, 26-12–26-13
Severe acute respiratory syndrome (SARS)
Canada, 4-3–4-4
emerging infectious diseases, 4-16–4-17
HCW, 4-17–4-18
national agencies response paradigms, 4-6–4-7
nosocomial diseases, 4-18
novel coronavirus (SARS-CoV), 4-16–4-17
personal protective equipment (PPE), 4-9–4-11
urine and feces, 4-18–4-19
SFGH, *see* San Francisco General Hospital
Sharps injuries prevention
Air Support Medical Sharps Pin Cushion™ container, 6-63–6-64
blood collection systems
Baxter/Edwards VAMP, 6-36
Becton Dickinson Vacutainer Systems, 6-59
Bio-Plexus Punctur-Guard, 6-59–6-60
Medical Safety Products Acci-Guard, 6-36–6-37
Ryan Medical Blood Adapter, 6-37–6-38
Viggo-Spectramed Safedraw, 6-60–6-61
catheter starters
Critikon ProtectIV, 6-23–6-24
Deseret Medical Intima, 6-24–6-25
Menlo Care Landmark, 6-25–6-26
Ryan Medical Shamrock, 6-26
disposable syringes
Becton Dickinson Safety-Lok Syringe, 6-38–6-39
Concord/Portex Needle-Pro, 6-61–6-62
NeedlePoint Guard Safety Syringe, 6-39–6-40
Sherwood Medical Monoject, 6-40–6-41
EPINet sharps injury data, 6-6–6-7
health care worker protection, 6-11

implement safety-engineered devices, 6-9–6-10
legislation and policy, 6-5
medication connectors
Baxter Healthcare InterLink IV Access System, 6-27–6-28
Baxter Healthcare Needle-Lock, 6-28–6-29
Burron Medical Safsite, 6-29–6-30
CU Medical Piggy Lock, 6-55–6-56
Edge Medical Safe-Draw multidose drug vial adapter, 6-58
ICU Medical Click Lock, 6-30–6-31
IMS Stick-Gard, 6-31–6-32
L&W Technology Safeport, 6-31–6-33
Pascall Medical SPIVE, 6-33
Ryan Medical Saf-T Clik, 6-33–6-34
Tri-State Hospital Supply Kleen-Needle, 6-34–6-35
needle guards
ICU Medical HR Needle, 6-40–6-42
North American Medical Products Safe-Site, 6-42
needleless medication/vaccine injectors, Bioject Biojector, 6-19–6-20
needle-recapping devices
J & T Products, 6-62–6-63
On-Gard Systems Recapper, 6-43–6-44
Terumo Safe-Guard Shield, 6-43–6-44
NSPA impact
injury rates, 6-8–6-9
market data, 6-7–6-8
occupational needlestick injuries
education and training, 6-17
needle-recapping controversy, 6-17–6-18
needlestick-prevention devices, 6-13–6-14, 6-44–6-51
preventive devices, 6-19
risk assessment, sources and management, 6-14–6-17
sharps disposal, 6-18–6-19
workers' compensation, 6-3–6-4
zero risk, 6-4
OSHA enforcement actions, 6-5
prefilled medication systems
Winthrop Pharmaceuticals Carpuject, 6-21–6-22
Wyeth-Ayerst Tubex, 6-22–6-23
safety devices efficacy, 6-9
Shiftwork, health and safety hazards, 21-1–21-2
checklist for reducing hazards, 21-15
hidden costs of shiftwork, 21-2–21-3
accidents, 21-7
effect on workers, 21-4–21-5
gastrointestinal function and, 21-9
health care industry, night work in, 21-3
heart disease and, 21-9–21-12
nurses, 21-5–21-7
psychological effects of, 21-8
reproductive function, 21-8–21-9
strategies
education, 21-14
hours limitation at night, 21-13–21-14
light control in the environment, 21-12
personal medical factors, 21-12–21-13
physical fitness and shift rotation frequency, 21-13
shift rotation and napping strategies, 21-12
Skin damage; *see also* Ethylene oxide (EtO)
EtO in, 13-2
laser beam exposure, 12-29
Skin protection, laser plume, 12-33
Skin testing, contact urticaria, 19-10
Sleep disorders, shiftwork workers, 21-4; *see also* Health care worker (HCW); Public safety; Shiftwork, health and safety hazards
Sliding boards, ergonomics, 9-48
Slip, trip, and fall (STF), 8-1–8-4
Small Particle Aerosol Generator® nebulizer, 16-22
Smoke evacuation systems
for plume control, 12-32
techniques for using, 12-33
SMR, *see* Standard mortality ratio
SOPs, *see* Standard operating procedures
Sorbent tubes, qualitative organic analysis; *see also* Health Hazard Evaluation Report, NIOSH
methodologies, 12-13
outcome, 12-17–12-19
SPAG-2® nebulizer, *see* Small Particle Aerosol Generator® nebulizer
Sputum induction booth, 11-86
Standard mortality ratio (SMR), 21-10–21-11
Standard operating procedures (SOPs), 12-25
Steam sterilization, infectious hospitals waste, 24-6; *see also* Medical waste
Sterilants, disinfectants and sanitizers, CDC and EPA classification schemes for, 14-29
STF hazard prevention
age group, employment length and gender, 8-3–8-4
body part and injury nature, 8-2–8-3
circumstances, 8-3–8-4
claim rates, age group and gender, 8-4
compensation claim rates, 8-3
evaluation, 8-6
friction characteristics
flooring testing and classification, 8-5–8-6
shoe–floor slip resistance, 8-4–8-5
full-time equivalent (FTE), 8-2
hazard assessments, 8-7
ice and snow removal, 8-10
identification, 8-6–8-7
incidents distribution, 8-2
lighting, 8-10
minimize tripping hazards, 8-9

operating rooms (OR), 8-9–8-10
stairs and handrails, 8-11
strategies
floors clean and dry, 8-7–8-8
housekeeping program, 8-7
slip-resistant shoes, 8-8–8-9
wet floors, 8-8
work-related injury, 8-2
Superfund Amendment and Reauthorization Act (SARA), 29-4
Surgical smoke
content and research, 12-2–12-3
standards and recommendations, 12-3
Surveillance systems, HCSA sector
Bureau of Labor Statistics (BLS), 2-14–2-15
data source limitations, 2-44
disease and cause-specific mortality data
industry subsector, 2-38–2-40
National Occupational Mortality Surveillance (NOMS), 2-38
proportionate mortality ratio (PMR), 2-38, 2-41–2-44
fatal occupational injuries, 2-31–2-32
industry level data, 2-29–2-31
infectious disease data
bloodborne infections, 2-37–2-38
HBV infections, 2-36–2-37
HCV, 2-37
sharps injuries, 2-34–2-36
tuberculosis, 2-37
injuries/illnesses, exposures and sources
assaults and violent acts, 2-27
back injuries, 2-26
floor and ground surfaces, 2-28
health care patient, 2-27
overexertion and repetitive motion, 2-25
slips, trips and falls, 2-26
sprains and strains, 2-24–2-25
worker motion, 2-28–2-29
key facts, 2-33
nonfatal occupational injuries and illness
incidence rates, 2-16–2-24
nursing and residential care facilities, 2-16–2-18
private and service-providing industries, 2-18
OSHA, 2-14
Survey of Occupational Injuries and Illnesses (SOII), 2-14
underestimate, 2-33–2-34

T

TCD, *see* Thermal conductivity detector
TDICT, *see* Training for Development of Innovative Control Technology Project
Telanar model 4000 dynamic head space concentrator, 12-13
Teletherapy, radiation therapy, 22-12–22-13; *see also* Hospitals, radiation safety procedures
Thermal conductivity detector (TCD), 17-12
Thermo luminescent dosimeters (TLDs), 22-9
Threshold limit values (TLVs), 14-1
Time-weighted average (TWA), 12-15, 28-4
Tissue biological effects, lasers, 12-24
TLDs, *see* Thermo luminescent dosimeters
TLVs, *see* Threshold limit values
Toluene biological effects, case study for, 28-1–28-4
Total particulate weights, determination of, 12-12; *see also* Health Hazard Evaluation Report, NIOSH
Training for Development of Innovative Control Technology Project (TDICT), 25-8
Training programs, laboratory safety, 27-4–27-5; *see also* Laboratory safety
Tuberculosis; *see also* Health care worker (HCW); Home health care
and aerosolized pentamidine treatment, association of, 15-6
and bronchoscopy, endotracheal intubation, and tracheostomy association, 15-7
in home health care, 25-8–25-9
Tuberculosis engineering controls
ClestraClean room air sterilizer, 11-77–11-79
health care settings, control options
health care worker (HCW), 11-84
HEPA bed tent, 11-85–11-86
recovery rate tests, HEPA, 11-87
RxAIR unit, 11-88
sputum induction booth, 11-86
implementing in county hospital
bronchoscopy, 11-106
emergency department (ED), 11-103–11-104
negative-pressure isolation room, 11-104–11-105
outpatient clinics, 11-106–11-108
pressure differential monitoring, 11-106
San Francisco General Hospital (SFGH), 11-103–11-104, 11-106, 11-108
sputum induction, centralization, 11-104
implementing quality assurance program
developing protocol, 11-110–11-111
negative pressure maintenance, 11-114
performing audits, 11-111
project evaluation, 11-111–11-113
project proposal, HCW, 11-109–11-110
survey results, 11-113
industrial hygiene approach, control
directional airflow, 11-11–11-13
hazard characterization, 11-3–11-4
patient isolation rooms, 11-21–11-25
regulations and guidelines, 11-4–11-6
research needs, 11-26
source characterization, 11-5–11-6
source control, 11-6–11-11

MICROCON HEPA air filtration system
controlling bioaerosols, emergency room, 11-74–11-77
isolation room investigation, 11-72–11-74
portable HEPA filtration, TB isolation
challenge data, 11-64–11-67
filtration mechanisms, 11-31–11-33
health care facilities, 11-33–11-34
in-room HEPA filtration, 11-35–11-37
mechanics of filtration, 11-31–11-33
respiratory isolation, TB, 11-34–11-35
respiratory protection, 11-38
source control, 11-38–11-39
reducing spread, tuberculosis
air quality survey, 11-48–11-50
move desks, chairs, and people, 11-56–11-57
natural ventilation, low-cost ways, 11-51–11-55
RS 1000 HEPA challenge test
air scrubber, 11-62
clearance rates/decay rates, 11-61
surgical mask, droplet nuclei TB, 11-89–11-90
surgical masks inefficiency, droplet nuclei TB
dust mist respirator, 11-90
mass median diameter (MMD), 11-89
TB prevention, workplace, 11-41–11-47
test method, 11-57–11-63
ultraviolet radiation
American Society of Heating, Refrigerating and Air Conditioning Engineers (ASHRAE), 11-93
effective factors, 11-92
germicidal UV lamps, 11-91
heating, ventilating, and air conditioning (HVAC) system duct, 11-94
inverse square law, 11-97
NIOSH exposure limit, 11-96
reciprocity law, 11-97
recommendations, 11-97
shielded UV fixtures, 11-92
technical details, 11-97
UV-C region, 11-94
ventilation
aerosol therapy cabinet, 11-98
ASHRAE standard, 11-100
calculation, number of room air, 11-100–11-101
dilution, 11-99
local exhaust, 11-98–11-99
make-up air supply, 11-101
room air distribution, 11-99–11-100
suggestions, 11-101
TWA, *see* Time-weighted average

U

UCSF health effects study, 15-15–15-17
UIHC, *see* University of Iowa Hospitals and Clinics
ULPA, *see* Ultralow penetration air
Ultralow penetration air (ULPA), 12-3
Ultraviolet germicidal irradiation (UVGI), 11-3, 11-19–11-21
Ultraviolet radiation, TB control
American Society of Heating, Refrigerating and Air Conditioning Engineers (ASHRAE), 11-93
duct irradiation, 11-20
effective factors, 11-92
germicidal UV lamps, 11-91
heating, ventilating, and air conditioning (HVAC) system duct, 11-94
HEPA filtration unit, 11-18–11-20
inverse square law, 11-97
NIOSH exposure limit, 11-96
reciprocity law, 11-97
recommendations, 11-97
shielded UV fixtures, 11-92
technical details, 11-97
ultraviolet germicidal irradiation (UVGI), 11-20
UV-C region, 11-94
Unipolar electrocautery unit, 12-5
United States and Canada
class, 4-7–4-8
emerging infectious diseases, 4-1–4-2
globalization and global warming effects, 4-7
national agencies response paradigms, 4-6–4-7
occupational health paradigms
centralization, 4-4–4-5
contagious airborne diseases, 4-3
health care worker (HCW), 4-2–4-4
health/protecting HCW arena, 4-2–4-3
HVAC systems, 4-4
severe acute respiratory syndrome (SARS), 4-3–4-4
surging population, 4-5
public health paradigms, 4-5–4-6
Universal control, laboratory safety, 27-2; *see also* Laboratory safety
Universal precautions
aim of, 17-20
of blood and body fluids, 25-8 (*see also* Home health care)
University of Iowa Hospitals and Clinics (UIHC), 24-3
University of Wisconsin Hospital, glutaraldehyde study, 14-25–14-26
U.S. Bureau of Labor Statistics (BLS), 26-7
Use test, procedure of, 19-10; *see also* Glove dermatitis; Latex allergy
U.S. Office of Management and Budget, 25-2
UVGI, *see* Ultraviolet germicidal irradiation

V

Vacuum scavenging systems, 17-7
Ventilation evaluation, ribavirin, 16-27–16-28
Ventilation, TB control
aerosol therapy cabinet, 11-98

ASHRAE standard, 11-100
calculation, room air change, 11-100–11-101
capture efficiency, 11-10
dilution, 11-99
empirical expressions, 11-8
flanges, 11-8
industrial ventilation manual, 11-10
local exhaust systems, 11-8, 11-98
make-up air supply, 11-101
room air distribution, 11-99–11-101
slot hood, 11-8
sputum induction booth, 11-9
suggestions, 11-101
tracer gas decay curve, 11-15
Veterans' Administration memorandum, glutaraldehyde elimination, 14-29–14-30
Violence, health care industry
barriers, 7-14–7-15
clinical history, 7-107–7-108
community, 7-109–7-110
community settings, 7-13–7-14
employee alarm systems, 7-44–7-45 history, 7-108–7-109
injury and illness prevention program
exceptions, 7-42–7-43
history, 7-44
steps record, 7-43
institutional settings
intervention types, 7-8
mental health and emergency departments, 7-8–7-9
psychiatric settings, 7-8
job assault
assault incident report form, 7-65
employer's responsibility, 7-50
factors, 7-49
grievance/petition form, 7-66
guidelines, 7-50–7-51
job survey, 7-64–7-65
posttraumatic stress disorder (PTSD), 7-51
prevention organization, 7-62–7-63
reasons and victims, 7-48
resource list, 7-54–7-62
sample contract language, 7-51–7-54
sexual harassment, 7-49
work environment, 7-49–7-50
work-related deaths, 7-48
joint labor/management experience
assaults *vs.* restraints, 7-73
Binghamton Psychiatric Center, 7-87–7-95
Buffalo Psychiatric Center, 7-74–7-87
health care facilities, 7-69
national profile, 7-71
New York State, 7-71–7-72
organizational culture and political will, 7-70–7-71
OSHA guidelines, 7-69–7-70
lawsuits, staff moral and bad publicity, 7-2
lean staffing, 7-4
magnitude and severity, 7-5–7-6
memo, 7-107
observations and comments, 7-105–7-107
prevention policy and programming
critical program components, 7-9
performance-based, 7-11
posttraumatic stress disorder (PTSD), 7-10
public sector health care facilities
authority, 7-46
PEOSH reporting requirements, 7-47
recommendations, 7-15–7-16
reported staff comments, 7-104–7-105
research, 7-7
risk factors, 7-6–7-7
security and safety
clinics and outpatient facilities, 7-30–7-31
emergency rooms and general hospitals, 7-31–7-32
employer cost, 7-22
evaluation, 7-37–7-38
hazard prevention and control, 7-27–7-28
home/field operations, 7-32–7-33
medical management, 7-33–7-34
prevention, 7-22–7-24
program development, 7-24–7-26
psychiatric hospital/in-patient facilities, 7-28–7-30
recordkeeping, 7-35
risk factors, 7-20–7-22
training and education, 7-35–7-37
worksite analysis, 7-26–7-27
staff survey, 7-103–7-104
state fund industries, 7-5
training and program development, 7-47 typology, 7-4–7-5
workplace prevention, 7-3
workplace study, 7-104

W

Waste anesthetic gases, sources of, 17-7–17-8
Weight loss surgery (WLS)
deep venous thrombosis (DVT), 3-48–3-49
Minnesota Multiphasic Personality Inventory-2 (MMMP-2), 3-48
psychosocial implications, 3-49–3-50
Roux-en-Y gastric bypass (RYGB), 3-48
wound care, 3-49
Wet chemistry (WC) analyzer
for formaldehyde monitoring, 14-6–14-7
for glutaraldehyde monitoring, 14-8
role of, 14-5
WHMIS, *see* Work Health Material Information System

Workers' compensation; *see also* Health care worker (HCW)
  and health care industry, 32-2–32-3
  in National Health Care System reforms, 32-9
  New York's training and education grant program by, 32-6
  occupational disease and, 32-4
  politics of, 32-3
  for tuberculosis, 32-4–32-5
  in workplace injuries and illness prevention, 32-6
Workers job performance, shiftwork, 21-4–21-5; *see also* Shiftwork, health and safety hazards
Work Health Material Information System (WHMIS), 14-20
Workplace hazardous drugs control, chemodrugs
  administering hazardous drugs, 10-11
  administrative controls, 10-15
  closed system devices, 10-14
  drug transportation, 10-11
  engineering controls, 10-12–10-13
  hazard assessment, 10-8–10-9
  medical surveillance, 10-16
  personnel exposure, safe levels, 10-17
  polices, programs and procedures, 10-9
  PPE, 10-14–10-15
  preparation areas, 10-10
  receiving and storage, 10-9–10-10
  spill management, 10-11–10-12
  training, 10-15–10-16
  ventilated cabinets, 10-13–10-14
Workplace health care facilities, ergonomics design
  dynamic simulation, future activity
    design concept development, 9-33–9-35
    enlarged plan, 9-36–9-38
    full-scale simulation, 9-38
    prognosis and follow-up, 9-39–9-41
    project data analysis, 9-27–9-29
    prototype, 9-38
    scenario activity, 9-29–9-33
    simulation props, 9-36
  functional building analysis
    macroscopic level, 9-25
    mesoscopic and microscopic level, 9-26
  problem situation, 9-24
  specific approach, 9-23–9-24
Workplace PF, definition of, 16-55
Work practices and procedure, EtO exposure prevention, 13-21–13-22
Work-related exposures
  to hepatitis B, 31-15–31-16
  to hepatitis C, 31-16

X

X-rays, properties of, 22-4; *see also* Radiation safety procedures, hospitals
Xylene biological effects, 28-4–28-7